The Royal Horticultural Society

Shorter
Dictionary
of
Gardening

THE ROYAL HORTICULTURAL SOCIETY

SHORTER
DICTIONARY
— OF —
GARDENING

MICHAEL POLLOCK
AND
MARK GRIFFITHS

MACMILLAN

First published 1998 by Macmillan
an imprint of Macmillan Publishers Ltd
25 Eccleston Place, London SW1W 9NF
and Basingstoke

Associated companies throughout the world

ISBN 0 333 65440 4

Designed by Robert Updegraff
Typeset by Florencetype
Printed and bound in Great Britain by The Bath Press, Bath

CONTENTS

Acknowledgements

The Publisher, Editors and the Royal Horticultural Society would like to
thank the following for their help in the preparation of this book

Illustrators
Vana Haggerty – horticultural techniques, pests, ornamental and edible crops
Camilla Speight – plant genera, species and cultivars

Photography
Andrew Lawson

Designer
Robert Updegraff

Consultants
Andrew Halstead
Ellen Henke
Chris Prior

And for their contributions and advice:
Roger Aylett, Jeffrey Brande, Michael Grant, Alan Leslie, Anthony Lord,
Stella Martin, Jonathan Pickering, Martin Rickard, Joyce Stewart, Debra
Whitehead, Brent Elliott and the Staff of the Lindley Library, the Staff at the
RHS Wisley, and Staff at the Royal Botanic Gardens Kew

Special thanks are due to Susanne Mitchell, Karen Wilson, Barbara Haynes
and Julian Ashby for help and support during the preparation of the book

Foreword

The New RHS Dictionary of Gardening was published to worldwide acclaim in 1992. This major work in four volumes brought together information on all aspects of gardening and was produced in collaboration with Macmillan Publishers Ltd with contributions from horticultural and botanical experts around the world. It still stands alone as the most comprehensive work available on horticulture.

At the beginning of 1995 it was decided that there was a need for a definitive single volume which would be a distillation of the original work aimed at a more popular readership. The three-and-a-half years taken to produce this volume is testimony to the work that has gone into the preparation, abridgement and verification of the text.

The RHS Shorter Dictionary of Gardening gathers in one volume all of the core horticultural matter to be found in *The New RHS Dictionary of Gardening* together with entries for over 1900 genera of ornamental and culinary plants.

The horticultural content covers the broadest possible spectrum of terms, techniques and processes. All the original entries have been reviewed and partially or extensively revised with the intention of meeting the needs and standards of dedicated amateurs and horticultural students.

The content of the plant entries focuses on garden merit and cultivation and includes a selection of the best species, hybrids and cultivars. In some instances - especially with the larger genera - this is tabulated for ease of reference.

Line illustrations of plants, techniques, implements, pests and botanical terms are used to clarify the text where it is helpful to do so to add more information. A section of colour illustrations demonstrates something of the diversity of plants and their ornamental value throughout the year.

The RHS in collaboration with Macmillan Publishers Ltd, is proud to be associated with the production of this comprehensive guide to horticulture. I have no hesitation in recommending it to the Society's members and to everyone interested in expanding their knowledge of this fascinating subject.

Sir Simon Hornby
PRESIDENT
THE ROYAL HORTICULTURAL SOCIETY

About this book

The RHS Shorter Dictionary of Gardening gathers into one volume all of the core horticultural matter to be found in the four volume *New RHS Dictionary of Gardening*. The original text has been substantially revised and much of it rewritten for this new work.

The horticultural and general entries have been written and compiled by Michael Pollock; these include the principles and practices of horticulture, gardening techniques, crops, horticultural science, botanical and horticultural terms and there is comprehensive coverage of commonly encountered pests, diseases and disorders.

The plant entries and tables have been written by Mark Griffiths; they include treatments of most genera that are widespread in cultivation. The plant entries move from a brief description of the genus, to considerations of garden merit and cultivation requirements. In many cases, there follows a descriptive paragraph detailing the best-known species, hybrids and cultivars.

Tables

Several of the largest and most familiar genera have been treated as tables for ease of reference. In these cases, the accepted name of the plant in question appears in bold. Beneath that name in the same left-hand column, synonyms and popular names may also appear. General information still appears in a prefatory paragraph. Individual cultivation requirements, climate zones and cultivars appear beneath each species as 'Comments'.

Climatic zones

Plant entries make mention of climatic zones. These may be given for a genus as a whole, or for individual species where their hardiness differs. The zones are an approximate indication of the coldest climatic band in which the plant will thrive. See Hardiness, p.339.

Plant Nomenclature

Botanical names adopted in this book have been adjudicated by members of the RHS Advisory Panel on Nomenclature and Taxonomy. For the most part, names follow those in *The RHS Plant Finder* 1997–98.

Synonyms are botanical names which, for whatever reason, no longer apply correctly to the plant in question. Many plants may be known to gardeners by names that are, strictly, synonyms. Where a name has changed within a genus, the synonym is listed after the currently accepted botanical name. For example, readers looking for *Helleborus corsicus* will find it listed now as *Helleborus argutifolius* (syn. *H. corsicus*). In other cases, a familiar garden plant may have been transferred to another genus and this is signified by a cross reference. Readers looking, for example, for the shrubby ANGELS' TRUMPETS are cross-referred from *Datura* to *Brugmansia*. Likewise (at least pending their return to the genus *Chrysanthemum*), the species familiar to gardeners and florists as CHRYSANTHEMUM can be found under *Dendranthema*.

Sometimes, however, a whole genus may be 'sunk' into synonymy. Familiar examples here include Coleus (= *Solenostemon*), Endymion (= *Hyacinthoides*), Gloxinia (= *Sinningia*), Montbretia (= *Crocosmia*). These synonyms, all vital cross references to well-known garden plants, are listed in a separate index at the end of this book, as are the plants' common or popular names.

Cultivars

Noteworthy cultivars are named and described in entries for ornamental plant genera and fruit crops. They are not usually included in entries for vegetable crops or within some of the main ornamental group entries (e.g. CHRYSANTHEMUM), because the very wide choice available to gardeners is best researched through the catalogues of seedsmen and nurserymen. The Royal Horticultural Society's Award of Garden Merit is awarded to plants judged to have excellent garden qualities. The lists of AGM plants are continuously expanding. For information on which plants have received the award readers are recommended to consult specialist plant and seed catalogues, *The RHS Plant Finder*, *The RHS Good Plant Guide* and the AGM list published periodically by the RHS.

Pests, Diseases, Disorders and use of Chemicals

The recognition and identification of pests, diseases and disorders are essential to the maintenance of good growth, flowering and cropping in cultivated plants. Information on frequently met problems is provided throughout the book by means of references in the plant entries, by marginal notes which refer the reader to main entries, and in the *Pests and Diseases Diagnostic Chart* which follows the main body of the text. There are many specific entries on pests, diseases and disorders providing information on identifying characteristics and, where appropriate, life cycles. Chemical control measures are rarely mentioned, emphasizing the need for utmost care in selection and application. References should be made to product manufacturers' recommendations for effective and safe use.

Abbreviations used in this book

cf. : compare	subsp./ssp. : subspecies
cv : cultivar	subspp. : subspecies (plural)
cvs : cultivars	syn. : synonym
f. : forma	var. : variety
q.v. : which see	x : hybrid
sp. : species	+ : graft hybrid (see *chimaera*)
spp. : species (plural)	

Conversions of Weights and Measurements

To convert	into	multiply by
millimetres	inches	0.039
centimetres	inches	0.394
metres	feet	3.281
metres	yards	1.094
square metres	square yards	1.196
grams	ounces	0.0352
kilograms	ounces	35.274
kilograms	pounds	2.205
litres	gallons (Imperial)	0.22
litres	gallons (US)	0.264
litres	pints (Imperial)	1.76
litres	pints (US)	2.113

THE ROYAL HORTICULTURAL SOCIETY

SHORTER
DICTIONARY
—— OF ——
GARDENING

A – Z

A

abaxial of the surface or part of a lateral plant organ turned or facing away from the axis and toward the plant's base, such as the underside of a leaf; cf. *adaxial*.

Abelia (for Clarke Abel (1780–1826), who discovered *A.chinensis*). Caprifoliaceae. East Asia; Mexico. 15 species, evergreen or deciduous shrubs with small funnel-shaped flowers through summer and early autumn. In some species the calyx is showy. Many members of this genus are not reliably hardy below zone 7, but can be grown against south-facing house walls and in sheltered, walled gardens. Plant in a sunny position, protected from harsh winds. Prune early-flowering species immediately after flowering, midsummer- and autumn-flowering species in early spring. Propagate from semi-ripe or ripe-wood cuttings in summer.

  A.chinensis (China; spreading, deciduous shrub to 1.5m with pairs of fragrant white flowers surrounded by rose-tinted calyces from summer to autumn); *A.* 'Edward Goucher' (semi-evergreen hybrid between *A.* × *grandiflora* and *A.schumannii*, with glossy dark green foliage bronze on emergence, and lilac-pink flowers with golden throats); *A.floribunda* (Mexico; somewhat tender evergreen with massed, tubular, cherry red flowers in midsummer); *A.* × *grandiflora* (semi-evergreen hybrid between *A.chinensis* and *A.uniflora* with vigorous, arching branches, dark green foliage emerging bronze-pink and turning deep purple-bronze in autumn; flowers mildly scented white, pink-tinged and surrounded by attractive, persistent, pink- or red-purple-tinted calyces; 'Francis Mason': leaves emerging in copper tones turning dark green with gold margins, flowers pale with red-bronze sepals; 'Goldsport' and 'Gold Strike': more solidly gold-variegated); *A.schumannii* (Central China; slender-branched, arching semi-evergreen to 1.5m tall with rose-pink flowers produced in great abundance throughout the summer; hardier than most other species); *A.triflora* (Himalayas; robust, deciduous shrub with an upright habit and sweetly fragrant shell-pink flowers surrounded by feathery red sepals).

Abeliophyllum (the foliage of this genus was thought to resemble that of *Abelia*). Oleaceae. Korea. 1 species, *A.distichum*, the KOREAN or WHITE FORSYTHIA, a deciduous shrub to 2m, with arching branches and rather dull, ovate leaves. The flowers appear in late winter and early spring on bare, maroon twigs. They resemble a forsythia's in general shape, but are sparkling white and sweetly scented (flushed pink in the cultivar 'Roseum'). Hardy in zone 7 if planted against a south-facing wall and given some support on wires or trellis. Cut back flowered shoots in mid spring. Propagate by simple or air layering in autumn or spring, or by half-ripe cuttings in a propagating case in summer.

*Abeliophyllum
distichum*

Abelmoschus (from Arabic, *abu-l-misk*, 'father of musk', alluding to the aromatic seeds of *A.moschatus*). Malvaceae. Old World Tropics and Subtropics. 5 species, annual and perennial, downy herbs with large, palmately lobed leaves, broadly funnel- or dish-shaped flowers close in shape to those of *Hibiscus* (a genus in which *Abelmoschus* is sometimes included) and pod-like fruit. *A.moschatus* is a vibrant half-hardy bedding plant for sunny, sheltered situations. Sow seed under glass in early spring. Plant out after the last frost.

  A.moschatus (SILK FLOWER, MUSK MALLOW; bushy annual to 60cm; flowers 8–10cm-long in shades of red and cerise, often stained white at the throat; colour selections include 'Mischief' (compact and bushy with bright red, white-eyed flowers), 'Pacific Light Pink' (to 40cm, bushy, flowers large, dark pink at rim fading to pale pink at centre), and 'Pacific Orange Scarlet' (flowers vivid scarlet, white-eyed).

Abies (from Latin, *abire*, to rise up, referring to the great height of some species). Pinaceae. North Temperate regions, Central America. FIR; SILVER FIR. Some 50 species, evergreen, coniferous trees with resinous bark and tiered branches in a whorled or candelabra-like arrangement. The needles are linear and flattened, often keeled and with two pale stripes of stomata beneath, and arranged radially, spirally or

Abelia grandiflora

pectinately on the branchlets. The male cones hang from leaf axils; the female cones (described in the table below) are ovoid to cylindric and appear in spring, standing more or less upright. These trees are mostly from cool, wet mountainous regions.

Avoid soils that dry out rapidly, and exposure to drying winds (*A.cephalonica* and *A.pinsapo* will tolerate drier and more alkaline soils than others). All species grow best in slightly acidic soils. Most resent atmospheric pollution, although *A.homolepis* and *A.veitchii* will survive in cities.

Plant young trees in autumn or spring on a well-drained site; select those that are 60–100cm tall with a single, straight leader (except, of course, for dwarf and low-growing cultivars). Keep the bore clear of weeds and mulch its surrounds. Prune to remove any surplus leaders, also the lowermost branch tiers as they die or attempt to resprout. *Abies* is attacked by adelgids and (in North America) by bagworms and caterpillars of the hemlock lopper. It usually fends off fungi, *Armillaria* included.

abiotic disorder a description of poor growth or malfunction in plants which is not caused by a living organism. It is commonly related to environmental factors or nutritional deficiency.

abortive of reproductive organs; undeveloped or not perfectly developed, and therefore barren: of seeds; failing to develop normally.

Abronia (from the Greek, *habros*, 'graceful'). Nyctaginaceae. Western North America, Mexico. SAND VERBENA. 35 species, annual or perennial, trailing herbs, often viscid, with rounded, fleshy leaves. Small tubular flowers are carried in stalked, terminal heads and are often highly fragrant, becoming more so at night. Grow in full sun on rather dry, light soils. In virtually frost-free climates they may become perennial, making leafy mats ideal for binding sand or for greening arid banks, retaining walls and rock gardens. In colder areas they are good basket plants for greenhouses and conservatories. For outdoor use as tender bedding plants, *Abronia* should be grown from seed each year or repropagated by cuttings taken from stock plants overwintered under glass. In frost-free regions, sow seed *in situ* in autumn; elsewhere under glass at 15°C/60°F in spring. Prior to sowing, soak the seed for 24 hours and remove the papery husk. Strike soft-wood cuttings in sand in spring.

A.latifolia (YELLOW SAND VERBENA; British Columbia south to California; thick-leaved, clammy trailer with bright yellow flowers; *A.umbellata* (PINK SAND VERBENA, BEACH SAND VERBENA; differs from *A.latifolia* in its narrower leaves, red-tinted stems and rose-pink flowers); *A.villosa* (resembles a hairy, sticky *A.umbellata*, with flowers of a deeper pink or rosy-purple; tolerates prolonged drought).

abscission, abscissing (abscising) of leaves, branches and fruits; a separating or falling away, caused by disintegration of a layer of plant tissues at the base of the organ (e.g. as the result of environmental conditions or pollination leading to hormonal action), with a subsequent development of scar tissue or periderm at the point of abscission.

Abutilon (from the Arabic name for mallow). Malvaceae. Tropics and subtropics. PARLOUR MAPLE. Some 150 species, annual and perennial herbs, or shrubs and trees with ovate and finely toothed to palmately lobed leaves, and nodding, solitary, bell- or bowl-shaped flowers with 5 showy, obovate petals and and a distinct staminal column. The most frequently grown species are evergreen shrubs, slender-branched and more or less glabrous in *A. × hybridum*, *A.megapotamicum*, *A. × milleri* and *A.pictum*; robust and hairy in *A.ochsenii*, *A.vitifolium* and their hybrid, *A. × suntense*. All have attractive flowers, complemented in some cultivars within the first group by handsomely variegated leaves.

A. × hybridum (syn. *A.globosum*; CHINESE LANTERN, FLOWERING or PARLOUR MAPLE; a varied group of garden hybrids, erect and spreading shrubs to 2m; their leaves ovate or palmately lobed, serrate, often variegated; flowers pendulous, to 7cm, white, pink, red, yellow, apricot and orange, sometimes conspicuously veined and almost always slightly inflated in appearance). *A.megapotamicum* (BRAZILIAN TRAILING ABUTILON; Brazil; slender-stemmed trailer or semi-climber to 3m; leaves dark green, cordate-lanceolate; flowers pendulous with pale yellow petals encased in a blood red calyx and with a red-black column; *A.* 'Variegatum'; leaves gold-mottled); *A.pictum* (syn. *A.striatum*; Brazil; shrub or small tree to 4m with maple-like leaves; flowers solitary, drooping, to 4cm with conspicuously veined, orange-yellow petals; 'Aureomaculatum': leaves gold-spattered; flowers dark-veined, coral-red; 'Thompsonii': orange-flowered with an intricate mosaic of emerald, lime and yellow leaf variegation); *A. × milleri* (*A.megapotamicum* × *A.pictum*; slender stemmed, rather loosely upright plants with flowers intermediate between parents; 'Variegatum': leaves marked with a gold, cream and jade green patchwork; calyx rosy; petals rich yellow; staminal column dark carmine).

Tender abutilons like those mentioned above are often grown as dot plants for summer bedding, or as half-hardy specimens for patio pots and baskets in cool temperate zones. They are also grown permanently in the cool greenhouse, conservatory or the home. *A.megapotamicum*, however, is hardy to −10°C/14°F and, if mulched, will usually regenerate from the base after hard frosts. It is well worth attempting outdoors as a wall shrub in sheltered situations in zones 7 and over.

The second group of popular abutilons comprises those robust, downy shrubs and small trees native to Chile and sometimes included in the genus *Corynabutilon*. These thrive on sheltered, sunny walls in zones 7 and over – *A.vitifolium* (to 8m, usually less; leaves large, palmately 3–5–7-lobed, coated in grey-white felt; flowers 5–8cm wide white or purple-blue; white-flowered plants include var.

ABIES

Name	Distribution	Habit	Leaves	Cones
A.alba A.pectinata SILVER FIR; CHRISTMAS TREE; EUROPEAN SILVER FIR	C & S Europe	large tree (to 60m); crown regular; bark grey; branches whorled	12–3cm long in 2 horizontal ranks, dark glossy green above, silvery- glaucous beneath	to 18 × 3.5cm, cylindric, bronze-green at first with exserted, reflexed bracts
Comments: Vulnerable to late spring frosts. Includes 'Pendula', medium-sized with weeping branches, and 'Pyramidalis', medium-sized and densely conical				
A.amabilis PACIFIC FIR; RED SILVER FIR; CASCADE FIR	W N America	large tree (to 35m); crown conic; bark silvery; branches tiered; buds very resinous	to 3.5cm long, crowded on upper side of branches, pectinate beneath, dark glossy green above, silver-white beneath, orange-scented if crushed	to 17 × 7cm, conic-ovoid, purple at first with concealed bracts
Comments: Will not tolerate dry and chalky soils. Includes 'Spreading Star' ('Procumbens'), low-growing (to 1m tall), branches wide-spreading, horizontal				
A.balsamea BALM OF GILEAD; BALSAM FIR	N America	medium-sized tree (to 25m); crown conic; bark grey; branches ascending	to 2.5cm long, 2-ranked, glossy dark green above with glaucous patch at tip, 2 grey-white bands beneath, strongly balsam-scented	to 8 × 2.5cm, oblong-cylindric, purple at first, gummy
Comments: Ill-suited to dry chalky soils and arid atmospheres. Often seen in the more tolerant, dwarf, flat-topped f. *hudsonia*, of which the dwarf, cushion-like 'Nana' is a good selection for a moist, part-shaded position on the rock garden.				
A.bracteata A.venusta BRISTLE CONE FIR; SANTA LUCIA FIR	California	large tree (to 40m); crown dense to open, brown; buds very long and slender	to 6cm, 2-ranked, glossy dark green, spine-tipped	to 8 × 0.5cm, round to ovoid, golder-green, bracts remarkably long, exserted, spine-tipped
Comments: Tolerates heat and drought.				
A.cephalonica GRECIAN FIR; GREEK FIR	Greece	to 40m; crown conic; branches low, spreading; dark grey	to 3cm, dark, glossy green, pale beneath, sharp-pointed, arranged around branchlets	to 18 × 2.5cm, cylindric, tapering, brown with bracts exserted
Comments: Disease-resistant and chalk-tolerant; plant where frosts will not damage early new growth. Includes 'Meyer's Dwarf' ('Nana') a broad, low-growing dwarf with short needles.				
A.concolor COLORADO FIR; WHITE FIR	W US	large, slow-growing tree to 50m; crown conic-columnar; bark pale grey; branches whorled, spreading	to 6cm, thick, scattered or in 2 ranks, blue-green	to 12 × 4cm, cylindric, tapering, pale-green to purple with a strong bloom
Comments: One of the finest large conifers – slower-growing in warm climates, favouring cool, airy humid conditions. Includes 'Candicans' with very bright silver-grey or 'blue' foliage, and 'Compacta' (to 2 × 2m) with an irregular, dense crown and grey-blue needles.				
A.concolor var. **lowiana** A.lowiana LOW'S FIR; PACIFIC WHITE FIR; SIERRA WHITE FIR	SW US	to 70m; crown narrowly conic; trunk thick; branches short	to 6cm, 2-ranked, grey-green	to 9 × 3cm, green at first
A.delavayi DELAVAYI'S FIR	China, Burma, India	to 25m; crown candelabriform; branches ascending	to 4cm, densely arranged around branchlets, dark green above, silvery beneath, revolute	to 12 × 4cm, cylindric, dark blue to violet ripening black
A.grandis GIANT FIR; GRAND FIR; LOWLAND FIR	W N America	fast-growing to 75m; crown conic to columnar; bark blistered, brown to ridged and grey-brown; branches arching to ascending	to 4.5cm, pectinately arranged, grooved and bright green above, with 2 glaucous bands beneath, very aromatic if crushed	to 10 × 3.5cm, cylindric, tapering, bright green at first
Comments: Somewhat more lime-tolerant than other species, very tall and fast-growing, favours abundant rainfall and humidity. Lower branches may be pruned out to allow underplanting. 'Aurea' has golden-yellow young growth.				
A.homolepis A.brachyphylla NIKKO FIR	Japan	to 40m; crown broadly conic; bark yellow to pink-grey, flaky	to 3cm, densely pectinately arranged, dark green above, with 2 white bands beneath, tip blunt or cleft	to 11 × 3cm, cylindric, smooth, gummy, purple at first
Comments: Pollution-tolerant. Includes 'Prostrata' (low, spreading); 'Scottiae' (dwarf); 'Shelter Island' (semi-prostrate, growth irregular, golden at first); 'Tomomi' (growth slender, sparse).				
A.koreana KOREAN FIR	S Korea	to 15m; crown broadly conic, compact, bark rough	to 2cm, crowded, arranged around branchlets, glossy dark green above, bright white beneath	to 8 × 2.5cm, cylindric, apex tapering, purple
Comments: Slow-growing and suitable for smaller gardens, where plants no taller than 1m will produce cones. Includes 'Aurea' (very small and slow-growing, leaves yellow-green); 'Piccolo' (dwarf, leaves very short, pointing forwards and upwards); 'Silberlocke' (leaves turned upwards showing silver-white undersides).				
A.lasiocarpa SUBALPINE FIR; ALPINE FIR	W N America	to 25m, narrowly conical to columnar, slow-growing; bark thin, smooth silver	1.5–4cm, densely 2-ranked or in a loose, brush-like arrangement, pale grey-green, apex blunt or notched	to 13 × 3cm, cylindric-conic, dark purple at first
Comments: Somewhat lime-tolerant; requires a fertile, damp soil. Often sold as an alpine. In drier conditions, very slow-growing and likely to lose its regular, narrow outline				
A.lasiocarpa ssp. **arizonica** A.arizonica CORKBARK FIR; CORK FIR	SW USA	bark thick, creamy-white, corky	silver-grey	
Comments: Attractive small tree with distinctive bark; makes a good bonsai. Includes 'Compacta', a slow-growing, compact, conical shrub with bright blue-grey leaves.				

ABIES

Name	Distribution	Habit	Leaves	Cones
A.magnifica CALIFORNIAN RED FIR; RED FIR; SILVER TIP FIR	Oregon, California	to 60m, crown narrow-conic; bark smooth, grey to red-brown, fissured; branches rather short, horizontal	to 4cm, grey-green to blue-green throughout, densely pectinate, curving upwards	to 25 × 9cm, broadly cylindric, purple at first
Comments: Very stately, performing best in cool, airy and humid conditions. Used as a Christmas tree in California. Includes 'Glauca' (leaves very strong blue green), and 'Prostrata' (dwarf, branches spreading).				
A.nordmanniana CAUCASIAN FIR; CHRISTMAS TREE; NORDMANN FIR	NE Turkey	to 55m; crown conic to columnar; branches whorled, downswept	to 3cm, glossy, dark green above, silvery beneath, densely arranged – forward-pointing and overlapping above on branchlets, pectinate beneath	to 18 × 4.5cm, red tinged green, scales long, reflexed
Comments: Resilient, disease-resistant species making a dense cone. Includes 'Golden Spreader' (dwarf, slow-growing, spreading, foliage light yellow) and 'Pendula' (semi-prostrate, branchlets pendulous).				
A.pinsapo SPANISH FIR; HEDGEHOG FIR	S Spain	to 35m; crown broadly conic; branches crowded, whorled; bark smooth then rough, rusty brown	to 1.8cm, often shorter, rigid and densely arranged around branchlets, dark green above, silvery beneath	to 16 × 3.4cm, green bloomed purple at first
Comments: Drought- and heat-resistant, chalk-tolerant and remaining dwarf for years. Includes 'Aurea' (slow-growing, weak, leaves golden); 'Glauca' (large, leaves blue-grey); var. **marocana** (*A.marocana*) the MOROCCAN FIR, leaves 2-ranked.				
A.procera *A.nobilis* NOBLE FIR; CHRISTMAS TREE	W USA	to 70m; crown conic; bark smooth and silver-grey at first then blistered and brown, ultimately fissured	to 3.5cm, arranged along sides and upper surface of branchlets, blue-green	to 23 × 7cm, oblong-cylindric, purple with long-exserted, reflexed bracts
Comments: Sometimes used as a Christmas tree; includes 'Glauca' (leaves blue-grey) and 'Prostrata' (dwarf, flat-topped bush, leaves blue-grey).				
A.spectabilis HIMALAYAN FIR	Himalayas	to 30m; crown conic, broad; branches red brown at first, spreading then ascending; bark rough, scaly; buds large, round	to 5cm, tough 4-ranked, dark green above, silvery beneath, tip cleft	to 15 × 5cm, cylindric to oblong-ovoid, blue-grey to violet
Comments: New growth badly affected by spring frosts.				
A.veitchii	Japan	to 35m; crown narrowly conic; branches short, spreading; bark grey, smooth above, fluted below	to 3cm, ranked along upper surface of branchlets, forward-pointing, glossy dark green above, silver beneath, apex truncate, cleft	to 7 × 2.5cm, cylindric, purple-blue, bract tips exserted
Comments: Pollution-tolerant.				

Abutilon pictum
'Thompsonii'

*Abutilon
megapotamicum*

album and 'Tennant's White'; 'Veronica Tennant': large, lavender blooms amid hoary foliage); *A.ochsenii* (differs from *A.vitifolium* in its 3-lobed leaves, usually less downy, leaves always with 3 lobes and in its smaller, pale lavender to violet flowers, often spotted at the base within; the hybrid between these species, *A.* × *suntense* includes the white-flowered 'Gorer's White' and 'White Charm', the dark purple 'Jermyns' and the violet 'Violetta'). Increase by softwood, greenwood or semi-ripe cuttings stripped of most of their leaves and inserted in a closed case, under glass.

Acacia (From Greek *akakia*, a plant name used by Dioscorides). Leguminosae. Subtropical and tropical regions. WATTLE; MIMOSA. Some 1200 species shrubs, trees and lianes (those below evergreen trees). The leaves are bipinnate, but are sometimes only present in juvenile stage and later replaced phyllodes. The flowers are small and numerous, usually yellow and packed in small, powder-puff-like arrangements. These in turn are crowded in globose or spiciform heads, themselves arranged in racemes or panicles.

Several species will tolerate frosts: *A.dealbata*, *A.longifolia*, *A.baileyana* to –10ºC/14ºF, *A.saligna* to –5ºC/23ºF. *A.decurrens*, *A.ulicifolia*, *A.melanoxylon*, *A.retinodes* and *A.podalyriifolia* have all been successfully cultivated outside in favoured regions of the UK, with the protection of a south- or southwest-facing wall. In cool temperate climates, the smaller *Acacia* species make beautiful specimens for the cool greenhouse or conservatory, often flowering in winter or early spring. *A.dealbata*, *A.melanoxylon*, *A.longifolia*, *A.pendula*, *A.retinodes* and *A.verticillata* are also amenable to tub cultivation under glass. Outdoors, grow in a well-drained, neutral to slightly acid soil in full sun. In zones at the limits of hardiness, plant near the base of a sheltered south- or southwest-facing wall.

Under glass, grow in direct sun in a free-draining, sandy, loam-based medium with leafmould or compost. When using fertilizers, it should be remembered that a number of species show sensitivity to high levels of phosphorus. Maintain a minimum winter temperature of 5ºC/40ºF and ventilate freely when conditions allow. Water moderately when in full growth, sparingly in winter. Move pot-grown specimens out-of-doors for the summer.

Propagate from seed or semi-ripe cuttings. Douse seed with scalding water and allow to soak in the cooled liquid for 12–48 hours before sowing; viable seeds should swell. Germination at 20–25ºC/68–77ºF takes 7–20 days. Strike semi-ripe heeled cuttings in a sandy propagating mix under mist or cover with gentle bottom heat, at 15–18ºC/60–65ºF.

Acacia baileyana

ACACIA

Name	Distribution	Height & Spread	Foliage	Flowers
A.aneura MULGA	Australia	to 7 × 4m	2–25 × 1cm, blunt, undivided, grey-green	golden-yellow in cylindric spikes in spring
Comments: Small graceful tree, will tolerate very dry and sunny conditions. Z8.				
A.baileyana BAILEY WATTLE; COOTAMUNDRA WATTLE; GOLDEN MIMOSA	SE Australia	5–10 × 2.5–8m	to 5cm, finely bipinnate glaucous grey- to blue-green	golden-yellow, strongly fragrant balls arranged in racemes in winter and spring
Comments: An exceptional small tree or multi-stemmed shrub with a compact habit and neat, feathery foliage (purple-tinted in 'Purpurea'); thrives on sunny walls in colder areas. Z8.				
A.cognata NARROW-LEAVED BOWER WATTLE	Australia	to 8 × 5m	to 10 × 0.3cm, undivided, linear-oblong, dark green	lemon yellow balls in spring
Comments: A small tree or shrub with graceful foliage on weeping branches. Z9.				
A.cultriformis KNIFE-LEAF WATTLE	New South Wales	to 4 × 4m	to 3.5 × 1.5cm, undivided, ovate to deltoid, thinly fleshy, silvery grey	clusters of golden yellow balls in spring
Comments: Multi-stemmed with arching branches; good for screening and dry banks in Z9, elsewhere for very sunny and sheltered sites. Z8.				
A.cyclops COASTAL WATTLE; WESTERN COASTAL WATTLE; ROOIKRANS	W Australia	to 4 × 4m	to 9 × 1.5cm, undivided, narrowly oblong, sharp-tipped and dark green	small, bright yellow balls, solitary or clustered, in spring
Comments: Resilient, dense and rounded shrub withstanding drought and coastal conditions, excellent for hedging or screening. The seeds are black with red rings. Z9.				
A.dealbata *A.decurrens* var. *dealbata* MIMOSA; SILVER WATTLE	SE Australia; Tasmania	to 30 × 25m	to 12cm, finely bipinnate, silver-grey to sea green	pale gold, intensely fragrant balls in large, branching sprays in late winter and spring
Comments: Very popular shrub or small to medium-sized tree; fast-growing with a narrowley erect, then tiered habit and spreading, feathery foliage. Requires a sheltered, warm position but will regenerate from root suckers if cut down by frost. Flowers are commonly sold as florist's mimosa. Z8				
A.decora GRACEFUL WATTLE; SHOWY WATTLE; WESTERN SILVER WATTLE	E Australia	2–5 × 1.5–3m	2–6 × 1cm, undivided, lanceolate, curving, blue-green	fragrant, bright gold balls massed in clusters in spring
Comments: Drought-resistant, useful for screening in warm areas. Z9.				
A.decurrens EARLY BLACK WATTLE; GREEN WATTLE	New South Wales	to 12 × 10m	to 7cm, finely bipinnate, dark green	yellow balls clustered in racemes in late winter and early spring
Comments: Resembles a deep green *A.dealbata*; roots and suckers sometimes invasive. *A.mearnsii* (BLACK WATTLE) differs in its branchlets which are not grooved and ribbed. Z8.				
A. 'Exeter Hybrid' *A.* 'Veitchiana'	garden origin	to 6 × 4m	undivided, long and narrow on arching branches	rich yellow, fragrant, in spikes in spring
Comments: (*A.longifolia* × *A.riceana*). A free-flowering container or border shrub for the large cool, conservatory. Z9.				
A.farnesiana MIMOSA BUSH; SCENTED WATTLE; SWEET ACACIA; SWEET WATTLE	Tropical America	to 7 × 5m	leaves bipinnate, feathery with segments to 1cm long	orange-yellow, fragrant balls throughout the year
Comments: A deciduous, thorny shrub or tree, possibly synonymous with *A.smallii*, a tree with which it is often confused and which is held to be hardier. Z8.				

ACACIA

Name	Distribution	Height & Spread	Foliage	Flowers
A.genistifolia A.diffusa SPREADING WATTLE	Australia, Tasmania	to 3 × 2m	to 5cm, undivided, rigid and pointed, dark green	pale or bright yellow balls in axillary clusters
Comments: A spreading, broom-like shrub with sharp phyllodes. Z9.				
A.longifolia SALLOW WATTLE; SYDNEY GOLDEN WATTLE	Australia, Tasmania	to 7 × 6m	to 15cm, undivided, oblong-lanceolate, dark green	pale to golden yellow spikes along branches in late winter and early spring
Comments: Fast-growing but short-lived large shrub or tree with a broad open head. Good for screening and soil binding; in coastal regions it sometimes becomes semi-prostrate. Drought- and lime-tolerant. Similar and sometimes treated as a variety of this species, **A.sophorae** (COASTAL WATTLE) is hardier and differs in its shrubby habit, and shorter, broader phyllodes. Z8.				
A.melanoxylon BLACKWOOD ACACIA; BLACK ACACIA	Tasmania, Australia	to 30 × 10m	to 14 × 2.5cm, undivided, oblanceolate, dark green	pale yellow balls in branching racemes in spring
Comments: Similar to A.longifolia, but with pinnate juvenile foliage. Fast-growing, erect tree; responds well to pruning. Z8.				
A.mucronata VARIABLE SALLOW WATTLE	Australia			
Comments: Similiar to A.longifolia, but with narrower phyllodes and looser inflorescences. Z8.				
A.paradoxa A.arreata HEDGE WATTLE; KANGAROO THORN	New South Wales	to 5 × 4m	1–3 × 1cm, o blong-lanceolate, dark green, on densely and viciously spiny branches	yellow, in solitary balls in late winter and spring
Comments: Flowers when young, making a fine, if thorny container shrub for terraces and conservatories. In warmer areas, a formidable hedge. Includes the weeping to straggling 'Pendula', and var. **angustifolia** with narrow phyllodes. Z8.				
A.pendula WEEPING ACACIA; WEEPING MYALL	E Australia	to 6 × 4m	to 10 × 1cm, undivided, lanceolate, blue-green	lemon-yellow in short, cylindrical panicles
Comments: A very beautiful small tree with weeping habit, silvery foliage and flowers produced throughout the year. Z9.				
A.podalyriifolia MOUNT MORGAN WATTLE; PEARL ACACIA; QUEENSLAND SILVER WATTLE	NE Australia	to 6 × 4m	to 5 × 2cm, undivided, rounded to ovate-elliptic, silvery grey, white-tomentose	bright yellow balls in branched racemes produced from early winter to early spring
Comments: Handsome large shrub or small tree with silvery foliage; can be trained and pruned. Good for large containers on terraces and in cool conservatories. Z8.				
A.pravissima OVEN'S WATTLE	SE Australia	3–8 × 2–6m	to 2 × 1.5cm, undivided, deltoid to cuneate, glaucous grey-green with a single spine below the broad apex on the underside	bright yellow, produced in small clusters in early spring
Comments: Small tree or shrub with arching, wand-like branches and unusually shaped phyllodes. Good for conservatories and sheltered walls and terraces. Z8.				
A.puchella PRICKLY MOSES	W Australia	to 3 × 2m	finely bipinnate with each segment to 0.5cm long	bright yellow balls clustered 2–3 together
Comments: Small to medium-sized shrb with prickly branches. Z8.				
A.redolens A.ongerup	W Australia	to 5 × 4m	2.5–7 × 0.5–1.5cm, undivided, narrow, grey-green, leathery and sweetly scented	yellow balls in short racemes or panicles in spring
Comments: Good for dry banks and impoverished soils in warm areas. 'Prostrata' is especially suitable for ground cover: low-growing and dense with yellow-green phyllodes. Z9.				
A.retinodes A.floribunda of gardens FLORIBUNDA ACACIA; MIMOSE DE QUATRE SAISONS; WATER WATTLE	S Australia, Tasmania	to 6 × 4m	to 18 × 1.5cm, undivided, linear, bright mid-green to grey-green	small, highly fragrant pale yellow balls freely produced in large panicles, principally in summer
Comments: A fairly hardy and lime-tolerant tree with an airy crown of weeping, narrow foliage. An excellent fast-growing, open screen. Flowers last well when cut. Z8.				
A.riceana RICE'S WATTLE	S Tasmania	2–10 × 1.5–6m	to 5.5 × 0.5cm, undivided, narrow, dark green and sharply pointed	pale yellow balls in dense racemes in spring
Comments: A large shrub with slender, arching branches. Distinguished from A.verticillata by the phyllodes, which are clustered, not whorled. Z8.				
A.saligna A.cyanophylla; A.cyanophyllodes BLUE-LEAVED WATTLE; GOLDEN WREATH WATTLE	W Australia	to 6 × 5m	15–30 × 1–3cm, undivided, linear-oblong to lanceolate, blue-green	deep yellow balls in clusters in spring
Comments: Multi-stemmed and spreading. Z9.				
A.stenophyllodes A.stenophylla DALBY MYALL; SHOE STRING ACACIA	Australia	to 10 × 6m	to 40 × 1cm, undivided, linear, leathery, pale green and drooping	pale yellow balls in spring
Comments: Fast-growing tree with a weeping, open crown and maroon-tinted new shoots. Z9.				

ACACIA Name	Distribution	Height & Spread	Foliage	Flowers
A.subporosa BOWER WATTLE; RIVER WATTLE	Australia	to 10m		
Comments: Close to *A.cognata*, but taller with broader phyllodes. 'Emerald Cascade' is low-growing with drooping, emerald green, narrow phyllodes and pale green flowers. Needs protection from coastal winds and burning sunlight. Z9.				
A.ulicifolia *A.juniperina* JUNIPER WATTLE	SE Australia	to 3 × 2m	to 2cm, undivided, needle-like, olive to dark green	solitary, creamy-yellow balls in spring
Comments: 'Brownii' is compact and bushy, to 60cm tall with large, rich yellow flowerheads. Z9.				
A.verticillata PRICKLY MOSES	SE Australia, Tasmania	to 6 × 4m	to 2 × 0.2cm, undivided, needle- like, whorled, dark olive green	pale yellow in short spikes in spring and early summer
Comments: Bushy shrub with dense, prickly growth. A good hedge if pruned, resisting intruders and saline winds. Unpruned, it becomes multi-stemmed and open with snaking branches and mats of gorse-like foliage. Z8.				

Acaena (from the Greek *akaina*, thorn, a reference to the spiny flower- and fruiting heads). Rosaceae. Southern Hemisphere (those below from New Zealand). NEW ZEALAND BURR; BIDI-BIDI; SHEEP'S BURRS. Some 100 species, small, evergreen, perennial herbs and subshrubs, carpet-forming with pinnate leaves and spring flowers packed in burr-like balls to 1.5cm diameter. Hardy to 5°C/–15°F or lower if protected with a dry mulch. Plant in light shade or full sun in sharply draining neutral to slightly alkaline soils. Increase by division, or by seed sown in a frame in autumn or spring.

A. 'Blue Haze' (vigorous spreader to 30 × 100cm with long, rooting stems leaves to 10cm long, leaflets widely spaced, 0.5–1cm long, bluntly toothed to lobed, smooth, grey-green to steel blue or pewter); *A.buchananii* (to 4cm, making close carpets of pea green to pale grey-green leaves; leaflets fewer and broader than in A. 'Blue Haze', downy beneath); *A.caesiiglauca* (taller than the last species to 5cm with larger leaflets of a pale, glaucous blue and rather downy); *A.microphylla* (leaves 1–3cm-long; leaflets of a pale olive to grey-green with a strong, red-brown tint, pronounced and persistent in the cultivar 'Kupferteppich').

Acalypha (from the Greek for nettle, *akalephe*: Linnaeus thought that the leaves of this genus resembled those of the stinging nettle). Euphorbiaceae. Tropics. Over 400 species, shrubs, trees and annual herbs. Provide a minimum temperature of 10°C/50°F, with filtered sunlight to medium shade, and high humidity. When in growth, water plentifully and liquid feed fortnightly. In winter, keep barely moist. Propagate from semi-ripe cuttings in late summer or from soft cuttings in early spring in a closed case with bottom heat.

A.godseffiana (New Guinea; to 1.6m; leaves narrowly ovate, bright green, usually deeply cut and fringed and edged, blotched or spattered with ivory, yellow and lime; flowers dull); *A.hispida* (CHENILLE PLANT, RED HOT CAT'S TAIL; New Guinea, Malaysia; to 4m, resembles a spectacular, woody love-lies-bleeding (*Amaranthus caudatus*) with 50cm-long, dark crimson to bright red tassels, and broad, sap green leaves); *A.wilkesiana* (COPPERLEAF, JACOB'S COAT, BEEFSTEAK PLANT; South Pacific; to 4.5m; leaves saw-toothed, elliptic to ovate, 20cm long, usually described as cop-per splashed with pink and red; includes forms with leaves bronze-maroon edged white to pink, or red with bronze and crimson markings, or red-bronze and very large, to broad and coppery black with twisted, crenate, green and red margins or green stippled and blotched orange and red).

Acanthocereus (from Greek *akantha*, thorn and *Cereus*, a related genus - a reference to the fierce spines). Cactaceae. Central and South America. 6 species, large cacti, shrubby, clambering or tree-like in habit with ribbed spiny stems and white, funnel-shaped, nocturnal flowers. Provide a minimum temperature of 10°C/50°F, with filtered sunlight to medium shade and high humidity. Plant in a loam-based, medium-fertility mix high in sand and grit and with a pH of 6–7.5. Shade in hot weather and maintain low humidity. Keep dry from mid-autumn until early spring, except for a light misting over on warm days in late winter. Propagate by detaching rooted lengths of stems.

A.colombianus (syn. *A.guatemalensis*, *A.pitajaya*; Colombia; differs from *A.tetragonus* in its erect habit and flowers with a perianth tube to 23cm long); *A.tetragonus* (syn. *A.acutangulus*, *A.pentagonus*; S US, W Indies, C America; gaunt shrub to 3m or more tall with arching stems; flowers to 25cm long with a perianth tube to 15cm).

Acantholimon (from Greek *akantha*, thorn, and *limon*, an allusion to the related genus *Limonium*). Plumbaginaceae. PRICKLY THRIFT. Mediterranean to Central Asia. 120 species, small, evergreen, tufted, perennial herbs and subshrubs with hummocks of needle-like leaves and short spikelets of flowers produced in summer. Suited to rock gardens and alpine troughs the following species are cushion-forming, dwarf shrublets. Plant in a warm, dry position in full sun on a light and sharply draining, gritty soil. Where wet and cold threaten, these plants succeed better in pans or tufa blocks in the alpine house or frame. Propagate by earthing up the bases of older plants: these may be divided once roots have formed.

A.glumaceum (Central Asia; rose flowers with white, downy calyces); *A.ulicinum* (Eastern Mediterranean to Asia Minor; pink flowers with white to purple calyces); *A.venustum* (Asia Minor; pink flowers with purple-veined, yellow calyces).

Acaena
'Blue Haze'

Acanthus (from Greek *akanthe*, thorn). Acanthaceae. Old World. BEAR'S BREECHES. 30 species, perennial herbs and shrubs with lanceolate to oblanceolate, toothed to pinnately lobed leaves, and tall spikes of tubular, hooded flowers carried in summer amid heavily veined bracts. All are fully hardy but prefer a warm and sunny position on a fast-draining soil. Increase by seed or division in early autumn, or take cuttings of the long, narrow roots in winter.

A.dioscoridis (Turkey, Iraq, Iran; leaves grey-green, laciniate to lobed; spikes 30–100cm with incised dull green bracts and purple-pink flowers); *A.hungaricus* (syn. *A.balcanicus*, *A.longifolius*; Balkans to Greece; leaves dark, glossy green, deeply pinntately cut, the lobes coarsely toothed but not spiny; spikes to 1m with spiny purple-red bracts and pale pink flowers); *A.mollis* (South Europe, Northwest Africa; leaves glossy dark green, broadly pinnately cut, the lobes wide, toothed and softly spiny; spikes to 1.5m with toothed, mauve-tinted bracts and white flowers; Latifolius Group includes 'Oak Leaf' with large, very broadly and shallowly lobed and scalloped leaves); *A.spinosus* (Mediterranean; leaves dark green, often with sparse silver hairs and obscurely silvered veins, 1–3 × pinnately cut, the lobes deep, narrow, rigid with sharply spiny margins; spikes to 1m tall with spiny, green-mauve bracts and white flowers; Spinosissimus Group-hybrids between this and the previous species with larger, softer, but deeply cut and sharply spiny leaves – most plants sold as *A.spinosus* belong here).

acaricide a chemical substance used to kill parasitic mites such as red spider mites.

acaulescent of a plant whose stem is absent or, more usually, appears to be absent, being subterranean or so short as to be inconspicuous.

Acca Myrtaceae. South America. 2 species, evergreen shrubs or small trees. *A.sellowiana* (syn *Feijoa sellowiana*), the PINEAPPLE GUAVA, ranges from Southern Brazil to Northern Argentina. It grows to 6m tall with peeling grey-brown bark and broadly oblong to elliptic, blunt, leathery leaves, dull green above, silvery beneath. Borne singly or in sparse clusters in summer, the flowers are some 4cm across and fleshy, with showy crimson stamens. The red-tinted, egg-shaped fruit is edible, as are the flowers. Hardy to –15°C/5°F given full sun, protection from harsh winds and a well-drained situation. Propagate by seed or semi-ripe cuttings in a closed case with bottom heat.

accelerator see *activator*.

accent plant a plant featured in garden design to focus attention on contrasting colour, texture or form, which may be natural, as with weeping willow (*Salix babylonica*), or achieved by trimming, as with yew (*Taxus baccata*).

accessory fruit a fruit or aggregate fruit in which the conspicuous and, usually, fleshy parts are distinct from the pistil, as in the enlarged torus of the strawberry.

acclimatization the gradual introduction of a plant to a different climate or environment ; most commonly used for the transfer of tender plants from a greenhouse to a frame or the open garden, a process often termed hardening off. Also refers to a late stage in the process of micropropagation, sometimes known as weaning.

accrescent growing together or becoming larger or longer with age, or after fertilization, as in the lantern-like calyx of *Physalis*.

accumbent (of cotyledons) lying face to face, the edges of one side lying against the radicle in the seed.

Acer (the Classical Latin name). Aceraceae. North and Central America, Europe, North Africa, Asia. MAPLE. Some 150 species, deciduous or evergreen trees and shrubs (those here deciduous), hardy and grown for their foliage which often colours beautifully in autumn, for their bark, which is sometimes attractively marked, and, less frequently, for their flowers and fruits. The leaves are most typically broad and palmately cut or lobed, but may also be ovate or elliptic to lanceolate, with three or no lobes. Small, usually yellow-green or red-brown flowers appear in arching to pendulous racemes or panicles in spring. The fruits hang on slender stalks and consist of a pair of winged nuts (samaras), fused at their ovules, with the wings spreading outwards.

The following maples are fully hardy. Most species require full sun or light shade, a fertile, well-drained soil, preferably neutral to slightly acid, and ample moisture. The Japanese Maples prefer dappled sun or shade, shelter and a moist, fertile soil rich in organic matter. They grow well in containers and small spaces. Propagate by seed sown when ripe, by softwood or semi-ripe cuttings, or by grafting.

Acca sellowiana

Acer (a) *A. davidii* (b) *A. griseum* (c) *A. palmatum* 'Rubrifolium' (d) *A. negundo* 'Flamingo'

ACER Name	Distribution	Habit	Bark	Leaves
A.buergerianum A.trifidum of gardens THREE-TOOTHED MAPLE; TRIDENT MAPLE	E.China; Korea; Japan	small, bushy tree with a low, spreading habit		to 8 × 10cm, acutely and palmately 3–5-lobed, dark green above, paler beneath with an entire or finely toothed margin, turning fiery red to orange- yellow in autumn
Comments: An attractive, small maple, useful for large containers, patios and bonsai. Z6.				
A.campestre FIELD MAPLE; HEDGE MAPLE	Europe; W Asia	variable shrub or small tree to 12m tall with a rounded, bushy crown or an open habit with twiggy, picturesquely gnarled branches	somewhat corky, grey-brown	to 8 × 10cm, bluntly and palmately 3–5-lobed, dull, dark green above, paler and somewhat hairy beneath, with an entire to toothed or lobed margin, turning clear yellow to amber in autumn; petioles contain milky sap.
Comments: Used in hedging, and a familiar feature of hedgerows. Includes 'Albovariegatum' (leaves blotched white), 'Elsrijk' (tree with a dense, broadly conical crown and small, dark green leaves, good for street planting), 'Nanum' (('Compactum') densely compact and shrubby with small, dark green leaves), 'Postelense' (small, shrubby, mop-headed tree or shrub, leaves yellow at first, then yellow-green with red-tinted petioles), 'Pulverulentum' (leaves blotched and finely flecked white), 'Schwerinii' (leaves blood-red at first, turning purple). Z4.				
A.capillipes	Japan	large shrub or small tree to 13m tall with a deep, broad crown of long, spreading, stem-like branches.	young shoots bright red, branches ultimately olive green with silvery white stripes.	to 11cm long, ovate and acutely 3-lobed with a tapering tip and serrate margins, emerging coral red, becoming jade green with a red-tinted petiole, turning orange-red in autumn.
Comments: Z5.				
A.cappadocicum A.laetum CAUCASIAN MAPLE; COLISEUM MAPLE	Caucasus; W Asia; Himalaya	medium-sized to large tree to 20m tall, often multi-stemmed.		to 12 × 14cm, broadly and palmately 5–7-lobed, entire, deep glossy green turning yellow in autumn; petioles containing milky sap.
Comments: Cultivars include 'Aureum' (leaves red-tinted at first then gold to yellow-green), 'Rubrum' (leaves blood-red at first turning dark green), and 'Tricolor' (leaves pink at first then pale green marked cream). ssp. *lobelii* (A.lobelii) grows to 18m tall with ascending branches in a compact, columnar to conical crown and large, dark green leaves with wavy margins. ssp. *sinicum* has red-brown, punctate shoots and conspicuously punctate petioles. Z6.				
A.carpinifolium HORNBEAM MAPLE	Japan	small to medium-sized tree to 10m tall.		to 11 × 4 cm, obovate with an acute tip, serrate margins and prominent, pinnate veins, fresh pale to mid green turning yellow to brown in autumn
Comments: Leaves strikingly similar to those of *Carpinus betulus* (COMMON HORNBEAM). Z5.				
A.caudatifolium A.morrisonense FORMOSAN MAPLE; MOUNT MORRISON MAPLE	China; Taiwan	fast-growing, large shrub or tree to 8m tall, erect then spreading	green striped white	to 10 × 7cm, ovate-oblong with three lobes, the midlobe longer and tapering. margins serrate, emerging red, expanding fresh mid-green with red petiole, turning red in autumn.
Comments: Needs moist, fertile soil. Z8.				
A.circinatum VINE MAPLE	W North America	to 12m tall, habit variable – in shade, a low-branching shrub with slender, sprawling and snaking stems and intricate twigs; in full sun, a single-trunked, small tree.		to 12cm wide, almost circular in outline with 5–11 lobes and serrate margins, emerging pink-tinted, hardening bright green and turning amber, flame and bright red in autumn.
Comments: Tolerates dry shade and is an excellent choice for underplanting in woodland gardens where the arching to clambering branches make airy bowers lit with brilliant autumn colour. The flowers (small, wine red) and fruit (small, deep crimson propellers) are also decorative. 'Monroe' is small and shrubby, the leaves with five distinct lobes with deeply dissected margins. Z5.				
A.cissifolium	China; Japan	small, spreading tree to 12m tall.		composed of 3 distinct leaflets to 8cm long, ovate and coarsely toothed, dark green turning red and yellow in autumn.
Comments: Needs light shade and a moist, neutral to slightly acid soil. ssp. *henryi* (A.henryi) has grey bark striped with blue-white, and larger, more or less untoothed leaflets, these turn brilliant crimson in autumn. Drooping catkins of pale gold flowers appear with the red-tinted new leaves. Z6.				
A.crataegifolium HAWTHORN MAPLE	Japan	small tree or large shrub to 10m tall with long, arching branches.	purple-red to grey-green with silver-white stripes.	variable, to 8cm long, ovate with 3–5, shallow lobes and serrate margins, fresh green to blue-green.
Comments: Foliage remarkably like that of the common hawthorn. 'Veitchii' has leaves splashed with lime and white, flushed pink at the base. Z6.				
A.davidii DAVID'S MAPLE; SNAKEBARK MAPLE	China	shrub or small to medium-sized tree to 15m tall, often multi-stemmed or low-branching with long, spreading branches	smooth, olive green striped silvery white	to 15cm long, ovate to oblong with a tapering tip, unlobed or shallowly 3-lobed with serrate margins, dark glossy green, turning bright yellow to orange and purple-red in autumn
Comments: Ornamental, red-tinted fruits hang in clusters along the branches in autumn. Includes the cultivars 'Ernest Wilson' (compact with ascending to arching branches and pale green leaves with pink petioles turning amber in autumn), 'George Forrest' (vigorous and open-crowned with long, spreading branches and large, deep green leaves with red petioles), and 'Madeleine Spitta' (columnar with erect branches, leaves turning orange in late autumn). One of the finest 'snakebarks' is HERS' MAPLE, ssp. *grosseri* (A.grosseri; A.hersii), to 9m tall with very decorative, white-marbled, grey-green bark and lobed or unlobed, ovate-cordate leaves to 7cm long; these turn a deep, glowing red in autumn. Z6.				
A.giraldii	China	medium-sized to large tree to 18m tall with a spreading crown	young branches flushed red to pink and variably striped white covered with grey-white bloom	to 11 × 15cm, with 3 broad and shallow lobes, entire to coarsely serrate, dark green above, pale blue-green beneath, petiole pink-tinted
Comments: Z6.				

ACER

Name	Distribution	Habit	Bark	Leaves
A.griseum PAPERBARK MAPLE	China	small to medium-sized tree to 12m tall with open, conical to columnar crown of spreading then ascending branches	twigs slender, dark brown and hairy; larger branches and trunk covered in smooth, glossy, red-brown bark that splits and peels in thin flakes and strips revealing pink-brown underbark	composed of 3 distinct leaflets to 5cm long, ovate-oblong and coarsely toothed, dark green above and glaucous, blue-green beneath, turning purple-red in autumn

Comments: One of the most valuable small trees – handsome in summer, glowing in autumn and outstanding in winter when its gaunt outlines and mahogany paper bark contrast especially well with hoar-frost and snow. Z5.

Name	Distribution	Habit	Bark	Leaves
A.heldreichii GREEK MAPLE	SE Europe	medium-sized tree to 16m tall		to 13cm wide, lobes 3–5, oblong-lanceolate and deeply toothed, the three terminal lobes cut almost to base, glossy deep green above, paler and somewhat glaucous beneath

Comments: From the Caucasus and SW Asia, RED BUD MAPLE, ssp. ***trautvetteri*** (*A.trautvetteri*) has red-brown branches with showy red buds in spring; the leaves are large and turn clear yellow in autumn; the fruits have conspicuous, red-tinted wings. From the Balkans, ssp. ***visianii*** has larger leaves and fruit. Z6.

Name	Distribution	Habit	Bark	Leaves
A.japonicum FULL MOON MAPLE; JAPANESE MAPLE	Japan	large shrub or small tree to 10m with a broad canopy		to14cm wide, almost circular in outline with 7–11, ovate to lanceolate, deep lobes, the margins toothed to incised, bright green gradually flushing wine red toward summer's end and colouring deep crimson in autumn

Comments: Appearing with the new leaves, the flowers are red with yellow stamens and hang in clusters. Includes the cultivars 'Aconitifolium' ('Laciniatum', 'Filicifolium') with deeply and finely cut leaves and good autumn colour, 'Green Cascade', with deeply divided, emerald leaves on arching branches, and 'Vitifolium' with broadly fan-shaped leaves composed of 10–12 lobes and colouring brilliantly in autumn.The shade-loving cultivar 'Aureum' (slow-growing with pale yellow leaves turning chartreuse green in summer) belongs to the species *A.shirasawanum*, another Japanese native differing from *A.japonicum* in its glaucous shoots. All favour a well-drained position sheltered from strong winds. Z5.

Name	Distribution	Habit	Bark	Leaves
A.macrophyllum BIG LEAF MAPLE; OREGON MAPLE	W N America	large, broad-crowned tree to 30m tall		to 28cm wide, palmately 3–5-lobed, the lobes irregularly toothed to cut, deep, lustrous green turning orange in autumn

Comments: Handsome large tree with drooping clusters of small yellow flowers in spring and tawny-bristly fruits. Armillaria-resistant. Z6

Name	Distribution	Habit	Bark	Leaves
A.maximowiczianum *A.nikoense* NIKKO MAPLE	Japan; C China	round-headed small tree to 15m tall		composed of 3 leaflets to 12 × 6cm,ovate to oblong and entire to toothed, dull green above, grey-green and hairy beneath, turning brilliant flame in autumn

Comments: Z5.

Name	Distribution	Habit	Bark	Leaves
A.mono *A.pictum*	China; Korea	medium-sized tree to 15m tall		to 15cm wide, palmately 5–7-lobed , the lobes ovate-deltoid, entire and apiculate, bright green and smooth, turning yellow in autumn

Comments: var. ***tricuspis*** has smaller, glossy, 3-lobed leaves. Z6.

Name	Distribution	Habit	Bark	Leaves
A.monspessulanum MONTPELLIER MAPLE	SE Europe to W Asia	small, rounded tree to 12m tall with a twiggy, dense crown, sometimes a shrub		to 8cm wide, with 3 deltoid-ovate lobes, margins variably toothed, smooth, dark green above, paler and hairless beneath, turning yellow-brown in autumn

Comments: Differs from *A.campestre* in its hairless leaves and the petioles which lack milky sap; ssp. ***turcomanicum***, however, has leaves with tawny-hairy undersides and blunt lobes. Forms from rather arid regions are small and shrubby with very neat foliage – var. ***microphyllum***, for example, is of a suitable scale for dry rock gardens. Z5.

Name	Distribution	Habit	Bark	Leaves
A.negundo ASH-LEAVED ELDER; BOX ELDER	N & C America	fast-growing, broad-crowned tree to 20m tall, often suckering	ultimately grey; slender young shoots are pithy, smooth, green and somewhat glaucous	to 20cm long, pinnately divided into 3–7, oblong leaflets with toothed to cut margins, bright, pale green above and pale beneath

Comments: Includes the cultivars 'Auratum' with golden leaves, 'Flamingo' with leaves edged and splashed pink to white, 'Elegans' with yellow-edged leaves, and 'Variegatum' with broadly white–margined leaves. var. ***violaceum*** has frosted, plum-coloured new shoots and tassels of pink-tinted flowers. Large trees are sometimes problematic especially in warmer regions - seeding and suckering freely, breaking easily and playing host to box elder bugs. Z4.

Name	Distribution	Habit	Bark	Leaves
A.oblongum EVERGREEN MAPLE	Himalaya; China	semi-evergreen or evergreen shrub or tree to 10m tall with spreading to upswept branches	peeling, grey	to 12cm long, oblong-obovate, usually entire, sometimes 3-lobed, tough-textured, glossy dark green above, glaucous beneath

Comments: New growth emerges bronze-pink. In var. ***concolor***, the leaves are pale green throughout. Z7.

Name	Distribution	Habit	Bark	Leaves
A.opalus ITALIAN MAPLE	S Europe	medium-sized, rounded tree to 15m tall		to 10cm wide, with 5, broad and shallow lobes, glossy dark green above and downy to glabrous beneath

Comments: var. ***obtusatum*** has larger leaves with 5–7, blunt and irregularly toothed lobes and grey-hairy undersides. Z5.

ACER Name	Distribution	Habit	Bark	Leaves
A.palmatum JAPANESE MAPLE	Korea; Japan	medium-sized to large shrub or small tree to 8m tall with a broad tiered crown	grey	to 10cm wide, rounded in outline, palmately and deeply 5–7-lobed, the lobes oblong to lanceolate, finely tapering and toothed to dissected, smooth thin-textured, pale, green, turning gold to bronze

Comments: Small trees and shrubs of great elegance and ideally suited to smaller gardens, especially where a Japanese effect is desired; they also make good tub and bonsai subjects. Perform best on moist and fertile, neutral to acid soils. Many cultivars developed ranging widely in habit, leaf shape and colour. They fall roughly into these categories: **Atropurpureum Group** with red-purple leaves turning scarlet in autumn; **Dissectum Group**, round-fiery red or deep headed shrubs with arching branches and leaves scarlet in autumn composed of 5–7–9, slender and finely dissected leaflets; **Elegans Group** (Heptalobum group, Septemlobum Group) with 7-lobed leaves, the lobes broader and less finely cut than in Dissectum Group; **Linearilobum Group** with 5–7, very deep slender and toothed lobes. Among the finest cultivars are 'Bloodgood' (leaves dark purple-red, turning blood red in autumn, fruit red). 'Butterfly' (leaves small, deeply cut, emerging pink, turning grey-green with cream margins). 'Osakazuki' ((Elegans Group) leaves green turning brilliant, fiery red in autumn), 'Senkaki' (young branches and twigs coral pink to sealing wax red, in winter especially, leaves mid-green turning amber in autumn). Z5.

A.paxii	C China	slow-growing, small, evergreen or semi-evergreen tree to 10m tall with a dense, compact crown	initially purple to grey-brown	to 7 × 4cm, obovate, usually 3-lobed, tough, lustrous light green above, pale and glaucous beneath

Comments: Z8.

A.pensylvanicum A.striatum GOOSEFOOT; MOOSEWOOD; SNAKEBARK MAPLE; STRIPED MAPLE	N America	small tree or shrub to 12m tall, often low-branching or multi-stemmed with long, spreading branches	young branches green to brown striped green and silver-white	to18cm long, obovate, base subcordate, apex broadly 3-lobed, mid green turning golden yellow in autumn

Comments: Favours neutral to slightly acid soils and grows well in semi-wooded locations. In the cultivar 'Erythrocladum' the young branches become bright coral red in winter. *A.* 'Silver Vein' (*A.laxiflorum* × *A.pensylvanicum* 'Erythrocladum') is vigorous with arching branches streaked green and white, and large, deep green leaves on long, red petioles turning bright yellow in autumn. Z3.

A.platanoides NORWAY MAPLE	N Europe to Caucasus, naturalized N America	broad-crowned, large tree to 30m tall	dark grey	to 18cm wide, broad with 5, ovate-deltoid lobes, the tips tapering with sparse but sharp, broadly angular teeth, thin-textured and smooth, mid to dark green, rather glossy above, turning clear yellow to amber in autumn

Comments: Racemes of bright yellow flowers appear before the leaves. Includes the cultivars 'Crimson King' (leaves dark purple-red to bronze-maroon), 'Crimson Sentry' (habit narrowly columnar, leaves deep purple-red), 'Drummondii' (leaves green deeply edged creamy white), 'Emerald Queen' (vigorous with lustrous, dark green leaves), 'Globosum' (small tree with a mop of short branches), 'Laciniatum' (EAGLE CLAW MAPLE, leaf lobes 5 with slender, talon-like tips), 'Lorbergii' (leaves with 5 lobes, their tips slender and pointing upwards) and 'Schwedleri' (young growth bright red, turning purple-green, then fiery orange in autumn). Z3.

A.pseudoplatanus SYCAMORE	Europe to W Asia, British Isles	small, untidy and suckering, or large, single-stemmed and broad-crowned tree to 40m	dirty grey to grey-brown and flaking	to 16cm wide, cordate in outline with 5, ovate and crenate-serrate lobes, dull dark to mid green above, grey-green beneath, turning yellow in autumn

Comments: Fast-growing tree, widely naturalized, often to the point of nuisance. Includes Atropurpureum' (leaves purple beneath), 'Brilliantissimum' (small, slow-growing tree, leaves emerge pink, turn yellow-green, then mid green), 'Leopoldii' (leaves yellow-pink turning green, splashed pink and creamy yellow), 'Prinz Handjery' (small and slow-growing, differs from 'Brilliantissimum' in its larger leaves tinted purple beneath), 'Purpureum' (leaves tinted purple beneath), and 'Simon Louis-Frères' (leaves pink at first, then green splashed white). Z5.

A.rubrum RED MAPLE; SCARLET MAPLE	N America	large tree to 40m tall	grey	to 10cm long, broadly cordate in outline, with 5–7, deltoid lobes, toothed and tapering, smooth dark green above, glaucous beneath, turning brilliant scarlet to yellow in autumn

Comments: Best autumn colour achieved on neutral to acid soils. Cultivars include 'Columnare' (habit broadly columnar, fiery red in autumn), 'October Glory' (autumn colour deep crimson, long-lasting), 'Red Sunset' (habit dense, erect, leaves bright red in autumn), 'Scanlon' (crown broadly ovate to conical, good autumn colour), 'Schlesingeri' (leaves to 12cm long with a truncate base and 3, large lobes, turning scarlet early in autumn). Z3.

A.rufinerve	Japan	medium-sized tree to 12m tall	green striped white; young branches glaucous blue-white	to 12cm long, rounded at base, with 3, spreading, biserrate lobes in apical half, dark green above, paler and red-veined beneath, turning vivid red, gold and amber in autumn

Comments: 'Albolimbatum' has leaves splashed white at margins and turning red and purple in autumn. Z6.

A.saccharinum A.dasycarpum; A.eriocarpum SILVER MAPLE	N America	fast-growing, broadly and loosely columnar tree to 40m tall	grey	to 14cm long, with 5 deeply cut lobes with tapering tips and toothed to cut margins, light green above, silvery beneath, turning yellow in autumn

Comments: Includes f. *laciniatum* with very deep, slender lobes, the margins finely cut, and 'Wieri' with weeping branchlets and finely dissected leaves. Z3.

A.saccharum ROCK MAPLE; SUGAR MAPLE	N America	broad-crowned tree to 40m tall	furrowed, grey	to 14cm wide, cordate in outline with 3–5, toothed and tapering lobes, thin-textured, smooth, dull green above, grey-green beneath, turning gold, fiery red and scarlet in autumn

Comments: Similar to *A.platanoides*. Sap source of maple syrup. Cultivars include 'Newton Sentry' (habit narrowly columnar with short, ascending branches and no distinct leader, leaves turning fiery red in autumn), and 'Temple's Upright' (habit narrowly conical to broadly columnar with ascending branches and a distinct leader). ssp. *nigrum* (*A.nigrum*) is the black maple to 40m tall with deeply fissured, black bark and usually 3-lobed leaves to 14cm long, deep, dull green above and grey-green and pilose beneath, turning yellow in autumn. Z3.

ACER

Name	Distribution	Habit	Bark	Leaves
A.tataricum TATARIAN MAPLE	SE Europe to SW Asia	large, spreading shrub or small tree to 10m tall	green obscurely striped white on young branches	to 10cm long, ovate and finely tapering, biserrate, bright to dull green above, paler and hairy beneath turning yellow in autumn; leaves on young shoots may be 3-lobed

Comments: From NE Asia, ssp. **ginnala** (*A.ginnala*) is the AMUR MAPLE, a spreading, large shrub or small tree with 3-lobed, toothed leaves to 8cm long turning bright red in autumn. Small, fragrant, yellow flowers are produced in clusters in spring and are followed by red fruit. This subspecies includes the cultivars 'Durand Dwarf' (dwarf with dense, red twiglets and small leaves), 'Flame' (to 7m tall with fiery autumn colour), and 'Pulverulentum' (leaves splashed and speckled white). Z4.

A.triflorum THREE-FLOWERED MAPLE	Manchuria; Korea	small, spreading tree to 12m tall	peeling, grey-brown	to 8cm long, composed of 3, ovate to oblanceolate leaflets, dark green above, glaucous beneath, turning fiery red in autumn

Comments: Differs from *A.maximowiczianum* in its peeling bark. A choice, slow-growing small tree with brilliant autumn colour. Z6.

A.truncatum SHANTUNG MAPLE	China; Korea	small, broad-crowned tree to 8m tall		to 10cm, base truncate, lobes 5, deltoid, deeply cut, bright green and smooth above, paler and hairy beneath

Comments: Attractive yellow-green flowers appear with the leaves. The cultivar 'Akikaze Nishiki' has leaves variably spotted, splashed and stained white. Z6.

A.ukurunduense *A.caudatum* ssp. *ukurunduense*	Japan; Korea	small tree		to 14cm wide, cordate in outline with 5–7 narrowly ovate, toothed lobes, veins impressed above, hairy beneath, colouring well in autumn

Comments: Dislikes chalk. Z6.

A.velutinum *A.insigne*	Caucasus; Iran	large tree		to 15cm wide, base subcordate, broadly and palmately 5-lobed, toothed, bright green above, paler and softly downy beneath

Comments: var. **vanvolxemii**: leaves to 30cm wide, blue-green beneath with hairs on the veins only. Z5

achene a dry, small, indehiscent fruit with a tight thin pericarp, strictly consisting of one free carpel as in Ranunculaceae, but sometimes applied to Compositae with more than one seed.

achenocarp any dry indehiscent fruit.

Achillea (for the Classical hero Achilles, said to have discovered the medicinal properties of these plants). Compositae. North temperate regions. YARROW; MILFOIL. 85 species, perennial herbs, often aromatic, with entire or finely divided and ferny foliage, and small, button-like flowerheads in loose clusters or crowded, flat-topped corymbs.

All are summer-flowering, fully hardy and prefer a sunny, well-drained position. The larger species and cultivars thrive in the herbaceous or mixed border; the flowerheads are attractive additions to dried flower arrangements. Smaller species are best used on the rock or dry garden, or planted on dry banks and walls. Propagate by division in early spring or autumn; take softwood cuttings in late spring.

A.clavennae (E Alps, W Balkans; to 12 × 20cm; carpeting with silky-silvery, pinnately lobed leaves to 8cm long; flowerheads to 1cm diam., bright white with golden centres in loose clusters); *A.clypeolata* (Balkans, Romania; to 40 × 30cm; semi-evergreen herbaceous perennial with finely divided, silvery leaves; flowerheads to 0.8cm diam., bright yellow, crowded in flat-topped heads; crossed with *A.filipendulina* to produce *A.*'Coronation Gold', vigorous, to 1m tall with deep golden flowerheads that dry well; and with *A.*'Taygetea' to produce 'Moonshine', with silver leaves and bright yellow flowerheads); *A.filipendulina* (Caucasus, Iran, Central Asia; semi-evergreen, herbaceous perennial to 100 × 60cm;

leaves green, hairy, deeply and finely pinnately lobed; flowerheads to 0.5cm diam., deep gold, crowded in broad, flat-topped to domed umbels on long stalks drying well; includes 'Altgold' with dark, burnished gold flowerheads, 'Gold Plate' with large, deep gold flowerheads on long, stout stalks; crossed with *A.ptarmica* to produce the creamy yellow-flowered 'Schwefelblüte'); *A.* × *kellereri* (*A.clypeolata* × *A.ageratifolia*; semi-evergreen perennial to 15 × 20cm with feathery, grey-green leaves and loose clusters of daisy-like flowerheads, white to cream with pale gold centres); *A.* × *lewisii* (*A.clavennae* × *A.clypeolata*; compact semi-evergreen perennial to 10 × 20cm, with feathery, grey-green leaves; the cultivar 'King Edward' has tight corymbs of tiny, buff-yellow flowerheads); *A.millefolium* (YARROW; erect perennial to 100 × 40cm with feathery dark green leaves and flat-topped corymbs to 15cm diam., crowded with small, white to pink flowerheads; includes 'Burgundy' with wine red flowerheads, 'Fanal', with snow-white flowerheads in a dense, rosette-like corymb, 'Forncett Beauty' with deep lavender flowerheads, 'Fire King' with bright pink flowerheads, the magenta 'Kelwayi', and the snowy 'White Beauty'); *A.ptarmica* (SNEEZEWORT; erect herbaceous perennial to 100 × 75cm, leaves dark green, lanceolate and toothed but not divided; flowerheads to 0.5cm diam., white in frothy corymbs;. includes 'Boule de Neige' with large, snow white, 'double' flowerheads); *A.* Summer Pastels (a seed race of herbaceous perennials with soft, grey-green foliage and domed corymbs in shades of pink, rose, buff, salmon, dull orange, purple and grey-lavender); *A.*'Taygetea' (erect perennial to 60 × 50cm with feathery, grey-green leaves and flat corymbs of lemon yellow flowerheads).

achenes

Achimenes longiflora

Achimenes (from Greek *chemaino*, to suffer from cold). Gesneriaceae. HOT WATER PLANT. Central and South America. 35 species, tender, bushy perennials with softly hairy, bushy growth arising annually from underground rhizomes clad with fleshy scales. They are grown for their showy, tubular to funnel-shaped flowers. They may be induced to flower at any time of year. Plant the dormant rhizomes in a fibrous potting mix. Grow in humid conditions, minimum temperature 18°C/65°F, and protect from bright sunlight. Water sparingly at first, increasingly as the new growth emerges. Pinch out the young shoots to encourage bushiness and support if necessary. When in full growth, *Achimenes* benefits from a weak, fortnightly feed and frequent misting over (the latter especially if grown in the home). As the flowers fade and the foliage deteriorates, water should be gradually witheld and humidity reduced. Once the top growth has died off, the rhizomes may be stored completely dry and in their pots at a minimum temperature of 7°C/45°F. Separate and clean the maggot-like rhizomes prior to replanting. These are fragile and their inevitable break-up is the principal method of propagation.

A.antirrhina (W. Mexico; to 30cm; flowers 4.5cm long, yellow marked maroon on the exterior with a red-stained throat to bright red-orange with a yellow throat); *A.erecta* (syn. *A.coccinea, A.pulchella*; the West Indies, Mexico, Panama; flowers bright scarlet with yellow throats); *A.grandiflora* (Mexico, Honduras; 45cm; leaves 4.5cm long, mauve to maroon with a pink, purple-spotted throat); *A.longiflora* (Mexico, Honduras; to 60cm; flowers to 6cm, violet, maroon, mauve, magenta or white with various markings). *Cultivars*: 'Ambroise Verschaffelt' (flowers large, white, heavily veined and blotched deep purple); 'Brilliant' (flowers scarlet): 'Glacier' (flowers snow white tinted icy blue); 'Little Beauty' (flowers deep pink with a yellow eye); 'Paul Arnold' (flowers dark purple, throat white tinged yellow with red dots); 'Peach Blossom' (habit trailing, flowers peach) ; 'Pearly Queen' (flowers cream tinged pink and lilac); 'Ruby' (flowers small, ruby red); 'Show Off' (flowers profuse, lilac-pink).

achlamydeous of flowers; lacking a perianth.

acicular of leaves; needle-shaped, and usually rounded rather than flat in cross-section.

acid in gardening mainly applicable to soil, potting mixtures and their ingredients, and water having a pH scale value less than 7.0; cf. *alkaline*. See *soils*; *pH*.

acinaceous of fruit; full of kernels.

acinaciform of leaves; shaped like a scimitar.

Aciphylla (from Greek *akis*, point, and *phyllon*, leaf, referring to the sharply pointed leaves). Umbelliferae. New Zealand; Australia. 40 species, evergreen perennials grown for their rosettes of tough, sword-shaped or compound leaves and spectacular, terminal inflorescences – usually a white or cream, pyramidal panicle composed of many compound umbels packed on candelabra-like branches. All of the species described here hail from New Zealand. Male plants are usually held to flower more reliably and impressively than females. They thrive in a sharply draining, gritty soil in full sunlight. In winter they require temperatures no lower than –10°C/14°F and protection from overwet conditions (e.g. a mulch of dry bracken or deep gravel around the crown). Propagate by seed sown under glass in late summer or early spring.

A.aurea (dense rosettes of sharply pointed, sword-shaped or pinnate leaves to 1m diam.; flower spikes over 1m); *A.colensoi* (COLENSO'S SPANIARD, WILD SPANIARD; swirling rosette of leaves to 50cm-long, coarse, glaucous and rigidly sword-like or pinnate with the lobes sharply tipped and ribbed in sealing wax red; flowers cream in metre-high towers); *A.scott-thomsonii* (GIANT SPANIARD; differs from the last species in its more frequently divided leaves to 1.5m long, emerging bronze and hardening pale grey green with the midribs faintly tinted red; flower spikes to 3m); *A.squarrosa* (SPEARGRASS, BAYONET PLANT; rosette to 1m across of rigid olive green leaves, sword-like or 2–3-pinnate, each segment ending in a vicious spine; flower spikes ivory, to 1m).

Acmena (named for Acmene, a nymph). Myrtaceae. 7 species, evergreen trees and shrubs grown for their flowers and fruit. *A.smithii* (syn. *Eugenia smithii*), the Australian LILLY-PILLY, is a large shrub or small tree with flaky bark and glossy ovate-lanceolate leaves 2–l0cm long. Small green-white flowers smother the branch tips in summer. The berries are lcm diam., edible and globose, in varying shades of purple, white and pink. *Acmena* grows well in full sun on moist fertile soils (minimum temperature 5°C/40°F). Propagation as for *Acca*.

Aconitum (the Classical Latin name, from Greek, *akoniton*). Ranunculaceae. Northern Hemisphere (temperate regions). MONK'S-HOOD. Some 100 species, biennial or perennial herbs, often tuberous, with erect to scrambling stems, palmately lobed or cut leaves and racemes or panicles of flowers in summer. The flowers consist of five petal-like sepals, the upper three large and forming a domed helmet and extending behind as a spur. All parts of these plants are highly toxic. The following are fully hardy and suitable for herbaceous and mixed borders, woodland gardens, and (the smaller species) for the rock garden. Taller plants may require staking. Grow *A.hemsleyanum* among shrubs or train on pea-sticks. Plant all in a moist but well-drained, fertile soil in full sun or part shade. Propagate by division in autumn.

A.anthora (Galicia; compact, tuberous perennial, erect to 60cm; leaves dark green, finely cut into numerous narrow segments; flowers yellow or purple-blue); *A. × cammarum* (syn. *A.bicolor*; large group of popular garden hybrids involving *A.variegatum* and *A.napellus*; to 1m; leaves to 10cm across and finely divided;

Aconitum lycoctonum subsp. *vulparia*

flowers some 4cm long, with high, beaked helmets in open, narrowly pyramidal panicles; cultivars include 'Bicolor' with white and indigo flowers, the tall, slender, violet-blue 'Bressingham Spire', the dull salmon-pink 'Carneum', 'Grandiflorum Album' with large, white flowers, the deep blue 'Newry Blue', and the violet-blue 'Spark's Variety'); *A.carmichaelii* (Asia, North America; erect, tuberous perennial to 1.5m; leaves 3–5-parted, lobed, toothed; flowers deep purple to blue or white, in a dense, erect and narrow panicle; cultivars include the late summer- to autumn-flowering, azure blue 'Arendsii'); A. *hemsleyanum* (syn. A *volubile* of gardens; Asia; stems slender, twining and scrambling with much-lobed leaves and racemes of mauve or lilac to green and blue flowers); *A.lycoctonum* (WOLF'S BANE, BADGER'S BANE; Europe, N Africa; tuberous perennial to 1.5m, leaves dark green, 5–7-parted, lobed and toothed; flowers purple to lilac, white or yellow with domed to narrowly cylindric helmets, carried on slender-branched, narrow panicles; subsp. *vulparia*: syn. *A.vulparia*, flowers pale creamy yellow); *A.napellus* (HELMET FLOWER, FRIAR'S CAP, BEAR'S FOOT, GARDEN WOLF'S BANE; Europe, Asia and N America; erect, tuberous perennial to 1.5m; leaves 5–7-parted, lobed and toothed; flowers blue, violet or lilac with domed, hooded helmets in tall, spire-like panicles in late summer; 'Album': flowers white).

Acorus (Latin name, from the Greek *akoron*, applied in ancient times to both *Acorus calamus*, the sweet flag, and to the superficially similar *Iris pseudacorus*, yellow flag). Araceae (Acoraceae). Northern Hemisphere. 2 species, hardy rhizomatous perennial herbs with sword-shaped leaves and inconspicuous flowers in club-shaped spadices. Both are fully hardy and suited to the bog garden or pond margin. Variegated forms of *A.gramineus* have long been used as bonsai and, more recently, as house plants. Plant in a moist or saturated, rich soil, or pot in the same. The roots and rhizomes may be covered by water. Grow in full or dappled sun. Increase by division.

A.calamus (SWEET FLAG, SWEET CALAMUS, MYRTLE FLAG, FLAGROOT; northern Hemisphere, deciduous or semi-evergreen; leaves to 150cm resembling those of a flag iris, fresh apple green, flushed purple-bronze in 'Purpureus', striped cream and yellow in 'Variegatus'); *A.gramineus* (E Asia; evergreen, to 50cm tall; leaves grassy or sedge-like, glossy dark green, tapering finely, in a distinct forward-leaning fan; 'Albovariegatus': dwarf, leaves striped white; 'Oborozuki': leaves vibrant yellow; 'Ogon': leaves variegated chartreuse and cream; 'Pusillus': very dwarf – seldom exceeds 8cm; 'Variegatus': leaves striped cream and yellow).

Acroclinium (Greek a*kros*, top, and *klinein*, to slope). Compositae. Australia; South Africa. 90 species, annual and perennial herbs, subshrubs and shrubs. *A.roseum* (syn. *Helipterum roseum*) the PINK AND WHITE EVERLASTING, ROSY SUNRAY, or PINK PAPER DAISY, is native to Southwest Australia. An annual to 60cm tall, with branching erect stems and narrow leaves, it is grown for its long-stalked, daisy-like flowerheads to 5cm diameter. These are papery, single or double, and range in colour from pure white to deep rosy pink. Grow in full sun and provide plants with ample feed and water at first. The flowerheads, if cut when still scarcely open and hung to dry in a cool shady place, retain their colour and can be used in dried floral arrangements. Sow seed *in situ* in spring, or (in cold regions) start under glass in early spring.

acropetalous of organs such as leaves or flowers produced, developing or opening in succession from base to apex; cf. *basipetalous*.

Actaea (from Greek *aktea*, the elder tree, a reference to the fruit and leaf shape). Ranunculaceae. Northern Temperate regions (those below from North America). BANEBERRY. 8 species, hardy herbaceous perennials grown for their berries and attractive, foliage. Long-stalked, ternately compound leaves arise in early spring from a thick buried rhizome; their segments are lanceolate-ovate and usually coarsely toothed. A wand-like stem to 80cm tall follows in spring and terminates in a spike of small white or cream flowers. The fruit is bead-like and toxic – pearly white with a blue-black 'pupil' in *A.alba* (syn. *A.pachypoda*), the WHITE BANEBERRY or DOLL'S EYES; bright scarlet in *A.rubra*, the RED BANEBERRY or SNAKEBERRY. Plant in the wild or woodland garden in a humus-rich, moist soil in part-shade. Increase by division in spring, or from fresh seed (its pulp removed) sown in a cold frame.

Actaea alba

Actinidia (from the Greek, *aktis*, a ray, alluding to the radiating styles). Actinidiaceae. East Asia. 40 species, largely dioecious, deciduous, woody climbers grown for their fruit, foliage and small, bowl-shaped, cream flowers. Most are familiar as ornamentals in cool temperate gardens. Strongly twining climbers, they are used as wall plants or allowed to scramble through old trees and over pergolas, where they provide a bold covering of handsome leaves—to 15cm long, broadly ovate, tapering and deep, smooth green in *A.arguta*, the TARA VINE or YANG-TAO; to 20cm long, broadly elliptic, blunt, tough and hairy in *A.deliciosa* (syn. *A.chinensis*), the CHINESE GOOSEBERRY or KIWI FRUIT; to 15cm long, ovate, tapering finely, emerging chocolate and lime green splashed cream and pink in *A.kolomikta*; and to 12cm long, ovate-oblong, tapering, emerging bronze, becoming deep green flecked or mottled silver or cream in *A.polygama*, the SILVER VINE.

In sheltered positions, and especially where rapid growth and wood ripening can be guaranteed by long, hot summers, all of the above will withstand winter lows of –17°C/1°F. They grow well in part shade, but will require ample sunlight for fruit ripening. Plant in deep and well drained, loamy soils, rich in organic matter (slightly alkaline for *A.kolomikta*; slightly acidic for *A.deliciosa*). Prune in winter to clean out tangled, dead or exhausted stems. Propagate by semi-ripe cuttings taken in a case with bottom heat at 20–25°C/70–77°F; also by layering in winter.

Cultivars of *A.arguta* and *A.deliciosa* grown for fruit production are usually selected for sex (i.e. female) and grafted on to seedling rootstocks. A small, male scion may then be grafted on to the established female scion as an in-built pollinator. See *kiwi fruit*.

actinomorphic of regular flowers possessing radial symmetry, capable of division in two or more planes into similar halves.

activator a nitrogenous substance such as sulphate of ammonia, poultry manure or a proprietary product, applied to plant waste in a compost heap to aid decay. Of particular value where non-leafy material is composted during periods of low temperature. Also known as an accelerator

active ingredient (abbrev. **a.i.**) the biologically active component of a pesticide which is toxic to the pest, pathogen or weed it is intended to control.

aculeate prickly, bearing sharp prickles.

aculeiform prickle-shaped.

acuminate of leaves and perianth segments with the tip or, less commonly, the base tapering gradually to a point, usually with somewhat concave sides.

acutangular of stems; sharply angular.

acute of the tips or bases of leaves or perianth segments where two almost straight or slightly convex sides converge to terminate in a sharp point, the point shorter and usually broader than in an acuminate leaf tip.

acyclic plant parts arranged spirally, not in pairs or whorls.

Ada (named after the queen of Caria in Asia Minor). Orchidaceae. South America. 15 species, epiphytic orchids. Native to the Colombian Andes, *A.aurantiaca* is cool-growing (winter minimum 10°C/50°F) and produces arching, 30cm sprays of 2.5cm-long flowers in late winter and early spring. These are narrow-petalled and appear half-closed, ranging in colour from scarlet to dark or bright orange to cinnabar red. Pot in a bark-based orchid mix. Water and feed frequently when in growth; reduce water supplies once the new pseudobulbs are fully formed, only increasing quantities again upon the emergence of the flower spikes. Shade from direct sunlight and provide a humid, cool and buoyant atmosphere in summer. Increase by division.

adaxial of the surface or part of a lateral plant organ turned or facing toward the axis and apex; sometimes used interchangeably with ventral; cf. *abaxial*.

adelgids (Hemiptera: Adelgidae). Conifer woolly aphids. Sap-feeding insects of conifers, about 1–2mm long, similar to aphids but with shorter antennae, and without the horn-like structure at the base of the abdomen. Many species occur across the northern Hemisphere.

The SPRUCE GALL ADELGID (*Adelges abietis*) is widespread in Britain, northern Europe and North America and causes small pineapple-like galls on shoots of *Picea* species during the summer. Infested branches become stunted and the dried up galls are unsightly. Winged adults emerge from the galls, and fly to infest other spruce trees, depositing a cluster of 20–50 yellow eggs on the needles. After egg laying, females die in such a position that their wings provide protection for the eggs. Hatched nymphs crawl to the bases of buds to overwinter, and in spring pass through three moults to become stem mothers, producing copious amounts of flocculent, white 'wool' amongst which 50 or more greenish eggs are deposited. Nymphs emerge after bud burst and crawl between developing needles, where their feeding induces pineapple-like galls.

Adelges viridis gives rise to similar galls on *Picea* species but emergent winged forms overwinter on *Larix* species and reinfest spruce in the following spring. *Adelges abietis* causes serious dieback and even death of the silver fir (*Abies alba*) and *Adelges piceae* (the BALSAM WOOLLY APHID of North America), which has a more conspicuous covering of white woolly wax, attacks *Abies alba*, *A.grandis* and *A.procera*. The stunted shoots of affected trees have swellings often referred to as gouty galls. Tufts of white woolly wax are also associated with LARCH ADELGIDS (*Adelges laricis*), and to a lesser extent with *Adelges cooleyi* on Douglas fir (*Pseudotsuga menziesii*) and the SCOTS PINE ADELGID *Pineus pini*. The bark of Weymouth pine (*Pinus strobus*) is most likely to be infested by *Pineus strobi*. It is not feasible to control adelgids on large specimen trees, where damage is rarely serious. Infestations on young conifers, especially Norway spruce (*Picea abies*), which is extensively grown for Christmas trees in the UK, may be prevented by spraying with suitable contact insecticides in February-early March.

Adenium (from the Arabic name for this plant, *oddaejn* (Aden)). Apocynaceae. Throughout Africa in arid places. MOCK AZALEA; DESERT ROSE; IMPALA LILY; KUDU LILY; SABI STAR. 1 species, *A.obesum*; a tender succulent, it produces a swollen caudex dividing into a few crabbed branches between 1m and 3m tall. The leaves are oblanceolate to obovate and are sometimes present only during a short growing season. The flowers are funnel-shaped with five spreading lobes, as much as 5 × 6cm, and are typically rosy pink or crimson with a white 'eye' and throat. Provide a minimum temperature of 15°C/60°F (these plants in fact thrive in fierce, dry heat). Pot in a gritty and sandy medium. Water freely during periods of high temperatures, scarcely at all otherwise. Grow in full sun; keep free of draughts and damp. Increase by stem cuttings or seed.

Adenium
obesum

Adenophora (from the Greek *aden*, gland, and *phoros*, bearing – the style has a glandular base). Campanulaceae. Eastern Europe to Japan. LADY-BELLS; GLAND BELLFLOWER. 40 species, hardy herbaceous perennials. They differ from the closely related *Campanula* in their thick, fleshy roots and the swollen disk at the base of the stamens. Rosettes of ovate to lanceolate leaves give rise to arching sprays of pendulous, bell-shaped flowers in summer. Hardy to –15°C/5°F, they thrive in full sun or part shade. Grow on well-drained, fertile and preferably slightly alkaline soils. Propagate by fresh seed sown when ripe or from basal cuttings in spring.

A.bulleyana (W. China; to 1m; flowers pale blue); *A.liliifolia* (Eurasia; to 50cm tall, hung with fragrant, white or pale blue flowers); *A.potaninii* (W. China; to 90cm arching sprays of lavender flowers); *A.triphylla* (to 90cm; flowers pale blue or violet flowers).

adherent plant parts usually free or separate (e.g. petals) but clinging or held closely together. Such parts are sometimes loosely described as united or, inaccurately, as fused, which is strictly synonymous with coherent. Some authors use this word to describe the fusion of dissimilar parts.

Adiantum (from Greek *adiantos*, unwetted or unwettable: the fronds repel water). Adiantaceae. Cosmopolitan. MAIDENHAIR FERN. 200 species, evergreen and deciduous ferns with short, creeping rhizomes and clumps of fronds. These are usually 1–5-pinnately compound with slender, dark stipes, delicate branches and finely textured, wedge- to fan-shaped pinnules. *A.pedatum* and *A.venustum* are among the most desirable hardy ferns for the cool rock garden or shady border. Plant in a moist soil rich in leafmould. Mulch the crowns with garden compost and dry bracken in winter. The remaining species are tender foliage plants for the home, greenhouse or conservatory. With the exception of the warm-growing *A.peruvianum* and *A.raddianum*, they need a cool, buoyant atmosphere and a minimum temperature of 10°C/50°F. Plant in an acid to neutral, well-drained mix high in coir, composted bark and leafmould; water throughout the year, but more sparingly in winter, Remove faded fronds. Propagate by division when repotting, ideally in spring just before growth begins.

A.capillus-veneris (MAIDENHAIR FERN, SOUTHERN MAIDENHAIR, VENUS' HAIR FERN; cosmopolitan; fronds to 70cm, arching, triangular, 2–3-pinnate, pinnules 1–2cm, cuneate to flabellate, emerging pale bronze turning bright lime green; stipe dark brown to black); *A.pedatum* (FIVE-FINGERED MAIDENHAIR FERN; N America, E Asia; fronds to 50cm tall, slender, black stipes crowned with five pedately arranged pinnae, pinnules emerging bronze-pink turning lime green, to 2cm, obliquely triangular; more crowded and blue-green in var. *subpumilum* (var. *aleuticum*), ALEUTIAN MAIDENHAIR); *A.peruvianum* (SILVER DOLLAR MAIDENHAIR; S America; fronds to 1m tall, arch-

ing, triangular, 1–3-pinnate; pinnules to 7cm, diamond-shaped to semi-circular, emerging soft silver-rose, turning pearly grey-green); *A.raddianum* (syn. *A.cuneatum, A.decorum*; DELTA MAIDENHAIR FERN; S America; fronds to 60cm, narrowly triangular, 3-pinnate, pinnules 1cm, cuneate to rhombic, lobed, pale green on very dark stipes; includes 'Fritz Luthii' with pale green fronds, and the much-divided and tasselled 'Grandiceps'); *A.tenerum* (BRITTLE MAIDENHAIR FERN; Warm Americas; fronds to 100cm, arching, broadly triangular, 3–5-pinnate, pinnules rhombic to fan-shaped, sometimes toothed or cleft, stipes black and glossy; includes 'Farleyense' FARLEY MAIDENHAIR, BARBADOS MAIDENHAIR, GLORY FERN, with large, fan-shaped, cut and crisped pinnules); *A.venustum* (EVERGREEN MAIDENHAIR; Asia; fronds to 80cm, usually smaller, arching, broadly triangular, 3–4-pinnate, pinnules to 1cm, ovate-cuneate, tip rounded, finely toothed, coppery at first, becoming fresh pale green and paler beneath, stipes dark, glossy).

Adiantum pedatum

adjuvant a substance added to a pesticide formulation or mixture to improve its action, by enhancing spreading, adherence or penetration, or by reducing phytotoxicity. Adjuvants include so-called spreaders, wetters and stickers besides surfactants and some antitranspirants. Many are based on refined mineral oil or vegetable oil, and although not pesticides, are subject to pesticide regulations.

Adlumia (for John Adlum (1759–1836), US soldier and grape breeder). Fumariaceae. Eastern N America, Korea. 1 species, *A.fungosa* (syn. *Fumaria fungosa*), CLIMBING FUMITORY, MOUNTAIN FRINGE and ALLEGHENY VINE, an herbaceous, biennial vine grown for its finely cut, pale green, ferny foliage and drooping panicles of small, spurred flowers. Produced in spring and summer, these are close to those of *Dicentra* in shape and a delicate shade of pink. Hardy to –17°C/–1°F. Sow seed in early spring in light shade on a cool, moist soil. Allegheny vine makes a low, bushy plant in its first year; thereafter it quickly climbs to 4m. Provide initial support with brushwood or pea sticks, then allow the stems to scramble through the surrounding vegetation. Self-sows freely.

Adlumia fungosa

adnate a plant feature attached by its whole length or surface to the face of an organ.

Adonis (in Greek mythology, the beautiful youth beloved of Aphrodite and killed by a wild boar: this flower was supposed to spring from his spilt blood). Ranunculaceae. Europe, Asia. 20 species, hardy annual or perennial herbs grown for their flowers produced in spring and summer. The stems arise from short, thick rhizomes and bear finely cut foliage beneath a large, solitary, bowl-shaped flower composed of oblong to obovate petals around a boss of anthers. Plant in moist but well-drained soil in sun or light shade with the crowns at a depth of 2–3cm. Propagate by fresh seed sown in pans in the cold frame in summer.

A.amurensis (E Asia; perennial to 15cm; flowers to 4cm diam., with 20–30, yellow petals, sometimes white, rose, or red-striped; includes the fringed, creamy 'Benten', 'Fukurokuju' with very large, double, golden flowers, and the double, yellow to green 'Pleniflora', in 'Ramosa', the flowers are double and bronze-red to brown); *A.annua* (PHEASANT'S EYE; Europe to Asia; annual to 40cm; flowers to 3cm diam., with scarlet to crimson petals, their black bases forming an 'eye'; *A.aestivalis* is similar and includes forms with yellow flowers); *A.brevistyla* (China, Tibet; perennial to 20cm tall; flowers to 3cm diam., with some 8 white petals tinted blue to lavender on the exterior); *A.vernalis* (Europe; perennial to 20cm; flowers to 8cm diam., with 12–20 yellow petals, white in 'Alba').

adpressed, **appressed** used of an organ which lies flat and close to the stem or leaf to which it is attached.

Adromischus (from Greek, *hadros*, thick, and *mischos*, stem). Crassulaceae. South Africa. 26 species, low-growing, frost-tender, succulent herbs and subshrubs grown for their attractive and often curiously shaped leaves. Plant in an open medium rich in grit and sand. When in growth, allow them to become just dry between waterings and feed with a fertilizer high in potassium and phosphorus but low in nitrogen. In the cooler, duller months, maintain a minimum temperature of 7°C/45°F, with low humidity and full sunlight; water only to prevent excessive shrivelling of the leaves. Propagate by leaf and stem cuttings (fallen leaves will often root and grow spontaneously).

A.cooperi (syn. *A.festivus*; PLOVER'S EGG PLANT; clump-former with thickly fleshy, 4–7cm-long leaves shaped like slightly compressed eggs, grey-green speckled red-brown); *A.cristatus* (leaves broadening to a squared tip and flattening toward a thin and strongly wavy apical margin); *A.maculatus* (leaves broadly oblanceolate to obovate to 10cm, rather thinner than in the above two species and grey-green to olive or dull brown with purple-brown blotches, the tip thin, horny-textured and usually wavy).

adsorption the attachment of an ion or molecule to the surface of a soil particle. See *base exchange*.

adult of leaves; the mature foliage of plants which bear leaves of distinctly different shape when young, such as eucalyptus, ivy and gorse; cf. *juvenile*.

adventitious, adventive growth which occurs at an unusual location on a plant, for example roots arising from a stem or leaf axil. Also used of non-native plants introduced accidentally or deliberately.

Aechmea (from the Greek, *aichme*, lance head, referring to the hard acute apices of the sepals) Bromeliaceae. C and S America. 170 species, tender, epiphytic, evergreen perennial herbs. Tough, strap-shaped leaves are carried in a stemless rosette, their overlapping and expanded bases forming a cup or funnel in which water collects. From this reservoir, the inflorescence emerges – a terminal spike or panicle clothed with showy bracts. Although small and often concealed, the flowers are usually brightly coloured.

Provide a minimum winter temperature of 10°C/50°F, and bright, filtered light. Plant in a very free-draining, bark-based mix which should be allowed to become almost dry between waterings. Alternatively, wrap the sparse and wiry roots in a pouch of well-rotted garden compost and sphagnum moss and attach the plant to driftwood, a bromeliad 'tree' or a bark slab; water by plunging and misting. Apply a weak liquid feed fortnightly in spring and summer. Propagate by detaching roots and offsets.

A.chantinii (AMAZON ZEBRA PLANT; leaves spiny-toothed, olive green brightly banded silver-grey beneath in an erect funnel; inflorescence 80cm tall, loosely branched and clothed with brilliant scarlet bracts, lowermost bent downwards; flowers waxy and orange-yellow); *A.distichantha* (robust to 1m with spreading, mid-green, saw-toothed leaves; inflorescence stout-stemmed, roughly pyramidal; branches short, covered in 2-ranked, overlapping, rose pink to lobster red bracts); *A.fasciata* (URN PLANT; funnel spreading with broadly strap-shaped leaves in sea green overlaid and banded with frosty scales; inflorescence 60cm tall, pyramidal and composed of many overlapping bracts, like a candy or flamingo pink torch, picked out with small flowers, shifting with age from pale blue to indigo then dark rose); *A.* Foster's Favorite Group (LACQUERED WINE CUP; vigorously offsetting, to 60×80cm with open funnels of glossy garnet leaves and pendulous spikes of small red and blue flowers followed by red berries); *A.fulgens* (CORAL BERRY; an open but deep rosette of mid-green, 35cm-long, oblong to ligulate leaves, glossy above, with grey scales beneath; inflorescence 15–20cm tall, branches slender, red clothed with thin, pink bracts; flowers small blue and violet turning deep pink and followed by bright red fruit); *A.nudicaulis* (rosette loose; leaves 30–80cm-long; black-toothed, strap-shaped and arching from a dark, urn-like base, olive above grey-banded beneath; inflorescence unbranched, to 25cm tall and cylinder-shaped with yellow flowers borne above brilliant red bracts); *A.racinae* (CHRISTMAS JEWELS; differs from *A.*'Foster's Favorite' in its paler, red-bronze leaves, red and yellow flowers and vivid ruby red fruit); *A.recurvata* (swollen-based, compact rosette of spiny-toothed, arching, narrowly triangular leaves, mid-green, sometimes flushing red toward their bases at flowering; inflorescence beacon-like, scarcely clearing the leaves and packed with bright red bracts and small, purple, pink or red and white flowers).

Aeonium (a plant name used by Dioscorides). Crassulaceae. Canaries, Madeira, North and East Africa (all those below from Atlantic Islands). 30 species, mostly perennial, tender shrubs and subshrubs with strap- to wedge-shaped, succulent leaves

Aechmea fasciata

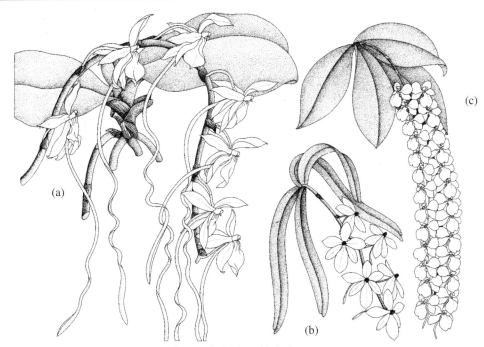

Aerangis (a) *A. kotschyana* (b) *A. luteoalba* var. *rhodosticta* (c) *A. citrata*

in wheel-like rosettes and small, yellow flowers packed in cymose, terminal inflorescences. Provide full sunlight and a gritty, loam-based medium that dries out between waterings. If quite dry, these plants may survive short, light frosts and will certainly tolerate a minimum of 5°C/40°F. Propagate by stem-tip or leaf cuttings.

A.arboreum (stoutly branched, tree-like perennial to 2m tall, each branch terminating in a loose rosette of leaves to 15cm long, obovate to oblanceolate, glossy, ciliate and mid-green, dark purple in 'Atropurpureum'; purple-black fading at the base to jade green in 'Schwarzkopf'; variegated white and cream in 'Variegatum'); *A.canariense* (CANARY ISLAND AEONIUM, GIANT VELVET ROSE; stemless with a short, unbranched stem; leaf rosette densely crowded, somewhat cupped/to flat leaves oblanceolate to 20cm long, mid-green, dewy with tiny, slightly sticky hairs); *A.haworthii* (PINWHEEL; bushy, to 1m tall with loose rosettes of obovate, acute, 5cm-long leaves, usually glaucous with red margins); *A.tabuliforme* (leaf rosettes to 40cm across, stemless, composed of many, overlapping, petal-like leaves, thinly succulent, obovate, acute, apple-green and ciliate).

Aerangis (from the Greek *aer*, air, and *angos*, vessel). Orchidaceae. Africa, Madagascar, Mascarene Islands, Sri Lanka. 50 species, epiphytic and lithophytic orchids with strap-shaped or elliptic leaves in a loose fan or in two ranks along elongated stems. White or ivory, moth-like flowers with long spurs are carried in arching racemes. Provide a minimum winter temperature of 10°C/50°F and semi-shaded, humid and buoyant conditions. Pot in a very open, bark-based orchid mix, or attach the plant to a pad of sphagnum moss and garden compost and tie it to

a suspended raft, cork block or fibre slab. Syringe daily during warm weather.

A.citrata (Madagascar; short-stemmed with broad, 10cm-long leaves and pendulous 25cm long racemes packed with rounded 2cm-diam. flowers, deep cream to very pale primrose and faintly lemon-scented); *A.ellisii* (Madagascar; large plant with distinct stems to 1m long; flowers pure white, nocturnally fragrant in two ranks along a horizontal raceme some 40cm long, each flower with a spur 15cm long pointing straight downwards); *A.fastuosa* (Madagascar; dwarf, very short-stemmed and broad-leaved, with 2–6cm-long racemes of sparkling white 5cm-wide flowers); *A.kotschyana* (Tropical Africa; medium-sized, short-stemmed plant with arching 40cm-long sprays of 5cm-wide, white flowers subtly tinted salmon pink with spiralling spurs to 20cm long) *A.luteoalba* (Ethiopia to Angola; dwarf, with a few, strap-shaped leaves to 10cm long – usually shorter – and arching racemes of disproportionately large flowers; var. *rhodosticta*: flowers with a red column).

aeration (1) the exchange of oxygen and carbon dioxide between the atmosphere and soil, which is dependent upon adequate soil pore spaces; cultivation, organic matter content, drainage, plant root and earthworm activity are all important influences. (2) the techniques used to relieve compaction, particularly on lawns. See *slitting*; *spiking* (q.v.).

aerator a specially designed hand-operated or mechanical device used to aerate soil or lawns by *slitting* or *spiking* (q.v.).

aerial roots roots borne wholly above ground, either adventitiously, as in *Hedera*, or from the rooting axis, as in many epiphytes.

Aerides odorata

Aerides (from Greek *aer,* air, due to their apparent ability to thrive on air alone). Orchidaceae. Tropical and subtropical E Asia. 40 species, tender epiphytic orchids grown for their pendulous racemes of fragrant, spurred flowers. They are tall with a single erect stem clothed with 2-ranked, strap-shaped leaves and giving rise to long, aerial roots. Most often grown is *A.odorata*, its flowers to 3cm wide and ranging in colour from pure white to mauve, but typically sparkling white or cream faintly and minutely spotted rose-purple on the lip and tipped magenta to purple on the tepals and spur. Provide a minimum temperature of 10°C/50°F. Plant in a coarse, bark-based mix in orchid pots or baskets; suspend in a brightly lit, well ventilated position with high humidity. Mist daily from late spring to autumn and water twice weekly by plunging. Reduce water in winter. Propagate by detaching rooted offsets or by cutting long and old stems into rooted sections when repotting.

aerobic used of conditions in which oxygen is available; or of organisms, especially bacteria, requiring oxygen for activity or life. Commonly met with in descriptions of managing a compost heap where regular turning encourages aerobic conditions and yields a superior product more rapidly than by anaerobic means; cf. *anaerobic*. See *compost heap*.

aeroplane house a type of greenhouse mainly used for commercial cropping, superseded by modern design. Of wooden construction, often on a low brick wall, 4.3–4.6m wide, with gutters 1.8–2m above ground level and a ridge height of approximately 3.9m.

Aeschynanthus (from Greek *aischyne*, shame, and *anthos*, flower, referring to the red flowers). Gesneriaceae. Indomalaysia. 100 species, tender, evergreen herbs and subshrubs, usually trailing or clambering with thinly leathery leaves and showy tubular flowers, the upper lip 2-lobed, and the lower lip 3-lobed. Fine plants for baskets and hanging pots in the home, greenhouses and conservatories; minimum temperature 18°C/64°F. Provide bright indirect light and medium to high humidity. Plant in a free-draining, fertile mix high in leafmould and composted bark. Keep moist throughout the year except during cool, dull weather when the plants should be kept rather dry. Root semi-ripe cuttings in spring and summer in a closed case with bottom heat.

A.marmoratus (syn. *A.zebrinus*; shrubby trailer with stems to 1m, leaves pale green with purple marbling, and stained purple beneath; flowers 3cm long, olive green flecked maroon); *A.pulcher* (ROYAL RED BUGLER; trailer with slender, cascading stems to 1m long; flowers deep scarlet with a yellow throat, 6cm long, calyx smooth, yellow-green flushed red above); *A.radicans* (LIPSTICK PLANT; trailer with cascading stems to 1m long, flowers deep scarlet, blotched yellow in throat, 7cm long emerging from

green to maroon calyx cups; *A.lobbianus* differs in its strongly purple-tinted stems, grey-green leaves and deep maroon calyx cups; the flowers are streaked maroon within); *A.speciosus* (syn. *A.splendens*; robust shrubby scrambler to 1.5m tall; flowers to 7cm, fiery orange-red above, orange-yellow below, calyx green); *A.* × *splendidus* (*A.parasiticus* × *A.speciosus*; differs from *A.speciosus* in its flowers marked with dark maroon).

Aesculus (Classical Latin name for an oak with edible acorns, applied to this genus by Linnaeus). Hippocastanaceae. N America; Europe; Asia. BUCKEYE; HORSE CHESTNUT. 15 species, hardy deciduous trees and shrubs with large, palmately compound leaves and showy flowers in slender or broad, terminal panicles and racemes in late spring and summer. The seeds are large and woody with a paler hilum and enclosed in a more or less spherical case with a thick, spiny, bumpy or smooth skin. Plant in full sun or light shade in a fertile, well-drained soil. Propagate by seed sown in autumn, or by grafting in late winter; also by budding in late summer. Physiological scorching of the foliage sometimes occurs in hot, dry summers – this can be alleviated by pruning to reduce total leaf area, and by irrigation. Some US species lose their leaves early in the season (i.e. mid- to late summer).

A.californica (CALIFORNIA BUCKEYE; California; small tree or spreading shrub to 12m tall, often multi-stemmed, with grey bark; glossy dark green leaflets, narrowly oblong, serrate; flowers in erect plumes, white to cream or pink-tinted with rose-mauve filaments, fragrant); *A.* × *carnea* (*A.hippocastanum* × *A.pavia*; RED HORSE CHESTNUT; broad-crowned tree to 25m tall; leaflets dark green and more or less smooth, toothed, oblong to obovate, often rather puckered and misshapen; flowers in loosely pyramidal panicles, rose to coral red or scarlet blotched orange-yellow; includes 'Briottii' has pale blood red flowers in large panicles); *A.chinensis* (CHINESE HORSE CHESTNUT; China; slow-growing, spreading tree to 30m tall; leaflets large, glossy dark green, toothed, oblong to obovate; flowers in erect spires, white); *A.flava* (syn. *A.octandra*; YELLOW BUCKEYE, SWEET BUCKEYE; North America; broad-crowned tree to 30m tall; leaflets glossy dark green turning scarlet in autumn, obovate; flowers yellow); *A.glabra* (OHIO BUCKEYE; C and E US; broad-crowned tree or large shrub to 30m tall; leaflets dark green and more or less smooth obovate to elliptic; flowers yellow-green with orange anthers); *A.hippocastanum* (HORSE CHESTNUT; Balkans; broad-crowned tree to 30m tall; leaflets mid- to dark green, coarse, with impressed veins, obovate; flowers in large, pyramidal panicles, white blotched rose or yellow, speckled red and fringed, spring; 'Baumannii': a vigorous slection with double, sterile flowers); *A.indica* (INDIAN HORSE CHESTNUT; Himalaya; broad-crowned tree to 30m tall; leaflets large, emerging bronze-pink, hardening smooth, dark green with paler undersides, broadly oblanceolate; flowers similar to *A.hippocastanum* but fewer on a longer, narrower panicle; includes 'Sydney Pearce' a vigorous selection with larger leaves, their petioles persistently red-tinted, and showy, pink-tinted flowers produced in great abundance); *A. × neglecta* (*A.flava* × *A.sylvatica*; SUNRISE HORSE CHESTNUT; SE US; spreading tree to 20m tall; leaflets emerging pink, hardening pale green, obovate; flowers yellow suffused red, early summer; 'Erythroblastos' has coral red new growth, amber autumn colour, and peach to pink flowers); *A.parviflora* (SE US; freely suckering, spreading shrub or low-crowned and multi-stemmed tree to 5m tall; leaflets large, emerging bronze, hardening dark green, smooth obovate; flowers in long, narrowly conical panicles, white with showy, pink-tinted stamens); *A.pavia* (RED BUCKEYE; N America; broad-crowned tree or large shrub to 5m tall; leaflets dark green and glossy, obovate, toothed; flowers carmine to red marked yellow, dark red in 'Atrosanguinea'; *A.turbinata* (JAPANESE HORSE CHESTNUT; Japan; broad-crowned tree to 30m tall; leaflets large, dark green with paler undersides obovate, toothed; flowers creamy white spotted red).

aestivation the arrangement of floral parts within the bud before flowering.

Aethionema (from the Greek, *aitho*, I scorch, and *nema*, thread or filament). Cruciferae. Europe, Mediterranean, W Asia. STONE CRESS. 40 species, short-lived, semi-evergreen or evergreen perennial shrublets usually with small and narrow, slightly fleshy leaves and dense, terminal racemes of 4-petalled flowers produced in spring and summer. Most are suited to the rock garden, although *A.grandiflorum* (the largest of the genus) will grace the foreground of herbaceous borders. Plant on a light, free-draining and somewhat alkaline soil in full sun. Hardy to –15°C/5°F. Take softwood cuttings in spring, or sow seed in autumn (with many, self-seeding is common).

A.armenum (Armenia; tufted shrublet 20cm; leaves glaucous; flowers small, pale pink; 'Warley Rose': very compact, leaves blue-grey, flowers rose pink; 'Warley Ruber': flowers deep rosy red); *A.grandiflorum* (syn. *A.pulchellum*; Iran, Iraq; PERSIAN STONE CRESS; loose subshrub to 45cm; leaves glaucous; flowers pale rose); *A.iberideum* (E Mediterranean to the Caucasus; compact subshrub to 15cm; leaves grey-green; flowers fragrant, white); *A.schistosum* (Turkey; erect to 10cm; stems densely branched; leaves leaden grey; flowers fragrant, rose-pink).

aetiology see *etiology*.

afoliate leafless.

agamospermy the development of seed without fertilization. A plant producing seed in this way is called an apomict.

Agapanthus (from Greek, *agape*, love, and *anthos*, flower). Liliaceae (Alliaceae). South Africa. AFRICAN LILY; BLUE AFRICAN LILY; LILY OF THE NILE. 10 species, evergreen or deciduous, perennial herbs with short, thick stems, fleshy roots and strap-shaped leaves. Bell- to trumpet-shaped, the flowers consist of six tepals that meet in a tube below and are outspread at their tips. They are slender-stalked and carried in summer in scapose, rounded umbels. Handsome free-flowering perennials suitable for herbaceous borders and containers, they are hardy in zones 6 and over. Grow in full or dappled sun; keep moist in spring and summer, drier in winter, when the crowns benefit from a dry mulch. Propagate by division in spring, but disturb as seldom as possible.

A.africanus (AFRICAN LILY; BLUE AFRICAN LILY; LILY OF THE NILE; to 60cm, flowers tubular, to 5cm long, deep violet blue; includes the small, white 'Albus Nanus', and the dark blue 'Sapphire'); *A.campanulatus* (to 100cm, flowers campanulate, to 3.5cm long, blue to white; includes the white var. *albidus* (Albus), the lavender-blue, large 'Isis', 'Profusion', free-flowering with pale blue flowers with darker stripes, and 'Variegatus' with cream-striped leaves); *A.inapertus* (to 180cm, flowers tubular, to 5cm long, bright blue to violet; includes the pale cream 'Albus'); *A.praecox* (to 100cm tall, flowers tubular to funnel-shaped, to 7cm long, bright to pale blue or white; subsp. *orientalis*: syn. *A.orientalis*, *A.umbellatus*, smaller and clump-forming with stout scapes and pale blue to indigo flowers to 5cm long; includes the white-flowered 'Albus', dwarf 'Nanus', and 'Variegatus' with white-striped leaves). *Cultivars* 'Alice Gloucester' (flowers white); 'Blue Moon' (flowers pale blue), the Headbourne Hybrids (to 1m tall, free-flowering and hardy with flowers in shades of blue), 'Lilliput' (to 45cm, small and fine with dark blue flowers), 'Loch Hope' (to 120cm tall, flowers large, dark blue, crowded in broad umbels), 'Peter Pan' (to 30cm, dwarf and free-flowering with mid-blue flowers), and 'Rancho White' (30–50cm tall, flowers white).

Agapetes (from Greek, *agapetos*, lovely or lovable, referring to the flowers). Ericaceae. East Asia, Pacific, Northern Australia. 95 species, evergreen, semi-scandent shrubs, many of them epiphytes with slender, arching branches and waxy, tubular to urn-shaped flowers, often beautifully marked and hanging from the axils of leathery leaves. Grow in lightly shaded and buoyant, humid conditions, with a minimum temperature of 5°C/40°F. Plant in small containers in an acid mix rich in coarse bark and leafmould. Water and syringe plentifully throughout the growing season, rather less as the days shorten and the flowers develop and open.

A.macrantha (NE India; gaunt, arching shrub to 2m; leaves to 12cm, laurel-like; flowers 5cm, pale candy pink, with dark red chevron markings); *A.rugosa* (syn. *Pentapterygium rugosum*; Khasi Hills; to 1m with rugose leaves to 10cm long; flowers white with purple marbling or red banding); *A.serpens*

Agapanthus praecox subsp. *orientalis*

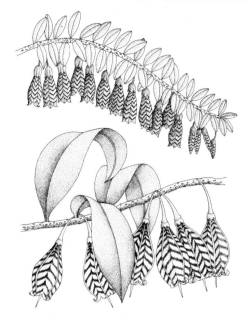

Agapetes (top) *A. serpens* (bottom) *A. macrantha*

(Nepal, Bhutan and North Assam; to 1m; leaves small, crowded; flowers 2cm long, bright scarlet or fiery red, ivory in 'Nepal Cream', with dark crimson or blood red chevrons; this species has been crossed with *A.*'rugosa* to produce *A.*'Ludgvan Cross', intermediate between the parents in habit and leaf form, with white-marked, carmine flowers).

Agastache (from Greek, *aga*, very much, and *stachys*, spike, referring to the abundance of flower spikes). Labiatae. N America; E Asia. GIANT HYSSOP; MEXICAN HYSSOP. 20 species, aromatic perennial herbs with clumped stems, toothed leaves and tubular, 2-lipped flowers whorled in spikes in summer. Hardy in zone 8. Plant in full sun in a warm, sheltered situation on a free-draining soil. Otherwise grow in the cool greenhouse or conservatory. Increase by cuttings in late summer.

A.barberi (Arizona to Mexico; to 60cm; leaves zestily aromatic, grey-green, toothed, to 5cm; flowers rose to purple, but usually glowing coppery orange with an undertone of pink); *A.mexicana* (syn. *Brittonastrum mexicanum, Cedronella mexicana*; Mexico; to 60cm, leaves toothed, ovate to lanceolate to 6cm; flowers rose to crimson).

Agave (from Greek, *agauos*, admirable, referring to the handsome appearance of the plants in flower). Agavaceae. Americas (those below from SW US to Mexico). CENTURY PLANT. Some 300 species, evergreen perennial herbs with rosettes of succulent leaves and erect spikes or candelabra-like panicles of tubular, cream to yellow-green flowers. Robust succulents grown for their handsome foliage and stately habit. The smaller species usually reach flowering size within a decade; larger species may require longer. Leaf rosettes tend to die after flowering. All require full sun and a fast-draining, gritty soil. Most

Agave parryi

need a minimum temperature of 7°C/45°F, although *A.americana* will tolerate lower temperatures and even light and short-lived frosts if dry. Propagate by basal offsets or by plantlets, which sometimes form on spent inflorescences.

A.americana (CENTURY PLANT; giant rosette of erect to spreading leaves each to 2m long, rigid, lanceolate, grey-green with hooked teeth and terminating in a sharp, dark spine; inflorescence candelabra-like, yellow-green and 4–10m tall; cultivars with white- and yellow-striped leaves are available); *A.attenuata* (clump of 1m-long, ovate to oblong leaves, grey-green with smooth margins; after some years, a short, stout trunk develops; flowers yellow-green, in arching racemes to 3m); *A.filifera* (compact with rosettes to 60cm wide; leaves narrow, rigid, dark green, marked with chalky white lines and edged with white fibres; 'Compacta': leaves crowded, to 10cm); *A.parryi* (low, compact and rounded rosette of broadly obovate leaves to 40cm long, glaucous blue-grey, with short, dark teeth and tipped with a brown spine); *A.victoriae-reginae* (ROYAL AGAVE; rosettes slow-growing to 60cm across, dense and domed; leaves tongue-shaped, dark green with smooth white margins and chalky superficial lines).

ageotropic applied to plant parts that are negatively geotropic, i.e. growing upwards against the influence of gravity as in the knee roots of the swamp cypress (*Taxodium distichum*). Also known as apogeotropic.

Ageratum (from Greek, *a-*, without, and *geras*, age, the florets retain their colour for a long time). Compositae. Tropical America. 43 species, annual and perennial herbs with button-like flowerheads. *A.houstonianum* is widely grown as a half-hardy annual suitable for bedding and containers. It is a bushy annual, 8–60cm tall with rounded, hairy leaves and a broad cap of tightly packed, powder-puff-like flowerheads in shades of blue, lilac and lavender. Cultivars include the taller 'Blue Horizon' and 'Blue Bouquet', the low, hummock-forming 'Blue Blazer', and 'Blue Danube', the pink 'Bengali' and 'Pinkie', and 'White Cushion'. Sow seed under glass in late winter; plant out in late spring, or sow seed *in situ* in late spring. Grow in a fertile, moist but free-draining soil in sun. Pinch out faded flowerheads regularly.

agglomerate crowded together in a head, as in the flowers of *Scabiosa* species.

aggregate (1) sand, grit or gravel used as an ingredient of rooting or growing media, or as the basal rooting layer in ring culture of tomatoes; (2) sand, gravel, or ballast used with cement to make concrete; (3) used of soil particles, called crumbs or peds.

aggregate flowers flowers gathered together, usually in a tight rounded bunch or head as in the family Dipsacaceae without being capitulate (as in Compositae).

aggregate fruits a collection of separate carpels produced by a single flower and borne on a common receptacle, as in the fleshy 'fruit' of *Rubus*.

Aglaonema (from Greek, *aglaos*, bright, and *nema*, thread, referring to the bright stamens). Araceae. Tropical Asia. 21 species, tender perenial herbs, evergreen, with erect to creeping stems, variegated, oblong to lanceolate leaves and inconspicuous flowers arranged in a club-shaped spadix and enclosed in a green or white, cowl-like spathe. Resilient and popular house plants. Grow in filtered sun or light shade, with a minimum temperature of 10°C/50°F. Propagate by stem cuttings, offsets or division.

 A.commutatum (to 1m; leaves 13–30cm long, emerald green banded or spotted yellow-white; includes 'Malay Beauty' with leaves to 30cm long, dark green spattered grey-green and zoned white, 'Treubii' with narrow leaves mottled grey-green or silver, and the heavily white-variegated 'White Rajah'); *A.nitidum* (to 1m; leaves to 45cm, unmarked or barred and blotched cream; the cultivar 'Curtisii' has leaves marked silver along the main veins; hybridized with *A.pictum* 'Tricolor' to produce 'Silver King' with large leaves heavily streaked silver-white, and 'Silver Queen' with narrower leaves marbled silvery white); *A.pictum* (30–50cm tall, with narrowly elliptic leaves to 16cm long, lustrous jade green blotched silver; includes 'Tricolor' with deep green leaves spattered silver and yellow-green).

AGM The Award of Garden Merit; conferred by The Royal Horticultural Society on plants judged to be of outstanding excellence for garden use. Criteria include retail availability of healthy stock. Awards are accompanied by a hardiness rating in the scale H1–H4, H4 being hardiest. The AGM was reconstituted in 1992 to replace the three-tier system of awards FCC, AM and HC established in 1921.

Agonis (from Greek, *agonos*, without angles, referring to the softly drooping and flowing branches of some species). Myrtaceae. Western Australia. 11 species, evergreen shrubs and small trees with fibrous bark, entire leaves and small, clustered flowers with five, rounded petals. Graceful trees for basically frost-free gardens or large, cool greenhouses and conservatories. Even if cut down by mild frosts, they will usually resprout. They tolerate all but wet and heavy soils and thrive in brilliantly sunny, rather dry conditions. Propagate as for *Eucalyptus*.

 A.flexuosa (WILLOW MYRTLE, WILLOW PEPPERMINT; to 10m; boughs snaking and arching, branchlets pendulous; leaves to 15cm long, peppermint-scented, linear silky at first; flowers white, 1cm-wide); *A.juniperina* (JUNIPER MYRTLE; more erect and shrubby than *A.flexuosa* with very slender, sharply pointed leaves).

Agrostemma (from Greek, *agros*, field, and *stemma*, garland). Caryophyllaceae. Mediterranean (naturalized Europe, N America). CORNCOCKLE. 4 species, slender annual herbs, erect and hairy with narrow leaves and broadly funnel-shaped, 5-petalled flowers in summer. Fully hardy. Plant in a sunny position on free-draining, rather poor soils. Deadhead regularly. Sow seed *in situ* in spring or late summer, or allow a few plants to set seed and naturalize themselves. *A.githago (*to 80cm; flowers white, rose or magenta; 'Milas': plum pink; 'Milas Rosea': pale lilac-pink).

Aichryson (Classical Greek name for a related plant). Crassulaceae. Atlantic Islands, Morocco, Portugal. 15 species, annual or short-lived perennial herbs or subshrubs with thinly fleshy leaves and small, star-shaped yellow flowers in panicles. Plant in a sharply draining, sandy medium, in full sun, with a minimum temperature of 7°C/45°F. Increase by stem cuttings. *A.* × *domesticum* (*A.tortuosum* × *A.punctatum*; YOUTH-AND-OLD-AGE. Canary Islands; subshrub to 30cm, leaves obovate to spathulate, to 1cm long; grey-green; 'Variegatum': leaves white-edged and solid creamy white on the same plant).

Ailanthus (Latinized form of the native name *ailanto*, meaning 'reaching for the sky', hence the popular name 'tree of heaven'). Simaroubaceae. E Asia to Australia. 5 species, deciduous shrubs and trees with pinnate leaves and small flowers in terminal panicles and followed in late summer by bunches of red-tinted, oblong samaras. A durable, very fast-growing and sometimes invasive tree. Plant on all but the driest or most saturated soils in zones 5 and over; provide some protection from harsh winds. Increase by rooted suckers and by seed sown in spring. Self-sown seedlings are common and may be a problem. *A.altissima* (TREE OF HEAVEN; China; to 30m; canopy broad, much-branched ; leaves to 1m long with red-tinted stalks and 15–30 mid green, lanceolate to ovate leaflets).

air-filled porosity (AFP) the percentage volume of potting compost which is occupied by air when the medium has been saturated and allowed to drain; a measure of aeration.

Ajuga (from Greek *a-*, not, and *zygon*, yoke: meaning obscure but perhaps to do with the calyx, which is not bilabiate). Labiatae. Europe; Asia. BUGLE. 40 species, annual or perennial herbs with low rosettes of obovate leaves, sometimes spreading by stolons, and giving rise in spring and summer to erect stems of tubular, 2-lipped flowers in whorls among leafy bracts. Excellent low ground-cover plants, fully hardy and thriving in sun, shade and a wide range of soils. Propagate by division or detaching rooted stolons.

 A.genevensis (BLUE BUGLE, UPRIGHT BUGLE; to 40cm tall, usually far shorter, not stoloniferous, with leaves to 12cm long; flowers bright blue, to 1.5cm long, amid violet to blue-tinted leafy bracts; 'Alba': flowers white; 'Rosea': flowers rosy pink; 'Tottenham': flowers lilac-pink; 'Variegata': leaves mottled creamy

aggregate fruits

Ajuga reptans

white); *A.pyramidalis* (PYRAMID BUGLE; creeping, stoloniferous dark green to 10cm long, flowers blue amid leafy, purple-tinted bracts in a pyramidal spike to 15cm tall; 'Metallica Crispa': leaves iridescent, bronze-maroon with crinkled margins); *A.reptans* (stoloniferous and carpeting leaves to 8cm long; flowers azure amid purple-blue-tinted bracts in spikes to 30cm tall; 'Atropurpurea'; leaves deep bronze-purple-tinted; 'Braunherz': lustrous dark, purple-black leaves, flowers deep blue; 'Burgundy Glow': leaves silvery variegated green cream, rose and deep red; 'Catlin's Giant': large, luxuriant purple-brown leaves, flowers blue on tall spikes; 'Jungle Beauty': leaves dark green to purple-tinted, wavy-margined; flowers indigo in spikes to 35cm; 'Multicolor': leaves dark green overlaid with red-bronze and splashed pink and cream; 'Rosea'; flowers pink).

Akebia (Latinized form of *akebi*, the Japanese name for these plants). Lardizabalaceae. China, Korea, Japan. 2 species, vigorous, woody perennial climbers with slender, twining stems, digitately compound leaves and unisexual flowers in short, drooping racemes. Plant in a moisture-retentive and well-drained soil. Grow in full sun or partial shade. Although hardy to –20°C/–4°F, they are sometimes cut down by hard frosts and late spring freezes will kill new growth. The flowers, also produced in spring, are usually unaffected. Increase by stratified seed or by stem cuttings under mist.

Akebia quinata

A.quinata (semi-evergreen or deciduous; leaves with five, oblong leaflets; flowers strongly vanilla-scented, deep plum purple to dark chocolate or maroon); *A.trifoliata* (syn. *A.lobata*; differs from the last species in being largely deciduous and possessing three leaflets per leaf their margins sometimes shallowly lobed rather than entire; flowers smaller, deep vinous purple, unscented); *A.* × *pentaphylla* (leaflets 4–5 (rarely 6 or 7) per leaf; flowers faintly scented).

alate winged; usually used of stems, petioles and fruits where the main body is furnished with marginal membranous bands.

Albizia (for Filippo degli Albizzi, naturalist who returned to Florence in 1749 from an expedition to Constantinople with many new seeds). Leguminosae. Tropics and subtropics. 150 species, deciduous to semi-evergreen trees, shrubs and lianes with feathery bipinnate leaves and mimosa-like flowerheads consisting of bundles of showy stamens in powder-puff-like balls or cylindrical spikes. *A.julibrissin* will grow outdoors in zones 6 and over provided the summers are long, hot and humid enough to ripen the new growth, or to spur a frosted stump into rapid resprouting. *A.lophantha* may survive outdoors against sheltered walls in zone 7, but is otherwise for summer bedding or the cool conservatory. Both flower from mid-spring to late summer and prefer full sun and a fertile, moist soil. Propagate by seed pre-soaked in tepid water for about twelve hours, or by semi-ripe cuttings; also by root cuttings in spring.

A.julibrissin (SILK TREE; Asia; domed or flat-crowned tree to 6m; flowerheads ivory to rosy pink, spherical, to 4cm wide); *A.lophanatha* (syn. *Paraserianthes lophantha*; Australia; tree to 4m; flower spikes to 6cm, cylindrical, lime green to gold).

Alcea (from Greek, *alkaia*, a kind of mallow). Malvaceae. Europe; Asia. HOLLYHOCK. 60 species, hardy hairy biennial or short-lived perennial herbs with broad, rounded to palmately lobed leaves and tall spikes of dish-shaped flowers with five, obovate to cuneate petals. *A.rosea* (syn. *Althaea rosea*) is a summer-flowering biennial, sometimes grown as an annual. An old favourite of cottage gardens, the hollyhock has been much-developed, resulting in cultivars that range in height from 1–3m, in flower form from the single to semi-doubles, full doubles and peony forms, some with ruffled or lacy petals. Colours run from white to darkest maroon-black through yellow, buff, pale rose, deep pink, and deep red. Full sun is best, together with shelter from strong winds. Sow seed *in situ* on well-drained soil in late summer or spring.

Alchemilla (from the Arabic name for this plant, *alkemelych*). Rosaceae. Cosmopolitan (chiefly Europe and Asia). LADY'S MANTLE. 300 species, hardy evergreen or semi-deciduous perennial herbs with clumps of slender-stalked, rounded leaves variously toothed, lobed or palmately cut. Produced in spring and summer, the flowers are very small, starry, green-yellow and carried in much-branched panicles. *A.mollis* tolerates sun or shade on all but the wettest or driest soils. It can be grown as a type of 'lawn', the foliage sheared regularly. *A.alpina* grows well in full sun on rock gardens and dry banks. Propagate by seed or by division in spring or autumn.

A.alpina (ALPINE LADY'S MANTLE; low mounds of small leaves with five to seven, toothed, lanceolate-obovate leaflets, dark green and smooth above, silvery-hairy beneath; in gardens, the name *A.alpina* is often misapplied to *A.conjuncta*, a larger plant with flowering stems to 30cm tall and leaves composed of 7–9, lanceolate to elliptic, toothed leaflets, these are smooth and mid-green above and silky-hairy beneath); *A.mollis* (LADY'S MANTLE; clump-forming perennial to 40 × 60cm with circular, toothed and shallowly lobed leaves to 15cm across; these are lime to sea-green and covered in soft, water-repellent down; the plant known as *A.mollis* 'Mr Poland's Variety', with more deeply lobed leaves that are hairless above is properly *A.venosa*).

algae primitive unicellular or multicellular photosynthetic plants living in water or moist environments; often a nuisance in ponds or irrigation systems or on damp surfaces.

algal spot a usually red-coloured spotting of foliage encountered in warm, wet regions, caused by algae, especially *Cephaleuros* species. In wet tropical climates or under protected cultivation with overhead watering, the surfaces of many plants are frequently colonized by superficial algae which are usually harmless except that they can reduce the amount of light available to the plant for photosynthesis. *Cephaleuros*, however, the usual cause of algal spot, is parasitic and grows within the tissues of its host plant, as well as on the surface in the form of raised orange or red hairy patches on foliage and shoots. It is sometimes confusingly referred to as 'red rust'. The diseases caused by *Cephaleuros* species can be unsightly on ornamentals such as *Anthurium*, *Camellia* and *Magnolia* and damaging to some crop plants including avocado, citrus, guava and tea. Generally, they are only a problem if other factors have weakened plants and predisposed them to infection. In serious cases control can be achieved by treatment with copper-containing preparations. Removal of infected leaves and twigs is advisable to reduce the overall level of inoculum.

algicide a chemical, such as dichlorophen, used to control algal diseases and weeds.

alginates seaweed derivatives; used as soil conditioners to enhance crumb stability and water and nutrient retention.

Alisma (Classical Greek name for this plant). Alismataceae. North temperate regions; Australia. 9 species, aquatic, perennial herbs making a clump of long-stalked leaves that remain partly submerged or stand clear of the water. In summer, many small, 3-petalled flowers are held above the plant on the slender, whorled branches of pyramidal panicles. WATER PLANTAIN, *A.plantago-aquatica*, is found in still or sluggish water and swamps throughout the Northern Hemisphere and Africa. The leaves are elliptic to ovate or lanceolate, ribbed, mid-green and up to 30cm long. The panicles stand 20–60cm high and swarm with white flowers on remotely whorled branches. Hardy in zone 4, used in ponds, lakes, ditches and streamsides planted in a loamy mix in submerged baskets, or directly into the banks and bed. Propagate by division, seed, or plantlets which form on the panicles.

alkaline in gardening, mainly applicable to soil and water having a pH scale value greater than 7.0; cf. *acid*. See *soils*; *pH*.

alkaloids nitrogenous compounds produced by plants, often as an end-product of nitrogen metabolism. Many have medicinal and/or toxic properties for example morphine, cocaine, nicotine, quinine, colchicine, strychnine.

Allamanda (for F. Allamand (1735–1795), Swiss physician and botanist who collected in Surinam and sent seed of this plant to Linnaeus). Apocynaceae. S America. 12 species, evergreen, usually scandent shrubs with leathery, oblong to lanceolate leaves and showy funnel-shaped flowers. Maintain a humid,

buoyant atmosphere; water and feed freely when in growth; keep barely moist in winter with a minimum temperature of 13°C/55°F. Shade only from the strongest summer sun. Tie in to support if necessary. Prune after flowering – this varies, but most plants in greenhouse cultivation flower from the late spring through to late winter. Propagate by soft or semi-ripe stem tip cuttings. *A.blanchetii* (syn. *A.purpurea*, *A.violacea*; PURPLE ALLAMANDA; erect or weakly climbing shrub with downy leaves and rich rosy-purple flowers to 9 × 6cm); *A.cathartica* (COMMON ALLAMANDA, GOLDEN TRUMPET; vigorous climber to 16m with glossy leaves and golden yellow flowers to 10 × 12cm with white markings in the throat; 'Grandiflora': flowers very large and freely produced; 'Hendersonii': flowers waxy, orange-yellow, tinged bronze on tube with white spots in throat; 'Nobilis': very large and robust with outsize pure gold flowers); *A.schottii* (syn. *A.neriifolia*; BUSH ALLAMANDA; resembles a more or less self-supporting *A.cathartica*, 1.5m; flowers, to 2.5 × 5.5cm, golden yellow strongly streaked or stained orange-red to chestnut brown within).

allée a walk or ride cut through dense woodland, or made by close plantings of shrubs and trees; a feature of formal French gardens since the 16th century.

allelopathy the usually harmful direct or indirect chemical effects of one plant on another. An example is the growth-inhibiting effect on other plants of toxic root exudate and leaf drip from the black walnut (*Juglans nigra*). Chemical diffusions from seeds or storage organs have been shown to be effective against micro-organisms. Allelopathic effect is difficult to confirm under natural conditions, and is of limited practical significance in gardens.

Allium (Latin, *allium*, the classical name for garlic, from the Celtic, *all*, meaning 'hot'). Liliaceae (Alliaceae). Northern Hemisphere. ONION. Some 700 species, bulbous perennial and biennial herbs with a characteristic 'onion' odour. Solitary or clustered bulbs produce leaves ranging in shape from grassy to narrowly tubular, to lanceolate, to elliptic or ovate. The flowers may be star-shaped or campanulate, consist of six tepals and are carried in scapose umbels usually in spring and summer.

The ornamental species are grown for their flowers and foliage, but *Allium* also numbers several important culinary herbs and vegetables – *A.cepa* (ONION, SHALLOTS), *A.fistulosum* (WELSH ONION, JAPANESE BUNCHING ONION, JAPANESE LEEK), *A.porrum* (LEEK), *A.sativum* (GARLIC), *A.schoenoprasum* (CHIVES), *A.scorodoprasum* (GIANT GARLIC, SPANISH GARLIC), *A.tuberosum* (CHINESE CHIVES, GARLIC CHIVES, ORIENTAL GARLIC).

The hardy, perennial ornamentals include tall species for the herbaceous and flower border, some of them excellent for cutting and drying, and many smaller species ideal for rock gardens, dry gardens, raised beds, pavement terraces and pot culture.

Allamanda cathartica

ALLIUM

Name	Distribution	Flowering time	Height	Leaves	Flowers
A.acuminatum	W N America	spring	10–30cm	2–4, linear, shorter than scape	deep rose pink to purple-pink or white, 10–30 per 4–6cm diam. umbel
A.aflatunense	C Asia	summer	80–150cm	6–8, 2–10cm across, ± glaucous, far shorter than scape	pale mauve to violet, small, in dense, 10cm diam. spherical umbels
Comments: Includes 'Purple Sensation' with intense violet umbels on 1m tall scapes.					
A.akaka	Turkey, Caucasus, Iran	spring	5–15cm	1–3, to 20 × 6cm, oblong-elliptic, grey- green, low-lying	white to pale lilac pink with red-tinted centre, very crowded in spherical 3–10cm diam., ± stalkless umbels
A.atropurpureum	S Europe	late spring–summer	4–100cm	3–7, linear, 15–35cm diam. umbel	dark purple to near-black in crowded heads, 3–7cm across
A.beesianum	W China	late summer	15–20cm	2–4, linear, 15–20cm, grey-green	bright to deep blue, 6–12, bell-shaped, nodding
Comments: Includes the white-flowered f. *album*					
A.caeruleum *A.azureum* BLUE ALLIUM	Siberia, Turkestan	summer	20–80cm	2–4, to 7cm, linear, 3-angled	cornflower blue, 30–50, small, packed in a 3–4cm diam. umbel on a slender scape
A.campanulatum	W US	summer	10–30cm	2–3, to 30cm, linear	pale pink to white, bell-shaped, 15–40 in a 2.5–7cm diam., hemispherical umbel
A.carinatum ssp. *pulchellum* *A.pulchellum*	C & S Europe; Russia; Turkey	summer	30–60cm	2-4, to 20cm, narrowly linear	purple, cup-shaped, to 30, nodding in a 2–5cm diam. umbel
Comments: The typical *A.carinatum* (KEELED GARLIC) differs in bearing bulbils in its umbels.					
A.cernuum LADY'S LEEK; NODDING ONION; WILD ONION	Canada to Mexico	summer	30–70cm	4–6, to 20 × 0.7cm, flat	deep pink to magenta, more rarely white or maroon, cup-shaped, nodding, 10–30 per lax, 3–5cm diam. umbel
A.christophii *A.albopilosum* STAR OF PERSIA	Iran, Turkey, C Asia	summer	15–50cm	2–7, 15–40 × 1–4cm, glaucous, downy, drooping	purple-violet, star-like, with metallic sheen, long- and slender-stalked, very many in a loose, spherical to hemispherical umbel to 20cm diam.
Comments: Inflorescences very attractive when dried. Includes 'Gladiator' (to 1.2m with large, lilac-mauve flowers), 'Globus' (*A.christophii* ×*A.giganteum*: to 35cm, flowers blue in large heads), 'Lucy Ball' (to 1.2m, flowers dark lilac in compact head), 'Rien Poortvliet' (to 1.2m, flowers lilac).					
A.cyaneum	China	summer	10–45cm	1–3, 15cm, thread-like, semi-cylindric	violet-blue to purple veined darker blue, bell-shaped, pendent, 5–18 in a small dense umbel
A.cyathophorum var. *farreri* *A.farreri*	China	summer	20–40cm	3–6, 18–24 × 0.1–0.5cm	maroon to dark mauve, bell-shaped, nodding, 6–30 in a loose umbel
A.flavum SMALL YELLOW ONION	C Europe to W Asia	summer	8–30cm	to 20cm, narrowly cylindric, glaucous	lemon-yellow, scented, bell-shaped, nodding, slender-stalked, 9–60 in hemispherical 1.5-3cm diam. umbel
Comments: Includes 'Blue Leaf' ('Glaucum') with strongly glaucous, blue-green leaves, and 'Minus' a smaller plant with contrasting purple stamens.					
A.giganteum GIANT ALLIUM	C Asia	summer	80–200cm	30–100 × 5–10cm, grey-green, basal	lilac, mauve or white, star-shaped, very many in a dense, 10–15cm diam. umbel
A.insubricum	N Italy	spring–summer	16–30cm	12–20 × 0.2–0.5cm, 3–4, flat, glaucous	purple-pink, bell-shaped, rather large and nodding, 3–5 per umbel
Comments: The very similar *A.narcissiflorum* differs in its fibrous bulb sheath.					
A.karataviense ALLIUM	C Asia Turkestan	late spring	10–25cm	2–3, 15–20cm, broad, oblong-elliptic to elliptic, glaucous, grey-blue tinted purple, low-lying	white to pale purple-pink, star-shaped, very numerous in a crowded spherical umbel to 20cm diam.; scape purple-tinted
A.macranthum	W China, Sikkim	summer	20–30cm	15–40 × 0.5cm, linear	deep purple, large, bell-shaped, long-stalked, to 20, nodding in a loose umbel
A.mairei	SW China	late summer to early autumn	10–40cm	to 25cm × 1cm, thread-like	white to pink, often spotted red, 2–20, bell-shaped, held erect in a narrow umbel on a wiry stalk
A.moly YELLOW ONION; LILY LEEK	S & SW Europe	summer	12–35cm	1–3, 20–30cm, lanceolate, glaucous	golden yellow, star-shaped, fairly large, 10–40 in a 4–7cm diam. umbel
Comments: Includes 'Jeannine', with bright yellow flowers in large umbels, 2 scapes produced per bulb					
A.narcissiflorum *A.pedemontanum*	N Italy, Portugal	summer	15–35cm	3–5, 9–18 × 0.2–0.6cm, flat, grey green	pure, bright pink to rosy purple, fairly large, bell-shaped, 5–8, nodding
Comments: See *A.insubricum*.					

ALLIUM

Name	Distribution	Flowering time	Height	Leaves	Flowers
A.neapolitanum *A.cowanii* DAFFODIL GARLIC; FLOWERING ONION; NAPLES GARLIC	Mediterranean, S Europe, Asia Minor, N Africa	spring	20–50cm	2, 8–35cm, linear-lanceolate	sparkling white, cup-or star-shaped, slender-stalked, small, to 40 per 5–11cm diam., open, hemispherical umbel
Comments: Includes 'Grandiflorum' with large, loose umbels of larger flowers, each with a dark eye.					
A.oreophilum *A.ostrowskianum*	Turkestan, Caucasus, C Asia	spring–summer	5–20cm	2, narrow	deep pink to rosy purple, broadly cup-shaped, proportionately large (to 2cm across), 10 or more in a loose, spherical umbel
Comments: Includes 'Zwanenberg' with deep carmine flowers.					
A.rosenbachianum	C Asia	late spring–summer	to 100cm	2–4, 1–5cm wide, strap-shaped	purple-pink to deep purple, star-shaped, small, to 50 crowded in a spherical umbel to 10cm diam.
Comments: Resembles a slightly smaller and earlier- flowering *A.giganteum*. Includes 'Album' (to 75cm, flowers white tinted green) and 'Purple King' (tall, flowers dark violet, stamens white).					
A.sativum GARLIC	wild origin unknown, widely cultivated and naturalized	spring–summer	25–100cm	strongly aromatic to 60cm, linear, flattened above	white to pink tinged purple-green, few in hemispherical 2.5–5cm umbels
Comments: Widely cultivated and long prized for its bulb, white or purple-skinned and consisting of bulblets (cloves). var. **ophioscorodon** has coiled stems.					
A.schoenoprasum CHIVES	Europe; Asia; N America; widely naturalized	spring–summer	10–60cm	strong onion-scented, 1–5, to 35cm, cylindric, hollow	white, pink-mauve or purple, small bell-shaped, crowded in 1.5–5cm diam. spherical umbels
Comments: Leaves are culinary chives, commonly grown in herb and cottage gardens. Variants include dwarf 'Schnittlauch' and curl-leaved 'Shepherd's Crook'.					
A.schubertii	E Mediterranean to C Asia	early summer	30–60cm	20–45cm, broadly linear to strap-shaped, wavy, glaucous	white tinted purple or wholly violet, small, narrowly star-shaped in spherical umbels to 5cm diam., but some on far longer, purple-tinted, radiating stalks
Comments: Remarkable for its purple flower heads with very unequal, spoke-like flower-stalks.					
A.scorodoprasum SAND LEEK; GIANT GARLIC; SPANISH GARLIC	E Europe; Caucasus; Turkey; N Iran	spring–summer	25–90cm	to 30cm, 2–5, linear	lilac to purple, small in hemispherical umbels, many replaced by purple bulbils
Comments: Culinary herb.					
A.senescens GERMAN GARLIC	Europe; N Asia	late spring–summer	7–60cm	4–30cm, 4–9, flat, narrowly strap-shaped	lilac, small, cup-shaped in a 2–5cm diam., hemispherical umbel
Comments: Includes var. **glaucum** with glaucous grey-green leaves growing in a low, swirling configuration.					
A.sikkimense *A.kansuense*	Himalaya; China; Tibet	summer	10–40cm	2–5, linear, flat, to 30cm	deep blue or purple, bell-shaped, to 15 in a small nodding umbel
A.sphaerocephalum DRUMSTICKS; ROUND-HEADED GARLIC; ROUND-HEADED LEEK	Europe; N Africa; W Asia	summer	to 90cm	2–6, 7–35cm, linear, cylindrical, hollow	purple-pink to maroon, small, bell-shaped, packed in a 2–3cm diam., spherical umbel atop a wiry scape
Comments: Flowers sometimes replaced by bulbils; white in sspp. **arvense** and **trachypus**.					
A.tanguticum LAVENDER GLOBE LILY	W China	summer	to 40cm	to 0.3cm wide, linear, flat	lavender with purple midveins to petals, many in a spherical umbel
A.unifolium	W US	early summer	to 60cm	solitary, grey-green, linear, flat	deep rosy pink, broadly star-shaped, 5–20 in a 5cm diam., hemispherical umbel

Species from the Western United States (e.g. *A. campanulatum*, *A. unifolium*) tend to be brightly coloured in shades of magenta, pink and red. They are suitable for hot, dry positions in well drained soils. The taller, Central Asian species (e.g. *A.christophii*, *A.giganteum*, *A.rosenbachianum*) usually produce large, spherical inflorescences atop long, leafless scapes as the basal foliage withers. They are outstanding planted among low-growing shrubs such as *Artemisia*, *Cistus*, *Lavandula*, *Salvia* and *Santolina*, which share their liking for arid, stony soils in full sun. To the fore, plant the TURKESTAN ALLIUM, *A.karataviense*, earlier-flowering with attractive, grey leaves and globes of purple to buff flowers. Smaller West Asian species (e.g. *A.carinatum* subsp. *pulchellum*, *A. flavum*) need a position at the front of the border. The smaller clump-formers are suitable for well-drained rock gardens in full sun, among them *A.mairei* and *A.oreophilum*. Himalayan natives like *A.beesianum* require rather damper, richer soils. Where the ground is heavy and winters wet and cold, some Californian species and dry-growing Asiatics

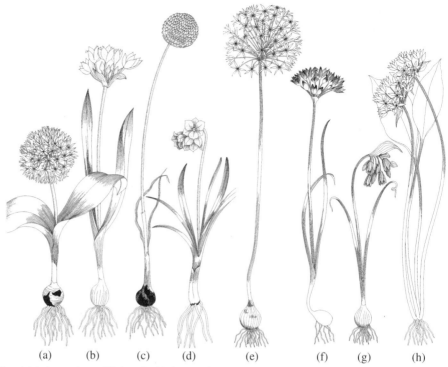

Allium (a) *A. karataviense* (b) *A. moly* (c) *A. scorodoprasum* (d) *A. insubricum* (e) *A. christophii* (f) *A. unifolium*
(g) *A. beesianum* (h) *A. ursinum*

almond

Pests
birds

Disease
peach leaf curl

like the spectacular *A.schubertii* are best grown in pans of gritty soil in full sun in a well-ventilated cool greenhouse, alpine house or bulb frame and kept dry in winter; otherwise, grow them in raised beds.

Plant bulbs in autumn or early spring at approximately one and a half times their depth. Clear the foliage once it has withered. If decorative, dead flower heads may be left until they break up. Propagate by offsets, removed when the plants are dormant, or by seed sown in spring at about 13°C/55°F. Divide spring-flowering clump-formers in summer, summer-flowering clump-formers in spring. Pests and diseases of *Allium* crops may also affect ornamentals. Pests include onion flies, stem eelworm and onion thrips. Diseases include bacterial soft rots, onion white rot, rust, smudge and smut.

allogamy see *cross-fertilization*.

allopolyploid a polyploid plant with sets of dissimilar chromosomes from parents of different species, and which is fertile as a result of a doubling of the parental chromosome numbers, for example swede (*Brassica napus* Napobrassica Group) (4×) derived from crossing turnip (*Brassica rapa* Rapifera Group) (2×) and rape (*Brassica napus*) (2×); cf. *autopolyploid*.

allotment (US, community garden) a piece of land let out by a public authority for cultivation by an individual or family. In the UK, usually a plot measuring 29 × 9.25m, contained within a group of

units. Their provision is catered for under numerous Allotment Acts.

almond (*Prunus dulcis*) a deciduous tree native of Asia Minor and parts of North Africa, closely allied to the peach. The almond is grown for its nuts or, in cool areas, the attractive pink blossom in spring. There are two distinct types: the sweet almond, whose nuts are prized for eating raw or cooked, and the bitter almond, which has a poisonous kernel and is used for flavouring and in medicines. In favourable situations some cultivars may reach a height of 10m. Its chilling requirement is 300–500 hours at temperatures below 7°C/45°F; early flowering predisposes the blossom and fruitlets to frost damage. Almond is insect pollinated and although most cultivars are self-fertile, cropping is improved where a pollinator is grown nearby. Propagation is usually by budding onto seedling almond or peach rootstock; cropping commences within three or four years. Almond is best trained and pruned as a bush peach, removing a proportion of the older wood in spring in favour of the production of fruit bearing one-year-old shoots.

Alnus (Classical Latin name). Betulaceae. Northern Hemisphere, south to Peru and Himalaya. ALDER. 35 species, hardy deciduous or semi-evergreen trees and large shrubs. The crown is typically ovoid to conical, the bark dark and sooty in appearance and the winter buds gummy. The leaves are ovate to orbicular and tough. The male flowers are small and carried in drooping catkins in winter and (more usually) early spring. Female flowers are borne in shorter, broader catkins that become small, woody 'cones'. Alders tolerate cold, wind and a wide range of soil types. All are able to thrive in wet, even waterlogged conditions, *A.glutinosa* especially, but *A.cordata* will also grow on rather dry, poor soils and is proving an excellent street tree. Sow seed in autumn, or propagate by budding in early autumn and cuttings in winter.

A.cordata (ITALIAN ALDER; Italy, Corsica; to 30m, crown ovoid to conic; shoots sticky, red-brown; leaves to 10cm long, broadly ovate to rounded, base cordate, margins finely toothed, leathery, deep, glossy green and smooth above, paler beneath, often persisting late into autumn; 'Purpurea': new growth bronze-purple); *A.glutinosa* (COMMON ALDER, EUROPEAN ALDER; Europe to C Asia, N Africa; to 20m, crown narrowly pyramidal; shoots smooth, sticky; leaves to 10cm long, obovate to rounded, base tapered, margins wavy to toothed, dark green and smooth; includes 'Aurea': new growth yellow ageing yellow-green; 'Imperialis': leaves deeply cut); *A.incana* (GREY ALDER; Europe; to 30m, crown conical; shoots grey-downy; leaves to 10cm long, ovate to oval, base rounded to cuneate, toothed, dull green above with corrugated veins, paler beneath with grey down; 'Aurea': shoots yellow becoming orange-red in winter, catkins are yellow-red; 'Ramulis Coccineis': winter shoots and buds red, catkins orange).

Alocasia (from Greek *a*-, without, and *Colocasia*, a genus from which this was split). Araceae. Southeast Asia. ELEPHANT'S EAR PLANT. 70 species, large, tender perennial herbs with tuberous or stoutly rhizomatous roots and long-stalked heart- to arrow-shaped leaves. The flowers are borne in dull, *Arum*-like inflorescences. Grow in a fertile medium rich in humus; water and feed frequently in spring and summer, providing high humidity and protection from full sunlight. Keep barely moist in winter (minimum temperature 10ºC/50ºF). Propagate by offsets, division of rhizomes or stem cuttings.

A.cuprea (leaves to 30cm, oblong to ovate, base cordate, green with a purple or coppery iridescent sheen between black-green zones around the sunken veins above, red-violet beneath); *A.macrorrhiza* (GIANT TARO; leaves to 125cm, ovate to cordate to sagittate, glossy green above with paler veins; widely grown for its edible rhizomes and shoots; ornamental cultivars include 'Variegata' with leaves blotched cream and dark green, and 'Violacea' with leaves tinged violet); *A.plumbea* (similar to *A.macrorrhiza* but with smaller leaves with wavy margins, dark green tinted purple to purple-red or violet-black with dark veins); *A.sanderiana* (KRIS PLANT; leaves to 40cm, sagittate with wavy margins, metallic violet-black or very dark green with silver-white margins and white veins); *A.veitchii* (leaves to 90cm, narrowly triangular to sagittate, dark green above, veined and edged grey, red-purple beneath; petiole with dark bands).

Aloe (from the Greek name, allied to Hebrew, *allal*, bitter). Liliaceae (Aloeaceae). Africa, Madagascar, Cape Verde Islands Some 325 species, evergreen perennial herbs, shrubs or trees with succulent leaves. The flowers are tubular and consist of six tepals; they are borne at various times of year in long-stalked, narrowly conical racemes or candelabra-like panicles. The larger species are fine plants for dry landscapes in frost-free gardens, or (where frosts are common) cool greenhouses and conservatories. The climbing *A.ciliaris* may be used to ornament pergolas and arbours. Several smaller species, notably *A.aristata* and *A.variegata*, are popular and resilient house plants, and *A.bakeri* might equally join their ranks. All prefer full sun, a minimum temperature of 7ºC/45ºF, a dry, buoyant atmosphere and a loamy soil high in grit and sand. Water liberally from late spring to early autumn, allowing the soil to dry out between waterings. At other times keep these plants more or less dry, watering only to prevent shrivelling of the leaves. Propagate by seed, stem cuttings or offsets.

Aloe variegata

| **ALOE** | | | | |
Name	Distribution	Habit	Leaves	Flowers
A.arborescens CANDELABRA ALOE; OCTOPUS PLANT; TORCH PLANT; TREE ALOE	S Africa; Zimbabwe; Mozambique; Malawi	erect shrub or tree to 6m tall with a stout trunk and a spreading canopy of candelabriform branches	in terminal rosettes, to 60cm long, grey-green and finely tapering with spiny margins	to 4cm long, scarlet tipped green-white on an unbranched spike to 1m tall in late winter and summer
Comments: Large shrub for virtually frost-free gardens or conservatories. Withstands drought, salt spray and very mild frosts. Includes 'Spineless' with grey-blue, spineless leaves, and 'Variegata' with leaves striped cream.				
A.aristata TORCH PLANT; LACE ALOE	S Africa	stemless forming clusters of dense rosettes.	to 10 × 1.5cm, lanceolate, tapering to a slender bristle, strongly incurved, green (red-brown in full sun on poor soils) with scattered white spots in bands and soft, white spiny teeth	to 4cm long, flame to bright red in a branched inflorescence to 50cm tall in winter and early summer
Comments: Popular dwarf species suitable for shallow containers in the home or glasshouse; used for edging and ground cover in dry, virtually frost-free regions.				
A.bainesii	S Africa; Swaziland; Mozambique	tree to 18m tall with a thick, forked trunk and forking candelabriform branches ending in leaf rosettes	60–90cm long, sword- shaped, green with scattered, brown-tipped, white marginal teeth	to 4cm long, rose tipped green on a branched inflorescence to 60cm tall, in winter
Comments: Stately, slow-growing tree for hot, dry climates.				

ALOE

Name	Distribution	Habit	Leaves	Flowers
A.bakeri	Madagascar	plant to 25cm tall with fairly slender, sprawling stems in a freely suckering clump	7–15cm long, linear-lanceolate and finely tapering with a curving tip and spiny margins, chocolate to green-brown banded brick red or dull pink with scattered green-white spots, or (in inferior forms or if grown in shade), olive tinted red-brown with scattered green-white streaks and spots	to 3cm long, apricot to scarlet grading through orange toward yellow-green mouth, on an unbranched spike to 30cm tall in winter

Comments: A fine, miniature aloe ideal for pot culture in the home where foliage and flowers provide interest throughout the year. Requires warm temperatures and is said to favour light shade although leaf colour is better in sun. Endangered in the wild, but its generously offsetting habit should guarantee its spread in cultivation.

Name	Distribution	Habit	Leaves	Flowers
A.brevifolia	S Africa (Cape Province)	stemless, forming low groups of compact rosettes	to 6 × 2cm, triangular-lanceolate, glaucous, grey-green with sparse, soft prickles beneath and toothed margins	to 3cm long, pale scarlet with a green mouth on unbranched spikes to 40cm tall in summer
A.ciliaris CLIMBING ALOE	S Africa (Cape Province)	slender, scrambling, leafy stems climbing to a height of 5m	to 15 × 1.5cm, linear-lanceolate, fresh green with white teeth	to 3cm long, scarlet to yellow-green at mouth on unbranched spikes to 30cm tall throughout the year

Comments: An unusual climber, tolerant of rather more shade and moisture than other aloes. Train on trellis or pillars in the glasshouse or in frost-free gardens.

Name	Distribution	Habit	Leaves	Flowers
A.distans JEWELLED ALOE	S Africa (Cape Province)	stems to 3m long, decumbent, branching and rooting with clumps of foliage	to 15 × 7cm, lanceolate, glaucous, with pale, raised spots, and yellow-white marginal teeth	to 3cm long, dull scarlet or lemon yellow marked orange, on a branched inflorescence to 60cm long in summer
A.ferox CAPE ALOE	S Africa (Cape Province)	stem erect, stout, unbranched, to 3m tall with a large teminal leaf rosette	to 100 × 15cm, lanceolate-ensiform, thickly fleshy, blue-green sometimes tinted red-brown with sharp, red-brown spines on sinuately toothed margins and sometimes covering surface	to 3.5cm long, scarlet to orange on a branched inflorescence to 60cm tall winter and spring

Comments: Provides a strong, vertical accent for xeriscapes and large succulent collections under glass.

Name	Distribution	Habit	Leaves	Flowers
A.humilis CROCODILE ALOE; HEDGEHOG ALOE; SPIDER ALOE	S Africa (Cape Province)	stemless, forming offsetting clumps of rosettes	to 10 × 1.8cm, ovate-lanceolate, tapering, glaucous, tuberculate with white marginal teeth and incurved tips	to 4cm, scarlet to orange on an unbranched spike to 35cm tall in spring

Comments: var. *echinata* (*A.echinata*) is smaller with soft prickles rather than tubercles on the leaf surfaces; 'Globosa' has blue-tinged leaves in a tight, spherical rosette.

Name	Distribution	Habit	Leaves	Flowers
A.marlothii	S Africa; Botswana; Swaziland	stem unbranched, 2–4m tall	to 150 × 25cm, lanceolate, grey-green with red-brown spines and marginal teeth	to 3.5cm long, orange to yellow-orange, tipped red-brown to mauve in a candelabra-like inflorescence to 80cm tall in winter

Comments: Slow-growing – small plants are often used in containers and dish gardens.

Name	Distribution	Habit	Leaves	Flowers
A.plicatilis FAN ALOE	S Africa (Cape Province)	much-branched shrub or small tree to 5m tall	to 30 × 4cm, in two flattened ranks making fans at the branch tips, strap-shaped and blunt, smooth blue-green	to 5cm long, scarlet tipped green on unbranched spikes to 50cm tall

Comments: Valued for its handsome habit: young plants have a cleanly symmetrical, icy-blue appearance.

Name	Distribution	Habit	Leaves	Flowers
A.saponaria *A.latifolia; A.macracantha; A.picta* SOAP ALOE	S Africa; Zimbabwe; Botswana	stemless, forming suckering clumps of broad rosettes	to 30 × 8–12cm, broadly lanceolate, pale to deep green with dull white, broken bands and brown marginal teeth	to 4cm long, yellow to salmon pink, orange or scarlet on a forking inflorescence 50–100cm tall

Comments: Fast-spreading and soon-congested, the rosettes may need frequent division.

Name	Distribution	Habit	Leaves	Flowers
A.striata *A.albocincta; A.hanburyana* CORAL ALOE	S Africa (Cape Province)	stem stout, to 1m long, decumbent, unbranched	to 50 × 20cm, ovate-lanceolate, tapering finely, untoothed, unspotted, blue-green with white striations (becoming red-tinted in strong sunlight) and pink to red margins	to 2.5cm long, peach-red to coral-red, on a branching inflorescence to 1m tall in winter and spring

Comments: 'Picta' (*A.striata* × *A.saponaria*) has dark green leaves striped with cream and banded yellow-green.

Name	Distribution	Habit	Leaves	Flowers
A.variegata *A.ausana* PARTRIDGE BREAST ALOE; TIGER ALOE	S Africa; Namibia	very short-stemmed, each plant composed of leaves tightly packed in an obscure spiral with their expanded bases overlapping	5–15 × 1.5–5cm, triangular in outline and 3-angled with margins folded upwards from a sharp keel beneath, thickly fleshy with a hard, smooth surface, dark green with white blotches arranged in bands, tip sharp, keel and margins horny, white	to 4cm long, flesh pink to dull scarlet on unbranched spikes to 30cm tall in winter

Comments: A very popular dwarf aloe widely grown as a houseplant.

| ALOE | | | | |
Name	Distribution	Habit	Leaves	Flowers
A.vera *A.barbadensis; A.perfoliata* BARBADOS ALOE; CURACAO ALOE; MEDICINAL ALOE	Cape Verde Islands; Canary Islands; Mediterranean Region; (introduced to Caribbean and S America)	stem short or absent, the plant a large, sprawling and suckering rosette	to 100 × 15cm, linear-lanceolate, tapering finely, outspread to incurved, fleshy and sappy, olive to dark grey-green, margins slightly pink-tinted with pale teeth	to 3cm, yellow on an unbranched spike to 90cm tall produced throughout the year

Comments: Long used in traditional medicine for its healing and cleansing properties – uses now adopted by the pharmaceutical and cosmetics industries.

Aloinopsis (from *Aloe* and Greek *opsis*, resemblance, on account of these plants' supposed similarity to *Aloe*). Aizoaceae. South Africa. 15 species, frost-tender, tuberous perennial herbs making dense, low clumps of succulent, wedge-shaped leaves. The stalkless flowers are daisy-like and yellow to pink. Grow in full sun in a dry, well-ventilated environment with a minimum temperature of 7°C/45°F. The compost should be high in sand and grit. Water sparingly from mid-spring to late summer; at other times merely mist over on warm, sunny days. Propagate by seed or division. *A.schooneesii* (leaves small, blue-green, broadly spathulate with thick, rounded tips; flowers to 1.5cm wide, yellow tinted red).

Alonsoa (for Alonzo Zanoni, Spanish Secretary for Santa Fe de Bogota about 1798). Scrophulariaceae. Tropical Americas. 12 species, small shrubs or perennial herbs with *Nemesia*-like flowers produced in summer and autumn. For use as a half-hardy annual in summer bedding, sow seed in early spring and under glass at 18°C/65°F. Alternatively, sow seed *in situ* in late spring. For winter flowering under glass, sow in late summer. Propagate also by softwood tip cuttings of non-flowering shoots in late summer. Under glass or in warm climates, treat as a perennial, pruning hard and repotting after flowering. *A.warscewiczii* (syn. *A.grandiflora*; Peru; MASK FLOWER; a bushy herb or shrub to 60cm tall; flowers 2cm-wide, scarlet).

Alopecurus (from Greek *alopekouros*, fox-tail). Gramineae. FOXTAIL GRASS. N Temperate regions, S America. 25 species, annual or perennial grasses, usually low to spreading with loosely clumped, short culms and narrow, flat leaf blades. The flowerspikes are narrow, hairy and tightly packed, resembling a fox-brush. The following species is fully hardy and will grow on most soils in sun or part shade. Its variegated form makes bright groundcover, especially if the flowerspikes are removed by shearing over. *A.pratensis* (COMMON FOXTAIL, FOXTAIL GRASS, LAMB'S TAIL GRASS, MEADOW FOXTAIL; Europe, Asia, N Africa; clump-forming perennial to 1m tall in flower, usually far shorter; 'Variegatus': syn 'Aureovariegatus', leaves bright green boldly edged and striped gold).

Alophia (from Greek *a*-, without, and *lophos*, crest). Iridaceae. Americas. 4 species, perennial cormous herbs with narrow, plicate leaves and terminal racemes of *Iris*-like flowers in summer. Cultivate as for *Homeria*. *A. drummondii* (syn. *A.caerulea*; US, Mexico; to 40cm; flowers to 6cm wide in shades of violet and indigo fading at the centre to white spotted brown).

Aloysia (for Maria Louisa (*d.* 1819), princess of Parma and wife of King Carlos IV of Spain). Verbenaceae. Americas. 37 species, glandular shrubs with opposite or whorled leaves and terminal spikes or racemes of small flowers. From Argentina and Chile, *A.triphylla* (syn. *Lippia citriodora*) is LEMON VERBENA. It grows to 3m tall, but usually makes a lower bush resembling a miniature tree with gnarled and fibrous bark. Lanceolate, to 10cm long and finely tapered, the leaves have an intense lemon perfume. Small, white flowers are produced in spikes and panicles in summer. Given good drainage with a warm position on a south-facing wall, and protection from harsh winds, it will tolerate temperatures to –10°C/14°F, regenerating from the stem base if given a deep, dry mulch. Grow in full sun. Prune out dead wood and any frosted top growth in late spring, cutting hard back into living wood. Root basal or nodal softwood cuttings in early summer in a closed case with gentle bottom heat.

alpine a plant indigenous to sites at high altitude, growing between the tree line and permanent snow line. Used more loosely of any plant suitable for a rock garden.

alpine house a greenhouse used for the cultivation of alpines and other perennial plants in containers or beds. Alpine houses are usually unheated, and always freely ventilated.

alpine meadow a garden feature simulating high-alpine meadow, in which alpine plants are grown among fine grasses. The term is often used to describe a piece of grassland where a variety of plants, especially small bulbs, are naturalized, in a form of wild gardening sometimes called meadow gardening. The grass is not mown until high summer so that the development of bulbous species is not repressed.

Alpinia (for Prospero Alpino (1553–1616), Italian botanist). Zingiberaceae. Asia; Australasia. Some 200 species, tender perennial herbs with stout, aromatic rhizomes and clumped, reedy stems with lanceolate leaves. Waxy and shell-like with a showy

Alpinia zerumbet

lip, the flowers are carried in terminal racemes, sometimes amid highly coloured bracts in summer. Provide a minimum temperature of 10°C/50°F and a humid, buoyant atmosphere in bright, filtered light. Water and feed freely in spring and summer, sparingly at other times. Increase by division.

A.purpurata (RED GINGER; South Pacific; to 3m tall with erect, 20–50cm-long, torch-like spikes of overlapping, purple-red bracts; flowers small, white; 'Eileen McDonald': bracts candy pink; 'Pink Princess': bracts rose; 'Tahitian Ginger': bracts deep scarlet); *A.vittata* (VARIEGATED GINGER; Solomon Islands; to 1.5m tall; leaves white- to cream-striped; 'Oceanica': smaller with narrow leaves flashed and striped pure white); *A.zerumbet* (PINK PORCELAIN LILY, SHELL GINGER; South East Asia; to 3m; flowers fragrant, waxy to 4cm long with pink-tinted, white petals and a ruffled golden lip veined wine red, inflorescence pendulous, lacking showy bracts; 'Variegata': leaves striped and zoned yellow and lime; *A.mutica*, the SMALL SHELL GINGER, or ORCHID GINGER, resembles a smaller *A.zerumbet* with more or less erect inflorescences).

Alstroemeria (for Baron Claus von Alstroemer (1736–1794), Swedish naturalist and pupil of Linnaeus). Liliaceae (Alstroemeriaceae). S America. LILY OF THE INCAS; PERUVIAN LILY. 50 species, hardy, fleshy-rooted perennial herbs with erect stems, scattered with ovate to lanceolate leaves and crowned with an umbel of slender-stalked, funnel-shaped flowers with six obovate tepals. Beautiful summer-flowering perennials hardy to −15°C/5°F, especially if grown in sheltered borders and thickly mulched in winter. Grow in a fertile, free-draining and moisture-retentive soil in sun or part-shade. Propagate by seed sown when ripe, or by division in late summer. Plants may also be grown under glass for winter flowering. They are excellent cut flowers.

A.aurea (syn. *A.aurantiaca*; to 1m; flowers 6–10cm across and bright orange to apricot or deep gold, the inner tepals spotted maroon to red); *A.* Ligtu Hybrids (to 1m; flowers ranging in colour from carmine to soft pink, salmon, orange or yellow, often with yellow zones and spotted or feathered maroon to red); *A.pelegrina* (syn. *A.gayana*; to 60cm; flowers to 10cm across, off-white flushed mauve or pink with a darker centre, and inner tepals stained yellow at base and flecked brown to maroon); *A.psittacina* (syn. *A.pulchella*; to 50cm; flowers to 8cm across, green tipped scarlet, overlaid with deep wine red, spotted and streaked dark maroon).

Alternanthera (from Latin *alternus*, alternate, and *anthera*, anther: alternate anthers are sterile in most species). Amaranthaceae. C and S America. JOSEPH'S COAT; CHAFF-FLOWER; JOYWEED; BROAD PATH; COPPERLEAF. 200 species, annual or perennial herbs, often with brightly coloured leaves and with spikes of small flowers. They are used as perennial foliage plants under glass, as half-hardy annuals outdoors in zones 9 and under, and, in the tropics and subtropics,

as permanent edging and groundcover. Grow in full or part sun in a moisture-retentive but well-drained soil. Take nodal cuttings in late summer to produce stock plants; overwinter at 13°C/55°F and water sparingly. Root cuttings in spring in a humid, closed case with bottom heat at 25°/77°F. Plant out in early summer.

A.bettzickiana (CALICO PLANT; annual or short-lived perennial to 1m tall; leaves olive green to yellow mottled or stained red to purple; includes dwarf, yellow-leaved 'Aurea Nana', and bright red 'Brilliantissima'); *A.ficoidea* (annual or short-lived perennial widespread in S America and erect to 50cm, or creeping and mat-forming; leaves green blotched red and purple or yellow, green, red and orange; 'Versicolor': leaves in copper, blood red and yellow; var. *amoena*: PARROT LEAF, SHOO-FLY, JOYWEED, mat-forming with lanceolate to elliptic leaves variously mottled and veined bronze, red, orange, yellow and purple).

alternate of leaves, branches, pedicels arranged in two ranks along the stem, rachis, with the insertions of the two ranks not parallel but alternating; cf. *paired*, *opposite*.

alternate host either of two plant species upon which a pest or disease organism spends part of its life cycle, for example the lettuce/poplar aphid (*Pemphigus bursarius*) and wheat/barberry rust (*Puccinia graminis*).

alternation of generations the occurrence of strongly differentiated, alternating sexual and asexual phases within the life cycle of organisms (i.e. the succession of gametophytes and sporophytes), as found in Pteridophytes such as ferns.

Alyogyne (from Greek *aluo*, to be at a loss, and *gyne*, woman, the stigma is virtually obscured by many whorled staminal filaments). Malvaceae. Australia. 4 species, tender evergreen shrubs grown for their *Hibiscus*-like flowers produced throughout the year. Grow in full sun in a well-drained loamy soil; avoid waterlogged conditions and keep almost dry in winter (minimum temperature 5°C/40°F). Increase by seed and semi-ripe cuttings. *A.huegelii* (LILAC HIBISCUS; hairy, upright shrub to 2.5m tall with palmately lobed leaves; flowers shimmering lilac to intense opal blue, to 15cm across, sometimes spotted purple-red at centre).

Alyssum (from Greek *a*, against, and *lyssa*, madness: the herb was used against madness and rabies). Cruciferae. Europe to W Asia. 168 species, annual and perennial herbs and subshrubs with small, 4-petalled flowers in head-like racemes in summer. Excellent plants for the rock garden, dry walls and banks, they are fully hardy and prefer full sunlight and a free-draining soil. Trim after flowering. Propagate by cuttings in spring or seed in autumn.

A.montanum (Europe; evergreen, tufted perennial to

20cm; leaves small, grey-hairy, flowers fragrant, bright yellow, to 0.6cm across, in round-headed racemes); *A.wulfenianum* (Asia Minor; white-downy, prostrate perennial to 20cm with small, grey leaves; flowers pale yellow to 1cm across in loose corymbs). *A.maritimum* (SWEET ALYSSUM) is now treated as *Lobularia maritima*, and *A.saxatile* as *Aurinia saxatilis*.

Amaranthus (from Greek, *a*-, without, and *maraino*, to wither: the dried flowers of some species are 'everlasting'). Amaranthaceae. Cosmopolitan. 60 species, annual herbs grown for their foliage and catkin-like racemes. Treat as half-hardy bedding or as pot plants. Sow under glass in early spring at 20°C/70°F and plant out after the last frost. In warm regions, sow *in situ* in spring.

A.caudatus (LOVE-LIES-BLEEDING, VELVET FLOWER, TASSEL FLOWER; 40–100cm; leaves green, stalks red-tinted; inflorescences tassel-like, strongly pendulous, to 30cm long, crimson to blood red – green in 'Viridis'); *A.hypochondriacus* (PRINCE'S FEATHER; 40–150cm, leaves green; inflorescence dense, composed of dark crimson spikes, erect and grouped together in a broadly pyramidal panicle; 'Erythrostachys': stems and leaves stained purple, spikes deepest blood red); *A.tricolor* (TAMPALA, CHINESE SPINACH; to 1m; leaves elliptic to ovate leaves to 20cm long; flower spikes small, red-tinted; a species with edible leaves, it also includes the ornamentals 'Joseph's Coat', with leaves flushed scarlet, gold, chocolate and green, and 'Molten Fire' with bronze to maroon leaves flushed scarlet).

× **Amarcrinum**. Amaryllidaceae. Garden origin. Intergeneric hybrids between *Amaryllis* and *Crinum*. × *A.memoria-corsii* combines the habit of *Crinum moorei* with the flowers of *Amaryllis belladonna*, bearing large heads of long-lived, fragrant and shell pink flowers in late summer and early autumn. The foliage persists year-round unless subjected to severe frost. Grow as for *Amaryllis*.

× **Amarygia**. Amaryllidaceae. Garden origin. Intergeneric hybrids between *Amaryllis* and *Brunsvigia*. × *A.parkeri* (*Amaryllis belladonna* × *Brunsvigia josephinae*) is the NAKED LADY LILY, intermediate between the parents, with fragrant flowers ranging from white to deep, clear rose, to carmine. × *A.bidwillii* (*Amaryllis belladonna* × *Brunsvigia orientalis*) differs in its shorter flowers with broader tepals. Grow as for *Amaryllis*.

Amaryllis (the name of a beautiful shepherdess, a character much-celebrated in pastoral poetry). Amaryllidaceae. South Africa. 1 species, *A.bella-donna* (BELLADONNA LILY, JERSEY LILY), a bulbous perennial herb with 2-ranked, strap-shaped leaves produced after the flowers. Carried in autumn atop stout maroon scapes, the funnel-shaped blooms are 10cm long, sweetly fragrant and glowing candy pink (white, white-eyed, rose and carmine cultivars are also available). Plant in a deep, fast-draining loamy

soil in a sheltered position in full sun. Keep moist, never wet, from winter until early summer; dry and baked prior to flowering. These bulbs are generally hardy in zones 6 and over, performing best beneath a wall and left undisturbed for many years. Apply a winter mulch of dry bracken or straw in cold areas. Propagate by offsets removed during dormancy. The popular tender 'amaryllis' is *Hippeastrum*.

Amberboa (from the Turkish name for *A.moschata*). Compositae. Mediterranean to Asia. 6 species, annual and biennial herbs grown for their large, colourful, cornflower-like flowerheads. An old favourite in cottage gardens, sweet sultan has a long flowering period and is suitable for cutting. Picked just as they open, the flowers dry well in borax. Sow seed *in situ* in full sun in autumn or spring. Protect autumn-sown plants where winter temperatures fall below –10°C/14°F. *A.moschata* (syn. *Centaurea moschata;* SWEET SUL-TAN; SW Asia; summer-flowing annual to 70cm tall; flowerheads, fragrant, long-stalked, to 7cm with a rich yellow disk crowned with outspread, finely cut florets in white, cream, yellow, gold-bronze, purple, carmine and pink).

Amelanchier (from amelanchier or amelancier, the Provencal name for *A.ovalis*). Rosaceae. North America; Europe; Asia. JUNEBERRY; SARVICEBERRY; SERVICEBERRY; SHADBUSH; SUGAR-PLUM. 25species, fully hardy deciduous shrubs or small trees with oval leaves. In spring, erect racemes of flowers composed of five, white, obovate to lanceolate petals appear; these contrast beautifully with the soft copper of the new growth. The fruits are small pomes, frosty blue-black to plum purple and offset by the fiery reds and oranges of the autumn foliage. Hardy to –30°C/–22°F. Grow in full sun or part shade on a neutral to acid soil (*A.alnifolia*, *A.asiatica*, and *A.ovalis* tolerate some lime). Propagate by layering, or by division of suckering species from autumn to early spring. Sow seed when ripe – *Amelanchier* hybridizes easily.

A.alnifolia (ALDER-LEAVED SERVICE BERRY; N America; suckering shrub or small tree to 4m tall; flowers fragrant, in erect racemes; 'Alta Glow': columnar with brilliant autumn colour; 'Regent': fiery autumn tints); *A.arborea* (shrub or tree to 20m tall; leaves red or yellow in autumn; flowers fragrant, in drooping racemes); *A.asiatica* (graceful shrub or small tree to 12m; leaves turning red-orange in autumn; flowers in downy racemes); *A.canadensis* (suckering shrub to 8m tall; leaves, turning orange-red in autumn; flowers in erect, downy racemes to 6cm long); *A.laevis* (ALLEGHENY SERVICEBERRY; erect shrub or small tree to 13m tall; leaves with fiery tints in autumn; flowers in nodding racemes to 12cm long); *A.lamarckii* (graceful, wide-branching shrub or tree to 10m tall; leaves turning brilliant orange-red in autumn; flowers in loose racemes to 8cm long; 'Rubescens': flowers soft pale pink).

ament see *catkin*.

Amaryllis belladonna

Amaranthus caudatus

Amelanchier canadensis

American garden a term used in 19th century European horticulture for a collection of trees, shrubs and herbs native to N America. The term also came to be used of moist acid borders.

ammonia a pungent nitrogen-containing gas (NH$_3$) produced as organic matter decomposes. It can be toxic to plants, but with special soil-injection equipment may be used on a farm scale as a fertilizer. Various ammonium salts are commonly used as fertilizers in gardening, for example ammonium nitrate, ammonium phosphate, ammonium sulphate.

ammonification the production of ammonia by soil organisms from organic nitrogen compounds. It is an intermediate step in the mineralization of nitrogen. See *nitrification*.

ammonium nitrate an inorganic fertilizer used as a top dressing, containing 35% nitrogen, half as nitrate and half as ammonium. It is soil acidifying, especially applied continually; where formulated with lime this effect is avoided.

ammonium phosphate an inorganic source of nitrogen and phosphate used in concentrated compound fertilizer and liquid feed. The forms most usually available are mono-ammonium phosphate (MAP) (NH$_4$H$_2$PO$_4$): 12% N, 61% P, and di-ammonium phosphate (DAP) ((NH$_4$)$_2$HPO$_4$): 21% N, 53% P. Both are water soluble. The pH of MAP is around 3.5 and of DAP around 8.0.

Ampelopsis glandulosa var. brevipedunculata 'Elegans'

ammonium sulphamate a non-selective inorganic contact herbicide, useful against annual, soft perennial and woody weeds. It is effective in destroying tree stumps when applied as crystals to cut surfaces. An interval of 8–12 weeks is necessary before replanting treated ground.

ammonium sulphate an inorganic fertilizer, commonly known as sulphate of ammonia, containing 21% nitrogen. It is quick acting as a top dressing, and is also used as an activator for compost heaps and as an ingredient of lawn sand. Its use gives rise to soil acidity, and application to chalky soils may result in loss of nitrogen as ammonia.

Amomyrtus. (Greek meaning 'not *Myrtus*') Myrtaceae. Southern S America. 2 species, evergreen shrubs or trees with flaking bark, leathery leaves, flowers of five, rounded petals with a boss of showy stamens, and small, black, edible berries. *A.luma* favours virtually frost-free conditions, growing well in areas of high humidity in full sun or light shade on rich, moist soils. For propagation, see *Myrtus*. *A.luma* (syn. *Myrtus luma*; LUMA; PALO MADRONO; Chile to 20m tall but more often a bushy shrub to 4m; leaves ovate to lanceolate, 3cm emerging coppery red before turning deep glossy green; flowers creamy white, 1cm across, produced in spring and summer).

Amorpha (from Greek *amorphos*, shapeless). Leguminosae. N America. 15 species, aromatic shrubs with glandular, pinnate leaves and pea-like flowers in terminal racemes in summer. Tolerant of poor soils, drought, and lows of –25°C/–13°F, quickly regenerating if cut down by frost. Plant in sun. Propagate by seed pre-soaked in hot water for two hours - germination may take three to eight weeks. Increase also by layering, rooted suckers and semi-ripe cuttings. *A.canescens* (LEAD PLANT; grey-downy shrub to 1m; flowers violet-blue); *A.fruticosa* (BASTARD INDIGO; to 4m tall with smooth or weakly hairy, glandular growth; flowers deep purple, indigo, pale blue or white).

Ampelopsis (from the Greek, *ampelos*, a grape, and *opsis*, resembling). Vitaceae. North America, Asia. 25 deciduous shrubs and woody vines. Those described here are fast-growing woody climbers, attaching themselves by tendrils which lack holdfasts (thus differing from *Parthenocissus*). The leaves are palmately lobed or divided or more elaborately compound. Small green-white flowers give rise to grape-like fruits by late summer. The following will tolerate winter lows of –25°C/13°F, especially if given a long, hot summer to ripen their wood. New growth is usually damaged by late frosts. Plant in a sheltered position in a deep, moderately fertile loamy soil in sun or shade. Provide support. Propagate by leaf bud cuttings in midsummer; treat with rooting hormone and root in a closed case with bottom heat. Greenwood or semi-ripe cuttings taken in summer may also root.

A.aconitifolia (N China, Mongolia; 10m; leaves to 12cm across, with 3–5, lanceolate to rhombic, deeply lobed and coarsely toothed segments in a palmate arrangement, more or less smooth and a fresh green becoming red-orange in fall; fruits small, sloe blue at first, turning orange-red in fine autumns); *A.glandulosa* var.*brevipedunculata* (BLUEBERRY IVY; China, Japan, Korea leaves palmately 3- or, rarely, 5-lobed, 5–15cm long, basically ovate-cordate in outline, toothed, turning from bright mid-green to red-bronze in autumn; fruits turquoise to amethyst; 'Elegans': stems pink-tinted, leaves splashed lime, white or cream); *A.megalophylla* (China; to 12m; leaves to 80cm, with ovate to oblong leaflets arranged pinnately towards the tip and bipinnately toward the petiole; fruit dark purple ripening black; *A.chaffanjonii* differs in leaves, claret-tinted, not glaucous, beneath, brilliant autumn colour and fruit red ripening black).

amphicarpic producing two different types of fruit; sometimes used of plants bearing two crops per season.

amphicarpogenous producing buried fruit, as in the peanut (*Arachis hypogaea*).

amphigeal producing two types of flower, one from the rootstock or stem base, the other from the upper stems.

amplexicaul of a leaf base or dilated petiole or stipule enlarged and embracing or clasping the stem.

Amsonia (for Dr Charles Amson, 18th-century physician and scientific explorer in North America). Apocynaceae. South Europe; Asia; North America. 20 species, perennial herbs and subshrubs with milky sap. Hardy to –20ºC/–4ºF or lower if mulched. Plant in full sun or light shade. Propagate by division in spring. *A.tabernaemontana* (BLUE STAR; SE US; perennial herb to 1m; flowers starry, 5-lobed pale to slate blue in terminal parnicles in summer).

Anacyclus (from Greek *an*, without, *kuklos*, ring, a reference to the flowerheads). Compositae. Mediterranean. 9 species, annual or perennial herbs with finely cut foliage and daisy-like flowerheads produced in summer. Hardy in zone 7. Grow in full sun on a gritty soil with excellent drainage. Remove spent flowerheads. Increase by seed sown *in situ* in spring, by cuttings, and by rooted stem sections. *A.pyrethrum* var. *depressus* (syn. *A.depressus*; Spain, Algeria, Morocco; prostrate, short-lived perennial to 5 × 20cm; leaves very fine, with silver down; flowerheads to 5cm across, ray florets white above, red-stained beneath, disc florets golden).

anaerobic used of conditions in which oxygen is absent, or of organisms that function in the absence of oxygen; cf. *aerobic*. See *compost heap*.

Anagallis (from *anagallis*, Greek word for this plant, from *anagelao*, to laugh; these herbs were said to dispel sadness). Primulaceae. PIMPERNEL. Cosmopolitan 20 species, hardy annual or perennial herbs with creeping to erect stems and ovate to elliptic leaves. Solitary, dish- to cup-shaped flowers with five broad lobes are produced on slender stalks from the leaf axils in spring and summer. Sometimes thought weedy and invasive, *A.arvensis* is often unwelcome in gardens, but is a bright and delicate groundcover for sunny positions on all but the wettest soils. *A.monellii* favours a warm, rather sandy site and is suitable for rock gardens and the edges of borders. *A.tenella*, should be allowed to colonize in the bog garden and at the margins of streams and ponds in the rockery. Propagate by seed sown *in situ* in spring or under glass in late summer; perennial species also by cuttings in spring.

A.arvensis (SCARLET or COMMON PIMPERNEL, POOR MAN'S WEATHER GLASS, SHEPHERD'S CLOCK; Europe; dwarf creeping annual; flowers 0.5–1.8cm across, typically bright scarlet, also white, pink, yellow and sky blue); *A.monellii* (BLUE PIMPERNEL; Mediterranean; erect perennial or biennial to 50cm; flowers 1–1.5cm across, rich blue above, tinted wine red beneath and at the centre or wholly blue or wine red; 'Phillippii': flowers large, royal blue); *A.tenella* (BOG PIMPERNEL; W Europe; creeping dwarf perennial; flowers 0.5–1cm across, sweetly scented and pale rose or white – deep pink in the cultivar 'Studland').

Ananas (from *nana*, the Tupi Indian word for the fruit of *A. comosus*). Bromeliaceae. S America. PINEAPPLE. 8 species, tender, perennial evergreen herbs. Dense, cone-like spikes of small flowers arise from the centre of rosettes of tough, rigid, sword-like leaves with spiny margins. The spikes terminate in a tuft of short, leafy bracts (coma), and swell as they develop, fusing to form an ovoid, fleshy syncarp (the pineapple fruit). Grow in full sunlight with a minimum temperature of 13ºC/55ºF. Water, syringe and feed liberally when in growth, especially in hot weather and during fruit development, but allow the soil to dry slightly between waterings. Keep virtually dry and do not mist in cool dull weather. Propagate by detaching rooted suckers, or by rooting the coma in a moist, sandy mix with bottom heat.

A.bracteatus (WILD or RED PINEAPPLE; to 1m; leaves olive; inflorescences 10–15cm-long, composed of lilac to red flowers and closely clothed in red-tinted bracts; var. *tricolor*: leaves striped cream and yellow and edged with bright red to pink teeth; fruit red-pink, topped with variegated coma); *A.comosus* (PINEAPPLE; seedless, edible cultigen taller than the last species with grey-green leaves; inflorescences 15–30cm-long; flowers violet-red, with yellow-tinted bracts; syncarp large, orange-yellow when ripe, sweetly fragrant; with 'Porteanus': leaves with a central yellow stripe; var. *variegatus*: leaves striped creamy yellow and sometimes red-pink).

Anaphalis (ancient Greek name for a similar plant). Compositae. Temperate regions. PEARLY EVER-LASTING. 100 species, white-hairy perennial herbs with clumps of spreading then erect stems terminating in clusters or flat-topped corymbs of flowerheads. These consist of small, yellow florets tightly encased by several series of papery, silver bracts. The following are hardy and flower between midsummer and early autumn. They are valued for their silvery effects and long-lasting flowerheads. The inflorescences dry well if cut just prior to opening. *A.margaritacea* and *A.triplinervis* are suitable for the herbaceous or mixed border. Smaller and more delicate, *A.nepalensis* requires a sunny position on a well-drained, gritty soil; var. *monocephala*, smaller still, is suited to rock gardens and alpine sink gardens. Increase by division in early spring.

A.margaritacea (N America, Asia and N Europe grows to 80cm; leaves grey-green above, silvery beneath; flowerheads clothed in pearly white bracts, 1cm wide, in a corymb to 15cm across; var. *yedoensis*: syn. *A. yedoensis*, leaves narrow, leaves and flowerheads somewhat tawny below); *A nepalensis* (syn. *A.triplinervis* var. *intermedia*; Himalaya, Western China; to 30cm tall; leaves grey-green with white-woolly undersides; flowerheads 1–1.5cm wide, silvery, 1–4 per stem; var. *monocephala*: syn. *A.nubigena*, low and tufted with white flowerheads, borne one per stem); *A.triplinervis* (temperate Asia; to 90cm tall; leaves grey-hairy; flowerheads silver-white 1cm wide, crowded in domed clusters).

anastomosing used of veins forming a network, united at their points of contact.

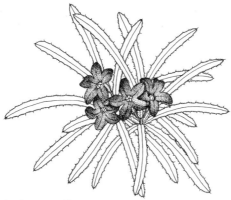

Anchusa caespitosa

Anchusa (from Greek, *ankousa*, alkanet, a skin paint – some species were used to make rouge). Boraginaceae. Europe; Africa; Asia. ALKANET; BUGLOSS. 35 species, hardy, coarsely hairy annual, biennial and perennial herbs grown for their flowers. Produced in spring and summer, these are funnel-shaped with five, broad and spreading lobes and are carried in terminal or axillary cymes. Grow *A.azurea* in a deep, fertile soil, well-drained but moisture-retentive, and in full sun. Deadhead promptly to maintain a tidy habit and promote a second flush of bloom. Increase by division or root cuttings in late winter. *A.caespitosa* requires a very sunny, well-drained position on a gritty soil, ideally on the rock garden or in an alpine sink. It is very intolerant of cold, overwet conditions. Propagate by offsets in summer. Treated as an annual, *A.capensis* is a fine summer bedding plant, sown *in situ* in spring. As a biennial, it should be overwintered in a frost-free greenhouse.

A.azurea (syn. *A.italica*; Europe, North Africa and Western Asia; spreading, clump-forming perennial to 1.5m tall; flowers 1cm across, deep blue to violet, in showy, terminal clusters; 'Dropmore': to 1m tall, flowers gentian blue; 'Little John': to 40cm: flowers small, bright blue; 'Loddon Royalist': to 1m tall, flowers bright blue flowers; 'Opal': to 1m tall, flowers sky blue). *A.caespitosa* (syn. *A.angustissima, A.leptophylla*; Crete; dwarf, tufted perennial to 8cm, making mounds of clumped, narrow and bristly leaves; flowers borne 1–3 together on short stalks among the leaf bases, 1.5cm wide, dark blue with a white throat); *A.capensis* (S Africa; CAPE FORGET-ME-NOT; erect, hairy biennial to 60cm flowers in terminal panicles, 1cm wide and bright blue with a white throat; 'Blue Angel': flowers ultramarine; 'Blue Bird': flowers vivid indigo; 'Pink Bird': flowers pink).

ancipitous describes stems and fruit with two sharp edges.

androdioecious having male and hermaphrodite flowers on separate plants of the same species, for example marsh marigold (*Caltha palustris*).

androecium the male component of a flower, the stamen or stamens as a whole.

Andromeda polifolia

androgynous hermaphrodite or, sometimes, monoecious.

Andromeda (for the daughter of Cepheus and Cassiope). Ericaceae. Arctic, temperate Northern Hemisphere. 2 species, dwarf evergreen shrubs, compact or spreading, with short, linear to lanceolate, revolute leaves and small, urn-shaped and 5-lobed flowers nodding in clusters at the branch tips in spring and summer. Fully hardy, *A.polifolia* needs dappled or full sun and a moist but porous, acid soil. Propagate by removal of rooted, creeping branches and suckers, and by simple layering. *A.polifolia* (MARSH ANDROMEDA, BOG ROSEMARY; N Europe, Alps, Carpathians and Russia; to 12cm forming dense mounds of spreading to erect stems; leaves to 2cm, dark green above with grey-white undersides; flowers white to pale pink, 0.5cm long in clusters to 3cm across; 'Compacta': compact, leaves grey-green, flowers pink; 'Grandiflora': leaves blue-green, flowers large, white tinted pale pink; 'Minima': dwarf, leaves needle-like, grey-green, flowers pale pink; 'Nana': vigorous, dwarf, flowers small, pink, profuse; 'Nana: Alba': small, compact, leaves silvery beneath, flowers small, white, profuse; 'Red Winter': tall, leaves red-bronze-tinted in autumn and winter).

andromonoecious having male and hermaphrodite flowers on the same plant, for example, horse chestnut (*Aesculus hippocastanum*).

Androsace (from Greek *aner*, man, and *sakos*, buckler, from the resemblance of the anther to an ancient buckler). Primulaceae. Northern temperate regions. ROCK JASMINE. 100 species, hardy annual, biennial or perennial herbs, those here summer-flowering, alpine perennials making dwarf rosettes or cushions of hairy foliage. The flowers are usually white or pink, with a short tube and five broad, outspread lobes; they may be solitary and short-stalked or stemless, or carried in long-stalked umbels. Grow in full sun, with an airy, cool atmosphere and on sharply draining, soils. *A.chamaejasme, A.imbricata* and *A.pyrenaica* grow best in a slightly acidic medium; the remaining species are calcicole. High-altitude alpines such as *A.cylindrica, A.hirtella, A.imbricata* and *A.pyrenaica* grow best in the alpine house or in raised beds that can be covered in cold wet weather. Lower-altitude species are more tolerant and may be grown in the open in troughs or on the scree and rock gardens; these include *A.carnea, A.chamaejasme, A.lanuginosa, A.primuloides* and *A.sempervivoides*. Propagate by rooted runners and rosettes of stoloniferous and freely offsetting species, the remainder by basal shoot cuttings taken in early summer.

A.carnea (W Europe; loosely tufted; leaves to 2cm, linear; scape to 6cm tall; flowers one to many, white or pink); *A.chamaejasme* (Northern Hemisphere; loosely tufted and shortly stoloniferous; leaves oblong to lanceolate; scape to 7cm; flowers 3–7, white or pink); *A.cylindrica* (Pyrenees; densely tufted and cushion-forming; leaves to 1cm long, oblong to lanceolate, persisting after death to form column-shaped shoots; scape to 2cm; flowers solitary, pink with yellow throat); *A.hirtella* (Pyrenees; densely tufted and white-hairy; leaves, linear; flowers solitary, very short-stalked, white); *A.imbricata* (syn. *A.vandellii* Alps, Pyrenees; densely tufted, forming rounded cushions; leaves, linear to lanceolate, white-hairy; flowers solitary, short-stalked, white with a yellow throat); *A.lanuginosa* (Himalaya; prostrate, white-hairy perennial; leaves, lanceolate to obovate; scape to 4cm; flowers many, pink); *A.primuloides* (Himalaya; stoloniferous; leaves lanceolate to linear, white-hairy; scape to 7cm; flowers, pink); *A.pyrenaica* (Pyrenees; dense, cushion-forming perennial; leaves, linear to oblong, overlapping in columnar shoots; flowers, solitary, very short-stalked, white with a yellow eye); *A.sarmentosa* (Himalaya; stolons long; leaves, lanceolate to ovate, silvery-hairy, scape to 10cm; flowers many, deep pink to brilliant carmine); *A.sempervivoides* (Kashmir, Tibet; leaves to 0.5cm, ovate to spathulate, densely overlapping in stoloniferous rosettes; scapes to 10cm; flowers, pink); *A.villosa* (Europe, Asia; densely tufted with cushion-forming, domed rosettes; leaves, oblong to lanceolate, white-hairy; scapes to 5cm; flowers two to three, white with yellow-green centres that turn red; var. *jacquemontii*: flowers purple-pink).

Anemone (From Greek *anemos,* wind, a name used by Theophrastus and reflected in the popular name 'windflower'). Ranunculaceae. North and South temperate regions. 120 species, hardy perennial herbs with rhizomatous, fleshy, fibrous or woody roots. The leaves are lobed, dissected or compound, rarely entire. Those on the stem are often arranged in a whorl below the inflorescence. Solitary or in cymes, the flowers are dish- to bowl-shaped and composed of numerous, oblong to oblanceolate segments.

The autumn-flowering species with a fibrous rootstock such as *A.hupehensis* and *A.tomentosa* occur predominantly in damp open woodland. This group, which includes *A. × hybrida,* is well suited to the autumn border and woodland garden. They are valued for the long succession of flowers that last well when cut. Plant in dappled shade or part-day sun, on a moist, fertile, loamy soil with ample organic matter. *A.hupehensis* will tolerate slightly drier and sunnier conditions, where it will bloom earlier although flowers will be smaller and of lower quality. Propagate by fresh seed sown ripe, by division or by root cuttings.

Tuberous and rhizomatous species include the spring-flowering *A.blanda, A.nemorosa* and *A.apennina.* They require moist but well-drained, humus-rich soils in the rock or woodland garden, but tolerate dry conditions during their summer dormancy. Also included in this category are taller species with a fibrous or slender-rhizomatous rootstock (e.g. *A.narcissiflora, A.sylvestris*). These prefer grittier, more freely draining soil. Light requirements vary from full sun (*A.narcissiflora*) to light shade (*A.sylvestris*). *A.nemorosa* is sufficiently robust for naturalizing in thin turf, where the first cut occurs after foliage die-back. Smaller species, notably *A.blanda* and its cultivars, are well suited to pot cultivation for the alpine display house. Propagate by fresh seed, rhizomatous species also by careful division.

Tuberous-rooted plants such as *A.coronaria, A. × fulgens* and *A.pavonina* flower from spring to early summer and are well suited to areas with dry hot summers, where they will grow in light sandy soils: shade in the hottest part of the day is appreciated. Plants of this type require dry dormancy after flowering. De Caen, Mona Lisa and St Brigid anemones are a commercial cut-flower crop in milder areas of Europe. Plant in a fertile but light-textured soil in sunny borders. In cold areas, mulch for winter protection or lift after flowering and overwinter dry tubers, plunged in sand under frost-free conditions. Under glass, give a winter minimum of 10°C/50°F and pot up tubers in a sandy loam-based mix of medium fertility for flowering 3–4 months later. *A.coronaria* may be flowered under glass almost all year with successive plantings; tubers of this and *A. × fulgens* deteriorate quickly and should be replaced every 2–3 years.

Susceptible to anemone and cucumber mosaic viruses; downy mildew, and anemone smut which produces swellings on the leaf stalks and leaves, bursting to release black spores. Seedlings also susceptible to flea beetle.

Anemone x *hybrida*

ANEMONE Name	Distribution	Flowering time	Habit	Leaves	Flowers
A.apennina APENNINE ANEMONE	S Europe	early spring	to 15cm, rhizomatous, spreading perennial	leaves divided into 3, toothed or lobed segments	solitary, to 3cm diam., blue or pink- or white-flushed, with 8–23 oblong tepals
Comments: Includes 'Petrovac' (flowers rich blue, many-tepalled), and 'Purpurea' (flowers soft purple-rose).					
A.biflora	Iran to Kashmir	spring	to 20cm, tuberous perennial	leaves divided into 3, lobed segments	2–3 together, nodding, 2–5cm diam., crimson, orange or yellow with 5, narrowly elliptic tepals
Comments: Requires a warm, dry summer rest.					
A.blanda	SE Europe to Cyprus and Turkey	early spring	to 18cm, tuberous perennial	leaves divided into 3, lobed segments	solitary, to 4cm diam., blue, mauve, white or pink with 9–15 narrowly oblong tepals
Comments: Includes 'Atrocaerulea' (('Ingramii') flowers deep blue), 'Blue Shades' (leaves very finely divided, flowers pale to deep blue), 'Bridesmaid' (flowers large, pure white), 'Bright Star' (flowers bright pink), 'Radar' (flowers magenta with a white centre), 'Violet Star' (flowers large, amethyst, exterior white), 'White Splendour' (flowers large, white, exterior pink).					

ANEMONE

Name	Distribution	Flowering time	Habit	Leaves	Flowers
A.coronaria POPPY-FLOWERED ANEMONE	S Europe, Mediterranean	spring	20–60cm, tuberous perennial	leaves divided into 3, lobed or toothed or cut segments	solitary, to 8cm diam., scarlet, blue, pink, white or bicoloured with 5–8, oval tepals

Comments: Many cultivars, dividing roughly into two groups: **De Caen Group** with single, large, poppy-shaped flowers in white, bright pink, red or purple with showy black anthers, and **St. Brigid Group** with semi-double to double, large flowers in bright pastels, strong reds and mauves, and bicolours. Excellent cut flower.

Name	Distribution	Flowering time	Habit	Leaves	Flowers
A. ×fulgens SCARLET WIND FLOWER	C Mediterranean	spring to early summer	to 25cm, tuberous perennial	leaves more or less divided into 3, lobed and toothed segments	solitary, to 6cm diam., brilliant scarlet with 15 oblong-obovate tepals, centre black

Comments: The offspring of *A.pavonina* × *A.hortensis*; includes 'Annulata Grandiflora' (flowers scarlet with a yellow centre) and 'Multipetala' (flowers scarlet, semi-double).

Name	Distribution	Flowering time	Habit	Leaves	Flowers
A.hupehensis *A.japonica* of gardens JAPANESE ANEMONE	China, Japan	late summer to autumn	to 70cm, fibrous-rooted, branching, erect perennial	leaves palmately 3-lobed to 3-parted with sharply toothed margins and thin hairs beneath	several together on long, uneven stalks in sparse umbels, to 5cm diam., white to pink or palest rosy mauve with 5–6 rounded tepals and golden anthers

Comments: This species is often confused with *A. × hybrida*; both have been called *A.japonica*. *A.hupehensis* differs from *A. × hybrida* in its fertile pollen and shorter stature. It includes 'September Charm' with silvery pink flowers. The true *Anemone japonica* is now treated as *A.hupehensis* var. *japonica*. Its flowers have up to 20, pink to rose-mauve tepals.

Name	Distribution	Flowering time	Habit	Leaves	Flowers
A. ×hybrida *A.japonica* of gardens JAPANESE ANEMONE	garden origin	late summer to autumn	to 150cm, fibrous-rooted, erect, branching perennial	very similar to *A.hupehensis* but with larger, very slightly hairy leaves	as for *A.hupehensis* but to 8cm diam. with 6–11(–15) tepals

Comments: Differs from *A. hupehensis* in its flowers, usually lacking fertile pollen and taller stature. Many cultivars ranging in height from 70–150cm, in habit from the bushy and compact to the tall and elegant with very long flowerstalks, in flower form from single to double and in colour from white to pale pink to edeep rosy mauve. They include 'Bressingham Glow' (to 90cm, compact, flowers semi-double, petals fluted, deep pink) and 'Prinz Heinrich' (stems slender, flowers deep rosy purple with numerous, flattened tepals).

Name	Distribution	Flowering time	Habit	Leaves	Flowers
A. ×lipsiensis *A. ×seemannii*	N and C Europe	early spring	to 15cm, rhizomatous, spreading perennial	leaves 3-parted in a ruff below flowers, deeply cut and sharply toothed	solitary, to 2cm diam., very pale yellow with bright gold and green stamens, tepals 5, ovate-elliptic

Comments: The offspring of *A.ranunculoides* × *A.nemorosa*, a naturally occurring hybrid.

Name	Distribution	Flowering time	Habit	Leaves	Flowers
A.narcissiflora	Europe, Asia, N America	summer	to 60cm, rhizomatous perennial	leaves 3–5-parted with narrow, deeply divided and coarsely toothed segments	up to 8 per umbel, to 4cm, white faintly tinted pink or blue beneath, with 5–6, obovate, spreading tepals

Name	Distribution	Flowering time	Habit	Leaves	Flowers
A.nemorosa WOOD ANEMONE; WINDFLOWER	N and C Europe	early spring	to 30cm, rhizomatous, carpeting perennial	leaves 3-parted with deeply cut, slender segments, some forming a ruff below flower	solitary, to 3cm cross, white, occasionally tinted pink or purple with 6–8 (5–12) oblong tepals and golden anthers

Comments: Many cultivars including 'Allenii' (flowers lilac or lavender), 'Plena' (flowers white, double), 'Robinsoniana' (flowers pearly grey to palest lilac), 'Vestal' (flowers white, centre with a button-like rose of petaloid segments), 'Vindobonensis' (flowers cream). 'Virescens' (flowers replaced by a pyramid of small, green, leafy bracts), 'Wilk's Giant' (flowers large, single, pure white).

Name	Distribution	Flowering time	Habit	Leaves	Flowers
A.pavonina	Mediterranean	early spring	to 30cm, tuberous perennial	leaves 3-parted with much-toothed and divided segments	solitary, to 10cm diam., scarlet, violet, purple, pink, white or yellow with a central boss of black stamens ringed with white, petals 7–9, broadly obovate

Comments: Includes the deep salmon-coloured 'Barr Salmon', and the **St. Bavo Group** with flowers in coral, rose, lavender and violet.

Name	Distribution	Flowering time	Habit	Leaves	Flowers
A.ranunculoides	N and C Europe	early spring	to 15cm, rhizomatous, spreading perennial	leaves 3-parted with segments much divided and toothed and held in a ruff below flowers	solitary, to 2cm diam., deep yellow, tepals 5–6, elliptic

Comments: Includes 'Flore Pleno', with semi-double flowers, 'Grandiflora' with larger flowers , and 'Superba' with olive-bronze leaves and bright yellow flowers

Name	Distribution	Flowering time	Habit	Leaves	Flowers
A.rivularis	N India, SW China	late spring, early summer	to 90cm, swollen-rooted erect, branching perennial	leaves 3, with 3-parted, lobed and toothed segments	2–5 per umbel on slender stalks, 1.5–3cm wide, white stained lavender beneath, with 5–8, oval tepals

Name	Distribution	Flowering time	Habit	Leaves	Flowers
A.sylvestris SNOWDROP WINDFLOWER	N India, SW China	late spring, early summer	to 90cm, swollen-rooted, erect, branching perennial	leaves palmately 3–5-parted, segments much toothed and divided	solitary, nodding, fragrant, 2.5–8cm diam., white, tepals 5+, elliptic

Comments: Includes 'Grandiflora' (flowers large, nodding and held well above plant, 'Elisa Fellmann' (('Plena') flowers semi-double), and 'Macrantha' (very vigorous, flowers large, nodding, strongly fragrant).

Name	Distribution	Flowering time	Habit	Leaves	Flowers
A.vitifolia	Asia	summer	to 90cm, rhizomatous, clump-forming perennial	leaves large, palmately 5-lobed and crenate, white-woolly beneath	a few together in lax umbels, 3.5–5cm diam., white sometimes tinted pink, with golden anthers, tepals 5–6, obovate-elliptic

anemone-centred applied to composite flowers, especially chrysanthemums and dahlias, where the disc florets are enlarged to form a central cushion, sometimes of a contrasting colour to the ray florets. Also used to describe peonies and camellias with petaloid stamens.

Anemonella (Latin diminutive form of the name *Anemone*, which this plant resembles). Ranunculaceae. 1 species, *A.thalictroides* (RUE ANEMONE), a hardy perennial herb native to Eastern North America. Finely divided, ferny foliage arises in spring from tuberous roots. In spring and early summer, 2–5, 2–3cm-wide flowers are carried on slender stalks atop a 10–20cm-tall stem and subtended by a compound, leafy ruff. They consist of 5–10 cupped, petal-like sepals, white to pale pink. In the cultivar 'Florepleno', the flowers are double and white, in 'Rosea' pale pink, and in 'Roseaplena' ('Schoaff's Double Pink') double and rosy pink. Plant in a moist, humus-rich sandy soil, in dappled sun or light shade. Mulch in winter. Propagate from fresh seed in summer, or by division of well-established plants.

Anemonopsis (*Anemone*, a related genus, and Greek, *opsis*, resemblance). Ranunculaceae. 1 species, *A.macrophylla* (FALSE ANEMONE), a rhizomatous, hardy perennial herb native to Japan. In summer, the ferny foliage is surpassed by slender stalks to 80cm tall and bearing delicately branched panicles. The nodding flowers are lavender, 2–4cm wide, and consist of 7–10 cupped, petal-like sepals and ten, far smaller petals. Plant in damp, cool and humus-rich soils in light shade or dappled sunlight. Propagate by division in spring.

Anemopaegma (from Greek *anemos*, wind, and *paigma*, sport, referring to the way in which the wind plays with the stems). Bignoniaceae. C and S America. YELLOW TRUMPET VINE. 40 species, evergreen, woody climbers grown for their showy, tubular to funnel-shaped flowers. The leaves consist of 1–2 pairs of smooth, leathery leaflets to 15cm long and often terminate in a coiling tendril. Plant in a fertile, loamy medium in large pots or tubs in the greenhouse or conservatory (minimum temperature 10°C/50°F), in full sun or part shade. Propagate by softwood cuttings.

A.chamberlaynii (Brazil; leaf tendrils usually 3-branched; flowers 5cm-long, rich golden yellow); *A.chrysoleucum* (Mexico to N Brazil; differs from the first species in its unbranched leaf tendrils and larger (6–10cm), fragrant flowers borne 1–3 (not 2–8) per axil; these are yellow and usually stained white with orange or purple-brown markings in the throat).

anemophily wind pollination

Anethum (from Greek *anethon,* dill). Umbelliferae. Old World. 2 species, annual and biennial herbs scented of anise with very finely divided leaves and small, yellow flowers in compound umbels. *A.graveolens* is the herb DILL (q.v.), native to south-west Asia, now widely naturalized in Europe and N America. It grows to 60 cm tall, with leaves to 30cm composed of many, thread-like segments.

anfractuose, anfractuous describes plant parts that are closely or tightly sinuous, or spirally twisted.

Angelica (from the ancient name, *Herba angelica*, 'angelic plant' – the many herbal properties of this plant were said to have been revealed by an angel). Umbelliferae. Northern Hemisphere. 50 species, biennial, monocarpic or perennial herbs, usually tall and robust with ternately 2–3-pinnate leaves and small, green-white flowers in large, compound umbels. *A.archangelica* is GARDEN ANGELICA, ARCHANGEL or WILD PARSNIP, a monocarpic herb to 2m tall, native to Europe and Asia. The 60cm-long leaves are smooth and dark green with toothed, oblong segments. The umbels are 12–25cm wide, massed with white or yellow-green flowers and carried on stout stalks.

Young stems of angelica are crystallized and used for confectionary decoration; they are also eaten fresh, cooked as for asparagus, or boiled with rhubarb to alleviate its tartness. Fresh or dried, the leaves are used in herbal tisanes. The flowers are a valuable nectar source for bees. A stately addition not only to the herb garden, but also to green/white borders and bedding schemes. *Angelica* will usually die after setting seed, but may perform as a perennial if the flowerheads are removed soon after fading in summer. Sow seed *in situ* when ripe.

angiosperms plants with their seeds enclosed in an ovary, i.e. flowering plants as opposed to gymnosperms, which are non-flowering plants with naked seeds.

Angraecum (Latinized form of the Malay name *anggrek* or *angurek*, used for epiphytic orchids with aerial roots). Orchidaceae. Tropical and South Africa; Indian Ocean Islands. 150 species, tender, epiphytic orchids usually with leathery leaves arranged in two ranks along erect stems and with abundant aerial roots. Waxy, star-like and long-spurred, in shades of white, cream and green, the flowers are produced in axillary clusters and sprays. Grow in light shade in a humid environment, minimum temperature 18°C/65°F. Plant in clay pots or baskets filled with a very open, bark-based medium. Water, feed and syringe frequently during warm weather; less often in winter. Increase by offsets and stem cuttings.

A.distichum (W Africa; stems to 20cm long concealed by the overlapping bases of two-ranked, short, folded leaves; flowers small, solitary, white); *A.eburneum* (Africa, islands in the Indian Ocean; to 1m tall with strap-shaped leaves and arching sprays of 6cm-wide flowers, green with upturned, shell-like white lips, heavily fragrant at night); *A.sesquipedale* (COMET ORCHID; Madagascar; leaves thick, strap-shaped; flowers starry, ivory, thickly waxy in texture, to 18cm across, spicily scented at night with a spur some 30cm long).

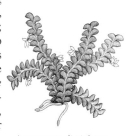

Angraecum distichum

*Anguloa
brevilabris*

angular, angulate of plant parts with laterally projecting angles, as in longitudinally ridged and angled stems.

Anguloa (for Don Francisco de Angulo (fl.1790), Spanish naturalist). Orchidaceae. South America. CRADLE ORCHID; TULIP ORCHID. 10 species, terrestrial or epiphytic orchids with large pseudobulbs and long, ribbed leaves. Borne in spring and summer, the flowers are large and solitary, arising on thick stalks alongside the new growth. They consist of broad and waxy, strongly cupped tepals which cradle a small, tongue-like lip. Most are strongly scented of chocolate and oil of wintergreen. Grow in humid, well-ventilated and lightly shaded conditions, minimum winter temperature 9°C/ 48°F. Pot annually in late winter in an open mix with sphagnum moss and perlite. Water and feed liberally while the new growth develops. Once the new pseudobulbs are complete (usually early autumn), allow the plants to dry out thoroughly between waterings and keep in cool, bright conditions; increase water and temperatures when the new growth appears. Increase by backbulbs and division.

A.brevilabris (flowers to 10cm deep, tepals olive, interior densely spotted dark blood red, lip short, pale barred blood red with small lateral lobes); *A.clowesii* (flowers to 10cm deep, tepals lemon to canary yellow); *A.hohenlohii* (syn. *A.ruckeri* of gardens; flowers to 12cm deep, tepals olive, interior densely spotted and stained dark blood red, lip red, larger than in *A.brevilabris*, with distinct, outspread lateral lobes); *A. × ruckeri* (*A.clowesii × A.hohenlohii*; flowers to 10cm deep, tepals dull yellow to olive, interior yellow spotted blood red); *A.uniflora* (flowers to 10cm deep, tepals white faintly flushed rose); *A.virginalis* (flowers to 8cm deep, tepals white, interior dotted and lined faintly with pale rose, lip with a kinked base, spotted red).

Anigozanthos (from Greek *anisos*, unequal and *anthos*, flower, referring to the unequal perianth lobes). Haemodoraceae. Australia. KANGAROO PAW; CAT'S PAW. 11 species, evergreen or semi-deciduous perennial herbs with grassy or iris-like leaves in basal fans or clumps and long-stalked spikes or panicles usually in spring and summer. The flowers are tubular, split on the curving lower side and with five short and acute lobes peeled back on the arching upper side. They are covered in dense and usually colourful felty hairs. Grow in full sunlight in a freely draining acid mix. Provide a rather dry, airy atmosphere and a minimum temperature of 5°C/ 40°F. Water freely during warm weather, sparingly in winter. Propagate by division in spring.

A.flavidus (TALL KANGAROO PAW; flowers 3–5cm long and typically lime green densely covered with sulphur yellow felt with orange anthers, arranged on the ascending branches of erect panicles that stand between one and two metres tall; red, pink, buff and orange forms also occur); *A.manglesii* (flowers arranged in two rows at the tip of a 30–100cm spike, 7–12cm long, yellow-green to emerald with bright

*Anigozanthos
manglesii*

red hairs toward the base); *A.rufus* (RED KANGAROO PAW; differs from *A.flavidus* in flowers, arranged on outspread branches, and basically olive, thickly covered in blood red felt with green anthers).

anion exchange capacity the ability of a soil to adsorb negatively charged ions known as anions. A measure of the retention and availability of such nutrients as phosphate, sulphate and molybdate.

Anisodontea (from Greek, *anisos,* unequal, and *odous*, tooth, referring to the unequal projections on the mericarps). Malvaceae. South Africa. 19 species, evergreen shrubs or shrubby perennial herbs with lobed leaves. Solitary, cymose or racemose, the flowers have five spreading, broadly obovate petals and are produced in spring and summer. Grow in full sun or light shade. Maintain a minimum winter temperature of 5°C/40°F with good ventilation, and water plentifully when in growth. Propagate by greenwood or semi-ripe cuttings in summer.

A.capensis (woody-based, bushy perennial to 1m tall; flowers 1.5–3cm wide, pale to deep magenta with darker veins and a basal blotch); *A. × hypomadarum* (hybrid of unknown parentage; slender-branched shrub to 3m tall long; flowers 2–3 cm wide and white or pale pink with deep purple veins merging in a basal blotch).

annual a plant that completes its life cycle of germination, flowering and seeding within one growing season. Represented, as examples, in the flower garden by cornflower (*Centaurea cyanus*) and in the vegetable garden by lettuce (*Lactuca sativa*).

annual rings clearly defined rings of secondary xylem evident in transversely cut woody stems and roots of plants growing in temperate zones. The rings are due to a seasonal rhythm in the growth of cambium, which is greatest in the spring and gradually falls off towards dormancy. The cycle is repeated each year, so that annual rings provide a means of calculating the age of trees especially. Also known as growth rings.

annual shoots shoots that live for one year only, produced by herbaceous perennial plants, most bulbs and some shrubs.

annular, annulate ring-shaped; used of organs or parts in a circular arrangement or forming rings.

annulus (1) the corona or rim of the corolla in Asclepiadaceae; (2) in ferns, an elastic ring of cells that partially invests and bursts the sporangium at dehiscence.

Anoda (from Greek *a-*, without, and Latin *nodus*, joint: the pedicels lack a node). Malvaceae. Tropical and subtropical Americas. 10 species, annual and perennial herbs and subshrubs with hastate, unlobed or lobed leaves and flowers with five, obovate petals carried one to two per axil in erect racemes in summer. Grow under glass (minimum temperature 7°C/45°F) in sun in a well-drained, gritty compost. Water moderately in winter. Alternatively, treat as a half-hardy annual grown in a sunny, sheltered situation. Propagate from seed sown *in situ* in late spring or, earlier, under glass. *A.cristata* (erect to sprawling annual or short-lived perennial herb to 1.5m tall; flowers 1.5–5cm wide, white, lavender or purple-blue; 'Opal Cup': flowers silvery lilac with an iridescent sheen and mauve veins; 'Snowdrop': flowers pure white).

Anomatheca (from Latin *anomalus*, abnormal, and *theca*, a case, referring to the warty capsule). Iridaceae. Central and South Africa. 6 species, cormous perennial herbs grown for their flowers. Cultivation as for *Babiana*. *A.laxa* (*Lapeirousia laxa*, *L.cruenta*; S Africa, Mozambique; leaves sword-shaped, exceeded in summer by a loose, erect to horizontal raceme some 30cm tall; flowers 2.5cm wide with a straight perianth tube and two groups of three segments, typically red with purple-red markings at the base of the lower three segments; warty seed capsules split to reveal red-brown seeds).

Anopterus (from Greek *ano*, upwards, and *pteron*, wing, referring to the winged seeds). Grossulariaceae. Australia; Tasmania. 2 species, evergreen trees and shrubs with leathery leaves and racemes of bell-shaped, 6-parted flowers. Grow in sheltered, humid locations in zone 8, or in a cool greenhouse and conservatory. Plant in a neutral to acid, sandy mix in bright, filtered light or part shade. Propagate by semi-ripe cuttings in summer. *A.glandulosus* (TASMANIAN LAUREL; shrub or small tree to 10m tall; leaves 6–18cm-long, lanceolate to elliptic leathery, toothed; flowers 2cm wide, waxy white to pale rose, in 10cm-long racemes in late spring).

Anredera (name derivation unknown). Basellaceae. South America (naturalized Southern Europe). 10 species, woody to herbaceous climbers arising from tuberous roots, with thinly fleshy leaves and small flowers in axillary spikes and racemes. Grow in full sun and with a minimum winter temperature of 5°C/40°F and sturdy support for the heavy climbing growth. Cut back the previous year's growth to just above ground in spring. Propagate by division of tubers in spring, and by tubercles which may form in the leaf axils. *A.cordifolia* (syn. *Boussingaultia cordifolia*, *B.gracilis* var. *pseudobaselloides*; MIGNONETTE VINE; MADEIRA VINE; vigorous, herbaceous vine; leaves to 6m fleshy, bright green and smooth, ʻheart-shaped; flowers fragrant white flowers in drooping spikes in late summer; plants named *A.baselloides* are often *A.cordifolia*. True *A.baselloides* differs in its simple, not 3-lobed style).

antemarginal used of veins, sori and other leaf features lying within, or extending just short of, the margin.

Antennaria (from Latin *antenna*: the pappus hairs of male flowers resemble insect antennae). Compositae. Northern Temperate Regions; Asia; South America. CATS' EARS; EVERLASTING; PUSSY TOES; LADIES' TOBACCO. 45 species, small, often woolly perennial herbs with leaves in basal rosettes and discoid, solitary or clustered flowerheads on erect stems in summer. Frost-hardy but intolerant of wet and cold combined. Plant in full sun. Propagate by division in spring. *A.dioica* (Europe, N America, N Asia; creeping, evergreen; leaves, spathulate, 0.5–4cm-long covered in silvery hairs; flowerheads to 0.5cm across in a terminal cluster resembling tiny everlastings, usually white with a silver-pink tint, but deep rose in the compact cultivar 'Nyewood Variety', rose-pink in 'Rosea' and deep blood red in 'Rubra').

anterior of the surface or part of an organ turned away from or furthest from the axis and projecting forward or toward the base or (in the case of a flower) any subtending bract; close to abaxial but broader in definition, meaning not only 'beneath' but also lower (of two, as in the lower lip of a bilabiate flower) or furthest as in the tip of an organ, cf. *posterior.*

Anthemis (the Greek name for this herb, from *anthemon*, flower). Compositae. Mediterranean to Western Asia. DOG FENNEL 100 species, hairy, aromatic perennial herbs and subshrubs with finely cut leaves and long-stalked, daisy-like flowerheads. Evergreen, hardy and summer-flowering. Plant in full sun. Propagate by division in autumn or spring, and by basal cuttings in spring or late summer.

A.punctata subsp. *cupaniana* (S Europe and N Africa; woody-based perennial making a dense and springy carpet of silvery leaves; flowerheads 6cm-wide, white with yellow centres); *A.tinctoria* (DYER'S CHAMOMILE, YELLOW CHAMOMILE; Europe, W Asia; erect, clump-forming, sparsely hairy to woolly perennial to 60cm tall; leaves ferny, mid-green; flowerheads to 3cm-wide, white to golden or pale cream with golden centres). For *A.nobilis*, see *Chamaemelum nobile.*

anther the pollen-bearing portion of the stamen, either sessile or attached to a filament.

anther cap in orchids, the cap-like case enclosing the pollinia.

anther sac a sac-shaped unit containing pollen bearing sacs disposed in two lobes; the tissues separating the members of each pair usually break down prior to anthesis, giving rise to a biloculate (two-celled) anther.

Anthericum (from the Greek *antherikos*, the flowering stem of the asphodel). Liliaceae (Anthericaceae). Southern Europe; Turkey; Africa (those below from Southern Europe). 50 species, fleshy-rooted, rhizomatous perennial herbs with clumps of grassy or strap-like leaves and slender, erect, simple to branching spikes of six-parted, starry white flowers. Fully hardy, plant in sun. Where temperatures fall far below –10°C/14°F, or winter conditions are wet, apply a dry mulch. Propagate by seed, or by division in spring. *A.liliago* (ST BERNARD'S LILY; to 90cm tall with usually unbranched racemes of 2cm-wide flowers in early summer); *A.ramosum* (slightly shorter than the first species with much-branched stems of flowers in summer; leaves grey-green).

anthesis the expansion or opening of a flower; the period in which the anthers and stigmas become functional, enabling fertilization to take place.

anthocorid bugs (Hemiptera: Anthocoridae). Small, usually predacious insects, about 2–4mm long, mostly black or brown and resembling small capsid bugs. *Anthocoris nemorum* preys upon aphids, scale insects, gall-midge larvae, small caterpillars, thrips and mites. Adults can consume an average of 50 fruit-tree red spider mites in a day or 500–600 aphids during the lifetime of one individual. *A.nemorum* and *A.nemoralis* are important predators of the aphid *Pemphigus bursarius*, which lives during the spring in galls on the petioles of poplar leaves before migrating to the roots of outdoor lettuce. Anthocorids enter the galls and a single nymph can rapidly kill the whole aphid colony, which averages 100 individuals.

anthocyanins pigments frequently found in cell sap, and responsible for red, blue and intermediate colours in flowers, foliage and stems.

anthracnose A group of fungal diseases usually caused by *Colletotrichum* species, characterized by black, sunken lesions on which pustules of mucilaginous spores are produced.

ANTIRRHINUM ANTHRACNOSE (*Colletotrichum antirrhini*) affects greenhouse-grown antirrhinums under damp growing conditions; it also occurs on plants outdoors at the end of summer. Leaf and stem lesions can result in the collapse of plants.

BEAN ANTHRACNOSE (*C.lindemuthianum*) is a major disease of haricot beans which can also affect scarlet runners, lima beans, cowpeas, broad beans and other legumes. Particularly damaging in wet conditions and under protected cultivation, it is readily spread by rain-splash, overhead irrigation and cultural operations. Symptoms are circular brown to black spots on pods developing a pinkish ooze of spores in damp weather, which leads to seed-borne infection, the usual route for introduction into gardens.

BRASSICA ANTHRACNOSE (*C.higginsianum*) is a seed-borne disease which can damage Chinese cabbage, mustard greens, turnips and other brassicas in warmer areas of the US.

CLIVIA ANTHRACNOSE (*C.cliviae*) may be the fungus responsible for the reddish spots commonly seen on *Clivia* foliage. No control is usually needed.

CUCURBIT ANTHRACNOSE (*C.lagenarium* or *C.orbiculare*) attacks all above-ground parts of cucurbits; it is particularly serious on protected cucumbers and on watermelons. Cucumber seedlings may suffer from damping-off. Pale green leaf spots occur later, which enlarge, dry up and may kill whole leaves. Sunken lesions are produced on stems and on developing fruits, causing cracking. Poorly ventilated conditions encourage spread of the disease. On watermelons, leaves develop small black spots which spread until the foliage becomes black and shrivelled. Fruits may be covered with numerous sunken lesions and also suffer sunburning as a side-effect of defoliation. The disease can become epidemic in conditions of high rainfall and temperatures of 24°C/75°F and over, and is a widespread problem on the Pacific Coast of the US, but rare in regions of low rainfall.

DIGITALIS ANTHRACNOSE (*C.fuscum*) causes damping-off of seedlings on foxgloves and other Scrophulariaceae. It may also cause serious damage to older plants as small purple-brown spots on leaves, and lesions on stems and petioles. It is seed-borne and may be prevented by hot-water treatment of seed.

HOLLYHOCK ANTHRACNOSE (*C.malvarum*) is a serious disease in North America, but less damaging in Europe. Seedlings are blighted and older leaves and stems show black spotting and blotching. *Abutilon*, *Lavatera* and *Malva* can also be infected.

STEM ANTHRACNOSE (*C.truncatum*) is a serious seed-borne disease of soy beans in the southern US. Reddish lesions occur on stems, leaves and pods, and if infection is heavy plants are stunted or killed.

Anthracnose of many plants is caused by *Glomerella cingulata* (anamorph *Colletotrichum gloeosporioides*). In tropical and subtropical areas it has an extremely wide host range as a pathogen, as a secondary fungus following other infections, injury or stress, and as a saprophyte. It is a common cause of leaf lesions, shoot dieback (withertip) and fruit spotting. In addition to anthracnose, the fungus causes cankers and fruit rots. The following are among the many anthracnose-type diseases caused by *G.cingulata*.

Fruit spotting of avocado, banana, guava, loquat, mango and papaya is a serious problem for which control measures include hot fungicidal dips, fruit wrapping on the tree, treatment with fungicidal waxes and gamma radiation. The natural microflora on fruit inhibit germination and penetration of spores, and encouraging these flora may lead to development of a biological control.

MANGO ANTHRACNOSE may infect fruits so badly that they drop before ripening or degenerate on the tree. It remains quiescent until fruit ripening after harvest, when it can cause serious losses during shipment and at point of sale. Fruit set may also be badly affected by earlier blossom blight. There is varietal resistance. Control with fungicide needs to be combined with hygiene measures, including the collection and destruction of diseased fruit and foliage.

In the US, PRIVET ANTHRACNOSE and TWIG BLIGHT are widespread and serious diseases on *Ligustrum* and the dogwoods. They are controlled by the use of resistant species and the destruction of diseased foliage.

Since spores are produced in a mucilaginous mass and dispersed by water, anthracnose diseases are encouraged by wet conditions and are particularly common in rainy climates, where control by spraying is difficult. Where chemical treatments can be used, copper-based and other general fungicides are usually effective. The level of disease can be reduced by crop rotation, the use of healthy seed, and careful disposal of crop residues, on which the fungus is able to continue to grow at the end of the season.

Anthriscus (from Greek, *anthriskos*, chervil). Umbelliferae. Europe; N Africa; Asia. 12 species, hardy annual, biennial and perennial herbs with fine, ternately divided leaves and in spring and summer, small, white to cream flowers in compound umbels. *A.cereifolium* is the herb CHERVIL, (q.v.). *A.sylvestris* is a beautiful addition to the wild garden, where it will naturalize itself, making banks of soft, foamy flower. It will grow on most soils in sun or part shade. Increase by seed sown *in situ*.

A.cereifolium (CHERVIL; annual to 60cm tall; leaves bright green, delicately scented of aniseed); *A.sylvestris* (COW PARSLEY, KECK; inedible biennial or perennial to 1m tall with spreading, ferny foliage and lacy flowerheads borne in spring; 'Moonlit Night': foliage solidly deep bronze-purple, flowers white; 'Ravenswing': foliage purple-black, flowers off-white – both cultivars come more or less true from seed).

Anthurium (from Greek, *anthos*, flower, and *oura*, tail, referring to the tail-like spadix). Araceae. Tropical Americas. FLAMINGO FLOWER; TAIL FLOWER. Perhaps 900 species, tender, evergreen, perennial herbs, many epiphytic, with climbing stems or, as is the case of those below, a short, rooting stem bearing a clump of long-stalked, leathery leaves. Produced throughout the year, the flowers are very small and packed in a slender spadix backed by a showy, shield-like spathe. Grow in dappled sun or shade with a minimum temperature of 15°C/60°F and medium to high humidity. Plant in small pots containing an open, neutral to acid, soilless mix high in coarse bark and leafmould. Keep moist at all times, never waterlogged. Apply a weak liquid feed monthly. Propagate by rooted offsets.

A.andraeanum (Colombia and Ecuador; to 60cm tall; spadix white to yellow, tapering, pointing downwards toward the tip of a 6–15cm-long, heart-shaped, puckered spathe in bright waxy red; cultivars with white, flamingo pink, golden, vermilion and scarlet spathes are available); *A.crystallinum* (Panama to Peru; leaves 30–50cm-long, broadly heart-shaped, soft, bronze-pink at first becoming deep, satiny emerald green overlaid with glistening, silver-white veins; spadix purple-brown backed by a narrow, purple-red spathe); *A.scherzerianum* (Costa Rica; to 30cm; spathe elliptic to ovate, glossy scarlet to orange-red; spadix to 12cm long, orange-red, spiralling; 'Atrosanguineum': spathe deepest blood red 'Rothschildianum': spathe red spotted: white, spadix yellow; 'Wardii': spathe large, deep burgundy).

Anthyllis (the classical name). Leguminosae. Mediterranean. 25 species, annual and perennial herbs or shrubs with pinnate leaves and heads of pea-like flowers in spring and summer. Hardy to at least –10°C/14°F, but resentful of cold *and* wet. Plant in full sun. Propagate by semi-ripe cuttings, or by detaching rooted branches.

A.hermanniae (domed, bushy perennial to 1m tall with a tangle of spreading, spiny-tipped, grey-hairy branches; leaves trefoil-like or reduced to a single, bright green leaflet; flowers yellow); *A.montana* (spreading, woody-based perennial to 30cm tall with grey-green to mid-green leaves composed of many leaflets; flowers pale pink marked red, in dense heads; 'Rubra': leaves grey, flowers crimson).

antibiotics substances produced by micro-organisms, especially fungi and actinomycetes, which are toxic to other micro-organisms. As a spin-off from medical research, streptomycin and terramycin have been used to control bacterial disease and downy mildew in plants, but this practice is questionable on public health grounds. Antibiotics specifically developed to control plant diseases include actidione (against powdery mildews, leaf spots and some rusts) and kasugamycin (against bacterial and fungal disease).

The use of antibiotics for the control of plant pathogens requires a thorough understanding of the implications for human health.

antidesiccant see *antitranspirant*.

Antigonon (from Greek, *anti*, against, and *gonia*, an angle, referring to the flexuous stems). Polygonaceae. Central America. 3 species, rampant perennial vines, usually tuberous-rooted, with axillary, tendril-tipped racemes of small flowers with colourful papery sepals. A vivid, fast-growing climber which, in frost-free gardens, will quickly cover any support or structure. Elsewhere, it should be grown in the cool greenhouse or conservatory or, in zones 8 and 9, outdoors in very sheltered positions as a deciduous perennial with a thick mulch in winter. Grow in full sun. Increase by division or seed. *A.leptopus* (CORAL VINE, CONFEDERATE VINE, MEXICAN CREEPER, CHAIN OF LOVE, ROSA DE MONTANA, QUEEN'S WREATH; Mexico; to 12m; flowers coral pink, white in 'Album', hot pink to crimson in 'Baja Red', produced chiefly in summer).

antipetalous opposite to or superposed on a petal, i.e. not alternate.

Anthurium scherzerianum

Antirrhinum (from Greek *anti*, opposite, and *rhis*, snout, referring to the corolla shape). Scrophulariaceae. SNAPDRAGON. Temperate regions. 40 species annual and perennial herbs and subshrubs with tubular, 2-lipped flowers solitary in the leaf axils or in terminal racemes; these have a billowing upper lip and a tongue-like lower lip with a conspicuous, bearded palate. The typical *A.majus* is short-lived with a woody base and rather ragged habit. Most cultivars, however, are treated as annuals or biennials and retain their vigour and bushiness; they range in height from tall (to 1m), to intermediate (to 45cm) and dwarf (to 30cm). The larger cultivars are suitable for cutting. In addition to these, more conventional snapdragons, new races of pendulous hybrids are becoming popular. These have grey-green foliage and a loose, cascading habit ideal for hanging baskets and containers. Examples include the deep rose and yellow 'Rose Pink' and 'Primrose Vein', white flushed yellow and rose, combed with red. Grow in full sun on a fertile, well-drained soil. Deadhead regularly. Sow seed *in situ* in late spring, or take cuttings in spring or early autumn.

A.majus (SNAPDRAGON; SW Europe, Mediterranean; erect, shrubby perennial to 1m; flowers in narrow, terminal racemes, tubular some 5cm long with two distinct lips and an enlarged palate; typically purple or pink, but the many seed races now available encompass plants with white, yellow, apricot, orange, flame, rose, red and purple flowers, some with mixtures of colour and differently toned lips or palates; flower forms range from small (3cm) to large (7cm), single and straight-tubed to penstemon- and trumpet-shaped, peloric and double).

antisepalous opposite to or superposed on a sepal, i.e. not alternate.

antitranspirant a chemical preparation applied to aerial parts of plants to reduce water-loss from foliage and shoots by sealing stomatal pores. Emulsions of plastic polymers are commonly used on newly transplanted evergreen woody plants to aid establishment. Some contribute a protectant effect against pests and diseases. Also known as an antidesiccant.

antrorse turned, curved or bent upward or forward, toward the apex; cf. *retrorse*.

ants (Hymenoptera: Formicidae) a very large family of small social insects, which live in nests containing wingless infertile females and at least one fertile female or queen. During summer, usually in hot humid weather, winged males and females swarm in mating flights; fertilized females shed their wings and choose a suitable site to establish a new nest. Some queens are capable of living for over 15 years. Ants feed on sugary substances such as nectar, as well as on oil-rich seeds, other ants and small invertebrates. Many species have a close association with aphids, scale insects and mealybugs, imbibing their excreted honeydew; for some sap-feeding insects this is an essential service. Ants assist aphids by warding off predators and parasites, and there is evidence that they are capable of transporting aphids to better feeding sites.

Plants are seldom attacked directly by temperate species of garden ants, but seed collecting is common, particularly oil-rich seeds such as *Buddleja*, *Meconopsis*, *Primula*, *Rhododendron* and *Viola*. Worker ants move along defined pathways between the nest and food source, laying down trail pheromones to mark the route. When nests are built under plants, extensive tunnelling and loosening of the soil around roots can cause plants to wilt, especially those growing in free-draining soils and in pots. Excavated soil is usually deposited on the surface, so that low-growing plants may become partially buried; this can be especially troublesome on lawns.

The COMMON BLACK ANT (*Lasius niger*), with workers about 4mm long, is prevalent in gardens throughout Europe and strains occur in North America. Nests are found beneath plants, stones, pavements and masonry, in banks and old tree stumps. The MOUND ANT (*L.flavus*) has similar-sized workers which are mostly pale yellow. This species produces ant hills which can reach the size of mole hills. PHARAOH'S ANT (*Monomorium pharaonis*) is a ubiquitous species, having been spread throughout the world by shipping. Workers are red-yellow and about 2mm long. *Pheidole* species are of comparable size, but workers are darker brown and the genus is notable in having a soldier caste, equipped with enlarged mandibles. RED ANTS (*Myrmica* species) are common in gardens; the workers are larger than those of *Lasius* and able to sting. LEAF-CUTTING ANTS (*Atta* species) occur in South America and the southern states of the US, where they are sometimes serious pests of tree crops and ornamentals; they cut leaf pieces from a wide range of plants and carry them back to their nests, sometimes virtually defoliating individual trees. A fungus on the decaying leaf fragments provides food for the ants.

Ants are best controlled by destroying the nest with boiling water or appropriate residual insecticides. Where nests are inaccessible, insecticides should be applied around the entrance holes and along pathways; proprietary baits may also be used, but these may need frequent renewal.

apetalous lacking petals.

apex the growing point of a stem or root; the tip of a plant organ or structure, most commonly used of a leaf tip.

Aphelandra (from Greek, *apheles*, simple, and *aner*, *andros*, man: the anthers are one-celled). Acanthaceae. Tropical America. 170 species, tender evergreen shrubs and subshrubs with tough, ovate to elliptic leaves and terminal spikes of tubular 2-lipped flowers amid overlapping bracts. Grow in bright, filtered light or light shade with a minimum temperature of 13°C/55°F. Pot in a soilless, free-draining mix; keep moist throughout the year and feed fortnightly when in growth. Both species flower at various times of year. The bracts of *A.squarrosa* are a long-lasting attraction. Cut back and repot after flowering. Root tip cuttings in a sandy mix in warm, humid conditons.

A.aurantiaca (Mexico, Colombia; shrub to 1m, the finest form is 'Roezlii' (FIERY SPIKE) with leaden, puckered leaves marked with silver veins; flowers brilliant scarlet, emerging from overlapping purple-tinted bracts in 4-sided spikes to 18cm long); *A.squarrosa* (ZEBRA PLANT, SAFFRON SPIKE; Brazil; shrub to 2m tall but usually sold and kept small, newly propagated, bushy plants; stems purple-tinted; leaves to 25cm long, glossy dark green with sunken herring-bone veins heavily outlined in white or gold; flowers maroon-flecked yellow, peeping from over-lapping, waxy, yellow bracts of 20cm-long 4-sided spikes; 'Leopoldii': leaves dark green, veined pure white, flowers yellow amid maroon bracts).

aphids (Hemiptera: Aphididae and other families) small, soft-bodied, sap-feeding insects, about 1–5mm long, with relatively long legs and antennae and a pair of horn-like structures (the siphunculi) on the back of the abdomen; GREENFLY, BLACKFLY, PLANT LICE, and BLIGHT are other descriptions for aphids. They may be black, green, yellow, pink or grey; in some, such as the WOOLLY APHID, the body is obscured by a covering of white waxy filaments.

Aphids feed by inserting sucking mouthparts into the phloem of plants: the resulting injury takes various forms including leaf curling, stem distortion, galling, malformed fruits, defoliation and, with root-feeding aphids, wilting and root splitting. Indirect damage is caused by the copious excretion of honeydew, which coats underlying foliage and becomes infected by fungal sooty moulds, so that affected leaves carry out their normal functions less efficiently.

Most aphids have a single host or a narrow range of host plants. Summer colonies usually consist of winged and wingless females that produce live young without fertilization. Winged aphids are weak fliers, but can be carried by air currents for long distances. Migratory species can find alternate hosts in this way and although mainly a minority group, are significant plant pests. They include the BLACK BEAN APHID (*Aphis fabae*), which overwinters as eggs on *Euonymus* and *Viburnum opulus* and migrates in late spring to beans and other hosts, and the LETTUCE ROOT APHID (*Pemphigus bursarius*), which migrates from poplar to the roots of lettuces and sow thistle. Like many other aphids, they produce sexual forms in autumn, and fertilized females lay shiny black eggs on the stems of woody plants. Some aphids overwinter as active stages.

Many aphids, particularly the PEACH-POTATO APHID (*Myzus persicae*), are vectors of viruses such as bean yellow mosaic, cauliflower mosaic, dahlia mosaic, lettuce mosaic, pea leaf-roll, plum pox, potato leaf-roll, tomato aspermy and tulip-breaking virus. *M.persicae*

overwinters either as eggs on peaches or nectarines, or as active stages under glass or outdoors in warmer countries. In summer, sparse colonies occur on a wide range of ornamentals and vegetables both outdoors and under glass. Two other *Myzus* species, the SHALLOT APHID (*M.ascalonicus*) and the VIOLET APHID (*M.ornatus*) also attack a wide range of host plants. Other common polyphagous species include the MOTTLED ARUM APHID (*Aulacorthum circumflexum*), a yellow aphid with a dark horse-shoe-shaped mark on the abdomen, which is widespread on greenhouse and house plants; the green-yellow greenhouse and POTATO APHID (*A.solani*), which occurs in Europe, Kenya, Japan, New Zealand, Peru and the US and attacks vegetables and ornamentals both outdoors and under glass; and the POTATO APHID (*Macrosiphum euphorbiae*), which is green or pink and attacks vegetables and ornamentals in Europe, Asia, Australia and the Americas.

Aphids that overwinter in the egg stage on deciduous trees and shrubs, especially fruit plants, can be controlled by the application of a winter wash when growth is dormant. In spring and summer aphids are readily susceptible to a number of contact and systemic insecticides, and early and repeated treatment is necessary. The instructions provided with proprietary products must be followed, especially with regard to persistence in edible crops. The choice of chemical sprays should have regard for useful predators and parasites. Aphids have many natural enemies including ladybirds, lacewing and hoverfly larvae, and braconid and chalcid wasps, some of which can be purchased for biological control. Aphids may be attacked by insect feeding fungi, and one of these, *Verticillium lecanii,* has been used to control certain species on greenhouse plants. See *blackfly, root aphids, woolly aphid.*

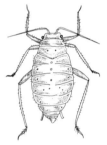

aphid

Aphyllanthes (from Greek, *aphyllos*, leafless, and *anthos*, flower, referring to the rush-like stems). Liliaceae (Aphyllanthaceae). SW Europe; Morocco. 1 species, *A.monspeliensis*, a perennial herb with a mop of wiry and chalky, leafless stems to 30cm long picked out in late spring and summer with 2cm-wide, 6-parted, lilac to bright blue flowers. Hardy to –5°C/23°F. Plant in sheltered sunny niches on the rock garden on a well-drained, humus-rich and sandy soil. Propagate by division.

aphyllous lacking leaves.

apical borne at the apex of an organ, farthest from the point of attachment; pertaining to the apex.

apical dominance suppression of the growth of lateral buds by a shoot tip, or apex. The dominance of the apical meristem accounts for the upward growth of most plants, which may be suppressed by natural plant growth processes or by stopping or pruning to produce branched, bushy or low growth. An apical bud is the terminal or topmost bud on a stem.

apicule a short sharp but not rigid point terminating a leaf, bract or perianth segment.

Apium (Classical Latin name for this plant). Umbelliferae. Europe; temperate Asia. 20 species, annual, biennial and perennial herbs. See *celeriac*, *celery*.

apogeotropic see *ageotropic*.

apomixis, apomict generally used to describe the process of reproducing without the fusion of gametes (asexually), which is also sometimes referred to as parthenocarpy. A plant in which seeds or seed-like structures develop from such unfertilized cells is termed an apomict, and its offspring are usually uniform. The phenomenon is common in plants such as *Rubus*, *Hieracium* and *Taraxacum*.

The term apomict is sometimes used to describe the product of other asexual (vegetative) means of increase, as an alternative to the term *clone*.

Aponogeton (from Aquae Aponi, the Latin name for the healing springs of Bagni d'Abano, plus Greek *geiton*, a neighbour. This name was originally applied to another aquatic, *Zannichellia*). Aponogetonaceae. Old World Tropics and Subtropics. 44 species, aquatic perennial herbs with bottom-rooting, tuberous rhizomes and submerged or floating, long-stalked, oblong to oval leaves. Small, often fragrant flowers are produced in summer on forked, spike-like racemes held just clear of the water. *A.disachyos* will tolerate lows to –5°C/23°F and needs fairly deep, still water in full sun. Plant in spring into the pond floor or in baskets of sticky loam plunged to a depth of 25–60cm. Grow *A.madagascariensis* in warm greenhouse ponds and aquaria that do not fall below 24°C/75°F. Plant in a mix of loam, sand and leaf-mould with a deep top-dressing of fine gravel. Propagate both species by division in spring.

A.distachyos (CAPE PONDWEED, WATER HAWTHORN; South African; leaves floating, bright green to 25cm long; flowers sweetly scented and with purple-brown to black anthers in two rows, one along each 10cm branch of a pinnately lobed, waxy white, forked raceme); *A.madagascariensis* (Madagascar; leaves 10–50cm long, lanceolate to oblong, fully submerged, leaf veins prominent, closely woven – the tissue that would normally connect them is absent or transparently thin, leaving a beautiful emerald latticework).

Aporocactus (from Greek *aporos*, vexatious, a reference to the complex classification of Cacti). Cactaceae. Mexico. 2 species, epiphytic or litho-phytic cacti with trailing or cascading, succulent, cylindrical stems and tubular flowers with a spreading limb. Plant in an open, acid mix rich in coarse bark, sand and grit. Choose a container or situation where the stems can cascade. Position in full sunlight with a minimum temperature of 5°C/40°F. Water freely in hot weather, scarcely at all

Aponogeton distachyos

apple

Pests
aphids
capsid bug
codling moth
sawfly
tortrix moth
winter moth

Diseases
brown rot
canker
crown gall
fireblight
powdery mildew
scab

Disorder
bitter pit

in winter. Propagate by rooting detached segments of stem. *A.flagelliformis* (RAT'S TAIL CACTUS; stems to 1m × 1.5cm with many short, yellow-tinted spines; flowers 5–8cm and scarlet to dark magenta).

appendage secondary part or process attached to or developed from any larger plant organ, for example, the leafy appendages terminating the lower inflorescence bracts of some *Heliconia* species or the whisker-like appendages of a *Tacca* inflorescence.

appendiculate furnished with appendages.

appendix a long, narrowed development of the spadix in Araceae.

applanate of plant parts, flattened.

apple (*Malus domestica*) The many cultivated forms are derived from hybridization between *Malus sylvestris* and *M.pumila,* both widespread in Southwest Asia, *M.sylvestris* (WILD CRAB) being common in Western Europe including Britain. There is evidence that apples were a human foodstuff as early as 6500 BC, and of their progressive introduction through the Middle East, Egypt, Greece and Western Europe, whence their cultivation spread throughout the world during the colonial period.

Good quality fruit production requires high summer temperature together with moderate, evenly distributed rainfall of 600–800mm per annum. Trees must be afforded relative freedom from frost, and shelter from strong wind. Fertile soil to a depth of 60cm is ideal, and free drainage essential. Natural hybridization has been an important source of good cultivars: the widely grown dessert apples 'Cox's Orange Pippin' and 'Golden Delicious' originated as chance seedlings. Many introductions result from he efforts of amateur selectors and breeders, but there are scientifically based breeding programmes throughout the world which continue to produce new cultivars by traditional methods and by relatively new techniques such as the irradiation of propagating material to induce genetic change. *Malus baccata* features in breeding initiatives to improve hardiness and disease resistance.

Propagation is by budding or grafting; the choice of rootstock influences the ultimate size of a tree. Commonly used are M27 (extremely dwarfing; an excellent choice for vigorous cultivars and container-grown trees; requires very good soil and permanent staking); M9 (very dwarfing and widely used; trees usually come into bearing within 3 or 4 years to produce a high percentage of good-sized fruit; requires permanent staking); M26 (dwarfing; a good choice for average soil conditions; trees usually come into bearing within 4 years); MM106 (semi dwarfing; suitable on most soils; often used for espalier and fan forms); and MM111 (vigorous, making large trees on good soils and moderately sized trees on poor soils).

Apples may be trained in various forms to suit garden size and layout; the commonest choice is the dwarf bush and cordon but dwarf pyramid, spindle bush, fan and pole are all suitable; standard and half standard are more challenging to manage in small gardens. Apple cultivars are not consistently self-fertile and depend on cross-pollination by insects and local advice should be sought on compatibility and flowering time prior to selection.

Plant in early winter at distances depending on rootstock, cultivar and local conditions. The following range is a general guide for garden purposes: bushes, 3–5m stations in rows 5–6m apart; cordons, 75cm apart in rows 1.8m apart; dwarf pyramids 1.2m apart in rows 2m apart; spindle bushes, 2m apart in rows 4m apart; fans and espaliers, 2.5m apart on dwarfing rootstocks and 3–4m apart on more vigorous ones.

Flowers and fruit are produced on shoots two or more years old and on short spurs on the older wood; fruit thinning is essential where heavy crops are likely. Established trees of trained forms such as cordons, espaliers and dwarf pyramids require pruning during late summer, cutting back laterals of the current season's growth to three leaves above the basal cluster where they arise direct from the main stem, or to one leaf in the case of shoots arising from existing side shoots or spurs. Secondary growth from later in the season is pruned in the winter, together with essential thinning and training. Free-growing forms such as bushes, spindle bushes and half-standards are winter pruned to stimulate growth the following season, and to remove dead and crossing shoots and maintain shape. Vigorously growing trees should be pruned more lightly than weak ones, which will benefit from tip pruning of branch leaders by up to a quarter of their length and the shortening of selected laterals. Successful pruning requires practice, and a basic understanding of growth habit and performance: continual hard pruning usually results in limited fruiting. Notable cultivars include: *Dessert, early*, 'Discovery', 'Epicure'; *midseason*, 'Ellison's Orange', 'James Grieve'; *late* 'Ashmead's Kernel', 'Gala'. Culinary, *early*, 'Emmeth Early', 'Grenadier'; *mid season*, 'Golden Noble', 'Peasgood's Nonsuch'; *late*, 'Bramley's Seedling', 'Lane's Prince Albert'.

appressed see *adpressed*.

approximate of plant parts drawn very closely together, sometimes confused with proximate.

apricot (*Prunus armeniaca*). Found in the wild throughout Central Asia but probably originating in China. Apricot cultivation in England is documented from around the mid 16th century, and trees were distributed during the 18th and 19th centuries to flourish in warmer climates, including Southern Europe, Australia and the US, where there are now considerable areas of commercial plantation. Apricots need a warm, sunny, sheltered position and in cool temperate areas do best where fan-trained against a tall south-facing wall. They may be similarly trained under greenhouse protection, and this is the most promising method for gardens in northerly latitudes. Since apricot is one of the earliest flowering fruits, protection from frost damage with temporary wall covers is desirable in cool areas. Even with care, cropping is usually variable in the UK.

Propagation is effected by grafting onto 'St Julien A' plum rootstock or seedling apricot or peach . The dwarfing rootstock 'Pixy' is a possibility but double working with 'St Julien A' is necessary to overcome incompatibility. Apricots are self-fertile but benefit from hand pollination by lightly touching the open flowers with a camel-hair brush or rabbit's tail around midday. Flowers and fruit are borne on shoots of the previous year and on spurs on older wood. To encourage spur formation, developing laterals should be pinched back to 7.5cm, or where growth is vigorous to 14cm; subsequent sublaterals should be shortened back to one leaf. Shoots needed to fill spaces must be tied in whilst still pliable.

Aptenia (Greek *apten*, wingless; the valves of the seed capsules lack wings). Aizoaceae. South Africa. 2 species, dwarf succulents with bushy, trailing stems and fleshy leaves. Small, daisy-like flowers are produced during warm and sunny spells throughout the year. Useful for edging, groundcover and baskets in frost-free gardens and, elsewhere, for outdoor summer display, and for the cool greenhouse. Pot in a sandy, loam-based mix; water moderately in spring and summer, very sparingly in winter. Position in full sun or light shade. Propagate by stem tip cuttings at any time of year. *A.cordifolia* (syn. *Mesembryanthemum cordifolium*; dwarf bush or cushion to 12 × 40cm; leaves mid to bright green, oval to heart-shaped, 0.5–1.5cm long; flowers 1cm across, purple-red; 'Variegata': leaves bordered cream).

aquatic a plant growing in water. Aquatics are often divided into floating plants, submerged or oxygenating plants, and marginals.

Aquilegia (from Latin *aquila*, eagle, alluding to the aquiline spurs). Ranunculaceae. COLUMBINE, MEETING HOUSES, GRANNY'S BONNETS, HONEYSUCKLE. N Hemisphere) 70 species, short-lived, evergreen perennial herbs, with a short, erect rhizome becoming woody, and a clump of rather ferny, ternately divided leaves, basically triangular in outline with wedge-shaped to rounded leaflets, their margins often cleft or cut. The flowers appear in summer, one to several atop a slender stem. They consist of five petal-like sepals, often spreading to reflexed, and five petals pointing forwards in a cup and projecting behind as slender, nectar-filled spurs. They are fully hardy, but species from the Southwest US prefer a warm and sheltered situation and are intolerant of overwet winter conditions – as are *A.alpina*, *A.jonesii*, and *A.viridiflora*, which are smaller plants suited to protected niches in the rock garden or pots in the alpine house. Grow in full sun on any fertile, free-draining soil. Sow seed *in situ* in spring or autumn. These plants hybridize freely and

apricot

Pests
birds
scale insects

Diseases
bacterial canker
(see bacterial diseases)
Eutypa canker
(see canker)
peach leaf curl
rust
silverleaf

Aquilegia canadensis

are in any case highly variable. Unless kept far apart or painstakingly pollinated, seed races and cultivars are unlikely to come true from seed, although a few, like *A.vulgaris* 'Nora Barlow', are reliable. Susceptible to mildew.

A.alpina (ALPINE COLUMBINE; Alps;15–80cm tall; flowers nodding, sepals bright blue to violet, to 4cm long, petals shorter and paler, sometimes white, spurs straight or curved, to 2cm); *A.atrata* (Alps, Apennines; to 80cm tall; flowers nodding, rich purple-violet to darkest maroon, sepals to 4cm, petals shorter, spurs to 1cm, hooked); *A.caerulea* (ROCKY MOUNTAINS COLUMBINE; W US; to 60cm tall; flowers erect, sepals to 4cm, blue, pink or white, petals shorter, white or cream, spurs to 5cm, straight or curved; includes the snow-white 'Candidissima', long-spurred blue and white 'Heavenly Blue', 'Helenae', a vivid blue cross with *A.flabellata*, the pale pink 'Coral', and 'Rostern' with carmine and cream flowers); *A.canadensis* (CANADIAN COLUMBINE, MEETING HOUSES, HONEY-SUCKLE; eastern N America; 30–60cm tall; flowers nodding; sepals to 1.5cm, not reflexed, red, petals to 1cm, yellow, spurs to 2.5cm, straight, red); *A.chrysantha* (GOLDEN COLUMBINE; SW US; 30–100cm tall; flowers held horizontally, sepals to 3cm, yellow sometimes tinged pink, petals shorter, yellow, spurs 3–7cm, curving outwards; includes the creamy white 'Alba', the double-flowered 'Florepleno', large, primrose yellow 'Grandiflora Sulphurea', and large, golden 'Yellow Queen'); *A.flabellata* (E Asia; 20–50cm tall; flowers nodding, sepals to 2.5cm, blue-purple to lilac or white, petals shorter, white or pale lilac tipped yellow-white, spurs to 2cm, curving; var. *pumila* 'Alba' is dwarf and white-flowered); *A.formosa* (WESTERN COLUMBINE; western N America; 50–100cm tall; flowers nodding, sepals to 2.5cm, red, petals shorter, yellow, spurs stout, straight, red; includes dwarf forms with paler, or white flowers, the double, red 'Rubraplena', and var. *truncata* (syn. *A.californica*), a tall plant with flame-coloured sepals and very short golden petals). *A.jonesii* (Rocky Mountains; to 10cm tall, flowers erect on short, slender stems, sepals to 1.5cm, blue to purple, petals shorter, blue, broad, spur to 1cm, incurved); *A.longissima* (SW US, to 100cm; flowers erect, sepals 2.5cm, yellow sometimes tinged red, petals same length as sepals, yellow, spurs to 15cm long, very slender, straight to drooping); *A.scopulorum* (W US; 10–20cm tall; flowers erect, sepals to 2cm, blue to white, sometimes red, petals shorter, white to pale yellow, spur 3–5cm, straight, blue to red); *A.viridiflora* (Siberia, China; 15–30cm; flowers to 3cm nodding, fragrant, sepals to 2cm, green, petals shorter, chocolate to maroon); *A.vulgaris* (COLUMBINE, GRANNY'S BONNETS; Europe, naturalized N America; 30–70cm tall; flowers nodding, blue to purple, red or white, sepals to 2.5cm, petals shorter usually white to yellow, petals shorter, spurs to 2cm, strongly hooked; many cultivars and seed races developed with yellow, white, pink, rose, red-purple, lilac, mauve, sky blue and dark blue flowers, some bicoloured, double, or lacking spurs; one of the most intriguing cultivars is 'Nora Barlow' – pom-pom flowers with ragged sepals

and petals in pale red or pink edged lime green and fading to cream; another group of cultivars, variously called 'Woodside' or Vervaeneana Group has leaves variegated lime and gold).

Arabis (from Greek *arabia*, Arabia). Cruciferae. North America, Europe, Asia. ROCKCRESS. 120 species, small, hardy, annual to perennial herbs with spathulate to obovate leaves, often toothed and hairy, and 4-petalled flowers in racemes in spring and summer. Suitable for the rock garden, dry walls and the alpine house. Grow in full sun on a fast-draining, loamy soil with added grit. Propagate by softwood cuttings in summer in a sandy mix in a shaded frame, or by division or seed in spring and autumn. Attacked by arabis midge: affected parts should be destroyed.

A.alpina (Europe; 5–40cm; flowers white or, rarely, pink; includes small, large, white, rose and variegated cultivars); *A. × arendsii* (a garden hybrid between *A.aubretoides* and *A.caucasica* with rose-red flowers; includes several cultivars with flowers ranging through pink, red, lilac and maroon); *A.blepharophylla* (California; to 10cm; flowers rose-purple; includes white and carmine cultivars); *A.caucasica* (syn. *A.albida*; South Europe; similar to *A.alpina*, with grey-green leaves; flowers white to pink; includes cultivars with variegated leaves and double flowers); *A.ferdinandi-coburgi* (Bulgaria; to 20cm tall; flowers white; includes cultivars with ivory-, lime- and gold-variegated leaves).

arabis mosaic virus a disease which affects many different plants but is actually insignificant on *Arabis*. It causes raspberry yellow dwarf disease, where leaves show yellow spots and yellow vein clearing; affected plants are stunted and produce few fruits. It is one of several viruses that affect rhubarb and strawberry with a yellow leaf mosaic and general stunting. Among susceptible ornamentals are *Forsythia*, jasmine, privet, *Sambucus* and *Syringa*. Most strains of the virus are spread by the root feeding of *Xiphenema diversicaudatum*, a free-living, plant-parasitic soil nematode. Nematodes acquire and transmit the virus very rapidly: they survive in soil for a long time, remaining infective, and cannot be eradicated easily. Infected plants must be removed; where appropriate, plant virus-free stocks, on new sites. See *virus*.

Arachnis (from Greek *arachnis*, spider, referring to the appearance of the flowers). Orchidaceae. SE Asia. SCORPION ORCHID. 7 species, tender evergreen orchids with tall, slender scrambling stems clothed with strap-shaped leaves. Carried in slender racemes and panicles throughout the year, the flowers are waxy and consist of five, narrowly oblanceolate, curving tepals and a small, tongue-like lip. The flowers are long-lived, last very well when cut and are common in floristry. Cultivation as for *Vanda* but with a minimum temperature of 15°C/60°F, full sun and high humidity. *A.flos-aeris* (flowers to 7cm across, dark green, cream, pale rust or dull, fiery orange with broad, maroon to fox red or chocolate bands).

Arachnis flos-aeris

arachnoid interlaced with a cobweb of fine white hair, as in the leaves of some tightly growing succulent rosette plants.

Aralia (from the French-Canadian name *aralie*). Araliaceae. Asia; Americas. 40 species, evergreen or deciduous trees, shrubs, climbers and suckering herbs, often with prickly stems and leafstalks and 1–3 × pinnately compound leaves. The flowers are small, green-white and produced in abundant umbels on large panicles in spring and summer. Plant in full sun or light shade in a sheltered postion on a freely draining but moist and fertile loam. The following species is fully hardy but will produce taller stems and more luxuriant foliage in warm, protected locations. Old or cold-damaged stems will die and should be cut out cleanly at the base. They will be replaced by new suckers. Propagate by division of clumps, removal of rooted suckers and root cuttings potted in early spring and rooted in heat. *A.elata* (syn. ANGELICA TREE; syn. *A.chinensis*; E Russia, China, Korea, Japan; clump-forming suckering, deciduous shrub to 10m; stems narrow, pithy, sparsely branched, erect and spiny; leaves to 1.5m long, triangular in outline, bipinnate with toothed mid to grey-green leaflets to 12cm long; inflorescence creamy-white, to 1m, assuming a pink tint as the summer advances; fruit glossy, purple-black; 'Albovariegata': syn. 'Variegata', leaflets edged white; 'Aureomarginata': leaflets edged yellow.)

araneous see *arachnoid*.

Araucaria (from the Arauco Indians of Central Chile, to whose territory *A.araucana* is native). Araucariaceae. SW Pacific; S America. 20 species, evergreen, coniferous trees with erect stems, resinous bark and whorled branches in a candelabra-like head, each branch with pinnately arranged, drooping to ascending branchlets clothed with scale- or needle-like, overlapping leaves. The following are widely grown and will withstand harsh winds, full sun and all but the wettest or driest soils. Only *A.araucana* is fully hardy. Young plants of *A.heterophylla* make attractive house and conservatory plants. Increase by seed.

A.araucana (syn. *A.imbricata*; MONKEY PUZZLE TREE, CHILE PINE; Chile, Patagonia; trunk columnar, dark, to 30m tall with a head of drooping then ascending branches with rope-like branchlets densely covered in 2–7cm-long, triangular, dark green, tough and sharp-tipped leaves); *A.bidwillii* (BUNYA BUNYA; Queensland; narrow to broadly domed tree to 45m tall with whorls of spreading branches hung with weeping branchlets; juvenile leaves 2–5cm long, oblong to lanceolate, sharp-tipped glossy and arranged in a spiral, adult leaves shorter, ovate, sharp-tipped very tough and overlapping); *A.cunninghamii* (Northern Australia, New Guinea; HOOP or MORETON BAY PINE; to 60m tall with spreading to upswept branches bare at their bases and tufted with drooping twigs at the tips; leaves 1–2cm-long leaves needle- to scale-like, olive to grey-green); *A.heterophylla* (syn.

A.excelsa; NORFOLK ISLAND or HOUSE PINE; Norfolk Islands; to 60m tall, strongly pyramidal; branches outspread, whorled, clothed with scale-like leaves to 0.6cm long; juvenile plants have soft, slender branches covered in needle-like, grey-green leaves, marked silver-white in 'Albospica', golden-striped in 'Aureovariegata' and blue-green in 'Glauca').

Araujia (from the Brazilian vernacular name for this plant). Asclepiadaceae. South America. 4 species, evergreen, woody-based perennial climbers grown for their flowers produced in summer. Provide a minimum temperature of 10ºC/50ºF. Plant in a moist, fertile soil; provide ample humidity and bright, filtered light. Propagate by tip cuttings. *A.sericofera* (syn. *A.sericifera*; CRUEL PLANT; climbing to 10m; flowers to 3cm diam., fragrant, salverform, white sometimes striped maroon within).

Araujia sericofera

arborescent with the habit or stature of a tree.

arboretum a garden or park chiefly devoted to a collection of trees and shrubs. Planted since ancient times, arboreta became popular in Europe in the late 18th century and have since served scientific as well as amenity purposes.

arboriculture the science and practice of tree growing, especially the establishment and maintenance of trees and shrubs in amenity and ornamental areas.

arbour a shady enclosure or retreat surrounded by trees or shrubs or trellis clad with climbing plants, often at the end of, or in conjunction with, a pergola. Known in the Middle Ages as herbers, they were then often extended to form long tunnel arbours. In 16th-century usage the term also meant a herb or flower garden, grassy sward or orchard, and in the 18th century included garden-houses.

Arbutus (the Classical Latin name). Ericaceae. N and C America, Europe, Asia Minor. MADRONE, MADROÑA, STRAWBERRY TREE. 14 species, evergreen trees and shrubs with flaking to peeling bark, tough-textured, dark green leaves and small, lily-of-the-valley-like flowers in racemes and panicles. The fruits are globose and fleshy, usually warm, glowing red and textured with small, prismatic bumps. *A.unedo* and *A. × andrachnoides* are hardy to –15ºC/5ºF; *A.andrachne* and *A.menziesii* to –10ºC/14ºF. All prefer full sun, an acid to neutral soil, and shelter from cold, drying winds. Propagate by basal cuttings taken in late winter; or by simple layering. Seed sown with the flesh removed and stratified for 4–6 weeks at or just above freezing point will usually germinate when brought under glass.

A.andrachne (GRECIAN STRAWBERRY TREE; SE Europe, Asia Minor; spreading tree or large shrub to 12m tall with peeling, red-brown bark; leaves to 10cm long, oval to oblong, entire, dark green above, paler beneath; flowers in late spring, dull white in erect panicles to 10cm long; fruit orange, granular, 1.5cm diam.; many plants cultivated under this name are *A.* × *andrachnoides*); *A.* × *andrachnoides* (a naturally occurring hybrid between the last species and *A.unedo*; spreading tree or large shrub to 10m with rusty red, peeling bark; leaves to 10cm long, oval to elliptic, finely serrate, dark green above, paler beneath; flowers in late autumn or spring, ivory to white in nodding panicles; fruit to 1cm diam., smoother than in *A.unedo*); *A.menziesii* (MADRONE; W US; tree to 30m with peeling red bark; leaves to 15cm long, elliptic to ovate, blunt, entire or sometimes serrate on young plants, dark green above, glaucous beneath; flowers in late spring, white in erect, pyramidal panicles to 15cm long; fruit to 1.5cm diam.); *A.unedo* (STRAWBERRY TREE; S Europe to Asia Minor, Eire; shrub or small, broad-crowned tree to 10m tall with coarse, flaking, red-brown bark; leaves to 7cm long, elliptic to obovate, serrate, glossy dark green above, paler beneath; flowers in late autumn, cream to white, pink or red-stained (f. *rubra*) in drooping panicles to 7cm long; fruit to 2cm diam., surface rough and granular, edible).

arcade a series of trained-plant arches forming a decorative garden feature; a pergola walk. Arcades are often constructed by training tree fruits to meet at the top.

Arctostaphylos (from Greek *arktos*, bear, and *staphule*, fruit: bears are said to eat the berries). Ericaceae. Northern Hemisphere. BEARBERRY; MANZANITA. 50 species, evergreen or semi-evergreen shrubs and small trees with small, urn- to bell-shaped flowers in terminal racemes and panicles and spherical, red to purple-black or brown berry-like fruit. The following are hardy and prefer full sun on a moist, acid soil. Propagate by simple layering in spring, or by severing rooted stems of trailing species; also by semi-ripe cuttings in summer.

A.alpina (syn. *Arctous alpinus*; ALPINE BEARBERRY, BLACKBEARBERRY Circumpolar regions; dwarf, creeping shrub to 15cm tall; leaves to 3cm, oblanceolate, finely serrate, fresh green in summer, red in autumn; flowers white flushed pink); *A.diversifolia* (syn. *Comarostaphylis diversifolia*; SUMMER HOLLY; California; erect shrub to 4m tall; leaves 3–8cm, oblong to broadly elliptic, toothed to entire, glossy; flowers white); *A.manzanita* (MANZANITA; California; shrub or tree erect to 6m, with purple-red bark and a gnarled habit; leaves to 5cm, broadly ovate, pale to grey-green; flowers white to pink in drooping panicles); *A.nevadensis* (PINE-MAT MANZANITA; W US; mat-forming, to 40cm tall; leaves to 3cm, lanceolate to obovate with a short, distinct tip, light green; flowers white to pink); *A.nummularia* (FORT BRAGG MANZANITA; California; mat-forming or rounded shrub; leaves to 1.5cm, elliptic to ovate, shiny; flowers white); *A.patula* (GREEN MANZANITA; California; erect shrub to 2m tall; leaves to 5cm, broadly oval to oblong, blunt and rounded, thick-textured; flowers pink, in broad panicles); *A.uva-ursi* (COMMON BEARBERRY,

HOG CRANBERRY, MOUNTIAN BOX; N America, N Europe, N Asia; mat-forming shrub with creeping and rooting stems; leaves to 3cm, obovate, dark green above, paler beneath, often turning red in winter; flowers white tinted pink; includes several ground-smothering cultivars with bright, glossy foliage and pink to white flowers, e.g. 'Point Reyes', 'Radiant', 'Vancouver Jade'; the California native, *A.hookeri*, MONTEREY MANZANITA, is sometimes treated as a subspecies of *A.uva-ursi*; it forms a dense mound 10–100cm high with shining bright green leaves to 2cm; white or pink flowers, one of the best ground-cover cultivars is 'Monterey Carpet').

Arctotheca (from Greek *arktos*, bear, and *theke*, chest). Compositae. South Africa. 5 species, stemless to creeping perennial herbs with lyrate to pinnately lobed leaves and daisy-like flowerheads. In frost-free regions, the following species is used for covering dry, sunny banks and rock gardens; in cooler zones, it is treated as an annual. Sow seed *in situ* in spring. *A.calendula* (CAPE WEED; fast-spreading groundcover; flowerheads to 5cm diam., bright yellow, on stalks to 50cm tall).

Arctotis (from Greek, *arktos*, bear, and *ous*, ear). Compositae. South Africa. AFRICAN DAISY. 50 species, annual or perennial herbs with low rosettes of grey-scurfy, pinnately lobed leaves and bright, daisy-like flowerheads in summer. Use as half-hardy bedding plants, repropagated annually by stem cuttings taken in late summer and overwintered in a heated greenhouse, or raised from seed in spring. Grow in full sun on a light, fast-draining soil.

A.fastuosa (syn. *Venidium fastuosum*; CAPE DAISY; leaves grey-white-hairy; flowerheads to 8cm diam., ray florets orange, purple-brown at base, disc florets maroon to black); *A.* × *hybrida* (syn. × *Venidioarctotis hybrida*; garden hybrids between *A. fastuosa* and *A.venusta*; leaves covered with thin, grey-white felt; flowerheads to 10cm diam., on stalks to 20cm tall, ranging in colour from grey-blue to white, cream, apricot, peach, yellow, orange, bronze, carmine and wine red, usually with darker centres); *A.venusta* (syn. *A.stoechadifolia*; BLUE-EYED AFRICAN DAISY; to 60cm tall; leaves with grey-white hair especially beneath; flowerheads to 7cm diam., white to purple-red with a blue to blue-red centre).

arcuate curved downwards, bow-shaped, usually applied to a smaller, more rigid structure than would be described as arching, for example the column of an orchid.

arcure an unusual and decorative system of training fruit trees, used especially for apples and pears, and involving bending branches to form tiers of spur bearing loops.

Ardisia (from Latin, *ardis*, a point, referring to the acutely pointed anthers). Myrsinaceae. Asia, Australasia, Americas. 250 species, evergreen trees,

arcure

shrubs and subshrubs with whorled, tough leaves, small 5-parted flowers in stalked, axillary clusters and berry-like fruit. *A.crenata* is grown as a potplant in the cool greenhouse, conservatory or home. It needs bright, filtered light. Grown as a standard, up to one metre tall with a clear stem, this species tends quite naturally to display its fruit in a ring of trusses below a mop of deep green growth, making a beautiful houseplant particularly at Christmas. *A.japonica* will tolerate mild frosts and makes handsome groundcover for lightly shaded, moist and fertile sites in zones 8 and 9. Trim over in spring. Propagate both by seed sown in spring, or by semi-ripe cuttings in summer.

A.crenata (syn. *A.crenulata*, *A.crispa*; Asia; CORALBERRY, SPICEBERRY; shrub to 2m tall; leaves 6–20cm long, elliptic to narrowly obovate, deep, glossy green with finely wavy-toothed margins, edged creamy-white in 'Variegata'; flowers white to pink in clusters at the end of slender stalks among and beneath the leaves; berries coral-pink to scarlet); *A.japonica* (MARLBERRY; China, Japan; low, spreading shrub with creeping and rooting stems and erect shoots to 40cm high; leaves 5–7cm-long, oval to ovate, glossy dark green, sharply toothed; flowers white to pale pink on short racemes; cultivars exist with leaves variously toothed and frilled and marked white, cream, gold and pink).

Arenaria (from Latin *arena*, sand – the plants grow in sandy places). Caryophyllaceae. Widespread, especially in northern temperate regions. SANDWORT. 160 species, low-growing, hardy perennial herbs with white, usually 5-petalled flowers. Suitable for sunny positions on fast-draining soils. Propagate by division, seed or (larger species) by softwood cuttings of basal shoots in early summer.

A.balearica (W Mediterranean Islands; matforming; flowers to 1cm diam., white); *A.montana* (SW Europe; grey-green and downy with prostrate shoots; flowers 2–4cm diam., white, similar to *Cerastium*, but with entire, not cleft petals); *A.purpurascens* (Pyrenees, N Spain; loosely tufted with branching stems ascending to 10cm; flowers to 2cm diam., pale purple, pink or white; *A.tetraquetra* (Pyrenees, N Spain; small, grey-green cushion-forming; flowers very short-stalked, to 1cm across, white, with four petals).

areole a small area or space, often between anastomosing veins; also a depression or elevation on a cactus stem, bearing spines.

Argemone (from Latin, *argema*, eye cataract: the genus was once used as a treatment for this condition). Papaveraceae. Americas. PRICKLY POPPY; ARGEMONY. 28 species, annual or perennial herbs and one shrub; the prickly stems contain yellow or orange latex. The leaves are glaucous and lobed with fiercely prickly margins. Poppy-like flowers are composed of usually six, broadly wedge-shaped petals and appear in summer. The seed pods are fiercely prickly. Sow under glass in early spring and plant out after the last frosts in a sunny, well-drained position. Seed may also be sown *in situ* in late spring. *A.mexicana* (DEVIL'S FIG, MEXICAN POPPY, PRICKLY POPPY; SW US, C America, W Indies; annual to 1m; leaves glaucous blue-green with silvery white flashes on the veins and spiny-lobed margins; flowers 5–7cm-wide bright to pale yellow, white or pale tangerine).

Argyranthemum (from Greek, *argyros*, silver, and *anthemos*, flower). Compositae. Canary Islands. Some 23 species, perennial, half-hardy subshrubs or shrubs with entire to dissected leaves and slender-stalked, daisy-like flowerheads from late spring to autumn. Marguerites are bright and trouble-free perennials for frost-free gardens or cool greenhouses and conservatories. They are equally among the finest half-hardy bedding or container plants for summer display in cooler regions. Grow in full sun on a well-drained soil. Deadhead regularly. Take stem cuttings in autumn and overwinter them in a frost-free greenhouse in zones 8 and under. Unless growing them to achieve height for training as standards, pinch out young plants to promote bushiness and plant them out in mid-spring.

A.frutescens (syn. *Chrysanthemum frutescens*; MARGUERITE; Canary Islands; a subshrub or shrub to 2m tall with finely pinnatisect, grey-green leaves to 10cm long and white flowerheads to 6cm across with bright yellow centre). *Cultivars:* 'Chelsea Girl' (to 70cm, leaves very fine, flowerheads white, single), 'Jamaica Primrose' (to 1m, leaves grey-green, coarsely cut, flowerheads large, single, golden), Nevada Cream' (flowerheads single, long-stalked, primrose to cream), 'Mary Cheek' (to 40cm, leaves fine, flowerheads double, pink), 'Peach Cheeks' (low and spreading, flowerheads single to double, peach pink to white with a yellow centre), 'Petite Pink' (to 40cm, compact, leaves fine and slightly fleshy; flowerheads single, pale pink with a golden centre), 'Powder Puff' (miniature with fine foliage and double, rather ragged-looking pink flowerheads), 'Rollason's Red' (to 40cm, leaves coarse, flowerheads single, ray florets crimson to scarlet, gold at the base, forming a bright ring around the dark red disc), 'Snowflake' (compact, leaves blue-grey, flowerheads small, double, pom-pom-like, pure white).

Argyreia (from Greek, *argyreios*, silvery, referring to the silver-hairy undersurfaces of the leaves of some species). Convolvulaceae. Tropical Asia to Australia. 90 species, perennial, evergreen, woody or woody-based climbers with large, rounded to heart-shaped leaves and showy funnel-shaped flowers. Provide a minimum temperature of 13°C/55°F, and full sun or bright, filtered light. Propagate by seed sown in spring in a heated case, or by softwood cuttings with bottom heat in spring. *A.nervosa* (syn. *A.speciosa* WOOLLY MORNING GLORY, India, Bangladesh; to 10m; leaves coated beneath in silver hairs; flowers 7cm-long flowers tubular to funnel-shaped, lavender with a deep mauve throat and silkily hairy).

Argyranthemum (top) *A.* 'Petite Pink' (bottom) *A.* 'Jamaica Primrose'

*Arisaema
sikokianum*

Argyrocytisus (from Greek, *argyros*, silver, and *Cytisus*). Leguminosae. Morocco. 1 species, an open-branched, semi-evergreen or deciduous shrub to 4m tall. *A.battandieri* (syn. *Cytisus battandieri*) is the PINEAPPLE BROOM or MOROCCAN BROOM, with large trefoil-like leaves covered in silvery down and tightly packed cone-shaped racemes of golden pea-like and pineapple-scented flowers carried through the summer. The cultivar 'Yellow Tail' is particularly fine, with chrome yellow racemes to 15cm long or more. It tolerates poor dry soils and hot dry conditions and, protected from wind and wet, will survive temperatures as low as –15°C/5°F. In colder locations, plant against a south-facing wall. Propagate from scarified or pre-soaked seed in a warm greenhouse in spring; also by semi-ripe cuttings.

Argyroderma (from Greek, *argyros*, silver, and *derma*, skin, referring to the silvery leaf surface). Aizoaceae. South Africa. 10 species, evergreen perennial herbs with very succulent, smooth, grey-green leaves in pairs with their bases fused. These leaf pairs may be borne on dwarf shrubs or in stemless clusters or be solitary, and are often partially buried. Daisy-like flowers appear in the cleft between the leaves. The following should be cultivated as for *Conophytum*.

A.delaetii (syn. *A.blandum*; stem largely subterranean; leaves semi-ovoid; flowers white, purple or yellow); *A. fissum* (syn. *A. brevipes*; mat-forming; leaves to 12cm long, yellow-green, finger-shaped; flowers with outer petals purple or yellow, inner petals white); *A.pearsonii* (syn. *A.schlechteri*; leaves semi-globose with a narrow fissure; flowers with outer petals purple, white or, rarely, white, inner petals yellow or white).

Ariocarpus (from *Aria*, a genus now included in *Sorbus*, and Greek, *karpos*, fruit, from a supposed resemblance between the fruits of these genera). Cactaceae. Texas, Mexico. 6 species, slow-growing cacti with thick rootstocks and unbranched, usually hemispherical to broadly conical stems. They are covered in large tubercles, usually spineless and bear openly funnelform flowers in autumn and winter. Grow in full sun with a dry, buoyant atmosphere and a minimum temperature of 5°C/40°F. Plant in a gritty compost with a pH of 6–7.5; water liberally but intermittently in spring and summer; keep dry from mid-autumn to early spring except for a light misting on warm days. Propagate by seed. Faster growth rates can be achieved by grafting of young plants. *A.fissuratus* (stem to 10cm diam., with prominent triangular tubercles, grey-green to grey-brown, each deeply fissured so as to resemble scales, and wrinkled with a woolly groove above; flowers to 4cm diam., magenta).

aril a generally fleshy appendage of the funiculus or hilum, partially or entirely enveloping the seed.

arillate possessing an aril; more loosely, any outgrowth or appendage on the testa.

Arisaema (from *Arum* and Greek *haima*, blood red, a reference to the red-blotched petioles of some species). Araceae. N America, Africa, Asia. 150 species, tuberous or thickly rhizomatous deciduous perennial herbs with trifoliate to palmately or pedately compound, long-stalked leaves. Produced in spring and early summer, the small flowers are packed on a spadix which bears a terminal appendage and is enclosed in a showy, pitcher-like spathe with a hooded to erect, ovate to lanceolate blade. They favour light shade and moist (but never wet) acid to neutral, soils rich in sand and leafmould. Those described here are hardy in zone 7 but may deteriorate when planted in regions with prolonged hard frosts and wet winters. Mulch thickly in winter. Propagate by seed sown when fresh in the cold frame, or by separating offsets and cormels produced by long-established plants.

A.candidissimum (China; spathe 8–15cm long, tube pale green striped white, limb white striped or zoned cool, candy pink, ovate to orbicular and slightly hooded with a tapered tip); *A.griffithii* (Himalaya; spathe blade large, broad, in pearly grey with deep purple interior and green, netted veins, arching forward and curving over itself, undersurface dull purple overlaid with an intricate network of green veins; spadix largely hidden, but ending in long, slender appendage); *A.sikokianum* (Japan; spathe 20cm long, tube dark, satiny maroon to black outside and snow white within, blade erect, ovate, acuminate, purple-green striped green-white within, far darker beneath and striped silvery white; spadix terminates just above the rim of the spathe in a bright white knob); *A.triphyllum* (JACK-IN-THE-PULPIT, INDIAN TURNIP; Eastern North America; spathe to 15cm long and yellow-green to purple, striped green or white with an erect to arching, tapering limb, the spadix narrowly club-shaped, green to purple and protruding from the tube).

Arisarum (from Greek *arisaron*, the name for *A.vulgare*). Araceae. Mediterranean; Southwest Europe; Atlantic Islands 3 species, tuberous or rhizomatous perennial herbs, deciduous or semi-evergreen with slender-stalked, arrow-shaped, dark green leaves borne close to the ground, and low-lying inflorescences produced in spring. These consist of a spathe, tubular below and hooded above with a finely tapering tip, the whole surrounding a slender spadix. They tend to lie beneath or just above the dense cover of foliage. Hardy to –10°C/14°F, or lower if mulched. *A.proboscideum* requires a moist soil in dappled sunlight or full shade. *A.vulgare* prefers a sunnier, slightly drier location on a gritty soil. Propagate by division when dormant (for *A.proboscideum* this is late summer; for *A.vulgare*, late spring).

A.proboscideum (MOUSE PLANT; Italy and Spain; rhizomatous; spathes 6–10cm long; white and bulbous at the base, chocolate-brown to maroon-black above where a hooded blade narrows sharply to a long and upturned tip (the mouse's 'tail'); spadix purple-brown, white-tipped, fully enclosed); *A.vulgare* (FRIAR'S

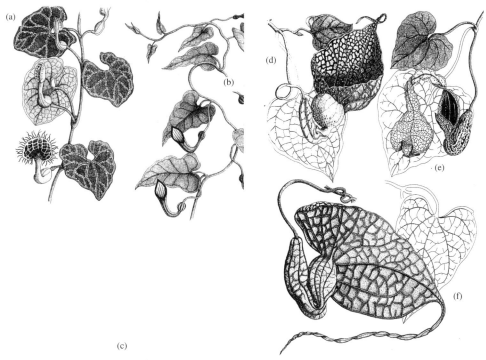

Aristolochia (a) *A. littoralis,* (b) *A. trilobata,* (c) *A. ringens,* (d) *A. gigantea,* (e) *A. cymbifera,* (f) *A. grandiflora*

COWL; Mediterranean, Canaries, Azores; tuberous; spathes 4–6cm-long, silver-white striped purple-brown at base, solid metallic maroon to violet-black above where the hooded blade expands and curves over like a cowl; spadix purple-green, protruding).

aristate of a leaf apex abruptly terminated in a very slender, pointed continuation of the midrib. It may also mean awned.

Aristea (from Greek *arista*, a point, referring to the leaves). Iridaceae. Africa; Madagascar (those below from South Africa). 50 species, rhizomatous, evergreen perennial herbs, clump-forming with erect, grassy to sword-like leaves and, in summer, tall, narrow panicles bearing spikes of short-lived, 6-parted flowers. Grow in a well-drained, acid medium. Position in part shade or bright, filtered light; provide a minimum temperature of 7°C/45°F, good ventilation and plentiful water when in growth. In mild locations in zones 7 and 8, they may survive frosts, especially if dry and mulched. Propagate by division after flowering, or by seed in sown spring at 20°C/70°F.

A.ecklonii (to 90cm tall with lax clusters of blue flowers to 3cm across); *A.major* (syn. *A.thyrsiflora*; to 1m tall with dense clusters of mauve to pale blue flowers to 5cm across).

Aristolochia (from Greek *aristos,* best, and *lochia*, childbirth – because of its foetus-shaped flowers, *A. clematitis* was used in medieval times as an abortifacient and birth-inducer). Aristolochiaceae. Cosmopolitan. BIRTHWORT; DUTCHMAN'S PIPE. 300 species, evergreen or deciduous woody or herbaceous, twining climbers and erect herbaceous peren-

nials. Produced in spring and summer, the flowers are axillary and pitcher-like with a swollen, inflated base which narrows to a slender tube curving upwards to an expanded limb. The hardy species are deciduous or semi-evergreen. Grow *A.clematitis* in full sun on a well-drained, fertile and free-draining soil. In exposed positions, the stems may require some support. Propagate by division or cuttings in early spring. *A.macrophylla* and *A.sempervirens* favour a sheltered position on a rich, moist soil, the first in dappled sun or semi-shade, the second in full sun. Support on trellis, walls, pergolas or tree trunks. The tender species are evergreen or briefly deciduous. *A.fimbriata* and *A.littoralis* require a moist, fertile soil in bright, filtered light with a minimum temperature of 10°C/50°F. Train on wires or pergola; cut back hard in early spring. The remaining species need a minimum temperature of 15°C/60°F, bright, filtered light and high humidity. Grow in moist and fertile, loamy soils and prune back hard after flowering. Most Aristolochias have malodorous flowers; although magnificent, the larger-flowered species can make a confined space almost unbearable. Propagate climbers by softwood cuttings – hardy species in midsummer, tender species in early spring using a rooting hormone and bottom heat. Multi-stemmed plants can also be divided. Seed needs three months stratification at 5°C/40°F, or may be sown when ripe in autumn after soaking in warm water for 48 hours. Surface sow under cover at 25°C/77°F.

A.clematitis (BIRTHWORT; Europe; E N America; hardy, erect herb to 1m tall; flowers to 4cm long, yellow-green, slender with an erect, triangular limb); *A.cymbifera* (Brazil; tender,woody climber to 10m; flowers to 20cm long, basically dirty white, base flask-like heavily mottled maroon, tube short and erect, limb divided into two lips, lower short and beak-like, marked maroon to violet, upper far longer and ruffled, hanging over the lower and expanding to a broad, blunt lobe intricately veined in purple or maroon); *A.fimbriata* (Brazil; tender, herbaceous or woody-based climber to 2m; leaves bright green veined yellow or silver; flowers to 6cm long, base swollen and rounded, tube olive, short, erect, expanding to a round or reniform limb with a deeply fringed margin and velvety maroon veined gold); *A.gigantea* (syn. *A.clypeata*; Panama; tender woody climber to 10m; flowers to 30cm long, base buff, flask-shaped and prominently veined, tube short and curving upwards to an inflated flesh-coloured sac, limb very broadly ovate to triangular with a shortly tapering, ribbon-like tip, held upwards and open like a scoop, white to ivory or yellow-orange, very heavily veined or blotched maroon to blood-red); *A.grandiflora* (syn. *A.gigas*; PELICAN FLOWER; Central America; tender woody climber to 15m; flowers to 300cm long (including tail), base flask-shaped, heavily veined, white to buff or yellow-green, curving upwards in a short, erect tube which becomes sac-like then expands into a broadly heart-shaped, downward-pointing limb to 50cm across and tipped with a spiralling, ribbon-like tail to 2m long; the limb is white to olive, buff or rust heavily veined in dark brown-red, maroon or purple); *A. × kewensis* (*A.labiata × A.trilobata*) garden hybrid intermediate between parents); *A.labiata* (syn. *A.brasiliensis*; ROOSTER FLOWER; South America; tender woody climber to 10m tall; flowers to 20cm long, cream to buff or brick red, mottled olive and yellow and veined red-brown to chestnut, base flask-shaped, tube short, slender, erect expanding in 2-lipped blade, lower lip beak-like, lanceolate and forward-pointing with the margins almost touching and rimmed, upper lip equal in length or slightly longer, broadly spathulate with margins ruffled and hanging downwards); *A.littoralis* (syn. *A.elegans*; CALICO FLOWER; South America, naturalized Central America and Southern US; tender, woody-based climber to 6m tall; flowers to 10cm long, base flask-shaped, white, narrowing and curving to white tube supporting a broad, heart-shaped to reniform limb which points upwards and is golden-throated and white overlaid with rich, velvety purple mottling); *A.macrophylla* (syn. *A.durior*; DUTCHMAN'S PIPE; E US; hardy, woody-based climber to 9m tall; flowers to 4cm long, green spotted purple-brown and yellow, base swollen, tube curving upwards like a pipe, limb with three spreading, rounded to acute lobes stained maroon); *A.ringens* (C America; tender woody climber; flowers to 28cm long, mottled red, yellow and green heavily veined red-brown to purple, base swollen and flask-shaped, tube slender, short and curving upwards, limb divided into two lips, upper erect and forward-pointing, spathulate with a frilled margin, lower longer and forward-pointing, chute-like, shaped like a tongue with its margins slightly inrolled); *A.sempervirens* (Mediterranean to Aegean;

hardy, woody-based climber to 5m; flowers yellow-green striped purple, to 5cm long, base swollen, tube slender, erect, expanding to a short, acute, erect limb); *A.trilobata* (syn. *A.macroura*; Central America; tender woody climber; flowers to 25cm long including tail, olive green veined and tipped brown to maroon, base swollen then narrowing and curving upwards to an erect tube with a distinct rim and erect lid tipped with a long, downward-spiralling tail).

Aristotelia (for Aristotle). Elaeocarpaceae. Australasia; S America. 5 species, evergreen or deciduous trees and shrubs with small flowers in axillary clusters or panicles followed by glossy berries. Hardy to −10°C/14°F. Plant in a slightly acid soil in sun; protect from winds. Propagate by ripewood cuttings, seed, or by simple layers.

A.chilensis (Chile; evergreen shrub or tree to 5m; leaves to 10cm long, ovate, toothed, glossy dark green, marked white in 'Variegata'; flowers green-white; fruit purple-black); *A.fruticosa* (New Zealand; MOUNTAIN WINEBERRY; small evergreen shrub with intricate, wiry branches, narrow leaves, very small flowers and red to purple-black fruit); *A.serrata* (New Zealand; NEW ZEALAND WINEBERRY; deciduous shrub or tree to 8m tall; leaves to 10cm long, ovate, deeply serrate; flowers pale pink in crowded panicles; fruit dark red ripening black).

aristulate bearing a small awn.

Armeria (a Latinized form of the old French vernacular name for sweet william, *armoires*, to which this genus was thought to bear some resemblance). Plumbaginaceae. Europe; Asia Minor; North Africa; Americas (Pacific coast). THRIFT; SEA PINK. Some 80 species, low-growing, tufted, evergreen perennial herbs or subshrubs with short-branching, cushion- to hummock-forming stems terminating in packed rosettes of very narrow, dark green leaves. Produced throughout the year but chiefly in spring and summer, the flowers are small, pink to white and 5-parted and carried amid chaffy bracts in dense heads atop slender, erect stalks. The most commonly grown species are *A.alpina*, *A.arenaria*, *A.juniperifola* (syn. *A.caespitosa*), *A.maritima* and *A.pseudarmeria*. The flowerstalks range in height from 6 to 30cm. They prefer a fast-draining sandy soil in full sun, and are suitable for exposed positions in the rock garden, on dry banks and walls, especially in coastal regions. Although fully hardy, they dislike cold, wet conditions, but do need ample moisture in spring. Propagate by seed or division in early spring.

armillaria root rot HONEY FUNGUS; BOOTLACE FUNGUS, a notorious lethal root rot of trees, shrubs and some herbaceous plants. *Armillaria mellea* is mostly responsible for infections of broad-leaved trees and shrubs in gardens; *A.ostoyae* is the main cause of death from this disease in conifers. First signs of infection may be a discoloration of foliage and progressive dieback of shoots, but symptoms only become visible when the fungus has invaded a significant part of the root system, and death can then follow rapidly. Other signs of infection may be an exceptional show of flowers, premature autumn colour, leaf-fall, and splitting of bark at or just above ground level, sometimes accompanied by a gummy exudate. If bark is lifted away from the collar and main roots, flat, often fan-shaped sheets of white fungal growth and/or flat black-brown strands may be present. The fungus also penetrates the wood, in which narrow black 'zone lines' occur.

In late summer or autumn toadstools may emerge in dense clusters around the base of dead tree stumps or other infected plants. These are extremely variable in size and colour, but the cap and stalk are typically honey-coloured at some stage of growth. The cap may be up to 23cm in diameter, the gills are white and there is a whitish ring attached to the upper part of the stalk. The fungus can spread by contact between diseased and healthy roots; it also survives in dead tree stumps and roots, from which it sends out long black rhizomorphs resembling bootlaces, which grow through the topsoil to infect new host plants. A large dead tree or stump can be a source of rhizomorphs for many years, possibly having died from other causes and later colonized by honey fungus. Spores from toadstools can theoretically infect newly exposed stumps, but this means of spread is unimportant. Trees and shrubs growing in gardens developed in or near to woodland, orchards or hedgerows are more likely to be attacked, particularly if growth is weakened for any reason.

The disease is incurable although some proprietary treatments are advertised for the protection of valuable plants. In woodland, control is neither practicable nor necessary. In parkland or arboreta, trees should not be planted within 9m of an infection site, and removal of the stump and root system of diseased trees is worthwhile. In gardens, orchards and other closely-planted areas disease control depends on the removal of all infected material. Measures need to be taken to prevent the disease spreading by removing all infected root and stump material. Dead, dying and adjacent plants should be removed with as much of the root system and surrounding soil as possible and the site refilled with soil from a non-woodland source. The use of chipping as a means of stump disposal goes some way towards eliminating the food source.

There is no evidence of infection resulting from the use of conifer bark mulches, which probably cannot act as a substantial food-base for rhizomorphs. Composted bark used for soil mixes is considered to be quite safe.

If a neighbouring source of infection cannot be removed, it may be possible to prevent the spread of rhizomorphs through the soil by means of a barrier made by burying heavy gauge plastic sheeting, vertically, at least 0.5m deep in light soil or 1m deep in clay soil. A narrow ditch, the bottom of which is cultivated, or a strip of ground which can be deeply cultivated are other means of isolation.

The following lists of plants notably susceptible or resistant to the disease are based on records kept at Wisley by the Royal Horticultural Society. Notably susceptible: *Acer* (except *A.negundo*) *Araucaria araucana, Betula, Ceanothus, Cedrus, Chamaecyparis, Cotoneaster, Crataegus,* × *Cupressocyparis leylandii, Cupressus, Forsythia, Hamamelis, Hydrangea, Juglans regia, Laburnum, Ligustrum, Malus, Paeonia, Picea omorika, Pinus* (except *P.patula*), *Prunus* (except *P.laurocerasus* and *P.spinosa*), *Rhododendron, Ribes, Rosa, Rubus, Salix, Sequoiadendron giganteum, Syringa, Thuja plicata, Ulmus, Viburnum, Vitis, Wisteria.*

Notably resistant: *Abelia, Abeliophyllum, Abies alba, A.grandis, A.procera, Abutilon, Acer negundo, Actinidia, Ailanthus altissima, Akebia, Albizia, Aloysia, Arbutus menziesii, Aristolochia, Aronia, Artemisia, Aucuba,* bamboos, *Buxus sempervirens, Callicarpa, Callistemon, Calycanthus, Caragana, Carpenteria, Carya, Castanea, Catalpa, Celastrus, Celtis, Ceratostigma, Cercidiphyllum, Cercis, Cestrum, Chaenomeles, Chionanthus, Choisya, Cistus, Clematis, Clethra, Colutea, Cornus, Corokia, Coronilla, Cotinus coggygria, Crataegus, Cunninghamia, Daboecia, Deutzia, Diervilla, Drimys, Elaeagnus, Embothrium, Enkianthus, Erica, Exochorda, Fabiana imbricata, Fagus sylvatica, Fothergilla, Fraxinus excelsior, Fremontodendron, Gaultheria, Gleditsia, Grevillea, Griselinia, Hakea, Halesia, Halimium, Hebe, Hedera, Helianthemum, Hippophae, Hoheria, Hypericum, Indigofera, Itea, Jasminum, Juniperus, Juglans hindsii, Kerria, Koelreuteria, Lagerstroemia, Larix, Laurus nobilis, Lavatera, Leptospermum, Leucothoe, Lindera, Liquidambar styraciflua, Lithospermum, Lomatia, Lonicera nitida, Maackia, Maclura, Mahonia aquifolium, Menziesia, Morus, Myrica, Myrtus, Nandina, Nothofagus, Nyssa, Oemleria, Ostrya, Paliurus, Parahebe, Parrotia, Passiflora, Paulownia, Phillyrea, Phlomis, Photinia, Phygelius, Phyllodoce, Physocarpus, Pieris, Pileostegia, Pinus patula, Pittosporum, Plagianthus, Podocarpus, Polygonum baldschuanicum, Prunus laurocerasus, P.spinosa, Pseudotsuga menziesii, Ptelea, Pterocarya, Pterostyrax, Punica, Quercus, Rhamnus, Rhus, Robinia, Romneya, Ruscus, Ruta, Sambucus, Santolina, Sarcococca, Sciadopitys, Shepherdia, Sophora, Stachyurus, Staphylea, Stephanandra, Stewartia, Styrax, Symphoricarpos, Symplocos, Tamarix, Taxus, Teucrium, Tilia, Tsuga, Ulex, Vaccinium, Yucca, Zelkova, Zenobia.*

In the US the following other trees and shrubs are reported to show resistance: French pear, fig, persimmon and the rootstock Myrobalan 29; *Acacia decurrens* var. *mollis, A. verticillata, Berberis darwinii, B. thunbergii, B. wilsoniae, Euonymus japonicus, Ilex aquifolium, Ligustrum japonicum, Prunus ilicifolia, P.lyonii, Pyracantha coccinea* 'Lalandei' and *Spiraea prunifolia.* The grading of plants as resistant or susceptible depends on disease records which are continually being added to, and these lists should not be regarded as definitive.

Armoracia (Classical name for horseradish). Cruciferae. Europe; Asia. 3 species, fully hardy perennial herbs with deep taproots, coarse, basal leaves and small, white flowers in racemes and panicles. *A.rusticana* is HORSERADISH (q.v.). The cultivar 'Variegata', with leaves splashed cream and white, is sometimes grown as an ornamental.

Arnebia (from the Arabian name). Boraginaceae. Europe; North Africa; Asia. PROPHET FLOWER; ARABIAN PRIMROSE. 25 species, bristly-hairy hardy annual or perennial herbs with 5-lobed, cup- to funnel-shaped flowers in simple or branched cymes in spring and summer. Plant in full sun or part shade in a sheltered position in a cool, moist but well-drained soil rich in humus and high in grit. Propagate by seed sown *in situ*.

A.*densiflora* (Greece, Turkey; perennial to 40cm tall, velvety and bristly with yellow flowers to 4.5cm long); *A.pulchra* (Caucasus, Iran, Asia Minor; a perennial to 40cm tall with flowers to 2.5cm long, rich yellow with purple-black spots on the lobes).

aroid A plant belonging to the family Araceae, characterized by an inflorescence made up of a spadix and spathe, for example *Arum* and *Zantedeschia*.

Aronia (from *aria*, the Classical Greek name for whitebeam, *Sorbus aria*). Rosaceae. North America. CHOKEBERRY. 2 species, hardy deciduous shrubs with elliptic to oblong or oblanceolate, wavy-toothed leaves, corymbs of small, white to pink flowers in summer, and red, purple or black berry-like fruit. Cultivate as for the smaller *Amelanchier* species.

A.*arbutifolia* (to 2m tall; leaves dull, turning scarlet in autumn; flowers in grey-hairy corymbs; fruit bright red; 'Brilliant': scarlet autumn colour and fruit); *A.melanocarpa* (differs from the last species in its shiny leaves turning red-brown in autumn, its glabrous corymbs of and its black fruit).

Arrhenatherum (from Greek *arren*, male, and *ather*, bristle, a reference to the awned male flowers). Gramineae. Europe; Africa; Asia. OAT GRASS 6 species, perennial grasses with slender, erect to spreading culms and narrow leaves, bearing airy panicles of papery, long-awned florets in summer. Somewhat invasive and not entirely ornamental, the typical *A.elatius* is seldom cultivated, although the panicles are attractive if cut and dried. A.'Variegatum' is, however, one of the purest-white of the variegated grasses. It requires regular division and replanting to maintain quality and vigour. Fully hardy, it will grow on any moderately fertile soil in sun or part shade. It suffers in regions with long, hot summers and may enter dormancy. A.*elatius* (*Avena elatior* is the FALSE OAT , FRENCH RYE; tussock-forming and to 1m tall with loose, purple-tinted and lustrous panicles; 'Variegatum': leaves edged and striped pure white).

Artabotrys (from Greek *artao*, to support, and *botrys*, bunch of grapes: a hooked tendril supports the fruit). Annonaceae. Old World Tropics. 100 species, tender shrubs climbing by woody hooks

(modified flowerstalks) and bearing intensely fragrant, 6-petalled flowers on long, woody stalks. Plant in a damp, semi-shaded position in the warm greenhouse or conservatory. Increase by semi-ripe cuttings or by air-layering. A.*hexapetalus* (CLIMBING ILANG-ILANG; S India, Sri Lanka; to 4m high; flowers ivory, to 2.5cm long, deliciously scented).

Artemisia (for Artemis, Greek goddess of chastity). Compositae. Northern Hemisphere; South America; South Africa. WORMWOOD; SAGEBRUSH; MUGWORT; LAD'S LOVE; OLD MAN. 300 species, aromatic annual, biennial and perennial herbs and subshrubs (those here usually perennial subshrubs and shrubs), with entire to toothed or lobed leaves often covered in short, silvery hair, and button-like, yellow and grey flowerheads carried in terminal racemes and panicles. The larger species and cultivars will grace any mixed border, but work especially well in warm sunny places often on poor soils. They prefer full sun and a fast-draining, gritty, neutral to alkaline soil. Clump-formers can be cut back in early spring; more shrubby types should be pruned only sparingly and then to remove frost-damaged or exhausted growth. Mulch at the limits of hardiness. Low-growing and truly prostrate plants, such as A.*caucasica*, A.*schmidtiana* and A.*stelleriana* 'Mori' grow well on sunny rock gardens, dry banks and walls. Propagate by division in spring or autumn, or (shrubby species) by softwood or heeled, semi-ripe cuttings in summer. The following are hardy in zone 7, or lower still if mulched in winter.

A.*abrotanum* (SOUTHERNWOOD; origin uncertain, naturalized in South and Central Europe; erect to 1m, stems clumped, more or less woody; leaves olive to dull grey-green, very finely cut into many slender lobes); *A.absinthium* (ABSINTHE, COMMON WORMWOOD, OLD MAN, LAD'S LOVE; Europe, Asia; bushy erect, woody-based, to 1m tall; leaves finely 2–3-pinnately cut, silver-hairy; the cultivar 'Lambrook Silver' is massed with very fine, silver-grey foliage); *A.arborescens* (Mediterranean; erect to sprawling shrub to 1.m tall with grey-green, finely cut foliage; 'Faith Raven': somewhat hardier with silver-white leaves); *A.armeniaca* (syn. *A.canescens*; Russia, Caucasus, Iran; erect, bushy, 30–100cm tall with silver-hairy, grey leaves cut into many filigree-like lobes); *A.caucasica* (syn. *A.assoana*, *A.pedemontana*; Spain to Ukraine; tufted, prostrate subshrub to 30cm tall with silver-white, ferny foliage); *A.frigida* (Russia; woody-based, tufted and mat-forming, to 50cm tall with finely pinnately to palmately cut leaves, grey-white with linear lobes); *A.lactiflora* (WHITE MUGWORT; China; vigorous, erect perennial with clumped, usually herbaceous stems and large, jaggedly cut dark green leaves; the flowerheads are grey-white and carried in massive panicles; 'Guizhou' is strongly flushed purple to maroon with white flowerheads;'Variegata' has grey-cream variegated leaves); *A.ludoviciana* (WESTERN MUGWORT, WHITE SAGE, CUDWEED; West US, Mexico; erect and bushy, woody-based perennial herb to 1m tall with

Artemisia ludoviciana

Artemisia schmidtiana 'Nana'

clumped stems and grey-silver, lanceolate leaves entire to jaggedly toothed or divided; includes the finely cut 'Silver Frost' and 'Silver King', the jagged 'Silver Queen' and 'Valerie Finnis'; var. *albula* has virtually white leaves, whilst var. *latiloba* has silver leaves with broad, sharp lobes); *A.pontica* (ROMAN WORMWOOD; Europe; erect, bushy perennial to 80cm tall with clumped stems and feathery, sage green to pewter grey leaves; 'Old Warrior': to 50cm. densely bushy and neat with very finely cut sage green to hoary grey foliage); *A.* 'Powis Castle' (garden origin, probably *A.arborescens* × *A.absinthium*; vigorous, erect to spreading shrub or subshrub to 1m with silver filigree leaves); *A.schmidtiana* (Japan; prostrate, carpet- or mound-forming subshrub to 60cm tall with silver-grey, finely cut, downy leaves; the dwarf, cushion-forming 'Nana' is an excellent rock or sink garden plant, likewise the low, hardy and bright grey-white 'Silver Mound'); *A.stelleriana* (BEACH WORM-WOOD, OLD WOMAN, DUSTY MILLER; NE Asia, E US; erect to arching, bushy, woody-based perennial to 60cm tall with silver-white-hairy leaves, deeply lobed, toothed or entire; 'Mori': compact, hardy and prostrate with off-white to silver-white leaves cut into fairly broad, dissected lobes, may be deciduous).

Arthropodium (from Greek *arthron*, joint, and *pous*, foot: the pedicels are jointed). Liliaceae (Asphodelaceae). Australasia. 12 species, evergreen or deciduous perennial herbs with narrow leaves in clumps or rosettes and starry, 6-parted flowers in racemes or panicles. In zones 7 and over, plant in full sun or part-shade on a fast-draining but permanently damp soil high in leafmould and grit. Mulch *A.cirrhatum* in cold winters. Increase by division.

A.candidum (deciduous; small clumps of finely grassy dull green leaves to 20cm long; flowers white, to 1cm wide, on thread-like stalks in a loose, erect panicle; 'Maculatum': smaller, leaves dull flesh pink mottled and spotted olive to bronze); *A. cirrhatum* (RIENGA LILY; ROCK LILY; evergreen; leaves narrowly lanceolate, to 60cm, white to glaucous beneath; flowers white, 2–4cm wide, on panicles to 30cm across standing on stout scapes to 1m tall).

artichokes a group of three types of perennial vegetable plant. The GLOBE ARTICHOKE (*Cynara cardunculus* Scolymus Group) is not known in the wild and probably derived from the cardoon (*C.cardunculus*). It is a herbaceous perennial grown in the vegetable garden mainly for the fleshy basal portions of the scales and the 'heart' or base of the flower, all of which are boiled. The young shoots, blanched by covering for six weeks with black polythene, may be used in salads. Globe artichokes are best replanted every three years by suckers taken from well-performing mother plants in spring; plant 90cm apart each way. Safeguard from frost damage by covering with a 30cm layer of straw.

JERUSALEM ARTICHOKE (*Helianthus tuberosus*), native to North America, is a herbaceous perennial 2–3m high, related to the sunflower (*H.annuus*). It is grown mainly for its edible tubers, harvested in autumn; they have the appearance of knobbly potatoes, 7–10cm long and 3–6cm in diameter, which are peeled and boiled. Rich in inulin rather than starch, the tubers are a suitable source of carbohydrate for diabetics. They are best replanted annually in spring, 8cm deep, at 40–50cm stations in rows 90cm apart. Plants are also grown in gardens as a temporary shelter screen.

CHINESE ARTICHOKE (*Stachys affinis*), a native of the Far East, is a herbaceous perennial 30–45cm high, grown for its small edible tubers produced on a creeping root system, with a flavour similar to Jerusalem artichoke. Plant tubers in spring 7-8cm deep and 30cm apart each way. They may be overwintered in the ground in all but the severest winters.

articulate, articulated of stems and stalks; jointed; possessing distinct nodes or joints, sometimes swollen at their attachment and breaking easily.

Arum (Latinized form of the Greek name for these plants, *aron*). Araceae. Europe: Asia. 26 species, hardy, tuberous perennial herbs with hastate to sagittate leaves and inconspicuous flowers arranged on a spadix terminating in a cylindric, club-shaped or tapering appendix and surrounded by a cup- or hood-like spathe. Most of the following species will tolerate cold to –5ºC/23ºF; *A.italicum* and *A.maculatum* are fully hardy. *A.pictum* and *A.rupicola* fare better in the cold greenhouse, alpine house or bulb frame. Most prefer a sheltered, warm position in full or dappled sun. Keep moist during the growing season (usually early autumn to late spring), but drier during the summer resting period. *A.italicum* and *A.maculatum* are easier to please and will grow in shade and permanent moisture. The first is an invaluable foliage plant for the winter garden, its brilliant fruiting heads giving way to tufts of marbled leaves. Sow ripe seed, cleansed of pulp, in a gritty medium in autumn, or divide plants after flowing. The berries are toxic.

Arthropodium candidum

A.creticum (Crete; spathe produced in spring, to 20cm, bright yellow to yellow-green or white, recurved, spadix tapering, yellow or, rarely, purple); *A.dioscoridis* (Cyprus to Turkey, Middle East; spathe produced in spring, to 40cm, green suffused and blotched deep, satiny maroon, more or less erect, spadix cylindric, purple-black); *A.italicum* (S and W Europe; leaves appearing in autumn and winter, to 35cm, glossy mid to dark green, often heavily marbled cream to white or yellow on veins—plants described as 'Marmoratum', or, erroneously, 'Pictum' – spathe produced in spring, to 40cm, yellow-green or green-white, faintly edged purple-red, erect, hooded, spadix club-shaped, yellow; fruiting head an oblong mass of showy orange-red berries); *A.maculatum* (LORDS-AND-LADIES, CUCKOO-PINT, JACK-IN-THE-PULPIT; Europe; leaves often spotted purple-black; spathe produced in early spring, to 25cm, pale, pearly green-white, often tinted purple-red especially on margins and with faint purple-red blotches, erect, hooded, spadix club-shaped, purple-

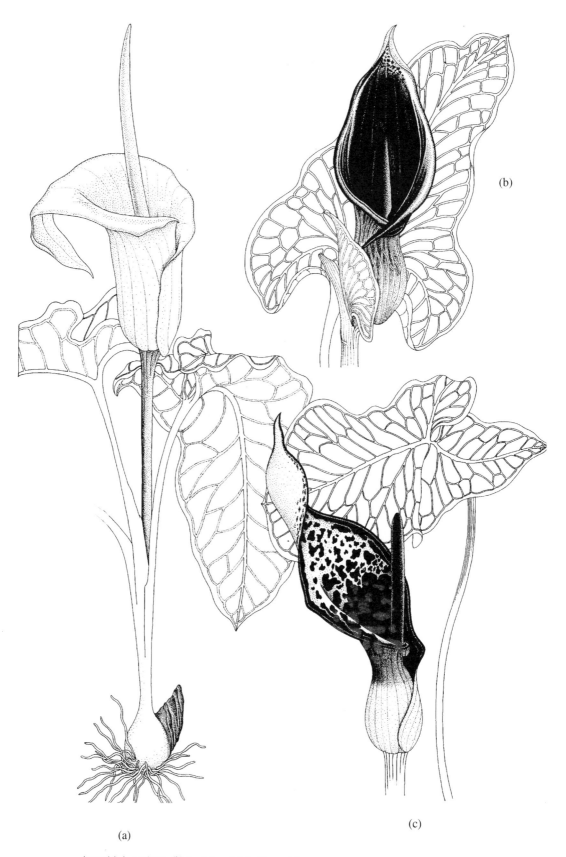

Arum (a) *A. creticum,* (b) *A. pictum,* (c) *A. dioscoridis*

red or, rarely, yellow; berries orange-red); *A.pictum* (W Mediterranean Is; leaves appearing in autumn, shimmering dark green with fine silver-green veins; spathe produced in autumn, to 25cm, dark, satiny purple-red, erect, hooded and stout, spadix club-shaped, maroon-black); *A.rupicola* (W Asia, SW Mediterranean; spathe produced in spring and early summer, to 40cm, green to purple-red or maroon, paler with at base, very narrow and erect, spadix tapering, green-white at base, maroon above). For *A.dracunculus*, see *Dracunculus vulgaris*; the white arum lily is *Zantedeschia aethiopica*.

Aruncus (Classical name for these herbs). Rosaceae. Northern Temperate Regions. GOAT'S BEARD. 2 species, hardy perennial herbs, clump-forming with slender-stalked, pinnate to ternately compound, ferny leaves composed of dark green, ovate and coarsely toothed leaflets. Small, off-white flowers are borne in frothy, plume-like panicles in summer. Hardy to –15°C/5°F, grow in moist soil in filtered sun or light shade. Propagate by division in autumn or early spring, or by seed in autumn. *A.dioicus* (*A.sylvester*; Europe to S Russia; to 2m tall smaller forms include the Japanese var. *astilboides*, and 'Kneiffii', a vigorous plant with narrow and gracefully cut leaflets and delicately branched, creamy panicles to 1m tall).

Arundinaria (from Latin *arundo*, reed). Gramineae. SE US. 1 species. See *Bamboos*.

Arundo (from Latin *arundo*, reed). Gramineae. Warm regions of the Old World. GIANT REED. 3 species, very large, rhizomatous perennial grasses, evergreen in mild climates, more or less deciduous elsewhere. Tall and cane-like, the stems bear two alternate ranks of sword-shaped leaves. Flowers are produced after long, hot summers in terminal, silky plumes. Plant in a fertile, moist but well-drained loamy soil in full sun. In frost-free regions, growth is fast, continuous and spreading. Elsewhere, the canes are usually killed or at least stripped by hard frosts – protect the crowns with a thick, dry mulch and sacking. Cut out dead or damaged canes in spring. Increase by division. *A.donax* (Mediterranean; canes to 6m, stout; leaves 60cm, rough-edged, grey-green, arching; flower spikes pearly pink becoming grey-white; 'Macrophylla' larger with purple-tinted stems and broader, blue-green leaves; var. *versicolor*: syn. 'Variegata', leaves striped and edged off-white).

Asarina (from a Spanish vernacular name for *Antirrhinum*, which it resembles). Scrophulariaceae. Mexico; SW US; S Europe. TWINING SNAPDRAGON. 16 species, sprawling or twining perennial herbs sometimes grown as annuals. Flowers produced in spring and summer, axillary, tubular to funnel-shaped with two lips and a distinct palate. Essentially tender, with a minimum temperature requirement of 5°C/40°F. In favoured locations in zone 8, they may, however, survive outdoors as

herbaceous perennials – this is certainly true of *A.procumbens* which overwinters successfully on sheltered rock gardens. Otherwise grow outdoors as half-hardy annuals. Provide support in the form of pea sticks, canes or surrounding vegetation. In the cool greenhouse or conservatory grow in pots and train on trellis or a wigwam of bamboo canes. Grow in full sun or part shade; ventilate, water and feed freely in summer; keep barely moist in winter; cut back in early spring. Increase by seed grown under glass in spring.

A.antirrhinifolia (VIOLET TWINING SNAPDRAGON; SW US to Mexico; glabrous twiner; flowers to 2.5cm long, mauve to deep violet, palate white marked yellow); *A.barclayana* (*Maurandya barclayana*; Mexico; glabrous climber; flowers to 7cm long, white to pink then deep purple, white within); *A.erubescens* (*Maurandya erubescens*; CREEPING GLOXINIA Mexico; grey-green climber; flowers to 7cm long, rose pink, white marbled pink within); *A.filipes* (*Antirrhinum filipes*; YELLOW TWINING SNAPDRAGON; SW US; glabrous, sometimes glaucous twiner; flowers to 1.5cm long, yellow spotted black on palate); *A.procumbens* (*Antirrhinum asarina*; SW Europe; woody-based, glandular hairy trailer; flowers to 3.5cm, white veined purple-rose, lips pale yellow or pale pink, palate yellow).

Asarum (from Classical Greek name *asaron*). Aristolochiaceae. Asia; N America; Europe. 70 species, low-growing, evergreen perennial herbs with spicily aromatic rhizomes and heart- to kidney- or arrow-shaped leaves borne close to the ground. The purple-brown flowers appear in spring. Solitary and short-stalked, they lie between the leaves and consist of a fleshy cup with three, tail-like lobes at its rim. The following are hardy in zone 6. Grow in the wild or woodland garden, or as a groundcover among shrubs. Plant in moist but porous soil rich in leaf-mould in dappled sunlight to deep shade. Increase by division of well-established clumps in spring.

A.arifolium (SE US; leaves 16–20cm, triangular to sagittate, blunt, marked light green or silver between veins); *A.asaroides* (Japan; leaves 8–12cm, broadly ovate, blunt, spotted); *A.blumei* (Japan; leaves 6–10cm, hastate to ovate, acute to blunt, dark green with white veins); *A. canadense* (WILD GINGER, eastern N America; leaves 6–10cm, broadly reniform, apex acuminate, dark green, usually lost in winter); *A.europaeum* (ASARABACCA, W Europe; leaves 2.5–10cm, reniform to cordate, blunt, dark glossy green); *A.hartwegii* (W US; leaves to 10cm, cordate to ovate, acute, dark green to bronze, mottled silver); *A.shuttleworthii* (SE US; leaves 5–10cm, orbicular to cordate, mid-green marked silver); *A.virginicum* (SE US; leaves 5–12cm, orbicular to cordate, dark green with purple or silver mottling).

Asclepias (from the Greek god of medicine, Asklepios, referring to the medicinal properties of some species). Asclepiadaceae. Americas; South Africa. SILKWEED; MILKWEED. 108 species, annuals

Asarum canadense

and perennials, herbs, subshrubs and shrubs, exuding milky sap if damaged, some tuberous-rooted. Carried in spring and summer in dense, flat-topped, umbel-like cymes, the flowers are small and starry with five, spreading and pointed lobes and a central corona – a crown of five, hooded lobes often differently coloured. The pods are slender and case-like or resemble a balloon; they disperse silk-tasseled seeds. *A.curassavica* and *A.physocarpa* should be grown in full sun with a minimum temperature of 7°C/45°F. Keep almost dry in winter. Short-lived and fast-growing, these shrubs are best replaced every two to three years with new plants grown from seed sown in late winter in a heated case. They are plagued by whitefly under glass. The remaining, hardy perennial species are handsome additions to the herbaceous border and the wild garden where they attract bees and butterflies. Plant in full sun. Increase by division in early spring.

A.curassavica (BLOOD FLOWER, SWALLOW WORT; S America; annual or short-lived perennial subshrub to 1.5m tall; flowers bright orange red with yellow centres); *A.hallii* (W US; tuberous perennial to 1m tall; flowers deep pink); *A.physocarpa* (syn. *Gomphocarpus physocarpus*; SWAN PLANT; S Africa; willowy, erect short-lived shrub to 2m tall; flowers cream to green-white; fruit to 6cm wide, rounded and inflated, pale, translucent green covered in soft spines and hanging from a curving stalk that resembles a swan's neck); *A.syriaca* (North America; tuberous perennial, 1–2m tall; flowers pink to mauve or white); *A.tuberosa* (BUTTERFLY WEED; INDIAN PAINTBRUSH; US; tuberous perennial, 1–2m tall; flowers flame red, vermilion, orange or gold, sometimes scented, sometimes pink or bicoloured).

aseptate of fruit; lacking partitions or divisions.

asexual lacking sexual characteristics or (of reproductive processes) occurring without the fusion of gametes. The description includes apomixis and mechanical methods of propagation such as grafting, cuttings and division.

ashes coke or coal ash is sometimes used to lighten heavy soils but may contain impurities; it is also used for surfacing standing ground or in plunge beds for pots. Fresh dry wood ash is a useful fertilizer. See *wood ash*.

Asparagus (Classical name for this plant). Liliaceae (Asparagaceae). Old World. 300 species, perennial herbs, sometimes shrubby or climbing, with thick rhizomes or root tubers and clumped, tough and slender stems. These are clothed with needle- or leaf-like cladophylls, subtended by the sharp, scale-like true leaves. With the exception of the two hardy species, all listed here are evergreen and hail from Africa. Small, green-white flowers are followed by glossy red to black berries.

EDIBLE ASPARAGUS (*A.officinalis*) has been used for over 2000 years, first as a medicine and then as a food. Grown for its edible shoots or 'spears'. This species and the ornamental *A.verticillatus* are hardy, requiring full sun and a freely draining, deep soil; *A.verticillatus* needs the support of surrounding vegetation or trellis.

Plant edible asparagus in spring as one-year old crowns, raised from seed; set 10cm deep at 30–40cm stations, in rows 60cm apart. Plants may be grown on the flat or, preferably, in ridges to aid drainage and weed control; earth-up in the autumn after planting. Plants are dioecious; males produce higher yield and do not give rise to troublesome seedlings; male plants can be selected in the second year in the seed bed, and there are all-male clones available through micropropagation. Close planting gives higher yield. Cutting can take place one year after planting, over a 3–4 week period; but it is better to leave beds uncut for at least two years from planting to build up strong crowns, after which harvest can be made over an 8–10 week period. Spears are cut about 20cm long; green when allowed to develop above the surface or mostly white when cut from more deeply earthed crowns. In autumn cut the tall stems of fern-like foliage down to within 2cm of ground level; mulch the rows with rotted organic matter. Beds can remain productive for many years.

The remaining species are tender. During the colder months, when growth virtually ceases, reduce water and withold fertilizer. Bright, filtered light is ideal, but most will tolerate light or even heavy shade. Provide a minimum temperature of 7°C/45°F. Propagate by division.

A.asparagoides (syn. *A.medeoloides, Medeola asparagoides*; stems erect to gently twining, 60–150cm tall; cladophylls to 3cm long, narrowly ovate with tapering tips, glossy green and long-lasting; foliage offered by florists under the name SMILAX); *A.densiflorus* (tuberous, making a clump of erect to sprawling stems to 1m tall, with whorled branches dense with linear cladophylls to 1cm long and bright emerald green; 'Myersii': syn. *A.meyeri*, *A.myersii*, FOXTAIL FERN, erect with short, compact lateral branches dense with cladophylls and creating soft, ferny cylinders; Sprengeri Group: syn. *A.sprengeri*, EMERALD FERN, looser, airy, emerald green foliage on gracefully arching stems); *A.drepanophyllus* (tuberous, making a dense clump of sprawling to arching, thinly woody stems 1–10m long; cladophylls tough, narrowly oblong, 2–7cm long and dark green); *A.falcatus* (SICKLE THORN; differs from the last species in its longer, slightly curving cladophylls with a distinct midrib); *A.officinalis* (EDIBLE ASPARAGUS; Europe, C Asia, N Africa; fleshy-rooted herbaceous perennial to 1.5m tall making a clump of finely branched, plume-like stems clothed with very fine, thread-like cladophylls); *A.setaceus* (erect to climbing evergreen with very slender stems ranked with horizontal branches of very fine, short, dark green cladophylls – the asparagus fern of florists); *A.verticillatus* (SE Europe to Siberia; differs from *A. officinalis* in its very long stems (scrambling to 2m) and black, not red berries).

asparagus

Pest
asparagus beetle

asperous, asperulous of plant surfaces; rough, coarse, sharp.

Asperula (from Latin, *asper*, rough). Rubiaceae. Europe; Asia; Australia. WOODRUFF. 100 species, annual and perennial herbs or dwarf shrublets with narrow leaves and terminal clusters of small, 4-parted flowers in summer. Fully hardy but intolerant of damp winter conditions, it should be grown on fast-draining soils in full sun on the rock garden, in sink gardens or in the alpine house. Increase by division, by seed or by soft cuttings in early summer. *A. arcadiensis* (syn. *A.suberosa* of gardens, (Bulgaria, Greece; perennial, making a tufted clump of frosted growth to 10cm tall; flowers bright pink clustered on an elongated spike to 5cm long). For *A.odorata* see *Galium odoratum*.

Asphodeline (related to *Asphodelus*). Liliaceae. (Asphodelaceae). Mediterranean to Caucasus. JACOB'S ROD. 20 species, more or less deciduous, rhizomatous perennial or biennial herbs producing an erect stem in spring and summer, clothed with slender leaves and terminating in a tall spike of 6-parted flowers among thin-textured bracts. Hardy to –15°C/5°F. Grow in full sun or dappled shade in a low- to med-ium-fertility, freely draining soil. Propagate by division or by seed sown in a sandy mix in the cold frame.

A.liburnica (C Europe, Balkans; perennial to 60cm tall; stem leafy below; flowers 3cm long and yellow striped green); *A.lutea* (KING'S SPEAR, YELLOW ASPHODEL; Mediterranean; perennial to 1m tall; stem clothed throughout in fine, grey-green leaves and ending in a spike of fragrant flowers each to 3cm long and golden yellow).

Asphodelus (from the Greek name for these plants, *asphodelos*). Liliaceae (Asphodelaceae). Mediterranean to Himalaya. 12 species, thickly rhizomatous hardy herbs, perennial or annual with slender, grassy leaves and spikes of starry, 6-parted flowers in late spring and summer. Of the following, the first is suited to scree or rock gardens, or to the alpine house, the others to the herbaceous border. Grow in full sun. Increase by seed.

A.acaulis (N Africa; low-growing perennial with a stemless spike of green-striped, white or pale pink flowers held low in the leaf rosette); *A.aestivus* (syn. *A.microcarpus*; Canaries, Mediterranean, Turkey; branching perennial to 2m tall with tan-striped, white flowers); *A.albus* (C and S Europe; erect perennial, sometimes branched, to 1m tall with white or pale rose flowers with deeper pink stripes).

Aspidistra (from Greek *aspidion*, a small, rounded shield, referring to the shape of the stigma). Liliaceae (Convallariaceae). E Asia. 8 species, tender, evergreen perennial herbs with thick, creeping rhizomes and erect, long-stalked, lanceolate to elliptic leaves. Cup-like and 6–8-lobed, the fleshy flowers arise along the rhizome and open at soil level. Grow in filtered sunlight to deep shade. Keep moist throughout the year; apply a weak feed fortnightly in spring and summer. Maintain a minimum winter temperature of 5°C/40°F. Propagate by division in spring. *A.elatior* (CAST IRON PLANT; BAR ROOM PLANT; BEER PLANT; China; leaves smooth, deep green, to 60cm tall; flowers 2–3cm-wide, cream mottled purple-red. 'Variegata': leaves long, striped creamy yellow; 'Variegata Exotica' leaves boldly streaked white).

Asplenium (from the Greek name, *asplenon*, derived from *a-*, without, and *splen,* spleen; these ferns were thought effective against complaints of the liver and spleen). Aspleniaceae. Cosmopolitan. SPLEENWORT. 700 species, evergreen ferns with short rhizomes and fronds in rosettes or tufts. Hardy species include *A.scolopendrium* and *A.trichomanes*, both of which will grow in a wide range of garden situations, although the former prefers some shade and moisture and the latter dislikes wet. Among tender *Asplenium* species are two very popular and resilient house plants, *A.bulbiferum* and *A.nidus*. The first will thrive in cool, airy and shaded conditions; the second prefers warmth and humidity, but will cope with minimum temperatures of 7-10°C/45-50°F and dry air. Both need a porous, soil-free potting mix, permanently damp but never wet, and protection from strong sunlight. Propagate all species by division; *A.bulbiferum* by plantlets; *A.scolopendrium* from spores.

Asplenium nidus

A. bulbiferum (HEN-AND-CHICKENS, MOTHER FERN; fronds to 60cm, olive to jade green, very finely cut and bearing plantlets); *A.nidus* (BIRD'S NEST FERN; fronds 30–80cm long, broadly oblanceolate, thin-textured, uncut and glossy light green with black midribs); *A.scolopendrium* (syn. *Phyllitis scolopendrium*; HART'S TONGUE FERN; fronds 20–40cm, lanceolate to narrowly spear-shaped, undivided and tough, light olive to dark green with a smooth, glossy texture; cultivars are available with fronds short or long, broad or narrow and with margins wavy, toothed, crested, crisped or cut, and sometimes forking at the tips); *A.trichomanes* (MAIDENHAIR SPLEENWORT; fronds 6–30 cm-long, with dark, wiry stalks ranked with small, broad and leathery, dark green leaflets).

Astelia (from Greek *a-*, without, and *stele*, a pillar: some species are epiphytic and thus unable to prop themselves up). Liliaceae (Asteliaceae). Australasia, Polynesia, Réunion, Mauritius, Falkland Islands. 25 species, evergreen perennial herbs with clumps of sword-like, ribbed leaves and inconspicuous flowers crowded in long-stalked panicles. Hardy in zones 7/8 if grown in a sheltered position and mulched in winter. Otherwise grow in containers and overwinter in a frost-free greenhouse. Plant in a gritty, acid soil. Position in full sun. Keep moist but avoid overwet winter conditions. Increase by seed or division. *A.nervosa* (New Zealand; leaves 50–200cm long linear to lanceolate, covered in brilliant silvery chaff; flowers green to purple-bronze; fruit orange to red).

Aster (from Greek *aster*, a star, referring to the shape of the flowers). Compositae. Americas; Europe; Asia; Africa. Some 250 species, mostly perennial herbs and subshrubs, rhizomatous or fibrous-rooted with erect, clumped stems and white to pink, red, lilac, mauve or blue, daisy-like flowerheads in summer and autumn. A large genus which includes low-growing montane natives for the rock garden to tall back border perennials, with a large number of the North American species, such as *A.acuminatus*, *A.cordifolius* and *A.divaricatus*, suited to naturalizing in the wild and woodland garden and other informal situations. Most asters thrive in sun or semi-shade in any moderately fertile, well-drained soil that retains sufficient moisture throughout the growing season. The rock garden species, generally flowering in early summer, include the very low-growing types such as *A.alpigenus*, *A.himalaicus*, *A.tongolensis* and *A.bellidiastrum* together with taller species such as *A.farreri*, *A.souliei* and *A.yunnanensis*. The many varieties of *A.alpinus*, with *A. × alpellus*, are also suited in scale to the larger rock garden and can be useful in smaller gardens. Where space is limited, the European and Asiatic species are slower growing and less rampant than their American counterparts. Selections from the *A.amellus* cultivars, in a height range of about 50–60cm, suit the front of sunny borders with well-drained, moderately fertile soils; plants benefit from division every third or fourth year in spring to avoid overcrowding and reduce losses from verticillium wilt. Hybrids of *A. × frikartii* and *A.sedifolius* are slightly taller. *A.thomsonii* 'Nanus' is more compact and is floriferous over exceptionally long periods; these appreciate similar treatment and conditions as for *A.amellus*.

The huge range of tall, clump-forming, late summer and autumn flowering asters, derived largely from the NEW ENGLAND ASTER, *A.novae-angliae* and the MICHAELMAS DAISY, *A.novi-belgii*, thrive in rich, fertile and moisture-retentive soil, in full sun or part shade. Top dress with garden compost after cutting back in later autumn after flowering. Flower quality and good health are maintained by frequent division and replanting of the more vigorous growth at the outside of the clumps, every second or third year in autumn or spring. Tall cultivars need staking in more carefully tended areas of the garden and this must be done early in the season, using peasticks or other grow-through support which will be obscured by the subsequent growth. *A.ericoides* and its cultivars, with slender-branched stems of small, starry flowerheads bring light effects to the sunny border with drier soils and have the advantage of being largely self supporting and rarely needing division.

A.divaricatus can be used to similar airy effect and is best left to flop gracefully over lower plants at the front of the border. *A.lateriflorus*, holding its minute flowerheads in horizontally branching sprays, suits similar soils and situations. *A.macrophyllus*, though its small flowers are not in the first order of merit, is particularly useful for naturalistic plantings in dry soils in shade, where the heart-shaped leaves and running root form good groundcover.

Propagation is usually by division, but also possible by soft cuttings in spring. A number of cultivars come approximately true from seed; most species can be raised from seed but *Aster* hybridizes freely and seedlings may not come true.

Aster may be affected by grey mould, leaf spots, sclerotinia rots, and aster wilt, powdery mildew may strike, especially in hot dry summers; modern varieties of *A.novi-belgii* are particularly susceptible. Selection of resistant cultivars is the preferred means of control, followed by treatment with appropriate fungicide at the first signs of infection, and regular treatment at fortnightly intervals thereafter as necessary. Remove and burn affected foliage in autumn.

Asteranthera (from Greek *aster*, star, and *anthos*, flower, referring to the flower shape). Gesneriaceae. 1 species, *A.ovata*, an evergreen, woody-based scrambler native to Chile and climbing to 4m. The stems are slender and picked out with small, downy leaves. The flowers appear singly or in pairs in summer and are bright red, to 6cm long, with a narrow tube widening to a 2-lipped limb to 3cm across, the lower lip marked yellow. Plant outdoors in zones 7 and over in a sheltered, sunny position on an acid soil. Keep moist but never waterlogged. Mulch in winter. Increase by semi-ripe cuttings, simple layering, or by division.

Astilbe (from Greek, *a-*, without, and *stilbe*, brightness, a reference to the dullness of the leaves of the type species). Saxifragaceae. E Asia; N America. 12 species, hardy, deciduous perennial herbs, rhizomatous and forming dense clumps, with slender-stalked, 1-3-ternately compound leaves composed of elliptic to ovate, toothed leaflets. Small, white, cream, pink or deep red, the flowers are crowded in dense, plume-like panicles. For woodland and water garden or moist borders. In some, the young, coppery foliage is handsome; flowering begins in early summer with *A.japonica* and hybrids, extending to late summer flowerers like *A.chinensis*. The many commonly available cultivars are usually grouped as *A. × arendsii* and range in height from 50 to 100cm, in colour, from white to darkest red, flowering from mid spring to late summer. Smaller species and forms like *A.chinensis* 'Pumila' and *A.simplicifolia* may be used in alpine troughs, beside rock garden pools or in pans. Plant in spring or autumn in rich, moisture-retentive soil (pH 5–7) and part-shade. Mulch with dry bracken in or straw in winter, and with leafmould and garden compost in early spring. Cut to ground level in late winter; lift and divide every three to four years in spring.

Astrantia (this name has two possible derivations – from Greek *aster*, referring to the starry flowerheads, or from the Mediaeval Latin *magistrantia*, 'masterwort', from *magister*, master). Umbelliferae. Europe to Asia. MASTERWORT. 10 species, fully hardy evergreen or semi-deciduous perennial herbs, with palmate or palmately lobed, toothed leaves arising from a short, woody rhizome. Produced in spring and summer, the flowers are

Astrantia major

small, white to green and carried on long-stalked umbels. These are surrounded by a collar of papery, green-white bracts with toothed margins, the whole resembling a larger flower. Grow in full sun or part shade. Increase by seed or division in early spring.

A.major (GREATER MASTERWORT; C and E Europe; 25 to 80cm; umbels to 3.5cm across, bracts lanceolate,white netted olive green or pale pink and coarsely toothed; subsp. *involucrata*: bracts larger, deeply cut; 'Shaggy': syn. 'Margery Fish', bracts deeply cut, pink-white; 'Sunningdale Variegated': leaves yellow-cream-variegated); *A.maxima* (S Europe, Caucasus; to 90cm tall; bracts pink-tinted, elliptic – white in the cultivar 'Alba', dark pink in 'Hadspen Blood', a very showy hybrid with *A.major*).

Astrophytum (from Greek *aster*, star, and *phyton*, plant). Cactaceae. US (Texas); Mexico. 4 species, tender, perennial cacti with an unbranched, dome- to ball-shaped stem, ribbed and sometimes spiny and marked with a pattern of chalky white scales. Provide a minimum temperature of 5°C/40°F and a dry, buoyant atmosphere in full sun. Plant in a gritty, low- to medium-fertility mix. Keep dry from mid-autumn to early spring except for a light misting on warm, sunny days. Water moderately in summer. Propagate by seed.

A.asterias (SEA URCHIN, SILVER DOLLAR CACTUS; stem circular, flattened to hemispherical, to 15cm wide with broad low, spineless ribs; flowers yellow sometimes tinted red); *A.myriostigma* (BISHOP'S CAP, BISHOP'S MITRE; stem rounded to domed, to 20cm wide with prominent and faceted, spineless ribs, the whole covered in minute, chalky scales; flowers gold with a yellow-orange heart); *A.ornatum* (stems spherical to column-shaped, 15–100cm tall with six to eight deep, wing-like and spiny ribs, dark green and lined with silvery white, dot-like scales; flowers, brilliant yellow).

asymmetric, asymmetrical (1) of a leaf margin or the shape of an inflorescence that is irregular or unequal in outline or shape, (2) of a flower incapable of being cut in any vertical plane into similar halves.

Atherosperma (from Greek, *ather*, bristle, and *sperma*, seed). Monimiaceae. Australia. 1 species, *A.moschatum* (AUSTRALIAN SASSAFRAS, TASMANIAN SASSAFRAS), a spreading, evergreen tree to 25m tall with nutmeg-scented, lanceolate leaves to 10cm long, glaucous to glossy above and white-downy beneath with toothed or entire margins. Creamy white, cup-shaped flowers to 2.5cm across appear in summer. Grow in a lime-free soil in full sun or part shade. Provide good ventilation and a minimum winter temperature of 3–5°C/37–40°F. Propagate by seed or semi-ripe cuttings in summer.

Athrotaxis (from Greek *athroos*, crowded, and *taxis*, arrangement, referring to the disposition of the leaves and cone scales). Cupressaceae. Tasmania. 3 species, evergreen conifers with fissured, flaking bark and small, subulate or scale-like leaves overlapping or spreading on erect branchlets. The cones are small and spherical. The following favour equable, rainy areas in zones 7–9. Plant in a moist, humus-rich soil in full sun or part shade with shelter from harsh winds. Propagate by cuttings.

A.selaginoides (KING WILLIAM PINE; to 30m tall with a broadly conical crown and red-brown bark falling away in shreds; leaves to 1cm, spreading and subulate, the outer surface bright green, the inner marked with two white lines); *A.cupressoides* (SMOOTH TASMANIAN CEDAR; to 16m; leaves scale-like; the hybrid offspring of these two is *A. × laxifolia,* SUMMIT CEDAR, hardier than either parent and with fresh, yellow-green new growth).

Athyrium (from Greek *athoros*, good at breeding, referring to the diverse forms of the sori). Athyriaceae. Cosmopolitan. 100 species, evergreen or semi-deciduous ferns with 2–3-pinnately divided fronds. *A.filix-femina* and *A.nipponicum* are popular garden plants. Both species prefer light shade and a humus-rich, moist and slightly acid soil; they will, however, tolerate some chalk and full sun, provided that the roots are cool and damp. Fully hardy, they nonetheless benefit from a dry winter mulch. Increase by division in early spring.

A.filix-femina (LADY FERN; northern temperate regions; fronds 20–100cm long, light-textured and a soft, pale green; numerous cultivars are available with fronds variously forked, bunched, crested, crisped and fringed – one of the most remarkable is 'Frizelliae' (TATTING FERN), with long, narrow fronds in which the segments are reduced to widely spaced balls of parsley-like growth); *A.niponicum* (syn. *A.goeringianum*; E Asia; fronds 10–30cm-long, broader than in *A.filix-femina*, of a heavier texture and dark green with a bronze-purple tint, especially when young; 'Pictum': PAINTED LADY FERN, JAPANESE PAINTED FERN, fronds emerging purple-bronze, turning dull aquamarine with silver-grey flush along the purple-tinted veins).

Atriplex (from the Greek name for these plants). Chenopodiaceae. Cosmopolitan. ORACH; SALT BUSH. 100 species, evergreen shrubs or annual herbs, often with a grey or silver-white mealy coating, and small flowers in dense panicles. *A.hortensis* is a fast-growing hardy annual with edible, spinach-like foliage. The red-flushed forms are very decorative. Sow seed *in situ* in spring. Unattended, these plants will seed themselves. *A.halimus* is fully hardy and tolerates saline conditions, drought and fierce winds. The remaining species are fine, silver shrubs for very dry, windy, saline or alkaline situations and are useful fire-retardants and windbreaks. All need a warm, dry and sheltered position on sharply draining soils. Increase by softwood cuttings in summer.

A.canescens (FOUR-WING SALTBUSH; to 2m tall, broadly spreading; leaves to 1.5cm long, linear to narrowly oblong, silvery grey); *A.halimus* (TREE PURSLANE; S Europe; semi-evergreen, bushy perennial shrub to 2m tall; leaves tough, ovate-

Astrophytum myriostigma

Athyrium nipponicum 'Pictum'

rhombic to hastate, pewter grey to silvery white); *A.hortensis* (MOUNTAIN SPINACH, ORACH; Asia, widely naturalized elsewhere; annual herb to 2m tall with arrow-shaped, leaves; var. *rubra*: leaves strongly purple-red-tinted – selections include 'Cupreatorosea' with red-tinted leaves burnished with copper, and 'Rosea' with pale rose leaves veined dark purple-red); *A.hymenelytra* (DESERT HOLLY; low, compact to 1m tall; leaves rhombic to oblong or bluntly hastate, coarsely toothed, silver-white, tough and holly-like); *A.lentiformis* (QUAIL BUSH; WHITE THISTLE; broadly spreading, to 3m tall, spiny; leaves oblong to ovate-deltoid, silver-grey; subsp. *breweri* lacks spines and is virtually evergreen).

attapulgite a clay mineral, which after calcining and crushing can be used as an ingredient of potting media. Its physical properties are similar to sand, but it has a higher buffering capacity.

attenuate of the apex or base of a leaf, sepal, petal, stem or bulb; tapering finely and concavely to a long drawn out point.

aubergine (*Solanum melongena*). EGGPLANT. a short lived perennial treated as an annual, it is 60–70cm high and is grown for its two-locular fruits. Of various shape and size, these are 5–15cm long and 3–6cm in diameter, yellow, black, white or purple in colour, and may be fried or roasted. Probably originating in India, but also found in Southeast Asia and China, the aubergine was introduced into Europe in the 13th century.

In warm climates, sow seeds in a sheltered place; transplant seedlings to ridges or raised beds either 50–60cm apart each way or spaced 60–75cm stations in rows 75–90cm apart. In temperate areas, raise seedlings in a greenhouse at a minimum temperature of 15°C/60°F; grow to maturity under protection in beds, growing bags or 25–30cm-diam. pots at a maintained temperature of 25–30°C/77-86°F. Stop to encourage bushy growth, and stake; fruit is harvested when fully coloured, 80–120 days from transplanting.

Aubrieta (for Claude Aubriet (1668–1743), French botanical artist). Cruciferae. Europe to Central Asia. AUBRETIA. 12 species, cushion- or mat-forming, evergreen, hardy perennial herbs with small leaves and clusters or short racemes of 4-petalled flowers throughout the year, but chiefly in spring and summer. Plant in a sunny, well-drained position, e.g. walls, dry banks, paving, edging and the rock garden. Shear off flowered growth. Increase by semi-ripe cuttings in early autumn or seed sown *in situ* in late spring.

A.deltoidea (Aegean; sprawling cushions of bristly, mid-green, rhomboid to ovate leaves, the margins entire or sparsely toothed; flowers to 1.5cm across, mauve to lilac, pink or purple-red; this species has silver-white- and gold-edged cultivars, some with minute leaves; *A.* × *cultorum* includes garden hybrids close to *A.deltoidea*, but ranging in flower shape and

Aubergine

aubergine

Pests
aphids
red spider mite
whiteflies

Disease
grey mould

auriculas

Pests
aphids
root aphids
woolly aphids
fungus gnats
red spider mite
slugs
vine weevil

Disease
root rot

size from the small and single to the large and semi-double, and in colour through deep violet to royal purple, mauve, lilac, magenta, carmine and rose pink).

auctoris, auctorum (abbrev. auct., auctt.) Latin 'of an author', 'of authors'. A formula used in full botanical citations to show when a name has been applied by authors to the wrong entity. It is used usually with the 'non' formula to contrast it with the true type of the plant as would be embodied by the correct author citation. For example, by *Betula platyphylla* auct. non Sukachev, we indicate that some authors call what is in fact *B.mandschurica* by a valid name belonging to a wholly different plant (the type named by Sukachev). This formula is sometimes used interchangeably with 'hort.' and 'misapplied'.

Aucuba (from the Japanese name for this plant). Cornaceae (Aucubaceae). East Asia. 3 species, evergreen shrubs and small trees with green branches, star-shaped maroon-green flowers and attractive berries. Plant in part-sun or shade on most soils. Male *and* female plants are needed for berrying. Once established, *Aucuba* will withstand hard frost, some drought and heavy atmospheric pollution. Prune to size in spring. Propagate from seed and suckers in spring or from basal cuttings of the current year's growth in autumn.

A.japonica (SPOTTED LAUREL; China, Japan; broad, rounded shrub to 4m tall; leaves 10–25cm long, elliptic to narrowly ovate, more or less toothed, thick and deep glossy green, unmarked in the cultivars 'Viridis', the narrow-leaved 'Salicifolia', and the dwarf 'Nana'; cultivars include 'Bicolor' (leaves with a central yellow blotch and large teeth), 'Crotonifolia' (female, leaves finely speckled and blotched gold), 'Gold Dust' (female, leaves speckled gold), 'Sulphurea' (leaves edged pale yellow), and 'Variegata' (female, leaves spotted and blotched yellow). The berries last from late summer to early spring, sometimes longer. Typically bright scarlet and about 1cm wide, they are also known in white and yellow forms (f. *leucocarpa*, f. *luteocarpa*), whilst the variegated selection 'Fructu-albo' has yellow-white to pale buff fruit).

auger a T-shaped metal hand tool about 1m long with the lower 15cm of the main stem twisted into a wide thread. It is used to collect cores or twists of soil from various depths for inspection or analysis.

aural screen a planting of trees and shrubs designed to reduce noise, especially traffic noise; the best effect is obtained with tall evergreen species planted more than one row deep.

auricle an ear-like lobe or outgrowth, often at the base of an organ (i.e. a leaf), or the junction of leaf sheath and blade in some Gramineae.

auriculas a group of distinctive primulas believed to arise from the natural hybrid *Primula* × *pubes-*

cens, the progeny of two variable high-alpine crevice plants *P.auricula* and *P.rubra*. The two pigments, flavone and hirsutin, produce a wide range of colours in the progeny, only true blue being absent. Auriculas have been documented since the mid 16th century, when they were valued for medicinal purposes and were introduced into the north of England by Huguenot silk weavers. Auricula societies flourished in England in the late 18th century, running shows with strict rules and displays of 'stage auriculas' shown in tiny theatres against a black velvet backdrop, often with mirrors. They are now divided into four classes, Shows, Alpines, Doubles and Borders, and are much grown for exhibition, from unheated greenhouses or frames. Hardy to –14°C/7°F, they require good drainage and a dry atmosphere in winter. Cultivars are propagated by offsets between spring and late summer and are hybridized by hand pollination. Fresh seed germinates best and most flower by the second spring from sowing. Grow outdoor types in raised beds rich in humus, tender cultivars singly in pots, protecting buds from frost and strong sunshine. Repot annually after flowering.

auriculate, auricled of the base of a leaf blade or perianth segment possessing two rounded, ear-shaped lobes that project beyond the general outline of the organ.

Aurinia (from Latin, *aureus*, golden, referring to the flowers of *A.saxatilis*). Cruciferae. Central Europe east to Ukraine and Turkey. 7 species, woody-based biennial and perennial herbs with sprawling to mound-forming branches, oblanceolate leaves and racemes or panicles of small, 4-parted flowers in spring and early summer. Fully hardy and ideal for rock gardens, dry walls and banks. Grow in full sun in well-drained, poor to medium-fertility soils. Increase by greenwood cuttings in early summer in a shaded cold frame, or from seed in autumn. Cut back after flowering. *A.saxatilis* (syn. *Alyssum saxatile*; evergreen perennial to 40 × 30 cm with downy, grey-green leaves and broad panicles of pale yellow flowers; 'Citrina': lemon yellow; 'Dudley Neville': rich cream tinted apricot; 'Gold Dust': deep gold; 'Variegata': grey-green leaves with yellow-white margins).

Austrocedrus (from Latin *auster*, the south wind, and *Cedrus*, cedar). Cupressaceae. Chile; Argentina. 1 species, *A.chilensis* (syn. *Libocedrus chilensis*), CHILEAN CEDAR, an evergreen coniferous tree to 15m tall. The crown is loosely ovoid to narrowly conical or columnar. Dark brown to orange-grey, the bark falls away in narrow shreds. Scale-like dark green leaves with white undersides overlap in four ranks along flattened, feathery branches. The cones are small, ovoid to oblong and mid-brown. Hardy in zones 7 and 8, the Chilean cedar prefers a cool, humid and sheltered location on a moist, sandy and humus-rich soil. It suffers in hot weather and drying winds. Increase by seed and cuttings.

author in taxonomy, the first worker validly to publish a new name or combination for a taxon. The author name, usually abbreviated, follows that of the taxon in a botanical citation. For example, *Physalis alkekengi* L., where L. stands for Linnaeus. The author citation can be qualified to denote a misapplied name (e.g. *Physalis alkekengi* hort. non L., which describes a name wrongly used by some growers for a plant more correctly named *P.alkekengi* var. *franchetii*), or used in a double citation to indicate that a name first described at one rank (here a species) has been relegated to the status of variety within another species, for example *Physalis alkekengi* var. *franchetii* (Mast.) Mak. In this case Makino, has reduced Masters's species *P.franchetii* to a variety of *P.alkekengi*. The convention is also used when a name is transferred to a different genus entirely.

autoecious completing the entire life cycle on one kind of host; cf. *heteroecious*.

autogamous self-fertilizing.

autopolyploid a polyploid with multiple chromosome sets derived from a single individual or species; cf. *allopolyploid*.

autumn colour Prior to discarding their foliage in autumn, deciduous trees and shrubs and most herbaceous perennials of temperate regions set in motion a process of transferring from their leaves useful materials such as carbohydrates for storage in stems or underground parts. To assist this function, accessory materials contained in the sap are called into use. Some of these materials are colourless but anthocyanin, which often features, is red where sap is very acid, blue in the absence of acidity, and violet in the intermediate condition. As chlorophyll degenerates it gives rise to yellow grains. Combinations of these and other materials impart a range of colour to the leaves of plants approaching dormancy. The timing and habit of autumn colour depends upon the species; its intensity and duration is influenced by soil and atmospheric moisture (for example, colours are rarely strong following a dry period), by temperature (early frost hastens leaf fall) and by light (transmission of colour is greatest in bright sunshine). Good autumn colouring species can be found among *Acer*, *Amelanchier*, *Berberis*, *Betula*, *Carpinus*, *Cornus*, *Euonymus*, *Liquidambar*, *Liriodendron*, *Parrotia*, *Sorbus* and *Viburnum*.

auxins see *growth regulators*.

available water capacity a measure of the water in soil available to plants; expressed as a percentage of the volume of soil; that which is held between field capacity and permanent wilting point.

awl-shaped of plant features; narrowly wedge-shaped and tapering finely to a point, subulate.

awn a slender sharp point or bristle-like appendage found particularly on the glumes of grasses.

awned see *aristate*.

axial relating to the axis of a plant.

axil the upper angle between an axis and any off-shoot or lateral organ arising from it, especially a leaf.

axillary situated in, or arising from, or pertaining to an axil. For example, an axillary inflorescence arises from the junction of a petiole and the stem, whereas a terminal inflorescence grows directly from the apex of the stem.

axis a notional point or line around or along which organs are developed or arranged, whether a stem, stalk or clump; thus the vegetative or growth axis and the floral axis describe the configuration and development of buds and shoots and flowers respectively, and any stem or point of origination on which they are found.

*Azolla
filiculoides*

AYR abbreviation for 'all year round'; applied to the continuous flower production of greenhouse chrysanthemums, especially by the manipulation of day length.

Azara (for J.N. Azara (1731–1804), Spanish patron of botany). Flacourtiaceae. South America (those below from Chile). 10 species of evergreen trees and shrubs grown for their attractive habit and foliage (usually neat, dark glossy green and gracefully disposed on arching or fan-like branches) and axillary clusters of small flowers with crowded yellow stamens. Free-standing, they nonetheless benefit from the protection of a south-facing wall or fence in zones 7 and 8: this will show their habit to best advantage and encourage free flowering and the production of attractive small berries. *A.petiolaris* and *A.microphylla* withstand temperatures to −15°C/5°F; *A.serrata* to −10°C/14°F; *A.lanceolata* to −5°C/23°F. Other species will not tolerate heavy frosts and the leaves of all species will scorch and drop if exposed to harsh cold winds. Plant in sun or light shade in a deep, humus-rich soil. Prune after flowering only to remove overcrowded, exhausted or badly placed branches. Propagate by semi-ripe cuttings in a closed case with bottom heat, or by simple layering of the past season's growth.

A.dentata (small tree to 2m; leaves elliptic, toothed, dark glossy to 3.5cm long; fruit yellow); *A.integrifolia* (tree-like shrub, usually to 5m tall with 4cm-long oval leaves more or less untoothed; flowers yellow; fruit glaucous black; var. *browneae*: leaves larger, obovate with a few apical teeth; 'Variegata': leaves smaller variegated with pale pink and creamy white); *A.lanceolata*: (shrub or small tree to 6m tall; leaves toothed, glossy dark green, lanceolate, on

slender arching branches; flowers ochre-yellow, fragrant; fruit violet-black); *A.microphylla* (tree to 8m tall with fan-like sprays of branchlets densely ranked with very small, rounded, dark green leaves – edged pale yellow in the cultivar 'Variegata' – flowers yellow-green, strongly scented of vanilla; fruit blood red); *A.petiolaris* (tree to 5m tall with glossy, 4cm-long, ovate to elliptic, toothed leaves on distinct stalks; flowers pale yellow, richly scented, carried on short cylindrical spikes in late winter and early spring; fruit purple-black with a plum-like bloom); *A.serrata* (shrub to 4m high with 6cm-long oblong to elliptic, toothed leaves of a deep, shining green; flowers small, golden, clustered, in each leaf axil; fruits black or white).

Azolla (from Greek, *azo*, to dry up, *ollymi*, to kill: these plants are killed by drought). Azollaceae. Cosmopolitan. MOSQUITO FERN; WATER FERN; FAIRY MOSS. 8 species, aquatic ferns. They are usually perennial, surviving cold periods as submerged fragments. At other times they float on the surface of standing water or colonize saturated mud. Their small bodies consist of forked or feather-shaped rhizomes from which hang thread-like roots. The 'fronds' are reduced to minute scales which overlap closely in two ranks along the rhizome branches. A. *filiculoides* (syn. *A. caroliniana*) is often grown in ponds and freshwater aquaria. Native to the warm Americas, it is now widely naturalized in the waterways, lakes and ditches of most temperate, subtropical and tropical regions in the Northern Hemisphere. Each plant body is some 1.5 × 1.8cm with water-repellent 'fronds' to 1mm long of a soft pea green. It will proliferate with remarkable speed, forming a velvety emerald skin that may become a nuisance. In hot sunshine, times of drought and with the onset of autumn, the fronds take on a rusty tint before becoming a deep fox- or blood red. *Azolla* can be grown outdoors in ponds in zones 7 and over. Where frosts occur, it will sink into winter dormancy, resurfacing in spring. By midsummer, it may be necessary to skim off surplus growth if other aquatics are not to be choked and the pond life below sealed in darkness.

Azorella (diminutive of Azores, application uncertain). Umbelliferae. South America; Antarctic Is; New Zealand. 70 species, dwarf, carpeting or hummock-forming perennial evergreen herbs with tufted rosettes of small leaves and, in summer, stalk-less umbels of small flowers. Hardy to −17°C/1°F, and suitable for perfectly drained but moist, sunny positions on the rock garden, scree, raised bed or sink garden. Propagate by division or seed. The following plant is sometimes confused with *Bolax gummifera* (syn. *Azorella glebaria*). *A.trifurcata* (Chile, Argentina; forming dense cushions or mats; leaves 0.6–1cm long, leathery and dark glossy green to grey-green with three distinct teeth at their tips; flowers yellow).

B

Babiana (Latinized form of the Afrikaans *bobbe-jaan*, baboon: baboons are said to eat the corms). Iridaceae. Africa (those below from South Africa). BABOON FLOWER. Some 60 species, cormous perennials grown for their brilliantly coloured flowers produced in spring and summer. They seldom exceed 20cm in height and produce a flattened fan of plicately ribbed, often hairy, lance-shaped leaves in spring. This gives rise to an erect, simple or branched spike of up to ten flowers, dish-shaped with six broadly elliptic tepals in the majority of species, irregular and tubular with narrow tepals in species formerly included in *Antholyza*.

Baboon flowers are usually described as frost-tender, requiring container-treatment much like that for *Freesia* and a minimum temperature of 10°C/50°F. They certainly respond well in such conditions and make attractive spring-flowering pot plants for the cool greenhouse, conservatory and home. Most can equally be grown outside where winter temperatures do not fall much below –5°C/23°F, in a warm, sunny position in a free-draining, light but fertile soil. Outdoors, plant deeply (i.e. to 10cm) and give a thick winter mulch of bracken litter or leaf-mould. Alternatively, lift corms in autumn and over-winter, cleaned and dry, in frost-free conditions. Replant or pot up in early spring in equal parts loam, leafmould and sand. Feed pot-grown plants prior to flower formation. Propagate by offsets.

B.plicata (to 20cm, flowers slightly irregular, fragrant, funnel-shaped, to 4cm diam., in shades of pale blue, violet and white with mauve and yellow blotches at the base of the lower tepals); *B.ringens* (syn. *Antholyza ringens*, to 20cm, flowers 5cm long in shades of red or maroon, closely 2-ranked on unbranched spikes to 20cm tall in summer); *B.rubro-cyanea* (to 20cm, flowers 4cm-diam. funnel- to dish-shaped flowers with spreading tepals, royal blue in their upper half, deep wine red below); *B.stricta* (to 20cm, flowers funnel- to disk-shaped with broad tepals in shades of purple, mauve, blue and yellow, sometimes mixtures thereof); *B.thunbergii* (syn. *Antholyza plicata*, to 60cm, flowers 5-6cm, tubular, curving in shades of bright red or deep pink stained green).

baccate of fruit; berry-like, with a juicy, pulpy rind.

Baccharis (for Bacchus, god of wine). Compositae. GROUNDSEL TREE. Americas. Some 350 species, hardy evergreen or deciduous shrubs adapted to dry exposed conditions with tough, small, toothed leaves, or few or no leaves and photosynthetic stems instead. Large clusters of small discoid flowerheads are followed by copious quantities of downy pappus. Grow in full sun in well-drained soil (minimum temperature –15°C/5°F). *B.halimifolia* will tolerate dry soils and saline, windy exposures; an excellent coastal shrub, it can be pruned to a hedge. *B.pilularis* is a fine open groundcover for binding impoverished soils in dry, sunny exposures, the cultivar 'Twin Peaks' especially. An annual shearing will improve its density. *B.sarothroides* is ideal for xeriscapes and for screening and erosion control in even the driest locations. Prune all species where necessary in early spring. Propagate by softwood or semi-ripe cuttings.

B.halimifolia, (GROUNDSEL-TREE or COTTON-SEED TREE; Eastern US to the West Indies; fast-growing, deciduous shrub to 3 × 3m; branches angled, scurfy; leaves 3–7cm, obovate, coarsely toothed; in autumn, flowerheads give way to masses of silver-white down); *B.pilularis* (COYOTE BUSH, DWARF CHAP-ARRAL-BROOM; California; fast-spreading ground-cover shrub to 50cm tall by 2m across; branches resinous, tangled; leaves 1–4cm-long obovate, saw-toothed, bright green and tough-textured; 'Centennial': a cross with *B.sarothroides*, heat-resistant and narrow-leaved; 'Pigeon Point': to 1 × 4m, fast-growing with larger, light green leaves; 'Twin Peaks': a small-leaved, slow-growing selection, male and thus free from the cotton down which can in this species become a nuisance); *B.sarothroides* (DESERT BROOM; to 2m, with no or very few leaves and bright green branches covered in snow white pappus in winter).

bacillus thuringiensis a bacterium used for biological control, mainly of caterpillars.

backbulb a dormant pseudobulb retained after active growth of usually only one season.

backcross a cross made between a hybrid plant and a plant genetically identical to one of the hybrid's parents. This procedure is used in breeding programmes for a number of seed raised generations with the object of concentrating desirable characters into an individual line.

backyard garden in the US a garden at the rear of a dwelling, usually enclosed with direct access from the living quarters. Also known as a yard garden.

Bacopa (from the South American name). Scrophulariaceae. WATER HYSSOP. 56 species, aquatic or semi-aquatic perennials with creeping to erect stems. Small, campanulate flowers are produced singly in the axils of thinly fleshy leaves. Plant in aquaria or at the margins of ponds and lakes. Cuttings of stem tips will root in warm water or wet loam, at 20°C/70°F or over.

B.caroliniana (N America; leaves 2.5cm, entire, ovate, scented of lemon; flowers blue); *B.monnieri* (Tropics; succulent, with obovate to spathulate leaves to 2cm long sometimes notched at the tips; flowers white to pale blue).

bacteria single-celled, microscopic organisms lacking a true nucleus and chlorophyll. Most bacteria are saprophytic and beneficial, breaking down organic matter in soil or fixing nitrogen, but some are pathogens.

bacterial diseases. Bacterial infections cause many different symptoms in plants, including seedling blights, leaf spots or blotches, shoot blight, wilting, scabs, galls, cankers and rots. The same pathogen may cause different symptoms on its host, depending on climatic conditions or the stage in the host's development at which infection takes place. In warm, moist conditions bacterial cells can multiply extremely rapidly and this accounts for the destructiveness of many bacterial diseases and their common occurrence in warm climates especially. Among the hundreds of kinds of bacteria known to cause plant diseases, some have a wide host range, for example, the CROWN GALL bacterium, (*Agrobacterium tumefaciens*) whilst others, such as that causing HYACINTH YELLOWS, (*Xanthomonas campestris* pv. *hyacinthi*), are restricted to a narrow range of host species.

The local spread of bacteria over short distances is effected principally by rain and wind. Bacteria on the surface of an infected plant disperse in rainwater, which is broken up into droplets and carried by the wind to other plants. Storms provide ideal conditions for the epidemic spread of bacterial diseases, as potential hosts will be likely to have waterlogged and wounded tissues which are highly susceptible to infection. Some diseases are transmitted by insects and other animals, including humans, and longer-distance spread is effected in this way. Many bacterial diseases are dispersed through infected plants and propagating material; many are seed borne or in accompanying debris; others are soil-borne but generally do not survive long in the absence of plant debris.

Bacterial blight is the common name for a bacterial disease causing widespread blackening of leaves and shoots. Examples include LILAC BLIGHT (*Pseudomonas syringae* pv. *syringae*), which also affects *Forsythia* and can cause blossom blight of pears. A number of plants can be attacked by other pathovars (pv.) of *P.syringae*, the diseases usually being referred to as bacterial leaf spots, such as blight of mulberries (*Morus* species) caused by *Pseudomonas syringae* pv. *mori*, and PEA BLIGHT (*P.syringae* pv. *pisi*), a serious seed-borne disease which has only recently become established in Britain. HALO BLIGHT (*Pseudomonas syringae* pv. *phaseolicola*) is a seed-borne disease of French and sometimes runner beans causing pale areas on leaves, stems and pods. Bacterial blight of walnuts (*Juglans regia*) is caused by *Xanthomonas campestris* pv. *juglandis*, which in the UK is mainly a disease of nursery stock but can also affect the fruits. An important bacterial blight affecting many rosaceous trees is FIREBLIGHT (*Erwinia amylovora*).

Bacterial canker manifests itself as sunken necrotic lesions of stem tissue. Cankers are often limited in size by a host reaction so there may be some malformation and outgrowth of the surrounding bark. *Pseudomonas syringae* pv. *mors-prunorum* is the cause of bacterial canker of *Prunus* species, especially cherries and plums, although almonds, apricots, peaches, nectarines and ornamental species are also susceptible. During autumn or winter, bacteria from the leaves infect stem tissue through wounds or leaf scars. Unfolding leaves are yellow, often narrow and curled, and may wither and die during the summer. Infection spreads in spring to produce shallow, depressed cankers; branches usually show a characteristic flattening on one side. Affected branches may be killed and gum may exude from the lesions. No new cankers are formed in summer, but by then the leaves have been attacked and develop circular brown spots. The centres of these spots often drop out to give a 'shot hole' symptom.

BACTERIAL CANKER OF POINSETTIA (*Curtobacterium flaccumfaciens* pv. *poinsettiae*) is a serious disease in which water-soaked streaks occur on stems and petioles, often extending to cause leaf-blotching and complete defoliation. BACTERIAL CANKER OF TOMATO (*Clavibacter michiganensis* subsp. *michiganensis*) is a vascular wilt disease in which deep cankers may be produced on the stems. The bacteria are seed- and soil-borne, and outdoor crops are most seriously affected. CITRUS CANKER (*Xanthomonas campestris* pv. *citri*) is a major disease thought to have originated in Asia, and has been subject to strict quarantine measures for many years. Eradication campaigns have been successful following outbreaks in Australia, New Zealand, South Africa and the US, but the disease has become established in several South American countries. Affected plants show branch cankering and the smaller shoots, foliage and fruit are also attacked. There are several strains of different pathogenicity to citrus cultivars. BACTERIAL CANKER OF POPLARS (*Xanthomonas populi*) produces a cream-coloured bacterial slime which oozes from cracks in young shoots in spring; large cankers may develop on branches and stems. Black poplars, including the Lombardy poplar (*Populus nigra* var. *italica*), are resistant to the disease.

Bacterial leaf and stem rot (*Erwinia chrysanthemi* pv. *dieffenbachiae*) is a widespread cause of brown leaf spotting and stem rot of *Dieffenbachia* species.

Bacterial leaf spot is caused by *Pseudomonas* and *Xanthomonas* species, and can give rise to leaf spots and other symptoms on various plants. LEAF SPOT OF ANTIRRHINUMS (*Pseudomonas syringae* pv. *antirrhini*) causes pale green leaf spots on seedling leaves which become brown and sunken in the centre and may coalesce, leading to total collapse. It is seed-borne and similar to diseases caused by *Pseudomonas* species on *Lobelia*, *Primula*, *Tagetes* and *Viola*; LEAF SPOT OF DELPHINIUMS (*P.syringae* pv. *delphinii*) results in large black blotches on leaves, stems and flower buds of some cultivars. ANGULAR LEAF SPOT OF CUCURBITS (*P.syringae* pv. *lachrymans*) results in small, irregular, water-soaked spots which enlarge to become angular in shape between leaf veins. Affected tissue, which may exude drops of liquid containing bacteria, dries out, causing holes in the leaves. Small spots may also develop on the fruit and cause it to rot. LEAF BLOTCH OF MAGNOLIAS may be caused by *P.syringae*. LEAF SPOT OF BERBERIS is caused by *P.syringae* pv. *berberidis*. LEAF SPOT OF BEGONIAS (*Xanthomonas campestris* pv. *begoniae*) gives rise to small translucent spots on the underside of leaves and the spots may enlarge and merge so that the leaves turn brown and die. LEAF SPOT OF IVY (*Xanthomonas campestris* pv. *hederae*) may be troublesome in greenhouses and during propagation when humidity is high and moisture persists on foliage.

Apart from copper preparations, most fungicides are ineffective against bacteria and control is best effected by an integrated programme of measures intended to break the cycle of disease, involving: (a) exclusion of the disease by the use of healthy seed and planting material, the destruction of infected tissue, plant debris and volunteer plants, general cleanliness and measures to ensure that contaminated tools, implements, and containers are not brought onto the site; (b) the prevention of damage to plants by cultural or harvesting operations, particularly during wet weather; (c) use of resistant selections if available; (d) control of possible insect or other animal vectors.

bacteriophage a virus that infects bacteria.

bagasse the fibrous residue of sugar cane processing, used as a ground mulch or as an ingredient of potting media.

bagging hook a sickle-shaped grass-cutting tool with a short handle.

bagworm (*Thyridopteryx ephemeraeformis*) a destructive garden insect familiar in N America, where its larvae devour the leaves of many trees and shrubs, especially conifers. They feed within spindle-shaped cases up to 50mm long, composed of plant foliage tied together with silk, in which the fully fed larvae pupate in late summer. Adult males, mostly black with almost transparent wings, fly to cocoons containing wingless females, which die within the cocoons after egg-laying. The larvae emerge in spring. Remove and destroy the bags containing eggs in winter or apply contact insecticide in spring just after the young larvae hatch.

Baileya (for L.H. Bailey (1858–1954), US botanist and horticulturist). Compositae. Western North America. 3 species, hairy annual or perennial herbs with daisy-like flowerheads carried from spring until early autumn. Grow in full sun. Sow seed *in situ* in late spring or, in cool climates, under glass in late winter for planting out in late spring. *B.multiradiata*, (DESERT MARIGOLD; annual to 35cm tall with entire or pinnately divided grey leaves and bright yellow flower heads to 5cm diam.).

ball (1) the mass of roots and growing medium at the base of plants lifted from the open ground or removed from pots; (2) a condition of rose flowers where the expanded bud decays before opening; (3) a class of dahlias with rounded flower heads larger than pompons.

ballbarrow a wheelbarrow with a large, usually pneumatic ball for the wheel.

balled and wrapped a description of plants lifted from the open ground where the root ball is wrapped in fabric for protection during transport. Also known as burlapped.

Ballota (from *ballote*, the Greek name for black horehound, *B.nigra*). Labiatae. Europe, Mediterranean (chiefly), West Asia. 35 species, perennial herbs, subshrubs and shrubs, mostly with paired leaves and whorled, two-lipped tubular flowers cupped by enlarged calyces in summer. Grow in sunny, sheltered situations on well drained soils in zones 7 and over. *Ballota* may require some protection where temperatures often fall below −10ºC/14ºF: frosts tends to kill off the soft top growth. Cut back or thin out dead wood in spring before new growth commences. Propagate by softwood or semi-ripe cuttings in spring and early summer.

B.acetabulosa (Eastern Mediterranean, Greece; stems woolly, erect, ranked with crenate, rounded to heart-shaped leaves to 5cm long; flowers lilac-pink, backed by a white-felted dish-like calyx to 2cm diameter; *B.pseudodictamnus* (S Aegean; smaller, more rounded than the last species, with ovate to cordate, virtually entire leaves and far smaller calyces; 'All Hallow's Green': smaller still, with finely downy foliage in soft lime green).

bamboos A group of long-lived grasses of the subfamily Bambusoidae, comprising about 90 genera and 1000 species; with natural habitats ranging from 50°N in Japan to 47°S in Chile, at altitudes from sea level to 4000m and therefore variously hardy in gardens. Bamboos differ from other grasses in having silica-rich culms and stalked leaves joining the leaf sheaths. They vary in height from a few centimetres to 40m. Flowering is uncommon. When it occurs plants may be seriously weakened. Seed production is correspondingly rare and propagation is best done by division. Outside the cool temperate West, bamboos are an important source of construction material, food and medicine. In gardens worldwide many species are grown as ornamental screens and groundcover. The ripened culms make excellent plant supports.

BAMBOOS

Name	Distribution	Habit	Culms	Leaves
Bambusa multiplex B.glaucescens; B.nana; B.argentea; HEDGE BAMBOO	S China	3–15m, densely clump-forming	1–4.5cm diam., arching with prominent nodes and many branches	2.5–15 × 0.5–1.5cm, in two rows, rather silvery beneath
Comments: Much used in subtropical and tropical Asia for hedging. Good bonsai.'Alphonse Karr': culms and branches striped orange-yellow and green, tinged pink when young. 'Golden Goddess': small, culms golden, leaves larger. 'Riviereorum' (CHINESE GODDESS BAMBOO): culms delicate, somewhat sinuous, leaves small. 'Silver Stripe': large, culms and leaves striped yellow or white. Z8/9.				
Bambusa oldhamii Sinocalamus oldhamii CLUMPING GIANT TIMBER BAMBOO; OLDHAM BAMBOO	China, Taiwan	6–15m, thickly clump-forming, erect with dense foliage	stout, 3–13cm diam., with white powder below nodes and short branches	15–30 × 3–6cm, tough and broad
Comments: Much used in warm places for screening. Z9.				
Bambusa ventricosa BUDDHA'S BELLY BAMBOO	S China	8–20m, densely clump-forming	3–5.5cm diam., smooth and green to yellow-green	10–20 × 1.5–3cm
Comments: If grown in containers or on poor soils, the culms seldom exceed 2m and develop obesely swollen internodes, a feature considered ornamental. Otherwise, they grow tall and straight. Z9.				
Bambusa vulgaris COMMON BAMBOO	so widely cultivated in S Asia that its natural habitat is obscure	5–25m, densely clump-forming	5–25cm diam., smooth with prominent nodes	10–25 × 1.8–4cm
Comments: Used for light timber, paper pulp and edible young shoots. 'Vittata': culms and leaves striped green and yellow. Z9.				
Chimonobambusa marmorea Arundinaria marmorea	long cultivated in Japan, native country uncertain	2–3m, running rhizomes making large, sometimes invasive stands of widely spaced culms	1–2cm diam., round in section, sheaths marbled purple or pink-brown with white spots when young, densely leafy with many branches	7.5–16 × 0.5–1.5cm
Comments: Z6.				
Chimonobambusa quadrangularis Arundinaria quadrangularis SQUARE BAMBOO	SE China, Taiwan, naturalized Japan	2–10m, running rhizomes with well-spaced, strongly erect culms and soft tiers of branches	1–3cm diam., green, square in section with 4 angles and thorn-like root protuberances at nodes	10–29 × 2–7cm, dark green.
Comments: Z6.				
Chusquea coronalis	C America	to 6m, clump-forming; culms erect to arching or sprawling	0.5–2cm diam., slender with rings of fine branches	2.3–7 × 0.3–0.8cm
Comments: Distinguished from C.culeou by the branches, which completely encircle the culms, and the leaves without tessellate veins. Z7.				
Chusquea culeou	Chile	4–6m, stately, large clumps with distinctively 'whorled' branches	1.5–3cm diam., with solid pith, pale papery sheaths and slender, rather short branches crowded in alternating semi-circles	6–13 × 0.5–1.5cm
Comments: Differs from C.coronalis in the branches on one side of each node only and the leaves with tessellate veins. 'Tenuis' (var. tenuis; C.breviglumis) grows to 2m with narrower culms spreading outwards at 45° from a narrow-based clump. Z7.				
Dendrocalamus giganteus Sinocalamus giganteus GIANT BAMBOO	SE Asia	15–40m, giant clump-former with a strong V-shaped outline from base and arching upper reaches hung with light branches	15–35cm diam., green to yellow-green, very stout with prominent nodes and large sheaths	20–55 × 3–12cm
Comments: A spectacular giant for subtropical and tropical gardens, large warm glasshouses and the atria of public buildings. Z10.				
Drepanostachyum falcatum Arundinaria falcata; Chimonobambusa falcata; Sinarundinaria falcata; Thamnocalamus falcatus	E Himalaya	2–6m, densely clump-forming and slow-growing	1.5–2.5cm diam., slender, white-powdery at first becoming glossy yellow with bushy branches at nodes	11–35 × 1–2.5cm
Comments: Intolerant of wet/cold conditions and temperatures below –6°C/21°F. Best grown in large tubs of fertile, sandy loam kept evenly moist in the cool conservatory. **D. khasianum** differs in its culms coloured a rich, polished coppery red. Z8.				
Fargesia dracocephala	China (Sichuan)	3–5m, makes a neat, fountain-like clump with arching top and branches	0.4–0.8cm diam. with numerous branches at first growing close to internodes, then spreading or drooping	5–17 × 0.5–2cm, held horizontally in a frond-like arrangement
Comments: Z6.				

BAMBOOS

Name	Distribution	Habit	Culms	Leaves
Fargesia murieliae *Arundinaria murieliae*; *Arundinaria spathacea*; *Fargesia spathacea*; *Sinarundinaria murieliae*; *Thamnocalamus spathaceus* UMBRELLA BAMBOO	C China	3–4m, clump-forming with culms gently arching outwards from a narrow base and making a light, airy, arching top	0.4–1cm diam., white-powdery at first, soon yellow-green, then yellow with pale brown sheaths gradually standing away from culms	6–15 × 0.5–1.8cm, apple green, tip finely tapering
Comments: Z5.				
Fargesia nitida *Arundinaria nitida*; *Sinarundinaria nitida* FOUNTAIN BAMBOO	C China	3–4m, clump-forming with graceful, slender culms bowed outwards from a narrow base and weighted with airy, drooping branches	0.4–0.8cm diam., grey-powdery at first then lined or tinted purple-brown with purple-green sheaths	4–11 × 0.5–1.1cm, dark green, flickering in the breeze
Comments: Includes 'Eisenach' (to 2m, culms strongly erect, black-tinted with prominent nodes), 'Ems River' (to 4m, culms almost black, very hardy), 'Maclureana' (to 6m, culms strongly arching, pearly grey-powdery), 'Nymphenburg' (culms dull grey-green, weeping). Z5/6.				
× *Hibanobambusa tranquillans* *Phyllostachys tranquillans*; *Semiarundinaria tranquillans* INYOU-CHIKUZOKU	Japan	2–5m, rhizomes running in warm, wet and fertile conditions, otherwise making open clumps with rather sparse branches	1–3cm diam., erect to arching at apex	15–25 × 3.5–5cm, dark green, tough, glossy
Comments: Possibly a hybrid between *Phyllostachys* and *Sasa*. Includes 'Shiroshima' with culms, sheaths and leaves striped white and yellow. Z8.				
Himalayacalamus falconeri *Arundinaria falconeri*; *Drepanostachyum falconeri*; *Sinarundinaria falconeri*; *Thamnocalamus falconeri*	C Himalaya	5–10m, makes a V-shaped clump of elegant slender culms and abundant feathery foliage	2–3cm diam., yellow-green with a purple ring below the nodes	7–16 × 0.8–2.7cm
Comments: 'Damarapa': (syn. *Arundinaria hookeriana* of gardens; *Sinarundinaria hookeriana* of gardens); culms emerging pink striped lime green then turning deep amber or red-brown with irregular, dark olive stripes. Requires a sheltered, warm site on a moist, fertile, sandy soil rich in leafmould. In cold regions, grow in tubs in the cool conservatory. Z8.				
Himalayacalamus hookerianus *Arundinaria hookeriana*; *Chimonobambusa hookeriana*; *Drepanostachyum hookerianum*; *Sinarundinaria hookeriana* BLUE-STEMMED BAMBOO	E Himalaya	6–9m, tall, V-shaped clump-former	1.5–3cm diam., blue-green, thickly white-powdered at first with a blue-grey ring below nodes, later yellow-green or purple-red	to 18 × 2cm, very graceful, appearing misty blue en masse and at a distance
Comments: The name *Arundinaria hookeriana*, properly a synonym of this species, is often misapplied to *Himalayacalamus falconeri* 'Damarapa' (q.v.). Z8.				
Indocalamus latifolius *Arundinaria latifolia*	E China	0.5–1m, rhizome running in warm, wet conditions, otherwise slowly building a spreading, mound-like clump	to 0.5cm diam., slender and sparsely branched	10–40 × 1.5–8cm, spreading, mid to deep green; veins not tessellated
Comments: Z7.				
Indocalamus tessellatus *Arundinaria ragamowskii*; *Sasa tessellata*; *Sasamorpha tessellata*	C China, Japan	1–2.5m, rhizomes running or growing slowly to make spreading, rather untidy, mounded clumps	0.5–1.5cm diam., slender and sparsely branched	to 60 × 10cm, spreading to drooping, mid to dark green with tessellated veins
Comments: The largest leaves of any bamboo listed here – an exotic and luxuriant screen or mounded, tall groundcover; also good for containers. Z7/8.				
Otatea acuminata *O.acuminata ssp. aztecorum*; *Arthrostylidium longifolium*; *Yushania aztecorum* MEXICAN WEEPING BAMBOO	Mexico to Nicaragua	2–8m, clump-forming with very graceful foliage on slender culms that arch almost to the ground	2–4cm diam., ultimately weeping	7–16 × 0.3–0.5cm, numerous, pendulous
Comments: Plant outdoors only in regions with short-lived frosts; otherwise suitable for the cool glasshouse or conservatory. Once established, will cope with drier conditions than most bamboos. Z9.				
Phyllostachys aurea FISHPOLE BAMBOO; GOLDEN BAMBOO	SE China, Japan	2–10m, strongly erect runner with dense foliage; may remain a clump-former in cold climates	2–5cm diam., smooth, green then brown-yellow with short, crowded internodes at base	5–15 × 0.5–2cm
Comments: Makes a good screen or hedge and will tolerate some drought. 'Albovariegata': culms slender, leaves striped white; 'Holochrysa': culms yellow sometimes striped green, leaves sometimes striped; 'Violascens': to 6m tall, culms stout, green striped purple, ultimately stained grey-purple to violet, leaves large, rather glaucous beneath. Z6.				
Phyllostachys aureosulcata YELLOW GROOVE BAMBOO	NE China	3–10m, vigorous, erect runner with dense foliage (may remain in clump in colder climates)	1–4cm diam., yellow-green with a distinct yellow groove, often flexuous below	5–17 × 1–2.5cm
Comments: Good hedging bamboo. 'Aureocaulis': syn. 'Spectabilis' culms yellow marked with a green groove. Z6.				
Phyllostachys bambusoides *P.mazellii*; *P.reticulata* GIANT TIMBER BAMBOO; JAPANESE TIMBER BAMBOO	China, Japan, widely grown elsewhere	3–30m, strongly erect and running, forming large groves of well-spaced culms decked with airy foliage. In colder climates it will remain in a clump	15–20 cm diam., stout, green, with heavily marked sheaths	9–20 × 2–4.5cm
Comments: Includes 'Albovariegata' (leaves marked white), 'Castillonis' (culms yellow with a green stripe), 'Castillonis Inversa' (culms green with a yellow stripe), 'Holochrysa' ('Allgold') (smaller and more open with golden-yellow culms sometimes striped green), 'Marliacea' (culms green, internodes strangely wrinkled). Z7.				

BAMBOOS

Name	Distribution	Habit	Culms	Leaves
Phyllostachys edulis *P.heterocycla* f. *pubescens*; *P.pubescens*; *P.mitis* of gardens MOSO BAMBOO	China, introduced Japan	3–27m, rhizomes running, forming large groves of stout, erect, grey-green culms with small feathery leaves; in colder climates, will remain in a clump	4–30cm diam., very thick, grey and velvety when young, ultimately green or almost orange with white powder below nodes, curved near base; sheaths heavily mottled brown	5–12 × 0.5–2cm

Comments: 'Bicolor': culms striped yellow; f. **heterocycla**: syn. *P.heterocycla*, TORTOISESHELL BAMBOO, lowest culm internodes short, distorted and bulging to one side. Z7.

Name	Distribution	Habit	Culms	Leaves
Phyllostachys flexuosa	China	2–10m, rhizomes running, making open groves of erect to arching culms with feathery foliage; remains clump-forming in colder climates	2–7cm diam., erect then arching, green-yellow becoming darker (almost black) with age, smooth with white powder below nodes	5–15 × 1.5–2cm

Comments: Z6.

Name	Distribution	Habit	Culms	Leaves
Phyllostachys nigra BLACK BAMBOO	E & C China	3–10m, running with fairly dense culms erect then arching above; remains in a tight clump in colder climates	1–4cm diam., green at first, typically turning glossy black in the second year with some grey-white powder below nodes	4–13 × 0.8–1.8cm, bright green, thin-textured

Comments: A beautiful and popular bamboo for sheltered borders and large containers. The blackest form is var. **nigra** with glossy, pure black culms; it includes the cultivar 'Othello' with tightly clumped culms that are jet black even in their first year. In var. **henonis** the culms are green to yellow-green. The culms are variably marked dark purple-brown on green in f. **punctata**; in 'Boryana' they are green blotched brown. Z7.

Name	Distribution	Habit	Culms	Leaves
Phyllostachys sulphurea	E China	4–12m, running, making groves of erect culms; remains clumped in colder climates	0.5–9cm diam., ultimately yellow, sometimes striped green, sheaths thick, spotted and blotched brown	6–16 × 1.5–2.5cm

Comments: var. **viridis** (*P.mitis*; *P.* 'Robert Young'; *P.viridis*) is a larger plant with green culms; it seldom thrives in cold climates, .Z7.

Name	Distribution	Habit	Culms	Leaves
Phyllostachys viridiglaucescens *P.elegans*	E China	4–12m, running, forming a dense stand of erect to arching culms; often remains clumped in cooler climates	1–5cm diam., smooth, with white waxy powder below nodes, often curved at base; sheaths large, with dark spots and blotches	4–20 × 0.6–2cm

Comments: Z8.

Name	Distribution	Habit	Culms	Leaves
Pleioblastus argenteostriatus *P.chino* var. *argenteostriatus*; *Arundinaria argenteostriata*; *Nippocalamus argenteostriatus*; *Sasa argenteostriata*	origin unknown, long cultivated in Japan	to 1m, running, forming erect stands of dense slender culms	0.5–1cm diam., thin, sparsely branched	14–22 × 1–2cm, pale to mid green striped creamy white

Comments: Attractive medium-height groundcover, best cut back every 1–2 years in spring. Z7.

Name	Distribution	Habit	Culms	Leaves
Pleioblastus auricomus *P.viridistriatus*; *Arundinaria auricoma*; *Arundinaria viridistriata*; *Sasa auricoma*	origin obscure, naturalized in Japan	1–3m, running but usually making a loose, mound-like clump of graceful stems	0.2–0.4cm diam., slender, reedy and sparsely branched	12–22 × 1.5–3.5cm, softly hairy, emerald green with brilliant yellow stripes

Comments: Very colourful small bamboo for ground cover or specimen planting. Cut hard back in spring every year. The leaves of f. **chrysophyllus** are a striking pure gold. Z7.

Name	Distribution	Habit	Culms	Leaves
Pleioblastus chino *P.maximowiczii*; *Arundinaria chino*; *Nippocalamus chino*	C Japan	2–4m, running and somewhat invasive making dense stands of lax, reedy stems	0.5–1cm diam., slender and sparsely branched	12–25 × 1–3cm

Comments: Includes the narrow-leaved, erratically white-striped f. **angustifolius** (*P.angustifolius*) and f. **gracilis** (*P.chino* 'Elegantissimus') which differs in its reliably white-striped, slender leaves, not hairy beneath. Z6.

Name	Distribution	Habit	Culms	Leaves
Pleioblastus gramineus *Arundinaria graminea*	Japan, E China	2–5m, running, making dense stands of slender culms	0.5–2cm diam., freely branches	15–30 × 0.8–2cm, narrow and finely tapering

Comments: A graceful, medium-sized bamboo with long narrow leaves. *P.linearis* differs in its hairy culm sheaths, truncate (not rounded) ligule and rather shorter leaves. Z7.

Name	Distribution	Habit	Culms	Leaves
Pleioblastus humilis *Arundinaria humilis*; *Nippocalamus humilis*; *Sasa humilis*; *Yushania humilis*	C Japan	1–2m, running, makes a dense stand of erect slender stems	0.2–0.3cm diam., reedy, branching throughout	10–25 × 1.5–2cm

Comments: Useful medium-height groundcover. Responds well to clipping or even mowing in spring and early summer. var. **pumilis** (*Sasa pumila*; *Arundinaria pumilia*; *Arundinaria gauntlettii*) is more sturdy and compact with very fresh green foliage. Z6.

Name	Distribution	Habit	Culms	Leaves
Pleioblastus pygmaeus *Arundinaria pygmaea*; *Nippocalamus pygmaeus*; *Sasa pygmaea* DWARF FERN-LEAF BAMBOO	origin obscure, long cultivated in Japan	10–20cm, running, making dense, short thickets of soft, slender culms	0.1–0.2cm diam.	2–4 × 0.2–0.5cm, arranged in two ranks, the tips sometimes withering and becoming papery white in hard winters

Comments: A charming miniature bamboo useful for containers, the margins of small ponds and for groundcover. The form most frequently grown is var. **distichus** (*P.distichus*; *Arundinaria disticha*; *Sasa disticha*), somewhat taller and very vigorous with leaves to 7cm long arranged in two distinct ranks at the culm tips. Both respond well to trimming or mowing. Z6.

BAMBOOS

Name	Distribution	Habit	Culms	Leaves
Pleioblastus simonii *Arundinaria simonii*; *Nippocalamus simonii* SIMON BAMBOO	C & S Japan	3–8m, running, but usually remaining confined to a clump of erect, elegant culms with arching branches and airy foliage	1.2–3cm diam., stout but thin-walled with white waxy powder below nodes and persistent sheaths that stand away from the culms as branches develop	13–27 × 1.2–2.5cm, tapering finely, somewhat glaucous beneath
Comments: Includes f. *variegatus* (*P.simonii* var. *heterophyllus*; *Arundinaria simonii* var. *albostriata*; *Arundinaria simonii* var. *striata*), less robust, with variable leaves, some narrow, some broad and many white-striped. Z6.				
Pleioblastus variegatus *P.fortunei*; *Arundinaria fortunei*; *Arundinaria variegata*; *Nippocalamus fortunei*; *Sasa variegata* DWARF WHITE-STRIPED BAMBOO	probably a selection of the green-leaved var. *viridis*, itself of unknown origin but widely cultivated in Japan	20–75cm, running, but usually making loose clumps of slender culms	0.2–1cm, typically very thin with a few erect branches	10–20 × 0.7–1.8cm, tapering finely, dark green to sea green with white to cream stripes of varying breadth
Comments: One of the finest and most widely grown variegated bamboos, a semi-dwarf making miniature stands of bright foliage. Once established, it responds well to trimming. Also suitable for bonsai treatment and use in cut flower arrangements. Z7.				
Pseudosasa amabilis *Arundinaria amabilis* LOVELY BAMBOO; TONKIN CANE	S China, widely cultivated throughout SE Asia	5–13m, running, making dense thickets of tall, erect then arching culms; in colder climates, more likely to remain in a clump	2–7cm diam., thick-walled with hairy, persistent sheaths and usually 3 branches per nodes	10–35 × 1.2–3.5cm
Comments: A very handsome bamboo producing long, rigid culms widely used for canes. It suffers in cold and windy conditions and favours moist, fertile and sheltered locations in zones 8 and 9. Z8.				
Pseudosasa japonica *Arundinaria japonica*; *Sasa japonica* ARROW BAMBOO	Japan, Korea	3–6m, running, making dense stands of erect then arching culms with resilient foliage, clump-forming in colder climates	1–2cm diam., thin-walled with persistent, thick sheaths and, usually, one branch per node	20–36 × 2.5–3.5cm, tapering finely, tough, two-thirds of undersurface glaucous (thus distinguished from *Sasamorpha borealis*)
Comments: One of the most commonly grown bamboos, excellent for large gardens and for screening; it resists high winds and will tolerate poor soils, dry or saturated. Z6.				
Qiongzhuea tumidinoda	China (Yunnan, Sichuan)	2–6m, running, erect to spreading above with outspread branches and graceful foliage	1–3cm diam., dull to glossy olive to sap green, basically narrow but bulging like a spinning top at each node; branches slender	5–15 × 0.5–2.5cm, finely tapering and thin-textured, mid green above, grey-green beneath
Comments: A very elegant bamboo valued for its unusual culms; requires a moist, fertile soil in a sheltered situation. Z6.				
Sasa kurilensis	Japan, Korea	1–4m, running, forming open thickets of slender culms	0.2–0.3cm diam., nodes glabrous, not prominent	15–30 × 1.2–4.8cm, dark shining green
Comments: Includes 'Shimofuri' with white-striped leaves. Z7.				
Sasa nipponica	Japan	50–80cm, running, forming loose stands or clumps of graceful culms	0.2–0.4cm diam., tinted purple, nodes glabrous, prominent, not branched	8–20 × 1.4–3cm, thin-textured
Comments: *Sasaella ramosa* is rather similar but distinctly branched. Z7.				
Sasa palmata *S.paniculata*; *S.cernua*; *S.senanensis*	Japan	1.5–4m, running, forming a thicket of slender culms topped with a canopy of arching, luxuriant foliage	0.2–0.5cm diam., often streaked purple, branches few	25–40 × 5–9.5cm, tapering finely, tough, bright shining green with a yellow-green midrib
Comments: A handsome medium-sized bamboo with a distinctive habit: a hand-like arrangement of broad leaves atop narrow, crowded culms. A rampant spreader. Z7.				
Sasa veitchii *S.albomarginata*	Japan	80–150cm, gently running, forming low thickets of outspread foliage	to 0.5cm diam., purple-lined, glaucous, more or less glabrous	15–25 × 3.2–6cm, broadly lanceolate-ovate, dark green with a yellow-green midrib, the margin withering and becoming papery-white
Comments: Excellent medium-height groundcover or a low, loose hedge. The broad, ribbed leaves are especially attractive in winter, when they develop ivory, parchment-like margins. Z7.				
Sasaella masumuneana *Arundinaria atropurpurea*; *Sasa masumuneana*	Japan	0.5–2m, running, forming low thickets of slender, branching culms topped with a graceful, nodding canopy	0.3–0.4cm diam., sheaths sparsely bristly	10–19 × 1.5–3.5cm, glabrous, sheaths lined purple
Comments: Includes f. *albostriata* with leaves striped white turning to yellow, and f. *aureostriata* with leaves striped yellow. Z7.				
Sasaella ramosa *Arundinaria vagans*; *Sasa ramosa*	Japan	1–1.5m, running and rampant, forming low thickets of broadly branching, slender culms topped with narrow, nodding leaves	0.3–0.4cm diam., sometimes purple, sheaths lacking bristles	10–20 × 1.4–3cm, hairy beneath, margins withering, becoming white and parchment-like
Comments: Z7.				

BAMBOOS

Name	Distribution	Habit	Culms	Leaves
Sasamorpha borealis *S.purpurascens*; *Arundinaria borealis*; *Sasa borealis*	Japan, China	1–3m, running, forming loose thickets	0.5–0.8cm, branching above, nodes hairy	10–20 × 2–3cm, dark green paler beneath, tinted purple toward base of stalk
Comments: Similar to *Pseudosasa japonica*. Z7.				
Semiarundinaria fastuosa *Arundinaria fastuosa*; *Arundinaria narihira*; *Phyllostachys fastuosa*	S Japan	3–12m, clump-forming with plum-coloured new growth and tall erect culms	2–7cm diam., hollow, lined purple-brown, glabrous with white powder below nodes; sheaths thick with shining, wine-coloured insides	9–21 × 1.5–2.7cm
Comments: Survives lows of –22˚C/–8˚F. Plant on moist rich soils in a sheltered position. Only the upper reaches of the culms are furnished with foliage, making this bamboo especially suitable for underplanting with low shrubs and perennials. Z7.				
Shibataea kumasasa *Phyllostachys kumasasa*; *Phyllostachys ruscifolia*	China, Japan	0.5–1.8m, running but appearing clumped and making a compact 'bush' of broad, bright green foliage	0.2–0.5cm diam., slender, green, with very short branches throughout	5–11 × 1.3–2.5cm, elliptic-ovate, dark to bright glossy green, borne along culms
Comments: Dwarf to medium-sized bamboo with a uniquely shrubby appearance and broad leaves, the whole not unlike *Ruscus*. Excellent high groundcover or low hedging; responds well to clipping over in spring and is even trimmed into rounded shapes in some Japanese gardens. Requires a moist, fertile, sheltered position. Z6.				
Sinobambusa tootsik	China	5–12m, running and sometimes rampant, forming light, elegant groves	2–6cm diam., smooth, with hairy sheaths	8–20 × 1.3–3cm, tapering finely
Comments: A graceful bamboo requiring a moist fertile soil, dappled sunlight and a humid atmosphere. Will not tolerate hard frosts and harsh winds. Z7.				
Thamnocalamus aristatus *T.spathiflorus ssp. aristatus*; *Arundinaria aristata*	NE Himalaya	3–10m, vigorously clump-forming, erect	1–6cm diam., yellow-green speckled brown with age, nodes with white waxy bloom below; branches red-lined, sheaths sometimes with dark markings	6.5–13.5 × 0.6–1.8cm, with hairy callus at base of stalk
Comments: An excellent hardy bamboo for screening or waterside plantings. Z7.				
Thamnocalamus crassinodus *Fargesia crassinoda*	Tibet	3–4m, erect, forming a v-shaped clump with very graceful, feathery foliage on spreading, fine branches	1–2cm diam., blue-green with thick white- or blue-grey bloom at first, later smooth, purple- or yellow-green, nodes swollen; sheaths large, parchment-like	4.5–9 × 0.5–1cm, grey-green
Comments: 'Kew Beauty', with very slender branches and small narrow leaves is one of the most elegant hardy bamboos. Z7.				
Thamnocalamus spathiflorus *Arundinaria spathiflora*	NW Himalaya			
Comments: Differs from *T.aristatus* in its slightly flexuous grey culms bloomed white at first and later flushed pink in strong sunlight; the leaves are grey-green with no callus at the base of the stalk. Z7.				
Thamnocalamus tessellatus *Arundinaria tessellata*	S Africa	2.5–7m, clump-forming or (in warm climates) running and invasive; pink-streaked shoots become tall, erect culms with dense whorls of short branches	1–6cm diam., pale green, nodes tinted purple-pink; sheaths large, papery, white; branches short, red-tinted, many per node	5–21 × 0.5–2cm, finely tapered, dark green
Comments: A striking and robust bamboo for mild climates or cool glasshouses. Z8.				
Yushania anceps *Arundinaria anceps*; *Arundinaria jaunsarensis*; *Chimonobambusa jaunsarensis*; *Sinarundinaria anceps*	C & NW Himalaya	2–5m, forms an erect to spreading clump with arching branches above	0.5–1.2cm diam., smooth; branches pendulous, not developing in first year; sheaths soon falling	6–14 × 0.5–1.8cm, mid-green, smooth, stalks sometimes purple-tinted
Comments: Popular medium-sized bamboo, resembles a larger, clump-forming *Fargesia*. 'Pitt White' (*Arundinaria niitikayamensis*) grows to twice the height of the type and is exceptionally graceful, if less hardy. Z7.				
Yushania maling *Arundinaria maling*; *Fargesia maling*; *Sinarundinaria maling*	NE Himalaya	3–10m, erect, clump-forming	2–5cm diam., rough when young, sheaths persistent	8.5–1.8 × 0.8–1.6cm, slightly rough-textured
Comments: Z7.				

Fargesia nitida

Shibataea kumasasa

Sasa veitchii

Pleioblastus variegatus

Phyllostachys aureosulcata

Phyllostachys nigra

Phyllostachys nigra 'Boryana'

Bambusa (from the Malay vernacular name). Gramineae. Tropical and subtropical Asia. 120 species, tender, clump-forming bamboos. See *bamboos*.

Banksia (for Sir Joseph Banks (1743–800), British botanist who first collected these plants). Proteaceae. AUSTRALIAN HONEYSUCKLE. Australia. 50 species, tender trees and shrubs, evergreeens with leathery leaves and, in spring and summer, long-styled flowers in dense terminal spikes which resemble brilliantly coloured, whiskery cones. Also decorative, the seeds ripen in flattened, woody capsules which gape like duck bills or oysters amid the bristly remnants of the flower cone. They thrive in sheltered locations in zone 9 and will tolerate short-lived frosts to −7°C/19°F if quite dry. In less favoured locations, they need a minimum temperature of 7°C/45°F and the protection of a greenhouse or conservatory. Grow in full sun in a free-draining sandy, slightly acid soil low in nutrients (especially nitrates and phosphates). Under glass ventilate liberally. Use slow-release fertilizer devoid of superphosphate. An application of sulphate of iron may wean shy-flowering plants. Water moderately when in growth, scarcely at all in winter. Prune only to remove dead growth or to cut flowers and seedheads for decoration. Increase by seed.

Banksia coccinea

Baptisia (from Greek, *bapto*, to dye, some species yield an indigo substitute). Leguminosae. US. FALSE INDIGO, WILD INDIGO. 17 species, hardy tap-rooted perennial herbs usually with three-parted leaves and erect, terminal racemes of pea-like flowers. Plant in a deep, well-drained, neutral to acid soil in full sun. Propagate from seed sown ripe in the cold frame, or by careful division in late winter (plants resent disturbance). *B.australis*, (BLUE FALSE INDIGO; summer-flowering perennial, 75–150cm tall; flow-ers pale violet to indigo marked white, to 1.5cm diam.; seed pods inflated grey, attractive in dried flower arrangements).

barb a hooked semi-rigid hair.

barbate bearded, with hairs in long weak tufts.

barbed of bristles or awns with short, stiff lateral or terminal hairs which are hooked sharply backward or downward.

barbellae short, stiff hairs, for example, those of the pappus in Compositae.

bare-root applied to plants which are lifted with little or no soil around their roots; as from nursery rows, where the specimens may be referred to as bare-root transplants.

bark the outermost protective tissue of woody plants. Shredded bark can be used as a soil conditioner, as a surface mulch, and as the sole or a constituent ingredient of growing media.

bark beetles (Coleoptera: Scolytidae). Small, dark-coloured, cylindrical beetles, up to 6mm long, adapted for burrowing in wood. Adults excavate tunnels, usually in the sapwood just beneath the bark, especially that of trees that are sickly or dying. Typically, a large number of beetles invade simultaneously in response to an aggregation pheromone. The FRUIT BARK BEETLES (*Scolytus rugulosus* and *S.mali*) are sometimes found in gardens, infesting fruit trees and hawthorn (*Crataegus monogyna*). *Ips* species are pests of forest conifers throughout the northern Hemisphere. In Europe and North America the ELM BARK BEETLES (*S.scolytus* and *S.multistriatus*) are serious pests of elm (*Ulmus* species) as they are vectors of Dutch elm disease. The SHOT-HOLE BORER (*Xyleborus dispar*), a European species, infests many trees and, in contrast to *Scolytus* species, tunnels extensively into the heartwood. The only feasible way of controlling bark beetles in gardens is to cut and burn infested trees and branches. In forests, traps containing aggregation pheromones may be used.

bark-bound used of trees with hard bark which prevents girth expansion. See *slitting*.

bark house a rustic garden house made of timber with bark still attached, featured in some Victorian gardens.

bark ringing excising with a sharp knife a narrow ring of bark around the trunk or limb of, usually, a fruit tree, to arrest vegetative growth.

Barleria (for Jacques Barrelier (1606–1673), French botanist). Acanthaceae. Old World Tropics and Subtropics. 250 species, tender evergreen herbs and shrubs grown for their tubular, 5-lobed flowers. In zones 9 and 10, these plants grow well in full sun or part shade planted in a moist, fertile, loamy soil. Under glass, provide a minimum temperature of 7°C/45°F. Propagate from semi-ripe or softwood cuttings in a closed case with bottom heat.

B.cristata (PHILIPPINE VIOLET; SE Asia; bristly shrublet to 1m; flowers 3-5cm, violet-blue to pink, borne amid spiny sepals); *B.lupulina* (HOP-HEADED BARLERIA; Mauritius; leaves with a white or pink midvein and hop-like spikes of yellow flowers); *B.obtusa* (S. Africa; spreading unarmed shrublet with elliptic leaves to 7cm long and loosely arranged axillary clusters of 3cm-long mauve flowers).

basal plate the compressed stem within a bulb from which the leaves arise as scales.

basal rot (*Fusarium oxysporum* f.sp. *narcissi*) a progressive rotting of narcissus bulbs after flowering caused by a specialized form of the widespread soil fungus *F.oxysporum*. The rot can be detected by softness, usually at the base of a bulb, and the dark coloration of the fleshy scales seen when the bulb is cut transversely. It should be distinguished from

BANKSIA

Name	Habit	Leaves	Inflorescence
B.ashbyi ASHBY'S BANKSIA	spreading tall shrub or tree to 8 × 3m	10–30cm, linear, wavy-serrate	bright orange, 15–20cm long
B.baueri WOOLLY BANKSIA	compact, low shrub to 2m	4–13cm, oblong cuneate, serrate, brown-green at first	orange-brown to grey, to 20 × 13cm, woolly
B.baxteri BAXTER'S BANKSIA	spreading shrub to 4 × 4m	7–17cm, rusty brown then bright green, divided into triangular lobes	globose 5–8cm diam., red-brown to green with yellow styles
B.blechnifolia	prostrate, spreading shrub to 50cm tall	25–45cm, erect, long-stalked and fern-like, appearing basal, deeply 8–22-pinnately lobed	cylindrical, to 16cm, red-pink, borne at ground level
B.brownii FEATHER-LEAVED BANKSIA	tall shrub to 4 × 2m	to 11cm, feather-like, soft-textured, whorled, white beneath with narrow, curving segments	15–20cm, oblong-cylindrical, flowers cream below, grey-brown above
B.coccinea SCARLET BANKSIA	shrub to small tree, 2–5 × 1–15.5m	to 9cm, oblong, dentate, velvety pink brown at first, then dark green above, silver beneath	to 20cm, globose, grey with scarlet styles
B.dryandroides DRYANDRA-LEAVED BANKSIA	dense, spreading shrub, to 1 × 1.5m	to 17cm, broadly linear with pinnate triangular segments	to 3cm diam., spherical, pale brown to dull orange
B.ericifolia HEATH-LEAVED BANKSIA	compact shrub to 5m tall	to 2cm, narrowly linear	to 30cm, cylindrical, orange-red to russet or grey with orange styles
B. 'Giant Candles' (B.ericifolia × B.spinulosa) BULL BANKSIA	compact shrub to 4m	to 4.5cm, linear, revolute, slightly toothed	to 40cm, deep orange
B.grandis	tall, robust tree to 10m	to 45cm, deeply lobed, pink-brown at first, then bright green	to 40cm, yellow, cylindrical
B.hookeriana HOOKER'S BANKSIA	multi-stemmed bushy shrub to 1.2m	to 20cm, narrow-linear, finely toothed, rusty brown at first	to 15cm, cylindrical, woolly white becoming orange
B.ilicifolia HOLLY-LEAVED BANKSIA	erect, stout tree to 10m	to 10cm, obovate to elliptic, glossy, entire to prickly toothed	to 20cm, flower buds pink tipped bright green opening cream, flowers ageing pink or red
B.laevigata TENNIS BALL BANKSIA	bushy shrub to 3.5m	to 14cm, broadly linear or narrowly obovate, toothed, rusty brown at first	tightly spherical, yellow and brown
B.laricina ROSE BANKSIA	rounded shrub to 1.2m	to 1.5cm, narrowly linear	spherical, yellow to red-brown
B.marginata SILVER BANKSIA	low shrub or tree to 12m	1.5–6cm, linear to cuneate, entire or serrate	5–10cm, cylindrical, yellow
B.media SOUTHERN PLAINS BANKSIA	dense shrub 2–5(–10)m	5–12cm, cuneate, toothed, pale brown at first	10–15cm, cylindrical, forming a bright yellow dome
B.nutans NODDING BANKSIA	compact, rounded shrub to 1m	1.2–2cm, linear, blue-green, revolute	cylindrical, nodding, green ageing pink-purple then rusty brown and onion-scented
B.occidentalis RED SWAMP BANKSIA	erect, bushy shrub to 3 × 3m	10–13cm, linear, whorled, silver beneath	to 14cm, cylindrical, red
B.pulchella TEASEL BANKSIA	small, spreading shrub to 1m	1cm, linear	to 2.5cm, cylindrical, orange-brown
B.quercifolia OAK-LEAVED BANKSIA	erect shrub to 3 × 2m	3–15cm, cuneate, serrate	to 10cm, narrowly cylindrical, red and green opening rusty brown
R.repens CREEPING BANKSIA	prostrate, creeping	to 30cm, erect, irregularly lobed, rusty brown at first	to 10cm, cylindrical, pink brown
B.serrata SAW BANKSIA	small tree, 2–15m	to 15cm, narrowly obovate, strongly serrate	to 15cm, cylindrical, green-yellow to cream, sometimes blue-grey in bud
B.speciosa SHOWY BANKSIA	large shrub or tree to 6m	to 40cm, linear, grey-green, silver beneath, deeply incised	to 12cm, squat, rounded, white opening yellow-green
B.verticillata GRANITE BANKSIA	spreading shrub or small tree to 4m	to 9cm, oblong, dark green, silver beneath	to 20cm, cylindrical, yellow
B.victoriae WOOLLY ORANGE BANKSIA	tall, rounded shrub to 5m	20–30cm, broadly linear with deep, sharply pointed lobes opening orange	to 12cm, cylindrical, white and hairy in bud,
B.violacea VIOLET BANKSIA	low shrub to 1.5m	to 2cm, linear	to 6cm diam., spherical, purple

damage by stem and bulb eelworm (*Ditylenchus dipsaci*), where the rotted tissue is in concentric rings. Cultivars vary in susceptibility. The fungus is capable of surviving in soil for many years in the absence of a narcissus crop. Infection can occur at any time but particularly via dying roots as temperatures rise in spring. Its progress is greatly influenced by temperature during growth, harvesting and storage, the most severe attacks occurring at about 25°C/77°F. Control measures in gardens should include rejecting diseased bulbs, rotating the planting site, lifting early and avoiding long exposure to sunshine, and storage at no more than 17°C/63°F. Post harvest dipping of bulbs in fungicide may be beneficial.

basal stem rot (*Pythium* species) a soil-borne pathogen of chrysanthemums, also transmitted by contaminated water and tools. Control is by strict hygiene, the use of mains water, sterilization of soil and compost, and the use of soil drench fungicide.

base dressing an application of organic manure or processed fertilizer incorporated into the soil prior to sowing or planting, with the aim of improving its fertility.

base exchange a process of importance in fertilizer application whereby a soil retains bases available as a salt solution in soil water while giving up an equivalent amount of bases which form new salts in solution. For example, the application of lime sets free insoluble potassium occurring naturally in clay soil in exchange for insoluble calcium which is retained. See *cation exchange capacity*.

base fertilizer see *base dressing*

basic slag a waste product from the lining of blast furnaces. Once an important insoluble phosphate fertilizer, containing between 7% and 22% phosphoric acid with calcium, magnesium and some trace elements.

basifixed of a plant organ; attached by its base rather than its back, as an anther joined to its filament by its base.

basil (*Ocimum basilicum*) an aromatic annual or short-lived perennial used as a herb, for its clove-like flavour, in salads, sauces and liqueurs. It reaches 20–60cm in height, and is suitable for container growing. In cool temperate climates basil is best treated as a tender annual, raised under protection in spring. Plant out after frost risk into a sheltered, sunny position, 15cm apart. For winter use indoors sow in early summer and pot up; and after light trimming bring indoors in late summer.

basing up building up a low ridge of soil in a complete circle around the base of a plant in order to retain applied water.

basil

basionym a taxonomic term; the 'base name' of a taxon that has been transferred by a later author from one rank to another within a species, or to a new species or genus altogether. For example, in 1964 Cullen and Heywood judged that *Paeonia arietina* (published by Anderson 146 years earlier) was in fact a subsp. of *Paeonia mascula*. *P.arietina* Anderson became *P.mascula* subsp. *arietina* (Anderson) Cullen and Heywood. The original name is now the basionym or base name of the new combination and synonymous with it. To make this change clear and to indicate that the type of Cullen's and Heywood's *arietina* is one and the same as Anderson's, the basionym author is cited in parentheses.

basipetalous developing from apex to base, as in an inflorescence where the terminal flowers open first, cf. *acropetal*.

basiscopic of ferns; directed towards the base of a frond; the first lateral vein or leaflet on a pinna branching off in a downwards direction.

bass, bast the strong inner fibrous bark of lime trees and similar fibres, once widely used as a horticultural tying material.

Bassia (for Ferdinando Bassi (1710–1774), Italian botanist). Chenopodiaceae. Warm temperate regions. 26 species, annual herbs or herbaceous perennials. *Bassia* quickly makes a dense miniature 'tree' ideal for bedding schemes as a dot or background plant or the centrepiece of container plantings. Sow seed under glass in early spring; plant out in sun after the last frost. Alternatively, sow seed *in situ* in late spring. *B.scoparia* (syn. *Kochia scoparia*; SUMMER CYPRESS, BURNING BUSH; annual 60cm, with a bushy spire of linear leaves turning from apple green to purple-red in late summer; f. *trichophylla*: particularly fine foliage, flushing deep blood red).

Bauhinia (for Johann Bauhin (1541–1613) and Caspar Bauhin (1560–1624), Swiss botanists. The two-lobed leaves characteristic of this genus symbolize the brothers' relationship). Leguminosae. Tropics and Subtropics. ORCHID TREE. 250–300 species, evergreen or deciduous shrubs, lianes and trees, often thorny and with large, rounded, 2-lobed leaves and showy, 5-petalled flowers with long stamens. The following are commonly grown outdoors in subtropical and tropical gardens and elsewhere under glass (minimum temperature 70°C/45°F). They flower at various times of year, depending on location; flowers are most often seen, however, from late autumn to spring. Use *B.corymbosa*, *B.galpinii* and *B.vahlii* as climbers or wall shrubs; the others as free-standing specimen shrubs and trees. Under glass, pot in tubs or beds and avoid overwet conditions in winter. *B.vahlii* may require some shading if its large leaves are not to fade and scorch. Prune to restrict size or to thin out congested growth after flowering. Propagate by seed, simple layering or grafting. Insert semi-ripe cuttings with the leaves removed in moist sand in a closed case with bottom heat.

B. × blakeana (*B.purpurea × B.variegata*; HONG HONG ORCHID TREE; S. China; deciduous or ever-green umbrella-shaped shrub or tree to 12m; flowers narrow-petalled mauve, rose or bright red, to 15cm across); *B.corymbosa* (syn. *B.glauca* of gardens; *B.scandens* of gardens; PHANERA VINE; S. China; liane, climbing by tendrils; flowers fragrant, white to pink); *B.forficata* (syn. *B.candicans*; S. America; evergreen to deciduous shrub or small tree to 10m tall with gnarled, leaning trunk and thorny branches; flowers narrow-petalled, 15cm-diam., cream to snow-white; will withstand some frost in zone 9 and very sheltered locations in zone 8); *B.galpinii* (syn. *B.punctata*; PRIDE OF THE CAPE, CAMEL'S FOOT or RED BAUHINIA; Africa; evergreen or semi-deciduous, spreading or climbing shrub to 3m; flowers orange or fiery red with ruffled petals); *B.purpurea* (SE Asia, confused with *B.variegata*, from which it differs in its oblanceolate, rosy purple petals which do not overlap one another); *B.vahlii* (MALU CREEPER; India; a robust liane, climbing to 35m; leaves 30cm. diam., kidney-shaped, downy; flowers ruffled, cream or white, to 5cm diam.); *B.variegata* (syn. *B.purpurea*; BUTTERFLY or ORCHID TREE, MOUNTAIN EBONY; E Africa; broadly spreading more or less deciduous shrub or tree to 12m; flowers pale pink to magenta or indigo or white).

beak a long, pointed, horn-like projection; particu-larly applied to the terminal points of fruits and pis-tils.

bean seed fly (*Delia platura*) a small fly similar to the cabbage root fly. The white, legless maggots, which are up to 9mm long, eat germinating French and runner bean seeds and seedlings. Crops may fail to emerge or be distorted and/or blind shoots are produced. Conditions unsuitable for rapid germina-tion increase the risk of damage. There may be three or four generations of maggots during the summer but damage is usually confined to the spring period. Control by incorporating insecticide along the seed row or by raising plants in cell modules.

beans a variable group of plants in the family Leguminosae, grown in most cases for their edible seeds and/or seed pods.

The BROAD BEAN (ENGLISH BEAN, EUROPEAN BEAN) (*Vicia faba*) is an erect hardy annual. The seeds, which may be eaten when immature, are produced in large fleshy green pods which may also be consumed whole at an early stage of develop-ment when 5–7.5 cm long. Sow selected cultivars in autumn for earliest cropping in the following year or successively from spring throughout summer. Sow direct to a depth of 5–7.5cm or in containers under protection for planting out. Rows should be 45cm apart with plants at 15cm final stations.

The FRENCH BEAN (HARICOT, DWARF, COMMON, KIDNEY, SNAP, STRING AND FLAGEOLET BEAN) (*Phaseolus vulgaris*) is a tender annual including bushy or climbing cultivars, mostly grown for their round or flat fleshy pods or their semi-mature (flageolet) or mature (haricot) seeds. Sow *in situ* to a depth of 5cm successively from late spring when soil temperature has reached 10°C/50°F and the risk of frost has passed Alternatively, sow in containers under protection for planting out in rows 45cm apart with plants at 30cm final stations. The optimum temperature range for growth is 16–30°C/ 60–85°F.

The RUNNER BEAN (SCARLET RUNNER, DUTCH CASE-KNIFE BEAN) (*Phaseolus coccineus*) is a tender climbing perennial treated as an annual in gardens and grown for its seed pods, which are edible when young. Sow *in situ* 5–7.5cm deep from late spring when the risk of frost has passed. Alternatively, sow in containers under protection for planting out in rows 60cm apart, with plants at 15cm stations. Plants can be trained up support structures such as 2.5m poles or canes erected in parallel rows and tied in pairs at the top or in groups of four to form wigwams. Strings or net may also be used. Non-climbing or dwarf cultivars are available, and climbing cultivars may be kept dwarf by pinching off developing shoots to produce an early low-yield crop.

Other cultivated beans include ASPARAGUS BEAN, YARDLONG BEAN (*Vigna unguiculata* subspecies *sesquipedalis*), CATJANG (*V.u.* subspecies *cylindrica*), CHERRY BEAN (*V.unguiculata*), MUNG BEAN (*V.radicata*), LIMA BEAN (*Phaseolus lunatus*), and FOUR-ANGLED or -WINGED BEAN (*Psophocarpus tetragonolobus*). Cultivation is described under the entries for these plants.

Bauhinia variegata

beard (1) an awn; (2) a tuft or zone of hair as on the falls of bearded irises.

Beaumontia (for Diana Beaumont (*d*.1831) of Bretton Hall, Yorkshire, England). Apocynaceae. East Asia. 9 species, evergreen twining shrubs with funnel- to bell-shaped, five-lobed flowers. *Beau-montia* requires hot, moist conditions in summer and cool, dry conditions in winter (minimum temperature 7°C/45°F). Grow in full light in a fertile but well drained medium. Root heeled, semi-ripe cuttings in a heated case in late summer. *B.grandi-flora* (EASTER LILY VINE, HERALD'S TRUMPET, MOONFLOWER, NEPAL TRUMPET FLOWER; to 5m; flowers fragrant, waxy white, produced in spring and summer, 8–13cm long and broadly funnel-shaped with wavy-margined lobes).

bedding the planting in outdoor beds of mostly green-house-raised flowering and foliage ornamentals for display. Half-hardy annuals are widely used as summer bedding plants, but also perennials; spring and winter bedding displays are possible using hardy plants. Carpet or tapestry bedding is a specialized form made popular in Victorian gardens. It involves high density planting of low growing species to create often intri-

beans

Pests
aphids, root aphids
bean seed fly
slugs
thrips
weevils

Diseases
halo blight
chocolate spot
(see botrytis*)*

cate patterns. Such beds may be raised and sloped and growth is contained by shearing or pinching.

bedeguar a moss-like gall of roses, also known as Robin's pincushion, caused by the larvae of the gall wasp *Rhoditis rosae.*

bed system the arrangement of usually closely-spaced plant rows, especially of vegetables, in narrow beds most conveniently 1.2m wide. For temperature and drainage advantage beds may be of raised construction, contained by timber or brick walls to a height up to 30cm and slightly wider at the base than at the surface.

bee plant a plant particularly attractive to bees, such as *Buddleja davidii, Cotoneaster horizontalis* and *Alcea rosea,* the hollyhock.

beetle mites (Acarina: Cryptostigmata) red-brown or black mites important in the breakdown of woodland litter. Several species feed on algae and lichen, and some are arboreal, such as *Humerobates rostrolamellatus.* They congregate on bark and may be mistaken for red spider mite eggs, but are larger and paler red. *Perlohmannia dissimilis* can damage the underground parts of potatoes, tulips and strawberries.

beetles (Coleoptera) Beetles constitute the largest order of the animal kingdom. They have forewings modified to form hardened wing cases (elytra), which are held flat over the body when the beetle is not in flight, protecting the membranous hindwings and abdomen. The immature larvae are generally soft-bodied grubs of very different appearance and habit to adults, and pass through a non-feeding pupal stage before becoming adult. Many beetles are major agricultural or horticultural pests, feeding on plants or on stored seeds; some are scavengers (e.g. dung beetles); others are useful predators on other pests; many more, including the fireflies, are of no economic importance. See *bark beetles*; *blister beetles*; *chafer beetles*; *Colorado beetle*; *flea beetles*; *ground beetles*; *Japanese beetle*; *ladybirds*; *leaf beetles*; *lily beetle*; *longhorn beetles*; *pollen beetles*; *rove beetles*; *sap beetles*; *wireworms.*

beetroot (*Beta vulgaris* subsp. *vulgaris*) a biennial grown as an annual vegetable for its edible swollen hypocotyl or 'root', which is typically bright red throughout. It is sometimes included in ornamental borders. Harvested 'seed' consists of a dried flowerhead or cluster which contains two or three true seeds; single-seeded or monogerm cultivars reduce the need for thinning. Sow seed 2cm deep. For the earliest crops, choose bolt-resistant cultivars, sown in rows 18cm apart with plants at 10cm final stations. Sow main crops in rows 20cm apart with plants at 5cm final stations. To advance the first harvest date, plants may be raised under protection in cells for transplanting or sown under cloches or plastic film. Mature roots may be stored in clamps or in boxes of sand, free from frost, for extended season use.

Begonia (for Michel Begon (1638–1710), Governor of French Canada, patron of botany). Begoniaceae. Tropics and Subtropics, especially Americas. Some 900 species, perennial herbs grown for their foliage and flowers. They may be fibrous- or tuberous-rooted, with thick, creeping rhizomes or low to trailing or erect, fleshy stems and a more or less bushy habit. Most are evergreen, although the tuberous species die-back in winter. The leaves range widely in shape, but are commonly asymmetric with one side slightly larger or more oblique than the other, especially at the base. They are thinly succulent, sometimes hairy, and often marked. Borne in axillary clusters and cymes or showy panicles, the flowers too are slightly fleshy; the males consist of two sets of two sepals, the females of sets of two to six.

For the most part, begonias are highly susceptible to frost and, indeed, to any sudden drop in temperature. Members of the Semperflorens and Tuberous groups are, however, familiar sights in cool temperate gardens, where they are used as half-hardy annuals in bedding, baskets and windowboxes. *B. grandis* subsp. *evansiana* has proved hardy year-round in zone 7, given a thick winter mulch and a sheltered position. All favour a fertile, porous medium, moist but never wet, and rich in leafmould, composted bark and sand. Feed actively growing plants fortnightly with a high-potash fertilizer. Avoid wetting foliage and sun scorch. In cultural terms, the genus can be divided into six groups—

1. *Cane-stemmed* Tall evergreens with fibrous roots and erect, branching stems, slender and cane-like with prominent nodes, and fleshy becoming thinly woody. The leaves are usually large and lanceolate to ovate with an obliquely heart-shaped base. Small but cheery flowers are carried in drooping, slender-stalked panicles. Excellent greenhouse and house plants – a common sight, for example, in office buildings. Provide a mimimum temperature of 10°C/50°F and bright light, not direct sunlight. New canes arise from the base of the plant and may require staking and some pinching to promote bushiness – they tend to last for two to three years and should then be cut out cleanly at the base. Propagate by tip cuttings in spring or by division of large clumps.

2. *Rex and rhizomatous* Low evergreens with thick, creeping to ascending rhizomes. The leaves are usually lanceolate to broadly ovate and often brilliantly coloured. Popular house plants. Provide a minimum temperature of 13°C/55°F with ample humidity during warm weather and light to heavy shade. Water carefully, especially in cool weather, and never allow water to rest on the crown or leaves. Propagate by division of rhizomes or leaf cuttings.

3. *Semperflorens* A group of hybrids with fibrous roots, low, clumped and bushy stems, rounded waxy leaves and clusters of small, brightly coloured flowers. These are the bedding begonias, essentially perennial but commonly treated as half-hardy annuals. Sow seed under glass in early spring at a temperature of 25°C/77°F; plant out in sunny or lightly shaded positions after the last frosts. Plants may be re-used if lifted in autumn, overwintered in dry, frost-free conditions, and pruned, divided and restarted in a moist, fertile medium at 15°C/60°F in spring. Especially vigorous or attractive examples might equally be potted up in autumn and brought into the home for winter display. Propagate by seed, division or stem cuttings.

4. Shrub-like Evergreens with fibrous roots and clumped, bushy stems that tend to remain rather succulent. Some, for example *B.metallica* and *B.scharfii*, make fine house plants. Grown for their foliage and flowers, the members of this group range widely in origin and appearance. Their requirements are basically the same as for the Cane-stemmed begonias, but with rather higher temperatures and humidity.

5. Tuberous Deciduous with a large, rounded tuber and annual, succulent stems, thick or slender, erect or pendulous. Grown mostly for their showy flowers. Grown under glass or treated as half-hardy 'annuals' (*B.* Tuber-hybrida Hybrids) planted out in late spring in dappled sun or light shade. Start tubers into growth in early spring on trays of moist coir in a shaded greenhouse at a temperature of 15°C/60°F. Once growth has begun, pot into the normal medium, potting on and staking as required. Flowerbuds develop three per axil – pinch out all until a strong head of foliage has been achieved. With large-flowered cultivars, pinch out the two outer buds in each truss of three: these are female and seldom so large and showy as the central male which should be encouraged to develop to its full potential. Foliage fades and flowering ceases in autumn, when these plants should be dried off. Once the dead topgrowth has fallen away, store the tubers dry in their pots or in boxes of sawdust at 7°C/45°F. Propagate in spring by detaching and rooting unwanted sideshoots, or by division of tubers (bisect longitudinally with a sharp, clean knife; dust the cut surfaces with flowers of sulphur); also by seed.

Not all tuberous begonias produce the distinctive, corm-like tubers of the type whose cultural regime is outlined above. These are described as semi-tuberous. In some, for example, *B.dregei* and *B.sutherlandii*, the tuber is a thickened stem and not all of the topgrowth is lost in winter. These plants and their hybrids (e.g. *B.* × *weltoniensis*) are usually grown year-round under glass or in the home. They require a dry winter rest in their pots and should be repotted in spring into a rather gritty medium. Propagate by stem cuttings and by bulbils which sometimes form in the growth axils.

6. Winter-flowering Evergreen and low-growing with slender stems and showy flowers produced from late autumn to early spring. This group is sometimes included among the Tuberous begonias, but can itself be split into Hiemalis types (with single to fully double flowers in a wide range of colours), Rieger begonias (improved plants of the Hiemalis type, widely grown as winter-flowering potplants for the home), and Cheimantha types with single, white or pink flowers and derived from *B.dregei* and *B.soco-trana* . Provide a miminum temperature of 15°C/60°F (not much higher in winter), with bright, indirect light, low humidity and a buoyant atmosphere. Keep consistently moist, never wet. Cut back flowered stems. Increase by stem cuttings or by rooted stems detached during repotting.

Clockwise from top left: Begonia rex 'Raspberry Swirl', *B.* 'Scherzo', *B.* 'Sophie Cecile', *B.* 'Little Darling', *B. listada*

BEGONIA

Name	Distribution	Habit	Leaves	Flowers
B.albopicta GUINEA-WING BEGONIA	Brazil	fibrous-rooted with clumped, erect, cane-like stems to 1m tall, succulent and simple at first, becoming much-branched and slightly woody	to 8cm long, obliquely ovate-lanceolate, tip acuminate, margins somewhat toothed and wavy, glossy dark green spotted silver-white above	small, green-white (pale rose in the cultivar 'Rosea')
Comments: Group 1.				
B.bowerae EYELASH BEGONIA	Mexico	creeping-rhizomatous, low and bushy	2.5–5cm long, oval, light emerald to lime green with irregular green-brown marginal markings and eyelash-like marginal hairs	small, palest pink
Comments: Group 2. Includes 'Major' with larger, rather faintly marked leaves, and 'Nigromarga' with emerald leaves marked dark brown, and pale marginal hairs.				
Cane-stemmed Group ANGELWING BEGONIAS	garden origin	fibrous-rooted with tall woody, cane-like stems, erect and branching with prominent nodes	10–30cm long, but usually large,obliquely ovate to lanceolate, base rounded to heart-shaped, apex acuminate, margins wavy to toothed, waxy, variously coloured and marked	small, pink to white in nodding panicles
Comments: Group 1. A large and varied group of hybrids, mostly large, sturdy and long-lived plants excellent for interiors and attractive in leaf and flower. Some of the finer cultivars include **B.** 'Edinburgh Brevimosa' (leaves large, very shiny, deep cerise with broad zones of maroon-black along impressed veins); **B.** 'Goody Plenty' (leaves large, satiny purple-black with lilac-pink spots and wavy margins); **B.** 'Lucerna' (leaves large, olive green spotted silver, flowers deep pink); **B.** 'Orpha C. Fox' (leaves medium-sized, dark silky green with zones of silver-grey between veins); **B.** 'Papillon' (leaves small to medium-sized, satiny maroon-black with wavy margins, contrasting beautifully with small, shell pink flowers); **B.** 'Pinafore' (leaves small to medium-sized, dark green and glossy above, red-tinted beneath, margins wavy, flowers deep salmon pink); **B.** 'Pink Spot Lucerne' (leaves medium-sized, slightly angled, deep olive with cerise spots and a pale margin); **B.** 'President Carnot' (leaves large, green with paler spots,flowers pink in large panicles); **B.** 'Snowcap' (leaves small, very dark olive green with dense silver-white spots).				
B.carolineifolia	Mexico; Guatemala	ascending to erect with very stout, fleshy and thin-barked cylindrical stems covered in prominent leaf scars	to 30cm across, palmately compound with 5–9 lanceolate to obovate leaflets in a circular arrangement, these with toothed or slashed, wavy margins, smooth, mid to deep green	small, pale fleshy pink in branching cymes
Comments: Group 2.				
B. × cheimantha BLOOMING FOOL BEGONIA; CHRISTMAS BEGONIA; LORRAINE BEGONIA	garden origin (*B.dregei* × *B. socotrana*)	fibrous-rooted with robust, branching stems to 35cm long	rounded with a cordate base and toothed margins, smooth bright green	large, pink in loose, pink-stalked cymes in winter
Comments: Group 6. Cultivars include 'Gloire de Lorraine' with rounded bright green leaves and white to pale pink flowers, the lilac-pink 'Love Me', ' Marjorie Gibbs' with a profusion of pale pink bloom offest by dark green, rounded leaves, and 'White Marina' with pale green leaves and large white flowers edged soft pink				
B.coccinea ANGEL-WING BEGONIA	Brazil	fibrous-rooted with robust, erect, cane-like stems	to 15cm long, obliquely oblong to ovate, serrate, deep, glossy green and somewhat glaucous above, red-tinted beneath	small,pink with a large, coral-red ovary, profuse, in spring
Comments: Group 1.				
B.corallina	Brazil	fibrous-rooted with cane-like stems erect to 3m	to 20cm long, lanceolate, glossy green with white spots above, red beneath	many, large, coral red, pendent
Comments: Group 1. Includes two cultivars with sweetly scented flowers, the white 'Fragrans', and the pale rose 'Odorata'.				
B.cubensis CUBAN HOLLY; HOLLY-LEAF BEGONIA	Cuba	stems erect, branched, brown-hairy	to 5cm long, obliquely ovate, acuminate, toothed, glossy dark green above, veins brown-hairy beneath	small, white
Comments: Group 4.				
B.dichroa	Brazil	roots fibrous;stems cane-like, erect to spreading, woody, branched, to 35cm long	15–30cm long,obliquely ovate-oblong,base cordate, apex acuminate, margins somewhat undulate,smooth, bright green with silver-white spots	orange to salmon pink with white ovaries in nodding umbels
Comments: Group 1.				
B.dregei GRAPE-LEAF BEGONIA; MAPLE-LEAF BEGONIA	S Africa	tuberous or semi-tuberous with succulent stems to 1m tall	to 8cm long, maple-like, i.e. ovate-rhombic and shallowly palmately lobed, toothed, pale green with purple veins and pewter spots above, purple-bronze beneath	small, white, pendent
Comments: Group 5. In the cultivar 'Macbethii', the leaves are smaller, much incised with deep green veins.				
B. × erythrophylla **B. × feastii** BEEFSTEAK BEGONIA; KIDNEY BEGONIA	garden origin (*B.hydrocotylifolia* × *B.manicata*)	creeping, rhizomatous, to 20cm tall	to 15cm long, almost round, base cordate but appearing peltate, thick, olive green above, purple-red beneath, margins and petiole with white hairs	small, pink, many in panicles held clear of foliage
Comments: Group 2. Includes 'Bunchii' (LETTUCE-LEAF BEGONIA) with bright green leaves with ragged and crested margins, and 'Helix' (WHIRLPOOL BEGONIA; POND-LILY BEGONIA) with the basal leaf lobes spiralled.				
B.foliosa FERN BEGONIA; FERN-LEAVED BEGONIA	Colombia; Venezuela	fibrous-rooted, stems to 1m tall, slender,cane-like, branching, erect to arching	1–3cm long, ovate-oblong, slightly toothed,glossy dark green, many in 2, close ranks	small, white
Comments: Group 1.				

BEGONIA

Name	Distribution	Habit	Leaves	Flowers
B.fuchsioides B.foliosa var. *miniata*; CORAZON-DE-JESUS; FUCHSIA BEGONIA	Venezuela	fibrous-rooted, stems to 1.5m tall, slender, cane-like, branching, erect to arching	2–5cm long, oblong, toothed, tinted pink at first later dark green, in 2 ranks	to 3cm across, bright rosy pink to coral red, hanging on pink stalks
Comments: Group 1.				
B.gracilis var. **martiana** B.martiana HOLLYHOCK BEGONIA	Mexico; Guatemala	tuberous, stem erect to 100cm, succulent, usually unbranched	small, orbicular to lanceolate, base cordate, succulent, crenate, pale green, with bulbils borne in axils	large, pink, fragrant
Comments: Group 5.				
B.grandis ssp. **evansiana** B.evansiana HARDY BEGONIA	E Asia	tuberous with erect to arching stems to 1m tall, these produced annually, red-tinted and lightly branched	to 15cm long, obliquely ovate, base cordate, apex acuminate, veins prominent, the upper surface rather corrugated and pale bronzy olive to lime green, the lower surface red-tinted	to 3cm wide, fragrant, white or pink, nodding in branching cymes
Comments: Group 5. Hardy outdoors in Z7 and over – plant in a sheltered, sunny or semi-shaded position in a fertile, humus-rich, moist but well-drained soil. In late summer and autumn, bulbils form at leaf axils and may be detached and grown on in the cool glasshouse. Mulch crowns thickly before the first frosts.				
B.heracleifolia STAR BEGONIA; STAR-LEAF BEGONIA	Mexico to Belize	rhizome stout, succulent, creeping	to 30cm across, orbicular in outline, base cordate, deeply palmately lobed, lobes 7–9, lanceolate, margins wavy, toothed and hairy, bronze-green above, red beneath with green blotches along veins and tufts of hair; petiole blotched red, hairy	pink, packed atop an erect, hairy stalk
Comments: Group 2. 'Sonderbruchii' has deeply cut leaves streaked bronze; var. *nigricans* has leaves edged dark olive to black; var. *punctata* has leaden, black-green leaves blotched black.				
B. × hiemalis WINTER-FLOWERING BEGONIAS	garden origin (*B.socotrana* × B. Tuberhybrida Hybrids)	fibrous-rooted to semi-tuberous with fleshy stems	obliquely oval	single or double, white to pink, yellow, orange or red, in winter or year-round
Comments: Group 6. Rots off easily – water from below. 'Krefeld' has mid green leaves and profuse orange to crimson flowers in winter				
B.listada	Brazil	fibrous-rooted with erect, branching stems to 30cm	to 6cm long, obliquely ovate-lanceolate, base unequally cordate, toothed, downy, velvety olive green with a lime to gold zone along midrib and red hairs beneath	few, white, to 5cm across, red-hairy
Comments: Group 4.				
B.manicata	Mexico	rhizomatous with fleshy, erect to decumbent stems to 60cm	obliquely ovate-cordate, to 16cm long, toothed and ciliate, glossy light green above, red-tinted and weakly hairy on veins beneath; leafstalk with a stiff collar of red hairs at its summit	small, pale pink with winged, pink ovaries, many together in panicles in early spring
Comments: Group 2. 'Aureomaculata' (LEOPARD BEGONIA; leaves blotched white or flushed red); 'Aureomaculata Crispa' (leaves pale green marbled yellow, margin ruffled, becoming pink in sun); 'Crispa' (leaves pale green with crisped and ruffled, hairy margins).				
B.masoniana IRON CROSS BEGONIA	New Guinea	low-growing with a creeping, fleshy rhizome	to 20cm long, obliquely ovate-cordate, margins toothed and hairy, surface covered in bumpy bristles, pale green with a broad, dark brown to black areas along the main veins forming a cross-like marking	small, green-white many together on an erect stalk in spring and summer
Comments: Group 2. A popular houseplant.				
B.mazae ·	Mexico	fibrous-rooted with trailing, fleshy stems	to 12cm, often smaller, rounded to heart-shaped, toothed, shiny green with red-bronze markings or veins above, red beneath	to 1.5cm across, white spotted red, fragrant, many together in winter and spring
Comments: Group 2. A good basket plant. Includes cultivars 'Nigricans' with dark and burnished leaves, and 'Stitchleaf' with pale green, heart-shaped leaves with dark purple 'stitches' along the margins				
B.metallica METAL-LEAF BEGONIA	Brazil	fibrous-rooted, silver-hairy, erect and branching, to 1.5m tall	to 15cm long, obliquely ovate, apex acuminate, base cordate, margins wavy and toothed, bronze-green with silver hairs and metallic purple veins	to 3.5cm across, pale pink to rose, many together from summer to autumn
Comments: Group 4.				
B.nelumbiifolia LILY PAD BEGONIA	Mexico to Colombia	rhizome short and stout, decumbent to ascending	to 30cm across, rounded, peltate, margin toothed and hairy, bright green, on long, hairy stalks	to 1.5cm across, white or pale pink in long-stalked clusters in winter and spring
Comments: Group 2.				
B.olsoniae	Brazil	fibrous-rooted, compact and shrubby, to 15cm tall	to 20cm across, obliquely rounded to heart-shaped, silky, bronze-green with pale green, white or pink veins	white to pale rose, one to a few together on long, arching stalks in winter
Comments: Group 4. A good basket plant				

BEGONIA

Name	Distribution	Habit	Leaves	Flowers
B.paulensis	Brazil	rhizome creeping to ascending, fleshy, downy	to 20cm, oblong, acuminate, peltate, margin toothed, surface puckered, glossy mid green with a network of white veins	3.5–5cm across, white with red hairs, many together on a tall stalk

Comments: Group 2.

Name	Distribution	Habit	Leaves	Flowers
B. 'Phyllomaniaca' CRAZY-LEAF BEGONIA	a Brazilian hybrid (*B.incarnata* × *B.manicata*)	stem erect to 60cm tall, branched and hairy with many buds and leafy outgrowths	to 20cm long, obliquely cordate,margin slightly lobed, toothed and hairy,upper surface glossy with many leafy outgrowths	to 2.5cm across, pink on erect stalks in winter

Comments: Group 4. Well-grown plants assume a remarkable, 'ragged' appearance, as if strewn with leaf fragments

Name	Distribution	Habit	Leaves	Flowers
B.prismatocarpa	Guinea	rhizomatous,stems creeping and rooting	to 5cm long, very obliquely ovate-cordate to rounded, margin wavy-toothed and 2–3-lobed, hairy	bright yellow, produced throughout the year

Comments: Group 2. 'Variegation' has leaves marked yellow-green and clear yellow flowers produced in great abundance

Name	Distribution	Habit	Leaves	Flowers
B.pustulata	Mexico; Guatemala	rhizomatous, low-growing	to 15cm long, often shorter, obliquely and broadly ovate, base cordate, apex acuminate, margin finely toothed, upper surface velvety with a dense pattern of small blisters, dark green faintly veined silver above, red-tinted beneath	small, rose pink, in summer

Comments: Group 2. var. *argentea* has very dark green leaves with a heavy pattern of silver flashes along the veins.

Name	Distribution	Habit	Leaves	Flowers
B.radicans *B.undulata* of gardens SHRIMP BEGONIA	Brazil	rhizomatous with scrambling, erect to pendulous, white-spotted stems to 2m long	to 7cm long, obliquely ovate to oblong, apex acuminate, margins wavy, red-tinted at first, becoming grey-green with white spots	to 2.5cm across, coral pink to brick red in dense, nodding clusters, usually in winter

Comments: Group 1. 'Purpurea' has persistently bronze-red-tinted leaves

Name	Distribution	Habit	Leaves	Flowers
B.rex KING BEGONIA; PAINTED LEAF BEGONIA	Assam	rhizome thick and fleshy, buried to creeping	to 25cm long, obliquely ovate, apex acute, base cordate, margins wavy with small, hairy teeth,surface puckered and sparsely hairy, rich metallic green with a broad silver zone above, purple-red beneath	to 5cm across,pale pink, a few together in winter and early spring

Comments: Group 2. A popular houseplant and the parent of many hybrids, most of which are sold as *B.rex*, but strictly belong to the *B.* Rex Cultorum Hybrids.

Name	Distribution	Habit	Leaves	Flowers
B. Rex Cultorum Hybrids *B.rex* of gardens BEEFSTEAK GERANIUM; REX BEGONIA	garden origin (*B.rex* crossed with related Asiatic species)	rhizomes thick, fleshy, creeping to erect	10–30cm long,obliquely ovate to oblong-lanceolate, base cordate to rounded, apex acute to acuminate, margins toothed to jaggedly lobed, smooth to hairy, dark green to wine red or bronze patterned or zoned silver-white, leaden grey, red, purple or black	white to pink or red in small clusters, seldom seen

Comments: Group 2. A large and popular group of hybrids widely grown as houseplants, includes 'Black Knight' (leaves large, black-red with rows of pink dots), 'Duartei' (leaves large with spiralling basal lobes, darkest green edged purple-black with silver streaks), 'Helen Lewis' (leaves medium-sized to large, silken dark purplewith silver bands), 'Fireworks' (leaves large, maroon marked white blending to purple between veins), 'Merry Christmas' (syn. *B.ruhrtal*, leaves large, deep green flecked pink with a large central cherry red zone fading to silver at its edges and deepening to burgundy at its centre, margins burgundy), 'Princess of Hanover' (leaves large with spiralling basal lobes, emerald green with silver bands edged dark red and covered in fine, pink hairs), 'Purple Curly Stardust' (leaves medium-sized with spiralling basal lobes and ruffled margins, purple with cream and pink mottling), 'Raspberry Swirl' (leaves medium-sized, rounded with overlapping basal lobes and jagged margins, silver edged purple-red with a purple-red centre radiating along veins as a raspberry red blotch), ' Roi des Roses' (leaves medium-sized, pale pink mottled silver), 'Silver Helen Teupel' (leaves long and deeply cut, silver with a radiating pink central zone),and 'Wood Nymph' (small, leaves rounded, rich brown speckled silver)

Name	Distribution	Habit	Leaves	Flowers
Rhizomatous Group	garden origin	low-growing plants with fleshy, creeping to erect rhizomes	3–25cm long, ranging widely in shape, colour and texture	white, pink, or red

Comments: Group 2. Large and disparate group of hybrids, making good houseplants. Includes 'Aries' (leaves medium-sized to large, broad and shallowly palmately lobed, maple-like, downy emerald green with chocolate markings toward margins), 'Beatrice Hadrell' (leaves medium-sized to large, incised, dark green with paler veins, flowers pale pink), 'Bethlehem Star' (leaves small,black with a fine cream, central star, flowers cream spotted pink), 'Bowkit' (leaves ovate, spirally twisted, yellow-green striped brown, flowers white flecked pink), 'China Doll' (compact, leaves cool green with red-brown mottling, flowers light pink), 'Curly Zip' (compact, leaves lime green with a red sinus and a central double corkscrew, flowers pink), 'Emerald Jewel' (leaves blistered, rich green, lightly marker silver and brown, flowers white), 'Little Darling' (leaves very small, ovate, acute, toothed and lobed, deep jade green patterned with darkest olive), 'Mac's Gold' (leaves small to medium-sized, star-shaped, yellow with chocolate brown marks), 'Munchkin' (leaves medium-sized, rounded, deep bronze,smooth, margins much-toothed,ruffled, crested and densely ciliate), 'Norah Bedson' (leaves medium-sized, rounded, bright green with dark brown markings, flowers pink), 'Oliver Twist' (leaves medium-sized, rounded, twisted and puckered,olive green, margins crested and covered in massed, red-tinted hairs), 'Scherzo' (leaves small,soft lime green with dark brown, stitch-like marks along margins), 'Silver Jewel' (leaves blistered, rich green streaked silver, flowers white)

Name	Distribution	Habit	Leaves	Flowers
B. × *ricinifolia* BRONZE-LEAF BEGONIA; CASTOR-BEAN BEGONIA; STAR BEGONIA	garden origin (*B.heracleifolia* × *B.peponifolia*)	rhizome thickly succulent and creeping to ascending	to 30cm long, round in outline with 5–7 lobes, dark bronze-green above, purple-red beneath	small, white to pale rose, many together on tall stalks

Comments: Group 2.

Name	Distribution	Habit	Leaves	Flowers
B.scharffii *B.haageana*	Brazil	fibrous-rooted,shrubby, stems erect,white-hairy, to 1m tall	to 25cm long, obliquely ovate, apex acuminate, base cordate, margin shallowly lobed, bronze-green with red veins above, red-tinted beneath	to 5cm across, pale pink with pink hairs in large, nodding clusters throughout the year

Comments: Group 4.

BEGONIA

Name	Distribution	Habit	Leaves	Flowers
B. Semperflorens-Cultorum Hybrids *B.semperflorens* of gardens; *B. × semperflorens-cultorum*; BEDDING BEGONIA; WAX BEGONIA	garden origin, hybrids derived from *B.cucullata* var. *hookeri* *B.schmidtiana*, and, sometimes, *B.fuchsioides*, *B.gracilis*, *B.minor* and *B.roezlii*	roots fibrous, stems clumped, fleshy and brittle, bushy, 15–40cm tall	5–10cm long, rounded, smooth and glossy, thinly fleshy, bright green to red-tinted or purple-bronze, (variegated in the Calla-lily begonias)	1–3.5cm across, single or double, white, pink or red, a few together in axillary clusters throughout year but chiefly in summer

Comments: Group 3. Large group of small, bushy and free-flowering perennials usually called *B.semperflorens* and used in gardens as half-hardy bedding annuals. Single-flowered selections include the green-leaved 'Derby' (flowers profuse, pale coral), Frilly Dilly Mixed (leaves bright green, flowers white, pink and red with frilly tepals), 'Red Ascot' (leaves emerald green, flowers dark crimson), and 'Viva' (flowers pure white). Bronze-leaved and single-flowered selections include the Cocktail Series ('Gin', white, 'Vodka', deep red, and 'Whisky', white). Double-flowered cultivars include 'Bo Peep' (leaves bronze, flowers dark pink, very double), 'Curly Locks' (leaves bronze, flowers sulphur yellow and pink), 'Dainty Maid' (leaves green, flowers white edged salmon pink), 'Old Lace' (leaves bronze, flowers rosy red, very double and lacy), and 'Pink Wonder' (leaves green, flowers pink and rounded, abundant). One of the best variegated cultivars is 'Charm' (leaves bright green blotched gold and cream, flowers single, pale pink, ever-blooming).

B.serratipetala	New Guinea	fibrous-rooted and shrubby with arching stems to 60cm long	to 4cm long, obliquely ovate, apex acuminate, margins lobed, sharply toothed and crisped, deep olive to metallic bronze-red with raised pink spots	rose pink to purple-red, sometimes with toothed tepals, throughout the year

Comments: Group 4.

Shrub-like Group		fibrous-rooted and bushy with erect to pendulous, branching, fleshy stems	10–30cm long, of various shapes and hues, hairy or smooth	white, pink or red

Comments: Group 4. Large and varied group making good houseplants, including 'Dancing Girl' (leaves variously shaped, spotted and streaked silver), 'Gloire de Sceaux' (leaves small, rounded, deep metallic bronze, flowers dark to pale pink, scented), 'Ingramii' (leaves small to medium-sized, mid green, flowers candy pink, produced in abundance), 'Midnight Sun' (outer, older leaves moss green, inner and younger leaves glowing pink to peach with green veins, flowers white), 'Thurstonii' (leaves medium-sized to large, obliquely ovate-lanceolate, metallic bronze-green with darker impressed veins), 'Tiny Gem' (leaves very small, crinkle-edged, deep green, flowers candy pink), 'Withlacoochee' (leaves small, oblong-ovate, very velvety, dark emerald to grey-green with white hairs, flowers pearly white).

B.solananthera	Brazil	rhizomatous, stems trailing to scrambling, to 80cm long	to 8cm long, broadly ovate, apex acuminate, margins angled, glossy deep green above, paler beneath	to 2cm across, white tinted rose, fragrant, many together in short, nodding clusters in winter and early spring

Comments: Group 6. A beautiful, winter-flowering basket plant for the warm glasshouse or home

B.stipulacea *B.angularis*; *B.zebrina*	Brazil	fibrous-rooted, stems cane-like, erect to spreading or pendulous, to 2.5m long	to 15cm long, obliquely ovate, apex acuminate, glossy with grey-white veins above, dull red beneath	to 1.5cm across, many, white or pink in winter or early spring

Comments: Group 1.

B.sutherlandii	S Africa	tuberous, stems red, slender, trailing	to 15cm long, obliquely lanceolate, margin toothed to lobed, bright green or sometimes olive with a red tint to veins and petiole	to 2.5cm across, orange to orange-red in pendulous bunches in summer

Comments: Group 5. An excellent basket plant. Somewhat hardier than many Begonias, it will overwinter in frost-free conditions if dry. 'Bulbils' may form in the axils of old stems and can be potted up and grown on.

B.partita	S Africa	shrubby, to 20cm tall from a swollen base	to 6cm long, variably 3-lobed, usually with small lateral lobes and a far longer, finely tapering midlobe, smooth pewter grey	small, white

Comments: Group 5. Requires a free-draining, rather gritty soil, full sun and good ventilation; can be given a dry winter rest

B. Tuberhybrida Hybrids *B. × tuberhybrida*; *B.tuberosa* of gardens TUBEROUS BEGONIAS	garden origin, hybrids between the Andean species, *B.boliviensis*, *B.clarkei*, *B.davisii*, *B.pearcei*, *B.veitchii* and, possibly, *B.frobelii* and *B.gracilis*	tubers usually large and circular, compressed with a central depression above, stems fleshy and branching, erect to arching or pendulous	6–20cm long, obliquely ovate to lanceolate, base rounded to cordate, apex acute to acuminate, margins toothed to cut, veins usually prominent, smooth to lightly hairy, pale green to dark green	4–20cm across, single to semi-double or double, tepals smoothly rounded to fringed, frilled, cut-edged or ragged, white, yellow, apricot, deep gold, orange, vermilion, salmon, scarlet, rose, crimson or mixtures and shades thereof

Comments: Group 5. A very wide range of hybrids grown in pots under glass or in the home, and outdoors in containers and borders for half-hardy summer display; the pendulous cultivars and some members of the Multiflora Group are excellent plants for hanging baskets. Cultivars include the pendulous 'Apricot Cascade' (flowers double deep apricot), 'Bridal Cascade' (flowers white edged pink), 'Crimson Cascade', 'Gold Cascade', and 'Orange Cascade'. The largest-flowered selections include 'Billie Langdon' (double and rose-like, pure white), 'Can-Can' (double, peach to apricot edged warm red), 'City of Ballarat' (double, deep orange), 'Masquerade' (double and ruffled, white edged rosy red), 'Midas' (double, pale gold), and 'Roy Hartley' (double, salmon pink to deep rose). The Multiflora Group covers smaller, bushy plants with numerous small to medium-sized flowers, among them 'Flamboyant' (leaves narrow, bright green, flowers single, scarlet), 'Helene Harms' (flowers bright yellow, semi-double), and 'Madame Richard Galle' (flowers small, double, apricot).

B.ulmifolia ELM-LEAF BEGONIA	Colombia; Venezuela	fibrous-rooted, stems erect, to 2m tall, 4-angled, pale green	7–12cm, ovate-oblong with toothed margins and a corrugated surface, light green to dull olive, coarsely hairy	to 1.5cm across, white, many in spring

Comments: Group 1.

BEGONIA

Name	Distribution	Habit	Leaves	Flowers
B.versicolor FAIRY-CARPET BEGONIA	China	rhizome short, creeping	7–12cm across, obliquely ovate to broadly oblong, margins finely toothed and hairy, emerald to apple green with silver-white markings and red veins above, red beneath	salmon pink or red in clusters on long, white-hairy stalks throughout the year
Comments: Group 2.				
B.vitifolia B.palmifolia	Brazil	fibrous-rooted, stems fleshy, smooth, erect to 1.2m tall	to 30cm, ovate to rounded in outline, hairy, margins toothed and broadly lobed, bright to deep green and shining	small, white on erect stalks
Comments: Group 1.				
B. × weltoniensis GRAPEVINE BEGONIA; MAPLE-LEAF BEGONIA	garden origin (*B.dregei* × *B.sutherlandii*)	tuberous or semi-tuberous with fleshy, branching stems, erect to 1m	to 8cm across, ovate, apex acuminate, margins toothed to shallowly lobed, dark glossy green tinted or veined purple-red	many, pink or white, in summer
Comments: Group 5.				
B.xanthina	Bhutan	stoutly rhizomatous, bushy, to 30cm tall	to 20cm long, obliquely ovate-cordate, margin finely toothed and hairy, glossy dark green above, purple-red, hairy beneath	to 3.5cm across, yellow to orange-yellow tinted red, pendent, several to many together in spring
Comments: Group 2. var. *pictifolia* has silver-marked leaves and pale yellow flowers.				

Clockwise from top left: Begonia masoniana , B. pustulata 'Argentea', *B.* 'Looking Glass', *B.* 'Snowcap',
B. x *tuberhybrida* 'Can-can', *B. semperflorens* 'Red Ascot', *B.* 'Munchkin'

Belamcanda (Latinized form of this plant's vernacular names, *balamtandam* (Malayalam) and *malakanda* (Sanskrit)). Iridaceae. E Europe; Asia. 2 species, rhizomatous perennial herbs very similar to *Iris* and grown for their flowers and attractive seeds. The leopard lily is hardy to –15°C/5°F, or lower still if given a dry winter mulch. Plant in sun or light shade. Propagate by seed sown in spring, or by division in spring or early autumn. *B.chinensis* (BLACKBERRY LILY, LEOPARD LILY; Russia to Japan; hardy, deciduous perennial to 1m; flowers short-lived, 4cm-wide, produced in succession from midsummer to autumn, ranging in colour from tawny yellow to deep orange spotted or blotched red, maroon or purple; seed capsules split to reveal blackberry-like clusters of glossy black seeds).

Bellevalia (for Pierre Richer de Belleval (1564–1632), who founded the Montpellier botanic garden in 1593). Liliaceae (Hyacinthaceae). Mediterranean to W and C Asia. 45 species, spring-flowering hardy bulbous perennials similar to *Muscari*, but often larger with looser racemes of tubular to campanulate rather than squatly urceolate flowers. Cultivate as for *Muscari*.

B.atroviolacea; (Russia and Afghanistan; spike conical, 8–30cm tall packed with nodding, violet-black flowers); *B.hyacinthoides* (syn. *Strangweja spicata*; Greece; raceme cylindric to 15cm tall; flowers pale blue, veined with darker blue); *B.paradoxa* (syn. *B.pycnantha*; *Muscari paradoxum*; *Muscari pycnanthum*; Russia and Turkey; to 40cm; raceme conical crowded with nodding, dark blue flowers).

bell glass a bell-shaped thick-glass cloche placed over plants to advance growth or sometimes as an aid to propagation. Developed in France in the 17th century, bell glasses are usually about 38cm high and 43cm wide. They are not commonly available.

Bellis (from the Latin *bellus*, pretty). Compositae. Europe; Mediterranean; W Africa. DAISY. 7 species, small, hardy annual or perennial herbs grown for their flowerheads. Although hardy to at least –15°C/5°F, the perennial cultivars of *Bellis perennis* are mostly grown as biennials. They are used in carpet bedding, for border and path edging and in containers, especially where winter colour is wanted. Plant in sun or part shade in soil that does not dry out in summer. Dead head often to prolong flowering. Sow seed in early summer to plant out in late summer or autumn. For winter flowers in pots, grow on in a low-fertility, loam-based mix in a cold frame and bring under glass in autumn; provide a winter temperature of 4–7°C/39–45°F and brisk ventilation.

B.perennis (ENGLISH DAISY; perennial often treated as a biennial; leaves 1–6cm, obovate to spathulate; scapes to 5cm tall; flowerheads 1–3cm-wide with yellow disc florets and white ray florets, often stained rose pink or red; cultivars range from the miniature 'Pomponette' and the compact but large-flowered 'Parkinsons Double White' and 'White Carpet' with pure white, 'double' flowerheads, to the larger, pom-pom-flowered types, some in very deep cherry red (e.g. 'Kito', 'Rob Roy'); in the cultivar 'Goliath', the flowerheads (white, pink, or red) may be as wide as 7cm; pink selections include 'Dresden China' (small flowerheads), 'Pink Buttons ('double' flowerheads) and 'Staffordshire Pink' (tall, i.e. to 12cm); 'Aucubifolia' has gold-variegated leaves, whilst those of 'Shrewsbury Gold' are pure, pale gold).

belvedere see *gazebo*.

bench, benching a narrow table-like structure of wood or metal upon which plants in containers are set out within a greenhouse. It may have a solid or slatted surface and is also referred to as staging. A potting bench may be of similar construction, with a solid surface used as a work table for potting and propagating operations.

beneficial insects Many species of insects are beneficial as predators, parasites or pollinators.

Most predatory insects are general feeders, preying as adults or larvae on several different types of insect and other small organisms, thereby contributing to natural pest control. They usually have fast-moving hunting skills and their mouthparts are adapted for biting and chewing or lacerating prey. Although most capsid bugs are plant feeders, some (especially the BLACK-KNEED CAPSID *Blepharidopterus angulatus*) are predacious, feeding significantly on fruit tree red spider mite. The related ANTHOCORID BUGS are also important predators. DRAGONFLIES (Odonata) take their prey on the wing, including mosquitoes, beetles and wasps. LACEWINGS (Neuroptera), some GALL-MIDGE LARVAE (Diptera: Cecidomyiidae) and HOVERFLY LARVAE (Diptera: Syrphidae) are important predators of aphids. Most of the remaining predators are beetles (Coleoptera), including carabid or GROUND BEETLES, and the staphylinid, or ROVE BEETLES, which are general predators of small insects and other small invertebrates. The GLOWWORM (*Lampyris noctiluca*) has larvae which feed on snails. Various species of LADYBIRD BEETLES are important predators of aphids, mealybugs, scale insects and spider mites, both as adults and larvae.

Most parasitic insects with plant-feeding hosts are more correctly called parasitoids. They are found in several families of Hymenoptera, and include ICHNEUMON WASPS, CHALCID WASPS and BRACONID WASPS. Only the larvae are parasitic, and they tend to be more host specific than predators.

Another beneficial function carried out by insects is pollination. Bees are the most valuable and abundant pollinators; other contributors include hoverflies and blowflies.

Berberidopsis (*Berberis* and Greek *opsis*, resemblance; the leaves of this plant were thought similar to those of *Berberis*). Flacourtiaceae. Chile. CORAL PLANT. 1 species, *B.corallina*, an evergreen woody-based perennial climber to 5m, with holly-like leaves. From midsummer to autumn, rounded, 1-2cm-wide coral-red flowers hang on slender stalks. Although hardy to –10°C/14°F, it suffers badly from wind scorch and prolonged frosts and may require protection in the form of a hessian drape and a thick mulch. Plant in a cool, semi-shaded and sheltered position in a moist but porous, neutral to acidic soil rich in decayed vegetable matter. Propagate by semi-ripe cuttings in a closed case with bottom heat, or by simple layering.

Berberidopsis corallina

Berberis darwinii

Berberis (from *berberys*, the Arabic name for the fruit). Berberidaceae. Europe, Asia, Africa, Americas. BARBERRY. 450 species, fully hardy evergreen or deciduous shrubs grown for foliage, flowers and fruit. The stems have thin grey-brown bark and yellow wood and are armed with sharp spines. The leaves are small, thin to tough and sometimes sharply toothed. Some deciduous species and cultivars colour brilliantly in autumn. Small, cream to yellow or fiery orange flowers are produced in axillary clusters, racemes or panicles, usually in spring and summer. The fruit is a berry ranging in colour from creamy yellow to pink, red, purple or blue-black.

Berberis species are widely used in gardens for hedging and specimen planting in lawns, borders and the rock garden. Plant in any well-drained soil in full sun or light shade. Prune out exhausted or congested stems in late winter; trim evergreen hedges after flowering, deciduous hedges in late winter. Propagate by seed cleansed of flesh and sown in a seedbed outdoors in late winter (offspring may be variable); or by heeled nodal and basal cuttings treated with a rooting hormone in a cold frame in early autumn. Treated with a rooting hormone, mallet cuttings will root under mist in the summer. Multi-stemmed plants can be earthed-up and encouraged to produce rooting stems which may be detached and grown on in autumn or spring. Sometimes affected by bacterial leaf spot. Some species are alternative hosts for rusts.

berceau a trellis arch covered by climbing plants.

Berchemia (for M. Berchem, 17th-century French botanist). Rhamnaceae. Asia; Africa; Americas. 20 species, mostly deciduous, woody climbers with twining stems and very small, white flowers in short panicles and racemes, followed by fleshy fruits in autumn. Hardy to –15°C/5°F. Grow in sun or light shade. Provide support or allow to scramble through trees and shrubs. Propagate from seed sown in autumn or spring, by semi-ripe stem cuttings or root cuttings in winter; also by simple layering.

B.racemosa (Japan and Taiwan; to 4m; leaves 6cm-long, ovate to cordate, splashed creamy white in 'Variegata'; fruit red to black); *B.scandens* (syn. *B.volubilis*; SUPPLEJACK or VINE RATTANY; US, Mexico, Guatemala; differs from the last species in its slightly larger, elliptic leaves marked cream to white in the cultivar 'Variegata'; oblong fruit turn from deep plum blue to black in autumn).

Bergenia (for Carl von Bergen (1704–1759, author of *Flora Francofurtana* (1750). Saxif-ragaceae. East Asia. SOW'S EARS. 8 species, hardy evergreen rhizomatous perennial herbs with broad and leathery leaves, and scapose cymes of 5-parted, cup-shaped flowers from late winter to spring. Once established, they will thrive in damp or rather dry situations, in light shade or full sun. Mulch annually in late winter with well-

BERBERIS

Name	Distribution	Habit	Leaves	Flowers	Fruit
B.aggregata B.geraldii	W China	densely branched, spiny, deciduous shrub to 2m tall	to 2.5cm long, oblong-ovate, base cuneate, apex obtuse, margin toothed, olive above, glaucous beneath, fiery red to crimson in autumn blue-white	pale yellow in erect, rounded panicles to 3cm long in late spring	to 0.7cm long, ovoid, pale red
Comments: parent of many hybrids including 'Crimson Bead' with dark carmine fruit in pendulous trusses, and 'Sibbertoft Coral' with coral red fruit bloomed. Z7.					
B.buxifolia B.dulcis; B.microphylla MAGELLAN BARBERRY	Chile	spiny, dense, evergreen or semi-evergreen shrub to 3m tall	to 2.5cm long, obovate, blunt, mucronate, tough, dark green above, glaucous beneath	deep orange-yellow, 1–2 per axil in late spring	to 0.8cm, globose, dark purple, grey-pruinose
Comments: includes 'Aureomarginata' with leaves edged gold, and 'Nana', a sparsely flowered dwarf form seldom exceeding 30 × 25cm. Z5.					
B.calliantha	Tibet	spiny evergreen shrub to 70cm tall; young stems red-tinted	holly-like, to 6cm long, oblong-elliptic, sharply toothed, deep, glossy green above, pale and glaucous beneath	relatively large, yellow, 2–3 per axil in late spring	to 1cm long, ovoid, purple-black, glaucous
Comments: Z7.					
B.candidula B.hypoleuca of gardens; B.wallichiana var. pallida	W China	evergreen, dense rounded shrub 50–100cm tall	to 3cm, elliptic-ovate, sparsely toothed, shiny dark green above, pale beneath	small, yellow stained red, 1 per axil in late spring	to 1cm long, ovoid, grey-pruinose
Comments: Z5.					
B. × carminea (B.aggregata × B.wilsoniae)	garden origin	deciduous shrubs, erect to domed, seldom exceeding 2m	to 3cm long, obovate, more or less toothed, pale beneath	yellow, in panicles to 5cm long in spring and early summer	to 0.8cm long, oblong-ovoid, pink to scarlet
Comments: vigorous garden hybrids valued for their colourful fruits borne in profusion in autumn; they include the cultivars 'Autumn Cheer' (leaves broad, fruit scarlet), 'Barbarossa' (stems to 2m tall, twigs red-tinted, leaves entire, fruit abundant, scarlet), 'Bountiful' (fruit large, rounded, crimson), 'Buccaneer' (leaves narrow, entire, more or less persistent, fruit large, rounded, cloudy white turning shiny red), 'Fireflame' (habit dense and tall, leaves small, fruit orange-vermilion), 'Pirate King' (vigorous, branches pendulous, fruit bright orange), and 'Sparkler' (fruit conic, tinted tangerine). Z6.					

BERBERIS

Name	Distribution	Habit	Leaves	Flowers	Fruit
B.coxii	E Himalaya	spiny evergreen shrub to 1.5m tall	to 6cm long, elliptic-ovate with incurved teeth, glossy dark green above, white and pruinose beneath	pale yellow, relatively large, to 6 per axillary cluster in late spring	to 1cm long, obovoid, glaucous blue-black

Comments: Z6.

Name	Distribution	Habit	Leaves	Flowers	Fruit
B.darwinii	Chile; Patagonia	spreading and suckering, evergreen, spiny shrub to 2m tall	holly-like, to 2cm long, obovate, tough, with 1–3 spiny teeth per side, glossy dark green above, paler beneath	orange-gold often tinted red in pendulous racemes to 10cm long in spring	to 0.7cm long, globose, blue-pruinose

Comments: includes cultivars 'Flame' (flowers vivid dark orange), 'Gold' (flowers vivid gold), 'Rubens' (flowerbuds red, flowers yellow tinted red), and 'Triumph' (hardy, dense and upright to 1m tall, leaves to 3cm long, tinted blue beneath). 'Blenheim', a hybrid between *B.darwinii* and *B.hakeoides*, is a large, erect shrub with clusters of sharply toothed leaves and deep golden flowers. The hybrid between *B.darwinii* and *B.valdiviana* is 'Goldilocks', a large, vigorous shrub with erect to arching branches, large, spiny, glossy dark green leaves and golden flowers in pendulous, red-stalked clusters. The cultivars 'Nana' (dwarf with small leaves and sparse flowers), and 'Prostrata' (dwarf and semi-prostrate with dull green leaves, red flower buds and profuse flowers) strictly belong to *B. × stenophylla*. Z7.

Name	Distribution	Habit	Leaves	Flowers	Fruit
B.empetrifolia B.revoluta	Chile; Argentina	evergreen, spiny, arching to semi-prostrate shrub to 30cm tall	to 1cm long, elliptic but strongly revolute, thus appearing narrow, mucronate, dark green above, pruinose beneath	small, deep yellow, paired, in late spring	to 0.7cm long, globose, glaucous blue-black

Comments: Z7.

Name	Distribution	Habit	Leaves	Flowers	Fruit
B.gagnepainii	W China	evergreen, spiny shrub with suckering, slender stems erect to 1.5m tall	to 4cm long, lanceolate, apex tapering finely, undulate with to 10 teeth per side, matt grey-green above, yellow-green beneath	bright yellow, to 7 per cluster in early summer	to 1cm long, ovoid, purple-black, pruinose

Comments: includes 'Fernspray' compact with long, narrow and undulate, pale green leaves, and 'Robin Hood', low-growing and rounded, to 1m tall, leaves tinted dark red with age. Z5.

Name	Distribution	Habit	Leaves	Flowers	Fruit
B. × hybrido-gagnepainii B. × media; B. × wokingensis; (*B.verruculosa × B.gagnepainii*)	garden origin	varied range of evergreen spiny shrubs	more or less ovate with spiny margins	close to *B.gagnepainii*	purple-black, grey-pruinose

Comments: a large complex of hybrids including 'Chenault' (*B.chenaultii*, to 2.5m tall, leaves to 3cm long, lanceolate, undulate, spiny shining above, pruinose beneath, flowers golden yellow), 'Park Juweel' (small, low, dense and thorny with more or less untoothed leaves colouring brilliantly in autumn and sometimes persistent), and 'Red Jewel' (differs from 'Park Juweel' in its broader leaves assuming deep, purple-red tints). Z5.

Name	Distribution	Habit	Leaves	Flowers	Fruit
B.jamesiana	SW China	erect, deciduous, spiny shrub to 2m tall	obovate, entire or slightly toothed toward apex, thick, olive green above, grey-green beneath, colouring vividly in autumn	pale orange in pendulous racemes to 10cm long	globose, opaque white ripening translucent pink

Comments: Z6.

Name	Distribution	Habit	Leaves	Flowers	Fruit
B.julianae	W China	evergreen, fiercely spiny, bushy shrub erect to 4m	to 10cm, obovate, serrate, tough, dark green above, paler beneath, tinted bronze-red when young	yellow, faintly fragrant in axillary clusters in early summer	to 0.8cm long, oblong, black, glaucous

Comments: good hedging shrub; includes 'Dart's Superb' (branches spreading, leaves small), 'Lombart's Red' (leaves tinted red beneath), and 'Nana' (dwarf). Z5.

Name	Distribution	Habit	Leaves	Flowers	Fruit
B.linearifolia	Chilean Andes	evergreen, erect to arching and rather spiny, rangy, shrub to 1.5m tall	to 5cm long, narrowly obovate-oblanceolate, entire with strongly revolute margins and a sharp tip, dark green above, pale beneath	orange in clusters in late spring and late summer	to 1cm long, ellipsoid, black, glaucous

Comments: includes the large-and free-flowering red-yellow 'Jewel' and the large-and orange-flowered 'Orange King'. Z6.

Name	Distribution	Habit	Leaves	Flowers	Fruit
B. × lologensis (*B.darwinii × B.linearifolia*)	garden origin	evergreen, spiny, medium-sized shrub	to 3cm long, obovate-spathulate, entire and/or toothed, tough, dark green	orange-yellow in umbel-like clusters	to 0.7cm long, ovoid, glaucous purple-blue

Comments: includes 'Apricot Queen' (growth erect flowers orange) and 'Stapehill' (flowers orange-yellow tinted fiery red). Z6.

Name	Distribution	Habit	Leaves	Flowers	Fruit
B. × mentorensis (*B.julianae × B.thunbergii*)	garden origin	deciduous, dense, erect and spiny shrub to 2m tall	to 4cm long, elliptic-obovate, entire, rigid or with 3 teeth toward the apex	solitary or paired, pale yellow in early summer	red-brown, ellipsoid

Comments: Z5.

BERBERIS

Name	Distribution	Habit	Leaves	Flowers	Fruit
B. × ottawensis (B. thunbergii × B. vulgaris)	garden origin	deciduous, erect shrub, very vigorous and suckering with large spines and (sometimes) expanded spiny stipules	2–6cm, obovate to elliptic-ovate, blunt to acute, soft, mid green with slightly paler undersides, colouring well in autumn	pale yellow to golden to drooping racemes	red

Comments: very robust and fast-growing, includes two purple-leaved forms often confused with each other – Purpurea', to 2m tall with deep and solid purple-red leaves, obovate-elliptic, entire and to 4cm long; 'Superba', to 3m tall with wine-red-tinted, pea green leaves, spathulate-orbicular, very finely serrate to entire and to 6cm long, subtended by large 3-parted thorns and spiny stipules. Z5.

B.poiretii	N China	deciduous or semi-evergreen shrub to 1m tall with angular, more or less spiny branches	to 4cm, narrowly oblanceolate, acuminate, with wavy margins and fine teeth	yellow flushed red in pendent racemes in late spring	narrowly oblong, pale red

Comments: *B.chinensis* differs in its broader leaves and darker fruit; *B.pallens* differs in its narrow leaves colouring well in autumn, its pale lemon flowers, and bright red fruit. Z5.

B.prattii	W China	deciduous, spiny shrub to 3m tall	to 4cm long, whorled, obovate, obtuse, entire to sparsely serrate, glossy olive green above, grey-pruinose beneath	yellow in erect or drooping panicles	to 0.6cm long, ovoid, coral pink

Comments: Z5.

B. × rubrostilla B. 'Rubrostilla'; (B.aggregata × B.wilsoniae)	garden origin,	deciduous, spiny, erect to arching shrub to 1.5m tall	to 2cm long, oblanceolate with to 6 teeth per side, grey-pruinose beneath	pale yellow in umbel-like clusters in late spring and early summer	to 1.5cm long, ovoid, scarlet, glossy

Comments: large-fruited and showy especially in autumn. Includes 'Chealii' (fruit garnet red), 'Cherry Ripe' (fruit pink-white ripening scarlet), ' Crawleyensis' (leaves larger, fruit large, scarlet). Z6.

B.sargentiana	W China	evergreen, spiny shrub to 2m tall	to 10cm long, elliptic, acute with to 25 teeth per side, thick, dark glossy green above with impressed veins, yellow-green beneath	pale yellow in clusters in early summer	to 0.6cm long, purple-black, oblong, glossy

Comments: Z6.

B. × stenophylla (B.darwinii × B.empetrifolia)	garden origin	evergreen shrub to 3m tall with slender, arching branches	to 2cm long, linear-lanceolate, mucronate, entire, tough, margin revolute, dark green above, pale and glaucous beneath	golden, solitary, in clusters, or in racemes in late spring and often again later in the year	to 0.7cm long, globose, purple-black

Comments: a resilient and varied group of hybrids used for specimen plantings, hedging, screening and high, impenetrable ground cover; popular cultivars include 'Compacta' (dwarf, compact), 'Corallina Compacta' (dwarf, compact, flowerbuds coral-tinted opening burnt orange), 'Crawley Gem' (low-growing with many slender and arching stems and red-tinted orange-yellow flowers), 'Cream Showers' (flowers cream-white), 'Etna' (stems many, arching, covered in dense, dark green leaves and clusters of brilliant, orange-red flowers), 'Irwinii' (syn. *B. × irwinii*, small and compact with golden yellow flowers), 'Pink Pearl' (leaves splashed ivory and rose, flowers ivory, coral and orange), 'Semperflorens' (flowers red-orange to tangerine, long-lasting and produced late in season). For 'Nana' and 'Prostrata' see *B.darwinii*. Z5.

B.thunbergii	Japan	deciduous, spiny suckering, bushy shrub to 1m tall	to 2cm long, ovate to obovate, blunt, olive-green above, grey-tinted beneath, scarlet to fiery red in autumn	yellow tinted red, clustered in umbel-like racemes	to 0.8cm long, ellipsoid, glossy red

Comments: compact shrub with superb autumn colour, excellent for low hedging. Includes cultivars 'Atropurpurea' (medium-sized shrub, foliage purple-purple-red), 'Atropurpurea Nana' (dwarf, foliage purple-red), 'Aurea' (leaves yellow, becoming lime green), 'Golden Ring' (leaves purple-red edged gold), 'Green Carpet' (low and spreading, leaves green turning red in autumn), 'Harlequin' (compact, leaves small, mottled pink), 'Helmond Pillar' (narrowly upright, leaves dark purple), 'Kelleriis' (leaves mottled silvery white, this colour flushing pink to dark rose in autumn; 'Silver Beauty' is similar but smaller), 'Kobold' (dwarf, dense and domed with green leaves and bright red fruit), 'Pink Queen' (leaves red-purple flecked grey and white), 'Red Pillar' (densely branched dwarf, erect shrub with purple-red leaves), 'Rose Glow' (small with new foliage purple mottled silver-white and rose pink). Z4.

B.verruculosa	W China	evergreen spiny shrub to 1.5m tall with arching, warty stems	to 2cm long, obovate-elliptic, more or less toothed, glossy dark green above, white beneath	golden, solitary or paired to 1cm long, ovoid or pyriform, purple-black, frosted	neat and slow-growing.

Comments: Z5.

B.vulgaris COMMON BARBERRY	North America; Europe; W Asia	deciduous, spiny suckering shrub to 2m tall	to 6cm long, obovate-elliptic, serrulate, matt green	small, yellow-white in pendent racemes to 6cm long	to 1cm long, oblong, bright to dull red, translucent

Comments: wood once harvested for tooth picks and, being bright yellow, held to be a cure for jaundice. The fruits are edible and especially sweet in the cultivar 'Dulcis'; 'Asperma' is seedless. 'Alba' is white-fruited, 'Lutea' yellow-fruited. Selections for leaf colour include 'Atropurpurea' with wine-red leaves and the silver-edged 'Marginata'. Z3.

B.wilsoniae	W China	deciduous, thorny mound-forming shrub to 1m tall	to 3cm long, obovate to linear-oblong, entire, mucronate, grey-green turning scarlet in autumn	small, pale yellow, to 6 per cluster	to 0.6cm diam., coral pink, globose

Comments: var. *stapfiana* has spathulate leaves with grey undersides and vivid scarlet fruit. Among the finest cultivars are 'Joke' (hardy and vigorous with a profusion of large, red-flushed pink fruit), and 'Orangeade' (semi-prostrate, spreading, fruit orange and pink). Wisley Hybrids closest to *B.wilsoniae* include 'Comet' (fruit rounded, scarlet), 'Coral' (fruit conic, scarlet), 'Ferax' (fruit oblong-ovoid, scarlet), 'Fireball' (fruit conic, vermilion), 'Stonefield Surprise' (shoots red-tinted, flowers and fruit small), and 'Tom Thumb' (small, shoots yellow leaves broad, persisting). Z6.

rotted garden compost mixed with a little slow-release organic fertilizer. Divide in autumn or spring.

B.ciliata (W Pakistan to Nepal; leaves to 35cm, hairy with ciliate margins, flushing deep red-bronze in cold weather; flowers nodding, rose-tinted white, on stalks to 30cm); *B.cordifolia* (Siberia and Mongolia; leaves to 40cm, smooth with sunken veins and wavy, toothed margins; flowers rose or magenta on 30–45cm stalks); *B.purpurascens* (E Himalaya; leaves to 25cm, dark, shining green turning deep purple-red in autumn and winter; flowers deep, rosy red on stalks to 40cm high); *B.stracheyi* (Afghanistan to Tibet; leaves to 20cm, smooth except for downy margins, flushing dark garnet in winter; flowers fragrant, pale rose to white in nodding clusters). *Cultivars* – 'Abendglut': to 40cm, with leaves large, tinted red; flowers dark red; 'Ballawley': glossy, purple-flushed in winter; flowers purple-crimson; 'Morgenröte': small; flowers large, cherry pink, produced twice a year; *B.× schmidtii* (*B.ciliata* × *B.crassifolia*: leaves smooth, bluntly toothed flowers, rose pink; 'Silberlicht': leaves large; flowers white, later tinted palest pink; 'Sunningdale': rounded, bronze-red leaves in autumn and winter; flowers deep lilac pink.

Berkheya (for M.J.L. de Berkhey, Dutch botanist). Compositae. South Africa. Some 80 species, perennial herbs, subshrubs and shrubs with daisy-like flowerheads. Plant on fertile, well-drained soils in a sheltered, sunny position in zones 8 and over. Propagate by division in spring. *B.macrocephala* (Natal; erect perennial to 1m; leaves deeply lobed, spiny; flowerheads to 6cm, ray florets rich yellow, disc florets orange-yellow, produced in summer in branched inflorescences).

berry a baccate, indehiscent fruit, one- to many-seeded; the product of a single pistil. Frequently applied to any pulpy or fleshy fruit.

Bertolonia (for A. Bertoloni (1775–1869), Italian botanist). Melastomataceae. South America. 14 species, tender perennial herbs with decorative, thinly succulent leaves and one-sided cymes of rosy-purple flowers. Plant in shallow pots, terraria or bottle gardens in a porous, soilless mix. Keep moist and humid in shade or filtered light; minimum temperature, 15°C/60°F. Draughts, cold and damp, and exposure of wet foliage to sun are all deleterious. Propagate by stem or leaf cuttings in a sandy propagating mix in a closed case. *B.marmorata* (Brazil; evergreen to 20cm; leaves ovate to oblong, upper surfaces downy and vivid green with streaks of silver-white, or overlaid with a copper lustre, or, again, satiny bottle green veined silver; their undersides are deep purple-red; flowers magenta).

Berzelia (for Berzelius, a Swedish chemist of the early 19th century). Bruniaceae. South Africa. 12 species, evergreen shrubs grown in the cool greenhouse or conservatory for their heather-like habit and small flowers in globose heads. Plant in a well-drained,

neutral to acid medium rich in organic matter. Position in full sun; ventilate freely. Keep the roots cool and moist in summer, drier in winter, with a minimum temperature of 7°C/45°F. Sow seed in spring or root semi-ripe cuttings in late summer. *B.lanuginosa* (to 2m; leaves downy, needle-like leaves; flowers minute, white in, summer).

Beschorneria (for Friedrich Beschorner (1806–1873), German botanist). Agavaceae. Mexico. 10 species, very large, evergreen, perennial herbs with tough, sword-shaped leaves far exceeded by scapose panicles of 6-parted, tubular flowers in late spring and summer. The following species is hardy in zone 7 if planted on a well-drained soil in a sheltered, sun-baked position. Propagate by division, or by rooted offsets in spring. *B.yuccoides* (leaves grey-green, to 1m long in an open rosette; panicle to 2m, stalk and bracts glaucous, coral to sealing wax red, flowers pendulous, 7cm-long, yellow-green).

Bessera (for W.S.J.G. von Besser (1784–1842), Austrian-Polish botanist). Liliaceae (Alliaceae). Mexico. 1 species, *B.elegans* (CORAL DROPS), a frost-tender, cormous perennial herb with narrow leaves. In summer, a loose umbel of 10–30 pendulous flowers is carried atop a slender scape, 100cm tall. The flowers are campanulate, to 4cm long, bright red to purple-red veined green on the exterior, ivory veined or edged scarlet on the interior. Plant in full sun in a sheltered postion. Where frosts are hard, mulch thickly in winter, or lift and store dry in a cool place. Alternatively, grow in a cool greenhouse or conservatory. Increase by seed or division in spring.

Betula (the Classical Latin name). Betulaceae. Temperate and Arctic N Hemisphere. BIRCH. Some 60 species, fully hardy deciduous trees and shrubs grown for their graceful habit, bark, foliage and catkins. The habit ranges from erect and conical or pyramidal to slender and weeping. The bark is usually papery, the thin older bark peeling to expose chalky white to glossy red new bark; the oldest bark (i.e. that on bole) becomes sooty grey and riven. The leaves are mostly ovate and thin-textured with toothed margins. Small tightly packed in male or female catkins and, the flowers appear in late autumn but do not open until spring (in the following descriptions, the catkins of the species below are female and measured at fruiting unless otherwise stated; produced in late winter and spring, the ornamental male catkins tend to be longer, narrower and pendulous with green-gold anthers).

Plant in full sun on any fertile, moist, but well-drained soil. Propagate by ripe seed; by side-veneer grafts in late winter in a cold greenhouse; by greenwood cuttings in early to midsummer; or by semi-ripe cuttings in late summer under mist or in a closed case. Attacked by bracket fungi (entering by bark wounds and resulting in older stems' snapping in their upper reaches); also by mildew and birch rust. *(See table overleaf)*

Bergenia ciliata

Beschorneria yuccoides

BETULA

Name	Distribution	Habit	Bark	Leaves	Catkins
(in the following descriptions, the catkins are female and measured at fruiting unless otherwise stated; produced in late winter and spring, the ornamental male catkins tend to be longer, narrower and pendulous with green-gold anthers)					
B.albosinensis CHINESE RED BIRCH	China	open-crowned, clear stemmed, medium-sized tree, 18–25m tall	glaucous grey-white at first then peeling, thin and coppery pink to orange-red	4–7cm long, narrowly ovate, acuminate, biserrate, glossy light green	3–4cm long, ovoid-cylindric, pendulous
Comments: var. ***septentrionalis*** differs in its orange-pink to red bark bloomed violet-white and peeling in shaggy rolls; the leaves are longer and narrower, dull green and drooping. Z7.					
B.alleghaniensis *B. lutea* YELLOW BIRCH	Eastern North America	open-branched tree, 20–30m tall, often multi-stemmed	thin, crisped or curled, tinged yellow or grey, peeling in translucent sheets and yellow-brown where exposed	6–12cm, ovate to ovate-oblong, acute, coarsely biserrate, dull green, silky hairy beneath, turning yellow in autumn	2.5–3cm long, thickly cylindrical, erect
Comments: Z3.					
B.davurica *B.dahurica* ASIAN BLACK BIRCH	NE Asia	medium-sized tree, 15–30m tall	thick, peeling in curling flakes, brown tinged grey, becoming silver-grey and rugged with age; twigs hairy at first, dark grey, resinous with white glands	5–10cm long, ovate-rhombic, serrate, acute, dark green and glabrous above, glandular-punctate beneath	2–2.5cm long, narrowly oblong, drooping
Comments: Z3.					
B.ermanii ERMAN'S BIRCH; GOLD BIRCH; RUSSIAN ROCK BIRCH	NE Asia	graceful, vigorous tree or large shrub to 25m tall, usually widely spreading	peeling, creamy yellow to white with pale brown lenticels, or red-brown, orange-brown to purple-brown on young branches; twigs hairless, glandular-warty when young	5–10cm, deltoid-cordate, coarsely serrate, apex cuspidate, with scattered glands	2–3cm long, ovate-ellipsoid, erect
Comments: 'Grayswood Hill' is a selected form of this species, sometimes offered under the name *B.costata*. Z2.					
B. 'Fetisowii'	C Asia (hybrid origin)	graceful, narrow-crowned tree	chalky white, peeling, colourful on trunk and branches		
Comments: Z4.					
B.fontinalis *B.occidentalis* AMERICAN RED BIRCH; WATER BIRCH	NW America	shrub to 6m tall, or tree to 12m tall	smooth, lustrous dark bronze	2.5–5cm long, rhomboid-ovate, apex acute, biserrate, dark dull green above, paler beneath, slightly hairy above, soon hairless beneath, turning yellow in autumn	to 3cm long
Comments: needs moist but not saturated conditions. Z3.					
B.lenta CHERRY BIRCH; SWEET BIRCH	Eastern North North	shrub or narrowly upright tree to 25m tall	polished red-brown to purple-black, fissured, not peeling, aromatic	6–12cm long, ovate-oblong, biserrate, acute to acuminate, glossy green above, paler beneath, golden in autumn	2–3.5cm long, oblong-ovoid, erect
Comments: Z3.					
B.maximowicziana MONARCH BIRCH	Japan	fast-growing tree to 30m tall with an open crown	thin, peeling in strips, tinged orange-brown, grey and pink-white; branches and twigs dark red-brown, shiny	8–14cm, broadly ovate, base cordate, apex acute to acuminate, dark green turning clear yellow in autumn sometimes with red veins	3–7cm, pendulous
Comments: resistant to bronze birch borer. Z6.					
B.nana DWARF BIRCH	subarctic regions	prostrate to ascending shrub or dwarf tree 0.5–1m tall	dull brown, rather hairy at first	0.5–1.5cm long, suborbicular to reniform, apex rounded, crenate, dark green tinged yellow or red in autumn	to 1cm long, ovoid, erect
Comments: charming dwarf, tree-like shrub, especially attractive in early spring when the twigs are bright with fresh new leaves and squat, yellow-brown catkins. Prefers moisture and will tolerate extreme cold; suitable for rock and bog gardens, acid borders, light woodland and wild gardens. Z2.					
B.nigra BLACK BIRCH; RED BIRCH; RIVER BIRCH	E US	fast-growing tree to 20m tall, with a pyramidal crown or multi-stemmed	thick, peeling in shaggy scales, pink-brown on young trees, on older trees very dark red tinged black	7–9cm long, ovate-rhombic, biserrate, glossy green above, pale and glaucous beneath, yellow in autumn	2.5–4cm long, cylindric, erect
Comments: prefers moist soil; resistant to bronze birch borer. 'Heritage' is a vigorous selection with peeling, light brown to cream bark and glossy, dark green leaves. Z4 .					
B.papyrifera CANOE BIRCH; PAPER BIRCH; WHITE BIRCH	N America; Greenland	open-branched, round-headed tree to 30m tall	bright white, dull white or dull grey-brown, smooth, peeling in papery layers; branchlets red-brown	4–9cm, broadly ovate, apex acute, biserrate, ciliate, with dark, glandular spots beneath	3–4cm long, cylindric, pendulous
Comments: similar to *B.pendula* but taller with a more open and less weeping crown. Z2.					

BETULA

Name	Distribution	Habit	Bark	Leaves	Catkins
B.pendula COMMON BIRCH; EUROPEAN WHITE BIRCH; SILVER BIRCH; WARTY BIRCH	Europe to W Asia	tree to 30m with a slender, open crown weeping branches and very slender, pendulous branchlets	thin, peeling and silvery white when young, thick, fissured and ashy black with age especially at base of trunk; twigs bumpy with glands	3–7cm long, broadly ovate- deltoid to rhombic, biserrate, thin viscid when young, smooth	1.5–3cm, cylindric, pendulous

Comments: Very widely planted and naturalized tree; popular cultivars include 'Dalecarlica' (SWEDISH BIRCH; tall, slender tree, branches weeping; leaves deeply cut); 'Fastigiata' (erect, medium-sized tree with closely upright branches; 'Obelisk' is similar but narrower still); 'Golden Cloud' (small tree with arching branches and golden leaves, requires shelter, light shade and good water supplies); 'Purpurea' (slow-growing, slender and weeping tree with white bark, purple-black twigs and purple-red leaves; favours cooler climates); 'Laciniata' (similar to 'Dalecarlica' and often grown under that name but more markedly weeping and with less deeply cut leaves); 'Tristis' (tall and graceful with a narrow crown and weeping branches); 'Trost's Dwarf' (dwarf shrub or tree seldom exceeding 1m, with arching branches and small, dissected leaves, good for bonsai, alpine troughs and rock garden); 'Youngii' (small, weeping tree with a dome-shaped crown). Z2.

B.platyphylla B.mandshurica MANCHURIAN BIRCH	Siberia to Manchuria, Korea, Japan	open-crowned tree to 20m tall	pure chalky white, not fissuring at base; twigs brown, warty and resinous	4–7cm long, triangular-ovate, apex shortly acuminate, coarsely serrate, hairless	to 3cm long, pendulous

Comments: differs from typical *B.pendula* in its unfissured, white bark and larger leaves, and from the solidly white-barked *B.pendula* var. *lapponica* in its acutely, not bluntly, toothed leaves. var. *japonica* (syn. *B.mandshurica* var. *japonica*, JAPANESE WHITE BIRCH), has larger leaves, sometimes finely hairy beneath and catkins to 5cm long. Z2.

B.szechuanica B.mandshurica var. szechuanica; B.platyphylla var. szechuanica	W China	tree to 25m tall	thin, peeling, chalky pure white; twigs shortly acuminate, dark grey to red brown with pale resinous warts	5–12cm long, ovate-cordate to deltoid, apex serrate, blue- green above, glaucous beneath	3–5cm long, pendulous

Comments: Z6.

B.utilis HIMALAYAN BIRCH	Himalaya	tree to 20m tall, usually single- stemmed with a broad, open crown	thin, peeling in horizontal papery flakes, pink to orange-brown, bloomed white; twigs red-brown in autumn	5–12cm long, ovate- acuminate, serrate, dark green above, paler beneath, tough, smooth above, somewhat hairy beneath, yellow in autumn	to 3.5cm long, cylindric, spreading to nodding

Comments: var. *jacquemontii* (syn. *B.jacquemontii*) differs in its pure, startling white bark; the shoots and buds lack resin (thus distinguished from var. *occidentalis*). It includes some of the finest silver-bark birches, medium-sized, relatively slow-growing and long-lived trees with clear trunks, ellipsoid to conical crowns and handsome leaves – 'Jermyns' has brilliant white bark and showy male catkins to 15cm long; 'Silver Shadow' has the brightest white bark and large, deep green and rather drooping leaves. Z7.

Biarum (name used by Dioscorides for a similar aroid). Araceae. Mediterranean; W Asia. 15 species, hardy or slightly frost-tender tuberous, perennial herbs with spathes in subtle or sinister tones, and foul-smelling, similarly coloured spadices. The inflorescence usually emerges in spring or autumn, unaccompanied by leaves and sitting directly on the soil. Plant in a light, fertile and very well drained soil in full sun (minimum winter temperature –5°C/23°F). Wet winter conditions are usually lethal. Flowering is unlikely unless a dry, warm summer dormancy can be ensured. Propagate by seed sown under cover in autumn, or by division when dormant.

  B.davisii (Crete, S Turkey; spathe squat, pitcher-like, to 8cm long, cream flecked pink-brown or dull mauve); *B.eximium* (S Turkey; spathe 10cm, recurved, oblong to ovate in velvety purple-black); *B.tenuifolium* (Portugal to Turkey; spathe lanceolate, recurved to 30cm long, dark purple).

Bidens (from Latin *bi-*, twice, and *dens*, tooth, referring to the biaristate fruit). Compositae. Cosmopolitan. TICKSEED; BEGGAR'S TICKS; STICK-TIGHT; BUR-MARIGOLD; PITCHFORKS; SPANISH NEEDLES. 200 species, hardy annual and perennial herbs and shrubs with simple or pinnate leaves and daisy-like flowerheads. Grow in any moderately fertile, moisture-retentive soil in sun. Propagate by seed sown *in*

situ in spring, or (perennial clones and 'Golden Goddess') by division in early spring. *B.ferulifolia* (syn. *B.procera*; Southern United States to Guatemala; bushy, annual or perennial herb to 1m tall; foliage finely divided; flowerheads yellow, to 5cm across, especially large and golden in the cultivar 'Golden Goddess', produced midsummer to late autumn).

biennial a plant that flowers and seeds in the second season following germination, such as Canterbury bells (*Campanula medium*). Some biennials are treated as annuals, for example carrot (*Daucus carota* subsp. *sativus*).

biennial bearing relating to fruit plants which produce little or no fruit in alternate years, as a result of a natural characteristic or an inducing factor.

bifid of leaves, bracts, sepals and petals that are cleft deeply, forming two points or lobes.

Bifrenaria (from the Latin *bi-*, twice, and *frenum*, bridle, a reference to the double band by which the two pollinia are connected). Orchidaceae. South America. 16 species, tender epiphytic orchids with fragrant and long-lasting flowers in spring and summer. Pot in a very open, bark-based mix. Grow in full sunlight and with high humidity. Water and feed

Bifrenaria
harrisoniae

copiously while in growth. Once pseudobulbs have developed fully, impose a cooler, drier atmosphere (minimum temperature 7°C/45°F) and reduce water to a daily misting. Recommence watering when new growth appears. Increase by division. *B.harrisoniae* (Brazil; flowers thickly waxy, 7cm across; tepals oval, ivory or rich cream cupping a frilled and hairy, dark rose or blood red lip).

big bud a condition of blackcurrants in which the buds become swollen and fail to develop, caused by the BLACKCURRANT GALL MITE (*Cecidophyopsis ribis*), which also transmits reversion disease.

bigeneric hybrid, bigener a plant resulting from cross-pollination between parents of different genera.

Bignonia (for the Abbé Bignon (1662–1743), Librarian to Louis XIV). Bignoniaceae. Southeastern US. 1 species, *B.capreolata*, the CROSS VINE, QUARTER VINE or TRUMPET FLOWER, a woody, evergreen climber grown for its flowers, produced in summer. The leaves consist of a pair of leaflets sometimes separated by a coiling tendril. The flowers are 4–5cm long, funnel- to trumpet-shaped, deep orange to scarlet (rich purple-brown to blood red in the cultivar 'Atrosanguinea'). Grow in a well-drained, fertile medium in full sun; minimum temperature –5°C/23°F. Water and feed generously when in growth and spray and ventilate freely under glass. Tub cultivation will restrict the roots and encourage flowering. Cut back the previous season's growth in spring. Propagate by layering, or by leaf bud cuttings rooted in a closed case with bottom heat.

Bignonia capreolata

bilabiate a flower possessing two lips, as in the corolla of Labiatae and Acanthaceae.

Billardiera (for J.J.H. de Labillardiere (1755–1834), French explorer and botanist). Pittosporaceae. Australia. 8 species, perennial, woody-based evergreen climbers grown for their pendulous flowers and glossy, oblong to globose berries in late summer and autumn. Plant in a moist neutral to acid medium with a cool rootrun and protection from scorching sunlight. Hardy in mild areas of zone 7 if planted on a sheltered, sunny wall and mulched thickly in winter. Train on wires or trellis. In colder regions, grow in a cool greenhouse or conservatory. Propagate by seed sown when fresh or stem cuttings in early summer; also by simple layering. *B.longiflora* (APPLEBERRY, BLUEBERRY, PURPLE APPLEBERRY; to 2m; leaves narrow, dark green; flowers slender, to 3cm long in shades of green-yellow, later dull pink to pale mauve; berries cobalt blue to indigo, pink, mauve or white).

Billardiera longiflora

Billbergia (for J.G. Billberg (1772–1884), Swedish botanist). Bromeliaceae. C and S America. 54 species, epiphytic, evergreen, perennial herbs with narrow, tough and saw-toothed leaves in a funnel-based rosette and tubular flowers in an erect or arching,

terminal raceme or panicle clothed with colourful bracts. Position in bright, filtered light or light shade; minimum temperature 7°C/45°F. Water and feed generously during warm weather; keep barely moist at other times. Increase by division or offsets.

B.chlorosticta (RAINBOW PLANT; Brazil; to 50cm tall with white-marked, red-brown leaves and an arching to nodding raceme of blue-tipped, sulphur flowers amid broad, lobster red bracts); *B.nutans* (FRIENDSHIP PLANT, QUEEN'S TEARS; Brazil to Northern Argentina; freely clump-forming, to 40cm tall with narrow, plain green leaves and an arching to nodding raceme of yellow-green flowers tipped slate-blue amid narrow, flamingo pink bracts); *B.pyramidalis*; E Brazil; to 60cm tall with lanceolate, purple-flushed leaves banded white beneath and an erect, pyramidal inflorescence with red or orange-pink, violet-tipped flowers amid bright rose bracts); *B.sanderiana* (Brazil; to 60cm tall with green leaves clouded with grey-white and a compound, nodding inflorescence of indigo-tipped, lime green flowers between bright rose bracts); *B. × windii* (ANGEL'S TEARS; *B.decora × B.nutans*; differing from the latter in its broader, spreading leaves obscurely banded grey beneath and larger floral bracts); *B.zebrina* (S Brazil to N Argentina; to 1m tall with an open rosette of bronze-tinted and silver-banded leaves and an arching to nodding raceme of sulphur flowers clothed with large, papery, pink bracts).

billhook, bill a heavy-duty trimming implement of variable design; basically a wide blade fashioned into a hook at the top. A long-handled version is called a slasher.

bindweed the popular name for two climbing perennials with funnel-shaped flowers that occur as invasive weeds in gardens. HEDGE BINDWEED (*Calystegia sepium)* is the stronger stemmed of the two producing large white, occasionally pink, flowers. FIELD BINDWEED (*Convolvulus arvensis*) bears small white and pink flowers and is more productive of seed. Bindweed is difficult to control because of its depth of rooting, and frequent and careful forking out and hoeing off is necessary to weaken and eradicate it. Glyphosate and glyphosate trimesium are effective as foliage-applied total weed killers.

bine a term sometimes used to describe climbing stems, especially of beans, peas and hops.

binomial a taxonomic term for the basic unit of naming in botany comprising a generic name and a species, cultivar, group or hybrid epithet describing and distinguishing the individual belonging to that genus. *Buddleja alternifolia*, *Buddleja* 'Petite Indigo', *Buddleja* Davidii Group and *Buddleja × weyeriana* are all examples of binomials.

biodynamic gardening a system akin to organic gardening, first advocated in the 1920s by Rudolph Steiner. The philosophy perceives plant life as essentially self-sustaining. It places emphasis on skilful handling of manures through composting, and on symbiotic relationships between plants and insects. The plant is considered as a whole and there is a holistic approach to its well-being. The concept of biodynamic gardening incorporates a theory of interplanetary influence on plant life.

biological controls strictly, the control of pests and weeds by the use of other living organisms such as parasites, predators or pathogens. More generally, the term includes specially devised cultural practices and host resistance. It embraces both the encouragement of natural pest control and the introduction of artificially bred beneficial organisms, including insects such as parasitoids and predators, predacious mites, insect-feeding fungi, bacteria, and insect viruses. Selected plant-feeding insects are being used increasingly for weed control. In refined biological control natural enemies are introduced from a pest's centre of origin and only relatively small numbers need to be released because, providing the environment is favourable, they should multiply and be self-perpetuating and reduce the pest to a non-damaging level. In common practice, biological control always requires repetitive action by humans; it involves the release of natural enemies at times when pest numbers rise to critical levels, or any method that avoids destroying natural enemies or provides more favourable conditions for them to flourish.

Biological control agents available to gardeners include chalcid wasps for the control of whitefly, mealybugs and some scale insects under glass; the ladybird beetle (*Cryptolaemus montrouzieri*), which is used against mealybugs; the gall-midge (*Aphidoletes aphidimyza*) whose larvae attack aphids; the predatory mite *Phytoseiulus persimilis*, which controls red spider mites under glass; and the bacterium *Bacillus thuringiensis*, used for controlling moth and butterfly caterpillars. Other biological controls against pests are the fungus *Verticillium lecanii* against greenhouse whitefly and aphids, and the parasitic nematodes *Steinernema bibionis* and *Heterorhabditis heliothis*, used against vine weevil grubs. Some diseases of plants can be controlled biologically by the introduction of fungi; for example, *Phlebia* (*Peniophora*) can be used as a competitive invader of cut tree stumps, and protects against later invasion by destructive decay fungi such as *Fomes* bracket fungi - a technique now well-established in forestry practice. *Trichoderma viride* works both as a parasite of the silver leaf pathogen (*Chondrostereum*) of plums and also as an 'antagonist' producing antibiotics which protect the host tree against infection. *Trichoderma* shows promise for controlling other tree diseases such as Dutch elm disease.

Most gardens, with a mixed flora of trees, shrubs, flowers, vegetables and fruit, encourage the establishment of natural pest enemies by providing a variety of food, shelter and reproductive sites, particularly if spraying with chemicals is restricted. Gardeners should be aware of the presence of natural predators in their gardens and take measures to encourage them. Sometimes it is possible to augment resources such as food, shelter and reproductive sites: hoverflies feed on the nectar of flowers; mulching with organic materials provides shelter for useful predators such as rove and ground beetles; and insectivorous birds such as tits can be encouraged with nest boxes. See *cultural controls*; *integrated pest management*.

biorational pesticides pathogens and parasites of pest species, naturally occurring chemicals, and pest growth regulators used in biological control or integrated pest management systems.

biotic disease a parasitic disorder caused by a living organism such as a virus, bacterium or fungus.

bipinnate of compound leaves where both primary and secondary divisions are pinnate.

birds Most birds are welcome in gardens but some cause a considerable amount of damage. The BULLFINCH (*Pyrrhula pyrrhula*) which for most of the year feeds on seeds from weeds and trees, often attacks the flower buds of top and bush fruits. Plum, pear, cherry and gooseberry are attacked from late autumn to early spring when the buds are dormant; apple and blackcurrant are at risk in spring after the buds have opened. Ornamental cherry, crab apple, almond, *Forsythia* and *Amelanchier* are also subject to attack. Barren shoots with flowers at the tip indicate that bullfinches have been at work. The HOUSE SPARROW (*Passer domesticus*) is usually more destructive in urban areas: plums, currants and gooseberries may be disbudded in late spring; and flowers of many plants damaged in spring and summer. *Crocus*, *Primula*, polyanthus, lettuce, peas, runner beans and freshly sown lawn seed are particularly vulnerable. WOOD PIGEONS and FERAL PIGEONS eat seeds of beans, peas and other crops and strip bush fruits. Brassicas are particularly subject to attack, both as young plants and during winter, when they may be completely defoliated. Other birds causing damage in gardens, especially to fruits, include TITS, BLACKBIRDS, STARLINGS and JAYS.

The only certain way to protect vulnerable crops is to cover them with 2cm-mesh netting or purpose-built fruit cages. Black cotton can be strung along seed rows or strands criss-crossed above new-sown lawns, crocuses and primulas in flower; it can also be wound among the branches of fruit trees to deter bullfinches. Nylon thread, which does not decay and can readily trap birds, must not be used. Strips of narrow plastic ribbon made taut above crops to vibrate in the wind may deter birds by the ultrasonic sounds produced. Scaring devices such as traditional scarecrows, glitter strips, windmills and imitations of predators can be effective initially; however, they need to be changed or repositioned frequently. Techniques such as distress-call transmissions are unsuitable in built-up areas; trapping, shooting or nest destruction are also not applicable to most gardens. Chemical repellents are restrictive, usually erratic in performance and subject to weathering.

hedge bindweed

field bindweed

biserial arranged in two rows.

biserrate doubly serrate, i.e. with toothed teeth.

bisexual of flowers, with both stamens and pistils; of plants, with perfect (hermaphrodite) flowers.

biternate with compound leaves ternately divided but with each division itself ternately compound.

bitter pit a disorder of apple and pear fruits, expressed as sunken spots in the skin with small brown areas below the lesions and scattered throughout the flesh. It results from induced calcium deficiency probably connected with water shortage, and may only become apparent in store. Its incidence can be reduced by spraying calcium nitrate, timely watering and mulching.

bitter rot most commonly used to refer to the round, brown, sunken lesions occurring on fruits of apple, pear and quince late in the season and after harvest; caused by the fungi *Glomerella cingulata* or *Pezicula* species. Concentric rings of white to pink pustules of spores develop on the lesions. The fungi are present in the small cankers which appear on branches. Spores from these are washed onto fruits and enter the lenticels, eventually causing rot as the fruits mature. Damaged fruits should not be stored; fruits in store should be frequently examined. *G. cingulata* also causes similar rot on cherries, cranberries, grapes (RIPE ROT), and peaches.

bizarre a type of carnation with patches of two or more colours on a differently coloured background.

blackberry

Pests
aphids
leafhopper
raspberry beetle

Diseases
spur blight
virus

blackberry (*Rubus fruticosus*) the true blackberry of Europe, as distinct from the North American dewberry (which describes trailing *Rubus* species native of the eastern states) and the upright-growing highbush or erect blackberries. They are grown for their glossy black aggregate fruits, picked with the plug intact. Most are self fertile and flowering is usually late enough in spring to avoid frost damage. Plant one- or two-year-old plants shallowly, 2.5–3m apart for erect cultivars, 3.5–4.5m apart for trailers; cut back to a bud 20–30cm above soil level after planting. Trailing blackberries require support, such as a line of upright posts bearing four parallel wires spaced 30–45cm apart, with the lowest 1m from ground level. Fruit is produced on second-year canes. New canes should be tied in away from the fruiting canes to reduce the spread of disease from the old ones. Notable cultivars include: *early*, 'Bedford Giant'; *midseason/late*, 'Fantasia', 'Loch Ness'.

blackcurrant

Pests
aphids
big bud mite
capsid bug

Diseases
American
gooseberry mildew
(see powdery
mildew)
grey mould
leaf spot
reversion

blackcurrant (*Ribes nigrum*) a deciduous shrub 1.2m high, grown for the juicy, many-seeded berries that develop in bunches from pendent racemes. It grows wild in central and eastern Europe and in northern and central Asia as far as the Himalayas. Most cultivars are self fertile.

The blackcurrant grows most satisfactorily in the cool moist climates of Europe, Canada and New Zealand, where the requirement of low temperature to break dormancy is best met, though the flowers are susceptible to spring frost. It is little grown in the US because the plant is an alternative host to white-pine blister-fungus.

Plant two-year-old certified bushes 1.5–1.8m apart each way depending on potential vigour, and cut down all shoots to one bud above ground level after planting. The best quality fruit is borne on one-year-old wood: the structure of established bushes should be thinned by about one third each year, pruning out two- and three-year-old shoots as near as possible to the base of the stool. Propagate from hardwood cuttings. Notable cultivars include: *early*, 'Boskoop Giant'; *mid-season*, 'Ben Lomond', Ben Sarek'; *late*, 'Jet'.

blackfly any species of *aphid* (q.v.) characterized by black colouring. It commonly refers to the BLACK BEAN APHID (*Aphis fabae*).

black knot (*Apiosporina morbosa*) a serious disease of *Prunus* species especially plum in North America. Black masses of fungus tissue form on infected stems, and knots up to four times the diameter of the host twigs produce spores. Control by cutting out knotted stems before spring; spray trees with copper or lime-sulphur just before bud break.

black leg a term used for certain plant diseases in which the base of the stem turns black. In black leg of potatoes, caused by the bacterium *Erwinia carotovora* subsp. *atroseptica*, inward curling and yellowing of leaflets is followed by wet black decay at the base of stems. Only healthy tubers should be saved as seed. Black leg of beetroot and sugar beet (*Pleospora bjöerlingii*) causes stems of young seedlings to become blackened and shrivelled; it is seed-borne but not usually a problem in gardens. Black leg of pelargoniums, in which cuttings or plants die from root and stem decay, is typically caused by *Pythium* species, but other fungi such as *Rhizoctonia solani* and *Thielaviopsis basicola*, like the bacterium *Xanthomonas campestris* pv. *pelargonii*, cause similar symptoms. Cuttings should be taken from healthy stock and plants grown in sterilized compost; fungicide soil drenches may be effective. BRASSICA CANKER (*Leptosphaeria maculans*) is also referred to as black leg.

black mildew parasitic fungi, mostly of warm climates, which form dark superficial mycelium on foliage. See *sooty mould*.

black root rot (*Thielaviopsis basicola*). a disease that attacks many different vegetables and ornamentals causing blackening of roots and often stem base infection. Affected seedlings are stunted and chlorotic, and may be killed outright. Drenches of systemic fungicide can give some control. Because the fungus is soil-borne it can best be avoided by growing seeds in sterilized compost and by careful hygiene. The fungus is also the cause of specific cherry replant disease, which results in poor growth and cropping; it can be overcome through soil fumigation. In black root rot of cucurbits the fungus *Phomopsis sclerotioides* causes black spots on roots which eventually rot, causing plants to wilt. It can be controlled by soil fumigation and systemic fungicide drenches.

black rot a term for several diseases including black rot of brassicas (*Xanthomonas campestris* pv. *campestris*). The latter is uncommon in the UK, but can occur in warm conditions if the seed is infected. Young plants may be killed in the seedbed following blackening of the cotyledons. On older plants, V-shaped yellow areas with dark veins develop at leaf margins and leaves may become desiccated. The petioles and stems, when cut, may show a black ring of vascular tissue. The disease is seed-borne and can also survive on crop debris. It is spread mainly by rain splash. Control by crop rotation, destruction of crop residues, and use of healthy seed. The same bacterium causes a similar disease of wallflowers. Black rot of carrots (*Alternaria radicina*) commonly causes damping-off of seedlings, leaf and stalk spots and a slow developing dry rot of stored roots. Celery, dill and parsley can also be affected by the same seed-borne fungus. GRAPE BLACK ROT (*Guignardia bidwellii*) causes shrivelled and blackened fruit. It overwinters on mummified berries from which spores are produced in spring, when they start to infect new growth. Red-brown dead spots, up to 5mm in diameter with minute black fruiting bodies develop on the foliage, purple-black lesions on shoots and tendrils. The rot can be of major importance particularly in humid regions. Control measures include the disposal of fallen mummified grapes and fungicide treatment.

black spot (*Diplocarpon rosae*) a common disease of roses. Black spots appear on leaves and radiating strands of mycelium beneath the cuticle give the characteristic fringed border to young spots. Spots may coalesce until large areas of the leaf are blackened and the rest turns yellow, resulting in premature leaf-fall. Warm moist conditions favour the disease and the spores overwinter on stem and bud scale lesions and fallen leaves. The ascospore-producing stage develops on fallen leaves in the US, but does not usually occur in Britain. Rose cultivars vary in susceptibility but because there are many races of the fungus none is immune, although among rose species, *R.bracteata* seems to be resistant to all races of the pathogen. Severe spring pruning to remove infected shoots is of more practical control value than leaf collection; fungicide sprays are effective if applied regularly throughout the growing season.

blade the thin, expanded part of a leaf or petal, also known as the lamina, excluding the petiole, stipe or claw; (in the strap-shaped leaves of certain monocots, i.e. grasses, orchids and bromeliads) the part of a leaf above the sheath.

blanching the practice of excluding light from parts of plants, so as to make vegetables, such as celery, leeks, chicory and seakale more palatable. It is also practised to induce root formation from shoots, for example in propagating plum and cherry rootstocks. Resultant growth is devoid of chlorophyll, and white or cream in colour.

blanketweed filamentous algae which form dense masses in ornamental pools. Their development is encouraged where water is shallow, rich in nutrients or supports imbalanced plant and animal life. Control measures include constant removal by hand, the immersion of loose bundles of barley straw at 10g per cubic metre, and approved algicides such as terbutryn.

Blechnum (from Greek *blechnon*, a fern). Blechnaceae. Cosmopolitan. 200 species, ferns with rather leathery, pinnate leaves in a rosette. The spore-bearing (fertile) fronds arise in midsummer and create a strong contrast with the sterile foliage, being longer and rigidly erect with modified, rust-brown pinnae. *B.penna-marina* is hardy in zone 6 and over and a charming groundcover for damp rock gardens and the drier fringes of the bog garden. *B.spicant* is hardy in zone 4, a superb strong green accent for stone walls and banks and the woodland garden, especially in winter; propagate by division in early spring. *B.tabulare* is rather frost-tender: in zones 7 and below, it needs a protected position and a deep mulch in winter. Plant all three in a sheltered situation in shade or diffused light; the soil should be damp and cool, neutral to acid and rich in garden compost, leafmould and grit. The tender *B.brasiliense* and *B.gibbum* need a minimum temperature of 10°C/50°F. Plant them in an open medium rich in bark and leafmould; keep away from draughts and bright sunlight; water and syringe throughout the year (the trunks especially). Propagate all species by offsets or division.

B.brasiliense (Brazil and Peru; trunk stout, erect, to 30cm tall crowned with a shuttlecock of oblong to lanceolate fronds to 90cm long with many linear to lanceolate pinnae to 15cm long, wavy, finely toothed, emerging a deep, burnished bronze, hardening glossy, dark green); *B.gibbum* (syn. *Lomaria gibba*; MINIATURE TREE FERN; Fiji; trunk narrow, black-fibrous; to 1m tall topped with rosettes of 30–90cm-long fronds, pinnae closely ranked, linear to oblong, undulate, to 10cm long, lime to mid-green); *B.penna-marina* (cool regions of S America and Australasia; dwarf fern seldom exceeding 15cm tall with loose and suckering rosettes of spreading, linear fronds with many bluntly ovate to oblong pinnae 0.5–1cm long); *B.spicant* (DEER FERN, HARD FERN; Northern Hemisphere; dense clumps of arching, dark emerald fronds to 60cm long, with slightly curving, lanceolate to oblong pinnae to 2cm long); *B.tabulare* (syn. *B.chilense*, *B.cordatum*, *B.magellanicum*; Southern Hemisphere; rhizomes thick, almost trunk-like, loose rosettes of lanceolate fronds to 1m long, these emerge pink-bronze and harden sea green with lanceolate, undulate and serrate pinnae).

Blechnum penna-marina

bleeding the continuous flow of sap from the cut surface of a plant, as occurs in grapevines (*Vitis* species) pruned in late winter.

bleeding canker a bark disease of non-coniferous trees, especially apple, beech, horse chestnut, lime and maple. It is usually caused by *Phytophthora cactorum* but also by *P.citricola* and *P.syringae*. The same fungi also cause phytophthora root rot. A gummy liquid, which dries to form a black shiny crust, oozes from lesions on the collar, trunk or branches; dieback occurs if stems are girdled. Affected tissue should be cut away down to healthy wood using a cutting tool sterilized with sodium hypochlorite (diluted to give 1% available chlorine), and a strip of healthy bark should be removed from around the edge of the original lesion. Resulting wounds can be treated with a recommended fungicide, and later with wound sealant.

Bletilla (diminutive form of the name *Bletia*, a related genus, named for Luis Blet (fl. 1794), a Spanish apothecary who kept a botanic garden at Algeciras). Orchidaceae. E Asia.10 species, deciduous, near-hardy terrestrial orchids with subterranean, corm-like pseudobulbs and narrowly lanceolate, pleated leaves. Produced in summer in slender loose racemes, the flowers consist of five oblong to oblanceolate tepals and a frilly lip. Hardy in zone 7. Plant in early spring in a soil rich in leafmould and garden compost. Keep moist and sheltered in sun or light shade. In winter protect the crowns with a thick mulch of coir and bracken or leaf litter, or lift the 'corms' and store frost-free in slightly damp sawdust. Propagate by division. *B.striata* (syn. *B.hyacinthina*; China, Tibet and Japan; to 60cm; flowers 2–3cm across, candy pink, rose, magenta or pale mauve, lip lined with pure white, frilly crests; white-flowered and variegated forms also occur).

bletting the process of over-ripening fruits such as medlars and figs to a preferred condition for harvest.

blight a term used loosely of plant disease or severe pest infestation causing withering and rapid death of plant parts. It is more correctly applied to disease infections such as POTATO BLIGHT caused by *Phytophthora infestans*; POTATO AND TOMATO EARLY BLIGHT (*Alternaria solani*); RASPBERRY CANE BLIGHT (*Leptosphaeria coniothyrium*); HALO BLIGHT of beans (*Pseudomonas syringae* pv. *phaseolicola*); FIREBLIGHT of pear and other rosaceous trees (*Erwinia amylovora*); and SPINACH BLIGHT, which is caused by the cucumber mosaic virus. See *bacterial blight*

blind, blindness used of a plant where growing points or flowers fail to develop. Blindness may be due to growth or environmental factors, or physical damage including pest or disease attack.

blister beetles (Coleoptera: Meloidae) soft-bodied beetles up to 20mm long with long legs and antennae; they range in colour from red-yellow to blue- or green-

black, some with white stripes or spots, and the head is wider than the thorax. Their name derives from the blister-forming fluid excretion containing cantharidin. They are most harmful in North America, where large numbers can defoliate vegetables and ornamentals over a short period in late summer. Examples include the STRIPED BLISTER BEETLE (*Epicauta vittata*), the SPOTTED BLISTER BEETLE (*E. maculata*) and the BLACK BLISTER BEETLE (*E.pennsylvanica*). Numbers can be reduced by spraying with contact insecticides.

blocking the technique of compressing growing media into variously shaped blocks in which to raise plants.

bloom (1) pertaining to flower or blossom; (2) the white or blue powdery or waxy coating on some leaves and fruits; (3) the rapid proliferation of algae in still water.

Bloomeria (for H.G. Bloomer (1821–1874), Californian botanist). Liliaceae (Alliaceae). SW US; Mexico. 3 species, hardy, bulbous, perennial herbs grown for their long-stemmed heads of star-shaped flowers produced in late spring. Plant in a well-drained soil in a sheltered, sunny place. *Bloomeria* becomes dormant in summer. Propagate by seed or division in late summer and autumn. *B.crocea* (syn. *B.aurea*; to 30cm tall; flowers 2.5cm-wide, dark-striped, golden).

blossom-end rot a common non-parasitic disorder of tomatoes and peppers. On tomato fruit it first shows as a small water-soaked bruise at the blossom end; as ripening progresses, the lesion darkens and becomes leathery and sunken, but not soft. The rest of the fruit is lighter coloured than normal at the stem end. In peppers, lesions are liable to develop fungal or bacterial rots. Blossom-end rot is caused by calcium deficiency in fruits but can be brought on by irregular watering. Plants most susceptible are those grown with ammonium rather than nitrate fertilizer, or grown in an acid medium, or subjected to periods of drought or waterlogging when fruit is starting to develop.

blossom fall also known as petal fall. See *bud stages*.

blossom wilt see *wilt*.

blown used of a flower beginning to wither.

blueberry (*Vaccinium* species) The most important in gardens is the HIGHBUSH BLUEBERRY developed from *Vaccinium corymbosum* and *V.australe*, natives to eastern North America. A deciduous shrub usually 1.5–1.8m high in cultivation, the highbush blueberry is grown for the blue-black, wax-covered, globose berries which may be up to 2cm across; they are borne in clusters and selectively picked. Plants are self-fertile and decorative both in flower and in autumn foliage, but the minimum cold requirement of 800 hours below 7.2°C/45°F limits the climatic range for successful

blueberry

Pests
aphids
birds

Disease
grey mould

growing. An acid soil is essential, the optimum pH range being 4.3–4.8. Their shallow rooting habit calls for regular watering during fruit development and for surface mulching, and fruit must be protected from birds. Highbush blueberries are grown as stooled bushes planted 1.5 m apart each way. The best fruits are produced on two- or three-year-old shoots. Pruning in the first three years should be light; thereafter, cut out the oldest unproductive wood, removing about one quarter of the bush annually. Propagate from summer softwood cuttings. Notable cultivars include: *early*, 'Earliblue', *mid-season*, 'Berkeley', *late*, 'Coville'.

blueing the process of making red or pink hydrangeas turn blue; achieved by lowering soil pH level to facilitate the uptake of aluminium or by applying aluminium sulphate or a proprietary product.

blue mould the visible growth of blue or blue-green fungus spores of those *Penicillium* species responsible for storage rots. *Penicillium corymbiferum* and other species affect many ornamental bulb species; infection occurs through small wounds while bulbs are in the soil. The rot develops rapidly in storage, especially under conditions of high humidity and lower temperature. Careful handling and the provision of adequate storage conditions, as well as dipping in a systemic fungicide are preventive measures. *Penicillium expansum* causes rotting of apples in storage. The store should be cool and well-ventilated, contact between individual fruits avoided, and damaged fruit removed. Citrus fruits are commonly affected by the post-harvest rots GREEN MOULD (*P.digitatum*) and the less serious BLUE MOULD (*P.italicum*). A devastating downy mildew disease of tobacco (*Peronospora tabacina*) is also referred to as blue mould.

bog garden an area in which the damp, often acidic conditions of natural wetlands is recreated, and where plants native to such places may be cultivated.

Bolax (from Greek *bolax*, clod). Umbelliferae. S America. 2 species, evergreen, cushion-forming perennials grown in rock gardens, alpine sinks, tufa and pans in the alpine house for their massed rosettes of very small, dark green leaves. Fully hardy, they require a moist, gritty soil in full sun. Propagate by detaching rooted rosettes. *B.gummifera* (syn. *Azorella glebaria*; South America and the Falklands; forms a dense cushion ultimately to 1m across, with jade green, 3-lobed leaves 2–6mm long).

bole the main stem of a tree or other arborescent plant.

bolting the premature production of flowers, especially of vegetables; usually triggered by environmental or cultivation factors.

Boltonia (for James Bolton (*c* 1758–1799), British botanist). Compositae. N America; Asia. FALSE CHAMOMILE. 8 species, hardy, perennial herbs with panicles of daisy-like flowerheads in autumn. Plant on any moderately fertile soil in full sun or dappled shade. Stake in exposed positions. Divide and replant regularly. *B.asteroides* (to 2m tall; flowerheads 2cm-wide with white, rosy mauve or purple ray florets and yellow disc florets; var. *latisquama*: flowerheads larger, darker ; 'Nana' and 'Snowbank': white, the former a dwarf selection, the latter attaining 2m).

Bomarea (for Jacques Christophe Valmont de Bomare (1731–1807), French patron of science). Liliaceae (Alstroemeriaceae). S America. 100 species, tender or near-hardy tuberous-rooted, twining perennial herbs with terminal umbels of brightly coloured, tubular to campanulate flowers to 6cm long. Grow with the support of wire or trellis, in a light, fertile, loam-based mix. Provide good ventilation, a minimum temperature of 5°C/ 40°F and bright, filtered light. Water and feed plentifully when in growth; keep almost dry in winter. Propagate by division in late winter, or by seed germinated at 23°C/75°F. Several species have proved hardy in sheltered sites in Zone 7, especially when thickly mulched in winter.

B.caldasii (syn. *B.caldasiana*, *B.kalbreyeri*; flowers orange-red to bright yellow flecked green, brown or red); *B.cardieri* (flowers with mauve to pink tepals tipped bright green and spotted darker mauve or green); *B.edulis* (flowers pink with the inner tepals yellow-green flecked purple-pink); *B.patacocensis* (flowers with orange to crimson outer tepals and inner tepals yellow to crimson tipped orange and flecked chocolate or violet); *B.racemosa* (flowers with brown-spotted, yellow to scarlet inner tepals); *B.shuttleworthii* (flowers with orange-red outer tepals spotted at their tips, and yellow inner tepals tipped green with a red midrib and dark spots above).

bone manure an additive for improving soil fertility, made from crushed animal bones, that typically contain 25% phosphate and 1–5% nitrogen, with some calcium. It should be sterilized to eliminate pathogens and handled with suitable precautions to avoid inhalation and other personal contact. Bone meal is a finely ground form.

Bongardia (for Heinrich Gustav Bongard (1786–1839), German botanist). Berberidaceae. W Asia. 1 species, *B.chrysogonum*, a tuberous, perennial herb. Pinnate leaves to 25cm long arise in early spring from a large, tuber; the grey-green leaflets are themselves lobed and marked purple-red. In late spring, a red-tinted panicle arches over the leaves, bearing golden flowers. Hardy in zone 8. Plant deeply in a very free-draining, sandy and gritty soil in full sun. Keep moist from winter to early summer; dry at other times. Sow seed in autumn: a minute tuber develops at first, pulling itself deep into the soil.

boning rods T-shaped wooden devices, typically 90cm long, used in land levelling. Sightings are made across three rods in a straight line, manoeuvred to establish a level.

bonsai the ancient Japanese art of container culti-vation incorporating miniaturization. 'Bonsai' means a plant in a pot, and originally described the prac-tice of taking specimens from the wild and main-taining them in the home or garden with the essence of their wildness preserved or recreated. Gradually potting, pruning, pinching and wiring became the rule and led to the cult of the miniaturized tree.

The aim of bonsai is to create on a miniature scale a picture of what one sees in nature using living plants with the container a key element. Most people identify woody plants most readily as bonsai, but some perennial herbs are of importance and may appear in advanced Western collections, among them terrestrial orchids, grasses, the smaller bamboos, ferns and mosses.

Among woody subjects many conifers are grown, and *Juniperus* is especially prized. Other trees and shrubs commonly chosen are *Acer palmatum, Pinus, Ulmus* and *Zelkova, Ginkgo, Lespedeza, Prunus* and *Wisteria.*

Woody bonsai are trained to one of the traditional styles. These imitate the growth forms of trees and shrubs in nature. Bonsai can be created from large well-established trees and shrubs cut down to the appropriate height and regrown and shaped into the desired style or young grown from seeds and cuttings and trained to the size required. To reduce a container-grown plant to a bonsai style can be achieved in two or three hours but take years of training to become good bonsai. Growing from seed gives maximum control but is slow. It will be 2–3 years before most young plants are large enough to be repotted and trained; usually the young tree will begin to look like a bonsai when it is five years old. Bonsai can be started from rooted cuttings which is quicker than seed and ensures that new plants will have the same characteristics as the parent. Aerial and ground layering are also suitable methods.

Pots and repotting Pots used for bonsai are usually made from frostproof ceramic materials such as stoneware. They must always have drainage holes. Dark containers complement a mass of foliage, a light colour a more delicate, open branch structure. A rugged tree looks good in a simple, angular container; an elegant bonsai may look better in a rounded and more elaborate pot. Shallow pots are usually used for trees with slender trunks, and deeper ones for trees with thicker and older trunks and for a bonsai that has part of its foliage falling below the line of the container. Tall pots are used for trees that hang downwards in the cascade style.

Young conifers usually need repotting every 2–3 years and older conifers every 3–5 years; young deciduous trees should be repotted annually and older deciduous bonsai every 1–2 years. Spring is generally the best time to repot although it may also be done in mid-autumn.

Potting media A suitable compost is made from one part leafmould or well-rotted garden compost, one part loam and one part sharp sand. Most bonsai prefer a slightly acid soil of pH 6.0 to 6.5. For *Juniperus* species and *Pinus* species the compost may contain as much as 75% sharp sand, and for fruiting and flowering plants and most of the decid-uous trees up to 50% loam may be used. For decid-uous trees, start feeding when buds open in spring right through to autumn. Evergreens should be lightly fed from mid-winter, increasing the rate from summer until mid-autumn. Liquid fertilizers can be used: a high-nitrogen product during the early part of the growing season and a high-potash one during the latter half. Fruiting and flowering trees should be fed extra potash, but hold back any feeding when the trees are in flower or fruit. Incorporating slow-release fertilizer into the potting compost is a suit-able method choosing the long-acting (8–9 months) and lower-nitrogen formulations.

Pruning is done to reduce the size of the trunk and branches, to obtain proportion and shape and to clean out congested growth. Special bonsai tools make pruning and trimming easy; always make a clean cut a little above a bud, and make large cuts at the rear of the tree so that the wound will not be visible. Pinching back, or soft pruning, encourages denser growth and foliage. Cutting the first flush of leaves will reduce the size of leaves in successive flushes and is often done for deciduous species. Vigorously growing species may need annual root pruning, and mature, old bonsai may require root pruning occasionally for re-invigoration. The best time to root-prune most trees is in early spring. About one third of the soil is removed from the outside of the root-ball and roots cut back with scissors before replanting in fresh soil in the same container.

Training generally involves wrapping wire around the trunk or branches to hold a position for several months. Copper wire is firmer, holds the shapes of branches better and is used for conifers and ever-green shrubs where it may take slightly longer for the position to set. Aluminium wire is more pliable and is more suitable for deciduous trees. Plastic-covered wire is available, also bare metal wire wrapped in thin paper. Wire conifers and evergreen shrubs in winter and deciduous trees in late spring. Hold back on watering prior to wiring and feed well for several weeks thereafter.

Ageing or *Jin* is an attempt to mimic storm damage or extreme old age by killing certain sections of branches or trunks and stripping them of bark. Select a branch, strip it of foliage and bark. A vari-ation on *Jin* is *sharimiki,* in which only a strip of bark is removed from live wood on a branch or a trunk, leaving the bonsai with a dramatic slash of silver-white on healthy wood.

General care Traditional bonsai are outdoor subjects and may be brought indoors for display but only for two days followed by six days outside. Indoors, protect from any heating source, mist regu-larly and provide good natural lighting. During winter bonsai may be protected in sheds or green-houses just above freezing point. Alternatively, the pots may be planted in sheltered parts of the open garden. Where neither is practicable, take precau-

Bonsai styles. Of between thirty and forty recognized styles, the following are popular: (a) *Formal upright*: a single bold trunk that tapers towards the top with branches evenly spaced and symmetrically balanced from all directions (b) *Informal upright*: a single trunk usually with one predominant curve; it may have a slight slant to it but remains balanced because the tip of the trunk is always directly over its base (c) *Twin trunk*: two trunks growing from a single base, with one taller and thicker than the other (d) *Multiple trunks*: three, four or more trunks growing from a single base (e) *Slanting*: the trunk grows at a pronounced slant, to the right or left, with branches emerging on both sides (f) *Cascade*: an arched trunk that cascades gracefully downwards, with its foliage below the bottom of the pot (g) *Semi-cascade*: similar to the cascade but the trunk is not as arched and the foliage may not spill down below the pot (h) *Windswept*: the trunk and all the branches grow at a slant, as it would if constantly blown by strong winds (i) *Broom*: a single upright trunk with numerous small branches fanning out from the top half (j) *Root-over-rock*: the tree is planted so that its roots may grow over a rock and then down into the soil (k) *Group or grove*: several different but usually related species of trees are grown together in a shallow tray to suggest a grove or forest (l) *Raft*: the trunk is buried horizontally and branches grow up vertically from it, resembling a line of trees.

tions to protect the pots from frost-damage and the plants from physiological drought.

Chrysanthemum bonsai A fully mature chrysanthemum bonsai can be achieved in 12 to 14 months, shaped into one of the traditional bonsai styles, especially informal upright and cascade. There should be many small flowers of the same size, one per stalk, blooming simultaneously. The bonsai may be kept for many years but is most commonly repropagated annually.

Cuttings are taken from the mature stems of very compact, small-flowered cultivars in early September and will root in two to three weeks. These are potted up in 10cm pots. In mid-winter select the two most vigorous shoots and prune back all others to the main stem. Prune back the original stem, 1.5cm at a time for the next month. Pot on about every 45 days, until the final 30cm diameter size is reached in late summer; root-prune as necessary. Top dress in late winter-early spring with a solid fertilizer and begin liquid feeding as the growing season progresses until colour can be seen on buds.

Training by wiring can begin in late winter using copper wire wrapped in paper to protect the delicate growth. In mid-spring and every 25 days or so thereafter pinch the tips of the shoots to control overall growth and shape the plant. A final overall pinching is done in late summer. Select one bud from each of the clusters and remove all others. Blooms open in mid-autumn.

Boophone (from Greek *bous*, ox, and *phone*, murder, a reference to the toxic foliage). Amaryllidaceae. S and E Africa. 6 species, deciduous, perennial herbs with large bulbs, strap-shaped leaves in two ranks and spherical umbels of slender-stalked flowers. The pedicels are long, becoming longer still in fruit, when the whole, globe-like inflorescence breaks free and is blown across the plains. Dried, these flowerheads make attractive decorations. Plant in early spring with the bulb's shoulders and neck exposed. Maintain a minimum winter temperature of 7°C/45°F with full sunlight and a dry, airy atmosphere. Water liberally from the emergence to the withering of the leaves; thereafter, keep dry in a hot, sunny place. Propagate by offsets or from ripe seed sown in sand at 20°C/70°F. *B.disticha* (CAPE POISON BULB; South African; leaves 30cm long, blue-green and wavy-margined; inflorescence short-stemmed, 30cm wide, composed of 5cm wide, pink flowers on rigid, ray-like pedicels to 10cm).

Borago (perhaps from Latin, *burra*, a hairy garment, referring to the bristly-hairy leaves). Boraginaceae. Europe. 3 species, annual or perennial herbs. BORAGE has a long history in herbalism. The 16th-century herbalist Gerard claimed a syrup of its flowers 'purgeth melancholy and quieteth the phreneticke and lunaticke person.' The cucumber-flavoured leaves are eaten raw in salads, added to cool drinks and cooked as for spinach. The flowers are also used in salads, or frozen in ice cubes for summer drinks. Sow seed *in situ* in spring in full sun. *B.officinalis* (BORAGE;

annual to 60cm tall, densely bristly-hairy; flowers star-shaped, sky blue or cobalt with a cone of purple-black scales at the centre).

Borassus (from Greek *borassos*, the immature spadix of the date palm). Palmae. Africa; SE Asia to New Guinea. 7 species, tall palms with stout, ringed, solitary stems and rounded to wedge-shaped, palmate leaves with spiny petioles and rigid, blue-green blades divided to at least half their depth into narrow segments. Provide a minimum temperature of 10°C/50°F. Sow seed on a moist, sandy mix at high temperatures. *B.flabellifer* (PALMYRA PALM, TODDY PALM, WINE PALM, DOUB PALM, LONTAR PALM; to 20m; trunk swollen at the base or middle; leaf blades to 3m long, divided into some 80 segments).

borax a salt containing boron and sodium; used for the correction of boron deficiency and sometimes as an insecticide or to kill persistent weeds.

Bordeaux mixture the first universal fungicide, the result of a serendipitous discovery by Pierre Millardet, a professor of botany at Bordeaux University. In 1882, while walking in the vineyards of the Medoc, where the newly introduced vine downy mildew (*Plasmopara viticola*) was devastating the crop, he noticed that vines near the path were free from disease. He found that it was the practice for growers to sprinkle a mixture of copper sulphate and lime (calcium hydroxide) on plants to discourage pilfering; those which had been so treated had escaped infection. Bordeaux mixture is a suspension of a flocculent precipitate of insoluble copper hydroxide and calcium sulphate. It continues to be useful against a wide range of pathogens, including bacteria and fungi, but it is not generally effective against powdery mildews or all rusts. It adheres well to plants, and these 'sticking' qualities are greatest when the mixture is used freshly prepared. For garden ornamentals use a 2:2:50 preparation, made by adding 80g copper sulphate dissolved in four litres of water to a second mixture of 40g hydrated (slaked) lime in six litres of water. Containers used for preparing the mixture and the spraying equipment must be able to resist its corrosive effect, and brass and plastic are suitable. Proprietary formulations of Bordeaux mixture are available with appropriate recommendations for use.

border a strip of cultivated ground, often backed by a wall or hedge, in which plants are grown. Commonly used of such an area containing herbaceous plants, but by no means exclusively so.

borecole an alternative name for curly or Scotch forms of kale (*Brassica oleracea* Acephala Group). See *kale*.

boron an essential trace element. Its deficiency is predisposed by high pH, and the development of hollow stems in brassicas is the symptom most commonly met with in gardens.

Borago officinalis

Boronia (for Francesco Borone, 18th-century Italian botanist). Rutaceae. Australia. Some 95 species, winter- and spring-flowering aromatic evergreen shrubs with small, 4-petalled, star- or bell-shaped flowers usually solitary and nodding from the axils of fine, simple or compound leaves. Plant in a free-draining, lime-free medium. Keep just moist in winter and maintain a minimum temperature of 7°C/45°F. Cut flowered stems hard back. Root semi-ripe cuttings in late summer under mist or in a case with bottom heat.

B.heterophylla (RED BORONIA, KALGAN; densely twiggy shrub to 2m tall, leaves needle- and feather-shaped; flowers fragrant cup-shaped, pink or scarlet); *B.megastigma* (SCENTED or BROWN BORONIA; slender, cypress-like shrub, 1.5–3m tall; leaves strongly aromatic, needle-like; flowers bell-shaped, chocolate brown on the exterior, primrose yellow on the interior, sweetly scented).

bosket, **boskage** a plantation or thicket of trees in a park or garden; sometimes containing walks and garden features. Often a backdrop to a parterre.

bossed see *umbonate*.

bostryx a uniparous helicoid cyme, i.e. a cyme producing one axis with each branching; often used interchangeably with helicoid cyme.

botrytis. Fungi of the genus *Botrytis* are responsible for many diseases of field crops, fruits, vegetables and ornamentals worldwide, mainly in cool temperate regions. They cause many and varied diseases including damping-off and rots of flowers, fruit, stems, roots, and storage organs. The most common disease is the ubiquitous GREY MOULD (q.v.) caused by *B.cinerea*. Other diseases are CHOCOLATE SPOT (*B.fabae*) of broad bean, SNOWDROP GREY MOULD (*B.galanthina*), HYACINTH FIRE (*B.hyacinthi*), LILY LEAF SPOT (*B.elliptica*), NARCISSUS SMOULDER (*B.narcissicola*), ONION NECK ROT (*B.allii*), PEONY BLIGHT (*B.paeoniae*) and TULIP FIRE (*B.tulipae).*

The relationships between the asexual fungal states classified as *Botrytis* and *Monilia*, their sexual states in the fungal genera *Botryotinia* and *Monilinia*, and the related genus *Sclerotinia*, are incompletely known. Many are important plant pathogens known predominantly or exclusively from their asexual or sclerotial states. The genus *Sclerotium* also contains sclerotium-producing plant pathogenic fungi, whose relationships are diverse and sometimes unknown. See *chocolate spot, neck rot.*

bottle garden a garden, usually of small slow-growing plants, made in a bottle, carboy or other glass container.

bottom heat warmth applied from below, in a frame or propagator, to advance growth or aid rooting and germination. Traditionally produced from decomposing organic matter but today usually provided through special electrically heated cables. See *hotbed.*

bottom-worked see *low-worked.*

botuliform sausage-shaped.

Bougainvillea (for L.A. de Bougainville (1729–1811), French navigator. He embarked on a voyage of discovery to the South Seas in 1768). Nyctaginaceae. South America. 14 species, mostly woody climbers, evergreen or briefly deciduous, many with stout thorns. The leaves are ovate to elliptic. Small, tubular, yellow-white flowers are produced amid clusters of brilliant, papery, leaf-shaped bracts. Although frost-tender, these spectacular climbers have become a common feature of gardens in the Mediterranean and in California where, once established in a sunny, sheltered spot, they survive freezing temperatures. In these and warmer regions, *Bougainvillea* is unsurpassed for covering trellis and pergolas; it can also be trained as a standard and, in the more shrubby cultivars derived from *B.glabra*, be planted as an informal hedge. In warm, wet climates, it remains evergreen.

B.glabra and its offspring tend to flower continuously and are best grown where they experience virtually no dry season, or can be watered and fed throughout the year. Within this group, very young plants will flower and are ideal for container cultivation. *B.spectabilis* and its hybrids require a dry season to induce flowering—under glass, water and feed plentifully at first to produce strong growth, then reduce supplies to promote flowering. In an open garden, grow in full sun on a rich, well-drained soil. Plants of the *B.spectabilis* alliance will hook themselves to trees, beams and rafters; otherwise, they require regular tying to a strong support framework. Under glass, they need a minimum night temperature of 7°C/45°F. Plant in a moderately fertile gritty, loam-based medium in pots, tubs or beds with a confined rootrun. Provide support. In early spring, spur prune established specimens (not *B.peruviana* which resents pruning) and remove any dead or exhausted wood. Pot-grown plants may be moved outside in warm summers. Root greenwood cuttings in late summer with bottom heat at 20°C/70°F. Air layer in spring.

B.glabra (PAPER FLOWER; The Brazil; thick-stemmed climber to 15m; inflorescence bracts white to deep magenta, sometimes outliving the flowers; perianth tube five-angled and covered in short hairs); *B.peruviana* (Peru; stoutly thorny with smaller bracts than *B.glabra* (to 10cm long), and the perianth tube, smooth and only obscurely five-angled; the hybrid between these two species, *B. × buttiana*, produces smaller bracts (to 5cm) in shades of rose, magenta and fiery orange; the perianth tube is angled, hairy and waisted below the middle); *B.spectabilis* (syn. *B.brasiliensis*; Brazil; robust climber with large, hooked thorns; inflorescence bracts purple-pink, to 6cm long; perianth tube obscurely angled, hairy); *B.* Spectoglabra Group (*B.spectabilis* × *B.glabra*, cultivars ranging in habit from low and spreading to bushy and erect, to tall and arching, leaves small,

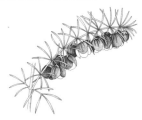

*Boronia
megastigma*

*Bougainvillea
'Mardi Gras'*

grey-green and downy or large, dark green and glossy, some cream-, white- and yellow-variegated forms; inflorescence bracts range from the small, persistent and densely clustered ('double') to large, crêpe-like and short-lived, in colour from white to lilac, shell pink, rose, deep cerise, garnet, purple-red, crimson, magenta, to bronze-red, brick red, coral, flame, vermilion, orange, apricot and pale yellow).

boulingrin a sunken grassed area at the centre of a bosket or dell.

bourse in apples and pears, the swollen stem at the base of a flower cluster, which may carry wood or fruit buds. Also known as a cluster base or knob.

Bouteloua (named for the brothers Boutelou, Claudio (1774–1842) and Esteban (1776–1813), Spanish botanists). Gramineae. US; Central America; West Indies. 39 species, annual or perennial grasses with one-sided panicles, the branches short, spicate and closely aligned along the underside of the summit, thus resembling the teeth of a comb. The flower heads are suitable for drying. Easily grown in zones 5 and over in full sun. Propagate by seed sown *in situ* or in pots under glass in spring, or by division. *B.gracilis* (syn. *B.oligostachya*; BLUE GRAMA GRASS, MOSQUITO GRASS; perennial; flowering stems to 60cm, the inflorescence occupying the topmost 5cm and angled at 45°, florets hanging (or seeming to hover) below the spikelets that clothe the 'comb' of branches).

Bouvardia (for Charles Bouvard (1572–1658), physician to Louis XIII and Keeper of the Jardin du Roi, Paris). Rubiaceae. Subtropical and tropical Americas. 30 species, perennial herbs or shrubs, frost-tender and grown for their clusters of narrowly tubular flowers, each with a limb of four spreading lobes, produced throughout the year. Provide a minimum winter temperature of 7°C/45°F, medium humidity and filtered sunlight. Propagate by softwood cuttings in spring, or by semi-ripe cuttings in summer.
    B.longiflora (syn. *B.humboldtii*; to 1m tall, with fragrant, pure white flowers to 9cm long); *B.ternifolia* (syn. *B.jacquinii*; weak-stemmed shrub to 1m tall with 3cm long flowers in red, pink, rose and coral); *B.triphylla* (to 1m; flowers orange-red, to 3cm, very close to *B.ternifolia*).

bower a shady recess in gardens. See *arbour*.

Bowiea (for James Bowie (c1789–1869), a Kew gardener who collected plants in South Africa and Brazil). Liliaceae (Hyacinthaceae). Southern and tropical Africa. 3 species, tender, bulbous herbs. The following species is a bizarre addition to the succulent collection, grown either in the cool greenhouse (minimum temperature 7°C/45°F) or indoors on a bright, draught-free windowsill. Plant in a very sandy and gritty, loam-based medium with the upper half to three quarters of the bulb exposed. Position in full sunlight in a dry, airy atmosphere. Water sparingly to encourage sprouting. As the new stem emerges, provide support in the form of a small trellis, wire mesh, or dead twigs. Water whilst the stem develops. Once growth and fruiting are finished, reduce the water, allowing the stem to die off. Keep virtually dry until the next growth flush begins, watering only to prevent excessive shrivelling of the bulb. Increase by seed. *B.volubilis* (CLIMBING ONION; bulb wide fleshy green; leaves narrow, dark green, usually present only in young plants; stem to 1m, slender, succulent, translucent green, twining and branching freely and intricately; flowers green-yellow, malodorous).

Boykinia (for Samuel Boykin (1786–1848), planter, physician and naturalist from Georgia, US). Saxifragaceae. North America; Japan (those below from Western N America). 9 species, hardy perennial herbs, mostly rhizomatous and hairy with rounded, long-stalked leaves and panicles of small, five-parted flowers in summer. Plant in damp, acid soils in light shade. Divide in late winter.
    B.aconitifolia (leaves 4–8cm long, rounded to kidney-shaped, lobed and toothed; stalks to 18cm long; flowers white in 15–80cm-tall panicles); *B.jamesii* (syn. *Telesonix jamesii*; differs from the last species in its smaller, toothed rather than lobed leaves and crimson flowers).

boysenberry see *hybrid berries*.

Brachycereus (from Greek *brachys*, short and *Cereus*). Cactaceae. Galapagos Islands LAVA CACTUS.1 species, *B.nesioticus*, a shrubby cactus to 60cm tall and spreading to 2m across in clumps of erect, ribbed and cylindrical, 3–5cm-wide stems covered in yellow spines. The large flowers are creamy-white and narrowly funnel-shaped. Plant in an acid, gritty and sandy, loam-based medium. Grow in full sun in a dry, airy environment, minimum temperature 10°C/50°F; water very sparingly in winter. Propagate by stem cuttings.

Brachychilum (from Greek, *brachys*, short, and *cheilos*, lip). Zingiberaceae. Java. 1 species, *B.horsfieldii* a rhizomatous, evergreen, tender perennial. To 2m tall, its slender, reed-like stems are ranked with narrow, deep green leaves and arise from a thick, aromatic rhizome. A loose, terminal inflorescence of 8cm-long flowers appears in summer; these are slender and tubular with twisted, pale gold petals. In late summer three-valved capsules split to reveal a bright orange interior encasing deep red seeds. Provide a minimum temperature 18°C/65°F and humid, shady conditions. Propagate by division in late winter.

Brachychiton (from Greek *brachys*, short, and *chiton*, outer garment, a reference to the seed coat). Sterculiaceae. Australia; Papua New Guinea. 31 species, tender, evergreen or deciduous trees and tall shrubs, often with swollen trunks, grown for their large panicles of small cup-shaped to tubular flowers produced in spring and summer. Large and woody, the beaked fruits are used by dried flower arrangers. Grow in an acid soil in full sun; minimum temperature 7°C/45°F, or cooler (–4°C/25°F) for short periods if well-established and dry. Propagate by seed or ripewood cuttings.

B.acerifolius (syn. *Sterculia acerifolia*; FLAME TREE, FLAME KURRAJONG; deciduous tree to 30m; leaves to 25cm, ovate to palmately lobed; flowers bright red or orange, on bare branches); *B.populneus* (syn. *Sterculia diversifolia*; BOTTLE TREE; KURRA-JONG; evergreen tree to 20m with a swollen trunk, poplar-like leaves to 12cm long, and clusters of small, green-white flowers).

Brachycome (from Greek *brachys,* short, and *kome*, hair, referring to the short pappus hairs). Compositae. Australasia. 70 species, annual and perennial herbs or shrublets with slender-stalked, daisy-like flowerheads in spring and summer. A tender perennial useful for edging, bedding and baskets and often treated as a half-hardy annual. Sow seed under glass in autumn, or *in situ* in spring; alternatively, take stem cuttings in early spring from stock plants cut back hard, potted up and overwintered in frost-free conditions. *B.iberidifolia* (SWAN RIVER DAISY; bushy, to 40cm with very finely cut foliage; flowerheads fragrant, 1cm-wide, sky to deep blue with a yellow centre, also white, pink, lilac and deep mauve).

Brachyglottis (from Greek *brachys*, short, and *glottis*, tongue, referring to the short ray florets). Compositae. New Zealand; Chatham Is; Tasmania. 30 species, evergreen trees, shrubs, climbers and perennial herbs, usually with grey-woolly foliage and panicles of yellow to cream, daisy-like flowerheads. Most were formerly included in the genus *Senecio*. Many *Brachyglottis* species are adapted to thrive in exposed, coastal situations, notably *B.buchananii, B.* 'Leonard Cockayne', *B.monroi, B.rotundifolia* and *B.* 'Sunshine'. The last is one of the most widely planted silver-leaved shrubs and seems tolerant of all but the wettest darkest spots. Smaller species, such as *B.bidwillii* and *B.compacta*, suit the large rock garden or silver border in a sheltered position. *B.hectoris* favours protected, semi-wooded places. All prefer full sunlight. Hardiness varies: *B.* 'Sunshine' withstands lows to –5°C/5°F with little damage. *B.monroi* will take –0°C/14°F and *B.compacta* and *B.rotundifolia* –°C/23°F. *B.kirkii,B.perdicioides* and *B.repanda* are frost-tender (minimum 3°C/37°F). Deadhead regularly. Propagate by seed or semi-ripe cuttings.

B.bidwillii (syn. *Senecio bidwillii*; compact, to 60cm; leaves to 2cm, elliptic to oblong, smooth above, silver-or buff-hairy beneath; flowerheads inconspicuous); *B.buchananii* (syn. *Senecio buchananii*; robust, to 1.5m; leaves 2.5–5cm, broadly oblong, smooth above, silver-hairy beneath; flowerheads yellow to cream); *B.compacta* (syn. *Senecio compactus*; to 1m; leaves 2–4cm, obovate to oblong, wavy-edged, white-hairy beneath; flowerheads yellow); *B.* Dunedin Group (hybrids between *B.compacta, B.greyi* and *B.laxifolia*,

shrubs to 2m tall with elliptic to oblong leaves to 5cm long, emerging silvery or white, hardening dull to lustrous sea or grey-green above, white-scurfy beneath; flowerheads bright yellow; the most commonly grown selection is 'Sunshine', syn. *Senecio greyi* of gardens, described above; 'Moira Read' has variegated leaves); *B.greyi* (to 2m; leaves 4–8cm, ovate to oblong, tough, silver-scurfy beneath; flowerheads bright yellow – the true plant is rare in gardens and the name is often misapplied to *B.* 'Sunshine'); *B.hectoris* (syn. *Senecio hectoris*; to 4m; leaves to 20cm, oblanceolate, coarsely toothed and pinnately cut at base, thinly white-hairy beneath; flowerheads white); *B.huntii* (syn. *Senecio huntii*; RAUTINI; compact shrub or tree to 6m; leaves to 10cm, lanceolate to oblong, tawny-hairy at first; flowerheads yellow); *B.kirkii* (syn. *Senecio kirkii; Senecio glastifolius*; KOHURANGI; to 3m; leaves 4–10cm long, obovate to oblong, wavy-toothed to entire, fleshy; flowerheads white); *B.laxifolia* (syn. *Senecio laxifolius*; resembles *B.greyi*, but with smaller, pointed leaves); *B.* 'Leonard Cockayne' (*B.greyi* × *B.rotundifolia*; spreading with wavy-edged, grey-green leaves to 15cm long, silver beneath; flowerheads yellow); *B.monroi* (syn. *Senecio monroi*; to 1m; leaves to 4cm, obovate to oblong, wavy and crenate, silver-felted beneath; flowerheads bright yellow); *B.perdicioides* (syn. *Senecio perdicioides*; RAUKUMARA; to 2m; leaves 2.5–5cm, oblong to elliptic, finely crenate, thin-textured, smooth, dull green; flowerheads to 1cm diam., yellow); *B.repanda* (syn. *B.rangiora; Senecio repanda*; PUKAPUKA; RANGIORA; to 6m; leaves to 25cm, broadly oblong to elliptic, with large, wavy teeth, mid to sea green above, silver-scurfy beneath; flowerheads inconspicuous, mignonette-scented; 'Purpurea': leaves purple above; 'Variegata': leaves stippled yellow, grey-green and cream); *B.rotundifolia* (syn. *Senecio reinoldii; Senecio rotundifolius*; MUT-TONBIRD SCRUB; to 6m, dense, rounded; leaves to 10cm, broadly ovate to rounded, tough, white-felted beneath; flowerheads yellow). For *B.* 'Sunshine' see above under *B.* Dunedin Group.

Brachysema (from Greek *brachys*, short, and *sema*, standard, referring to the very short standard of the flower). Leguminosae. Australia. 15 species, shrubs with clusters of pea-like flowers in spring and summer. Plant in a free-draining, loamy and sandy soil in full sun; minimum temperature 7°C/45°F; water moderately in spring and summer, scarcely at all in winter. Avoid fertilizers containing phosphorus. Propagate by seed in spring, or from semi-ripe cuttings in a closed case; also by layering. *B.celsianum* (syn. *B.lanceolatum*; SCIMITAR FLOWER, SWAN RIVER PEA; spreading to prostrate evergreen shrub to 2m, stems flattened, green and ribbon-like; leaves lanceolate; flowers scarlet to 6cm-long with a sharply curved keel).

Brachystelma (from Greek *brachys*, short, and *stelma*, garland, referring to the dense inflorescence). Asclepiadaceae. Tropical and South Africa. 100 species, perennial herbs usually with swollen, turnip-

Brachyglottis 'Sunshine'

like roots or caudices bearing short-stalked clusters of five-parted flowers. Many species are grown in succulent collections; the best-known is probably the South African *B.barbariae* with a flattened caudex to 10cm wide. The flowers are produced before the leaves in a dense, hemispherical umbel to 12cm wide. They are malodorous and a dirty purple-brown with yellow centres and tawny spots. Very slender, the 2.5cm-long petals stand erect and joined at the tips in a cage- or lantern-like structure. Plant in a gritty medium; do not bury the caudex. Place in full sunlight in dry, airy conditions with a minimum temperature of 8 °C/46°F. Water sparingly when in growth; keep dry when dormant. Propagate by seed.

bracing The provision of artificial support to tree branches, using cables or metal rods.

bracken (*Pteridium aquilinum*) an invasive fern which spreads as a weed by means of its thick, fleshy, brown underground stems. Persistent removal of frond growth will weaken the stand, and is best done by cutting in early summer and again as regrowth emerges a few weeks later. Glyphosate and glyphosate trimesium are effective herbicides against bracken. Fronds can be composted, used as a surface mulch or for insulating tender plants against winter cold.

bracket fungi (Aphyllophorales: Basidio-mycetes). The fruiting bodies of various saprophytic or parasitic fungi, which form bracket-like projections on infected trees or on deadwood. The structures may be annual (and usually produced in autumn) or perennial.

braconid wasps (Hymenoptera: Braconidae). small wasps with relatively long antennae and wings, and a distinct dark area about half-way along the front margin of the forewing. Closely related to ichneumon flies. Females parasitize many insects, particularly caterpillars of butterflies and moths, and lay eggs in their bodies. *Cotesia glomerata* is a familiar parasitoid of caterpillars of the large cabbage white butterfly (*Pieris brassicae*), and up to 150 fully fed larvae emerge from the collapsed skin of the host and pupate together in conspicuous sulphur-yellow cocoons. Several braconids species are used for biological control of pests, for example, some *Aphidius* and *Praon* species infest aphids.

bract a modified protective leaf associated with the inflorescence (clothing the stalk and subtending the flowers), with buds, and with newly emerging shoots and stems. Bracts exhibit varying degrees of reduction, from the leafy inflorescence bracts of *Euphorbia pulcherrima* to the navicular bracts of *Heliconia*, or the thin scale-like stem bracts of many herbaceous spcies. In the developing shoots of members of the last category, a progression from bract to true leaf can be seen as the shoot develops.

bracteate possessing or bearing bracts.

*Brassavola
flagellaris*

*Brassavola
cucullata*

bracteole a secondary or miniature bract, often borne on a petiole or subtending a flower or inflorescence.

bract scale in conifers, the scale of a female cone that subtends the seed-bearing scale.

Brahea (for Tycho Brahe (1546–1601), Danish astronomer and pioneer of science). Palmae. C America. (those listed below are from Mexico and Baja California, 12 species, slow-growing palms with stout, columnar trunks clothed in fibres and large, fan-shaped leaves cut into many, narrow segments. The inflorescences are large, drooping and much-branched, bearing many, small, cream flowers. The following two species thrive on parched, rocky soils in full sunlight and will tolerate several degrees of frost if dry and well-established. Increase by seed.

B.armata (BLUE HESPER PALM, MEXICAN BLUE PALM, SHORT BLUE HESPER, BLUE FAN PALM, GREY GODDESS; to 10m; foliage silver-blue; inflorescence to 4m long, very showy); *B.brandegeei* (SAN JOSÉ HESPER PALM; to 30m with pale, grey-green leaves to 1.5m long).

Brassavola (for A.M. Brasavola (1500–1555), Venetian botanist. His name is spelt with one 'S', that of the genus with two). Orchidaceae. Central and South America; West Indies. 30 species, evergreen, epiphytic and lithophytic orchids with more or less pendulous, pencil-thin pseudobulbs terminating in a single, tough, narrow leaf and solitary or clustered flowers, often nocturnally scented and in shades of white, cream and pale green; the tepals are slender, the lip is funnel-shaped. Provide a minimum temperature of 7°C/45°F, with bright light and a buoyant, humid atmosphere. Plant in baskets or small pots containing a very open, bark-based medium, or attach to rafts and bark slabs against pads of moss and leafmould. While in growth, water and feed frequently, allowing the medium to dry between drenchings; syringe daily in hot weather. Except for misting to prevent the shrivelling of leaves and pseudobulbs, withhold water for one to two months on completion of new pseudobulbs. Recommence watering when root and shoot activity resumes. Propagate by division.

B.cordata (flowers 4.5cm-wide, vanilla-scented, tepals lime green to ivory, lip bright white); *B.cucullata* (flowers fragrant with long and drooping tepals, to 7cm, very narrow and ivory, lip cream, fringed at its base, tapering narrowly at the tip); *B.flagellaris* (leaves and pseudobulbs pendulous, long and whip-like, picked out with 6cm-wide flowers, the tepals spreading and pale cream, the lip white stained lime green in the throat); *B.nodosa* (LADY OF THE NIGHT; flowers to 7.5cm wide with nodding, green, ivory or white tepals and a white lip stained green and spotted maroon at the throat). For *B.digbyana* and *B.glauca* see *Rhyncholaelia*.

Brassia (for William Brass, 18th-century botanist who collected in Guinea and South Africa). Orchidaceae. Central and South America. SPIDER

ORCHID. 25 species, tender epiphytic orchids with ovoid to cylindrical pseudobulbs, strap-shaped leaves and racemes of flowers with very long, slender tepals and a shorter, trowel-shaped lip. Pot in a very open, bark-based mix. Grow in cool to intermediate conditions (minimum winter temperature 7°C/45°F) with high humidity. Admit full sunlight except in the hottest months, when the plants should be shaded, freely syringed and ventilated. Water and feed liberally when in growth; reduce temperatures and keep almost dry once the new pseudobulbs are developed. Increase by division.

B.caudata (tepals to 18cm, pale yellow-green to yellow-orange marked red-brown, lip paler and spotted red-brown); *B.lawrenceana* (tepals to 7cm, green-yellow, spotted maroon at the base, lip white to pale lime); *B.verrucosa* (tepals to 12cm, pale lime green spotted chocolate or black at the base, lip white, spotted red-brown at the base and scattered with green warts).

brassicas a term commonly used as a collective description for the important vegetable crops Brussels sprouts, cabbage, cauliflower, kale, kohlrabi, swede and turnip, members within the genus *Brassica* of the family Cruciferae.

× **Brassocattleya** (*Brassavola* (or *Rhyncholaelia*) × *Cattleya*). Orchidaceae. Garden hybrids ranging widely in habit and size from the dwarf and tufted to the tall and rangy, with the ellipsoid to cylindrical pseudobulbs and tough oblong to elliptic leaves characteristic of *Cattleya*. The flowers are *Cattleya*-like, but have very ruffled lips; they vary from small and neat to exceptionally large and frilly, in shades of white, pink, magenta, purple, red, orange, apricot, yellow and lime green. Many grexes are available. Cultivation as for *Cattleya*.

× **Brassolaeliocattleya** (*Brassavola* (or *Rhyncholaelia*) × *Laelia* × *Cattleya*). Orchidaceae. Garden hybrids ranging in habit from dwarf to medium-sized or large, squat, robust or reedy. Flowers slender and starry to buxom and lacy in shades of white, rose, magenta, mauve, red, orange, apricot, yellow, green or cream, often flushed, veined or marked a different colour. Many grexes are available. Cultivation as for *Cattleya*.

break, breaking (1) the growth made from an axillary bud, especially in chrysanthemums and carnations, as a result of pinching out the growing point or as a natural occurrence; (2) a stage in the opening of buds; see *bud stages*; (3) a mutation or sport seen in flowers, often characterized by streaks or flecks, and possibly virus induced.

breastwood shoots growing outwards from espalier or other wall-trained trees.

Breynia (for J.P. Breyn, 17th-century German botanist). Euphorbiaceae. Tropical Asia; Pacific Islands; Australia. 25 species, evergeen shrubs and trees with small, usually 2-ranked leaves and inconspicuous flowers. Plant in any moderately fertile, soilless or loam-based mix; position in dappled sunlight or light shade; minimum temperature 7°C/45°F. Propagate by soft cuttings in early summer. *B.nivosa* (syn. *B.disticha*; SNOW BUSH; shrub to 1.5m; leaves 1.5–2cm long in two ranks giving the branches a ferny appearance, densely mottled, spattered and zoned white, often with the veins picked out in emerald green; 'Atropurpurea': leaves rich purple-red; 'Roseapicta': leaves mottled pink and red).

Brassia verrucosa

Briggsia (for Munro Briggs Scott (1889–1917), botanist) Gesneriaceae. India to China. 23 species, rhizomatous, near-hardy perennial herbs grown for their stalked cymes of tubular, two-lipped flowers produced in summer. Suitable for the alpine house or frame, or for very sheltered positions on the rock garden in zone 8. Plant in pans, raised beds or rock crevices in a neutral to acid soil rich in leafmould. Keep moist from spring to autumn. Provide full sun except where likely to scorch or dry the plants excessively (i. e. under glass in summer). Propagate from seed. *B.muscicola* (China, Bhutan and Tibet; leaves woolly in a stemless rosette; flowers pale yellow, the interior orange-yellow marked purple).

Brimeura (for Maria de Brimeur, a 16th-century Dutch aristocrat who loved flowers). Liliaceae (Hyacinthaceae). Mediterranean; SE Europe; Aegean. 2 small, perennial, bulbous herbs grown for their flowers which are produced in spring and resemble miniature bluebells (*Hyacinthoides*). Plant in full sun or light shade in any well drained soil; minimum temperature –15°C/5°F. Propagate by division of established clumps as the foliage withers in early summer. *B.amethystina* (flowers bright blue, indigo or white); *B.fastigiata* (flowers, lilac, lilac-pink or white).

brindille a thin lateral shoot of a fruit tree, 7.5–30cm long and usually terminating in a fruit bud.

Briza (Classical Greek name for a grass, probably rye). Gramineae. Cosmopolitan (mainly northern temperate regions). QUAKING GRASS. 12 species, annual or perennial hardy grasses with erect, open panicles, their branches very fine and hung with ovate spikelets, the lemmas large, papery and overlapping in two ranks giving the whole spikelet the appearance of a rattlesnake's tail. Among the most graceful of ornamental grasses, *Briza* has a pearly, translucent quality. The parchment flowerheads are used in dried flower arrangements. Grow in full sun in any moderately fertile soil, chalkland included; ample moisture is required until the flowerheads are developed, rather drier conditions thereafter. Sow seed *in situ* in early spring, or divide *B.media*.

B.maxima (syn. *B.major*; GREAT QUAKING GRASS, PUFFED WHEAT; annual to 60cm; spikelets on thread-like branches, each to 2.5cm and olive to straw-coloured tinged red-brown or purple); *B.media* (COM-

Bromelia balansae

MON QUAKING GRASS, COW QUAKES, DODDERING DICKIES, LADY'S HAIR GRASS, PEARL GRASS and TOTTER; perennial to 1m tall; spikelets to 1cm, ovate to deltoid, often tinted purple, in erect panicles to 18cm); *B.minor* (LESSER QUAKING GRASS; annual to 60cm with spikelets to 0.5cm long, remarkably airy and delicate in the cultivar 'Minima').

broadcasting the scattering of seeds evenly over a seed bed rather than sowing in drills.

broadleaf, broad-leaved a tree or shrub with broad flat leaves as opposed to needles. Commonly used to mean a non-coniferous tree.

broad mite see *tarsonemid mites*.

broccoli

Pests
aphids
cabbage
caterpillars
cabbage root fly
flea beetle
turnip gall weevil
(*see* weevils)

Diseases
clubroot
leaf spot

broccoli (*Brassica oleracea* Italica Group). a name most commonly used in the UK to describe purple and white sprouting broccoli; green sprouting broccoli is usually known as calabrese. Purple and white sprouting broccoli differ from cauliflower in their more divided leaves and the terminal and axillary flowerbuds, that are the edible part, form less dense heads.

In parts of Britain the term broccoli is sometimes incorrectly used to refer to large-headed white cauliflower (Botrytis Group), maturing in winter and which should properly be known as winter cauliflower. In North America the name broccoli refers both to calabrese, which there includes sprouting and asparagus broccoli (Italica Group), and to the large white-headed winter-flowering cauliflower (Botrytis Group) which can only be grown in favourable parts.

Raise in a seedbed or in modules under protection from mid April-late May sowings; plant out June or July 60cm apart each way. Plants may require staking. The main season is usually from February to late May, when the heads in tight bud are snapped off selectively over a period of 6–8 weeks, during which time small heads develop from side shoots. Early maturing cultivars are available. Purple forms are hardier and more prolific than white forms. See *calabrese*.

Brodiaea (for James Brodie, Scottish botanist). Liliaceae (Alliaceae). W US. 15 species, deciduous, cormous perennial herbs with narrow leaves often withering before the inflorescence expands in early summer. The flowers are funnel- or bell-shaped and carried on slender stalks in lax, scapose umbels. Hardy in zone 8. Grow in sun on light, fertile, sandy loams. Keep moist when in growth; dry and warm after flowering. Propagate by division of established colonies, removal of offsets or seed sown when ripe. The HARVEST BRODIAEAS, *B.coronaria* (syn. *B.grandiflora*, *B.howellii*) and *B.elegans* produce 2–12, 2–4cm-long, purple-blue to deep mauve flowers, the first on a scape to 30cm, the second to 50cm. *B.minor* produces pink to deep violet blue flowers to 2.5cm long on 7–30cm-tall scapes. See also *Dichelostemma* and *Triteleia*.

bromeliads members of the family Bromeliaceae, e.g. *Aechmea*, *Ananas*, *Billbergia*, *Bromelia*, *Cryptanthus*, × *Cryptbergia*, *Dyckia*, *Fascicularia*, *Guzmania*, *Neoregelia*, *Nidularium*, *Pitcairnia*, *Puya*, *Tillandsia*, *Vriesea*.

Bromelia (for Olaf Bromel (1639–1705), Swedish botanist). Bromeliaceae. Central and South America. 47 species, perennial herbs, usually terrestrials with spreading rosettes of rigid and spiny, sword-like leaves and showy flowerspikes in spring and summer. Grow in full sun, minimum temperature 7°C/45°F, in a very open medium. Water, feed and syringe liberally in warm weather; keep virtually dry in winter. Propagate by offsets. *B.balansae* (HEART OF FLAME; leaves flushing brilliant scarlet at the heart of the rosette in the flowering season; inflorescence torchlike, densely branched, to 1m, with white-edged, rosy flowers guarded by outward-curving, scarlet bracts).

Bromus (from Greek *bromos*, oat). Gramineae. BROME, CHESS. N Temperate Regions. 100 species, annual, biennial or perennial grasses with loose erect culms, narrow leaves and long-awned spikelets in graceful panicles in summer. The following are fully hardy. With the exception of the shade-loving *B.ramosus*, they will grow in most situations but prefer sun and a free-draining soil. Increase by seed or by division. The flowerheads of all except *B.madritensis* are suitable for drying.

B.briziformis (Europe, Temperate Asia; annual or biennial to 60cm, with loosely pyramidal panicles hung with *Briza*-like spikelets); *B.inermis* (Europe, Temperate Asia; creeping perennial erect to 1m; panicles drooping; 'Skinner's Gold': leaves striped gold, seed heads golden); *B.madritensis* (COMPACT BROME, STIFF BROME, WALL BROME; Mediterranean; annual erect to 60cm with very fine, feathery arching panicles tinted pink to wine-red); *B.ramosus* (WOOD BROME, HAIRY BROME; Europe, N Africa, SW Asia; clump-forming perennial, 1–3m tall with sea green foliage and fine, purple-tinted spikelets in gracefully arching panicles).

Broussonetia (for Pierre Marie August Broussonet (1761–1807), Professor of Botany at Montpellier). Moraceae. East Asia; Polynesia. 8 species, deciduous trees or large shrubs with large, entire to lobed leaves, small, male flowers in catkins and female flowers in globose heads. Tolerant of dry, poor soils, heat and pollution, *Broussonetia* will thrive where temperatures do not often fall below −5°C/25°F; elsewhere it will need a sunny, sheltered site and may be killed to the ground by frost (it suckers freely). Propagate by semi-ripe, heeled cuttings in late summer under mist or in a closed case, or from hardwood and root cuttings in winter, or by layering in spring.

B.papyrifera (PAPER MULBERRY; China and Japan; broad-crowned tree to 15m tall, or, a spreading shrub to 3m; leaves 7–20cm long, ovate, often toothed, 'tattered' or 3-lobed and variably so on the same plant; female plants produce orange-red ball-like fruits; 'Billardii': leaves are much dissected; 'Laciniata': leaves reduced sometimes to small remnants of blade hanging on the tip of the midvein; 'Variegata': leaves are splashed white or yellow; 'Leucocarpa': fruit white.

Browallia (for Johan Browall (1705–1755), Bishop of Abo, Sweden, a defender of Linnaeus' system of plant classification). Solanaceae. South America. AMETHYST FLOWER; BUSH VIOLET. 3 species, tender annual and perennial herbs or shrubs with trumpet-shaped flowers. Grow in a shaded, humid environment (minimum temperature 10°C/50°F). For summer flowers, sow seed in early spring; for winter display, sow seed in late summer or cut back, repot and feed the strongest, existing plants. Take short, soft cuttings of cultivars in spring or late summer. *Browallia* is usually treated as an annual, but may persist for some years if regularly pruned, fed and repotted, *B.speciosa* especially.

B.americana (syn. *B.elata*; bushy, annual or short-lived perennial to 60cm tall; flowers blue to violet or white to 2.5cm; includes the cultivars 'Caerulea' (flowers sky blue), 'Compacta' (habit small, compact), 'Grandiflora' (*B.grandiflora*) and 'Major' (flowers larger), and the dwarf 'Nana'); *B.speciosa* (syn. *B.gigantea*; shrubby perennial to 1m; flowers to 4cm in shades of blue and violet with a white eye; many cultivars developed in shades of white, pale blue, royal blue, indigo, violet and lilac; 'Major': SAPPHIRE FLOWER, flowers large, purple-blue); *B.viscosa* (annual to 60cm tall; flowers to 3cm violet-blue with a white eye; 'Alba': flowers white; 'Sapphire': sapphire blue).

Browningia (for W.E. Browning, Director of the Instituto Ingles at Santiago, Chile). Cactaceae. South America. 7 species, shrubby to tree-like cacti with spiny, strongly ribbed, cylindric stems and nocturnal, tubular to funnelform flowers. Large but slow-growing, this cactus is usually offered as an unbranched stem to 1m tall. Provide a minimum temperature of 10°C/50°F. Plant in a neutral to slightly acid medium high in grit and sand. Shade in hot weather and maintain low humidity. Keep cool and dry from mid-autumn to early spring except for occasional mistings on warm days. Propagate by seed or cuttings of detached branches. *B.hertlingiana* (syn. *Azureocereus hertlingianus*; Peru; erect and tree-like, ultimately to 8m; stems to 30 cm wide, glaucous blue with 18 wavy ribs; spines 5cm, tawny; flowers white to pink).

brown coal see *lignite*.

brown rot a major fungal disease of stone and pome fruit, mainly caused by three similar species of *Monilinia*. There are differences in biology and world distribution of these fungi, but the general pattern of disease development is the same. Most damage is caused to fruit while it is still on the tree, although post-harvest rotting can also be a serious problem. Blossoms of fruit trees may also be attacked by the disease, which subsequently causes cankers on fruit spurs and smaller branches, resulting in crop loss and providing a source of inoculum for further infection. Symptoms of infection on fruit are soft brown rotten areas which spread quickly, and on blossom, brown spotting of petals followed by shrivelling of the whole flower.

M.fructicola, which is native to North America and is particularly damaging to peaches and other stone fruit, is the most important of the brown rots; it now occurs also in South America, Australia and New Zealand. Infected fruits form abundant tufts of powdery ashy-grey spores, sometimes in concentric rings. European stone fruit growing could be at risk from this disease. *M. fructigena*, primarily a disease of apples and pears, is of Eurasian origin and is still mainly restricted to the Old World. It sporulates abundantly on affected fruit, forming raised masses of fungal tissue in concentric rings covered with buff-coloured spores. There is a danger of introducing this species into American and Australian orchards.

M.laxa is thought to have originated in Eurasia but now has a worldwide distribution. Its fawn-grey spores form less luxuriantly and it is mainly a disease of *Prunus*, causing BLOSSOM BLIGHT and canker rather than fruit rot; a form of this species is particularly pathogenic to apples and pears. Fruits infected before maturity shrivel on the tree and are the principal means whereby the brown rot fungus, having overwintered in them, is able to infect blossom in spring. Mummified fruits infected by *M.fructicola* give rise to the ascospore-producing sexual stage of the fungus in spring; this provides an additional opportunity for dispersal of the disease, as well as the possibility of genetic change in the fungus' virulence and fungicide tolerance. The sexual stage of the life cycle is rare in the other two species of *Monilinia*.

Control measures include: (a) reducing the over-wintered inoculum by disposing of mummified fruits, dead spurs and branches; (b) a fungicidal spray programme from blossom time where the disease is established, especially on peaches; (c) the prevention of fruit damage by insects; (d) avoiding harvesting in wet weather; (e) careful handling of fruits.

Other brown rot fungi are *M.aucupariae*, which causes a fruit rot and foliage blight of *Sorbus commixta* in Europe and Japan; *M.johnsonii*, CRATAEGUS LEAF BLOTCH and BLOSSOM BLIGHT, which is widespread; *M.kusanoi*, blight of flowering cherry in Japan; *M.linhartiana*, QUINCE FRUIT ROT in Turkey; *M.oxycocci*, CRANBERRY HARD ROT; and *M.vaccinii-corymbosi*, BLUEBERRY BROWN ROT or MUMMY BERRY, in North America. Other diseases also referred to as brown rot are (*Pseudomonas solanaceanum*) bacterial wilt of potatoes; ORCHID LEAF SPOT disease such as *Erwinia cypripedi*; and decay caused by certain wood-rotting fungi.

Bruckenthalia (for S. von Bruckenthal (1721–1803), Austrian nobleman). Ericaceae. SE Europe; Asia Minor. 1 species, *B.spiculifolia*, SPIKE HEATH, a hardy, heath-like, mat-forming shrublet to 25cm tall, with very small, dark, and narrow leaves, and erect, 3cm-long bell-shaped racemes of small magenta to candy pink flowers in late spring and summer. Plant in a moist, acid soil in full sun. Propagate by semi-ripe cuttings with a heel in summer inserted in a closed case; also by detaching rooted stems and by seed.

Brugmansia (for Sebald Justin Brugmans (1763–1819), Professor of Natural History at Leiden). Solanaceae. South America. ANGEL'S TRUMPET. 5 species shrubs or small trees. Produced in summer, the large, pendulous, trumpet-shaped flowers widen to a pleated, 5-pointed limb. The following species were formerly included in the genus *Datura*, a genus now limited to short-lived or annual shrubby herbs with very similar but less markedly pendulous flowers. Grow in full sun or light shade in a fertile, loamy medium. Water and feed generously when in growth. In winter, they should be kept frost-free and almost dry. Prune hard in early spring. Propagate by semi-ripe, heeled cuttings in late summer.

B.arborea (syn. *Datura cornigera*; to 4m; flowers to 15cm, white, only slightly pendent; this name is often misapplied to *B.* × *candida*, which has far larger, pendulous flowers); *B.aurea* (to 10m; flowers to 24cm, nocturnally fragrant, white to golden yellow); *B.* × *candida* (*B.aurea* × *B.versicolor*; fast-growing to 5m; flowers to 32cm, fragrant, creamy yellow becoming pure white, rarely pink; double-flowered, pure white forms include 'Double White' and 'Knightii'); *B.* × *insignis* (*B.suaveolens* × *B.versicolor*; resembles *B.suaveolens* but with longer flowers (to 38cm) in white, pink, salmon or apricot with narrower corolla lobes); *B.* × *rubella* (*B.arborea* × *B.sanguinea*; to 3m; flowers to 15cm, red-flushed); *B.sanguinea* (RED ANGEL'S TRUMPET; to 10m; flowers to 25cm, yellow-green at base, then yellow deepening to scarlet or orange-red at limb); *B.suaveolens* (to 5m; flowers to 30cm, strongly nocturnally fragrant, white to pale yellow or pink); *B.versicolor* (to 5m; flowers to 50cm, white sometimes becoming orange or peach; often mislabelled *B.arborea*; the popular 'Grand Marnier' with soft apricot flowers is derived from this species, as are the salmon 'Charles Grimaldi' and the cream to pink 'Frosty Pink').

Brunfelsia (for Otto Brunfels, 16th-century botanist and physician). Solanaceae. Tropical America. 40 species, shrubs and small trees. Borne in summer, the flowers are fragrant, tubular with five broad lobes. Grow in a humid, lightly shaded environment with a minimum temperature of 13°C/55°F. Plant in a slightly acid mix; water and feed generously during periods of warm weather and active growth, sparingly at other times. Propagate from greenwood cuttings in a case with bottom heat in spring and summer.

B.lactea (JAZMIN DEL MONTE; Peru; shrub to 3m tall with intensely fragrant, waxy ivory flowers to 7.5cm across); *B.pauciflora* (syn. *B.calycina*; Brazil; bushy shrub to 3m tall; flowers purple with a white eye ringed in lavender blue to 7cm across; 'Floribunda': YESTER-DAY-TODAY-AND-TOMORROW; flowers violet with small white centre, fading to lavender then white over a period of three days; 'Macrantha': probably synonymous with *B.grandiflora*, a more luxuriant shrub, with flowers to 5 × 8cm in deep purple with a white to lavender throat and also fading through lilac to white over a period of days).

Brugmansia arborea

Brussels sprouts

Pests
aphids
cabbage caterpillars
cabbage root fly
flea beetle
turnip gall weevil
    (*see* weevils)
whiteflies

Diseases
clubroot
leaf spot

Brunnera (for Samuel Brunner (1790–1844), Swiss botanist). Boraginaceae. Eastern Europe to Siberia. 3 species, hardy, rhizomatous perennial herbs with broad, coarsely hairy leaves and, in spring and early summer, erect, terminal panicles of small forget-me-not-like flowers. Plant in a humus-rich, moist soil in dappled sun or full shade with some protection from harsh winds. Propagate by division in autumn or root cuttings in early spring. *B.macrophylla* (to 50cm; leaves ovate to cordate to 15cm long; flowers clear blue; 'Dawson's White' and 'Variegata': leaves variegated white; 'Hadspen Cream': leaves splashed cream; 'Aluminium Spot' and 'Langtrees': leaves spotted silver; 'Betsy Baring': flowers white).

Brunsvigia (for Karl Wilhelm Ferdinand, Duke of Brunswick-Luneburg (1713–1780), patron of the arts and sciences). Amaryllidaceae. South Africa. 20 species, deciduous perennial herbs with large bulbs and broadly strap-shaped leaves to 90 cm long emerging after the flowers. Produced in late summer, the flowers are funnel-shaped and carried in spherical umbels on stout, red-tinted scapes. After flowering, the individual flower stalks extend after the fashion of *Boophone*. Cultivation as for *Boophone*, although these bulbs have proved slightly more hardy, sometimes surviving in very sunny, sheltered positions outside in zone 8, protected in winter with a deep mulch or framelights.

B.josephinae (syn. *B.gigantea*; JOSEPHINE'S LILY; scape to 90cm tall with many, 6cm-long, crimson flowers each on slender, ray-like stalk); *B.orientalis* (syn. *B.multiflora*; differs in its hairy rather than glabrous leaves, its shorter scape and flower stalks and the flowers to 7cm long and pink or scarlet).

brushwood killer a herbicide used on woody perennial weeds.

Brussels sprouts (*Brassica oleracea* Gemmifera Group) a tall biennial grown mostly as a winter vegetable for its small cabbage-like axillary buds. It has a loose head of large leaves which may also be harvested like cabbage. It is widely grown in Britain, and mainly in California within the US.

Sow 1–2cm deep in a seedbed from mid March to mid April, transplanting to 60cm each way from mid May to the end of June. Stopping the plants by removing the growing tip was once recommended for advancing maturity and improving sprout quality but is no longer necessary because of the improvement in cultivars, particularly F_1 hybrids.

brutting a partial breaking of hazelnut or fruit tree shoots, leaving the ends hanging, to prevent late summer growth. See *hazelnuts*.

bryobia mite see *red spider mites*.

bryophytes small multicellular lowly plants, belonging to the division Bryophyta; mostly living on land and represented by mosses and liverworts.

bubbler a water-feature device whereby a low pressure jet of water is propelled through the centre of a millstone, cobblestones etc. to create a gentle flow.

buccae the lateral sepals (wings) in the flowers of *Aconitum* and allies.

bud blast (*Pycnostysanus azaleae*) a fungal disease of *Rhododendron* species characterized by drying and silvering of buds. It is associated with the rhododendron leafhopper. The removal of infected buds is a useful control measure, but only practicable on a small scale. The incidence of the disease can be limited by the use of recommended insecticides, applied at 10–14 day intervals from early August, to control the leafhopper.

bud burst see *bud stages*.

bud drop the shedding of flower buds, particularly by tuberous begonias, lupins, sweet peas, runner beans and tomatoes. It is most often caused by low night temperature or waterlogging, or draught under glass.

budding a technique of *grafting* (q.v.) using a single bud, for propagating a wide range of woody plants.

budding knife a knife designed for bud-grafting, commonly termed budding. It is usually made with a part of the blade or handle modified for lifting bark.

Buddleja (for Adam Buddle (1600–1715), amateur botanist and cleric). Loganiaceae. Asia; America; Africa. 100 species, evergreen or deciduous shrubs and trees, usually downy or felty, with narrow, pith-filled branches, lanceolate to ovate leaves and panicles of 4-lobed, tubular flowers from late spring to autumn.
    B.alternifolia, B.davidii, B.globosa, B. 'Lochinch', *B.* 'West Hill' and *B.* × *weyeriana* are hardy to −15°C/5°F and, if cut to the ground, will usually regenerate. *B.colvilei, B.crispa, B.fallowiana* and *B.lindleyana* are usually damaged by temperatures below −10°C/14°F and grow best against a sheltered, south-facing wall in zones 7 and 8. *B.asiatica, B.* × *lewisiana* and *B.madagascariensis* are frost-tender and should be grown in cool greenhouses and conservatories. Grow all in fertile, well-drained soils in full sun and with protection from strong winds. Plant tender species in tubs or borders, ideally near a wall to provide support for their wand-like branches and with good ventilation and abundant water supplies except in cool weather.
    Much is made of the differing pruning requirements of buddlejas. If the plants are growing and flowering well, there is really little need to prune them; since they do, however, produce a good deal of short-lived wood and are easily damaged by freezing weather, the following guidelines may be of use – prune *B.asiatica, B.crispa, B.davidii, B.fallowiana, B.* × *lewisiana, B.lindleyana, B.* 'Lochinch', *B.madagascariensis* and *B.* 'West Hill' in spring, cutting back either to near ground level oʀ to within 5cm of a permanent woody framework established in the first few years after

Buddleja globosa   *B. alternifolia*   *B. davidii*

planting by cutting back selected stems to half their length. *B.alternifolia* and *B.colvilei* flower on the previous season's growth and should be pruned after flowering: cut back old, flowered shoots to the desired shape and remove dead, diseased and congested branches. *B.globosa* and *B.* × *weyeriana* flower on growth which emerges from terminal buds on the previous year's wood: unless a major assault is needed, remove only weak or exhausted branches in early spring. Propagate by heeled, semi-ripe cuttings in the cold frame, or, *B.alternifolia,* hardwood cuttings in autumn. Pinch out new plants to encourage branching.
    B.alternifolia (China; deciduous shrub or small tree to 9m; branches slender, weeping; leaves narrow, dark green, to 7cm long, silvery in the cultivar 'Argentea'; flowers fragrant, lilac-purple flowers to 1cm long, in dense clusters on the previous year's wood); *B.asiatica* (E Asia; evergreen, winter-flowering shrub or small tree; branches long, spreading and woolly; leaves narrowly lanceolate, with white-felted undersides; flowers white, fragrant, to 0.6cm long, in dense, drooping, narrow panicles); *B.colvilei* (Himalayas; summer-flowering evergreen, shrub or small tree to 6m tall with large, dark green leaves; flowers 2.5cm wide, showy, bell-shaped, gathered in lax, drooping trusses to 20cm long, rose-purple to crimson or maroon or deep red in the cultivar 'Kewensis'); *B.crispa* (Northern India; bushy, deciduous shrub to 3m tall with thickly white-felted braches and grey-woolly, wavy-toothed, triangular to ovate-lanceolate leaves; flowers fragrant, lilac with orange or white throats are in bluntly pyramidal panicles); *B.davidii* (BUTTERFLY BUSH, SUMMER LILAC; China, Japan; evergreen, spreading shrub or tree to 4m tall; leaves lanceolate, dull to dark green above and felty beneath; flowers fragrant, 1cm-long, in dense, narrowly conical panicles; typically deep lilac with an orange-yellow throat, there are many cultivars ranging in colour through white ('Peace','White Bouquet', 'White Cloud', 'White Profusion') to pink ('Charming', 'Pink Delight', 'Pink Pearl'), magenta, ('Dartmoor'), lilac ('Fascinating', 'Fortune'), mauve ('Amplissima', 'Veitchiana'), darkest purple ('Black Knight', 'Dubonnet'), violet ('Empire Blue', 'Fromow's Purple', 'Ile de France', 'Magnifica') and purple-red ('Border Beauty', 'Royal Red'), usually with an orange-yellow throat; in 'Harlequin', the

leaves are variegated ivory and the flowers a deep purple-red; variety *nanhoensis*: low-growing, elegant shrub with slender branches, smaller and narrower leaves and panicles of fine blue-mauve flowers (white in 'Alba') includes 'Nanho Blue' with powder blue flowers and 'Nanho Purple' with orange-eyed, violet flowers); *B.fallowiana* (Burma, Western China; robust, deciduous shrub to 5m tall with stout, white-woolly branches and crenate, ovate to lanceolate leaves grey above with white-woolly undersides; flowers fragrant, 1cm-long, lavender or white with orange throats; crossed with *B.davidii*, producing 'Lochinch', a grey-downy, vigorous, bushy shrub with fragrant,violet flowers with a deep orange eye, and 'West Hill' with narrow, arching panicles of pale lilac flowers with a gold to bronze eye, both flowering in late summer and autumn); *B.globosa* (ORANGE BALL TREE; Chile, Argentina, Peru; semi-evergreen erect shrub or tree to 6m; leaves wrinkled, glossy, dark green; flowers honey-scented, yellow-orange in tight, 2cm-wide balls); *B. × lewisiana* (*B.asiatica* × *B.madagascariensis*; includes 'Margaret Pike', a large shrub with long, felty branches and dense racemes of pale yellow flowers in winter); *B.lindleyana* (evergreen shrub to 2m with slender, arching branches, pale green, elliptic to lanceolate leaves to 12cm long and narrow spikes of curving, 1.5cm-long, purple-violet flowers in summer); *B.madagascariensis* (Madagascar; evergreen sprawling to semi-scandent shrub producing 25 cm-long panicles of orange-yellow flowers in winter); *B. × weyeriana* (*B.globosa* × *B.davidii*; intermediate between parents; includes 'Golden Glow', orange-yellow flushed lilac, and 'Sungold', deep orange).

bud scales the coverings of a bud.

bud stages the phases through which flower buds and flowers progress. These have established descriptions in fruit plants, for example, breaking, bud burst, and have relevance to the timing of sprays or other treatments for the control of pests and diseases.

budwood, bud stick a plant shoot from which buds are obtained for budding.

Buglossoides (*Buglossum* and *-oides*, 'resembling'). Boraginaceae. Europe; Asia. 15 species, hardy, bristly annual or perennial herbs with erect to decumbent stems and cymes of 5-lobed, blue or white flowers in spring and summer. Grow on a moist but well-drained, neutral to alkaline soil in part-shade. Mulch in very cold areas. Propagate by division. *B.purpureocaerulea* (syn. *Lithospermum purpureocaeruleum*; W Europe to Iran; rhizomatous perennial, 25–70cm tall with narrow leaves to 8cm long; flowers to 1.5cm, emerging red-purple, becoming bright blue).

bulb a modified bud, usually subterranean, consisting of a short thickened stem, serving as a storage organ. There are two principal kinds, (1) naked, composed of free, overlapping scaly modified leaves, all of them fleshy, for example, *Lilium*; (2) tunicated, with thin,

large narcissus fly and larva

small narcissus fly and larva

bulbs (clockwise from top right) Cardiocrinum; Fritillaria; Notholirion; Hyacinthus; Lilium *(stoloniferous);* Narcissus; Lilium *(non-stoloniferous)*

membranous, fibrous or reticulated outer and fleshy concentric inner layers, for example, *Allium*. The 'solid bulb' is a corm.

bulb fibre a growing medium for bulbs in containers, traditionally made from 6 parts peat, 2 parts crushed oyster shell and 1 part crushed charcoal, by volume.

bulb flies (Diptera: Syrphidae) species of plant-feeding hoverflies, of which the most injurious is the LARGE NARCISSUS FLY (*Merodon equestris*). First recorded in England in 1869, it is thought to have originated in southern Europe and has now spread to many parts of the world, including America. Its principal host is *Narcissus*, but it also attacks other members of Amaryllidaceae, and some bulbs of other families. The adult is a stout hairy fly, about 15mm long, resembling a small bumble bee, but with one pair of wings and short antennae. Fertilized females produce about 40 eggs, normally deposited singly on the neck of host plants. The legless grub crawls down the outside of the bulb, enters through the basal plate and eats out the central portion of the bulb. When fully fed, the cream-white maggots are up to 18mm long. The bulbs are usually killed, although daughter bulbs often remain unharmed and produce thin leaves known as 'grass'.

SMALL NARCISSUS FLIES (*Eumerus* species) are widespread in Europe and North America. They are smaller, about 6mm long, with three pairs of crescent-shaped, white markings on a black, hairless abdomen. The legless larvae are 8mm long when fully fed, and there are usually several per bulb. In most cases damage by *Eumerus* species is secondary.

Infested bulbs should be removed and burnt. Fill in the holes left at the soil surface as the foliage begins to die down to deter flies from reaching the necks of the bulbs. Dusting the soil surface with contact insecticide at the same time is helpful. Covering bulbs in June-early July with a fine-mesh floating mulch will exclude egg-laying females.

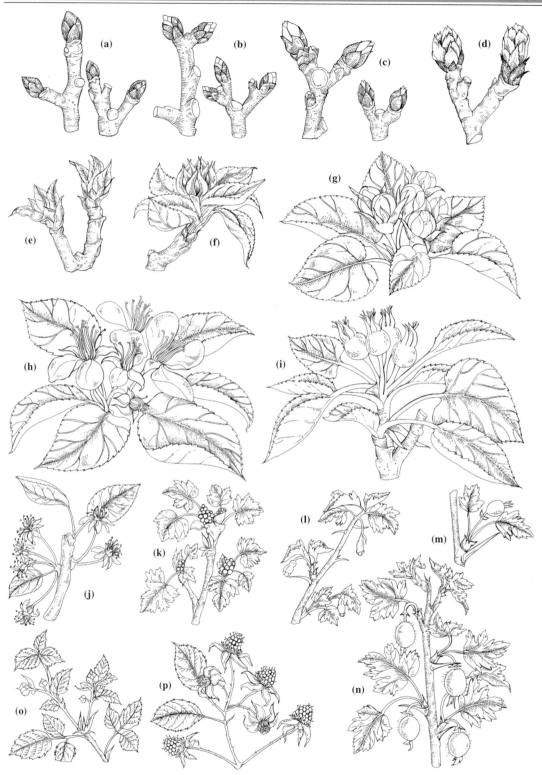

Bud Stages. Apple, Pear, Plum and Cherry: (a) *Dormant*, no sign of growth (b) *Swelling*, bud scales just beginning to part (c) *Breaking*, bud scales opening (d) *Burst*, leaf tips apparent at top of the bud (e) *Mouse ear*, leaves fully defined (f) *Greencluster*, small green flower buds visible amongst expanding leaf clusters (g) *Pink bud*, unopened flowers within pink coloured buds (= *White bud* in Pear, Plum and Cherry) (h) *Petal fall,* most flowering over (i) *Fruitlet*, small swelling fruits visible (= (j) *Cot split* in plums, accompanied by a splitting off of the yellow calyx). **Blackcurrant:** *Dormant, Breaking, Burst*, similar to Apple, etc. (k) *Grape*, leaves about 25mm across, flower clusters in unopened green bud, reminiscent of bunches of grapes. **Gooseberry:** *Dormant, Breaking*, similar to Apple etc. (l) *Early flower*, flower buds just showing (m) *Fruit set*, flowers fully open (n) *Fruit swelling*, three weeks after fruit set. **Raspberry:** *Dormant, Burst*, similar to Apple etc. (o) *Greenbud*, leaves opened and green flower buds showing (p) *Fruitlet,* green fruits swelling.

bulb frame an all-glass frame used by specialist growers to provide a dry environment for bulbs in their resting period.

bulbiferous bearing bulbs or bulbils.

bulbiform in the shape of, or resembling, a bulb.

bulbil, bulblet a small bulb or bulb-like growth, arising from a leaf axil or in place of flowers in an inflorescence.

bulbiliferous bearing bulbils.

Bulbocodium (from Greek *bolbos*, bulb, and *kodya*, capsule). Liliaceae (Colchicaceae). S and E Europe. 2 species, hardy, perennial, cormous herbs. Apparently stemless flowers with 6, erect to spreading oblong to lanceolate tepals emerge in early spring before or with narrow leaves. Cultivate as for the smaller *Colchicum* species. *B.vernum* (flowers 1–3, 5cm-wide with spreading, rose to magenta tepals; *B.versicolor* differs in its smaller flowers).

Bulbophyllum (from Greek *bylbos*, bulb, and *phyllon*, leaf, referring to the pseudobulbs which usually bear a single leaf). Orchidaceae. Tropics and subtropics. 1,200+ species, evergreen, epiphytic and lithophytic orchids, small plants with creeping rhizomes carrying rounded to ovoid pseudobulbs each terminating in single or paired, oblong to elliptic, thinly fleshy leaves. Produced throughout the year, the flowers vary greatly in size, colour and shape – they may be solitary or borne on racemes or slender-stalked umbels. The dorsal sepal and petals are usually short, ovate to lanceolate and can be fringed or tufted with hairs and ribbons, the lateral sepals longer, linear to lanceolate and often twisted downwards and together; the lip is small, fleshy and sometimes hinged.

Grow in an open, bark-based mix in small, shallow pots or baskets, or attach to rafts, branches and cork slabs against pads of moss. Provide bright, humid conditions (minimum temperature 10°C/50°F). Water and mist freely when in growth; at other times, merely mist to prevent shrivelling of pseudobulbs. Propagate by division.

B.barbigerum (W and C Africa; flowers to 2cm across, several in erect racemes, tepals red-green, lip deep burgundy, long and projecting forwards with a dense beard of glossy hairs); *B.guttulatum* (India; flowers to 4cm across, several in an umbel, tepals broad, yellow or pale green speckled purple, lip short, pale purple; *B.umbellatum* differs in larger flowers in cream or ivory overlaid with fine lines of rose-purple spots, lip rose-purple); *B.lobbii* (India to Philippines; flowers solitary, to 10cm across, waxy, red to dull ochre or pale yellow sometimes speckled or veined red-brown, lateral sepals large and curving outwards like horns or handle-bars). *B.longiflorum* (Africa, Asia, Australasia, Polynesia; flowers produced very freely, to 2cm long in a semi-circular umbel, dorsal sepal and petals small, pink to brown or olive speckled deep red, tipped with maroon hairs, lateral sepals larger, cream to olive, striped or speckled rose, wine red or maroon, curving forward and together in a shoe-like form, lip very small, maroon); *B.longissimum* (SE Asia; flowers large in an umbel, dorsal sepal and petals short, cream lined purple-red, lateral sepals to 20cm long, white to cream striped rosy purple, curving inwards and together, tapering very finely, lip short, tongue-like); *B.medusae* (Southeast Asia; flowers in a dense umbel, white to ivory lightly spotted rose or maroon at the centre with very long (to 15cm), finely tapering lateral sepals drooping in a ghostly mop).

bulb planter a tool consisting of a hollow cylinder on a handle; used to remove plugs of soil prior to planting bulbs. It is sometimes called a transplanter or, in the US, a turf plugger.

bulb scale a component of a bulb.

bulb scale mite see *tarsonemid mites*.

bulb scaling a method of propagating bulbs, such as lilies, from the outer scales of a mother bulb; see *cuttings*.

bulky organic high-volume animal or plant waste, usually low in nutrient content, used to maintain or improve fertility and moisture holding in soils through incorporation or surface mulching.

bullace (*Prunus institia*) a type of plum producing small rounded fruits varying in colour from almost black through red to pale green-yellow. Native to Britain and across Europe to Asia Minor; closely allied to the damson but with more rounded fruits that are not as distinctively purple. Plums and damsons have so evolved that they have surpassed the bullace in cultivation, and it is now most widely represented by its near relative the small-fruited Mirabelle plum of France. Notable cultivars are 'White Bullace' and 'Mirabelle de Nancy'.

bullate where the surface of an organ (usually a leaf) is 'blistered' or puckered (i.e. with many superficial interveinal convexities).

bumble bees (*Bombus* species) widely distributed, fat-bodied, hairy insects up to 25mm long, with membranous wings and waisted abdomens. Females have heavily fringed hind tibiae for carrying pollen. They are basically dark in colour, with horizontal bands or patches of yellow or orange, and sometimes white hairs. They are social insects, building annual colonies in underground cavities, and are valuable pollinators. Unlike hive bees, they are able to forage in less favourable weather conditions. All bumble bees are able to pollinate flowers with a short corolla but only those with long tongues, such as *Bombus hortorum*, can pollinate flowers with long corolla tubes. Short-tongued bumble bees often bite

bullace

Pests
aphids

Diseases
bacterial canker
sliver leaf

Bulbophyllum medusae

a hole at the base of the corolla of such plants as beans, sweet peas, antirrhinums and aquilegias to obtain nectar. The timing of pesticide sprays should take account of bumble bee activity.

Buphthalmum (from Greek *bous*, ox, and *opthalmos*, eye, referring to the wide, eye-like flowerheads). Compositae. Europe; W Asia. 2 species, hardy, perennial herbs grown for their daisy-like flowerheads produced in summer. Plant in medium- or low-fertility soils in full sun in the herbaceous border or wild garden. Stake in exposed or over-fertile conditions. Propagate by division. *B.salicifolium* (OX EYE DAISY, YELLOW OX EYE; forming a clump of metre-tall willowy stems with slender, 10cm-long, oblong to lanceolate leaves; flowerheads 5–7cm wide, typically golden yellow).

Bupleurum (from Greek *boupleuros*, 'ox-rib', a name given to another plant). Umbelliferae. Europe; Asia; Northern US; Canary Islands; South Africa. HARE'S EAR; THOROW-WAX. 100 species, annual and perennial herbs and shrubs with entire, leathery leaves and, in summer, small yellow flowers in slender-rayed, compound umbels. Sow seed of herbaceous species in spring on well-drained soils in sunny borders, meadow gardens and the fringes of woodlands and shrubberies. Thorow-wax will naturalize itself in favourable locations. *B.falcatum* and *B.petraeum* are suitable for sunny banks, large rock gardens and dry borders on very well-drained soils in zones 7 and over. They suffer in cold, damp conditions. Propagate by division or seed in early spring. In a sheltered, sunny position, the shrubby *B.fruticosum* will survive temperatures as low as –15°C/5°F. It tolerates poor, rather dry soils and is an excellent evergreen for training on walls and allowing to trail over walls and banks. It grows well on the coast. Prune out exhausted or frost-damaged growth in spring. Propagate by semi-ripe cuttings in summer.
  B.falcatum (SICKLE-LEAVED HARE'S EAR; woody-based, metre-high perennial with scimitar-like leaves to 10cm long); *B.fruticosum* (SHRUBBY HARE'S EAR; slender-branched, evergreen shrub, to 2 × 2m, with erect to sprawling branches clothed with dull or sea green, oblanceolate leaves to 8cm long with grey undersides); *B.petraeum* (ROCK HARE'S EAR; woody-based perennial to 30cm tall, making low rosettes of narrow, glaucous, 20cm-long leaves); *B.rotundifolium* (THOROW-WAX; annual or short-lived perennial to 60cm tall; leaves rounded, clasping to perfoliate and pink-tinted at first becoming glaucous sage green; flowerheads lime green to gold; 'Green Gold': bright green leaves, bracts lime, flowers golden yellow).

Burgundy mixture a fungicide prepared in a similar way to Bordeaux mixture, except that sodium carbonate is used instead of calcium hydroxide, giving a precipitate of basic copper carbonate. It is not widely used. See *Bordeaux mixture*.

burlap see *hessian*.

burlapped see *balled and wrapped*.

burr, bur a prickly fruit, usually adapted for distribution on the coats or feet of animals.

bush (1) a low shrub with branches arising initially from ground level or from a short leg; (2) used of tomatoes which have a determinate habit.

bushel An outmoded measure of volume equivalent to approximately $0.04m^3$. Horticulturally significant in the formulae of John Innes Seed and Potting composts. A bulk of about one bushel is contained in a box of dimensions 25cm × 25cm × 55cm.

bush fruit a group description of currants, gooseberries and blueberries. c.f. *cane fruit*.

bush tree a trained form of top fruits, usually having a stem 60–90cm high with the main branches arising fairly close together to form the head.

Butia (the vernacular name in Brazil). Palmae. S America. YATAY PALM; JELLY PALM; PINDO PALM. 12 species, palms with short to columnar trunks, large pinnate leaves and spikes of small yellow flowers followed by egg-shaped, edible fruit. Plant on any well-drained soil in full sun or light shade with plentiful water. Once established, it will take periods of drought and low temperatures to -2°C/28°F. Propagate by seed. *B.capitata* (to 6m tall with a stout trunk armoured with husky petiole bases; leaves 1–2m long and erect to strongly arching, segments 75 cm-long, glaucous, grey-green; fruit yellow to bright red).

Butomus (from Greek *bous*, ox, and *temno*, to cut, a reference to the sharp leaf margins which render this plant unusable as cattle fodder). Butomaceae. Europe; Asia. 1 species, *B.umbellatus* (FLOWERING RUSH, WATER GLADIOLUS, GRASSY RUSH), a hardy, summer-flowering deciduous, aquatic perennial to 1m tall with rush-like leaves and scapose, umbels of slender-stalked, rosy pink flowers each to 3cm diam. Pot in a rich, sandy and loamy mix and cover with up to 25cm depth of water in an open, sunny pond, lake or a half barrel. Propagate by division in spring.

butterflies (Lepidoptera). Butterflies generally differ from moths in having clubbed antennae and slender bodies, and in their diurnal activity, although these distinctions are not definitive. Most are welcome visitors to gardens with the exception of CABBAGE WHITE BUTTERFLIES (*Pieris* species), the larvae of which cause extensive damage to the foliage of brassicas and garden nasturtiums (*Tropaeolum* species) (See *cabbage caterpillars*). Butterflies contribute to pollination whilst visiting flowers as nectar feeders and find *Buddleja* species and *Sedum spectabile* particularly attractive. Caterpillars of the RED ADMIRAL (*Vanessa atalanta*) and TORTOISESHELL (*Aglais urticae*) feed upon stinging nettle (*Urtica dioica*).

bush tree

C

cabbage a collective name for several types of plant within the genus *Brassica* grown as vegetables for their edible leaves.

CABBAGE (*Brassica oleracea* Capitata Group) probably originated along the Mediterranean coast, where wild forms still occur, and was known to the ancient Greeks for its medicinal properties. A biennial grown as an annual, it forms round or pointed compact heads; the leaves are green or red, smooth or wrinkled. Cabbage is the most widely cultivated of the brassicas in the UK, but selection and breeding have led to types suited to conditions worldwide. In temperate climates, fresh harvest is possible year round by growing cultivars of different maturity groups and with successional sowing.

Cultivars are conveniently grouped according to season of maturity, but there is much overlap: spring cabbage (April–May, often harvested before heading as greens); summer cabbage (June–August, including red cabbage); autumn cabbage (September–October); winter cabbage (November–March, including January King type, very hardy savoys with crinkly leaves, and Dutch winter white). All have similar cultivation requirements and are usually sown 1–2cm deep (in July–August for spring cabbage, March–May for the other crops) in rows 15cm apart. Transplant to 10–40cm final spacings in rows 30–60cm apart, as recommended for the chosen cultivar. Raising in modules under protection maximises use of a growing site and allows for manipulation of harvest date.

CHINESE CABBAGE (CHINESE LEAVES, PEKING CABBAGE) (*Brassica rapa* Pekinensis Group) is a biennial grown as an annual. It has large, deeply veined leaves with a prominent, edible midrib, forming a dense, compact head. The typical barrel type is 20–25cm high, with tight heads; the cylindrical type is 38–45cm high, and greener, with a loose habit. Most cultivars are best sown *in situ* in early summer for maturing in early autumn, or raised earlier in modules under protection for planting out. They can be grown under cover to extend the season. Space 30–40cm apart each way. Chinese cabbage has a tendency to bolt where raised at low temperature, and where transplanted or deprived of water.

CHINESE WHITE CABBAGE (PAK-CHOI) (*Brassica rapa* Chinensis Group) is a biennial grown as an annual, closely related to Chinese cabbage. It is upright and does not form a compact head; the leaves are spoon-shaped with thick, broad, white stems. There are many different cultivars, some 10–15cm high, which should be closely planted. They are tolerant of a wide range of climatic conditions, but must have a regular supply of water.

PORTUGUESE CABBAGE (*Brassica oleracea* Tronchuda Group) is an annual non-heading, transitional type between kale and ordinary cabbage; the leaves are large with prominent white midribs. Grow as for cabbage, raising from a seedbed in spring; plant 50–60cm square.

Other *Brassica* species, sometimes referred to as cabbages, include TEXSEL GREENS (ABYSSINIAN or ETHIOPIAN CABBAGE) (*Brassica carinata*), fast-growing, with shiny leaves with a flavour reminiscent of spinach. The young leaves and stems are used as a cooked vegetable, the flower buds and young leaves eaten raw, and the seeds crushed for use as a condiment. Sow *in situ* 13mm deep, successively from spring to late summer, in rows 30cm apart; thin to stand 2.5cm apart. Alternatively, seed may be broadcast, or sown for transplanting 50cm square. Plants will regenerate if cut, and may be grown under cover to extend the season. See *oriental greens*.

cabbage caterpillars the larvae of various pests that commonly attack all leafy brassicas. The CABBAGE MOTH (*Mamestra brassicae*) has green or brown larvae up to 35mm long; there is one generation, active from July to September. The caterpillars often bore into cabbage hearts, but also feed on other vegetable and ornamental plant foliage. They pupate in the soil. The LARGE CABBAGE WHITE BUTTERFLY (*Pieris brassicae*) has hairy yellow larvae up to 45mm long, with black markings; there are two generations, active from June to July and August to October. They feed in groups on outer leaves at first, but later tend to separate. The SMALL CABBAGE WHITE BUTTERFLY (*Pieris rapae*) has green larvae up to 30mm long, with a velvety covering of hairs. There are two generations, active from May to June and from July to September, and the caterpillars often bore into cabbage hearts. Butterflies of both species also attack other Cruciferae, and pupate

cabbage

Pests
aphids
cabbage caterpillars
cabbage root fly
whiteflies
flea beetle
turnip gall weevil
 (see weevils)

Diseases
clubroot
downy mildew

small cabbage white butterfly

on plant stems, fences and in sheltered places. Caterpillars of the DIAMOND-BACK MOTH (*Plutella xylostella*) are often troublesome.

Inspect plants regularly for infestation. Control by hand-picking. Alternatively, use a bacterial spray of *Bacillus thuringiensis*, or a chemical insecticide. Caterpillars of *P.brassicae* are sometimes parasitized by the braconid wasp (*Cotesia glomerata*).

cabbage root fly (in the US, CABBAGE MAGGOT) (*Delia radicum*) a widely distributed fly, similar in size to a house fly, the larvae of which are destructive pests of cabbage, calabrese, cauliflower, Brussels sprouts, radish, turnip, swede and kohlrabi, the ornamentals *Matthiola*, *Alyssum*, *Arabis*, *Draba* and *Lunaria*, and cruciferous weeds. The legless white larvae, which grow up to 8mm in length, feed on roots near the taproot causing collapse; they may tunnel through leaf midribs and the buttons of Brussels sprouts. There are two or three generations per year in temperate regions, four in warmer zones. The off-white, oval eggs are deposited at the base of the host, just below the soil surface; they hatch in 3–7 days.

Control by placing 10cm-diameter discs of thin, durable material or proprietary mats around stem bases; alternatively, dust seed rows, or the soil around the base of transplants, with a recommended insecticide.

Cabomba (native name in Guiana). Cabombaceae. Tropical and subtropical America. 7 species, aquatic perennial herbs with clumps of long, submerged to floating stems and cleft to finely divided leaves in whorls or pairs. Grow in pools or freshwater aquaria where the temperature does not fall below 5°C/40°F. Propagate by stem cuttings inserted into the pool or tank bed in spring and summer. *C.caroliniana* (FANWORT, FISH GRASS, WASHINGTON GRASS; SE US; leaves 2–7cm, bright green, fan-shaped, divided into thread-like segments).

cactus (1) any member of the family Cactaceae, which is the best known of over 40 families belonging to the range of plants known as succulents. Cacti are natives of hot, dry regions of the New World, with the one exception of *Rhipsalis baccata*, which also occurs spontaneously in Africa, Madagascar, and Sri Lanka. They are predominantly of cylindrical or globose habit and bear spines, but a few genera, such as *Schlumbergera*, have spineless leaf-like stems while the genus *Pereskia* has barely succulent stems and large, 'leafy' leaves. Apart from flower characters, they most are reliably distinguished from other succulent plants by cushion-like growths called areoles, from which branches, small, succulent leaves, spines and flowers emerge; (2) a class of dahlias. See *dahlia*.

Caesalpinia (for Andrea Cesalpini (1519–1603), Italian botanist, philosopher and physician to Pope Clement VIII, author of *De Plantis* (1583) and other botanical works). Leguminosae. Tropics and subtropics. 70 species, trees, shrubs and perennial herbs, some climbing, some thorny. The bipinnate leaves are composed of many small leaflets; they persist in warm, humid conditions, but usually fall in cool, dry winters. In spring, summer, or throughout the year, the showy flowers are borne in terminal racemes and panicles and consist of five, rounded petals and a cluster of long, protruding stamens.

With the exception of *C.decapetala* var. *japonica*, these shrubs are tender. Killed to the ground by frost, they may regenerate if dry, mulched and sheltered, and might be attempted outdoors in very favoured locations in zones 8 and 9. Otherwise, grow under glass in large tubs or borders with a minimum winter temperature of 7°C/45°F. Water and feed generously in spring and summer; keep almost dry in winter. Outdoors, plant on well-drained soils in full sun and away from strong winds. *C.decapetala* needs to be trained against a wall or pergola; the other species can be used as free-standing specimens, wall shrubs, screens or stooled back. Cut down in spring, *C.pulcherrima* responds quickly in hot summers, with several, unbranched and free-flowering shoots from a single rootstock. Propagate by seed, pre-soaked in warm water for 24 hours, or by softwood cuttings in a case with bottom heat.

C.decapetala (MYSORE THORN; prickly climber or shrub to 3m tall; flowers pale yellow, red-spotted; stamens slender, downy, pink or red; includes var. *japonica*, with 4cm-wide, golden flowers and bright scarlet stamens); *C.gilliesii* (syn. *Poinciana gilliesii*; POINCIANA, BIRD-OF-PARADISE; angular shrub or small tree to 4m; flowers bright yellow; stamens scarlet, 7cm long); *C.mexicana* (MEXICAN BIRD-OF-PARADISE; thornless shrub or small tree to 10m; flowers in dense clusters 3cm wide, petals pale gold marked orange-red at the base; stamens yellow); *C.pulcherrima* (FLAMBOYANT TREE, BARBADOS PRIDE, PARADISE FLOWER, PEACOCK FLOWER BAR-BADOS FLOWER FENCE; shrub or small tree to 10 × 10m; flowers 4cm wide in apricot, orange and flame; stamens bright red, to 8cm long). Royal poinciana is *Delonix regia*.

caespitose tufted, growing in small, dense clumps.

caisse de Versailles a large wooden container for a tender tree, intended to be overwintered in an orangery and moved outside in summer.

caked see *cap*.

calabrese (*Brassica oleracea* Italica Group) a low-growing brassica up to 60cm tall, more dwarf than purple or white sprouting broccoli and grown for the edible blue-green terminal and axillary flowerheads, which are harvested in bud. In the UK, it is regarded as distinct from broccoli. It is best sown *in situ*, successively from late March to early July, for summer–autumn harvest; thin to 15cm spacings in rows 30cm apart, or raise in modules under protection. Late summer sowings may be overwintered for spring maturing under protection, or grown outdoors in areas with very mild winters. See *broccoli*.

cactus

Pests
fungus gnats
mealybug
red spider mite
scale insects
vine weevil

Disease
rots

calabrese

Pests
aphids
cabbage caterpillars
cabbage root fly
flea beetle

Diseases
clubroot
leaf spot

Caladium (Latinized form of the Malay name *kaladi*). Araceae. ANGEL'S WINGS,ELEPHANT'S-EAR. Tropical S America. 7 species, tuberous perennial herbs with large, slender-stalked leaves, often beautifully veined and patterned, and dull green, *Arum*-like inflorescences. Used in frost-free climates for year-round bedding, elsewhere for summer bedding and for display in the greenhouse and home. It is usually bought as a pot plant in full leaf and will remain in glory for six to eight months after which the foliage tends to die back. Store dormant tubers almost dry in their pots or sawdust at a minimum temperature of 13°C/55°F. Check for signs of decay; if necessary, dust with fungicide. Bring the tubers into growth in early spring at 23°C/75°F, having repotted them into a porous but moisture-retentive soilless medium (that said, *Caladium* seems to thrive in wet and heavy soils in hot climates or hot seasons such as the New York summer). Grow in bright, indirect light with high humidity and a minimum temperature of 18°C/65°F. Water, syringe and feed frequently in warm weather. For summer bedding, place outdoors in early summer. Reduce water as the leaves begin to fade. Propagate prior to repotting by separation of small tubers or by cutting the parent tubers into sections (dust cut surfaces with a fungicide).

C.bicolor (to 90cm tall with 20–50cm-long, heart- to arrow-shaped leaves of a soft, slightly quilted and chalky in texture, oriented downwards on the petioles and variably spotted, veined, blotched, zoned or solidly stained white, rose pink or blood red on a lime or grey-green background; now incorporated in this species are *C.marmoratum*, with leaves marbled white, ivory and grey, and *C.picturatum*, with rather narrow, spotted leaves; the many hybrids between these former species were grouped under the names *C. × hortulanum* and *C. × candidum*, now also synonyms of *C.bicolor*—they range in height from 15 to 100cm; in leaf size from 8 to 30cm; in leaf shape from narrowly lance-shaped to broadly heart-shaped; and in colour from almost pure white through cream and grey to pink and red to the deepest burgundy, and combinations thereof in washes, spots, veins, tints and suffusions).

Calamagrostis (from Greek *kalamos*, a reed, and *agrostis*, a kind of grass). Gramineae. REED GRASS. Temperate northern Hemisphere. 250 species, clump-forming, perennial grasses with reed-like stems, slender leaves and silky flowers in plume-like panicles in summer. Hardy in zone 4. Grow in full sun or light shade on any moist but well-drained soil. The flower-heads can be used in dried-flower arrangements. Increase by division. *C. × acutiflora* (*C.arundinacea × C.epigejos*; FEATHER REED GRASS; erect to 2m with lax, arching foliage and soft, silvery to pearly pink or purple-bronze panicles; 'Karl Foerster': tall erect, with upright, red-bronze to buff panicles; 'Overdam': to 1, arching, leaves white-striped).

Calathea (from Greek *kalathos*, basket: the flowers are surrounded by bracts, as if in a basket). Maran-taceae. Tropical America. Some 300 species, evergreen perennial herbs with long-stalked, ovate to oblong-lanceolate and often colourful leaves, mostly 20–35cm long. The flowers are small, usually white to cream with a pink to mauve tint, and carried among bracts on a spike or raceme. Provide a minimum temperature of 15°C/60°F and light to deep shade in a humid, draught-free environment. Plant in a soilless mix high in bark, leafmould and sand. Keep evenly moist, never wet, throughout the year. Increase by division.

C.crocata (ETERNAL FLAME; Brazil; leaves dark green with a grey feathered pattern above, purple beneath; flowers orange-red in spikes with showy orange bracts); *C.ecuadoriana* (RED ZEBRINA, TIGRINA; Ecuador; leaves dark green with midrib and veins pale lime to yellow above, purple-red beneath; flowers yellow-orange amid green bracts); *C.lancifolia* (RATTLESNAKE PLANT; Brazil; leaves narrow, wavy, pale yellow-green with a dark midrib and margins and alternating, small and large, dark blotches along veins above, purple-tinted beneath); *C.lindeniana* (Brazil, Peru; leaves dark green with olive, feathered zones along the midrib and on margins above, purple-tinted beneath); *C.makoyana* (PEACOCK PLANT, CATHEDRAL WINDOWS, BRAIN PLANT; Brazil; leaves pale green, feathered cream with dark, oblong blotches along veins and a dark border above, purple beneath); *C.majestica* (syn. *C.ornata*; Brazil; leaves glossy green above, dull purple beneath; includes 'Albolineata', with fine white lines along lateral veins, 'Roseolineata', with fine pink lines along veins, and 'Sanderiana', with broader, glossy leaves, olive green with white and pink markings); *C.zebrina* (ZEBRA PLANT; Brazil; leaves deep velvety green with broad chartreuse stripes along lateral veins above, purple-tinted beneath; includes several cultivars with silver-grey markings).

calcarate furnished with a spur.

calcareous used of alkaline soils that contain chalk or limestone.

Calceolaria (from Latin *calceolus*, a slipper, referring to the shape of the flower). Scrophulariaceae. SLIPPER FLOWER, SLIPPERWORT, POCKETBOOK FLOWER, POUCH FLOWER. C and S America. 300 species, annual, biennial and perennial herbs, subshrubs and shrubs, generally with hairy stems and leaves. The slipper-like flowers are usually borne in stalked cymes in spring and summer (winter if grown under glass). They consist of a rounded and inflated corolla with a short, rim-like upper lip and a larger, puffy and pouch-shaped lower lip.

The low-growing, alpine species need full sun, a cool, buoyant atmosphere and a sharply draining acid soil high in grit. Although frost-hardy, they are intolerant of wet, cold conditions and are usually short-lived outdoors unless planted in a sheltered rock garden niche or sink and given cloche protection in winter. Alternatively, grow in the alpine house or cold greenhouse. The larger perennial species

Caladium bicolor

Calceolaria
darwinii

(*C.integrifolia, C.pavonii*) require a fertile, lime-free medium in the cool greenhouse or conservatory with good light and ventilation; keep almost dry in winter. Sow seed of the showy garden hybrids (C. Fruticohybrida Group, C. Herbeohybrida Group) under glass in summer or spring and grow on with a minimum temperature of 10°C/50°F. For bedding and window box displays, plant out in late spring. Propagate all by seed, basal cuttings or division.

C.arachnoidea (Chile; rhizomatous, tufted perennial herb to 60cm tall, covered in silver down; flowers to 1cm in diameter, dull violet-purple, very rounded, clustered on slender stalks); *C.biflora* (Chile, Argentina; perennial herb with low-lying rosettes of hairy leaves; flowers 2cm in diameter, bright yellow spotted red within and carried or more two together on stalks to 20cm tall); *C.darwinii* (S Patagonia, Tierra del Fuego; dwarf perennial herb with low-lying tufts of small, glossy leaves; flowers solitary to 3cm long and basically yellow-ochre, the upper lip short and hooded, the lower lip large, triangular and with a white band on its inner rim and stained and spotted ruby red; *C.*'Walter Shrimpton' is a hybrid between this species and *C.fothergillii*, of greater vigour and with rather longer, sac-like flowers in rich ochre, extensively spotted red-brown with a broad white band like a clerical collar); *C.fothergillii* (Falkland Islands, Patagonia; dwarf perennial herb with low-lying leaf rosettes; flowers solitary, to 3.5cm long sulphur yellow spotted or streaked ruby red below); C. Fruticohybrida Group (a popular group of rather shrubby hybrids, usually grown under glass or treated as half-hardy annuals and differing from the next group in their smaller, less puffy flowers, usually carried in dense, stalked clusters; many cultivars and seed races are available, ranging in colour from yellow to orange and bronzy red, often spotted or stained); C. Herbeohybrida Group (a group of popular hybrids grown under glass and often treated as annuals or biennials; they have a soft and bushy growth habit with broad leaves and large heads of showy, strongly inflated flowers in tones of pale yellow, gold, tangerine, scarlet ruby red, maroon and bronze, often spotted or stained in a darker shade and sometimes scented); *C.integrifolia* (Chile; erect to sprawling perennial subshrub to 1.5m, covered in clammy hairs; flowers small, yellow, clustered in short, branched cymes; var. *viscosissima*: larger, with very sticky leaves); *C.pavonii* (Ecuador, Peru; scandent, buff-hairy, perennial subshrub to 1.5m; leaf bases broadly winged and connate; flowers yellow sometimes marked purple-red in large cymes); *C.polyrhiza* (Chile, Patagonia; prostrate perennial herb with small, rounded, hairy leaves, long-stalked flowers solitary, to 3cm long and yellow with purple spots); *C.tenella* (Chile; creeping to mat-forming perennial herb; leaves small; flowers relatively large, rounded and open, golden yellow with a few orange-brown flecks, in short-stalked clusters).

calceolate slipper-shaped; resembling a round-toed shoe in form.

calcicole a plant that favours or tolerates a soil with a high lime content, for example, *Syringa*; sometimes called a calciphile.

calcifuge a plant that will not grow normally in soil with a high lime content, for example, *Rhododendron*; commonly called a lime-hater or sometimes a calciphobe.

calcium a macronutrient that is essential for plant growth. Calcium is found in, or applied to, soils mainly as chalk, limestone, magnesian limestone or gypsum. Its concentration in soil determines acidity or alkalinity, which is measured on the pH scale.

calcium carbonate the chemical constitutent of chalk and limestone. See *lime*.

calcium hydroxide hydrated or slaked lime, a relatively expensive but quick-acting form of lime.

calcium nitrate an inorganic fertilizer combining calcium and ammonium nitrate (15.5% nitrogen). A non-acidifying source of nitrogen, it is recommended for the treatment of bitterpit of apples and pears.

calcium oxide quicklime, made by heating limestone. It is caustic and dangerous to handle. See *lime*.

Calendula (from Latin, *calendae*, the first day of each month – a reference to the long flowering period of these plants). Compositae. Mediterranean, Macaronesia. MARIGOLD. 20 species, annual and perennial herbs, often glandular and aromatic, with flat, wide flowerheads with strap-shaped ray florets. Treated as a hardy annual, *Calendula* is used for bedding, borders and window boxes. Sow *in situ* in spring or autumn in a sunny position on free-draining soils. For prolonged blooming, deadhead or sow successionally.

C.officinalis (RUDDLES, COMMON MARIGOLD, SCOTCH MARIGOLD, POT MARIGOLD; hardy, muskily scented annual or perennial of obsure origin, now widespread through cultivation, to 70 × 40cm, with flowerheads to 7cm wide in shades of orange and yellow, produced from late spring to autumn; many seed races are available—dwarf races (to 30cm) include Bon Bon (early-flowering, vivid orange), Dwarf Gem and Fiesta; among the taller races (to 60cm) are the pompon-centred Kablouna and the quilled, 'cactus-bloomed' Radio; colour ranges from the bright yellows of Sun Glow and Lemon to the vivid orange Orange Prince and the red-tinted Indian Prince; larger-flowered seed races include the rich orange, 45cm-tall Mandarin Flower Hybrids and the mixed orange, yellow and apricot of Pacific Beauty Mixed and Art Shade Mixed; in Japan, the marigold has been developed as a cut flower, giving rise to tall, strong-stemmed and fully 'double' selections like 'Early Nakayasu'; three old and dependable cultivars are 'Chrysantha', with 'double' bright gold flowerheads, 'Prolifera', with proliferous flowerheads, and 'Variegata', with leaves variegated yellow).

Calla (from Greek *kallos*, beauty). Araceae. Temperate regions. 1 species, *C.palustris* (BOG ARUM, WATER ARUM), a perennial, marginal to aquatic herb with creeping, green rhizomes and smooth, heart-shaped leaves to 8cm across. Minute, green flowers are carried in spring on a creamy, club-shaped spadix. This is loosely surrounded by a white spathe to 6cm long. Decorative, red berries follow in late summer. Hardy in climate zone 4. Plant in full sun at the muddy margins of ponds and lakes. Allow the rhizomes to creep toward the water. Propagate by division in early spring.

Calliandra (from Greek *kalos*, beautiful, and *aner*, man, a reference to the beautiful stamens). Leguminosae. POWDER PUFF TREE. Tropics and subtropics. 200 species, trees, shrubs and perennial herbs with bipinnate leaves and stalked, terminal and axillary flowerheads. The corolla is small and scarcely distinguishable, but the stamens are many, colourful and slender, making, *en masse*, a powder-puff-like ball. The following species are evergreen and autumn- to spring-flowering, although they may become at least semi-deciduous in arid seasons outdoors and may flower sporadically throughout the year under glass. Grow in full sun in a loamy medium with extra sand and grit. Provide a minimum winter temperature of 5°C/40°F. Water and syringe freely in summer. Keep rather dry in winter. Prune if necessary, after flowering. An excellent choice for indoor bonsai. Propagate from seed soaked in warm water for 24 hours then sown in a moist, sandy mix at 23°C/75°F.

C.eriophylla (MOCK MESQUITE, MESQUITILLA, FAIRY DUSTER; California to Mexico; spreading shrub to 1 × 1.5m with 2.5cm-long leaves composed of many tiny leaflets; flowerheads 2–3.5cm wide, with red to rosy pink or white stamens); *C.haematocephala* (PINK POWDER PUFF; S America; flat-topped, spreading shrub or tree to 6m, usually less in cultivation; leaflets to 7cm, emerging bronze-pink and hardening mid-green; flowerheads 6–9cm wide, white at the base of the stamens deeping through pink to vivid red at their tips); *C.tweedii* (TRINIDAD, MEXICAN or BRAZILIAN FLAMEBUSH; warm Americas; spreading, shrub to 2m tall with very fine, ferny foliage; flowerheads 7cm wide, with bright crimson stamens).

Callicarpa (from Greek *kalos*, beautiful, and *karpos*, fruit). Verbenaceae. Subtropics and tropics. 140 species, trees and shrubs, usually deciduous and covered with felty or scurfy hair. The leaves are oblong to elliptic-lanceolate and the flowers small and dull, borne in compact, axillary cymes. They are followed in late summer by clusters of brilliantly coloured, bead-like berries; these last well into early winter. Hardy to −18°C/0°F, they will usually resprout if cut down by hard frosts. Plant in full sun and away from strong winds on fertile, well-drained soils, with ample moisture in spring and summer. Propagate by seed or short, heeled cuttings taken in early spring.

C.americana (southern N America, West Indies; much-branched shrub to 3m tall with red-purple, hairy twigs; leaves colouring bronze, pink and flame in autumn; fruit 6mm-wide in shades of rose-pink, red-violet and blue-mauve – off white in 'Lactea'); *C.bodinieri* var. *giraldii* (BEAUTYBERRY; China; 3m-tall shrub with slender, spreading branches clothed in tawny hair; leaves turning purple-red to amber in autumn; fruit 2–3mm-wide, shining amethyst purple or deep mauve – abundant and violet in 'Profusion').

Callisia (from Greek *kallos*, beauty). Commelinaceae. Subtropical and tropical America. 20 species, tender perennial and annual herbs (those listed here are perennial). Trailing and rooting, thinly succulent stems are clothed in slightly fleshy, ovate to lanceolate leaves. Small, white to pink and three-petalled flowers are produced throughout the year. *C.elegans* and *C.repens* are fine plants for baskets and hanging pots and should be treated as for tender *Tradescantia* species. The remaining two species prefer a gritty, loamy medium, full sun and cool, dry winters (minimum temperature 7°C/45°F). Propagate by detaching rooted lengths of stem or plantlets formed on stolons; increase also by stem cuttings.

C.elegans (stems to 30cm long; leaves loosely 2-ranked, 2.5–10 cm long and dull green, finely striped silver-white above, purple-tinted beneath); *C.fragrans* (bold rosette of 30cm-long, mid-green leaves – striped in the cultivar 'Melnickoff' – giving rise to leafy stolons bearing miniature plants); *C.navicularis* (growth of two types – succulent, keel-shaped leaves to 3cm long overlapping in short, bulb-like shoots, or more remotely disposed along creeping stolons, leaves smooth, mid-green above and striped or stained purple beneath); *C.repens* (syn. *Tradescantia* 'Little Jewel'; mat-forming trailer with dense, glossy, deep green, ovate leaves to 2.5cm long).

Callistemon (from Greek *kallos*, beauty, and *stemon*, stamen). Myrtaceae. Australia. BOTTLEBRUSH BUSH. 25 species, shrubs and small trees, evergreen with slender, arching branches and narrow leaves; these emerge pink and silky and harden leathery, smooth and dark green. Produced in summer, the flowers consist of bundles of long and spreading stamens. They are held flush with the branch on one- to two-year-old wood in a cylindrical arrangement – the whole resembling a bottlebrush. The following are hardy in zone 7, or in favoured locations in zone 6. *C.sieberi* will thrive in most regions in zone 6. Plant in full sun on a fast-draining soil, ideally with the shelter of a south-facing wall. Avoid pruning if possible. Propagate by semi-ripe cuttings in summer.

C.citrinus (CRIMSON BOTTLEBRUSH; medium shrub to small tree; leaves lemon-scented if crushed; flowers bright crimson with dark anthers; 'Splendens': dense spikes of carmine flowers); *C.linearis* (NARROW-LEAVED BOTTLEBRUSH; medium-sized shrub; leaves narrow; flowers dull red with

Callistemon citrinus

gold anthers); *C.pallidus* (LEMON BOTTLEBRUSH; medium to tall shrub; flowers cream to yellow-green with yellow anthers); *C.rigidus* (STIFF BOTTLE-BRUSH; small- to medium-sized shrub with hard-edged leaves; flowers deep red); *C.salignus* (PINK TIPS, WILLOW BOTTLEBRUSH, WHITE BOTTLEBRUSH; tall shrub to medium-sized tree with papery white bark and narrow leaves; flowers green-white, pink, red or rosy mauve); *C.sieberi* (ALPINE BOTTLEBRUSH; erect to spreading, tall to dwarf shrub, with dense, short, pointed leaves; flowers cream to yellow); *C.speciosus* (ALBANY BOTTLEBRUSH; erect, medium-sized shrub with prominently veined, rigid and spreading leaves; flowers deep red or white with golden anthers); *C. subulatus* (arching shrub to 1.5m with hard-pointed, narrow leaves; flowers crimson); *C.viminalis* (WEEPING BOTTLEBRUSH; small shrub or tree with cascading branches; flowers bright red; 'Red Cascade': strongly weeping, flowers rose-red).

Callistephus (from Greek *kallos*, beauty and *stephos*, crown, referring to the fruit appendages). Compositae. China. 1 species, *C.chinensis* (CHINA ASTER), an erect and bushy annual herb to 80cm tall with coarsely toothed, ovate leaves to 8cm long. Produced in summer and autumn, the flowerheads resemble large daisies or chrysanthemums. To 12cm across, they are typically white to pale mauve or violet with a central disc of yellow. Seed races range in flower form from single to double, anemone-centred to quilled, in colour from white to cream, palest pink, flamingo pink, deep rose, magenta, apricot, peach, salmon, scarlet, crimson, lilac and deep purple, and in habit from dwarf and compact to tall and long-stalked. Sow seed under glass in early spring or outdoors in late spring. Plant out after the last frosts in full sun on a well-drained, fertile soil. Stake taller cultivars. Deadhead regularly.

Calluna (from Greek *kallunein*, to beautify, referring to its use as a brush or broom). Ericaceae. N America, Azores, N and W Europe to Siberia. 1 species, *C.vulgaris* (LING, SCOTS HEATHER), a low and bushy evergreen shrub to 60cm tall with many, tiny, scale-like leaves overlapping on the narrow, upright branches. Produced in summer and autumn, the flowers are shortly tubular to bell-shaped, to 0.5cm long and borne at the leaf axils, forming long and slender racemes on the branch tips. They differ from the flowers of *Erica* in the calyx, which is typically pink and virtually conceals the corolla. This plant is immensely variable, with over a thousand cultivars recorded. They range in height and habit from low and creeping, to compact and domed to tall and cypress-like, in flower colour from pure white to pale pink, rose, lilac-pink, magenta, salmon, crimson and mixtures thereof. Many are excellent foliage plants, colouring especially well in winter and early spring, with leaves ranging through deep emerald green to yellow, gold, orange, flame, bronze, rusty red and purple-red. Fully hardy. Plant on sandy,

well-drained, acid soils in full sun. Further requirements as for hardy *Erica* species.

callus an abnormal or isolated thickening of tissue, either produced in response to wounding or abscission or as a surface feature of leaves and perianth segments. It is important in propagation by cuttings and in one method of tissue culture (see *micropropagation*). Callus tissue is also found as protuberances on the lips of many orchid flowers.

Calocedrus (from Greek *kalos*, beautiful, and *kedros*, cedar). Cupressaceae. Western N America, E Asia. 3 species, large evergreen, coniferous trees with flattened sprays of small, scale-like leaves and rounded to oblong, red-brown cones. Fully hardy. Plant in any but the wettest and heaviest soils in full sun. Water young plants in dry climates. Once established, this a remarkably drought-tolerant conifer. It will also survive harsh winds and poor soils. Propagate as for *Chamaecyparis*. The genus is Armillaria-resistant.

   C.decurrens (syn. *Libocedrus decurrens*; INCENSE CEDAR; western N America; broadly conical to pyramidal tree to 45m tall with deeply fissured red-brown bark and fans of rich green foliage; cones 2.5cm long, oblong, carried erect on pendulous shoots; size and habit vary according to climate—dense and fastigiate in warm, moderately dry summers and cool or cold winters, dense and broadly ovoid in cool, wet summers and mild winters, broadly conic and open with level branches in cold wet winters and hot, dry summers; 'Aureovariegata': medium-sized, foliage spattered with yellow-cream; 'Compacta': habit globose to columnar, densely branched; 'Depressa': dense, globose, dwarf, tinted bronze in winter; 'Glauca': foliage glaucous, tinted blue-grey; 'Intricata': dwarf, compact, with tortuous branches, foliage tinted brown in winter; 'Riet': globose, dwarf, seldom exceeding 75cm).

Calocephalus (from Greek *kalos*, beautiful, and *kephale*, head). Compositae. Australia. 18 species, annual and perennial herbs and small evergreen shrubs, usually densely covered in white wool, with entire leaves and florets packed in small, globose heads. In cool-temperate gardens, *Calocephalus* is often damaged or killed by cold, wet winters. For this reason, it tends to be seen in the form of young, rather sparse plants, used for silver effects in bedding, baskets and patio pots and repropagated annually. In milder climates or in very sheltered positions in zone 8, it will survive, forming broad, impenetrable hummocks; these thrive in gravelly, dry places and withstand severe buffetings and salt spray. Plant in a very free-draining soil in full sun. Propagate by semi-ripe cuttings in late summer and overwinter in a cool well-ventilated greenhouse. Plant out after frosts have passed. *C.brownii* (shrublet to 50 × 80cm; branches wiry and intricate, clothed with scale-like leaves and in silvery white to grey scurf; flowerheads resembling silver buttons).

Calochortus (from Greek *kalos*, beautiful, and *chortos*, grass, referring to the grassy leaves). Liliaceae (Calochortaceae). N America to Guatemala (those listed here are from the SW US, principally California). 60 species, bulbous perennial herbs with grassy leaves and erect, branched stems bearing showy flowers in spring and summer. They fall into three groups: GLOBE TULIPS or FAIRY LANTERNS with several, nodding, globose to campanulate flowers; STAR TULIPS, with erect, campanulate to cup-shaped flowers, usually with the petal margins rolled outwards and (in *C.tolmiei*) covered in hairs; and MARIPOSA LILIES, with large, cup-shaped flowers on tall stems. *Calochortus* flowers consist of three, narrow sepals and three broad petals with a semi-circular zone in their lower half that may be differently coloured, hairy and marked with a dark, glandular patch.

Although happiest in climatic zone 9, *Calochortus* may survive outdoors in favoured locations in zones 7 and 8. Plant in a deep, freely draining, sandy soil in a sheltered position in full sun. Keep just damp in autumn and winter, moist in late winter and spring. After flowering, keep dry until mid-autumn, allowing the foliage to wither and the soil to bake. In many places, these conditions are more easily achieved in a raised bed with covers, or in pots and pans in the alpine house or bulb frame. Leave the bulbs in the ground or in their containers during dormancy. Propagate by ripe seed in the cool greenhouse or frame, by offsets when dormant, or from bulbils which sometimes form in the leaf axils.

C.albus (WHITE GLOBE LILY, FAIRY LANTERN; 20–80cm; flowers to 6cm in diameter, petals broad, white, bearded with a deep red-brown spot); *C.amabilis* (GOLDEN FAIRY LANTERN; 10–30cm; flowers to 5cm in diameter, petals golden yellow, fringed, bearded); *C.amoenus* (PURPLE GLOBE TULIP; like a shorter *C.albus*, with nodding, deep rose or mauve flowers); *C.barbatus* (to 60cm; flowers nodding, campanulate, to 5cm in diameter, petals yellow, often tinted rose or mauve, bearded and fringed); *C.clavatus* (50–100cm; flowers erect, cup-shaped, 8–10cm in diameter, bright yellow, lower half of petals covered in hairs and bounded by a dark brown arc); *C.kennedyi* (to 50cm; flowers campanulate, erect, 5–10cm in diameter, petals rich orange to vermilion with a glossy, black gland and hairs toward the base); *C.luteus* (similar to *C.venustus* with erect, campanulate dark yellow flowers with dark brown markings and hairs toward base of petals); *C.splendens* (LILAC MARIPOSA; 20–60cm; flowers erect, campanulate, 5–10cm in diameter, petals pale pink with fine teeth and hairs); *C.superbus* (similar to *C.venustus*, flowers campanulate, petals white, cream or yellow, streaked purple at base usually with a brown or maroon spot surrounded with yellow); *C.tolmiei* (syn. *C.maweanus*; CAT'S EARS, PUSSY EARS; 5–30cm; flowers open-campanulate, erect or spreading, 2–5cm in diameter, white or cream often tinged rose or purple, petals hairy and fringed); *C.uniflorus* (syn. *C.lilacinus*; STAR TULIP; stem very short, flowers 3–6cm in diameter, petals pale lilac, sometimes toothed and bearded); *C.venustus* (WHITE MARIPOSA TULIP; 20–60cm; flowers 5–10cm in diameter, white to yellow or purple or dark red, petals bearded below, with a dark red-brown spot); *C.vestae* (similar to *C.venustus*, the petals white to purple-tinged streaked with red to purple near base and each with a central brown spot edged with yellow); *C.weedii* (30–60cm; flowers erect, campanulate, petals to 3cm, deep yellow with purple dots or margins, bearded and fringed).

Calochortus venustus

Calomeria (from Greek, *kalos*, beautiful, and *meris*, part, in reference to the capitula). Compositae. Africa, Madagascar, Australia. 14 species, annual to perennial herbs with small, button-like flowerheads in large, much-branched panicles in summer. A spectacular biennial grown in the cool greenhouse (minimum temperature 7°C/45°F) or in subtropical bedding schemes. Sow seed in summer under glass. Grow on in individual pots in a fertile, loam-based mix in a well-ventilated, sunny greenhouse. Water moderately when in growth, scarcely at all in winter. Pot on to flower in large containers, or plant outdoors in a warm and sunny position in late spring. Support with canes. *C.amaranthoides* (syn. *Humea elegans*; INCENSE PLANT, PLUME BUSH; Australia; robust and intensely aromatic biennial or perennial to 2m or taller; flowerheads small, red-brown to pink, massed on the cascading branches of a broad panicle).

Calothamnus (from Greek *kalos*, beautiful, and *thamnos*, bush or shrub). Myrtaceae. Western Australia. NET BUSH. 25 species, small to large evergreen shrubs with wand-like branches clothed in papery bark and rigid, narrow, terete or flat leaves. The flowers grow directly from the old wood and below the new growth, forming a stemless, crowded spike or cluster composed of long, brilliantly coloured stamens, the whole resembling a one-sided *Callistemon* inflorescence. The following species should be cultivated as for *Callistemon* but kept rather dry and frost-free in winter.

C.quadrifidus (COMMON NET BUSH, 2–3m, upright to spreading; flower spikes to 20cm long, stamens to 2.5cm, of equal length, deep red with golden anthers); *C.sanguineus* (BLOOD RED NET BUSH; much-branched, dwarf to medium-sized shrub; inflorescence to 10cm, stamens to 2.5cm, not equal, deep red with golden anthers); *C.validus* (BARRENS CLAW FLOWER; erect, rounded, medium-sized shrub (i.e. to 2m); inflorescence to 5cm long, stamens to 4cm long, of equal length, rich red with golden anthers); *C.villosus* (SILKY NET BUSH; downy, upright to spreading, small to medium-sized shrub; branches hairy; inflorescence to 10cm, stamens more or less equal, bright red with golden anthers).

Caltha (name of a yellow flower, used by Virgil and Pliny). Ranunculaceae. Temperate regions. 10 species, perennial herbs with long-stalked, rounded to reniform leaves and cymes of buttercup-like

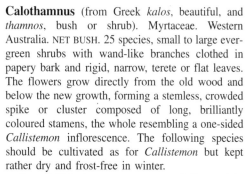

Calochortus clavatus

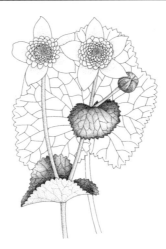

Caltha palustris 'Florepleno'

flowers composed of broad, petal-like sepals in yellow or white. These fully hardy, spring-flowering perennials will grow in sun or part shade in moist or saturated soils (neutral to acid for *C.leptosepala*). They thrive especially at the margins of ponds, streams and lakes and may even be partially submerged. The smaller cultivars are suitable for damp positions in the rock garden. Propagate by division in spring, or from fresh seed sown in late summer in the cold frame.

C.leptosepala (N America; to 30cm tall; flowers white with a golden boss of stamens); *C.palustris* (KINGCUP, MARSH MARIGOLD, MEADOW BRIGHT; northern temperate regions; variable, from 10–80cm tall, low and compact to tall and rangy; flowers rich golden yellow sepals; includes creeping, alpine and very large forms; 'Florepleno' has fully double flowers).

Calycanthus (from Greek *kalyx*, calyx and *anthos*, flower – the sepals and petals are alike). Calycanthaceae. China, US. ALLSPICE, SPICE BUSH. 6 species, spicily aromatic, deciduous shrubs, to 4m tall, with dull to glossy dark green, elliptic to obovate leaves and solitary flowers in summer. These are composed of numerous, similar sepals and petals, oblong to oblanceolate and spreading. Fully hardy, they will thrive on most fertile, moist, but well-drained soils in sun or part shade. Increase by seed harvested from the distinctive, goblet-shaped capsules in autumn, or by softwood cuttings in summer.

C.fertilis (SE US; leaves more or less smooth beneath; flowers to 5cm in diameter, purple-red to maroon, slightly fragrant; includes the dwarf 'Nanus', and 'Purpureus', with leaves tinted purple beneath); *C.floridus* (syn. *C.sterilis*; CAROLINA ALLSPICE, STRAWBERRY SHRUB; SE US; leaves hairy beneath; flowers to 5cm in diameter, dark maroon, strawberry-scented); *C.occidentalis* (CALIFORNIAN ALLSPICE, SPICE BUSH; distinguished from *C.fertilis* by the summer buds that are not concealed by petiole bases; flowers to 7cm in diameter, maroon fading to yellow, fragrant).

calyptra a hood or cap-like structure terminating a circumscissile calyx or pyxis.

calyx a collective term for the sepals, whether separate or united, that form the outer whorl of the perianth or floral envelope.

Camassia (from *quamash*, so-named by Native Americans for whom the bulbs of *C.quamash* were an important foodstuff). Liliaceae (Hyacinthaceae). Americas (those listed here are from western N America). CAMASS, QUAMASH, CAMOSH. 6 species, deciduous, bulbous perennial herbs with narrow leaves and, in late spring and early summer, tall spikes of white, blue or violet, star-like flowers with six, linear to oblong tepals. Plant in a deep, moisture-retentive and fertile soil in full sun or light shade. Where temperatures fall below –15°C/5°F for long periods, mulch the crowns thickly. Propagate by seed, by offsets or by division of established clumps in autumn or early spring.

C.cusickii (raceme dense, standing to 1m; flowers 5cm wide, pale blue, larger and deep blue in the cultivar 'Zwanenburg'); *C.leichtlinii* (flowers 4–8cm across, creamy white, blue or violet, carried on a raceme to 120cm tall; in this species, the tepals always twist together as they wither: this never happens in *C.cusickii* and only sometimes in *C.quamash*; includes "Blauwe Donau', with dark blue flowers, 'Electra', with rich blue flowers to 10cm across, 'Plena' and 'Semiplena', with creamy white flowers, double and semi-double respectively); *C.quamash* (syn. *C.esculenta*; flowers 2–7cm wide, white to pale violet or blue, carried loosely on racemes to 60cm tall; includes 'Orion', with a denser raceme and flowers that are deep purple in bud, opening dark, steely blue).

cambium a secondary or inner meristem that increases the girth of a plant stem or root by adding vascular tissue.

Camellia (for Georg Kamel [Camellus] (1661–1706), Moravian Jesuit who travelled in Asia and wrote a history of the plants of Luzon published in 1704 by John Ray). Theaceae. Asia. About 250 species, evergreen shrubs or trees with elliptic to lanceolate, tough and shiny leaves. The flowers are solitary or clustered, bowl-shaped and composed of 5–12, broad, silky petals and a central boss of golden stamens. Camellias need cool winters and many will tolerate moderate frosts but they must have good drainage, plenty of organic matter in the soil, and a pH between 5 and 7. *C.japonica* prefers the more acid end of this range, most other species nearer to neutral. They need warmth and some sunlight in late summer to form flower buds which open during the following autumn, winter or spring. In cold climates, the flowers and young growth may be damaged by cold winds and late frosts. Partial shade tempers late frosts and hot sun; this is especially important for young plants. If their roots are cool, established bushes of many cultivars of *C.japonica* and most of *C.reticulata* and *C.sasanqua* will flourish in full sun in all but very hot climates. *C.japonica* is the hardiest species. Damage can occur in some forms of *C.japonica* at –13°C/9°F. At –22°C/–8°F considerable damage can be expected to established plants; smaller ones are liable to 'bark-split' and can be cut to the ground; such plants should be pruned back to just above ground level; most will re-grow. If severe frost persists, and especially if it is accompanied by strong winds, even large plants will be killed. As the term 'camellia house' would suggest, most will thrive in an unheated glasshouse, given a cool root-run, excellent ventilation and protection from scorching sun. Free drainage is very important. Where soils are very impermeable or alkaline, raised beds can be made of slightly acid sand, sandy compost or light topsoil, topped up with a moderately acid mulch. If necessary, prune in spring after flowering and before growth starts. Neglected large plants can be rejuvenated by hard pruning, although a season's flowering will be lost. Increase by semi-ripe cuttings of the current season's growth. Insert them into a moist, fine mix under mist, glass or polythene, after being wounded and dipped into a hormone rooting powder. Some cultivars, particularly those of *C.reticulata*, dislike root disturbance and are best rooted directly into small individual pots. Side or cleft grafting is also used for these and especially for promising new cultivars. Most new cultivars are raised from seed. Fresh seed germinates easily, especially with bottom heat.

Camellias are affected by a number of pests. Scale insects are often the most important and the sooty mould they engender is a serious disfigurement. Diseases include the soil-borne *Phytophthora cinnamomi*. Plants grafted onto C. *sasanqua* are virtually immune. Camellia dieback, *Glomerella cingulata*, is chiefly a problem with C. *sasanqua* and C. × *williamsii* on very acid soils. In areas where dieback is endemic, spray after pruning and treat cuttings and grafts before use by soaking in a fungicidal solution. Widespread in the Far East and North America, flower blight, *Ciborinia camelliae*, is spread by affected flowers. The recommended treatment is to remove lower branches, clear all vegetation beneath the bushes, apply a dry mulch and spray with fungicide.

CAMELLIA Name	Distribution	Habit	Leaves	Flowers	Flowering time
C.chrysantha	S China	open shrub or tree to 5m	10–17cm, oblong-lanceolate, acuminate, dark green, tough	cup-shaped, single; petals 2–5cm, 9–11, pure yellow	spring
C.cuspidata	W China	bushy shrub, erect to 3m	5–7.5cm, elliptic to lanceolate, acuminate to caudate, serrulate, bronze at first, later dark, glossy green	cup-shaped, single; petals 6–7, to 3cm, white	early spring
C.granthamiana	Hong Kong	open shrub to 3m	to 10cm, elliptic to oblong-elliptic, bluntly acuminate, bluntly toothed, glossy deep green with impressed veins	fragrant, saucer-shaped, single; petals 8, to 7cm, white, broad and cleft at tips	late autumn
C. × *hiemalis* (*C. japonica* × *C. sasanqua*)		erect, bushy shrub or tree to 3m	6–9cm, oblong-lanceolate, serrulate	fragrant, cup-shaped, semi-double; petals 7, to 3.5cm, white, pink or red	late autumn to winter
C.honkongensis	Hong Kong	bushy shrub or tree to 10m	to 12cm, oblong-elliptic, tip shortly tapered, iridescent bronze at first, hardening dark, glossy green	cup-shaped, single; petals 6–7, 3cm, broad and blunt or cleft, deep crimson with velvety undersides	late spring
C.japonica *C.hozanensis*	Japan	shrub or small tree to 10m, usually less in gardens	5–10cm, broadly elliptic to elliptic-oblong, apex briefly acuminate, finely toothed, rigid and glossy dark green	saucer-shaped, single to double; petals typically 5–6, to 4cm, red, broad and rounded to cleft	late winter to early spring
Comments: The common camellia of gardens, with over 2000 cultivars varying greatly in habit, hardiness and leaf shape and colour. The flowers range in type from formal to semi-doubles, peony-, rose- and anemone-forms, and in colour from pure white to pale pink, deep rose, crimson and velvety scarlet, including bicolours, striped and picotee forms. The stamens are typically golden and showy, but may be replaced to varying degrees by petaloid growths. The Higo Japonicas are an ancient group of cultivars developed in Japan and characterized by large and single, flat flowers with bosses of golden anthers or petaloids. The SNOW CAMELLIA is ssp. *rusticana* (syn. *C. rusticana*), with small, widely spreading flowers.					
C. × *maliflora* (parentage obscure)		erect, bushy and densely leafy shrub to 2.5m	3.5–5cm, oblong-elliptic to broadly lanceolate, apex bluntly acute to acuminate, finely toothed, thin-textured, mid-green	dish to cup-shaped, semi-double; petals and petaloids numerous, to 2cm across, blush rose, often white at centre with deeper rose margins	spring
C.oleifera	China, Indochina	bushy shrub or small tree to 7m	4–9cm, elliptic to oblong-acuminate, apex acute or acuminate, finely toothed, dark green	fragrant, cup-shaped, single; petals 5–7, to 3.5cm, obovate with rounded, cleft tips, white, sometimes flushed pink	early spring
Comments: FORTUNE'S YELLOW CAMELLIA is the cultivar 'Jaune', with white petals and a dense boss of yellow petaloids					

CAMELLIA

Name	Distribution	Habit	Leaves	Flowers	Flowering time
C.reticulata	China	loose-branched shrub or tree to 15m	8–12cm, broadly elliptic to oblong-elliptic, apex acute to shortly acuminate, finely toothed, tough, dark green	saucer-shaped, single; petals 5–6, to 5cm, broad and often cleft, crimson to rose pink or white (in the typical plant , now known as 'Captain Rawes', the flowers were semi-double, very large and carmine)	spring

Comments: Over 400 cultivars and hybrids of this species have been recorded. They range in flower form from cupped to bowl-shaped, single to semi-double and double, some with peony centres or irregular to wavy petals and petaloids. Colours vary from white to pale rose, deep pink, crimson, dark cherry red and combinations thereof.

Name	Distribution	Habit	Leaves	Flowers	Flowering time
C.rosiflora	China	spreading shrub to 1m	4–8cm, elliptic, bluntly acuminate, finely toothed, dark green	saucer-shaped, single; petals 6–9, to 1.8cm, rounded, sometimes notched, soft, pale rose	spring
C.saluenensis	W China	bushy shrub to 5m	3–6cm, oblong-elliptic, apex acute to blunt, finely toothed, glossy dark green with conspicuous veins	cup-shaped, single; petals 6–7, to 3cm, broadly oval, notched, shell pink to deep rose, or white	early spring

Comments: Includes cultivars with semi-double and rose bud-like flowers, some with ruffled or irregular petals, in shades of white and pale pink tinted or veined with darker rose, also solid pale pink and coral

Name	Distribution	Habit	Leaves	Flowers	Flowering time
C.sasanqua	Japan	erect, bushy shrub or small tree to 5m	3–8cm, elliptic to oblong, acute to cuspidate, finely toothed, bright glossy green	fragrant, dish- to cup-shaped, single to semi-double; petals 6–8, to 3.5cm, rounded to notched, white to rose	autumn

Comments: Many cultivars are available including plants with a compact and densely bushy habit and handsome foliage, especially when newly emerged. The flowers range from small to large, tightly cupped to spreading, single to double, in tones of white, pink, salmon, cerise and red, often with one of these colours tinting, edging or overlaying a paler background.

Name	Distribution	Habit	Leaves	Flowers	Flowering time
C.tsaii	Burma, N Vietnam, W China	shrub to 3m	6–9cm, oblong-lanceolate, apex acuminate or caudate, finely toothed and somewhat undulate, dark glossy green assuming bronze tints with age	cup-shaped, single; petals 5, 2cm, broad, entire to cleft, white	spring
C. × *vernalis* (*C.japonica* × *C.sasanqua*)		erect shrub to 3m	3.5–7.5cm, elliptic-oblanceolate, apex bluntly tapered, finely toothed, bright, glossy green	fragrant, dish- to cup-shaped, single to semi-double, white sometimes suffused pink or solid rose to red	late winter
C. × *williamsii* (*C.saluenensis* × *C.japonica*)		erect to spreading shrub to 4m	7–10cm, elliptic to oblong-lanceolate, shallowly and finely toothed, bright, shining green	cup-shaped, single, 5–12cm diam., white flushed rose or pale rose with deeper centres	early spring

Comments: Many cultivars are available, including singles, semi-doubles and doubles, anemone- and peony-forms, in shades of white, pink, rose and red, sometimes tinted with a deeper shade. 'Golden spangles' has deep pink flowers and yellow-green-variegated leaves.

Campanula (diminutive form of the Latin *campana*, a bell, alluding to the shape of the flower). Campanulaceae. BELLFLOWER. Temperate N Hemisphere. Some 300 species, annual, biennial or perennial herbs varying greatly in habit and size. The flowers are typically bell-shaped and produced in spring and summer. With very few exceptions, *C.isophylla*, *C.pyramidalis* and C.*vidalii* among them, most are reliably hardy to between –15 and –20°C/5 to –4°F, although several alpine species dislike a combination of winter cold and wet. The larger perennials are invaluable in the herbaceous and mixed border. These include lower-growing species for the foreground of borders, such as *C.carpatica* and the slightly taller and invasive *C.glomerata*. *C.persicifolia*, reaching almost 1m in height, is especially elegant. The clump-forming, deeply rooting *C.latifolia*, and the undemanding *C.lactiflora* are among the tallest hardy species. They may need staking in exposed positions, but otherwise need only a moderately fertile and well-drained soil in sun or light shade. Many of the long-stemmed perennials make beautiful cut flowers. Magnificent but slightly frost-tender, *C.pyramidalis* is more safely grown in the cool greenhouse or conservatory in cool temperate regions, and

may be moved out of doors for the summer months. This species,*C.isophylla* and *C.vidalii* need a moist but freely draining, medium-fertility, loam-based potting mix, bright indirect light and good ventilation with a winter minimum of 7°C/45°F. *C. medium* is a biennial, usually sown in summer for early summer bloom in the following season; plants are overwintered *in situ* or lifted and overwintered in the cool glasshouse. Alternatively, seed sown under glass in late winter/early spring will flower in its first year.

Many of the smaller species suit rock gardens, troughs and dry stone walls. Most make few demands other than a gritty, well-drained, moisture-retentive, neutral or limy soil, in sun or light shade. *C.morettiana* is suitable for tufa rock plantings and deserves protected cultivation in the alpine house. Other delicate species requiring protection from winter wet include *C.betulifolia*, *C. raineri* and *C. zoysii*. These need a lean and gritty alpine mix with plentiful water in spring and summer, and to be kept dry but not arid in winter. Propagate biennials by seed; perennials by seed, division and basal cuttings; smaller alpines also by removal of rooted offsets. (*See table overleaf*)

CAMPANULA

Name	Distribution	Habit	Leaves	Flowers
C.alliariifolia	Caucasus, Asia Minor	hairy, erect perennial to 70cm	to 8cm, cordate, hairy	campanulate, blue or white with bearded lobes, nodding in erect racemes
C.barbata BEARDED BELL-FLOWER	Norway, Alps	tufted, shortlived perennial or biennial to 50cm	lanceolate to strap-shaped, hairy	campanulate with ciliate lobes, lavender to white, nodding in one-sided racemes
C.betulifolia	Armenia	perennial with erect then arching, purple-tinted stems to 30cm	ovate to lanceolate, toothed	tubular-campanulate, erect to nodding, white to pink with a deep pink flush on the exterior
C.carpatica CARPATHIAN BELLFLOWER	Carpathian Mountains	clump-forming perennial to 40cm	heart-shaped to ovate, crenate	broadly campanulate, blue or white, large, erect, solitary and long-stalked
Comments: Includes cultivars with flowers in shades of white, pearl, lilac, lavender and blue				
C.cochleariifolia *C.pusilla* FAIRIES' THIMBLES	Europe	tufted and creeping dwarf perennial	small, broadly heart-shaped to ovate, toothed	campanulate, solitary, nodding, blue to pearl or white
Comments: Includes cultivars ranging in flower colour from pale to deep blue, silver-blue to pure white, some double				
C.elatines ADRIATIC BELLFLOWER	Adriatic Coast	tufted, compact perennial to 15cm	rounded to heart-shaped, finely toothed	campanulate to rotate and starry, with deflexed lobes, blue or white in loose spikes or panicles
C.garganica	Italy	close to *C.elatines*	ovate to cordate	star- to bell-shaped, in shades of lavender blue
Comments: Includes cultivars with white, pale lilac, lavender and deep mauve blue flowers; 'Dickson's Gold' has golden leaves and blue, star-shaped flowers; 'W.H. Payne' has erect, starry flowers, strong lavender blue with white centres				
C.glomerata CLUSTERED BELLFLOWER	Europe, Temperate Asia	bristly-hairy perennial erect to 75cm	oblong to ovate or lanceolate, wavy-toothed	violet-blue to white, broadly tubular, in dense, terminal heads subtended by bracts
C. × *haylodgensis* *C.carpatica* × *C.cochleariifolia*	garden origin	sprawling, tufted perennial to 15cm	rounded to ovate or heart-shaped, toothed	campanulate, large, pale blue or white
Comments: Includes double-flowered cultivars				
C.isophylla ITALIAN BELLFLOWER, FALLING STARS	N Italy	dwarf, trailing or creeping perennial	broadly ovate to cordate, toothed	campanulate to star-shaped, violet to pearly grey or white, large, in erect, loose clusters
Comments: Includes cultivars with white-variegated and grey-downy leaves and flowers in shades of blue and white				
C.lactiflora	Caucasus	erect perennial with branching stems to 1.5m	ovate to oblong, toothed	broadly campanulate, milky blue fading to a white centre, erect in a broad, leafy panicle
Comments: Includes cultivars with white, lilac and violet flowers				
C.latifolia	Europe to Asia	erect, unbranched, finely hairy perennial to 1m	ovate to oblong with a heart-shaped base, toothed, hairy	cup- to bell-shaped, pale blue to lavender, in axils or terminal clusters
Comments: Includes cultivars with white, indigo, purple and lilac flowers				
C.latiloba	Siberia	differs from *C.persicifolia* in its broadly lanceolate leaves on narrowly winged stalks		deep blue, lavender or white
C.medium CANTERBURY BELLS, CUP AND SAUCER	S Europe	hairy, erect, much-branched biennial to 90cm tall	elliptic, hairy more or less toothed	campanulate to urceolate with the lobes forming a curved-back rim, blue, lavender, mauve, pink, pale rose, cream or white, sometimes with an enlarged calyx - the 'saucer' to the corolla 'cup', borne in terminal racemes and panicles
Comments: Many seed races and cultivars				
C.morettiana	N Italy, Alps	dwarf, tufted, hairy perennial	heart-shaped, crenate, more or less smooth	campanulate to funnel-shaped, violet-blue to white, solitary, erect
C.persicifolia WILLOW BELL, PEACH BELLS	Europe, N Africa, W Asia	erect, glabrous perennial to 70cm	lanceolate to narrowly oblong, crenate, dark green	broadly campanulate, blue or white, nodding in a slender, terminal raceme
Comments: Includes many cultivars - among them plants with flowers ranging from pure white to pink, lilac, indigo and darkest blue, some double				

CAMPANULA

Name	Distribution	Habit	Leaves	Flowers
C.portenschlagiana	Europe	tufted, downy perennial with sprawling to ascending stems to 25cm long	cordate to rounded, toothed	funnel-shaped to campanulate with short, deflexed lobes, deep lavender in loose panicles
Comments: Includes cultivars with white, pale violet and vivid blue flowers				
C.poscharskyana	Balkans	similar to *C.portenschlagiana* but less coarse with more finely tapering and toothed leaves	funnelform to star-shaped, lavender to violet	
Comments: Includes cultivars with white, pink, lavender and deep violet flowers				
C.pulla	E Europe	short-lived, tufted perennial to 12cm tall, spreading by underground runners	very small, rounded to spathulate, toothed, glossy, in rosettes	funnelform to campanulate, dark violet, solitary, nodding on slender stalks
C.pyramidalis CHIMNEY BELLFLOWER	Europe	biennial or short-lived perennial attaining 2m in flower with a stout base and a spire-like or narrowly pyramidal heavily branched inflorescence	heart-shaped	broadly campanulate, fragrant, pale blue to white
Comments: A spectacular half-hardy plant best grown under glass or moved outdoors in summer to a very sheltered location				
C.raineri	Switzerland, Italy	tufted perennial to 8cm, spreading by underground runners	obovate, toothed, grey	broadly campanulate, large, erect, solitary, pale lavender blue
C.rotundifolia BLUEBELL, HAREBELL	N Hemisphere	tufted perennial erect to 45cm	rounded to cordate, wavy-edged	campanulate, white to deep blue or indigo, slender-stalked, solitary or few
C.takesimana	Korea	erect perennial to 60cm	ovate to cordate, toothed	tubular to campanulate, large, pendulous and lantern-like, lilac to white spotted dark mauve-pink within on an arching, branched inflorescence
C.trachelium NETTLE-LEAVED BELLFLOWER, THROAT-WART	Europe, N Africa, Siberia	erect, bristly-hairy red-tinted perennial to 1m	ovate to cordate, crenate, bristly	tubular to campanulate, blue-purple to lilac in dense panicles
Comments: includes white and blue, double-flowered cultivars				
C.vidalii *Azorina vidalii*	Azores	evergreen subshrub to 50cm tall	narrow, glossy dark green	campanulate to urceolate, waxy, dusky pink to white, hanging in arching panicles
Comments: Tender				
C.zoysii	Italian Alps	tufted perennial to 10cm	obovate to rounded	cylindric to urceolate, pendulous, clear blue to pale mauve

campanulate of a corolla; bell-shaped; a broad tube terminating in a flared limb or lobes.

Campsis (from Greek *kampsis*, bending: the stamens are curved). Bignoniaceae. E Asia, N America. TRUMPET CREEPER. 2 species, deciduous vines clinging by means of aerial roots. The leaves are pinnate with saw-toothed, ovate to lanceolate leaflets. Borne in summer in terminal cymes or panicles, the flowers are large and funnel-shaped with five short, spreading lobes. Plant in full sun or light shade in a warm, sheltered position on a medium- to high-fertility, loam-based soil with plentiful water in spring and summer. *C.grandiflora* is hardy to –10°C/14°F, *C.radicans* to –20°C/–4°F. Hardiness is improved where summers are long and hot enough to ripen new wood fully. Train on walls, fences, trellis and trees. Propagate by simple layering or by semi-ripe to hardwood cuttings in a cold frame in autumn.

C.grandiflora (syn. *C.chinensis*; China, Japan; to 6m tall with few or no aerial roots; flowers to 5 × 6cm, orange outside and deep yellow inside; includes the hardier 'Thunbergii', with orange flowers produced later in the season); *C.radicans* (SE US; differs from the last species in its abundant aerial roots and flowers to 7 × cm, pale orange with a yellow interior and flame to scarlet lobes; includes f.*flava*, with large flowers of a solid rich yellow, 'Praecox', with scarlet flowers, and 'Speciosa', shrubby and weakly climbing, with small, flaming orange flowers); *C.* × *tagliabuana* (hybrid offspring of *C.grandiflora* and *C.radicans*; the best-known is 'Madame Galen', with deep apricot to salmon flowers to 8cm across and dark green foliage).

camptodromous describing venation in which the secondary veins curve toward the margin without forming loops.

canal a narrow, rectilinear stretch of water. Canals are a feature of western and Mogul formal gardens and are often associated with fountains.

canaliculate channelled with a long, concave groove, like a gutter.

Canarina (from the habitat of *C.canariensis*). Campanulaceae. Canary Islands, tropical E Africa. 3 species, fleshy-rooted, scrambling, perennial herbs differing from *Campanula* in their 6-parted flowers and berry-like fruit. Provide a minimum temperature of 7°C/45°F, full sun and an airy atmosphere. Water moderately during the growing and flowering period (late autumn to mid-spring) and provide support. Keep just dry during the summer dormancy and remove faded growth. Increase by seed, division or basal cuttings. *C.canariensis* (syn. *C.campanula*; CANARY ISLAND BELLFLOWER; to 1.5m; leaves hastate, serrate to sinuate; flowers large, pendulous, waxy, orange-yellow striped blood red to maroon, bell-shaped, with an upturned rim).

candelabriform of branching patterns and stellate hairs; candelabra-like, with tiered whorls or ranks of radiating or divergent branches.

cane the slender stem of a clump-forming perennial, produced annually and usually completing its development within a year or two. Canes are typically rather hard and jointed and may be hollow and silica-rich as in bamboos, woody as in *Rubus*, or fleshy as in *Dendrobium*.

cane fruit a collective term for raspberry, blackberry and hybrid berries. cf. *bush fruit*.

canescent hoary, or becoming so; densely covered with short, grey-white pubescence.

cane spot (*Elsinoe veneta*) a fungus causing spotting of the main stems, leaves and flower stalks of cane fruits, especially in wet seasons. The spots are purple, later turning grey with purple margins, becoming sunken with the outer tissue cracked. Dieback of canes and fruit infection may occur. Severely infected canes should be cut out and burnt; where regularly a serious problem a recommended fungicide spray should be applied repeatedly, commencing when the buds are early developed.

canker a general term for a sunken necrotic lesion of a root or stem, usually of a woody plant, often with malformation and outgrowth of the surrounding bark. It is caused by a variety of fungi and bacteria.

APPLE AND PEAR CANKER (*Nectria galligena*) is a fungus that kills twigs and fruit spurs and often forms deep cankers on older branches. The cankers elongate and become elliptical in shape as they grow, the bark shrinks in concentric rings, and the branch usually swells; the stem may become girdled and die back. Spores infect bark through leaf scars or wounds caused by frost, pruning, or pests and diseases. Fruits are sometimes infected: the rot starts around the eye and extends. Cut off and burn small affected branches and spurs; on larger branches, pare back to healthy tissue using a knife sterilized in sodium hypochorite; burn the parings and protect the wounds with a proprietary canker paint. A systemic fungicide programme for the control of scab and mildew should also keep canker in check. Wet, heavy soil aggravates canker, especially on susceptible apple cultivars such as 'Cox's Orange Pippin', 'James Grieve', 'Worcester Pearmain' and 'Spartan'. *N.galligena* can also affect ash, beech, hawthorn, poplar and *Sorbus* species.

Gibberella baccata f.sp. *mori* causes die-back of young shoots on mulberry; cut out affected parts.

GARDENIA CANKER (*Phomopsis gardeniae*) enters through small wounds. Swollen brown lesions occur near soil level and may girdle the stem, resulting in death. Obtain cuttings from healthy plants and raise in a sterile medium.

ROSE CANKER is commonly caused by *Leptosphaeria coniothyrium*, but *Cryptosporella umbrina*, *Clethridium corticola* and *Coniothyrium wernsdorffiae* can be involved. They enter through pruning wounds. Protect larger cuts with canker paint, and spray with Bordeaux mixture at pruning time.

CYPRESS CANKER (mainly *Seiridium cardinale*) is recorded throughout the world, and is usually referred to as CORYNEUM CANKER. Members of the Cupressaceae vary in susceptibility: the Monterey cypress (*Cupressus macrocarpa*), Italian cypress (*Cupressus sempervirens*) and Leyland cypress (× *Cupressocyparis leylandii*) are highly susceptible; the Lawson cypress (*Chamaecyparis lawsoniana*) and *Thuja plicata* are relatively resistant; *Juniperus sabina* and *Thuja occidentalis* are resistant. In Britain, the disease is only a problem on the Monterey cypress. The first symptoms are fading and death of old foliage at branch ends in spring, followed by the death of new growth and dying back of branches. Infection of bark and sapwood results from injury to twigs, branches or the main stem. Cankers may be 10–15cm long and 25mm wide; they become sunken as the rest of the stem continues to grow. On young, vigorous trees, resin exudes from the margins of lesions. Young trees with main-stem cankers can be killed within a year; multiple cankers on older trees can result in death within 5–10 years. The canker is spread locally by water splash or during cultural operations. All infected branches should be cut out well below the cankers, and badly affected trees removed.

CANKER STAIN of *Platanus* (*Ceratocystis fimbriata* f.sp. *platani*) is a lethal disease of the London plane (*P.acerifolia*). Infection enters through wounds in the bark or sapwood, and can also be introduced from diseased trees on saws, ropes and even wood-protection paints. Several species of Nitidulid sap beetles have been reported as carriers. Symptoms are darkly stained lesions in underlying layers of bark and sapwood. Infected sapwood dies and bark develops cankers. Vertical spread is rapid, up to three metres a year. External symptoms are long, shrunken, often cracked cankers following the grain of the wood. Infected tissues are killed and rapidly invaded by secondary wood-rotting fungi. Fully grown trees decline following infection, with thinning and yellowing of foliage over a period of years, which results in dieback and eventual death. Young trees of pole size may die within two years. There is no cure for infected trees. Remove diseased trees as soon as they are identified; protect healthy trees from injury; and disinfect pruning tools.

EUTYPA CANKER and die-back (*Eutypa lata*), also known as DEAD ARM or DYING ARM, is a serious disease of the grapevine (*Vitis* species) and of apricot (*Prunus armeniaca*) in Europe, Australia, North America and South Africa. *Phomopsis viticola* is a common secondary invader of dead arm infection. Several strains of *E.lata* occur, all pathogenic to the grapevine but of different degrees of virulence to apricot. The fungus is widespread as a secondary organism on the dead twigs of woody plants such as apple, ash, hawthorn and ivy, which act as a constant source of infection. The fungus is not usually seen on vines less than seven years old. First symptoms are weak growth of the new season's shoots in mid-spring; the leaves are small and distorted and may become yellowed and spotted. The vascular tissue of dying branches shows dead areas beneath pruning wounds; death of these branches (dead arm) occurs the following year. Symptoms on apricots are cankers on branches, usually with gumming, resulting in dead foliage that remains attached to the tree until winter; death of the branch follows. Remove and burn dead wood during dry spells; treat pruning wounds with fungicide or antagonistic fungus immediately after pruning. Pneumatic secateurs with a spraying attachment combine pruning and treatment operations.

BOX CANKER and BLIGHT is caused by *Pseudonectria rousseliana*. Affected plants are slow to start into growth in the spring, and foliage becomes straw-coloured; diseased leaves and branches are covered with pink fungal spore pustules. Cankers develop on main stems and branches, often following winter injury. Remove all cankered stems and dead foliage, and apply fungicide.

ELAEAGNUS CANKER (*Phomopsis elaeagni*) is a serious fungal disease of *Elaeagnus angustifolia*, which is widely planted in North America. It is prevalent on nursery plants. Other species of fungi also cause cankers of *Elaeagnus* in the US.

CANKER OF BRASSICAS (also known as BLACK LEG) is caused by *Leptosphaeria maculan*. Symptoms include damping-off as well as stem canker. It is common on oilseed rape (*Brassica napus*) and often affects broccoli (*B.oleracea* Italica Group) and Brussels sprouts (*B.oleracea* Gemmifera Group), especially when grown from home-saved seed. Elongated, sunken lesions dotted with black fungal bodies occur at ground level, and small, circular brown spots on leaves. On swedes (*B.napus* Napobrassica Group) and turnips (*B.rapa* Rapifera Group), the canker causes a dry rot on roots. The fun-

gus is seed-borne; commercial seed will normally be derived from healthy crops and treated with fungicide.

BLACK CANKER of parsnips is caused by *Itersonilia pastinacae* and *Phoma* species. Lesions on roots spread and develop into black rot around the shoulder and crown. On black fen soils, canker may also be caused by *Mycocentrospora acerina*, where the rot is purple rather than black. An orange-brown canker with roughened skin is caused by other weakly parasitic soil-inhabiting fungi. Spores of *Itersonilia* arise from leaf spots and are washed into the soil to cause original infection. Earthing up roots, rotation and close spacing are useful control measures. The cultivar 'Avonresister' is resistant to all three types of canker. See *bleeding canker*.

cankerworm see *winter moths*.

Canna (a Classical name for a type of reed). Cannaceae. New World subtropics and tropics. 9 species, tender or semi-hardy perennial herbs, deciduous to evergreen with fleshy rhizomes, broad leaves, and, in late spring and summer, erect to arching, terminal racemes and panicles of showy flowers with three sepals, three petals, and one to four staminodes, some of them petal-like.

Cultivars of *C. × generalis* are excellent plants for the cool greenhouse or, outdoors, in tubs and bedding displays, massed or as dot plants. Plant in a rich, loamy, moist soil in full sun. Where frosts are common and hard, lift the rhizomes once the top growth has withered and overwinter them, cool but frost-free, in barely damp wood shavings or vermiculite. In early spring, start the rhizomes into growth in pots in warm, humid conditions. Plant out in late spring. In favoured districts in zones 7 and over, plants might be risked outside all year in a sunny, sheltered position with a thick mulch in winter. *C.indica* can be similarly treated, but prefers a warmer and damper site and will take some shade. It seldom survives heavy frosts. Under glass, it is an attractive plant for large containers and beds, remaining in leaf and flower throughout the year if encouraged to form large clumps and given a minimum temperature of 10°C/50°F. *C.iridifolia* grows best in a humid greenhouse with a minimum temperature of 18°C/65°F. It needs plenty of headroom and a moist, fertile, loamy medium. At high temperatures, this species thrives in saturated conditions and associates well with other large, tropical and subtropical marginals, e.g. *Cyperus* and *Thalia*. Propagate all by division in spring. Soak seed of species in warm water for 24 hours or make a small nick in the seed coat prior to sowing in a soilless medium at 20°C/70°F.

C. × generalis (syn. *C. × orchiodes*; CANNA LILY; gardenhybrids derived from a number of species and much interbred, ranging from 30–130cm and from low and clump-forming to tall and single-stemmed; leaves 30–80cm long, broadly ovate to narrowly oblong, blue-green to jade striped gold, from bronze to purple-red to chocolate veined fiery orange;

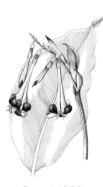

Canna iridifolia

flowers large and ruffled (to 8cm in diameter) or small and spidery, white, cream, yellow, orange, peach, coral pink, vermilion, scarlet, blood red, or combinations thereof in patches, spots and streaks); *C.indica* (syn. *C.musifolia*; INDIAN SHOT; C and S America; 0.5–2m; smooth, mid-green leaves to 50cm long, usually with red-purple stained sheaths or midribs; flowers, to 6cm across, with narrow segments in shades of scarlet and orange, marked yellow and held among purple-red bracts; includes 'Purpurea', to 2m tall, with large, paddle-like purple-bronze leaves and small, scarlet flowers); *C.iridifolia* (Peru; 1–3m leaves to 1m broadly oblong, blue-green; flowers to 6cm, with curving, narrowly lanceolate segments in shades of deep pink or orange).

Cantua ('cantua' is the Quecha name for *C.buxifolia*). Polemoniaceae. Temperate S America. 6 species, evergreen shrubs with tubular, 5-lobed flowers in pendulous, terminal corymbs in late spring and summer. In Mediterranean-type climates, *Cantua* will grow in the open border or year round in tubs on the terrace. In zones 6 to 8, it requires the protection of a sheltered, south-facing wall; alternatively, grow in borders and tubs in the cool greenhouse or conservatory. Under glass grow in sun, water plentifully in spring and summer, sparingly in winter. Root semi-ripe cuttings in a closed case with bottom heat. *C.buxifolia* (SACRED or MAGIC FLOWER OF THE INCAS; Peru, Bolivia, Chile; erect to spreading shrub 1–5m tall, flowers 6–8cm long, narrowly trumpet-shaped, the tube pink to purple with yellow stripes, the spreading lobes cerise to crimson or scarlet).

cap, **capping** (1) the formation of an impermeable crust on a soil surface, usually due to the impact of rain drops on weakly structured, fine-textured soils. Capping can reduce water infiltration and aeration and form a barrier to emerging seedlings. A capped surface may alternatively be referred to as caked. (2) a layer of soil or plastic sheeting placed over a compost heap.

cape gooseberry (*Physalis peruviana*) a half-hardy perennial up to 1m in height, native to Tropical South America; grown as an annual for its edible, golden, cherry-sized fruits that are borne within persistent lantern-shaped calyces. It may be grown in sheltered sites outdoors, or under greenhouse protection in border soil or pots. Sow seeds 6mm deep, in pots or trays, in gentle heat during spring. Seedlings are pricked out into small pots or other modules, and planted out when established, 75cm apart in ground beds or into 25cm pots. Support individual plants with canes 75cm out of the ground. Plants may be pinched when 30cm high to induce branching. Water sparingly and apply liquid fertilizer as for tomatoes after the first flowers are set. Fruits on plants outdoors may be slow to ripen; they are ready to harvest when golden yellow, and the sweet, distinctive flavour improves where left for a week or two on the plant. The cultivar 'Goldenberry' produces large fruits.

capillarity the movement of liquid as a result of surface tension. Capillarity is of horticultural relevance to the availability of soil moisture and the functioning of certain watering systems.

Capillary watering applies to the irrigation of container-grown plants, utilizing the property of upward movement of water from a wetted mat or sand bed. A *capillary bed* is a specially constructed ground-level bed, usually of fine sand, used especially for the capillary watering of container-grown ornamentals. A *capillary bench* is a carefully levelled greenhouse bench covered by a capillary mat (usually made of synthetic fibre or wool) or sand, used for capillary watering of pot plants.

capillary slender and hair-like; much as filiform, but even more delicate.

capitate (1) arranged in heads, as in the inflorescence of Compositae; (2) terminating in a knob or somewhat spherical tip.

capitulum a head of densely clustered and sessile or subsessile flowers or florets on a compressed axis, characteristically found in members of the daisy family, Compositae.

caprification cross-pollination in the fig (*Ficus carica*), accomplished by the FIG WASP (*Blastophaga psenes*), which breeds inside the wild form or caprifig. On emergence, with pollen adhering to its body, the wasp is attracted to the inflorescences of female figs and transfers pollen on contact. Caprification is a necessary process for certain cultivated figs, such as the ancient 'Smyrna' grown for drying, but modern cultivars are self-fertile.

Capsicum (from Greek *kapto*, to bite, referring to the bitingly hot taste of chilli). Solanaceae. Tropical America. PEPPER; GREEN PEPPER, BELL PEPPER, CHILLI PEPPER. 10 species, shrubby, annual to perennial herbs grown for their edible fruits – these are large, chambered berries with thick, sweet to bitingly pungent rinds. The two principal crop species are *C.annuum* (CAPSICUM, RED PEPPER, SWEET PEPPER, PAPRIKA), and *C.frutescens* (HOT PEPPER, SPUR PEPPER, TABASCO PEPPER), both much-developed and highly varied. *See peppers.* Some ornamental (and inedible) cultivars are grown, usually as annuals for container display in the home and cool greenhouse, and for summer bedding. They are listed below. Sow seed in spring with a minimum temperature of 10°C/50°F. Grow in a fertile, loamy medium with plentiful water in full sun. Syringe in summer to promote fruit set and discourage red spider mite. *C.annuum* – 'Holiday Cheer' (bushy with a dense head of round fruits turning from green to cream, yellow and red), 'Holiday Time' (bushy with a dense head of conical fruits turning from green to yellow, red, and, sometimes, purple), 'Piccolo' (slender and tree-like in outline, with small leaves variegated cream, dark and pale green and

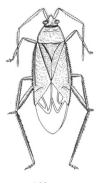

capsid bug

purple, small, mauve flowers and round, purple-black fruits); *C.frutescens* – 'Chameleon' (low and bushy with fruits turning from yellow to purple, then red), 'Long Gold Spike' (fruits conical, gold to brick red, erect in a spire-like, branching head to 1m long; a fine plant for cut displays in autumn).

capsid bugs (Hemiptera: Miridae) insects up to 6mm long, with relatively long probosci, legs and antennae. Adults have hardened forewings that lie flat and cover the membranous hindwings and they vary in colour from green to red-brown. All are elusive and deceptively common. The cylindrical curved eggs are either deposited on the surface of plants or inserted into plant tissue. Some species overwinter in the egg stage on woody plants, others pass the winter as adults in sheltered situations. Emergent nymphs resemble the adults but are smaller, sexually immature and wingless. Most capsids are plant feeders; some are largely predacious, including the BLACK-KNEED CAPSID (*Blepharidopterus angulatus*) and FLOWER BUGS (*Anthocoris* species). Plant-feeding species infest a wide range of fruit, vegetable, woody and herbaceous plants. Shoots become distorted or blind; leaves become puckered with irregular-shaped holes developing around the punctures; blooms may be malformed or aborted; and fruits of apples and pears develop raised corky patches.

The principal European species are the TARNISHED PLANT BUG (*Lygus rugulipennis*), which is green mottled with red-brown, and mostly attacks ornamentals; the COMMON GREEN CAPSID (*Lygocoris pabulinus*), which is bright green with darker markings, and is particularly injurious to soft and top fruit, but also attacks ornamentals; the POTATO CAPSID (*Calacoris norwegicus*), which is similar to the common green capsid in colour and found chiefly on potatoes and a range of Compositae; and the APPLE CAPSID (*Plesiocoris rugicollis*), which is similar to the common green capsid in colour, and mainly attacks apples and soft fruits.

In North America, the name tarnished plant bug is applied to *Lygus lineolaris*, a capsid similar in biology and appearance to the European species but which has 3–5 generations each year, depending on latitude. It attacks a range of garden plants, particularly apples, peaches and strawberries, which become misshapen. The FOUR-LINED PLANT BUG (*Poecilocapsus lineatus*) is yellow-green with four dark stripes running lengthwise down the back. There is one generation each year, and it overwinters in the egg stage. It attacks woody and herbaceous plants, especially currants. Weeds serve as alternative hosts to capsids and should be destroyed. Using a recommended insecticide, apples and pears should be sprayed after petal fall, and susceptible ornamental plants at the first sign of capsid presence. The level of damage to vegetables does not usually warrant control.

capsule a dry, dehiscent seed vessel.

Cardamine pentaphyllos

septicidal capsule

Caragana (Latinized form of karaghan, Mongolian name for *C.arborescens*). Leguminosae. E Europe, C Asia. PEA TREE, PEA SHRUB. 80 species, very hardy deciduous shrubs and small trees, with neat, pinnate leaves, the leafstalks often sharply tipped and persisting as spines. Small, pea-like flowers are borne singly or in clusters in spring. They will survive almost any conditions, except wet and shade. Top-grafted on stocks of the typical plant, the weeping filigree-leaved cultivars of *C.arborescens* are beautiful small trees, although more prone to cold and wind damage. Propagate from seed sown when ripe or pre-soaked in warm water, or take softwood and semi-ripe cuttings in summer.

C.arborescens (SIBERIAN PEA TREE; tree-like shrub to 6m tall with dark green leaves composed of eight to twelve leaflets; flowers pale yellow; includes 'Lorbergii', with very slender, ribbon-like leaflets); *C.frutex* (RUSSIAN PEA SHRUB; upright, suckering shrub to 3m tall; leaves with four leaflets; flowers yellow; includes 'Globosa', more rounded and compact).

Caralluma (from *carallum*, name given to *C.adscendens* by the Telugas of Coromandel, India). Asclepiadaceae. Spain, Africa, India, Burma. Some 80 species, clump-forming, leafless perennial herbs with erect to sprawling, succulent stems, angled and with rows of teeth or tubercles. Often malodorous, the flowers are fleshy and star- to cup-shaped with five, acute lobes in spring and summer. Cultivation as for *Stapelia*.

C.adscendens (India, Burma, Sri Lanka; stems to 60 cm; flowers outspread to bell-shaped, erect to pendulous, to 2cm wide, green to tawny or chestnut, spotted or striped purple, often hairy); *C.europaea* (N Africa, S Europe; stems to 15cm; flowers to 2cm wide, spreading to campanulate, yellow to red-brown with purple bands, more or less hairy); *C.joannis* (Morocco; stems to 10cm; flowers to 2.5cm, bell- to star-shaped, olive-yellow, spotted red, lobes marked purple with small hairs on margins).

carbon/nitrogen ratio the relative proportion in plant remains of carbon (C) and nitrogen (N). The C/N ratio influences the respective rates of decomposition of carbon and nitrogen in soil or compost heaps. Where the ratio is high, as in woody debris or straw, decomposition is slow, nitrogen is depleted and a supplementary dressing is advisable.

carbonate of lime see *calcium carbonate*.

Cardamine (diminutive form of Greek *kardamomon*, a name used by Dioscorides for cress). Cruciferae. N Hemisphere. BITTERCRESS. 150 species, perennial or annual herbs (those listed here are hardy perennials) with racemes or panicles of 4-parted flowers in spring and early summer. Plant in semi-wooded locations and herbaceous borders (*C.enneaphyllos*, *C.pentaphyllos*), or in shaded rock gardens, borders and woodland (*C.trifolia*), or naturalize in long, damp turf (*C.pratensis*). They favour a moist, cool and fertile soil. Increase by division or seed in autumn; *C.pratensis* also by leaf cuttings.

C.enneaphyllos (syn. *Dentaria enneaphylla*; W Carpathians, E Alps, S Italy; 10–35cm tall, with a scaly, fleshy rhizome; leaves pinnately lobed, each leaflet ovate to lanceolate, toothed; flowers to 3cm across, white to cream, nodding in a congested raceme); *C.pentaphyllos* (syn. *Dentaria pentaphylla*; mountains of W and C Europe; 15–60cm tall, with a scaly, creeping rhizome; leaves digitately lobed with five, lanceolate to ovate, toothed leaflets; flowers to 3cm across, rose to pale lilac or white in congested racemes); *C.pratensis* (LADY'S SMOCK, CUCKOO FLOWER; widespread in Europe; 6–30cm tall, clump-forming; leaves in a loose rosette, pinnate with small, broad leaflets; flowers to 2cm in diameter, white veined to flushed pale rose to lilac, in slender, erect racemes; includes the double-flowered 'Florepleno'); *C.trifolia* (C and S Europe; 10–30cm tall, with a creeping rhizome; leaves neat, composed of three, broad and toothed, dark green leaflets; flowers to 2cm in diameter, white or pink, in racemes atop slender, bare stalks).

Cardiocrinum (from Greek *kardia*, heart, and *krinon*, lily). Liliaceae (Liliaceae). E Asia. GIANT LILY. 3 species, spectacular perennial herbs, with single, stout stems arising from annually produced bulbs, glossy, deep green heart-shaped leaves and, in summer, large, funnel-shaped, white flowers with six tepals. Suited to the woodland garden or a shady, moist border among shrubs, the giant lily is hardy to –10°C/14°F, but fares better if thickly mulched in winter. Plant in a deep, nutrient- and humus-rich soil that remains cool and moist. The bulbs perish after flowering but are replaced by daughter bulbs that take three to five years to reach flowering size. The seed capsules are large and attractive, bursting in autumn to release many papery seeds – these will germinate where they fall or in a cool, shaded frame. Seed-grown plants may take seven years to flower, but tend to be taller and more robust than vegetatively propagated plants.

The most popular species is *C.giganteum* from Himalaya, Burma and China. It grows to 4m tall with a clump of basal leaves to 45cm long, and stem leaves diminishing in size toward its summit. The fragrant flowers are regularly funnelform, to 20cm long and marked maroon inside. The Chinese *C.cathayanum* is shorter than *C.giganteum* (to 1.5m) and also differs in having leaves absent from the lower part of the stem, but scattered in its upper two thirds. The flowers are 10cm long, irregularly funnelform and cream tipped and spotted purple. The Japanese *C.cordatum* grows to 2m tall, with leaves arranged in a loose whorl in the lower half of the stem and 15cm-long, irregularly funnelform, cream flowers blotched yellow and spotted red-brown at the base within.

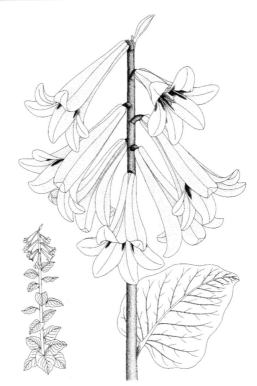

Cardiocrinum giganteum

Cardiospermum (from Greek *kardia*, heart, and *sperma*, seed: the seed is marked with a heart-shaped white spot). Sapindaceae. Tropical Africa, America, India. 14 species, slender-stemmed herbs or shrubs climbing by tendrils. The leaves are usually biternately compound with toothed or pinnately cut leaflets. In summer, small, 4-petalled flowers are followed by inflated capsules. They require frost-free conditions, although *C.halicacabum* is often grown as a half-hardy annual, the seed sown under glass in early spring or *in situ* in early summer. Grow in full sun or shade in a fertile, loam-based medium with ample moisture; provide support. Both species can also be propagated by softwood cuttings in spring.

C.grandiflorum (HEARTSEED; tropical America and Africa; perennial, herbaceous vine climbing to 8m; leaves to 20cm long, hairy beneath; flowers fragrant, cream, to 1cm wide; fruit smooth, triangular-ovoid, to 7 × 4cm); (*C.halicacabum*; BALLOON VINE, HEART PEA; tropical Asia, Africa and America, naturalized southern US, Australia; vine to 3m; differs from the first species in its smooth leaves to 12cm, smaller, white and scentless flowers and very inflated capsules, bladder-like, bristly and to 2.5 × 2.5cm).

cardoon (*Cynara cardunculus*) a herbaceous perennial, to 2m high, closely related to the GLOBE ARTICHOKE (*Cynara cardunculus* Scolymus Group) and grown for its edible hearts and fleshy petioles which are sometimes blanched. Serve raw or boiled. Sow seed 3–4cm deep in rows 1.5m apart; thin to final spacings of 50cm.

cardoon

Pests
aphids
slugs

Carex (from Greek *keiro*, to cut, alluding to the sharp leaf margins). Cyperaceae. Cosmopolitan. SEDGE. Some 1000 species, evergreen, rhizomatous or tufted, perennial herbs with short or distinct and slender stems, usually long, narrow and grass-like leaves and very small, grain-like flowers in male or female spikes arranged together in a panicle. All prefer a moist, neutral to slightly acid soil with a high organic content. The very hardy weeping sedges, *C.pendula* and *C.pseudocyperus*, grow well at the waterside or even partially submerged, and will soon self-sow. *C.atrata*, *C.grayi* and *C.riparia* can be used to colonize the margins of water and bog gardens. The broad-leaved sedges, *C.plantaginea* and *C.siderosticha*, are handsome groundcover for damp and shaded borders. The narrow-leaved Japanese sedge grasses (*C.hachijoensis*, *C.morrowii* and *C.oshimensis*) thrive in similar conditions, but will also take more light and less water. *C.fraseri* naturalizes well in light, acid woodland and the bog garden, making a carpet of white, button-like flowerheads in early spring. The bronze and copper New Zealand sedges favour a sheltered position in zones 7 and 8, with full sunlight, on acid, gritty, moist soils. *C.atrata* and the more tender *C.baccans* produce attractive fruiting heads (jet black and brick red respectively). The palm sedges, *C.muskingumensis* and *C.phyllocephala*, make dense stands of reedy stems, hardy in zone 7 and ideal for stark modern landscapes. The variegated *C.phyllocephala* 'Sparkler' is an excellent container plant for the cool greenhouse, likewise *C.brunnea* 'Variegata'. Propagate all by division in early spring.

C.atrata (BLACK SEDGE, JET SEDGE; Europe; clump-forming, to 60cm tall; stems dark below; leaves slender, pale blue-green; female spikes to 5cm, rounded to oblong, purple-black); *C.baccans* (CRIMSON-SEEDED SEDGE; India to China; loosely tufted to 1m tall; leaves to 1cm wide, leathery, flat, deep green; in fruit, female spikes resemble tight clusters of orange-red berries); *C.berggrenii* (New Zealand; broad, low and tufted with short, linear leaves that may be metallic grey-brown, bronze,

glaucous and, in some selections, distinctively narrow); *C.brunnea* (Japan, Australia; stems slender, tufted, 30–90cm tall; leaves to 0.5cm wide, yellow- to bright green, robust; spikes nodding; includes 'Variegata', small, with leaves broadly edged yellow); *C.buchananii* (LEATHERLEAF SEDGE; New Zealand; densely tufted, 10–70cm tall; leaves very slender, straight and erect to outward-leaning and arching, often curling and whip-like at tips, pink-orange to red-green below to copper or bronze above); *C.comans* (NEW ZEALAND HAIRY SEDGE; New Zealand; densely tufted, 6–40cm tall; leaves very slender, arching to drooping in a pale green mophead, sometimes taking on red-bronze tints in cold weather; includes forms with deep bronze leaves, and 'Frosted Curls', with very narrow, hair-like, curling and tangled leaves, bronze-green to bleached grey-green at tips in winter); *C.conica* (Japan, S Korea; tufted, to 50cm tall; leaves to 0.4cm wide, dark green, rigid, flat, rough; includes 'Snowline', small, neat, with arching leaves dark green edged white, and 'Variegata', with leaves variegated white); *C.elata* (TUFTED SEDGE; Europe; tussock-forming, 20–100cm tall; leaves fine, to 0.6cm wide, folded in section, glaucous; includes 'Aurea', with leaves edged yellow, 'Bowles Golden', with golden yellow leaves edged green, and 'Knightshayes', with yellow leaves); *C.firma* (Europe; to 20 cm tall, densely tufted, mat-forming; leaves to 10cm, narrow, rigid, pointed, blue-green; includes 'Variegata', with leaves striped creamy yellow); *C.flagellifera* (New Zealand; resembles a taller, coarser *C.buchananii*; leaves whip-like, pale bronze to tawny brown, arching to form a low dome); *C.fraseri* (syn. *Cymophyllus fraseri*; FRASER'S SEDGE; US; tufted, to 35 cm tall; leaves to 5cm wide, smooth, rigid, finely toothed; spikes ball-like, bright white); *C.grayi* (GRAY'S SEDGE, MACE SEDGE; N America; clump-forming, 30–100cm tall; leaves flat, to 1cm wide, pale green, margins rough; female spikes to 3cm wide, round and spiky, resembling a mace); *C.hachijoensis*

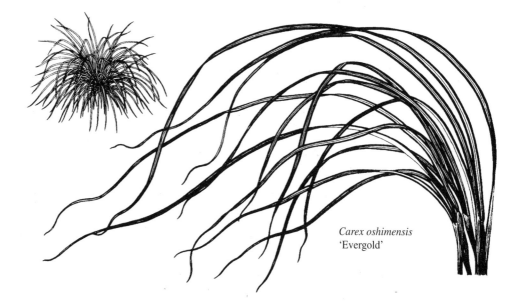

Carex oshimensis
'Evergold'

(Japan; differs from *C.morrowii* in narrower, softer leaves; for the cultivars 'Evergold' and 'Variegata', see under *C.oshimensis* and *C.morrowii* respectively); *C.morrowii* (Japan; resembles *C.oshimensis* but with slightly wider leaves with rough margins; includes 'Fisher's Form', large, with leaves striped cream, and 'Variegata', with leaves narrowly striped white near margins – this cultivar is often included in *C.hachijoensis*); *C.muskingumensis* (PALM SEDGE; N America; stems narrowly cane-like, erect, tufted, to 1m tall, ranked with spreading to arching, slender, bright green leaves; includes 'Wachtposten' syn. 'Sentry Tower', a particularly fine form, with long stems and shining, deep green leaves); *C.oshimensis* (JAPANESE SEDGE GRASS; Japan; to 30cm tall, densely tufted, making a spilling, hair-like tussock; leaves fine and narrow, smooth, deep, glossy green, typically with narrow white margins; includes 'Aureavariegata', with leaves with a broad, central, golden stripe, 'Evergold', with leaves with a creamy yellow central stripe – this cultivar is often included in *C.hachijoensis* – and 'Variegata', with leaves with a central white stripe); *C.pendula* (WEEPING SEDGE, PENDULOUS SEDGE; Europe; clump-forming, to 1.5m tall; leaves to 2cm wide, long and finely tapering, coarse; spikes to 8cm long, cylindrical, pendulous on slender-branched, arching panicles); *C.petriei* (New Zealand; resembles a smaller *C.buchananii*, with narrower, pinker leaves and stouter spikes); *C.phyllocephala* (Japan; stems tufted, cane-like, to 45 cm tall, with long, narrowly strap-shaped leaves at their summit; includes 'Sparkler', with lime to dark green leaves striped and edged white); *C.pilulifera* (PILL SEDGE; dwarf, tufted, to 30cm tall; leaves very fine; spikes in a small head; includes 'Tinney's Princess', with very small leaves striped creamy yellow); *C.plantaginea* (N America; tufted, to 35cm tall; leaves broadly strap-shaped, ribbed; spikes purple); *C.pseudocyperus* (CYPERUS SEDGE; temperate regions; resembles a smaller, softer *C.pendula* with coarser spikes; to 80cm tall; leaves flat, to 1cm across, bright green; spikes to 5cm long, green-brown pendulous on arching stems); *C.riparia* (Northern Hemisphere; to 1m tall, fast-spreading; leaves flat, to 1.5cm wide; spikes to 10cm, cylindrical; includes 'Variegata', with striped or almost wholly white leaves); *C.siderosticha* (to 10cm tall, with smooth, lax, strap-shaped leaves in clumps, emerging pink turning bright green; includes 'Variegata', with leaves edged and striped white); *C.uncifolia* (New Zealand; spreading or loosely tufted, to 10cm tall; leaves very narrow, curving, with curly tips, pink to red-brown).

carina keel, (1) the anterior petals of a papilionaceous flower; (2) the keel of the glume in flowers of Gramineae; (3) the midvein of a leaf, petal or sepal, prominent to ridged beneath.

carinate of a leaf, bract or perianth segment, boat-shaped or, more usually, keeled, with a line or ridge along the centre of its dorsal surface.

Carissa (based on an Indian name). Apocynaceae. Old World tropics and subtropics. 20 species, evergreen shrubs and small trees with intricate, often spiny branches, glossy, ovate to oblong leaves and clusters of star-shaped, fragrant flowers. These have a slender corolla tube expanding to a spreading limb of five, oblong to obovate lobes; they are followed by colourful, succulent fruit. Grow on most well-drained, moderately fertile soils in full sun or light shade. Tolerant of light frost, wind, salt spray and short-lived drought, it makes an excellent specimen planting, screen, hedge or groundcover. In zones 8 and below, grow in the cool greenhouse or conservatory. It responds well to frequent clipping and hard pruning. Propagate by heeled, semi-ripe cuttings of side shoots rooted in a warm case.

C.macrocarpa (syn. *C.grandiflora*; NATAL PLUM; S Africa; shrub or tree to 9m tall; branches tipped with formidable forked spines and clothed with glossy green leaves to 6cm long; flowers to 5cm wide, white and jasmine-scented; fruits plum-shaped and bright red to purple-black, 2–5cm-long, edible; 'Horizontalis': NATAL CREEPER, dense and trailing, leaves very small, fruit vivid scarlet).

Carissa macrocarpa

carnation (*Dianthus caryophyllus*) The wild ancestor of the border and perpetual flowering carnations originated in the Pyrenees and has been cultivated in western Europe since the 15th century. The modern classification of show carnations includes Selfs (all colours except blue); White-, Yellow- and Apricot-ground Fancies; Other Fancies; White- and Yellow-ground Picotees; Other Picotees; Cloves. Fancies are striped, flaked, or suffused with one or more contrasting or blended colours.

Border carnations are grown as annuals or perennials for cutting or garden decoration and are 40–60cm high; they flower prolifically in midsummer and prefer a cool temperature climate. Grow unstopped but staked. Propagate perennial beds by layering every three or four years, soon after flowering. Plant out in autumn or early spring 38–45cm apart each way. When grown as annuals, sow seed in early spring under protection at 16°C/60°F.

Perpetual flowering carnations are grown primarily as fragrant cut flowers. They require light, airy, greenhouse protection for good quality and a winter minimum of 4–7°C/40–45°F. They are raised from cuttings, potted into 8cm-diameter containers and stopped at eight or nine pairs of leaves. Selected side shoots are stopped thereafter. Grow on in 15cm-diameter pots for the first year, then pot up into 21cm-diameter pots and stand on a deep gravel base. Support with 1.2m canes and wire hoops. Disbud for large blooms, leaving only the crown bud; for sprays, remove only the crown bud.

American spray carnations are an important commercial cut flower of comparatively recent introduction, originating in Connecticut from selected crosses. Culture is as for perpetual flowering carnations except that the crown buds are always removed to induce sprays.

carnation

Pests
aphids
carnation
    tortrix moth
scale insects
thrips

Diseases
grey mould
rust
wilt
virus

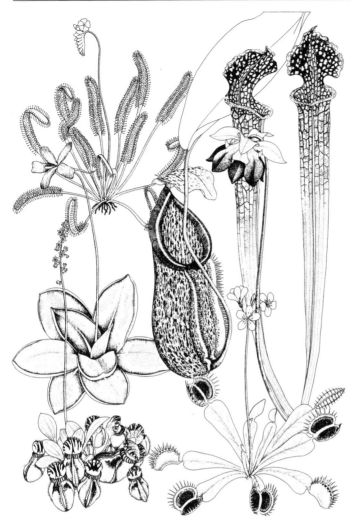

carnivorous plants (clockwise from top right) *Sarracenia leucophylla, Dionaea muscipula, Cephalotus follicularis, Pinguicula moranensis, Drosera capensis,* (centre) *Nepenthes* x *hookeriana*

Malmaison carnations The name alludes to the resemblance of the blooms to the Bourbon rose 'Souvenir de la Malmaison'. They were first raised in France in 1857 from a seed mutant, and bud sports followed. Cultivation is as for perpetual flowering carnations but they require more shade in summer. Disbud to leave 3–4 buds per stem. Propagate by cuttings in winter or by layers in summer. See *pinks*.

Carnegiea (for Andrew Carnegie (1835–1919), steel magnate, philanthropist and patron of science). Cactaceae. 1 species, *C.gigantea* (SAGUARO), from the southwestern US and northern Mexico. This giant cactus grows to 16m tall with a trunk some 75cm thick. For many years it remains unbranched, as a deeply ribbed column with grey-brown spines to 7cm long. Produced at the column's summit in spring and summer, the white, funnel-shaped flowers are some 10cm long and open at night. Edible, red-green fruit split jaggedly from top to bottom, revealing bright

red pulp. When they finally appear, the branches are few and erect, resembling the main stem. Grow in a frost-free garden or a heated greenhouse (minimum temperature 10ºC/50ºF) in a gritty and sandy, loam-based medium, pH 6–7.5. Under glass, admit full sunlight except in very hot weather and maintain low humidity and good air circulation. Keep dry from mid-autumn to early spring, except for a light misting on warm days in late winter. Branch cuttings will root but are seldom available. In view of its endangered status in the wild, saguaro is best raised from seed sown on a fine, sandy medium in a heated case. It is very slow-growing.

carnivorous plant a plant adapted to attract, capture and digest small creatures for its own nutritional benefit. Recognized in 450–500 species within the families Sarraceniaceae, Nepenthaceae, Cephalotaceae, Droseraceae, Byblidaceae, Lentibulariaceae, Dioncophyllaceae, Roridulaceae and Bromeliaceae, carnivorous plants thrive in moist acidic conditions. The adaptation serves to provide a protein source of nitrogen in habitats where the nutrient is otherwise deficient. Their prey is predominantly insects (hence the common description, insectivorous plant) but may include protozoa, spiders, crustacea, amphibians and even small mammals.

There are four trapping mechanisms: (1) the snap-trap, with hinged leaves resembling an open jaw, as in VENUS' FLY TRAP (*Dionaea muscipula*), to which the prey are attracted by pigment and secretion; (2) suction, as in *Utricularia* species, where the trap is a small bladder with appendages and mouth covered by a trap door; a partial vacuum arising in the closed bladder keeps the trap shut. The finely balanced mechanism is disrupted by the prey, which is then drawn into the open receptacle in a flow of air or water; (3) pitfall traps, as in *Sarracenia*, where modified leaves form pitchers into which insects are lured and then trapped by surface hairs and wax; (4) flypaper traps, as in the SUNDEWS (*Drosera* species), where insects are attracted to glistening glands on leaf and stem surfaces; they are overwhelmed by copious secretions of sticky fluid and then digested. For cultivation details refer to the above genera and *Byblis, Cephalotus, Darlingtonia, Drosophyllum, Heliamphora, Nepenthes, Pinguicula.*

carotenoids yellow, orange, or red pigments, including carotene and xanthophyll, that are associated with chlorophyll and responsible for these colours in many plant parts.

carpel a female sporophyll, a simple pistil or one element of a compound pistil bearing an ovule.

Carpenteria (for Professor William Carpenter (1811–1848), Lousiana physician). Hydrangeaceae. California. 1 species, *C.californica* (TREE ANEMONE), an evergreen shrub to 2m tall with glossy, elliptic to oblong leaves to 10cm long. Produced in summer, the fragrant flowers are 5–7cm across, pure white, with five to seven, spreading petals and a central boss of golden. Plant in full sun with shelter from strong winds on a well-drained soil. It will tolerate occasional lows to –15°C/5°F if grown on a warm, south-facing wall and mulched in winter. It withstands dry conditions. Propagate by seed in spring and autumn, or by basal ripewood cuttings in spring and greenwood cuttings in summer (both rooted in a case with bottom heat).

carpet bedding see *bedding*.

Carpinus (Classical Latin name). Betulaceae. Europe, E Asia, N and C America. HORNBEAM. 35 species, mostly deciduous trees with slender branchlets and ovate to cordate leaves, usually toothed and with ribbed veins. Inconspicuous female flowers are carried in pendulous, hop-like catkins clustered with narrow, leafy bracts; these persist and are an attractive feature of the tree. *C.betulus* can be used for formal hedging, informal screening, and for pleaching. It and the remaining species are fine specimen trees, flourishing on most soils, especially chalk. Although all are fully hardy, late frosts may damage the new growth. Leave newly planted hornbeam hedges to grow for two seasons, pruning hard thereafter in late winter and trimming in late summer. Sown in autumn in an open bed, ripe seed will germinate the following spring. Taken in late summer, wounded and treated with a rooting agent, cuttings will root in a closed case or under mist.

C.betulus (COMMON HORNBEAM; Europe, Asia Minor; broad, pyramidal- to irregular-crowned tree to 20m with a fluted trunk and grey bark; leaves 6–12cm long, dentate and bright to mid-green, becoming amber to clear yellow in autumn; 'Columnaris': similar to 'Fastigiata', but slow-growing, dense, compact and columnar; 'Fastigiata': medium-sized, narrowly pyramidal and erect, becoming broader with age; 'Frans Fontaine': similar to the last but retaining narrow, fastigiate habit; 'Incisa': with small, narrow leaves with one or a few deep indentations; 'Pendula': dwarf and domed with strongly weeping branches; 'Purpurea': leaves tinted purple at first; 'Variegata': leaves sporadically marked creamy white); *C.caroliniana* (syn. *C.americana*; AMERICAN HORNBEAM, BLUE BEECH; eastern N America and C America; grey-barked, spreading tree to 13m tall with arching branches; leaves to 20cm long, toothed and dark to glaucous blue-green, assuming red-orange tints in autumn); *C.japonica* (JAPANESE HORNBEAM; widely spreading tree or small shrub to 15m tall with scaly, grey bark; leaves 5–10cm long, downy, with small teeth and deeply impressed veins giving a corrugated appearance, emerging with a red tint and opening deep green); *C.turczaninovii* (China, Japan, Korea; small, shrubby tree to 12m tall with slender branches and red-tinted new growth; leaves, to 6cm long, and serrate).

Carpobrotus (from Greek *karpos*, fruit, and *brota*, 'edibles': some species have fleshy edible fruit). Aizoaceae. South Africa, Australia, Mexico, Chile, widely naturalized elsewhere. ICE PLANT. 30 species,

mat-forming succulents with creeping, slightly woody stems, narrow, cylindrical to 3-sided, thickly fleshy leaves, and flushes of short-stalked, daisy-like flowers opening *en masse* and for a short time in hot, bright weather. These succulents will thrive on most well-drained soils (sand included) in full sun. They withstand drought, mild frost, high temperatures and salt spray and are excellent groundcover for dry and impoverished areas in Mediterranean and subtropical gardens. Their fleshy, carpeting growth has made them an invaluable fire break in some dry regions. In cold-winter regions, grow in beds, tubs or large pots in the greenhouse and move outdoors for the summer months; avoid cool, wet conditions. Propagate by detaching rooted lengths of stem.

C.chilensis (syn. *Mesembryanthemum aequilaterale*; western US to Mexico and Chile; leaves 5cm-long, straight and 3-sided; flowers faintly fragrant, 6cm wide, rosy purple); *C.edulis* (syn. *Mesembryanthemum edule*; HOTTENTOT FIG, KAFFIR FIG; South Africa; leaves to 8cm, dull, dark green, tapering and curving toward tip; flowers 7–9cm wide, pale yellow, becoming flesh-coloured then bright pink; fruit fleshy, edible, top-shaped).

carrier an inert material used to facilitate the application of a pesticide.

carrot (*Daucus carota* subsp. *sativus*) a biennial, widely grown in temperate and tropical regions as an annual for its edible root. Widely distributed through cultivation, it probably originated in Afghanistan from a dark-rooted type, spreading to China by the 13th or 14th century, and to Western Europe by the 15th century. Yellow and white mutants arose; the familiar orange, due to the presence of beta carotene, derives from selections made in the Netherlands around 1600.

Carrot is a cool-season crop, best at a mean temperature range of 16–18°C/60–65°F; temperatures above 29°C/84°F give rise to reduced top growth and increased root flavour, while below 16°C/60°F roots become long, tapered and of poor colour. Sow successionally *in situ* from March, 2cm deep in rows 30–37cm apart with roots thinned to 2–5cm apart; early crops can be sown in rows 15cm apart.

Earliest maturity is achieved from sowings made under glass, plastic cloches or fleecy or perforated polythene film; remove covering at six true leaves. Late supplies can be maintained by storing in the sown site, which needs to be well-drained and covered with straw in severe winter weather; alternatively, store in clamps outdoors or cover with sand in boxes kept in a cool dry place.

Cultivars are grouped on the basis of maturity date, size, and root shape, as follows: Amsterdam Forcing (early, small to medium size, slender, cylindrical); Nantes (mid-season, medium size, broader and longer than Amsterdam Forcing, cylindrical); Chantenay (maincrop, medium size, broader and shorter than Nantes, conical); Berlicum (late, large, cylindrical); Autumn King (very late, very large and long, conical).

Carpenteria californica

carrot

Pests
aphids
carrot fly

Disease
virus

carrot fly (in the US, CARROT RUST FLY) (*Psila rosae*) a shiny-black fly, up to 10mm long, with yellow legs; its larvae are a pest of the carrot, principally, but also of celery, parsnip and parsley. Taproots and lateral roots are damaged, foliage yellows, and plants wilt and may die; there are often extensive rust-coloured tunnels in the taproot.

There are two generations per year, three in a favourable climate. Oval white eggs are laid in the soil surface around the host plant and hatch within seven days, and the colourless larvae become creamy white and up to 10mm long when fully fed.

Avoid damage from the first generation by sowing after mid-May; damage by the second generation can be avoided by lifting an early-sown crop before late August. Other preventive measures include the use of resistant cultivars, such as 'Flyaway' and 'Sytan', open sites, avoidance of the need to thin, and planting in blocks surrounded by a 60cm-high polythene barrier to deflect egg-laying females. Control chemically by incorporating insecticide along seed rows before sowing; for late-lifted roots, apply insecticide as a soil drench.

Carthamnus (from Arabic, *qurtom*, and Hebrew, *qarthami*, to paint, referring to the dye extracted from the florets of *C.tinctorius*). Compositae. Mediterranean, Asia. 14 species, annual and perennial herbs with spiny to pinnately cut leaves and terminal heads of small florets surrounded by spiny bracts. *C.tinctorius*, SAFFLOWER or FALSE SAFFRON, is a W Asian annual with erect, spiny-leaved stems to 1m tall. Produced in late summer and autumn, the flowerheads are 3–4cm wide and composed of bright orange-yellow florets in a thistle-like involucre surrounded by leafy, spiny bracts. The flowerheads are good for cutting and dry well if taken when still in bud. Sow seed *in situ* after the last frosts.

cartilaginous hard and tough in texture, but flexible.

Carum (Latin form of the Greek name, *karon*: *C.carvi* was extensively grown in Caria, a district in Asia Minor). Umbelliferae. Temperate and subtropical regions. 30 species, perennial and biennial, tap-rooted herbs with finely 2–3-pinnately divided leaves and compound umbels of small flowers in summer. CARAWAY, *C.carvi*, is a Eurasian biennial to 60cm tall, cultivated in the herb garden for its aromatic leaves, its roots (sometimes prepared as for parsnip) and, principally, for its small, narrow, liquorice-flavoured seeds. It will grow in full sun on any well-drained, deep and moderately fertile soil, including heavy clay. Harvest seed in mid- to late summer as the flowerheads begin to dry off and the seed darkens. Sow seed *in situ* when ripe or in early spring.

Carya (from Greek *karya*, a nut-bearing tree). Juglandaceae. HICKORY, PECAN. US, China. 25 species, large deciduous trees, with smooth, brown to grey twigs and pinnate leaves. Produced in spring, the flowers are inconspicuous – the males borne in branched, drooping catkins; the females in a terminal spike. The fruit is a thick-skinned, green drupe, ovoid to rounded and splitting into four valves containing sometimes edible nuts. The following species are fully hardy and valued for their majestic habit, autumn colour and fruit (which needs long, hot summers to ripen). Plant in full sun or light shade, on a deep, moist and fertile soil. All but the youngest saplings will suffer if transplanted. Propagate by seed sown ripe or stratified over winter for spring-sowing. Increase edible nut-bearing cultivars by splice or veneer grafting on stocks of *C.cordiformis*.

C.cordiformis (BITTERNUT, SWAMP HICKORY; NE US; to 28m tall; bark smooth then fissured, young branches rusty-hairy; leaflets 7–13cm, 5–9, ovate to lanceolate, mid-green turning bright yellow in autumn; fruit rounded, to 3cm, nuts bitter); *C.glabra* (PIGNUT, SMALL-FRUITED HICKORY, BROOM HICKORY; NE US; to 30m tall; bark grey, furrowed; leaflets to 15cm, 3–7, ovate to lanceolate, dark green turning yellow and orange in autumn; fruit rounded to pear-shaped, to 2cm, nuts bitter); *C.illinoinensis* (PECAN; E US and Midwest to Texas and Mexico; to 50m; leaflets to 15cm, 11–17, oblong to lanceolate; fruit to 7cm, oblong with a thin, brown shell and edible nuts; many commercial cultivars are available, bred for fruit, suitability as pollinators, and disease- and climate-tolerance); *C.ovata* (SHAGBARK HICKORY; E US; to 40m; bark grey, shaggy; leaflets to 15cm, 5, oblong to obovate, acuminate, dark green turning golden-yellow in autumn; fruit to 5cm, thin-shelled, nuts edible; numerous cropping cultivars are available).

caryopsis of Gramineae; a one-celled, one-seeded, superior fruit in which the pericarp and seed wall are adherent.

Caryopteris (from Greek *karyon*, nut, and *pteron*, wing, referring to the winged fruit). Verbenaceae. E Asia. BLUE BEARD. 6 species, deciduous herbs and shrubs with aromatic leaves and clusters of small flowers in late summer and autumn. Grow in full sun on any light, well-drained and moderately fertile soil. Where temperatures fall below –10°C/14°F for prolonged periods, provide a deep winter mulch and the protection of a sheltered, south-facing wall. Cut back hard to a woody framework in spring (plants growing in areas with severe winters are often killed to the ground and should be pruned that low, effectively becoming 'herbaceous' perennials). Propagate by semi-ripe or green wood cuttings rooted in a sandy mix in a case with bottom heat.

C. × clandonensis (syn. *C.incana*, *C.mongholica*; slender-branched, bushy shrub, to 1.5 × 1m; leaves to 10cm, toothed, sage- to grey-green above, silver to white beneath; flowers lavender to deep blue; cultivars include: 'Arthur Simmonds', very widely grown, resilient, with compact spikes of bright blue flowers; 'Dark Knight', of low and spreading habit, with silvery leaves and deep blue flowers; 'Ferndown', with very deep blue-violet flowers;

Caryopteris
x *clandonensis*

'Heavenly Blue', with deep blue flowers; 'Kew Blue', with silvery leaves and flowers of a deep Cambridge to lavender blue; 'Longwood Blue', to 60cm tall, with silvery leaves and sky blue flowers; 'Worcester Gold', with golden leaves and deep lavender flowers);*C.incana* (syn. *C.mastacanthus*; COMMON BLUE BEARD,BLUE SPIRAEA; China, Japan; to 1.5m tall, grey-felted shrub; leaves toothed; flowers blue to mauve, white in the cultivar 'Candida'); *C.mongholica* (N China, Mongolia; smaller than *C.incana* and weaker in appearance; flowers deep blue).

Caryota (from Greek *karyon*, nut). Palmae. SE Asia to the Pacific Islands and Australia (both species listed here are from SE Asia). FISHTAIL PALM. 12 species, medium-sized to large palms with single or clump-forming trunks and large, bipinnate leaves; these are triangular in outline and consist of many wedge- to diamond-shaped leaflets that have a distinctive ragged or cleft edge (hence 'fishtail'). Large, branched inflorescences ultimately bear plum-shaped, irritant fruit. Provide a humid environment, protection from scorching sunlight and a minimum temperature of 10°C/50°F. Plant in a medium containing composted bark, leafmould, and clay or polystyrene granules and well-rotted manure; keep moist and regularly syringed, except in cool weather. Propagate by seed surface-sown on damp sand in a heated case, or, for *C.mitis*, by offsets.

C.mitis (BURMESE FISHTAIL PALM, CLUSTERED FISHTAIL PALM, TUFTED FISHTAIL PALM; clumped stems, 3–12m tall; leaves 1–4m long, erect to spreading, with 10–30cm-long, deltoid, light green segments); *C.urens* (WINE PALM, JAGGERY PALM, TODDY PALM; differs in its taller and stouter, solitary stem (to 12m), outspread to arching leaves and drooping, dark green segments).

cascade (1) a small waterfall, especially one of a series. Cascades are important artificial features in formal gardens, often in the form of a long series of steps and usually in combination with fountains and other waterworks; (2) a form of chrysanthemum; (3) a training style in bonsai.

Cassia (from Greek *kasia* in Dioscorides). Leguminosae. Tropics. SHOWER TREE. 535 species, semi-deciduous or evergreen trees, shrubs and herbs with pinnate leaves and broadly 5-petalled flowers with long stamens in racemes or panicles principally in spring and summer. Tender. Plant in full sun or light shade on a free-draining, moderately fertile, sandy loam. They prefer infrequent but deep watering and will tolerate some drought. Under glass, these are attractive flowering shrubs, blooming when young and responding well to hard pruning in early spring. Provide a minimum temperature of 10°C/50°F, filtered sunlight and a fertile, sandy medium. Water and feed generously in warm weather; keep almost dry in winter. Propagate by seed or by semi-ripe cuttings in sand in a closed case with gentle bottom heat.

C.fistula (syn. *C.excelsa*; GOLDEN SHOWER TREE, PURGING CASSIA, INDIAN LABURNUM, PUDDING PIPE TREE; Asia, the warm Americas, Australia and the Pacific; tree to 20m; flowers 3–6cm wide, vivid yellow, sweetly scented, carried in dense drooping racemes to 40cm long); *C.grandis* (PINK SHOWER, HORSE CASSIA, APPLEBLOSSOM CASSIA; tropical Americas; to 30m; flowers to 2.5cm wide, pink fading to orange-pink, peach or light yellow, white or cream, in erect racemes to 25cm long); *C.javanica* (PINK SHOWER, RAINBOW SHOWER; SE Asia; to 25m tall; flowers to 7cm wide, crimson to pale pink or buff, in rigid, corymbose racemes to 10cm long); *C.leptophylla* (GOLD MEDALLION TREE; Brazil; to 15m; flowers deep yellow, 3cm wide, carried in densely packed, terminal racemes to 25cm long); *C.moschata* (BRONZE SHOWER; tropical Americas; to 20m; flowers to 3cm wide, bronze or golden yellow, often veined red in loose racemes to 30cm long). Many ornamental *Cassia* species are now included in the genus *Senna*.

Cassiope (for Cassiope, wife of Cepheus, King of Aethiopia, and mother of Andromeda). Ericaceae. Europe, Asia, N America. 12 species, hardy, dwarf evergreen shrubs, with slender, creeping to erect stems clothed with small overlapping, scale-like leaves, thus appearing heath- or clubmoss-like in appearance. In spring, the solitary flowers nod on thread-like stalks; they are small and bell- to urn-shaped with four to six, pointed lobes. Plant in a moist, acid soil, rich in sand and leafmould. They prefer full sunlight but will suffer in scorching conditions. A shallow winter mulch of ground leafmould and garden compost will encourage the sprawling stems to layer and rejuvenate themselves and will protect young shoots. Propagate by semi-ripe cuttings in summer; also by detaching rooted stems.

C.fastigiata (Himalaya; to 30 × 20cm; flowers bell-shaped, white 8mm across with curving lobes, on red or green stalks); *C.lycopodioides* (NE Asia to Alaska; to 10cm tall; flowers 8mm long, white or pink and bell-shaped with slightly curving lobes on thread-like, red-tinted stalks); *C.mertensiana* (WHITE HEATHER; N America; 15–30cm tall; flowers 6mm-wide, bell-shaped, white, on slender, red-tinted stalks); *C.selaginoides* (Himalaya; to 25 × 15cm; flowers bell-shaped, white or pink-tinged to 1cm wide with 5, curving lobes); *C.tetragona* (Arctic Circle, and N Europe; to 30cm; flowers pendulous, bell-shaped, 6–8mm across, white or pink-tinged and 4–5-lobed; *C.wardii* (Tibet; flowers white, urn-shaped, some 8mm long with five lobes, tinted red inside at the base, on short stalks).

Castanea (the Classical Latin name for these trees, after Castania, a town in Thessaly famous for them). Fagaceae. N America, Europe, Asia. SWEET CHESTNUT, CHINKAPIN. 12 species, hardy, deciduous trees and shrubs with grey to black, smooth to furrowed bark, oblong to lanceolate or oval, usually toothed leaves and small, white to yellow-green,

Cassiope lycopodioides

muskily scented flowers in long catkins in summer. The fruit are ovoid to triangular, glossy brown, edible nuts, one to several enclosed in a spiny case (burr) that splits open in late summer.

The genus divides into two groups on the basis of habit – true chestnuts are large trees, usually single-stemmed unless damaged by the elements or grazing animals or coppiced by man or chestnut blight; chinkapins (chinquapins) are smaller and shrubby. *C.crenata*, *C.dentata*, *C.mollissima* and *C.sativa* belong to the first group, while chinkapins include *C.alnifolia*, *C.pumila* and *C.ozarkensis*.

Chestnuts are attractive, drought-tolerant and hardy trees, grown for their large, bright green leaves (yellow-brown in autumn), their pleasing, if rather rank-smelling, catkins, their nuts, eaten roasted or candied, and for timber. Where fruit is the main objective, only *C.crenata* and *C.sativa* are well-adapted to cool-summer regions, but all grow and fruit well in climates such as those of the eastern US. Individual trees are usually self-sterile and two or more clones should be near each other to ensure a good seed set. Grow in well-drained, neutral to slightly acid deep loams (dry and highly acidic, sandy soils included). Propagate by seed in autumn (most species hybridize freely), or by budding in summer or grafting in winter on to stocks of *C.crenata* and *C.sativa*.

The greatest threat to chestnuts is the chestnut blight. Introduced accidentally to the US at the turn of the century, it has devastated the native *C.dentata*, a species now often to be seen only in the form of suckering stands that have arisen from infected mature trees and that in turn await death through reinfection. *C.sativa* is also susceptible but survives in the cooler summers of northern Europe and has yet to be affected in the British Isles. Species from E Asia (the blight's place of origin) are largely resistant, as are hybrids between them and the western species (e. g. Dunstan Hybrid Chestnuts). Fruit are sometimes affected by chestnut weevils, which lay their eggs on the developing fruit in summer. Having bored into the burr and fed on the developing kernel, the weevils pupate in the soil beneath the trees and it is at this stage that they are most easily treated. The two-lined chestnut borer tunnels under the bark and may cause severe damage to small or weak trees.

Chinkapins are hardy and useful in large gardens, grown for their foliage and spreading habit. They will withstand wet and acid conditions, adapting well to the wild or woodland garden and to lakesides, and are an excellent, soil-enriching understorey for impoverished and sandy coniferous woodland.

C.alnifolia (BUSH or TRAILING CHINKAPIN; SE US; a suckering, spreading shrub to 1m tall; leaves oblong to elliptic, toothed, 12cm long, glossy dark green above, tawny-hairy beneath with short-spiny burrs); *C.crenata* (JAPANESE CHESTNUT; Japan; spreading, short-trunked, to 20m tall; leaves to 15cm, oblong, glabrous above, grey-downy beneath with small, bristly teeth; fruit to 4cm, 1–4 per 4–8cm-wide, slender-spiny burr; crossed with *C.dentata* to

produce blight-resistant clones such as 'Essate-Jap' and 'Sleeping Giant'); *C.dentata* (syn. *C.americana*; AMERICAN SWEET CHESTNUT; eastern US; to 45m tall, but often smaller and multi-stemmed due to chestnut blight; leaves to 20cm long, oblong to lanceolate, dull green, glabrous throughout with large, glandular teeth on margins; fruit to 3cm, 2–3 per long-spiny, 5–7cm burr); *C.mollissima* (CHINESE CHESTNUT; China, Korea; to 20m tall, broad-crowned and stout-trunked; leaves to 20cm, oblong to lanceolate to elliptic, downy beneath at first, with bristly teeth; fruit to 3cm, 1–4 per 5–8cm, slender-spiny burr; large-fruited selections include 'Nanking', and the small, sweet-fruited 'Kelsey': chestnut-blight-resistant); *C.ozarkensis* (OZARK CHINKAPIN; differs from *C.pumila* in its smooth, not hairy shoots and its larger leaves – to 23cm long); *C.pumila* (ALLEGHENY CHINKAPIN; eastern US; a suckering shrub or small tree, 4–10m tall; leaves oblong to ovate, to 15cm long, smooth above and pale-hairy beneath with short teeth and 4cm, slender-spiny burrs; var. *ashei*: COASTAL CHINKAPIN, leaves densely hairy beneath, burrs with short spines); *C.sativa* (SWEET CHESTNUT, SPANISH CHESTNUT; Europe, N Africa, W Asia; broad to columnar, to 40m tall; leaves to 25cm, oblong to lanceolate, deep, glossy green, smooth above, initially hairy beneath, with remote, bristly teeth; fruit to 3cm long, 1–5 per 4–7cm, slender-spiny burr; includes the ornamental cultivars 'Asplenifolia' and 'Laciniata', with finely cut leaves, 'Albomarginata', with white-edged leaves, 'Variegata', with yellow-edged leaves, 'Heterophylla' with leaves of varying shapes, sizes, teeth and lobing, and 'Holtii', with a narrowly conic crown; the finest fruting clone is generally thought to be 'Marron de Lyon', bearing large, sweet fruit even when young).

Castanopsis (*Castanea*, and Greek, *opsis*, resemblance). Fagaceae. E Asia. Some 110 species, evergreen trees and shrubs with scaly bark and leathery leaves. Inconspicuous flowers are borne in catkins and followed by nuts contained by prickly cupules. Hardy in climate zone 7, but thriving in regions with long, hot and humid summers; elsewhere, plant it in a sheltered and warm location. Grow in full sun on a fertile, moist and acid to neutral soil. Increase by seed. *C.cuspidata* (SE China, S Japan; tree to 25m with spreading to pendent branches and dark grey bark; leaves to 9cm, ovate to oblong with a long, drawn-out apex, thick, tough, entire, glossy dark green above, bronze beneath).

castor meal a by-product of the manufacture of castor oil, used as an organic fertilizer with an analysis of 5–6% nitrogen, 1–2% phosphate and 1% potash.

Casuarina (the long, drooping branchlets were thought to resemble the tail feathers of the cassowary (*Casuarius*)). Casuarinaceae. Australia, Pacific. BEEFWOOD, BULL OAK, WHISTLING PINE, AUSTRALIAN PINE, SHE OAK. Some 70 species, trees and shrubs, usually with slender, weeping branches and many, jointed and whip-like branchlets. These give young plants the look of a tamarisk or a broom; in older trees, they hang from the branches like very long, fine pine needles. The true leaves are reduced to tiny withered scales. Woody, cone-like fruit add to the pine-like appearance. Grown in regions that experience little or no frost, where they make fast-growing and elegant street and specimen trees and are well-suited to dry, windy and saline conditions. Elsewhere, they are grown in borders and tubs in the cool greenhouse or conservatory. Plant in full sun in a moisture-retentive but free-draining soil with a high sand or grit content. Provide a minimum temperature of 5°C/40°F and avoid cold, overwet conditions. Propagate from seed or by semi-ripe cuttings in a closed case with bottom heat.

C.cunninghamiana (RIVER SHE OAK; to 35m tall; branchlets 7–25 cm × 0.4–0.7mm, 6–9-ridged, glabrous, dark green; cones to 1.5cm, dull grey); *C.equisetifolia* (HORSETAIL TREE, SOUTHSEA IRONWOOD; to 35m tall; branchlets 7–20cm × 0.5–1mm, 6–8-ridged, hairy, grey-green; cones to 3cm, grey-brown); *C.torulosa* (to 20m tall, usually less; branchlets 5–15cm × 0.4–0.5cm, 4–5-ridged, bronze green; cones to 3cm, glossy mahogany to black-brown; forms are grown with very fine, purple-bronze foliage); *C.verticillata* (syn. *C.stricta*; MOUNTAIN SHE-OAK, DROOPING SHE-OAK, COAST BEEFWOOD; to 10m tall; branchlets to 10cm × 1.5mm, 9–13-ridged, somewhat hairy, dark green; cones 2–4cm, glossy dark brown to black).

Catalpa (from 'catawba', the vernacular name of this plant among the Native Americans of Carolina and Georgia). Bignoniaceae. N America, Cuba, E Asia. 11 species, hardy deciduous trees with large, ovate to cordate leaves, and, in midsummer, pyramidal panicles of showy, trumpet-shaped flowers edged with five, wavy lobes. Long, inedible fruit resemble narrow beans and hang, blackened and thinly woody, through the winter and spring. Hardy to –15°C/5°F, or lower where long, hot summers ensure wood ripening. Grow in any deep, fertile, moisture-retentive soil in an open, sunny site with shelter from strong winds. Sow seed outdoors in autumn or in cold frames in spring after stratification for three weeks at 1°C/34°F. Taken in late spring before the leaves are fully developed, softwood cuttings will root in a closed case with bottom heat.

C.bignonioides (COMMON CATALPA, EASTERN CATALPA, CATAWBA, INDIAN BEAN, BEAN TREE, INDIAN CIGAR; SE US; broad-crowned tree to 15m tall; leaves 10–20cm-long, thinly hairy, pale to deep green, usually unlobed and malodorous when crushed; flowers 3–5cm long, white with purple spots and yellow stains in the throat; includes 'Aurea', with large, yellow to lime leaves, and 'Nana', UMBRELLA CATALPA, often misnamed *C.bungei*, with a dense, broad mop of small leaves, usually top grafted for use as a closely pruned, ball-headed standard); *C.bungei* (N China; differs from *C.bignonioides* in its more triangular, 15cm-long leaves that taper to narrowly pointed lobes; flowers fewer, and white to pale rose with dark maroon spots and yellow streaks); *C. × erubescens* (*C.bignonioides* × *C.ovata*; intermediate between the parents with entire and lobed leaves together on the same tree; includes 'Purpurea', with purple-black shoots and leaves to 30cm, bruised purple to faintly violet-tinted dark green); *C.fargesii* (W China; broad-crowned tree to 20m tall; leaves 8–14cm long; flowers 3cm long, rose to mauve with yellow and maroon markings; includes f. *duclouxii*, with 3-lobed leaves and purple-rose flowers); *C.ovata* (China; to 10m tall with 10–20cm-long leaves that taper finely and have one to five lateral lobes; flowers to 2.5cm, white stained yellow and marked purple within; includes 'Flavescens', with paler green leaves and flowers stained dull, pale yellow); *C.speciosa* (WESTERN CATALPA, HARDY CATALPA, SHAWNEE WOOD; differs from *C.bignonioides* in its height (to 30m), its leaves, to 30cm long and unscented when crushed, and in having fewer and more heavily marked flowers; it occupies a more westerly range, from southern Illinois to Arkansas, and flowers some two to three weeks before *C.bignonioides*).

Catananche (from Greek *katanangke*, a strong incentive – this plant was used in love philtres). Compositae. Mediterranean. 5 species, annual and perennial herbs with long-stalked, daisy-like flowerheads in summer. *C.caerulea* (CUPID'S DART or BLUE CUPIDONE) is a perennial herb to 90cm tall with grassy, grey-green foliage. The flowerheads are 5cm wide and carried on slender stalks to 30cm long, each consisting of strap-shaped florets in a deep lavender blue and encased by shining, papery bracts. In the cultivars 'Alba' and 'Perry's White', the flowerheads are white; in 'Bicolor', they are white with a deep blue centre; in 'Blue Giant' and 'Major', they are large, profuse and cornflower blue. A hardy perennial suitable for flower borders and for drying – air-dried, the flowerheads tend to close up, making a silvery 'bud'; dried in silica or borax, the florets retain their open attitude and colour. Grow in full sun on any well-drained soil. Sometimes short-lived, especially on wet and heavy soils. Propagate by root cuttings or by seed sown under glass in early spring.

*Catalpa
bignonioides*

cataphyll any of the several types of reduced or scarcely developed leaves produced either at the start of a plant's life (i.e. cotyledons) or in the early stages of shoot, leaf or flower development (e.g. the bract-like rhizome scales of some monocots – those subtending and enveloping pseudobulbs, those found in the crowns of cycads alternating with whorls of developed leaves, and the imperfect leaf forms found, progressing up the stem from the most reduced scales, in the basal portions of the stems of many herbs).

catch crop a quickly maturing crop grown in the interval between harvesting one main crop and sowing or planting another.

catenary

catenary an ornamental support for climbing or rambling plants, especially roses. It consists of chains or ropes hanging loosely, but evenly, from the tops of vertical posts set out in a continuous line.

caterpillars (Lepidoptera) the larvae of butterflies and moths. They have biting mouthparts and pairs of stumpy abdominal legs (prolegs) bearing hook-like crotchets. Most lepidopterous larvae have five pairs of prolegs. Those of the family Geometridae have only two pairs and because of this they progress in a series of looping movements, hence the common name LOOPER (in North America they are called inchworms, cankerworms and spanworms). The larvae of SAWFLIES (*Hymenoptera*) have six to eight pairs of prolegs devoid of crotchets.

All parts of plants are subject to attack by lepidopterous caterpillars: the roots and underground organs on herbaceous plants may be damaged by the caterpillars of SWIFT MOTHS (*Hepialus* species); stems may be girdled or severed at ground level, especially on young vegetables, by CUTWORMS (*Noctua* and *Agrotis* species); leaves may be tied together or rolled by the caterpillars of TORTRIX MOTHS (e.g. the FRUIT TREE TORTRIX MOTH, *Archips podana*); foliage may be eaten on deciduous trees and shrubs by gregarious web-formers, such as the LACKEY MOTH (*Malacosoma neustria*), and by gregarious non web-formers, such as the BUFFTIP MOTH (*Phalera bucephala*); foliage may also be eaten by solitary larvae, such as those of the DIAMOND BACK MOTH (*Plutella xylostella*) on the cabbage tribe, those of the ELEPHANT HAWK MOTH (*Deilephila elpenor*) on *Fuchsia* and *Impatiens* especially, and those of the MAGPIE MOTH (*Abraxus grossulariata*), a looper caterpillar, on *Ribes* species. Caterpillars that bore into stems include the LEOPARD MOTH (*Zeuzera pyrina*) on fruit and ornamental woody plants; those that damage fruit include the CODLING MOTH (*Cydia pomonella*) on *Malus* species; those that mine leaves include the AZALEA LEAF MINER (*Caloptilia azaleella*).

Various residual contact insecticides, many of which also act as stomach poisons, are available for caterpillar control, particularly for those feeding openly on foliage. Infestations by tunnellers and borers may be prevented or reduced by timely application. Pheromone traps can be used to monitor pest presence and ensure optimum timing for insecticide application. A bacterial spray containing *Bacillus thuringiensis* is effective against most caterpillars and harmless to other types of animals.

See *bagworm, butterflies, cabbage caterpillars, clearwing moths, codling moth, cutworms, larvae, leaf roller, moths, slugworms, swift moth, tent caterpillar, tortrix moth, winter moth*.

Catharanthus (from Greek *katharos*, pure, or unblemished, and *anthos*, flower). Apocynaceae. Madagascar. 8 species, bushy annual or perennial herbs with *Vinca*-like flowers chiefly in summer. In frost-free regions, this species can be grown

Catharanthus roseus

year round outdoors and is a colourful bedding and tub plant, tolerating a wide range of conditions once established, drought included. In zones 8 and under, *Catharanthus* is used for summer bedding or grown under glass and in the home as a bright and free-flowering perennial. Grow in full sun or part shade in a fertile, moisture-retentive but freely draining medium. Water and feed liberally in spring and summer, sparingly in winter; maintain a minimum temperature of 7°C/45°F. Cut back in early spring to promote fresh, bushy growth. Sow seed in late winter or early spring at 20°C/70°F. Root greenwood or semi-ripe cuttings in summer in a closed case with bottom heat. *C.roseus* (syn. *Vinca rosea*; MADAGASCAR PERIWINKLE, ROSE PERIWINKLE, CAYENNE JASMINE; evergreen to 60 cm tall, perennial sometimes treated as an annual, with glossy leaves and white to candy pink to rose-purple 5cm-wide flowers with a dark purple-red or pink eye).

cation exchange capacity (CEC) the ability of a soil to adsorb positively charged ions, known as cations, on the surface of clay particles and organic matter. It is a measure of the retention of such nutrients as potassium, magnesium, calcium and ammonium. See *base exchange*.

catkin a cylindrical, bracteate spike or spike-like inflorescence composed of single flowers or cymules, the flowers usually apetalous and unisexual. It is also known as an *ament*.

cats (*Felis catus*) Cats foul newly cultivated soil and thereby disturb seedbeds and newly sown lawns. They are attracted to the extra warmth provided by cloches and greenhouses and prefer dry soil. Keep vulnerable areas well-watered and exclude cats by covering sites and access with wire or plastic netting. Pepper dust and other repellents have some deterrent effect but are not long-lasting. Cats will eat certain plants for therapeutic reasons; as well as catmint, *Actinidia kolomikta*, and similarly pungent plants, they will chew *Chlorophytum* and *Cyperus* leaves indoors. Spraying plants with quassia will act as a deterrent.

Cattleya (for William Cattley (died 1832), English horticulturist and patron of botany). Orchidaceae. C and S America. 45 species, tender, evergreen, perennial herbs grown for their spectacular flowers. Epiphytes and lithophytes, they produce club- or cane-like pseudobulbs, 10–50cm tall, along a creeping rhizome. The pseudobulbs terminate in one or two, tough, oblong to elliptic leaves, from the base of which a papery sheath emerges and splits to free a stem of one to ten thickly textured flowers. These consist of three, lanceolate to oblong sepals, two, usually broader and larger petals and a funnel-shaped lip with a spreading, colourful midlobe.

The many *Cattleya* hybrids are among the most popular orchids. They include miniatures, medium-sized and large plants. Flowers range in shape and size from the small and star-like to the very large (to 18cm in diameter), lush and lacy; in colour, through white, pale rose, magenta, mauve, burgundy, velvety red, scarlet, vermilion, orange, gold, lemon yellow, bronze and green, often with a differently coloured lip or blotches or veins. The range is greatly extended by intergeneric hybrids with *Brassavola, Epidendrum, Laelia, Rhyncholaelia* and *Sophronitis*. Some are languorously scented; some last well when cut. The species generally possess more character than the hybrids and are regaining popularity. The principal flowering time is winter to spring, but this may vary from year to year with the same plant and even a small collection should provide some colour at all times. Most of the species are grown by orchid specialists, and several at least are easy and reliable plants for the greenhouse (minimum temperature 10°C/50°F) or the home. Grow in very bright, humid and buoyant conditions with some protection from full, midday sun. Plant in an open, bark-based medium in small pots, pans or baskets. Water, syringe and feed frequently when in growth. Once the new growth is fully developed, water only to prevent shrivelling of pseudobulbs and impose cooler, drier conditions. As temperatures rise in spring, mist over with increasing frequency; return to a full watering regime when the new shoots and roots appear. Propagate by division, taking one active growth and two to three backbulbs per division. *Cattleya* resents disturbance and is quite happy to grow over the edge of the pot for a season or two.

C.aurantiaca (tall plants with clusters of 3cm-wide, starry flowers in scarlet, vermillion or pale orange); *C.bicolor* (small to medium-sized plants with several flowers to 10cm wide, tepals bronze, lip tongue-like, bright purple-red); *C.bowringiana* (medium to tall, grey-green plants with clusters of 5cm-wide, spreading flowers, tepals rose to magenta, lip stained deep mauve or rosy purple); *C.dowiana* (medium-sized plants with one to three, 15cm-wide flowers, tepals golden yellow, petals broad and frilly, lip large stained deep, velvety red); *C.forbesii* (small plants with a few, 8cm-wide, delicately scented flowers, tepals pale olive to buff or bronze, lip cream stained yellow and flesh pink); *C.guttata* (tall plants with clusters of 7cm wide flowers, tepals olive to lime green spotted rosy purple to maroon, lip white with a bright magenta or deep garnet midlobe); *C.labiata* (medium-sized plants with one or few broad and frilly, fragrant flowers to 18cm wide, tepals white to pale rose, magenta or mauve, lip very large and showy, rose-mauve stained yellow and overlaid with mauve or red-purple); *C.loddigesii* (medium-sized to tall plants with clusters of spreading, fragrant flowers to 10cm wide, tepals palest rose to lilac, lip shell-like, white to cream tinted yellow and stained rose to magenta); *C.mossiae* (medium-sized plants, flowers one or few, to 18cm wide, very broad and lacy, fragrant, tepals white to pale rose or magenta or lilac, lip white to rose or lilac with a central orange patch and purple-red veins and tip); *C.schilleriana* (medium-sized plants with a few, fragrant flowers to 10cm wide,

tepals narrow and wavy, deep olive to bronze spotted chocolate to maroon, lip white densely combed with rosy purple); *C.skinneri* (medium-sized to tall plants with clusters of spreading, rounded flowers to 12cm wide, rose to deep magenta, lip with a white or cream throat); *C.walkeriana* (small plants, flowers to 10cm wide, one or few, fragrant, tepals pale pink-lilac to bright rose-purple, petals rounded and spreading, lip with lateral lobes held flat and open and midlobe kidney-shaped, stained white to yellow at base, deep magenta to mauve at tip).

caudate of a leaf or perianth segment apex, tapering gradually into a fine tail-shaped point, tip or appendage.

caudex strictly, the basal axis of a plant, comprising both stem and root; sometimes applied to the aerial stems of palms and superficially palm-like plants. The term is most often used in connection with plants with stout, swollen or succulent unbranched stems, usually crowned with narrower branches, leaves or inflorescences.

caudiciform resembling or possessing a caudex, encountered in the phrase 'caudiciform succulents', a disparate group of succulent and semi-succulent plants of horticultural importance and defined by its members' exhibiting a caudex.

caulescent producing a well-developed stem above ground.

cauliflory the production of flowers directly from older wood, i.e. springing from the trunk or branches as in the Judas tree (*Cercis*), cocoa (*Theobroma*), and *Goethea*. Such plants are termed cauliflorous.

cauliflower (*Brassica oleracea* Botrytis Group) Closely related to cabbage, cauliflower is grown for its edible inflorescence made up of thick, fleshy flower stalks compacted to form a spherical head of edible curd, usually white. Most are annual but some biennial and perennial forms also exist. A native of the Mediterranean region, cauliflower has been widely cultivated since the 18th century and is an important crop in France, Italy and the UK.

Curd development is governed by temperature; with annual cultivars in temperate zones, it occurs after the formation of a prerequisite number of leaves, at around 17°C/63°F; above this temperature, curds may not form or be of poor quality. With biennial and perennial cultivars, curds are formed after exposure to 10°C/50°F or below, depending partly on plant size. In situations where the temperature fluctuates around a threshhold at critical times, vegetative growth may take place, resulting in curds with leafy bracts. Cultivars vary in winter hardiness, and may be grouped according to maturity time: early summer; summer; autumn; winter-heading; spring-heading overwintered.

cauliflower

Pests
aphids
cabbage caterpillars
cabbage root fly
flea beetle

Diseases
clubroot
leaf spot

For early summer-heading curds, sow in October 1cm deep under protection, into a seedbed or, for best results, in pots. Overwinter with careful ventilation and plant out in March or as soon as suitable conditions prevail. For summer- and autumn-heading curds, sow from mid-March to mid-May under a cold frame or cloches, or in a sheltered outdoor seedbed. For winter- and spring-heading cultivars, sow in late May in an outdoor seedbed. Plant out summer-heading cultivars from early March to mid-May, 55cm square or at 45cm spacings in rows 60cm apart; plant out autumn-heading cultivars in mid- to late June, 55–64cm each way or at 50–60cm spacings in rows 60–65cm apart; winter- and spring-heading cultivars should be planted in mid- to late July, 70cm each way. Mini cauliflowers can be produced over 13–18 weeks during summer months from sowings made from April onwards, with the plants spaced 15cm each way. There is interest in green curd cultivars, grown for autumn maturity, and in the perennial cultivar 'Nine Star Perennial', sown in spring and planted 1m square. See *broccoli, calabrese*.

cauline attached to or arising from the stem.

Ceanothus (from Greek *keanothos*, name used by Theophrastus for spiny plants). Rhamnaceae. W N America, Mexico. CALIFORNIA LILAC. Some 60 species, hardy evergreen or deciduous shrubs and small trees, often with tough, small dark green leaves and grown for their dense panicles of small, blue, mauve, pink or white flowers produced in spring and summer. The following are hardy in climate zone 6, but should be planted against a south-facing wall in areas that experience prolonged hard frosts and cold winds. Many have a spreading habit, which benefits from some support and is shown to best advantage on a wall; an exception is *C.thyrsiflorus* var. *repens*, an excellent groundcover. Plant in full sun on a free-draining, light soil. Prune deadwood from evergreen species in spring; trim over after flowering. Reduce long, flowered shoots of deciduous species and hybrids to within a few buds of the main framework in spring. Increase by softwood or semi-ripe cuttings in summer.

C.arboreus (FELTLEAF CEANOTHUS, CATALINA MOUNTAIN LILAC; evergreen shrub or tree 3-7m; leaves 3–8cm, broadly ovate to rounded, densely white-tomentose beneath; inflorescence to 12cm, pyramidal; flowers pale blue, fragrant; includes 'Trewithen Blue', spreading, with dark green, broadly oval leaves and deep blue flowers); *C.* 'Burkwoodii' (*C.dentatus* var. *floribundus* × *C.* × *delilianus* 'Indigo'; dense, evergreen, spreading shrub to 1.5m; leaves 1.5–3cm, elliptic, toothed, dark green and glossy above, pale-hairy beneath; inflorescence 3–6cm, flowers bright blue with darker stamens); *C.* 'Burtonensis' (*C.impressus* × *C.thyrsiflorus*; spreading evergreen shrub to 2m; leaves small, circular, dark green and glossy, appearing crinkled; inflorescence to 2cm in diameter, dense, rounded; flowers dark blue); *C.* 'Concha' (evergreen,

to 2m, spreading; leaves to 2.5cm; flowers dark blue, red-tinted in bud); *C.* 'Delight' (*C.papillosus* × *C.rigidus*; evergreen shrub to 5m; leaves glossy pale green; inflorescence a long panicle; flowers deep blue); *C.* × *delilianus* (FRENCH HYBRID CEANOTHUS; *C.coeruleus* × *C.americanus*; evergreen shrub to 5m; leaves 4–8cm, elliptic to oblong, dark green above, paler and downy beneath; flowers blue in large panicles; includes the popular 'Gloire de Versailles', a large, semi-deciduous and free-flowering form, with a rangy habit, pale to mid-green leaves and fragrant, powder blue flowers); *C.* 'Dark Star' (evergreen, arching to 2 × 3m; leaves ovate, toothed, dark green; flowers dark purple-blue, honey-scented); *C.dentatus* (CROPLEAF CEANOTHUS; dense, evergreen shrub to 1.5m; leaves 0.5–2cm, elliptic to narrowly oblong or linear, dark green above, paler beneath, margins revolute and glandular-papillate; flowers dark blue in short, dense clusters); *C.gloriosus* (POINT REYES CREEPER; prostrate to decumbent, evergreen shrub, 2-3m across; leaves 2–5cm, broadly oblong to rounded, toothed, tough, dark green; inflorescence umbel-like; flowers dark blue to violet to purple); *C.impressus* (SANTA BARBARA CEANOTHUS; low, evergreen shrub to 1.5m; leaves 0.6–1.2cm, elliptic to suborbicular, veins sunken above, margins wavy, dark glossy green; flowers dark blue in a narrow inflorescence to 2.5cm long); *C.incanus* (COAST WHITETHORN; erect evergreen shrub to 4m, twigs thorny, bloomed white; leaves 2–6cm, broadly ovate to elliptic, grey-green above, paler beneath; flowers white in 3–7cm-long, dense panicles); *C.* 'Italian Skies' (a hybrid with *C.foliosus*; evergreen shrub to 1.5m tall, spreading, densely branched; leaves to 1.5cm, oval to oblong, glossy dark green with small teeth; inflorescence to 7cm, conical; flowers intense blue); *C.* × *lobbianus* (*C.dentatus* × *C.griseus*; erect, evergreen shrub to 1m; leaves to 2.5cm, elliptic, leathery, dark green, toothed; flowers bright blue in rounded, axillary clusters with blue-haired stalks); *C.* × *pallidus* 'Marie Simon' (*C.* × *delilianus* × *C.ovatus*; much-branched, deciduous shrub to 1.5m tall; leaves broadly oval; flowers soft pink in conical, dense and terminal panicles); *C.papillosus* (WART LEAF CEANOTHUS; evergreen shrub to 5m; leaves 1–5cm, oblong to elliptic, margins revolute with glandular teeth, dark green and hairy, upper surface strongly papillose to pustular; flowers pale to deep blue or lilac-mauve in short, narrow spikes); *C.prostratus* (SQUAW CARPET; prostrate evergreen shrub forming dense mats; leaves 1–2cm, cuneate to obovate, margins undulate, tough, pale, glossy green; flowers pale lavender to dark blue or white in umbel-like, terminal panicles); *C.* 'Puget Blue' (evergreen, to 3m, dense; leaves elliptic to oblong, glandular-toothed; flowers pale to dark blue); *C.rigidus* (MONTEREY CEANOTHUS; evergreen tree to 2m, or low and spreading; leaves 0.6–1.5cm, cuneate to broadly obovate, leathery, sometimes toothed, tough, glossy dark green; flowers fragrant, lilac to dark blue or white in sparse, umbel-like clusters); *C.* 'Southmead' (a hybrid involving *C.* × *lobbianus*;

evergreen dense, bushy shrub to 1.5m tall; leaves small, oblong, glossy dark green; flowers very dark, rich blue, in rounded inflorescences); *C.thyrsiflorus* (BLUEBLOSSOM, BLUE BRUSH; large, evergreen shrub or small tree to 6m; leaves 1–1.5cm, oblong-ovate to broadly elliptic, toothed, dark, glossy green; flowers pale to dark blue or white in rounded, short panicles; includes var. *repens* (CREEPING BLUE BLOSSOM), prostrate, giving dense cover, with small, dark, glossy leaves and dark blue flowers – the best selection is 'Blue Mound'); *C.* × *veitchianus* (*C.griseus* × *C.rigidus*; robust evergreen shrub to 3cm; leaves to 2cm, obovate to ovate, toothed, glossy green; inflorescence 2–5cm, dense, oblong to capitate; flowers intense, dark blue).

CEC see *cation exchange capacity*.

Cedrus (from the Greek name, *kedros*). Pinaceae. E Mediterranean, N Africa, W Himalaya. CEDAR. 2 species, hardy evergreen coniferous trees with a pyramidal to tiered habit and needle-like foliage. Oblong to cylindric cones stand erect on outspread branches. These stately conifers are hardy in climate zone 6, thrive in most well-drained soils and will tolerate some drought once established. Plant young trees in mid-spring; feed and stake them in the early years and prune to remove any forked leaders and weak lower branches. Propagate by seed, except in the case of cultivars, which must be grafted (*C.libani* subsp. *atlantica* Glauca Group will, however, come more or less true from seed).

C.deodara (DEODAR, HIMALAYAN CEDAR; W Himalaya; 35–60m tall; crown broadly conic, leader nodding, branches tiered with drooping tips; leaves 3–6cm, rich green to grey-green, somewhat 4-sided; cones 8–14cm, glaucous blue-grey to brown, composed of scales with ridged apices; includes dwarf, bushy, fastigiate, blue-green, yellow and very hardy cultivars); *C.libani* (CEDAR OF LEBANON; Lebanon, Syria, Turkey; to 45m; crown conic, becoming flat-topped and irregular with age, branches tiered and horizontal to slightly ascending; leaves 1–3cm, mid-green, often blue-tinted, 4-sided; cones 8–12cm, dull green becoming brown, composed of scales without ridged apices; includes dwarf, spreading, weeping, gold and variegated cultivars; subsp. *atlantica*: syn. *C.atlantica*, ATLAS CEDAR, to 50m tall, crown ovoid to conic at first, ultimately flat-topped, with roughly 4-sided, 1–3cm-long, blue-green to blue-grey leaves (Glauca Group) and grey-bloomed cones; includes fastigiate, weeping, pyramidal, white-variegated, blue-grey and silver cultivars).

Celastrus (from *kelastros*, Greek name for *Phillyrea*). Celastraceae. Widespread. BITTERSWEET. 30 species, evergreen or deciduous shrubs. Those listed here are deciduous vines with ovate to elliptic, toothed leaves and bright, orange to red seeds held in woody, 3-valved and spherical capsules in autumn and winter. Rampant vines hardy in climate zone 4 and thriving on most soils in sun or semi-shade. They are best left to make dense tangles and luxuriant cascades of growth among trees or along walls and fences in the wilder parts of the garden. Cut back hard in spring to remove dead stems and keep within bounds. Root cuttings in late winter in a cold frame. Plants of both sexes are usually required for fertile seed.

C.orbiculatus (syn. *C.articulatus*; ORIENTAL BITTERSWEET, STAFF VINE; NE Asia, naturalized N America; to 12m; fruit interior yellow, seeds coral pink to red); *C.scandens* (WAXWORK, AMERICAN BITTERSWEET, STAFF TREE, STAFF VINE; N America; to 7m; fruit interior orange-yellow, seeds carmine to scarlet).

celeriac (*Apium graveolens* var. *rapaceum*) a biennial, usually grown as an annual for its edible celery-flavoured root, which is swollen and knobbly and is shredded for use in salads, or boiled. Celeriac requires a long season of uninterrupted growth. Sow in February/March under protection, at around 15°C/60°F, in seed boxes or modules, or in March/April in a cold frame. Pot up and harden off for planting outside in late May, 30–38cm square; keep well-watered. Harvest from October. The crop is best left *in situ* and covered with straw in severe winter weather.

celeriac

celeriac

Pest
carrot fly

celery (*Apium graveolens* var. *dulce*) a biennial, grown as an annual for its crisp, edible, fleshy petioles. Native to damp sites in Europe and Asia, it has been cultivated for over 2000 years and does best in temperate zones with a mean monthly temperature of 16–21°C/60–70°F. Above this temperature range, leaf growth is poor, below it, plants may bolt. Celery matures in autumn/winter until the first frosts, but the season can be extended by greenhouse protection or by covering plants outdoors.

Two main divisions may be distinguished by cultivation method. Trench celery, with white-, pink- or red stemmed cultivars, is traditionally blanched by planting in a trench for gradual earthing up; the same result can more easily be achieved with thick

celery

Pests
carrot fly
celery fly
leaf miner
slugs

Diseases
heart rot
leaf spot

paper collars. For self-blanching celery, with its white- or cream-stemmed cultivars, close planting and straw packing are used to produce good-quality plants. Trench celery is usually sown under protection in February/March at around 15°C/60°F, and potted up and hardened off for planting outdoors in late May, 30–45cm square as double rows in a prepared trench, or single rows 90cm apart with plants spaced 25cm apart. Self-blanching celery should be raised similarly and planted out 25 cm square in block formation. American green-stemmed cultivars should be treated in the same way. All celery requires an uninterrupted growing season, with generous watering.

celery fly CELERY LEAF MINER (*Euleia heracleii*) a small fly, about 5mm long, brown with mottled irridescent wings. The legless maggots are up to 8mm long, white or very pale green. Celery, celeriac, and parsnips are subject to attack, the maggots mining within the leaves and giving rise to blotch-like blisters, eventually causing shrivelling. There are two or three generations per season, and early summer attacks are most damaging. Control with recommended insecticide spray if attack threatens to be severe.

cell (1) the primary unit of plant tissue, microscopic in size, consisting of cytoplasm and usually a single nucleus, which contains the code of plant characteristics, and is bounded by a thin wall of cellulose; (2) the lobe of an anther or locule of an ovary or pericarp; (3) a small individual container for raising plants, manufactured of plastic or polystyrene in the form of a multi-celled tray; it is often called a module. See *module*.

cellulose a polymer of glucose molecules, the major constituent of plant cell walls.

Celmisia (for Celmisius, son of the nymph Alciope, after whom a related genus was named). Compositae. Australasia (those listed here from New Zealand). 60 species, more or less hardy perennial herbs and subshrubs with silvery-hairy leaves and slender-stalked, white, daisy-like flowerheads produced in spring and summer. Hardy in climate zones 7 and over, they are suited to the rock garden, scree, raised beds, alpine troughs and peat terrace. Small and cushion-forming species grow best in pans in the alpine house. Grow in sun, with shelter from cold, drying winds, in a gritty, humus-rich, lime-free soil. Water freely when in growth and keep roots cool. Keep just moist in winter. Propagate by cuttings and division in early summer, or by fresh seed germinated in cool, moist conditions in a mix high in silver sand and leafmould.

C.bellidioides (creeping, mat-forming herb; leaves to 1.5cm, oblong-ovate to spathulate, dark green above, grey-floccose beneath; flowerheads to 2cm in diameter); *C.coriacea* (clump-forming herb with lanceolate to oblong leaves to 60cm long, tough and ribbed, very lustrous and silvery; flowerheads to 8cm in diameter); *C.ramulosa* (small procumbent to ascending shrub or subshrub to 20cm; leaves to 1cm, tough, linear to oblong, grey-tomentose; flowerheads to 2.5cm in diameter); *C.traversii* (clump-forming herb with oblong to lanceolate leaves to 40cm, dark green with cream-felted undersides and red-purple margins and midrib; flowerheads to 5cm in diameter, on a scape to 50cm tall); *C.walkeri* (sprawling, much-branched shrub or shrublet to 2m; leaves to 5cm, narrowly oblong, leathery, white-hairy beneath; flowerheads to 4cm in diameter).

Celosia (from Greek *keleos*, burning: the inflorescence of some species is flame-like). Amaranthaceae. Asia, Africa, Americas. COCKSCOMB, WOOL-FLOWER. 50 species, annual and perennial herbs with rather fleshy stems and small, pearly flowers packed in showy terminal, plume-, spire- or coral-like heads. Sow seed under glass at 18°C/65°F in late winter. Grow on in a fertile, loamy mix at 15°C/60°F in direct or filtered sunlight, watering and feeding freely. Reduce water and feed to induce and prolong flowering. For half-hardy bedding, harden off carefully and plant out after the last frosts in a well-drained, fertile soil in full sun. *C.argentea* (RED FOX; widespread, intensively developed annual, 0.25–2m tall with lanceolate to ovate, mid-green to maroon leaves and flowers in shades of red, crimson, purple, magenta, pink, scarlet, salmon, apricot, orange and yellow; inflorescence types range from the broad, terminal cockscomb with brain-like folds of the Cristata Group (syn. *C.cristata*) to the tapering, torch-like spikes of the Plumosa Group).

Celtis (from *keltis*, a Greek name applied to another tree). Ulmaceae. Tropics, Europe, N America. NETTLE TREE, HACKBERRY. 70 species, trees or shrubs, deciduous or evergreen. Those listed here are deciduous, hardy trees grown for their broad-headed, spreading habit, ovate to lanceolate, toothed leaves and small, berry-like fruits which hang on slender stalks and follow inconspicuous green flowers in summer. Although tolerant of harsh winters, these trees require long, hot summers for good growth and fruit. Plant on any fertile, deep soil in full sun. Propagate by fresh seed collected in autumn and stratified in moist sand for two to three months with the flesh removed.

C.australis (MEDITERRANEAN HACKBERRY, EUROPEAN NETTLE TREE, LOTE TREE; Mediterranean, Middle East; to 20m; leaves to 15cm, roughly hairy and dark green above, paler and downy beneath; fruit red ripening brown); *C.occidentalis* (COMMON HACKBERRY; N America; to 25m; leaves to 12cm, smooth, glossy green above, slightly hairy beneath; fruit yellow or red ripening deep purple); *C.sinensis* (JAPANESE HACKBERRY; E Asia; to 15m; leaves to 8cm, tough, smooth, glossy dark green above, duller beneath; fruit dark orange).

celtuce see *lettuce*.

Centaurea (from Greek *kentauros*, centaur). Compositae. Mediterranean, Asia, N America, Australia. KNAPWEED, STAR THISTLE. 450 species, annual and perennial herbs, fully hardy and grown for their thistle-like flowerheads. Produced from late spring through summer, these consist of a scaly, ovoid involucre crowned with showy outspread florets, tubular and deeply lobed. Grow in full sun on any well-drained soil. Annual species, such as *C.cyanus* and its cultivars, are excellent seasonal fillers; their bright blooms are also suitable for cutting and drying. Sow seed *in situ* in spring. Successional sowings will give flowers throughout the summer. Perennial species are fine plants for the herbaceous or mixed border; propagate by seed or division in autumn or spring (*C.montana* may also be increased by root cuttings).

C.cyanus (CORNFLOWER, BLUE-BOTTLE; northern temperate Regions; annual to biennial to 90cm tall; leaves lanceolate, entire to toothed, pale-floccose beneath; flowerheads deep violet-blue to azure, sky blue, mauve, pink, red and white); *C.dealbata* (Caucasus; perennial to 1m tall; leaves pinnatisect, long-stalked, grey-hairy; flowerheads bright pink to lilac-mauve); *C.hypoleuca* (W Transcaucasia, Turkey, Iran; perennial to 50cm tall; leaves lyrate to pinnatisect, grey-hairy beneath; flowerheads pink to bright magenta-rose in the cultivar 'John Coutts'); *C.macrocephala* (Caucasus; perennial to 1m; leaves lanceolate, entire to deeply cut, green; florets yellow to yellow-orange); *C.montana* (PERENNIAL CORN-FLOWER; European mountains; perennial to 80cm; leaves ovate to lanceolate, entire to lobed, floccose beneath; outer florets typically blue, inner florets violet to mauve); *C.pulcherrima* (Transcaucasia; perennial to 40cm; leaves lanceolate, entire to lyrate or pinnatisect, grey-hairy beneath; outer florets bright rose-purple, inner florets paler) *C.margaritae*, *C.moschata* and *C.suaveolens* are now treated as *Amberboa moschata*.

centipedes (Chilopoda) small soil inhabiting animals with segmented bodies, They have one pair of legs on each trunk segment, in contrast to millipedes which have two pairs; and, depending on species, they may have 15 to over 100 pairs of legs. The legs on the first segment behind the head are modified into a pair of claw-like appendages, each with a poison gland used to seize and immobilize prey. Centipedes are mostly carnivorous, feeding on small slugs, worms and insects and they are generally beneficial in the garden.

The most common species belong to the orders Geophilomorpha and Lithobiomorpha. Geophilomorphs which are adapted to subterranean life, are very long and slender, and have 31–177 pairs of legs. They are often brought to the surface during cultivation and writhe and twist with little progress. The most common species is *Necrophlaoephagus longicornis*, which is yellow, and 80mm long. Lithobiomorphs are surface dwellers. They are found under stones and in other sheltered places. They have flattened bodies and adults have 15 leg-bearing segments and relatively long and stout legs, allowing for rapid progress over the soil surface. One of the most common species is *Lithobius forficulatus*, which is chestnut brown and about 25mm long.

central cell a binucleate cell in the centre of the embryo sac from which the endosperm develops. See *polar nuclei*.

Centranthus (from Greek *kentron*, spur, and *anthos*, a flower, from the spurred flower). Valerianaceae. Europe, Mediterranean. 12 species, hardy annual and perennial herbs, with clumps of erect stems decked with broad, packed cymes of small flowers in late spring and summer. Hardy in zone 6. Grow in full sun on a fast-draining, poor and preferably alkaline soil in the herbaceous or mixed border, or in the crevices of old walls. Shear over after flowering. Propagate from seed in autumn or spring. *C.ruber* (RED VALERIAN, JUPITER'S BEARD, FOX'S BRUSH; perennial to 80cm tall; flowers crimson to pale red, held above foliage in branched heads from late spring to early autumn; white, pink and deep red forms are common in cultivation).

centre leader a central stem carrying branches and/or fruiting spurs.

centrifugal progressing or extending from the centre towards the margin.

centripetal developing or progressing towards the centre from the margin.

cephalanthium the capitulum or flowerhead of Compositae.

Cephalaria (from Greek *kephale*, head: the flowers are in round heads). Dipsacaceae. 65 species, annual and perennial herbs with pinnately toothed to lobed leaves and, in summer, small flowers in rounded heads surrounded by bracts and carried on slender stalks. Fully hardy, plant in a sunny, well-drained position in a large herbaceous border or the wild garden. Increase by division in spring, or by seed in autumn. *C.gigantea* (syn. *C.tatarica*; GIANT SCABIOUS, YELLOW SCABIOUS; Caucasus, Siberia; stately perennial to 2m tall with long-stalked, 5cm-wide heads of ivory to yellow flowers).

cephalium a woolly growth bearing flowers, terminating the stem of certain Cactaceae.

Cephalocereus (from Greek *kephale*, head, and Latin, *Cereus*, referring to the head-like, differentiated flowering zone). Cactaceae. Mexico. 3 species, tender cacti with tall, usually unbranched column-like stems that are ribbed and spiny. Tubular to campanulate flowers open at night amid wool and bristles. Provide a minimum temperature of 10°C/50°F, low humidity

Centaurea montana

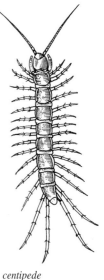

centipede

and shade from the hottest summer sun. Plant in an acid to neutral soil high in grit and sand. Keep dry from mid-autumn until early spring, except for a light misting over on warm days in late winter. *C.senilis* (OLD MAN CACTUS; stem grey-green, growing extremely slowly to 12×0.4m, usually grown as far smaller, narrower plants, covered in a mass of white, shaggy hairs; flowers off-white marked pink).

Cephalophyllum (from Greek *kephalos*, head, and *phyllon*, leaf, referring to the dense, head-like clumps formed by the leaves of some species). Aizoaceae. Namibia, South Africa. 30 species, low-growing, tender perennials with succulent, cylindric to triangular leaves and large, daisy-like flowers opening on warm, sunny days. Cultivate as for *Conophytum*.

C.*alstonii* (leaves to 7cm, grey-green; flowers dark red); C.*pillansii* (leaves 2.5–20cm, dark green; flowers yellow with a red centre).

Cephalotaxus (from Greek *kephale*, head, and Latin *taxus*, yew, referring to the head-like inflorescences of these yew-like trees). Cephalo-taxaceae. N India, China, Taiwan, Japan. PLUM YEW. 9 species, evergreen coniferous trees with 2-ranked, yew-like leaves and fleshy, egg-shaped fruits. Hardy in climate zone 5, but better in milder regions. It favours a moist, sandy soil, a cool, humid atmosphere and semi-shade, and may be grown as a specimen, an understorey shrub, or as a screen or hedge. Propagate by greenwood cuttings of terminal shoots, under mist or in a cold frame, in late summer or early autumn. Sow seed in a cold frame when ripe; alternatively, stratify seed and sow in spring (germination may be delayed until the second spring). *C.harringtonia* (COW'S TAIL PINE, JAPANESE PLUM YEW; Japan; shrub or small tree to 5m tall; leaves to 6cm, linear to falcate, dark green above, with two grey bands beneath; fruit 2–3cm, ripening purple-brown; dwarf, shrubby, fastigiate, procumbent and semi-weeping forms exist).

Ceratostigma plumbaginoides

Cephalotus (from Greek *kephalotos*, with a head: the stamens are capitate). Cephalotaceae. W Australia. AUSTRALIAN PITCHER PLANT. 1 species, *C.follicularis*, a small, tender perennial herb, carnivorous with leaves of two types: the first ovate to spathulate in a central tuft, and the second (the traps) pitcher-like and low-borne. The pitchers are some 1–5cm long, squat and green with bristly wings, the rim heavily ribbed, waxy and tinted blood red and the lid veined purple-red and zoned with paler, semi-lucent patches. Small, green-white flowers are produced in slender-stalked spikes in summer. Plant in deep clay pots containing a mix of silver sand, peat and live sphagnum; maintain a minimum temprature of 10°C/50°F and position in full sun with high humidity. From mid-spring to autumn, stand the pot to one third of its depth in rain water; during the winter months, reduce the water depth keeping the medium constantly moist but not sodden. Propagate by division or leaf cuttings in late spring.

Ceratostigma willmottianum

Cerastium (from Greek *keras*, horn, alluding to the shape of the fruit). Caryophyllaceae. Widespread. MOUSE-EAR CHICKWEED. Some 100 species, hardy annual or perennial herbs, often tufted or mat-forming with small, hairy leaves. Produced in summer, the flowers are white and consist of five, obovate petals with notched tips; they are carried one or a few together on slender stalks. Grow *C.alpinum* in the rock garden or in the alpine house and *C.tomentosum* as groundcover to the fore of borders, in the interstices of paving, or on dry walls. Both are hardy in climate zone 4 and prefer full sun, minimal winter wet and a rather dry, poor and neutral to alkaline soil. Shear off dead flowerheads. Increase by division or by seed in spring.

C.*alpinum* (ALPINE MOUSE EAR; Arctic, European mountains; mat-forming perennial with very small, grey-white hairy and obovate to elliptic leaves; flowers to 2cm in diameter, on stalks to 8cm); C.*tomentosum* (SNOW-IN-SUMMER; mountains of Europe and W Asia; mat-forming perennial with grey to white-hairy, linear to lanceolate to elliptic leaves, to 2cm long; flowers to 2.5cm in diameter, on stalks to 15cm).

Ceratophyllum (from Greek *keras*, horn, and *phyllon*, leaf). Ceratophyllaceae. Europe, Africa, Asia, US. HORNWORT. 30 species, submerged aquatic herbs with slender floating stems whorled with brittle, thread-like leaves. Grow the following species in unheated freshwater aquaria or ponds, pushing the stem bases into the substrate. Increase by stem cuttings or by spontaneously offsetting winter buds. *C.demersum*, (Africa, Mediterranean, Eastern Europe; fully hardy with stems to 1m long encircled by forked, dark green leaves).

Ceratopteris (from Greek *keras*, horn, and *pteris*, fern). Parkeriaceae. Tropics. WATER FERN, FLOATING FERN, FLOATING STAG'S HORN FERN. 4 species, tender aquatic or semi-aquatic ferns with free-floating or anchored rosettes of smooth, pinnately lobed leaves filled with air spaces – these give the leaves a succulent appearance. Three rather similar species, *C.cornuta*, *C.pteridioides* and *C.thalictroides*, are grown in aquaria, pools and lakes (minimum temperature 10°C/50°F). They make rosettes to 30cm across, composed of rather fleshy to spongy, bright green fronds; the fronds range from oblong and shallowly toothed to finely 2–3-pinnately lobed. Float the rosettes on soft, still water rich in nutrients, or bed into the saturated mud of shallows and margins. Increase by offsets.

Ceratostigma (from Greek *keras*, horn and *stigma*, alluding to the horn-like excrescences on the stigmas). Plumbaginaceae. Asia, Africa. PLUMBAGO. 8 species, hardy, woody-based perennial herbs or small shrubs with ovate to obovate, more or less bristly leaves and terminal heads of blue flowers in late summer and autumn. Plant in a sheltered position in full sun on a well-drained soil. Where temperatures fall below –15°C/5°F for prolonged periods, these plants benefit from a dry winter mulch and may perform as herbaceous perennials. Propagate by removal of rooted suckers or spontaneous layers in spring, or by softwood or semi-ripe cuttings.

C.griffithii (Himalaya; evergreen, procumbent to creeping, evergreen shrub forming mats to 0.3 × 1.5m; leaves 1–3cm, tough and bristly, becoming red-tinted in hard weather; flowers bright blue); *C.plumbaginoides* (W China; multi-stemmed, more or less erect and semi-deciduous woody-based herb or subshrub to 50cm; leaves to 9cm, undulate, thin-textured, bright green slightly bristly, flushing rusty red in autumn; flowers dark to brilliant blue); *C.will-mottianum* (CHINESE PLUMBAGO; W China to Tibet; deciduous shrub, broad, bushy and erect to 1m, with ribbed and bristly, twiggy branches; leaves 2–5cm, bristly, tinted red in autumn; flowers pale to deep sky blue amid red-brown calyx lobes).

Cercidiphyllum (from *Cercis*, and Greek *phyllon*, leaf). Cercidiphyllaceae. W China, Japan. KADSURA, KATSURA TREE. 1 species, *C.japonicum*, a hardy deciduous tree to 30m tall, with an erect, sometimes multi-stemmed habit, grey, furrowed bark and slender, glossy shoots. The winter buds are small, red and claw-like, often appearing on short spurs. Broadly ovate to kidney- or heart-shaped, the leaves are 5–10cm long and finely crenate. Soft pea-green at first, they shift in autumn through shades of amber, yellow and scarlet, becoming candy-scented. The small red flowers are inconspicuous. A form with pendulous branches is sometimes grown. Plant in sun or part shade on a moist, humus-rich and preferably acid soil. Propagate by seed cuttings taken in late spring and rooted in humid, heated case.

Cercis (from Greek *kerkis*, a weaver's shuttle: a name given by Theophrastus to describe the large, woody, flattened fruits). Leguminosae. Asia, Europe. 6 species, hardy deciduous trees and shrubs with broadly heart- to kidney-shaped leaves and small, pea flowers in clusters or short racemes, often sprouting directly from the older wood in spring. Plant in full sun on a fertile, well-drained soil. Most are grown for their flowers, but *C.* 'Forest Pansy' is an exceptional foliage plant, especially when young, or if stooled back in late winter. Propagate by seed or (cultivars), by semi-ripe cuttings in a closed case with bottom heat; increase also by budding.

C.canadensis (EASTERN REDBUD, REDBUD; N America; spreading tree to 14m; leaves 3–12cm, smooth to shining above, puberulent beneath; flowers to 1cm, crimson, lilac or white; 'Forest Pansy': leaves darkest wine red at first, ultimately purple-red); *C.siliquastrum* (JUDAS TREE, LOVE TREE; E Mediterranean; spreading tree or shrub to 10m, usually smaller; leaves 6–10cm, matt grey- to mid-green, glabrous; flowers 1–2cm, pale rose to magenta; includes white and dark-flowered forms).

Cereus (from Latin *cereus*, a wax torch taper, referring to the shape of some species). Cactaceae. West Indies, S America. 40 species, tender, shrubby or tree-like cacti with erect, ribbed and spiny, columnar stems and large, funnel-shaped flowers composed of many tepals and usually opening at night. Grow in full sun with low humidity and a minimum temperature of 10°C/50°F. Plant in an acid to neutral medium high in grit and sand. Water moderately in summer, very sparingly in winter. Increase by rooting sections of stem.

C.uruguayensis (syn. *C.peruvianus*; Brazil to Argentina; tree to 6m; stems 8–12cm in diameter, glaucous, with 8 deep, rounded ribs and spines to 2cm long; flowers to 18cm, white with green outer tepals edged in red; 'Monstrosus': stems distorted, compressed to fan-shaped); *C.validus* (syn. *C.forbesii*; Argentina; tree to 6m; stems to 12cm in diameter, blue- to grey-green, with 4–7 deep, thin ribs and spines to 4.5cm long; flowers to 20cm, white with red-tinted outer tepals).

ceriferous wax-producing.

Cerinthe (from Greek *kerinos*, waxy, bees were thought to obtain wax from the flowers). Boraginaceae. HONEYWORT. Europe. 10 species, annual, biennial or perennial herbs (those here annual or, rarely, biennial). They are loosely branching and erect to spreading with oblong to broadly obovate, sessile leaves. Produced in spring and summer, the tubular flowers nod amid colourful bracts in arching, terminal cymes. The following species are among the most striking of all hardy annuals. They prefer a well-drained, sandy or gritty soil, full sun and shelter, consorting brilliantly with blue-grey grasses and Mediterranean-type plants. As the season progresses, they tend to become bare and leggy and may need support from pea sticks or dead twigs. Sow seed under glass in early spring for planting out after the last frosts, or sow *in situ* in late spring. In the right situation, they will self-sow; otherwise, collect the black, shot-like seed from late summer onwards. Pinch out new plants to establish a bushy, free-flowering habit. Any shoots so removed make good cuttings, rooting quickly under glass.

C. major (S Europe; to 60cm; leaves blue-green with irregular grey spots; bracts sea green tinted bronze-purple, becoming violet then petrol blue with age; flowers gold at base, progressing through copper, wine red and violet to the indigo tip; 'Purpurascens': a robust plant with especially fine colouring – maroon then blue-purple); *C.retorta* (S Europe; to 50cm; leaves pale to mid-green with two rows of dull silver blotches and a purple-brown stain near the apex; bracts yellow-green suffused maroon to violet; flowers yellow tipped dark maroon or violet).

cernuous nodding, usually applied to flowers with curved or drooping pedicels attached to a straight or erect inflorescence axis.

Ceropegia linearis
subsp. *woodii*

Ceropegia (from Greek *keros*, wax, and *pege*, fountain, alluding to the appearance of the waxy flowers). Asclepiadaceae. Old World tropics and subtropics. 200 species, tender perennials, tuberous or caudiciform with slender, twining to cascading stems and small, succulent leaves, or erect and shrubby with candelabra-like, succulent and virtually leafless stems. The flowers are short-stalked and solitary or clustered; they consist of a bladder-like base, a narrow, upward-curving tube and a 5-lobed limb, the lobes narrow, hairy and outspread or (more characteristically) directed inwards in a lantern or cage. Provide a minimum temperature of 10°C/50°F with low to medium humidity and full sun. Plant in a gritty, loam-based medium; water moderately in spring and summer, very sparingly in winter. Increase by tip cuttings of mature shoots with two to three nodes or, for *C.linearis*, by detaching lengths of stem on which tubers have developed. *C.linearis* subsp. *woodii* is a popular houseplant, an excellent choice for baskets and hanging pots.

C.haygarthii (South Africa; stem stout, slightly succulent, twining, to 2m; flowers 1–3cm long, white to buff, tube pitcher-shaped, widening above and spotted purple, the lobes narrowing greatly and joining across aperture of pitcher, then twisting together to form an erect tail tipped with rounded, hairy appendage); *C.linearis* subsp. *woodii* (syn. *C.woodii*; HEARTS ENTANGLED, SWEETHEART VINE, HEARTS ON A STRING; South Africa; tuberous with many slender, pendulous stems and small, thinly succulent heart-shaped leaves, tinted maroon beneath and marbled grey-green above; flowers with flesh pink tube and dark 'cage' covered in black hairs); *C.sandersonii* (PARACHUTE PLANT, UMBRELLA PLANT; Mozambique, South Africa; stems to 2m, thickly succulent, twining; flowers 3–5cm long, green marked dark green and covered with white hairs above, the tube flaring from a narrow base, the lobes fusing above in a lantern-like canopy – the 'umbrella' – to 2.5cm in diameter).

Chaenomeles
x *superba*

certified stock plants that have been examined by an officially authorized inspector and found to conform to the standards of a certification scheme regarding freedom from pathogens (especially viruses), vigour, and trueness to type. Certified stock is the best-quality planting material available to gardeners, and particularly important as regards new stocks of strawberries, raspberries and blackcurrants.

cespitose see caespitose.

Cestrum (from the Ancient Greek name *kestron*, for similar but unrelated plants). Solanaceae. S America. BASTARD JASMINE, JESSAMINE. Some 175 species, evergreen or deciduous trees and shrubs, many frost-tender with spreading to semi-scandent branches and lanceolate to elliptic leaves. Borne in panicles or corymbs, tubular to funnel-shaped flowers appear from early summer to autumn. *C.elegans*, *C.fasciculatum*, *C.parqui* and *C.roseum*

will survive mild frosts, especially if planted on a sheltered, south-facing wall, as suits their semi-climbing habit. Plants cut down by frost will usually resprout if thickly mulched in winter. The remaining species need a minimum temperature of 7°C/45°F. Grow in full or filtered sun on a fertile, well-drained soil. Feed and water container-grown plants freely when in growth, sparingly at other times. Provide support for scandent species. Prune in spring to remove weak or dead growth. Increase by semi-ripe or greenwood cuttings in summer.

C.aurantiacum (Guatemala; evergreen shrub to 2m, sometimes climbing; flowers bright orange); *C.elegans* (Mexico; evergreen shrub to 3m, bushy with arching branches; flowers deep pink to purple-red); *C.fasciculatum* (Mexico; evergreen shrub to 2m with arching branches; flowers carmine to purple-red); *C.*'Newellii' (a possible hybrid with *C.elegans*, with crimson flowers and dark green leaves); *C.nocturnum* (LADY OF THE NIGHT, NIGHT JESSAMINE; West Indies; evergreen shrub to 4m with spreading to scandent branches; flowers pale green to ivory, deliciously scented at night); *C.parqui* (WILLOW-LEAVED JESSAMINE; Chile; deciduous shrub to 3m; flowers yellow-green to yellow, or violet in the cultivar 'Cretian Blue'); *C.roseum* (Mexico; evergreen shrub to 2m; flowers rose to purple, rosy pink in the cultivar 'Illnacullin').

Chaenomeles (from Greek *chainein*, to gape, and *melon*, apple, referring to the incorrect belief that the fruit is split). Rosaceae. E Asia. FLOWERING QUINCE, JAPANESE QUINCE, JAPONICA. 3 species, deciduous, fully hardy shrubs or small trees with spiny, slender branches, toothed, oval to lanceolate leaves and, from late winter to early summer, showy flowers composed of five, broad petals. Dull yellow and aromatic, the quince-like fruits are used to make jellies and conserves, often combined with apple and pear. Plant on a well-drained soil in full sun. These shrubs are grown to best advantage on walls, tied to trellis, wires or wall nails, with their lateral shoots reduced to spurs of two or three buds after flowering (or fruiting) and any weak shoots that grow outwards removed over the spring and summer. Increase by softwood or greenwood cuttings in summer.

C.cathayensis (China; shrub or small tree to 6m, with tortuous branches armed with spiny spurs; flowers to 4cm in diameter, white, sometimes flushed pink); *C.japonica* (MAULE'S QUINCE; Japan; spreading, thorny shrub to 1m; flowers 4cm in diameter, orange-red, scarlet or crimson; includes white, dwarf and brilliant orange-red forms); *C.speciosa* (China; spreading, thorny shrub to 3m; flowers to 4.5cm in diameter, scarlet to crimson; includes cultivars with single, semi-double and double flowers in tones of blood red, scarlet, salmon pink, crimson, rose, palest pink and pure white – for example, the shell pink-tinted cream 'Apple Blossom'; *C.* × *superba* (*C.speciosa* × *C.japonica*; erect to spreading, spiny shrub to 1.5m; leaves close to *C.japonica*; flowers white, pink, crimson to orange and orange-scarlet; includes

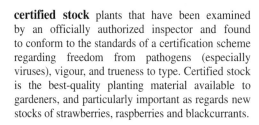

ABOVE Japanese irises thrive in sunny, damp sites like this streamside. Equally at home is the giant *Phormium*, a plant more usually associated with parched coastal gardens, but here providing a suitably grassy, bronze foil for glowing candelabra primulas.

RIGHT By late spring, the bog garden or waterside becomes a place of rich colour and formal contrasts: here the hot tones of candelabra primulas contend with the foamy, fragrant plumes of *Rodgersia aesculifolia* and *Carex elata* 'Aurea', Bowles's golden sedge.

Woodland gardens

ABOVE Secreted among the cool creams and yellows of *Helleborus orientalis* and *Erythronium* 'Pagoda' are two plants with subtle, silver-mottled leaves – wine red *Trillium sessile*, and *Pulmonaria saccharata* 'Argentea' with flowers passing through pink, mauve and blue. Informal 'woodland' plantings like this one are easily achieved in shady corners and under shrubs. They flourish in late winter and spring and are virtually self-maintaining.

RIGHT An imaginative reworking of the woodland theme, this shade planting combines elements of the bog garden – candelabra primulas, *Rodgersia* – with Japanese maples and azaleas, naturalized narcissus and bluebells and dramatic *Phormium*.

'Crimson and Gold', with crimson petals and golden anthers, 'Etna', small with deep fiery red flowers, the pink 'Hever Castle', orange-scarlet 'Knap Hill Scarlet' and the large, dark crimson 'Rowallane').

chafer beetles (Coleoptera: Scarabaeidae) stout, medium to large beetles with antennae terminating in a club consisting of 3–7 flat, leaf-like segments. The larvae live in soil or decaying wood and are white or grey and C-shaped, with a heavily sclerotized yellow or brown head. They have three pairs of thoracic legs, and the last segment of the abdomen is enlarged.

The larvae of many species feed on plant roots. They are abundant in grassland, particularly on light soils, and attack lawns and the roots of a wide range of fruit, vegetable and ornamental plants. Adults feed on leaves, buds, flowers and fruits of deciduous trees and shrubs. Chafer beetles are on the wing in summer and the females deposit eggs in the soil. The life cycle varies: larger species, such as the COCKCHAFER, have larvae that live for two-and-a half years before pupating in the soil; smaller species, such as the GARDEN CHAFER and the JAPANESE BEETLE, complete the cycle in one year.

The COCKCHAFER (*Melolontha melolontha*) is up to 30mm long, with a black head and thorax and red-brown wing shields. In both sexes the abdomen terminates in a spade-like point. Males are much attracted to light during the midsummer flight period. The larvae are up to 60mm long and take two-and-a-half years to become fully fed.

The GARDEN CHAFER (*Phyllopertha horticola*) is up to 11mm long, with a metallic green head and thorax and brown wing shields. The larvae, up to 18mm long, complete development in one year.

The SUMMER CHAFER (*Amphimallon solstitialis*) is uniformly brown and up to 18mm long. It is seldom injurious in gardens, although it swarms occasionally on elms and poplars.

The ROSE CHAFER (*Cetonia aurata*) is metallic golden-green, up to 20mm long; it is locally common and may feed on flowers, buds and leaves of roses from June to August.

Chafer beetles are not easy to control. Apply a soil insecticide around the roots of vulnerable plants; treat infested lawns similarly and roll them in late spring to aid the killing of pupae and emerging adults.

See *Japanese beetle*.

chalcid wasps (Hymenoptera: Chalcidoidea) an important group of beneficial insects. They are no more than 1 or 2mm in length and although abundant in gardens, are usually overlooked. They have elbowed antennae and the wings are held flat when at rest. Many have a metallic green or blue body sheen. Some larvae feed on seeds or within galls (e.g. *Megastigmus* species, and *Torynus varians*, which can damage apple fruits in North America).

Several chalcid wasps are utilized in biological control and many are potential candidates. The majority are parasites of insects, mites and spiders; a few parasitize parasites and should be excluded from the breeding phase of biological-control organisms. *Encarsia formosa* is widely used for the biological control of greenhouse whitefly (*Trialeurodes vaporariorum*); *Aphelinus* species parasitize aphids and scale insects; and the apple woolly aphid parasite (*A.mali*) occurs widely throughout the world. Several other cosmopolitan genera, including *Aphycus*, *Aphytis*, *Physcus*, *Aspidiotiphagus* and *Prospantella*, parasitize scale insects. *Pteromalus* species parasitize the pupae of cabbage white butterflies and related insects. *Trichogramma* species are bred in large numbers in many countries for mass-release programmes to attack the eggs of various injurious species of moth. Unlike many parasitoids, they have a wide range of hosts; as many as 20 individuals may develop in a single host egg.

chalk a soft white form of limestone, composed of calcium carbonate. It is incorporated in loam-based seed and potting composts.

Chamaecyparis (from Greek *chamai*, lowly, dwarf, and *kuparissos*, cypress). Cupressaceace. E Asia, N America. CYPRESS. 8 species, fully hardy, evergreen coniferous trees and shrubs with small, awl- to scale-like leaves, aromatic and covering finely branched sprays or fans of growth; the cones are small and usually globose. Pollution-tolerant, moderate- to fast-growing conifers suitable for screening, hedging and lawn specimens; dwarf cultivars are used for rock gardens, island beds and containers. They succeed best in full sun on moist, well-drained, acid to neutral soils. Propagate by seed or, for cultivars, by heeled greenwood cuttings in late summer in a humid cold frame or by softwood cuttings under mist in summer. All species are susceptible to phytophthora root rot (*Phytophthora cinnamomi*). *Phytophthora lateralis* is a serious problem on *C.lawsoniana*, especially in parts of the western US where this most popular and versatile of conifers is now gravely threatened.

C.lawsoniana (LAWSON'S CYPRESS, OREGON CEDAR, PORT ORFORD CEDAR; W US; to 60m in wild, crown narrowly conic then columnar; branches level, branchlets pendulous; leaves to 2mm, keeled, bright green or tinted blue-grey with pale lines beneath; includes many cultivars ranging in habit from dwarf and rounded to columnar in tones of green, blue-green, grey, blue, lime and yellow; there are also some white-variegated forms and others with thread-like or feathered foliage; the most popular cultivar is 'Ellwoodii', erect to 3 × 1.5m with blue-grey leaves); *C.nootkatensis* (NOOTKA CYPRESS, YELLOW CYPRESS; western N America; to 40m, slender or broadly conic, becoming domed with age; branches upswept, branchlets pendulous; leaves to 4mm, deep green; includes cultivars with yellow to moss green or blue-grey foliage and dwarf, compact or tall and weeping habit); *C.obtusa* (HINOKI CYPRESS; Japan; to 40m, slow-growing with a broadly conic crown becoming columnar; branchlets in flattened, layered fans; leaves to 2mm, blunt and closely overlapping, dark green

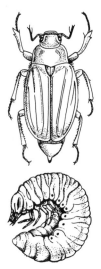

cockchafer beetle and larva

*Chamaedorea
elegans*

above, silver-white beneath; includes dwarf, tiered, congested, filiform and rounded cultivars with very deep green, cream to gold or grey foliage – one of the most popular is the dwarf, dark green 'Nana'); *C.pisifera* (SAWARA CYPRESS; Japan; to 40m, broadly conic to irregularly conic; bark rusty brown and peeling; leaves to 3mm, mossy green above, silvery beneath, tapered and pointed; cultivars range in size from the dwarf and slow-growing to medium-sized trees, in habit from flat-topped to rounded and bushy and in foliage colour from steel blue (e.g. 'Boulevard', 'Squarrosa') to yellow-green or white-variegated); *C.thyoides* (WHITE CYPRESS, WHITE CEDAR; E US; to 30m, slow-growing, crown narrowly conic to irregularly ovoid; leaves to 3mm, dark grey-green or glaucous, in fan-shaped sprays; includes dwarf forms with blue-grey, gold and bronze foliage, the leaves sometimes needle-like; the most popular is 'Andelyensis', a bluntly conic shrub to 3m with fan-like branchlets of linear, blue-green leaves).

Chamaecytisus (from Greek *chamai*, lowly, dwarf, and *Cytisus*). Leguminosae. Europe, Canary Islands. 30 species, more or less evergreen trees, shrubs and subshrubs, usually with small, trifoliolate leaves and pea-like flowers in spring and summer. The following species are hardy in climate zones 6 and over. Cultivate as for *Cytisus*.

C.albus (syn. *Cytisus albus*; PORTUGUESE BROOM; SE and C Europe; erect to 80cm; leaflets often hairy; flowers white or yellow in terminal clusters); *C.purpureus* (syn. *Cytisus purpureus*; SE Europe; to 60cm, bushy and low-growing; leaflets dark green, smooth; flowers pale pink to lilac flushed ruby or crimson; white, opal, pink and dark red cultivars are grown); *C.supinus* (syn. *Cytisus supinus*; S and C Europe to Ukraine; to 1m, ascending to prostrate; leaflets smooth or thinly hairy; flowers yellow in short, terminal racemes); *C. × versicolor* (*C.purpureus × C.hirsutus*; to 50cm, erect to diffuse with long, slender branches; leaflets hairy beneath; flowers pale yellow with a pink-violet keel; includes 'Hillieri', a low-growing plant with arching branches and large yellow flowers later becoming bronze- to pink-tinted).

Chamaedaphne (from Greek *chamai*, lowly, dwarf, and *Daphne*). Ericaceae. Northern temperate regions. LEATHERLEAF, CASSANDRA. 1 species, *C.calyculata*, a spreading evergreen shrub, fully hardy and seldom exceeding 1 × 1.5m. To 5cm long, the leaves are oval to lanceolate, tough and rusty-scaly beneath. Pure white, urn-shaped flowers appear in short racemes in spring. Grow as for *Andromeda*.

Chamaedorea (from Greek *chamai*, lowly, near the ground, and *dorea*, a gift: the fruits are easily reached). Palmae. C and S America. 100 species, small, tender palms, often with cane-like stems and pinnate to fish-tail-shaped leaves. Small, golden to creamy flowers are carried in freely branched. Provide a minimum temperature of 15°C/60°F, with filtered sun to light

shade and medium to high humidity. Grow in a porous, soilless mix kept consistently moist. Feed and syringe frequently in summer. Increase by careful division of stems when repotting. These conditions are ideal, but *C.elegans* has proved itself capable of surviving (and often receives) far harsher treatment. *C.elatior* may need some support with age.

C.elatior (stems solitary or clumped, slender, cane-like, green, to 4m, ultimately weakly climbing; leaves to 3m, pinnate, with narrow drooping leaflets); *C.elegans* (syn. *Neanthe bella*; PARLOUR PALM; stems solitary or clumped, to 2m × 3cm, erect; leaves to 30cm, pinnate; a very popular pot plant, especially when young); *C. microspadix* (syn. *C.metallica* of gardens; stems to 1m, cane-like; leaves to 25cm, fish-tail-shaped, ribbed and puckered, metallic sea green).

Chamaemelum (from Greek *chamai*, lowly, and *melon*, apple). Compositae. Europe, Mediterranean. CHAMOMILE. 4 species, annual or perennial frost-hardy herbs with finely and pinnately cut foliage and yellow-centred, white, daisy-like flowerheads. Plant in full sun on any loamy soil with a high sand content and ample moisture, but avoid sites where there is risk of water-logging. Flowerheads of the typical plant are used to make chamomile tea. The non-flowering cultivar, 'Treneague', is used in lawns and walks and on seats and banks. Chamomile lawns become bare and patchy unless regularly sheared and diligently replanted wherever thin. Increase by seed in autumn or, for 'Treneague', by division or rooted stem cuttings in spring. *C.nobile* (W Europe; strongly aromatic, decumbent perennial with finely 2-3-pinnatisect mid- to grey-green leaves and small flowerheads; 'Treneague': non-flowering, carpeting cultivar suitable for lawns).

chamaephyte a low-growing plant with growth buds above the soil surface; usually bushy and of mountainous or arid origin.

Chamaerops (from Greek *chamai*, lowly, dwarf, and *rhops*, a bush). Palmae. Mediterranean. 1 species, *C.humilis*, a palm with shaggy, solitary to clumped trunks 0.25–3m tall. To 1m long, the palmately divided leaves consist of a coarsely toothed petiole and a fan of linear to lanceolate, scurfy grey segments. This palm will tolerate short frosts in climate zone 9 and favoured locations in zone 8, especially if grown in a sunny, sheltered position and kept dry in winter with its upper trunk and crown wrapped in sacking. Else-where, grow in full sun in a frost-free greenhouse or conservatory. Plant in a fast-draining, gritty and low- to medium-fertility soil. Keep moist in summer, virtually dry in winter. Increase by offsets or seed.

Chamelaucium (name derivation obscure). Myrtaceae. SW Australia. WAXFLOWER. 21 species, strongly aromatic, evergreen, tender shrubs with soft, needle-shaped foliage and bright flowers composed of five, broad and spreading petals. Grow in full sun with low to medium humidity and a minimum winter temperature of 7°C/45°F. Plant in tubs or borders of sandy, acid soil. Water, feed and ventilate liberally in spring and summer; keep almost dry in winter. The following species usually flowers in late winter and spring and is a popular cut flower. Increase by seed or by semi-ripe cuttings. *C.uncinatum* (GERALD-TON WAX; shrub to 3m tall; leaves strongly lemon-scented; flowers to 2.5cm in diameter, red, purple, mauve, pink or white).

charcoal the black, porous residue from wood burnt slowly under conditions of limited air supply. Charcoal is of no nutrient value, consisting wholly of carbon. It is used in crushed form to lighten texture and absorb surplus moisture in potting mixtures, or powdered to reduce bleeding from cuttings of latex-bearing plants, such as *Ficus* species.

chards see *swiss chard*.

charmille a smoothly-clipped hedge, usually of HORNBEAM (*Carpinus betulus*). It is used in the French garden to line allées.

chartaceous of leaf and bract texture; thin and papery.

Chasmanthe (from Greek *chasma*, chasm, and *anthos*, flower: the perianth is gaping). Iridaceae. South Africa. 3 species, cormous perennial herbs with 2-ranked linear to lanceolate leaves and, from autumn to spring, spikes of showy flowers with curving perianth tubes and six, unequal segments. Cultivate as for *Babiana*.
C.aethiopica (40–70cm; spike unbranched, flowers to 8cm, red, maroon in throat, tube with yellow stripes); *C.floribunda* (50–150cm; spike branched, flowers to 8.5cm, orange-red or yellow).

Chasmanthium (from Greek *chasma*, chasm, and *anthos*, flower: the spikelets gape widely). Gramineae. US, Mexico. 6 species, hardy perennial grasses with linear to lanceolate leaves and loosely clumped, slender culms that terminate in arching panicles of large, flattened spikelets. Hardy in climate zone 5. Grow in sun or dappled shade on a rich, moisture-retentive soil with shelter from strong winds. Increase by division in spring. The inflorescences are suitable for drying. *C.latifolium* (syn. *Uniola latifolia*; NORTH AMERICAN WILD OATS, SPANGLE GRASS, SEA OATS; to 1m tall; panicle weeping; spikelets 2–4cm, oblong to lanceolate to broadly ovate, with overlapping lemmas, like a large, flat *Briza* and changing by late summer from olive-bronze to straw yellow).

Cheilanthes (from Greek *cheilos*, lip, and *anthos*, flower, referring to the clustered marginal sori). Adiantaceae. Cosmopolitan. LIP FERN. Some 180 species of ferns, those described below are small, half-hardy and evergreen with short to creeping rhizomes and pinnately compound fronds. Grow in a cold greenhouse or frame in pans containing a humus-rich medium with a little added dolomitic

limestone. Water sparingly in summer, avoiding foliage; keep almost dry in winter. Provide a dry, cool and buoyant atmosphere. Admit bright, dappled light in summer, full sun in winter. Increase by spores or by careful division.
C.lanosa (N America; fronds 15–30cm, blades broadly ovate to lanceolate, tripinnatifid, pinnules oblong, soft green, on black, hairy stipes); *C.pteridioides* (syn. *C.fragrans*; S Europe; fronds 3–15cm, blades subtriangular to linear to lanceolate, bipinnate, pinnules oblong to suborbicular, toothed, dark green and sweetly scented, on shiny, chestnut-coloured stipes).

Cheiridopsis (from Greek *cheiris*, sleeve, and *opsis*, resemblance: the older leaf bases enclose the new growth in a dried sheath). Aizoaceae. South Africa. 80 species, tender perennial herbs forming low clumps of thickly succulent, semi-cylindrical leaves usually with flat tops. Daisy-like flowers open on bright days in summer. Cultivate as for *Conophytum*.
C.denticulata (syn. *C.candidissima*; leaves 8-10cm, erect, more or less 3-sided, white-grey or blue-green sometimes with a reddened tip; flowers 7cm in diameter, yellow to white); *C.purpurea* (syn. *C.purpurata*; leaves to 6cm, bluntly 3-sided, grey to blue-grey, becoming purple-red-tinted; flowers to 4.5cm in diameter, yellow; the name *C.purpurea* is sometimes applied in horticulture to a plant with rose-magenta flowers).

chelate see *sequestrol*.

Chelidonium (from Greek *chelidon*, swallow: the flowers were said to appear with the swallows in spring, and to die on their departure). Papaveraceae. Europe, W Asia, naturalized E US. 1 species, *C.majus*, GREATER CELANDINE or SWALLOW WORT, a hardy biennial or perennial with irritant, orange sap. The leaves are more or less lanceolate with rounded tips and wavy to toothed, or lobed. To 2.5cm in diameter, the flowers consist of four orange to yellow petals (more in 'Florepleno') and are carried on branching stems to 1m tall in summer. Hardy in climate zone 6 and suitable for the wild or woodland garden on most soils in sun or shade. Sow seed *in situ* in spring.

Chelone (from Greek *chelone*, tortoise: the flower is said to resemble a turtle's head). Scrophulariaceae. N America. SHELLFLOWER, TURTLEHEAD. 6 species, hardy erect perennial herbs with tubular, hooded and 2-lipped flowers in dense, terminal spikes in summer and early autumn. Hardy in climate zone 6. Plant in a moist, fertile soil in full sun or part-shade. Increase by division, or by soft tip cuttings in summer. *C.obliqua* (to 60cm; flowers to 2cm, deep rose pink to lilac or purple-red, with a sparse yellow beard).

chemical controls see *pesticides*.

chemotropism, chemonasty growth or bending towards a chemical stimulus, particularly seen in fungi and in the pollen tubes of flowers.

Chasmanthium latifolium

cherry

Pests
aphids
birds
slugworm
winter moth

Diseases
bacterial canker
shot-hole
silver leaf

cherry The edible fruit of selected *Prunus* species. Three types are grown in gardens: SWEET CHERRY, derived from *P.avium*, which is widespread from Europe, including Britain, to western Asia (its seedlings are often known as GEANS or MAZZARDS); ACID CHERRY, derived from *P.cerasus*, which has a distribution similar to *P.avium*; and the DUKE CHERRY, a hybrid of the two. Edible fruits are also produced by *P.pseudocerasus*, *P.tomentosa* and *P.besseyi*. There is evidence that cherries were a Bronze and Iron Age food in Britain, and trade in cultivated forms flourished from the 16th century. Much breeding work was undertaken in England during the 19th century, with significant introductions from British Columbia.

For gardens, acid cherries, especially in the form of fans or bushes, are the most suitable choice because they are the smallest of the three types. The self-fertile 'Morello' is the most successful, and is suitable for north- or east-facing walls. It is propagated by bud grafting on to semi-vigorous 'Colt' rootstock, although in the US selected seedlings of *P.avium* are preferred. The wide choice of cherry cultivars includes those with fruit ranging in colour from creamy white to purple-black; many are self-incompatible and require a cross pollinator.

Prune all cherries in summer. Since acid cherries carry fruit almost exclusively on young shoots of the previous year, aim for replacement, thinning new shoots 5–7.5cm apart along the ribs of the fan; after cropping, cut fruited laterals back to a trained replacement shoot. Sweet cherries form fruit buds at the base of young laterals of the previous year and on older wood: all shoots except those required as replacement shoots should be pinched at 5–6 leaves, and then to three leaves in late summer. The cherry flowers later than other stone fruits and is therefore less vulnerable to frost damage, but all types require protection from birds. Harvest with scissors.

chervil (*Anthriscus cereifolium*) a rapidly growing annual, up to 60cm high and parsley-like in appearance; it is used as a herb for its aniseed flavour. Sow *in situ* in a sheltered but unshaded site outdoors or under protection, or raise in containers for transplanting. Sow from February to April for a summer crop, or in August for cropping between autumn and spring. Seed may take 2–3 weeks to germinate.

Cheshunt compound a chemical preparation for the prevention and treatment of damping-off diseases (especially *Pythium* species) in seedlings. It was devised at the former Cheshunt Experimental Station, England, and is a powdered mixture consisting of two parts by weight of copper sulphate and eleven parts of ammonium carbonate. It is dissolved in warm water at the rate of 3g to one litre and is used to water seed boxes and seedbeds at the time of sowing. Repeat treatment when pricking out or transplanting seedlings.

Chiastophyllum (from Greek *chiastos*, arranged diagonally or crosswise, and *phyllon*, leaf). Crass-ulaceae. Caucasus. 1 species, *C.oppositifolium*, a hardy perennial herb with creeping stems and low rosettes of rounded and toothed, succulent leaves, each to 8cm long. In spring and summer, arching panicles to 15cm tall bear numerous small, golden flowers. In the cultivar 'Frosted Jade', the leaves are marked white to cream and tinted pink in winter. Hardy to –15°C/5°F. Grow in cool crevices in the rock garden in well-drained but moisture-retentive soil. Increase by side-shoot cuttings in early summer.

chicory (*Cichorium intybus*) WITLOOF CHICORY a native of Europe including Britain, where it grows wild, and used for centuries as a cooked or salad vegetable of distinctive slightly bitter flavour. Chicory may be grown similarly to lettuce or more suitably forced in darkness to produce tight clusters of mild-flavoured, folded leaves known as chicons. The taproot is roasted and ground to flavour coffee. Sow from May to early June lcm deep in rows 30cm apart; thin to 20cm spacings. Summer and autumn leaves may be eaten but are bitter. Lift the roots, which should be 5–9cm in diameter, in late October to December; trim the leaves hard and the thongs to 20cm. Close pack in deep boxes or large pots, in light soil or sand, to crown level. Place in a dark cellar or shed, or cover potted roots with an upturned pot, with light totally excluded. The optimum temperature range for forcing is 10–18°C/50–65°F. Harvest 3–4 weeks from housing. On light or sandy soils, forcing can be done outside *in situ* by earthing up trimmed plants to 15-18cm depth.

RED CHICORY (RADDICHIO) has red, sometimes variegated leaves; it forms a small crisp heart, the inner leaves part white and part coloured. It is grown and used as lettuce, though rather more bitter. Sow from April through to August, thinning plants to 20–35cm apart each way.

SUGAR LOAF CHICORY resembles cos lettuce; the inner leaves are tightly packed, white and sweetened, but of sharper flavour than lettuce. It is mainly grown for autumn maturity as a salad ingredient, from June-July sowings, as for witloof chicory. Alternatively, grow as a cut-and-come-again salad crop from early-season sowings onwards.

Chilean potash nitrate see *potassium nitrate*.

chilling the exposure of plants to low temperature, usually during dormancy. It is necessary to initiate growth or flowering in certain plants, for example, rhubarb *Rheum* × *cultorum*.

chimaera, chimera a plant in which two genetically different sorts of tissue coexist. Outer and inner tissues may be different in character, causing variegation or no obvious difference. Chimaeras may arise as a result of a natural mutation, or through grafting as in the case of + *Laburnocytisus* 'Adamii', which bears parts characteristic of both *Laburnum anagyroides* and *Chamaecytisus purpureus*.

Chimonanthus (from Greek *cheimon*, winter, and *anthos*, flower). Calycanthaceae. China. 6 species, deciduous shrubs to 3m tall with dull green, ovate to lanceolate leaves to 10cm long. Waxy and bell-shaped, the flowers consist of numerous sepals and petals, are deliciously scented and appear in short-stalked clusters in late winter on bare twigs. The following species is one of the finest winter-flowering shrubs. Hardy in climate zone 7, it nonetheless performs better if grown against a sheltered, sun-soaked wall. Plant in a fertile, moist but well-drained soil in full sun. Prune only to limit size or to remove weak or damaged growth. Propagate by seed, layers or semi-ripe cuttings in summer. New plants may take some years to reach flowering size. *C.praecox* (syn. *C.fragrans*; WINTERSWEET; flowers to 2.5cm long, nodding, dull yellow to buff with the inner petals marked maroon to blood red; cultivars with pure yellow, large, small and more heavily marked flowers are available).

Chimonobambusa (from Greek *cheimon*, winter, and Latin *Bambusa*: referring to the late appearance of the new shoots in some areas). Gramineae. S and E Asia. Some 20 species, small to medium-sized bamboos with running rhizomes; see *bamboos*.

Chinese gooseberry see *kiwi fruit*.

chinoiserie a European interpretation of the Chinese style of garden design, favouring an irregular and asymmetrical layout and incorporating pagodas, bridges and ornaments. It has been a major influence in English landscape gardening; surviving examples of garden buildings include the Pagoda at Kew Gardens, built in 1761.

Chionanthus (from Greek *chion*, snow, and *anthos*, flower). Oleaceae. E Asia, Pacific, E US. FRINGE TREE. 100 species, hardy deciduous trees and shrubs with elliptic to ovate leaves and, in summer, lacy panicles of flowers composed of four, white, oblong petals. These are followed by purple-blue, fleshy fruits. Hardy in climate zone 5. Plant in full sun on a moist but free-draining, fertile soil (acid to neutral for *C.virginicus*). Propagate by seed sown when ripe, or by layering in autumn or spring.
 C.retusus (CHINESE FRINGE TREE; Taiwan; to 3m; leaves to 10cm, pale green above; panicles erect); *C.virginicus* (E US; leaves to 20cm, dark green above, turning pale gold in autumn; panicles drooping).

Chionodoxa (from Greek *chion*, snow, and *doxa*, glory: they flower amid the melting snows). Liliaceae (Hyacinthaceae). W Turkey, Crete, Cyprus. GLORY OF THE SNOW. 6 species, bulbous perennial herbs, usually small or dwarf with linear leaves and erect racemes of shortly tubular, 6-tepalled flowers in spring. Cultivation as for hardy *Scilla*.
 C.forbesii (syn. *C.luciliae* of gardens; W Turkey; raceme 8–30cm tall with 4–12 slightly drooping flowers, deep blue with white centres, or snow white

or pink; includes Siehei Group, syn. *C.siehei* and 'Tmoli', syn. *C.tmolusii*); *C.luciliae* (W Turkey; raceme 5–14cm, with 1–2(–3), erect flowers, soft violet blue with white centres, or pure white or pink; includes Gigantea Group, syn. *C.gigantea*); *C.sardensis* (W Turkey; raceme 10–40cm with 4–12 slightly drooping or outward-facing flowers, of pure rich blue).

× **Chionoscilla** (*Chionodoxa* × *Scilla*). Liliaceae (Hyacinthaceae). 1 species, × *C.allenii*, a natural hybrid between *Chionodoxa forbesii* and *Scilla bifolia*, a hardy, spring-flowering bulb to 8cm tall, with oblanceolate leaves and erect racemes of starry blue or lilac-pink flowers with paler or white centres. Cultivate as for hardy *Scilla*.

chip budding a modified form of bud grafting used for propagating a wide range of woody plants. See *grafting*.

chipping (1) the puncturing of a hard seed coat before sowing by nicking with a knife, or filing; it is a form of scarification; see *seeds and seed sowing*; (2) a method of propagating tunicate bulbs, especially *Narcissus* species, by slicing them into portions. See *cuttings*: bulb scaling.

chitting (1) the sprouting of potato tubers before planting; (2) the germination of seeds before sowing, especially as practised in fluid drilling.

chive (*Allium schoenoprasum*) a hardy perennial about 25cm high, grown as a herb for its onion-flavoured leaves, which are used chopped in salads. Sow in modules under protection in spring; plant out clusters of 4 or 5 seedlings 25cm apart. Established clumps can be divided in autumn or spring.

chlamydia (1) bud scales; (2) perianth envelopes.

Chlidanthus (from Greek *chlide*, a costly ornament, and *anthos*, flower). Amaryllidaceae. Peru. 1 species, *C.fragrans*, a tender or half-hardy, bulbous perennial herb with linear, grey-green leaves. It bears fragrant, trumpet-shaped flowers, to 7cm across and in shades of yellow, red, cinnabar and green striped red atop 30cm-tall scapes. Grow in a sandy, loam-based medium in full sun. Water liberally from spring until flowering occurs in mid- to late summer; thereafter, keep frost-free and dry. Increase by seed or by offsets in spring.

chlorenchyma parenchymatous tissue containing chloroplasts; the type of tissue in which photosynthesis occurs.

chlorine a micronutrient or trace element rarely deficient in plants due to its almost universal occurrence in soil.

Chlorogalum (from Greek *chloros*, green, and *gala*, milk, an allusion to the colour of the sap). Liliaceae (Hyacinthaceae). Western N America. SOAP

Chimonanthus praecox

Choisya ternata

PLANT, AMOLE. 5 species, hardy bulbous perennial herbs with linear leaves and, in summer, erect panicles of small, 6-parted flowers that open in the afternoon. Cultivate as for *Camassia*. *C.pomeridianum* (SOAP PLANT, WILD POTATO; 60–150cm; flowers with purple- or green-striped, white tepals to 2cm long).

chlorophyll a mixture of two green pigments occurring in most plants, mainly in leaves but also in some stems; it is largely responsible for the conversion of light energy to sugars in the process of photosynthesis.

Chlorophytum (from Greek *chloros*, green, and *phyton*, plant). Liliaceae (Anthericaceae). Widespread in tropics and subtropics. 215 species, more or less tender, evergreen perennial herbs with fleshy roots. Linear to strap-shaped or ovate to lanceolate, the leaves are smooth and borne in basal rosettes or clumps. Throughout the year, slender-stalked panicles bear small, white flowers and sometimes plantlets. The following species is one of the most ubiquitous and long-suffering of foliage house plants. Notwithstanding, it prefers a minimum temperature of 7°C/45°F, bright indirect light, a fertile, free-draining, loam-based medium and ample supplies of water and feed throughout the year. Increase by division or plantlets. *C.comosum* (SPIDER PLANT, RIBBON PLANT; South Africa; clumped or rosette-forming with linear, arching leaves, green usually boldly striped white or cream; panicles loose-branched, bearing small, white flowers and numerous plantlets).

chloroplasts usually spherical or discoid bodies found in the cells of leaves and young stems; they contain the chlorophyll of green plants.

chlorosis a condition in which plants become abnormally pale green or yellow due to a partial or complete loss of chlorophyll. It may be caused by inadequate light, mineral deficiency (especially on high lime soils), or diseases such as those caused by viruses and mycoplasma-like organisms.

chocolate spot (*Botrytis fabae*) a fungal disease of the broad bean (*Vicia faba*); numerous small brown spots appear on leaves. The disease may become aggressive in wet seasons, causing plant death. Control by optimizing growing conditions and by the use of fungicide spray.

Choisya (for Jacques Denis Choisy (1799–1854), Swiss botanist). Rutaceae. SW US, Mexico. 9 species, evergreen, hardy aromatic shrubs with digitately compound, glossy leaves and, in summer and late autumn to winter, terminal clusters of fragrant white flowers composed of 4 or 5, spreading, obovate petals, each to 1.5cm long. *C.ternata* is hardy in climate zone 6 but benefits from a sheltered position; the remaining two species are less tolerant, thriving only in very favoured locations in zones 8 and 9. Something of their finesse is, however, shared by the hardy *C.*'Aztec Pearl'. All prefer a fertile, moist but fast-draining soil and full sun. Increase by semi-ripe cuttings in late summer.

C.arizonica (syn. *C.dumosa* var. *arizonica*; Arizona; bushy shrub to 1m tall; leaves with 3 or 5, linear, crenulate leaflets to 5cm long); *C.dumosa* (SW US, Mexico; bushy shrub to 2m; leaves with 8–13, linear to oblong, cre-nulate leaflets to 4.5cm long); *C.ternata* (syn. *C.grandiflora*; MEXICAN ORANGE BLOSSOM; Mexico; bushy shrub to 2m; leaves with 3, oblong to obovate, entire leaflets to 8cm long; 'Lich': syn 'Sundance', pale gold to lime green foliage; 'Aztec Pearl': hybrid with *C.arizonica*, with slender leaflets).

choripetalous see *polypetalous*.

Chorizema (from Greek *choros*, a dance for joy, and *zema*, a drinking vessel: the botanist Labillardière was close to exhaustion and dehydration when he came across this plant and water in the same place). Leguminosae. Australia. FLAME PEA. 18 species, tender, evergreen shrubs, sometimes climbing, with entire to sharply toothed leaves and racemes of colourful, pea-like flowers in spring and summer. Provide a minimum temperature of 7°C/45°F, full sun and a rather dry, buoyant atmosphere. Plant in a sandy, acid to neutral mix; water and feed moderately in spring and summer, sparingly in winter. Provide support. Sow seed in spring after immersing it in hot water and allowing it to cool and soak for 24 hours; root semi-ripe cuttings in summer.

C.cordatum (HEART-LEAVED FLAME PEA, AUSTRALIA FLAME PEA, FLOWERING OAK; dense upright shrub to 3m, sometimes climbing; leaves sometimes minutely spiny-toothed or lobed; flowers with red, orange and yellow standard and magenta or mauve keel); *C.ilicifolium* (HOLLY FLAME PEA; scrambling herb or shrub to 1m; leaves sinuously toothed and holly-like, tough, dark; flowers with orange-red and yellow standard and rose-mauve keel).

chromatid one of the two strands of a replicated chromosome, prior to their separation in mitosis or meiosis.

chromoplast, chromoplastid microscopic granules in plant cells containing other colouring than chlorophyll.

chromosome a microscopic body made up of DNA and protein, found in the nucleus of a plant cell; it contains the genes, or units of inheritance, that transmit a plant's characteristics. For each species, chromosomes are distinctive in shape and their number is always even.

Chrysalidocarpus (from Greek *chrysalis*, and *karpos*, fruit, alluding to the resemblance of the fruits to butterfly pupae). Palmae. Madagascar, Comoros Islands. 20 species, tender palms, often multi-stemmed with smooth, erect trunks, and graceful pinnate leaves. Grow in a moisture-retentive, coir-based mix, in bright, indirect light, with a winter minimum temperature of 15°C/60°F and plentiful supplies of water and liquid feed in summer. Propagate by detaching rooted suckers. *C.lutescens* (syn. *Areca lutescens*; YELLOW PALM, ARECA PALM, BUTTERFLY PALM, CANE PALM, GOLDEN YELLOW PALM; stems to 9m × 12cm, clumped, resembling giant canes; leaves to 2m, pinnae bright green, erect to arching, narrow, rachis, petioles and leaf sheaths yellow-green to golden covered in wax; young plants – 30–100 cm tall, stemless or multi-stemmed and light-textured – are commonly grown as pot plants).

chrysalis the pupa or transitional non-feeding stage between larva and adult that is represented in all advanced orders of insects. It is usually concealed in a cell or cocoon. There are variations in form, for example, the appendages and rudimentary wings are free from the body in Coleoptera and Hymenoptera, whereas in Lepidoptera they are tightly fixed to it.

Chrysanthemum (from Greek *chrysos*, gold, and *anthemon*, flower). Compositae. Europe, N Africa. 5 species, hardy annual herbs with entire to pinnately lobed and toothed leaves and large, daisy-like flowerheads. Sow seed *in situ* in spring on any well-drained, fertile soil in full sun. *C.segetum* is often sold in wildflower mixtures and will naturalize.

C.carinatum (Morocco; to 1m; flowerheads to 10cm in diameter, ray florets white, yellow, often tinged red or with a very dark red-brown basal stripe and white zone, disc florets purple; includes red, pink, orange, yellow, maroon, bronze and white zoned red or orange cultivars); *C.coronarium* (CROWN DAISY; Mediterranean, Portugal; to 80cm; flowerheads to 5cm in diameter, yellow; includes

'double' and deep yellow cultivars); *C.segetum* (CORN MARIGOLD; E Mediterranean, W Africa, naturalized throughout Europe; to 80cm; flowerheads to 6.5cm in diameter, primrose to golden yellow). For the florists' chrysanthemum see *chrysanthemums* and *Dendranthema*.

chrysanthemums FLORISTS' CHRYSANTHEMUM a perennial grown for cut flowers, garden decoration and exhibition. The origin of the florists' chrysanthemum has been variously credited to China, Japan and Tibet. It probably derived from *Dendranthema indicum*, and has been grown in Britain since the late 18th century; many thousands of cultivars have been progressively introduced.

Chrysanthemums are classified into 30 sections according to form, flowering time, colour and size. The major exhibition forms are Incurved, Reflexed, Intermediate, Anemone-centred, Single and Pompon. Charms are dwarf, bushy plants bearing hundreds of small, single flowers; they are grown in pots and form spherical or hemi-spherical plants up to 90cm in diameter. Cascades have similar flowers, 2.5cm in diameter, and are trained as pillars, fans and pyramids, and, especially, in cascading forms.

Chrysanthemums may be disbudded, with one bloom per stem or lateral, or grown as sprays with many blooms per stem. Propagate from 5cm spring cuttings taken from boxed stools, which should be overwintered in a well-ventilated, frost-free cold frame or cold greenhouse and moved to a warm greenhouse five weeks before cuttings are taken. Cuttings should be taken 14–15 weeks before the recommended stopping date for a particular cultivar and rooted in small pots or seed trays at an optimum temperature of 16°C/60°F. Pot up rooted plants, using a loam-based compost, individually into 9cm pots

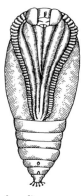

chrysalis

chrysanthemum

Pests
aphids
capsid bug
earwigs
eelworm
leaf miner
red spider mite
whiteflies
thrips

Diseases
powdery mildew
rust

chrysanthemums clockwise from top left: Incurved, Reflexed, Intermediate, Anemone-centred, Charm, Pompon, Single

or box into deep trays, and protect from frost. Plants should be watered regularly, and liquid fertilizer applied at 7–10 day intervals between bud appearance and first colouring. For garden decoration or cut flowers, plant singly in groups in the border, supporting with 1.2m canes. For late flowering, from earlier rooting dates, pot on successively through 15cm, 20cm and 23–25cm sizes of pot.

The purchase of healthy stock plants is of great importance and vigilance to allow for timely chemical sprays.

The stopping date for individual cultivars needs to be established. Stop plants by removing the growing tips; the resulting laterals, or breaks, are further stopped and the flower buds removed according to choice of single large blooms or sprays per stem. Charms, cascades and exhibition plants require special training. For quality blooms, protection is needed from autumn. Chrysanthemum blooms can be produced year round by manipulating day length with the aid of artificial lighting and shading.

chrysanthemum stool miner (*Psila nigricornis*). An insect pest, related to the carrot fly, which attacks the roots of chrysanthemums (*Dendranthema* species). It is particularly damaging to stools of chrysanthemums brought into a heated greenhouse to produce cuttings, and the shoots may be killed. Control by treating the growing medium with an insecticide.

Chrysogonum (from Greek *chrysos*, gold, and *gone*, joints: the golden flowers are borne at the nodes). Compositae. E US. GOLDEN KNEE. 4 species, *C.virginianum*, a hardy perennial herb to 20cm tall with downy, ovate to lanceolate and toothed leaves, and a succession of 3cm-wide, starry yellow flowers from late spring to autumn. Hardy to –15°C/5°F or lower; use as groundcover in dappled shade on moist, humus-rich soils. Propagate by ripe seed sown when fresh, or by division in spring.

Chrysosplenium (from Greek, *chrysos*, gold, and *splen*, spleen, hence *splenium*, a herb with medicinal properties). Saxifragaceae. GOLDEN SAXIFRAGE. Temperate regions. 55 species, low perennial herbs with creeping stems, rounded leaves and small yellow flowers in terminal heads surrounded by yellow-tinted, leafy bracts in spring. Both species listed here are suited to damp, shady positions on acid soils rich in leafmould. They thrive in the woodland garden and as groundcover in the bog garden and fern border. The first is hardy in zone 7, but may be damaged by late frosts. The second is hardy in zone 5. Increase by division or cuttings of new shoots.

C.davidianum (Himalaya; creeping; hairy; leaves to 5cm, dark green, toothed to lobed; flowerheads to 7cm diam., rich gold to lime green); *C.oppositifolium* (Europe; tufted to creeping, sparsely hairy; leaves to 2cm, bright, pale green, lobed; flowerheads to 3cm diam., pale gold to lime green).

Cissus rhombifolia

Cissus discolor

Chusquea (from the vernacular name used in NE S America). Gramineae. Mexico to Chile. Some 120 species, clump-forming bamboos with pith-filled culms. See *bamboos*.

Cicerbita (Italian name of Mediaeval origin for the sow thistle). Compositae. Northern temperate regions. 18 species, hardy perennial herbs with milky sap, runcinate to lyrate-pinnatisect leaves and thistle-like flowerheads on branching stems in summer. Hardy to at least –15°C/5°F, and suitable for naturalizing in the wild garden. Grow in moist but well-drained, humus-rich, neutral to acid soils. Deadhead to prevent seeding (conversely, propagate by seed).

C.alpina (syn. *Lactuca alpina*; MOUNTAIN SOW THISTLE; Arctic and Alpine Europe; 50–250cm; flowerheads pale blue); *C.bourgaei* (syn. *Lactuca bourgaei*; Asia Minor; 1.5–3m; flowerheads lavender or lilac).

Cichorium (Latinized form of the Arabic name). Compositae. Europe, Africa, Asia, naturalized elsewhere. 8 species, hardy annual or perennial herbs with milky latex, runcinately to pinnately lobed leaves and short-lived, daisy-like flowerheads on slender, erect stems in summer. *C.endivia* is ENDIVE, grown for its edible leaves, as is *C.intybus*, CHICORY. The latter is also ornamental, hardy in climate zone 4 and an excellent plant for summer meadows where it contrasts brilliantly with ripe grasses and wild poppies. Grow in full sun on a free-draining soil. Once established, this plant will withstand drought. Increase by seed in autumn or spring. *C.intybus* (CHICORY, WITLOOF; Europe, W Asia, N Africa, naturalized N America; perennial to 1m tall with a long, thick taproot; flowerheads sky blue). See *chicory*, *endive*.

ciliate bearing a marginal fringe of fine hairs.

Cimicifuga (from Latin *cimex*, a bug, and *fugo*, flee, from the shoo-fly properties of *C.foetida*). Ranunculaceae. Northern temperate Regions. BUGBANE, RATTLETOP. 15 species, hardy perennial herbs with long-stalked, ternately compound leaves composed of toothed leaflets and, in late summer, wand-like racemes of small, white flowers. The following species are hardy in climate zone 5. Grow in sun or light shade on a moist humus-rich soil. Increase by division.

C.racemosa (BLACK COHOSH, BLACK SNAKEROOT; N America; 80–150cm; raceme to 90cm, erect, slender, branched, flowers fragrant); *C.simplex* (E Russia, Mongolia, Japan; 60–120cm; raceme to 30cm, erect to bowed; includes Atropurpurea Group, with forms with very dark, purple-bronze leaves, for example 'Brunette').

cinereous ashy grey.

circinate of an unexpanded leaflet, frond or frond segment, rolled up in a close coil with the tip at its centre, like a crozier.

circumscissile a form of dehiscence in which a pod opens along a line parallel with its circumference, allowing the top to come off like a lid. See *calyptra*.

cirrhous, (cirrose) of the apex; terminating in a curled, coiled or spiralling continuation of the midrib.

Cissus (from Greek *kissos*, ivy). Vitaceae. Tropics and subtropics. GRAPE IVY, TREEBINE. 350 species, herbs, shrubs and vines. Those listed here are shrubby with handsome, evergreen foliage on slender stems climbing by tendrils. The flowers are small and followed by inedible, dry berries. *C.antarctica* and *C.rhombifolia* are popular pot plants, especially in offices and other public buildings where, grown on trellis and canes, they withstand years of abuse. They prefer a minimum temperature of 10°C/50°F, light shade and moderate humidity. *C.discolor* is suited to hanging baskets and trellis. It requires a minimum temperature of 15°C/60°F, shade and high humidity. *C.striata* will tolerate light frosts and bright light. Propagate all by stem cuttings in late spring rooted in a heated case; increase also by layering.

C.antarctica (KANGAROO VINE; Australasia; leaves 7–10cm, ovate to oblong, apex long-acuminate, margins entire to sinuate or irregularly dentate, glossy, dark green, smooth, tough); *C.discolor* (REX BEGONIA VINE; SE Asia to Australia; leaves 6–25cm, ovate-oblong to lanceolate, quilted dark velvety green above with a double row of grey-green, silvery-white or pale pink blotches, dark wine red beneath); *C.rhombifolia* (VENEZUELA TREEBINE; tropical America; leaves composed of 3 leaflets, each 2.5–10cm long, rhombic to ovate, glossy dark green above, remotely and sharply toothed; 'Ellen Danica': leaflets deeply incised); *C.striata* (IVY OF PARAGUAY, MINIATURE GRAPE IVY; Chile, Brazil; leaves composed of 5, palmately arranged, obovate to oblanceolate leaflets, each 1.5–3cm long, smooth and thin-textured and with gland-tipped teeth at their apices; leaves usually dropped in cold weather). For *C.bainesii*, see *Cyphostemma bainesii*; for *C.juttae* see *Cyphostemma juttae*; for *C.voinieriana*, see *Tetrastigma voinierianum*.

Cistus (from *kistos*, the Greek name for this plant). Cistaceae. S Europe, N Africa. ROCK ROSE. 20 species, small to medium-sized, more or less hardy evergreen shrubs with gummy to downy lanceolate to ovate leaves. Produced in abundance on bright summer days, the flowers are short-lived and composed of five, broadly obovate and crêpe-like petals. The following species are hardy to −12°C/10°F, especially if planted in a sheltered position and given a thick, dry mulch (or complete covering) in exceptionally hard winters. They need a fast-draining, light soil, full sun and protection from strong winds. Increase by softwood or heeled cuttings taken in late summer and rooted in a sandy mix at 15°C/60°F. Overwinter newly rooted plants in frost-free conditions.

C. × *aguilarii* (*C.ladanifer* × *C.populifolius*; leaves to 10cm, lanceolate, undulate, bright green above, paler beneath; flowers to 3.5cm in diameter, white; bases of petals purple-blotched in 'Maculatus'); *C.albidus* (to 2m; leaves to 5cm, oblong to ovate, covered in white down; flowers to 4cm in diameter, rosy pink to lilac with yellow basal spots); *C.* × *cyprius* (*C.ladanifer* × *C.laurifolius*; to 2m; leaves to 10cm, viscid, aromatic, oblong to lanceolate, undulate, dark grey-green and glabrous above, grey-hairy beneath; flowers to 7cm in diameter, white, often with bold, carmine basal spots); *C.* × *dansereaui* (syn. *C.* × *lusitanicus*; *C.ladanifer* × *C.hirsutus*; to 60cm; leaves to 6cm, oblong to lanceolate, dull green; flowers to 7cm in diameter, white with crimson basal spots); *C.* × *hybridus* (syn. *C.* × *corbariensis*; *C.populifolius* × *C.salviifolius*; spreading, to 1m tall; leaves to 5cm, ovate, undulate, dark green above, paler and hairy beneath; flowers to 3.5cm in diameter, white, usually with yellow basal spots); *C.incanus* (to 1m; leaves 1.5–7cm, ovate to elliptic, green or silver-hairy with sunken veins; flowers to 6cm in diameter, pink-purple; subsp. *creticus*: syn. *C.creticus*, leaves undulate-crispate, flowers rose pink to purple, to 6cm in diameter, with yellow basal spots); *C.ladanifer* (COMMON GUM CISTUS, LADANUM; viscid, aromatic, to 2m; leaves 4–10cm, linear to lanceolate, dark leaden green above, paler beneath; flowers to 7–10cm in diameter, white, usually with deep maroon basal spots); *C.laurifolius* (to 1.5m; leaves 3–9cm, ovate to lanceolate, undulate, dark green and smooth above, white-hairy beneath; flowers to 6cm in diameter, white with yellow basal spots); *C.monspeliensis* (MONTPELIER ROCK ROSE; to 1.5m; leaves 1.5–5cm, linear to lanceolate, rugose, dark green above, thinly grey-hairy beneath; flowers to 2.5cm in diameter, white); *C.parvifolius* (to 1.5m; leaves to 3cm, ovate, grey-hairy; flowers to 3cm in diameter, pink); *C.*'Peggy Sammons' (*C.albidus* × *C.laurifolius*; to 1m; leaves to 6cm, oval, soft sage green; flowers to 5cm in diameter, pink); *C.populifolius* (to 2m; leaves 5–9cm, ovate to cordate, glabrous, mid-green, clammy; flowers to 5cm in diameter, white, often with yellow basal spots); *C.* × *purpureus* (*C.ladanifer* × *C.incanus* subsp. *creticus*; to 1m; leaves to 5cm, oblong to lanceolate, grey-hairy beneath; flowers to 7cm in diameter, pink with dark red basal spots); *C.salviifolius* (to 60cm, spreading; leaves 1.5–4cm, ovate to oblong, grey-green and rugose above, paler beneath; flowers to 4cm in diameter, white with yellow basal spots); *C.*'Silver Pink' (to 50cm, low, spreading, bushy; leaves to 7cm, ovate to lanceolate, pale sage green to grey above, silvery beneath; flowers to 7cm in diameter, pale, silky rose); *C.* × *skanbergii* (*C.monspeliensis* × *C.parviflorus*; to 1m; leaves to 5cm, oblong to lanceolate, downy beneath; flowers pale pink).

CITES an acronym for the Convention on International Trade in Endangered Species, established in 1972 to monitor and control international trade in species actually or possibly threatened with

Cistus 'Silver pink'

Cistus ladanifer

extinction from their natural habitat. Around 126 states throughout the world are signatories to the Convention, which is concerned with the movement of animals and plants across international boundaries and not internal trade within countries. The Convention operates through a system of export and import licences with regard to three lists of species which comprise CITES Appendices I, II and III.

× **Citrofortunella** Rutaceae. Garden hybrids between *Citrus* and *Fortunella*, these are small, tender, evergreen trees or shrubs with intensely citrus-scented foliage, fragrant, waxy white flowers and small and tart, orange-like fruit. The following hybrid is commonly offered as the dwarf ornamental orange. Largely decorative, the fruit is painfully sharp, but will add something to marmalade, duck and cocktails. Grow in full sun with a minimum temperature of 7°C/45°F and ample ventilation. Keep barely moist in cool weather; water, feed and syringe freely in the growing season. Hand-pollination with a fine brush is sometimes helpful for plants grown indoors and flowering in winter. Sudden fluctuations of temperature and water supplies will cause fruit drop. Prune only to remove exhausted, bare growth. Propagate by seed or by greenwood or semi-ripe cuttings in summer; increase also by grafting. × *C.microcarpa* (syn. *Citrus mitis Citrus reticulata* × *Fortunella margarita*; CALAMONDIN ORANGE; small shrub or miniature tree to 2m with a rounded crown; leaves, elliptic to broadly ovate, dark glossy green; fruit 2.5–3.5cm, globose, bright orange, peel loose, strongly orange-scented, pulp sour; includes 'Tiger', with leaves streaked white, and 'Variegata', with white and grey leaves).

Citrus (from *Mala citrea*, ancient name of *Citrus medica*). Rutaceae. Asia, Pacific Islands, widely cultivated and naturalized elsewhere. 16 species, small, aromatic evergreen trees and shrubs with thorny branches, ovate to elliptic leaves and fragrant waxy white flowers. The edible fruit (hesperidium) is composed of juicy pulp encased in a thick rind dotted with oil glands. The genus includes the following important fruit crops, some (e.g. *C.meyeri*) grown as ornamentals under the same regime as × *Citrofortunella*, but requiring rather more room: *C.aurantifolia* (LIME); *C.aurantium* (SEVILLE ORANGE); *C.bergamia* (BERGAMOT); *C.limon* (LEMON); *C.meyeri* (DWARF LEMON, MEYER LEMON); *C.× paradisi* (GRAPEFRUIT); *C.reticulata* (MANDARIN ORANGE, TANGERINE, CLEMENTINE, SATSUMA); *C.sinensis* (ORANGE, SWEET ORANGE); *C.× tangelo* (TANGELO, UGLI FRUIT).

cladophyll, cladode a flattened or acicular branch that takes on the form and function of a leaf, arising from the axil of a minute, bract-like, and usually caducous true leaf, for example in *Ruscus*.

Cladrastis (from Greek *klados*, branch and *thraustos*, fragile, referring to the brittle twigs).

Leguminosae. E Asia, N America. 5 species, hardy deciduous trees with large pinnate leaves and pea-like flowers in terminal racemes and panicles in summer. Hardy to –30°C/–22°F, the following will grow on most, fertile, well-drained soils in full sun, thriving on chalk. Prune young plants in late summer to encourage a strong, clear leader – the branches of this tree are brittle and easily brought down by strong winds. Propagate by seed, scarified and stratified at 5°C/40°F for two to three months; increase also by root cuttings in winter. *C.lutea* (KENTUCKY YELLOW-WOOD; round-headed tree to 15m; leaves dark green turning yellow in autumn; flowers white, fragrant, in drooping panicles).

clair-voyée an aperture cut into a garden wall to allow a view of the land beyond; typically placed at the end of a walk or allée.

clamp in vegetable growing, a carefully constructed conical pile of root vegetables covered with layers of straw and soil to enable outdoor storage over the winter.

Clarkia (for Captain William Clark (1770–1838), explorer in North America). Onagraceae. Western N America, S America. FAREWELL TO SPRING, GODETIA. 33 species, annual herbs with linear to lanceolate or elliptic leaves and spikes or racemes of flowers with four, obovate and clawed, silky petals. Sow *in situ* in spring (also in autumn in more or less frost-free climates) on any well-drained soil of low to moderate fertility in full sun. Pinch young plants to encourage bushiness. Larger cultivars may need some support. *C.amoena* (SATIN FLOWER; California; fast-growing, slender-stemmed, erect or decumbent, 20–100cm tall; flowers 4–6cm in diameter, cupped to outspread, single or double with smooth or ruffled petals in shades of deep crimson, magenta, pink, lavender, scarlet, apricot, salmon and white, sometimes with paler margins or bases or darker basal blotches; many cultivars and seed races are available).

clasping partially or wholly surrounding an organ, as for example, in the case of a leaf base which clasps a stem; such leaves are termed *amplexicaul*

class the principal category of taxa intermediate in rank between Division and Order, for example Monocotyledonae.

clathrate latticed, pierced with holes and windows, as for example, in the leaves *Aponogeton madagascariensis*.

clavate, claviform shaped like a club or a baseball bat; thickening to the apex from a tapered base.

claviculate bearing hooks or tendrils.

claw the narrowed petiole-like base of some sepals and petals.

clay the finest texture of soil, comprising particles of less than 0.002mm in diameter and consisting mainly of alumino-silicate minerals. See *soil*.

Claytonia (for John Clayton (1686–1773), American botanist). Portulacaceae. Americas, Australasia, New Zealand. PURSLANE. 15 species, small, rosette-forming herbs with thinly succulent, smooth leaves and cup-shaped, 5-petalled flowers in spring. Both of the following are hardy in climate zone 6. *C.megarhiza* requires a deep, sharply draining soil, high in grit and leafmould and moist in spring and summer, but almost dry in winter. It thrives in full sun in airy conditions and is perhaps best-suited to culture in long clay pots in the alpine house or frame. *C.virginica* occurs on damp and cool, humus-rich soils in wooded or semi-wooded places. It is suited to similar conditions in the garden. Propagate by seed, offsets, or division.

C.megarhiza (western N America; taprooted; leaves 5–15cm, spathulate to obovate; flowers white to pink, white to deep pink); *C.virginica* (SPRING BEAUTY; eastern N America; cormous; leaves 6–12cm, linear to oblanceolate; flowers white tinted pink, some forms yellow veined red).

clearwing moths (Lepidoptera: Sesiidae) moths differing from most other Lepidoptera in having wings largely devoid of scales. They fly in the daytime and resemble wasps or ichneumon flies. The CURRANT CLEARWING MOTH (*Synanthedon tipuliformis*) is a destructive pest of blackcurrants and, to a lesser extent, red- and whitecurrants and gooseberries. It occurs widely over Europe and parts of Asia and has been introduced into North America, Australia and New Zealand. The moths have a wingspan of up to 20mm, and a black abdomen with three or four yellow bands, terminating in a prominent tuft of black hairs. The larvae tunnel into shoots, causing wilting and death or weakness of the stems. Remove and burn affected shoots; where the moths are troublesome, spray a residual contact insecticide in late spring, when the moths are first seen.

In North America, the crown and main roots of peach may be attacked by the PEACH TREE BORER (*Synanthedon exitiosa*), whose adults, up to 25mm long, are steel blue with yellow markings. In addition, flowering dogwoods may be attacked by the DOGWOOD BORER (*S.scinula*); raspberries by the RASPBERRY CROWN BORER (*Pennitsetia marginata*), which hollows out the crowns; and grapes by the GRAPE WOOD BORER (*Vitacea polistiformis*), whose larvae bore into the roots, reducing vigour.

cleft of a flat organ (e.g. a leaf or perianth segment); cut almost to the middle.

Cleistocactus (from Greek, *kleistos*, closed, and *cactus*, referring to the unexpanded perianth). Cactaceae. S America. Some 30 species, shrubby or tree-like cacti with cylindric stems and narrowly tubular flowers in summer. Provide a minimum temperature of 5°C/40°F, full sun and low humidity. Plant in a fast-draining mix high in grit and with a pH of 6–7.5. Keep dry from mid-autumn to early spring, except for light misting on warm days in late winter. At other times, water sparingly. *C.strausii* (SILVER TORCH; Bolivia; stems 1–3m × 4–8cm, erect and branching below, with many ribs and dense, fine, white spines; flowers to 8cm, dark red).

cleistogamy, cleistogamous self-pollination and self-fertilization within an unopened flower. The process often follows unsuccessful insect pollination on a plant and occurs late in the season, as in sweet violet (*Viola odorata*).

Clematis (from Greek *clema*, a tendril; name used by Dioscorides for several climbing plants). Ranunculaceae. VIRGIN'S BOWER, LEATHER FLOWER, VASE VINE. N and S Temperate regions, mountains of Tropical Africa. Over 200 species, deciduous or evergreen, semi-woody to woody climbers or woody-based, erect to sprawling perennials. The flowers are campanulate to bowl-shaped, solitary, clustered or borne in panicles, and composed of petal-like sepals and numerous stamens. In some species, petal-like staminodes are also present. The fruit consists of numerous achenes with long, persistent, often plumose styles.

Clematis have long been cultivated in Japan, and have been cultivated in Europe since the 16th century. Japanese cultivars, mainly those of the Patens Group, were introduced into European gardens in the mid-19th century. Nurserymen and plant breeders took the opportunity of using these earlier-flowering cultivars in a vast breeding programme, crossing them with European species and cultivars and other large-flowered clematis introduced from China, mainly during the period 1855–80. Many of the cultivars raised at that time are still popular today.

Clematis species and cultivars have many uses in the modern garden. The vigorous species such as *C.montana* are ideal for growing over buildings and through large trees. The herbaceous species are suitable for borders. Less vigorous climbers, like *C.alpina* and *C.macropetala*, are superb against north- or east-facing walls or fences. *C.viticella* and its colourful cultivars are best-suited for growing over shrubs or small trees and even over medium-height groundcover plants. The winter-flowering evergreen species like *C.cirrhosa* are more suited to the sheltered garden or, in cold districts, the conservatory or cool glasshouse. The spectacular evergreen *C.armandii* is far hardier in cool temperate areas than is often believed and will thrive on a north-facing wall even in climate zone 7.

The large-flowered clematis cultivars fall into two basic groups: those that flower on ripened stems of the previous season and those that flower on the current season's growth from midsummer onwards. Compact and free-flowering, the early large-flowered types can, for purposes of cultivation, be split again into sections. The very early flowering types (typical of

Clematis florida 'Sieboldii' *(left), C. armandii (right)*

the *C.*Patens Group) such as *C.*'Miss Bateman', are ideal for growing in containers or small gardens. The double and semi-double, large-flowered cultivars, such as *C.*'Vyvyan Pennell', are more suited to growing through wall-trained shrubs such as *Pyracantha* because of the weight of their flowers. The later, early summer-flowering cultivars with large flowers and vigorous habit, such as *C.*'Marie Boisselot', fare better grown through large free-standing shrubs or small trees. The second group of large-flowered cultivars, such as the popular *C. × jackmanii*, which flowers on the current season's growth, will grow well through wall-trained shrubs, climbing or rambler roses or free-standing species roses.

In addition to the climbing species and cultivars, there are several herbaceous species which are particularly useful for the border. These include *C.heracleifolia*, *C.integrifolia*, *C.recta* and *C.stans*. The slender-stemmed taller species, such as *C.recta* may need the support of twigs or sticks. The herbaceous species and their cultivars flower on the current season's stems. *C. × jouiniana* might be termed semi-herbaceous, producing a sprawling mass of vigorous stems. It is best used for covering paving, breaking up the outlines of borders and over tree-stumps and drain-covers.

The New Zealand species and their cultivars are fine plants for the conservatory or frost-free glasshouse in cool-temperate areas; they may also be grown successfully in virtually frost-free gardens in a sunny well drained site. When grown in containers, the potting medium should be gritty, well-drained and never wet during the winter months. Some cultivars make ideal plants for the alpine house, such as *C. × cartmanii* 'Joe'. All are evergreen and flower on the previous season's ripened stems. They therefore require only the removal of dead and weak stems after flowering each year.

In the garden, *Clematis* will grow in most well-drained, fertile soils in sun or part-shade. To avoid damage by animals and total infection by clematis wilt, all climbing clematis should be planted with the top of the root ball buried some 8cm below the soil surface. This will help to build up a good root crown of growth buds. Newly planted clematis should be pruned hard the first spring after planting, especially those planted in containers. All growth should be removed down to a strong pair of leaf axil buds 30cm above soil level. Most clematis climb by their leaf petioles, which twist themselves around the support, whether it be wire, trellis or another plant. They may also be tied on to supports.

For the purpose of pruning, climbing clematis fit into three basic groups according to whether they flower on old or previous season's wood or that of the current year. Of species it can, however, be remarked that pruning can be limited to the removal of excess, exhausted or congested growth after flowering.

GROUP 1. Types which produce short flower-stalks direct from the previous season's leaf axil buds; including the spring-flowering species *C.alpina*, *C. macropetala* and *C.montana*. Remove any dead or damaged stems after flowering and, as the growing season progresses into summer and autumn, train new growth into the support if it can be reached, keeping stems within their allocated space.

GROUP 2. The early large-flowered cultivars, the double and semi-double and the large-flowered types which flower before midsummer (i.e. Florida, Lanuginosa and Patens Groups). These produce stems of varying lengths with a single flower at the end. The stems grow away from axillary buds which will have ripened the previous season. Remove any dead or damaged stems down to a strong pair of buds during late winter or very early spring. Leave all healthy growth. Tie in the old growth and new growth as it appears so that it can support the large flowers.

GROUP 3. These produce all their flowers in clusters or panicles at the end of the current season's growth (i.e. Jackmanii, Texensis and Viticella Groups). Remove all top growth down to just above the base of the previous season's stems during late winter and very early spring. New growth may be visible at this time; prune to just above where it commences. All top growth should be trained and tied in to its support during the growing season.

In addition to these groups, evergreen clematis like *C. armandii*, *C. cirrhosa* and some New Zealand species may be divested of old and exhausted stems after flowering. Herbaceous species can be cut near the ground between late autumn and early spring.

Species can be successfully propagated from seed sown soon after harvesting in a cold frame. Cultivars should be reproduced by layering, cuttings or grafting. Layering can be carried out during late winter or early spring with old stems, or in early summer with current season's growth. Herbaceous clematis can be increased by division. *C.heracleifolia* and its cultivars increase easily by layering. Take cuttings before flowering during late spring or, from the very early spring-flowering types, in early summer. Internodal cuttings of soft to semi-ripe wood are best; one of the pair of leaves should be removed. Insert the cuttings in a light, sandy potting mixture in a cold frame or a propagation unit can be used. Nodal cuttings of *C.armandii* will root in a cool mist unit. Graft large-flowered cultivars on to rootstocks of *C.viticella* or *C.vitalba* during early spring. A propagation unit or cold frame can be used; a cleft-graft is generally preferred. Pot grafted plants with the union well below soil level.

Clematis wilt causes the plant to collapse while in full growth and the growing tips to turn black. All top growth should be removed down to soil level and burnt. If the clematis was planted deeply enough, it should start to grow again from below soil level. The wilt is caused by a fungus and can be controlled by spraying with a fungicide at intervals.

Clematis cirrhosa var. *balearica* 'Freckles' *(left)*, *C. tangutica* 'Bill Mackenzie' *(right)*

CLEMATIS

Name	Distribution	Habit	Leaves	Flowers
C.afoliata *C.aphylla*	New Zealand	sprawling to climbing, almost leafless shrub to 2m	where present, trifoliolate with minute leaflets	2.5cm diam., sepals green-white, 4ate, spreading

Comments: Spring

C.alpina *Atragene alpina*	N Europe & mts of C & S Europe, N Asia	deciduous climber to 2.5m	biternate, leaflets 9, to 5 cm, ovate-lanceolate, serrate	pendulous, star- to bell-shaped, sepals 4, to 4cm, blue or mauve, oblong-lanceolate, staminodes petaloid, pale blue to white, spoon-shaped

Comments: Spring. See also *C. macropetala*. 'Bluebell': as 'Pamela Jackman', blue. 'Burford White': large, clear white, bell-shaped, tepals pointed; vigorous. 'Columbine': pale lavender, almost campanulate; sepals long and acute. 'Columbine White': white, on upright stems. 'Frances Rivis' ('Blue Giant'): large, rich sky blue. 'Grandiflora': large, mauve-blue. 'Helsingborg': pale to deep indigo, staminodes petaloid, paler. 'Inshriach': small, on slender, swan-necked pedicels, sepals narrow, sometimes twisted, dark lilac to pale mauve, staminodes green-white tipped mauve, many in a tight central boss. 'Pamela Jackman': to 8cm diam., dark azure, staminodes tinted blue.'Pink Flamingo': white blushed pale rose, staminodes petaloid. 'Ruby': soft rosy red, staminodes cream. var. *ochotensis* indigo. var. *sibirica* : yellow-white, rarely flushed blue; 'Gravetye': large, milky yellow, very early-flowering; 'White Moth': semi-double, white.

C.armandii	C & W China.	evergreen climber to 9m	trifoliolate, leaflets to 15 cm, oblong-lanceolate to ovate, entire, emerging bronze-pink, soon becoming rich glossy green, tough and prominently veined	to 6cm diam., fragrant, sepals 4, pure white or cream-white, later rose

Comments: Spring. 'Apple Blossom': leaves strongly tinted bronze when young; flowers cup-shaped, pink in bud, fading to white. 'Snowdrift': leaves large, new foliage tinted copper; flowers large, white, waxy, very fragrant.

C.barbellata	W Himalaya	climber to 4m, differing from *C.montana* in flower colour		campanulate, sepals 4, ovate, to 4cm, downy, dull purple to brown-violet

Comments: Spring. 'Pruiniana': flowers plum, lantern-shaped, stamens petaloid, white

C.campaniflora	Portugal	deciduous climber to 6m	5- or 7-ternate, leaflets to 7.5cm, narrowly lanceolate, ovate or oval, simple or sometimes lobed	bowl-shaped, pendulous, sepals white tinged violet-blue, 4, to 2cm, oblong, recurved, downy

Comments: Summer

C.cartmanii (*C.marmoraria* × *C.paniculata*.)	garden origin	bushy dwarf evergreen with finely dissected dark green leaves and large panicles of white flowers		

Comments: 'Joe': tufted to mat-forming, branches purple-tinted, flowers to 4cm diam., white, profuse

C.chrysocoma	China	deciduous climber to 2.5m	trifoliolate, tawny-hairy, leaflets to 4.5cm, broadly ovate or rhomboid, trilobed, coarsely toothed	4.5cm diam., solitary, sepals white tinged pink, 4, broadly oblong, downy beneath

Comments: Summer-autumn.

C.cirrhosa	S Europe, Mediterranean.	evergreen climber to 4m	to 5cm, simple and toothed, jaggedly cut or trilobed, or ternate or biternate with toothed or lobed leaflets, glossy dark green	pendulous, campanulate, to 7cm diam., each long-staked and subtended by a small cup-like pair of bracteoles, sepals ovate, to 2.5cm, cream, sometimes spotted red

Comments: Winter. 'Wisley Cream': flowers cream; var. *balearica* : sepals pale cream, always spotted and flecked red-maroon within. 'Freckles': flowers large, cream, intensely flecked and spotted maroon to violet within, on very long, slender pedicels.

C.colensoi	New Zealand	evergreen climber, much-branched	7 cm, triangular, trifoliolate and pinnately cut	sepals green, 6, narrowly lanceolate, downy

Comments: Spring-summer.

C. **County Park Hybrids** (*C.petriei* × *C.marmoraria*)	garden origin	dwarf, usually evergreen shrubs, trailing and bushy or weakly climbing, with dark, much-dissected foliage and dense panicles of pale flowers		

Comments: Spring-summer. 'Pixie': trailing, flowers pale yellow-green in a congested panicle, male selection. 'Fairy': a female selection with silky golden styles and showy fruiting heads where pollinated

C.crispa BLUE JASMINE, MARSH CLEMATIS, CURLY CLEMATIS, CURLFLOWER	SE US	climbing deciduous shrub to 2.5m	pinnately 3-, 5- or 7-foliolate, leaflets to 7cm, often trifoliolate or lobed, thin, lanceolate to ovate	campanulate, pendulous, sepals to 5cm, fused below, distinct and spreading at tips, lavender, almost white at margins, thin and wavy

Comments: Summer-summer.

CLEMATIS

Name	Distribution	Habit	Leaves	Flowers
C.douglasii	W N America	herbaceous, non-climbing perennial to 60cm	simple below, 2-pinnate above, leaflets oblong, lanceolate or ovate, downy	tubular or campanulate, nodding on long, swan-necked stalks, sepals 4, deep mauve to violet, exterior paler, to 3cm, oblong with reflexed, tapered tips

Comments: Spring-summer. var. *scottii*: leaves grey-green, flowers urceolate, lavender to pale magenta, downy; includes 'Rosea' with rose pink flowers

Name	Distribution	Habit	Leaves	Flowers
C.× durandii (*C. × jackmanii × C.integrifolia*)	garden origin	robust, erect to weakly climbing shrub to 180cm	to 15cm, simple, ovate, glossy green	to 12cm diam., sepals deep violet-blue, usually 4, to 6, outspread or reflexed, obovate, slightly wavy

Comments: Summer-autumn

Name	Distribution	Habit	Leaves	Flowers
C. × eriostemon (*C.integrifolia × C.viticella*)	garden origin	sprawling subshrub to 3m	pinnately divided, leaflets usually 7, elliptic, deep green, entire or slightly lobed	sepals 4, dark violet to lavender, spathulate, 4cm, reflexed

Comments: ' Bergeronii' : to 3m; leaves thick, large; flowers almost campanulate, mauve; 'Hendersonii': to 2.5m; flowers to 6.5cm diam., violet; 'Intermedia': as 'Hendersonii', but stems thicker; flowers similar to *C.integrifolia*, violet blue.

Name	Distribution	Habit	Leaves	Flowers
C.fargesii	China	deciduous climber to 6m	bipinnate, leaflets to 5 cm, ovate, coarsely serrate, sometimes trilobed, dull green, silky	to 6.5cm diam., sepals 6, obovate, to 2cm diam., pure white above, yellow and downy outside

Comments: Summer-autumn

Name	Distribution	Habit	Leaves	Flowers
C.flammula	S Europe, N Africa, W Syria, Iran, Turkey	deciduous subshrub climbing to 5m	3- or 5-foliolate, leaflets narrowly lanceolate to rounded, bright green, smooth, rather tough	very fragrant, to 2.5cm diam., sepals pure white, 4, to 1.2cm, oblong, obtuse

Comments: Summer-autumn

Name	Distribution	Habit	Leaves	Flowers
C.florida	China, Japan	deciduous or semi-evergreen climber to 4m	to 12.5cm, ternate, each division trifoliolate, leaflets to 5cm, ovate to lanceolate, entire or toothed, shiny , downy beneath	to 7.5cm diam., sepals white or cream, sometimes lined or tinted green, 4, deep violet, to 2.5cm

Comments: Summer. 'Albaplena': flowers double, staminodes and sepals white flushed green, exterior with central green stripe. 'Sieboldii' ('Bicolor'): flowers white with many purple staminodes forming a rose-like centre.

C. Florida Group. Woody climbers, 2.5–3m, flowering spring to summer on previous year's wood; flowers usually semi-double or double, or single later in season, 15–22cm across, white to lilac or deep violet. 'Duchess of Edinburgh' (double, small, to 15cm diam., rose-shaped, white), 'Haku Ookan' (flowers semi-double, rich violet, stamens white), 'Kathleen Dunford' (large, to 22cm diam., semi-double, rosy purple, stamens gold), 'Miss Crawshay' (small, to 15cm diam., semi-double, mauve-pink, stamens gold), 'Proteus' (semi-double, deep coral pink), 'Sylvia Denny' (double, white, stamens pink).

Name	Distribution	Habit	Leaves	Flowers
C.forsteri	New Zealand	evergreen climber	trifoliolate, leaflets to 7cm, lanceolate to broadly ovate, smooth, entire to toothed or pinnately cut, tough, bright green	sepals 5, white to yellow

Comments: Summer. 'Tempo': male selection, very vigorous; flowers to 4cm diam., pale green becoming white

Name	Distribution	Habit	Leaves	Flowers
C.fusca	NE Asia	climber to 3m	pinnate, leaflets to 6cm, ovate, smooth or downy beneath	urn-shaped, nodding, sepals 4, to 2.5cm, dull purple to violet inside, densely hairy outside

Comments: Summer. var. *violacea* : flowers violet.

Name	Distribution	Habit	Leaves	Flowers
C.glauca	W China to Siberia	climber to 4m	pinnate or bipinnate, leaflets to 5cm, elliptic to lanceolate, 2–3-lobed, blue-green	sepals to 2cm, yellow

Comments: Summer

Name	Distribution	Habit	Leaves	Flowers
C.graveolens	Himalaya	climber to 4m; close to *C.orientalis* but with more finely divided glaucous leaves, and heavily scented flowers with spreading sepals		fragrant, sepals 4, yellow, ovate to obovate, often notched at apex, 3cm

Comments: Summer

Name	Distribution	Habit	Leaves	Flowers
C.heracleifolia	C & N China	woody-based perennial herb, erect to 1.5m	trifoliolate, leaflets to 6.5cm, rounded to ovate, coarsely toothed, with pale hairs	tubular to narrowly campanulate, to 2.5cm, sepals 4, deep blue, tips recurved, exterior hairy

Comments: Summer-autumn. 'Azur': hyacinth blue. 'Jaggards': deep blue. var. *davidiana* : fragrant, indigo-blue. 'Manchu': flowers deep blue. 'Wyevale': leaves broad, much divided; flowers small, deep blue, hyacinth-like, fragrant, clustered.

CLEMATIS

Name	Distribution	Habit	Leaves	Flowers
C.hookeriana	New Zealand	differs from *C.paniculata* in leaflets deeply lobed and green-yellow to light yellow flowers		
C.integrifolia	C Europe, SW Russia, W & C Asia	erect, non-climbing herbaceous perennial or subshrub to 1m.	to 9cm, simple, ovate to lanceolate, entire, glabrous above, downy beneath, prominently veined	bell- to star-shaped, nodding on long, erect stalks, sepals dark violet or blue, rarely white, usually 4, lanceolate, to 5cm, spreading with recurved tips and slightly wavy margins

Comments: Summer. 'Alba': flowers white. 'Olgae': flowers clear light blue, sepals long and recurved, sweetly scented. 'Pangbourne Pink': flowers pale rose. 'Pastel Blue': flowers delicate powder blue. 'Pastel Pink': flowers soft light pink. 'Rosea': flowers sugar pink, underside darker, scented. 'Tapestry': flowers large, mauve to rich red, sepals mauve.

Name	Distribution	Habit	Leaves	Flowers
C.×jackmanii (*C.lanuginosa* × *C.viticella*)	garden origin	climber to 4m	simple to trifoliolate, leaflets to 12cm, broadly ovate, deep green and glabrous above, paler and downy beneath	rich velvety violet-purple, numerous, to 12cm diam., sepals 4–6, broadly ovate

Comments: Summer. 'Alba': flowers grey-white; 'Purpurea Superba': flowers deep violet. 'Rubra': flowers red to plum, double, or single later in season.

C. Jackmanii Group. 2–6m, flowering summer to autumn on new shoots; flowers 12–20 cm diam.; sepals usually 4, wide to narrow and pointed, shell pink, red or blue to purple. 'Comtesse de Bouchard' (flowers almost circular, satin pink flushed lilac, abundant, stamens cream), 'Gipsy Queen' (flowers to 15cm diam., rich purple with 3 red stripes, abundant), 'Hagley Hybrid' (flowers cup-shaped, sepals pointed, shell pink, abundant, stamens brown), 'Madame Baron Veillard' (sepals pointed, lilac-pink, profuse, long-lasting, stamens white), 'Madame Grange' (sepals incurving, purple-tinted bronze with brown midstripe, silky pink beneath), 'Mrs Cholmondeley' (flowers to 22cm diam., sepals long and narrow, lavender, profuse, anthers brown; long-flowering), 'Niobe' (sepals pointed, dark velvety ruby, darker in bud, stamens gold), 'Perle d'Azur' (flowers sky blue, sepals slightly corrugated, semi-pendent, profuse, stamens green), 'Rouge Cardinal' (low; flowers to 18cm diam., 6 recurving and blunt sepals deep crimson, darker in bud, stamens brown), 'Star of India' (flowers to 12cm diam., plum with red midstripe, abundant, stamens yellow).

Name	Distribution	Habit	Leaves	Flowers
C.×jouiniana (*C.heracleifolia* var. *davidiana* × *C.vitalba*)	garden origin	vigorous, sprawling, semi-evergreen, woody-based perennial or semi-scandent shrub to 4m	leaflets 3 or 5, to 10cm, ovate, coarsely toothed, weakly hairy	fragrant, sepals 4, white to ivory deepening to opal then lilac or sky blue toward tips, strap-shaped, 2cm, tip expanded and somewhat recurved

Comments: Autumn. 'Mrs Robert Brydon': small, off-white, somewhat tinted blue. 'Oiseau Bleu': mauve to pink. 'Praecox': pale blue flushed silver.

Name	Distribution	Habit	Leaves	Flowers
C.ladakhiana	India , Tibet	deciduous or semi-evergreen climber to 3m, stems tinged brown-purple	pinnate to bipinnate, leaflets narrow	sepals spreading, narrowly elliptic, yellow to orange-yellow, usually with darker markings or staminodes

Comments: Summer

Lanuginosa Group. Woody climbers, 2.5–5m, flowering on short side-shoots on current year's growth; flowers very large, loosely arranged, summer to autumn, appearing consecutively, single or double, 15. 'Bracebridge Star' (flowers lavender with crimson midstripe, stamens plum), 'Edith' (clear white, stamens dark red), 'Fair Rosamond' (flowers small, to 15cm diam., white flushed blue with red midstripe, stamens purple; scented), 'General Sikorski' (flowers blue, edges crenulate, stamens gold), 'Henryi' flowers, to 20cm diam., creamy white, stamens coffee-coloured; long-flowering), 'Horn of Plenty' (cup-shaped, rosy purple with darker midstripe, stamens plum), 'Lady Caroline Nevill' (semi-double, soft lilac with darker midstripe, stamens beige), 'Marie Boisselot' ('Madame le Coultre'); (vigorous, to 5m; sepals to 6, overlapping, flat, palest pink fading to pure white; anthers pale yellow to pale brown), 'Nelly Moser' (large, to 22cm diam., sepals pointed, palest lilac with carmine midstripe, anthers rusty-red), 'Silver Moon' (palest pearly lilac, stamens yellow), 'William Kennet' (vigorous; sepals to 8, edges crimped, lavender blue with darker midstripe, abundant).

Name	Distribution	Habit	Leaves	Flowers
C.macropetala	N China , E Siberia, Mongolia	deciduous climber, to 1m	biternate, leaflets to 4cm, ovate to lanceolate, coarsely toothed and lobed	to 10cm diam., pendulous, sepals blue or violet-blue, 4, oblong to lanceolate, to 5cm, exterior downy, staminodes numerous, outer violet-blue, inner blue-white

Comments: Spring-summer. See also *C. alpina*. 'Ballet Blanc': very double, small, white. 'Blue Bird' (*C.macropetala* × *C.alpina* var. *sibirica*): semi-double, large, deep lavender. 'Jan Lindmark': pale purple. 'Maidwell Hall': semi-double, to 5cm diam., deep blue. 'Markham's Pink': strawberry pink, nodding. 'Rosy O'Grady' (*C.macropetala* × *C.alpina*): to 7cm diam., semi-double, deep bright pink, sepals long and pointed. 'White Swan' (*C.macropetala* × *C.alpina* var. *sibirica*): to 12cm diam., pure white. 'Snow Bird': white, late-flowering.

Name	Distribution	Habit	Leaves	Flowers
C.marmoraria	New Zealand	prostrate, suckering evergreen shrub to 6cm tall, the smallest *Clematis* species, loosely resembling a rigid clump of parsley	leaflets deeply and closely divided, glossy green, thick	to 2cm diam., sepals 5 green-white becoming creamy white; stamens many, filaments green, anthers cream
C.montana	C & W China, Himalaya	vigorous, climber, to 8m	trifoliolate, leaflets to 10cm, ovate to lanceolate, toothed to trilobed	to 5cm diam., lantern- to star-shaped, sepals white or pink, 4, rarely 5, elliptic, exterior downy

Comments: Spring-summer. 'Alba': flowers off white; 'Alexander': flowers large, perfumed, cramy white with yellow stamens; 'Peveril': to 8cm diam., pure white, stamens long and shimmering. 'Vera': pink, fragrant. f. *grandiflora*: very vigorous, flowers white, abundant. var. *rubens*: young leaves tinged purple-bronze, flowers pink-red. 'Elizabeth': flowers palest pink, vanilla-scented. 'Freda': less vigorous; leaves bronze; flowers cherry pink edged crimson. 'Lilacina' (*C.montana* f. *grandiflora* × *C.montana* var. *rubens*): flowers pale mauve. 'Marjorie': vigorous; flowers semi-double, creamy pink tinted orange, staminodes petal-like, salmon. 'Mayleen': leaves tinted bronze; flowers large, to 7.5cm diam., pink, stamens gold. 'Odorata': flowers palest pink, sweetly scented, abundant. 'Perfecta' (*C.montana* f. *grandiflora* × *C.montana* var. *rubens*): vigorous; branches deep brown tinted red; flowers to 9cm diam., white flushed lilac. 'Picton's': flowers small, deep satin pink, anthers gold. 'Pink Perfection': vigorous; flowers profuse, small, deep pink, sepals round. 'Superba' (possibly *C.montana* var. *rubens* × *C.* 'Mrs Geo. Jackman'): vigorous; flowers large, white. 'Tetrarose': tetraploid; stems tinted ruby; leaves bronze; flowers large, to 10cm diam., rich rosy mauve. 'Undulata': strong-grower; flowers white, flushed mauve, sepals wavy.

CLEMATIS

Name	Distribution	Habit	Leaves	Flowers
C.orientalis	Aegean, Ukraine, SE Russia, Iran, W Himalaya, W China, Korea	deciduous vine or scrambler to 8m	pinnately 5, trilobed, entire or toothed, grey-green	to 5cm diam., sepals yellow or green-yellow, 4, oblong or elliptic, outspread, later recurved, thick and fleshy, stamens maroon

Comments: Summer-autumn

C.patens	Japan, China	deciduous climber	pinnately 3–5-lobed	to 15cm diam., sepals 6, cream white, violet or bright blue, elliptic to obovate, to 8cm, anthers purple-brown

Comments: Spring-summer. 'Standishii': flowers to 14cm diam., pale lilac with a metallic lustre, centre lilac-rose

C. Patens Group. Woody climbers, 2–3.5m, flowering in spring on old wood; flowers with pointed sepals, normally single, 15–25cm across, sepals wide and overlapping to pointed, flat to wavy, white to purple, often with a darker midstripe. 'Barbara Dibley' (small; sepals long, deep red to violet, midstripe darker, stamens plum), 'Barbara Jackman' (small, to 16cm diam., 6 broad, acuminate sepals, violet with dark pink to red midstripe, anthers cream), 'Bees Jubilee' (pinky mauve, midstripe deep carmine stamens brown), 'Captain Thuilleaux' ('Souvenir de Captain Thuilleaux'); (sepals pointed, cream with broad strawberry midstripe, anthers brown), 'Countess of Lovelace' (double, lilac-blue, rosette-form), 'Daniel Deronda' (semi-double, violet-blue, stamens cream), 'Dawn' (compact, flowers large, pearly pink sepals overlapping, stamens carmine), 'Doctor Ruppell' (rose madder with carmine midstripe, stamens gold), 'Elsa Spath' ('Xerxes') (flowers large, to 20cm diam., sepals to 6, deep violet with dark purple midstripe, stamens plum), 'Gillian Blades' (flowers large, to 22cm diam., pure white, sepals flat with frilled edges), 'H.F. Young' (to 3.5m; sepals pointed and overlapping, Wedgwood blue, stamens cream, early-flowering, abundant), 'Lasurstern' (to 3.5m; to 22cm diam., sepals narrow, wavy-edged, rich blue, anthers white; long-flowering), 'Lincoln Star' (to 2m; sepals pointed, raspberry pink, centre plum, anthers burgundy), 'Lord Nevill' (dark violet blue, crenulate, stamens red), 'Miss Bateman' (flowers small, to 15cm diam., stellate, 8 overlapping sepals creamy white, stamens rusty red), 'Mrs George Jackman' (semi-double, sepals overlapping and elliptic, creamy white, abundant, anthers brown), 'Prins Hendrik' (to 25cm diam., sepals pointed and wavy-edged, lavender, stamens purple), 'Richard Pennell' (flowers saucer-shaped, sepals overlapping, rosy purple, anthers red and gold), 'The President' (to 18cm diam., saucer-shaped, 8 sepals, purple with paler midstripe, stigma and filaments off-white), 'Vyvyan Pennell' (double, large, deep violet, later single and bluer, stamens gold), 'Wada's Primrose' ('Moonlight', 'Yellow Queen'); (flowers small, to 15cm diam., primrose with yellow anthers), 'Walter Pennell' (double, deep pink flushed mauve, stamens buff).

C.petriei	New Zealand	sprawling woody vine to 4m	bipinnate or tripinnate, leaflets to 3cm, tough, ovate to oblong, entire or bluntly toothed	yellow-green, sepals 6, to oblong, 2cm, downy

Comments: Summer. 'Limelight': male selection, leaves purple, flowers lime. 'Princess': female selection, similar to 'Limelight' but with smaller flowers and showy seed heads.

C.phlebantha	W Nepal	erect to trailing or sprawling bush to 1.5m	to 7.5cm, pinnate, leaflets 1.5cm, white-woolly beneath, lobed	to 4.5cm diam., sepals white, 5veined, elliptic to obovate, cupped to spreading, exterior densely woolly, anthers yellow

Comments: Spring-summer.

C.×pseudococcinea (*C.×jackmanii × C.texensis*)	garden origin			

Comments: Similar to *C.texensis* but with campanulate flowers. 'Countess of Onslow': pink tinted mauve, sepals with a central. red bar. 'Duchess of Albany': deep pink, erect, centre tinted brown, ventral side striped white. 'Duchess of York': pink, midstripe darker. 'Grace Darling': pale carmine pink, profuse. 'Sir Trevor Lawrence': bright crimson edged light violet, stamens cream. See also *C.* Texensis Group.

C.ranunculoides	China	erect , non-climbing perennial herb to 50cm or climber to 2m	trifoliolate or pinnately 5-foliolate, or simple and trilobed, leaflets rounded to obovate or ovate, to 5cm, coarsely toothed, hairy	pendulous, sepals purple to pink, 4, oblong, 1.5cm, spreading, reflexed, downy

Comments: Spring to autumn.

C.recta	S & C Europe	erect, non-climbing perennial to 1.5m, densely branched	to 15cm, pinnate, leaflets to 9cm, oval to lanceolate, entire, glabrous and deep blue-green above, paler beneath	to 2cm diam., erect, in many-flowered panicles, sepals milky white, 4, narrowly ovate or oblong, spreading

Comments: Summer. 'Grandiflora': flowers large, very abundant. 'Plena': flowers double. 'Purpurea': branches and leaves flushed deep purple-bronze, becoming green in time.

C.rehderiana	W China	deciduous, woody-stemmed climber to 7.5m	pinnate, leaflets to 7cm, broadly ovate, often trilobed, coarsely toothed, hairy	campanulate, pendulous, sepals soft primrose yellow or pale green, 4, to 2cm, reflexed, downy outside

Comments: summer

C.spooneri *C.chrysocoma* var. *sericea*	China	deciduous climber to 6m	trifoliolate, leaflets to 7cm, oval or ovate, coarsely toothed, yellow-silky beneath	to 6cm diam., sepals pure white, 4, rounded to oval or obovate, to 2.5cm diam., silky-hairy outside

Comments: Spring

C.stans *C.heracleifolia* var. *stans*	Japan	deciduous erect shrub or subshrub or climber to 180cm	trifoliolate, leaflets to 15 cm, broadly ovate, coarsely toothed, with distinct, hairy veins	tubular, sepals white, 1cm, recurved, blue inside, white and tomentose outside

Comments: Summer-autumn

CLEMATIS

Name	Distribution	Habit	Leaves	Flowers
C.tangutica *C.orientalis* var. *tangutica*	Mongolia, NW China.	climber to 3m	pinnate or bipinnate, leaflets to 8cm, oblong to lanceolate, lobed and toothed, bright green	campanulate to lantern-shaped, pendulous, sepals golden yellow, 4, oval to lanceolate, to 4cm, apex tapering to spreading, exterior hairy

Comments: Summer-autumn. 'Bill Mackenzie' (*C.tangutica* × *C.tibetana* ssp. *vernayi*): vigorous; flowers very large, yellow, nodding, long-lasting. 'Corry' (*C.tangutica* ssp. *obtusiuscula* × *C.tibetana*): flowers large, very open, lemon. 'Drake's Form': flowers large. 'Lambton Park': flowers large, to 7cm diam., yellow, nodding. ssp. *obtusiuscula* : free-flowering and vigorous, leaflets with fewer teeth, flowers with blunt, more widely spreading sepals.

Name	Distribution	Habit	Leaves	Flowers
C.texensis	SW US (Texas)	climbing subshrub or erect to scrambling perennial to 2m	soft grey-green, pinnate, leaflets to 8cm, ovate to rounded	to 3 cm, much narrowed towards mouth, pendulous, sepals scarlet-red or carmine, thick, narrowly ovate, somewhat reflexed

Comments: Summer. 'Major': flowers scarlet with white or yellow interior; 'Passiflora': flowers to 2cm, scarlet throughout.

C. Texensis Group. Non-climbing, erect to sprawling or weakly climbing shrubs or woody-based semi-herbaceous perennials, flowering abundantly on young shoots over a long summer period; flowers urceolate to narrowly campanulate. 'Duchess of Albany': flowers tubular, nodding, bright pink to lilac at margins; summer-early autumn. 'Etoile Rex': flowers to 5cm, nodding, bell-shaped, cerise to mauve, margin silver-pink; summer-early autumn. 'Gravetye Beauty': flowers cerise to scarlet, tubular-campanulate, then outspread. 'Lady Bird Johnson': flowers dusky red, later edged purple, stamens creamy yellow. 'The Princess of Wales': flowers deep vivid pink, stamens creamy yellow.

Name	Distribution	Habit	Leaves	Flowers
C.tibetana	N India.	differs from *C.orientalis* in the finely cut, smooth and glaucous foliage and thickly textured pale yellow flowers		

Comments: ssp. *vernayi* : leaves glaucous; flowers to 5cm, solitary or in threes, nodding, narrowly campanulate then outspread, sepals green-yellow to burnt orange, thickly fleshy, somewhat wrinkled, stamens dark purple – easily distinguished by its remarkably thick sepals - these have led to its being given the popular name 'Orange Peel Clematis', which has also been applied as a cultivar name to a selection of this subspecies with thick sepals of a rich yellow-orange.

Name	Distribution	Habit	Leaves	Flowers
C.× triternata (*C.flammula* × *C.viticella*)	garden origin	climber to 4m	simple or bipinnate, leaflets entire	3cm diam., in terminal panicles; sepals lilac, 6

Comments: 'Rubromarginata': flowers white, edged wine red, profuse.

Name	Distribution	Habit	Leaves	Flowers
C.× vedrariensis (*C.chrysocoma* × *C.montana* var. *rubens*)	garden origin	closely resembles *C.montana* var. *rubens*, vigorous evergreen climber to 6m	leaflets to 6cm, ovate, often 3, trilobed, coarsely toothed, dull purple- green and pale- hairy above, paler and more densely so beneath	to 6.5cm, sepals rose, lilac or pink, 4–6, occasionally rounded to oval, outspread, stamens yellow

Comments: Spring-summer. 'Rosea': flowers large, pale pink.

Name	Distribution	Habit	Leaves	Flowers
C.viticella	S Europe	deciduous semi-woody climber to 3.5m	to 12.5cm, pinnately trifoliolate, leaflets to 6.5cm, lanceolate to broad-ovate, often 2–3- lobed	nodding, sepals blue, purple or rose-purple, 4, oblong to obovate, apex acuminate, to 4cm, undulate, exterior silky

Comments: Summer-autumn. 'Nana': bushy, non-climbing. 'Plena': flowers double, violet. 'Purpurea': flowers plum. 'Purpurea Plena Elegans': flowers to 8cm diam., very double, deep violet. 'Rubra': flowers deep carmine. 'Rubra Grandiflora': flowers large, carmine red. f. *albiflora* : flowers pure white.

C. Viticella Group. Woody climbers, 2.5–6m, flowers abundant, appearing consecutively over a short season, to 15cm diam., single to double, white to red and deep purple with coloured midstripe or veins. 'Alba Luxurians' (creamy white, sepals tipped green, centres dark; vigorous), 'Ascotiensis' (large, to 20cm diam., bright blue, sepals pointed, stamens green), 'Duchess of Sutherland' (small, claret with lighter midstripe, stamens gold), 'Ernst Markham' (to 15cm diam., sepals blunt-tipped, vibrant petunia red, profuse, stamens gold), 'Etoile Violette' (flowers deep purple, sepals blunt-tipped, stamens gold), 'Huldine' (vigorous, to 6m; flowers to 10cm diam., pearly white with lilac midstripes beneath, profuse), 'Kermesina' (flowers wine red, stamens brown), 'Lady Betty Balfour' (to 6m, vigorous; deep violet-blue, stamens yellow. late-flowering), 'Little Nell' (to 5cm diam., white with creamy midstripe shading to lavender at margins, abundant), 'Madame Julia Correvon' (to 8cm diam., ruby red, sepals twisted and recurved), 'Margot Koster' (to 2.5m; flowers to 10cm diam., deep lilac-pink, sepals spaced, abundant), 'Mary Rose' (flowers small, double, spiky, dusky amethyst, profuse), 'Minuet' (large, cream, sepals edged lavender; stems long and erect), 'Mrs Spencer Castle' (small, double, heliotrope pink, stamens gold), 'Royal Velours' (flowers deep velvety purple), 'Venosa Violacea' (sepals boat-shaped, violet with paler centre, veins dark purple), 'Ville de Lyon' (vigorous; sepals wide and rounded, carmine-red edged darker, profuse; stamens gold), 'Voluceau' (petunia red, stamens yellow).

Cleome hassleriana

Cleome (name used by Theophrastus). Capparidaceae. Tropics and subtropics. 150 species, annual and perennial herbs with palmately lobed leaves and dense, terminal racemes of flowers in spring and summer. These consist of four, narrowly ovate and clawed petals and a bundle of very long, slender stamens. Sow under glass some two months before the last frosts or *in situ* in late spring. Grow in a warm, sunny position on light, fertile and preferably sandy soils. Propagate by seed. *C.hassleriana* (syn. *C.spinosa* of gardens; SPIDER FLOWER; S America; annual, 50–150cm tall; flowers white to pink or purple with petals to 3cm long and stamens to 10cm).

Clerodendrum (from Greek *kleros*, chance, and *dendron*, tree, alluding to the variable medicinal properties of some species). Verbenaceae. Tropics and subtropics. Some 400 species, trees and shrubs, mostly tender and many climbing, with ovate to elliptic or obovate leaves and, in summer, panicles or corymbs of flowers composed of a campanulate to tubular calyx, a slaverform 5-lobed corolla and exserted stamens. *C.bungei* and *C.trichotomum* will tolerate temperatures as low as –15°C/5°F; the former may be killed to the ground, but will resprout from the roots if thickly mulched. Both require a sheltered, sunny site on a well-drained, deep and fertile soil. Remove deadwood in spring. The remaining species need a minimum temperature of 10ºC/50ºF, full sun except on the hottest summer days and moderate to high humidity. Water and feed freely in spring and summer; keep barely moist in winter. Support climbing species. Cut back after flowering and then only if it is necessary to alleviate overcrowding, remove weak or dead growth or keep the plant within bounds. Long-established specimens of climbing species may, however, need hard rejuvenation pruning from time to time – this should be done in early spring. Propagate by seed in spring or by softwood and semi-ripe cuttings rooted in a sandy mix in a heated case; increase also by root cuttings in spring and by removal of rooted suckers.

C.bungei (GLORY FLOWER; E Asia; suckering, semi-evergreen erect shrub to 2.5m tall; leaves foetid; flowers white to rose or purple-red, sweetly scented in rounded terminal heads to 15cm long); *C.paniculatum* (PAGODA FLOWER; SE Asia; erect evergreen shrub to 1.5m; flowers scarlet in a terminal, much-branched panicle to 30cm tall); *C.philippinum* (syn. *C.fragrans*; GLORY BOWER; E Asia; semi-evergreen, erect shrub to 3m; flowers pink, white, or sometimes tinged lilac, often double, sweetly fragrant, in dense, domed heads); *C.speciosissimum* (syn. *C.fallax*; erect, evergreen shrub to 4m, often grown as a smaller, pot plant; flowers bright scarlet in erect, conical panicles to 45cm); *C. × speciosum* (*C.splendens × C.thomsoniae*; JAVA GLORY BEAN, PAGODA FLOWER; differs from *C.thomsoniae* in its pale red or pink calyx and deep purple-rose corolla). *C.splendens* (tropical Africa; evergreen climber to 3m; flowers bright red to scarlet in crowded, terminal panicles); *C.thomsoniae* (BAG FLOWER, BLEEDING HEART VINE; tropical W Africa; evergreen climber to 4m tall; flowers in drooping, terminal panicles, calyx large, white, corolla deep red); *C.trichotomum* (Japan; deciduous, broad-crowned shrub or tree to 7m tall; leaves foetid; flowers fragrant in broad, nodding cymes, calyx red-tinted, corolla white; fruit turquoise to petrol blue, among persistent, red calyces; includes var. *fargesii*, a hardy and free-fruiting form;); *C.ugandense* (BUTTERFLY BUSH; tropical Africa; erect to sprawling, evergreen shrub to 3.5m tall; flowers in terminal panicles to 15cm, calyx crimson-tinted, corolla pale blue to violet).

Clethra (from Greek *klethra*, alder). Clethraceae. Asia, Americas, Madeira. WHITE ALDER, SUMMER-SWEET. Some 30 species, trees and shrubs, usually deciduous with obovate to lanceolate, serrate leaves and fragrant, 5-petalled flowers, small, bell-shaped and white, in racemes from summer to autumn. *C.alnifolia* is the hardiest species, surviving in climate zone 4; the Asian species will suffer in zones below 5 or 6, whilst *C.arborea* is tender and should be grown in the cool greenhouse or conservatory in zones 8 and under. All prefer a humus-rich, acid soil, moist but well-drained, and light shade to dappled sunlight. Propagate by seed in spring or by heeled cuttings of lateral shoots in mid- to late summer in a

heated frame. *C.alnifolia* produces suckers that may be detached and grown on once rooted.

C.alnifolia (SWEET PEPPER BUSH, SUMMER SWEET, BUSHPEPPER; eastern N America; tree or shrub to 4m; flowers in erect, cylindrical panicles; includes pink-flowered forms); *C.arborea* (LILY-OF-THE-VALLEY TREE; Madeira; evergreen shrub or tree to 8m; young stems tinged red; flowers nodding in slender racemes); *C.barbinervis* (China to Japan; shrub to 10m with peeling bark; leaves turning red and yellow in autumn; flowers with fringed petals in hairy, spreading racemose panicles to 15cm); *C.delavayi* (W China; tree to 13m; flowers nodding in a raceme to 25cm long, white to cream emerging from pink sepals).

Clerodendrum thomsoniae

Cleyera (for Dr Cleyer, 18th-century Dutch botanist). Theaceae. E Asia, C America. 17 species, evergreen or deciduous trees and shrubs with flowers composed of five, obovate, white to creamy yellow petals produced in summer. Hardy in favoured locations in climate zone 7. Plant in sun or part shade with protection from strong winds, in a moist, fertile and acid soil. Increase by semi-ripe cuttings in summer. *C.japonica* (SAKAKI; China, Korea, Japan; evergreen, glabrous shrub to 4m; leaves 7–10cm, narrowly oblong to ovate, blunt, glossy, dark green above, paler beneath; flowers to 2cm in diameter, cream, fragrant, produced in early summer; includes two very similar cultivars, 'Fortunei' (syn. *C.fortunei*), with pointed leaves variegated cream and rose toward margins, and 'Tricolor', with pointed leaves tinged pink at first, marbled grey and cream and edged cream to yellow-green – both are sometimes confused with *Eurya japonica* 'Variegata').

Clianthus (from Greek *kleos*, glory, and *anthos*, flower). Leguminosae. Australia, New Zealand. 2 species, evergreen trailing to clambering shrubs and subshrubs with pinnate leaves and pea-like flowers characterized by a billowing, reflexed standard to 5cm long and a finely tapered, claw-like keel to 7cm long. *C.formosus* requires a minimum temperature of 5ºC/40ºF, full sunlight and a dry, buoyant atmosphere. Grow in baskets or pots of a loam-based mix high in sand and grit; keep just moist in spring and summer, virtually dry in winter. This species is best watered from below and should not be syringed. Propagate from seed sown in spring at 18ºC/65ºF. The life of this often short-lived perennial can be prolonged in cultivation by grafting scions of 6–12-month-old seedlings on to stocks of *Colutea arborescens* of the same age; use an apical wedge graft. *C.puniceus* will survive outdoors on sunny and sheltered walls in favoured regions of climate zones 7 and over. Plant in full sun on a fast-draining, gritty soil. Tie in to wall, if necessary, and remove any dead or exhausted growth in spring. Propagate by seed or by semi-ripe cuttings in summer in a heated case.

C.formosus (syn. *C.dampieri*, *C.speciosus*; DESERT PEA, GLORY PEA, STURT'S DESERT PEA; Australia; silky-hairy, prostrate subshrub; flowers in

Clianthus puniceus

erect, long-stalked, umbel-like racemes, vivid scarlet, the standard with a glossy black, bulging blotch); *C.puniceus* (GLORY PEA, PARROT'S BEAK, LOBSTER CLAW; New Zealand; erect to clambering shrub to 5m with spreading to arching branches; flowers in drooping racemes, satiny scarlet, vermilion, coral-pink or white).

click beetles see *wireworms*.

climacteric in fruit cultivation, the point of maximum rate of respiration in stored fruit, such as apples. It coincides with the attainment of ripeness.

climatic zone *see hardiness*.

climber any plant that climbs, or has a tendency to do so, by means of various adaptations of stems, leaves or roots.

clinandrium in orchids, the part of the column containing the anther.

cline a population of plants exhibiting continuous morphological variation therefore presenting problems in plant classification, for example the orchid genus *Ophrys*.

Clitoria amazonum

Clitoria (the keel of these flowers resembles the clitoris). Leguminosae. Tropics. 70 species, tender herbs and shrubs, many climbing, with trifoliolate or pinnate leaves and pea-like flowers. These are usually carried with the large and showy standard held downwards and the small, incurved keel uppermost. Provide a minimum winter temperature of 15°C/60°F, medium to high humidity and bright, filtered light. Grow in a fertile, loam-based medium. Water, feed and syringe generously in summer; keep barely moist in winter. Train on wires, trellis or canes. Increase by seed in spring, or by spontaneous layers and softwood cuttings in summer.

    C.amazonum (Brazil; woody climber to 4m; flowers to 8cm, white or pale pink lined with dark rose veins); *C.ternatea* (BLUE PEA, BUTTERFLY PEA; tropical Asia; soft-stemmed, woody-based climber to 3m; flowers to 5cm, bright blue fading to white at centre with a yellow stain).

Clitoria ternatea

Clivia (for Charlotte Florentina, Duchess of Northumberland, née Clive, d. 1866, who was first to flower the type species). Amaryllidaceae. KAFFIR LILY. S Africa. 4 species, tender perennial, evergreen herbs with glossy, dark green, strap-shaped leaves in two ranks, their bases overlapping and bulb- or stem-like. Appearing throughout the year, the flowers are waxy, funnel-shaped, composed of six, oblong to obovate segments and carried in umbels atop stout scapes. Provide a minimum winter temperature of 7°C/45°F and bright indirect light. Grow in a fertile, free-draining mix. Water and feed freely in spring and summer; keep barely moist in winter. *C.miniata* is an excellent house plant. It benefits from occa-

sional leaf sponging. Propagate by seed in late winter, or by division after flowering.

    C.miniata (leaves to 60 × 7cm; scape to 30cm, flowers erect to spreading, each to 7cm long and scarlet to orange with a yellow throat; includes flame-, peach-, yellow-, and large-flowered cultivars, and forms with leaves striped white, yellow or cream); *C.nobilis* (differs from *C.miniata* in its narrower, drooping flowers to 5cm long, red edged yellow and tipped green).

cloche a low portable unit constructed of glass or rigid-plastic panes on a wire frame; used for the protection of plants and to advance growth. The term is also applied to plastic film stretched over wire hoops, a construction alternatively known as a low continuous polythene tunnel.

clod a consolidated mass of soil particles, which can form at the surface of heavy soil after inappropriate cultivation.

clone the genetically identical offspring of a single parent, produced by asexual reproduction or vegetative propagation.

close case, closed case see *propagation frame*

clubroot (*Plasmodiophora brassicae*) FINGER AND TOE, CLUBBING, CLUB FOOT ANBURY, a fungal disease affecting most crucifers, including weeds such as charlock and shepherd's purse, and ornamentals such as candytuft, stocks and wallflowers; it is especially severe on brassicas. Radish and certain cultivars of calabrese, Chinese cabbage and kale have been found to be resistant.

    Affected plants become stunted with discoloured leaves, wilt in dry weather, and roots become abnormally swollen. Symptoms vary according to the habit of the host plant: on plants with a slender taproot and numerous laterals, such as cabbage, there may be clusters of small galls or a single large club-shaped swelling; on swedes and turnips, warty outgrowths occur mainly at the base of the natu-rally swollen roots. Symptoms can be confused with those caused by the turnip gall weevil, but such galls either contain the maggots or have conspicuous exit holes.

    Plasmodiophora is a slime mould; it survives in soil as minute resting spores which can remain viable for at least 20 years. In the presence of susceptible roots, spores germinate to form swimming zoospores which eventually penetrate root hairs.

    Clubroot is most damaging on acid soils and can be much reduced by liming to bring the soil up to pH7. Rotation is advisable, but alone is not effective because of the longevity of the resting spores. Avoid brought-in brassica plants, which are a common source of infection. On infected sites, raise plants in pots of sterilized compost and dip the rootball into a recommended fungicide at planting time. In the early stages of infection, a foliar feed may help plant establishment.

cluster cups a descriptive term for the fruiting bodies of rust fungi.

Clytostoma (from Greek *klytos*, glorious, and *stoma*, mouth, referring to the flowers). Bignoniaceae. Tropical America. 9 species, tender, summer-flowering evergreen vines. The leaves are composed of two or three leaflets and terminate in a tendril. Produced in late spring and summer, the 5-lobed flowers are funnel-shaped and borne in pairs or short panicles. Cultivate as for *Bignonia* but with a minimum temperature of 10°C/50°F. *C.callistegioides* (syn. *Bignonia callistegioides*, *B.speciosa*, *B.violacea*; ARGENTINE TRUMPET VINE, LOVE CHARM; Brazil, Argentina; fast-growing woody climber to 5m; flowers to 7cm long, throat creamy yellow veined lilac, lobes mauve-pink fading to rose).

CO₂ enrichment the introduction of extra carbon dioxide into a greenhouse to maintain a maximum rate of photosynthesis in intensively cultivated crops, especially tomatoes. It may be generated by burning propane, natural gas, or paraffin or derived from liquid CO_2.

Coade stone an artificial stone made of fired clay with added ingredients; invented and manufactured by Eleanor Coade in the late 18th century and used for a range of garden and architectural ornaments.

Cobaea (for B. Cobo (1572–1659), Jesuit priest and naturalist active in Mexico and Peru). Polemoniaceae (Cobaeaceae). C and S America. 20 species, herbaceous or shrubby perennial, evergreen climbers, tender and sometimes treated as annuals. The leaves are pinnate and tendril-tipped. Solitary flowers are produced in great abundance from late spring to autumn. They are usually fragrant and composed of a leafy, spreading calyx and a 5-lobed, campanulate corolla. In frost-free regions, the following species will thrive in a fertile, moist soil in full sun, soon covering its support. Cut back and feed plants grown as perennials in spring. Elsewhere, grow year round in a cool greenhouse or use as a half-hardy annual, sowing seed in early spring at 20°C/70° for planting out in warm and sheltered locations in early summer. *C.scandens* (CUP AND SAUCER VINE; MEXICAN IVY; VIOLET IVY; Mexico; rampant, woody-based climber to 10m, usually treated as a half-hardy annual; flowers with campanulate corolla (the 'cup') to 5 × 4cm, green-cream and muskily fragrant at first, later violet, then deep purple and sweetly scented; the corolla sits on a leafy pale green calyx, the 'saucer').

cobnut see *hazelnuts*.

Coccoloba (from Greek *kokkos*, a berry, and *lobos*, a pod, referring to the fruits). Polygonaceae. Americas. 150 species, trees, shrubs and lianes, tender and evergreen with large, leathery leaves, small, green-white flowers and fleshy, grape-like fruit. Young specimens make good pot plants grown in sun or part shade with a minimum temperature of 10°C/50°F. Plant in a fertile, sandy mix. Keep moist throughout the year, and syringe and feed in summer. In the tropics and subtropics, this species is often to be seen as a large shrub, tree or hedge, showing remarkable tolerance of scorching sun, poor soils and saline winds; it is an excellent coastal plant. Propagate by greenwood cuttings. *C.uvifera* (SEA GRAPE; coastal tropical America; stoutly branched, erect shrub or tree to 10m, usually smaller; leaves to 20cm wide, broadly oblong to obovate, base more or less cordate, apex rounded, glossy olive to dark green, often with bronze to yellow or red veins and wavy margins).

Cocculus (diminutive form of the Greek *kokkos*, berry, referring to the small fruits). Menispermaceae. Asia, Africa, America. 11 species, deciduous or evergreen trees, shrubs and lianes with small, green-white flowers and globose fruits containing horseshoe-shaped seeds. Hardy to −15°C/5°F and suited to woodland gardens, trellises and pergolas in light shade or dappled sun. Plant in a moist, humus-rich soil. Prune in early spring to keep within bounds and remove dead or congested growth. Propagate by seed, semi-ripe cuttings and root cuttings.

C.carolinus (CAROLINA MOONSEED, CORAL BEADS, SNAILSEED; E US; deciduous twiner to 4m; leaves to 10cm, ovate to cordate, sometimes obscurely 3–5-lobed with 3–7 principal veins; fruit bright red); *C.orbiculatus* (E Asia; differs from *C.carolinus* in having leaves sometimes shallowly 3-lobed with 3–5 principal veins, and pruinose black fruits).

coccus part of a schizocarp or lobed fruit.

cochlea a tightly coiled legume.

cochleate coiled like a snail's shell.

cochlear, cochleariform spoon-shaped.

Cochlioda (from Greek *kochlion*, a little snail, referring to the curiously shaped callus). Orchidaceae. Peru. 6 species, compact, tender, evergreen perennial herbs with clumped pseudobulbs and strap-shaped leaves. They are grown for their flowers composed of five, lanceolate to elliptic tepals and a showy, 3-lobed lip.

Cobaea scandens

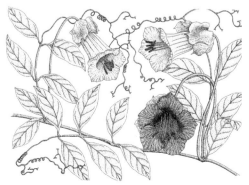

Cochlioda rosea

Provide a minimum temperature of 7°C/45°F with full sun in winter, and bright, indirect light and high humidity when in growth. Plant in an open, bark-based mix. Water and feed freely in spring and summer; keep almost dry in winter, apart from occasional mistings on warm days. Increase by division.

C.noezliana (flowers vivid scarlet, lateral lobes of lip rounded); *C.rosea* (flowers deep rose to crimson, lateral lobes of lip rhombic).

cockroaches (Dictyoptera: Blattidae) active insects up to 40mm in length, with long, many-jointed antennae and thickened forewings lying horizontally and overlapping the hindwings on a flattened body. The indigenous European species are small and of no economic significance; injurious species have, however, become widespread, mainly by shipping from warmer regions. They can breed throughout the year in a heated greenhouse. Eggs are laid in batches of about 15 in a tough purse-shaped capsule which may be carried around by the female for some days before being deposited in soil or in a crevice. Over a period of 3–12 months, a female may produce as many as 30 capsules. After 4–12 weeks (or more, depending on temperature), the young hatch inside the capsule, which then splits open. The nymphs resemble the adults but are wingless; they moult 6–12 times, according to species, before attaining their adult form.

The largest of the four cosmopolitan species is the AMERICAN COCKROACH (*Periplaneta americana*), which is up to 38mm long, and has red-brown wings extending beyond the abdomen. The AUSTRALIAN COCKROACH (*P.australasiae*) is similar, but about 30mm long, with a narrow yellow band behind the head. The COMMON COCKROACH (*Blatta orientalis*) is up to 25mm long and dark brown, almost black. The wings of the male do not reach the end of the abdomen and those of the female are rudimentary. The GERMAN COCKROACH (*Blattella germanica*) is up to 12mm long, and is dull yellow with two dark stripes on top of the thorax; the wings extend beyond the abdomen. The adults and nymphs of these four species are omnivorous. Greenhouse plants constitute only a small part of their diet, and all plant parts are subject to attack. Cinerarias, chrysanthemums, pot cyclamen, nicotianas, *Schizanthus* and the aerial roots of orchids are especially vulnerable. Cockroaches are mostly active at night and are more prevalent in old greenhouses, which provide suitable hiding places.

Control by attention to hygiene and the elimination of daytime resting sites. Apply contact insecticides to floors, walls and heating pipes, or use proprietary baits. Treatment should be continuous because eggs may take a year to hatch, especially at lower temperatures. Where resistance to chemicals is suspected, pitfall traps may be used.

codling moth and caterpillar

cocoa husks the husks of cacao nuts (*Theobroma cacao*), used as a ground mulch, soil conditioner or occasionally as a potting medium ingredient. The husks should be aged before use to reduce nitrogen and salt levels.

coconut fibre see *coir*.

Cocos (from Portuguese and Spanish *coco*, a grinning face, referring to the face-like marks on the base of the nut). Palmae. 1 species, *C.nucifera*, the COCONUT, a palm widespread throughout the tropics, especially in coastal regions. It grows to 30m tall with an unbranched trunk, often grey, fissured and ringed and tapering upwards from a swollen base. The leaves are pinnate and spreading to arching, mid-green and to 6m long. The fruits are rounded, to 35cm long and covered in a smooth, green to orange coat; the 'coconut' is the endocarp, woody and fibrous and lined with white, oily endosperm. Widely grown and naturalized in the tropics, especially in coastal regions. Elsewhere, grow in beds, large pots or tubs in a fertile and free-draining, sandy medium rich in lime and potash. Water plentifully, and maintain high humidity and a minimum temperature of 18°C/65°F. Shade from full summer sun. Fresh seed germinates readily at 27–30°C/85–90°, and it is as a seedling, the 'nut' still attached, that this palm is usually offered for sale.

Codiaeum (from the native Ternatean name, *kodiho*). Euphorbiaceae. SE Asia, Pacific Islands. 6 species, tender evergreen trees and shrubs grown for their leathery, often brightly coloured leaves. Small, creamy white flowers are carried in slender racemes. Provide a winter minimum of 13°C/55°F, with medium to high humidity and protection from strong direct sunlight, draughts and temperature fluctuations. Grow in a moist, fertile and lime-free potting mix; water, feed and mist freely when in full growth. Prune leggy plants hard in late winter, dressing the wounds with charcoal. Propagate by air layering or root greenwood stem tip cuttings in a heated case – control bleeding by dipping in charcoal. *C.variegatum* (syn. *Croton pictum*; CROTON; SE Asia; shrub to 2m; leaves 30cm, obovate to elliptic to oblong or lanceolate, tough, glossy dark green; many cultivars are grown and are particularly popular as house plants: their leaves range in shape from entire to jaggedly lobed, near-round to ribbon-like, blade- to cup-shaped, and in colour from deep emerald to lime or gold, variously stained, spotted and veined yellow, orange, pink, red, maroon and purple-black).

codling moth (*Cydia pomonella*) a small moth with a 19–22mm wingspan and grey-brown forewings with a copper-coloured patch at the tips. Its larvae are a serious pest of mature apple fruits (dessert cultivars are particularly susceptible) and also attack pear, quince, walnut and some ornamental *Malus* species. In northern Europe, there is normally only one generation during summer, and sometimes a second; in hot countries, there are regularly three. Eggs are laid on fruits and foliage, and after hatching, the caterpillars tunnel towards the core of the fruit. Damage often goes unnoticed until the fruit is mature; damaged fruits may ripen prematurely and drop early. The fully fed white-pink caterpillar leaves the fruit either through the eye end or by tunnelling through the side of the apple. The skin around the exit hole may be dark red and the tunnel is filled with caterpillar excrement.

The caterpillar overwinters under loose bark and pupates the following spring, so that bands of sacking or corrugated cardboard fitted around the trunk or larger branches of a tree can serve as artificial over-wintering sites; these may be removed and destroyed during winter. Many caterpillars are eaten by birds. Fruit can be protected effectively by spraying with a suitable insecticide at egg-hatching times in early summer. In order to assess more accurately the most effective time for spraying, place a pheromone trap in the tree to indicate flight periods. On isolated trees, pheromone traps may give some direct control by depleting the male moth population.

Codonopsis (from Greek *kodon*, bell, and *opsis*, appearance, alluding to the shape of the corolla). Campanulaceae. Asia. 30 species, hardy perennial herbs, often muskily scented and scandent, grown for their 5-lobed, campanulate flowers produced in spring and summer. Hardy to –10°C/14°F. Grow in a light, fertile, humus-rich and well-drained soil in sun or dappled shade. Keep roots cool and moist in summer; mulch in winter. Allow climbers to twine through pea sticks or surrounding shrubs. Propagate by division or by seed in spring.

C.clematidea (Russia to Himalaya; erect scrambler to 80cm; flowers to 2.5cm, bell-shaped, pale blue with tangerine and black markings at base within); *C.convolvulacea* (Himalaya to SW China; twiner to 3m; flowers to 4cm in diameter, azure to violet-blue with distinct, spreading, ovate lobes; *C.vinciflora* is similar); *C.ovata* (W Himalaya; erect to 30cm, rarely twining; flowers funnelform to campanulate, pale blue with darker veins and a near-black base within).

Coelogyne (from Greek *koilos*, hollow, and *gyne*, woman, referring to the deep stigmatic cavity). Orchidaceae. Asia. 200 species, tender perennial herbs with globose to flask-shaped pseudobulbs crowned with paired, lanceolate to elliptic leaves. Produced in spring and summer, the flowers consist of five linear to elliptic tepals and a showy lip and are carried in terminal clusters or racemes. Maintain a minimum winter temperature of 10°C/50°F for all species except *C.pandurata* and *C.speciosa*, which require a minimum of 15°C/60°F. Grow in a humid, buoyant atmosphere with filtered light in summer and full sun in winter. Pot in an open, bark-based mix. Water, feed and syringe freely from mid-spring to early autumn; in winter, water and syringe only to pre-vent severe shrivelling of pseudobulbs and then only on warm days. Propagate by division after flowering.

C.cristata (India; flowers to 6cm in diameter, often fragrant, lacy, snow white marked yellow on crested lip, in nodding racemes to 15cm long); *C.dayana* (Borneo; similar to *C.tomentosa*, but flowers slightly larger and a deep parchment colour, heavily edged and blotched chocolate brown on the white-veined lip); *C.nitida* (syn. *C.ochracea*; India to Laos; flowers to 4cm in diameter, sweetly fragrant, sparkling white marked orange-yellow and red on lip, in erect to nodding racemes to 20cm long); *C.ovalis* (India to Thailand; flowers to 4cm in diameter, pale tan marked dark brown on fringed lip, solitary or a few together on a very short, erect raceme); *C.pandurata* (SE Asia; flowers to 8cm in diameter, sometimes fragrant, jade to lime green, veined black on lip, on an erect to arching raceme to 30cm tall); *C.speciosa* (SE Asia; flowers to 8cm in diameter, olive to buff or pale salmon, marked brown to buff on the white-tipped, fringed lip, solitary or a few and nodding on an erect raceme to 12cm tall); *C.tomentosa* (syn. *C.massangeana*; SE Asia; flowers to 4cm in diam-eter, pale yellow, ivory or tan, veined brown and white on pale yellow lip, many together in strongly pendulous racemes to 45cm long).

Coelogyne cristata

Coffea (from the Arabic name). Rubiaceae. Asia and Africa, widespread in tropics through cultivation. COFFEE. 40 species, tender evergreen shrubs and small trees with leathery, ovate to elliptic leaves and clus-ters of fragrant, creamy white flowers. These give rise to green-red berries; the seeds are roasted as coffee beans. Grow in neutral to slightly acid, loam-based mix. Provide bright filtered light, high humidity and a minimum temperature of 15°C/60°F. Water and feed freely when in growth. Prune in spring to encourage the formation of several erect leaders in young plants and to limit the flower-bearing lateral branches to one per node in established plants. Under glass, hand-polli-nation may be needed for fruit set. Propagate by seed sown in spring or by cuttings of upright shoots with a snag of old wood. *C.arabica* (COFFEE, ARABIAN COFFEE; Ethiopia, Sudan; shrub to 3m; berries to 1.5cm long and green ripening red, yellow or purple; dwarf forms are sometimes grown as ornamentals).

Codonopsis convolulacea

coherent of parts usually free or separate, fused together, as in a corolla tube. The term is sometimes used to describe the adhesion of similar parts; cf. *adherent*.

coir fibre from the outer husk of the coconut (*Cocos nucifera*), once widely used in plunge beds and as a rooting medium; now the subject of renewed interest as a peat alternative in growing composts. Special care is required in the watering of coir; nutrient replenishment is also essential.

Coix (from Greek *koix*, a reedy plant). Gramineae. Tropical Asia. 6 species, annual and perennial herbs with slender leaves and gracefully arching panicles in summer, the female flowers are enclosed in distinc-tive, bead-like utricles which hang like tear drops. In zones 8 and under, sow seed in late winter under glass (minimum temperature 15°C/60°F) and pot several plants together in a rich loamy mix. Protect from fierce summer sun. Alternatively, grow outdoors as half-hardy annuals; raise under glass and plant out in a warm, sunny position in a moist but well-drained,

Coelogyne tomentosa

fertile soil in late spring. In frost-free regions with long, hot summers, sow seed *in situ* in spring. *C.lacryma-jobi* (JOB'S TEARS; SE Asia; annual to 1.5m tall; utricles to 1.5cm in diameter, white to grey tinged blue or brown).

colchicine an alkaloid derived from the AUTUMN CROCUS (*Colchicum autumnale*); used in plant breeding to produce growths with more than the normal number of chromosomes.

Colchicum agrippinum

Colchicum (of Colchis). Liliaceae (Colchicaceae). E Europe to Asia, N Africa. AUTUMN CROCUS. 45 species, hardy perennial herbs with large corms and linear to elliptic leaves. Produced mostly in late summer and autumn, the flowers appear before the leaves. Virtually stalkless, they are campanulate to funnelform or star-shaped and consist of six, lanceolate to oblong lobes. Plant in full sun on a fertile, well-drained soil. The larger species and cultivars are suitable for naturalizing in turf; the smaller species suffer in dull and overwet condtions and may need to be grown in a bulb frame or in a sheltered rock garden. Increase by seed or by division in autumn.

C.agrippinum (Greece to SW Turkey; flowers funnel-shaped, segments to 5cm, chequered rose-purple and white); *C.autumnale* (AUTUMN CROCUS, MEADOW SAFFRON; W and C Europe; flowers funnel-shaped to campanulate, segments to 6cm, purple-pink to pale pink or white); *C.bivonae* (syn. *C.bowlesianum*, *C.sibthorpii*; S Europe to Turkey; flowers campanulate, segments 4–8cm, rosy purple, chequered); *C.byzantinum* (Turkey, Syria, Lebanon; flowers funnel-shaped, segments to 5cm, pale mauve-pink); *C.cilicicum* (Turkey, Syria, Lebanon; flowers funnel-shaped to campanulate, segments 4–7cm, pale lilac to deep rose-purple, sometimes obscurely chequered); *C.luteum* (C. Asia to China; flowers funnel-shaped to narrowly campanulate, segments to 3cm, pale to deep yellow); *C.speciosum* (Turkey, Iran, Caucasus; flowers funnel-shaped to campanulate, segments to 8cm, pale pink to deep rose-purple or white, sometimes white in throat); *C.* 'The Giant' (flowers large, lilac pink and faintly chequered with a white throat); *C.variegatum* (Greece, SW Turkey; flowers broadly funnel-shaped, segments to 5cm, purple-red, strongly chequered, white in throat); *C.* 'Waterlily' (flowers large, double, deep rose-pink with a white throat).

cold frame an unheated frame.

cold house an unheated greenhouse.

cold store a refrigerated room or container used to store plants, propagating material or produce at low temperature.

coleoptile a sheath protecting the plumule of a germinating grass seedling.

coleorhiza a sheath protecting the radicle of a germinating grass seedling.

collar sometimes used to describe the junction point of the stem and roots of a plant.

collar rot rotting of the stem at, or just above, soil level. The original damage may be caused by careless weeding, animals, insects, waterlogging or heating up under densely packed basal mulch. It provides an entry for bacteria and fungi. Collar rot sometimes refers to the post-emergence damping-off of seedlings and may be applied to the black-leg symptoms caused by the bacterium *Erwinia carotovora* on potato, the fungi *Leptosphaeria maculans* on brassicas and *Pythium* species on pelargonium cuttings. When the rot extends below soil level, the symptom may be described as foot rot. See *black leg*.

collenchyma a strengthening tissue of young plant organs, composed of living cells with irregularly thickened walls.

Colletia (for Philibert Collet (1643–1718), French botanist). Rhamnaceae. Temperate S America. 17 species, hardy shrubs, leafless or with very small and short-lived leaves, and green, photosynthesizing stems and branches. These are covered in ferocious spines. White to cream or pale flesh pink and scented, small flowers cluster along the spines in winter and spring. They give rise to fleshy berries. Hardy to –5°C/23°F, or lower if planted against a warm and sheltered wall. Grow in full sun on a fast-draining, lean soil. Increase by heeled semi-ripe cuttings of short side shoots in a heated case.

C.hystrix (syn. *C.armata*; S Chile; to 2m; spines dark green, to 2.5cm, thinly hairy); *C.paradoxa* (syn. *C.cruciata*; Uruguay, S Brazil; to 2m; spines chalky blue-green, thickly glaucous, to 5cm, tapering from a broad, flattened base); *C.spinosissima* (syn. *C.infausta*; S Chile; to 2.5m; spines olive green, to 2.5cm, smooth).

Collinsia (for Zaccheus Collins (1764–1831), vice president of the Philadelphia Academy of Natural Sciences). Scrophulariaceae. N and C America. 25 species, half-hardy annual herbs with racemes of tubular, 2-lipped flowers in summer. Cultivate as for *Alonsoa*.

C.bicolor (CHINESE HOUSES, INNOCENCE; California; to 60cm; flowers to 2cm, upper lip white, lower lip rose-purple; white and multicoloured forms are available); *C.grandiflora* (BLUE LIPS; western N America; to 35cm; flowers to 2cm, upper lip pale purple to white, lower lip dark blue or violet).

colloid a substance in a colloidal state where particles or droplets are held in suspension by reason of their very small size. Colloids have a high ratio of surface area to volume, are very reactive chemically and physically and usually possess electrical charge; they attract oppositely-charged ions and water. Some clay particles and fractions of humus are colloidal. Emulsions, aerosols and smokes are other colloids important in horticulture.

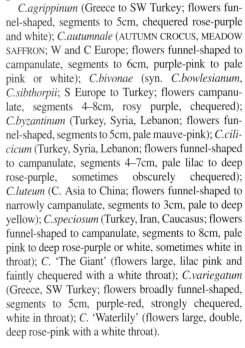

Colocasia (from Arabic *kolkas* or *kulcas*). Araceae. Tropical Asia (some widespread through cultivation). 6 species, large, tender, evergreen, perennial herbs. Long-stalked, sagittate to cordate leaves arise from a turnip-shaped tuber, massive and edible in *C.esculenta*. The inflorescences resemble those of *Arum* and are usually dull yellow-green. Cultivate as for *Alocasia*.

C.esculenta (COCOYAM, TARO, DASHEEN; leaves to 60cm, mid-green and smooth, on stalks to 1m tall; includes the cultivars 'Fontanesii', with dark purple-red to violet-black leafstalks and leaf blades veined and stained violet-black, and 'Illustris', the IMPERIAL TARO or BLACK CALADIUM, with purple-brown leafstalks and bright green leaf blades stained violet-black between the veins).

colonnade a row of columns or sometimes trees, either forming part of a loggia or leading from one group of buildings to another. Colonnades are a feature of formal gardens.

Colorado beetle (*Leptinotarsa decemlineata*) (Coleoptera: Chrysomelidae) a beetle with a stout, oval, strongly convex body up to 10mm long and yellow wing cases with five black stripes. Adults overwinter in soil and orange-yellow eggs are laid in masses, mainly on the foliage of potato, but also on tomato and tobacco. There are up to two generations per year. The larvae are soft and red-orange, with three pairs of short, stumpy thoracic legs. Both stages feed openly on foliage with devastating effect. The Colorado beetle is a native of the western US but widespread in North America. Recorded in Europe since 1876, it was established in France by 1920 and thereafter in most European countries except Britain and Ireland. Control with contact insecticide.

Colquhounia (for Sir Robert Colquhoun (d. 1838), a patron of the Calcutta Botanic Garden who collected plants in Kumaon). Labiatae. Asia. 3 species, hardy or slightly frost-tender evergreen shrubs covered in felty white down. The leaves are ovate to lanceolate with toothed margins. Showy, tubular and two-lipped, the flowers appear in whorls in late summer. Hardy to –7°C/19°F or lower if grown against a warm and sheltered wall and thickly mulched in winter. Plant in full sun on a fast-draining but fertile soil. Prune hard in spring to remove frosted top growth and to establish a low, woody framework from which tall and free-flowering shoots will arise by summer. Root softwood cuttings in late summer in low humidity. *C.coccinea* (Himalaya to China; to 3m; flowers to 2cm, flame to scarlet marked yellow within).

columella the central axis of a multi-carpelled fruit.

column (1) a solid ornamental upright or pillar, as used in a colonnade or other formal garden feature; (2) an upright form of fruit tree producing tightly packed spurs on a central trunk. Also known as a pole tree. In apples columnar trees may be geneti-

cally induced. The term is occasionally used for weak-growing pears; (3) a feature of orchids where the style and stamens are fused together; (4) the tube-like configuration formed by the fusion of staminal filaments in some Malvaceae.

Columnea (for Fabio Colonna (1567–1640), author of the first botanical book with copperplate illustrations). Gesneriaceae. Tropical Americas. 160 species, tender perennial, evergreen herbs and subshrubs, usually with long stems cascading from a low crown, small, paired, downy leaves and tubular flowers with a hood-like upper lip, two spreading lateral lobes and an entire lower lip. Cultivate as for *Aeschynanthus*.

C. × banksii (*C.schiedeana × C.oerstediana*; stems to 80cm, pendent; leaves to 4cm, ovate to oblong, smooth above, dark green, tinted purple-red beneath; flowers to 8cm, bright red marked yellow within); *C.crassifolia* (Guatemala, Mexico; stems to 30cm, ascending; leaves to 10cm, narrowly elliptic, smooth above, mid-green; flowers to 10cm, brilliant scarlet); *C.gloriosa* (GOLDFISH PLANT; C America; stems to 1m, cascading; leaves to 3cm, ovate, hairy; flowers to 8cm, scarlet with a yellow throat); *C.microphylla* (Costa Rica; stems to 60cm, cascading; leaves to 1cm, ovate to orbicular, covered in soft, purple-red hairs; flowers to 8cm, scarlet with a yellow blotch in throat; includes dwarf forms with very small leaves, and 'Variegata', with grey-green leaves edged cream).

column foot a basal platform found in the column of some Orchidaceae, to which the lip is attached.

Colutea (name used by Theophrastus). Leguminosae. Europe, Africa, Asia. 26 species, hardy deciduous shrubs or small trees, sometimes spiny, with pinnate or trifoliolate leaves, pea-like flowers in summer and strongly inflated, translucent, bladder-like pods. Hardy to –20°C/–4°F. Grow in full sun on a well-drained soil, with protection from fierce winds. Prune in spring to keep within bounds and remove exhausted growth. Propagate by seed sown in gentle heat in early spring, or by semi-ripe, heeled cuttings rooted in a sandy mix in midsummer.

C.arborescens (BLADDER SENNA; S Europe; shrub to 5m; flowers to 2cm, yellow faintly marked red on standard; pods to 8cm, bladder-like, pale green becoming red-tinted, then pearly grey-green); *C. × media* (*C.arborescens × C.orientalis*; shrub to 3m; flowers to 3cm, yellow to coppery orange stained bronze-red or orange); *C.orientalis* (S Russia, N Iran; shrub to 3m; flowers to 1.5cm, orange-red with coppery veins and yellow markings).

coma (1) a tuft of hairs projecting from a seed, as in Asclepiadaceae; (2) a tuft of leaves or bracts terminating an inflorescence or syncarp, for example the crown of a pineapple; (3) a leafy head in the crown of a tree or shrub.

Combretum (name used by Pliny for a climbing plant). Combretaceae. Tropics. 250 species, tender

Columnea gloriosa

Colorado beetle and larva

trees and shrubs, many of them climbers. The showy flowers have long stamens and appear in dense racemes and panicles in summer. Cultivate as for *Petrea*.

C.grandiflorum (N Africa; evergreen, twining shrub to 6m; flowers comparatively large and red in one-sided spikes to 10cm long); *C.microphyllum* (FLAME CREEPER, BURNING BUSH; small deciduous or semi-evergreen tree or scrambler to 8m; flowers bright red and downy in short racemes).

comfrey *Symphytum officinale*, a robust, coarsely hairy and large-leaved perennial herb in the Family Boraginaceae. Native to Europe and Western Asia, it has a long history of medicinal use (hence such vernacular names as BONESET). Today it is grown for ornament (see *Symphytum*) and for its mineral-rich foliage which makes an excellent liquid manure if harvested fresh and added in bulk to water.

comfrey

Commelina (for Jan Commelin (1629–1692) and Caspar Commelin (1667–1731), Dutch botanists). Commelinaceae. Mostly tropics and subtropics. WIDOW'S TEARS, DAYFLOWER. Some 100 species, annual and perennial herbs (those listed here are usually treated as annuals), sometimes tuberous, with slender stems, ovate to lanceolate leaves and clusters of 3-, but apparently 2-petalled flowers enclosed in a boat-like spathe. *C.coelestis* is treated as half-hardy in cold winter regions, its tuberous roots lifted in autumn, stored in frost-free conditions, brought into growth in gentle heat in late winter and replanted outdoors in late spring. *C.benghalensis* will thrive in sun or shade in moist, frost-free conditions; a fast-growing and dependable groundcover, it can be useful under staging. *C.communis* is similar but will grow outdoors in cold-winter areas with long, hot summers (e. g. the east coast of the US). Propagate the *C.coelestis* by division; the others by spontaneous layers or stem tip cuttings.

C.benghalensis (tropical Asia, Africa; stems to 30cm, creeping and rooting; flowers to 1.5cm, deep blue to violet; 'Variegata': leaves creamy white-edged); *C.coelestis* (C and S America; erect to 1m; flowers bright blue to indigo; includes cultivars with white flowers and variegated leaves); *C.communis* (China, Japan, naturalized S Europe, E US; stems sprawling and rooting, to 70cm long; flowers pale to bright, deep blue; 'Aureostriata': leaves cream-striped).

common name a popular or vernacular name for a plant, as opposed to a botanical name, for example, daisy for *Bellis perennis*.

community plot, community garden see *allotment*.

compaction the loss of pore spaces in soil caused by excessive cultivation or surface traffic. Wet soils of heavy texture are particularly susceptible, and deterioration of structure results.

companion, companionate planting the practice of growing plants reputed to have beneficial effects on their neighbours in close proximity to other species. Such plants may be held to have growth-enhancing or pest-suppressing properties, for example, *Tagetes patula* is claimed to be an insect pest deterrent.

compatibility (1) the ability of one plant to fertilize another; of particular relevance in top fruit growing; (2) the ability of a stock and scion to be combined successfully in a graft; (3) the suitability of two or more pesticides for mixing together.

complanate flattened, compressed.

complete of flowers, possessing all four whorls (sepals, petals, stamens and carpels).

compost (1) rotted organic matter (usually plant refuse) that is absorbent, humus-rich, and, where well-made, friable; it is added to soil to improve fertility, structure and water-holding capacity; low in nutrients; (2) in the UK, the term is also used to describe seed and potting mixtures or media.

compost heap a stack or bin of organic waste kept moist and, ideally, aerated to encourage its breakdown by micro-organisms.

A compost heap should be established either on a soil base, which encourages worm activity, or with a floor of widely spaced bricks covered with strong wire mesh to aid air flow. The containing walls should be constructed of concrete blocks, straw bales or thick timber, and a composting unit should preferably comprise two or more compartments, each about 1–1.5m square. The front wall of each bin should be made from removable boards, added as the height of the heap increases, with a detachable waterproof cover provided where rainfall is high. To ensure good drainage, the soil base should be forked over and a 5cm layer of brushwood put down. Organic matter is added in layers about 15cm deep, with different sorts mixed together; thickened material should be cut or shredded. Layers of chopped, damp straw incorporated throughout the heap will aid aeration. Where practicable, filling the bin in one operation gives the best results.

For good microbial activity, the carbon/nitrogen ratio should be 30:1, carbon occurring mostly in roots and stems, and nitrogen in green leaves. During autumn and winter, the application of an activator is beneficial. The compost heap is filled to near the top of the bin and the waterproof covering insulated with straw, sacks or other similar heavy fabric.

High temperature is necessary to ensure decompostion and destruction of weed seeds; in a well-prepared heap, up to 80°C/180°F can soon be attained. The compost heap can be kept active by regularly turning the contents of one compartment into another, moving the outermost material to the centre. See *activator*.

compound made up of two of more similar parts. For example, the compound leaves of clover (*Trifolium* species), the compound inflorescences of dandelion (*Taraxacum* species) and the compound fruits of mulberry (*Morus* species).

compound fertilizer a manufactured fertilizer providing more than one nutrient.

compressed flattened, usually applied to bulbs, tubers and fruit and qualified by 'laterally', 'dorsally' and 'ventrally'.

concatenate linked as in a chain.

conceit any fanciful, grotesque, or exaggerated garden feature, such as a folly, grotto, ruin or elaborate waterworks. The term is sometimes applied to a whole garden or landscape having a conglomeration of such features.

concolorous of a uniform colour.

conductivity a measure of the electrical conductivity of soil or liquid feed, which determines the salinity or concentration of dissolved mineral salts, some of which are plant nutrients. The higher is the salt concentration the lower is the resistance to electrical flow, which means the higher is the conductivity. High salt concentration may inhibit water uptake by roots.

conduplicate of leaves; folded once lengthwise, so that the sides are parallel or applied; of cotyledons, folded in this manner, with one cotyledon enclosed within the other, and enclosing the radicle in its fold.

cone (1) in gymnosperms, an assemblage of bracts and sporophylls, usually densely crowded and overlapping in an ovoid, spherical or cylindrical structure on a single axis and (in the female cone) persisting to the seed-bearing stage. In the families Gnetaceae and Podocarpaceae the bracts or scales may be very few so that the whole scarcely resembles a cone; but the term is retained here, as is strobilus, to avoid the misleading use of 'inflorescence'; (2) in flowering plants, a compact spike or raceme with conspicuous and congested bracts suggesting a true cone, as found in Zingiberaceae.

conferted closely packed or crowded.

conglomerate tightly clustered, usually into a ball.

conifer a tree or shrub, usually evergreen and many with needle-like leaves, belonging to the Division Gymnospermae which is characterized by the production of ovules not enclosed in an ovary. The seeds are often aggregated in woody cones, for example, *Pinus*, *Sequoia*, *Cupressus* or with fleshy coats as in the drupes of *Cephalotaxus*, or with fleshy arils as in *Taxus*, or swollen receptacles as in *Podocarpus*. Deciduous conifers include *Larix*, *Metasequoia* and *Taxodium*. 'Softwood' is a loose alternative term for conifers.

conifer spinning mite see *red spider mites*.

conjugate coupled, a linked pair, as in the leaflets of a pinnate leaf.

conk the hard fruiting body of a wood-rotting fungus, found on the branches, trunks and stumps of trees. It is also called a bracket fungus.

connate united, usually applied to similar features when fused in a single structure.

connate-perfoliate where opposite sessile leaves are joined by their bases, through which the axis appears to pass.

connective the part in a stamen that connects the lobes of an anther at the point of attachment to the filament.

connivent converging, and even coming into contact, but not fused; of petals, gradually inclining toward the centre of their flower.

conoid resembling or shaped like a cone.

Conophytum (from Greek *konos*, cone, and *phyton*, plant, referring to the conical plant bodies). Aizoaceae. South Africa. Some 80 species, tender, succulent perennials, dwarf with the fleshy leaves more or less fused in pairs; these form the plant 'bodies', low-lying, solitary or clumped, globose to cylindric, round- to flat- or sunken-topped, and notched to deeply cleft. Daisy-like flowers open at their cleavages in bright sunlight. Provide a minimum temperature of 7°C/45°F, full sunlight and a dry atmosphere. Grow in shallow pots or pans containing a loam-based medium very high in sand and grit and top-dressed with gravel or pebbles. Water sparingly from late summer to early autumn, very occasionally during sunny spells from mid-autumn to early winter; give a few waterings again during early spring, then withhold water from late spring to midsummer, when the old plant bodies dry up completely and the new leaves form within them. Propagate by seed germinated and grown on for a year in warm, moist and lightly shaded conditions, prior to gradual acclimatization to the harsher, adult regime. Increase also by careful division of clumps when repotting in late summer. *Conophytum* is soon killed by overwet and draughty conditions.

C.bilobum (clump-forming, plant bodies to 7cm, sea green to blue-grey, narrowly wedge-shaped with a rounded apex and a deep, wide cleavage making 2, long and distinct lobes sometimes spotted or edged red-brown; flowers golden yellow); *C.meyeri* (clump-forming, plant bodies to 2cm, obovoid, apex slightly bilobed and keeled, dull green to pale white-green, smooth to somewhat velvety and sometimes

Conophytum obcordellum, C. truncatum and C. bilobum

edged or marked red-brown; flowers white to yellow); *C.notabile* (plant bodies to 3cm, clumped, obovoid, with a rounded apex and slight fissure, grey-green and faintly spotted; flowers orange-red); *C.obcordellum* (syn. *C.nevillei*; plant bodies to 2cm, clumped, squatly top-shaped with a flattened or dimpled apex, more or less circular if viewed from above, silvery pale green to sea green, dotted and sometimes stained red-brown; flowers scented, white, yellow or pink); *C.truncatum* (plant bodies to 3cm, squatly top-shaped to rounded with a round-edged, flattened upper surface, dimpled and kidney-shaped if viewed from above, sea-green with fine, darker dots; flowers creamy yellow, spidery).

conservation measures taken to safeguard species from extinction, and to protect or restore natural habitats and gardens.

conservatory originally synonymous with green-house, now usually a greenhouse structure attached to a dwelling; used for relaxation in the surroundings of plants.

Consolida (Latin name for a wound-healing herb). Ranunculaceae. Mediterranean to C Asia. LARKSPUR. Some 40 species, annual herbs related to *Delphinium*, with slender, erect stems, finely and palmately cut leaves and showy, spurred flowers in summer. Sow in spring *in situ* on a fertile, moist but well-drained soil in full sun. Many cultivars and hybrids of the following species provide excellent cut flowers, harvested just as the lowermost buds open. To retain their form and colour as *dried* flowers, cut when almost fully open and hang in bundles in a cool, dry place. Taller cultivars may need staking. *C.ambigua* (syn. *C.ajacis*, *Delphinium consolida*; Mediterranean; stems to 80cm tall, branching above; flowers to 2.5cm in diameter, in racemes terminating branches, sky blue to azure, violet, lilac, pink or white, or white freaked and splashed with pink and blue, some-times semi- or fully double and with toothed to cut petals).

conspecific belonging to the same species; often used of highly similar populations scarcely worthy of recognition as separate species.

contact in horticulture, applied to (1) insecticides that enter insects through the intact cuticle; (2) herbi-cides that damage plant tissue by direct contact.

container-grown applied to plants established in containers, from which they can be planted out with minimal root disturbance.

continuous an uninterrupted symmetrical arrange-ment. The term is sometimes used as a synonym of decurrent.

contractile roots roots that contract in length and pull parts of a plant further into the soil, for example the corms of *Crocus*.

Convolvulus althaeoides

Convolvulus cneorum

controlled release see *slow release.*

Convallaria (from Latin, *convallis*, a valley, alluding to the habitat). Liliaceae (Convallariaceae). Northern temperate regions. 3 species. *C.majalis* is LILY OF THE VALLEY, a hardy, rhizomatous herba-ceous perennial to 20cm tall, with one to four, ovate to elliptic leaves on a short shoot and, in spring, an arching raceme of fragrant white to ivory flowers. These are 0.5–1cm deep and spherical to urceolate to campanulate, edged with six, short reflexed segments. They give rise to glossy, scarlet berries. Hardy to at least –20°C/–4°F. Grow in a moderately fertile, humus-rich and moisture-retentive soil in part shade, ideally in the woodland garden or under shrubs, where the creeping rhizomes will build up extensive colonies. For forced blooms, lift strong, 2–3-year-old 'crowns' (lengths of rhizome termi-nating in a bud) in the autumn and pot into a fertile, loamy mix with sand and leafmould; overwinter in the cold greenhouse or frame, increasing water and heat as growth proceeds. Propagate by division.

convolute rolled or twisted together longitudinally, with the margins overlapping, as leaves or petals in a bud.

Convolvulus (from Latin *convolvere*, to twine). Convolvulaceae. Cosmopolitan. 250 species, annual and perennial herbs and subshrubs, bushy or vigor-ously twining, with short-lived funnel-shaped flow-ers. The perennial species listed below are hardy to –5°C/23°F. They need a warm and sunny, sheltered position on a well-drained, low to medium fertility soil. Most benefit from a dry winter mulch, and some, like *C.althaeoides*, may be killed to the ground by cold. *C.sabatius* is often used as a half-hardy annual. Support climbing species, or allow them to sprawl and trail. Propagate by softwood cuttings in early summer. In warm regions, the annual *C.tricolor* should be sown *in situ* in autumn or spring; else-where, sow in early spring under heated glass and plant out in a sunny, warm position after the last frosts.

C.althaeoides (S Europe; slender-stemmed, sprawling or climbing perennial to 1m tall; leaves grey-green, sagittate to ovate-cordate, often deeply and finely palmately to pinnately cut; flowers to 4cm in diameter, pink to purple-pink); *C.cneorum* (SILVERBUSH; Italy and Balkans to N Africa; erect or spreading bushy shrub or woody-based perennial to 50cm; leaves oblanceolate to linear, covered in bright, silver, silky hairs; flowers to 2.5cm in diame-ter, palest shell pink with darker zones along folds); *C.sabatius* (syn. *C.mauritanicus*; Italy, N Africa; woody-based perennial to 50cm with bushy to trail-ing habit and slender stems; leaves oblong to rounded, mid- to dark green, sometimes silvery; flowers to 2.5cm in diameter, pale to deep blue, lilac or pink); *C.tricolor* (syn. *C.minor*; Portugal to Greece and N Africa; erect and bushy annual to 60cm; leaves oblanceolate to ovate, mid-green; flowers to 5cm in diameter, pale to deep blue with a white throat and golden eye; includes low and very deep blue cultivars, e.g. 'Blue Flash' and 'Royal Ensign', and some with paler flowers, e.g. 'Cambridge Blue'; because of its name, this species is commonly confused with *Ipomoea tricolor*, the climbing morning glory, another popular annual).

cool house a greenhouse where heating is maintained to provide frost protection or, more usually, a minimum winter temperature of 7°C/45°F.

coping the top course of masonry in a wall; usually sloping and sometimes providing an overhang to protect wall-trained fruit trees.

copper a micronutrient or trace element. Copper deficiency occurs in soils high in organic content and lime, and is manifest in arrested growth or dieback of shoot tips.

copper sulphate a compound that is 1) applied to soil for the correction of copper deficiency, 2) used as an ingredient of the fungicide known as Bordeaux mixture.

coppicing the practice of regularly cutting down young growth arising from the base of a tree or shrub, thereby giving rise to numerous vertical shoots. It is mainly a forestry practice but is also used on certain ornamentals to obtain colourful bark (e.g. *Cornus* species), juvenile foliage (e.g. *Eucalyptus* species) or very large foliage (e.g. *Paulownia* species). It is also known as stooling.

Coprosma (from Greek *kopros*, dung, and *osme*, odour, the plants are often malodorous when crushed). Rubiaceae. SE Asia, Australasia, Pacific Islands. 90 species, dioecious, evergreen shrubs and small trees, erect to prostrate or scrambling with leaves ranging from small and narrow to large, broad and glossy. Inconspicuous flowers give rise to fleshy berries. The hardiest species listed here is *C.brunnea*, which will tolerate lows to −5°C/23°F if planted on a fast-draining, acid soil in full sun, with shelter from winter wet and harsh winds. The others are essentially frost-tender, growing best in cool greenhouse conditions, in a moist, acid to neutral, fertile soil and with good ventilation and protection from the hottest summer sun. Increase by ripe wood cuttings in autumn, or by seed sown in spring in a cold frame with the berry pulp removed.

C.brunnea (syn. *C.acerosa*; SAND COPROSMA; New Zealand; stems wiry, creeping to ascending; leaves to 1.5cm, linear; fruit translucent pale blue; includes forms with bronze foliage and cobalt blue to violet fruit, some of them hybrids with *C.petriei*); *C. × kirkii* (*C.brunnea × C.repens*; prostrate to spreading or ascending shrub to 1.5m; leaves to 3cm, linear-oblong to narrowly obovate, leathery and dark green; fruit white faintly flecked red; includes cultivars with bronze, white-edged and yellow-splashed leaves); *C.repens* (LOOKING-GLASS PLANT, MIRROR PLANT; New Zealand; spreading to erect shrub to 8m; leaves to 8cm, broadly oblong to obovate, apex blunt, base tapered, glossy dark green; fruit yellow-orange to red; includes cultivars with leaves splashed or edged white or gold and some with coppery tones).

coral spot (*Nectria cinnabarina*) a fungal disease producing pink to orange pustules and groups of red fruiting bodies scattered over bark, dead twigs and branches. It grows commonly as a saprophyte on woody debris, but can also enter living tissue through wounds to become parasitic, and cause dieback. Although generally regarded as a secondary fungus causing minor diseases, coral spot can be serious, especially where infection follows grafting operations. *Acer*, *Aesculus*, apple, beech, *Cercis*, currants, *Elaeagnus*, elm, fig, gooseberry, hornbeam, lime, *Magnolia*, mulberry, *Pyracantha*, *Ribes* and walnut are particularly affected. Trained forms of trees and shrubs requiring frequent pruning are vulnerable. Prune off dead shoots and branches to a point well below the diseased area; burn prunings and affected woody debris. Treat posts, fences and other wooden structures with wood preservative. Badly affected soft fruit bushes should be dug up and burned.

coralloid resembling coral in structure. It is used of certain types of root adapted to house blue-green algae or saprophytic fungi. It is also applied to the heavily cristate inflorescence of some *Celosia* cultivars, the body shapes of some cacti and to the calli and stigmatic surfaces of some monocots.

coppicing

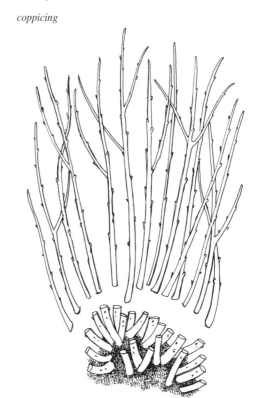

oblique cordon

U cordon

step over cordon

cordate heart-shaped; applied to leaves and leafy stipules that are usually ovate-acute in outline, with a rounded base and a deep basal sinus or notch where the petiole is inserted.

cordon (1) a trained form of tree or bush fruit, comprising one stem, usually planted obliquely, or two or three stems in the form of a vertical U or double U shape. Step-over cordon refers to forms with one or two shoots trained horizontally along a low level wire; (2) a system of growing sweet peas where each plant is restricted to one shoot in order to produce large flowers for exhibition.

Cordyline (from Greek *kordyle*, club, referring to the thickened roots). Agavaceae. SE Asia, Australasia, Polynesia. 15 species, evergreen shrubs and small trees, at first with simple, cane-like stems, later more tree-like and branched above. Leaves oblong to broadly lanceolate to linear, in dense mops and rosettes at branch ends. Flowers small, cream, massed in very large, dense panicles. *C.australis* is hardy outdoors in favoured locations in zone 7; *C.indivisa* in zone 8. Much can be done to aid their survival in terms of siting, selection of cultivar, tying up of leaves and wrapping of crown base in winter. Where very hard winters threaten, they are perhaps best brought into a frost-free greenhouse. Plant both in a warm, sunny position on a fertile, fast-draining soil. *C.australis* is a good coastal and city plant for milder regions. Younger specimens of coloured cultivars are among the most popular dot and patio container plants. The TI TREE, *C.fruticosa* is equally well-loved in the tropics and subtropics and, elsewhere, as a house plant. It needs a minimum winter temperature of 10°C/50°F, bright, indirect light and a moist, fertile, potting mix. Propagate all species by seed or suckers in spring. Increase *C.fruticosa* also by 5–8cm cuttings of mature stem laid flat on sand in a heated frame; pot on as the new shoots appear.

C.australis (CABBAGE TREE; New Zealand; to 20m, becoming branched above with age; leaves to 100 × 2cm, arching, linear, light green to olive; includes cultivars with leaves flushed deep purple-bronze to copper and ruby red, some striped white, cream, yellow or pink, or with a red midvein); *C.fruticosa* (syn. *C.terminalis*; TI TREE; SE Asia, Australia, Polynesia; to 4m, shrubby; leaves to 60 × 10cm, oblong to lanceolate, narrowing to a grooved petiole, glossy green, often tinged red; widely grown with many cultivars available, ranging from dwarf to tall, and with leaves brightly stained, striped and zoned in shades of scarlet, crimson, pink, purple, bronze, copper, lime, yellow, cream and white); *C.indivisa* (syn. *Dracaena indivisa*; New Zealand; to 8m, scarcely branched above; leaves to 200 × 10cm, narrowly lanceolate, somewhat glaucous with the midvein often tinted red; purple- and bronze-leaved cultivars are grown).

core flush a physiological disorder of apples stored for long periods. It is more serious at low temperature and where carbon dioxide level is high and water loss from the fruit excessive. Symptoms are discoloration of the core between the seed cavities, often extending into the flesh.

Coreopsis (from Greek *koris*, a bug, and *opsis*, like, alluding to the appearance of the fruit). Compositae. Americas. TICKSEED. 80 species, annual or perennial herbs with entire to toothed to finely pinnately lobed leaves and bright, daisy-like flowerheads in summer. Fully hardy, *Coreopsis* species prefer full sun and a fertile, moist but free-draining soil. Propagate by seed or, perennials, by division in spring or from softwood cuttings in summer.

C.auriculata (SE US; erect perennial to 1.5m; leaves entire or 1–2-lobed; flowerheads to 5cm in diameter, long-stalked, yellow; includes the vivid gold 'Cutting Gold', the dwarf 'Nana', with orange-yellow flowerheads, and 'Superba', with bright yellow flowerheads tinted purple); *C.grandiflora* (C and SE US; erect annual or perennial to 60cm; leaves entire to 3–5-parted; flowerheads to 6cm in diameter, long-stalked, ray florets yellow, disc florets tinted orange; includes 'Badengold', a short-lived perennial to 100cm tall with golden flowerheads, 'Ruby Throat', with a red-stained centre, and the deep yellow, 'double' 'Sunray'); *C.lanceolata* (C and SE US; perennial to 60cm; leaves usually entire; flowerheads to 6cm in diameter, golden, long-stalked, ray florets usually with jagged margins; includes the dwarf 'Goldfink', and cultivars with double and red-tinted flowerheads); *C.tinctoria* (N America; erect, bushy annual to 1.2m; leaves 1–2-pinnate; flowerheads to 3cm in diameter, long-stalked and borne several together, ray florets yellow marked brown-red at base, disc florets dark red); *C.verticillata* (SE US; bushy perennial, erect, 20–90cm tall; leaves very finely bipinnate; flowerheads to 5cm in diameter, golden yellow or pale lemon yellow in 'Moonbeam').

coriaceous leathery or tough but smooth and pliable in texture.

coriander (*Coriandrum sativum*) a hardy annual to 45cm high, grown as a culinary herb for its seed and aromatic leaves, which are used especially in oriental cooking. Sow successively from spring to late summer *in situ*, 1cm deep in rows 30–45cm apart, thinning to 15cm final spacings. For winter supply, late autumn sowings can be made under cover.

Coriaria (from Latin *corium*, hide or leather: some species are used in tanning). Coriariaceae. Europe, Asia, Americas, New Zealand, Pacific Islands. 30 species, herbs, shrubs and small trees with angular stems and arching, frond-like branches ranked with small, ovate leaves. At first the flowers are small, dull and green, but as the fruits mature, the petals persist, becoming enlarged, fleshy and colourful. Hardy in favoured areas of zone 7, especially if grown against a warm, sheltered wall and mulched in winter. Plant in full sun on a moist, fertile soil. Prune back exhausted or frost-damaged growth in spring. Increase by seed

coriander

sown in autumn, by semi-ripe cuttings or by rooted suckers. *C.terminalis* (Himalaya to W China; deciduous suckering subshrub or shrub to 1.5m; leaves red in autumn; fruiting raceme to 15cm long, cylindrical with enlarged corollas to 1cm across and glossy black; includes f. *fructorubro*, with translucent red corolla, and var. *xanthocarpa*, with translucent yellow corolla).

cork a layer of protective tissue, usually elastic and spongy, that replaces the epidermis in the older superficial parts of some plants. It is also known as phellem.

cork bark the thick corky bark of trees, such as *Quercus suber*, which is often used by gardeners to sustain epiphytic bromeliads, ferns and orchids. The round sections can be filled with sphagnum moss or other substrate and the plant established through a hole in the bark.

cork cambium a layer of meristematic tissue from which cork is derived. It is also known as phellogen.

corm a solid, swollen, subterranean, bulb-like stem or stem-base. Corms are annuals, the next year's corm developing from the terminal bud or, in its absence, one of the lateral buds.

cormel, cormlet a small corm developing from and around the mother corm.

cormous bearing corms.

Cornell mixes soilless growing media based on peat and vermiculite or perlite, developed by Cornell University, New York.

corniculate horned, shaped like a new moon, bearing or terminating in a small horn-like protuberance.

Cornus (Latin name for *C.mas*). Cornaceae. DOGWOOD, CORNEL. Temperate regions of the N hemisphere; also S America and Africa. Some 45 species, perennial, semi-woody herbs, or, more usually, shrubs and small trees, mostly deciduous. They bear small, pale flowers in flat-topped cymes, panicles or heads, often subtended by bracts. These give rise to small, colourful drupes. *C. alba* and *C.stolonifera* are invaluable for their winter beauty, having brilliant autumn foliage, handsome fruits and colourful stems which range from yellow through brilliant crimson to almost black-purple. The 'flowering dogwoods' such as *C. florida*, *C.kousa* and *C.nuttallii* also have stunning autumn colour, but it is for their large, white or pink floral bracts, produced in spring and early summer, that they are celebrated. *C.mas* is valued as an early spring-flowering shrub. The only non-woody species commonly grown is *C.canadensis*, which forms an attractive carpet of creeping stems, with white-bracted flowers in summer and red fruits in autumn. Most are hardy to between –10°C and –25°C/14–13°F, although *C.capitata* is hardy only where the temperature does not fall below –5°C/23°F. *C. florida* and *C.nuttallii* will tolerate low temperatures, but need long, hot humid summers to ripen their wood. Late frosts may also damage their decorative bracts. Dogwoods thrive in a wide range of soil types from acid through to thin chalky soils. Many of the coloured-stemmed dogwoods will grow in poorly drained soils and are ideal for waterside plantings. For the best winter colour, cut back *C.alba* and *C. stolonifera* to near ground level in spring, or grow them as a low pollard or stub. This promotes fresh, rod-like growths that will colour brilliantly the following autumn and winter. Most other species have gracefully tiered habits that should be left untouched. Increase by seed or cuttings; also, *C. canadensis* by detaching rooted lengths of stem.

Cornus kousa   *Cornus nuttallii*

CORNUS

Name	Distribution	Habit	Bark	Leaves	Inflorescence	Fruit
C.alba RED-BARKED DOGWOOD, TARTARIAN DOGWOOD	Siberia to Korea	deciduous, erect to spreading, suckering shrub to 3m	new shoots turn sealing wax red in winter	4–8cm, ovate-elliptic with short, pointed tip, mid to dark green above, slightly glaucous beneath	flowers small, cream in flattened, 5cm-wide cymes in late spring	small, white to blue

Comments: Cultivars with variegated leaves include 'Argenteomarginata' (edged white), 'Aurea' (soft yellow-green), 'Elegantissima' (grey-green edged white to cream), 'Gouchaltii' (pink-flushed, edged yellow), and 'Spaethii' (edged yellow, bronze-flushed at first). Cultivars with exceptional winter stem colour include 'Atrosanguinea' (dwarf with crimson stems), 'Kesselringii' (stems purple-black), and 'Sibirica' (stems bright coral red).

Name	Distribution	Habit	Bark	Leaves	Inflorescence	Fruit
C.alternifolia PAGODA DOGWOOD, GREEN OSIER	E N America	deciduous shrub or tree to 8m with a flat-topped crown and tiered branches		6–12cm, ovate-elliptic, dark green above, somewhat glaucous beneath	flowers small, yellow-white in 5cm-wide cymes in early summer	small, black, rarely yellow

Comments: Includes 'Argentea' with white-marked leaves, 'Corallina' with dull red stems, the yellow-fruited 'Ochrocarpa' and the strongly tiered 'Umbraculifera'.

Name	Distribution	Habit	Bark	Leaves	Inflorescence	Fruit
C.canadensis *Chamaepericlymenum canadense* BUNCHBERRY, CREEPING DOGWOOD, DWARF CORNEL, CRACKERBERRY	Greenland to Alaska	creeping perennial with rooting, ground-smothering stems to 8cm tall		2–4cm, ovate-lanceolate, in a whorl at the summit of erect stems	green to red-violet, clustered in a tight umbel surrounded by 4–6, ovate, white bracts each to 2cm long in early summer	bright red berries
C.capitata *Dendrobenthamia capitata* BENTHAM'S CORNEL	Himalaya, China	evergreen or semi-evergreen, spreading tree to 16m		5–12cm, ovate to lanceolate, dark above, pale beneath	flowers small, white in clusters to 2cm across, subtended by obovate, cream to white bracts to 8cm long in early summer	a fleshy, strawberry-shaped, scarlet aggregate to 2.5cm diam.
C.controversa GIANT DOGWOOD	Himalaya, China, Japan	deciduous tree to 17m tall with horizontal, tiered branches		7–15cm, ovate to broadly elliptic, cuspidate, dark green above, paler beneath, flushed red in autumn	small, white in flat-topped cymes to 10cm diam.	

Comments: 'Variegata' (WEDDING CAKE TREE) has an uneven white-yellow margin

Name	Distribution	Habit	Bark	Leaves	Inflorescence	Fruit
C. 'Eddie's White Wonder' (*C.florida* × *C.nuttallii*)	garden origin	deciduous tree, erect and pyramidal to 6m tall with drooping branches		to 15cm, ovate-elliptic, mid-green turning orange-brown and purple-red in autumn	small and green, surrounded by 4 (–6) large, rounded, white bracts in late spring	
C.florida EASTERN FLOWERING DOGWOOD, COMMON WHITE DOGWOOD	E N America	deciduous tree or shrub to 10m tall with many, spreading then ascending branches		7–15cm, oval to ovate, dull green above, pale beneath, turning orange-red in autumn	small, green in a button-like head surrounded by 4 bracts each to 4cm long, obovate with curled margins and white to cream or pink, in late spring	scarlet

Comments: Includes many cultivars with bracts ranging in size, number and in colour from pure white to bright flamingo pink and deep rose. 'Welchii' has rather distorted leaves variegated red to pink and yellow-white.

Name	Distribution	Habit	Bark	Leaves	Inflorescence	Fruit
C.kousa KOUSA	China, Korea, Japan	deciduous, erect tree or shrub to 7m tall		5–9cm, ovate-acuminate, undulate, dark green above, paler beneath, turning red in autumn	small, green in rounded heads surrounded by 4, large, lanceolate, creamy white bracts in summer	a fleshy, scarlet, strawbery-like aggregate to 2cm diam.

Comments: Includes the larger var. *chinensis* with longer, more pointed bracts turning from green to white and assuming pink tints

Name	Distribution	Habit	Bark	Leaves	Inflorescence	Fruit
C.macrophylla	Himalaya, China, Japan	deciduous tree or shrub to 15m spreading and loosely tiered branches		10–17cm, elliptic-ovate, apex finely tapering, dark green above, pale beneath	small, yellow-white in broad cymes to 15cm diam., in summer	small, blue
C.mas CORNELIAN CHERRY	Europe, W Asia	deciduous, spreading and bushy shrub or tree to 5m tall		4–10cm, ovate-elliptic, mid- to dark green	small, bright yellow in rounded cymes to 2cm diam., produced before leaves in late winter and early spring	small, oblong, bright red

Comments: Includes cultivars with variegated leaves and white, yellow and violet-blue fruit

Name	Distribution	Habit	Bark	Leaves	Inflorescence	Fruit
C. 'Norman Hadden' (*C.kousa* × *C.capitata*)	garden origin	deciduous, erect to spreading tree to 8m			small and green surrounded by showy white bracts that assume pink tints with age	a bright red, large, strawberry-like aggregate

CORNUS

Name	Distribution	Habit	Bark	Leaves	Inflorescence	Fruit
C.nuttallii MOUNTAIN DOGWOOD, PACIFIC DOGWOOD	W N America	deciduous tree to 30m tall with a broadly conical crown		8–12cm, oval to obovate, mid- to dark green, downy at first	small, green in tight heads surrounded by 4–8, broadly oval to obovate bracts, each to 8cm long and green-white turning pure white, later assuming a pink flush	
C.sanguinea BLOOD TWIG, COMMON DOGWOOD, DOGBERRY	Europe	deciduous, erect and freely suckering shrub to 4m tall	new shoots green flushed purple-red especially in winter	4–10cm, broadly elliptic to ovate, dull green flushed purple-red in autumn	small, off-white, muskily scented in dense cymes to 5cm diam. in summer	small, purple-black
Comments: Includes cultivars with variegated leaves and green stems, also 'Winter Beauty' – compact with bright orange-yellow shoots assuming fiery tints in winter; the leaves turn amber in autumn.						
C.stolonifera AMERICAN DOGWOOD, RED OSIER	E N America	deciduous shrub to 3m, freely suckering with spreading to erect, slender branches	young branches dark purple-red to deep red especially in winter	5–10cm, oval to oval-lanceolate with a finely tapering tip, dark green above, glaucous beneath	small, dull white in cymes to 5cm across in late spring	small, white
Comments: Includes 'Flaviramea' with yellow shoots in winter, 'Kelseyi', dense and dwarf with red-tipped, yellow-green winter shoots, and 'White Gold' with white-edged leaves.						

Corokia (from the Maori vernacular name *korokia-taranga*). Cornaceae. New Zealand. 3 species, evergreen shrubs and small trees with slender, twiggy branches, often spiralling or zig-zagging and becoming congested. These are silver to ash grey at first and ultimately covered in thin, black-brown bark. The leaves are tough and lanceolate to spathulate, bronze to olive above, silvery beneath. Clusters of small yellow flowers in late spring are followed by orange-red berries. Hardy in climate zone 7. Plant in full sun on a fertile, well-drained soil that is moist in spring and summer, but rather dry in winter. For best results, place within the shelter of a south-facing wall. Root short, heeled cuttings of the previous season's growth in a sandy mix in a frame or case in autumn.

C.buddleioides (to 4m; branching pattern fairly straightforward; leaves to 10cm, lanceolate); *C.cotoneaster* (WIRE NETTING BUSH; to 2m; branching pattern intricate and tortuous; leaves 0.5–2cm, orbicular to spathulate, sometimes shallowly and sparsely toothed); *C.* × *virgata* (*C.buddleioides* × *C.cotoneaster*; to 3m; branching pattern somewhat twisted and intricate to erect and straightforward; leaves 1–4cm, spathulate to oblanceolate).

Corokia cotoneaster

corolla the interior perianth whorl; a floral envelope composed of free or fused petals. Where the petals are separate, a corolla is termed choripetalous or polypetalous; where fused, the corolla is termed gamopetalous or sympetalous – in such cases, the petals may be discernible only as lobes or teeth on the rim of a corolla tube, cup or disc.

corona a crown- or cup-like appendage or ring of appendages. It may be a development of the perianth (as in *Narcissus*), of the staminal circle (as in *Asclepias*), or located between perianth and stamens (as in *Passiflora*).

Coronilla (from Latin *coronilla*, a little crown, alluding to the radiating golden umbels of most species). Leguminosae. Europe, Asia, Africa. CROWN VETCH. 20 species, evergreen herbs and small shrubs with pinnate or trifoliolate leaves and pea-like flowers in stalked, circular umbels at various times of year, including winter (*C.valentina*). Hardy in zone 7, but best given the shelter of a south-facing wall. Grow in full sun on a light, fast-draining soil. Support or tie-in to walls if necessary. Prune after flowering to remove any weak growth. Older plants may need harder, rejuvenative pruning at this time. Root greenwood cuttings in summer.

C.emerus (syn. *Hippocrepis emerus*; SCORPION SENNA; C and S Europe; shrub to 2m; leaves to 6cm, pinnate, dark green; flowers fragrant, yellow); *C.valentina* (S Portugal, Mediterranean, N Africa, Aegean; shrub to 1.5m; leaves to 5cm, pinnate, leaflets bright green above, glaucous beneath; flowers golden yellow, fragrant; subsp. *glauca*: syn. *C.glauca*, leaves glaucous, blue-green, flowers lemon to golden; 'Variegata': foliage and young shoots striped and zoned creamy yellow to white).

Coronilla valentina subsp *glauca*

Correa (for Jose Francesco Correa de Serra (1750–1823), Portuguese botanist). Rutaceae. Australia. AUSTRALIAN FUCHSIA. 11 species, evergreen shrubs and trees with small, leathery leaves and pendulous tubular to campanulate flowers year-round, usually waxy and with four, short, reflexed lobes. Most species tolerate short-lived frosts to about –5°C/23°F (*C.decumbens* lower still), but in colder zones need greenhouse protection. Plant in a fast-draining, gritty and sandy, acid medium. Position in full or dappled sun in an airy, rather dry atmosphere. Water moderately when in growth, sparingly in winter. Root softwood or semi-ripe cuttings in a closed case with bottom heat.

C.backhouseana (dense, spreading shrub to 2m; flowers to 2.5cm, creamy yellow to pale green); *C.decumbens* (prostrate shrub to 1m; flowers to 2.5cm, red tipped green); *C.* × *harrisii* (*C.pulchella* × *C.reflexa*; flowers to 3cm, scarlet); *C.pulchella* (more or less prostrate shrub to 1.5m; flowers to 2.5cm, orange to pink-red or, rarely, white); *C.reflexa* (syn. *C.speciosa*; erect or prostrate shrub to 3m; flowers to 4cm, white to green, pink or red, sometimes bright red tipped green).

corrugate crumpled or wrinkled but more loosely so than in rugose. Corrugate aestivation describes the irregular and apparently crumpled folding of a perianth in bud, as for example happens with the petals of many members of the poppy family, Papaveraceae.

Corydalis claviculata

Cortaderia (from *cortadera*, Argentine name for these plants). Gramineae. S America, New Zealand, New Guinea. 24 species, large, evergreen perennial grasses with very long and slender, sharp-edged leaves making dense tussocks. The flowers are carried in tall, plume-like panicles in late summer and autumn. Hardy in climate zone 5. Plant in full sun on a well-drained but moisture-retentive soil. These grasses accumulate a good deal of dry, dead material within their dense and massive crowns – this should be cleared each spring, along with spoilt plumes and any damaged leaves (great care being taken to avoid their razor-like margins). The short cut of setting light to the whole clump and incinerating growth dead or alive is a little harsh and possibly dangerous. Increase by division in late spring. *C.selloana* (PAMPAS GRASS; temperate S America; leaves arching, making tussocks to 1.5 × 2m; panicles to 120cm, oblong to pyramidal, pearly white or cream, sometimes flushed purple, red or rose; cultivars range from dwarf to very tall (i.e. 3m) and include selections with white- or yellow-striped foliage and panicles in tones of silver-white, pale violet and rose).

cortex bark or rind; stem tissue situated between the stele and epidermis.

corticate covered with thick bark, with a hard coating, but softer centre; corky.

Cortusa (for J.A. Cortusi, 16th-century Italian botanist). Primulaceae. C Europe to N Asia. 8 species, hardy perennial herbs with rosettes of long-stalked, rounded to heart-shaped, toothed and hairy leaves and funnel-shaped, 5–lobed flowers in long-stalked umbels in summer. Hardy in zone 7 and suitable for lightly shaded areas of the rock garden and for planting beneath shrubs. It favours a moist, humus-rich soil and a mulch of leafmould and garden compost in autumn. Propagate by seed sown as soon as ripe, by division in early spring or autumn, or by cuttings of thick, mature roots in late summer. *C.matthioli* (flowers pink to magenta, nodding, on scapes to 30cm tall).

Corydalis (from Greek *korydalis*, crested lark: the flowers resemble the head of this bird). Fumariaceae. Northern temperate regions, South Africa. Some 300 species, annual or deciduous perennial herbs with ferny, biternately compound foliage and racemes of flowers in spring and summer; these are tubular and angled horizontally, the outer petals recurved at the tip and spurred at the base, the inner petals beak-like. Hardy in climate zones 6 and 7. Plant in full sun or light shade in a humus-rich, well-drained soil that is cool and moist in spring and summer. Propagate the perennial species by division when dormant, or – if tuberous – by detaching offsets. Sow seed of annual species *in situ* in spring.

C.cashmeriana (Himalaya; fibrous-rooted perennial to 20cm; leaves bright green; flowers bright blue with darker tips); *C.cava* (syn. *C.bulbosa*; C Europe; tuberous perennial to 15cm; leaves pea-green; flowers dull purple or white); *C.cheilanthifolia* (China; fleshy-rooted perennial to 30cm with very fern-like, soft olive green leaves; flowers bright yellow); *C.claviculata* (CLIMBING CORYDALIS; Europe; slender, climbing annual to 1.5m tall with sea green, tendril-tipped leaves; flowers cream to pink, tipped dark red); *C.diphylla* (Himalaya; tuberous perennial to 15cm; leaves grey-green; flowers lilac-white, tipped purple); *C.flexuosa* (Himalaya; fleshy-rooted perennial to 20cm; leaves bright green to grey-green; flowers sky blue to deep cobalt; includes various cultivars with flowers in a range of blues and foliage marked or tinted bronze to deep maroon); *C.fumariifolia* (syn. *C.ambigua* of gardens; Japan; tuberous perennial to 10cm; leaves glaucous; flowers azure); *C.lutea* (Europe; bushy perennial to 30cm with mid-green leaves, their undersides glaucous; flowers bright yellow); *C.nobilis* (Siberia; fleshy-rooted perennial to 50cm with glaucous leaves; flowers yellow, tipped purple-brown); *C.ochroleuca* (SE Europe; similar to *C.lutea* but with white to cream flowers tipped pale yellow); *C.popovii* (C Asia; tuberous perennial to 15cm with blue-green leaves and maroon-lipped, white flowers); *C.solida* (FUMEWORT; N Europe, Asia; tuberous perennial to 20cm with dark-tipped, pale mauve-pink to light red flowers; includes 'George Baker', with deep, rosy red flowers); *C.wilsonii* (China; fleshy-rooted perennial to 30cm; leaves glaucous, blue-green; flowers bright yellow, tipped green).

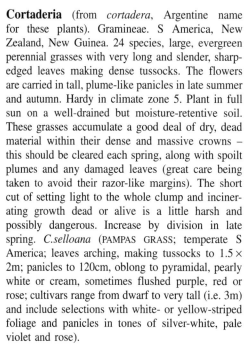

Corylopsis (from Greek *korylos*, hazel, and *opsis*, appearance, the leaves resemble those of *Corylus*). Hamamelidaceae. Asia. WINTER HAZEL. 30 species, hardy, deciduous shrubs and small trees with slender branches, and broadly ovate leaves with impressed veins. Small, fragrant and yellow, the bell-shaped flowers appear before or alongside the leaves and hang in narrow racemes amid translucent green bracts. Hardy to −15°C/5°F. Grow in a sheltered, partially shaded position on a well-drained but moisture-retentive, acid to neutral soil. Mulch in dry summers. Allow space for the strongly spreading habit of *C.pauciflora* and *C.spicata*. Propagate by seed in autumn, by softwood or greenwood cuttings rooted in a closed case in late spring, or by simple layering.

C.glabrescens (Japan, Korea; to 5m; leaves glaucous beneath, veins in to 12 pairs; inflorescence to 2.5cm, flowers very pale yellow, anthers yellow or purple); *C.pauciflora* (BUTTERCUP WITCH-HAZEL; Japan, Taiwan; to 3m; leaves to 6.5cm, glaucous beneath, veins in to 9 pairs; inflorescence to 3cm, flowers relatively large (i.e. to 1.8cm across), sweetly fragrant, primrose yellow, anthers yellow); *C.sinensis* (syn. *C.willmottiae*; China; to 5m; leaves glaucous and downy beneath with veins in 7–12 pairs, deep purple-bronze-tinted in 'Spring Purple'; inflorescence to 8cm with lemon yellow flowers with yellow anthers); *C.spicata* (SPIKE WITCH-HAZEL; Japan; to 3m; leaves glaucous and downy beneath with veins in 6–7 pairs; inflorescence to 6cm, flowers bright yellow, anthers red-brown).

Corylus (Classical Latin name). Betulaceae. N Temperate regions. HAZEL. 15 species, hardy deciduous shrubs or small trees with large and coarse, ovate leaves, their veins impressed and margins toothed. The tiny, yellow male flowers hang in catkins in early spring; the female flowers resemble small buds with protruding red styles. The fruits are woody nuts enclosed by two, leafy, lobed bracts. For habit and foliage colour respectively, *C.avellana* 'Contorta' and *C.maxima* 'Purpurea' are exceptional shrubs, whilst *C.colurna* is proving to be a tough but distinguished street tree. All are hardy to at least −30°C/−22°F. Plant in full or dappled sun on any fertile, free-draining soil. Prune to restrict size and remove suckers from grafted plants. Propagate by seed sown immediately upon ripening, or, for cultivars by simple layering and grafting. Hazel is attacked by several leaf-eating moths, and also by the hazel sawfly, nut gall mite, nut weevil, hazelnut aphids and grey squirrels.

C.avellana (HAZELNUT, COBNUT; Europe; multi-stemmed shrub to 6m, or tree to 10m; leaves to 10cm, with short, stiff hairs; fruit edible, to 2cm, amid jagged bracts to 1cm; includes 'Contorta', CORK-SCREW HAZEL, HARRY LAUDER'S WALKING STICK, a slow-growing shrub to 3m with strongly spiralling shoots and contorted leaves); *C.colurna* (TURKISH HAZEL; SE Europe, Asia Minor; tree to 30m with a pyramidal crown; leaves to 12cm, lustrous deep green; fruit to 2cm amid thickened, glandular and deeply cut bracts to 3cm); *C.maxima* (FILBERT; SE Europe, Asia Minor; differs from *C.avellana* in its larger leaves, usually biserrate, and in bracts, to twice the length of the fruit, deeply lobed and forming a loose tube; includes 'Purpurea' (syn. 'Atropurpurea'), with dark purple-red foliage).

corymb an indeterminate flat-topped or convex inflorescence, where the outer flowers open first; cf. *umbel.*

corymbose, corymbiform resembling or forming a corymb.

Coryphantha (Greek, *koryphe*, summit, and *anthos*, flower – the flowers arise near the stem apex). Cactaceae. SW Mexico, US. 45 species, cacti with single or clustered globose to cylindric stems, usually spiny, and funnel- to bell-shaped flowers. Provide a minimum temperature of 5°C/40°F, with full sun and low humidity. Plant in a mix containing at least 50% grit. Keep dry from mid-autumn until early spring, except for a light misting on warm days in late winter. Water thoroughly in spring and summer, but allow the soil to become dry between waterings. Increase by seed.

C.cornifera (stem to 12cm, pale green with pyramidal tubercles and spines in flattened circles covering stem; flowers yellow); *C.radians* (differs from *C.cornifera* in its woolly tubercles and flowers with red-tinged outer tepals). For *C.vivipara*, see *Escobaria vivipara.*

Cosmos (Greek *kosmos*, ornament). Compositae. Tropical and warm Americas. 26 species, annual and perennial herbs and subshrubs with simple to lobed or pinnate leaves and long-stalked, daisy-like flower-heads in summer. *C.atrosanguineus* is a perennial, developing root tubers much like those of *Dahlia*. Once established, it will survive frosts provided that the crown is thickly mulched in winter. Alternatively, lift the tubers in late autumn and store as for *Dahlia*, or root basal cuttings in late spring and summer – these can be overwintered under glass and treated as half-hardy annuals. Sow seed of the annual *C.bipinnatus* and cultivars in a cool greenhouse in autumn or early spring for planting out in a sunny position on a fertile, well-drained soil after the last frosts. In areas with hot, humid summers, sow *in situ* in late spring.

C.atrosanguineus (CHOCOLATE COSMOS; Mexico; perennial, 30–100cm tall, ultimately tuberous-rooted; leaves pinnate; flower heads to 5cm in diameter, deep velvety black-red to darkest maroon, chocolate-scented); *C.bipinnatus* (Mexico, S US; annual, 30–200cm tall; leaves finely bipinnate; flowerheads to 10cm in diameter, carmine to rose-pink or lilac with a yellow centre; many seed races and F_1 hybrids have been developed, including plants dwarf and tall, with single, semi-double and double flowerheads in shades of red, pink, orange, yellow and white, sometimes marked and usually with contrasting centres).

costa a single pronounced midvein or midrib; the term is sometimes used to describe the rachis of a pinnately compound leaf.

costapalmate used of 1) a pinnate leaf where the pinnae are congested in a radial, fan-like order and perhaps to some degree united, the whole therefore appearing palmate; 2) a palmate leaf where the petiole continues through the blade as a distinct midrib.

Costus (name used by Theophrastus and Pliny). Zingiberaceae. Tropics. 90 species, tender perennial herbs, mostly evergreen with creeping, aromatic rhizomes and clumped, cane-like stems. Showy flowers with ruffled, crepe-like lips appear in cone-shaped, terminal heads. Cultivate as for *Alpinia*. *C.afer* (SPIRAL GINGER; W Africa; 1–4m; leaves lanceolate, smooth, spirally arranged; flowers white marked yellow); *C.malortieanus* (syn. *C.zebrinus*, *C.rumphii*; STEPLADDER PLANT; Brazil, Costa Rica; to 1m; leaves broadly obovate to obcordate, thick, hairy, striped or mottled dark green to purple-brown; flowers yellow marked red-brown); *C.speciosus* (CREPE GINGER, MALAY GINGER; SE Asia to New Guinea; to 3m; leaves elliptic to lanceolate, striped creamy white in 'Variegatus'; flowers white to pink stained yellow-orange, amid red-tinted bracts); *C.spiralis* (SPIRAL FLAG; West Indies, Ecuador; to 1.5m; leaves obovate to oblong, spirally arranged; flowers rose to flesh pink or red, bracts sometimes red-tinted).

cot split a bud stage of plums, when the yellow calyx splits off small green fruitlets. See *bud stages*.

Cotinus (from Greek *kotinos*, wild olive). Anacardiaceae. Northern temperate regions. SMOKEWOOD, SMOKE BUSH. 3 species, deciduous trees and shrubs with elliptic to obovate leaves and very small flowers in large, finely branched and feathery panicles in summer, many of these abort with the result that the bare panicle branches assume a ghostly or smoke-like appearance. Fully hardy. Plant in sun on a well-drained, low to medium-fertility soil. Prune in late winter to remove deadwood and straggling growth – contact may cause dermatitis. Root heeled, ripewood cuttings in early autumn in a cold frame.

C.coggygria (syn. *Rhus cotinus*; SMOKE TREE, VENETIAN SUMAC; S Europe to C China; shrub to 5m; leaves to 7.5cm, soft mid-green, often tinted pink to bronze at first; inflorescences to 20cm, purple-buff becoming smoky grey; cultivars and hybrids with the next species include plants with a fog of pink to wine red flowers and purple or maroon leaves, sometimes edged scarlet and usually colouring richly in autumn); *C.obovatus* (AMERICAN SMOKEWOOD, CHITTAMWOOD; S US; tree or shrub to 10m, differs from *C.coggygria* in its leaves with cuneate (not rounded or truncate) bases, sea green often tinted purple when young, becoming gold, amber and scarlet in autumn).

Cotoneaster (from Latin *cotoneum*, quince and *-aster*, a suffix indicating incomplete resemblance; the name refers to the loose resemblance of some species to the quince, *Cydonia*). Rosaceae. Europe, N Africa, E Asia, Siberia, Himalaya. Over 70 species, deciduous or evergreen shrubs or small trees. Produced in spring and summer, the flowers are small, solitary, clustered or carried in flat-topped corymbs. White or pink-tinted, they consist of five spreading, rounded petals and numerous stamens. They are followed by small, red or black pomes; these are highly ornamental and often persist on the shrub long into the winter.

Cotoneasters are grown for their profuse flowers, which are particularly attractive to bees, and for the rich colours of their autumn foliage and berries; these are most attractive in autumn and winter, and a useful food source for birds.

Low-growing and spreading species and cultivars are excellent groundcover on banks. They are also useful for underplanting large shrubs and trees, or for trailing over rocks and walls.

The evergreen *C.franchetii*, *C.conspicuus* and *C.lacteus* and the deciduous *C.bullatus* and *C.hupehensis* can be used in the shrub border or for woodland edge plantings; the evergreens also making good windbreaks. *C.simonsii*, *C.franchetii*, *C.lacteus*, *C.divaricatus* and *C.frigidus* are often used in informal hedging. A number of dwarf species, such as the ground-hugging *C.adpressus* and *C.microphyllus* var. *thymifolius* are well-suited to the rock garden. *C.horizontalis* and *C. microphyllus* can be wall-trained with little or no support and will thrive even on north- and east-facing walls, where they will provide bright flowers, rich autumn colour and vivid fruit.

Cotoneasters will tolerate atmospheric pollution, and all except the boggiest soils. Most are hardy to –20°C/–4°F, although the foliage of evergreen species and cultivars may be scorched or killed at very low temperatures. Grow in full sun or semi-shade. For hedging, use young plants of *C.franchetii* or *C.simonsii* set 30–35cm apart; or plants of *C.frigidus*, *C.lacteus* and *C. × watereri* 60–90cm apart.

Pruning is usually necessary only to restrict size and promote bushy growth; it should be undertaken in late winter. Remove outward-facing growths on wall-grown plants and tie in new growth. Trim informal hedges selectively, taking out badly placed branches. Propagate evergreen species and cultivars by semi-ripe heel cuttings in early autumn. Increase deciduous species by cuttings in mid to late summer, in a closed case with bottom heat. Sow seed outdoors in autumn or stratify in damp sand over winter and sow under glass in spring. They hybridize freely, and seed may not come true.

Susceptible to fireblight, armillaria root rot and silver leaf. Many of the pests of apples also attack *Cotoneaster*. They may also become infested with the hawthorn webber moth.

(see table overleaf)

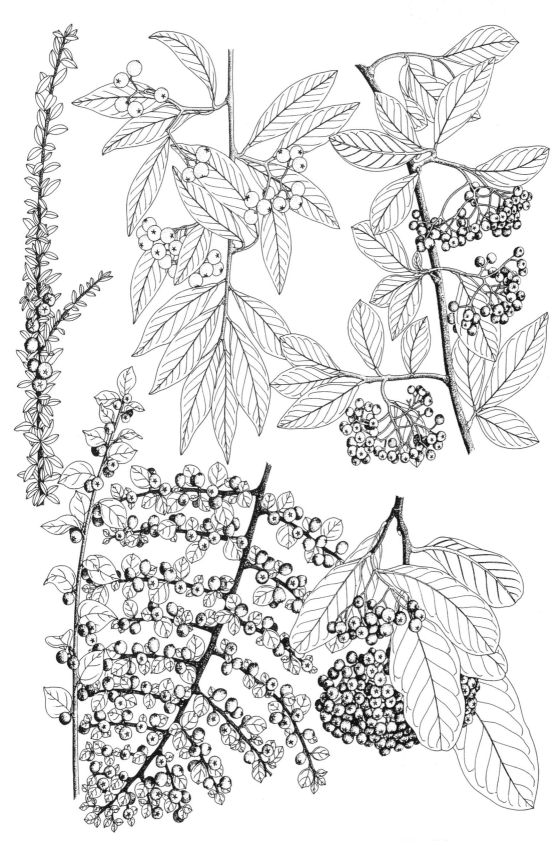

Cotoneaster, *top row left to right: C. conspicuus* 'Decorus', *C.* 'Rothschildianus', *C.* 'Cornubia',
bottom row left to right: C. simonsii , C. horizontalis, C. x *watereri*

COTONEASTER

Name	Distribution	Habit	Leaves	Flowers	Fruit
C.adpressus	China	scandent, spreading or prostrate deciduous shrub	to 1.5cm, broadly ovate or obovate, dull green, scarlet in autumn	red with some white, solitary or paired	red
Comments: 'Little Gem': growth weaker, cushion-forming with little autumn colour and no fruit					
C.bullatus	China	deciduous shrub to 4m	to 7cm, ovate to oblong or elliptic, apex tapering, dark green and bullate with fiery tints in autumn	white tinted red in clusters to 5cm across	currant red
Comments: 'Firebird': spreading shrub to 3m, fruit dark orange or red in dense clusters Comments: var. *floribundus*: flowers abundant, fruit large, densely clustered					
C.cashmiriensis *C.cochleatus* of gardens, *C.microphyllus* var. *cochleatus*	Kashmir	slow-growing, low, dense, dwarf shrub close to *C.microphyllus*	0.5-1cm, ovate to elliptic, usually notched, deep glossy green	white	small, bright red
C.congestus	Himalaya	congested, spreading, scandent or prostrate, evergreen or semi-evergreen shrub or subshrub	to 1cm, often in two rows, obovate to elliptic or oblong, tough, dull green	white	carmine
C.conspicuus	Tibet	spreading scandent or prostrate evergreen or semi-evergreen shrub to 1m	to 2cm, oblong or lanceolate, tough	white, solitary	shiny, scarlet
Comments: 'Decorus': spreading and low-growing with abundant flowers and fruit; 'Tiny Tim': mat-forming, fruit orange washed red					
C. 'Cornubia'	garden origin	semi-evergreen, erect to arching shrub to 5m	to 12cm, elliptic to obovate or broadly lanceolate, tough, mid to dark green with sunken veins	white in flat-topped clusters	bright red in large, drooping clusters
C.dammeri	China	scandent to prostrate evergreen shrub with long shoots	to 3cm, elliptic to obovate or oblong, dark glossy green	white, usually solitary	red
Comments: 'Coral Beauty': very low-growing with abundant orange-red fruit; 'Donnard Gem': low, spreading, with tiny grey-tinted leaves and persistent red fruit; 'Skogholm': branches rooting, spreading to 3m across, leaves elliptic, large, dull dark green, flowers dull red					
C.dielsianus	China	arching to pendulous, semi-evergreen shrub to 3m	to 2.5cm, rounded to oval or obovate, tough, yellow- to grey-hairy beneath	pink, clustered	blood red to currant red (coral in the round-leaved var. *elegans*)
C.divaricatus	China	spreading deciduous shrub to 2m with distinctly divaricating branches	to 2cm, ovate to elliptic, glossy dark green, red in autumn	white tinted red, clustered	deep red
C. 'Exburiensis'	garden origin	evergreen or semi-evergreen arching shrub to 5m, similar to *C.* 'Rothschildianus'	lanceolate, large, glossy bright green and rugose	white	apricot yellow, tinged pink in winter
C.franchetii	China, Tibet, Burma	evergreen slender shrub to 3m with arching branches	to 3cm, oval, grey-green above, felted with grey-white or mustard-coloured hairs beneath	white tinged rose pink, clustered	orange-scarlet
C.frigidus	Himalaya	spreading, deciduous shrub or tree to 18m	to 12cm, elliptic to oblong or obovate, dull green, woolly at first beneath	white, clustered	light red
Comments: includes cultivars with pendulous branches and creamy yellow fruit					
C.glaucophyllus	China	evergreen shrub to 4m	oval, glaucous white beneath, with conspicuous veins, covered in tawny hairs when young	white, in clusters	orange to deep red
C. 'Gnom'	garden origin	evergreen, prostrate shrub to 20 x 200cm	narrowly oblong to lanceolate, dark green	white, clustered	light red

COTONEASTER

Name	Distribution	Habit	Leaves	Flowers	Fruit
C.horizontalis	China	deciduous shrub to 1.75m with arching branches and branchlets 2-ranked and flattened in a distinctive herring bone pattern	0.5-1.5cm, rounded to elliptic, glossy dark green, usually tinted deep bronze to fiery red in autumn	solitary or in pairs, white tinted pink or red	orange-red to blood red

Comments: Includes cultivars with white or cream-variegated leaves and of dwarf, low and spreading habits

Name	Distribution	Habit	Leaves	Flowers	Fruit
C.hupehensis	China	deciduous shrub to 2m with arching branches	to 3.5cm, elliptic to ovate, dark green above, grey and somewhat hairy beneath	white, clustered	large, bright crimson
C. 'Hybridus Pendulus'	Garden origin	evergreen shrub, with narrow, spreading to creeping branches, but usually grown as a standard with a domed and weeping crown	oblong to lanceolate, dark green	white, clustered	bright red, abundant
C.lacteus	China	evergreen shrub to 4m	oblong to obovate or elliptic, tough, mid- to dark green above with impressed veins, grey-green beneath	milky white in broad clusters	dull scarlet to deep brick red, persistent
C.linearifolius *C.microphyllus* f. *thymifolius*	Nepal	dwarf, evergreen shrub with prostrate branches	very small, narrow, glossy dark green	white, in small clusters	deep pink to red
C.microphyllus	Himalaya	low, spreading, evergreen shrub to 1 x 2m with short, stiff branches and neat foliage	0.5-1cm, oval, dark, glossy green	white, solitary or in small clusters	carmine red

Comments: See also *C.cashmiriensis* and *C.linearifolius*

Name	Distribution	Habit	Leaves	Flowers	Fruit
C.prostratus *C.rotundifolius*	Himalaya	evergreen, arching to prostrate shrub	1-1.5cm, broadly oval to rounded, dark green, glossy above, pale beneath	white, solitary	crimson to deep ruby red
C. 'Rothschildianus'	garden origin	evergreen shrub to 5m with arching branches	lanceolate to elliptic, bright to deep green	creamy white in broad clusters	creamy yellow to gold
C.salicifolius	China	evergreen shrub to 5m with a graceful, arching habit	oblong to lanceolate, tapering narrowly, dark green	white in clusters	bright red

Comments: Includes numerous cultivars and involved in hybrids with erect and stongly arching to low or prostrate habits, most with narrow, dark foliage and fruit in shades of orange, flame, scarlet and brick red

Name	Distribution	Habit	Leaves	Flowers	Fruit
C.simonsii	Himalaya, Nepal, Sikkim	erect, twiggy, deciduous or semi-evergreen shrub to 4m	1-3cm, ovate to broadly elliptic, dark, shiny green turning red-bronze in autumn	solitary or in small clusters, white tinted pink-red	orange-red to scarlet, persistent

Comments: An excellent hedging plant

Name	Distribution	Habit	Leaves	Flowers	Fruit
C.sternianus *C.franchetii* var. *sternianus; C.wardii* of gardens	Tibet, Burma	evergreen or semi-evergreen shrub to 3m, with spreading to stiffly arching branches	to 3.5cm, elliptic, sage green above, with white, downy undersides	white tinted red in small clusters	orange-red, abundant
C. × *watereri* varied group of hybrids between *C.frigidus*, *C.henryanus* , *C.rugosus* and *C.salicifolius*	garden origin	large shrubs or small trees, evergreen or semi-evergreen, often with arching branches	usually large, narrow and dark green	white in broad clusters	red, abundant in large, showy clusters

Comments: Covers many clones – low and near-prostrate to tall and arching with fruit in shades of scarlet, brick red, flame, orange, salmon, yellow-pink, yellow and cream; among them some of the most popular cotoneasters, including *C.* 'John Waterer', a strong-growing shrub to 5m with arching branches, large, dark green lanceolate leaves and abundant red fruit.

cottage garden a small, country garden, traditionally planted with flowers such as sweet williams (*Dianthus barbatus*), hollyhocks (*Alcea rosea*), pinks (*Dianthus plumarius* derivatives) and pansies (*Viola* species), all growing in a colourful jumble. The flowers are often interspersed with vegetables, and fences, walls and porches are clothed with honeysuckle (*Lonicera* species) and rambler roses (*Rosa* species).

Cotula (from Greek *kotyle*, a small cup). Compositae. Cosmopolitan. 80 species, hardy annual or perennial herbs, rhizomatous, with pinnately compound leaves and stalked, button-like flowerheads in spring and summer. The first species is hardy in zone 7 and will grow in full sun at the water's edge or in the bog garden. The second needs full sun, a fast-draining, gritty medium on the rock garden or in an alpine trough, and protection from wet, cold conditions. Propagate by division or by seed in spring.

C.coronipifolia (BRASS BUTTONS; South Africa; annual or perennial with rather fleshy, ascending to decumbent stems to 30cm long; leaves to 1.2cm, linear, entire, toothed or pinnately lobed; flower heads to 1cm in diameter, yellow or cream); *C.hispida* (S. Africa; perennial with dense, tufted stems to 4cm; leaves very finely cut, covered in silver hair; flowerheads small, golden on slender stalks).

Cotyledon (from Greek *kotyle*, a small cup or cavity, referring to the leaves of some species). Crassulaceae. Africa. 9 species, small, tender evergreen shrubs and subshrubs with large and thickly succulent leaves and nodding, bell-shaped to tubular flowers in dense panicles. Cultivate as for the tender *Crassula* species.

C.ladysmithiensis (South Africa; to 30cm; leaves to 6cm, oblong-obovate, rounded, toothed, thickly fleshy and covered in short, yellow-brown hairs; flowers red-brown); *C.orbiculata* (South Africa; to 1m; leaves to 16cm, obovate to linear, terete to flattened, grey-green often edged red; flowers orange to red); *C.undulata* (SILVER CROWN, SILVER RUFFLES; South Africa; to 1m; leaves to 15cm, broadly obovate and grey-green with a strongly wavy apical margin; flowers orange).

cotyledon the primary , storage, either solitary (in monocots), paired (in dicots) or whorled as in some conifers. It may remain within the seed coat or emerge and become green during germination.

couch grass (*Elymus repens*) TWITCH, SCUTCH a perennial, invasive grass-weed of gardens, which spreads rapidly by means of underground stems. Control by forking out. Glyphosate and glyphosate trimesium are effective as foliage-applied total weedkillers; apply when growth is active, especially in spring, at 10–15cm high.

courgette (*Cucurbita pepo*) ZUCCHINI the immature fruit of some cultivars of marrow (mainly hybrids), grown as a vegetable throughout the summer in temperate regions; the fruits are pale or dark green or yellow, sometimes striped. Sow 2.5cm deep in small pots under protection at 13°C/55°F minimum; transplant 4–5 weeks later after hardening off, and only when danger of frost is passed. Space 90cm apart each way. Seedlings can be advanced outdoors under cloches or film. In mild climates, courgettes

courgette

Pests
aphids
red spider mite
whiteflies

Diseases
virus
downy mildew
grey mould

may be sown direct outdoors. Plants produce separate male and female flowers and are insect-pollinated; under cool conditions, hand pollinate by removing male flowers and dabbing pollen into the female flowers, which can be recognized by the immature ovary below the perianth. Harvest fruits regularly, when 10cm long.

coursonne a small fruiting lateral of apples or pears.

cover cropping see *green manuring*

crab apple (*Malus* species) hardy, deciduous trees bearing pome fruits, found throughout temperate zones of the northern Hemisphere. *M.sylvestris* is native of Europe. The fruits of most species and cultivars are small and unacceptably sour for eating, though some selections are suitable for jam. *M.floribunda* and *M.* 'Aldenhamensis', *M.* 'Golden Hornet' and *M.* 'John Downie' are very effective as pollinators of apple. Crab apples make ornamental contributions to a garden, and are grown for blossom (white in *M.hupehensis*, *M.florentina* and *M.sargentii*, pink in *M.floribunda* and *M.spectabilis*, deep rose in *M.* × *purpurea*); for fragrance (violet-scented in *M.angustifolia* and *M.coronaria* 'Charlotte'); for decorative fruits, ranging from yellow to deep red/purple; and for autumn colour (e.g. *M.coronaria*, *M.glaucescens* and *M.kansuensis*). Prune, as for apples, in winter to remove congested and diseased growth. Propagate species by seed, cultivars by budding or grafting on to seedling rootstocks. For detail on ornamental species, see *Malus*.

Crambe (Greek name for cabbage). Cruciferae. C Europe to Asia, Africa. 20 species, annual or perennial herbs (those listed here are perennial), evergreen to deciduous, with a stout, woody rootstock and leaves that are tough, broadly elliptic to ovate, puckered and wavy. Appearing in late spring and summer, the flowers are small, fragrant, four-petalled, white and in large panicles. *C.cordifolia* is a magnificent perennial for the large herbaceous border. It prefers a fast-draining, alkaline soil in full sun. In areas with very hard or dull, wet winters, mulch the crowns once the foliage has died away. *C.maritima*, needs a sandy and stony, free-draining soil, again in full sun, and is at home in coastal regions. It is also cultivated for its edible stems, blanched and harvested in late winter and early spring. Both species are hardy in climate zone 5. Propagate by seed sown in spring or autumn in the cold frame or cold greenhouse; alternatively, increase by root cuttings or division in spring.

C.cordifolia (Caucasus; leaves to 30cm, dark glossy green, much puckered; panicle to 1.5m tall, pyramidal, with many fine, spreading branches and a mass of flower); *C.maritima* (SEA KALE; coastal N Europe, Baltic Sea and Black Sea; leaves to 30cm, fleshy, wavy and crisped to lobed, blue-grey to purple-tinted, glaucous; panicle to 50cm tall, dense and corymbose). See also *sea kale*.

cranberry a dwarf shrub producing small, dark red acid berries of culinary use. The AMERICAN CRANBERRY (*Vaccinium macrocarpon*) is a dwarf, evergreen shrub, native to eastern North America. It has been cultivated since the early 18th century but is seldom grown in gardens. The fruits are acid, rich in vitamin C, and are harvested for cooking in autumn, picked over several times. The native BRITISH CRANBERRY (*V.oxycoccus*) is not grown commercially; the related LINGEN or LINGBERRY (*V.vitis-idaea*) is grown on a small scale in Europe.

American cranberries require a very acid soil (pH4–5) and a controlled soil-water level in order to produce bog conditions and afford frost protection. Plant from cuttings in late spring, 15–45cm apart each way, according to local practice; fruits are produced in the second or third year. No pruning is necessary.

crane flies see *leatherjackets*.

Crassula (diminutive of Latin *crassus*, thick or swollen). Crassulaceae. Some cosmopolitan, most restricted to South Africa. 300 species, annual, biennial and perennial herbs and small shrubs (those listed here are perennials from South Africa) with succulent leaves and small, usually white, star-like flowers in clusters or panicles. All except *C.sarcocaulis* (which is hardy in zone 7) need a minimum winter temperature of 7°C/45°F and a dry atmosphere. Most need full sun, although some species with thinner-textured leaves prefer bright, indirect light. Grow in a free-draining, gritty, loam-based medium. Water frequently in summer, but allow the soil to become almost dry between times. From late autumn to mid-spring, water only to prevent shrivelling of leaves. Propagate by stem or leaf cuttings.

C.arborescens (SILVER JADE PLANT, SILVER DOLLAR, CHINESE JADE; 1–2m tall, shrubby and tree-like with stout branches; leaves 2–5cm, obovate to round, tip rounded, base tapering, blue-green and grey-bloomed with a red-brown tinted margin); *C.deceptor* (syn. *C.deceptrix*; to 15cm tall, branches covered in compressed, 4–ranked leaves forming a squat column, each leaf to 1cm long, ovate to wedge-shaped, thickly fleshy, grey and covered in dots); *C.lactea* (FLOWERING CRASSULA, TAILOR'S PATCH; to 20cm, stems horizontal to scrambling, thick; leaves 2–7cm, terete, oblanceolate, base tapering, apex pointed, dull, deep green; flowers cream tinted pink, produced in great abundance); *C.multicava* (to 40cm tall, shrubby; leaves to 5cm, oblong to ovate, mid- to grey-green; flowers cream tinted rose, massed in stalked panicles, sometimes giving rise to plantlets); *C.muscosa* (syn. *C.lycopodioides*; MOSS CYPRESS, PRINCESS PINE, TOY CYPRESS, WATCH-CHAIN CYPRESS; to 60cm, resembling a miniature cypress or clubmoss; stems slender, concealed by very small, olive green, scale-like leaves which overlap in four ranks); *C.ovata* (DOLLAR PLANT, JADE PLANT, JADE TREE; differs from *C.arborescens* in glossy mid- to dark green leaves sometimes edged with red-brown – one of the most popular *Crassula* species, and widely grown as a house plant); *C.perfoliata* var. *falcata* (syn. *C.falcata*; AIRPLANE PLANT, PROPELLER PLANT, SCARLET PAINT BRUSH, SICKLE PLANT; stout-stemmed and erect to 50cm; leaves thick, grey, broadly lanceolate, paired and twisted in opposite directions from their fused bases; flowers fragrant, bright red, in large clusters); *C.sarcocaulis* (shrubby, to 50cm tall with a broad, dense and twiggy crown; leaves 0.5–2cm, semi-terete, elliptic to lanceolate, pointed, red-tinted; flowers pink, malodorous, in massed clusters); *C.schmidtii* (tufted, mat-forming, to 15cm tall; leaves to 3cm, linear to lanceolate, green with dark spots, tinged red-brown, ciliate; flowers pink to magenta); *C.socialis* (cushion-forming, to 10cm; stems densely clothed in 4–ranked leaves, these small, triangular, thickly fleshy and mid-green; flowers white in long-stalked clusters).

+**Crataegomespilus** Rosaceae. Garden origin. Graft hybrids between *Crataegus* and *Mespilus*, differing from *Mespilus* in the smaller flowers clustered at the branch tips, and from *Crataegus* in the *Mespilus*-type fruit. Cultivate as for *Mespilus*. Increase by grafting. +*C.dardarii* (*Crataegus monogyna* + *Mespilus germanica*; BRONVAUX MEDLAR; deciduous tree to 6m with spiny branchlets; leaves to 15cm, narrowly oblong to elliptic, dark green above, woolly beneath; flowers to 1.5cm in diameter, white; fruit to 2cm in diameter; includes 'Jules d'Asnières', with flowers tinted rose).

Crataegus (from Greek *kratos*, strength, referring to the wood). Rosaceae. N Hemisphere. HAWTHORN. Some 200 species, hardy deciduous trees and shrubs, often thorny, with entire, toothed, lobed or pinnatisect leaves and corymbs of flowers in late spring, usually white and headily scented. The fruits are small, red to orange or brown pomes and often persist in decorative clusters late into autumn and winter. The following species are fully hardy and will grow on all but the wettest soils. They prefer full sun, but will tolerate light shade. Several make excellent hedges and windbreaks, and most cope with harsh winds, coastal exposures and urban pollution. Sow seed of species in autumn; increase cultivars by budding.

C.crus-galli (COCKSPUR THORN; E US; wide-spreading, flat-topped tree to 10m with vicious, ultimately branching thorns; leaves 2.5–10cm, obovate, toothed, dark glossy green becoming orange-red in autumn; flowers with pink anthers; fruit to 1cm, subglobose, deep red); *C.ellwangeriana* (E US; tree to 6m tall with sparse spines; leaves to 8cm, ovate to oval, shallowly lobed, serrate, dark green and rugose above, hairy beneath; flowers with pink anthers; fruit 1.5cm, oblong, red); *C. flava* (SUMMER HAW, YELLOW-FRUITED HAW; N America; tree to 6m with spines to 2.5cm; leaves to 5cm, obovate to oblong, apex with 3, biserrate lobes, tough, pale green; flowers with purple anthers; fruit to 1cm, rounded to pyriform, green-yellow, edible); *C. laciniata* (syn. *C.orientalis*; SE Europe, W Asia; tree to 7m tall with sparse thorns and downy shoots; leaves to 5cm, triangular to rhombic with 5-9, deep, toothed lobes, deep green

Crassula perfoliata var. *falcata* (above)

C. deceptor (below)

with ashy grey down; flowers large with maroon anthers; fruit to 1.5cm, globose, downy, orange-red); *C.laevigata* (syn. *C.oxyacantha*; ENGLISH HAWTHORN, QUICK-SET THORN, WHITE THORN, MAY; thorny shrub or small tree, 2.5–6m tall; leaves to 5cm, obovate with 3-5, shallow and blunt, serrate lobes; flowers white to pink or crimson with pink-purple anthers; fruit to 1cm, ovoid to globose, deep red; there are many cultivars including forms with semi- and fully double flowers, pure white to deep crimson, variegated leaves and dark blood red to yellow fruit); *C. × lavallei* (*C. stipulacea × C. crus-galli*; the most popular cultivar is 'Carrierei', a spreading tree to 7m tall with elliptic to oblong and irregularly toothed leaves that turn from dark green to deep red in autumn; the flowers have pink anthers and the long-lasting fruit are orange-red, ellipsoid and to 1.8cm); *C.macrosperma* (N America; thorny tree to 8m tall; leaves ovate to elliptic with five, broadly triangular and sharply toothed lobes; flowers with red anthers; fruit to 1.5cm, obovoid to oblong, red; includes 'Acutiloba', with broad and jaggedly serrate leaves); *C.mollis* (RED HAW; C US; spreading tree to 13m with thorny, white-hairy shoots; leaves to 10cm, broadly ovate, with 4–7 pairs of shallow, glandular-toothed lobes, hairy becoming rugose; flowers with yellow anthers; fruit to 2.5cm, globose-pyriform, red, downy); *C.monogyna* (COMMON HAWTHORN, ENGLISH HAWTHORN, MAY; W Europe; thorny shrub or tree to 5m tall; leaves to 6cm, broadly ovate to rhombic with 3–7 lobes, glossy dark green; fruit to 1cm, ovoid to ellipsoid, red; widely grown as a hedge; there are many cultivars including forms with tightly compact to low and spreading or narrowly erect habits, pendulous to twisted branches, variegated leaves, pink flowers and yellow fruits; one of the most celebrated is 'Biflora', the GLASTONBURY THORN, which produces flowers and a few leaves in winter as well as spring); *C.pedicellata* (SCARLET HAW; E US; slender-branched, thorny tree to 7m; leaves 5–10cm, broadly ovate with 4–5 pairs of shallow, biserrate lobes, dark green and rugose, turning fiery red in autumn; flowers with pink-red anthers; fruit to 1cm, pyriform, bright red); *C.phaenopyrum* (WASHINGTON THORN; SE US; broad-crowned tree to 10m with slender branchlets and long, sometimes branched thorns; leaves to 7cm, triangular to broadly ovate, 3–5-lobed toward base, toothed, bright glossy green turning orange-red in autumn; flowers with pink anthers; fruit to 0.5cm, subglobose, bright red, lasting well); *C. × prunifolia* (*C.crus-galli × C.succulenta* var. *macracantha*; tree to 6m with a dense, spreading crown and large, stout thorns; leaves to 8cm, broadly ovate to oval, coarsely serrate, dark green turning rich purple-red in autumn; flowers with pink anthers; fruit to 1.5cm, globose, bright red); *C.tanacetifolia* (TANSY-LEAVED THORN; W Asia; erect shrub or tree to 10m; branchlets felty at first and often thornless; leaves 2.5–6cm, oval to rhombic with 5–7, narrow, pinnately cut and toothed lobes, hairy at first; flowers with red anthers; fruit to 2.5cm, globose, yellow-red, reminiscent of apples in scent and flavour).

crateriform shaped like a bowl or goblet, in the form of a concave hemisphere slightly contracted at the base.

crazy paving paths or terraces made of broken paving slabs fitted closely together in a random arrangement.

creeper strictly, a plant that spreads over the soil surface, rooting as it grows, for example, *Vinca* species. The term is often loosely applied to climbers and other wall plants, for example, *Vitis* species.

cremocarp a dry, seed-like fruit consisting of two one-seeded carpels with an epigynous calyx. The carpels separate into mericarps on ripening.

crenate scalloped, with shallow, rounded teeth. Where the teeth themselves have crenate teeth, the term used is bicrenate.

crenulate minutely crenate.

Crepis (from Greek *krepis*, name of a plant in Theophrastus). Compositae. N Hemisphere. HAWK'S BEARD. 200 species, annual or perennial herbs with rosettes of entire to pinnatifid leaves and long-stalked, dandelion-like flowerheads in summer and autumn. Fully hardy. Grow in full sun or part shade on a well-drained soil. Increase perennials by root cuttings or division, annuals and biennials by seed in autumn. All will self-sow.

C.aurea (Alps; glabrous perennial to 30cm; flowerheads orange to deep yellow, tinted maroon beneath); *C.incana* (PINK DANDELION; Greece; grey-hairy perennial to 15cm; flowerheads bright pink to magenta); *C.rubra* (more or less hairy annual or short-lived perennial to 60cm; flowerheads usually rose, sometimes red or white).

cress a group name for several low-growing salad herbs, including AMERICAN LAND CRESS (*Barbarea verna*), ITALIAN CRESS (*Eruca sativa*) and GARDEN CRESS (*Lepidium sativum*), all members of the Cruciferae. Garden cress is the most commonly grown, the young seedlings being cut and used as a salad garnish. Sow seed in succession thickly on to a water-soaked, fibrous mat lining a shallow dish, or in trays or pots; maintain a temperature of 10°C/50°F and harvest after 10–14 days. For salad greens, all three may be sown outdoors in early spring in rows 30cm apart at 2–3 week intervals. Indoor cress is often grown with milder flavoured MUSTARD (*Sinapis alba*), raised similarly.

crest an irregular or dentate elevation or ridge; generally applied to an outgrowth of the funiculus in seeds but also found on the summit of some organs and on the lip of some orchids.

crickets (Saltatoria: Gryllidae and other families) insects related to grasshoppers with long thin antennae and two pairs of wings folded horizontally over a broad flat abdomen. Adult males produce a characteristic chirping sound by rubbing forewings against hindwings. Most lay eggs in the soil, but some bush crickets place them on plants. The eggs hatch into worm-like larvae; these moult almost immediately to form the first nymphal stage, which resembles the adult, and gradually increase in size with subsequent moults. They are mostly omnivorous, and are common in hot dry places. Plants are attacked occasionally, but only in frames and heated greenhouses in temperate countries.

The HOUSE CRICKET (*Acheta domesticus*) may eat the tops of young seedlings and attack older plants at ground level. Control by eliminating daytime hiding places. Alternatively, apply contact insecticide or employ baits; pitfall traps may also be used. The MOLE CRICKET (*Gryllotalpa gryllotalpa*) is robust, up to 45mm long; it has expanded forelegs and is armed with teeth modified for digging. It is rare in Britain, but common in southern Europe, and has been introduced into North America. It burrows in soil to eat the roots and stems of many plants, especially potatoes, carrots, tomatoes and sweetcorn. Control by applying an appropriate soil insecticide.

BUSH CRICKETS (Tettigoniidae), also known as LONG-HORNED GRASSHOPPERS, mostly frequent trees and shrubs, and feed mainly on small insects, although some species may occasionally eat plants. Females can be recognized by their prominent sword-shaped ovipositor.

Crinodendron (from Greek *krinon*, a kind of lily, and *dendron*, tree). Elaeocarpaceae. Chile. 2 species, hardy evergreen shrubs or small trees with leathery leaves and solitary, pendulous flowers in late spring. Hardy to –10°C/14°F, especially if given the shelter of a wall or surrounding trees and shrubs; a cool root-run is, however, important. Grow in partial shade or dappled sun on a moist but well-drained, fertile, humus-rich and acid soil. Propagate by softwood or semi-ripe cuttings in a closed case with gentle bottom heat.

C.hookerianum (syn. *Tricuspidaria hookeriana*; CHILE LANTERN TREE; tree to 9m, or shrub to 4m;

Crinodendron patagua (above)

C. hookerianum (below)

flowers scarlet to carmine, to 3cm, lantern- or urn-like, fleshy); *C.patagua* (tree to 14m, or shrub to 4m; flowers white, to 2.5cm, resembling a fringed bell).

Crinum (Greek *krinon*, lily). Amaryllidaceae. Tropics and subtropics. 130 species, perennial bulbous herbs, evergreen or deciduous, the bulb often with an elongated, stem-like neck. The leaves are strap- or sword-shaped. Large, funnel-shaped and 6-lobed, the flowers are often fragrant and arise in scapose umbels in summer and autumn. More or less hardy, the South African species and hybrids prefer a rich but well-drained soil in full sun, that is moist in spring and summer, but rather dry in winter; *C. × powellii*, for example, will thrive at the foot of a south-facing wall in zone 7. *C.americanum* requires virtually frost-free conditions and a permanently moist, acid soil. It will tolerate part shade. Tropical Asiatic species need light shade, medium to high humidity and a minimum temperature of 13°C/55°F. Grow these in large pots or beds of moisture-retentive, humus-rich soil. All species resent disturbance. Propagate by offsets in spring, or from ripe seed sown in spring at 20°C/70°F.

C.americanum (FLORIDA SWAMP LILY, SOUTHERN SWAMP CRINUM; S US; scape to 75cm; flowers to 10cm, creamy white); *C.asiaticum* (POISON BULB; tropical Asia; scape to 60cm; flowers to 15cm, white, fragrant); *C.bulbispermum* (syn. *C.longifolium*; S Africa; scape to 90cm; flowers to 20cm, green-white tinted or streaked pink to red); *C.macowanii* (PYJAMA LILY; South Africa; scape to 90cm; flowers to 20cm, pink or white striped deep pink along lobes); *C.moorei* (South Africa; scape to 90cm; flowers to 12cm, fragrant, pale to deep pink or white); *C. × powellii* (*C.bulbispermum × C.moorei*; garden origin; leaves to 1m; scape to 120cm; flowers to 18cm, fragrant, pale to candy pink to soft rose or white).

crispate curled and twisted extremely irregularly, used either of a leaf blade or, more often, of hairs.

crispy-hairy of hairs; wavy and curved, in dense short ringlets.

cristate crested or, less commonly, crest-like. The term is sometimes used to denote the presence of crests (cristae) (e.g. *Coelogyne cristata*), but more often to describe a perceived abnormality of growth (e.g. closely sculpted brain- or coral-like ridges, as found in *Celosia*, or the ruffled and usually forked tips of fern fronds).

crocks fragments of broken clay pots placed in the base of a growing container to prevent loss of the growing medium through the drainage hole, and to create an area of free drainage and air circulation at the base of the rootball. Crocking is beneficial for soil-based and soilless composts and essential for the finer bark and leafmould mixes used for certain orchids, ferns and aroids. Also known as shards.

Crocosmia (from Greek *krokos*, saffron, and *osme*, smell, referring to the smell of saffron given off by the dried flowers when immersed in water). Iridaceae. South Africa. MONTBRETIA. 7 species, deciduous perennial herbs, cormous with narrow, sword-shaped and ribbed leaves. The flowers are red, orange or yellow, tubular and curving and produced in summer on slender-stalked, erect to angled racemes or sparsely branched panicles. Hardy in zone 6. Grow in full sun on a fertile, well-drained soil. In areas with very hard winters, plant in a warm and sheltered site, or lift the corms and store them over winter in dry and frost-free conditions. Increase by division of corms in spring.

C.aurea (to 1m tall; flowers pale to burnt orange, straight-sided, to 5cm long, opening to broad lobes); *C.* × *crocosmiiflora* (*C.aurea* × *C.pottsii*; a garden hybrid differing from *C.aurea* in the curved and slightly inflated upper part of the perianth tube and the narrower lobes); *C.masoniorum* (to 1m; flowers vermilion, to 5cm, narrowed to base, widening above to outspread lobes; includes 'Firebird', with fiery orange-red flowers tinted green in the throat); *C.paniculata* (to 1m; flowers to 6cm, deep orange, curving downwards, somewhat inflated with lobes spreading); *C.pottsii* (to 90cm; flowers to 3cm, widening sharply from a narrow base, orange flushed red, lobes to 1 × 0.5cm). *Cultivars:* 'Bressingham Blaze', with widely funnel-shaped, fiery orange-red flowers; 'Citronella', with yellow flowers with red-brown markings at centre; 'Emberglow', with glowing orange-brown flowers, arching upwards;'Jackanapes', with small flowers, bicoloured yellow and deep orange-red; 'Lucifer', with large, flame flowers; 'Solfatarre', with apricot-yellow flowers and smoky bronze leaves; and 'Star of the East', with large, soft apricot-yellow flowers with a paler throat and darker lobes.

Crocus (Greek *krokos*, saffron, an ancient plant name that shares its derivation with the Hebrew *karkom* and Arabic *kurkum*). Iridaceae. Europe, N Africa, Middle East, Central Asia. Some 80 species, small, cormous perennial herbs. The leaves are grassy, often with a pale green or silver-white central stripe. The leaf bases dilate to form a cap-like corm tunic which is papery or composed of parallel or reticulate fibres. Produced in autumn, late winter and spring, the flowers are funnel- to goblet-shaped, narrowing below into stem-like tube and expanding above into 6 lobes. The style is divided at the tip into 3 or more branches.

Most are hardy to –20°C/ –4°F, if the ground does not freeze for long periods.They are suitable for a range of garden situations, with the most vigorous species, such as *C.speciosus* and *C.vernus*, used to great effect in the wild garden, naturalized beneath deciduous trees and shrubs, or in grass where the first cut does not occur until the crocus foliage has died back. On light, freely draining soils where turf is sufficiently fine, *C.tommasinianus*, *C.flavus*, *C.nudiflorus*, *C.serotinus*, *C.sieberi* and the more robust cultivars of *C.chrysanthus* can be used to similar effect. These species also suit beds and borders, and will tolerate low groundcover that will mask the fading foliage. Cultivars of *C.vernus* are useful for indoor decoration in spring, gently forced in pots plunged beneath 15cm of sand for 6–8 weeks in the cold frame and then brought indoors to a temperature of 10–12°C/50–54°F. The smaller species are suited to the rock garden, raised bed, or for troughs and tubs, planted in a gritty loam-based mix. More delicate species, and those whose perfection may be marred by rain and wind in winter, can be grown in the alpine house.

Although most will thrive in a sunny position, sheltered from wind, on a moderately fertile well-drained soil, in cultural terms, crocuses may be split into three groups – Members of the first group need moisture to complete their growth cycle, but undergo a baking during dormancy in the long hot summer. They need a sheltered, sunny situation and a fast-draining soil. Although hardy, in some regions, these species are generally more safely grown with cloche protection, or in the bulb frame or alpine house, where they can be protected from excessive rainfall in summer. They include *C aerius*, *C. biflorus*, *C. cartwrightianus*, *C. dalmaticus*, *C. minimus*, *C. niveus*, *C. olivieri*, *C. tournefortii*.

The second group is more tolerant of damp and can be grown in the open garden, given a warm, sunny position and good drainage. They include *C.ancyrensis*, *C.angustifolius*, *C. chrysanthus*, *C. etruscus*, *C. flavus*, *C. gargaricus*, *C. imperati*, *C. malyi*, *C. medius*, *C. pulchellus*, *C. sativus*, *C. serotinus*, and, if given the perfect drainage of a raised bed, *C. cancellatus*, *C. goulimyi*, *C. laevigatus* and *C. longiflorus*. The third group includes plants from climates that experience intermittent rainfall throughout the year, with a peak in autumn and winter. This group does not share the requirement for a warm dry dormancy and needs at least some moisture to prevent the dormant corms from shrivelling. They also prefer some shade from the hottest sun in summer. They include *C. banaticus*, *C. cvijicii*, *C. nudiflorus*, *C. tommasinianus* and *C.vernus*, and, given a sheltered situation in well-drained but retentive soil, *C. kotschyanus* and *C. speciosus*.

Plant corms 5–8cm deep. Propagate by removal of cormlets, by division of established colonies or by seed sown when ripe in a sandy propagating mix. Plants raised from seed will bloom in their third or fourth year. A number of species, notably *C.imperati* and *C.tommasinianus*, will self-seed freely where conditions suit. Rodents and pheasants may dig up and eat the corms; squirrels may eat the emerging shoots. Planting under fine mesh may reduce losses. Birds may damage the flowers. Tulip-bulb aphids sometimes infest corms in storage and persist to form colonies which damage young shoots. *Crocus* may be affected by blue mould, gladiolus dry rot, hard rot, scab, hyacinth black slime, narcissus basal rot, tulip grey bulb rot and the rust fungus, *Uromyces croci*.

(see table overleaf)

Crocus (a) *C. korolkowii*, (b) *C. chrysanthus*, (c) *C. sieberi*, (d) *C. nevadensis*, (e) *C. biflorus*, (f) *C.vernus*

CROCUS

Name	Distribution	Flowers
C.aerius *C.biliottii*	N Turkey	tube to 9cm, deep blue, often flushed, royal blue from base of lobes, overlaid with darker feather veins; style trifid, vermilion to scarlet

Comments: Mid–late spring. For *C.aerius* of gardens, see *C.biflorus* ssp. *pulchricolor*

Name	Distribution	Flowers
C.ancyrensis	Turkey	tube yellow or mauve; throat yellow, glabrous, lobes 1–3cm, vivid yellow to pale orange; style trifid, deep orange

Comments: Spring. 'Golden Bunch': up to 7 flowers per corm

Name	Distribution	Flowers
C.angustifolius *C.susianus* CLOTH OF GOLD CROCUS	SW Russia	throat yellow, glabrous or minutely hairy, lobes yellow, exterior flushed or veined maroon; style trifid, deep yellow to vermilion

Comments: Spring

Name	Distribution	Flowers
C.banaticus *C.byzantinus* ; *C.iridiflorus*	N Romania, former Yugoslavia, SW Russia	lilac-mauve, inner lobes to 3cm, erect, outer lobes to 5cm, spreading or deflexed, darker; style slender, multifid, feathery, violet

Comments: Autumn

Name	Distribution	Flowers
C.biflorus	S Europe, Asia Minor	throat white to golden yellow, glabrous, lobes to 3cm, white, lilac or pale blue with dark mauve veins on exterior; style trifid, deep yellow

Comments: Spring and autumn. ssp. **alexandri** : white, exterior suffused deep violet . ssp. **melantherus**: flecked white and feathered grey-purple on exterior; throat pale yellow, anthers damson purple to black. ssp. **pulchricolor**: syn. *C.aerius* of gardens, flowers indigo throughout, colour most intense at base of lobes, throat bright yellow; this subspecies crossed with *C.chrysanthus* has given rise to the cultivars 'Blue Pearl', 'Blue Bird' and 'Advance'. ssp. **weldenii**: white flushed pale lilac at base or, rarely, throughout undersides of lobes, throat white or pale blue.

Name	Distribution	Flowers
C.boryi *C.ionicus* , *C.cretensis*	W & S Greece, Crete	tube 5–15cm, ivory, exterior sometimes veined or flushed mauve, lobes to 5cm, often smaller, obovate; style multifid, burnt orange

Comments: Autumn

Name	Distribution	Flowers
C.cancellatus	former Yugoslavia, Greece, Turkey, W Syria, Lebanon, N Israel	white to opal or lilac with purple veins, throat yellow, sometimes hairy, lobes to 5.5cm; style multifid, deep orange

Comments: Autumn. ssp. **mazziaricus** : flowers white to lilac sometimes stained yellow at base, usually veined violet and more cupped.

Name	Distribution	Flowers
C.candidus *C.kirkii*	NW Turkey	tube to 7cm, white, tube stained purple-maroon, throat yellow, glabrous, outer lobes flushed and spotted violet-grey or blue; style 6-branched, deep yellow

Comments: Late winter, early spring

Name	Distribution	Flowers
C.carpetanus *C.lusitanicus*	C & NW Spain, N Portugal	tube 6–11cm, white flushing to lilac usually with fine mauve veins, throat white with yellow tint; style ivory or lilac, trifid, flattened and ruffled at tips

Comments: Spring

Name	Distribution	Flowers
C.cartwrightianus *C.sativus* var. *cartwrightianus* WILD SAFFRON	Greece	fragrant, throat hairy, lobes 1.5-3cm, white, lilac or mauve with darker veins and a stronger purple flush at base, albino forms also occur; style trifid, vermilion, branches club-shaped

Comments: Autumn–early winter

Name	Distribution	Flowers
C.chrysanthus *C.annulatus* var. *chrysanthus* ; *C.croceus*	Balkans, Turkey	tube 4.5cm, occasionally striped or stained bronze-maroon; style trifid, deep yellow

Comments: Late winter: large and rounded, pure white, exterior grey-blue. 'Blue Pearl': pearly blue, base bronze, interior silver-blue with a rich orange stigmata. 'Blue Peter': soft blue inside with a golden throat, exterior rich purple-blue. 'Brassband': apricot yellow outside with a bronze green-veined blotch, inside straw-yellow with a tawny glow. 'Cream Beauty': rounded, on short stalks, creamy yellow. 'E.A. Bowles': rounded, old gold, throat dark bronze. 'Elegance': violet outside, edges paler, inner tepals with a deep blue blotch, inside pale violet with a bronze centre. 'Gipsy Girl': large, golden yellow, outer tepals striped and feathered purple-brown. 'Ladykiller': outer tepals rich blue, narrowly edge white, inner tepals white, blotched with inky blue, silver-white inside. 'Snow Bunting': pure white with golden throat, exterior feathered purple. 'Snow White': pointed, frost-white with a gold base. 'Zenith': glossy violet-blue, inside silver-blue, contrasting with gold anth. and prominent gold throat. 'Zwanenburg Bronze': fls golden yellow, exterior dark bronze.

Name	Distribution	Flowers
C.cvijicii	S former Yugoslavia, N Greece, E Albania	solitary, tube 3.5–7.5cm, dellicate yellow-golden, rarely white, tube to 3.5cm, white or pale yellow, sometimes tinted purple, throat white, sometimes stained yellow, hairy; style trifid, yellow-orange, branches widened at tips

Comments: Late spring

Name	Distribution	Flowers
C.cyprius	Cyprus	1–2, tube, 4.5–9cm, scented, white-lilac, exterior flushed mauve, darkest at base of lobes and on tube, throat yellow; style trifid, orange, tips enlarged

Comments: Late winter

Name	Distribution	Flowers
C.dalmaticus	SW former Yugoslavia, N Albania	differs from *C.sieberi* in its single flowers, their exteriors pearly grey-mauve to biscuit and faintly striped purple

Comments: Late winter–early spring

CROCUS

Name	Distribution	Flowers
C.etruscus	N Italy	delicate lilac, tube white, 4.5–7cm, throat hairy, sulphur yellow, lobes 3–4cm, the outer whorl ivory, pearly or tawny, feathered with purple below
Comments: Late winter–spring. 'Zwanenburg': flowers pale blue, but appearing a stronger, almost pure blue if planted *en masse*		
C.flavus *C.luteus* ;*C.aureus*	S former Yugoslavia, C & N Greece, Bulgaria, Romania, NW Turkey	tube 7–18cm, scented, light golden yellow to apricot-yellow, tube exterior and base of lobes sometimes striped or stained brown; style trifid, yellow
Comments: Spring		
C.gargaricus	NW Turkey	tube 5.6–10.5cm, yellow-orange, lobes yellow, style deep yellow, trifid and widening, fimbriate at tips
Comments: Spring		
C.goulimyi	S Greece	8–20cm, throat hairy, white, lobes 1.5–4cm, rounded, lilac-pale mauve, inner whorl paler than outer; style trifid, white deepening to orange
Comments: Autumn. 'Albus': pure white		
C.hadriaticus *C.peloponnesiacus*	W & S Greece	tube 3–9cm, white, yellow, maroon or mauve, throat hairy, yellow, sometimes white, lobes 2–4.5cm, elliptic, white suffused yellow, buff or lavender at base, occasionally tinted lilac throughout; style trifid, orange
Comments: Autumn		
C.imperati *C.neapolitanus*	W Italy, Capri	tube 3–10cm, white, occasionally expanding yellow or mauve, throat deep yellow, lobes 3–4.5cm, bright purple within, exterior tawny with to 5, feathered violet stripes; style orange to coral red, trifid
Comments: Late winter–early spring. 'de Jager': lilac inside, with buff and purple stripes on the exterior. ssp. *suaveolens*: flowers smaller, each outer lobe marked with 3 purple stripes.		
C.korolkowii	Afghanistan, N Pakistan, Russia	scented, tube 3–10cm, golden, sometimes marked bronze or maroon, throat yellow to metallic buff, lobes to 3cm, elliptic, golden yellow, outer lobes marked dark brown or maroon below; style trifid, orange, tips widened, papillose
Comments: Late winter–early spring		
C.kotschyanus *C.zonatus*	C & S Turkey, NW Syria, C & N Lebanon	scented, tube to 13cm, white, throat hairy, white tinted yellow with 2 golden splashes at base of each lobe, sometimes fusing, lobes 3–4.5cm, pale lavender with darker, parallel veins running almost to apex; style trifid, ivory to pale yellow, tips expanded and subdivided
Comments: Autumn. var. *leucopharynx*: blue-lavender, veined blue, with central white zone, yellow markings absent; style pale cream. ssp. *suworowianus*: unscented, throat glabrous with pale yellow markings, lobes held nearly erect, cream with mauve veins, tone suffusing where veins converge		
C.laevigatus *C.fontenayi*	Greece, Crete	fragrant , tube 2–8cm, white tinted yellow or mauve toward throat, throat yellow, lobes 1–3cm, obovate to elliptic, white or pale rose-lilac above, outer lobes white, mauve, bronze or yellow, with up to 3 broad dark purple or maroon feathered lines below, markings sometimes absent or exterior wholly purple; style multifid, orange
Comments: Autumn–early spring		
C.longiflorus *C.odorus*	SW Italy, Sicily, Malta	scented, tube 5–10cm, pale yellow, striped violet; throat yellow, glabrous or thinly hairy lobes 2–4.5cm, lilac, exterior with darker veins, sometimes shaded bronze; style trifid, vermilion, tips widened, crenate, sometimes subdivided
Comments: Autumn		
C.malyi	W former Yugoslavia	tube 4–9cm, white, occasionally flushed yellow, bronze or mauve, throat hairy, yellow, lobes 2–4cm, white, often with grey-blue or bronze stain at base; style trifid, orange, tips widening, notched
Comments: Spring		
C.medius	NW Italy, SE France	tube 8–20cm, white to mauve at apex, throat white veined violet, lobes 2.5–5cm, obovate, pale blue-mauve, veined darker at base; style multifid, vermilion, branches notched and recurving
Comments: Autumn		
C.minimus *C.insularis*	Sardinia, Corsica, Isles Sanguinaires	tube 4-11cm, white flushed mauve at apex, throat white sometimes tinted pale yellow; lobes 2-3cm, oblanceolate, light rosy mauve to deep purple, the undersides of the outer tepals striped, veined or shaded darker purple to brown on a bronze background; style trifid, orange, tips, flattened, crenate-lobed
Comments: Late winter–early spring		
C.niveus	S Greece	tube 9–18cm, ivory to dull yellow, throat golden, hairy, lobes 3–6cm, obovate, white, sometimes stained lilac; style trifid, scarlet, tips flattened,

Crocus (a) *C. speciosus*, (b) *C. tournefortii*, (c) *C. sativus*, (d) *C.banaticus*, (e) *C.pulchellus*, (f) *C. niveus*

CROCUS

Name	Distribution	Flowers
C.nudiflorus C.multifidus; C.aphyllus; C.fimbriatus;C.pyrenaeus	SW France, N & E Spain	tube 10–22cm, white flushed mauve toward apex, throat white-mauve, lobes 3–6cm, pale mauve to amethyst; style multifid, orange
Comments: Autumn		
C.ochroleucus		tube 5–8cm, white, throat hairy, pale yellow-golden, spreading to base of lobes, lobes 2–3cm, elliptic, ivory; style trifid, golden yellow
Comments: Autumn–winter		
C.olivieri	Balkans, Turkey	tube 5–7cm, yellow or maroon, throat yellow; lobes 1.5–3.5cm, yellow-orange, or striped bronze-maroon below and on tube; style divided × 6, pale orange
Comments: Late winter–spring. ssp. **balansae** : flowers striped or stained brown bronze or maroon below.		
C.pulchellus		differs from C.speciosus in its golden yellow throat, gracefully incurving, opal-blue lobes with lilac veins
Comments: Autumn–early winter. Turkey. 'Zephyr': flowers very large, pale, pearly blue.		
C.sativus C.officinalis SAFFRON	widespread through cultivation	ancient sterile cultigen derived from C.cartwrightianus, from which it differs chiefly in perianth lobes to 5cm and style branches to 3cm
C.serotinus	Portugal , Spain, Gibraltar, N Africa	scented, tube 2–5cm, white to mauve; throat hairy, white or ivory; lobes 2.5–4cm, obtuse, pale mauve to lilac blue, sometimes veined purple; style multifid, burnt orange
Comments: Autumn to winter. More often seen is ssp. **salzmannii** with lilac-blue flowers to 10cm long, these sometimes have yellow throats.		
C.sieberi	Greece, Crete	scented, tube 2.5–5cm, white or mauve, usually deep yellow toward throat; throat glabrous, golden yellow or orange; lobes 2–3cm, obtuse, white within, tinted purple, exterior of outer whorl striped, barred or suffused mauve; style trifid, yellow-orange, branches short, the tips divided or ruffled
Comments: Spring–early summer.'Bowles White': pure white with an orange throat; stigma scarlet. 'Firefly': outer tepals nearly white, inner violet, base yellow. 'Hubert Edelsten': deep purple, blotched white at tip. 'Violet Queen': rounded, violet-blue. ssp. **atticus**: pale lilac to violet with an orange style. ssp. **nivalis** : lilac, throat yellow. ssp. **sublimis** : throat yellow, tepals blue-mauve, darker at tips, sometimes white at base. ssp. **sublimis** f.**tricolor** : flowers with 3 distinct bands of colour: lilac, pure white and golden yellow.		
C.speciosus	Russia, Iran, Turkey	scented, tube 5–20cm, white or pale mauve; throat glabrous, off-white, rarely, tinted yellow, lobes 3–6cm, mauve-blue above, exterior veined or spattered purple, sometimes on a paler, opal ground; style multifid, orange
Comments: Autumn. 'Aitchisonii': large, pale lilac with feather venation; 'Albus': white; 'Artabir': pale lilac with strong venation; 'Cassiope': large, pale violet with yellow throat; 'Globosus': lavender-blue, appearing somewhat inflated; 'Pollux': large, pale mauve with pearly exterior; 'Oxonian': indigo; 'Trotter': large, white.		
C.tommasinianus	Balkans, Hungary, Bulgaria	tube 3.5–10cm, white, throat white, thinly hairy, lobes 2–4.5cm, pale mauve to violet, occasionally white or rose, darker at apex, sometimes with silver or bronze hue below; style trifid, orange, flattened and fimbriate at tips
Comments: Late winter. 'Barr's Purple': outer lobes tinged grey, inner lobes rich purple-lilac. 'Pictus': outer lobes marked mauve or purple at apex. 'Roseus': pink-purple. 'Ruby Giant': probably a hybrid with C.vernus; dark red-purple, especially in upper part and base of lobes. 'Whitewell Purple': exterior purple-mauve, interior silver-mauve.		
C.tournefortii C.orphanidis	S Greece, Crete, Cyclades	sometimes scented, tube 3–10cm, white, rarely suffused purple, throat glabrous or downy, yellow to ivory, lobes 1.5–3.5cm, pale lilac, often veined darker at base of lobes, rarely white throughout; style multifid, orange-crimson
Comments: Autumn–Winter		
C.vernus DUTCH CROCUS	Italy, Austria, E Europe	tube 5–15cm, mauve or white, throat white, often suffused mauve, lobes 3–5.5cm, white, violet or blue-mauve, sometimes striped white or violet, usually with a darker, v-shaped marking near tips; style trifid, yellow-orange, widened and fimbriate at tips
Comments: Spring–early summer. 'Early Perfection': violet-purple to blue with dark edges. 'Enchantress': light amethyst with silvery floss, dark base. 'Pickwick': white with deep lilac stripes, base deep purple. 'Remembrance': large, rounded, silver-purple, base flushed purple. 'Vanguard': light blue, exterior grey; blooms more than 2 weeks in advance of other cultivars. 'Glory of Limmen': very large, rounded, white with short purple stripes at base. 'Haarlem Gem': small, silvery-lilac externally, pale-lilac-mauve within. 'Jeanne d'Arc': large, white, with 3 thin purple stripes, tinged violet at base. 'Kathleen Parlow': large, white, base and tube purple. 'Little Dorrit': very large, rounded, pale silver-lilac. 'Paulus Potter': large, glossy red-purple. 'Purpureus Grandiflorus': large, deep glossy rich purple. 'Queen of the Blues': fls large, lavender-blue. 'Striped Beauty': pale silver-grey, striped deep mauve, base violet-purple. ssp. **albiflorus**: usually white, sometimes marked purple		
C.versicolor	S France to W Italy, Morocco	tube 6–10cm, white or purple-striped, throat lemon yellow or ivory, lobes 2.5–3.5cm, white-mauve, exterior usually with darker stripes, sometimes with bronze-yellow hue; style trifid, pale orange, tips widened, indented
Comments: Late winter–spring.'Picturatus': flowers white on the exterior, feathered violet		

crome, cromb a dragfork. See *fork*.

crop rotation see *rotation*.

cross a naturally occurring or induced hybrid between two plants.

cross-fertilization fertilization resulting from cross-pollination; cf. *self-fertilization*.

cross-pollination the transfer of pollen from one flower to another flower on a different plant; often used to imply cross-fertilization; cf. *self-pollination*.

Crossandra (from Greek *krossoi*, fringe, and *aner*, male, referring to the fringed stamens). Acanthaceae. Africa, Madagascar, Indian subcontinent. 50 species, tender evergreen shrubs and subshrubs with 5-lobed, narrowly tubular flowers in terminal spikes clothed in green, overlapping bracts. Cultivate as for *Aphelandra*.
   C.infundibuliformis (FIRECRACKER FLOWER; S India, Sri Lanka; to 1m tall; flowers to 3.5cm across, bright orange to salmon pink with fan-shaped lobes, in spikes to 10cm long); *C.nilotica* (tropical Africa; to 50cm; flowers to 2.5cm across, brick-red to orange or apricot with obovate lobes, in spikes to 7cm long).

Crotalaria (from Greek *krotalon*, rattle: the seeds rattle in the dried pods). Leguminosae. Widespread in tropics and subtropics. RATTLEBOX. 600 species, herbs or shrubs with simple to trifoliolate leaves and pea-like flowers. The following species will survive mild frosts if thickly mulched in winter. Grow in full sun on a free-draining, fertile soil. Cut back stems after flowering. Increase by semi-ripe cuttings in summer, or by seed in spring. *C.agatiflora* (CANARY BIRD BUSH; E Africa; shrub to 6m; leaves trifoliolate, grey-green; flowers to 7cm across, pale lemon yellow tinted olive green, in erect terminal racemes to 40cm long).

crotch the junction of primary branches with the centre stem or trunk of a tree.

crown (1) a corona; (2) a collective term for the main stem divisions and foliage of a tree or shrub and the branching pattern and overall habit that they assume (i.e. domed, spreading, narrowly conical); (3) the basal portions of a herbaceous plant, usually where root or rhizome and aerial stems or resting buds meet; (4) a length of rhizome with a strong terminal bud, used for propagation as, for example, with *Convallaria*; (5) the head of a single-stemmed tree-like plant or shrub bearing a distinct apical whorl, rosette or flush of foliage; (6) the leaves and terminal buds of a low-growing plant when arranged in a fashion resembling that of the larger plants mentioned under (5), for example, many ferns.

crown bud a flower bud at a shoot tip, surrounded by other usually smaller buds. It is of special importance in the training of carnations and chrysanthemums.

crown gall (*Agrobacterium tumefaciens*) a widespread bacterial disease. It is an important problem in woody plant nurseries, especially in warm climates, but is not of major importance in temperate gardens. Tumour-like galls are usually found at ground level or on roots, and occasionally on stems well above ground, as soft, white, rounded swellings with uneven surfaces that become brown with age; they vary in size from that of a pea to a huge ball. On woody plants, the galls become hard and persist; on herbaceous plants, they are soft and disintegrate.

Crown gall affects monocot and dicot woody and herbaceous plants. Common hosts among fruits and nuts are apple, particularly the rootstocks M9 and M2, almond, apricot, blackberry, cherry, fig, grape, gooseberry, loganberry and other *Rubus* fruit, nectarine, pecan, plum, peach, pear, raspberry and walnut. Other hosts include *Begonia*, chrysanthemums, *Dahlia*, *Daphne*, elm, *Euonymus*, *Gladiolus*, hollyhock, honeysuckle, hawthorn, *Libocedrus*, juniper, lupin, marigold, marrow, *Phlox*, poplar, privet, *Rhododendron*, rhubarb, rose, runner bean, swede, sweetpea, tomato, wallflower and willow. Several strains affect only certain host plants.

Crown gall is soil-borne and capable of living saprophytically in soil for long periods. A wound parasite, it can only enter injured roots and stems. On entry, it stimulates the cells of the host plant to grow and multiply abnormally. Where infection takes place during rapid growth, galls develop within days or weeks, depending on the host plant. Where a perennial plant is infected late in the growing season, the bacterium may not produce galls until the following spring. Secondary galls remote from the original gall are sometimes formed as chains along roots or on stems. Aerial galls are most frequently found on roses, *Daphne* species and raspberry and blackberry canes. The bacterium is more likely to persist and spread on wet soils.

Preventive measures include the provision of good drainage; the inspection of new stocks to avoid infecting the soil; a long rotation plan of non-susceptible crops on affected sites; and avoidance of injury to roots. Where shrubs and fruit show aerial galls, the affected parts should be cut out.

crown lifting the removal of the lower branches of, usually, ornamental trees, carried out mainly to admit light or prevent obstruction of passage or a view.

crown rot used of an infection at soil level girdling and killing the host.

crownshaft a characteristic of some palms in which the overlapping leaf bases form a tight cylindrical or vase-shaped shaft distinct from the, usually, bare stem below.

crown thinning the selective removal of tree branches to reduce crowding and admit more light and air into the crown.

cruciate, cruciform cross-shaped.

crumb a small rounded, porous aggregate of soil particles found in top soil where the level of organic matter is high. Also generally called a ped.

Cryptanthus (from Greek *kryptos*, hidden, and *anthos*, flower). Bromeliaceae. Brazil. EARTH STAR. 20 species, tender, evergreen perennial herbs, stemless with narrowly triangular to lanceolate, tough and finely saw-edged leaves in a low-lying rosette. Small, white flowers appear at the rosette's centre. Grow in small containers in a coir and bark mix with added sand and polystyrene granules; provide a minimum temperature of 13°C/55°F and bright, indirect light or dappled sun. Water freely in summer, avoiding foliage; keep barely moist in winter. Propagate by removal of rooted offsets after flowering. Small and colourful, these bromeliads make resilient house plants and are commonly used in mixed container plantings, bottle gardens and Wardian cases.

C.acaulis (GREEN EARTHSTAR, STARFISH PLANT; leaves to 10cm, undulate, in a small, low, flat rosette, typically grey-green, but cream-striped to red-flushed or bronze in some cultivars and hybrids); *C.bivittatus* (leaves to 25cm, in a dense, low, spreading rosette, dark green with 2 broad, longitudinal, white or pink bands above; red-flushed and banded red in var. *atropurpureus*); *C.bromelioides* (RAINBOW STAR; stoloniferous with stalked rosettes; leaves 10–20cm, lanceolate, base narrowed, olive green or variously striped white, cream, pale green and rose); *C.zonatus* (ZEBRA PLANT; leaves to 20cm in a spreading rosette, undulate, dark olive to chocolate, irregularly banded silver-grey to cream or white; chocolate banded silver in the more robust and erect 'Zebrinus').

Cryptanthus bivittatus

× **Cryptbergia** Bromeliaceae. Garden origin. Hybrids between *Cryptanthus* and *Billbergia* with linear-lanceolate, sharply pointed leaves to 30cm long in freely suckering, narrow, funnel-shaped rosettes; the small, white flowers are rarely produced. Cultivate as for *Billbergia*. × *C.meadii* (*Cryptanthus beuckeri* × *Billbergia nutans*; leaves mottled pink); × *C.rubra* (*Cryptanthus bahianus* × *Billbergia nutans*; leaves dark bronze-red to brilliant blood red).

cryptogams flowerless and seedless plants that nonetheless increase by means of sexual fusion, producing spores. (i.e. neither angiosperms nor gymnosperms, but pteridophytes, bryophytes.)

Cryptogramma (from Greek *kryptos*, hidden, and *gramme*, a line, referring to the lines of sori concealed before maturity). Cryptogrammataceae. Widespread in temperate and subtropical regions. ROCK BRAKE, PARSLEY FERN. 4 species, dwarf, deciduous ferns with dense tufts of deeply and much-divided fronds that are said to resemble parsley. The following species are fully hardy and suitable for the rock garden, scree, wall crevices and small pans in the alpine house. Grow in an acid, gritty mix in full sun. Keep moist until established. Increase by division.

C.acrostichoides (AMERICAN ROCK-BRAKE; N America; fronds to 12cm, stipes green, blades bright green, 2–3-pinnate with crowded, ovate to oblong, crenate pinnules); *C.crispa* (EUROPEAN or MOUNTAIN PARSLEY FERN; fronds to 20cm, stipes straw yellow to pale green, blades bright pale green, 2-pinnate with cuneate to elliptic, dentate pinnules, thick-textured).

Cryptomeria (from Greek *kryptos*, hidden, and *meris*, a part, referring to the hidden floral parts). Cupressaceae. Japan, China. JAPANESE CEDAR, SUGI. 1 species, *C.japonica*, a hardy, evergreen, coniferous tree to 60m tall with a conical crown and a thick bole covered in fibrous, peeling, red-brown bark. The leaves are soft, awl-shaped and some 1–1.5cm long, turning from sea green to deep, glossy green, sometimes purple-bronze or rust in cold weather. They clothe crowded, drooping branchlets on tiered boughs. Many cultivars are available including dwarf, rounded, conical, pendulous and contorted types and forms with cristate, filamentous, persistently juvenile, variegated, purple-flushed and sea green growth. *C.japonica* is hardy in climate zone 6. Plant in full sun with shelter from harsh, drying winds. A deep, moisture-retentive and acid to neutral soil is preferred, as is a mild and humid climate. Propagate as for *Chamaecyparis*.

cryptophyte a plant whose growth buds are located below the soil surface; cf. *chamaephyte*.

Cryptotaenia (from Greek, *cryptos*, hidden, and *tainia*, a fillet or band). Umbelliferae. Temperate Regions. 4 species, perennial herbs with short clumped stems bearing lobed leaves, and forked sprays of small flowers in summer. Hardy in zone 6, *C.* 'Atropurpurea' promises to become one of the most desirable, purple-leaved foliage plants. It prefers a moist, rich soil and will grow in full sun or part shade. Increase by division in early spring. *C.japonica* (E. Asia; to 50cm; leaves to 12cm diam., smooth, divided into 3 broad, shallowly toothed lobes; flowers pink; 'Atropurpurea': foliage maroon to chocolate brown).

Ctenanthe (from Greek *kteis*, comb and *anthos*, flower, referring to the arrangement of the bracts). Marantaceae. Costa Rica, Brazil. 15 species, tender, evergreen perennial herbs, clump-forming with long-stalked, oblong to lanceolate leaves, often beautifully marked and coloured, and small, white and purple flowers in racemose cymes. Cultivate as for *Calathea*.

C.lubbersiana (Brazil; to 40cm; leaves to 18cm, deep green with yellow-green variegation above, pale green beneath); *C.oppenheimiana* (Brazil; to 1m; leaves to 40cm, green with silvery-grey feathering above, wine-red to purple beneath; includes 'Tricolor', NEVER-NEVER PLANT, with leaves blotched cream).

cubiform dice-shaped, cubic.

cuckoo spit the frothy mass surrounding the larvae of *froghoppers* (q.v.) or spittlebugs.

cucullate furnished with or shaped like a hood.

cucumber *cucumber* (*Cucumis sativus*) a frost-susceptible annual, native of Sino-Himalaya, with a vigorous trailing habit; it is grown as a salad vegetable for the immature, edible fruits. Grown in the open in tropical and subtropical areas; in temperate areas selections giving the best quality and yield require protection for all of the life cycle.

In frost-prone zones, sow singly in pots under protection at 20°C/70°F, reducing to 15°C/60°F on germination. Transplant into well-manured beds, growing bags or 22–25cm pots, spaced 60cm apart; maintain a temperature of 20°C/70°F, with high humidity, and support with canes. Train the main shoot vertically and stop at 1.3m. All-female cultivars bear fruits on the main stem, so remove laterals; other cultivars require horizontal wire supports, and laterals are stopped two leaves beyond a female flower.

RIDGE CUCUMBERS are hardier and can be transplanted outdoors in early summer. Space 60cm apart each way on ridged beds and stop at 6–7 leaves; where trained up trellis or netting, space 45cm apart and stop when at top of support.

culm the stems of Gramineae.

cultigen a plant found only in cultivation or in the wild having escaped from cultivation. Cultigens include many hybrids and cultivars, some exotic ornamentals whose wild origins have become lost or were never recorded, and several ancient and important crops that have undergone development and distribution at man's hands for so long a time that, although they may have 'wild' distributions, details of their original forms and native places are lost or confused.

cultivar (abbreviation cv.) a 'cultivated variety'; a distinctive plant or plants arising and/or maintained in cultivation. This can happen either as a result of hybridisation, or through some type of mutation, or within the natural variation of a species. When reproduced (whether by sexual or asexual means according to the particular type of plant and cultivar), such plants

retain their distinguishing characteristics. A cultivar is named with a cultivar (or fancy) epithet, a word or words in a vernacular language (unless published prior to 1959, or a botanical (Latin) epithet already established for a taxon, like a variety or forma that has since been judged to be a cultivar). The epithet is printed in roman characters, takes a capital first letter and is either enclosed in single quotation marks or prefixed by the abbreviation cv. Many of the most popular garden plants are cultivars. These include plants selected for their foliage (often named 'Argentea', 'Aurea', 'Dissecta', 'Variegata' and so forth), and for their flowers ('Alba', 'Florepleno', etc). Examples of cultivars with more recent, vernacular names include *Mahonia* 'Charity' and Rose 'Peace.

cultivator an implement for cultivating soil. The term is applied in particular to powered machines with rotating blades or tines, but is also used for various hand tools, especially those with recurved prongs, and also sometimes for those with blades fixed behind a wheel.

cultrate, cultriform knife-shaped, resembling the blade of a knife.

cultural controls the suppression of plant pests and diseases through good husbandry methods. These include careful management of the soil; the production of healthy well-grown plants; the use of resistant cultivars, for example root-aphid resistant lettuces; general hygiene, including careful disposal of infected debris; the use of healthy seed and planting stock; frequent inspection and rogueing; rigorous elimination of weeds; crop rotation which is especially useful in deterring eelworms; cultivation of the soil to expose soil pests such as leatherjackets and wireworms to predation by birds; and manipulation of sowing or planting dates of vegetable crops to avoid times when adult pests (e.g. the carrot fly) are egg-laying. cf. *pesticides*. See *biological controls*.

Cuminum (from the Greek name for this plant, *kyminon*). Umbelliferae. Mediterranean to C Asia. 2 species, annual herbs with fine, biternately divided foliage and small white to pink flowers in compound umbels. *C.cyminum* is grown for its aromatic seeds, harvested when dry and ground to make the spice cumin, which is so important in the cuisine of the Middle East and Indian subcontinent. Cultivate as for *dill* (q.v.).

cuneate, cuneiform inversely triangular, wedge-shaped.

Cunninghamia (for James Cunningham, who collected *C.lanceolata* in China about 1701). Taxodiaceae. China, Taiwan. 2 species, hardy, coniferous evergreen trees to 50m tall. The crown is conical, the bark red-brown, fibrous and deeply fissured and the branches whorled. Narrowly ovate to lanceo-

---

cucumber

Pests
aphids
red spider mite
whiteflies

Diseases
virus
downy mildew
grey mould
gummosis
wilt

late, sharp-pointed, glossy and pliable, the leaves are arranged in two ranks along drooping, 2-ranked branches. Hardy to –25°C/–13°F, although shelter from harsh winds and a mild, humid climate are preferred. Plant in a rich, moist and acid to neutral soil in sun or part shade. Propagate as for *Chamaecyparis*. *C.lanceolata* (syn. *C.sinensis*; CHINA FIR; C China; leaves 3–7cm long, lanceolate, lustrous bright green above with 2 broad, white bands beneath).

cup the corona of *Narcissus*, where it is shorter than the perianth segments. A large cup is more than one third of the length of the perianth segments, a small cup less than one third their length. See *trumpet*.

Cuphea (from Greek *kyphos*, curved, referring to the curved capsule). Lythraceae. SE US to Brazil. 260 species, small, bushy, tender annual or short-lived, evergreen perennial herbs and subshrubs with neat foliage and brightly coloured tubular flowers in summer. The following species need a minimum temperature of 7°C/45°F, full sun and a fertile, free-draining medium that is moist in spring and summer and only barely so in winter. *C.ignea* is often grown as a half-hardy annual, raised from cuttings taken in late summer and overwintered under glass or as house plants. Pinch new plants to promote bushiness and set out in a warm, sunny position in late spring. Increase all by greenwood cuttings in spring or summer.

C.cyanea (Mexico; glandular-hairy subshrub to 2m; flowers to 2.5cm, orange-red with yellow-green apex bearing purple-black petals); *C.hyssopifolia* (FALSE HEATHER, ELFIN HERB; Mexico, Guatemala; downy subshrub to 60cm; flowers to 1cm, silky, pale lilac veined lavender, or pink, or white); *C.ignea* (FIRECRACKER PLANT, CIGAR FLOWER; Mexico, Jamaica; more or less glabrous shrublet to 60cm; flowers to 3cm, scarlet, apex black edged white; includes 'Variegata', with leaves flecked lime and gold).

× **Cupressocyparis** Cupressaceae. Garden hybrids between *Cupressus* and *Chamaecyparis*, fast-growing, coniferous evergeen trees with dense, pinnate or plumose sprays of growth covered in small, overlapping, scale-like leaves. The following hybrid is fully hardy and will tolerate all but very wet or dry soils. It responds well to pruning and shearing and is widely grown as a hedge and windbreak. It does, however, grow very fast – many towering, back-yard avenues were originally intended to be orderly, head-high hedges. Increase by cuttings. *C.* × *leylandii* (*Cupressus macrocarpa* × *Chamaecyparis nootkatensis*; LEYLAND CYPRESS a vigorous, fast-growing tree to 35m or more with a dense, columnar to pyramidal crown, smooth to fissured green-brown bark and pendent branchlets; the leaves are dark green tinged grey and carried in flat sprays; cultivars include: 'Castlewellan Gold', slower-growing with plumose sprays of yellow-green leaves, tinted bronze in winter; 'Harlequin', with grey-green leaves in plumose sprays broken with patches and flashes of white to ivory; the fast-growing, dense and narrowly columnar 'Leighton Green'; and the compact and conical 'Robinson's Gold', with lime green to golden foliage with bronze tints in winter and early spring).

cupressoid with *Cupressus*-like foliage, i.e. narrow branchlets clothed with scale-like leaves, for example *Hebe armstrongii*, *Tamarix*.

Cupressus (from Greek *kuparissos*, a word taken from an unknown near Eastern language, allied to the Hebrew *gopher*, the name of *C.sempervirens*). Cupressaceae. N and C America, N Africa to China. CYPRESS. 20 species, evergreen, coniferous trees with peeling or flaking bark, sprays of aromatic growth covered in overlapping, scale-like leaves, and spherical cones composed of resinous, peltate scales. Most species are from mountains, some of these in or around desert. All are good for low rainfall areas on hot sites except *C.torulosa*, which prefers cooler, wetter areas. *C.arizonica* var. *glabra* thrives on any soil, from chalk to highly acid sand; *C.macrocarpa* and *C.sempervirens* are less happy on chalk but thrive on limestones; others prefer soils at pH6 or below. All but *C.arizonica* var. *glabra* have been killed by –25°C/–13°F, and *C.lusitanica* by –20°C/–4°F. Winds of –10°C/14°F will scorch most species. *C.torulosa* 'Cashmeriana' is not hardy and should be grown in the cool greenhouse or conservatory in zones 8 and under. For propagation and pests, see *Chamaecyparis*.

C.arizonica (ARIZONA, CYPRESS, SMOOTH CYPRESS; Arizona to Texas and N Mexico; to 25m, crown broadly ovoid-conic; shoots dense, branching angular, leaves very finely serrate; includes var. *glabra* (syn. *C.glabra*) with smooth, purple to red bark exfoliating in papery layers and bright blue-green leaves with white resin glands – dwarf, globose, fastigiate, blue-, silver- and gold-leaved cultivars are available, one of the best is 'Blue Ice', narrow and erect with cool, blue-grey foliage); *C.lusitanica* (MEXICAN CYPRESS, PORTUGUESE CYPRESS, CEDAR OF GOA; to 45m, crown conical, bark brown, furrowed, exfoliating in strips; leaves aromatic, dark to blue-green, spirally arranged; includes cultivars with whip-like and pendulous branchlets and grey and variegated foliage); *C.macrocarpa* (MONTEREY CYPRESS; California; to 45m, bark purple-brown or grey, thick, ridged; crown columnar to conical becoming flat-topped with age with tiered branches; leaves slightly fleshy, bright green to yellow-green with paler margins; includes dwarf, globose, narrowly columnar, prostrate and golden cultivars); *C.sempervirens* (ITALIAN CYPRESS, MEDITERRANEAN CYPRESS; Crete, Rhodes, Turkey to Iran, naturalized elsewhere in Mediterranean; to 40m, bark grey-brown, shallowly ridged, scaly; crown narrowly conical with dense more or less ascending branchlets; leaves dark green, aromatic;

Cuphea ignea

nodal cutting

dwarf, open and tiered and very tightly columnar cultivars are grown, as are golden and steely blue-grey forms); *C.torulosa* (HIMALAYAN CYPRESS; Himalaya; to 40m, bark brown, furrowed, curled, peeling in long ribbons; crown conical; branches horizontal, branchlets whip-like, more or less pendulous, young shoots maroon; leaves pale yellow-green, incurved; includes cultivars with congested foliage, leaves turning brown in winter, and open, bushy habit; 'Cashmeriana': syn. *C.cashmeriana*; KASHMIR CYPRESS; Bhutan; to 40m tall, crown broadly pyramidal with an open, tiered habit, rusty red bark exfoliating in strips, and long, weeping feathery branchlets of a vivid blue-grey).

cupule a cup-shaped involucre of hardened, coherent bracts, subtending a fruit or group of fruits, as in the acorn. See *cyathium*.

Curcuma (from Arabic *kurkum*, yellow: *C.longa* is turmeric). Zingiberaceae. Tropical Asia. 40 species, tender perennial herbs, evergreen or deciduous with reed-like stems clothed with lanceolate to oblong leaves; these arise from a thick and aromatic rhizome and clumped, swollen roots. The flowers are small and thin-textured with a conspicuous, tubular lip; they protrude from the colourful, overlapping bracts of a scapose or terminal, cone-like inflorescence. Cultivate as for *Brachychilum*.

C.petiolata (QUEEN LILY; Malaysia; to 60cm tall; inflorescence to 15cm, terminal, bracts violet to magenta, flowers yellow and white); *C.roscoeana* (HIDDEN LILY; Malaysia; to 90cm tall; inflorescence to 20cm, terminal, bracts orange-red, flowers yellow).

curd the tight, unexpanded flowering portion, or head, of cauliflowers and broccoli.

currants a collective name for BLACKCURRANT (*Ribes nigrum*), REDCURRANT (*Ribes rubrum* Redcurrant Group), WHITECURRANT (*Ribes rubrum* Whitecurrant Group), and PINKCURRANT (*Ribes rubrum* Pinkcurrant Group).

cushion the enlarged area of tissue above, beneath or to either side of an insertion or at the axis of two branches or a stem and a branch; also known as a pulvinus.

cushion plant a plant with a low, dense and rounded habit; for example many high alpines.

cuspidate of apices, terminating abruptly in a sharp, inflexible point, or cusp.

cuticle the outermost, multi-layered skin of waxy tissues covering the epidermis and containing cutin, fatty acids and cellulose.

cutin a polymer of fatty acids forming a major part of the cuticle.

cuttings the most widely used technique of vegetative propagation by which identical characteristics of plants are replicated. It involves selecting and detaching portions of root, stem, leaf or buds and inserting them in a growing medium, within a suitable environment, in order to induce the production of roots and eventually an independent plant. The ease of regenerating from cuttings varies widely and is influenced by genetic potential, stock plant conditioning through nutrition, hard pruning, manipulation of temperature and light, and time of taking.

Stem cuttings Successful rooting by this method depends on stage of shoot maturity, and stem cuttings may be considered as follows:

SOFTWOOD CUTTINGS are generally taken from spring into summer, selecting from the first flush of new growth and continuing until it becomes lignified, but cuttings may be available all the year round from greenhouse plants such a *Fittonia*, *Pilea*, chrysanthemums and *Fuchsia*. Cuttings should be prepared with a sharp knife and the cut made usually below a node. Nodal *or* internodal tip cuttings are perfectly satisfactory for easy-rooting plants, including shrubby subjects such as *Buddleja davidii*, *Forsythia × intermedia*, *Fuchsia* species and *Spiraea × bumalda*; leaf-bud stem sections are suitable for climbers such as *Lonicera* and *Vitis* species.

For plants that are more difficult to root, softwood heel cuttings may be obtained by using a light downward pressure to carefully pull the shoot from the parent plant so that it breaks away from the main stem with a basal plate or 'heel' of tissue attached. Plants that benefit from this treatment include cultivars of *Cotinus coggygria*, *Acer palmatum*, deciduous azaleas and plants that produce bulky leafy growth or long vigorous internodal shoots, for example, cultivars of *Actinidia chinensis*, *Sambucus nigra*, *Jasminum officinalis* and climbing *Lonicera*.

The rooting medium should be sterile and open, as can be achieved with peat or coir and sand mixes. Whether mist, fog, a glass frame or a plastic film cover is used as the propagating system, the aim must be to minimize water and energy loss from the time of taking cuttings through to rooting. This can be achieved by ensuring adequate humidity and temperature around the tops of cuttings; a good air/water ratio at the rooting zone; the appropriate pH for the rooting media (a pH of 7 and over may encourage hard callus tissue, thus restricting root emergence); uniform basal heat, which may be between 15°C/60°F and 25°C/77°F; and sufficient light. Rooting time varies with species and rooting environment. Root initials will develop on some chrysanthemums after four days, with root emergence taking between 7–10 days; this contrasts with deciduous azaleas, where emergence of significant roots may take up to 6 weeks.

GREENWOOD CUTTINGS are a transitional condition of stem cutting, between softwoods and semi-ripe or semi-hardwoods. A 'firming up' of shoots or a change in stem and foliage colouration can be taken as an indication that shoots are of a suitable maturity, for example, stems of some cultivars of hybrid *Clematis* will change from a light green to a reddish-purple colour, and the leaves will become deep green. A second category of greenwood cuttings may be found, for example, in *Acer palmatum* cultivars, where extension growth is produced in usually three distinct flushes, each punctuated by a short period of bud dormancy lasting 2–3 weeks. Observation of these growth patterns is important for judging correctly the appropriate time for taking cuttings.

The speed of root development from greenwood cuttings is slower than from softwoods, and problems sometimes occur in overwintering late-struck cuttings. The method is suitable where the softwood stage has been missed and it may be more effective with some plants, for example, *Magnolia* × *soulangeana* and its cultivars.

SEMI-RIPE CUTTINGS As the season's growth slows down and there is development of terminal and axillary buds, stems become firmer and may be described as being in a semi-ripe or semi-hardwood condition. This stage is important for the propagation of broad-leaved evergreens and conifers.

Semi-ripe cuttings can be used for easy-to-root deciduous plants, such as *Forsythia*, *Weigela* and *Berberis thunbergii* cultivars. Judge the timing of taking so as to achieve rooting well before the onset of winter, and if cold frames are used, winter covering should be provided. For the propagation of evergreens, systems under protection with basal heat are normally used, although cold frames or low plastic tunnels are suitable and desirable for some plants, including *Berberis* × *stenophylla*, *Lavandula spica* cultivars, *Salvia officinalis* cultivars and *Santolina chamaecyparissus*. Such plants are best propagated from late August to early October and root initiation should take place before December. *Ceanothus*, *Cistus*, *Elaeagnus* and *Pyracantha* will generally root best in August to October, followed by *Camellia*, *Ilex* and *Mahonia* in October to November. *Osmanthus*, *Phillyrea* and *Ulex* often root best after a cold spell, bringing the programme to December and possibly into February.

The cutting making procedure is similar to that described for softwood cuttings, but an effective variation is the mallet cutting which is particularly useful for woody plants with hollow or pithy stems, and for those producing very short sideshoots, for example some spiraeas. Mallet cuttings are made by cutting a detached stem of the previous year's growth into sections each containing a vigorous side shoot, with a portion of the mother stem, about 2.5cm in length, retained as the base of the new cutting. Thick mallets are best slit lengthwise.

CONIFER CUTTINGS Most conifers are propagated by semi-ripe cuttings, although some species, including *Picea glauca* var. *albertiana* 'Conica' and the deciduous *Metasequoia glyptostroboides*, can be increased from softwood and greenwood cuttings. Selected forms of *Abies*, *Cedrus*, *Larix*, *Picea* and *Pinus* are difficult to root from stem cuttings and are normally raised by grafting.

Semi-ripe propagation of evergreen conifers can be undertaken from August right through to March, commencing with *Cupressus* in August/September and *Chamaecyparis* and *Thuja* from September through to November. *Juniperus* and *Taxus* forms root best after a period of low temperature and propagation of these may be delayed until January. Selection of cutting material is important: three-dimensional terminal shoots should be selected that are neither weak nor taken from sublateral branches low down. Conifers with a normal upright habit can be made to produce horizontal growth where rooted from horizontal terminal cuttings, for example, *Sequoia sempervirens* 'Adpressa'.

Suitable cutting length varies according to the vigour of the plant. Cuttings of *Chamaecyparis pisifera* 'Boulevard' may be no longer than 5cm, contrasting with those from *C.lawsoniana* 'Stewartii' at 170mm.

Effective propagation facilities range from enclosed environments with basal heat, such as intermittent mist, fog or plastic-covered frames, to basic cold frames out-of-doors.

WINTER HARDWOOD CUTTINGS This technique is generally reserved for deciduous woody plants, with cuttings taken during the dormant season when material is fully mature and may be cut into lengths using secateurs, a guillotine or a power-driven bandsaw. In most cases, it is a cheap reproduction method requiring modest facilities. Plants normally considered include *Clematis montana*, *Cornus alba*, *Deutzia scabra*, *Forsythia* × *intermedia*, *Kerria japonica*, *Ligustrum ovalifolium*, *Lonicera fragrantissima*, *Philadelphus* species, *Physocarpus opulifolius*, *Platanus* × *hispanica*, *Populus* species, *Ribes* species, *Salix* species, *Spiraea thunbergii*, *Stephanandra incisa*, *Symphoricarpos orbiculatus*, *Tamarix* species, *Viburnum* × *bodnantense* and *Weigela* species. To provide sufficient suitable cutting material in quantity, the establishment of stock plants as hedges or stools is appropriate, established preferably from an approved Clonal Selection Scheme to ensure trueness to type, health and vigour.

Taking hardwood cuttings at leaf fall will improve rooting, although with easy-rooting plants, such as willows and poplars, they may be taken any time during the dormant period. Hardwood cuttings may vary from a short stem 3.5cm long with one bud at the apex (a vine eye cutting), to a branched stem 2.5m long (e.g. willow 'sets' planted direct into the final position). For the majority of species, lengths of 15–20cm are suitable. Cut the base horizontally and the top slanting away from the apex bud; it is not necessary to cut just below a node except with hollow- or pithy-stemmed plants such as *Forsythia* and *Sambucus*.

Some cuttings need to be modified to encourage a 'leg' (e.g. gooseberries and red currants) or to discourage basal suckering (e.g. *Rosa rugosa* and *R.multiflora* selections for rootstocks); this is achieved by cutting out the lower buds on the stem, retaining just two or three at the top.

To improve rooting performance, hardwood cuttings may be subjected to basal stem wounding, dipping in rooting hormones, and pre-callusing tech-

hardwood cutting

niques. Basal heat treatment between 18°C/65°F and 20°/70°F for no more than three months will speed up rooting in some subjects, such as *Malus* clonal rootstocks, *Tilia*, *Platanus*, *Ailanthus* and *Laburnum*.

The open ground is suitable for easily rooted hardwood cuttings and those pre-treated with basal heat. Plant vertically in well-defined rows, with the apical bud just showing above the surface. Cold frames are useful for rooting hardwood cuttings in exposed localities, with the bed prepared by blending equal parts of granulated pine bark and crushed washed grit; slow-release fertilizer may also be added. Greenhouse protection is a convenient additional facility but there may be difficulties in preventing shoot growth advancing too quickly before root development.

Rooting enhancement techniques wounding the base of stem cuttings can enhance rooting, especially in semi-ripe and hardwood types.

SLICE WOUNDING is a speedy and effective method, where a sliver of wood 2–5cm long is removed at the base of the cutting, just exposing the cambium. With conifers, stripping the lower foliage by hand so that a rough scar is left exposing the cambium can be effective.

SPLIT WOUNDING hardwood cuttings with a pair of secateurs encourages well-distributed root formation. Cut through the centre of the base of the cutting to leave a split 3–5cm long.

SLIT OR INCISION WOUNDING is suitable for cuttings prone to basal rots or those that are too thin or soft to make an effective slice wound. Normally, two or three slits are made by carefully drawing the point of a knife along the base of the stem for 2–5cm to penetrate the cambium layer. Roots appear from new cambium formed along the slits.

SCRAPING the base of the stem for 2–5cm with a knife blade is sometimes advocated for cuttings that are thin-stemmed or have a thin epidermal layer, for example, *Ceanothus impressus*.

SYNTHETIC HORMONES Dip treatments increase the chances of successful rooting, promoting quicker root initiation and general improvement in quality and quantity of roots. The substances are often synthetic relatives of natural auxins. Most used are IBA (4-indol-3-yl-butyric acid), which has stable qualities and wide effectiveness. NAA (2-(1-napthyl)acetic acid) is often blended with IBA, and certain plants, such as large-flowered hybrid rhododendrons, are considered more responsive to it. Fungicides are often incorporated, as is boron, which seems to further aid root development. The most appropriate concentration of powder preparations for softwood and easy-rooting greenwood cuttings is 0.1% IBA; for other greenwood and easy-rooting semi-ripe cuttings, 0.3% IBA is suitable; for difficult semi-ripe and hardwood cuttings, 0.8% IBA is recommended. Powders are applied to the base of the cutting to a depth of 1cm. After use, the product should be re-sealed and stored in a refrigerator at between 3°C/37°F and 5°C/40°F; contamination from plant debris, water, long exposure to air, light and

root cutting

warmth will accelerate the breakdown of the active ingredient. Liquid preparations are more effective than powders, but less widely available.

The concentrations used for a quick dip are 500–1000ppm for softwood cuttings;1000–2500ppm for greenwood, semi-ripe and easy-rooting hardwood cuttings; and 2500ppm–5000ppm for difficult semi-ripe and hardwood cuttings. The use of correct concentrations for dipping is important; too much rooting hormone will cause distortion, leaf drop and basal rots. An alternative, safer option is a long-soak method from two to 24 hours, using low concentrations ranging from 20ppm for softwood cuttings to up to 250ppm for more lignified material. Recommendations come with the product, which may be supplied in tablet form for dissolving.

Root cuttings Among the limited range of plants that have the capacity to be increased by this means are those that naturally produce adventitious shoots from the root zone, for example, *Aesculus parviflora*, *Acanthus mollis*, *Ailanthus altissima* and *Aralia elata*.

Most root cuttings need to be taken during late winter; healthy stock plants between three- and five-years-old provide the most suitable material. The roots should be washed free of soil and one-year-old roots of uniform size severed from the stock plant as close to the crown as possible.

Vigorous hardy plants producing long fleshy roots can be planted vertically, direct in the open ground. Cuttings should be 10–15cm long with a minimum diameter of 5mm, and regeneration may take up to 16 weeks. Where cold frame or greenhouse protection is available, cuttings 5–6cm long may be used and the regeneration period reduced to 8–10 weeks; if basal heat is available at 18°C/65°F, the cutting size can be reduced to 2.5–3cm long, and the regeneration period may be as short as 4–6 weeks.

Plants with roots that regenerate easily, for example *Geranium sanguineum* and *Primula denticulata*, can be placed horizontally in the planting situation and covered; difficult ones, such as *Koelreuteria paniculata* and *Romneya coulteri*, respond more favourably when planted vertically. Open ground beds are suitable for root cuttings of vigorously rooting plants that are capable of tolerating adverse weather, such as *Ailanthus altissima*, *Cochlearia armoracia* and *Rhus typhina*.

Cold frames provide a suitable protected environment for root cuttings taken from alpines (e.g. *Anchusa*, *Erodium*, *Geranium*, *Primula denticulata*); herbaceous perennials (e.g. *Acanthus*, *Eryngium*, *Limonium*, *Papaver*, *Phlox*, *Verbascum*); subshrubs (e.g. *Romneya*); shrubs (e.g. *Aesculus parviflora*, *Aralia chinensis*, *Chaenomeles speciosa*, *Clerodendrum trichotomum*, *Rhus glabra*, *R.typhina*, *Rubus biflorus*, *R.thibetanus*) and trees (e.g. *Ailanthus altissima*, *Catalpa bignonioides*, *C.bungei*, *C.ovata*, *Paulownia tomentosa*, *Populus alba*, *P.tremula*). The frame can consist of a raised bed covered with plastic film supported by hoops, or one constructed with fixed surrounds and covered with glazed lights. A base dressing of a general fertilizer should be applied and the pH maintained at between 5.5–6.5. The prepared cuttings should be planted vertically in rows, with the tops just below the soil surface, or spaced horizontally in a shallow drill 50mm deep and covered with the frame soil and a 12mm layer of 5mm washed lime-free crushed grit.

Alternatively, root cuttings may be planted in trays or pots, and placed in the cold frame. A rooting mixture of equal parts peat, 3mm washed lime-free crushed grit and sterilized loam is suitable, with the container filled to within 12mm of the rim. Cuttings planted vertically should be placed so that the top ends are flush with the surface of the mixture, while horizontally planted cuttings should be spaced over the surface. To help aeration, the container is topped off with a covering, such as 5mm washed lime-free crushed grit.

Greenhouse facilities extend the attributes of the cold frame, especially where basal heat is available. In addition to the plants listed under cold-frame propagation, the following can be propagated in such an environment: climbing plants (e.g. *Bignonia capreolata*, *Campsis radicans*, *Eccremocarpus scabra*, *Passiflora caerulea*); trees and shrubs (e.g. *Albizia julibrissin*, *Daphne genkwa*, *Embothrium coccineum*, *Kalopanax pictus*, *Koelreuteria paniculata*, *Lagerstroemia indica*, *Paulownia tomentosa*, *Phellodendron amurense*, *Zanthoxylum piperitum*). It is important to keep the air temperature in the greenhouse cool but above freezing, while the basal temperature should not exceed 18°C/65°F. Potting-up can usually be done during early summer, and with further protective growing-on facilities, substantial plants can be produced by the end of the growing season.

Using root cuttings as a source of stem cutting material is a technique that greatly enhances rooting capabilities on plants such as *Albizia julibrissin* and *Populus* and *Alnus* species, when compared to conventional stem cuttings.

Leaf cuttings Some plants can be induced to regenerate from appropriately treated leaves, in particular species of the families Begoniaceae, Crassulaceae, Gesneriaceae and Melastomataceae. Leaves of *Paulownia tomentosa* will differentiate and produce buds and roots; and leaf rooting is possible with *Clematis*, *Cyclamen*, *Hedera*, *Jasminum* and *Mahonia*, although these will not develop into fully composite plants. A range of monocotyledonous plants will also regenerate from leaf cuttings, including *Endymion*, *Galanthus*, *Heloniopsis*, *Hyacinthus*, *Lachenalia*, *Leucojum* and *Sansevieria*.

The most suitable rooting environment is a closed case shaded from direct sunlight, with high humidity maintained and a basal heat between 18–20°C/65–70°F. Plants responsive to leaf propagation can usually be propagated at any time of the year. Leaves selected should be fully expanded and healthy, with no signs of ageing. Succulent leaves are prone to rotting, and propagation facilities, potting medium and leaves should be kept as clean as possible.

LEAF-PETIOLE CUTTINGS This is a suitable technique for plants that produce small leaves, for example, *Begonia* 'Gloire de Lorraine', *Peperomia*, *Ramonda* and *Saintpaulia*. The leaf petiole should be trimmed to leave a stalk about 4cm long; this is then inserted into a soft, open rooting medium, for example, of equal parts moss peat and perlite – a suitable mix for all leaf cutting systems.

MIDRIB LEAF CUTTINGS Plants such as *Gesneria*, *Sinningia*, and in particular *Streptocarpus*, will produce plantlets from veins severed along the length of their leaves. Detached leaves should be placed upside down on a clean piece of glass and the central midrib cut out. The two remaining leaf pieces are then placed in the surface of the rooting medium, with the cut tissue horizontal and all cut ends of the lateral veins just buried. Plantlets should develop from each lateral vein within 6–8 weeks.

LATERAL VEIN-LEAF CUTTINGS The narrow, strap-like leaves of *Streptocarpus*, *Sansevieria* and *Heloniopsis* can be cut into a series of 2.5–3.5cm-wide sections across their width, and the cuttings planted vertically with the basal ends inserted into the rooting compost. When propagating *Lachenalia*, *Galanthus* and *Leucojum* by this method, extra care must be taken to keep the leaf sections turgid and, since foliage dies down after flowering, timing of propagation is critical.

LEAF SLASHING Plants with large broad leaves with reticulated veining, such as *Begonia rex*, can be induced to produce plantlets by slashing the reverse side of selected leaves. Cuts about 10mm long are made across the main veins and repeated every 2.5cm or so. The leaf is then pegged down, top side up, on the compost surface. Buds will form at the incisions to produce roots and plantlets.

LEAF SQUARES are a means of rapid multiplication of any plant that will regenerate from leaves, but are generally reserved for large-leaved plants, such as some *Begonia* species. Selected leaves are placed upside down on a clean piece of glass, and cut up into approximately 2cm squares. Sound leaf pieces are then laid on the rooting medium face upwards, each secured by a small clean stone resting on its surface. Alternatively, they may be inserted vertically, with care taken to ensure that the end originally nearest the leaf stalk is buried.

Bulb scaling The scaly and fleshy specialized leaves that form the bulbs of *Lilium* and *Fritillaria* species will readily form bulblets. Firm clean scales are snapped off the base of the bulb. These are mixed with moist sharp sand, perlite or vermiculite to avoid dehydration and placed in trays or sealed in polythene bags. Callus tissue forms at the base of the scale and in a few weeks one or more bulblets develop from it; these are prised off to be grown on in a sterilized medium. With broad scales, the number of progeny can be increased by nicking the base edge to a depth of 5mm every 5mm of its width. Outermost scales are the best choice and dipping in rooting hormone increases yield.

Asiatic, Oriental, Caucasian, Martagon, European and Eastern North American lilies should be

bulb scaling

scaled between flowering time and spring. They benefit from being stored in a dark place at 18–24°C/65–75°F and usually produce bulblets within 12 weeks. Most Asiatic sorts will grow away freely, but others first require a period of 6–8 weeks cool storage in a refrigerator. Western American lilies should be scaled in the autumn/early winter and stored in a dark place at 3–10°C/37–50°F. The bulblet-bearing scales are potted up 1cm apart in 15cm pots to grow on.

TWIN-SCALING is a variation used for bulbs such as daffodils, snowdrops, hippeastrums, nerines and hyacinths. Here, the outer brown scales are removed, the roots trimmed without damaging the basal plate, and the nose sliced off. The bulb is upended and divided into up to 16 segments, each with a section of basal plate attached. The segments are then separated into pairs of scales, each retaining a fragment of basal plate. From a bulb cut into eight segments, up to 40 twin-scales can be produced.

CHIPPING is an alternative system where the mother bulb is prepared similarly, but the portions are not divided into pairs of scales. Large bulbs may be divided into 16 chips, small bulbs into 4 or 8 chips.

Good hygiene is vital and at all stages equipment must be kept scrupulously clean. Twin-scales and chips should be treated with fungicide, then mixed with moist vermiculite, in 30 micron sealed polythene bags, and stored in a dark place at 20°C/70°F for about 12 weeks. Bulblets should be hardened off before planting in a cool, frost-free environment. Flowering size is usually reached in three years. See *scooping, foliar embryos.*

cutworms (Lepidoptera: Noctuidae) the larvae of various noctuid moths, so-called because they attack plants at ground level, severing young plants completely and often girdling the stems of older plants. Roots, corms, tubers and leaves are also eaten. Most damage occurs on light soils during a dry summer. Young vegetables, especially lettuces, onions and brassicas, are particularly susceptible; carrots, celery, beet, potatoes and strawberries are also vulnerable, as are China asters, chrysanthemums, dahlias, marigolds, primulas, zinnias and nursery trees and shrubs.

The principal European species are the YELLOW UNDERWING MOTH (*Noctua pronuba*), which has brown forewings, yellow hindwings and a wingspan of up to 60mm; the TURNIP MOTH (*Agrotis segetum*), which varies in colour from grey or brown to almost black, with a wingspan of up to 40mm; and the HEART AND DART MOTH (*A.exclamationis*), which is similar to *A.segetum* but with forewings bearing characteristic dark markings. The habit of North American species is similar, but in warmer areas the number of generations increases.

The moths are on the wing from June until August depending on species. Up to 1000 eggs are deposited in clusters on stems or leaves of plants. The newly hatched caterpillars feed on the foliage, but within a week or so descend to the soil. The plump larvae

large yellow underwing

are various shades of brown yellow or green and up to 50mm long. They live most of their lives in the soil near to the surface.

Control weeds that encourage egg-laying, and cultivate soil in autumn and winter to expose the larvae to predators such as birds. In susceptible areas, treat the bases of young plants with insecticide. Small infestations can be controlled by locating and destroying caterpillars in the soil near recently damaged plants.

Cyananthus (Greek *kyanos*, blue and *anthos*, flower). Campanulaceae. Himalaya, China. TRAILING BELLFLOWER. 30 species, evergreen, tufted and carpet-forming perennial herbs with small leaves and solitary flowers in late summer. Fully hardy, the following species thrive in pockets of moist, humus-rich soil in light shade or dappled sun in the rock garden. Propagate by softwood cuttings in spring.

C.lobatus (Himalaya; a mat of 1cm-long, obovate, toothed to lobed leaves; flowers purple, blue or white, 3cm, narrowly funnelform with five broad lobes); *C.microphyllus* (differs in its smaller, oblong to elliptic leaves and more broadly funnelform, violet-blue flowers with ovate to lanceolate, not obovate lobes).

Cyanotis (Greek *kyanos*, blue and *ous*, ear). Commelinaceae. Old World tropics. 30 species, tender perennial herbs, usually with hairy, thinly succulent leaves overlapping in two ranks along creeping stems that radiate from initial rosettes. The flowers are small and three-petalled. Cultivate as for the tender *Tradescantia* species but with less water in winter.

C.kewensis (TEDDY BEAR VINE; India; stems to 15cm long; leaves lanceolate, 2–5cm-long, dark green above, purple beneath and velvety; flowers purple-pink); *C.somaliensis* (PUSSY EARS; Somalia; stems to 25cm long; leaves oblong to linear, densely white-hairy, 2–10cm long; flowers purple-blue).

Cyathea (from Greek *kyatheion*, little cup, a reference to the appearance of the sori). Cyatheaceae. Tropics. TREE FERN, SAGO FERN. 600 species, evergreen tree ferns with erect, rather slender, smooth stems marked with the scars of fallen frond bases (thus easily distinguished from *Dicksonia*). The large fronds form a spreading or shuttlecock-like crown; they are basically triangular in outline and finely bipinnately divided. *C.australis* and *C.cooperi* will survive lows of –7°C/19°F with some foliage loss: they might be attempted outdoors in favoured, maritime climates in zone 8. Elsewhere, grow in a frost-free conservatory or greenhouse. *C.arborea* and *C.medullaris* need a minimum temperature of 10°C/50°F. All prefer a porous planting medium, ideally soilless, and rich in garden compost, bark and charcoal. Position in dappled sunlight or shade, sheltered from strong winds and in a humid, buoyant atmosphere. Water and syringe freely in warm weather, less in winter but never allow to dry out.

C.arborea (WEST INDIAN TREE FERN) reaches a height of 10m. Its canopy consists of fronds to 3.5m long, divided into fresh green pinnules that are themselves divided into 25–32 pairs of 3mm-wide, serrate, oblong-falcate segments. The stipes arch strongly, are tinged yellow and clothed below with creamy scales. The AUSTRALIAN or ROUGH TREE FERN, *C.australis*, and the very similar *C.cooperi*, attain heights between 3 and 18m. The 1–4m fronds spread gracefully and are lime to mid-green with entire, serrulate or crenate, 10cm-long pinnules. The stipes are covered with brown scales which become darker as they descend toward the stem. *C.medullaris* (BLACK TREE FERN) is native to Australia, New Zealand and the Pacific Islands. The largest species listed here, its black stem may be over 15m tall. It is crowned with arching fronds to 6m long, dark, glossy green above, paler beneath. The stipes are stained black and covered below with black scales.

cyathiform shaped like a cup, similar to urceolate but without the marginal contraction.

cyathium an inflorescence-type characteristic of *Euphorbia*, consisting of a cupule (an involucre of small bracts sometimes furnished with glands and subtended by petaloid bracts or appendages) enclosing several 'stamens', each equivalent to a single male flower, and a stalked trilocular 'ovary' (a 3-carpelled pistil equivalent to the female flower). In dioecious species, the male or female is lacking according to the sex of the individual.

cycad a primitive gymnosperm belonging to any one of the families Cycadaceae, Zamiaceae or Stangeriaceae. Although allied to conifers, cycads very superficially resemble ferns and palms. Having dominated the world's flora 200 million years ago, they now survive, more or less unchanged, scattered throughout the tropics and subtropics. See *Cycas, Dioon, Encephalartos, Macrozamia, Zamia*.

Cycas (from *koikos*, the Greek name for the doum palm (*Hyphaene*)). Cycadaceae. Old World tropics and subtropics. CYCAD, SAGO, FALSE SAGO, FERN PALM. Some 20 species, tender, evergreen perennials. The trunks are stout and column-like, usually unbranched and covered with scars and the remains of leaf bases. They terminate in a crown of large, pinnate leaves with many tough-textured, narrow segments. Male plants produce large, yellow cones composed of closely packed, diamond-shaped scales; the female cones are looser with egg-shaped seeds borne on much reduced, felty, leaves. Handsome foliage plants for subtropical and tropical gardens and for interiors. The most commonly cultivated species, *C.revoluta* is a resilient pot or landscape plant, tolerating neglect and even mild frosts provided the leaf rosette is tied upwards and wrapped with sacking in winter. The remaining two species require a minimum winter temperature of 10°C/50°F. Plant in large pots, tubs or borders in a loamy mixture rich in garden compost, bark and sand. Provide *C.circinalis* and

C.rumphii with a shady, humid environment. *C.revoluta* will accept brighter, drier conditions. Water, feed and syringe frequently in spring and summer, but keep barely moist at cooler, duller times of year. Propagate by seed, surface-sown on a moist, sandy mixture at 15–29°C/60–85°F, or by detaching the bud-like offsets that sometimes develop at the base and along the trunks of older plants.

C.circinalis (FALSE SAGO) is native to southeast India. It makes a trunk to 5m tall, some 30cm wide. The leaves are 1.5–3m long; the segments to $30 \times$ 1.5cm, linear to lanceolate and somewhat curved, smooth, mid- to pea green and of a rather supple, leathery texture. *C.rumphii* may be synonymous with this species, differing only in its larger and more luxuriant habit, its softer and broader leaf segments, sometimes a deep sea green, and certain cone details. It is widespread in coastal southeast Asia, and extends to East Africa, Madagascar and the Pacific Islands. The Japanese *C.revoluta* is a smaller and stockier plant, to 3m tall. Its 1–2m-long leaves are held erect to spreading and are crowded with narrow, linear segments, each to 15×0.5cm, rigid, and glossy dark green with a distinct midrib and revolute margins.

Cyclamen (from the Greek *kyklos*, circle, referring to the rounded tubers). Primulaceae. SOWBREAD, PERSIAN VIOLET. Europe, Mediterranean to Iran, Somalia. 19 species, tuberous perennial herbs. The tubers are usually circular and somewhat flattened. They bear tufts of stalked, circular to heart-shaped leaves. The flowers are solitary and nod on curving stalks. The five petals are rounded and twist upwards.

With the exception of *C. persicum* and its cultivars, the following are hardy. *C. hederifolium, C. coum* and *C. trochopteranthum* will survive in climate zone 6. The remainder prefer rather higher temperatures and protection from wet winter conditions. They should be grown outdoors in very sheltered positions in zones 7 and over, and elsewhere in the alpine house or frame. Plant in sun or part-shade on a gritty, fertile and humus-rich soil. Keep moist when in growth, drier when at rest (i.e.summer except for *C. purpurascens*). Pot-grown plants can be dried out completely during dormancy. The hardiest species, however, cope well with conditions in the open garden and may be naturalized under trees and shrubs and even in short grass without suffering. Grow *C. persicum* and its cultivars in the cool greenhouse or in the home (minimum temperature 5°C/40°F). Plant the tubers in a fertile, porous, soilless medium with their upper surface exposed. Position in bright, filtered light in well-ventilated conditions. Water and feed when in growth. Dry off after flowering, when the leaves will wither. Tubers may be started into growth at any time of year. Increase by seed. Old tubers may develop short, woody 'stems' at their growing points. These may be removed and rooted. *C. persicum* and its cultivars are affected by black root rot.

(see table overleaf)

Clockwise from top: *Cyclamen persicum, C. trochopteranthum, C. hederifolium* (with different leaf forms), *C. coum*

CYCLAMEN

Name	Distribution	Leaves	Flowers
C.africanum	Algeria	to 10cm, circular to reniform or cordate, crenate to serrate, dark green lightly marked silver-grey, lustrous, pale green beneath	to 2.5cm, violet-scented, petals slightly reflexed, ovate-lanceolate to lanceolate, pale pink to deep rose with a deep red patch at base; autumn
C.cilicium	SW Turkey	to 4cm, suborbicular, serrate to crenate, grey-green to dark green with silver-grey and dark blotches, purple-red beneath	to 2cm, petals obovate-acute, twisted, strongly reflexed, white to deep pink with a dark crimson blotch at the base suffusing petals; autumn

Comments: Includes the white-flowered 'Album'

Name	Distribution	Leaves	Flowers
C.coum C.hiemale, C.orbiculatum, C.vernale	Caucasus, Turkey, Lebanon, Israel	2.5–6cm diam., orbicular-reniform, dark to mid green marked with lighter spots or silver-grey zone or band above, purple-red-tinted beneath	to 2cm, petals ovate-elliptic, strongly reflexed, somewhat twisted or folded, white to pale rose or deep carmine with a dark crimson to maroon blotch at base; late winter-early spring

Comments: Includes many cultivars ranging in colour from pure white to deepest magenta; members of the Pewter-leaved Group have leaves almost wholly overlaid with solid, dull silver to pewter; ssp. *caucasicum* (syn. *C.caucasicum*) differs in its cordate leaves with silver markings, and flowers with a pale mauve rim and acute petals

Name	Distribution	Leaves	Flowers
C.creticum	Crete	to 4cm diam., cordate, dentate, dark grey-green, usually spotted or patterned silver-grey above, purple-red beneath	to 2.5cm, fragrant, petals lanceolate, white to pale pink; spring
C.cyprium	Cyprus	to 3.5cm diam., broadly cordate to ovate-lanceolate, apex tapered, toothed to shallowly lobed, olive green blotched light green near margin with a hastate inner pattern marked with grey-green, stained red beneath	to 2.5cm, petals sharply reflexed, folded near base, twisting and toothed toward apex, pure white or pale pink with a v-shaped, purple-pink marking near base; autumn-early winter
C.graecum	Greece, S Aegean Is, S Turkey, Cyprus	to 6cm diam.,cordate-acute, toothed, velvety dark green with a moss green to silver-grey zone at centre, veined lime green or silver grey above, purple-red-tinted beneath	to 2cm, petals strongly reflexed, oblong-elliptic, pale pink to deep carmine with a darker stain at base; autumn-early winter
C.hederifolium C.neapolitanum	S Europe to Turkey	ivy-like, 4–8cm, rounded to cordate, bluntly hastate or lanceolate-oblong, entire to toothed or lobed, very variable in colour – dark to pale green or grey-green above, variously mottled, veined or zoned pale green to silver-grey	to 2.5cm, petals ovate-lanceolate, acute, reflexed, constricted at base, pale to deep pink, usually with a dark carmine stain at base, sometimes pure white, rarely fragrant; late summer to early winter

Comments: Many cultivars available covering the range of variation described here

Name	Distribution	Leaves	Flowers
C.libanoticum	Syria, Lebanon	to 8cm diam., rounded to cordate, dull blue-green with darker blotches, red-tinted beneath	to 5cm, petals oblong-elliptic, white to palest pink in lower half with a dark carmine blotch at base, clear pink in upper half;winter-early spring
C.mirabile	SW Turkey	to 3.5cm diam., suborbicular to reniform, minutely toothed, blunt, dark green, zoned or marbled grey, pink or red above, purple beneath	to 3cm, petals oblong to obovate, more or less toothed toward apex, pale pink with a dark carmine blotch at base; autumn
C.parviflorum	E Turkey	to 2.5cm, suborbicular, base cordate, margins entire, dull dark green	to 1cm, very short-stalked, petals broad, spreading then abruptly reflexed and slightly twisted, mauve-pink with a dark purple blotch at base; winter
C.persicum	E Mediterranean, Rhodes, Crete, Libya	5–12cm diam., ovate-cordate, finely toothed, dark green variably veined, mottled, spotted or zoned pale green or silver-grey, often red-tinted beneath	2–8cm, often fragrant, petals oblong-lanceolate to obovate,strongly reflexed and twisted, white to pale rose, deep pink, magenta or cerise with a darker blotch at base; winter to spring, but bred and grown to flower throughout year

Comments: This is the florist's *Cyclamen* with many cultivars ranging in habit from the large and robust to the dwarf and compact, in leaf from the large, dark and fleshy to the small and exquisitely marked, and in flower from small, fragrant and 'species-like' to very large, in tones of white, pink, salmon, cerise, scarlet and red, the petals small and slender to broad and heavy, ruffled and 'double'

Name	Distribution	Leaves	Flowers
C.pseudibericum	S Asia Minor	to 6.5cm, obcordate, toothed, dark green marbled silver-grey above, red-maroon beneath	to 2.5cm, fragrant, petals broadly lanceolate to elliptic, mauve to bright magenta or pink with a purple-brown blotch at the base streaking across the white rim; winter-spring
C.purpurascens C.europaeum, C.fatrense	C & E Europe	to 8cm, circular-ovate, base cordate, finely toothed, dark green veined or marbled silver-grey above, tinted red beneath	to 3cm, very fragrant, petals oblong-elliptic, reflexed and slightly twisted,lilac-pink to deep magenta or purple-red, darker at base; late summer-autumn
C.repandum	C & E Mediterranean	6–10cm, broadly cordate, shallowly lobed and toothed, dark green with a paler inner pattern bordered silver-grey, sometimes with paler or silver spots	to 3cm, fragrant, petals linear-oblong, white to deep carmine with a darker stain at rim; spring
C.rohlfsianum	Libya	to 10cm, reniform to orbicular, margins shallowly lobed, toothed, deep green with a darker central zone bordered with a jagged region of grey-green to silver	to 3cm, usually fragrant, petals lanceolate-obovate, strongly reflexed and twisted, pale rose to lilac pink, darker toward base; autumn
C.trochopteranthum C.alpinum	SW Turkey, Cilician Taurus	2.5–5cm, reniform to broadly ovate, base cordate, apex blunt, dark green zoned silver above, tinted purple-red beneath	to 2cm, scented, petals broadly obovate, twisted and spreading, propeller-like, pale pink to magenta or deep carmine with a dark basal

cyclic bearing the production of fruit or seed crops at intervals of longer than one year. It commonly occurs in some apple cultivars as biennial bearing.

Cydonia (for the town of Cydon in Crete). Rosaceae. W Asia, naturalized throughout temperate Europe, Mediterranean, Middle East, S America. QUINCE. 1 species, *C.oblonga*, a deciduous shrub or small tree with a spreading crown, to 6m tall. The leaves are 10cm long, ovate to elliptic, deep green above and woolly grey beneath. The solitary flowers appear in spring and early summer. Some 5cm wide, they consist of five, rounded, white or pale pink petals and are followed in late summer and autumn by broadly pyriform fruits to 10cm long. These are golden-yellow, downy, scented like a spicy pear with a bouquet of pine; they are used for jellies, preserves, added to pies and souffles, stewed with spirits and spices, served with meat and game and put to many herbal uses. The seeds are toxic.

Hardy in zones 4 to 7, quince nonetheless needs a long, warm summer to ripen its fruit. When grown for fruit, it is treated as a bush or standard tree with an open centre, or, in less favoured districts, as a fan or espalier against a south-facing wall. *C.oblonga* provides a dwarfing rootstock for pears. *C.* 'Vranja' and *C.* 'Bereczki' are vigorous plants and generous bearers of well-flavoured fruit, suited to fan training. The fruits of *C.* 'Champion', also a heavy cropper, and of *C.* 'Portugal' are milder in flavour. The apple-shaped fruit of *C.* 'Maliformis' ripens well in colder climates. The portuguese quince, *C.oblonga* 'Lusitanica', produces an abundance of large, pale rose flowers.

Grow in full sun, in deep, moist, fertile loam; on light soils, apply an organic mulch to conserve water. Prune in winter to establish a fruiting framework, cutting back leaders by one third of the previous season's growth to an outward-facing bud. Fruit is borne on spurs and on the tips of the previous summer's growth: once a framework is established, little pruning is required save to relieve overcrowding and to remove badly placed branches. Unless frosts are heavy, wait until fruit is fully ripe for harvest in mid- to late autumn. In a cool place, fruit will keep for 2–3 months. For propagation, see *apples*. For the Japanese or flowering quince, see *Chaenomeles*, several species of which were once included in *Cydonia*. See also *quince*.

Cymbalaria (from Greek *kymbalon*, cymbal, referring to the leaf shape of some species). Scrophulariaceae. W Europe. 10 species, perennial herbs, often short-lived with slender, trailing and rooting stems and small, shallowly lobed, reniform to rounded leaves. From late winter to early autumn, many solitary, 2-lipped flowers appear; these resemble a miniature snapdragon or *Linaria*. Use the following species for baskets, cracks in walls and paving, and small-scale carpeting effects in sun or shade on damp or (once established) rather dry and impoverished soils. *C.aequitriloba* may not survive temperatures below –17°C/1°F; *C.muralis* is hardy in zone 6 and becomes invasive given the least

encouragement. Sow seed *in situ* in spring or in a frame in late winter; otherwise, take rooted sections of stem in spring or summer.

C.aequitriloba (dense, mossy mat of entire or bluntly 3–5-lobed leaves; flowers 8– 13mm-long, purple); *C.muralis* (KENILWORTH IVY, COLISEUM IVY, PENNYWORT, IVY-LEAVED TOADFLAX rampant short-lived trailer, often infesting old walls; leaves reniform to semi-circular with five to nine, blunt to acute lobes; flowers lilac to mauve, white or violet and with a yellow palate, 9–15mm long).

Cymbidium (diminutive form of the Greek *kymbe*, boat – a reference to the hollow recess in the lip). Orchidaceae. Asia to Australia. 50 species, epiphytic, lithophytic and terrestrial orchids with clustered, small to large pseudobulbs sheathed by long, narrow leaves. Produced in basal racemes usually in winter and spring, the flowers are long-lasting and waxy with similar, ovate to lanceolate tepals and a showy lip with two, incurved lateral lobes and a projecting, platform-like midlobe.

Cymbidium is the most popular orchid genus, once widely sold as a cut flower, today so easily produced that it is simpler to sell whole plants in full bloom. These are likely to be hybrids, hundreds of which have been registered. They range in habit from the miniature (i.e. 80cm tall with many, 5cm-wide flowers) to very large (i.e. to 1.5m tall with up to 30, 12cm-wide flowers). The foliage is generally grassy and the flower spikes may be erect, arching or pendulous. The flowers are in shades of white, rose, bright pink, wine red, chestnut brown, amber, ochre, yellow and green, often marked with red or brown, especially on the lip, which may be a different ground colour altogether with a ruffled margin, prominent, sometimes golden crests and hairy.

The following species are also grown. They need a minimum winter temperature of 5°C/40°F. Pot in a bark-based mix with added slow-release fertilizer and a moisture-retentive medium such as rockwool or coir. Water and feed frequently when in growth and provide a humid, bright and warm environment. By midsummer, the new growth should be well-developed and will benefit from a ripening period of hot, sunny days and cool nights – easily achieved by placing the plants outdoors until autumn. Once the new pseudobulbs are plump and fully formed, place in a cool, airy situation and reduce water to a minimum (*C.ensifolium* and *C.goeringii* are exceptions, requiring a permanently moist medium and cool, shaded conditions). The flower spikes should now begin to form and may require staking and training. Those of *C.devonianum* and *C.elegans* may bury themselves in the pot unless guided over the edge. Grown in too consistently warm an atmosphere, *Cymbidium* will usually fail to flower, producing luxuriant foliage instead – the most common problem with plants bought in flower and grown in the home.

Increase by division or from backbulbs taken when repotting (best carried out after flowering and as seldom as possible – these plants flower better if pot-bound). *Cymbidium* is affected by fungal rots and cymbidium mosaic virus. Plants infected by the latter should be discarded, and care taken to sterilize blades used in division, to avoid its transmission.

C.aloifolium (India, China, Java; leaves fleshy; inflorescence to 90cm, pendulous; flowers to 5cm wide, tepals cream-yellow streaked red-brown, lip cream or white stained yellow, veined maroon; spring- to summer-flowering); *C.atropurpureum* (Thailand; Malaysia, Philippines; differs from last species in flowers that are more solid maroon and coconut-scented); *C.bicolor* (Indochina to Malaysia; inflorescence to 75cm, strongly arching; flowers to 4.5cm wide, fragrant, tepals yellow-cream striped maroon, lip white stained yellow dotted or blotched maroon; spring- to summer-flowering); *C.canaliculatum* (Australia; inflorescence to 45cm, arching; flowers to 4cm wide, tepals green to pale bronze streaked red-brown, lip ivory spotted purple-red; autumn-and winter-flowering); *C.dayanum* (N India to Japan, Malaysia; inflorescence to 35cm; flowers to 5cm wide, fragrant, tepals white with a central maroon stripe, lip white stained yellow, marked maroon; summer-flowering); *C.devonianum* (N India to Thailand; inflorescence to 40cm, strongly pendulous; flowers to 3.5cm wide, tepals olive streaked blood red, lip garnet to deep maroon; spring-flowering); *C.eburneum* (India to China; inflorescence to 36cm, few-flowered; flowers to 12cm wide, fragrant, tepals white, sometimes tinged pink, lip ivory stained yellow; winter- to spring-flowering);

Clockwise from top right: Cymbidium ensifolium, C. devonianum, C. goeringii, C. aloifolium, C. atropurpureum, C.elegans

C.elegans (syn. *Cyperorchis elegans*; N India to China; inflorescence to 60cm, pendulous; flowers ivory, crowded, narrowly campanulate; autumn- to winter-flowering); *C.ensifolium* (India, China, Philippines; habit grassy; inflorescence to 70cm, few-flowered; flowers to 5cm wide; tepals slender, straw yellow to olive green streaked red-brown, lip cream to white spotted red; spring- and summer-flowering); *C.erythrostylum* (Vietnam; inflorescence to 60cm, few-flowered; flowers to 12cm wide, tepals white to palest pink, lip cream veined and blotched deep red; summer-flowering); *C.finlaysonianum* (SE Asia; inflorescence to 1m, pendent; flowers to 6cm wide, tepals olive to yellow-green tinted pink, lip white marked red; summer-to autumn-flowering); *C.floribundum* (syn. *C.pumilum*; S China, Taiwan; inflorescence to 40cm, arching; flowers 3–4cm wide, tepals rusty brown or yellow green, lip white marked red to maroon; winter- to spring-flowering); *C.goeringii* (China, Japan; plants small, grassy; flowers solitary on stalks to 10cm, 5cm wide, tepals slender, curving, olive, yellow green, or rusty red, lip white marked red; winter- to spring-flowering); *C.hookerianum* (N India, China; inflorescence to 70cm, arching; flowers to 14cm wide, fragrant, tepals apple green spotted red below, lip cream edged green, spotted maroon; winter- to early spring-flowering); *C.insigne* (Thailand, Vietnam, China; inflorescence to 1.5m arching, flowers to 9cm wide, tepals white to pale pink sometimes spotted red, lip white stained yellow spotted and veined red; late autumn- to spring-flowering); *C.iridioides* (syn. *C.giganteum*; N India, Burma, S China; inflorescence to 1m; flowers to 10cm wide, scented, tepals olive green with broken stripes of rusty red, lip yellow marked red; summer-flowering); *C.lowianum* (Burma, China, Thailand; inflorescence to 1m, arching; flowers to 10 cm wide, tepals olive to lime green irregularly veined rust to blood red, lip white to yellow spotted red brown; winter- to summer-flowering; flowers unmarked in 'Concolor'); *C.madidum* (syn. *C.iridifolium*; Australia; inflorescence to 60cm, pendent; flowers to 3cm wide, tepals pale bronze to yellow-green, lip primrose blotched red-brown; summer- to winter-flowering); *C.parishii* (Burma; resembles a larger *C.eburneum* with highly scented flowers with the callus not velvety on midlobe; summer-flowering); *C.sanderae* (Vietnam; inflorescence to 40cm, few-flowered; flowers to 8cm wide, tepals white flushed pink and spotted red at base, lip cream stained yellow, marked maroon; winter- to late spring-flowering); *C.tigrinum* (NE India, Burma; flowers to 5cm wide, fragrant, tepals olive to yellow, faintly lined or dotted purple-red, lip white lined, spotted and edged purple-red; spring- to summer-flowering); *C.tracyanum* (China, Burma, Thailand; inflorescence to 1.2m; flowers to 12.5cm wide, tepals olive to tawny yellow, heavily streaked chestnut to rusty brown, lip cream to yellow lined brown or maroon; autumn- to summer-flowering).

cymbiform boat-shaped, attenuated and upturned at both ends with an external dorsal ridge, as in the keel of many papilionaceous flowers.

Cymbopogon (from Greek *kymbe*, boat, and *pogon*, beard). Gramineae. Old World tropics and subtropics. 56 species, largely tender, perennial grasses, many with aromatic foliage. All parts of the following species are strongly scented of lemon. Sections of culm and lengths of leaf are cut and used fresh, whole, grated or chopped as a flavouring in southeast Asian cooking. Grow in a moisture-retentive, loamy soil in full sun; maintain a moderately humid atmosphere with a minimum temperature of 10°C/50°F. Water plentifully when in growth. Propagate by division. *C.citratus* (LEMON GRASS; southern India, Sri Lanka, widely naturalized elsewhere in Asia; to 1.5m tall; culms cane-like leaves slender, grey-green).

cyme a more or less flat-topped and determinate inflorescence, the central or terminal flower opening first.

cymose arranged in or resembling a cyme or bearing cymes.

cymule a small and generally few-flowered cyme.

Cynara (from Greek *kyon*, dog – the phyllaries are shaped like dog's teeth). Compositae. Mediterranean, NW Africa, Canary Islands. 10 species, hardy perennial herbs with bold, pinnately cut leaves and discoid, thistle-like flowerheads in summer; these are large and goblet-shaped and composed of thick and densely overlapping phyllaries that surround a boss of purple-blue florets. The following are grown for the kitchen (blanched young leaves of *C.cardunculus,* flowerheads of *C.cardunculus* Scolymus Group), but are equally ornamental. *C.cardunculus* is a magnificent foliage plant for any large border, but especially valuable in the white and silver garden. If the globe artichoke lacks the stately, frosted appearance of the cardoon, its flowerheads nonetheless have a *Protea*-like quality welcome in late summer. Although tolerant of impoverished, stony soils and heavy clay, both species prefer a moist but well-drained, loamy soil in full sun with some protection from strong winds. Where hard frosts or cold, wet weather are common, mulch *C.cardunculus* in winter. Propagate by seed or by division in spring, or by root cuttings in sandy soil in a cold frame in early spring.

C.cardunculus (CARDOON; Mediterranean, N Africa; to 1.75m; leaves 50–100cm, arching, deeply and jaggedly cut and densely overlaid with fine silver wool; flowerheads spiny, 6cm wide and packed with purple florets; Scolymus Group: syn. *C.scolymus;* GLOBE ARTICHOKE; Mediterranean; to 2m; differs from *C.cardunculus* in its duller and less spiny foliage and flowerheads broader and clothed with thickly leathery to fleshy phyllaries; the florets are lilac to mauve). See *artichoke, cardoon*.

Cynoglossum (from Greek *kyon*, dog and *glossa*, tongue, referring to the shape of the leaves). Boraginaceae. Temperate regions. HOUND'S TONGUE. 55 species, annual, biennial and perennial herbs with coarse, hairy foliage and cymes of forget-me-not-like flowers in spring and summer. Fully hardy, hound's toungue prefers a fertile, well-drained soil with ample moisture in spring and early summer. Annual and biennial species grow well in full sun in the flower border. The perennials will also thrive in semi-wooded locations. Sow seed of annual and biennial species *in situ* in autumn or spring. Divide perennials in early spring.

CHINESE FORGET-ME-NOT, *C.amabile*, is an east Asian biennial to 60cm tall with 0.5cm-wide flowers in blue, pink and white. The cultivar 'Firmament' has grey-tinted leaves and azure flowers. *C.cheirifolium* from southwest Europe is biennial, to 40cm tall, with white-hairy leaves and 0.8cm-wide flowers which turn form pale mauve to indigo. WESTERN HOUND'S TONGUE, *C.grande*, attains a height of 60cm, with trusses of 1cm-wide, deep blue flowers. This North American native usually dies back to a stout rhizome after flowering. HIMALAYAN HOUND'S TONGUE, *C.nervosum*, is a perennial growing to 80cm with sprays of intense blue flowers held over narrow leaves. From North America, *C.virginaticum* is an unbranched perennial to 90cm tall with ovate, 20cm-long leaves and 1cm-wide flowers in shades of white, mauve and pale blue.

Cypella (from Greek *kypellon*, a goblet or cup, referring to the shape of the flowers). Iridaceae. C and S America. 15 species, small, summer-flowering bulbous perennials resembling *Iris*. Planted in a free-draining soil at the base of a sunny, sheltered wall, *C.herbertii* may survive frosts, especially if given a thick, dry mulch in winter and watered well in late spring and summer. Otherwise, lift the bulbs and store cool and dry in winter. Under glass, maintain a minimum temperature of 10°C/50°F; water carefully but plentifully when in growth. Store in sand and keep dry and frost-free when dormant. Sow ripe seed into a sandy potting mix, or detach bulbils and offsets, growing them on in moist, cool conditions until near flowering size. *C.herbertii* (Brazil, Uruguay, Argentina; panicle to 30cm high; flowers 4–6cm-wide, short-lived but produced in succession, three large outer segments are tawny yellow to golden, often with a purple-brown central stripe, ovate and waisted at the middle, the inner segments far smaller, pale yellow heavily marked purple to maroon).

Cyperus (the classical name for this plant, from the Greek for sedge). Cyperaceae. Warm and temperate regions worldwide. 600 species, mostly perennial, rhizomatous, evergreen herbs from marshy places. Where present, the basal leaves are grassy. Inconspicuous, the flowers are packed in small spikelets at the tips of slender rays that form spreading umbels atop tall, clumped stems; a collar of leaf-like bracts subtends the umbel in some species. All of the following are grown for their graceful inflorescences, produced or persisting throughout most of the year. The first two species are attractive greenhouse or house plants, the variegated cultivars especially. Plant in a fertile, loamy mix; keep constantly moist or even stand or submerge

in water. *C.haspan* will grow outdoors in sheltered locations as a marginal or submerged aquatic. *C.papyrus* requires a large tub *in* or a saturated bed *beside* ponds under glass. Inflorescences of this species and *C.alternifolius* are used as cut flowers. Growth is rapid and rhizomes may need to be reduced or rejuvenated every few years. Propagate by division in spring, or by stem cuttings of the umbels cut just below the bracts and rooted in water.

C.albostriatus (syn. *C.diffusus*, *C.elegans*) is a South African native, 30–50cm tall. Its leaves and inflorescence bracts can be distinguished by their prominent veins (boldly striped white in the cultivar 'Variegatus'). The umbel rays are 2.5–10cm long, shorter than the 6–9, spreading bracts. From Madagascar, the UMBRELLA PLANT, *C.alternifolius* is taller than the last species (40–150 cm) and leafless. The leafy inflorescence bracts are longer than the 6cm umbel rays. The cultivar 'Variegatus' is heavily striped or zoned white. *C.haspan* from central Asia differs from the last two species in having umbel bracts shorter than the 4–12cm-long rays. It is leafy, 10–50cm tall and hardy in zone 8. The others listed here are tender (minimum temperature 7°C/45°F). *C.papyrus* is the PAPYRUS or EGYPTIAN PAPER REED, a majestic African native 2–5m tall with thick, pithy stems which carry mops of countless, nodding rays. It lacks leaves and the umbel bracts are reduced to papery spathes. A dwarf (i.e. 1m tall) form of this species is grown under the names 'Nanus' and *C.prolifer*.

Cyphomandra (from Greek *kyphos*, a tumour or swelling, and *aner*, man, referring to the humped shape formed by the anthers). Solanaceae. S America. 30 species, herbs, shrubs, vines and trees similar to *Solanum*, but differing in their large, juicy fruit. *C.crassicaulis* (*C.betacea*) is the TREE TOMATO or TAMARILLO, a hairy, sparsely branched shrub, 1–3m tall. It is grown for its dull red, ovoid to ellipsoid, 8cm-long fruit, in the subtropics, tropics and elsewhere under glass (minimum temperature 10°C/50°F). In flavour, the fruit resembles a tart tomato and is good stewed with a little sugar. Plant in a fertile, loamy medium; position in part shade or full sun in a sheltered, humid environment. Water and feed liberally when in growth and fruiting, scarcely at all after fruiting when growth will pause and much of the foliage may drop. Raise from seed.

Cyphostemma (from Greek *kyphos*, curved or swollen, and *stemma*, wreath). Vitaceae. Africa, Madagascar (both species listed here are from Namibia). 150 species, tender, deciduous shrubs usually with greatly swollen, succulent trunks covered in peeling, papery bark and forking into short, thick branches or producing soft, climbing stems. The leaves are large and fleshy. Packed in flat-topped cymes, the small, yellow-green flowers are followed by colourful berries. These splendid caudiciform succulents require a minimum temperature of 10°C/50°F, full sunlight, a dry, buoyant atmosphere and a loam-based potting medium high

*Cyperus
alternifolius*

in sand and grit. Water only when in growth and then sparingly, and avoid wetting the foliage and stems. They deteriorate quickly in cool, damp conditions. Propagate by stem cuttings or by seed.

C.bainesii (syn. *Cissus bainesii*) makes a flask-shaped trunk to 60cm tall with few, if any, short branches at its summit. The leaves are downy and divided into 3 leaflets, these are ovate to elliptic, some 12cm long and coarsely toothed. *C.juttae* (syn. *Cissus juttae*) develops a massive bole to 2m tall, often with several thick, short limbs. The leaves are simple to 3-parted, with each segment 10–35cm long, oval, coarsely toothed, smooth and waxy blue-green.

*Cypripedium
reginae*

Cypripedium (from Greek *Kypris*, Venus, and *pous*, foot). Orchidaceae. Northern temperate regions. SLIPPER ORCHID, LADY'S SLIPPER ORCHID, MOCCASIN FLOWER. 45 species, deciduous hardy or near-hardy orchids, rhizomatous with broad, ovate to elliptic leaves and one to many flowers produced in spring and early summer. The flowers consist of an ovate dorsal sepal, two narrower and spreading petals and a showy, inflated, slipper-like lip backed by the fused lateral sepals. With the exception of *C.calceolus* which tolerates lime (although not the variety *parviflorum*), the following species require a slightly acid soil rich in decayed leaf matter. They associate well with ferns, trilliums and dwarf, bog-loving Ericaceae in damp, semi-wooded locations and the drier reaches of the bog garden. They are frost-hardy but benefit from a thick dry winter mulch. In very exposed locations, they fare better in pots in the alpine house, *C.japonicum*, *C.macranthum* and the otherwise scarcely visible *C.debile* especially. *Cypripedium* resents disturbance and is slow to increase; it may, however, be divided in early spring, taking some of the old soil with the new plants.

C.acaule from eastern North America produces a pair of basal leaves between which arises a 10–30cm-high stalk carrying a single flower to 8cm across. The drooping tepals are deep olive to chocolate brown, the lip rose pink. The European *C.calceolus* carries two to four leaves on a 15–50cm stem which terminates usually in a single, 5–8cm-wide flower. The tepals are olive to chocolate or maroon, the lip clear yellow. Its variety *parviflorum* from eastern North America has smaller flowers with spiralling petals; those of the Asian var. *pubescens* are larger and a uniform yellow flecked purple-brown. The largest species listed here, *C.californicum* from the western US produces metre-high stems hung with up to twelve, 4cm-wide flowers, their tepals yellow-green, the lip snowy white. The oriental *C.debile* seldom exceeds 10cm, a dwarf species with glossy green, heart-shaped leaves paired midway on the stem and a single, nodding flower to 3cm wide, pale green with blood red stippling on the lip. Also from Japan and China, *C.japonicum* (syn. *C.formosanum*) can be readily identified by its large, pleated and fan-shaped leaves. The single flower is 10cm wide with green-white, spotted tepals and a rosy pink lip which is bowl-shaped and puffy. *C.macranthum* extends from eastern Russia to Japan. It bears three to four large leaves and a solitary, purple-red-stained, 8cm-wide flower. From North America, *C.reginae* (syn. *C.album*) is a robust species with broad, ribbed leaves and two to four, 6–9cm-wide flowers toward the summit of its metre-high stems; the tepals white, the lip pink or white. Several hybrids are available and are proving hardy and vigorous. One of the best is *C.* Gisela Frosch, a cross between *C. calceolus* var. *parviflorum* and *C. macranthum*, with chocolate tepals and a puffy rose-purple lip. Many species of *Paphiopedilum* were formerly included in this genus.

cypsela an achene invested with an adnate calyx, as in the fruit of Compositae.

Cyrilla (for Dominico Cirillo (1734–1790), physician and professor of botany at Naples, executed by the Bourbons for his radical views). Cyrillaceae. SE US to Brazil. LEATHERWOOD, BLACK TITI, MYRTLE. 1 species, *C.racemiflora* (syn. *C.parviflora*), a highly variable tree or shrub, 1–9m tall, deciduous or evergreen and fully frost-hardy or slightly tender according to provenance. Leatherwood is grown for its 4–10cm-long, elliptic to oblong leaves which are glossy dark green turning deep red in autumn, and for its small, white flowers. These appear in late summer, crowded in whorled and slender racemes to 15cm long. Plant in full sun on a damp, acid soil rich in leafmould. Propagate by semi-ripe cuttings.

Cyrtanthus (from Greek *kyrtos*, curved, and *anthos*, flower, referring to the curved perianth tube). Amaryllidaceae. Tropical Africa, South Africa (those listed here are from South Africa). FIRE LILY. 47 species, bulbous perennial herbs with linear or strap-shaped leaves and tubular to funnel-shaped flowers borne on scapes.

C.breviflorus (syn. *Anoiganthus breviflorus*, with two to ten, yellow or white, 3cm-long, open flowers on a scape to 30cm tall) is hardy to –5°C/23°F in warm, sheltered postitions in well drained soils. *C.falcatus* (with six to fifteen, 6cm-long, flesh and yellow-green flowers narrow and nodding atop a 30cm-tall scape) will withstand light frosts if perfectly dry when dormant and mulched with its own dead foliage. *C.angustifolius* (FIRE LILY, to 40cm tall, with four to ten, nodding, 5cm-long, scarlet flowers), *C.brachyscyphus* (DOBO LILY, to 25cm tall, with six to eight, narrow, 2.5cm-long flowers in coral red, flame or scarlet) and *C.mackenii* (IFAFA LILY, to 30cm tall, with four to ten, 2.5cm-long, slender, white to cream flowers with a sweet fragrance) will take the mildest of frosts, growing well outdoors in southern Europe and California.

Most other species are not hardy but are well suited to pot cultivation under glass in zones 8 and below (minimum temperature 5°C/40°F). Some species, notably the SCARBOROUGH or GEORGE LILY, *C.elatus* (syn. *C.purpureus*, *Vallota purpurea*, *Vallota speciosa*, to 1m with six to nine, 10cm-long, broadly funnel-shaped flowers in scarlet, pink or white), and *C.sanguineus* (syn. *Gastronema sanguineum*, to 25cm tall, with one to three, 10cm-long, broadly funnelform, scarlet flowers) make excellent house plants. Grow in a fibrous, loam-based mix with added leafmould and sharp sand in full sun or bright indirect light. Water plentifully and apply dilute liquid feed weekly when in growth. Dry off deciduous species gradually after flowering; water evergreen species very sparingly in winter. Top-dress with fresh growing medium as growth recommences; repot as seldom as possible. Propagate by seed or offsets in spring.

Cyrtomium (from Greek *kyrtos*, curved or arched, referring to the growth habit). Dryopteridaceae. Cosmopolitan. 20 species, evergreen ferns with loose rosettes of pinnate leaves, the segments usually large and firm-textured. Grow the following species in the home, or under glass (just frost-free or warmer), or in very sheltered postions outdoors, where it will tolerate lows of –10°C/14°F for short periods if mulched. Plant in a neutral medium, soil-less for containers, humus-rich for borders. Provide a cool, humid environment in light to deep shade and keep moist throughout the year; feed in late spring and summer. Spores soon give rise to fast-growing new plants. *C.falcatum* (JAPANESE HOLLY FERN, FISHTAIL FERN; Far East, South Africa, Hawaii; fronds 20–60 cm long, semi-erect to spreading, broadly lanceolate in outline, segments in three to eleven pairs, each 7–13cm long, deep, glossy green, ovate to oblong, tapering and curving above, broad and rounded below toothed or shallowly lobed).

cyst nematodes see *eelworms*.

Cystopteris (from Greek *kystis*, bladder, and *pteris*, fern, alluding to the shape of the indusium). Woodsiaceae. Cosmopolitan. BLADDER FERN. 18 species, small, deciduous ferns with creeping rhizomes and lacy, fresh green, 2–3-pinnate fronds. Fully hardy, the following species will colonize gulleys and crevices in the rock garden or stony ledges by ponds, favouring neutral, porous soils that never dry out, light shade and shelter from strong winds. Remove faded foliage in late winter and mulch the rhizomes thinly with a mixture of garden compost, leafmould and a little bonemeal. Propagate by division in early spring or, *C.bulbifera*, by bulbils.

The North American *C.bulbifera* (BERRY BLADDER FERN, BULBLET BLADDER FERN) produces 15–70cm-long fronds, broadly lanceolate in outline with small bulbils in the axils of the upper pinnae. *C.fragilis* (BRITTLE BLADDER FERN, BRITTLE FERN) is found throughout the temperate northern Hemisphere and extends south to Chile. Its 6–45cm-long fronds lack bulbils (although the sori are characteristically bladder-like), are very finely divided and cut, and a pale emerald which contrasts well with the brittle, dark stipes.

Cytisus (from *kytisos*, a name applied by the Ancient Greeks to woody Leguminosae). Leguminosae. N Africa, W Asia, Europe, Canary Islands. BROOM. 33 species, evergreen or deciduous trees and shrubs, usually with slender, ribbed green branches and small, simple to trifoliolate leaves. Pea-like flowers are borne on short axillary shoots or terminal racemes in spring and summer. Most will tolerate temperatures as low as –15°C/5°F, often far lower given perfect drainage and a sheltered location. *Cytisus* resents root disturbance. Prune lightly to remove dead or exhausted growth. Hard pruning seldom succeeds. In prostrate and decumbent species, pruning is rarely necessary, although deadheading is beneficial. Propagate by heeled semi-ripe cuttings of lateral shoots in summer; treat these with a hormone rooting compound and root in a sandy propagating mix in a case with gentle bottom heat. Increase also from ripewood cuttings in late summer in a cold frame.

Cytisus x *praecox*

C.ardoinoi (SW Alps, S France; dwarf, prostrate, deciduous shrub to 40cm tall with ridged and winged young branches; flowers rich yellow, solitary or a few together in clusters); *C.* × *beanii* (*C.ardoinoi* × *C.purgans*; low-growing, deciduous shrub to 40cm with many, slender erect to arching branches; flowers golden yellow, borne in profusion along branches); *C.* × *dallimorei* (*C.multiflorus* × *C.scoparius*; erect shrub to 2.5m; flowers solitary or clustered, yellow stained ruby red to purple-red or lilac-pink); *C.* × *kewensis* (*C.ardoinii* × *C.multiflorus*; arching to prostrate shrub to 30cm with very small, 3-foliolate leaves; flowers solitary or clustered, pale yellow or cream); *C.multiflorus* (WHITE SPANISH BROOM; Portugal, Spain; erect, much-branched shrub to 2.5m; flowers white, solitary or in clusters; includes cultivars with pale yellow-cream, pure white and pink-tinted flowers); *C.nigricans* (SE and C Europe; erect shrub to 1m; flowers yellow in terminal racemes); *C.* × *praecox* (*C.multiflorus* × *C.purgans*; WARMINSTER BROOM; to 1m tall, similar to *C.multiflorus* but with denser branchlets; flowers pale yellow to deep gold, massed along ascending branches; includes gold-, cream- and white-flowered cultivars); *C.scoparius* (COMMON BROOM, SCOTCH BROOM; Europe; erect, many-branched shrub to 2m tall with yellow flowers; includes the dwarf, low-growing subsp. *maritimus* with large golden flowers, the showy, red-blotched 'Andreanus', the ivory and gold 'Diana', and the large, rich red 'Windlesham Ruby'). The following popular *Cytisus* species have been transferred to other genera: *C.albus* (= *Chamaecytisus albus*), *C.battandieri* (= *Argyrocytisus battandieri*); *C.canariensis* (= *Genista canariensis*); *C.purpureus* (= *Chamaecytisus purpureus*); *C.* × *spachianus* (= *Genista* × *spachiana*).

cytokinins a class of plant hormones involved in cell division and organ differentiation and causing delay in senescence.

cytology the branch of biology that concerns itself with the structure, functions, generation and life history of cells.

D

Daboecia (named after St Dabeoc). Ericaceae. W Europe to the Azores. 2 species, summer-flowering low-growing and compact evergreen shrubs differing from *Erica* in their broader leaves and relatively large, glandular flowers which fall away rather than wither and persist on the plant. For cultivation see Heaths and heathers.

   D. azorica (Azores; leaves dark above and silvery beneath; flowers ruby red, bell-shaped);*D. cantabrica* (CONNEMARA HEATH, ST DABEOC'S HEATH; W Europe; taller than the last species, forming a straggling bush to 50cm high; flowers urn-shaped, rose-purple; selections available in a range of colours including white ('Alba', 'Snowdrift'), pale pink ('Donald Pink'), dark pink ('Praegerae', dwarf and spreading with narrow flowers), magenta ('Porter's Variety'), dwarf and narrow-flowered). deep rose-purple ('Atropurpurea') and 'Bicolor', with pink, white and striped flowers on the same raceme; hybrids between the two species are collectively termed the Scotica group, among them the ruby-flowered 'Jack Drake' and free-flowering crimson 'William Buchanan', combining the colour and compactness of *D. azorica* with the hardiness of *D. cantabrica*).

Dacrydium (from Greek *dakrudion*, small tear, referring to the resin drops these trees exude). Podocarpaceae. SE Asia, W Pacific Islands, New Zealand. Some 30 species, evergreen, coniferous trees or shrubs, with scale-like, dark green leaves closely overlapping on narrow branchlets and inconspicuous cones. RED PINE or RIMU, *D.cupressinum*, is a New Zealand native to 40m tall with a narrowly conic to rounded crown hung with long and slender, weeping branchlets. This exceptionally graceful, slow-growing conifer is marginally hardy in zone 8. In colder regions, grow as a container specimen in the unheated greenhouse or conservatory. It requires a moist, acid soil, full sun, warm, humid summers and mild winters. Propagate by seed or cuttings of stem tips under mist.

dactyloid, dactylose finger-like.

Daboecia cantabrica (left)
D. azorica (right)

Dactylorhiza (from Greek *daktylos*, finger, and *rhiza*, root, referring to the finger-like tuber lobes). Orchidaceae. Europe, Asia, N Africa, N America. 30 species, deciduous, terrestrial orchids with lobed tubers and lanceolate to strap-shaped leaves sometimes spotted maroon. Produced in spring and summer in dense, terminal racemes, the flowers consist of five ovate tepals and a broadly wedge-shaped to rounded, shallowly 3-lobed lip. Hardy in climate zone 6. The southern species may be damaged by prolonged frosts, frost-heave and cold, wet conditions, but will usually survive if mulched in winter. In very cold regions, grow them in pans in the alpine house. Plant in sun or part-shade in a cool, moist soil rich in sand and leafmould. Several species will thrive in damp conditions and are suited to the bog garden. These include *D. elata*, *D. incarnata*, *D. majalis*, *D. praetermissa*, *D. purpurella* and *D. traunsteineri*. *D.fuchsii* and *D. maculata* will naturalize in grassland. Most species prefer a circumneutral soil, although *D. maculata* favours acid conditions whilst *D. fuchsii* and *D. traunsteineri* are lime-loving. Increase by division in early spring, or by seed allowed to self-sow in areas unlikely to be disturbed (hybrids may arise).

   D. aristata (Japan, Korea, N America; 10-40 cm tall; leaves sometimes spotted; flowers in a loose spike, purple-pink to maroon, lip spotted dark purple); *D. elata* (ROBUST MARSH ORCHID; S Europe, Mediterranean, especially Algeria; 30-100 cm tall; leaves unspotted; flowers large, in a dense, long spike, deep rose to rich magenta with darker stippling on lip; clones with yellow-variegated leaves are sometimes offered); *D. fuchsii* (COMMON SPOTTED ORCHID; N, W, C and E Europe, east to Mongolia; 20-60cm tall; leaves spotted, tips rather broad and blunt, thus differing from *D. maculata*; flowers in a dense spike, white to pale pink or pale lilac-mauve, spotted or streaked blood red or maroon, especially on lip); *D. incarnata* (EARLY MARSH ORCHID; Mediterranean; 15-80cm tall; leaves unspotted; flowers in a short spike, pale rose, lilac, pink, cream or rich rosy mauve, lip spotted red); *D. maculata* (SPOTTED ORCHID; N, W and C Europe east to Asia; 15-60cm tall; leaves usually spotted,

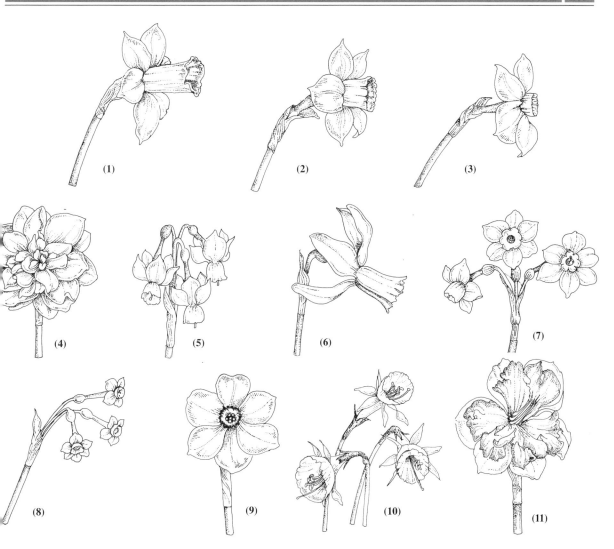

Daffodil Divisions: Div. 1 Trumpet: Div. 2 Large cup: Div. 3 Small cup: Div. 4 Double: Div. 5 Triandrus:
Div. 6 Cyclamineus: Div. 7 Jonquilla: Div. 8 Tazetta: Div. 9 Poeticus: Div. 10 Bulbocodium: Div. 11 Split corona

tips tapering, thus differing from *D. fuchsii*; flowers in a dense spike, white, pale lilac or rose to mauve,with dark red or magenta spots and lines especially on lip); *D. majalis* (BROAD-LEAVED MARSH ORCHID; S and C Europe east to Russia and W Asia; 20-75cm tall; leaves plain green or spotted, broad; flowers in a dense spike with leafy bracts, lilac to magenta, lip blotched white and patterned with dark purple-red); *D. praetermissa* (SOUTHERN MARSH ORCHID; W Europe including S UK; 20-70cm tall; leaves unspotted; flowers in a dense spike, deep magenta to rosy purple, lip spotted and streaked purple); *D. purpurella* (NORTHERN MARSH ORCHID; NW Europe including N UK; similar to *D. praetermissa*; to 40cm tall; leaves sometimes spotted purple; flowers in a dense spike, dark rosy purple to wine red, spotted or ringed dark purple on lip); *D. traunsteineri* (IRISH MARSH ORCHID; N and C Europe; 15-50cm tall; leaves sometimes spotted; flowers in a loose spike, bright magenta, lip spotted purple).

daffodil a popular name for *Narcissus* (q.v.), especially used of cultivars. Confusingly, in the cut-flower trade and some bulb catalogues, the term is reserved for cultivars with long trumpets (coronas), those with short cups (coronas) being referred to as narcissi.

The Royal Horticultural Society is the International Registration Authority for the genus *Narcissus*, and the current (1998) system identifies 13 divisions. (1)*Trumpet*, one flower per stem; trumpet at least as long as the petals (perianth segments). (2) *Large cup*, one flower per stem; cup more than one third but less than equal to the length of the petals. (3) *Small cup*, one flower per stem; cup not more than one third the length of the petals. (4) *Double*, one or more flowers per stem; with doubling of the petals or cup or both. (5) *Triandrus*, characteristics of *N.triandrus* clearly evident; usually two or more pendent flowers per stem; petals reflexed. (6) *Cyclamineus*, characteristics of *N.cyclamineus* clearly evident; usually one flower per stem; petals reflexed; the flower at an

daffodil

Pests
bulb scale mite
bulb flies
eelworm

Diseases
basal rot
virus

acute angle to the stem with a very short neck. (7) *Jonquilla*, characteristics of *N.jonquilla* group clearly evident; usually 1–3 flowers per rounded stem; leaves narrow, dark green; petals spreading not reflexed; flowers fragrant. (8) *Tazetta*, characteristics of *N.tazetta* group clearly evident; usually 3–20 flowers per stout stem; leaves broad, petals spreading and not reflexed; flowers fragrant. (9) *Poeticus*, characteristics of *N.poeticus* group without admixture of any other; usually one flower per stem; petals pure white; cup usually disc-shaped, with a green or yellow centre and a red rim; flowers fragrant. (10) *Bulbocodium*, characteristics of Section Bulbocodium clearly evident; usually one flower per stem; perianth segments insignificant compared with the dominant corona; anthers attached more or less centrally to the filament; filament and style usually incurved. (11) *Split corona*, corona split rather than lobed, usually for more than half its length. (12) *Miscellaneous*, all daffodils not falling into one of the other divisions. (13) Daffodils distinguished solely by botanical name.

Within each division, daffodils are further distinguished by code letters for colours: white or whitish = W; green = G; yellow = Y; pink = P; orange = O; red = R. The colour code consists of two letters or groups of letters separated by a hyphen, letters before the hyphen describing petals (perianth segments) and letters after the hyphen describing the trumpet or cup (corona). For purposes of colour coding, the petals and trumpets or cups are divided into three zones, none of which need to be in specific proportions. Where petals are substantially of one colour, one letter is used; where of more than one colour, two or three letters are used, the outer zone being described before the mid-zone and/or base. Where the trumpet or cup is of one colour, one letter is again used; otherwise, three letters are always used, the eye-zone being described before the mid-zone and rim. In double daffodils, the code letter(s) before the hyphen describe not only the outer whorl of petals but also any extra petals of the same colour interspersed with the trumpet or cup segments. The full classification consists of the division followed by the colour code; for example, a large cup cultivar with flowers of mainly yellow petals, white at the base, and with the cup having a white eye-zone, white mid-zone and yellow rim would be coded 2YW-WWY.

The leaves and flowers are tolerant of severe frost, except those of division 8. Generally, plant bulbs one and half times their depth but deeper in light-textured soils. Shallow planting results in greater increase and smaller bulbs. Plant in late summer or early autumn, in random groups, with bulbs 5cm apart; planting should be earliest for plants in division 9, which have virtually no resting period.

Daffodils may be grown for decoration or exhibition in deep, 22–25cm-diameter pots; a mass of flowers can be obtained by planting one layer half way down the pot and another layer immediately above. Daffodils are particularly attractive where established in grass; remove turf temporarily to allow planting, or use a bulb-planting tool. Turf plantings will usually thrive for many years, but both the grass and the daffodil foliage should be left uncut for four to six weeks after the flowers have faded. In borders, lift bulbs after three years to maintain vigour in the stock, but only after the leaves turn yellow. Dry the bulbs, clean off old tunics, discard defective bulbs and, where not replanted immediately, store in a cool dark place. Daffodils are usually propagated from offsets or, for rapid multiplication, by chipping or twin-scaling; see *cuttings*; they are sometimes increased by micropropagation (q.v.). Seed is used for hybridizing and for increasing many species.

The development stages or forms of daffodil bulbs are variously described. Mother bulbs have two or three new bulbs attached at the base and can be expected to produce one or two flower stems in the first year and to increase. Offsets are new bulbs developed to the stage of detachment from the mother bulb, either naturally or by being pulled apart when loose enough; they are usually flattened on one side and produce leaves in the first year, but rarely flower until at least the second year. Chips are very small offsets (the term should not be confused with that used to describe a system of propagation; see *chipping*). Rounds are evenly rounded bulbs with a single growing point or 'nose' at the top, developed from offsets, chips or seed and usually producing one flower when of suitable size. Double-nosed bulbs consist of a round bulb with a large offset still attached, so that there are two associated growing points or 'noses' at the top, each of which will produce a flower.

daffodil bulb forms

double-nosed   *round*   *offset*   *mother bulb*

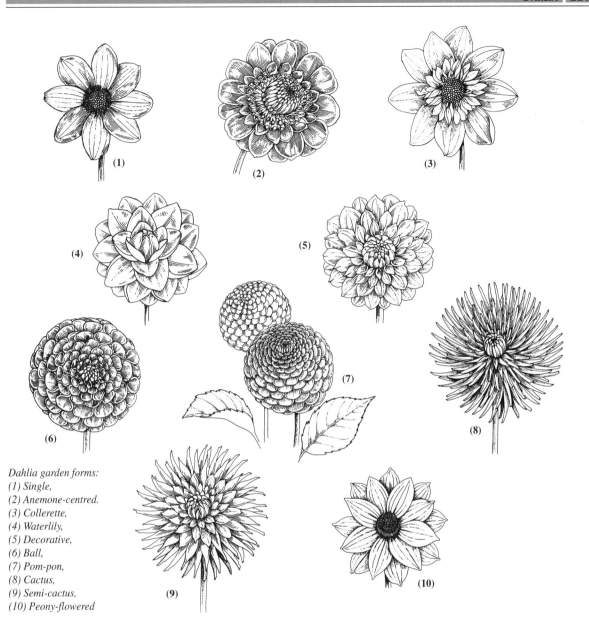

Dahlia garden forms:
(1) Single,
(2) Anemone-centred.
(3) Collerette,
(4) Waterlily,
(5) Decorative,
(6) Ball,
(7) Pom-pon,
(8) Cactus,
(9) Semi-cactus,
(10) Peony-flowered

Dahlia (for Anders Dahl (1751-1789), Swedish botanist). Compositae. C America. 30 species, perennial, tuberous-rooted herbs and subshrubs with pinnately compound leaves and long-stalked radiate flowerheads produced in summer and autumn. In cultivars, the disc florets are often replaced by ray florets, giving a 'double' (sometimes spherical) effect. The florets of cultivars range from strap-shaped, to rolled and quill-like, to broadly petal-like. The species are undeservedly rare in cultivation. They are fine plants for the cool glasshouse and conservatory, and for subtropical bedding schemes. The taller species should be encouraged to climb, supported on wires and trellis.

Garden dahlias derive especially from *Dahlia coccinea* and *D.pinnata* and cultivation of the genus dates from at least the 16th century. Dahlias were introduced into England in 1789 and their popularity spread throughout Europe and the US in the mid-19th century. 20,000 cultivars are now listed in the International Register of Dahlia Names maintained by the Royal Horticultural Society. Dahlias are frost tender but prolific and long-flowering with strongly fragrant sap. Seed-raised types include dwarf forms, suitable for bedding and containers. Border types, which are perpetuated by tubers, require staking, and they are often disbudded.

Ten groups are classified: (1) *Single-flowered*: open-centred blooms with one or two complete rows of outer florets surrounding a disc; blooms usually about 10cm diameter; most plants to 0.3m high. (2) *Anemone-flowered*: one or more rows of ray florets surrounding a dense group of upward pointing tubular florets; blooms 7.5cm diameter; plants 0.6–0.9m high. This group consists of very few cultivars. (3) *Collerette* (US Collarette), open centre blooms, sur-

dahlia

Pests
aphids
capsid bug
earwigs
red spider mite
slugs and snails
thrips

Diseases
crown gall (see gall)
grey mould
virus

rounded by an inner ring of short florets, known as the 'collar' and one or two complete outer rows of usually flat ray florets; blooms 10–15cm diameter; plants 1.1m high. (4) *Water lily or Nymphaea-flowered*, fully double blooms with generally sparse flat ray florets, having slightly incurved or recurved margins; blooms 10–12cm diameter; plants 0.9–1.2m high. (5) *Decorative*, fully double blooms with no central disc, and broad, flat or slightly involute ray florets, sometimes slightly twisted with the apex obtuse; plants 0.9–1.5m high. (6) *Ball*, ball-shaped or globose blooms, sometimes slightly flattened at the top with ray florets blunt or rounded at their tips and cupped for more than half their length; blooms 5–15cm diameter; plants 0.9–1.1m high. (7) *Pompon*, sometimes referred to as drumstick dahlias, and similar to Ball group cultivars but more globose with florets involute for their entire length; miniature sized blooms up to 5cm diameter; plants up to 0.9m high. (8) *Cactus-flowered*, blooms fully double with no central disc, and long pointed ray florets, finely quilled for more than half the length; plants 0.9–1.1m high. In the US, this group is subdivided into *straight cactus types*, with straight slightly incurved or recurved rays, and *incurved types* where pointed rays turn towards the centre of the bloom. (9) *Semi-cactus flowered*, blooms fully double with slightly pointed ray florets broad at their base and revolute for more than half their length, either straight or incurving; plants 1.2m high. (10) *Miscellaneous*, a grouping of relatively small disparate classes, including Orchid, Star, Chrysanthemum and Peony cultivars, the latter considered a distinct group in the US, all growing to 0.3–1.1m high. Decorative and Cactus flowered groups are further divided in the US to give twelve groups. Subdivisions of groups are made according to bloom diameter and colour. Those made according to size are the perogative of national societies on account of the effect of varying cultural conditions.

In the UK, five subdivisions of bloom size apply: giant, 254mm or more; large, 203–254mm; medium, 152–203mm; small, 102–152mm; miniatures, up to 102mm. Subdivisions according to colour are: white; yellow; orange; bronze; flame; red; dark red; light pink; dark pink; lilac, lavender or mauve; purple, wine or violet; blends, two or more colours intermingled; bicoloured, ground colour tipped with another colour; variegated, several colours striped or splashed in one bloom. A cultivar annotated 5a Lt.Pk is a light pink giant-flowered decorative type.

Sow bedding types in late winter/early spring at 16°C/60°F, and harden off, planting out after any risk of frost. Propagate border types from tubers or rooted cuttings. In frost-prone areas, lift tubers in mid- to late autumn, or immediately after the first frost, and place upside down until the stems are dry. Treat with fungicide and store in dry sand at 5°C/40°F min. Overwintered tubers may be planted out direct, six weeks before the last frost is expected, 15cm deep. Alternatively, raise plants from 7.5 cm-long cuttings taken from tubers forced at 15–18°C/60–65°F in late winter. Pot up and harden off

daisy grubber

to plant out after any risk of frost, 60–90cm apart, according to ultimate height. Stop plants from rooted cuttings at 38cm high. On all plants, restrict the number of side shoots according to the size of blooms required: 4–6 for giant/large, 7–10 for medium/small. Support each plant with three canes, placed 15cm apart, and surrounding string ties. For high-quality blooms, remove the pairs of buds from axils behind the terminal bud.

D. coccinea (Mexico to Guatemala; to 3m, usually tinged purple; ray florets scarlet, orange or deep maroon); *D. excelsa* (Mexico; to 6m; ray florets lilac, solitary or a few clustered together); *D. imperialis* (Guatemala to Colombia; to 9m; ray florets yellow sometimes tipped red); *D. pinnata* (Mexico; to 2m, tinged purple; ray florets pale purple stained yellow or pink at base); *D. tenuicaulis* (Mexico; to 4m; ray florets lilac, many clustered together).

Dais (from Greek *dais*, a torch, referring to the shape of the inflorescence). Thymelaeaceae. South Africa, Madagascar. 2 species, deciduous or semi-evergreen shrubs with small leaves and fragrant, tubular to star-shaped flowers in terminal heads. Provide a minimum winter temperature of 5°C/40°F in full sun. Plant in a gritty mix; water plentifully when in growth, very sparingly when dormant. Propagate by semi-ripe cuttings, or by seed sown in spring. *D.cotinifolia* (leaves to 7cm, obovate to oblong, blue-green; flowers fragrant, to 1.5cm in diam., pale lilac).

daisy grubber a short-handled tool designed for extracting thick-rooted weeds, especially from lawns. It consists of a narrow forked blade, bent into a U-shape at a point along its length to facilitate leverage.

damping DAMPING DOWN is the wetting of greenhouse floors and staging in order to increase humidity, especially when temperatures are high. It is important as an aid to pollinating tomatoes. DAMPING OVERHEAD is the spraying of greenhouse plants with water to reduce transpiration loss. It is usually performed only in warm houses, and out of bright sunlight.

damping-off a soil-borne fungal disease that destroys sprouting seeds, or causes stem rotting of young seedlings at soil level. It commonly affects bedding plants, herbaceous perennials, newly sown grass, vegetables and tree seedlings. Several fungi may be involved including *Pythium* species, (which thrive in wet conditions and in soils with a high pH), *Rhizoctonia solani* (especially in drier and more acid conditions), *Thielaviopsis*, *Botrytis*, *Fusarium* and *Phytophthora*. Pre-emergence damping off is encouraged by overwatering or poor drainage, and by temperatures that are too low for quick germination. In post-emergence damping-off, there will often be spreading patches of disease. The latter condition is prevalent in wet, warm conditions where seed has been sown too thickly and ventilation is poor.

The disease is difficult to control. To prevent its

occurrence, avoid the conditions mentioned above and use sterilized or otherwise pathogen-free growing media. Disinfect containers prior to use and employ clean water in raising. Post-emergence damping-off spreads rapidly but early treatment with fungicide can check its progress. Cheshunt compound and copper oxychloride have been successfully used in the UK. Seed treatment and the application of chemicals to the soil prior to planting, are recommended in some cases.

damson a type of plum derived from *Prunus institia*. The fruits are small with a pronounced neck, oval, purple and initially acid sour. The dwarf compact trees have leaves much smaller than plums; closely allied to the bullace. Probably known in Syria in pre-Roman times; the name damson or 'damascene' derives from Damascus. The trees succeed in wet and harsh conditions; notable self-fertile cultivars are 'Prune Damson' for flavour and 'Merryweather Damson' for good size.

Danaë (for Danaë, daughter of King Acrisius of Argos). Liliaceae (Ruscaceae). W Asia. ALEXANDRIAN LAUREL. 1 species, *D.racemosa*, an evergreen perennial herb, related to *Ruscus* but with flowers in short terminal racemes. The stems grow from 50 to 100cm tall in shrubby clumps, emerging as asparagus-like shoots and hardening in one year. The true leaves are scale-like and are replaced by luxuriant, dark, glossy green cladophylls. These are 3–7cm long, ovate to lanceolate and taper finely to the tip. The small, green flowers are sometimes followed by orange-red berries. It is hardy in zone 7. Grow in sun or part shade on a moist, fertile soil. Propagate by division, or by seed in autumn, although growth from seed is slow.

Daphne (for Daphne, the nymph changed by Apollo into *Laurus nobilis*, the bay tree). Thymelaeaceae. Europe, Asia, N Africa. Some 50 species, small evergreen and deciduous shrubs, with heads or clusters of flowers usually in late winter and spring. These are tubular with four spreading lobes. In some species, they are followed by berries.

These shrubs are grown for their beautiful and intensely fragrant blooms, sometimes followed by brightly coloured fruits. In some, the foliage is also handsome, laurel-like (*D.odora*, *D.laureola*) or variegated, as in cultivars of *D.odora*, or the gold-edged *D.* 'Carol Mackie'. Most are hardy to at least –15°C/5°F. *D.mezereum*. *D.petraea* and *D.cneorum* withstand temperatures down to –30°C/–22°F. Evergreen species may become semi-deciduous in less favoured climates. Plant in a well-drained but moisture-retentive soil. For woodland species, incorporate additional organic matter, and provide shade during the hottest part of the day. *D.arbuscula*, *D.bholua*, *D.blagayana*, *D.genkwa*, *D.* × *hybrida*, *D.mezereum* and *D.pontica* tolerate partial shade, *D.laureola* will grow in deep shade. Most thrive in soils ranging from slightly acid to alkaline, but *D.arbuscula*, *D.blagayana*, *D.genkwa*,

D.pontica, *D.retusa*, and *D.tangutica*, demand lime-free conditions.

Propagate species from seed sown ripe, after removal of the fleshy exterior. Stand in cool, shaded position (heat inhibits germination). Retain the seed pan for a second flush of seedlings in the following year. Progeny may be variable, but is more likely to be virus-free, and for this reason seed is the most sensible means of increase for *D.mezereum*. Select seed from a free-flowering parent with good flower colour where possible. Propagate cultivars by semi-ripe heel cuttings, in a closed case. With trailing plants, propagate by simple layering in summer. Increase also by whip-grafting in late winter: deciduous species on to *D.mezereum*, and evergreens on to *D.laureola* understock. Plants suffering sudden die-back may be under stress or infected with virus diseases. Cucumber mosaic and other viruses cause mottling and distortion of the leaves. The fungus *Marssonina daphnes* causes a spotting at the base of the leaves and on the petioles which may result in defoliation. *(see table overleaf)*

Daphniphyllum (from *Daphne* and Greek *phyllon*, leaf.) Daphniphyllaceae. Asia. 15 species, evergreen shrubs or trees with stout branches, tough leaves and small flowers in dense clusters in early summer – these are green in female plants, purple-red in male plants. The following species is hardy to –20°C/–4°F, but must be sheltered from cold, drying winds. Grow in cool, humid borders or light open woodland, preferably on neutral soils that are humus-rich and moisture-retentive. Propagate from seed in spring, or by greenwood cuttings under mist or in a closed case with gentle bottom heat. *D.himalense* var. *macropodum* (syn. *D.macropodum*; Japan, Korea, China; shrub or tree to 15m; young branches red-brown; leaves 8–20cm, oblong, somewhat deflexed, leathery, dark glossy green above, glaucous beneath, petiole and midrib usually tinted red).

dard a short lateral shoot of a fruit tree, not more than 7.5cm long and usually terminating in a fruit bud.

Darlingtonia (for Dr William Darlington of Philadelphia, botanist). Sarraceniaceae. US (N California, S Oregon). 1 species, *D.californica*, the COBRA LILY or CALIFORNIA PITCHER PLANT, an evergreen, carnivorous perennial herb. It forms a clump of insect-trapping leaves, each to 35 cm long. These are pitcher-like (i.e. tubular and filled with a digestive brew of water and bacteria). They spiral gently upwards from a narrow base to a broader, bulbous apex, which resembles a hooded cobra. The pitcher mouth is small and opens on the underside of the hood, guarded by a forked tongue-like appendage. The pitchers are usually bright green laced with translucent white windows; with age and bright sunlight, the traps assume purple-red tints. White flowers nod on tall stalks in summer. Grow in full sunlight in a buoyant but humid environment, with

damson

Pests
aphids
birds
plum moth (*see* moths)
red spider mite

Diseases
bacterial canker
  (*see* bacterial
  diseases)
brown rot
silver leaf

DAPHNE

Name	Distribution	Habit	Leaves	Flowers	Fruit
D.alpina	S & C Europe	deciduous, erect or prostrate to 45cm	1-4cm, lanceolate to oblanceolate, grey-green, hairy	fragrant, white in terminal heads in spring or early summer	red to orange-yellow, downy
D.arbuscula	Hungary	evergreen, procumbent, to 20cm	to 2cm, linear to oblanceolate, obtuse, glossy, tough	very fragrant, rose pink, in terminal clusters in summer	grey-white, not fleshy
D.bholua	E Himalaya	evergreen or deciduous, erect or spreading, to 4m	5-10cm, elliptic to oblanceolate, undulate to obscurely toothed, dull, deep green, leathery	very fragrant, white to purple-pink, in terminal clusters in winter	black
Comments: Includes the hardy, deciduous 'Gurkha', and the hardy, evergreen 'Jacqueline Postill' with large, deep pink flowers					
D.blagayana	E Europe, Balkans, Greece	evergreen, procumbent, to 30cm	3-5cm, oblong to obovate, blunt, smooth	fragrant, creamy white in dense terminal heads in spring	pink-white
D. × *burkwoodii* (*D.caucasica* × *D.cneorum*)	garden origin	vigorous, erect, semi-evergreen to deciduous, to 1.7m	2-4cm, oblanceolate, bluntly apiculate, dull green, more or less smooth	fragrant, white flushing rose pink to mauve-red in terminal clusters in spring	
Comments: Includes the vigorous 'Albert Burkwood', with white flowers ageing pink then rose-mauve, and 'Carol Mackie' with leaves edged bright, pale gold					
D.cneorum GARLAND FLOWER	C & S Europe	low-growing, evergreen, to 40cm	1-2.5cm, oblanceolate, deep green above, blue-green beneath	fragrant, pale to deep rose in dense terminal clusters in spring and sometimes again in late summer	yellow to brown
Comments: 'Eximia' has dark green leaves and, in late spring, white flowers with pink to red exteriors and throat, opening from rosy buds					
D.genkwa	China	deciduous, erect, slender, sparsely twiggy, to 1m	3-6cm, lanceolate to ovate, thinly downy at first	subtly fragrant, amethyst, lilac, rose-purple or white, in sparse clusters before leaves in spring	
D.giraldii	China	erect, deciduous, to 1.2m	3-6cm, oblanceolate	fragrant, golden yellow, sometimes tinted purple in bud, in terminal clusters in early summer	red
D. × *houtteana* (*D.laureola* × *D.mezereum*)	garden origin	erect, bushy, semi-evergreen, to 1m	to 9cm, oblanceolate to elliptic, glossy dark green strongly flushed purple-red or maroon at first	deep to pale lilac-rose, in clusters below leaves in spring	
D. × *hybrida* (*D.sericea* × *D.odora*)	garden origin	bushy, evergreen to 1.5m	3-7cm, oblong to ovate, glossy, leathery	very fragrant, deep purple-pink in terminal clusters throughout the year	
D.jasminea	SE Greece	erect or semi-prostrate evergreen to 30cm	to 1cm, oblong to lanceolate, blue-grey, smooth	highly fragrant, white sometimes with a red-purple exterior, in sparse terminal clusters in spring	
D.laureola SPURGE LAUREL	Europe, W Asia	evergreen, erect to 1.5m	5-8cm, obovate to lanceolate or oblong, leathery, glossy	fragrant, yellow-green, in clusters below leaves in late winter and early spring	black
Comments: ssp. *philippi* is semi-prostrate and compact, the flowers smaller, lime green and sometimes tinted purple					
D. × *mantensiana* (*D.* × *burkwoodii* × *D.retusa*)	garden origin	evergreen, dwarf and rounded habit	to 3.5cm, glossy dark green	very fragrant, deep rose-purple to lilac in terminal clusters in spring, summer and autumn	
Comments: 'Manten', the clone usually seen, is described here					
D.mezereum MEZEREON, FEBRUARY DAPHNE	Europe	deciduous shrub to 1.5m with numerous, erect, rod-like branches arising from a short trunk-like stem	3-8cm, oblanceolate, soft sea green, thinly downy at first	fragrant, rose to deep purple-red in clusters below leaf buds or leaves in winter and early spring	bright red, persistent
Comments: Includes selections with white, palest pink and deep crimson flowers, also with variegated leaves					

DAPHNE

Name	Distribution	Habit	Leaves	Flowers	Fruit
D.×napolitana (*D.sericea* × *D.cneorum*)	garden origin	bushy erect evergreen to 75cm	2-3.5cm, oblanceolate to narrowly ovate, shiny above, glaucous beneath	very fragrant, rose lilac , in terminal clusters in spring with further flushes in summer and autumn	
D.odora WINTER DAPHNE	China, Japan	evergreen, erect to procumbent to 2m	5-8cm narrowly ovate to lanceolate or elliptc, leathery, glossy	sweetly fragrant, rose tinted purple-red or white with a rosy purple exterior, in terminal clusters in late winter and spring	red
Comments: 'Aureomarginata' : leaves deeply edged creamy yellow to pale gold					
D.petraea	N Italy	evergreen, mat-forming, to 30cm	1cm, narrowly oblanceolate to spathulate, tough, dark green	fragrant, rose pink in terminal clusters in spring and summer	green-brown, downy
Comments: 'Grandiflora': flowers larger, deeper rose to magenta					
D.pontica	SE Europe, Caucasus, Asia Minor	evergreen to 1.5m	3-10cm, obovate to oblong, pointed, deep, glossy green	fragrant, pale yellow-white or yellow-green in pairs usually below new growth in spring and summer	black
D.sericea *D.collina*	E Mediter-ranean	small, many-branched evergreen to 1m	1-5cm, oblanceolate to obovate or narrowly elliptic, glossy above, downy beneath	strongly scented, dark rose ageing buff in clusters in spring, sometimes with a second flush in autumn	red or orange-brown
Comments: The name *D.collina* applies in gardens to a very hardy form with dense, dark green leaves and small, purple-pink flowers					
D.tangutica	NW China	evergreen, erect and open or rounded and densely branched to 1.75m	3.5-6cm, oblanceolate to oblong or elliptic, more or less acute glossy, leathery, deep green	fragrant, exterior rose-purple, interior white flushed rose-purple in tube, in dense, terminal clusters in spring and again in late summer	red
Comments: Often treated as synonymous with this species, *D.retusa* differs in its slow-growing, dwarf and rather stiff habit and its shorter, retuse leaves					
D. 'Valerie Hillier' (*D.cneorum* × *D.longilobata*)	garden origin	dwarf., spreading evergreen	to 5cm, narrowly oblong to elliptic, glossy green	fragrant, pink-purple fading to white edged pink from spring to autumn	

Daphne 'Albert Burkwood'

Daphne odora 'Aureomarginata'

a minimum winter temperature of 5°C/40°F. Plant in a mix of live sphagnum, moss peat and silver sand. In spring and summer, stand the pot in rainwater or distilled water to half its depth and water or mist overhead daily with the same. In winter, keep permanently moist, not sodden. In the home, grow in large terraria. Propagate by division in spring.

Darmera (for Karl Darmer, 19th-century Berlin horticulturist). Saxifragaceae. W US. UMBRELLA PLANT, INDIAN RHUBARB. 1 species, *D.peltata* (syn. *Peltiphyllum peltatum*), a perennial herb with fleshy, creeping and flattened rhizomes. The rounded leaves are some 30–60cm across, with toothed to lobed margins and attached to hairy stalks some 1–2m tall. Produced in spring, the flowers are small, white to pink, and borne in domed cymes on stout and hairy, purple-tinted scapes to 40cm tall. Fully hardy, it is an excellent perennial for damp gardens and the margins of ponds, lakes and streams. Plant in full sun or part shade on a humus-rich medium that never dries out. Once established, it will tolerate saturated soils when in growth. Mulch the networks of stout, exposed rhizomes in winter. Propagate by division in spring.

Darwinia (for Erasmus Darwin (1731–1802), naturalist and writer). Myrtaceae. W Australia. 60 species, dwarf to medium-sized shrubs with aromatic leaves. Produced in spring, the flowers are small and arranged in terminal heads surrounded by a pair of conspicuous bracts. Provide a minimum temperature of 7°C/45°F with full sun and a buoyant, rather dry atmosphere. Grow in a perfectly drained, sandy, acid to neutral medium low in nitrogen. Water moderately when in growth, at other times, sparingly. Trim back after flowering to maintain compactness. Propagate by seed in spring or by semi-ripe cuttings in summer. *D.citriodora* (LEMON-SCENTED MYRTLE; dwarf to small shrub; leaves to 2cm, glaucous, lemon-scented; flowerheads to 3cm across, erect to nodding, bracts orange-red and green, flowers orange-red).

Dasylirion (from Greek *dasys*, thick, and *leirion*, lily, owing to the thickness of the stem). Agavaceae (Dracaenaceae). S US to Mexico. SOTOL, BEAR GRASS. 18 species, evergreen perennials, sometimes with a short, stout trunk at whose summit the linear, spiny-edged leaves emerge in a fountain-like rosette. Off-white, bell-shaped flowers are borne in large, narrow panicles in summer. Grow in full sun with a minimum temperature of 5°C/40°F and in a dry, airy atmosphere. Plant in a fast-draining, loam-based medium high in sand and grit. Water moderately in spring and summer, very sparingly in winter. Propagate by seed in spring. The following may prove hardy in very sheltered locations in zones 7 and 8. Plant in a gritty soil near a sun-baked wall. Provide a dry mulch in winter.
D.acrotrichum (Mexico; trunk to 1.5m; leaves to 100 × 1–2cm, pale green, finely toothed with hooked, pale yellow spines with fibrous tips; inflorescence to 4m); *D.glaucophyllum* (Mexico; trunk absent or to 30cm; leaves to 120 × 1-2cm, glaucous, margins very finely toothed and bearing hooked, yellow spines; inflorescence to 6m); *D.texanum* (TEXAS SOTOL; Texas, N Mexico; trunk to 5m, often buried; leaves to 100 × 1–1.5cm, glossy green, marginal spines yellow becoming brown and with fibrous tips; includes the glaucous-leaved 'Glaucum').

Datura (from an Indian vernacular name). Solanaceae. N and C America, naturalized elsewhere. THORN APPLE. 8 species, annual or short-lived perennial shrubby herbs with entire or sinuately toothed leaves and large flowers in summer. These are broadly funnel-shaped with five, pointed folds. The fruit is a spiny, 2-chambered capsule. Woody perennial species with pendulous, not erect flowers, are now included in *Brugmansia*. Grow in full sun in moisture-retentive but well-drained, fertile and preferably calcareous soil. Propagate by seed sown *in situ* in spring or earlier under glass. Set out seedlings after danger of frost has passed. The foliage is extremely susceptible to viruses affecting other solanaceous plants and may act as a host; do not grow in close proximity to potato crops. All parts are extremely toxic.
D.ceratocaula (C and S America; purple-tinted annual to 90cm; flowers fragrant, to 18cm, white, marked or flushed red-mauve); *D.innoxia* (DOWNY THORN APPLE, INDIAN APPLE, ANGEL'S TRUMPET; C America, naturalized Old World; hairy annual to 1m tall; flowers to 10cm, white to pink or lavender); *D.metel* (HORN OF PLENTY, DOWNY THORN APPLE; S China, widely naturalized in the tropics; glabrous annual to 1.5m, stems sometimes purple-tinted; flowers to 20cm, white to creamy yellow or deep purple, sometimes bicoloured, often double); *D. stramonium* (JIMSON WEED, JAMESTOWN WEED, COMMON THORN APPLE; Americas, naturalized Europe; glabrous to short-hairy annual to 2m; flowers to 10cm, white or purple).

Daucus (from *daukos*, ancient Greek name). Umbelliferae. Cosmopolitan. 22 species, tap-rooted annual and biennial herbs with finely cut, 2–3-pinnate leaves and small, white to cream or mauve flowers in compound umbels. A native of Europe and Central Asia, *D.carota*, WILD CARROT, is a summer-flowering, hardy biennial, useful in cultivation for meadow areas of the wild garden. It is a food plant for the caterpillars of the swallow-tail butterfly, and a source of nectar and pollen for bees. Sow seed *in situ* in autumn or early spring. Subspecies *sativus*, with a fleshy, orange taproot, is the edible *carrot* (q.v.).

Davallia (for E. Davall (1763–98), Swiss botanist). Davalliaceae. Warm temperate, subtropical and tropical Old World. HARE'S FOOT FERN. Some 40 species, deciduous or semi-deciduous frost-tender ferns with finely pinnate or pinnatifid fronds, deltoid to ovate in outline and delicately cut into lacy segments. The genus is most readily recognized by the tough, creeping rhizomes. Covered in chestnut scales, they branch and ramble freely, cascading from pots, encasing baskets and sometimes forming impenetrable tangles. Grow in bright indirect light or shade, with a minimum winter temperature of 7°C/45°F. Provide an airy, humid atmosphere. Water and syringe generously when in full growth, but scarcely at all during dormancy and cold weather. Well-established plants will withstand periods of drought and extremes of temperature, but may drop all their fronds. Plant the rhizome bases in an open, soilless medium with added bark; thereafter, allow the rhizomes to follow their own path – often through the air and outside the pot. Repot as seldom as possible. These ferns were often grown as 'balls' – apparently self-supporting spheres of dense foliage hung in lightly shaded, humid places. To create such a 'ball', make a globe of wire mesh lined with moss and filled with potting mix. Poke rooted lengths of rhizome into the holes or pin them to the surface. Keep moist, warm and shaded until established. Water large 'balls' by plunging. Push fresh compost between the tangle of rhizomes at least once a year. The best species for this treatment are *D.bullata* and *D.canariensis*. Increase all by detaching rooted pieces of rhizome.

D.bullata (HARE'S FOOT FERN; tropical Asia; fronds to 30cm, 3–4-pinnate, long-stalked and rather soft-textured, bronze at first becoming dark olive to jade green, fertile fronds distinctly bullate above, segments linear, acute); *D.canariensis* (CANARY ISLAND HARE'S FOOT FERN, DEER'S FOOT FERN; W Mediterranean to Atlantic Islands; fronds to 45cm, 4-pinnate, emerald to dark green, rather tough, fertile fronds bullate above, segments ovate to rhomboid); *D.fejeensis* (RABBIT'S FOOT FERN; Fiji; fronds to 45cm, 4-pinnate, fine and feathery, segments linear, sometimes cleft at apex); *D.mariesii* (SQUIRREL'S FOOT FERN, BALL FERN; E Asia; fronds to 20cm, 3–4-pinnate, finely cut, bright green, segments lanceolate to oblong, entire to lobed); *D.trichomanoides* (SQUIRREL'S FOOT FERN, BALL FERN; S E Asia; fronds to 30cm, 3–4-pinnate, very finely cut, dark green, segments cleft to toothed and horned).

Davidia (for the French missionary Armand David, who introduced the tree from China in the 1890s). Nyssaceae (Davidiaceae). SW China. DOVE TREE, GHOST TREE, HANDKERCHIEF TREE. 1 species, *D.involucrata*, a deciduous tree to 20m, with flaking grey-brown bark and a broadly pyramidal crown. The leaves are broadly ovate, to 16 cm, sharply serrate, with a cordate base and an acuminate apex. The small purple flowers appear in spring and summer, forming a 2cm-wide ball-like head that hangs on a slender stalk. The head is subtended by two, drooping and fluttering, pure white bracts, the larger to 20cm. Hardy in climate zone 6. Plant on deep, moist loam in a sheltered site. In cold areas, protect when young with burlap or sacking during the winter. Propagate by semi-ripe cuttings under mist in summer or from heeled cuttings in a cold frame; alternatively, propagate by seed sown in a cold frame in spring while still moist from fruit (drying may inhibit germination).

day-degrees a measure of accumulated heat units; calculated by progressively adding the difference between the daily mean temperature and a prescribed base temperature. The measure is used to determine optimum sowing or harvest dates for certain vegetable or fruit crops.

day length, day neutral see *photoperiodism*.

dead-heading the removal of faded blooms from flowering plants. It is carried out to improve the plant's appearance and to prevent seed from setting, thereby in many cases encouraging further flowering.

dealbate whitened, covered with a white powder.

deblossoming the removal of flowers from fruit trees, and sometimes from late-planted strawberries, in the first season after planting, as an aid to establishment. It is also a means of discouraging biennial bearing in top fruits.

Decaisnea (for Joseph Decaisne (1807–82), Director of the Jardin des Plantes, Paris.) Lardizabalaceae. E Asia. 2 species, deciduous shrubs with pinnate leaves and, in early summer, flowers composed of six, narrow sepals and hanging in slender, pendulous racemes. These are followed by large, bean-like fruit containing black seeds embedded in white pulp. Hardy to at least –15°C/5°F, but often hit by late frosts. Grow in a moist, fertile, loamy soil in sun or partial shade. Propagate from seed sown fresh in the cold frame in autumn, or under glass in early spring, although germination may be slow and irregular by the latter method. *D.fargesii* (W China; erect to 3m with several, rod-like stems; leaves to 80cm, composed of up to 25 leaflets, each to 14cm long, ovate to elliptic and glaucous beneath; flowers to 4cm, yellow-green in panicles to 40cm long; fruit to 10cm long, thickly cylindrical, pendulous, leaden blue).

deciduous (1) falling off when no longer functional, as with non-evergreen leaves, or the petals of many flowers; (2) of a plant that sheds its leaves annually, or at certain periods; cf. *evergreen*.

deck a terrace made of wood boarding, common in the US. Popular as an alternative to stone or concrete paving, decking is also suitable for raised viewing platforms and linking across water, to provide an effective contrast of form. However, deck surfaces can be slippery when wet.

declinate bent or curved downward or forward.

decompound a compound leaf with two or more orders of division, for example bipinnate, tripinnate, triternate.

Decumaria (from Latin *decimus*, tenth, referring to the number of flower parts). Hydrangeaceae. W China, S US. 2 species, climbing deciduous or semi-ever-

Decaisnea fargesii

Davidia involucrata

*Decumaria
barbara*

green shrubs, similar to *Hydrangea anomala* and *Schizophragma*, from which they differ in their inflorescences, which are wholly composed of small, fertile florets. Fully hardy; cultivate as for *Schizophragma*.

D. barbara (CLIMBING HYDRANGEA, WOOD VAMP; SE US; to 10m; leaves to 10cm, ovate-oblong, apex tapering, margin toothed; flowers creamy white, sweetly fragrant, in broad panicles to 10cm);*D. sinensis*(leaves to 7cm, broader than in *D. barbara* and often persistent; flowers muskily scented).

decumbent of a stem, lying horizontally along the ground, but with the apex ascending and almost erect.

decurrent of a leaf blade, where the base extends down and is adnate to the petiole (if any) and the stem.

decurved curved downwards.

decussate of leaves, arranged in opposite pairs with adjacent pairs at right angles to each other, thus forming four longitudinal rows.

deep bed method a method of cultivation in beds constructed above soil level, usually to a height of about 30cm. Ideally, the beds should be about 1.2m wide so that all necessary work can be carried out from the pathways, avoiding compaction through treading. The method is frequently associated with deeply dug beds and high organic-matter input, and is sometimes referred to as the French intensive method. See *bed*, *raised bed*.

deer any mammal of the family *Cervidae*, of which two species, the ROE DEER (*Capreolus capreolus*) and FALLOW DEER (*Cervus dama*), are most likely to invade gardens in Europe; other species such as SIKA and MUNTJAC may do so in some areas. Deer attack trees and shrubs by browsing foliage and stripping bark. They may also fray bark by rubbing their newly formed antlers against it to remove the velvet covering. Low-growing plants are eaten to ground level. They mostly feed at dusk and dawn so are seldom seen, but hoof marks are usually apparent in soft soil. Since deer lack incisor teeth in the upper jaw, shoots of damaged trees and shrubs are cleanly cut on one side with a characteristic frayed edge on the other. A wide range of plants is subject to attack but the extent to which a particularly garden is affected will depend on alternative woodland sources of food, deer population and weather conditions. Control is best effected by erecting strong fencing around a garden to a height of 2m or by protecting trees with trunk guards. Repellents and scaring devices are seldom effective in the long term.

deficiency a condition in plants, soil or growing media where essential nutrients are absent or insufficient to sustain healthy growth. Symptoms of deficiency – variously expressed in leaves, shoots and fruits, according to species and the nutrient concerned – commonly arise as a result of the inability of plants to absorb or transport available nutrients. This may be due to soil acidity or alkalinity, waterlogging or drought. See *nutrients*.

deflexed bent downward and outwards; cf. *reflexed*.

deflocculation the disintegration of aggregated clay particles, as may occur on soils low in calcium and high in sodium. It leads to the loss of soil structure.

defoliant a chemical substance which when applied to a plant induces abscission, causing the leaves to fall.

defoliation the removal of leaves from plants. It may occur as a natural response to seasonal change in the case of deciduous species, or as a result of pest or disease attack, drought, waterlogging, or other disorder in the case of evergreen species also. It may be a deliberate operation for cultural advantage, especially in the case of tomatoes, where lower leaves are progressively removed as a means of improving air circulation and extracting diseased tissue.

defruiting see *fruit thinning*.

degree-days see *day-degrees*.

dehiscence the mode and process of opening (i.e. by valves, slits, pores or splitting) of a seed capsule or anther to release seed or pollen.

dehiscent of a seed capsule; splitting open along definite lines to release seeds when ripe; cf. *indehiscent*.

dehorning the severe cutting back of a selection of large branches in an established tree, especially a fruit tree, to envigorate, contain or improve shape. The process is less drastic than heading back.

Deinanthe (from Greek *deinos*, strange, and *anthos*, flower, referring to the unusually large flowers). Hydrangeaceae. E Asia. 2 species, robust, erect perennial herbs with ovate to elliptic, toothed or lobed leaves, borne toward the stem apex. Produced in summer in terminal clusters or panicles, the *Hydrangea*-like flowers are numerous, large and nodding. Hardy to about –15°C/5°F. Grow in sheltered shade on a cool, humus-rich, moisture-retentive but freely draining lime-free soil. Propagate by division or by ripe seed sown when fresh. *D.caerulea* (China; to 50cm; flowers pale mauve to lilac-blue with blue stamens).

delayed open-centre a form of fruit tree where the lower main branches arise from a central trunk over a distance of 1.2–1.8m, above which the head opens out to produce a goblet shape.

Delonix (from Greek *delos*, conspicuous, and *onyx*, claw, referring to the showy, clawed petals). Leguminosae. Old World tropics. 10 species, evergreen or deciduous trees with bipinnate leaves and racemes of showy flowers, each composed of five clawed petals and a bundle of long stamens. The following species is widely planted in tropical and subtropical regions as a fast-growing and long-flowering street tree. In temperate zones, it will grow in large tubs in the greenhouse or conservatory, with a minimum winter temperature of 10°C/50°F. In frost-free zones, grow in deep, well-drained soils with full sun and shelter from cold winds. Propagate as for *Caesalpinia*. *D.regia* (syn. *Poinciana regia*; FLAMBOYANT, PEACOCK FLOWER, FLAME TREE, ROYAL POINCIANA; Madagascar; to 10m with spreading crown; flowers to 10cm across, scarlet to flame or orange, variably marked or striped gold to cream or pure gold).

Delosperma (from Greek *delos*, visible, and *sperma*, seed, referring to the lidless seed capsules). Aizoaceae. Africa. Over 150 species, perennials or, rarely, annuals or biennials, with succulent leaves and daisy-like flowers. Cultivate as for *Lampranthus*. *D.tradescantioides* (South Africa; dwarf and creeping perennial with ovate to cylindrical, 3-angled leaves to 3cm long and white flowers to 2.5cm wide).

Delphinium (from Greek *delphis*, dolphin; *delphinion* was a name used by Dioscorides). Ranunculaceae. North temperate regions. 250 species, annual, biennial or perennial herbs with palmately or digitately lobed and cut leaves in basal clumps and on tall, erect stems. Produced in summer in narrow, sometimes branching racemes, the flowers are spurred and consist of five petal-like sepals and two pairs of smaller petals. When the latter are distinctively coloured and downy, they are termed the 'bee'. The smaller species are suited to the rock garden and sunny borders on free-draining, gritty soils. Those from the Western United States may need winter protection in cold areas.

The tall spectacular hybrids of gardens, which are available in varying shades of blue, and purple, mauve, pink, cream and white are derived mainly from *D.elatum*, *D.grandiflorum* and *D × belladonna*.

The Pacific Hybrids bred in California to flower as annuals and similar English hybrids which are suitable perennials, each with *D.elatum* in their parentage, produce large, erect and dense inflorescences which may reach 2.5m in height. Belladonna hybrids generally produce more but smaller flowers on less dense inflorescences which attain up to 1.5m in height. They produce side growths which make the plants suitable for cut flowers as well as garden decoration, although the colour range is narrower than in the taller hybrids. There is a wide choice of hybrid cultivars.

Delphiniums thrive in any fertile soil which is not waterlogged in winter. Perennial types can be planted from spring to early autumn and annual types in the spring; adequate watering is essential. Hybrids and cultivars do not come from seed. They may be propagated by division in early autumn or early spring or, preferably, from 5–10cm shoot cuttings, with solid stems, taken in the spring. Rooting can be achieved in a mixture of peat and sand, or in perlite or in water.

Where appropriate, seed should be packaged and refrigerated to preserve viability, before sowing during late winter or early spring under protection at a temperature of 15°C/60°F. In the US delphiniums are not usually grown as perennials.

Tall garden hybrids require staking, and for high quality flower spikes shoots should be thinned to 5–7 per established plant. Watering is essential in dry spells, and cutting the stems to ground level immediately after flowering will produce a second flush, but supplementary feeding should be applied.

D. × *belladonna* (*D. elatum* × *D. grandiflorum*; garden hybrids with tall, branching stems to 1.75m with single or semi-double flowers in shades of sky blue, dark blue, gentian and indigo, often with a white 'eye'); *D. brunonianum* (Himalaya; perennial to 50cm tall with pale purple-blue, dark-spurred flowers); *D. cardinale* (California; short-lived perennial to 60cm with spikes of scarlet flowers with yellow 'eyes'); *D. elatum* (Southern and Central Europe to Siberia; perennial to 2m tall with dense spikes of single, semi-double or double flowers in shades of blue, purple and white and with contrasting 'eyes'); *D. grandiflorum* (syn. *D. chinense*; Siberia, E Asia; perennial often treated as an annual, to 1m tall with blue, purple or white flowers in loose racemes, usually with contrasting dark and white 'eyes'); *D. nudicaule* (California; short-lived perennial, 20-60cm tall with loose racemes of yellow-eyed, orange to fiery red flowers); *D. tatsienense* (China, Tibet; short-lived perennial, 20-60cm tall with loose spikes of bright blue or white flowers). For *D. ajacis* and *D. consolida* see *Consolida ambigua*.

deltoid, deltate resembling an equilateral triangle attached by the broad end rather than the point; shaped like the capital form of the Greek letter, delta Δ.

demersed of a part constantly submersed under water.

Dendranthema (from Greek *dendron*, tree, and *anthemon*, flower). Compositae. Europe, C and E Asia. About 20 species, muskily aromatic perennial herbs, sometimes woody at base, with pinnatifid or entire leaves and daisy-like flowerheads in loose corymbs produced in late summer and autumn. *D.pacificum* (syn. *Ajania pacifica*) is native to Japan, a perennial to 40cm tall with clumped stems of lobed leaves, their undersides bright silver, and white to yellow flowerheads. Fully hardy, grow in sun. Increase by division. The florist's chrysanthemum is *D. grandiflorum*, an ancient cultigen probably raised in gardens from the yellow-flowered, Japanese *D.indicum* (see *chrysanthemum*).

dendritic, dendroid branching finely at the apex of a stem, in the manner of the head of a tree.

dendrology the botanical science concerned with the natural history of trees.

Delphinium nudicaule

delphinium

Pests
aphids
slugs and snails

Diseases
powdery mildew
virus

Dendrobium (from Greek *dendron*, tree, and *bios*, life, referring to the epiphytic habit of most species). Orchidaceae. Asia, Polynesia, Australia. At least 1000 species, mostly epiphytic, perennial herbs. They range widely in habit from dwarf plants with clumps of fleshy leaves to tall species with cane-like pseudobulbs. The flowers consist of five spreading tepals and a showy lip. They are produced at various times of year, singly or in clusters, racemes or panicles. Plant in small containers, preferably clay pots or baskets, containing a very open, bark-based mix. In very humid conditions, plants can also be mounted on slabs of bark or dead branches. Grow in bright, filtered sun with a buoyant, humid atmosphere. Provide a minimum temperature of 10°C/ 50°F. Water, mist and feed generously when in growth. When resting, they should be watered only to prevent the pseudobulbs from shrivelling. *D. nobile*, *D. bigibbum* and their respective hybrids make excellent houseplants. Increase by division; also by detaching and growing on the plantlets that sometimes develop along the cane-like pseudobulbs.

DENDROBIUM

Name	Distribution	Flowering time	Hardiness	Habit	Flowers
D.aggregatum	India	spring	cool-growing	evergreen with squat, spindle-shaped pseudobulbs	to 4cm diam., often scented, in arching racemes, golden yellow, throat orange
Comments: Grow in pans; give cool dry rest in winter					
D.aphyllum *D.pierardii*	India	late winter	cool-growing	deciduous with narrow pendulous canes	to 5cm diam., in clusters along canes, translucent white to pink, lip cream-yellow marked purple at base
Comments: Mount on bark or raft to allow for pendulous habit					
D.bigibbum COOKTOWN ORCHID	Australia	year round	intermediate- to warm-growing	evergreen with erect canes and leathery dark green leaves	to 6cm diam., in long-stalked sprays, pink, purple, magenta or white, lip spurred, often with a white throat
Comments: Includes ssp. ***phalaenopsis*** with broader petals and strong colour. A common cut flower with many hybrids and cultivars					
D.brymerianum	Burma	spring	warm-growing	evergreen with tall erect canes	to 6cm diam., in terminal racemes, glossy yellow, lip marked orange with long branching threads on margin
D.canaliculatum ANTELOPE ORCHID	Australia	autumn	intermediate-growing	evergreen with squat pseudobulbs and narrow leaves	to 2.5cm diam., fragrant, in tall dense racemes, white tipped yellow to brown, lip white marked purple
D.chrysanthum	India	spring	cool-growing	deciduous, canes long, slender, pendulous	to 4cm diam. fragrant, in short clusters, golden yellow, lip fringed, with a dark red-brown blotch
Comments: Grow in a basket to allow for pendulous habit					
D.crumenatum DOVE ORCHID	Burma, Malaysia	summer	intermediate-growing	evergreen; canes sometimes branched	to 4cm diam., along slender leafless stem tips, fragrant, white, lip marked yellow, sometimes veined pink
Comments: Flowers very short-lived but produced in great abundance					
D.cucumerinum	Australia	summer	intermediate-growing	dwarf creeping plant with succulent, warty, gherkin-shaped leaves	to 1.5cm diam., in short racemes, green-white to cream marked red-brown
Comments: Grown for its curious habit; must be mounted on a raft or bark					
D.densiflorum	India	spring	cool-growing	evergreen, medium-sized; canes club-shaped	to 5cm diam, fragrant in dense pendulous racemes, yellow, lip orange in throat, finely fringed
D.fimbriatum	India, Burma, Thailand, China, Malaysia	spring	cool-growing	deciduous, canes long, narrowly spindle-shaped	to 5cm diam., in pendulous racemes, deep orange-yellow, lip golden with maroon blotch, margin intricately fringed
D.infundibulum	Burma, Thailand	late spring	intermediate-growing	evergreen; canes erect, with sparse black felt	to 8cm diam., in clusters, sparkling white, lip funnel-shaped, frilled with yellow stain at base
D.kingianum	Australia	year round	cool-growing	medium-sized evergreen with dense clumps of tapering pseudobulbs and tough leaves	to 3cm diam., in erect terminal racemes, white cream, pink or purple-red
Comments: Very vigorous plant flowering profusely if left to form large pot-bound clumps					
D.loddigesii	Laos, China	spring	intermediate-growing	semi-evergreen; short sprawling canes with fleshy leaves	to 5cm diam. solitary along stems, lilac-purple, lip orange banded white, edge purple, fringed
Comments: Grow in pans					

DENDROBIUM

Name	Distribution	Flowering time	Hardiness	Habit	Flowers
D.nobile	India, China, Thailand	winter-spring	cool-growing	deciduous or semi -evergreen; canes tall, erect	to 8cm diam., scented, in short clusters, white flushed mauve to pink lip white edged rosy-pink, blotched velvety crimson to maroon in throat
Comments: Very resilient, popular plant with many colour forms including whites, purples and yellow-marked. Parent of many hybrids					
D.moschatum	India to Thailand	summer	cool-growing	evergreen; canes erect	to 7cm diam., in drooping racemes, pale yellow tinted or veined pale rose, lip hairy, slipper-shaped with 2 dark purple blotches
D.speciosum	Australia	year-round	intermediate-growing	evergreen; pseudobulbs stoutly club-shaped; leaves leathery	to 4cm diam., fragrant, waxy, in dense terminal racemes, narrow, not opening fully, white to cream, lip marked purple
Comments: Highly floriferous if allowed to become pot-bound and placed outdoors in full sun from mid to late summer					
D.spectabile	New Guinea	summer	warm-growing	evergreen; pseudobulbs tall, cane-like; leaves fleshy	to 8cm diam., in erect racemes, tepals narrow, wavy and twisted yellow-green striped purple, lip paler veined purple-red narrowing to a beak-like tip
D.strebloceras	Molucca	summer	warm-growing	evergreen; pseudobulbs tall, cane-like; leaves fleshy	to 8cm diam., fragrant in lax racemes, green-yellow marked purple, pet. narrow, erect and strongly twisted
D.wardianum	Assam, Burma, Thailand	spring	cool-growing	semi-evergreen; canes long, narrow, arching	to 10cm diam., fragrant, solitary or clustered, white tipped magenta, lip with orange stain and dark crimson blotches in throat
Comments: Requires high temperatures and plentiful water and feed in growth, cool dry conditions at rest					

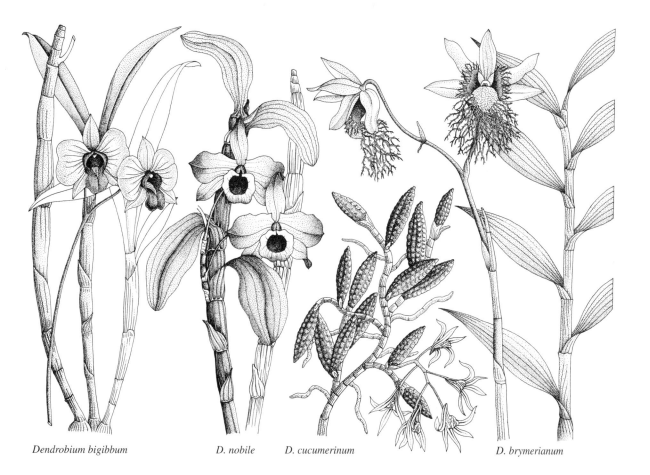

Dendrobium bigibbum   *D. nobile*   *D. cucumerinum*   *D. brymerianum*

Dendromecon (from Greek *dendron*, tree, and *mekon*, poppy). Papaveraceae. SW US, Mexico. TREE POPPY, BUSH POPPY. 1 species, *D.rigida*, a summer-flowering, evergreen shrub or small tree, 3–6m tall, with tough, ovate to lanceolate leaves to 10cm long. The solitary flowers are 2–7cm in diameter, fragrant and composed of four yellow petals and numerous stamens. It is hardy in zone 8 if planted near a sunny, south-facing wall and on a well-drained loamy soil, with added sharp sand and rubble. Protect in winter with a dry mulch. Propagate by ripe wood cuttings in summer, with moderate heat, in a very sandy mix.

denitrification the reduction of nitrates to nitrites, ammonium and gaseous ammonia through microbial activity in anaerobic conditions. See *nitrogen*.

Dennstaedtia (for August Wilhelm Dennstedt (1776–1826), German botanist and physician). Dennstaedtiaceae. Tropics and subtropics, eastern N America. Some 70 species, evergreen or deciduous ferns with creeping rhizomes and more or less deltoid, 1–4-pinnate or -pinnatifid fronds. The following species is hardy to at least –30°C/–22°F. Plant in the woodland garden in full or part shade on a moist and acid, humus-rich soil. Propagate by division or spores. *D.punctiloba* (HAY-SCENTED FERN; eastern N America; deciduous, fronds to 75 × 20cm, bright green, deltoid to lanceolate, finely 2–3-pinnate; pinnules ovate to rhomboid, acute, notched to lobed; scented of hay when dried).

dentate toothed; of the margins of leaf blades and other flattened organs, cut with teeth. Strictly, dentate describes teeth that are shallow and represent two sides of a roughly equilateral triangle, in contrast to serrate (saw-toothed), where the teeth are sharper and curved forwards, and crenate, where the teeth are blunt and rounded. Less precisely, the term is used to cover any type of toothed margin.

denticulate minutely dentate.

depauperate reduced in stature, number or function as if starved and ill-formed. Species with organs described thus are, however, usually perfectly healthy but adapted to cope with a particular ecological factor.

dependent hanging downward as a result of its weight, as in a flower- or fruit-laden branch.

deplanate flattened or expanded.

depressed sunken or flattened, as if pressed from above.

descending tending gradually downwards.

Deschampsia (for Louis Deschamps (1765-1842), French naturalist). Gramineae. HAIR GRASS. Temperate Regions. 50 species, perennial or, rarely, annual grasses forming dense tussocks. They flower in summer, bearing many small spikelets in finely branched, long-stalked panicles. Both species described below are perennial, fully hardy and will grow in sun or light shade. They prefer moist (not wet) conditions, but will tolerate some drought, especially from late summer onwards. *D. cespitosa* will grow in heavy, poorly drained soils. *D. flexuosa* needs a neutral to acid soil, ideally rich in leafmould and sand. Clumps of *D. flexuosa* may become tired and bald within a few years. Increase both species by division or seed.

D. cespitosa (syn. *D. caespitosa*; HASSOCKS, TUFTED HAIR GRASS, TUSSOCK GRASS; northern temperate regions; evergreen to semi-deciduous, densely tufted, forming large, erect to spreading clumps; leaves 10-30 × 0.25-0.5cm, linear, glossy dark green, ridged and channelled above, rough to touch; inflorescence 20-100cm, loosely pyramidal to cylindrical, branches occupying up to half length of inflorescence, sometimes weeping to one side, spikelets small, green or golden suffused silver to bronze or purple, pale blonde when dry and persisting well into autumn; var. *parviflora*: a small, rare and rather fine variety; var. *vivipara*: spikelets develop into plantlets instead of seed; 'Bronzeschleier': syn. 'Bronze Veil', inflorescence large, tinted bronze; 'Goldhaenge': syn. 'Golden Showers', late-flowering, tall with golden inflorescences; 'Tautraeger': syn. 'Dew Carrier': small, compact, inflorescence tinted blue); *D. flexuosa* (WAVY HAIR GRASS, COMMON HAIR GRASS; northern temperate regions; deciduous to semi-evergreen, densely tufted forming low, mop-like clumps; leaves to 15 × 0.1cm, thread-like, terete, bright green, smooth; inflorescence 20-100cm, loosely pyramidal, branches occupying uppermost quarter of inflorescence, very fine, somewhat wavy, glistening, coppery red to bronze-green, spikelets small, shimmering, bronze to copper-red, silver or yellow-green; 'Tatra Gold': leaves pale gold to lime green, inflorescence soft bronze, comes true from seed).

Desfontainia (for René Louiche Desfontaines (1752–1833), French botanist). Loganiaceae. Chile, Peru. 1 species, *D.spinosa*, a bushy, evergreen shrub to 3m with holly-like leaves. Produced in late summer and autumn, the solitary flowers are 1.5–9cm long, tubular to funnelform, crimson to scarlet-orange and edged with five yellow, shallow lobes. With shelter from cold, drying winds, it is hardy to –10°C/14°F, but may also be grown in the cold greenhouse or conservatory. Plant in a humus-rich, moisture-retentive, acid soil in partial shade. Mulch established plants and water plentifully in dry weather. Propagate from seed surface-sown on a soil-less propagating mix, or in summer by semi-ripe cuttings rooted under mist or in a closed case with bottom heat.

deshooting the removal of soft young shoots from trained fruit trees, especially those growing towards or out from the wall on wall-trained trees.

desiccant a chemical substance that causes the aerial parts of plants to dry out. It is used to facilitate the harvesting of commercially grown potatoes. In gardens, lawn sand has such an effect on some plants growing in turf.

Desmodium (from Greek *desmos*, band, referring to the stamens which are united in a band). Leguminosae. Widespread. BEGGARWEED, TICK-TREFOIL. Over 300 species of deciduous shrubs, subshrubs, herbs or perennials with 3-lobed to trifoliolate or pinnate leaves and pea-like flowers in racemes or panicles in late summer. The genus is hardy in zone 6. Cultivate as for *Indigofera*. *D.elegans* (syn. *D.tiliifolium*; shrub to 1.5m; leaves with three leaflets, each to 6cm long, obovate to rounded, dark green above, grey-downy beneath; flowers rose to carmine or pale lilac, in racemes to 20cm long).

determinate (1) of inflorescences such as cymes where a central or terminal flower opens first, thus ending extension of the main axis; (2) of tomato cultivars of bush habit, where side branches develop instead of a continuous main stem.

Deutzia (for Johann van der Deutz (1743–88), friend and patron of Thunberg, the namer of this genus). Hydrangeaceae. Asia, C America. 60 species, usually deciduous shrubs with pith-filled branches and peeling bark. The leaves are ovate to lanceolate or elliptic, sometimes toothed and more or less hairy. Often fragrant and in tones of white, cream and rose to purple-red, the flowers appear in spring and summer in racemes, cymes, panicles or corymbs. They consist of five elliptic to lanceolate or rounded petals and ten stamens. Most species will tolerate cold to at least −15°C/5°F, although a number, including *D.gracilis*, *D.grandifora*, *D.× lemoinei* and *D. × rosea*, are susceptible to damage by late frosts. They perform best in moist but well-drained, fertile and humus-rich loamy soils, with shade from early morning sun in regions that suffer late spring frosts, and with protection from hot afternoon sun. After flowering, remove old and overcrowded growth at the base. Flowers are carried on the short laterals produced by the previous season's growth and are produced more generously on young wood. Propagate species by seed, cultivars and hybrids by soft nodal cuttings in late spring – treat cuttings with 0.4% IBA and root in a sandy mix in a closed case with bottom heat. Alternatively, propagate by semi-ripe cuttings and hardwood cuttings in a cold frame, by removal of suckers or by layering for low-growing plants such as *D.compacta*, *D.scabra*, *D. × rosea*.

D.compacta (China; to 1m; leaves to 7cm, lanceolate, finely toothed; flowers to 1.2cm across, white, in small, compact corymbose panicles); *D. × elegantissima* (garden origin; to 1.5m; leaves ovate to oblong-ovate, sharply toothed above; flowers to 2cm across, pink, in loose cymes; includes 'Elegantissima', with upright growth, dark green leaves and pale pink flowers edged carmine, 'Fasciculata', an erect, rounded

shrub with mid-green leaves and pale pink flowers, and 'Rosealind', with deep carmine flowers); *D.gracilis* (Japan; erect to 2m; leaves lanceolate to ovate, serrulate; flowers pure white, to 2cm across, in racemes or narrow panicles to 8cm long; includes 'Grandiflora', a hybrid with *D.scabra* that is larger in all respects, and forms with yellow-mottled leaves and pink flowers); *D.longifolia* (China; to 2m; leaves lanceolate, finely tapering, serrulate; flowers to 2.5cm across, white, tinted purple-pink outside, in broad cymes to 8cm; includes 'Elegans', with drooping branches and purple-pink-tinted flowers, and 'Veitchii', with narrow leaves and purple-tinted flowers); *D. × magnifica* (garden origin; to 2m, robust and upright; leaves ovate to oblong, sharply and finely toothed; flowers white, single or double, in dense panicles to 6cm long; includes 'Staphyleoides', with large, drooping panicles of white, fragrant flowers with reflexed petals); *D.monbeigii* (China; to 1.5m, with slender branches; leaves ovate to lanceolate, serrate; flowers to 1.4cm across, white, in corymbose inflorescences to 6cm); *D.pulchra* (Philippines, Taiwan; to 4m; leaves lanceolate to narrowly ovate, entire to toothed, thick-textured, very hairy beneath; flowers white, sometimes tinted pink, in a slender and sparse, nodding panicle to 12cm); *D. × rosea* (garden origin; dwarf and arching to 1m; leaves elliptic to oblong or lanceolate, serrate; flowers white, tinted pink, in short, broad panicles; includes 'Carminea', with red-pink flowers); *D.scabra* (Japan; to 2.5m; leaves broadly ovate, coarsely serrate; flowers to 1.5cm across, white, honey-scented, in loose, broadly pyramidal panicles; 'Plena': flowers double, pale pink); *D.setchuenensis* (China; to 2m; leaves ovate, finely tapering, serrulate; flowers white, to 1cm across in loose, broad corymbs).

dewpoint as temperature falls during calm clear nights, the capacity of air to hold moisture is reduced until saturation is reached, and this is known as the dewpoint. Condensation and the formation of dew consequently occurs on the cool surfaces of plants.

di-ammonium phosphate see *ammonium phosphate*.

diadelphous of stamens, borne in two distinct bundles, or with several stamens united and a further solitary stamen set apart from the others, as in many Leguminosae.

dialypetalous see *polypetalous*.

diandrous possessing two perfect stamens.

Dianella (diminutive of Diana). Liliaceae (Phormiaceae). Tropical Africa, Asia, Australasia. flax lily. 25–30 species, evergreen, fibrous-rooted, perennial herbs with erect, slender stems, each bearing a terminal fan of grass-like to narrowly sword-shaped leaves. To 2cm across, the star-like flowers consist of six tepals and nod in spreading, loose pan-

Deutzia longifolia

icles. They are followed by deep purple-blue berries. Grow in the cool greenhouse or conservatory, with a minimum temperature of 5–7°C/40–45°F. They will also survive outdoors in sheltered sites where temperatures do not drop far below zero for long periods. Grow in any moderately fertile soil in sun or dappled shade. Water plentifully and feed when in full growth; keep just moist in winter. Propagate by seed sown under glass in spring or by division. *D.caerulea* (stem to 60cm tall; flowers to 1.5cm across, blue to white with pale yellow anthers, in panicles to 30cm); *D.tasmanica* (stem to 120cm tall; flowers to 2cm across, pale blue with brown anthers, in panicles to 60cm).

Dianthus superbus

Dianthus (from Greek *dios*, and *anthos* - 'flower of the gods'). Caryophyllaceae. CARNATION, PINK. Mostly Europe and Asia. 300 species, perennial, biennial and, rarely, annual herbs. They range in habit from woody-based mat-formers to erect herbs. Produced in spring and summer, the flowers are solitary or borne in terminal clusters. They consist of a cylindric calyx and five, spreading, clawed petals. These are usually obovate, toothed to cut at the outer margin and sometimes bearded below. Many are perfumed, most characteristically of cloves.

The perennial species listed here are fully hardy and suited to well-drained, sunny positions – for example, border edges, cracks in paving, wall crevices and the rock garden. The smaller species may fare better in alpine sinks or raised beds. They need a gritty soil, neutral to alkaline in all cases except the acid-loving *D. pavonius*. They will deteriorate in wet conditions. Increase by cuttings or rooted offsets in spring and summer. *D. barbatus* and *D. chinensis* are usually treated as annuals or biennials, sown under glass in spring and planted out in autumn or (in severe climates) the following spring, having been overwintered in a cold frame. They need a well-drained soil and full sun. See also *carnation* and *pinks*.

D. alpinus (Alps; tufted perennial, 5-15cm; leaves linear to oblong-lanceolate, shiny; flowers solitary, deep pink with crimson purple spots on a white 'eye', petals 1-2cm, toothed and bearded; includes white- and salmon-flowered cultivars); *D. armeria* (DEPTFORD PINK; Europe, W Asia; hairy biennial or annual, 20-40cm; leaves narrowly oblong; flowers in dense, involucrate heads, pink with pale dots, petals to 0.5cm, toothed and bearded); *D. barbatus* (SWEET WILLIAM; S Europe, widely naturalized; glabrous, short-lived perennial, 30-70cm; leaves lanceolate-elliptic; flowers in dense, broad, involucrate heads, red-purple with pale dots or a white band at centre, petals to 1cm, bearded; the parent of numerous cultivars, hybrids and seed races including short and tall plants, some bushy, others willowy, with dense heads of flowers in shades of white, pale rose, bright pink, cerise, crimson and blood red, many of them bicoloured, for example, a white-fringed flower with a bold red centre); *D. carthusianorum* (S and C Europe; glabrous perennial, 20-60cm; leaves linear,

grassy; flowers in slender-stalked heads, red-purple, pink or white, petals to 1.5cm, toothed, bearded); *D. chinensis* (INDIAN PINK; China; biennial or short-lived perennial, 20-70cm, usually somewhat hairy; leaves lanceolate; flowers in loose clusters, lilac-pink with a purple 'eye', petals to 1.5cm, deeply toothed or cut; includes cultivars with flowers in shades of pink, red and white, many treated as annuals, among them 'Heddewigii' – compact, very free-flowering plants in a range of colours); *D. deltoides* (MAIDEN PINK; Europe, Asia; tufted, mat-forming perennial, 15-45cm, sometimes hairy, often glaucous; leaves narrowly oblanceolate to linear; flowers solitary, pale to deep pink with pale spots in a dark 'eye', petals to 0.8cm, toothed and bearded); *D. gratianopolitanus* (syn. *D. caesius*; CHEDDAR PINK; W and C Europe; compact, mat-forming perennial, 5-20cm; leaves linear, glaucous; flowers solitary, pink or red, very fragrant, petals 1cm, toothed, sometimes bearded); *D. haematocalyx* (Balkans; tufted perennial, 10–30cm; leaves linear, glaucous, with thickened margins; flowers solitary or a few per cluster, deep purple-pink with buff exteriors and a purple-red calyx, petals 0.5-1cm, toothed, sparsely bearded); *D. microlepis* (Bulgaria; dwarf, cushion-forming perennial to 5cm; leaves linear, glaucous; flowers solitary, clear pink or white, petals to 0.7cm, toothed and bearded); *D. monspessulanus* (S and C Europe; loosely tufted, mat-forming perennial, 30–60cm; leaves linear, grassy, sometimes glaucous; flowers solitray or a few grouped together, strongly fragrant, pink, pale lilac or white, petals 1–2cm, finely cut into narrow segments); *D. myrtinervius* (Balkans, Macedonian border of Greece; densely mat-forming perennial to 5cm; leaves very small, elliptic to narrowly oblong; flowers solitary, very short-stalked, bright pink, sometimes with a dark-bordered white 'eye', petals 0.5cm, toothed, sparsely bearded); *D. pavonius* (syn. *D. neglectus*; SW Alps; densely tufted, mat-forming perennial, 5–15cm; leaves linear, usually grey-green; flowers solitary, pale pink to crimson, buff on exterior, petals 1cm, toothed and bearded); *D. superbus* (Europe, Asia; mat-forming perennial, 20-60cm; leaves linear to narrowly lanceolate, pale green; flowers solitary, very fragrant, pink to pale lilac or purple-pink, petals 1-3cm, finely and deeply cut into many narrow lobes).

diapause the state of reduced metabolic activity in which insects may overwinter in temperate climates.

Diapensia (ancient Greek name). Diapensiaceae. Circumboreal. 4 species, low-growing, tufted, evergreen subshrubs with glabrous, entire leaves and solitary, terminal flowers in early summer – these are tubular with five, spreading lobes. In cool, damp climates, *D.lapponica* is suited to north-facing crevices in the rock garden, shady scree beds or troughs. Plant in part shade or good indirect light on a moist, lime-free gritty soil, enriched with leafmould. Increase by careful division in spring; alternatively, propagate by soft tip cuttings taken after flowering. *D.lapponica* (Alaska, N Asia; creeping subshrub to 7.5cm; leaves to 1cm, spathulate to linear, leathery turning red in winter; flowers solitary, white, to 2cm across).

Diascia (from Greek *dis-*, two, and *askos*, sac, referring to the two sacs marked by yellow and maroon patches on the corolla and known as 'windows'). Scrophulariaceae. South Africa. 50 species, erect to decumbent annual and perennial herbs (those listed here are perennial). The leaves are small, elliptic to ovate or heart-shaped, serrate and often glandular. Bright flowers, carried from late spring to autumn in terminal racemes, have rounded petals and lateral spurs containing dark glands. Most will survive temperatures of –5°C/23°F on very well-drained soils. Take cuttings to ensure against winter losses, planting out on a sunny, sheltered site when danger of frost is past. These plants may also be treated as half-hardy annuals and are especially suited to baskets and containers. Deadhead to prolong flowering. Under glass, grow in pots of low-fertility loam-based medium; water plentifully in summer, and feed fortnightly with a dilute liquid fertilizer, when in active growth. Propagate by softwood cuttings in summer, and from lightly covered seed sown in cold frames or under glass, at 15°C/60°F, in autumn or early spring; overwinter autumn-sown plants at 10°C/50°F. Alternatively, increase by division in spring.

D.barberae (syn. *D.cordata*; TWINSPUR; stems erect or sprawling, to 30cm, glandular-hairy above; flowers in loose racemes, rose-pink with a yellow blotch and maroon-black glands; includes 'Ruby Field', with profuse, salmon-pink flowers); *D.rigescens* (stems erect or sprawling, to 50cm; leaves becoming red-brown-tinted with age; flowers in dense racemes, rose-pink with dark glands); *D.vigilis* (sprawling, to 50cm with small, rounded, fleshy leaves; flowers in loose racemes, pale pink with dark glands).

dibber, dibble a tool used to make holes in soil or containerised growing media for planting or for inserting seed, seedlings or cuttings. Dibbers for outdoor use take the form of a wooden implement about 25cm long, as thick as a standard spade handle, with a T- or D-shaped hand grip and sometimes a steel tip. Dibbers for greenhouse use are pen-sized and made of wood or plastic. Both forms have tapered ends.

Dicentra (from Greek *dis*, two, and *kentron*, spur, referring to the 2-spurred flowers). Fumariaceae. Asia, N America. 19 species, annual or perennial deciduous herbs (those listed here are perennial), with ferny, ternately decompound leaves. Borne in spring and summer in panicles, racemes or corymbs, the pendulous flowers are heart-shaped in outline and consist of four petals in two pairs, the outer pair spurred at the base and usually pouched, the inner pair tongue-shaped, with convex inner faces and crested apices forming a hood covering the anthers.

The genus is fully hardy. Grow in humus-rich, moist and acid to neutral soil, ideally with light shade. *D. formosa*, *D. peregrina* and *D. spectabilis* will grow in a sunny border on a cool, moist soil. *D. cucullaria* naturalizes well in woodland and shrubbery. Propagate from seed sown in spring at 15°C/60°F; also, divide in spring or after foliage dieback. Root cuttings of *D.spectabilis* will succeed in a sandy mix, placed in a cold frame.

D.cucullaria (DUTCHMAN'S BREECHES; N America; to 30cm; leaves soft, ferny, blue-green; flowers white to palest pink, tipped yellow); *D.eximia* (STAGGERWEED, TURKEY CORN; US; to 60cm; leaves finely cut, mid-green above, glaucous beneath; flowers magenta to pink, rarely white); *D.formosa* (WILD BLEEDING HEART; western N America; to 40cm; leaves finely cut, blue-green to silver grey with a purple-pink cast; flowers red to coral, salmon, rose or pure white; very glaucous plants are sometimes named subsp. *oregana*, one of the best silver-leaved white-flowered cultivars is 'Langtrees'.); *D.peregrina* (E Asia; to 15cm; leaves glaucous, finely and deeply cut; flowers white to magenta or purple, tipped purple); *D.spectabilis* (BLEEDING HEART; to 100cm; leaves mid- to dark green, loosely but deeply divided; flowers large, with heart-shaped, deep red to pink outer petals and white inner petals; includes pure white forms).

dichasium a type of determinate inflorescence; the basic structure has three flowers (members), one terminating the primary axis, the other two carried on more or less equal secondary branches arising from beneath the primary member in a false dichotomy. The secondary members may themselves be dichasia. Such a structure is known as a dichasial cyme or a compound dichasial cyme, although the former term has been applied by some authors to the basic dichasium itself.

dichogamy the maturing of a flower's anthers and stigmas at different times, preventing self-pollination. See *protandry, protogyny*.

Dichorisandra (from Greek *dis*, two, *choris*, apart, and *aner*, man, referring to the way two stamens stand apart from the others in some species). Commelinaceae. Tropical America. 25 species, evergreen perennial herbs, with erect, rather stout stems and large smooth, lanceolate to elliptic leaves. Produced in clusters in summer, the flowers consist of three, showy, rounded petals. Provide humid conditions, with a minimum winter temperature of 12°C/55°F and shade from direct sunlight. Grow in a well-drained, fertile and moist medium. Propagate by division or cuttings.

D.reginae (Peru; stem to 30cm; leaves to 18cm, striped or flecked silver above, flushed purple beneath; flowers white edged violet-blue); *D.thyrsiflora* (SE Brazil; stem to 200cm, bare below; leaves to 30cm, deep green above, faintly purple-tinted beneath; flowers violet, white at base).

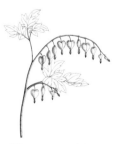

Dicentra spectabilis

dichotomous branching regularly by forking repeatedly in two, the two branches of each division being basically equal.

Dicksonia (for James Dickson (1738–1822), British botanist). Dicksoniaceae. America, SE Asia, Australasia, Polynesia. Some 25 species, large tree ferns with erect, column-like trunks – these are clothed with dead frond bases, shaggy hairs and fibrous roots and crowned with spreading rosettes of 1–4-pinnately compound, lanceolate to deltoid fronds. *D.antarctica* is hardy to –5°C/23°F and may be grown outside in milder areas in zone 7. Water copiously during the summer growing months. Plants growing outside should have their trunks hosed down daily during hot and dry weather to encourage rooting through the stem. Under glass, provide bright indirect light, plentiful water and frequent dampings down during hot weather (temperatures greater than 32°C/90°F are harmful). Grow in a slightly acid medium, moist but porous and rich in leafmould. Specimens in large containers or those planted out should be top-dressed or mulched with fresh compost or organic matter annually. During the winter, water moderately and ensure that maximum light reaches plants growing under glass. Those growing outside need to be protected with bracken or sacking if severe weather threatens. Propagate by spores. Although slow-growing, young plants so raised are usually quick to establish and will cope better with life in the garden than the ancient, wild-collected trunks recently imported into Britain in such numbers. *D.antarctica* (AUSTRALIAN TREE FERN, SOFT TREE FERN; Australia, Tasmania; trunk to 300 × 40 cm, densely fibrous, fox red to dark brown, with large scars and remnants left by fallen fronds; fronds 150–250cm, lanceolate-triangular, light to dark green and coarsely textured; pinnae to 45cm, lanceolate; pinnules linear; segments oblong, toothed or lobed).

*Dicksonia
antarctica*

dicotyledons (abbreviation dicot(s)) one of the two major divisions of the angiosperms, the other being the monocotyledons. Dicots are usually characterized by having two cotyledons, the presence of cambium in many species and floral parts commonly occurring in fours and fives.

Dictamnus (from Greek *diktamnon*, dittany). Rutaceae. SW Europe to E Asia. DITTANY, BURNING BUSH. 1 species, *D.albus*, a woody-based, perennial herb to 80cm tall with glandular, trifoliolate to pinnate leaves. Borne in summer in erect, terminal racemes, the flowers are some 4cm across and consist of five, elliptic to lanceolate white, pink, red or lilac petals, sometimes streaked or dotted with red. Hardy in zone 4. plant in a sunny or partly shaded position on any well-drained, fertile soil. Propagate from seed sown in a cold frame in spring (seedlings may take several years to attain flowering size), or by division of large, mature clumps in spring or autumn.

didymous twinned; (in pairs) the two parts similar and attached by a short portion of the inner surface.

didynamous possessing two pairs of stamens, the two pairs being of unequal lengths.

dieback (1) a general term for disorder in which shoots or branches die from the tip backwards, usually relatively slowly when compared to a wilt; (2) a non-specific symptom of a number of diseases or disorders, particularly affecting woody plants.

Dieffenbachia (for Herr Dieffenbach, gardener in 1830 at Schonbrünn, Austria). Araceaea. Tropical America. DUMB CANE, MOTHER-IN-LAW'S TONGUE, TUFTROOT. Some 25 species, evergreen, erect, perennial herbs with stout, rather fleshy green stems and oblong or lanceolate to elliptic leaves. These are smooth and commonly variegated in tones of grey, white, cream and yellow-green. The flowers are inconspicuous – packed in a club-shaped spadix and surrounded by a dull, green-white spathe. Popular foliage plants for tropical and subtropical gardens; elsewhere, they are suited to the home, interior landscapes and the intermediate to warm greenhouse (minimum winter temperature 10°C/50°F). Grow in bright, indirect light to shade, ideally with medium to high humidity. Pot in a porous, fertile, soil-less medium. Water and feed freely in warm weather and when in full growth, more sparingly at other times. Propagate by tip cuttings, air layering, division of basal shoots and slightly dried 5–8cm stem cuttings, in warm, humid conditions. The sap is poisonous.

D.amoena (to 2m, very robust with thick stems; leaf blades to 50cm, elliptic to oblong, dark green with creamy-white zones along lateral veins); *D.maculata* (to 1m; leaves to 25cm, oblong to lanceolate, acuminate, often heavily spotted white; includes many cultivars and hybrids, among them, 'Exotica', compact, with ovate dark green heavily zoned leaves veined white or green-white and to 25cm long, and 'Rudolph Roehrs' with broad leaves heavily spattered or almost wholly zoned lime green to yellow and edged and veined dark jade green); *D.× memoria-corsii* (leaves to 30cm, elliptic to oblong, dark green heavily blotched silver-grey and spotted white); *D.seguine* (to 1.2m; leaves to 45cm, oblong to elliptic or lanceolate, glossy deep green; several cultivars and hybrids variously spotted, blotched, zoned, and veined white, cream or emerald green.

Dierama (from Greek *dierama*, funnel, referring to the shape of the flowers). Iridaceae. Tropical and South Africa. ANGEL'S FISHING ROD, WAND FLOWER, AFRICAN HAIRBELL. Iridaceae. 44 species evergreen, cormous, perennial herbs, with clumps of narrow, grey-green grassy leaves. The flowers are bell- to funnel-shaped and composed of six, silken tepals; they hang on the thread-like, weeping branches of tall and slender, arching panicles in summer. All species are reliably hardy to between –5°C and –10°C/23–40°F, although well-established clumps of *D.pulcherrimum* have survived temperatures of –17°C/1°F. Grow in a sunny, sheltered site in deep, rich, moist but free-draining soil. Propagate by division in spring or by seed in autumn or spring.

D.dracomontanum (syn. *D.pumilum*; 30–100cm tall; flowers 2–3cm, pale to deep rose pink, mauve, purple-pink, or coral); *D.pendulum* (100–200cm; flowers 3–5cm, widely bell-shaped, pale rose to magenta; includes dwarf forms with rose flowers); *D.pulcherrimum* (90–180cm; flowers 3–5cm, narrowly bell-shaped, pale pink to deep magenta or purple-red, rarely white).

Diervilla (for M. Dierville, a French surgeon who travelled in Canada 1699–1700, and introduced *D.lonicera*). Caprifoliaceae. N America. bush honeysuckle. 3 species, low, often suckering, deciduous shrubs with ovate to lanceolate leaves and, in summer, tubular, honeysuckle-like flowers in small clusters. Hardy to –30°C/–22°F. Plant in sun or part shade on any well-drained soil. Prune in spring before growth begins. Propagate from suckers in spring, or from seed.

D.lonicera (eastern N America; to 1m; branches smooth; flowers green-yellow, 3 per axil or 5 per terminal cyme); *D.sessilifolia* (SE US; to 1m; branches hairy; leaves tinted red at first, colouring well in autumn; flowers sulphur yellow, 2 per terminal cyme; has crossed with *D.lonicera* to produce *D.× splendens*).

Dietes Iridaceae. Africa (1 species Lord Howe Island). 6 species, rhizomatous perennial herbs with fans of sword-shaped leaves and short-lived flowers on branching stems in summer. The flowers are composed of six segments, the outer three larger and outspread. Even the hardiest species, *D.bicolor*, will not tolerate temperatures much below –5°C/23°F, and should be kept dry as temperatures fall and given a dry mulch in winter. Other species will not stand prolonged freezing. Grow in full sun or light dappled shade in an open but moisture-retentive, humus-rich soil; apply liquid feed and water if neces-

sary during full growth, but reduce both water and feeding during the summer months after flowering. Propagate by seed or by division after flowering (divisions may not re-establish readily and should be given a period of protected cultivation).

D.bicolor (South Africa; 60–100cm; flowers 6–8cm in diameter, yellow, blotched brown and orange at centre); *D.iridioides* (syn. *D.vegeta* of gardens; tropical E Africa to South Africa; 30–60cm; flowers 4–7cm in diameter, white stained yellow, mauve or blue and spotted deep yellow or brown at centre).

diffuse spreading widely outwards.

diffusion the movement of a substance towards equilibrium from a high to low concentration. It is an important function in plant nutrient uptake and also relevant to soil aeration.

Dierama pulcherrimum

digamous with two sexes in the same flower cluster.

digging the process of turning over soil with a spade or fork in preparation for sowing or planting crops. Digging can improve soil structure and fertility by breaking down clods, either directly or as a result of exposure to weathering, and by facilitating the incorporation of organic matter and lime. It can also improve drainage and aeration and is a means of burying weeds and exposing pests to predators. Some exponents of organic gardening are satisfied that equal or better results can be obtained by not digging soils of naturally good structure. During digging, the spade or fork is forced vertically into the soil to the full blade or prong depth and the handle pressed down and backwards to lift the soil and turn it upside down, thereby burying weeds and crop residue.

Single digging cultivates down to blade or prong depth, which is referred to as the top spit. A trench about 30cm wide is taken out at one end of the plot and the soil removed by barrow to the other end and dumped evenly just outside the boundary. Large plots may be equally divided down the longest dimension and a starting trench taken out across just half the width, the resultant soil being placed off the plot, along the adjacent untrenched boundary. Digging proceeds by turning the soil into the first trench, thereby creating a new trench. This operation is repeated progressively, the last trench being filled with soil originally barrowed to the far end. Where a large plot has been divided, the end trench is filled with soil taken from a new trench started across the undug half. The digging of the second half then proceeds in the opposite direction to the first half until the last trench is filled with soil originally dumped off the plot boundary.

Ridging is a variation aimed at exposing more surface area. Strips of ground of three spades width are dug at a time, following the single digging method described above. The left and right spadesful are thrown forward onto the central spadeful, and in this manner wide ridges are formed down the plot. These are levelled out by forking prior to sowing or planting.

Double digging (double spading, bastard trenching) follows the same principle of operating as single digging, except that the working trench is 60–75cm wide and more definitely maintained. As digging progresses, the base of the trench is forked to one spit depth.

Trenching involves cultivating to three spits deep, and requires a working trench 1m wide. The top and second spits are turned over and the trench base forked to one spit depth. This more complex operation requires the trench to be divided along its length and each half to be dug in sequence to avoid mixing of the layers. The full-width top spit of the first trench and the second spit of one half width of the trench are initially removed in the manner described for single digging.

It is best to dig heavy, textured soils in late autumn to benefit from the action of frost in breaking up clods. Light-textured soils are best left until spring to preserve their structure. Annual digging is usually necessary, but on most sites double-digging should only be required every three years and trenching only where serious compaction is identified. Organic manure can be worked in during any of the digging operations, and there are advantages in applying it so that it becomes mixed throughout the top spit. On lighter-textured soil, the manure may be spread over the ground in autumn – earthworm activity will draw much of it into the surface layer before spring digging. If lime is needed, apply it over the soil surface in the autumn, before digging, and always fork in several weeks ahead of any dressing of animal manure.

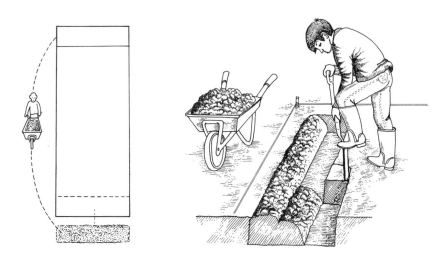

single digging: planning and method

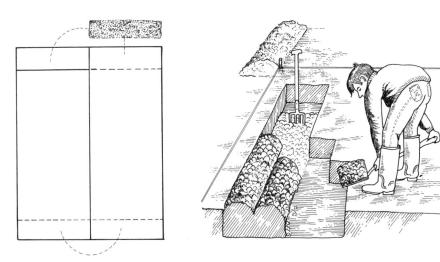

double digging: planning and method

Digitalis (from Latin *digitus*, finger, referring to the shape of the flower). Scrophulariaceae. Europe, NW Africa, to C Asia. FOXGLOVE. Some 20 species, biennial or perennial herbs, with leaves occurring in a basal rosette and on the erect flowering stems. Tubular to deeply bell-shaped and weakly 2-lipped, the flowers are carried in a one-sided spike in summer. Grow in freely draining, humus-rich soils in partial or dappled shade, or in sun where soils are reliably moisture-retentive throughout the growing season. Propagate by seed, which requires light for germination; perennials may also by increased by division.

D.ferruginea (RUSTY FOXGLOVE; S Europe, W Asia; biennial or perennial to 1m; flowers to 3cm, golden brown tinted rusty red with darker veins inside); *D.grandiflora* (LARGE YELLOW FOXGLOVE; Europe; biennial or perennial to 1m; flowers to 5cm, cream to pale yellow or apricot, with darker veins or spots within, broadly funnel-shaped); *D. lanata* (GRECIAN FOXGLOVE; E Europe, Aegean, Turkey, Siberia; biennial or short-lived perennial to 1.5m; flowers to 2cm, white with nutmeg spots and maroon veins); *D.lutea* (STRAW FOXGLOVE; Europe, NW Africa; perennial to 100cm; flowers to 2cm, creamy white to pale yellow, narrow); *D. × mertonensis* (*D.grandiflora* × *D.purpurea*; garden origin; perennial to 75 cm; flowers to 5cm rose-mauve to strawberry pink or coppery red); *D.purpurea* (COMMON FOXGLOVE; W Europe; biennial or perennial, 80–180cm tall; flowers to 5cm, purple, red, pink or white, the interior usually paler, hairy and heavily marked with white-edged, deep maroon spots; many cultivars and seed races are offered, including short or tall plants, with 'double' and otherwise distorted flowers, ranging in colour from deepest purple-red to bronzy gold and pure white).

digitate with leaflets palmately arranged resembling the fingers of an outspread hand.

digonous two-angled, as in the stems of some cacti.

digynous with two separated styles or carpels.

dill (*Anethum graveolens*) a strongly aromatic annual up to 60cm high, grown for its distinctively flavoured leaves and seeds, which are used in salads, fish cooking, sauerkraut and pickling-spice mixtures. The leaves are best used fresh, but may be dried or frozen. Sow successionally *in situ* from spring to midsummer. Seeds are harvested on turning brown, and the seed heads dried between single layers of paper at low temperature.

diluent an inert material used to reduce the concentration of active ingredient in a pesticide formula.

dilutor a device used in liquid-feeding systems for dispensing a concentrated solution of fertilizer into a water pipeline. It usually functions on a diaphragm or water displacement principle. Dilutors are particularly identified with commercial nurseries, where they may also be used to distribute certain pesticides.

Dimorphotheca (from Greek *dis*, two, *morphe*, form, and *theke*, case: the fruits of the ray and disc florets differ in shape). Compositae. South and tropical Africa. SUN MARIGOLD. 7 species, herbs or subshrubs, with entire to pinnately lobed leaves and long-stalked, daisy-like flowerheads in summer. Cultivate as for *Arctotis*. *D.pluvialis* (syn. *D.annua*; WEATHER PROPHET; South Africa; aromatic annual to 40cm; leaves to 10cm, obovate to oblanceolate, toothed to pinnately lobed, deep green, hairy; flowerheads to 6cm in diameter, ray florets white above with dark purple bases, blue-violet beneath, disc florets purple-brown).

dioecious having male and female flowers on separate plants.

Dionaea (from Dione, mother of Aphrodite, alluding to the plant's beauty). Droseraceae. SE US (coastal N and S Carolina). VENUS' FLY TRAP, VENUS' MOUSE TRAP. 1 species, *D.muscipula*, a low-growing, carnivorous, perennial herb. To 10cm long, the leaves are semi-erect or lie close to the ground in a basal rosette. They consist of a winged, spathulate petiole, bearing at its tip a blade of two semi-circular lobes that are hinged at the middle and edged with lash-like teeth. The inner surface of these lobes is glossy, glandular and turns a deep, meaty red in bright sunlight. Each lobe bears three fine trigger hairs. When alighting insects touch these hairs, they cause the lobes to snap shut. The blade then becomes a digestive purse, leaching the imprisoned prey of its vital juices. In spring and summer, white, 5-petalled flowers are carried in umbel-like cymes atop slender scapes. Plant the bulb-like rosette bases in a porous mixture of vermiculite, peat and sphagnum moss, either singly in fairly deep pots or massed in pans or trays. Keep constantly moist by standing containers in soft water and/or by frequent overhead watering with a fine rose. Tolerant of temperatures just above freezing, at which point growth will cease and even die back to resting bulbs: water should be reduced accordingly. Temperatures between 13°C/55°F and 27°C/80°F will ensure continuous growth. Fresh air and full sunlight will promote a low-growing rosette habit and the most highly coloured traps. Propagate by seed (sown under cover on damp sphagnum and bottom-watered), division or by leaf cuttings of the petiole.

Dionysia (for Dionysos, Greek god of wine). Primulaceae. W and C Asia. 41 species, tufted or cushion-forming alpine subshrubs. The stems are branched and in old plants become woody and covered by withered leaf remains. The leaves form rosettes and are very variable in shape and size, from large and *Primula*-like to minute and scale-like; they are usually hairy and glandular on one or both surfaces, and farinose beneath. Narrowly tubular with a limb of five, spreading lobes, the flowers are virtually stalkless and solitary, or carried in scapose and sometimes tiered umbels. Alpine-house conditions are essential and

dill

plants grow best if double-potted or if single-potted and placed in a plunge bed. Plant cushions may be wedged between small pieces of rock on the pot surface or grown in tufa to ensure that the neck zone remains dry. Use a gritty, loam-based potting mix. Keep just moist, never wet, and position in sun. Propagate by softwood cuttings in summer.

D.aretioides (N Iran; leaves to 0.7cm, oblong to narrowly spathulate, hairy, farinose, bluntly toothed, forming low, dense, grey cushions; flowers to 1.5cm in diameter, bright yellow, solitary, stemless, scented, the petals more or less toothed); *D.lamingtonii* (Iran; leaves to 0.3cm, oblong to spathulate, entire, hairy, forming dense, compact cushions; flowers to 0.5cm across, bright yellow, stemless, the petals shallowly cleft); *D.tapetodes* (W Asia; leaves to 0.4cm, obovate to spathulate, entire to crenulate, minutely glandular, farinose, in tight cushions; flowers to 1cm in diameter, solitary, stalkless, bright yellow, the petals more or less rounded).

Dioon (from Greek *dis*, two, and *oon*, egg – the seeds are borne in pairs). Zamiaceae. C America. 10 species, long-lived and slow-growing evergreen cycads, superficially resembling tree ferns or palms. The stems are erect, usually unbranched, stoutly columnar and clothed with the scars or remains of leaf petioles. Tough, pinnate leaves are produced in terminal rosettes. Massive, ovoid cones are borne in the woolly crowns on plants of separate sex. Grow *D.edule* in full sun with a medium high in sand and grit, rather dry and airy conditions and a minimum temperature of 7°C/45°F. *D.spinulosum* prefers part shade, humidity, a moist, humus-rich soil and a minimum temperature of 15°C/60°F. Propagate by seed.

D.edule (Mexico; stem to 1.8m, stoutly columnar; leaves to 1.5m, semi-erect; pinnae closely set, to 13 × 1.5cm, linear to lanceolate, tip narrowly acuminate and sharply pointed, base decurrent, rigid, grey-green, margins usually entire but spiny-toothed in some juveniles and varieties); *D.spinulosum* (Mexico; stem to 10m, narrowly columnar; leaves to 2m, arching; pinnae well-spaced, to 15 × 4cm, lanceolate, tip broadly acuminate with a soft but distinct point, base decurrent, rigid but more flexible than in *D.edule*, glaucous sea-green at first, hardening mid-to dark green, margins always spiny-toothed).

Dioscorea (for Dioscorides, 1st-century Greek physician and herbalist, author of *Materia Medica*). Dioscoreaceae. Tropics and subtropics. YAM. Some 600 species, twining perennial herbs, often with tuberous roots. The flowers are yellow-green and inconspicuous. *D.discolor* needs part to full shade, high humidity and a minimum winter temperature of 13°C/55°F. Plant in a rich, coir-based mix with added leafmould and sand. Water and feed liberally when in growth and during warm weather; at other times, keep just moist. Grow *D.elephantipes* in full sun with a dry, buoyant atmosphere and a minimum winter temperature of 7°C/45°F. Plant in a deep, fertile, loam-based mix high in grit. Leave the caudex two thirds exposed. Water and feed generously once the

Dionysia aretioides

Dioscorea elephantipes, D. discolor (top)

new shoots appear at the summit of the caudex; provide support as they develop. Withhold water once the growth turns yellow and dies off, except for very occasional soakings to prevent the caudex from withering. Dormancy may last for six months; after two or three, it may be deliberately broken by raising the temperature and by watering. Propagate by seed.

D.discolor (tropical South America; stems vigorously climbing to 4m; leaves to 30cm, ovate to cordate with a finely tapering tip, velvety emerald above with broken zones of deep chocolate-maroon and irregular flashes of silver along midrib and principal veins, maroon beneath); *D.elephantipes* (syn. *Testudinaria elephantipes*; ELEPHANT'S FOOT, HOTTENTOT BREAD; South Africa; caudex to 1m in diameter, woody, fissured and facetted; stems to 2m, of annual duration; leaves 2–6cm in diameter, broadly heart-shaped to reniform, smooth, bright green).

Diosma (from Greek *dios*, divine, and *osme*, smell, referring to the pleasant fragrance of the crushed leaves). Rutaceae. South Africa. Some 28 species, evergreen shrubs with aromatic, glandular leaves and small, 5-petalled flowers in winter and spring – these are usually fragrant and white or red-tinted. Grow in full sun, with a minimum winter temperature of 5°C/40°F. Plant in a well-drained acid mix; reduce water in winter. Prune after flowering. Increase by seed in spring or semi-ripe cuttings. *D.ericoides* (BREATH OF HEAVEN; to 80cm tall, rounded and many-stemmed; leaves to 0.6cm, aromatic, oblong, crowded; flowers white, very fragrant).

Diospyros (from Greek *dios*, divine, and *pyros*, wheat or grain, alluding to the edible fruits). Ebenaceae. Europe, Americas, Africa, Asia. Around 475 species, deciduous or evergreen trees or shrubs (those listed below are deciduous trees) with hard, often black wood. Small flowers in summer give rise to large, fleshy berries. Plant all species in deep, fertile and loamy soils, in sun or light shade. The following species are hardy in zone 6, but may be susceptible to frost, especially when young. *D.kaki*, which is grown for its fruit, should be protected from wind. For commercial production, *D.kaki* is generally restricted to zone 8, although as a garden fruit, where high yields may not be as important, it shows more adaptability. In cooler temperate areas (to zone 5), provide the protection of a warm south-facing wall, where the fruit may ripen over long hot summers. Plants of both sexes are needed for fruit to set. Propagate from seed sown when ripe and stratified, by layers in spring, or from semi-ripe cuttings in a closed case/under mist in summer. Commercial cultivars are whip-grafted just below soil level, or shield-budded just before dormancy, on to seedling rootstock.

D.kaki (KAKI, PERSIMMON, JAPANESE PERSIMMON; wild origin unknown; to 14m; leaves to 20cm, ovate to obovate, deep, glossy green above, turning orange-red in autumn; flowers pale yellow; fruit to 7.5cm in diameter, globose to oblong, waxy yellow to orange or red with a persistent, accrescent calyx); *D.lotus* (DATE PLUM; temperate Asia; to 25m; leaves to 12cm, lanceolate to elliptic, tough, falling when green; flowers yellow-buff to brown; fruit to 2cm in diameter, globose to ovoid, yellow, red or blue-black, pruinose); *D.virginiana* (AMERICAN PERSIMMON, POSSUMWOOD; E US; to 20m; leaves to 12cm, elliptic to ovate or oblong, deep glossy green turning red in autumn; flowers yellow-white; fruit 2.5cm in diameter, orange).

Dipelta (from Greek *di-*, two, and *pelte*, shield, referring to the form of the bracts). Caprifoliaceae. C and W China. 4 species, deciduous shrubs allied to *Weigela*. The flowers are tubular to campanulate, large and solitary or clustered; they hang on slender stalks, encircled by two to four shield-like bracts which become larger after fertilization of the flower. The following species will withstand winter temperatures down to –20°C/–4°F. Cultivate as for *Weigela*.

D.floribunda (to 4.5m; flowers fragrant, to 3cm, white sometimes flushed shell pink at base, marked or stained deep yellow in throat, subtended by shield-like, green-white bracts); *D.yunnanensis* (2–4m; flowers to 2.5cm, fragrant, cream stained pale pink, throat marked orange, subtended by cordate, thin-textured bracts).

Diphylleia (from Greek *dis*, double, and *phyllon*, leaf, alluding to the deeply lobed leaf). Berberidaceae. Eastern N America, Japan. 3 species, perennial, rhizomatous herbs with large, peltate and long-stalked leaves, their blades rounded and deeply 2-lobed. Small, white flowers appear in spring in terminal cymes. These are followed by colourful berries. Fully hardy. Grow in damp, humus-rich soil in part shade. Propagate by division in spring or by seed. *D.cymosa* (UMBRELLA LEAF; E US; to 1m; leaves 30–60cm across, cleft, each segment 5–7-lobed and toothed; flowerstalks becoming red in fruit; berries to 1.2cm across, blue).

Diplarrhena (from Greek *diploos*, double, and *arrhen*, male – the flowers have two perfect stamens). Iridaceae. SE Australia, Tasmania. 2 species, rhizomatous perennial herbs with sword-like leaves in tufts or fans. The fragrant flowers are carried in slender-stalked clusters in summer and consist of six segments, the outer three broad and rounded, the inner three smaller. In frost-free zones or areas where temperatures only occasionally fall to –10°C/14°F, *Diplarrhena* is suitable for the herbaceous border or rock garden. In colder regions, plant in pots or large tubs in the cool greenhouse or conservatory. Grow in full sun in a freely draining, sandy, humus-rich and acid soil. Propagate by seed, division, or by plantlets. *D.moraea* (flowering stem to 65cm; flowers to 4cm in diameter, fragrant, white, the inner segments suffused or veined mauve-blue or yellow toward apex).

diploid of a plant, having the normal number of chromosomes for the species, i.e. twice the number borne in the pollen and ovule cells; commonly annotated 2 ×; cf. *haploid, polyploid, triploid, tetraploid.*

Dipsacus (name used by Dioscorides). Dipsacaceae. Europe, Asia, Africa. TEASEL. 15 species, biennial or short-lived perennial herbs, with tall, erect stems and toothed to pinnately cut lanceolate to oblong leaves in pairs, with their bases more or less fused in a cup. The small flowers are packed in heads which resemble finely textured, prickly cones. These are surrounded by erect and rigid, spine-like bracts. The flowers may be dried and used in everlasting arrangements. Harvest for drying when in bloom. *Dipsacus* is fully hardy. Grow in sun on any moderately fertile soil. Allow to self-sow, or sow seed *in situ* in autumn or spring.

D.fullonum (COMMON TEASEL; Europe, Asia, naturalized N America; biennial to 2m; basal leaves oblanceolate, crenately toothed, spiny-pustular; flowers lilac in an oblong to ovoid inflorescence with unhooked bracts); *D.sativus* (FULLER'S TEASEL; Europe, Asia, N Africa, naturalized N America; differs from *D.fullonum* in its entire basal leaves and cylindrical inflorescence with hooked bracts).

Dipteronia (from Greek *di-*, double, and *pteron*, wing, referring to the fruit). Aceraceae. C and S China. 2 species, deciduous trees and large shrubs with pinnate leaves, small, 5-parted green-white flowers in erect panicles, and clusters of paired, winged fruits – these turn red as they mature in late summer. Cultivate as for the smaller *Acer* species. *D.sinensis* (tree or bushy shrub to 10m; leaves to 25cm; leaflets 7–11, to 9cm, ovate to lanceolate, serrate).

Disa (alluding to the mythical Queen Disa of Sweden who came to the King of the Sveas wrapped in a fishing-net: the dorsal sepal of *D.uniflora* is net-veined). Orchidaceae. Africa, Madagascar. About 130 species, terrestrial orchids with tuberous roots and linear to lanceolate leaves in basal rosettes and on the erect flowering stems. The flowers are borne in slender racemes or a few together. Each consists

Dipelta floribunda

Dipsacus sativus

Disa uniflora

of a large, erect and hood-like dorsal sepal with a basal spur, small petals and lip, and two larger, outspread lateral sepals. Grow in the cool greenhouse or conservatory (minimum temperature 7°C/45°F) in pans of neutral to acid fibrous mix, enriched with leafmould; protect from scorching sunlight and avoid wetting foliage. Keep plants moist throughout the growing season, watering with rainwater. Decrease water supplies slightly after flowering. Repot every two years. Propagate by division or seed.

D.uniflora (South Africa; 15–60cm; flowers 2–6cm long, lateral sepals and lip usually carmine red, hood orange within, veined red, petals light carmine at base, the blade yellow with red spots; yellow and pink forms also occur). *Disa* grexes include 'Betty's Bay' (tall plants with bright orange flowers), 'Diores' (tall plants with many flowers in shades of pink, red and orange), 'Kewensis' (small flowers in shades of pink and orange), 'Kirstenbosch Pride' (tall spikes with many small, bright orange-red flowers), 'Langleyensis' (tall spikes of pale pink flowers), and 'Veitchii' (tall spikes of bright orange flowers).

Disanthus (from Greek *dis*, twice, and *anthos*, flower: the flowers are paired). Hamamelidaceae. China, Japan. 1 species, *D.cercidifolius*, a deciduous, bushy shrub to 4 × 2m. To 10cm wide, the leaves are ovate to orbicular with a cordate base. Blue-green above and paler beneath, they turn a rich wine red in autumn, finally assuming fiery tints. Small and spidery, the flowers are dark purple-red and appear in autumn and winter. Hardy in zone 7. Plant in a sheltered position in dappled shade on an acid soil. Cultivate otherwise as for *Hamamelis*.

disbudding (1) the removal of buds, carried out to allow those remaining to develop top-quality flowers, especially where intended for exhibition. It is of special relevance in the cultivation of carnations, chrysanthemums, dahlias and roses; (2) a technique for restricting side growths on vines and on wall-trained fruits, although in the latter case the technique is more accurately described as deshooting.

disc, disk (1) a fleshy or raised development of the torus which may occur within the calyx, within the corolla and stamens, or surrounding the pistil, in which case it is composed of coalesced nectaries or staminodes; (2) (in Compositae) the central part, or sometimes whole, of the capitulum bearing short tubular florets as opposed to the peripheral ray florets; (3) the central part of the lip in Orchidaceae, often elevated and callused or crested; (4) a circular flattened organ, for example the disc-like tendril tips of some plants climbing by adhesion; (5) the basal plate of a bulb around which scales are arranged.

Discaria (from Greek *diskos*, disc, due to the fleshy nectariferous disc of flowers). Rhamnaceae. Temperate S America, Australia, New Zealand. 12 species, deciduous shrubs or small trees, with long, slender, green thorns, small and often obsolete leaves, and dense clus-

ters of minute 4-parted flowers in spring. The following species is suitable for sunny, south-facing walls on any well-drained soil. It will survive winter temperatures as low as –15°C/5°F. Prune, if necessary, after flowering. Propagate by seed or by semi-ripe cuttings. *D.toumatou* (WILD IRISHMAN; New Zealand; shrub or small tree to 5m with green, flexuous twigs and many spines to 5cm long; flowers green-white).

disc floret a flower with a tubular corolla, often toothed, found in the centre of a radiate capitulum or occupying the whole of a discoid capitulum. Found in the family Compositae. See *ray floret*.

discoid (1) of a leaf, with a round fleshy blade and thickened margins; (2) more commonly, pertaining to the capitula of some Compositae, composed entirely of disc florets.

disease any plant ailment caused by pathogenic organisms, such as fungi, bacteria or viruses, affecting the viability, quality or economic value of a plant. The term is occasionally used more loosely to denote any ailment of a plant, including malfunctions caused by faulty cultivation or unsatisfactory growing conditions, which are more accurately termed disorders.

Disporum (from Greek *di-*, twice, and *spora*, seed, referring to the usually 2-seeded fruits). Liliaceae (Colchicaceae). US, E Asia. 10–20 species, perennial, rhizomatous herbs, with leafy, sparingly branched stems and ovate to lanceolate leaves. Nodding, bell-shaped to tubular flowers appear in spring and are followed by orange to red berries. Cultivate as for *Tricyrtis*. Propagate by division in spring or by seed.

D.hookeri (NW US; 30–100cm; leaves 3–14cm, lanceolate to ovate, apex acuminate, base cordate, stem-clasping; flowers to 2cm, green-white; fruit scarlet); *D.sessile* (Japan; to 60cm; leaves 5–15cm, oblong to lanceolate; flowers to 3cm, white tipped green; fruit blue-black; includes 'Variegatum', with leaves striped cream).

dissected cut in any way, a general term applicable to leaf blades or other flattened organs that are incised, lacerate, laciniate, pinnatisect or palmatisect.

distal the part furthest from the axis, thus the tip of a leaf is distal; cf. *proximal*.

distant of leaves on a stem, stipes on a rhizome or flowers on a floral axis, widely spaced, synonymous with remote; cf. *proximate*.

distichous of leaves, distinctly arranged in two opposite ranks along a stem or branch.

Distictis (from Greek *di-*, twice, and *stiktos*, spotted: the seeds are very flat and resemble two rows of spots in the fruit). Bignoniaceae. Central America,

W Indies. 9 species, slender shrubs climbing by tendril-tipped leaves. Produced in spring and summer, the flowers are tubular to funnel-shaped. Cultivate as for *Bignonia* but with slightly higher temperatures and ample heat and light to ripen the new growth.

D. buccinatoria (Mexico; flowers to 8cm, rich purple-red, throat golden yellow);*D. lactiflora* (W Indies, Puerto Rico, St Domingo; flowers to 3cm, white, throat yellow); *D. laxiflora* (Mexico, Nicaragua; flowers to 8cm, rich purple in bud, openingmagenta, fading to lilac-white); *D.* 'Rivers' (flowers dark mauve with a golden throat).

distinct with similar parts that are not united.

Distylium (from Greek *dis*, twice, and *stylos*, style). Hamamelidaceae. E Asia. 12 species, small, evergreen trees and shrubs with leathery, dark green foliage and, in spring and early summer, short racemes of flowers with tinted calyces, no petals and colourful anthers. Cultivate as for *Hamamelis*. *D.racemosum* (ISU TREE; Japan; broadly spreading shrub or small tree to 25m in wild; leaves to 7cm, elliptic to obovate; flowers in racemes to 7.5cm long, stamens red).

diurnal of activity taking place only during daylight.

divaricate broadly divergent and spreading, a term usually applied to branching patterns where the branches spread 70° to 90° outwards from the main axis.

divergent broadly spreading from the centre.

Division the highest rank of the principal categories in the plant kingdom. The names of Divisions end in *phyta* ('plants').

division a method of increasing plants, suitable for those that produce a mass of closely knit shoots or buds in a clump or crown of growth which can be separated into pieces. Many herbaceous perennials can be readily propagated in this way. The best time for division is when roots and shoots have started to form in the spring. It is best to use portions from the periphery of the crown, and depending on species these may be pulled off by hand or prised off by inserting two border forks back to back and levering the handles. Plants with tough compacted crowns are most easily divided by cutting through lifted, washed crowns.

Many alpines can be divided to produce new plantlets annually. These divisions should be made about ten days after main flowering ceases, when new root and shoot growth commences. However, in the case of late summer- and autumn-flowering plants, division should be delayed until the spring.

Herbaceous plants with tough, grassy or sword-like leaves crowded in dense clusters should be divided with a spade or hatchet, in time to establish well before winter. Some shrubs produce a suckering base or crown or layer themselves, providing young shoots may then be separated.

DNA abbreviation for deoxyribonucleic acid; a complex molecule found in chromosomes, and the chemical basis of genes.

Docynia (the name is an anagram of *Cydonia*, a close relation). Rosaceae. E Asia. 2 species, evergreen or deciduous shrubs or small trees with clusters of fragrant, 5-petalled flowers; these are white, opening from pink-tinted buds in spring, and are followed by downy, ovoid, quince-like fruit in autumn. *Docynia* species are not reliably frost-hardy and in areas at the limits of their hardiness (zones 8 and under), are best positioned against a south-facing wall. Plant in full sun on a well-drained soil. Propagate by budding in summer, grafting in winter, or by seed in autumn. *D.delavayi* (S China; tree to 10m; leaves persistent, oblong, entire, white-hairy beneath; fruit yellow, 4cm in diameter).

Distictis buccinatoria

Dodecatheon (from Greek *dodeka*, twelve, and *theos*, god – Pliny applied the name to a spring flower, probably the primrose, protected by the twelve gods). Primulaceae. N America, Bering Straits. SHOOTING STAR, AMERICAN COWSLIP. 14 species, perennial herbs forming clumps or tufts of oblanceolate to obovate leaves. Produced in spring and summer, the *Cyclamen*-like flowers nod in slender-stalked umbels. Hardy in zone 6 and suitable for the woodland or damp rock garden. Grow in full sun or part shade on a cool, moist but well-drained soil rich in grit and leafmould. Propagate from ripe seed or by division.

D.dentatum (W US; to 10cm; flowers white, anthers yellow, tinted black); *D.hendersonii* (SAILOR-CAPS, MOSQUITO-BILLS; California; to 40cm; flowers mauve-pink with a cream and yellow basal rim, anthers violet-black); *D.meadia* (SHOOTING STAR, AMERICAN COWSLIP; E US; to 50cm; flowers purple-pink or pink or white, with a cream and yellow basal rim and purple-red anthers); *D.pulchellum* (western N America, Mexico, naturalized E US; to 35cm; flowers deep carmine to rose-purple or lilac, with a cream to gold basal rim and purple-black anthers).

Dodecatheon meadia

Dodonaea (for Rembert Dodoens (1516–85), Dutch physician and herbalist). Sapindaceae. Tropics and subtropics, especially Australia. Some 50 species, evergreen shrubs and trees with tough, glossy foliage and showy winged fruits. In a sheltered, warm and sunny spot. *D.viscosa* will tolerate short-lived, mild frosts and withstand drought, salt, wind and pollution. In hot, arid zones, it makes a good windbreak or hedging plant, responding well to light clipping. Plant in any well-drained soil in full sun. Propagate from seed sown in spring or by greenwood cuttings in summer. *D.viscosa* (NATIVE HOPS; South Africa, Australia, Mexico; light-textured, tree-like shrub to 2m with erect, resinous foliage; leaves to 13cm, narrowly elliptic to oblanceolate, wavy, glossy pale green; seed capsules clustered, to 2cm in diameter, flushed purple-red; 'Purpurea': leaves strongly flushed red-bronze to purple).

dolabriform hatchet-shaped.

division

doleiform barrel-shaped.

dollar spot (*Sclerotinia homoeocarpa*) a fungal disease of turf causing round discoloured patches. In the US, it occurs on a wide range of grasses including *Agrostis*, *Poa* and *Stenotaphrum*. In the UK, it is common only on CREEPING RED FESCUE (*Festuca rubra*) in some locations. Infected leaves are white or straw-coloured, arising in patches up to 5cm in diameter (hence dollar spot), which may merge to form larger areas of infection. The disease is associated with poor fertility, especially nitrogen deficiency, which should be rectified. Other control measures are selection of resistant types and spraying with a recommended fungicide.

Dombeya (for Joseph Dombey (d. 1794), French botanist). Sterculiaceae. Africa and Madagascar to Mascarene Islands. About 200 species, deciduous or evergreen shrubs or small trees, with large and hairy, ovate to cordate leaves. The flowers consist of five petals and are candy- or toffee-scented. They hang in pompon-like umbels on slender, drooping stalks in spring and summer. Grow in bright light or partial shade, providing protection from the hottest sun in summer, in a well-drained, fibrous loam-based medium, with additional leafmould and sharp sand. Water plentifully when in growth, reducing supplies as light levels and temperatures fall in winter; maintain a minimum temperature of 10°C/50°F. Prune after flowering to confine to allotted space. Propagate by seed or semi-ripe cuttings in a closed case with gentle bottom heat. *D.burgessiae* (syn. *D.mastersii*; central and South Africa; shrub or tree to 4m; leaves to 22 × 18cm; flowers to 5cm in diameter, white to candy pink, sometimes with darker veins).

dominant describing one of the forms taken by components of a single gene, which preferentially exhibits its characteristics when paired with another form in cross-pollination; cf. *recessive*.

Doritis (from Greek *dory*, spear, referring to the spear-shaped lip). Orchidaceae. SE Asia. 2 species, epiphytic, evergreen orchids with abundant aerial roots and short-stemmed growths of 2-ranked, leathery leaves. The flowers are borne at various times in rigidly erect and sometimes branching racemes. The tepals are obovate to elliptic and the lip 3-lobed, with the lateral lobes rounded and erect and the midlobe arrow-like and forward-pointing. Cultivate as for *Phalaenopsis*. *D.pulcherrima* (flower spikes 20–60cm, flowers to 4cm in diameter, deep magenta with lip midlobe mauve, lateral lobes red, disc lined white).

dormancy a condition in living plants where vegetative activity ceases (true dormancy) or is brought to an absolute minimum (quiescence) for a prolonged period. It may be due to physiological factors or seasonal rhythm, especially temperature, or it may be a response to environmental conditions unfavourable to growth. Dormancy is exhibited in deciduous species during winter, when plants rest in a leafless condition, and in bulbs, corms, tubers and seeds, which are a means of succession from one season to another, or over a number of years. In many species a period of dormancy is a suitable time for transplanting. Dormancy can be artificially broken through special techniques, most commonly through exposure to controlled temperature.

dormant oil see *winter wash*.

Doronicum (from the Arabic name for this plant, *Doronigi*). Compositae. Europe, Asia. LEOPARD'S BANE. 35 species, perennial tuberous or rhizomatous herbs with clumps of ovate to cordate leaves and yellow, daisy-like flowerheads on slender stalks in spring and summer. Fully hardy. Grow in any moderately fertile, well-drained but moisture-retentive soil in part shade or in good light, with protection from the strongest sun in summer. Propagate by division in autumn.

D.orientale (SE Europe, Caucasus, Lebanon; to 60cm; flowerheads 2–6cm in diameter; includes 'Finesse', with long-stalked flowerheads and slender, bright yellow florets, 'Frühlingspracht', with double, golden yellow flowerheads, 'Gerhard', with lemon yellow flowerheads and a green disk, 'Goldenkranz', with large deep yellow, double flowerheads, the dwarf, golden 'Goldzwerg', and 'Magnificum', with large flowerheads); *D.pardalianches* (syn. *D.cordatum*; GREAT LEOPARD'S BANE; W Europe to SE Germany; to 90cm; flowerheads 3–5cm in diameter); *D.plantagineum* (W Europe; to 80cm; flowerheads to 5cm in diameter; cultivars derived from this species include the tall, large-flowered 'Excelsum' and the free-flowering, mildew-resistant 'Strahlengold').

Dorotheanthus (for Dorothea, the mother of G.F. Schwantes, the botanist who named this plant, and Greek *anthos*, flower). Aizoaceae. South Africa. 6 species, succulent annuals, with fleshy leaves more or less covered in crystalline papillae. Brightly coloured, daisy-like flowers open on warm, sunny days. Sow seed thinly during late winter and early spring. Grow under glass with a temperature of 15–20°C/60–68°F; shade from direct sunlight. Prick out seedlings into trays or small unit containers. Grow on in full sunlight at a temperature of 10–15°C/50–60°F and with plenty of ventilation. Harden the plants off gradually during late spring, ready for planting outside in full sun on a well-drained soil in early summer. *D.bellidiformis* (syn. *Mesembryanthemum criniflorum*; LIVINGSTONE DAISY; leaves 2.5–7cm, tongue-like, thinly fleshy, glistening; flowers to 4cm in diameter, white, pale pink, magenta, red, orange, apricot or white edged red; many seed races available).

dorsal pertaining to the back of an organ, or to the surface turned away from the axis, thus abaxial; cf. *ventral*.

dorsifixed of an organ, attached by its dorsal surface to another.

Doronicum pardalianches

dorsiventral flattened, and having separate dorsal and ventral surfaces, as with most leaves and leaf-blades.

Doryanthes (from Greek *dory*, spear, and *anthos*, flower, in reference to the tall flowering stems). Agavaceae. E Australia. SPEAR LILY. 3 species, very large, evergreen perennial herbs, with sword-shaped leaves. The flowers are carried in short spikes compressed into a large terminal globose head or oblong thyrsus and held aloft on a tall scape. They are commonly replaced by bulbil-like plantlets. Grow in a humus-rich, well-drained soil in full sun with a minimum temperature of 10°C/50°F. Water plentifully when in growth; keep just moist in winter. Propagate by seed, suckers or by bulbils. *D.palmeri* (basal leaves to 2m; scape to 2.5m; flowers to 5cm, rich orange-red, paler within, among red-flushed bracts).

dot plant a bold specimen plant used to give contrast of foliage colour, form and height in a bedding scheme, for example, *Eucalyptus globulus* or standard fuchsias.

double, semi-double any flower with many more than the basic number of petals, sepals, florets or coloured bracts. Fully double blooms have densely packed rounded heads in which stamens are totally obscured or absent and the flower usually sterile. Semi-double blooms have two or three times the basic number of petals, sepals or florets, arranged in several rows with stamens conspicuous. A double form may be designated 'flore pleno' (abbreviation fl.pl.) 'plenus' or 'pleniflorus'. Hose-in-hose is a term describing a form of doubling in tubular flowers where one perfect corolla develops within another, frequently found in primulas and rhododendrons.

double-digging, double spading see *digging*.

double working the grafting of an intermediate cultivar or interstock between scion and rootstock to overcome incompatibility, to confer disease or cold resistance, or to control vigour.

dovecote a pigeon-house. Originally substantial, functional buildings, dovecotes later became popular as decorative garden features, often in the form of a small wooden house on a long pole.

downy mildew diseases caused by fungal parasites belonging to the family Peronosporaceae including *Bremia*, *Peronospora*, *Plasmopara* and *Pseudoperonospora*. The fungi penetrate deeply into plant tissue and their aerial growth can be seen as a glistening downy white or purple growth, usually on the under-surface of leaves. Yellow patches develop beneath the visible fungus growth and affected leaves turn brown and eventually die. In some downy mildews, there is marked distortion of the host due to swelling and curling of the infected tissues. Spread is by spores during the growing season, but a resistant sexual stage may overwinter inside plant tissues. Downy mildews are favoured by damp conditions. The following plants are common sufferers: crucifers (affected by *Peronospora parasitica*), onions (*P.destructor*), pea (*P.viciae*), rose (*P.sparsa*, mainly on greenhouse plants), spinach and beet (*P.farinosa*), umbellifers (*Plasmopara nivea*), grapevine (*P.viticola*) and cucurbits (*Pseudoperonospora cubensis*). Other plants sometimes affected include *Anemone*, *Antirrhinum*, *Campanula*, chrysanthemums, *Clarkia*, *Digitalis*, *Geum*, *Hebe*, *Helleborus*, *Laburnum*, *Mesembryanthemum*, *Myosotis*, *Papaver*, *Primula* and *Viola*.

Control by good ventilation, removing infected plant debris and, where essential, fungicide application. Copper-containing chemicals are effective and generally recommended for garden use. Many broad-spectrum organic fungicides also give good protection, including, for example, the dithiocarbamates mancozeb and zineb.

downy see *pubescent*.

Draba (name used by Dioscorides for another crucifer, *Lepidium draba*). Cruciferae. N Hemisphere, mountains of S America. 300 species, dwarf, annual or perennial, cushion-forming herbs with tiny leaves in tight rosettes and 4-petalled, usually yellow flowers in short racemes in late spring and summer. Fully hardy. Grow in the rock garden, in troughs or raised beds or in the alpine house. All dislike winter wet, especially those with woolly foliage – these, if grown in the open, should be protected with a cloche or propped pane of glass and with a topdressing of chippings or small stones. Grow in a gritty, well-drained and moderately fertile soil in full sun. In the alpine house, grow in pans of a mix of equal parts loam, leafmould and sharp sand, with a wedge of tufa or layer of limestone chippings around the neck of the plant. Water moderately during growth and flowering, avoiding the cushions; otherwise water very sparingly. Remove dead or dying material from the rosettes to avoid rot. Propagate from seed sown fresh in early spring. Alternatively, increase by careful division, by small offsets in spring or by soft cuttings in summer.

D.aizoides (C and S Europe, Great Britain; leaves to 1cm, linear, sparsely ciliate, incurved in dense, compacted cushions; flowers bright yellow, 4–18 on smooth flowering stems to 10cm tall); *D.hispanica* (Spain; differs from the first species in having slightly broader, roughly hairy leaves, shorter, bristly flowering stems and pale yellow flowers); *D.mollissima* (Caucasus; leaves to 0.6cm, oblong, blunt, white-hairy, in tight cushions; flowers golden yellow in tight, short racemes); *D.polytricha* (Armenia, Turkey; leaves to 0.5cm, oblong to spathulate, with long white hairs, in tight cushions; flowers bright yellow on hairy stems); *D.rigida* (Armenia, Turkey; leaves to 0.6cm, broadly linear, ciliate, dark green in tight, mossy rosettes; flowers bright yellow on smooth stalks; includes var. *bryoides* with very small, incurved dark green leaves and more or less stemless flowers).

*Dracaena
fragrans
'Massangeana'*

Dracaena (from Greek *drakaina*, a female dragon, referring to the resinous red gum exuded from the stems of *D.draco*). Agavaceae (Dracaenaceae). Old World, 1 species America. Some 40 species, evergreen shrubs or trees, sparingly to much-branched, the bark smooth and ringed to rough and fissured. The leaves vary in shape from narrow and sword-like to lanceolate or elliptic, smooth and tough. Creamy, star-shaped and nocturnally fragrant flowers are borne in large panicles at various times of year. *D.draco* prefers full sun, a rather dry, lean soil and a minimum winter temperature of 5°C/40°F. In Mediterranean-type climates, it becomes a large, much-branched tree. The other species listed here are often grown in the home, or in interior landscapes, and are valued for their resilience and decorative foliage. Grow them in a fertile, porous medium in light shade or dappled sun, with a minimum winter temperature of 13°C/55°F. Water, feed and syringe liberally in warm weather; keep barely moist at other times. Sponge over leaves of plants grown indoors. Propagate by seed, stem tip cuttings or air-layering.

D.cincta (origin obscure; shrub to 1.8m, stem slender, usually unbranched; leaves to 38 × 1.5cm, linear, mid- to deep green, edged red to brown; some plants cultivated as *D.marginata* may belong here; 'Tricolor': leaves striped cream and edged red); *D.concinna* (Mauritius; shrub to 1.8m, stem slender, red-tinted; leaves to 90 × 7.5cm, linear to ensiform, dull green edged purple-red; some plants cultivated as *D.marginata* may belong here); *D.deremensis* (tropical E Africa; shrub to 4m, stem slender, sparsely branched; leaves to 70 × 5cm, linear to lanceolate, narrowing toward base, spreading and clothing much of stem, dark to grey green; includes several popular variegated cultivars, for example 'Bausei' with a central white band, 'Lemon and Lime', lime green with a yellow central band and margins, 'Roehrs Gold', with a pale gold central band and margins, and 'Warneckii', grey-green with a pure white central stripe and margins); *D.draco* (DRAGON TREE; Canary Islands; tree to 10m, ultimately with a very thick, grey-barked trunk exuding red resin if wounded and developing a massively branched, broad crown; leaves to 60 × 4cm, linear to lanceolate, glaucous, tough); *D.fragrans* (Sierra Leone to Malawi; shrub or tree to 15m, branching above, stems thick; leaves to 100 × 10cm, lanceolate, narrowing below, outspread with tips curving, glossy pale to dark green; includes the popular 'Massangeana', CORN PLANT, with pale green leaves broadly striped yellow); *D.hookeriana* (SE Africa; shrub to 1.8m, sometimes branched; leaves to 75 × 11cm, lanceolate, semi-erect with curving tips and narrow base, margins translucent white or red-tinted; includes 'Gertrude Manda', with broad and thick leaves with translucent edges, and 'Variegata', with white-striped leaves); *D.reflexa* (syn. *Pleomele reflexa*; SONG OF INDIA; Madagascar, Mauritius; graceful shrub to 5m, freely branching with slender, rather sprawling stems; leaves to 20 × 2–5cm, linear to lanceolate or narrowly elliptic, tapering finely at both ends, dark green; includes forms with narrow,

white-edged and yellow-striped leaves); *D.sanderiana* (BELGIAN EVERGREEN, RIBBON PLANT; Cameroon; erect, sparingly branched shrub to 1.5m; stems slender; leaves to 20 × 1.5–4cm, lanceolate, base and apex tapering, spreading and somewhat wavy, clothing stems, pale to grey-green broadly edged or striped cream-white); *D.surculosa* (syn. *D.godseffiana*; GOLD DUST DRACAENA, SPOTTED DRACAENA; shrub, 1–4m with slender, twiggy branches; leaves to 20cm, elliptic, apex acute, base shortly tapered, deep glossy green, spotted to flecked or blotched gold to cream). For *D.indivisa*, see *Cordyline indivisa*; for *D.godseffiana*, see *D.surculosa*; for *D.marginata*, see *D.cincta* or *D.concinna*; for *D.terminalis*, see *Cordyline terminalis*.

Dracocephalum (from Latin *draco*, dragon, and Greek, *kephale*, head, referring to the shape of the flowers). Labiatae. Europe, Africa, Asia, N America. About 45 species, annual or perennial herbs and dwarf shrubs with whorls of tubular, 2-lipped flowers in terminal or axillary spikes or racemes. The following species is perennial and hardy to –15°C/5°F. Grow in a sunny position in a fertile well-drained soil. Propagate by division in spring or autumn, or from basal cuttings of young growth in spring. *D.ruyschianum* (C Europe to Japan; erect to 60cm; flowers to 3cm, violet-blue, rarely pink or white).

Dracula (from the diminutive form of Latin, *draco* – the name does not commemorate the Transylvanian Count, although both names mean 'little dragon'; one of the most splendidly undead-looking species also rejoices in the name *D. vampira*). Orchidaceae. Central and South America. Some 100 species, evergreen perennial herbs. They are similar to *Masdevallia*, the flowers consisting of 3 sepal-segments more or less fused in a triangular or 3-pointed shield or cup, with each segment drawn out to a long, tail-like tip (the descriptions below measure the flowers from tip to tip). Unlike *Masdevallia*, the flowers hang on slender stalks and have a distinctly clawed, usually pouch-shaped lip. They are produced in abundance in spring and summer.

Small orchids with spectacular, if rather sinister flowers, they need a temperature range of 10-24°C/50-75°F, shade, high humidity and a buoyant atmosphere. Plant in small baskets or perforated pots (flowerstalks often burrow through the medium before finding an outlet) in a fine, bark- or perlite-based mix with chopped sphagnum moss. Keep moist and syringe frequently. Feed whilst the new growth develops. Increase by division after flowering.

D. bella (syn. *Masdevallia bella*; Colombia; flowers to 20cm across, basically triangular with fused segments and long, slender tips, buff or dull olive densely spotted oxblood with darker tips and a white or pale lip); *D. chestertonii* (syn. *Masdevallia chestertonii*; Colombia; flowers to 7cm across with three distinct segments, each with a slender tip, sulphur green or grey-white, dotted and edged with purple to grey-black warts, lip ochre lined with red radiating gills,

Dracula bella

resembling an upturned mushroom); *D. chimaera* (syn. *Masdevallia chimaera*; Colombia; flowers to 30cm across, segments triangular to ovate with very long, slender tips, buff to olive flecked and stained maroon to violet-black, hairy and warty, lip cream to flesh pink, buff to orange at base); *D. erythrochaete* (syn. *Masdevallia erythrochaete*; Guatemala, Nicaragua, Costa Rica, Panama; flowers to 12cm across, segments more or less fused, forming a broad cup with long, slender tips, cream to grey-pink densely spotted flesh pink to maroon and hairy, with maroon tails, lip white or buff); *D. platycrater* (syn. *Masdevallia platycrater*; Colombia; flowers to 25cm across with distinct and outspread, narrow segments and slender tips, white to rose dotted purple-red, lip pink); *D. vampira* (Ecuador; flowers to 30cm across with ovate to triangular segments and very slender, long tips, white to yellow green densely overlaid and veined purple-black, with golden sunburst-like markings at centre, tips maroon-black, lip white flecked violet).

Dracunculus (name used by Pliny for a plant with a curved rhizome, from Latin diminutive of *draco*, a dragon). Araceae. Mediterranean, Madeira, Canary Islands. 3 species, tuberous herbaceous perennials with pedately compound leaves – their mottled sheaths and stalks forming a pseudostem – and, in spring and summer, large and malodorous, *Arum*-like inflorescences. The following species will establish itself in most sheltered, well-drained positions in zones 7 and over, tolerating full sun or dappled shade. It becomes dormant in summer, when the soil should be warm and dry. In very wet or cold areas, protect the corms with a dry mulch in winter. Increase by offsets in late summer, or seed in autumn. *D.vulgaris* (syn. *Arum dracunculus*; DRAGON ARUM; C and E Mediterranean; stem 40–100cm, mottled maroon; spathe to 100 × 20cm, wavy, deep red-purple, spadix black-red).

drag brush a wide-headed brush-like hand tool, designed to be pulled across turf to work in top dressings.

drag, dragfork see *fork*.

drainage the removal by gravity of excess surface or soil water to a lower level. Good soil drainage is very important in gardens, providing or improving conditions for healthy plant growth and preventing flooding of surfaces. Imperfect drainage may be indicated by surface puddling after rain, and by the presence of moss in lawns. It may be due to a high water table, or more often to compacted layers which arise naturally on some soils or as a result of treading or cultivating in unfavourable conditions.

Drainage status can be assessed by digging 30cm-deep holes across a site shortly after heavy rain and examining them a few hours later for water presence. Drainage defects are more apparent in soils of heavy texture, and on sites where a problem is identified it is advisable to dig inspection holes to at least 60cm depth. Where these are kept covered from rainfall for 24 hours and are then found to have an accu-

mulation of water, the existence of a high water table is indicated and skilful site management will be needed. Alternatively close examination of the side walls of the inspection holes will reveal whether compaction is a problem.

Soil drainage can be improved by careful cultivation, and, where compaction is a problem, double digging or trenching (see *digging*) may be desirable. Incorporation of lime and rotted organic matter serves to improve soil structure, and in some situations grit or sand additives can aid drainage.

Where fundamental site characteristics cause impeded drainage, it may be worth laying down a drainage system. However, such a project is expensive, both in terms of labour and materials and must be carefully surveyed and executed. The first essential is to plan for disposal of collected water from the lowest point of the site. Ideally, the water should be directed to a ditch or other authorized outlet or, alternatively, to a soakaway pit, constructed at least 1.8m square and filled with rubble. Purpose-made drain-pipes of earthenware or plastic, 5–7.5cm in diameter, should be laid on a firmed 5cm bed of coarse gravel in trenches 35–40cm deep. The trenches must follow a natural slope or a constructed one with an ideal fall of 1:40. A herringbone design is appropriate, with the feeders meeting main lines at an angle of 45 degrees. The drain pipes should be overlaid with a shallow layer of coarse gravel. As an alternative to drainpipes trenches may be filled with a 10–15cm layer of coarse gravel, or for a shorter serviceable period with tight bundles of brushwood, all conforming to the same principles of depth, layout and outfall. On heavy clay soils, narrow drainage tunnels, known as mole drains, may be formed at intervals across a site by using a flat-bladed plough to draw a metal cylinder through the soil is such a way that it penetrates below cultivation depth. Mole drains are usually constructed in conjunction with a pipe system and will remain functional for a number of seasons.

drawn describing seedlings or plants that become abnormally tall and thin as a result of poor light or crowding.

Dregea (for Johann Franz Drege (1794–1881), botanist). Asclepiadaceae. Old World (warm regions). 3 species, woody-based, evergreen climbers with large leaves and umbels of starry and fragrant, *Hoya*-like flowers in summer. The following species is hardy in zone 8. Plant on a well-drained, moist and fertile soil in sun or part shade. Provide support. Increase by seed in spring or stem cuttings in summer. *D.sinensis* (syn. *Wattakaka sinensis*; China; evergreen climber to 3m tall; leaves to 10cm, broadly ovate to cordate, grey-felted beneath; flowers to 1.5cm in diameter, white marked pink, in long-stalked, downy umbels to 10cm across).

drench, drenching the liberal application of a pesticide solution to the soil around the roots of a plant; usually to control a soil-borne pest or disease by contact, but also for systemic action, where the pesticide is taken up by roots and transported throughout the plant to destroy pests feeding upon aerial parts.

drepanium a sickle-shaped cyme.

Drepanostachyum (from Greek *drepanon*, sickle, and *stachys*, spike, in token of the sickle-shaped inflorescence.) Gramineae. Himalaya. A genus of clumping bamboos; see *bamboos*.

dressing (1) any bulky organic manure or fertilizer applied to the soil in a solid as opposed to a liquid state, in preparation for cropping or as an application around established plants (see *base dressing*, *side dressing*, *top dressing*); (2) any fungicidal or insecticidal substance applied to seeds, bulbs, corms, or tubers to prevent disease or pest attack.

dribble bar a perforated metal or plastic bar fitted to a watering can or sprayer for the application of insecticide, fungicide or weedkiller. Also known as a sprinkler bar.

dried blood a quick-acting organic fertilizer containing 10–12% nitrogen, used either as a dry, base- or top-dressing or as a liquid feed when mixed with water. The sterilized product should be handled with suitable precautions to avoid inhalation and other personal contact.

drift (1) any part of a pesticide spray that has been carried beyond its intended target. Drift may cause damage to neighbouring plants with which it comes into contact. Particularly at risk are tomatoes, in which even very low concentrations of hormone-type weedkillers, carried as drift over long distances, can cause serious distortion of growth; (2) a block planting of young trees.

drill a shallow furrow or groove made in soil for the sowing of seeds, or for transplants.

Drimys (from Greek *drimys*, acrid, sharp). Winteraceae. S America, Malaysia, Australasia. Some 30 species, evergreen trees or shrubs with leathery, aromatic leaves and clusters of star-shaped flowers in late spring. *Drimys* is hardy in climate zone 8, and in sheltered areas in zone 7. Plant in semi-shade on light, lime-free soils which are well drained but moisture-retentive. Propagate by greenwood cuttings in late summer, on a mist bench or in a closed case, by softwood basal cuttings in spring, or by simple layering.

D.lanceolata (Australia; PEPPER TREE; shrub or small tree to 3m; branchlets dull red to purple-red; leaves to 12cm, lanceolate to oblong, subcoriaceous; flowers pale brown to green-white; fruit black); *D.winteri* (Chile, Argentina; tree or shrub to 20m, often multi-stemmed; bark aromatic; leaves 3–20cm, elliptic to oblong or lanceolate, coriaceous, glaucous and pale beneath; flowers white to cream, fragrant; fruit glossy black).

drip irrigation see *trickle irrigation*.

drip line (1) the actual or predicted line on the soil surface below a tree canopy where its leaves shed rain drips. It is used as a guide for fertilizer application; (2) a term used to describe a small-bore irrigation pipe in trickle systems.

drip-point, drip-tip a leaf tip, either acuminate, caudate or aristate, from which water readily drips in wet conditions.

drop layering, dropping see *layering*.

dropsy see *oedema*.

Drosanthemum (from Greek *drosos*, dew, and *anthemon*, flower, referring to the glistening, dew-drop appearance of the papillae). Aizoaceae. South Africa. 90 species, succulent perennial shrubs with fleshy, usually papillose leaves and bright, daisy-like flowers opening on warm and sunny days. Cultivate as for *Lampranthus*.

D.hispidum (freely branching shrub to 60cm tall and over 1m across, stems rooting, with rough white hairs; leaves 1.5–2.5cm, cylindric, obtuse, light green to red-tinged with large, transparent papillae; flowers to 3cm in diameter, deep purple-red); *D.speciosum* (shrub to 60cm, stem papillose, with rough spots; leaves 1–1.6cm, semicylindric, curved, obtuse, covered in crystal-like papillae; flowers to 5cm in diameter, deep orange-red with a green centre).

Drosera (from Greek *droseros*, dewy, from the glands on the leaf surface). Droseraceae. Cosmopolitan. DAILY DEW, SUNDEW. Some 100 species, highly varied carnivorous herbs, mostly perennial and all trapping their prey by sweet and glistening, red-tinted tentacles. These cover the leaf surfaces and bend slowly inwards to ensnare victims. The leaf itself may roll inwards, enclosing the prey and forming a pocket in which the digestive process takes place. The dish-shaped flowers have five (sometimes four or eight) rounded petals and may be solitary, racemose or paniculate.

All *Drosera* species require full sun and a humid atmosphere. The hardy species will endure hard frosts, dying back to resting winter buds. These include *D.anglica*, *D.intermedia* and *D.rotundifolia*. Since, however, it is almost impossible to recreate the sphagnum bog they inhabit in the open garden, these sundews tend to be grown in the cool greenhouse or in the home. The tender (remaining) species need a minimum temperature of 7°C/45°F. Plant all species in small clay pots containing a mix of coarse peat, sharp sand and fresh sphagnum. In spring and summer, stand the pots in rainwater up to one third of their depth. In autumn and winter, keep just moist. Tuberous species will die back during this period. In hot, dry weather, syringing with soft water can be beneficial, together with light shading to prevent scorching. In the home, it may be necessary to grow these plants in a case, or at least to stand them in large bowls or trays of water. Propagate by seed or leaf and root cuttings. Alternatively, increase by division or by gemmae (tiny bud-like propagules produced during the resting period) sown on damp compost.

D.adelae (Australia; leaves 10–25cm, narrow-lanceolate, thin-textured with a distinct midrib, pale green to olive, sparsely red-glandular; flowers to 0.3cm in diameter, cream to beige or red-brown); *D.anglica* (GREAT SUNDEW, ENGLISH SUNDEW; N Europe, N Asia, N America; leaves to 1.3cm, linear-oblanceolate, pale green, glands bright red; flowers to 0.5cm in diameter, white); *D.binata* (Australia, New Zealand; leaves to 30cm, the blade deeply forked into 2–14 linear to filiform spreading lobes with long, red tenatcles; flowers to 2cm in diameter, white to pink); *D.capensis* (S Africa; leaves 3–6cm, narrowly oblong to spathulate-linear, with red tentacles; flowers to 2cm in diameter, rosy pink); *D.capillaris* (PINK SUN-DEW; subtropical and tropical Americas; leaves to 0.6cm, broadly spathulate to orbicular; flowers to 1cm in diameter, pink or white); *D.cistiflora* (South Africa; leaves to 10cm, linear; flowers 5cm in diameter, scarlet, violet or white); *D.filiformis* (DEW-THREAD SUNDEW, THREAD-LEAVED SUNDEW; US; leaves 10–25cm, filiform with purple-red tentacles; flowers to 3cm in diameter, rose-pink); *D.gigantea* (W Australia; leaves to 0.6cm in diameter, cup-shaped, peltate, with red tentacles, on slender scrambling stems; flowers to 1.5cm in diameter, white); *D.glandulifera* (COMMON SCARLET SUNDEW; Australia; leaves to 0.5cm in diameter, transversely oval, concave, with red tentacles; flowers to 0.8cm in diameter, red with black centres); *D.intermedia* (LOVE NEST SUNDEW; N Europe and Asia, N and C America; leaves to 0.5cm, obovate, covered in red tentacles; flowers 1cm in diameter, white); *D.prolifera* (Australia; leaves to 1.5cm, reniform with pale tentacles; flowers to small, pink to red, on scapes which in time bend over and bear plantlets, coming to resemble proliferous stolons); *D.pygmaea* (Australia, New Zealand; leaves to 0.2cm in diameter, round, concave; flowers to 0.3cm in diameter, white); *D.regia* (GIANT SUNDEW; S Africa; leaves to 70cm, linear to lanceolate; flowers to 3.5cm in diameter, pale pink to purple); *D.rotundifolia* (ROUND-LEAVED SUNDEW; northern temperate regions; leaves 0.5–1cm in diameter, round with bright red tentacles; flowers to 0.6cm in diameter, white to pink); *D.spathulata* (SPOON-LEAF SUNDEW; NE Asia to New Zealand; leaves 1–2cm, oblong to spathulate; flowers to 0.8cm in diameter, white to pink); *D.whittakeri* (Australia; leaves to 3.5cm, spathulate to obovate, pale green with bright red tentacles; flowers to 3cm in diameter, white, fragrant).

Drosophyllum (from Greek *drosos*, dew, and *phyllon*, leaf, alluding to the glistening appearance of the glandular leaves). Droseraceae. Portugal, S Spain, Morocco. 1 species, *D.lusitanicum*, a carnivorous perennial to 30cm. It resembles *Drosera*, but its sticky leaf tentacles are shorter and do not move to enmesh prey. The plant develops a slender, usually trailing, woody stem bearing a terminal rosette of thread-like leaves. These are some 20cm long and densely covered on their undersides by glandular hairs which secrete adhesive and digestive fluid. Withered leaves persist in a shaggy, grey-brown ruff. The flowers consist of five broad, bright yellow petals and are carried in a corymbose cyme in spring and summer. Provide a minimum winter temperature of 7°C/45°F, full sun and an airy, rather dry atmosphere. It resents disturbance and is best grown from seed in a small clay pot containing a mix of two parts moss peat, two parts loam-based compost and one and a half parts sand. As the plant increases in size, plunge the smaller pot within a larger pot of the same mix. The compost in the larger pot should be moist at all times (use rainwater) but that in the smaller pot (i.e. around the crown of the plant) should be kept dry. The roots will descend into the large pot below. Good results have also been achieved by drilling through a piece of pumice stone on tufa. This is placed on a pan containing the medium described above. A young plant may then be established in the hole in the stone (also containing the medium). This will eventually root into the pan, which should be kept just moist by watering from beneath.

drought the depletion of soil moisture due to lack of rainfall or insufficient watering. Physiological drought occurs when soil salinity is so high that it restricts the absorption of water by plant roots. Either form of drought causes flagging and various degrees of damage in plants not adapted to grow in such conditions.

drupaceous resembling, or pertaining to a drupe; with a juicy seed coat.

drupe an indehiscent, one- to several-seeded fruit, in which the endocarp is osseous or cartilaginous, and is contained within a soft, fleshy pericarp. Stone fruits, such as mangoes or plums, are drupes.

drupe

drupelet a small drupe, generally a part of an aggregate fruit such as a blackberry (*Rubus*).

dry rot a general term for diseases that cause a dry decay, with discoloration and shrivelling of tissue, of various plant parts including fruits and underground storage organs, for example *Fusarium solani* rot of potatoes.

dry set a condition in which fruits appear to set but fail to swell. It is especially common in greenhouse tomatoes grown in hot dry conditions, and is largely preventable by regular damping down.

dry wall a stone wall built without mortar, used in gardens to stabilize vertical banks and terrace edges. Dry walls are suitable for planting vigorous alpines and trailing plants.

Dryandra (for Jonas Dryander (1748–1810), Swedish botanist.) Proteaceae. W Australia. Some 60 species, evergreen shrubs or small trees with leathery, toothed to lobed leaves, often white-downy beneath. Small, tubular flowers are carried in dense, terminal heads subtended by persistent bracts in spring and summer. Cultivate as for *Banksia*. Chlorosis may be treated with chelated iron. *D.formosa* (SHOWY DRYANDRA; W Australia; shrub to 2m; leaves 10–20cm, soft, triangular-lobed, tomentose beneath; flowerheads large, golden-orange).

Dryas (after Dryas, a wood nymph to whom the oak was sacred – the leaves of *D.octopetala* resemble those of the oak). Rosaceae. Arctic and alpine regions. MOUNTAIN AVENS. 5 species, evergreen, creeping shrublets, forming extensive, loose mats of slender, woody branches clothed with tough, rugose leaves. Produced in early summer, the flowers are solitary, borne on long stalks and consist of 7–10, rounded petals and numerous stamens. They give rise to snow white, feathery seed heads. The following species are suitable for the rock garden, making good light cover for spring bulbs, and for planting in paving joints. Hardy to –15°C/5°F. Plant in gritty, well-drained, neutral to slightly alkaline soils. Grow in an open position in sun. Propagate by ripe seed sown fresh, by semi-ripe, heeled cuttings, rooted in sharp sand; also by division of rooted stems.

Dryas octopetala

D.drummondii (N America; leaves to 4cm, elliptic to obovate, pale green above, white-tomentose beneath or throughout, margins coarsely crenate; flowers cream or white tinted yellow, nodding); *D.octopetala* (MOUNTAIN AVENS; northern temperate regions; leaves to 4cm, oblong to ovate, dull green and glabrous above, white-tomentose beneath, margins crenate; flowers white, erect); *D.* × *suendermanii* (a vigorous hybrid between the two preceeding species; flowers slightly nodding, cream in bud, opening white).

Dryopteris (from Greek *dryas*, oak, and *pteris*, fern – some species are found in oak woods). Dryopteridaceae. Cosmopolitan. WOOD-FERN, BUCK-LER FERN, SHIELD FERN, MALE FERN. 150 species, ferns with erect to creeping rhizomes and shuttlecock-like to loose and spreading crowns of fronds – these are triangular to oblong-lanceolate in outline and 1–3-pinnate. The following species are hardy down to –30°C/–22°F if well-established and mulched during cold periods. Most are tolerant of a wide range of pH. Grow in part to full shade or dappled sun (some species, for example, *D.erythrosora*, will tolerate full sun if cool and moist at the roots, but their fronds will be of poorer quality). Plant in a moist but porous soil, rich in leafmould and well-rotted garden compost. Water generously during the first spring and summer. Propagate by spores or by division of mature plants with multiple crowns.

D.affinis (GOLDEN SHIELD FERN; Europe, Asia; stipes covered with golden to orange-brown scales; frond blades 30–125cm, elliptic to narrowly elliptic, pinnules oblong, lobed, toothed, deep, glossy green; includes dwarf, congested and crested forms); *D.carthusiana* (NARROW BUCKLER FERN; Europe; stipes covered with pale brown scales; frond blades 6–30cm, stiffly erect, lanceolate to ovate-lanceolate, pinnules pinnate or pinnately lobed, lime green, glandular-hairy beneath); *D.dilatata* (BROAD BUCKLER FERN; Europe; stipes covered with pale brown scales; frond blades 7–100cm, triangular-ovate, pinnules ovate to oblong, dentate or pinnately lobed, dark green, sparsely glandular beneath; includes cultivars with crisped pinnules or crested fronds); *D.erythrosora* (JAPANESE SHIELD FERN, COPPER SHIELD FERN; E Asia; stipes red-brown to deep copper with brown or black scales; frond blades 30–70cm, broadly ovate-triangular to oblong, pinnules narrowly oblong to linear-lanceolate, dentate or pinnately lobed, emerging deep copper-pink, turning gold to lime with a bronze tint, ultimately deep olive flushed copper); *D.filix-mas* (MALE FERN; N America, Europe, N Asia; stipes with brown scales; frond blades 30–90cm, lanceolate-triangular, apex finely acuminate; pinnae oblong-lanceolate, finely tapering with oblong, serrate segments, dull mid- to deep green, more or less hairy; includes cultivars with crested, crisped and much-reduced fronds); *D.goldieana* (GIANT WOOD FERN, GOLDIE'S WOOD FERN; NE America; stipes covered with dark, glossy brown scales; frond blades to 1m cm, ovate, abruptly acuminate, pinnae oblong-lanceolate, acuminate, segments oblong, serrate); *D.intermedia* (FANCY FERN; N America; stipes straw-coloured, covered in dense scales; frond blades to 70 × 30cm, ovate-lanceolate, abruptly acuminate, pinnules with segments oblong, obtuse, minutely spiny-toothed, bright green, glandular); *D.marginalis* (N America; stipes covered in pale brown scales; frond blades 25–75cm, lanceolate, pinnae oblong-lanceolate, segments oblong, blunt, entire to crenate, coriaceous); *D.oreades* (MOUNTAIN MALE FERN; Caucasus; stipes with pale brown scales; blades 30–80cm, oblong-lanceolate to narrowly lanceolate, pinnae linear-lanceolate, attenuate, pinnules oblong, bluntly toothed; includes wavy to crested forms).

Duchesnea (for Antoine Nicolas Duchesne (1747–1827), French horticulturist, who published in 1766 *Histoire naturelle des fraisiers*, an account of the history of the strawberry). Rosaceae. Asia. INDIAN STRAWBERRY, MOCK STRAWBERRY. 2 species, stoloniferous perennial herbs. They closely resemble strawberries (*Fragaria*), differing in their yellow flowers and small, rather dry and inedible fruit. Hardy to –20°C/–4°F, *D.indica* is a vigorous and fast-growing carpeting plant. It favours a moist, sandy but fertile soil in dappled sun. Increase by offsets freely produced on runners. *D.indica* (syn. *Fragaria indica*; India to Japan, naturalized N America; leaflets to 7cm, obovate, crenate, silky-hairy beneath; flowers to 2cm in diameter; fruit bright red).

Dudleya (for William Russel Dudley (1849–1911), first professor of botany at Stanford University, California). Crassulaceae. Western N America. 40 species, succulent perennials with fleshy, ovate to linear leaves in basal rosettes. The flowers are tubular or star-shaped with five, acute lobes; they are borne in spring and summer in panicles. Grow in full sun, in a perfectly drained, gritty medium, with a minimum winter temperature of 7°C/45°F. Water very sparingly when plants are dormant. Increase by stem cuttings.

D.brittonii (leaves 7–25cm, crowded, finely pointed, dusted white, in a solitary rosette; flowers yellow to green); *D.lanceolata* (leaves 10–15cm, lanceolate, tip tapering narrowly, crowded in a solitary rosette, glaucous; flowers red-yellow); *D.pulverulenta* (CHALK LETTUCE; leaves 7–30cm, oblong to spathulate, apex tapering and pointed, thickly white-glaucous in a large, solitary rosette; flowers red); *D.saxosa* (leaves 5–15cm, pointed, semi-terete, glaucous, in few or solitary rosettes; flowers yellow to red); *D.virens* (ALABASTER PLANT; leaves 4–25cm, linear, pointed, glaucous; flowers white).

Duranta (for Castor Durantes, early-18th-century physician and botanist in Rome). Verbenaceae. Tropical America. Some 30 species, trees or shrubs with 5-lobed, salverform flowers in slender racemes or panicles in summer and fleshy, globose fruit. Cultivate as for *Lantana. D.erecta* (syn. *D.plumieri*, *D.repens*; PIGEON BERRY, SKY FLOWER, GOLDEN DEWDROP; tropical America; shrub or small tree to 6m, sometimes spiny; racemes to 15cm, flowers white, lilac, blue or purple; fruit to 1cm diameter, yellow).

dust, dusting the application to plants or soil of a pesticide formulated as a fine powder. Dusts have good adherence to plants but can be unsightly. They are best administered by means of a special applicator.

Dutch elm disease (*Ophiostoma novo-ulmi* (syn. *Cyratocystis ulmi*); *O. ulmi*) a serious disease of *Ulmus* and *Zelkova* species, first detected in Holland in 1918 and in the eastern US in the 1920's. The major epidemic in the UK resulted from the introduction of *O. novo-ulmi* from the US on infested elm wood. This more virulent species has largely replaced *O. ulmi* which causes a less serious form of the disease. Dutch elm disease has been responsible for devastating losses to the elm throughout the British and American countryside. Trees of all ages are liable to infection. Leaves in the crown wilt, become yellow and die during summer months; notching appears in twig crotches and the twigs may become crookshaped. Longitudinal streaks can be seen under the bark. Symptoms recur and spread in succeeding seasons, and the tree is killed within a few years, although growth may regenerate from the base.

The disease is mainly spread by bark beetles, especially the ELM BARK BEETLE (*Scolytus scolytus*), which emerge from brood galleries in dead elm bark and transmit fungal spores as they feed in the crotches of twigs and bore into older bark. Transmission through root contact also occurs. The only practical means of control is to fell and burn affected trees.

Dutch gardening in Britain, a style of gardening broadly dating from the reign of William and Mary and reflecting a combination of French and contemporary Dutch influence. It is characterized by an elaborate layout with statuary, topiary and low box hedges, orangeries, trees in tubs, and the use of bulb flowers.

Dutch light a single sheet of glass, usually 1422 × 730mm, fitted into a wooden framework. Dutch lights are used to cover garden frames and sometimes mounted to form a Dutch-light greenhouse, with sloping sides and a pitched roof.

Duvalia (for H.A. Duval (1777–1814), author of *Plantae succulentae in horto alenconio* (1809)). Asclepiadaceae. Arabia, Africa. 19 species, small, succulent, perennial herbs with fleshy, 4–6-angled and toothed stems. The short-lived leaves are minute and scale-like. The fleshy flowers are deeply bell-shaped with five, recurved lobes. Cultivate as for *Stapelia. D.corderoyi* (South Africa; stems to 5cm tall, usually prostrate, purple-tinted, thick; flowers to 5cm in diameter, olive green edged red-brown with deep purple hairs).

dwarf pyramid a trained form of fruit tree, with tiers of successively shorter horizontal branches forming a conical shape. Also known as a fuseau.

dwarf, dwarfing a small or slow-growing form of a plant, which may arise by mutation, breeding or, as in the case of many dwarf conifers, by the propagation of witches brooms. Dwarfing rootstocks are particularly valued for predicting the ultimate size of fruit trees. Dwarfing can also be achieved by cultural methods, as in bonsai, or by the use of growth inhibitors on pot plants.

Dyckia (for Count Salm-Dyck (1773–1861), German botanist). Bromeliaceae. C and S America. 104 species, stemless, rosette-forming perennial herbs with rigid, sword-shaped leaves covered with grey scales and edged with spiny teeth. Small flowers are crowded in erect, branching, scapose inflorescences. Grow in full sun in a dry, airy atmosphere (minimum temperature 7°C/45°F); pot in a very gritty, open medium. Keep virtually dry in winter. *D.remotiflora* will tolerate several degrees of frost if dry. Propagate by rooted offsets.

D.brevifolia (S Brazil; leaves to 20cm; flowers yellow); *D.* 'Lad Cutak' (*D.brevifolia* × *D.leptostachya*; leaves 15–25cm, maroon-flushed; flowers orange-yellow in a spike to 1m); *D.leptostachya* (throughout range; leaves 40–100 cm, bulbous below, covered in ash-grey scales; flowers red-orange); *D.remotiflora* (S Brazil, Uruguay; leaves 10–25cm, dark green; flowers dark orange).

Dudleya pulverulenta

dwarf pyramid

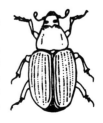

bark beetle

E

e-, ex- prefix meaning 'without' or 'deprived of', for example estipellate, without stipels.

ear (1) an auricle; (2) the spikelet of some grasses.

earthing up drawing up of soil around plants, usually with a draw hoe or drag fork. It is carried out on potato crops to prevent tuber-greening and infection from blight; on brassicas to prevent wind-rocking; on leeks and celery to blanch stems; and in layering and stooling fruit-tree rootstocks to encourage the formation of rooted shoots.

earthworms (Annelida: Oligochaeta) worms characterized by long, soft, multi-ringed or segmented cylindrical bodies, which have pairs of bristles on the underside to aid movement. They are widely distributed in soils. Mature worms have a swollen glandular area about half way along the body which, depending on species, is either saddle-shaped or completely encircling; its main function is to produce a yellow, lemon-shaped cocoon in which 20–30 eggs are deposited. Young worms emerge after 4–24 weeks, and take 35–70 weeks to reach maturity. Unless the weather is extremely cold or dry, earthworms are active within 50cm of the surface.

The earthworms' burrowing improves drainage and aerates the soil, and bacterial decomposition is hastened by their digestion of organic matter. The small earthworms known as MANURE WORMS or BRANDLING, particularly *Eisenia foetida*, are useful in compost heaps, where they accelerate the decomposition of vegetable matter.

Worms feed by pulling organic matter lying on the surface into their burrows and by extracting the organic matter content from soil ingested during burrowing activities. This soil is ejected in a semi-liquid state and used for lining burrows. Excess excreted soil is deposited as worm casts above or below the surface, where it can be a problem on lawns.

Most earthworms cannot thrive in acid soils, so that any measures to reduce pH will lower their numbers. Where control of earthworms is necessary recommended chemical treatment may be used.

earwigs (Dermaptera: Forficulidae) insects up to 25mm long, with large semi-circular membranous hindwings folded under shortened leathery wing sheaths, leaving most of the abdomen exposed. The end of the abdomen is characterized by a pair of horny pincers – these are curved in males and almost straight in females. Of most importance to gardeners is the COMMON EARWIG (*Forficula auricularia*), which occurs in Europe, North America, northern Africa, western Asia, Japan, Australia and New Zealand. The female lays eggs in a soil cavity and remains with them until they hatch in early spring; the nymphs resemble adults but are smaller. A second brood of young may be produced in midsummer. Adults and nymphs feed on small insects such as aphids and attack plant tissue, preferring multi-petalled flowers such as chrysanthemums and dahlias. They also make irregularly shaped holes in leaves and attack young buds, causing distorted shoots, and often create secondary damage in cavities in fruits, especially peaches and apples.

Disposing of rubbish and fallen leaves helps to reduce the earwig population by limiting shelter, as does cutting bamboo stakes close to a node. Pieces of sacking, corrugated cardboard or flower pots stuffed with straw and inverted on canes attract earwigs, and these serve as traps that can be removed and destroyed at intervals. Susceptible plants may be sprayed with residual contact insecticides, applied at dusk on warm evenings.

Eccremocarpus (from Greek *ekkremes*, pendent, and *karpos*, fruit). Bignoniaceae. Chile and Peru. GLORY FLOWER. 5 species, evergreen or herbaceous perennial vines, the bases and older branches sometimes thinly woody. The leaves are bipinnate or twice pinnatisect, with terminal tendrils. Produced in summer in terminal racemes, the flowers are tubular, swelling toward the tip, then abruptly contracting at the mouth before expanding again in a small, entire or 2-lipped limb. The following species is a valuable climber for walls, fences and pergolas, and will scramble through shrubs, bearing a succession of exotic and brightly coloured blooms throughout the summer until the first frosts. In a warm and sheltered

Eccremocarpus scaber

position with good drainage, it will resprout from the base, even where temperatures have fallen to –5°C/23°F and below. In colder zones, grow in a cool greenhouse or conservatory. Remove overcrowded growth and frost-damaged stems in spring. Propagate from seed under glass in late winter, or by soft tip cuttings in early summer. *E.scaber* (Chile; to 4m; flowers to 3cm, typically scarlet to orange-yellow, also pink and deep crimson in cultivars and hybrids).

Echeveria (for Atanasio Echeverria, one of the botanical draughtsmen employed on *Flora Mexicana* (1858)). Crassulaceae. C and S America. Around 150 species, succulent, evergreen perennial herbs and subshrubs with short, usually unbranched stems and fleshy leaves in terminal rosettes. Small and urn-shaped, the 5-lobed flowers are carried on slender, erect to arching racemes. Grown for their attractive foliage, as house or greenhouse plants, most will withstand temperatures down to 5°C/40°F. Some are used in carpet bedding and may tolerate light frosts if dry. Grow in full sun, with good ventilation, in a gritty medium. Water generously in summer; keep dry but not arid in winter. Propagate by leaf cuttings, seed, division and offsets.

E.agavoides (Mexico; leaves 3–8cm, ovate-triangular, sharply pointed, waxy, with transparent margins, sea green with a red tip; flowers orange-pink outside, yellow within); *E.derenbergii* (Mexico; leaves 2–4cm, cuneate to obovate, thick, bristle-tipped, pale green with a red margin; flowers yellow tipped red); *E.elegans* (MEXICAN SNOW BALL, WHITE MEXICAN ROSE, MEXICAN GEM; Mexico; leaves 3–6cm, thickly spathulate-oblong, bristle-tipped, icy blue-grey in rounded rosettes; flowers pink outside, yellow-orange within); *E.gibbiflora* (Mexico; leaves 15–30cm, obovate to spathulate, pointed, tinted purple with wavy margins; flowers red outside, buff within); *E.harmsii* (Mexico; hairy throughout; leaves 2–5cm, oblanceolate, pointed, downy pale green with red margins; flowers red with yellow interiors); *E.pulvinata* (PLUSH PLANT, CHENILLE PLANT; Mexico; leaves 2–6cm, spathulate-obovate, finely pointed, minutely hairy, grey- to mid-green with red margins; flowers yellow marked red); *E.secunda* (Mexico; leaves 2.5–7.5cm, spathulate-cuneate, blunt, bristle-tipped, glaucous, margin often red; flowers red outside, yellow within); *E.setosa* (MEXICAN FIRECRACKER; Mexico; leaves 4–5cm, oblanceolate, pointed, white-hairy, bristle-tipped; flowers yellow with red markings).

Echinacea (from Greek *echinos*, hedgehog or sea-urchin, referring to the prickly phyllaries). Compositae. E US. CONE FLOWER. 9 species, tall perennial herbs producing large daisy-like flowerheads in summer and autumn, the disc florets usually raised in a cone-like mass. Grown in herbaceous and cut-flower borders and in native plant collections, they will withstand heat with high humidity, as well as drought, partial shade and temperatures as low as –15 to –20°C/5 to –4°F. Give a deep dry mulch of bracken litter in zones at the limits of their hardiness. Cut back and feed plants that have flowered in midsummer for a second flowering. Propagate by seed, division or root cuttings. *E.purpurea* (syn. *Rudbeckia purpurea*; to 1.5m tall; ray florets to 8cm, magenta to red-purple, disc florets orange and dark red-brown; includes cultivars with pale pink, deep carmine and burnt orange and pure white flowerheads).

echinate covered with many stiff hairs or bristles, or thick, blunt prickles.

Echinocactus (from Greek *echinos*, hedgehog, and Latin, *cactus*). Cactaceae. SW US, Mexico. 5 species, disc-shaped, globose or shortly columnar cacti with strongly ribbed stems, broad woolly crowns and stout spines. The flowers are shortly funnel-shaped to campanulate. Grow in full sun in a cool frost-free greenhouse (minimum temperature 2–7°C/36–45°F) with low humidity. Plant in a very gritty, acid to neutral compost. Keep dry from mid-autumn until early spring, except for light misting on warm days in late winter. Propagate by seed or offsets.

Echeveria agavoides

E.grusonii (GOLDEN BARREL CACTUS; C Mexico; stem to 130cm, globose, green with many raised ribs with golden to pale yellow spines; flowers to 6cm, yellow); *E.platyacanthus* (syn. *E.ingens*; C and N Mexico; stem to 3 m, glaucous and sometimes banded purple when young with many distinct ribs, spines, yellow or brown; flowers 3–6cm, yellow).

Echinocereus (from Greek *echinos*, hedgehog, and Latin, *Cereus*). Cactaceae. SW US, Mexico. 45 species, low-growing shrubby cacti with simple or clustering stems – these are self-supporting, procumbent or rarely clambering, globose to cylindric and ribbed. The flowers are funnel-shaped. Some can withstand freezing temperatures when dry and may be hardy outdoors in sheltered districts in zone 8. *E.reichenbachii* for example may be grown in a cold greenhouse or out-of-doors with winter protection from rain. The remaining species need a cool greenhouse (minimum temperature 5°C/40°F). Grow all species in a gritty, neutral compost in full sun with low humidity. Withhold water from mid-autumn until early spring, except for light misting on warm days in late winter. Propagate by stem cuttings, seed or offsets.

E.cinerascens (E Mexico; stems sprawling, cylindric, 10–30 × 1.5–12cm, spines to 4.5cm; flowers to 10cm, pink to magenta with paler throat); *E.leucanthus* (syn. *Wilcoxia albiflora*; NW Mexico; stems ascending, cylindric, to 30 × 0.6cm, spines minute; flowers 2–4cm, white with darker throat and sometimes pale pink stripes); *E.pectinatus* (SW US, N Mexico; 8–35 × 13cm, globose to cylindric, spines to 2cm; flowers 5–16cm diameter, white to yellow or pink to lavender); *E.pentalophus* (E Mexico, S Texas; stems cylindric, sprawling to erect, 20–60 × 1–6cm, spines very short or to 6cm; flowers 8–12cm, bright pink or magenta with white or yellow throat, rarely pure white); *E.reichenbachii* NE Mexico, SW US; globose to cylindric, erect, to 40 × 10cm, short spines; flowers 5–12cm, pale pink to purple or

*Echium
wildprettii*

crimson); *E.schmollii* (syn. *Wilcoxia schmollii*; E
Mexico; stem ascending, cylindric, to 25 × 1cm,
spines to 0.7cm, hair-like, giving the whole a woolly
appearance; flowers to 5cm, bright pink); *E.trigloch-
idiatus* (SW US, N Mexico; stems ovoid to cylindric, 5–30 × 5–15cm, woolly, spines 1–7cm; flowers
3–9cm, brilliant scarlet or pink).

Echinops (from Greek *echinos*, sea-urchin or
hedgehog, and *opsis*, resemblance). Compositae.
Europe to Asia. GLOBE THISTLE. About 120 species,
perennial, biennial or rarely annual herbs (those listed
below perennial) with entire to pinnately cut leaves,
the margins spiny-toothed, the undersides white-
felted. The flowerheads are globe-like and composed
of many, spiky tubular florets, grey to blue or white
and with blue to grey anthers. They stand above
foliage on slender stalks in summer. Fully hardy.
Grow in full sun in any well-drained soil of low to
moderate fertility. Propagate by seed sown under
glass or *in situ*, division or by root cuttings in winter.
E.bannaticus (E Europe; 50–120cm; flowerheads
to 5cm diameter, grey-blue or deep blue); *E.ritro* (C
and E Europe to C Asia; 20–60cm; flowerheads
3–5cm diameter, pale to deep blue or, rarely, white);
E.sphaerocephalus (S and C Europe to Russia;
50–200cm; flowerheads grey-blue or white).

Echinopsis (from Greek *echinos*, hedgehog, and
opsis, appearance). Cactaceae. Central S America.
50–100 species, shrubby, tree-like or columnar cacti
with simple, branched or clustering, cylindric to
globose, ribbed stems with few to many spines.
Produced in spring and summer, the flowers are large
and bell- to funnel-shaped. Provide a minimum
temperature of 2–7°C/36–45°F. Grow in full sun with
low humidity in a gritty, neutral medium. Keep dry
from mid-autumn until early spring, except for light
misting on warm days in late winter. Increase by
seed, stem cuttings and offsets.
E.aurea (syn. *Lobivia aurea*; Argentina; stem
globose to elongate, 10–40 × 4–12cm, spines to 3cm;
flowers 5–9cm, yellow); *E.eyriesii* (S Brazil to N
Argentina; stem globose to shortly cylindric, to 15 ×
10cm, spines very short and dark; flowers 20–25cm,
white); *E.oxygona* (syn. *E.multiplex*; S Brazil, N
Argentina; stem globose, to 30 × 15cm, spines to
2.5cm, yellow-brown; flowers 25 × 10cm, pink);
E.rhodotricha (Paraguay, NE Argentina; stem
globose to shortly columnar, to 80 × 9cm, spines to
2.5cm pale yellow tipped brown; flowers to 15cm
diameter, white).

Echium (name used by Dioscorides). Boraginaceae.
Canary Islands, Europe, Africa, W Asia. Some 40
species, bristly annual, biennial or perennial herbs or
shrubs. The flowers are produced in summer in
terminal cymes which usually form a panicle. In
zones that are frost-free or almost so, the giant, spire-
like and monocarpic species (*E.candicans E.simplex*,
E.wildpretii, *E.pininana* and *E.giganteum*) are
impressive and beautiful focal plants for the large

mixed or herbaceous border; in cooler climates, grow
in the cool greenhouse or conservatory. Since they
are easily raised from seed, it is worth attempting
these noble plants outdoors even where frosts are to
be expected: they can always be given light protection at the harshest times. Hybrids between *E.pini-
nana* and *E.wildpretii* have shown cold tolerance to
temperatures of at least −10°C/14°F: overwinter
seedlings from summer sowings in the cool greenhouse at 7°C/45°F and plant out in spring to flower
in the following season. Under glass, provide direct
sunlight and water plentifully when in growth. Keep
rather dry in winter and maintain a minimum temperature of 5°C/40°F, with good ventilation. The hardy
E.vulgare is well-suited to the wild garden and is
attractive to bees and butterflies; it also serves as a
fast-growing annual filler for the mixed border or for
bedding. Grow in an open position in sun on low-
to moderate-fertility, well-drained soils. Propagate by
seed. Sown under glass or (*E.vulgare*) *in situ*.
E.candicans (PRIDE OF MADEIRA; Madeira; biennial shrub to 2.5m; leaves lanceolate, silvery; flowers
red-purple in bud, opening blue-white in cylinder-
shaped panicles to 30cm); *E.giganteum* (Canary
Islands; giant, monocarpic herb with a single, spire-
like stem to 2.5m tall, bearing many flowering
branches; leaves to 20cm, lance-shaped, bristly;
flowers white); *E.pininana* (Canary Islands; as for
E.giganteum, but to 4m tall with slender, silvery-
hairy leaves and blue and rose flowers); *E.simplex*
(PRIDE OF TENERIFE; Canary Islands; as for *E.gigan-
teum*, but leaves densely silver-hairy); *E.vulgare*
(VIPER'S BUGLOSS; Europe; biennial, bristly herb,
25–60cm; leaves to 15cm, oblong to lance-shaped;
flowers blue to violet, or rosy pink or white in a
loosely branching panicle); *E.wildpretii* (syn. *E.bour-
gaeanum*; Canary Islands; as for *E.giganteum*, but
with many overlapping and downward-pointing, narrow leaves covered in white hairs; flowers pale red).

ecology the study of the relationships between
organisms and their environment.

ecosystem the unit composed of a community and
its environment.

ecotype a variant of a species adapted to a particular environment.

edaphic pertaining to soil.

edema see *oedema*.

Edgeworthia (for M.P. Edgeworth (1812–81),
English botanist employed by the East India
Company). Thymelaeaceae. Nepal to China, introduced Japan. PAPERBUSH. 3 species, deciduous or semi-
evergreen shrubs to 2m, with thick branches covered
with papery bark and tough, lanceolate to oblong
leaves to 10cm long. Fragrant silky-hairy flowers are
borne in dense, stalked, axillary heads in spring. They
each consist of a white to yellow calyx with a golden

or pale orange stain at the base of the shortly cylindrical tube; the limb is divided into four lobes. With shelter from cold winds and in a well-drained soil, they will survive temperatures to −15°C/5°F without injury. Grow in full sun or part shade. In areas at the limits of hardiness, grow against a south-facing wall, or in the cool greenhouse. Propagate from semi-ripe cuttings in summer, or from seed in autumn.

E.chrysantha syn. E.papyrifera; PAPERBUSH; China; flowers yellow to orange-red); *E.gardneri* (Nepal, Sikkim; flowers white stained yellow).

edging (1) the practice of trimming lawn edges; (2) the use of solid materials, such as long narrow strips of plastic or thin metal, paving slabs or edging tiles, to demarcate lawn edges and prevent grass growing into beds and borders; (3) a continuous planting along the boundaries of beds or borders, often employing low, shrubby species, such as box or lavender.

edging iron, edging knife a long-handled tool, consisting of a half-moon-shaped blade fixed in the same plane as the handle, used for slicing lawn edges or turf. Wheeled versions are known as turf racers.

edging shears long-handled shears designed for trimming the edges of lawns. The blades may be fashioned to lie in the same plane as the handles or at right angles to them.

Edraianthus (from Greek *hedraios*, sitting, and *anthos* flower – the flowers are sessile). Campanulaceae. Mediterranean to Caucasus. GRASSY BELLS. Some 24 species of tufted perennial herbs, similar to *Campanula*, with bell-shaped, 5-lobed flowers in late spring and early summer. Suited to the sunny rock garden, troughs and dry stone walls. Grow in light, freely draining, humus-rich, calcareous soils, in sun or with some shade at the hottest parts of the day. They are hardy to about −15°C/5°F, but the resting buds are susceptible to winter wet. They may perform better in the alpine house, watered plentifully when in growth and kept almost dry in winter. Propagate by seed sown in a cold frame in spring or by softwood cuttings of side shoots in late spring/early summer.

E.dalmaticus (Balkans; to 7cm; leaves linear; flowers several together, blue to violet or white); *E.pumilio* (Balkans; dwarf with very slender, grey-hairy leaves and solitary, violet to blue flowers); *E.serpyllifolius* (Balkans; to 5cm; leaves spathulate; flowers solitary, deep violet, exceeding blunt bracts).

EDTA ethylene diamine tetra-acetic acid; an organic substance used in the chelating of trace elements to maintain their availability in soils. See *sequestrol*.

eelworms NEMATODES worm-like animals, some of which are predators or parasites; many feed on bacteria or fungi. Those that are plant pests are usually less than 1mm long and have pin-shaped mouthparts with which they pierce plant cells in order to feed. Their flexible bodies enable them to swim in an eel-like manner in thin water films between plant cells or soil particles. At least two species, *Steinernema bibionis* and *Heterorhabditis heliothis*, have been used for the biological control of insect pests, such as vine weevil grubs. Eelworms reproduce by laying eggs which hatch and pass through four immature stages, each of which resembles the adult form. The following are the principal types of eelworm that damage plants.

CYST EELWORMS (*Heterodera* species, *Globodera* species) develop inside the roots of their host plants; when fully mature, the females' bodies swell and burst through the root surface, where they can be seen as tiny spherical or lemon-shaped brown cysts. These are packed with eggs which can remain viable in the soil for many years; when host plants are grown the eggs are stimulated to hatch by chemicals secreted from the roots. Heavy infestations disrupt the uptake of water and nutrients, resulting in poor growth and the early death of plants. In gardens, only potatoes and tomatoes are commonly damaged by cyst eelworms.

ROOT KNOT EELWORMS (*Meloidogyne* species) are mainly pests of tropical and subtropical areas but elsewhere can be troublesome in greenhouses. They develop inside the roots of many plants causing swellings; heavy infestations can result in plant death through disruption of water and nutrient uptake.

STEM AND BULB EELWORM (*Ditylenchus dipsaci*) appears to be a single species, occurring in a number of biological races capable of attacking different plants. The races commonly found in gardens typically attack *Narcissus*, onion and *Phlox*, and sometimes tulip. This nematode lives within stems, bulbs and foliage; symptoms are distinctive forms of distorted growth. See *stem and bulb eelworm*.

LEAF AND BUD EELWORMS (*Aphelenchoides* species) live within the leaf tissues and/or between the embryonic leaves in buds. Two species are common in gardens and attack an extremely wide range of plants, including chrysanthemums, ferns, *Begonia*, *Anemone* and strawberry. Typical damage consists of brown infested areas within the leaf, sharply separated from healthy tissue by the larger leaf veins.

Various types of virus vectors such as *Trichodorus* species and *Longidorus* species live in the soil and feed on plant roots. They are particularly associated with virus diseases, including arabis mosaic, cherry leaf-roll and tobacco rattle, which affect strawberry, cane fruits and some ornamentals.

Infestations can be limited by burning affected plants and by crop rotation. Care should be taken to acquire only healthy plant material. Some potato cultivars have resistance to, or are tolerant of, cyst eelworms. *Narcissus* bulbs and chrysanthemum stools can be essentially freed of eelworms by immersion in hot water for set periods (see *hot water treatment*). Effective chemical treatment against eelworms is too toxic for garden use.

Egeria (Egeria, a water nymph of Roman mythology who married Numa). Hydrocharitaceae. S America, naturalized elsewhere. 2 species, evergreen, perennial aquatic herbs. The submerged stems are long and densely furnished with opposite or whorled, linear to oblong leaves. Cymes of small, white to yellow flowers are subtended by a tubular spathe and held clear of the water in summer. The following species is an excellent oxygenator for the cold or tropical aquarium, and offers cover for spawning and young fish. Provide good light and a minimum temperature of 7°C/45°F. Trim regularly to keep fresh appearance. Propagate by stem cuttings in spring. *E.densa* (syn. *Anacharis densa*, *Elodea densa*; S America, widely naturalized elsewhere; leaves to 2 × 0.3cm, dark green).

Ehretia (for Georg Dionysus Ehret (1708–1770), German botanical artist who worked in England). Boraginaceae. Africa, Asia, Americas. 50 species, deciduous or evergreen trees or shrubs with oblong to elliptic leaves. Borne in summer in terminal panicles, the flowers are small, star-shaped and sometimes fragrant. The corolla is tubular-campanulate with five spreading or reflexed lobes. The fruits are subglobose, glabrous drupes, to 2cm and yellow or orange to red. Given a sheltered position with good drainage, *E dicksonii* will tolerate temperatures to about –10°C/14°F. Grow in full sun. Propagate by softwood or semi-ripe cuttings. *E.dicksonii* (China, Taiwan, Ryukyu Islands; deciduous tree to 12m; leaves to 18cm, oblong to elliptic, serrate, leathery and shiny, bristly above, velvety beneath; flowers to 0.6cm, white tinted yellow, fragrant).

Eichhornia (for J.A.F. Eichhorn (1779–1856), Prussian politician). Pontederiaceae. Tropical S America, widely naturalized in other tropical and subtropical waterways. WATER HYACINTH, WATER ORCHID. 7 species, perennial, aquatic herbs. The leaves are borne in stoloniferous, floating rosettes. They consist of a swollen, spindle-shaped petiole filled with spongy, buoyant tissue, and a glossy, rounded to heart- or kidney-shaped blade. Composed of six, broad petals, the flowers are produced in summer in dense, erect spikes. Grow on still freshwater in sun with shelter from wind. Where frosts occur, collect the strongest-looking plants in autumn and overwinter them on trays of wet loam, leafmould and sand, with a minimum temperature of 10°C/50°F and good light. Propagate by detaching young plants produced freely on runners. *E.crassipes* (leaves 5–15cm, shining pale to dark green; flowers to 4cm, pale lilac to violet blue, often with paler or deeper feathering and a central violet and yellow blotch).

Elaeagnus (from Greek *elaia*, olive, and *agnos*, the Greek name for *Vitex agnus-castus*. The name was applied by Theophrastus to a willow). Elaeagnaceae. Asia, S Europe, N America. OLEASTER. About 45 species, deciduous or evergreen shrubs or trees, often spiny. The leaves are lanceolate to ovate or oblong and often covered with minute silvery or brown scales, especially beneath. Small and often fragrant, the tubular to campanulate flowers are carried in

Elaeagnus pungens
'Maculata'

clusters in late winter and spring. They are followed by small, fleshy fruits. These shrubs will thrive in a range of soil types provided they are free-draining (the deciduous species are particularly tolerant of poor soils). They are useful in coastal gardens or as shelter belts in exposed areas. They respond well to hard pruning where required, quickly producing young shoots from the old wood. Propagate from fresh seed sown in a cold frame – a stratification period of three months at 4°C/39°F is otherwise required. *E.commutata* can be propagated easily from suckers. Root cuttings, hardwood cuttings (*E.angustifolia*) and leafy summer cuttings (*E.pungens*) are also successful.

E.angustifolia (RUSSIAN OLIVE, OLEASTER, WILD OLIVE, SILVER BERRY; W Asia; deciduous shrub or small tree to 7m; leaves 5–9cm, narrow, covered in pale bronze to silver scales at first; flowers yellow to silver, fragrant; fruit yellow covered in silvery scales); *E.commutata* (SILVER BERRY; N America; deciduous suckering shrub to 5m; leaves to 6cm, broad, lustrous, silvery; flowers fragrant, silvery yellow; fruit yellow, covered in silvery scales); *E. × ebbingei* (*E.macrophylla* × *E.pungens*; shrub to 3m, usually evergreen; leaves to 10cm, broad dark, glossy green or metallic sea green above, silvery beneath; flowers intensely fragrant, white to cream; includes cultivars with gold-edged and yellow-green variegated leaves); *E.macrophylla* (Japan, Korea; evergreen shrub to 3m; leaves 5–12cm, broad, glossy deep green above, silver-scaly beneath; flowers cream, scaly, highly fragrant; fruit red, scaly); *E.pungens* (Japan; evergreen shrub to 4m; leaves 4–8cm, broad, wavy, deep smooth green above, dull silvery white and brown-scaly beneath; flowers fragrant, silvery white; fruit brown becoming red; includes several popular cultivars with leaves edged, splashed or zoned yellow, grey-green, pale green and darkest emerald, the most popular being 'Maculata'); *E.*'Quicksilver' (*E.angustifolia* × *E.commutata* (?); tall, deciduous suckering, more or less pyramidal shrub with narrow, grey leaves overlaid with silver scales); *E.umbellata* (Himalayas, China, Japan; deciduous shrub or small tree to 10m; leaves to 10cm, wavy, bright green above with silver and some brown scales beneath; flowers yellow-white, scented; fruit silver-bronze, ripening red; includes cultivars with pale lime and silvered foliage and amber or deep red fruit).

Elaeocarpus (from Greek *elaia*, olive, and *karpos*, fruit: the fruit resembles an olive.) Elaeocarpaceae. E Asia, Indomalaysia, Australasia, Pacific. 60 species, evergreen trees or shrubs with small, fragrant flowers in axillary racemes in summer. These have 3–5, fringed petals and are followed by small, spherical fruits. Where frosts are light and short-lived, provide a sheltered position against a south- or southwest-facing wall; otherwise, grow in the cool greenhouse or conservatory with a minimum temperature of 5–7°C/40–45°F. Under glass, admit full light but ensure shade from the hottest summer sun; water plentifully when in full growth, less in winter. Cut back the previous season's growth in late winter.

Propagate by seed sown under glass in spring or by semi-ripe cuttings in a sandy propagating medium with gentle bottom heat. *E.cyaneus* (syn. *E.reticulatus*; Australia; tree to 18m tall, more usually a shrub in cultivation; leaves to 15cm, oblong, toothed; flowers small, ivory; fruit blue).

electronic leaf a device for controlling the frequency of water droplet application in a mist propagation unit. It comprises a plastic sensor, in which two electrodes are implanted. When a film of moisture connects the two electrodes, an electric current flows and the unit is programmed 'off'. When the current is interrupted as a result of the 'leaf' drying out, a switch is activated and a fine water mist is emitted over the bench.

Eleocharis (Greek, *helodes*, marshy or of marshy places, and *charis*, grace). Cyperaceae. More or less cosmopolitan. SPIKE RUSH. Around 150 species, annual or perennial grass-like herbs with very narrow leaves and 3–4-angled stems. Borne in summer, the flowers are minute and arranged in solitary, brown and bristly spikelets. *Eleocharis* is fully hardy. Cultivate as for *Carex* in bog gardens and at pond margins with a slightly acid pH. *E.acicularis* (NEEDLE SPIKE RUSH, SLENDER SPIKE RUSH, HAIR GRASS; widespread in N America, Europe, Asia; perennial, 5–30cm tall).

Eleutherococcus (from Greek *eleutheros*, free, and *kokkos*, pip, here referring to the fruit structure). Araliaceae. S and E Asia, Japan, the Ryukyu Islands, Taiwan and the Philippines (Luzon); most diverse in C and W China. About 30 species, generally deciduous and often bristly or prickly shrubs or trees some more or less scandent or sprawling with digitately compound leaves. Small, 5-parted flowers are carried in umbels or heads in terminal, simple or compound inflorescences. They are followed by small, drupaceous fruits that turn black or purple-black at maturity. In positions sheltered from north and east winds, *E.sieboldianus* will usually survive temperatures from –10° to –15°C/14–5°F and withstands poor soils and pollution. Often used for hedging in cool and warm temperate areas, it seems to thrive on urban pollution, and one or two heavy trimmmings per year. That said, for best results, grow on any well-drained, humus-rich soil in full sun. Specimen shrubs may be pruned to thin crowded wood and to shorten ungainly growth in late winter. Propagate by ripe seed in autumn or by root cuttings in late winter; alternatively, increase by suckers and semi-ripe cuttings, with bottom heat, in summer. *E.sieboldianus* (syn. *Acanthopanax sieboldianus*; E China, Japan (introduced); arching to scandent shrub to 3m with slender, cane-like stems, often spiny; leaflets usually five per leaf, toothed; flowers green-white; fruit black; 'Variegatus': leaves edged creamy-white).

ellipsoid elliptic, but 3-dimensional.

elliptic, elliptical ellipse-shaped; midway between oblong and ovate but with equally rounded or narrowed ends, with two planes of symmetry and widest across the middle.

elongate lengthened, as if stretched or extended.

Elsholtzia (for John Sigismund Elsholtz (1623–1688), Prussian horticulturist and physician). Labiatae. C and E Asia, NE Africa. 38 species, aromatic shrubs or perennial and annual herbs with small, tubular and 2-lipped flowers in dense, whorled spikes in late summer and autumn. Although the shrubby growth is usually hardy to –10°C/14°F, it may be killed by prolonged frosts and is in any case best cut hard back each year in spring. Mulch in winter. Propagate from seed sown under glass in spring or by cuttings of new growth, under mist or a warm cloche, in early summer. *E.stauntonii* (China; sub-shrub 1.75m; leaves to 15cm, ovate, toothed flowers dark pink or white).

Elymus (from Greek *elymos*, classical name for millet). WILD RYE; LYME GRASS. Gramineae. Northern temperate Asia. Some 150 species of tufted or rhizomatous perennial grasses valued for the often bright blue-greens of the foliage clumps, topped by flowering spikes in shades of buff and parchment. *E.magellanicus* is one of the most intensely blue of all grasses. They are tolerant of temperatures to at least –10°C/14°F. They thrive in any gritty, moisture-retentive soil in sun. Propagate by division.

E.hispidus (syn. *Agropyron glaucum*; Europe, Asia; to 50cm, erect, tightly clump-forming; leaves intense grey-blue, hispid); *E.magellanicus* (syn. *Agropyron magellanicum*; S America; to 30cm, loosely clump-forming; leaves arching, brilliant blue-grey, smooth or finely grooved).

emarginate the apex of a leaf or other organ shallowly notched, the indentation (sinus) being acute.

emasculation the removal of anthers from a flower before the pollen is ripe; carried out by plant breeders to prevent self-fertilization.

Embothrium (from Greek *en*, in, and *bothrion*, a little pit, in reference to the sunken anthers). Proteaceae. C and S Andes. 8 species, evergreen shrubs or trees with leathery leaves. Produced in clusters or racemes in late spring and early summer, the flowers are narrowly tubular and split into four, narrow and twisted lobes, parted by a slender style. Hardy to –15°C/5°F, given a warm and sheltered site with protection from cold drying winds. Plant in a well-drained, moist, neutral or acid soil in sun, against a south-facing wall or in light, open woodland. Propagate by seed sown under glass in spring, from suckers and from root cuttings in the cold frame; also by nodal cuttings of the current season's growth, under mist or in a closed case with gentle bottom-heat, in summer. *E.coccineum* (CHILEAN FIREBUSH,

CHILEAN FLAMEFLOWER; Chile; shrub or tree to 10m with clumped, stems; leaves to 12cm, oblong; flowers to 4.5cm, sealing-wax red to brilliant scarlet).

embryo the rudimentary plant within the seed.

embryo culture a form of micropropagation using excised plant embryos.

emersed raised out of and above the water.

Emilia (probably a commemorative name, but for whom it is unknown). Compositae. India, Polynesia, tropical Africa. About 24 species, annual herbs with lanceolate to oblong or lyrate leaves. Small, pompon-like flowerheads are carried in summer in long, branching sprays. Grown for its brightly coloured blooms suitable for mixed bedding, cutting and drying. Treatment as for *Acroclinium*. *E.coccinea* (syn. *E.flammea*, *E.javanica*; TASSEL FLOWER; tropical Africa, Asia; to 60cm; flowerheads scarlet, fiery orange or rich yellow).

EMLA an acronym for two eminent British horticultural research stations (East Malling in Kent and Long Ashton in Somerset), both formerly concerned solely with fruit research. Their collaboration was particularly important in the re-selection of *Malus* clonal rootstocks. See *apple*.

Emmenopterys (from Greek *emmenes*, enduring, and *pteryx*, wing – one of the calyx lobes becomes enlarged and wing-like). Rubiaceae. China, SE Asia. 2 species, deciduous trees with entire, somewhat leathery, elliptic to oblong leaves. Produced during hot summers in terminal panicles, the flowers are funnel- or bell-shaped with five, spreading, ovate and downy lobes. Of the five calyx lobes, one is sometimes enlarged, ovate to oblong and white. Hardy to –25°C/–13°F; grow in a sunny site sheltered from cold winds, in a deep, fertile and moisture-retentive loam. Propagate from softwood cuttings in early summer. *E.henryi* (to 26m; young growth red-bronze; leaves to 20cm dark green above, pale beneath, petioles red to purple; panicles to 18cm, corolla white, to 3cm, enlarged calyx lobe white, to 5cm with a claw to 5cm).

Encephalartos (from Greek *en-*, within, *kephale*, the head, and *artos*, bread – the inner tissues of the upper trunk are starchy and, with much careful preparation, edible). Zamiaceae. South and C Africa (those listed below are from South Africa). Some 25 species, evergreen cycads, stemless or with short and squat to tall and columnar trunks. Long and leathery, the leaves are pinnate and borne in terminal flushes or rosettes. The leaflets are tough and two-ranked. Large cones are carried on plants of separate sex. Species with glaucous, woolly or narrow pinnae will tolerate low temperatures (minimum 5°C/40°F), full sunlight and dry conditions: these include (leaflets glaucous) *E. horridus*, *E.lehmannii* and *E.trispinosus*, and (leaflets woolly or very narrow) *E. ghellinckii*, *E.frid-*

erici-guilielmi. Grow in a gritty, sandy, neutral to slightly acid mixture; water freely when in growth (i.e. when bract, leaf or cone activity is visible in the crown) and apply a weak, fortnightly feed. Water very sparingly in winter. The remaining species require a minimum temperature of 13°C/55°F, at least partial shade, high humidity, plentiful water supplies throughout the year and a more open, bark-based mix rich in garden compost and leafmould. Propagate by seed sown under glass in warmth, or by offsets.

E.altensteinii (4–7m; leaves to 3.5m, leaflets to 15cm, linear to oblong, rigid, fresh green, with 1–3 spiny teeth along each margin); *E.ferox* (to 1m; leaves to 1.8m; leaflets to 12cm, oblong to ovate, glossy dark green with 2–4 broad, sharp and spiny teeth along each margin); *E.friderici-guilielmi* (to 4m; leaves to 1.5m, leaflets to 18cm, narrowly linear, entire, tip sharply pointed, densely woolly at first, later dark green); *E.ghellinckii* (to 3m; leaves to 1m, spirally twisted, leaflets to 14cm, linear but with strongly revolute margins thus appearing needle-like, covered in grey wool at first, later rigid bright green); *E.horridus* (stem very short; leaves to 1m, leaflets to 10cm, broad and deeply cut with 2–3 large, spine-tipped and revolute lobes, rigid, bright glaucous blue-grey); *E.lebomboensis* (to 4m; leaves 1–3m, leaflets to 17cm, lanceolate, bright green with 2–4 teeth along each margin); *E.lehmannii* (to 2m; leaves to 1.5m, leaflets to 18cm, oblong to lanceolate, usually entire, glaucous blue-grey); *E.longifolius* (to 4m; leaves to 2m, leaflets to 20cm, lanceolate, dark glossy green, entire or with 1–4 teeth along lower margin); *E.natalensis* (to 4m; leaves to 3.5m, leaflets to 25cm, broadly lanceolate, dark green, entire or with 1–4 teeth along one or both margins); *E.transvenosus* (to 13m; leaves to 2.5m, leaflets to 25cm, broadly lanceolate, with 2–5 small teeth along upper margin, 1–3 along lower margin, glossy, dark green); *E.trispinosus* (differs from *E.horridus* in the narrower, less strikingly glaucous leaflets with one or two strongly recurved and spine-tipped lobes on lower margin only).

enchytraeid worms (Annelida: Oligochaeta) many-segmented, small, thread-like worms, up to 2mm long, translucent white in colour and often abundant in leafmould, compost and manure, where they feed on decaying organic matter. They are commonly known as POT WORMS, ASTER WORMS, ROOT WORMS or WHITE WORMS. Although frequently found near plant roots, they do not feed on living plant tissue and are harmless.

Encyclia (from Greek *enkyklo*, to encircle, referring to the characteristic of the lateral lobes of the lip which encircle the column). Orchidaceae. Florida, West Indies, C and S America. Some 150 species, epiphytic and lithophytic orchids, closely allied to and formerly included in *Epidendrum*. They vary greatly in size and habit, but generally have clumped, rounded to flask- or spindle-shaped pseudobulbs topped with oblong-lanceolate to linear leaves. The flowers consist of five, obovate to linear tepals and a showy lip; they are borne usually in spring and summer in terminal racemes or panicles.

Grow in pans or baskets of very open bark-based mix in sunny, cool to intermediate conditions (minimum temperature 7°C/45°F). Water and feed freely in spring and summer and maintain high humidity. Once the new pseudobulbs are complete and ripened, water and syringe only to prevent their shrivelling. Resume watering when new growth starts. These guidelines apply to most species, with the following exceptions—The dwarf *E.tripunctata* should be mounted on fibre blocks on pads of moss and kept just moist and in filtered sunlight throughout the year. *E.vitellina* requires cool, buoyant, rather dry conditions in full sunlight with a decided dry winter rest. *E.mariae* and *E.citrina* must be mounted 'upside-down' on cork or bark slabs and suspended in a cool, bright, dry environment. Plunge often and syringe daily when in growth. Once growths are complete, lower temperatures (to near-freezing, if possible) and withhold water completely. With lengthening days, start back into growth by raising temperatures slightly and misting roots – this will encourage development of flower spikes for spring blooming. Increase all by division when repotting in spring.

E.adenocaula (syn. *E.nemoralis*, *Epidendrum adenocaulum*, *Epidendrum nemorale*; Mexico; flowers to 9cm diameter, slender, nodding, pale rose-pink to magenta, the lip with darker markings, several together on an arching, warty panicle to 100cm); *E.alata* (syn. *Epidendrum alatum*; Mexico to Nicaragua; flowers to 6cm diameter, tepals pale green or yellow-green marked purple-bronze or red-brown, the lip white veined maroon, many together on a minutely warty panicle to 80cm); *E.brassavolae* (syn. *Epidendrum brassavolae*; Mexico to Panama; flowers to 7cm diameter, slender, spreading, tepals green to olive or pale tan, lip cream tipped rose-purple, many on an erect raceme to 30cm); *E.citrina* (syn. *Cattleya citrina*, *Epidendrum citrinum*; Mexico; flowers fragrant, to 8cm long, strongly pendulous and not opening widely, waxy lemon yellow, the lip veined deep gold, solitary or paired on a short, hanging stalk); *E.cochleata* (syn. *Epidendrum cochleatum*; CLAMSHELL ORCHID, COCKLE ORCHID; widespread in C America; flowers to 6cm diameter, tepals slender, lime to olive, lip held uppermost, shell-like, white, tipped and veined dark purple to purple-black, many in an erect raceme to 30cm); *E.cordigera* (C America; flowers to 7cm diameter, tepals chocolate brown to maroon or bronze-green, lip large, wavy, cream streaked or flushed rose to magenta); *E.fragrans* (syn. *Epidendrum fragrans*; C America, northern S America, West Indies; flowers highly fragrant, to 4cm diameter, tepals white to ivory, lip held uppermost, striped maroon); *E.mariae* (syn. *Epidendrum mariae*; Mexico; similar to *E.citrina*, but with narrower, lime to olive tepals and a white lip); *E.prismatocarpa* (syn. *Epidendrum prismatocarpum*; Costa Rica to Brazil; flowers to 5cm diameter, fragrant, tepals yellow-green spotted or banded dark maroon to black, lip lilac-purple, base olive, margins white, in tall racemes); *E.radi-ata* (syn. *Epidendrum radiatum*; Guatemala, Honduras, Mexico; similar to *E.fragrans* but with smaller flowers, the lip rounded rather than heart-shaped, with a cleft and wavy, not smooth and pointed tip); *E.tripunctata* (syn. *Epidendrum tripunctatum*; Mexico; flowers to 1.5cm diameter, a few borne together in short racemes, tepals lime to olive green, lip white, column dark purple with yellow teeth); *E.vespa* (syn. *Epidendrum vespa*; tropical America; flowers to 2cm diameter, tepals olive to yellow or cream blotched purple to maroon or dark chocolate, lip white or yellow marked pink, many in a raceme to 35cm); *E.vitellina* (syn. *Epidendrum vitellinum*; Mexico, Guatemala; flowers to 5cm diameter, tepals vermilion to scarlet, lip orange to deep gold, several in an erect raceme to 40cm).

endemic occurring naturally in one locality only.

endive (*Cichorium endivia*) an annual, native of Asia and Europe, of similar appearance to lettuce, forming a rosette of leaves with a loose heart. It is used in salads or braised, the green outer leaves being sharper flavoured than the light-coloured, sweeter inner leaves. The curled or staghorn-leaf endive is better adapted to higher temperatures than the broad or Batavian forms. The main season is autumn to early winter outdoors, later under protection. Sow in cell modules for transplanting or *in situ* for thinning to a final spacing of 38 × 38cm for broad-leaf types and 30 × 30cm for curled-leaf types. Sow during April in heat for early summer cropping and successionally outdoors from May to July for summer to autumn maturity. Latest crops of broad-leaf cultivars are best raised under glass or cloches from an August sowing. Endive is useful for cut-and-come again harvest, and leaves of broad-leaf types especially can be made sweeter by blanching. Either bunch and tie the rosettes together or cover with an upturned container or dinner plate to exclude light, for about 10 days in summer and twice as long in winter.

endo- prefix meaning 'within', 'inner', for example endodermis, the inner layer of skin in a periderm.

endocarp the innermost layer or wall of a pericarp, enclosing the seed, and either membranous or bony or cartilaginous.

endosperm a nutritive tissue of developing seed.

Enkianthus (from Greek *enkyos*, swollen or pregnant, and *anthos*, flower – the type species has basal swellings on the flowers). Ericaceae. Himalaya to Japan. 10 species, usually deciduous shrubs with whorled branches and obovate to elliptic, toothed to entire leaves. Carried in spring in drooping, terminal umbels or racemes, the flowers are waxy, campanulate to urceolate, and 5-lobed. Suitable for the peat terrace and shrub border, for open situations in the woodland garden and for moist and peaty pockets in the rock garden. Hardy to –15°C/5°F. Grow on a moisture-retentive, lime-free soil, enriched with leafmould, in full sun or dappled shade; mulch in hot, dry weather. Propagate from ripe seed, by semi-ripe cuttings in summer, or by simple layering.

E.campanulatus (Japan; variable, deciduous shrub to 5m tall; leaves becoming vivid red in autumn; flowers pale cream to pale pink with faint salmon, peach or rust veins, deeper and more solid red in some forms, campanulate, in drooping racemes); *E.cernuus* (Japan; deciduous shrub to 3m; flowers white or deep red, broadly campanulate with jagged, unequal lobes, in drooping, hairy racemes); *E.perulatus* (Japan; deciduous shrub to 2m; leaves becoming yellow and red in autumn; flowers white, urceolate, in drooping umbels).

Ensete (from the African Amharic name *anset*) Musaceae. Old World tropics. 7 species, massive evergreen perennial herbs, close to *Musa*, but with a more shuttlecock-like growth habit and unpalatable fruit. They are august foliage plants for subtropical and tropical gardens and, elsewhere, for tubs and borders in large greenhouses and interiors. Fast-growing, young plants can be overwintered under glass and planted out in late spring to make striking foci for 'tropical' bedding schemes in temperate gardens. Grow in full sun or shade in a humid, sheltered environment with a minimum temperature of 10°C/50°F. A free-draining, highly fertile medium is essential, as are plentiful supplies of water and feed in the growing season. Stems will deteriorate after flowering and should be cut down once the replacement suckers are well-established. Increase by rooted suckers and by seed (see *Musa*).

E.ventricosum (syn. *Musa arnoldiana*, *Musa ensete*; ABYSSINIAN BANANA, ETHIOPIAN BANANA; Africa; pseudostem 1–12m; leaves 2–6 × 1–1.5m, olive green with a maroon midrib); *E.gilletii* (syn. *Musa gilletii*; differs from *E. ventricosum* in its narrower leaves with sealing-wax red midribs).

ensiform sword-shaped, straighter than lorate, and with an acute point.

entire continuous; uninterrupted by divisions or teeth or lobes. Thus, 'leaves entire' describes leaves with margins that are neither toothed nor lobed.

entomophily the transfer of pollen by insects.

Eomecon (from Greek *eos*, the dawn or east, and *mekon*, poppy, referring to its distribution). Papaveraceae. E China. SNOW POPPY. 1 species, *E.chionantha*, a perennial, rhizomatous herb to 40cm tall with branched stems containing orange sap. To 10cm across, the leaves are heart-shaped to arrow-shaped, leathery, slightly succulent, grey-green and wavy-edged. Produced in summer, the pure white flowers are 5cm across with four, broad petals and many stamens. With good drainage *Eomecon* will tolerate temperatures to −10°C/14°F. Plant in shade or sun on well-drained, humus-rich, fertile soils that remain cool and moist in summer. The rhizomes resent disturbance but may spread extensively and become

a nuisance. Propagate in spring from seed sown *in situ*, or by division.

Epacris (from Greek *epi*, upon, and *akris*, a summit; alluding to the high-altitude habitat of some species). Epacridaceae. E and SE Australia, Tasmania, New Zealand and New Caledonia. AUSTRALIAN HEATH. 35 species, heath-like, evergreen shrubs with small leaves and cylindric to campanulate, 5-lobed flowers in erect, terminal racemes. Grow in moist and cool, humus-rich but minerally poor neutral to acid soils in dappled shade or sun. Under glass, maintain good ventilation; water moderately when in full growth, less in autumn and winter. Overwinter at about 5°C/40°F. Propagate by seed in spring under glass, or by semi-ripe cuttings. *E.impressa* (COMMON AUSTRALIAN HEATH; S Australia, Tasmania; spindly erect shrub to 1.5m; leaves to 1.5cm, narrow, pointed; flowers to 1.8cm, white to rose or purple-red, nodding in erect terminal racemes).

Ephedra (from a Greek name used by Pliny for the mare's tail, *Hippuris*, which *Ephedra* very loosely resembles). Ephedraceae. S Europe, N Africa, Asia, subtropical America. JOINT FIR. Some 40 species, erect to scrambling shrubs with slender, green and jointed stems, the leaves reduced to small and withered scales. The 'inflorescences' of these primitive conifer allies are inconspicuous, yellow-green cones. Where male and female plants are grown together, the female may bear fleshy, berry-like fruit. Those named here are very similar in general appearance and grow to about 2 × 3m. They are drought- and lime-tolerant and good for groundcover on dry soils, for rock gardens, dry walls and raised beds. Some have proved useful in stabilizing sand. Of those most commonly seen in temperate gardens, *E.distachya* and *E.gerardiana* are hardy to − 15°C/5°F, while *E.americana*, *E.fragilis* and *E.viridis* are frost-tender. Plant in a porous, freely draining soil in full sun. Propagate by seed sown under cool glass in autumn, by division or by simple layering.

ephemeral short-lived; strictly, of flowers lasting for one day only, but also used to describe plants that survive for only a few weeks and may have several generations in one season.

epi- prefix meaning 'upon', 'uppermost' as epigynous, growing upon or arising from the ovary.

epicalyx an involucre of bracts surrounding a calyx; a false calyx.

epicarp see *exocarp*.

epicormic of branches or buds, developing on or from the trunk of a tree. The growth can be latent or adventitious.

epicotyl the portion of the stem of an embryo or seedling above the cotyledons.

Epidendrum (from Gk *epi*, upon, and *dendron*, tree, referring to the epiphytic habit of most species). Orchidaceae. Tropical Americas. Some 500 species, perennial herbs: the tall cane-growing species are often, at least initially, terrestrial; the smaller caned species and pseudobulbous plants are epiphytic and lithophytic. Pseudobulbous species (e.g. *E.ciliare*) should be cultivated as for *Laelia* or *Rhyncholaelia*, potted tightly into an open bark-based mix and grown in cool to intermediate conditions with full sun and a cool dry winter rest. The tall-growing cane species (e.g. *E.ibaguense*) are amongst the most flexible of all orchids: in frost-free sunny conditions, they will scramble freely and flower throughout the year. Provide a foothold in a bark mix with plenty of well-rotted farmyard manure. Syringe and foliar-feed regularly. Propagate by adventitious plantlets. The smaller cane-type species (e.g. *E.pseudepidendrum*) require intermediate conditions, high humidity, dappled sunlight and only a semi-dry winter rest. The magnificent *E.parkinsonianum* should be treated as a large *Brassavola*, but mounted on bark or a raft and suspended in full sunlight.

E. ciliare (widespread in C and S America; pseudobulbous; flowers in erect racemes to 30cm, fragrant, white to cool green or pale yellow with a white lip, each to 15cm across with slender tepals and a 3-parted lip, the midlobe narrow, the outer lobes broader and finely cut and fringed; *E. oerstedii* differs in its lip with entire, not fringed lateral lobes); *E. conopseum* (GREENFLY ORCHID; SE US; stems short, slender, clumped; flowers in loose, erect racemes to 16cm, fragrant, lime green to grey green, usually tinted maroon or bronze, each to 2cm across); *E. ibaguense* (widespread in C and S America; stems erect, cane-like, to 2m, rooting only at or near base; flowers in erect, long-lasting racemes to 30cm, each to 3cm across, vermilion to fiery scarlet, rarely rose to magenta, lip cross-shaped and fringed, often held uppermost; cf. *E. radicans*); *E. nocturnum* (Tropical Americas; habit variable, stems stout or reedy, short or tall; flowers in short racemes, fragrant, to 7cm across, tepals very slender, white to parchment or yellow-green, lip white, resembling that of *E. ciliare* but with entire lateral lobes); *E. parkinsonianum* (C America; stems short, on creeping rhizomes; leaves large, thick, tough, sickle-shaped and pendulous; flowers similar to those of *E. nocturnum*; *E. falcatum* differs in its pink-tinted flowers with a yellow throat); *E. peperomia* (S America; dwarf, forming mats of creeping, bright green growth; flowers to 2.5cm across, solitary, short-stalked, insect-like, tepals lime green, narrow, lip large and broad, glossy chestnut brown);*E. pseudepidendrum* (Costa Rica, Panama; stems erect, cane-like; flowers in loose, erect to arching racemes to 15cm, each to 6cm across, tepals narrow, apple green, lip large, glossy orange to orange-red); *E. radicans* (widespread in C and S America; differs from *E. ibaguense* in its scrambling stems rooting freely throughout their length, the column is also longer and arching); *E. stamfordianum* (Mexico to Colombia; pseudobulbs narrowly spindle-shaped; flowers many,

in erect to arching racemes or panicles to 60cm long, fragrant, each to 4cm across, tepals yellow-green spotted red-brown or purple, lip white with a yellow-green midlobe, often spotted or stained purple to rose).

epidermis the outermost layer of cells which serves to protect plant tissues.

Epigaea (from Gk *epi*, on, and *gaia*, earth, referring to the lowly habit). Ericaceae. N America, W Asia, Japan. 3 species, small, creeping, evergreen shrubs with bristly branches and tough, ovate to oblong or elliptic, hairy leaves. Produced in spring in short clusters or racemes, the scented flowers are funnel- to bell-shaped with five, rounded lobes. Suitable for shaded pockets in the rock garden, peat terrace or woodland garden, or for pot cultivation in the shaded alpine frame. Hardy to at least –30°C/–22°F. The flower buds need a period of chilling before they will open. Grow in acid moist but well-drained soils. Protect young plants from prolonged frost with bracken litter or evergreen branches. Germinate ripe seed in shade, watering with lime-free water from below. Overwinter seedlings in their first year at 13–15°C/55–60°F. Increase also by semi-ripe cuttings rooted in a closed case, or by layers.

E.asiatica (Japan; leaves 3.5–10cm; flowers to 1cm long, white tinted pale rose); *E.gaultherioides* (syn. *Orphanidesia gaultherioides*; NE Turkey; leaves 8–12cm; flowers to 5cm diameter, soft pink); *E.repens* (TRAILING ARBUTUS, MAYFLOWER; N America; leaves 2–8cm; flowers fragrant, to 1.5cm long, white to pale pink or deep rose).

epigeal describing a type of germination where the cotyledons emerge from the ground to form the first green foliage leaves of the plant, as occurs, for example, in French beans; cf. *hypogeal*.

epilithic growing on rocks.

Epilobium (from Greek *epi*, upon and *lobos*, pod: the petals surmount the pod-like ovary). Onagraceae. Temperate regions. 200 species, perennial, rarely annual herbs and subshrubs, with shortly tubular to funnelform, 4-petalled flowers. Spreading by running white stolons and prolific air-borne seed, *E.angustifolium* is usually thought a weed. Its white form is less rampant, and an attractive plant for the herbaceous border. In the rock garden, the lower-growing species make carpets of colour. These small plants should be given shade part of the day in hot climates; they require a cool, moist and gritty root-run. In zones 7 and under, they may suffer in harsh winters. Propagate from seed sown in spring or by division, autumn to spring; low-growing species also by softwood cuttings from autumn to spring.

E.angustifolium (syn. *Chamaenerion angustifolium*; GREAT WILLOWHERB, FRENCH WILLOW, ROSEBAY WILLOWHERB, FIREWEED, WICKUP; northern Hemisphere; perennial with erect, clumped stems to 2m tall; leaves long, narrow; flowers pale pink to

Epidendrum pseudepidendrum

purple-pink, rarely white); *E.chloraefolium* (New Zealand; branching, clump-forming and bushy perennial to 40cm tall; leaves small, often tinted bronze at first; flowers white to pink); *E.glabellum* (New Zealand; similar to *E.chloraefolium* but larger, the leaves not tinted bronze; flowers white to rose violet, pink or pale yellow). For *E.canum* see *Zauschneria*.

Epimedium
x *versicolor*
'Sulphureum'

Epimedium (from *epimedion*, the ancient Greek name for a wholly unrelated perennial). Berberidaceae. Mediterranean to E Asia. BARRENWORT, BISHOP'S HAT; BISHOP'S MITRE. Around 25 species, more or less evergreen perennial herbs. They make dense mats of creeping, thin but tough rhizomes, which give rise to slender-stalked, ternately compound leaves, the leaflets rhombic or lanceolate, thinly leathery and sometimes finely toothed. Produced in spring, the delicate flowers nod on slender stalks along erect to arching racemes. They consist of two sets of four sepals, the inner set petal-like and spreading, and four, spurred petals. Hardy to –10°C/14°F, or lower. Grow in partial shade in any fertile, humus-rich, moist soil. Carefully remove faded foliage in early spring to show flowers to best advantage. Propagate by division in spring or by fresh, ripe seed in late summer in the cold frame.

E.alpinum (S Europe; leaflets to 13cm, bright green, at first tinged red; flowers with dull red inner sepals and yellow petals to 0.4cm); *E.grandiflorum* (NE Asia; leaflets to 13cm, light green tinted bronze at first; flowers white, violet or deep rosy red, the inner petals usually darker than the petals); *E. × perralchicum* (*E.perralderianum × E.pinnatum*; similar to *E.pinnatum* except leaves occasionally simple, bronze when young; flowers large bright golden yellow); *E.perralderianum* (N Africa; similar to *E.pinnatum* except leaves bronze when young, sepals green not brown); *E.pinnatum* subsp. *colchicum* (N Iran; leaflets to 8 cm; flowers yellow sometimes veined red); *E.pubigerum* (Turkey, SE Europe; leaf undersides hairy; flowers with inner sepals pale rose or white and petals bright yellow); *E. × rubrum* (*E.alpinum × E.grandiflorum*; leaflets to 14cm, bright red-tinted at first; flowers with bright crimson inner sepals and yellow petals); *E. × versicolor* (*E.grandiflorum × E.pinnatum* subsp. *colchicum*; differs from *E.grandiflorum* in the smaller flowers with the petals only just exceeding or shorter than the sepals; includes several cultivars with copper- or red-mottled or tinted young growth and flowers in various shades of yellow); *E. × warleyense* (*E.alpinum × E.pinnatum* subsp. *colchicum*; leaflets to 13cm, hairy beneath, sparsely spiny-toothed; flowers with yellow to brick red inner sepals and yellow petals); *E. × youngianum* (*E.diphyllum × E.grandiflorum*; flowers white or rose to purple-red).

Epipactis (the Greek name for this genus). Orchidaceae. Northern temperate zone, tropical Africa, Thailand, Mexico. Some 24 species, terrestrial herbaceous orchids with mat-forming rhizomes and clumps of erect stems. The leaves are linear to elliptic and often pleated. Borne in summer in lax terminal racemes, the flowers consist of five, ovate to elliptic tepals and a showy, boat-shaped lip. The following species is fully hardy and may be grown in full sun or light shade. It needs a fertile soil, rich in leafmould and fast-draining but never dry. Increase by careful division of the rhizomes in autumn or spring when the new growing points are just visible. *E.gigantea* (N America; 20–90cm; flowers to 3cm diameter, tepals green, tinted or veined rose-purple, lip cream to yellow with purple-red to brown veins and spots and an orange-yellow midlobe).

epipetalous growing on or arising from the petals.

Epiphyllum (from Greek *epi*, upon, *phyllon*, leaf: the flowers are borne on leaf-like stems.) Cactaceae. Warm Americas. ORCHID CACTUS. About 15 species, mostly epiphytic cacti with green and succulent, laterally flattened stems, their margins wavy to notched or deeply lobed. The flowers are tubular to funnel-shaped and consist of numerous, spreading, silky tepals. Provide a minimum temperature of 15°C/60°F, plentiful water, humidity in warm weather and protection from burning sun. Plant in a fast-draining and acid, bark-based mix. Reduce water in winter and rest winter-flowering species in late summer. Orchid cacti are popular in cultivation and have spawned numerous cultivars, mostly with large, broadly funnel-shaped flowers in shades of white, cream, pink, rose, red and orange-red.

E.anguliger (FISHBONE CACTUS; S Mexico; stems flattened, to 100 × 4–8cm, deeply pinnately lobed; flowers 15–18, scented, outer tepals lemon or golden, inner tepals white); *E.crenatum* (S Mexico to Honduras; stems flattened to 50 × 3.5cm, rather thick, crenate; flowers 20–29 × 10–20cm, outer tepals pale yellow, inner tepals white); *E.oxypetalum* (S Mexico to Honduras, widely cultivated in the tropics and subtropics; stems flattened, to 30 × 10–12cm, undulate-crenate; flowers 25–30cm, outer tepals red, inner tepals white).

epiphyte a plant that grows on another plant, but is not parasitic and does not depend on it for nourishment.

Epipremnum (from Greek *epi*, upon, and *premnon*, a trunk, referring to the climbing and epiphytic habit). Araceae. SE Asia to W Pacific. 8 evergreen perennial lianes, climbing by adventitious, adhesive roots. The leaves are leathery, smooth, and entire to pinnately lobed, or occasionally perforated. They tend to occur in markedly different juvenile and adult phases, the juveniles usually smaller, more heart-shaped and entire than the adults. (Plants in cultivation are generally juveniles.) Inconspicuous flowers are carried in club-shaped spadices surrounded by cowel-like spathes. The following species favours a minimum winter temperature of 10°C/50°F, semi-shade, moderate to high humidity and an open, bark-based mix with plentiful water and feed. That said, *E.aureum* and its cultivars are often to be seen in the home, in offices and in public buildings, contending bravely with dehydrated moss-poles and neglected hydroponic planters. Propagate by rooted lengths of stem. *E.aureum* (syn. *Pothos aureus, Rhaphidophora aurea, Scindapsus aureus*; GOLDEN POTHOS, DEVIL'S IVY, HUNTER'S ROBE; Solomon Islands; stem climbing or sprawling to cascading, green striped yellow or white; juvenile leaves 6–30cm, bright green, variegated yellow or white; adult leaves to 80cm, pinnately cut, marked yellow or cream; includes 'Marble Queen', with moss green leaves boldly streaked bright white on white petioles, and white-streaked stems, and 'Tricolor', with leaves variegated cream to white on off-white petioles and stems).

Episcia (from Greek *episkios*, shaded referring to the natural habitat of these plants). Gesneriaceae. S America. 6 species, evergreen, creeping herbs with hairy ovate leaves and bell- to funnel-shaped, 5-lobed flowers at various times of year. Of cascading habit, they are suited to hanging baskets. Grow in an open, bark-based mix. Keep moist and humid in spring and summer with bright, indirect light to shade; in winter, keep barely moist, position in full sun and maintain a minimum temperature of 15°C/60°F. Increase by stem cuttings and offsets.

 E.cupreata (Colombia, Venezuela, Brazil; leaves to 8cm, puckered with fine, impressed veins, dark emerald green to purple-green, bronze or dark chocolate brown above, often with a metallic tint or silver blotches, purple-red beneath; flowers scarlet often marked or ringed yellow or purple); *E.dianthiflora* (syn. *Alsobia dianthiflora*; LACE FLOWER VINE; Costa Rica, Mexico; leaves to 4cm, thinly fleshy, green, downy; flowers sparkling white, lobes deeply fringed); *E.lilacina* (Costa Rica; leaves to 5cm, dark green, purple-tinted beneath; flowers lilac).

Epithelantha (from Greek *epi*, upon, *thele*, nipple, and *anthos*, flower, referring to the position of the flowers). Cactaceae. SW US, N Mexico. 1 species, *E.micromeris* (syn. *Mammillaria micromeris*), a dwarf cactus with globose to obovoid stems, 1–7.5cm in diameter. These are covered in radiating, white to pale grey or pale yellow spines. The flowers are 3–12mm in diameter and arise in the woolly stem-apex. They are campanulate, almost white or pale orange to pink, and are followed by red and juicy, club-shaped fruit. Grow in full sun with low humidity and a minimum temperature of 7°C/45°F. Plant in a gritty, neutral compost. Keep dry from mid-autumn until early spring, except for light misting on warm days in late winter. Increase by offsets.

epithet any word in a binomial which is neither the generic name nor a term denoting rank. The epithet qualifies a generic name or a name of lower rank, for example, *Primula vulgaris* subsp. *sibthorpii*, where *Primula* is the genus name, *vulgaris* the specific epithet, and *sibthorpii* the subspecific epithet.

Epsom salts see *magnesium sulphate*.

equitant when conduplicate leaves overlap or stand inside each other in two ranks in a strongly compressed fan, as in many Iridaceae.

erect of habit, organ or arrangement of parts, upright, perpendicular to the ground or point of attachment.

Eranthemum (from Greek *eros*, love, and *anthemon*, flower). Acanthaceae. Tropical Asia. Some 30 species, perennial shrubby herbs and shrubs with tubular, 5-lobed flowers amid bracts in spikes or panicles. Grow in bright filtered light or with part-day shade. Water plentifully but carefully when in full growth, allowing the surface of the medium to dry out between waterings, and maintain a moderately humid but buoyant atmosphere. Provide a minimum winter temperature of 13°C/55°F. Propagate by softwood cuttings in a closed case with bottom heat in late spring/early summer. Cut back flowered stems of mature specimens by about half their length after flowering. *E. pulchellum* (syn. *E.nervosum*; BLUE SAGE; evergreen shrub to 1.25m; flowerspikes to 7cm; bracts green feathered with white; flowers to 3cm, deep blue).

Eranthis (from Greek *er*, spring, and *anthos*, flower). Ranunculaceae. S Europe (also naturalized W and N Europe and N America), Asia (Turkey to Japan). WINTER ACONITE. 7 species, perennial herbs with small, rounded tubers. The basal leaves are petiolate and deeply palmately lobed. The three stem leaves form a ruff below the flower. Solitary, terminal and yellow to white, the flowers are cupped, short-stalked and consist of two whorls of ovate to elliptic perianth segments. Valued for the bright carpet of flower colour they provide in late winter and early spring. They may be naturalized under deciduous trees or shrubs or in light grassland. Plant in summer or early autumn in humus-rich, heavy soil. Propagate by division in spring after flowering. Like *Cyclamen* and *Galanthus*, *Eranthis* can be difficult to establish from dry tubers: purchase 'in the green' (i.e. bare-root but in leaf) where possible. Sow seed as soon as ripe (hybrids are sterile).

 E.hyemalis (S France to Bulgaria; to 8cm; flowers 2–3cm diameter, bright yellow, produced in late winter; Cilicica Group: to 5cm; leaves tinged bronze at first; flowers shiny golden yellow, larger and emerging later; Tubergenii Group: leaves tinged bronze at first; flowers large, golden, appearing later; includes 'Guinea Gold', with deeply and narrowly cut bracts and fragrant, long-lasting, deep golden flowers).

Eranthis hyemalis

Ercilla (for Don Alonso de Ercilla (1533–95), of Madrid). Phytolaccaceae. Americas, South Africa. 2 species, evergreen climbers attaching themselves by disc-like holdfasts. The leaves are leathery and entire, and the flowers small and borne in spring in dense spike-like racemes. These are sometimes followed by glossy berries. Tolerant of temperatures to between –5 and –10°C/23–14°F. Grow in a well-drained lime-free soil in sun to part shade. Provide support. Propagate by nodal stem cuttings in midsummer. *E.volubilis* (syn. *E.spicata*; Peru, Chile; to 6m; leaves to 5cm, ovate to oblong, fleshy, glossy dark green; flowers purple or green with white stamens; fruit dark purple).

Eremurus (from Greek *eremia*, desert, and *oura*, tail, referring to the appearance of the flower spikes in their desert or semi-desert habitats). Liliaceae (Asphodelaceae). W and C Asia. DESERT CANDLE, FOXTAIL LILY. 40–50 species, fleshy-rooted perennial herbs with strap-shaped leaves forming tufts or rosettes. Small, starry and 6-parted, the flowers are packed in summer in long, tapering racemes atop erect scapes. *Eremurus* produces a fleshy rooted crown with a central growing-point which may rot if too wet in winter. Where possible, grow on raised beds or banks where water drains quickly; alternatively, set the crown on sharp sand and cover it with same when planting. They require cold in winter to flower well, and are unsuited to cultivation in completely frost-free zones. Grow in rich, well-drained, sandy soil in full sun. Without shelter from wind, taller species and hybrids may require staking. Protect from winter wet with a mulch of sharp sand or ashes. In regions that experience late frosts, give young shoots the protection of dry bracken litter or straw. Propagate from ripe seed in autumn, or by careful division when the foliage dies back in late summer. *E.himalaicus* (to 2.5m tall, flowers white); *E.* × *isabellinus* (*E.stenophyllus* × *E.olgae*; to 1.5m tall, flowers orange, pale yellow, gold, pink, white to copper; includes the Shelford Hybrids and Highdown Hybrids); *E.robustus* (to 1m; flowers pink); *E.spectabilis* (to 1m; flowers pale yellow suffused green); *E.stenophyllus* (to 1.5m tall; flowers orange-yellow to clear yellow ageing topaz).

erianthous woolly flowered.

Erica (from the Greek name *ereike*). Ericaceae. HEATH, HEATHER. Africa, Madagascar, Middle East, Europe, Atlantic Is. Some 735 species, evergreen shrubs and shrublets with small, narrow leaves. Tubular to urn- or bell-shaped, the flowers are carried in heads at the branch tips or disposed along the branches in a raceme- or panicle-like arrangement. For cultivation, see *heaths and heathers*.

Erica canaliculata (top)
Erica lusitanica (below)

ERICA Name	Distribution	Habit	Leaves	Flowers
E.arborea TREE HEATH	SW Europe, Mediterranean, N Africa	tree-like, to 4m	to 5mm, dark green, in whorls of 4	to 4mm, white tinged grey or cream, campanulate, fragrant in large, pyramidal panicles
Comments: Late spring. Z7. Lime-tolerant. Includes clones with bright gold to lime green foliage				
E.australis SPANISH HEATH	W Iberian Peninsula, Tangier	erect to 2m	to 5mm, dark green, in whorls of 4 or 3	to 9mm, red-pink, tubular to campanulate with reflexed lobes; on short branches
Comments: Spring-summer. Z8.				
E.canaliculata	S Africa	erect to 2m	4-10mm, bright to dark green, in whorls of 3	to 3.5mm, cup-shaped, purple-pink to pale rosy pink with conspicuous dark purple-red to black anthers; in whorls at ends of branchlets
Comments: Year-round. Z 9, sometimes surviving in sheltered places in zone 7.				
E.carnea *E.herbacea* WINTER HEATH	C & E Europe	dwarf, with procumbent to ascending stems to 25cm	to 8mm, dark green in whorls of 4	to 6mm, purple-pink, cylindric, often with dark anthers; in terminal, somewhat 1-sided racemes
Comments: Late winter-early spring. Z5. Lime-tolerant. Many cultivars, ranging in habit from the tall and open to the dwarf and dense; in foliage colour from dark green to emerald, sea-green, gold and lime to copper and bronze; in flower colour from pure white to palest shell pink, rose, purple-pink, deep magenta and carmine				

ERICA

Name	Distribution	Habit	Leaves	Flowers
E.ciliaris DORSET HEATH	SW Europe	spreading, to 80cm	to 4mm, ovate to lanceolate, ciliate	to 12mm, deep pink-red, swollen-tubular to urn-shaped; in terminal racemes
Comments: Late summer-autumn. Z 6. Includes cultivars with foliage turning red-bronze in winter, and flowers in tones of white, pink and rose, also bicolours – E.'David McClintock', for example, has white to blush flowers tipped with purple-pink				
E.cinerea BELL HEATHER	W Europe	low, bushy, to 60cm with hairy young twigs	to 5mm, dark green, revolute, in whorls of 3	to 7mm, pink, purple-red or white, urn-shaped; in terminal racemes or umbels
Comments: Summer. Z5. Many cultivars – in habit, prostrate to low and spreading, dome-shaped or rounded to erect and narrow, densely bushy to loose and open; in foliage colour, dark green to bronze-red or gold, some changing from yellow to deep rusty orange in winter, or with bright copper new growth; in flower colour, pure white, pale rose, pink, magenta, crimson and deep ruby				
E.× darleyensis (E.erigena × E.carnea)	garden origin	as for E.carnea, but more vigorous and bushy, to 60cm	to13mm	white to rosy pink with dark anthers
Comments: Winter-spring. Z6. Lime-tolerant. Cultivars range in foliage from dark green to lime or golden yellow, some with cream- or red-tipped new growth, or turning bronze-red in winter; in flower colour from silvery white to rose, pink, magenta and lilac				
E.erigena E.hibernica; E.mediterranea of gardens IRISH HEATH	Tangier, Iberian Peninsula, SW France, Eire	as for E.carnea, but erect to 2m		
Comments: Late winter-early spring. Z7. Cultivars range in habit from low, slow-growing and compact to erect and rangy; in leaf colour from deep green to grey-green and gold, some tinted purple-red or copper in winter or when new; in flower colour from white to shell pink, rose, deep crimson, magenta and lilac, some becoming darker with age, some scented				
E.gracilis	S Africa	compact, to 50cm tall	2-4mm, in whorls of 4	to 4mm, pale to deep pink, ovoid to rounded, in terminal whorls of 4
Comments: Summer. Z9. Includes cultivars with glowing deep red flowers				
E.× hiemalis FRENCH HEATHER; WHITE WINTER HEATHER	origins and parentage obscure	to 60cm	to 5mm, bright green, in whorls of 4	to 15mm, white suffused pink with dark anthers, tubular; in short, terminal racemes
Comments: Winter. Z8. Includes cultivars with long, tubular flowers in shades or purple-pink and rose, often with paler or white lobes and opening				
E.lusitanica PORTUGUESE HEATH	W Iberian Peninsula to SW France, naturalized SW England	close to E.arborea, but to 3m tall and not so dense and tree-like	to 7mm	to 5mm, pink in bud, opening white, narrowly campanulate
Comments: Spring to early summer. Z7/8. Requires shelter. E.'George Hunt': leaves yellow, flowers white				
E.mackayana	Spain, Ireland	low and spreading to erect, to 50cm	to 6mm, oblong to lanceolate, revolute, glandular, in whorls of 4	to 7mm, pink, squatly urn-shaped, in terminal umbels
Comments: Mid-summer to early autumn. Z4. Includes clones with white, rose and deep purple-pink flowers, some semi-double or double				
E.pageana	S Africa	erect to 50cm, downy	5-8mm, closely packed, in whorls of 4	6-8mm, yellow, downy, tubular to campanulate with a green calyx; in terminal clusters of 3-4
Comments: Spring. Z 4				
E.scoparia BESOM HEATH	SW Europe, N africa, Canary Is	slender and erect, to 6m	to 7mm, glossy, in whorls of 3-4	to 3mm, green tinged red-brown, campanulate; in narrow leafy racemes
Comments: Late spring to early summer. Z7/8. Includes small, neat clones with bright green, shiny leaves				
E.× stuartii E.× praegeri ; (E.mackayana × E.tetralix)	W Ireland	close to E.mackayana, distinguished by hairy ovary and calyx		
Comments: Summer to autumn. Z6. Includes clones with pink to pale mauve flowers, some tipped with a deeper tone, maroon, for example. Several have new foliage tinted yellow, orange or coral				
E.terminalis CORSICAN HEATH	SW Mediterranean	to 2.5m, erect and tree-like	dark green, in whorls of 4	to 7mm, rosy pink to deep mauve-pink fading to rusty red, urceolate; in umbel-like clusters
Comments: Summer to early autumn. Z8. Lime-tolerant				

ERICA

Name	Distribution	Habit	Leaves	Flowers
E.tetralix CROSS-LEAVED HEATH	Iberian Peninsula, France, UK	dwarf and spreading, to 25cm	to 6mm, lanceolate to narrowly oblong, grey-green, ciliate, in whorls of 4	to 9mm, pale pink, swollen tubular to urceolate; in terminal umbels
Comments: Summer to early autumn. Z5. Includes clones with white to pale pink and dark rosy red flowers, sometimes arranged in a tight, radiating cluster; some have very fine silvery grey foliage				
E.umbellata DWARF SPANISH HEATH	W Iberian Peninsula, Tangier	dwarf, to 80cm	to 4mm, in whorls of 3	to 6mm, rose pink to purple with dark brown anthers, campanulate to ovoid; in umbels
Comments: Late spring to early summer. Z7				
E.vagans CORNISH HEATH	W France, Spain, Cornwall, Ireland	decumbent to erect, to 80cm	to 10mm, mid-green, in whorls of 4 or 5	to 4mm, cylindric to campanulate, lilac pink to white, with dark maroon to brown anthers; in leafy racemes
Comments: Mid-summer to late autumn. Z5. Somewhat lime-tolerant. Includes clones with white, cream, pink, cerise, deep ruby and purple flowers; some with yellow foliage, or foliage taking on colourful tints in winter. In *E.*'Viridiflora', the flowers are replaced by green bracts.				
E.× veitchii (*E.arborea × E.lusitanica*)	garden origin	resembles a more vigorous *E.arborea*	similar to *E.lusitanica*	cylindric to near-spherical, white to cream with pink stamens, fragrant
Comments: Spring. Z7/8. 'Gold Tips': young shoots golden, fading to lime or pale green				
E.× watsonii (*E.ciliaris × E.tetralix*)	garden origin	compact, hairy, to 30cm	mid-green, often colourful in spring	broadly campanulate to urceolate, pink
Comments: Summer. Z5. Includes clones with white, rose and deep crimson flowers; some with new foliage flushed gold, yellow-orange, copper or red-bronze				
E. × williamsii (*E.vagans × E.tetralix*)	garden origin	resembles *E.vagans* but usually with yellow to rich gold new growth		campanulate, rose pink with dark brown anthers
Comments: Spring. Z5				

ericaceous (1) a plant belonging to the family Ericaceae; (2) planting media with low pH suitable for growing calcifuge plants; (3) resembling the genus *Erica* in general habit.

Erigeron (from Greek, *eri*, early, and *geron*, an old man, either alluding to the hairy pappus, or to the hoary appearance of the leaves of some species in spring). Compositae. Cosmopolitan, especially N America. FLEABANE. Some 200 species, annual or, more commonly, perennial herbs with daisy-like flowerheads in spring and summer – the ray florets range widely in colour, the disc florets are yellow. The following species are fully hardy and suitable for the herbaceous border and rock garden; *E.karvinskianus* is an exquisite colonizer of damp old walls and paving. Grow the others in sun on any moderately fertile, freely draining soil that remains moist in the growing season. Provide the support of pea-sticks for taller perennials and deadhead to produce a second flush of bloom later in the season. Cut back in autumn. Propagate by seed sown *in situ*, also by division and by basal softwood cuttings in spring.
    E.alpinus (Europe; hairy perennial herb to 40cm tall; flowerheads to 3.5cm diameter, ray florets lilac); *E.aurantiacus* (ORANGE DAISY; Turkestan; mat-forming, downy perennial herb to 30cm; flowerheads to 5cm diameter, ray florets bright orange-yellow); *E.aureus* (N America; hairy perennial herb to 20cm; flowerheads to 3cm diameter, ray florets yellow); *E.glaucus* (BEACH ASTER, SEASIDE DAISY; W US; tufted perennial herb to 50cm, stems sprawling; leaves grey-green; flowerheads with lilac to violet ray florets; includes the white-flowered 'Albus', and 'Elstead Pink' with deep lilac-pink flowerheads); *E.karvinskianus* (syn. *E.mucronatus*; C America; woody-based perennial with slender, sprawling stems to 30cm; leaves small; flowerheads to 2cm diameter, slender-stalked, white or pink, becoming red-purple with age); *E.speciosus* (NW US; perennial herb to 80cm; flowerheads to 5cm across, ray florets violet to blue or white).

Erinacea (from Latin *erinaceus*, resembling a hedgehog, referring to the overall appearance of the plant). Leguminosae. W Mediterranean to E Pyrenees. 1 species, *E.anthyllis* (syn. *E.pungens*; HEDGEHOG BROOM, BRANCH THORN, BLUE BROOM), an evergreen, clump-forming shrublet, to 30cm tall with thorn-like branchlets. The leaves are small and short-lived. Pale blue, indigo or mauve, pea-like flowers appear in clusters in summer. Hardy to –15°C/5°F. Grow in a hot position on the rock garden, at the base of a sunny wall or on raised beds or low walls. Plant in gritty free-draining soils, with shelter from cold, drying winds. Propagate from seed: cold-stratify overwinter and germinate in warmth in early spring. Alternatively, increase by heel cuttings in early autumn.

erinous prickly, coarsely textured with sharp points.

Erinus (Greek name *erinos*, used by Dioscorides for a basil-like plant). Scrophulariaceae. N Africa, S and C Europe. 2 species, tufted perennial herbs with tubular, broadly 5-lobed flowers in erect racemes in late spring and summer. A hardy perennial alpine, although shortlived, it seeds itself freely and will carpet large areas. Grow on light, well-drained soils in sun or part-shade on rock gardens, walls, raised beds, scree, troughs or in tufa; it is also useful for sowing in spring directly into paving joints. A mulch of shingle or grit provides a good seedbed and protection against winter rots around the neck. Keep moist but never waterlogged. Propagate from seed sown as soon as ripe in cold frames or *in situ*. Small softwood cuttings will also succeed. *E.alpinus* (ALPINE BALSAM, FAIRY FOXGLOVE, LIVER BALSAM; S and C Europe; to 30cm, carpet-forming small leaves; flowers 0.6–1cm, typically purple, white in 'Albus', red in 'Carmineus', crimson in 'Dr Hähnle', pink in 'Roseus', violet in 'Lilacinus').

Eriobotrya (from Greek *erion*, wool, and *botrys*, a cluster of grapes, referring to the woolly, clustered panicles). Rosaceae. E Asia. Some 10 species of evergreen shrubs or trees with stout branches and large, leathery leaves. Small, white flowers are produced in late winter and spring in broad and often densely woolly panicles. The fruit is an obovoid to globose pome, with persistent calyx teeth at the apex and a few large seeds. *E.japonica* is extensively cultivated in subtropical and warm-temperate climates for its edible fruits – these are eaten raw, poached, stewed as preserves or pickled with sweet spices. In Japan over 800 cultivars are known. The fruit, which resembles an apricot, has a firm flesh and slightly acid, sweet aromatic flavour; the seeds are slightly poisonous. In temperate zones, it makes a handsome ornamental, with bold and luxuriant foliage. The flowers, borne intermittently from early winter to spring, are produced in quantity only after hot summers. The fruit ripens in spring, but is likely to be rendered inedible by hard frosts in cold areas. It will survive temperatures as low as –17°C/1°F, and tolerates dry soils and maritime exposures. Plant in fertile, well-drained soils, in full sun or light shade with shelter from cold winds, which will scorch the foliage. In regions at the limits of its hardiness, grow against a south-facing wall. Propagate by seed sown under glass when ripe, by semi-ripe cuttings with a heel, or by air layering. *E.japonica* (LOQUAT, JAPANESE LOQUAT, JAPANESE MEDLAR, NISPERO; China, Japan; large and spreading shrub or tree to 7m tall; branches stout, covered in tawny hairs at first; leaves to 30cm, broadly oblanceolate to elliptic, leathery and corrugated; fruit to 4cm diameter, globose to pyriform, yellow to apricot).

Eriogonum (from Greek *erion*, wool, and *gonia*, a joint or knee, referring to the downy nodes). Polygonaceae. N US, mostly western. WILD BUCKWHEAT; UMBRELLA PLANT; ST. CATHERINE'S LACE. 150 species, herbs and subshrubs (those listed below are perennial), usually with woolly leaves and small,

pale flowers in terminal heads or umbels. Hardy to between –10 and –15°C/14–5°F. Most are also tolerant of exposed sites, although those with woolly leaves need protection from winter wet. Grow in a gritty, perfectly drained soils in full sun. Water sparingly in spring and autumn, moderately when in full growth and keep almost dry in winter. They resent disturbance once established. Propagate by ripe seed sown in a cool greenhouse in autumn or in spring; alternatively, increase by heeled greenwood cuttings in midsummer or by division in early spring.

E.arborescens (California; shrub to 1.5m; leaves 2–3cm, linear to oblong, white-hairy beneath; flowers white to pale rose); *E.crocatum* (SAFFRON BUCKWHEAT; California; herb or subshrub to 20cm; leaves 1.5–3cm, ovate to elliptic, white-felted; flowers sulphur yellow); *E.giganteum* (ST CATHERINE'S LACE; S California; shrub to 2.5m; leaves 3–10cm, oblong to ovate, grey- to white-hairy; flowers white fading to rusty red); *E.ovalifolium* (W US; silver-hairy, mat-forming and small to dwarf; leaves 0.5–1.5cm, broadly oblanceolate to elliptic; flowers white to pink or rose-purple); *E.umbellatum* (SULPHUR FLOWER; western N America; herb or subshrub, low and spreading, rarely exceeding 30cm; leaves to 2cm, spathulate to obovate, tomentose, in rosettes; flowers cream to sulphur yellow, sometimes titnted red with age).

Eriophyllum (from Greek *erion*, wool, and *phyllon*, leaf). Compositae. Western N America. WOOLLY SUNFLOWER. About 12 species, hairy, annual to perennial herbs and shrubs, with entire to toothed or pinnatifid leaves and daisy-like flowerheads in late spring and summer. Grow in full sunlight on a well-drained, preferably sandy soil. Clip over after flowering. Propagate by seed sown under glass in autumn, or by division in spring. *E.lanatum* (hairy herb, 20–60cm tall; leaves to 15cm, spathulate to narrowly oblanceolate, entire to pinnately lobed; flowerheads 2–5cm in diameter, yellow).

Eritrichium (from Greek *erion*, wool, and *trichos*, hair, the plants are woolly). Boraginaceae. Temperate northern Hemisphere. ALPINE FORGET-ME-NOT. Some 30 species, low-growing, tufted, alpine perennial herbs, with small, downy leaves and forget-me-not-like flowers in short cymes in early summer. Hardy to about –15°C/5°F, but intolerant of winter wet and perhaps best grown in the alpine house in a lean and gritty alpine mix. Provide a deep collar of grit around the neck and/or a prop of small stones to keep foliage clear of the soil surface. Alternatively, grow in tufa rock half-buried in the pot. Water plentifully when in growth, avoiding the foliage, and keep just moist in winter. Sow seed in a mix of loam, leafmould and sharp sand, with additional parts of crushed slate and tufa dust; expose to frost but protect from rain, and germinate in the cold greenhouse; increase also by softwood cuttings in summer.

E.elongatum (W US; to 7.5cm, densely tufted, mat-forming; leaves to 0.7cm, oblanceolate, woolly;

flowers to 0.7cm diameter, blue with yellow crests); *E.nanum* (Alps; to 5cm; densely tufted; leaves to 1cm, elliptic to linear, silver-hairy; flowers to 0.8cm diameter, sky blue with yellow crests).

Erodium cheilanthifolium

Erodium (from Greek *erodios*, heron, the carpels resemble the head and beak of a heron). Geraniaceae. Europe, C Asia, temperate Australia, tropical S America. STORKSBILL, HERON'S BILL. 60 species, mostly perennial, but some annual herbs and subshrubs (those listed below are small, compact perennial herbs). *Erodium* differs from *Geranium* in having only 5 fertile stamens (*Geranium* has 10). The leaves are usually lobed or pinnately divided with toothed margins. Produced from spring to late summer, the flowers are 5-petalled, and are carried in stalked umbels or rarely solitary. The fruit resembles a slender, bird's bill and, when ripe, splits lengthways into five 1-seeded parts. Plant the smaller species on the rock garden, in a sunny, well-drained spot with a humus-rich, preferably alkaline, soil. The taller species are less demanding as to soil type (although drainage must be good). The following are hardy in zone 7, but the more southern natives will suffer from winter damp and are best protected with a propped pane of glass in the open garden during the winter months. Alternatively grow in pans in the alpine house. Propagate from seed sown under glass, or by division in spring.

E.cheilanthifolium (syn. *E.petraeum* subsp. *crispum*; S Spain, Morocco; leaves to 5cm, white-hairy, ferny; flowers, white to pale pink veined red and with dark stains); *E.chrysanthum* (Greece; leaves to 3.5cm, ferny, silver-hairy; flowers creamy yellow to pale sulphur); *E.corsicum* (Corsica, Sardinia; leaves silvery-downy ovate, crumpled, crenate; flowers rose pink or white veined red); *E.foetidum* (syn. *E.petraeum*; Pyrenees, S France; leaves ferny, softly hairy; flowers veined red); *E.glandulosum* (syn. *E.petraeum* subsp. *glandulosum*; Pyrenees; leaves ferny, hairy, aromatic; flowers lilac with dark stains or rose stained purple); *E.manescaui* (Pyrenees; leaves to 30cm, ovate to lanceolate, pinnate; flowers purple-red with darker patches); *E.reichardii* (syn. *E.chamaedryoides*; ALPINE GERANIUM; Majorca, Corsica; leaves cordate, crenate, dark green; flowers white with rose veins); *E.* × *variabile* (*E.corsicum* × *E.reichardii*; intermediate between parents; includes white, bright pink and double-flowered cultivars).

erose irregularly dentate, as if gnawed or eroded.

Eryngium (from Greek *eyringion*, name used by Theophrastus). Umbelliferae. Cosmopolitan. ERYNGO, ERINGOE, SEA HOLLY. 230 species, perennial, biennial and annual herbs (those listed below are perennial unless otherwise stated) with mostly basal leaves ranging in shape from sword-like to heart-shaped, the margins toothed to intricately cut, lobed and spiny. Produced in late spring and summer, the flowers are small, generally blue-grey or white and packed in a cone-like, hemispherical to cylindrical head, subtended by a ruff of spiny to lacerate, decorative bracts. Most are hardy to –15°C/5°F; South American species such as *E.agavifolium* and *E.bromeliifolium* are less reliable where temperatures fall far below freezing point, but have been successfully cultivated in sheltered niches with perfect drainage. Grow in full sun on moderately fertile, well-drained soils. Propagate by seed in autumn in the cold frame; also by root cuttings in winter or by careful division in autumn or spring.

E.agavifolium (Argentina; to 1m tall; leaves sword-shaped, spiny-toothed; flowerheads to 5cm, green-blue with entire to spiny-toothed bracts to 1cm long; often confused with *E.bromeliifolium*); *E.alpinum* (Europe; to 60cm; leaves ovate to heart-shaped, spiny-toothed; flowerheads to 4cm, steel blue to sky blue, violet, lilac or white with softly spiny, pinnately lobed bracts to 6cm); *E.amethystinum* (Italy, Sicily, Balkans; to 70cm; leaves obovate to finely palmately divided, the segments pinnately lobed and spiny; flowerheads to 2cm, electric blue to amethyst on silver stems, with spiny bracts to 5cm); *E.bourgatii* (Spain, E Mediterranean; to 40cm; leaves 3-parted, each segment narrowly pinnately lobed, spiny and silver-veined; flowerheads to 3cm, sky blue with electric blue and silver, toothed bracts to 5cm); *E.bromeliifolium* (Mexico; differs from *E.agavifolium* in the white-blue flowerheads to 2.5cm exceeded by many rigid, sharp bracts); *E. campestre* (C and E Europe; leaves pinnately divided; flowerheads small, massed, blue-grey); *E.eburneum* (S America; differs from *E.agavifolium* in the entire but spiny-margined leaves and taller flowering stems to 2m, with ivory flowerheads to 2cm and spiny-tipped bracts); *E.giganteum* (MISS WILLMOTT'S GHOST; Caucasus; biennial to 1.5cm; leaves heart-shaped to triangular, crenate to spiny and incised; flowerheads to 4cm, electric blue to pale green or white with spiny, grey-white or silver-blue-tinted bracts to 6cm); *E.maritimum* (SEA HOLLY; coasts of Europe, naturalized east coast of US; short-lived perennial 20–60cm tall with tough, spiny, glaucous blue-grey foliage; flowerheads to 5cm, pale blue to lilac with bracts to 4cm resembling leaves); *E.* × *oliverianum* (*E.alpinum* × *E.giganteum*; to 100cm; lower leaves heart-shaped to ovate, slightly 3-lobed, spiny-toothed with conspicuous veins; flowerheads to 4cm, vivid blue with linear, spiny-toothed and purple bracts); *E.planum* (C and SE Europe; to 75cm; leaves ovate to oblong, toothed to palmately lobed, spiny and tinted blue; flowerheads to 3cm, iridescent azure; lilac to pink forms are also grown); *E.* × *tripartitum* (parentage unknown; to 1m; leaves 3-lobed and spiny-toothed; flowerheads to 1cm metallic blue, exceeded by narrow, blue-tinted bracts); *E.variifolium* (N Africa; to 45cm; leaves rounded to heart-shaped, margins toothed, dark green marbled white; flowerheads to 2cm diameter, silver-blue exceeded by spiny, white-blue to steel grey bracts); *E. yuccifolium* (SW US; similar to *E. agavifolium* but usually deciduous with blue-green glaucous leaves and green-white flowerheads).

Erysimum (name used by Hippocrates, from Greek *eryo*, to draw out: some species produce blisters). Cruciferae. Europe to W and C Asia, N Africa and N America. Around 80 species, annual, biennial or perennial, evergreen herbs, usually bushy with clumped, woody-based stems and oblong to linear leaves. Carried in terminal racemes, the flowers are more or less fragrant and consist of four, spreading, obovate petals.

Derived from *E.cheiri*, most wallflowers are grown for spring and early summer flowering and most commonly used for bedding, cutting and for containers. Selection has yielded dwarf, intermediate and tall forms, and colours that include pastels and range from white and pale lemon yellows through orange-yellow to deep crimson, brown and purple; many are scented. They are short-lived perennials, treated as biennials. Fully hardy, they need a moist but well-drained, neutral to slightly alkaline soil in full sun. *E.* × *allionii*, the SIBERIAN WALLFLOWER, is used similarly but is less commonly grown, having a colour range restricted to orange, golds and apricots. Nevertheless, it has a number of advantages, being denser in habit than *E.cheiri*, spicier of scent, more tolerant of cold, and flowering slightly later.

The genus includes several perennials, some of them shrubby, including the bushy, two-tone *E.bicolor*, the shrubby, silver and mauve *E.*'Bowles Mauve', the low and tufted *E.kotschyanum*, found on high scree, the violet, bushy *E.linifolium* and *E.purpureum* with beautiful ash-grey leaves and purple flowers. These are hardy in zone 6 given sharply draining, rather poor soils in full sun, with protection from harsh winds and prolonged cold. Repropagate and replace regularly; overwinter small, new plants in a cold greenhouse.

Biennials should be sown in late spring/early summer; grow on in a nursery bed, and transfer to flowering positions in early autumn. Where winter temperatures fall consistently below –15°C/5°F to –20°C/–4°F, overwinter in the cold frame and plant out in early spring. Propagate perennials from seed in spring and early summer, or from heel cuttings in summer, rooted in a sandy propagating medium in the cold frame; alternatively, increase by division.

E. × *allionii* (syn. *Cheiranthus* × *allionii*; SIBERIAN WALLFLOWER; perennial grown as a biennial; flowers large, bright orange); *E.bicolor* (Madeira, Canary Islands; subshrub or shrub to 90cm; flowers lightly fragrant, cream to pale yellow or bronze turning pale lilac; *E.mutabile* differs in its mauve-pink flowers turning orange-bronze); *E.*'Bowles Mauve' (shrub or subshrub to 75cm; leaves grey-green; flowers small, rich mauve); *E.cheiri* (syn. *Cheiranthus cheiri*; WALLFLOWER; S Europe; perennial subshrub cultivated as a biennial, 25–80cm tall; flowers bright orange-yellow, usually striped or spotted purple-red); *E.helveticum* (syn. *E.pumilum*; C and E Europe; clump-forming, tufted perennial to 10cm; flowers small, fragrant, bright yellow in flat heads opening from purple-tinted buds); *E.hieraciifolium* (syn. *E.alpinum*; biennial or perennial, 30–120cm tall;

flowers yellow with hairy undersides to petals; sometimes treated as a synonym of *E.* × *allionii*); *E.kotschyanum* (Turkey; densely tufted perennial to 10cm; flowers orange-yellow to yellow); *E.linifolium* (syn. *Cheiranthus linifolius*; Spain, Portugal; woody-based perennial 12–70cm tall with numerous slender stems; flowers mauve or violet); *E.purpureum* (Turkey, Syria; erect, woody-based perennial to 30cm; leaves grey-green; flowers violet-mauve).

Erythrina (from Greek *erythros*, red, referring to the colour of the flowers). Leguminosae. Pantropical. CORAL TREE. 108 species, deciduous or evergreen shrubs, trees and perennial herbs, usually with blunt, conical thorns or recurved prickles, and trifoliolate to pinnate leaves. Produced in late summer in erect racemes, the flowers are large, pea-like and usually red with a large, cowl- or boat-like standard petal.

E.crista-galli is usually grown in the cool greenhouse or conservatory in temperate zones. It will, however, grow outdoors with the shelter of a south-facing wall, and will tolerate temperatures as low as –10°C/14°F provided the stem bases are thickly mulched with leaf litter or sawdust and covered with bracken. In harsh climates, the stems are usually killed by frost to ground level. They are best left until mid- to late spring when they can be cut down and removed along with the bracken cover. By this time new shoots should be in evidence; these will develop quickly, flowering by late summer.

E. × *bidwillii* has a thick woody rootstock and shows similar tolerances. The roots of this species are sometimes lifted and stored in frost-free conditions over winter. In warm temperate zones, grow in full sun in moderately fertile, well-drained soils. Under glass, grow in direct sun in a freely draining, high-fertility, loam-based mix; water plentifully when in growth, gradually withdrawing water in autumn to overwinter in cool, dry conditions with a minimum temperature of 5–7°C/40–45°F. Propagate from seed sown in gentle heat in spring; seed-grown plants usually flower in their third or fourth year. Alternatively, increase by semi-ripe cuttings in summer, or by heeled cuttings of young growth in spring.

E. × *bidwillii* (*E.crista-galli* × *E.herbacea*; woody-based perennial herb or shrub to 4m with dark red flowers to 5cm long); *E.crista-galli* (CORAL TREE, COCKSPUR CORAL TREE; S America; prickly erect shrub or small tree to 9m; flowers to 7cm long, bright scarlet, in large, leafy, terminal racemes).

Erythronium (from Greek *erythros*, red, more or less the colour of *E.dens-canis*). Liliaceae (Liliaceae). N US, Europe and Asia. DOG'S-TOOTH VIOLET, ADDER'S TONGUE, TROUT LILY, FAWN LILY. Some 20 species, perennial herbs with yellow-white bulbs, which are elongated and tooth-like. The leaves are basal, ovate to elliptic, and mottled white, pink, chocolate or maroon, or plain. Produced in spring, the nodding flowers are borne singly or several together on an erect scape. Deeply bell- to hat-shaped, they consist of six, ovate to lanceolate perianth segments, their tips sometimes strongly reflexed. Hardy to –15°C/5°F and below. Most species are suited to shady pockets in the rock garden and for naturalizing in the shrub and mixed border, but are perhaps seen at their best in the dappled sunlight of the woodland garden. Grow in partial shade in a well-drained, humus-rich and moisture-retentive soil. *E.dens-canis* is sturdy enough to naturalize in thin grass. Mulch established clumps annually with leafmould. Propagate by division of established clumps as leaves fade, and re-plant immediately to avoid desiccation of the bulbs. Increase also by ripe seed sown fresh in a moist and humus-rich propagating mix.

E.americanum (YELLOW ADDER'S TONGUE, TROUT LILY, AMBERBELL; eastern N America; leaves mottled brown and white; flowers solitary, yellow, exterior marked brown or purple, interior spotted); *E.californicum* (FAWN LILY; California; leaves mottled brown-green; flowers 1–3, creamy white with a central ring of orange-brown); *E.dens-canis* (DOG'S TOOTH VIOLET; Europe, Asia; leaves mottled pink and chocolate; flowers solitary, rose to mauve, sometimes pale pink, lilac, plum purple or pure white); *E.grandiflorum* (AVALANCHE LILY; W US; leaves plain green; flowers 1–3, yellow); *E.hendersonii* (SW Oregon, NW California; leaves dark green, mottled, margins crisped; flowers to 10, lilac-pink with a purple centre); *E.oregonum* (Oregon to British Columbia; differs from *E.californicum* in its flowers with yellow centres); *E.revolutum* (N California, Vancouver Islands; leaves mottled, margins crisped; flowers 1–3 per stem, deep pink with a yellow centre; includes 'Pagoda', with bronze-marbled leaves and pale sulphur flowers with brown centres).

Escallonia (for Señor Escallon, a Spaniard who travelled in South America in the late 18th century). Grossulariaceae (Escalloniaceae). S America. Some 50 species, aromatic and more or less viscid, evergreen shrubs and small trees with glossy, dark green leaves and racemes or panicles of shortly funnel-shaped flowers, each with five, spreading lobes, from late spring to early autumn. In cool and exposed maritime gardens, many of the taller species are useful screens and windbreaks, and in milder regions can be clipped as hedging for more formal situations. Most are valuable additions to the shrub border, grown for their small, usually dark glossy leaves and for their flowers – these are often carried late in the season, from summer into early autumn. Where temperatures fall much below –10°C/14°F, most species need the protection of a south-facing wall. If cut back by frost they will usually regenerate from the base or from older wood if thickly mulched at the roots. The Donard Hybrids (*E. × langleyensis*), especially 'Slieve Donard' and 'Apple Blossom', and *E. rubra* cultivars are among the hardiest, surviving temperatures to –20°C/–4°F. *E.*'Iveyi' is almost as hardy, being cut back at temperatures at about –15°C/5°F. Grow in full sun

Escallonia
'Slieve Donard'

to light shade in deep, moisture-retentive and well-drained soils. Summer-flowering species bloom on wood made in the previous season and can be cut back after flowering to remove old, weak and overcrowded growth; those that flower late in the season are best cut back in spring. Propagate species by seed. For cultivars and hybrids, increase by softwood cuttings in early summer, by nodal, basal or heeled semi-ripe cuttings in summer, or by hardwood cuttings in autumn or early winter.

E.'Edinensis' (arching shrub to 2m tall; flowers deep pink); *E.*'Iveyi' (erect shrub to 3m; flowers fragrant, white opening from pink buds); *E. × langleyensis* (hybrids between *E.rubra* and *E.virgata*, including 'Apple Blossom', erect to 1.2m; flowers profuse, pale pink with white 'eyes', 'Donard Scarlet', erect and narrow to 2m, with scarlet flowers, and 'Slieve Donard', compact to 1.5m, with rose flowers); *E.leucantha* (Chile, Argentina; large shrub or small tree to 12m; flowers white in racemes to 5cm long); *E.rubra* (Chile; shrub to 4m tall; flowers pink to red; includes var. *macrantha*, a dense and vigorous shrub to 4m tall with darker flowers, and its cultivars 'Crimson Spire', erect to 2m, with very dark, glossy leaves and crimson-red flowers, and 'Woodside', dense, dwarf and spreading with intense red flowers); *E.virgata* (Chile, Argentina; deciduous shrub to 2m, with slender, arching branches; flowers small, white or palest pink).

escape (1) a garden plant that has become established in the wild; (2) a plant capable of surviving in adverse environmental conditions through adaptation of its life cycle; (3) a susceptible plant which fails to succumb to a prevalent disease.

Eschscholzia (for Johann Friedrich Eschscholz (1793–1831), German physician and naturalist). Papaveraceae. Western N America, naturalized in Europe. Some 10 species, annual or perennial herbs with glaucous and deeply dissected leaves. Solitary and long-stalked, the flowers are broadly funnel-shaped and consist of four to eight petals that fold together in parasol-fashion in dull weather. They are brilliantly colourful, the best blooms and strongest colours being obtained on poor, well-drained soils, where the plants may self-sow and overwinter after blooming. Sow seed *in situ*. Remove faded blooms to prolong flowering.

E.caespitosa (TUFTED CALIFORNIA POPPY; C California; tufted annual to 25cm, branching freely below; leaves more finely and narrowly divided than in *E. californica*, blue-grey; flowers to 5cm across, bright yellow); *E.californica* (CALIFORNIA POPPY; W US, naturalized elsewhere; annual or short-lived perennial to 60cm tall; leaves glaucous grey-green; flowers to 7cm across, typically yellow to orange; cultivars and seed races include dwarf and large plants, with single, semi-double or double flowers in shades of white, cream, yellow, orange, vermilion, flame, scarlet, salmon, rose, cerise, carmine and purple-bronze).

Escobaria (for the Mexican brothers Romulo and Numa Escobar, early 20th century). Cactaceae. W US, S Canada, N Mexico and Cuba. About 17 species of low-growing cacti, the stems simple or clustering, squatly globose to cylindric, tuberculate and spiny; the flowers are shortly tubular. Provide a minimum temperature of 5–7°C/40–45°F. Grow in a gritty compost of low to neutral pH, in full sun and low humidity; keep dry from mid-autumn until early spring, except for light misting on warm days in late winter. Increase by seeds or offsets. *E.vivipara* (syn. *Coryphantha vivipara*; S Canada, US, N Mexico; stems 6–15cm, densely grey-spiny; flowers 2–6cm, pink, purple or yellow).

esculent suitable for human consumption.

espalier strictly, a system of supports for training plants, especially fruit trees, consisting of horizontal wires stretched along a wall, or between upright posts (a variation correctly described as *contre espalier*). The wires are spaced 40–50cm apart. Conventionally, 'espalier' refers to the trained woody plant structure itself, which usually consists of a vertical stem with three or more horizontal tiers of branches, all in the same plane.

Espostoa (for Nicholas E. Esposto, Peruvian botanist, early 20th century). Cactaceae. S America (S Ecuador, Peru, Bolivia). 10 species, shrubs or small trees, the stems are succulent, columnar, many-ribbed and spiny; the flowers are tubular to campanulate, usually nocturnal, and develop at a distinct cap or patch of wool at the stem apex (cephalium). Grow in a gritty, slightly acid to neutral compost, providing a minimum temperature of 10–15°C/50–60°F. Shade in hot weather and maintain low humidity. Keep dry from mid-autumn until early spring, except for light misting on warm days in late winter. Seedlings are readily raised, and young plants of all species are valued for their attractive, very dense covering of fine spines and/or hairs. *E.lanata* (Ecuador, Peru; stems 4–7m × 15cm with 20–30 ribs, covered in long, silky hairs, spines yellow to black; flowers 4–8cm in diameter, white to purple, from a white to tawny cephalium).

estrade a type of trained tree in which the branches form successive horizontal tiers around the trunk.

ethnobotany the study of man's use of and relationships with plants, especially within a given community or race.

ethylene a colourless flammable gas (C_2H_4), of horticultural relevance because of its association with ripening fruits. For example, stored apples emit ethylene as their respiration rate increases, encouraging other fruits in store to ripen prematurely.

etiolated describing pale and/or spindly, elongated plant growths which are due to the exclusion of light.

Etiolation is employed to advantage in the process of blanching.

etiology (aetiology) the study of the causes of disease.

Etlingera (for the botanist A.E. Etlinger, (fl. 1774). Zingiberaceae. Sri Lanka to New Guinea. 57 species, rhizomatous, perennial herbs with cane-like stems. Scapose inflorescences arise from the rhizome: they consist of many, waxy, overlapping bracts, the lowermost spreading to reflexed like the tepals of a water lily or an overblown *Magnolia*, the uppermost forming a dense and torch-like cone. These bracts virtually conceal small flowers, each with a tubular, frilly lip and three petals. Grow as for *Alpinia*. *E.elatior* (TORCH GINGER, PHILIPPINE WAXFLOWER; syn. *Alpinia elatior*, *Alpinia magnifica*, *Nicolaia elatior*, *Nicolaia speciosa*, *Phaeomeria magnifica*; to 5m tall; inflorescence to 30cm across, standing 1m tall, bracts pale coral to deep flamingo pink; flowers white or yellow with a dark crimson lip).

Euanthe (from Greek *euanthes*, blooming well, referring to the spectacular inflorescence). Orchidaceae. Philippines. 1 species, *E.sanderiana* (syn. *Vanda sanderiana*), a robust epiphytic orchid to 1m tall. It closely resembles *Vanda*, with two ranks of strap-shaped leaves along an erect stem. Borne in an axillary raceme, the flowers are 6–10cm in diameter and flat with broad sepals: the dorsal sepal is rose, tinged white and variably spotted blood red; the lateral sepals are tawny-yellow, net-veined or flushed brick-red to rose; the petals are ovate to rhombic and dotted red; the short, fleshy lip is honey-coloured, and streaked, veined or stained purple-red. Cultivate as for *Vanda*.

Eucalyptus (from the Greek, *eu*, well, and *kalypto*, to cover, referring to the calyptra or lid that covers the flowers before opening). Myrtaceae. GUM. Australia, Malesia, Philippines. Over 500 species, evergreen trees and shrubs, usually oily and intensely aromatic. The bark flakes or peels, exposing the pale new surface beneath. Often tough and glaucous, the leaves vary greatly in form, from the circular to the narrow and willowy. Moreover, many have distinct juvenile phases where the leaves are quite different from those of the adult plant. The flowers consist of numerous stamens set on a top-shaped or cup-like hypanthium. Prior to opening, this is sealed by a lid or *calyptra*, derived from the calyx and corolla. The fruits are cylindrical to squatly spherical, woody, and may persist for many years. The following need full sun, shelter from harsh winds and a freely draining soil. They cope well with drought. *E. globulus*, *E. gunnii* and *E. pauciflora* subsp. *niphophila* are hardy in sheltered places in climate zone 6, and will usually regenerate if cut down by hard frosts. *E. coccifera*, *E. dalrympleana* and *E. perriniana* will survive outdoors in zone 8, or in favoured positions in zone 7. The remaining species need a cool greenhouse or conservatory. Increase by seed.

espalier

EUCALYPTUS

Name	Habit	Bark	Juvenile leaves	Adult leaves	Flowers
E.camaldulensis RIVER RED GUM	tree to 45m	smooth, white, grey, brown or red	ovate to broadly lanceolate, grey- to blue-green	8-30cm, narrowly lanceolate, thick, grey-green	cream, small
Comments: Z8/9					
E.citriodora LEMON-SCENTED GUM	tree, 25-50m	white, powdery, sometimes pink, red or blue-grey	narrowly to broadly lanceolate	8-16cm, lanceolate, lemon-scented when crushed	white
Comments: Z9					
E.coccifera TASMANIAN SNOW GUM	shrub or tree to 10m	smooth, white-grey, yellow or pink when fresh and peeling	broadly elliptic to rounded, somewhat glaucous	5-10cm, elliptic or lanceolate, pointed to scimitar-like, thick, grey-green, peppermint-scented	white
Comments: Z8					
E.dalrympleana MOUNTAIN GUM	tree to 40m	blotched white and grey to yellow-white, peeling to expose pink, green to olive patches	sometimes stem-clasping, rounded to ovate, light green to glaucous	10-20cm, narrowly lanceolate, tapering finely, shiny	white
Comments: Z8					
E.eremophila TALL SAND MALLEE	shrub to 4.5m	smooth, light brown or yellow-brown to grey-white	alternate, ovate	to 8cm, narrowly lanceolate, acuminate	white
Comments: Z9					
E.ficifolia FLOWERING GUM; RED-FLOWERING GUM	tree to 10m	rough, fibrous, shaggy	ovate to broadly lanceolate, bristly	8-15cm, broadly lanceolate to ovate, tapering, thick, glossy, dark green	showy, crimson
Comments: Z9					
E.forrestiana FUCHSIA GUM	tree or shrub to 5m	rough, grey, flaky to smooth and grey, exposed new bark grey-brown	alternate, ovate	to 9cm, lanceolate, tapering finely, shining deep green	white or red giving rise to red pear-shaped fruit
Comments: Z9					
E.globulus TASMANIAN BLUE GUM; BLUE GUM	tree to 70m	white to cream, yellow or grey, smooth then peeling in ribbons	ovate to lanceolate, grey-green, glaucous to resinous, opposed in stem-clasping pairs	12-25cm, lanceolate to falcate, glossy to dull mid-green	white
Comments: Z8. The most popular gum in British gardens after *E.gunnii* – very fast-growing with bolder, duller, rather rank foliagE.Young plants are often used in subtropical bedding schemes.					
E.gunnii CIDER GUM	tree to 25m	smooth, brown grey-green, peeling to expose pink to grey-white areas	neat, ovate to rounded, blue-grey to grey-green, glaucous, perfoliate or opposed in stem-clasping pairs	to 8cm, broadly elliptic to ovate, grey-green, glaucous	white
Comments: Z7. The most hardy and popular gum in British gardens; pruned hard or even stooled in spring, it produces a mop of fresh blue, juvenile growth – an ideal complement for silver-grey or purple foliage and pastel flowers in shrubberies and mixed borders					
E.macrocarpa	shrub to 5m, spreading or sprawling	smooth, grey	opposite, broadly elliptic to rounded, grey-green, glaucous	to 12cm, opposite, broadly ovate, glaucous	showy, red, pink or cream
Comments: Z9					
E.nicholii NARROW-LEAVED PEPPERMINT	tree to 15m, willowy with a light, airy crown	rough, fibrous, yellow-brown	small, linear, grey-green, crowded	6-13cm, narrowly lanceolate and tapering finely, blue- to grey-green, peppermint-scented	white
Comments: Z 8					
E.pauciflora GHOST GUM	tree or suckering shrub to 20m	smooth, white to ghostly grey or brown-red, falling away in patches and strips giving a mottled appearance	large, ovate to elliptic, thick, grey-green with a white bloom	to 20cm, lanceolate to falcate, blue-green	white
Comments: Z7. More common in gardens is the SNOW GUM, ssp. *niphophila* (syn. *E.niphophila*), to 6m tall with silvery grey, tough, ovate juvenile leaves and ovate to lanceolate, blue-green adult leaves, often with red margins and stalks. Hardier than the type, this gum has striking bark – a patchwork of white, ivory and pale rusty pink					

Clockwise from top left: *Eucalyptus macrocarpa, E. globulus, E. eremophila, E. pyriformis, E. pauciflora, E. forrestiana* (middle)

EUCALYPTUS Name	Habit	Bark	Juvenile leaves	Adult leaves	Flowers
E.perriniana SILVER DOLLAR GUM; SPINNING GUM	tree or multi-stemmed shrub to 8m	white to grey with brown patches	circular, perfoliate, bright silver-grey	to15cm, lanceolate to scimitar-shaped, pendulous, blue-grey	white
Comments: Z8. Especially showy when juvenile; sometimes hardy in sheltered areas					
E.pyriformis DOWERIN ROSE; PEAR-FRUITED MALLEE	shrub to 4.5m	smooth, grey to light brown	alternate, ovate-lanceolate to rounded	to 8cm, ovate to lanceolate, light green	red, pink, yellow or ivory from large pear-shaped buds
Comments: Z9					
E.viminalis MANNA GUM; RIBBON GUM	tree to 50m	smooth, grey, white or yellow-white and rough, fibrous and peeling on upper branches	lanceolate to cordate, sometimes stem-clasping	to 20cm, narrowly lanceolate, tapering finely, dark green	white
Comments: Z 9					

Eucharis (from Greek *eu*, good, and *charis*, attraction, in reference to the blooms). Amaryllidaceae. W Amazon and adjacent E Andes (Guatemala to Bolivia). 17 species and 2 natural hybrids, summer-flowering, bulbous evergreen perennial herbs, with dark green, shiny, ovate to broadly lanceolate stalked leaves. The flowers are carried in a scapose umbel; they are usually white and highly fragrant with six, spreading, ovate tepals. The six stamens are broadened at their bases and fused to form a cup. Grow in dappled sunlight or shade, with a minimum temperature of 10°C/50°F, in a fibrous loam-based medium, with added leafmould and sharp sand. Keep dry but not arid until growth commences early in the year, then water moderately, applying a balanced liquid fertilizer weekly when in full growth. Propagate by offsets.

E.amazonica (Ecuador, N Peru, S Colombia; scape to 70cm; flowers to 10cm diameter, white); *E.* × *grandiflora* (AMAZON LILY, EUCHARIST LILY, STAR OF BETHLEHEM; a natural hybrid from Colombia involving *E.amazonica* and very similar to it, usually differing in its slightly smaller flowers).

Eucomis (from Greek *eukomos*, lovely-haired, referring to the crown of bracts at the apex of the inflorescence). Liliaceae (Hyacinthaceae). PINEAPPLE LILY. Africa (those listed below are from South Africa). Some 15 species, robust, bulbous perennials with strap-shaped to broadly lanceolate, smooth leaves. The flowers are small, usually green, and star-shaped. They appear in late summer, packed in dense, cylindrical spikes; these are carried atop stout scapes and crowned with a tuft of leaf-like bracts. The following species favour a minimum temperature of 4–10°C/39–50°F, but may be tried outdoors in mild regions of zones 7 and over, and either lifted in winter or grown in a very sheltered site and protected with a heavy dry mulch. Plant about 15cm deep in autumn, in a sunny position, in rich, well-drained soil. In pots,

grow one bulb to a 12.5cm pot or three bulbs to a larger-sized pot, with bulb tips just showing. Give bright, filtered light and good ventilation. Water sparingly when dormant, plentifully while in growth. Propagate from seed sown in spring, or by offsets.

E.autumnalis (syn. *E.undulata*; to 50cm; leaves with wavy margins; flowers white or pale olive at first becoming deeper green, bracts wavy); *E.bicolor* (to 60cm; leaves with wavy margins; scape flecked maroon, flowers edged purple, bracts wavy, edged purple); *E.comosa* (to 70cm; leaves with wavy margins, spotted to striped purple beneath; scape spotted purple; flowers purple-tinted, bracts sometimes edged or spotted purple); *E.pallidiflora* (to 75cm; leaves with crispate margins; flowers green-white, bracts crispate).

Eucommia (from the Greek *eu*, well, and *kommi*, gum, from the rubber-like latex produced by this tree). Eucommiaceae. C China. 1 species, *E.ulmoides*, the GUTTA-PERCHA TREE, a deciduous tree to 20m tall. The leaves are 7–15cm long, narrowly ovate to elliptic and finely toothed. Hairy at first, they soon become glossy, dark green and will exude threads of latex if broken. The flowers lack petals and are borne in axillary clusters with or before leaves; the males consist of numerous red-brown anthers. Hardy to −15°C/5°F or lower. Grow in full sun with shelter from cold winds. Propagate from seed sown in autumn, or from softwood cuttings.

Eucryphia (from Greek *eu*, well, and *kryphios*, covered; the sepals form a cap over the flower bud). Eucryphiaceae. Chile, SE Australia (Tasmania, New South Wales, Victoria). 6 species, evergreen trees or shrubs (*E.glutinosa* is deciduous in cultivation), with tough and often glossy, simple or pinnate, finely toothed leaves. Often fragrant, the solitary flowers are white and appear in summer and early autumn.

Eucryphia glutinosa

Generally, they consist of four broad petals and numerous fine stamens with pink anthers. They are best grown in a moist, but well-drained acid soil; *E.cordifolia* and *E.* × *nymansensis* will grow on alkaline soils. Ideally, site among other trees and shrubs which will shade the base from hot sun and keep the soil cool and moist, but allow the crown to grow into the light. Most species are hardy in climate zone 7, although protection from cold winds is essential. Pruning is generally not required. Propagate by semiripe cuttings in late summer, by seed sown when ripe or by layers or suckers.

E.cordifolia (ROBLE DE CHILE, ULMO; C and S Chile; columnar tree to 20m; leaves to 7cm, simple oblong; flowers to 5cm diameter); *E.glutinosa* (NIRRHE; C Chile; upright tree or shrub to 10m; leaves to 5cm, leaflets 3–5, elliptic to oblong, colouring well in autumn if deciduous; flowers to 6cm diameter); *E.lucida* (LEATHERWOOD, PINKWOOD; Tasmania; erect shrub or tree to 8m; leaves to 5cm, juveniles with 3 leaflets, adults simple, oblong to lanceolate; flowers 3–5cm diameter, nodding; includes pink- and cream-flowered cultivars); *E.milliganii* (Tasmania; differs from *E.lucida* in its narrowly columnar habit, smaller leaves and cup-shaped flowers to 2cm diameter); *E.* × *nymansensis* (*E.cordifolia* × *E.glutinosa*; dense, vigorous, columnar tree to 15m tall; leaves simple or with 3 leaflets to 8cm long; flowers to 7.5cm diameter).

Euonymus (from the ancient Greek name *euonymon dendron*, hence to Latin meaning 'of good name' – an ironic allusion to its toxicity). Celastraceae. Asia, Europe, N and C America, Madagascar, Australia. Over 170 species, deciduous or evergreen shrubs or trees with inconspicuous, green-white to yellow flowers. These give rise to clusters of 3–5-valved, shell- or envelope-like fruit, splitting to reveal bead-like, white to coral or orange-yellow seeds, enclosed in fleshy, colourful arils. Evergreen species are hardy in climate zone 6, deciduous species in zone 4. Plant all in well-drained, moist and fertile soils in full sun or light shade. Evergreens used for hedging or edging may be trimmed in spring. Give support to climbing cultivars where required. Increase by nodal cuttings taken between summer and autumn or by fresh seed sown under glass or in the cold frame.

E.alatus (WINGED SPINDLE TREE; N Asia; much-branched, deciduous shrub to 3m; branches with 4, deep and corky wings; leaves 2–7cm, ovate to elliptic, serrate, pale green turning crimson to brilliant scarlet in autumn; fruit pale red with orange arils); *E.americanus* (STRAWBERRY BUSH; E US; erect, deciduous shrub to 3m; branches 4-angled; leaves to 8cm, narrowly ovate to lanceolate, tough, persisting well into autumn; fruit pink, warty, aril bright red); *E.europaeus* (COMMON SPINDLE TREE; Europe to W Asia; deciduous shrub or small tree to 7m; branches with corky stripes; leaves to 8cm, ovate to oblong, mid-green, finely toothed, turning scarlet in autumn; fruit pink to bright red, arils orange; includes dwarf and large cultivars, some with leaves pale yellow-green, variegated or turning very deep purple-red in autumn, some with orange-red fruit); *E.fortunei* (China; evergreen, slender-stemmed shrub, procumbent or climbing to 5m; leaves to 6cm, ovate to elliptic, finely toothed, leathery and glossy dark green; fruit ivory to yellow, arils orange to pink; a variable and much-developed species including selections with erect and bushy, vigorously climbing, dwarf and compact or densely ground-smothering habits, the leaves ranging in size from 1 to 6cm, in colour from dark green assuming purple-red autumn tints, to white-, cream-, or yellow-variegated); *E.hamiltonianus* (Himalayas to Japan; deciduous shrub or small tree, similar to *E.europaeus* but with smooth branches and longer, narrower and thicker leaves; fruit pink, arils red; includes cultivars with good autumn colour and bright coral fruit); *E.japonicus* (China, Korea, Japan; erect, evergreen shrub or small tree to 8m; leaves to 7cm, elliptic to oblong, bluntly and finely toothed, tough, smooth, glossy deep green; fruit pink, arils orange; a popular and much-developed species often used for hedging and edging, it includes many cultivars ranging in habit from the dwarf and box-like to the tall and shrubby, in leaf form from very small and narrow to large and broad, in colour from grey-green to deep emerald, edged, zoned or splashed white, cream or yellow); *E.obovatus* (RUNNING STRAWBERRY BUSH; N America; slender-stemmed, deciduous shrub, prostrate and rooting, or climbing if supported; leaves to 6cm, obovate to elliptic, crenate, pale green; fruit carmine, warty, aril red; 'Variegatus' has white-marked leaves).

Eupatorium (for Eupator, King of Pontus, who used one species as an antidote for poison). Compositae. E US, Eurasia. JOE PYE WEED, THOROUGHWORT, BONESET. About 40 perennial herbs, subshrubs or shrubs with many, button-like flowerheads packed in terminal panicles or corymbs in late summer and autumn. Fully hardy. Grow in any moderately fertile, well-drained but moisture-retentive soil in sun or part shade. Propagate by seed or division. *E.purpureum* (JOE PYE WEED, TRUMPET WEED; E US; purple-tinged, perennial herb 1–3m tall; leaves whorled, to 25cm, ovate to lanceolate, vanilla-scented when bruised; flowerheads pale pink to green-yellow or rose-purple).

Euphorbia (named for Euphorbus, physician to the King of Mauretania). Euphorbiaceae. SPURGE. Cosmopolitan. At least 2000 species, ranging greatly in habit, from small, weedy annuals, to herbaceous perennials, shrubs and succulents. All have milky, toxic sap and bear small flowers in cyathia. A cyathium is a small cup composed of bracts. It is sometimes furnished with glands and subtended by other, larger bracts. Inside the cyathium are found several 'stamens', each equivalent to a male flower, and a stalked ovary equivalent to the female. The cyathia themselves are often disposed on radiating branches and surrounded by showy bracts.

The succulent species need a minimum temperature of 10°C/ 50°F, full sun and low humidity. Grow in a gritty and sandy, loam-based medium. Water and feed moderately in warm weather; keep more or less dry at other times, watering only to prevent shrivelling of stems.

The herbaceous perennial species are hardy in climate zone 6. Evergreens and semi-evergreens with glaucous, fleshy leaves need a sunny, sheltered site on a fast-draining, gritty soil. They are excellent plants for rock, gravel and silver gardens. Other perennials will grow in a range of situations, from sun to light shade, and prefer a damp and fertile, humus-rich soil. That said, *E. amygdaloides* var. *robbiae* and *E. cyparissias*

will tolerate full sun and impoverished, drought-stricken clay, often spreading to the point of nuisance. The semi-shrubby *E. characias* is fully hardy and will grow in sun or shade. The shrub *E.mellifera* is tender in climate zone 7 unless grown in a sheltered position.

Annuals and biennials should be sown *in situ* in spring on well-drained soil in sun. *E. lathyris* is unlikely to be welcome in any garden, but *E. marginata* is highly desirable, a fine cut flower and an excellent filler for white borders.

E. fulgens and *E. pulcherrima* (poinsettia) need a minimum temperature of 5°C/40°F, full sun and a dry, buoyant environment. Both will become sizeable, if rather sparse, shrubs in time, but they are usually grown as pot plants, repropagated every few years. Keep poinsettias moist and well-fed during growth and flowering. After flowering (usually late winter), reduce the water until leaves and bracts drop. Cut back the stems. Keep the plants more or less dry and in a cool, shaded place. Repot in midsummer. Increase light, temperature, water and feed as the nights grow longer. Plants should be in flower by Christmas.

Increase succulents, glaucous-leaved perennials and shrubs by stem cuttings; herbaceous perennials by division in early spring; annuals and biennials by seed. The milky sap is toxic and irritant: care should be taken when handling all species.

EUPHORBIA Name	Distribution	Habit	Leaves	Inflorescences
E.abyssinica	Ethiopia, Eritrea, Somalia	tree to 10m tall with candelabra-like branches each thickly succulent with several deep, wavy and spiny wings	absent	dense, inconspicuous
Comments: Z10				
E.amygdaloides WOOD SPURGE	Europe, W Asia	evergreen perennial herb to 50cm with clumped and suckering stems	5-12cm, spathulate to narrowly obovate, dark green, more or less hairy at first, then smooth	lime green to bright yellow in an erect to nodding, terminal spike amid leaf-like bracts in early spring
Comments: Z5. includes 'Rubra', a compact and bushy plant with stems and foliage flushed ruby red to deep purple-red, especially in winter and early spring, and bright lime green flowers; also, the popular subsp. *robbiae* (syn. *E.robbiae*), a freely spreading plant with broader, smooth and dark green leaves in a ruff-like arrangement and tall, loose spikes of lime to pale gold flowers				
E.candelabrum	NE Africa to S Africa	tree to 20m with candelabra-like branches, each succulent and with deep, spiny and dark-toothed wings	small, short-lived	yellow-green
Comments: Z10				
E.caput-medusae MEDUSA'S HEAD	S Africa	succulent perennial to 30 x 80cm with a stout, more or less buried base from which radiate numerous low-lying branches, each cylindrical, snaking and covered in cone-shaped tubercles	narrow, small, short-lived	each with a white fringe, packed in terminal heads
Comments: Z10				
E.characias	W Mediterranean, Portugal	evergreen perennial herb to 1.5m with numerous clumped stems arising from a woody base	10-22cm, linear to oblong or narrowly oblanceolate, sea green to grey-blue, softly and minutely downy	small, bell-shaped, yellow-green with deep maroon, eye-like glands and sometimes rims, in a dense and branching terminal head from spring to summer
Comments: Z5. Includes the taller ssp. *wulfenii* with grey-green leaves and bright lime flowerheads with yellow glands; 'Burrows Silver', smaller with white-variegated leaves and creamy flowerheads and 'Portuguese Velvet', compact with crowded, short and broad, velvety leaves				

EUPHORBIA

Name	Distribution	Habit	Leaves	Inflorescences
E.cyparissias CYPRESS SPURGE	Europe, naturalized N America	perennial herb, fast spreading with erect and branching annual stems to 30cm	1.5-3cm, linear to oblong, soft sea green or bright green, giving the branches the appearance of feathery cylinders; autumn colour yellow to fox red	small, lime green to bright gold in frothy, branching heads in spring

Comments: Z5. 'Fens Ruby' : small, bushy plant with new growth tinted copper red to soft bronze

Name	Distribution	Habit	Leaves	Inflorescences
E.dulcis	Europe	perennial herb to 40cm tall, with erect, branching annual stems arising from a thick rootstock	3-8cm, oblong to oblanceolate, mid green, smooth, soft-textured	yellow in broad, branching heads, glands and bracts sometimes tinted red or purple, in early summer

Comments: Z6. 'Chameleon': leaves and stems flushed solid dark chocolate to purple-red; self-seeds freely and comes true from seed

Name	Distribution	Habit	Leaves	Inflorescences
E.fulgens SCARLET PLUME	Mexico	evergreen shrub to 2m with slender, arching branches	to 16cm, narrowly lanceolate, dark green	small, 'flower'-like with rounded 'petals', bright scarlet in dense axillary clusters throughout the year

Comments: Z10. Includes forms with white, cream, yellow, orange and pink flowers. A popular cut flower.

Name	Distribution	Habit	Leaves	Inflorescences
E.griffithii	E Himalaya	clump-forming perennial herb with erect, branching stems to 1m	8-18cm, narrowly oblong to lanceolate, dark green with a paler midrib often tinted pink, upper leaves and bracts flushed pink to glowing orange-red	copper to fiery red or orange, in dense terminal heads in spring

Comments: Z5. Includes 'Dixter', a compact plant with luxuriant, dark foliage tinted bronze and pink and orange inflorescences, and 'Fireglow', a tall plant with bronze-pink new foliage and bright fiery red inflorescences

Name	Distribution	Habit	Leaves	Inflorescences
E.lathyris CAPER SPURGE, MYRTLE SPURGE, MOLE PLANT	cosmopolitan	biennial herb to 1.5m tall; stem erect, unbranched below, extensively branched above in a broad inflorescence	6-18cm, oblong to lanceolate, sea green with paler midribs, smooth, angled downwards in four distinct ranks along stem	yellow-green with triangular bracts on the forking stems of a large, terminal head in summer

Comments: Z5. A widespread weed, sometimes encouraged for the striking regularity of its growth habit and its curious inflorescences. Also said to deter moles. The fruits resemble capers (but are poisonous).

Name	Distribution	Habit	Leaves	Inflorescences
E.mammillaris CORN COB, CORKSCREW	S Africa	succulent shrub or shrublet to 30cm with 5-sided, dense and erect branches to 6cm thick and covered in tubercles and spines	absent	small and green-maroon, the stalks of dead flowerheads persisting as spines to 1cm long

Comments: Z10. 'Variegata': stems marbled green-white, bright green and tinted pink when young

Name	Distribution	Habit	Leaves	Inflorescences
E.marginata SNOW ON THE MOUNTAIN, GHOST WEED	N America	softly and minutely hairy annual, erect to 1m tall	to 8cm, ovate to oblong, grey-green, upper leaves and inflorescence bracts edged bright white	white, surrounded by wholly white or white-edged bracts in terminal heads from spring to autumn

Comments: Z5. A good cut flower

Name	Distribution	Habit	Leaves	Inflorescences
E.martinii (E.amygdaloides × E.characias)				

Comments: A natural hybrid including plants intermediate between both or favouring either of the parents. The plants are usually compact with ruffs of narrow leaves and columns of flowerheads; these tend to be cupped as per E.characias, but smaller and bright yellow or lime, often with dark wine red glands and margins. Z5.

Name	Distribution	Habit	Leaves	Inflorescences
E.mellifera	Madeira, N Tenerife, La Palma	evergreen shrub to 2m or tree to 3m with stout, olive, candelabra-like branches and a dense, broad crown	8-18cm, oblong to bluntly lanceolate, smooth, mid to deep green with a white to pale green midrib	honey-scented, yellow-green with maroon glands in crowded heads in early summer

Comments: Z7. A handsome if rather tender shrub almost extinct in the wild and richly deserving a sheltered, sunny position in the garden

Name	Distribution	Habit	Leaves	Inflorescences
E.meloformis MELON SPURGE	S Africa	dwarf succulent to 10cm tall with a more or less spherical stem with 8 to 12 facets; dark olive with obscure transecting bands of deep green and purple-green	inconspicuous and soon lost	small, yellow-green, borne on very short stalks at the sunken stem apex

Comments: Z9. An easy succulent, often confused with E.obesa, from which it differs in its persistently spherical stem with a sunken top.

Name	Distribution	Habit	Leaves	Inflorescences
E.milii E.splendens CROWN OF THORNS	Madagascar	erect to scrambling shrub to 1m tall with grey, viciously spiny branches	2-6cm, obovate to oblong, dark green	slender-stalked with cupped, scarlet bracts and yellow glands, produced throughout the year

Comments: Z9. A popular potplant with numerous cultivars, the inflorescences in shades of scarlet, flame, peach, pink, yellow, ivory and white

EUPHORBIA

Name	Distribution	Habit	Leaves	Inflorescences
E.milii *E.splendens* CROWN OF THORNS	Madagascar	erect to scrambling shrub to 1m tall with grey, viciously spiny branches	2-6cm, obovate to oblong, dark green	slender-stalked with cupped, scarlet bracts and yellow glands, produced throughout the year
Comments: Z9. A popular potplant with numerous cultivars, the inflorescences in shades of scarlet, flame, peach, pink, yellow, ivory and white				
E.myrsinites	S Europe, Asia Minor	evergreen perennial herb with a stout stock from which arises a crown of succulent trailing stems, each to 30cm long	to 3cm, obovate with a short, pointed tip, fleshy, blue-grey, glaucous, densely and spirally arranged	sulphur-yellow, on branching stems in dense terminal heads from late winter to mid spring
Comments: Z6. A superb plant for sunny situations on the rock and dry garden. *E.rigida* and *E.seguieriana* are similar but have narrower leaves.				
E.nicaeensis	S, C, E Europe	evergreen perennial to 80cm with clumped, red-tinted stems arising to 80cm from a woody stock	to 8cm, oblong to lanceolate, grey-green, tough to thinly fleshy, blunt	yellow-green in broad heads in summer
Comments: Z6.				
E.obesa GINGHAM GOLFBALL	S Africa	succulent perennial to 30cm tall with a 6-facetted stem, spherical at first, becoming stoutly cylindrical with a bluntly pointed apex, dull green with obscure, transecting bands of pale olive, grey or dull purple	inconspicuous, short-lived or absent	small, more or less stalkless, yellow-green
Comments: Z9. A popular succulent; cf *E.meloformis*.				
E.palustris	Europe	deciduous perennial resembling a taller, smoother, cleaner *E.polychroma*, with clumped stems sparsely branching above and erect to 1.25m	to 8cm, elliptic to oblong or lanceolate, mid green with a paler midrib, upper leaves sometimes turning rusty red or orange in autumn	bright sulphur yellow or pale gold in large, domed heads in spring
Comments: Z5. A robust, handome spurge suitable for damp, even waterlogged sites.				
E.polychroma *E.epithymoides*	C & SE Europe, Asia Minor	deciduous perennial herb with a bushy clump of erect, unbranched or sparsely branched stems to 40cm tall	3-6cm, obovate to elliptic or oblong, soft and minutely hairy at first, bright green often with a purple blush, becoming mid green	sulphur yellow to bright gold in broad, cap-like heads
Comments: Z5. Includes cultivars with bright gold and purple-tinted inflorescences.				
E.portlandica PORTLAND SPURGE	W Europe	neat, deciduous or evergreen biennial or perennial herb with clumped, bushy stems erect to 30cm	0.5-2cm, oblong to oblanceolate, bluntly pointed, thinly fleshy, sea green to blue-green sometimes flushed red	in loose terminal clusters with triangular, sea-green bracts and yellow-orange glands
Comments: Z5. Provide full sun and a well-drained soil.				
E.pulcherrima POINSETTIA	Mexico	deciduous shrub to 3m, typically gaunt and flowering on bare branches, but usually seen as a potplant, short, leafy and bushy	10-20cm, ovate to elliptic with jagged or sinuate, pointed lobes or teeth	small and green-yellow in terminal clusters surrounded by very showy, leaf-like bracts to 20cm long and ranging in colour from dark scarlet to flamingo pink, peach, cream or white
Comments: Z9. A popular houseplant, grown for display at Christmas and available in a range of colours and sizes				
E.rigida *E.biglandulosa*	S Europe to Caucasus	evergreen perennial herb resembling an erect *E.myrsinites* to 40cm tall	to 5cm, lanceolate, tapering to a sharp point, tough, fleshy, pale grey-blue, glaucous, sometimes tinted rose	golden yellow sometimes with maroon glands, in broad terminal heads at various times of year; in full sun and with age, the bracts become amber or scarlet with glowing orange-red tints
Comments: Z7. Plant in full sun on a sharply drained soil				
E.schillingii	Nepal	deciduous, erect, clump-forming perennial to 1m tall	to 13cm, oblong to bluntly lanceolate, fresh emerald green with a white midrib, pale veins and a thin, clear red margin	bright yellow in broad, rounded terminal heads in spring and summer
Comments: Z6				
E.seguieriana	SW to C and E Europe, W Asia	evergreen perennial herb resembling *E.myrsinites* but with looser, narrower and less rigid foliage	to 7cm, narrowly linear to oblong, finely tapering, thinly fleshy, blue green, glaucous, forward-pointing	bright sulphur yellow in broad, branching terminal heads, usually from mid summer
Comments: Z7. subspecies ***niciciana*** is a smaller, free-flowering plant with dense, spreading leaves; these are usually linear, sharply pointed and to 2cm				

EUPHORBIA

Name	Distribution	Habit	Leaves	Inflorescences
E.sikkimensis	E Himalaya	differs from *E.griffithii* in its coral pink, ruby-veined new growth and the infloresecnce bracts, which are yellow, not orange or red		
Comments: Z6				
E.wallichii	Himalaya	deciduous perennial herb with erect, clumped stems to 50cm	to 12cm, narrowly elliptic to oblong, dark jade green with a white midrib and fine, purple-tinted margins	terminal, yellow-green with large, green glands
Comments: Z6				

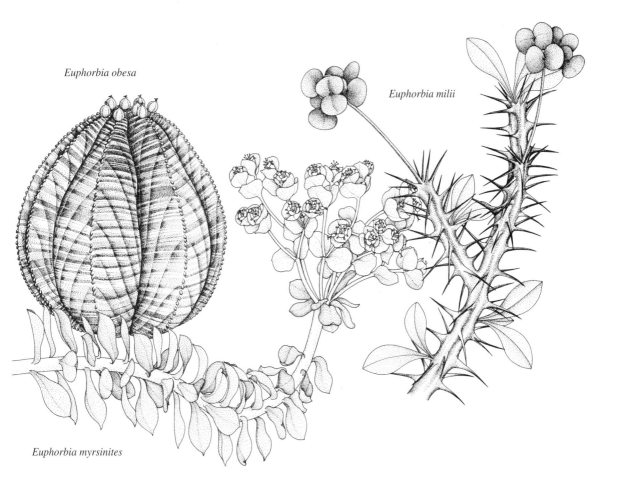

Euphorbia obesa

Euphorbia milii

Euphorbia myrsinites

Euptelea (from Greek *eu*, well, and *ptelea*, elm, referring to the edible fruit). Eupteleaceae. E Asia. 3 species, deciduous trees and shrubs, usually with broadly pyramidal crowns. The leaves are toothed and often tinted red on emergence. Small and lacking sepals and petals, the flowers are produced before the leaves in clusters along shoots and consist of numerous stamens with red anthers. Hardy in zone 6 and suitable for open situations in the woodland garden or shrub border. Propagate by fresh seed sown as soon as it becomes ripe, or by layers.

E.pleiosperma (W China; to 10m; leaves to 12cm, broadly ovate, apex tapering, with shallow, regular teeth, colouring red in autumn); *E.polyandra* (Japan; to 7m; leaves to 12cm, broadly ovate with deep, irregular teeth, colouring red and yellow in autumn).

Eurya (from Greek, *eury*, broad, perhaps referring to the broad petals). Theaceae. S and E Asia, Pacific Islands. Some 70 species, mostly evergreen trees and shrubs, usually with a herringbone branch pattern and tough, dark green and crenately toothed leaves.

Small, 5-petalled flowers with many stamens appear in summer. These are sometimes followed by small berries. Hardy in zone 8, given a protected site. Where winters are more harsh, grow in the cool greenhouse or conservatory. Plant in sun or part shade, with shelter from cold, drying winds. Protect from frost when young. Under glass, water plentifully when in full growth. Keep just moist in winter with a minimum temperature of 5°C/40°F. Propagate by seed sown fresh or in spring, or by semi-ripe cuttings in late summer.

E.emarginata (Japan; shrub or small tree; leaves 2–3.5cm, obovate to oblong, emarginate; flowers yellow-green; fruit purple-black); *E.japonica* (E Asia; shrub or tree to 10m; leaves 3–8cm, oval to obovate, apex bluntly acute; flowers white; fruit black; includes 'Variegata', with leaves edged creamy white, and 'Winter Wine', with leaves turning burgundy in autumn; sometimes confused in cultivation with *Cleyera japonica*).

Euryops (from Greek *euryops*, having large or wide eyes, referring to the broad discs of some species). Compositae. Arabia, Africa (those listed below are from South Africa). About 100 species, evergreen shrubs and subshrubs to perennial or annual herbs with yellow daisy-like flowerheads in summer.

The following species are suitable for rock gardens, containers and raised beds. They need excellent drainage and some shelter (for example, a south-facing wall). They will withstand temperatures as low as –13°C/9°F. Grow in full sun on a gritty, neutral to slightly acid soil. Trim over after flowering if necessary. Propagate by seed or by greenwood or semi-ripe cuttings with a heel.

E.acraeus (syn. *E.evansii* of gardens; dense rounded shrub to 1m tall; leaves to 3cm, narrowly oblong, tip 3-toothed, blue-green, glaucous); *E.pectinatus* (vigorous shrub 1–2m tall, often grey-hairy; leaves to 10cm, pinnately lobed, silvery).

Eustoma (from Greek *eu*, good, and *stoma*, mouth, referring to the beautiful corolla throat). Gentianaceae. Southern N America to northern S America. 3 species, annual, biennial or perennial herbs usually with glaucous, ovate to linear or lanceolate leaves in opposite pairs. The long-stalked flowers are solitary or in loose, erect panicles. The corolla is funnel-shaped to campanulate with five to six, oblong or obovate and satiny lobes. Beautiful plants for the cool greenhouse or conservatory, where they need a minimum winter temperature of 5°C/40°F. Treat as biennials. Raise from seed in autumn and prick out into a sandy, soilless mixture. Grow on in bright, well-ventilated conditions. Cool temperatures with good light and fresh air are vital; overheating at the roots is disastrous. The slender stalks may require support. Protect blooms from strong sunlight.

E.exaltatum (S US, Mexico, West Indies; glaucous annual or short-lived perennial to 60cm; flowers to 4cm deep, white, blue, pink or purple); *E.grandiflorum* (syn. *Lisianthius russellianus*;

*Eustoma
grandiflorum*

PRAIRIE GENTIAN; US, Mexico; erect, glaucous annual or perennial to 60cm; flowers to 10cm deep; a popular cut flower available in shades of white, cream, pale yellow, peach, pink, deep rose, purple, indigo and blue, often with a darker or far paler central patch or margins).

eutrophication the increase of algal and bacterial growth in a pond or other receptacle of water, which occurs as a result of nutrient enrichment.

evapotranspiration the combined loss of water from evaporation and transpiration.

evergreen a plant that retains green leaves through more than one growing season, as opposed to a deciduous plant.

everlasting flowers that retain their form for a very long period after cutting, especially if carefully dried, for example *Helichrysum* species. Many belong to the Compositae, with papery petals or bracts; grasses and prepared foliage such as beech are also included in the description. Also referred to as immortelle

evolute unfolded.

Exacum (from Latin *ex*, out, and *agere*, to drive, referring to the purgative properties of members of the genus). Gentianaceae. Old World tropics and subtropics. 25 species, annual, biennial or perennial herbs with small, salverform to rotate flowers in cymes. A free-flowering pot plant for the home or conservatory, where it needs a minimum winter temperature of 10°C/50°F. Sow seed in late summer, at 15°C/60°F, and prick out into a soilless mix. Keep moist but never wet; provide good light and a weak liquid feed. Increase also by cuttings in autumn. *E.affine* (GERMAN VIOLET, PERSIAN VIOLET; S Yemen, naturalized elsewhere; densely bushy annual or short-lived perennial, 20–60cm; leaves small and neat; flowers 0.8–2cm diameter, fragrant, sky blue to pale violet or rich purple with yellow anthers; includes dwarf and large cultivars and flowers in various shades of mauve, blue, pink and white).

excrescence an outgrowth or abnormal development.

excurrent (1) projecting beyond the margin or apex of its organ, for example a midvein terminating in an awn; (2) a growth habit where the primary axis remains dominant and recognizable throughout life, with the branches secondary to it – as in a strongly single-stemmed tree with a remote, tiered branching pattern.

exfoliating peeling off in thin layers, shreds or plates.

exo- prefix meaning outward.

exocarp the outermost wall of a pericarp.

Exochorda (from Greek *exo*, outside, and *chorde*, a cord, referring to the fibres outside the placenta). Rosaceae. China. PEARLBUSH. 4 species, deciduous shrubs with entire or serrate, pale green, and usually softly hairy leaves. Showy, white flowers are produced in spring in erect to spreading racemes; they consist of five, obovate petals and numerous stamens. Fully hardy. Grow in sun or dappled shade. Prune after flowering if necessary, to reduce overcrowding or to remove weak growths. Propagate by simple layering, by removal of rooted suckers, or by seed sown in autumn. Alternatively, increase by semi-ripe cuttings in a closed case, or basal softwood cuttings in spring – these should be wounded and treated with hormone rooting preparation, and grown under mist or in a closed case with bottom heat.

    E.giraldii (China; to 3m, spreading; flowers to 2.5cm diameter in racemes to 8cm); *E.* × *macrantha* (*E.korolkowii* × *E.racemosa*; erect to 2m; flowers to 3cm diameter; includes the popular cultivars 'Irish Pearl' (syn. 'The Pearl'), with very long, spreading flowering shoots, leaves pale green above and tinted blue-grey beneath, and 'The Bride', compact with arching to spreading branches, pea green leaves and large flowers produced in great abundance); *E.racemosa* (China; to 3m; flowers to 4cm diameter, in racemes to 10cm long).

exotic a plant introduced from its natural habitat into another. It is often loosely used of showy trop-ical specimens, but the term is not necessarily linked to hardiness.

explanate spread out, flat.

explant in micropropagation, a tiny portion of a plant which is inserted in an aseptic growing medium.

exserted obviously projecting or extending beyond the organs or parts surrounding it, as with stamens that stick out from the corolla; cf. *included*.

extension shoot, extension growth a shoot, especially on a fruit tree, which has been selected to extend the branch framework. It is more often called a leader.

extra- prefix meaning outside; for example extra-floral, outside the flower proper.

extrorse turned or facing outwards, often used of the dehiscence of an anther; cf. *introrse*.

eye (1) a latent growth bud, especially on tubers such as potatoes; (2) the centre of a flower when it is coloured differently from the rest, as in many pinks, pansies, auriculas and delphiniums; (3) the centre of primula flowers, which may be described as pin-eyed or thrum-eyed according to whether the stigma or stamens are visible at the top of the corolla tube. See *heterostyly*.

F

F$_1$ the first generation progeny of a cross. The F$_1$ hybrid relates to seed derived from two highly selected, inbred parent lines and is of particular value in vegetables and flowers, for the desirable characteristics of uniformity, vigour and high yield. F$_1$ hybrids will not breed true from their own seed.

F$_2$ the progeny from self- or cross-pollinated F$_1$ plants. F$_2$ hybrids are less uniform and vigorous than their parents, but often have desirable characteristics.

Fabiana (for Francisco Fabian y Fuero (1719-1801), Spanish botanist). Solanaceae. Warm temperate S America. Some 25 species, heath-like, evergreen shrubs. Borne in early summer, the flowers are tubular with a limb of five, short lobes. Hardy to about -5°C/23°F if given a sheltered position at the base of a warm, south-facing wall. Plant in sun on light, well-drained soils. Prune in spring to shape straggly older plants. Propagate from seed sown in a sandy medium in spring, or by greenwood cuttings rooted with gentle bottom heat in summer. *F.imbricata* (Chile; erect to 2m with many short, downy lateral branches and minute, dark green, overlapping leaves; flowers to 1.5cm long, white to rose – pale mauve to blue-violet in f. *violacea*).

facultative saprophyte a pathogen that can grow on dead organic matter as well as on living hosts; cf. *obligate parasite*.

Fagus (the Classical name for this tree). Fagaceae. Northern temperate regions. beech. 10 species, deciduous trees, usually with smooth, grey bark, slender young shoots and narrowly ellipsoid buds. Oblong to elliptic or ovate, the leaves are alternate, 2-ranked, entire to dentate and shiny. Small, male flowers are borne in slender-stalked heads. The female flowers give rise to cup-like, thinly woody casings, often covered in soft spines and splitting open in four sections to release small, chestnut brown triangular nuts.

The following species are fully hardy and tolerant of most fertile, well-drained soils and sites. They perform poorly, however, in coastal exposures and drought-stricken areas. They make fine specimen trees for large gardens; *F.sylvatica* is also an excellent hedge plant, and will retain its withered foliage in winter if juvenility is simulated by pruning. The US and Chinese species may grow only slowly in the cool summers of Britain while doing well in the eastern US. Propagate as for *Quercus*, although the seeds are a little more tolerant of drying, and the seedlings do not produce a taproot. Plant out when 2–4 years old and 30–60cm tall. Seed should be from extensive single-species stands or wild-collected if identity is to be guaranteed. Cultivars are grafted on seedlings of *F.sylvatica*. Young trees are very shade-tolerant and benefit from shelter for the first few years; in open ground, they can be damaged by frost.

F.grandifolia (syn. *F.americana*; AMERICAN BEECH; eastern N America; to 35m tall; leaves 6–15cm, obovate to oblong, serrate, blue-green above, paler beneath, turning golden in autumn); *F.orientalis* (ORIENTAL BEECH; SE Europe to W Asia; to 40m; leaves 8–17cm, elliptic-oblong to obovate, dentate, dark green turning yellow in autumn); *F.sylvatica* (COMMON BEECH, EUROPEAN BEECH; Europe; to 48m; leaves 5–11cm, elliptic-ovate, wavy-edged, pale and softly downy at first, becoming deep, glossy green and turning yellow to rusty brown in autumn; includes f. *laciniata*, CUT- OR FERN-LEAF BEECH, with deeply cut leaves, f. *pendula*, WEEPING BEECH, with strongly weeping branches, and f. *atropurpurea*, COPPER BEECH, with purple-bronze leaves; cultivars are available which display these traits in combination, for example 'Rohanii' with deeply cut, purple leaves; two further important trends in variation are represented by clones with yellow new foliage, for example 'Aurea Pendula', 'Zlatia', GOLDEN BEECH, and by clones with strongly fastigiate habit, for example 'Dawyck', DAWYCK BEECH, the purple-leaved 'Dawyck Purple' and the purple-, cut-leaved 'Rohan Obelisk').

fairy ring a ring of toadstools formed in lawns by fruiting bodies of one of a number of fungi. In the UK, fairy rings are most commonly formed by the fungus *Marasmius oreades*. The ring of dead grass at the edge of the circle is usually surrounded by rings of darker green grass, these benefit from nutrients produced by fungal activity. Fungicides may be applied after spiking and de-thatching the turf. Removal of infected soil is also effective.

Fagus sylvatica

falcate strongly curved sideways, resembling a scythe or sickle.

fall in irises, the outer petaloid perianth segments, which may be pendulous or horizontal in presentation. It is a prominent feature of bearded irises.

fallow the practice of leaving an area of ground clear of any crop for a period, either short-term in the course of natural rotation, or long-term in preparing a site. A fallow period benefits soil structure by allowing weathering and can reduce the level of soil-borne pests through the action of predators which because there is no cover have more ready access to their prey; it is also an important opportunity for weed removal.

Fallugia (for Abbot Virgilio Fallugi of Vallombrosa (fl. *c*.1850), Italian botanist). Rosaceae. SW US, Mexico. apache plume. 1 species, *F.paradoxa*, a deciduous shrub to 2.5m. The slender branches are white and woolly at first and later covered with buff and white, flaking bark. The leaves are 0.5–2cm long and feathery. Solitary or clustered, the 2cm-wide flowers appear in summer; they consist of five, white and spreading petals and a boss of green-yellow stamens. They are followed by feathery, purple to smoky red-grey seedheads. It is hardy to –5°C/23°F, but not in combination with winter wet. Grow in full sun, in perfectly drained soil, with shelter from cold winds. Propagate by softwood cuttings or by seed.

family the principal category of taxa intermediate between order and genus, for example, Rosaceae, the rose family. The names of most families derive from that of the type genus (in the above example, *Rosa*) and are plural nouns ending in *aceae*. Among the exceptions to this rule are Palmae, Gramineae, Leguminosae, Umbelliferae; these too have modern equivalents in Arecaceae, Poaceae, Fabaceae and Apiaceae, respectively.

family tree in fruit cultivation, a tree comprising two or more cultivars grafted on to a common framework.

fan a form of trained tree with branches radiating from a short stem like the ribs of a fan. This method of training is most often used for peaches, nectarines and apricots, and sometimes for plums, cherries and bush fruits suitable for planting against a wall.

fancy (1) of flowers, especially carnations, that have variegated markings, as opposed to those of a single colour known as selfs; (2) a 19th-century term denoting a particular pursuit, especially the art of breeding and showing specific plants – the specialist being called a fancier; (3) a term used of names published in accordance with the *International Code of Nomenclature for Cultivated Plants* and applied to cultivars, grexes and hybrids. It is invariably formed in a vernacular language, not Latin, for example *Lupinus* 'Loveliness', *Heuchera* 'Palace Purple'.

Farfugium (name used by Pliny). Compositae. E Asia. 2 species, rhizomatous, evergreen perennial herbs with long-stalked leaves in clumps, their blades rounded to fan-shaped and broad. Daisy-like flowerheads are borne in loose corymbs. Hardy to –10°C/14°F, and a fine foliage plant for pond and stream margins, Japanese gardens and partly shaded terraces. Plant in full or dappled sun on a moist, fertile soil. Increase by division in early spring. *F.japonicum* (syn. *Ligularia kaempferi*, *Ligularia tussilaginea*, *Tussilago japonica*; Japan; leaves to 30cm across, long-stalked and rounded to kidney-shaped, toothed to entire, thick-textured, dark green; flowerheads yellow, to 6cm across, inflorescences to 75cm tall; includes 'Argenteum' with leaves stippled and blotched grey-green and cream on deep green, and 'Aureomaculatum', with leaves spotted yellow).

Fargesia (for Paul Guillame Farges (1844–1912), French missionary and naturalist in Central China). Gramineae. China, Himalaya. Some 40 species of clumping bamboos; see *bamboo*.

farina a powdery or mealy coating on the stems, leaves, and sometimes flowers of certain plants.

farinose, farinaceous floury, having a mealy, granular texture.

fasciation see *malformation*.

fascicle a cluster or bundle of flowers, racemes, leaves, stems or roots, almost always independent but appearing to arise from a common point.

Fascicularia (from Latin *fasciculus*, a small bundle, alluding to the growth habit). Bromeliaceae. Chile. 5 species, low, evergreen, perennial herbs with rigid, narrow and spiny-toothed leaves in a dense, flat and stemless rosette. In summer and autumn, small flowers are carried in a dense, head-like spike – these are embedded in the rosette centre and surrounded by spiny and often colourful bracts. Hardy in climate zone 8, with full sun and a warm, perfectly drained position as can be found in raised borders on south-facing walls, dry, sheltered rock gardens and xeriscapes. Plant in a slightly acid soil rich in sand and grit. Increase by rooted offsets in late spring. *F.bicolor* (leaves to 50cm, deep green covered with silver scales, those at the centre of the rosette turning bright scarlet at flowering time; flowers small, blue, amid red to white bracts).

fastigiate describing the habit of trees and shrubs that have a strongly erect, narrow crown, and branches virtually parallel with the main stem.

father plant the pollen parent of a hybrid.

× **Fatshedera** (*Fatsia japonica* 'Moseri' × *Hedera hibernica*). Araliaceae. Garden origin. × *F.lizei*, is an evergreen shrub of loose sprawling habit, to

fan

1.2m tall with palmately and deeply 5-lobed, dark lustrous green leaves to 20cm across. Small and sterile, the green-white flowers are borne in umbellate panicles in late summer and autumn. Cultivars include 'Annemieke' with yellow-variegated leaves, 'Pia', with wavy leaves, and the cream-marked 'Variegata'. Hardy to about –15°C/5°F. Grow in sun or shade. Although not self-clinging like its *Hedera* parent, it does show a slight tendency to climb and may be tied to trellis, pillars or other supports; it is also useful as a conservatory or houseplant. Specimens with single, straight and clean stems make excellent stocks for standard ivy 'trees' – take an upright, 40–100cm tall × *Fatshedera*, remove leaves, branches and apex and cut deeply and cleanly crosswise into the tip; insert two or four scions of *Hedera*, bind the wound and grow on in warm, sheltered conditions. Increase by greenwood cuttings in summer.

Fatsia (from the Japanese name of *F.japonica*, *Fatsi*). Araliaceae. E Asia (Japan, Taiwan, Bonin Islands), introduced Hawaii. 3 species, evergreen shrubs or small trees with stout, usually sparsely branched stems and foliage crowded towards branch tips. The leaves are large, leathery and palmately lobed. Small, 5-petalled flowers are borne in umbels on the branches of a large panicle, and are followed by small, drupaceous fruits. Hardy to –5°C/23°F, although in cold areas it should be given the protection of a south- or west-facing wall. It is an excellent choice for sheltered city gardens. The more compact and variegated cultivars are slightly less frost-hardy and demand a richer soil and part shade in the garden. In colder areas, it is also grown in the home or conservatory, where it prefers cool, humid conditions. Grow in sun or part shade. Prune to remove lax growth or leggy stems. Propagate from seed sown at 15–20°C/60–68°F in autumn, by air-layering or by greenwood cuttings (with all but one leaf removed) rooted with bottom heat in summer. Root cuttings will also succeed. *F.japonica* (syn. *Aralia japonica*; JAPANESE ARALIA; Japan; shrub to 5m, with widely spreading, sparsely branched stems becoming bare below; leaves to 50 × 40cm, with 7–11, oblong-elliptic, toothed lobes, dark glossy green above, long-stalked; flowers creamy white in rounded umbels along the branches of a loosely pyramidal panicle to 80cm long, produced in autumn; fruits small, black; includes 'Aurea', with gold-variegated leaves, 'Marginata', with grey-green leaves edged off-white, 'Moseri', compact and vigorous, and the white-edged 'Variegata').

Faucaria (from Latin *fauces*, jaw: the toothed leaves resemble gaping jaws). Aizoaceae. TIGER JAWS. South Africa. 30 species, succulent, almost stemless perennials. The fleshy leaves are decussate and spread outwards; in outline they are rhombic, spathulate or broadly lanceolate, often with a flattened upper surface fringed with ferocious-looking marginal teeth. The underside is deeper, resembling the keel and prow of a ship. Large, daisy-like flowers open during the afternoon from late summer to late autumn. Cultivate as for *Conophytum*, although these plants are more robust and will take sporadic and sparing waterings throughout the spring and summer; the leaves usually persist for several seasons. *F.tigrina* (leaves 3–5cm, margins with 10 hair-tipped teeth, grey-green with numerous, small dots; flowers yellow).

faveolate, favose honey-combed.

feathered (1) of a maiden or first-year tree with a slender upright trunk bearing a number of lateral growths or feathers; hence, 'feathered maiden'; (2) in flowers, especially tulips, used of feather-like markings of one colour upon a different ground colour; (3) of leaf surfaces where the veins or colour markings spring from a central rib; hence, feather-veined.

feather-veined describing leaves whose veins all arise pinnately from a single mid-rib.

fedge a composite word from fence and hedge, denoting a garden feature which consists of a fence or net-clad structure up which mostly climbing or scrambling plants are grown to form a screen.

Felicia (for Herr Felix (d. 1846), German official). Compositae. Tropical & South Africa, Arabia. BLUE MARGUERITE, BLUE DAISY, KINGFISHER DAISY. 83 species, annual to perennial herbs, dwarf subshrubs and shrubs with long-stalked, daisy-like flowerheads in summer, the ray florets blue, mauve, pink or white, and the disc florets yellow. Suitable for bedding, baskets, containers and rock gardens. Plant in full sun on well-drained soil. Both species listed here are soon killed by hard frosts and overwet conditions. Remove dead flowerheads and trim back any weak growth. Increase by seed sown under glass in spring, or, for *F.amelloides*, by cuttings taken in late summer or autumn and overwintered under glass.

F.amelloides (BLUE DAISY, BLUE MARGUERITE; South Africa; bright green, perennial subshrub to 60cm; stems slender, trailing to erect; flowerheads to 4cm diameter, long-stalked, ray florets pale blue; includes 'Santa Anita', with large, clear blue flowerheads, and the cream-blotched 'Santa Anita Variegata'); *F.bergeriana* (KINGFISHER DAISY; South Africa; hairy, grey-green annual herb to 25cm tall, forming bushy mats; flowerheads to 3cm diameter, short-stalked, ray florets bright blue).

felted-tomentose tomentose, but more woolly and matted, the hairs curling and closely adpressed to the surface.

fence a barrier constructed of wood, or sometimes plastic, in contrast to walls of brick or stone. There is a wide choice of fence designs including close-boarded, interwoven, weather-boarded, split chestnut, and planed paling, rustic, wattle and trellis. Some are available as prefabricated panels 1.8m wide.

Fenestraria (from Latin *fenestra*, window, referring to the translucent leaf tips). Aizoaceae. Namibia. 2 species, succulent, evergreen perennial herbs. Smooth and fleshy, the leaves are 2-4cm long, mid- to dark green and stand erect in stemless clusters. Each leaf is club-shaped, expanding upwards to a flattened or domed tip with a translucent zone, or 'window'. Daisy-like flowers appear in late summer and autumn. Cultivation as for *Conophytum*. In habitat, these plants grow with only their leaf tips showing, but they are best cultivated with their leaves largely exposed.

F.aurantiaca (flowers 3–7cm in diameter, golden yellow); *F.rhopalophylla* (BABY'S TOES; flowers 1.8–3cm in diameter, white).

fenestrate irregularly perforated by numerous openings or translucent zones (windows).

fennel (*Foeniculum vulgare*) a native of Europe, including coastal parts of Britain. COMMON FENNEL (*F.vulgare*), is a perennial grown for its aniseed-flavoured leaves and leaf sheaths, which are used in salads, meat and fish dishes, and for its seeds, which are used in baking, fish dishes and herbal tea. Sow *in situ* outdoors after the risk of frost has passed, or divide existing plants in spring and set out at final spacings of 30–35cm. *F.* 'Purpureum' is an attractive border plant. *F.vulgare* var. *azoricum*, FLORENCE FENNEL or FINNOCHIO, is cultivated as an annual. It is smaller than *F. vulgare* and is grown for the succulent, swollen, aniseed-flavoured bases of the leaf stalks, which are served fresh in salads or cooked; the foliage can be used as flavouring. Sow May–June *in situ* for thinning, or raise in cell modules for transplanting no later than the 3–4-leaf stage to final spacings of 30cm, in rows 50cm apart. For transplanting under glass to extend the cropping season, choose bolt-resistant cultivars and sow in mid-July–early August.

ferns any of a group of plants belonging to the division Pteridophyta, prominent in the Earth's vegetation for millions of years; their fossilized form is the origin of the vast coal deposits that were laid down in the Carboniferous period. Ferns are established widely in temperate zones, are prolific and dominant in the tropics and are absent only in extremely cold or dry regions. They occur at altitudes from sea level up to 4400m, the majority in wet climates from humid lowland jungle to highland cloud forests. Their lightweight spores are carried very long distances on air currents and local adaptations arise. Ferns have long been used for medicinal purposes including disorders of the spleen, hence the common name spleenwort.

Exotic ferns have been grown in England for display since 1628, when specimens were imported by John Tradescant from Virginia to his garden in Lambeth, London. Interest in their cultivation increased in the late 18th century with the development of reliable techniques for raising fresh spores, giving rise to a spectacular popularity which peaked in the Victorian 'Pteridomania'. This in turn gave rise to widespread plundering of native habitats, particularly in Britain. The fashion for fern cultivation diminished from the late 19th century, and between the two world wars collections declined, with an estimated 60% loss of old collected variants.

There exist between 230 and 250 genera, including 10–12,000 species, from greatly diverse environments. Ferns are grown for the elegant and delicate symmetry of their fronds, for textural contrasts, and for colour – for example, the beautifully coloured emerging croziers of *Adiantum* and *Blechnum* species, the autumn colour of *Osmunda regalis*, and the metallic sheen on the emergent fronds of *Pityrogramma*, and *Cheilanthes* species. Variegated sports are established in cultivation (e.g. *Pteris* cultivars), as are cristate forms. Aromatic fronds are produced in some *Pityrogramma* and *Dryopteris* species. Hardy ferns are valuable for shady areas and are undemanding except for requiring moisture; some, like *Dryopteris filix-mas* and *Blechnum penna-marina*, are useful groundcover plants. In New Zealand *Dicksonia squarrosa* is used to stabilize roadside banks. They have few culinary uses since many ferns contain carcinogens, but the steamed croziers of *Matteuccia struthiopteris* and *Osmunda cinnamomea* are eaten for their asparagus-like flavour, and the soft pith of *Cyathea* species may be baked or roasted.

Successful fern cultivation requires an understanding of the three types of natural habitat. (1) Woodland or rainforest ferns are native to the herbaceous understorey and will grow in dappled, sometimes total shade with medium to high levels of humidity. Rooting is usually shallow. These include taller forest plants such as the tree ferns *Dicksonia* and *Cyathea* and epiphytes such as *Platycerium*, as well as climbers (e.g. *Lygodium* species), hemi-epiphytes and many of the hardy, woodland species familiar to gardeners. (2) Ferns of rocky, more exposed conditions or of subalpine meadows, such as *Cryptogamma* and *Woodsia* species, may be treated as alpine subjects either in an alpine house or on walls and dry rockeries. The group includes high-altitude species naturally subject to snow cover and long periods of freezing, and subjects which encounter scree, rock crevice, wet rocky grassland or stream and waterfall overhangs. Their light requirements may be higher but a moist, shaded root run is essential. (3) Natives of arid or semi-arid zones with limited seasonal rainfall include the 'Resurrection Ferns' of the New World, such as *Cheilanthes* species, which regenerate from shrivelled growth. Such species may grow on exposed ground, rock ledges, scree, gullies and crevices; in cultivation, they require alpine house or rock garden conditions.

Growing regimes reflect these different habitats. In cool temperate zones, tropical ferns should be treated as greenhouse or indoor subjects, requiring a minimum temperature of 18°C/64°F, high humidity with frequent damping down in the summer months, indirect sunlight from mid-spring to mid-

fennel

ferns

Pests
aphids
scale insects

Disease
rust

Osmunda regalis
(royal fern)

autumn, and direct sunlight at other times. Warm-temperate ferns should be grown in the glasshouse or indoors, with a minimum temperature of 13°C/55°F, medium humidity and bright filtered sunlight. Cool temperates should be grown in cool greenhouses (minimum temperature 7°C/45°F), indoors or in a sunless position outdoors, with medium humidity. Alpine, non-hardy ferns require cold, frost-free greenhouse conditions: damp down and ventilate when the temperature exceeds 20°C/68°F, and shade and water plentifully in summer. Moderate watering and frequent misting are essential if grown indoors. In centrally heated rooms, tropical and subtropical species are the best choice. For a cool room, choose cool temperate types, mist occasionally and water moderately in summer and sparingly in winter. Ferns are well adapted to thrive indoors in low light levels placed away from draughts. Repot every other year and feed at 14 day intervals throughout the growing season with an organically based fertilizer.

Potting media should be free-draining and enriched with organic matter. Most species prefer a slightly acid medium of pH 6–6.5, though there are other preferences, such as a very alkaline mix for the limestone polypody, *Gynnocarpium robertianum*, For terrestrial temperate and tropical ferns, a suitable mix is 1 part loam: 2 parts sharp sand: 3 parts leafmould: 1 part medium grade forest bark, with hoof-and-horn added. For epiphytic ferns, use 1 part fine-grade bark: 1 part charcoal: 1 part perlite, and mount to establish on slabs of cork or tree-fern fibre with a pad of sphagmum moss. Aquatic ferns are grown in ponds, aquaria or slow-moving water, where potted plants should be in a medium low in organic matter to avoid fermentation. Hardy ferns for the garden include *Adiantum*, *Asplenium*, *Athyrium*, *Blechnum*, *Dicksonia*, *Dryopteris*, *Gymnocarpium*, *Matteuccia*, *Onoclea*, *Osmunda*, *Polypodium*, *Polystichum* and *Woodsia*.

Some ferns are propagated from active or dormant bulbils and plantlets (e.g. *Asplenium bulbiferum*), or from offsets (e.g. *Dicksonia* species); others by air-layering (e.g. *Davallia* species) and by stem propagation (e.g. *Lycopodium* species).

Most ferns, however, are propagated by division or from spores. Division of plants should take place in spring. Ensure that basal roots are included, shallowly potted and kept in warm, moist conditions. Reproduction by spores entails a cycle in which two generations (sporophyte and gametophyte) lead an independent existence, each requiring different cultural conditions. The mature fern plant (sporophyte) develops specialized spore-producing structures on the underside of its leaves. Spores germinate under damp shady conditions to produce a flattened plate of cells called a prothallus (gametophyte), on the underside of which male and female structures arise to facilitate fertilization. The cross-fertilization of these organs results in the development of new sporophytes – the 'ferns' we know and grow.

Spores of temperate and hardy ferns are generally long-lasting and those of tropical habitats short-lived.

Dryopteris filix-mas
(male fern)

Successful raising from spores requires the careful sterilization of the sowing area before the spores are sown. Use a sterile mixture of equal parts peat and sharp sand, and surface sow in sterilized clay pots covered with a sheet of glass. Hardy and temperate species are maintained at 15–16°C/59–61°F and tropical species at 21°C/70°F. The pots should be stood in shade in 1cm depth of water to maintain a moist medium for the prothalli – these will usually form after a few weeks. They resemble fragile liverworts or mosses. If very crowded, they may be separated into small blocks using a sterile scalpel and replanted with space around them in larger pans. Again, keep moist and covered and grow on in shade. Young sporophytes will develop within a period ranging from two weeks to several months. Once they fill the pot, they may be carefully moved on to a fine, soilless medium and raised in shade with humidity.

fern-case a glazed container, sometimes of bell-glass form, used to protect ferns grown indoors from dry air and draughts. It is akin to a *Wardian case* (q.v.).

fernery a place devoted to the cultivation of ferns. It may be a glasshouse, fern-case or, as featured in Victorian gardens, an outdoor area where ferns are placed among rocks, tree stumps and roots.

fern-leaf a descriptive term for the narrow and elongated leaves of tomatoes infected with tobacco or cucumber mosaic viruses.

Ferocactus (from Latin *ferox*, fierce, and Cactus). Cactaceae. SW US, Mexico. BARREL CACTUS. 23 species, large cacti with simple or clustered, squat and rounded to barrel-shaped or cylindric stems covered in ribs and fierce spines. The flowers are bell- to funnel-shaped. Provide a minimum temperature of 2–5°C/36–40°F. Grow in a gritty, slightly acidic to neutral medium in full sun with low humidity. Keep dry from mid-autumn until early spring, except for light misting on warm days in late winter. Propagate by seed. *F.chrysacanthus* (NW Mexico; stem globose to shortly cylindric, to 100 × 30cm, spines to 5cm, yellow or red, flattened and twisted; flowers yellow to orange with red-tinted midveins to tepals); *F.cylindraceus* (syn. *F.acanthodes* of gardens; NW Mexico, SW US; stem to 3m × 50cm, cylindric or barrel-shaped, spines 7–17cm long, red, orange or yellow, sometimes hooked; flowers tinged green yellow or red); *F.hamatacanthus* (syn. *Hamatocactus hamatacanthus*; N Mexico, SW US; stems to 60 × 30cm, hemispheric to cylindric, spines to 8cm, some hooked; flowers to 10cm, yellow); *F.latispinus* (C & S Mexico; stems 10–40 × 16–40cm, squatly globose or flattened; spines flattened; flowers purple-pink or yellow); *F.wislizeni* (SW US, NW Mexico; stem 1.6–3m × 45–80cm, globose to barrel-shaped or cylindric, tapering above, spines to 10cm, brown to grey, the lowermost usually flattened and hooked; flowers to yellow, orange or red).

Ferraria (for Giovanni Battista Ferrari (1584–1655), Italian botanist). Iridaceae. Tropical and South Africa. 10 species, small, cormous perennial herbs with sword-shaped leaves. Produced in spring and summer, the flowers are short-lived, often malodorous and composed of six, radiating and ovate to lanceolate segments. Hardy to about –5°C/23°F. Plant 15cm deep in a warm, sunny and sheltered site with a protective mulch of bracken litter in winter. Where grown under glass they are best planted directly into the greenhouse border, or else in deep pots. Grow in direct sunlight in a well-drained potting mix with additional sharp sand; maintain a minimum temperature of 10°C/50°F in spring and summer. Reduce watering after flowering until leaves wither; keep completely dry and frost-free when dormant. Propagate by seed sown in late summer or autumn, or by offsets in autumn. *F.crispa* (syn. *F.undulata*; South Africa; to 45cm; flowers to 5cm diam., deep, velvety chestnut with cream to white markings at centre, spotted and lined in a paler shade, or yellow to tan spotted and lined dark maroon, margins wavy to crisped).

ferrous sulphate a constituent of lawn sand and other moss-killer products. It is sometimes recommended as a corrective treatment for iron deficiency in plants.

ferruginous brown-red, rust-coloured.

fertile (1) producing viable seed; used of anthers containing functional pollen, of flowers with active pistils, of fruit bearing seeds and of spore-bearing fern fronds; (2) used to describe a flower-bearing shoot.

fertility (1) of soil; its suitability to support good plant growth. Soil fertility is influenced by drainage, physical structure and the presence of organic matter, lime and nutrients; (2) of plants; the capacity to produce viable seeds, in which an important factor is self- or cross-compatibility of pollen.

fertilization (1) the union of male and female sexual cells (gametes), resulting in a unicellular zygote from which an embryo develops; see *cross-fertilization*, *pollination*, *self-fertilization*; (2) (in the US) the use of fertilizers.

fertilizer correctly, any material that provides plant nutrients in relatively concentrated form, as opposed to bulky manure, although the terms fertilizer and manure are often used loosely and interchangeably. Fertilizers can be of organic or inorganic origin, produced naturally or synthetically, and may be formulated as powder, granules, or liquid. The concentration of nutrients is stated on containers as a percentage; for compounded products containing the three primary nutrients, this will be expressed in code form, thus 7–7–10 indicates 7% nitrogen, 7% phosphate and 10% potassium. The higher the percentage, the greater the concentration of nutrient per unit of weight; see *compound fertilizer*; *slow-release*; *straight fertilizer*.

Ferula (Latin name for fennel). Umbelliferae. Mediterranean to C Asia. GIANT FENNEL. 172 species, robust perennial herbs arising from a thick rootstock. The leaves are finely divided into very narrow segments. Small, yellow or yellow-white flowers are carried in large, compound umbels in late spring and summer. Suitable for the large herbaceous and mixed border, or for waterside plantings and hardy to at least –5°C/23°F. Grow in moist, deep and fertile soils in sun. They do not transplant well and should be given permanent positions when young. Mulch thickly in winter. Propagate by fresh seed in late summer, or by careful division. This is not the edible fennel, for which see *Foeniculum*. *F.communis* (GIANT FENNEL; Mediterranean; 2–3m tall; leaves mid-green with thread-like segments; flowers yellow; giant, bronze- and purple-leaved forms are sometimes grown).

festooning the bending of shoots or branches of fruit trees into a more horizontal position by means of weights or ties; it is also practised on some ornamentals. Festooning is done to control vigour or stimulate flowering, and is employed especially in the arcure form of training.

Festuca (from Latin *festuca*, stalk or stem; in Pliny the wild oat growing among barley). Gramineae. Cosmopolitan (mainly cold and temperate regions). FESCUE. Some 300 species, rhizomatous or tufted perennial grasses up to 2m, usually very much shorter, with flat, folded or involute leaves. The species described below are ornamental, valued for the attractive, often brilliant blue-greens of the low, hair-like foliage tussocks. Most are tolerant of cold to at least –15°C/5°F, although many do not tolerate heavy snow cover or prolonged winter wet. Plant in light, well-drained soil in full sun. Clip over in spring to encourage dense fresh foliage; divide and replant older plants. Propagate by division or seed, although species hybridize readily and may not come true from seed.

F.eskia (Europe; densely tufted, forming a low tussock or thatch of very slender dark green leaves to 8cm long); *F.glauca* (BLUE FESCUE, GREY FESCUE; Europe; densely tufted, forming a tussock, cushion or mop of very slender leaves to 12cm long; these are typically glaucous blue-green; a varied and confused group with many cultivars, some of which doubtless belong to other species, for example, *F.cinerea*, *F.ovina*, its blue-grey subspecies *coxii*, and *F.rubra* – they range in habit from relatively tall, loose and spilling to dense and compact, and in colour from deep green to bronze-red-tinted, sea green, and blue-green to silver-grey).

fibrillate finely striated or fibrous.

fibrillose with thread-like fibres or scales.

fibrous (1) of a thread-like, woody texture; (2) used of plant roots produced as a mass of fine growth, in contrast to fewer coarse or fleshy roots; (3) used of loam with a high content of fibre derived from dead grass roots and rotted turf.

Ficus (Latin name for the edible fig). Moraceae. Tropics and subtropics. FIG. About 800 species, deciduous or evergreen trees, shrubs and woody root-climbing vines (those listed below are usually ever-green in cultivation, except *F.carica*). Their growth habits and foliage vary enormously from species to species, although all show the characteristic protec-tive stipule which encloses the new growth in a rolled, pointed tube. This falls away as the leaves emerge, leaving a ring-like scar. The minute flowers are embedded on the inner surface of a fleshy recep-tacle (the 'fig', or syconium). This may be large or small, rounded, pear-shaped or oblong, but is usually – if tautologously – 'fig-shaped'. Pollination is effected by various genera of wasps, which are specially adapted to enter the syconium through the tight aperture at its apex and lay their eggs within.

With the exception of *F.carica*, the following species require a minimum temperature of 10°C/50°F, bright, indirect light and medium to high humidity. Some, however, survive in less than ideal conditions and show remarkable tolerance of drought, dry air and harsh light, especially when pressed into service as houseplants and staples of the interior land-scaper's art; most notable among these are *F.benjamina*, *F.binnendykii* 'Amstelveen', *F.deltoidea*, the universal rubber plant, *F.elastica*, and *F.lyrata*. The climbing species listed below make excellent house-plants when young and still creeping – small speci-mens of *F.pumila* are very commonly used to cascade over the rims of mixed planters. Plant in a fertile, porous, soil-free medium. Water, syringe and feed generously during summer or when temperatures are high. At other times, keep just moist and protect from draughts. Propagate by stem cuttings, air-layering, or, for climbing species, by detaching rooted lengths of stem. For cultivation of *F. carica* see FIG below.

F. benghalensis (BANYAN TREE; S Asia; ever-green tree to 30m, ultimately with many stem buttresses and massive aerial roots; leaves to 25cm, broadly ovate, base rounded to cordate, apex blunt or briefly tapered, leathery, mid to dark green; 'Krishnae': leaves cup-like with inrolled, fused margins); *F. benjamina* (WEEPING FIG; S Asia to Australia and Pacific; graceful evergreen shrub or tree to 30m, branches weeping; leaves to 13cm, ovate-lanceolate, cuspidate, thinly leathery, dark green; includes dwarf, strongly weeping and bushy clones with foliage ranging from large and long to short and tufted, some variegated white, cream, grey or gold); *F. binnendykii* (NARROW-LEAF FIG; SE Asia; evergreen shrub or tree to 20m, branches somewhat weeping; leaves to 20cm, pendulous, narrowly lanceolate, tip finely pointed, tough; the cultivar usually grown is 'Amstelveen' with narrow, glossy dark green leaves hanging dagger-like); *F. carica* (BROWN TURKEY FIG, COMMON FIG; Cyprus, Turkey, Middle East, W Asia; deciduous shrub or tree to 10m; leaves to 30cm with three to five broad lobes, coarsely textured and somewhat downy; fruit to 5cm diam., dull green to brown or deep, glau-cous maroon, edible; for cultivars see FIG below);

F. deltoidea (MISTLETOE FIG; SE Asia; evergreen shrub or small tree to 7m; leaves to 8cm, broadly spathulate to deltoid, apex rounded, base cuneate, tough, olive to dark green, dull or somewhat glossy; fruits small, inedible, abundant even on young plants); *F. elastica* (RUBBER PLANT, INDIA RUBBER TREE; evergreen tree to 60m with many buttresses and aerial roots, usually seen as a smaller, sparsely unbranched shrub; stipules large, dull coral pink to scarlet; leaves to 40cm, oblong to elliptic, tip acuminate or apiculate, tough, somewhat rigid, glossy dark green with a prominent, pale or red midrib; a very popular houseplant, it includes tall, dwarf, rarely branched and bushy clones with leaves ranging from small and broad to long and narrow, some variegated white, grey or creamy yellow or flushed wine red); *F. lyrata* (FIDDLE-LEAF FIG; Tropical Africa; evergreen tree to 12m; leaves to 50cm, obovate to lyrate or pandurate, apex rounded, very tough but flexible, glossy mid-green with corrugated upper surface); *F. macrophylla* (MORETON BAY FIG, AUSTRALIAN BANYAN; Australia; evergreen tree to 55m, sometimes devel-oping massive buttresses and aerial roots; leaves to 25cm, oblong to elliptic or ovate, tough, glossy dark green above, paler beneath; young plants superficially similar to *F. elastica* but with leaves less leathery);*F. microcarpa* (syn. *F. retusa* of gardens, *F. nitida* of gardens; INDIAN LAUREL, CURTAIN FIG, GLOSSY-LEAF FIG; E Asia to Australia and Pacific; evergreen tree to 30m, ultimately producing many buttresses and festoons of aerial roots; leaves to 12cm, elliptic to obovate, tough, glossy dark green; includes var. crassifolia with broad, thickly textured leaves, and dwarf clones with variegated leaves); *F.pumila* (CREEPING FIG, CLIMBING FIG; E Asia; evergreen or semi-decid-uous creeping or climbing shrub with matted, slender branches clinging by dense, short aerial roots; leaves 1-5cm, broadly oblong to elliptic or ovate, thinly leathery with a puckered, papery texture, dark green to bright mid-green; a very popular pot plant with numerous cultivars, the leaves ranging from very small and rounded to broadly toothed or strongly wavy, some variegated white or gold); *F. religiosa* (SACRED FIG; BO TREE, PEEPUL; fast-growing, deciduous or semi-deciduous tree to 10m, sometimes producing buttresses and aerial roots; leaves to 18cm, broadly triangular with a finely tapering tip, thinly leathery, rather papery, dull to deep green); *F. rubiginosa* (RUSTY FIG; Australia; evergreen tree to 12m; leaves to 18cm, oblong to elliptic or ovate, tough, flushed red at first and with rusty hairs beneath, later dark green and smooth; includes cultivars with variegated leaves); *F. sagittata* (E Asia; similar in habit to *F. pumila*, but more robust and with larger, cordate to sagittate leaves, the surface not puckered; includes variegated clones);*F. sarmentosa* (E Asia; differs from *F. pumila* in its larger leaves, lobed in juve-nile plants, lanceolate in adults; includes variegated cultivars).

FIG 289

field capacity the maximum amount of water a freely drained soil will hold before it is lost to drainage by the force of gravity.

field-immune used of plants that are not infected by a pathogen under normal growing conditions despite being experimentally susceptible.

fig (*Ficus carica*) a deciduous bush or tree, with large, lobed leaves, grown for its small, rounded, green, maroon or brown coloured fruits. One of the oldest fruits known in cultivation. The fig is believed to originate in the Tajik mountains of Transcaucasia, but has long been native to southwest Asia and the eastern Mediterranean, growing wild in southern Europe and northern Africa. Its cultivation is recorded from 2700 bc. It is valued for its nutritious fruits, which can be dried and stored, and for its use as a shade tree.

The fig was introduced into Britain by the Romans but there is little evidence of its later cultivation until the 16th century. In the 18th century new cultivars were introduced from Europe including 'Brown Naples', later known as 'Brown Turkey' and still the most widely grown cultivar in the UK. By the late 18th century 75 cultivars were grown by the Horticultural Society at Chiswick, the fruit being found to thrive against relatively warm walls and in glasshouses, some with heat. Fig cultivation increased progressively into the 20th century but

Fig

declined with the great country estates after World War II. Today the fig is grown widely in commercial plantings in Calfornia, Turkey, southern Europe and northern Africa, principally for its dried fruit. The ancient cultivar 'Smyrna' is most famous because of its high quality and suitability for drying; in common with older selections, it relies on cross-pollination by the small fig wasp (*Blastophaga psenes*) in the process known as caprification. Modern cultivars are self-fertile.

In warm climates, goblet-shaped bushes are successful, spaced 6–9m apart with five or six branches quite severely pruned and thinned. In cooler climates, such as the UK, trees are best planted against a south-facing wall as fan-trained specimens: wall protection aids fruit ripening, shields the plant from rain and provides some safeguard against severe winter weather. Roots should be restricted by constructing a sealed brick-, concrete block- or slab-lined bed, 75cm long and 60cm wide and deep, with the soil base covered by a 30cm layer of rubble. Propagate from suckers, by cuttings or layering. One- or two-year-old specimens are best planted in March or April, when there is no longer a risk of severe frost; position 22cm from the wall with 3.6m between plants. Train the main ribs so that they are no less than 30cm apart.

Figs can also be grown successfully in 25–37cm pots, provided the base is well-crocked and firmly filled with a soil-based medium. Small bushy trees up to 1.3m should be trained and repotted every second year when dormant, with the old compost carefully replaced with a fertile mix. Renew the surface compost in the intervening year, and apply high-potash liquid feed every ten days during summer.

In cooler climates, only one harvest is possible per season outdoors; the crop arises from small pea-sized fruitlets formed towards the end of short laterals from late summer. Frost damage to young fruitlets can be prevented by covering the plants with bracken or some other insulating material after leaf fall, or by moving pot-grown figs under glass. Fruits ripen in late summer or early autumn; those fruitlets formed early on new shoots of the current year will not ripen but turn yellow, and should be removed. Regular light pruning and thinning is necessary to contain growth. In late spring cut out old damaged wood and any shoots that do not lie on the same plane as the wall; a proportion of the remainder should be cut back near to the main branch to stimulate new growth. Encourage suitable short growths from around midsummer by pinching shoots at the fourth or fifth leaf no later than late June. Water in dry summers, particularly pot- or bed-grown trees, but not near ripening time. Fruit thinning may be necessary in warm climates where a crop may be produced from both overwintered fruits and those initiated early in the current year. Every other fruit may be removed at swelling for improved quality. Fruits are best picked warm, fully ripe and hanging downward, particularly where showing a longitudinal split and a drop of nectar at the eye. Netting against birds is usually necessary.

fig

Pest
scale insects

Diseases
grey mould
virus

filament (1) a stalk bearing an anther at its tip, the two forming a stamen; (2) a thread-like or filiform organ, hair or appendage.

filamentous composed of or bearing filaments.

filbert see *hazelnut*.

filial generation a cross-bred generation; hybrid offspring, that of the first crossing being denoted by the symbol F_1, the second by F_2.

filiform of leaves, branches etc., filament-like, i.e. long and very slender, rounded in cross-section.

Filipendula (from Latin *filum*, thread, and *pendulus*, drooping, referring to the root tubers of some species, which hang together with threads. Rosaceae. Northern temperate regions. MEADOWSWEET; DROPWORT. 10 species, rhizomatous perennial herbs, sometimes tuberous. The leaves are basal or alter-nate on erect stems and pinnate, with pairs of large leaflets alter-nating with smaller ones. The terminal leaflet is largest and coarsely toothed to palmately cut. White or pink and often with contrasting stamens, the small, 5-petalled flowers are borne in great abundance in frothy, cymose corymbs in spring and summer. Fully hardy, they are suitable for streamside and bog garden plant-ings, for the herbaceous and mixed border, and for moist, cool woodland or other naturalistic plantings. Plant on well-drained but moisture-retentive soils in sun or semi-shade. With variegated cultivars, removal of young flower shoots will result in successive flushes of fresh, colourful foliage. Propagate by division in autumn or winter, or by seed in autumn or spring.

F.camtschatica (Kamchatka, Korea, Japan; to 3m tall, hairy; flowers white to pale pink); *F.palmata* (Siberia to China and Japan; to 1m tall; flowers white with red anthers; includes pure white-, pink-, red- and purple-red-flowered cultivars); *F.purpurea* (Japan; to 1.3m tall, more or less glabrous and purple-tinted; flowers purple-red or pink; includes cultivars with white flowers and with strongly purple-tinted foliage); *F.rubra* (QUEEN OF THE PRAIRIE; US; to 2.5m tall; flowers deep peach to rose); *F.ulmaria* (MEAD-OWSWEET, QUEEN OF THE MEADOWS; Europe, W Asia; 1–2m tall, more or less hairy; flowers creamy white in frothy corymbs; includes double- and pink-flowered cultivars and plants with leaves flushed gold or consis-tently striped or zoned creamy-yellow – 'Aurea' is one of the finest pure gold foliage plants); *F.vulgaris* (DROPWORT; Europe, N & C Asia; to 80cm tall; flow-ers white, sometimes tinted red or purple; includes cul-tivars with large, double and pale pink flowers).

fillis a strong soft string of various ply, usually coloured brown or green. It is specially made for tying plants.

fimbriate bordered with a fringe of slender outgrowths, usually derived from the lamina rather than attached as hairs.

fir a common name applied to trees of the genus *Abies*, but also used of the genera *Pseudotsuga* and *Cunninghamia*.

fireblight (*Erwinia amylovora*) a bacterial disease principally affecting pome fruits in the family Rosaceae, including apple, *Amelanchier*, *Chaen-omeles*, *Cotoneaster*, hawthorn, loquat, *Photinia*, *Pyracantha*, *Sorbus* and especially pear. It has been known in North America since 1780, where it has occasionally also infected other rosaceous plants, including apricot, cherry, *Kerria*, plum and rose. Identified in New Zealand in 1920 and in England 1956–57, it has spread to other parts of western Europe, the eastern Mediterranean and the Far East.

Fireblight was the first plant disease shown to be caused by a bacterium, and the name refers to the fact that the flowers, leaves and twigs are blackened as if burnt by fire. It overwinters in cankers, from which it exudes as a slime at about the same time as the flowers open in spring. The bacteria are disseminated by wind, rain or insects and infect the shoots of blossoms (especially in pears), from where they spread down the branches. In pears, the disease can spread rapidly and kill the trees, but in apple and other hosts the main branches and trunk are not usually affected. In gardens and orchards, hawthorn hedges are often a source of infection.

Badly affected trees should be removed com-pletely; in other cases, branches may be cut back to a point at least 60cm below where there is any sign of the disease beneath the bark. Pruning tools should be immersed in strong disinfectant after each cut, and the severed surfaces treated with a wound sealant and all infected tissues burnt. The disease is no longer subject to statutory control in the UK.

firmer an oblong or round implement, usually consisting of a flat piece of wood with an additional smaller piece fixed to one surface as a handle; used for levelling and firming compost in a seed box or pot.

Firmiana (for Karl Josef von Firmian (d. 1782), governor of Lombardy and patron of the Padua botanic garden). Sterculiaceae. E Asia, 1 species E Africa. 9 species, trees or shrubs with entire or palmately lobed leaves. The flowers are borne in axillary and terminal panicles or racemes. Usually yellow, they lack petals but consist of a showy, bell-shaped and 5-lobed calyx with a prominent, stamen bearing pistil. Suitable for outdoor cultivation in sheltered gardens where temperatures seldom fall below freezing. Otherwise, grow in the cool glasshouse in full light or part-shade. Water plenti-fully when in growth. Maintain good ventilation, with a winter minimum temperature of 2–5°C/36–40°F. Prune only when necessary to confine to allotted space. Propagate by seed sown under glass as soon as it ripens. *F.simplex* (syn. *F.platanifolia*, *Sterculia platanifolia*; CHINESE PARASOL TREE, CHINESE BOTTLETREE, JAPANESE

VARNISH TREE, PHOENIX TREE; E Asia; deciduous tree to 20m; leaves 30–40cm across, with 3–7 deep lobes, deep green; flowers to 4cm across, lemon yellow; 'Variegata': leaves white-mottled).

fish manure, fish meal, fish guano an organic compound fertilizer made by drying and pulverizing fish waste. It contains 7–10% nitrogen, 5–15% phosphate, and 2–3% potash.

fistulose, fistular hollow and cylindrical like a pipe.

Fittonia (for Elizabeth and Sarah May Fitton, authors of *Conversations on Botany* (1823)). Acanthaceae. S America. 2 species, downy evergreen perennial herbs with low or creeping and rooting stems. The leaves are oblong to broadly and bluntly elliptic, soft-textured and overlaid with a network of metallic or brightly coloured veins. Small, white and two-lipped, the tubular flowers are carried in slender spikes clothed with overlapping, green bracts. Striking, small foliage plants for the home and glasshouse (minimum temperature 15°C/60°F). Plant in shallow pots or pans containing a porous, soilless mix. Water and feed freely during warm weather; keep barely moist at other times. Position in light shade, with moderate to high humidity and protection from draughts. The flower spikes are often pinched out to encourage bushy and luxuriant growth. Propagate by stem cuttings or spontaneous layers. *F.albivenis* (Argyroneura Group: syn. *F.argyroneura*, SILVER NET-LEAF, SILVER THREADS, with silver-white-veined, emerald green leaves; 'Pearcei': SNAKESKIN PLANT, with larger and thinner-textured leaves in sage green, net-veined carmine; Verschaffeltii Group; syn. *F.verschaffeltii* MOSAIC PLANT; to 8cm; leaves 5–10cm, soft-textured and finely downy, olive green to purple-bronze with an intricate network of ruby to scarlet veins).

Fitzroya (for Robert FitzRoy RN (1809–65), Captain of HMS *Beagle*). Cupressaceae. C Chile, N Patagonia. 1 species, *F.cupressoides* (syn. *F.patagonica*), PATAGONIAN CYPRESS, an evergreen, coniferous tree to 45m or shrubby at altitude in the wild. The thick trunk is clothed with fissured and peeling, rusty red bark. Slender and pendulous, the branchlets bear overlapping leaves, each to 0.5cm long, oblong and dark green with a paler midrib. An enormous and very long-lived but slow-growing tree in the hills of Chile and Argentina. It favours mild, damp maritime climates, growing well on the Pacific coast of North America from central California to Vancouver. Its needs are similar to those of *Abies*, although it requires extra shelter, humidity and a moist, acid soil. Propagate by seed sown in a cold frame or by cuttings.

flabellate, flabelliform fan-shaped, with a wedge-shaped outline and sometimes conspicuously pleated or nerved.

flag, flagstone a flat slab of stone used for paving. Various simulated products are available in concrete.

flagellate with whip-like runners; sarmentose.

flaked describing bicoloured flowers, especially carnations and tulips, where one colour overlies a ground colour in large splashes.

flamed describing flowers, especially tulips, where in the centre of each petal there is a band of solid colour in combination with fine feathering.

flame gun a large, hand-held or wheel-mounted blow-torch, used principally to destroy surface weeds. It produces a long flame with intense local heat from the ignition of paraffin vaporized under pressure.

flat a shallow tray with open-jointed base strips, often used for plant raising. See *seed box, seed tray.*

flatworms (*Platyhelminthes*: *Turbellaria*) Many flatworms live in ponds, lakes and waterways where they prey on small animals or feed on detritus. Some live on land, especially in tropical and subtropical areas. They have flat, ribbon-like unsegmented bodies and some species are up to 20cm long when extended. The New Zealand flatworm (*Artioposthia triangulata*) and the Australian flatworm (*Australasion sanguinea*) have become established in the UK and Ireland. Both feed on earthworms and in some areas they have reduced earthworms to a very low level. They reproduce by laying egg capsules that contain 6–10 eggs. There are no suitable means of controlling flatworms.

flea beetle (Coleoptera: Chrysomelidae) small beetles, up to about 3mm long and mostly black, although some species have a broad yellow stripe on their wing cases. The common name refers to their ability to jump when disturbed. Five species of *Phyllotreta* are of most concern in Europe: the SMALL STRIPED FLEA BEETLE (*P. undulata*), the LARGE STRIPED FLEA BEETLE (*P.nemorum*), and the TURNIP FLEA BEETLES (*P.atra*, *P.cruciferae* and *P.nigripes*). Other *Phyllotreta* species occur in North America and their habits are similar.

Adults overwinter in such places as hedge bottoms, under loose bark, or in grass tussocks; on emergence in spring, they are attracted to brassicas and related plants. Eggs are deposited on or near the plants and the resulting larvae feed on the roots or (in the case of *P.nemorum*) tunnel the leaves, but consequent injury is seldom serious. The adult beetles attack emerging seedlings, biting small holes in leaves and stems causing severe damage, especially in dry seasons when seedling establishment is slow. Cruciferous plants are particularly susceptible, but other plants attacked include beet, anemones, godetias, and irises.

Damage may be reduced by keeping seedlings well-watered to hasten growth and by clearing away plant debris from beneath hedge bottoms to reduce adult hibernation sites. Seed dressings and recommended contact insecticides are available as control measures.

Fittonia argyoneura
Verschaffeltii Group

Another flea beetle, the CABBAGE STEM BEETLE (*Psylliodes chrysocephala*), occurs in Europe and has been recorded in Canada. The adults are metallic green or blue and up to 5mm long with reddish legs. The larvae tunnel inside the stems and leaf veins of brassica plants causing a growth check and often death. The pest can be controlled with contact insecticide applied at setting out.

In North America, the GRAPE FLEA BEETLE (*Altica chalybea*), with metallic blue-green adults up to 5mm long, attacks the buds and unfolding leaves of grape and the larvae skeletonize the leaves. Other plants subject to attack include Virginia creeper, plum, apple, pear, quince, beech and elm. Infestation can be controlled with appropriate contact insecticides.

fleecy film a soft fibrous material of spun plastic, used to cover the soil surface to advance warming, or placed over plants to extend their season of maturing and provide frost protection. The film allows air and rain to penetrate and is best laid over shallow V-shaped drills of sown crops such as carrot, or over cell-raised plants, such as lettuce, after they have been set out. Cover for no more than four weeks in the spring, longer in autumn. Where carefully anchored, the film provides a barrier against aphids, root flies, bulb flies, moths and butterflies. Fleecy film can also be used for greenhouse insulation and shading.

flexible paving paving slabs bedded close together on sand, without mortar.

flexuous of an axis, zig-zag, bending or curving in alternate and opposite directions.

flies one of the largest groups of insects, classified in the order Diptera. They are widely distributed and include many serious pests of cultivated plants. True flies are easily recognised as the adult insects have only one pair of wings, the hind pair being modified into club-like structures used as sensory and balancing organs. The larval stages are legless maggots which often have no obvious head or other external features. Damage to plants is usually caused by the larvae which in most species are concealed in the soil, in galls or other plant tissue. See *bean seed fly*; *bulb flies*; *cabbage-root fly*; *carrot fly*; *celery fly*; *chrysanthemum stool miner*; *frit fly*; *fruit flies*; *fungus gnat*; *gall midge*; *hoverfly*; *leaf miners*; *leatherjacket*; *onion fly*; *sawfly*.

floating mulch a temporary ground or crop covering made of plastic fabric, described as 'floating' on account of its loose but secure fitting. It is used to advance soil temperature, extend the season of vegetable crops and, in some cases, provide protection against frost and insects. The most usual kinds are perforated polythene sheet film, close-woven net and fleecy film.

floccose possessing dense, woolly hairs that fall away easily in tufts.

flocculation the process whereby fine soil particles join together into larger aggregates as a result of the neutralizing of clay colloids. Lime, gypsum, humus and other soil conditioners are used for the flocculation of clay soils to improve structure and, hence, drainage, aeration and root penetration.

flocculent, flocculose slightly floccose, woolly.

flood bench a level, water-tight bench which is periodically flooded and then pumped dry. It is one form of irrigation for pot plants.

floral diagram a diagram providing a condensed and simplified view of the flower from above, with the parts of the flower shown in one plane and symbols representing each part. The organs of a flower displayed in a floral diagram (starting from the centre) are the ovary, stamens, corolla and calyx, with the bracts and bracteoles, if present, shown beyond, in relation to their position beside the flower. The ovary in cross-section shows the arrangement of carpels, and the side on which the anthers open to release pollen is indicated. An absent organ, in terms of expected symmetry, is indicated by a dot or an asterisk. If one part is attached to another, lines are drawn between them, and the separation or linkage of petals and sepals is also made clear.

To supplement a diagram and section a floral formula can be used, in which K = calyx, C = corolla, A = androecium, G = gynoecium. To these symbols are appended figures showing the numbers of each part, brackets around a number showing that the parts are united. If the number of a part is large and indefinite the symbol ˙ is used; if a part is absent, a zero (0). Where perianth whorls are linked, a bridging line or square brackets indicate this. A bar above or below the number for the gynoecium indicates an inferior or superior ovary, a feature visible in the flower section. The floral diagram does not of course indicate the overall shape of the flower. This can be rectified by accompanying it with a realistic half-flower section which at once indicates its actual form.

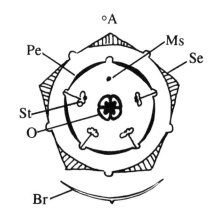

Floral diagram of Lamium album:
A= axis; Br= bract; Se= sepal; Pe= petal; St= stamen; Ms= missing stamen; O= ovary
Floral formula: K(5) C(5) A(4) G2

flore pleno (abbrev. fl.pl.) see *double*.

floret (1) a very small flower, generally part of a congested inflorescence; (2) in grasses, a collective term embracing various structures of the flower head.

floricane a biennial stem which fruits and flowers in its second year, as found in, for example, *Rubus* species.

floriferous bearing flowers freely.

florigen a hypothetical plant hormone, responsible for initiating flowering.

florilegium a collection or selection of flowers; occasionally, also used of a book or selected list of plants.

flower a structure bearing one or more pistils or stamens, or both, usually with some floral envelope. If surrounded by a perianth of calyx and corolla, it is termed perfect.

flower bud see *fruit bud*.

flower gatherer a pair of secateurs or scissors specially designed to hold a flower stem as it is cut.

fluid drilling, fluid sowing the process of sowing pre-germinated seeds mixed with a specially prepared gelatinous carrier. It was developed to advance the establishment of plants in instances where soil is too cold for direct sowing, and also to economize on seed.

In fluid drilling, seeds are sown under protection on damp absorbent paper at a suitable temperature, and when the roots are no more than 5mm long the germinated seeds are mixed throughout the gel. The mixture is transferred to a polythene bag; one corner of the bag is then cut off, and the gel mixture 'piped' along prepared soil drills, which must be kept moist.

flush a surge in the production of flowers, fruits, or fruiting bodies, especially used of roses and mushrooms.

fluted of a trunk with long, deep, rounded grooves running vertically.

fly speck a superficial growth of the fungus *Schizothyrium pomi* on apple, citrus, pear, plum and possibly other fruits. It is often associated with sooty blotch and occurs as groups of black, circular dots, the individual dots being 0.2–0.4mm in diameter. No damage is caused.

Fockea (for Gustav Woldemar Focke (early 19th century), German physician and plant physiologist). Asclepiadaceae. Southern Africa (Angola to the Karroo). 10 species, perennial succulent shrubs, developing a massive, woody, turnip-like caudex crowned with slender, twining to sprawling branches. The leaves are oblong to ovate. To 4cm diameter, the starfish-like flowers are borne singly or several together in leaf axils. Plant in deep pots or large pans containing a fast-draining, gritty, loam-based medium. Maintain a minimum winter temperature of 10°C/50°F in full sun with low humidity. Water and feed generously during the spring and summer growing season but allow the plants to dry out between waterings. Keep dry from mid-autumn to early spring except for occasional mistings or drenchings to prevent withering and collapse of caudex. Propagate by seed sown under glass or by stem cuttings in spring (plants gained from cuttings seldom develop the attractive caudex).

F.crispa (syn. *F.capensis*; to 4m; leaves to 3cm, finely hairy, oval and wavy-margined; flowers grey-green with brown blotches); *F.edulis* (differs from the last in its smooth, dark green leaves without wavy margins).

Foeniculum (the Latin name for fennel). Umbelliferae. Europe, Mediterranean. 1 species, *F.vulgare*, FENNEL, an aromatic, glaucescent perennial or biennial herb to 2m. The leaves are to 50cm long, triangular in outline, 3–4-pinnate, and very finely cut, with thread-like segments to 5cm long. Small, yellow flowers are borne in compound umbels in summer. The cultivar 'Purpureum' (syn. 'Purpurascens') has especially fine, purple-bronze-flushed leaves and is a beautiful foil for pastel flowers and silvery foliage in the herbaceous border. Hardy in zone 6. Grow in a well-drained site in a warm sunny position. Propagate by seed or by dividing established plants in the spring as soon as fresh growth appears; seeds may be sown outside once the risk of frost has passed. See *fennel*.

fogger, fogging nozzle a device for producing very small droplets of water or pesticide solution.

fogging the application of a pesticide carried in droplets so fine that visibility within an enclosed treated area is reduced. The pesticide particles slowly settle, well distributed mainly on the upper surfaces of leaves. Formulations are usually oil-based and are applied with special equipment to produce an aerosol.

foliaceous resembling a leaf, in appearance or texture.

foliage plant a plant grown predominantly for the colour, markings, shape or texture of its leaves.

foliar embryo an embryonic plant that arises on the leaves of mature specimens of a species, and provides a means of increasing a stock. This may be a natural occurrence in the case of plants such as *Tolmiea menziesii*, *Bryophyllum pinnatum* and the ferns *Asplenium bulbiferum* and *Cystopteris bulbifera*.

The production of foliar embryos at specific locations on a leaf can be stimulated in certain plants. Sedums will produce small plantlets only at the point where the leaf was detached from its parent and these may be encouraged to root by pressing the leaf base lightly into the surface of a sandy compost, which should be kept dry. *Tiarella* species, *Mitella* species and *Cardamine trifoliata* produce plantlets at the junction of the leaf petiole and leaf blade. In these species, fully expanded leaves are removed, retaining about 20mm of petiole, and inserted into a rooting compost with the leaves lying flat on the surface; these are then placed in a humid, shaded environment at 20°C/68°F until plantlets develop. *Kalanchoe blossfeldiana* will produce plantlets only around the indented leaf-margin; suitable leaves may be removed, anchored on the surface of a sandy medium and kept dry until plantlets have formed. Also known as proliferation and vivipary.

foliar feeding the application of a dilute solution of fertilizer to the leaves of plants; useful as an emergency treatment for correcting deficient trace elements but also of value for supplementary feeding. The absorption of liquid fertilizer is greatest where leaf cuticles are thin, as on the undersurfaces of leaves or where they are just expanding. Liquid fertilizer should not be applied in bright sunlight because foliage may be scorched.

foliate bearing leaves.

foliolate bearing leaflets.

follicle a dry, dehiscent, one- to many-seeded fruit, derived from a single carpel and dehiscing by a single suture along its ventral side.

folly a garden structure designed to catch the eye, but with no functional purpose, often taking the form of a pyramid, pinnacle, tower or mock ruin.

Fontanesia (for R.L. Desfontaines (1750–1833), French botanist). Oleaceae. Near East, China. 1 species, *F.phillyreoides*, a deciduous shrub to 3m. The leaves are 2–10cm long, ovate to lanceolate, glossy, glabrous and serrulate to entire. In early summer, small, 4-petalled flowers are borne in dense axillary racemes or terminal panicles. It is fully hardy. Grow in a well-drained soil in sun or partial shade. Propagate by semi-hardwood cuttings in late summer in a cold frame, or by hardwood cuttings outside in autumn; sow seed in a cold frame in spring.

Fontinalis (from Latin *fontalis*, meaning of or belonging to a spring or fountain). Fontinalaceae. Cosmopolitan. WATER MOSS. Some 55 species, aquatic mosses forming clumps of long and slender, submerged stems covered in small, dark green leaves. The following species grows rooted to rocks and tree roots submerged in cold, fast-flowing rivers and streams, in ditches and ponds and on cliff faces in water seepage. It will grow as an oxygenator in ponds or cold-water aquaria, providing useful cover for spawning fish and acting as host to the small invertebrates they feed upon. Grow in lime-free water in indirect light. Propagate by division of large clumps in spring. Take stems with part of the anchorage attached or tie gently to new anchorage in the form of rough stone or brick. *F.antipyretica* (Northern Hemisphere; stems to 80cm, slender, mossy and feathery, dark emerald to olive green).

foot rot a rot of the lower part of the stem-root axis. Where used generally to describe a base rot disease symptom, it often overlaps with the terms collar rot, damping-off and root rot. Foot rot is also used more specifically to refer to particular diseases, such as those caused by the soil-borne fungi *Fusarium solani* on peas and beans, *Mycosphaerella pinodes* on peas and *Phytophthora cryptogea* on tomatoes.

forcing the acceleration of vegetative growth and especially flowering or fruiting, usually by temperature and light manipulation. Rhubarb and hyacinths are commonly forced. Commercial forcing is carried out in specially designed greenhouses or sheds, often with additional bottom heat; but in the domestic garden it is usually improvised in greenhouses and frames or achieved with the use of forcing pots.

Forestiera (for Charles le Forestier, French physician and naturalist). Oleaceae. N & C America. 20 species, usually deciduous trees or shrubs. The leaves resemble those of *Ligustrum*. Small, cream or green-white flowers are borne on axillary clusters or in racemes. They are followed by small, black, fleshy fruits. Hardy in climate zone 6. Plant on a well-drained soil in a sunny site. Propagate by half-ripe cuttings in summer or by seed when ripe in a cold frame.

F.acuminata (US; shrub or small tree to 2m with smooth branches; leaves 3–10cm, ovate-oblong to lanceolate, tip serrate; flowers green-white; fruit to 1.2cm, purple); *F.neomexicana* (SW US; shrub to 3m with somewhat armed branches; leaves to 5cm, ovate to lanceolate, crenulate; flowers yellow-white; fruit to 0.5cm, purple, pruinose).

forest transplant a young tree, usually of forestry or native species, that is raised from seed and especially suitable for planting on a large scale or in difficult sites. A three-year-old transplant decribed as 2+1 indicates two years in a seed bed and one year in a nursery row.

fork (1) a hand implement comprising a steel shaft with a head of three or four narrow prongs; the shaft is attached to a handle usually made of wood. Handgrips may be D-, T- or Y-shaped, or occasionally absent. A standard digging fork has four tines about 30cm long, which are round or square in section, with the shaft mounted on a handle about 60cm long; there are curved-tine and long-handled versions for various purposes. A border or lady's fork differs in its shorter tines and narrower head. A flat-tined, spading or potato fork has broad flat tines, suitable for digging heavy-textured soil.

A dragfork is a long-handled fork without a handgrip, the tines bent at right angles to the shaft. Dragforks come in various sizes and designs, and are used as cultivating tools or for pulling garden refuse or manure. A hollow-tine fork has tubular tines, designed to remove cores of soil from lawns to aid drainage, aeration and top dressing. A pitchfork has only two tines, which are slightly curved upwards and fixed to a straight handle. It is suitable for handling straw or very long cut grass. A handfork has three or four flat tines about 10cm long on a handle about 12.5cm long or longer, without a handgrip; it is used for cultivating small areas.

(2) a point on a tree where two branches of similar size meet.

form, forma (abbrev. f.) an infraspecific taxon, subordinate to subspecies and variety. The lowest rank in the taxonomic hierarchy, forms are usually distinguished only by minor characteristics. These characteristics may be of importance to the gardener, however, and include such features as habit and leaf and flower colour, e.g. *Hosta crispula* f. *viridis*.

formaldehyde a chemical substance (H-HCO) effective as a fungicide; it is used for sterilizing bulbs, soil, containers and tools, and for greenhouse hygiene. Formaldehyde volatilizes rapidly and is toxic to plants in active growth. It is a severe irritant to the eyes and nose and its use is limited by regulations and subject to precautions.

formal garden a garden of geometric, usually symmetrical design, typically incorporating architectural features, clipped trees and water.

Forsythia (for William Forsyth (1737–1804), superintendent of the Royal Gardens, Kensington). Oleaceae. E Asia, C Europe. 6 species, deciduous shrubs with golden-green branches covered with lenticels. The leaves are simple or three-parted with entire to toothed margins. The short-stalked, yellow flowers are borne before the leaves, singly or in clusters and usually nodding; tubular to bell-shaped, they consist of four, oblong lobes. Among the most colourful and floriferous of early spring-flowering shrubs, they are fully hardy. They will tolerate part shade but bloom more prolifically in sun. Cut back after flowering to within a few centimetres of the old wood. Older plants may be rejuvenated by thinning by one third to one half every two to three years. The slender branches of *F.suspensa* should be allowed to develop in weeping profusion, given some support but otherwise unhindered. For hedging, choose vigorous, erect cultivars. Propagate by half-ripe cuttings in summer in a case or cold frame, or by hardwood cuttings outside in autumn. Susceptible to leaf spot fungi, forsythia gall (which may disfigure but not damage) and bacterial blight.

F. × *intermedia* (*F.suspensa* × *F.viridissima*; garden origin; to 3m, erect to spreading; flowers in clusters, deep yellow; includes 'Arnold Giant', with large, nodding, golden flowers, 'Beatrix Farrand', large, with deep gold flowers with broad lobes, 'Karl Sax', vigorous and free-flowering, with deep yellow flowers marked orange in the throat and foliage turning purple-red in autumn, 'Lynwood', free-flowering, vigorous and erect, and the erect and vigorous 'Spectabilis', with densely packed, deep yellow flowers with 4–5–6, twisted lobes); *F.ovata* (KOREAN FORSYTHIA; Korea; to 1m, domed, compact; flowers small, bright yellow, sparse, solitary or paired; includes the hardy, erect and profusely flowering 'Ottawa'); *F.suspensa* (China; semi-scandent to erect shrub to 3m tall; leaves often 3-parted; flowers clustered, nodding, pale to deep yellow with somewhat twisted lobes; includes var. *fortunei*, erect to arching with deep yellow flowers, and var. *sieboldii*, slender-stemmed and weeping to semi-scandent with clear, pale yellow flowers, f. *atrocaulis*, with maroon-tinted branchlets and pale flowers, and 'Nymans', which is similar to f. *atrocaulis* but produces larger, brighter flowers later in the season, i.e. mid- to late spring); *F.viridissima* (China; erect to 3m with persistently green, 4-angled branches; leaves flushing claret to plum-purple in autumn; flowers bright yellow stained green with narrow lobes).

Fothergilla (for Dr John Fothergill (1712–80), who cultivated many American plants at Stratford-le-Bow, Essex). Hamamelidaceae. E US. 2 species, deciduous, low-growing shrubs, to 3m tall with widely spreading branches. Obovate to oval and coarsely toothed, the leaves are pale green when first opening, mature to mid-green in summer and turn a glorious crimson or orange-yellow in autumn. The fragrant flowers open before the leaves in bottle brush-like, terminal heads or spikes. They lack petals, but are made up of about 24 stamens with white filaments, each to 2.5cm long. Fully hardy and among the finest shrubs for autumn colour, they are fairly low-growing and need a cool, moist, acid soil in sun. Otherwise, cultivation as for *Hamamelis*.

F.gardenii (syn. *F.alnifolia*, *F.carolina* *F.parvifolia*; SE US; leaves to 6.5cm, toothed in upper half; flowers with white filaments to 2.5cm long; includes compact, blue-green-leaved cultivars and forms with pink flowers, possibly hybrids with the next species); *F.major* (*F.monticola*; Allegheny Mountains; leaves to 10cm, more or less entire or sparsely toothed in upper half; flowers with pink-tinged, white filaments to 2cm long).

foundation planting in garden design, the basic structural planting of trees and hedges.

fountain a jet or spout of water, produced artificially by pumping and/or gravity. Fountains are an important feature of formal gardens.

foveolate with shallow small pits or precise indentations.

fractiflex in intermittent zig-zag lines.

Fragaria (the Latin name for the strawberry, from *fragrans*, fragrant, referring to the aroma of the fruit). Rosaceae. Northern temperate zones, Chile. STRAW-BERRY. Some 12 species, stoloniferous, perennial herbs with 3-parted, toothed leaflets. Made up of five, rounded to obovate petals and numerous stamens, the flowers are carried in cymes in spring and summer. The fruit consists of many small achenes scattered on the surface of an enlarged, conical, fleshy, red receptacle. Grow in fertile, well-drained, moisture-retentive soil in sun. When grown on very rich soils, variegated cultivars may produce green crowns. Propagate from seed in early spring, or by removal of runners (plant-bearing stolons). Among the edible strawberries, several ornamental cultivars can be found. *F.* × *ananassa* (*F.chiloensis* × *F.virginiana*), the GARDEN or CULTIVATED STRAW-BERRY, the WILD STRAWBERRY, *F.vesca* and its variant 'Semperflorens', the ALPINE STRAWBERRY or FRAISE DU BOIS; all have variegated cultivars. *Fragaria* 'Pink Panda' is a pleasing, pale-tomentose plant with large, candy pink flowers. For *F.indica*, see *Duchesnea indica*. See also *strawberry*.

Frailea (for Manuel Fraile, (b. 1850 at Salamanca Spain), curator of cacti at the US Department of Agriculture, Washington DC). Cactaceae. E Bolivia, S Brazil, Paraguay, Uruguay and N Argentina. About 15 species, dwarf cacti with tufted or solitary, squatly globose to cylindric stems, usually weakly ribbed or tuberculate. Produced amid wool and bristles, the yellow, shortly funnelform flowers open only briefly, if at all. Provide a minimum temperature of 10–15°C/ 50–60°F, full sun and low humidity. Plant in an acidic medium, high in grit. Water very sparingly in winter (to avoid shrivelling). Most species of *Frailea* can be raised to maturity from seed in less than three years. They need ample watering during the warmer months and may flower intermittently over the summer and autumn; except in the brightest weather, the flowers often fail to develop fully or to open, but fruits will still be formed. *F.pygmaea* (syn. *F.pulcherrima*; Argentina, S Brazil, Uruguay; stems 1–7 × 1–2.5cm, globose to squatly cylindric, dark green to purple-red, with small tubercles, woolly areoles and short yellow to white spines; flowers small, yellow).

frame a structure with a removable or hinged glass or plastic cover resting on a low framework of masonry, wood or metal; it is used for propagation, forcing, hardening-off or winter protection. It is usually unheated, when it is known as a cold frame, but is sometimes provided with a source of bottom heat. See *hotbed*, *soil warming*.

frameworking a form of grafting fruit trees in which scions of a different cultivar are inserted on the pre-pared framework of an established tree. See *grafting*.

Francoa (for F. Franco MD, who encouraged the study of botany in 16th-century Spain). Saxifrag-aceae. Chile. BRIDAL WREATH. 5 species, perennial evergreen herbs with a clump of obovate to broadly lanceolate leaves pinnately cut into blunt lobes. Abundant small flowers are produced in summer in long-stalked racemes; they consist of four, oblong petals, usually white and with a dark pink spot at base of claw. Plant in a well-drained, fairly moist, sunny position. In colder areas, overwinter in frost-free con-ditions and plant outdoors the following spring. Although frost-hardy, these plants are commonly grown indoors in cool, bright conditions. Propagate from seed in early spring or by division.

F.appendiculata (inflorescence to 68cm, sparingly branched, compact; flowers shell pink, sometimes spotted); *F.ramosa* (inflorescence to 90cm, much-branched; flowers white); *F.sonchifolia* (leaves with broadly winged petioles; inflorescence to 60cm, simple to sparingly branched; flowers pink with red spots).

Franklinia (for Benjamin Franklin (1706–90), American statesman). Theaceae. SE US (Georgia, now extinct in the wild). 1 species, *F.alatamaha* (syn. *Gordonia alatamaha*), a deciduous tree or shrub, erect to 10m tall with smooth bark. To 15cm long, the leaves are obovate to oblong, sparsely toothed and dark glossy green, turning bright red in autumn. Produced in late summer, the cup-shaped, solitary flowers are fragrant, 8cm across and composed of five, rounded, white petals and a boss of golden stamens. This tree needs long hot summers to flower well and for wood to ripen – specimens perform exceptionally only in continental climates, where they prove hardy to at least 15°C/5°F. In less favoured climates, plant in a sheltered south-facing niche with good drainage and full sun. Grow on moisture-reten-tive but well-drained, neutral to slightly acid soils. Propagate by fresh seed sown ripe, by softwood cut-tings or by semi-ripe cuttings in mid-summer; alter-natively, increase by simple layering of low branches.

Fraxinus (the Latin name for the ash). Oleaceae. Temperate Europe, Asia, N America, a few in tropics. ASH. 65 species, medium-sized to large trees, mostly deciduous (those listed below are deciduous) with pinnate leaves and panicles of small flowers borne before or alongside the leaves. The fruits are 1-seeded samaras, each with a long, flattened wing. Fully hardy, the following species should be planted on moist but well-drained soils in full sun. Propagate from stratified seed in spring. Graft cultivars in spring or bud in summer on seedling stock of the same species under glass. Susceptible to ash heart rot and ash bark beetle.

F.americana (syn. *F.alba*, *F.juglandifolia*; WHITE ASH; NE US; tree to 40m tall, crown spreading; leaflets to 9, to 15cm, oblong to lanceolate, entire or serrate, slightly downy beneath; fruit wing oblong, 3–5cm; includes 'Acuminata', with long-tapering, entire leaflets, dark green above and almost white beneath, turning purple-red in autumn, 'Autumn Applause', with a rounded, dense crown and deep red to mahogany autumn colour, and 'Autumn Purple', with

Fraxinus 'Raywood'

leaves persisting late into autumn and turning purple to chocolate mottled mauve and bronze); *F.angustifolia* (syn. *F.oxycarpa*; NARROW-LEAVED ASH; S Europe, N Africa; tree to 25m; leaflets to 13, to 7cm, oblong to lanceolate, finely serrate, dark green above, paler beneath, glabrous; fruit wing elliptic-oblong, to 4cm; includes the striking 'Raywood', with an elegant, feathery crown, the leaflets narrow and dark green, assuming a hazy, metallic appearance toward summer's end and turning deep plum to claret in autumn – an exceptional street tree); *F.excelsior* (COMMON EUROPEAN ASH; Europe to Caucasus; tree to 40m; leaflets to 11, to 12cm, obovate, serrate, dark green above, paler beneath, glabrous; fruit wing oblong, to 4cm; includes f. *diversifolia*, with only one or three leaflets per leaf, 'Jaspidea', with yellow new growth turning gold in autumn, and 'Pendula' with a gnarled, spreading crown and branches weeping to the ground); *F.nigra* (BLACK ASH; N America; tree to 25m; leaflets to 11, to 12cm, oblanceolate, serrulate, veins brown and downy beneath; fruit wing bluntly oblong, to 4cm; includes 'Fallgold', with leaves turning gold in autumn; *F.mandshurica*, MANCHURIAN ASH, differs in its more distinctly toothed leaflets with veins sunken beneath; *F.ornus* (FLOWERING ASH, MANNA ASH; S Europe, Asia Minor; broad-crowned tree to 8m; leaflets to 7, to 7cm, obovate, serrulate, dark green above, paler beneath with a somewhat downy midrib; flowers white and fragrant in showy, lacy panicles; fruit wing narrow-oblong, to 2.5cm); *F.pennsylvanica* (RED ASH; N America; tree to 18m; leaflets to 9, 7–15cm, lanceolate, entire or serrulate toward tip, olive green; fruit wings spathulate to lanceolate; includes 'Patmore', vigorous and strongly erect with good pest-resistance and dark green foliage lasting well into autumn, 'Summit', with a broadly pyramidal crown and gold autumn tints, and 'Urbanite', tolerant of city pollution and sun scorch, with a broadly pyramidal crown and thick, glossy leaves becoming bronze in autumn); *F.sieboldiana* (syn. *F.mariesii*; Japan, China; compact-crowned tree to 8m; leaflets to 7, to 7cm, ovate, serrate or entire, dark green, glabrous except for downy midrib; flowers cream, fragrant in panicles to 15cm); *F.velutina* (ARIZONA ASH; SW US; shrub or small tree; leaflets 3–5–7, lanceolate to elliptic, coarsely serrate, dull green above, more or less hairy beneath, tough-textured; includes 'Fan-Tex', fast-growing and well-proportioned with large, deep green and leathery leaves).

Freesia

Freesia (for Friedrich Heinrich Theodore Freese (d. 1876), a pupil of the botanist Ecklon). Iridaceae. Sout Africa. 11 species, cormous perennial herbs with a fan of lightly pleated, narrow leaves and erect to horizontal, simple or branched flowering stems in winter and spring. Often fragrant, the flowers are more or less funnel- or goblet-shaped, sometimes 2-lipped, and consist of six tepals.

Frost-tender bulbs, grown in the cool glasshouse for winter and early spring flowers, or, using prepared bulbs, in the open garden for flowers in summer. Grow indoor bulbs in a medium-fertility sandy mix, at 5–8cm spacings in prepared beds, boxes or pots; plant in succession from late summer through to winter. Water and keep in the frame or cold glasshouse (5°C/40°F) and cover with a 3cm layer of coir and a top dressing of grit. Gradually increase water as growth resumes, and raise the temperature to 10°C/50°F; provide full light and good ventilation, and feed fortnightly with a dilute liquid feed as flower buds show. Provide support. Temperatures above 15°C/60°F will result in spindly plants with quickly fading blooms. Reduce water as plants die back and keep completely dry when dormant; lift and store cool and dry until the following season. In mild, essentially frost-free areas, corms may be planted directly into the open ground in late summer/early autumn into fertile sandy soils in a sunny and sheltered position. Propagate by pre-soaked seed in the cool glasshouse or cold frame, or by offsets in autumn.

F.alba (stem 12–40cm; flowers 2.5–6cm long, sweetly scented, white, sometimes flushed purple outside and lined with purple in throat, lowest tepals sometimes marked with yellow); *F.corymbosa* (syn. *F.armstrongii*; stem to 50cm with several branches, spikes 3–10-flowered; flowers 2.5–3.5cm, scented or not, ivory or pale yellow with the lower tepals marked bright yellow, or pink with a yellow throat); *F.leichtlinii* (stem 8–20cm; flowers 2.5–4cm long, scented, cream or pale yellow, the lower tepals orange-yellow, the upper tepals sometimes flushed with purple-brown on the outside); *F.refracta* (close to *F.alba* and *F.leichtlinii*, but with spicily scented, two-lipped flowers in pale yellow, sometimes flushed green or purple, the lower tepals marked orange and the throat veined purple). The florist's freesias are complex hybrids involving the above species; they have been grouped under various names – *F.* × *hybrida*, *F.* × *kewensis*, *F.* × *ragioneri*, and *F.* × *tubergenii*. They range in height from 10cm to 30cm. The flowers may be single or 'double', scarcely to sweetly scented, and vary in colour from silvery white to ivory, yellow or bronze and from soft pink to red and mauve, blue and indigo. These colours are frequently combined, for example as bronze, pearly or lavender flushing on the exterior, or as a pale yellow or white throat.

Fremontodendron

Fremontodendron (for Major-General John Charles Frémont (d. 1890), US explorer, botanist and horticulturist). Sterculiaceae. Southwest N America. FLANNEL BUSH; CALIFORNIA BEAUTY. 2 species, evergreen shrubs and trees covered in short, felty hair. The leaves are unlobed or palmately 3-, 5-, or 7-lobed. Produced from late spring to early autumn, the solitary, short-stalked flowers consist of a cupped, showy calyx with five, ovate to elliptic, petal-like lobes and a staminal tube divided into 5 parts toward the tip and ending in a filiform style. Hardy to −15°C/5°F. Grow in full sun against a sheltered, south-facing wall. They prefer poor, dry soils. It may be necessary to tie in branches to a framework of trellis or wire. Cut out frost-damaged growth in

Fremontodendron californicum (left), *F. mexicanum* (right).

spring. Pruning is otherwise inadvisable (the countless short hairs are, in any case, extremely irritating). Propagate by seed or by softwood or semi-ripe cuttings in a sandy propagating mix with gentle bottom heat.

F.californicum (syn. *Fremontia californica*; shrub to 7m; leaves unlobed to 3-lobed, sparsely hairy above, densely so beneath; flowers opening all at once, 3.5–6cm in diameter, shining golden yellow); *F.mexicanum* (syn. *Fremontia mexicana*; differs from the first species in its 5–7-lobed leaves and flowers opening in succession, 6–9cm in diameter, deep gold to orange yellow, becoming stained brick-red at the base; this species has crossed with the first, producing *F.*'California Glory', a vigorous and free-flowering plant hardier than either parent with lemon yellow flowers tinged red on the exterior).

French garden the grand garden style developed by Le Nôtre during the reign of Louis XIV in which a garden was centred on the house and designed with a series of axes, often combined with water features in the form of canals. Also incorporated were elaborate parterres, formal lakes, clipped trees and hedges along straight walks, geometric woods and circular open spaces from which avenues radiated in a fan pattern.

French intensive method see *deep bed method*.

friable used of soil that is of good crumb structure, and suitable for working down to a seed bed or tilth.

frill girdling the formation of downward-sloping overlapping cuts around the trunk of a tree, near soil level, using a light axe or billhook. Frill girdling is done in preparation for the application of brushwood killer or ammonium sulphamate, either of which is deposited in the cuts as a treatment to kill the tree.

frit a formulation of fertilizer that is incorporated into a melted glass matrix, which is then shattered into 'frits' and ground up to produce an insoluble nutrient additive. Frits are often used to supply trace elements, particularly in potting composts.

frit fly (*Oscinella frit*) a very small fly, the maggots of which are about 3mm long and may attack sweet corn during May and June. The growing points of seedlings are bored into and there is distortion, wilting and death. Seedlings are most susceptible up to the five to six leaf stage. Where the fly is a troublesome pest raise in cell modules, or spray emerging shoots outdoors with a recommended insecticide.

Frithia (for Frank Frith, collector of South African succulents and discoverer of this plant). Aizoaceae. South Africa. 1 species, *F.pulchra* (FAIRY ELEPHANT'S FEET), a dense, stemless succulent. It resembles *Fenestraria*, the leaves to 2cm long, densely clustered and finger- to club-shaped, with a flat tip bearing a 'window' and the other surfaces covered in minute, rough warts. Crimson to mauve with a paler centre, or pure white, the daisy-like flowers appear in summer. Cultivate as for *Conophytum*.

Fritillaria (from Latin *fritillus*, a dice-box, referring to the shape of the capsules, or to the chequered pattern on the flowers of some species, reminiscent of the traditional decoration on a dice-box). Liliaceae (Liliaceae). Temperate regions of northern hemisphere, particularly Mediterranean, SW Asia (mountainous regions), western N America. Some 100 species, bulbous perennial herbs with alternate to whorled, oblanceolate to linear leaves. The flowers are usually nodding and broadly bell-shaped. The six tepals are commonly marked with tessellations or have dark nectaries at the base on their inner surfaces.

The species described here are spring-flowering and fully hardy unless otherwise specified. The smaller species will, however, suffer in cold, wet winters and dull damp summers. For this reason, they tend to be grown in the alpine house, bulb frame or raised bed, with cloche protection in winter. Most of the smaller species require a sharply draining, fertile soil, full sun, ample moisture during the growing and flowering season and a dry (but not arid) rest in summer. *F. meleagris* prefers permanently damp sites and is suitable for naturalizing in long grass, providing that the first lawn cut does not occur until the leaves have died back in summer. *F. camschatcensis* needs a lightly-shaded position on a cool damp and humus-rich soil. The larger *F.imperialis* and *F.persica* are suited in scale to the mixed border, and will thrive in full sun on well-drained, fertile loams. In climate zones 6 and over, they require no additional winter protection.Increase by seed sown when ripe or in spring, in a loam-based propagating mix with additional sharp sand; germination may take six months. Propagate also by bulbils. *Fritillaria* is attacked by lily beetle and, in the case of *F.meleagris*, greedy pheasants.

FRITILLARIA

Name	Distribution	Height and foliage	Flowers
F.acmopetala	W Asia, E Mediterranean	15–45cm tall; leaves linear	1–3, broadly campanulate, tepals 2.5–4cm, lanceolate to oblanceolate, green stained brown
F.aurea syn. *F.bornmuelleri*	Turkey	4–15cm tall; leaves lanceolate to ovate-lanceolate, glaucous	solitary, broadly and deeply campanulate, tepals 2–5cm, ovate-oblong, yellow with orange or red-brown tessellations
F.bucharica	C Asia, NW Afghanistan	10–35cm tall; leaves lanceolate to ovate	1–10, cup-shaped, tepals 1.5–2cm, lanceolate, white or off-white with green veins
F.camschatcensis BLACK SARANA	NE Asia to NW N America	15–75cm tall; leaves broadly lanceolate, whorled below	1–8, broadly campanulate to cup-shaped, tepals 2–3cm, oblong-ovate, shining purple-brown to black with finely grooved and ridged inner surfaces
F.cirrhosa	E Himalaya, China	20–60cm tall; leaves linear, whorled below, the upper leaves or bracts very slender with coiling, tendril-like tips	1–4, broadly campanulate, tepals 3.5–5cm, narrowly elliptic, basically green or yellow-green, often tinted or tessellated purple-brown
F.crassifolia	Anatolia	6–20cm tall; leaves lanceolate	1–3, broadly campanulate, tepals 1.8–2.5cm, ovate, green with brown tessellations
Comments: var. **kurdica** has yellow-green tepals heavily suffused with maroon and a central yellow stripe			
F.imperialis CROWN IMPERIAL	S Turkey to Kashmir	50–150cm tall, muskily scented; leaves lanceolate, whorled, bright green	3–5, campanulate, nodding in a terminal umbel crowned with leafy bracts, tepals 4–5cm, ovate-oblong, orange to red
Comments: Includes cultivars with yellow, tangerine, deep orange and red flowers, 'Crown on Crown' with two whorls of flowers, one above the other, and selections with yellow- and white-edged leaves and bracts			
F.meleagris SNAKE'S HEAD FRITILLARY, GUINEA-HEN FLOWER, CHEQUERED LILY, LEPER LILY	Europe	12–30cm tall; leaves linear	solitary or paired, broadly campanulate, nodding, tepals 3–4.5cm, oblong-elliptic, white or pink heavily chequered purple-red
Comments: A variable species, chequered to varying degrees, some flowers pure white, some with the faintest tessellations, others darkest purple-red with a leaden cast			
F.michailovskyi	NE Turkey	6–24cm tall; leaves lanceolate	1–4, broadly campanulate, clustered together, tepals 2–3cm, ovate-elliptic, purple-brown to green with yellow tips
F.pallidiflora	E Siberia, NW China	10–80cm tall; leaves broadly lanceolate, glaucous	2–5, broadly and deeply campanulate, tepals 2.5–4.5cm, oblong-elliptic, pale buttery yellow, sometimes tinted green or faintly chequered
F.persica *F.eggeri, F.libanotica*	Middle East, W Asia	20–90cm tall; leaves lanceolate, glaucous grey-green, many alternating on stems	7–20, nodding in an erect, terminal raceme, narrowly campanulate; tepals 1.5–2.5cm, dull purple-black to chalky maroon or green
Comments: 'Adiyaman': to 1m tall with many dark maroon flowers			
F.pontica	Balkans, N Greece, NW Turkey	8–30cm tall; leaves linear-lanceolate	solitary or paired, broadly campanulate, nodding, tepals 2.5–4.5cm, oblong-obovate, green faintly marbled and edged brown within and with dark, visible nectaries

Fritillaria meleagris (left),
F. imperialis (centre),
F. camschatcensis (right).

FRITILLARIA

Name	Distribution	Height and foliage	Flowers
F.pudica YELLOW FRITILLARY	W N America	8–30cm tall; leaves linear to narrowly lanceolate	1–6, campanulate, nodding, tepals 1–2.5cm, elliptic-obovate, deep yellow sometimes suffused with warm orange or red
F.pyrenaica	SW France, NW Spain	15–30cm tall; leaves linear to lanceolate	solitary or paired, broadly campanulate, nodding; tepals 2.5–3.5cm, oblong-elliptic, the tips recurved, yellow chequered brown and heavily suffused dark purple-brown on exterior
F.raddeana	NW Iran, Turkmenia	50–150cm tall; leaves lanceolate, whorled and alternate, dark green	6–20, broadly campanulate in a terminal umbel crowned with leafy bracts, tepals 4–5cm, oblong to rhombic, white to palest yellow suffused green
F.recurva SCARLET FRITILLARY	California, S Oregon	20–90cm tall; leaves linear-lanceolate, grey-green, whorled	3–12, narrowly campanulate, nodding in a raceme, tepals 2–3cm, ovate to oblong-lanceolate, the tips recurved, orange-red to scarlet with yellow tessellations
F.sewerzowii *Korolkowia sewerzowii*	C Asia, NW China	15–20cm tall; leaves lanceolate	4–12 in a raceme or solitary, narrowly campanulate with a broadly flared mouth, tepals 2.5–3cm, oblong-obovate, green to livid purple without and yellow to brick red at base and within
F.stenanthera	E Russia	10–30cm tall; leaves linear to ovate-lanceolate	1–10 in a raceme, narrwoly campanulate and flared at mouth, tepals 1.5–2cm, obovate-lanceolate, pale pink
F.tubiformis *F.delphinensis*	SW Alps	4–35cm tall; leaves linear to lanceolate, glaucous	solitary, broadly campanulate to turbinate, tepals 3.5–5cm, ovate-elliptic, incurved, exterior purple-pink to maroon, chequered and suffused grey, interior with yellow tessellations
F.verticillata	C Asia, W Siberia	20–60cm tall; leaves narrowly lanceolate to linear, whorled, the upper leaves or bracts with tendril-like tips	1–5, broadly campanulate, nodding, tepals to 4.5cm, oblong-obovate, white, suffused and faintly chequered green to brown

froghoppers (Hemiptera: Cercopidae) sap-sucking insects, up to 6mm long, which are so named because of the adults' frog-like shape and ability to jump. The US common name of SPITTLEBUG refers to the characteristic white froth or 'cuckoo spit' which surrounds the pale-coloured nymphs when they feed. The COMMON FROGHOPPER (*Philaenus spumarius*) varies in colour from yellow through shades of brown to almost black; in the US, it is known as the MEADOW SPITTLEBUG. The adult female lays eggs in the autumn, and nymphs hatch in spring, feeding on plants until reaching maturity in midsummer. A wide range of plants may be attacked, including blackberries, raspberries, perennial asters, campanulas, chrysanthemums, *Coreopsis*, geums, lavender, *Lychnis*, *Phlox*, roses, rudbeckias, solidagos and willows, but insecticides are seldom necessary and a powerful jet of water will normally remove cuckoo spit and nymphs.

frost the condition occurring when air temperature falls below 0°C/ 32°F, the freezing point of water. When atmospheric moisture freezes to produce ice crystals on the ground or on other surfaces, the condition is known as white or hoar frost. Temperatures between 0° and –3°C/32–27°F are generally referred to as slight frost, between –4°C/25°F and –6°C/21°F as keen frost, between –7°C/19°F and –10°C/14°F as hard frost, and below –10°C/14°F as severe frost. The susceptibility of plant tissues to frost damage varies according to species, provenance, acclimatization and site exposure. Some parts of a plant may be more susceptible than others: in fruit trees, for example, the blossom more so than the bark. The effects of frost may depend on its duration as well as its severity: repeated freezing and thawing or very rapid thawing can be particularly damaging.

The effects of frost can be lessened by choosing frost-resistant plants, by protecting plants with insulating materials including glass, and by artificial heating. Protection from freezing wind is desirable, and for frost-tender species sites likely to accumulate cold air behind barriers or in hollows should be avoided. Frost can benefit soil structure, especially on clay soils where the freezing of soil water causes the shattering of large clods.

frost heave the lifting upwards of soil and plants resulting from expansion when ice forms in the soil.

frost pocket a site where cold air accumulates behind a barrier or in a hollow, giving rise to frequent frosts.

fruit the fertilized and ripened ovary of a plant, together with any adnate parts.

fruit bud, blossom bud, flower bud a bud which on opening gives rise to flowers (or flowers and leaves), with potential for fruit production. It is distinguished from a vegetative bud by its larger size and more rounded shape.

fruit cage a wood or metal framework, to about 1.8m high, clad with plastic or wire netting, and used especially to protect soft fruits from birds.

fruit flies (Diptera: Tephritidae) small, greyish-yellow flies with mottled wings, which are distributed worldwide but mostly in the tropics. Their white maggots feed on fleshy fruits and the flowerheads of Compositae, and also mine the leaves or make galls. The CHERRY FRUIT FLY (*Rhagoletis cerasi*), which is found in Europe but not the UK, in parts of Asia and in North America, attacks cherries, honeysuckle and *Prunus* species; as many as six larvae may invade a single fruit. The APPLE FRUIT FLY (*R.pomonella*) is a major pest of apple in North America; the maggots make winding tunnels in the flesh of the fruit and also attack plum, pear, cherry, blueberry, huckleberry and hawthorn. In North America, currant and gooseberry fruits may be attacked by the CURRANT FRUIT FLY (*Epochra canadensis*). WALNUT HUSK FLIES (*Rhagoletis* species) occur throughout the US; the larvae burrow into the green husks of walnuts, turning them black and slimy. The MEDITERRANEAN FRUIT FLY (*Ceratitis capitata*) attacks peach, citrus, plum and various other fruits in the subtropics. European species include the BERBERIS SEED FLY (*Anomoia purmunda*), which damages berries of *Berberis*, *Pyracantha* and *Cotoneaster*, and the ROSE HIP FLY (*Rhagoletis alternata*), which feeds on the flesh and seed of rose hips. Of the leaf-mining species, only the celery fly (*Euleia heraclei*) is a serious pest in Europe.

Chemical control methods must be aimed at the adults. In North America, pheromone traps and biological controls are used.

fruit house a glasshouse designed for growing fruit. Typically a tall, narrow lean-to structure with fruit trees trained on the back wall. Also known as an orchard house.

fruitlet an immature fruit. See *bud stages*.

fruit rot see *blue mould*.

fruit set the open flower condition in gooseberry fruit development. See *bud stages*.

fruit setting the development of fruit following the pollination and fertilization of flowers, especially used of top and soft fruits, tomatoes and cucurbits. Fruit setting can be artificially induced by certain plant hormones. See *dry set*.

fruit swelling an early development stage of top and soft fruits. See *bud stages*.

fruit thinning the removal of a proportion of immature fruitlets from fruit plants in order to improve the size and colour of those remaining, or to prevent overcropping, or as a means of discouraging biennial bearing. It is also referred to as de-fruiting.

Fuchsia fulgens

Fuchsia procumbens

Fuchsia (for Leonhart Fuchs (1501–66), German herbalist and botanical illustrator). Onagraceae. C & S America including Mexico and Tierra del Fuego, also New Zealand, Tahiti. LADY'S EAR DROPS. 105 species, evergreen to deciduous shrubs and trees. Often drooping on slender pedicels, the flowers consist of a tubular to bell-shaped calyx with four, ovate to lanceolate, outspread lobes and four or no petals – these are often shorter and broader than the sepals, differently coloured and rolled together or spreading. The stamens and long stigma are usually exserted. Globose to oblong, berry-like fruit follow in late summer and autumn.

Most fuchsia cultivars are the result of breeding between many species, much of it undocumented. *F.magellanica* and possibly *F.coccinea* arrived at Kew around 1788. *F.magellanica* and *F.fulgens* appear to be the original parents of plants resembling modern cultivars, though perhaps 10 other species may have been involved later. *F.magellanica* has a few variants and hybrids, and *F.triphylla* probably crossed with *F.boliviana* and then with *F.fulgens* and *F.splendens*, giving rise to several cultivars grouped as Triphylla Hybrids, which typically have flowers in clusters with very long tubes and very short sepals. There are also some cultivars and hybrids among the small-flowered species.

Fuchsias are of great value for summer bedding, windowboxes, hanging baskets and containers on terraces; they will grow in any aspect except deep shade. Hardy species and cultivars are useful in permanent shrub borders and as ornamental hedges up to 2m tall, flowering from mid-summer until late autumn. The tender cultivars and species are lifted in late autumn and stored in a frost-free environment during the winter months. Garden plants will grow in any fertile, free-draining soil. Some protection from frost should be given in winter, either by earthing up the crowns or by covering them with sand or leaf litter.

Plant outdoor fuchsias in late spring and early summer; they are best planted deeply to afford the rootstock some protection and a chance of regeneration in areas where frosts are severe. Summer-bedding fuchsias are planted out in early summer. Garden plants should be watered during hot, dry weather. Prune in spring once new growth has started, cutting away all old, flowered wood.

All fuchsia species and cultivars may be grown in the glasshouse, either in pots or planted directly in the ground. A minimum winter temperature of 7°C/45°F is required for species and Triphylla types, but cultivars can be successfully overwintered at 1°C/34°F. Maintain humidity during the summer months by damping-down and misting, except when plants are in bud or flower. Water moderately in spring and plentifully in summer. Shade from direct sunlight. Under glass, pot in either a medium fertility loam-based potting mix (high-fertility for baskets) or a soilless medium with added grit (this will need more frequent feeding). Repot when growth commences in spring. Feed with a high-nitrogen fertilizer in spring to encourage growth, changing to a high-potash fertilizer in mid-season to encourage

FUCHSIA

Name	Distribution	Habit	Leaves	Flowers
F.arborescens TREE FUCHSIA	Mexico	shrub or small tree, 3–8m tall	10–20cm, oblong-elliptic to obovate-lanceolate, entire, dark green above, paler beneath	to 1cm, pink to pale purple with contrasting stamens, in dense, erect panicles
Comments: Z10				
F. × bacillaris *F.cinnabarina*; (*F.microphylla* × *F.thymifolia*)	garden origin	erect or spreading, twiggy, small shrub, 30–200cm tall	0.5–2cm, ovate to broadly lanceolate, entire to serrate, bright green	to 1.5cm, rose-pink to red, solitary
Comments: Z9. includes 'Reflexa' with minute leaves and very small, cerise flowers				
F.boliviana syn. *F.cuspidata*	N Argentina to Peru	erect shrub or small tree, 2–4m tall	5–20cm, narrowly elliptic to broadly ovate, finely toothed, upper surface deep green and smooth to downy, often with red-tinted veins, undersurface paler and downy	5–8cm, pale pink to vermilion with a scarlet limb, narrowly tubular, pendulous in short racemes or panicles
Comments: Z10. 'Alba' has pale green leaves and pure white flowers				
F.denticulata	Peru and Bolivia	erect to scrambling or sprawling shrub, 4–10m tall	4–17cm, elliptic to lanceolate, denticulate, firm, dark green and smooth above, paler or red-tinted beneath	5–7cm, pink to pale red tipped green-white, petals orange to vermilion, tubular, clustered toward branch tips
Comments: Z10				
F.excorticata	New Zealand	shrub or small tree 1–2m, with peeling and papery, rust-coloured bark	5–10cm, ovate-lanceolate, more or less entire	2–3cm, green marked maroon with blue anthers, solitary, long-stalked
Comments: Z9				
F.fulgens	Mexico	tuberous-rooted shrub, 0.5–3m tall	9–15cm, ovate-cordate with fine red teeth, glandular, pale green above, flushed red beneath	5–8cm, pink to orange or fiery scarlet, tubular, pendulous , in racemes
Comments: Z10				
F.magellanica *F.globosa*, *F.gracilis*	Chile, Argentina, widely naturalized elsewhere	erect shrub to3m tall	1.5–5cm, ovate-elliptic, wavy-toothed, mid green, sometimes tinted red	2–4cm, deep crimson to pink or white with purple to pink or white petals, solitary or paired, pendent
Comments: Z6. A variable and hardy species; includes var. *eburnea* ('Alba'), with white flowers, var. *gracilis*, a slender and highly floriferous shrub with small, scarlet and violet flowers, and var. *molinae*, with the palest shell pink flowers. Variegated cultivars include 'Variegata', with leaves edged cream, and 'Versicolor' ('Tricolor') with foliage tinted red or pink at first, becoming grey-green tinted silver and pink and irregularly edged creamy white. One of the most commonly grown cultivars is 'Riccartonii', a robust, suckering and hardy shrub with dark calyx and broader sepals.				
F.microphylla *F.minutiflora*	Mexico to Costa Rica and Panama	bushy or scrambling, densely twiggy and small shrub, 0.5–5m tall	0.6–4cm, broadly lanceolate to oblanceolate, entire or toothed, bright green	0.5–1cm, white to pink, magenta, cerise or deep red with similarly coloured, or paler, or purple petals, solitary
Comments: Z10				
F.procumbens	New Zealand	wiry-stemmed, prostrate to climbing or cascading shrub to 1m tall	0.6–2cm, more or less rounded with a cordate base	1–1.8cm, gold to pale orange suffused green and tipped with maroon, with blue anthers, erect, solitary
Comments: Z9				
F.splendens	Mexico to Costa Rica	erect shrub, 0.5–2.5m tall	3.5–13cm, ovate to cordate, serrate, smooth or hairy, often flushed red and hairy beneath	3–6cm, rose pink to orange-red tipped green, solitary, pendent to spreading
Comments: Z9				
F.thymifolia	Mexico to Guatemala	erect to spreading, twiggy shrub, 0.5–3m tall	0.8–6cm, elliptic to ovate, entire or toothed, pale, bright green	0.5–1cm, green-white to pale pink, the colour deepening with age, solitary, pendent
Comments: Z9				
F.triphylla	Hispaniola	erect shrub or subshrub, 0.3–2m tall	2.5–10cm, elliptic to oblanceolate, entire to finely toothed, mid- to deep green, hairy and often tinted bronze above, flushed purple beneath	4–6cm, orange-red to coral pink, narrowly tubular with a bulbous base, in erect to nodding racemes
Comments: Z10				

flowering and ripen the growth. Greenhouse plants may be set outside in autumn to ripen the wood before they are overwintered in frost-free conditions. They must not be allowed to dry out completely during this period.

For plants grown under glass, little pruning is needed other than in spring, when the previous season's plants are started into growth and are cut back to 2–3 nodes from the main branch, primarily to shape the plant. During the early part of the growing season, pinch out the growing tips to encourage the plant to bush out. When training standards, leave the leading shoot intact and pinch out the side shoots until the main stem has reached two-thirds of the desired height of the finished standard. The next five or six pairs of side shoots are left and the growing tip then removed. Side shoots are pinched out at every 2–3 pairs of leaves until the head of the standard is formed. For weeping standards, train laterals downwards to a wire framework.

Propagate from seed or by cuttings.

full blossom an early development stage of top fruits. See *bud stages*.

fumigation the application of a pesticide to an enclosed growing area in the form of vapour or smoke. On a large scale, for example, in stores or commercial greenhouses, volatile substances are vaporized to produce toxic gases. Similar formulations may be injected into the soil or applied under plastic sheeting for soil sterilization. In the smaller greenhouse, fumigation with smokes is a suitable method.

fungicide a substance used to control diseases caused by fungi. Fungicides may eradicate established infections, prevent spread, or both. Most have contact action, a few are systemic. Fungicides are most usually applied as sprays or dips, but seed dressings can be very effective. The term fungistat may be used to describe a substance capable of preventing growth of a fungus without killing it.

Commonly used fungicides are derived from a wide range of chemicals. Those available for garden use are covered by regulations designed to ensure efficacy, while protecting users, consumers and the environment.

The effectiveness of fungicides may depend on environmental conditions, and always on timely application. They should only be used where essential. Careful husbandry and good management of the growing environment, including rotation, attention to drainage and general crop hygiene, will all mitigate against fungal attack. Over use may lead to the build-up of resistant fungus strains.

fungus an organism belonging to a group distinct from plants and animals. They are incapable of photosynthesis, and obtain sustenance either as parasites of other living organisms or as saprophytes of dead organic material. Fungi reproduce by spores and include such diverse types as yeasts, lichen fungi, moulds, mushrooms and pathogens of plants and animals.

fungus gnats (Diptera: Sciaridae) midge-like flies, up to 4mm long, mostly black or dark grey with relatively long legs. They are often seen moving rapidly over the soil surface on pot plants and in cucumber crops, occasionally making short slow flights. Fungus gnats occur in fungi, leafmould, rotting wood, manure and compost heaps, and soils containing any of these are readily infested; they are often abundant in peat compost. Under favourable conditions, such as in greenhouses where temperatures are high and on

moist media with high organic content, breeding continues throughout most of the year.

Most are harmless, but some may damage seedlings, cuttings and young plants. *Lycoriella auripila* and *L.solani* are important pests of mushrooms. The larvae burrow into small buttons, and the stalk and cap of mature specimens. In greenhouses, avoid accumulations of the plant debris on which flies breed; where the gnats are troublesome, insecticides can be used, either pre-mixed with the growing medium or watered as a drench. Non-chemical control is possible with sticky traps and with the pathogenic eelworm *Steinernema bibionis* or the predatory mite *Hypoaspis miles*.

In North America, the POTATO SCAB GNAT (*Pnyxia scabiei*) can be a serious pest of potatoes. The larvae invade tubers and continue to develop after harvest. Clean seed and crop rotation provide good control. Also known as SCIARID FLIES.

funicle, funiculus the stalk or thread bearing the ovule or seed and attaching it to the placenta.

funnelform funnel-shaped, e.g. a corolla tube that widens gradually from its base.

furcate forked; the terminal lobes prong-like.

Furcraea (for Antoine François de Fourcroy (1755–1809), French chemist). Agavaceae. Mexico, C America. Some 20 species, evergreen, perennial herbs with large rosettes of sword-like, succulent leaves. Tubular, fragrant flowers composed of six, oblong tepals are carried in very tall, scapose panicles. Cultivate as for *Agave*, with a minimum temperature of about 6°C/43°F. *Furcraea* rarely sets seed but frequently produces large numbers of bulbils. *F.foetida* (syn. *F.gigantea*; MAURITIUS HEMP, GREEN ALOE; C America; leaves to 180cm, sword-shaped, in a basal rosette, with entire to spiny-toothed margins; inflorescence to 7m tall; flowers strongly scented, milky white inside; var. *mediopicta*: spineless, dark green leaves, striped cream and brushed with pale to blue-green).

fusarium patch disease (*Monographella nivalis*) a common and damaging fungal disease of lawns and sports turf, especially *Poa* species in the UK. It causes patches of dead grass, dark brown or orange at the edges, and wet and slimy, with white or pink fungal bodies matting dead leaves. It is favoured by wet conditions, high nitrogen and dense thatch. Control is by good management or by a recommended fungicide applied in the growing season.

fuseau see *dwarf pyramid*.

fusiform spindle-shaped; swollen in the middle and tapering narrowly to both ends.

FYM farmyard manure; cow, pig or horse dung, usually mixed with straw and valuable for improving soil fertility. See *manure*.

G

Gagea (for the amateur botanist Sir Thomas Gage (1761–1820), of Hengrave Hall, Suffolk). Liliaceae (Liliaceae). Europe, C Asia, N Africa. About 50 species, small, bulbous perennial herbs with linear leaves. Borne singly or in umbels or racemes in spring, the flowers are cylindric to campanulate or rotate and consist of six segments. The following species need shallow soils and hot, dry conditions in summer for bulb ripening. They are best grown in shallow pans in a sunny bulb frame, watered copiously when in growth, and dried out when dormant, with just sufficient moisture left to prevent shrivelling. Outdoors in zones 7 and over, grow them in well-drained, sandy soils in full sun, in situations that are dry but not completely arid in summer. Provide adequate moisture when in growth. Propagate by division of bulbils or by seed in spring, under glass.

G.graeca (syn. *Lloydia graeca*; E Mediterranean, Greece, Crete; flowers 1–1.5cm across, white with purple stripes); *G.peduncularis* (N Africa, Balkans, Aegean; flowers 1–2cm across, yellow).

Gaillardia (for M. Gaillard de Charentoneau, French patron of botany). Compositae. S US, Mexico, S America. BLANKET FLOWER. About 30 species, annual, biennial and perennial herbs with entire, toothed or pinnatifid, hairy leaves. Daisy-like flowerheads are produced in summer, the ray florets yellow to red, tipped with yellow or red-purple, the disc florets purple to red-bronze. Grow in sun in any moderately fertile, well-drained soil; they are hardy but generally perform best on poor soils and in hot dry conditions. Cut back after flowering. In zones experiencing heavy snows, protect the crowns of perennial species with a dry mulch covered with pea sticks or brushwood. Propagate by seed, cultivars by root cuttings in winter also by division.

G.aristata (N America; perennial to 70cm; flowerheads to 10cm across, ray florets yellow or yellow stained red, especially at base); *G. × grandiflora* (*G.aristata × G.pulchella*); like *G.aristata* but larger and more vigorous; includes many cultivars with flowerheads ranging from pure yellow to burnished gold, dark orange, bronze-red or blood red, in pure shades or bicolours); *G.pulchella* (INDIAN BLANKET, FIRE FLOWER, FIRE WHEELS; US, Mexico; annual or short-lived perennial to 60cm; flowerheads to 6cm across, ray florets red with yellow tips or bands, or entirely yellow or red; largely represented in cultivation by 'Lorenziana', with many enlarged and tubular disc florets in red, yellow or bicolours).

Galanthus (from Greek *gala*, milk, and *anthos*, flower). Amaryllidaceae. W Europe (*G.nivalis* often naturalized) to Iranian Caucasus and Caspian Sea. SNOWDROP. About 15 species, dwarf to small bulbous perennials ultimately forming clumps. Strap-shaped to elliptic, the leaves emerge from a slender, tubular sheath alongside the flowers. Solitary flowers appear in autumn, winter and early spring, nodding atop slender scapes. They are bell-shaped with three larger outer petals, usually oblong to obovate and white, and three smaller, inner petals which form a cup and are marked green.

Grow in dappled sun on moist but well-drained, humus-rich soils that do not dry out excessively in summer; a mulch of leafmould is beneficial. Dried snowdrop bulbs do not establish easily. Plants are better if bought and planted 'in the green', that is, after flowering and with fully developed leaves. *G.reginae-olgae* requires more sun and protection from cold winds than other species. In pots, grow in a mix of equal parts loam, leafmould and sharp sand, in a cool shaded frame in summer, bringing into the alpine house in early winter. Most species are reliably hardy to at least –15°C/5°F. *G.ikariae* is more tender and will suffer below –5°C/23°F. To bring into bloom in mid-winter, pot up in spring after flowering and move from the frame to cold greenhouse in early autumn when bulbs are dormant. Increase by division of clumps after flowering or sow ripe seed sown in a cool shaded frame.

Grey mould is indicated when infected shoots first appear above the ground covered with the grey mycelium and spores of the fungus; the rot extends into the bulbs. Small black sclerotia develop on affected bulbs and contaminate the soil, so that any bulbs bearing these resting stages of the fungus should be discarded.

G.caucasicus (Caucasus; winter-flowering; leaves oblong, glaucous; scape to 14cm, outer flower segments obovate, 2-3cm, inner segments marked green at tips; var. *hiemalis*: smaller, flowering in late autumn and early winter; 'Lady Beatrix Stanley': leaves glaucous, flowers double with narrow, claw-like outer segments and inner segments spotted green at tips, probably a hybrid; 'Straffan': vigorous with two scapes per growth); *G.elwesii* (Balkans, W Turkey; winter-flowering; leaves broadly oblong to elliptic, glaucous, hooded at tip; scape to 10cm, flowers honey-scented, outer segments broadly obovate to rounded, usually flared, to 3cm, inner segments marked green at tip and base); *G.gracilis* (syn *G.graecus*; Bulgaria, Greece, Turkey; winter-flowering; leaves narrow, twisted, glaucous; scape to 10cm, flowers sometimes violet-scented, outer segments broadly obovate and rounded, to 2.5cm, inner segments marked green at apex and base); *G.ikariae* (syn. *G.latifolius*; Aegean Is., Turkey, Caucasus; winter-flowering; leaves strap-shaped to broadly oblong, bright, glossy green; scape 7-15cm, outer flower segments oblong to spathulate to 3cm, inner segments notched and marked green at tips); *G.nivalis* (COMMON SNOWDROP; Europe; winter- and early spring-flowering; leaves linear to narrowly strap-shaped, more or less glaucous; scape to 10cm; flowers sometimes faintly scented, outer segments oblong, to 2cm, inner segments marked green at tip; 'Atkinsii': tall with large, elongated flowers;

'Florepleno': flowers double with many, sometimes ruffled inner segments; 'Lutescens': small and delicate with a yellow-green scape, yellow ovary and yellow markings on the inner segments; 'Pusey Green Tips': flowers with a long, ear-like bracteole, long, arching, green-tipped outer segments and numerous ruffled inner segments; 'Scharlockii': flowers with two long, ear-like bracteoles and outer segments marked green at tips); *G.plicatus* (E Europe; winter-flowering; leaves broadly strap-shaped with recurved margins and loosely folded upper surface, glaucous; scape 10-20cm, outer flower segments oblong to narrowly elliptic, to 2.5cm, inner segments marked green at tips; subsp. *byzantinus*: inner segments marked green at tips and base); *G.reginae-olgae* (Sicily, Greece, SW Turkey; typically autumn-flowering; leaves linear with a glaucous stripe, produced after flowers; scape to 10cm, flowers faintly scented, outer segments to 2.5cm, inner segments marked green at tips; subsp. *corcyrensis*: plants flowering in late autumn with leaves developed; subsp.*vernalis*: plants flowering in winter and early spring with leaves developed); *G.rizehensis* (Turkey; late winter- and early spring-flowering; leaves linear, dull deep green; scape 10-20cm, outer flower segments oblong to oval, to 2cm, inner segments marked green at tips).

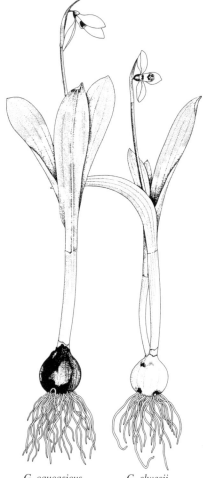

Galanthus plicatus   *G. nivalis*   *G. caucasicus*   *G. elwesii*

Galax (from Greek *gala*, milk; meaning obscure). Diapensiaceae. SE US. WANDFLOWER, WAND PLANT, BEETLE WEED, GALAXY, COLTSFOOT. 1 species, *G.urceolata* (syn. *G.aphylla*), a tufted, evergreen perennial herb with a matted, creeping rootstock. The leaves are basal and slender-stalked with blades 2.5–7.5cm in diameter, circular to cordate and toothed. Glossy dark green, they often become bronze in cold weather or bright sunlight. Small, white flowers are carried in spike-like racemes to 45cm tall in late spring and early summer. Fully hardy and suitable for the woodland garden and for underplanting in shrubberies. Excellent groundcover, the foliage also lasts well when cut. Grow in moist, leafy, lime-free soils with part shade, and mulch annually in spring with acidic organic matter, such as pine needles. Propagate by division in early spring, or by seed sown in early spring or in autumn and kept moist in the shaded cold frame.

Galega (from Greek *gala*, milk; once thought to increase milk flow). Leguminosae. S Europe to Asia Minor, tropical E Africa. Some 6 species, erect, perennial herbs with pinnate leaves. The pea-like flowers are carried in axillary racemes in summer. A vigorous perennial for the large herbaceous and mixed border and for more informal plantings in the wild garden, it is hardy at least to –15°C/5°F. Grow in sun or light shade on any moisture-retentive soil. Propagate by division or by seed soaked for 12 hours before sowing. *G.officinalis* (GOAT'S RUE; C and S Europe, Asia Minor; to 1.5m; flowers to 1cm long, white to lavender; includes the Hartlandii Group, syn. *G.* × *hartlandii*, with flowers in shades of white, palest lavender, lilac blue and mauve-pink, also combinations thereof).

Galium (from Greek *gala*, milk; referring to former use of *G.verum* in curdling milk for cheese manufacture). Rubiaceae. Subcosmopolitan (temperate). BEDSTRAW, CLEAVERS. Some 400 species, annual or perennial herbs with clumped, erect to climbing stems and whorled leaves. Starry, 4–parted flowers are borne in cymes in summer. *G.odoratum* develops its persistent, sweet coumarin fragrance on drying and is an ingredient of pot-pourri and sachets, traditionally used as a moth-deterrent. It has been collected and cultivated since the Middle Ages at least, to flavour wines, brandy, apple jellies, sorbets, creams and fruit salads. As a tisane (allow to wilt slightly before infusing), it has a fresh grassy flavour and a gently sedative effect. Harvest before flowering, and dry in a cool dry and shaded place. As an ornamental, it provides useful groundcover for cool shade, under shrubs or in the woodland garden. It is hardy to at least -15°C/5°F. Propagate from seed sown fresh in late summer. Divide in spring or at any later time during the growing season.

G.odoratum (syn. *Asperula odorata*; WOODRUFF, SWEET WOODRUFF; Europe, N Africa; creeping to ascending perennial to 45cm with slender, 4–angled stems, the dried shoots strongly scented of hay; leaves whorled, 1–5cm, dark to bright green; flowers white in terminal clusters).

gall a localized swelling or outgrowth resulting from an abnormal increase in the number of plant cells or from individual cells becoming enlarged. Galls can be caused by bacteria (for example, the beneficial nitrogen-fixing root nodules on legumes caused by *Rhizobium* species, or the damaging CROWN GALL (*Agrobacterium tumefaciens*); by fungi (for example, CLUBROOT of brassicas, caused by *Plasmodiophora brassicae*); by eelworms (for example, ROOT KNOT EELWORM, *Meloidogyne* species, on tomatoes); by viruses, which may cause LEAFY OUTGROWTHS referred to as enations; by mites (for example, BIG BUD of blackcurrants, caused by *Cecidophyopsis ribis*); or by insects (for example, RED LEAF BLISTER of red and white currants caused by the aphid *Cryptomyzus ribis*, OAK APPLES caused by the gall wasp *Biorhiza pallida* and ROOT GALLS on brassicas caused by the TURNIP GALL WEEVIL, *Ceutorhynchus pleurostigma*).

gall midges (Diptera: Cecidomyiidae) small fragile flies, up to 3mm in length, mostly with delicate hairy wings. The legless larvae, up to 4mm long, have diverse feeding habits depending upon the species to which they belong. Some are predators, mainly of aphids or, sometimes, mites; others feed on decaying organic matter; some feed on fungi, especially rusts and mildew; many are plant parasites, sometimes inducing plant galls.

Several gall midges are pests of top and soft fruit, including two common in both Europe and North America. The PEAR MIDGE (*Contarinia pyrivora*) has yellow or white larvae which infest pear fruitlets, causing them to become blackened and fall prematurely; the PEAR LEAF MIDGE (*Dasineura pyri*) has white larvae which live and feed within the upwardly rolled margins of pear leaves. Other species that attack fruit in Europe include the BLACKCURRANT LEAF MIDGE (*D.tetensi*), the RASPBERRY CANE MIDGE (*Resseliella theobaldi*), and a related species, the RED BUD BORER (*R.occuliperda*), which is a pest of newly-budded fruit trees and roses.

European species responsible for damage to vegetables include the SWEDE MIDGE (*Contarinia nasturtii*), whose yellow-white larvae feed on young leaves and shoots of various brassicas, often causing blindness (where growing points fail to develop), and the PEA MIDGE (*C.pisi*), whose larvae feed on the flowers, leaves and pods of peas.

Gall midges common on ornamental plants in both Europe and North America include the honey locust or GLEDITSIA GALL MIDGE (*Dasineura gleditchiae*), which constructs leaflets into pod-like structures. European species targetting ornamentals include the YEW GALL MIDGE (*Taxomyia taxi*), which produces galls resembling globe artichokes on the tips of young shoots. The ROSE MIDGE (*Dasineura rhodophaga*) is an important pest of roses in North America, its white to light orange larvae deforming shoot tips with malformation and death of flower buds.

Control by hand-picking and destruction of the galls. Residual contact insecticides can be effective if applied repeatedly from the onset of symptoms.

gall mites (Acarina: Eriophyidae) very small mites, up to 0.25mm long, with white or yellow, elongated, maggot-like bodies; they differ from other mites in having two pairs of legs rather than four. Gall mites are capable of moving only short distances and are mainly dispersed by wind. All are plant feeders and most attack a single plant species or closely related plants. Often there is no discernible damage but some mites are associated with characteristic galls on the leaves of woody plants, such as the NAIL GALL MITE of Europe (*Eriophyes tiliae*), which makes small, pointed red galls on the upper surface of lime (*Tilia* species) leaves. Several species common to Europe and parts of North America live in leaf buds causing them to swell, including the YEW GALL MITE (*Cecidophyopsis psilaspis*). Amongst the rust mite species that feed openly on leaves, the APPLE RUST MITE (*Aculus schlechtendali*) is widespread in Europe and North America; it causes severe russeting of leaves and fruitlets.

The BROOM GALL MITE (*Aceria genistae*) invades the vegetative buds of *Cytisus*, which as a result develop into cauliflower-like growths. Several species of gall mite have been associated with witches' brooms, which commonly occur on birch and willows, although the implication of the mites has never been confirmed.

Gall mites are difficult to control in gardens due to the restricted availability of effective chemicals. Most damage is only cosmetic and can be tolerated; where damage is more severe, it may be necessary to replace plants.

gall wasps (Hymenoptera: Cynipidae) small, inconspicuous insects, up to 8mm long, with brown or black bodies and two pairs of wings. They lay eggs in plant tissue, the majority infesting oaks and others living mostly on wild roses. As the eggs hatch, the larvae feed on the plant tissue which swells up around them to form galls. Within the life cycle of oak gall wasps, two different types of gall may be produced on the host plant to accommodate sexual and asexual generations. Gall wasps seldom cause significant damage to warrant control measures; see *bedeguar.*

Galtonia (for Sir Francis Galton (1822–1911), anthropologist, explorer and geneticist). Liliaceae (Hyacinthaceae). Eastern South Africa. 3 species, bulbous perennial herbs. The leaves are rather fleshy and narrowly lanceolate to strap-shaped. Long-stalked and cylindric to conical, the flower spikes arise in late summer and autumn. White or green, the flowers are showy, funnel- to bell-shaped and nodding. *G.candicans* is the hardiest species. With deep mulch protection, it may be grown where winter temperatures fall to –15°C/5°F; the remaining species are slightly less cold-tolerant, suffering at temperatures between –5 to –10°C/23–14°F. Grow in a sunny, sheltered position on any well-drained soil that does not dry out when plants are in leaf. In harsh climates, lift the bulbs in winter and store them in dry, frost-free conditions prior to replanting in spring. Propagate by ripe seed sown in spring, or by offsets.

G.candicans (SUMMER HYACINTH; to 1m tall; leaves 50–100 × 5cm; flowers fragrant, to 3cm, white faintly tinted green at base); *G.princeps* (differs from *G.candicans* in its flowers, strongly tinted green at base and on lobes, these are carried in shorter, broader spikes); *G.viridiflora* (to 1m; flowers 2–5cm, pale green with white-edged lobes).

gamete a fertile reproductive cell of either sex, which unites with another gamete at fertilization to form a zygote.

gametophyte the sexual stage in the life-cycle of a plant showing alternation of generations, when the gametes are produced on a specially developed form of the plant.

gamopetalous with petals united by their margins, in the form of a tubular or funnelform corolla, at least at their bases.

gamophyllous with leaves or leaf-like organs united by their margins.

gamosepalous with sepals united by their margins.

garden line a length of thin cord or string wound directly or by a spool on to a stout wooden or metal pin, with the free end tied to a second pin. Pulled taught between two marker canes, it is used to lay out straight lines for seed drills or planting rows.

gardenesque a style of gardening promoted by J.C. Loudon in the mid-19th century and based on the principal of attributing equal importance to carefully cultivated exotic plants and their setting. It aimed to make the garden appear as a work of art rather than of nature. Later writers used the term derogatorily when referring to Victorian gardens, which they felt lacked a sense of design.

Gardenia (for Dr Alexander Garden (1735–91), botanist of South Carolina and correspondent of Linnaeus). Rubiaceae. Tropical Old World. Some 250 species, shrubs or trees with leathery leaves and axillary or terminal flowers. Typically white to cream and produced at various times of year, the flowers are either solitary or in few-flowered cymes and usually fragrant. They consist of a cylindric to campanulate corolla tube and a spreading limb of 5–12 lobes. Grow in a humus-rich, acidic, soilless medium, with high humidity and bright indirect light. Water freely when in growth and moderately at other times. Maintain a minimum temperature of 15°C/60°F in winter. In areas where tap water is alkaline, use rainwater or apply sequestered iron at approximately fortnightly intervals. Propagate in spring by greenwood cuttings. Alternatively, take ripewood cuttings, with a heel, in late summer or autumn, and root these in a sandy mix in a closed case with a bottom heat.

G.augusta (syn. *G.jasminoides;* GARDENIA, JASMIN, CAPE JESSAMINE; E Asia; evergreen shrub or tree to 12m tall, usually far smaller in cultivation; leaves glossy dark green; flowers 5–10cm across, white to ivory, intensely fragrant, rose-like in shape and usually double with a shortly cylindric corolla tube to 3cm long; includes several cultivars, among them 'Fortuniana', with large, heavily perfumed flowers turning from pure white to ivory); *G. thunbergia* (WHITE GARDENIA; South Africa; evergreen shrub or tree to 5m; leaves glossy; flowers white to cream, fragrant, with a limb of overlapping segments each to 3cm long, and a narrow corolla tube to 7cm long).

garlic (*Allium sativum*) a hardy onion-like perennial, probably of Central Asian origin but cultivated in Egypt before 2000 BC; grown for its strongly aromatic oily bulbs, consisting typically of a number of offsets, or cloves. The cloves are used, either raw or cooked, as a savoury flavouring, and the young green leaves may be finely chopped in salads; also for medicinal purposes.

Bulb formation is encouraged by increased day-length and higher temperatures, but garlic is hardier and more easily grown than often thought. Cool storage of dormant cloves at 0–10°C/32–50°F hastens bulbing, whilst high temperature delays or prevents it.

Plant only healthy individual cloves, of at least 13mm diameter or about 10g in weight, in early spring (from cold store) or in autumn. Space 15cm apart in rows 30cm apart, and cover with soil to a depth of 25mm above the clove tip. Garlic can be raised in small pots or cell modules prior to setting out, although a cold period out-of-doors is necessary. Bulbs become dormant in summer, and should be lifted when the leaves die down. After drying, they can be stored as ropes through to spring, or for longer in cold storage. Pink and white forms exist and some store better than others.

garotting a method of restricting growth and improving the yield of fruit trees by twisting a metal wire tightly around a branch or the main trunk. It is similar in principle to bark ringing but less drastic in effect.

Garrya (for Nicholas Garry (d. l830) of the Hudson Bay Company, who helped David Douglas on his plant collecting expeditions in western North America). Garryaceae. W US, Mexico, West Indies. Some 18 species, dioecious, evergreen trees or shrubs with leathery leaves. Inconspicuous flowers are carried amid overlapping, paired bracts along pendulous, tassel-like catkins - these are more showy on male plants. Hardy in zone 7, especially if grown as a wall shrub. Plant in spring on any well-drained soil, in sun or shade. Propagate in late summer by cuttings of semi-ripe side shoots with a heel, or, in late winter, by hardwood basal cuttings rooted under mist, or in a case. Prune only to establish shape or to remove deadwood; do this immediately after flowering since catkins are borne at the tips of the previous season's wood. Purple blotches or black-

ening of the leaves are symptoms of frost damage and wind scorch. *G.elliptica* (SILK TASSEL, TASSEL TREE; W US; shrub or small tree to 4m tall; leaves to 8cm, oblong to elliptic, wavy-edged, tough, glossy grey-green to matt dark green; male catkins to 20cm long, grey narrow, hanging in clusters in winter and spring; includes 'James Roof', with deep sea green leaves and silver-grey to mushroom-coloured catkins, to 20cm long and in abundant clusters).

Gasteria (from Greek *gaster*, belly, alluding to the inflated base of the perianth tube). Liliaceae (Aloeaceae). South Africa. 14 species, compact, more or less stemless perennial succulents. The fleshy 'leaves are strap-shaped and distichous or spirally arranged in compact rosettes. Nodding on a lax, simple or sparingly branched raceme, the flowers are tubular and pink to vermilion tipped green. Cultivate as for the smaller *Aloe* species.

G.bicolor (syn. *G.caespitosa*; leaves 3–22cm, dark green spotted white in obscure crossbands, tip pointed; var. *liliputana:* syn. *G.liliputana*, miniature with leaves to 6cm long, blotched white); *G.carinata* (leaves 3–12cm, keeled beneath, dark green covered in pale tubercles; var. *verrucosa:* syn. *G.verrucosa*, leaves not keeled but densely covered in white tubercles).

Gaultheria (for Dr Gaultier, a mid-18th-century physician and botanist from Quebec). Ericaceae. Americas, Japan to Australasia. (Including *Pernettya*.) Around 170 species, evergreen shrubs and shrublets with tough-textured leaves and small, urn-shaped flowers in spring and early summer. These are followed by spherical fruits, often vividly coloured, lasting well and with a scent of winter-green. Fully hardy. Grow in sun or semi-shade in moist, humus-rich, lime-free soils. Prune only when necessary to remove deadwood and weak growth. Vigorous species such as *G.shallon* withstand close clipping. Species formerly included in *Pernettya* are particularly useful when mass-planted as medium-height groundcover. Species such as *G.mucronata* and *G.pumila* may require one male pollinator to 5–10 fruiting plants to fruit well. Propagate by greenwood cuttings in summer, by division in early spring or autumn, or by seed sown in early spring.

G.cuneata (W China; dwarf, compact shrub to 30cm; leaves 1–3cm, ovate to obovate, glandular-serrate, leathery; flowers white; fruit to 0.6cm, white); *G.forrestii* (SW China; rounded bush to 1m; leaves 5–9cm, narrowly ovate to oblong or oblanceolate, bristly-serrate, leathery; flowers to 0.5cm, fragrant, milky white; fruit to 0.6cm, blue); *G.miqueliana* (Japan; shrub to 30cm; leaves 2–4cm, ovate to bluntly obovate, glandular-serrate; flowers to 0.4cm, white; fruit to 1cm, white or pale pink); *G.mucronata* (syn. *Pernettya mucronata*; Chile, Argentina; rigidly branched, suckering shrub to 1.5m; leaves 0.8–1.8cm, oval to elliptic, sharply pointed, entire to toothed, rigid, dark green; flowers to 0.6cm, white to pink; fruit to 1.2cm, white, rose or lilac to crimson or purple-black; includes

Gaultheria shallon

Garrya elliptica

numerous cultivars); *G.nummularioides* (Himalaya; prostrate shrublet; leaves 0.6–1.5cm, ovate to elliptic, rounded, bristly-serrate, rugose, dull green; flowers to 0.6cm, white to pink or red-brown; fruit to 0.8cm, blue-black); *G.procumbens* (WINTERGREEN, CHECKERBERRY, TEA BERRY, MOUNTAIN TEA; eastern N America; creeping shrublet to 30cm; leaves 2–5cm, elliptic to obovate, finely crenate to bristly serrate, otherwise smooth, mid- to deep green; flowers to 0.7cm, white to pale pink; fruit to 1.5cm, dark pink to deep red, flesh white to pink, strongly scented of wintergreen); *G.pumila* (syn. *Pernettya pumila*; southern S America, Falkland Islands; prostrate shrub; leaves to 0.6cm, ovate to lanceolate, blunt to rounded, leathery, glossy, smooth; flowers to 0.6cm, white; fruit to 2cm, white or red-tinted; 'Harold Comber': fruit large, deep pink); *G.shallon* (SALAL, SHALLON; Alaska to California; low-growing, suckering and spreading shrub to 60cm; leaves 5–10cm, broadly ovate, apex acute, bristly serrate; flowers to 0.5cm, white tinged pink, in drooping racemes; fruit to 1cm, red ripening with a black tinge); *G.tasmanica* (syn. *Pernettya tasmanica*; Tasmania; mat-forming shrub; leaves to 0.8cm, oblong, acute, more or less crenate, leathery and glossy; flowers to 0.6cm, white; fruit to 1cm with a vivid red, enlarged fleshy calyx); *G.trichophylla* (W China, Himalaya; wiry-stemmed, cushion-forming shrub; leaves to 1cm, elliptic or ovate to narrowly oblong, bristly-serrate; flowers to 0.4cm, red to pink or white; fruit to 1cm, pale blue); *G. × wisleyensis* (*G.shallon × G.mucronata*; syn. × *Gaulnettya wisleyensis*; vigorous, suckering shrubs to 1m, forming dense thickets of small, tough, dark green leaves, the flowers white to pink or red-tinted, the fruit showy and white to purple-red or ruby; includes numerous cultivars).

Gaura (from Greek *gauros*, superb, referring to the flowers). Onagraceae. N America. 21 species, annual, biennial or perennial herbs. Borne in spikes in summer and autumn, the short-lived flowers are tubular to funnel-shaped and consist of four, clawed petals. A graceful hardy plant for herbaceous borders, the wild garden or collections of native plants. Plant in sun on any well-drained soil. Propagate from seed in spring or early summer. *G.lindheimeri* (Texas, Louisiana; willowy perennial to 1.5m tall; flowers to 3cm across, white suffused with pink in loose, erect spikes).

Gaylussacia (for J.L. Gay-Lussac (1778–1850), French chemist). Ericaceae. Americas. Some 40 species, deciduous or evergreen shrubs. The flowers are small and tubular, urceolate or campanulate with five lobes. They are carried in axillary racemes in late spring and give rise to berry-like fruit. Fully hardy. Plant on moist, sandy, acid soils in the shrub border, open woodland or rock garden, in part shade or full sun. Propagate by softwood cuttings in summer, by division of rooted suckers or by seed in autumn. *G.baccata* (BLACK HUCK-LEBERRY; E US; erect, deciduous shrub to 1m tall; leaves pale green turning red in autumn, resinous; flowers dull red; fruit black, glossy, edible).

Gazania (perhaps named for Theodore of Gaza (1398–1478), who translated the botanical works of Theophrastus into Latin). Compositae. Tropical and southern Africa. TREASURE FLOWER. About 16 species, perennial or annual herbs. They produce showy, daisy-like flowerheads on slender stalks in summer. In dry climates that are frost-free or almost so, the following are grown as annuals or perennials in flower borders and rockeries. In cool temperate gardens, they are grown as half-hardy annuals, but, in warm sunny positions with good drainage, they may overwinter successfully. They are drought-tolerant and will withstand salt-laden winds. Propagate by seed in early spring under glass or *in situ* after the last frost. Root heeled cuttings in summer; in sand in a closed case with bottom heat; overwinter new plants at 10°C/50°F. *G.rigens* (TREASURE FLOWER; S Africa; decumbent perennial to 30cm tall; leaves oblong to spathulate, entire to pinnately lobed, dark green above, silvery beneath with margins incurved; flowerheads to 8cm across, ray florets orange with black basal spots, disc florets orange-brown; var. *uniflora*: syn. *G.uniflora*, smaller flowerheads with bright yellow ray florets; many cultivars and hybrids have been developed, including plants with very silvery, or yellow-variegated leaves, and flowerheads in tones of white, yellow, tangerine, orange, red, pink and purple-bronze).

gazebo a roofed structure designed to command a view. A gazebo may take the form of a turret or lantern on a housetop or, often, a separate garden house. It is also known as a belvedere. The term gazebo is sometimes used of a projecting window or balcony.

gazon coupe a form of parterre in which shapes are cut from turf and filled with coloured sand or gravel.

Gelsemium (from Italian, *gelsomino*, jasmine). Loganiaceae. Americas, SE Asia. YELLOW JESSAMINE, CAROLINA JASMINE. 3 species, perennial, evergreen, twining shrubs with clusters of sweetly fragrant flowers in late spring and summer - these are funnel-shaped and five-lobed. In zones that are more or less frost-free, the following species is suitable for training on trellis and pergolas or for growing on steep banks. Elsewhere, it will need the shelter of a warm, south-facing wall. In such conditions and where long, hot summers ensure ripening, it may tolerate temperatures to –10°C/14°F or below. Otherwise, grow in the cool greenhouse or conservatory. Plant in full sun or light shade. Under glass, this climber flowers more profusely if restricted at the roots. Prune hard after flowering. Propagate by seed in spring at 18–20°C/65–70°F, or by semi-ripe cuttings in a closed case with gentle bottom heat. *G.sempervirens* (FALSE JASMINE, EVENING TRUMPET FLOWER; S US, C America; to 6m tall; leaves glossy; flowers sweetly scented, to 3cm, yellow with an orange centre).

gel sowing see *fluid drilling*.

BELOW LEFT A limited palette of pink, mauve, purple-grey and lime brings ravishing results in this mixture of *Lavandula stoechas*, *Allium* 'Purple Sensation', fennel and *Euphorbia amygdaloides*. The fennel and lavender have the added advantage of aromatic foliage. The *Allium* is one of a number of bulbs that flower in high summer and bring vivid accents to monochromatic silver plantings. All three will take full sun and dry soils.

RIGHT Pastels are not the only colours produced by or capable of enhancing grey-leaved plants. Here *Malva sylvestris*, *Lychnis coronaria* and *Knautia macedonica* weave brilliantly together. Other 'silvers' with glowing red flowers include *Hedysarum coronarium*, *Lotus berthelotii*, *Glaucium corniculatum* and *Papaver triniifolium*.

BELOW RIGHT Two popular grey-leaved plants, *Echinops* and *Stachys*, frame this rich, high summer melange of *Lilium* 'Pink Perfection', *Lavatera*, purple-leaved *Sedum* and variegated *Phlox*.

Roses

FACING PAGE, TOP Climbing roses deck metal arches interwoven with perennial sweetpeas and the burgundy foliage of *Vitis vinifera* 'Purpurea'.

FACING PAGE, RIGHT A hybrid tea climber, 'Guinée' is possibly the darkest-red rose of all, sometimes appearing almost black. After a first flush, it continues to produce a few, richly scented blooms throughout the summer.

FACING PAGE, LEFT Hybrid musk 'Penelope' is a vigorous shrub rose. Its dark foliage provides a perfect backdrop for a summer-long succession of sweetly scented flowers – orange-pink in bud and opening a delicate shade of creamy pink.

ABOVE LEFT 'Canary Bird' is a hybrid between two forms of the Chinese *Rosa xanthina*. It makes a bush of cascading, slender branches clothed with ferny foliage. Muskily scented, golden flowers appear in profusion in late spring, and sometimes again in autumn.

Rosa moyesii is a large, multi-stemmed species from Western China with small, pinnate leaves and single flowers ranging in colour from pink to magenta and scarlet. The commonest clone in gardens is 'Geranium' (*right*) with vivid pillar box red flowers. This species produces an abundance of brightly coloured, flask-shaped hips. Its influence can be seen in *Rosa* 'Highdownensis' (*top right*), a pink-flowered seedling of *R. moyesii* with arching stems and a profusion of brilliant fruit.

Climbers and wall shrubs

ABOVE Two slightly tender shrubs, pineapple broom, *Argyrocytisus battandieri* and the white-flowered tree anemone, *Carpenteria californica* thrive on this warm wall. Although neither is a true climber, both appreciate some support and shelter from cold winds. Planted around them are *Campanula*, *Armeria* and *Allium*.

LEFT One of the best-loved of all climbers, pale mauve *Wisteria* contributes to this artful monochrome, with *Ceanothus* and *Geranium* x *magnificum*.

RIGHT Suited to containers in warm sunny spots, the half-hardy annual climber *Ipomoea lobata* bears flowers that shift with age from flaming red to rich yellow and creamy white. Each inflorescence shows these stages at once, hence its popular name, Spanish flag.

BELOW The flowers of this large, wall-trained hybrid tea rose 'Shot Silk' are mirrored by the soft pink leaf tips of *Actinidia kolomikta* – a woody climber related to the Chinese gooseberry, with roots and stems as attractive to cats as its foliage is to humans.

BELOW RIGHT Two climbers, sweet pea 'Blue Velvet' and *Clematis viticella* 'Étoile Violette' entwine the slats of this wooden obelisk in a cloud of inky mauves. A fast-growing hybrid of *C. viticella*, 'Étoile Violette' performs well in large containers and should be pruned hard back in mid-spring to give it a few weeks headstart before the sweet pea seedlings are added to the planting.

Another combination of wall shrub and climber – here *Clematis* 'Étoile Rose' grows through *Carpenteria californica*. This *Clematis* was raised in 1903 by Lemoine at Nancy; like its parent, *C. texensis*, it favours a warm, sheltered site in full sun.

geminate paired.

gemma an asexual reproductive body, bud-like and detaching itself from the parent plant.

gemmation a form of secondary growth in potato tubers, occurring especially after spasmodic development as a result of a drought period. Characteristically, areas near the eyes swell into smooth knob-like protuberances.

gemmiparous bearing vegetative buds.

gene a segment of chromosome which encodes for a specific characteristic; a unit of inheritance in plants.

genetic engineering a laboratory process whereby particular genes are identified, extracted and transferred from one plant cell to another, including that of a different species. By this means, desirable characteristics can be implanted and perpetuated by propagation.

genetics the study of heredity and variation; loosely used to describe the practice of plant breeding.

geniculum a knee-like joint or node where an organ or axis is sharply bent.

Genista (name used by Virgil). Leguminosae. Europe, Mediterranean to W Asia. BROOM, WOAD-WAXEN. Some 90 species, shrubs, or small trees, sometimes spiny, mostly deciduous, but appearing evergreen due to green, flattened branchlets. Pea-like, yellow flowers are carried in terminal racemes or heads in spring and summer. Grow in full sun on well-drained and not too-fertile soils. The low-growing and compact species are suited to warm, sunny ledges in the rock garden. *G.sagittalis* thrives on the tops of dry-stone walls, or trailing over the edges of raised beds. *G.hispanica* is hardy to about −20°C/−4°F and one of the best shrubs for clothing dry sunny banks. Most species are cold-tolerant to at least −15°C/5°F, including the graceful weeping tree broom, *G.aetnensis*. *G.tinctoria* can show cold tolerance to −35°C/−30°F and below. *Genista* does not shoot freely if cut hard back. Increase by softwood or semi-ripe cuttings or by seed in autumn.

G.aetnensis (MOUNT AETNA BROOM; Sardinia, Sicily; slender shrub or tree to 6m with fine, weeping branches; leaves linear, sparse or absent; flowers golden yellow); *G.canariensis* (syn. *Cytisus canariensis*; Canary Islands; bushy shrub to 2m; leaves trifoliolate, downy, each leaflet to 1cm; flowers bright yellow, fragrant); *G.cinerea* (SW Europe, NW Africa; erect shrub to 3m with long, slender branches; leaves to 1cm, narrowly elliptic, downy beneath; flowers bright yellow); *G.hispanica* (SPANISH GORSE; S France, N Spain; spiny, erect shrub to 70cm; leaves to 1cm, ovate-oblong, downy beneath, present only on flowering branches; flowers golden yellow); *G.lydia*

(E Balkans, Syria; prostrate shrub to 30cm, the branches ascending with prickly tips, grey-green; leaves to 1cm, linear to elliptic; flowers golden yellow); *G.pilosa* (W and C Europe; prostrate to erect shrub to 45cm, branches ascending, downy; leaves to 1.5cm, oblanceolate, silky beneath, dark green above; flowers golden yellow; includes prostrate and mound-forming cultivars); *G.sagittalis* (syn. *G.delphinensis*; S and C Europe; prostrate shrub to 15cm, branches green with broad wings, giving the whole a leafy, evergreen appearance; leaves to 2cm, lanceolate; flowers golden yellow); *G. × spachiana* (*G.stenopetala × G.canariensis*; syn. *Cytisus × spachianus*, *G.fragrans*; evergreen shrub 3–6m tall; leaves trifoliolate, leaflets to 1.5cm, obovate, rounded, dark green above, silky beneath; flowers yellow); *G.tenera* (Madeira, Tenerife; resembles *G.cinerea*, but with shorter, stouter branches, becoming twiggy; leaves grey-green; flowers yellow; 'Golden Showers': flowers bright gold in profusion); *G.tinctoria* (DYER'S GREENWEED; Europe, Asia Minor, Ukraine; variable, more or less erect shrub to 2m; leaves to 5cm, elliptic to oblanceolate, bright green, glabrous; flowers golden yellow; 'Royal Gold': profuse, conical panicles of rich yellow flowers).

genotype the genetic constitution of an organism; cf. *phenotype*.

Gentiana (named for Gentius, ancient King of Illyria, said to be the discoverer of medicinal properties of gentian roots). Gentianaceae. Cosmopolitan except Africa, usually in mountainous regions. GENTIAN. About 400 species, annual, biennial or perennial herbs (those here perennial). The flowers are rotate to funnel-shaped, the corolla tube sometimes pleated, and the limb divided into five to seven short and pointed lobes.

Gentiana sino-ornata

The smaller alpine species do well on scree beds with underlying moisture, especially *G.verna*. Dwarf gentians (*G.ornata*, *G.sino-ornata* and *G.acaulis*) are well-suited to moist sites in the rock garden. Although they should not be planted where they will suffer tree drip, they do appreciate some protection from fierce sunlight. *G.saxosa* is best cultivated in the alpine house, in a 2:2:1 mix of loam, sand and leafmould or coir, with a collar of sharp grit around the neck. *G.acaulis*, *G.farreri*, *G.verna* and their hybrids may also be accommodated in the alpine house, where their flowers will not be damaged by rain and their winter requirements (for cool and relatively dry conditions) can be met more easily.

Some species lend themselves to pocket plantings in paving where, given a gritty substrate, they will appreciate the coolness at their roots. These include *G.acaulis*, *G.septemfida* and *G.sino-ornata*. For border plantings, two types are useful: those forming low-growing mats, such as *G.sino-ornata*, *G.acaulis*, *G.farreri* and *G.septemfida*, suitable for drifts and groups at the border's edge, and taller-growing subjects, such as *G.asclepiadea* and *G.lutea*. These last two are shade-tolerant and may naturalize in the wild, woodland and bog gardens, on moist, acid soils.

All require perfect drainage and ample moisture at the roots during their growing season.

Sow seed as soon as ripe, thinly and combined with fine, dry sand into a light, porous, sandy mix, and plunge pots in the cold frame. All species require light for germination. Take 4–6cm cuttings of those species that do not form a central rosette and taproot; cut cleanly with a sharp knife, close to the crown, and insert firmly into a sandy sterilized mix. Place in a cool, shaded frame and keep moist. Mat-forming species with fibrous root systems may be divided in early spring. Layering, which often occurs naturally, can be induced by pegging down shoots in spring; these may be detached in the following season. Rust fungus gives rise to brown pustules on the leaves and stems.

G.acaulis (syn. *G.excisa*, *G.kochiana*; Europe, Spain to Balkans; tufted; leaves in a basal rosette; flowers solitary, to 5cm, campanulate, dark blue, spotted green within tube); *G.angustifolia* (Europe, particularly SW Alps, Jura, Pyrenees; tufted; leaves in a rosette; flowers solitary to 5cm, funnel-shaped, deep sky blue, paler and spotted green within tube); *G.asclepiadea* (WILLOW GENTIAN; Europe, Asia Minor; clump-forming with erect to arching stems to 30cm; flowers in clusters, to 3.5cm, narrowly campanulate, azure, spotted purple within white-striped throat; sky blue and white-flowered forms are also grown); *G.clusii* (C and S Europe; tufted; leaves in rosettes; flowers solitary, to 5cm, funnelform or campanulate, deep azure, paler and spotted olive green within); *G.farreri* (NW China; stems slender, branched, trailing; leaves tufted or paired; flowers solitary, to 6cm, narrowly funnel-shaped, Cambridge blue tinged green, tube white within, striped deep violet-blue and white on exterior); *G.gracilipes* (NW China; stem to 25cm, branched, decumbent; flowers solitary, long-stalked, 3.5cm, narrowly campanulate, deep purple-blue, exterior green); *G.lutea* (GREAT YELLOW GENTIAN; Europe; robust, erect to 2m, stems thick; leaves paired and opposite; flowers yellow in axillary clusters, to 2.5cm, rotate to campanulate); *G.* × *macaulayi* (*G.sino-ornata* × *G.farreri*; vigorous with low, branching and rooting stems; flowers solitary, to 6.5cm, widely funnel-shaped, deep blue, striped violet and panelled green-white outside; includes the mid-blue, late-flowering 'Edinburgh', the large, deep blue 'Kidbrooke Seedling', and the pale blue 'Wells Variety'); *G.ornata* (C Nepal to SW China; prostrate, to 10cm; flowers solitary, to 3.5cm, broadly campanulate, pale blue striped purple-blue with white panels outside); *G.saxosa* (New Zealand; tufted, stems prostrate then ascending; leaves fleshy; flowers solitary or two to five per cyme, to 2cm, white veined purple-brown, campanulate, with blunt, oblong lobes); *G.septemfida* (W and C Asia; erect or ascending, 15–30cm, clumped; leaves paired; flowers in terminal clusters, to 3.5cm, narrowly campanulate, deep blue with pale spots, paler within); *G.sino-ornata* (W China, Tibet; prostrate with sprawling and rooting stems; leaves paired; flowers solitary, to 5.5cm, funnel-shaped, deep blue, paler within, exterior with stripes of purple-blue, panelled green-

Geranium phaeum

white); *G.verna* (SPRING GENTIAN; low, tufted, flowering stems to 10cm; flowers solitary, 1–3cm in diameter, rotate, brilliant blue, sometimes pale blue, or pink to purple-red, throat white).

genus (pl. genera) the principal rank in the taxonomic hierarchy between family and species. The genus represents a single highly distinctive species (monotypic or monospecific genus) such as *Ginkgo* or, more often, a number of species united by a common suite of distinctive characters. The genus or generic name is italicized and takes a capital letter.

geocarpy the subterranean ripening of fruits developed from flowers borne above ground, as in the peanut (*Arachis hypogaea*).

geophyte a plant growing with a stem or tuber below ground; usually applied to tuberous or bulbous species from arid or semi-arid lands.

geotropism directional growth of a plant in response to gravity. Positive geotropism is demonstrated by roots growing downwards, negative geotropism by shoots growing upwards. See *stimulus movements*.

Geranium (from Greek, *geranos*, crane, referring to the long beak of the fruit; the name *geranion* was used by Discorides). Geraniaceae. Temperate regions. CRANE'S BILL. Some 300 species of mostly perennial, but also annual herbs and some subshrubs, not to be confused with *Pelargonium*, which is popularly known as geranium. Those here are perennial herbs, often hairy and to varying degrees glandular. Their leaves are typically rounded in outline and palmately divided, the segments themselves may be lobed, forked and toothed. Produced in spring and summer, the flowers have 5, usually obovate petals.

G.canariense, *G.incanum*, *G.maderense* and *G.traversii* var. *elegans* are not fully hardy. In climate zones 8 and under, they need a warm and sheltered site, fast-draining soil and protection from overwet winter conditions. They might equally be grown in the cold greenhouse or conservatory. These species are, however, easily propagated from seed and well-worth risking outdoors. The remaining species are hardy.

Several of the smaller plants thrive on the rock garden, old walls, among paving stones and in raised beds and sinks. They include *G.cinereum*, *G.dalmaticum*, *G.orientalitibeticum*, *G.pylzowianum* and *G.sanguineum* var. *striatum*. They prefer full sun and a free-draining soil. Increase by division, rooted lengths of stem and runners. Two of the smaller species, *G.renardii* and *G.sessiliflorum*, have differing needs – the first seems happier in the dry garden among pinks, artemisias and silver sages; the second favours a cool, moist but gritty soil, still in full sun.

The others described here are superb hardy perennials for the herbaceous and mixed border, woodland garden and for underplanting shrubberies. They tolerate a wide range of conditions, from full sun to shade, and will usually withstand periods of drought.

Some, like *G.endressii* and *G.macrorrhizum* are near-invincible groundcover. *G.pratense* is a lovely plant for the wild or meadow garden. *G.phaeum* and *G.sinense* are among the most richly and subtly coloured of all hardy plants; whilst the vibrant lime and magenta of the scrambling *G.* 'Anne Folkard' is almost too bright to behold. All of these should be propagated by division in early spring.

G. 'Ann Folkard' (*G.procurrens* × *G.psilostemon*; garden origin; leaves emerging pale gold, later lime green; inflorescence to 1m, sprawling and scrambling, much-branched, flowers produced in succession over a long period, to 3cm across, deepest magenta veined black); *G.canariense* (Canary Is; short-lived, to 40cm tall; leaves to 30cm across in a spreading rosette atop a short 'trunk', stalks stained purple-red, blades cut into 5 pinnately lobed and toothed segments; flowers to 5cm across, deep pink); *G.cinereum* (Pyrenees; rosette-forming, to 15cm; leaves to 5cm across, round, cut into 5–7 wedge-shaped, lobed segments, grey-green, softly hairy; flowers to 2.5cm across, white to pink with purple or white veins); *G.clarkei* (Kashmir; to 50cm; leaves with 7 pinnately lobed and toothed segments; flowers to 4cm across, deep purple, pink or white with purple-pink veins); *G.dalmaticum* (Balkans, Albania; trailing dwarf to 15cm; leaves to 4cm across, round, glossy green with 5–7 toothed to lobed segments; flowers to 3cm across, bright pink); *G.endressii* (Pyrenees; hairy, clump-forming, to 50cm; leaves to 15cm across, with 5 lobed and toothed segments; flowers 2–4cm across, pale rose becoming candy pink with darker veins; 'Wargrave's Pink': a popular cultivar with pale salmon flowers, good weed-smothering groundcover); *G.farreri* (China; dwarf, tap-rooted, rosette-forming; leaves to 5cm across, round to reniform, cut into 7 segments, each 3-lobed and toothed, stalks and margins tinted red; flowers to 3.5cm across, pale mauve-pink with blue-black anthers); *G.himalayense* (syn. *G.grandiflorum*; Himalaya; carpet-forming, hairy, to 30cm; leaves to 20cm across, cut into 7 lobed and toothed segments; flowers to 6cm across, deep violet-blue, fading to magenta then white at centre); *G.ibericum* (Caucasus, Turkey, Iran; clump-forming, to 60cm; leaves to 10cm across, hairy, cut into 9–10, overlapping, lobed and toothed segments; flowers to 5cm across, rich lavender blue with feathered purple veins); *G.incanum* (S Africa; bushy, branching, to 1m; leaves aromatic, cut into 5 narrow, toothed and lobed segments, grey- to white-hairy; flowers to 2cm across, deep magenta with a white 'V' at the base of each petal and darker veins); *G.* 'Johnson's Blue' (*G.himalayense* × *G.pratense*; garden origin; clump-forming, to 30cm; leaves to 15cm across, cut into 5–7 segments each deeply lobed and toothed; flowers to 5cm across, deep lavender blue); *G.macrorrhizum* (S Europe; clump-forming and carpeting, to 40cm, muskily scented and clammy; leaves to 20cm across with 5–7 broad, lobed and toothed segments, colouring well in autumn; flower stalks and calyces tinted purple-red to brown, flowers to 3cm across, magenta to pale pink or white; an excellent, weed-smothering

groundcover); *G.maculatum* (NE US; clump-forming, hairy, to 75cm; leaves to 20cm across, deeply cut into 5–7 widely spaced and deeply toothed segments, colouring well in autumn; flowers to 4cm across, pink or white); *G.maderense* (Madeira; to 1.5m; leaves to 60 cm across, deeply lobed and finely cut and toothed, dark green, in a rosette atop a short, stout 'trunk', stalks tinted red-brown; flowers to 4cm across, deep magenta with paler veins); *G.* × *magnificum* (*G.ibericum* × *G.platypetalum*; garden origin; flowers rich violet-purple with darker veins); *G. nodosum* (Pyrenees to Balkans; clump-forming and spreading, to 50cm; leaves 5–20cm across, glossy, broadly and sparsely lobed and shallowly toothed; flowers to 4cm across, lilac-pink with darker markings at centre); *G.orientalitibeticum* (syn. *G.stapfianum* 'Roseum'; China; dwarf tuberous carpeter to 20cm, differs from *G.pylzowianum* in its larger, marbled leaves and petals lacking claws, these are bright purple-pink with a white base); *G.* × *oxonianum* (*G.endressii* × *G.versicolor*; garden origin; vigorous clump-forming and carpeting, to 65cm; leaves sometimes marked; flowers pink with notched petals and dark veins; 'Claridge Druce': leaves dark and rather glossy, flowers rose pink with darker veins; 'Walter's Gift': new growth brown tinted orange, later bronze-green, flowers pale pink, heavily veined); *G.palmatum* (Madeira; differs from *G.canariense* in developing a trunk only with age, in its stalked, central leaf segments, and in its purple-pink flowers with cream, not red, anthers); *G.phaeum* (BLACK WIDOW; Europe; clump-forming, to 70cm; leaves to 20 cm across, rounded, shallowly lobed and toothed, usually with maroon markings near the base; flowers to 2cm across, silky purple-black to maroon, deep mauve-pink, lilac or white; includes numerous cultivars, one of the finest is 'Samobor', with broad rounded leaves clearly marked with a chocolate horseshoe, and flowers fading from maroon to cloudy mauve); *G.pratense* (MEADOW CRANE'S BILL; Europe, Asia; clump-forming, to 70cm; leaves to 20cm across, with 7-9 deeply cut, lobed and toothed segments; flowers to 4.5cm across, violet to blue or white, sometimes with paler, or pink veins; includes cultivars with white, opalescent, sky blue, deep blue, violet, pink and mauve flowers); *G.procurrens* (Himalaya; clump-forming, glandular-hairy with trailing and rooting stems; leaves to 10cm across with 5 segments, each 3-lobed and toothed; flowers to 3cm across, deep purple-pink with black veins and markings at the centre); *G.psilostemon* (Turkey; clump-forming, erect, to 100cm; leaves to 25cm across with 7 segments and deep, sharp lobes and teeth, colouring well in autumn; flowers to 3.5cm across, intense magenta, veined and stained black at centre); *G.pylzowianum* (China; 5-20cm, tuberous, carpeting and spreading by runners; leaves to 5cm across, cut into toothed and lobed, wedge-shaped segments, dark green; flowers to 4.5cm across, pink, centre white to green, veins darker); *G.renardii* (Caucasus; clump-forming, to 20cm; leaves 5–10cm across, rounded, shallowly and bluntly lobed, dull grey to sage green with fine, pale down and impressed wrinkled veins;

flowers to 3.5cm across, white to palest lavender with dark purple, branching veins); *G.sanguineum* (BLOODY CRANE'S BILL; Europe; spreading, hummock-forming, to 25 cm; leaves 5–10cm across, rounded, deeply cut into 5–7, toothed and lobed segments; flowers to 4cm across, deep purple-red to bright carmine, pink or white, the centre white and the veins usually darker; includes numerous cultivars; var *striatum*: syn. var. *lancastriense*, a low-growing plant suitable for the rock garden with clear pink flowers marked with darker veins); *G.sessiliflorum* (New Zealand; dwarf, clump-forming, to 8cm; leaves to 2.5cm across, rounded, shallowly lobed; flowers to 1cm across, white to dull cream; selections such as 'Nigricans' and 'Porter's Pass' have pale bronze to deep chocolate leaves turning orange to coral red as they die); *G.sinense* (China; clump-forming, to 40cm; leaves to 20cm across, deep green, with 5–7 deep, toothed segments; flowers to 3cm across, velvety dark maroon to chocolate brown with a pink centre and red anther filaments);*G.sylvaticum* (Europe; clump-forming, glandular-hairy, to 70cm; leaves to 20cm across, deeply cut with 7–9, toothed and lobed segments; flowers to 3cm across, blue to violet with a white base, sometimes lilac, white or rose; 'Mayflower': flowers rich violet-blue with a white centre); *G.traversii* var. *elegans* (Chatham Is.; low-growing and spreading, silvery throughout; leaves to 10cm across, with 5–7 lobed and toothed segments; flowers to 3cm across, cloudy pink with darker veins); *G.wallichianum* (Himalaya; clump-forming, low-growing, to 45cm; leaves with 3–5, deep, sharply lobed and toothed segments; flowers to 3.5cm across, purple-pink with a white centre; 'Buxton's Variety': syn. 'Buxton's Blue', dense, carpeting, leaves emerald green flecked with pale green or silvery white, flowers porcelain blue with a bright white centre); *G.wlassovianum* (E Russia to China; clump-forming, bushy, softly hairy, to 40cm; leaves to 15cm across, rounded, shallowly cut into 7 pinnately lobed segments; flowers to 4.5cm across, dark to pale purple-violet with a white centre and feathered, deep violet veins).

Gerbera (for the German naturalist Traugott Gerber (d. 1743), who travelled in Russia). Compositae. Africa, Madagascar, Asia, Indonesia. TRANSVAAL DAISY, BARBERTON DAISY. About 40 species, hairy, stemless, perennial herbs with entire to toothed or pinnately lobed leaves and showy, long-stalked, daisy-like flowerheads. The following species and its hybrids provide exceptionally long-lasting cut flowers, notable for their range of radiant colours, from softer pastels to deep shades of red and orange. Scald the bases of the flower stalks and stand them in cool deep water before arranging. In essentially frost-free zones, grow permanently in the open garden in full sun on a free-draining soil; otherwise, grow in the cool, well-ventilated greenhouse, in the border or in pots. Dwarf hybrids such as 'Happipot' are particularly well-suited to pot cultivation and may be brought into the house for temporary display when in bloom. Under glass, provide a freely

draining, medium-fertility, loam-based mix and ensure some shade from the hottest summer sun; water plentifully and liquid feed fortnightly when in growth. Overwinter at 5–7°C/40–45°F, and keep just moist. Divide and repot in spring. *G.jamesonii* (BARBERTON DAISY, TRANSVAAL DAISY; flowerheads to 15cm across, on stalks to 70cm tall; many cultivars and seed races are available with single or double flowerheads, some very large, and in shades of yellow, orange, vermilion, deep scarlet, brick red, salmon, crimson, magenta, pale pink and white).

germination a complex sequence of physiological and structural changes which occurs as a seed starts into growth.

germination test any of various methods for determining the viability of seeds, including germinating a sample of seed under optimum conditions, studying a sample's reaction to chemical treatment, or germinating excised embryos. Viability is expressed as a percentage of seeds successfully tested.

Gesneriads a group name for plants of the family Gesneriaceae, including *Achimenes, Aeschynanthus, Alloplectus, Asteranthera, Briggsia, Chirita, Codonanthe, Columnea, Conandron, Episcia, Eucodonia, Gesneria, Gloxinia, Haberlea, Jankaea, Kohleria, Lysionotus, Mitraria, Nautilocalyx, Nematanthus, Niphaea, Paraboea, Petrocosmea, Phinaea, Ramonda, Rehmannia, Saintpaulia, Sinningia, Streptocarpus*.

The family includes small shrubs and trees, epiphytic vines and almost stemless rosette-forming plants, of tropical, sub-tropical, Asian and European origin. The majority have fine fibrous root systems (for example, *Columnea, Ramonda*), in some cases producing rhizomes bearing scales of modified leaves (for example, *Achimenes, Gloxinia*). Others form tubers with fine roots arising on the upper surface (for example, *Sinningia*). Fibrous rooters grow continuously, whilst rhizome and tuber formers have a period of dormancy which varies according to species.

Cool-temperate genera are native of high altitudes in Europe, Asia, and the Americas and can be grown in the rock garden or alpine house. For outdoor cultivation in temperate zones, well-drained positions such as rocky crevices are necessary to avoid excess winter wet. Plant rosette-formers such as *Haberlea* and *Ramonda* in vertical crevices; provide support for shrubby/climbing genera such as *Asteranthera* and *Mitraria* to keep the plants above wet ground. For outdoor cultivation in warmer zones, plant in moisture-retentive soils. It is important to identify soil pH requirements; for example, *Conandron* and *Ramonda serbica* are acid lovers, while *Haberlea, Jankaea*, and *Ramonda myconi* are lime lovers. Protect from prolonged hard frost, and hot sunshine. In the alpine house, pot in shallow containers of well-drained soil-based compost. Water sparingly during the winter resting phase; water moderately and position in bright filtered light in summer.

Subtropical and tropical genera can be grown indoors

in temperate areas at temperatures ranging from 7–30°C/45–85°F, with shrubs and terrestrial species (e.g. *Lysionotus*, *Petrocosmea*) at the lower end of the range, epiphytes (e.g. *Aeschynanthus*, *Columnea*) at the higher end and others (e.g. *Saintpaulia* and *Streptocarpus*) in the mid range of 15–20°C/60–68°F. In temperate zones, position plants in bright filtered light during summer. They are day-length responsive so, for optimum results, supplement by artificial means to provide 12–14 hours lighting. Outdoors in long-day warmer zones, provide shade, especially for *Episcia* and *Aeschynanthus*. Water plentifully but carefully, avoiding wetting the leaves. Store dormant rhizomes and tubers free of frost and repot each season.

In their active growth stage, all gesneriads prefer medium to high humidity and, in high-temperature areas, outdoors misting twice a day is beneficial. Some genera, (e.g. *Episcia*, *Gesneria*, and *Phinaea*), require high humidity to grow well in temperate zones, as can be obtained in a closed case. Plants benefit from modest, balanced supplementary feeding during their growing period; nitrogen should be omitted during flowering. Pot into shallow containers of peat-based compost or equal parts sterilized loam, sphagmum peat, perlite and vermiculite; ensure a pH range of 6.5–7.0, except for *Phinaea* which requires 7.5. Grow large vining and epiphytic sorts in hanging mesh baskets or pots, and high-humidity lovers in terraria.

Gesneriads may be propagated from seed or cuttings. Sow freshly harvested seed, without covering, at a temperature of 20–25°C/68–77°F to achieve germination in 4 weeks. Some genera (e.g. *Streptocarpus*) suspend development and should be pricked out very early. Leaf cuttings should be set into 5–6.5cm pots, and placed in a propagator, at 20–25°C/68–77°F, or in a warm position under a polythene-bag cover. Increase large-leafed types, (e.g. *Streptocarpus*), by leaf cuttings made as 5cm-long, wedge-shaped segments. Stem cuttings, with or without growing tips, may be taken, cutting below leaf nodes, into portions 5–10cm long; the lowest leaves should be removed. This method can be used for shoots arising from tubers.

Scaly rhizomes can be broken into segments 1cm long and several inserted into individual 9cm pots to a depth of 1cm. Tubers may be increased by cutting into individual segments, each possessing a bud; dust cut surfaces with fungicide and allow to dry for a while before potting with the bud eye just proud of the surface.

Geum (name used by Pliny). Rosaceae. Europe, Asia, New Zealand, N and S America, Africa. AVENS. Some 50 species, perennial herbs and subshrubs (those listed below are perennial herbs). The leaves are lyrate or pinnate, usually hairy, in basal clumps and on the erect flowering stems. Bell- to bowl-shaped and produced in spring and summer in stalked cymes, the flowers consist of five, persistent calyx lobes and five broad petals. The following species are hardy to at least -15°C/5°F, but in areas of prolonged low temperatures, should be protected in winter with evergreen branches or a dry mulch of bracken litter or leafmould. Grow

in well-drained, moisture-retentive soils rich in organic matter. Propagate cultivars by division in autumn or spring and divide plants every three to four years to maintain vigour. Species may be propagated by seed, sown in autumn or spring, although species hybridize freely and seed may not come true.

G. × *borisii* (*G.bulgaricum* × *G.reptans*; clump-forming, to 50cm; flowers bright yellow or, in the forms usually cultivated, orange-scarlet); *G.chiloense* (Chile; to 60cm; flowers scarlet to yellow, washed deep fiery orange, in erect cymes); *G.coccineum* (Balkans; to 45cm; flowers to 3cm in diameter, cupped to outspread, deep fiery red to orange or apricot); *G.* × *heldreichii* (*G.coccineum* × *G.montanum*; to 30cm; flowers large, orange-red); *G.montanum* (ALPINE AVENS; C and S Europe; to 30cm; flowers to 3cm in diameter, pale yellow); *G.rivale* (WATER AVENS, INDIAN CHOCOLATE, PURPLE AVENS, CHOCOLATE ROOT; Europe; to 35cm; flowers to 2.5cm in diameter, nodding, calyx dark purple-brown, corolla campanulate, cream to dusky pink or brick red to pale orange streaked purple-red).

Gevuina (the native name). Proteaceae. Chile. CHILE NUT, CHILEAN HAZELNUT. 1 species, *G.avellana*, an evergreen shrub or tree to 12m, with rusty-tomentose new growth. The leaves are pinnate or bipinnate, 20–42cm long, coriaceous, sap-green and coarsely toothed. Appearing in late summer in narrow racemes, the flowers are spidery and cream to pale buff. These are followed by fleshy-coated fruits that turn from red to black. Usually considered tender, *G.avellana* is nevertheless known to tolerate temperatures to –10°C/14°F in a sheltered woodland environment. It may also be grown in tubs in the cool greenhouse or conservatory. Grow in semi-shade on a cool, moist soil rich in sand and leafmould. Shield from harsh winds. Propagate by semi-ripe cuttings or by seed.

Gibbaeum (from Latin *gibba*, hump, referring to the hump-shaped plant bodies of many species). Aizoaceae. S Africa (W Cape). 20 species, highly succulent, more or less stemless, clump-forming perennial herbs, with paired and fleshy leaves united at their base to a greater or lesser extent. Daisy-like flowers open at the junction of the leaves on bright days. Provide a minimum winter temperature of 7°C/45°F, with full sun and a dry, draught-free environment. Water sparingly during periods of sunny weather when in growth and not at all when dormant. Plant in a perfectly drained, gritty medium. Feed during periods of active growth in sunny weather, using a very weak, low-nitrogen, liquid fertilizer. Propagate as for *Conophytum*.

G.petrense (leaves to 1cm, triangular, smooth, pale grey-green; flowers pink to red); *G.velutinum* (leaves to 6cm, narrowly triangular, pale grey-green, velvety; flowers white or pink).

gibberellins a group of naturally occurring plant hormones that influence growth, especially as stimulators. They are synthesized as gibberellic acid. See *growth regulators*.

gibbous swollen on one side or at the base.

Gilia (for Filippo Gilii (1756–1821), an astronomer in Rome and author, with Caspar Xuarez, of *Osservazioni fitologiche*). Polemoniaceae. SW US, southern S America. 25 species, annual, rarely perennial herbs. The leaves are usually pinnately lobed, rarely entire, and often downy or glandular. Carried in summer and autumn in panicles or clusters atop slender stems, the flowers are tubular to funnelform. Suitable for annual and cut-flower borders, or for pots in the cool greenhouse where late winter/early spring blooms are required. Grow in full sun on a very well-drained, fertile soil; provide support in exposed situations. Sow seed *in situ* in spring or early autumn. For late-winter flowering under glass, sow in late summer and overwinter in well-ventilated conditions with a minimum night temperature of 7°C/45°F.

   G.achilleifolia (S California; glandular-hairy annual to 70cm; flowers 1–2cm, violet-blue in a dense, fan-like cluster of cymes); *G.capitata* (western N America; smooth to glandular-hairy annual to 80cm; flowers to 0.8cm, lavender, in small, crowded terminal heads).

Gillenia (for Arnold Gille, 17th-century German botanist, who had a garden at Cassel). Rosaceae. N America. 2 species, rhizomatous perennial herbs with erect to arching, branching stems and trifoliolate leaves. Five-petalled flowers are carried in loose panicles in summer. Hardy to at least –15°C/5°F. Grow in sun or semi-shade in a lime-free, moist but porous soil. Provide support by staking. Propagate by division, or by seed sown in autumn or spring. *G.trifoliata* (BOWMAN'S ROOT, INDIAN PHYSIC; NE US, Canada; erect, slender, finely branched; flowers to 2cm across with narrow, white petals).

Ginkgo (from Chinese *yin-kuo*, silver apricot, via Japanese pronunciation, *ginko*.) Ginkgoaceae. China. 1 species, *G.biloba*, (MAIDENHAIR TREE) a deciduous, dioecious tree to 40m. The trunk is erect, and the crown basically open and conical with several, large and upswept stems from a low fork. The branches are spreading with weeping tips. The leaves are fan-shaped with an irregular upper margin often shallowly cut into two lobes. They are 5–12cm across, tough, yellow-green above and paler beneath, turning pale gold to rich butter yellow in autumn. Male flowers are carried in short, yellow-green catkins in spring. The female inflorescences consist of two, cap-shaped ovules at the head of a short, drooping stalk. The fruit matures the following autumn, and is fleshy-coated, spherical and to 2.5cm long. Pale olive at first, it ripens to purple-black and smells of rancid butter. A remarkable tree, the *Gingko* is grown for its beautiful maidenhair-like foliage, elegant crown, and its curiosity value as a 'living fossil'. It resists many pests, diseases and urban pollution, and is particularly suitable as a street tree (especially the cultivar 'Fastigiata' and similar clones); the autumn colour is

Ginkgo biloba

brief but spectacular. It does best in climates with hot summers. Male clones are often grown in the US and Europe, for the fruit is malodorous and messy when ripe. Female plants are more often seen in Japan and China, where ginkgo nuts are eaten and used in medicine. Plant on any well-drained soil in a fairly sheltered site. Hard pruning will result in the production of long, soft watershoots, and felled trees will coppice. Old specimens may become hollow-trunked. Propagate from seed stratified and sown in autumn. Take half-ripe cuttings in summer or hardwood cuttings in winter. Graft scions of desirable clones and males (side-veneer) on to seedling stock in spring, providing intermediate greenhouse conditions, medium light and occasional syringing.

girdling (1) the removal of a circle of bark around a tree trunk or branch, often the result of the activity of rabbits or deer but sometimes due to other mechanical damage. It usually results in the death of the tree or limb; see *bark ringing*; (2) (US) the growth of roots around a tree trunk, or other larger roots, so tightly as to prevent normal expansion.

glabrate, glabrescent (1) nearly glabrous, or becoming glabrous with age; (2) minutely and almost invisibly pubescent.

glabrous smooth, hairless.

gladiate sword-like.

Gladiolus (from Latin *gladiolus*, diminutive of *gladius*, a sword, referring to the leaf shape). Iridaceae. Africa, mainly South Africa, Madagascar, Europe, Arabia, W Asia. About 180 species, cormous perennial herbs with linear to sword-shaped leaves. Produced in spring and summer in narrow spikes, the flowers consist of six equal or unequal perianth lobes. These may be broad or narrow, ruffled or pointed and closed or spreading, their bases forming a long or short, usually curved tube.

   The species will flower in late winter or spring if grown under glass. In the open garden, they flower in summer and early autumn, as do the hybrids. The species are generally rather tender and may not survive the winter in climate zones 7 and below. *Gladiolus communis* subsp. *byzantinus* and *G.italicus* are, however, reliably hardy in climate zone 7, while *G.atroviolaceus*, *G.callianthus* and *G.papilio* may last for many years in a sheltered, sunny position. The remaining species should be planted outdoors in spring and lifted for dry, frost-free storage in autumn. Otherwise, keep them under glass, planted in autumn for spring flowering – in which case, dry the corms off when the new growth is complete and store them dry in their pots until repotting in the autumn. All need full sun and a sandy or gritty soil that dries out thoroughly once the foliage has withered. The hybrids should be planted out in spring, 10–15cm apart and deep, in fertile, well-drained soil in full sun. Water in dry weather and stake taller cultivars if necessary. Although many of the hybrids will survive the winter outdoors if protected from slugs and wet, it is safe practice to lift the corms in the autumn and store them in dry, frost-free conditions. Increase by cormlets removed from the dormant parent corms. Affected by rots when in storage (dust with fungicide). Gladiolus scab produces lesions and blotches on the foliage. Foliage streaked with yellow and dying prematurely probably indicates gladiolus yellows. Destroy infected plants.

G.atroviolaceus (Greece, Turkey, Iraq, Iran; 35–70cm; flowers deep violet-purple, almost black); *G.callianthus* (syn. *Acidanthera bicolor*; E Tropical Africa; 50–100cm; flowers scented, white with dark purple-red marks or staining in throat); *G.cardinalis* (WATERFALL GLADIOLUS, NEW YEAR LILY; S Africa; 60–100cm; flowers crimson or scarlet with a white or cream, mauve-edged marking on each of the lower perianth lobes); *G.carneus* (syn. *G.blandus*; PAINTED LADY; S Africa; 20–100cm; flowers white, cream, mauve or pink, usually with yellow, red or purple markings on the lower perianth lobes or with dark blotches in throat); *G. × colvillei* (*G.tristis* × *G.cardinalis*; syn. *G.nanus*; a garden hybrid and parent of the Nanus or Miniature Hybrids); *G.communis* subsp. *byzantinus* (syn. *G.byzantinus*; BYZANTINE GLADIOLUS; S Spain, Sicily; 50–100cm; flowers deep purple-red with narrow, paler marks bordered with purple on the lower perianth lobes; white, pink and vivid cerise forms are also grown); *G.dalenii* (syn. *G.natalensis*, *G.primulinus*, *G.psittacinus*; Tropical Africa, S Africa; 50–100cm; flowers green, yellow, orange, red, pink or purple, sometimes striped or mottled with another colour); *G.italicus* (syn. *G.segetum*; FIELD GLADIOLUS; S Europe; 40–100cm; flowers purple-pink to magenta, lower perianth lobes with a narrow pink blotch outlined in purple); *G.papilio* (syn. *G.purpureo-auratus*; S Africa; 50–90cm; flowers off-white to dull yellow suffused cloudy purple to dove grey especially on exterior, lower lobes with buff to golden, wedge-shaped patches outlined with mauve); *G.tristis* (MARSH AFRIKAANER; S Africa; 40–100cm; flowers white, pale cream or yellow, often tinged with green, usually dotted, lined or flushed dark green, brown or purple). *Cultivars.* The best-known gladioli are garden hybrids, of which there are many. They fall into four groups – MINIATURE HYBRIDS (Nanus Hybrids): 50–90cm tall with closely arranged flowers each to 5cm across and usually frilled; examples include 'Amanda Mahy' (orange-pink, lower lobes flecked pale mauve), 'Charm'(pink with yellow-green blotches), 'Hot Sauce' (light red with yellow markings, heavily frilled), 'Nymph'(white-pink blotched cream and edged pale rose). BUTTERFLY HYBRIDS: to 1.2m with densely arranged flowers each to 10cm across and conspicuously blotched on the throat; examples include 'Chartres' (purple with darker blotches), 'Georgette' (cherry red with a rich yellow centre), 'Madame Butterfly' (shell pink with a salmon and red throat). PRIMULINUS HYBRIDS: to 90cm with loosely arranged flowers each to 7.5cm across with the uppermost perianth lobe hooded and arching over the stigma

and anthers; examples include 'Lady Godiva' (white), 'Pegasus' (yellow tipped red), 'Red Star' (starry, vivid purple-orange with a paler centre). LARGE-FLOWERED HYBRIDS (Grandiflorus Hybrids): 90–150cm; with closely arranged flowers 6–14cm across; examples include 'Applause' (pink), 'Blue Conqueror' (deep violet-blue with paler centres), 'Ebony Beauty' (velvety black-red with white stamens), 'Hunting Song' (orange-red), 'Jacksonville Gold' (bright yellow), 'Minuet' (white), 'Vesuvius' (vivid scarlet).

gland any cell or cells secreting a substance or substances, such as oil, calcium or sugar, for example nectaries.

Gladiolus callianthus

glandspine in some Cactaceae, a short spine borne in the upper part of an areole, which is nectariferous in its first year.

glandular-pubescent either covered with intermixed glands and hairs, or possessing hairs terminated by glands.

glass a widely used cladding material for greenhouses, frames and cloches, which has the advantages of being long-lasting, with very high levels of sunlight transmission and good heat retention. Its disadvantages are fragility, and weight and cost compared with substitutes such as acrylic, polycarbonate, polyester, polyethylene or polyvinyl chloride. For garden purposes, horticultural glass at least 3mm thick should be chosen; the most common pane size is 60cm square. For conservatory glazing, thicker toughened glass is essential, often in double-layered sheets for improved heat retention.

glass case any of a wide range of glazed structures for protecting plants. Long wall cases were formerly a common feature of large gardens, used for protecting trained forms of peaches, pears and cherries. Within greenhouses, glass cases may be used to contain ferns or orchids, and to house cuttings and germinating seeds; such units are sometimes referred to as close cases. Occasionally, the term is used to refer to a bell glass, which is a type of cloche.

glaucescent slightly glaucous.

Glaucidium ('resembles *Glaucium*' – the general shape of the flowers is similar). Paeoniaceae. Japan. 1 species, *G.palmatum*, a perennial, rhizomatous herb to 40cm tall, and white-pubescent when young. Rigid stems bear 2 leaves at their summits, each to 20cm, kidney-shaped to rounded, palmately lobed and toothed to incised. Produced in late spring, the solitary, terminal flowers are to 8cm wide and consist of four broad, petal-like sepals; these are mauve to pale lilac, (rarely white), with a boss of yellow stamens. Hardy to –15°C/5°F and below. It requires shelter from cold drying winds, protection from summer heat and drought and is well-suited to the woodland garden, peat terrace or north-facing herba-

Gladiolus papilio

ceous border. Grow in a cool, moisture-retentive, humus-rich soil in partial shade. Propagate by seed or careful division in spring.

Glaucium (from Greek *glaukos*, grey-green, referring to the leaf colour). Papaveraceae. Europe, N. Africa, C and SW Asia. HORNED POPPY. Some 25 species, annual, biennial or perennial herbs with orange latex. The oblong to oblanceolate leaves are pinnately lobed or dissected and toothed. Produced in summer in loosely and broadly branched terminal panicles, the flowers are cup-shaped, short-lived and consist of four petals. They are followed by long, narrowly cylindric and sharply tapered (horn-like) seed pods. Hardy in climate zone 6 and an excellent plant for silver, dry and coastal gardens. Grow in a warm, sunny position on a fast-draining soil. Sow seed *in situ* in spring, or in the cold frame in summer.

G.flavum (YELLOW HORNED POPPY; Europe, N Africa, W Asia; biennial or short-lived perennial to 1m tall, usually shorter; leaves bright grey-green to cool blue-green; flowers to 5cm across, deep golden yellow; fruit to 30cm; *G.corniculatum* has darker flowers, from deep orange to scarlet).

glaucous coated with a fine bloom, whitish, or blue-green or grey, and easily rubbed off.

Glechoma (from Greek *glechon*, name for a plant in PseudoDioscorides). Labiatae. Europe, naturalized N America. Some 12 species, evergreen perennial herbs, most with creeping and rooting stems. Small, two-lipped, tubular flowers are carried in whorls in summer. Fully hardy and tolerant of sun, shade and most soils, the following perennial species is usually cultivated in its variegated form and is a very useful trailing plant for hanging baskets and containers. It makes equally dependable, if invasive, groundcover. Increase by detaching rooted lengths of stem. *G.hederacea* (GROUND IVY, ALEHOOF, FIELD BALM, GILL-OVER-THE-GROUND, RUNAWAY ROBIN; Europe to Caucasus, naturalized N America; flowering stems ascending, 5–30cm tall, non-flowering stems creeping to trailing, mat-forming and rooting; leaves to 3cm, broadly ovate, toothed; flowers to 2cm, violet to mauve or lilac, sometimes white or pink; includes the pink-flowered 'Rosea', and 'Variegata', with leaves with broken edges and zones of white and silver-grey).

Gleditsia (for Johann Gottlieb Gleditsch (1714–1786), German botanist, director of Berlin Botanical Garden, and friend of Linnaeus). Leguminosae. C and E Asia, N and S America, Iran, tropical Africa. HONEY LOCUST. 14 species, deciduous trees, usually with stout, simple or branched thorns on the trunks and branches. The leaves are pinnate and/or bipinnate. Small, green-white flowers are carried in spring and summer in axillary racemes and may give rise to large seed pods. Honey locusts are widely used in landscaping and frequently grown in parks and gardens. Some

species are used as hedges, and some for fences and as resilient street trees – most notably *G.triacanthos* in the US, which tolerates large-scale container cultivation in northeastern cities. Where fallen pods would cause litter problems, the sterile and unarmed *G.triacanthos* 'Moraine' may prove useful. They withstand temperatures down to at least –30°C/–22°F. Grow in full sun. Propagate species by scarified seed, cultivars by grafting or by budding on to seedling stocks of *G.triacanthos*. In the US, *Gleditsia* is affected by mimosa webworm. The *G.triacanthos* cultivars, 'Green Glory' and 'Moraine', show some resistance.

G.caspica (CASPIAN LOCUST; N Iran, Transcaucasia; to 12m, trunk thorns to 15cm, branched, slightly flattened; leaves to 25cm long, pinnate, with shining green leaflets, each to 5cm, ovate to oval and finely crenate); *G.japonica* (Japan, China, introduced E US, California; to 20m, trunk thorns to 8cm, branched, slightly flattened; leaves to 30cm, bipinnate or pinnate with oblong to lanceolate leaflets to 4cm, glossy, entire to sparsely crenate); *G.triacanthos* (HONEY LOCUST; N America, introduced temperate Old World; to 45m, trunk and branches armed with stout, sharp, simple or branched, flat thorns; leaves to 20cm, bipinnate or pinnate with oblong to lanceolate, crenulate, deep green leaflets to 4cm long; includes thornless cultivars and others with golden to lime green foliage (e.g. 'Sunburst'), or dark bronze red foliage (e.g. 'Rubylace'), some weeping, and some with dark green foliage persisting well into autumn – for example, 'Shademaster').

gley used of soils that have developed under impeded drainage conditions. Typically, the soil has a comparatively shallow surface horizon and grey colouring, with a thin coating to granular components and frequently rust-coloured discolouration from ferric oxide deposit.

Globba (from *galoba*, native name in Amboina, Indonesia). Zingiberaceae. SE Asia. Some 70 species, rhizomatous perennial herbs with clumped, reed-like stems. Produced in late spring and summer, the inflorescence is a pendulous raceme with showy, reflexed bracts. The delicate flowers consist of three petals, petal-like staminodes and one fertile stamen with a slender, curling filament. The lower flowers are sometimes replaced by bulbils. During growth, maintain a minimum temperature of 18°C/65°F, with high humidity, plentiful water and bright indirect light. In the winter months, keep virtually dry, provide full sun and a minimum temperature of 10°C/50°F. Propagate by division in spring, or by removing bulbils. *G.winitii* (Thailand; to 1m; inflorescence to 15cm, with pink to magenta or mauve bracts and yellow flowers).

globose spherical, sometimes used to mean near spherical.

globular composed of globose forms, as in the aggregate fruits of *Rubus*.

Globularia (from Latin *globulus*, a small round head, from the form of the flowerhead). Globulariaceae. Cape Verde Islands, Canary Islands, Europe and Asia Minor. GLOBE DAISY. 22 species, usually evergreen perennial herbs or shrublets. They form low hummocks or mats of neat foliage, giving rise in spring and early summer to fluffy, ball- or button-like flowerheads. Hardy in climate zone 6 and well-suited to the rock garden. Grow in gritty, well-drained, neutral or slightly alkaline soil in full sun. Propagate by division, by softwood or semi-ripe cuttings or by seed sown when ripe.

G.cordifolia (Europe; dwarf, creeping and mat-forming shrublet with spathulate, notched leaves and button-like, sky blue to lavender (rarely white or pink) flowerheads); *G.meridionalis* (syn. *G.cordifolia* subsp. *meridionalis*; Europe; more robust than *G.cordifolia* with oblanceolate, rarely notched leaves and mauve-blue to lilac-pink flowerheads).

glochid a barbed spine or bristle, often tufted, on the areoles of cacti.

glomerate aggregated in one or more dense or compact clusters.

glomerule a cluster of capitula or grouped flowers, usually subtended by a single involucre.

gloriet in Spanish gardens, a pavilion or arbour placed in a central open space from which paths may radiate.

gloriette a hill-top focal point in a landscape garden.

Gloriosa (Latin *gloriosus*, glorious). Liliaceae (Colchicaceae). Tropical Africa and Asia. GLORY LILY, CLIMBING LILY, CREEPING LILY. 1 species, *G.superba*, a tuberous, perennial herb. Its slender stems climb to 2m by means of tendrils at the leaf tips. Solitary flowers are produced in summer and autumn on long pedicels in the leaf axils. The flowers are usually angled downwards and consist of six spreading tepals, each 4–10cm long, lanceolate, bowed at centre and gently but distinctly reflexed at tip. They range in colour from yellow to red or purple or bicoloured, with the margins often incurved, undulate or crisped. The most popular named variants are the lemon yellow-flowered 'Citrina', 'Rothschildiana', to 2m tall, with wavy flowers to 18cm across in scarlet or ruby stained and edged golden yellow, and 'Simplex', a shorter plant with broader, if smaller, tepals in deep orange and yellow.

Beautiful climbers for the warm greenhouse or home. Grow in full sun in a freely draining, fertile mix. Support the climbing stems on pea sticks, a cane frame or a wire hoop. Water plentifully and feed fortnightly with dilute liquid feed when in full growth. Withhold water as growth dies back in late summer/early autumn; store the long, narrow tubers dry in their pots at a minimum temperature of 10°C/50°F. Repot every one or two years in early

spring. Propagate by seed sown in a sandy propagating mix in late winter, with gentle bottom heat, by careful division in spring, or by offsets.

Glottiphyllum (from Greek *glotta*, tongue, and *phyllon*, leaf, referring to the thick, fleshy, tongue-like leaves). Aizoaceae. South Africa. More than 50 species, highly succulent perennials with low, branching stems giving rise to clumps of thickly fleshy leaves. Solitary and virtually stalkless, the daisy-like flowers open on bright days in spring and summer. Provide a minimum winter temperature of 7°C/45°F, low humidity and full sun. Grow in a low-fertility potting mix with high proportions of sand and grit. Water very sparingly during periods of very warm weather. Keep dry from autumn to spring. Propagation as for *Conophytum*.

G.nelii (leaves to $5 \times 2 \times 1$cm, upper surface flat, lower surface obscurely carinate, light green with translucent angles; flowers golden-yellow); *G.semicylindricum* (leaves to $5 \times 0.6 \times 0.6$cm, semi-cylindric, fresh glossy green with faint dots and small, tooth-like projections on margins; flowers yellow).

glume a small, dry, membranous bract found in the inflorescences of Gramineae and Cyperaceae and usually disposed in two ranks.

glutinose, glutinous see *viscid*.

Glyceria (from Greek *glykys*, sweet, referring to the flavour of the seeds). Gramineae. Northern temperate regions, Australia, New Zealand, S America. SWEET GRASS, MANNA GRASS. 16 species, perennial grasses, often found in wet and marshy places. They form extensive colonies of pithy, reed-like stems with narrowly strap-shaped leaves. The flowers are borne in summer in plume-like, terminal panicles. Fully hardy, the following species thrives in fertile, moisture-retentive soil in the herbaceous and mixed border, but is more often used as an aquatic or marginal. Grow in full sun. *Glyceria* spreads vigorously in watery places and should be confined in a tub or basket in smaller pools. Propagate by division in early spring. *G.maxima* (syn. *G.aquatica*; REED SWEET GRASS, REED MEADOW GRASS; Europe, Asia; to 2.5m tall in flower; includes var. *variegata*, with leaves striped green and cream, flushed pink on emergence, and 'Pallida', with leaves boldly striped off-white).

Gloriosa 'Rothschildiana'

goblet a form of fruit tree or bush in which the main branches are trained to encircle an open centre.

Gomesa (for Dr Bernadino Antonio Gomes, Portuguese botanist and physician). Orchidaceae. Brazil. Some 20 species, epiphytic or lithophytic orchids with clumped pseudobulbs and strap-shaped leaves. Small flowers are carried in arching racemes in spring and summer. Provide a minimum temperature of 10°C/50°F. Pot in an open mix of fern fibre, medium-grade coarse bark and sphagnum. Water, feed and syringe freely when in growth, and keep in buoyant, humid conditions in light shade. On completion of growth, impose a cooler, brighter regime, watering only to prevent shrivelling of pseudobulbs. Propagate by division when repotting. *G.crispa* (flowers to 2cm across, scented, tepals pale, dull yellow to lime-green, wavy, lip sometimes marked rusty-red or burnt orange).

Gomphrena (name used by Pliny for a kind of Amaranth). Amaranthaceae. Tropical America, Australia. 90 species, annual, rarely perennial herbs. Produced in spring and summer, the individual flowers are inconspicuous, but are carried in dense ball- or button-like heads amid highly coloured, chaffy bracts. In cool temperate zones, *Gomphrena* is treated as a frost-tender annual in beds and borders, for cutting and drying and as a pot plant for the greenhouse or conservatory. It is grown for its 'clover-like' flowerheads in strong colours that are retained on drying. Grow in full sun, in a sheltered sunny position in a deep, fertile, well-drained soil. Sow *in situ* in spring in warm temperate areas; elsewhere, sow under glass in early spring, hardening off before planting out after the last frosts. *G.globosa* (GLOBE AMARANTH, BATCHELOR'S BUTTONS; Panama, Guatemala; annual, usually erect, to 60cm tall; flowerheads 2–6cm, globose to broadly cylindric, in shades of white, pink, scarlet, deep crimson, purple-red or mauve; includes dwarf, compact and rounded and decumbent cultivars).

Gongora (for Don Antonio Caballero y Gongora, Bishop of Cordoba, late 18th century). Orchidaceae. W Indies, Mexico to Peru and Brazil. Some 50 species, epiphytic orchids. Waxy, intricate and fragrant, the flowers resemble insects in flight and are suspended along chain-like pendulous racemes in spring and summer. Grow in baskets where the flowers can be shown to advantage. Cultivate as for *Stanhopea*, but with rather more shade, moisture and higher temperatures when in growth.

G.armeniaca (flowers to 4cm long, salmon-pink, yellow or orange, sometimes spotted purple-brown); *G.galeata* (flowers to 5cm, yellow-brown to creamy green); *G.quinquenervis* (flowers to 4.5cm, pale yellow spotted and banded red-brown).

Gongora galeata

gooseberries

Pests
aphids
birds
capsid bug
caterpillars
sawflies

Diseases
American
    gooseberry
    mildew
        (*see* powdery
        mildew)
grey mould
leaf spot

gooseberry (*Ribes uva-crispa* var. *reclinatum*) a hardy, deciduous, spiny shrub, grown for its ovoid berries. The many hundreds of European cultivars derive from *R.uva-crispa*, which is a native of the Mediterranean region, North Africa, the Caucasus and northern China. North American cultivars, which have smaller fruits on less compact or upright bushes, derive from the native *R.hirtellum*, which is often crossed with European hybrids and shows resistance to mildew. Recently introduced into Europe, genes of the American species *R.divaricatum* confer resistance to American gooseberry mildew fungus.

The first recorded date of cultivation in England is 1276, but there was no general development until the introduction of gooseberries and other fruits from the Continent in the early 16th century. Much hybridizing followed throughout the 18th and 19th centuries, and gooseberry clubs flourished in the north of England during the 19th century, of which a few survive, members competing to produce heavy fruits.

Fruits are available green for cooking from late spring through to summer when ripe for dessert. The cool, moist climates of northern Europe, the northern US and Canada favour fruit development, although frost can damage flowers and young fruits. Wind shelter is desirable.

In gardens, the gooseberry is usually grown as a bush on a 15cm leg, or trained as a cordon or fan; commercially, often as a suckering bush or hedgerow. Bushes of European cultivars should be planted 1.5–2m apart, and American cultivars 3m apart. Plant single cordons 0.5m apart, double ones 1m apart. Select two- or three-year-old plants for bushes, and one-year-olds to train against canes as cordons. In all cases, remove stem suckers.

Propagate from 30–40cm long hardwood cuttings, made from the current year's growth; taken just before leaf fall, with all but the topmost four buds removed. Leaving some basal buds will improve rooting, but it then becomes necessary to remove the consequent shoots at planting time. In North America and Germany, mound layering is practised. After planting, cut main branches by one half and lateral shoots to one bud. Prune in the early years to establish a bush framework of six to eight branches radiating from the top of the stem. Fruit is borne on one- and two-year-old shoots. Annual pruning aims at maintaining a supply of new shoots and removing congestion: basically, shorten leaders by half, cutting to an upward bud on spreading cultivars, and reduce laterals to 8cm. Cordons are treated as a branch of a bush, where old spurs require thinning with age. Summer pruning of laterals to five buds at green-fruit stage encourages spur formation, light and air penetration, and removes mildew-infected tips. Gooseberries require netting as protection from birds which depredate buds in winter. Successful cultivars for the UK include the earlies 'May Duke' (red) and 'Golden Drop' (yellow); mid-season, 'Invicta' (green) and 'Jubilee' (green); lates, 'Lancashire Lad' (red) and 'Lancer' (green). Of the American gooseberries 'Captivator', 'Josselyn', 'Oregon Champion' and 'Welcome' are notable.

gootee see *layering*.

gopher a small burrowing rodent that with other ground squirrels are occasional garden pests in western North America, burrowing into garden areas with resultant damage to plants. The usual control is to place poisoned bait in the burrows, but this causes painful death and is not permitted in some states. Trapping is an alternative method. Advice from local pest-control officials should be sought.

Gordonia Theaceae. SE Asia, warm N America. 70 species, evergreen trees and shrubs with saucer-shaped, solitary flowers composed of broad petals and a showy boss of stamens. Cultivation requirements are mainly as for *Franklinia*, although it requires higher temperatures (ideally frost-free conditions) and hot humid summers.

G.axillaris (syn. *G.anomala*; China, Taiwan; shrub or tree to 12m; leaves to 17cm, oblong, leathery, dark glossy green; flowers to 15cm across, with five to six, creamy white petals and orange yellow stamens from early winter to spring); *G.lasianthus* (LOBLOLLY BAY; SE US; shrub or tree to 20m; leaves to 20cm, obovate to lanceolate, deep glossy green; flowers 6–8cm across, with five white petals and yellow stamens in summer).

gourds hard-shelled fruits of the family Cucurbitaceae, some of which are grown for ornament or utensils, others primarily as edible crops. Often the immature fruit is edible and the mature fruit hard-shelled, for example *Lagenaria siceraria*. Some species provide both hard-shelled, inedible fruits and distinct edible ones whose shells are not durable, for example *Cucurbita pepo*, which includes summer squashes. All gourds are frost-sensitive and in temperate climates should be raised in a greenhouse at temperatures between 18–20°C/65–68°F. Sow in containers for transplanting outside when the plants are 10–15cm tall, and have been hardened off. In warm areas seeds may be sown direct into the open ground. A spacing of 2.5 × 2.5m is appropriate for most cultivars, and 1m apart for small-fruited kinds. A rich, well-drained soil is needed with liberal amounts of water applied throughout the growing season. Supports are necessary for some gourds, while others naturally trail on the ground. Pruning the vines of most species is recommended, reducing the main shoots to 2.5m and the laterals to 4–6 leaves.

Ornamental gourds In Africa and South and Central America the shells of gourds are traditionally carved, decorated, used as kitchen utensils and made into musical instruments and ceremonial masks

In the garden, ornamental gourd plants may be trained on poles, fences or trellises to form screens,

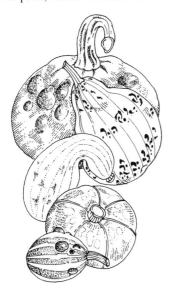

and they provide decorative display over arbours and pergolas. Plants may be allowed to trail over terraces or steep banks as a decorative feature, and in some situations to help stabilize soil.

Ornamental gourds include attractive cultivars of: *Benincasa hispida* (BOTTLE GOURD, CHINESE MELON) which in temperate climates requires warm greenhouse conditions and produces 25cm long fruits, covered in fine hairs and a waxy secretion; *Cucurbita ficifolia* (MALABAR MELON, FIG-LEAVED or SIAMESE GOURD), is the hardiest of the ornamental gourds suitable for training over pergolas and producing large, green fruits resembling a watermelon. *C.maxima* (TURK'S CAP and large yellow forms) which produces exceptionally large fruits and requires stout supports. *C.moschata* which has red, green, white and butternut forms, and is adapted for training, shows resistance to virus diseases and vine borer. *C.pepo* (CUSTARD, CROOKNECK and ORANGE GOURDS) and *C.pepo var. ovifera*, (YELLOW-FLOWERED GOURD) are popular ornamental gourds; *Lagenaria siceraria* (WHITE-FLOWERED CALABASH or BOTTLE GOURD) includes the MINIATURE BOTTLE GOURD, WARTED BOTTLE GOURD, POWDER HORN, CLUB and DOLPHIN types, but does not readily fruit in temperate areas; *Luffa cylindrica* (SMOOTH LOOFAH, DISHCLOTH GOURD) and *L.acutangula* (ANGLED LOOFAH, RIDGED GOURD) can be grown outdoors in subtropical areas or under greenhouse conditions in temperate climates. The bathroom loofah is the bleached, fibrous interior of *L.aegyptiaca*; When grown under warm greenhouse conditions *Momordica charantia* (BITTER GOURD) produces highly decorative fruits, light green, ripening to bright red, and covered with prominent warts which burst open when ripe; *Trichosanthes cucumerina* produces gourds up to 60cm long, which are bright green when young, becoming bright orange when mature and often twisted and curled in shape.

After harvest, gourd fruits are usually washed with a mild disinfectant, dried for 4–6 weeks and waxed to ensure preservation. Fruit shape can be modified by tying plastic-coated wires around the young developing fruits, or confining them in a rigid container. They may be stored for weeks or months if hung in a dry, well-ventilated room.

Edible gourds. In tropical and subtropical climates, *Benincasa*, *Coccinia*, *Lagenaria*, *Luffa*, *Momordica*, *Telfairia* and *Trichosanthes* are fairly widely grown, and in temperate zones many of these can be grown successfully under greenhouse conditions.

The most important species of edible gourds cultivated in warm-temperate areas are *Cucumis anguria* (WEST INDIAN GOURD or GHERKIN); *C.melo* (CANTALOUPE, HONEYDEW, MUSK or SWEET MELON); *Cucurbita moschata*, which includes the WINTER SQUASHES; *C.pepo*, which includes PUMPKINS, SUMMER SQUASHES and VEGETABLE MARROWS; also the vegetable spaghetti, with flesh that separates into spaghetti-like strands when boiled; *C.maxima*, which includes WINTER SQUASHES and the SQUASH GOURD; *C.mixta*, the WINTER SQUASH and pumpkin. See *cucumber; melons; pumpkins and squashes.*

gourmand see *water shoot.*

graft (1) a plant resulting from grafting; (2) loosely, a plant shoot, or scion, suitable for grafting; (3) an old word for a spade depth of soil; (4) a type of spade with a long slightly concave blade, used for digging drains; it is also called a grafting tool.

graft hybrid see *chimaera.*

grafting a method of vegetative propagation which normally involves joining together two separate plants in such a way that they eventually function as one. Grafts may arise naturally or be purposely made. Closely clipped hedges such as BEECH (*Fagus sylvatica*) and FIELD MAPLE (*Acer campestre*) will often form an interconnecting network of natural grafts. Similarly, where trees with thin bark, such as *Eucalyptus*, *Fagus*, *Fraxinus*, *Pinus*, *Platanus* and *Populus*, are growing close together, natural bridge grafts may form between individuals of the same species.

Grafting is a means of replicating selected plants which do not come true from seed or which are difficult to propagate by other vegetative methods. Bench grafting refers to any grafting method performed at a bench, where the rootstocks may be bare-root plants, unrooted cuttings or young plants established in pots.

Grafted plants often produce a larger plant in a shorter period of time, and the selection of an appropriate rootstock can improve the plant's tolerance of soil type, pH and temperature, and provide resistance to certain pests and diseases. Specified rootstocks can induce vigorous or dwarfing growth response, as exploited in the propagation of most tree fruits. To overcome incompatibility, preserve cold or disease resistance or control vigour, it is sometimes necessary to graft an intermediate variety or interstock, a technique known as double working.

Grafting may also be used to produce curiosities, such as a prostrate form grafted on to an upright stock, or family fruit trees, with different cultivars grafted on to a single plant. Grafting can promote the early development of features that would take many years if plants were raised from seed; for instance the distinctive bark of *Betula ermanii* or the flowering of *Wisteria*. Dioecious plants such as *Ilex aquifolium* may be grafted with a scion of the opposite sex, to improve the probability of fruiting, while bridge grafting can be used to to overcome an incompatible graft union or to combat the results of the girdling of stems by vandalism, machinery or rodents.

Grafted plants of the same species will generally form a compatible union although occasionally (in particular with conifers) the graft may either fail to form a union or the union may fail after ten years or so. Some species will form compatible unions with a different species belonging to the same genus, as with *Sorbus hupehensis* on *S.aucuparia*. Two different genera can form compatible unions, for instance *Amelanchier* on *Crataegus*; such successful intergeneric grafts will invariably be between members of the same family. Long-term compatibility of scions of one genus upon a rootstock of another is however unreliable.

There are two main grafting techniques.

Approach grafting (inarching) This technique is a means of propagating plants that are difficult to graft by other methods, and also of saving valued dying trees. Here, the scion continues to grow on its own roots until the union with the rootstock is made. The plant providing the scion may be grown in a container, and sometimes the stock plant also. The plants are arranged so that the stems are alongside each other, and, in the simplest method, a sliver of bark and wood about 50mm long is taken off each stem, on the facing sides; other methods include the use of tongues and inlays. The two cut surfaces are then bound together with plastic film, rubber strips or wax tape; in some cases, the whole area may need to be covered with grafting wax to form a waterproof seal. When the two cut surfaces have grown together, the scion plant is cut just below the point of union and the upper part of the stock removed.

A similar method may be used for providing disease-resistant rootstocks for the production of tomatoes and cucumbers in infected border soil. One type of inarching can be used to change the rootstock of an established orchard tree, and involves planting at least three plants of the new choice of rootstock close to the trunk of the tree and grafting the top of each into the trunk.

Detached scion grafting This is the most widely used grafting method and involves removing a scion from a stock plant and working it on to a rootstock. The timing of the operation is crucial, as is the physiological condition of both rootstock and scion: the rootstock should preferably be more advanced in growth than the scion, thus ensuring that the healing of the union occurs well before the scion breaks into growth. Water loss from the scion and graft union must be kept to a minimum, and for this reason the graft should be made quickly, firmly secured and covered with wax or another waterproofing substance. Detached scion grafting may be carried out in the open field or under protection, usually between the time the sap begins to rise in the spring and before it ceases in autumn; except in the case of bud grafting, it is advisable to avoid midsummer. Evergreen and tender plants should be grafted under cold frames or in cool or heated greenhouses with high humidity and shade.

The two principal methods of detached scion grafting are apical grafting and side grafting.

APICAL GRAFTING, where the top of the rootstock is decapitated and replaced by a scion.

Splice grafting (whip grafting), a simple method suitable for bench grafting, which is the process of joining a root stock with a scion before potting. The rootstock and scion are held together by hand when tying and must be of similar diameter. The scion cut should overshoot the top of the rootstock by about 3–5mm to ensure that the developing callus forms a bridge between the two components.

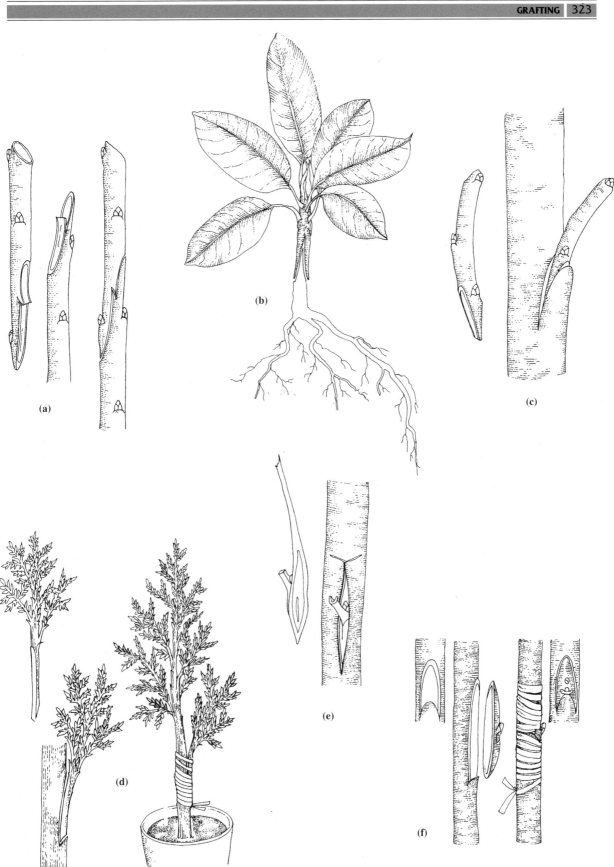

Grafting: (a) Whip and tongue graft (b) Saddle graft (c) Side cleft graft (d) Spliced side graft (e) Shield or T-budding (f) Chip budding

Whip and tongue grafting is widely used to produce young plants in the field, notably where bud grafts have failed to take. The rootstock is trimmed of all side growth and the top cut to within 10cm of the ground. The scion is prepared from a stout one-year-old shoot and a piece selected that contains at least three or four buds, the lowest being about 3–4cm above the bottom end. The tip should be neatly trimmed back just above the adjacent top bud. The tongue is made in the top half of the slanting cut surface, pointing downwards. The slanting cut on the rootstock must match the dimensions of that on the scion, allowing for the scion cut to overshoot the top of the rootstock by 3–5mm. An equivalent facing tongue is made near the top of the rootstock cut, allowing the scion to be fitted. The graft is tied and made watertight.

Saddle grafting is a more difficult apical graft, traditionally used for large-flowered hybrid rhododendrons. Rootstock and scion must be of the same diameter. The cut rootstock is prepared by removing the top to leave a clear stem of about 10cm, the top end of which is cut on both sides to form a central apex about 3–4cm long. The base of the scion is cut by removing a wedge of tissue matching the shape of the cut rootstock, so that the scion may neatly straddle it. The cambium layers should exactly match. The graft is securely tied and sealed.

Wedge or cleft grafting is a simplified inverted saddle graft, achieved by splitting the top of the rootstock through the centre. It is normally practised on easy subjects like *Hibiscus syriacus* cultivars and evergreen azaleas. The base of the scion is cut in a V-shape and pushed into the split rootstock. The union may be secured with a simple clip like a small plastic clothes peg.

Rind or crown grafting is a method of overcoming the problem of oversized rootstocks, and is also used for topworking mature fruit trees to change them from one cultivar to another. A slanting cut is made at the base of the scion, and a further incision, about 1–2cm long, made through the rind of the decapitated rootstock, allowing the rind to be peeled back. The scion is pushed behind the rind to the required depth, leaving an arch of a few millimetres visible on the cut face of the scion at the top of the rootstock to allow the callus to form a bridge between the two components. Since the scions are held firmly under the rind, tying is not required, but the unions must be made waterproof with a suitable sealant.

SIDE GRAFTING involves inserting scions without beheading the rootstock and is useful for propagating and for supplying an extra branch in any part of a tree; a spliced side graft (veneer side graft) is used for a wide range of evergreen and deciduous trees and shrubs where one- or two-year-old rootstocks have been established in pots. *Betula*, *Fagus* and *Rhododendron* are best grafted in late winter and *Acer palmatum*, *Hamamelis* and *Cedrus* in summer. The foliage should be retained on evergreens, but removed on summer-grafted deciduous plants to reduce water loss, with the leaf petioles retained to allow for natural abscission and to prevent bleeding.

The scion is prepared with a 45° cut at the bottom, followed by a slice-cut down the stem, about 2.5–3.5cm long, which meets the sharp end of the 45° basal cut. The stem of the rootstock should be cleaned of shoots and a matching cut made as low down as is practical. Various modifications of this graft are used.

Side cleft grafting is where two slanting cuts are made on opposite sides of the base of the scion to form a wedge about 2–3mm long; one of the cuts should be slightly longer than the other. An angled incision of about 20° is made in the side of the rootstock, of sufficient depth to accommodate the scion. The incision is sprung open and the scion pushed home, ensuring that the longer cut is opposite the inner face of the rootstock and in line with the cambium. The graft is tied and sealed.

FRAMEWORKING AND TOP GRAFTING or topworking are procedures for changing the cultivar in mature trees, especially apples. In frameworking, the main structure is retained and branches headed back ('dehorned') by about one metre; only strong healthy laterals are retained and these are cut back to about 5cm. The remaining shoots are grafted by whip and tongue or rind methods and the framework supplemented with side grafts set into the branches. Topworking involves cutting the main branches right back to near the crotch and inserting a number of scions, usually as rind grafts, around each cut. Topworked trees take longer to come into bearing and may be more prone to infection from diseases such as silverleaf.

ROOT GRAFTING covers any method of grafting shoots upon pieces of root. It can be used for a wide range of woody and herbaceous perennials, including *Clematis*, *Gypsophila*, *Hibiscus syriacus*, *Wisteria* and *Campsis*, and is recorded as successful with apples and rhododendrons. When there are large fleshy roots and thin scion wood, as in tree peonies, an inlay-type graft is normally used. Where the root and scion are of approximately the same diameter, splice, wedge or saddle grafts are suitable. Nurse grafting is a form of root grafting useful for propagating difficult stem cuttings or for plants where other sorts of grafting are likely to produce excessive rootstock suckers. Normal root grafting methods are used except that the scion/cutting is encouraged to root and the nurse root is removed once this has occurred. Plants that are nurse grafted include *Clematis* cultivars, using a tailed-scion graft on to roots of *C.vitalba* or *C.viticella*.

CUTTING GRAFTING is practised on plants that are difficult to root: the graft is made on to a stem or root cutting, which is easily regenerated. The method is used commercially on a limited range of plants, including *Salix caprea* 'Pendula', large-flowered hybrid rhododendrons, and grape vines.

BUDDING is a detached side-grafting technique, which employs a single bud to propagate fruit and ornamental trees and shrubs. The methods most widely used are shield-budding (T-budding) and chip budding, normally carried out between midsummer and early autumn; also practised is patch budding of walnuts. In all cases the rootstock should be well-established and actively growing, with selected buds dormant, well-developed, uniform in size and vegetative.

In shield or T-budding the scion buds, which occur in the axils of leaves, are removed from shoots of the current year's growth by means of a shallow slicing cut. This commences about 12mm below the bud, passing well under it and being completed some distance above it, so that there is a tail of rind which is convenient for handling. The removal of a sliver of wood at the back of the bud is generally required, especially for roses. The prepared bud forms the shield, and this is inserted beneath the rind of the rootstock, which is prepared with a T-shaped cut made down to the wood. The bud graft is tied with a suitable binding to secure the components in close contact, to minimise dessication and to make a watertight seal.

Chip budding is used extensively and involves a single bud graft that does not rely on lifting the rootstock bark. It is essentially a cleft graft, with the bud fitting exactly on to the chip of stem tissue removed from the rootstock. Both cambiums should be in line with each other to produce a neat speedy union with minimal callus tissue. Chip buds need to be tied more securely than T-buds. Plastic tape, 25mm wide, is used to completely cover the stem tissue, with pressure points above and below the bud, which may itself be covered. Union will take between 5–8 weeks, depending on the plant, time of budding, and weather conditions. Hot dry weather encourages speedy unions. The bud-take can be assessed the following February and the rootstock headed back to the top of the live bud. A slanting cut of about 22°, away from the bud, should be made. Clip and stake devices can be used initially to support the extension growth and keep it straight.

Grafting and sealing materials For tying grafts, raffia has been largely superseded by plastic materials but it is still used for delicate operations, such as the split scion on *Clematis*, and rind grafts on miniature roses. For bench grafts, suitable tying materials are waxed cotton (or waxed fillis for large diameter scions), 15cm strips of 5mm-wide rubber bands, or paraffin wax tape, which is self-sealing and gradually disintegrates. Paper and cotton tapes may be used for outdoor grafts, but the union must be made rainproof with a sealant. Plastic grafting tapes are available in widths of 12mm or 25mm, and unions can usually be made watertight without the use of a sealant.

Proprietary sealants include bitumastic emulsions, applied cold to cut ends of scions and rootstocks where the union has been made rainproof by tying. Resin-based mastics, formulated with grease, mineral oil, sythetic wax or ethyl alcohol, are good alternatives to hot resin-based sealants, which require a portable heat source when used outdoors. Paraffin wax is widely used for bench grafting, and rubberised latex compounds are suitable for both bench and outdoor grafts, applied cold to set and forming a flexible rainproof covering.

grafting knife a strong folding knife designed for grafting, comprising a blade about 7.5cm long, with an angled point at its end, set in a handle 10–12.5cm long.

grafting tool see *graft*.

graft union the point on a grafted plant at which the scion and stock are joined.

graminaceous grassy, resembling grass in texture or habit.

granules a pesticide formulation, in which a chemical substance is combined with a carrier to form small pellets, usually for application to the soil. Granules facilitate even distribution through special applicators. They are suitable for the continuous release of an active chemical and for band treatment of insecticide along crop rows, and for broadcasting herbicide substances.

grape marc the seed, skin and stem residue from grape pressing, which can be composted and used in growing media. The compost is saline, with a high potassium content, and the residual seeds have properties similar to very coarse sand.

grape any of several species of *Vitis*, cultivated for their fruit. The grape grown in Europe and Asia is derived from *Vitis vinifera*, the EUROPEAN WILD GRAPE. Several *Vitis* species indigenous to America (in particular the SUMMER GRAPE, *V.aestivalis*, the MUSCADINE, *V.rotundifolia*, and the FOX GRAPE, *V.labrusca*) have also been the sources of a range of cultivars.

Ancient vine growing is indicated by grape seeds found in Early Bronze Age sites at Jericho and Lachish. It spread west and north and also into Egypt, and figures prominently in Greek and Roman literature and in the Bible. Viticulture was introduced to Britain by the Romans, and by the time of the Domesday Book 38 vineyards were recorded in the south of England from Kent to Gloucestershire. Imports of superior wine gradually undermined the importance of the home-grown product, and vinegrowing in Britain has ebbed and flowed with changing climatic conditions.

In North America, early efforts to cultivate vines in the eastern states foundered through the ravages of *Phylloxera vastatrix*, but this pest was not present in California and vineyards flourished there. Resistant Californian grape vines became the rootstocks on to which *Phylloxera*-susceptible European vines were grafted in the late 19th century, thereby saving the great vineyards from extinction. At about the same time, European vines were cross-pollinated with American cultivars, producing *Phylloxera*-resistant hybrids which are widely grown today.

Grapes under glass A greenhouse needs to provide adequate space and ventilation and sufficient root run into a good growing medium, with surface access to it for routine top-dressing and watering. A

grapes

Pests
birds
mealy bug
phylloxerids
red spider mite
scale insects
wasps

Diseases
black rot
downy mildew
grey mould
powdery mildew

Disorder
shanking

lean-to structure is most suitable, the back wall offering residual heat. The soil must be fertile and well-drained, and the vine can be planted inside or outside the greenhouse; if outside, the stem or rod is led through an opening in the supporting wall of the structure. Grapes for greenhouse cultivation are classified as (1) Sweetwater: early-ripening cultivars, which are the best choice for unheated houses, for example 'Black Hamburgh' and 'Foster's Seedling'. (2) Muscats: the finest flavoured grapes normally requiring heat, for example 'Muscat of Alexandria' and 'Madresfield Court'. (3) Vinous: the latest to ripen, requiring adequate heat for example 'Alicante' and 'Gros Colmar'.

PLANTING AND TRAINING. Planting of bare-roots plants is carried out in November or December, but container-grown specimens may be planted at any time. In small houses, planting at one end and training the vine up and along the ridge is suitable, with fittings inserted to support wires that should hold the vine 35cm away from the glass, although in a small greenhouse 25cm will suffice. Where there is space, suitable extra main shoots are trained along wires spaced 1.2m apart.

The simple rod and spur training method is commonly used in greenhouse culture. After planting, shorten the stem to about 7.5cm above soil level or the graft union; always prune by the New Year to prevent bleeding of sap. One leading shoot (also called rod or stem) should be allowed to grow up and tied loosely to the wire during the first summer. Laterals (also called rods or canes) growing out from the leading shoot are shortened to five leaves and any sub-laterals to one leaf. Competing leaders should be removed in the first winter, the leading shoot reduced by two-thirds of its length and the laterals pruned to one bud from the leading shoot stem.

In the second summer, the leading shoot extends and is tied to the wires, and the treatment of laterals and sub-laterals repeated. Any flower trusses must be pruned off to prevent fruiting before the third year.

In subsequent winters the leading shoot is shortened and restricted to the available space; laterals are repeatedly shortened to one bud, and with annual cuts spurs are formed which eventually require thinning with a pruning saw. Where laterals are trained at an upward angle, they should be cut loose in winter and allowed to hang, or retied horizontally, until growth begins, to ensure even growth at each spur; the laterals are then returned to their normal positions.

As growth starts in spring, only two fruiting laterals must be retained at each spur. The weaker of the two is pinched to two leaves and is kept as insurance against any accident to the lateral which will normally carry the crop. As this develops, flower trusses and usually also vine leaves will appear up to 60cm along its length. The tip of the fruiting lateral should be pinched out two leaves beyond the best flower truss, other trusses being cut out. If no flower truss appears, such shoots should be pinched at six to eight leaves. Any sub-laterals are pruned to one leaf and tendrils removed.

In the third summer, just one or two bunches of grapes are allowed to develop, gradually increasing the number of bunches in subsequent summers. In general, at least 30cm of stem should be left between each bunch.

MANAGEMENT. Most vines are self-fertile but tapping the laterals around midday to distribute pollen will assist setting in many cultivars. The Muscats, especially, benefit from cupping hands gently and passing them over the flower truss from one vine to another; lightly dusting flower trusses with a soft brush is an alternative procedure.

Glasshouse ventilators must be opened to provide a cold environment until late winter after which they should be progressively closed to maintain uniform temperature for the encouragement of early growth. Ventilation should be restricted during pollination but given freely after fruit set to reduce risk of disease. Ventilators may need to be fitted with covers of fruit netting if birds are troublesome.

Damping down is advantageous on sunny days except during flowering and fruit ripening, and mist-spraying of the foliage in early morning is advisable, until the first sign of fruit ripening.

Fruit thinning is essential and special grape thinning scissors are used to remove smaller, crowded fruits within the bunch, a forked stick being used to manipulate the bunch.

Winter treatment involves pricking the soil over lightly with a fork in order to remove the top 4–5cm layer, which is replaced with a fresh, rich, soil-based mixture. Following the first waterings, a mulch of well-rotted farmyard manure or compost should be applied. Once active growth is evident, feeding at two-to three-week intervals with a high-potash fertilizer should be given, until the fruits show colour.

Bunches of grapes should be harvested with a small piece of the spur attached to act as a handle to avoid damaging the bloom on the fruit surface.

Grapes in containers. If space is limited grapes can be grown in pots, tubs or other containers with a minimum diameter of 30cm. Drainage holes are essential, and should be covered with a layer of broken crocks and fibrous turf. The container is filled with a fertile soil-based medium in preference to a peat-based mix.

Where practical, a containerized vine should be moved in and out of a greenhouse or conservatory onto a sheltered terrace to aid growth and ripening. Cultivation requirements are similar to those of greenhouse vines, involving development of a permanent rod which is spur-pruned each winter. A suitable support for the rod is essential to a height of about 1.75m and only five or six bunches should be retained per plant. A mulch of rotted farmyard manure and regular feeding as for greenhouse crops is beneficial. Thorough daily watering is essential once growth begins, being reduced as the grapes begin to colour. Repot in mid-winter every two or three years.

Grapes on outdoor walls. Sheltered, warm sunny walls are suitable for grape growing, the vines being trained and pruned as for greenhouse culture. Control of dense growth by thinning and pinching and by annual pruning in winter is essential for success. The soil should be thoroughly and deeply prepared and only where the soil is poor should small amounts of rotted manure or compost be incorporated; otherwise excessive growth and delayed cropping will result. A 1–2 year-old vine should be planted 20–25cm away from the wall, in winter in milder climates and in early spring in regions with cold winters. Feed, renew the top soil and mulch as for the greenhouse crop. Early-ripening cultivars are the best choice.

Where phylloxera or nematodes are a problem grapes must be propagated onto resistant rootstocks. For other situations most vines can be propagated from hardwood cuttings taken in early winter and rooted outdoors. Muscadine cultivars do not root readily from cuttings and are usually layered. For other cultivars green softwood cuttings are also a suitable means of increase.

grape scissors scissors with long narrow tapering blades, specially designed for thinning bunches of grapes. Also known as vine scissors.

grape stage see *bud stages.*

Graptopetalum (from Greek *graptos*, painted or written upon, and *petalon*, petal). Crassulaceae. Paraguay, Mexico to Arizona. Around 12 species, succulent perennial herbs and shrubs. The fleshy leaves are usually borne in rosettes. Small, 5–petalled flowers are carried in spring and summer in lax cymes. Cultivate as for *Echeveria.*

G.amethystinum (JEWEL-LEAF PLANT; Mexico; subshrub to 30cm tall; leaves 3–7cm in loose rosettes, oblong to ovate, blunt, margin rounded, blue-green to lavender or purple-red); *G.paraguayense* (GHOST PLANT, MOTHER OF PEARL PLANT; Mexico; decumbent herb to 30cm; leaves 3–5cm in open rosettes, obovate to spathulate, thick, short-pointed, emerging pale mauve, becoming thickly grey-white-glaucous).

Graptophyllum (from Greek *graptos* to write, and *phyllon*, leaf, alluding to the beautifully marked leaves). Acanthaceae. Australasia, SW Pacific. 10 species, evergreen shrubs, often with colourfully marked leaves and producing racemes of two-lipped, tubular flowers in summer. The following species is grown primarily for its handsome foliage. It will tolerate poor light levels and drier atmospheres in the office and home. Provide a minimum temperature of 10°C/50°F, with protection from draughts, and full sun. Water and feed freely when in growth, but keep barely moist in winter. *G.pictum* (CARICATURE PLANT; habitat obscure, widepread as an escape in SE Asia; to 2m; leaves to 15cm, oval, glossy dark green blotched or marbled cream in central zone; flowers crimson to purple).

grass hook see *scythe*

grasshoppers (Orthoptera: Acrididae) mostly large insects, up to 50mm long or more, usually brown or green, with biting mouthparts and hind legs modified for jumping. Most feed predominantly on grasses but some have a wider range. Locusts are grasshoppers that have both a solitary phase in their life cycle and a gregarious one when they swarm. Grasshoppers and locusts lay their eggs in the soil.

Locusts are important pests in Africa and Asia. In Australia, the AUSTRALIAN PLAGUE LOCUST (*Chortoicetes terminifera*) damages pastures, field crops and vegetables. The SPUR-THROATED LOCUST (*Austacris guttulosa*) and the SMALL PLAGUE LOCUST (*Austroicetes cruciata*) are less serious pests. In North America, grasshoppers are more important pests than locusts. The CLEARWING GRASSHOPPER (*Camnula pellucida*) attacks cereals and other crops and several *Melanoplus* species are pests of a wide range of plants. Control of locusts often requires national or international action on a concerted scale; like grasshoppers, they are susceptible to residual contact insecticides or suitable baits containing stomach poisons.

grassing down the sowing of grass; especially around fruit trees to slow down the rate of growth through competition for water and nutrients, and thereby encourage fruiting.

grease banding a technique used mainly on fruit and ornamental trees to control winter moths and related species; it involves encircling the trunk with a 100mm band of greaseproof paper smeared with grease. Female WINTER MOTHS are wingless and gain access to trees by climbing the trunk where they become trapped in the greaseband. The device should be applied in mid-autumn and regularly maintained throughout the winter. A band of vaseline around the stems of chrysanthemums can help to prevent CHRYSANTHEMUM EELWORM (*Aphelenchoides ritzema-bosi*) from ascending the stems in humid weather.

grecian saw a pruning saw with a narrow curved blade, with the teeth set for pulling action.

greenback a condition of tomato fruits in which an area around the stalk remains hard and green whilst other parts ripen. It is associated with hot sunshine and an inadequate supply of potash. There are differences in susceptibility between cultivars. Prevention is by greenhouse shading and careful feeding and watering.

green bud see *bud stages.*

green cluster see *bud stages.*

greenfly see *aphids.*

greenhouse a structure in which plants are cultivated for the production of a harvested crop or for display. It may be covered with glass, rigid plastic sheeting or plastic film. In the UK, the terms greenhouse and glasshouse are interchangeable but to

grecian saw

avoid confusion arising from the introduction of plastic alternatives to glass, there is international agreement to use the term 'greenhouse' in all technical and scientific publications.

Glass-clad greenhouses, regardless of design, are often described according to the temperature level normally maintained: COLD HOUSE, without heating; COOL HOUSE, heated to maintain frost protection only or a minimum of about 7°C/45°F; TEMPERATE HOUSE, heated to maintain about 13°C/55°F minimum; HOT HOUSE (also warm house or stove house), heated to maintain about 18°C/65°F minimum.

The design of greenhouses must take into account five main requirements – maximum light transmission, minimum obstruction of the growing area, structural integrity, energy efficiency and cost. Light transmission is affected by the proportion of opaque structural material, orientation and roof shape. The use of steel and aluminium extrusions gives increased strength for size and reduces the amount of opaque material. An east-west orientation increases light transmission as does a high south-facing eave, with the side wall inclined.

Cladding materials Glass is the best choice in terms of light transmission, heat retention, durability and cost-effectiveness. Panes usually measure 600 × 600mm, but may be smaller for some wooden greenhouses; Dutch-light panes are 1422 × 730mm. A thickness of 3mm is adequate for panes up to 600mm wide, except in situations where there are particular safety risks; special types are available for double glazing.

Where timber glazing bars are used, the glass is normally bedded on putty and secured by brass pins; the bars may be capped with aluminium strip and soft mastic. With aluminium bars, the glass is bedded on mastic or into profiled polyvinyl chloride (PVC) or neoprene extrusion and secured with stainless-steel clips. Dutch-light panes are normally dry glazed, each sheet being held in grooves machined along the frame members.

Glass alternatives are available in the form of rigid plastic sheeting and plastic film. Acrylic and acrylic-coated polycarbonate are the most satisfactory sheet plastics, having optical properties similar to glass and being available in twin-walled form for improved insulation; acrylic can be formed into curved profiles. Very tough and durable film plastics can also be used to make tensioned glazing panels.

Polythene film cannot be regarded as a direct replacement for glass but it provides suitable covering for low-cost tunnel greenhouses. The risk of it degrading in sunlight is greatly reduced where the product contains an ultraviolet light inhibitor, and the incorporation of the copolymer, EVA, improves heat retention characteristics. The film is stretched over galvanized-steel tubular arches anchored in the soil, to form walk-in tunnels. Single-span structures are available in widths from 4.3m to 10m, and multi-span forms, up to 2.4m high, with connecting gutters, in spans from 6.4m to 10m. These structures are more susceptible to wind than glasshouses so the arches must be well-secured, the structure braced and the cover applied as a single sheet secured to a base rail or dug into the ground. Polythene tunnels are mainly used for the production of salad crops, celery, peppers and strawberries, and for the protection of container-grown nursery stock.

Heating and ventilation Where supplementary heating is provided, the range and season of plants can be extended. However, it must be remembered that the cost of heating escalates dramatically as the minimum temperature requirement is raised. Commercial greenhouses are usually heated either by hot water circulated through fixed pipes, or by fan-propelled hot air distributed through large-diameter polythene tubes. More common and convenient for garden greenhouses are portable units fuelled by electricity, paraffin or gas.

Electric fan heaters are a suitable choice for the small greenhouse, circulating air quickly and providing a source of cool air in summer. Electric tubular heaters, fixed about 10cm from the walls, can maintain a higher temperature in large greenhouses, and they give more even distribution of heat. Both sorts must be thermostatically controlled for efficiency.

Paraffin heaters are relatively cheap to buy and operate, although they cannot be thermostatically contolled. They emit water vapour, which can encourage disease, and at high settings or where inadequately maintained may produce harmful gases; they are best regarded as a stand-by where only frost protection is required.

Flueless gas-fired heaters employing natural or bottled gas are available, and these may be thermostatically controlled and will enhance atmospheric carbon dioxide level. Both types of fuel-burning heaters require regular maintenance, and a little ventilation should be provided.

The required output capacity of greenhouse heaters can be calculated, each based on temperature lift, which is the degree difference between the required minimum inside the glasshouse and the likely minimum outside. Another factor to take into account is the co-efficient of heat loss through the cladding, which necessitates calculating the area of all exposed surfaces and dividing that for brick walling by two because of its better insulating properties. A suitable formula for calculating required electric-heater capacity in kilowatts is then:

$$\frac{8 \times \text{equiv. surface area (m}^2) \times \text{temp. lift (°C)}}{1000}$$

Greenhouse ventilation limits the rise of air temperature, provides cooling, controls the rise of humidity and maintains the ambient level of carbon dioxide. Ventilators built into the ridge and low down on the sides of the structure are recommended for good air circulation; and their total area when fully open should be equivalent to at least one sixth of the greenhouse floor area. Extractor fans give positive air movement but result in uneven temperature distribution. They should be fitted high in the gable-end away from the door, and can be thermostatically controlled. See *bench, shading*.

green manuring the practice of growing certain quick-maturing plants and subsequently digging them in to improve soil fertility. Examples of green manures are RAPE (*Brassica napus*), WHITE MUSTARD (*Sinapis alba*), CLOVER (*Trifolium* species), WINTER FIELD BEANS (*Vicia faba*) and GRAZING RYE GRASS (*Secale cereale*). Usually sown broadcast in close rows, green manures must be dug in whilst lush.

Green manuring provides a source of humus, thereby improving soil structure; where leguminous nitrogen-fixing species are grown, it is also a means of adding nitrogen to the soil. However, green manuring is essentially a technique supplementary to manuring and fertilizing and should not be solely relied upon on poor soils. On light-textured soils, it provides protection from winter erosion. Also known as cover cropping, sheet composting.

green tip see *bud stages*

greenwaste collected plant residues, which following composting are a valuable source of organic matter for soil amelioration and mulching.

Grevillea (for Charles F. Greville (1749–1809), a founder of the Royal Horticultural Society, once vice president of the Royal Society). Proteaceae. Australia, New Caledonia. SPIDER FLOWER. Some 250 species, evergreen trees or shrubs. The leaves range in shape from small, rigid and needle-like to broad, soft and pinnately lobed. Tubular, curving flowers are borne in racemes or panicles usually in spring and summer. Many species are known to tolerate frost to about –5°C/23°F, if given a perfectly drained soil and a sheltered position against a south-facing wall. Many of the cultivars show a greater tolerance than their parents. *G.alpina*, *G.juniperina* and *G.rosmarinifolia* are among the hardiest species, the last having survived temperatures to 10°C/14°F. The genus is grown primarily for its nectar-rich flowers, although some species also have very attractive foliage – for example, *G.robusta*, a popular plant for the home or office, grown either as a compact, rooted stem cutting or as a specimen shrub. Most thrive given full sun and good drainage in a neutral or slightly acid soil. Propagate from fresh seed in spring, or by semi-ripe cuttings. *G.robusta* is resistant to root-rotting fungus and is used as a rootstock for grafting western Australian species and as a stock for standard and weeping specimens.

G.alpina (E Australia; prostrate to erect shrub to 2m; leaves 1–3cm, linear to rounded; flowers red, pink, green, yellow and white, crowded in short, spidery racemes); *G.banksii* (RED-FLOWERED SILKY OAK; E Australia; shrub to 4m; leaves to 25cm, pinnatifid, lobes linear, silky grey beneath; flowers red or white in cylindrical racemes to 10cm); *G.* 'Canberra Gem' (*G.juniperina* × *G.rosmarinifolia*; vigorous, rounded shrub to 2.5m; leaves to 3cm, linear, sharply pointed; flowers bright pink, waxy, in clusters sometimes throughout the year); *G.juniperina* (E Australia; dense rounded shrub to 2m tall; leaves 1–2cm, narrowly lanceolate to linear; flowers green to yellow pink or red); *G.* Poorinda Hybrids (a group of garden hybrids including 'Poorinda Constance', a large, rounded shrub to 4m with sharply pointed, oblong leaves to 3cm and red flowers in spidery clusters, the purple-red 'Poorinda Peter', with bronze-tinted, pinnate leaves, and the apricot-flowered 'Poorinda Queen'; *G.robusta* (SILKY OAK; E Australia; fast-growing tree to 35m; leaves to 30cm, bronze at first, becoming olive, broadly oblong, pinnatifid, segments deeply cut, silky; flowers orange); *G.rosmarinifolia* (E Australia; dense shrub to 2m tall; leaves to 3cm, linear, rigid, bright to grey-green; flowers pink to red with cream, sometimes pure creamy yellow).

grex a group name for all plants derived from the crossing of the same two or more species. The grex (herd or hybrid swarm) is close in some senses to the *group*, but relates more specifically to an established parentage or lineage. The term has become limited to orchid hybrids. It is printed in roman with a capital initial. The grex may be combined with a cultivar or clonal name which is printed within quotation marks, for example, *Paphiopedilum* Maudiae 'The Queen'.

grey bulb rot (*Rhizoctonia tuliparum*) a soil-borne disease of ornamental bulbs in cool temperate regions. Poor emergence is an early indication of the disease, and lifted plants have rotted shoots with soil adhering even though the roots may appear healthy. The fungus shows as greyish mycelia around rotted bulbs and between bulb scales, with numerous large flattened white or black fruiting bodies on the bulbs and in the surrounding soil; by means of these fruiting bodies, the fungus can survive for several years. It is most serious on tulips, but *Colchicum*, *Crocus*, hyacinths, *Iris*, *Ixia*, lilies and *Narcissus* can also be infected. Control by removing all affected bulbs and the surrounding soil as soon as symptoms are noticed. The affected area should not be used for bulb growing for at least five years.

Greyia (for Sir George Grey (1812–1898), governor general of Cape Colony). Greyiaceae. South Africa. NATAL BOTTLEBRUSH. Some 3 species, shrubs or small trees. The toothed leaves are white-woolly when young. Produced in spring and summer, the small flowers have showy stamens and are carried in dense, bottlebrush-like spikes. Plant in a well-drained, sandy medium in full sun. Under glass, provide good ventilation, maintain a minimum temperature of 7–10°C/45–50°F, and water moderately when in growth. Keep almost dry in winter. Propagate by seed in spring, by basal cuttings of new growth, by semi-ripe cuttings, or by removal of rooted suckers. *G.sutherlandii* (to 4.5m; leaves to 7cm, rounded to heart-shaped, toothed; flower spikes to 10 × 5cm, red).

grey mould the visible growth of the ubiquitous fungus *Botrytis cinerea*. It is found on many different hosts and causes leaf, flower and bud rots, die-back of woody plants and rots of fruits and vegetables. Affected organs are covered with a fluffy grey mould from which clouds of spores are released, rapidly spreading the disease. The fungus is common on decaying plant material and is a weak parasite which mainly attacks damaged or stressed plants growing under cool, damp, overcrowded conditions.

Control depends on good hygiene, involving the removal of crop debris and damaged or affected tissues, good ventilation, especially in greenhouses, and the avoidance of damp or overcrowded conditions. Chemical control is possible with recommended systemic and protectant fungicides. However, control by the continual use of systemic fungicides alone may lead to the build-up of resistant strains of the fungus, or development of infection by other fungi, such as *Alternaria* and *Stemphylium*, which are not controlled by the group of fungicides in use.

Although a major disease of vines, requiring costly control measures during the growing season, grey mould can also be beneficial to certain wine-producers. *Botrytis* is the agent causing 'noble rot' (*pourriture noble*) of ripe grapes, the condition in which the berries lose moisture and thereby yield juice of a higher sugar content. The infection is essential to the production of some wines, such as Sauternes and Tokay.

greywater see *irrigation*.

Grindelia (for David Hieronymus Grindel (1776–1836), Russian botanist). Compositae. Western N and S America. GUM PLANT, TARWEED, ROSIN-WEED. About 60 species, resinous annual or perennial herbs, rarely shrubs. They produce large, daisy-like flowerheads in summer. Suitable for the larger rock garden, for the warm sunny border, and for dry, sunny banks, especially on poor dry soils. Hardy to –15°C/5°F. Grow in any well-drained soil in full sun. Propagate by seed, or by semi-ripe cuttings in late summer. *G.chiloensis* (syn. *G.speciosa*; Patagonia; clammy, shrubby perennial to 1m; flowerheads deep golden yellow).

Griselinia (for F. Griselini (1717–1783) Italian botanist). Cornaceae (Griseliniaceae). New Zealand, Brazil, Chile. 6 species, evergreen trees and shrubs with leathery leaves and inconspicuous flowers in racemes or panicles. Where temperatures do not fall below –5°C/23°F, *G.littoralis* may be grown in the open garden; it will thrive in colder areas still (to –12°C/10°F) if given a sheltered, south-facing niche and good drainage; it is suitable for screens and hedging in coastal areas. *G.lucida* is frost-tender, but may be grown in the conservatory in cold areas. Grow in sun or part shade. Trim hedges in summer. Propagate from nodal cuttings in late summer/early autumn, in a sandy propagating mix in the cold frame.

G.littoralis (New Zealand; shrub or tree to 8m; leaves to 9cm, ovate to oblong, olive to apple green, glossy, slightly fleshy, more or less undulate; includes 'Dixon's Cream', with leaves splashed creamy white,

and 'Variegata', with leaves blotched or zoned white); *G.lucida* (New Zealand; shrub to 4m; leaves to 18, mid-green, glossy; includes the large- and round-leaved var. *macrophylla* and the yellow-cream-marked 'Variegata').

grit very small stone fragments, larger than sharp sand, used as an ingredient of certain growing media, or on the surfaces of pots and sinks for alpines especially.

grotto an artificial garden feature made to resemble a small, picturesque cave. Grottos have been known since Roman times and have been either excavated or constructed, often from tufa rock, and ornamented with stones, shells or attractive minerals. They sometimes incorporate water works.

ground bed usually a temporary cold frame, comprising low board or block walls covered by glass lights or film plastic.

ground beetles (Coleoptera: Carabidae) a very large and widely distributed family of small to medium-large beetles. Most are black, with long bead-like antennae, biting mouthparts and slender legs adapted for running. Many species are flightless. Normally, ground beetles are active at night and spend the day under stones, leaves, logs etc., especially in damp places. Adults and larvae feed on small worms, slugs, snails and insect larvae, and the species *Bembidion lampros* and *Trechus quadristriatus* are predators of the eggs and larvae of the cabbage root fly (*Delia brassicae*). Some ground beetles, including the strawberry seed beetle (*Harpalus rufipes*) and the strawberry ground beetles (*Pterostichus madidus* and *P.melanarius*), may damage ripening fruits of strawberries.

groundcover selected species planted between shrubs to cover the ground and conceal all signs of the soil beneath. Provided that perennial weeds are removed before planting, groundcover will smother other weeds and can provide an attractive garden feature in itself, as well as enhancing plants beneath which it grows. It may be useful as a mulch for conserving moisture, but where well-established it can compete with main plantings for moisture and nutrients.

Plants used as groundcover are usually less than 30cm high, although in large areas some shrubs can be grown for the purpose. Species should not be so vigorous as to become weeds, and the majority of plants should be evergreen. However, individual species that spread over an area of about 60cm diameter and die back in the autumn are useful for underplanting spring-flowering dwarf bulbs. For groundcover, flower qualities are secondary to foliage effect.

Herbaceous perennials are valuable groundcovering plants and among these are the evergreen 'carpeters' *Saxifraga umbrosa* (London Pride), *Ajuga reptans*, (bugle), *Potentilla alba* and *Waldsteinia ternata*, all of which will spread over a wide area from a single plant.

Larger plants, making clumps which eventually join together, include *Alchemilla mollis*, *Geranium endressii* and *G.* × *magnificum*, *Tellima grandiflora*, *Brunnera macrophylla*, *Bergenia* species, *Pachyphragma macrophylla* and *Pulmonaria* species. Suitable taller hardy perennials which are effective in expansive colonies are *Hosta*, *Hemerocallis*, *Helleborus orientalis* and *Nepeta*. Effective small-scale groundcover of dense spreading growth is provided by the alpine campanulas and phloxes, *Dryas octopetala*, *Polygonum vacciniifolium*, *Asarum europaeum* and the smaller cultivars of *Geranium sanguineum*.

Low-growing shrubs, such as *Cornus canadensis*, *Cotoneaster dammeri*, *Euonymus fortunei*, *Lonicera pileata*, *Mahonia repens*, *Pachysandra terminalis*, *Rubus pentalobus* and *Vinca* species, can be effective. Suitable taller shrubs include lavender, hydrangeas, *Prunus laurocerasus* 'Otto Luyken', *Sarcococca* species, *Viburnum davidii*, and, for acid soil, *Arctostaphylos*, *Gaultheria* and *Vaccinium* species. Heaths and heathers are some of the most valuable of all groundcover. *Erica carnea* and *E.erigena* hybrids flourish on alkaline soil, *E.vagans* on neutral ground, and all other species of *Erica* and *Calluna* in soils free of lime.

Some climbing plants make effective groundcover, eventually covering large areas and being therefore relatively inexpensive in outlay. These include the less rampant forms of *Hedera*, *Hydrangea petiolaris* and *Lonicera japonica* 'Halliana'. Certain ferns can be used, two British natives, the male fern, *Dryopteris filix-mas*, and the hart's tongue, *Asplenium scolopendrium*, being among the most valuable. Others are *Athyrium filix-femina*, *Blechnum chilense*, *Cystopteris bulbifera*, *Matteuccia struthiopteris* and *Polypodium vulgare*.

Prostrate and semi-prostrate conifers are worth consideration, especially those with coloured foliage, for example, the procumbent, grey selections of *Juniperus*. Ornamental grasses and sedges, among them *Festuca glauca*, *Luzula* species, *Phalaris arundinacea*, *Stipa calamagrostis* and *Carex* species, can also be used.

All of the foregoing are satisfactory in most gardens in the UK and the eastern states of North America. In temperate regions like the Mediterranean and the western seaboard areas of North America, where frost in winter is unlikely and the summer is characterized by long dry periods, dense-growing low shrubs can be selected from the natural range of *maquis* and *garigue* plants, while low creepers can be found among such plants as *Acaena* species, *Armeria*, dwarf *Ceratostigma* species, *Convolvulus cneorum*, spreading *Chamaecytisus* and *Genista*, *Liriope*, mesembryanthemums, *Osteomeles*, the dwarf saxifrages, *Sedum* species and *Teucrium* species.

Subtropical areas, such as parts of Florida, can support perennials which are often grown as tender summer bedding in northern parts, for example *Chlorophytum*, coleus, *Iresine*, *Helichrysum petiolare* and *Tradescantia*. Other plants for full sun include *Asystasia gangetica*, *Hemigraphis alternata* and *Lantana montevidensis*. Ferns provide important groundcover plants for shade in such climates, together with species such as *Episcia*, *Fittonia*, *Peperomia* and *Pilea nummularifolia*. *Asparagus sprengeri* grows in sun or shade. See *mulch*.

ground elder (*Aegopodium podagraria*) an invasive perennial weed of uncultivated ground, borders and sometimes lawns. It occasionally establishes itself from seed but is mainly transported by fleshy underground stems. These form a dense network in the top layer of soil and give rise to thick leaf cover which can stifle all but the most vigorous of plants. Left unchecked, ground elder can spread up to a metre per year. Control by forking is difficult because of the need to remove all root portions, which readily become established amongst the roots of garden plants. The best recourse is to lift infested plants for renewal whilst the site is thoroughly cleared either by meticulous cultivation, total weed killer or, where practicable, by smothering with a very deep organic mulch or thick polythene sheet, left down for several growing seasons. Glyphosate and glyphosate trimesium are effective as total weedkillers.

ground elder

groundkeeper used particularly of a potato tuber that is unharvested and gives rise to a plant in the following season. It may be of significance as a source of disease infection in the new season. Also known as a volunteer.

groundwork the ground level of a bedding scheme comprised of low growing plants.

group a category intermediate between species and cultivar, used to distinguish: (1) an assemblage of two or more similar cultivars and/or individuals within a species or hybrid; (2) plants derived from a hybrid of uncertain parentage. Group names are written in roman script, take a capital initial letter, no quotation marks and are always followed by the word Group, with an upper case initial.

growing bag a sealed, plastic bag, about 32cm wide and 1m long, filled with 50–60 litres of compressed soilless growing medium, usually fertilized peat or composted bark. Growing bags are used for the cultivation of tomatoes and other plants where it is beneficial to avoid rooting into greenhouse border soil, and for growing on concrete surfaces. Two or three planting holes are cut into the top of the module and 5cm-drainage slits made in the sides, about 2.5cm from the base. Regular watering and supplementary feeding are essential.

growing point the tip of a stem.

growing room, growth chamber a windowless room, usually within an existing building, in which light, temperature, and humidity are controlled. Commonly used for raising bedding plants, tomatoes, cucumbers and lettuce in winter and early spring, and for research purposes. Lighting is provided by static or mobile rigs of 2.4m-long 125W fluorescent tubes, giving a visible irradiance of $40W/m^2$ or $20W/m^2$ depending on 12- or 24-hour cycles of operating.

Growmore a standard, compound, balanced fertilizer, made to a formula first recommended by the UK Ministry of Agriculture during World War II, and marketed as National Growmore. It contains 7% nitrogen: 7% phosphate: 7% potash.

growth cycle periods of active growth in plants alternating with dormancy or slowed growth brought about by low rainfall or changed levels of temperature or day length. In temperate zones, it is illustrated by deciduous trees losing their leaves in autumn and by evergreen growth slowing during cold spells and in poor light conditions.

growth regulators chemicals that influence plant growth. They either occur naturally in association with plant tissue or are manufactured and applied to achieve a desired effect.

Ethylene gas is produced by ripening apples and hastens maturity in unripened fruits and wilting in cut flowers. Various complex chemicals, often referred to as hormones, occur within plants. Among these are auxins, which cause stem elongation and root formation on cuttings, gibberellins, which break dormancy and induce flowering and stem elongation, cytokinins, which stimulate cell division and abscisic acid which inhibits growth.

Synthetic imitations of plant hormones are developed to produce abnormal growth as in the weedkillers 2,4–D and mecoprop. Other manufactured growth regulators include maleic hydrazide, which suppresses sprout development in potatoes and sucker growth in trees and shrubs, dikegulac, which retards growth of hedges, chlormequat, which restricts stem growth in ornamentals, and IBA (4–indol–3–yl-butyric acid), which promotes the rooting of cuttings.

growth rings see *annual rings*.

grubber see *mattock*; *daisy grubber*.

grubs a popular term for the larvae of beetles (Coleoptera); also sometimes applied to those of bees, wasps and ants (Hymenoptera). Beetle larvae have biting mouthparts, a well-developed, hardened head capsule and usually three pairs of thoracic legs. The larvae of ants, wasps and bees are legless with the head only slightly hardened and less prominent.

guano originally, sea-bird droppings used as a rich natural fertilizer. The term is now loosely applied to other concentrated organic ferilizers, especially those that are made from fish and known as fish guano.

guard cells two cells that surround each stoma (which are the pore-like openings distributed over the epidermis of the aerial parts of plants), and that serve to open and close it.

gumming the exudation of gum-like resins by plants, particularly *Prunus* species and conifers, often in response to injury or disease, but sometimes as a reaction to poor soil conditions or insect attack. In stone-fruit trees, gumming is a physiological condition triggered by stress; plum fruits may form gum around the stone.

gummosis the exudation of gum from sunken spots on fungus-infected cucumber fruits, upon which dark mould may develop. The condition is encouraged in a wet, cool environment with limited ventilation.

Gunnera (for Ernst Gunnerus (1718–1773), Norwegian bishop and botanist). Gunneraceae (Haloragidaceae). Australasia, South Africa, S America, Pacific region north to Hawaii. Some 50 species, dwarf or giant rhizomatous, perennial herbs. The small species form interlacing mats, the large ones, clumps or colonies. In either case, the leaves are broadly ovate to rounded, basal and long-stalked. In spring and summer, inconspicuous flowers are borne massed in panicles and give rise to dry or fleshy fruit. The two giants described here (*G.manicata* and *G.tinctoria*) are magnificent foliage plants for the waterside, wet ditches and bog gardens. In climate zones 6 and over, they thrive on moist or wet, rich soils in sun or part shade. In autumn protect them from frost by covering crowns with the current season's foliage (break the leafstalks and bend the leaves down to cover the growing point), or with straw or bracken. *G.magellanica* is suitable for moist areas in the rock garden or peat bed, and for pans in the alpine or cool greenhouse. Grow in full sunlight or partial shade in a moist medium, rich in organic content. Increase all species by division in spring.

G.magellanica (southern S America, Falkland Islands; dwarf, mat-forming; leaves 2–9cm in diameter, blades reniform, somewhat cupped, deep green, crenate, stalks 2–15cm; fruit a cluster of orange-red berries); *G.manicata* (GIANT RHUBARB; S Brazil,

Gunnera manicata

Colombia; rhizomes massive and shaggy; leaves 1–2m in diameter, blades rounded to reniform, margins broadly lobed and jaggedly toothed, upper surface dull green and bullate, lower surface with very prominent, prickly veins, stalks 1.5–2.5m tall, stout, prickly; flowers minute, rust to fox red, massed in a conical to cylindrical panicle to 1.5m tall); *G.tinctoria* (syn. *G.chilensis*, *G.scabra*; Chile; differs from *G.manicata* in forming smaller clumps with shorter leafstalks and blades to 1.5m wide, these more sharply toothed and lobed; the flowers and fruit, and sometimes the leaf veins, have a strong wine red tint).

guttation the exudation of drops of water, mostly from the margins or tips of leaves, seen for example in *Tropaeolum majus*, *Alchemilla mollis* and lawn grass. It usually takes place under humid atmospheric conditions which give rise to low transpiration when water absorption is occurring freely and root pressure develops.

guying the use of guy wires to support trees. The wires are fixed high up around the trunk, and attached to pegs driven into the ground.

Guzmania (for Anastasio Guzman (d. 1802), Spanish naturalist and apothecary). Bromeliaceae. C and S America. 126 species, perennial, epiphytic herbs. The leaves are strap-shaped and smooth and arranged in a basal rosette, their bases forming a cup or vase. From its centre, the flower spike emerges at various times of year, branched or unbranched and usually covered in waxy, colourful bracts. The flowers are small and broadly tubular. Cultivate as for *Aechmea*.

G.lingulata (C and S America; to 45cm; leaves pale to mid-green; flowers white to yellow amid orange to red or yellow bracts); *G.monostachya* (syn. *G.tricolor*; C and S America; to 40cm; leaves pale green, often with brown stripes or tint at base; flowers white, lower bracts green striped brown-black, upper bracts bright scarlet); *G.sanguinea* (C and S America; to 30cm; leaves mid to dark green, flushing red at centre when in flower; flowers yellow amid red to orange-yellow bracts); *G.vittata* (Brazil, Colombia; to 60cm; leaves dark green with paler bands and sometimes banded purple beneath; flowers white amid purple-spotted, green bracts).

Gymnocalycium (from Greek *gymnos*, naked, and *kalyx*, bud). Cactaceae. Brazil, Paraguay, Bolivia, Uruguay, Argentina. Perhaps 50 species of low-growing cacti. Their stems are globose and mostly unbranched with prominent ribs and spines. The flowers are funnelform to campanulate. Provide a minimum winter temperature of 7°C/45°F. Grow in a neutral compost containing more than 50% grit. Maintain low humidity; keep dry from mid-autumn until early spring, except for light misting on warm days in late winter. Slight shading during the summer months is desirable in greenhouses exposed to full sun. The flowering season is spring to early summer, although a few species may produce a few flowers well into autumn. Increase by seed, stem cuttings or grafting.

G.andreae (N Argentina; stems to 4.5cm in diameter, globose, lustrous dark blue- or black-green, spines dull white or dark brown, needle-like, often curved; flowers to 3cm, yellow); *G.gibbosum* (Argentina; stems to 20 × 15cm, glaucous, globose becoming club-shaped, brown-green, spines brown to grey, straight or slightly curved; flowers to 6cm, pure white or outer tepals tinted pale pink); *G.mihanovichii* (N Paraguay; stems to 8cm in diameter or more, squatly globose to shortly cylindric, dark olive green with cross banding, spines dull yellow tipped brown; flowers to 4.5cm, yellow-green tipped red; cultivars exist that lack chlorophyll, having stems wholly and brightly coloured yellow, red or pink; to survive, these clones must be grafted); *G.quehlianum* (N Argentina; stem to 3.5 × 7cm, flattened to globose, turning bronze in full sun, spines adpressed, somewhat bristly, horn-coloured, red-brown toward base; flowers to 4cm, white with a red throat); *G.schickendantzii* (N Argentina; stem, to 15 × 30cm, broadly flattened to globose, turning bronze in full sun, ribs somewhat spiralled, spines flattened and recurved, grey-red to horn-coloured, often darker-tipped; flowers white tinged pink or red-green).

Gymnocladus (from Greek *gymnos*, naked, and *klados*, a branch). Leguminosae. US, China. Some 5 species, deciduous trees with bipinnate leaves. The flowers are small, green-white and borne in short terminal panicles or racemes in early summer. The fruit is a large, woody pod. Hardy to –30°C/–22°F and tolerant of salt, drought and alkaline soils. Propagate from seed sown fresh or in spring, after chipping or soaking in hot water. Alternatively, take root cuttings in winter. *G.dioica* (KENTUCKY COFFEE TREE; central and eastern N America; to 25m, bark coarsely fissured, branchlets thick; leaves 30–80cm, emerging pale pink, turning dark green, yellow in autumn; flowers dull green-white, downy, in racemes; fruit to 15 × 4cm, brown or maroon, thick, succulent becoming woody).

gymnosperm a plant in which seeds are borne naked on a sporophyll, rather than enclosed in an ovary, as in angiosperms; the ovules develop without the enclosing pericarp found in angiosperms. Gymnosperms include conifers, cycads, *Welwitschia*, *Ephedra*, and *Gnetum*.

Gymnospermium (from Greek *gymnos*, naked, and *sperma*, seed). Berberidaceae. Europe to C Asia. 4 species, deciduous perennial herbs with tuberous rhizomes. The ferny leaves are ternately divided into pinnately lobed leaflets. The flowers loosely resemble those of *Berberis* but are borne in slender-stalked, terminal racemes. Cultivate as for *Bongardia*.

G.alberti (C Asia; to 25cm; flowers to 2cm across, yellow with red-brown veins on sepals and a stalked ovary); *G.altaicum* (Black Sea; differs from *G.alberti* in having plain yellow sepals and an ovary that is not stalked).

*Gypsophila
paniculata*

Gynandriris (from Greek *gyne*, female, *andros*, male, and *Iris* – referring to the united stamens and pistil). Iridaceae. Southern Africa, Mediterranean, east to Pakistan. 9 species, cormous perennial herbs. Short-lived, *Iris*-like flowers are produced in spring. Hardy to about –5°C/23°F, the following species needs plentiful moisture when in growth but requires a dry summer baking to ensure flowering. Plant deeply, at about 15cm, in a well-drained, gritty soil in a warm sunny position. Sow seed in a sandy propagating mix in early spring at about 10°C/50°F; increase also by offsets. *G.sisyrinchium* (Portugal, Mediterranean to SW Asia; 10–40cm; leaves slender, usually prostrate and curling; flowers produced in succession, 1–6 per cyme, faintly scented, violet-blue to lavender, the falls with a white or white and orange patch).

gynodioecious having female and hermaphrodite flowers on separate plants of the same species, as occurs in, for example, ground ivy (*Glechoma hederacea*).

gynoecium the female element of a flower, comprising the pistil or pistils.

gynomonoecious having female and hermaphrodite flowers on the same plant, as occurs in, for example, *Syringa × persica*.

gynophore the stalk of a pistil, which raises it above the receptacle.

gynostegium the staminal crown in Asclepiadaceae.

gynostemium a single structure combining androecium and gynoecium, as in the column of Orchidaceae.

Gynura (from Greek *gyne*, woman, and *oura*, a tail: the stigma is long and rough). Compositae. Old World tropics. VELVET PLANT. About 50 species, perennial herbs and subshrubs with entire to pinnately dissected leaves and orange-yellow button-like flowerheads. The following are striking foliage plants suitable for pot or basket cultivation in the greenhouse or home. Grow in a fertile and well-drained potting mix in bright indirect light, ensuring shade from strong summer sun. Maintain a minimum temperature of 15°C/60°F. Keep evenly moist. Propagate by softwood or semi-ripe cuttings in a shaded, closed case with gentle bottom heat. *G.aurantiaca* (VELVET PLANT, PURPLE VELVET PLANT, ROYAL VELVET PLANT; perennial herb, erect to sprawling or clambering, to 60cm, all parts covered in short purple hairs; leaves coarsely toothed, dark green with a velvety purple sheen from the hairs; includes the popular 'Purple Passion', syn. *G.sarmentosa* of gardens, a decumbent plant, the leaves narrow with wavy-lobed and toothed margins and thickly covered in velvety purple hairs).

Gypsophila (from Greek *gypsos*, chalk, and *philos*, loving, referring to the preference of some species for limy soils). Caryophyllaceae. Eurasia, particularly common in SE Europe. BABY'S BREATH. About 100 species, annual or perennial herbs of varying habit. The leaves are often somewhat fleshy and glaucous. Usually numerous and small in spreading panicles, rarely large and solitary, the flowers are borne in spring and summer. They are broadly funnel-shaped and consist of five, obovate petals.

Gypsophila is represented in cultivation by the small montane species, such as *G.repens*, and larger species, such as *G.paniculata*, from dry, stony or sandy habitats. The perennial *G.paniculata* and the fast-growing annual *G.elegans* are particularly useful in the flower border. They are grown for the light airy masses of tiny flowers, carried in summer on slender much-branched stems. Both species are excellent cut flowers. The following are fully hardy, but favour a warm and sheltered position with protection from excessive winter wet. Grow in full sun on well-drained soils – sandy, poor and stony for the alpine species, deep and fertile for the larger species. Avoid disturbing established plants. Cut back or trim over after flowering. Propagate perennials by softwood cuttings in summer, by seed in spring, or (*G.paniculata* cultivars) by grafting in winter. Sow seed of *G.elegans* under glass in early spring for planting out after the last frosts, or *in situ* in mid-spring.

G.cerastioides (Himalaya; loose, grey-hairy, mat-forming perennial; flowers to 2cm diam., white or lilac veined pink, in loose corymbs); *G.elegans* (Asia Minor, Caucasus, S Ukraine; glabrous annual to 50cm, branching above; flowers to 3cm diam., white veined pink or purple, on long, slender stalks in a finely but loosely branched panicle; a popular cut flower, rather coarser and larger than the *G.paniculata*; cultivars include the large, pure white 'Covent Garden' and 'Giant White', the pale pink 'Rosea', the vinous 'Purpurea' and the carmine 'Red Cloud'); *G.paniculata* (BABY'S BREATH; C and E Europe, C Asia; stoutly rhizomatous, diffusely branched perennial to 120cm, usually smooth and glaucous; flowers to 0.8cm diam., white or pink, many in a finely branched, cloud-like panicle; includes cultivars with large or dwarf habit, and single or double flowers); *G.repens* (C and S Europe; mat-forming, grey-glaucous perennial with flowering stems ascending to 20cm; flowers to 1.5cm diam., white, pink or purple-pink in loose panicles; cultivars include pale to deep rose-flowered plants and forms with single or double flowers, one of the finest being 'Dorothy Teacher', a neat, compact, blue-green plant with profuse flowers turning from white to soft pink).

gypsum hydrated calcium sulphate (sulphate of lime), used to improve the structure of clay soils without affecting their alkalinity. It is valuable for land reclamation after flooding by sea water.

gyrate curving in a circular or spiral fashion.

H

ha-ha a deep wide ditch separating a garden or park from the surrounding countryside, designed to exclude animals without creating a visual barrier. Originating in the early 18th century, the device became a characteristic of the English Landscape style of 'Capability' Brown (1716–83) and his followers.

Haageocereus (for F. Haage, and *Cereus*). Cactaceae. Peru, N Chile. To 10 species, shrubby or arborescent cacti with ribbed and spiny stems. Mostly white or dull red, the tubular to funnelform flowers bloom at night and, often remain open through the following morning. Grow in a cool frost-free greenhouse (minimum temperature 2–7°C/36–45°F). Use 'standard' cactus compost, with a moderate to high inorganic content (more than 50% grit) and a pH of 6–7.5. Provide full sun and low air humidity. Keep dry from mid-autumn until early spring, except for light misting on warm days in late winter. Increase by stem cuttings.

H.decumbens (S Peru; stems to about lm × 4–8cm, decumbent or pendent, spines very stout, dark; flowers white); *H.multangularis* (syn. *H.chosicensis*; C Peru; stems to 1.5m × 6–10cm, erect or ascending, spines very stout; flowers red, pink or white); *H.versicolor* (N Peru; stems slender, erect, to 1.5m x 5cm, spines variable in colour; flowers white).

Haberlea (for Karl Konstantin Haberle (1764–1832), Professor of Botany at Budapest). Gesneriaceae. Balkans. 2 species, tufted perennial herbs. Borne in low rosettes, the leaves are obovate to broadly oblong, coarsely crenate and more or less hairy. Produced in late spring and summer in loose scapose umbels, the flowers are broadly tubular with five rounded lobes. Cultivate as for *Ramonda*.

H.ferdinandi-coburgi (C Bulgaria; similar to *H.rhodopensis* except leaves more or less glabrous above; flowers lilac, tube darker above, throat yellow-spotted); *H.rhodopensis* (C and S Bulgaria, NE Greece; leaves softly hairy; flowers 1.5–2.5cm pale blue-violet; includes the white-flowered 'Virginalis').

habit (1) the characteristic growth form of a plant; for example the spreading, columnar or weeping habit of trees; (2) loosely used to describe plant type; for example herb, shrub or tree.

habitat the natural environment in which a species lives, characterized by climate, elevation, soil type and association with other organisms.

Habranthus (from Greek *habros*, graceful, and *anthos*, a flower). Amaryllidaceae Temperate S America, especially Argentina, Uruguay and adjacent regions of Brazil and Paraguay. 10 species, perennial bulbous herbs with narrowly strap-shaped leaves and solitary, scapose flowers – these are funnel-shaped and 6-lobed. Somewhat tender bulbs grown for their flowers. *H.brachyandrus* may be grown at the base of a south- or southwest-facing wall in sheltered gardens with little or no frost. *H.tubispathus* is reliably hardy in zone 8, especially if established in a dry, sheltered and sunny position. Grow in a fibrous loam-based mix with additional leafmould, planting firmly with neck and shoulders above soil level. Propagate by offsets and also from fresh ripe seed.

H.brachyandrus (S Brazil; to 30cm, flowers to 9cm, brilliant or pale pink, tube red-black at base); *H.tubispathus* (syn. *H.andersonii*; Argentina, S Brazil, Uruguay, S Chile; to 15cm, tinged red, flowers to 7cm, burnt orange, yellow or golden-bronze above, grey-pink with darker stripes beneath).

Hacquetia (for Balthasar Hacquet (1740–1815), author of *Plantae Alpinae Carniolicae*). Umbelliferae. Europe. 1 species, *H.epipactis*, a clump-forming, perennial herb to 7cm tall (usually shorter). The emerald green leaves are near-circular in outline and palmately cleft and toothed. Small yellow flowers are produced in early spring in a dense, rounded umbel, subtended by a ruff of leaf-like bracts – the whole arrangement gives the impression of a single, green-petalled flower with a boss of golden stamens. A slowly spreading perennial for the rock and woodland garden and underplanting among winter- and spring-flowering shrubs. Hardy to –15°C/5°F. Grow in any moist, humus-rich soil, in part shade. Propagate by fresh seed, by careful division of well-established plants in early spring, and by root cuttings.

Haemanthus (from Greek *haima*, blood, and *anthos*, flower, referring to the colour of the flowers.) Amaryllidaceae. South Africa, Namibia. 21 species,

Habranthus tubispathus

deciduous or evergreen bulbous herbs with broadly strap-shaped 2-ranked leaves and small flowers packed in scapose umbels and subtended by several bracts. The fruits are coloured berries. Plant with the bulb neck at or just above soil level, in well-crocked pots of humus-rich loam-based potting medium, with added coarse grit and well-rotted manure. Maintain a winter minimum temperature of 10°C/50°F; grow in bright filtered light, but move into part shade as the flower buds develop to preserve colour and prolong flowering. Dry off deciduous species as leaves yellow; keep evergreen species dry but not arid during rest. The plants flower better when under-potted and resent disturbance. Propagate by offsets, removed as new growth begins and grown on in a closed case until established; or from ripe seed.

H.albiflos (South Africa; leaves sometimes spotted white, usually downy; scape to 35cm, green, glabrous or pubescent, spathe white veined green, ciliate, flowers white; fruit white, orange or red); *H.cocci-neus* (CAPE TULIP; South Africa; leaves sometimes barred maroon, green or white, glabrous or downy; scape to 37cm, cream to pale red streaked dark red, umbel to 10cm across, spathe 6–9-valved, coral, ver-million or scarlet, flowers coral to scarlet with white markings; fruit white to deep pink); *H.sanguineus* (South Africa; leaves scabrid above, glossy beneath, margin red or transparent; scape to 27cm, claret red, spathe red-pink, flowers red to salmon pink marked white; fruit glassy white to claret).

haft the base of an organ when narrow or constricted; most often applied to the haft of the falls in *Iris* flowers.

hair an outgrowth of the epidermis; unicellular or comprising a row of cells, and conforming to one of several types (e.g. dendritic, stellate, scale-like, peltate) according to branching, form, grouping and attachment.

Hakea (for Baron Christian Ludwig von Hake (1745–1818), German patron of botany). Proteaceae. PINCUSHION TREE. W Australia. Some 110 species, shrubs or small trees, usually hairy and with leathery to rigid leaves. Borne in short axillary racemes or clusters, the flowers are tubular with slender lobes and a protruding style. Cultivate as for *Banksia*.

H.lissosperma (syn. *H.sericea* of gardens; NEEDLE BUSH; E Australia; shrub or small tree to 6m tall; leaves to 10cm, terete, sharply pointed, curved upwards; flowers white in clusters); *H.suaveolens* (West Australia; erect shrub, 1.5–2m; leaves 7–20cm, pinnate, segments erect, sharply pointed; flowers cream in dense racemes).

Hakonechloa (from Hakone in Honshu, Japan). Gramineae. Japan. 1 species, *H.macra*, a perennial grass 30–75cm tall. The stems are slender and clumped, bending outwards in a graceful 'shaggy' crown. The leaves are narrowly lance-shaped and fresh green, becoming rust- and orange-tinted in autumn. In the cultivar 'Aureola', the leaves are striped lime and creamy yellow to gold; in 'Albo-aurea', they are lime green, gold and cream, acquiring pink to maroon stripes and tints in bright sun and in autumn. Given shelter from cold drying winds, they are hardy to between –15 and –20°C/5 to –4°F and easily grown in any moderately fertile, moisture-reten-tive soil with additional leafmould. Light shade will enhance leaf colour. Plants spread slowly once estab-lished to form substantial clumps; also excellent for container cultivation. Propagate by division in spring.

Halesia (for Dr S. H. Hales (1677–1761), author of *Vegetable Staticks* (1727)). Styracaceae. SILVERBELL TREE, SNOWDROP TREE. China, eastern N America. 5 species, deciduous trees or shrubs. In late spring and early summer, the flowers hang in clusters from the previous year's wood. White or pale rose, they are broadly bell-shaped and four-lobed, and are followed by winged fruit. Plant in deep, humus-rich, lime-free soils that are moist but well-drained, in sun with shel-ter from wind. Propagate by simple layering in spring, or, in the same season, by softwood basal cuttings, treated with hormone-rooting powder, under mist or in a closed case. Overwinter young plants in the cold frame. Seed exhibits a complex dormancy: sow fresh in autumn, and maintain at 14–25°C/57–77°F for 60–100 days, followed by cold stratification at 0–5°C/32–40°F for a further 2–3 months. Bring into equable conditions when germination starts.

H.monticola (syn. *H.carolina* var. *monticola*; SE US; tree to 28m; leaves 8–16cm, elliptic to obovate, serrate, soon glabrous except veins beneath; flowers white, 1.5–2.5cm; f. *rosea* with pale rose flowers, and var. *vestita* with leaves white-tomentose beneath when young and flowers sometimes tinted rose); *H.tetraptera* (syn. *H.carolina*; S US; small tree or shrub, to 6m; leaves 5–6cm, ovate to lanceolate, serrulate, glabrous above, grey-downy beneath; flowers white, 1–1.5cm; includes 'Meehanii', with broader, wrinkled leaves and smaller flowers with more deeply cut lobes, and var. *mollis*, with broader, downy leaves and larger flowers).

half hardy applied to plants which, in a given climatic zone, may be grown outdoors after the risk of frost has passed. Used particularly of bedding plants, and other annuals. See *hardiness*.

half-pot see *pots and potting*.

× **Halimiocistus** Cistaceae. S Europe. Hybrids between *Cistus* and *Halimium*, small to medium-sized evergreen shrubs with small, narrow and downy leaves and lax cymes of five-petalled flowers in summer. Cultivate as for *Cistus*.

× *H.*'Ingwersenii' (*Halimium umbellatum* × *Cistus hirsutus*; Portugal; to 50cm; flowers white in sparse, long-stalked branching cymes); × *H.sahucii* (*Hali-mium umbellatum* × *Cistus salviifolius*; S France; to 1m; flowers 3–5 per cyme, white); × *H.wintonensis* (*Halimium ocymoides* × *Cistus salviifolius*; garden origin; to 60cm; flowers white, often yellow at centre with maroon blotches in long-stalked, 2–4-flowered cymes; includes 'Merrist Wood Cream', with flowers deep cream to primrose blotched dark maroon).

Halimium (from Greek *halimos* belonging to the sea). Cistaceae. Mediterranean, SW Europe, N Africa, Asia Minor. Some 12 species, evergreen shrubs and subshrubs, differing from the closely related *Cistus* in their predominantly yellow flowers. Grey-leaved summer-flowering plants, hardy in zone 7 and suited to areas with mild winters and warm dry summers. Cultivate as for *Cistus*.

H.lasianthum (SW Europe; to 70cm; flowers to 4cm in diameter, petals golden with a dark basal spot; includes 'Concolor', with unspotted flowers, 'Sandling', with the flowers conspicuously blotched maroon, and subsp. *formosum* with larger flowers, the basal spot higher up each petal); *H.ocymoides* (SW Europe; to 1m; flowers golden, typically with a wide maroon spot at the base of each petal, in erect, terminal panicles forming a broad inflorescence; includes the compact, broad-leaved 'Susan'); *H.umbellatum* (Mediterranean; to 40cm; flowers white stained yellow at base of petals, to 2cm in diameter, 3–6 per umbel-like panicle).

halo blight a bacterial disease of beans. See *bacterial diseases*.

halophyte a plant adapted to, or tolerant of, saline soil, such as the glassworts (*Salicornia* species) found on salt marshes.

Hamamelis (from Greek for a pear-shaped fruit). Hamamelidaceae. WITCH-HAZEL. N America, Europe, E Asia. 5 species, deciduous shrubs or small trees with tomentose branches and buds and ovate to obovate, toothed leaves, most turning yellow in autumn but also showing tints of red and orange. Produced in late winter in axillary clusters, the scented, spidery flowers consist of four slender petals arising from the cup of a glossy, chestnut-coloured calyx.

Most species are hardy to –20°C/–4°F. Grow with shelter from cold drying wind, in fertile, neutral or preferably slightly acid, well-drained and humus-rich soils; incorporate additional organic matter, such as leafmould or well-rotted compost, at planting time. Deep soils overlying chalk are acceptable; lime-induced chlorosis may be treated with sequestered iron. All species prefer moisture-retentive soils. Pruning is necessary only to remove deadwood. Grafted specimens may produce suckers which should be removed at the base as they appear. Cultivars may be propagated by side-grafting in gentle heat in late winter/early spring or by budding in late summer on to seedling understock of *H.virginiana*. Layering in autumn is useful where small numbers are required. Take softwood cuttings in summer, treat with 0.8% IBA and root under mist or in a closed case with bottom heat; overwinter in the first year in the cold greenhouse or frame. Propagate species by ripe seed

sown fresh in the open cold frame; some germination may occur in the first spring, but seed usually requires a second winter in the cold frame before germinating. Dormancy may be broken by warm treatment, for a period of nine weeks at 20°C/70°F, followed by 14 weeks at 5°C/40°F.

H. × intermedia (*H.japonica × H.mollis*; garden origin; shrub to 4m; flowers to 3cm in diameter, petals crumpled; cultivars include: 'Advent', the earliest flowering, with bright yellow flowers; the medium-sized 'Allgold', with ascending branches, leaves that are yellow in autumn, and butter-yellow flowers with narrow, twisted petals and red calyces; 'Arnold Promise', with densely clustered, deep sulphur-yellow flowers, with red-green calyces; the late-flowering 'Carmine Red', with narrow, twisted, pale bronze petals to 2cm, with tips suffused copper; 'Diane', with yellow and scarlet foliage in autumn and coppery carmine flowers with purple-bronze calyces; 'Hiltingbury', with large, rounded leaves turning red, orange and copper in autumn, and pale copper-red flowers with maroon calyces; the large ascending shrub 'Jelena' – also known as 'Copper Beauty' – with leaves turning bronze, red and orange in autumn and densely clustered, yellow flowers suffused deep copper-red, with twisted petals and burgundy calyces; the erect medium-sized shrub 'Magic Fire' – also known as 'Feuerzauber' or 'Fire Charm' – with leaves turning yellow in autumn and copper-orange petals, suffused red and twisted; the medium to large shrub 'Moonlight', with circular leaves turning yellow in autumn, and highly fragrant, densely clustered, pale sulphur-yellow flowers, with a burgundy basal blotch; 'Nana', with deep yellow petals to 3cm; 'Orange Beauty', with golden yellow to orange-yellow flowers; 'Primavera', with flowers densely clustered and sickle-shaped, primrose yellow petals to 2cm; 'Ruby Glow' – also known as 'Rubra Superba' or 'Adonis' – with leaves turning orange, bronze and scarlet in autumn, and copper-red petals to 2cm; the early flowering 'Sunburst', with sulphur yellow flowers; and 'Winter Beauty', with flowers as 'Orange Beauty' but larger, and petals brown-red at base); *H.japonica* (JAPANESE WITCH-HAZEL; Japan; shrub or tree to 3m; flowers yellow with petals to 2cm and calyces green or purple within; cultivars include: the vigorous 'Arborea', to 5m, with yellow, wavy petals and maroon calyces; 'Sulphurea', with sulphur yellow petals to 1.5cm; 'Zuccariniana', with erect to spreading branches, leaves turning yellow in autumn, flowers with twisted, wavy, sulphur yellow petals to 1.2cm); *H.mollis* (CHINESE WITCH-HAZEL; W China; shrub or small tree to 5m; flowers sweetly fragrant, with calyces red-purple within and straight petals to 1.7cm, golden yellow and tinged red at base; cultivars include the early flowering 'Brevipetala' with orange-yellow petals to 1cm; 'Coombe Wood', with golden yellow flowers, suffused red at base; the large shrub 'Goldcrest', with golden petals to 2cm, each with a wavy tip and a burgundy basal blotch, and calyces with a purple-red interior; 'Pallida', with flowers covering branches, sulphur yellow; and the late-flowering 'Westerstede', with pale yellow flowers); *H.vernalis* (Southern Central US; upright suckering shrub to 2m, similar to *H.virginiana*, except calyx interior red, petals yellow to red; cultivars include 'Carnea', with pale fleshy pink flowers; 'Lombart's Weeping', with pendulous branches, blue-green leaves and pale red flowers; 'Red Imp', with claret petal bases and copper tips; 'Sandra', with violet-purple young foliage, becoming green, flushed purple beneath, turning red, scarlet and orange in autumn, with cadmium yellow flowers; and 'Squib', with cadmium yellow petals and green calyces); *H.virginiana* (VIRGINIAN WITCH-HAZEL; E US; shrub or small tree, to 5m; leaves yellow in autumn, calyces green or brown inside, petals 17mm, crinkled, golden yellow; includes 'Rubescens', with petal bases suffused red; *H.macrophylla* from the SE US has larger leaves with superb autumn colour and small, crinkled, pale yellow flowers, opening in late autumn and winter).

hamate hooked at the tip.

handbarrow a wooden platform, about 120cm long and 70cm wide, with the side bearers extended to form handles, and usually having a short leg at each corner of the platform. It is used mostly to transport plants in containers, with one person standing between each pair of handles for holding and lifting.

handlight see *bellglass*; *cloche*.

hanging basket a hemispherical container of galvanized or plastic-coated wire suspended on chains to display ornamental plants or, sometimes, for growing pendulous dwarf tomatoes. Square baskets made of wooden slats are used for orchid-growing. Hanging baskets are commonly used for summer decoration outdoors but make valuable additions to the large greenhouse or conservatory.

To retain the growing medium, the basket should be lined with moss (where it can be obtained from approved sources), plastic sheeting or netting, coconut fibre, foam plastic, impregnated cardboard or wool waste. Soilless composts are suitable, and the incorporation of moisture-absorbing polymers and slow-release fertilizer is advantageous. Dead-heading flowers, regular watering and supplementary feeding are essential practices for the maintenance of a good display.

hapaxanthic with only a single flowering period.

haploid describing the basic number of chromosomes borne in the unpaired pollen and ovule cells of a diploid plant. It is commonly annotated **x**; c.f. *polyploid*.

haptonasty, haptotropism the response of plants to contact or touch, as demonstrated by the tendrils of climbing plants and by some carnivorous plants. See *stimulus movements*.

Hardenbergia (for Countess von Hardenberg, sister of the celebrated traveller Baron von Hügel, who collected plants in western Australia in 1833). Leguminosae. Australia, Tasmania. 3 species, evergreen vines. The leaves consist of one to five, ovate to lanceolate leaflets. Small, pea-like flowers are produced in spring and summer in long, axillary racemes. Where winter temperatures do not fall much below –5°C/23°F, they are well-suited to growing over arches, fences or pergolas, or trailing over a retaining wall. In cooler zones, they may be pot-grown in the cool greenhouse or conservatory, with a minimum winter temperature of 7–10°C/49–50°F. Grow in moist, well-drained, lime-free soils in sun or light dappled shade. Water moderately and feed monthly with dilute liquid feed when in growth; keep just moist in winter. Container-grown plants with integral support may be plunged outdoors in summer. Propagate by seed sown in spring; pre-soak for 24 hours prior to sowing at 20°C/70°F and cover with fine grit. Alternatively, root tip cuttings in late spring in moist sand or vermiculite, in a closed case with bottom heat.

H.comptoniana (WESTERN AUSTRALIA CORAL PEA; W Australia; evergreen vine to 3m; leaves with 3–5 leaflets; flowers blue to purple, standard with a green-spotted, white blotch at base); *H.violacea* (syn. *H.monophylla*; VINE LILAC, PURPLE CORAL PEA, FALSE SARSPARILLA; E Australia, Tasmania; evergreen vine to 2m+; leaves with only one leaflet; flowers purple, white, pink, or lilac, standard spotted yellow or green at centre).

hardening off the process of acclimatizing plants to more rigorous conditions. It is achieved by gradually increasing exposure to lower temperature and air movement, often by transferring containerized plants to unheated frames. It is usually practised with half-hardy or tender plants raised under glass and intended for planting outside.

hardiness a complex quality attributed to plants that are capable of withstanding climatic rigours without greenhouse protection. In tropical and subtropical climates, the term may extend to drought tolerance. Natural adaptations among plants to ensure hardiness include a reduced surface area, a deciduous or herbaceous habit, seed dormancy, an accelerated life cycle, hairiness, thickened cell walls, thick bark, procumbent growth, and a lack of fleshy parts (except in xerophytes). In cells, the presence of oils and crystals are effective against freezing.

Hardiness may be much affected by macroclimate. For example, plants overwintering with their metabolic processes naturally at a near standstill may succumb to periods of unusually persistent rainfall. A seasonal rise in the temperature of a locality promotes rapid growth and full ripening of wood and flower buds, which are therefore more tolerant of wind and very low winter temperatures. These factors are influenced by the prevailing climate, which in turn may be affected by topography.

Whereas little can be done to alter the effect of macroclimate on plant hardiness, there are opportunities for improving the microclimate in the garden: drainage may be improved by building raised beds, soil structure can be ameliorated, shelter and surface insulation can be provided with layered mulches and by wrapping up frost-susceptible plants and protecting them with glass or plastic. Natural selection and hybridization can contribute to plant hardiness, and seeds and plants should be chosen with their provenance in mind.

Low-temperature hardiness. In the northern Hemisphere, injury to plants from wind frosts and sustained low temperatures is usually associated with bursts of cold air from the north or northeast, most likely during the winter months from December to March. A period of sub-zero temperatures often causes the ground to freeze to some depth and evergreen trees and shrubs may suffer particularly since water from the soil is difficult to extract and they cannot transpire; severe wind chill can 'burn' or kill green shoots and stems. Many plants develop a hardier constitution as they mature, and stock grown in containers under protection must be hardened off. Where this is not done, it is advisable to delay planting out until the spring, or to protect evergreens from severe weather. Most slightly tender species may rejuvenate if cut hard back after frost injury.

Radiation frost injury is more common and can be sudden, the most serious damage occurring in late spring or early summer. Soft growth of trees and shrubs as well as fruit crops is very vulnerable, as are half-hardy or tender bedding display plants. Damage is severe after a mild spell has stimulated new growths or early flowers. Remedial measures include locating frost pockets and draining off cold air from behind walls and hedges by creating suitable openings with permeable fences; tender plants should not be planted out until the main risk of frost is past.

Hardiness zones Hardiness zones based on winter isotherms are a means of expressing the cold tolerance of garden plants. The isotherm is a line on a map connecting places with the same mean temperature at a given time. Alfred Rehder in his *Manual for Cultivated Trees and Shrubs* (1972) developed a system using eight zones to cover the cold-temperate areas of the US, with five-degree Fahrenheit bands based on the lowest mean temperatures of the coldest months. In 1960 the US Department of Agriculture (USDA), in cooperation with the American Horticultural Society, published a map based on ten-degree Fahrenheit differences in the annual minimum temperature, on which the US State boundaries were superimposed. The map was revised in 1990 to include Alaska, Canada, Mexico and Hawaii and has 11 zones, using data from 124,500 weather stations. A new zoning system modelled on that of the USDA was introduced for western Europe in *The New Royal Horticultural Society Dictionary of Gardening* (1992). Zone maps have been published for South Africa, Australia, New Zealand, Japan and China.

Maps based on isotherms represent meteorological patterns, not hardiness zones. Hardiness is a complex phenomenon, a highly relative quality of the individual plant – not of its projected site – and is affected by various linked factors of which cold is only one.

The Royal Horticultural Society's Award of Garden Merit to plants designates hardiness limits for Britain as follows – H1: heated glass; H2: unheated glass; H3: hardy in some regions or particular situations; H4: generally hardy throughout the British Isles. *The European Garden Flora* has a more elaborate classification – G2: requires a heated greenhouse even in south Europe; G1: requires a cool greenhouse even in south Europe; H5: hardy in favourable areas, withstands 0–5°C/32–40°F minimum; H4: hardy in mild areas, withstands –5° to –10°C/23°–14°F minimum; H3: hardy in cool areas, withstands –10° to –15°C/14°–5°F minimum; H2: hardy almost everywhere, withstands –15° to –20°C/5° to –4°F minimum; H1: hardy everywhere, withstands –20°C/–4°F and below. These guides to plant hardiness must be regarded as flexible and therefore of general use only.

In this volume, the climatic zone numbers given in individual plant entries follow those adopted by the *New RHS Dictionary of Gardening* and the USDA, they can be summarized as follows:

Zone	°F	°C
1	<–50	<–45.5
2	–50 to –40	–45.5 to –40.1
3	–40 to –30	–40.0 to –34.5
4	–30 to –20	–34.4 to –28.9
5	–20 to –10	–28.8 to –23.4
6	–10 to 0	–23.3 to –17.8
7	0 to +10	–17.7 to –12.3
8	+10 to +20	–12.2 to –6.7
9	+20 to +30	–6.6 to –1.2
10	+30 to +40	–1.1 to +4.4
11	≥+40	≥+4.4

The zones are an approximate indication of the coldest temperature band in which a plant will thrive.

harding a vernacular term for a plant that does not thrive.

hardwood (1) a common name for a broad-leaved tree, and descriptive of its timber; cf. *softwood*; (2) used to describe cuttings made from mature shoots of woody plants.

hastate arrow-shaped; triangular, with two equal and approximately triangular basal lobes, pointing laterally outward rather than toward the stalk; cf. *sagittate*.

Hatiora (anagram of *Hariota*, an earlier generic name commemorating Thomas Hariot (1560–1621), English scientist and historian, mentor to Sir Walter Raleigh). Cactaceae. SE Brazil. 4 species, epiphytic or lithophytic cacti with freely branching, cylindric, angled, winged or flat stems composed of short segments. These are unarmed or bear soft, bristly spines. The flowers are campanulate with a short tube and a spreading perianth of intense yellow, pink or red. Small, obovoid fruits follow. Grow in an intermediate greenhouse (minimum temperature 10–15°C/50–60°F) in 'epiphyte' compost, consisting of equal parts organic/inorganic matter and a pH below 6. Provide shade in summer and maintain high humidity. Reduce watering in autumn/winter until flower-buds begin to develop. *H.gaertneri*, often incorrectly called *Schlumbergera gaertneri* and the hybrid *H. × graeseri* (*H.gaetrneri* × *H.rosea*) are floriferous and popular house plants, flowering in the northern Hemisphere at about Eastertide. Increase by stem cuttings.

H.gaertneri (EASTER CACTUS; E Brazil; stem segments mostly flat, oblong or elliptic, truncate, 4–7cm, shallowly crenate; flowers to 4.5cm. intense scarlet, rose, magenta, ruby and orange to gold in some cultivars and hybrids; fruit oblong, red); *H.salicornioides* (BOTTLE CACTUS, BOTTLE PLANT, DRUNKARD'S DREAM; SE Brazil; small shrub, much-branched, erect or pendent, stem-segments whorled, bottle-shaped, 1.5–5cm; flowers to 1.3cm, yellow to orange; fruit globose, white).

haulm the stems of potatoes, peas and beans, especially.

haustorium (plural haustoria) a sucker attachment in parasitic plants penetrating the host.

haw (1) fruit of the hawthorn (*Crataegus monogyna*); (2) an archaic term for a hedged enclosure or a hedge.

Haworthia (for Adrian Hardy Haworth (1765–1833), English plantsman, amateur botanist and entomologist, author of books on succulent plants and bulbs). Liliaceae (Aloeaceae). South Africa, extending to Namibia, Swaziland and Mozambique. Over 70 species, dwarf, perennial, evergreen, rosette-forming succulents, distinguished from *Aloe* by their smaller, usually tufted habit and 2-lipped white or off-white flowers. They require a winter minimum of 4–7°C/39–45°F, a porous, well-drained but nutritious soil and good ventilation and lighting, although the softer-leaved species appreciate partial shade. As house plants, they are among the few succulents to accept a sunless, north-facing windowsill. Annual repotting is recommended, as the thick fleshy roots die back and are renewed each growing season. During the colder winter months, impose a dry rest; water freely during warm, bright spring andsummer days. Increase by offsets and division of clumps.

H.arachnoidea (COBWEB ALOE; Cape Province; stemless, clustering; rosettes 2–13cm in diameter, leaves arachnoid, ascending to incurved to 7cm, oblong to lanceolate, green, soft and juicy, semi-translucent flecked, aristate, longitudinally lined, margins and upper part of keel with white to pale brown teeth); *H.attenuata* (Cape Province; stemless or short-stemmed with basal offsets; leaves erect or slightly curving, to 8cm, dark green, flexible, narrowly triangular, acuminate, rough from white tubercles in bands on underside; includes 'Variegata', with variegated leaves; *H.clariperla*, a widely grown plant with orange, pearl-dotted leaves, is a variant of this species; *H.fasciata* (ZEBRA HAWORTHIA; South East Cape; similar to *H.attenuata* but with shorter leaves that are smooth on the upper side and with bands of white tubercles beneath); *H.maughanii* (Western Cape; plant body below soil level with only the flat, translucent window tip of each leaf exposed); *H.truncata* (Western Cape; similar to *H.maughanii* but with flatter, broader leaves in two series like the pages of an open book).

hazelnuts the fruits of the hazel (*Corylus*), of which the most important are the COBNUT (*C.avellana*) and the FILBERT (*C.maxima*) both of which are natives of Europe; in the US, the name filbert is commonly applied to both these species. The widespread use of *C.avellana* dates back to prehistoric times, since the hardened shoots were suitable for roadmaking and for thatching, wattle, stakes and utensils. Selected forms of hazelnuts were grown by the Greeks and Romans, and by the 16th-century the distinction between cobs and filberts was recognised in cultivation. Named cultivars resulting from breeding programmes were introduced in the early 19th century, and the twin-shelled 'Cosford Cob' remains famous, together with 'Lambert's Filbert', erroneously known as 'Kentish Cob'.

Hazelnuts are unsuited to small gardens because trees may reach 4.5m in width, and ensuring good pollination requires more than one cultivar. The male flowers produce attractive yellow catkins in late winter on the same plant as the inconspicuous red female flowers. Unlike cobnuts, filbert fruits bear a husk longer than the nut and usually completely enclosing it. Selections with frilled and twisted husks are known as frizzled filberts; others produce ornamental purple leaves.

Plant bushes about 4.5m apart with 8–12 main branches trained to form an open centre. Pruning should restrict height to 2m, with the leading shoot of a one-year-old specimen pruned at 45cm to induce

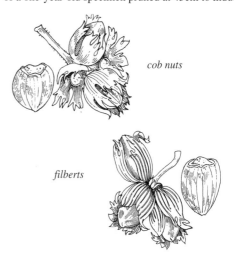

cob nuts

filberts

hazelnuts

Pests
sawflies
squirrels

branch leaders from the top 15cm, and any below this level removed. In the following winters, branch leaders are shortened by about one-third and side-shoots crowding the centre of the tree removed together with basal shoots; any suckers must be cut out with a spade to avoid proliferation. Once the desired height has been reached, branch leaders should be cut back to a weak side-shoot and any strong upright shoots removed.

On an established bush, summer pruning is important, using a technique known as 'brutting'. This consists of breaking by hand stronger side shoots at about half their length in August and allowing the broken part to hang by the rind, the shoot not being completely severed. This reduces vigour and encourages the formation of female flowers, as well as allowing greater light and air penetration to improve the quality of the maturing crop. The brutted shoots are cut back to 3–4 buds in winter, the pruning best timed to coincide with flowering so that the disturbance of shoots assists pollination.

The nuts are harvested as the husks yellow, laid out in trays in an airy room and turned to ensure even drying. For prolonged storage, they may be packed in layers in earthenware jars, intermingled with coconut fibre mixed with a little salt.

Artificial hybrids include the TRAZEL, *C. × colurnoides* (*C.avellana × C.colurna*), with large long nuts; the HAZELBERT or MILDRED FILBERT (*C.avellana × C.americana*), with large nuts resistant to Eastern filbert blight; and the FILAZEL (*C.avellana × C.cornuta*), which is hardy into zone 4.

head (1) a dense cluster or spike of flowers; (2) the part of a tree above the clear trunk; (3) the harvestable part of hearted leaf and stem vegetables, such as lettuce, cabbage and celery, when alternatively it may be referred to as a heart. It is also used for the edible part of a cauliflower.

heading back drastic cutting back of trees or shrubs by shortening for all or most of the main branches. It is done in preparation for topwork grafting of fruit trees, for reshaping, and for inducing vigour in ornamentals.

heart (1) see *head*; (2) used to describe soil, which is said to be in 'good heart' when it is adequately aerated, irrigated and otherwise fertile.

heart rot (1) a general term describing rot of the heartwood of trees, usually caused by bracket fungi. These mostly enter through wounds, and rotting may not be evident until a tree is blown over or felled. Pruning and treating wounds with a fungicidal sealant may help prevent the spread of such fungi, but little can be done once the main trunk heartwood is infected. (2) used to describe celery heart rot, a wet, slimy brown rot at the centre of plants, caused by the bacterium *Erwinia carotovora*; heart rot of swede results from boron deficiency.

heartwood the inner, non-functional xylem layers of a woody stem, which are harder and usually darker than the outer sap wood.

heaths and heathers a common group name for species of *Erica*, *Daboecia* and *Calluna* (q.v.) that are native of Europe and Africa. For most of the group, a maximum soil pH of 4.5–5.0 is required for good growth, although *E.erigena*, *E.terminalis* and cultivars of *E.carnea* and *E. × darleyensis* will tolerate alkaline soils. The term heath strictly describes species of *Erica* and *Daboecia*, and these occur on poor soils with little shelter, and, in Europe, where rainfall is high. Cape heaths are southern African species, which do not survive winter temperatures below –7°C/20°F. Heather refers to the great number of cultivars of *Calluna vulgaris*, also called ling, which thrive on moor and mountain. Grow in full sun on a moist but free-draining soil. Some will tolerate drought once established.

Propagate from cuttings of semi-ripe shoots, 3–4cm long, taken in summer and inserted in a peat/sand mixture under plastic film. Layering in September/October is possible and especially useful for certain dwarf cultivars of *Calluna*. Trimming keeps plants in good shape. Summer-flowering kinds should be trimmed in March to just below the old flower heads but not into old wood. Winter and spring flowerers are treated similarly, immediately after flowering. The tree heaths should be pruned only when there is a need for rejuvenation or containment, in April or May. Spring planting is generally preferable; choose well-hardened plants and apply an annual mulch of pulverized bark or similar.

heavy used of soils containing a high proportion of clay, which therefore need careful and timely management. See *soil*.

Hebe (from *Hebe*, goddess of youth and cup-bearer to the Gods). Scrophulariaceae. New Zealand, Australia, temperate S America (those here from New Zealand unless otherwise stated). Some 75 species, small to medium-sized evergreen shrubs ranging in habit from dwarf and procumbent to erect and bushy, to tree-like. The leaves vary from needle- or scale-like and overlapping on whipcord- or cypress-like branchlets, to rounded or lanceolate and more openly arranged on stouter shoots. Small, four-lobed flowers are carried in axillary or subterminal racemes or heads usually in spring and summer. Hardy to –15°C/5°F. Plant in a position sheltered from cold winds, in full sun and on well-drained soils. Propagate by heeled, semi-ripe cuttings in a gritty medium in the cold frame in late summer; whipcord types are more prone to rotting off and need careful and judicious watering.

H.albicans (compact spreading or rounded shrub to 60cm; leaves to 2 × 1.5cm, spreading to imbricate, ovate to oblong, glaucous, fleshy; racemes to 6cm, flowers white, anthers purple; cultivars include the

Hebe
'Autumn Glory'

low, domed shrub 'Cranleigh Gem', to 60cm, with grey leaves, flowers in short broad racemes, and dark anthers; the very compact 'Pewter Dome', with small, dull grey-silver leaves; 'Red Edge', with leaves edged red and mauve flowers; the hardy 'Sussex Carpet', to 30cm, spreading, with glaucous leaves and a short inflorescence; and 'Trixie', with green leaves and white flowers); *H.*'Alicia Amherst' (robust with deep green, lustrous leaves and showy racemes of strong purple flowers); *H.* × *andersonii* (*H.salicifolia* × *H.speciosa*; spreading shrub to 1.6m; leaves to 10 × 3cm, ovate to lanceolate, obtuse, deep green; flowers crowded on spikes to 10cm, violet; includes the vigorous 'Variegata', with dark green leaves, mottled or graded with grey-green and irregularly edged white to ivory, and lilac flowers); *H.armstrongii* (erect shrub to 1m; branches whipcord, spreading then ascending, terete, yellow-green particularly in winter; leaves to 1.8mm, closely adpressed; flowers white, in erect terminal clusters; differs from *H.ochracea* in its yellow-green (not ochre) branches and less arching habit); *H.*'Autumn Glory' (to 80cm; stems flushed dark red; leaves to 3cm, broadly obovate, dark green edged red when young; flowers violet-blue in long racemes); *H.buchananii* (shrub to 20cm, forming clumps to 90cm across; branches tortuous, black; leaves to 0.7cm, spreading, ovate and concave, dull dark green; flowers white, crowded; includes the smaller and more compact 'Minor', forming a hummock of minute green leaves); *H.canterburiensis* (low shrub to 1m; leaves to 1.7cm, loosely overlapping, distichous, elliptic to obovate, subcoriaceous, margins pubescent, apex acute; inflorescence to 2.5cm, crowded, flowers white; includes the procumbent 'Prostrata', with internodes flushed red and bright green leaves, twisted back and overlapping); *H.*'Carnea' (dense, glabrous, to 1m; leaves to 6cm, narrowly oblong to lanceolate, green faintly tinted red at margins and midrib; racemes 6–8cm, flowers pink fading to white, giving a 2-tone appearance to inflorescence); *H.carnosula* (shrub to 40 cm, forming leafy tufts of erect shoots; leaves to 2cm, closely overlapping, decussate, obovate, concave, thick, somewhat glaucous; flowers white, in terminal rounded inflorescence to 1.5cm diameter, anthers purple; commonly mistaken for *H.pinguifolia*, from which it can be distinguished by its less glaucous leaves); *H.cupressoides* (cypress-like whipcord with dense, rounded or open habit, to 1.5 × 2m; branchlets sometimes glutinous and aromatic; juvenile leaves linear, sometimes lobed, mature leaves to 0.15cm, scale-like, closely adpressed but not hiding stems, narrow-ovate to triangular, ciliate; flowers very pale blue, anthers red-brown; includes 'Boughton Dome', to 30 × 50cm, forming a broad and densely branched dome of fine grey-green branchlets clothed with both juvenile and adult leaves); *H.elliptica* (New Zealand, Tierra del Fuego, Falkland Islands; shrub to small tree to 5m, usually about 1.5m; leaves to 2.4cm, spreading, often oriented in one direction, elliptic to oblong, coriaceous, lustrous dark green, ciliate; flowers clustered at branch tips, white to violet or pink);

H.'Fairfieldii' (compact, upright to 60cm, close to *H.hulkeana* but with branches rigidly erect, duller leaves, shorter, stout inflorescence and larger lilac flowers); *H.* × *franciscana* (*H.elliptica* × *H.speciosa*.; shrub to 1.3m, habit rounded, dense; leaves to 6 × 2.5cm, larger, duller than *H.elliptica*, distichous, obovate to elliptic, fleshy, dark green, apiculate; inflorescence to 7.5cm, cylindrical; flowers purple tinged pink; includes 'Blue Gem', with red-violet flowers and violet anthers, and 'Variegata', with leaves mottled and margined yellow at beginning of season); *H.*'Great Orme' (similar to *H.*'Carnea' but more robust, internodes dark, leaves to 9cm, oblong-lanceolate, somewhat falcate); *H.*'Hagley Park' (to 40 × 60cm; leaves obovate to spathulate, serrate; flowers in terminal panicles to 30cm, similar to *H.hulkeana* but a stronger lilac-pink); *H.hectoris* (whipcord to 75cm; branches crowded, or spreading to decumbent with age, yellow-green; leaves to 0.4cm, scale-like, overlapping, closely adpressed, concealing stems, glossy, yellow-green; flowers white to pink in congested terminal heads); *H.hulkeana* (NEW ZEALAND LILAC; sprawling shrub to 1m, usually slender, habit loose, internodes tinted purple-red; leaves to 5 × 3cm, spreading, elliptic to rounded, subcoriaceous, glabrous, deep lustrous green above, pale beneath, margin often flushed red, serrate; flowers lavender to white in panicles to 30cm; includes 'Averil', with pale green leaves not edged red and more compact inflorescences); *H.*'La Séduisante'(young leaves lustrous, tinted purple; flowers red-purple in long showy racemes in late summer); *H.lycopodioides* (rigid and narrowly erect whipcord to 60cm; resembles *H.hectoris* but branchlets tetragonal; juvenile leaves linear-subulate, mature leaves to 3mm, scale-like, adpressed, ribbed or striped yellow, margins yellow; flowers white with lilac-blue anthers); *H.*'Midsummer Beauty' (robust, soon attaining 2m or more; leaves oblong to lanceolate or broadly elliptic, deep green tinted purple-red when young; flowers pale blue to lilac or magenta, often fading to pale rose, in long slender racemes); *H.ochracea* (whipcord to 60cm; branches spreading, rigid, dark, bowed outwards; branchlets olive green to ochre, arising from upper side of branches; leaves to 1.5mm, ochre to olive, tightly adpressed; flowers small, white, in short spikes; often confused with *H.armstrongii*, this species is the more commonly cultivated and can be distinguished by its ochre branches and strongly outspread to arching habit; includes 'Greensleeves', erect to 60cm, with rigid, green leaves, and the very compact 'James Stirling'); *H.pimeleoides* (densely branched low-growing shrub to 45cm; leaves 6–15mm, elliptic to broadly obovate, sometimes concave, spreading to recurved, glaucous, blue-grey, margins occasionally red; flowers white flushed pale lilac to pale violet; cultivars and varieties include: the very small, creeping 'Minor', with flowers tinted lilac blue; 'Quicksilver', with small, recurved, especially glaucous leaves; var. *glaucocaerulea,* erect to 35cm becoming decumbent, with larger, broadly elliptic, concave, strongly glaucous leaves, edged red in summer, and violet-blue flowers; 'County Park', to 20 × 50cm, decumbent, with grey-green leaves edged red and flushed red-purple in cold weather, and violet flowers; 'Wingletye', close to 'County Park' but lower-growing, with strongly glaucous leaves that do not flush red in cold weather, and lilac blue flowers); *H.pinguifolia* (procumbent or ascending to 1m; leaves to 1.5cm, overlapping to spreading, obovate to elliptic, concave, thick, glaucous, blue-green, often edged red, obtuse to subacute; flowers white, anthers blue; often confused with *H.carnosula*, which has broader leaves and a glabrous ovary and style; includes the popular cultivar 'Pagei', with grey, oblong to obovate, scarcely concave leaves, and white flowers in dense short spikes); *H.rakaiensis* (shrub to 90cm; branchlets slender with rough leaf scars; leaves to 2cm, spreading, obovate to oblong, entire, subcoriaceous, glossy, bright green; racemes to 4cm, simple, covering plant, flowers white; often labelled *H.subalpina* or *H.buxifolia* in gardens); *H.recurva* (spreading shrub to 1m; leaves to 5cm, spreading to deflexed, narrowly lanceolate, glabrous, grey-green; flowers white – often pink in bud – in spikes to 6cm long; selections of *H.recurva* are available with longer, plain green leaves, or particularly strong grey tones, and in some cases, mauve flowers fading to white); *H.salicifolia* (New Zealand, Chile; shrub to 5m, usually shorter; branchlets green; leaves to 15cm, suberect to spreading, lanceolate to oblong, apex acuminate, entire or denticulate, glossy dark green; racemes to 20cm, narrowly conical to cylindric, flowers white, usually tinted pale lilac); *H.speciosa* (strong-growing shrub to 2m; branches stout, somewhat angular; leaves to 10cm, broadly elliptic to obovate, glossy, margin cartilaginous; flowers dark red to magenta, crowded in conic-cylindric racemes to 7cm long; a highly variable species giving rise to many cultivars and hybrids); *H.tetragona* (erect whipcord to 1m, much-branched; branchlets distinctly tetragonal, yellow-green; leaves to 3.5mm, decussate, adpressed, deltoid to subulate); *H.*'Youngii' (to 15cm, procumbent, compact; leaves to 0.75cm, elliptic, dark green, faintly edged red at times; flowers deep violet fading to white in short racemes).

Hedera (Latin name for the plant.) Araliaceae. Europe, Asia, N Africa. IVY. 11 species of evergreen, woody, climbing or creeping plants with distinct sterile, juvenile and fertile, arborescent stages. The stems of the juvenile stage climb by means of aerial rootlets. Juvenile leaves are conspicuously lobed or cordate, those of arborescent stage are more nearly entire; all are glabrous and often waxy and shining above, stellate-hairy or scaly beneath. Small, cream to yellow-green flowers are carried in globose umbels. They are followed by small, spherical drupes; these are usually black, sometimes orange, yellow or cream.

Ivies are valued as coverage for walls, sheds and even unsightly tree stumps; ivy-clad buildings have considerable charm. For large expanses of wall the larger-leaved kinds such as *H.colchica* and *H.hibernica* are the most suitable; for smaller areas, the diversity of leaf shape and variegation of the many cultivars of *H.helix* can be used to advantage. Variegated ivies look particularly fine against red-brick backgrounds and add immensely to the winter beauty of the garden. The yellow-green leaved clones such as *H.helix* 'Buttercup' associate well with purple-leaved shrubs or blue-flowered plants, while against white walls the winter purple of the leaves of *H.helix* 'Atropurpurea' can make a dramatic contrast.

Its creeping habit and preference for shade make ivy an excellent groundcover plant and one whose variegated cultivars can be used with great effect to brighten dark corners. The sulphur-cream *H.hibernica* 'Sulphurea' underplanted with autumn crocus (*Colchicum* species) makes an attractive colour combination and will guard the flowers against mud splash.

Concern is sometimes expressed as to the effects of ivy on trees or buildings. Ivy is not parasitic and, while the growth of ivy on trees of value or beauty is to be deprecated, an unloved tree may well be given up to it. When such a tree becomes ivy-laden, it makes a most pleasing winter picture and a refuge for birds. On sound buildings and walls, ivy is harmless but can become unwieldy and harbour insects. To obviate this, it should be clipped over in late spring every other year or so. Walls and fences that are very weak may fall if a mass of ivy becomes heavy with snow or rain; clipping prevents this.

The non-climbing forms of *H.helix*, 'Congesta' and 'Erecta', and the slow-growing, rock-hugging 'Conglomerata' make good feature plants for rock garden or patio. In completely shaded town gardens, ivy will grow perfectly well and here the variegated kinds can be used to brighten a dark site. 'Standard' ivies can be produced through grafting small-leaved, trailing kinds on the single stems of × *Fatshedera,*

Increase by nodal cuttings of juvenile growth with one or two leaves. Woody shoots of the adult stage are less easy and may require mist facilities. Cuttings are best taken between June and October when the young growth has hardened sufficiently. Inserted in sharp sand in lightly shaded frames, these root rapidly and may then be potted into a medium-fertility soil-based mix. When well rooted and growing, plant out or, if specimen or conservatory plants are required, pot on into larger pots and then train either to canes in the form of a trellis or on moss poles. In the conservatory or home, ivies should never be placed in direct sunlight or be allowed to dry out. During summer, a balanced liquid feed every 14 days is beneficial.

Soil preparation for ivies growing outside should be generous, as for any permanent planting. In non-lime areas, the inclusion of chalk, lime or mortar rubble can be advantageous but is not essential. Plants intended to climb need to be guided by being tied to a cane leading to their support. If this is not done the ivy will initially creep along the ground, possibly ascending at a point not envisaged by the planter. *(See table overleaf)*

HEDERA

Name	Distribution	Leaves
H.azorica	Azores	9–11 × 10–12cm, with 5–7 bluntly acute lobes, centre lobe only slightly longer than laterals, sinuses shallow, light matt green, hairs stellate, 3–5-rayed, white and so extensive as to give young stems and leaves a felted appearance
H.canariensis *H.algeriensis* ALGERIAN IVY	Canary Is., N Africa	juvenile leaves unlobed or 3-lobed, 10–15 × 8–12cm, bluntly acute, base cordate, margin entire, mid-green and glossy, stems purple-red; adult leaves elliptic-lanceolate, hairs scale-like, with 1–15 rays

Comments: 'Gloire de Marengo': leaves with areas of silvery-grey and creamy yellow; a popular houseplant but an outdoor climber in favoured districts. 'Marginomaculata': leaves extensively mottled creamy yellow. 'Montgomery': leaves 3-lobed, stems and petioles wine-red; sometimes listed as *H.algeriensis*. 'Ravensholst': with large, mostly 3-lobed leaves, centre lobe acute and broad, some leaves unlobed and elliptical; dark glossy green. 'Striata': vigorous and a darker green with a very slight yellow variegation at the leaf centre.

Name	Distribution	Leaves
H.colchica PERSIAN IVY; BULLOCK'S HEART IVY; COLCHIS IVY	Caucasus to Asiatic Turkey	ovate, generally unlobed, 6–12 × 6–8cm, apex acute, base cordate, margin entire, dark green, resin-scented when crushed; adult leaves lanceolate; hairs scale-like, with 12–20 rays

Comments: 'Dentata' (ELEPHANT'S EARS): leaves pendent, larger than the type, 15–20 × 15–18cm, base auriculate, the margins carrying widely spaced fine teeth, light green. 'Dentata Variegata': similar to 'Dentata' but leaves smaller, light green variegated with areas of grey-green and irregular marginal areas of cream-yellow. 'Sulphur Heart' ('Paddy's Pride'): similar to 'Dentata Variegata' but leaves a deeper green with an irregular central splash of yellow

Name	Distribution	Leaves
H.helix COMMON IVY; ENGLISH IVY	Europe and E Russia	juvenile leaves 3–5-lobed, 5–8cm, leathery, entire with cordate base and pale venation, young stems and leaves with white-stellate, 4–6-rayed hairs; leaves of arborescent stage elliptic, entire, less than 5cm wide

Comments: 'Angularis': leaves 5-lobed, centre lobe prolonged and acuminate, laterals bluntly acute, dark green in summer, colouring purple in winter. 'Atropurpurea': leaves 3-lobed, centre lobe prolonged and acuminate, laterals bluntly acute, dark green in summer, purple in winter. 'Buttercup': leaves 5-lobed, apex acute, base cordate, light green, butter yellow in sunlight. 'Chrysophylla' ('Spectabilis Aurea'; 'Aurea Spectabilis'): leaves 3-lobed, lobes acute, leaf base truncate, mid to deep green, slightly suffused yellow. 'Congesta': non-climbing, erect; leaves 3-lobed, dark green, in a distichous arrangement. 'Conglomerata': creeping but producing short upright shoots; leaves unlobed, margins undulate and waved. 'Digitata': leaves 5-lobed, deeply divided, cente lobe only slightly longer than laterals. 'Erecta': non-climbing, erect; leaves 3-lobed; stems green; larger in all parts than 'Congesta'. 'Eva': leaves small with grey-green centre, splashed deep green, margined cream. 'Fantasia': leaves 5-lobed, bright green heavily mottled cream-yellow. 'Goldchild': leaves 3–5-lobed, base cordate, mid-green with bright yellow, mostly marginal variegation. 'Goldheart': leaves 3-lobed, dark green with prominent yellow splash in centre. 'Glacier': leaves 3-lobed, apex acute, mid-green overlaid grey-green with occasional cream variegation. 'Glymii': leaves unlobed, ovate-acuminate, glossy, becoming deep purple in winter. 'Ivalace': leaves 5-lobed, margins strongly undulate, convolute at the sinus giving a crimped and laced effect, dark green and glossy. 'Kolibri': leaves 3–5-lobed, light green with flecks of grey-green and areas of cream variegation. 'Little Diamond': leaves unlobed, diamond-shaped, grey-green, white variegation slight and mainly at the edge. 'Manda's Crested': leaves 3–5-lobed, lobes convolute with blunt, downward-pointing apices giving a curly effect. 'Maple Leaf': leaves 5-lobed, apex acuminate, centre lobe 1.5 times length of laters, edges irregularly indented, mid-green. 'Midas Touch': leaves unlobed, deltoid or triangular, apex bluntly acute, base cordate, dark green, irregularly variegated bright yellow. 'Minor Marmorata' (SALT AND PEPPER IVY): leaves 3-lobed, small (3–5cm), dark green, spotted and splashed cream-white. 'Parsley Crested': leaves unlobed, ovate to almost circular, margins crimped and crested, light green. 'Pedata' (BIRD'S FOOT IVY): leaves 5-lobed, centre lobe prolonged and narrow, basal lobes back-pointing, dark green, veins grey-white. 'Professor Friedrich Tobler': leaves with 3–5 narrow lobes divided almost to the centre vein, dark green. 'Shamrock': leaves 3-lobed, centre lobe broad, lateral lobes sometimes folded in pleated fashion alongside centre lobe, dark green. 'Spetchley': leaves 3-lobed, very small (0.5–2cm), centre lobe prolonged, dark green. 'Tricolor': leaves unlobed, triangular, small (2–4cm), grey-green with cream-yellow edge that is slightly pink in summer, intensifying to purple in winter. 'William Kennedy': leaves 3-lobed but laterals often reduced to give a single-lobed small (1.5–3cm) leaf. var. *poetarum* (syn. *H.poetica*; *H.helix* var. *poetica*): POET'S IVY; ITALIAN IVY; leaves 5-lobed, 5–7 × 6–8cm, the two basal lobes much reduced giving a somewhat 'square' appearance to the light green leaf; grown for the novelty of its dull orange berries.

Name	Distribution	Leaves
H.hibernica *H.helix* 'Hibernica' IRISH IVY	Atlantic coast of GB, Ireland, W France, Spain, Portugal	juvenile leaves 5-lobed, 5–9 × 8–14cm, lobes triangular and bluntly acute, central lobe larger, base cordate, dark green, with stellate hairs tending to lie in parallel fashion rather than the haphazard fashion of *H.helix*; adult leaves ovate-elliptic

Comments: 'Deltoidea' (SWEETHEART IVY): leaves sometimes 3-lobed but more often deltoid and unlobed, base strongly cordate with overlapping lobes giving an almost heart-shaped leaf, dark green. 'Hamilton': leaves 5-lobed, centre lobe only slightly longer than laterals, mid green, edge thickened. 'Helford River': leaves 5-lobed, centre lobe prolonged and acuminate, mid-green with grey-white veins. 'Lobata Major': leaves 3-lobed, lateral lobes at right angles to the long-acuminate central lobe. 'Rona': leaves 5-lobed, mid-green with freckled and diffused yellow variegation; less green leaf intrusion than with 'Variegata'. 'Rottingdean': leaves 5-lobed, digitate, centre lobe slightly longer than laterals, sinuses narrow and convolute at the clefts, dark green. 'Sulphurea': leaves edged and splashed sulphur yellow. 'Variegata': some leaves entirely yellow or parti-coloured.

Name	Distribution	Leaves
H.maderensis	Madeira	juvenile leaves 3-lobed, occasionally unlobed, 4–5 × 6–7cm, sinuses very shallow or absent, lobes obtuse, base truncate or slightly cordate, mid to dark green, very glossy at first; hairs scale-like with 10–16 rays
H.maroccana MOROCCAN IVY	Morocco	3-lobed, 7–9 × 7cm, centre lobe twice as long as laterals, sinuses shallow, apex acute, base cuneate or truncate, dark green; hairs scale-like, rays 6–8; petioles purple-red

Comments: 'Spanish Canary': leaves very large; hardy and fast-growing

Name	Distribution	Leaves
H.nepalensis *H.himalaica, H.cinerea* HIMALAYAN IVY	Himalaya to Kashmir	juvenile leaves ovate to lanceolate, 6–10 × 4–5cm, obscurely lobed, lobes often little more than 3–6 marginal indentations, apex acute, base truncate, matt olive-green, hairs scale-like, 12–15-rayed; adult leaves elliptic-lanceolate, unlobed

Comments: Readily identified by the dull orange berries. 'Suzanne': differs from type in 5-lobed leaves, the short basal lobes back-pointing to give a 'bird's foot' effect, dark green. In var. *sinensis* from Northern China, the juvenile leaves are usually unlobed and ovate-acuminate with a cuneate base, they emerge maroon, rapidly maturing to matt mid-green.

Name	Distribution	Leaves
H.pastuchovii	Russia, Iran	juvenile leaves narrowly ovate, 4–9 × 3–4cm, unlobed, apex acuminate, base slightly cordate, dark green, glossy, slightly leathery, hairs scale-like with 8–12 rays, sparse
H.rhombea *H.japonica* JAPANESE IVY	Japan	juvenile leaves ovate to triangular, unlobed, 2–4 × 4–5cm, apex acute, mid-green, leathery, stems green, hairs scale-like, rays 10–18

Comments: 'Pierrot': leaves smaller, stems thin and wiry. 'Variegata': leaves slightly smaller, finely edged with white

Hedera: (a) *H. hibernica,* (b) *H. nepalensis,* (c) *H. colchica* 'Sulphur Heart', (d) *H.*'Conglomerata Erecta', (e) *H.* 'Manda's Crested', (f) *H.* 'Pedata', (g) *H.* 'Atropurpurea', (h) *H.* 'Parsley Crested', (i) *H.* 'Dentata Variegata', (j) *H.* 'Spetchley', (k) *H.* 'Goldheart'

hedge a continuous row of trees or shrubs, usually low-growing and closely planted to provide a boundary line, screen or ornamental feature. There is no hard and fast definition of a hedge by height, but living barriers more than 5m tall are usually referred to as shelter belts, and continuous plantings less than 50cm high as edging. Hedges may be formally clipped or informal, and a very wide range of species are suitable subjects, for example: FORMAL EVERGREEN *Aucuba, Berberis, Cotoneaster, Escallonia, Ilex , Lonicera nitida, Quercus ilex, Taxus baccata*; FORMAL DECIDUOUS *Alnus glutinosa, Carpinus betulus, Crataegus monogyna, Fagus sylvatica, Prunus cerasifera, Salix, Sambucus nigra, Sorbus intermedia*; INFORMAL EVERGREEN *Brachyglottis, Camellia japonica, Choisya ternata, Cotoneaster, Mahonia aquifolium, Rhododendron ponticum, Rosmarinus, Viburnum tinus*; INFORMAL DECIDUOUS *Amelanchier lamarckii, Cornus stolonifera, Deutzia scabra, Forsythia*; *Fuchsia magellanica*; *Hydrangea macrophylla, Kerria*; *Rosa* species.

Hedychium (from Greek *hedys*, sweet, and *chion*, snow: *H.coronarium* has fragrant white flowers). Zingiberaceae. Tropical Asia, Himalaya and Madagascar. GINGER LILY, GARLAND LILY. About 40 species, perennial herbs with stout rhizomes and reed-like stems bearing two-ranked, oblong to lanceolate leaves. Fragrant flowers are carried in dense, terminal spikes in summer. They consist of three petals, a showy lip, petal-like staminodes and long, slender filaments. The Himalayan species, particularly *H.coccineum* and *H.densiforum* and its cultivars, are hardy to –2°C/28°F if grown at the foot of a south-facing wall or given a mulch as winter protection. *H.gardnerianum* is a fine plant for the cool conservatory or subtropical bedding scheme. *H.coronarium* and *H.flavescens* need a minimum of 18°C/65°F with high humidity, and plentiful water. Liquid feed plants in pots throughout growth. Reduce water considerably in winter and allow plants to become dormant, cutting out old spikes. Propagate by division in spring or by seeds sown fresh at 18°C/65°F.

H.coccineum (RED GINGER LILY, SCARLET GINGER LILY; Himalaya; to 3m; flowers pale to deep red or orange, lip 1–2cm across, red and yellow to mauve; includes 'Tara', with bold spikes of orange flowers, and var. *angustifolium*, with narrower leaves and orange flowers); *H.coronarium* (BUTTERFLY LILY, GARLAND FLOWER, WHITE GINGER; India; to 3m; flowers very fragrant, white, lip 5cm across, white with yellow-green centre; 'F. W. Moore': *H.coronarium* × *H.coccineum*, showy spikes and fragrant, amber yellow flowers, with a base of segment blotched orange-yellow; var. *maximum*: larger flowers and filaments tinged pink); *H.densiflorum* (Himalaya; 1–5m; flowers vermilion, in dense, cylindric spikes; lip 0.7cm across; filaments red; includes 'Assam Orange', with shorter stems and very fragrant, vivid burnt orange flowers); *H.flavescens* (YELLOW GINGER; Bengal; to 3m; flowers in dense 20cm spike; corolla pale yellow with linear, green lobes, lip red-yellow at base, staminodes yellow, filaments yellow); *H.gardnerianum* (KAHILI GINGER; N India, Himalaya; 1–2m+; flowers fragrant, white and yellow, in dense 25–35cm spike, lip 1–2cm across, yellow, filaments bright red).

Hedysarum (the Greek name *hedysaron*, from *hedys*, sweet). Leguminosae. Northern temperate regions. Some 100 species, perennial herbs, subshrubs or shrubs with racemes of pea-like flowers in summer. Suitable for the herbaceous border; the flowers also last well when cut. Grow in full sun, in any moderately deep and fertile, well-drained soil. Hardy to –15°C/5°F. Propagate by seed, or by careful division in spring. *H.coronarium* (FRENCH HONEYSUCKLE; S Europe; herb, 30–100cm; leaflets 1.5–3.5cm, 7–15, elliptic to orbicular; flowers intensely fragrant, bright red to purple-red; 'Album': flowers white).

heel a small strip of wood and bark retained at the base of a cutting at the time of removal from the main shoot. The presence of a heel often improves rooting potential.

heeling in the temporary insertion of plants in soil, usually prior to permanent planting, in order to prevent deterioration due to drying out.

Helenium (from Greek *helenion*, for an Old World plant (perhaps *Inula helenium*), said to have been named for Helen of Troy). Compositae. Americas. SNEEZEWEED. About 40 species, annual, biennial or perennial herbs with linear to lanceolate leaves and daisy-like flowerheads with a prominent, domed disc produced from midsummer to autumn. Fully hardy. Grow in full sun in fertile and moisture-retentive soil, enriched with well-rotted organic matter; apply balanced fertilizer in spring and provide support for taller plants. Lift and divide every second or third year, and discard less vigorous portions of the clump. Provide a winter mulch where low temperatures are prolonged. Propagate also by seed or basal cuttings.

H.autumnale (SNEEZEWEED; N America; perennial to 1.5m; flowerheads to 5cm diameter, clustered in a corymb, ray florets yellow to bright yellow; includes 'Grandicephalum', to 150cm, with large, yellow flowerheads, the dwarf 'Nanum Praecox', with yellow flowerheads produced early, 'Praecox', with yellow, brown and red flowerheads, produced early, 'Rubrum', with dark red flowerheads, and 'Superbum', to 150cm tall, with large, bright yellow flowerheads with wavy ray florets; see below for further cultivars); *H.bigelovii* (California, Oregon; perennial, to 90cm; flowerheads long-stalked, to 6cm diameter; ray florets yellow; includes 'Aurantiacum', with gold ray florets). *Cultivars* Taller selections include the 150cm yellow 'Sonnenwunder', the gold, brown-centred 'Goldrausch', 'Zimbelstern', with wavy ray florets, and the brown-red, 120cm-tall 'Wonadonga' and 'Margot', and the 120cm 'Flammenrad' and 'Riverton Beauty', yellow with red edges and eyes; medium cultivars include the gold 'Butterpat',

heel cutting

'Waltraut', 'Bressingham Gold' and the bronze to velvety wine red 'Moerheim Beauty', 'Goldlackzwerg', 'Baudirektor Linne', 'Bruno' and 'Wonadonga'. Smaller clones include the yellow 'Aurantiacum', the gold and copper 'Wyndley', the red-brown 'Kupferzwerg' and the soft mahogany 'Crimson Beauty'.

Heliamphora (either from Greek *helios* the sun, or from *helos* a marsh, and *amphora*, a jug or pitcher, referring to its appearance and habitat). Sarraceniaceae. Guyana Highlands, Guyana and Venezuela. SUN PITCHER. 6 species, carnivorous perennial herbs. Borne in a basal clump, the leaves are pitcher-shaped and bright green, assuming a red tint in full sun. Stoutly tubular in the lower half, they then narrow in a short bottle-neck before expanding above in a flattened blade; this terminates in a spoon-shaped cap. White flowers nod in loose, arching racemes in spring and summer. *Heliamphora* requires a very open but absorbent compost, such as perlite and living sphagnum and perfect drainage (that said, these plants sometimes thrive in terraria). Under glass, air should circulate freely and high humidity be maintained by regular overhead spraying. Provide bright light but shade from fierce sun. Maintain winter temperatures of between 5.5–20°C/42–70°F, and summer temperatures not exceeding 25°C/77°F. In the open greenhouse, drench with rainwater once or twice daily in warm weather; keep moist and humid at all times. Dilute foliar feed or slow-release fertilizers are beneficial but must be carefully applied.

H.heterodoxa (pitchers cauline, to 30cm, upper part of pitcher interior with velvety hairs or glabrous, cap large); *H.minor* (pitchers constricted above and below middle, usually not exceeding 7.5 × 1cm in cultivation, pubescent in upper portion of interior, cap very small, or absent); *H.nutans* (pitchers basal, constricted above and below middle, to 20cm, green with red margins, hairy in upper portion of interior, cap medium-sized, i.e. to 1.5cm across).

Helianthemum (from Greek *helios*, sun, and *anthemon*, flower – the flowers open only in bright sunshine). Cistaceae. Americas, Europe, Mediterranean, N Africa, Asia Minor to C Asia. ROCK ROSE, SUN ROSE. 110 species, evergreen or semi-evergreen shrublets with small, more or less grey-downy, oblong to linear leaves. Borne on one-sided, racemose cymes in spring and summer, the short-lived flowers are dish-shaped and consist of five, crêpe-like petals and numerous golden stamens. Suitable for rock gardens (in company with similarly tough and fast-growing species), for the front of sunny borders and for wall crevices or gravel beds. All are hardy to at least -10°C/14°F and may be overwintered with a covering of bracken or sacking in colder areas. Provide a sunny open site and a light well-drained soil. Trim back after flowering for a second flush of flowers in late summer. Plants are short-lived, and may require replacement when too leggy and sparse. Propagate by softwood or heeled cuttings in late summer, rooted in a sandy mix; overwinter in a cold frame for planting out in spring.

H.apenninum (Europe, Asia Minor; to 50 cm; flowers to 2.8cm in diameter, petals white, yellow at base; var. *roseum*: flowers pink); *H.nummularium* (Europe, Asia Minor; to 50cm; flowers 2–5cm in diameter, golden yellow, pale yellow, pink, white or orange, rarely cream; includes the dwarf, compact 'Amy Baring', with deep yellow flowers and subsp. *grandiflorum*, with orange-yellow flowers); *H.oelandicum* (Europe; to 20cm; flowers to 0.8cm in diameter, yellow); *H.* × *sulphureum* (*H.apenninum* × *H.nummularium*; low shrub; flowers pale yellow). *Cultivars* The rock roses most frequently encountered in gardens tend to be hybrids with *H.apenninum*, *H.nummularium*, and *H.croceum* involved in their parentage. They range in flower colour from white to cream, pale yellow, gold, orange, tangerine, flame, scarlet, wine red, crimson, rose pink and blush and have grey, sage green or silvery foliage.

Helianthus (from Greek *helios*, sun, and *anthos*, flower). Compositae. N and S America. SUNFLOWER. About 70 species, annual to perennial herbs with showy daisy-like flowerheads in summer. *H.tuberosus* is the JERUSALEM ARTICHOKE (see *artichoke*). *H.salicifolius*, is a hardy, spreading perennial, well-adapted to dry, sunny situations in the garden. The annual *H.annuus* is available in a wide range of cultivars, including dwarf (30–60cm), intermediate (100–120cm) and tall forms (3m and more). Smaller cultivars make fine cut flowers. Sow seed under glass in early spring for planting out in early summer, or sow *in situ* in late spring. Grow in full sun in any fertile well drained soil. Mulch perennials annually in spring with well-rotted farmyard manure. Stake tall cultivars and provide a deep winter mulch where prolonged low winter temperatures are not accompanied by snow cover. Lift and divide perennials every third or fourth year to maintain vigour. Propagate by seed and (perennials) by division or basal cuttings.

H.annuus (COMMON SUNFLOWER; US; stout, often giant, annual with an erect, more or less unbranched stem to 5m; leaves to 40cm, broad-ovate to cordate, acute, coarsely toothed, roughly hairy or bristly, dark green to silver; flowerheads terminal, one or few, to 60 cm across, ray florets yellow, occasionally tinged red, bronze or purple, disc florets red or purple; includes numerous cultivars and seed races – dwarf (50cm) to giant (6m), flowerheads 15–60cm in diameter, 'single' or 'double', with florets in pale primrose to deep gold or copper red or with purple-bronze tints); *H.* × *multiflorus* (*H.annuus* × *H.decapetalus*; perennial, to 2m, with flowerheads to 12cm in diameter, florets golden yellow, disc florets occasionally replaced by ray florets; includes numerous cultivars, to 180cm tall, 'single' or 'double', pale primrose to deep gold); *H.salicifolius* (Southern Central US; perennial, to 3m; leaves to 20cm, narrowly linear to lanceolate, slightly hairy, arching; flowerheads numerous, to 8cm in diameter, ray florets golden yellow, disc florets purple, rarely yellow).

Helichrysum (from Greek *helios*, sun, and *chrysos*, golden). Compositae. Warm Old World, especially South Africa and Australia. EVERLASTING FLOWER. About 500 species, annual, biennial and perennial herbs, subshrubs and shrubs, often hairy and glandular and producing button- or daisy-like flowerheads, usually in summer. Almost without exception *Helichrysum* species require a sheltered position in sun with well-drained soil, although they are suited to a diversity of situations in the garden depending on size and habit; some also have flowers suitable for drying. The low-growing perennials and dwarf shrublets, (for example, *H.bellidioides*) are suitable for a sunny position in gritty, freely draining soils in the rock garden, but will require at least some protection from excessive wet in winter. Cushion-forming types, such as *H.milfordiae*, are more reliable when given the protection from winter wet afforded by the alpine house; *H.milfordiae* is also suited to tufa plantings. Many species have beautiful foliage, extremely useful for colour and textural contrast in the herbaceous border and invaluable in the silver border; these include taller aromatic species, such as *H.italicum*, var. *serotinum*. In regions where winters are severe, protect with a mulch of bracken litter and cut back damaged growth in spring. *H.petiolare* is frequently grown as a half-hardy annual in cool temperatures zone for edging, groundcover and hanging baskets. The species most often grown for cutting and drying, *H.bracteatum*, comes in a wide range of heights and colours. It is easily grown in any moderately fertile and moisture-retentive soil in full sun and may be sown *in situ* in spring, but better harvests are obtained by starting earlier under glass, especially in regions where summers are short. Cut the flowers before they are fully open; dry upside down and away from direct light in small bunches; strip foliage from stems to avoid rotting. This species is usually treated as an annual, but may live for several years if protected from cold, wet conditions. Propagate shrubby species by heeled semi-ripe cuttings in summer; perennials by division or seed in spring. Remove rosettes of *H.milfordiae* and root in a sandy propagating mix in the cold frame.

H.bellidioides (New Zealand; perennial herb to 20cm, stems prostrate; leaves to 6mm, orbicular, silvery-arachnoid; flowerheads to 3cm in diameter, solitary, phyllaries linear, papery white); *H.bracteatum* (syn. *Bracteantha bracteata*; STRAWFLOWER, GOLDEN EVERLASTING, YELLOW PAPER DAISY; Australia; annual or perennial herb to 1.5m; leaves to 12cm, oblong to lanceolate, acuminate, glabrous or subglabrous, sometimes downy and silver-grey as in the copper- to golden-flowered 'Dargan Hill Monarch'; flowerheads solitary, to 7cm in diameter; seed races include dwarf to tall plants with large or small flowerheads in shades of sulphur, golden yellow, orange, bronze, copper, red, rose, purple and white); *H.italicum* (syn. *H.angustifolium*; S Europe; bushy aromatic subshrub or small shrub to 1m; leaves to 6cm, narrowly linear, pale-tomentose to glabrescent, revolute; flowerheads yellow, 2–4mm in diameter in

a cluster to 8cm across; includse subsp. *serotinum,* the CURRY PLANT, to 60cm, with silver, slender, strongly curry-scented foliage); *H.milfordiae* (syn. *H.marginatum*; South Africa; subshrub to 15cm, cushion-forming, prostrate; leaves to 1.5cm, obovate to subspathulate, densely tomentose, apex scarious; flowerheads 4cm in diameter, solitary, phyllaries glossy white, usually crimson- or brown-tipped); *H.petiolare* (syn. *H.petiolatum*; LIQUORICE PLANT; South Africa; climbing or sprawling subshrub or shrub to 1m; stems slender, branched, thinly grey-tomentose; leaves to 3.5cm, rounded to broadly ovate, silver-grey-tomentose; flowerheads to 7mm in diameter, in loose corymbs; includes 'Limelight' (syn. 'Aureum') with gold to lime-green leaves, 'Roundabout', dwarf and compact to 15cm, with variegated leaves, and 'Variegatum', with variegated grey and cream leaves); *H.splendidum* (E and South Africa; shrub to 1.5m; stems branched, leafy, thinly tomentose; leaves to 6cm, linear to lanceolate, tomentose beneath, revolute; flowerheads to 5mm in diameter, in crowded or loose, terminal corymbs, bright yellow, rarely orange). For *H.coralloides*, see *Ozothamnus coralloides*; for *H.rosmarinifolium*, see *Ozothamnus rosmarinifolius*; for *H.selago*, see *Ozothamnus selago*.

Heliconia (for Mount Helicon in Greece, the home of Muses, a pun reflecting the closeness of this genus to *Musa*). Heliconiaceae. C and S America, S Pacific Islands, to Indonesia. FALSE BIRD OF PARADISE, WILD PLANTAIN. About 100 species, large, evergreen perennial herbs with clumped pseudostems and paddle-shaped leaves. Borne amid waxy, brilliantly coloured boat-shaped bracts in erect to pendulous spikes, the flowers are white to yellow or orange, often tipped or striped with blue-black. Magnificent perennials for the warm greenhouse or moist, semi-shaded positions in subtropical and tropical gardens, *Heliconia* species are grown for their stately, banana-like habit (sometimes to 6m tall), handsome foliage (beautifully marked in *H.indica*) and their superb waxy inflorescences. These may be erect or pendulous, flat or spiralling, and range in colour from yellow to pink to sealing-wax red and combinations thereof. All are suitable for cutting. They need an open, moisture-retentive bark mix, rich in leafmould and well-rotted manure. Plant in tubs or borders (dwarf cultivars of *H.stricta* are excellent potplants) in semi-shaded, humid conditions (minimum temperature 15°C/60°F); syringe and foliar-feed frequently, except during flowering period and at the coolest times of the year. Cut away flowered growths once new suckers have formed. Propagate by division.

H.acuminata (Brazil to S Venezuela and SE Peru; stem 0.5–3cm; leaves 15–70cm; inflorescence erect, 50–100cm, bracts distichous, 6–25cm, 4–6, red, orange or yellow, glabrous to hairy, flowers white, orange or yellow, usually with dark green 'eye-spots' toward apex, or olive-green with white or pale orange tips); *H.bihai* (WILD PLANTAIN, MACAW FLOWER, BALISIER; C and S America; stem 0.5–5m; leaves 0.25–2m; inflorescence erect, 45–110cm, bracts 3–15,

Heliconia caribaea

distichous or alternate, sides red, keel yellow, margins green and/or yellow, more or less glabrous, flowers white, apex pale green); *H.caribaea* (WILD PLANTAIN, BALISIER; Jamaica and E Cuba to St. Vincent; stems 2–5.5m; leaves 60–130cm; inflorescence 20–40cm, erect, bracts to 25cm, 6–15 in 2 overlapping rows, broadly triangular, red or yellow, keels and apices sometimes green or yellow, flowers white, apex green); *H.chartacea* (Amazon Basin; stems to 4m; leaves to 1.2m; inflorescence to 1m, pendent, bracts to 40cm, 4–28, slender, loosely spiralling, carmine to blush pink edged lime green, somewhat glaucous, flowers dark green; includes the candy pink and deep rose-magenta 'Sexy Pink' and 'Sexy Scarlet'); *H.hirsuta* (Antilles and Belize to Bolivia, N Paraguay and E Brazil; stems 0.5–5m; leaves 15–50cm; inflorescence 20–70cm, erect, often hairy or cobwebby, bracts 3–9 slender, upcurved, loosely 2-ranked, orange to red, flowers yellow to orange or red, with dark green eyespots near apex); *H.indica* (Papua New Guinea; stems to 5m, stout; leaves to 2.5m; inflorescence to 1m, erect, bracts to 35cm, distichous, tough, carinate, spreading, often with leafy appendages, green, flowers green; includes 'Spectabilis', with leaves green above, purple-bronze beneath or purple-green throughout, often with fine rose to white markings, and 'Striata', with green leaves with green-yellow or white markings); *H.metallica* (Honduras to NW Colombia and Bolivia; stems 0.5–4m; leaves 0.25–1.4m, emerald-green above, metallic purple beneath; inflorescence 30–90cm, erect, bracts 5–14cm, 3–8, distichous, linear to lanceolate, green to red, cobwebby, flowers rose pink to purple tipped white); *H.psittacorum* (PARROT'S FLOWER, PARAKEET FLOWER, PARROT'S PLANTAIN; E Brazil to Lesser Antilles; stems 0.5–2m; leaves 10–50cm; inflorescence 12–70cm, erect, bracts 4–16cm, 2–7, pink, orange or red, sometimes edged yellow or green, upcurved, glabrous, waxy, not overlapping, slender, flowers yellow-green, orange or red with a dark green apical band and tipped white); *H.stricta* (S Venezuela and Surinam to Bolivia and Ecuador; stems 0.5–4m; leaves 40–150cm; inflorescence 20–30cm, erect, bracts 7–11cm, distichous, overlapping, broadly carinate, slightly inflated, red or orange, keel and margins yellow, tipped green, flowers white or pale yellow at base and apex, with a bright green band toward apex); *H.zebrina*(Central E Peru; stems 1–2m; leaves to 75cm, dark green, often with dark green or purple bands and blotches, purple beneath; inflorescence 20–50cm, erect, bracts 6–21cm, 6–9 distichous, narrow, spreading, red, glabrous, flowers pale yellow-green).

helicoid spirally clustered, in the shape of a spring or a snail-shell.

helicoid-cyme see *bostryx.*

Helictotrichon (from Greek *heliktos*, twisted, and *thrix*, *trichos*, hair or bristle, referring to the shape of the awn). Gramineae. OATGRASS. Temperate northern Hemisphere, South Africa. 60 species of perennial grasses with slender, clump-forming culms and fine, arching foliage. *H.sempervirens* grows to 1m tall and is one of the finest of the blue and silver grasses, making a fountain of blue-grey foliage overtopped by weeping, ash-blond flowerspikes. Grow in well-drained soil in sun; give mulch protection where winter temperatures fall much below −15°C/5°F. Remove dead foliage regularly to retain form. Propagate by division or seed in spring.

Heliopsis (from Greek *helios*, sun, and *opsis*, appearance, referring to the showy flowerheads). Compositae. N America. OX-EYE. About 13 species, loosely branched, erect perennial herbs with daisy-like flowerheads in late summer. Noted for ease of cultivation, extreme cold-hardiness, longevity, vigour and the warm colours of the large daisy flowers that appear from mid- to late summer and autumn. Best grown on a fertile and moisture-retentive soil in full sun. Propagate by division. *H.helianthoides* (eastern N Amercia; to 1.5m; flowerheads to 7.5cm in diameter, ray florets pale to deep yellow, apex toothed, disc florets yellow; cultivars range from 1–5m high with 1- or many-ranked ray florets ranging in colour from light to dark gold).

heliotropism see *phototropism.*

Heliotropium (from Greek *helios*, sun, and *trope*, turning, from the long-disproved idea that the inflorescence turned with the sun). Boraginaceae. Tropical and temperate regions. HELIOTROPE; TURNSOLE. Some 250 species, annual or perennial herbs, subshrubs or shrubs with small, 5-lobed flowers in scorpioid spikes or racemes in summer. The most commonly cultivated plants are those hybrids derived from *H.arborescens*; when treated as frost-tender annuals, these are suitable for borders, pots and window boxes. When container-grown, they may be trained as pyramids or standards, for summer bedding or for permanent positions in the cool greenhouse or conservatory, where their delightful perfume may be more easily appreciated. Grow in moist but well-drained, moderately fertile soils in full sun. For potplants, use a medium-fertility, loam-based mix and water plentifully when in growth. Maintain a winter minimum temperature of 5–7°C/40–45°F and ventilate freely as temperatures rise to 13°C/55°F. Propagate from seed sown in early spring at 16–18°C/6l–65°F. Many cultivars are available. Propagate these by greenwood cuttings in summer or by semi-ripe cuttings in late summer or autumn. Alternatively, take soft tip cuttings in late winter from stock plants grown on under glass over winter at 13°C/55°F. Standards are usually grown on through the winter at 10–16°C/50–60°F from late summer semi-ripe cuttings; training for standards and pyramids follows procedures similar to those used for *Fuchsia.*

H.arborescens (syn. *H.peruvianum*; HELIOTROPE, CHERRY PIE; Peru; hairy, perennial shrub to 2m; flowers sweetly scented, violet or purple to white; cultivars include 'Chatsworth', with rich purple-blue, very fragrant flowers; 'Florence Nightingale', with pale mauve flowers; 'Grandiflorum', with large flowers; 'Iowa', bushy and erect, with dark green leaves, lightly tinted purple, and dark purple, fragrant flowers in large clusters; 'Lemoine', to 60cm, with deep purple, fragrant flowers; 'Lord Roberts', small, with soft violet flowers; 'Princess Marina', with deep violet flowers; 'Marine', bushy and compact, to 45cm high, with dark green, wrinkled leaves and deep violet, fragrant flowers in large clusters; 'Regal Dwarf', dwarf, with dark blue, fragrant flowers in large clusters; 'Spectabile', compact, to 1.2m high, with pale violet, scented flowers; and 'White Lady', small and shrubby, with white flowers, tinged pink in bud).

Helipterum (from Greek *helios*, sun, and *pteron*, wing, in reference to the plumose pappus). Compositae. Australia, South Africa. EVERLASTINGS, STRAWFLOWERS. About 50 species, annual or perennial herbs or shrubs, more or less covered in white hairs. Papery, daisy-like flowerheads appear in summer and dry well. Cultivate as for *Helichrysum bracteatum*. *H.splendidum* (SHOWY SUNRAY, SPLENDID EVERLASTING, SILKY-WHITE EVERLASTING; West Australia; tufted annual to 50cm; leaves linear, grey-green; flowerheads to 6cm in diameter, phyllaries cream-white, often with a purple band).

Helleborus (name used by Theophrastus for medicinal herbs, probably not of this genus, but later always associated with it and with *Veratrum*). Ranunculaceae. Europe, especially Italy and Balkans, Turkey to Caucasus, W China. HELLEBORE. 15 species, rhizomatous perennial herbs for the most part evergreen. The leaves are trifoliolate to palmate or pedately compound. They may be basal, or, in caulescent species, borne on the flowering stem. The flowers may be solitary or, more usually, produced a few together in loose cymes subtended by leaflike bracts; in some (caulescent) species, many flowers will be carried in a paniculate cyme. Usually nodding, they are dish- to bell-shaped and consist of five broadly ovate to elliptic segments, numerous shortly tubular nectaries and many stamens.

But for *H.argutifolius*, *H.lividus*, their offspring and *H. thibetanus*, the following are hardy in zone 5. They should be planted on a moist but well-drained and fertile soil – heavy clay is tolerated but over wet conditions may encourage leaf spot. Site in sun or part-shade with protection from harsh winds. Feed in spring and summer. Most flower from late winter to mid-spring. With stemless species bearing all-basal leaves (e.g. *H.orientalis*), it is advisable to remove the old foliage once the flowerbuds are visible, the better to display the flowers. *H.argutifolius*, *H.lividus* and their hybrid offspring need full sun, shelter and a fast-draining gritty soil in zones 6 and over. Where cold, wet weather threatens, protect *H.lividus* with a dry mulch, a cloche, or grow in the cold greenhouse. *H. thibetanus* is still quite new to gardens, but it seems likely to need a shaded, humid position on a cool, permanently moist soil rich in leafmould. It may die back by midsummer, is somewhat tender and should be mulched in cold winters.

Division is possible for most species (not *H.foetidus*). Fresh seed sown in cold frames in early summer is also worthwhile and seedlings grow quickly into good plants, flowering in 2-3 years and often producing attractive colour forms. In suitable conditions, they will seed freely; the self-sown seedlings may be carefully transplanted in early spring. *H.argutifolius*, *H.foetidus* and *H.lividus* may also be propagated using cuttings of young vegetative shoots. All (especially those with softer leaves such as *H.niger*) are susceptible to leaf spot.

H.argutifolius (syn. *H.corsicus*; Corsica, Sardinia; stems to 1m, erect or sprawling, soon becoming naked below; leaves cauline, trifoliate, coriaceous, leaflets to 25cm, elliptic to lanceolate, glabrous, olive to grey-green, spiny-toothed; inflorescence terminal, many-flowered, flowers pale green, cup or bowl-shaped, to 5cm across); *H.atrorubens* (NW Balkans; to 35cm; leaves basal, deciduous, dark green, pedate, glabrous, central leaflet, 10–20cm, lateral leaflets divided into 3–5 lobes (to 7 segments in all), coarsely serrate; inflorescence to 30cm, flowers produced before leaves, 2–3, deep purple suffused green within, pendent, flat or shallowly saucer-shaped, 4–5cm across); *H.foetidus* (STINKING HELLEBORE, SETTERWORT, STINKWORT, BEAR'S FOOT; W and C Europe; stems to 80cm, erect, hard, dying after fruiting; leaves cauline, coriaceous, dark green or grey-green, pedate; segments 7–10, linear to lanceolate or narrowly-elliptic, tapering finely, to 20cm, coarsely serrate or nearly entire; inflorescence terminal, to 30cm, branches many-flowered, flowers pendent, green, usually flushed red-purple at apex, campanulate, to 2×2.5cm, sometimes fragrant; includes Wester Flisk Group, with stems, petioles and inflorescence branches tinted red and flowers yellow-green edged maroon); *H.lividus* (Majorca; to 45cm+; stems erect or sprawling, dying after fruiting; leaves cauline, trifoliolate, coriaceous, leaflets 10–20cm, obliquely elliptic, deep green to grey-green with silvery white veins above, suffused pink-purple beneath, serrate or entire; inflorescence terminal, flowers to 10, creamy-green suffused pink-purple, flat to bowl-shaped, 3–5cm across); *H.multifidus* (C Balkans; leaves basal, usually deciduous, coriaceous, pedate, leaflets deep green much-divided, segments 20-45, linear to narrowly lanceolate, to 13cm, serrate; inflorescence exceeding leaves, branched, flowers 3–8, conic to cup-shaped, nodding, green, to 4.5cm diameter, scented); *H.niger* (CHRISTMAS ROSE; C Europe; to 30cm, evergreen; leaves basal, leathery, dark green, pedate, segments 7–9, oblong or oblanceolate, to 20cm, toothed towards apex; flowers solitary or 2–3 on stout peduncle, waxy white, tinged green at centre, flushed pink, peach or dull purple with age, saucer- to bowl-shaped, to 8cm across; includes numerous cultivars with large and well-rounded flowers, often in very pure tones of white, but some

Helleborus orientalis subsp. *guttatus*

ivory or near-yellow, others assuming rose, tawny or apricot tints, especially as the flowers age, or when the plants are grown on very lean soils in full sun; clones with variegated leaves are sometimes offered); *H.niger* × *H.lividus* (leaves pedate to trifoliate, deep green with pale veins; flowers large, flat, white suffused pink-brown or pink-purple, turning dull purple with age); *H.niger* × *H.* × *sternii* (very variable; leaves somewhat pedate with some of the attractive veining of *H.lividus*; flowers large, white, tinged pink-brown or green outside, colouring more deeply with age); *H.* × *nigercors* (*H.niger* × *H.argutifolius*; stem short, leafy, with terminal clusters of flowers, but some leaves and flowers basal; leaves large, grey- to dark green, semi-rigid, trifoliate to pedate with 7 segments, dentate or spinulose; flowers large, saucer-shaped, waxy white sometimes tinged green); *H.odorus* (Balkans, S Hungary, S Romania; to 30cm; leaves basal, coriaceous, pedate, central leaflet undivided, laterals with 3–5 lobes, elliptic or oblanceolate, to 20cm, hairy beneath, coarsely serrate; inflorescence branched, short at first; flowers fragrant, usually 3–5, saucer-shaped, facing outwards, to 6cm in diameter, green to yellow); *H.orientalis* (LENTEN ROSE; NE Greece, N and NE Turkey, Caucasian Russia; to 45cm; leaves basal, coriaceous, deep green, pedate with central leaflet entire, laterals divided giving 7–9 elliptic or oblanceolate segments, to 25cm, serrate; inflorescence to 35cm, branched, flowers 1–4, saucer-shaped, 6–7cm across, white to cream or cream tinged green, flushed purple or pink particularly below and at centre, green or purple-green following fertilization; includes subsp. *abchasicus*, syn. *H.abchasicus*, with flowers suffused red-purple, often finely spotted, and subsp. *guttatus*, syn. *H.guttatus*, with white or cream flowers, spotted red-purple. Members of the *H.orientalis* complex will hybridize freely, producing plants with flowers in shades of white, pale pink, rose, plum and deep purple, often flushed with another of these tones or spotted or streaked purple-red. Hybridization also occurs between *H.orientalis* and other species. The influence

of *H.purpurascens* will produce dark pruinose purple shades, for example 'Queen of the Night'. Crossing with *H.torquatus* introduces strong, glaucous, purple-black tones, for example 'Pluto', wine-purple outside, green and slaty purple inside, with purple nectaries, and 'Ballard's Black', with dark purple-black flowers. Hybrids involving *H.odorus*, *H.cyclophyllus* or *H.multifidus* exhibit shades of yellow and green, e.g. 'Citron', clear pale yellow; 'Orion', cream tinged green, purple stain in centre and purple nectaries, and 'Yellow Button', with small, deep yellow flowers); *H.purpurascens* (east central Europe; to 20cm at flowering; leaves basal, deciduous, coriaceous, palmate with 5 leaflets, each divided into 2–6 lanceolate segments, to 20cm, hairy beneath, coarsely serrate; inflorescence appearing before leaves, branched or simple, flowers 2–4, nodding, cup-shaped, to 7cm across, externally purple-violet or red-purple to brown or green, glaucous, internally sometimes green); *H.* × *sternii* (*H.argutifolius* × *H.lividus*; leaves pewter to dark grey-green, obscurely veined green-white or silver, suffused purple to pale rose, entire or spiny-toothed; flowers lime green to green flushed pink or purple-brown; includes 'Ashwood Strain' with silver, jagged leaves, the compact 'Boughton Beauty', with grey-green leaves marked and sometimes tinted rose-purple and flowers and floral axis suffused rose; Blackthorn Group, with leaves leaden grey-green, conspicuously veined, and flowers slightly flushed with pink, and 'Bulmer's Strain' with neat grey foliage, marbled as in *H.lividus*); *H. thibetanus* (SW China; to 30cm; leaves similar to *H. orientalis*, but softer-textured, more finely toothed, emerging bronze-green turning pea green; flowers nodding, bowl- to bell-shaped, to 3cm diam., palest rose with textured veins, or white with pale pink veins); *H.torquatus* (Balkans; to 40cm; leaves basal, deciduous, emerging with or after flowers, coriaceous, pedate with 10–30 narrow segments to 18cm, tapered, dark green above, puberulent beneath, coarsely serrate; inflorescence with apical leafy bracts, flowers 1–10, nodding or facing outwards, saucer- to cup-shaped, to 5.5cm across, externally dark plum purple, pruinose, internally same colour or blue-green; includes 'Aeneas', with double, predominantly green flowers, with exterior shaded maroon, and 'Dido', with double flowers, with a lime green interior and dark purple exterior); *H.vesicarius* (S Turkey, N Syria; to 60cm, becoming dormant in summer; leaves basal, thin, bright green, pedate, with 3 primary divisions each to 18cm, the leaflets lobed and toothed; inflorescence loosely cymose with leafy bracts, flowers bell-shaped, green tipped maroon or chocolate; fruit to 7.5cm, pale yellow-green, inflated with distinct wings); *H.viridis* (W Europe; to 40cm; leaves deciduous, thin, glossy, pedate, segments 7–13, oblong-lanceolate to narrow-elliptic, to 10cm, pubescent beneath, serrate; inflorescence emerging with leaves, long-branched; flowers 2–4, nodding, saucer-shaped or rather flat, to 5cm across, green).

helminthoid worm-shaped.

Helonias (from Greek *helos*, swamp, the natural habitat of the plant). Liliaceae (Melanthiaceae). N America. SWAMP PINK. 1 species, *H.bullata*, an evergreen tuberous perennial. Bright green and glossy, the leaves are to 45 cm long, oblong to lanceolate and form a basal rosette. Starry and scented, the flowers are pink, with pale blue anthers, and appear in spring in dense, conical racemes atop scapes to 45cm tall. Suited to bog garden and waterside plantings. Hardy at least to –15°C/5°F. Grow in moist, fertile, humus-rich, acidic soils in sun or light, dappled shade. Propagate by division after flowering.

Heloniopsis (from *Helonias* and Greek *opsis*, resemblance). Liliaceae (Melanthiaceae). Japan, Korea, Taiwan. 4 species, rhizomatous, perennial, evergreen herbs with oblong or lanceolate leaves in a basal rosette. In spring, one or a few flowers nod in a loose, scapose, umbellate raceme; these differ from those of *Helonias* in the larger and broader perianth segments and the purple-blue to dark violet anthers. Hardy to between -10 and -15°C/1F5°F. Grow in a moisture-retentive, humus-rich soil in light shade, with some protection from cold winds. Propagate by division of established clumps or by seed sown when ripe or in spring.

H.orientalis (syn. *H.japonica*; Japan, Korea, Sakhalin Islands; leaves 8–10cm, tinged red-brown toward apex; scape to 20cm; flowers to 4cm diam., tepals pink or violet; anthers blue-violet).

helophyte a plant that grows in permanent or seasonal mud, under water or around the water's edge.

*Hemerocallis
lilio-asphodelus*

Hemerocallis (from Gk *hemere*, day, and *kallos*, beauty; the flowers only live for one day.) Liliaceae (Hemerocallidaceae). E Asia, Japan, China. DAYLILY. Some 15 species, rhizomatous, clump-forming, perennial herbs with linear, arching leaves and short-lived flowers in summer; these are produced in succession on tall, more or less branching stems, and are funnel-shaped with 6 ovate to lanceolate, spreading segments. They will grow in most fertile garden soils including sand and heavy clay, but do best at a pH of 6–7. Abundant blooms will be produced if they are planted in a soil liberally enriched with well-rotted manure. Copious water is required to maximize both the size and number of blooms, but the ground must be free-draining. If grown in waterlogged ground or rough grass, daylilies produce more foliage and fewer flowers. Most daylilies require long periods of full sun. Some species and older hybrids (*H.*'Corky', *H.*'Golden Chimes', *H.*'Lemon Bells') will tolerate part shade, but the palest, pastel forms, especially pink-toned cultivars, may not open fully unless exposed to long periods of strong sunlight. Some dark red and deep purple cultivars benefit from part shade in tropical and subtropical climates. Unless naturalized, plants should be lifted and divided every three years to maintain vigour and promote flowering. Divide cultivars in spring, or after flowering. Most species, except some forms of *H.fulva*,

produce fertile seed, although the seedlings may not resemble the parent.

Young foliage is particularly susceptible to slug and snail damage, but daylilies are also prone to attack by the gall midge indicated by the presence of small white maggots in the flower bud, the daylily aphid which leaves white flecks on the foliage in spring. Daylilies can be affected by 'spring sickness', which rots the new inner leaves. This is more likely to be a problem in temperate climates and will prevent flowering for a least a year. It is seldom a problem where continental winters are the norm. Crown rot is also a problem in temperate regions. It is encouraged by the ground's heaving at times of alternate freezing and thawing. It can be prevented by thick winter mulching.

Daylilies hybridize readily and some 20,000 cultivars are now registered with the International Registration Authority; this figure is being increased by some 400–800 each year. The American Hemerocallis Society has categorized daylily cultivars in several useful ways. Plants are classed as evergreen, semi-evergreen or dormant. Evergreen sorts are best suited to zones 6–9; many will not thrive in cold climates, especially if there is no snow cover. Dormant sorts are suited to zones 2–6 and may not perform well in warmer climates. Semi-evergreen sorts are best suited to temperate climates but some can do well in hotter or colder regions. Some evergreens will acclimatize to colder conditions as they mature. It is a matter of trial and error. Leaves vary from graceful and grasslike (*H.minor*) to coarse and strap-like (*H.fulva*) and range from 30cm (*H.*'Eenie Weenie') to 1.5m (*H.altissima*). Mature foliage colour ranges from blue-green to mid- or dark green, sometimes yellowish, although excessive yellowness can denote an unhealthy plant. A few cultivars have leaves with unstable longitudinal variegation. Daylilies are further categorized according to height of flower stems which, for garden value, should be proportional to height of leaf-mound. Dwarf: below 30cm; low: 30–60cm; medium: 60–90cm; tall: over 1m. Daylilies are also categorized by flower size expressed as a diameter. Miniature: less than 7.5cm; small-flowered: 7.5–11.5cm; large-flowered: over 11.5cm. They are further classified by flower colour and pattern as selfs, blends, polychrome, bi-tone, reverse bi-tone, bicolour, reverse bicolour, banded, eye-zoned, edged, haloed, tipped or watermarked. Throat and eye colour can be an attractive contrast. The perianth consists of six segments arranged in threes, the outer, narrower sepals lying below the petals. The midribs of the segments may differ in colour from the segment; the stamens may also differ in colour from the segments giving added garden value. Flowers of heaviest substance are found in tetraploid cultivars, where the colours are usually brighter and the blooms more prolific. Some of the species' characteristic gracefulness can be lost in tetraploid cultivars. Bloom texture varies from smooth, satiny to velvety and creped, and may have an overlay known as diamond dusting. The edges may be ruffled. Natural flower form is trumpet-shaped but the flowers of modern hybrids may be circular, triangular, star-shaped, flat or spider-like. Those with more than six segments are known as doubles or camellia forms. Flower colour in the species ranges from lemon-chrome to tawny-orange. Hybridizers have raised cultivars in almost every colour except true blue and pure white, although some daylilies appear virtually white if exposed to strong sunlight. All daylilies have a yellow undertone to a greater or lesser degree and the colour range usually encountered runs from pale citron yellow to glowing orange and gold overlaid with darkest blood red. Heavy rain can cause the pigment to scatter in some rich red cultivars.

Most species are diurnal. Nocturnal species (e.g. *H.citrina*) open their flowers in late afternoon and remain open all night. *H.lilio-asphodelus* is an extended bloomer, the flower sometimes remaining open up to 16 hours. Both diurnal and nocturnal types can also have this characteristic. *H.lilio-asphodelus* and *H.citrina* are very strongly fragrant, a characteristic inherited by some of their offspring. The season of blooming is categorized as extra early, early, early-midseason, midseason, late-midseason, late and very late. Some are remontant, either flowering early and repeating in autumn, or flowering over a quick succession of bloom periods. Repeat blooming is often influenced by growing conditions.

H.citrina (China; flowering stems erect, branched above, taller than foliage; flowers 20–65, 9–12cm, fragrant, pale lemon, tinged brown beneath, opening at night, perianth tube to 4cm, tepals narrow); *H.fulva* (origin uncertain, perhaps Japan or China; flowering stems taller than foliage, forked; flowers 10–20, 7–10cm, widely funnelform, rusty orange-red, usually with darker median zones and stripes, perianth tube 2–5cm); *H.lilio-asphodelus* (syn. *H.flava*; flowering stems branched above, weak, ascending, taller than foliage; flowers 8–12, 7–8cm, fragrant, shortly funnelform, yellow, perianth tube to 2.5cm); *H.minor* (Japan, China; flowering stems taller than foliage, forked or shortly branched above; flowers 2–5, 5–7cm, shortly funnelform, lemon yellow, tinged brown outside, tepals narrow, perianth tube less than 2cm).

hemicryptophyte a plant whose latent buds are located at or just above the soil surface, and which is shallow rooted and poorly anchored.

Hemigraphis (from Greek *hemi*, half, and *graphis*, brush – a reference to the hairy outer filaments). Acanthaceae. Tropical Asia. Some 90 species, annual or perennial herbs and subshrubs, usually with toothed or crenate leaves. Slender and 5-lobed, the tubular, usually white flowers appear intermittently in terminal spikes with conspicuous bracts. Suited to hanging baskets, terraria and as groundcover in the tropical greenhouse or in gardens in the humid tropics and subtropics. Grow in bright filtered or indirect light. Otherwise cultivate as for *Eranthemum*.

H.alternata (syn. *H.colorata*; RED IVY, RED FLAME IVY; India, Java; evergreen, downy, prostrate perennial herb; leaves to 9cm, cordate to ovate, crenate, bullate, silver-grey above, flushed purple beneath); *H.repanda* (Malaysia; evergreen, prostrate, perennial herb, more or less hairless, flushed red or maroon; leaves to 5cm, linear to lanceolate, bluntly toothed to lobed, satiny leaden-grey, deeply flushed red or purple).

hen and chickens see *malformations*.

Hepatica (from Greek *hepar*, liver: the leaves were thought to resemble the liver in outline and colour and thus, by the doctrine of signatures, to cure liver disease). Ranunculaceae. Northern Temperate Regions. Some 10 species, small perennials. The leaves are basal, 3–5-lobed, entire to crenately toothed and often dull grey-green flushed purple, particularly beneath and during dormant season, when they usually persist. Solitary, scapose flowers appear before the new growth in spring. A calyx-like involucre of three reduced leaves subtends a bowl-shaped flower of 5–12 petaloid sepals. Fully hardy. Grow in cool, moist, but well-drained humus-rich soil in part shade. Apply a mulch of organic matter in autumn or late spring; annual dressings of slow-release fertilizer will improve flowering. Propagate by division in late spring or autumn. Fresh seed sown in early summer will germinate the following spring.

Hepatica nobilis

H. × media (*H.transsilvanica* × *H.nobilis*; intermediate between parents; includes 'Ballardii', with deep blue flowers and many sepals, and 'Millstream Merlin', with semi-double, deep blue flowers); *H.nobilis* (syn. *H.triloba*; Europe; leaf lobes 3, ovate, tinged purple and hairy beneath, petioles 5–15cm; flowers 1.5–2.5cm in diameter, blue-purple to white or pink; includes 'Ada Scott', with double, indigo flowers, and 'Alba', with white flowers, 'Barlowii', with very rounded, sky blue flowers, 'Caerulea', with blue flowers, 'Little Abington', with double, deep blue flowers, 'Marmorata', with leaves spotted white, 'Plena', with double, blue flowers, 'Rosea', with pink flowers, 'Rubra', with red-pink flowers, and 'Rubra Plena', with double, red flowers); *H.transsilvanica* (syn. *H.angulosa*; Romania; similar to *H.nobilis* but leaf lobes crenately toothed, flowers 2.5–4.5cm in diameter, pale blue or opal-white, rarely tinted pink).

herb (1) a plant grown for its medicinal or flavouring properties, or for its scented foliage (see *herbs*). The term *pot herb* refers to those grown for culinary puposes (2) strictly, a plant without persistent stems above ground, often confined to perennials with annual stems or leaves arising from a persistent subterranean stem or rootstock. More generally, any non-woody plant.

herbaceous (1) describing a plant without persistent stems above ground; (2) pertaining to soft green leaves.

herbaceous border a garden planting primarily devoted to the cultivation of herbaceous perennials. It was first introduced in the early 1800s and significantly developed around the end of the century under the influence of Gertrude Jekyll and William Robinson, when it became an essential feature of British gardens. Today, herbaceous perennials are often grown with annuals and bedding plants, bulbs and shrubs, in what is correctly referred to as a mixed border.

herbal originally, a book describing plants in general; nowadays referring to one that covers only plants grown for medicinal or culinary purposes.

herbarium a collection of preserved plant specimens used for identification and classification, and typically maintained in a botanic garden. Originally, a herbarium was a small garden or lawn; it later became synonymous with an arbour.

herber see *arbour.*

Herbertia (for Dr William Herbert (1778–1847), a dean of Manchester and expert on bulbous plants). Iridaceae. Temperate S America. 6 species of perennial cormous herbs, allied to *Tigridia* and bearing 1–2 short-lived flowers, each composed of 6 segments, the 3 outer broadly triangular to ovate and spreading or reflexed, and the 3 inner segments, smaller rounded and erect. In zones that are frost-free or almost so, they are suited to the warm and well-drained sunny border; where temperatures only occasionally fall to about –5°C/23°F, they can be grown outside with a deep protective mulch of bracken litter. Otherwise, cultivate as for *Homeria.*

H.amoena (Argentina, Uruguay; scapes to 30cm, flowers violet, outer perianth segments to 2cm); *H.lahue* (S Chile, Argentina; scape 8–15cm; outer perianth segments 1–1.5cm, violet stained blue near base, inner perianth segments violet); *H.pulchella* (southern S America to S Brazil; scape 8–15cm; outer perianth segments 2.5–2.8cm, blue to lilac or purple tinted pink, often with a central white stripe, claw white and bearded, flecked or flushed violet, inner segments mauve).

herbicide a chemical substance that destroys plants. Herbicides are used especially to control those plants regarded as weeds and are usually applied in spray or granular form.

herbs plants grown primarily for their culinary or medicinal properties or aromatic foliage, and often planted in combination for ornamental effect.

A herb garden can be an attractive garden feature, with simple designs being the most effective, for example, those based upon circles, squares or rectangles surrounding a central feature, such as a statue, sundial, fountain or beehive. Cultivation in raised beds can be both practical and decorative, and for further labour-saving, herbs may be grown successfully in gravel. When planted in a border in groups of three or five individual plants herbs make an effective contribution of contrasting colour, texture and leaf shape. Herbs are ideal for patios, balconies, window boxes, troughs and other containers; those

especially suitable for containers include basil (*Ocimum basilicum*), borage (*Borago officinalis*), chamomile (*Chamaemelum nobile*), lemon balm (*Melissa officinalis*), rosemary (*Rosmarinus officinalis*), rue (*Ruta graveolens*), sage (*Salvia officinalis*), great burnet (*Sanguisorba officinalis*), tarragon (*Artemisia dracunculus*), parsley (*Petroselinum crispum*), mint (*Mentha* species), thyme (*Thymus* species), chives (*Allium schoenoprasum*) and marjoram (*Origanum* species). Herbs often produce a better essential oil when grown in poor or dry soil, and it is best to mulch annually with garden compost, rather than to apply manures or fertilizers.

Herbs must be harvested when the oils are at their most concentrated, usually just as the flowers reach maturity. Dry, clean material should be selected. Plants with roots containing essential oils should be harvested in early autumn, then rubbed or washed free of soil and dried off. Examples include horse-radish (*Armoracia rusticana*), liquorice (*Glycyrrhiza glabra*), and orris (*Iris florentina*). For seed crops, such as anise (*Pimpinella anisum*), caraway (*Carum carvi*) and coriander (*Coriandrum sativum*), enfold the seed heads in paper bags and suspend these in a dry, warm, shaded place for at least three weeks to facilitate shedding.

Drying may be carried out in a constantly warm airing cupboard, with the herbs spread out on trays to allow air to circulate and turned occasionally. Alternatively, the herbs can be dried on racks in an oven at a temperature of not more than 32°C/90°F; for really fast drying, they may be heated in small quantities in a microwave oven for about one-and-a-half minutes. Some herbs, such as parsley (*Petroselinum crispum*), dill (*Anethum graveolens*) and fennel (*Foeniculum vulgare*), do not dry well and are better frozen without blanching soon after picking.

Hermannia (for Paul Hermann (d. 1695), German botanist and professor of Botany at Leiden). Sterculiaceae. Tropical and South Africa. HONEY-BELLS. Over 100 species, herbs and subshrubs, generally hairy. Solitary or in cymes, the flowers are campanulate to rounded and consist of five obovate to oblong petals, their bases often narrowed into a claw and more or less spirally twisted. Suited to sunny banks and rock gardens in essentially frost-free Mediterranean-type climates. In cooler zones, grow in the cool to intermediate greenhouse. Plant in sun, with some shade during the hottest parts of the day, in any well-drained soil. Under glass, water moderately when in growth, less at other times, and maintain a minimum winter temperature of 7–10°C/45–50°F. Repot and prune leggy specimens before growth resumes in late winter/early spring. Propagate by seed or softwood cuttings in spring. *H.incana* (syn. *H.candicans*; South Africa; grey-hairy perennial herb to 2m; flowers to 2cm diam., honey-scented, yellow, petals broadly rounded or truncate, hairy).

hermaphroditic bisexual, having both pistils and stamens in the same flower.

Hermodactylus Iridaceae. S Europe, from France through Balkans, Greece, Turkey, and Israel. 1 species, *H.tuberosus*, (syn. *Iris tuberosa*), MOURNING WIDOW IRIS, a winter- to spring-flowering tuberous-rooted perennial herb. It grows to 30cm and resembles a large Reticulata iris. The flowers are scented, to 5cm diam. and pale olive to apple green with velvety brown-black falls. It is undemanding in cultivation, well-suited to dry chalky soils, and will form large clumps where conditions suit. Although hardy to −15°C/5°F, it is sometimes given cold greenhouse protection in regions where early spring weather may spoil the flowers. Grow in sun in any well-drained soil. The flowers last fairly well when cut and kept in a cool room. Propagate by division in late summer.

Hesperaloe (from Greek *hesperos*, western, and *Aloe*, from the plant's habitat and appearance). Agavaceae. SW US, Mexico. 3 species, evergreen perennial herbs, forming grassy clumps of linear leaves with white marginal fibres. Short-lived, the narrowly campanulate flowers are carried in sparsely branched racemes in summer and autumn. Cultivate as for *Nolina*. *H.parviflora* (syn. *H.yuccifolia*; SW Texas; leaves to 1.3m, leathery, margins with fine white threads; panicle slender, to 1.2m; flowers dark to light red, golden yellow within, not opening fully; 'Rubra': flowers bright red).

Hesperantha (from Greek *hespera*, evening, and *anthos*, flower). Iridaceae. Subsaharan Africa, with most species in South Africa. 55 species, cormous perennial herbs with linear leaves and spikes of starry to saucer-shaped flowers opening after dusk in spring and summer. Cultivate as for *Ixia*. *H.cucullata* (syn. *H.buhrii*; South Africa; to 30cm; flowers scented, to 4cm diam., white with outer tepals stained red).

hesperidium a berry, pulpy within with a tough rind, for example the fruits of *Citrus*.

Hesperis (from Greek *hespera*, evening, when the flowers are most fragrant). Cruciferae. Europe, N and W Asia. 60 species, biennial or perennial, erect herbs with lax racemes of 4-petalled flowers in late spring and summer – these are especially fragrant in the evening. A plant for the herbaceous border and wild garden. It tends to become woody, tired and less floriferous with age: new stock should be propagated every 2–3 years. Double-flowered cultivars are especially good for cutting. Hardy to at least −15°C/5°F. Plant in moist but well-drained neutral to alkaline soils, in full sun or light shade. Propagate cultivars from basal cuttings in spring. Sow seed *in situ* in autumn or spring; the white-flowered cultivars come approximately true, some individuals being tinted pale pink. *H.matronalis* (DAMASK VIOLET, DAME'S VIOLET, SWEET ROCKET; S Europe to Siberia; biennial or perennial, 60–100cm; flowers scented, especially at evening, petals 1–2cm, white to lilac; includes cultivars with single and double flowers in shades of white, pink, rose, purple and lilac).

Hesperocallis (from Greek *hesperos*, western, and *kallos*, beauty). Liliaceae (Agavaceae). SW US (S California, W Arizona). 1 species, *H.undulata*, the DESERT LILY, a bulbous perennial herb to 30cm tall, with wavy-edged linear leaves and terminal racemes of fragrant, funnelform flowers, to 7cm long and white striped with green. Native to desert and semi-desert, where the bulbs are often found at great depth, *Hesperocallis* is especially useful for dry, frost-free gardens. In cool temperate zones, it may be grown in the cool greenhouse or conservatory. Plant in a sandy and gritty medium. Position in full sun with a dry, buoyant atmosphere. Water moderately when growth; keep dry when dormant. Propagate by seed and offsets.

hessian a strong, coarse cloth, also known as burlap, used in gardens to insulate frames or cloches from frost, to form windbreaks and to wrap root balls after lifting trees or shrubs.

Heterocentron (from Greek *heteros*, different or variable, and *kentron*, spur). Melastomataceae. C and S America. Some 27 species, tender perennial herbs and low-growing shrubs with 4-petalled, bowl-shaped flowers borne singly or in small panicles from late summer to winter. In frost-free regions, grow in sun in any fertile and well-drained soil. Under glass, admit full sun and use a medium-fertility loam-based mix. Water plentifully and feed weekly with dilute liquid fertilizer when in full growth; maintain a winter minimum temperature of 7°C/45°F. Propagate by seed, by softwood or stem tip cuttings and by division. *H.elegans* (SPANISH SHAWL; Mexico, Guatemala, Honduras; carpet-forming subshrub; flowers solitary; petals to 2.5cm, magenta to mauve).

heteroecious requiring two unrelated plant hosts to complete a life cycle, as in many rust fungi; cf. *autoecious*.

heterogamous (1) bearing two types of flower, as in the ray and disc florets of Compositae; (2) with sexual organs abnormally arranged or developed, or with the sexual function transferred from flowers of one sex to another.

Heteromeles (from Greek *heteros*, different, and *mela*, the apple tree). Rosaceae. US (California). CHRISTMAS BERRY, TOLLON. 1 species, *H.arbutifolia* (syn. *Photinia arbutifolia*), an evergreen, tree-like shrub to 9m tall. To 10cm long, the leathery leaves are lanceolate to obovate, thick-textured and sharply toothed. Small and 5-petalled, the flowers are white and borne in large, flattened panicles in summer. They give rise to bright-red, berry-like fruit. Its general requirements are as for *Photinia*, but with a warm and sheltered site if grown in zones 8 and lower.

heteromorphic describing a plant that assumes different forms or shapes at different stages in its life history cf. *polymorphic*.

Hermodactylus tuberosus

heterophyllous used of a plant bearing two or more different forms of leaf on the same plant.

heterosis the improvement in vigour, fertility or yield potential exhibited by F_1 hybrids. It is also known as hybrid vigour.

heterosporous producing spores of two sexes, which develop into male and female gametophytes.

heterostylous of species whose flowers differ in the existence or number of styles.

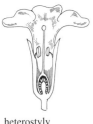

heterostyly condition in which the styles and stamens of flowers are of different lengths in individual plants of the same species. Heterostyly is a natural adaptation to ensure cross-pollination, found, for example, in species of *Primula* and *Linum*. In dimorphic heterostyly, there are two types of flower: 'thrum-eyed', where the style is short and the stamens long, and 'pin-eyed', where the opposite condition exists. In some plants, such as *Oxalis* and *Lythrum salicaria*, trimorphic heterostyly exists, with three variations of style and stamen length.

heterostyly
(pin-eyed)

heterozygous describing a plant that bears both dominant and recessive genetic components for a particular character, giving rise to a hybrid. It is used of cross-pollinated species or cultivars; cf. *homozygous*.

Heuchera (for Johann Heinrich Heucher (1677–1747), a professor of medicine at Wittenberg). Saxifragaceae. N America. ALUM ROOT, CORAL BELLS. Some 55 species, evergreen perennial herbs with a semi-woody rootstock and basal clumps of slender-stalked rounded to heart-shaped leaves. Small, bell-shaped flowers are produced in spire-like panicles or racemes in spring and summer. Evergreen ground-cover for the woodland garden and damp, sunny, border edges, valued for foliage and graceful flowering spikes in subtle shades of coral pink and green. *H.* 'Palace Purple' is one of the most striking small herbaceous perennials: its dark maroon foliage and delicate white flowers commend it as a contrast plant for mixed borders, preferably on slightly damp soils; it has also been used in bedding schemes, propagated from seed in the year prior to display. A new generation of smaller cultivars offers still more exciting foliage colour – for example, *H.*'Stormy Seas', with cloudy pewter leaves, shot with silver and ruby, or the satiny, liver red *H.*'Cascade Dawn', a marvellous companion for *Millium effusum* 'Aureum'. Plant in well-drained but moisture-retentive soil, in sun or part shade, with the crowns just above soil level. Mulch well against soil heave in areas with cold winters. Reduce and replant or mulch if the crowns lift in subsequent years. Propagate in autumn by division or from seed in spring.

heterostyly
(thrum-eyed)

H. × *brizoides* (a range of hybrids with *H.sanguinea* as part of the parentage; the precise application of this name is obscure, but it is also likely to embrace hybrids involving *H.micrantha* and *H.americana*; includes Bressingham Hybrids, with graceful inflorescences and flowers ranging in colour from pink through rose, carmine and red with some whites, 'Green Ivory', with a robust inflorescence and green and white flowers, 'Scintillation', with silver marbled leaves and flowers tipped vivid pink, 'Snowflake', to 60cm with white flowers in a lax panicle, and 'Taff's Joy', with leaves variegated cream and tinted pink); *H.cylindrica* (flowers cream, tinged green or red; includes 'Greenfinch', vigorous with sulphur-green flowers, 'Hyperion', with coral pink flowers on 6cm spikes, 'Siskiyou Mountains', dwarf with scalloped leaves and cream flowers in a 10cm raceme); *H.micrantha* (stem and petioles white-pilose at base, leaves shallowly lobed; often assigned to this species but likely to be of hybrid origin is the popular *H.* 'Palace Purple', all parts of which are deep glossy purple-bronze except for its small white flowers); *H.sanguinea* (New Mexico, Arizona; to 60cm; leaves 2-5.5cm, broadly reniform to ovate, near pentagonal in outline, glandular, veins hairy beneath; flowers bright red, 6–12mm; includes 'Alba', with white flowers, 'Maxima', with burgundy flowers, 'Splendens', with rich red flowers, 'Variegata', with variegated leaves, and 'Virginalis', with white flowers).

× **Heucherella** (*Heuchera* × *Tiarella.*) Saxifragaceae. Garden origin. Evergreen, perennial herbs to 45cm tall with rounded leaves to 13cm in diameter, the margins shallowly lobed and toothed, the surfaces bristly-hairy and light green (mottled brown when young, becoming dark green with age and turning bronze in autumn). In spring and summer, small pink flowers are carried in loose, slender panicles to 40cm tall. Cultivate as for *Heuchera*. Propagate by division.

× *H.alba.* (*Heuchera* × *brizoides* × *Tiarella wherryi*; close to × *H.tiarelloides*, but lacking stolons and normally white-flowered; includes 'Bridget Bloom', with shell pink flowers in dense spikes); × *H.tiarelloides* (*Heuchera* × *brizoides* × *Tiarella cordifolia*; stoloniferous; leaves cordate to suborbicular, coarsely toothed and lobed, crenate; scapes tinted brown, flowers pink).

× **Hibanobambusa** (Bamboo of Mount Hiba). Gramineae. Japan. 1 species. See *Bamboos*.

Hibbertia (for George Hibbert (d. 1838), a patron of botany). Dilleniaceae. Madagascar, Australasia, Polynesia. BUTTON FLOWER, GUINEA GOLD VINE. Some 125 species, evergreen shrubs, many of them climbers. Chiefly produced in summer, the flowers are large, solitary and terminal, consisting of five, broad and spreading petals and numerous stamens. Given the shelter of a warm south- or southwest-facing wall, the following may tolerate very light and short-lived frosts. In cooler regions, grow in the cool greenhouse or conservatory. Plant in sun or part shade. Provide support. Under glass, admit bright filtered light, ensuring shade from the hottest summer sun. Water plentifully and feed fortnightly when in full growth, reducing water as light levels and temperatures fall. Maintain a minimum winter temperature of 10°C/50°F. Prune to thin out congested growth in early spring. Propagate by semi-ripe cuttings in a sandy propagating mix, or by layers. *H.scandens* (SNAKE VINE, GOLD GUINEA PLANT; Australia; shrub to 12m, procumbent or twining; leaves 5–10cm, obovate to lanceolate, silky-hairy; flowers to 5cm diameter, malodorous, petals golden yellow, obovate to triangular, with ragged tips).

hibernaculum the winter-resting body of some plants; usually a bud-like arrangement of reduced leaves.

Hibiscus (Greek name for mallow). Malvaceae. Warm temperate, subtropical and tropical regions. MALLOW, ROSE MALLOW, GIANT MALLOW. About 220 species, annual or perennial herbs, shrubs, subshrubs and trees. Bowl- to funnel-shaped, the showy flowers consist of five, more or less obovate petals with, at their centre, a long, and protruding staminal column. The hardiest here include *H.sinosyriacus*, tolerant of temperatures to –l5°C/5°F, and *H.syriacus*, with many exceptional cultivars, most tolerating cold to –20°C/–4°F. *H.mutabilis* is frost tender, but will regenerate from the woody base if cut back by light frost. In subtropical and warm temperate, essentially frost-free zones, *H.rosa-sinensis* is used for hedging and screening and, as with *H.schizopetalus*, makes a fine specimen for large containers. Species from moist habitats, such as *H.coccineus* are suited to waterside plantings. *H.moscheutos* is a vigorous perennial for similar situations where winter temperatures do not fall much below –5°C/23°F. Several are grown as frost-tender annuals in cool-temperate regions, including *H.acetosella* and *H.sabdariffa*. *H.trionum* has naturalized as a weed in parts of North America. The F$_1$ hybrids *H.* Southern Belle Group and 'Dixie Belle' are treated as annuals but may survive several degrees of frost, especially if protected with bracken litter.

Grow hardy species in a well-drained, humus-rich, fertile soil in full sun, with mulch protection at the roots and the shelter of a warm south-facing wall in zones at the limits of hardiness. Little regular pruning is necessary; long shoots may be tipped back immediately after flowering and overgrown specimens may be cut hard back and thinned out in spring. Grow tender species in a medium- to high-fertility loam-based mix or soilless medium. Water plentifully when in growth and, unless high winter temperatures of about 16°C/60°F can be maintained, reduce water in winter to keep almost dry until growth resumes in spring. Maintain a minimum temperature of 7–10°C/45-50°F for *H.rosa-sinensis*, *H.mutabilis* and *H.coccineus*; tropical species such as *H.schizopetalus* require higher temperatures 16°C/60°F. Ventilate at 20°C/70°F and above. Tip prune young plants to encourage bushiness, and cut hard back in spring.

Propagate annuals from seed sown under glass in early spring or *in situ* in late spring. Propagate shrubs by greenwood or semi-ripe cuttings in a shaded case with bottom heat, or by simple layering in midsummer/early autumn. Increase perennials by division.

H.acetosella (E and C Africa; annual or woody-based perennial herb to 1.5m, usually smooth, tinted red; leaves unlobed or 3- or 5-lobed; flowers solitary, petals purple-red or yellow, deep purple at base; includes 'Red Shield' with brilliant maroon leaves); *H. × archeri*. (*H.rosa-sinensis × H.schizopetalus*; similar to *H.rosa-sinensis,* but leaves more coarsely toothed, petals red, laciniate or crenate); *H.cannabinus* (INDIAN HEMP, BIMIL, DECCAN HEMP; origin unknown, probably East Indies; annual or short-lived woody-based perennial to 3.5m; leaves 3-, 5-, or 7-lobed; flowers axillary, forming racemes, petals 4–8cm, pale yellow or less commonly pale purple, with red-purple spot at the base); *H.coccineus* (SE US; tall, hemp-like, woody based, perennial herb to 3m; leaves glaucous, palmately lobed; flowers solitary, petals to 8cm, deep red); *H.moscheutos* (COMMON ROSE MALLOW, SWAMP ROSE MALLOW; S US; robust, woody-based perennial herb to 2.5m; leaves 8–22cm, broadly ovate to lanceolate, unlobed or 3- or 5-lobed, downy; flowers solitary, petals 8–10cm, white, pink, rose, base sometimes banded   crimson, or petals wholly or edged crimson); *H.mutabilis* (COTTON ROSE, CONFEDERATE ROSE MALLOW; China; shrub to 3m or small tree to 5m; leaves 8–17cm, palmately 3-, 5- or 7-lobed, hairy beneath; flowers solitary or in clusters, petals 5–7cm, white, pink or scarlet, with darker base); *H.rosa-sinenis* (CHINESE HIBISCUS, HAWAIIAN HIBISCUS, ROSE-OF-CHINA, CHINA ROSE, SHOE BLACK; probably Asian tropics; shrub to 2.5m or tree to 5m, sparsely puberulent to glabrate; leaves to 15cm, ovate to broadly lanceolate, toothed; flowers solitary, petals 6–12cm, very variable in colour, but commonly red to deep red, darker towards the base. Cultivars selected for foliage include: 'Cooperi', leaves narrowly lanceolate, olive green marbled red, pink and white, flowers rose; 'Lateritia Variegata', leaves pointed and irregularly lobed, heavily variegated off-white, flowers gold; 'Snow Queen', leaves broadly ovate, marbled white, and grey-green. Those selected for flower colour and form include 'Aurora', flowers, pom-pom shape, blushing pink; 'Bridal Veil', flowers large, single, crêpe-textured, pure white; 'Crown of Bohemia', flowers fully double, gold with flaming orange throat; 'Fiesta', flowers large, single, crinkled, deep apricot orange with pink-red eye; 'Kissed', petals reflexed, vivid red; 'Percy Lancaster', flowers single, petals narrow, palest pink washed apricot, eye russet; 'Ruby Brown', flowers single, large, brown tinted orange with dark red throat; 'Sunny Delight', flowers single, large, brilliant yellow, throat white); *H.sabdariffa* (ROSELLE, JAMAICA SORREL, RED SORREL, SORREL; Old World tropics, widely cultivated; annual or robust, woody-based perennial herb or subshrub to 2.5m; leaves 8-15cm, ovate, undivided or palmately 3- or 5-divided; flowers solitary or in short leafy raceme; petals 4–5cm, light yellow, purple-red at base); *H.schizopetalus* (JAPANESE HIBISCUS, JAPANESE LANTERN; Kenya, Tanzania and N

Hibiscus syriacus 'Blue Bird'

Mozambique; shrub to 3m; leaves to 12cm, ovate, serrate; flowers on long slender pedicels; petals to 7cm, deeply and irregularly laciniate, pink or red; staminal column long-exserted; *H.sinosyriacus* (C China; similar to *H.syriacus*, but leaves large; includes 'Autumn Surprise', flowers white, base feathered cherry, 'Lilac Queen', flowers white, lightly tinted mauve, base clear burgundy, profuse, 'Red Centre', stem with silver sheen when mature, flowers white, centre red, 'Ruby Glow', flowers white, base clear burgundy); *H.syriacus* (Old World, widely cultivated; shrub or small tree to 3m; leaves 3–7cm, 3-lobed, coarsely toothed; flowers solitary or paired, petals 3.5–7cm, white, red-purple or blue-lavender, with crimrose base. Cultivars selected for foliage include 'Meehanii': habit low, leaves edged yellow, flowers single, lavender; 'Purpureus Variegatus': leaves variegated white. Those selected for flower colour and form include 'Admiral Dewey': flowers double, wide, snow white, abundant; 'Ardens', habit broad, flowers densely double, purple tinted blue; 'Blue Bird': habit erect, flowers wide, very large, sky blue with small red eye; 'Coelestis': pale violet, rose at base, single; 'Diana': habit upright, leaves trilobed, flowers single, white, very wide, crinkled edges, profuse; 'Duc de Brabant': deep rosy purple, double; 'Hamabo': flowers single, light pink, crimson markings and eye; 'Hélène': flowers single, large, white with dark red edge and surrounding veins, abundant; 'Jeanne d'Arc': white, semi-double; 'Lady Stanley': white suffused shell pink, maroon at base, semi-double; 'Lucy': flowers double red; 'Monstrosus': white with maroon eye, single; 'Snowdrift': white, large, single; 'Violet Clair Double': vinous purple, red-purple at centre, double; 'William R. Smith': pure white, single, very large; 'Woodbridge': flowers single, large, rich pink to carmine at centre); *H.trionum* (FLOWER OF-AN-HOUR, BLADDER KETMIA: arid Old World tropics; erect or ascending hispid annual or short-lived perennial herb to 1.2m; leaves to 7cm, deeply 3- or 5-lobed or -parted, incised or entire; flowers solitary, calyx with purple veins, becoming inflated, petals to 4cm, white, cream or yellow, with red base).

Hidalgoa (for Miguel Hidalgo, an eminent Mexican of the early 19th century). Compositae. Mexico, C America. CLIMBING DAHLIA. 5 species, semi-woody perennial climbers. The leaves are usually pinnately divided and toothed. Solitary and brightly coloured, the daisy-like flowerheads appear in summer. Cultivate as for *Delairea*.

H.ternata (Mexico; to 50cm; flowerheads to 5cm diameter, ray florets orange); *H.wercklei* (Costa Rica; to 1m+; flowerheads to 6cm diameter, ray florets, scarlet above, yellow beneath) .

Hieracium (from the classical Greek name *hierakion* used for a range of yellow-flowered Compositae). Compositae. Europe, to N and W Asia, NW Africa and N America HAWKWEED. Some 10,000 species, perennial, more or less hairy herbs, with milky sap. The leaves are arranged in basal rosettes or

on the flowering stems. Yellow, daisy-like flowerheads are carried in long-stalked and slender-branched panicles in spring and summer. The following are hardy in zone 6 and suitable for the large rock garden or the foreground of well-drained borders. Easily grown in low-fertility, gritty neutral or alkaline soils in sun. Propagate by seed or division.

H.lanatum (SE France, W Switzerland, NW Italy; 10-15cm; leaves covered with dense, white, woolly hairs , to 10cm, elliptic to lanceolate); *H.maculatum*. (W and C Europe, sporadically elsewhere; 20-80cm; leaves spotted or blotched brown-purple, to 4.5cm, ovate to lanceolate).

hilum the scar on a seed at the point at which the funicle was attached.

hip, hep the fleshy developed floral cup, with enclosed achenes, that forms the fruit of a rose.

Hippeastrum (from Greek *hippeus*, a rider, and *astron*, a star; reference unclear). Amaryllidaceae. Americas. AMARYLLIS; KNIGHT'S STAR LILY. About 80 species, bulbous perennial herbs with strap-shaped leaves and flowers atop a hollow scape. These consist of a tubular to funnelform perianth tube often, dilating at the throat, and six, spreading lobes, the inner 3 sometimes narrower than the others.

Grow under glass in bright filtered light, in a free-draining, medium-fertility, loam-based mix (e.g. 3:1:1 loam, leafmould, sharp sand). Plant firmly with the neck and shoulders of the bulb above soil level, winter-flowering species and cultivars in late summer/early autumn, summer-flowering species in early spring. Maintain a minimum temperature of 13–16°C/55–60°F and water sparingly until growth commences; at this point, temperatures may be increased to 16–18°C/60–65°F for earlier flowers, although flowering is prolonged at lower temperatures. When in active growth, water plentifully, syringe the leaves with soft water, and feed weekly with a dilute liquid fertilizer. Leaves continue their development after flowering, and feeding and watering should be continued until growth ceases and leaves yellow – usually by midsummer. Reduce water gradually and withhold completely when dormant, keeping bulbs dry and warm (not below 5–7°C/40–45°F) until growth resumes. *Hippeastrum* dislikes root disturbance: repot only once every three of four years, before the new cycle of growth begins.

Propagate by offsets, and by seed sown when ripe at 16–18°C/60–65°F; if grown on without dry rest during their first year, seedlings mature and flower more quickly, and may flower in their second or third year from sowing. Red spot disease is caused by the fungus *Stagonospora curtisii* – watery red lesions occur on the bulbs, leaves and flower stalks, and red scabs form on the outer scales of the bulbs. Emerging foliage and flower buds may be sprayed with a copper-based or dithiocarbamate fungicide.

H.aulicum (LILY OF THE PALACE; Brazil, Paraguay; flowers usually 2, crimson, to 15cm, throat green,

hip

lobes obovate, pointed, the 2 upper inner ones much broader than the others); *H.papilio* (S Brazil; flowers 2, pale green streaked and stained dark red, perianth segments oblanceolate, to 14 × 5cm); *H.reginae* (MEXICAN LILY; Mexico to Peru and Brazil, West Indies, W Africa; flowers 2–4, drooping red, perianth tube to 2.5cm, lobes to 13cm, obovate, bright red with a large green-white star in the throat); *H.striatum* (Brazil; flowers 2–4, to 10cm, crimson, keeled green to halfway up the lobes); *H.vittatum* (Peruvian Andes; flowers 3–6, to 12cm diameter, perianth tube to 2.5cm, lobes obovate to oblong, to 4cm across, keel white, margins irregular and white, striped red in between). *Cultivars* 'AppleBlossom', white suffused soft pink; 'Beautiful Lady', pale mandarin red; 'Best Seller', scape short, flowers cerise; 'Bouquet', salmon; 'Byjou', soft burnt apricot; 'Cantate', milky deep red; 'Christmas Gift', white; 'Dazzler', pure white; 'Dutch Belle', opal rose; 'Ludwig's Goliath', large, bright scarlet; 'Lydia', pale salmon; 'Oskar', rich deep red; 'Orange Sovereign', pure orange; 'Picotte', white rimmed red; 'Red Lion', dark red; 'Royal Velvet', deep velvety red; 'Star of Holland', scarlet with white star at throat; 'Susan', large, soft pink; 'United Nations', white striped vermillion; 'Valentine', white with pink veins, heavier towards edges; 'White Dazzler', pure white.

hippocrepiform shaped like a horseshoe.

Hippophaë (from Greek *hippophaes*, a kind of spurge). Elaeagnaceae. Europe, Asia. SEA BUCK-THORN. 3 species, thorny, deciduous, dioecious trees or shrubs. Lanceolate to linear, the leaves are more or less covered in somewhat metallic, scale- or chaff-like hairs, as is the new-growth. Inconspicuous flowers appear before the leaves and, where male and female plants are grown together, give rise to long-lasting and brightly coloured berries. The genus is primarily used for its ornamental qualities, especially in maritime plantings, although *H.rhamnoides* is also used in shelterbelts and in soil stabilization, especially on coastal dunes, and may subsequently naturalize. Plant in full sun or part shade. Prune only to remove straggling shoots in later summer. Propagate by layering in autumn, from suckers, by semi-ripe cuttings in summer, or by hardwood cuttings in late autumn. Alternatively, sow seed fresh in autumn, or in spring following stratification at 4°C/39°F over three months. *H.rhamnoides* (W Europe to China; thorny shrub, 1–9m; twigs rigid, metallic-scaly then grey, thorny; leaves to 6cm, linear to lanceolate, silver; fruit to 6mm diameter, orange, clothing bare branchlets in winter).

hirsute with long hairs, usually rather coarse and distinct.

hirsutullous slightly hirsute.

hispid with stiff, bristly hairs; not so distinct or sharp as when *setose*.

hispidulous minutely hispid.

hoary densely covered with white or grey hairs.

hoe a hand implement used for surface cultivation and destroying weeds, typically consisting of a flat, usually oblong, metal plate, or blade, attached to a handle about 1.5m long. The two most common types are the Dutch hoe (also known as the push hoe or scuffle hoe), primarily used for surface hoeing, and the more versatile draw hoe, which is useful for cultivation, drill drawing, thinning seedlings and earthing up. The Dutch hoe has the blade set on the same plane as the handle, with the operator walking backwards to leave the cultivated ground untrodden. The draw hoe has the blade set at right angles to the handle, with the operator adopting a more stooping posture and walking forward over the cultivated ground. An onion hoe is a type of draw hoe, fixed to a handle about 30cm long, and used for cultivating amongst closely spaced plants.

hoggin a fine gritty aggregate, often with clay content, used for bedding paving slabs and providing a firm undersurface for gravel.

Hoheria (Latinized form of the Maori name for these plants). Malvaceae. New Zealand. LACEBARK. 5 species of evergreen or deciduous shrubs and small trees with thinly leathery leaves. Solitary or clustered, the flowers are produced in the leaf axils in summer. They consist of five, obovate, white, cream or ivory petals and a boss of many stamens. Given shelter from cold, drying winds, most species tolerate temperatures to –15°C/5°F. Deciduous species will regenerate freely from the base if cut back by frost. *H.populnea* is more tender and prone to damage at –5°C/23°F; in severe conditions, *H.sexstylosa* may defoliate, but usually recovers in spring. The soft growth generated on over-rich soils is more prone to frost damage and, in addition, may suffer sudden die-back for other, physiological reasons. Grow in sun or light dappled shade, on a deep, well-drained, humus-rich and moderately fertile soil. In regions at the limits of hardiness, provide the shelter of a warm south- or southwest-facing wall, and a protective mulch of organic matter in winter. Prune in spring: removing deadwood and weak or overcrowded growth. Propagate by semi-ripe cuttings or by layers. Ripe seed sown in autumn germinates freely; *H.sexstylosa* in particular self-seeds in profusion. Seed-raised specimens will pass through a juvenile stage.

H.angustifolia (evergreen tree to 10m with slender flexible branches; leaves 2–5cm, narrowly obovate, oblanceolate to lanceolate, coarsely spinulose, dentate to serrate; flowers to 2cm diameter, petals snow white, narrowly oblong, notched); *H.glabrata* (deciduous tree to 10m; leaves 5–14cm, glabrate, broadly ovate to ovate-lanceolate, acuminate, crenate to dentate; flowers to 4cm diameter; petals white to cream, obovate, anthers purple); *H.* 'Glory of Amlwich' (*H.glabrate* × *H.sexstylosa*; small tree; leaves to 9cm, ovate and slender-pointed, serrate, pale green; flowers to 4cm diam-

Hippeastrum papilio

Hoheria lyallii

eter, snow white, profuse); *H.lyallii* (LACEBARK; deciduous tree to 6m; leaves 5–10cm, cordate to ovate, grey-green, felty white-pubescent; flowers 2–3cm in diameter, snow white, petals obovate, anthers purple); *H.populnea* (LACEBARK; evergreen tree to 10m, branchlets slender, bark exfoliating in fine ash-grey strips; leaves 7–14cm, broadly ovate to elliptic, acuminate, serrate; flowers 2.5–3cm in diameter, pure white; includes 'Albavariegata', with leaves broadly edged white, 'Purpurea', with leaves flushed and veined maroon beneath, and 'Variegata', with pale yellow-green leaves edged dark green); *H.sexstylosa* (RIBBON-WOOD; evergreen tree to 6m; bark glossy red-brown, overlaid with grey exfoliating strips; leaves 5–15cm, lanceolate to ovate, glossy; flowers 2–2.5cm in diameter, pure white, fragrant, petals oblong, notched, anthers white; includes 'Crataegifolia', with small, coarsely toothed juvenile leaves).

Holboellia Lardizabalaceae. (for Frederik Ludvig Holboell (fl. 1847), a superintendent at the Copenhagen Botanic Garden) N India to China. About 5 species, evergreen, twining shrubs with palmately compound leaves and unisexual flowers in spring and summer. These are small, waxy and bell-like. They are grown for their handsome, evergreen foliage and small, delicately coloured flowers, and may be used to drape a wall or cover a tree stump, or to climb small trees, a trellis or a tripod system in a border. The cream to shell-pink *H.coriacea* is considered to be the hardiest species, but may suffer damage during prolonged spells below –5°C/23°F. Cut back weaker growths and dead wood in spring. *Holboellia* succeeds on most garden soils, either in sun or shade. Increase by seeds, cuttings and layering.

Holcus (from Greek *holkos*, a kind of cereal). Gramineae. Europe, temperate Asia, N and South Africa. 8 species, annual or perennial grasses with flimsy, tufted culms and linear leaves. The flowers are borne in spike-like, dense or open panicles in summer. Fully hardy. Grow in sun in well-drained but moisture-retentive soil. Clip over H.'Albovariegatus' after flowering. Propagate by division, and, for species, also by seed. *H.mollis* (CREEPING SOFT GRASS; Europe; perennial to 45cm; leaves to 1cm wide, grey-green, glabrous to slightly hairy; flowers in narrowly oblong to ovate, dense to loosely branched panicles to 12cm; 'Albovariegatus': with leaves broadly edged white with a narrow green stripe).

hollow heart a disordered condition of potatoes, most usually found in large tubers, where a sound cavity is apparent when the potato is cut across. It is the result of checked development, most likely due to a drought period being followed by abundant rain, which causes rupturing of the central tissue.

Holmskioldia (for Theodor Holmskiold (1732–94), Danish botanist). Verbenaceae. Tropical Africa, Asia. Some 10 species, evergreen climbing shrubs. Produced in autumn and winter in panicles or racemes, the flowers consist of a bowl- to bell-shaped calyx with a short tube and a circular, spreading rim, and a tubular, two-lipped corolla. Once established, *Holmskioldia* is highly drought-tolerant and extremely useful for plantings in poor sandy soils in tropical and subtropical gardens. In cooler climates, it needs warm greenhouse protection (minimum temperature 15°C/60°F), and treatment as for climbing *Clerodendrum*. *H.sanguinea* (CUP AND SAUCER PLANT, MANDARIN'S-HAT, CHINESE-HAT PLANT; Himalaya; calyx to 2.5cm diameter, brick red to orange, net-veined, corolla scarlet).

Holodiscus (from Greek *holos* entire, and *diskos*, disc, referring to the entire disc of the flower). Rosaceae. Western N America to Colombia. 8 species, deciduous shrubs with pinnately lobed and toothed leaves and small, creamy white flowers massed in cascading sprays in summer. Hardy to –15°C/5°F. Grow in full sun or light shade in moist, fertile soil that does not dry out in summer. Prune after flowering or in early spring, only to remove old or overcrowded branches. Increase by layering; also, by semi-ripe heeled cuttings. *H.discolor* (CREAMBUSH, OCEAN-SPRAY; western N America; to 5m; flowers in a 30cm plume-like panicle).

Homeria (from Greek *homereo*, to meet together: the filaments are united in a sheath around the style). Iridaceae. South Africa. 31 species, cormous perennial herbs with linear to sword-shaped leaves. The flowers are short-lived but produced in long succession over the summer. They consist of six perianth segments in a broadly cup-shaped arrangement. Grow in the cool greenhouse in a medium-fertility, loam-based mix with additional leafmould and sharp sand; maintain good ventilation and a minimum temperature of 5–7°C/40–45°F. Water plentifully when in growth and dry off gradually as flowers fade. Store dormant corms in a cool dry place. Propagate by seed in autumn, or by division or offsets. *H.ochroleuca* (to 80cm; flowers to 8cm diameter, pale yellow, occasionally stained orange at centre, muskily scented).

homogamous having hermaphrodite flowers, or flowers of the same sex.

Homogyne (from Greek *homos*, similar, and *gyne*, female, alluding to the similarity of the male and female florets). Compositae. Europe. 3 species evergreen, rhizomatous perennial herbs with basal, broadly cordate to rounded and toothed leaves and button- to daisy-like flowerheads. Attractive low groundcover for shaded niches in the rock garden and as path edging in the woodland garden. Grow in gritty but moisture-retentive and well-drained humus-rich soils in shade or in good north light. *H.alpina* prefers acid soils. Propagate by division or by ripe seed, sown fresh. *H.alpina* (ALPINE COLTS-FOOT; Europe; leaves 2–4cm, reniform, leathery, glabrous, petiole to 10cm, hairy; flowerheads on scapes to 40cm, phyllaries purple, florets purple-red).

homologous of organs or parts that resemble each other in form or function.

homomorphic having the same shape.

homonym the same scientific name given to two or more distinct plants. Under the rules of nomenclature, only one plant may assume the name, and this is usually the first plant to have received it.

homosporous (1) producing only one kind of seed; (2) developed from only one kind of spore.

homozygous describing a plant bearing the same genetic components at a given position on a chromosome, so that there can be no genetic segregation. It is used of self-pollinated species or cultivars; cf. *heterozygous*.

honey bee (Hymenoptera: Apidae) (*Apis mellifera*) a widely distributed social insect of benefit to gardening. Thought to be of tropical origin, it has been introduced into most countries of the world for its honey. A colony consists of three well-defined castes: a queen, worker bees (sterile females) and drones (males). In summer, large colonies may contain up to 80,000 workers and several hundred drones. Workers leave the hive to forage for water, nectar, pollen or propolis (a resinous plant exudate collected from buds and other parts of trees, used to fill unwanted gaps in the hive). Nectar is the main substance collected, and this is imbibed by the bees; pollen and propolis are collected in the pollen baskets on the hind legs.

In visiting flowers honey bees are of great value as pollinators; most fruits, vegetables and ornamental plants relying on insect pollination. To avoid harming honey bees, careful timing of pesticide application is crucial, choosing to spray during their periods of least activity.

honeydew a sugary fluid excreted by sap-feeding insects such as aphids, whiteflies, mealybugs and scale insects, which creates a sticky coating on leaves and stems. It is most markedly seen under *Tilia* species in urban areas, where paving is often covered with the sticky residue. Honeydew, which is sweet, is eagerly sought and imbibed by ants; it is often colonized by the sooty mould fungus. Any sign of stickiness on a plant indicates the possibility of attack by one of these sap-feeding pests, either of the plant or one nearby.

honey fungus see *armillaria root rot*.

hooded cucullate or, more loosely, referring to inarching parts enclosing others.

Hoodia (for Mr Hood (fl. 1830), a grower of succulent plants). Asclepiadaceae. South Africa, Namibia. Some 17 species, succulent, perennial, leafless herbs. The stems are grey-green, angled and covered in conical tubercles and hard, thorn-like teeth. Saucer- or cup-shaped with five, short and pointed lobes, the large flowers appear near the stem apex at various times of year. Cultivate as for *Stapelia. H.bainii* (Cape Province; stem to 15-angled, tubercles compressed, spiralled, spines pale brown; flowers 2–7cm in diameter, deeply cup-shaped to flat, pale yellow to buff, veins darker, with a dark red-brown corona).

hoof-and-horn a slow-release organic fertilizer made from the ground-up hooves and horns of cattle; it contains 7–15% nitrogen and 1–10% phosphate, with some calcium. It should be sterilized to eliminate pathogens, and handled with suitable precautions to avoid inhalation and other personal contact.

hops, hop manure matured spent hops from breweries, which can be used to improve soil structure although the nutrient content is low. Hop manure is made by adding fertilizers to matured hops, and the analysis varies considerably according to the manufacturer. Spent hops can be applied at any time, but hop manure is best applied in spring.

Hordeum (Latin *hordeum*, barley, a corruption of *horridus*, bearded with bristles, referring to the bristle-like, long-awned glumes and vestigial spikelets). Gramineae. Temperate northern Hemisphere and S America. BARLEY. Some 20 species of annual or perennial grasses. Grown for graceful, feathery flowerheads, which dry and dye well. *H.jubatum* is a highly ornamental barley, with delicate, long, straight and silky awns faintly tinged crimson. Pick immature heads and air dry. Sow seed *in situ* in autumn or spring, on light, well-drained, soils in full sun. *H.jubatum* (SQUIRRELTAIL BARLEY, FOX-TAIL BARLEY; N America, NE Asia; annual or perennial to 60cm; flowers in dense, finely bristly, silky, nodding, pale green to purple-tinged spikes to 12cm).

Horminum (from Greek *hormao*, I excite, referring to the use of the plant as an aphrodisiac). Labiatae. Pyrenees, Alps. PYRENEAN DEAD-NETTLE, DRAGON MOUTH. 1 species, *H.pyrenaicum*, a rhizomatous perennial herb to 45cm. Borne in basal rosettes, the dark green leaves are to 7 × 5cm, ovate to cordate and toothed. Tubular, 2-lipped and violet, the flowers nod in whorls held clear of the foliage in summer. Hardy to –20°C/–4°F and suitable for the front of the border or for cultivation in the rock garden. Grow in moderately fertile, well-drained soil in full sun. Propagate by division in spring or autumn, or from seed sown in a cold frame in winter for germination in spring.

hormone see *growth regulators*.

hornet see *wasps*.

horny hard and brittle, but with a fine texture and easily cut.

horseradish (*Armoracia rusticana*) a fleshy-rooted perennial, also known as RED COLE, which is native to temperate regions of Europe and Asia and naturalized in North America. It has been used since ancient times as a medicinal and condiment herb. The grated root is used in a traditional English sauce to accompany meat or fish, or (in small quantities) as a salad garnish; young leaves may be added to salads.

Since horseradish regenerates from its spreading roots, it is potentially a pernicious weed and in gardens its territory must be contained and all portions of root removed on lifting. Propagate by pencil-thick roots, or thongs, 15–20cm long, taken from mother-plants to overwinter; store in bundles buried in sand under cool conditions. Plant sprouted thongs in spring 5–7cm deep at 45cm spacings in rows 60cm apart. Harvest roots in autumn and store in darkness. Quality can be improved by removing the top soil when plant growth reaches 20cm high and rubbing off side roots. Replant in beds annually, or at intervals of no more than three years, to maintain quality.

horsetail *Equisetum arvense* is the COMMON HORSE-TAIL, a flowerless perennial with leafless, branching rough-textured stems. Unchecked, it can be a serious weed spreading by tuber-bearing underground stems which may penetrate to a depth of two metres. Stalked shoots bearing fruiting bodies appear in early spring, but propagation by spores is less common.

Complete eradication of dense established stands is unlikely. Repeated application of approved herbicides will at least weaken plants, and glyphosate and glyphosate trimesium are effective foliage-applied total weedkillers, suitable for use on uncropped ground. Vigilance is needed to avoid the introduction of horsetail; light infestations should be tackled early with meticulous cultivation of the ground to remove all parts of the plants. Where horsetail is established in neglected lawns, it can usually be suppressed by regular mowing.

common horsetail

hort., horti, of the garden; **hortorum** of gardens; **hortulanorum,** of gardeners; all three are abbreviated hort.; never capitalized. The last two indicate either the misapplication of a name by gardeners or denote a name of confused authorship or of no standing. The citation *horti.* usually precedes a grower's name and indicates a plant name published in a nursery catalogue, for example hort. Lemoine.

horticulture the art, practice and science of garden cultivation; in common parlance, it includes the commercial production of fruit, vegetables and ornamentals.

hortus conclusus 'enclosed garden' – a mediaeval garden within a garden, enclosed by a hedge, fence, or wall, and containing a lawn and often a fountain, roses, arbours, and turf seats.

hortus siccus 'dried garden' – a collection of dried specimens comprising a herbarium.

hose-in-hose see *malformations*.

hose reel a device for rolling up a garden watering hose. The through-feed type is the most convenient, with a special valve enabling the hose to be wound or unwound whilst the water supply remains connected.

host an organism upon which another lives and may feed.

Hosta (for Nicolaus Tomas Host (1761–1834), physician to the Austrian Emperor). Liliaceae (Hostaceae, Funkiaceae). Japan, China, Korea. PLANTAIN LILY. 40 species, clump-forming herbaceous perennials, fleshy-rooted with thick, short rhizomes and stout resting buds. Emerging in mid-spring, the leaves form a basal mound or rosette. They are broadly heart-shaped to narrowly lanceolate in outline, glossy to glaucous, thin and pliable to thick and tough, smooth to ruggedly textured. Produced in summer and autumn, the funnel-shaped flowers are ranked along one side of a slender raceme, or gathered among bracts at the top of a scape. The following are among the most valuable hardy foliage plants. Their flowers and papery, wind-swept seed heads are also beautiful. They may be grown as single specimens, massed for dense groundcover, or used in containers. All will grow in full sun given adequate moisture, but leaf quality is usually better in dappled sun or shade and especially so in the case of plants with blue-grey, lime green and golden leaves. Hostas are hardy in climate zone 5 if thickly mulched in winter. Plant in a fertile, open soil rich in leafmould and compost. Keep moist throughout the year; avoid freezing, waterlogged conditions in winter and early spring. Increase by division in spring. Martyrs to slugs and snails.

H.crispula (Japan; leaves to 25 × 15cm, ovate-lanceolate, undulate, dull olive green, deeply and irregularly edged white; flowers pale mauve); *H.decorata* (Japan; leaves to 16 × 12cm, ovate to rounded, dull dark green, leathery, narrowly edged white; flowers rich violet or white); *H.fortunei* (garden origin; leaves to 30 × 20cm, heart-shaped to rounded or broadly lanceolate, heavy-textured, mid to deep green, sometimes glaucous and rugose above; flowers mauve to pale violet; 'Albopicta': leaves unfurling pale creamy yellow with a green margin, ultimately pale green with a hint of variegation; 'Aureamarginata': leaves edged gold; 'Marginato-alba': leaves edged white); *H.hypoleuca* (Japan; leaves to 45 × 30cm, broadly ovate-cordate, somewhat leathery, pale green, glossy or glaucous above, powdery white beneath; flowers pale lavender to milky white); *H.lancifolia* (garden origin; leaves to 17 × 3cm, lanceolate, finely tapering, thin-textured, glossy, deep olive green; flowers deep mauve); *H.montana* (Japan; leaves to 30 × 18cm, broadly ovate to heart-shaped, dull or shiny mid to deep green or glaucous, veins deeply impressed; flowers grey-mauve to white; 'Aureomarginata': leaves narrower, broadly and irregularly edged gold, undulate); *H.plantaginea* (AUGUST LILY; China; leaves to 28 × 20cm, ovate-cordate, glossy olive green with sunken veins; flowers white, fragrant, opening at dusk; sometimes confused with *H.* 'Thomas Hogg'); *H.rectifolia* (Japan; leaves to 15 × 5cm, ovate to ellliptic, erect, margins somewhat inrolled, dull mid- to dark green; flowers violet); *H.sieboldiana* (Japan; leaves to 50 × 30cm, ovate to cordate or rounded, heavily textured, puckered, matt glaucous grey-blue to blue-green or olive above, paler beneath; flowers pale lilac-grey fading to white clouded with pale lilac; var. *elegans*: leaves heavily textured, strongly puckered, thickly glaucous blue-grey, flowers pale pearly lilac; 'Frances Williams': leaves yellow-edged); *H.sieboldii* (Japan, Sakhalin Is; leaves to 15 × 6cm, lanceolate to elliptic, sometimes wavy, somewhat puckered, matt mid-green above edged pure white; flowers pale mauve, striped violet and lined white); *H.tardiflora* (garden origin; leaves to 15 × 7cm, erect, lanceolate to narrowly elliptic, semi-rigid, thick-textured, glossy deep olive green above; flowers clear mauve, produced in autumn); *H.tokudama* (garden origin; leaves to 30 × 15cm, cordate to rounded, heavily textured, puckered, glaucous blue-grey, flat to concave above; flowers pale grey-mauve to off-white; 'Aureonebulosa': leaves clouded yellow at centre; 'Flavocircinalis': leaves edged bright yellow); *H.undulata* (garden origin; leaves to 15 × 6cm, ovate to lanceolate-elliptic, very wavy and twisted, glossy mid green with a creamy white centre, later breaking into white streaks; flowers pale purple; var. *erromena*: leaves wholly mid-green; var. *univittata*: leaves large, twisted with narrow, white centres); *H.ventricosa* (China; leaves to 24 × 18cm, broadly ovate to cordate, slightly wavy, thin-textured, glossy dark green; flowers deep mauve; 'Aureomaculata': leaves bright yellow in centre, fading by midsummer); *H.venusta* (Korea, Japan; leaves to 3 × 2cm, lanceolate-elliptic to ovate-cordate, dull to glossy olive to dark green; flowers pale violet). *Cultivars* There are many hosta cultivars and hybrids, ranging in size from the very large and robust to dwarf plants suitable for rock and container gardens. The leaves vary in shape from the narrowly lanceolate to almost round, in texture from thick and puckered to thin and smooth, glossy to glaucous. Leaf colour runs through frosty blue-grey to aquamarine, olive and dark jade green, including some solid golds and limes. Variegation may be marginal, central or throughout, and take the form of white, cream, gold or lime flashes, stripes, patches and clouding. There follows a selection of popular cultivars – 'August Moon' (leaves large, ovate, pale golden yellow, faintly glaucous, corrugated, flowers near-white); 'Big Daddy' (leaves very large, round, thickly textured, puckered, glaucous blue, flowers near-white); 'Blue Moon'(leaves small, heart-shaped, glaucous blue, flowers mauve-grey); 'Ginko Craig' (leaves small, low, dark green edged clear white, flowers violet); 'Gold Edger' (small, low, neat, leaves ovate-cordate, cool lime turning soft gold,

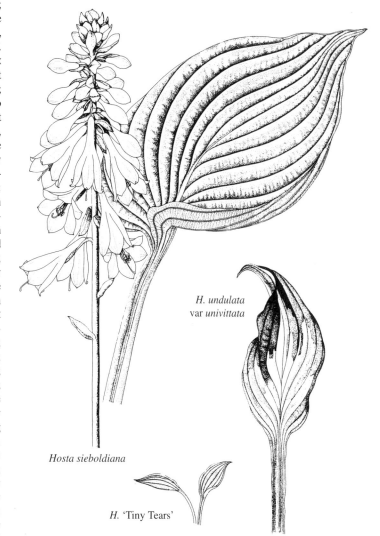

H. undulata
var *univittata*

Hosta sieboldiana

H. 'Tiny Tears'

flowers pale mauve); 'Gold Standard'(leaves large, ovate-cordate, chartreuse green turning pale yellow, edged dark green, flowers lavender); 'Halcyon' (leaves medium-sized, heart-shaped, glaucous blue-grey, flowers lavender-grey); 'Kabitan' (leaves narrow, lanceolate, glossy, thin-textured, yellow-green with a narrow, dark green, wavy margin; flowers deep mauve); 'Popo' (dwarf, leaves very small, cordate, blue-grey); 'Royal Standard' (leaves large, ovate, bright green, flowers white); 'Shining Tot' (dwarf, leaves very small, dark, glossy green, flowers lilac); 'Sum and Substance' (leaves very large, smooth, thickly textured, pale lime green to yellow, flowers pale lavender); 'Thomas Hogg' (syn. *H.undulata* 'Albomarginata', leaves ovate to elliptic, apex rounded, margin slightly wavy, mid-green edged creamy white, flowers pale purple); 'Tiny Tears' (dwarf, leaves very small, narrow, dark green, flowers violet); 'Wide Brim' (leaves large, heavily puckered, dark green with a wide, irregular cream margin, flowers mauve); 'Zounds' (leaves very large, heavy-textured and puckered, gold with a metallic sheen, flowers pale lavender).

host plant any plant that provides sustenance, or other support, for another organism, such as a plant, virus, bacterium, fungus or insect.

hotbed a specially prepared growing bed producing a warm-growing medium to advance the growth or maturity of plants and crops. A hotbed is particularly valuable for growing cucumbers and melons, advancing certain root and salad vegetables, and for rooting cuttings or providing bottom-heat for seed germination. It may be constructed in a frame or greenhouse or freestanding outdoors.

Traditionally, hot beds were based on fermenting horse manure, although tanning bark was also used, either alone or with animal manure. A common practice involved mixing equal parts fresh stable litter and fallen leaves into a well-watered heap, which was turned several times within a week. The substrate was then stacked on to a well-drained growing site, and up to 15cm of growing medium added. As a result of the relative scarcity of stable manure, hotbeds were bottom-heated through hot-water pipe systems, a method now generally superceded by electric warming cables.

hot house see *greenhouse*.

hot water treatment a pest-control treatment that involves immersing plants in warm water, exploiting the difference between the thermal death point of certain pests (especially eelworms) and plants. It is usually carried out when the plants are dormant. Temperature and immersion time are critical and vary with the subjects; a few degrees temperature variation may kill or seriously damage plants or, alternatively, leave pests unharmed. Pre-warming of some stock improves safety or efficacy. Home-improvised tanks may be used, but sensitive thermostatic control and a means of water circulation are essential for success.

STEM EELWORM (*Ditylenchus dipsaci*), BULB FLIES (*Merodon equestris* and *Eumerus* species) and the TARSONEMID BULB SCALE MITE (*Steneotarsonemus laticeps*) in narcissus can be controlled by immersion of dormant bulbs for three hours at 44–45°C/111–113°F. Dormant forms of stem eelworm, known as 'eelworm wool', may survive unless a wetting agent is added to the water. Hyacinths may be similarly treated but should be immersed for four hours, and tulips, which are easily damaged, must be dry stored at 34°C/93°F for three days, followed by a three-hour pre-soak in cold water. Iris bulbs infested with POTATO TUBER EELWORM (*Ditylenchus destructor*) require immersion for three hours at 44–45°C/111–113°F, preceded by 10–14 days' dry storage at 30°C/86°F.

Lilies are subject to attack by the LEAF EELWORM (*Aphelenchoides fragariae*) and those with white bulbs may be disinfested by treating at 41°C/106°F for two hours, following storage for 8–10 weeks at 2°C/36°F to prevent sprouting; cultivars with darker bulbs for example, *Lilium speciosum* and *L.regale* should be similarly pre-treated, followed by hot water treatment at 39°C/102°F for two hours. Ferns

and other plants, such as gloxinias and begonias, infested with the same eelworm can be treated after soil removal at 49°C/120°F for five minutes. Chrysanthemum stools, trimmed and washed clean of soil, can be disinfested by immersion for five minutes at 46°C/115°F or 20–30 minutes at 43.5°C/110°F. Phlox can be prepared in a similar way and treated at 43.5°C/110°F for one hour to rid the stools of STEM EELWORM (*Ditylenchus dipsaci*). Strawberry runners attacked by *D.dipsaci* and *Aphelenchoides fragariae* may be disinfested by immersion at 46°C/115°F for 10 minutes, a treatment which also controls tarsonemid mites.

Hot water treatment can be used to treat mint runners infected with RUST, and also to protect chrysanthemum stools against WHITE RUST.

Hottonia (for P. Hotton, Dublin botanist). Primulaceae. Northern Hemisphere. 2 species, aquatic, perennial herbs, rooting in mud or floating. The leaves are submerged and finely dissected. Salverform with five, spreading lobes, the flowers stand clear of the water in whorled racemes in summer. Vigorous aquatics for ponds and slow-moving streams, where their ferny leaves are submerged below water level while the spikes of white to lilac flowers rise in spring above the surface. They are also useful as shallow-water oxygenators, especially if fish are present. They produce winter-resting buds which sink to the bottom until the following spring. Plant in spring by throwing divisions/cuttings into water or by rooting into submerged marginal mud. Propagate also from seed sown in spring in pans standing in water.

H.inflata (AMERICAN FEATHERFOIL; E Europe; leaves 1-5cm, ovate or oblong, lobes narrow-linear; flowers 0.5–0.8cm, white); *H.palustris* (WATER VIOLET; Europe, W Asia; leaves 2–13cm, lobes linear, sometimes further divided; flowers to 2.5cm diameter, violet, throat yellow).

house plant a plant that will grow satisfactorily in the temperature, humidity and light of a dwelling house. Suitable subjects include a range of evergreen foliage plants, such as *Monstera*, *Sansevieria* and *Peperomia*, as well as flowering plants such as *Cyclamen*, *Hippeastrum*, and chrysanthemums. Greatest success is achieved where the recommended environment is maintained without wide fluctuation, with careful attention to watering and feeding.

Houstonia (for Dr William Houston (1695–1733), British botanist and writer on American plants). Rubiaceae. N America, Mexico. Some 50 species, hardy perennial herbs or, rarely, subshrubs with small, broadly funnel-shaped and 4-parted flowers in spring and summer. Low-growing, stem-rooting perennials, grown for their loose mats of small shining leaves and for their dainty flowers in clear blue, white or blue-violet. They are suitable for crevices in the rock garden, for the alpine house or frame, and for other shady situations in the garden. Hardy to –20°C/–4°F. Grow in a cool, sheltered site, in moist, well-drained, leafy soil. Propagate by division in autumn or spring, or from seed sown in a soilless propagating mix in the cold frame in spring.

H.caerulea (BLUETS, INNOCENCE, QUAKER-LADIES; N America; perennial herb to 20cm, mat-forming; leaves small; flowers few, solitary on slender stalks to 6cm, corolla salver-shaped, to 1cm, pale blue to lilac or violet, or white, and yellow-eyed; includes 'Alba', tightly mat-forming with white flowers, and 'Fred Millard', mat-forming with deep blue flowers); *H.longifolia* (LONGLEAF BLUETS; E US; erect perennial herb to 30cm; flowers numerous in cymes, corolla white or purple, funnel-shaped to 8mm); *H.purpurea* (MOUNTAIN HOUSTONIA; C and S US; erect perennial herb to 50cm; flowers in dense cymes or clusters, corolla pale purple, lilac or white, funnel-shaped, to 8mm); *H.serpyllifolia* (SE US; perennial herb to 8cm, stem prostrate; leaves very small; flowers on erect stalks to 4cm, corolla violet-blue or white, salver-shaped, to 1cm).

Houttuynia (for Martin Houttuyn, 18th-century Dutch naturalist). Saururaceae. E Asia. 1 species, *H.cordata*, a semi-evergreen perennial herb of damp places. All parts of this plant have a peppery aroma. The clumped stems are 15–40cm tall and spread widely and often somewhat unpredictably by means of subterranean suckers. The leaves are to 6 × 3cm, heart-shaped, smooth and dark green (brilliantly painted red and cream in the cultivar 'Chameleon'). Small, green flowers are carried in small, squat, cone-like spikes subtended by white bracts. An attractive groundcover in the bog garden or damp border, *Houttuynia* can also be grown as a pond marginal in up to 10cm of water. It can be invasive but is less vigorous on drier soils. Hardy to at least–15°C/5°F, it should be grown on moist soil in full sun to light shade. Propagate by division in spring.

Hovenia (for David Hoven, an Amsterdam senator). Rhamnaceae. E and S Asia. 2 species, deciduous shrubs or trees, usually with large leaves. Small, fragrant and 5-petalled, the flowers are borne in cymes in summer. With age, the fleshy flower stalks become red and edible. Grow in fertile, sandy loam soils in full sun. Hardy to at least −15°C/5°F. Propagate by softwood cuttings in summer, by hardwood cuttings or by seed. Seed germinates freely if sown fresh; stored seed may need acid scarification before sowing. Poorly ripened wood is likely to be more susceptible to coral spot; application of high-potash fertilizer in midsummer may help. *H.dulcis* (RAISIN TREE, CHINESE/JAPANESE RAISIN TREE; E Asia; tree to 20m; leaves 10-20cm broadly ovate, apex acuminate, base cordate, coarsely toothed, downy beneath; flowers to 6mm in diameter, green-yellow, pedicels fleshy becoming red and edible after frost).

hoverflies (Diptera: Syrphidae) flies that are easily recognizable by their habit of remaining stationary in flight before darting to another position; they frequent flowers to feed on nectar or, sometimes, pollen. Many are brightly coloured with black and yellow stripes, resembling wasps; their legless larvae may be scavengers, living in decaying organic matter, dung or stagnant water. Some other species are injurious to plants, including the bulb flies (see *bulb flies*). A third group, including *Syrphus ribesii*, are beneficial in gardens because their larvae feed upon aphids.

Howea (native to Lord Howe Island, in the S Pacific). Palmae. Lord Howe Island. 2 species, large palms, ultimately with solitary, erect stems. The leaves are pinnately compound with fibrous sheaths, unarmed petioles and linear to lanceolate leaflets. *Howea* species are among the most elegant of feather palms. Suitable for outdoor cultivation in frost-free temperate climates. Elsewhere, they show remarkable tolerance of neglect, low light, low temperatures and of the dry atmospheres that generally pertain in the home. Among the most widely used palms for interior decoration, they are slow-growing and exceptionally beautiful as juveniles, and require infrequent repotting. Propagate by seed.

H.belmoreana (BELMORE SENTRY PALM, CURLY PALM; to 7m; trunk markedly ringed, usually swollen at base; pinnae to 2.5cm across); *H.forsteriana* (SENTRY PALM, FORSTER SENTRY PALM, KENTIA PALM, THATCH LEAF PALM; to 18m but grown as a stemless juvenile; trunk obliquely ringed, straight; pinnae to 1m long, 3–4cm across, dark green).

Hoya (for Thomas Hoy, head gardener to the Duke of Northumberland at Syon House in the late 18th Century). Asclepiadaceae. WAX FLOWER, PORCELAIN FLOWER. Asia, Polynesia, Australia. Some 200 species, evergreen shrubs and subshrubs, some climbing with slender, rooting stems, others bushy. The leaves are tough, leathery and, in some cases, thinly succulent. Borne at various times of year in long-lasting pendulous umbels, the flowers are waxy and star-shaped. They consist of five, pointed lobes surrounding a fleshy and often differently coloured corona. The flowers are sometimes sweetly fragrant and dripping with nectar.

Provide a minimum temperature of 10°C/50°F. Grow in dappled sun or light shade. Plant in an open mix rich in organic matter (that suitable for the orchids *Paphiopedilum* will suit). Water, syringe and feed freely in warm weather; water sparingly in cool conditions. Train climbing species on wires, trellis and moss poles. Shrubby species like *H.lanceolata* are excellent plants for baskets. Increase by semi-ripe cuttings (the wound allowed to dry) in a heated case; also by layering.

H.carnosa (WAX PLANT; India, Burma, S China; climber to 6m; leaves to 10cm, elliptic-ovate, smooth, thick; flowers to 1.5cm across, white to fleshy pink with a glistening, papillose texture, corona purple-pink to pale red; includes cultivars with leaves variegated cream and yellow, or splashed with silver; *H.compacta*, also known as *H.carnosa*

Hoya carnosa

Hoya lanceolata subsp. *bella*

'Hindu Rope', differs in having leaves tightly folded together with only their lower surfaces exposed); *H.imperialis* (SE Asia; climber to 8m; leaves to 20cm, elliptic-oblong, smooth, thick; flowers to 7cm across, chocolate to wine red or magenta, corona creamy white); *H.lanceolata* (including *H.bella* and *H.sikkimensis* of gardens; MINIATURE WAX PLANT; Himalaya to N Burma; shrubby, to 45cm, with arching branches; leaves to 3cm, ovate-lanceolate, very thinly fleshy, often pale green; flowers highly fragrant, to 1cm across, sparkling white, corona purple-red); *H.linearis* (Himalaya; stems drooping to trailing; leaves 2–5cm, linear, more or less narrowly cylindric to terete, grooved, downy; flowers to 1cm across, faintly scented, ivory, corona yellow tinged pink); *H.longifolia* (STRING BEAN PLANT; Himalaya, SE Asia; stems trailing or weakly climbing; leaves to 10cm, angled downwards, linear to narrowly lanceolate, thick, dark green, smooth, with a sunken midvein; flowers 1–4cm across, white sometimes flushed pink, corona rose pink to red); *H.polyneura* (FISHTAIL HOYA; Himalaya, S China; shrubby to 1m, branches arching; leaves to 10cm, broadly ovate-rhombic, smooth, glossy deep green with darker veins; flowers to 1.5cm across, white to cream, corona purple-red).

Huernia (for Justus Huernius (1587–1652), a Dutch missionary and the first collector of plants from the Cape of Good Hope). Asclepiadaceae. DRAGON FLOWER. South Africa to Ethiopia and Arabia (1 species), W Africa. Some 70 species, succulent, perennial herbs. The stems are clumped, fleshy, and grey-green with toothed angles. The leaves are rudimentary and short-lived. Malodorous and fleshy, the flowers are bell-shaped with five short and pointed lobes. Cultivate as for *Stapelia*.

H.boleana (Ethiopia; stem to 8 × 1.5cm, 5 angled, teeth soft to 6mm, grey-green to green mottled purple; corolla 17mm in diameter, lobes deltoid, inner surface blood-red, covered with large yellow papillae); *H.macrocarpa* (Eritrea, SW Saudi Arabia, S Yemen; stem to 9cm, 4-angled, thick, ridges sinuate, teeth to 8mm; corolla 11–20mm in diameter, papillose to smooth, exterior yellow green, interior pale yellow with concentric maroon bands or wholly purple to crimson, lobes broadly triangular, attenuate); *H.pillansii* (COCKLEBUR; Cape Province; stem to 4cm, sub-globose when young, with conical tubercles in many rows, apices softly setose; corolla to 4mm in diameter, cream to pink with red spots, lobes recurved, attenuate, interior pale yellow with crimson spots, densely red-papillose); *H.primulina* (Cape Province; stem to 8cm, sometimes spotted red, ridges sinuate, with dark, hooked teeth; corolla 2.5cm in diameter, pale to sulphur-yellow, smooth, fleshy and waxy, sometimes tinged red, lobes triangular-attenuate); *H.zebrina*. (OWL-EYES, LITTLE OWL; N Transvaal, Botswana, Namibia; stem to 8cm, 5-angled, teeth 4–5mm, conical; corolla 3.5–4cm in diameter, yellow blotched red-brown, lobes triangular, lime or sulphur-yellow, interior downy, banded red-brown).

humidity the amount of water vapour in the atmosphere, measured on a scale of relative humidity (RH): RH 0% = totally dry air, RH 100% = saturated air, at a given temperature. Excepting the extremes of xerophytes and hygrophytes, most species grow satisfactorily within the range of 65–80% RH. Below 65%, growth is progressively and adversely affected; above 80%, there are increasing problems with fungal and bacterial diseases. Above 90%, transpiration begins to be restricted and the movement of nutrients, particularly calcium, is reduced to the extent that deficiency symptoms may appear.

Humidity can be reduced by applying heat and ventilation or, on a large scale, with dehumidification systems that employ heat exchangers or refrigerant cooling. Damping down, manually with a hose, watering can or overhead sprinkler, will raise humidity. 'Fogging' systems are available for expansive greenhouses; these produce free-floating droplets of less than 5 microns with no excess water.

humification the breakdown of organic matter in soil leading to the formation of humus.

Humulus (from Latin *humus*, soil, referring to the creeping habit). Cannabidaceae. Temperate Europe, N America, C and E Asia. HOP. 2 species, twining, herbaceous perennials with more or less palmately lobed leaves and pendulous, cone-like inflorescences with papery bracts. Rapidly growing twiners, *Humulus* species are especially useful as summer screening plants on trellis or wires, and for scrambling through hedging and large open shrubs; they are valued for their foliage. In autumn, *H.lupulus* also produces fragrant straw-coloured inflorescences, used in brewing and, when air-dried, to make swags and garlands. The young shoots are edible in salads or cooked as for spinach. *H.lupulus* is hardy at least to –15°C/5°F. *H.japonicus* is usually grown as annual in cool temperate climates, rapidly forming a green screen that makes an admirable backdrop for more brightly coloured flowers. Grow in any moderately fertile, well-drained soil in sun or part shade; *H.lupulus* 'Aureus' colours best in sun, when few foliage climbers can rival the beauty of its fresh gold to lime leaves. Propagate cultivars of *H.lupulus* by greenwood leaf-bud cuttings in midsummer, taken from material on which leaves have fully expanded. Apply 0.2% IBA and root in a closed case with bottom heat at 20°C/70°F. Progeny from seed is highly variable. Sow *H.japonicus in situ* in spring, or earlier under glass at 15–18°C/60–65°F. Lower the temperature on germination, and plant out in late spring.

H.japonicus (JAPANESE HOP; temperate E Asia; stem very rough; leaves to 20cm, 5–7-lobed, strongly serrate, petioles longer than blades; includes 'Lutescens', with pale gold to lime green foliage, and 'Variegatus', with foliage blotched and streaked white); *H.lupulus* (COMMON HOP, EUROPEAN HOP, BINE; northern temperate regions, widely naturalized; stem rough; leaves 3–5-lobed, coarsely toothed, petioles usually shorter than blades; 'Aureus': leaves golden yellow).

humus a complex mixture of compounds resulting from the microbiological decomposition of organic matter in soil; it forms a black or dark brown amorphous colloid. Humus affects the retention and availability of nutrients by increasing cation exchange capacity, and slowly releases nitrogen and phosphate. Humus influences soil structure and, on clay soils, will assist in bonding particles; on sandy soils especially, it increases available water.

Hunnemannia (for John Hunneman (d. 1839), botanist and plant collector). Papaveraceae. MEXICAN TULIP POPPY, GOLDEN CUP. Mexico. 1 species, *H.fumariifolia*, a perennial, glaucous herb to 1m tall with finely, ternately divided leaves. Produced in summer, the yellow flowers are solitary, to 8cm in diameter and composed of four, rounded petals. Usually grown as an annual, although perennial in frost-free climates. Sow *in situ* in spring, or in late winter in the greenhouse at 13–15°C/55–60°F, in a sandy medium. Plant out in late spring, taking care not to disturb roots, in a sunny, well-drained site.

hurdle see *wattle*.

husk the outer layer of certain fruits, for example *Juglans*, *Physalis*, generally originating from the perianth or involucre.

Hyacinthella (diminutive of *Hyacinthus*). Liliaceae (Hyacinthaceae). E Europe to W Asia and Israel. Some 20 species of small, bulbous, perennial herbs with slender leaves and bell-shaped flowers in scapose racemes in spring. Grow in a well-drained, sunny position in the rock garden. They may also be grown in the alpine house or bulb frame where they can be given protection from wet when dormant in summer. Otherwise, cultivate as for *Muscari*.

H.glabrescens (central S Turkey; flowers to 6mm, tubular to campanulate, deep violet-blue, on distinct pedicels); *H.nervosa* (SE Turkey, through Syria, Jordan and Iraq to Israel; flowers 1cm, tubular to narrowly campanulate, sessile, pale blue).

Hyacinthoides (from *Hyacinthus* and Greek *-oides*, resembling). Liliaceae (Hyacinthaceae). BLUEBELL. W Europe, N Africa. Some 4 species, perennial, bulbous herbs with narrowly strap-shaped leaves and scapose racemes of bell-shaped flowers with spreading perianth lobes in spring. Fully hardy. Grow in any humus-rich, moisture-retentive soil in semi-shade; flower colour will bleach in sun and these are, in any case, bulbs associated by most with woodland carpets and dappled glades. They tolerate a range of soil types, but prefer heavier soils. Propagate by division of established colonies (bulbs will be found quite deep in the soil), or by seed sown *in situ* when ripe or in spring in the shaded cold frame.

H.hispanica (syn. *Endymion hispanicus*; SPANISH BLUEBELL SW Europe, N Africa; close to *H.non-scripta*; flowers 6–8, unscented, widely campanulate, borne in a rather loose, not 1-sided raceme, tepals spreading, not curved at tips, anthers blue; plants cultivated as *H.hispanica* are often hybrids between this species and *H.non-scripta*. Cultivars include: 'Alba', with white flowers; 'Danube', with abundant, dark blue flowers; 'Excelsior', tall, with large, blue violet flowers, with a marine blue stripe; 'La Grandesse', with pure white flowers; 'Mount Everest', with white flowers, in a broad spike; 'Myosotis', with porcelain blue flowers, with a sky blue stripe; 'Queen of the Pinks', with deep pink flowers; 'Rosabella', with soft pink flowers; 'Rose', with violet pink flowers, in a large spike; and 'White City', with white flowers); *H.italica* (SE France, NW Italy, Spain, Portugal; scape 10–40cm; flowers 6–30, borne in dense, conical, erect, not 1-sided raceme, tepals 5–7mm, spreading widely, blue-violet, occasionally white, anthers blue); *H.non-scripta* (syn. *Endymion non-scriptus*; BLUEBELL; W Europe; scape 20–30cm, flowers 6–12, fragrant, tubular with tepals curved at tips in loose, 1-sided, drooping raceme, tepals 1.5–2cm, oblong to lanceolate, violet-blue, occasionally pink or white, anthers cream; includes 'Alba', with white flowers, and 'Rosa', with pink flowers).

Hyacinthus (an ancient Greek name used by Homer, the flowers being said to spring from the blood of the dead Hyakinthos). Liliaceae (Hyacinthaceae). HYACINTH. W and C Asia. 3 species, bulbous perennial herbs with linear-lanceolate to broadly strap-shaped leaves and cylindrical, scapose racemes of tubular to campanulate flowers with spreading to recurved perianth lobes in spring.

Smaller species, such as *H.orientalis* and *H.litwinowii,* are suited to the sunny rock garden or bulb frame and to sandy, well-drained soils. Most commonly grown are those highly fragrant and showy cultivars derived from *H.orientalis*, the Dutch hyacinths, and the looser-flowered Roman hyacinths, from *H.orientalis* var. *albulus*, which, with successional planting and selection may be had in bloom from Christmas until late spring. The Multiflora Group, producing several slender stems with individual blooms more loosely set, are also suitable for indoor and outdoor use. These cultivars are available in a wide colour range and are used in traditional spring bedding, in beds and borders and in tubs and other containers.

Outdoor bulbs are planted in autumn in an open and sunny position in any well-drained, moderately fertile soil. Prepared bulbs for forcing are potted in autumn, usually into a proprietary bulb mix, and require a period of about 8–10 weeks, in cool, dark, moist conditions, to develop a good root system; at this stage, temperatures should not exceed 7–10°C/45–50°F. When necessary to apply water, avoid wetting the shoot, as this may cause the bulb to rot. On emergence of the shoot tips, increase temperatures to 10°C/50°F; thereafter, increase light and temperature as growth progresses. Alternatively, on shoot emergence, place pots in a dark cupboard, at 18–20°C/65–70°F, and water plentifully. Bring into subdued light as the bud emerges (usually when the shoot is between 6–8cm high) and increase

Hyacinthus orientalis

Hyacinthus litwinowii

light levels as growth progresses and the stem elongates. Forced hyacinths may be planted out after blooming in the open garden, and will flower in subsequent years, although the blooms will decline in quality. Bulbs forced in a bulb glass are best discarded after flowering.

Propagate species by ripe seed sown thinly in a loam-based propagating mix in the cold frame, so that seedlings may remain *in situ* during their first season's dormancy. Increase cultivars by cutting out a cone of tissue from the bulb base, or by cutting crosswise slits through the base – both treatments result in the formation of large numbers of bulbils.

Hyacinth fire, caused by the fungus *Botrytis hyacinthi*, was so-called by Dutch growers because it could spread through the fields 'like fire'. Beginning at the tips, the leaves become brown and shrivelled; in damp conditions, a grey mould develops on affected parts and the flowers may be destroyed. If detected early enough, the disease can be controlled by fungicide sprays. In the disease known as black slime, the leaves turn yellow and fall over just after flowering; they can be pulled away easily because they are rotten at the base. The bulb scales are dry and dark grey but, under moist conditions, a white, fluffy mycelial growth develops on them. Black sclerotia, sometimes up to half an inch across, are formed between and within the scales and can remain in the soil to perpetuate the fungus. The aphid-transmitted virus diseases, hyacinth mosaic and ornithogalum mosaic, can cause a conspicuous yellow or grey leaf mottling; affected plants should be destroyed to prevent the diseases from spreading.

H.litwinowii (Central Asia, E Iran; 10–25cm; flowers 3–13, not fragrant, perianth 1.8–2.5cm, green-blue, tube constricted above ovary, lobes longer than tube, spreading and recurved above, filaments longer than anthers); *H.orientalis* (COMMON HYACINTH; C and S Turkey, NW Syria, Lebanon; to 30cm; flowers 2–40, waxy, heavily scented, perianth 2–3.5cm, pale blue to deep violet, pink, white or cream, tube as long as or exceeding lobes, constricted above ovary, lobes spreading or recurved, anthers longer than filaments. Cultivars and subspecies include: 'Amethyst', with lilac flowers; 'Anna Marie', with light pink flowers; 'Appleblossom', miniature, with shell pink flowers; 'Ben Nevis', with double, large, ivory white, compact flowers; 'Blue Jacket', with navy flowers, striped purple; the late-flowering 'City of Haarlem', with primrose flowers; 'Delft Blue', with soft blue flowers; 'Distinction', with deep burgundy to purple flowers; 'Gipsy Queen', with pale salmon orange flowers; 'Hollyhock', with double, crimson, compact flowers; 'Jan Bos', with cerise flowers; 'Lord Balfour', miniature, with claret flowers tinted violet; 'Multiflora White', with multiple stems and sparse, white flowers; 'Myosotis', with palest blue flowers; 'Ostara', with large, purple-blue flowers; 'Pink Pearl', with deep pink flowers, with paler edges; 'Sunflower', miniature, with bright yellow flowers; subsp. *orientalis*, with 2–12 flowers and a 2–3cm perianth, pale violet-blue at base shading to white above, with lobes to four-fifths the length of tube).

hybrid berries

Pests
leaf hoppers
raspberry beetle

hyaline transparent, translucent; usually applied to the margins of leaves and bracts.

hybrid a plant produced by the cross-breeding of two or more genetically dissimilar parents which generally belong to distinct taxa. A natural hybrid is one arising in nature; a spontaneous hybrid is one arising without the direct intervention of man, usually in gardens; an artificial hybrid is a cross deliberately made by man. Specific or interspecific hybrids arise between species of the same genus; intergeneric hybrids arise between taxa of different genera – the number of genera involved being usually denoted by a prefix, for example bigeneric, trigeneric.

hybrid berries the large group of hybrids produced by crossing a wide range of *Rubus* species; often the species crossed are far apart botanically. The cultivation and propagation of hybrid berries should follow the same practices for the blackberry.

The LOGANBERRY was raised at the end of the 19th century in California by Judge Logan, reportedly from *Rubus ursinus* 'Auginbaugh' × 'Red Antwerp' raspberry. The phenomenal berry was raised in 1905 in California by Luther Burbank as a second-generation selection from a cross between the 'Auginbaugh' and 'Cuthbert' raspberries. The VEITCHBERRY was introduced in England in 1902 as a cross between *Rubus rusticanus* × raspberry 'November Abundance', and is important as a parent of the earliest-fruiting blackberry 'Bedford Giant'. The YOUNGBERRY, introduced in 1926 as a cross between the phenomenal berry and a dewberry ('Austin Mayes'), has sweeter and better-coloured fruits than loganberry and is widely grown in South Africa as South African loganberry.

BOYSENBERRY was introduced in 1935 and is similar to the youngberry, but of unknown origin. Grown widely in New Zealand and the north western states of the US, it has red-purple fruits and is moderately vigorous. The NECTARBERRY is probably a sport of boysenberry, and another selection is 'Riwaka', grown in New Zealand. MARIONBERRY resembles boysenberry though its origin is unrecorded. It has large, black, well-flavoured fruit on vigorous plants, and yields over a longer period than boysenberry.

NESSBERRY was introduced in 1921 in Texas, from breeding work at the Texas Experimental Station. It combines the raspberry flavour of the fruit with the drought resistance of the dewberry, and has been important in breeding several erect blackberries, suited to commercial cultivation in the southern US.

TAYBERRY and TUMMELBERRY were introduced by the Scottish Crop Research Institute in 1979 and 1983 respectively. The parentage of tayberry is 'Aurora' blackberry and a tetraploid raspberry; tummelberry, which ripens a week later than tayberry, results from a cross between tayberry and a sister seedling. Both have larger fruits than loganberry and are less acid.

SUNBERRY is a loganberry-type fruit with good flavour, introduced from the East Malling Research

Station, Kent, in 1981 as a cross between *Rubus ursinus* × raspberry 'Malling Jewel' seedling; SILVANBERRY is an Australian raised hybrid of very mixed parentage; it is very vigorous and produces large, dark red, sweet fruits; JAPANESE WINEBERRY (*R.phoenicolasius*) has red bristly stems and produces bright red conical fruits. See *jostaberry*.

hybrid vigour see *heterosis*.

hydathode a water-secreting gland on the surface or margin of a leaf, usually situated at the end of a vein and often surrounded by a coneretion of white salts. It is similar to a stoma but with functionless guard cells, and is sometimes termed a lime-dot.

Hydrangea (from Greek *hydor*, water, and *aggeion*, vessel: an allusion to the cup-shaped fruit). Hydrangeaceae. China, Japan, the Himalaya, Philippines, Indonesia and N and S America. 100 species, deciduous or evergreen shrubs, small trees or climbers. Numerous small flowers are borne in panicles or corymbs. Many species bear larger, sterile flowers, usually at the outer edges of the corymbs; in these flowers,the conspicuous, coloured part is the enlarged calyx.

Hydrangea species are grown primarily for their flowers, although a number of species also have handsome foliage, including *H.quercifolia*, *H.aspera* and *H.sargentiana*; those with foliage colouring well in autumn include *H.serrata*, *H.quercifolia* and *H.heteromalla*. Flowers show two characteristic forms: the large, conspicuous and round-headed form composed of sterile ray florets, as exemplified by the Hortensia group of garden hybrids, and the 'lacecap' type, comprising the majority of species, where a central boss of small, insignificant, fertile florets is accompanied or surrounded by larger sterile ray florets. In *H.paniculata* and its clones, the flowers are carried in elegant, densely flowered, conical panicles.

Hydrangea species make most attractive specimens for the shrub or mixed border. Many, such as *H.quercifolia* and *H.aspera*, are effective in more informal situations in the open woodland garden (in general, large-leaved species require a position in semi-shade, with *H.sargentiana* thriving in part to full shade). *H.paniculata* is amenable to training as a single-stemmed standard; with hard pruning, it makes a multi-stemmed shrub, noted for its generosity in bloom.

The many named garden hybrids, largely derived from *H.macrophylla* and *H.serrata*, fall into two (horticultural) groups – the round-headed 'Hortensia' types, and the slightly hardier 'lacecaps'. These two groups are well-suited to seaside gardens. Although tolerant of alkaline soils, they may become chlorotic on shallow chalk. They include several cultivars suitable for tub cultivation or for use as house plants (usually for a single season). As with some of the species, the flowers are often dried for winter decoration. Tub cultivation has the additional advantage of permitting greater control over flower colour than in open ground. Colour is dependent on cultivar and on the availability of aluminium, which is largely determined by soil pH. On acid soils at pH 4.5–5.5, flowers may achieve an intense deep blue ('Marechal Foch', 'Nikko Blue'), this changing to shades of pink as soil pH rises; with careful selection of cultivars, blooms may become a clear warm rose pink ('Queen Elizabeth', 'King George') or deeper shade (as in 'Hamburg' and 'Ami Pasquier') on soils at pH 7.4. In containers and on near-neutral soils, blue coloration may be maintained or enhanced by a weekly or fortnightly application of blueing compound (aluminium sulphate) at 85gm per 13.5 litres of water.

The climbing species, such as *H.anomala* and *H.serratifolia*, will climb up and through trees, although they may need initial support until they develop their aerial roots. *H.petiolaris* is particularly useful for shaded walls, usually showing good tolerance of atmospheric pollution.

Some hydrangeas are extremely hardy, such as *H.paniculata* and *H.arborescens* subsp. *discolor* and their forms. *H.aspera* and its allies and the Hortensia and 'lacecap' cultivars are generally winter hardy; in Europe, most appear to tolerate temperatures down to −20°C to −25°C/-4°F to -13°F. An undeserved reputation for a lack of hardiness is due to their long growing season, which makes early shoots and unripened autumn growth particularly vulnerable to late spring and early autumn frosts.

Grow in fertile, moist but well-drained soils, if necessary incorporating well-rotted organic matter to improve fertility; an annual mulch of the same in spring, with a top dressing of balanced fertilizer, is beneficial. Site in sun or part shade and, when in flower, provide some protection from cold winds and from full sun to preserve colour. *H.involucrata* and *H.paniculata* will tolerate full sun, provided there is sufficient moisture in the growing season. In containers, use a medium-fertility, loam-based mix (omitting lime for blue flowers). For early flowers in pots, take leaf bud or stem cuttings in early spring, root in a closed case with bottom heat, pot on individually and pinch out in early summer. Overwinter under glass, at 5–10°C/40–50°F, watering sparingly; in late winter/early spring, gradually raise the temperature to 13°C/55°F, growing on in full sun with plentiful water.

A minimum of pruning is generally required. Vigorous species that flower at the tips of the current season's growth (e.g. *H.paniculata*, *H.arborescens*) may have the previous season's growth cut back in late winter/early spring to 2–3 pairs of buds, in order to obtain extra large panicles of flowers for a more formal effect. Without pruning, such species will make large shrubs with a larger number of smaller flowers, a result which may be preferred. With clump-flowering types that produce shoots from the base (e.g. *H.serrata* and *H.macrophylla* cultivars), flowers are carried occasionally on these shoots but in the main are formed terminally on the previous season's wood. Pruning of these types should be confined to the removal of old flower heads (leave this until spring to help frost protection) and weak, congested or exhausted growth at the base; this allows light and air to the centre of the bush, encouraging good ripening

Hydrangea quercifolia

and flowers. Shoots should not be simply shortened back since, in most cases, the removal of the terminal bud will cause the flowers to be lost for that season.

Plants may, after many years, be rejuvenated by cutting hard back to the base. The crowded sheaf of shoots that results must be thinned as it grows to allow a few strong shoots to ripen fully. Climbing species, which flower on lateral shoots, require minimal pruning, other than the removal of unwanted extension growth in summer and the restriction of the outward spread of wall-grown plants. Cut back outward-growing laterals to a suitable bud in spring; this may best be done over several seasons to minimize loss of bloom.

Propagate species by seed sown under glass in early spring (offspring may be variable). Increase climbing species by basal softwood cuttings of non-flowering shoots; treat these with 0.8% IBA and root in a closed case with bottom heat at 20°C/70°F, or use the warm bench with plastic method. Propagate other species by hardwood stem or tip cuttings, or from leaf bud cuttings of the current season's growth; root these in a closed case with bottom heat at 13–15°C/55–60°F.

H.anomala (Himalaya, China; deciduous climber to 12m; shoots becoming rough and peeling; leaves 7.5–13cm, ovate, apex shortly acuminate, cordate at base, coarsely toothed, downy tufts in the vein axils beneath; corymbs fairly flat 15–20cm across, with few white, peripheral sterile flowers, each 1.5–3.7cm across, and numerous, small, cream fertile flowers); *H.arborescens* (E US; loose, open, deciduous shrub, 1–3m; shoots downy at first; leaves 7.5–17.5cm, broadly ovate, acuminate, with coarse, teeth, shiny, dark green above, paler and downy beneath on veins; corymbs fairly flat, much-branched, 5–15cm across, with 0–8 long-stalked, creamy white sterile flowers, each 1–1.8cm across, and numerous, small, dull white fertile flowers); *H.aspera* (Himalaya, W and C China, Taiwan, Java, Sumatra; spreading deciduous shrub, or small tree, to 4m; shoots at first with spreading hairs, later hairless and peeling; leaves 9–25cm, lanceolate to narrowly ovate, acute or acuminate, rounded or tapered to the base, serrate, densely downy beneath, sparsely hairy above; corymbs fairly flattened, to 25cm across, with few to many white to pale pink or purple, darker-veined, sterile flowers with 4 rounded sepals, each to 2.5cm across, fertile flowers small, numerous, white-purple or pink; includes subsp. *strigosa*, with short, stiff hairs beneath leaves); *H.heteromalla* (Himalaya, W and N China; deciduous shrub to 3m; shoots hairy at first later glabrous; leaves 9–20cm, narrowly ovate, rounded, crenate or some-times cordate at base, toothed and bristly at the margins, pubescent beneath; corymbs flattened, 15cm across, with few white or ivory sterile flowers, each 2.5–5cm across, and numerous, small, white fertile flowers; includes 'Bretschneiden', with peeling bark, leaves white beneath, and white 'lacecap' flowers); *H.involucrata* (Japan, Taiwan; open, deciduous shrub to 2m; shoots bristly at first; leaves 7.5–15cm, broadly ovate to oblong, acuminate, finely toothed, bristly; corymbs irregular, 7.5–12.5cm across, enclosed by

broadly ovate bracts covered with flattened white hairs, with few, long-stalked, pale blue or faintly pink sterile flowers, each 1.8–2.5cm across, and numerous, small, blue fertile flowers; includes 'Hortensis', with more numerous, double, pink-white, sterile flowers); *H.longipes* (C and W China; loose, spreading, deci-dous, shrub, 2–2.5m; shoots loosely downy at first; leaves 7.5–17cm, rounded to ovate, apex abruptly acuminate, cordate at base, sharply toothed, bristly; corymbs fairly flat, 10-15cm across, sterile flowers 8-9, white or faintly purple, to 2cm across, fertile flowers small, white, numerous); *H.macrophylla* (Japan; spreading deciduous shrub to 3m; shoots ± glabrous; leaves 10–20cm, broadly ovate, acute or acuminate, coarsely toothed, glabrous; corymbs flattened, much-branched, with few pink lilac or blue sterile flowers, each 3–5cm across, and numerous small blue or pink fertile flowers; the wild type, sometimes called *H.macrophylla* var. *normalis* (syn. *H.maritima*) is a maritime plant with a corymb of the 'lacecap' type; it is thought to be the ancestor of many cultivars of both 'lacecap' and 'mophead' types); *H.paniculata* (E and S China, Japan, Sakhalin; large deciduous shrub or small tree to 4m; shoots at first pubescent; leaves 7.5–15cm, ovate, acuminate, toothed, sparsely bristly; panicle conical or pyramidal, 15–20cm, with few white-pink sterile flowers, each 1.75–3cm across, and numerous yellow-white fertile flowers. Cultivars include: 'Everest', with dark green leaves and all-sterile, white to pink florets in large heads; 'Floribunda', with a long, narrow inflorescence and largely sterile, packed flowers; 'Grandiflora', with small, sterile, white flowers, turning pink-red, in very large panicles; 'Greenspire', with sterile, green flowers, becoming red-tinged; 'Kyushu', erect, with dark, shining, tapering leaves and many sterile flowers; 'Pink Diamond', with sterile flowers becoming pink in large panicles; 'Praecox', early-flowering, with toothed sterile flowers in small pani-cles; 'Tardiva', late-flowering; and 'Unique', with a very large inflorescence with many white sterile flowers); *H.petiolaris* (Japan, Sakhalin, Korea, Taiwan; deciduous climber to 20m; shoots becoming rough and peeling; leaves 3.5–11cm, ovate to rounded, shortly acuminate, ± cordate at base, finely toothed, sometimes pubescent beneath; corymbs flat, 15–25cm across, with to 12 white, peripheral sterile flowers, each 2.5–4.5cm across, and numerous, small, off-white fertile flowers); *H.quercifolia* (SE US; loose, rounded, deciduous shrub, 1–2.5m; shoots thick, red-downy, then hairless and flaky; leaves 7.5–20cm, ovate to rounded and deeply 5–7-lobed, minutely toothed, bristly, turning red-bronze in winter; panicles conical-pyramidal, 10–25cm, with numerous long-stalked, white sterile flowers, each 2.5–3.5cm across, and numerous, small, white fertile flowers; includes 'Snow Flake', with predominant, double, green sterile florets, turning white as they mature); *H.sargentiana* (China; loose, spreading, deciduous shrub to 3m; shoots with small, erect hairs and translucent bristles; leaves 10–25cm, broadly ovate, rounded at the base, velvety above, densely bristly beneath; corymb fairly flattened, 12.5–22.5cm across, with a few, pink-white peripheral sterile flowers to 3cm across, and numerous, small, pale purple fertile flowers); *H.scandens* (S Japan, E Asia; spreading or almost pendulous shrub to 1m; shoots glabrous or very finely pubescent; leaves 5–9cm, lanceolate or oblong to ovate, shortly toothed, finely pubescent on the veins beneath; corymbs fairly flattened, to 7.5cm across, with a few white-blue sterile flowers, each 1.75–3.75cm across, fertile flowers numerous, white, with small, clawed petals); *H.seemanii* (Mexico; evergreen climber or creeper with dark green, leathery, elliptic, acuminate leaves and white flowers; differs from *H.serratifolia* in its inflorescence – a single terminal corymb (not many) – and the presence of large, sterile flowers); *H.serrata* (Japan and Korea; spreading deciduous shrub to 2m; shoots at first finely pubescent; leaves 5–15cm, lanceolate, acuminate, glabrous above, veins beneath with short hairs; corymbs flattened, 5–10cm across, with to 12 pink or blue sterile flowers, each 1–1.5cm across, and numerous, small, pink or blue fertile flowers; 'lacecap' cultivars derived from *H.serrata* include 'Bluebird', 'Blue Deckle', 'Diadem' and 'Miranda', blue, compact, tending to be hardier than those of *H.macrophylla* origin and some of the best for general garden use); *H.serratifolia* (Chile, Argentina; evergreen climber, to 30m; shoots at first finely pubescent; leaves 5–15cm, elliptic, acuminate, base often cordate, usually entire, leathery; inflorescence to 15 × 9cm, composed of numerous small corymbs arranged one above the other, each at first enclosed by 4 papery bracts, flowers small, fertile, white, some variants exist with 1 or few white sterile flowers). *Cultivars* In most cultivars, the colour of the corymbs is influenced by the presence of aluminium in the soil. Some 'lacecap' cultivars, such as 'Bluebird', are often associated with *H.macrophylla;* they belong, however, to *H.serrata*, which some authorities treat as a subspecies of *H.macrophylla.*

Mophead hydrangeas (Hortensias) These bear predominantly sterile flowers, which may be single or double and produced throughout the flowerhead; the inflorescences are compact to loose, usually rounded. They include: the semi-dwarf 'Ami Pasquier', with crimson to plum, non-fading flowers; 'Ayesha' – also known as 'Silver Slipper' – with glossy green leaves and cupped lilac-like florets in misty lilac to pink flattened heads; 'Générale Vicomtesse de Vibraye', tall, with light aqua blue flowers in dense heads; 'Hamburg', large, with long-lasting, deep vivid pink to purple flowers and deeply serrated sepals; 'Heinrich Seidel', of stiff growth, with cherry red to purple flowers in dense heads; 'Holstein', with abundant, pink to sky blue flowers and serrate sepals; 'Joseph Banks' – also known as 'Hortensis' – 2.25m, with flowers opening green, ageing pink to blue; the original introduction from China; 'Madame Emile Mouillère – also known as 'Sedgwick's White' – stem unspotted, with white flowers with a pink to blue eye; the compact, late 'Mathilde Gutges', with intense blue to purple flowers; 'Niedersachen', tall, with very pale blue or pink flowers, reliable; 'Nigra', with an almost black stem and

rose to blue flowers; 'Nikko Blue', with pink to bright blue flowers; 'Otaksa', with rounded leaves and pink to blue flowers; an early 'Hortensia' from Japan; 'Pia', dwarf, to 30cm, with pink to red flowers in an irregular inflorescence; 'Preziosa', a hybrid with *H.serrata*, tinged purple-red throughout, with rose to purple red flowers; 'Vulcain', dwarf, with deep pink-purple or orange and green flowers; and 'Westfalen', with red to purple flowers, nearest to crimson on alkaline soils.

Lacecap hydrangeas These bear sterile flowers that are largely peripheral to the flowerhead with pedicels to 5cm; the inflorescences are flattened. 'Lacecaps' include: 'Geoffrey Chadbund', with red to purple flowers; 'Lanarth White', compact, with pink to blue fertile flowers and white ray florets; 'Lilacina', strong growing, with pink through lilac to blue flowers and serrate florets; 'Mariesii', with pale pink flowers, reluctantly blue, the inflorescence including both large and small ray florets; 'Blue Wave', with blue flowers, edged by blue to pink single ray florets; 'Quadricolor', with leaves mottled cream, yellow, pale and dark green; 'Sea Foam', with blue flowers edged by white ray florets; arose as a branch-sport on a plant of 'Joseph Banks'; 'Tricolor', with leaves mottled cream, yellow, pale and dark green; 'Veitchii', with dark green leaves and white flowers ageing pink; and 'White Wave', with pink to blue fertile flowers, surrounded by pearly white ray florets.

Hydrocharis (from Greek *hydor*, water, and *charis*, graceful, referring to the appearance of these aquatic plants). Hydrocharitaceae. Europe, W Asia, N Africa. FROGBIT. 2 species , aquatic, perennial herbs with slender, stolon-like stems, usually floating and rooting, and rosettes of small, stalked leaves with smooth, rounded to reniform leaves. Produced in spring and summer, the flowers are solitary or clustered and consist of three, broad petals. For sunny garden ponds. In cold areas, a sufficient depth of water must be given to prevent freezing in winter. Plants are free-floating; alkaline water is preferred. Resting buds are formed in autumn each year and sink to overwinter at the bottom, rising with increasing light levels in early summer to produce young plants. Increase in spring by the introduction of runners thrown on to the water surface. Alternatively, propagate from seed sown in pots of a loam-based medium; stand these in shallow trays of water. *H.morsus-ranae* (FROGBIT; Europe, W Asia; leaves small, rounded to heart-shaped, bright green, flowers to 2cm across, white stained yellow).

hydroculture a synonym of hydroponics, but often reserved for the cultivation of plants in nutrient-enriched water using sterile aggregate to support plant roots. As such it is a valuable technique for interior landscaping, where ornamental plants are grown in shallow tanks. See *hydroponics.*

hydroleca see *clay granules.*

hydrophyte a plant that grows partly or wholly submerged in water.

hydroponics, hydroponic culture the cultivation of plants in nutrient-enriched water. The Nutrient Film Technique (NFT), which was devised in the UK for commercial production in 1975, overcomes the constraints of poor aeration. Plants are raised in small blocks of rockwool and stood in polythene-film troughs containing a circulated solution less than 5mm deep. A prepared slope of 1:75 is required from one end of the greenhouse to the other for the continuous troughs, the edges of which are clipped together between plants. Dilute nutrient solution is pumped from a catchment tank to the feeder end of the troughs, from where there is a flow rate along the troughs of about 2 litres per minute. Electronic devices monitor pH and conductivity of the solution and inject nutrients and acid. Typical nutrient concentrations in parts per million for tomatoes would be 150–300 calcium, 40–50 magnesium, 3–6 iron, 0.5–1.0 manganese, 0.1 copper, 0.1 zinc, 0.3–0.4 boron and 0.05 molybdenum. See *hydroculture*.

hydroseeding the sowing of imbibed seed in a stream of water, practised with grass seed for large areas or on steep slopes.

hydrotropism growth in the direction of moisture, demonstrated especially by the roots of seedlings. See *stimulus movements*.

hygrometer an instrument for measuring atmospheric humidity, such as a wet and dry bulb thermometer, or electrolytic or dial gauge.

Hygrophila (from Greek *hygros*, wet, and *-philos*, loving, referring to the aquatic habitats of some species). Acanthaceae. Tropics. Some 100 species, aquatic or marginal perennial herbs. The leaves may range in form from entire to finely dissected, according to the degree of submersion. The flowers are small, tubular and 2-lipped, and arranged in whorl-like clusters in terminal racemes. Suited to the pool margin in the tropical greenhouse or aquarium; the submerged foliage is almost invariably more attractive. Leaf form in *H.difformis is* extremely variable, the more desirable, finely pinnate forms depending on high light intensity for their development; maintain a minimum water temperature of 20°C/68°F. In similar well-illuminated conditions, the undemanding *H.polysperma* may develop a red cast to the leaves and, where water temperatures remain at or above 15°C/60°F, will remain evergreen. Detached leaves will root if allowed to float freely. *Hygrophila* does not show marked preferences for substrates.

H.difformis (WATER WISTERIA; SE Asia; evergreen; stem to 60cm, slender, soft, rooting; immersed leaves to 10cm, pinnatifid, translucent, brittle, emersed leaves smaller, thicker, lanceolate to ovate, crenate; flowers to 2cm, lilac or violet marked purple on lower lip); *H.polysperma* (India, Bhutan; deciduous; stem to 50cm, submerged, rather woody; leaves to 4cm, oblong to lanceolate, pale green; flowers to 0.8cm, white to sky blue or pale lilac, downy).

hygrophyte a plant that grows under permanently moist conditions.

hygroscopic expanding when water is present, contracting in its absence; hygroscopic water is water retained in an essentially dry soil and unavailable to plants because of high surface tension forces.

Hylocereus (from Greek *hyle*, wood, and *Cereus*). Cactaceae. C America, West Indies, Colombia and Venezuela. Some 16 species, epiphytic, climbing or scrambling cacti. Segmented and slender, the stems are usually 3-winged or angled, with short spines and aerial roots. Large, funnel-shaped flowers open at twilight and last for one night only. Provide a minimum temperature of 10°C/50°F. Grow in an acid mix rich in bark. Shade in summer, and maintain high humidity. Reduce watering in winter. *H.undatus* (epiphytic or climbing to 5m or more; flowers 25–30 × 15–25cm, white; fruit to 15cm in diameter, globose, edible).

Hylomecon (from Greek *hyle*, wood, and *mekon*, poppy). Papaveraceae. Japan, Korea, E China. 1 species, *H.japonica*, a deciduous, perennial herb, to 30cm. To 25cm long, the leaves are pinnately divided, toothed and bright green. Produced in spring and summer, the long-stalked, solitary flowers are 5cm across with four, rounded, golden yellow petals. Fully hardy and suitable for damp, cool and humus-rich soils in the wild garden, especially in light woodland, and for the shady rock garden. Propagate from seed sown in spring, or by careful division in early spring.

Hymenanthera (from Greek *hymen*, membrane, and Latin *anther*, anther, referring to the membrane joining the anthers together). Violaceae. E Australia, New Zealand, Norfolk Island. Some 7 species, mostly dioecious, evergreen, twiggy shrubs with small, solitary or clustered flowers in spring and summer, and spherical to ellipsoid, berry-like fruits. Hardy to about -10°C/14°F. Grow in moderately fertile, well-drained but moisture-retentive soils in sun. Under glass, grow in a loam-based mix with additional sharp sand and leafmould; water plentifully when in full growth, moderating as light levels and temperatures fall in winter. The distinctive branch systems of *Hymenanthera* make it difficult to prune without spoiling its habit and form. Propagate by seed sown in spring in the cold frame, or by semi-ripe cuttings in sand in a propagating case with bottom heat. *H.crassifolia* (New Zealand; spreading to erect shrub to 2m; branches divaricate, hairy, spiny-tipped; leaves spathulate to obovate, thickly leathery, to 2cm, blunt to retuse, revolute; fruit purple).

Hymenocallis (from Greek *hymen*, membrane, and *kalos*, beauty, referring to the attractive corona). Amaryllidaceae. Subtropical and tropical Americas. SPIDER LILY. 30-40 species, bulbous perennial herbs with elliptic to strap-shaped leaves, their bases stalk-like or sheathing. In spring and summer, stout scapes are topped with umbels of fragrant white or ivory flowers. The perianth is tubular to funnel-shaped, with six spreading lobes. The six stamens are fused at their bases into a showy cup (corona) inserted into the top of the perianth tube. Provide a minimum temperature of 10°C/50°F. Grow in a light, well-drained potting mix of equal parts loam, leafmould and sharp sand, in bright filtered light. Water plentifully when in growth and continue watering and liquid-feeding after flowering, as for *Hippeastrum*. Keep just moist when resting. Propagate from offsets or by seed.

H. × *festalis* (*H.longipetala* × *H.narcissiflora*; scape 10cm, flowers pure white; perianth tube curved, 4cm, lobes to 11.5 × 1.3cm, curved: staminal cup to 5 × 6.5cm, with reflexed teeth, free ends of filaments to 4cm; style far exserted); *H.latiflora* (CAYMAN ISLANDS SPIDER-LILY, CHRYSOLITE LILY; flowers 6–12cm, perianth tube to 16cm, lobes to 10cm, linear, arching); *H.* × *macrostephana* (*H.narcissiflora* × *H.speciosa*; scape 30–45cm; perianth tube to 8.5cm, green below, white above, segments linear to lanceolate, 9–11cm, white to pale green-yellow, outer whorl with thickened green tips; corona funnel-shaped, to 6 × 7.5cm, bluntly toothed, white; free part of filaments 2.5cm, style far exserted); *H.narcissiflora* (BASKET FLOWER, PERUVIAN DAFFODIL; scape equalling leaves; perianth tube funnelform above, to 10cm, spreading, green, lobes to 10 × 1.2cm, lanceolate; staminal cup, to 5cm long and more than 5cm in diameter, striped green, with rounded spreading toothed processes, filaments to 12mm, pointing inwards; style exceeding limb); *H.palmeri* (ALLIGATOR LILY; scape to 25cm; perianth tube 8cm, yellow-green, lobes to 10cm, filiform, linear, spreading from base; staminal cup funnelform, with erect dentate margins, to 3–5cm, filaments 4cm); *H.speciosa* (scape to 40cm; flowers 7–12, green tinged; perianth tube to 9cm, lobes to 15cm; staminal cup funnelform, to 5cm, margin toothed, filaments nearly erect to 5cm).

Hymenosporum (from Greek *hymen*, a membrane, and *sporos*, seed, the seed has a membranous wing). Pittosporaceae. Australia. 1 species, *H.flavum*, AUSTRALIAN FRANGIPANI, an evergreen, narrow-crowned tree or large shrub, to 12m tall. The glossy, deep green leaves are 7–15cm long, oval to oblong and entire with a narrowly tapering apex. Produced in spring and summer in lax panicles to 20cm across, the flowers are intensely fragrant and creamy white, ageing golden-yellow. To 2.5cm in diameter, they consist of five, obovate petals, united in a short tube below and spreading above. Grown for its fragrant creamy flowers and dark glossy foliage, *Hymenosporum* makes an attractive lawn specimen in frost-free climates. It is also suited to tub cultivation in the cool greenhouse or conservatory. Propagation as for *Pittosporum*.

Hyophorbe (from Greek *hys*, pig, and *phorbe*, food: the fruits are eaten by pigs). Palmae. Mascarene Islands. BOTTLE PALM, PIGNUT PALM. 5 species, medium-sized to large palms with a single trunk often strongly swollen at the base or middle, then tapering

to a slender neck. The leaves are pinnate and arching with linear-lanceolate leaflets and smooth petiole bases forming a distinct crownshaft. Beautiful feather palms from the coastal forests of the Mascarene Islands, where they are threatened with extinction. In the coastal gardens of the low-altitude tropics and subtropics, they make curious and elegant specimens, with arching foliage emerging from the prominent crownshaft above a more or less swollen stem; they are sometimes used there in avenue plantings. When available, they are fine palms for large pots in the warm greenhouse or conservatory. Propagate by seed.

H.lagenicaulis (BOTTLE PALM; Round Island; trunk to 6m, closely ringed, grey, with vertical cracks, flask-shaped, expanded base to 70cm in diameter; leaves to 2m, pinnae to 70 each side of rachis, 17–60cm, with 2 pairs of lateral veins prominent above and beneath); *H.verschaffeltii* (SPINDLE PALM; Rodrigues Island; trunk to 5m, grey, markedly ringed, to 25cm in diameter, straight becoming tapered; leaves to 1.75m, pinnae to 80 each side of rachis, glossy green above, dull green beneath, midrib prominent above, no veins prominent beneath).

hypanthium cup-, ring- or tube-like structure formed by the enlargement and fusion of the basal portions of calyx, corolla and stamens, together with the receptacle on which these parts are borne, as in Rosaceae. It is sometimes termed the floral cup or, inaccurately, the calyx tube.

Hypericum (Greek name for this plant, used by Dioscorides and earlier authors, from *hyper*, above, and *eikon*, image). Guttiferae. Cosmopolitan except for tropical lowlands, arctic, high altitude and desert regions. Over 400 species, small trees, shrubs or herbs, evergreen or deciduous and dotted with pale and dark glands. The stems are often ridged or angled at first, and the rounded to oblong to ovate leaves are paired or sometimes whorled. Solitary or cymose flowers are produced usually in summer. They are typically dish- or bowl-shaped with five (rarely four) yellow petals and numerous stamens in distinct bundles or a single boss. The fruit is a fleshy, spherical to ellipsoid capsule sometimes ripening from dull red to purple-black.

In general terms, the following are hardy in climate zone 6, possibly lower. They favour full sun or part shade and a fertile, well-drained soil. *H.calycinum* is rampant and resilient groundcover in shade or sun, but may become a nuisance in smaller gardens. *H.*'Rowallane' and *H.*'Hidcote' are useful additions to informal shrub and mixed borders. The colourful fruits of *H.*'Elstead' extends the season of interest into autumn; they are also used by flower arrangers, lasting well in water.

Grown in a sunny, sheltered spot, *H.hookerianum*, its hybrids and cultivars are amongst the most attractive free-standing shrubs in the genus. Many of the alpine, Mediterranean and Southwest Asian shrublets are found on chalk or limestone soils and will benefit from neutral to alkaline soil in the garden. These

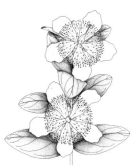

*Hypericum
calycinum*

include *H.cerastioides*, *H.coris*, *H.olympicum* and *H.reptans*. They require open, well-drained sites in full sun on the rock garden, gravel garden, on scree or between paving stones. *H.empetrifolium* may succumb to wet and cold unless grown in the alpine house. The more choice, Asiatic shrubs and the alpine shrublets may be cut to the ground at temperatures below –10°C/14°F, but will usually regenerate from the base if the rootstock has been mulched. *H.calycinum* is hardy to at least –20°C/–4°F.

Prune *H.calycinum* by shearing back almost to ground level every second year. Cut out deadwood for other species in spring or stool them back annually to within a few centimetres of the previous season's growth. Plants with long, gracefully arching stems should be pruned only by removal of old shoots at the base. Propagate from seed sown in spring; by heeled, semi-ripe cuttings in a cold frame, or (alpine species) by basal softwood cuttings in early summer. Where creeping rhizomes or rooting stems are produced, propagate by division of layers. Affected by the rust fungus *Melampsora hypericorum*, which produces pustules on the undersurfaces of leaves. Leaves may also be tied together by the caterpillars of the carnation tortrix moth.

H.acmosepalum (China; shrub, 0.6–2m, branches erect, gradually outcurving; leaves 1.8–6cm oblong to elliptic, apex obtuse to rounded, paler or glaucous beneath, thinly coriaceous to thickly papery; flowers 3–5cm, diameter, 1–3, terminal, star-shaped, deep yellow, sometimes red-tinged); *H.balearicum* (Balearic Islands; evergreen shrub or tree, 0.6–2m, spreading; stem and leaves glandular; leaves 0.6–1.5cm, ovate to oblong, apex rounded, somewhat undulate; flowers 1.5–4cm in diameter, solitary, star-shaped, golden, exterior faintly red-tinged); *H.beanii* (China; shrub 0.6–2m, bushy, robust, erect or arching; leaves 2.5–6.5cm, narrowly elliptic or oblong to lanceolate or ovate, paler to glaucous below, thickly papery to thinly coriaceous; flowers 3–4.5cm in diameter, 1–14, terminal, star-shaped to cupped, golden yellow); *H.bellum* (N India, China, Burma; shrub 0.3–1.5m, erect to arching; leaves 1.5–7.8cm, oblong-lanceolate to subcircular, apex obtuse to rounded or indented, often apiculate, paler or glaucous beneath, thickly papery; flowers 2.5–6cm in diameter, 1–7, terminal, cupped, golden to butter-yellow or pale yellow); *H.calycinum* (Bulgaria, Turkey; shrub 0.2–0.6m tall, evergreen, with creeping branching stolons and erect stem; leaves 4.5–10.4cm, oblong to elliptic or narrowly ovate, apex obtuse or apiculate, pale beneath, leathery; flowers 1 (rarely 2–3), star-shaped, bright yellow); *H.cerastioides* (Bulgaria, Greece, Turkey; perennial herb, 7–27cm, often shrubby at base, decumbent or ascending; leaves 1–3cm, oblong to elliptic or ovate, apex rounded, downy, slightly paler beneath, thickly papery; flowers 2–4.5cm in diameter, star-shaped, 1–5, bright yellow); *H.coris* (SE France, Switzerland, N and C Italy; dwarf subshrub or perennial herb, 10–45cm; leaves 0.5–1.8cm, in whorls, linear, apex shortly apiculate to rounded, revolute, glaucous beneath, coriaceous; flowers 1.5–2cm in diameter, star-shaped, 1-to 20, in a broadly pyramidal to shortly cylindric inflorescence, golden yellow, sometimes veined red); *H.* × *cyathiflorum* (probably *H.addingtonii* × *H.hookerianum*; shrub to about 1.5m, widely spreading; leaves 3–7.5cm, lanceolate, apex acute to apiculate, paler beneath, thickly papery; flowers 4–5cm in diameter, cupped, golden yellow; introduced as *H.patulum* 'Gold Cup'). *H.empetrifolium* (Albania, Greece, Crete, Turkey, Cyprus, Libya; shrublet to 60cm erect, tufted and narrowly branched, or cushion-like, sometimes procumbent, rooting at nodes; leaves 0.2–1.2cm, in whorls, linear or narrowly elliptic, revolute, paler beneath, coriaceous; flowers 1–2cm diameter, star-shaped, 1–40 in cylindric to slender pyramidal inflorescence, golden yellow); *H.forrestii* (SW China, NE Burma; shrub, 0.3–1.5m, bushy ± erect; leaves 2–6cm, lanceolate or triangular-ovate to broadly ovate, apex obtuse to rounded, paler beneath, thickly papery; flowers 2.5–6cm in diameter, 1–20, usually deeply cupular, golden); *H.*'Hidcote' (probably *H.* × *cyathiflorum* 'Gold Cup' × *H.calycinum*; shrub to 1.75m, bushy; branches arching to spreading; leaves 3–6cm, triangular to lanceolate, apex acute to obtuse or slightly mucronate, thickly papery; flowers 3.5–6.5cm in diameter, produced over a long time, forming subcorymbiform inflorescence, cupped, golden yellow); *H.hookerianum* (India, N Thailand, Bangladesh; shrub 0.3–2m, bushy, round-topped erect to spreading; leaves 2.5–8cm, narrowly lanceolate to oblong, apex acute to rounded, paler or glaucous beneath, thickly papery; flowers 3–6cm in diameter, 1–5, terminal, deeply cupped, deep to pale yellow); *H.* × *inodorum* (*H.androsaemum* × *H.hircinum*; deciduous shrub 0.6–2m, bushy, erect; leaves 3.5–11cm, oblong-lanceolate to broadly ovate, apex acute to rounded, somewhat paler beneath papery; flowers 1.5–3cm in diameter, star-shaped or cupped, golden yellow; widely naturalized; yellow-leaved and variously variegated cultivars include 'Summergold', 'Ysella', 'Goudelsje', 'Hysan' and 'Beattie's Variety'; 'Elstead' is a selection (or back-cross to *H.androsaemum*) with large fruit that flush rosy-red, not the usual cerise, during ripening); *H.kouytchense* (China; shrub 1–1.8m, bushy; branches arching or pendulous; leaves 2–5.8cm, elliptic to ovate or lanceolate, apex acute to obtuse or rounded to apiculate, paler beneath, thickly papery; flowers 4–6.5cm in diameter, star-shaped, bright golden yellow); *H.* × *moserianum* (*H.patulum* × *H.calycinum*; shrub 30–70cm, semi-evergreen, spreading or arching; leaves 2.2–6cm, oblong to lanceolate or ovate, acute to apiculate, paler beneath, thickly leathery; flowers 4.5–6cm in diameter, 1–8, forming subcorymbiform inflorescence star-shaped or slightly cupped, bright yellow; 'Tricolor': leaves variegated cream, pink and green); *H.olympicum* (Greece, Balkans, Turkey; dwarf shrub, 0.5–4cm erect to creeping; leaves 5–38mm, oblong to elliptic or ovate to linear, apex acute to obtuse, thinly glaucous, thinly coriaceous; flowers 2–6cm in diameter, star-shaped, golden to lemon yellow, sometimes tinted red; 'Sunburst': robust, flowers large; 'Citrinum': flowers lemon yellow. 'Sulphureum': leaves narrow, flowers pale yellow); *H.patulum* (China, introduced Taiwan, Japan, shrub 0.3–1.5m, bushy, arching to spreading, sometimes weakly frondose; leaves 1.5–6cm, lanceolate or oblong-lanceolate to ovate, glaucous beneath, thickly papery; flowers 2.5–4cm in diameter, forming subcorymbiform inflorescence, cupped, golden yellow; *H.reptans* (Himalaya; shrublet, prostrate or ascending to 0.3m, forming clumps or mats; leaves 0.7–2.2cm, elliptic to obovate, obtuse to rounded, paler beneath, leathery; flowers 2-3cm in diameter, solitary, ± deeply cupped; petals, deep golden yellow, sometimes tinged red); *H.* 'Rowallane' (probably *H.leschenaultii* × *H.hookerianum* 'Charles Rogers'; shrub to 3m; branches erect, gradually outcurving; leaves 2.7–6.7cm, ovate to oblong-lanceolate, paler or glaucous beneath, thinly coriaceous to thickly papery. Flowers 5–7.5cm in diameter, 1–3, terminal, shallowly cupped).

hypertrophy abnormal development of plant tissue due to an increase in cell size which may result from various causes including disease infection.

hypertufa an artificial material with water absorbent properties similar to tufa, made by mixing one part cement, one part sand, and one or two parts finely granulated peat, or a similar material, with water. It is used for the cultivation of alpines, and for coating glazed sinks to resemble troughs of natural stone.

hypocotyl the axis of an embryo lying beneath the cotyledons and, on germination, developing into the radicle.

Hypoestes (from Greek *hypo*, under, and *estia*, house: the calyx is enclosed by bracts). Acanthaceae. South Africa, Madagascar, SE Asia. Some 40 species, evergreen perennial herbs, subshrubs and shrubs with clustered or solitary, tubular, two-lipped flowers. The following species is a popular houseplant, valued for its colourful foliage. Cultivation as for *Justicia*. *H.phyllostachya* (POLKA DOT PLANT, MEASLES PLANT; evergreen subshrub to 1m; leaves to 5cm, oval, dark green flecked or blotched white to pink or purple-red; flowers magenta to lilac).

hypogynous borne beneath the ovary, generally on the receptacle; used of the calyx, corolla and stamens of a superior ovary.

Hypoxis (from Greek *hypo*, beneath, and *oxys*, sharp, referring to the base of the capsule). Hypoxidaceae. N America, Africa, Australia, tropical Asia. STAR GRASS. About 150 species, perennial cormous herbs. The clumped, basal leaves are narrowly sword- or strap-shaped, sometimes ridged beneath and often hairy. Produced singly or in sparse racemes in spring and summer, the flowers are starry and consist of six, outspread, ovate to oblong tepals. Grow in a light, freely draining soil in full sun or in light dappled shade with a minimum temperature of 7°C/45°F.

Keep well-ventilated as the shoots appear; water sparingly until the roots are growing well, and thereafter moderately. Keep moist when in full growth and then gradually dry off and store dry in a shaded and frost-free position. Propagate by seed and offsets.

H.angustifolia (South Africa; leaves 10–15cm, weakly hairy or smooth; flowers to 2cm across, bright yellow, usually paired); *H.capensis* (WHITE STAR GRASS; South Africa; leaves 10–30cm, smooth; flowers 2-6cm across, white or yellow with purple basal spots, solitary).

Hypsela (from Greek *hypselos*, high, referring to the occurrence of the type species in the Andes). Campanulaceae. S America, New Zealand, Australia. 4 species, prostrate perennial herbs, with small, solitary flowers tubular below and spreading above in a 5-lobed, 2-lipped limb. Creeping plants forming dense low mats of ground cover, with a mass of tiny flowers in spring and summer. Well-suited to humus-rich and moisture-retentive gritty soils in the rock garden, in part to full shade. Hardy to between –5 to –10°C/23–14°F. Propagate by division in spring. *H.reniformis* (syn. *H.longiflora*; S America; stem to 5cm, creeping; leaves to 1cm, elliptic to reniform; flowers small, white suffused pink, veined carmine, dotted yellow at base; includes 'Greencourt White', with bright green leaves and white flowers).

Hyssopus (identified, possibly mistakenly, with a plant described by Dioscorides). Labiatae. S Europe to C Asia. Some 10 species, strongly aromatic perennial herbs or dwarf shrubs with narrow leaves and spikes of tubular and 2-lipped flowers in summer. Grow individually in borders or pots; alternatively, use *H.officinalis* as a low hedge in the herb garden. Plant in well-drained soils on sunny sites. Pinch to give bushy plants. Propagate from seed or by softwood cuttings in spring or early summer. *H. officinalis* (HYSSOP; S and E Europe, widely naturalized in Europe and US; erect to 60cm; flowers violet or blue, occasionally white; includes 'Albus', with white flowers, 'Grandiflorus', with large flowers, 'Purpurascens', with deep red flowers, 'Roseus', with rose pink flowers, and 'Sissinghurst', dwarf and compact).

hysteranthous with leaves developing after the flowers; cf. *precocious* and *synanthous*.

Hystrix (from Greek *hystrix*, porcupine, referring to the spiky panicles). Gramineae. N America, N India, China, New Zealand. BOTTLE-BRUSH GRASS. 6 species of perennial grasses. Grown for its flowerheads, *H.patula* produces slender grey stems and swaying heads of long-awned, horizontally spreading spikelets. The flowerheads should be picked when immature for air drying. Hardy in zone 6, it grows best on moisture-retentive but well-drained soil in sun or light dappled shade, but also tolerates dry shade, and is therefore ideal for naturalistic woodland gardens. Propagate by division or by seed in spring.

Hyssop

I

IAA indol-3-yl-acetic acid; an auxin capable of producing a wide range of growth effects in plants. Where synthesized as a product to aid the rooting of cuttings, it has the disadvantage of breaking down quickly before use in the presence of sunlight and where contaminated with plant or soil debris.

IBA 4 indol-3-yl-butyric acid; an auxin widely used in synthesized form as an aid to rooting cuttings. It is more extensively used than IAA because of its relative stability and effectiveness on many different plants.

Iberis (from Iberia, where most of the species occur). Cruciferae. S Europe, W Asia. CANDYTUFT. Some 30 species, annual or perennial herbs and subshrubs bearing small, white or mauve, 4-petalled flowers in racemes or corymbs in spring and summer. Hardy to –15°C/5°F. Grow in well-drained and rather poor, neutral to alkaline soil in full sun. Propagate annuals by seed sown *in situ* from early spring to summer. For greenhouse blooms, sow in autumn and mid-winter at 15°C/60°F, growing on in light, well-ventilated conditions, with a maximum temperature of 10°C/50°F. Propagate perennials by semi-ripe cuttings.

I.amara (W Europe; annual; erect, bushy, 15–40cm tall; leaves lanceolate, entire to toothed; flowers white or purple-white, fragrant); *I.saxatilis* (SW Europe; evergreen perennial subshrub to 15cm tall; leaves subcylindrical, fleshy, linear to needle-like; flowers white, often tinged purple with age); *I.sempervirens* (S Europe; spreading evergreen, perennial subshrub to 30cm; leaves oblong to spathulate; flowers white; 'Snowflake': mound-forming, dark green leaves, flowers sparkling white); *I.umbellata* (S Europe; bushy annual to 30cm tall; leaves linear to lanceolate; flowers white, pale purple-pink or mauve, sometimes bicoloured, in umbels).

ichneumon wasps (Hymenoptera: Ichneumonidae) small to medium-sized insects which are potentially beneficial to the gardener as a biological control of pests. They have two pairs of membraneous wings, the forewings having a dark area half way along the front margin. The first segment of the abdomen is constricted to form a narrow waist. The adults are active in warm sunny weather frequenting flowers such as umbellifers. Their larvae are all parasitic, especially of butterfly and moth caterpillars, whilst some parasitize sawfly larvae. Others show preference for beetles and to a lesser extent fly larvae, and a few parasitize spiders, aphids and lacewings. The red-brown *Ophion* species are familiar, being often attracted to artificial lights at dusk, and these parasitize various noctuid moths. *Rhyssa persuasoria* is about 25mm long and the female has an ovipositor twice this length with which it parasitizes the larvae of wood-boring wood wasps (*Urocerus* species).

Idesia (for Eberhard Ysbrant Ides (fl. 1720), Dutch traveller in Asia). Flacourtiaceae. S Japan, Taiwan, C and W China, Korea. 1 species, *I.polycarpa*, a dioecious, deciduous tree to 12m tall with stout and spreading branches. The leaves are to 20cm long and cordate to ovate with toothed margins, dark green above and glaucous beneath. The petioles are red and to 20cm long. Small, fragrant and yellow-green, the flowers are produced in summer in a pendulous panicle to 30cm long. Orange-red berries persist in crowded trusses after leaf fall. Hardy to –15°C/5°F. Grow in free-draining but moist, neutral to slightly acid soils in sun or light shade. Plants of both sexes are needed for fruit-set. Propagate by seed, or by softwood or semi-ripe cuttings in summer.

Ilex (from its resemblance to the leaves of *Quercus ilex*, the evergreen oak). Aquifoliaceae. Cosmopolitan, temperate and tropical. HOLLY. Over 400 species, evergreen and deciduous trees, shrubs and climbers with tough, entire, toothed to spiny-margined leaves. Small and yellow-green to white, the flowers are borne in the axils singly, in clusters, or in cymes. They are followed by shiny and sometimes highly coloured berries.

Hollies are excellent, hardy plants for woodland or specimen planting, while some make fine hedges and windbreaks They will tolerate all but waterlogged soils. Propagate by grafting in spring on seedlings of the species. Alternatively, increase by cuttings in late summer; reduce the number of leaves, dip in fungicide, put in a sandy mix and place in mild heat. Cuttings will also root in an outside frame. Seed can take 2–3 years to germinate.

I. x *altaclarensis* (*I. aquifolium* x *I. perado*; large evergreen trees and shrubs to 20m, differing from *I. aquifolium* in their more vigorous growth, larger, broader and more variable leaves and larger fruit; includes numerous cultivars, among them selections with broad and dense to narrow, erect habits, leaves ranging in form from large, broad and spineless to puckered and prickly, in colour from darkest green to lime and pure gold, some variegated yellow, white and grey-green, some female clones are very free-fruiting, with fruit ranging in colour from scarlet to deep brick red);*I. aquifolium* (COMMON HOLLY, ENGLISH HOLLY; S & W Europe, N Africa, W Asia; evergreen shrub or tree to 25m; leaves 5-10cm, elliptic to ovate, dark glossy green, undulate, entire or spiny; fruit red; many cultivars, broadly conforming to the range described for *I.* x *altaclarensis*, but embracing a far greater variety of habit, from low and dense to tall and columnar; the berries may be dark blood red, scarlet, orange-red or yellow; the leaves range from small to large, narrow to broad, entire to viciously spiny; leaf colouration can include gold, lime, white and silvery grey splashes tints and variegation, and, when young, bright pink flushes); *I.* x *aquipernyi* (*I. aquifolium* x *I. pernyi*; evergreen shrub or small tree to 6m; leaves similar to *I. pernyi*, to 4cm, apex prolonged, margins wavy-spiny); *I. ciliospinosa* (W China; evergreen shrub to 6m; leaves elliptic to ovate, to 5cm, acute, weakly spined, dull dark green; fruit red); *I. cornuta* (CHINESE HOLLY, HORNED HOLLY; China, Korea; dense, rounded evergreen shrub to 4m; leaves rectangular, 5-8cm, dull green, spines variable; fruit red; includes cultivars with dense, compact habits, leaves spiny to more or less spineless, some variegated yellow, others narrow and twisted); *I. crenata* (BOX-LEAVED HOLLY, JAPANESE HOLLY; Japan, Korea, Sakhalin Is.; evergreen shrub or small tree to 5m; leaves ovate to elliptic, 1.5-3cm, acute, crenate, dark green; fruit glossy black; includes many cultivars, ranging in habit from near-prostrate to dwarf and compact, some densely bushy, taller and suitable for low hedging; the leaves range from small to miniature broad to narrow, toothed to spineless, flat to puckered, they may be bright green, golden, variegated, dark green, purple-flushed or almost black; the best female clones produce an abundance of glossy black berries); *I. dipyrena* (HIMALAYAN HOLLY; W China, E Himalaya; evergreen tree to 15m; leaves 5-11cm, oblong or elliptic, acute or briefly tapering, entire or spiny in juveniles, dull mid-green; fruit red); *I. fargesii* (China; evergreen shrub or small tree to 12m; leaves 6-12cm, oblanceolate to oblong-elliptic, apex tapering, serrate or entire near base, leathery, dull dark green); *I. georgei* (Burma, Yunnan; wide-spreading, dense shrub to 5m; leaves 3.5-5cm, lanceolate to ovate, apex acuminate, weakly spined, leathery, dark glossy green; fruit red); *I. glabra* (GALLBERRY, INKBERRY; E N America; evergreen, erect shrub to 3m; leaves 2-5cm, narrowly obovate to oblanceolate, entire or with a few teeth near the apex, glossy green; fruit black; includes cultivars with white berries, also

dwarf and slow-growing, hardy selections, some with foliage flushed dark wine red in winter); *I. integra* (Japan, Korea, Taiwan; evergreen shrub to 7m; leaves 5-8cm, obovate to elliptic, blunt, entire, dark glossy green; fruit dark red); *I. kingiana* (syn. *I. insignis*; E Himalaya, Yunnan; evergreen tree to 5.5m; leaves 11-22cm, oblong, narrowly tapering, very spiny when juvenile, adults slightly serrate or entire, leathery, glossy green; fruit bright red); *I.* x *koehneana* (*I. aquifolium* x *I. latifolia*; evergreen tree to 7m; leaves 8-15cm, oblong to elliptic, acute, glossy mid-green; fruit bright red; cultivars include free-fruiting, lime-green, bright green and black-green-leaved clones); *I. laevigata* (SMOOTH WINTERBERRY; E US; differs from I. verticillata in its glabrous leaves and longer-stalked, orange-red to yellow fruit); *I. latifolia* (TARAJO; Japan, China; erect, evergreen shrub to 7m; leaves 8-18cm, oblong to oblong-ovate, blunt or briefly tapered, serrate, leathery, dark glossy green; fruit orange-red); *I. macrocarpa* (S China; deciduous tree to 17m; branchlets short, spur-like; leaves7-11cm, ovate to elliptic, tapered, finely toothed; fruit very large, black); *I.* x *meserveae* (*I. aquifolium* x *I. rugosa*; evergreen shrub to 2m, resembling a small-leaved *I. aquifolium*; leaves glossy, often blue-green, spiny; fruit red; cultivars include clones with improved hardiness, slow-growing, compact habit, purple-tinted stems and new growth, foliage ranging in colour from mid-green to dark, glossy blue-green); *I. opaca* (AMERICAN HOLLY; E & C US; evergreen tree to 15m; leaves 5-12cm, oblong to elliptic, entire or spiny, usually tipped with a spine, dull matt green above, yellow-green beneath; fruit crimson; many cultivars varying in hardiness and rate of growth, some tall and open or conical to densely columnar, others small and compact to near-dwarf; there are also clones with exceptionally fine foliage and fruit, some yellow-berried; some, like 'Mrs Santa' are suitable for hedging; 'St Mary' is offered as a Christmas potplant in N America); *I. pedunculosa* (Japan, China, Taiwan; evergreen tree to 10m; leaves 4-8cm, ovate, acuminate, entire, dark glossy green; fruit bright red, long-stalked); *I. pernyi* (C & W China; evergreen shrub or small tree to 8.5m; leaves 1.5-3cm, triangular, margins 5-spined, apex tapered, base square, dark glossy green; fruit dark red; 'Jermyns Dwarf': habit low, arching); *I. serrata* (JAPANESE WINTERBERRY; Japan, China; deciduous shrub to 4m; leaves 4-8cm, elliptic, apex acute to narrowly tapered, finely toothed, somewhat downy at least at first; fruit red; includes yellow- and white-fruited cultivars); *I. verticillata* (BLACK ALDER, WINTERBERRY; E N America; deciduous, suckering shrub to 2m; leaves 4-10cm, obovate to lanceolate, acuminate, serrate, bright green, hairy beneath; fruit red; includes cultivars with foliage turning rich bronze in autumn and berries ranging from bright scarlet to orange-red and yellow; leafless, fruiting branches of some are sold in the cut-flower trade); *I. yunnanensis* (Burma, China; evergreen shrub to 4m; leaves 2-3.5cm, ovate to lanceolate, crenate to serrulate, glossy dark green; fruit red).

I. crassifolia

I. 'Golden Milkboy'

I. 'Indian Chief'

I. 'Lawsoniana'

I. 'Sentinel'

Illicium (from Latin *illicium*, allurement, alluding to the attractive odour of these plants). Illiciaceae. SE Asia, SE US, West Indies. ANISE TREE. Some 40 species, evergreen, aromatic shrubs and small trees with thick-textured, glossy leaves. Produced in late spring and summer, the solitary flowers are borne in the leaf axils and consist of numerous, oblong, petal-like segments. Strongly anise-scented, the fruit is star-shaped with thinly woody 1-seeded carpels. The following are hardy in zone 7, but prefer a sheltered situation, or wall protection. Grow in sun or part shade, in lime-free, humus-rich soils that are well-drained but never dry. Propagate from simple layers, or by semi-ripe cuttings.

I.anisatum (STAR ANISE; Japan, Taiwan; highly aromatic shrub or small tree to 8m; flowers to 3cm in diameter, segments to 30, yellow-green to white); *I.floridanum* (PURPLE ANISE; SE US; shrub or small tree to 3m; flowers to 5cm in diameter, nodding, segments to 30, dark red, purple or maroon); *I.henryi* (C and W China; shrub or small tree to 7m; flowers to 2cm in diameter, cupped, segments 10–14, copper to dark red).

imago the adult form of an insect.

imbibition the absorption of water by seeds in preparation for germination.

imbricate of such organs as leaves or bracts, overlapping.

immersed (1) entirely submerged in water; cf. *emersed*; (2) features embedded and sunken below the surface of a leaf blade.

immortelle see *everlasting*.

immunity absolute and specific resistance of plants to disease or, less commonly, to pest infestation. Immunity is due to metabolic or morphological factors; it should not be confused with resistance or tolerance (two terms often used interchangeably). Immunity is a possible aim of plant breeders, and successful products are known as immune cultivars. The advantages of immune cultivars should be seen against one potential disadvantage: that, unlike resistant cultivars, they present fungal, bacterial and viral pathogens with a threat to their existence, and this encourages genetic variation in the attacking organism. A consequent breakdown in immunity of the host species can have catastrophic effects.

imparipinnate where a pinnately compound leaf terminates in a single leaflet, pinna or tendril; cf. *paripinnate*.

Impatiens (from Latin *impatiens*, impatient, referring to the elastic valves of the seed pod, which violently discharge the ripe seed). Balsaminaceae. Cosmopolitan except S America, Australia and New Zealand. PATIENCE PLANT, BALSAM, BUSY LIZZIE, SULTANA. 850 species, annual or perennial herbs or subshrubs with more or less succulent stems. Borne in spring and summer, the flowers consist of five petals and a curving spur. The fruit is an explosive capsule which, when touched, suddenly splits and rolls inwards, flinging the seeds in all directions. *I.walleriana*, *I.hawkeri* and *I.balsamina* make good house plants, but are more usually grown as summer-flowering half-hardy bedding for borders and containers; they are particularly useful in shady sites. Species like *I.niamniamensis* and *I.repens* are grown in the intermediate to hot greenhouse – keep the larger perennials fairly dry during winter dormancy when leaf loss may occur. *I.glandulifera* and *I.capensis* will survive and self-sow in the open where winter temperatures fall to –15°C/5°F and below. They are suitable for naturalizing in the cool, shady wild garden, but may become invasive. Sow seed of these hardy species *in situ* in late spring.

For frost-tender species, sow seed under glass in spring and prick out into individual pots of a loam-less or medium-fertility loam-based mix; grow on in sunny, airy but frost-free conditions, and pinch and turn frequently to create bushy plants. For bedding, plant out after the last frosts, and keep well-watered. Regular pinching of the succulent stems is necessary to obtain compact, attractive specimens. For the greenhouse or home, grow on in a low-fertility loam-based mix and give cool conditions and filtered light in summer, with a direct light and a minimum temperature of 10°C/50°F in winter. If the side-shoots and first flower buds of *I.balsamina* are removed, tall flowering spikes result. Propagate good colour forms by softwood cuttings in spring; root in a soilless propagating mix in a closed case with gentle bottom heat. Seed of many of the warmer-growing species is short-lived. This is not the case however with *I.balsamina*, *I.hawkeri* and *I.walleriana* and their cultivars.

I.balsamina (GARDEN BALSAM, ROSE BALSAM; SE Asia; more or less unbranched annual, erect to 75cm; flowers 2.5–5cm in diameter, saccate, white, creamy yellow, pink, lilac, deep crimson, scarlet or purple, some bicoloured, some double or semi-double); *I.capensis* (JEWELWEED, LADY'S EARINGS, ORANGE BALSAM; N US, naturalized Europe; annual to 1m, erect to spreading; flowers to 2.5cm, saccate, orange-yellow with red-brown spots); *I.glandulifera* (HIMALAYAN BALSAM, POLICEMAN'S HELMET; Himalaya, naturalized Europe, N America; stout annual erect to 2m; flowers to 4cm, fragrant, deeply saccate, pink to rose-purple, to lavender or white with a yellow-spotted interior); *I.hawkeri* (New Guinea to Solomon Islands; branched perennial herb to 10cm; stems thick, red-tinted; leaves tinged bronze; flowers 6–8cm in diameter, red, crimson, purple, pink or white; includes cultivars with large, double and frilled flowers, some with very deep bronze-red leaves, or variegated emerald, yellow and red); *I.niamniamensis* (E tropical Africa; perennial to 90cm, more or less erect and unbranched; flowers to 5cm, deeply saccate and compressed, sepals red, petals yellow-green; 'Congo Cockatoo': flowers large, red and bright yellow-green); *I.repens* (India, Sri Lanka; creeping and freely branching perennial; flowers to 4cm, saccate, yellow, spur downy); *I.walleriana* (syn. *I.sultanii*; BUSY LIZZIE, PATIENCE PLANT, SULTANA; perennial treated as an annual, 15–60cm tall, bushy with translucent, succulent branches; flowers 2.5–5cm diameter, flat, bright red, crimson, orange, pink, white or multicoloured; numerous cultivars and seed races include dwarf and large plants, with single and double flowers in a wide colour range, and leaves variegated yellow or tinted red or purple-bronze).

Imperata Gramineae. Cosmopolitan. 8 species, clump-forming to running perennial grasses with linear to lanceolate leaves. Silky flowers are carried in spike-like panicles from late summer to autumn. Grow in damp but well-drained, fertile soils, preferably in full sunlight. Hard frosts and winter wet may cause young plants to deteriorate or perish – their crowns may need a dry mulch in zones 6 and under. Increase by division in early spring. *I.cylindrica* (Japan; leaves to 50cm erect to arching, bright green; the plant usually grown is 'Red Baron', (syn. 'Rubra'), JAPANESE BLOOD GRASS; it seldom exceeds 30cm, with new growth flushed bright ruby red at first, becoming brilliant blood red to deep garnet in autumn).

imperfect when certain parts usually present are not developed. Imperfect flowers are unisexual.

impressed a feature sunken into the surface of, for example, a leaf or stem.

inarching see *grafting*.

inbreeder a genetically uniform cultivar that reproduces uniformly by self-pollination and self-fertilization, for example lettuce, tomato, pea.

inbreeding the process of fertilizing the flowers of a plant with its own pollen; resultant plants may be referred to as an inbred line.

inbreeding depression the loss of vigour, fertility, yield, or other desirable characteristics which may result from inbreeding.

Incarvillea (for Pierre d'lncarville (1706–57), French missionary in China, and botanical correspondent of the botanist Bernard de Jussieu). Bignoniaceae. C and E Asia, Himalaya. 14 species, annual or perennial herbs (those listed below are perennials), usually with basal clumps of pinnate or pinnatisect leaves. Produced in summer in terminal racemes or panicles, the flowers are tubular to campanulate with a limb of five, broad and spreading lobes. The following species are hardy to –15°C/5°F, but where the ground is likely to be frozen for long periods, protect the crowns with a deep, dry mulch. Grow them in deep, fertile soils neither waterlogged in winter nor dry in summer. Position in full sun but with some protection at the hottest part of the day. Propagate by seed in autumn or spring and prick out into individual pots. Division is possible but difficult: established plants resent root disturbance.

I.delavayi (China; to 60cm; corolla tube to 6cm, purple-pink and yellow outside, yellow lined purple inside, lobes to 3cm, rounded, wavy, purple-pink); *I.mairei* (SW China; to 30cm; corolla tube 3–6.5cm, deep rose outside, white to grey or yellow inside, lobes to 3cm, rounded, wavy, crimson to deep carmine).

inchworm see *caterpillars*.

incised dissected, but cut deeply and irregularly with the segments joined by broad lamina.

included enclosed within; as of a grass floret by its glume, or stamens within a corolla.

incompatibility (1) the inability of one plant to fertilize another, a factor which is of particular relevance in top fruit growing; (2) the inability of a given rootstock to form a long-term union with a scion.

incomplete flowers lacking one or more of the four whorls of the complete flower.

incrassate thickened.

incumbent folded inward and lying on or leaning against another organ. It is used of cotyledons when the radicle lies on one side instead of along the edge, and of an anther when it lies against the inner face of its filament.

incurved, incurving applied to one of the major exhibition forms of chrysanthemum in which the florets turn upwards and inwards, either tightly to form a compact globose flower (incurved), or loosely (incurving).

indehiscent not splitting open to release its pollen or, more usually, seeds or spores. It is generally applied to fruit types, such as achenes, berries, drupes and pomes; cf. *dehiscent*.

indeterminate (1) describing an inflorescence that does not terminate in a single flower, and in which the spike continues to grow as lower flowers open; (2) used of tomato cultivars where the main stem continues to grow to great length unless stopped; cf. *determinate* or *bush*.

Indianesque a garden style reflecting elements of Indian gardening, particularly in its use of ornaments and architectural detail. Occasionally practised in England by Humphry Repton and his followers in the late 18th and early 19th centuries.

indigenous strictly native; not exotic, naturalized or introduced.

Indigofera (from Latin *indigus* indigo, and *-ferus*, bearing). Leguminosae. Widespread in tropical and subtropical regions. Some 700 species, small trees, shrubs, or annual or perennial herbs with pea-like flowers in racemes or spikes. Although liable to be cut back to the ground in cold winters, they will tolerate temperatures to about –15°C/5°F. Grow in full sun, with the shelter of a warm wall and a deep winter mulch in zones at the limits of hardiness. Prune out dead and frosted growth and rejuvenate old, leggy specimens by cutting back to ground level in spring. Where growth above ground survives several winters, cut the latest season's growth back to framework of older wood in spring. Propagate by semi-ripe cuttings of lateral shoots rooted in the cold frame; alternatively, increase by root cuttings or by removal of suckers.

I.decora (Japan, C China; shrub to 80cm; racemes erect, to 20cm, flowers to 2cm, standard white suffused pale crimson at base, wings pink); *I.dielsiana* (SW China; shrub to 150cm; racemes to 15cm, erect, flowers to 1.5cm, pale red-pink, downy); *I.heterantha* (syn. *I.gerardiana*; NW Himalaya; shrub to 2.5m; leaves grey-downy; racemes dense, erect, 7–15cm; flowers to 1.5cm, pale pink, rosy purple or crimson); *I.pseudotinctoria* (Taiwan, Japan, C China; subshrub to 90cm; racemes 4–10cm, packed; flowers to 0.5cm, pale red to pale pink or white tinted rose); *I.tinctoria* (SE Asia; racemes slender, shorter than leaves, standard pale red to rose, wings and keel of a somewhat deeper hue).

Indocalamus (from *indo*, Indian, and Greek *kalamos*, reed). Gramineae. China, Japan, Malaysia. Some 25 species of relatively small, running bamboos; see *Bamboos*.

indumentum a covering of hair, scurf or scales, most often used in the general sense of hair.

induplicate of a leaf, sheath, sepal or petal, folded inwards.

indurate hardened and toughened.

indusium the epidermal tissue that covers or surrounds the spore-bearing region of a fern frond.

inerm, inermous unarmed; without spines, prickles or teeth.

inferior beneath.

infertile (1) of a plant unable to produce viable gametes or seed, and fruit; (2) sometimes used of soils which are acid or low in nutrients, or of poor structure or with impeded drainage.

infiltration the passage of rain or irrigation water into the soil.

inflexed bent inwards towards the main plant axis.

inflorescence the arrangement of flowers and their accessory parts on an axis.

infra- as a prefix, denotes below, as in infrastipular, meaning below the stipules.

infrageneric describing a taxon below the rank of genus, i.e. subgenus, species, subspecies, variety, forma, cultivar.

infraspecific describing a taxon below the rank of species, i.e. subspecies, variety, forma, cultivar.

infructescence fruiting stage of the inflorescence.

infundibular, infundibuliform funnel-shaped.

injection the placement of a pesticide beneath the soil surface, for example, as a possible control for eelworm; or, occasionally, into the branches, trunks, or roots of trees. In most cases, special equipment is necessary to ensure correct dosage.

inoculum (1) potentially infective material present in the garden environment, for example fruiting bodies of black spot (*Diplocarpon rosae*) overwintering on fallen rose leaves; (2) material containing microorganisms used to inoculate a plant or substrate.

inorganic denoting a substance devoid of carbon; of neither plant nor animal origin. Inorganic fertilizers, whether in natural or synthesized form, are of mineral origin. The identification of organic/inorganic origins is of particular relevance in considerations of organic gardening.

insects an extremely diverse class of animals represented by more than 800,000 described species, which constitute about 70 per cent of all known animal species. Adult insects have a hard external skeleton with three pairs of jointed legs; the body consists of a head, thorax and abdomen. The head has compound eyes, a pair of antennae and mouthparts, which may be adapted for biting and chewing or for sucking liquids. The thorax consists of three segments, each of which has a pair of legs, and in most insects there is a pair of wings on each of the front two segments. Some insects, such as worker ants, have no wings while others, such as female winter moths, have tiny non-functional wings. The abdomen is made up of 11 segments, although some may be very small and modified to form part of the genitalia.

Most insects reproduce by laying eggs but some groups, such as the aphids, can give birth to live young. The immature stages in the life cycle may be very different in appearance to the adult form and may have entirely different dietary requirements. A typical example is the cabbage white butterfly, the larvae of which are caterpillars that feed on brassica leaves; the larval stage is followed by a non-feeding resting stage called a pupa from which the adult butterfly emerges to suck nectar from various flowers. Lacewings, moths, flies, beetles, ants, bees, wasps and sawflies also go through this form of development. Other insects, including earwigs, termites, aphids, scale, mealybugs, whiteflies, capsid bugs and thrips, go through incomplete metamorphosis, producing immature stages which are rather similar in appearance to the adult form. The length of life cycle may be as little as two to three weeks, as shown by aphids breeding under warm conditions, while the larvae of some wood-boring insects need several years to reach maturity.

Virtually all cultivated plants are attacked by at least one insect pest. Thrips and the various insects which belong to the Hemiptera have needle-like mouthparts used to suck sap. Most feed on foliage and stems but some species of aphid and mealybug take sap from roots. Many aphids, whiteflies, scale insects, suckers and mealybugs excrete a sugary liquid called honeydew. This makes the foliage sticky and allows the growth of black sooty moulds. Some aphids, leafhoppers and thrips are efficient vectors of plant virus diseases, carrying viral particles on their mouthparts or in their saliva when they transfer to a new plant. The remaining plant pests have mouthparts used for biting and chewing plant tissues and may be root feeders, stem borers, defoliators, leaf miners, leaf rollers, gall formers, flower feeders or fruit borers. Some moth caterpillars and beetle larvae specialize in feeding on dry plant material such as seeds and stored food.

Not all insects are plant pests. The vast majority are of no economic importance and some are of definite benefit in gardens. Predators and parasites help to keep pest numbers down (see *beneficial insects, biological control*) and inconsiderate use of pesticides can lead to a resurgence of certain pests once their natural enemies have been eliminated.

insecticide a chemical used to kill or control insect pests. Insecticides are sometimes given names that describe the type of pest or stage of its lifecycle controlled for example acaricide (mites), nematicide (nematodes or eelworms), molluscicide (slugs and snails) and ovicide (eggs).

Most insecticides work by contact action, i.e. it is necessary for the pest to be either covered by the chemical or to walk over a treated surface in order for it to be affected. Some insecticides are absorbed into the plant and are effective against sap-feeding pests such as aphids. Others are stomach poisons deposited on plant surfaces to be eaten by pests such as caterpillars and beetles; a few insecticides are volatile and work by vapour action. In practice, many chemicals act on pests in more than one way. Insecticides are most usually applied as sprays, but also as dusts, granules, smokes, baits and aerosols.

Plant-derived insecticides Plant-derived insecticides are naturally occurring, chemically unrelated substances, which include some of the earliest pesticide discoveries. Pyrethrum, derived from the flowerhead of *Tanacetum cinerariifolium*, was used as early as the 18th century in Persia. Derris, or rotenone, is produced from the roots of *Derris* and *Lonchocarpus* species; its insecticidal properties were developed in the 1920s from the practice of South American native peoples using crushed roots to paralyse fish. Nicotine is extracted from tobacco waste, which emanates from *Nicotiana* species. Soft soaps are derived from palm oil as well as other natural fats. All of these substances will provide some control of pests but they have very short persistence. Their harmful effects on beneficial insects are limited and when diluted they are of low toxicity to man and most other vertebrate animals, except fish.

Synthesized insecticides These fall into the following main groups.

ORGANOCHLORINES A group of insecticides, developed during the period 1939–60, which combine relatively low mammalian toxicity with a high degree of insecticidal activity. Many are very persistent; those such as DDT, aldrin, dieldrin and chlordane have been banned in many countries because of their accumulation within the environment and harmful effects on birds and animals. Lindane, also known as gamma-HCH or BHC, is of moderate persistence and still used against a wide range of pests. Dicofol is an acaricide used against tarsonemid and red spider mites, although resistance to it is now widespread among the latter.

ORGANOPHOSPHATES A group of insecticides, developed from 1950 onwards, which contain many compounds of short to medium persistence and which break down to harmless chemicals, and are therefore less damaging to the environment than organochlorines. They include contact insecticides, such as malathion, fenitrothion, trichlorfon, dichlorvos, diazinon and pirimiphos-methyl, and systemics, including dimethoate, heptenophos and oxydemeton-methyl, which are effective against sap-feeding pests and leaf miners.

CARBAMATES Among carbamates, developed in the mid 1950s for use as insecticides, are carbaryl and pirimicarb; the latter is a selective insecticide that controls aphids but leaves most other pests and beneficial insects unharmed.

PYRETHROIDS A series of synthetic pyrethroid compounds, available as a result of changes made to the molecular structure of natural pyrethrum. Some of the early products introduced during the 1970s, such as resmethrin and bioresmethrin, have the low mammalian toxicity and short persistence of natural pyrethrum, but greater insecticidal activity. Permethrin is a pyrethroid of moderate persistence, which has low toxicity to mammals and birds. Some of the later pyrethroids used by professional growers, such as deltamethrin and cypermethrin, are more toxic and have greater persistence.

Resistance to insecticides Various pests of fruit, vegetables and greenhouse crops have developed resistance to at least one group of pesticides; these include red spider mites, whiteflies, aphids, some root flies of vegetables, Colorado beetle, diamond back moth, codling moth and pear psyllids. There is no evidence that pesticides induce mutations but instead that they act as a selective agent, killing only individuals with susceptible genetic characteristics and leaving those with genes that provide them with immunity.

Three factors are conducive to the development of resistance. First, the frequency of the occurrence of resistant genes in the original population: some resistant strains are very localized, indicating, in such cases, that the resistant genes are not universally distributed. Second, the intensity of selection, which is dependent upon the magnitude of the population exposed to the pesticide and the proportion killed. Third, the number of generations passed through each year.

In the development of resistance, there are essentially two types of mechanism involved. In one, enzymes undergo genetically-based changes which enhance their ability to break down the pesticide to a non-toxic product; the other, also under genetic control, makes the organisms insensitive to the pesticide. Although such mechanisms usually develop in response to the selective pressure of a single pesticide, the system when established, is invariably capable of affording protection against a group of related pesticides. Once resistance has been established, effective chemical pest control can only be obtained by switching to a chemically unrelated group of insecticides. However, this practice often proves of short-lived value because many insects and mites have the capacity to develop additional mechanisms that provide multiple resistance.

Resistance develops through intensive and widespread use of insecticides, so alternation of substances or, where appropriate, the use of non-chemical methods are valuable in helping to avoid the problem. See *beneficial insects, biological control, integrated pest management, organic gardening*.

insectivorous plant see *carnivorous plant.*

inserted attached to or placed upon.

insertion the point or mode of attachment for a body to its support.

instar any stage between moultings in the growth of an insect larva or nymph.

insulation a means of conserving heat in greenhouses, frames and garden buildings, and around plants, in order to maintain a required growing temperature and/or to protect tender plants from frost. Insulation of heated garden structures results in economy of fuel usage.

On greenhouse structures, double-glazing and double-skin cladding with rigid or film plastic are effective forms of insulation. Polythene sheet fitted to the inside surfaces of a greenhouse provides useful insulation, while products made by fusing two sheets of film to enclose air bubbles are very efficient in heat retention. In large greenhouses, mechanized installations for drawing an insulating sheet horizontally across the width of the greenhouse at about the height of the eaves are worthwhile; these systems are known as thermal screens. During cold spells, frames may be insulated by covering with a polythene sheet, hessian, old carpet or similar. All of these methods of insulation involve some loss of light.

Covering plants with glass or plastic cloches, or horticultural fleece provides varying degrees of protection from heat loss. Bracken or straw wrapped around tender plants outdoors is a valuable means of insulation but some form of outer cover is necessary in order to prevent waterlogging. Polythene screens are suitable for some wall plants. Mounding with soil or any organic mulching material will conserve heat in vegetable clamps and around the roots of plants that are not reliably hardy.

integrated pest management (IPM) The term IPM initially referred to the control of insects and other pests, but has been extended to encompass disease and weed control. The description 'integrated control' was first used in the US in 1956, and in 1967 the Food and Agriculture Organization defined the concept as 'a pest management system that utilizes all suitable techniques and methods, in as compatible a manner as possible, to maintain the pest population at levels below those causing economic damage'. Although in its conceptual sense IPM involves the scientific study of targeted organisms, in practice its implementation is usually concerned with supervisory control. Its broad objectives are an increase in biological control and conservation of natural enemies present, together with a reduction in the use of pesticides and utilization of any secondary measures of control that are available.

Gardeners may not be in a position to practise full IPM, but some of its principles can be followed by increasing the use of non-chemical control methods. For example, the use of greasebands on tree trunks against winter moths reduces the need to use insecticides, as does the placing of discs around the base of brassicas to protect them against root fly (*Delia brassicae*). Correct timing often means that a single application of a non-chemical control method can replace several protective sprays; this would apply, for example, when using pheromone traps for codling moth (*Cydia pomonella*).

Many pesticides can be applied as spot treatments to individual plants or in band treatments along plant rows; also as seed dressings, granules and baits. Not only are such methods less wasteful and therefore more economic but, most important, they target the application effectively, with reduced risk to beneficial insects. Pesticides with a narrow-spectrum and short persistence should be preferred.

The garden is an ideal environment for both plants and pests, and gardening skills are needed to tip the balance in favour of plant health. Healthy plants produced by sound husbandry are less likely to fall prey to pests and diseases. See *beneficial insects, biological control*.

inter- as a prefix, denotes between or among, as in interstaminal, meaning between the stamens.

intercropping, intersowing growing a relatively rapid maturing crop between rows of a slower-growing one in order to make maximum use of available ground. For example, lettuce or radishes can be sown and harvested between widely spaced rows of establishing Brussels sprouts.

intergeneric hybrid the offspring of a cross between plants of different genera. It is designated in written descriptions by a multiplication sign before the hybrid genus name, for example the common conifer × *Cupressocyparis leylandii*, which originates from the hybridization of *Cupressus macrocarpa* with *Chamaecyparis nootkatensis*.

intermediate (1) applied to one of the major exhibition forms of chrysanthemum in which flowers are intermediate in form between Incurved and Reflexed; (2) a hybrid plant showing characteristics intermediate between both parents.

internal rust spot a condition of potatoes which is only apparent when affected tubers are sectioned. It typically shows as small rust-coloured patchy spots distributed across the tuber; these are often more concentrated towards the surface. The condition is most prevalent in potatoes grown on sandy, infertile soils, especially those low in organic matter, and where plants encounter drought stress. There are differences in cultivar susceptibility.

internode the length of plant stem between nodes.

interspecific hybrid the offspring of a cross between two species of the same genus. It is designated in written descriptions by a multiplication sign between the genus name and the hybrid epithet, for example *Forsythia × intermedia*, the hybrid offspring of *F.suspensa* and *F.viridissima*.

interstock the intermediate portion of stem that is grafted to the scion and rootstock of a double-worked tree. Interstocks are used as a means of overcoming incompatibility. For example, the pear cultivar 'Beurré Hardy' is commonly used as an interstock between a quince rootstock and some scion cultivars such as 'Packham's Triumph' and most clones of 'William's Bon Chrétien'. See *grafting*.

intra- as a prefix, denotes within or inside, as in intrastaminal, meaning inside the stamens or the circle of stamens.

intraspecific describes hybrids that have occurred between two distinct varieties of the same species.

introduction a non-indigenous plant, usually purposefully introduced.

introrse turned or facing inwards, towards the axis, as with an anther that opens towards the centre of its flower.

intumescence a small, localized proliferation of plant tissue, on leaves and fruit especially, caused by a humid growing environment or other situation where water uptake is greater than the rate of loss by aerial parts. Similar lesions which are more widespread, and often larger, are referred to as *oedema* (q.v.).

Inula (name used by Pliny for a plant similar to *Helenium* (Greek *helenion*), perhaps *I.helenium*). Compositae. Temperate and warm Old World. About 90 species, mostly perennial herbs or subshrubs, rarely annual or biennial (those listed below are herbaceous perennials), usually hairy, with daisy-like flowerheads composed of slender, golden-yellow ray florets, in summer. Fully hardy. Grow in sun in any moderately fertile and well-drained soil. Propagate by seed or division.

I.acaulis (Asia Minor; to 20cm, stem tinted purple; leaves 3–6cm, oblanceolate to spathulate, entire, ciliate, minutely glandular; ray florets to 1.5cm); *I.ensifolia* (E Europe; to 60cm; leaves to 9cm, linear to lanceolate, entire, ciliate, glabrous; ray florets to 2.2cm); *I.hookeri* (Himalaya; to 60cm; leaves to 13cm, oblong to lanceolate, minutely toothed, hairy; ray florets 2.5cm); *I.magnifica* (Caucasus; to 2m, stem tinted purple-black; leaves to 25cm, elliptic to ovate, dentate, dark green, glabrous above, villous beneath; ray florets to 2cm); *I.oculus-christi* (E Europe, Caucasus, Tukey, Iraq, Iran; to 15cm; leaves to 14cm, oblong to lanceolate or elliptic, entire or denticulate; ray florets to 2cm).

invalid applied to botanical names published without, for example, the necessary formalities of description, diagnosis, type designation, or not published in a valid (i.e. botanically acceptable) publication.

involucel a secondary or diminutive involucre.

involucre a single, highly conspicuous bract, a bract pair, or a whorl of small bracts or leaves subtending a flower or inflorescence.

involute rolled inward, toward the uppermost side.

Iochroma (from Greek *ion*, violet, and *chroma*, colour, referring to the flowers of some species). Solanaceae. Tropical America. Some 20 species, shrubs or small trees with clusters of tubular to narrowly trumpet-shaped, 5-lobed flowers in summer. Provide a minimum temperature of 10°C/50°F, full or bright indirect light and a buoyant atmosphere. Grow in a well-drained, medium-fertility, loam-based mixture; water plentifully in summer, sparingly in winter. In cool temperate zones, pot-grown specimens may be moved outside for the summer. Prune in spring to restrict or rejuvenate. Propagate by greenwood cuttings, rooted with bottom heat in a case in summer, or by seed sown in spring. *I.cyaneum* (syn. *I.tubulosum*; northwestern S America; shrub to 3m; flowers to 3.5cm, deep purple-blue).

Ionopsidium (from Greek *ion*, violet, and *opsis*, appearance, alluding to the faint resemblance to tufted violet species). Cruciferae. Mediterranean. 5 species, tufted, annual herbs with leaves in rosettes and small, 4-petalled flowers in summer and early autumn. Suitable for the rock garden, as edging and for crevices in paving, where it may self-seed. Sow seed *in situ* in moist, fertile, well-drained soils in spring. *I.acaule* (VIOLET CRESS; Portugal; to 8cm; flowers lilac or white tinged violet).

Ipheion (from *iphyon*, a Classical Greek name used in a number of different forms for an assortment of plants, including bulbous genera). Liliaceae (Alliaceae). S America. 10 species, tufted, bulbous perennials, most parts smelling of garlic if bruised, the leaves especially. The bulbs are small and clumped and the leaves narrow and smooth, arising in loose tussocks. Produced in spring and summer, the flowers are salverform with a spreading limb of six, ovate-oblong tepals; they are carried singly or in pairs on slender scapes. *I.uniflorum* is hardy in zone 6, but may need mulch protection where frosts are prolonged. Plant in a well-drained gritty soil in dappled or full sun in a sheltered position. Propagate by seed in spring or by offsets in late summer as the foliage dies back. *I.uniflorum* (syn. *Triteleia uniflora, Milla uniflora, Brodiaea uniflora, Beauverdia uniflora*; SPRING STAR-FLOWER; Uruguay, Argentina; leaves to 25cm, narrowly strap-shaped, glaucescent; scapes 5–20cm, numerous, flowers to 4cm in diameter, pale blue to deeper lilac, with a white or cream suffusion in throat and darker stripes along the midvein of each segment; includes the pure white 'Album', 'Froyle Mill' with deep violet flowers, the clear blue 'Rolf Fiedler', and 'Wisley Blue', with pale blue flowers with deeper tips).

Ipheion uniflorum

Ipomoea (from Greek *ips*, a worm, and *homoios*, resembling). Convolvulaceae. Widely distributed in tropical and subtropical regions. MORNING GLORY. 450–500 species, annual to perennial herbs, shrubs or small trees, woody to herbaceous and usually climbing. The leaves are entire to lobed. Solitary to cymose or paniculate, the flowers are deeply funnel-shaped to tubular with a circular, entire rim or a limb of five lobes.

Apart from *I.tricolor*, the ubiquitous morning glory, other popular species include *I.alba*, the spectacular moonflower, a woody-based perennial with huge white blooms, and *I.coccinea, I.quamoclit* and *I. × multifida* – vigorous vines, though delicate in appearance, with small flowers in brilliant tones of crimson and scarlet. The last two, with finely divided leaves, are especially attractive. The spanish flag, *I.lobata*, is a weakly twining annual or short-lived perennial with erect spikes of flowers which pass with age through scarlet, yellow and cream. Subtropical gardens and large heated conservatories might accommodate the two robust, semi-woody species, *I.mauritiana* and *I.horsfalliae*, the first with narrowly funnel-shaped deep red or purple-pink flowers, the second with wider blooms in shades of pink with a darker 'eye'.

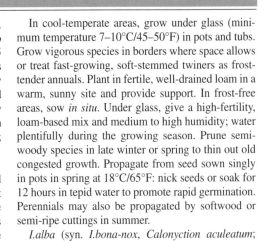

In cool-temperate areas, grow under glass (minimum temperature 7–10°C/45–50°F) in pots and tubs. Grow vigorous species in borders where space allows or treat fast-growing, soft-stemmed twiners as frost-tender annuals. Plant in fertile, well-drained loam in a warm, sunny site and provide support. In frost-free areas, sow *in situ*. Under glass, give a high-fertility, loam-based mix and medium to high humidity; water plentifully during the growing season. Prune semi-woody species in late winter or spring to thin out old congested growth. Propagate from seed sown singly in pots in spring at 18°C/65°F: nick seeds or soak for 12 hours in tepid water to promote rapid germination. Perennials may also be propagated by softwood or semi-ripe cuttings in summer.

I.alba (syn. *I.bona-nox, Calonyction aculeatum*; MOONFLOWER, BELLE DE NUIT; Pantropical; woody-based, herbaceous perennial climber, 5–30m; leaves to 15cm, ovate to heart-shaped, entire to 3-lobed, smooth; flowers to 15 × 14cm, trumpet-shaped, white faintly striped green, fragrant at night); *I.coccinea* (syn *Quamoclit coccinea*; RED MORNING GLORY, STAR IPOMOEA; US; annual herb climbing to 4m; leaves to 10cm, ovate to heart-shaped, entire or toothed; flowers to 2.5 × 2cm, tubular, scarlet with a golden throat, fragrant); *I.hederacea* (syn. *Pharbitis hederacea*; S US to Argentina; hairy, annual herb climbing to 3m; leaves broadly ovate to heart-shaped, usually 3-lobed; flowers to 4 × 3cm, funnel-shaped, blue, purple or cerise with a white tube); *I.horsfalliae* (W Indies; perennial, woody climber to 8m; leaves to 20cm, palmately 3–5-lobed, smooth; flowers to 4 × 5cm, trumpet-shaped, cerise to deep purple-red); *I.lobata* (syn. *I.versicolor, Mina lobata, Quamoclit lobata*; SPANISH FLAG; Mexico; annual or short-lived pernnial herb climbing to 5m; leaves ovate to heart-shaped, toothed to deeply cut or lobed; flowers in 1–sided racemes, each to 4 × 1.5cm, broadly tubular, curving and narrowing toward tip and appearing closed, scarlet in bud, opening yellow tipped scarlet, fading to cream); *I.mauritiana* (Pantropical; woody, perennial climber to 5m; leaves to 15cm, palmately lobed; flowers to 6 × 6cm, funnel-shaped, pink to deep rosy red with a darker stain in throat);. *I. × multifida* (syn.. *I. × sloteri, Quamoclit × sloteri*; *I.coccinea × I.quamoclit*; CARDINAL CLIMBER; garden origin; annual climber to 1m; leaves finely pinnately cut; flowers to 5 × 2.5cm, trumpet-shaped, scarlet to crimson); *I.nil* (syn. *Pharbitis nil*; Pantropical; hairy annual herb climbing to 5m; leaves to 15cm, broadly ovate, entire or 3-lobed; flowers to 6 × 5cm, funnel-shaped, blue, purple or red; includes numerous cultivars with flowers large to small, single or semi-double, in shades or combinations of white, blue, cerise, scarlet, mauve, violet and maroon, some non-climbing, others with variegated foliage;. *I. × imperialis*, IMPERIAL JAPANESE MORNING GLORY, differs in its larger, double flowers often with fringed or fluted limbs); *I.purpurea* (syn. *Convolvulus major* of gardens, *Pharbitis purpurea*; COMMON MORNING GLORY; Mexico, widely naturalized elsewhere in Tropics; annual herb, usually hairy, climbing to 2.5m;

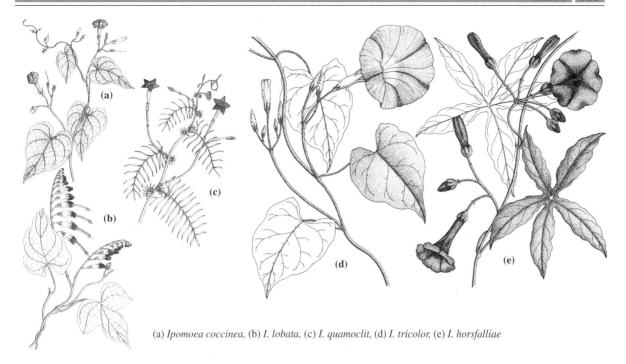

(a) *Ipomoea coccinea*, (b) *I. lobata*, (c) *I. quamoclit*, (d) *I. tricolor*, (e) *I. horsfalliae*

leaves to 10cm, ovate to rounded, entire or 3-lobed; flowers to 5 × 5cm, funnel-shaped, blue, purple, pink, red, white, or with stripes of these colours on a white background; includes cultivars with silver-variegated leaves and deep violet, double flowers); *I.quamoclit* (syn. *Quamoclit vulgaris*, *Quamoclit pennata*; CYPRESS VINE, STAR-GLORY, INDIAN PINK; Tropical America; annual climber to 3m; leaves 1–9cm, finely pinnately cut; flowers to 3cm, narrowly trumpet-shaped with a small, star-like limb, scarlet; includes cultivars with very fine, ferny foliage and flowers in shades of plum, scarlet, flame, crimson, pink and white, some doubles); *I.tricolor* (syn. I. *rubro-caerulea*; MORNING GLORY; Central America, widely naturalized elsewhere; glabrous annual or perennial climbing herb to 4m; leaves ovate to heart-shaped, tip pointed; flowers to 6 × 7cm, funnel-shaped, bright sky blue, often with faint lavender-pink stripes at least at first, tube white, golden at base within; includes cultivars with very large flowers in shades and combinations of white, sky blue, cobalt, lavender, pink and carmine, the most famous selection is probably the large, sky blue 'Heavenly Blue'; this species should not be confused with *Convolvulus tricolor*).

Ipomopsis (from Greek *ipo*, to impress, and *opsis*, appearance, due to the showiness of many species of this genus). Polemoniaceae. US (Pacific coast to Florida and South Carolina), S Argentina (1 species). 24 species, annual to perennial herbs with entire to pinnately lobed leaves in basal rosettes and on stems. The flowers are tubular or hypocrateriform and appear in spring and summer in terminal, cymose racemes. Grow in a sandy, fertile medium in the conservatory or cool greenhouse in full sun with good ventilation. Keep moist but never wet. Sow seed under glass in early spring.

I.aggregata (SCARLET GILIA, SKYROCKET; western N America; glandular-hairy biennial, erect to 80cm; leaves finely pinnate; flowers to 4cm, pink to magenta or bright red mottled yellow, rarely pure yellow or white); *I.rubra* (STANDING CYPRESS; SE US; more or less glabrous, unbranched biennial or perennial erect to 2m; leaves finely pinnate; flowers to 2.5cm, scarlet, interior yellow spotted red).

Iresine (from Greek *eiros*, wool, referring to a harvest garland bound with wool). Amaranthaceae. Americas, Australia. BLOODLEAF. 80 species, annual to perennial herbs or subshrubs with entire, often brilliantly marked or coloured leaves and spikes of inconspicuous, white or green flowers. Frost-tender plants, valued for their colourful foliage. The following species are grown outdoors in the tropics and subtropics, and used for summer bedding or as pot plants elsewhere. They need a minimum temperature of 10°C/50°F. Grow in sun or light shade. Water and feed freely in summer; keep just moist in winter. Pinch out young plants to promote a bushy habit. Increase by seed sown under glass in early spring at a minimum temperature of 15°C/60°F, or by stem cuttings in spring.

I.herbstii (syn. *I.reticulata*; BEEF PLANT, BEEFSTEAK PLANT, CHICKEN GIZZARD; Brazil; annual erect to 1m with glossy and thinly succulent, purple- or red-tinted stems; leaves 2–6cm, ovate with variegation ranging from green with yellow veins or mottling to deep purple-red with pink veins; includes 'Aureoreticulata', with red-tinted leaves netted with yellow veins, 'Brilliantissima', with rich crimson leaves, and 'Wallisii', dwarf with purple-black veins); *I.lindenii* (Ecuador; erect perennial to 1m; leaves 4–10cm, ovate to lanceolate, glossy deep blood red with prominent red veins; includes 'Formosa', with yellow leaves veined with crimson).

Iris (from the Greek messenger Iris who came to earth via a rainbow). Iridaceae. North temperate zones. FLAG, FLEUR-DE-LIS, SWORD LILY. About 300 species, rhizomatous, fleshy-rooted or bulbous perennial herbs with grassy to sword-shaped leaves. The flowers consist of a tubular perianth with six segments. Each of the three outer segments ('falls') is narrowed toward the base into a 'haft', sometimes with a beard of coloured hairs or crested or ridged, and often with a patch of different colour (the 'signal'); the apex of the fall is expanded, showy and hangs downwards. Each of the inner three segments ('standards') is narrowed toward the base into a claw and usually expanded toward the tip; they tend to stand erect between the falls, but are sometimes much-reduced or bristle-like. The style has three petal-like arms, which arch over the falls. The fruit is a leathery capsule. Irises fall into several, natural groups with distinctive characteristics and cultural requirements.

Rhizomatous irises (1) BEARDED OR POGON IRISES (subgenus Iris, section Iris). Several species are good plants for the mixed border and of these *I.germanica* and its white-flowered var. *florentina* and *I.pallida*, with its two variegated-leaf forms 'Argentea' and 'Aurea', are well known. Some of the smaller species such as *I.lutescens* and *I.pumila* are suitable for the rock garden. There is, in addition, an enormous number of named hybrid cultivars of bearded iris. These are classified as follows: Miniature Dwarf Bearded (MDB) up to 20cm, the first to flower; Standard Dwarf Bearded (SDB), 21–40cm; Intermediate Bearded (IB), 40–70cm; Miniature Tall Bearded (MTB), 40–70cm; Border Bearded (BB), also 40–70cm, and the Tall Bearded (TB), over 70cm. The MTB and BB cultivars flower with the TBs. Many iris species will occasionally flower for a second time later in the year and this trait has been developed by the hybridizers, particularly of the bearded cultivars, to create cultivars that bloom again, often repeatedly, until the first frost. To sustain this additional growth in one season, remontant irises may require a further application of fertilizer after the first flowering and should be watered in dry summers.To flower well, bearded irises require a sunny situation and well-drained soil. The soil should ideally be in the pH range 6.0–7.5; they do not require lime, but ground limestone should be worked in if the soil is acid. They will tolerate a higher pH and can grow well on chalky soils. Some humus should be incorporated but avoid fresh manure or green compost, which may cause rhizome rot. Raise the soil level in wet places or if the soil is heavy. Although iris rhizomes can be transplanted at any time, for the best results move only when absolutely necessary and then after flowering during summer dormancy. For planting on sandy soils, dig a hole allowing the roots to fall straight down, push the nose of the rhizome into the side of the hole and fill in with soil, ensuring that no space is left beneath the rhizome. On heavy soil, place the rhizome on a mound or ridge within the planting hole, with the roots spread out, cover with soil and firm in. Under most conditions leave the top surface of the rhizome

exposed, but cover thinly in hot areas, to avoid sun scorch. Shorten the leaf fan to prevent wind-rocking. Water in and keep moist until new growth is seen. Apply low-nitrogen compound fertilizers during soil preparation and subsequently in early spring and autumn. Do not trim off leaves of established plants in summer. After 3–4 years, when the clump becomes crowded, dig up and separate it into single rhizomes; discard the oldest, leafless and exhausted portions. Replant in groups with noses together, in fresh soil. Propagate species by seed or by detached rhizomes. Sow seeds thinly in a soil-based medium with a quarter extra grit; cover with coarse grit and leaves exposed to the weather. Cultivars are propagated by vegetative means only. Principal diseases of bearded iris are rhizome soft rot and scorch. With rot, remove infected tissue, cutting back to firm rhizome, and allow to dry, then dust the new surface with a fungicide. If the infection is grave (affecting more than half the rhizome), discard altogether. With scorch, the leaves suddenly turn red-brown as if burnt; the rhizome remains firm but roots wither and the plant dies. This can occur at any time; it is not believed to be contagious, but affected rhizomes should be removed and destroyed. Leaf spots are either bacterial or fungal, usually occurring after flowering.

(2) ONCOCYCLUS IRISES (subgenus Iris, section Oncocyclus) occur in areas of dry, often hot summers and dry, cool or cold winters. The winter months may be spent under snow or in desert conditions; next comes a period of growth and flowering in spring. This is followed by a long summer dormancy. The flowers are large, often in vivid colours with striking patterns. Except in places where the climatic conditions resemble their habitat, these irises cannot be grown in the open garden. Instead, they are grown either in raised beds outdoors and covered with glass or plastic to protect them from winter and summer rain, or in the cold greenhouse in deep pots. Soil should be gritty, free-draining and with pH above 6.5, with a low-nitrogen fertilizer added. Allow full sun and ventilation at all times. Water moderately to initiate growth between late winter and early spring; continue to water and feed until the foliage begins to die down after flowering. Thereafter keep dry. Germination of seed is often slow and erratic. They are very susceptible to virus diseases.

(3) REGELIA IRISES (subgenus Iris, section Regelia) differ from the last group in having a beard on the standards *and* the falls. They grow on open rocky hillsides and also have a long summer dormancy. They are best suited to bulb frame culture like the Oncocyclus irises, but are easier to grow; some species, such as *I.hoogiana*, can be grown outside in temperate climates in a well-drained sunny site if protected from heavy winter rain. Many crosses have been made between Oncocyclus and Regelia irises; classified as Regeliocyclus Hybrids, these are easier to grow than either parent yet retain their strange beauty. They are good plants for the alpine house. Hybrids between Oncocyclus or Regelia irises with bearded iris species or cultivars are called Arilbred Irises and are suitable for garden use in temperate areas with low rainfall.

Iris reticulata (left), *I. magnifica* (middle),
I. iberica subsp *elegantissima* (right)

(4) EVANSIA IRISES (subgenus Limniris, section
Lophiris) are mostly creeping to spreading plants
with branched rhizomes and fans of foliage.
I.confusa is frost-tender. In climate zones 7 and
below, it should be grown in large pots of a fertile,
neutral to acid and gritty medium. Keep moist
throughout the year; feed in summer and place
outside in full sun. Bring under cover in autumn.
The remaining species are more or less hardy in zone
6. *I.cristata* is suitable for the rock garden or peat
bed. *I.gracilipes* enjoys a cool, peaty soil in dappled
shade. *I.japonica* will grow in semi-shade in most
good soils, but needs a warm and sheltered spot.
I.tectorum prefers a gritty but moist soil against a
sunny wall. Increase by division.

(5) SIBERIAN IRISES (subgenus Limniris, series
Sibiricae) are easy garden plants, hardy in climate
zone 4 and requiring a fertile soil which does not dry
out during the spring and early summer growth
period. Although often grown beside streams and

pools, they are not true waterside plants and will do equally well in a mixed border. Hybrids between Siberian and Pacific Coast Irises (Cal Sibs) are also known. Many selected hybrids are available, both diploid and tetraploid, the latter having stouter, more upright foliage and larger flowers of heavier substance with broad flaring falls in clear rich colours. Increase by division or seed. In the US, Siberian irises are attacked by the iris borer and the iris weevil.

(6) PACIFIC COAST IRISES (subgenus Limniris, series Californicae) occur in California, Oregon and Washington, growing in neutral or slightly acid, well-drained soil in lightly wooded areas. All are very variable in colour and hybridize freely. They are easy to grow as garden plants, but they develop chlorosis and lose vitality in alkaline soils. They are usually hardy in climate zone 6 and prefer a sunny to lightly shaded position on a moist but porous, humus-rich soil. They resent disturbance. Propagate by careful division when the plant is showing renewed growth some months after flowering, and keep moist and shaded until established. They are readily raised from seed in a soil-based medium, but the offspring are likely to be hybrids. There are many cultivars available in a wide range of colours.

(7) WATER IRISES (subgenus Limniris, series Laevigatae) are plants of wet places, ditches and swamps and are excellent candidates for the margins of ponds or streams. They are hardy in climate zone 4. Included in this group are *I.pseudacorus* and *I.versicolor*. *I.laevigata* is often grown in containers set on shelves in garden ponds. Propagation is by division of clumps and is best carried out in early autumn. *I.ensata* has been cultivated in Japan for centuries and many new forms, including tetraploids, are now being raised in the US and Europe. The flowers are large, sometimes double, with spreading standards, in a wide range of colours. They flower late in the iris season, often into mid-summer. Japanese irises are often grown as marginal plants in containers, and are probably best removed from the water and kept drier in winter. Japanese irises will develop chlorosis if grown in alkaline soils but some of the Higo strain or the CaRe (calcium resistant) cultivars will survive up to pH 7.4. Members of this group are occasionally attacked by the larvae of the iris sawfly. If the plants are in or near a pond take care to use a chemical control that is not harmful to pond life.

(8) LOUISIANA IRISES (subgenus Limniris, series Hexagonae) are showy, late-flowering waterside irises from the swamps of the southern US. They are robust plants with branched flower stems, often in a zig-zag form, and are popular garden plants in the southern US and Western Australia, where the high summer temperatures necessary for successful growth can be attained. Many superb hybrids have been raised. Although hardy in climate zone 4, they seldom thrive in cool-temperate regions, needing long, hot and humid summers. That said, the bright blue *I.brevicaulis* and coppery red *I.fulva* will flower in Western Europe if planted in sun on a humus-rich, neutral to acid soil and kept moist and well-fed while in growth. The hybrid between them, I. × *fulvala*, will also grow under the same conditions.

(9) THE TRIPETALAE SERIES (subgenus Limniris, series Tripetalae) includes the iris with the widest distribution, *I.setosa*. In this series the standards are tiny and the flower appears to have only three petals. Ranging from Japan to Alaska and eastern Canada, *I.setosa* exists in many variants, the smaller of which are good for the rock garden. Fully hardy. Increase by seed or division.

(10) SPURIA IRISES (subgenus Limniris, series Spuriae) are very diverse in habit and size, ranging from 25–150cm tall. The taller species are herbaceous and good plants for the back of a mixed border on deep soils with ample humus. Some of the smaller species such as the yellow *I.kerneriana* are rock-garden plants requiring good drainage, while the plum-scented *I.graminea* is happy in a humus-rich soil in part shade. There are two general groups, one typified by *I.orientalis*, which becomes dormant after flowering, and the other including *I.spuria* itself, which retains its leaves until winter. Many excellent hybrid plants have been raised in a range of bright colours, both self- and multi-coloured. Propagation is by division in autumn or early spring, planting the rhizome below the surface in a rich deep soil and keeping moist until established. New plantings may be slow to flower but can remain undisturbed for many years.

(11) IRIS FOETIDISSIMA (subgenus Limniris, series Foetidissimae) is a widespread native of Western Europe and North Africa which will grow almost anywhere or in any soil. It seems most at home in semi-wooded or lightly shaded situations on humus-rich soil in climate zones 6 and over. This hardy, evergreen iris is much prized by flower arrangers for the scarlet seeds that are displayed when the capsules split in autumn; forms with white or yellow seeds are known. Slow-growing white- and cream-variegated forms occur. *I.foetidissima* var. *citrina* produces larger, clear topaz-yellow flowers. This species is very susceptible to virus disease.

(12) THE ALGERIAN OR CRETAN IRIS, *I.unguicularis* (subgenus Limniris, series Unguiculares) flowers from early autumn to mid-spring. Hardy in climate zone 6, it needs a warm sunny position, ideally against a south-facing wall on a sharply draining, neutral to alkaline soil. In summer, it should become dry and baked. A topdressing of potash or bone meal in early autumn and spring will promote growth and flowering. Propagate by division in spring and keep watered until established. This species is, however, better left undisturbed for many years, indeed, new plants may take years to flower, preferring cramped, impoverished conditions. Seed capsules will sometimes be found in the clumps at ground level; seed will germinate if sown in a soil-based medium by the following autumn. Larvae of the angleshades moth may eat the ends of flower buds while still at ground level.

Bulbous irises (13) XIPHIUM IRISES (Subgenus Xiphium) occur in Western Europe and North Africa, usually on alkaline soils. As there is a dormant period after flowering, several of the species need bulb frame culture in wet summer areas. *I.xiphium* occurs over a wide range of habitats. Both the typical plant and the many named colour forms called Spanish irises are reliable garden plants. Their leaves appear in autumn. Hybrids between *I.xiphium* and *I.tingitana*, known collectively as Dutch irises, are also good garden plants, and are widely used for the cut-flower market throughout the year; colours range from white and yellow to deep purple. *I.latifolia*, erroneously named the English iris in the 16th century and still so-called, is an exception, growing in damp grassy places in the Pyrenees. An easy plant for the border, it prefers a soil which does not dry out. Many colour forms are available in blue, violet, purple and white.

(14) JUNO IRISES (Subgenus Scorpiris) comprise a group of over 55 species. Most of them are difficult in temperate areas. They occur in western Asia ranging from Turkey to central Russian Asia and south to Jordan with the one exception of *I.planifolia* found in Southwest Europe and North Africa. They are fleshy-rooted, bulbous plants with channelled leaves in one plane. The flowers are attractive, the falls large and brightly coloured, the standards very small, often reduced to bristles. The majority of the Junos grow in areas of hot dry summers and cold winters with snow cover, and unless this can be achieved naturally then overhead protection from summer and winter rain is essential. Grow in deep pots in an unheated, well-ventilated alpine house or plant out in a bulb frame. A free-draining soil with added dolomitic limestone is recommended; feed with balanced fertilizer in the growing period

from mid-autumn until the leaves turn yellow after flowering. Never wet the foliage, as water trapped in the channelled leaves may cause rot. Some of the Junos are more tolerant of damp, for example *I.aucheri*, *I.bucharica* and *I.magnifica* will thrive in the open garden in cool-temperate areas in well-drained loam. The smaller forms of *I.bucharica* are excellent plants for the rock garden, flowering in early spring. Propagate from small young bulbs carefully separated when replanting, or from seed. Artificial cross-pollination may be necessary as many species are self-sterile.

(15) RETICULATA IRISES (subgenus Hermodactyloides) are dwarf irises, their bulbs covered in netted (reticulate) tunics. They occur in Turkey, Iran, Central Asia and south to Jordan. In temperate gardens both *I.histrioides* and *I.reticulata* will flourish in a sunny position in well-drained soil, and can be left undisturbed for years. There are many colour forms available and also some named hybrids. The yellow *I.winogradowii* is a high mountain plant which likes plenty of moisture during the growing season and, unlike the other reticulatas, does not require a summer dry period. A hybrid with *I.histrioides* named *I.*'Katharine Hodgkin' is a hardy bulb which increases rapidly in the open garden or a raised bed. *I.danfordiae* will flower well in the first season but then breaks up into many tiny bulblets and should be replaced annually (flowering colonies only seem to develop when this species is planted deeply in a bulb frame). Increase by bulb offsets or from seed, which germinates freely. Affected by ink disease.

IRIS

Name	Distribution	Group	Habit	Leaves	Flowers
I.acutiloba	Transcaucasus	[2]	stem 8–25cm, unbranched	narrowly sickle-shaped, 2–6mm wide	solitary, 5–7cm diam., white-yellow, strongly veined and lined brown-grey; falls with one central and one rust-black spot, bearded, beard hairs sparse, long, purple-brown, falls and standards more or less pointed
Comments: Spring-early summer					
I.aphylla	C & E Europe to W Russia and N Caucasus	[1]	to 30cm, deciduous; flowering stems branched from middle or near base, bracts 3–6cm, inflated, sometimes purple-tipped	arranged in fans, outer leaves curved, inner leaves upright, 0.5–2cm wide	flowers 1–5, 6–7cm diam., pale to deep purple to blue-violet, beard hairs blue-white, tips yellow
Comments: Spring					

Aril Irises: [group 3]. Regeliocyclus Hybrids. 'Ancilla': flowers white veined purple, falls flaked and netted purple. 'Chione': Standards white veined lilac blue, falls veined grey, blotched black-brown. 'Clotho': Standards deep violet with black beard, falls brown, veins and blotch black. 'Dardanus': Standards tinged lilac, falls cream blotched matt purple. 'Theseus': Standards violet with darker veins, falls ivory marked deep violet. 'Thor': flowers grey veined purple, blotch bright purple. 'Vera': flowers chocolate-brown tinged purple with blue beard. Aril-Median Irises: (Aril × Miniature Tall Bearded irises). 'Canasta': to 28cm; standards pale violet, falls rich beige with deep red signal and veining, the latter paler at edges, beard brown-bronze. 'Little Orchid Annie': to 30cm; standards pale amethyst veined and shaded green-gold, ruffled, midrib green, falls orchid edged yellow-green, signal and veining red, beard dark yellow. Arilbred Irises (Aril × Tall Bearded irises). 'Lady Mohr': to 75cm; standards light mauve-blue, falls yellow and green with deep red markings, beard brown. 'Loudmouth': to 25cm; fls deep red, falls with black signal. 'Nineveh': Standards purple shaded pink, falls violet-red, beard dark brown. 'Saffron Charm': fls in blend of deep yellow, grey and lavender.

I.aucheri	SE Turkey, N Iraq, N Syria, Jordan, NW Iran	[14]	to 40cm; roots fleshy	25×2.5–4.5cm, completely concealing stem until fruiting	3–6, blue, rarely almost white; tube to 6cm; fall blade with central yellow ridge, margin wavy, haft broadly winged; standards 2–3.5cm, horizontal to deflexed, obovate
Comments: Late winter–spring.Crossed with *I.persica* to produce *I.*'Sindpur'					

IRIS Name	Distribution	Group	Habit	Leaves	Flowers
I.bakeriana	SE Turkey, N Iraq, W Iran	[15]	to 15cm; bulb tunics netted	cylindric, 8-ribbed, otherwise close to *I.reticulata*	solitary, white, tipped violet on falls with a cream ridge surrounded by violet spots and veins, haft spotted and veined violet, standards and style branches lilac

Bearded (Pogon) Irises: [group1]. (1) *Border Bearded* (stems to 70cm; fls large). 'Brown Lasso': Standards butterscotch, falls pale violet and brown. 'Carnival Glass': red-brown, occasionally tinged blue or red. 'Impelling': bright pink and yellow with an orange beard. 'Impetuous': sky-blue, ruffled, beard white. 'Marmalade Skies': apricot-orange. 'Whoop Em Up': vivid golden-yellow, falls clear chestnut-red, rimmed with bright yellow. (2) *Intermediate Bearded* (stems to 70cm; abundant). 'Annikins': deep violet blue. 'Black Watch': very dark velvety black-purple. 'Golden Muffin': Standards ruffled, yellow, falls amber-brown edged yellow. 'Indeed': Standards clear lemon-yellow, falls translucent white edged bright yellow with a white beard. 'Silent Strings': clear blue. 'Why Not': bright apricot-orange with darker beard. (3) *Miniature Dwarf Bearded* (stems to 20cm, to 2 flowers per stem). 'April Ballet': Standards light blue with a violet-blue spot. 'Egret Snow': Standards pure white, falls occasionally flecked blue. 'Jasper Gem': rust-red bicolor. 'Orchid Flare': pink with white beard. (4) *Miniature Tall Bearded* (stems to 70cm; flowers 8 or more). 'Carolyn Rose': cream-white lined rose-pink. 'Dappled Pony': Standards dark blue, falls white heavily flecked violet. 'Smarty Pants': yellow with red stripes on falls. 'Surprise Blue': lavender blue with white hafts. (5) *Standard Dwarf Bearded* (stems to 40cm; flowers 3 or 4 per stem). 'Austrian Sky': blue with darker markings on falls. 'Blue Denim': blue, flaring with a white beard. 'Gingerbread Man': deep brown with a bright blue beard. 'Green Halo': pale olive-green with a darker 'halo'. 'Melon Honey': melon-orange with a white beard. 'Toots': velvet burgundy with a yellow beard. (6) *Tall Bearded* (over 70cm high; flowers large, usually ruffled). 'Broadway': Standards golden-yellow, beard orange, falls white. 'Going My Way': white striped violet-purple. 'Jane Phillips': large, pale blue. 'Love's Allure': Standards grey-lilac flecked purple, falls edged sandy gold. 'Red Lion': deep brown-burgundy. 'Vanity': light pink flashed white with a pink-red beard.

I.bucharica *I.orchioides* of gardens	Russia, NE Afghanistan	[14]	to 40cm	to 20×3.5cm at flowering, later elongated on expanded stem, shiny green, channelled, margins white	2–6 in upper lf axils; perianth tube to 4.5cm, golden yellow to white, fall with central ridge yellow surrounded by green to dull purple suffusion or markings, standards spreading, trilobed or lanceolate

Comments: Spring

I.chrysographes	W China, NE Burma	[5]	to 50cm; rhizome stout	50–1.5cm	1–4 on hollow, branched or unbranched stems, fragrant, 5–10cm diam., dark red-purple streaked gold on falls; falls deflexed; standards spreading

Comments: Late spring–summer. Hybridizes with *I.forrestii* and *I.sibirica*. 'Black Night': flowers indigo-violet. 'Inshriach': flowers almost black. 'Margot Holmes': flowers purple-crimson lined yellow at throat. 'Rubella': flowers burgundy.

I.clarkei	E Himalaya	[5]	rhizome stout; stems to 60cm, 1–3-branched	to 2cm wide, glossy green above, glaucous beneath	2 per branch, 5–10cm diam., blue-violet to red-purple; falls with large, white, central signal patch, veined violet; haft slightly yellow; standards bent to horizontal

Comments: Late spring–early summer

I.confusa	W China	[4]	vigorous, clump-forming, to 1.5m; stems green, cane-like, procumbent and branching, then ascending, with an apical fans of foliage	bright green, sword-shaped, to 5cm across	to 5cm diam., many borne on branching stems to 90cm long, short-lived, white spotted yellow or mauve; falls 2.5×5cm, crest yellow, signal patch yellow

Comments: Spring

I.cristata	NE US	[4]	rhizome small, much-branched	to 15×1–3cm at flowering, linear	1–2 per spathe, subsessile, 3–4cm diam., pale lilac to purple; tube to 10cm; falls obovate, blunt, reflexed, 1×1cm, central patch white, crest of 3 crisped yellow ridges; standards erect, oblanceolate, shorter and narrower than falls

Comments: Early summer. 'Alba': fls white. 'Caerulea': fls blue.

I.crocea	Kashmir	[10]	rhizome woody, roots wiry	to 75×1.5–2cm, sword-shaped, leathery	flowers 12–18cm diam., in terminal clusters on branched stems to 1m, dark golden yellow; falls oblong, margins crisped; standards oblanceolate, erect, margins slightly crisped; style branch lobes narrow-triangular

I.danfordiae	Turkey	[15]	to 15cm; bulb tunic netted, tiny, producing offset bulblets	1–15cm, taller than flowers, quadrangular	solitary, to 5cm diam., yellow; tube to 7.5cm; falls lightly spotted green in centre and lower part, ridge deeper yellow-orange; standards reduced, bristle-like, 3–5mm

Comments: Early spring.

I.douglasiana	W US (S Oregon, California)	[6]		to 100×2cm, dark green, base stained red, ribbed	2–3, 7–10cm diam. on branching 15-70cm stems, lavender to purple, veins darker; tube to 3cm; falls yellow in centre; standards erect, lanceolate, clawed

Comments: Summer. 'Alba': flowers white. 'Southcombe Velvet': flowers deep violet.

IRIS

Name	Distribution	Group	Habit	Leaves	Flowers
I.ensata *I.kaempferi* JAPANESE WATER IRIS	Japan, N China, E Russia	[7]	stoutly rhizomatous, aquatic or marginal	20–60 × 0.4–1.2cm, leathery, midrib prominent	3 or 4 per branch on flowering stems to 90cm, 8–15cm diam., purple to red-purple; tube 1–2cm; fall haft yellow, blade elliptic- ovate, also suffused yellow; standards smaller, erect

Comments: Summer. A large number of cultivars, many of which originate from Japan, with a few (such as some of the Higo Hybrids) beginning to enter western gardens; flowers to 20cm in shades of white, pink, blue and violet often mottled or flecked; single to double; often with spreading standards. 'Alba': flowers pure white. 'Blauer Berg' ('Blue Mountain'): flowers bright sky blue, abundant; middle to late season. 'Gei Sho Ne': habit unusually short; flowers violet; middle to late season. 'Good Omen': growth strong; flowers double, violet tinged red, abundant; early season. 'Major': habit tall to 46cm; flowers darker. 'Moonlight Waves': flowers white with lime-green centre. 'Peacock Dance': falls white, veins red-violet, signal patch yellow; standards dark red-purple; style branches purple. 'Prairie Love Song': flowers white, signal patch yellow. 'Raspberry Rimmed': flowers white, margins red-pink, signal patch yellow. 'Returning Tide': flowers pale blue, signal yellow. 'Sorcerer's Triumph': flowers double, white, veins red-purple, signal orange. 'Stranger in Paradise': flowers white, margins pink, signal patch yellow. 'Summer Storm': flowers to 18cm diam., violet tinged red, veined blue; late flowering. 'The Great Moghul': flowers black-purple. 'Unschuld' ('Innocence'): flowers clear white, perianth seg. undulate; early flowering. 'Variegata': leaves striped white; flowers purple. 'Aichi-no-Kagayaki' (*I.ensata × I.pseudacorus*): flowers large, bright yellow, veins brown. 'Chance Beauty' (*I.ensata × I.pseudacorus*): flowers large, bright yellow, veins red-brown.

Name	Distribution	Group	Habit	Leaves	Flowers
I.foetidissima ROAST BEEF PLANT; GLADWYN; GLADDON IRIS; STINKING GLADWYN	S & W Europe, N Africa, Atlantic Is.	[11]		evergreen, sword- shaped, in basal fans, deep green, to 2.5cm wide	to 5 per branch, 5–7cm diam., malodorous, short-lived, on short-branching stems to 80cm; falls obovate, 2cm wide, blade blue- lilac to topaz or yellow, veins purple, centre fading to white, haft bronze to brown, winged; standards smaller, oblanceolate, erect, flushed lilac

Comments: Summer-flowering. Grown chiefly for its handsome foliage and scarlet, bead-like seeds persisting in the capsules from late summer through winter. 'Fructo-alba': seeds white. 'Variegata': leaves striped cream. var. *citrina:* flowers pale yellow and mauve. var. *lutescens:* flowers pure clear yellow.

Name	Distribution	Group	Habit	Leaves	Flowers
I.forrestii	W China, N Burma	[5]		narrow-linear, shiny above, glaucous below, shorter than stem	2, on an unbranched stem to 45cm, scented, 5–6cm diam., yellow lined purple-brown on fall haft; falls oblong-ovate, blade 5cm; standards erect, oblanceolate

Comments: Summer.

Name	Distribution	Group	Habit	Leaves	Flowers
I.fosteriana	NE Iran, Turkmenistan, NW Afghanistan	[14]	10–15cm; bulb long, slender, tunics dull olive	to 17×0.4–0.8cm, lanceolate, silver-edged, tips acuminate	1–2, to 5cm diam.; tube to 4cm; fall and blade orbicular, to 5cm, pale yellow, crest conspicuous, yellow, sometimes on deeper yellow, patch brown-veined; standards obovate, deflexed, purple; style branches pale yellow
I.fulva	C US (Mississippi Valley)	[8]	to 90cm; rhizome slender, green	1.5–2.5cm wide, green, ensiform	to 6.5cm diam., bright red to rust, orange or rarely deep yellow; tube to 2.5cm; falls and standards hanging downwards, falls oblanceolate, to 2.5cm wide, blunt, narrowed to a claw, standards broad, truncate, notched, to 5×2cm

Comments: Summer.

Name	Distribution	Group	Habit	Leaves	Flowers
I. × fulvala (*I.fulva ×* *I.brevicaulis*)	garden origin	[8]			purple-red
I.gatesii	N Iraq, SE Turkey	[2]	40–60cm	linear, straight to curved, 0.5–1cm wide	solitary, 13–20cm diam., very showy and heavily textured, ground colour cream, opal blue or, more usually, pearl grey, heavily overlaid with maroon or blue-mauve veins and stippling, beard yellow or maroon, signal patch absent or small and dark

Comments: Summer

Name	Distribution	Group	Habit	Leaves	Flowers
I.germanica	widely naturalized; either a Mediterranean native or an ancient fertile hybrid	[1]	rhizome stout, horizontal, branching; flowering stem 60–120cm, with 1–2 short branches and thick, purple- tinted bracts	2-ranked, equitant, 30– 40×2.5–4.5cm, grey- green	9–10cm diam., in shades of blue, violet, white, beard yellow, standards sometimes paler than falls

Comments: Late spring. 'Amas': falls deep blue-purple, standards paler blue, prominent blue-white beard tipped orange. 'Askabadensis': tall, flowering later; falls red-purple, veins on hafts yellow-brown, beard tipped yellow, standards paler blue. 'Karput': lvs edged red-purple; falls narrow, black-purple, standards paler red-purple. 'Nepalensis': falls and standards both dark purple-red tipped orange; beard blue-white grading to white. 'Sivias': blue; beard white, hairs scarcely tipped yellow. var. *florentina:* smaller ; flowers scented, palest blue-white; veins on haft of falls yellow, beard yellow.

Name	Distribution	Group	Habit	Leaves	Flowers
I.gracilipes	China, Japan	[4]	10–15cm; dwarf, clump- forming, stoloniferous; stems slender, branching	grassy, ensiform, to 30×0.5–1cm	3–4cm diam., pink to blue-lilac veined violet; tube 1.5cm, falls obovate, to 2.5×1.5cm, notched, centre white, crest wavy, yellow and white

Comments: Early summer. 'Alba': fls white

IRIS

Name	Distribution	Group	Habit	Leaves	Flowers
I.graebneriana	Russia	[14]	to 40cm	2-ranked, shiny green above, grey-green beneath, margins white	4–6, 7cm+ diam., blue sometimes tinged violet; tube to 6cm, fall blade deeper blue, centre ridge white on veined, white ground, standards obovate, to 2.5cm, tips pointed
Comments: Spring					
I.graminea *I.colchica*	NE Spain to W Russia, N & W Caucasus	[10]	stem 20–40cm, strongly flattened or 2-winged	borne on stem, uppermost leaves taller than flowers	1–2, fruit-scented, 7–8cm diam.; falls rounded, 1.2cm wide, violet, blade white at centre veined violet, haft winged, occasionally tinged green or yellow; standards erect, purple, 0.5cm wide; style branches purple, base tinged green or brown
Comments: Summer.					
I.histrio	S Turkey, Syria, Lebanon	[15]	bulb offsetting, tunic netted	30–60cm, 4-angled	1 per spathe, 6–8cm diam., falls lilac, streaked and spotted, ridge yellow surrounded by a pale, bluer spotted area, standards lilac, oblanceolate, unmarked
Comments: Winter–early spring					
I.histrioides	Central N Turkey	[15]		4-angled in section, not or scarcely developed at flowering time, later extending to 50cm	6–7cm diam., blue, fall blade pale spotted blue with a yellow ridge
Comments: Early spring . 'Golden Harvest': yellow. 'Imperior': dark blue. 'Major': purple-blue, spotted white on falls. A plant with much larger, dark violet blue flowers with almost horizontal falls has also been offered as *I.histrioides* 'Major'. 'White Excelsior': large white. 'White Perfection': large, white. var. *sophenensis:* flower segments narrower, dark violet-blue, lightly spotted; fall ridge yellow. See also under *I.winogradowii*.					

I. × *hollandica* – The Dutch Irises – *I.xiphium* × *I.tingitana*, possibly with the influence of *I.latifolia*. See *I.xiphium*.

Name	Distribution	Group	Habit	Leaves	Flowers
I.hoogiana	Russia	[3]	40–60cm; rhizome stout, producing long stolons	erect, almost straight, tinged purple, to 50×1.5cm - 2–3 per pair of bracts, 7–10cm diam., scented, plain grey-blue; tube to 2.5cm; falls and standards bearded yellow	late spring. 'Alba': white, with a faint overlay of pale lavender-blue. 'Bronze Beauty': grey-violet, falls deep rich violet, edged cinnamon-brown. 'Noblense': blue-violet, crest golden yellow. 'Purpurea': deep purple.
I.iberica	SW Asia	[2]	to 20cm, rhizome compact	to 6mm diam., grey-green, linear, curving	solitary, to 6.5cm diam., tube to 3.5cm, falls reflexed from base, white densely veined brown and spotted, signal patch black-brown, velvety, beard hair purple-brown, standards to 8.5cm, erect, white to pale lilac, obscurely veined
Comments: Late spring. ssp. *elegantissima*: falls cream to ivory spotted and veined maroon, blade deflexed almost vertically, standards white, sometimes veined brown at base					
I.innominata	W US	[6]	to 25cm	slender, 2–4mm wide, dark green, base tinged purple	1–2 per stem, 6.5–7.5cm diam., cream-yellow to orange or pink-lilac to dark purple, veins darker; tube to 3cm; falls to 6.5mm, margins of falls and standards frilly, standards shorter, style lobes broad, margins toothed
Comments: Summer. Used in hybridizing, see Pacific Coast hybrids. 'Lilacina': lavender. 'Lutea': yellow. 'Spinners': soft brown-yellow, veined and marked rich brown					

Japanese Irises: See *I.ensata, I.laevigata*

Name	Distribution	Group	Habit	Leaves	Flowers
I.japonica	C China, Japan	[4]	45–80cm, rhizome stoloniferous; stems erect, branched	to 45cm, in fans, ensiform, shiny green	3–4 per branch, 4–5cm diam., appearing flattened due to spreading perianth lobes, margins fringed, white to light blue-lilac; falls 2cm wide, blade with orange frilly crest on purple blotched area; standards lilac
Comments: Spring. 'Ledger's Variety': frilled, white, marked purple, crest orange. 'Rudolph Spring': pale purple-blue marked orange. 'Variegata' ('Aphrodite'): leaves striped white and marked purple					
I.kerneriana	N & C Turkey.	[10]	rhizome slender, with fibrous remains of old lf bases	linear, to 5mm wide, slightly shorter than flowering stems	2–4 per branch, 7–10cm diam., deep cream to pale yellow; fall blade elliptic, to 2cm wide, with deep yellow central blotch, strongly recurved, tip almost touching stem, haft narrow, 5mm; standards erect, margins wavy, notched
Comments: Summer					
I.korolkowii	Russia, NE Afghanistan	[3]	40–60cm; rhizome thick, stoloniferous, with fibrous leaf base remains	linear, 0.5–1cm wide, base tinged purple	2–3 per spathe, 6–8cm diam., appearing elongated, cream-white lightly veined deep maroon, falls oblong, to 4×2.5cm, acute, deflexed, signal patch dark green to black-brown, beards inconspicuous, standards acute, erect
Comments: Early summer. 'Concolor': bright blue-purple. 'Violacea': cream veined red-violet.					

IRIS

Name	Distribution	Group	Habit	Leaves	Flowers
I.laevigata	E Asia	[7]	aquatic or marginal, to 45cm; similar to *I.ensata* but stouter	1.5–4cm, without prominent midrib	2–4, 8–10cm diam., blue-violet, tube to 2cm; fall haft pale yellow, standard erect, smaller than falls

Comments: Summer–autumn. 'Alba': flowers white, style branches mauve. 'Albopurpurea': falls purple, mottled white around edges, standards white. 'Atropurpurea': red-purple. 'Colcherensis': flowers white, fall centre dark blue. 'Lilacina': light blue. 'Midnight': very deep blue lined white. 'Montrosa': very large, deep blue, white at centre. 'Mottled Beauty': white, falls spotted pale blue. 'Regal': red-purple. 'Rose Queen': Soft rose. 'Snowdrift': double, white, style branches light violet. 'Variegata': leaves striped green and white; flowers pale blue.

I.latifolia *I.xiphioides* ENGLISH IRIS	Spain and Pyrenees	[13]	to 80cm.	appearing in spring, to 65×0.8cm, channelled, grey-white above	1–2 per bract pair, 8–10cm diam., violet-blue, tube 5mm, falls with central yellow blotch, broad, ovate-oblong, to 7.5cm, haft winged, standards to 6cm, oblanceolate

Comments: Late spring–summer. 'Almona': flowers light blue-lavender. 'Blue Giant': Standards blue-purple, speckled darker, falls deep blue. 'Isabella': flowers pink-mauve. 'La Nuit': flowers dark purple-red. 'Mansfield': flowers magenta-purple. 'Mont Blanc': flowers white. 'Queen of the Blues': flowers rich purple-blue.

Lousiana Hybrids: [group 8]. Hybrids of the Series *Hexagonae* (including *I.fulva*) are becoming increasingly popular in the US. They are mostly tender, and bear flowers in a wide range of colours including shades of yellow, pink, red, brown, purple and blue. 'Black Gamecock': flowers blue-black with line signal. 'Mme. Dorothea K. Williamson' (*I.fulva × I.brevicaulis*): hardy; flowers very large, plum-purple. 'Gold Reserve': flowers golden-orange veined red. 'May Roy': Standards pale pink, falls pink tinged purple. 'Roll Call': to 90cm; flowers violet with green styles. 'Sea Wasp': to 110cm; flowers blue with yellow line signal.

I.lutescens *I.chamaeris*	NE Spain, S France and Italy	[1]	fast-growing, to 30cm; rhizome stout	30×0.5–2.5cm, equitant, straight to curved	1–2, 6–8cm diam., yellow, violet, purple, white or bicoloured, tube to 3.5cm, falls oblong-spathulate, to 7.5×2cm, beard yellow, standards oblong, erect or converging, to 7.5×2.5cm, margins crisped

Comments: Early to mid-spring 'Campbellii': dwarf; flowers bright violet-blue, falls darker. 'Jackanapes': flowers blue and white. 'Nancy Lindsay': dwarf; flowers pale yellow.

I.magnifica	Russia	[14]	30–60cm; bulb tunics papery	many, remote, 3–5cm wide, channelled, pale green, lustrous	3–7, to 8cm diam., pink-lilac, tube 4–4.5cm, falls yellow at centre, crest white, fall haft widely winged, to 2.5cm wide, standards obovate, to 3cm, horizontal to deflexed

Comments: Late spring

I.missouriensis	W US	[7]	to 75cm	leaves to 7mm wide, absent in winter, usually taller than flowers	2–3 per bract pair, 5–8cm diam., on pedicels to 20cm, white, lilac, lavender or blue; falls obovate, deflexed, veined, signal patch yellow; standards oblanceolate, upright

Comments: Late spring-summer. 'Alba': flowers white.

I.orchioides	Russia	[14]	bulb tunics papery; roots fleshy	18×1–3cm, straight to curved, pale green, channelled, concealing stem until after flowering	3–4, 5cm diam., light yellow suffused mauve, tube 3–6cm, fall haft broadly winged, 2cm wide, crest deep yellow, toothed, on dark yellow ground veined green or mauve, standards to 1.5cm, trilobed or linear

Comments: *I.orchioides* of gardens is usually *I.bucharica*.

I.orientalis	NE Greece, W Turkey	[10]	to 90cm, branched	to 90×1–2cm	2–3 per pair of papery bracts, 10cm diam., white; fall blade rounded, signal area large, yellow, haft narrow, slightly pubesc., standards to 8.5cm, erect

Comments: Summer. For *I.orientalis* of gardens, see *I.sanguinea*.

Pacific Coast or **Californian Irises:** [group 6]. (*I.douglasiana × I.innominata*, etc.). 'Arnold Sunrise': flowers white flushed blue, fall marked with orange. 'Banbury Fair': Standards off-white, falls pale lavender flecked lavender at centre. 'Blue Ballerina': flowers white, falls marked with purple. 'Broadleigh Rose': flowers marked with shades of pink. 'Lavender Royal': flowers lavender with darker flushes.

I.pallida DALMATIAN IRIS	former Yugoslavia	[1]	stems branched, to 120cm	glaucous, 20–60×1–4cm, usually evergreen	2–6, fragrant, 8–12cm diam., soft lilac-blue, bracts silvery, papery; beard yellow

Comments: Late spring-early summer. 'Argenteavariegata': leaves striped blue-green and white. 'Aureavariegata' ('Aurea', 'Variegata'): leaves striped yellow. ssp. *cengialtii*: 30–45cm; leaves usually deciduous; flowers deep blue-purple, beard hairs white, tipped orange or yellow.

I.pseudacorus YELLOW FLAG	Europe to W Siberia, Caucasus, Turkey, Iran, N Africa	[7]	marginal or aquatic; stems branched, to 2m; rhizome stout	to 90×3cm, grey-green, midrib prominent	4–12, 5–12cm diam., bright yellow veined brown or violet; tube to 1.5cm, falls rounded, to 4cm wide, blotched deeper yellow; standards oblanceolate, erect, to 3cm

Comments: Early–midsummer. 'Alba': creamy white, with brown veins near tips. 'Gigantea': large, golden yellow. 'Golden Fleece': deep yellow without darker veins. 'Mandschurica': matt yellow. 'Variegata': leaves striped yellow at first. var. *bastardii*: flowers pale yellow; fall blade not blotched deeper yellow.

I.pumila *I.aequiloba, I.taurica*	SE & EC Europe to Urals	[1]	dwarf, to 15cm	almost straight, grey-green, to 15×1.5cm, dying away in winter	usually solitary, rarely 3 in some forms, mostly purple-violet, some white yellow or blue; tube 5–10cm, beard blue or yellow

Comments: Spring. A parent of the dwarf bearded cultivars.

IRIS Name	Distribution	Group	Habit	Leaves	Flowers
I.reticulata	N & S Turkey, NE Iraq, N & W Iran, Russia.	[15]	bulb pear-shaped with netted tunic and producing bulblets after flowers	4-angled, very narrow, to 30×0.2cm, synanthous or hysteranthous	solitary, dark violet blue to paler blue to red-purple; tube 4–7cm; falls 5cm, often yellow-ridged, haft 2.5cm; standards erect, oblanceolate, 6cm

Comments: Early spring.

Reticulata Hybrids [group 15]. (derived from *I.histrioides* and *I.reticulata* and hybrids between them). 'Alba': flowers white. 'Cantab': flowers very pale Cambridge blue, crested yellow. 'Clairette': Standards sky-blue, falls deep blue marked white. 'Edward': flowers dark blue marked orange. 'Gordon': flowers light blue with orange blotch on white ground lightly striped blue. 'Harmony': Standards blue, falls royal blue blotched yellow and white. 'Ida': Standards light blue, falls paler, blotch pale yellow on white ground slightly spotted blue. 'Jeannine': flowers violet, falls blotched orange with white-and-violet striped patches; fragrant. 'Joyce': Standard lavender-blue, falls deep sky-blue with yellow markings and grey-brown stripes. 'J.S. Dijt': flowers dark red-purple; fragrant. 'Natascha': flowers white tinged blue, falls veined green with golden-yellow blotch. 'Pauline': flowers purple-violet with dark purple falls blotched blue and white. 'Purple Gem': flowers violet, falls black-purple blotched purple and white. 'Royal Blue': Standard deep velvet blue, falls blotched yellow. 'Spring Time': Standards pale blue, falls dark blue tipped white, with purple spots and yellow midrib. 'Violet Beauty': Standards velvet-purple, falls deep violet with orange crest.

I.rosenbachiana	C Asia	[14]		bright glossy green, to 5cm at flowering, later to 25cm, tips curving outwards	1–3, light purple; falls blotched deep purple, crest orange

Comments: Winter–early spring. var. **baldschuanica** is smaller, with pale primrose flowers veined and blotched brown-purple.

I.sanguinea *I.orientalis* of gardens	SE Russia, Korea, Japan	[5]		similar to *I.sibirica* but leaves as tall as or taller than the 75cm, unbranched flowering stems	2, blue-purple marked white

Comments: Early summer. 'Alba': white. 'Kobana': narrow, white. 'Snow Queen': ivory.

I.serotina	SE Spain	[13]	40–60cm	narrow, channelled, appearing in autumn, dying before flowering	1–2 on an unbranched stem, violet-blue, tube to 1cm, fall centre yellow, standard reduced, to 1cm, extremely narrow

Comments: Late summer.

I.setosa	Northeastern N America, E Russia, N Korea, Japan	[9]	15–90cm	rhizome stout, clad with old leaf bases; stems usually 2–3-branched to 50×2.5cm, deciduous, base suffused red	to 15, 5–9cm diam., tube to 1cm, falls orbicular, 2.5cm diam., light blue-purple to purple, haft narrow, palest yellow, veins blue-purple, standards greatly reduced so that flowers appear to have only 3 petals, bristle-like, erect

Comments: Late spring–early summer. 'Kosho-en': white. 'Nana': Small, purple. f. **serotina:** flowers solitary, sessile.

I.sibirica	C & E Europe, NE Turkey, Russia	[5]	stems 1–2-branched, 50–120cm, taller than leaves	to 4mm wide	to 5, to 7cm diam.; falls blue-purple, veined and marked white and gold, haft paler but darker-veined

Comments: Late spring–summer. 'Alba': white.

Siberian Hybrids: [group 5]. (*I.sibirica* × *I.sanguinea*). Over 150 cultivars with flowers in a wide range of colours from white and yellow through pink and red to violet. 'Ann Dasch': flowers dark blue, falls marked yellow. 'Anniversary': flowers white with yellow hafts. 'Butter and Sugar': flowers white and yellow. 'Caesar's Brother': flowers dark pansy purple. 'Ego': flowers rich blue. 'Ewen': flowers burgundy, tetraploid. 'Helen Astor': flowers dark plum tinged rosy red, conspicuous white veins near throat. 'Mrs Rowe': flowers small, grey-pink. 'Papillon': flowers pale blue. 'Ruffled Velvet': flowers red-purple marked yellow. 'Sparkling Rose': flowers rose-mauve, falls flecked blue. 'Tropic Night': flowers blue-violet. 'Wisley White': flowers white.

I.spuria	Europe, Asia, Algeria	[10]	to 50cm	30×1.2cm	violet-blue, yellow or white, fall blades rounded, to 2.5cm diam., standards oblanceolate

Comments: Summer. ssp. *spuria*: to 80cm; flowers 6–8cm diam., lilac or blue-violet, veins violet; falls to 6cm, striped yellow in centre, haft exceeding blade. ssp. *carthaliniae*: to 95cm; flowers 4–5, sky blue or white, veined deeper blue. ssp. *halophila*: 40–85cm; flowers 4–8, 6–7cm diam., white, dull light yellow to bright yellow, veined deeper, falls to 6cm, haft longer than blade. ssp. *maritima*: 30–50cm; flowers to 4, bract green, falls to 4.5cm diam., veins dense, purple, blade deep purple, haft longer than blade, centrally striped green. ssp. *musulmanica*: syn. *I.klattii*, *I.musulmanica*, *I.violacea*; 40–90cm; flowers light violet to dark lavender-violet, veins darker, fails to 8cm, blade striped yellow, base tinged yellow, haft equalling or exceeding blade. ssp. *notha*: 70–90cm; flowers violet-blue, fall haft striped yellow, exceeding blade.

Spuria Group. [group 10]. Many cultivars with flowers in shades and combinations of white and yellow through orange, red and brown to blue. 'Connoisseur': lavender blue. 'Elixir': saffron. 'Imperial Bronze': deep yellow veined brown. 'Protégé': Standards blue, falls white veined blue. 'Red Oak': brown-purple. 'Shelford Hybrid': blue.

I.susiana MOURNING IRIS	Origin unknown, possibly Lebanon	[2]			to 12cm diam., pale lilac-grey heavily veined deep purple, falls and standards similarly sized and shaped, blade round, to 8cm wide, signal patch velvety purple-black, beard deep brown-purple

Comments: Late spring

I.tectorum ROOF IRIS	C & SW China, possibly Burma, naturalized Japan	[4]	rhizome stout; stems sometimes branched, to 40cm	thin, ribbed, glossy dark green to 30×2–2.5cm	2–3 per spathe, to 10cm diam., somewhat flat, blue-lilac, veined and patched darker, falls 2.5cm wide, crest well-divided, frilly, white, spotted darker; standards and falls spreading, margins wavy

Comments: Early summer. 'Alba': white, sparsely veined yellow. 'Variegata': leaves striped and streaked cream.

IRIS

Name	Distribution	Group	Habit	Leaves	Flowers
I.tenax	NW US	[6]	to 30cm	tinged pink at base	1–2, 7–9cm diam., palest yellow to lavender or red-purple; tube short, to 1cm, falls lanceolate, 2.5cm wide, reflexed, with a white or yellow central patch of purple, standards lanceolate, 0.6cm wide

Comments: Early summer. Used in hybridizing; see Pacific Coast hybrids.

I.unguicularis *I.stylosa*	Algeria, Tunisia, W Syria, S & W Turkey	[12]	stemless; rhizome tough, branching	evergreen, in tufts, linear-ensiform, to 60×1cm	solitary short-stalked, fragrant, tube very long, to 20cm, falls obovate, reflexed, 2.5cm wide, white veined lavender, central band yellow, haft linear, veins dark, standards size and shape as falls, erect, lilac; style branches yellow-glandular above

Comments: Winter-spring. ssp. *cretensis* (syn. *I.cretensis, I.cretica*): dwarf; leaves very narrow and grassy; flower segments to 5.5cm, purple-blue, fall blades and hafts white veined violet, blade striped orange in centre. 'Alba': flowers white, falls with central green-yellow line. 'Ellis's Variety': leaves narrow; flowers bright violet-blue. 'Marginata': flowers lilac, margins white. 'Mary Barnard': flowers violet-blue. 'Oxford Dwarf': flowers deep blue, falls white veined purple, with central orange line, tips lavender. 'Speciosa': leaves short, narrow; flowers fragrant, deep violet, central yellow stripe. 'Starker's Pink': dwarf; leaves shorter, narrower; flowers pink-lavender. 'Variegata': flowers mottled and streaked purple on lavender ground. 'Walter Butt': flowers large, fragrant, pale silver-lilac; late autumn-winter.

I.variegata	C & SE Europe	[1]	20–50cm, stems branched	dark green, sword-shaped, ribbed, to 30×3cm	3–6, 5–8cm diam., bright yellow to white, tube to 2.5cm, falls obovate, 2cm wide, reflexed, veined red-brown, standards oblong, erect

Comments: Spring–summer. Forms with brown-red falls occur. Used in hybridizing; parent plant of the miniature tall bearded hybrids.

I.verna	SE US	[8]	to 6cm at flowering	equitant, sword-shaped, to 15×1cm after flowering, glaucous, purple at base	to 5cm diam., bright blue-lilac, tube 2–5cm, falls obovate, 4×1cm, centre striped orange with brown spots, standard obovate, erect

Comments: Spring.

I.versicolor	E US	[7]	marginal; rhizome stout, creeping; stems 20–80cm, branched	35–60×1–2cm	several per branch, 6–8cm diam., violet to red-purple, falls wide-spreading, blade oval, 8–2.5cm, blotched green-yellow, surrounded by white veined purple, haft white, purple-veined, standards oblanceolate, smaller, 4cm, paler, erect

Comments: Summer. 'Gerald Derby' (*I.versicolor × I.virginica*): flowers large, purple-blue. 'Kermesina': flowers red-purple. 'Rosea': flowers pink

I.virginica SOUTHERN BLUE FLAG	E US	[7]	similar to *I.versicolor* but flowering stems usually unbranched, often curved	1–3cm wide, soft, tips drooping	1–4, blue, fall centre yellow-hairy (not bearded)

Comments: Summer. 'Alba': flowers white. 'Giant Blue': flowers large, blue. 'Wide Blue': leaves blue-green; flowers blue.

I.warleyensis	Russia	[14]	20–45cm; roots slightly thickened	channelled, spaced along stem, to 20×3cm	to 5, 5–7cm diam., pale to dark violet to blue-purple, tube 5cm, fall blade orbicular, margins white, crest white to yellow, on yellow ground, dissected, haft unwinged, standard deflexed, linear to trilobed, to 2cm

Comments: Spring. 'Warlsind' (*I.warleyensis × I.aucheri*): to 25cm; falls yellow, blade blotched purple-blue, ridge yellow, standards and styles white

I.wendelboi	SW Afghanistan	[14]	dwarf, to 10cm; bulb tunics papery	well-developed at flowering, to 20×1cm, glaucous, arching, concealing stem until after flowering	1–2, to 5.5cm diam., deep violet, tube 3cm; falls not winged, crest bright golden yellow, frilly, standards much reduced, to 0.5cm

Comments: Spring.

I.winogradowii	Russia	[15]	to 15cm; differs from *I.histrioides* in producing rice-grain-like bulblets and pale primrose fls spotted green on the fall haft and centre of blade		

Comments: Early spring. 'Frank Elder' (*I.winogradowii × I.histrioides*): flowers pale blue slightly shaded yellow. 'Katherine Hodgkin' (*I.winogradowii × I.histrioides*): flowers yellow veined blue and faintly tinged pale blue.

I.xiphium *I.hispanica* SPANISH IRIS	Iberian peninsula, Mediterranean	[13]	stems to 40–60cm	20–70×0.3–0.5cm, channelled, appearing in autumn	1–2, usually blue or violet, sometimes white, yellow or mauve, tube 1–3mm, fall blade centre usually orange or yellow

Comments: Spring–early summer. Hybridized with *I.tigitana* to produce Dutch irises. 'Battandieri' (syn. *I.battandieri*): flowers white, ridge on fall blades orange-yellow. 'Blue Angel': flowers bright blue, central falls marked yellow. 'Bronze Queen': flowers golden brown, suffused purple and bronze. 'Cajanus': flowers yellow. 'King of the Blues': flowers blue. 'Lusitanica' (syn. *I.lusitanica*): flowers yellow. 'Praecox' (syn. *I.xiphium* var. *praecox*): flowering earlier; flowers large blue. 'Professor Blaauw': flowers bright violet-blue. 'Queen Wilhelmina': flowers white. 'Taitii' (syn. *I.taitii*): flowers pale blue. 'Thunderbolt': falls bronze-brown, blotched yellow, standards purple-brown.

Irishman's cutting a portion of a plant with pre-formed roots, which has been detached from the mother-plant to establish as an individual. Commonly, the term is used to refer to any very small piece of a plant taken as a division, or to a naturally rooted layer.

iron an essential trace element in plants necessary for the formation of chlorophyll. Plants are especially susceptible to iron deficiency on alkaline soil, but also through waterlogging, low temperature, excessive phosphate in soil or, occasionally, heavy metal contamination. Iron deficiency is expressed as interveinal or overall yellowing of young leaves, which progresses down the shoots; the condition is referred to as chlorosis. Application of iron seque-strol (chelate) and mulching with organic matter will alleviate iron deficiency.

irregular zygomorphic; asymmetrical.

irrigation the application of water, sometimes containing nutrients in solution, to support plant growth. The frequency of need and most effective timing and rate of application vary with species, growing media and situation, as well as with levels of rainfall and atmospheric conditions, which are in turn affected by climate and season. Careful consideration must be given to these factors in order to ensure that irrigation is of maximum benefit, to conserve water as a valuable natural reserve and to avoid harmful waterlogging.

Incorporating organic matter into the soil improves water availability, and the application of a mulch to the surface of moist soil reduces water loss through evaporation. Damping down or misting around growing plants cools leaves and reduces transpiration, and similar benefits can be obtained by careful ventilation and shading in greenhouses, and by providing wind-shelter for plants outdoors. Many ornamental plants desirable in gardens are adapted to conditions of low water availability; most established trees and shrubs have a great capacity to extract water from depth; and many fruit and vegetable crops have different water requirements at particular stages of growth. All of these considerations are important in defining the real need for irrigation, which is a laborious, expensive and potentially wasteful undertaking.

Methods of irrigation

WATERING CANS are convenient for small areas outdoors and for plants grown under protection; a capacity of 9 litres is suitable. Galvanized cans are durable but heavier than plastic ones and, whatever the type, a long spout is advisable to enable careful placement between plants. A fine oval-shaped rose is necessary for watering seedlings, while the round coarser type of rose is a good choice for irrigating established plants because of the larger droplet size and faster delivery.

HOSEPIPES provide a convenient means of continuously delivering water over a distance from the tap source, but distribution is imprecise without special fittings. Most hosepipes are available in PVC, commonly 13mm-diameter bore, and may be reinforced to reduce kinking and interruption of the flow. Control of water output, distribution pattern and droplet size can be effected with a fitted control nozzle or handgun, or by squeezing the hosepipe by hand at or near its outlet. Flat hosepipes are available, reducing storage space required, and all types are best stowed on a reel, some designs of which allow the water to flow evenly when the hose is only partly unwound.

SPRINKLERS automatically deliver a spray of coarse droplets over a broadly defined area. In large greenhouses, these may be mounted overhead, but in gardens they are most commonly used as single static, rotating or oscillating units, placed on the ground and fed through a hosepipe. Travelling sprinklers are available for large lawns, and in some circumstances it may be appropriate to have a permanent grid of distribution pipes below ground level into which sprinklers can be connected. Irrigation by sprinkler can be wasteful because much water may be lost to evaporation or blown off target, and puddling may occur where plant foliage obstructs distribution. Most sprinklers work efficiently only above 30 psi, and a low trajectory is preferable. On average a sprinkler may discharge 900 litres of water per hour.

TRICKLE SYSTEMS are especially useful for plants suited to individual watering, such as pot plants and tomatoes; they can also be used in outdoor borders and on a capillary bench, which is an effective means of irrigating pot plants. Trickle irrigation will work at lower flow pressure than sprinklers, but output needs to be closely monitored and outlets kept unblocked and clean. Water distribution may be via nozzles fitted directly into the line, or through spurs of flexible tubing of 2–5mm diameter, which can be fitted singly or in multiples at each regularly spaced outlet.

SEEPHOSES, or soakhoses, are an excellent means of low-level watering because they dispense water exactly where it is required. Such lines may consist of small-bore plastic delivery pipes, continuously stitched along their length, or of compressed rubber with micro-pores 5–8 microns in diameter; with both systems water oozes along the length of run. Some types of seephose are recommended for burying below the surface, thereby further reducing evaporation loss. Lay flat polythene tube, which swells under water pressure, is another form similar to seephose, with water emitted through small regularly spaced perforations. As the water pressure is increased, the water is discharged in thin sprays. Seephose systems of garden irrigation are both effective and economic in water use.

AUTOMATIC DEVICES are available to control water flow for indoor and outdoor irrigation systems and these make an excellent contribution to economies.

Water supply Irrigation water must be free of grit or biological matter, including algae, which can block small outlets. Filtration and the covering of stagnant storage vessels are important. Care must be taken not to allow water sources to become infected with plant disease organisms as might happen where plant debris or soil drainage water contaminates a low-level tank.

High concentration of dissolved mineral salts can affect plant growth and block irrigation nozzles; over 300ppm calcium can cause lime-crusting. Gardeners do not have recourse to acidifying treatments such as nitric acid used by commercial growers, and domestic water softeners are unsuitable in view of phytotoxic residues. The use of ionized water is rarely a practical proposition, and the best means of avoiding high-lime content is to collect rainwater.

WATER BUTTS fitted with lids and low-level taps are useful for collecting roof water, although small debris will usually be washed down; the volume of storage is also limited and quickly used up during periods of little rainfall. Supply from wells, boreholes, rivers and streams has relevance to a limited number of sites, and for most gardeners mains supply is the norm. Mains water is filtered at the supply source and biological and chemical content is controlled to within limits suitable for plant growth. Its availability may be subject to metering, or surcharge for certain garden uses, and also to restrictions at times of drought. Irrigation systems adapted for special uses, such as automatic injection of liquid feed, must be fitted with devices to ensure that back-siphonage into the mains is not possible. 20mm taps usually discharge about 900 litres per hour from a mains supply.

DOMESTIC WASTE WATER from washing, sometimes called greywater, is often suggested for irrigation use during drought. The problem is the variability of content, and while it is unlikely that moderate use of domestic waste water would have a damaging effect on soil structure, soil organisms or plant growth, heavy applications loaded with soap detergent and other household cleaners, or water softening additives, can give rise to an increase in soil pH, phosphate content and the total salinity of soil. The safest course is to avoid repeated irrigation of the same site with waste water and to reserve its use for established plants.

Need, timing and quantity Soil moisture loss from a growing medium arises through evaporation from the surface and transpiration from the aerial parts of plants, collectively referred to as evapotranspiration. Plant requirements are met from soil reserves, which are replenished by rainfall or irrigation. By the end of March in most years soil outdoors will be fully charged with water as a result of winter rain. It is then described as being at field capacity, the level at which further added water passes to drainage. As the season progresses, evaporation and transpiration result in water loss from the soil and where the loss is greater than rainfall or irrigation, a soil moisture deficit (SMD) arises.

Calculation of irrigation requirements for outdoor crops balances input of water from rainfall with losses through evapotranspiration. Rainfall can be measured with a rain gauge and weekly evapotranspiration rates can be predicted from tables of local averages, or more accurately through adjusted data from weather centres.

Average monthly values for SMD in southern and Midland England (in mm) are April 50; May 75; June 100; July 100; August 75 and September 50. In the north of England, the SMD is about one third less. A reasonable target is to apply 25mm of water each time the SMD reaches 25mm, but woody plants in general may tolerate a SMD of 50–75mm. 25mm of rainfall is equivalent to an application rate of 20 litres of water per square metre.

Whether or not irrigation is practised with reference to collected or provided data, it is important to ensure that high SMD levels do not build up. Water shortage in soils should be anticipated in order to maintain steady growth.

In terms of irrigation needs, garden plants that should be given priority are seedlings and transplanted crops and specimen plants, together with leafy vegetables and those with swelling edible parts such as peas, beans, sweetcorn, courgettes and tomatoes. Raspberries and strawberries benefit from irrigation as the fruits colour and begin to swell respectively. A good yield of early potato tubers depends on moisture availability, whereas established leeks, carrots, parsnips and onions are less responsive.

Irrigation techniques should be selected with the aim of ensuring that plants are thoroughly watered. Light sprinkling is wasteful and ineffective. Crops growing under greenhouse protection need special care to meet water requirements, particularly those in containers. Irrigation need can be assessed by handling the surface of the soil, or by making a judgement based on the weight of those containers that are small enough to be easily lifted. Tapping clay pots to test for a hollow 'ring' is effective. Proprietary moisture deficit indicators are available; and saucers to provide reservoirs of water, and capillary benching are useful aids to greenhouse irrigation.

Isatis (Greek name for woad). Cruciferae. Europe, Mediterranean to C Asia. WOAD. Around 30 species, annual to perennial herbs with loose racemes of 4-petalled flowers in summer. The following species is hardy to −15°C/5°F. Grow in a sunny position, on a well-drained fertile soil. Sow seed *in situ* in autumn or spring. *I.tinctoria* (DYER'S WOAD, WOAD; Europe; biennial to 1m; flowers yellow).

island bed a large bed usually devoted to herbaceous perennials, heathers or dwarf conifers, in which the display can be seen from all sides. It may be irregular in outline and often deep and curving, and may be set into a lawn or surrounded by hard standing.

Isoplexis (from Greek *isos*, equal, and *plexis*, plait, as the upper corolla segment and the lip are of equal length). Scrophulariaceae. Canary Islands, Madeira. 3 species, evergreen subshrubs with toothed leaves and erect, terminal racemes of tubular, 5-lobed and 2-lipped flowers in summer. Grow in full sun in a medium-fertility loam-based mix, to which sand and leafmould have been added. Maintain a minimum temperature of 7°C/45°F and ventilate freely whenever possible; water plentifully in summer, moderately in winter. Propagate from seed in a sandy seed mixture in spring or by greenwood cuttings in late summer. *I.canariensis* (syn. *Digitalis canariensis*; Canary Islands; to 1m; raceme to 30cm, flowers to 3cm, orange-yellow to brick red or orange-brown).

Isopyrum (from Greek *isos*, equal, and *pyros*, grain: the fruit resembles grains of wheat). Ranunculaceae. Northern Hemisphere. FALSE RUE ANEMONE. 30 species, tufted perennial herbs with ferny, ternately lobed leaves and small, cup-shaped flowers, each composed of five, petal-like sepals, nodding on fragile stalks in spring. Suitable for woodland, peat beds and the cool, shaded rock garden. Cultivate as for *Anemonella*. *I.thalictroides* (Europe; to 25cm; flowers white).

Italian garden, Italianate the style of the great medieval and Renaissance formal gardens of Italy, where stonework, statuary, and waterworks played the most important role. Plants were treated formally as hedges, topiary, knots, and parterres and trimmed in pots. It was actively copied in Britain between 1830 and 1860.

Itea (from Greek *itea*, willow, after the pendulous catkins). Grossulariaceae (Iteaceae). NE Asia, eastern N America. 10 species, evergreen or deciduous shrubs and trees with toothed leaves and small, green to white flowers in racemes or panicles in summer. *I.ilicifolia* is a fine small shrub suitable for well-drained, fertile soils in warm and sunny positions. In zones 7 and under, it will succumb to hard frosts, overwet soils and cold winds, and tends therefore to be planted on south-facing walls and, at first, protected with dry bracken or evergreen clippings. *I.virginica* is altogether hardier. It favours semi-shade or full sun and moist, slightly acid positions where it can spread freely – lake or pond margins and islands, for example. It is excellent as tall ground-cover. Propagate both by soft stem-tip cuttings in late spring, inserted in mist or under polythene, or, in late winter, by heeled semi-hard cuttings in a sandy mix in the cold frame. Air-layering will also succeed.

I.ilicifolia (W China; erect, evergreen shrub to 5m; shoots smooth, slender; leaves 5–12cm, elliptic, spiny-toothed, glossy olive to dark green, tough; raceme slender, pendulous with cream to gold or pale lime flowers); *I.virginica* (SWEETSPIRE, VIRGINIA WILLOW, TASSEL-WHITE; E US; arching deciduous shrub to 3m; leaves 3–10cm, narrowly elliptic to oblong, rich red in autumn; racemes erect, to 15cm, with cream flowers).

Itea ilicifolia

Ixia (from Greek *ixia*, the name of a plant noted for the variability of its flower colour). Iridaceae. South Africa (Cape Province). 45–50 species, perennial herbs with small corms, grassy foliage, and, in summer, slender spikes of tubular flowers with six, spreading, ovate to lanceolate segments. Where temperatures seldom fall below freezing, *Ixia* can be planted in a sunny, south-facing border or under a south wall, about 10–15cm deep, with a winter covering of bracken litter or leafmould. Plant in autumn to bloom in spring and summer, or in spring for summer blooms. For potted specimens, set 6–8 corms firmly in a 12cm pot in a mix of sandy loam and leafmould. Plunge in a cool frame and water very sparingly during winter until the flower spikes appear, then give full light and good ventilation. Continue to water until foliage begins to die back, then dry off gradually and store dry in cool but frost-free conditions. Offsets usually flower in their second year. Sow seed in autumn, and let seedlings remain in the same pans for their first year. Seed-raised plants will usually flower in their third or fourth year.

I.maculata (18–50cm; flowers 2.5–5cm in diameter, orange, yellow-orange or white with a dark brown, black or purple centre); *I.monadelpha* (15–40cm; flowers 3–4cm in diameter, white, pale pink, lilac, mauve, purple, violet, blue, or claret, usually with a brown or red-brown central marking); *I.viridiflora* (50–100cm; flowers 2.5–5cm in diameter, pale aquamarine with a purple-black centre).

Ixiolirion (from *Ixia* and Greek *lirion*, lily, for its resemblance to *Ixia*). Amaryllidaceae. SW and C Asia. 4 species, bulbous perennial herbs. Produced in a rosette, the leaves are linear to lanceolate, and usually persist in winter. Carried in spring and summer in umbels or loose racemes, the starry flowers consist of six segments, either free or united toward the base in a short tube. Hardy to –15°C/5°F, they need a warm and sheltered position in a rich and perfectly drained soil, where they can be baked in summer and protected from excessive wet in winter. They are sometimes grown in the cool conservatory or greenhouse. Plant 15cm deep in late summer/early autumn. A winter mulch of bracken litter or leafmould is beneficial. Propagate by removal of offsets after flowering, or by seed in autumn. *I.tataricum* (syn. *I.montanum*; SW and C Asia, Kashmir; to 40cm; flowers in a loose umbel, 2–5cm, pale to mid blue or violet).

Ixora (from the name of a Malabar deity to whom the flowers were offered). Rubiaceae. Tropics (Old and New World). Some 400 species, evergreen shrubs or trees bearing paniculate or corymbose cymes of small, narrowly tubular flowers in summer, each with four spreading lobes. Grow in warm greenhouse conditions (minimum temperature 15–18°C/60–65°F), with high humidity and bright indirect light, in a fibrous, soilless medium. Water plentifully. Prune as necessary after flowering. Propagate from short-jointed semi-ripe cuttings of non-flowering shoots, potted singly and plunged in a shaded closed case with bottom heat.

I.coccinea (JUNGLE GERANIUM, FLAME OF THE WOODS, JUNGLE FLAME, RED IXORA; tropical Asia; shrub to 2.5m; flowers red, orange, apricot, pink or yellow); *I.javanica* (COMMON RED IXORA, JAVANESE IXORA; Java; shrub or tree to 4m; flowers red or, rarely, pink to orange); *I thwaitesii* (WHITE IXORA; Sri Lanka; shrub or tree to 6m; flowers white to cream).

J

Jaborosa (from Arabic, *jaborose*, for the closely related *Mandragora*). S America. Solanaceae. 20 species, low-growing, perennial herbs with clumped, elliptic leaves and low-borne, solitary flowers in summer – these are tubular to campanulate with a starry limb of five, narrow lobes. Plant in a sandy, well-drained soil in full sun or part shade. Although hardy in climate zones 7 and 8, where hard frosts are common and prolonged, the following species should be grown in a warm, sunny position at the base of a south-facing wall. Propagate by division in spring. *J.integrifolia* (S Brazil, Uruguay, Argentina; leaves to 16cm, dark green; flowers green to white, to 6cm diameter).

Jacaranda (from the vernacular Brazilian name for these plants). Bignoniaceae. Tropical America: Belize, Guatemala, West Indies to Argentina. Some 45 species, deciduous or evergreen trees, usually with bipinnate leaves. Tubular to campanulate and 5-lobed, the flowers are carried in terminal or axillary panicles in spring and summer. *J.mimosifolia* is widely planted in the subtropics and warm temperate regions as a specimen, street and avenue tree. In zones 8 and under, it is grown in the cool greenhouse or conservatory, predominantly as a foliage plant, although it may begin to bloom at about 1.5m in a large tub. It is also used, when young, as a foliage plant in summer bedding. Grow in a well-drained, high-fertility, loam-based mix in full sun. Water plentifully when in growth, sparingly at other times. Maintain a minimum temperature of 7°C/45°F, with good ventilation. Prune plants grown for their foliage hard back in late winter and repot or topdress at the same time. Propagate by seed or by heeled semi-ripe cuttings. *J.mimosifolia* (syn. *J.ovalifolia*; Argentina, Bolivia, widely cultivated elsewhere; tree to 15m tall with a rounded crown; leaves deep green, ferny; flowers 2.4–5cm, blue-purple with a white throat; 'Alba' is white-flowered).

Jamesia (for Dr Edwin James (1797–1861), who discovered the plant on Major Long's expedition to the Rocky Mountains in 1820). Hydrangeaceae. W US. CLIFFBUSH, WAXFLOWER. 1 species, *J.americana*, a deciduous, rounded shrub to 1.5m tall, with peeling, papery bark. The downy leaves are 3–7cm long, ovate, toothed and prominently veined. To 1cm across, the 5-petalled flowers are white (pink in 'Rosea') and carried in terminal panicles to 6cm in early summer. Fully hardy. Grow in any freely draining, moderately fertile soil in full sun. Prune after flowering to remove old, weak or overcrowded growth. Propagate by ripe seed, semi-ripe cuttings or by layering in spring.

Jankaea (for Victor Janka von Bulc (1837–1900), Austrian soldier and botanist). Gesneriaceae. C Greece (Mount Olympus). 1 species, *J.heldreichii*, a perennial, evergreen herb. The leaves are 2–4cm long, bluntly obovate, hairy and carried in low rosettes. Bell-shaped with four to five rounded lobes, the flowers are lilac and carried in small clusters atop stalks to 5cm tall. This plant resents overwet conditions, and especially water on its foliage or around the crown. For this reason, it is often grown in the alpine house, in tufa or with a collar of stones. It will grow outdoors in climate zones 7 and over, but needs the enhanced drainage of a scree or sink garden, and cloche protection in winter. Plant in a fast-draining, gritty, alkaline medium rich in leafmould. Grow in a dry, buoyant atmosphere with full sun. Water moderately in spring and summer, scarcely at all in winter. Propagate by division.

Jankaea heldreichii

japanese beetle

Japanese beetle *(Popillia japonica)* (Coleoptera: Scarabaeidae) a beetle that is bluntly ovoid in shape and about 11mm long, with the head, thorax and legs metallic green and the hardened wing cases dull coppery brown. The larvae have creamy-white C-shaped bodies with a swollen last segment and well-developed brown heads, legs, and biting mouthparts. When fully fed, they are about 25mm long. The larval stage lasts about 10 months, but may be longer in colder areas where the complete life cycle can extend to two years.

Japanese beetle is native to the Far East but is widespread in North America since its introduction in 1916, and poses a threat to other areas including Europe, Africa and western Asia. It is mainly a pest of permanent pasture and ornamental grasses, but its larvae also feed widely on plant roots, often causing serious damage to nursery stock. The adults fly in sunny weather, attacking especially apple, cherry, plum, elm, horse chestnut, lime, lombardy poplar, willow, raspberry, strawberry, dahlia, hollyhock, rose, Virginia creeper and zinnias. They eat holes in leaves and flowers, and in severe attacks may cause complete defoliation. Deep holes are made in ripe fruits. In North America biological control organisms have been used successfully, including the eelworm *Neoplectana hoptha*, two hymenopterous parasites, and the bacterium *Bacillus popilliae* which causes 'milky disease' of the grubs.

Japanese beetle is very similar in appearance to the garden chafer *(Phyllopertha horticola)*, which is common in Europe, but may be readily distinguished by its green legs.

Japanese gardening an ancient art originating around 300 AD with elaborate landscape creations. It evolved to reflect aspects of Japanese traditions, symbolism and faith. Gardens became places for contemplation, sometimes with plantings reduced to a minimum and shaped trees, moss, rocks, water or raked sand predominating.

jardin anglais the English garden; a European name for the landscape garden, the fashion for which resulted in the destruction of formal gardens around Florence and elsewhere in Europe in the late 18th and 19th centuries.

jardinière an ornamental stand or container for plants, normally used indoors.

Jasione (name given by Theophrastus to *Convolvulus*). Campanulaceae. Temperate Europe, Mediterranean. SHEEP'S BIT. Some 20 species, annual, biennial or perennial herbs, usually with clumped, oblong to lance-olate leaves, and 5-parted flowers in terminal heads or panicles, surrounded by one or more rows of spiky involucral bracts. The following species is fully hardy and suitable for the rock garden. Grow in well-drained sandy soils in sun. Propagate by division or seed. *J.laevis* (syn. *J.perennis*; SHEPHERD'S SCABIOUS, SHEEP'S BIT; tufted perennial, to 15cm tall; inflorescence bracts ovate to deltoid, toothed; flowers blue).

Jasminum (Latinized form of Persian *yāsmiñ*). Oleaceae. Old World tropics, subtropics and temperate regions. JASMINE, JESSAMINE. About 200 evergreen or deciduous shrubs and small trees, many sprawling, scandent or twining. The leaves are pinnate, trifoliolate or with only one leaflet, thus 'simple'. Often fragrant, the flowers are solitary or clustered in axils, or in cymes or cymose panicles – they consist of a tubular corolla with 4 or more spreading lobes. Grown for their often heavily fragrant flowers, the genus *Jasminum* comprises species for a number of garden situations, from *J.parkeri*, with tiny yellow flowers in early summer, which forms low, dense and evergreen hummocks suitable in scale for the rock garden, to the free-standing semi-evergreen *J.humile*, a fine wall shrub, bearing large and fragrant flowers from early spring to late autumn. The twining climbers, such as *J.officinale* and the tropical species, *J.rex*, will scramble over trelliswork, shrubs, trees or other supports.

The rambling or scandent shrubby species, which require tying with shoots carefully trained and spaced, can be used as wall shrubs, or trained over pergolas, arches, arbours and fences. *J.nudiflorum* is also useful as a soil-stabilizer on steep banks; it will cascade over a retaining wall and is tolerant of urban pollution. A sunny site is best but *J.officinale* and *J.nudiflorum* will tolerate the shade of a north-facing wall. *J.mesnyi*, with spectacular semi-double flowers, and *J.polyanthum*, vigorous and intensely fragrant, for south- or southwest-facing walls in areas with little or no frost, make excellent conservatory plants in colder zones. *J.polyanthum* may also be grown as a house plant trained on wire hoops or canes.

Cultivate frost-tender species in a cool greenhouse or conservatory border (minimum temperature 5°C/40°F) or in a medium-fertility loam-based mix, with wires or other support. (*J.polyanthum* requires a temperature drop to induce flowering.) Water plentifully in summer, sparingly in winter, and ventilate freely when possible. *J.rex* requires a minimum temperature of 18°C/65°F, and succeeds best with part-day shade. Tie in leaders of scandent species in a fan formation after planting. Cut back flowered growth of *J.nudiflorum* immediately after flowering; remove weak growth and tie in as necessary. Thin, but do not shorten, shoots of *J.officinale* after flowering. Other species require no regular pruning, other than thinning of overgrown plants. Propagate by semi-ripe cuttings in late summer, overwintering at 7–10°C/45–50°F. Take cuttings of *J.mesnyi* and *J.polyanthum* with a heel, and root with bottom heat of 15°C/60°F. Also by layers, and seed sown when ripe.

JASMINUM

Name	Distribution	Habit	Leaves	Flowers
J.azoricum J.fluminense	Azores	evergreen climbing shrub	leaflets 3, ovate, acute, undulate, cordate at base, terminal leaflet larger	fragrant, long-stalked, borne in terminal panicles, corolla white, tube to 0.5cm, lobes to 6, 1cm long
Comments: Late summer. Z9.				
J.beesianum	China	deciduous or semi-evergreen, sprawling shrub	simple, ovate-lanceolate, 2–5cm, olive green, slightly downy	1–3 per axil, small, fragrant, pale pink to deep rose
Comments: Early summer. Z7. Produces abundant shining black berries which persist in winter. A form of *J.beesianum* with longer leaves and larger, pale pink flowers is often seen, this may be a backcross with *J.* × *stephanense*. True *J.beesianum* has smaller, darker flowers.				
J.floridum	China	erect to arching, semi-evergreen shrub	leaflets 3–5, 1–3.5cm, oval to ovate, acuminate, glabrous, shining green above, pale beneath	yellow, profuse, in cymes
Comments: Late summer. Z9.				
J.fruticans	Mediterranean, Asia Minor.	erect, evergreen or semi-evergreen shrub to 1.25m	leaflets 1–2cm, 3, tough, narrow-oblong, obtuse, dark green, glabrous, minutely ciliate	to 5 per terminal cyme; corolla yellow, tube to 1.25cm, lobes 5, obtuse, to half length tube
Comments: Summer. Z8.				
J.grandiflorum CATALONIAN JASMINE; ROYAL JASMINE; SPANISH JASMINE	S Arabia, NE Africa, widely grown in warm and temperate regions elsewhere	semi-evergreen or evergreen climbing shrub	leaflets 5–7, to 2cm (terminal leaflet longer), ovate	in terminal, branching cymes, intensely fragrant; corolla white, exterior sometimes tinted red, tube to 2cm, lobes 5–6, oblong
Comments: Z10. Source of jasmine oil.				
J.humile ITALIAN JASMINE	Middle East, Burma, China	evergreen or semi-evergreen shrub to 6m, erect, occasionally tree-like	leaflets to 7, to 4cm, ovate-lanceolate	sometimes scented, in umbel-like clusters; corolla yellow, tube to 2cm, lobes to 1cm, rounded, spreading
Comments: Summer. Z6. Includes 'Revolutum' (syn. *J.revolutum, J.reevesii, J.triumphans*): semi-evergreen, robust shrub, leaflets 3–7, terminal leaflet to 6cm, laterals to 4cm, flowers 2.5cm wide, yellow, fragrant, 6–12 together, stamens slightly protruding; f. *farreri* (syn. *J.farreri, J.giraldii*): evergreen, spreading shrub to 1.5m, leaflets 3, oval-lanceolate, distinctly tapering, terminal leaflet to 10cm, dull green, coarse above, hairy beneath, inflorescence hairy, flowers yellow, to 12 per cyme.				
J.laurifolium	Assam, Bangladesh	evergreen slender climber	narrowly elliptic-lanceolate	starry, white
Comments: Summer. Z 9. Includes f. *nitidum* (syn. *J.nitidum, J.gracile* var. *magnificum*): ANGEL HAIR JASMINE; ANGEL WING JASMINE; WINDMILL JASMINE; STAR JASMINE; CONFEDERATE JASMINE, leaves to 8cm, simple, elliptic-lanceolate to ovate-lanceolate, glossy bright green above, flowers fragrant, corolla white, often tinted red in bud and on exterior, tube to 2cm, slender, lobes to 1.5cm, 9–11, linear-lanceolate.				
J.mesnyi J.primulinum PRIMROSE JASMINE	W China	evergreen rambling to cascading shrub to 3m	leaflets 3, 2.5–7cm, lanceolate, dark glossy green	3–5cm wide, solitary in axils, corolla usually semi-double, bright yellow, lobes to 1cm across, obtuse
Comments: Summer. Z8.				
J.multiflorum J.pubescens STAR JASMINE	India	evergreen pubescent climber	to 5cm, simple, ovate, base rounded to cordate	large, fragrant, in sparse to crowded clusters; corolla white, lobes oblong-lanceolate, 5–8, to half length of tube
Comments: Summer. Z10				
J.nudiflorum J.sieboldianum WINTER JASMINE	N China.	deciduous, rambling to cascading shrub, to 3 × 3m, with long, slender branches	leaflets 3, 1–3cm, oval-oblong, dark green, glabrous, ciliate	borne before leaves, singly in axils of previous year's growth; corolla to 3cm wide, yellow, 6-lobed
Comments: Winter–early spring. Z 6.				
J.officinale COMMON JASMINE; TRUE JASMINE; JESSAMINE	Asia Minor, Himalaya, China.	deciduous climbing shrub to 10m, shoots twining, slender	leaflets 5–9, 1–6cm, elliptic, acuminate	highly fragrant, in branched cymes; corolla white, tube to 2cm, lobes 4–5, shorter than tube, +/– triangular-ovate
Comments: Summer–early autumn. Z7. 'Affine': flowers larger, more profuse, exterior pink; corolla lobes broader; 'Aureum': leaves blotched golden yellow; 'Inverleith': leaves 2–5cm; leaflets 3–7, lanceolate, terminal leaflet often lobed, flowers to 1.5cm long, cylindrical-funnelform, exterior strongly tinted pink-red, lobes 5, to 1cm, broadly ovate, apiculate, pure white above, tinted red beneath, especially in bud.				
J.parkeri	NW India	evergreen twiggy dwarf shrub, to 30cm, forming a dense mat or mound	leaflets 3–5, 0.3–0.6cm, ovate-acuminate	usually borne singly, yellow, 1.5cm wide
Comments: Summer. Z7.				
J.polyanthum J.blinii	China	deciduous or evergreen climbing and twining shrub to 8m	leaflets 5–7, lanceolate, narrow-acuminate at apex, terminal leaflet to 8cm, coriaceous, 3-nerved	in axillary panicles, highly fragrant; corolla to 2cm, white within, exterior flushed pink, lobes obovate
Comments: Summer. Z9.				
J.rex	Thailand	glabrous climber	simple, to 16cm, dark glossy green, broadly ovate, acuminate	on slender, drooping stalks, 2–3 per axillary cyme, unscented; corolla white, salverform, tube to 2.5cm, lobes 8(–9), ovate-oblong
Comments: Summer. Z10				

JASMINUM

Name	Distribution	Habit	Leaves	Flowers
J.sambac ARABIAN JASMINE	widespread through long cultivation; may originate in India	evergreen twiner, shoots pubescent, angular	simple, semi-rigid, shiny, glabrous or hairy, broad-ovate to 8cm, acute, conspicuously veined	in clusters, highly fragrant; corolla waxy white, pink with age, tube 1.2cm, lobes to 1.5cm, 6–9, oblong

Comments: Flowers continuously. Z10. 'Grand Duke of Tuscany'('Florepleno'): flowers double, lobes rounded, the whole resembling a small *Gardenia*.

J. × stephanense (*J.beesianum* × *J.officinale*)	garden origin, also said to occur wild in W China	fast-growing climber, twining, to 5m, stems often yellow-green or cream at first	simple or 3–5-lobed, leaflets pale olive green, pubescent beneath.	pale pink, small, in sparse cymes

Comments: Summer. Z7.

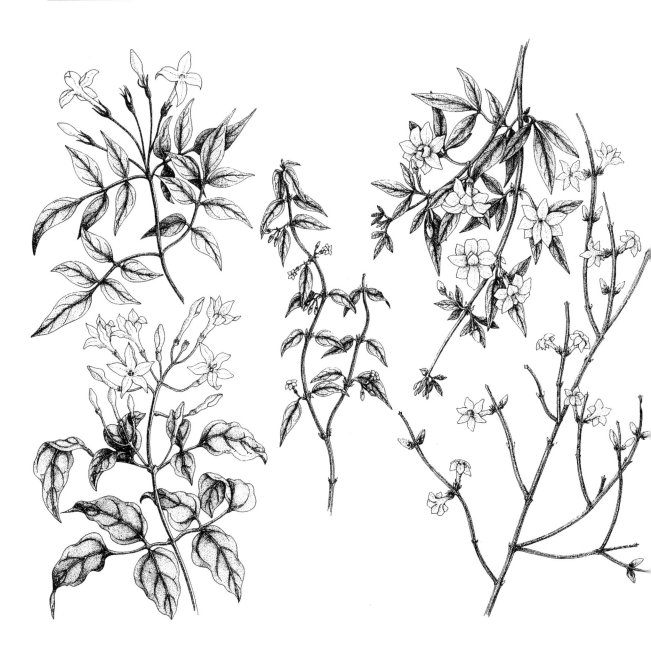

Clockwise from top left: *Jasminum officinale, J. mesnyi, J. nudiflorum, J. azoricum, J. beesianum* (middle)

Jatropha (from Greek *iatros*, physician, and *trophe*, food, alluding to the medicinal qualities). Euphorbiaceae. Tropics and warm temperate regions, mainly S America. TARTOGO, GOUT PLANT, BARBADOS NUT, PHYSIC NUT, CORAL PLANT, PEREGRINA, JICAMILLA. 170 species, perennial herbs and shrubs, with poisonous sap. The leaves are simple or palmately lobed or cut. Small flowers are produced in dense, flat-topped, thick-stalked and many-branched cymes. Grow in a minimum temperature of 10°C/50°F, with full sun and a dry atmosphere. Plant in a well-drained, loam-based mix with additional leafmould and sharp sand. Water moderately when in full growth. Keep virtually dry during the winter rest period, when the plants may become leafless. Propagate by seed or by semi-ripe cuttings. All parts are poisonous.

J.multifida (CORAL PLANT, PHYSIC NUT; tropical America; shrub or tree to 7m; leaves to 30cm diameter, dark green, palmately and deeply 7–11-lobed, lobes finely dissected and toothed, petioles sometimes finely spotted purple-brown, as are stems of young plants; flowers scarlet); *J.podagrica* (GOUT PLANT, TARTOGO; Guatemala, Panama; stem to 2.5m, usually thickly gouty with branches only at its summit and a large central swelling; leaves to 30cm diameter, ovate to orbicular, with 3–5, broad and shallow lobes; flowers coral red).

Jeffersonia (for Thomas Jefferson (1743–1826), third president of the United States, author, architect and horticulturist). Berberidaceae. N America, E Asia. TWIN LEAF. 2 species, perennial herbs with radical, 2-lobed, rounded and toothed leaves peltately attached to long stalks. Carried on slender scapes in spring, the solitary flowers are cup-shaped with five to eight, oblong petals. Fully hardy. Grow in light shade or dappled sun, in an acid to neutral humus-rich soil. Keep the root run cool and moist. Sow fresh ripe seed in the cold frame or cool greenhouse, or divide well-established plants in early spring.

J.diphylla (RHEUMATISM ROOT; N America; leaves to 15cm across, glaucous beneath; scapes exceeding petioles; flowers white); *J.dubia* (NE Asia; leaves to 10cm across, glaucous, tinged mauve-grey; scapes shorter than petioles; flowers lavender to pale blue).

John Innes base a compound fertilizer containing 2 parts hoof and horn (73% nitrogen), 2 parts superphosphate (18–21% phosphate), and 1 part potassium sulphate (48% potash) by weight. It is formulated for use in John Innes compost.

John Innes compost any of various soil-based seed, cutting and potting media made to formulae developed by the John Innes Horticultural Institute in Norwich, UK. The recommended basic ingredients are: partially sterilized, fibrous, clay-loam of pH 5.5–6.5, riddled through a 9mm sieve; coarse sand with particles up to 3mm diameter; dust-free moss or sedge peat of moderately coarse texture.

John Innes seed compost comprises 2 parts (by volume) loam: 1 part peat: 1 part sand. Added to each cubic metre of mix is 0.6kg of ground limestone and 1.2kg superphosphate. John Innes cutting compost comprises 1 part (by volume) loam: 2 parts peat: 1 part sand, with no added fertilizer.

John Innes potting composts comprise 7 parts (by volume) loam: 3 parts peat: 2 parts sand. They are devised in three grades, JIP-1, JIP-2, and JIP-3, which contain an increasing quantity of base fertilizers to suit varying degrees of plant vigour and seasonal change in growth rate. For JIP-1, 0.6kg ground limestone, 1.2kg hoof and horn meal, 1.2kg superphosphate and 0.6kg potassium sulphate is added per cubic metre. All fertilizer ingredients except ground limestone are doubled and tripled for JIP-2 and JIP-3 respectively.

Batches of John Innes compost will vary due to differences in the quality of loam especially. Storage life is affected by moisture content, temperature and organic matter and ideally the composts should be used soon after making. JIP-1 will store longer than JIP-2 or JIP-3, and for best results the fertilizer should be withheld until just before the compost is required for use.

For lime-hating plants an acid, or ericaceous, compost known as JIS-A is suggested, having the same ingredients as John Innes seed compost but with the ground limestone replaced by flowers of sulphur at the same rate.

jostaberry (*Ribes × culverwellii*) a bush fruit hybrid between blackcurrant and gooseberry, bearing small clusters of dull berries which resemble blackcurrants. The plant is vigorous, upright, spineless and self-fertile. Its flowers are frost-susceptible and carried on old and young wood. It may be cultivated and propagated as a blackcurrant and grown either as a stooled bush, or a half standard, in which case it will require staking. Jostaberry is resistant to American gooseberry mildew, blackcurrant leaf spot and gall mite. Also known as yostaberry.

Jovellana (for Dr C.M. de Jovellanos (1744–1811), student of the Peruvian flora). Scrophulariaceae. Chile, New Zealand. Some 6 species, tender herbs, subshrubs and shrubs, resembling *Calceolaria* in the strongly saccate corolla, but having two almost equal corolla lobes (lips) with the margins flat or involute. The following species is usually grown in the cool greenhouse or conservatory, but will tolerate mild frosts if planted in a warm, sheltered position and mulched in winter. Grow in full sun in a free-draining loam-based medium. Avoid excessive moisture in winter. Propagate by seed or by semi-ripe cuttings. *J.violacea* (Chile; puberulent shrub to 2m; flowers to 1.5cm diameter, pale violet spotted purple, throat blotched yellow).

Jovibarba (from Latin *Ioves*, of Jupiter, and *barba*, beard). Crassulaceae. Europe. 5 species, perennial herbs with monocarpic leaf rosettes. They are close to and sometimes included in *Sempervivum* but distinguished from that genus by their secondary leaf rosettes and campanulate flowers. Cultivate as for *Sempervivum*.

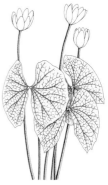

Jeffersonia diphylla

J.hirta (syn. *Sempervivum hirtum*; C and SE Europe; rosettes 5-7cm across; leaves lanceolate to obovate, ciliate, green tipped brown to red-brown or flushed deep red; flowers pale yellow-brown); *J.sobolifera* (syn. *Sempervivum soboliferum*; C and SE Europe; rosettes 2–4cm across; leaves obovate to oblong, bright green tipped red with age, ciliate; flowers yellow).

Juanulloa (for Jorge Juan (1713–73) and Antonio Ulloa (1716–95), Spanish travellers in S America). Solanaceae. C and S America. Some 10 species, epiphytic or climbing and stem-rooting shrubs. Produced in clusters or racemes, the flowers consist of a large, campanulate calyx, often fleshy and ridged, with oblong to lanceolate lobes, and a tubular, 5-lobed corolla. Grow in a freely draining, medium-fertility loam-based mix in the intermediate greenhouse (minimum temperature 13°C/55°F). Keep moist, humid and well-fed in spring and summer with some protection from full sunlight; position in full sun in winter and keep rather drier. Propagate by seed or cuttings. *J.mexicana* (syn. *J.aurantiaca*; Peru, Colombia, C America; stems to 2m, arching; calyx to 1.5cm diameter, waxy, pale orange, corolla to 4cm, waxy, brilliant orange).

Jubaea (for King Juba of the ancient African kingdom of Numidia). CHILEAN WINE PALM, HONEY PALM, SYRUP PALM, COQUITO PALM, LITTLE COKERNUT. Palmae. Chile. 1 species, *J.chilensis*, a palm to 25m tall, the stem to 1.3m diameter at base, occasionally swollen in middle, becoming bare, with oblique scars and vertical cracks. The leaves are 1–5m long and pinnate, with linear, rigid, grey-green segments to 50cm long. A large but slow-growing and elegant feather palm, *Jubaea* is notable for its tolerance of relatively cool conditions. It is suitable for outdoor cultivation as a specimen or in avenue plantings in dry, Mediterranean-type climates and may tolerate several degrees of short-lived frost. In colder climates, it may be grown in the cool to intermediate greenhouse or conservatory and is particularly handsome as a juvenile specimen. Plant in a loamy mix high in grit and sand; keep moist in warm weather, rather dry in winter. Admit full sun and ventilate freely. Propagate by fresh seed, which may take up to six months to germinate.

Juglans (from Latin *Jovis glans*, 'Jove's acorn'). Juglandaceae. N and S America, SE Europe, Asia. WALNUT (q.v.). About 15 species, deciduous trees with furrowed bark and pinnate leaves. Inconspicuous male flowers are borne in axillary catkins; the female flowers in terminal spikes. The females give rise to 2–4-celled drupes with thick, furrowed pericarps and 2–4-lobed seed.

The young growth of many species is prone to damage by late spring frosts, although in the genus as a whole spring flushing is relatively late. Substances produced by the roots and fallen leaves of many species, notably *J.regia* and *J.nigra*, appear to have toxic effects (a form of allelopathy) on a variety of plants.

All species produce edible nuts but quality is highest in cultivars of *J.regia*. Bearing commences between 5 and 15 years of age, and may be irregular due to a number of factors. It seems likely that flower initiation depends on suitable conditions the previous summer, and flowers and young growth are destroyed by even short periods (hours) with temperatures of between –2°C/28°F and –3°C/26°F. Although walnuts are self-fertile, in some individuals male and female flowers mature at different times and pollination may not occur in solitary trees. When ripening is poor in cool damp climates, walnuts can nevertheless yield a good harvest of nuts for pickling, provided the nuts are gathered before the development of the hard shell. However, walnuts are cultivated as far north as Warsaw, and in recent years Polish cultivars (e.g. the *J.regia* Carpathian Group) have yielded promising results for cold temperate zones. Increased interest is being shown in Australia, where 'Franquette', a comparatively old French cultivar (on *J.nigra* rootstock), which is reliable, late-flowering, and of excellent quality, is widely planted.

Plant in deep, well-drained but moderately moist, fertile and slightly alkaline soils. Prune only when quite necessary, in late summer/early autumn when the tree is in full leaf, or else when it is completely dormant, otherwise wounds bleed profusely. For fruit production, *J.regia* may be pruned to a vase or cup-shaped framework when young, and, in the smaller garden, trees may be kept relatively dwarfed by judicious light pruning.

Propagate from seed sown ripe, or stratified in autumn and sown in spring, in a sandy, humus-rich propagating medium. Seed dormancy is broken at temperatures of 5–6°C/40–43°F. Protect from rodents and, since walnuts develop deep taproots which make transplanting difficult, move seedlings as soon as possible to nursery beds and then on to final positions. Cultivars will not come true from seed; propagate by splice or veneer grafting on to seedling stock of *J.regia* under glass in winter, or by budding on to nursery-grown seedling stock in summer. For heavier soils, *J.nigra* is the preferred rootstock.

Bacterial leaf blight causes small angular black spots on the leaves and can also attack fruit. Trees may be attacked by two species of walnut aphids. Blister-like growths on the leaves are caused by the walnut blister mite, which is widespread in Europe and Asia and has been introduced into Australia. Additional North American pests include the walnut caterpillar, walnut husk fly, a fruit fly with larvae which tunnel into the green husks turning them black and slimy; and two species of weevil, the black walnut curculio and the butternut curculio, with larvae that burrow into the nuts.

J.ailanthifolia (HEARTNUT, JAPANESE WALNUT; Japan; tree to 15m; leaves to 50cm, leaflets 11–17, to 15 × 5cm, oblong to elliptic, acuminate, serrate, grey-tomentose to glabrous above, glandular-hairy beneath; fruit globose to ovoid, sticky-hairy; includes var. *cordiformis*, syn. *J.cordiformis*, with narrower leaflets); *J.cathayensis* (CHINESE WALNUT, CHINESE BUTTERNUT; shrub or tree to 23m; leaves to 80cm, viscid; leaflets 11–19, to 15 × 8cm, oblong to lanceolate, finely dentate, hairy; fruit ovoid to 4.5cm); *J.cinerea* (BUTTER NUT, WHITE WALNUT; NE US; to 30m; young branches glandular-glutinose; leaves 25–50cm, leaflets 11–19, 6–12cm, oblong to lance-olate, acuminate, serrate, hairy, glandular beneath; fruit to 10cm, glandular-glutinose); *J.microcarpa* (syn. *J.rupestris*; SW US, Mexico; shrub or small tree to 7m; young shoots grey to brown; leaves with 15-23 leaflets, finely serrate); *J.nigra* (BLACK WALNUT; E US, naturalized C Europe; to 45m; leaflets 15–23, to 11cm, ovate to oblong, pubescent beneath; fruit globose, about 4cm diameter, hairy); *J.regia* (ENGLISH WALNUT, PERSIAN WALNUT, MADEIRA WALNUT; SE Europe to Himalaya and China, C Russia, naturalized US; to 30m; leaflets 5–9, rarely more, 6–12cm, elliptic to obovate, entire to serrate, with tufted hairs in vein axils beneath; fruit subglobose, smooth, green, 4–5cm across).

Juncus (from Latin *iuncus*, a rush, derived from *iugere*, to join, from the use of the stems in tying). Juncaceae. Cosmopolitan, but rare in the tropics. RUSH. About 225 species, rhizomatous grassy herbs growing in wet places. The stems and leaves are clumped and terete to narrowly sword-like and folded. The flowers are small, chaffy, green or brown and carried in dense, rounded, terminal or lateral cymes. Fully hardy and suitable for ponds and boggy ground. In smaller ponds, they can be grown in plunged containers. Propagate by division in spring.

J.effusus (COMMON RUSH, SOFT RUSH; virtually cosmopolitan; stems and leaves indistinguishable, to 1m, dark green, slender, smooth, terete; inflorescence a dull brown lateral cyme near apex of stem; includes 'Spiralis', CORKSCREW RUSH, with spirally twisted leaves and stems, 'Vittatus' with stems and leaves banded ivory, and 'Zebrinus' with leaves banded white; similar to *J.effusus* but smaller and finer is *J.decipiens,* whose cultivar 'Curly Wurly' has tightly spiralling foliage); *J.ensifolius* (western N America; 20–80cm, stems clumped, clothed with sea green, overlapping, sword-shaped leaves that resemble a miniature flag iris; flowers purple-black, glossy, in small, spherical heads).

June drop the natural shedding in early summer of a proportion of the immature fruits of apple, pear, plum, cherry and citrus. Fruits fall over a short period, and the extent of the occurrence appears to be affected by soil moisture status. Planned fruit thinning should be done immediately after June drop.

Juniperus (Latin name for the tree). Cupressaceae. Northern Hemisphere from Arctic to C America, Himalaya, Taiwan; to 20°S in E Africa. JUNIPER. Perhaps 60 species, evergreen, coniferous trees or shrubs. The older bark is usually grey-brown and stringy, exfoliating in flakes or long vertical strips and exposing red-brown, smooth bark beneath. The leaves are sharp and needle-like or small, scale-like and overlapping. The cones are globose, ovoid or lumpy, bloomed with glaucous wax, fleshy and inde-hiscent, with the scales marked only by the finest lines; in colour they are green, ripening to orange, red, brown, purple or black.

The following are fully hardy and will adapt to most garden situations, although full sun and a well-drained soil are important. *J.drupacea* prefers lime. *J.recurva* favours a humid, sheltered, cool and damp site. Many of the glaucous species will tolerate poor, chalky and stony soils, drought and sun. Both *J.conferta* and *J.horizontalis* (especially its selection 'Bar Harbor') endure salt spray and residue from salt-treated roads. Many species abound in cultivars and one at least can usually be fitted to a particular garden need. The prostrate cultivars of several species, for example, are resilient, impenetrable groundcover. Dwarf cultivars are excellent rock garden and heather/conifer bed subjects. *J.chinensis* and *J.sargentii* are among the best conifers for bonsai treatment. Propagation as for *Chamaecyparis*. Affected by needle blight and stem die-back. The causal fungi enter through wounds or weakened tissue, especially in wet weather. Control by fungicidal spray. Junipers can be attacked by rusts; these cause the branches to swell and erupt with yellow-brown bodies. They are also attacked by phytophthora root rot. Pests include tent caterpillar and the juniper webber, which feeds *en masse* amid a complex of webbing. Infested foliage turns brown and dies, usually in early summer. Insecticides will work if applied in late summer or early autumn, when the caterpillars are young. Also afflicted by juniper scale, the conifer spinning mite and the bagworm.

J.chinensis (CHINESE JUNIPER; China, Mongolia, Japan; tree or shrub to 20m; crown conic; bark peeling in strips; foliage pungently scented, juvenile leaves subulate, dark green with 2 pale bands beneath, adult leaves rhombic, blunt, scale-like, overlapping; fruits violet to brown with a white bloom; includes golden, grey-blue and narrowly columnar to pyramidal cultivars; see also under *J. × media*); *J.communis* (COMMON JUNIPER; Europe, Asia; dense shrub or tree, 1–20m tall; crown erect to spreading, typically bushy and erect; bark flaking or peeling; leaves subulate to needle-like, deep green with a single pale band on inner surface; fruit blue-grey to purple-black, glaucous; includes very dwarf to tall cultivars ranging in habit from tightly erect to pendulous and prostrate); *J.conferta* (SHORE JUNIPER; Japan, Sakhalin; prostrate to ascending shrub; bark red-brown; leaves needle-like, dark green, inner surface with a pale, glaucous band; fruit shiny dark purple; includes cultivars with blue-green and emerald green foliage, some turning yellow-green to bronze in winter; all are tolerant of coastal conditions); *J.davurica* (DAHURIAN JUNIPER; Siberia, Mongolia, China; semi-prostrate shrub; bark grey, flaking; juvenile leaves needle-like, adult leaves scale-like, diamond-shaped, pointed, sage green; fruit purple-brown; includes cultivars with foliage variegated or stained golden yellow); *J.drupacea* (SYRIAN JUNIPER; Greece, Syria, Turkey; columnar tree to 15m in the wild, bark red-brown, peeling in long strips; leaves to 2.5cm, needle-like, whorled, hard, sharply pointed, shiny green; fruit red-brown, violet-

Juncus ensifolius

bloomed); *J.horizontalis* (CREEPING JUNIPER; N America; low to prostrate, spreading shrub to 0.6 × 3m; leaves needle- or scale-like, grey-green or grey-blue; fruit dark blue with a pale bloom; includes cultivars with very low, feathery or dense foliage, some very resistant of salt spray, others ranging in colour from dull green to bright steely blue, the best turning purple-bronze in winter); *J.* × *media* (*J.sabina* × *J.chinensis*; spreading shrub to 2m, differs from *J.sabina* in its thicker shoots and spiky crown with long, horizontal shoot tips; bark scaly, red-brown; leaves yellow- to grey-green, needle- or scale-like; fruit dark purple bloomed pale blue; cultivars include compact and low to robust and loosely tiered clones, some sports juvenile, they range in colour from dull olive green to blue-grey and gold, some are yellow-variegated; one of the best-known is 'Pfitzeriana', a spreading, grey-green flat-topped shrub to 3 × 5m); *J.procumbens* (BONIN ISLES JUNIPER; S Japan; to 75cm, procumbent; bark maroon-brown, smooth; leaves in whorls of 3, linear, sharply pointed, pale green, the inner face with 2 grey-green bands; fruit dark brown; includes dwarf, prostrate and compact cultivars, some with gold-tipped foliage); *J.recurva* (DROOPING JUNIPER, HIMALAYAN WEEPING JUNIPER; SW China, E Himalaya; tree or shrub to 20m, crown conical, branches weeping at least at tips; bark red-brown, peeling in strips; leaves subulate to lanceolate, sharply pointed, rigid, rather dry to touch, grey-green with a pale grey inner surface; fruit dark green-brown to black; var. *coxii*: COFFIN JUNIPER; small tree or large shrub, crown spreading, branches distinctly weeping); *J.rigida* (TEMPLE JUNIPER; Japan, Korea, N China; tree to 15m, or shrub with an open crown and pendulous branches; bark brown to yellow-brown, peeling; leaves in whorls of 3, needle-like, sharply pointed, rigid, bright green, inner face with a single, blue-grey band; fruit dark maroon to purple-black); *J.sabina* (SAVIN; C & S Europe, C Asia, NW China; spreading shrub or tree to 4m; bark rust-brown, flaking; leaves very aromatic, needle-like to scale-like, grey-green to blue-green; fruit very dark blue with a white bloom; cultivars include dwarf to robust clones of prostrate to tiered habit with compact to loosely feathered, dark green, grey or blue-grey foliage, sometimes tipped yellow); *J.sargentii* (Japan, China; prostrate shrub to 1 × 3m; branches layered to sweeping; bark red-brown; leaves needle-like to scale-like, grey-green or dark green, camphor-scented; fruit blue-black); *J.scopulorum* (ROCKY MOUNTAIN JUNIPER; Rocky Mts; shrub or tree to 15m; crown open; bark red-brown, fissured; leaves scale-like, sharply pointed, yellow-green to dark green; fruit blue-black covered in a pale bloom; includes cultivars with dense or lax foliage in shades of grey-green and bright blue-grey, short or tall with procumbent, conical and very narrow habits; one of the best-known is 'Skyrocket', a very slender, columnar tree to 6m tall with feathery, strongly ascending blue-grey foliage, sometimes listed under *J.virginiana*); *J.squamata* (FLAKY JUNIPER; NW Asia to China; prostrate to sprawling or erect shrub to 8m; bark rusty brown, flaky; leaves in whorls of 3, subu-late, rigid, aromatic, deep grey-green to silvery blue-grey with a bright, blue-white band; fruit glossy black; cultivars include dwarf, prostrate, columnar and bushy plants, many with bright silver-blue leaves, some assuming lilac tints in winter; one of the best-known is 'Meyeri', a broadly spreading shrub with dense, spiky boughs of steely blue foliage and old, dead growth rusty and persistent); *J.virginiana* (PENCIL CEDAR, EASTERN RED CEDAR; E N America; tree to 20m, crown conic to columnar, branches spreading; bark rusty brown, peeling in shreds; leaves needle- and scale-like, grey-green marked with white, scented weakly of soap or white spirit; fruit purple-maroon thickly bloomed; includes numerous cultivars – tall to low, narrowly erect to tiered and prostrate, pale to deep green or grey-green, sometimes turning purple-bronze in winter; this species is one of the parents of *J.* 'Grey Owl', a shrub to 2 × 3m with spreading branches and feathery, silver-grey foliage).

Justicia (for J. Justice, 18th-century Scottish gardener). Acanthaceae. Tropical and subtropical northern and southern Hemispheres, and temperate N America. WATER WILLOW. Some 420 species, perennial herbs, subshrubs or shrubs with jointed stems and oblong to ovate leaves. The flowers are borne in axillary or terminal spikes, cymes or panicles, usually subtended by bracts; they are tubular and curving, with two lips, the uppermost narrowly hood-like. Grow in a well-drained, medium- to high-fertility, loam-based mix in full sun or bright filtered light, with protection from the hottest summer sun and good ventilation. Water plentifully and feed weekly when in full growth. Keep just moist in winter, with a minimum temperature of 7°C/50°F. Propagate by semi-ripe stem cuttings in spring.

J.adhatoda (India, Sri Lanka; shrub to 3m; flowers to 3cm, white veined red or purple on lower lip); *J.brandegeana* (syn. *Beloperone guttata*, *Drejerella guttata;* SHRIMP PLANT, FALSE HOP; Mexico; downy shrub to 2m; flowers to 3cm, white marked purple-red in 'shrimp-' or 'hop-like,' arching spikes to 8cm long, clothed with overlapping, downy, yellow, or more usually dusty pink to brick red or dull brown bracts); *J.californica* (CHUPAROSA HONEYSUCKLE; southern N America; subshrub to 1.5m; branches downy, arching; flowers red); *J.carnea* (syn. *Jacobinia carnea*; BRAZILIAN PLUME, FLAMINGO PLANT, PARADISE PLANT; northern S America; shrub to 2m; flowers to 5cm, fleshy pink to magenta or purple-red in dense, erect, plume-shaped spikes to 15cm with green bracts); *J.rizzinii* (syn. *J.floribunda*, *J.pauciflora*; Brazil; small, rounded, downy shrub; flowers to 2cm, scarlet tipped yellow in nodding clusters); *J.spicigera* (syn. *J.ghiesbreghtiana*, *Jacobinia spicigera*; Mexico to Colombia; shrub to 1.8m; flowers to 4cm, orange or red in a secund raceme).

juvenile (1) used of the sexually immature stage of a plant; (2) used of the immature foliage of those plants that bear leaves of a distinctly different shape when young, such as *Cryptomeria* and *Eucalyptus*; cf. *adult*.

K

Kadsura (Japanese name for *K.japonica*). Schisandraceae. E and SE Asia. 22 species twining, evergreen shrubs with entire or obscurely toothed leaves and small flowers usually borne singly at the leaf axils in summer. These are followed by fleshy berries. Hardy in climate zone 7. In colder areas, grow in the greenhouse or conservatory. Outdoors, plant in a sheltered position, in sun or part shade, on moderately fertile, well-drained neutral to acid soils. Plants of both sexes are required for fruit set. Increase by semi-ripe stem cuttings. *K.japonica* (Japan, Korea; climbing to 4m; flowers to 1.5cm wide, ivory; fruit scarlet).

Kaempferia (for Engelbert Kaempfer (1651–1716), German physician). Zingiberaceae. Tropical Asia. About 50 species, rhizomatous perennial herbs with aromatic tubers and foliage Those here are low-growing plants with clumps of broad leaves. Produced in spring and summer in short spikes, the flowers consist of three petals, large, petal-like staminodes and a showy, deeply lobed lip. Grow in semi-shade with a minimum temperature of 15°C/60°F. Plant in a porous, fertile medium rich in leafmould. Keep warm, humid and moist when in growth; during the winter resting period, keep cool and dry. Propagate by division of tubers.

K.atrovirens (PEACOCK PLANT; Borneo; leaves bronze with a metallic sheen above; flowers white and lavender, pink or violet); *K.pulchra* (Thailand, Malaysia; leaves pale green suffused and feathered with dark purple-green or bronze-green feathered silver-green; flowers lilac); *K.roscoeana* (PEACOCK LILY, DWARF GINGER LILY; Burma; leaves deep green marked paler green above, flushed red beneath; flowers white); *K.rotunda* (RESURRECTION LILY; SE Asia, widely cultivated; leaves green marked silver above, tinted purple beneath; flowers white with a lilac lip).

kainit a natural mineral containing potassium chloride together with sodium chloride, magnesium sulphate and impurities. Formerly an orchard fertilizer, it is now rarely used on account of its variability and the risk of phytotoxicity caused by its impurities.

Kalanchoe (from the obscure native name for a Chinese species). Crassulaceae. Tropics, especially Old World. 125 species, perennial, occasionally annual or biennial, succulent shrubs, herbs and climbers (those below perennials). The leaves are fleshy and the flowers, which are bell-shaped to tubular, are carried in racemes and panicles at various times of year. Many of the smaller species, notably *K.blossfeldiana* and its hybrids, carry exotic and vibrantly coloured flowers in late winter and spring. They are commonly grown as house plants. The trailing to weeping cultivars are useful for hanging pots and baskets. Grow in a medium-fertility loam-based mix with added grit, in sun or light shade, with a minimum temperature of 10°C/50°F. Water plentifully from spring to autumn, sparingly and occasionally in winter. Propagate from stem or leaf cuttings. For large-leaved species such as *K.beharensis,* the midrib is cut across in several places, and pinned to the surface of a sandy propagating mix. New plants should form within a month. A few species, such as *K.daigremontiana*, produce large numbers of tiny plantlets at the tips and margins of mature leaves; these are easily detached and root quickly when placed on the soil surface. Surface sow seed of all species in early spring at 20°C/70°F.

K.beharensis (Madagascar; densely felty shrub to 6m; leaves to 35cm, spear-shaped to triangular, toothed, stained brown, silvery beneath; flowers green-yellow marked violet within); *K.blossfeldiana* (FLAMING KATY; Madagascar; bushy herb to 60cm; leaves to 7cm, oblong to ovate, glossy green, tip blunt, base rounded, crenate; flowers red, orange, pink or yellow in erect, crowded, domed to flat-topped bunches); *K.daigremontiana* (DEVIL'S BACK-BONE; Madagascar; robust, erect perennial to 1m; leaves to 20cm, lanceolate, fleshy, smooth, marbled brown-purple beneath, margins toothed, bearing plantlets; flowers violet-grey); *K.delagonensis* (syn. *K.tubiflora*; Madagascar; perennial shrublet to 2m; leaves 2–12cm, narrow, terete, grey-green, tip toothed and bearing plantlets; flowers magenta to orange); *K.fedtschenkoi* (Madagascar; dense shrublet

Kalanchoe tomentosa

to 50cm; leaves 2–6cm, oblong to ovate, glaucous blue, toothed near apex; flowers purple-red; includes 'Variegata', with cream-marked leaves); *K.pumila* (Madagascar; sprawling shrublet to 30cm; leaves to 3cm, chalky, grey-white, obovate, toothed toward tip; flowers pink with purple lines); *K.*'Tessa' (bushy, sprawling to cascading perennial with neat, elliptic, smooth and deep green leaves and large, nodding, tubular flowers in deep orange-red); *K.tomentosa* (PANDA PLANT; Madagascar; erect perennial to 1m, sparingly branched; leaves 2–9cm, oblong to obovate, grey-green and covered in dense, felty, pale hairs except for the red-brown-stained tip; flowers yellow-green stained purple and covered in red-brown hairs); *K.uniflora* (Madagascar; creeping perennial; leaves 0.5–4cm, oblong to rounded, convex, sparsely crenate; flowers pendent, purple-red); *K.* 'Wendy' (more or less erect, bushy perennial to 40cm; leaves elliptic to obovate, smooth, bluntly toothed; flowers large, urn-shaped, nodding, magenta with yellow lobes).

kale

Pests
cabbage root fly
flea beetle

Disease
clubroot

kale (*Brassica oleracea*, Acephala Group) a biennial member of the Cruciferae, closely related to cabbage; it is grown as an annual for its edible shoots and young leaves, which are harvested through the winter until spring. One of the hardiest brassicas, it is capable of withstanding winter temperatures as low as –15°C/5°F, but is also tolerant of high summer temperatures. It is widely distributed in both temperate and tropical regions.

Transplants may be raised from a spring sowing and within 6–8 weeks will be large enough to plant firmly, at 45–75cm spacings according to cultivar. For an early crop of spring greens, grow at closer spacings sowing seed broadcast under cloches during the latter part of the winter, or in the open during early spring. Thin to 7–10cm spacings. Plants may be harvested as spring greens when no more than 15cm high, and the remaining stems allowed to resprout for a further harvest.

Kale forms may be classified into the curly or Scotch kales (borecole) and the broader smooth-leaved types. A number of kales grow extremely tall, including the so-called Jersey kale (also known as tree cabbage, walking-stick cabbage, Jersey longjacks and *chou cavalier*). Though normally reaching around 150cm, Jersey kale has been recorded to 540cm. The straight, relatively slender stems are strong enough to dry and make into walking sticks; the 75cm leaves are decorative and can be cooked when young.

RAPE or SIBERIAN KALE is similar in appearance to the curly kales but is a different species (*Brassica napus*). It will not withstand transplanting and must be sown *in situ*.

CHINESE KALE (*Brassica oleracea* Alboglabra Group) is a perennial grown as an annual and long established in tropical Asia. It is well adapted to high temperature, although some cultivars will also tolerate low temperatures. The whole plant is normally used for cooking, including the inflorescence. Raise in a seedbed or alternatively propagate from cuttings, transplanting 30cm apart. Harvest 60–70 days from planting out; useful laterals may arise after cutting of the terminal shoot. In temperate areas, cultivars suited to low temperatures may be raised under glass at 20°C/70°F.

Kalmia (for Peter Kalm (1716–1779), Swedish botanist and student of Linnaeus). Ericaceae. US, Cuba. 7 species, evergreen shrubs with oblong to elliptic, entire, leathery leaves and small flowers in corymbs, umbels or fascicles in summer. The flowers are bell-shaped to broadly tubular, with five lobes. Suitable for planting in open woodland or at the woodland edge and in association with *Rhododendron* and other large Ericaceae. Hardy to –30°C/–22°F. Grow in full sun or dappled shade, in moisture-retentive, sandy soils that are lime-free and rich in organic matter. Sow seed in late winter, lightly covered, in the warm greenhouse. Propagate from wounded hardwood cuttings in late winter, in a closed case with bottom heat, or by layering. Alternatively, propagate by softwood cuttings in summer.

K.angustifolia (SHEEP LAUREL, PIG LAUREL, LAMBKILL; E US; to 1.5m; flowers 0.7–1.2cm in diameter, scarlet-pink, pink, deep rosy purple or white in axillary corymbs); *K.latifolia* (MOUNTAIN LAUREL, CALICO BUSH; E US; 3–10m; flowers 2–2.5cm in diameter, white to pale pink, deep rose or crimson, in large terminal corymbs).

Kalmiopsis ('resembling *Kalmia*'). Ericaceae. US (Oregon). 1 species, *K.leachiana*, an evergreen dwarf shrub to 30cm. To 3cm long, the leaves are oval or obovate, bright green with large, yellow glands beneath. Produced in spring in terminal racemes, the bell-shaped flowers are 1–2cm across, rose-purple and 5-lobed. A choice shrub, suitable for the rock garden or alpine house and frame; hardy to –15°C/5°F. Grow in moist, peaty, acid soil, in semi-shade or in sun, with some protection at the hottest part of the day. Propagate by surface sown seed, by layering, or by semi-ripe cuttings.

Kalopanax (from Greek *kalos*, beautiful, and *Panax*). Araliaceae. China, Korea, Sakhalin, Kuriles, Japan, Ryukyu Islands. 1 species, *K.septemlobus* (syn. *K.pictus*, *K.ricinifolius*, *Acanthopanax ricinifolius*), the TREE ARALIA, a round-headed, sparingly branched tree to 30m. The branches and trunk are prickly, especially in young growth. To 35cm across, the leaves are palmately lobed, finely toothed, dark green above and paler beneath. They are divided into five to seven, triangular to ovate lobes and carried on petioles to 50cm long. Small white flowers are borne in paniculate umbels in summer and followed by small, blue-black drupes. Hardy in zone 6. Plant in deep, fertile, moisture-retentive soil. Give sun or semi-shade. On young plants, unripened wood is susceptible to frost damage. Propagate from seed in autumn or semi-ripe cuttings under mist or with bottom heat in summer. Young plants are slow-growing.

kaolinite a clay mineral with relatively low cation exchange and water-holding capacities.

keel (1) a prominent ridge, like the keel of a boat, running longitudinally down the centre of the under-surface of a leaf, petiole, bracts, petal or sepal; (2) the two lower united petals of a papilionaceous flower.

kelp any of various types of very large seaweed, especially *Laminaria* species. The term is sometimes used more generally to describe any seaweed harvested as bulky organic manure.

Kennedya (for Lewis Kennedy (1775–1818), a founding partner of Lee and Kennedy, nurserymen of Hammersmith, London). Leguminosae. Australia, Tasmania, New Guinea. CORAL PEA. Some 16 species, woody or herbaceous perennial climbers. The leaves are usually trifoliolate. Showy, pea-like flowers are produced in trusses or racemes in late spring and summer. In frost-free regions, *Kennedya* species are grown as climbers or for groundcover. They are ideal for covering dry, sunny banks and warm walls. In cooler zones, they are suited to the cool greenhouse or conservatory, trained up pillars and along rafters or scrambling through shrubs. Grow in a sandy soil in full sun in a warm, sheltered position. Maintain a winter minimum temperature of 5–10°C/40–50°F. Water moderately when in growth, and keep virtually dry in winter. Prune after flowering to remove overcrowded growth. Propagate by seed sown in spring at 20°C/70°F. Immerse seed briefly in boiling water and soak in tepid water for 12 hours before sowing. Take semi-ripe cuttings in summer.

K.nigricans (BLACK CORAL PEA, BLACK BEAN; W Australia; woody climber to 6m; flowers violet-purple to true black blotched golden yellow); *K.rubicunda* (DUSKY CORAL PEA; New South Wales, Victoria; woody-based climber to 3m; flowers deep coral to scarlet with a paler blotch at base of standard).

kernel the inner part of a seed or the whole body within the seed coat.

Kerria (for William Kerr (d. 1814), Kew gardener who collected in China, Java and the Philippines). Rosaceae. China, Japan; widely cultivated in Japan. JEW'S MALLOW; JAPANESE ROSE. 1 species, *K.japonica,* a deciduous shrub to 2.25m. The branches are green and rod-like, branching in their second season and suckering freely. To 7cm long the leaves are oval and saw-toothed. They are usually mid- to dark green, but are edged and splashed white to cream in the cultivar ('Variegata') 'Picta'. Borne in spring, the buttercup-like flowers are to 5cm in diameter, pale to deep golden yellow, with five ('Simplex') or more ('Pleniflora') ovate to orbicular petals and numerous stamens. Hardy to –20°C/–4°F. Grow in a sunny position. Prune after flowering to remove old and overcrowded stems. Propagate by semi-ripe cuttings or by suckers.

key a device for identifying plants, often by means of a series of couplets of mutually exclusive statements, each of which leads to a further pair or to the identification (a dichotomous key).

kieserite see *magnesium sulphate.*

Kirengeshoma (from Japanese *ki*, yellow, and *rengeshoma*, native name for *Anemopsis*, which this plant is held to resemble). Hydrangeaceae. Japan, Korea. 1 species, *K.palmata*, a graceful perennial herb. The stems are 60–120cm tall, erect to arching and purple- or red-tinted. The pale green leaves are 10–20cm long, broadly ovate, toothed to palmately incised, and with a tapering tip and a heart-shaped base. Borne in late summer, the flowers are narrowly bell-shaped, to 3cm long, and slender-stalked. They consist of five, overlapping, pale yellow to apricot petals. Hardy in zone 6. Grow in deep, moist, humus-rich, lime-free soils, in semi-shade or with part-day sun. Mulch in winter. Propagate by division or by seed, which may take up to 10 months to germinate.

Kirengeshoma palmata

Kitaibela (for Paul Kitaibel (d. 1817), Hungarian botanist). Malvaceae. Balkans. 1 species, *K.vitifolia*, a perennial herb to 2.5m tall. To 18cm long, the long-stalked leaves are rhombic to rounded, with five to seven triangular and coarsely toothed lobes. Produced in summer, the broadly funnel-shaped flowers are solitary or carried in axillary cymes; they consist of five cuneate petals, to 2.5cm long and white or rose. Fully hardy, *Kitaibela* will grow in any fertile, moisture-retentive but well-drained soil in sun. Propagate by seed in autumn or spring, or by division or cuttings.

kitchen garden a plot of ground predominantly devoted to edible produce but often including flowers for cutting, and commonly greenhouses for growing crops out of season. Traditionally, large kitchen gardens were entirely walled in, the walls themselves being used for training fruit trees.

kiwi fruit (*Actinidia deliciosa*) CHINESE GOOSE-BERRY a hardy perennial climber, very vigorous with shoots up to 10m in length, bearing large heart-shaped leaves. It is native to China, and grown for its edible, hispid fruits that are up to 10cm long and 4.5cm across. This rampant climber can become highly invasive if not pruned with care. The young blossom-bearing shoots are susceptible to frost damage so that fruiting plants require a sheltered site. Propagate from seed or preferably from cuttings. The plants should be spaced 3–4.5m apart, with a support system provided as for blackberries and hybrid berries, or by means of a pergola. Under greenhouse protection they may be trained in a similar manner to grapes. After planting the plants should be cut back to 30cm height. One male plant is required for up to 6 female plants, and pollination by hand is beneficial under greenhouse conditions. Young plants are tip-pruned on reaching the top wire to encourage lateral growths, which should be trained along the support wires, pinching out the growing tips when the shoots reach the end posts. Thereafter fruiting sub-laterals should be pinched back to six or seven leaves beyond the last fruit; non-flowering shoots should be pinched back to five leaves throughout the growing season. Three-year-old fruiting laterals are best pruned hard back to encourage replacement growth. The fruit is autumn ripening and should be stored in an airy shed or greenhouse for up to 6 weeks after picking.

Knautia (for Christian Knaut (1654–1716), German botanist at Halle). Dipsacaceae. Europe, Caucasus, Siberia, Mediterranean. 60 species, annual and perennial herbs with small flowers borne in summer in long-stalked, flat-topped heads subtended by leafy bracts. Suited to the wild flower garden, to areas managed as meadow, and to the cottage garden. Both are fully hardy if grown in a warm sunny site on any well-drained soil. Propagate by seed or basal cuttings.

K.arvensis (BLUE BUTTONS, FIELD SCABIOUS, SCABIOUS; Europe, Mediterranean, Caucasus, naturalized N America; perennial to 1.5m; flowers pale purple-blue); *K.macedonica* (syn. *Scabiosa rumelica*; C Europe; perennial to 0.8m; flowers deep purple).

knee a spongy outgrowth from a horizontal subterranean root projecting above the soil surface in saturated conditions, as in *Taxodium*.

knife-ringing see *bark-ringing*.

Kniphofia (for J.H. Kniphof (1704–1756), German botanist). Liliaceae (Aloeaceae). REDHOT POKER. Africa (those species listed below are from South Africa). 68 species, evergreen perennial herbs forming clumps of grass-like to strap-shaped leaves. The flowers are tubular to narrowly funnel-shaped and point downwards in dense, cylindric to conical, long-stalked spikes in summer and autumn. Unopened or recently opened flower buds may be of a far deeper or wholly different colour, giving the spike a bicoloured appearance.

Redhot pokers are best known for their range of hot reds and oranges, although numerous cultivars are available in softer colours, for example, the creamy 'Snow Maiden', the glowing yellows of 'Maid of Orleans', and the cool green of 'Green Jade'. Most species are cold-tolerant to temperatures of –15°C/5°F, although at these low temperatures, plants require a warm and sheltered position and should be mulched deeply with leafmould or bracken litter. In the northern US, plants are sometimes lifted and over-wintered in a moist soilless medium in frost-free conditions. However, most resent root disturbance and mulch protection is preferable wherever possible. Grow in full sun in deep, fertile and freely draining, preferably sandy soil, with sufficient organic matter to ensure an adequate supply of moisture when in growth. Propagate by seed or division; increase cultivars by division only. Thrips occasionally cause mottling of the foliage.

Kniphofia galpinii

K.angustifolia (to 60cm; leaves grass-like; flower spike graceful, loose, flowers to 3cm, white to yellow, orange and coral); *K.caulescens* (1.5–2m; robust, forming clumps of short, stout stems; leaves thick, keeled, glaucous grey-green; flower spike dense, broadly cylindric, flowers to 2.5cm, creamy white tinted peach, flushed flame in bud); *K.galpinii* (to 80cm; leaves slender; flower spike small, dense, red at apex shading to orange-yellow as flowers open below, flowers to 3.5cm; sometimes confused with *K.triangularis*); *K.triangularis* (to 1m; leaves slender; flower spike small, dense, flowers to 3.5cm, coral to orange with flared perianth lobes); *K.uvaria* (1–2m; leaves narrowly strap-shaped; flower spike oblong to ovoid-conical, dense, flowers 3–4cm, brilliant red to green-yellow). *Cultivars* dwarf (to 80cm): 'Ada' (primrose yellow), 'Bressingham Comet' (orange tipped red), 'Bressingham Glow' (deep orange to flame, late-flowering), 'Canary Bird' (yellow), 'Candle Light' (clear yellow to cream, spikes slender and graceful), 'Little Maid' (ivory tipped soft yellow), 'Tubergeniana' (soft primrose), 'White Fairy' (white); medium to tall (80–150cm): 'Atlanta' (orange-yellow), 'Green Jade' (bright green), 'Nobilis' (deep, fiery orange), 'Percy's Pride' (green-yellow), 'Royal Standard' (bright yellow below, vermilion above), 'Samuel's Sensation' (deep flame shading to yellow at base, late-flowering with very long spikes), 'Spanish Gold' (rich dark yellow).

knock down a term sometimes applied to contact insecticides that kill only those pests actually covered during the application of the material. Such substances are quick acting and usually non-persistent.

knot, knot garden a bed laid out in an intricate pattern, sometimes resembling a maze, using low-growing, clipped plants such as dwarf box *(Buxus)*, cotton lavender *(Santolina),* and thrift *(Armeria)* . The area around and between the lining plants may be filled with flowering plants, which are replaced in season; in Tudor times especially, such areas were often left bare or covered with colourful materials, such as sands, chalk or coal. Knots were especially popular in the 16th and 17th centuries, and were originally designed to be seen mainly from above, from terraces or windows. The name was derived from the resemblance of the pattern to a knot of rope arranged on the ground.

Koeleria (for G.L. Koeler (1765–1806), botanist specializing in grasses). Gramineae. Temperate regions, tropical Africa. Some 25 species, annual and perennial grasses with clumped, erect stems and narrow leaf blades. The flowers appear in summer in narrowly cylindric panicles held over the foliage. These dry a parchment tone and can be used in dried arrangements. Fully hardy. Grow in full sun in any not too-fertile, well-drained soil, including calcareous soils and shallow soils over chalk. Propagate by seed or division in autumn or spring.

K.glauca (C Europe, Siberia; perennial to 60cm; leaves slender strongly glaucous blue-green to grey); *K.vallesiana* (W Europe; perennial to 40cm; leaves to 10cm, very narrow, blue-green to silver).

Koelreuteria (for Joseph G. Koelreuter (1733–1806), professor of Natural History at Karlsruhe). Sapindaceae. China, Taiwan. GOLDEN RAIN TREE, SHRIMP TREE, CHINESE RAIN TREE. 3 species, flat-topped, deciduous shrubs or trees. The leaves are simple or bipinnate. Yellow, 4-petalled flowers are borne in large terminal panicles in summer. They are followed by conspicuous, bladder-like seed pods. *K.paniculata* remains of interest throughout the season, with red-flushed young growth, yellow autumn tints and inflated fruits in warm, earthy colours. It will withstand winter temperatures of at least –10°C/14°F, but flowering and autumn colour are much improved where a long, hot growing season, dry autumn and cold winter are common. Plant in sun. It is tolerant of poor, dry soils, pollution and exposure to wind (not salt-laden), growing quickly to a specimen of reasonable size; it resents severe pruning. Young plants should be kept well-watered until they are three or four years old. Propagate from seed, sown in autumn or stratified and sown in spring. Alternatively, increase by root cuttings in winter. *K.paniculata* (PRIDE OF INDIA, VARNISH TREE, GOLDEN RAIN TREE; N China, Korea; tree to 15m; leaves to 40cm, ferny, emerging pink-red, falling bright yellow; flowers in large, pyramidal panicles; fruit to 5cm, oblong to ovoid, splitting into three brown, papery segments and persisting well after leaf fall).

Kohleria (for Michael Kohler, 19th-century teacher of natural history at Zurich). Gesneriaceae. Tropical America. 50 species, perennial herbs and shrubs with ovate to elliptic, hairy leaves. Nodding in summer on stalked cymes, the flowers are tubular and somewhat swollen, with a narrowed throat and a limb of five, small, spreading lobes. Provide a minimum temperature of 15°C/60°F with light shade, medium to high humidity and protection from draughts. Grow in a porous, soilless medium; keep moist when in growth, virtually dry during the winter resting period. Increase by division of rhizomes.

K.amabilis (Colombia; to 60cm tall, white-hairy; leaves sometimes with silver markings, often flushed red; flowers to 2.5cm, deep rose with brick red bars and stripes, limb rose with maroon bars and spots); *K.bella* (Costa Rica; to 40cm tall, pale red-hairy; flowers to 4cm, red tinged yellow below, limb yellow with purple spots); *K.bogotensis* (Colombia; to 60cm tall, hairy; leaves with pale green or white markings; flowers to 2.5cm, red fading to yellow at base, limb yellow spotted red); *K.eriantha* (shrubby perennial herb to 1m, red-hairy; flowers to 5cm, orange to scarlet or cinnabar red, yellow-white with red spots); *K.warscewiczii* (syn. *K.digitaliflora;* Colombia; to 60cm tall, covered in white hairs; flowers to 3cm long, deep rose to magenta, white below, limb green with deep purple spots).

kohlrabi

kohlrabi (*Brassica oleracea,* Gongylodes Group) TURNIP CABBAGE a biennial member of the Cruciferae, grown as an annual for its spherical, swollen stem which is trimmed of leaves and boiled. It has a flavour similar to turnip for which it is often used as a substitute, being better able to withstand drought and high temperature.

Kohlrabi is most suited to temperate regions but can be grown in warmer climates. It is sensitive to low temperature and a week at 10°C/50°F will cause plants to bolt: the recommended range is 18–25°C/64–77°F with an optimum of 22°C/72°F. For continuity, sow at regular intervals from early spring to late summer in drills 30cm apart, later thinning seedlings to 25cm apart. Alternatively, seedlings may be transplanted when no more than 5cm tall. Kohlrabi is best harvested when young as the swollen stems become woody; it can be stored for winter use either *in situ* or in boxes of sand.

Kolkwitzia (for Richard Kolkwitz (b. 1873), professor of botany in Berlin). Caprifoliaceae. BEAUTY BUSH. C China. 1 species, *K.amabilis,* a deciduous shrub to 3.5m. Produced in spring and early summer, the flowers are paired in terminal corymbs. The slender sepals are densely hairy to bristly. To 2cm long, the bell-shaped corolla is white stained rose pink outside (strongly so in the cultivars 'Pink Cloud' and 'Rosea'), with deep yellow to orange markings in the throat and a limb of five, rounded lobes. Grown for its masses of beautiful bell-shaped flowers, which resemble those of the closely allied *Abelia, Kolkwitzia* makes a valuable addition to the shrub border. Its exfoliating bark is attractive in winter. It thrives in chalk soils and is hardy to –30°C/–22°F, although it may suffer die-back of shoot tips in very cold weather. Plant in full sun. Prune out old, damaged or weak shoots from the base after flowering. Propagate by softwood cuttings or from suckers, which are freely produced.

kohlrabi

Pests
cabbage root fly
flea beetle

Disease
clubroot

L

labellum a lip, especially the enlarged or otherwise distinctive third petal of an orchid.

labiate possessing a lip or lips; describing corollas of the sort found in many Labiatae.

labium one of the lip-like divisions of a labiate corolla or calyx.

Lablab (from a Hindu plant name). Leguminosae. Tropical Africa, widely cultivated in India, SE Asia, Egypt, Sudan. 1 species, *L.purpureus* (syn. *Dolichos lablab*; DOLICHOS BEAN, HYACINTH BEAN, BONAVIST, LUBIA BEAN, SEIM BEAN, INDIAN BEAN, EGYPTIAN BEAN), a perennial twining herb, to 6m. The leaves are trifoliolate, composed of ovate to triangular or rhombic leaflets to 15cm, tinted purple and colouring well in autumn. White, purple or bright rose, the pea-like flowers are carried in erect racemes to 40cm long in summer. Grow in well-drained soils in full sun; maintain a minimum winter temperature of 7–10°C/45–50°F; water plentifully when in growth. Train on canes or trellis. Apply liquid feed weekly when growing strongly. Alternatively, treat as half-hardy annuals grown as for runner beans (see *beans*). Increase by seed in spring.

+ Laburnocytisus (*Laburnum* combined with *Cytisus*). Leguminosae. Garden origin. +*L.adamii*, a graft hybrid between *Laburnum anagyroides* and *Chamaecytisus purpureus*, is a tree to 7.5m tall, resembling *Laburnum* in overall habit, but with a mixture of sporadic, true growths and flowers of both *Laburnum anagyroides* and *Chamaecytisus purpureus* where the chimaera breaks down. Otherwise, the growth is intermediate between the parents, thus the racemes are *Laburnum*-like but with smaller flowers in dull yellow tinged purple, appearing fleshy pink or pale bronze. Cultivate as for *Laburnum*. Propagate by grafting carefully chosen scions of the intermediate form on to *Laburnum anagyroides*.

Laburnum (classical Latin name). Leguminosae. SC and SE Europe, W Asia. BEAN TREE. 2 species, small to medium-sized deciduous shrubs or trees with trifoliolate leaves. Produced in spring and early summer, the flowers are pea-like, yellow, and packed in graceful, pendulous racemes. All parts are poisonous, the pulse-like seeds especially. Fully hardy, laburnums are used as lawn and border specimens, and may be fan-trained as screens and living fences. They are perhaps at their most beautiful when grown in walkways over arches and pergolas. Plant in any moderately fertile, well-drained soil in sun. Pruning of free-standing specimens is rarely necessary, but may best be done in late summer. To make a laburnum arch, support young plants with wooden stakes – these will eventually rot by which time trunks will be self-supporting. Space plants at about 2m and train side branches along rigid wires between the stakes and arched over the top. As branches meet at the top of the structure, they can be grafted together. Propagate species from seed sown ripe under glass or in the frame. Take hardwood cuttings of the past season's growth in late winter; treat with rooting hormone and insert into open ground, with sharp sand around the base. Propagate hybrids by budding on to seedling stock of *L.anagyroides* in summer. A leaf miner, *Leucoptera laburnella*, makes spiral mines within the leaves; although rarely fatal, heavy infestations disfigure small plants.

L.alpinum (SCOTCH LABURNUM, ALPINE GOLDEN CHAIN; south central Europe; twigs glabrous; inflorescence to 35cm+, denser than in *L.anagyroides*; pedicels equalling flowers; corolla 1.5cm+, bright yellow; includes the low-crowned, weeping 'Pendulum' and the erect 'Pyramidale'); *L.anagyroides* (COMMON LABURNUM, GOLDEN CHAIN; C and S Europe; twigs grey-green, weakly hairy; inflorescence to 20cm; pedicels shorter than flowers; corolla 2cm, lemon to golden yellow; includes the weeping 'Pendulum', the yellow-leaved 'Aureum', and 'Autumnale' ('Semperflorens') with a second flush of flowers produced in autumn); *L.* × *watereri* (*L.alpinum* × *L.anagyroides*; Tyrol, S Switzerland; closer to *L.anagyroides* but only shoot tips hairy; inflorescence to 50cm; flowers fragrant; includes 'Alford's Weeping', with a widely spreading, weeping crown, and 'Vossii', exceptionally floriferous, with long racemes).

lacebugs (*Stephanitis* and other species.) (Hemiptera: Tingidae) small, flattened, sap-sucking insects, up to 6mm in length, with lace-like forewings. Of around 700 recorded species, only a few are of garden importance. They include *Stephanitis rhodo-dendri*, the RHODODENDRON BUG of Europe and North America, which inserts eggs into the midrib of rhododendron leaves in late summer and autumn. These hatch in spring to produce grey-black, spiny-bodied nymphs, which feed on the underside of leaves; adults appear four weeks later. Lower leaf surfaces become fouled with brown excrement and pale yellow spots appear on the upper surfaces, giving a mottled appearance. Small infestations are best controlled by pruning and destroying affected branches, otherwise contact or systemic insecticides should be applied from midsummer onwards.

lacecap a flat-topped inflorescence in which the outer flowers are sterile and enlarged, as in, for example many *Hydrangea* cultivars.

laced used to describe flowers of certain cultivars of garden pinks, which have a patch of contrasting colour at the edge of each petal forming a contin-uous, narrow, scalloped band around the flower, thereby bearing some resemblance to lace-work.

lacerate irregularly, and more or less broadly, shal-lowly and sharply cut, as if torn.

lacewing flies (Neuroptera) insects up to about 25mm long, with long thread-like antennae and two pairs of similar sized transparent lace-like wings, held like a roof over the back when at rest; they are useful predators in gardens.

The familiar GREEN LACEWINGS (Chrysopidae) are often attracted to artificial light and have yellow-green bodies and wings, and iridescent eyes. Eggs are laid on stalks, separately or in bundles, and the bristly larvae are mostly pale green or yellow with darker markings. They feed on aphids, other small insects and sometimes mites, and typically cover their backs with the empty skins of their prey. Most over-winter in a spun cocoon before pupating in the spring.

The BROWN LACEWINGS (Hemerobiidae) are mostly smaller, with brown or grey-brown wings; their eggs are not laid on stalks. Many frequent conifers where the larvae attack aphids, and some also attack thrips and mites on other trees and herba-ceous plants.

The POWDERY LACEWINGS (Coniopterygidae) are small fragile insects, their body and wings covered with white powdery wax; they resemble whiteflies except that the wings are not held flat. Their larvae have plump bodies tapering at the hind end with relatively long legs; they feed on small aphids and mites. Of these *Conwentzia* species feed on the fruit tree red spider mite, *Panonychus ulmi* – the larvae have been recorded as consuming up to 35 of these mites a day and the adults 30 mature female mites in an hour.

Lachenalia (for Werner von Lachenal, late 18th-century Swiss botanist). Liliaceae (Hyacinthaceae). South Africa. Some 90 species, bulbous perennial herbs with strap-shaped to circular leaves and erect spikes of tubular to urn-shaped flowers. These consist of six perianth segments in two whorls: the outer whorl shorter, forming a fleshy tube or cup, each segment often with a marked apical swelling; the inner whorl protruding, usually broader and more showy, with lips coloured and recurved. Most flower in winter and spring. Grow in a medium-fertility, sandy loam-based mix, in full sun; maintain a minimum temperature of 7°C/45°F. Plant in late autumn. Water sparingly as growth commences, plentifully when in full growth. Dry off as leaves wither and keep almost completely dry when dormant. Propagate by ripe seed when available; seedlings may flower in the season following first dormancy. Offsets and bulbils are also produced.

L.aloides (syn. *L.tricolor*; leaves strap-shaped, blotched green or purple; raceme to 28cm, scape often mottled and tinted red-brown; flowers pendulous, tubular to funnelform, outer segments fleshy, lemon yellow to apricot or white, sometimes flushed orange, scarlet or blue-green from the base, apical swellings bright green, inner segment 2–3.5cm, tipped cinnabar red, magenta, scarlet or green; includes 'Pearsonii', with leaves spotted red-brown above and scape to 18cm, mottled red-brown, outer segments to 1.5cm, apricot, apical swelling lime green, inner segments to 3cm, apricot to gold, tips stained red to maroon, and 'Nelsonii', with leaves spotted purple and bright yellow flowers tipped green); *L.contaminata* (WILD HYACINTH; leaves, grass-like; scape mottled; inflorescence 6–25cm, flowers campanulate, white, apical swelling on outer segments maroon, inner segments striped maroon near tips); *L.mutabilis* (leaf solitary, sometimes spotted or banded maroon, often crispate; inflorescence 10–45cm, outer perianth segments pale blue to white, apical swelling dark brown, inner segments yellow with brown markings near tip); *L.orchioides* (leaves strap-shaped, sometimes spotted; inflorescence 8–40cm, flowers fragrant, oblong to cylindrical, outer segments pale blue at base shading to green-yellow or cream, apical swellings green; includes var. *glaucina*, with outer perianth segments blue at base, shading to purple, and apical swelling dark purple, sometimes entirely blue with dark blue swelling); *L.rubida* (leaves strap-shaped, often spotted; inflorescence 6–25cm, flowers pendulous, cylindrical, outer segments bright pink to ruby red, or pale yellow spotted red, apical swelling yellow-green or pink-red, inner segments tipped purple, marked white).

laciniate irregularly and finely cut, as if slashed.

lacrimiform, lachrymaeform tear-shaped.

Lactuca *see* lettuce.

lacuna a cavity or depression, especially an air hole in plant tissue.

Lachenalia aloides

lacunose pitted with many deep depressions or holes.

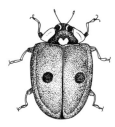

ladybird and larva

ladybirds (Coleoptera: Coccinellidae) the common species are medium-sized beetles with oval or rounded bodies and bright colouration. Most adults and larvae feed principally on aphids, scale insects, mealybugs, thrips and mites, but some, such as the 22-spot and 16-spot ladybirds, feed on fungal spores, especially those of powdery mildews. Typically, the adults hibernate in sheltered situations beneath leaves, loose bark or in buildings, and in spring the females deposit clusters of pale yellow spindle-shaped eggs, usually on leaves. The resulting larvae, mostly leaden grey or grey-blue spotted with yellow, orange or white, wander at random in search of prey. When fully fed, after about three weeks, they attach themselves to a leaf or other support to change into pupae. Adult ladybirds appear about ten days later.

Common species include the TWO-SPOT (*Adalia bipunctata*), usually having red wing cases with two black spots but very variable; the red and black SEVEN-SPOT (*Coccinella septempunctata*); the red and black TEN-SPOT (*Adalia decempunctata*), which is very variable in colour; and the 14-spot (*Propylea 14-punctata*) and 22-spot (*Psyllobora 22-punctata*), which are both yellow with black spots. Common North American species include the CONVERGENT LADYBIRD (*Hippodamia convergens*) and the SINUATE LADYBIRD (*H.sinuata*), as well as *Adalia bipunctata*.

These and many other ladybirds provide natural methods of controlling various pests and some feature in biological control programmes, including the VEDALIA LADYBIRD (*Rodolia cardinalis*), which was introduced from Australia at the end of the last century to control the fluted scale insect (*Icerya purchasi)* in California and Florida. *Cryptolaemus montrouzieri*, also from Australia, has been used to control mealybugs (*Pseudococcus* species) under glass in temperate countries, and *Coccinella septempunctata* is released annually in Maine, US, to control aphids on potatoes.

A small number of ladybirds are plant feeders. These include the MEXICAN BEAN BEETLE (*Epilachna varivestis*), which is up to 8mm long, has yellow, coppery or bronze wing cases with 16 black spots, and is principally a pest of soya beans in North America; and the SQUASH BEETLE (*E. borealis*), which mainly infests squash and pumpkins. Both adults and larvae skeletonize the leaves of plants and can be controlled by appropriate contact insecticides.

Laelia (for Laelia, one of the vestal virgins). Orchidaceae. C and S America, from West Indies south to Brazil. About 70 species, evergreen epiphytic or lithophytic herbs with globose to pear- or spindle-shaped pseudobulbs and leathery, ovate to linear leaves. The flowers basically resemble those of *Cattleya* but range widely in size and shape from the small and spidery to the large and lacy. Cultivate as for *Cattleya*.

L.anceps (Mexico, Honduras; flowers 3–6, fragrant, 8–10cm in diameter, rose-lilac or magenta, lip deep purple tinged with pink and yellow in the throat, some-times wholly white); *L.autumnalis* (Mexico; flowers 3–6, scented, to 10cm in diameter, rose-purple, lip rose-white with purple apex and yellow centre; *L.gouldiana* is a stouter plant with broader, solid magenta blooms, possibly a cross between *L.anceps* and *L.autumnalis); L.cinnabarina* (Brazil; flowers 5–15, about 5cm in diameter, deep orange-red, slender); *L.crispa* (Brazil; flowers 4–7, to 12cm in diameter, white, lip usually mainly purple with some yellow marks); *L.flava* (Brazil; flowers 3–10, canary yellow, about 6cm in diameter); *L.harpophylla* (Brazil; flowers 3–7, 5–8cm in diameter, vermilion, lip with a paler margin and yellow in centre); *L.milleri* (Brazil; flowers 6cm in diameter, few, orange-red, tinged with yellow in throat); *L.pumila* (Brazil; flowers solitary, to 10cm in diameter, flat, slightly drooping, rose-purple, rarely white, lip deep purple with yellow throat, sometimes edged and veined amethyst); *L.purpurata* (Brazil; flowers 15–20cm in diameter, 3–7, tepals white tinged with pink, sometimes pure white, lip white at apex, purple towards base, yellow with purple veins in throat, rarely deep violet); *L.speciosa* (syn. *L.majalis*; Mexico; flowers 12–18cm in diameter, 1–2, tepals rose-lilac, rarely white, lip white spotted with deep lilac, margin pale lilac); *L.tenebrosa* (Brazil; flowers about 16cm in diameter, 4, tepals copper-bronze, lip purple, darker in throat). There are many cultivars and grexes of *Laelia*, either slender or stout in habit and with large or small flowers in shades of pink, mauve, yellow, orange, red and white.

× **Laeliocattleya** (*Laelia* × *Cattleya*). Orchidaceae. A wide range of colourful hybrids. The plants are usually robust, upright, with one or two leaves at the apex of each pseudobulb. The inflorescence is terminal, with few to many flowers of various size, always brightly coloured and conspicuous. Cultivate as for *Cattleya*.

laevigate appearing smoothly polished.

Lagarosiphon (from Greek *lagaros*, narrow, and *siphon*, tube, alluding to the female perianth). Hydrocharitaceae. Africa, introduced to Europe and New Zealand. CURLY WATER THYME. 9 species, perennial, aquatic herbs with submerged long, branching stems and linear to lanceolate, spiralling to opposite or whorled leaves. The genus is fully hardy, but also a good aquarium oxygenator. *L.major* (South Africa, introduced to Europe and New Zealand; stem to 3mm in diameter; leaves 6.5–25 × 2–4.5mm, thick, opaque).

lageniform flask-shaped.

Lagerstroemia (for the Swedish merchant, Magnus von Lagerstrom (1691–1759) of Göteborg, friend of Linnaeus). Lythraceae. Tropical Asia to Australia. 53 species, deciduous shrubs or trees, usually with peeling bark and leathery elliptic to obovate leaves. The flowers have six or more, crumpled and clawed, silken petals and many stamens; they are carried in summer in pyramidal to cylindric panicles. These are shrubs for warm subtropical climates or temperate regions with hot summers. They are frequently used in Mediterranean areas and in southern California and Florida for specimen, hedge, screen and avenue plantings. They have the additional interest of muted autumn colours and a smooth, fluted, and shredding bark. In colder areas (with winter temperatures down of about –12°C/10°F), *L.indica* will succeed given the shelter of a warm south-facing wall. It will tolerate still harsher winters where hot, humid summers promote rapid new growth from bases cut back by hard frosts. *L.speciosa* is tender and requires a minimum temperature of 10°C/50°F.

Grow in any well-drained soil in a sheltered, sunny spot. Unless pollarding or coppicing is to be carried out, prune annually in late winter only to remove congested older wood and to open up the plant centre; tip back long shoots as required. *Lagerstroemia* does not transplant well and must be moved with a large root-ball in late spring. Overwinter container-grown specimens at a minimum temperature of 13°C/55°F and water very sparingly. Propagate by semi-ripe cuttings rooted under mist or with bottom heat in summer; alternatively, increase by hardwood cuttings in winter or from seed sown in spring, although the flower colour of seedlings is often inferior.

L.indica (CRAPE MYRTLE, CREPE FLOWER; China, Indochina, Himalaya, Japan; tree or shrub to 6m; panicles to 20cm, petals suborbicular, to 1cm, pink, purple or white; includes numerous cultivars ranging from dwarf shrubs to trees some 10m tall and varying widely in hardiness, bark quality and autumn colour; the flowers may be small or large, smooth-petalled or ruffled, and in tones of white, cream, pink, lavender, deep rose, pale red and darkest crimson); *L.speciosa* (QUEEN'S CREPE MYRTLE, PRIDE OF INDIA, PYINMA; tropical Asia; tree to 24m; panicles to 40cm, petals suborbicular, to 3 × 2cm, purple or white).

Lagunaria (for Andres de Laguna (d.1560), Spanish botanist.) Malvaceae. Australasia (Queensland, Norfolk Island, Lord Howe Island). 1 species, *L.patersonii* (NORFOLK ISLAND HIBISCUS, QUEENSLAND PYRAMID TREE, COW ITCH TREE), a columnar to pyramidal evergreen tree or shrub to 15m. Somewhat leathery and scurfy beneath, the leaves are to 10cm long, ovate to broadly lanceolate and entire with a cuneate base and blunt apex. *Hibiscus*-like, pale pink to magenta flowers, to 6cm in diameter, appear in the upper leaf axils in summer. Cultivate as for tender *Hibiscus*.

Lagurus (from Greek *lagos*, hare, and *oura*, tail, referring to the fluffy panicles). Gramineae. Mediterranean. HARE'S TAIL. 1 species, *L.ovatus*, an annual grass to 60cm. In spring and summer, it produces a spike-like, ovoid to oblong panicle, to 6 × 2cm, densely and softly hairy and light green or tinged mauve. These may be hare's-tail flowerheads used fresh or dried in floral arrangements. Grow in light, sandy, freely draining soils in sun; cut or pull flowerheads for drying before maturity. Seed may be sown *in situ* in spring, or in pots in autumn and overwintered in well-ventilated frost-free conditions for planting out in spring.

lambourde sometimes used to describe a fruit spur.

lamellate composed of one or more thin, flat scales or plates.

lamina see *blade*.

Lamium (Latin name used by Pliny). Labiatae. Mediterranean. DEAD NETTLE. Some 50 species, perennial or annual herbs, with stoloniferous, creeping stems. Produced in spring and summer in whorls, the flowers are tubular, two-lipped and hooded. Useful, fully hardy groundcover, tolerant of a wide range of soil types, climates and light conditions. Generally, cultivars with pale foliage (e.g. *L.* 'Cannon's Gold' and 'White Nancy') prefer shade, or, if grown in sun, a cool, moist root run. *L.orvala* is sometimes damaged by late frosts. Propagate by removal of rooted running stems of most species, by division of clumps.

L.galeobdolon (syn. *Galeobdolon argentatum*; *Lamiastrum variegatum*; *L.luteum*; YELLOW ARCH-ANGEL; Europe to W Asia; vigorous perennial to 60cm, erect to widely creeping, occasionally stoloniferous; leaves toothed; flowers yellow, flecked brown; cultivars include: 'Florentinum', with large leaves, splashed silver, and, in winter, purple; 'Hermann's Pride', mat-forming, with narrow, toothed leaves, streaked and spotted silver; 'Silberteppich' (syn. 'Silver Carpet'), slow growing, clump-forming, with silver leaves; 'Silver Angel', prostrate, fast-growing, with leaves marked silver; and 'Variegatum', with smaller, oval leaves, mid-green marked silver; *L.maculatum* (Europe and N Africa to W Asia; perennial to 80cm, ascending to trailing, stoloniferous, pubescent; leaves toothed, white-mottled or -striped; flowers pink to red or purple, rarely white; cultivars include: 'Album', with leaves blotched silver and white flowers; 'Aureum' ('Golden Nuggetts'), with gold leaves with a white centre, and pink flowers; 'Beacon Silver', habit low, with silver leaves, with a thin green edge, and pink flowers; 'Cannon's Gold', with pure gold leaves and purple flowers; 'Pink Pewter', with leaves tinted silver and rich pink flowers; 'Sterling Silver', with pure silver leaves and purple flowers; and 'White Nancy', with silver leaves with a thin green edge, and white flowers); *L.orvala* (GIANT DEAD NETTLE; S C Europe; perennial to 100cm, erect to spreading; flowers large, purple-pink with a mottled lip and throat; 'Album': flowers white).

Lammas growth a term usually applied to fruit trees to describe the growth of shoots occurring in late summer i.e. around Lammas, the first day of August.

Lampranthus (from Greek *lampros*, bright, brilliant, and *anthos*, flower, referring to the brilliant flowers). Aizoaceae. South Africa. 200 species, subshrubs with succulent, terete or 3-angled leaves and bright, daisy-like flowers. Most make excellent summer bedding or container plants in temperate regions; a few are frost hardy to –1°C/30°F or –2°C/28°F and thus can be grown outside the year round in sheltered favourable sites in temperate regions. They require a position in full sun and a low-fertility, well-drained soil. Plant as soon as frosts are over and leave outside until the first autumn frosts; the plants should then be dug up and potted, or cuttings taken and kept frost-free over the winter. Under glass, grow in full sun with low humidity and a minimum temperature of 7°C/45°F; plant in a loam-based medium rich in grit and sand. Water sparingly in summer, scarcely at all in winter.

L.aurantiacus (leaves 2–3cm, grey-pruinose; flowers orange); *L.haworthii* (leaves light green, densely light grey-pruinose; flowers light purple); *L.multiradiatus* (syn. *L.roseus*; leaves 2.5–3cm, green to grey-green, glaucous with translucent dots; flowers pale pink to rose-red); *L.spectabilis* (leaves flushed red; flowers purple-red to magenta).

lanate, lanose woolly, possessing long, densely matted and curling hairs.

lanceolate lance-shaped; 3–6 times as long as broad, and with the broadest point below the middle; tapering to a spear-like apex.

land drain purpose-made drainpipes of earthenware or plastic, usually 7.5cm in diameter and 45cm long, although plastic types often come in longer lengths. Land drains are used to improve water drainage from soil. See *drainage*.

landscape gardening the art of laying out an estate to blend with the surrounding countryside, dating from the early 18th century when there was a move against formality and enclosure in gardening. Landscape gardening declined in the 19th century as the number of privately owned large properties contracted but the principles of landscaping have persisted. The term is now commonly used to cover garden design, construction and maintenance on a small scale and of variable standards, but there remain many well-trained and gifted professionals continuing the art.

lanuginose as lanate, but with the hairs shorter; somewhat woolly or cottony.

Lantana (from the superficial similarity to *Viburnum lantana*). Verbenaceae. Tropical Americas and Africa, a widespread weed elsewhere in the tropics and subtropics. SHRUB VERBENA. Some 150 species, aromatic shrubs, subshrubs and perennial herbs with small, salverform, 4–5-lobed flowers in compact terminal heads in spring and summer. Suitable for hedging and specimen plantings in frost-free zones, and for bedding out or for the greenhouse or conservatory in cool temperate climates. *L.camara* is easily trained as a standard. The trailing *L.montevidensis* is suited to groundcover in the conservatory border or to hanging baskets. Grow in full sun. Water plentifully when in full growth; keep just moist in winter with a minimum temperature of 10°C/50°F. Propagate by seed in spring or by semi-ripe cuttings in summer.

L.camara (tropical America, widespread weed in tropics and subtropics; shrub to 2m; flowers yellow to orange or red often with a brighter eye; many cultivars ranging in habit from dwarf to tall, with flowers mostly bicolours, in shades ranging from white through to yellow and salmon to red); *L.montevidensis* (syn. *L.sellowiana*; S America; herb or shrub to 1m, trailing or scandent; flowers rose-lilac to violet).

Lapageria (for Joséphine Tascher de la Pagérie (1763–1814), wife of Napoleon Bonaparte and an enthusiastic patron of gardening). Liliaceae (Philesiaceae). Chile. CHILEAN BELLFLOWER, CHILE BELLS, COPIHUE. 1 species, *L.rosea*, an evergreen climber to 10m with wiry stems clothed with sharp scale-like bracts. The leathery leaves are ovate or ovate to lanceolate and glossy dark green. Campanulate flowers hang from the axils of upper leaves in summer. They consist of six waxy tepals, 6–9cm long, oblong to elliptic and luminous rose to pale salmon, often stained, spotted or streaked deeper crimson. Given the protection of a partially shaded and sheltered wall, *L.rosea* will grow outside where temperatures seldom fall below –5°C/23°F; otherwise, it makes a fine specimen for the borders of the cold greenhouse or conservatory. Grow in cool, well-drained but moist soils, humus-rich and lime-free, in light shade or where direct sunlight penetrates for short periods only. Propagate by layering in spring or autumn or by seed, pre-soaked for about 48 hours before sowing.

Lardizabala (for Miguel de Lardizabel y Uribe, 18th-century Spanish naturalist). Lardizabalaceae. Chile. 2 species, evergreen climbers with compound leaves and, in winter, slender spikes of small, curiously coloured flowers followed by attractive berries. Hardy to –10°C/14°F; plant in a shady protected site on well-drained soils rich in decayed matter. Propagate by seed in spring or by stem cuttings in spring or autumn. *L.biternata* (Chile; twining to 4m; leaves ternate to biternate, dark green, leaflets 5–10cm, ovate; male inflorescence to 10cm, flowers green edged dark brown-purple; fruit 6cm, dark purple).

Larix (Classical Latin name). Pinaceae. Northern Hemisphere. LARCH. Some 14 species, deciduous, coniferous trees. The crown is usually conic in young trees, becoming domed or irregular with age. The branches are horizontal or somewhat drooping. Soft, needle-like leaves are arranged in loose spirals on long shoots and in dense, whorl-like spirals on short shoots.

Lapageria rosea

In cool temperate maritime climates, only two species, *L.decidua* and *L.kaempferi* (and *L. × marschlinsii*, the hybrid between them), are wholly successful, making healthy trees, 25–50m high, in most areas. All tolerate acid and infertile soils; *L.decidua* will also grow on chalk. Dead branches should be removed since they are very brittle and can be dangerous. *L.decidua* is best choice for specimen planting, although *L. × marschlinsii* may be preferred on less fertile sites for more rapid growth. They and *L.kaempferi* make fine shelterbelts, summer screening and copse groups, and are covered in small purple-pink cones in early spring, leafing out bright green, then turning gold to russet in autumn. Pests common in Europe and North America include the larch sawfly, with defoliating olive-green larvae up to 25mm long, and larch adelgids made conspicuous by a covering of white woolly wax. Caterpillars of the larch casebearer make holes and mines in the foliage and cover themselves with a cigar-shaped case made from pieces of needles. A needle cast disease can cause serious damage to young larch trees in the nursery. The needles turn brown from the tips and fall off prematurely.

L.decidua (EUROPEAN LARCH; Alps, Carpathians; to 50m; crown slender, conic, branches horizontal, branchlets drooping; leaves 2–4(–6.5)cm, soft, flat or slightly keeled beneath, pale green; cones 1.5 to 5cm, conic to cylindric, brown, scales 40–50, straight or incurved, sometimes short-hairy on outer surface; about 20 cultivars are described, mostly varying in habit, from dwarf ('Compacta', 'Corley', 'Repens'), erect ('Fastigiata'), pendulous ('Pendula', 'Viminalis') to twisted ('Cervicornis', 'Tortuosa') and sparsely branched ('Virgata'); in 'Alba', the immature cones are pale green, almost white; subsp. *polonica*: POLISH LARCH, W Poland and Ukraine, to 30m, with a very slender crown, drooping branchlets, and smaller, blunter cones, 1–2.8cm, with concave scales); *L.kaempferi* (JAPANESE LARCH; Japan; to 45m; crown broadly conic, branches and shoots horizontal or slightly ascending; leaves 2–3.5cm, slightly keeled beneath, grey-green, with stomatal bands throughout; cones 2–3, rarely 4cm, ovoid, scales truncate to emarginate, apex reflexed; about 20 cultivars are described, mainly varying in habit, and mostly dwarf ('Blue Dwarf', 'Nana', 'Varley', 'Wehlen') or with pendulous branchlets ('Georgengarten', 'Inversa', 'Pendula'), or both ('Hanan'); in 'Blue Haze' the leaves are brightly glaucous, likewise the semi-fastigiate 'Pyramidalis Argentea'); *L.laricina* (TAMARACK, AMERICAN LARCH; northern N America; to 20(–30)m; crown conic, branches horizontal, branchlets drooping; leaves 2–3cm, keeled, pale green with 2 stomatal bands beneath; cones ovoid, 12–24mm, green or purple, becoming straw-brown, scales 15–20, nearly circular, margins slightly incurved; includes 'Arethusa Bog', dwarf, with short, narrow shoots and arching branches, 'Aurea', with leaves gold when young, later light green, and 'Glauca', with metallic blue leaves); *L. × marschlinsii* (*L.decidua* × *L.kaempferi*; DUNKELD LARCH, HYBRID LARCH; vigorous; similar to *L.decidua* but shoots and leaves faintly glaucous; cones conic, scale tips slightly reflexed); *L.occidentalis* (WESTERN LARCH; western N America; to 50m; branches horizontal; leaves 2.5–5cm, keeled beneath, slightly convex above, acute, blue-green to grey-green, with 2 stomatal bands beneath; cones 2.5–4.5 × 1cm ovoid, scales rounded, to 12mm broad, reflexed on opening, to 2.5cm broad); *L. × pendula* (*L.decidua* × *L.laricina*; WEEPING LARCH; to 30m; branchlets pendent; leaves similar to *L.decidua* but blunter; cones 2–3.3cm, ovoid, scales 20–30, exterior base downy; includes 'Contorta', with shoots twisted when young, and 'Repens', with horizontally creeping branches); *L.russica* (syn. *L.sibirica*; SIBERIAN LARCH; Russia; to 30m; crown narrowly conic; branches horizontal to upswept; leaves 2.5–4cm, clustered, very narrow, soft, bright green above, with 2 stomatal bands beneath and scattered stomata above; cones 2.5–3.5cm, rarely 4cm, scales thick, pubescent, margins incurved; includes the columnar and compact 'Fastigiata', 'Glauca', with glaucous needles, 'Longifolia', with longer needles, 'Pendula', with weeping branches, and the more vigorous 'Robusta').

larva the immature stage in advanced insects, which does not resemble the adult form and which needs to pass through a chrysalis or pupal stage, undergoing complete metamorphosis, before developing into an adult. Larvae vary in form and habits. See *caterpillars*, *grubs*, *maggots*.

latent bud a bud that remains inactive and concealed until stimulated into growth by the removal of adjacent shoots through natural damage or pruning. It may also be referred to as dormant.

lateral on a tree or shrub, a side shoot or branch as distinct from a leader. On woody fruit plants, laterals bear fruit buds, and may be described as one-year-old or maiden laterals, or two- or three-year-old laterals. The identification and understanding of types of laterals is important to successful pruning.

latex a milky fluid or sap, usually colourless, white or pale yellow, found in plants such as *Asclepias* and *Euphorbia*.

lath house a structure clad with spaced wooden or plastic laths or trelliswork to provide plants with shelter from strong sunlight, wind and light frost. Probably originating in the southern US, lath houses are used for the acclimatization of nursery stock and by some growers of bonsai and orchids.

lath light a frame covered by laths, used for the same purposes as a lath house.

lath screen a portable wooden unit, usually 3m long and 1m high, clad with 2.5cm-wide laths, nailed vertically and spaced 2.5cm apart; plastic variants are available. It is used for wind protection around low-growing plants.

Lathyrus vernus
(narrow-leaved
form)

Lathyrus (name used by Theophrastus, from the Ancient Greek name for the pea or pulse, combining *la-*, very, and *thoures*, a stimulant: the seeds were said to have excitant or irritant properties.) Leguminosae. Eurasia, N America, mountains of E Africa and temperate S America. VETCHLING, WILD PEA. 110+ species, annual or perennial herbs, often climbing by leaf tendrils, with pinnate leaves and racemes of showy pea-like flowers.

The species described below have handsome flowers, are often fragrant and most will tolerate temperatures below –15°C/5°F. Those climbing by means of tendrils are useful for clothing trellis and pergola or for trailing unsupported over walls, slopes and embankments. *L.nervosus*, with periwinkle blue flowers, may be grown outdoors where winter temperatures do not fall much below zero for prolonged periods; it thrives best in zones with cool moist summers. The non-climbing, clump-forming *L.vernus* is especially delightful in late winter when its light bunches of emerging stems, packed with rose, blue and mauve blooms, bring relief to the woodland garden and the foreground of herbaceous borders. *L.sylvestris* is eminently suited to wild gardens and native plant collections. *L.latifolius*, the everlasting sweetpea, is a favourite plant for cottage gardens. *L.odoratus* is the SWEETPEA (q.v.).

Grow in any moderately fertile, well-drained soil in sun, or light dappled shade. Provide appropriate support (i.e. canes, trellis or host-shrubs for climbers; for semi-scandent or erect perennials, a few birch twigs pushed into the ground around the crowns should suffice). Dead-head throughout the season and cut back perennials in autumn to ground level. Feed yearly on very poor soils.

Propagate from pre-soaked seed sown under glass in early spring or *in situ* in spring; alternatively, increase perennials by division in early spring. This last method is most successful with *L.vernus* and the non-climbing species; many of the climbing perennials do not transplant well and even seed-grown specimens are best planted out when young and then left undisturbed. See also *Sweetpea*.

L.grandiflorus (TWO-FLOWERED PEA, EVERLASTING PEA; S Italy, Sicily, S Balkans; perennial climbing to 2m; stems angled; leaves terminating in a 3-branched tendril, leaflets to 5cm, 1 pair (rarely 3 pairs), ovate; flowers to 3cm, standard violet, keel pink, wings mauve); *L.latifolius* (PERENNIAL PEA, BROAD-LEAVED EVERLASTING PEA; C and S Europe, naturalized N America; perennial climbing to 3m; stem winged; leaves terminating in a 3-branched tendril, leaflets to 15cm, 1 pair, linear to elliptic, somewhat blue-green; flowers to 3cm in diameter, magenta-purple, pink, or white; 'Albus': white; 'Blushing Bride': white flushed pink; 'Pink Beauty': dark-purple and red; 'Red Pearl': carmine red; 'Rosa Perle': 'Pink Pearl', vigorous, pink, long-lasting; 'Splendens': deep-pink flowers); *L.nervosus* (LORD ANSON'S BLUE PEA; S America; perennial climbing to 60cm; leaves terminating in a 3-branched tendril, leaflets to 4cm, 1 pair, ovate to oblong; flowers to 2.2cm, indigo); *L.odoratus*

(SWEETPEA; Crete, Italy, Sicily; annual climbing to 2m; leaflets to 6cm, 1 pair, oval to oblong; flowers to 3.5cm, typically purple; now highly developed resulting in a vast range of cultivars with flowers clustered or on long racemes, large or small, sometimes ruffled or 'double', sweetly to heavily scented, in shades of red, rose, mauve, purple, purple-black, white, opal, blue, lilac, peach, cream, pale yellow and variously mottled or veined); *L.rotundifolius* (PERSIAN EVERLASTING PEA; E Europe, W Asia; perennial climbing to 1m, stem angular; leaves terminating in a 3-branched tendril, leaflets to 6cm, 1 pair, ovate to orbicular; flowers to 2cm, deep pink); *L.sylvestris* (FLAT PEA, NARROW-LEAVED EVERLASTING PEA; Europe; perennial climbing to 2m; stem angular, winged; leaves with a branched tendril, leaflets to 15cm, 1 pair, linear to lanceolate; flowers to 2cm, purple-pink mottled purple and green; 'Wagneri': deep red); *L.vernus* (SPRING VETCH; Europe except extreme N; bushy, non-climbing perennial to 60cm, usually shorter; leaves terminating in a point, not a tendril, leaflets 3–10cm, in 2–4 pairs, oval to lanceolate, acuminate, dark green; flowers 1–2cm, pink to red-violet turning green-blue; includes 'Albiflorus', with blue-white flowers, 'Alboroseus', with rose-white flowers, and 'Roseus' with rose to lavender blue flowers; also included here are very graceful forms with narrow leaflets and lilac to opal blue flowers, these may belong to other, lesser-known species).

lattice another word for trelliswork; often referring to that made of metal.

Laurelia (from Spanish *laurel*, local name for *L.sempervirens*). Monimiaceae. New Zealand, Chile, Peru. 2 species, evergreen, aromatic trees and shrubs with leathery, dark green leaves and inconspicuous flowers in panicles or racemes in summer. Hardy to about –5°C/23°F but susceptible to cold winds, the following species needs a warm sheltered position in zones at the limits of its hardiness. Grow in sun or part shade on moderately fertile soils that do not dry out in summer. Propagate by seed, by semi-ripe cuttings in a closed case, or by layering. *L.sempervirens* (syn. *L.serrata*; CHILEAN LAUREL; Chile, Peru; to 30m; leaves 6–9cm, elliptic to ovate, serrate, glossy green).

Laurus (Latin name for these plants). Lauraceae. S Europe, Canary Islands, Azores. 2 species, aromatic evergreen trees or shrubs with tough, ovate to elliptic or oblong to lanceolate leaves and small, 4-parted, white to creamy buff flowers in axillary clusters. *L.nobilis* is widely grown for its foliage both as an ornamental and culinary herb. Tolerant of clipping, it is used for screening and hedging and for tubs and large pots, trained as a standard or pyramid. Given perfect drainage and a warm, sheltered site it will withstand occasional lows to –15°C/5°F. *L.azorica* has similar requirements but prefers a warm wall and responds badly to pruning and shearing. Grow in a moisture-retentive but well-drained fertile soil in full sun. In containers, use a high-fertility loam-based mix and liquid feed fortnightly when in full growth. Trim formal shapes in summer, using secateurs rather than shears to avoid unsightly damage to the foliage. Propagate by semi-ripe cuttings in summer or by basal hardwood cuttings of the previous season's growth in mid- to late winter; root in a closed case with bottom heat.

L.azorica (CANARY LAUREL; Canary Islands, Azores; tree to 10m; young branchlets downy; leaves 5–12cm, broadly lanceolate to elliptic, dark green, glabrous above, paler and pubescent, especially on midrib beneath); *L.nobilis* (TRUE LAUREL, BAY LAUREL, SWEET BAY, BAY TREE; Mediterranean; small tree or shrub 3–5m; young branchlets glabrous; leaves 5–10cm, narrowly elliptic to oblong to ovate, often undulate, dark green, glabrous beneath; includes 'Aurea', with leaves tinged yellow, f. *angustifolia* (syn. 'Salicifolia'), with narrow leaves, and 'Crispa' (syn. 'Undulata'), with wavy leaf margins).

Lavandula

Lavandula (from Latin *lavo*, I wash, referring to lavender water, made from oil of lavender). Labiatae. Atlantic Islands, Mediterranean, N tropical Africa, W Asia, Arabia, India. LAVENDER. 28 species, aromatic, evergreen shrubs and subshrubs with linear to oblong, entire to toothed or pinnate leaves. Small, shortly tubular and 2-lipped flowers crowd the upper portions of long-stalked spikes in summer. Aromatic shrubs grown for ornament and perfume. A few (*L.angustifolia* and cultivars), are frost-tolerant to –10°C/14°F. Others (*L.stoechas* and *L.dentata*), are hardy to at least –5°C/23°F. These slightly tender species are nonetheless worth attempting in climate zone 7 in sheltered, sunny sites on well-drained soils. Trim back after flowering. Propagate from seed sown under glass in spring, from cuttings of new growth in early summer, or from semi-ripe cuttings put into a cold frame in late summer.

L.angustifolia (syn. *L.spica*; ENGLISH LAVENDER; Mediterranean; shrub, 1–2m; leaves 2–6cm, entire, lanceolate, oblong or linear, revolute, tomentose, grey when young, green with age; spike 2–8cm, flowers dark purple or blue; 'Alba': flowers white; 'Rosea': flowers pink; 'Atropurpurea': flowers very dark purple; 'Dutch White': tall, with profuse white flowers in small heads; 'Hidcote': to 30cm, dense, with lanceolate, grey leaves, lilac flowers, deep purple calyx in dense spikes; 'Hidcote Giant': very tall, with deep purple flowers; 'Hidcote Pink': as 'Hidcote' but with pale pink flowers; 'Loddon Pink': to 45cm, with soft pink flowers; 'Munstead': to 45cm, with small leaves, large blue-lilac flowers; 'Nana Alba': dwarf, white-flowered; Old English Group: to 50cm, with leaves to 7cm, branching flowering spikes, to 115cm, and pale lavender to violet flowers; 'Vera': DUTCH LAVENDER, tall, robust, with rather broad grey leaves and lavender flowers); *L.dentata* (Spain, N Africa; shrub to 1m; leaves 1.5–3cm, oblong, linear or lanceolate, wavy-toothed to pinnatifid, grey-tomentose; spikes 2.5–5cm, bracts usually tinged purple, apical bracts sometimes in a coma, flowers powder-blue to dark purple; var.

candicans: with most parts white-tomentose; 'Silver Form': leaves soft, silver, flowers large, blue); *L.lanata* (S Spain; shrub to 1m, white-woolly throughout; leaves 3–5cm, linear to spathulate or oblong, scarcely revolute; spike 4–10cm, flowers lilac to deep indigo); *L.stoechas* (FRENCH LAVENDER; Mediterranean region; shrub 30–100cm; leaves 1–4cm, linear to oblong or lanceolate, entire, usually grey-tomentose; spike 2–3cm, sterile bracts along spike or in a coma, 1–5cm, oblong-obovate, erect, petal-like, red-purple, rarely white, flowers purple, white or pale pink; 'Alba': flowers white; subsp. *pedunculata*: inflorescences long-stalked with rich mauve sterile bracts tinged red, rarely white, long and erect, and forming a showy coma; 'James Compton': flowers deep purple in a spike with large, pale purple bracts; 'Papillon': with very long, narrow, bright purple sterile bracts).

Lavatera

Lavatera (for J. R. Lavater, 16th-century physician and naturalist in Zurich). Malvaceae. Macronesia, Mediterranean to NW Himalaya, C Asia, E Siberia, Australia, California and Baja California. TREE MALLOW. 25 species, annual, biennial and perennial herbs or soft-wooded shrubs, usually downy to thickly hairy, with palmately angled or lobed leaves, and, in summer, broadly dish- or funnel-shaped flowers composed of five obovate to cordate petals. The following species are hardy in climate zone 6. Grow all species in full sun on a light, well-drained, moderately fertile soil. Give shrubby perennials protection from cold drying winds and a winter mulch. *L.oblongifolia* needs a warm, sheltered site, and a rather dry, gritty soil. Propagate annuals and biennials by seed sown *in situ*: annuals in spring, biennials in late summer. Increase perennials by cuttings of basal shoots in spring, and shrubs and perennials also by softwood cuttings in spring and early summer.

Lavendula dentata

L.arborea (TREE MALLOW; Europe, Mediterranean, Macaronesia, naturalized California and Baja California; tree-like biennial or perennial to 3m; leaves 8–18cm, orbicular-cordate, 5-, 7- or 9-lobed, pilose, crenate; flowers with petals 1.5–2.5cm, lilac or purple-red, with darker veins at the base; includes 'Ile d'Hyéres', to 3.5m, with large, palmate leaves and small, magenta flowers, in leaf axils, in early summer, and 'Variegata', of vigorous habit, with large, marbled white leaves); *L.assurgentiflora* (MALVA ROSA; SW US; deciduous shrub to 6m; leaves 8–15cm, 5-or 7-lobed, lobes triangular, coarsely toothed, white-pubescent; flowers with petals 4cm, red-purple, with darker veins); *L.cachemiriana* (Kashmir; annual or short-lived perennial herb to 2.5m, downy; leaves downy beneath, cordate-orbicular, crenate, 3- or 5-lobed; flowers with petals to 4cm, pink, deeply bifid); *L.oblongifolia* (Spain; erect, evergreen shrub to 1.5m, branches twiggy, covered in tawny hairs; leaves ovate to lanceolate, silky white-hairy; flowers small, sometimes not opening, rose pink with mauve markings); *L.thuringiaca* (syn. *L.olbia* of gardens; TREE

LAVATERA; C and SE Europe; perennial herb, subshrub or shrub to 1.8m, softly grey-tomentose; leaves to 9cm, cordate to rounded, 3- or 5-lobed; flowers with petals 1.5–4.5cm, purple-pink; 'Barnsley': to 2m, with lobed leaves and flowers opening white with red eye, fading to pale pink with a rose eye; 'Bressingham Pink': to 1.8m, with pale pink flowers; 'Candy Floss': with leaves softly tinted grey and bright pink flowers, with white stamens; 'Ice Cool': to 1.8m, with pale green leaves and pure white flowers, occasionally fading to pink; 'Rosea': to 2m, with downy leaves, tinted grey, and pale rose-mauve flowers); *L.trimestris* (Mediterranean; annual herb to 1.2m, sparsely hairy; leaves 3–6cm, rounded to cordate, slightly 3-, 5- or 7-lobed; flowers with petals 2.5–4.5cm, white, rose, pink or red; 'Loveliness': to 1.2m, with large, trumpet-shaped flowers, of a striking deep rose; 'Mont Blanc': dwarf and compact with dark green leaves and pure white flowers; 'Pink Beauty': dwarf and bushy habit, to 60cm, with large, delicate pale pink flowers with violet veins and eye; 'Silver Cup': dwarf and bushy habit, to 60cm, with large, glowing pink flowers, to 12cm in diameter; 'Splendens': flowers large, white or red).

lawn an area densely covered with grass or certain other low-growing plants, and usually closely mown. As a meadow spangled with flowers, it was first widely known in medieval Europe, although the word 'lawn' is not recorded until 1674, and its contemporary meaning, as a closely mown cover, not until 1733.

Lawns serve as decorative features in themselves, or to enhance plantings and other facets of garden design. They provide a stable surface for access and recreation.

Suitable species Turfgrasses may be divided into two main groups according to their temperature tolerance.

COOL-SEASON TURFGRASSES Cool-season or temperate grass species with a temperature optimum of 15–24°C/60–75°F are employed for lawns in Britain, northern Europe parts of North America and other temperate regions. Most commonly used are FESCUES (*Festuca rubra* subsp. *rubra* and *F.rubra* subsp. *commutata*), BENTS (*Agrostis tenuis, A. castellana, A.stolonifera*), meadowgrasses, which are also known as BLUE GRASSES (*Poa pratensis, P.trivialis, P.nemoralis*), and PERENNIAL RYEGRASS (*Lolium perenne*). ANNUAL MEADOWGRASS (*Poa annua*) is a widely found invasive weed-grass of temperate lawns and may often make up the greatest proportion of a lawn, although it is never intentionally sown.

These grasses are usually sown as mixtures, although in temperate areas of the US smooth-stalked meadowgrass or KENTUCKY BLUEGRASS (*Poa pratensis*) and its cultivars is sometimes sown as a single species.

For the finest lawns, a mixture of bents and fescues is used. BROWNTOP and HIGHLAND BENT (*Agrostis tenuis* and *A.castellana*) are mixed with CHEWINGS FESCUE and slender CREEPING RED FESCUE (*Festuca rubra* subsp. *commutata* and *F.r* subsp. *litoralis*) in a ratio of 20–30% bent to 70–80% fescue by weight.

In Britain, the major wear-resistant species is perennial ryegrass (*Lolium perenne*); it is usually mixed with red fescue, smooth-stalked meadowgrass/Kentucky bluegrass, and browntop or highland bent. Pure swards of smooth-stalked meadowgrass/ Kentucky bluegrass (*Poa pratensis*) are extensively used in the Northeast, Midwest and other temperate regions of the US.

For lawns subjected to shade, WOOD MEADOWGRASS (*Poa nemoralis*) is often sown along with ROUGH-STALKED MEADOWGRASS (*Poa trivialis*). TURF TIMOTHY (*Phleum bertolonii*) may be used in place of perennial ryegrass on moist areas.

WARM-SEASON TURFGRASSES Warm-season species come from tropical and subtopical areas including Asia, South America, Africa and southern China, and these have a temperature optimum of 26.5°–35°C/80°–95°F. The major species used for lawns are BERMUDA GRASS (*Cynodon* species), *Zoysia* species and ST AUGUSTINE GRASS (*Stenotaphrum secundatum*).

Bermuda grass forms a dense, uniform, high-quality turf which tolerates heavy wear. Three of the zoysia grasses are used for lawns: JAPANESE LAWN GRASS (*Zoysia japonica*), MANILA GRASS (*Z.matrella*) and MASCARENE GRASS (*Z.tenuifolia*). Zoysias are often used in warm humid regions and areas where the climate is subtropical only in the summer months; they form a fine if rather slow-growing turf. St Augustine grass is by nature a coarse-textured, invasive species. Despite this, it is an acceptable species for general-purpose lawns which do not receive heavy wear. Other grass species for low-maintenance coarser lawns include CENTIPEDE GRASS (*Eremochloa ophiuroides*), CARPET GRASS (*Axonopus* species), and BAHIA GRASS (*Paspalum notatum*). Their coarse-textured leaves restrict their use to low-quality lawns.

Warm-season grasses are not usually mixed because they have a strong creeping habit and do not integrate well. They become dormant and lose their colour at temperatures below 10°C/50°F.

OTHER LAWN PLANTS Broad-leaved species are sometimes used for special effect but they are unsuitable for lawns subjected to heavy wear. These include fragrant CHAMOMILE (*Chamaemelum nobile*), DICHONDRA (*Dichondra micrantha*), PENNYWORT (*Hydrocotyle novae-zealandiae*) and even various species of moss, as in some Japanese gardens. BRASS BUTTONS (*Cotula dioica, C.pulchella* and *C. maniototo*) makes a fine, feathery, close-knit surface of some durability. It will tolerate drought and damp and has been used for bowling greens. Historically, CLOVER (*Trifolium* species), MEDICK (*Medicago* species) and SAINFOIN (*Onobrychis* species) were used for lawns in Europe.

Site preparation The site of a lawn should be cleared of all perennial and annual weeds and, where necessary, improvements made to soil drainage.

A base dressing of fertilizer should be applied to the prepared surface. On most soils, a compound containing 10% nitrogen, 15% phosphate and 10% potash, applied at 100–150g/m^2 and lightly raked into the soil surface, will be suitable.

For most grasses, satisfactory growth can be achieved at pH levels between 5.5 and 7.0. Fine fescues and bents have an optimum range of 5.5–6.5 while perennial ryegrass, the meadowgrasses and many of the warm-season grasses grow better at pH levels between 6.0 and 7.0.

Lawn establishment Lawns can be established from seed, turf (sod in the North America), or by vegetative propagation. Most cool-season lawns are produced from seed or turf, while those using warm-season species are generally produced by vegetative means.

SEED Establishing a lawn from seed is by far the cheapest method and enables the gardener to sow specific grass species and cultivars; however, seed takes much longer to establish than turf. It is only possible to sow seed during warm moist periods unless irrigation is available. Ideally, cool-season species are best sown in late summer or early autumn, with spring the second-best option. Grass seed should be sown at a rate of 25–30g/m^2 for most seed mixtures. Pure bent lawns should be sown at a rate of between 8–10g/m^2, while pure perennial ryegrass and smooth-stalked meadowgrass will require higher seed rates. Sowing too thickly may encourage seedling diseases such as damping-off.

After sowing, the seed should be lightly raked into the surface and sandy soils rolled with a light (50–100kg) roller to consolidate the surface. Heavy soils should not be rolled until after germination, at which time a second rolling should be carried out on sandy soils. The grass seedlings will take 7–14 days to emerge depending on species, temperature, and water availability. The lawn should be regularly irrigated in dry weather. When seedlings are 2.5–3.8cm high, they should be mown using a rotary mower and any excess clippings raked up and removed. If sown in the late summer or autumn, the lawn should be mown as necessary through the late autumn period, before gradually lowering the height of cut the following spring to that required for summer use. Lawns from seed should be subjected to as little wear as possible during their first season.

TURF Turf comes from established grass sward which is lifted intact and transplanted to a new site. It gives an immediate visual effect and can usually be walked on within a short period of laying. Turf is much more expensive than seed, and may carry pests, diseases and weeds.

Different types of turf are available, including meadow turf, treated meadow turf, sea-marsh turf, custom-grown turf and seedling turf, and careful inspection of the offered product is essential.

Meadow-turf, grown initially for agricultural purposes, is likely to contain coarse grasses, including annual meadowgrass and broad-leaved weed species such as the daisy (*Bellis perennis*). Treated meadowgrass has had selective herbicide applied to eliminate broad-leaved weed species.

Sea-marsh turf was once considered the finest turf available in the UK because it contains fine fescues and bents. However, it often grows in river estuaries where silt and clay are deposited and care must be taken not to cover a free-draining soil with a silty impermeable barrier.

Seedling-turf, 6–15 weeks old, is raised on a shallow soilless medium and can be quickly grown to order. It is light and easy to transport but does not have the maturity of older turf and needs protection from heavy use until established.

Traditionally in the UK, high-quality turf was sold in 30cm squares and lesser-quality turf in 90 × 30cm strips. Most turf sold by large producers is lifted by tractor-mounted machines and cut into one square metre pieces.

Turf should be laid with an alternate bond, brick fashion, so that each piece tightly meets the next. Moist soil encourages rapid rooting and, after laying, turf should be lightly rolled to eliminate air pockets between the turf and the soil. In wet conditions, the rolling should be left until the turf has knitted together. Finally a light topdressing of sand or sandy loam is applied over the surface and brushed into any gaps. Regular irrigation that penetrates the turf and soil beneath to a depth of 10–15cm will be needed in dry conditions.

Turf can be laid almost year round in Britain, although in the more extreme climates of continental Europe and North America the operation will be restricted by very hot or very cold conditions.

VEGETATIVE PROPAGATION Vegetatively propagated lawns require less material than turf for a given area, but there is no instant cover and about two months is required for rooting and spread. Species commonly propagated by this method are the strong, creeping, warm-season grasses, together with cool-season creeping bent in North America. Vegetative propagation is not often practised in Britain.

For stolonizing, the soil is prepared as for seeding and the stolons spread at a rate of around 2.25–5 litres/m^2 and pressed into the soil surface.

Mowing Ideally no more than one third of the leaf growth should be removed at one mowing if growth is not to be adversely affected. For all lawns, infrequent heavy mowing will cause the greatest damage to the grasses.

Fine fescues and bents used in high-quality lawns can be cut as low as 6mm in the summer and 12mm in winter, and these will need cutting every other day in summer so as not to remove more than one-third of the growth.

Ryegrasses and other species used in utility lawns grow better at heights between 1.2–2.5cm in summer and 2.5–3.7cm in winter. In addition, the greater amount of grass will protect the surface against wear. These lawns will only need cutting once or twice a week in summer if no more than a third of the grass is to be removed.

Grass mowing will be least frequent in the cold of winter and during summer drought conditions.

The two most commonly used types of mower are the cylinder or reel mower, and the rotary mower which includes the hover-type. The cylinder mower gives the best finish, although the rotary type gives perfectly acceptable finishes for utility lawns.

For high-quality lawns, grass clippings should be removed (boxed off) to discourage earthworms, lessen the spread of annual meadowgrass and other seeding weeds, reduce the accumulation of organic matter and leave a better surface finish.

Fertilizers On heavy-textured soils, where clippings are returned and rainfall or irrigation rates are low, only small amounts of fertilizer will be required; $30g/m^2$ of a fertilizer containing 10% nitrogen will be a suitable annual dressing. This may be applied as one application in mid- to late spring, or in split applications in early summer and after autumn renovations. Phosphorus and potassium are likely to be needed only every 3–5 years.

On light-textured soils where the clippings are removed and high rainfall or irrigation is plentiful, as much as $150g/m^2$ of a fertilizer containing 10% nitrogen per annum may be needed. Such a large amount of fertilizer should be applied in at least three applications, in late spring, in summer and after autumn renovations. Phosphorus and potassium will also be required at a quarter and three-quarters the rate of nitrogen respectively per annum.

Iron is included in some compound fertilizers, and also widely applied as a constituent of lawn sand, which is used to control moss and encourage grass growth. Lawn sand usually contains 3 parts by weight ammonium sulphate, 1 part ferrous sulphate and 20 parts fine sand. It should be applied at $136g/m^2$, in spring or summer outside of hot dry spells.

Fertilizer must be applied evenly over the entire surface and, unless heavy rain is expected, it should be watered in to prevent scorch.

Irrigation Where ornamental effect is essential and water usage can be afforded, enough water should be applied at times of drought to ensure that the soil is moistened to a depth of 100–150mm. Shallow watering encourages detrimental surface rooting, and heavy infrequent irrigation is in general better than light frequent applications. Well-established lawns are better able to recover from drought conditions that cannot be offset due to unavailability of irrigation water or restrictions in supply.

Maintenance and weed control Common problems on well-used lawns are soil compaction and thatch accumulation.

Compaction restricts root growth and prevents rapid drainage, while lawns that have become acidic and from which earthworms have been eliminated can build up organic debris or thatch on the soil surface.

The techniques of scarification, hollow tining or coring, spiking and slitting are used to remove thatch and relieve soil compaction.

A powered scarifier has vertical blades that rotate at a high speed and penetrate the lawn up to 1.2cm in depth. Light scarification can be undertaken using a wire rake. It is advisable to scarify in two directions, one at 30 degrees to the other.

Hollow tining or coring removes cores of grass, thatch, and soil 6–18mm in diameter and 7.5–10cm in depth. It can be undertaken with a special hand tool or by machine. The resulting holes may be left open or filled with a suitable sandy topdressing, which will prevent the hole from closing and improve the soil.

Spiking with a purpose-made tool or hand fork relieves soil compaction and allows air and water into the soil. For maximum effect, the spikes are pulled back after entering the soil, thereby lifting the turf and creating cracks and fissures

Slitting involves penetration of the soil with flat, knife-like blades to a depth of 7.5–10cm.

Heavy scarification and hollow tining should be undertaken in early autumn when warm, moist conditions permit rapid recovery. Light scarification, spiking and slitting can be undertaken throughout the season except in dry conditions.

Bulky topdressing is also beneficial after autumn work since it can aid thatch breakdown, fill core holes, and correct surface irregularities. A topdressing containing medium/fine sand, sieved soil and peat or well-rotted compost, in a ratio of 6:3:1, is suitable for most lawns; apply at $1–3g/m^2$ on a dry day and lute or brush well into the surface and core holes.

Rolling the lawn in spring resettles the surface but excessive rolling can cause compaction.

Brushing dew from grass, especially in the autumn, helps prevent fungal disease spread.

Soil excreted by earthworms and deposited as worm casts on the surface of lawns can be a problem in making mowing difficult or unsightly, damaging the cutting edges of cylinder mowers. When flattened, worm casts smother grass, provide sites for weed germination and make a lawn slippery. Treatments to raise acidity, such as the application of sulphate of ammonia or sulphate of iron, discourage earthworms. Brushing with a birch broom will disperse worm casts.

Broad-leaved weeds should be treated with hormonal herbicide or weed-and-feed fertilizer. Small infestations of plants, such as daisy, dandelion and plantain, can be removed with a daisy grubber or knife, or killed by herbicide applied with a 'spot-weeder'.

Moss thrives in poorly drained lawns of low fertility and high acidity. It should be treated by cultural improvements and chemical moss killer or lawn sand. See *chafer beetles, dollar spot, earthworms, fairy ring, fusarium patch disease, leatherjacket, ophiobolus patch, red thread, snow mould.*

lax loose, for example flowers widely spaced in an open inflorescence.

laxpendent hanging loosely.

layering (1) a training technique used in the cordon system of growing sweet peas, and in large-scale tomato growing; (2) a method of vegetative propagation whereby stems are induced to form adventitious roots while still attached to the parent plant, to be eventually severed and grown on as separate individuals.

Plants, such as *Campsis*, *Hedera* and *Rubus*, have a natural propensity to layer themselves. Artificial methods of layering have been devised, utilizing inherent growth characteristics of plants together with varying degrees of shoot manipulation, including pruning, blanching, etiolation and constriction. Such strategies, besides providing shoots with a high capacity to produce roots, also cause natural hormones and carbohydrates to accumulate, thus encouraging quick rooting at specific locations on the stem. Layering is less widely used than formerly, mainly because of the advances made in rooting clonal material from cuttings. However, on certain plants, such as *Corylus maxima* 'Purpurea', clonal *Malus* rootstocks and certain rhododendrons, all of which are difficult to root by cuttings, layering is still a prime technique of production. For best results, the soil should be light textured, deep, well-drained, moisture-retentive, stone-free and of a pH appropriate to the species.

Simple layering Simple layering provides a means of propagating a wide range of woody plants. Vigorous one-year old shoots are encouraged to develop close to the soil, mainly by hard pruning, and are buried in such a way that the tip of the shoot is exposed well above soil level.

Many woody plants respond to this treatment, such as *Carpenteria californica*, *Chimonanthus praecox*, *Corylopsis pauciflora*, *Corylus avellana* cultivars, *C.maxima* cultivars, *Cornus florida*, *Daphne cneorum*, *Disanthus cercidifolius*, *Kalmia latifolia* cultivars, *Magnolia* species and cultivars, *Rhododendron* species and cultivars and *Syringa vulgaris* cultivars. The time of year can influence the success rate; thus, late spring is recommended for *Disanthus cercidifolius*, when shoot growth has started, and midsummer for *Chimonanthus praecox*, when shoots are supple but not easily bruised.

Rooting is stimulated in stems blanched by covering them with soil and by constriction. A shoot is selected and pulled down to the soil to ascertain the position where the stem needs to be buried and constricted. There are a number of ways of constricting stems: by bending, twisting to rupture the tissue, cutting a tongue about 3cm long and exposing the cambium, girdling by removing a ring of bark about 5mm wide, or by twisting copper wire into the stem tissue. A trench about 15cm deep is made with one side sloping towards the parent plant and a vertical side opposite. The constricted shoot is then bent down into the trench with the shoot tip placed against the vertical side. The tip should be exposed above the soil surface by 10–15cm, secured with a peg and covered with soil; it may need to be tied to a small cane to ensure vertical growth. Layers will take one or two years to root and when sufficiently established, they can be severed from the parent plant, lifted during the dormant season and subsequently grown on. Layering is also the preferred method for increasing border carnations; a tongue is made near the leafy end of the shoot and pegged down into a depression that is later refilled with soil.

Compound or serpentine layering This method is a modification of simple layering, utilizing the trailing shoots of climbing plants; it is suitable for subjects that do not root easily from stem cuttings, such as *Wisteria, Clematis armandii, Lapageria rosea* and *Vitis davidii*. Stem blanching and constriction techniques are applied at intervals along the trailing stems, and to avoid restricting the flow of sufficient sap, cutting a shallow tongue in the stem is probably the safest method.

During early spring, vigorous shoots are positioned on the soil surface in a circular pattern. Starting at the basal end, the first constriction is made and the shoot carefully bent into a hole about 10cm deep and secured with a peg before filling back with soil. This procedure is repeated along the stem, ensuring that vegetative buds are looped above the soil between each layer. New growth developing from the buds should be pinched back once or twice during the growing season to avoid excessive extension growth at the expense of root development. When sufficiently rooted, layers are severed close to the base of the parent plant. During the dormant season, these may be lifted and sectioned into separate plants.

Tip layering This method of layering is suitable for rooting plants of the genus *Rubus*, including blackberry, loganberry, boysenberry and tayberry. Healthy parent plants are established for one growing season, and in the spring of the second season are cut back to about 15–20cm above ground level. When new growth is 60cm in length, each shoot should be tipped to stimulate side laterals. Once side laterals have produced 1–1.5m of growth, shoots may be pulled down into a 15cm deep trench, made with a sloping side towards the parent plant and a vertical side opposite; this will encourage the shoot to grow vertically. The shoot tip is placed so that the tip is bent upwards at the vertical side and fixed with a wire staple before being covered with soil. Rooting takes place within three weeks and rooted layers can be lifted during December.

Air or Chinese layering This layering method is valuable in creating new plants from the upper growth of specimens. It is especially useful for plants that have become too tall, as often occurs with house plants, such as *Ficus elastica* cultivars, or with *Mahonia aquifolium* cultivars. Marcottage, gootee, and circumposition are other names for this method.

A woody shoot, preferably not more than two years old is selected, and either cut to remove a ring of bark about 1cm wide, or slit shallowly upwards for about 2.5cm length of stem and a sliver of wood pushed in to keep the wound open. A rooting

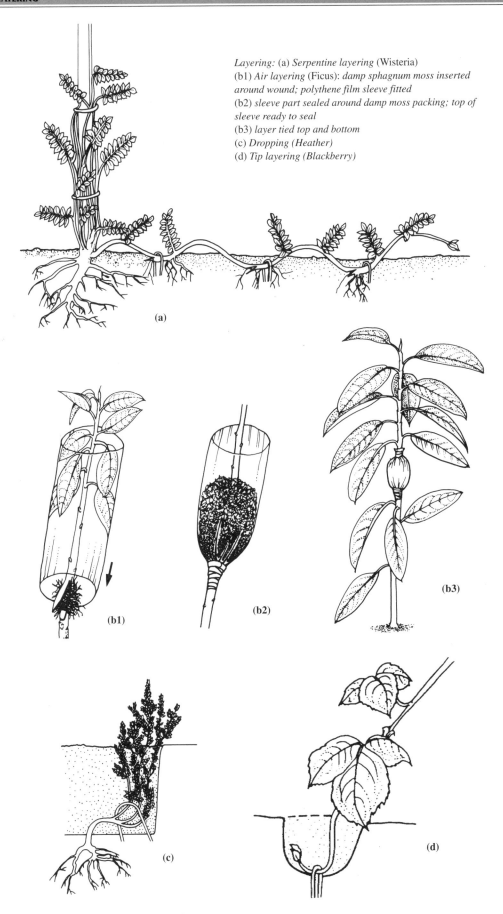

Layering: (a) *Serpentine layering* (Wisteria)
(b1) *Air layering* (Ficus): *damp sphagnum moss inserted around wound; polythene film sleeve fitted*
(b2) *sleeve part sealed around damp moss packing; top of sleeve ready to seal*
(b3) *layer tied top and bottom*
(c) *Dropping (Heather)*
(d) *Tip layering (Blackberry)*

hormone should be applied over a 2cm area around the wound. A handful of moist sphagnum moss is bound in and around the cut with a few turns of thread, and the area wrapped in a sleeve of black plastic film (30 microns), which should overlap and be tied firmly top and bottom. To avoid potentially lethal temperatures (35–40°C/95–104°F) building up inside the black plastic through radiation, the use of co-extruded plastic film (black with white reverse) is advised, with the white side outward-facing. Alternatively, aluminium foil or aluminized plastic may be used. When roots have developed sufficiently, cut the stem below the moss ball and grow the plant on. The old stem should be cut back to vegetative buds to prevent stem die-back and to encourage new growth.

Stooling or mound layering This is a widely used commercial method, especially for the production of clonal rootstocks. Roots are stimulated by blanching developing shoots; this is done by heaping soil around their bases. Plants suitable for this technique must be able to tolerate hard annual pruning down to ground level, and be able to root effectively from blanched stems only; they include clonal apple rootstocks such as M9, MM106, rootstocks for plums like 'St Julien A' and ornamentals such as *Prunus cerasifera* 'Nigra', *P. tenella* 'Fire Hill' and *Tilia × euchlora*.

New stool beds are established by obtaining young parent plants that are free from viruses and true to name. Planting should be done during the dormant period and the young plants grown on over one season to become well-established. During March of the following year, the stems should be cut back (stooled) about 2.5–5cm above soil level. Each stool should produce 2–6 new shoots, and once the shoots attain 12–15cm in height, moist, finely cultivated soil is carefully mounded up around them leaving the shoot tips visible. This procedure is normally carried out three times during the growing season, the final mounding up being done soon after midsummer. Maintenance work required during the growing season will include irrigation and weed, pest and disease control. By late autumn, layers will be well-rooted, and they should be harvested when fully dormant. The soil mounds are broken down to expose the rooted layers, which are gathered by cutting back to the original stool. There will be an annual increase in production up to about five years, but a well-maintained stool bed will last at least fifteen years.

French or continuous layering The technique used here is similar to stooling, except that selected shoots are pegged down so that their buds produce new vertical shoots and these are then stimulated to root by blanching. Plants most suitable for this method should be capable of rooting effectively after blanching, produce long unfeathered shoots and be tolerant of hard pruning; they include *Cornus alba* cultivars, *Cotinus coggygria* cultivars, *Hydrangea paniculata* and *Viburnum plicatum*.

Mother plants are established for one growing season in well-prepared soil at 2–3m-square spacings depending upon vigour. In the following spring, the plants are cut back to 2.5–5cm above soil level and will produce 2–6 shoots, which should be left to develop fully.

During late winter in the third season after planting, long, healthy and ideally unfeathered shoots are pegged down on to the surface of the soil, arranged like the spokes of a wheel. Thin, weak shoots and any surplus to requirements should be cut hard back. In spring, growth will develop at right angles to the pegged shoots and from the centre of the mother plant. The centre shoots should be encouraged to develop fully as replacement layers for next season. Growth from the pegged shoots are mounded to a depth of 15cm.

Rooted layers should be ready by early winter, and the layered shoots are severed close to the mother plant, lifted, sectioned and grown on.

Etiolation or trench layering This method is generally reserved for plants that need to be propagated on their own roots, such as *Prunus avium* F12/1 rootstocks and walnuts. Etiolation layering differs from stooling inasmuch as buds are encouraged to develop and extend on layered shoots in darkness, without foliage and virtually devoid of pigmentation. Such shoots, in which the cells of the buried sections become elongated, have a remarkable capacity to produce adventitious roots.

Mother plants are established from single shoots about 60cm long; these are planted at a 45-degree angle, at 60cm spacings, in rows 1.5–3m apart. Plants are allowed to establish for one growing season, and during the early spring of the second year, a trench about 5cm deep and 30cm wide is taken out along the row. A little soil is removed just beside the base of each angled stem, allowing the stem to be easily bent horizontally into the trench, where it is held firmly in place with stout pegs.

Before the buds have begun to swell, the shoots are covered with soil 2.5cm deep and just as shoots appear above ground level a further covering of the same depth is made. This is repeated up to three times, after which the shoots are allowed to attain about 15cm in height and then mounded up with the tips visible.

Rooted layers will be ready for harvesting by early winter. One strong shoot is retained every 30cm along the row to continue the sequence of production.

Dropping This relatively simple layering technique involves disposing of the parent plant once rooted layers are removed. It is normally carried out on small bushy woody plants, such as *Berberis × stenophylla* 'Irwinii', *Calluna* and *Erica* species and cultivars, and dwarf *Rhododendron* species and cultivars.

Young bushy plants are selected during spring and lower leaves removed where rooting is expected to take place. A deep trench is dug into which the plants are placed with their shoots evenly spaced out. The soil is replaced, ensuring that the shoot tips are above the soil level. When the shoots have rooted, between one to two years later, the parent plants are lifted and the rooted layers removed and grown on.

layer planting used of groups of plants in a floral bed that are arranged to provide successional flowering; also to achieve variable height for improved display.

Layia (for George Tradescant Lay (d. *c*1845), botanist to Captain Beechey's voyage in the *Blossom*, which visited California in 1827). Compositae. W US. About 15 species, annual herbs with daisy-like flowerheads from summer to early autumn. They are frost-tender annuals for dry, sunny banks, and do not thrive where summers are hot and humid. Sow seed *in situ* in spring (or earlier under glass), in any well-drained soil in sun. *L.platyglossa* (syn. *L.elegans*; TIDY TIPS; California; to 30cm, glandular, hairy; leaves linear to narrowly oblong, dentate to pinnatifid; ray florets yellow with white tips, anthers black).

laying a traditional and effective method of training hedges by bending partially severed shoots to an angle of about 45 degrees and weaving them around wooden stakes driven into the hedge line at intervals. Hedges can be layed at any time between October and May but the work is easiest if done in spring when sap is rising.

laying-in the process of tying in young shoots against a wall, carried out during the training of a fruit tree.

lazy-bed a system of growing potatoes, in which tubers are placed on the soil surface and covered with soil from trenches dug between the rows or bed.

leaching the drainage of water through soil or other growing media, removing soluble chemicals, especially nitrogen. It has the deleterious effects of potential waste and/or pollution of watercourses. It can be beneficial in diluting a build-up of soil salts which often follows regular supplementary feeding, as for tomatoes; and leaching is induced in such situations by flooding the soil. See *nutrients* and *plant nutrition*.

leader the main shoot of a woody plant, often referred to as the central leader; also, the dominant terminal shoot that extends the growth of a branch. In pruning, leaders are usually cut lightly (tipped) in comparison to laterals.

leaf a lateral member borne on the shoot or stem of a plant, generally differing from the stem in form and structure.

leaf beetles (Coleoptera: Chrysomelidae) one of the largest families of beetles, the adults of which are mostly oval-shaped, drab to brightly coloured and ranging in size from 1–12mm long. The larvae are soft, fleshy grubs with three pairs of short, stumpy thoracic legs. Adults and larvae often feed openly on the foliage of plants, or the larvae mine leaves or feed on the roots. Species attacking vegetables include the FLEA BEETLES (*Phyllotreta* species) and related species in North America; the COLORADO BEETLE (*Leptinotarsa decemlineata*) on potatoes in North America and Europe (not the UK); the ASPARAGUS BEETLE (*Crioceris asparagi*), with conspicuous yellow and black adults up to 6mm long and grey black humped larvae, both of which feed on the foliage of edible asparagus; and, in North America, the BEAN LEAF BEETLE (*Ceratoma trifurcata*), a yellow and black marked beetle up to 7mm long, which feeds on the foliage of beans and other legumes and has root-feeding larvae. In North and South America, cucurbits can be severely attacked by CUCUMBER BEETLES (*Acalymma* and *Diabrotica* species), with pale yellow green adults, marked with spots and bands of black and up to 8mm long, which feed on foliage and whose larvae tunnel into roots; some of these larvae are known as CORN ROOTWORMS because they can seriously damage the roots of maize.

Fruit in North America is attacked by rootworms, particularly the GRAPE ROOTWORM (*Fidia viticida*), with chestnut brown hairy adults up to 6mm long and root-feeding larvae which damage the roots of grape vines, and the STRAWBERRY ROOTWORM (*Paria fragariae*), a glistening brown beetle with black markings, up to 3mm long, which feeds on foliage and whose larvae infest the roots and crowns of strawberries, raspberries, blackberries and some other plants, such as rose.

The WATER LILY BEETLE (*Galerucella nymphaeae*) is small and brown, up to 6mm long, and has soft-bodied brown-black larvae which feed on the upper surface of leaves, making furrows and holes. The SUNFLOWER BEETLE (*Zygogramma exclamationis*) has adults resembling the Colorado beetle which feed, along with their hump-backed larvae, on the foliage of sunflowers in North America.

The adults and larvae of several species feed openly on the foliage of trees. These include the ELM LEAF BEETLE (*Pyrrhalta luteola*), with yellow or green adults up to 6mm long, which together with their larvae cause defoliation of elm trees in Europe (not the UK) and North America; various small, metallic blue, red or green beetles, mostly *Phyllodecta* species up to 5mm long, attack the foliage of willows and poplars; the EUROPEAN VIBURNUM BEETLE (*Pyrrhalta viburni*), is a dun brown beetle, up to 6mm long, with creamy yellow larvae with black markings.

Leaf beetles that feed openly at both adult and larval stages can be controlled with recommended contact insecticides. Where larvae are root feeders, measures must be aimed at eliminating the adults before egg-laying commences. See *Colorado beetle, flea beetle, lily beetle*.

leaf blister a descriptive term for foliage diseases caused by species of the fungus *Taphrina*, where limited infected leaf areas become swollen. The effect is similar to, but less severe than, that caused by PEACH LEAF CURL (*Taphrina deformans*) (q.v.). *T. caerulescens* can be damaging in the US and fungicide treatment of the dormant buds is recommended, as for peach leaf curl; *T. bullata* causes brown blisters on pears, similar to those caused by leaf blister mites; *T. caerulescens* causes grey-pink blisters on oaks, sometimes resulting in severe defoliation; *T. populina* causes blisters on poplars, which are bright yellow on the underside of leaves when the fungus is sporulating; *T. sadebeckii* forms yellow blisters on alder; and *T. ulmi* forms small green blisters on elm, which turn brown. Leaf blisters of various ferns are also caused by *Taphrina* species.

leaf-cutter bees (*Megachile* species) (Hymenoptera: Megachilidae) small solitary bees, up to 10mm long, which resemble honey bees but have a stouter build and are more hairy. The females use their mandibles to remove uniform oval or circular portions of leaves to construct thimble-shaped cells; these are filled with a mixture of pollen and nectar before an egg is deposited and the cell capped with a circular piece of leaf. Up to around 20 cells are built end on end in situations such as tunnels made in decaying wood, old brickwork or dry soil. Larvae feed within the cells during the summer and overwinter there, pupating in the spring and emerging as adults in early summer. Commonly found in gardens is *Megachile centuncularis*, which attacks the foliage of roses, and other species cause similar injury to many ornamental plants, but control measures are seldom necessary.

leafhoppers (Hemiptera: Cicadellidae) sap-feeding insects, mostly up to 5mm long, similar to froghoppers but generally less robust, and with a slender tapering body and a double row of spines along the hind legs. Adults are widespread on plants in summer, jumping, flying briefly and then resetting on leaves. Some species transmit plant diseases. Typically, they feed on the underside of leaves, causing small white spots to appear on the upper surface. In severe attacks, the spots coalesce and the whole leaf becomes chlorotic, before withering. The underside of infested leaves is invariably littered with discarded white moultskins.

In gardens, the ROSE LEAFHOPPER (*Edwardsiana rosae*) is common in both Europe and North America on roses and apple; similar species attack hornbeam and beech hedges. The GREENHOUSE LEAFHOPPER (*Hauptidia maroccana*) is commonly found in greenhouses and gardens in temperate countries, attacking a wide range of plants. The RHODODENDRON LEAF-HOPPER (*Graphocephala fennahi*), common in Europe and North America, does no direct damage but by depositing eggs in the flower buds facilitates spread of the fungal disease known as bud blast. These leafhoppers are conspicuous, up to 9mm in length with red oblique stripes on the green wing cases. Apples and potatoes are subject to attack by several different species of leafhopper; grapes by *Erythroneura* species; *Rubus* by *Macropsis* species; and strawberry by *Aphrodes* and *Euscelis* species, which transmit the virus causing green petal disease. Recommended contact or systemic insecticides are effective, where control is necessary.

leaflet one of the leaf- or blade-like ultimate units of a compound leaf.

leaf miners the leaf-mining larvae of several different groups of insects, including the Lepidoptera (moths), Coleoptera (beetles), Diptera (flies) and Hymenoptera (some sawflies). Apple is attacked by some species, including the SPOTTED TENTIFORM LEAF MINER (*Phyllonorycter* species), which is a more serious pest in North America than in Europe; the larvae live in raised blister-like mines. In Europe, the APPLE LEAF MINER (*Lyonetia clerkella*) is more widespread on apple, and also occurs on cherry, ornamental *Prunus*, hawthorn and birch, but the serpentine mines mostly become numerous in late summer and autumn when it is too late for the tree to be adversely affected. Other moths in Europe and North America with leaf-mining larvae include the AZALEA LEAF MINER (*Caloptilia azaleella*), which causes blotch mines on evergreen and some deciduous azaleas, and the LILAC MINER (*C. syringella*), which causes blotch mines on lilac and privet; the larvae of both these species vacate the mines when they are about half-grown to feed within a turned-over leaf margin, which is held in place by silken threads.

The CHRYSANTHEMUM LEAF MINER (*Phytomyza syngenesiae*) causes serpentine mines on the leaves of chrysanthemums, cinerarias, lettuces and some other cultivated plants; the HOLLY LEAF MINER (*P. ilicis*) causes blotch mines on *Ilex* species; the columbine leaf miners *P. aguilegiae* and *P. miniscula* cause blotch and serpentine mines respectively on *Aquilegia* species; and the TOMATO LEAF MINER (*Liriomyza solani*) is responsible for serpentine mines on the leaves of tomatoes. Other *Liriomyza* species also attack chrysanthemums, peas and other vegetables in both Europe and North America. The CELERY FLY (*Euleia heraclei*), a serious pest of celery in Europe, has larvae that live in communal blotch mines, which subsequently become desiccated, brown and papery giving the foliage a scorched appearance. In Europe, the larvae of the CARNATION FLY (*Delia cardui*) cause extensive mining of carnations, pinks and sweet williams. The larvae of the BEET LEAF MINER (*Pegomya hyoscyami*), common to both continents, cause blotch mines on beetroot, spinach beet and related plants.

Light infestations of leaf miners can be controlled by hand-picking and destroying affected leaves; once larvae become established in mines, they are not easily killed by spraying, although systemic insecticides have proved effective against several species.

leafmould the residue of decayed leaves, which forms a dark brown, friable material, rich in humus but low in nutrients. It is valuable as a soil improver and surface mulch; also as a compost ingredient, for which it was formerly widely used until superseded by peat but it remains a candidate alternative. Leafmould occurs naturally in woodland, and can be prepared by collecting fallen leaves and depositing them in a compost bin or small wire-mesh compound. Leaves decay more slowly than mixed garden waste, but with turning a suitable leafmould should be available for use after 18 months. Oak, beech and hornbeam yield the best quality leafmould. Thick leaves, such as sycamore and horse chestnut, take longer to decay, and hard leaves, such as holly and laurel, are best mixed with general plant waste for composting. Pine needles decay over a long period to produce a useful medium for acid-loving plants.

leaf mould (*Fulvia fulva*) a serious worldwide disease of tomato plants grown under conditions of high humidity; it is rare on outdoor plants. Pale yellow patches appear on the upper surface of older leaves as a result of velvety green-grey fungal growth produced on the undersides. The disease spreads up the plant and affected leaves progressively turn brown and shrivel; flowers and fruit may also be infected. Plant vigour and yield can be seriously affected, and the fungus can carry over from one crop to the next as spores or fruiting bodies on debris or seed.

Many cultivars of tomato have resistance to leaf mould, although different races of the fungus may overcome resistance. Preventative measures include close attention to ventilation, aiming to keep the greenhouse relative humidity level below 85%. Avoidance of prolonged wetness of foliage is also beneficial. Weekly sprays of a recommended fungicide will check the spread of leaf mould. All infected plant debris should be removed and empty greenhouses disinfected by burning sulphur.

leaf rollers the larvae of several groups of insects, and some mites, are associated with partial or complete leaf rolling. The rolling may be tight or loose, or in a downwards or upwards direction, but the damage is always unsightly and reduces the normal functions of affected leaves.

The larvae of tortricid moths typically feed within a rolled or folded-over portion of a leaf, which is held in place by strands of silk. Common in Europe and North America are the FRUIT TREE TORTRIX MOTH (*Archips podana*), which attacks various deciduous trees and shrubs, conifers, and especially apple, raspberry and blackberry; the STRAWBERRY TORTRIX (*Acleris comariana*) on strawberry, geums, potentillas and azaleas; the LEAF TIER (*Cnephasia longana*) on chrysanthemums, irises, strawberry, fruit trees, shrubs, the new growth of conifers and various vegetables; and the STRAWBERRY LEAF ROLLER (*Ancylis comptana*) on strawberry, potentillas, thymes and *Dryas octopetala*.

Several adult weevils roll one or both sides of a leaf to form a tube in which the larvae live and feed; they include the HAZEL LEAF ROLLER WEEVIL (*Byctiscus betulae*) on hazel, birch and other deciduous trees, and the POPLAR LEAF ROLLER (*B. populi*) on poplars and aspens. The leaf-rolling habit also occurs among gall midges: the orange larvae of the VIOLET GALL MIDGE (*Dasineura affinis*) live in the upward-rolled, swollen leaves of violets, and the white larvae of the PEAR LEAF MIDGE (*D. pyri*) infest the upward-rolled margins of pear leaves. One common European sawfly, the LEAF-ROLLING ROSE SAWFLY (*Blennocampa pusilla*), causes a tight downward roll of both leaf margins of roses. Hand-picking

and crushing the larvae is usually an effective control measure control, as is burning affected leaves. Some mites are associated with leaf rolling, especially tarsonemid mites, such as the BROAD MITE (*Polyphagotarsonemus latus*) on *Gerbera* and *Cyclamen*. See *tarsenomid mites, tortrix moths*.

leaf scar scar tissue on a shoot indicating the site of leaf abscission.

leaf scorch see *scorch*.

leafsoil a soil containing a high proportion of leafmould.

leaf spot spots on plants usually caused by fungal pathogens, affecting a wide range of plants, particularly when the foliage is ageing. Although leaf spots can be conspicuous and unsightly on ornamentals, the majority cause only slight damage in garden plants. However, leaf-spot diseases of crops, such as banana and celery, can seriously reduce yield.

Typical leaf spots are scattered randomly and have definite edges, often with darker margins. On net-veined leaves, they are usually circular, while on leaves with parallel veins, they are elongated along the direction of the veins. The dead centres of spots are often lighter in colour and may show the spore-bearing structures. Some leaf-spot fungi, such as *Alternaria* and *Cercospora*, have spores which are dispersed in dry conditions, but the majority, including *Ascochyta* and *Septoria*, have sticky spores and these are adapted to spread in water.

Leaf spots may also be caused by bacterial pathogens (see *bacterial leaf spot*), and some virus diseases produce (usually) chlorotic spots. Air pollution, pesticide injury, nutrient or water deficiency, pest attack and physical damage by hail, can all produce leaf spots.

As many fungal leaf-spot diseases are spread by spores produced on the spots, a degree of control can be obtained by picking off affected leaves as soon as symptoms appear and destroying fallen ones. Avoiding overhead watering and the handling of plants when they are wet are other practical means of control. Chemical control with copper-based fungicides can be effective, and celery leaf spot can be controlled by seed treatment. See *anthracnose, black spot, chocolate spot*.

leaf trace a vascular bundle supplying a leaf.

leaf weevils (*Phyllobius* species) (Coleoptera: Curculionidae) small weevils, up to 10mm long, dark brown to black and covered with dense scales which often appear metallic gold or green-bronze. Mostly present in early summer, the adults bite small holes in leaves and occasionally the blossom of trees and shrubs, including top fruits, birch, lime, flowering cherry, crab apples, poplars and rhododendrons. The larvae live in the soil and feed on roots without causing significant damage.

The principal species are the BROWN LEAF WEEVIL (*Phyllobius oblongus*), the SILVER-GREEN LEAF WEEVIL (*P. argentatus*) and the COMMON LEAF WEEVIL (*P. pyri*). Control is seldom necessary but can be achieved by spraying with a recommended contact insecticide.

leafy gall (*Corynebacterium fascians*) a bacterial disease causing leafy galls around the collar of infected plants, blindness, absence of flower buds, and fasciation. Susceptible plants include carnations, chrysanthemums, *Dahlia*, *Gladiolus*, *Heuchera*, nasturtiums, *Pelargonium*, *Petunia*, *Phlox*, *Schizanthus*, sweetpeas, sweet williams and *Verbascum*. When associated with the eelworms *Aphelenchoides ritzemabosi* or *A. fragariae*, the bacterium causes cauliflower disease of strawberry, a severe stunting and swelling of the leaf and flower stalks. Leafy gall is not usually serious but affected plants should be destroyed and contaminated land kept free of susceptible plants.

lean used loosely to denote soil that is free-draining and low in organic matter and nutrients.

lean-to a greenhouse or conservatory with a single pitch roof, usually built against a wall and therefore often used to accommodate trained fruits. Lean-tos are usually relatively economic to heat and if suitably sited readily trap solar heat. See *greenhouse*.

leatherjackets (Diptera: Tipulidae) larvae of the large, fragile, spindly-legged daddy-long-legs, or crane fly. The legless, grey-brown larvae are up to 45mm long, with leathery skins and inconspicuous heads. Adult females lay up to 300 eggs, singly or in clusters depending upon species, just below the soil surface in grassland, including lawns, or in cereal crops, where the resulting larvae feed upon roots.

Leatherjackets can be serious lawn pests, causing brown patches in midsummer; their presence can be confirmed by soaking the area with water and covering it with black polythene, under which they will emerge within 24 hours. Affected lawns may be further damaged by birds searching for leatherjackets. The underground portions of plants, such as young brassicas, lettuces, strawberries and various ornamentals, may be damaged by leatherjackets, especially in newly cultivated gardens.

One of the most injurious species is *Tipula paludosa*, a native of Europe but also established in North America, particularly in British Columbia and Washington. Other common species are *T. oleracea* and the spotted crane fly (*Nephrotoma maculata*).

To reduce leatherjackets on lawns, attract them to the surface in the manner described above and then remove by hand-picking. Recommended soil insecticides are more effective if used in late autumn when the larvae are young.

Ledebouria (for Carl Friedrich von Ledebour (1785–1851), German botanist). Liliaceae (Hyacinthaceae). South Africa. Some 16 species, bulbous, perennial herbs with thinly fleshy, narrowly lanceolate leaves, often striped or spotted red or green. They produce spikes of small, bell-shaped flowers with recurved perianth segments. Grow in full sun with low humidity, good ventilation and a minimum temperature of 7°C/45°F. Pot in pans containing a gritty, loam-based medium and with the bulbs for the most part exposed. Water moderately in spring and summer, very sparingly in winter. These bulbs associate well with succulents and cacti. Propagate by offsets.

L.cooperi (syn. *Scilla cooperi*; leaves to 25cm, somewhat fleshy, oblong to ovate or linear, sometimes striped brown; flowers pale purple sometimes with green keels); *L.socialis* (syn. *Scilla socialis*; leaves to 10cm, slightly fleshy, lanceolate, pearly grey-green with some dark green blotches above, green or deep pink-purple beneath; flowers pale purple with green keels).

Ledum (from Greek *ledon*, mastic, a name used for *Cistus*). Ericaceae. Northern temperate regions. 3–4 species of erect or diffuse evergreen shrubs with aromatic foliage and small, white, 5-petalled flowers in umbel-like corymbs in spring and summer. Fully hardy and suited to the bog or damp woodland garden in shade or semi-shade and to moist, humus-rich, acid soils. Propagate by semi-ripe cuttings in a closed case with bottom heat, by simple layering or by seed. *L.groenlandicum* (LABRADOR TEA; northern North America, Greenland; 0.5–2m; leaves 2–6cm, linear to oblong, revolute, dark green and somewhat hirsute above, rusty-woolly beneath; flowers white in clusters to 5cm in diameter; 'Compactum': dwarf).

Leea (for James Lee (1715–1795), London nurseryman). Leeaceae. Old World tropics. 34 species, shrubs and small trees with simple or 1–3-pinnate leaves, small flowers in umbels and berry-like fruit. Provide bright filtered light or shade from direct sun in high summer, and grow in a high-fertility loam-based medium. Maintain high humidity, water plentifully, and feed fortnightly with a dilute liquid feed when in full growth; reduce water in winter when a minimum temperature of 16°C/60°F is appropriate. Propagate by stem cuttings, air-layering.

L.amabilis (Borneo; evergreen shrub to 2m; leaves 1–3-pinnate, leaflets to 6cm, lanceolate, sparsely toothed, lustrous bronze above with a broad white central stripe, claret beneath with a translucent, central, green stripe; includes 'Splendens', the whole of which is flushed dark bronze-red); *L.coccinea* (WEST INDIAN HOLLY; Burma; evergreen shrub to 2m; leaves 2–3 pinnate, leaflets 5–10cm, elliptic to obovate, revolute, sometimes toothed; flowers pink).

leek (*Allium porrum*) a monocotyledonous biennial with long green flat leaves, white at the base, which form a tight cylinder and normally not a distinct bulb. Wild relatives occur in the eastern Mediterranean region and into western and southern Russia.

leather jacket

leek

Pests
eelworm
onion fly

Diseases
rust
white rot

The leek is adapted to a temperate climate and is more cold-tolerant than the onion, although it also is vernalized by low temperature and consequently bolts to become unusable. Cool conditions favour growth, and day temperatures greater than 24°C/75°F may reduce yields.

Leeks are mostly grown as a hardy winter vegetable but they can be harvested from late summer until late spring; they are used as a cooked vegetable and in soups. They grow best in a deep soil worked with well-rotted organic manure and with a pH of 6.5–7.5. Germination is optimum at soil temperatures between 11–23°C/52–73°F but drastically reduced below 7°C/45°F and above 27°C/80°F.

Earliest crops are raised in gentle heat under protection during early spring, either direct sown into cell modules or pricked out, and hardened off before planting. Early outdoor sowings can be made under cloches or in unprotected seedbeds from March to early May, in drills approximately 2cm deep. Plant out seedlings when 10–15 weeks old and about 20cm tall, in rows 30cm apart, with 15cm between plants. Higher-density plantings will give similar yields of small, more slender leeks. Seed can be sown *in situ* but weed control is more difficult and the plants are likely to have a reduced portion of blanched stem. Leaves and roots can be trimmed slightly to facilitate planting, with some loss of yield.

The length of blanched leaf base is influenced by cultivar and planting method. The simplest procedure is to drop individual plants into 15cm-deep holes made with a dibber, and water immediately. The holes will gradually fill up as the season progresses, increasing the proportion of leaf excluded from light. Alternatively, plants can be set into V-shaped drills 7.5cm deep which are later filled in, or they may be established on the flat and subsequently earthed up several times during growth. Leeks should be kept well-watered and lifted as required. Later cultivars can remain outside during the winter until late spring. Cultivars with short, thick stems are known as pot leeks and are of particular interest to exhibitors.

leg a short stem from which the branches of a shrub are trained, as on a gooseberry bush.

legume (1) the one-celled seed pod of plants of the pea family Leguminosae; (2) more generally, used to describe any plant of that family, especially peas and beans. Leguminous plants support nitrogen-fixing bacteria that give rise to root nodules, thus harnessing atmospheric nitrogen for the benefit of the host.

Lemboglossum (from Greek *lembos*, boat or canoe, and *glossa*, tongue, referring to the boat-shaped lip). Orchidaceae. C America, Mexico. Some 14 species, evergreen, epiphytic perennials formerly included in *Odontoglossum* Showy flowers are borne in racemes or panicles at various times of year. Provide cool to intermediate conditions (minimum temperature 10°C/45°F), in light shade with humid, buoyant air. Water frequently when in growth, less so at other times. See also *Odontoglossum*.

L.bictoniense (Mexico, Guatemala, El Salvador; raceme erect, to 80cm, flowers often fragrant, tepals usually pale green or yellow-green banded or spotted red-brown, sepals to 2.5cm, elliptic to oblanceolate, petals smaller, lip to 2cm, subcordate, crisped or crenulate, white to rose or magenta-tinted; white, golden and lime-green self-coloured forms occur); *L.cervantesii* (syn. *Odontoglossum cervantesii*; Mexico, Guatemala; raceme 5 to 32cm, flowers fragrant, tepals white to rose irregularly banded brown-red in basal half, sepals to 3.5cm, narrowly ovate to oblong, petals broader, lip white to rose, striped purple at base, to 2.5cm, lateral lobes erect, midlobe broadly cordate, toothed, callus yellow); *L.cordatum* (syn. *Odontoglossum cordatum*; Mexico, Guatemala, Honduras, Costa Rica, Venezuela; raceme or panicle to 60cm, tepals yellow blotched and barred deep red-brown, sepals to 5cm, elliptic to lanceolate, petals shorter, lip to 2.5cm, usually white spotted red-brown, cordate, margins slightly erose); *L.rossii* (syn. *Odontoglossum rossii*; Mexico, Guatemala, Honduras, Nicaragua; raceme to 20cm, tepals white, pale yellow or pale pink, the sepals and lower portions of petals mottled and spotted chocolate to rust, sepals to 4.5cm, oblong to elliptic or lanceolate, broader, crisped to undulate, lip to 3cm, rounded to cordate, undulate, callus deep yellow spotted red-brown); *L.stellatum* (syn. *Odontoglossum stellatum*; Mexico, Guatemala, El Salvador; raceme to 8.5cm, tepals yellow-bronze barred brown, to 3cm, linear to lanceolate, petals sometimes yellow-white, lip white or pink marked mauve, to 2cm, ovate to triangular or suborbicular, lacerate to dentate).

lemma the lower and stouter of the two glumes that immediately enclose the floret in most Gramineae; also referred to as the flowering glume.

lenticel an elliptical and raised cellular pore on the surface of bark, or the suberous tissue of fruit, through which gases can penetrate.

lenticellate possessing lenticels.

lenticular, lenticulate lens-shaped; almost flattened and elliptical, but with both sides convex.

lentiginous minutely dotted, as if with dust.

Leonotis (from Greek *leon*, lion, and *ous*, ear, alluding to the appearance of the hair-fringed corolla lip). Labiatae. 1 species pantropical, the others Southern Africa. LION'S EAR. 30 species, aromatic, annual or perennial herbs and subshrubs with tetragonal stems, toothed leaves and bearded, tubular and 2-lipped flowers in terminal whorls from late summer to early winter. *L.leonurus* tolerates temperatures down to freezing. In cool climates, it is cultivated as a conser-

Lemboglossum bictoniense

Lemboglossum cervantesii

vatory specimen or treated as a tender biennial; plants in their second year make striking specimens for the summer border. Grow in full sun, in a high-fertility loam-based mix. Water pot-grown plants plentifully when in full growth, sparingly at other times. Overwinter in bright, cool but frost-free conditions. Cut back in early spring. Propagate by seed sown in early spring under glass, or from greenwood cuttings in late spring. *L.leonurus* (South Africa; downy, evergreen shrub to 2m; flowers to 6cm, orange-red to scarlet; includes 'Harrismith White', with white flowers).

Leontopodium (from Greek *leon*, lion, and *pous*, foot, in reference to the shape of the flowerheads). Compositae. Eurasia (especially mountains), possibly also Andes. EDELWEISS. About 35 species, perennial herbs, usually low-growing and forming clumps of narrow, grey-hairy leaves. In spring and summer, they produce erect, slender-stalked cymes of felty, button-like flowerheads surrounded by a star-shaped arrangement of lanceolate, usually whitewoolly bracts. Hardy alpines, suitable for the rock garden, trough and raised bed. Grow in full sun in any gritty, perfectly drained, not too fertile, circumneutral or alkaline soil; in areas with damp winters, provide shelter from wet and winds. Propagate by seed sown fresh or by division.

L.alpinum (EDELWEISS; mountains of Europe; loosely mat-forming perennial; inflorescence stalks to 20cm tall, crowned with an irregular star of spreading, linear-oblong, densely white-woolly bracts, the whole to 10cm in diameter); *L.stracheyi* (Himalaya; inflorescence stalks to 50cm tall, crowned with one or two irregular stars to 6.5cm in diameter, with many, grey-hairy, or green and glandular bracts).

lepidote covered with tiny, scurfy, peltate scales.

Leptospermum (Greek *leptos*, slender, *sperma*, seed). Myrtaceae. Mostly S Australia; 1 species common in Tasmania is widespread in New Zealand, 2 are found in SE Asia. TI-TREE, TEA TREE. 79 species, evergreen shrubs or trees with smooth and flaking, fibrous or papery bark, tough and sometimes aromatic leaves and axillary flowers with five, rounded, speading and silken petals. Hardy to about −10°C/14°F but best grown in a sunny, sheltered position on a freely draining, humus-rich, neutral or acid soil. Where prolonged and hard frosts are common, grow under glass in a fertile, well-drained but moisture-retentive ericaceous medium. Water moderately when in growth, reducing as temperatures and light levels fall to keep just moist in winter with a minimum temperature of 5°C/40°F. Maintain good ventilation in winter; plunge pot-grown specimens outdoors in a sunny , warm place in summer months. Propagate by semi-ripe or softwood cuttings in summer.

L.lanigerum (Tasmania, S Australia, N, S and W Victoria; shrub or tree to 5m; leaves 0.2–1.5cm, oblong, to narrowly oblanceolate, usually grey-

downy; flowers white, 1.5cm in diameter; 'Silver Sheen': leaves silver-grey, later flowering than *L.lanigerum*, i.e. summer, not spring, hardy in zone 7, probably attributable to *L.myrtifolium*); *L.scoparium* (MANUKA, TEA TREE; N, S and W Victoria, Tasmania, New Zealand; shrub 2–4m; leaves 0.7–2cm, elliptic, or broadly lanceolate or oblanceolate, often silvery-hairy at first; flowers white or, rarely, pink or red, 0.8–1.2cm in diameter; numerous cultivars are grown with single or double flowers in shades of white, cream, pink, rose, red or cerise; these vary in habit from dwarf and bushy to tall and spire-like, and from tender to frost-hardy, with foliage ranging in shape from needle-like to broad and glossy, and in colour from darkest blood red to emerald green).

Leschenaultia (for Leschenault de la Tour, botanist of the voyage of discovery under Captain Nicolas Baudin in 1802). Goodeniaceae. Australia. Some 24 species, herbs, subshrubs or shrubs with a heath-like habit and linear leaves. Produced in small clusters in spring and summer, the flowers consist of a short tube and five, erect or spreading lobes, often winged and with 2-lobed tips. Grow in full light but with some protection from the strongest summer sun, in an acid, soilless mix low in phosphates and nitrates. A topdressing of grit or gravel is beneficial in pots or in the open garden. Water carefully and moderately in growth, sparingly at other times. Maintain good ventilation and a winter minimum of 7–10°C/45–50°F. Trim after flowering. Propagate by seed, or by greenwood or semi-ripe cuttings in sand, in a closed case with gentle bottom heat. *L.floribunda* (W Australia; shrubby erect, annual to 1m; flowers to 1.6cm blue or white).

Leontopodium alpinum

lesion a defined area of diseased tissue such as is caused by a leaf spot or canker.

Lespedeza (for Vincente Manuel de Cespedes, the Spanish governor of eastern Florida during the late 18th century at the time of the botanist Michaux's travels in that region; the name of the governor was rendered Lespedez in Michaux's *Flora*). Leguminosae. E US, E and tropical Asia, Australia. BUSH CLOVER. Some 40 species, shrubs or herbs with trifoliolate leaves and pea-like flowers in racemes, axillary fascicles or panicles in summer. Cultivate as for *Indigofera*.

L.bicolor (EZO-YAMA-HAGI; E Asia, naturalized E US; to 3m, woody-based perennial; flowers 1–12cm, purple-rose or rose-violet, loosely packed in axillary racemes or in terminal clusters; includes 'Summer Beauty', of spreading habit, to 1.6m tall, with long-lasting flowers, and 'Yakushima', dwarf, to 30cm tall, with small leaves and flowers); *L.thunbergii* (MIYAGINO-HAGL; Japan, China; perennial herb or subshrub, 1–2m; flowers to 2cm, rose-purple, in numerous pendulous racemes to 15cm grouped in terminal panicles to 80cm; includes 'Alba', with white flowers).

lettuce (*Lactuca sativa*) A cultigen possibly derived from *Lactuca serriola*, it originated in the region encompassing Asia Minor, Transcaucasia, Iran and Turkestan. It was first used as a medicinal plant and treated as a food plant as early as 4500 BC. Headed types did not appear until the 16th-century. Lettuce is widely cultivated in both temperate and tropical regions as an annual salad vegetable; in cooler regions, it is important as a greenhouse crop. It is successfully grown where mean temperature is in the range of 10–20°C/50–68°F; higher temperatures prevent heading, cause bolting and also produce a bitter leaf flavour.

The ideal site is an open area with light, well-drained and fertile soil of around 6.0 pH. The range of cultivars enables lettuce to be produced in the open throughout most of the year, and some are hardy enough to overwinter unprotected in mild areas. Cultivars can be divided into hearting and non-hearting groups. Hearting types include BUTTERHEADS (LOOSEHEADS in US), which have soft-textured, flat leaves forming a dense firm head, and CRISPHEADS, which are usually larger, later maturing, with less tendency to bolt, and crisp-textured, wrinkled leaves. The term 'Iceberg' lettuce refers to crisphead cultivars producing large heads, from which the outer leaves are removed after harvest to leave the white solid hearts, which have good keeping quality. 'Webbs' lettuce refers to some crisphead cultivars with a less dense heart and more usable outer leaves. Butterhead and crisphead cultivars may be referred to as cabbage lettuce. A third type of hearting lettuce is the COS lettuce with upright, long thick leaves usually forming a loose heart and taking longer to mature than the cabbage types. Some cultivars are described as semi-cos, producing compact, crunchy hearts of sweet flavour.

Non-hearting lettuces are mainly the so-called SALAD BOWL types, which form a rosette of leaves suitable for picking individually; alternatively, the head may be cut for use, with re-sprouting usually occurring from the stump. These cultivars are the least prone to bolting, and include highly decorative sorts with deeply indented or frilly leaves, which may be in various shades of green or red according to cultivar. CUT-AND-COME-AGAIN cultivars are sometimes referred to separately in the non-hearting category. They include old European, loose-leaf types which may be smooth or curly leaved.

It is possible to modify the growth habit of lettuce by planting it at high density to encourage the formation of leaves and inhibit head formation. The leaves can then be cut 2–3cm above soil level; the remaining stumps will produce a second crop that will be ready to harvest 4–7 weeks later. Cos lettuce are best suited for leaf production.

Lettuce may be sown direct, or in seed trays or cell modules under protection for subsequent transplanting. They do not establish well in hot weather so summer raisings are best direct-sown. Seed should be sown in drills 2cm deep at two-week intervals for continuity of supply. Sowing may take place from early spring until autumn, but germination, particu-larly of the butterhead types, is often erratic when soil temperatures are greater than 25°C/77°F. Watering will reduce soil temperature and improve germination, and other helpful devices include covering the seedbed with a reflective white material after sowing or sowing after the heat of daytime has passed. Except during midsummer, thinnings from direct sowings can be used to provide a further crop maturing approximately 10 days later.

Later crops will benefit from protection under cloches during the autumn, and for a protected winter crop, autumn sowings can either be made in the open or under glass for transplanting to unheated frames or greenhouses. Cold-hardy cultivars can be overwintered outdoors to provide an early crop during late spring; these should be sown *in situ* during late summer and thinned to approximately 8cm apart in the autumn. A final thinning should be done when growth recommences in spring to leave 30cm between plants. Overwintered crops benefit from a topdressing of nitrogen fertilizer during early spring.

Spacing of lettuce plants varies according to cultivar, but for general guidance, standard butterheads can be grown at 25cm spacings in rows 30cm apart; crispheads 30–38 × 38cm apart; and cos and 'salad-bowl' types 35 × 35cm apart. 'Cut-and-comeagain' types may be spaced as close as 2.5cm, in rows 12cm apart.

For heading lettuce, the time from planting to maturity varies from 60–80 days during the summer to 90–145 days during the cooler period of the year. Plants should be kept well-watered, particularly during the later stages of growth.

INDIAN LETTUCE (*Lactuca indica*) is an erect perennial up to 1.3m, grown for its edible leaves; it is of Chinese origin but now grown in India, Malaysia, Indonesia, the Philippines and Japan. In warm climates, seeds are sown in nursery beds or containers and transplanted at 30 × 30cm spacings. Propagation by root cuttings is possible and axillary buds may develop from the base to produce edible shoots. In temperate areas, seeds may be sown early in the year under greenhouse conditions, at temperatures ranging from 20–25°C/68–77°F; seedlings are transferred to 25–30cm pots when 10–15cm high. The first leaves may be harvested about 60 days from sowing or planting.

STEM LETTUCE (CELTUCE, ASPARAGUS LETTUCE, CHINESE LETTUCE) (*Lactuca sativa* var. *asparagina*) is widely grown in its native China, and cultivated in tropical southeast Asia. It is a non-heading form of lettuce whose immature stems and leaves are used as a cooked vegetable. The mature leaves are large, coarse and inedible and the thickened stems may grow to 1m high. In temperate climates, the crop can be grown under greenhouse conditions, with a temperature range of 20–25°C/68–77°F, or outdoors in warm sheltered sites. Seeds may be sown early in the year and seedlings transferred to 25–30cm pots, or spaced 30–35cm apart in the border, when at the 3–4-leaf stage. Harvest at 3–4 months, when the stems are 2.5cm in diameter and the plants 30cm high.

lettuce

Pests
aphids, root aphids
cutworms
leatherjackets
slugs and snails

Diseases
downy mildew
grey mould
virus

Disorder
tipburn

Leucadendron (from Greek *leukos*, white, and *dendron*, tree, alluding to the best-known species, *L.argenteum*, the silver tree). Proteaceae. South Africa. Some 80 species, shrubs or trees with tough foliage and cone-like, terminal inflorescences surrounded by showy, leaf-like bracts. Cultivate as for *Protea*.

L.argenteum (SILVER TREE; tree to 10m; leaves to 15cm, lanceolate with adpressed brilliant silvery hairs; inflorescence 12–20cm in diameter, bracts silver-grey, downy without, shining within); *L.* 'Safari Sunset' (*L.salignum* × *L.laureolum*; vigorous shrub to 2.5m; leaves to 9cm, deep green, flushed red particularly towards branch tips; inflorescence 10–20cm in diameter, bracts light copper red becoming deep wine red fading to golden yellow, smooth, glossy; foliage used in floristry).

Leucanthemella (diminutive of Greek *leukos*, white, and *anthemon*, flower). Compositae. SE Europe, E Asia. 2 species, perennial herbs with branched stems and daisy-like flowerheads in summer. Cultivate as for *Tanacetum*. *L.serotina* (SE Europe; to 1.5m, branched above; leaves to 12cm, lanceolate to oblong, base 2–4-lobed, toothed; ray florets white or red; includes 'Herbstern', with clear white ray florets and tinted yellow disc florets).

Leucanthemopsis (from *Leucanthemum*, and Greek *opsis*, appearance). Compositae. European mountains and N Africa. 6 species of dwarf, tufted perennials with pinnately lobed or dissected leaves and daisy-like flowerheads. Hardy in zone 7. Grow in sun on a free-draining soil. Protect from cold, wet conditions. Increase by stem tip cuttings or seed. *L.alpina* (Europe; tufted or mat-forming to 15cm; leaves to 4cm, ovate to spathulate, crenate to pinnatifid or palmatifid, grey-hairy or glabrous; flowerheads to 4cm in diameter, ray florets white, sometimes becoming pink, disc florets orange-yellow; includes subsp. *tomentosa*, dwarf, tomentose).

Leucanthemum (from Greek *leukos*, white, and *anthemon*, flower). Compositae. Europe, N Asia. About 25 species, annual or perennial herbs with daisy-like flowerheads in summer. The hardy border perennials grouped under *L.* × *superbum* are greatly valued for the profusion of blooms carried over long periods in summer; they are good for cutting. Grow in full sun, on moderately fertile, well-drained and moisture-retentive soils; division of clumps every second or third year will maintain flower quality, although a number of the cultivars are easily raised from seed and flower abundantly in their first year.

L. × *superbum* (*L.maximum* × *L.lacustre*; SHASTA DAISY; garden origin; perennial erect to 1m+; flowerheads to 10cm in diameter, ray florets pure white, disc florets yellow; over 50 cultivars, mostly white; the original single flowers are now mostly superseded by fringed, semi-double and double cultivars; notable single-flowered cultivars include the old 'Phyllis Smith' and the feathery 'Bishopstone'; notable semi-doubles include 'Aglaia' and the 90cm

'Esther Read'; anemone-centred doubles include 'Wirral Supreme' and 'T.E. Killin'; fully double flowers include 'Cobham Gold', with yellow central florets, and the tall, to 100cm, 'Fiona Coghill' and 'Starburst'; dwarf cultivars include 'Powis Castle' and the 45cm 'Little Silver Princess'); *L.vulgare* (OX-EYE DAISY, MOON DAISY, MARGUERITE; temperate Europe and Asia; perennial herb; to 1m, simple or branched; flowerheads 2.5–9cm in diameter, solitary or clustered, ray florets white, occasionally short or absent, disc florets yellow; includes 'Hofenkröne', erect to 60cm, with 'double', white flowerheads, and 'Hullavington', with stem and leaves blotched yellow).

Leucocoryne (from Greek *leukos*, white, and *koryne*, club, referring to the white staminodes). Liliaceae (Alliaceae). Chile. 12 species, bulbous perennial herbs, many with the characteristic smell of garlic. The grass-like leaves often wither before flowering. Funnel-shaped flowers with six segments are carried in scapose umbels in spring and early summer. Grow in direct sun, in a medium-fertility loam-based mix containing plentiful sharp sand. Provide a minimum winter temperature of 5°C/40°F. Water sparingly in spring when growth resumes. After flowering impose a dry rest period. Propagate from seed sown when ripe, or under glass in spring. Alternatively, increase by offsets. *L.ixioides* (GLORY OF THE SUN; to 45cm; flowers fragrant, tepals to 1.5cm, free portions white or deeply edged lilac to violet-blue, tube white, stamens and staminodes yellow-white).

Leucogenes (from Greek *leukos*, white, and *genea*, race). Compositae. New Zealand. 2–3 species, woody-based, tomentose perennial herbs with dense, overlapping leaves and button-like flowerheads in a dense cluster, subtended by a collar of leaves. Both species are extremely desirable for the alpine house and rock garden, with densely silver-white woolly foliage, and each flowerhead surrounded by a ray of white felted bracts. They suffer in cold, wet conditions, but may survive outdoors in protected locations in climate zone 8. Plant in crevices in the rock garden on a lean, gritty and moisture-retentive medium, in bright light but with shade from hot sun. In the alpine house, grow in well-crocked pots in a mix of equal parts loam, leafmould and coarse sand. Keep in cool and well-ventilated conditions, avoiding hot sun. Water plentifully and carefully when in growth; keep only just moist in winter. Repot as growth recommences in spring. Propagate by softwood cuttings in moist sand in early summer, or from fresh seed sown ripe.

L.grandiceps (SOUTH ISLAND EDELWEISS; leaves to 10 × 4cm, obovate to cuneate, blunt, tomentose; flowerheads subtended by a collar of some 15, silvery-white tomentose leaves); *L.leontopodium* (NORTH ISLAND EDELWEISS; leaves to 2 × 0.5cm, linear to oblong, acute, silver-white- to yellow-tomentose; flowerheads subtended by a collar of some 20, white leaves).

Leucojum trichophyllum (left), L. vernum (middle), and L. aestivum (right)

Leucojum (from Greek *leukos*, white, and *ion*, violet, a name first applied by Theophrastus to *Matthiola* and a white-flowered bulbous plant). Amaryllidaceae. W Europe to Middle East, N Africa. SNOWFLAKE. About 10 species, bulbous perennials, resembling and related to *Galanthus*. In autumn or spring, they bear clusters of nodding, bell-shaped flowers on slender scapes. The following are fully hardy. *L.vernum* and *L.aestivum* are suited to naturalizing in damp rough grass, or for growing in damp pockets in the rock garden; the latter thrives in moist, rich, heavy soils, especially by ponds or streams, and tolerates waterlogged conditions. Grow in partial shade, or in sun where soils remain permanently moist. *L.autumnale* is suitable for the sunny border front, path edging and for warm sunny locations on the rock garden; it is one of the least demanding of the smaller species for growing in the open, given a well-drained sandy soil. Although sometimes grown in the open in warm and perfectly drained situations, *L.roseum* is also suited to cultivation in the alpine house or bulb frame, where its delicate beauty can be appreciated at close quarters, and where its requirements for hot dry dormancy in summer and protection from winter wet are more easily met. Propagate by division in spring or autumn, after flowering, or by seed sown when ripe.

L.aestivum (SUMMER SNOWFLAKE, LODDON LILY; Eurasia; scape to 60cm, flowers produced in spring, 2–7 per scape, faintly chocolate-scented, perianth segments broadly oblong, to 2cm, white, marked green just below apex); *L.autumnale* (SW Europe, N Africa; scape slender, green- to red-brown, usually about 15cm, flowers produced in late summer and autumn, 1–4 per scape, perianth segments oblong, 1cm, crystalline white, often flushed pink at base); *L.nicaeense* (S France; scape to 15cm, usually less, flowers 1–3 per scape, produced in spring, perianth segments oblanceolate, to 12mm, white, spreading); *L.roseum* (Corsica, Sardinia; scape to 15cm, flowers 1 per scape produced in late summer and autumn, perianth segments oblanceolate, to 1cm, pale pink, darker along median line); *L.trichophyllum* (S Portugal, SW Spain, Morocco; scape slender, to 25cm, flowers 2–4 per scape produced in winter and spring, perianth segments oblanceo-late to oblong, white or flushed pink to purple); *L.vernum* (SPRING-SNOWFLAKE; France to east central Europe, naturalized Great Britain; scape stout, to 20cm, deep green, flowers 1(–2) per scape, appearing in late winter, perianth segments broadly oblong, to 2cm, white, variably marked green or yellow just below apex).

Leucospermum (from Greek *leukos*, white, and *sperma*, seed). Proteaceae. South Africa, Zimbabwe. 46 species, small, evergreen trees or shrubs with tough leaves and showy, terminal flowerheads. These are cone-like and composed of tubular flowers with reflexed segments and long, deflexed styles. Cultivate as for *Protea*.

L.catherinae (CATHERINE'S PINCUSHION, CATHERINE-WHEEL LEUCOSPERMUM; shrub to 2.5m; inflorescence to 15cm in diameter, flowers pale orange, deepening to red-gold); *L.cordifolium* (syn. *L.nutans*; shrub to 2m; inflorescence spherical, to 12cm in diameter, flowers yellow, orange or crimson); *L.reflexum* (small erect shrub to 2m; inflorescence 4.5–6cm in diameter, flowers crimson).

Leucothoë (for the daughter of Orchamur, King of Babylon and beloved of Apollo). Ericaceae. US, S America, Himalaya, E Asia, Madagascar. Some 44 species, deciduous or evergreen shrubs with leathery leaves and small, white, tubular to ovoid flowers in terminal or axillary racemes or panicles in spring. Hardy in climate zone 6, *Leucothoë* is suitable for the peat terrace, for open situations in the woodland garden and other naturalistic areas, and for moist sites in the rock garden. *L.keiskei* makes attractive groundcover, and is also amenable to pot cultivation in the alpine house. Grow in dappled shade in moist, humus-rich, lime-free soils, and mulch annually with leafmould. Prune when necessary to restrict growth after flowering, cutting out old and weak growth at ground level. Propagate by semi-ripe cuttings in late summer in a closed case with bottom heat. Alternatively, increase by division of suckering species, or by layering.

L.axillaris (SE US; evergreen shrub about 1.5m; leaves 5–11cm, ovate to oblong, dentate in apical half, lustrous green, glabrous above, paler and sparsely hirsute beneath; 'Compacta': compact); *L.fontanesiana* (DOG-HOBBLE, DROOPING LAUREL, SWITCH IVY; SE US; evergreen shrub to 2m; branches arching, tinged red; leaves 6–16cm, oblong to lanceolate tapering finely, glabrous, ciliate, dentate, dark green, lustrous above; 'Nana', compact, low-growing; 'Rainbow', young growth crimson, leaf variegation pink and cream, later white and green; this species was treated by many as *L.walteri*); *L.keiskei* (Japan; evergreen shrub; branches erect or prostrate, slender, young shoots red, glabrous; leaves 4–9cm, narrowly ovate to broadly lanceolate, apex long and slender, margins shallowly dentate, sparsely setose beneath); *L.racemosa* (FETTER BUSH, SWEET BELLS; E US; deciduous shrub, 1–2.5m; leaves 2.5–6cm, oblong to ovate or elliptic, acute, shallowly round-toothed and finely serrate).

levelling (US: grading) a technique for producing level surfaces on a garden site by excavating banks and ridges and filling hollows; also, a method of determining heights, often for the same purpose, using a surveyor's level. Levelling land is a demanding operation which should only be undertaken where essential for the siting of an amenity, and where the slope cannot be exploited as a design feature. Terracing is a compromise option in some situations, but if the gradient is steep expensive retaining walls will be required.

Much of the levelling necessary in the course of gardening can be assessed and verified by eye, but there may sometimes be a need to use levelling devices. On sites where the variation is only slight, a spirit level, wooden pegs and a long plank are sufficient. The pegs are driven in progressively across the plot to such depths as allow the bridging plank to lie dead level, when checked with a spirit level. When all of the peg tops are in position, the soil is moved until level with the tops or with predetermined regularly spaced marks on each peg.

On steeper sites, the same principle is employed using boning rods. These are wooden stakes, each about 90cm long, with a pointed end and a short horizontal crosspiece fixed at the top to form a T-shape. The first rod is driven into the soil at the lowest point of the slope, and the second a few feet away and manipulated in such a way that its top is level with that of the first rod when checked with a bridging plank and spirit level. More boning rods can then be sighted into level positions by eyeing across three horizontal members. On the steepest gradients trenching will be necessary to accommodate insertion of the rods to the desired level. Working datum points for soil-moving can be fixed by marking the individual boning rods at equal distances from the top.

Steps should be taken to avoid burying topsoil and exposing subsoil. To this end, wherever possible, temporarily remove topsoil and carry out levelling on the subsoil.

Lewisia (for Captain Meriwether Lewis (1774–1809) of the Lewis and Clark expedition across America). Portulacaceae. Western N America. 19 species glabrous, low-growing, perennial herbs with rosettes of thinly succulent leaves and, in spring and summer, cymes or panicles of broadly funnel-shaped flowers with five or more petals. Fully hardy and suitable for the rock garden, scree, alpine house and frame. Although cold-tolerant, they will rot away in wet winter conditions and may need to be grown in the alpine house or frame. Outdoors, grow deciduous species in perfectly draining, gritty and humus-rich soils on sloping sites in sun. Site evergreen species in a north- or northeast-facing niche, ensuring shade from the hottest sun, in a circumneutral or slightly acid medium; provide a collar of grit at the neck, or grow in a vertical crevice where excess moisture will not accumulate in the rosette. Despite their alpine origins, *Lewisia* species need a fairly fertile soil to perform well and the addition of bonemeal or finely graded, well-rotted manure to the medium is recommended. In the alpine house, use well-crocked pots with an open mix of equal parts loam, leafmould and sharp sand; top-dress with a collar of grit. Water moderately and carefully when in growth, allowing the medium to become almost dry between waterings; withhold water from deciduous species as they enter dormancy and

keep warm and dry. Continue watering evergreens until late summer, then reduce water to keep almost dry in winter. Propagate by fresh seed sown in the shaded cold frame. Propagate evergreen species also by offsets in summer; establish these in a shaded frame in a gritty propagating mix before potting on.

L.columbiana (evergreen; leaves 2–10cm, narrowly oblanceolate or linear, fleshy; flowers 1–2.5cm in diameter, off-white veined pink to pink or deep pink-magenta); *L.cotyledon* (evergreen; leaves 3–4cm, spathulate, oblanceolate or obovate, slightly glaucous, fleshy; flowers 2–4cm in diameter, pink-purple with pale and dark stripes, sometimes white, cream with pink-orange stripes, apricot or yellow); *L.nevadensis* (deciduous; leaves 4–15cm, narrowly linear or linear-oblanceolate, fleshy; flowers 2–3.5cm in diameter, white or rarely pink-white); *L.rediviva* (BITTERROOT; deciduous; leaves many, 1.5–5cm, linear or clavate, subterete, withered at flowering; flowers 5–7cm in diameter, rose-pink, purple-pink or white); *L.tweedyi* (evergreen; leaves 4–8cm, loosely tufted, green, often with purple suffusion, broadly oblanceolate or obovate; flowers 4–7cm in diameter, pink-peach to yellow or rarely white).

Leycesteria (for William Leycester, chief justice of Bengal *c*1820). Caprifoliaceae. W Himalaya to SW China. 6 species, deciduous or semi-evergreen suckering shrubs with erect, clumped stems. Five-lobed, funnelform flowers are produced in pendent racemes in late spring and summer and followed by berries. *L.formosa* tolerates urban and maritime conditions, limey soils and windswept locations, and is hardy to about –15°C/5°F. *L.crocothyrsos* will tolerate temperatures to –5°C/23°F with protection from cold winds. Plant both in rich moisture-retentive soil in sun or part shade. Sow fresh seed in autumn in the cold frame, or store in damp peat and sand for sowing in early spring. Divide established clumps of *L.formosa* in autumn. Propagate *L.crocothyrsos* by semi-ripe cuttings of lateral shoots with a heel. The hollow stems of *L.formosa* are liable to be cut back by frost. It benefits anyway from being cut to the ground in spring.

L.crocothyrsos (Himalaya (Assam), N Burma; to 2.5m; flowers rich yellow in whorls of 6 on arching racemes to 10cm; fruit translucent, yellow-green); *L.formosa* (HIMALAYA HONEYSUCKLE; Himalaya, W China, E Tibet; to 2m; flowers white tinted red-purple, whorled on pendulous racemes to 10cm with wine red bracts; fruit bead-like, sea-green becoming maroon then purple-black, ripening at different speeds giving a multicoloured effect).

Leymus (from Greek *elymos*, a kind of millet). Gramineae. Northern temperate regions, 1 species Argentina. Some 40 species, rhizomatous, perennial grasses with slender, sharply pointed, glaucous leaves and narrow, straw-coloured inflorescences in summer. *L.arenarius* is often used as a sand-binder with marram grass, *Ammophila arenaria*, but is valuable in gardens as (invasive) groundcover, especially in association with other coastal natives and blue-

Lewisa rediviva

grey and silver-leaved plants. Shear over in spring for a flush of fresh blue-green growth. Fully hardy, it needs full sun and a sandy or gritty, free-draining soil. Remove unwanted shoots to prevent its rapid spread. Increase by division or rooted offsets. *L.arenarius* (syn. *Elymus arenarius*; BLUE LYME GRASS, BUNCH GRASS, SAND WILD RYE, LYME GRASS, SEA LYME GRASS, EUROPEAN DUNE GRASS; N and W Europe, Eurasia; culms to 1.5m clumped and spreading by stolons; leaf blades linear, flat, to 60 × 1.5cm, tough-textured, glaucous blue-grey, margins becoming rolled in dry conditions; spikes to 25cm, slender, becoming parchment-coloured).

liana, liane a woody climbing vine.

Liatris (derivation of name obscure). Compositae. N America. BUTTON SNAKE ROOT, GAY FEATHER, BLAZING STAR, SNAKE ROOT. About 35 species, cormous or tuberous perennial herbs with clumped, narrow leaves and many, small and feathery flowerheads arranged in erect and leafy-stemmed, narrowly cylindrical spikes in summer. Fully hardy and suited to the herbaceous and cut-flower border and to more informal situations. Flowering is prolonged by removing spent blooms at the tip of the spike (they flower from the top down). Most species are tolerant of poor dry soils in cultivation but perform best on fertile, well-drained and moist soils. Tubers are prone to rot in soils that are excessively wet in winter and also make a desirable food source for various rodents. Propagate by division or seed sown ripe in autumn.

L.pycnostachya (BUTTON SNAKEROOT; SE US; to 1.5m; flowerheads red-purple, rarely white, crowded in very dense spikes to 30 × 3cm); *L.scariosa* (SE US; to 80cm; flowerheads purple, few to many, in spikes; includes 'Alba', with white flowerheads, 'Magnifica', with very large, white flowerheads, 'September Glory', to 125cm, with deep purple flowers, and 'White Spire', with flowers in long white spikes); *L.spicata* (BUTTON SNAKEWORT; E US; to 1.5m; flowerheads red-purple, clustered in a dense spike, to 70cm; includes 'Alba', with white flowerheads, 'Blue Bird', with vivid mauve-blue flowers, 'Kabold' (syn. 'Goblin'), dwarf to 40cm, with bright violet flowerheads, and 'Snow Queen', to 75cm, with snow-white flowerheads).

Libertia (for Marie A. Libert (1782–1865), Belgian student of liverworts). Iridaceae. Australia, temperate S America. 20 species, rhizomatous perennial herbs, with grassy to sword-like leaves in basal tufts or fans and panicles of small, saucer-shaped flowers in summer. Hardy to −10°C/14°F. Grow in sun or dappled shade in a moist but well-drained, slightly acid soil. In regions at the limits of its hardiness, mulch in winter with bracken litter or other organic matter. Propagate by division in spring or by seed.

L.grandiflora (New Zealand; differs from *L.ixioides* in its outer perianth segments with an olive to bronze keel, and inner segments to 1.5cm); *L.ixioides* (New Zealand, Chatham Islands; leaves 20–40cm,

linear, rigid; inflorescence a broad panicle composed of umbellate 2–10-flowered clusters, outer perianth segments white tinted brown or green, inner segments to 1cm, white); *L.peregrinans* (New Zealand; leaves to 70cm, olive-bronze to dull orange with red-tinted veins and margins; flowers to 2cm, white).

lichen a composite organism formed from the symbiotic relationship of a fungus and an alga, the green cells of the latter establishing amongst thread-like growths of the fungus. Lichens take on many forms and are commonly found as coloured crusty growths on rocks, stone walls, roofs, tree trunks and soil. In moist rural climates, they may develop abundantly on the branches of trees, but rarely thrive in industrial areas. Their presence in an area therefore provides some indication of freedom from pollution.

Lichens are slow-growing and persistent. Although attractive on stonework, they are horticulturally undesirable where occurring on lawns as gelatinous patches or as dog lichens, and where dense colonies develop on fruit trees. Control can be obtained by spraying cresylic acid on paths and tar oil on dormant deciduous trees and woody shrubs; infested lawns should be aerated and treated with lawn sand containing dichlorophen. All of these treatments should be carried out according to product recommendations.

lifting (1) the removal of a plant from the ground, prior to transplanting, division, or winter storage; (2) used of the harvesting of root and tuber vegetable crops where the use of a fork is required; (3) the heaving up of plants out of the ground following severe frost.

light (1) an essential requirement of green plants, providing energy for the manufacture of organic compounds through the process of photosynthesis. Plants vary in their need and tolerance of light intensity, and growth and especially flowering in some is influenced by the duration of daylight; see *photoperiodism*, *supplementary lighting*; (2) a movable glazed cover of a frame. The traditional English light is multi-paned and usually measures 1.8 × 1.2m; the Dutch light measures approximately 1.4 × 0.75m and is formed of one sheet of glass; (3) describing soils containing a high proportion of sand. Such soils are naturally prone to drying out and early exhaustion of nutrients, but have the advantages of being early warming and relatively easy to cultivate.

ligneous, lignose woody in texture.

lignin a deposition within cell walls of woody plants that is formed from phenolic polymers and which give rise to thickening and strengthening; the basis of woody plant structure.

lignite a kind of coal which retains abundant traces of fibrous vegatation, representing a mid stage in the conversion of peat to bituminous coal. It is sometimes added in crushed form to potting media to increase the cation exchange capacity. Lignite has the physical characteristics of sand, and its effectiveness is influenced by particle size, fine grades leading to reduced drainage and aeration. Problems may arise from its use, due to salinity. Also known as brown coal.

lignotuber the swollen woody base of some shrubs adapted to withstand fire and drought.

Ligularia (from Latin *ligula*, a little tongue, referring to the tongue-shaped ray florets). Compositae. Temperate Eurasia. LEOPARD PLANT. About 180 species, perennial herbs with long-stalked, basal leaves and, in late summer, small, button- or daisy-like flowers in tall spikes or scapose clusters. Fully hardy, *Ligularia* species demand deep, moist, fertile and humus-rich soils in full sun or dappled shade. Propagate by division.

L.dentata (syn. *L.clivorum*; China, Japan; to 1m; leaves to 30 × 40cm, reniform to orbicular, deeply cordate, dentate; flowerheads few to many, in broad, scapose heads, ray florets bright orange; includes 'Desdemona', to 120cm, with purple-bronze leaves and deep orange flowerheads, 'Othello', to 120cm, with rounded, long-stemmed, dark purple to maroon-black leaves and deep orange flowerheads, and 'Sommergold', with rich gold flowers); *L.przewalskii* (N China; to 2m; stems dark purple; leaves to 40cm in diameter, deeply palmately lobed, segments lobed or toothed; flowerheads yellow, many, small, in long, narrow racemes); *L.stenocephala* (China, Japan, Taiwan; to 1.5m; stems dark purple; leaves to 35 × 30cm, hastate to cordate or triangular, apex tapering; flowerheads yellow, many, in a long, slender raceme; includes 'The Rocket', to 180cm, with near-black stems and leafstalks, and 'Weihenstephan', to 180cm, with large, gold flowerheads).

ligulate (1) possessing a ligule; (2) strap-shaped, usually more narrowly so than in lorate.

ligule (1) a strap-shaped body, such as the limb of ray florets in Compositae; (2) the thin, scarious, sometimes hairy projection from the top of the leaf sheath in Gramineae and some other families, for example Palmae, Zingiberaceae; (3) any strap-shaped appendage; (4) an envelope sheathing the emerging foliage of Palmae.

Ligustrum (Latin name for privet). Oleaceae. Europe, N Africa, E and SE Asia, Australia. PRIVET. Some 50 species, shrubs and small trees, deciduous or evergreen, with smooth oblong or ovate leaves, panicles of small, white to cream flowers in summer and dark, fleshy fruit. The genus includes tough shrubs or small trees useful for hedging or informal screens, even in part shade and on city sites. *L.ovalifolium* and its cultivar 'Aureum' have largely replaced *L.vulgare* as a popular hedging plant since they are more likely to remain evergreen, although very severe winters may kill hedges back to ground

level. More choice species, such as *L.lucidum* and *L.sinense*, are grown as specimen shrubs or trees in sunny, sheltered sites; they are valued for their foliage, soft, plume-like spires of cream or white, late-summer flowers and attractive fruits. Although drought- and shade-tolerant, hedges of *L.ovalifolium* and *L.vulgare* will make a healthier backdrop given sufficient moisture, sun and an annual application of fertilizer. Prune vigorous, well-established hedge plants two to three times annually. Old, straggly hedges may be rejuvenated by pruning hard back and watering and feeding thoroughly. Propagate evergreens from greenwood cuttings in spring or summer in a closed case or under mist, and deciduous species from hardwood cuttings in winter. Alternatively, sow seed in spring. Privet is susceptible to privet thrips, privet aphid, leaf spot, and lilac leaf miner.

L.japonicum (Japan, Korea; evergreen shrub to 4m; leaves to 10cm, glabrous, dark green, broadly ovate, often with red-green margins and midrib; flowers scented, in pyramidal panicle to 15cm; 'Revolutum': narrow, erect, to 1m, with small, narrow leaves; 'Rotundifolium': leaves round or broadly ovate, to 4cm; 'Variegatum': leaves stippled and edged white); *L.lucidum* (CHINESE PRIVET, WHITE WAX TREE; China, Korea, Japan; evergreen shrub or small tree to 10m; leaves to 10cm, ovate to elliptic, apex tapering; panicle 10–20cm; fruit blue-black; 'Aureovariegatum': leaves yellow-variegated; 'Compactum': dense, with waxy, dark green leaves; 'Latifolium': leaves very large, glossy dark green; 'Macrophyllum': leaves large; 'Tricolor': leaves marked white and yellow, flushed pink when young); *L.ovalifolium* (CALIFORNIA PRIVET, OVAL-LEAVED PRIVET; Japan; semievergreen shrub to 4m, upright; leaves 3–7cm, elliptic to ovate, blunt to acute, glossy, dark above, yellow-green beneath; flowers off-white, malodorous, many, in panicle to 10cm; 'Albomarginatum': leaves edged white; 'Argenteum': leaves margined silver; 'Aureum': leaves yellow with broad, golden margins; 'Compactum': dense and slow-growing; 'Tricolor': new growth flushed pink and leaves ultimately variegated yellow-white; and 'Variegatum', with leaves stippled pale yellow); *L.sinense* (China; deciduous shrub to 4m; leaves to 7cm, elliptic to oblong, olive green above, paler beneath, midrib downy; flowers off-white, fragrant, in panicles to 10cm; fruit claret-coloured; 'Variegatum': leaves variegated soft grey-green and white); *L. × vicaryi* (*L.ovalifolium* × *L.vulgare*; golden origin; deciduous, spreading; leaves golden); *L.vulgare* (COMMON PRIVET; N Europe, Mediterranean, N Africa, Asia Minor; distribution enlarged and confused by cultivation; deciduous shrub to 5m, upright, dense; leaves oblong to ovate or lanceolate, glabrous, dark green; flowers off-white in dense, erect panicles to 5cm, malodorous; fruit black, glossy; 'Argenteovariegatum': leaves speckled white; 'Aureovariegatum': leaves marked golden yellow; 'Aureum': leaves golden; 'Buxifolium': leaves persistent, small, ovate; 'Leucocarpum': fruit white-green; 'Xanthocarpum': fruit bright yellow; var. *italicum*: syn. 'Sempervirens', leaves evergreen).

Lilium (Latin form of Greek *leirion*) Liliaceae (Liliaceae). LILY. Temperate Northern Hemisphere. About 100 species, bulbous perennials. The bulbs are composed of fleshy scales and produce erect stems clothed with narrow leaves. The flowers are carried toward the summit of these stems in late spring and summer. They consist of 6 tepals and are erect and cup-shaped, horizontal and funnel- or bowl-shaped, or pendulous and shaped like a bell or 'turk's-cap' (i.e. tepals outspread with recurved tips).

In general terms, the following are hardy and will grow in sun or dappled shade in a well-drained, humus-rich soil. Most lilies grow in acidic soils, but some grow on alkaline substrates. The following are considered to be lime-loving or lime-tolerant: *L.amabile*, *L.bulbiferum*, *L.canadense*, *L.candidum*, *L.cernuum*, *L.chalcedonicum*, *L.davidii*, *L.hansonii*, *L.henryi*, *L.lankongense*, *L.longiflorum*, *L.martagon*, *L.monadelphum*, *L.pardalinum*, *L.pyrenaicum*, *L.regale*. Hybrids derived from these species are most probably lime-tolerant. Only a few lime-loving species, such as *L.henryi* and *L.candidum*, actually appear to deteriorate when grown on acid soils. With the exception of *L.candidum*, lilies like to be planted at a depth of about 8–10cm and on a deep, fertile soil. Bulbs may be planted on a layer of sharp sand where the subsoil is heavy. Moisture should be plentiful in late winter and spring, easing gradually in summer so that, following flowering, the soil is still moist enough for the bulb to replenish itself. The soil should never be wet. In cool-temperate areas, plant bulbs outdoors in autumn. If this is not possible, planting should be delayed until spring, the bulbs having been potted up and kept cool and moist. If the bulbs show signs of rot or other moulds, they should be washed clean and then immersed in a fungicide solution for 20 minutes.

The often-quoted advice to keep the roots in the shade while allowing the heads to be in the sun is generally sound. Many of the Asiatic species, like *L.pumilum*, will accept full sun. Most of the Oriental species and hybrids will appreciate dappled shade for some of the day; this applies also to forms of *L.martagon* and Western American species. The Caucasian lilies will take full sun as in the wild but must have adequate moisture, and as will *L.candidum* and other southern European species used to the heat of sun at lower latitudes. Most lilies, other than those which are stoloniferous or stoloniform (e.g. *L.nepalense*), may be grown in pots – the Asiatic and Oriental and hybrid trumpet lilies are particularly suitable. Space the bulbs at least 2cm apart and 1cm away from the sides of the pot. The deeper the pot, the better. Loam-based mediums with good aeration, drainage and a high humus level are best, but soil-less mediums can be used if compaction is avoided, adequate moisture is always provided, and the appropriate liquid feeding carried out.

In regions with prolonged, hard winters in zones lower than 7, certain lilies should be grown under glass. These include *L.candidum*, *L.formosanum*, *L.longiflorum*, *L.nepalense**, *L.speciosum* and *L.wallichianum**. Those marked with an asterisk are stoloniferous and prefer to be grown in greenhouse borders or large tubs and boxes so that their stolons can wander. Lilies may be increased by seed, from bulbils formed in the leaf axils and bulblets formed below ground. Bulbs can also be increased by division and scaling (see *cuttings*). Affected by botrytis, basal rots and viruses. Attacked by lily beetle.

LILIUM Name	Distribution	Height	Leaves	Flowers
L.amabile L.fauriei	Korea	to 90cm	to 9cm, scattered, lanceolate, 3-veined, absent on lower part of stem, numerous towards top	1–5, turk's-cap, in a raceme, malodorous; tepals 5.5cm, red with dark purple spots, 2–3 fleshy papillae on upper surface; pollen red; stigma red-brown
Comments: Summer. var. *luteum*: flowers yellow				
L.auratum GOLDEN-RAYED LILY OF JAPAN, MOUNTAIN LILY	Japan	60–200cm	to 22×4cm, numerous, scattered, lanceolate, dark green, 5–7-veined, petiolate	6–30, 30cm wide, bowl-shaped, horizontal or slightly pendulous, in a raceme, very fragrant; tepals 12–18×4–5cm, white with yellow or crimson central streaks and crimson-spotted, fleshy papillae on basal part; pollen chocolate to red
Comments: Short-lived, requiring lime-free soil in full sun with root shade provided by low-growing plants. Summer–autumn. Crossed with *L.speciosum*, *L.japonicum* and *L.rubellum* to give rise to a wide range of cultivars - 'Apollo': central band and spots ruby-red. 'Crimson Beauty': cherry red, margin white. 'Praecox': heavily spotted; early flowering. 'Rubrum': tepals banded crimson. 'Tom Thumb': dwarf. var. *platyphyllum*: to 2.3m; leaves broadly lanceolate; flowers larger with fewer spots on tepals. var. *rubrovittatum* : tepals banded yellow at base, deep crimson at apex. var. *virginale:* to 2m; flowers white, streaked yellow, spotted pale yellow or pink; includes 'Album': flowers pure white.				
L.bakerianum L.lowii	N Burma, SW China	60–90cm	10×1.5cm, scattered on upper part of stem, linear to lanceolate, 1–3-veined	1–6, bell-shaped, pendulous; tepals white, interior spotted red-brown, exterior tinged green, 7×1.5cm; anthers orange
Comments: Summer				
L.bulbiferum L.chaixii FIRE LILY	S Europe	40–150cm	10×2cm, lanceolate, scattered, 3–7-veined	to 5, occasionally to 50, erect, cup-shaped, tepals 6–8.5×2–3cm, bright orange, tips darker, interior spotted maroon; anthers brown; pollen orange
Comments: Grows in well-drained soil in summer shade. Summer. var. *croceum*: syn. *L.croceum;* tepals orange				

LILIUM

Name	Distribution	Height	Leaves	Flowers
L.canadense MEADOW LILY, WILD MEADOW LILY	Eastern N America	to 1.5m	15×2cm, lanceolate to oblanceolate, mostly in whorls, veins 5–7	10–12, bell-shaped, pendulous, in an umbel; tepals 5–7.5×1–2.5cm, yellow, basally spotted maroon; pollen yellow to red-brown

Comments: Plant in moist but well-drained soil. Summer. 'Chocolate Chips': tepal exteriors and tips crimson, fading to orange then yellow toward the interior base, spots large. 'Fire Engine': tepals slightly recurved, exterior and interior tips crimson, fading to orange then yellow near the base; pollen burnt-orange. 'Melted Spots': flowers yellow-orange, spots dense on the basal two-thirds. 'Peaches and Pepper': flowers peach-orange, finely spotted. var. ***coccineum:*** syn. *L.canadense* var. *rubrum;* flowers red, throat yellow

Name	Distribution	Height	Leaves	Flowers
L.candidum MADONNA LILY, WHITE LILY	Balkans, E Mediterranean	1–2m	basal lvs 22×5cm, 3–5-veined, produced in autumn, retained in winter, stem leaves scattered, 7.5×1cm, lanceolate	5–20, funnel-shaped, fragrant, in a raceme; tepals 5–8×1–4cm, white, interior base yellow,; pollen yellow

Comments: Summer

Name	Distribution	Height	Leaves	Flowers
L.cernuum NODDING LILY	Korea, NE Manchuria, Russia	to 60cm	8–15×0.1–0.5cm, scattered, mostly crowded on centre third of stem, 1–3-veined, sessile	1–14, turk's-cap, 3.5cm wide, fragrant, in a pendent raceme; tepals lilac, purple, pink or occasionally white, spotted purple; pollen lilac

Comments: Summer

Name	Distribution	Height	Leaves	Flowers
L.chalcedonicum *L.heldreichii* SCARLET TURK'S-CAP LILY, RED MARTAGON OF CONSTANTINOPLE	Greece, S Albania	45–150cm	to 11.5×1.5cm, edged silver, spirally arranged, sessile, 3–5-veined	to 12, pendulous turk's-cap, 7.5cm wide, slightly scented; tepals scarlet, without spots; pollen scarlet

Comments: Summer. 'Maculatum': scarlet, spotted black.

Name	Distribution	Height	Leaves	Flowers
L.× dalhansonii (*L.hansonii*×*L.martagon* var. *cattaniae*)	garden origin	1.5–2m		turk's-cap, 3cm, malodorous; tepals, dark maroon, to pink, white or yellow, spotted darker; style purple

Comments: Summer. 'Backhouse hybrids': cream, yellow, pink to maroon with darker spots. 'Damson': plum-purple. 'Destiny': yellow, spotted brown. 'Discovery': rose-lilac, base white, tinged pink, spotted deep crimson, tips darker, exterior pink with a silvery sheen.

Name	Distribution	Height	Leaves	Flowers
L.davidii *L.thayerae*	W China	1–1.4m	6–10×0.2–0.4cm, linear, acute, dark green, 1-veined, margins finely toothed and inrolled	5–20, turk's-cap, unscented, pendulous, in a raceme, buds hairy; tepals to 8cm, vermilion spotted purple; pollen orange or scarlet

Comments: Summer. var. ***unicolor:*** syn. (*L.biondii;* stems shorter, to 1m; bulbs not stoloniferous; leaves more numerous and longer, very crowded; flowers paler, spots red, mauve or absent. var. ***willmottiae:*** syn. *L.sutchuenense, L.chinense*; stems to 2m, arching; bulbs stoloniferous; Preston Hybrids, Patterson Hybrids, Fiesta Strain and North Hybrids developed from *L.davidii*.

Name	Distribution	Height	Leaves	Flowers
L.duchartrei *L.farreri* MARBLE MARTAGON	W China	60–100cm	10.5×1.5cm, dark green above, pale green beneath, lanceolate, scattered, sessile, margins rough, veins 3–5	to 12, turk's-cap, pendulous, fragrant, in an umbel; tepals white, spotted deep purple, tube green at base, exterior white flushed purple; pollen orange

Comments: Summer

Name	Distribution	Height	Leaves	Flowers
L.formosanum	Taiwan	30–150cm	7.5–20×1cm, scattered, dark green, oblong-lanceolate, margins revolute, 3–7-veined	1–2, sometimes to 10, fragrant, funnel-shaped, horizontal; tepals 12–20cm, white, exterior flushed purple, tips recurved; anthers yellow to purple; pollen brown or yellow

Comments: Sometimes treated as a biennial. Summer–autumn. var. ***pricei:*** stems 30–60cm; flowers 1–2, more deeply coloured.

Name	Distribution	Height	Leaves	Flowers
L.grayi	E US	to 1.75m	to 5–12×1.5–3cm, lanceolate or oblong-lanceolate, sessile, in whorls	1–8, bell-shaped; tepals 6.25×0.5cm, interior light red with yellow base, spotted purple, exterior crimson, darker towards base; anthers yellow; pollen orange-brown

Comments: Summer. 'Gulliver's Thimble': bright crimson.

Name	Distribution	Height	Leaves	Flowers
L.hansonii	E Russia, Korea, Japan	to 120cm	18×4cm, dark green, oblanceolate to elliptic, 3–5-veined	3–12, turk's-cap, fragrant, pendulous; tepals 3–4×1.5cm, thick, oblanceolate, deep orange-yellow, spotted purple-brown toward base; anthers purple; pollen yellow

Comments: Summer

Name	Distribution	Height	Leaves	Flowers
L.henryi	China	1–3m	8–15×2–3cm, shiny, scattered, upper leaves ovate, sessile, basal leaves lanceolate, petiolate, 3–5-veined, crowded below flowers	4–20, turk's-cap, pendulous, in a raceme; tepals 6–8×1–2cm, orange, spotted black, lanceolate; anthers deep red

Comments: Summer. 'Citrinum': fls pale lemon-yellow, spotted brown

Name	Distribution	Height	Leaves	Flowers
L.japonicum *L.krameri, L.makinoi* BAMBOO LILY, SASA-YURI	Japan	40–90cm	15×2.5cm, scattered, lanceolate, 3–5-veined, margins rough; petiole to 1cm	1–5, fragrant, funnel-shaped, horizontal; tepals 10–15×2–3.5cm, oblanceolate to oblong, pink, occasionally white; anthers brown; pollen red or brown

Comments: Summer. 'Albomarginatum': leaves edged creamy-white. 'Album': flowers white. var. ***platyfolium:*** more vigorous, with broader leaves.

LILIUM

Name	Distribution	Height	Leaves	Flowers
L.kelloggii	W US	30–125cm	to 10×2cm, in whorls of 12 or more, sessile, lanceolate or oblanceolate, acute, 1–3-veined	to 20, pendulous, turk's-cap, fragrant, in a raceme; tepals 5.5×1cm, mauve-pink or white with dark purple spots and central yellow stripe toward base; anthers and pollen orange
Comments: Summer				
L.lancifolium **L.tigrinum** DEVIL LILY, KENTAN, TIGER LILY	E China, Japan, Korea	60–150cm	12–20×1–2cm, numerous, lanceolate, scattered, 5–7-veined, margins rough	to 40, in a raceme, pendulous, turk's-cap, to 12.5cm wide; tepals 7–10×1–2.5cm, lanceolate, interior orange, spotted deep purple; anthers orange-red to purple; pollen brown
Comments: Summer–early autumn. 'Florepleno': flowers double. 'Yellow Tiger': flowers lemon-yellow, spotted dark purple. var. *flaviflorum*: flowers yellow. var. *fortunei*: stems to 2m, densely woolly; flowers 30–50, orange-red. var. *splendens*: flowers larger, bright orange-red, spotted black.				
L.lankongense **L.forrestii**	W China	to 120cm	10×0.8cm, numerous, sessile, scattered, crowded near base, oblong-lanceolate, acute, 3–7-veined, margins rough	to 15, 5cm wide, in a raceme, pendulous, turk's-cap, fragrant; tepals 4–6.5×1–2.5cm, rose-pink, spotted purple, green central stripe; anthers purple; pollen brown
Comments: Summer				
L.leichtlinii	Japan	to 120cm	15×1cm, numerous, scattered, linear-lanceolate, 1–3-veined, margins rough	1–6, in a raceme, unscented, turk's-cap, buds white-hairy; tepals to 8×2cm, lemon yellow spotted maroon, lanceolate; anthers and pollen red-brown
Comments: Summer. var. *maximowiczii* : syn. *L.maximowiczii*, *L.pseudotigrinum*, *L.leichtlinii* var. *tigrinum;* to 2.5cm; flowers 1–12; tepals bright orange-red, spotted purple-brown; anthers and pollen red.				
L.longiflorum EASTER LILY	S Japan, Taiwan	to 1m	to 18×1.5cm, scattered, lanceolate or oblong-lanceolate	to 6, funnel-shaped, scented, in an umbel; tepals to 18cm, white; pollen yellow; stigma green
Comments: Summer. 'Albomarginatum': leaves edged white. 'Gelria': flowers white; tepals slightly recurved at apex; pollen yellow. 'Holland's Glory': flowers white, to 20cm. 'White America': flowers white, tips and nectaries green, pollen deep yellow. 'White Europe': flowers white, tepals strongly recurved, throat veined green, exterior green-white below, venation cream above, pollen lemon-yellow. var. *takeshima:* stems taller, purple-brown, flowers flushed purple outside, pollen orange.				
L.mackliniae	NE India	to 40cm	3–6×0.4–1cm, spiral, horizontal, linear-lanceolate or elliptic-lanceolate	1–6, in a raceme, pendulous, bell-shaped; tepals 5×2cm, interior rose-pink, exterior purple-pink; anthers purple; pollen yellow-orange or brown
Comments: Late spring–summer				
L.maculatum **L.elegans,** **L.thunbergianum,** **L.fortunei,** **L.batemanniae,** **L.atrosanguineum**	Japan	to 60cm	5–15×1.5cm, lanceolate to elliptic, scattered, 3–7-veined	cup-shaped, erect; tepals 8–10cm, yellow, orange or red, variably spotted
Comments: Summer. Considered by some to be *L.dauricum* × *L.concolor.* 'Alutaceum': deep apricot, spotted purple-black. 'Aureum': orange-yellow, spotted black. 'Bicolor': brilliant orange, margins bright red, spots few, faint. 'Biligulatum': deep chestnut-red. 'Sanguineum': orange-red. 'Wallacei': apricot with raised maroon spots.				
L.× marhan (*L.hansonii*×*L.martagon* var. *album*)	garden origin	to 1.5m		pendulous, turk's-cap, 5–6cm wide; tepals thick, orange-yellow, spotted red-brown
Comments: Summer				
L.martagon MARTAGON, TURK'S-CAP	NW Europe, NW Asia	to 2m	to 16×6.5cm, oblanceolate, 7–9-veined, in whorls of 8–14	to 50, in racemes, pendulous, turk's-cap, 5cm wide, fragrant; tepals 3–4.5 × 0.6–1cm, dull pink, spotted maroon; pollen yellow
Comments: Summer. var. *albiflorum*: white with pink spotting. var. *album*: white, not spotted. var. *cattaniae*: maroon, unspotted. var. *caucasicum*: lilac-pink. var. *sanguineopurpureum:* dark maroon, spotted.				
L.monadelphum **L.szovitsianum,** **L.colchicum** CAUCASIAN LILY	NE Turkey, Caucasus	100–150cm	12.5×2.5cm, lanceolate or oblanceolate, spirally arranged, 9–13-veined	1–5, occasionally to 30, fragrant, pendulous, turk's-cap; tepals 6–10×1–2cm, yellow, interior spotted purple or maroon, exterior flushed purple-brown; pollen orange-yellow
Comments: Summer				
L.nanum	Himalaya, W China	6–45cm	to 12×0.5cm, linear, scattered, 3–5-veined	solitary, pendulous, bell-shaped, scented; tepals to 1–4×0.3–1.6cm, pale pink to purple with fine dark purple or brown mottling; anthers yellow-brown
Comments: Summer. var. *flavidum:* flowers pale yellow.				
L.nepalense	Bhutan, Nepal, N India	to 1m	to 14×3cm, oblong-lanceolate, 5–7-veined	1–3, slightly pendulous, funnel-shaped, with a musky nocturnal scent; tepals to 15cm, tips twisting, green-white to green-yellow, base red-purple to dark maroon; anthers purple; pollen orange-brown
Comments: Summer. Stoloniferous and requiring a large, cool and moist rootrun in mild, shady conditions.				
L.occidentale EUREKA LILY	W US	70–200cm	to 13×1.5cm, in whorls, linear-lanceolate to lanceolate, 7-veined	1–5, occasionally 20, in a raceme, pendulous, turk's-cap, to 7cm wide; tepals 3.5–6×1.5cm, crimson with green-yellow throat, or vermilion with orange throat and brown spots; anthers purple; pollen orange-red
Comments: Summer.				

Lillium pyrenaicum (top left), *L. nepalense* (middle),
L. canadense (right)

LILIUM

Name	Distribution	Height	Leaves	Flowers
L.pardalinum LEOPARD LILY, PANTHER LILY	W US	2–3m	to 18×5cm, in whorls of to 16, elliptic or oblanceolate, glabrous, to 3-veined	to 10 in a raceme, unscented, pendulous, turk's-cap, to 9×9cm; tepals lanceolate, orange-red to crimson, spotted deep maroon toward base, some spots outlined yellow; anthers red-brown; pollen orange
Comments: Summer. 'Californicum': deep orange, spotted maroon, tips scarlet. 'Johnsonii': tall; flowers finely spotted. var. *giganteum*: to 2.5m; flowers crimson and yellow, densely spotted.				
L.parryi LEMON LILY	SW US	to 2m	to 15cm, oblanceolate, 3-veined, margins slightly rough	to 15, sometimes many more, fragrant, horizontal, funnel-shaped; tepals to 7–10×0.8–1.2cm, oblanceolate, tips recurved, lemon-yellow, base sparsely spotted maroon; anthers orange-brown; pollen red
Comments: Summer				
L.philadelphicum *L.montanum* WOOD LILY	Eastern N America	to 1.25m	5–10×1.5cm, mostly in whorls of 6–8, oblanceolate, margins sometimes rough, 3–5-veined	1–5, cup-shaped, erect, in an umbel; tepals to 7.5cm, oblanceolate, orange to vivid orange-red, spotted purple, edges revolute; pollen deep red
Comments: Summer				
L.primulinum OCHRE LILY	W China, N Burma, Thailand	to 2.4m	to 15×4cm, scattered, lanceolate, 1–3-veined	2–8, occasionally 18, fragrant, funnel-shaped, pendulous; tepals 6–15×1–5cm, oblong-lanceolate, yellow, sometimes marked purple-red; pollen brown
Comments: Summer–autumn . var. *ochraceum*: flowers red-purple toward base.				
L.pumilum *L.tenuifolium,* *L.linifolium* CORAL LILY	N E Asia	15–90cm	to 10×0.3cm, scattered, sessile, linear, 1-veined	to 7–20, scented, pendulous, turk's-cap, in a raceme; buds woolly; tepals 5×5cm, oblong-lanceolate, scarlet, base sometimes dotted black; pollen scarlet
Comments: Summer. 'Golden Gleam': apricot yellow.				
L.pyrenaicum	Europe to NE Turkey and Georgia	15–135cm	to 12.5×2cm, linear-lanceolate, acute, margins sometimes silver or ciliate	to 12, pendulous, turk's-cap, 5cm wide, in a raceme; tepals green-yellow, streaked and spotted dark maroon; pollen orange-red
Comments: Summer. 'Aureum': deep yellow. ssp. *pyrenaicum*: flowers yellow, interior lined and spotted dark purple; f. *rubrum*: orange-red. ssp. *carniolicum*: to 120cm; flowers yellow, orange or red, sometimes spotted purple. var. *albanicum*: to 40cm; flowers yellow. var. *jankae*: to 80cm; flowers yellow. ssp. *ponticum*: 15–90cm; flowers deep yellow or deep orange, interior red-brown, spotted purple toward base. var. *artvinense*: flowers deep orange.				
L.regale REGAL LILY	W China	50–200cm	5–13×0.4–0.6cm, scattered, sessile, linear, 1-veined	1–25, fragrant, horizontal, funnel-shaped, in an umbel; tepals 12–15×2–4cm, white, interior of tube yellow, exterior tinted purple; anthers and pollen golden
Comments: Summer. 'Album': flowers almost pure white; anthers orange. 'Royal Gold': flowers yellow.				
L.rubellum	Japan	30–80cm	10×3.5cm, scattered, petiolate, ovate-lanceolate to oblong-elliptic, 5–7-veined	1–9, very fragrant, horizontal, funnel-shaped; tepals 7.5cm, oblanceolate to oblong, rose-pink, base sometimes spotted maroon; pollen orange-yellow
Comments: Summer				
L.speciosum	China, Japan, Taiwan	stems 120–170cm	to 18×6cm, scattered, lanceolate, petiolate, 7–9-veined	to 12, fragrant, pendulous, broadly turk's-cap, 15cm wide, in a raceme; tepals to 10×4.5cm, white, base flushed carmine, spotted pink or crimson, ovate-lanceolate, margins wavy; pollen dark red
Comments: Late summer.'Grand Commander': deep pink tinted lilac, edged white, spotted red. 'Krätzeri': white, exterior with central green stripe. ' Melpomene': deep carmine, segments edged white. 'Uchida': brilliant crimson, spotted green, tips white. var. *album*: white. var. *magnificum*: rose, spotted deep crimson, margins paler; pollen red. var. *roseum*: rose. var. *rubrum*: carmine.				
L.superbum TURK'S-CAP LILY	E US	1.5–3m	3.5–11× 0.8–2.8cm, lanceolate or elliptic, 3–7-veined, in whorls of 4–20 and scattered	to 40, pendulous, turk's-cap, in a raceme; tepals orange flushed red, spotted maroon toward green base; anthers red; pollen orange-brown
Comments: Summer				
L.× testaceum *L.isabellinum,* *L.excelsum,* *(L.candidum ×* *L.chalcedonicum)* NANKEEN LILY	garden origin	100–150cm	5–10cm, scattered, linear, margins ciliate, veins pubesc. beneath	6–12, scented, pendulous, turk's-cap; tepals to 8cm, yellow to pale orange, interior spotted red; anthers red; pollen orange
Comments: Summer				
L.tsingtauense	NE China, Korea	to 120cm	to 15×4cm, glabrous, oblanceolate, petiolate, often in 2 whorls	2–7, in a raceme, unscented, erect, bowl-shaped; tepals to 5×1.5cm, orange or vermilion with purple spotting; anthers and pollen orange
Comments: Summer. var. *carneum*: flowers red, unspotted. var. *flavum*: flowers yellow, spotted red.				
L. wallichianum	Himalaya	to 2m	25×1.2cm, scattered, linear or lanceolate	1–4, scented, horizontal, funnel-shaped, 20cm wide with basal tube to 10cm; tepals 15–30cm, interior creamy white, exterior tinged green; pollen yellow
Comments: Summer–autumn				
L.washingtonianum	W US	to 2m	to 15×3.5cm, in whorls, oblanceolate, margins wavy	to 20, scented, to 11cm wide, bowl-shaped, horizontal to ascending. in a raceme; tepals 8×2.5cm, oblanceolate, white, spotted purple at base; pollen yellow
Comments: Summer. var. *purpurascens*: flowers opening white, becoming pink then purple.				

lily beetle (*Lilioceris lilii*) 8mm-long beetle with bright red thorax and wing cases, and black head and legs; its larvae are rounded, humped and covered in wet black excrement, reaching 8mm long when fully fed. It is widely distributed in Europe, but its occurrence is localized in the UK being mostly reported in SE counties. Most *Lilium* species are susceptible, also *Fritillaria imperialis* and *F. meleagris*; other plants sometimes attacked are *Polygonatum* species, *Convallaria* species and *Nomocharis saluenensis*. Adults and larvae cause defoliation and, in heavy infestations, also feed on flowers, seed capsules and stems. The larvae normally feed in groups from the tips, while the adults create irregular holes at random.

Adults overwinter in the soil and emerge mainly in March/April to lay red eggs in clusters on the undersides of leaves. All stages can be found in summer due to progressive emergence from the soil. Control limited infestations by hand-picking adults, eggs and larvae; otherwise, spray recommended contact insecticides, and apply repeatedly.

limb (1) one of the larger branches of a tree; (2) (bot.) a broadened, flattened part of a plant organ extending from a narrower base, as in the flared upper part of a tubular or funnelform corolla.

lime any of several substances containing calcium which are used in the garden to improve fertility and reduce acidity. Strictly, the term refers to calcium oxide or quicklime, produced by heating limestone or chalk in a kiln, but this form is caustic and dangerous to handle and unsuitable for garden use. Calcium hydroxide, available as slaked or hydrated lime, is quicker-acting and safer to handle, although it can damage plant foliage. Calcium carbonate, in the form of crushed chalk or ground limestone, is widely used, and is the bulkiest and cheapest of the three forms. One kilogram of calcium oxide is as effective as 1.5kg of calcium hydroxide or 2kg of calcium carbonate, depending on the size of particles. Magnesian or dolomitic limestone provides both calcium and magnesium nutrients. Some fertilizers, such as calcium nitrate, contain lime.

Lime is important in plant nutrition as a source of calcium, and also as a means of neutralizing soil acidity, or producing an alkaline condition where this is desirable. It improves the structure of clay soils by a chemical process known as flocculation, which gives rise to the aggregation of soil particles into large, stable crumbs providing good aeration. By neutralizing soil acidity, lime dressing facilitates the activity of those soil-inhabiting bacteria that convert ammonium salts into nitrates, and also increases the availability of important plant nutrients. The beneficial activities of earthworms and nitrogen-fixing bacteria are encouraged, whilst clubroot disease and some soil-inhabiting plant pests are discouraged.

Most vegetable crops actually do best on a slightly acid soil, although for large groups of ornamentals within the family Ericaceae acid conditions are essential, and these have little or no requirement for calcium. Some plants, for example, many alpines, thrive in soils high in lime, but generally overliming can lead to nutrient deficiency in plants, especially of phosphorus, iron, manganese and other trace elements.

The degree of soil acidity is indicated by the pH level, and this should be regularly checked, particularly in vegetable gardens. Most crops do best at a pH of 6.5 on mineral soils and a pH of 5.8 on peat soils. Clay soils have a high buffering capacity and require large quantities of lime to affect pH, whereas sandy soils require less but the effect may be short-lived.

Indications of lime requirement gained from soil-testing need adjustment with respect to soil texture, organic matter content, the degree of acidity, rainfall and the evenness of incorporation. However, for garden purposes, the following general guidlines are suggested for using hydrated lime to reduce acidity to around 6.5 pH on three different soil types.

	g/m^2 of hydrated lime		
Original pH	Sandy or gravelly	Medium loam	Peat or Clay
4.5	480	690	860
5.0	300	490	590
5.5	170	280	350
6.0	100	140	300
6.5	0	0	0

Where there is a need for lime dressing in excess of $300g/m^2$, it should be applied over several seasons. Even then, split dressings are advisable for best effect, with each incorporated to a depth of 15cm. Autumn or winter dressing is recommended, and there should be an interval of at least several weeks between applications of animal manure and lime to avoid strong release of ammonia, with lime always being applied after manure. See *pH*.

lime-dot see *hydathode*.

lime-induced chlorosis interveinal or uniform chlorosis of young leaves, often followed by stunting, caused by deficiency, unavailability, or immobility of nutrients such as manganese and especially iron in soils that are high in lime.

limestone a sedimentary rock consisting chiefly of calcium carbonate, although other constituents such as magnesium carbonate and silica may also be present. It is a source of lime for application to garden soils. In water-worn form, it has been used in the making of rock gardens, a practice now condemned on the grounds of conserving the natural environment.

lime-sulphur a complex mixture of calcium polysulphides, prepared by boiling together sulphur and calcium hydroxide. It was first made at Versailles in 1851, and was an important fungicide and acaricide for many years. It has been replaced by non-phytotoxic forms of sulphur, and other fungicides.

Limnanthes (from Greek *limne*, marsh, and *anthos*, flower, referring to the habitat). Limnanthaceae. W US. MEADOW FOAM. 7 species, annual herbs with pinnately divided, leaves and dish-shaped, 5-petalled flowers in summer. A hardy annual, sowing itself in profusion where conditions suit. It is suitable for path and border edging, for the interstices of paving, for the base of walls and for other situations offering a cool, moist root run. The lightly scented flowers, carried over long periods in summer, are a good nectar source for bees. It is easily grown in sun in any moderately fertile, moisture-retentive soil. Sow seed *in situ* in spring or, in mild winter areas, in autumn. *L.douglasii* (POACHED EGG FLOWER; California, S Oregon; to 30cm; leaves pinnately cut, thinly succulent, bright green; flowers fragrant, to 2.5cm in diameter, petals shiny yellow with white margins, or entirely white; includes var. *sulphurea*, with yellow petals).

Limonium (from Greek *leimon*, meadow – salt meadows are a common habitat of these plants). Plumbaginaceae. Cosmopolitan. SEA LAVENDER, MARSH ROSEMARY, STATICE. About 150 species, perennial herbs or shrubs, rarely annuals, usually with rosettes of entire or pinnatifid leaves. The flowers are small and papery with a shortly tubular, 5-lobed corolla; they are carried in summer and autumn in spreading panicles composed of short spikelets. *L.bellidifolium* is hardy in climate zone 7 and suited to sunny, perfectly drained sites in alpine sinks and the rock garden. More robust perennials, such as *L.latifolium*, are hardy to –20°C/–4°F. As with many members of this genus, it is well-adapted to coastal gardens and dry soils. Grow in any deep, well-drained soil in full sun. Good textural contrasts may be achieved by naturalizing in gravel or shingle. The flowers of most species can be dried for decoration; they range in form and colour from the subtle and delicate sprays on fine sinuous stems, as found in *L.latifolium*, to the more extravagant and densely flowered panicles of cultivars of *L.sinuatum*, frequently seen as a florist's flower. All species are easily raised from seed. Division is possible but difficult and re-establishment may be slow. *L.latifolium* and other large perennials may be increased by root cuttings in a sandy propagating mix in late winter/early spring.

L.bellidifolium (Mediterranean and Black Sea to E England; perennial herb, 10–30cm, woody at base; flowering stem branching from base, tuberculate; flowers about 5mm, blue-violet; includes 'Filigree', with purple-blue flowers, and 'Spangle', to 1m, with small, pale blue flowers, in loose sprays); *L.latifolium* (SE and C Europe; woody-based perennial herb to 80cm; panicle to 60cm, domed; flowers about 6mm, pale violet; includes 'Violetta', with violet-blue flowers); *L.perezii* (Canary Islands; perennial subshrub to 70cm; panicle downy; flowers with a purple to blue, downy calyx and yellow corolla); *L.sinuatum* (Mediterranean; hairy perennial to 40cm, often treated as an annual; flowering stem winged, inflorescence dense; flowers with downy, white or pale violet calyx and papery, white or pink corolla, becoming purple; includes many cultivars and seed races with flowers in shades of white, yellow, gold, apricot, rose, deep pink, red, carmine, purple, lilac, sky blue and deep blue).

Linaria (from Latin *linum*, flax, referring to the flax-like habit and foliage of some species). Scrophulariaceae. Northern temperate regions, especially Europe. TOADFLAX, SPURRED SNAPDRAGON. About 100 species, annual to perennial herbs with more or less erect and clumped stems clothed with narrow leaves. Tubular, 2-lipped, spurred, snapdragon-like flowers are carried in terminal racemes in summer. Hardy in climate zones 6 and over. Grow in full sun on well-drained, rather poor and sandy soils. *L.purpurea*, with long-lived and graceful spires of delicate flowers, is suitable for the flower border, *L.alpina* for scree, rock gardens, raised beds and wall crevices. The taller species such as *L.maroccana* make good cut flowers. Mediterranean species such as *L.triornithophora* and *L.tristis* are slightly more tender, tolerating temperatures to -5°C/23°F, but may be treated as annuals or protected in winter with a mulch of bracken or evergreen prunings. Perennial species tend to be short-lived, but will self-seed, proving most persistent in perfectly drained, sunny sites. To flower *L.maroccana* in pots, overwinter autumn-sown plants in a low fertility, loam-based medium, in sunny, well-ventilated, frost-free conditions. Propagate annuals and *L.alpina* from seed sown *in situ* in spring or, in mild climates, in autumn. Sow perennials in early spring under glass or in the cold frame; some cultivars, notably *L.purpurea* 'Canon Went', come true from seed. Alternatively, increase perennials by division, or from basal or softwood cuttings in spring in the cold frame.

L.alpina (ALPINE TOADFLAX; C and S Europe; glaucous, dwarf annual, biennial or perennial; stem 5–25cm, decumbent or ascending; inflorescence short; flowers violet with yellow palate, rarely wholly yellow, white or pink; includes 'Alba', with white flowers, 'Rosea', with rose-pink flowers with an orange-yellow palate); *L.bipartita* (CLOVEN-LIP TOADFLAX; NW Africa, Portugal; annual; stem to 40cm, slender; inflorescence lax; flowers violet, lips widely diverging, palate orange; includes 'Alba', with white flowers, 'Queen of Roses', with pink flowers, and 'Splendida', with deep purple flowers); *L.genistifolia* (SE and C Europe, Asia Minor; perennial, 30–100cm, erect, branched above; leaves rigid; flowers lemon-yellow to orange, palate orange-bearded; 'Nymph': flowers cream); *L.maroccana* (Morocco, naturalized NE US; annual to 45cm, erect, branched; flowers brilliant violet-purple, palate orange to yellow with smaller paler patch; 'Carminea': bright rosy carmine; 'Diadem': large, rich violet with a white eye; Excelsior Hybrids: from white to yellow and beige, to salmon, rose-carmine and crimson, to purple and blue; 'Fairy Bride': white; 'Ruby King': deep blood red); *L.purpurea* (C Italy to Sicily, naturalized elsewhere including Great Britain; glaucous perennial; stem 20–60cm, ascending to erect, often branched above;

inflorescence slender; flowers violet tinged purple; includes 'Canon J. Went', tall, with tiny, pale pink flowers, and 'Springside White', with grey-green leaves and white flowers); *L.reticulata* (PURPLE-NET TOADFLAX; Portugal, N Africa; glaucous annual: stem 60–120cm; inflorescence short, downy; flowers deep purple, palate coppery orange or yellow with purple striations; 'Aureopurpurea': rich purple with orange or yellow palate; 'Crown Jewels': small, maroon, red, gold and orange); *L.triornithophora* (THREE-BIRDS-FLYING; W Spain, N and C Portugal; perennial glabrous and somewhat glaucous; stem 50–130cm, erect or diffuse; inflorescence lax with flowers 3(–4) per whorl, pale lavender striped purple in tube with yellow palate, tube inflated; 'Rosea': pink flowers); *L.tristis* (DULL-COLOURED LINARIA, SAD-COLOURED LINARIA; S Spain and S Portugal, NW Africa, Canary Islands; glaucous perennial; stem 10–90cm, decumbent to ascending; flowers olive to yellow tinged dull purple to brown); *L.vulgaris* (COMMON TOADFLAX, BUTTER-AND-EGGS, WILD SNAP-DRAGON; Europe except for extreme N and much of Mediterranean; perennial; stem 15–90cm, erect, simple or branched, glandular-pubescent above; inflorescence dense; flowers pale to bright yellow, palate coppery).

Lindera (for Johann Linder (1676–1723), Swedish botanist). Lauraceae. Temperate and tropical E Asia (Himalayas to Malaysia, China, Japan), N America. 80 species, aromatic, evergreen or deciduous shrubs or trees grown principally for their foliage. In spring, they produce small, yellow flowers. Where plants of both sexes are present, small drupes are produced. Grow in part shade or dappled sunlight in a moisture-retentive, fertile, lime-free soil enriched with leafmould. *L.benzoin* tolerates temperatures as low as –25°C/–13°C, and *L.obtusiloba* as low as –15°C/5°F. Position plants where they have some protection from late spring frosts. Prune to remove deadwood in spring; old leggy specimens may be rejuvenated by cutting hard back to the base if necessary, although this is best carried out over several seasons. Propagate by fresh seed sown when ripe – seed has short viability and should not be allowed to dry out before sowing. Alternatively, increase by semi-ripe cuttings in a closed case or by layering.

L.benzoin (SPICE BUSH, BENJAMIN BUSH; E US; deciduous, highly aromatic shrub to 4m; leaves 6–15cm, obovate, thin, glabrous above, entire, ciliate; fruit bright red; 'Xanthocarpa': fruit yellow); *L.obtusiloba* (Korea, China, Japan; deciduous shrub or small tree to 10m; leaves 6–12cm, ovate, apex often bluntly 3-lobed, glaucous beneath, pale gold in autumn; fruit black).

Lindheimera (for Ferdinand Jacob Lindheim (1801–1879), German botanist). Compositae. S US. 1 species, *L.texana*, TEXAS STAR, an annual herb to 65cm with pinnatifid to entire leaves. To 4cm across, the daisy-like flowerheads are clustered in corymbs in late summer and autumn; they consist of white ray florets and yellow disc florets. An undemanding, hardy annual for the cut-flower border and for native plant collections. Grow in sun in any well-drained soil. Sow *in situ* in spring or earlier under glass.

line a term used by plant breeders to describe a more or less uniform assemblage of individuals; equivalent to a cultivar.

linear slender, elongated, the margins parallel or virtually so.

lineate see *striated*.

line out to plant out young plants or hardwood cuttings in straight rows to be grown on for sale.

liner a young hardy nursery-stock plant in its first pot.

lingulate resembling a tongue.

Linnaea (for Carl Linnaeus (1707–1778), botanist). Caprifoliaceae. Circumpolar regions. TWIN-FLOWER. 1 species, *L.borealis*, a creeping, dwarf, evergreen shrub with slender, trailing and rooting branches and glossy green, rounded leaves, seldom more than 0.5cm long. Fragrant flowers appear in summer, paired and nodding, atop a delicate, red-flushed stalk to 8cm tall. They are campanulate to funnelform, 1–2cm long and white or pale candy-pink with deeper markings within the five, rounded lobes. A charming and fully hardy, if rather fragile-seeming groundcover for alpine sinks, peat beds, and the rock or woodland garden. Plant in a cool, moist but well-drained acid soil in dappled sun. Propagate by rooted runners in spring.

Linnaea borealis

Linum (Latin for flax). Linaceae. Temperate northern Hemisphere. FLAX. Some 200 species, annual, biennial or perennial herbs or subshrubs with short-lived, dish-shaped, 5-petalled flowers in terminal racemes in summer. Grow in a sheltered position in sun in any light, well-drained, moderately fertile and humus-rich soil. Sow seed of annual species *in situ* in early spring, or in autumn for early flowers in pots under glass. Overwinter at 10°C/50°F, watering carefully and sparingly until light levels and temperatures rise in spring; as roots fill their pots, water moderately and liquid feed until flower buds form. Propagate perennials by seed, cultivars by cuttings of basal shoots in spring. Perennials tend to be short-lived and should be propagated frequently. Increase shrubby species by semi-ripe cuttings in a shaded cold frame.

L.arboreum (S Aegean; glabrous perennial shrub to 1m; flowers yellow); *L.flavum* (GOLDEN FLAX; C and S Europe; erect, woody-based perennial, 30–40cm; flowers golden yellow); *L.* 'Gemmels Hybrid' (*L.campanulatum* × *L.elegans*; to 15cm; flowers rich yellow); *L.grandiflorum* (FLOWERING FLAX; N Africa; erect glabrous annual to 75cm; leaves narrow; flowers rose with a dark centre; 'Bright Eyes': ivory with a deep brown centre, 'Caeruleum': blue-purple; 'Coccineum': scarlet; 'Roseum': rose pink; and

'Rubrum': bright red); *L.narbonense* (Mediterranean; glaucous perennial to 60cm; leaves narrow; flowers azure with white eye; 'Heavenly Blue': clear vivid blue; 'Six Hills': bright blue); *L.perenne* (PERENNIAL FLAX; Europe; erect perennial, to 60cm; leaves narrow; flowers pale blue; 'Album': white; 'Caeruleum': sky blue; subsp. *alpinum*: to 30cm, with smaller flowers); *L.suffruticosum* (Spain to N Italy; woody-based perennial, 25–40cm, flowering stem procumbent; leaves narrow; flowers white, veined purple with a violet or pink centre; subsp. *salsoloides*: leaves narrower, flowers pearly-white; 'Nanum': low-growing); *L.usitatissimum* (FLAX; Asia, widespread as long-established escape in Europe and N America; erect annual, to 120cm; leaves narrow; flowers blue).

lip (1) in, for example, Labiatae, one of the two distinct corolla divisions, one, the upper, often hooded, the other, lower, often forming a flattened landing platform for pollinators; (2) a staminode or petal modified or differentiated from the others.

Liquidambar (from Latin *liquidus*, liquid, and *ambar*, amber, referring to the fragrant resin). Hamamelidaceae. N America, Eurasia, China. SWEET GUM. 4 species, medium-sized to large deciduous trees with palmately lobed leaves superficially similar to *Acer* but in alternate, not opposite, arrangement. They are shiny dark green, turning maroon, bright orange and red in autumn. The flowers are inconspicuous. *L.styraciflua*, the most commonly cultivated species, is a stately tree with a conical head and handsome, maple-like foliage; it is suitable for planting in parks, avenues and large gardens, and is hardy to at least –15°C/5°F. The best selections produce brilliant autumn tints. Other species are often not hardy below –5°C/23°F. *L.formosana* Monticola Group is slightly more cold-tolerant and shows attractive purple-hued foliage in spring. Plant on deep, fertile, well-drained but moisture-retentive soils in full sun, giving plenty of room for development. *Liquidambar* resents transplanting and, if this is essential, prepare by root pruning a year in advance. Propagate from seed sown in autumn into outdoor seed beds or from stratified seed in spring; the latter may take up to two years to germinate. Increase also by softwood cuttings in summer and by layering. Protect young plants from frost and plant out after the second year.

L.formosana (FORMOSAN GUM; S China, Taiwan; straight-trunked tree to 40m; leaves to 15cm in diameter, 3-lobed, sometimes with subsidiary lobes at base, base cordate to truncate, finely serrate, glabrous above, usually downy beneath; includes Monticola Group, cold-hardy, with large, glabrous, 3-lobed leaves, colouring beautifully in autumn); *L.orientalis* (ORIENTAL SWEET GUM; Asia Minor; slow-growing, bushy-headed, small tree to 30m in the wild; leaves usually 5-lobed, 5–7cm in diameter, lobes oblong, deep, glabrous, coarsely toothed and glandular); *L.styraciflua* (SWEET GUM, AMERICAN SWEET GUM, RED GUM; E US; tall, narrowly pyramidal tree, to 45m in the wild; leaves 5- or 7-lobed, 15cm in diameter,

lobes triangular, shining and glabrous above, hairy in vein axils beneath; 'Aurea': leaves mottled and striped yellow; 'Burgundy': deep wine red autumn colour, turning later and persisting longer than in other forms; 'Festival': erect with autumn leaf tints of yellow, peach, pink; 'Golden Treasure': with golden yellow leaf margins; 'Lane Roberts': autumn colour deep black crimson-red; 'Moonbeam': leaves variegated cream and vivid autumn colour; 'Palo Alto': fiery orange-red autumn colour; 'Pendula': branches more or less weeping from an upright bole; 'Rotundiloba': leaf lobes rounded; 'Variegata': leaves mottled yellow).

liquid feeding the use of a dilute solution of fertilizers in water for feeding plants, used especially for those grown in containers. The solution may be applied by watering can or through an irrigation system using a dilutor. Liquid feeding allows for the composition and concentration of nutrients, and the frequency of feeds, to be readily changed.

Concentrated liquid fertilizers are available as proprietary products containing various percentages of nitrogen and potassium to suit the condition of crops. Alternatively, they can be obtained by collecting liquid seeping from cowsheds or manure heaps, or by steeping a permeable sack of animal manure in a container of water. Infusions of comfrey serve a similar purpose. Liquid feeds are readily absorbed by plants and therefore quick-acting.

Liriodendron (from Greek *leirion* lily, and *dendron* tree). Magnoliaceae. Eastern N America, China, Indochina. 2 species, large deciduous trees with broadly oblong leaves, the apex cleft, truncate or retuse, the base broadly 1–2-lobed on each side. Produced in summer, the vase-shaped, solitary flowers consist of nine tepals, the outer three sepal-like, the inner six petal-like and in 2 whorls of 3, surrounding numerous, spirally arranged stamens. A graceful, upswept branch system and foliage that turns butter yellow in autumn, place *Liriodendron* in the first rank of specimen and avenue trees for parks, open spaces and larger gardens. The delicately scented, pale green flowers are usually obvious only at close quarters, but are beautiful when cut and displayed where their elegant form and bright internal markings may be appreciated. Although tolerant of temperatures to at least –15°C/5°F, flowering is more profuse in hot summer climates. Plant in a sunny, open position on deep, moisture-retentive, fertile soils. Limit young trees to a single leader as two or more are likely to incur structural weakness. Trees planted in woodland, as they would grow in habitat, tend to attain considerable heights with very straight and virtually clear trunks. Propagate from seed (a large proportion of which may be infertile) in the autumn, or, in spring, by grafting.

L.chinense (CHINESE TULIP TREE; C China, Indochina; similar to *L.tulipifera*, but to 16m; leaves more deeply lobed; flowers smaller, inner tepals to 4cm, green, veined yellow); *L.tulipifera* (TULIP TREE, YELLOW POPLAR, TULIP POPLAR, CANARY WHITE-

WOOD; eastern N America; tree to 50m+; leaves to 12cm, bright green, paler beneath; flowers pale yellow-green, banded orange near base, inner tepals 6 × 3cm; includes 'Aureomarginatum', with yellow or green-yellow leaf margins, 'Contortum', with contorted leaves and twiglets, 'Fastigiatum', of narrow-pyramidal habit, with fastigate branches, and 'Integrifolium', with leaves lacking side lobes).

Liriope (after the wood nymph Liriope). Liliaceae (Convallariaceae). Japan, China, Vietnam. LILY TURF. Some 5 species, perennial, evergreen, tufted herbs with grassy leaves and small, grape-like flowers in elongated spikes or racemes from late summer to late autumn; these are followed by black, berry-like fruits. Fairly drought-tolerant evergreen groundcover, hardy to –15°C/5°F, possibly lower with shelter from cold drying winds. Cultivation as for *Ophiopogon*.

L.muscari (China, Taiwan, Japan; to 45cm, tufted, tuberous; leaves to 60cm, narrowly strap-shaped, glossy dark green; flowers dark mauve, densely clustered on a spike-like raceme; includes numerous cultivars, tall or short, compact and clumped or freely spreading, with pale to dark green, narrow to strap-like leaves, some striped or edged silvery white to cream or gold, and with white to lilac, violet or dark mauve flowers, the racemes long and slender or congested and resembling a bunch of grapes); *L.spicata* (China, Vietnam; to 25cm, rhizomatous; leaves to 35cm, grass-like, minutely serrate; flowers pale mauve to nearly white, more distinctly tubular than in *L.muscari* and on a mauve to brown stem; includes cultivars with white and pale mauve flowers, dark green or white-striped leaves and white-marbled fruit).

Lisianthius (from Greek *lysis*, loosening, and *anthos*, flower; meaning uncertain, but referring perhaps to the slow unfurling of the silken flowers). Gentianaceae. C and S America, Caribbean. 27 species, glabrous herbs and shrubs with opposite, sometimes amplexicaul, ovate to lanceolate leaves. Showy, funnel-shaped and 5-parted flowers are produced in spring and summer in cymes, corymbs or umbels. The florist's *Lisianthius* is usually *Eustoma grandiflorum*. In temperate zones, *Lisianthius* requires the protection of the greenhouse or conservatory, with a winter minimum of 15°C/60°F. In exceptionally favoured areas, *L.nigrescens* may survive in a sheltered niche. Its trumpet-shaped, purple-black blooms last well in water and are useful in flower arrangements. Grow in a well-drained medium of 2:1:1 loam, coir and sand. Pots may be moved to a sheltered spot outdoors in summer. Sow seed in spring for flowers the following year. *L.nigrescens* (FLOR DEL MUERTO; C America; shrub to 2m; flowers to 5cm, violet to rich purple or blue-black, nodding). For *L.russellianus*, see *Eustoma russellianum* and *E. grandiflorum*.

Lithocarpus (from Greek *lithos*, stone, and *carpos*, fruit, alluding to the hard-shelled fruit). Fagaceae. SE Asia and Indonesia, 2 in Japan and 1 in western N America. TANBARK OAK. 300 species, oak-like evergreen trees with leathery, mostly entire leaves and hard-shelled, acorn-like nuts in calyx cupules. *Lithocarpus* is similar to *Quercus* but differs in the male flowers being in erect spikes (pendulous in *Quercus*) and the acorns being borne few to many, often densely, on stout, stiff spikes. Cultivate as for the smaller, evergreen *Quercus* species.

L.densiflorus (TANBARK OAK; N America; tree to 30m, leaves 5–13cm, elliptic to oblong, acute, stiff and leathery, toothed, initially downy above, white-tomentose beneath); *L.henryi* (C China; tree to 20m+; leaves 10–25cm, elliptic to oblong, finely tapering, entire, glossy above, paler beneath).

Lithodora (from Greek *lithos*, stone, and *doron*, gift). Boraginaceae. SW Europe to Asia Minor. Some 7 species, perennial shrublets and subshrubs, usually with bristly leaves and 5-lobed, funnel- to bowl-shaped flowers in cymes in spring and summer. Given good drainage in a raised bed, *L.diffusa* will survive temperatures to about –15°C/5°F; the remaining species will tolerate cold to between –5 and –10°C/14–23°F. Grow in full sun in perfectly drained but moisture-retentive soils (acid to neutral for *L.diffusa*). Trim over after flowering to maintain compactness. In the alpine house, use well-crocked clay pots with a mix of equal parts loam, leafmould and coarse sand. Water moderately when in growth and keep almost dry in winter. Propagate by seed in autumn, by soft stem cuttings, rooted in a closed case with bottom heat, or by semi-ripe cuttings.

L.diffusa (syn. *Lithospermum diffusum*; NW France to SW Europe; stem to 60cm, procumbent or straggling, bristly; flowers to 2cm, blue, sometimes purple, exterior pubescent, throat with a dense ring of long white hairs; includes white, pale blue and low-growing, deep azure cultivars, for example 'Heavenly Blue'); *L.oleifolia* (syn. *Lithospermum oleifolium*; E Pyrenees; stem to 45cm, slender, diffuse, ascending; flowers to 1.5cm, pink becoming blue, or sky blue, exterior silky, interior glabrous); *L.zahnii* (syn. *Lithospermum zahnii*; S Greece; stem to 60cm, tufted, much-branched, erect or ascending, silky-hairy; flowers to 1.2cm, white or blue, interior glabrous).

Lithophragma (from Greek *lithos* stone and *phragma*, fence or screen, referring to its habitat). Saxifragaceae. Western N America. WOODLAND STAR. 9 species, perennial herbs with rosettes of long-stalked, rounded to trifoliolate leaves arising from subterranean bulbils. Carried in erect racemes in spring, the flowers consist of five spreading petals that are basically obovate, clawed and entire to 5-lobed. Hardy woodland natives for well-drained peaty soils in light shade. They may also be grown in the alpine house: keep moist while in growth, rather dry at other times. The foliage is visible only for a short growing season, withering in summer. Propagate in early spring by division, bulbils or from seed. *L.parviflorum* (leaves 3-lobed to trifoliate; flowering stem to 50cm, petals white-pink 7–16mm, 3-cleft).

lithophyte a plant that grows on rocks or stony soil, deriving nourishment from the atmosphere rather than the soil.

Lithops (from Greek *lithos*, stone, and *opsis*, appearance). Aizoaceae. South Africa, Namibia. LIVING STONES, FLOWERING STONES, PEBBLE PLANTS. 35 species, small, succulent, stemless perennials. They consist of solitary or clumping plant bodies each formed by the fusion of two, fleshy lobes with a marked central fissure. The upper surface of each lobe, (the face), may be elliptic to reniform, concave, flat or convex, with a remarkable range of stone-like characters – cracks, pitting, wrinkles, dots, mossy hieroglyphs – or with a central panel of a different colour and usually semi-translucent, either clear or occluded to varying degrees by island-like markings or incursions of marginal coloration and veining. Solitary, daisy-like flowers are produced in summer and autumn.

Provide a minimum temperature of 10°C/50°F, full sun and a dry atmosphere at all times, possibly with light shading in the height of summer if ventilation is at all inadequate. Plant in pans or half pots containing a low-fertility, loam-based medium with large proportions of sand and fine grit. A topdressing of pea gravel and pebbles flush with the upper leaf surfaces will help maintain soil temperature and moisture, and, of course, draw attention to the plants' powers of mimicry. Water very sparingly from early summer to late autumn; keep dry during the winter and merely spray over lightly on warm spring days. Watering should only begin in the summer when the old pair of leaves is almost totally shrivelled. In the autumn, water only during periods of dry sunny weather – plants that remain too wet for too long at this time of year can split their bodies or rot. Repotting is done when growth starts in late spring. A weak low-nitrogen fertilizer can be given to well-established plants. Propagation as for *Conophytum*.

L.aucampiae (body 20–32mm, face elliptic to reniform, brick to sandy brown or ochre with sienna to green-brown dots, often joined by slender fissures, sometimes forming a semi-lucent green-brown panel with blotches and a broken margin; flowers yellow); *L.bromfieldii* (body 15–32mm, lobes equal, face elliptic to reniform, buff to grey with wrinkles sunken, irregular, dark red edged grey-green, brain-like; flowers yellow); *L.dorotheae* (body 20–30mm, convex, face rounded, dark beige or buff, panel translucent grey-green or olive with jagged to lobed edges and large blotches, sometimes reducing it to a pattern of hieroglyphs, with red dots; flowers yellow); *L.hookeri* (body to 20–30mm, face flat, reniform to elliptic, buff to pale glaucous brown, panel rich brown, filled by interlocking fleshy figures with deep dark grooves, forming a vermiculate pattern; flowers yellow; includes var. *susannae*, body 15–20mm, face pale grey, sometimes with an obscure, darker panel, markings dark, sunken, forming a broken network); *L.julii* (body 20–30mm, face reniform, flat to slightly concave, dove to dark grey, panel dark brown to olive with red dots, eroded

edges and broad blotches, leaving a pattern of loose reticulation, often with a brown stain on the inner lip of each lobe; flowers white; subsp. *fulleri*: body 12–14mm, face buff to dull grey, panel translucent grey, sometimes with pale blotches, edges eroded and lobed, with dark brown or red spots or dashes in sinuses; var. *brunnea*: body 12–14mm, face slightly convex, buff, panel olive to chocolate-brown with few, minute or many, distinct markings, edges shallowly to deeply eroded with red-brown flecks in sinuses); *L.karasmontana* (body 30–40mm, face elliptic to reniform, flat, concave or convex, rugose, dull grey or beige, panel dark brown, with an impressed pattern of faint dendritic lines, or, in some variants, pale beige with face rugose and entirely suffused pale brick-red; flowers white; var. *lericheana*: body 15–20mm, lobes near equal, buff, face rounded, rugose, pink-tinted, pale olive to dull jade-green, markings few, large, edges eroded with lobes irregular; var. *tischeri*: body 20–25mm, face broadly reniform, flat, rugose, capped pale ginger, panel dark olive to chocolate, so obscured by markings and irregular marginal incursions as to be an open pattern of dull, impressed 'veins'; subsp. *bella*: body 25–30mm, face convex, grey to buff, panel dull olive reduced to an open, jagged-edged network by broad markings and marginal incursions); *L.lesliei* (body 20–45mm, face elliptic to reniform, flat to convex, grey-green to buff to pale terracotta, sometimes capped pale gold, panel pale to dark olive, with eroded edges and dense, dotted, irregular markings forming a fine, mossy, dendritic pattern; flowers yellow; var. *mariae*: body 20–30mm, lobes grey-buff, face sandy gold, panel olive, finely and densely marked appearing minutely green-gold-speckled; subsp. *burchellii*: body pale grey or buff, panel charcoal-grey with many fine markings creating an intricate, spreading network, or virtually unmarked, characterized by radiating marginal lines with expanded tips); *L.marmorata* (body 28–30mm, face narrowly reniform, flat to convex, pale grey or beige sometimes tinted green or lilac, panel translucent, dark grey or grey-green, spreading outward from an arch on inner lip, closely and deeply marked with jagged edges, resembling a grey fissured stone; flowers white); *L.olivacea* (body to 20mm, face flat, reniform with a straight inner margin, pale grey or beige, panel translucent, olive, edge slightly eroded, clear or with a few, scattered, raised markings; flowers yellow with white centre); *L.otzeniana* (body 25–30mm, face reniform to elliptic, convex, buff, panel dull olive-brown, semi-translucent with edges appearing deeply eroded and a few broad, intramarginal markings; flowers yellow with white centre); *L.pseudotruncatella* (body 25–30mm, face broadly reniform, flat to slightly concave, grey to buff, panel olive-brown, obscured and appearing as a fine network of dots and mossy 'veins', terminating in dull red dashes, sometimes staining edges pale copper; flowers yellow; var. *elisabethae*: slightly smaller, body grey tinted lilac-blue or pink, markings dark grey, marginal dashes bright red; subsp. *archerae*:

face reniform, pale grey with a slightly darker central zone, faint, delicate radial fissures, obscure spotting and, sometimes, marginal red dots and dashes; subsp. *dendritica*: face grey with a regular network of fine, dark hieroglyph-like markings, radiating from a distinct, straight line of colour at inner margin); *L.schwantesii* (body 30–40mm, face oblong to reniform, rugose, grey or buff capped pale ginger, panel olive-grey with a network of cinnamon lines; flowers yellow; subsp. *gesneri*: body 20–25mm, lobes apex, grey-tan, capped pale brown with a network of red-brown lines; subsp. *steineckeana*: body 15mm, face semicircular, convex, grey-white suffused dirty cream, usually with a few grey-green dots, panel small, opaque, seldom present, possibly a garden hybrid; subsp. *terricolor*: body 20–25mm, face oblong-reniform, slightly convex, buff to tan, closely spotted olive or mid-green, flowers yellow, occasionally with a white centre).

Littonia (For Dr S. Litton, professor of Botany in Dublin in the mid-19th century). Liliaceae (Colchicaceae). South Africa, Arabia. 8 species tuberous, perennial climbing herbs with slender stems, tendril-tipped leaves and orange, campanulate, nodding flowers in summer. Cultivate as for *Gloriosa*. *L.modesta* (South Africa; stem to 12m, slender; leaves linear to lanceolate; tepals to 5cm, orange-yellow, lanceolate).

littoral, litoral growing on the sea-shore.

Livistona (for Patrick Murray, 17th-century Baron of Livingston, whose collection became the basis of the Edinburgh Botanic Garden). Palmae. Asia, Australasia. 28 species, shrubby or arborescent palms. The stems are solitary, erect and clothed with leaf sheaths, becoming bare and ringed or covered with leaf bases; the leaves are palmate. Elegant fan palms, they are grown outdoors as specimens in frost-free gardens or in deep pots in the greenhouse or conservatory (minimum temperature 7°C/45°F). Plant in full sun or light shade in a fertile, well-drained, acid to neutral soil. Water pot-grown specimens only sparingly in winter. Propagate by seed sown in deep containers.

L.australis (GIPPSLAND PALM, CABBAGE PALM, AUSTRALIAN PALM, AUSTRALIAN FAN PALM; east coastal Australia; stem to 25m × 30cm, spiny-fibrous at first, becoming bare; leaf blades to 1.75m, glossy, segments to 70, with drooping, cleft tips; fruit to 2cm, red-brown to black, waxy); *L.chinensis* (CHINESE FAN PALM; S Japan, Ryukyu Islands, Bonin Island, S Taiwan; stem to 12m × 30cm, swollen at base; leaf blades dull; fruit to 2.5cm, deep blue-green to grey-pink, glossy).

Lloydia (for Edward Lloyd (1660–1709), Welsh antiquary and naturalist, and keeper of the Ashmolean Museum, Oxford). Liliaceae (Liliaceae). Temperate northern Hemisphere. About 12 species, bulbous perennials with narrowly linear leaves. Produced in summer on slender stalks, the flowers consist of six spreading perianth segments. Fully hardy. Grow in light shade or bright indirect light, in a gritty, perfectly drained but moisture-retentive soil. Propagate by seed sown when ripe in the cold frame or in spring. *L.serotina* (SNOWDON LILY; temperate northern Hemisphere; to 15cm; flowers to 1.5cm, white, pale yellow at base, veined red-purple).

loam a term widely used imprecisely in horticultural literature to describe soil considered of excellent quality for plant growth; 'medium loam' generally refers to a fertile soil that is rich in fibrous organic matter, moisture retentive, easily worked, and without predominant proportions of silt, clay, sand or stones.

The term 'loam' is used more specifically in classification systems of soil texture, where it does not imply fertility but serves to describe soils with conspicuous physical characteristics, such as silt-loam, clay-loam, sandy-loam and fibrous loam.

Formulae for soil-based growing media, such as the John Innes seed and potting composts, prescribe loam as an ingredient. In this instance, it refers ideally to fibrous clay-loam soil taken, after obtaining the necessary permission, from the top 40–60cm of long-term grass sward. The turf is lifted intact and stacked, grass side down, to mature in a heap about 3m square and 2m high.

loamless compost see *soilless media*.

lobed divided into segments; the segments separated from each other by sinuses that do not reach to the base of the organ.

Lobelia (for Matthias de l'Obel (1538–1616), botanist and physician to James I of England). Campanulaceae. Tropical to temperate climates, particularly America. Some 365 species, annual or perennial herbs, shrubs and treelets. Produced in summer, usually in racemes, the flowers are tubular and bilabiate, with the lower three lobes large and spreading, and the upper 2 lobes small and recurved.

The 'annual' lobelias are small, tender and half-hardy perennials commonly grown as edging and in window boxes, with trailing kinds well-suited to hanging baskets. Valued for their (typically) rich, deep blue flowers, modern selections have extended the colour range to include pure white (*L.erinus* 'White Lady'), carmine pink (*L.* 'Rosamond') and pale lilacs, as in *L.* 'Lilac Fountain'. The hardy *L.siphilitica*, *L.cardinalis*, the more tender *L.fulgens*, and the hybrids derived from them, (*L.* × *speciosa*), include some of the most beautiful of garden perennials, bearing tall strong spikes of bloom in luminous colours in late summer. They are suited to the herbaceous border and to moist soils at stream and pond side. Although sometimes grown as marginal water plants, they are generally short-lived when submerged. The hardy *L.dortmanna* needs to grow in small, fairly shallow ponds, or at lake margins.

Lloydia serotina

L.tupa is a spectular perennial. Hardy in zone 7, it needs a sunny, warm and sheltered situation, a moist but free-draining soil and mulch protection in winter. Increase annuals by seed in spring, perennials by seed or division in spring.

L.cardinalis (CARDINAL FLOWER, INDIAN PINK; N America; short-lived perennial herb, to 90cm, strongly tinged purple-bronze; leaves to 10cm, glossy, often purple-red, narrowly ovate to oblong or lanceolate, toothed; flowers to 5cm, bright scarlet, in narrow spikes; includes 'Alba', with white flowers, and 'Rosea', with pink flowers); *L.dortmanna* (WATER LOBELIA; N America, W Europe; aquatic, glabrous, perennial herb to 60cm; leaves oblong; flowers to 2cm, pale mauve, few, pendulous, in a raceme standing clear of the water); *L.erinus* (EDGING LOBELIA, TRAILING LOBELIA; South Africa; small perennial herb, branches slender, sprawling then ascending or cascading; leaves small, dark green, glabrous, ovate to linear, toothed; flowers to 2cm, throat yellow to white, tube and limb blue to violet, in short, lax spikes; there are many cultivars and seed races including dwarf and vigorous plants, some bushy, some trailing, with single or double flowers ranging in colour from violet through blue to pink and white); *L.fulgens* (syn. *L.splendens*; Texas, Mexico; close to *L.cardinalis* but slightly hairy; leaves to 15cm, linear to lanceolate; inflorescence usually one-sided, flowers scarlet, exterior more or less downy; 'Illumination': flowers scarlet in large spikes); *L.* × *gerardii* (*L.cardinalis* × *L.siphilitica*; robust perennial, to 1.5m; flowers violet tinged with pink, to purple, lower lip with two white marks, in large racemes; 'Rosenkavalier': pure pink; 'Vedrariensis': dark violet); *L.siphilitica* (GREAT LOBELIA, BLUE CARDINAL FLOWER; E US; perennial herb to 60cm; stem erect, very leafy; leaves to 10cm, ovate to lanceolate, somewhat downy, toothed; flowers to 2.5cm, blue, in dense spikes; 'Alba': white flowers; 'Nana': compact); *L.* × *speciosa* (*L.cardinalis* × *L.fulgens* × *L.siphilitica*; erect, smooth to hispid perennial; flowers to 3cm, red or mauve tinged with purple or violet, in tall spikes; cultivars include 'Bees Flame', with bronze-red leaves and scarlet flowers, 'Brightness', with brilliant red flowers, 'Dark Crusader', with green to dark bronze leaves and blood-red flowers, 'Queen Victoria', with beetroot-red leaves and vivid red flowers, and 'Russian Princess', with tinged red leaves and bright purple flowers). *L.tupa* (Chile; perennial to 2m; stem upright, robust; leaves to 30cm, lanceolate, pale sea green, finely downy; flowers large, brick red to blood red, in terminal spikes).

lobulate possessing or bearing lobules.

Lobularia (Latin *lobulus*, referring to the small fruit). Cruciferae. North temperate regions. 5 species, annual or perennial hairy herbs with narrow leaves and small, scented, 4-petalled flowers in compact terminal racemes, in spring and summer. Hardy, low-growing, bushy annuals and short-lived perennials, commonly used in bedding and edging, as temporary groundcover, and on dry walls. They are tolerant of maritime conditions. Grow in full sun, on well-drained soils. Sow seed under glass in later winter/early spring, and harden off in the cold frame before planting out in spring. Alternatively sow thinly *in situ* in spring. Deadheading, by trimming with scissors, prolongs flowering. Plants may be attacked by crucifer downy mildew, crucifer white blister and club root. *L.maritima* (SWEET ALISON, SWEET ALYSSUM; S Europe, widely naturalized; low-growing bushy annual or short-lived perennial, 10–55cm; leaves 1–3cm, linear to ligulate, somewhat silvery-grey; flowers fragrant in compact heads, petals to 3mm, rounded, usually white; many cultivars and seed races available, including compact and taller forms and plants with small or large flowers in tones of white, pink, purple-red and mauve).

locular, loculate furnished with locules; divided into separate chambers or compartments.

locule a cavity or chamber within an ovary, anther or fruit.

loculicidal a form of dehiscence, where the capsule splits longitudinally and dorsally, directly through the capsule wall.

locus the position of a gene on a chromosome.

locusts see *grasshoppers*.

lodicule one of two, or sometimes three, minute extrastaminal scales, adpressed to the base of the ovary in most Gramineae.

loganberry see *hybrid berries*.

loggia a covered sitting or walking place attached to a building, with one side open to the air. The term is often applied to garden rooms or house extensions, even those that are completely enclosed by glass.

Lomatia (from Greek *loma*, edge, referring to the winged edge of the seeds). Proteaceae. Australasia, S America. 12 species, evergreen shrubs with entire, toothed, pinnatifid or pinnately compound leaves. Produced in summer in racemes, the flowers are obliquely tubular, with a limb splitting into 4 narrow twisted lobes. Cultivate as for *Embothrium*, but with rather more shelter – for example, at the fringes or in clearings of woodland gardens and mature shrubberies, on moist, acid soil. *L.ferruginea* (Chile, Patagonia; to 7m; leaves to 20cm, leathery, olive-green above, tomentose beneath, initially rusty-tomentose; flowers scarlet fading to olive green at tips).

loment a legume (q.v.) that is contracted between the seeds it contains, drying and splitting transversely into one-seeded segments at maturity.

long-arm pruner a pruning tool consisting of cutting blades mounted on a long pole and operated by a lever attached as a handle at the opposite end; it is used for pruning parts of plants inaccessible from ground level. It is often fitted with a saw or fruit-picking attachment, and is sometimes called a lopper.

long-day plant see *photoperiodism.*

longhorn beetles (Coleoptera: Cerambycidae) family of beetles with antennae often longer than the body. Many species are brightly coloured, and some up to 100mm long, though most are between 10–60mm in length. The larvae are soft white or yellow grubs with very small or absent legs, mostly tunnellers into dead or decaying wood.

Longhorn beetles are more common in the tropics than elsewhere but two species that are pests occur in gardens in North America. Larvae of the APPLE-TREE BORER tunnel the cambium and heartwood of trees such as apple, pear, quince, plums, cherry, mountain ash and hawthorn at ground level; the adults are up to 20mm long and olive-brown with conspicuous white stripes. The pest can be controlled by routine orchard sprays. Larvae of the RASPBERRY CANE BORER (*Oberea bimaculata*) first invade shoot tips of plants, such as raspberry, blackberry and sometimes roses, but later burrow downwards to ground level. The adults are slender, up to 12mm long, with two black spots on the yellow thorax. Damage can be reduced by cutting out and burning affected canes or shoots.

long tom see *pots and potting.*

Lonicera (for Adam Lonitzer (1528–86), German naturalist, author of a herbal (*Kreuterbüch*) much reprinted between 1557 and 1783). Caprifoliaceae. HONEYSUCKLE, WOODBINE. N Hemisphere. About 180 species, deciduous or evergreen shrubs, some bushy, others twining vigorously. Borne in clusters in leaf axils or in terminal heads, the flowers are tubular to campanulate; the 5-lobed corolla limb may be bilabiate (with a 4-lobed upper lip) or regular. The fruit is a many-seeded berry.

Lonicera has a wide range of uses in the garden. The climbing and twining species are used for covering trelliswork, walls, fences, pergolas and old tree stumps. Although they will grow in sun or part shade, in habitat the climbers tend to grow with their roots in shade and their shoots reaching up to the sun. The shrubby honeysuckles will grow in full sun or light shade. Deciduous shrubby honeysuckles include several sweetly fragrant winter-flowering species and some colourful shrubs such as *L.tatarica.* Among the evergreen shrubby species, *L.pileata* makes good, loose groundcover, while *L.nitida* is familiar as a hedging plant. Those listed below are fully hardy except for the showy climber *L.hilde-brandtiana,* which needs a minimum temperature of 7°C/45°F. Honeysuckles like a fertile, well-drained soil. Prune young plants to encourage a good framework of branches. Thereafter prune after flowering to keep within bounds and thin out old wood. *L.nitida* may be sheared throughout the growing season. Propagate by semi-ripe cuttings in summer, or by hardwood cuttings in late autumn, also by simple layering in late autumn. Attacked by aphids; many species affected by mildew.

longarm pruner

Lonicera japonica 'Halliana' *(left), L. henryi (middle), L. nitida (right)*

LONICERA

Name	Distribution	Habit	Leaves	Flowers	Fruit
L.×americana (*L.caprifolium* × *L.etrusca*)	S & SE Europe, Balkans	climber resembling *L.caprifolium*		fragrant, yellow, tinted maroon, in crowded whorls from axils of connate leaves, to 5cm, tube slender	red

Comments: Summer. 'Atrosanguinea': flowers deep red outside. 'Quercifolia': leaves lobed, sometimes with yellow margin or red-striped. 'Rubella': flowers light purple outside, buds more deeply coloured. Z6.

Name	Distribution	Habit	Leaves	Flowers	Fruit
L.×brownii (*L.sempervirens* × *L.hirsuta*) SCARLET TUMPET HONEYSUCKLE	garden origin	deciduous climber to 3m	elliptic, blue-green, somewhat hairy beneath, uppermost leaf pairs connate	orange-scarlet in heads	

Comments: Spring–summer. 'Dropmore Scarlet': vigorous; flowers long trumpet-shaped, bright scarlet, midsummer-early autumn, long lasting. 'Fuchsioides': flowers orange-scarlet. 'Plantierensis': flowers large, coral-red, tipped orange. 'Punicea': flowers orange-red outside; slow-growing. 'Youngii': flowers deep crimson. Z5.

| *L.caprifolium* ITALIAN WOODBINE; ITALIAN HONEYSUCKLE | Europe, W Asia | deciduous climber to 6m | to 10cm, obovate or oval, glaucous, especially beneath, terminal pair fused | yellow-white, pink-tinged, to 5 cm, fragrant, in 4–10-flowered whorls | orange-red |

Comments: Spring–summer. 'Pauciflora': flowers to 3cm, purple or rose outside, off-white inside. 'Praecox': leaves grey-green; flowers cream, often tinted light red, later turning yellow, early-flowering. Z5.

| *L.etrusca* | Mediterranean region | semi-evergreen climber to 4m | to 8cm, oval or obovate, glaucous, blue-green, upper leaves connate | closely packed in whorls, yellow tinted red, becoming deep yellow, to 5cm | red |

Comments: Summer. 'Donald Waterer': flowers red and white, yellowing with age, fragrant. 'Michael Rosse': flowers cream and pale yellow, later darkening, slightly fragrant. 'Superba': shoots flushed red; flowers cream, later orange, in large terminal panicles; strong-growing. Z7.

| *L. fragrantissima* WINTER-FLOWERING HONEYSUCKLE | China | erect to spreading, evergreen to deciduous shrub to 3m tall; young shoots glabrous and pruinose, thus differing from *L. standishii* | to 7.5cm, oval-elliptic, dull green above, somewhat glaucous beneath | to 2cm, cream to white, intensely fragrant, paired in axils | dull red |

Comments: Winter - early spring. Z 5.

| *L.×heckrottii* (*L.sempervirens* × *L.×americana*) | garden origin | loosely climbing to sprawling deciduous shrub | to 6cm, oblong or elliptic, glaucous beneath, uppermost pairs connate | fragrant, rich pink outside, yellow inside, 4cm, in whorls on terminal spikes | red |

Comments: Summer. Sometimes called *L.* 'American Beauty'. 'Goldflame': leaves dark green; flowers yellow inside, flushed strong purple. Z5.

| *L.henryi* | W China | evergreen or semi-deciduous climber | to 9cm, oblong to lanceolate , deep, shiny green | 2cm, maroon or yellow, usually paired | purple-black |

Comments: Summer. Z4.

| *L.hildebrandtiana* GIANT BURMESE HONEYSUCKLE; GIANT HONEYSUCKLE | China, SE Asia | evergreen climber, sometimes semi-deciduous, to 25m | to 12cm, oval or rounded, deep green above | to 16cm, creamy white, changing to gold then amber, fragrant | |

Comments: Summer. Z9.

| *L.japonica* JAPANESE HONEYSUCKLE; GOLD-AND-SILVER FLOWER | Japan, Korea, Manchuria, China; naturalized SE US | evergreen or semi-evergreen climber to 9m | to 8cm, usually smaller, oblong to elliptic, incised or lobed on juvenile or very vigorous shoots | white, becoming yellow, paired, intensely fragrant, to 4cm | blue-black |

Comments: Spring–summer. 'Aureoreticulata' ('Reticulata'): leaves small, bright green, with golden netted veins, sometimes lobed. 'Halliana': leaves rich green; flowers white, sometimes tinted red, later yellow. 'Hall's Prolific': habit climbing, to 6m high, to 3m wide; leaves ovate; flowers white, later cream to yellow. 'Purpurea': leaves tinted purple; flowers maroon outside, white inside. 'Variegata': leaves variegated yellow. Z4.

| *L. ledebourii* | W US | erect, deciduous shrub to 2m | to 12cm, ovate to oblong | to 2cm, deep orange, in clusters subtended by purple-tinged, heart-shaped bracts | black, amid persistent red bracts |

Comments: Summer. Z6

| *L. maackii* | E Asia | deciduous, erect shrub to 5m | to 8cm, narrowly obovate to broadly lanceolate, dark green | to 2cm, white then yellow, paired | dark red or black |

Comments: Spring-summer. 'Erubescens': flowers suffused pink, fruit red. Z3.

LONICERA

Name	Distribution	Habit	Leaves	Flowers	Fruit
L. morrowii	Japan	deciduous shrub to 2m	to 5cm, oblong to elliptic	to 1.3cm, downy, white turning yellow, paired	dark red, shiny
Comments: Spring-summer. Z 3					
L. nitida	SW China	twiggy evergreen shrub to 3.5m	0.5-1.5cm, ovate to elliptic or rounded, dark, glossy green	to 1cm, cream to white, paired	glossy blue-purple
Comments: Spring-summer. Commonly used for hedging and groundcover, with many cultivars, ranging in habit from tall, erect and dense to low and arching, with very small leaves in tones of dark jade, emerald, lime and gold. Z6.					
L.periclymenum WOODBINE; HONEYSUCKLE	Europe, Asia Minor, Caucasus, W Asia	deciduous climber to 4m	to 6.5cm, ovate, oval or obovate, uppermost pair separate	fragrant, red and yellow-white, to 5cm, in 3–5-whorled terminal spikes	bright red
Comments: Summer. 'Belgica': flowers white, flushed purple outside, later yellow, scented, in large clusters; fruit large, red, abundant. 'Graham Thomas': flowers large, white, later yellow tinted copper, long-lasting. 'Quercina': leaves 'oak-like', sinuately toothed to lobed, sometimes variegated white. 'Serotina', LATE DUTCH HONEYSUCKLE: leaves narrow; flowers dark purple outside, later fading, yellow inside, profuse; fruit red.					
L. pileata	China	low-growing, evergreen or semi-evergreen shrub to 1.5m	to 3cm, oblong, dark green, glossy	to 0.8cm, yellow-white, paired	translucent purple-blue
Comments: Spring-summer. Includes cultivars with bright emerald foliage, some very low-growing and excellent groundcover. Z 5					
L. x *purpusii* (*L. fragrantissima* × *L. standishii*) WINTER-FLOWERING HONEYSUCKLE	garden origin	erect, semi-deciduous shrub to 3m with arching, smooth to bristly branchlets	to 10cm, ovate to elliptic, dark green above, paler beneath	creamy white, to 2.5cm, paired or clustered, highly fragrant	red
Comments: Winter to early spring. Z6.					
L.sempervirens TRUMPET HONEYSUCKLE; CORAL HONEYSUCKLE	E & S US	vigorous evergreen climber	to 8cm, oval, deep green above, blue-green beneath, uppermost pair of leaves connate	to 5 cm, rich scarlet-orange outside, more yellow inside, in whorls	bright red
Comments: Spring–autumn. 'Magnifica': red outside, interior yellow. 'Sulphurea': bright yellow, long- and late-flowering. 'Superba': bright scarlet. Z3.					
L. standishii WINTER-FLOWERING HONEYSUCKLE	China	erect, deciduous or semi-evergreen shrub to 2m with warty and bristly branchlets	to 10cm, broadly oblong-lanceolate, downy	to 2.5cm, creamy-white to palest pink, intensely fragrant, paired	red
Comments: Winter-spring. Z6.					
L. tatarica	Russia, C Asia	erect, deciduous shrub to 4m	to 6cm, ovate to lanceolate, glaucous beneath	to 2.5cm, pink, paired	scarlet to yellow-orange
Comments: Spring-summer. Includes cultivars of tall to dwarf and compact habit, with flowers in shades of deep rosy red, pink and white, and scarlet to golden fruit. Z3.					
L.× *tellmanniana* (*L.sempervirens* × *L.tragophylla*)	garden origin	vigorous, deciduous climber	to 10cm, ovate to oblong, deep green above, pruinose beneath, upper pair fused	to 4.5cm, rich orange to coppery gold, whorled	
Comments: Summer. Z5.					
L.tragophylla	China	deciduous climber	to 14cm, oblong, glaucous, uppermost 1–3 pairs fused	to 8cm, orange to yellow, often tinted red above, whorled	red
Comments: Summer. Z6.					
L. × xylosteoides (*L.tatarica* × *L.xylosteum*)	garden origin	erect, deciduous shrub to 2m	to 6cm, broadly elliptic to obovate, blue-green, hairy	small, light red, hairy with a swollen base	yellow to red
Comments: Spring-summer. Includes low-growing and compact cultivars, some with white flowers. Z 6					
L.xylosteum FLY HONEYSUCKLE	Europe, NE Asia	erect, deciduous shrub to 3m	3-7cm, oblong-ovate to obovate, downy	to 1cm, white often tinted red or pink, carried on middle or lower half of branches	red or yellow
Comments: Summer. Z3.					

Lonicera tragophylla

L. periclymenun

L. fragrantissima

Lopezia (for Tomas Lopez (*c*1540), Spanish botanist who studied South American plants). Onagraceae. C America. 21 species, annual or perennial herbs with terminal racemes of zygomorphic, 4-petalled flowers likened to colourful insects in flight. In frost-free areas, plant in well-drained soil on a sunny site or in part-day shade. In cold areas, plant in a medium-fertility loam-based mix and grow under glass, maintaining a minimum night temperature of 10°C/50°F; water moderately, and pinch tips for bushier specimens. Propagate from seed, sown with a little bottom heat, and by cuttings in spring. *L.racemosa* (MOSQUITO FLOWER; Mexico, El Salvador; variable annual or perennial to 1.5m; petals to 1cm, white to palest lilac, or purple, pink to red or vermilion, lowermost obovate to rounded, uppermost linear to oblong).

Lophomyrtus (Greek *lophos*, crested, and *Myrtus*). Myrtaceae. New Zealand. 2 species, evergreen shrubs or small trees with leathery leaves and white, axillary flowers with four spreading petals. Cultivate as for *Myrtus*.

L.bullata (RAMA RAMA; to 5.4m; leaves to 5cm, broadly ovate to suborbicular, strongly bullate, glossy, often tinged purple-red; *L.obcordata* (ROHUTU; to 4.5m; leaves to 1.25cm, obcordate, cuneate, emarginate); *L. × ralphii* (*L.bullata × L.obcordata*; intermediate between parents; leaves puckered; flowers white tinted pink; 'Purpurea': leaves deep purple-red).

Lophophora (from Greek *lophos*, crested, and *phoros*, bearing, referring to the tufted areoles). Cactaceae. E and N Mexico and S Texas. 2 species, cacti with a turnip- or carrot-like rootstock and low, depressed-globose stems, weakly ribbed and lacking spines. Campanulate flowers arise in the woolly, sunken stem apex. The fruit are cylindric to clavate, pink or red and juicy when ripe. Grow in a gritty, circumneutral compost, in a cool, frost-free greenhouse with full sun and low humidity. Light frosts are tolerated if these plants are perfectly dry. Keep dry from mid-autumn until early spring, except for light misting on warm days in late winter. *L.williamsii* (MESCAL, PEYOTE; N and NE Mexico, S Texas; stem 3–6 × 4–11cm, blue-green, ribs obscurely tuberculate, areoles small, white-woolly; inner tepals 2.5–4mm wide, pink with pale to white margins).

lopper see *long-arm pruner*.

lopping shears a pruning tool with short stout blades on medium length handles, used for cutting stout branches. They are sometimes called loppers.

lorate strap-shaped.

Lorette pruning a method of pruning fruit trees, especially pears, developed by Louis Lorette in the early 19th century. All pruning is carried out from spring to late summer by severe cutting of side laterals to stimulate a maximum amount of fruit bud from the base of shoots. The system was developed

as relevant to vigorous seedling rootstocks, and it has little advantage on selected types which are best pruned by the renewal method.

Loropetalum (from Greek *loron* strap, and *petalon*, petal: the petals are long and narrow). Hamamelidaceae. Himalaya, China, Japan. 2 species, evergreen, downy shrubs or small trees, to 3m tall, with ovate to elliptic leaves and terminal clusters of spidery, 4-petalled flowers in winter and spring. Although tolerant of temperatures down to –5°C/23°F, *Loropetalum* flowers well only where temperatures seldom drop below 5°C/40°F. In colder regions, it may be grown in the intermediate greenhouse in a well-drained but moisture-retentive, high-fertility, loam-based medium to which additional organic matter has been added. Propagation as for *Hamamelis*. *L.chinense* (syn. *L.indicum*; India, China, Japan; shrub to 3m; leaves to 4cm, dark green, bristly-white, downy; petals 2cm, white, strap-shaped and wavy, giving a feathery appearance to whole inflorescence; 'Roseum': flowers pink).

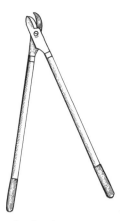

lopping shears

Lotus (from the Greek name *lotos*, used by Dioscorides and Theophrastus for certain Leguminosae). Leguminosae. Mediterranean Europe, south to Sahara Desert, W Asia; W US; Australia; South Africa. Some 100 species, annual or perennial herbs and subshrubs with palmate or pinnate leaves and axillary racemes of pea-like flowers in spring and summer. The shrubby *L.hirsutus* is hardy to –15°C/5°F. Grow in full sun on a fast-draining soil. Prune in spring. The herbaceous *L.jacobaeus* and the softly shrubby, scrambling *L.maculatus* and *L.berthelotii* are suitable for planting outdoors in gardens that are frost-free or almost so. In zones 8 and lower, these species make beautiful specimens for the cool greenhouse or conservatory (minimum temperature 5°C/40°F). Alternatively, they can be used as half-hardy bedding plants, planted out after the last frosts, and overwintered in frost-free conditions as stock plants or newly rooted cuttings. With cascading stems of filigree foliage and fiery

Lotus berthelotii

flowers, *L.maculatus* and *L.berthelotii* are among the finest plants for hanging baskets, patio containers and groundcover planting in silver and grey gardens. Grow in full sun in a sheltered site in a fast-draining soil. Keep moist when in full growth, but rather dry and in cool, airy conditions in winter. Trim after flowering. Propagate by scarified seed in spring, or by semi-ripe cuttings in summer.

L.berthelotii (CORAL GEM, PARROT'S BEAK, PELICAN'S BEAK; Canary Islands, Cape Verde Islands, Tenerife, naturalized US; cascading, climbing or prostrate perennial subshrub, to 1m; branches slender, ash-grey; leaves very short-stalked, leaflets 1cm, linear, silver-grey; flowers 2–4cm, typically scarlet but also orange-red or maroon-black, keel long, slender, beak- or claw-like; includes 'Kew Form', with silver leaves and red flowers; *L.maculatus* differs in its burnished bronze-gold and fiery red flowers); *L.hirsutus* (syn. *Dorycnium hirsutum*; HAIRY CANARY CLOVER; Mediterranean, S Portugal; densely silver-downy subshrub to 50cm; leaflets to 2cm, obovate; flowers cream flushed shell pink in umbels to 6cm; seed heads dark, attractive); *L.jacobaeus* (Cape Verde Islands; erect perennial, 30–90cm, more or less grey-downy; flowers in clusters, dark maroon-black to velvety brown except for olive-yellow standard).

low-worked used of plants grafted at a point near ground level. Also known as bottom-worked.

Ludisia (derivation of name obscure, possibly after the subject of an Ancient Greek elegy written by her widower). Orchidaceae. SE Asia, China, Indonesia. 1 species, *L.discolor* (syn. *Haemaria discolor*), an evergreen, low-growing perennial herb with creeping, fleshy rhizomes and subcordate to broadly elliptic or ovate leaves to 9cm. These are thinly fleshy velvety-papillose and chocolate to bronze to black, typically with shimmering gold to copper-red veins. Small, sparkling white and yellow flowers are carried in erect racemes to 30cm tall at various times of year. Place the rhizomes, half-buried, in a mix of leafmould, coir, coarse bark and charcoal in pans or half pots. Water freely, allowing a slight drying between each watering. Mist with soft water during warm weather. Provide medium to high humidity, temperatures in excess of 15°C/60°F (if lower, reduce watering accordingly) and shade. Repot and divide after flowering, when the flowered growths will deteriorate: increase by division of sprouting sections of rhizome.

Luma (native Chilean name). Myrtaceae. Argentina, Chile. 4 species, evergreen shrubs or small trees with leathery leaves, white, 4-petalled flowers and dark-purple berries. Hardy in sheltered locations in zone 7. Cultivate as for *Myrtus* but with protection from strong winds and on a moist but gritty, acid to neutral soil.

L.apiculata (Argentina, Chile; shrub or small tree to 10m; bark cinnamon, peeling to expose ash-grey wood; leaves 1–4.5cm, elliptic to suborbicular, apiculate; 'Glanleam Gold': leaves deeply edged creamy yellow, tinged pink at first; 'Penwith': leaves grey-green edged cream-white, tinted red-pink in winter); *L.chequen* (Chile; shrub or small tree to 9m; bark grey-brown; leaves 0.5–2.5cm, elliptic, ovate or lanceolate, rarely suborbicular, acute).

Lunaria (from Latin *luna*, moon, alluding to the shape of the pod and the moonlight-like quality of the septum). Cruciferae. C and S Europe. 3 species, biennial or perennial herbs with erect, branching stems, toothed, ovate to cordate leaves and terminal racemes of 4-petalled flowers in spring and summer. The fruit is a silique, very compressed and oblong or elliptic to nearly circular in outline. Stripped of its casing and seeds, all that is left of the fruit is a translucent, pearly white septum. These persist on the bleached branches of the inflorescence, and are popular with dried flower arrangers. Hardy to –15°C/5°F. Grow in moist fertile soils, in part shade or sun. Propagate from seed sown in autumn or spring; *L.rediviva* may also be increased by division. Both will naturalize. For dried arrangements, harvest as pods turn brown; hang to dry and allow the casings and seed to fall away.

L.annua (syn. *L.biennis*; HONESTY, SILVER DOLLAR, PENNY FLOWER; biennial to 1m; flowers unscented, to 3cm across, purple to white, sometimes white freaked or flecked purple; fruit 2–7cm, oblong to round, apiculate; includes 'Alba', with white flowers, 'Haslemere', with leaves variegated off-white and purple flowers, and 'Variegata', with leaves variegated and edged cream); *L.rediviva* (PERENNIAL HONESTY; similar to *L.annua* except perennial, flowers fragrant, fruit tapering at base and apex).

lunate crescent-shaped.

Lupinus (from Latin *lupus*, wolf, referring to the belief that these plants ravage the land, exhausting the fertility of the soil). Leguminosae. Western N America, Mediterranean, S America, S Europe, N Africa. LUPIN. Some 200 species, annual or perennial herbs or shrubs with palmate leaves and pea-like flowers in long, erect and cylindrical racemes in late spring and summer.

The most commonly cultivated perennials are those derived mainly from *L.polyphyllus* and its crosses with *L.arboreus* and several annual species. Some of the finest of these are the Russell Lupins, raised by George Russell of York over a period of 25 years and introduced in 1937. Named selections are chosen for colour and height, from dwarf races such as Garden Gnome, Minarette and Dwarf Gallery, which seldom exceed 45–60cm, to *L.* 'Band of Nobles', which may achieve 150cm. Colours range from cream and white shades through yellows in *L.* 'Chandelier', orange reds in *L.* 'Flaming June', carmine in *L.* 'The Pages' to rich violets in *L.* 'Thundercloud'; they include strong primary colours and bicolours, as in *L.* 'The Châtelaine' and *L.* 'The Governor'. More subtle shades are found in *L.* 'Blushing Bride' (ivory white), *L.* 'George Russell' (creamy pink), and *L.* 'Wheatsheaf' (golden-yellow flushed pink). Hybrids sold as mixed shades are at their most splendid when planted in groups of five or more, where fine contrasts can be achieved. Most hybrids and species are not long-lasting when cut, but will survive longer if the cut stem is filled with water before arranging. *L.poly-phyllus* and hybrids are hardy in climate zone 5. Grow in full sun, in deep, moderately fertile, well-drained soils that are slightly acid to neutral.

L.arboreus and the rather finer *L.chamissonis* are useful in the shrub border, herbaceous border and for naturalizing on poor dry soils and in maritime areas. Given good drainage and shelter from cold winds, they will tolerate temperatures to –15°C/5°F.

Sow seed in spring in pots and plant out when small. Lupins are generally intolerant of root distur-bance. Germination is quicker if seeds are pre-soaked for 24 hours in warm water. Increase also by basal cuttings, taken with a small part of the rootstock attached, and rooted in sand in the cold frame.

The fungus *Pleiochaeta setosa* causes small black-purple spots which may enlarge so that the leaves shrivel and die; spots may also occur on stems and pods. The disease can be controlled by copper-based fungicides. Lupins can also be affected by black root rot, crown gall, powdery mildew and sclerotina rot. Lupins may become severely infested with the lupin aphid. In North America, plants may be attacked by the capsid bug.

L.arboreus (TREE LUPIN; W US; evergreen shrub to 3m; leaflets lanuginose beneath, grey-green; flowers sulphur-yellow, sometimes blue or lavender, in erect, lax, terminal raceme to 25.5cm); *L.chamis-sonis* (SW USA; evergreen, slender-branched shrub to 1.5m, leaflets narrow, silver-hairy; flowers blue and cream to yellow in racemes to 10cm); *L.poly-phyllus* (California to British Columbia; stout peren-nial to 1.5m, usually unbranched; flowers blue, purple, pink or white, verticillate, in a somewhat dense raceme to 60cm; includes 'Moerheimii', with white and rose flowers); *L.subcarnosus* (TEXAS BLUEBONNET; SW US; decumbent annual to 40cm, branched at base; flowers bright blue, standard white at centre, turning purple, crowded, in several-flow-ered raceme to 12cm long). For hybrid lupins and cultivars, see above under Russell Lupins.

lute an implement used for working in topdressings, for filling hollows in lawns, and for levelling soil. Usually made in one of two forms: a length of wood about 1m wide, fixed to a handle in the fashion of a rake, or a series of shorter crosspieces, made into a frame, and fixed by a hinged bracket to the handle.

Luzula (from Italian *lucciola*). Juncaceae. Cosmopolitan, especially temperate Eurasia. WOOD-RUSH. 80 species, perennial or, rarely, annual herbs (those here perennial) with basal clumps and rosettes of grassy or sedge-like leaves, sometimes with long wavy white hairs. Brown, green or white, the incon-spicuous flowers are borne in spring amid small, chaffy bracts in stalked umbel-like corymbs. Fully hardy groundcover for moist and shaded conditions. Cultivars of *L. sylvatica* will brighten the darkest

places; they are also useful for stabilizing banks of heavy soil. The graceful *L.nivea* is less invasive and appreciates a sunny aspect. Propagate by division between autumn and spring, or by seed sown in spring and summer.

L.nivea (SNOW RUSH; Alps, C Europe; to 60cm, loosely tufted; leaves linear, to 30 × 0.4cm, flat, dark green, hairy; inflorescence loose with up to 20 clusters of off-white flowers; includes 'Little Snow Hare', with fine, bright white flowerheads with long hairs, and 'Snow Bird', with snow white flowerheads); *L.sylvatica* (syn. *L.maxima*; GREATER WOODRUSH; S, W and C Europe; to 80cm, loosely tufted, in large tus-socks; basal leaves to 30 × 2cm, channelled, smooth or with silky hairs; inflorescence spreading, with many red-brown flowers in groups of two to five; includes 'Aurea', with broad, golden-yellow leaves, 'Hohe Tatra', with smooth, golden to lime green leaves, the colour especially bright in winter, 'Marginata', with deep green leaves, edged white and brown, 'Tauernpass', with very broad, low-lying leaves, and 'Woodsman', with bright sap green leaves).

Lycaste (possibly named for Lycaste, daughter of King Priam). Orchidaceae. Mexico, C America, W Indies, S America. Some 45 species, perennial, epiphytic, terrestrial or lithophytic herbs with large, ovoid pseudobulbs bearing bold, plicate, lanceolate leaves. The flowers appear with or shortly before the new growth and are borne singly on sheathed stalks arising from the base of the pseudobulbs. They are large, waxy and fragrant, with cupped to spreading lanceolate to elliptic tepals; the lip is trilobed and the midlobe spreading to decurved, pubescent, entire

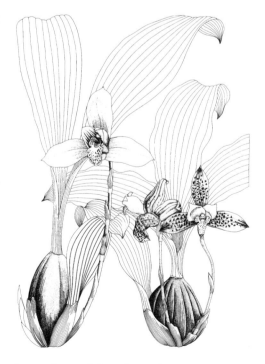

Lycaste skinneri (left), *L. deppei* (right).

to fimbriate or undulate. Pot in a mix of coarse bark, charcoal, perlite, sphagnum and leafmould. When in growth, water and feed freely and maintain humid, buoyant conditions with light shade (avoid leaf scorch); aim to promote the largest and firmest possible pseudobulbs. Once growth is complete, reduce temperature (minimum 10°C/50°F) and watering, and increase light. Increase by division.

L.aromatica (Mexico, Honduras, Guatemala; flowers scented, sepals to 4×2cm, green-yellow, petals to 3.5×2cm, deep yellow, lip to 3cm, golden yellow, dotted orange); *L.cochleata* (Guatemala, Honduras; similar to *L.aromatica*; flowers to 5cm in diameter, sepals green-yellow, petals deep orange, lip fimbriate); *L.cruenta* (Mexico, Guatemala, Costa Rica, El Salvador; flowers spicily scented, sepals to 5×2.5cm, yellow-green, petals to 4×2.5cm, bright yellow to yellow-orange, lip yellow, dotted maroon, spotted or zoned crimson at base, saccate, midlobe pubescent); *L.deppei* (Mexico, Guatemala; sepals to 6×2.5cm, pale green flecked or faintly lined oxblood to red, petals to 4.5×2cm, white, flecked red at base, lip bright yellow with red dots, base striped red, strongly veined, crenate); *L.longipetala* (Ecuador, Peru, Colombia, Venezuela; tepals yellow-green, tinged brown, sepals to 8.5×3cm, petals smaller, lip red-brown to violet-purple, denticulate or fimbriate); *L.macrophylla* (Costa Rica, Panama, Colombia, Venezuela, Peru, Bolivia; sepals olive-green, edged pink-brown, to 4×2cm, petals white, spotted rose-pink, to 3.5×2cm, lip white, margins dotted rose, midlobe spreading, ciliate); *L.skinneri* (syn. *L.virginalis*; Guatemala, Mexico, Honduras, El Salvador; sepals white to pink to violet-rose, to 8×3.5cm, petals pink to red-violet, often marked crimson, to 7.5×4cm, lip white to pale rose, flecked red-violet, to 5cm, disc pubescent, fleshy).

Lychnis (From Greek *lychnos*, lamp; the name *Lychnis* was used for these plants by Theophrastus, and refers to the use of the grey-felted leaves of *L.coronaria* as lamp-wicks). Caryophyllaceae. Widespread in the northern temperate zone; several species grown in gardens. CATCHFLY. About 20 species, mostly perennial herbs, differing from *Silene* only in the combination of 5 styles and a capsule opening with 5 teeth (most *Silene* species have 3 styles and a capsule opening with 6 teeth). *Lychnis* species range through northern temperate regions in diverse habitats, from fenland, damp meadow and moist woodlands (*L.flos-cuculi*) to rocky or alpine meadow habitats (*L.alpina*). Woolly or downy-leaved species such as *L.flos-jovis, L.coronaria* and *L.viscaria* are suited to dry borders or raised beds. *L.coronaria* self-seeds freely in gravelly soils. *L. × haageana* and *L.chalcedonica* require more moisture to thrive; the latter is sometimes used in bog- garden or streamside plantings with the roots above water level. *L. × haageana* flowers in its first year from seed and may be grown as an annual. *L.alpina* is a dwarf, mat-forming plant for the rock garden. Most species tolerate temperatures below –15°C/5°F, although

L.flos-jovis and *L. × haageana* may not survive where temperatures fall below –10°C to –15°C/14–5°F. Propagate perennials by division in autumn or spring, or from seed sown when ripe or in spring.

L.alpina (subarctic regions and mountains of northern Hemisphere; glabrous, tufted perennial to 15cm; inflorescence dense, more or less capitate, of 6–20 flowers, petals usually pale purple, deeply bifid; includes cultivars with white to rose flowers); *L.chalcedonica* (MALTESE CROSS; Europe, Russia; erect, hispid perennial to 50cm; inflorescence capitate, 10–50-flowered, petals to 1.5cm, bifid, bright scarlet; includes cultivars with white to rose, salmon and deep red flowers, some double); *L.coronaria* (DUSTY MILLER, ROSE CAMPION; SE Europe, but naturalized locally elsewhere from widespread cultivation; biennial or short-lived perennial to 80cm, covered in silver-white to grey, felty hairs; inflorescence few-flowered, flowers long-stalked, petals to 1.2cm, entire or shallowly 2-toothed, purple-red; includes cultivars with white, rose, deep carmine and double flowers); *L.flos-cuculi* (RAGGED ROBIN; Europe, also Caucasus and Siberia; sparsely hairy perennial with erect flowering stems to 75cm; flowers on slender stalks in loose cymes, petals to 1.5cm, pale purple, deeply and narrowly 4-lobed; a dwarf variant is sometimes grown in rock gardens; includes cultivars with white, pink, red and double flowers); *L.flos-jovis* (European Alps, naturalized locally elsewhere; white-tomentose perennial with erect, usually unbranched stem to 80cm; inflorescence more or less capitate 4–10-flowered, petals to 0.8cm, scarlet, bifid with broad, often cut lobes; includes 'Alba', with white flowers, and 'Nana' ('Minor'), dwarf, to 25cm, with red flowers); *L. × haageana* (short-lived, glandular-hairy perennial to 60cm; inflorescence few-flowered, petals to 2cm, broadly obovate, bifid, with a narrow tooth on each side and toothed on margins, scarlet or rich orange-red; name applied to a group of garden hybrids of uncertain parentage but clearly involving Far Eastern species, especially *L.fulgens* and *L.sieboldii*; selections from crosses between these hybrids and *L.chalcedonica*, known as *L. × arkwrightii*, include 'Vesuvius' with dark foliage and very large orange-scarlet flowers; these hybrids are best treated as summer annuals; they include 'Grandiflora', with large, red flowers, 'Hybrida', with red flowers, and 'Salmonea', with salmon pink flowers); *L.viscaria* (Europe, W Asia; glabrous or sparsely hairy perennial with stiff stem to 60cm, simple or slightly branched above, sticky below upper nodes; panicle narrow, interrupted, spike-like, petals to 1cm, entire or shallowly bifid, usually purple-red; includes cultivars with white, rose, magenta and bright red flowers, some double; subsp. *atropurpurea*: flowers deep purple).

Lycianthes (from *Lycium*, and *anthe*, flower). Solanaceae. Tropical America, E Asia. About 200 species, perennial shrubs or vines with showy, rotate and 5-lobed flowers in summer. Cultivate as for *Brugmansia*. *L.rantonnetii* (syn. *Solanum rantonnetii*; BLUE POTATO BUSH; Argentina to Paraguay;

shrub to 2m; flowers 1–2.5cm in diameter, dark blue or violet, paler blue or tinged yellow in centre; fruit to 2.5cm, red; 'Royal Robe': flowers fragrant, violet blue with yellow centres).

Lycium (from Greek *lykion*, from Lycia, Asia Minor; name of a species of *Rhamnus* transferred to this genus by Linnaeus). Solanaceae. Cosmopolitan, temperate and subtropical regions. BOXTHORN, MATRIMONY VINE. About 100 species, deciduous and evergreen shrubs, often spiny with slender branches, erect, spreading or scrambling. Small, dull white, green or purple flowers are borne in the leaf axils in spring and followed by bright red berries. Tolerant of maritime conditions and hardy to –23°C/–10°F. They will often succeed trained as wall plants or espaliers against a framework of wire. They also make good informal hedges, responding well to shearing, and providing a thorny and impenetrable barrier. Other garden uses include shrub plantings for stabilizing banks and clothing of unsightly tree stumps and retaining walls with their tangled and cascading stems. Prune in late winter or early spring to confine to the space allocated, or to thin old, crowded or weak wood; shear hedging plants two to three times in the growing season. Espaliers should be pruned after fruiting in winter or early spring. Propagate from seed sown in spring and by rooted suckers. Alternatively, increase by hardwood cuttings in autumn or spring, layering, or by semi-ripe cuttings rooted with gentle bottom heat in summer. *L.barbarum* (syn. *L.chinense*, *L.europaeum*, *L.halimiifolium*;· COMMON MATRIMONY VINE, DUKE OF ARGYLL'S TEA TREE; SE Europe to China; deciduous, erect or spreading, to 3.5m, usually spiny; leaves to 5cm, narrowly oblong to lanceolate, greygreen to dull mid-green; flowers to 1cm, in clusters of 1–4, dull lilac; fruit to 2cm, orange-red or yellow).

Lycopersicon *see* tomato.

Lycoris (for a beautiful Roman actress, the mistress of Mark Anthony). Amaryllidaceae. China, Japan. 10–12 species, bulbous perennial herbs with strapshaped leaves. The flowers are produced in late summer and autumn in scapose umbels; they have six, slender, reflexed and wavy tepals and prominent stamens. *Lycoris* species will tolerate temperatures to at least –15°C/5°F, but need a period of dry warmth during their summer dormancy and minimal winter wet. They are also amenable to pot cultivation in the cool greenhouse or conservatory. They resent root disturbance, and may take several years to become well-established; topdressing is preferable to repotting, and plants may remain in the same pot for up to 4–5 years. Propagate by offsets or from ripe seed. *L.aurea* (GOLDEN HURRICANE LILY, GOLDEN SPIDER LILY; China, Japan; scape to 60cm, flowers 9.5–10cm, golden-yellow, funnelform, lobes recurved, wavy); *L.radiata* (SPIDER LILY, RED SPIDER LILY; Japan; scape to 50cm, flowers 4–5cm, rose-red to deep red, lobes strongly reflexed, wavy; includes

'Alba', with white flowers tinged yellow at base, and 'Variegata', with crimson flowers, edged white as they fade); *L.squamigera* (MAGIC LILY, RESURRECTION LILY; Japan; scape to 70cm, flowers fragrant, 9–10cm, pale rose-pink flushed or veined lilac or purple, funnelform, lobes recurved, slightly wavy).

Lygodium (from Greek *lygodes*, flexible, twining, referring to the climbing habit). Schizaeaceae. Cosmopolitan. CLIMBING FERN. About 40 species, climbing ferns. The roots and rhizomes form a dense clump from which arise long and slender vining fronds with twining, wiry rachises, clasping rachillae and 2–3-pinnately or palmately compound pinnae. Both species listed here prefer a minimum winter temperature of 7°C/45°F, although *L.palmatum* may survive outdoors in climate zone 6, losing its fronds in winter. They need a moist, porous and fertile medium rich in leafmould and garden compost (pH 5 or lower for *L.palmatum*). Provide a cool, buoyant and humid atmosphere in light shade with support from wires, trellis, canes or surrounding vegetation. Water copiously throughout the growing season, and feed at two week intervals. Water moderately in winter. Cut out faded fronds and mulch in early winter. Propagate from spores or by division. Serpentine layering is also a possibility: an actively growing frond is pinned to the soil surface at each node.

L.japonicum (Japan to Australia; fronds bright green, divisions finely pinnately divided and 'ferny', to 20cm long); *L.palmatum* (E US; fronds dark green, divisions palmately lobed, to 10cm).

Lyonia (for John Lyon (died before 1818), enthusiastic collector of American plants, who introduced many into England). Ericaceae. US, E Asia, Himalayas, Antilles. Some 35 species, evergreen or deciduous shrubs or small trees with leathery leaves, conspicuously angled branches and campanulate to urceolate flowers densely packed in axillary clusters or short racemes in summer. Cultivate as for *Leucothoë*.

L.ligustrina (MALE BERRY, HE HUCKLEBERRY, MALE BLUEBERRY, BIG BOY; E US; deciduous shrub to 4m; leaves 3–7cm, oblong to elliptic or lanceolate, entire or finely serrate; flowers densely packed in downy, terminal panicles 8–15cm long, oval-urceolate to subspherical, off-white, downy); *L.ovalifolia* (Himalaya, W China, Japan, Taiwan; deciduous shrub or small tree to 12m; leaves 6–10cm, ovate to elliptic, or oblong, finely downy, somewhat setose beneath; flowers in downy racemes 3–6cm long, white, tubular to urceolate, downy).

lyrate pinnatifid, but with a large, rounded terminal lobe and smaller lateral lobes diminishing in size toward the base of the whole.

Lysichiton (from Greek *lysis*, a loosening, and *chiton*, cloak, referring to the way in which the spathe opens as the fruit ripens). Araceae. NE Asia, western N America. SKUNK CABBAGE. 2 species, robust, clump-forming, deciduous, perennial herbs of marshy places, with thick rhizomes and very large leaves, ovate to oblong, muskily scented and usually appearing with or after the inflorescence. In early spring, minute, green flowers are packed on a thick club-like spadix enclosed within a large, cowel-like spathe. Both species will tolerate cold to at least −15°C/5°F. They are ideal for the bog garden, streams and the pond or lakeside; grow in wet or damp, deep humus-rich soils in sun or part shade. They may be slow to establish and flower. When used as a waterside marginal, prepare the planting site with rich loam to a depth of at least 30cm, built up to water level. Propagate by division in late winter.

L.americanus (western N America; leaves 50–125 × 30–80cm; spathe to 40cm, bright yellow); *L.camtschatcensis* (NE Asia; close to *L.americanus*, but more compact in all parts, spathe white, not yellow).

Lysimachia (from Greek *lysis*, releasing, and *mache*, strife: supposed to possess soothing qualities). Primulaceae. N America, Eurasia, S Africa. LOOSESTRIFE. 150 species, erect or procumbent herbs, rarely dwarf shrubs (those listed here are perennial herbs), with 5-parted, rotate flowers in spring and summer. Fully hardy and easily grown in moist borders in sun or part shade or at the waterside and in bog gardens. Rampant *L.nummularia* is a rapidly spreading, evergreen groundcover. *L.clethroides* makes a graceful show of curving white racemes in late summer and does well in the semiwild garden or in woodland. The glaucous-leaved *L.atropurpurea* and *L.ephemerum* are non-invasive perennials for damper parts of the grey border. *L.ciliata* is a handsome, erect and free-flowering hardy perennial for damp borders – its chocolate and bronze-leaved cultivars providing outstanding foils for copper, bronze, gold and lime ferns, grasses and sedges. Propagate by division or, *L.nummularia*, by rooted stem lengths. *L.atropurpurea* can be short-lived and may need to be propagated from seed collected every two to three years.

L.atropurpurea (Balkans; erect, clump-forming, to 80cm; leaves narrow, grey-green, wavy; flowers small, darkest wine red to purple-black in tight, narrow spikes); *L.ciliata* (N America, naturalized Europe; stem to 1m, erect, glabrous; leaves to 14cm, ovate to lanceolate, ciliate, strongly flushed chocolate to bronze in 'Firecracker'; flowers solitary or paired, in axils of upper leaves, to 1cm in diameter, yellow with red blotches); *L.clethroides* (GOOSENECK LOOSESTRIFE; China, Japan; stem to 1m, simple, erect; leaves to 13cm, ovate to lanceolate, sparsely hairy, glandular-punctate; flowers crowded in slender, nodding, terminal racemes, to 1cm in diameter, white); *L.ephemerum* (SW Europe; stem to 1m, simple, erect; leaves to 15cm, linear or lanceolate to spathulate, glaucous grey-green; flowers in narrow terminal racemes, to 1cm in diameter, white with purple-tinted calyces); *L.nummularia* (CREEPING JENNY, MONEYWORT; Europe, naturalized eastern N America; evergreen, glabrous, stem to 50cm, prostrate, fast-creeping and rooting; leaves to 2cm, broadly ovate or cordate to circular; flowers solitary, rarely in pairs in leaf axils, 1–2cm in diameter, yellow; 'Aurea': leaves golden yellow); *L.punctata* (C Europe, Asia Minor, naturalized N America; stem to 90cm, erect; leaves to 7cm, lanceolate to elliptic, ciliate, puberulent, dotted with glands beneath; flowers whorled, to 1.5cm in diameter, yellow).

Lythrum (from Greek *lythron*, blood, referring to the colour of the flowers). Lythraceae. N America, Europe, Old World. LOOSESTRIFE. 38 species, perennial or annual herbs or small shrubs with 4-angled stems (those listed here are clump-forming hardy herbaceous perennials). In summer, flowers with four to eight spreading, obovate petals are borne in slender, erect, terminal spikes. Fully hardy perennials for the bog garden, water margin or sunny, moisture-retentive borders. The long, upright wands of crumpled magenta flowers last for many weeks in late summer. Autumn foliage colour is a further bonus. Propagate by seed or division in spring.

L.salicaria (PURPLE LOOSESTRIFE, SPIKED LOOSESTRIFE; Old World, naturalized NE US; erect, downy perennial to 120cm; flowers in whorled clusters; petals to 1cm, pink-purple; 'Atropurpureum': flowers dark purple; 'Brightness': to 90cm, with deep rose-pink flowers; 'Feuerkerze' (syn. 'Firecandle'): to 90cm, with slender, rose-pink spires; 'Happy': dwarf to 60cm, with small leaves and red flowers; 'Lady Sackville': to 100cm, with deep pink spires; 'Morden Pink': magenta; 'Purple Spires': vigorous, with abundant, rosy-purple flowers; 'Red Gem': to 100cm, with long, red spikes; 'Robert': to 90cm, of neat habit, with bright pink spikes); *L.virgatum* (Europe and Asia, naturalized New England; similar to *L.salicaria* but leaves glabrous; 'Dropmore Purple': flowers purple; 'Modern Rose': compact, with rose-red flowers; 'Rose Queen': to 90cm, with pink flowers from purple buds; 'The Rocket': to 75cm, with vivid rose-pink flowers).

Lysimachia punctata

M

Maackia (for Richard Maack (1825–1886), Russian naturalist). Leguminosae. E Asia. 8 species, deciduous shrubs or trees with pinnate leaves and pealike flowers in erect, dense and branching racemes in summer. Hardy to –15°C/5°F. Grow in a warm, sunny position in any fertile well-drained soil. Restrict pruning to young plants or to the removal of small branches; large wounds do not heal quickly. Sow seed in autumn after pre-soaking in hot water for 24 hours. Propagate also by root cuttings. *M.amurensis* (E Asia; tree to 20m, often a shrub in cool climates; leaves to 20cm, leaflets to 8cm, 7–11, ovate; flowers to 1.2cm, white or cream to yellow, then brown; fruit to 5cm, winged).

Macadamia (for John Macadam, M.D., secretary of the Philosophical Institute of Victoria). Proteaceae. Madagascar to Australia. 10 species, evergreen trees or shrubs with whorled leaves, racemes of inconspicuous flowers in spring, and edible nuts. Although *Macadamia* species may survive slight frosts, growth is optimal between 20–25°C/70–77°F, ceasing below 10°C/50°F and above 30°C/86°F; cold weather can result in loss of the entire fruit crop. They are liable to wind injury when young, and on shallow soils are easily uprooted in storms. In temperate regions, grow them as foliage ornamentals in the cool to intermediate greenhouse, with a minimum winter temperature of 10°C/50°F. Plant in a well-drained, medium-fertility mix; keep evenly moist when in growth, drier at other times. *Macadamia* may be pruned and repotted or top-dressed in spring. It may be put outside in summer. Propagate by fresh husked seed with bottom heat of 25°C/77°F. *M.integrifolia* (MACADAMIA NUT; E Australia; tree to 20m; leaves to 14cm, oblong to obovate, serrate; flowers white, in pendent racemes to 30cm; fruit spherical, to 3.5cm in diameter).

Macfadyena (for James MacFadyen (1795–1850), Scottish botanist, author of Flora Jamaica (1837)). Bignoniaceae. Mexico, West Indies to Uruguay. 4 species, woody climbers. The leaves consist of two leaflets with a tendril between them. Yellow, tubular to campanulate and flared flowers are produced in spring and summer. Plant outdoors in regions that are frost-free or almost so. Elsewhere, a handsome plant for the cool greenhouse or conservatory. Cultivate as for *Bignonia*. *M.unguis-cati* (flowers 5–12cm long, yellow striped orange in throat).

Mackaya (for James Townsend Mackay (1775–1862), Scottish-born botanist and gardener). Acanthaceae. South Africa. 1 species, *M.bella*, an evergreen, erect shrub to 1.75m with glossy green, elliptic leaves. Borne throughout the year in terminal, loose spikes, the flowers are 6cm long, tubular to campanulate and pale violet with five, purple-veined and flared lobes. Cultivate as for *Justicia*.

Macleania (for Mr John Maclean of Lima, an English merchant and patron of botany). Ericaceae. Tropical C and S America. Some 40 species, evergreen shrubs with slender, often pendulous branches, sometimes scrambling. The leathery leaves are tinged red when young. Tubular, 5-lobed flowers hang in clusters or racemes in summer. Grow in a freely draining, humus-rich, neutral to acid medium, in bright filtered light or part shade. Water moderately when in full growth; reduce water supplies as light levels and temperatures fall, keeping plants just moist in winter, with a minimum temperature of 10°C/50°F. Provide support. Cut back in winter or after flowering to shape and restrict shoot length as necessary. Propagate by seed, by semi-ripe cuttings or by simple layering. *M.insignis* (S Mexico, Honduras, Guatemala; to 1.8m; flowers to 3cm, orange to scarlet).

Macleaya (for Alexander Macleay (1767–1848), colonial secretary for New South Wales). Papaveraceae. E Asia. 3 species, perennial herbs with creeping, sometimes invasive rhizomes, glaucous foliage and yellow sap. The leaves are broad and rounded, toothed and palmately lobed. Small, feathery flowers are carried in terminal, plume-like inflorescences in summer. Fully hardy and easily established in most soils in sun or part shade. They are especially attractive at the rear of herbaceous borders, complementing soft pinks, blues or mauves with their fine, bronze-grey foliage and smoky flowers. They will, however, become invasive. Remove flowerheads after blooming and cut plants down to ground level in autumn. Propagate by division, in the dormant season, or by transplanting suckers in spring.

M.cordata (PLUME POPPY, TREE CELANDINE; China, Japan; to 2.5m; leaves to 20cm across, rounded to cordate, obtusely toothed and lobed, grey to olive-green above, downy white beneath; flowers beige to cream-white, in plumed racemes to 1m, stamens 25–30; 'Alba': flowers white, 'Flamingo': foliage grey-green, flowers buff pink; var. *thunbergii*: leaves distinctly glaucous beneath); *M.microcarpa* (C China; as for *M.cordata* except flowers beige, flushed pink below with 8–12 stamens; 'Coral Plume': flowers pinker; 'Kelway's Coral Plume': flowers deep coral).

Maclura (for William Maclure (d. 1840), American geologist). Moraceae. America, Asia, Africa (mostly warmer regions). OSAGE ORANGE, BOW WOOD. 12 species, spiny, dioecious shrubs, treelets or climbers. Small, yellow-green flowers are followed (where plants of both sexes are grown) by large, wrinkled and fleshy fruits. The primary horticultural interest of *M.pomifera* lies in its unusual but inedible fruits, about the size of an orange. It also has attractive, ridged, deep orange-brown bark, clear yellow autumn colour, and, given space, develops an open and rounded head. It is valued as a thorny and impenetrable hedge, able to withstand hard shearing and harsh wind. It fares best in areas with hot, dry summers. Plant in any well-drained soil, in full sun. It is hardy to –21°C/–6°F, although young growth may be cut by frost. When necessary, prune in winter; the milky sap may cause dermatitis. Sow fresh seed in an open seed bed in autumn, or stored seed, stratified for two months at 4°C/39°F. Increase also from semi-ripe cuttings in a closed frame under mist in summer. *M.pomifera* (E US; tree to 18m, deciduous; branchlets green, thorns to 2.5cm+; leaves 5–15cm, ovate, shiny above, tomentose beneath, turning yellow in autumn; fruit 8–12cm in diameter, globose, with a deeply wrinkled, shiny surface, initially green, orange-yellow when ripe, inedible; includes the thornless 'Inermis', 'Pulverulenta' with powdery, white leaves, and the hardy 'Fan d'Arc' with large, dark leaves).

macronutrients see *nutrients* and *plant nutrition*.

Macrozamia (from Greek *makros*, great, plus *Zamia*). Zamiaceae. Australia. 12 species, cycads, large, 'palm-like' evergreens with stout, buried to exposed and column-like trunks and tough, finely pinnate leaves in a terminal rosette. They are dioecious and produce massive cones. If pollinated, the female cones develop large, egg-shaped seeds encased in a fleshy, brightly coloured coating. Suitable for tubs or landscaping outdoors in subtrop-

ical and tropical regions; elsewhere, they are subjects for containers and beds in the cool or intermediate greenhouse (minimum temperature 10°C/50°F), in the home and for interior landscaping. Cultivation requirements are as for *Cycas circinalis*, although *Macrozamia* thrives in full sunlight and tolerates some drought.

M.communis (BURRAWONG; New South Wales; to 3m; stem largely subterranean; leaves to 2m, pinnae to 130, to 25 × 1cm, rigid, widely spaced, linear, pungent, pale beneath); *M.moorei* (Queensland, New South Wales; to 9m; stem columnar, exposed to 7m tall; leaves to 3m, many, smooth, semi-erect, pinnae to 50 pairs, linear-lanceolate to 30 × 1cm, pungent, somewhat glaucous); *M.pauli-guilielmi* (Queensland; stem swollen, ovoid, often subterranean; leaves to 1m, few, woolly then smooth, grey-green, spiralling, pinnae to 120, erect, narrow-linear to filiform, to 20 × 0.5cm, rachis flattened, very twisted, woolly at base); *M.spiralis* (New South Wales; stem subterranean; leaves to 2m, slightly twisted, pinnae to 25 × 1cm, linear-falcate to narrowly lanceolate, dark, glossy green, often with a small, orange-pink callus at base).

maggots the common name given to the legless larvae of flies (Diptera). Some garden pest maggots have a well-developed head, such as fungus gnats. In others, including leatherjackets and gall midge larvae, the head is reduced and partly retracted into the thorax. Maggots of more advanced flies, such as hoverflies, fruit flies, leaf miners and vegetable root flies, have vestigial heads. Garden pest maggots cause damage by burrowing into and feeding on plant tissues.

magnesian limestone see *lime*.

magnesium a major (macro-) nutrient; an essential constituent of chlorophyll and of importance in the transport of phosphate. Deficiency can occur on acid sandy soils, and may be induced by unbalanced fertilizer application, for example, very high dressings of potassium; soil compaction, waterlogging and water stress aggravate the disorder. Symptoms of deficiency are interveinal chlorosis on older leaves where the margins may remain green; in severe cases, the affected tissue dies. Mature plants are more susceptible. On acid soils, dressings of magnesian limestone (9–11% Mg) may be applied, otherwise kieserite (16–17% Mg). Foliar sprays of magnesium sulphate (10% Mg) are beneficial, applied every 14 days as Epsom salts at 210gm to 10 litres of water.

magnesium carbonate the chemical constituent of magnesian or dolomitic limestone.

magnesium sulphate a magnesium fertilizer, obtainable in two forms: kieserite (16–17% Mg), which is relatively insoluble and suitable for soil application; and Epsom salts (10% Mg), which is soluble and more expensive and therefore used mainly for liquid and foliar feeding.

Magnolia (for Pierre Magnol (1638–1715), Professor of Botany and Director of Montpellier Botanic Gardens, France). Magnoliaceae. Japan, Himalaya, W Malaysia (to Java), eastern N America to Tropical America. Some 125 species, evergreen or deciduous shrubs or trees with large, solitary flowers in spring and summer. These are star-like to vase-shaped, and consist of numerous ovate to strap-like segments. The seeds have colourful coats and are borne in curiously-shaped cones.

The following species are fully hardy. Given adequate moisture at the roots, most magnolias enjoy full sun. However, the late spring- and summer-flowering species, such as *M. wilsonii*, *M. sieboldii*, *M. sieboldii* subsp. *sinensis*, prefer light dappled shade or some shade for at least part of the day. All prefer a sheltered position. *M. grandiflora* and *M. delavayi*, grow best on a warm wall in colder regions. Magnolias are tolerant of urban pollution and make excellent choices for town gardens and street plantings.

They thrive in a wide variety of soil types. Provided there is an adequate supply of moisture and drainage is good, most will at least tolerate alkaline conditions, given adequate reserves of humus in the soil. The ideal soil is a neutral to acid loam, with plenty of organic matter. *M. virginiana* will grow in damp or waterlogged conditions.

Pruning is generally a matter of shaping the plant at planting time by removing weak and badly placed growth and tipping back long shoots. Routine pruning is usually restricted to removing deadwood and watershoots. Where a specimen outgrows its allotted space, it can be pruned hard back immediately after flowering for deciduous species and in early spring for evergreens. It is important when removing larger branches to preserve the branch/bark ridge, leaving the branch collar intact.

The early-flowering species and hybrids can be grown from softwood cuttings with mist and bottom heat at 20–24°C/70–75°F or, with careful management, from semi-ripe cuttings using the warm bench and polythene system or a closed case with bottom heat. Light wounding and treatment with 0.8% IBA are advantageous. Cuttings will take 6–8 weeks to root in trays or pots and should be overwintered *in situ*, leaving potting on until growth has started in the following spring. They should be kept barely moist over winter and cool but frost-free. Simple layering in early spring is often a good alternative using standard procedures, although shrubby species like *M.stellata* may not produce suitable material and, where specimens are large and high-branching, it may be awkward to effect. In commercial nurseries, it is common practice to stool stock specimens grown on good fertile soil to force them to produce long flexible growth from the base; stems tongued and pegged in spring will be ready to sever in the following spring, to be lined out in late autumn/early winter. Autumn layers of the current season's wood with the leaves removed are treated in a similar fashion, but will take two years to root. An alternative for the amateur gardener – although not for the timid – involves cutting a single stem hard back in early spring to force new growth suitable for layering in the following spring. Magnolias generally show good regeneration and often produce strong growth even from old wood.

When available, seed is an excellent method of propagation for the species and will yield strong young plants that grow away very quickly, although they will show a range of variation and may take several years after first flowering to achieve good quality blooms. Seed should be sown fresh, or vernalized for spring-sowing: store cleaned seed for about 100 days at 2–4°C/3–39°F. Dried seed may germinate after a year's dormancy but frequently fails to do so.

Chlorosis can be caused by excessive alkalinity and by both excesses and deficits of potash. It should not be confused with the natural pallor of the young leaves on many species. Lime-induced chlorosis is quickly remedied by foliar applications of iron chelates (sequestered iron). On alkaline soils, care must be taken to ensure that mulching materials such as leafmould and garden compost are lime-free. A black bacterial leaf spot is probably caused by *Pseudomonas syringae*; it can be controlled by copper-based fungicidal sprays. Pale irregular spots on the leaves of *M.grandiflora* are caused by the fungus *Phyllosticta magnoliae* but its effects are usually cosmetic. In the US, magnolias are sometimes attacked by magnolia scale.

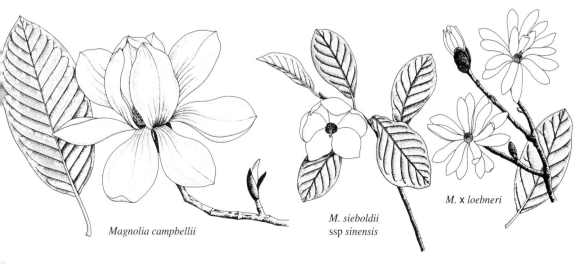

Magnolia campbellii

M. sieboldii ssp *sinensis*

M. x *loebneri*

MAGNOLIA

Name	Distribution	Habit	Leaves	Flowers
M.acuminata CUCUMBER TREE	EN America	deciduous tree to 30m, crown conical, later candelabriform	to 24cm, ovate to elliptic or oblong-ovate, apex acute, base cuneate, dark green above, downy or glaucescent beneath	cup-shaped, erect, segments to 9×3cm, 9, grey-green tinged green to yellow-green, sometimes flushed maroon, oblanceolate to oblong-spathulate

Comments: Spring-summer. 'Variegata': leaves stippled bright gold. f. *aurea*: inner segments golden yellow. var. *subcordata*: YELLOW CUCUMBER TREE, flowers faintly scented, lime-green to clear yellow. 'Elizabeth' (*M.acuminata* × *M.denudata*): flowers precocious, yellow fading to cream. 'Yellow Bird' (*M.acuminata* × *M. × brooklynensis*): flowers yellow

Name	Distribution	Habit	Leaves	Flowers
M.campbellii	Himalaya to China	robust, deciduous tree to 30m	to 23cm, elliptic-ovate to oblong-lanceolate, acute, base rounded, unequal, coriaceous, dark sea green above, paler, sometimes silky beneath	precocious, segments to 16, to 14×6cm, broad, fleshy, concave, white or crimson to rose-pink, paler above, inner whorls erect, outer whorls often reflexed

Comments: Early spring. var. *alba*: flowers white. 'Maharaja': flowers large, white, purple at base. ssp. *mollicomata*: flowers produced earlier, usually bright pink beneath, white above. 'Lanarth': habit fastigiate, flowers opening deep red. 'Charles Raffill': flower buds stained claret, segments purple-pink beneath, white stained pink at margin above. 'Kew's Surprise': flowers darker pink than in 'Charles Raffill'.

Name	Distribution	Habit	Leaves	Flowers
M.delavayi	SW China	evergreen tree, to 10m, often cut back by hard winters	to 50×20cm, ovate to oblong, sinuate, base round to cordate, coriaceous, glabrous, deep sea-green above, grey-green, glaucescent beneath	to 20cm across, slightly fragrant, inner segments 6–7, to 10×5cm in 2 whorls, fleshy, ivory to topaz, cupped

Comments: Late summer.

Name	Distribution	Habit	Leaves	Flowers
M.denudata M.heptapeta, M.yulan YULAN, LILY TREE	E and W China	deciduous tree to 15m, crown broadly pyramidal	to 15cm, obovate to ovate, softly pubescent beneath	to 15cm diameter citrus-scented, precocious, goblet-shaped, erect, segments 9, to 7.5cm, white to ivory, sometimes stained purple-pink at base

Name	Distribution	Habit	Leaves	Flowers
M.fraseri EAR-LEAVED UMBRELLA TREE	SE US	erect deciduous tree to 16m	to 27×18cm, whorled at branch tips, obovate, base deeply cordate or auriculate, glabrous light green, sometimes glaucous beneath	fragrant, tulip-shaped, expanding to 20cm diameter; segments 9, to 12×5cm, white tinted yellow at first, later ivory, spathulate to obovate

Comments: Spring

Name	Distribution	Habit	Leaves	Flowers
M.grandiflora LARGE-FLOWERED MAGNOLIA, BULL BAY, SOUTHERN MAGNOLIA	SE US	evergreen tree to 30m	to 20cm, elliptic to ovate, stiffly coriaceous, glossy dark green above, rust-red pubescent beneath	to 25cm diameter, fragrant, erect, segments 9–12 to 12×9cm, creamy white, obovate or spathulate, fleshy, cupped

Comments: Summer-autumn. 'Exmouth': habit conic; leaves narrow, acuminate, finely pubescent beneath, margins recurved; flowers to 25cm diameter. 'Ferruginea': erect and compact, leaves rust-brown tomentose beneath, glossy dark green above. 'Goliath': habit bushy; leaves short, broad, obtuse, blistered, glabrous beneath. 'Praecox': early- and long-flowering. 'Praecox Fastigiata': early and long-flowering, narrow and erect in habit

Name	Distribution	Habit	Leaves	Flowers
M.hypoleuca	Japan	deciduous to 30m, crown pyramidal	to 40cm, obovate, grey-green, in spreading whorls	to 20cm diameter, fragrant, cup-shaped or spreading, segments to 12, ivory flushed yellow-pink with age, fleshy, stamens crimson to blood-red

Comments: Summer. fruiting cones red, showy

Name	Distribution	Habit	Leaves	Flowers
M.kobus M.praecocissima	Japan	deciduous shrub or tree to 30m, spreading, conical or domed	to 19cm, oblong-elliptic, pubescent beneath	to 10cm diameter, precocious, erect, segments usually 6, 7×3.5cm, creamy white, often stained wine-red or pink, spathulate to obovate

Comments: Winter-spring

Name	Distribution	Habit	Leaves	Flowers
M.liliiflora M.quinquepeta MULAN, WOODY ORCHID	China	deciduous shrub to 4m	to 20cm, elliptic to obovate, acute, sparsely pubescent, sap-green above, paler beneath	erect, goblet-shaped, segments 9, inner 6 to 7.5cm, obovate-oblong, obtuse, concave tip, thick, white often flushed pink to claret

Comments: Spring. 'Nigra': flowers claret to amethyst above, paler below

Name	Distribution	Habit	Leaves	Flowers
M. × loebneri (*M.kobus* × *M.stellata*)	garden origin	shrub or small tree, habit as for *M.stellata* but twigs velutinous not silky	15cm, narrow-obovate	with 12 spathulate segments each to 8×3cm, white, occasionally tinged pink

Comments: 'Leonard Messel': flowers to 13cm diameter, lilac-pink. 'Merrill': flowers larger than in *M.stellata* semi-double

Name	Distribution	Habit	Leaves	Flowers
M.macrophylla LARGE-LEAVED CUCUMBER TREE, GREAT-LEAVED MACROPHYLLA, UMBRELLA TREE	SE US	large deciduous shrub or tree to 20m	to 95cm, in dense whorls, elliptic, oblanceolate to oblong-ovate, base cordate to auriculate, pale green, glabrous above, downy white beneath	fragrant, to 30cm diameter, segments 6–9, 20×14cm, thick, ivory or cream, inner 3 spotted or tinged purple toward base, outer segments narrow-spathulate, tinged green

Comments: Fruiting cones rose-pink with red seeds

Name	Distribution	Habit	Leaves	Flowers
M.rostrata	China, Burma	medium-sized rather skeletal deciduous tree	to 50cm, broadly obovate, prominently veined, covered at first in tawny velvet-like hair	fleshy, cream to pink

Comments: Summer. fruiting cone long, pink, showy

MAGNOLIA

Name	Distribution	Habit	Leaves	Flowers
M.salicifolia ANISE MAGNOLIA, WILLOW-LEAFED MAGNOLIA	Japan	tree or shrub to 12m; twigs scented of lemon or anise	to 12cm, narrow-oval to lanceolate, glabrous, dull green above, paler beneath, sometimes downy	precocious, inner 6 segments to 9×4cm, lanceolate to spathulate, white, occasionally tinged green or flushed pink at base
Comments: 'Wada's Memory' (*M.kobus* × *M.salicifolia?*): small tree, habit upright, compact, leaves elliptic, flushed red-brown at first; flowers to 17cm diameter, fragrant, white, segments 6, at first borne horizontally, then drooping and fluttering				
M.sargentiana	W China	deciduous tree to 25m	to 18cm, elliptic to obovate, occasionally oblong or suborbicular, deep glossy green above, paler, grey-pubescent beneath	fragrant, precocious, pendent, segments 10–14, to 9×3.5cm, spathulate or oblong-oblanceolate, white to pale purple-pink above, purple-pink below
Comments: Spring. 'Caerhays Belle' (*M.s.* variety *robusta* × *M.sprengeri* 'Diva'): segments 12, broad, large, pale carmine to salmon pink				
M.sieboldii	E Asia	deciduous shrub or tree to 10m	12cm, oblong to ovate-elliptic, deep green, subglabrous above, glaucescent, pubescent beneath	to 10cm, fragrant, cup- then dish-shaped, held horizontally, not nodding as in *M.wilsonii* and *M.sieboldii* ssp. *sinensis*, segments to 12, white, obovate to spathulate, to 6×4.5cm, stamens crimson
Comments: Spring-summer. ssp. *sinensis* (syn. *M.sinensis*): CHINESE MAGNOLIA; broadly branching shrub or tree to 6m; young shoots silky; leaves 7.5—21cm, oval to oblong, obovate or suborbicular, glabrous above, glaucescent and, at first, velutinous beneath; flowers to 13cm diameter, nodding, segments 2.5—5cm across, white obovate, to oblong-spathulate, stamens bright crimson				
M. × *soulangiana* (*M.denudata* × *M.liliiflora*) SAUCER MAGNOLIA, CHINESE MAGNOLIA	garden origin	deciduous shrub or tree to 10m; habit of *M.denudata* but more slender	to 16.5cm, broadly elliptic to suborbicular, shiny green above, often puberulent beneath	precocious, erect, segments 9, cupped, oblong-obovate, concave at tip, white, marked rose-pink to violet-purple beneath to 11×7cm
Comments: Spring. The most widely cultivated magnolia, its many cultivars are differentiated by colour, ranging from white to claret, and whether early- or late-flowering				
M.sprengeri	China	deciduous tree to 20m; young shoots yellow-green, glabrous	to 13.5cm, obovate to lanceolate-elliptic, glabrous, above, pale, glabrous to villous beneath	precocious, fragrant, dish-shaped, segments 12- 14, 12×5.5cm, spathulate to oblong-ovate, white, occasionally tinged red to pale pink
Comments: Spring. 'Burncoose': rose-purple. 'Claret Cup': claret beneath, fading to white above. 'Diva': rosy red				
M.stellata *M.tomentosa* STAR MAGNOLIA	Japan	as for *M.kobus* but smaller, to 7.5m; twigs silky-pubescent, bark muskily scented at first	13.5cm, narrow-oblong to obovate	star-like, segments 12—33, spreading, 6.5×1.5cm, snow white, sometimes faintly flushed pink
Comments: spring-summer. Includes cultivars with semi-double to double flowers in pure white to rose flowering early to late in season				
M.tripetala UMBRELLA TREE, ELKWOOD; UMBRELLA MAGNOLIA	EN America	deciduous tree, to 12m; crown broadly spreading	50×26cm+, in whorls at ends of branch, oblanceolate, conspicuously veined, pale pubescent beneath at first	creamy-white, erect, vase-shaped, muskily scented, segments 9–12, oblong-spathulate, fleshy
Comments: Spring-summer. Fruiting cones to 10cm, red-pink; seeds red. 'Silver Parasol' (*M.tripetala* × *M.hypoleuca*): leaves large, in showy, umbrella-like whorls, silver beneath; flowers like *M.hypoleuca*				
M. × *veitchii* (*M.denudata* × *M.campbellii*)	garden origin	deciduous tree, to 30m	22cm, oblong to obovate, dark green veins pubescent above, often tinted purple at first	precocious, vase-shaped, erect, segments 7–10, obovate to spathulate, 12.5×6cm, white tinged violet to pink beneath, pale pink above
M.virginiana SWEET BAY, SWAMP BAY, SWAMP LAUREL	E US	evergreen or semi-deciduous shrub or tree, to 30m+ often multi-stemmed and straggling	11cm, narrow-oblong to suborbicular, shiny green above, glaucous to silver-velvety beneath	rounded to cup-shaped, 6cm diameter, white to ivory, highly fragrant, segments 6–15, 5×2cm, obovate or suborbicular
Comments: Summer-autumn				
M. × *wiesneri* (*M.hypoleuca* × *M.sieboldii*)	garden origin	tree or shrub to 7m+	appearing whorled, obovate, undulate, veins pubescent above, green, glaucous beneath with velvety midvein	spicily fragrant, cup-shaped, erect, segments 9, 6×3.5cm, obovate, creamy white, stamens crimson
Comments: Distinguished from *M.sieboldii* by its larger flowers and broader, tougher leaves				
M.wilsonii	W China	deciduous shrub or tree to 8m+	to 16cm, lanceolate to oblong-ovate, matt green above, velvety beneath	fragrant, nodding, cup-shaped becoming saucer-shaped, segments 9–12, inner segments 6.5×4.5cm, white, spathulate to narrow-lanceolate, incurved, stamens rose-purple
Comments: Closely related to *M.sieboldii*				

Mahonia (for Bernard McMahon, American horticulturist, d. 1816). Berberidaceae. Asia, N and C America. OREGON GRAPE, HOLLY GRAPE. 70 species of evergreen shrubs and small trees closely related to *Berberis* but lacking stem thorns and with tough, sharply toothed pinnate or trifoliolate leaves. Small, sometimes fragrant flowers are borne in slender racemes, clusters or congested panicles near or at the stem summit from late autumn to spring. The fruits are berry-like and plum-red to pruinose purple-black.

The robust, taller-growing Asian species are usually native to damp woodlands, while the shorter, spreading, American species are plants of forest edges or dry scrub. Most are tolerant of a wide range of sites, given a free-draining soil. The larger Asiatics, notably *M.japonica*, *M.bealei*, *M.napaulensis*, *M.lomariifolia* and *M. × media*, are particularly valued for their bold foliage and winter blooming. They are best used as specimen plants in a semi-shaded position in a woodland garden or mixed shrubbery. Provide a damp, slightly acid to neutral soil with a high humus content. Prolonged exposure to harsh winds and temperatures below –15°C/5°F is liable to cause leaf scorch and flower damage. *M.lomariifolia* requires a slightly sheltered site although its hybrids, *M. × media*, are more resilient; *M.napaulensis* suffers in gardens where temperatures fall regularly below –10°C/14°F. *M.japonica* will tolerate –20°C/–4°F. Thin out only leggy, exhausted stems in spring.

Most North American species tolerate a wider range of soils and conditions than members of the first group. They may sometimes become invasive, and should be held in check by the removal of unwanted offshoots. Older portions of these species may lose vigour and become bare, in which case it may be necessary to reduce and replant them. *M.nervosa* is unusual in its very short or absent stems and suckering, ground-smothering habit. It favours a neutral to acid, humus-rich soil in full sun or light shade. Glaucous-leaved species from the southern US and Central America require dry, perfectly drained sites in full sun. They will suffer from wind scorch, winter wet and exposure to temperatures below –10°C/14°F. Plant in gritty, slightly acid soil, preferably on a south-facing wall.

Propagate by fresh seed; clean off the fleshy exterior and surface sow on a sandy propagating mix under glass. Division of most species is possible where offshoots or basal branches have rooted; to this effect, stooling may be advantageous. Take stem section cuttings of members of the first group in late winter. The suckering species may be readily divided in spring. Mallet cuttings of the warmer-growing American species, taken in late winter may root in a cold frame; alternatively, they may be grafted on to stocks of *Berberis thunbergii*. For diseases and pests, see *Berberis*.

M.acanthifolia (Himalaya; erect shrub to 5m; leaves to 60cm, leaflets 17–27, 5–10cm, oblong-ovate to oblong-lanceolate, apex acute to acuminate, base truncate, thinly leathery, dull to slightly glossy above with veins impressed, sinuately toothed, teeth 3–7 on lower margin, 2–5 on upper margin; raceme to 30cm, spreading, rather thick and crowded, flowers deep yellow; 'Maharajah': fine form, with deep yellow flowers); *M.aquifolium* (OREGON GRAPE; western N America; shrub to 2m, sparingly branched, suckering freely; leaflets 5–13, to 8 × 4cm, ovate, with to 12 spiny teeth per side, glossy, dark green turning purple-red in winter; flowers golden, on dense, erect racemes to 8cm; cultivars include: 'Apollo', low-growing, expansive, to 60cm, with larger, dull leaves and good winter colour; 'Atropurpurea', with red-purple leaves in winter; 'Orange Flame', low-growing, to 60cm, with leaves becoming bronze, tinted with orange-red after first season, dark red in winter; and 'Smaragd', spreading, with red-bronze winter colour and profuse flowers); *M.bealei* (syn. *M.japonica* Beale: Group; W China; erect, scarcely branching shrub to 2m; leaves to 50cm; leaflets to 10 × 6cm – terminal leaflet larger and broader –, 9–15, obliquely and broadly ovate acuminate, coarsely and sinuately spiny-toothed (3–4 teeth per side), rigid, sea green to olive green; flowers scented, pale yellow in short, crowded, more or less erect racemes; a hybrid between this species and *M.napaulensis* has been offered under the name *M.japonica* 'Trifurca', an upright shrub with bold ruffs of outspread leaves, the leaflets broad and very sharply and coarsely toothed, the flowers pale yellow in short, erect, clustered racemes; a hybrid between *M.bealei* and *M.lomariifolia* has been recorded: *M.* 'Arthur Menzies' has an upright, compact habit, leaves to 45cm, leaflets to 19, deep sea green, with 3–4 spines per side, and lemon-yellow flowers in erect racemes); *M.confusa* (China; small shrub; leaflets to 20, narrow-lanceolate, spinose, sea-green to leaden grey above, silvery beneath, rachis purple-tinted; flowers pale yellow in erect, slender racemes); *M.fremontii* (SW US; to 4m; leaves to 10cm, leaflets 3–7, to 6cm, oblong to lanceolate, sinuately spiny, bright glaucous blue; flowers in clusters, pale yellow); *M.gracilipes* (China; erect, loosely branching, to 1.5m; leaves to 50cm, leaflets 5–13, 8–20cm, oblong to obovate, apex acute to acuminate, spiny, sea green above, silver beneath; flowers maroon and white); *M.japonica* (Japan, in cultivation only – possibly originating in China or Taiwan; to 3m; leaves 30–40cm, leaflets 7–19, 5–11cm, obliquely ovate-lanceolate to oblong-lanceolate, apex acuminate, base rounded with 4–6 sinuate, spiny teeth per side, very leathery, dull dark green to sea green; raceme 10–20cm, slender, spreading to ascending or pendulous, flowers sulphur yellow, scented of lily of the valley; often confused with *M.bealei*, especially forms with broader leaflets; includes 'Hiemalis' (syn. 'Hivernant'), with leaves to 50cm, narrower, red-tinted leaflets, and racemes, to 35cm, flowering profusely in mid-winter); *M.× lindsayae* (*M.japonica* × *M.siamensis*; includes 'Cantab', a medium-sized, spreading shrub, with

Mahonia nervosa

long, arching leaves, leaflets to 15, remote, with to 6 spines per side, petiole tinted red in cold weather, and large, fragrant, lemon yellow flowers on spreading or drooping racemes); *M.lomariifolia* (Burma, W China; stems erect to 4m, seldom branching above base; leaves to 60cm, outspread, leaflets to 7 × 1cm, 18–41, crowded, overlapping toward apex, oblong-ovate to oblong-lanceolate, with to 7 spines per side, sharply acuminate, subfalcate, dark green; flowers pale yellow, crowded on erect spikes to 15cm); *M.* × *media* (*M.japonica* × *M.lomariifolia*; tall, vigorous hybrids with spreading ruffs of rigid dark green leaves and long racemes of yellow flowers; includes 'Buckland', with an infloresence to 65cm in diameter composed of some 14 racemes, and fragrant, pale yellow flowers; 'Charity', erect to 5m, with ovate-lanceolate, acuminate leaflets, with to 3 teeth on upper margin and to 4 on lower, and deep yellow flowers in erect racemes to 35cm; 'Faith', similar to *M.lomariiflora*, but with broader leaflets and paler flowers; 'Hope', with dense, bright pale yellow flowers; 'Lionel Fortescue', with a fragrant, erect inflorescence; 'Underway', of branching, bushy habit, with flowers in autumn; and 'Winter Sun', with horizontal racemes and topaz-yellow flowers opening in autumn and winter); *M.napaulensis* (Nepal, Sikkim, Assam; differs from *M.acanthifolia* in having shorter leaves with fewer, rather narrower leaflets, usually more closely toothed and glossy, and faintly scented flowers produced in later winter); *M.nervosa* (NW US; dwarf, more or less stemless shrub to 25cm, suckering and forming dense, leafy colonies; leaves to 25cm, resembling a miniature *M.japonica* in shape, sea green turning a rich bronze to maroon in cold weather; flowers pale gold in racemes to 8cm long); *M.nevinii* (California; erect to 2.25m; leaflets to 3cm, 3–7, narrow-lanceolate, with to 6 spiny teeth per side, grey-blue, glaucous; flowers in loose, broad racemes); *M.pinnata* (California; distinguished from *M.aquifolium* by its more erect, rigid habit, and more finely serrate leaflets; to 2.5m; leaflets 5–9, to 6cm, ovate-lanceolate, sinuate with to 13 slender, forward-pointing spines per side, dull sea green above, tinted red in winter, pruinose beneath; racemes to 6cm, flowers primrose yellow; plants grown under this name are often *M.* × *wagneri*, particularly the clone 'Pinnacle'); *M.repens* (western N America; suckering, semi-prostrate shrub to 0.5m; leaves to 25cm, leaflets usually 5, ovate, dull green above, papillose beneath, glaucous at first, with 9–18 spiny teeth per side; flowers deep yellow, in racemes to 8cm; includes the very low-growing 'Rotundifolia', with larger, broadly ovate, entire or serrulate leaves); *M.siamensis* (N Thailand to Burma, China; to 4m; leaves to 70cm, leaflets 11–17, obliquely lanceolate to ovate, 14–17cm, sinuately toothed, dull green very leathery with veins impressed; raceme 12–25cm, 6–10-fascicled, ascending to erect, flowers deep yellow, very fragrant); *M.trifoliolata* (Mexico; erect to 2.25m; leaflets 3, to 6.8cm,

narrowly lanceolate, rigid, undulate with 1–4 lobe-like marginal spines per side, grey to blue-green, glaucous; flowers pale yellow in short corymbs); *M.* × *wagneri* (*M.pinnata* × *M.aquifolium* cultivars include 'Aldenhamensis, vigorous and erect to 1.5m, with large, sea-green leaves, tinted blue beneath, bronze young leaves and yellow flowers; 'Fireflame', to 1.25m, with 7 leaflets, green or glaucous blue above, grey-green beneath, blood red to bronze in winter; 'Moseri', erect to 80cm, with young leaves tinged bronze-red, later dark green, and yellow flowers; 'Pinnacle', syn. 'Pinnata', erect to 1.5m, vigorous, with young foliage and winter leaves tinted copper, otherwise bright green; 'Undulata', erect to 1.5m, vigorous, with very glossy leaves with undulate margins, young and winter leaves tinted red-bronze; and 'Vicaryi', broadly erect to 1m, with small leaves, some tinted red in autumn, and yellow flowers in small dense racemes).

Maianthemum (from Greek *maios*, the month of May, and *anthemon*, blossom). Liliaceae (Convallariaceae). Northern temperate regions. 3 species, low-growing perennial herbs with creeping rhizomes. They produce slender stems bearing two to three, smooth, heart-shaped leaves and, in late spring, small white flowers in terminal racemes. These are followed by red berries. Fully hardy. Plant in humus-rich, slightly acid soils in moist and shaded locations in the wild and woodland garden where they may soon form large colonies. They suffer in hot, dry summers. Propagate by division.

M.bifolium (FALSE LILY OF THE VALLEY; W Europe to Japan; stem 5–10cm; leaves 3–8cm, broadly cordate-ovate, thinly-textured, short-stalked with a deep, open sinus); *M.canadense* (TWO-LEAVED SOLOMON'S SEAL; N America; stem 5–10cm; leaves 5–10cm, broadly ovate-cordate, more or less sessile with a narrow sinus).

maiden mainly used to describe a tree or shrub, especially a fruit tree, in its first year after grafting or budding and before any pruning has taken place. A feathered maiden has lateral growths. It is often also used to describe newly rooted strawberry plants, and sometimes other plants in their first year after propagation.

Maihuenia (from *maihuen*, a local name for the plant). Cactaceae. Argentina and Chile. 2 species, low, shrubby cacti with clumped, cylindrical stems bearing small, terete leaves at their summits and slender spines. Funnel-shaped, yellow flowers appear in summer. Grow in an unheated greenhouse, or outdoors in zones 7 and over, with winter protection from rain. Provide a gritty, more or less neutral medium and full sun. Keep dry from late autumn until early spring. *M.poeppigii* (S Argentina, S Chile; dwarf shrub, forming low mounds; stem segments shortly cylindric, 1.2–3cm; leaves 4–6mm, central spine 1–1.5cm; flowers 3–4.5cm, yellow).

maincrop used to distinguish those vegetable crops that do not mature particularly early or late in the season; applied especially to potatoes and carrots.

Malcolmia (for William Malcolm, late 18th-century London nurseryman). Cruciferae. Mediterranean to Afghanistan. MALCOLM STOCK. 35 species, annual to perennial herbs with branching, often prostrate stems and entire or pinnately lobed, downy leaves. Produced from spring to autumn in loose racemes, the flowers consist of four, linear or obovate petals. A hardy annual, grow in well-drained, medium- to low-fertility neutral to alkaline soils, in sun or part-day shade. Plants suffer in hot, humid summers. Sow seed thinly *in situ*, from early spring until autumn. *M.maritima* (VIRGINIA STOCK; Mediterranean; annual to 35cm; flowers fragrant, petals 12–25mm, red-purple, rarely violet, notched, long-clawed; Compacta Mixed: compact, to 40cm, with small, fragrant, pink, white and red flowers).

Malephora (from Greek *male*, armpit, and *phorein*, to bear). Aizoaceae. South Africa, Namibia. 15 species, erect or creeping shrubby perennials with succulent, semicylindric leaves and daisy-like flowers, golden, yellow or pink and to 5cm in diameter. Cultivate as for *Lampranthus*. *M.crocea* (stem stout, gnarled; leaves 2.5–4.5cm, more or less 3-sided, pruinose; flowers 3cm in diameter, golden yellow inside, tinted red outside; var. *purpureocrocea*: flowers brilliant red).

male sterility the failure of flowers to form fertile pollen.

malformations to a greater or lesser extent, all plant species show variation from the condition usually accepted as normal. Adaptation to habitat is held to account for many such variations. Frequently, however, malformations or monstrosities in plants arise from other factors and may be seen as excessive or arrested growth, or other abnormal development in form or position of parts.

Imbalance of auxin and other non-specific hormones accounts for many instances of malformation. Parasitism by mites or viruses can influence the production of hormones and stimulate the formation of galls or witches' broom growths. Occasionally, an abundant supply of mineral nutrients, particularly nitrogen, can lead to malformation.

Genetic change is also a potential cause of abnormal growth, and minor variations brought about in this way or through environmental influence are held by some to account for reversion to an ancestral condition. Chromosome changes and abnormality can produce unusual growth forms, for example, autotetraploids are frequently larger than diploids of the same species, for example the grape cultivar 'Perlette', *Clematis montana* 'Tetrarose' and the majority of *Narcissus* cultivars. The growing points (apical meristems) of a plant contain immature tissue and physical or chemical damage can influence their future behaviour. The accidental division of the apex into two independent meristems can lead to two shoots developing from one region. Breakdown in growth coordination can take place in the apex, often induced by hormonal imbalance and bringing about abnormal multiplication of shoots and bushy development, or unusual leaf arrangement (phyllotaxy), or clusters of leaves produced instead of flowers. Plant parts sometimes develop out of place and inflorescences can revert to vegetative shoots; such malformations may be of hormonal origin or due to changes in temperature or day length at critical times in the plant's growth cycle.

One of the most striking abnormalities is FASCIATION, which results from unusual activity in the apical meristem. Familiar examples occur in *Ranunculus*, *Delphinium* and *Forsythia* with wide, laterally branching stems giving the impression of several normal stems fused together. Sometimes, an abnormal number of flowers is produced on fasciated stems, as may be seen in lilies. In 'ring fasciation', a ring of subsidiary flowerheads is produced around a normal central flower. This occurs, for instance, in members of the Compositae, such as the daisy, where the abnormality is referred to as 'hen-and-chickens'. Sometimes lateral buds develop on a thickened stem, and branches of normal dimension and appearance grow from it.

Some fasciated forms are cultivated, for example, *Salix udensis* 'Sekka', whose flattened, recurving stems are encouraged to form by hard pruning. Another example is *Celosia argentea* var. *cristata*, the cock's comb, a tetraploid cultigen derived from *C.argentea*, which can be seed-propagated; and there are cristate forms among cacti, forming fan-shaped or densely convoluted growths. Many fern cultivars have crested ends of fronds and are given names like 'Cristata' or 'Monstrosa'.

In some plants, leaves may fail to separate properly from stems, or parts of flowers may appear to be stuck together. In double flowers, it is quite common to find PETALOID STAMENS.

Another interesting variation is the rearrangement of flower parts, probably the outcome of a small genetic change and well-documented in *Antirrhinum*. This phenomenon is known as PELORY or PELORIA and a frequently met example is the foxglove *Digitalis purpurea* 'Monstrosa', in which a large, open, regular flower tops the spike of normal 2-lipped, bell-shaped flowers; this is heritable from seed. In the abnormality of flowers known as HOSE-IN-HOSE, one perfect corolla develops within another, in some cases giving the appearance of a double bloom and in others clearly separated. This condition is found particularly in primulas, but also in rhododendrons and *Mimulus*.

Freak plants produced artificially by grafting are often regarded as teratological, but this is not strictly so. Unusual colour mutants, lacking chlorophyll, can sometimes be perpetuated by grafting, for example, *Gymnocalycium* in the Cactaceae. Unsuitable matching of stock and scion may bring about peculiar forms, for example, the selection of a vigorous scion for a restricting rootstock in *Fraxinus* can lead to the scion bulging above the graft. The reverse condition is commonly seen in *Prunus*.

Malformations (a) Spiral fasciation (Asparagus), *(b) Flat fasciation* (Asparagus), *(c) Multiple fasciation* (Pinus pinaster), *(d) Flat fasciation* (Salix udensis *'Sekka'), (e) Peloric foxglove* (Digitalis purpurea), *(f) 'Hen-and-Chickens' proliferation* (Calendula), *(g) Flat fasciation with multiple flowers (florist's chrysanthemum), (h) Proliferation (rose)* , *(i) Gamopetaly (*Papaver bracteatum) , *(j) Petaloid calyx (*Mimulus), *(k) Sepals becoming leafy (*Primula vulgaris), *(l) Foliar proliferation above inflorescense (*Polyanthus Primula), *(m) Phyllody – bracts replacing inflorescence (*Plantago major), *(n) Bifurcation of frond (*Asplenium scolopendrium), *(o) Cristation of frond (*Dryopteris filix-mas) , *(p) Cristate inflorescence (*Celosia argentea *'Cristata'), (q) Cristate stem (*Mammillaria vaupelii *'Cristata'), (r) Monstrous growth (*Pachycereus schottii *'Monstrosus'),*

mallet cutting see *cuttings, semi-ripe*

Malope (name of Greek origin, used by Pliny for a mallow). Malvaceae. Mediterranean to W Asia. 4 species, annual and perennial herbs with unlobed or palmately lobed leaves. Produced in summer, the flowers are funnelform and consist of five, obovate petals. *Malope* performs best in regions with cooler summers and may fail in very hot, humid conditions. Grow in moderately fertile and humus-rich, sandy soils in full sun. Deadhead regularly. Propagate from seed sown under glass in early spring, or, later, *in situ*. *M.trifida* (W Mediterranean; annual to 1.2m; flowers 4–6cm, purple-red; 'Alba': white; 'Grandiflora': large, dark rose; 'Rosea': rose red; 'Vulcan': large, bright red).

Malus (Latin for apple-tree). Rosaceae. Europe, Asia, US. APPLE, CRAB APPLE. Some 35 species, deciduous trees and shrubs, many with thorny or short, spur-like lateral shoots. Produced in corymbs in spring and summer, the flowers consist of five, outspread, rounded to obovate petals and numerous stamens. They are followed by pomes, often highly coloured, and ranging from small 'crabs' to large apples. *Malus* provides ornamental interest throughout the year, especially in spring when in blossom. The decorative fruit, with a colour range from clear yellow through bright reds to deep red-purple, often persists well into winter. The foliage of some colours well in autumn. Taller species, such as *M.floribunda, M.hupehensis* and *M.prattii*, when grown as standards, are suitable as lawn specimens, at the woodland edge, or in large borders. Smaller species or those grown as half-standards can be selected for the smaller garden or border. *M.tschonoskii* is a small conical tree, bearing few fruits but with brilliant autumn colours in orange, purple and scarlet, and is well-suited to planting in public areas.

Malus tolerates light, dappled shade, but autumn colour and fruiting are better in sun. All are hardy to at least -15°C/5°F and most tolerate a range of fertile soils. Grow in a sunny position, in moisture-retentive but not boggy soil. Prune in winter, to remove dead, diseased, damaged or crossing wood. On grafted specimens, it may be necessary to remove suckers. Standard and half-standard trees grown for ornamental purposes are often trained in the nursery with an open, wide-branching crown; for fruiting purposes, it may be preferable to obtain a feathered maiden to form a central leader (spindle-bush) tree, with a well-spaced framework of laterals.

Propagate species by seed, although this may not come true. *M.hupehensis* produces seed apomictically, and seedlings of this species will come true. Sow seed in spring, following three months of cold stratification in a damp medium at 1°C/34°F. Temperatures above 15–20°C/60–70°F in the sowing medium may induce secondary dormancy. Cultivars are T-budded, or whip and tongue grafted on to clonal rootstocks. Pests and diseases of crab apples are similar to those affecting orchard apples. See *apples*.

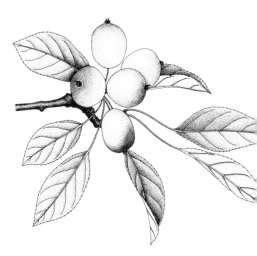

Malus 'Golden Hornet'

MALUS Name	Distribution	Habit	Leaves	Flowers	Fruit
M. × arnoldiana (*M.baccata* × *M.floribunda*)	garden origin	shrub to 2m, resembling *floribunda*; shoots arching, soon glabrous	5–8cm, elliptic to ovate, long- tapered, biserrate, ultimately glabrous except for veins beneath	carmine-red in bud, later pink becoming white	1.5cm, globose, yellow- green
Comments: 'Linda': flowers single, carmine becoming light pink; fruit to 3cm diameter, carmine					
M. × atrosanguinea (*M.halliana* × *M.sieboldii*)	garden origin	spreading, with slightly pendulous shoots	ovate, serrate, lobed at base on strong shoots, waxy	single, deep carmine	1cm, globose, red or yellow flushed red
M.baccata. SIBERIAN CRAB	NE Asia	tree or shrub to 5m, often confused with *M.robusta*; shoots slender, glabrous	3–8cm, ovate, long-tapered, serrate, glabrous	3–3.5cm diameter, white	1cm, globose, yellow and red
Comments: 'Columnaris': habit narrowly upright to 8m; flowers snow white; fruit yellow with red cheek. 'Gracilis': habit shrub-like, slow-growing, branches slender, dense; leaves small, narrow; buds pink, flowers white, stellate; fruit small, red. 'Lady Northcliffe': buds light pink, flowers profuse, white; fruit to 1.5cm diameter, dark orange. 'Macrocarpa': fruit large, to 3cm, diameter, glossy yellow stained red. 'Orange': habit large, shrub-like; buds light pink, flowers white; fruit large, to 4cm diameter; red washed orange. 'Spring Snow': habit upright; flowers large, white; fruitless; seedling of 'Dolgo'. var. *himalaica*: leaves wide, elliptic, roughly serrate, glabrous, veins on lower leaf surfaces downy; flowers 3cm diameter, pink in bud; fruit 1–1.5cm, yellow flushed red. variety *jackii*: crown broader, branches stout; leaves broadly elliptic; flowers 3cm diameter, white; fruit 1cm, red, waxy. var. *mandschurica*: leaves broadly elliptic, margins with few serrations, lower surface downy at first; flowers to 4cm diameter, white, scented; fruit 12mm, ellipsoid, bright red, ripening early.					

MALUS

Name	Distribution	Habit	Leaves	Flowers	Fruit
M.coronaria	NE United States	small tree to 7m wide, spreading, shoots with numerous short-thorned laterals, becoming glabrous	5–10cm, ovate-oblong, pointed, serrate, slightly lobed, becoming glabrous and scarlet-orange in fall	to 4cm diameter, pink	4cm, depressed-globose, green, base slightly ribbed

Comments: 'Charlottae': leaves red stained orange in autumn; flowers double, soft pink, fragrant. 'Nieuwlandiana': habit shrub-like, to 3m; flowers in large hanging clusters, vivid pink, very fragrant, fruit blue pruinose. var. *dasycalyx*: leaves paler green beneath; flowers smaller, pink, highly fragrant; fruit 4cm, yellow-green.

Name	Distribution	Habit	Leaves	Flowers	Fruit
M.floribunda JAPANESE CRAB	Japan	shrub to 4m, or tree to 10m, crown dense, shoot tips slightly pendulous, downy	4–8cm, long-tapered, deeply serrate, sometimes lightly lobed	2.5–3cm, diameter, deep pink in bud, opening pale pink, white inside	0.5cm, yellow

Comments: 'Peachblow': habit upright; buds dark pink, flowers white; fruit to 1cm diameter, red

Name	Distribution	Habit	Leaves	Flowers	Fruit
M.hupehensis	India	tree or shrub, 5–7m, shoots spreading, soon glabrous	5–10cm, ovate to oblong, long-tapered, deeply incised, slightly downy beneath	4cm diameter, pink fading to white	1cm, rounded, yellow-green, flushed red

Comments: 'Rosea': flowers profuse, cherry blossom pink; fruit yellow, sometimes tinted red

Name	Distribution	Habit	Leaves	Flowers	Fruit
M. × magdeburgensis (*M.pumila × M.spectabilis*)	garden origin	shrub or small tree, resembling *M.spectabilis*; crown globose	6–8cm, elliptic, acuminate, downy beneath	4.5cm diameter, bright red in bud, opening deep pink	3cm, rounded, green-yellow flushed red
M.prattii	W China	erect shrub or small tree to 7m, shoots downy at first	6–15cm, elliptic-oblong, long-tapered, finely biserrate, veins downy beneath, turning red and orange in autumn	2cm diameter, white	1–1.5cm, rounded-ovate, red to yellow, white-pitted
M.prunifolia	NE Asia	small tree to 10m, shoots downy at first	5–10cm, elliptic or ovate, sinuate, downy beneath	3cm diameter, pink in bud opening white	2cm, ovoid-conical, yellow-green to red

Comments: 'Cheal's Crimson': habit upright to broad, to 5m; buds pink, flowers single, white; fruit prolific, to 2.5cm diameter, light orange with scarlet cheek. 'Cheal's Golden Gem': flowers single, white; fruit rounded, gold. 'Fastigiata': habit narrow upright, later spreading; flowers white; fruit yellow and red. ' 'Hyslop': habit upright to spreading; crown broad; fruit globose, to 4cm diameter, yellow stippled and blotched light red, edible. 'Pendula': branches weeping.

Name	Distribution	Habit	Leaves	Flowers	Fruit
M. × purpurea (*M.astrosanguinea × M.pumila* 'Niedzwetzkyana')	garden origin	large shrub or small tree, bark red-black	8–9cm, ovate, scalloped, occasionally lobed, brown-red at first, shiny	3–4cm diameter, purple-red, soon fading	1.5–2.5cm, rounded, purple-red

Comments: 'Aldenhamensis': habit shrub-like, low-growing, to 3.5m; leaves bronze red; buds deep red, flowers profuse, maroon ; fruit red tinted brown. 'Eleyi': habit shrub-like, slightly spreading; leaves darker; flowers single, maroon; fruit ovoid, purple. 'Eleyi Compacta': habit compact, shrub-like; branches short; leaves, flowers and fruit purple. 'Hoser': buds dark purple; flowers pink; fruit pruinose purple. 'Jadwiga': habit pendulous; crown broad; buds maroon, flowers pink; fruit conic, to 5cm diameter, blackcurrant purple. 'Lemoinei': habit shrub-like, upright, vigorous; leaves dark purple, later bronzed or dark green; flowers single to semi-double, maroon; fruit to 1.5cm diameter, dark purple.

Name	Distribution	Habit	Leaves	Flowers	Fruit
M. × robusta *M.baccata × M.prunifolia*	garden origin	vigorous, conical shrub or small tree	8–11cm, elliptic, margins scalloped, bright green	3–4cm diameter, white, occasionally pink.	1–3cm, ellipsoid to rounded, yellow or red, long-stalked, sometimes pruinose

Comments: 'Alexis': buds light pink, flowers single, white; fruit to 3cm diameter, vivid pink-red, pruinose. 'Beauty': habit narrowly upright; flowers single, white; fruit to 3cm diameter, scarlet. 'Erecta': habit narrowly upright when young, later spreading; flowers single to semi-double, white stained light pink; fruit small, to 2cm diameter, yellow with red cheeks, pruinose. 'Red Sentinel': flowers single, white, early, fruit glossy scarlet, persistent. 'Red Siberian': fruit bright red. 'Yellow Siberian': fruit yellow. var. *persicifolia*: large shrub, branches slender; leaves 5–10cm, oval-lanceolate, finely scalloped; flowers 4cm diameter, numerous, pink in bud opening white; fruit 2cm, abundant, rounded or oblong, red

Name	Distribution	Habit	Leaves	Flowers	Fruit
M.sargentii	Japan	shrub to 2m, thorny	5–8cm, ovate, serrate with 3 lobes, orange in fall	2.5cm diameter, along whole length of shoot, buds pale pink, opening white	1cm, globose, dark red

Comments: 'Rosea': buds dark pink, flowers fading to white

Name	Distribution	Habit	Leaves	Flowers	Fruit
M.sieboldii	Japan	shrub to 4m, shoots arched, black-brown	3–6cm, ovate-elliptic, tapered, dentate and 3-5-lobed, downy, red or yellow in fall	2cm diameter, deep pink in bud, eventually white	small, globose, red to yellow-brown, persistent

Comments: 'Fuji': habit low, spreading; flowers anemone-centred, white stained green or purple; fruit small, orange. 'Seafoam': somewhat pendulous; flowers carmine, fading to white flushed with pink. 'White Angel': upright, later nodding; flowers in clusters, snow-white; fruit prolific, small, scarlet. 'Wintergold': shrub-like, upright to spreading; flowers pink in bud fading to white; fruit gold, persistent.

Name	Distribution	Habit	Leaves	Flowers	Fruit
M.spectabilis ASIATIC APPLE	China	large shrub or small tree to 8m, pyramidal at first, later spreading; shoots becoming red-brown	5–8cm, elliptic-oblong, short-tapered, dentate, downy beneath	4–5cm diameter, single or semi-double, dark pink in bud, opening pale pink	2–3cm, rounded, yellow

Comments: 'Blanche Amis': habit shrub-like, upright; buds carmine, flowers semi-double, white inside, carmine outside; fruit globose, small, yellow. 'Plena': flowers profuse, double, pink, soon fading; fruit globose, to 2cm diameter, yellow. 'Riversii': flowers double, pink, fruit large, to 3.5cm diameter, yellow.

MALUS

Name	Distribution	Habit	Leaves	Flowers	Fruit
M.sylvestris EUROPEAN APPLE, WILD CRAB	C Europe	tree or shrub to 7m, slightly thorny	4–8cm, oval to rounded, scalloped or incised, subglabrous	to 4cm diameter, pink-white outside, pinker within	2–4cm, rounded, yellow-green flushed red
Comments: 'Plena': flowers double, soft pink, fading to snow white; fruit scarlet, sometimes blotched yellow; often listed as *M.spectabilis alba plena*. var. *domestica*: ORCHARD APPLE; the variety giving rise to edible apples					
M.toringoides	W China	shrub or small tree to 8m, soon glabrous	3–8cm, ovate and crenately lobed or simple, veins downy beneath	2cm diameter, white	1.5cm, globose to pyriform, yellow flushed red, persistent
M.transitoria	NW China	closely resembles *M.toringoides*, smaller, crown narrower; shoots felted when young	2–3cm, broadly ovate, deeply narrow-lobed, downy	to 2cm diameter, white	1.5cm, red
M.trilobata	W Asia	erect shrub or small tree	5–8cm, 3-lobed, shiny, soon glabrous, intense red in fall	3.5cm diameter, white	2cm, ellipsoid, red
M.tschonoskii	Japan	tree to 12m, pyramidal then spreading; shoots felted	7–12cm, ovate-elliptic, serrate to slightly lobed, white-felted at first, orange-red in fall	3cm diameter, white.	2–3cm, rounded, yellow-green flushed red
M.yunnanensis	W China	compact, erect tree to 10m, shoots felted at first	6–12cm, broadly ovate, roughly biserrate, occasionally with 3–5 pairs of broad lobes, felted beneath, red-orange in fall	1.5cm diameter, usually in crowded corymbs to 5cm, white	1–1.5cm, red, pitted
Comments: var. *veitchii*: leaves slightly lobed, soon glabrous; flowers to 1.2cm diameter; fruit to 1.3cm, red with white pits					
M. × zumi (*M.baccata* var. *mandshurica* × *M.sieboldii*)	Japan	small pyramidal tree, shoots more or less downy	5–9cm, ovate, long tapered, scalloped to lobed, downy beneath at first	3cm diameter, pink in bud, fading to white	1cm, rounded, red
Comments: var. *calocarpa*: more spreading; leaves smaller, always entire on fruiting wood, distinctly lobed on strong shoots; flowers smaller, white; fruit 1–1.3cm, crowded bunches, scarlet. 'Golden Hornet': flowers white; fruit gold. 'Professor Sprenger': fruit orange					

Malva (from Latin *malva*, mallow, from the Greek *malache*, probably derived from Greek *malachos*, soothing, referring to its medicinal properties). Malvaceae. Europe, Asia, Africa; widely introduced into temperate and tropical regions. 30 species, annual or perennial herbs and subshrubs, with rounded to reniform or cordate leaves, often toothed or lobed. The flowers are broadly funnel- to bowl-shaped with five, obovate petals and are carried in loose terminal spikes in summer. The following are hardy to −15°C/5°F. Grow in any reasonably drained and moderately fertile soil, in full sun. Staking with twiggy sticks may be necessary, especially on moist rich soils. Cut down flowered stems in autumn. Sow seed in early spring in the cold frame or under glass at 15°C/60°F; alternatively, increase also by soft-wood cuttings in spring. *Malva* is prone to rust: collect and burn all affected material in autumn and spring.

M.moschata (MUSK MALLOW; Europe, NW Africa, naturalized US; aromatic perennial erect to 1m; leaves shallowly 3-lobed to deeply divided; petals 2–3cm, white or pink; includes 'Alba', to 60cm, with deeply cut leaves and silky white flowers, and 'Rosea', with rose flowers tinted purple); *M.sylvestris* (TALL MALLOW, HIGH MALLOW, CHEESES; Europe, N Africa, SW Asia, US as garden escape; perennial to 1m; leaves broadly cordate or suborbicular, lobed; petals 1.5–2cm, mauve, paler and hairy at base, notched at apex; includes 'Alba', with pure white flowers, 'Brave Heart', upright, to 90cm, with large, pale purple flowers with a dark eye, 'Cottenham Blue', to 75cm, with pale mauve-blue flowers with darker veins, and 'Primley Blue', low-growing, with soft violet flowers with darker veins).

Malvaviscus (from Latin *malva*, mallow, and *viscum*, glue, referring to the sticky seeds). C and S America. 3 species slender-stemmed, evergreen shrubs, sometimes vinelike, with unlobed or palmately lobed, toothed leaves. Produced in summer, the flowers resemble those of *Abutilon*, with the petals often incurved in a bell- or turban-like arrangement. From the centre protrudes a long, staminal column. In frost-free, warm temperate regions, grow outdoors in full sun on any well-drained soil. In cool temperate zones, grow under glass with a minimum winter temperature of 10°C/50°F, in direct sunlight, in a loam-based mix. Water plentifully when in growth; keep just moist during the winter months. Cut flowered stems hard back in winter to maintain shape. Propagate by soft-wood or semi-ripe cuttings or by seed in spring. *M.arboreus* (WAX MALLOW; Mexico to Peru and Brazil; to 4m; leaves lobed, downy; flowers long-stalked, petals 2.5–5cm, rich red; var. *mexicanus*: TURK'S CAP, leaves almost glabrous, more or less unlobed).

mamillate, mammillate furnished with nipple-like prominences.

Mammillaria (from Latin *mammilla*, little nipple, referring to the tubercles). Cactaceae. SW US, Mexico, C America, Caribbean region, Colombia, Venezuela. Some 150 species, low-growing cacti with globose to squatly cylindrical stems covered in prominent, spiny tubercles. Campanulate to funnelform flowers are borne freely at various times of year. The fruit is berry-like, oblong or clavate, and often bright red. These small, undemanding cacti are popular as houseplants, producing an abundance of small, colourful flowers. Provide a minimum temperature of 7°C/45°F, full sun and low humidity. Plant in a gritty and sandy, loam-based mix. Water moderately in spring and summer. Keep dry from mid-autumn until early spring, except for light misting on warm days in late winter. Increase by seed and offsets.

M.bocasana (C Mexico; freely clustering; stem more or less globose, almost hidden by spines; axils naked or with fine hairs or bristles; central spines 1(–5), 5–10mm, 1–2 hooked, red or brown, radial spines 25–50, 8–10(–20)mm, hair-like, white; flowers creamy white, outer tepals with pale pink stripe); *M.candida* (NE Mexico; simple or clustering; stems to 14cm in diameter, globose to stoutly cylindric; tubercles cylindric; areoles with scant wool; axils with fine white bristles; central spines 8–12, to 1cm, white or tipped pink or brown, radial spines numerous, to 1.5cm, white; flowers rose pink); *M.densispina* (C Mexico; usually simple; stem to 12 × 6(–10)cm, globose to short-cylindric; central spines 5–6, 10–12(–15)mm, straight, black-tipped, radial spines about 25, unequal, 8–10mm, yellow; flowers pale yellow); *M.elongata* (C Mexico; clustering; stem 1–3cm in diameter, cylindric, elongate; tubercles short, central spines 0–3, to 15mm, pale yellow to brown, radial spines 14–25, 4–9mm, yellow; flowers pale yellow or tinged pink); *M.geminispina* (C Mexico; soon clustering and forming mounds; individual stem 6–8cm in diameter, becoming cylindric; axils with wool and short bristles; central spines 2 or 4, uppermost 15–40mm, white, dark-tipped, radial spines 16–20, 5–7mm, chalky white; flowers deep pink); *M.gracilis* (east C Mexico; offsetting very freely, stem to 13 × 1–3cm, cylindric; tubercles short, obtusely conic; central spines 0–2(–5), up to 10–12mm, white or dark brown, radial spines 11–17, 3–8mm, bristly, white; flowers pale yellow with pink or brown midline); *M.haageana* (syn. *M.elegans*; SE Mexico; stem usually simple, reaching 15 × 5–10cm in larger variants: tubercles small, crowded; axils often woolly and sometimes with bristles; central spines usually 2, sometimes 1 or 4, to 15mm, brown, radial spines 15–25, 3–6mm, white; flowers small, deep purple-pink); *M.hahniana* (C Mexico; simple or clustering; individual stem to 20 × 12cm; tubercles very numerous, triangular-conic, small; axils with long white bristles; central spines 1(–4), up to 4mm, white tipped brown, radial spines 20–30, 5–15mm, hairlike, white; flowers deep purple-pink); *M.magnimamma* (C Mexico; usually clustering and forming mounds 50cm or more in diameter; individual heads to 10–12cm in diameter; tubercles pyramidal-conic; axils woolly in flowering zone; spines

3–6, usually 1 longer and stronger than the others and curved; flowers pale yellow to deep purple-pink); *M.microhelia* (C Mexico; simple or clustering; stem to 25 × 3.5–6cm, cylindric; tubercles shortly conic; axils soon naked; central spines 0–4(–8), to 11mm, dark red-brown, radial spines (30–)50, 4–6mm; yellow, often slightly recurved; flowers creamy white to red or purple); *M.plumosa* (NE Mexico; clustering to form mounds to 40cm or more in diameter; stem to 6–7cm in diameter, globose, hidden by spines; tubercles cylindric; axils woolly; spines up to 40, 3–7mm, feathery, white; flowers small, creamy white or tinged brown-pink); *M.prolifera* (NE Mexico, SW US, Cuba, Hispaniola; forming dense clumps; stem to about 9 × 4.5cm, globose, to cylindric-clavate; tubercles terete; axils naked or with fine white hairs; central spines 5–12, 4–9mm, straight, puberulent, white to red-brown, sometimes dark-tipped, radial spines 25–40, 3–12mm, fine, straight to tortuous, white; flowers creamy yellow, or tinged pink, outer tepals brown-striped); *M.rhodantha* (C Mexico; usually simple, occasionally offsetting or branching; stem to 40 × 12cm, cylindric; tubercles obtuse-conic to cylindric, axils woolly at first and with a few bristles; central spines 4–9, to 18(–25)mm, usually recurved, typically red-brown, radial spines typically 17–24, 4–9mm, glassy white to pale yellow; flowers purple-red); *M.schiedeana* (clustering; stem to 10 × 6cm, depressed-globose; tubercles tapering, cylindric-terete; axils with long white woolly hairs; central spines 0, radial spines very numerous, 2–5mm, adpressed, minutely pubescent, white, pale yellow towards base, golden yellow at base, the tip usually hair-like; flowers creamy white); *M.sempervivi* (east central Mexico; simple at first, sometimes later dividing apically or offsetting; stem to 10cm in diameter, depressed-globose; tubercles angled-conic, dark blue-green, axils woolly; central spines usually 2, sometimes 4, to 4mm, brown or black, later grey, radial spines rudimentary, on juvenile stem only; flowers nearly white or pale pink); *M.zeilmanniana* (C Mexico; freely clustering; stem to 6 × 4.5cm; tubercles subcylindric, central spines 4, the upper 3 straight, the lowest hooked, slightly longer, all red-brown, radial spines about 15–18, finely bristly, pubescent, white; flowers violet-pink or purple, rarely white).

Mandevilla (for H.J. Mandeville, British Minister at Buenos Aires, who introduced *M.laxa*). Apocynaceae. C and S America. Some 120 species, woody, evergreen lianes with funnelform to salverform, 5-lobed flowers at various times of year. Grow in a coarse, well-drained but moisture-retentive medium rich in organic matter, with shade from the strongest summer sun. Provide a minimum temperature of 7–10°C/45–50°F. Young plants grow well and flower freely in well-lit, warm rooms in the home and are often offered for sale as houseplants trained on wire hoops. Water plentifully and liquid feed occasionally when in growth and flower. Provide support and mist over frequently, especially in warm bright weather or if grown in the home. Reduce water as light levels

Mandevilla splendens

Mandevilla × amabilis

and temperatures drop in autumn; keep just moist in winter. Prune in late winter/early spring to thin old and crowded growth and cut back remaining branches to short spurs. Propagate by semi-ripe or softwood nodal cuttings.

M. × amabilis. (*M.splendens* × *?*; climber to 4m; flowers 9–12cm in diameter, rose pink, rosy crimson at centre, throat yellow); *M. × amoena* (*M. × amabilis* × *M.splendens*; a backcross with more abundantly produced, darker flowers to 10cm in diameter; includes *M.* 'Alice du Pont' and *M.* 'Splendens Hybrid'); *M.boliviensis* (Bolivia, Ecuador; WHITE DIPLADENIA; slender climber to 4m; flowers to 5 × 8cm, white, throat golden yellow); *M.laxa* (syn. *M.suaveolens*; CHILEAN JASMINE; Argentina; climber to 4m; flowers highly fragrant, to 5cm in diameter, white to ivory); *M.sanderi* (Brazil; climber to 5cm; flowers rose-pink, throat to 4.5cm; 'Rosea': BRAZILIAN JASMINE, flowers to 8cm, salmon pink with throat and tube yellow within); *M.splendens* (SE Brazil; climber to 6m; flowers 7.5–10cm in diameter, rose pink; includes 'Rosacea', with rose-pink flowers, flushed and edged deeper rose and with a tube ringed bright rose at throat and yellow within).

Mandragora (Classical Greek name used by Hippocrates). Solanaceae. Mediterranean to Himalaya. 6 species, perennial herbs with stout, forked taproots and oblong to lanceolate leaves in a basal rosette. The short-stalked, solitary flowers are bell-shaped, 5-lobed and appear in spring. They are followed by toxic, tomato-like fruit. Hardy in climate zone 7. Grow in a warm, sheltered position, in deep, light, well-drained but moisture-retentive, circumneutral soil; it will tolerate some shade. Propagate by seed sown ripe or in spring, or by root cuttings in winter. *M.officinarum* (MANDRAKE, DEVIL'S APPLES; N Italy, W Balkans; leaves to 30cm, somewhat puckered and wavy; flowers to 2.5cm, blue to purple, lobes triangular; fruit to 3cm in diameter, yellow).

Manettia (For X. Manetti (b. 1723), keeper of the botanic garden at Florence). Rubiaceae. Tropical America. Some 80 species, soft-stemmed or woody-based, evergreen twiners with brightly coloured, tubular to funnel-shaped flowers at various times of year. Grow in bright, indirect light, on trellis or other support, in the greenhouse or conservatory, with a minimum temperature of 10°C/50°F. Water plentifully when in growth; reduce watering as temperatures fall in autumn. Avoid persistent wetting of foliage, especially in bright sunlight or cold weather. Cut back if necessary in spring. *Manettia* species can also be grown as half-hardy annuals, planted outside for summer display, in humus-rich soils with part shade. Most plants will degenerate after a few years, making frequent repropagation desirable: take softwood cuttings of new growth in spring or semi-ripe cuttings in summer.

M.cordifolia (FIRECRACKER VINE; Bolivia to Argentina, Peru; twining herb, to 4m; leaves heart-shaped, smooth; flowers to 5cm, vivid red or

dark orange fading to yellow at lobes, tube distended above); *M.luteorubra* (syn. *M.bicolor*, *M.inflata*; BRAZILIAN FIRE CRACKER, TWINING FIRE CRACKER, FIRECRACKER VINE; Paraguay, Uruguay; perennial, twining herb or shrub to 4m; leaves ovate to elliptic, roughly but minutely hairy; flowers to 5cm, tube cylindric, bright red, downy, lobes yellow).

manganese an essential minor (micro-) nutrient, required as a constituent of enzymes that are important in respiration and protein synthesis. Symptoms of deficiency occur in plants grown on high lime soils, especially those rich in organic content; however, symptoms may be due to the element being unavailable rather than absent. Deficiency is commonly expressed as leaf chlorosis, commencing at the margins with colour left in the finest veins, thus giving a distinct netted appearance. Although commencing in shoot tips, older leaves appear more affected as new leaves grow out of the symptoms. Characteristic spotting occurs on potatoes and beet; and also on the cotyledons of peas and some beans where it causes the condition known as marsh spot, found when the seeds are split open. The avoidance of overliming will prevent deficiency and sprays of manganese sulphate are an effective treatment. Manganese toxicity can occur on soils of a pH below 5.0.

Manglietia (from the Malay name for one of the species). Magnoliaceae. Malaysia to S China, E Himalaya. Some 25 species, shrubs or trees, resembling *Magnolia*, but with leaf petiole bases appearing swollen, 9 flower segments in 2 whorls, occasionally more, and ovules to 6 per carpel (*Magnolia* has 2 ovules per carpel). Cultivate as for *Michelia*. *M.insignis* (W China, Himalaya, Burma; tree to 12m; young shoots grey-downy; leaves to 20cm, oblong to elliptic, glossy above, glaucous beneath, tough; flowers to 7.5cm in diameter, white tinged pink or carmine).

manure is a term often used to refer to any substance added to soil as a source of plant nutrients, although it is most appropriately used as a description of bulky organic material derived from animals or plants. Manure provides nutrients and humus that benefit soil structure. The type most commonly referred to is farmyard manure (FYM), which is cow, pig or horse dung mixed with straw or litter. The nutrient content varies according to the type of animal and its feeding, the proportion of litter and the duration and method of storage. Exposure to the elements rapidly lowers nutrient levels, especially of nitrogen. The average analysis of FYM is 0.5% nitrogen (N), 0.25% phosphate (P) and 0.5% potash (K).

Horse dung is usually the richest and driest, and has the best effect on soil structure. It is also very suitable for hot beds and mushroom beds. Pig and cow manure have a close texture and are better used on light soils than on heavy ones. Cow manure will dry rapidly and can be crumbled for use in potting mixtures.

FYM is a possible source of weed seeds and occasionally plant pests and diseases. It should be

Manettia luteorubra

composted for six months before use, preferably stacked on a concrete surface under cover.

Poultry manure has a higher nitrogen and phosphate content than farmyard manure at about 2.5% N and 1% P, which rises on drying. Pigeon manure has the highest nutrient content of all, at about 14% N, 11% P and 3% K. Both can be used fresh as long as care is taken not to exceed the nitrogen requirements of a crop, but they are best added in small quantities to a compost heap, or dried and mixed with soil or sand for top-dressing.

Sewage sludge provides organic matter for the improvement of soil structure but it is of variable nutritional content and it is important to assure freedom from heavy metal contamination. It is more suitable for use on agricultural land than in gardens.

Organic manures can be worked into the soil at autumn or spring digging or alternatively spread over the surface, 8–12cm deep, during autumn or winter, for incorporation by earthworms prior to spring digging; incorporate throughout the top 18cm at rates of not less than 5.5kg/m^2. See *compost, green manuring, guano, fish manure, leafmould, liquid feeding*.

Maranta (for B. Maranti (fl. 1559), Venetian botanist). Marantaceae. Tropical C and S America. Some 32 species, evergreen herbaceous perennials with short, clumped stems arising from a starchy rhizome. The leaves are long-stalked with obovate to elliptic blades and exhibit sleep movements – standing erect as if in prayer at night. Small flowers are carried in sparsely branched, bracteate spikes throughout the year. Cultivate as for *Calathea*. The following species is one of the most widely grown foliage houseplants. *M.leuconeura* (PRAYER PLANT, TEN COMMANDMENTS; Brazil; low-growing; leaves to 12cm, elliptic, lustrous dark green, zoned grey or maroon, veined silver, red or purple above, grey-green or maroon beneath; flowers white stained or spotted violet; var. *erythroneura*: HERRINGBONE PLANT, leaves velvety black-green with prominent red veins and a lime green zone along midvein; 'Massangeana': leaves blue-tinted, dull rusty-brown toward centre with a jagged silver band along midrib and silver lines along lateral veins; var. *leuconeura*: leaves broadly elliptic, dark green above with a pale, comb-like central zone and silver feather veins; and var. *kerchoviana*: RABBIT'S FOOT, leaves grey-green with a row of purple-brown to dark olive blotches on each side of midrib).

marcescent withered but persisting, for example, the leaves of young beech plants or those trimmed as hedges in late summer.

marcottage, marcotting see *air layering*.

marginal plant a plant grown either in shallow water or in moist soil around pool edges and banks of watercourses.

Margyricarpus (from Greek *margaron*, pearl, and *karpos*, fruit, referring to the white fruit). Rosaceae. S America (Andes). 1 species, *M.pinnatus*, PEARL FRUIT, an evergreen, densely branched subshrub to 60cm with pinnate leaves to 2cm long divided into many fine and silky leaflets. Inconspicuous flowers are followed in summer by fleshy, rounded fruits to 0.7cm in diameter and white with purple tints. Grow in well-drained, moisture-retentive, preferably lime-free soils in an open position, but with some shade during the hottest part of the day. Hardy to –5°C/23°F – protect from prolonged winter frosts. In the alpine house, grow in pans and water freely when in growth, keeping dry but not arid in winter. Propagate by softwood cuttings in early summer, by simple layering, or from seed sown in autumn.

marionberry see *hybrid berries*.

marjoram (*Origanum* species) a culinary herb native of mountainous Mediterranean areas and known to have been cultivated in Ancient Egypt. SWEET or KNOTTED MARJORAM (*O. majorana*) is grown as an annual, and reaches a height of 60cm. Sow under protection in March or a little later outdoors, planting or thinning to 20cm stations. POT MARJORAM or OREGANO (*O.vulgare*) is hardier and grown as a bushy perennial, up to 90cm high. Raise from seed, cuttings or division, and trim regularly. There are attractive variegated forms of MARJORAM, usually less distinctly flavoured. Leaves can be dried satisfactorily or specimens potted to overwinter productively under protection.

marjoram

marl a calcareous clay soil. Marl is sometimes applied to sandy soils to improve structure, fertility and water retention.

marrow (*Cucurbita pepo*) VEGETABLE MARROW a bush or trailing annual producing large, cylindrical, edible fruits; immature marrows of some cultivars are known as courgettes or zucchini. Originating in northern Mexico and the southern US, the marrow is now widely distributed in many tropical and subtropical regions and can be cultivated outdoors in temperate regions during the summer period.

Marrows require a fertile, moisture-retentive soil achieved by incorporating well-rotted organic matter. A soil temperature of a least 13°C/55°F is necessary for germination, and plants are best sown in small pots under protection in the spring; set out 4–5 weeks later, after hardening off and when risk of frost has passed. Later direct sowings can be made in the open or under cloches, placing two or three seeds per site at a depth of 2.5cm; thin later to leave only the strongest seedling. Space 90cm apart each way for bush types and 1.2m for trailing types. Marrows must be kept well-watered particularly during flowering, and mulching of the soil surface is benefical.

Separate male and female flowers are produced and these are insect-pollinated; under cool conditions, hand-pollinate by removing male flowers and

marrow

Pests
aphids
red spider mite
whiteflies

Diseases
virus
downy mildew
grey mould
gummosis
powdery mildew

Masdevallia macrura

Masdevallia elephanticeps

Masdevallia veitchiana

dabbing pollen into the female flowers – these can be recognized by the immature ovary below the perianth. Regular harvesting extends the production period and increases total yield. Courgettes should be cut when about 10cm long and marrows when between 15–20cm long. Marrows developed to their full size can be stored for winter use. See *courgette, pumpkins* and *squashes*.

Martynia (for John Martyn (1699–1768), Professor of Botany at Cambridge). Pedaliaceae. C America and West Indies, in highlands; naturalized in India and New Caledonia. UNICORN PLANT, DEVIL'S-CLAW, ELEPHANT-TUSK, PROBOSCIS FLOWER. 1 species, *M.annua*, an annual or perennial herb, 1–2m, clammy and hairy with ovate to broadly deltoid, shallowly lobed leaves to 25cm long. Produced in summer, the flowers are 5cm long, tubular, 2-lipped and cream to maroon, with the tube usually yellow, orange- or red-spotted within and the lobes blotched deep purple. They give rise to remarkable fruit – an oblong capsule, 2.5–3.5cm long with two long, curved horns. Sow seed in early spring under glass at 20°C/70°F, and prick out into individual pots. Harden off before planting out after the last frosts, into any well-drained, fertile soil in full sun.

Mary garden a medieval enclosed garden with features of religious significance, including flower-spangled turf and plants having Christian symbolism, especially the Madonna lily.

Masdevallia (for José Masdevall (d. 1801), Spanish botanist and physician). Orchidaceae. Mexico to Brazil and Bolivia. Some 350 species, small evergreen perennial herbs with clumps of obovate to oblanceolate leaves borne on short stalks. The flowers are solitary or racemose, thinly fleshy and essentially triangular in outline due to the showy, expanded sepals. These are ovate to triangular, fused below in a narrow or cup-shaped tube, then spreading and terminating in a tail. The petals and lip are far smaller and virtually hidden within the sepaline cup. Pot in a mix of fine bark, charcoal and sphagnum in small pots. Position in light shade in a buoyant, humid, cool environment – growth will deteriorate where temperatures exceed a day maximum of 25°C/75°F or a night range of 5–10°C/40–50°F. Although these plants should never be allowed to dry out, they are susceptible to damping off, and should therefore be watered carefully. Propagate by division.

M. barleana (Peru; inflorescence to 25cm, 1-flowered; sepaline tube narrow, scarlet, campanulate-cylindric, slightly decurved, dorsal sepal to 4cm, red-orange with red lines, ovate-triangular to subquadrate, lateral sepal connate for two-thirds, bright carmine shaded scarlet, with scarlet lines, tails to 14mm, sometimes crossing); *M.caudata* (Colombia, Venezuela, Ecuador; inflorescence 1-flowered, flowers slightly fragrant, to 15cm across, usually smaller; sepaline tube short, cup-like, dorsal sepal lime to buff, spotted and lined lilac, free portion to 2.5cm, obovate, con-cave, tails to 6.5cm, slender, yellow, lateral sepal buff flushed or spotted lilac or rose, free portion to 1.5cm, ovate, tails to 5cm, pale green, deflexed); *M.coccinea* (Colombia, Peru; inflorescence to 40cm, 1-flowered; flowers waxy, variable in colour, sepals deep magenta, crimson, scarlet, pale yellow or cream-white, sepaline tube to 2cm, campanulate to cylindric, dorsal sepal to 8cm, triangular or linear, tail slender, erect or recurved, lateral sepals ovate-triangular, large, fused, with ridges and short tails); *M.ignea* (syn. *M.militaris*; Colombia; inflorescence to 40cm, 1-flowered; flowers to 8cm across; sepals scarlet to orange, often tinged crimson, sepaline tube to 2cm, cylindrical, hooded, curved, dorsal sepal free portion to 1cm, triangular, tail to 4cm, lateral sepals broadly falcate-ovate, tails short); *M.macrura* (Colombia, Ecuador; inflorescence usually equalling leaves, 1-flowered; flowers to 25cm across; sepals red to dull brown-yellow, with maroon warts, sepaline tube to 1.5cm, cylindrical or flattened, ribbed, dorsal sepal free portion to 15cm, lanceolate to triangular, tail long, yellow-green, lateral sepals to 12.5cm, ovate to oblong, connate for basal 4cm, tails long, strongly decurved); *M.mooreana* (syn. *M.elephanticeps*; Colombia, Venezuela; inflorescence to 10cm, 1-flowered; flowers to 9cm across; sepaline tube broadly cylindrical, sepals long, tapering, forward-pointing, dorsal sepal yellow-white streaked purple at base, triangular, tail yellow, to 5cm, lateral sepals crimson to purple, interior with black-purple papillae, triangular, tail yellow toward apex); *M.schroederiana* (Peru; inflorescence to 21cm, 1-flowered to 8cm across; sepaline tube short-campanulate, white, ribbed, expanding as a flattened oblong platform formed by fusion of the lateral sepals, pearly white flushed ruby red in upper portions and at dorsal sepal, tails long, yellow, lateral tails held horizontally, dorsal tails erect); *M.tovarensis* (syn. *M.candida*; Venezuela; inflorescence to 18cm, 1–4-flowered; flowers to 3.5cm across, pure white, with tails cream or pale jade; sepaline tube cylindrical, to 6mm, dorsal sepal free portion to 4 × 0.6cm, filiform, erect, lateral sepals to 4 × 1cm, shaped like a lyre, connate for half of length, ribbed, tails short, often crossing); *M.veitchiana* (Peru; inflorescence to 45cm, 1-flowered; flowers to 8cm across, sepal interior shinning vermilion covered with many iridescent purple papillae, sepaline tube to 3cm, campanulate-cylindrical, dorsal sepal free portion to 3cm, triangular-ovate, tail, to 3.5cm, lateral sepals larger than dorsal sepal, broadly ovate or triangular, acuminate, curving downwards, connate for 3cm, tails short, often forward-pointing or overlapping); *M.wageneriana* (Venezuela; inflorescence 1-flowered to 5cm, slender; flowers to 6cm light green-yellow or cream, orange-yellow toward base, spotted and streaked violet, sepaline tube short, dorsal sepal free portion broadly ovate-oblong, to 1cm, concave, tail slender, to 5cm, sharply recurved, lateral sepals similar to dorsal sepal, connate for more than half of length). For *M.bella*, see *Dracula bella*; for *M.chimaera*, see *Dracula chimaera*; for *M.erythrochaete*, see *Dracula erythrochaete*; for *M.vampira*, see *Dracula vampira*.

mat (1) the layer of dead grass and debris accumulating near soil level in turf; (2) a thick cover of coir or other fabric used to protect frames against frost; (3) a coir or chain mat used to work top-dressings into turf. See *capillary mat*.

Matteuccia (for C. Matteucci (1800–1868), Italian physicist). Dryopteridaceae (Athyriaceae). Temperate N America, Europe, E Asia. OSTRICH FERN. 4 species, deciduous perennial ferns with erect, stout stocks and fast-spreading rhizomes. The sterile fronds are lanceolate to oblong in outline and 2-3 × pinnately compound; they stand semi-erect in shuttlecock-like rosettes. The fertile fronds are erect and smaller with darker, much-reduced pinnae. Fully hardy and ideal for damp situations, *Matteuccia* spreads by underground rhizomes to form colonies of fresh lime 'shuttlecocks'. The foliage is a delightful feature in the garden in spring but can become rather ragged late in a dry season. The fertile fronds persist throughout the following winter. Plant in moist or wet soils rich in decayed vegetable matter, in shade or dappled sun. *M.struthiopteris* can be invasive. Propagate by offsets – these may surface at some distance from the mother plant. *M.struthiopteris* (OSTRICH FERN, SHUTTLECOCK FERN; Europe, E Asia, eastern N America; usually deciduous, spreading by buried runners; sterile fronds 12–100, soft, bright green, pinnae 30–70, narrowly lanceolate, pinnatifid).

Matthiola (for Pierandrea Mattioli (1500–1577), Italian physicist and botanist). Cruciferae. W Europe, C Asia, South Africa. STOCK, GILLYFLOWER. 55 species, annual or perennial herbs, occasionally subshrubs, with terminal racemes of 4-petalled flowers in summer. These are highly fragrant, especially at night, and are excellent for cutting.

Grow in sun in a fertile, well-drained, neutral or slightly alkaline soil. Sow seeds of annuals in early spring under glass, at 15°C/60°F. Grow on in well-ventilated conditions at 10°C/50°F; water moderately, allowing the plants to become almost dry between waterings. Harden off and plant out after the last frost; stake tall cultivars. In zones that do not experience hot humid summers, seed may be sown *in situ* in spring. To ensure double flowers, seed stock must be from a selected race and germinated in warmth. On emergence of the first pair of leaves, the temperature is dropped to at least below 10°C/50°F. The genetic difference expressed in cotyledon colour should then be quite distinct: those with yellow green or pale coloured cotyledons will be double-flowered, and dark green seedlings may be discarded. With the Trisomic seven-week strains, after pricking out and growing on in the usual way, doubles should become apparent at four-leaf stage, although the difference is less obvious – the sturdier seedlings are likely to be double. Stocks of the Stockpot type, a dwarf strain grown for bedding, cutting and pot plants, exhibit notched seed leaves in double-flowered plants. For greenhouse flowers, sow from late summer in succession until mid-winter, spaced 15–20cm in the greenhouse border or given a final pot size of 10–15cm. Use a porous, medium-fertility, loam-based propagating mix. Maintain a temperature of about 10°C/50°F and keep well-ventilated. Liquid feed weekly, as the roots fill their pots. Sow biennials in summer in a seedbed or lightly shaded frame; overwinter in the cold greenhouse or frost-free frame, and plant out in early spring. Provide cloche protection if overwintering outdoors in all but the mildest areas. A bacterial leaf spot, leaf rot and stem rot, canker, black root rot, club root, downy mildew and the cabbage root fly can affect stocks.

M.incana (BROMPTON STOCK; S and W Europe; woody-based biennial, 30–80cm; leaves lanceolate, grey-green, hairy; flowers purple, sometimes pink or white; includes 'Annua', the fast-growing TEN WEEKS STOCK, grown as an annual); *M.longipetala* (NIGHT-SCENTED STOCK; Greece to SW Asia, S Ukraine; annual, 8–50cm; leaves oblong to linear, simple or pinnatisect, entire or toothed; flowers yellow, green-white or pink). *Cultivars.* Modern selections of garden stocks are available in a wide range of sizes, from the 20cm dwarf Stockpots to the Column types up to one metre tall, in colours from pastel pinks to mauve-blues and carmines, and shades of yellow, copper and gold. The annuals are the largest group, and include the Ten-week and dwarf, bushy Trisomic seven-week races. Annuals fall into the following groups – (a) *Ten-week Dwarf Large-flowering*: compact, free flowering, to 30cm, excellent for bedding purposes. (b) *Ten-week Excelsior* (*Column*): of upright habit, to 80cm, producing the single columnar densely flowered spike valued by florists; for the greenhouse or outdoors and excellent for cutting. The Pacific seed race is also of this columnar type, and better suited to outdoor cultivation. (c) *Ten-week Giant Imperial*: of bushy and branching habit, to 60cm, with long well-shaped spikes; for cutting and bedding. (d) *Ten-week Giant Perfection*: of erect and bushy habit, to 70cm, with long spikes of large flowers; for bedding and cutting and exhibition. (e) *Ten-week Mammoth* or *Beauty*: erect, bushy and compact, to 45cm, producing several spikes of large flowers; for bedding, cutting and for greenhouse cultivation for winter bloom. (f) *Perpetual-flowering*, or *All the Year Round*: dwarf, vigorous plants, with large spikes of pure white flowers, excellent for cutting, and pot culture. Other major divisions are Brompton stocks, erect and bushy biennials, to 45cm, used for bedding, and Intermediate or East Lothian stocks which, treated as annuals, flower to follow on from the Ten-week stocks, and are also grown as biennials flowering in spring and early summer. Virginia stock is *Malcolmia maritima*.

mattock a short-handled tool for breaking up hard ground and grubbing up trees; similar to a pickaxe, with one side of the head pointed and the other fashioned into a broader chisel-like blade. In the *grubbing axe*, one blade is made in the same plane as the handle for use as an axe. The *grubber* or *grubbing mattock* has a broad blade at right angles to the handle, and the other blade pointed or chisel-like.

Maxillaria
tenuifolia

Maxillaria (from Latin *maxilla*, jaw, referring to the supposed resemblance of the column and lip to the jaws of an insect). Orchidaceae. Tropical America from West Indies and Mexico to Brazil, with 1 species in Florida. Over 600 species, evergreen, epiphytic perennial herbs. Those listed here have pseudobulbs and slender leaves. Produced in spring and summer, the flowers are solitary with petals similar to sepals but usually smaller; the lip is fleshy, concave and entire or trilobed. Cool or intermediate-growing (minimum temperature 10°C/50°F), *Maxillaria* species require an open bark mix, a brief winter rest and plentiful water, high humidity and light shade in spring and summer.

M.cucullata (Mexico; flowers dingy dark brown, tepals 25–28mm, lanceolate, lip 23mm, dark purple-red); *M.grandiflora* (Ecuador, Peru, Colombia, Venezuela, Guyana; flowers nodding, milk-white, scented, fleshy, dorsal sepals 35–45mm, ovate to oblong, lateral sepals slightly longer and wider, petals shorter and narrower; lip to 25mm, ovate); *M.lepidota* (Venezuela, Colombia, Ecuador; tepals yellow, marked red at base, lip creamy yellow with maroon marks, sepals to 60mm, lanceolate, petals to 45mm, curved-lanceolate, lip 20mm, fleshy, midlobe covered with yellow-green farina); *M.meleagris* (Mexico, Guatemala, Panama; flowers scented of coconut, buff-orange, buff-olive or flesh-coloured, marked with dark red, lip dark blood red, sepals 12–30mm, elliptic, petals shorter, lip 7–16mm, trilobed, recurved, midlobe fleshy, tongue-like); *M.picta* (Brazil; flowers large, golden yellow inside, cream-yellow outside, banded and flecked purple-brown, lip yellow-white or cream spotted red, sepals about 30mm, oblong, petals slightly shorter and narrower; midlobe tongue-shaped); *M.sanderiana* (Peru, Ecuador; flowers large, fleshy, tepals white flecked violet-purple, lip dull yellow with red markings, dark purple on outer surface, dorsal sepal 60–75mm, oblong to lanceolate, lateral sepals wider and longer, petals somewhat shorter, lip 30mm, ovate, obscurely trilobed, margin crisped); *M.tenuifolia* (Mexico to Honduras and Nicaragua; flowers with a strong coconut scent, deep red mottled with yellow towards base of tepals, lip yellow, marked dark red, sepals spreading 20–25mm, lanceolate or ovate, petals shorter, projecting forwards, lip about 16mm, obscurely trilobed, midlobe tongue-shaped, fleshy); *M.venusta* (Colombia; flowers nodding, tepals milk-white, lip yellow with 2 red spots, margins of lateral lobes red, sepals to 75mm, lanceolate, dorsal sepal concave, lateral sepals curved, petals slightly shorter, lip fleshy, trilobed, midlobe triangular, recurved).

maypoling the practice of supporting heavily laden branches of a fruit tree with strong ties in order to avoid breakage of limbs. The ties may be attached to the top of a pole which is lashed to the trunk of the tree or to a central stake.

Maytenus (from Chilean vernacular name). Celastraceae. Tropical and subtropical regions. Some 255 species, evergreen shrubs or small trees. Produced in spring, the flowers are inconspicuous, but the fruits split to reveal seeds with colourful arils. *M.boaria* will tolerate temperatures to –10°C/14°F, but needs a warm and sheltered site in sun or semi-shade. The soil should be well-drained but never dry. Root semi-ripe cuttings under mist and with gentle bottom heat. *M.boaria* (syn. *M.chilensis*; MAYTEN; Chile; evergreen shrub or tree to 24m with a broad, weeping crown; leaves 2.5–5cm, lanceolate to narrowly elliptic, finely serrate; fruit yellow; seeds 2, aril scarlet).

maze a layout of paths with numerous alternatives, dead ends, and confusing turnings. The earliest garden mazes were two-dimensional, made up of paths cut in turf to form a knot garden. The hedge-maze, properly a labyrinth, is the later more popular concept formed with evergreens, such as box, holly or yew, designed to confuse the user, with the hedges usually trained to a height that is impossible to see over.

Mazus (from Greek *mazos*, breast, referring to the protuberances in the corolla throat). Scrophulariaceae. SE Asia, China, Taiwan, Malay Archipelago, Australasia. Some 30 species, annual or perennial herbs, usually creeping or prostrate, with tubular flowers in spring. These are 2-lipped, the upper lip shorter and hooded, the lower lip large, 3-lobed and colourfully marked in the throat. A charming dwarf groundcover for sheltered, slightly acid positions in the rock garden, *Mazus* is hardy in climate zone 8. Protect in winter with evergreen branches or overwinter stock plants in frost-free conditions. Increase by division in spring. *M.reptans* (Himalaya; creeping perennial; leaves to 1cm; flowers purple-blue, lower lip blotched white, yellow and red-purple; 'Albus': flowers white).

meadow gardening see *alpine meadow*.

meal (1) a powdered fertilizer produced by drying and grinding organic materials, such as bones or fish; it is easy to apply and the nutritional content is quickly available to plants; (2) a natural surface covering of the leaves of some plants, known as farina.

mealybugs (Hemiptera: Pseudococcidae) sap-feeding insects, up to 5mm long, with soft, flesh-coloured bodies fringed with fine waxy filaments and covered with a powdery white secretion or 'meal'. Males are winged and resemble very small flies but some species have females only and reproduce by parthenogenesis. Mealybugs congregate in leaf axils, sheaths and under loose bark, covering themselves and their eggs with a fluffy, white wax, which is often the first sign of infestation. Established colonies debilitate plants and deposit honeydew on the foliage; this acts as a substrate for unsightly sooty mould growth.

ROOT MEALYBUGS seldom exceed 3mm in length and are plainly visible in pockets and dry patches in the root balls of infested plants knocked out of their pots. Poor growth followed by wilting is the most obvious aerial symptom.

Most mealybugs are natives of tropical and subtropical regions. In temperate zones, they are mainly a problem on species grown under protection, particularly *Bougainvillea*, *Codiaeum*, *Cycas*, *Ficus*, *Gardenia*, *Hippeastrum*, *Hoya*, *Passiflora*, *Plumbago*, *Saintpaulia*, citrus plants, palms, and cacti and succulents. Because of their hidden feeding sites and waxy covering, mealybugs are difficult to control but approved insecticides may be applied either by immersion, use of a drenching spray or by direct application with a fine brush. Woody deciduous subjects can be sprayed or painted with a tar-oil wash during dormancy, after removal of loose bark and dead leaves. Biological control may be achieved using a ladybird beetle (*Cryptolaemus montrouzieri*) and/or certain parasitic wasps (*Leptomastix dactylopii*, *Leptomastidea abnormis* and *Anagrus pseudococci*).

meat meal, **meat guano** an organic fertilizer made by processing meat refuse into powdered form. Where made only from meat, it is purely nitrogenous, providing up to 10% nitrogen. The inclusion of ground bone gives an analysis of around 6% nitrogen and 20% phosphate, and also adds trace elements. Both products are relatively slow-acting. They should be sterilized to eliminate pathogens, and handled with suitable precautions to avoid inhalation and other personal contact.

Meconopsis (from Greek *mekon*, poppy, and *opsis*, appearance). Papaveraceae. Himalaya to W China, W Europe. ASIATIC POPPY, BLUE POPPY. 43 species, annual, biennial or perennial herbs, often monocarpic, with hairy to bristly leaves in basal rosettes and, in spring and summer, poppy-like flowers on erect, simple or branching stems.

Hardy and semi-hardy biennials and perennials which, with one exception (*M.cambrica*), are native to the Himalayas and mountains of China and Tibet. Many are monocarpic, although they may take between two and four years before flowering. They are cultivated for their beautiful nodding poppy-like flowers, although the basal leaf rosettes of some species are also very handsome. These are plants for the woodland garden, peat terrace and shady herbaceous border. Although tall, many are suitable for the rock garden if planted in moist crevices with perfect drainage out of direct sun. They are hardy in climate zone 7, but need cool, moist summers and relatively mild winters. Nearly all grow best in part shade, and this is important to ensure and preserve good flower colour. They require shelter from summer heat and strong winds, and a moist but not stagnant root run in soil that is fast-draining. Even the hardiest species will rot if water accumulates around the collar. In very wet areas, protect evergreen species with open cloches or panes of glass. The exception is *M.cambrica*, which will grow in virtually any soil, in any position. Plant in spring in a humus-rich, lime-free soil. Deadhead to prolong flowering and cut down perennial species in autumn. Lift and divide perennial species every three years to prevent deterioration. Some, such as *M.grandis*, *M.* × *sheldonii*, *M.quintuplinervia*, can be divided in spring or autumn. All species can be seed-propagated. Sow ripe seed in late summer in equal parts peat and sand. Prick out into boxes and overwinter in a well-ventilated cool greenhouse or frame. Plants may suffer from downy mildew (*Peronospora*), which is grave at seedling stage and should be treated with appropriate fungicide. The perennial species are thought to be longer-lived if they develop several crowns before being allowed to flower – for example, *M.betonicifolia* will be monocarpic if flowered as a biennial but may flower over several seasons if raised slowly (in cool conditions) and stopped in the first two years to induce offsetting.

M.betonicifolia (BLUE POPPY; China; short-lived perennial, 30-200cm; leaves to 35cm, stalked, oblong to ovate, serrate, sparsely bristly; flowers to 6, in somewhat drooping cymes, petals to 5cm, mauve-pink to bright sky blue; 'Alba': white); *M.cambrica* (WELSH POPPY; W Europe; perennial to 60cm; leaves pinnately lobed, to 20cm, glabrous to hairy; flowers solitary, in axils of upper stem leaves; petals to 3cm, yellow; includes var. *aurantiaca*, with orange flowers, 'Frances Perry', with scarlet flowers, and 'Florepleno', with semi-double, yellow or orange flowers); *M.grandis* (Himalaya; perennial to 120cm; leaves to 30cm, narrowly oblanceolate to elliptic to oblong, serrate, rusty-bristly; petals to 6cm, purple or deep blue); *M.horridula* (Himalaya to W China; short-lived perennial, to 80cm; leaves elliptic to narrowly obovate, to 25 × 3cm, grey-green with purple or yellow warty and spiny bristles; petals to 4cm, cobalt blue to violet or white); *M.integrifolia* (Tibet, Upper Burma, W China; short-lived perennial, to 90cm, covered with downy, orange-red hairs; leaves entire, oblanceolate to obovate or linear, to 37cm, densely hairy; petals to 3cm, yellow or white); *M.napaulensis* (SATIN POPPY; C Nepal to SW China; perennial, to 2.6m; leaves to 50cm, pinnately cut, with dirty gold to brown or tawny hairs and bristles; petals to 4 × 3cm, red or pink to purple or blue, rarely white); *M.paniculata* (E Nepal to NE Assam; short-lived perennial, to 2m; leaves pinnately cut, white- to tawny-shaggy and bristly; petals to 5cm, yellow); *M.quintuplinervia* (NE Tibet, China; perennial, to 30cm; leaves to 25cm, golden to rusty-bristly, obovate to narrowly oblanceolate; petals to 3cm, pale mauve-blue or light violet, rarely white); *M.* × *sheldonii* (*M.betonicifolia* × *M.grandis*; leaves oblong to lanceolate, bristly, serrate; petals to 3cm, blue; includes 'Slieve Donard', to 1m, with aquamarine flowers).

Meconopsis napaulensis

Medicago (from *Medike*, name given by Dioscorides to a grass). Leguminosae. Europe, Mediterranean, Africa, Asia. MEDIC, MEDICK. Some 56 species, annual or perennial herbs or small shrubs with trifoliolate leaves and racemes of small, pea-like flowers, usually in summer. The fruit is a legume, but often spiny and spiralling like a snail's shell. The shrubby *M.arborea* is hardy to about

−10°C/14°F. In regions at the limits of its hardiness, it requires the protection of a south wall, but is resistant to salt spray and wind. Plant in full sun on a fast-draining soil. Prune out dead or weak growth in spring. Sow seed in autumn or spring (stratification or scarification may improve germination); increase also by division of softwood or semi-ripe cuttings in summer. *M.arborea* (MOON TREFOIL; S Europe; grey-downy shrub 1–2m; flowers 12–15mm, yellow; fruit in spiral of 1–2 turns).

Medinilla (for J. de Medinilla de Pineda, Governor of the Marianne Islands in 1820). Melastomataceae. Tropical Africa, SE Asia, Pacific. Some 150 species, evergreen shrubs, usually with bold, leathery leaves and small flowers whorled amid large waxy bracts in showy panicles. Provide a minimum temperature of 15–20°C/60–70°F. Grow in bright filtered light, ensuring shade from the hottest summer sun, in a fertile, open and porous medium with leafmould and bark. Provide high humidity, syringe frequently, water plentifully and feed fortnightly when in full growth. During the winter months, water only to prevent wilting and lower temperatures slightly; this will help promote flowering. Propagate by semi-ripe cuttings in a sandy propagating mix in a humid closed case with bottom heat; pinch out young plants to encourage branching. *M.magnifica* (Philippines; robust epiphytic shrub to 3m; branches stout, tetragonal to winged, jointed; leaves 20–30cm, broadly ovate, deep glossy green; panicle to 50cm, strongly pendulous, bracts very large, candy pink, paired, concave, flowers to 2.5cm in diameter, pink to coral-red).

medium a seed or potting compost mixture or other material in which plants may be grown or propagated.

medlar

Diseases
brown rot
powdery mildew

medlar (*Mespilus germanica*) a fully hardy deciduous tree or shrub to 5m high, grown for its small flattened apple-shaped fruits which have a wide, open eye and horny calyces. The MEDLAR is indigenous throughout Europe and probably originated in Transcaucasia; although recorded in Ancient Greece, Rome and Asia Minor, it has never been of great horticultural importance. It is occasionally found wild in southern England and was introduced by the French into North America in the 17th-century; it is sometimes found as a hedging subject in Florida. The MEDLAR forms a decorative tree with arching branches, large pink and white flowers, and orange-gold tinted foliage in autumn. Its fruits have a characteristic flavour which only develops when part-decayed (bletted) after late autumn harvest and storage with calyces downwards, on slatted trays; they are also used to make preserves.

MEDLARS are grown as bushes or half standards; they are self fertile and have a chilling requirement of between 100–450 hours below 7°C/45°F. Trees are usually grafted onto QUINCE A rootstocks, or for half standards onto SEEDLING PEAR. Pruning and training is similar to that employed for apple.

medulla parenchymatous tissue found in the centre of stems of many dicotyledonous plants. It is also called pith.

medullary consisting of pith; spongy.

Meehania (for Thomas Meehan (1826–1901), writer and nurseryman). Labiatae. Asia, N America. JAPANESE DEAD NETTLE. 6 species, creeping stoloniferous perennial herbs with toothed leaves and small, tubular and 2-lipped flowers whorled in leafy spikes in spring and summer. *Meehania* species form large patches of groundcover in dappled shade or in exposed but sunless areas, and are hardy to −15°C/5°F. They flower from mid- to late summer, giving a display of lilac or blue flowers on rather untidy running stems which may form dense clumps. Propagate by rooted stolons, or by seed sown in a cold frame in spring. *M.urticifolia* (Japan, NE China, Korea; hairy; 15–30cm; leaves ovate to cordate, coarsely toothed; flowers 4–5cm, blue-purple, lower lip spotted dark purple).

megaspore the larger of the two types of spore produced by heterosporous plants; they develop into female gametophytes.

megasporophyll in non-flowering plants, a sporophyll bearing megaspores; in angiosperms, a carpel.

Melaleuca (from Greek *melas*, black, and *leukos*, white, in allusion to the often black trunk and white branches). Myrtaceae. Mainly Australia; also New Caledonia, New Guinea and Malaysia. HONEY MYRTLE, BOTTLEBRUSH, PAPER BARK. Over 150 species, evergreen trees and shrubs, usually with papery bark, tough leaves dotted with aromatic oil glands and rounded, 5-petalled flowers with showy stamens. Provide a minimum temperature of 5°C/40°F, full sun and a dry, buoyant atmosphere. Plant in a fast-draining, nitrogen-poor medium. Water moderately in spring and summer, sparingly

medlar

in winter. Increase by seed in spring, or by semi-ripe cuttings in summer.

M.armillaris (BRACELET; New South Wales, Victoria, South Australia; to 8m; leaves 1.4–2.8cm, linear or narrowly elliptic to compressed or semiterete; inflorescence 20–70-flowered, stamens white to mauve); *M.elliptica* (W Australia; to 3m; leaves to 0.6cm, opposite, ovate to suborbicular, close, leathery; inflorescence in oblong or cylindrical spikes, stamens red); *M.hypericifolia* (New South Wales; to 6m; leaves 1-4cm, opposite, elliptic to obovate; inflorescence a many-flowered, dense axillary spike, stamens crimson-red); *M.leucadendra* (RIVER TEATREE, WEEPING TEATREE, BROAD-LEAVED PAPERBARK, PAPERBARK, BROAD-LEAVED TEATREE; N Australia, W Australia, Queensland, also in Malaysia and New Caledonia; tree to 30m; leaves scattered, 8–23cm, narrowly ovate to elliptic, thinly leathery; inflorescence a many-flowered, open spike, stamens white); *M.nesophila* (WESTERN TEA MYRTLE; W Australia; to 7m+; leaves to 2.3cm, alternate, obovate-oblong to oblong-cuneate; inflorescence in terminal dense heads, stamens lavender or rose-pink); *M.squarrosa* (South Australia, SE Australia; to 15m; leaves 0.5–1.5cm, decussate, ovate, pointed; inflorescence a dense terminal spike, stamens white or pale yellow); *M.viridiflora* (New Guinea, New Caledonia, Australia; to 25m; leaves 5–22cm, scattered, obovate, elliptic or broadly ovate; inflorescence a many-flowered spike, stamens white, yellow, green, pink or red; var. *rubriflora*: leaves thinner, shorter, stamens red).

Melasphaerula (from Greek *melas*, black, and *sphaerula*, a Latinized diminutive of a sphere, referring to the corms). Iridaceae. South Africa. 1 species, *M.ramosa*, a perennial, cormous herb to 60cm with lanceolate, grass-like leaves and small, white to cream, purple-veined flowers nodding on wiry panicle branches. Cultivate as for *Ixia*.

Melastoma (from Greek *melas*, black, and *stoma*, mouth; the mouth is stained black by the berries). Melastomataceae. SE Asia. Some 70 species, evergreen shrubs with broad, distinctly veined and tough-textured leaves and 5-petalled flowers in cymes in summer. Cultivate as for *Medinilla*. *M.candidum* (Taiwan and Ryukyu Islands to SE Asia and Philippines; 1.2–2.5.m, hairy; flowers 5–8cm in diameter, fragrant, pink or white).

Melia (Greek name for the ash). Meliaceae. Old World tropics, Australia. 5 species, deciduous or semi-evergreen trees or shrubs with pinnate or 2-pinnate leaves. The flowers appear in spring and summer in loose, axillary panicles; they consist of five to six, spreading, oblong petals with the stamens united in a tube. Drupaceous fruits follow in autumn. Hardy in climate zone 8 or below where summers are long and hot, *M.azederach* is a rapidly growing tree, often used in its native regions for reforestation. It is valued for its dense, rounded crown and fragrant blooms; the yellow berries are toxic. It tolerates a range of well-drained soils and hot, dry conditions, self-seeding in profusion where conditions suit. Propagate by seed. *M.azederach* (CHINABERRY, PERSIAN LILAC, PRIDE OF INDIA, BEAD TREE; N India, China; deciduous tree to 15m; leaves to 80cm, leaflets to 5cm, ovate to elliptic, serrate; flowers 2cm in diameter, lilac, in panicle 10–20cm long, staminal tube violet; fruit 1–5cm in diameter, yellow; 'Umbraculiformis': TEXAN UMBRELLA TREE, branches radiating in a spreading head).

Melianthus (from Greek *meli*, honey, and *anthos*, flower – the calyces are filled with nectar). Melianthaceae. South Africa; *M.major* naturalized in India. 6 species, shrubs with robust, sparsely branched stems, often following a semi-herbaceous habit in colder locations, otherwise evergreen, becoming woody, especially the basal portions, and scented in all parts. The leaves are pinnately compound with winged petioles, conspicuous stipules and large, jaggedly toothed leaflets. Tubular, hooded and richly nectariferous flowers are borne in erect racemes in summer. *M.major* is a magnificent plant grown for its blue-green, architectural foliage in sheltered locations in zones 7 and over. It is also used as a temporary dot plant in bedding schemes, overwintered in a cold greenhouse. Grow in full sun or light shade in moisture-retentive but freely draining soils; fertile soils give good foliage effects, although flowering is better on poor soils. Mulch with a thick layer of coir or bracken litter in autumn in areas where frosts are prolonged – the top growth will be cut back by frost, but the base will resprout in spring. When siting young plants, bury the older, slightly woody stem portion to protect the rootstock and promote offsetting. Propagate by seed in spring, by greenwood stem tip cuttings in summer, or by removal of rooted suckers in spring. Red spider mite may be a problem under glass. *M.major* (HONEY FLOWER; to 2.25m; stem sea-green and thickly glaucous at first, becoming woody below; leaves to 50cm, glaucous, sea-green to grey-blue, leaflets 5–13cm, 9–11, ovate-oblong, coarsely serrate; raceme to 80cm, flowers brown-red).

Melica (from *melike*, the Greek name for a grass). Gramineae. Temperate regions (except Australia). MELIC. 70 species, perennial grasses with clumped culms, linear to lanceolate leaves and small but ornamental spikelets hanging from the delicate branches of slender, arching panicles in spring and summer. These are followed by rice-like grains, pearly in most species but jet-like in *M.altissima* 'Atropurpurea'. Hardy in zone 6. Grow in sun or shade on a light, moisture-retentive, loamy soil. Most species perform well on calcareous soils. Propagate by seed or division. *M.altissima* (SIBERIAN MELIC; C and E Europe; to 150cm; leaf blades to 25cm, scabrous; panicle erect, 1-sided, dense, to 25×2.5cm; spikelets oblong, to 1cm; includes 'Alba, with very pale green spikelets, and 'Atropurpurea', with deep purple-black spikelets, sweeping downwards and overlapping).

Melicytus (from Greek *meli*, honey, and *kytos*, a hollow vessel, referring to the staminal nectaries). Violaceae. New Zealand, Norfolk Island, Fiji, Solomon Islands. 4 species, dioecious shrubs or trees with small flowers and showy berries. Hardy to about –5°/23°F given wall protection and shelter from cold drying winds. Grow outdoors or in a cool, well-ventilated greenhouse, in a moderately fertile, loam-based mix, with direct sun or bright filtered light. Water moderately when in growth, sparingly in winter. Propagate by seed sown when ripe or in spring. *M.ramiflorus* (WHITEY WOOD; tree, to 10m+; bark grey-white; leaves to 15cm, toothed; flowers to 4mm in diameter, yellow-green; fruit glossy, lavender to violet, dark blue or purple).

Meliosma (from Greek *meli*, honey, and *osme*, scent, alluding to the honey-scented flowers). Sabiaceae. Tropical America, tropical and temperate Asia. 25 species, deciduous or evergreen trees or shrubs with pinnate leaves and small, starry and fragrant flowers in large, pyramidal panicles in spring and summer. Hardy in zone 7. Grow in sun on deep, moisture-retentive, moderately fertile, circumneutral soils. Propagate by seed, layers or from softwood cuttings under mist.

M.pinnata (syn. *M.oldhamii*; Korea, China; deciduous, 20–40m; leaves 17.5–35cm, leaflets 5–13, lowermost 2.5 × 2cm, orbicular to obovate, terminal leaflet 7.5–14cm, dentate, pubescent above when young; panicle 15–30cm; flowers pure white); *M.veitchiorum* (W China; deciduous to 50m; leaves 45–75cm, leaflets 8.5–17.5cm, 7–9, ovate or oblong, entire or sparsely toothed, midrib pubescent beneath; panicle to 45cm; flowers cream-white).

Melittis (from Greek *melissa* or *melitta*, bee: the plant is attractive to bees). Labiatae. BASTARD BALM. Western central and southern Europe to Ukraine. 1 species, *M. melissophyllum*, a perennial herb to 50cm with erect, hairy stems. The leaves are oval, to 8cm, crenate, hispid and honey-scented when fresh. Tubular and 2-lipped, the flowers are white, pink or purple or with a large purple blotch on the lower lip. They appear in whorls in early summer. Found in open woodlands, *Melittis* is useful for the front of the herbaceous border and for the herb garden. It is fully hardy. Plant in any moisture-retentive soil with added leafmould, preferably in light, woodland shade. Propagate by division in spring. When dried, *Melittis* retains its fragrance for a long time.

Melocactus (from Latin *melo*, melon, and Cactus, referring to the shape). Cactaceae. Tropical America. About 30 species, cacti with globose, strongly ribbed and spiny stems. The flowers are small, tubular, red to pink and sunken in a woolly apical cephalium. The fruit is a juicy berry, usually clavate and red, pink or white. Grow in full sun with low humidity and a minimum temperature of 10°C/50°F. Plant in a gritty, acid medium. Water very sparingly in winter (i.e. only to avoid shrivelling). Increase by seed or offsets.

M.bahiensis (E Brazil; stem 9.5–21 × 11–21cm, depressed-globose to conic; ribs 8–14, acute to rounded; central spines 1–4, 1.7–5cm, radial spines 7–12, 2–6cm; fruit red to crimson); *M.curvispinus* (syn. *M.oaxacensis*; widespread in C and S America, Cuba; stem 6–30 × 8–27cm, depressed-globose, globose or ovoid; ribs 10–16, acute; central spines 1–5, to 5.2cm, radial spines 7–11(–15), 1.6–4.2cm, curved, rarely straight; fruit bright red); *M.intortus* (syn. *M.communis*; West Indies; stem to 60(–90) × 30cm, globose to tapered-cylindric; ribs (9–)14–27, rounded, thick; central spines 1–3, radial spines 10–14, 2–7cm, stout, straight; cephalium growing tall; fruit pink to red); *M.matanzanus* (N Cuba; resembling *M.violaceus* and *M.neryi*, but with 8–9 ribs and a cephalium with dense orange-red bristles); *M.neryi* (Venezuela to N Brazil; resembling *M.violaceus*, but with a dark blue-green stem and curved, sometimes longer spines); *M.violaceus* (E Brazil; stem 5–18 × 6–17m, depressed-globose to subpyramidal; ribs 8–15, acute; central spine 0–1, ascending, shorter than radials, radials 5–9(–11), to 1–2.4cm, straight; fruit pink or white).

melon, sweet (*Cucumis melo* subsp. *melo*) an annual, prostrate herb grown for its edible fruits, which are usually consumed fresh. Assumed to have originated in Africa, melons were known to the Romans, and in the 15th century were brought from Turkish Armenia to the papal estate of Cantaluppe near Rome, and distributed from there to the rest of the Mediterranean area, western Europe, and on to North America by Columbus on his second voyage.

Melons are a warm-season crop and, according to cultivar, need 85–125 days without frost from sowing to the fruits reaching maturity. In colder climates, with a shorter frost-free period, melons need protection for all or part of the growing season: they take 3–4 months to reach maturity, and warm dry conditions are essential for ripening.

There are many cultivars, grouped according to fruit character. The Cantalupensis group, the cantaloupes, including Charentais types, have ovoid or round fruits with a ridged rind; they include the well-kown 'Ogen'. The Reticulatus group, the musk melons, have net markings on the rind, orange flesh and are widely grown in the US, in the Mediterranean area and under glass in the north. The Inodorus group, winter melons, includes honeydew melons with a harder rind which facilitates longer storage.

Cultivars either have all male or all female flowers on the same plant or they may have male and hermaphrodite flowers on the same plant, as is the case with most American cultivars. Assisted pollination is recommended on a garden scale (see *marrow*).

The most suitable soil is a well-drained fertile loam (pH 5.5–7.0), with a minimum depth of 40cm. Where grown outdoors, soil preparation is started in the autumn before planting, with organic manure thoroughly incorporated to maintain soil structure and improve moisture retention. In some areas, melons are grown on ridges. Before sowing in rows

melon

Diseases
collar rot
grey mould
root rot
virus

or ridges, the planting site may be covered with a plastic mulch to raise soil temperature.

Plants are best raised in pots under protection for transplanting when two or three true leaves have formed, after hardening off and when risk of frost has passed. They may also be direct-sown outdoors, up to 3.7cm deep. Plants should be spaced 0.6–1.0m apart in rows on the flat, or 1.2–2.0m apart on ridges. The minimum soil temperature for germination is 15°C/60°F and the optimum for seedling growth about 30°C/85°F. In cooler climates, young plants grown outdoors may be protected by glass cloches, translucent paper or low polythene-clad tunnels.

Water demand is high when the plants are growing fast and fruits are swelling, but should be reduced just before the fruits start to mature; trickle or seep-hose irrigation is the most efficient system. Fruits that are allowed to ripen develop the best flavour and picking should be carried out at regular intervals. In most musk melon cultivars, fruit is ready for picking when it separates easily from the stem; in cantaloupes, ripeness is indicated by a change in rind colour or, in certain cultivars, by the development of scent and fine circular cracks at the base of the stem. Melons can only be stored for relatively short periods. Musk melons can be kept for 6–12 days at 3–4°C/37–39°F (relative humidity 85–90%), and honeydew melons for 14–90 days at 10–15°C/50–60°F.

In greenhouses, melons may be grown in beds or in raised troughs and boxes (30cm deep, 46cm wide) in a loam compost rendered highly fertile with well-rotted manure. The usual time for sowing is spring. Seeds are sown singly in pots or in soil blocks at a temperature of 16–18°C/60–65°F; after seedlings have emerged, the temperature is reduced to 13–15°C/55–60°F. When true leaves appear, melons are transplanted into the greenhouse bed, 45–60cm apart, along the side walls. When about 10–15cm tall, the growing point is pinched out, stimulating the production of two shoots to be trained up wires on the side walls. A temperature of 16–27°C/60–80°F is maintained, and ventilation provided when the temperature rises above this. When each shoot has produced five leaves, the growing tip is again removed to stimulate side shoots on which the flowers are produced. When three leaves are produced on each of these side shoots, the tips are pinched out. Six to eight flowers are selected per plant and hand-pollinated; once set, four fruits of uniform size are selected (not more than one per lateral) for growing on, and the others removed.

Plants should be kept well-watered, and given a weekly liquid feed from the stage when the fruits are about the size of a walnut. Stop the fruit-bearing laterals at the first joint beyond the fruit when it begins to swell. As the fruits grow, support with nets tied to the wires. Leaves and laterals are thinned out as necessary. When the fruit has reached full size, humidity levels are lowered for ripening, feeding is stopped, and watering reduced so that the medium is kept just moist enough to avoid rind splitting.

For early crops on hot beds in frames, the growing medium must be of sufficient depth to maintain a temperature of about 18°C/65°F. Two plants are positioned in the centre of each frame and two shoots from each led to the corners of the frame. When each shoot has produced 6–8 leaves, the growing tips are cut back to four leaves. The side shoots thus stimulated are next stopped at three leaves and flowers hand-pollinated, eventually allowing two fruits to develop per plant for main crops, and one for early and late crops. A flat stone may be placed under each fruit to keep it clean and to aid uniform ripening.

membranous (membraneous), membranaceous thin-textured, soft and flexible.

Menispermum (from Latin *menis*, a tiny half-moon-like device inscribed at the opening of books, and *spermum*, seed, referring to the shape of the seed). Menispermaceae. Eastern N America, E Asia. MOONSEED. 2 species, woody-based, dioecious twiners with slender stems and rounded, obscurely lobed leaves. Inconspicuous, yellow-green flowers are borne in axillary, stalked racemes or panicles in summer. Where plants of both sexes are grown together, the female bears small dark berries, toxic and hanging in grape-like bunches. Hardy to –10°C/14°F; plants will re-sprout from the base in spring if cut down. Grow in partial shade or full sun in any moist moderately fertile soil. Provide trellis, fence or tree support. Propagate by seed, by removal of rooted suckers, or by ripewood cuttings in the cold frame or cool greenhouse in autumn. *M.canadense* (YELLOW PARILLA; eastern N America; stem to 6m; leaves 5–20cm, ovate to cordate or rounded, entire or obscurely 5–7-lobed, dark green above, paler beneath; fruit pruinose purple-red to blue-black).

Mentha (the classical Latin name, from Greek *minthe*). Labiatae. Eurasia, Africa. MINT. 25 species, aromatic herbaceous perennials or, occasionally, annuals with creeping rhizomes and small, 2-lipped, tubular or campanulate flowers in dense whorls. Nearly all mints are invasive creeping plants. They are grown for their aromatic foliage and have long been cultivated for their culinary, antiseptic and aromatic qualities. They will tolerate a wide range of soils and habitats, thriving in hot, well-drained places where plenty of moisture is available to them. Commonly grown for culinary use are spearmint (*M.spicata*) and applemint (*M.suaveolens*), also *M. × villosa*. See *mint*. These and all others listed here are fully hardy, except the slightly tender *M.requienii*, a tiny mat-forming plant requiring cool, damp shade. Propagate by division in early spring.

M.aquatica (WATERMINT; Eurasia; subglabrous to tomentose perennial; stem 15–90cm, often red-purple; leaves 2–6cm, ovate to ovate-lanceolate, serrate; flowers lilac in terminal heads comprising 2–3 verticillasters); *M. × gracilis* (*M.arvensis × M.spicata*; GINGERMINT, RED-MINT; N Europe, including Great Britain, widely cultivated; variable perennial, 30–90cm; stems erect, usually glabrous, often tinged red; leaves 3–7cm, ovate to lanceolate, oblong or elliptic, entire or remotely serrate, glabrous or sparsely pubescent; flowers lilac in distant verticillasters; includes 'Aurea', to 30cm, with gold, strongly fragrant leaves, and 'Variegata', with leaves flecked gold, with a fruity scent and ginger flavour); *M.* × *piperita* (*M.aquatica* × *M.spicata*; PEPPERMINT; Europe; perennial, 30–90cm, usually glabrous, often tinged red-purple; leaves 4–9cm, ovate to lanceolate, petiolate, serrate, apex acute, usually glabrous or thinly hairy; flowers lilac-pink in congested verticillasters forming a terminal oblong spike; cultivars include 'Candymint', with sweetly fragrant leaves; 'Citrata', LEMON MINT, with lemon-scented leaves; 'Crispa', with crinkled leaves; 'Lime Mint', with leaves scented of lime; and 'Variegata', with leaves deep green mottled cream with a peppermint flavour); *M.pulegium* (PENNYROYAL; SW and central Europe to Great Britain; glabrous to tomentose perennial, 10–40cm; stem procumbent or upright; leaves 5–30mm, narrowly elliptic to oval entire or toothed, usually glabrous above, pubescent beneath; flowers lilac in distant verticillasters; 'Cunningham Mint': low, with oval, light green leaves); *M.requienii* (CORSICAN MINT; Corsica, Italy, Sardinia; glabrous or sparsely hairy procumbent perennial; stems 3–10cm, thread-like, creeping, forming cushions or mats; leaves 2–7mm, ovate orbicular, entire or sinuate; flowers lilac); *M.spicata* (SPEARMINT; S and C Europe; perennial, 30–100cm; leaves 5–9cm, lanceolate to ovate, smooth or rugose, serrate, glabrous to hairy; flowers lilac, pink or white in cylindrical spikes; includes 'Chewing Gum', with deep mahogany leaves scented of bubble gum, 'Crispa', with strongly curled leaves, and 'Kentucky Colonel', with large, ruffled, richly scented leaves); *M.suaveolens* (APPLEMINT; S and W Europe; perennial, 40–100cm; stem white-tomentose, apple-scented; leaves 3–4.5cm, rugose, downy, serrate; flowers white or pink in congested verticillasters forming a terminal spike 4–9cm, often interruped and branched; often found under the name of *M.rotundifolia*; 'Variegata': PINEAPPLE MINT: leaves streaked white, with a pineapple aroma); *M.* × *villosa* (*M.spicata* × *M. suaveolens*; garden origin; a variable hybrid; leaves 4–8cm, broadly ovate or orbicular, softly hairy, serrate; flowers pink).

mentum a chin-like extension of the flower, formed by the association of the bases of the lateral sepals with the foot of the column in many orchids.

Menyanthes (possibly derived from *menanthos*, 'moonflower', a name used by Theophrastus for a plant growing on lake Orchomenos, or a reference to the month-long flowering period of this plant). Menyanthaceae. BOG BEAN, BUCK BEAN, MARSH TREFOIL. 1 species, *M.trifoliata*, a perennial aquatic or marginal herb, often rooting at water margins and with rhizomes spreading across the water surface. The leaves are glabrous and trifoliolate with oblong to obovate leaflets. Produced in spring and summer in erect racemes to 20cm tall, the flowers are some 2.5cm across, white flushed pink, with five, heavily fringed

and bearded petals. *M.trifoliata* is fully hardy. Plant in submerged pots and baskets in ponds or establish at the margins where the rhizomes can spread freely across the water surface. Increase by division.

Menziesia (for Archibald Menzies (1754–1842), botanist and naval surgeon). Ericaceae. US, E Asia. 7 species, deciduous shrubs with drooping, bell- to urn-shaped flowers in terminal umbels or racemes. Hardy to –20°C/–4°F. Grow in dappled shade in moist but free-draining soil that is lime-free and humus-rich; maintain an acidic organic mulch and protect from cold dry winds and late frosts. Remove fading flowerheads, but prune only to remove dead-wood after flowering. *Menziesia* blooms on the previous season's wood. Propagate from greenwood cuttings in summer in a closed case with bottom heat, or from seed sown in autumn at about 13°C/55°F. *M.ciliicalyx* (Japan; erect shrub, 30cm–1m; flowers to 1.7cm, yellow or green-white, purple-tipped, tubular to urceolate, white downy inside, lobes 4 or 5, small; var. *multiflora*: flowers many, purple-tinted; var. *purpurea*: flowers purple).

Merendera (from *quita meriendas*, Spanish for Colchicum.) Liliaceae (Colchicaceae). Mediterranean, N Africa, W Asia, outliers in Middle East and NE Africa. Some 10 species, perennial, cormous herbs with linear to lanceolate leaves partially developed at flowering. Produced in autumn, the stalkless flowers are funnelform to spreading with six linear to narrowly obovate tepals. Hardy outdoors in zone 7, but perhaps more safely grown in the bulb frame or in pans in the alpine house where the flowers and dormant bulbs can be protected from rain. Grow in full sun in a gritty soil. Propagate by seed and offsets.

M.montana (syn. *M.bulbocodium*; Iberian Peninsula, central Pyrenees; flowers pale magenta to rosy purple, tepals to 26 × 3–11mm, narrowly elliptic to narrow oblong-oblanceolate); *M.robusta* (C Iran, S Russia, N India; flowers deep pink to pale lilac or white, fragrant, tepals 18–14 × 2–9mm, oblong to narrowly elliptic).

mericarp a one-seeded carpel; one of a pair split apart at maturity from a syncarpous or schizocarpous ovary.

mericlone a clone produced by meristem culture.

meristem undifferentiated plant tissue found at sites of active cell division and from which new cells are produced.

meristem culture a form of micropropagation using meristematic tissue. It is valuable for the production of virus-free plants.

-merous a suffix denoting number and disposition of parts; often combined simply with an arabic number, thus a 3-merous or trimerous flower has parts arranged in threes.

Merremia (for the German naturalist, Blasius Merrem (d. 1824)). Convolvulaceae. Pantropical. 600 species, perennial, herbaceous or woody climbers similar to *Ipomoea*, with showy, campanulate or funnel-shaped flowers. Cultivation requirements are as for the tender, perennial species of *Ipomoea*, although *Merremia* is tolerant of full sun and thrives in rather dry situations. Maintain a minimum winter temperature of 7°C/45°F, and water very sparingly at this time. *M.tuberosa* (WOOD ROSE, YELLOW MORNING GLORY, SPANISH WOODBINE; Mexico to tropical S America, introduced in many other tropical regions; woody, tuberous climber to 20m; leaves to 15cm, 7-lobed; flowers 5–6cm, campanulate, bright yellow).

Mertensia (for Francis Karl Mertens, (1764–1831), Professor of Botany, Bremen). Boraginaceae. Northern temperate regions. Some 50 species, perennial herbs with tubular to campanulate, 5-lobed flowers nodding in scorpioid cymes in spring. *M.echioides* is suited to cool, moisture-retentive places in the rock garden; plant in light, neutral to slightly acid soils with good light but shade from hot sun. *M.maritima* and *M.simplicissima* thrive in sun on nutritionally poor, gravelly or sandy soils and are best-suited to maritime gardens or perfectly drained sites in the dry and silver garden. From moist woodland, *M.pulmonarioides* prefers deep, moist but well-drained soils rich in organic matter. It is well-suited to naturalizing in open woodland in shade or semi-shade. Its foliage dies back in summer. All are hardy to at least –15°C/5°F. Propagate from seed sown ripe in the cold frame. Increase also by careful division in early autumn or spring; *M.maritima* and allies resent root disturbance and are more reliably propagated by seed.

M.echioides (Pakistan, Kashmir, Tibet; stem to 30cm, erect or decumbent, pubescent; leaves to 9cm, lanceolate to elliptic; flowers deep blue to blue-purple); *M.maritima* (OYSTER PLANT; Alaska to Massachusetts, Greenland, Eurasia; stem to 100cm, decumbent or spreading; leaves to 10cm, fleshy, spathulate to oblong-ovate, glaucous sea green; flowers pink becoming blue; similar but superior to this species is *M.simplicissima*, syn. *M.asiatica*, a sprawling herb with thinly fleshy silver-blue chalky leaves and small blue flowers); *M.pulmonarioides* (syn. *M.virginica*; BLUEBELLS, VIRGINIA BLUEBELLS, VIRGINIA COWSLIP, ROANOKE BELLS; N America; stem to 70cm, erect; leaves to 20cm, elliptic to ovate, soft blue-green; flowers blue, sometimes pink or white; includes 'Alba', with white flowers, and 'Rubra', with pink flowers).

Meryta (from an artificial Greek word *merytos*, meaning growing together – the male flowers in some species form a knob or head). Araliaceae. Pacific Islands (Micronesia through Vanuatu to New Zealand and SE Polynesia). 250 species, small to medium evergreen trees with large and leathery, spirally arranged leaves, inconspicuous, umbellate flowers in compound inflorescences and small drupes. Provide a minimum temperature of 10°C/50°F. Plant in beds or large tubs in a medium fertility, loam-based mix with additional organic matter. Give strong, filtered to indirect light and water moderately. Propagate from seed in autumn, by semi-ripe cuttings rooted with bottom heat, or by air layering. *M.sinclairii* (PUKA; New Zealand; round-headed tree to 8m; leaves oblong to obovate, to 50cm; inflorescence to 45cm; flowers green-white, in dense umbels; fruit black, fleshy, to 13mm).

mesh pot a plastic pot with sides consisting mainly of large rectangular perforations, used in orchid culture.

mesocarp the middle layer of a pericarp, often fleshy, occasionally fibrous, rarely membranaceous or spongy.

mesophyte a plant that grows in a habitat where the soil is not continuously waterlogged and where extreme drought is not regularly experienced.

Mespilus (the Latin name for this fruit). Rosaceae. Europe, Asia Minor. MEDLAR. 1 species, *M.germanica*, a deciduous tree or shrub, to 5m, usually shorter. The bark is grey-brown and fissured to flaking, the branches arching and divaricate. To 12cm long, the oblong to lanceolate leaves are dull green above, tomentose beneath, with serrate margins. They turn yellow, amber and rusty brown in autumn. To 5cm wide, the white flowers are usually solitary with 5, rounded petals and numerous stamens. The fruit is turbinate, to 2.5cm, brown when ripe, and crowned with 5 narrow calyx lobes around the transecting ends of the bony seed vessels. See *medlar*.

metamorphosis the change in appearance, life style, and/or feeding habits that distinguishes the juvenile and adult forms of insects and amphibians, such as frogs. It may be radical (complete metamorphosis) or slight (incomplete metamorphosis).

Metasequoia (from Greek *meta*, with or beside (i.e. 'related to'), and *Sequoia*). Cupressaceae. W China (E Sichuan, W Hubei). 1 species, *M.glyptostroboides*, DAWN REDWOOD, a deciduous conifer, fast-growing, to 45m+. The bole is to 2m in diameter, deeply ridged and buttressed at base with fox red, furrowed and fibrous bark flaking in long shreds. The crown is conic, becoming columnar. Bright green, linear-oblong leaves to 4cm long are borne in two ranks on the feathery shoots; they turn yellow, red and finally rusty brown in autumn. Fully hardy. Avoid dry sites and frost pockets where first growth will be scorched. Like *Taxodium*, it can be planted beside or even in shallow standing water. Warm summers are needed for good growth. Pruning basal branches is said to decrease irregularities in the trunk, although for some, these flutes, ridges and pockets lend the tree its unique character.

*Mertensia
pulmonarioides*

*Metasequoia
glyptostroboides*

Metrosideros (from Greek *metra*, middle, and *sideros*, iron, in reference to the hard wood of the genus). Myrtaceae. South Africa, Malaysia, Australasia, Pacific Islands. 50 species, aromatic evergreen shrubs, trees or woody climbers with tough foliage. The flowers are showy with five, spreading petals and a central boss of long, colourful stamens. In favoured maritime climates that are frost-free or almost so, they make beautiful specimens and may also be used as hedging. Blooming even when young, they are excellent shrubs for large pots and tubs in the cool greenhouse or conservatory in cooler temperate zones, and may be moved outdoors for the summer months. Plant in a free-draining but moisture-retentive, medium- to high-fertility potting mix with additional leafmould. Water plentifully when in growth, less at other times. Under glass, provide a winter minimum temperature of 7°C/45°F, protect from the hottest summer sun and maintain good ventilation with moderate humidity. Prune immediately after flowering to restrict size and to remove old weak and overcrowded growths. Propagate by seed sown in spring, or root semi-ripe cuttings in summer in a heated case.

M.diffusa (SMALL RATA VINE; New Zealand; slender liane to 6m+; leaves 7–8mm, oblong to ovate-lanceolate, somewhat downy when young; flowers white to pink with white to pink stamens); *M.excelsa* (CHRISTMAS TREE, POHUTUKAWA; New Zealand; tree to 20m; leaves 5–10cm, elliptic to oblong, coriaceous, thick, white-tomentose beneath; flowers crimson with crimson stamens); *M.perforata* (FLOWERY RATA VINE; New Zealand; liane to 15m+; leaves 6–12mm, broadly ovate to oblong or rounded, glabrous above, pale beneath and more or less bristly; flowers white or pink with white or pink stamens); *M.robusta* (NORTHERN RATA; New Zealand; tree to 25m+; leaves 2.5–5cm, elliptic to ovate-oblong, glabrous, coriaceous; flowers red with red stamens).

mice small, widely distributed mammals that damage bulbs and corms, especially lilies, crocuses and tulips, and dig up and eat seeds, particularly peas and beans. The main culprit is the WOOD- or LONG-TAILED FIELD MOUSE (*Apodemus sylvaticus*), which measures up to 100mm from nose to base of tail, which is about as long again. Adults are brown with white underparts and have relatively large ears. They live in burrows in fields, hedgerows and banks, and, especially in winter, may enter greenhouses and other buildings. Spring traps, baited with carrot, apple or chocolate and set during autumn or early winter, help to control mice before damage occurs. Seedbeds and bulbs may be protected with 10mm-mesh netting. Bulbs and corms are more vulnerable where newly planted. Pesticide-treated seeds are unpalatable to mice. Proprietary mouse poisons are effective, but special care must be taken to prevent other animals gaining access to them.

Michelia (for P.A. Michel (1679–1737), Florentine botanist). Magnoliaceae. SE Asia, India, Sri Lanka. Some 45 species, evergreen or deciduous shrubs or trees, resembling *Magnolia* but with axillary flowers.

Michelia species thrive best in frost-free gardens (although *M.doltsopa* and *M.figo* have survived outdoors in sheltered situations in southern Britain). Elsewhere, grow in the cool greenhouse or conservatory. Plant in humus-rich, well-drained, neutral to alkaline soils, in direct light or partial shade. Water pot-grown specimens plentifully during the growing season, sparingly in winter. Prune only if necessary to restrict growth. Propagate from seed sown ripe in autumn or spring, or by greenwood cuttings in summer, either under mist or with bottom heat in a closed case.

M.doltsopa (SW China, E Himalaya, E Tibet; tree, to 12m; leaves to 18cm, oblong to lanceolate, glossy above, grey-silky beneath; flowers to 5 × 4cm, ovoid to globose, fragrant, white); *M.figo* (BANANA SHRUB; China; evergreen shrub to 4.5m, densely branched; leaves 5–10cm, elliptic to oblong, dark green; flowers to 3cm in diameter, cupulate, scented of banana, ivory tinged yellow, base and margins tinged rose to purple).

Microbiota (from Greek *mikros*, small, and *Biota* (a synonym of *Platycladus*), the closest ally of this conifer). Cupressaceae. SE Siberia. 1 species, *M.decussata*, an evergreen, spreading, semi-prostrate coniferous shrub, 40(–70)cm high, and to 2m or more across. The leaves are small, subulate or scale-like, pale green at first becoming bronzed red or purple to dull gold in hard winters; they overlap on flat sprays. A valuable prostrate evergreen for rockeries and dwarf conifer collections, *Microbiota* is exceedingly hardy and site-tolerant. Increase by cuttings.

microclimate the environmental conditions of a specific locality or garden area, which may work to the advantage or disadvantage of cultivated plants. Conditions within a particular geographical area may be affected by topographical features, such as high ground, or in a garden by hedge planting or the construction of water features, raised beds or a greenhouse. Wind, rainfall and soil and air temperature may thus be different to the conditions prevailing outside of the defined area.

Micromeria (from Greek *mikros*, small, and *meris*, part, referring to the usually very small flowers). Labiatae. Mediterranean, Caucasus, SW China. Some 70 species, perennial, rarely annual, herbs or dwarf shrubs, often aromatic and bearing spikes of whorled, 2-lipped, tubular flowers in summer. Hardy and suitable for rock gardens on open, sunny sites and in poor, well-drained soils. Propagate by softwood or semi-ripe heel cuttings in spring and summer. *M.juliana* (SAVORY; Mediterranean; dwarf, downy shrub, to 40cm; leaves to 1.3cm, ovate to narrowly lanceolate, revolute; flowers to 4mm, purple).

micronutrients see *nutrients* and *plant nutrition*.

micropropagation the propagation of plants from the culture of very small portions of tissue grown *in vitro* (literally 'in glass'), under aseptic conditions, on a specially-formulated gelatinous or liquid medium.

Micropropagation, sometimes referred to by horti-culturists as tissue culture, evolved from techniques conceived by research biologists in the early part of the 20th century as a means of studying the functioning and development of tissues, and ultimately cells, separately from the whole organism. The technology was first used horticulturally in the 1920s for the germination of orchid seeds; the method still forms the basis of seed-raising in the orchid industry.

The discovery that small pieces of tissue from shoot tips, approaching a meristem in size, are often free of virus infection led to the use of meristem culture for the production of virus-free individual plants, an importants benefit of micropropagation. Development work on this process, including the modification of the medium on which these plantlets were grown *in vitro*, principally by the addition of plant-growth regulators, showed that axillary buds could be stimulated to give multiple-shoot plantlets. Such multiple shoots could then be divided and transferred to a fresh medium at regular intervals to achieve the rapid multiplication in plant material that is characteristic of micropropagation.

Micropropagation is used in horticulture for the production of a wide range of ornamental, fruit and vegetable crops, and as a means of rescuing rare and endangered species. Variations of the technique are used to assist in plant-breeding procedures, and as a means of introducing new cultivars rapidly.

There are a number of routes to *in vitro* micro-propagation, depending on species. Plantlets are commonly established from meristems or shoot tips, but leaf or stem tissue, bulb scales or even anthers and pollen are also used. Natural or artificially-induced callus tissue can be a source of shoot development, either directly or following procedures for the separation of cells. Choosing the most appropriate method of micropropagation requires a clear objective and good knowledge of the plant to be propagated.

An essential consideration for success is freedom from contamination by microorganisms, which can arise on or within the mother plant or from any part of the working environment. Plant material used for micropropagation is subjected to prescribed steps of sterilization involving washing and immersion; the growing medium, glassware and operating instruments are sterilized most effectively with the aid of an autoclave. The sterilization of plant tissue and all micropropagation operations must be carried out in a sterile environment, created by means of a laminar flow cabinet in which a non-turbulent airstream, filtered to remove contaminants, is created across the working area.

The growing medium must supply all the plant tissues' requirements for nutrients and moisture to sustain growth and development *in vitro*. Variations in nutrient levels and pH have been developed to suit particular groups of plants. In most cases, plant tissue culture *in vitro* requires the incorporation of plant growth hormones, usually cytokinins or auxins for shoot and root growth respectively. Manipulating the ingredients and levels of hormonal substances added is a means of influencing plant development.

The control of temperature and artificial light level is essential, and must be matched to the prescribed requirements of particular groups of plants. Rapidly-growing cultures require division and transfer to a fresh medium at regular intervals, the process being referred to as subculturing. Typically, many plants require subculturing on a four-week cycle, but some herbaceous plants may need processing as frequently as every two weeks, while for some woody plants every ten weeks is adequate.

Micropropagation is a specialist technique of immense current and potential value to the practice of horticulture. Successful implementation demands a knowledge of the protocols developed for specific plants, and careful attention to the provision and maintenance of aseptic working conditions and procedures.

Besides the challenges posed by cultural practice *in vitro* and the skill required to wean plantlets into growing compost, certain limitations of the technique should be understood. There is a risk, albeit rare, of permanent off-types occurring as a result of induced change in genetic make-up. The risk is highest where callus tissue is liberally produced, and/or with high concentrations of cytokinins or auxins. Transient variation may occur in micropropagated offspring due to a change in environmental conditions, and as a quality-control measure in commercial situations, a proportion of propagules need to be grown on to test trueness to type before release.

microspore the smaller of the two types of spore produced by heterosporous plants; they develop into male gametophytes.

midrib, midvein the primary vein of a leaf or leaflet, usually running down its centre as a continuation of the petiole or petiolule.

Mikania (for J.G. Mikan (1743–1814), Professor of Botany at Prague, or his son J.C. Mikan (d. 1844), plant collector in Brazil). Compositae. Tropics, especially New World. About 300 species, evergreen, perennial, woody or herbaceous lianes producing clusters of small, button-like flowerheads in late summer and autumn. Cultivate as for *Delairea*. *M.scandens* (WILD CLIMBING HEMP-WEED, CLIMBING HEMP-VINE; tropical America; herbaceous; leaves to 10cm, triangular to hastate, entire or toothed, glabrous to hairy; flowerheads strongly vanilla-scented, white, tinged yellow, or lilac to purple).

mildew in horticulture and plant pathology, used to describe certain diseases in which the pathogen is visible on the plant surface. DOWNY MILDEWS are caused by fungi of the family Peronosporaceae and POWDERY MILDEWS by those of Erysiphaceae. The SOOTY MOULDS caused by fungi of the families Capnodiaceae and Meliolaceae are sometimes called dark or black mildew. Mildew is also a former name for honeydew. See *downy mildew*, *powdery mildew*, *sooty moulds*.

Milium effusum
'Aureum'

Milium (Latin *milium*, millet). Gramineae. Eurasia, N America. 6 species, annual or perennial grasses with tussocks of short culms, linear, arching leaves and loose, airy panicles of small spikelets in spring and summer. Hardy to –20°C/–4°F and below. Plant in fertile, moist but well-drained soil in sun or shade (ideally dappled sunlight). In hot, humid summers, these grasses may enter summer dormancy. Propagate by seed and division; *M.effusum* 'Aureum' comes true from seed and will self-sow. Often affected by mildew. *M.effusum* (WOOD MILLET; loosely tussock-forming perennial to 30cm; leaf blades to 30 × 1.5cm, soft-textured, arching; panicle to 30 × 20cm, basically pyramidal in outline, loose and airy, light green or tinged purple, branches delicate, whorled, spike-lets small; includes 'Aureum' (BOWLES' GOLDEN GRASS), with pale golden yellow to lime green leaves and spikelets, and the weak-growing 'Variegatum', with bright green leaves striped white).

Milla (for Juliani Milla, gardener to the Spanish court in Madrid, 18th century). Liliaceae (Alliaceae). S US, C America. 6 species, cormous perennial herbs with linear leaves and, in summer, tubular, 6-lobed flowers in scapose umbels. Cultivate as for *Ipheion*. *M.biflora* (to 30cm; flowers campanulate to funnelform, 1.5–3.5cm, white sometimes tinted lilac, usually striped or veined green on exterior). For *M.uniflora*, see *Ipheion uniflorum*.

millepedes (*Diplopoda*) small insect-like animals characterized by having two pairs of legs on each trunk segment, except for the first few segments behind the head and some at the hind end; there may be as many as 50 segments.

Millepedes thrive in moist soil of high organic content and feed principally on decaying or soft vegetable matter. They lay eggs in batches in the soil and the young may take several years to reach maturity. Only a few species are harmful to garden plants. The SPOTTED SNAKE MILLEPEDE (*Blaniulus guttulatus*) usually causes the most serious damage; it is creamy white, up to 20mm long, with orange-red dots along each side of the body. It attacks seedlings and soft growth, such as strawberry fruits and mushrooms, as well as enlarging damage to potato tubers and bulbs that has been initiated by other pests such as slugs. The FLAT-BACKED MILLEPEDES (*Brachydesmus superus* and *Polydesmus angustatus*) may damage seedlings and other plants outdoors; in temperate countries, the TROPICAL GREEN-HOUSE MILLEPEDE (*Oxidus gracilis*), which is up to 40mm long and varies from white to brown, can be injurious to greenhouse plants, particularly cucumbers. Good hygiene and thorough cultivation help reduce numbers and can provide sufficient control, but for persistent attacks, dust seedlings, bulbs and corms with recommended contact insecticide.

millipede

Miltonia (for Lord Fitzwilliam Milton (1786–1857), landowner and orchidologist). Orchidaceae. Brazil, one species in Peru. Some 20 species, epiphytic, evergreen perennial herbs with ovoid to ellipsoid

Miltonia clowesii

pseudobulbs, strap-shaped leaves and racemes of showy flowers at various times of year. The flowers consist of ovate to oblong, spreading tepals and a larger, simple or pandurate, oblong to rounded lip. Cultivate as for *Odontoglossum*.

M.candida (Brazil; resembles *M.clowesii* except tepals oblong-obtuse, lip shorter than other segments, clasping column at base); *M.clowesii* (Brazil; inflorescence to 45cm; tepals yellow, heavily blotched and barred chestnut-brown, sepals to 4 × 1cm, lance-olate, acuminate, undulate, petals to 3.5 × 1cm, lip white at tip, deep purple at base, to 4 × 2cm, subpandurate, caudate, callus white or yellow); *M.regnellii* (Brazil; inflorescence to 40cm; tepals cream suffused rose to lilac or pale amethyst, sepals to 3 × 1cm, oblong-lanceolate, recurved, petals wider; lip to 3.5 × 3.5cm, pale rose streaked lilac or amethyst, margins white, rotund to obovate, obscurely trilobed, callus of several radiating yellow lines); *M.spectabilis* (Brazil; inflorescence to 25cm; flowers solitary, sepals to 4 × 1.5cm, white, often tinged rose at base, lanceolate-oblong, petals similar with white patch at base, lip rose-violet, margins white or pale rose, to 5 × 4.5cm, obovate-orbicular, callus yellow with pink lines). There are many cultivars and grexes in *Miltonia* (some strictly assignable to *Miltoniopsis*) with pale green foliage and flowers in shades of white, pink, red and lilac, among them the classic pansy orchids with broad blooms in deep velvety red. For *M.phalaenopsis* see *Miltoniopsis phalaenopsis*; for *M.roezlii*, see *Miltoniopsis roezlii*; for *M.superba*, see *Miltoniopsis warscewiczii*; for *M.vexillaria*, see *Miltoniopsis vexillaria*.

Miltoniopsis (Miltonia and Greek *opsis*, resemblance). Orchidaceae. Pansy orchid. Costa Rica, Panama, Venezuela, Ecuador, Colombia. 5 species of epiphytes or lithophytes closely allied to *Miltonia*, generally with larger, pansy-like flowers. Cultivate as for *Odontoglossum*.

M.phalaenopsis (syn. *Miltonia phalaenopsis*; Colombia; raceme to 20cm; tepals pure white, sepals to 2 × 1cm, elliptic-oblong, acute, petals wider, elliptic, obtuse, lip to 2.5 × 3cm, white blotched and streaked purple-crimson or pale purple, 4-lobed, callus with yellow spots); *M.roezlii* (syn. *Miltonia roezlii*; Colombia; raceme to 30cm; sepals to 5 × 2cm, white, ovate-oblong, acuminate, petals to 5 × 2.5cm, white blotched wine-purple at base, lip to 5 × 5.5cm, white, orange-yellow at base, widely obcordate); *M.vexillaria* (syn. *Miltonia vexillaria*; Colombia, Ecuador; raceme to 30cm; tepals rose-pink or white flushed rose-pink, sepals to 3 × 1.5cm, obovate-oblong, recurved above, petals slightly wider, margins white, lip much larger than sepals, white or pale rose, deep rose on disc, suborbicular, cleft, callus yellow); *M.warscewiczii* (syn. *Miltonia superba*; Costa Rica; raceme to 30cm; flowers cream-white, each segment with a rose-purple blotch at base, tepals 3 × 1.5cm, ovate-elliptic, acute, lip to 3.5 × 3.5cm, widely pandurate, basal lobes 2, small, rounded, midlobe emarginate, callus yellow).

mimicry resemblance of a plant to an insect or object; a feature that serves to attract pollinators or provide camouflage, for example bee orchids (*Ophrys* species), which attract male bees by reason of the flowers similarity to females, and pebble plants (*Lithops* species), which are not readily distinguished from the rocks and stones amongst which they grow.

Mimosa (from Greek *mimos*, mimic; the leaves of many species resemble animals in their ability to move). Leguminosae. C and S America, S US, Asia, tropical and E Africa. Some 400 species, usually spiny herbs, shrubs or small trees. The leaves are bipinnate with numerous, oblong to linear leaflets and a swelling at the base of the stalk. When touched, or at nightfall, or in times of drought, the leaflets fold rapidly together and the leafstalks wilt – a trick with the dual functions of alarming would-be grazers and conserving water. The flowerheads are typically *Acacia*-like, rounded, stalked and packed with small, 'fluffy' flowers. Grow in pots or beds, in bright filtered light with a minimum temperature of 10°C/50°F. Water plentifully and apply liquid feed to established plants weekly when in growth; water very moderately in winter. Although leggy plants can be pruned hard prior to repotting in spring, they are best replaced after a year or two. Propagate by seed, pre-soaked in hot water and sown uncovered in early spring at 18–20°C/65–70°F. *M.pudica* (SENSITIVE PLANT, TOUCH-ME-NOT, SHAME PLANT, LIVE-AND-DIE, HUMBLE PLANT, ACTION PLANT; S Mexico to C America, West Indies, Hawaii, Fiji, Australia; prostrate to semi-erect herb, subshrub or shrub, 0.3–0.8m; pinnae (1–)2 pairs, to 10cm, leaflets 10–25 pairs, mid- to grey-green, soft-textured, oblong; flowers pale pink to lilac, in heads to 2cm in diameter).

mimosiform, mimosoid possessing flowerheads resembling those of the genus *Mimosa*, in which calyx and corolla are inconspicuous or reduced and the stamens are showy.

Mimulus (diminutive form of the Latin *mimus*, a mimic actor, referring to the resemblance of the markings of the corolla to a grinning face). Scrophulariaceae. Mostly America, also South Africa and Asia. MONKEY FLOWER, MUSK. 150 species, annual or perennial herbs, or rarely shrubs, with racemes of showy, tubular flowers in spring and summer. These are 2-lipped, the throat open or closed by a raised and hairy palate. The soft-stemmed perennial species are hardy in climate zone 7 and thrive in full sun or light shade on moist, or saturated, fertile soils; many of these will grow in water and are excellent marginals. The shrubby species (e.g. *M.aurantiacus*) are frost-tender and should be grown under glass in cold areas or, where winters are mild, in the open garden against a warm wall in a dry, sunny spot. Where winters are severe, the herbaceous *Mimulus* are best treated as annuals for summer-bedding display or as flowering pot plants for windowsills and cool greenhouses. Where grown in the open garden in hard winter areas, take cuttings in summer as an insurance policy or protect plants in the open with cloches. In late winter prune pot-grown specimens; repot them, and start into growth at 13°C/55°F, ventilating freely whenever possible. Water plentifully when growing strongly and plunge in the open during the summer months. Feed established plants fortnightly with a dilute liquid fertilizer. Sow seed in early spring at 13–15°C/55–60°F under glass and cover very lightly with grit or sieved soil. For winter/spring-flowering annuals in pots, sow in autumn and overwinter under glass. Seed may also be sown *in situ* after the last frosts. Increase by division in spring or by softwood cuttings in spring and summer. Propagate shrubby species by semi-ripe cuttings in late summer, rooted in a closed propagating case.

M. aurantiacus (syn. *M. glutinosus*; BUSH MONKEY FLOWER, ORANGE BUSH MONKEY FLOWER; S Oregon to California; clammy shrub, 60–120cm; flowers 3.5–4.5cm, orange or deep yellow); *M.guttatus* (COMMON LARGE MONKEY FLOWER; Alaska to Mexico; largely glabrous annual or biennial herb; stem 40–100cm, erect or decumbent; flowers 3–3cm, bright yellow, throat with pair of red to purple-brown-spotted hairy ridges; includes 'A.T. Johnson', to 15cm, with yellow flowers spotted red); *M. × hybridus* (*M.luteus* × *M.guttatus*; flowers large, colour variable; cultivars include 'Andean Nymph', with pink flowers with cream-tipped petals; 'Bee's Dazzler', with brilliant crimson-red flowers; 'Brilliant', with deep purple-crimson flowers; Calypso Hybrids, with large, burgundy, scarlet and gold flowers; 'Duplex', with enlarged, colourful calyx; 'Fireflame', with bright flame red flowers; 'Highland Pink', with rich pink flowers; 'Highland Red', with deep red flowers; 'Highland Yellow', with yellow flowers; 'Hose in Hose', with soft tan flowers, one sitting inside the other; 'Inschriach Crimson', dwarf, carpet-forming, with crimson flowers; 'Leopard', with yellow rust-spotted flowers; 'Malibu', to 25cm, of strong growth, with orange flowers; 'Mandarin', with abundant, bright orange flowers; 'Nanus', of dwarf habit, with glossy leaves and bright orange to gold flowers with a red throat; 'Old Rose', to 15cm, with bright red flowers; 'Queen's Prize', dwarf, with large flowers in bright colours, many spotted and blotched; 'Royal Velvet', to 20cm, with large velvety, mahogany red flowers with gold throats speckled mahogany; 'Röter Kaiser' ('Red Emperor'), to 20cm high, with crimson to scarlet flowers; 'Shep', with large, yellow flowers splashed brown; 'Whitecroft Scarlet', with vermilion flowers; and 'Wisley Red', to 15cm, with velvety blood red flowers); *M.luteus* (MONKEY MUSK, YELLOW MONKEY FLOWER; Chile, naturalized elsewhere; glabrous perennial herb; stem 10–30cm, decumbent to ascending, hollow; flowers 2–5cm, yellow, with deep red or purple spots); *M.moschatus* (MUSK FLOWER, MUSK PLANT; British Colombia to Newfoundland, S to W Virginia and California; perennial herb, glandular-pilose, sometimes aromatic; stem 10–40cm, decumbent; flowers 1.8–2.5cm, pale yellow, throat tubular, finely brown-spotted and streaked with black).

mineralization see *nutrients*

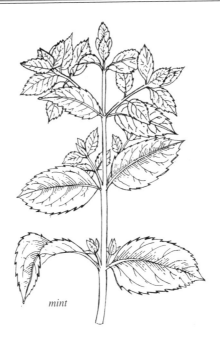

mint

mint

Disease
rust

mint (*Mentha* species) a group of creeping, invasive perennials grown as culinary herbs for their aromatic foliage; there are many types, with variation in flavour. Commonly grown are SPEARMINT (*M. spicata*) and APPLEMINT (*M.suaveolens*), natives of central, south and west Europe, and also BOWLES' MINT (*M. villosa alopecuroides*), of garden origin. Propagate in spring from division of young shoots with roots, or by 5cm long pieces of root planted horizontally 5cm deep and about 23cm apart. Mint beds are best re-established every three years. For shoot and leaf production in winter and spring, roots can be lifted in autumn and planted in a large container within the greenhouse. Mints may need containment with physical barriers to control the plants' invasive habit.

Mirabilis (from Latin *mirabilis*, wonderful). Nyctaginaceae. UMBRELLAWORT, MARVEL OF PERU. SW US, S America. Some 50 species, annual or perennial herbs with large tuberous roots. The stems are regularly forked, the leaves ovate to elliptic or lanceolate. Produced in summer in axillary corymbs, the flowers are tubular with a spreading limb of five lobes. Hardy to –5°C/23°F, given a warm, sheltered position in very well-drained soil. Alternatively, lift and overwinter tubers in a frost-free place as for *Dahlia*. Grow in full sun or part-day shade. Propagate from seed sown at 15°C/60°F in spring: harden off young plants before planting out. Increase also by division of tubers in spring. *M.jalapa* (MARVEL OF PERU, FOUR-O'-CLOCK, FALSE JALAP; Peru, garden escape in N America and S Europe; erect perennial, to 1m; flowers fragrant, opening in the late afternoon or evening, 3–5 × 1–2.5cm, purple, crimson, apricot, yellow or white, several shades often present on one plant, often striped or mottled).

Miscanthus sinensis 'Zebrinus' *(left), and (right)* 'Cabaret'

Miscanthus (from Greek *mischos*, stalk, and *anthos*, flower, referring to the stalked spikelets). Gramineae. Old World tropics, South Africa, E Asia. 17 species, deciduous perennial grasses with clumped, reed-like culms clothed with slender, arching leaves. Silky-hairy spikelets are borne on the arching branches of terminal panicles in summer and autumn. Hardy in climate zone 6 and tolerant of a wide range of sites, these grasses prefer fertile, moist soils in full sun. Most are dry and dormant by mid-winter, although the windswept skeletons of the inflorescences remain beautiful, especially when rimed with frost. The flowering time ranges from early summer to early autumn. Although the flowers of some selections open with a reddish cast and others open silver, all take on a silver-white appearance upon drying. They are superb for cut or dried arrangements. Clumps eventually tend to die out in the centre, and may require lifting and dividing after 5–7 years. Annual maintenance generally consists of cutting back by mid-spring before new growth begins, or in late autumn if winter interest is not a consideration. Species are easily grown from seed. The cultivars should be propagated by division. Division is safest done in spring, but may succeed in an autumn followed by a mild winter. The same holds true for transplanting times. The mealybug, *Pilococcus miscanthi*, now poses a serious threat to *Miscanthus sinensis* and possibly to other species. The eggs hatch and crawlers appear in late spring. This mealybug is big and prolific, and can reduce even the largest clump of *Miscanthus* to a stunted, chlorotic mess that will cease to flower. By autumn, the lower culms and insides of the leaf sheaths may be white-caked with mealybugs. It attacks all parts of the plant. including the roots, so above-ground mechanical or chemical methods are not sufficient for control. For those not fond of using pesticides in the garden, the best strategy is probably to remove and destroy any plants that show infestation and to be cautious when purchasing new plants.

M.sacchariflorus (AMUR SILVER GRASS; Asia, escaped in US; to 3m; leaves to 90 × 3cm, flat, stiff, smooth, margins rough, midrib pale; inflorescence to 40 × 13cm, very pale green, tinged red or purple, racemes to 30cm, flimsy, beige, with long silky hairs; 'Aureus': to 1.5m, leaves striped gold; 'Robustus': vigorous, to 2.2m; 'Variegatus': leaves narrowly striped white); *M.sinensis* (EULALIA; E Asia; to 4m; leaves to 120 × 1cm, flat, glabrous to puberulous above, blue green; inflorescence to 40 × 15cm, obpyramidal, pale grey tinted brown or mauve, racemes to 20cm, with long silky hairs; there are many cultivars: (*habit*) 'Goliath': large, to 2m; 'Gracillimus': more slender than type, leaves with a white midrib; 'Grosse Fontaine': large fountain-like, early-flowering; 'Strictus': as for 'Zebrinus', but smaller, very erect, with leaves banded yellow; 'Yaku Jima': to 85cm, with narrow, arching leaves; (*leaves striped*) 'Cabaret': leaves wide with a broad central creamy white band and dark green edge, flushed pink in autumn, panicles copper; 'Goldfeder': leaves edged and striped yellow; 'Morning Light': leaves narrow, arching, neatly edged white, panicles red-bronze; 'Sarabande': leaves with a wide central silver stripe; 'Variegatus': leaves with white or cream-white stripes; (*leaves banded*) 'Tiger Tail': leaves banded cream; 'Zebrinus': ZEBRA GRASS, leaves and culms banded white and yellow; 'Graziella': leaves narrow, somewhat arching, burgundy and orange, with bronze tints in autumn, panicles large, white, nodding; 'Purpureus': syn. 'Purpurascens', small, leaves with a pink-tinted central vein, turning shades of brown and orange-red in autumn, panicles oatmeal flushed pink; (*flowers*) 'Flamingo': panicles tinted flamingo pink; 'Kleine Fontaine': small, with graceful, arching leaves and early flowers; 'Malepartus': robust, with burgundy leaves in autumn, and large, copper to silver panicles tinted mauve; 'Rotsilber': panicles red-tinted; 'Silberfeder': upright, with narrow, silver-veined leaves, slender stems and loose, silver panicles occasionally flushed pink); *M.tinctorius* (Japan; tufted perennial to 1m; leaves flat, to 40×1.2cm, margins scabrous, mostly glabrous, pilose towards base; inflorescence erect, with a very short axis, racemes to 15cm; 'Variegatus': to 30cm, leaves streaked white to cream or pale yellow).

mist propagation a method of rooting cuttings under protection by means of a fine spray of water applied at frequent intervals to prevent the cuttings from wilting. A mist propagation unit consists of a bench or ground bed layered with a rooting medium, along the length of which spray nozzles are fitted on risers. The misted water application is controlled by an electronic leaf, which simulates the regularity with which cutting leaves dry out. See *electronic leaf*.

mist-sprayer, **mister** a small specialized sprayer used to deliver a fine mist of water around house-plants to increase humidity.

Mitella (diminutive of Latin *mitra*, a cap or mitre – thus a small cap or mitre, alluding to the shape of the fruit). Saxifragaceae. N America, NE Asia. BISHOP'S-CAP, MITREWORT. 20 species perennial, minutely hairy herbs making clumps of long-stalked, ovate-cordate, lobed and crenate leaves. Small, green or yellow flowers are carried in narrow, erect spikes in summer. An attractive, dainty groundcover for moist woodland or pockets in the rock garden. Fully hardy, it prefers a cool, damp, humus-rich soil and shade. Increase by division in spring. *M.breweri* (British Colombia to California and Idaho; to 10cm; leaves 4–10cm broad; flowers green-yellow, petals 1–2mm, divided into 5–9 segments).

mites a group of very small plant pests with needle-like mouthparts which they insert into plant tissues in order to suck sap. Typically, they have four pairs of jointed legs, with bodies consisting of a head and a fused thorax and abdomen. There are three important groups of mite pests: TETRANYCHIDS or RED SPIDER MITES; TARSONEMID MITES; and ERIOPHYID or GALL MITES.

BRYOBIA MITES are black with pink legs, and feed from the upper leaf surface of plants such as primulas, ivy and gooseberry.

FALSE RED SPIDER MITES (Acarina: Tenuipalpidae) are Tenuipalpidae) small, bright red, slow-moving mites, up to 2mm long, with an oval body; they are related to the red spider mite (Tetranychidae). Most species feed on the undersides of leaves near to veins, and those which are occasionally significantly injurious are mostly in the genus *Brevipalpus*, which includes the SCARLET TEA MITE (*B.californicus*), the ORCHID MITE (*B.oncidii*), the CITRUS FLAT MITE (*B.lewisi*), the privet mite (*B.obovatus*) and the RED CREVICE TEA MITE (*B.phoenicis*). All infest hosts such as citrus, tea, coffee, peach, rubber, coconut, apple, pear, olive, fig, walnut, grape, date palm, privet, orchids, buddleias, fuchsias and many other ornamental plants. They feed by sucking the sap and cause a fine pale mottling of the upper leaf surface. These mites are more tolerant of acaricides than are red spider mites.

The BRYOBIA MITES (*Bryobia* species) and the FALSE RED SPIDER MITES (*Brevipalpus* species) are closely related to the tetranychids.

See *gall mites*, *red spider mites*, *tarsonemid mites*.

Mitraria (from Latin *mitra*, cap or mitre, referring to the shape of the fruit). Gesneriaceae. Chile. 1 species, *M.coccinea*, a climbing or straggling, evergreen perennial herb, eventually becoming somewhat woody, with dense, tangled branches. To 2cm long, the leaves are ovate, dentate, lustrous dark green, coriaceous and hairy. In late spring and summer, solitary flowers hang from the leaf axils, each to 3cm long, tubular to goblet-shaped, downy and scarlet to orange-red. Cultivate as for *Asteranthera*.

mixed border see *herbaceous border*.

MLO see *phytoplasmas*.

module a small individual container or growing unit for plant raising, mainly used for direct single or multi-seed sowing to each unit, which removes the need for pricking out and offers the possibility of economizing on seed. Modules are of particular value in reducing the effects of growth check at planting time, and also in facilitating early establishment under protection to allow for planting out at a favourable or convenient time. Modules take the form of small pots made of various materials: moulded plastic, polystyrene, peat or fibre cellular trays of various sizes (the smallest units of which are known as plugs), and soil blocks.

Mogul (**Mughal**, **Mughul**) **gardens** formal gardens of India created by Persian Mogul invaders in the 16th and 17th centuries. They are based on water and fountains, with pavilions, lawns, flower-beds and shade trees.

mole drain see *drainage*.

moles (*Talpa europaea*) small subterranean mammals, up to 165mm long, with short tails, dense dark brown fur and forelegs adapted for burrowing. Moles construct a system of tunnels that may reach 1m in depth, and their activity damages plants indirectly by undermining roots; excavated soil is thrown up into surface molehills which can disfigure lawns. Moles are territorial and live for up to 3–4 years, feeding principally on earthworms.

Signs of activity in a small garden may be the work of an individual since moles are solitary animals except in the breeding season. The young are born usually four to a litter in spring, in an underground nest chamber or, where there is a high water table, in an unusually large molehill. They disperse above ground to new territories in early summer.

Specially designed mole traps are available for control, and fumigant smokes can be used, but the most lethal poisons may only be used by pest-control contractors. Electronic repellent devices have been used with varying degrees of success. Once established, control is difficult and recolonization of vacant tunnel systems can occur from adjoining infested sites.

Molinia (for J.I. Molina (1740–1829), writer on the natural history of Chile). Gramineae. Eurasia. 2 species, perennial grasses with tufted culms and slender leaves. Small and shimmering, the florets are loosely borne on the thread-like branches of erect, spreading panicles in summer. Fully hardy. Grow in moisture-retentive, lime-free soils in sun. Propagate by division in early autumn and spring.

M.caerulea (PURPLE MOOR-GRASS; to 120cm; leaf blades to 45 × 1cm; panicle dense, to 40 × 10cm, purple to olive-green; spikelets to 1cm; 'Dauerstrahl': stems tall, yellow-tinted 'Heidebraut': stems soft straw yellow, to 1.5m, spikelets yellow-tinted, forming a glistening cloud; 'Moorhexe': narrowly upright, with slender stems, to 50cm, and dark flowers in straight, narrow panicles; 'Moorflamme': leaves colouring well in autumn, inflorescence dark; 'Overdam': to 60cm, with strong stems and fine leaves; 'Variegata': to 60cm, leaves striped dark green and cream-white; subsp. *arundinacea*: to 2.5m, with leaves to 12mm across and panicles with long, spreading branches; cultivars of subsp. *arundinacea* – 'Altissima', to 1m, with golden-yellow leaves in autumn; 'Fontane', with inclining stems, forming a fountain; 'Karl Foerster', tall, with leaves to 80cm; 'Skyracer', tall, with leaves to 1m, clear gold in autumn; 'Transparent', with sparse spikelets, giving the whole inflorescence a light, spacious, transparent quality; 'Windspiel', with slender, swaying stems and gold-brown, dense plumes).

molluscicide a pesticide active against slugs and snails (molluscs), such as metaldehyde and methiocarb.

Moltkia (for Count Joachim Gadake Moltke of Denmark, d. 1818). Boraginaceae. N Italy to N Greece, SW Asia. Some 6 species, perennial herbs or subshrubs usually coarsely hairy and with massed clusters of funnel-shaped flowers in summer. Fully hardy, *Moltkia* prefers neutral to alkaline soils but otherwise has garden uses and cultural requirements similar to those of *Lithodora*. Propagate also by layering.

M. × *intermedia* (*M.petraea* × *M.suffruticosa*; closely resembles *M.suffruticosa* except larger and more shrubby with broader leaves and deep bright blue flowers, in spreading heads; includes 'Froebelii', with azure-blue flowers); *M.petraea* (Balkans to C Greece; white-bristly, slender shrublet to 40cm; leaves to 5cm, oblanceolate to linear, revolute; cymes compact, flowers to 8mm, blue or violet-blue); *M.suffruticosa* (N Italy; loosely branched, bristly shrublet to 25cm; leaves to 15cm, linear, sometimes revolute; cymes short, dense, flowers to 17mm, blue).

Moluccella (from the original collection region, the Moluccas). Labiatae. Mediterranean to NW India. 4 species, glabrous, annual or short-lived perennial herbs with erect, scarcely branching stems and wavy-toothed leaves. The flowers are borne in summer in whorls forming a tall, erect spike. They are remarkable for the calyx which enlarges with age becoming a green, net-veined and prickly-edged dish or funnel, far exceeding the small, tubular and two-lipped corolla. A distinguished annual for the flower border and, later, for drying. Treat as half-hardy, sowing seed under glass in early spring at 15°C/60°F, or, later, *in situ*, in sunny well-drained borders. *M.laevis* (BELLS OF IRELAND; W Asia; annual, erect, 30–100cm; calyx light green, with pale or white net markings, campanulate or saucer-shaped, margin more or less spiny, corolla white to pale lilac).

molybdenum an essential plant micronutrient, required for enzyme activity. Deficiency, which is predisposed by soil acidity, is rare in the UK, but more common in Australia and New Zealand. It is seen particularly in brassicas, lettuce, tomatoes and pointsettia; in cauliflower, it produces blind plants with reduced and twisted lamina, or only the midribs persisting, in a condition known as whiptail (which may occur on soil up to pH 7). Death of growing points and chlorotic mottling and cupping or rolling of leaves are symptoms common to other molybdenum-deficient plants. Legumes may show signs of nitrogen deficiency because the function of nitrogen-fixing bacteria is dependent on the availability of molybdenum.

monadelphous stamens united by fusion of their filaments into a single group or bundle.

Monarda (for Nicholas Monardes (1493–1588), Spanish botanist and physician, author of the first medicinal flora of North America (1571) translated 1577 as *Joyfull Newes*). Labiatae. N America. WILD BERGAMOT, HORSEMINT, BEE BALM. 16 species, aromatic annual or perennial herbs, more or less hairy with clumped, erect stems. Hooded, tubular and 2-lipped, the flowers are borne in late spring and summer in dense clusters or heads or in an interrupted spike, and subtended by leafy bracts. Fully hardy. Plant in full sun. Propagate by division of established clumps during the dormant period. Various mildew fungi can be a nuisance, especially in hot dry summers.

M.didyma (BEE BALM, BERGAMOT, OSWEGO TEA; Canada, US; strongly aromatic perennial herb, 70–120cm; flowers bright crimson, 3–4.5cm; includes 'Adam', to 1m, with cerise flowers, 'Alba', with white flowers, 'Cambridge Scarlet', with red flowers, 'Croftway Pink', with rose-pink flowers, 'Salmonea', with salmon pink flowers, and 'Burgundy', with dark purple-red flowers); *M.fistulosa* (Canada, US, Mexico; perennial herb, 35–120cm; flowers lavender, lilac or pink, 2–3cm; differs from *M.didyma* in having bluntly 4-angled stem, usually only 1 flower cluster (not 2 superimposed) and fewer flowers with bracts tinted purple-pink, not red, and densely, not sparsely, hairy calyx throat). *Cultivars* selections or hybrids of *M.didyma* and *M.fistulosa*: 'Beauty of Cobham', tall, with lilac-pink flowers; 'Blue Stocking', tall, with brilliant deep violet flowers; 'Capricorn', with purple flowers; 'Croftway Pink', with clear rose pink flowers; 'Kardinal', tall, with red flowers tinted purple; 'Loddon Crown', with maroon flowers; 'Mahogany', tall, with deep wine red flowers; 'Morning Red', with dark salmon flowers; 'Pisces', with strong pink flowers with a green calyx; 'Prairie Night', with rich violet flowers; and 'Snow Queen', low, with creamy white flowers.

moniliform a regularly constricted organ, giving the appearance of a string of beads or a knotted rope.

mono-ammonium phosphate see *ammonium phosphate*.

monocarpic dying after bearing flowers and fruit only once.

monochasial cyme as a dichasial cyme, but with the branches missing from one side.

monochlamydeous with only one perianth whorl.

monocotyledons (monocots) one of the two major divisions of the angiosperms, the other being the dicotyledons. Monocots are characterized by the single cotyledon in the seed, usually the absence of cambium and thus of woody tissue and, in many cases, parallel leaf venation. The monocotyledons include Gramineae, Liliaceae, Orchidaceae, Iridaceae, Bromeliaceae, Palmae.

monoecious having separate male and female or bisexual flowers on the same plant, cf. *dioecious*.

monogeneric a family or grouping of higher rank containing a single genus.

monogerm seed seed that contains single embryos and is produced mechanically or genetically. It is available for beet species, the seed of which otherwise comprises a cluster of embryos, requiring resultant seedlings to be thinned.

monopetalous strictly, with one petal; often loosely used for gamopetalous.

monopodial of a stem or rhizome in which growth continues indefinitely from the apical or terminal bud, and which generally exhibits little or no secondary branching.

monotypic having only one component, for example a genus with one species.

Monstera (from Latin *monstrum*, a marvel, possibly alluding to curious shape of leaves). Araceae. Tropical America. SWISS-CHEESE PLANT, WINDOW-LEAF. 22 species, evergreen, epiphytic, perennial lianes with freely rooting, clambering stems and large, smooth, heart-shaped leaves, often pinnately lobed or perforated. Small flowers are borne on a dense, club-like spadix surrounded with a thick-textured spathe. The fruit is cone-shaped, composed of tightly packed white berries. In temperate zones, *Monstera* species are most commonly grown as houseplants or in warm greenhouses. Under such conditions, poles of sphagnum moss are often used in imitation of natural supports, dampened to improve atmospheric humidity and to provide a moist medium for the running aerial roots. If the necessary draught-free conditions with filtered light and high humidity are not provided, the leaf perforations typical of well-grown adult specimens of many species may fail to develop. Too much or too little water will cause leaf yellowing. Plants benefit from frequent sponging of leaves. Propagate by internodal cuttings, or from cuttings of growing tips with one leaf attached, rooted with bottom heat in a sandy propagating mix. Alternatively, increase by air-layering.

M.deliciosa (CERIMAN, SWISS-CHEESE PLANT; Mexico to Panama; robust climber with stout, green stems and copious, thick aerial roots; leaves 25–90 × 25–75cm, orbicular to ovate, base cordate, margins pinnately lobed in adult foliage, segments curved, oblong, usually perforated with elliptic to oblong holes; spathe to 30cm, spadix to 25 × 3cm, swelling in fruit, becoming cream-coloured and tasting of banana and pineapple when mature; includes 'Albo-variegata', with large, partly rich deep green leaves with other sections a contrasting creamy-white, and 'Variegata', with leaves and stem irregularly variegated cream or yellow); *M.dubia* (C America; robust, bushy climber with thick aerial roots; adult leaves to 130 × 60cm, oblong, pinnately cut, segments 12–20; leaves linear in juvenile forms and silver-variegated); *M.obliqua* (northern S America; slender-stemmed climber; leaves to 20 × 7cm, elliptic to oblong-lanceolate, entire or perforated with holes covering much of the surface area).

monstrous, **monstrosity** describing plants showing some abnormality of growth or flower, due to genetic make-up or pest or disease. Monstrous plants are sometimes prized by gardeners. See *malformations*.

montmorillonite a primary type of clay mineral having a high capacity for cation exchange and water-holding.

moongate a circular doorway in a wall; Chinese in origin and used in Western gardens from the 18th century.

mor a type of humus found as an unincorporated surface layer on mineral soils, typically as leaf litter on a forest floor where decomposition and incorporation are restricted by acidity, cf. *mull*.

Moraea (for Robert More, 18th-century botanist and natural historian. The name was originally spelt *Morea*, but changed to *Moraea* by Linnaeus in 1762, possibly in honour of his father-in-law Johan Moraeus, a Swedish physician). Iridaceae. Subsaharan Africa from Ethiopia to South Africa. About 120 species, perennial, cormous herbs with erect, simple or branched stems bearing terminal clusters of flowers with six tepals. Species which grow in winter and flower in spring need protection from frost in cool temperate climates. They are well-suited to cultivation in the cool greenhouse or conservatory. Species which flower in summer, are frost-hardy, withstanding winter temperatures between –5 and –10°C/23–14°F. All need full sun and a fertile, sandy and well-drained medium. Provide plentiful moisture when in growth, but keep dry when dormant. Increase by seed.
M.huttonii (to 1m; flowers scented, yellow with darker yellow marks, style crests with brown or purple blotch; outer tepals to 5cm, inner tepals to 4cm, lanceolate, erect); *M.polystachya* (to 80cm; flowers pale blue or violet, outer tepals 3–5cm with yellow or orange nectar guides, inner tepals 3–4cm, erect or reflexed); *M.spathulata* (to 90cm; flowers yellow, outer tepals 3.5–5cm with darker yellow nectar guides, inner tepals erect, 3–4cm).

moraine in nature, the accumulation of rocky debris, sand and silt at the snout or sides of a glacier, produced by grinding action on the bed rock. Being freely drained, it supports a specialized flora of rosette- or cushion-forming plants with long taproots. Moraine conditions can be imitated in the garden by the provision of specially constructed beds. See *rock plant gardening*.

Morina longiflora

Morina (for Louis Morin (1636–1715), French botanist). Morinaceae. E Europe to Asia. 4 species, aromatic perennial, prickly herbs. The leaves are mostly basal, tufted, lanceolate to oblong, undulate and spiny-toothed, progressively reduced on flowering stems to whorled, leafy and spiny bracts. The flowers are borne in whorls on an erect spike in summer. They are tubular and 5-lobed, with a spreading, somewhat 2-lipped limb. Hardy to –17°C. Grow in any fertile, humus-rich, well-drained but moderately moisture-retentive, sandy or gritty soil in sun. Fresh, ripe seed germinates freely in an open, gritty propagating mix and is best sown in individual pots to avoid damage to the taproot. Overwinter young plants in a well-ventilated cold frame. Increase also by root cuttings. *M.longiflora* (WHORLFLOWER; Himalaya; to 1.3m; corolla to 2.5cm, white, flushing shell pink, then bright crimson, particularly the limb).

morphology the study of the external form and relative position of plant organs; one of the most important divisions of botanical science.

Morus (from Latin *morus*, mulberry). Moraceae. Western N America and S Europe east to Japan, south to lowland tropics in C Africa. MULBERRY. About 12 species, deciduous trees and shrubs. The crown is usually dense and rounded with rugged, gnarled branches, the bark brown, rough, and often burred. Young shoots exude white latex when cut. The leaves are generally ovate to cordate and toothed, turning yellow in autumn. The inflorescences are green and catkin-like, the females developing as fruits that superficially resemble a large raspberry but differ by consisting of closely packed but separate drupes. These are green at first, ripening through orange or white to red or deep purple-black by late summer to mid-autumn of the same year.
M.alba (WHITE MULBERRY; China; tree to 16m; fruit 1–2.5cm, green-white ripening pink to dark red; cultivars include 'Aurea', with yellow leaves and bark; 'Laciniata', with leaves deeply lobed and toothed; 'Macrophylla', with large leaves; Multicaulis Group, suckering shrubs, with leaves to 35cm and fruit that is nearly black when ripe; 'Nana', dwarf, shrubby, rounded, with regularly lobed leaves; 'Nigrobacca', with fruit ripening purple-black; 'Pendula', producing a weeping crown if grafted on tall stocks of the typical mulberry; and 'Pyramidalis', with a pyramidal crown); *M.nigra* (BLACK MULBERRY; probably originating in SW Asia, now widespread through cultivation; tree to 15m; fruit 2–2.5cm, green ripening through orange and red to deep purple); *M.rubra* (RED MULBERRY; E US, SE Canada; tree to 15m, rarely 20m; fruit 2.5–3cm, green ripening through orange and red to purple; includes 'Nana', dwarf, slow-growing, with smaller, 3–5-lobed leaves). See *mulberry*.

mosaic leaf discoloration with numerous discrete lighter and darker areas, for example yellow or pale greens on darker greens, often angular and tending to be delineated by the veins. Mosaic is a characteristic virus disease symptom; cf. *mottle*. See *virus*.

moss primitive non-flowering, plants belonging to the Bryophyta. Delicate in form, they posses stems and leaves but no true root. Mosses generally grow in damp places and, like ferns, exhibit alternation of generations, except that the conspicuous generation is the sexual one which can live independently.

The common peat or bog moss of the genus *Sphagnum* is of horticultural significance: harvested peat is formed in bogland from its incompletely decomposed remains, and it is used in its living form as a medium for supporting epiphytes and for lining hanging baskets and slatted containers. The use of sphagnum moss and sphagnum peat conflicts with wildlife conservation objectives. Peat and moss alternatives are available for most horticultural uses, and moss, such as Highland moss (*Pseudoscleropodium purum*), is obtainable from sustainably harvested sources.

Some species of moss may be cultivated or maintained for ornamental purposes, but naturally growing plants are most commonly met in gardens as undesirable invaders of lawns.

moss house a rustic building featured in some Victorian gardens, with interior walls formed of wooden slats into which mosses collected from the wild were pressed.

moss pole a wooden pole on to which sphagnum moss is bound, or a cylinder of wire or plastic netting filled with moss; used to provide support for climbing plants with aerial roots.

mother plant a plant from which young plantlets or propagating material are obtained; often stock plants exclusively reserved for the purpose and kept free of pathogens, especially for the production of certified stock.

moths Together with butterflies, moths make up the insect order Lepidoptera, none of which are pests as adult insects: some do not feed at all, others suck nectar or similar sweet liquids. They lay eggs on suitable host plants. These hatch into caterpillars with three pairs of jointed legs at the head end and up to five pairs of claspers or prolegs on the abdomen; geometrid larvae, known as loopers or inchworms, have only two pairs of prolegs. Caterpillars pass through several stages before becoming fully fed, and often leave the food plant to pupate in the soil before emerging as adults. All caterpillars have mouthparts adapted for biting and chewing, and most feed on foliage. Some feed exposed, while others, for example, many of the tortrix moths, feed hidden from view by binding two leaves together. Some moths, such as the GYPSY MOTH, BROWN TAIL MOTH, LACKEY MOTH, SMALL ERMINE (*Yponomeuta* species) and COTONEASTER WEBBER, have larvae that feed gregariously and cover infested shoots with dense silk webbing; these can defoliate shrubs and small trees completely. Some of the smallest moths have larvae that feed as leaf miners within the foliage of plants such as apple, oak, lilac and laburnum.

The roots and stem bases of many plants are damaged by soil-dwelling caterpillars, known as cutworms, which are the larvae of various species of noctuid moth. SWIFT MOTH LARVAE (*Hepialus* species) also feed in a similar manner. Annual flowers, lettuce, small herbaceous plants and root vegetables are most at risk.

The larvae of CLEARWING MOTHS are stem borers, mainly of woody plants such as currant, apple, willow and poplar. Caterpillars of the LEOPARD MOTH attack the branches and trunk of apples and other trees, and pine trees have caterpillar pests that bore into buds killing or distorting the leading shoot.

A number of moths have caterpillars that feed on ripening fruits or seeds, including the CODLING MOTH, which is a worldwide pest of apples and pears, the PLUM MOTH, the PEA MOTH and the ORIENTAL FRUIT MOTH, which attacks peach and many other fruits. Some moth caterpillars can be very destructive in seed stores.

Most moth larvae cause insignificant amounts of damage and some have been used successfully in the biological control of weeds such as prickly pear and ragwort. Where control measures against caterpillars are required, the young larvae should be targeted with contact insecticides since these are more effective than systemic types. Biological control is possible for some pests, such as cabbage caterpillars, using a bacterial spray containing *Bacillus thuringiensis*. See *clearwing moths, codling moth, swift moths, tent caterpillars.*

motile hairs, appendages, lobes and spores capable of movement.

mottle leaf discoloration with small chlorotic areas of irregular shape without sharply defined edges and in a pattern that is not related to the vein network. Mottle is a common virus disease symptom; cf. *mosaic*. See *virus*.

mould the visible surface growth of fungal spores and mycelium, which is common on organic matter of plant or animal origin with a high moisture content. BLUE MOULD is the spores of *Penicillium* species, which cause storage rots of fruit and bulbs. GREY MOULD describes the fungus *Botrytis cinerea*, which causes various forms of die-back and rots in many plants, and rots fruits and vegetables. *Rhizopus stolonifer*, which can cause fruit rot, is sometimes called PIN MOULD, and LEAF MOULD is a disease of tomatoes caused by the fungus *Fulvia fulva*. SOOTY MOULD fungi grow on plant surfaces that are covered with honeydew excreted by insects. SLIME MOULDS are organisms related to fungi which ingest bacteria, fungus spores and particles of organic matter.

mount an artificial hill within a garden. Mounts were constructed in Europe, especially Britain, from the 14th to early 18th century to provide a viewpoint over high garden walls. Mounts could be square or circular, with stairways or a spiral ramp leading to a flat viewing terrace, with perhaps an arbour, pleasure house, or temple on top.

mouse ear see *bud stages*.

mowing see *lawns*.

mucilage a viscous substance or solution.

mucilaginous slimy.

mucro an abrupt, sharp, terminal spur, spine or tip.

mucronate an apex terminating suddenly with an abrupt spur or spine developed from the midrib.

mucronulate diminutive of mucronate.

Mucuna (Brazilian name for these plants). Leguminosae. Widespread in the tropics and subtropics of both Hemispheres. Some 100 species, evergreen, woody climbers and erect shrubs with trifoliolate leaves and pea-like flowers in axillary clusters or racemes in spring and summer. *M.bennettii* is a spectacular flame-flowered climber for tropical gardens or the warm greenhouse. Its cultural requirements are as for *Strongylodon*. *M.pruriens* and *M.sempervirens* are less showy, more intriguing, with racemes and clusters respectively of purple-black or mauve to white flowers. In climate zones lower than 9, they perform best in cool greenhouses and conservatories, but might be attempted outdoors in favoured regions.

M.bennettii (NEW GUINEA CREEPER; New Guinea; perennial woody climber to 20m; flowers to 8.5cm, vivid scarlet or flame-coloured, in a short inflorescence); *M.pruriens* (syn. *M.cochinchinensis*; VELVET BEAN; Asia, naturalized elsewhere; annual or perennial semi-woody, climbing herb to 4m; racemes to 30cm, flowers 3–4cm, dark damson coloured to pale purple or white; var. *utilis*: VELVET BEAN, FLORIDA BEAN, BENGHAL BEAN, flowers with a purple standard flushed green and wings of a dirty red); *M.sempervirens* (China; perennial evergreen climber to 12m; raceme short, nodding, flowers waxy, bruised purple-black, malodorous).

Muehlenbeckia (for H.G. Muehlenbeck (1798–1845), Swiss physician and student of the flora of Alsace). Polygonaceae. S America, New Zealand, Australia, New Guinea. 15 species of largely dioecious, evergreen climbing or procumbent subshrubs and shrubs with wiry stems, small leaves, inconspicuous white flowers and fruit surrounded by the enlarged, fleshy white perianth. Grown for its intricate habit, minute foliage and small but very sweetly fragrant flowers, *M.complexa* is suitable for covering tree stumps and rocky banks, or for scrambling

Muehlenbeckia complexa

through shrubs. It is also good for hanging-baskets, forming a mass of intricately tangled dark wiry stems. This species is sometimes treated as a standard, grown on without support in open beds to promote a number of long stems, then potted in tubs of rich loamy soil with the stems gathered and tied to a central cane. Plants grown in this way produce a weeping crown of fine branchlets. Despite its delicate appearance, *M.complexa* can match other climbing Polygonaceae in vigour and has become a troublesome invader in warmer regions. Plant in a position sheltered from cold, drying winds in climate zones 7 and over. Grow in well-drained soil in sun to part shade. Prune only to restrict to allotted space. Propagate by semi-ripe cuttings in summer. *M.axillaris* (Australia, New Zealand; small, prostrate or straggling shrub to 1 × 1m; branches delicate, intricate, black; leaves 0.5–1cm diam., oblong – pandurate to orbicular, dark green above, grey beneath; 'Nana': dwarf, leaves pandurate); *M.complexa* (MAIDENHAIR VINE, WIRE VINE, MATTRESS VINE, NECKLACE VINE; New Zealand; fast-growing semi-deciduous liane, creeping or climbing to 5m, stems slender, dark and wiry, forming dense tangles; leaves 5–20mm, dark green above, purple to silver beneath, oblong to circular or pandurate, rounded or cordate at base; includes var. *microphylla*, a dense shrub to 60cm, with few, rounded, very small leaves, and var. *triloba*, with pandurate, 3-lobed leaves).

Muilla (anagram of the closely related *Allium*.) Liliaceae (Alliaceae). SW US, Mexico. 5 species, cormous herbaceous perennials, resembling *Allium* but lacking its characteristic odour. Cultivate as for *Allium*. *M.maritima* (syn. *M.serotina*; California, Mexico; to 50cm; flowers 4–70 per umbel, tepals 3–6mm, white, tinged green, midrib brown, anthers purple).

Mukdenia (from the Chinese city, Mukden). Saxifragaceae. N China, Korea. 2 species, deciduous perennial herbs. The long-stalked leaves are peltate, round to reniform, cordate at the base, palmately lobed and toothed. In spring, paniculate racemes of numerous small, white flowers stand taller than the foliage. Deciduous groundcover for woodland fringe and other shaded locations. Cultivate as for *Mitella*. *M.rossii* (N China, Manchuria, Korea; to 60cm; leaves bronze-green; flowers small, white).

mulberry (*Morus* species) Probably originating in the Caucasus or further east, the BLACK MULBERRY (*Morus nigra*) is prized for its fruit which superficially resembles a large raspberry. There are biblical references to the mulberry and its spread can be traced from ancient Greece and Rome across Europe to Britain and thence to other parts of the world. Mulberries are grown as either standard or half standard trees; they are slow growing but may eventually reach 10m in height, forming ornamental specimens with attractive gnarled bark, and a

mulberry

spreading head. They may survive for several hundred years. Mulberries require a sheltered, sunny situation, and in the UK do best in southern parts. Small trees can be maintained in large tubs.

The flowers, which are borne on inconspicuous catkins, are unisexual and self-fertile; the trees have a high chilling requirement and are therefore late into leaf and flower, thus avoiding frost damage.

Fruits ripen in late summer, changing gradually from green to deep crimson when they acquire a sweetly acidic flavour; the juice stains strongly and is persistent. Either pick the fruit from the trees or gather when fallen onto grass or a collecting sheet; use immediately for dessert, preserves or wine making, or for freezing.

Prune young trees in winter to establish a framework and minimally thereafter to maintain shape; where cut or damaged between spring and autumn the wounds will bleed profusely and should be cauterized with a hot iron. Propagation may be by simple layering in autumn, or by air layering, but usually by 18cm long hardwood cuttings taken from two year-old wood, with a heel, in autumn or spring.

WHITE MULBERRY, *M.alba* is not usually grown for its fruits, but in China, its leaves have provided food for silkworms for over 5000 years.

mulch a covering placed on the surface of culti-vated soil. The benefits of a mulch include: conserving soil moisture by reducing evaporation; insulating the soil against extreme cold in winter, whilst keeping it relatively cool in summer; in some instances, encouraging early soil warming in spring; stimulating beneficial soil bacteria; deterring some soil-inhabiting pests; smothering weeds; keeping some fruits and vegetables off the soil surface; providing nutrients; and improving soil structure.

Mulches usually consist of bulky organic mate-rials, such as rotted animal manure, chopped straw, spent mushroom compost, spent brewery hops, leaf-moulds, chipped bark and garden compost. Lawn mowings can be used provided they do not form too dense a layer, as can chopped bracken.

Unless incorporated into the surface layer, decaying organic mulches have little depleting effect on soil nitrogen. Certain types, such as farmyard manure and garden compost, may be a source of weeds if they have not been prepared or stored care-fully. Some spent mushroom compost offered for sale may include lime, which will adversely affect pH where ericaceous plants are grown. Assurance should be obtained that straw for mulching does not contain residual hormone weedkiller, which might have a harmful effect upon plant growth.

Non-organic mulching materials include pea gravel and stone chippings, which are visually acceptable for some garden situations, and black polythene sheet for weed suppression and moisture retention around fruit and vegetable crops and newly planted trees. Clear polythene and spun fibre fleece aid soil warming and seed germination, as well as moisture retention. Film or sheet covers are often referred to as 'floating mulches', and some plants, for example, strawberries, may be planted through holes in the mulch; all such covers must be firmly buried at the edges to avoid removal by wind. Perennial weeds should be cleared from a site before mulches are put down.

The best time for mulch application is in spring when winter rain will have fully charged the soil with water. Organic materials should be spread to a depth of at least 10cm. It is important to remember that, in times of prolonged drought, sufficient water must be applied to penetrate the mulch layer; sheet mulches should have irrigation lines installed beneath the covers at the time of laying down.

In very cold conditions, the temperature immedi-ately above a mulch tends to be lower than over bare soil; where the temperature falls below freezing, shoots and flowers, for example, of strawberries, should be protected with additional straw, bracken or a cloche.

Hoeing provides some of the insulating benefits of mulching, but only if it is done to a shallow depth.

mule a sterile hybrid of two plant species.

mull a type of humus consisting of a dark-coloured mixture of mineral and organic material, found in the surface layer of soil, typically on a forest floor where decomposition and incorportion of organic matter occurs freely; cf. *mor*.

multifid as bifid, but cleft more than once, forming many lobes.

multifoliate many-leaved.

multigeneric describing hybrids with more than two genera involved in their immediate ancestry, for example *Potinara* (*Brassavola × Cattleya × Laelia × Sophronitis*). Multigeneric hybrid names are formed either by combining the generic names of the parents, as for example with *Sophrolaeliocattleya*, or, as in the case above, by creating a commemorative per-sonal name using the suffix *-ara*.

mulberry

Diseases
canker
powdery mildew

multiple fruit a fruit derived from an entire inflorescence, i.e. from the parts of a number of flowers.

mummified fruit dry and shrivelled fruit infected by fungi, especially that of apples, grapes, peaches and plums. Mummified fruit can often carry over disease and should be collected and burnt.

municipal waste see *town refuse*.

muriate of potash see *potassium chloride*.

muricate rough-surfaced; furnished with many short, hard, pointed projections.

muriculate slightly muricate.

Murraya (for John Andrew Murray (1740–1791), pupil of Linnaeus and Professor of Medicine and Botany at University of Gottingen). Rutaceae. Asia (India and China south to Australia). 5 species, aromatic evergreen trees and shrubs with pinnate leaves and fragrant, orange blossom-like flowers in spring and summer. Grown in their native regions as border specimens, as screens or as hedging, in cool temperate climates *Murraya* species are handsome specimens for tubs or large pots in the greenhouse or conservatory. Young plants of *M.paniculata* are also used as houseplants. Plant in a fertile, moisture-retentive but well-drained medium in full sun or part shade. Water plentifully and feed fortnightly when in full growth, moderately at other times. Maintain a winter minimum of 10°C/50°F. Prune if necessary in early spring. Propagate by seed in spring or by semi-ripe cuttings.

M.koenigii (CURRY LEAF; Asia; evergreen tree, 4.5–6m; leaves pungently aromatic, leaflets 2.5–4cm, oblong-lanceolate to ovate, membranous; flowers white to ivory; fruit dark blue tinged black); *M.paniculata* (ORANGE JESSAMINE, SATIN-WOOD, COSMETIC BARK TREE, CHINESE BOX; China and India south to Australia; evergreen tree or shrub, 4.5–7.5m, evergreen; leaves glabrous and glossy, leaflets 2.5–4cm, cuneate-obovate to rhombic; flowers white; fruit orange to red).

Musa (from Arabic *mouz*, banana, or for Antonius Musa, physician to the Emperor Augustus). Musaceae. Old World tropics and subtropics, east to Japan, south to Queensland. BANANA, PLANTAIN, MANILA HEMP. About 40 species, giant, evergreen perennial herbs with thickened rhizomes and swollen and suckering rootstocks. The leaves are large and paddle-shaped with a distinct, sunken midrib. The sappy and fibrous petiole bases stand one within the other, forming a trunk-like pseudostem. The inflorescence is a terminal spike, pendulous or erect, with tubular flowers in clusters subtended by large and fleshy to waxy, coloured bracts. The fruit is the banana, most familiarly finger-like and gently curving, but globose to ellipsoid in some species. After flowering/fruiting, the main growth dies, but is soon replaced by offsets.

Magnificent and fast-growing foliage and flowering ornamentals for tropical and subtropical gardens and, in cooler climates, for large containers or borders in the greenhouse, conservatory or home. Careful note should be taken of eventual size when buying, although some 'dwarf' selections of *M.acuminata* and *M.× paradisiaca* are now widely offered as houseplants and seldom make the home uninhabitable. With the exception of *M.basjoo*, those described here need a minimum temperature of 10°C/50°F, but are largely unfussy as to light and humidity, provided they are grown in a moist and richly fertile medium. Water and feed generously when in growth. Remove the blades of faded leaves and cut out flowered growths once they begin to deteriorate. The Japanese *M.basjoo* is virtually hardy in sheltered areas of zone 7 and can certainly be used as a dot plant in bedding schemes or as a patio container plant. Overwinter it in a cold greenhouse with leaves removed and the rootstock wrapped in scarcely damp burlap or plunged in coir or sawdust. If grown outdoors year round, this banana should be sited with protection from wind which quickly shreds the leaves. Wrap the stem with sacking in autumn and mulch the base thickly. Where the stems are killed by frost, replacement shoots will usually appear by early summer. Propagate all species by seed germinated in spring at 20°C/70°F, or by removing suckers with a portion of root-bearing rhizome attached. See also *Ensete*.

M.acuminata (syn. *M.cavendishii*; BANANA, PLANTAIN; SE Asia, N Queensland; pseudostems to 7m or more, green blotched brown; leaves to 3 × 0.7m, sometimes dark-flecked above, green or purple beneath; inflorescence pendulous, bracts bright red to purple or yellow; fruit to 20 cm, yellow, pulp white to yellow; includes the BLOOD BANANA, *M.sumatrana*, with leaves blotched wine red above, stained red beneath – colouring strong and persistent in 'Rubra'); *M.basjoo* (syn. *M.japonica*; JAPANESE BANANA; Japan, Ryukyu Islands; to 3m, stoloniferous; pseudostem green to red-tinted; leaves 2 × 0.7m, oblong-lanceolate, bright green, midrib sometimes red-tinted; inflorescence horizontal to pendulous, bracts yellow-green or tinged purple-brown; fruit to 6 × 2.5cm, yellow-green with white pulp; 'Variegata': leaves banded or flecked lime green, cream and white); *M. × paradisiaca* (EDIBLE BANANA, FRENCH PLANTAIN; *M.acuminata × M.balbisiana*; Tropics; to 8m; leaves to 2.5 × 0.7m, oblong, green; infloresence pendulous, to 1.5m, bracts purple-red; fruit usually seedless, often yellow, pulp white; there may be as many as 300 cultivars of edible banana worldwide; two ornamental cultivars are 'Koae', with leaves laterally striped white and pale green on dark green, and 'Vittata' with leaves variegated pale green and white, the midrib white, edged red); *M.uranoscopus* (syn. *M.coccinea*; Indochina; pseudostem to 1m, glossy red-green; leaves to 75 × 30cm, oval to elliptic, glossy above, somewhat waxy beneath, midrib green to rose-pink; inflorescence to 0.75m, erect, bracts magenta to scarlet; fruit oblong to cylindric, to 5 × 2.5cm, red-purple to pink-green). For *M. ensete*, see *Ensete ventricosum*; for *M.superba*, see *Ensete superbum*.

Muscari (from Greek *moschos*, musk, alluding to the scent of some species). Liliaceae (Hyacinthaceae). Mediterranean, SW Asia. GRAPE HYACINTH. 30 species, bulbous perennial herbs with slender, rather fleshy leaves and erect spikes packed with small, bell- to urn-shaped flowers in spring. Hardy in climate zone 7. Plant in full sun on a free-draining soil. Top-dress established clumps with bonemeal in spring; lift and divide when overcrowded, incorporating fresh top soil before replanting. Propagate by offsets after flowering or by seed sown when ripe.

M.armeniacum (SE Europe to Caucasus; raceme 2.5–7.5cm, flowers crowded, fertile flowers 3.5–5.5mm, obovoid to urceolate, azure, sometimes flushed purple, rarely white, lobes paler or white, sterile flowers few, smaller and paler; includes white-, pale blue- and white-rimmed cultivars); *M.aucheri* (syn. *M.tubergenianum*; Turkey; raceme dense, ovoid or cylindric, fertile flowers 3–5mm, subspherical or ovoid, bright azure, rarely white, lobes paler or white, sterile flowers paler); *M.azureum* (E Turkey; raceme 1.5–3cm, dense, fertile flowers 4–5mm, campanulate, bright blue with a darker stripe on the lobes, sterile flowers smaller and paler; includes white- and pale blue-flowered cultivars); *M.botryoides* (C and SE Europe; raceme dense then loose and cylindric, fertile flowers 2.5–5mm, subspherical, scented, azure or white, lobes white; includes white, bright blue and pink cultivars); *M.comosum* (TASSEL HYACINTH; S and C Europe, N Africa, SW Asia, naturalized Europe; raceme loose, fertile flowers 5–9mm, oblong-urceolate, brown-olive, lobes yellow-brown, sterile flowers subspherical or obovoid, rarely tubular, bright violet on violet, ascending pedicels; includes 'Plumosum', syn. 'Monstrosum', with a sterile, mauve-blue, densely branched inflorescence); *M.latifolium* (S and W Asia; raceme dense then loose, fertile flowers 5–6mm, oblong-urceolate, tube strongly constricted, deep violet, sterile flowers blue); *M.macrocarpum* (Aegean Islands, W Turkey; raceme loose, fertile flowers 8–12mm, fragrant, oblong-urceolate, blue-violet then yellow, expanding to form brown or yellow corona, sterile flowers tinged purple, few or none); *M.massayanum* (E Turkey; raceme dense, cylindric, fertile flowers to 1cm, pink to violet at first, later pale green- or yellow-blue, lobes dark brown, sterile flowers pink or violet on ascending pedicels, forming a dense tuft); *M.muscarimi* (syn. *M.moschatum*; SW Turkey; flowers muskily scented, fertile flowers 8–14mm, narrowly urceolate, purple then pale green to ivory, strongly contracted then expanded to form a brown corona, sterile flowers purple-tinged, rarely present); *M.neglectum* (syn. *M.racemosum*; COMMON GRAPE HYACINTH; Europe, N Africa, SW Asia; raceme dense in flower; fertile flowers 1.5–3.5mm, ovoid to oblong-urceolate, strongly constricted, deep blue-black, lobes white, sterile flowers smaller and paler blue). For *M.paradoxum*, see *Bellevalia pycnantha*.

muscariform in the shape of a broom or brush.

mushroom compost spent compost from commercial mushroom production, used as a soil improver, a mulch, or an ingredient of potting media. It is the product of composting straw with horse manure or nitrogenous ferilizers, and, depending on the type of casing ingredients used for spawning mushrooms, may be alkaline in nature; hence, regular applications can alter soil pH. Where incorporated, it improves soil structure, provides some nutrients and can enhance nutrient and water retention. Its physical properties are determined by age: for example, with decomposition, aeration decreases and moisture-holding increases.

mushrooms Edible fungi have been collected from the wild since ancient times but few species are in cultivation due to the complex microbial processes and associations involved in their growth, and to the difficulties involved in preparing substrates on which they will establish and fruit. Mushroom cultivation is unique in that it provides the only profitable process for converting waste lignocellulose from agriculture and forestry into a useful, edible product. Mushrooms cultivated on a significant scale worldwide are *Agaricus bisporus* and *A.bitorquis*, SHIITAKE MUSHROOM (*Lentinus edodes*), CHINESE or STRAW MUSHROOM (*Volvariella volvaceae*), WINTER MUSHROOM (*Flammulina velutipes*), OYSTER MUSHROOM (*Pleurotus* species), NAMEKO or VISCID MUSHROOM

mushrooms

Pests
eelworms
sciarid flies

Diseases
virus
bacteria

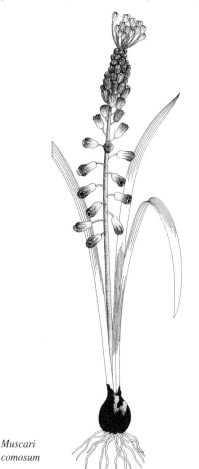

*Muscari
comosum*

(*Pholiota nameko*), JEW'S EAR (*Auricularia* species) and SNOW MUSHROOM (*Tremella fuciformis*).

Agaricus accounts for two thirds of the world production of cultivated fungi; it dominates production in the west and is the only species of significance cultivated in the British Isles.

Agaricus bisporus is cultivated on substrates prepared from cereal straw, preferably wheat. In the first phase of preparation, straw or strawy horse manure is wetted and mixed with activators, such as chicken manure or molassed malt waste; these provide a source of nitrogen and soluble carbohydrate to promote microbial breakdown of the substrate when stacked. Fermentation develops and the stack is turned at intervals of 2–4 days over a period of 7–14 days to maintain aeration and ensure uniformity. In the second phase, the fermenting compost is transferred to a facility in which it is pasteurized at 55–60°C/130–140°F over a period of 4–8 days in a process known as 'peak heating'. At the end of this period, the compost is cooled at 25°C/77°F and inoculated with a pure culture of *Agaricus* mycelium (spawn) grown on cereal grain. The spawned compost is maintained at 25°C/77°F for 10–14 days during which it is colonized by the fungal mycelium (spawn-running). The upper surface of the colonized compost is then covered with a layer of peat/chalk (casing) and the temperature lowered to 16–18°C/60–65°F. The mycelium continues to colonize the compost and also invades the casing layer. After 18–21 days, fruit bodies are initiated (pinning) and these develop into the typical gilled caps. The crop is produced in flushes at intervals of 7–10 days over a period of 30–35 days. Production is then terminated by heating to at least 60°C/140°F with live steam (cooking out), the spent compost removed and the facility cleaned in preparation for the start of the next cycle. The production cycle takes 12–14 weeks.

Commercial crops may be cultivated in purpose-built sheds, in insulated 'tunnels' covered with plastic film, or in caves. The substrate may be contained and the crop produced in ridged beds on the floor, in tiered wooden trays, on metal shelves, in deep troughs or in open-topped plastic bags. The processes involved in preparation and handling of compost, filling of trays and shelves, spawning, casing, watering and clearance of the used compost are all highly mechanized.

Cultivation requires accurate control of the environment and particularly of temperature, CO_2 level, ventilation rate and evaporative capacity of the air at different stages in the production cycle. It is therefore difficult for private gardeners to prepare composts on which mushrooms can be successfully grown, but there are mushroom-growing kits available, consisting of prepared compost already spawned, together with a bag of casing material. The packaging may be a plastic bag, a water-resistant cardboard or rigid plastic box, or a rigid plastic tub, which doubles as a growing container. The purchaser simply follows the instructions enclosed.

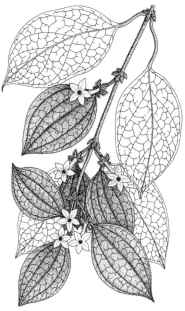

Mussaenda erythrophylla

Mussaenda (from the Sri Lankan name). Rubiaceae. Tropical Old World. Some 100 species, shrubs, subshrubs and herbs, often scandent. They bear 5-lobed, tubular to funnel-shaped flowers in axillary or terminal, panicles or cymes. With some flowers in each inflorescence, a calyx lobe is greatly enlarged as a highly coloured, leaf-like limb. Grow in the hot greenhouse (minimum temperature 15–18°C/60–65°F), in direct sunlight, in a pH neutral mix of equal parts loam, coir and leafmould with added sharp sand. Water plentifully when in growth, sparingly in winter. Provide support for climbing specimens, and thin out crowded stems or prune to a framework in spring. Propagate by semi-ripe cuttings or air layering in summer, or by seed in spring. *M.erythrophylla* (Tropical Africa; erect or climbing, evergreen or deciduous shrub to 8m; flowers in dense panicles or cymes, pink to red and white to yellow, many with an enlarged calyx lobe to 10cm long, oval and, typically, scarlet; includes 'Queen Sirikit', with deep pink to ivory inflorescences).

mustard, white (*Sinapis alba*) a cruciferous annual, native of the Mediterranean region, C Asia, and N Africa. Of horticultural importance as a salad plant and for green manuring, it is also grown for its seed, which is an ingredient of table mustard and curries, and as a fodder crop. Mustard is often grown with common garden cress (*Lepidium sativum*) for harvesting as dense young seedlings with elongated stems about 4–5cm high. It should be sown indoors two days later than cress for even development of the two, either on a soilless medium like blotting paper or any other moisture-retentive substrate. Mustard is milder in flavour than cress and is sometimes substituted with rape (*Brassica napus* subsp. *oleifera*).

White mustard can also be grown as a cut-and-come again salad vegetable, either sown in seed trays under protection or broadcast or in drills out-of-doors, especially in spring and autumn. Other vegetable crops known as mustards are SPINACH MUSTARD (*Brassica rapa* Perviridis Group) and MUSTARD GREENS (*Brassica juncea*) see *oriental greens*, and TEXSEL GREENS (*Brassica carinata*) see *cabbage*.

mutation a spontaneous or induced genetic change in a plant that gives rise to characteristics such as leaf variegation and colour change or distribution in flowers. Many mutations are perpetuated as desirable forms in gardens. A plant arising from such a change is known as a mutant, sport or break.

muticous blunt, unpointed.

Mutisia (for José Celestino Mutis of Cadiz (1732–1808), teacher of anatomy and student of South American plants). Compositae. S America. About 60 species, evergreen subshrubs or shrubs, many of them sprawling or twining, often with tendril-tipped leaves. Brightly coloured daisy-like flowerheads are produced in late spring and summer. Given the protection of a south-facing wall and good drainage, *M.oligodon* will tolerate temperatures to about –15°C/5°F and *M.decurrens* to about –10°C/14°F. Grow in deep, well-drained, moderately fertile soils, with shade at the roots but where top growth will receive full sun; a large rock or deep organic mulch placed over the roots will provide the necessary protection and help conserve moisture when in growth. Provide support by means of wire or trellis, or plant at the foot of a wall shrub in a sheltered, sunny position. Under glass, water moderately when in growth, and maintain good ventilation with a minimum winter temperature of 7–10°C/40–50°F. Prune to remove weak and overcrowded growth after flowering or in spring in cool climates. Propagate by stem cuttings in summer, by simple layering, or by removal of rooted suckers; increase also by seed in spring.

M.decurrens (Chile, Argentina; much-branched, rhizomatous subshrub to 2m; leaves to 10cm, lanceolate, with a long tendril at the apex and decurrent base, margins entire or sharp-toothed; flowerheads to 12cm in diameter, brilliant orange); *M.oligodon* (S Chile, S Argentina; straggling shrub or liane to 1m; leaves to 3.5cm, oblong to elliptic, apex acute or obtuse and sparsely toothed, sometimes terminating in a tendril, base cordate; flowerheads to 7cm in diameter, bright red).

mycelium typically, the vegetative body of a fungus made up of a mass of microscopic filamentous branches, known as hyphae.

mycology the study of fungi.

mycoplasmas (MLOs) see *phytoplasmas*.

mycorrhiza a symbiotic association between the root cells of a plant and fungal mycelium, whereby the absorption of nutrients is enchanced to the benefit of both organisms. The relationship is common in plants such as orchids, trees and heaths.

Myoporum (from Greek *myo*, to close, and Latin *porum*, a pore, due to the densely glandular-punctate leaves). Myoporaceae. Australia and New Zealand to E Asia and Mauritius. About 30 species, glandular-punctate evergreen trees and shrubs, often shrubby and heath-like. Small, 5-lobed and campanulate to hypocrateriform flowers are borne singly or in clusters in summer and are followed by colourful drupes. In frost-free, Mediterranean-type climates, grow *Myoporum* species in the rockery or on sunny banks; they thrive in dry soils and maritime conditions. In colder zones, grow in a sandy loam-based mix in the cool, well-ventilated greenhouse or conservatory. Propagate by simple layers, by semiripe cuttings or by seed.

M.laetum (New Zealand; tree to 10m; leaves 4–10cm, lanceolate to obovate, somewhat fleshy, crenate to sinuate; flowers to 1cm in diameter, campanulate, white with purple spots; fruit pale to dark maroon); *M.parvifolium* (Australia; low, spreading, glabrous shrub to 50cm; leaves 1–2.5cm, linear; flowers honey-scented, to 1cm wide, white).

Myosotidium (from *Myosotis* and Greek *oides*, resembling – the flowers resemble those of *Myosotis*). Boraginaceae. New Zealand (Chatham Islands). 1 species, *M.hortensia*, the CHATHAM ISLANDS FORGET-ME-NOT, a perennial or biennial herb to 1m tall with long-stalked, thinly fleshy, reniform to broadly ovate or ovate-cordate leaves to 30cm in diameter and glossy above with deeply impressed veins. In summer, it produces broad, stalked heads of large, dark to sky blue forget-me-not flowers. Hardy in climate zone 7. In gardens with cool summers and little frost, it is suitable for the front of borders or the rock garden. Where hard frosts occur, protect with evergreen branches or dry bracken litter, otherwise grow in the cold greenhouse. Plant in a cool position in semi-shade, sheltered from wind, in free-draining, humus-rich soils. Propagate from seed sown ripe in autumn or by careful division.

Myosotis (from Greek *mys*, mouse, and *ous*, ear, referring to the appearance of the leaves). Boraginaceae. Temperate distribution, mainly Europe, New Zealand. FORGET-ME-NOT, SCORPION GRASS. Some 50 species, hairy annual, biennial or perennial herbs. They bear small, rotate or salverform flowers with five rounded lobes and distinct faucal scales (the 'eye') at various times of year and usually in paired cymes. *M.sylvatica* and *M.alpestris* occur in damp woodland and meadow. Their many cultivars are grown as hardy annuals or biennials (tolerating winter temperatures of at least –15°C/5°F); these are available as compact dwarf cultivars, such as 'Ultra-

Myosotis scorpioides

marine', with deep indigo blue flowers, 'White Ball', compact with white blooms, and the taller rose pink flowered 'Rose'. Traditionally used in spring bedding and as border edging, they are also suited to window boxes and to pot cultivation in the cold greenhouse for winter and early spring blooms. They will self-seed. *M.scorpioides*, *M.laxa*, and *M.palustris*, found in wet habitats at water margins and in marshy ground, are suited to pond edge and other damp sites, in mud or very shallow water, especially in wilder areas of the garden. *M.palustris* is vigorous and invasive, sometimes occurring as a lawn weed; it may also be used in the wildflower lawn. *M.australis* occurs at higher altitudes in exposed, moist and rocky habitats. Grow it in gritty, moderately fertile and perfectly draining soils in sun or in light, partial shade – scree plantings, for example. Propagate by seed or division in early spring. Powdery mildew may seriously reduce flowering of annual/biennial species. *Myosotis* smut shows as off-white spots on the leaves, and often occurs on plants allowed to self-sow.

M.alpestris (Europe, Asia, N America; tufted perennial to 30cm; flowers bright or deep blue, to 9mm in diameter); *M.australis* (Australia, Tasmania, New Zealand; perennial to 30cm; corolla white or yellow); *M.caespitosa* (Europe, Asia, N Africa, N America; perennial to 60cm, sparingly covered with white hairs; flowers blue to 4mm in diameter); *M.laxa* (Europe, N America; annual to perennial; stem to 50cm, decumbent, adpressed-pubescent; flowers bright blue with a yellow eye, to 5mm in diameter); *M.palustris* (Europe, Asia, N America; close to *M. caespitosa*, except flowers to 8mm in diameter; includes 'Alba', with white flowers, and

'Mermaid', with dark green leaves and deep blue flowers with a yellow eye); *M.scorpioides* (FORGET-ME-NOT; Europe; perennial; stem to 50cm, angled, mostly glabrous; flowers bright blue, with white, yellow or pink eye, to 8mm in diameter; includes 'Sapphire', with bright blue flowers, 'Semperflorens', of dwarf habit, to 20cm, and 'Thuringen', with sky blue flowers); *M.sylvatica* (GARDEN FORGET-ME-NOT; N Africa, Europe, W Asia; biennial to perennial; stem to 50cm, hairy; flowers bright blue, purple or white-blue, varying to pink, with a yellow eye to 8mm in diameter; cultivars include 'Blue Ball', small and compact, with rich indigo flowers; 'Blue Bird', tall, with deep blue flowers; 'Carmine King', erect, with rosy carmine flowers; 'Compacta', dense and low; 'Royal Blue', tall, early-flowering, with abundant, indigo flowers; 'Robusta Grandiflora', vigorous, with large flowers; 'Rosea', tall, with soft pink flowers; and 'White Ball', small and compact, with large, white flowers).

Myriophyllum (from Greek *myrios*, many, and *phyllon*, leaf, referring to the much-divided leaves). Haloragidaceae. Cosmopolitan. MILFOIL. Some 45 species, aquatic or terrestrial herbs (those listed here are perennial aquatics with slender, floating to emergent stems clothed with many whorled leaves, entire or slightly dentate when emergent, usually finely pinnatifid when submerged). Inconspicuous flower spikes are produced in summer. Useful oxygenators for freshwater pools and aquaria, as well as providing fish shelter. *M.verticillatum* will survive in ponds where winter temperatures fall as low as –15°C/5°F; *M.aquaticum* and *M.hippuroides* will not tolerate temperatures below freezing. Frost-tender species may be overwintered as rooted pieces in a moist medium in a frost-free greenhouse. Propagate by stem cuttings rooted directly into the growing medium or by division.

M.aquaticum (PARROT FEATHER, DIAMOND MILFOIL; S America, Australia, New Zealand, Java; stem to 2m; leaves more or less uniform, in whorls of 4 or 5, to 4cm, pinnatifid, segments short, bright yellow-green or blue-green); *M.hippuroides* (WESTERN MILFOIL; SW US; stem to 60cm; leaves in whorls of 4 or 5, pale green, emergent leaves linear to lanceolate, entire to serrate, submerged leaves to 2cm); *M.verticillatum* (MYRIAD LEAF; N America, Europe, Asia; stem to 1m; emergent leaves to 1cm, pectinate to pinnatifid, submerged leaves to 4 × 4cm, in whorls of 4–6, with 8–16 pairs of opposite segments).

myrmecophytes, myrmecophilous plants plants that live in symbiosis with ants, forming hollow structures in which the insects nest in exchange for benefits, such as carbon dioxide enrichment and protection from predators and competition by other plants. Examples include, *Myrmecodia*, *Lecanopteris*, *Dischidia* and *Cecropia*.

Myrrhis (name used by Dioscorides.) Umbelliferae. Europe. 1 species, *M.odorata*, SWEET CICELY, an

These cultivars of two keynote early autumn perennials – *Chrysanthemum* and *Anemone* x *hybrida* – share the same palette, cool pinks, pale gold and lime green, bringing freshness and brightness to the border at a time more usually identified with the colours of decay.

Grasses

Glaucous, grey-blue
tones are well-
represented among
ornamental grasses,
particularly in the
genera *Festuca*,
Helictotrichon,
Koeleria, *Leymus*
and *Sesleria*. The
most intensely blue
grass of all,
however, is *Elymus
magellanicus*.
Needing full sun
and excellent
drainage, it is the
perfect complement
for old-fashioned
pinks, *Cerinthe* and
silver foliage plants.

The genus *Deschampsia* contains at least two hardy species of special interest to gardeners, *D. flexuosa*, the slender-leaved wavy hair grass, and the broader-leaved tufted hair grass, *D. cespitosa*. In late summer and autumn, the light-textured panicles of *D. cespitosa* 'Goldehange' become a haze of pale gold which persists well into winter.

An ideal choice for containers, Japanese gardens and cool, moist situations, *Hakonechloa macra* 'Aureola' has perhaps the most to offer of all the smaller variegated grasses. Its gently spreading and spilling habit resembles a miniature bamboo, while its graceful leaves – cream to bright gold and striped with fresh green – assume wine red tints in late summer before flushing rusty pink in winter.

Freshwater cord grass, *Spartina pectinata* forms spreading colonies on damp soils. Its graceful flowerheads and autumn colour make it an attractive choice for the margins of large ponds and lakes. The cultivar 'Aureomarginata' is less invasive and will grow on any soil that is not too dry, forming stands of richly variegated, rush-like foliage. Other marginal and aquatic grasses and grass-like plants include *Acorus*, *Eriophorum*, *Glyceria*, *Juncus*, *Phragmites*, *Scirpus* and *Typha*.

Fountain grasses, members of the genus *Pennisetum*, produce feathery and beautifully coloured cylindrical panicles in summer. Several are not fully hardy and are often treated as annuals; these include *P.* 'Burgundy Giant', a magnificent grass with purple-bronze to wine red foliage. Shown here, *Pennisetum orientale* is a reliably hardy perennial with soft-textured flowerheads that turn from pure white to pearly pink in late summer.

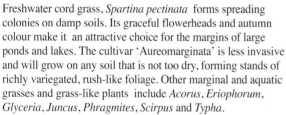

The aptly named squirrel or foxtail barley, *Hordeum jubatum* is one of the best grasses for cutting and drying, producing masses of long-awned, pink-tinged flowerheads. Other grasses suitable for drying include *Briza*, *Chasmanthium*, *Cortaderia*, *Hystrix*, *Lagurus* and *Miscanthus*. In all cases, it is best to harvest the flowerheads just before the florets are fully developed.

Autumn-flowering perennials

ABOVE In this autumn border, the dark foliage of a *Dahlia* creates the ideal foil not only for its own, golden-orange flowerheads, but also for dusky pink *Sedum spectabile* and the slender mauve flower spikes of *Verbena bonariensis*.

LEFT Among the families of flowering plants, none is better represented in the autumn border than the daisy family, Compositae. Here we see *Helenium*, *Aster*, *Chrysanthemum* and *Dahlia*.

BELOW The genus *Aster* contains some of the most dependable hardy perennials for the autumn garden. New England asters (*A. novae-angliae*) and Michaelmas daisies (*A. novi-belgii*) are available in a wide variety of bold colours.

aromatic, perennial herb to 2m with bipinnate leaves, the segments 1cm, oblong to lanceolate, pinnatifid or deeply toothed, pale beneath and often blotched white. Small, white flowers are carried in compound umbels in early summer. In the herb garden, it is valued for its culinary and medicinal virtues. It is also well-suited to naturalizing in woodland walks and for plantings in the shrub border. If cut back periodically in the herbaceous border, the fresh green foliage makes an admirable foil for other plants. It is hardy to at least −15°C/5°F and self-seeds freely. Propagate from ripe seed, by careful division, or by root cuttings, in spring or autumn.

Myrsine (name given by Dioscorides to the myrtle, *Myrtus communis*). Myrsinaceae. Azores to China and New Zealand. 5 species, evergreen trees and shrubs with leathery, aromatic leaves, inconspicuous flowers and, where plants of both sexes are present, small drupes. *Myrsine* species are grown for their often aromatic foliage and attractive fruits. In warm mediterranean-type climates, they are sometimes used as hedging and as specimens for the shrub border. Given good drainage and shelter from cold drying winds, the aromatic *M.africana* may tolerate several degrees of short-lived frost and is suitable for the sunny rock garden; in cooler zones, young plants are grown as houseplants. Grow in full sun or part shade in any fertile, well-drained, circumneutral soil. Propagate by seed or semi-ripe cuttings. *M.africana* (CAPE MYRTLE, AFRICAN BOXWOOD; Azores, mountains of E and South Africa, Himalaya, China; shrub 0.5–1.5m, young shoots softly hairy; leaves 0.6–2cm, narrowly obovate to elliptic, apex rounded or truncate, sparsely toothed; flowers pale brown, 3–6 per cluster; fruit 6mm in diameter, blue-lilac).

Myrtillocactus (so-named because *Vaccinium myrtillus* has similar-looking fruits). Cactaceae. Mexico, Guatemala. 4 closely related species of arborescent or shrubby cacti with numerous, ascending, few-ribbed and spiny branches. The flowers are small and rotate, the fruit small, edible and berry-like. Plant in a circumneutral compost high in grit. Provide a minimum temperature of 10°C/50°F. Shade in hot weather and maintain low humidity; keep dry from mid-autumn until early spring, except for light misting on warm days in late winter. *M.geometrizans* (C and S Mexico; tree 4–5m,

with short trunk; branches numerous, upcurving, 6–10cm in diameter, blue-green, ribs 5–6, central spine, 1, 1–7cm, dagger-like, almost black, radial spines 5–9, 2–10mm, red-brown then grey; flowers to 2×2.5cm, creamy white; fruit 1–2cm in diameter, dark red or purple).

Myrtus (Greek name for this plant). Myrtaceae. Mediterranean, N Africa. 2 species, aromatic, evergreen shrubs with neat, glossy leaves and, from spring to summer, fragrant, solitary flowers with four broad and spreading petals and many showy stamens. In mild areas of zones 7 and over, *M.communis* is grown as specimen and sometimes for hedging. In colder areas, it tends to be grown as a standard or pyramid in containers and moved into the cold greenhouse for the winter months. Given shelter from harsh winds and good drainage, it will, however, tolerate mild frosts; the smaller-leaved cultivars are hardier still. Grow in full sun in a moderately fertile, well-drained soil. In pots, use a free-draining, medium-fertility, loam-based mix. Under glass, water moderately when in growth; keep just moist in winter, and at a temperature of around 5°C/40°F. Prune in spring to maintain size and shape and to remove damaged growth. Semi-ripe nodal cuttings will root in a frame or a case with bottom heat.

M.communis (MYRTLE; Mediterranean and SW Europe, widely cultivated since ancient times, native range uncertain; much-branched erect shrub to 5m; leaves to 5cm, ovate to lanceolate, coriaceous, lustrous green, aromatic; flowers white to ivory or pink-white, to 3cm in diameter, fragrant; fruit to 1cm, blue-black or red-black to dark purple-red: 'Albocarpa': fruit white; 'Buxifolia': leaves neat, box-like; 'Compacta': dwarf, dense; 'Florepleno': flowers double, white; 'Leucocarpa': fruit white; 'Microphylla': dwarf and densely leafy, with very small and narrow leaves; 'Microphylla Variegata': leaves striped white; 'Minima': dwarf, with small leaves; 'Variegata': small, leaves with white to cream margins or blotches; var. *italica*: narrowly upright, with 3×1cm, oval to lanceolate leaves; var. *latifolia*: leaves $2–3 \times 1–1.5$cm, oval to oblong, acuminate; var. *romana*: leaves $3–4.5 \times 1–1.5$cm, broadly ovate, strongly acuminate, light green, in whorls of 3–4; subsp. *tarentina*: compact, rounded habit, with small, narrow, dark leaves, pink-tinted, cream flowers, and white fruit).

N

NAA 2-(l-napthyl) acetic acid; an auxin used in synthesised form as an aid to rooting cuttings, usually blended with IBA in powder form. It is considered to be especially useful in the rooting of large flowered hybrid rhododendrons. NAA is also used in preparations available for the control of suckers or water shoots in raspberries and pome and stone fruits.

Nandina (from Japanese name, *nandin*). Berberidaceae. HEAVENLY BAMBOO. India to Japan. 1 species, *N.domestica*, an evergreen or semideciduous shrub to 2m with erect and clumped, slender stems. The leaves are bipinnate to tripinnate, to 90cm long and held semi-erect or horizontal. They are composed of elliptic to rhombic leaflets to 7cm long. Smooth and thinly leathery, these emerge slightly glaucous and lime green tinted rose, harden to sap-green and turn bright red to purple in autumn. Small, white flowers are borne in terminal panicles in summer and followed by red berries. Hardy in climate zone 6. Plant in a humus-rich, moist but well-drained soil, in a cool but sunny position, sheltered from cold winds. Remove damaged or exhausted shoots by cutting them right back to the base. Propagate by single node cuttings, taken in midsummer and rooted in a closed case with bottom heat in a sandy mix. Seeds do not germinate freely but may succeed if sown when ripe.

nappe a smooth sheet of water falling over a weir or cascade.

Narcissus (Greek name derived from *narke*, numbness, torpor, from its narcotic properties: the youth Narcissus fell in love with his own reflection in a pool and was turned into a lily by the gods). Amaryllidaceae. DAFFODIL. S Europe and Mediterranean; also N Africa, W Asia, China and Japan. About 50 species, bulbous perennial herbs with grassy to strap-shaped leaves. The flowers are characteristically yellow or white, sometimes fragrant, and solitary or in an umbel borne on a leafless scape. The perianth is tubular at the base, with 6 segments and a conspicuous corona in the form of a trumpet or a smaller ring or cup. For cultivation see *daffodils*.

Narcissus triandrus

nastic movements growth movements in plants that take place in response to an environmental stimulus. Unlike tropism, the direction of growth is not influenced by the location of the stimulus. For example, the opening and closing of some flowers, induced by changes in light or temperature.

National Growmore see *Growmore*.

naturalizing the practice of introducing plants into a setting where they give the appearance of having established naturally. It may involve plants generally within their native habitat or exotic ones.

Naturalizing bulbs in grass is effective in gardens, and randomly planted drifts of daffodils are popular. Crocuses, snakeshead fritillary, *Fritillaria meleagris*, and *Scilla*, *Chionodoxa* and *Leucojum*, together with *Dactylorhiza* orchids, combine well in grass areas, which must be mown only after leaves of the bulbous species have died down; where wild flowers are also established in grass, mowing should be left until late summer to allow seed ripening. Many plants, including *Cyclamen*, *Galanthus*, hellebores and ferns, are well-suited to naturalizing in woodland areas, and spring-flowering bulbs established around the base of deep-rooted and lightly-shaded trees can be attractive.

Nautilocalyx (from Greek *nautilos*, sailor, and *calyx*, probably referring to the boat-shaped floral bracts). Gesneriaceae. Tropical America. About 38 species, perennial evergreen herbs and subshrubs with fleshy stems, elliptic to lanceolate, toothed leaves and solitary or cymose axillary flowers. These are tubular with a limb of five lobes. Provide a minimum temperature of 15°C/60°F, with moderate to high humidity and light shade. Pot in a sandy soil-less mixture. Keep moist when in growth, but almost dry when at rest during the cooler, duller times of year. Increase by stem cuttings in summer.

N.bullatus (Peru; to 6m; leaves bullate, purple beneath; flowers downy, pale yellow); *N.lynchii* (Colombia; to 60cm, erect; leaves flushed dark purple or red-brown, glossy; flowers pale yellow, exterior covered with red hairs, interior flecked purple).

NARCISSUS

Name	Distribution	Height	Leaves	Flowers
N.assoanus *N.requienii,* *N.juncifolius, N.pallens* RUSH-LEAVED JONQUIL	S France, S & E Spain	7–25cm	to 20×0.2cm, cylindric, spreading or prostrate, green, terete, smooth	2–3, to 2.2cm diam., horizontal or slightly ascending, yellow, fragrant; perianth tube 1.2–1.8cm, straight; segments obovate, 0.7–1×0.7cm spreading to incurved, corona cup-shaped, conical, 0.5×1.1–1.7cm, crenate, deeper yellow than segments
N.asturiensis *N.minimus*	N Portugal, NW & NC Spain	7-14cm	8×0.6cm, glaucous-green, spreading, channelled	solitary, to 3.5cm across, usually drooping, soft yellow; perianth tube to 0.8 cm, green-yellow; segments to 1.4×0.4cm, usually twisted, deflexed; corona 1.7cm, widened below, constricted at middle, mouth spreading, fimbriate

Comments: 'Giant': larger in all parts.

N.bicolor *N.abscissus*	Pyrenees and Corbières	to 35cm	30–35×1.1–1.6cm, green or glaucous, erect, flat	solitary, horizontal or ascending; perianth tube 1cm, green-to orange-yellow, broad; segments 3.5–4cm, cream or pale sulphur-yellow, spreading or deflexed, corona to 4cm, yellow, 1.5–2cm diam., mouth lobed or dentate
N.broussonetii	Morocco	to 40cm	28×0.9cm, glaucous, erect	1–8, ascending, white, to 3.5cm diam., fragrant; perianth tube to 2.8cm, funnel-shaped, white; segments 1.6×1.2cm, spreading or incurved; corona rudimentary

Comments: Autumn

N.bulbocodium HOOP PETTICOAT DAFFODIL, PETTICOAT DAFFODIL	W France, Spain, Portugal, N Africa	2.5-20cm	10–30 cm×1–5mm, semi-cylindric, erect, ascending or prostrate, dark green	solitary, horizontal, pale yellow to deep golden-yellow, often green-tinged, to 4.5cm diam.; perianth tube 0.6–2.5 cm, yellow, often tinged green, especially below; segments much shorter than corona, 0.6–2cm×0.5–5mm, often tinged green; corona funnel-shaped, 0.9–3.2×0.7–3.4cm, yellow, margin of mouth spreading or incurved, entire to dentate or crenate

Comments: var. ***bulbocodium***: plants usually dwarf; flowers golden-yellow. var. ***conspicuus***: plant robust; flowers dark yellow to citron; var. ***citrinus***: a name loosely applied to large-flowered pale yellow plants.

N.cantabricus N.clusii WHITE HOOP PETTICOAT DAFFODIL	S Spain, Morocco, Algeria	5-10cm	ascending or spreading, to 15cm×1mm, semicylindrical, slightly channelled	solitary, ascending, to 40mm diam., pure- or milk-white, fragrant; perianth tube to 1.2cm, funnel-shaped, white, green below; segments 1.2×0.5cm, white, virtually outspread; corona to 1.5×4cm, entire or crenate or undulate
N.cyclamineus	NW Portugal, NW Spain	to 20cm	12–30cm×4–6mm, bright green, spreading, keeled	solitary, drooping or pendent, deep yellow; perianth tube 2–3mm, green; segments sharply reflexed, corona 2cm, slightly constricted just below flared margin, lobed or fimbriate
N.dubius	S France, SE Spain	15-25cm	50cm×7mm at flowering, spreading, dark green, inner face flat, outer striate	2–6, ascending, white; perianth tube 1.6×0.2cm, green below; segments 7×6mm, spreading; corona 4×7mm, cup-shaped, crenate
N.elegans	W & S Italy, Sicily, Corsica, Sardinia, Morocco to Libya	20cm +	12–25cm×3–5mm, erect, glaucous, striate on outer surface	2–7, horizontal, 2.5–3.5cm diam., fragrant; perianth tube 1.6×0.2cm, green; segments 1.5× 0.3–0.7cm, white, spreading, becoming twisted with age, corona 1×2cm, green, becoming dull orange

Comments: Autumn

N.jonquilla JONQUIL	S & C Spain, S & E Portugal, naturalized elsewhere	to 40cm	erect to spreading, to 45×0.8cm, channelled at base, cylindric towards apex, striate, green	1–6, to 3cm diam., ascending; perianth tube to 3cm, slightly curved, pale green, segments elliptic, to 1.3cm, spreading, yellow; corona cup-shaped, 7–10×2–4mm, margin shallowly lobed or somewhat crenate
N.minor *N.nanus,* *N.pumilus,* *N.provincialis,* *N.lobularis*	Pyrenees, N Spain	14-20cm	erect, 8–15×0.4–1cm, sage-green or glaucous, flat or channelled	solitary, to 3.7cm diam., horizontal or ascending; perianth tube 1–1.8cm, yellow or green-yellow; segments ovate-lanceolate, 1.5–2.2cm, somewhat twisted, drooping, yellow, often with deeper median streak; corona 1.7×2.5cm, plicate, dilated at mouth, margin frilled
N.×odorus *N.calathinus,* *N.campernellii:* *(N.jonquilla ×* *N.pseudonarcissus)* CAMPERNELLE JONQUIL	garden origin; naturalized S Europe	to 40cm	to 50×0.8cm, strongly keeled, bright green	1–4, ascending, bright yellow, very fragrant, perianth tube to 2cm; segments to 2.5×1.3cm; corona to 1.8×2cm, lobed to more or less entire

Comments: 'Rugulosus': larger than the typical plant . 'Plenus': flowers double.

N.papyraceus *N.tazetta* ssp. *papyraceus* PAPER-WHITE NARCISSUS	S Europe	to 40cm +	to 30×1.7cm, erect, keeled, glaucous	2–20, 2.5–4cm, diam., ascending, fragrant; perianth tube to 1.5×0.3 cm, green below, white above, segments white, to 1.8cm, ovate; corona cup-shaped, 3–6×8–11mm, entire or slightly notched, white

Comments: Winter–spring. ssp. ***polyanthus***: syn. *N.polyanthus*; flowers 3–12(–20), 2.5–4cm diam.; corona pale yellow when young, becoming white.

NARCISSUS

Name	Distribution	Height	Leaves	Flowers
N.poeticus *N.ornatus* POET'S NARCISSUS, PHEASANT'S-EYE NARCISSUS	France to Greece	35-50cm	to 45×0.6–1cm, erect, channelled, green or somewhat glaucous	solitary, 4.5–7cm diam., horizontal to ascending, fragrant; perianth tube cylindric, 2.5×0.4cm, green; segments suborbicular to cuneate, to 3×2.2cm, spreading white, yellow at base; corona flat, to 2.5×14mm, yellow with red, frilled margin
Comments: Late spring. var. *recurvus:* PHEASANT'S EYE NARCISSUS; perianth segments strongly reflexed, pure white.				
N.pseudonarcissus WILD DAFFODIL, LENT LILY, TRUMPET NARCISSUS	W Europe	12-90cm	8–50×0.5–1.5cm, erect, strap-shaped, usually glaucous	usually solitary, occasionally 2–4, horizontal to drooping, fragrant; perianth tube 1.5–2.5cm; segments 1.8–4cm, spreading, sometimes twisted, white to deep yellow; corona 1.5–4.5cm, white to deep yellow, subentire to 6-lobed
Comments: ssp. *pseudonarcissus:* 20–35cm; flower solitary, to 6.5cm diam.; perianth tube yellow, usually tinged green; segments 2–3.5cm, twisted, deflexed, white to sulphur yellow. ssp. *major:* syn. *N.major, N.hispanicus, N.confusus, N.maximus:* to 50cm; flower solitary, to 9.5cm diam., deep yellow; perianth tube 1.8cm, green-yellow; segments 1.8–4cm, twisted, inner deflexed, outer reflexed; corona 2–4cm, margin expanded. ssp. *moschatus:* syn. *N.moschatus:* 15–35cm; flower solitary, 5–6cm diam., horizontal or drooping, uniform sulphur-white; perianth tube 8–15mm, green; segments 2–3.5cm twisted; corona 3–4cm, slightly flanged at margin. ssp. *nevadensis:* syn. *N.nevadensis:* to 30cm; flowers 1–4, 5cm diam., ascending; perianth tube 1.5cm, green-yellow; segments 1–2×1cm, deflexed, not twisted, white with yellow central streak; corona 1.5–2.5cm, subcylindric, margin slightly expanded, yellow. ssp. *nobilis:* syn. *N.nobilis;* 15–30cm; flower solitary, horizontal or ascending, 8–12cm diam.; perianth tube to 2.5cm, bright yellow; segments 3–4cm, twisted, white with yellow mark at base on reverse; corona 3–4cm, margin expanded, deeply dentate. ssp. *obvallaris:* syn. *N.obvallaris:* TENBY DAFFODIL; to 20cm; flower solitary, 4cm diam., horizontal; perianth tube 1cm, yellow with green stripes; segments to 3cm, slightly twisted, yellow; corona to 3.5cm, margin dilated, 6-lobed, sometimes reflexed. ssp. *pallidiflorus:* syn. *N.pallidiflorus:* to 30cm; flower solitary, horizontal or drooping, 7.5cm diam.; perianth tube 2.5cm, green, with yellow streaks; segments 3–4cm, twisted, pale yellow with darker median streaks; corona 3–4cm, margin expanded, recurved, pale yellow, of a slightly deeper colour than segments.				
N.romieuxii *N.bulbocodium* ssp. *romieuxii*	N Africa	10-20cm	to 20×0.1cm, erect or spreading, dark green, weakly striate	solitary, 2.5–4cm diam., horizontal or ascending, yellow; perianth tube to 2.5cm, green at base, yellow above; segments to 1.3×0.4cm, spreading; corona 1.5×3cm, margin 6-lobed and crenate, spreading to flattened
Comments: ssp. *albidus:* flowers white				
N.rupicola *N.juncifolius* ssp. *rupicola*	Spain, Portugal	14-23cm	18×0.3cm, erect, 2-keeled, glaucous	solitary, to 3cm diam., ascending; perianth tube 2.2cm, green or green-yellow; segments to 1.5×1.1cm, spreading, yellow; corona 3–5×6–18mm, conic or reflexed, deeply 6-lobed to crenate or subentire, yellow
Comments: ssp. *watieri* : syn. *N.watieri:* flowers white.				
N.serotinus	Mediterranean	13-30cm	10–20×0.1–0.5cm, erect or spreading, dark green, sometimes with longitudinal white stripes	solitary or occasionally 2–3, to 3.4cm diam., ascending, fragrant; perianth tube to 2cm, dark green; segments oblong-lanceolate, to 1.6×0.7cm, spreading or recurved, twisted, white; corona minute, to 1.5×4mm, 6-lobed, dark yellow to orange
Comments: Autumn				
N.tazetta BUNCH-FLOWERED NARCISSI, POLYANTHUS NARCISSUS	S Portugal, Mediterranean, east to Iran, probably introduced further east, where fully naturalized in Kashmir, China and Japan	20-45cm	20–50×0.5–2.5cm, erect, twisted, keeled, glaucous	1–15, 4cm diam., horizontal, fragrant; perianth tube cylindric, 2cm, pale green; segments broad-ovate, 0.8–2.2cm, patent, incurving, white; corona cup-shaped, 0.5×1cm, bright to deep yellow
Comments: ssp. *aureus:* syn. *N.aureus, N.bertolonii, N.cupularis:* perianth segments deep yellow to golden yellow; corona deep yellow to orange. ssp. *corcyrensis:* syn. *N.corcyrensis:* flowers 1–2; perianth segments narrow, sometimes reflexed, pale yellow; corona yellow. ssp. *italicus:* syn. *N.italicus:* perianth segments cream or very pale yellow; corona deeper yellow. 'Canaliculatus': syn. *N.canaliculatus* of gardens; to 20cm; leaves narrow, erect, glaucous, striate; flowers small; segments white, corona ochre yellow; sometimes shy-flowering.				
N.triandrus ANGEL'S TEARS	Spain & Portugal, NW France	20-30cm	15–30×1.5–5mm, keeled or striate, flat or channelled, erect or decumbent, sometimes curled at tip, green, or slightly glaucous	1–6, pendulous, white to bright yellow; perianth tube 1.5cm, green below, yellow above; segments sharply reflexed, lanceolate to linear-oblong, 1–3cm, often with deeper median streak; corona cup-shaped, 0.5–1.5×0.7–2.5cm, entire, somewhat undulate
Comments: var. *concolor:* flowers deep yellow: the name is loosely applied in horticulture to any yellow-flowered example of this species.				
N.willkommii *N.jonquilloides*	S Portugal, SW Spain	to 18cm	to 37×0.3cm, erect, flattened at base, rounded above, glaucous dark green	usually solitary, 3cm diam., horizontal; perianth tube to 1.6cm, green-yellow; segments broad-elliptic, 0.6–1.3×0.7cm, spreading or reflexed and curving inwards, slightly imbricate, yellow; corona cup-shaped, 0.6×1cm, deeply 6-lobed, yellow

navicular, naviculate shaped like the keel of a boat.

neck used of the upper part of a bulb, where the leaves and flower stem emerge.

neck rot generally, decay at the neck of a bulb; more specifically, an onion disease caused by the fungus *Botrytis allii*. The term also refers to a disease known as gladiolus scab and neck rot, caused by the bacterium *Pseudomonas gladioli*.

Neck rot of *Narcissus* is usually caused by *Fusarium oxysporum* f.sp. *narcissi*, better known as the agent of basal rot, but other fungi, including *Botrytis* and *Trichoderma* species, may be involved. Infection of the dying flower stalk spreads down into the bulbs and may lead to extensive rotting in store. It is usually necessary to cut across the neck of the bulb in order to see symptoms of the disease.

ONION NECK ROT, *Botrytis allii*, is very common in stored onions. A rot spreads from brown sunken lesions around the neck, and a fluffy growth of grey mycelium and small black fruiting bodies may develop on the bulb. The disease is seed-borne and invades the cotyledons and leaf bases, although there may be no obvious symptoms during the growing season. Crop rotation should be practised, crop residues removed, and onions properly ripened and dried before being stored.

GLADIOLUS SCAB and NECK ROT affects crocuses, freesias and other members of the Iridaceace, as well as gladioli. Circular or elongated dark brown spots appear on the leaves, especially at the base where they may coalesce and cause the leaves to turn brown and dry up. Pale yellow spots develop on the corms and in severe attacks lead to large cavities. There is a copious exudation of gum containing bacteria from the corm lesions; this may dry up, leaving a lacquer-like surface. Affected corms should be destroyed and the contaminated site kept free of Iridaceae for as long as possible.

necrosis the death of plant cells, especially when the affected tissues become dark in colour. It is a common symptom of fungus infection. A necrotic reaction to tissue infection is one in which areas of tissue, usually on leaves, die as in NECROTIC RINGSPOT of *Prunus*. See *pedicel necrosis*.

nectar sweet, liquid secretion attractive to pollinators.

nectarberry see *hybrid berries*.

nectary a gland, often in the form of a protuberance or depression, which secretes and sometimes absorbs nectar.

nectarine see *peaches*.

Nectaroscordum (from Greek *nektar*, nectar, and *skorodon*, garlic, referring to the large nectaries on the ovaries of these relations of the onion). Liliaceae (Alliaceae). S Europe, W Asia, Iran. 3 species, bulbous perennials closely resembling *Allium*. The leaves are linear to narrowly strap-shaped. The bell-shaped flowers nod in scapose umbels in spring and summer. The flowerstalks stand erect after flowering, and remain attractive. Hardy to at least –13°C/5°F. Grow in any light, well-drained soil that is neither excessively dry nor waterlogged, in sun or part shade. Propagate by seed in early spring, or by offset bulbils. In favourable conditions, these bulbs will naturalize themselves. *N.siculum* (syn. *Allium siculum*; SICILIAN HONEY GARLIC; France, Italy; flowers 10–30, bell-shaped, pendulous; tepals to 1.5cm, cream, flushed flesh-pink to caramel brown, stained dark red within, green toward base; includes subsp. *bulgaricum*, syn. *N.bulgaricum*, *N.dioscoridis*, from East Romania, Bulgaria, Turkey, and the Crimea, with white to yellow tepals, flushed pale pink and green above, edged white, and flushed green below).

needle a linear stiff leaf; as found in conifers.

needle blight term used to describe some foliage diseases of conifers, in which the leaves are infected and die but do not fall. *Didymascella thujina* is a serious form of needle blight affecting young *Thuja plicata* in North America and Europe.

needle cast the premature shedding of conifer leaves, which may be due to diseases such as *Hypodermella laricis* on larch, *Lophodermella sulcigena*, SWEDISH PINE CAST, on pines, *Lophodermium setisiosum* on Scots pine, *Meria laricis* on European larch, and *Rhabdocline pseudotsugae* on Douglas fir. Symptoms show as yellow spotting of needles in the spring, followed by brown discolouration before falling. In severe attacks, only current year foliage remains to produce a very thin canopy. Control by fungicide sprays should be combined with collecting and burning of fallen needles to reduce disease carry-over.

Neillia (for Patrick Neill (1776–1851), Scottish naturalist, secretary of the Caledonian Horticultural Society). Rosaceae. E Himalaya to China and W Malaysia. Some 10 species, subshrubs or, more often, arching shrubs with flexuous branches and alternate leaves. These are usually 3-lobed and irregularly serrate. Produced in spring and early summer in spreading to nodding racemes, the flowers are cylindric to campanulate with five rounded lobes. Hardy to –20°C/–4°F. In areas where low temperatures are prolonged, protect the roots with a deep mulch. Grow in moist, well-aerated soil, in sun to part-shade. Prune after flowering, cutting out old stems at ground level. Propagate by semi-ripe cuttings in a closed case with bottom heat, by softwood cuttings in early summer, by removal of suckers in autumn, or by seed.

N.sinensis (C China; to 3m; branchlets glabrous; leaves to 8cm, oval to oblong, incised-serrate and lobed, light green; flowers white to pink in nodding racemes to 6cm); *N.thibetica* (W China; to 2m; branchlets finely downy; leaves to 8cm, ovate, finely downy beneath; flowers pink to white in 8–15cm racemes).

Nelumbo (name used in Sri Lanka for *N.nucifera*). Nelumbonaceae. Eastern N America, warm Asia to Australia. LOTUS. 2 species, perennial aquatic herbs, evergreen in warm climates, with spreading, cylindrical, spongy rhizomes. The leaves are large and usually emergent, the circular, glaucous blades peltately attached and held aloft on tall stalks. Solitary and long-stalked, the showy flowers are chalice-like and composed of numerous oblong-ovate petals and many golden stamens. The ovaries are sunken in pits in the flat upper surface of a top-shaped receptacle. These persist long after the flowers and are handsome in dried flower arrangements.

The sacred lotus is usually treated as a tender aquatic below zone 9. Some of its cultivars, especially those of Far Eastern provenance, prove perfectly hardy, however, in outdoor pools in places as far north as New York City. Deep-water cultivation, where the leaf stalks will reach their potential height of 2.5m, is generally practicable only in warm subtropical climates or in the tropical house pool. In cooler regions (climate zones 7–9), shallow water heats up more quickly, providing the warmth necessary for good growth and flowering. Grow in tubs, baskets or beds in heavy loam, enriched with well-rotted farmyard manure or compost; set rootstocks horizontally at about 2.5cm deep and gradually increase the depth of water from 5 to 40cm as growth proceeds. Remove any fading foliage during the growing season. In outdoor temperate cultivation, gradually reduce the water level and mulch deeply, with leafmould or compost, as temperatures begin to fall in late summer and autumn, or lift, wash and store the rhizomes in moist sand in a cool frost-free place for the winter.

The lotus will also grow in tubs and half barrels on terraces: establish rhizomes on a loamy rich mix in the lower two-thirds of wooden or glazed tubs, and fill the remaining space with water. As frost kills foliage, remove tubs to shelter (a cold greenhouse or garden shed) and keep just moist. Move containers outside, feed and replenish water in late spring. Propagate by division of the rhizomes. Chip seed before sowing in small pots of rich loamy medium in early spring; set in an aquarium or warm greenhouse, covered with water, and maintain a temperature of 25–30°C/77–85°F. Pot on and gradually acclimatize to greater depths of water, setting out in late spring/early summer after the last frost. Seed-grown plants will flower in their third year.

N.lutea (WATER CHINQUAPIN, AMERICAN LOTUS, YANQUAPIN; eastern N America; leaves to 2m above water surface, leaf blades 50cm across; flowers 10–25cm across, pale yellow; includes 'Flavescens', with leaves splashed red in centre and small flowers); *N.nucifera* (SACRED LOTUS; Asia from Iran to Japan, south to Australia; leaves to 2m above water surface, leaf blades to 80cm across; flowers 10–30cm across, fragrant, single or double, palest pink to rose, salmon or cerise, or white to creamy yellow; numerous cultivars are grown).

Nematanthus (from Greek *nema*, thread, and *anthos*, flower). Gesneriaceae. S America. Some 30 species, climbing or trailing subshrubs. The leaves are usually small and thinly fleshy. Produced at various times of the year in small, compact clusters, the flowers consist of a green, orange or red- or purple-tinged calyx, enclosing a tubular, 5-lobed corolla. Cultivate as for *Aeschynanthus*, although *Nematanthus* species are generally bushier.

N.gregarius (syn. *Hypocyrta nummularia*; *N.radicans*; CLOG PLANT; eastern S America; climbing or pendent, to 80cm; leaves glabrous; flowers to 2.5cm, pouched, bright orange with a purple-brown stripe leading to each lobe; 'Variegatus': leaves green with yellow centre); *N.strigilosus* (eastern S America; climbing to 1.5m; leaves hairy beneath; flowers to 2cm, tube orange, limb yellow).

nematicide a pesticide active against nematodes.

nematodes see *eelworms*.

Nemesia (name used by Dioscorides). Scrophulariaceae. South Africa. About 65 species annual or perennial herbs or subshrubs with shortly tubular, 2-lipped flowers in terminal racemes in summer. Commonly used as frost-tender annuals for summer bedding, mixed borders and pot-plant display in cool greenhouses. Plant out after last frosts in a sunny position. After the first flush of flowers, trim back to promote a second display. Water well in dry weather. Sow seed in autumn or spring at 15°C/60°F; harden off before planting out into the open garden.

N.barbata (glabrous annual herb to 50cm; flowers to 1.5cm, blue, upper lip pale blue within, with white margin and purple lines, white without, with short obtuse lobes, lower lip longer than upper, entire or nearly so, deep blue, lower part white striped purple); *N.strumosa* (more or less glandular-hairy annual herb, 15–60cm, erect; flowers yellow or purple to white, often veined purple without, throat yellow with darker markings, lower lip 2–3cm broad, notched at apex, bearded within); *N.versicolor* (more or less glabrous annual herb, to 50cm; flowers to 1.2cm, blue, mauve, yellow or white, lips often different colours, lower lip broad, obtuse, shortly bilobed, palate broad; includes f. *compacta*, more compact, with profuse, white, rose, violet and blue flowers). *Cultivars and hybrids.* Mostly derived from *N.strumosa* and *N.versicolor*, these are the nemesias popular for summer bedding. They include: 'Blue Gem', of bushy habit, to 20cm, with a cloud of sky-blue flowers; Carnival Hybrids, compact and bushy, to 30cm, and very floriferous; 'Funfair', to 25cm, with flowers in brilliant colours; 'Grandiflora', with large flowers; 'Mello Red and White', with raspberry red and white flowers; 'Mello White', with white flowers; 'Nana Compacta', of dwarf habit; 'Suttonii', to 50cm, with irregularly shaded flowers, each with a broad lip in front and pouch at base and ranging from carmine, through yellow and pink to white; and 'Tapestry', to 25cm, of upright habit, with flowers richly coloured.

Nemopanthus (from Greek *nema*, thread, *pous*, foot, and *anthos*, a flower, referring to the very slender pedicels.) Aquifoliaceae. Eastern N America. MOUNTAIN HOLLY. 1 species, *N.mucronatus*, a deciduous, stoloniferous shrub to 3.5m, differing from *Ilex* in the much reduced calyx and free petals. The leaves are oblong to ovate, to 6cm long, entire or slightly serrate. Small, green-yellow flowers are followed by spherical, red fruit. Cultivate as for the smaller *Ilex* species.

Nemophila (from Greek *nemos*, grove, and *philos*, loving, in reference to the habitat of some species). Hydrophyllaceae. Western N America. 11 species of annual glabrous, downy or hispidulous herbs with spreading, brittle and slightly succulent stems and pinnately lobed leaves. Produced in summer, the flowers are campanulate, cylindrical or rotate with 5 rounded petals. Fast-growing annuals, used for border edgings, window boxes and other containers and suitable for north-facing sites with good light. They are sometimes grown in pots in the cool greenhouse or conservatory. Grow in moisture-retentive soils in sun or part shade with shelter from wind. Sow *in situ* in spring or autumn.

N.maculata (FIVE SPOT; C California; to 30cm, glabrous to hairy, erect to spreading; flowers to 4.5cm in diameter, petals white, each with a deep violet blotch near or at the apex, sometimes faintly veined or tinted mauve-blue); *N.menziesii* (syn. *N.insignis*; BABY-BLUE-EYES; California; to 12cm, hirsute, spreading-procumbent; flowers to 4cm in diameter, petals white to sky-blue, often with a white or yellow-stained centre, or spotted or stained darker blue or purple-black; cultivars include 'Alba', with white flowers with a black centre; 'Crambeoides', with pale blue flowers veined purple, unspotted; 'Coelestis', with white flowers edged sky-blue; 'Grandiflora', with pale blue flowers with a white centre; 'Insignis', with pure blue flowers; 'Marginata', with blue flowers edged white; 'Occulata', with pale blue flowers with a purple-black centre; 'Pennie Black', with deep purple-black petals edged bright white; 'Purpurea Rosea', with purple-pink flowers; var. *atromaria*, with white flowers spotted black-purple; and var. *discoidalis*, with bronze-purple flowers edged white).

Neolloydia (for Professor F.E. Lloyd (1868–1947), American botanist). Cactaceae. E and NE Mexico and SW Texas. 14 species, low-growing or dwarf cacti with depressed-globose to shortly cylindric, tuberculate and spiny stems. The flowers are small and shortly funnelform. Plant in a gritty compost with a pH of 6–7.5. Provide a minimum temperature of 7°C/45°F, full sun and low humidity; keep dry from mid-autumn until early spring, except for light mistings on warm days in late winter. Increase by offsets. *N.conoidea* (E and NE Mexico, SW Texas; stem simple or clustering, 5–24 × 3–6cm, globose to ovoid or shortly cylindric, grey to slightly yellow-green; flowers 2–3cm, magenta).

Neomarica (from Greek *neon*, new, plus *Marica*). Iridaceae. Tropical America, W Africa. 15 species, rhizomatous, herbaceous perennials with tough, sword-shaped leaves in a basal fan. The short-lived, *Iris*-like flowers are carried in clusters atop tall stems. Provide a minimum temperature of 10°C/50°F. Grow in bright filtered light or full sun, in a well-drained, fertile, loam-based mix with additional organic matter and sharp sand. Water moderately when in growth; keep almost dry in autumn and winter. Propagate by division or by seed. *N.caerulea* (Brazil; to 60cm; flowers 8–10cm in diameter, outer perianth segments blue to lilac, inner segments deep blue, the claws and blade bases of all segments yellow-white banded brown and orange-yellow).

Neoporteria (for Carlos Porter, Chilean entomologist). Cactaceae. Chile, S Peru and W Argentina. Some 25 species, mostly rather small cacti with globose to short-cylindric, ribbed and tuberculate stems and felted, depressed areoles. Funnelform or campanulate flowers arise at the apex in summer. Cultivate as for *Neolloydia*.

N.chilensis (Chile; stem globose to short-columnar, woolly at apex; ribs 20–21, spines glassy white; flowers pink or pale yellow); *N.napina* (Chile; stem to 10 × 5cm, globose to ovoid, grey-green or tinged red, ribs about 14, spines almost black; flowers pale yellow or tinged pink); *N.nidus* (Chile; stem to 30 × 5–9cm, globose to short-cylindric, ribs 16–18, spines weak and tortuous, interlaced, brown, yellow or almost white; flowers pink); *N.subgibbosa* (Chile; stem to 1m × 10cm, globose to short-cylindric at first, green to grey-green, ribs 16–20, spines yellow, brown or nearly black at first or amber-yellow; flowers pink, paler towards the throat); *N.villosa* (Chile; stem to 15 × 8cm, short-cylindric, grey-green, becoming tinged purple-black, ribs 13–15, spines bristly, dark and hair-like, pale brown or off-white; flowers pink with white throat).

Neoregelia (for E.A. von Regel (1815–92), German botanist and director of the Imperial Botanic Gardens, St Petersburg, 1875–92). Bromeliaceae. S America. 71 species, mostly terrestrial, evergreen perennial herbs with strap-shaped leaves in a dense, funnel-shaped or tubular rosette. Small flowers are borne in a low, dense head of tightly packed bracts, usually held deep with the rosette. Cultivate as for *Aechmea*.

N.carolinae (BLUSHING BROMELIAD; Brazil; leaves 40–60cm, innermost with scarlet bases when in flower, sparsely scaly beneath, closely toothed; inflorescence bracts bright red, papery, flowers lavender-blue; f. *tricolor*: leaves with yellow, white and green stripes); *N.concentrica* (Brazil; leaves to 40cm, mid-green often with obscure, darker spots, spiny teeth black, inner leaves flushing purple-pink at base when in flower; inflorescence bracts yellow-white flushed violet or purple, flowers white or blue); *N.fosteriana* (Brazil; leaves to 30cm, copper-red with a few green

Nemophila maculata

spots, pale grey-scaly, apex dark red, margins laxly toothed; inflorescence compound, red, flowers red); *N.marmorata* (MARBLE PLANT; Brazil; leaves to 60cm, laxly toothed, sheaths purple spotted pale green, blades darkly mottled or spotted, particularly at base, apex sometimes with a bright red spot; flowers white tinged pink); *N.princeps* (S Brazil; leaves 20–50cm, laxly toothed, bases large, green, densely scaly, blades green, grey-scaly beneath, the innermost flushed bright red in flower; flowers white tipped dark blue); *N.spectabilis* (PAINTED FINGERNAIL, FINGERNAIL PLANT; Brazil; leaves 40–50cm, grey-scaly and banded white beneath, outer leaves green with a bright red apical spot, inner sometimes wholly red, often edged purple, subentire or denticulate; inflorescence bracts red or purple, flowers violet-blue).

Nepenthes (name used by Homer, meaning grief-assuaging, applied because of supposed medicinal properties). Nepenthaceae. Madagascar, Seychelles, tropical Asia to Australia (N Queensland). PITCHER PLANT, TROPICAL PITCHER PLANT. 70 species, climbing or scrambling, evergreen perennial shrubs. They are carnivorous with each tough, oblanceolate leaf terminating in a tendril which in turn supports a pitcher-like trap. These pitchers are cylindric to rounded, usually green with red spots or tints. The mouth of the pitcher has a thickened, ribbed and often colourful rim, and a fixed lid, standing erect or projecting over pitcher-mouth. Along the front of the pitcher run two wings or ridges, usually toothed or fringed. Small, dull flowers are carried in dense spikes at various times of year. Provide a minimum winter temperature of 18°C/65°F. The compost must be well-drained; two parts bark, two parts perlite and one part moss peat is a standard compost. Ample moisture and high humidity are required but free drainage is essential; lattice or basket pots are recommended for this reason. Dilute liquid fertilizer is beneficial, applied to the roots or as a foliar feed. Old plants benefit from heavy pruning in spring. Increase by air layering.

N.alata (Philippines, Malaysia, Borneo, Sumatra; pitchers weakly dimorphic – lower 6.5–13cm, cylindric above, constricted at centre, inflated at base, light green with red flecks, or heavily suffused red, lip green or occasionally red, lid elliptic with glandular crest at base; wings prominent, fimbriate; upper pitchers elongate); *N.ampullaria* (Malaysia, Borneo, Sumatra to New Guinea; pitchers to 5cm, rounded, squat, produced low on plant, green spotted and blotched deep red, or entirely green or deep red, lip with narrow rim, lid to 3.5cm, narrow, wings broad, widely spreading, strongly toothed); *N. × hookeriana* (*N.rafflesiana × N.ampullaria*; natural hybrid, Malaysia, Sumatra, Borneo; pitchers dimorphic – lower to 11cm, ovoid, pale green with dark red spots, or sometimes heavily blotched red, lip broad, descending into pitcher, green, lid flat, obovate, wings broad, fimbriate; upper pitchers to 12.5cm, funnel-shaped); *N.mirabilis* (MONKEY CUP; Malaysia, Sumatra, Borneo, S China, SE Asia to New Guinea and Queensland; pitchers cylindric or slightly inflated at base, to 18cm, pale green with red spots, or wholly red, lip broad, flattened, striped red, lid orbicular to ovate, wings present in lower pitchers, toothed); *N.rafflesiana* (Malaysia, Sumatra, Borneo; pitchers dimorphic – lower to 12.5cm, ventricose, rounded at base, green, heavily spotted red, lip broad, crimson, ribbed, narrowing upwards to form elongate process to lid, spiny above, lid oblong-orbicular, wings to 2.5cm broad, teeth incurved; upper pitchers to 23cm, funnel-shaped); *N.rajah* (KING MONKEY CUP; Borneo; pitchers dimorphic – lower to 35 × 15cm, rounded, green lightly spotted red, or entirely red to purple externally, spotted red and purple-black within, lip broad, outer margin projecting, undulate, ribbed, crimson with darker bands, lid large, wings narrow, toothed; upper pitchers funnel-shaped).

Nepeta (named used by Pliny, probably after Nepi in Italy). Labiatae. Eurasia, N Africa, Tropical Africa. About 250 species usually perennial, aromatic herbs with toothed leaves and small, tubular and 2-lipped flowers whorled in spikes in summer. The following species are fully hardy. Grow them in well-drained, light soils in sun. Propagate by division in spring or autumn, or by stem tip or softwood cuttings in spring or summer. Sow seed *in situ* in autumn.

N.cataria (CATNIP, CATMINT; Europe, where it is widely naturalized, SW and C Asia; to 1m, erect; leaves 3.5–8cm, ovate, cordate at base, serrate, grey-tomentose beneath; inflorescence spike-like, flowers white spotted blue-violet; includes the lemon-scented 'Citriodora'); *N. × faassenii* (synonyms *N.mussinii* of gardens, *N.racemosa* of gardens; *N.racemosa × N.nepetella*; to 60cm; leaves to 3cm, oblong to ovate, truncate at base, crenate, silver-grey; flowers pale lavender with darker spots; 'Six Hills Giant': tall, vigorous with lavender-blue flowers in large sprays; 'Snowflake', low and spreading, with leaves tinted grey and flowers snow white; 'Superba', spreading, with leaves tinted grey and abundant, dark blue flowers); *N.govaniana* (W Himalaya; erect; leaves large, ovate to oblong or elliptic, rounded at base, crenate; raceme elongated, lax, flowers yellow); *N.grandiflora* (Caucasus, eastern Central Europe; erect to 40–80cm; leaves to 10cm, ovate, cordate at base, crenulate, glabrous; spikes elongate, interrupted, flowers blue); *N.nepetella* (SW Europe to S Italy; stem to 80cm; leaves to 4cm, oblong to lanceolate, truncate at base, crenate to dentate, pubescent to lanate, green to glaucous; inflorescence usually branched, calyx often tinged pink or blue, corolla pink or white; includes subsp. *amethystina*, from North Africa and the Iberian Peninsula, with blue-violet flowers); *N.nervosa* (Kashmir; to 60cm; leaves to 10cm, linear to lanceolate, entire or somewhat dentate, strongly veined; raceme dense, cylindric, flowers blue or yellow).

Nephrolepis (from Greek *nephros*, kidney, and *lepis*, scale, alluding to the common form of the indusia). Oleandraceae. Pantropical. LADDER FERN, SWORD FERN, BOSTON FERN. Some 30 species, epiphytic or terrestrial ferns with spreading rhizomes and prolific, wiry stolons. The fronds are pinnate with linear to oblong blades and numerous leaflets. *N.exaltata* and *N.cordifolia* are remarkably tolerant house plants. The many *N.exaltata* cultivars have arisen from the original 'Boston Fern' (*N.exaltata* 'Bostoniensis'), a 19th-century mutation with markedly drooping fronds. Many of these cultivars, particularly those with thickly layered or lacy foliage (e.g. 'Childsii' and 'Smithii'), resent very moist conditions and are more susceptible to a range of pests and diseases. Provide a minimum temperature of 7°C/45°F. Grow in a coarse, fast-draining, soilless mix with sand and bark. Give bright filtered light and a moist but buoyant atmosphere, ventilating whenever possible. Syringe in hot weather. Water sparingly in winter. Propagate by division in late winter or spring, and by young plantlets formed on the stolons.

N.cordifolia (ERECT SWORD FERN, LADDER FERN; pantropical; frond blades to 60 × 5cm, erect, arching or pendent, lanceolate to linear, acute or acuminate, sterile pinnae to 20 × 9mm, fertile pinnae 30 × 5mm, to 70 pairs, short-stalked, oblong to linear, apex toothed, base unequal and cordate to obtuse, entire or dentate; 'Duffii': syn. *N.duffii*, DUFF'S SWORD FERN, rachis usually forked, pinnae orbicular, crowded, attached in more than one plane; 'Plumosa': pinnae lobed); *N.exaltata* (BOSTON FERN; pantropical; frond blades linear, 50–250 × 6–15cm, pinnae numerous, 2–8 × 0.7–1.3cm, apex acute or subacute, base subcordate with auricle overlapping rachis, bluntly serrulate to crenate; 'Bostoniensis': syn. *N.bostoniensis*, fronds erect to pendent; this is the 'original' Boston Fern, which has produced, among others, the following sports – 'Childsii', fronds approximate and overlapping, to 4-pinnate, deltoid; 'Elegantissima', fronds 2-pinnate; 'Fluffy Ruffles', fronds dense, to 3-pinnate, deltoid; 'Smithii', fronds finely 3-pinnate, lace-like).

Nerine (from Greek Nereis, the name of a sea nymph, an allusion to the probably apocryphal tale that these bulbs were washed ashore on Guernsey following a shipwreck). Amaryllidaceae. South Africa, Lesotho, Botswana, Swaziland. 30 species, bulbous perennials with strap-shaped or thread-like leaves appearing with or soon after the flowers. Numerous flowers are produced generally in autumn in scapose umbels. These consist of six, slender and spreading, wavy segments and long, declinate stamens. In cool-temperate zones, most species are cultivated in the cool greenhouse, since they do not tolerate temperatures below freezing. *N.masonorum*, a dainty species for warm sheltered pockets in the rock garden in areas with mild winters, is well-suited to cultivation in the alpine house in cooler zones. *N.bowdenii* will grow where winter temperatures drop to −15°C/5°F, although it needs protection from winter wet: in regions at the limits of its hardiness, grow at the base of a south-facing wall in perfectly drained soil, with a dry mulch of bracken litter or leafmould. In northern Britain, large and well established clumps are most commonly seen at the base of south-facing house walls, in the rain shadow of

the eaves. *N.sarniensis* may be treated similarly, but is rather more susceptible to cold and wet.

N.bowdenii (leaves strap-shaped; scape to 45cm, tepals to 7cm, wavy, candy pink to deep rose or white); *N.filifolia* (leaves very slender; scape to 30cm, tepals to 2.5cm, crisped, white, rose, magenta or crimson); *N.flexuosa* (leaves narrowly strap-shaped; scape to 1m, flexuous, tepals to 3cm, crisped, pale pink or white); *N. masonorum* (leaves thread-like; scape to 30cm, tepals to 1.5cm, wavy, pale to deep rose with a darker pink midstripe); *N.sarniensis* (GUERNSEY LILY; leaves strap-shaped; scape to 45cm, tepals to 3.5cm, slightly crisped, pink, pale rose, crimson, scarlet, mauve-pink, orange-red or white); *N.undulata* (leaves very narrow; scape to 45cm, tepals to 2cm, pale candy pink or rose, strongly crisped).

Nerium (name used by Dioscorides). Apocynaceae. Mediterranean to W China, widely naturalized. OLEANDER, ROSE BAY. 1 species, *N.oleander*, a glabrous evergreen shrub to 2.6m tall with spreading to erect branches and tough, dark green, lanceolate leaves to 20cm long. Produced in summer in broad, terminal cymes, the flowers are white to rose, salmon, deep pink, cerise, peach or yellow, often fragrant and funnelform with a 5-lobed, spreading limb to 5cm across. All parts are extremely toxic. In cool climates, *Nerium* is commonly grown in tubs in the greenhouse or conservatory, to be moved outdoors for the summer months. However, given a sheltered sunny position, it may tolerate temperatures to −10°C/14°F for short periods. Grow in full sun. Keep almost dry in winter. Prune established specimens in late winter to shape and restrict size. Propagate cultivars by semi-ripe cuttings of terminal shoots, or by stem sections in summer after flowering.

Nertera (from Greek *nerteros*, lowly, alluding to the habit). Rubiaceae. S China and SE Asia to Australia, New Zealand, Polynesia, Antarctic and Hawaii to C and S America, Tristan da Cunha. Some 15 species, diminutive, prostrate or creeping herbs with very small leaves, solitary and inconspicuous flowers and bright, bead-like berries. Grow in sandy/gritty, moisture-retentive and freely draining soils, with shelter from cold winds in a semi-shaded situation. Maintain a minimum temperature of 5°C/40°F. Propagate by careful division, tip cuttings or by seed in spring in a sandy propagating mix in the shaded cool greenhouse or frame. *N.granadensis* (syn. *N.depressa*; BEAD PLANT, CORAL MOSS, ENGLISH BABYTEARS; S America, Taiwan and SE Asia, Australia and Tasmania, New Zealand; forming dense mats and mossy patches to 40cm wide or more; leaves to 0.8cm, usually far smaller, rounded, bright green; fruit 0.5cm wide, globose or ovoid, orange to scarlet or dark red; easily confused with the unrelated *Soleirolia*, whose fruits are quite insignificant).

nervose, nerved furnished with ribs or veins.

nessberry see *hybrid berries*.

Nerine bowdenii

netted reticulate, net-veined.

neuter, neutral sterile, asexual; used of flowers, lacking both pistils and stamens.

NFT see *nutrient film technique*.

Nicandra (for Nikander of Colophon, (*c.* 150 AD), Greek botanist and medical writer). Solanaceae. Peru. 1 species, *N.physaloides*, SHOO-FLY, APPLE OF PERU, a glabrous annual herb, to 130cm tall. The flowers are to 3.5×4cm, bell-shaped with a white throat and 5-lobed, with a lilac-purple to blue and white limb. A sturdy, freely-branching annual, grown for its bell-shaped flowers produced freely throughout late summer and autumn, although opening fully for only a few hours around midday. The attractive fruits are enclosed in purple calyces and resemble those of *Physalis*. It may self-sow freely in the garden. Sow *in situ* after the last frosts on to any rich, well-drained soil in full sun.

niche a recess in a garden wall, often at the end of a vista, in which a statue, urn or stone vase is placed.

nicking removing a small crescent of bark, or slitting the bark on a shoot below a dormant bud, as a means of inhibiting its growth; cf. *notching*.

Nicotiana (for Jean Nicot (1530–1600), French consul in Portugal, who introduced *Nicotiana* to France). Solanaceae. Tropical America, Australia, Namibia. TOBACCO. 67 species, mostly clammy, annual or perennial herbs and shrubs. Produced in summer in terminal panicles, the flowers are tubular to funnel-shaped with a 5-lobed limb. Their fragrance becomes especially intense at evening. In cool-temperate areas *Nicotiana* species and cultivars are generally treated as half-hardy annuals, although some species are perennial. In sheltered gardens, *N.alata* and *N.sylvestris* may survive temperatures down to about –5°C/23°F, shooting in spring from dormant buds on their thick rootstocks. *N.glauca* is almost as hardy. Plant out in late spring in rich, moisture-retentive, but well-drained soil. Deadhead bedding plants to prolong flowering. Surface sow the fine seed in spring at 18°C/65°F, planting out in late spring or early summer. Harden off before planting out. Prone to viruses.

N.alata (syn. *N.affinis*; JASMINE TOBACCO, FLOWERING TOBACCO; S Brazil to NE Argentina; viscid, sparsely branched perennial, to 1.5m; flowers green-white, interior white, tube to 10cm, pubescent, limb to 2.5cm diameter); *N.glauca* (TREE TOBACCO, MUSTARD TREE; S Bolivia to N Argentina, naturalized in United States; glaucous sunshrub or shrub to 6m+; flowers cream yellow-green, tube to 4.5cm, limb to 0.4cm in diameter, lobes short, throat slightly inflated); *N.langsdorffii* (Brazil; erect, branched, viscid-pubescent annual to 1.5m+; flowers lime green, tube to 2.5cm, viscid limb small, pleated); *N.× sanderae* (*N.alata* × *N.forgentiana*; shrubby, viscid-pubescent annuals to 60cm+; flower tube green-yellow at base, limb to

4cm+ in diameter, red, occasionally white or green to rose or purple; includes 'Breakthrough', early-flowering, dwarf, compact, with fragrant flowers in a range of colours; 'Crimson King', with deep crimson flowers; 'Daylight Sensation', with day-blooming flowers in shades of lavender, purple, white and rose; 'Dwarf White Bedder', of low habit and bushy, with pure white, fragrant flowers; 'Fragrant Cloud', with large, pure white flowers, fragrant at night; 'Grandiflora', with large flowers, with a large, widely dilated throat; 'Lime Green', with flowers bright sulphur yellow tinged lime green; 'Nana', dwarf; Nikki Hybrids, bushy and hardy, with flowers in a range of colours including white, shades of pink, red and yellow; 'Rubella', with rose red flowers; Sensation Hybrids, with flowers in a range of colours including pink, red and white, fragrant; 'Sutton's Scarlet', with a dark red corolla; and 'White', with heavily scented, white flowers); *N.tabacum* (TOBACCO; COMMON TOBACCO; NE Argentina, Bolivia; annual or biennial, viscid-pubescent, to 120cm, stem sometimes becoming woody at base; flowers green-white to rose, tube inflated, to 5.5cm, limb to 1.5cm in diameter; var. *macrophylla*: leaves large, flowers rose to carmine red; 'Variegata': leaves to 30cm, variegated cream with green, flowers white tinged pink).

Nidularium (from Latin *nidulus*, a little nest, referring to the inflorescence bracts). Bromeliaceae. E Brazil. 23 species, evergreen perennial, terrestrial or epiphytic herbs. The leaves are usually strap-shaped, finely toothed and arranged in a flat rosette with a central reservoir. Small flowers are carried in a compound inflorescence surrounded by showy bracts. Cultivate as for *Aechmea* but avoid direct sunlight and mist frequently in warm weather.

N.fulgens (leaves to 40cm, pale green with darker mottling, pointed, slightly scaly beneath, laxly toothed; inflorescence crowded, domed, on a very short scape, bracts cerise, coarsely toothed, sepals red, petals white, tipped dark blue, with white margins); *N.innocentii* (leaves 20–60cm, purple- to blood-red beneath or throughout, shiny, toothed; scape short; inflorescence bracts red tipped green, large, toothed, sepals white or pink, petals white with green bases; var. *lineatum*: leaves pale green with fine white lines, inflorescence bracts green with brick red tips; var. *striatum*: leaves green striped white, inflorescence bracts carmine); *N.procerum* (Brazil; leaves 0.4–1m, broadly acuminate, tough, finely toothed, pale waxy green tinted cooper; inflorescence bracts tinted red to strong red, flowers vermilion tipped blue).

Nierembergia (for John Eusebius Nieremberg (1595–1658), Spanish naturalist). Solanaceae. S America. CUPFLOWER. Some 23 species, annual or perennial herbs or subshrubs. Produced in summer, the flowers are cupped to tubular with a spreading limb of five lobes. Hardy to –12°C/10°F. In colder areas, they are used as bedding annuals or pot plants for summer colour, flowering in the first year from seed. Plant out in a sunny position after the last frosts. Clip back

N.hippomanica lightly after flowering. Propagate from seed, division or by heeled cuttings in summer. Harden off new plants in a cold frame before planting out.

N.hippomanica (Argentina; erect, white-hairy herb to 30cm+; flowers blue tinged violet, tube slender, yellow; var. *violacea*: syn. *N.caerulea*, flowers violet blue; 'Purple Robe': flowers purple-blue); *N.repens* (WHITECUP; Andes, warm temperate S America; procumbent herb; flowers white, tinged yellow or rose pink at base, tube slender; 'Violet Queen': flowers rich purple-blue).

Nigella (from diminutive of Latin *niger*, black, referring to the seed). Ranunculaceae. Eurasia. FENNEL FLOWER, WILD FENNEL, LOVE-IN-A-MIST, DEVIL-IN-A-BUSH. Some 14 annual herbs, with erect, usually unbranched stems. The leaves are finely pinnately cut with slender segments. Produced in spring and summer, the flowers are solitary and consist of five petal-like sepals and many smaller petals. In some species, the flowers are subtended by a ruff of conspicuously veined leaves, each terminating in hair-like divisions. The fruit is a large capsule, ultimately inflated with persistent horn-like styles. Hardy annuals, valued for their fine foliage, flowers and seed capsules, which are used in dried-flower arrangements. Sow *in situ* in autumn or spring in full sun. Established plants will withstand dry, infertile conditions. Propagate by seed – self-sowing is common. *N.damascena* (LOVE-IN-A-MIST; S Europe, N SAfrica; to 5–50cm; flowers white, rose pink, pale blue or purple-blue, 3.5–4.5cm in diameter, subtended by a finely divided, collar-like involucre; includes numerous cultivars and seed races ranging from tall to dwarf and single, semi-double and double, in shades of deep blue, sky blue, violet, lavender, lilac, mauve, purple-red, rose, and white).

Nipponanthemum (from Nippon and Greek *anthemon*, flower). Compositae. Japan. 1 species, *N.nipponicum*, a perennial herb or subshrub to 1m. The leaves are to 9cm long, spathulate, crenately toothed, dark and lustrous. Produced in late summer, the long-stalked, daisy-like flowerheads are to 8cm in diameter, with white ray florets and yellow disc florets. Grow in full sun in very well-drained and moderately fertile soil. Where winter temperatures fall much below −10°C/14°F, protect roots with a deep mulch of leafmould or bracken litter, and foliage with a covering of evergreen branches. Plants cut back to the ground in winter may regenerate from the roots. Propagate by seed or division.

nitrate see *nitrogen*.

nitrate of potash see *potassium nitrate*.

nitrate of soda see *sodium nitrate*.

nitrification see *nitrogen*.

Nitro chalk see *ammonium nitrate*.

nitrogen (N) a major (macro-) nutrient, essential to, and highly influential on, the rate of plant growth. As a constituent of protein, and therefore all protoplasm, it is needed in great quantity by most plants. It is mainly absorbed in the form of nitrate or ammonium ions.

Nitrogen can be stored long-term in soils as organic matter. Such organic nitrogen becomes available to plants through its mineralization to inorganic ions as a result of the activity of soil-inhabiting microorganisms in the process of nitrification. Organic nitrogen is reduced to ammonia, nitrite and then nitrate. This process depends upon the presence of water, oxygen, other nutrients, temperature and pH. Organic matter relatively high in carbon can be slow to break down, and may in fact take up free nitrogen from the soil in the process of mineralization. These considerations make it necessary to add nitrogen compounds to compost heaps incorporating woody waste.

Nitrogen-fixing bacteria and blue-green algae short-cut mineralization by obtaining nitrogen directly from the air. Some of these organisms live in the soil; others, such as *Rhizobium* bacteria, form nodules on roots of leguminous plants. *Actinomyces*, a bacteria-like organism, has a similar symbiotic relationship with alders and *Elaeagnus*.

Nitrogen deficiency can occur on soils that are low in organic matter, very dry, or of low temperature; where impeded drainage or other factors create anaerobic conditions; when leaching occurs following winter rain or excessive irrigation; and where large quantities of organic matter, relatively high in carbon, are added to the soil. It also occurs as a result of exhaustion of growing media in pots, or where plants are too closely spaced. Symptoms are feeble growth, pale or yellow-green foliage, often developing red or purple tints, and premature defoliation.

Besides bulky manures, organic sources of nitrogen include dried blood, hoof and horn, shoddy and urea. Inorganic sources include ammonium nitrate, ammonium phosphate and ammonium sulphate, as well as calcium nitrate, potassium nitrate and sodium nitrate.

nitrogen fixation see *nitrogen*.

noble rot see *grey mould*.

nocturnal of flowers, opening or fragrant only during the evening and night.

node the point on an axis where one or more leaves, shoots, whorls, branches or flowers are attached.

nodose possessing many closely packed nodes; knobbly.

nodule a small rounded growth on a root or other plant part. Nodules commonly occur on roots and are particularly associated with nitrogen-fixing bacteria and certain eelworm infestations.

Nigella damascena
'Miss Jekyll'

Nolana (from Latin *nola*, a little bell, alluding to shape of corolla). Nolanaceae. Chile to Peru, Galapagos Islands. 18 species, glandular annual or perennial herbs or subshrubs. The flowers are bell- to funnel-shaped with five lobes. Annuals, easily grown outdoors in any moderately fertile soil in sun, or as pot plants in the cool greenhouse. Sow seed *in situ* in spring.

N.humifusa (syn. *N.prostrata*; Peru; to 15cm, often decumbent; flowers to 1.7cm, lilac with throat white, streaked violet or purple; 'Shooting Star': trailing, with lilac to lavender flowers, streaked dark purple); *N.paradoxa* (Chile, Peru; to 25cm, usually decumbent; flowers to 3.5cm, bright dark blue, throat yellow or white; 'Blue Bird': deep sky-blue flowers with white throats; 'Cliff Hanger': trailing with cornflower blue flowers with pale yellow throats; subsp. *atriplicifolia*: leaves spathulate to linear, flowers blue, violet or white flowers, with yellow or white tubes).

Nolina (for P.C. Nolin, 18th-century French agriculturalist). Agavaceae. S US, Mexico to Guatemala. Some 24 species, evergreen shrubs and trees. The stem is usually sparsely branched and grossly swollen at the base. Tough, linear leaves are arranged in terminal tufts and rosettes. Small, creamy white flowers are borne in great abundance in large panicles. For outdoor cultivation only in frost-free desert and semi-desert regions. Elsewhere, they are usually grown as part of a greenhouse succulent collection or as a feature of interior landscapes. Grow in a well-drained, gritty medium; provide a minimum temperature of 10°C/50°F. Water plentifully in summer and sparingly in winter. Propagate in spring by seed or offsets.

N.gracilis (syn. *Beaucarnea gracilis*; differs from *N.stricta* in glaucous leaves with rough margins); *N.hookeri* (syn. *Calibanus hookeri*; EC Mexico; caudex massive, turnip-like, mostly subterranean, lacking aerial branches; leaves clumped, glaucous, grassy); *N.recurvata* (syn. *Beaucarnea recurvata*, *N.tuberculata*; BOTTLE PALM, ELEPHANT FOOT TREE, PONY TAIL; SE Mexico; trunk flask-shaped, to 8m, swollen to 2m in diameter at base, narrowing and sparingly branched above; leaves in fountain-like terminal rosettes, to 1.8 × 2cm, linear, channelled, arching, dark green); *N.stricta* (syn. *Beaucarnea stricta*; Mexico; differs from *N.recurva* in its rigid, straight, pale to mid-green, dagger-like leaves).

Nomocharis (from Greek *nomos*, meadow, and *charis*, loveliness, referring to the plant's habitat and beauty). Liliaceae (Liliaceae). W China, SE Tibet, Burma, N India. 7 species, bulbous perennial herbs, resembling small lilies (*Lilium*), with flat to bowl-shaped, white to pink or pale yellow flowers, often spotted purple or maroon. Usually carried in summer in loose terminal racemes, the flowers consist of six ovate to elliptic segments, their margins entire to toothed or fringed. Hardy in climate zone 7. They require acid, cool and shady conditions and are well-suited to the peat terrace and woodland garden. Grow in dappled light or semi-shade in well-drained, moist peaty or leafy soils. Propagate by seed.

*Nomocharis
pardanthina*

N.pardanthina (China; to 90cm; flowers 5–9cm diameter, nodding to erect, flattened, outer segments white to pink blotched purple, dark maroon at base, inner segments more densely blotched, with fringed tips); *N.saluenensis* (China, Burma; to 85cm; flowers 6–9cm diameter, horizontal or drooping, saucer-shaped, outer segments rose-pink with a dark maroon patch and spots at base, inner segments entire).

nonvascular without vascular tissue, for example algae, lichens, fungi.

Nopalxochia (from the ancient Mexican name, meaning cactus with scarlet flowers). Cactaceae. S Mexico. 3–4 species, epiphytic cacti with flattened, wavy-edged stems and funnel-shaped flowers. Provide a winter minimum temperature of 10°C/50°F, high humidity and shade from direct sun. Plant in an acid compost rich in sand, leafmould and coarse bark. Water freely in summer, very sparingly in winter. Increase by stem cuttings.

N.ackermannii (S Mexico; stem to 20–70 × 5–7cm; flowers 12–14cm, orange-red); *N.phyllanthoides* (S Mexico; stem to 40 × 0.6cm; flowers 8–10cm pink).

notching removing a small crescent of bark on a shoot above a dormant bud to stimulate it into growth; cf. *nicking*.

Nothofagus (from Greek *nothos*, false, and *Fagus*, to which it is related). Fagaceae. Temperate southern S America, New Zealand, E Australia, and tropical high altitude New Caledonia and New Guinea. SOUTHERN BEECH. About 40 species, evergreen and deciduous trees and shrubs resembling beech trees (*Fagus*). The leaves are oblong to ovate with wavy to toothed margins and distinct veins. The nuts are surrounded by a thick, splitting involucre, often covered with sticky protuberances. These trees favour cool, wet summers in maritime temperate climates. *N.antarctica* is hardy in zones 7 and over. All other species are killed or injured at –20°C/–4°F and some at –10°C/14°F. They are not suitable for frost-prone, inland areas and should only be planted on sites with good air drainage. Regions with hot, humid summers should also be avoided. Plant in full sun on moist but well-drained soils with a pH of 5–7. Increase by seed.

N.alpina (syn. *N.procera;* RAULI BEECH; Chilean Andes; deciduous tree to 40m; leaves 4–15cm, ovoid to lanceolate, matt green and thinly pubescent, margin slightly scalloped, finely crenate-serrate, veins 15–22 pairs, impressed above, veins ending in a slight sinus with 4 to 5 unequal intervening teeth); *N.antarctica* (NIRE; Tierra del Fuego to Chile; deciduous tree or shrub to 17m; leaves 1.5–4cm, oblong, finely but irregularly toothed, with 4 pairs of veins, glossy rich green above, paler beneath, often sweetly aromatic, glabrous except for a few hairs on veins; 'Benmore': syn. 'Prostrata', a low-spreading form with interlacing branches in a dense mound); *N.betuloides* (COIGUE DE MAGELLANES, GUINDO BEECH; Chile, W

Argentina; evergreen tree to 25m, or shrub in exposed sites; leaves 1–3cm, ovate, crenate, glossy dark green, slightly sticky, finely freckled with white glands beneath, veins often pink); *N.dombeyi* (COIGUE; Chile, Argentina; evergreen tree to 50m; leaves 2–4cm, ovoid to lanceolate, irregular-serrate, glossy dark green above, matt and pale beneath, minutely freckled black); *N.menziesii* (SILVER BEECH; New Zealand; evergreen tree to 30m; leaves 1–1.5cm, orbicular to broadly ovoid, margin double-crenate, silvery grey on young trees, glossy dark green on older trees, coriaceous, glabrous except for hair-filled vein axils at base of blade); *N.obliqua* (ROBLE BEECH; Chile, W Argentina; deciduous tree to 40m; leaves 3–8cm, ovoid to oblong, mid-green above, paler beneath, glabrous, margin double-toothed, veins 7–12 pairs, impressed above, each ending in a large lobe-like tooth, itself 1–3-toothed each side).

Notholirion (from Greek *nothos*, false, and *leirion*, lily; these plants have been placed in *Lilium* and *Fritillaria* but are now considered separate from both). Liliaceae (Liliaceae). Afghanistan to W China. 4 species, bulbous perennials with linear to lanceolate leaves produced in autumn and winter. Lily-like, trumpet-shaped flowers are carried in racemes in summer; they consist of six, spreading segments with recurved tips. Leafing early in the season, *Notholirion* is prone to frost damage; for this reason, in cool temperate zones, it is commonly cultivated in large containers in the cool greenhouse. Use a loam-based mix with silver sand and leafmould; grow in bright, filtered light and water plentifully when in full growth. Reduce water when dormant. Increase by seed or bulbils. *N.campanulatum* (N Burma, W China; to 80cm; flowers to 20, pendulous, to 5cm, crimson to maroon tipped green).

Notospartium (from Greek *notos*, southern, and *spartion*, broom). Leguminosae. New Zealand (South Island). PINK BROOM, SOUTHERN BROOM. 3 species, shrubs and trees. Adult plants are mostly leafless with a mop of slender, flattened, weeping, branchlets. In spring and summer, pea-like flowers are carried in pendulous lateral racemes. Hardy to about –10°C/14°F and suitable for a warm, sheltered border or for the base of a sunny wall. Grow in a well-drained but moisture-retentive, humus-rich soil. Provide support for older plants and protect young plants from prolonged winter frost with bracken litter or evergreen branches. Propagate from seed sown in the cold frame or greenhouse in spring, or by semi-ripe cuttings in summer, rooted in a closed case with bottom heat.
N.carmichaeliae (shrub to 5m; branchlets rush-like, grooved, olive; racemes to 5cm, sericeous, flowers light purple to pink); *N.glabrescens* (round-headed tree to 9m; branches slightly grooved, grey-green to olive; racemes to 5cm, glabrescent, flowers magenta with standard paler and stained dark purple-red).

novirame a flowering or fruiting shoot arising from a primocane.

NPK the combined chemical symbols for the three major plant nutrients, nitrogen, phosphorus, and potassium; commonly used to describe the make up of a basic balanced fertilizer.

nuclear stock plants propagated vegetatively from a single parent. The term is primarily used of plants grown from a mother plant that has been tested for trueness to type and freedom from viruses, other diseases, and pests, thus providing a basis for the production of certified stock.

nucleus the membrane-bound body containing the genetic material of plant and animal cells.

Nuphar (from *nufar*, Arabic name for *Nymphaea*). Nymphaeaceae. Temperature regions of northern Hemisphere. COW LILY, SPATTERDOCK, YELLOW POND LILY, WATER COLLARD. Some 25 species, perennial, aquatic herbs with floating to emerging large, ovate to orbicular leaves. Held above the water surface, the flowers are solitary and consist of a shallow cup of broad sepals and numerous smaller petals. These surround numerous stamens and a large ovary. The fruit is large, green and flask-shaped. Fully hardy; grow in deep water and full sun or light shade. Plant in containers or on the bed itself in still or slow-moving water. For propagation see *Nymphaea*.
N.advena (Europe and C US, Mexico, West Indies; leaves to 33cm; flowers to 4cm in diameter, yellow tinged red); *N.lutea* (YELLOW WATER LILY, BRANDY BOTTLE; E US, West Indies, N Africa, Eurasia; leaves to 40cm; flowers to 6cm in diameter, bright yellow).

nursery, nursery garden a site devoted to the propagation and raising of young plants.

nursery bed an area reserved for the rearing of young plants, where they can remain until large enough to be transferred to more permanent positions. It is sometimes also called a reserve border.

nut (1) an indehiscent, one-celled and one-seeded, hard, ligneous or osseous fruit, for example, the acorn of *Quercus*; (2) more loosely applied to a drupe with a thin exocarp and a large nutlet.

nutant nodding; usually applied to a whole inflorescence or stem.

nutlet a small nut or a small stone of a drupaceous fruit. It is similar to an achene, but with a harder and thicker pericarp.

nutrient film technique (NFT) see *hydroponics*.

nutrients nutrients are mineral ions used by plants to form the proteins, fats and other compounds needed for growth. Essential plant nutrients are divided into the macronutrients (needed in relatively large amounts of 10s or 100s of kg/ha), which are nitrogen (N), phosphorus (P), potassium (K), magnesium (Mg), calcium (Ca) and sulphur (S), and the micronutrients or trace elements (needed in relatively small amounts of g/ha), which are iron (Fe), manganese (Mn), copper (Cu), zinc (Zn), boron (Bo), molybdenum (Mb) and chlorine (Cl). Other trace elements may be needed, but in such small amounts that it is effectively impossible to prove the requirement.

Nutrients are not always beneficial and problems can arise if levels are too high. There can be harmful effects of salinity from excessive concentrations; this problem is most likely to occur when soluble NPK fertilizers are used. Some micronutrients, notably zinc, copper and boron, can prove directly toxic at very much lower levels than those producing salinity effects.

Excess nitrogen may lead to a delay in flowering and fruiting, enhanced vegetative growth, a reduced root–shoot ratio, decreased winter-hardiness, and greater pest and disease susceptibility.

With some sensitive species, it is possible to see visual signs of damage at what would normally be regarded as relatively low levels of a given nutrient. For example, many Australasian plants are adapted to soils that are extremely deficient in phosphorus, and these suffer toxicity when planted in more normal substrates.

Some trace elements are capable of improving the growth of particular plants, the most commonly encountered of which is sodium. This can usefully be applied to plants from the beet and cabbage families, which have developed from seaside habitats. However, sodium can be damaging to soil structure and should not normally be applied as a fertilizer without good reason.

Nutrients in soils   Very few soils are actually deficient in total amounts of plant nutrients; deficiency symptoms more commonly arise due to shortfall in the manner or rate in which nutrients are made available. Nutrients are taken up by plants as mineral ions dissolved in soil water, and for prolonged growth and vigour, there must be a constant supply. They originate in the following ways: WEATHERING OF SOIL MINERALS   Mineral-rich rocks are broken down by the action of frost, water and organic acids and are the ultimate source of most soil-borne nutrients, although nitrogen is an important exception. BREAKDOWN OF ORGANIC MATTER   Organic matter from dead plant and animal remains is gradually broken down by soil organisms to release constituent mineral nutrients in a process known as mineralization; this is the source of a regular supply of nitrogen ions, in particular. INPUTS FROM THE ATMOSPHERE   Many nutrients fall to earth dissolved in rain and this low input is of importance in some ecosystems; inputs are greater in areas of high atmospheric pollution. Nitrogen-fixation in plants can greatly increase soil fertility by converting nitrogen from air within the soil structure into organic form. Nutrient ions in solution are prone to leaching and the most important mechanisms of temporary storage are: CATION EXCHANGE CAPACITY (CEC) Positively charged ions (cations) are held by electrostatic charge

on the surface of clay particles and organic matter. These stored nutrients are in equilibrium with those held in the soil solution so that, within broad limits, as the concentration in solution falls, more stored ions are released. CHEMICAL COMPLEXES BETWEEN THE IONS   Nutrient ions can combine with other nutrients or other soil minerals to form relatively insoluble salts. These can in turn break down to release ions if the equilibrium with the soil solution changes. The relative solubility of different mineral elements changes dramatically with soil pH, so that attention to lime status in a soil is very important.

Any problem that interferes with or slows the normal movement of water through the soil, such as drought or compaction, can severely limit the availability of nutrients to a plant. Water shortage can also limit nutrient uptake by restricitng root growth. *Nutrient losses*   The most obvious means by which nutrients are lost from soil is by uptake into vegetation; they can be locked up in the biomass of large long-lived plants, especially trees, and it can be centuries before they enter the soil again. When plants are harvested, grazed or mown, the nutrients removed can represent a loss from the system unless replaced by rotted matter.

Nutrients can be lost from a growing medium by leaching, the process of drainage water removing dissolved mineral salts and clay particles. Leaching is a natural process intensified by heavy rainfall or excessive irrigation and the effects are only serious if the mineral ions move out of the root zone. Leaching into lower soil levels can actually be beneficial if it allows plants to continue nutrient uptake when there is surface drought. It is also advantageous in reducing soil salt concentrations which may arise from the continual use of supplementary fertilizers. Some elements are more mobile than others, for example, nitrogen ions are prone to leaching, whereas phosphorus moves extremely slowly in soil. Leaching is worse on coarse-textured soils where the drainage rate is high and CEC is low. Leaching rates can therefore be reduced by increasing CEC or by reducing the rate of water movement down the profile by improving structure and water retention; adding clay or organic matter helps to meet both these objectives. Vegetation restricts leaching by recovering and recycling ions and by reducing water drainage, whereas leaving land fallow can cause depletion of readily available nutrients. Reduced leaching is one of the benefits obtained by using green manures on unused land. On sandy soils, soluble fertilizers, such as ammonium nitrate, can be lost in just a few days of wet weather, making this an ineffective and uneconomic fertilizer. Slow-release fertilizers are often much better value on sands, even though they appear to be more costly. In general, thoughtful and timely application of manures and fertilizers helps to reduce waste and the risk of ground water pollution.

Fire causes nutrient loss; if vegetation is burnt, certain elements, such as nitrogen and sulphur, form gases that join the atmosphere, although potassium remains in the ashes.

Soil nitrogen can be lost to the atmosphere by the process of denitrification, which occurs following bacterial action in waterlogged soils. A similar loss can occur by ammonia volatilization, when fertilizers or manures rich in ammonium compounds are applied to well-aerated soils which have a high pH or have recently been limed.

Nutrient antagonisms Not only are the absolute levels of a given nutrient important, but also the relative proportions of two or more elements can affect plant growth and yield and influence the availability of other minerals. For example, adding extra nitrogen to a soil already deficient in phosphorus can increase the severity of the deficiency. Similarly, high potassium levels aggravate magnesium deficiency. Other nutrient antagonisms include N/K; K/Mg; Mg/K; P/K; P/Zn; Na/Ca; K/Ca; Mg/Ca; Ca/Mn; Cu/Fe; Zn/Fe; Co/Fe and Ni/Fe.

The opposite phenomenon is seen when the addition of one nutrient improves the uptake or efficiency of use of another. These are called synergisms and include N/Mg and K/Fe.

Identifying nutrient deficiencies In the majority of cases and particularly at low levels of deficiency, visual clues can vary according to the availability of other nutrients and the type of plant; confusion may also result due to problems such as virus infection, physical root damage, waterlogging and herbicide damage which can cause similar symptoms.

Deciding which nutrient is causing a deficiency problem can be difficult; it is important to note whether the symptoms are affecting old or young leaves and leaf margins or leaf veins. A pictorial reference book is invaluable. The major problem with using visual symptoms to identify a deficiency is that growth and yield can be affected long before there are obvious signs. Annual dressings of fertilizers provide some insurance but can be expensive and may contribute to a build up of soil salts harmful to plant growth, and to pollution.

If more precision is required, an assessment of the current nutrient status of the soil can be made by soil analysis, achieved by laboratory tests, by the use of portable electrodes or with soil-testing kits. Most soil analysis relies on the use of chemical extractants which remove nutrients at levels that are roughly equivalent to the amounts taken up by plant roots. Although a broad indication of fertility can be obtained, there are drawbacks relating to the interpretation of results. A soil is only deficient when the plant in question cannot obtain enough nutrients: both the demands and the root systems of a mature tree will be very different from those of a small vegetable seedling. In addition, the more unusual the soil type to be tested, the less certainty there is with regard to the accuracy of the tests. Levels of nitrogen are particularly difficult to interpret, as the amount available in the soil after a cold wet winter indicates nothing about the likely levels in summer. In addition, most soil-testing kits will only give readings for nitrogen as nitrate ions. For vegetables grown at a high pH this will be adequate, but it will be of no value for plants growing on acid ground where ammonium is important. Foliar analysis and sap testing can be more reliable techniques, but are impracticable for most gardeners.

Assessment of the nutrient status of soil on a new site is desirable, and determining levels of phosphorus, potassium and magnesium, as well as pH, may be worthwhile. When fertility is established, combining moderate levels of regular feeding, with an eye for possible deficiency symptoms, usually means that repeated complicated tests of garden soil are unnecessary.

The natural response of many gardeners faced with a nutrient deficiency is to add fertilizer. However, if the real problem is linked to poor nutrient availability rather than low levels, fertilizing may be of limited or no benefit and alternative feeding strategies may be needed. For example, iron is present in huge amounts in soil, yet deficiencies can result from a high pH. This can be corrected by foliar feeding, using special sequestrene compounds that remain soluble at high pH, or by adding bulky organic manures, which slightly acidify the soil and also produce their own natural sequestrenes.

The nutrient status of crops can also be modified by various cultural practices: mulching or irrigating reduces drought and improves nutrient movement; mulches or cloches increase soil temperature and bacterial activity early in the year; cultivation aerates the soil and increases the rate of organic matter breakdown, and also influences the rooting depth and extent of the plants. Marling a sandy soil or adding organic matter can improve structure, water retention and CEC.

The choice of fruit rootstock affects the extent of rooting and hence the likelihood of nutrient deficiencies, while the presence of grass in an orchard can influence nutrient availability. Above all, modifying the soil pH is one of the most effective means of altering nutrient availability in gardens.

It is not always desirable for soils to have high nutrient levels. In nature, very fertile soils are dominated by a few aggressively competitive plants. The decline of wildflowers in countryside areas can often be directly attributed to the increased use of fertilizers, particularly in species-rich grassland where the more attractive plants have been replaced by vigorous grasses.

nyctinasty the night-time folding up of leaflets, as occurs, for example, in some *Oxalis* species.

nymph the immature stage of those more primitive insects that undergo incomplete metamorphosis; the offspring resemble the parents except that they are smaller, sexually immature and develop wings gradually. Nymphs are usually found in similar places to adults, and consume the same type of food. Familiar plant pests with nymphal stages include springtails, earwigs, grasshoppers and crickets; others include leafhoppers, aphids, scale insects, mealybugs, whiteflies and capsid bugs.

Nymphaea (from Greek *nymphe*, a water-nymph, referring to the habitat). Nymphaeaceae. Cosmopolitan distribution. WATER LILY. Some 50 species, aquatic, perennial herbs with stout rhizomes or tubers and long-stalked floating leaves, their blades usually rounded and with a deep basal sinus. The flowers are solitary, floating or emergent and fragrant, with many oblong petals in a bowl-like arrangement around a boss of stamens.

Species from cooler regions are suitable for permanent positions in the pools and lakes of temperate gardens, one of the most important factors for success being the selection of appropriate species and cultivars for the given depth of water. Depth requirements range from 15 to 30cm for *N.tetragona* and its hybrids, making them suitable for even the smallest stretches of water, including half barrels in the patio or courtyard garden. The majority of *N.* Laydekeri Hybrids are grown at a depth of 30–40cm, and the *N.* Marliacea Hybrids at 90cm, although the tiny 'Mary Patricia' requires only 15–30cm, and the more robust 'Carnea' and 'Rosea' need up to 1.2m of water above the crown. Deep-water species, at depths of 2–2.5m for larger-scale plantings, include *N.alba* cultivars and hybrids such as 'Gladstoniana', with semi-double, starry white flowers. The tropical species and hybrids are fine plants for large pools in the heated greenhouse (minimum temperature 15°C/60°F).

Plant all in baskets or on the pond floor in a loamy medium in water with a pH of 6.0–7.0. Position in full sun away from disturbance by waterfalls, fountains or pumps. Feed when necessary with a slow-release fertilizer placed in the baskets or in the vicinity of root systems. At planting time, remove adult leaves and damaged roots and treat cut surfaces with charcoal. Plant the rhizomes or tubers vertically with the fibrous roots spread out beneath. The crown must be just at the surface of the medium. Top dress with a layer of pea gravel. Lower the container into the water slowly; gradually increase the plant's depth to the appropriate level as growth progresses. In larger expanses of water, plant roots in a packet of soil wrapped in hessian, place carefully in water and allow to sink. Temperate species are generally cold hardy where temperatures fall to –15°C/5°F and below, but there must be sufficient depth of water above the crown (about 22–30cm) to protect it from freezing. Alternatively, tubers can be lifted and stored in cool conditions if kept moist and protected from vermin. When plants become overcrowded, flowers are smaller and fewer and foliage stands clear of the water. It may be necessary to remove overcrowded foliage in summer by pulling away the older, outer leaves. Propagate by division or by removal of pieces of rhizome with a sprouting eye. Increase viviparous tropical types by the removal of young plantlets borne in the leaf blade sinus.

Waterlily beetle can be a destructive pest – adults overwinter among pondside vegetation, laying eggs on upper leaf surfaces in late spring. The emerging larvae feed on foliage, eating holes in the leaves and flowers. The water lily aphid attacks many soft-leaved aquatics during the summer, distorting foliage and flower stems and discolouring flowers. Eggs overwinter on *Prunus*, especially plum and blackthorn, but also on flowering cherries; large-scale landscape plantings with these trees at the waterside will almost certainly make it difficult to maintain a healthy population of *Nymphaea* species. Waterlily leaf spot shows as dark rotting patches on leaves that eventually disintegrate. Remove and destroy affected parts. With crown rot, the crown suffers a putrid rot and rapidly collapses.

N.alba (EUROPEAN WATER LILY; Eurasia, N Africa; leaves to 30cm in diameter, entire, dark green above, red-green to yellow beneath; flowers to 20cm, diameter, white, opening diurnally, floating, faintly fragrant, stamens yellow to orange); *N.caerulea* (BLUE LOTUS; N and tropical Africa; leaves to 40cm in diameter, entire or undulate, green above, spotted purple beneath; flowers to 15cm in diameter, pale blue, emergent, opening diurnally, stamens yellow); *N.capensis* (CAPE BLUE WATER LILY; S and E Africa, Madagascar; leaves to 40cm in diameter, dentate to undulate, green, spotted purple beneath when young; flowers to 20cm in diameter, bright blue, opening diurnally, sweetly fragrant, emergent, stamens golden; cultivars and varieties include 'Eastoniensis', with serrate leaves and steel blue flowers; var. *zanzibariensis*, with smaller leaves, tinged purple beneath, and deeper blue flowers to 30cm in diameter; 'Azurea', with light blue flowers; 'Jupiter', with large, scented, dark violet blue flowers; and 'Rosea', with leaves tinted red beneath and pale pink flowers flushed red); *N.* × *daubenyana*. (*N.caerulea* × *N.micrantha*; leaves viviparous; flowers 5–18cm in diameter, emergent, azure blue, opening diurnally); *N.flavovirens* (Mexico, S America; leaves to 45cm in diameter,

Nymphaea alba

subentire to deeply sinuate, sometimes red beneath; flowers to 20cm in diameter, white, strongly fragrant, opening diurnally, emergent, stamens deep yellow; cultivars include 'Astraea', with star-shaped, blue flowers shading to white at centre, and yellow stamens; 'Mrs C. W. Ward', with rosy red flowers and golden yellow stamens tipped pink; 'Purpurea', with vivid purple flowers and gold stamens; 'Stella Gurney', also known as 'Pink Star', with large, pale pink flowers; and 'William Stone', with large, dark blue flowers, violet at centre, and stamens gold tipped blue); *N.gigantea* (AUSTRALIAN WATER LILY; tropical Australia, New Guinea; leaves to 60cm in diameter, undulate, dentate, tinged pink to purple beneath; flowers to 30cm in diameter, sky blue to blue-purple, opening diurnally, emergent, stamens bright yellow); *N. × helvola;* syn. *N.* 'Pygmy Yellow'; *N.tetragona × N.mexicana*; leaves to 6cm, red blotched brown; flowers to 5cm in diameter, canary-yellow, stamens orange); *N.* Laydekeri Hybrids (syn. *N. × laydekeri*; hybrids involving *N.alba* var. *rubra*, *N.tetragona* and others; hardy garden hybrids lacking the vigour of the Marliacea Hybrids; usually with brown-mottled or -tinted leaves and flowers 6–8cm in diameter; cultivars include 'Fulgens', with leaves mottled brown, and crimson-magenta flowers to 9cm in diameter, with red stamens; 'Liliacea', with fragrant, lilac-pink flowers, with orange-red stamens; and 'Purpurata', with green leaves and crimson flowers, with orange-red stamens). *N.lotus* (EGYPTIAN WATER LILY, LOTUS, WHITE LILY; Egypt to tropical and SE Africa; leaves to 50cm in diameter, undulate to serrate, dark green above, green or brown and usually hairy beneath; flowers to 25cm in diameter, white, sometimes tinged pink, opening diurnally or nocturnally, slightly fragrant, emergent); *N.* Marliacea Hybrids (syn. *N. × marliacea*; MARLIAC HYBRIDS, involving *N.alba*, *N.odorata* var. *rosea*, *N.mexicana* and others; hardy, robust garden hybrids with large flowers held above the water; cultivars include 'Albida', with fragrant, white flowers, exterior tinted pink below, with yellow stamens; 'Carnea', with fragrant, flesh-pink to deep rose flowers; 'Chromatella', with leaves mottled purple-bronze, and chrome yellow, semi-double flowers; and 'Rosea, with leaves flushed purple at first, and rosy red flowers); *N.odorata* (FRAGRANT WATER LILY, POND LILY; E US; leaves to 25cm in diameter, entire, dull dark green above, usually purple and rough beneath; flowers to 15cm in diameter, white, usually floating, sweetly fragrant, opening diurnally, stamens gold; cultivars include 'Eugene de Land', with scented, pale orange-pink flowers held above water; 'Exquisita', with small, star-shaped, rose flowers; 'Helen Fowler', with large deep pink, very fragrant flowers held above water; 'Roswitha' with rich rose-red flowers; 'Sulphurea Grandiflora', *N.odorata × N.mexicana*, with dark green, marbled leaves and very large, stellate flowers of a bright rich yellow; 'William B. Shaw', with large, flat, creamy pink flowers, with an internal zone of dark red; var. *rosea*,

with deep pink, strongly scented flowers to 10cm in diameter; and 'Prolifera', with abundant flowers); *N.tetragona* (syn. *N.pygmaea*; PYGMY WATER LILY; NE Europe, N Asia to Japan, N America; leaves to 10cm, entire, blotched brown when young above, dull red beneath; flowers to 5cm in diameter, white, sometimes faintly lined purple, floating, slightly fragrant, anthers golden yellow; cultivars include 'Alba', with small, oval leaves, purple beneath, and white flowers to 2.5cm in diameter; 'Hyperion', with dark amaranth flowers; 'Johann Pring', with rich pink flowers – inner ring of stamens light orange, outer ring dark pink – to 5cm in diameter; 'Rubis', with deep red flowers, lacking white dots on outer petals; f. *rubra*, with leaves tinted purple, red beneath, and dark red flowers, with orange stamens).

nymphaeum a grotto or building usually containing statuary and waterworks, a feature of Roman and Italian Renaissance gardens.

Nymphoides (resembling *Nymphaea*). Menyanthaceae. Cosmopolitan. FLOATING HEART. 20 species, aquatic perennial herbs with creeping rhizomes and elongate stems. The leaves are floating, long-stalked and rounded to heart-shaped. The flowers are borne on slender stalks, the flowering nodes sometimes with clusters of short spur roots, and the corolla rounded with five entire or fringed lobes. A hardy, floating aquatic for smaller ponds or containers. Cultivate as for hardy *Nymphaea* in full sun. Propagate by division in spring or by separation of young plantlets in autumn. Damaged, or killed, by formalin, when this is used against algae. *N.peltata* (YELLOW FLOATING HEART, WATER FRINGE; Europe, Asia, naturalized US; freely stoloniferous; leaves 5–10cm in diameter, mottled; flowers to 2.5cm in diameter, golden yellow, fringed).

Nyssa (for Nyssa, a water nymph: *N.aquatica* grows in swamps). Nyssaceae. TUPELO. Eastern N America, E Asia. 5 species, deciduous trees with inconspicuous green flowers and leaves colouring beautifully in autumn. The following are fully hardy and perform best in full sun or light shade on moist or wet, acid to neutral soils. *N.sylvatica* is tolerant of pollution, maritime conditions and shade. Avoid root disturbance. Sow seed, following stratification at 5°C/40°F for three months, or increase by layers or semi-ripe cuttings in late summer.

N.sinensis (CHINESE TUPELO; C China; similar to *N.sylvatica* but 10–15m and more spreading; leaves to 20cm, oblong to lanceolate, sparsely hairy and tinged red when young; autumn colour usually finer than that of *N.sylvatica*); *N.sylvatica* (BLACK GUM, PEPPERIDGE, SOUR GUM; eastern N America; broadly columnar to conical, 20–30m; leaves to 15cm, ovate to elliptic or obovate, downy on the veins beneath, flame-coloured in autumn; fruit ovoid, 1cm, blue-black; var. *biflora*: SWAMP TUPELO, SE US, to 15m, with base of trunk swollen when growing in water, and more leathery, oblanceolate leaves).

O

ob- a prefix, meaning inverted; thus, oblanceolate means lanceolate, but with the broadest part furthest from the base.

obelisk a pointed, tapering shaft of masonry, usually square in section, often used in landscape gardens as an eye-catcher.

obligate essential; unable to exist without; for example, true parasites are said to be obligate.

oblique (1) of a base, with sides of unequal angles and dimensions; (2) of direction, extending laterally, the margin upwards and the apex pointed horizontally.

obtuse apex or base terminating gradually in a blunt or rounded end.

Ochna (from Greek *ochne*, name used by Homer for the wild pear, the leaves of which are said to resemble those of some species in this genus). Ochnaceae. Old World tropics. BIRD'S EYE BUSH. 86 species, deciduous or evergreen trees and shrubs with leathery, glossy and toothed leaves. Produced in spring and summer, the flowers give rise to fruits with enlarged and fleshy calyces and glossy drupes attached to a swollen, lobed receptacle. Grow in full sun; water plentifully and feed fortnightly when in full growth; maintain a winter minimum temperature of 10°C/50°F. Prune in early spring if necessary to shape or confine to bounds. Propagate by seed or semi-ripe cuttings. *O.serrulata* (MICKEY-MOUSE PLANT, BIRD'S EYE BUSH; South Africa; to 2.25m; flowers solitary or clustered, calyx lobes bright red in fruit; drupes glossy, black).

Ocimum (plant name used by Theophrastus). Labiatae. Old World tropics. 35 species of aromatic annual and perennial herbs, subshrubs and shrubs with small, two-lipped flowers whorled in spikes. *O. basilicum*, is one of the most important culinary herbs; the leaves are used in salads, casseroles, sauces and liqueurs. Although more or less perennial, basil is best-treated as a tender annual in cool temperate climates and sown under glass in a soilless compost in spring; pot on under glass in a similar medium and plant after the last frosts. In hotter climates, it is possible to obtain two or more crops per year – sow *in situ*. *O.basilicum* (COMMON BASIL, SWEET BASIL; tropical Asia, now widespread through cultivation; annual or short-lived perennial, 20–60cm; leaves 1.5–5cm, narrowly ovate to elliptic, entire to serrate; cultivars include 'Citriodorum', with lemon-scented leaves; 'Crispum', with leaves curled around the edges; 'Minimum', BUSH BASIL, GREEK BASIL, 15–30cm, bushy with very small leaves; 'Purple Ruffles', with purple leaves, curled around the edges; 'Purpureum', also known as 'Dark Opal', with red-purple, clove-scented leaves; and 'Spicy Globe', compact and globose, with small leaves and white flowers).

ocrea, ochrea a tubular or inflated sheath formed by a pair of coherent stipules, as in Polygonaceae.

Odontadenia (from Greek *odous*, tooth, and *aden*, gland; referring to the 5-toothed glands). Apocynaceae. Old World tropics. 30 species, evergreen climbing shrubs. Large, yellow and funnel-shaped flowers are borne in large, loose cymes in spring and summer. Cultivate as for *Allamanda*. *O.macrantha* (syn. *O.grandiflora*, *O.speciosa*; Costa Rica to Peru and N Brazil; flowers scented, bright yellow shaded orange, to 8cm diameter).

× **Odontioda** Orchidaceae. Garden hybrids between *Odontoglossum* and *Cochlioda*. Plants consist of a group of compressed pseudobulbs with two or more strap-like leaves. The flowers arise at various times of year in simple or branched racemes. Most have a rounded shape and wide sepals and petals, the lip large or small depending on the ancestry. They come in many colours, often conspicuously marked in a variety of contrasting tones; the bright red of so many of the crosses is inherited from the *Cochlioda* parent. Most of these hybrids are 'cool' growers (minimum temperature 10°C/50°F), although a few are tolerant of warmer conditions and may thrive in the home; see *Odontoglossum* for general requirements.

× **Odontocidium**. Orchidaceae. Garden hybrids between *Odontoglossum* and *Oncidium*. The flowers arise on simple or branched racemes and are very varied – rounded in shape, or with long, narrow sepals and petals, the lip large or small depending on the ancestry; many colours are represented, often marked in contrasting tones. Most of these hybrids are 'cool' growers (minimum temperature 10°C/50°F), although a few are tolerant of warmer conditions; see *Odontoglossum* for general requirements.

Odontoglossum (from Greek *odous, odontos,* tooth, and *glossa,* tongue, referring to the tooth-like processes of the lip). Orchidaceae. S America. Some 60 species, evergreen perennial herbs, for the most part epiphytes or lithophytes with pseudobulbs and strap-shaped leaves. The flowers are carried in erect to arching racemes or panicles and vary greatly in size, shape and colour. The sepals are more or less equal and spreading, the petals similar to the sepals but often shorter. The lip is simple or 3-lobed, its basal portion often claw-like, the lateral lobes spreading or erect, the midlobe entire or cleft, and the disc crested, bumpy or with keels. These plants are cool-growing (winter minimum temperature 10°C/50°F) and demand a buoyant, freely ventilated and humid atmosphere in light shade. Their roots are slender: a fine bark mix is best, containing rockwool or sphagnum. Water and syringe freely when in growth, and impose drier cooler conditions when at rest. Never syringe in cold weather. Propagate by division.

O.cirrhosum (Ecuador, Peru, Colombia; raceme or panicle to 60cm, arching; flowers to 8cm across, tepals slender, wavy, with narrow, curved tips, white, more or less blotched red-brown, lip with a narrow, pointed midlobe, white stained yellow in throat and marked red-brown); *O.crispum* (Colombia; raceme to 50cm, arching; flowers to 9cm across, tepals broad, spreading, toothed and lacy, pure white, sometimes tinted pale rose or with a few red spots, lip short, broad, white stained yellow at base with a few red spots); *O.cristatum* (Colombia, Ecuador, Peru; raceme to 50cm, arching; flowers to 8cm across, tepals ovate, tapering narrowly, creamy yellow to olive with chestnut blotches, lip finely pointed, olive blotched chestnut with a white throat); *O.harryanum* (Columbia, Peru; raceme erect to 1m, sometimes branched; flowers to 10cm across, tepals broadly oblong, wavy, yellow to olive overlaid with chestnut to chocolate brown, lip large, heart-shaped, white heavily veined with red); *O.lindleyanum* (Colombia, Venezuela, Ecuador; raceme or panicle to 30cm; flowers to 6cm across, tepals narrowly oblanceolate, golden blotched and barred chestnut brown, lip narrow, forward-pointing, white with a spear-shaped midlobe progressing from yellow at the tip through burnt orange to fuchsia pink at the base); *O.luteopurpureum* (Colombia; raceme to 1m; flowers to 10cm across, tepals ovate to lanceolate, pointed, wavy, creamy yellow heavily blotched and overlaid with glossy maroon, lip axe-shaped, finely fringed, white blotched maroon at base); *O.pendulum* (syn. *Cuitlauzina pendula, O.citrosmum*; Mexico; raceme to 50cm, pendulous; flowers to 7cm across, lemon-scented, tepals broadly obovate, pure white sometimes flushed pale rose or lilac, lip broadly hatchet shaped, mauve-pink, yellow at base); *O.pulchellum* (syn. *Osmoglossum pulchellum*; LILY OF THE VALLEY ORCHID; Mexico, Guatemala, El Salvador; raceme to 50cm, erect to arching; flowers to 3cm across, sweetly scented, tepals obovate, cupped, sparkling white tinted rose, lip white); *O.wallisii* (Colombia, Venezuela, Peru; raceme or panicle to 50cm, arching; flowers to 5cm across, tepals narrowly oblanceolate, golden blotched red-brown, lip white spotted rose-purple near apex). For *O.bictoniense, O.cervantesii, O.cordatum, O.rossii* and *O.stellatum* see under *Lemboglossum*; for *O.grande* see *Rossioglossum*.

Odontoglossum pendulum

oedema the formation of small blister-like swellings on leaves or stems, which may become brown and corky. The result of increased intracellular water, oedema occurs where intake exceeds the rate of transpiration. It is favoured by a humid atmosphere and heavy watering. Tomatoes and pelargoniums especially are prone to oedema. Also known as edema or dropsy.

Oemleria (for Augustus Gottlieb Oemler, 19th-century German-born pharmacist and naturalist working in Savannah, Georgia, US). Rosaceae. Western N America. OSO BERRY, OREGON PLUM. 1 species, *O.cerasiformis* (syn. *Osmaronia cerasiformis*), a deciduous, slender-stemmed shrub to 5m. The leaves are to 10cm long, oblong to oblanceolate, glossy dark green above, grey and downy beneath. Small, white, fragrant flowers are carried in nodding racemes in early spring and are followed by small, sloe-black fruits. Hardy to –20°C/–4°F. Grow in well-drained but moisture-retentive neutral to acid soils in shade. Prune after flowering to thin out old shoots and to relieve overcrowding. Old plants may be rejuvenated by cutting hard back. Propagate by rooted suckers.

Oenanthe (from Greek *oinanthe,* inflorescence of the grapevine). Umbelliferae. Western N America. 30 species of glabrous perennial herbs, many of them highly toxic. The stems are procumbent to ascending and often rooting. The leaves are pinnately compound or decompound with toothed leaflets. Small, white flowers are borne in tight compound umbels. Hardy in climate zone 6, *O.javanica* 'Flamingo' is a brightly variegated groundcover plant ideal for wet situations or use in containers and bedding. It will also thrive in shallow water. Propagate by stem tip cuttings, division or by simple layering. *O.javanica* (syn. *O.japonica*; India to Japan, Ryukyu Islands, Taiwan, Malaysia, N Australia; stems 20–40cm, procumbent to erect; leaf segments 1–3cm, ovate to narrowly obovate, toothed, sometimes lobed; 'Flamingo': leaves splashed and zoned pink, cream and white).

Oenothera
'Siskiyou'

Oenothera (name used by Theophrastus for another plant). Onagraceae. N and S America, many naturalized elsewhere, usually on disturbed ground. EVENING PRIMROSE, SUNDROPS, SUNCUPS. 124 species, annual, biennial or perennial herbs. Produced in spring and summer, the flowers are usually short-lived but showy, either solitary in leaf axils or gathered into corymbose, racemose or spicate inflorescences. Bowl- to funnel-shaped, they consist of four, silky, obovate petals and open at dawn or dusk. The following species are hardy in climate zone 6 and fare best in full sun on well-drained, low- to medium-fertility soils. Their longevity varies, influenced by climate and soil type: certain perennial species are better treated as annuals or biennials when gardening on heavy soils or in cold areas. The tall-growing species are prolific seeders useful for naturalizing in wild areas (gravelly banks and near-pure sand included). Low-growing species make good front-of-border or scree plants. Sow seed from spring to early summer. Propagate perennials by division, or by softwood cuttings taken in spring before flowering.

O.biennis (COMMON EVENING PRIMROSE, GERMAN RAMPION; eastern N America, naturalized Europe and elsewhere; erect, annual or biennal, 10–150cm; inflorescence spicate; petals 1.8–2.5cm, yellow, ageing gold); *O.deltoides* (DESERT EVENING PRIMROSE; California; annual, 5–25cm, branching from base; flowers solitary, white, ageing pink, 4–8cm diameter); *O.fruticosa* (SUNDROPS; eastern N America; biennial or perennial, 30–80cm, hairy, tinged red; petals 1.5–2.5cm, toothed, deep yellow; cultivars include 'Fireworks', with leaves tinged purple and yellow flowers; 'Golden Moonlight', to 80cm high, with large, bright yellow flowers; 'Highlight', with yellow flowers; 'Illumination', with leaves, tinged bronze, and large, deep yellow flowers; 'Silvery Moon', to 80cm high, with large, pale yellow flowers; 'Sonnenwende', to 60cm, with leaves red in autumn, and red-orange flower buds; 'Yellow River', with stem brick red and bright canary yellow flowers; 'Youngii', to 50cm, with large, bright yellow flowers; subsp. *glauca*, syn. *O.tetragona*: leaves broader, tinted red when young); *O.macrocarpa* (syn. *O.missouriensis*; OZARK SUNDROPS; south central US; perennial, rather short-stemmed, branched from base, decumbent to erect, pubescent; flowers yellow, to 10cm in diameter; includes 'Greencourt Lemon', with lemon yellow flowers); *O.perennis* (syn. *O.pumila*; SUNDROPS; eastern N America; perennial, 10–50cm; stem slender, simple or few-branched; flowers in a loose spike, petals 0.6–0.8cm, yellow); *O.rosea* (Texas to Peru, naturalized S Europe; erect, strigulose, perennial or annual, 15–60cm; inflorescence spicate, petals 0.4–1cm, pink to red-violet); *O.speciosa* (WHITE EVENING PRIMROSE; SW US to Mexico; rhizomatous perennial, erect, 30–60cm, strigose; petals 2–2.5cm, white ageing rose, sometimes pink when young; 'Rosea': to 30cm, with pale pink flowers opening during the day; 'Siskiyou': habit neat, bushy, flowers rose-pink with white centres veined pink, the very centre and anthers golden).

officinal strictly meaning 'of apothecaries' shops'. Pertaining to commercial value, especially as sold in shops; applied to plants with real or presumed medicinal properties, hence the specific epithet *officinalis*.

offset, offshoot a young plant produced naturally by vegetative means; it is attached to its parent and easily detached from it. The term is commonly applied to bulbs, corms and many rosette-forming plants, such as sempervivums; it is also used of fibrous-rooted plants that produce a number of loosely joined groups of shoots, such as perennial asters. See *daffodils*.

oilstone a fine-grained sharpening stone lubricated with oil, used for sharpening knives and secateurs.

okra (*Abelmoschus esculentus*) LADY'S FINGER A tender annual, up to 2m in height, grown for its edible pods used when immature as a cooked vegetable, or matured, dried and finely ground for use as a flavouring. The pods are 10–25cm long, angled, beaked and bristly and there are green, red or yellow skinned cultivars. In warm temperate and sub-tropical climates OKRA can be grown outdoors. An ideal temperature range for good growth is 20–30°C/68°–86°F, so that greenhouse protection may be required. Pre-soak seed and germinate at a minimum temperature of 16°C/61°F. In suitable climates seed can be sown *in situ* outdoors in rows 60–70cm apart at 20–30cm stations. Under protection grow in pots, growing bags or border soil allowing at least 40cm between plants. Plants require staking and the growing points should be pinched out at 60cm height. The maintenance of high humidity is beneficial. Harvest from 8–12 weeks after sowing, cutting the pods off with a knife.

Olea (from Greek *elaia*, the olive). Oleaceae. Temperate and tropical Old World. OLIVE. Some 20 species of long-lived evergreen trees and shrubs with leathery leaves, small, white or off-white flowers in summer, and ovoid or globose drupes. Although badly damaged by temperatures lower than –10°C/14°F *O.europaea* will grow in zone 8, if given the shelter of a wall and full sunlight. However, good fruiting can be expected only in warm-temperate regions with moist winters and hot, dry summers. Plant in a deep fertile soil, with perfect drainage. The weighting or arching of branches will encourage fruiting. Pruning may stimulate the overproduction of non-fruiting water shoots. Container-grown plants thrive in cool, well-lit greenhouses and conservatories and may be moved outdoors in summer. Propagate by seed sown in gentle heat in spring, by semi-ripe cuttings in summer. *O.europaea* (COMMON OLIVE, EDIBLE OLIVE; to 7m; leaves to 8cm, elliptic to lanceolate, grey-green above, silver-scurfy beneath; fruit ripening red to purple-black, to 5cm; widespread throughout the Mediterranean region, var. *europaea* is the fruit- and oil-yielding cultigen; it includes numerous cultivars; the fruit may range in size from 1 to 5cm, in shape from near-spherical, to egg- to spindle-shaped, in colour from purple-red to blue-black or yellow-green, in quality from firm and waxy to smooth and yielding, and from sour to sweet).

oleaginous oily, though of a fleshy nature.

Olearia (for Johann Gottfried Olschlager (1635–1711), German horticulturist – the family name was Latinized Olearius). Compositae. Australasia. DAISY BUSH. About 130 species, herbs, evergreen shrubs or small trees with leathery leaves and daisy-like flowerheads in spring and summer. Most are not reliably hardy where temperatures fall much below –5°C/23°F, unless given warm wall protection. They are valuable in mild maritime gardens, showing good resistance to salt-laden winds and maintaining a dense and compact habit when grown in full sun and an exposed position. Several species of particularly dense habit are suitable for hedging, including *O.macrodonta*, *O.avicenniifolia* and *O. × haastii* (the two last also tolerate urban pollution). With *O.nummularifolia*, *O.virgata*, *O.mollis* and *O.ilicifolia*, these three are among the hardiest, tolerating temperatures to –15°C/5°F; *O. × scilloniensis* and O. 'Talbot de Malahide' are almost as hardy, to about –10°C/14°F. Grow in any well-drained soil in full sun. *Olearia* responds well to hard pruning, breaking freely from older wood. Propagate by semi-ripe cuttings in summer or by heeled ripewood cuttings of lateral shoots in the cold frame in early autumn. Softer-leaved types, such as *O.phlogopappa*, can be increased by softwood cuttings.

O.avicenniifolia (New Zealand; shrub or small tree to 7m; leaves 5–10cm, ovate to lanceolate, tapered at base and apex, entire, grey-white, glabrous above, white- or pale yellow-tomentose beneath; flowerheads white); *O.chathamica* (New Zealand; shrub, to 2m; leaves 3–8cm, oblanceolate to elliptic, acute, tapered at base, regularly and obtusely dentate, glabrous above, midrib raised beneath; flowerheads white or tinged purple, with dark purple centres); *O. × haastii* (*O.avicenniifolia × O.moschata*; New Zealand; shrub, 1–3m; leaves 1–2.5cm, crowded, oval or ovate, entire, dark glossy green, glabrous above, white-tomentose beneath; flowerheads white with centres yellow); *O.* 'Henry Travers' (*O.chathamica × O.semidentata*; medium-sized shrub; leaves lanceolate, grey-green, silvery beneath; flowerheads large, lilac, deep mauve centres); *O.hookeri* (Tasmania; shrub to 1m; leaves 3–6cm, narrow, grooved above, revolute; flowerheads indigo, centres yellow); *O.ilicifolia* (New Zealand; tree or shrub to 5m, with musky fragrance; leaves 5–10cm, linear to oblong or lanceolate, acute to acuminate, coriaceous to glabrous above, white- to yellow-tomentose beneath, undulate, serrate-dentate; flowerheads fragrant, white); *O.lacunosa* (New Zealand; shrub to 5m; leaves 7–16cm, linear to oblong, acute to acuminate, coriaceous, rugose, midrib yellow above, minutely sinuate-dentate; flowerheads white);

O.macrodonta (NEW ZEALAND HOLLY; New Zealand; shrub or tree to 6m; leaves 5–10cm, ovate to oblong, acute to acuminate, dark glossy green above, silver-tomentose beneath, musk-scented when crushed, undulate, acutely dentate-serrate; flowerheads white with red-tinted centres); *O. × mollis* (*O.ilicifolia × O.lacunosa*; New Zealand; shrub to 3m; leaves to 10cm, lanceolate, spinulose, rounded at base, veins sunken above, very prominent beneath, densely white- to pale yellow-tomentose beneath, revolute; flowerheads white; 'Zennorensis': shrub to 2m, with narrow, acute, sharply dentate leaves, 10 × 1–1.5cm, dark olive green above, white-tomentose beneath); *O.nummulariifolia* (New Zealand; shrub to 3m; leaves 0.5–1cm, dense, ovate to suborbicular, glabrous above white- to buff- or yellow tomentose beneath, revolute; flowerheads fragrant, cream or pale yellow); *O.phlogopappa* (Tasmania, Victoria, New South Wales; aromatic shrub to 3m; leaves 1.5–5cm, oblong or narrowly obovate, sinuous or shallowly toothed, dark dull green above, white- or grey-tomentose beneath; flowerheads white, occasionally pink, mauve or blue, with centres yellow). *O. × scilloniensis* (*O.lirata × O.phlogopappa*; shrub to 3m; leaves to 11cm, elliptic to oblong, obtuse, deep green and reticulate above, pale green, tomentose beneath, sinuate; flowerheads white with yellow centres; 'Master Michael': leaves tinted grey, flowers blue); *O.semidentata* (Chatham Island, New Zealand; rounded shrub to 3.5m; leaves 3.5–7cm, linear to lanceolate, serrate above, rugose, glabrous above, white-tomentose beneath; flowerheads pale purple with darker violet-purple centres; for *O.semidentata* of gardens, see *O.* 'Henry Travers'); *O.solandri* (New Zealand; shrub or small tree to 4m; leaves to 1.5cm, midrib sunken, linear to spathulate or narrowly obovate, subcoriaceous, glabrous above, white- to yellow-tomentose beneath; flowerheads yellow; includes 'Aurea', with leaves strongly tinged gold); *O.* 'Talbot de Malahide' (syn. *O.albida*; very similar to *O.avicenniifolia* except leaves blunter and more rounded); *O.virgata* (New Zealand; shrub to 5m; leaves 0.5–2cm, narrowly obovate, glabrous above, white-tomentose beneath; flowerheads yellow to white; var. *lineata*: leaves narrow, strongly revolute).

Olsynium (from Greek meaning 'hardly united', referring to the stamens). Iridaceae. Americas. About 12 species, perennial herbs with fibrous roots and basal fans or tufts of linear to lanceolate leaves. Produced in summer in terminal clusters, the flowers are bell-shaped and consist of six tepals. Graceful perennials for cool, semi-shaded positions in acid pockets on the rock garden or the peat bed. They will survive frosts to –20°C/4°F and become dormant in late spring, but nonetheless require permanent moisture. Increase by seed or division. *O.douglasii* (syn. *O.grandiflorum*, *Sisyrinchium douglasii*, *Sisyrinchium grandiflorum*; GRASS WIDOW, PURPLE-EYED GRASS; western N America; 15–30cm high; flowers pendent, wine-red, purple-pink or white).

Omphalodes (from Greek *omphalos*, navel, and *oides*, like, the nutlets resemble a navel). Boraginaceae. Europe, N Africa, Asia, Mexico. NAVELWORT, NAVELSEED. Some 28 species, annual, biennial or perennial herbs, with long-stalked, oblong to ovate leaves. Produced in spring and summer in cymose sprays, the small, bowl-shaped flowers have five, rounded lobes and, usually, a paler eye. *O.luciliae* requires a sunny position with its roots in light, moist, sharply drained soil, with additional limestone. Hardy to about –5°C/23°F, it dislikes winter wet and may be more safely grown in the alpine house. The annual *O.linifolia*, with white blooms which last well when cut, is sown *in situ* in spring, in sun and well-drained soil. *O.verna* is hardy in climate zone 6 and suitable for cool positions in the rock garden or for naturalizing in light, open woodland. Increase by division in early spring.

O.cappadocica (Asia Minor; perennial to 28cm, erect or ascending; flowers bright blue with a white eye, to 5mm in diameter; 'Bright Eyes': flowers white edged bright blue; 'Cherry Ingram': to 25cm, with deep blue flowers); *O.linifolia* (SW Europe; annual to 40cm, erect; flowers to 12mm in diameter, white or blue); *O.luciliae* (Greece, Asia Minor; perennial to 25cm, tufted, glabrous; flowers to 8mm in diameter, rose becoming blue); *O.verna* (CREEPING FORGET-ME-NOT; Europe; perennial to 20cm, stoloniferous; flowers to 12mm in diameter, blue; 'Alba': flowers white).

Omphalogramma (from Greek *omphalos*, navel, and *gramma*, line). Primulaceae. Himalaya, W China. 15 species, perennial, usually rhizomatous, herbs. The solitary flowers are bell- to funnel-shaped with six to eight lobes and are produced atop glandular-hairy scapes in spring. *O.vinciflorum* is hardy in climate zone 7 and suited to the shady rock garden, with gritty, moist, lime-free soil. Otherwise, cultivate in the alpine house or cold frame, in pans of a gritty alpine mix with added leafmould. Propagate from seed or by division in spring. *O.vinciflorum* (China; flowers or scapes to 20cm, deep indigo-blue or purple, 3–5cm in diameter).

Oncidium (from Greek *onkos*, a lump or tumour, referring to the fleshy, protuberant calli on the lips of many species). Orchidaceae. Tropical and subtropical America. Some 450 species, evergreen perennial herbs, for the most part epiphytes or lithophytes. They vary greatly in habit and size, from dwarf species to 'giants', many with pseudobulbs, some lacking these but with tough, overlapping leaves or merely one or two very large and fleshy leaves. Produced in racemes or panicles, the flowers are generally small and often in bright shades of yellow barred or mottled with brown. The tepals are oblong to obovate, the lip larger and entire to 3-lobed with a warty or crested callus. A large genus of cool and intermediate-growing orchids. The smaller species and those with distinct pseudobulbs need dappled sunlight and a humid, buoyant atmosphere in the cool house (winter minimum 10°C/50°F). In the growing season, water and feed freely; at other times, water only to prevent shrivelling. Pot in a fine- to medium-grade bark-based mix. The larger species, with reduced pseudobulbs and massive, leathery, 'mule's ear' leaves, should be mounted on bark or planted in baskets and afforded maximum light and humidity throughout the year. They need less water than the first group but favour a warmer and steamier situation. Increase by division or seed.

O.bifolium (Brazil, Uruguay, Argentina; plants small to medium-sized, pseudobulbous; raceme to 40cm, flowers to 3cm across, tepals small, yellow marked brown, lip far larger with a broad, kidney-shaped midlobe, rich golden yellow with red callus); *O.cavendishianum* (Mexico, Honduras, Guatemala; pseudobulbs very small or absent; leaves solitary, very large, tough and fleshy; panicles to 1.5m, flowers to 4cm across, tepals rounded, yellow to olive spotted chocolate or red-brown, lip larger, 3–parted with a notched, fan-shaped midlobe, rich yellow with red dots on callus); *O.crispum* (Brazil; plants medium-sized, pseudobulbous; panicle to 1m, erect to arching; flowers to 8cm across, tepals obovate, the petals far larger and wavy, yellow to buff, heavily veined and overlaid with chocolate brown, lip with two small lateral lobes and a large, wavy, fan-shaped midlobe, yellow to tan mottled or veined brown especially near margin); *O.flexuosum* (Brazil, Argentina, Paraguay, Uruguay; plants small to medium-sized, pseudobulbous; panicle wiry, to 1m, flowers to 2.5cm across, tepals small, recurved, yellow barred red-brown, lip far larger, broadly fan-shaped, bright gold, sometimes spotted red-brown, callus marked red-brown); *O.fuscatum* (syn. *Miltonia warscewiczii*; *Miltonioides warscewiczii*; Peru, Colombia, Costa Rica; plants medium-sized, pseudobulbous; panicle to 50cm, flowers to 5cm across, tepals oblong to oblanceolate, wavy, brown edged yellow-white, lip large, oblong to shield-shaped, purple to crimson with a deep white margin and a shiny red-brown central patch); *O.globuliferum* (Colombia, Venezuela, Costa Rica, Panama; plants small to medium-sized, pseudobulbous; raceme to 35cm, flowers to 3cm across, tepals small, oblanceolate, wavy, golden yellow spotted or barred red-brown at base, lip far larger, kidney-shaped, deep gold spotted red-brown at base); *O.hastatum* (Mexico; plant medium-sized, pseudobulbous; panicle to 1.5m; flowers to 2.5cm across, tepals lanceolate to ovate, yellow to olive blotched brown, lip narrowly arrow-shaped, lateral lobes white, midlobe yellow to flame red with a crimson callus); *O.jonesianum* (Brazil, Paraguay, Uruguay; plants small to medium-sized; pseudobulbs very small or absent; leaves clumped, pendulous, fleshy, narrowly cylindrical and pointed; raceme to 50cm, flowers to 7.5cm across, tepals obovate, wavy, yellow to olive blotched red-brown, lip fan-shaped, white with a red-spotted, yellow base); *O.ornithorhynchum* (C America; plant small to medium-sized, pseudobulbous; panicle to 50cm, arching to pendulous, crowded; flowers to 2cm across, lacy, pale lilac-pink to rich rose with a yellow callus); *O.pulchellum* (syn. *Tolumnia pulchella*; Jamaica, Guianas, Cuba, Hispaniola; plants dwarf to small, pseudobulbs absent; leaves tough, narrow, folded and keeled, in a fan-like growth; panicle to 50cm, flowers to 2.5cm across, tepals small, white tinted rose to rich magenta, lip large, broadly fan-shaped, white tinted rose to deep magenta, yellow at base); *O.pusillum* (syn. *Psygmorchis pusilla*; C and S America, W Indies; plant dwarf, short-lived, pseudobulbs absent; leaves folded and keeled, in a fan-like tuft; flowers to 3cm diam., borne singly in succession on short, wiry stalks, tepals very small, yellow blotched red-brown, lip large, frilly, golden yellow with a red-spotted, white callus); *O.sphacelatum* (Mexico to El Salvador; plants medium-sized to large, pseudobul-bous; panicle to 1.5m, flowers to 2.5cm across, tepals ovate to lanceolate, pointed, yellow barred and blotched chestnut brown, lip large with rounded lateral lobes and a broadly kidney-shaped midlobe, golden blotched orange-yellow at base with an orange-spotted, white callus); *O.superbiens* (Colombia, Venezuela, Peru; plant medium-sized to large, pseudobulbous; panicle to 2m, slender and snaking, flowers to 7cm across, tepals broadly obovate, long-clawed, sepals chocolate brown tipped yellow, petals yellow or white barred chocolate, lip small, purple-brown, trowel-shaped); *O.tigrinum* (Mexico; plants medium-sized to large, pseudobulbous; raceme or panicle to 1m, erect, flowers to 6cm across, tepals lanceolate to ovate, wavy, yellow barred dark brown, lip far larger with a broad, kidney-shaped midlobe, bright yellow); *O.varicosum* (Brazil; plants medium-sized, pseudobulbous; panicle to 1.5m, loose, arching, flowers to 5cm across, tepals small, olive to yellow barred chestnut brown, lip far larger, fan-shaped, golden yellow with a red-spotted callus).

onion (*Allium cepa*) a biennial grown as an annual for its edible bulb which, usually, has a strong pungent flavour; the leaves and immature bulbs are also edible. The onion probably originated in the regions of Iran and western Pakistan and its use can be traced back to 3200 BC, when it was an important food crop for the Egyptians.

Hot dry summers are ideal for bulb maturation but there are many cultivars suited to a range of growing conditions, including cool-season production in tropical regions. For high yield, cool weather is desirable during the early stages of growth. The crop is tolerant of frost, but prolonged exposure to temperatures below 10°C/50°F will cause plants to bolt. Optimum seedling growth occurs in the range 20–25°C/68–77°F and declines rapidly at temperatures above 27°C/80°F. Bulb formation takes place in response to long-day conditions and, in temperate climates, most cultivars bulb up during early summer for late summer harvest. Some cultivars produce bulbs over a long-day range from 12 to more than 16 hours, making it possible to manipulate crop supply. Bulbing is also advanced by raised temperatures and nitrogen deficiency. Bulb

dormancy may be prolonged by storing at 1°C/34°F, with low relative humidity.

With choice of cultivars and by storage, onions can be continuously available from summer until the following spring. Immature salad onions can be harvested throughout most of the year. Most onion cultivars are brown- or yellow-skinned, with some red-skinned; they may be globe-shaped or flat.

Bulb onions can either be propagated from seed, or vegetatively using small bulblets referred to as sets. Maincrop onions should be sown in a well-prepared seedbed from February to April. Germination is poor in cold wet soils and cloches should be used to warm the soil for early sowings in cold areas. Seed, preferably treated with a fungicidal dressing, should be sown thinly in 2cm-deep drills in rows 30cm apart, with seedlings thinned to 4cm spacings for medium-sized bulbs, or up to 10cm spacings for large bulbs. Early summer maturing crops can be obtained by raising transplants under protection for early planting in spring, or by sowing suitable cultivars outside in autumn.

Transplants are best raised in multi-seeded cell modules or blocks of up to six seedlings; sow at 10-16°C/50–61°F in January and, after hardening off, plant out 25-30cm apart each way. For outdoor autumn sowings, specially selected cultivars should be chosen. The sowing date is critical and depends on local climatic conditions. If sown too early, plants may develop to a size where they are vernalized by low winter temperature and will bolt in spring; if sown too late, they may be too small to survive the winter. Sow seeds 2.5cm apart in rows spaced 30cm apart, at times based on local recommendations and aiming for seedlings to be 15–20cm high before the onset of winter.

Onions raised from sets have the advantages of being less prone to disease and onion fly attack; they yield reasonably in poorer soil conditions, and mature earlier than seeded crops. Cultivars are available for both spring and autumn planting and should be planted firmly, 5cm apart, with the tips at soil level, in rows spaced 25cm apart. The risk of bolting is reduced with small sized sets, and with heat-treated ones which should not be planted before late March.

For salad use, thinnings of maincrop onions can be used, but for a continuous supply sow specific cultivars, usually referred to as spring onions (or scallions in the US), from spring to midsummer at intervals of two to three weeks. Over-wintered sowings of extra-hardy cultivars can be harvested in spring.

Weed control is important and during dry spells watering may be necessary, but not as the crop reaches maturity. Onions are ready to harvest when the leaves start to die back and fall over; after drying in the sun or in a cool dark shed, they may be tied in ropes and hung up or stored in net bags or shallow trays.

Pickling onions are bulb onion cultivars which tolerate poorer, drier soils than bulb or spring onions. Sow in March 6mm apart, since competition ensures small size, in rows 30cm apart. Harvest in late summer.

onions

Pests
eelworm
onion fly

Diseases
downy mildew
neck rot
white rot

SHALLOTS (*Allium cepa.* Aggregatum Group) have a distinct flavour, and are available in yellow- or red-skinned forms. Raise from sets, about 2cm in diameter; plant December to March to tip level, 15cm apart, in rows 20cm apart. Bulb clusters are harvested in late summer.

WELSH ONION, PERENNIAL GREEN ONION, or CIBOULE (*Allium fistulosum*), is a hardy clump-forming type used as a salad onion; it grows to 60cm tall. Sow in spring or August in rows 30cm apart, thinning to 23cm apart. Lift and replant clumps every 2–3 years. JAPANESE BUNCHING ONIONS (*Allium fistulosum*) are a much shorter improved form, also available as a single-stemmed type; they are best sown as annuals or biennials in rows 30cm apart, and thinned to 7.5cm spacings.

The EVERLASTING ONION, a cultivar of *A.cepa*, is similar in habit to the Welsh onion but with thinner, flatter leaves and a mild flavour. It should be propagated periodically by division.

The EGYPTIAN or TREE ONION (*Allium cepa* Proliferum Group) produces clusters of very small aerial bulbs, borne on tall stems in place of flowers. They are hardy and can be used throughout the year. Plant 25cm apart in spring or autumn; the plants are often self-perpetuating as the bulb-laden stems fall over.

onion fly (*Delia antiqua*) a small grey fly similar in appearance to a housefly, with white legless maggots up to 8mm long, which attack roots of seedling onions, and sometimes shallots and leeks. Early signs of attack are a yellowing and wilting of the outer leaves; symptoms are worse in hot weather on dry soils. There are two generations and damage usually occurs during the period June–August; second-generation maggots may tunnel inside the bulbs as well as attacking roots.

Spring-sown seedlings and sets can be protected by treating the soil with a recommended insecticide at sowing or planting time. Plants overwintered from a late-summer sowing will require a second treatment in May/June. Heavily infested plants should be removed and burnt.

onion fly and larva

Onoclea (from Greek *onos*, vessel, and *kleio*, to close, referring to the closely rolled fertile fronds). Athyriaceae. Eastern N America, E Asia. 1 species, *O.sensibilis*, the SENSITIVE FERN or BEAD FERN, a deciduous perennial fern with extensively creeping rhizomes. The fronds are 30-100cm long, more or less erect and pinnately divided with pinnae in 8-12 pairs, each pinna to 8cm wide, deeply lobed to sinuate or entire. The fronds emerge a soft coppery pink and become bright lime green. The rachis is broadly winged. Hardy in climate zone 5, *Onoclea* is most usually cultivated in bog and woodland gardens on damp leafy soils, or at the margins of ponds and lakes, where the spreading rhizomes may actually grow outwards covering the water's edge with a dense undulating blanket. Plant in part or full shade on permanently damp, acid to neutral soils rich in organic matter. Propagate by division in early spring.

Onoclea sensibilis

Ononis (name used by Theophrastus). Leguminosae. Canary Islands, Mediterranean, N Africa, Iran. RESTHARROW. Some 75 species, annual or perennial herbs or dwarf shrubs. The leaves are usually trifoliolate, and glandular-hairy. Pea-like flowers are produced in panicles, spikes or racemes in summer. The following species are hardy to −15°C/5°F. Grow in well-drained, neutral to alkaline soils in full sun, giving *O.fruticosa* a sheltered position. Propagate by scarified seed sown in spring or when ripe in autumn. Take care not to damage the developing taproot when pricking out. Alternatively, take softwood cuttings in early summer and root in a closed case with bottom heat. *O.fruticosa* (C and E Spain, central Pyrenees, SE France; dwarf shrub to 1m; flowers pink); *O.natrix* (S and W Europe, N Spain; dwarf shrub, erect, many-branched, to 60cm; flowers yellow, often veined violet or red); *O.rotundifolia* (SE Spain to E Austria and C Italy; perennial dwarf shrub, to 50cm, erect; flowers pink or white).

Onopordum (also spelt *Onopordon*, Latinized form of a Greek name, *onopordon* from *onos*, ass, and *pordos*, flatulence, referring to its effect on these animals when eaten). Compositae. Europe, Mediterranean, W Asia. About 40 species, biennial herbs, usually with spiny-winged stems and pinnately lobed and toothed leaves. Erect, branching stems bear large, thistle flowerheads in summer. Hardy in climate zone 6. Grow in any fertile, well-drained or heavy, preferably slightly alkaline soil, in full sun or light shade. Unless deadheaded, *Onopordum* will self-seed. Propagate by seed in autumn or spring, sown *in situ* or in the cold frame. *O.acanthium* (GIANT THISTLE; SCOTS THISTLE; W Europe to C Asia; stately, giant biennial to 3m, covered in grey-white felt; flowerheads large, scurfy and spiny with white or purple florets).

Onosma (from Greek *onos*, ass, and *osme*, smell: the plant is said to be liked by asses). Boraginaceae. Mediterranean to E Asia. Some 150 species, bristly-hairy biennial and perennial herbs, often woody-based. The tubular to funnel-shaped flowers are carried in terminal cymes in summer. Hardy in climate zone 7. Plant on the rock garden, gravel garden or dry borders in full sun on a light, fast-draining soil. Increase by softwood cuttings in summer, or by seed in autumn.

O.alborosea (SW Asia; perennial to 25cm; leaves grey-green, white-bristly; flowers nodding, white to pink-purple, becoming purple or violet-blue); *O.stellulata* (Balkans; perennial to 25cm, shortly hairy and bristly; flowers light yellow).

Oophytum (from Greek *oon*, egg, *and phyton*, plant, referring to the egg-shaped plant bodies). Aizoaceae. Africa (Cape Province). 2 species, highly succulent perennials they resemble *Conophytum*, forming dense mats of closely packed, spherical to ovoid bodies. The flowers are daisy-like. Cultivate as for *Conophytum*.

O.nanum (bodies 2×0.7cm, minutely papillose; flowers white with reddened tips); *O.oviforme* (bodies to 2×1cm, olive-green, often flame-coloured, glossy-papillose; flowers white below, purple-pink above).

open-centre a form of fruit tree in which the main branches are trained to form a cup or goblet shape, leaving the centre clear. See *delayed open centre*.

open pollination pollination by natural means; as opposed to closely controlled crossing undertaken by a plant breeder.

operculum the lid or cap of a capsule with circumscissile dehiscence.

ophiobolus patch (*Gaeumannomyces graminis*) a disease of lawn grasses, very prevalent in the northwest US, but uncommon in the UK. It especially affects creeping bent grass, *Agrostis stolonifera*; rings of yellow or orange grass occur up to 1m in diameter, with weeds and resistant grasses sometimes creating a green centre. Dark brown hyphae can be found on infected roots. Wet conditions and surface alkalinity favour the disease, and management of these conditions provides more effective control than fungicide application.

Ophiopogon (from Greek *ophis*, serpent, and *pogon*, beard). Liliaceae (Convallariaceae). Asia. 4 species, evergreen, perennial herbs, with clumped, narrowly strap-shaped, grassy leaves. Small and bell-shaped, the flowers are borne in congested spikes in spring and summer. The berry-like fruits are usually dark violet blue and persist late into autumn. Excellent plants for edging, groundcover and containers. *O. planiscapus* 'Nigrescens' is a remarkable plant, striking dramatic contrasts with its purple-black foliage and ink black fruit. All are hardy in climate zone 6, although foliage may be damaged in hard winters. Grow in sun, where soils remain moist during the growing season, or in partial shade, in any moderately fertile, well-drained soil. Propagate by ripe seed sown fresh in a sandy propagating mix in the cold frame, or by division.

Ophiopogon planiscapus 'Nigrescens'

O.intermedius (China; leaves to 60×0.5cm, dark green; flowers white to lilac, numerous, in a loose raceme; 'Argenteomarginatus': leaves edged white); *O.jaburan* (Japan; close to *O.japonicus*, but more robust; roots not tuberous; leaves to 60×0.5cm; flowering stem to 60cm, flowers tinted lilac or white in a dense 7.5–15cm raceme; fruits violet-blue; 'Aureovariegatus': leaves striped yellow; 'Crow's white': leaves variegated white; 'Vittatus': leaves pale green striped and edged creamy white; 'White Dragon': leaves boldly striped white, with almost no green); *O.japonicus* (Japan; roots tuberous; leaves to 40×0.3cm, rather rigid, dark green, somewhat curved; flowering stem 5–10cm, flowers white to light lilac in a loose short raceme; fruit blue, 0.5cm in diameter; 'Compactus': miniature and dense, to 5cm; 'Minor': compact, to 8cm, with curling, black-green leaves; 'Silver Dragon': leaves variegated white); *O.planiscapus* (Japan; roots thickened; leaves to 35×0.3–0.5cm, arching forward, dark green; flowers white or lilac, in 6.5cm racemes; fruit dull blue; 'Nigrescens': to 15cm, with curving, purple-black leaves, lined silver-green at base when young, flowers tinted pink to lilac, fruit black).

Ophrys (from Greek *ophrys*, meaning 'eyebrow' and probably alluding to the hairy lip of some species). Orchidaceae. Europe, N Africa, W Asia. 30 species, perennial herbs with paired, spherical tubers. The leaves are ovate to narrowly lanceolate and produced in winter in low rosettes, giving rise in spring and summer to slender, erect racemes. The flowers are remarkably insect-like, with three, ovate to elliptic sepals and narrow, antenna-like petals. The lip resembles the insect body, entire and rounded to slender and 3-lobed, variously covered in felt- or bristle-like hairs. In many species, the base or centre of the lip is marked with a sparkling to glossy, iridescent patch – the speculum. The following are hardy in climate zone 6, although the southernmost species suffer in prolonged frosts and all dislike cold, wet winter conditions. The familiar and durable *O.apifera* will naturalize in undisturbed grass. *O.fusca* and *O.insectifera* tolerate some shade, growing well in woodland clearings and fringes and especially among pines. The remaining species prefer alpine house conditions, or culture in raised beds where they can be covered in winter. Grow in full or dappled sun in a loamy medium rich in sand and fine leafmould. The addition of limestone chippings to the mix benefits most species. Keep moist when in growth and flower. After flowering, allow the plants to dry out. Repot the tubers in autumn and recommence watering. Increase by division when repotting, and by seed.

O.apifera (BEE ORCHID; W & C Europe; 10–50cm; flowers to 2.5cm across, sepals pink to pale lilac, lip broad, rounded, velvety chestnut brown with 2 yellow-white spots at tip, speculum red-brown with a double, horse-shoe-shaped, yellow-white border); *O.ciliata* (MIRROR OPHRYS; syn. *O.speculum*, *O.vernixia*; Mediterranean, N Africa; 10–50cm; flowers to 2.5cm across, sepals green tinted pink and

Ophrys ciliata

striped purple-brown, lip distinctly 3-lobed, olive green fringed with dense red-brown hairs, speculum covering midlobe, glossy blue thinly edged yellow); *O.fusca* (Mediterranean, SW Romania; 10–40cm; flowers to 2.5cm across, sepals green, lip obovate, shallowly 3-lobed, dark velvety brown with a narrow, yellow-green margin and a 2-lobed, purple-blue speculum); *O.holoserica* (syn. *O.arachnites*, *O.fuciflora*; LATE SPIDER ORCHID W, C & S Europe, Mediterranean, Russia; 10–50cm; flowers to 3cm across; sepals broad, white to pale rose or magenta, lip broad, square, with a forward-pointing, green tip, velvety red-brown to maroon or yellow, speculum H- to W-shaped, mauve to blue bordered white); *O.insectifera* (FLY ORCHID; Europe, including Scandinavia; 10–50 cm; flowers to 2cm across, sepals green, lip tongue-shaped with 2 shorter, distinct lateral lobes, dark, velvety brown, speculum dark blue to violet); *O.lutea* (YELLOW BEE ORCHID; Mediterranean; 8–30cm; flowers to 3cm across, sepals pale green, lip rounded, broad, 3-lobed, bright yellow with a central, velvety brown patch and a 2-lobed, rich blue speculum); *O.scolopax* (WOODCOCK ORCHID; S Europe; 10–40cm; flowers to 2.5cm across, sepals green to white or dusky pink, lip with 2 narrow, projecting, hairy lateral lobes, midlobe rounded with a green, forward-pointing tip, dark velvety brown, speculum H- or ×-shaped, bordered white); *O.tenthredinifera* (SAWFLY ORCHID; Mediterranean; 10–45cm; flowers to 3cm across, sepals broad, pale rose to magenta, lip broad, rounded to square, hairy, green to yellow with a large, brown central patch, a small blue, W-shaped speculum and a tuft of brown hairs at the tip).

Oplismenus (from Greek *hoplismenos*, 'armed', referring to the awned spikelets). Gramineae. Tropics, subtropics. Some 6 species, trailing annual or perennial grasses, with slender, rooting stems and lanceolate, finely tapering leaves. These tender grasses are grown in temperate areas in hanging baskets and as low edging in the warm greenhouse or conservatory. Grow in bright indirect light or filtered light. Water plentifully when in growth. Maintain a minimum temperature of 15°C/60°F. Propagate by division of rooted stems.

O.burmannii (tropics; annual glabrous to pubescent, to 60cm; leaves to 6cm; includes 'Albidus', with white leaves with a pale green median stripe); *O.hirtellus* (tropical America, Africa, Polynesia; evergreen perennial, to 90cm+; leaves to 5cm, somewhat hairy; 'Variegatus': syn. 'Vittatus', leaves striped white, sometimes tinted pink).

Oplopanax (from Greek *hoplon*, a weapon, and *Panax*, referring to the spiny stems). Araliaceae. NE Asia, N America. 3 species, deciduous, densely spiny, coarse suckering shrubs. Small, green-white flowers are carried in umbel-like clusters arranged in narrowly conical panicles. The fruits are red, round drupes. Hardy to at least –15°C/5°F, although the young growth is likely to be cut back by spring frosts. On cool moist soils, it forms tall, impenetrable thickets of bold, rather maple-like foliage, adorned with attractive red fruits. Plant in sun or part-shade. Propagate from seed in autumn, or by suckers or root cuttings. *O.horridus* (DEVIL'S CLUB; central and western N America, especially Pacific coast; leaves to 40cm in diameter, shallowly or deeply 5–13-lobed, serrate; inflorescence to 20cm; fruit to 7mm).

opossum (*Procyon lotor*) a species of small marsupial animals. The adults are 60–110cm long, with a distinctive black 'mask' across the eyes region of the face and a long tail ringed with dark and pale fur. In parts of the US opossums may invade gardens and feed upon a wide range of fruits, vegetables and buds. They are nocturnal, climbing animals, and shooting or trapping are the only effective controls.

opposite used of two organs at the same level, or at the same or parallel nodes, on opposite sides of the axis.

Opuntia (a pre-Linnaean name for some kind of spiny plant associated with the ancient Greek town of Opus, or the surrounding region, known as Eastern or Opuntian Locris, between Thermopylae and Thebes). Cactaceae. Throughout America, from S Canada to Patagonia. PRICKLY PEAR, TUNA. Over 200 species, stoutly succulent trees and shrubs, some low and clumped or spreading. The stem segments are cylindric, club-shaped, subglobose, or more or less flattened. Where present, the leaves are terete or subulate, usually small and short-lived. Most species have spines cushioned by tufted glochids. The flowers are broadly funnel- to bowl-shaped with numerous silky tepals and many stamens. The fruit resembles a more or less prickly pear and is edible in some species.

A few species from the central and eastern United States are frost-hardy, and some from lowland tropical areas require warm temperatures, but the great majority can be grown in a cool greenhouse (minimum temperature 7°C/45°F). Plant in a gritty, sandy, loam-based medium with a pH of 6–7.5. Provide full sun and low humidity. Keep dry from mid-autumn until early spring, except for light misting on warm days in late winter. Avoid contact – the numerous, minutely barbed glochids are easily detached from the plant but difficult to see and remove from the skin, where they will cause irritation. When repotting, hold the plants with a collar of folded newspaper. Increase by rooting stem segments.

Opuntia microdasys

O.brasiliensis (eastern S America; tree-like, to 6–9m or more, with cylindric, unjointed trunk and branches; stem segments flat and somewhat leaf-like, obovate to oblong or lanceolate, to 15 × 6cm, 4–6mm thick; leaves small, subulate, spines to 15mm; flowers pale yellow); *O.cholla* (Baja California; small tree, to 5m; stem segments cylindric, 5–30 × 2–3cm, tuberculate, glochids numerous, yellow, spines 3–25mm, yellow then grey; flowers purple-pink); *O.compressa* (east and central US; shrub, forming clumps or mats 10–30cm × 2m or more, stem segments elliptic to obovate to orbicular, 5–12.5 × 4–10cm, often tinged purple, leaves subulate 4–7mm, spines absent or to 2.5cm; flowers yellow, often with red centre); *O.ficus-indica* (INDIAN FIG, BARBARY FIG, PRICKLY PEAR; Mexico, widely naturalized; large shrub or small tree to 5m with a trunk sometimes 1m in diameter; stem segments obovate to oblong 20–60 × 10–40cm, spines to 2.5cm, white or off-white; flowers yellow; fruit 5–10 × 4–7cm yellow, orange, red or purple in different cultivars); *O.microdasys* (C and N Mexico; shrub, forming thickets 40–60cm; stem segments oblong, obovate or suborbicular, 6–15cm, green, velvety, glochids yellow, or white, spines absent or very short; flowers yellow, outer segments often tinged red); *O.robusta* (central Mexico; shrub to 2m or more; stem segments orbicular or nearly so, massive, to 40cm or more, waxy pale blue, spines to 5cm, white, pale brown or yellow below; flowers yellow); *O.tunicata* (Mexico, SW US, and naturalized in parts of S America; shrub, to 60cm, densely branched; spines to 5cm, yellow or off-white; flowers yellow); *O.verschaffeltii* (Bolivia, N Argentina; low shrub, forming clumps; stem segments usually elongate in cultivars, 6–20cm, with low tubercles, leaves terete, to 3cm, persistent, spines 1–3cm and bristle-like or absent; flowers orange to deep red).

orangery originally, a sheltered south-facing area within a garden set aside for the cultivation of orange trees. During the 16th century, such areas were enclosed by structures for winter protection, and by the 17th century the term was used more to describe the elaborate structures than the garden area. The orangery was the forerunner of the greenhouse and conservatory.

Orbea (from Latin *orbis*, a disc, referring to the thickened annulus in the flowers). Asclepiadaceae. S & E Africa. 20 species, succulents very similar to *Stapelia*, whose cultural requirements they share. *O.variegata* (syn. *Stapelia variegata*; TOAD CACTUS, STARFISH CACTUS; Cape Province; stems to 20cm, sprawling to erect, leafless, with toothed ridges; flowers 5–9cm across, star-shaped, flat, fleshy, wrinkled, malodorous, pale yellow to sulphur, blotched, spotted, lined and ringed maroon to dark brown).

orbicular perfectly circular, or nearly so.

orchard a garden or field area mainly devoted to fruit trees.

orchard house see *fruit house*.

orchids a large group of highly developed flowering plants belonging to the family Orchidaceae. Orchids are not difficult to cultivate if the correct environmental conditions are provided. The growing requirements of orchids are usually designated 'warm', 'intermediate', 'cool', and 'hardy' according to the species' natural habitat.

'Warm-growing' orchids comprise those species, and their hybrids, occurring in tropical regions at sea level or low altitude. They require a minimum greenhouse night temperature of 20°C/68°F throughout the year. Included in this group are *Phalaenopsis, Vanda, Catasetum*, the larger *Angraecum* species, some *Paphiopedilum* and larger *Dendrobium* species.

'Intermediate-growing' orchids come from mid-range altitudes. They require a minimum greenhouse night temperature of 14°C/57°F during winter, slightly higher in summer. A wide range of orchids can be grown under these conditions, including *Cattleya* species and their hybrids, *Laelia, Brassavola, Oncidium, Brassia*, some *Coelogyne* species, and many *Paphiopedilum, Dendrobium* and *Bulbophyllum* species.

'Cool-growing' orchids originate at mid-range or high altitudes in the tropics. They require a minimum greenhouse night temperature of 10°C/50°F during winter, slightly higher in summer. This group includes many kinds within *Cymbidium, Paphiopedilum, Coelogyne, Masdevallia, Odontoglossum, Lycaste, Laelia, Oncidium*, and *Maxillaria*, and many epiphytes from eastern Africa.

Most orchids are grown in containers placed on greenhouse staging, or suspended from the walls on a special framework. Clay or plastic pots are mostly used, but wire or wooden slatted baskets are suitable for some kinds with pendulous growth or inflorescences. Many epiphytes also grow well mounted on slabs of cork-oak bark or firmly attached to chunks of tree-fern fibre. At first, they need to be tied to the surface, with coarse nylon thread, plastic tape or copper wire, but new roots will soon grow and attach the plants to the mount.

Composts and potting Orchid plants need repotting when the growing compost is exhausted, or as they begin to grow over the edge of the pot, or where vigorous root growth causes the whole plant to rise out of the pot. However, the presence of adventitious roots is quite normal in many genera, particularly *Phalaenopsis* and the *Laelia* alliance, and such overgrown plants are best left undisturbed for many seasons. Some orchids resent disturbance and take a long time to re-establish after repotting. For some of the larger orchids in the *Vanda* alliance, including *Angraecum* species, repotting is rarely necessary where plants are well-managed. Large plants of most genera will eventually need rejuvenating, but plants grown in rock wool require repotting much less frequently.

The growing medium should be a well aerated, moisture-retentive, a source of nutrients, and, since it is replaced only after long intervals, must decompose slowly. Suitable growing composts are made from natural materials, such as conifer bark, coarse grit, fibrous peat, or chopped dried sphagnum mixed with inert materials like perlite, perlag, charcoal and sometimes polystyrene fragments. These may have bone meal, dried blood or hoof and horn meal base dressing added, or, alternatively, have liquid fertilizer applied on a regular basis.

A mix containing large particles will have large air spaces and need watering more frequently than a close mix. There are two successful basic compost formulae. One is for terrestrial orchids and comprises 3 parts fibrous peat: 2 parts coarse perlite: 2 parts coarse grit: 1 part horticultural grade charcoal; the other is for epiphytic orchids and comprises 3 parts washed, medium grade bark chips: 1 part coarse perlag: 1 part horticultural grade charcoal: 1 part fibrous peat or dried leaves or chopped sphagnum. Rock wool mixed with perlite provides a substrate very similar to mossy tree branches in tropical forests. A carefully controlled system of watering and feeding is necessary with this medium, so it is inadvisable to use both rock wool and other composts in a mixed collection. Pine bark requires soaking and all other ingredients, including perlite and rock wool, should be made moist before use.

Choose a pot size to allow space for at least one year's growth – many orchids do best in smaller pots. To ensure good drainage, fill one third of the container with polystyrene fragments, vertically packed pot crocks, or large stones (these are particularly useful in plastic pots to provide stability).

When potting, the base of the pseudobulbs, or other part from which roots will emerge, should be just below the level of the pot rim and plants may need staking until they become established. After potting, the compost should be thoroughly watered for two or three days; newly potted plants are then kept dry for two or three weeks, during which period they must be kept in a humid place and misted over frequently.

Watering Orchids variously adapt to a weekly or daily watering or misting regime; they need to dry out periodically, some growing best with a long moist season followed by a dry period of several months. New growers need to study a plant's natural habitat and try to imitate it in greenhouse management. Rainwater is the best choice, preferably stored in a covered tank within the greenhouse.

Once established and with new root growth visible, a regular watering regime should be followed, with plants being heavily and thoroughly watered on each occasion. The need for water can be assessed by checking the weight of the pots; as a general rule, this should be done once or twice a week. Daily watering will be necessary in summer, whereas in winter once every two to four weeks may be sufficient. Avoid moisture lodging in leaf axils or the apex of new shoots overnight, since this can encourage waterborne disease. Misting early in the day and repeatedly during periods of sunshine helps ensure high humidity and lowers leaf temperature; however do not wet leaves or flowers in strong sunshine.

Feeding It is beneficial to feed orchid plants regularly during their active growth season, and this is essential where they are potted in rock wool, coarse bark or perlite only. Dilute liquid fertilizer is most convenient; proprietary products are suitable but should be diluted to a quarter or half the strength recommended for other pot plants.

Resting Many terrestrial orchids have pseudobulbs or tubers by which means the plants survive a long dormant season that may be too hot or too cold for active growth. Similarly, many epiphytic orchids are structurally adapted with pseudobulbs or succulent rigid leaves to withstand drought. The nature of these essential resting periods must be understood so that appropriate cultural treatment can be given, especially as regards restricted watering. For example, in the Indian *Dendrobium* species, dormancy may last

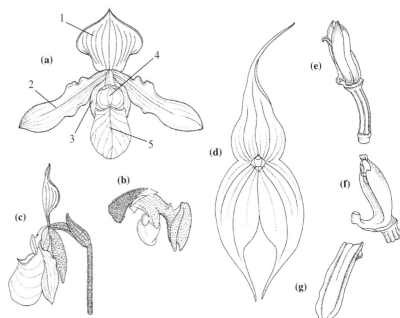

The orchid flower
(a) Paphiopedilum
(1) dorsal sepal (2) petal
(3) synsepalum formed
by fusion of lateral
sepals(4) staminode
(5) slipper-like lip
(b) dissection showing
apex of ovary, pollinia
and staminode
(c) side view
(d) Masdevallia – *flower*
with fused sepals
(e) dissection with
sepals removed
revealing small petals,
lip and ovary
(f) column (g) lip

only a few weeks, while in many high altitude Mexican orchids, it continues over several winter months, during which period the plants often flower.

Flowering and staking   Most orchids flower once a year but *Phalaenopsis* hybrids flower twice a year and some hybrids, such as *Odontoglossum* species, will flower at ten-month intervals. Many genera produce long inflorescences bearing numerous flowers, but most plants of *Paphiopedilum* and *Lycaste* species and hybrids bear flowers singly.

For attractive presentation, nearly all inflorescences need training; this is best done by placing a cane or wire support adjacent to each flower spike soon after it appears and tying the spike in when it reaches about 10–15cm in height.

Propagation   Orchids can be increased by division of mature plants, by activating dormant buds on 'backbulbs', by taking stem or inflorescence cuttings, and from seeds.

DIVISIONS AND 'BACKBULBS'   Large plants can be divided easily into two or more portions, and many may readily fall apart when removed from their pots. Other plants have pseudobulbs joined to each other by a tough rhizome, which needs to be cut with a sterilized knife or secateurs. For most orchids, two or three growths should be retained in each division; *Cattleya* and its allies need at least one leafy pseudobulb on the back growth, with a prominent dormant bud at its base. For *Lycaste* and members of the *Odontoglossum* alliance, two or three pseudobulbs are usually separated as a clump. *Cymbidium* species are the easiest to propagate from old leafless pseudobulbs, which, after removal of loose sheaths and drying of the cut base, can be immersed up to one third their height in sharp sand, and kept moist and cool. Within two or three months, a new shoot will appear above the surface; after a similar period, the shoot will bear its own roots. New plants propagated by this means should reach flowering size in two or three years.

CUTTINGS   Pseudobulbs, stems and inflorescence stalks of some orchids provide suitable cutting material. Each section must contain one or more dormant buds; when detached from the rest of the plant and laid on a bed of damp moss or inserted in a pan of moist grit, the buds will produce new plantlets after a few months in humid conditions. Suitable for this method are the cane-like stems of *Epidendrum* and *Dendrobium*, and the inflorescence stalks of *Phalaenopsis, Phaius* and *Calanthe*. On many pseudobulbs, such as those of *Pleione*, dormant buds may spontaneously develop plantlets, sometimes known as 'keikis'.

SEEDS   Orchid seeds are minute and the food reserves in the embryo are inadequate for the early development of new plants without some supplement. In nature, most orchid seeds develop a symbiotic partnership with a soil- or bark-inhabiting fungus: on invading the seed, the fungus is digested and nutrients obtained from it. This process can be simulated most successfully in a specialised laboratory by sowing sterilized seeds with a fungus culture on to agar, with the addition of porridge oats to sustain the fungus. An

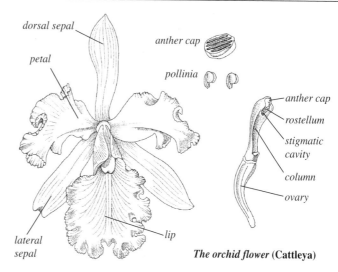

The orchid flower (**Cattleya**)

alternative method is to use a medium containing all that the germinating seed requires in terms of nutrients and thus dispense with the intermediary fungus. All these techniques must be carried out in sterile conditions, which the amateur can achieve by using a domestic pressure cooker to prepare glassware and media; a large polythene bag can be used to provide isolation for the sowing operation, which should be carried out as speedily as possible.

Conical flasks or sterile bottles containing the newly sown seeds are kept under controlled conditions while the embryo grows out to form a rounded protocorm, covered in rhizoids, and then a small plantlet. In some instances, the containers need to be kept in the dark for the first few months, but epiphytic orchids develop green protocorms almost immediately and are kept under artificial light for 12-16 hours per day. After a few months, the plantlets must be transferred to freshly prepared medium in a new container, a process that should again be carried out under sterile conditions. Six to twelve months after sowing, the plantlets are large enough to be removed from the flask, washed carefully and potted up in a fine compost mix. Extra warmth and humidity should be provided for the first few weeks. Species and hybrids can grow to flowering size in 18-20 months but four to six years is average for most orchids.

MERISTEM PROPAGATION   Techniques for the culture and multiplication of the apical meristem of a young shoot, and the various forms of tissue culture, are valuable for rapid multiplication. The processes are similar to those used for propagation from seed, but since the starting material is already mature, the protocorms develop into new, flowering-size plants far more quickly. Sterilized equipment and plant material is used, and a warm growth chamber is necessary. After eventual transfer to an orchid compost, the propagules grow away rapidly.

USE OF COLCHICINE   The poisonous alkaloid colchicine can be used to change the chromosome constituents in cells of orchid protocorms. The concen-

orchids

Pests
aphids
cockroaches
red spider mites
scale insects
vine weevil

Diseases
leaf spot
virus

tration and duration of the chemical application to plantlets must be carefully and accurately monitored or monstrosities may arise. There has been success with many of the free-flowering *Cymbidium* hybrids, and some of the *Paphiopedilum* protocorms. The resultant adult plant will usually have larger leaves and flowers of superior shape and there may be improvements in fertility. However, plants reach flowering size more slowly and may have fewer flowers.

Registration and nomenclature The Royal Horticultural Society is the International Registration Authority for Orchid Hybrids, and publishes bi-monthly lists of new registrations in its journal *The Orchid Review*. Collated lists of registrations for a given period of years are published occasionally under the title *Sander's List of Orchid Hybrids*. The complete International Register now lists over 100,000 hybrids.

Orchid hybrids are named at two or three levels, for example in the hybrid *Cattleya* Bow Bells 'White Wings', the name of the genus is *Cattleya*, the name of the grex is *Cattleya* Bow Bells and the name of the cultivar is *Cattleya* Bow Bells 'White Wings'.

The term grex is unique to orchids and is the equivalent to the cultivar-group of the International Code of Nomenclature for Cultivated Plants – 1995. It denotes a group of individual plants derived from an artificial hybrid, and the grex name is applied to all progeny directly raised from the same two named parent taxa, whenever that cross is made.

A cultivar in orchids is a clone – a genetically similar assemblage of individuals derived originally from a single seedling following vegetative propagation. A grex epithet should always be used in conjunction with a generic name, and both a generic name and grex epithet should accompany the epithet of any cultivar.

Hardy species A number of orchids from the mountains of sub-tropical regions and temperate areas are amenable to cultivation. They may require frost-free conditions or a completely dry period during dormancy, and are most easily accommodated in a cold frame or unheated greenhouse. Common examples are the species and hybrids of *Pleione*, which grow best in shallow pans of fast-draining but moisture-retentive compost; keep under warm, humid conditions in summer, dried off in the autumn, and very cool during the winter. European orchids, such as *Orchis*, *Ophrys* and *Dactylorhiza*, can also be grown easily in pots in an alpine house, as can some species of *Cypripedium*, and *Bletilla*.

Some terrestrial genera are deciduous, coming into growth in spring and flowering during late spring or midsummer. They may be grown outdoors in the garden and include some species of the lady's slipper genus, *Cypripedium*, *Dactylorhiza* and *Orchis* .

Wild orchid plants are specially protected in most countries and should not be collected without permission. CITES controls the import and export of orchid species. When purchasing terrestrial orchids, it is important to be sure that stocks have originated as nursery-propagated plants.

Orchids for the windowsill Many of the tropical orchids which in the past have been cosseted in greenhouses, will do equally well on a windowsill if a few basic rules are observed. Of these, the most important are ensuring that there is enough light for the plants, without too much heat, and that there is adequate humidity in the immediate surroundings.

During the winter months, a south-facing windowsill or a table placed within a bay window, is a good situation; in summer, an east-facing windowsill which receives morning sunshine is suitable. A deep windowsill is ideal, fitted with a polypropylene tray containing about 2–3cm of a clean, moisture-retentive material; this should be kept moist but not waterlogged in order to maintain a humid atmosphere around the pots. Orchids with 'intermediate' or 'warm' temperature requirements do best in centrally heated homes. Many *Phalaenopsis* and *Paphiopedilum* species and hybrids give pleasure with their long-lasting flowers, and a wide variety of species can be grown successfully with patience and understanding.

Orchis (from Greek *orchis*, testicle, referring to the shape of the tubers). Orchidaceae. Temperate N Hemisphere. 35 species, herbaceous perennials with spherical tubers and lanceolate leaves. Produced in spring and summer in erect spikes, the flowers are spurred with a hood of ovate tepals and a showy, entire to 3-lobed lip. *O.mascula*, *O.morio* and *O.purpurea* are hardy in zone 6; *O.laxiflora* in zones 7–8; *O.papilionacea* in zones 8–9. All may be grown in the alpine house, the last especially. Grow in a circumneutral, loamy mix rich in sand and leafmould. *O.mascula* will grow in light shade and is suited to the woodland garden; the remainder prefer full or dappled sun. Keep moist when in growth and flower; rather dry when at rest (usually high summer to autumn). *O.laxiflora* needs persistently damp condtions. Increase by division before growth recommences, or by seed.

O.laxiflora (Continental Europe, Mediterranean, Channel Is; 20–100cm; raceme loose, long; flowers to 2.5cm across, tepals curved upwards, lilac-mauve to rose or rich purple-red, lip fan-shaped, very shallowly 3-lobed, rose to purple-red, often with a white blotch); *O.mascula* (EARLY PURPLE ORCHID; Europe; 10–50cm; leaves usually spotted maroon; raceme dense, narrow; flowers to 2cm across, tepals curved upwards, rosy mauve to purple-red, lip fan-shaped, shallowly 3-lobed, white edged and spotted pink to purple-red); *O.morio* (GREEN-WINGED ORCHID; Europe, Mediterranean; 7–30cm; raceme rather loose, cylindrical; flowers to 1.8cm across, tepals pointing forward in a hood, white to pale rose or purple-red with distinct green-veins, lip distinctly 3-lobed, rose to purple-red with midlobe white blotched purple-pink; white- yellow- and deep purple-red-flowered forms occur); *O.papilionacea* (BUTTERFLY ORCHID; Mediterranean; 15–40cm; raceme fairly loose, somewhat pyramidal; flowers to 3cm across, tepals forward-pointing, hooded, purple-red with darker veins, lip broad, fan-shaped, rose to purple-red with darker spots and lines); *O.purpurea* (LADY ORCHID; Europe, Mediterranean; 30–80cm; raceme dense, somewhat conical; flowers to 2.5cm across, tepals forward-pointing, hooded, purple-brown, lip white spotted purple-red, deeply 3-lobed, the lateral lobes narrow and outspread, resembling arms, the midlobe itself deeply cleft into two parts, 'legs').

ordo, order the principal category of taxa intermediate in rank between class and family, for example Commelinales, the order containing the families Rapateaceae, Xyridaceae, Mayacaceae and Commelinaceae. Until the present century, botanists used *ordo* to denote the rank of family and *cohors* for the modern order.

organic gardening a system of cultivation that makes minimal use of manufactured chemical substances, placing emphasis on the use of manures and fertilizers of plant and animal origin to establish and maintain soil fertility; control of pests, diseases and weeds is also pursued chiefly by non-chemical methods, such as crop rotation, hygiene, ground mulching, hand-picking and trapping, cultivar resistance, and the use of natural predators and parasites. Recycling of organic residues through composting and green manuring are other important features of the system. Non-cultivation or organic-surface cultivation, also known as the no-dig system, entails placing organic residues on the soil surface, from where they are drawn into the top profile by the action of earthworms and microorganisms.

These approaches stem from a broader philosophy which advocates a holistic approach to gardening, recognizing the interdependence of life forms. Considerations of conserving wildlife and natural resources, and the avoidance of pollution, are therefore central to the concept of organic gardening.

The principles of organic agriculture were promoted by Rudolf Steiner in the 1920s, and his influence persists today in parts of Germany and Switzerland. The Rodale Organization was founded in the US in the early 1940s, and in 1954 Lawrence Hills established the Henry Doubleday Research Association (HDRA) in the UK. The HDRA is the world's largest specialist organic-gardening organization for amateurs.

The pursuit of organic gardening requires understanding of garden ecology, dedication, and at least the same level of input as less stringent regimes. The incidence of pests, diseases and other disorders is less likely where plants' requirements for steady growth are met; this applies to the choice of site, soil preparation and maintenance, nutrition and watering. In the garden situation, considerations of maximum yield and perfect quality are usually less important than in commercial growing, and the principles of organic gardening are both attainable and commendable. See *biodynamic gardening*.

organochlorines, organophosphates see *insecticides*.

oriental greens a collective name for a number of fast-growing brassicas, cultivated for eating cooked or raw. CHINESE CABBAGE or CHINESE LEAVES (*Brassica rapa* Pekinensis Group) and CHINESE

WHITE CABBAGE or PAK-CHOI (*Brassica rapa* Chinensis Group) may also be considered under this heading (see *cabbage*).

LOOSE HEADED CHINESE CABBAGE (*Brassica rapa* Pekinensis Group) is a faster-growing type of Chinese cabbage that produces open heads of light green leaves which may be serrated or wavy, some cultivars having very loose tops with pale centres. It is grown in the same way as Chinese cabbage for harvesting as mature heads, or from spring sowings as a cut-and-come-again crop, first harvested five weeks after sowing.

KOMATSUNA or SPINACH MUSTARD (*Brassica rapa* Perviridis Group) is closely related to turnip, producing large leaves on vigorous plants. The various forms are tolerant of drought and a wide temperature range; they will overwinter outdoors but yield better quality where protected. Sow midsummer for autumn/winter harvest outdoors, or in late summer where they are to be overwintered under cover; space seedlings 5–30cm apart depending on required stage of maturity. Cut-and-come-again supplies can be obtained from successional spring/early summer sowings outdoors.

MIZUNA GREENS (*Brassica rapa* var. *nipposinica*) produces highly decorative rosette heads of much-divided, glossy green leaves. Young leaves are suitable for salad use, the older ones as a cooked vegetable. Grow as komatsuna.

MUSTARD GREENS (*Brassica juncea*) are a robust, hardy group of brassicas producing coarse leaves, some cultivars being coloured purple and most spice-flavoured. Grow as Komatsuna, spacing on average 30cm apart.

CHINESE BROCCOLI or CHINESE KALE (*Brassica oleracea* Alboglabra Group) is an annual grown for its succulent, well-flavoured flowering shoots. It crops best from midsummer sowings in rows 10cm apart; thin to 10–15cm spacings for harvesting whole young plants, or transplant 25–38cm apart for large plants to be cut continually. The main flowering shoot should be harvested first and sideshoots as they develop. See *cabbage*.

Origanum (from Greek *oros*, mountain, and *ganos*, joy). Labiatae. Mediterranean to E Asia. MARJORAM, OREGANO. Some 20 species, subshrubs or perennial herbs, often pubescent, generally aromatic. Small, 2-lipped flowers appear in summer, whorled amid conspicuous, overlapping bracts in terminal spikes. Plant in full sun on lean, fast-draining soils. *O.rotundifolium* favours lime-free soil; other species prefer slightly alkaline conditions. Trim back after flowering. Mulch in cold winter areas in zones 7 and under. Harvest for freezing or drying just before flower buds open. Propagate from seed sown under glass in early spring, or later in the open ground. Alternatively, take softwood cuttings in early summer; pot on individually, overwintering in frost-free conditions for planting out the following spring. Winter supplies for the kitchen may be taken as cuttings in summer and overwintered at a minimum of 10°C/50°F.

oriental greens

Pests
aphids
cabbage caterpillars
cabbage root fly
whiteflies

Disease
clubroot

O.amanum (Mediterranean, Turkey; stem to 20cm, hirsute to scabrous; leaves to 1.9cm, cordate; flowers pink, bracts purple); *O.dictamnus* (DITTANY OF CRETE, HOP MAJORAM; Crete; dwarf shrub to 30cm; leaves 1.3–2.5cm, woolly-white, ovate to round; flowers amid hop-like, rose-purple bracts); *O.laevigatum* (Turkey, Cyprus; stem to 70cm, glabrous; leaves 3cm, ovate to elliptic, smooth; flowers purple; 'Hopley's': to 75cm, with large pink flowers and bracts); *O.majorana* (SWEET MARJORAM, KNOTTED MARJORAM; originally Mediterranean and Turkey, now widespread escape in Europe; annual, biennial or perennial herb to 60cm, stem glabrous or tomentose; leaves to 2cm, ovate, grey, downy; flowers white to mauve or pink; bracts grey-green); *O.onites* (POT MARJORAM; Mediterranean; mound-forming shrublet to 60cm; leaves to 2cm, ovate, rounded to cordate at base, bright green, aromatic; flowers mauve or white; 'Aureum': leaves gold); *O.rotundifolium* (Armenia, Georgia, Turkey; subshrub to 30cm, spreading by rhizomes; leaves to 2.5cm, rounded or cordate, stem-clasping, blue-grey; inflorescence nodding, hop-like, bracts pale green tinged purple-pink, flowers white or pale pink); *O.vulgare* (WILD MARJORAM, OREGANO, POT MARJORAM; Europe; rhizomatous, woody, branched perennial herb to 90cm, strongly aromatic; leaves to 4cm, round to ovate, entire or slightly toothed; bracts purple or green, flowers purple; var. *album*: to 25cm, bushy, with light green leaves and white flowers; 'Aureum': to 30cm, spreading, with small, gold leaves and lavender flowers; 'Compactum': to 15cm, compact and cushion-forming, with small, round, dark green leaves; 'Gold Tip': leaves tipped yellow; 'Nanum': dwarf, to 20cm, with purple flowers; 'Roseum': flowers pink; 'Thumble's Variety': to 35cm, with large, pale yellow leaves, later turning yellow-green, and soft white flowers).

Ornithogalum (name used by Dioscorides, from Greek *ornis*, bird, and *gala*, milk). Liliaceae (Hyacinthaceae). South Africa, Mediterranean. Some 80 species, bulbous perennial herbs with linear to lanceolate or obovate leaves, sometimes with a silver-white median stripe above. The flowers are 6-parted and starry, each with a prominent ovary; they are produced in spring and summer in long-stalked, pyramidal to subcylindric racemes or corymbs. The hardiest species, such as *O.umbellatum*, withstand temperatures to between –15 and –20°C; *O.montanum* and *O.nutans* are almost as tolerant, surviving temperatures at the warmer end of this range. *O.narbonense* and *O.oligophyllum* are hardy where temperatures seldom fall below –10°C/14°F. *O.arabicum* suits warm sunny borders, or, where temperatures fall much below freezing, containers of medium-fertility, loam-based mix in the cold greenhouse. The South African species are generally frost-tender and are planted temporarily in summer borders of cool temperate gardens or grown in pots in the cool greenhouse. They include a number of species which provide long-lived cut flowers, notably *O.thyrsoides*. Grow in any moderately fertile, well-drained soil in sun. Propagate by seed sown when ripe in the cold frame. Increase also by offsets when dormant.

O.arabicum (Mediterranean; scape 30–80cm, raceme cylindric to rounded or conical, flowers fragrant, 3–5cm across, white or cream with a black or purple-black ovary; very free-flowering, large-flowered forms are sometimes called *O.corymbosum*); *O.dubium* (S Africa; scape to 30cm, raceme broadly and loosely pyramidal, flowers 2–4cm across, orange, red, yellow or white, often stained brown at centre); *O.montanum* (Europe, W Asia; scape 10–60cm, raceme broadly cylindric, flowers drooping, 2–5cm across, pearly translucent white striped green); *O.narbonense* (Mediterranean, Caucasus, Near East; scape to 90cm, raceme loose, many-flowered, flowers to 5cm across, milky white striped green); *O.nutans* (Europe, W Asia, naturalized E US; scape to 60cm, raceme cylindric, one-sided, flowers nodding, 3–5cm across, translucent white striped green); *O.oligophyllum* (syn. *O.balansae*; Balkans, Turkey Georgia; scape to 15cm, raceme broadly pyramidal to rounded, flowers 2–3cm across, white to ivory edged pure white, striped yellow-green on exterior); *O.saundersiae* (GIANT CHINCHER-INCHEE; S Africa; scape 30–100cm, raceme pyramidal, flowers to 3cm across, white to cream, ovary dark green or black); *O.thyrsoides* (CHINCHERINCHEE, WONDER FLOWER; S Africa; scape to 50cm, inflorescence broadly pyramidal, flowers 2–4cm across, white to ivory, stained green at centre; a long-lasting cut flower, it includes cultivars with snow white, topaz and golden flowers); *O.umbellatum* (STAR OF BETHLEHEM; Europe, N Africa, Middle East; scape 10–30cm, raceme broadly pyramidal to rounded, flowers 3–4cm across, shiny white striped green on exterior).

Orontium (from *orontion*, the Greek name for an aquatic growing in the Syrian river Orontes). Araceae. E US. GOLDEN CLUB. 1 species, *O.aquaticum*, an aquatic, perennial herb with thick rhizomes rooting into the muddy bed, pond margins or (in cultivation) a water lily basket. Floating or emergent, the leaves are to 25cm, oblong to narrowly elliptic with glaucous silvering above and often purple-tinted beneath. Breaking the water's surface in spring and summer, the inflorescence consists of a white-stalked spadix to 18cm long; this is bright yellow and resembles a slender, upcurved club. Hardy to about –15°C/5°F. Plant rhizomes in fertile, loamy soil and submerge in still water in full sun. Propagate by division in spring.

Orostachys (from Greek *oros*, mountain, and *stachys*, spike). Crassulaceae. N Asia to Europe. 10 species, biennial succulent herbs with fleshy, spine-tipped leaves in a dense, rounded rosette. Small, starry flowers are carried in a dense terminal raceme in the summer of the second year. Grow in a free-draining, gritty medium in full sun, with good ventilation and a minimum temperature of 8°C/47°F. Keep near-dry in cool weather. Plants usually die after flowering but set seed freely and are easily raised by this means.

O.chanetii (China; similar to *O.fimbriata* except leaves linear with a small, hard spine; flowers white

Ornithogalum dubium

Ornithogalum thyrsoides

to pink); *O.fimbriata* (Tibet, Mongolia, Japan, China; to 15cm; leaves 2.5cm, oblong, tip with a long spine; flowers tinged red).

Oroya

Oroya (after Oroya, Peru). Cactaceae. Peru. 2 species, low-growing cacti with squatly globose to shortly cylindric, many-ribbed stems bearing elongate areoles and pectinate spines. Shortly funnel- or bell-shaped, the flowers open in spring and summer. Grow in a cool greenhouse (minimum temperature 2–7°C/36–45°F) in a gritty, acid to neutral, free-draining medium. Position in full sun with low humidity. Keep dry from mid-autumn until early spring, except for light misting on warm days in late winter. Increase by seed or offsets. *O.peruviana* (to 40 × 20cm; spines to 2cm; flowers to 3cm, pale to deep rosy red usually with a yellow centre).

Orthrosanthus (from Greek *orthros*, morning, and *anthos*, flower; the flower opens early in the morning). Iridaceae. Tropical America, Australia. 7 species, rhizomatous, evergreen perennial herbs with basal clumps and fans of narrow, rigid or grassy leaves. The *Iris*-like flowers are short-lived but produced in succession in a slender-stalked panicle in summer. Grow in the border or in pots in the cool greenhouse or conservatory (minimum temperature 7°C/45°F). Plant in a humus-rich and well-drained soil in sun. Water carefully and moderately when in growth; keep almost dry at other times. Propagate by careful division of established plants or by seed, which germinates readily, although seedlings develop slowly.

O.chimboracensis (Mexico to Peru; flowers lavender-blue, to 4cm in diameter); *O.violaceus* (China; flowers violet, to 6cm in diameter).

Osmanthus (from Greek *osme*, fragrance, and *anthos*, flower). Oleaceae. N America, Asia, Pacific. DEVIL-WOOD, SWEET OLIVE, CHINESE HOLLY. 30 species, evergreen shrubs or small trees with dark green, smooth and leathery leaves, the margins entire to sharply toothed. Clustered in the leaf axils or, rarely, in terminal panicles, the flowers are small, highly fragrant and bell-shaped to tubular with four lobes. The fruit is an oval, blue-black to purple drupe. All species here will grow in sun or light shade on a range of fertile, well-drained soils, preferably neutral to acid, although some lime is tolerated and *O. × burkwoodii*, *O.delavayi* and *O.yunnanensis* thrive in chalky conditions. *O.decorus* and *O.heterophyllus* are very hardy. The remaining species will grow in climate zone 7, but need a site sheltered from freezing winds. In regions of prolonged winter frost, *O.fragrans* will need the protection of a cool, airy greenhouse in winter (minimum 4°C/39°F). Remove straggly or winter-damaged growth. Propagate in late summer by semi-ripened cuttings in a propagating case with bottom heat or root ripe growths in autumn in a cold frame. Alternatively, increase by layering in autumn or spring, or by seed sown in a cold frame on ripening.

O.armatus (W China; shrub, 2–4m; leaves 7–14cm, oblong to ovate, strongly toothed, spiny, sometimes entire; flowers off-white; fruit deep violet); *O. × burkwoodii* (syn. × *Osmarea burkwoodii*; *O.delavayi* × *O.decorus*; compact shrub to 2m; leaves 2–4cm, ovate to elliptic, serrate; flowers white, fragrant); *O.decorus* (Caucasus, Lazistan; broad shrub, to 3m; leaves to 12cm, entire, oblong apex acuminate, yellow beneath; flowers white; fruit blue-black; 'Baki Kasapligil': slow-growing, hardy, with narrow leaves); *O.delavayi* (W China; stocky shrub to 2m; leaves to 3cm, ovate, finely toothed; flowers white, sweetly fragrant); *O. × fortunei* (*O.fragrans* × *O.heterophyllus*; expansive shrub, to 3m; leaves to 10cm, oval, acuminate, teeth large, triangular, or leaves entire; flowers white, fragrant); *O.fragrans* (FRAGRANT OLIVE, SWEET TEA; Himalaya, Japan, China; shrub or small tree to 12m; leaves to 10cm, oblong to lanceolate, entire or finely toothed; flowers white, highly fragrant; f. *aurantiacus*: flowers orange); *O.heterophyllus* (syn. *O.ilicifolius*; HOLLY OLIVE, CHINESE HOLLY, FALSE HOLLY; Japan, Taiwan; erect dense shrub, to 5m; leaves to 6cm, elliptic to oblong, entire or with large lobe-like teeth – both forms found together on same branch; flowers white, fragrant; fruit purple-blue; 'Goshiki': leaves prickly, cream- and bronze-variegated, young growth rose pink; 'Latifolius Variegatus': leaves large, variegated cream; 'Myrtifolius': leaves ovate, entire; 'Purpureus': new growth purple-red); *O.serrulatus* (W China; shrub, to 3m; leaves resembling those of *O.yunnanensis*, but broader with finer, forward-pointing teeth or sometimes entire, flowers white); *O.yunnanensis* (syn. *O.forrestii*; W China; similar to *O.serrulatus* but leaves with longer, more spiny teeth; tall shrub to 5m; leaves to 20cm, ovate to lanceolate, tip finely tapering, spotted black beneath with to 30 sharp teeth on either side, or entire; flowers waxy, off-white to light yellow, very fragrant; fruit dark purple, pruinose).

Osmunda (opinions vary as to the name derivation – possibly for the Nordic god Thor, also called Osmunder; or for Osmundus – Asmund – an 11th-century Scandinavian writer of runes and Christian apologist; or from the Latin *os*, mouth, and *mundare*, to clean). Osmundaceae. Asia, Americas. FLOWERING FERN. About 12 species, large terrestrial ferns with profuse, matted and fibrous roots and more or less erect, bipinnate or bipinnatifid fronds in large clumps or crowns. The spores are produced on specialized, narrow pinnules, either separately on entirely fertile fronds or in the middle or at the ends of more regular (i.e. 'sterile' – type) fronds. The following species are hardy in climate zone 5 and will grow in full sun (*O.regalis*) or dappled to deep shade (*O.cinnamonea*, *O. claytoniana*). They need a cool, acid mix rich in leafmould. They must never be dry; indeed *O.regalis* will grow at the water's edge and make islands of its own dense root growth in shallow pools and streams. Propagate by spores or by division in spring.

O.cinnamonea (CINNAMON FERN, FIDDLEHEADS, BUCKHORN; N and S America, West Indies, E Asia; whole plant densely rusty-hairy at first; sterile fronds

60–90cm, pinnate, pinnae 8–10 × 2–2.5cm, ligulate to lanceolate, cut almost to rachis; fertile fronds distinct, much smaller, becoming cinnamon brown, pinnae lanceolate); *O.claytoniana* (INTERRUPTED FERN; America, Himalaya, China; foliage pink-downy at first; fronds 30-60cm, pinnate to pinnatifid, sterile pinnae 10-15 × 2.5cm, lanceolate, cut almost to rachis, fertile pinnae borne at centre of regular fronds, far smaller with dense cylindric pinnules); *O.regalis* (ROYAL FERN, FLOWERING FERN; cosmopolitan; rhizome becoming erect and mound-like, to 1m tall, massive, with tough fibrous roots; fronds 60–180cm, bipinnate, sterile pinnae 15–30cm, pinnules 2.5–5 × 1–2cm, oblong, blunt, minutely serrate, fertile pinnae cylindric, forming large rusty panicle at frond apex; 'Crispa': pinnules crisped; Cristata Group: pinnules finely crested; 'Purpurascens': growth purple when young rachis, purple throughout season).

osmunda fibre dried and coarsely chopped roots of the royal fern, *Osmunda regalis*, formerly an ingredient of potting media for orchids. In the interests of conserving the royal fern, it is now no longer recommended for this use, and there are satisfactory alternatives. See *orchids*.

Osteomeles (from Greek *osteon*, bone, and *melon*, apple, referring to the fruit). Rosaceae. China to Hawaii and New Zealand. 2 species, evergreen shrubs or trees with small, finely pinnate leaves and white, 5-petalled flowers in small, terminal corymbs in late spring and early summer. These are followed by small, red to blue-black pomes. Hardy to –10°C/14°F, but best given the shelter of a south-facing wall. Grow in sun on a well-drained soil. Propagate by semi-ripe cuttings, or by layering. Sow stratified seed in spring. *O.schwerinae* (SW China; to 3m; branchlets slender, pendulous; leaves to 7cm, grey-downy, leaflets elliptic to obovate-oblong, to 1.2cm, cuspidate; flowers white; var. *microphylla*: leaves smaller and finer, less woolly).

Osteospermum (from Greek *osteon*, bone, and Latin *spermum*, seed). Compositae. Africa (those listed below are from South Africa). About 70 species of shrubs, subshrubs or annual to perennial herbs with entire, toothed, pinnatifid or pinnatisect leaves and large, daisy-like flowerheads. Brilliantly coloured perennials grown in containers and as half-hardy summer bedding. Given a warm, sunny position with perfect drainage and shelter from cold winds, the following may survive where temperatures fall to between –5 and –10°C/23–14°F. Alternatively, take cuttings in summer and overwinter young plants under glass prior to planting out the following spring.
  O.barberiae (spreading, rhizomatous perennial to 50cm, glandular-pubescent; ray florets magenta above, usually light orange-brown beneath, disc florets deep purple or yellow; 'Compactum': to 10cm, with deep pink flowerheads with a dark purple reverse); *O.ecklonis* (robust subshrub to 1m; ray florets white above, indigo often with white margin beneath, disc

Osteospermum 'Whirligig'

florets bright blue; 'Deep Pink Form': to 30cm, with numerous, narrow, dark pink florets; 'Giant Mixed'; to 35cm, with flowers in cream, orange and salmon pastels; 'Starshine': to 75cm, with snow-white flowers with blue eyes; 'Weetwood': to 25cm, with white flowers, olive green below); *O.jucundum* (perennial herb, to 50cm; ray florets red on both surfaces, disc florets black-purple at apex). *Cultivars*: shorter cultivars (to 25cm) include the white- and pink-backed 'Cannington Roy' and the pink 'Hopleys' and 'Langtrees'; taller cultivars (to 60cm) include the white-flowered, variegated-leaved 'Silver Sparkler', the pink 'Pink Whirls' and 'Bodegas Pink' with variegated leaves, and 'Buttermilk' with pale yellow flowers. 'Whirligig' has grey-green foliage, powder-blue to chalky grey ray florets, strongly contracted with margins inrolled above mid-point, then expanded again at tip, and dark blue-grey disc florets. The Cannington Hybrids include pink, white- and purple-flowered forms 15–30cm tall. Seed races, such as the 25cm Dwarf Salmon and the 45cm Tetra Pole Star, are also offered.

Ostrowskia (for Michael Nicolayevich von Ostrowsky, minister of Imperial Domains and Russian patron of botany in the late 19th century). Campanulaceae. Turkestan. 1 species, *O. magnifica*, GIANT BELLFLOWER, a perennial herb erect to 1.8m, with whorled, grey-green, ovate leaves to 15cm. Produced in summer in terminal racemes, the bell-shaped flowers are to 5cm long with broadly ovate lobes, lilac suffused with white or opal blue with darker mauve veins. Hardy to at least –15°C/5°F. It dies back completely in midsummer and then requires a dry rest until late autumn. Grow in a warm, sunny and sheltered position on deep, fertile, freely draining but moisture-retentive soil. Mulch in winter with leafmould or dry bracken. Propagate by seed or root cuttings. It resents disturbance.

Ostrya (from *ostrys*, the Greek name for these trees). Betulaceae (Carpinaceae). Europe, Asia, America. Some 9 species, deciduous trees with leaves alternating in 2 rows, their veins parallel and the margins dentate or serrate. The male inflorescence resembles that of *Carpinus*, but develops in autumn. The female catkins are made up of leafy bracts and take on a hop-like appearance. Cultivate as for *Carpinus*. *O.carpinifolia* (HOP HORNBEAM; S Europe, Asia Minor; to 20m; young shoots downy; leaves ovate, rounded at base, acute at apex, to 10 × 5cm, lustrous dark green, hairy between veins above, paler beneath with sparse hairs on veins, veins in 15–20 pairs, margins double-dentate); *O.virginiana* (EASTERN HOP HORNBEAM, IRONWOOD; eastern N America; to 20m; young shoots glandular-hairy; leaves ovate to lanceolate, rounded to cordate at base, apex long-acuminate, 7–12 × 3–5cm, dark green, hairy on midrib and between veins above, pale, pubescent beneath, veins in 11–15 pairs).

Otatea (from the Aztec name for these bamboos). Gramineae. Mexico to Nicaragua. 2 species; see *bamboos*.

Othonna (from Greek *othone*, linen, in reference to the soft leaves – a name used by Dioscorides and Pliny for the same or similar plants). Compositae. Mostly South Africa. About 150 species, perennial herbs or small shrubs, usually glabrous and glaucous. The leaves are entire to variously dissected, lobed or toothed, membranous, leathery or fleshy. Daisy-like flowerheads appear in summer. The following species is hardy in zone 7, given a sheltered position in full sun. Plant on a lean and gritty soil and protect from winter wet. Propagate by softwood cuttings in early summer. *O.cheirifolia* (syn. *Othonnopsis cheirifolia*; Algeria, Tunisia; spreading evergreen subshrub or shrub, to 40cm; leaves to 8cm, paddle-shaped, glaucous blue-grey, thinly succulent; flowerheads yellow).

Ourisia (for General Ouris (d. 1773), governor of the Falkland Islands). Scrophulariaceae. Andes, Antarctic S America, New Zealand, Tasmania. Some 25 species, low, usually dwarf perennial herbs or subshrubs. Produced in late spring and early summer, the flowers are salverform to funnel-shaped with a spreading and slightly zygomorphic limb of five broad lobes. Hardy in climate zone 7. Grow in raised beds or rock gardens on acid soils that are cool and moist but contain plenty of sand and grit. Shelter from scorching sun and drying winds. Propagate from fresh seed, or by division in spring.

　O.caespitosa (New Zealand; stems creeping; leaves 0.4–0.8cm; flowers white with a yellow throat, to 16mm in diameter); *O.macrocarpa* (New Zealand; stems erect, 10-60cm; leaves to 1.5cm; flowers 15–25mm in diameter, white sometimes with yellow throat); *Cultivars* include 'Loch Ewe' (*O.coccinea* × *O.macrophylla*), to 20cm, slowly spreading in tight rosettes, with shell pink flowers, and 'Snowflake' (*O.macrocarpa* × *O.caespitosa*), to 10cm, with dark, glossy leaves and white flowers.

outbreeder a plant that reproduces predominantly by cross-pollination.

ovate egg-shaped in outline; rounded at both ends but broadest below the middle, and 1.5–2 times as long as broad.

ovicide a pesticide used to kill the eggs of pests, for example tar oils applied to dormant fruit trees against aphid and sucker eggs.

oviparity the laying of fertilized eggs by insects; cf. *viviparity.*

ovoid egg-shaped, ovate but 3-dimensional.

Oxalis (from Greek *oxys*, acid, sharp, sour, referring to the sour taste of leaves). Oxalidaceae. Cosmopolitan but with centres of diversity in South Africa and S America. SORREL, SHAMROCK. 800 species, annual or perennial herbs and shrubs. They range greatly in habit, from succulent subshrubs to gaunt, truly woody shrubs, and from annual herbs

Oxalis adenophylla

to bulbous and tuberous perennials. The leaves are typically palmately lobed. The flowers are bowl- to funnel-shaped or salverform with five, obovate petals. Relatively few of the desirable species are reliably frost-hardy; those that are include *O.adenophylla* and *O.enneaphylla* (particularly valued for their neatly pleated glaucous foliage, suited to the sunny raised bed and large trough or to well-drained, humus-rich, sandy niches on the rock garden), *O.depressa* (for the well-drained interstices of paving or partially shaded bases of walls and rockwork), and *O.acetosella* (for naturalizing in the wild and woodland garden in moisture-retentive, humus-rich soils with shade or dappled sunlight). These will tolerate temperatures as low as −15°C/5°F, and thrive in temperate climates where summers are relatively cool, or where they can be given shade from the hottest sun in summer. Other species will tolerate short-lived frosts, especially in a sheltered rock garden or where their roots are protected by paving or gravel; these include *O.bowiei*, *O.hirta*, *O.laciniata*, *O.latifolia*, *O.lobata*, *O.purpurata* and *O.tetraphylla*; otherwise, these may be easily grown in the alpine house in well-crocked pans or pots in a gritty, leafy alpine mix. The excellent purple-leaved house plant *O.triangularis* is proving hardy outdoors in sheltered regions of climate zone 7, if thickly mulched in winter. Tropical shrubby species, such as *O. hedysaroides*, need a fertile, moist and soil-free medium, protection from draughts and full sun, moderate to high humidity and a minimum temperature of 10°C/50°F. Propagate by ripe seed, division, or offsets. Soft-stemmed species may be increased by cuttings in sand, in a shaded closed case with gentle bottom heat.

　O.acetosella (WOOD SORREL, CUCKOO BREAD; N Temperate Regions; rhizomatous, creeping, to 8cm; leaves to 3cm across, pale green; flowers white veined pink; var. *purpurascens*: flowers rose veined purple; f. *rosea*: flowers rich pink); *O.adenophylla* (Chile, Argentina; bulbous, to 8cm; leaves to 3cm across, leaflets grey-green, folded; flowers lilac-pink, throat white veined pink, maroon at base); *O.bowiei* (syn. *O.purpurata* var. *bowiei*; S Africa; bulbous, to 30cm; leaves to 10cm across, purple beneath; flowers bright rose red to pink); *O.chrysantha* (Brazil; mat-forming perennial to 5cm tall; leaves to 1.5cm across, green, downy; flowers golden yellow with red markings in throat); *O.depressa* (syn. *O.inops*; S Africa; bulbous; to 10cm tall; leaves to 2cm across, grey-green, sometimes with dark markings; flowers bright pink to rich, rosy mauve or white, throat yellow); *O.dispar* (Guyana; gaunt shrub to 1.5m; leaves to 15cm across; flowers golden yellow, sweetly scented); *O.enneaphylla* (SCURVY GRASS; Falkland Is, Patagonia; tufted, rhizomatous with bulbils; to 5cm tall; leaves to 2cm across, leaflets grey-green, folded, narrow, somewhat fleshy; flowers white to deep rose or red, scented; crossed with *O.laciniata* producing *O.* 'Ione Hecker' with narrower, wavy leaflets and purple-blue flowers pale at the edges, dark in the throat); *O.hedysaroides* (C America; erect, bushy subshrub to 1m tall; leaves to 5cm across; flowers yellow; 'Rubra': stems and foliage strongly flushed maroon); *O.hirta* (S Africa; bulbous with erect or trailing stems to 30cm; leaves to 3cm across, leaflets narrow; flowers purple-red with a yellow throat); *O.laciniata* (Patagonia; rhizomatous with chains of tiny bulbils; 5–10cm tall; leaves 1–2cm across, leaflets grey-green, folded, with purple, wavy margins; flowers fragrant, steel blue, lilac, violet or crimson with darker veins);*O.lobata* (Chile; rootstock bulbous; to 10cm tall; leaves to 1cm across, bright green, smooth, sometimes marked with purple; flowers bright golden yellow); *O.tetraphylla* (syn. *O.deppei*; Mexico; LUCKY CLOVER, GOOD LUCK PLANT; bulbous; 10–30cm tall; leaves 4–12cm across, mid green usually with maroon bands or stains; flowers red to pink or lilac with a white-edged, yellow throat); *O.triangularis* (Brazil; rootstock swollen; to 15cm tall; leaves to 10cm across, dark purple to beetroot red sometimes with darker or paler markings; flowers white tinted purple-pink).

Oxydendrum (from Greek *oxys*, sharp, and *dendron*, tree; the foliage tastes sour). Ericaceae. E US. 1 species, *O.arboreum*, SORREL TREE or SOUR-WOOD, a deciduous shrub to 9m or tree to 20m, slender and sometimes multistemmed, with rusty-red to grey bark and branches forming a fine canopy in older specimens. The leaves are 8-20cm long, oval, elliptic or oblong to lanceolate, lustrous, and pale to deep green, colouring vivid red in autumn. White and urn-shaped, the small flowers are produced in summer on finely branched panicles to 25cm long; the panicle branches persist in autumn and impart a pale, rather ghostly appearance to the whole tree. Hardy in climate zone 5, especially if planted in a sheltered position with the protection of surrounding trees. Grow on moist but well-drained,

acid soils, in light dappled shade or full sun. Propagate from seed sown in autumn or spring, or by softwood cuttings in summer – treat these with rooting hormone and root in a closed case with bottom heat.

oxygenator a plant grown submerged in a pool or aquarium and producing oxygen for the benefit of fish and other plants and life forms, for example Canadian pondweed, *Elodea canadensis*.

oyster shell the protective shell of sea molluscs; formerly widely used in crushed or ground form as an ingredient of potting media, especially bulb fibre, and valued for its high calcium content (50%) and water-holding property. However, it is not commonly available and has been replaced by materials such as perlite, vermiculite and rock wool.

Ozothamnus (from Greek *ozos*, branch, and *thamnos*, shrub). Compositae. Australasia. About 50 species, evergreen shrubs and woody perennial herbs with small and often heath-like leaves. Neat, button-like flowerheads are produced in dense, domed clusters in summer. The following will tolerate temperatures to –10°C/14°F, and probably lower, given good drainage and shelter from cold drying winds. They thrive in warm, sunny situations on a sharply drained but moist and slightly acid soil. Cut back leggy specimens in spring before growth commences. Propagate by semi-ripe cuttings in summer.

O.coralloides (syn. *Helichrysum coralloides*; New Zealand; shrub to 50cm, branches spreading, densely tomentose; leaves closely adpressed, to 5mm, scale-like, thick and leathery, silver-hairy on inner surface, smooth dark green on outer surface); *O.ledifolius* (syn. *Helichrysum ledifolium* KEROSENE WEED; Tasmania; strongly aromatic, inflammable shrub to 1m, young shoots downy, older leaves and branches viscid, producing yellow gum; leaves to 15mm, oblong to linear, obtuse, spreading, leathery, strongly revolute, glabrous, above, downy beneath, red-tinted at first); *O.rosmarinifolius* (syn. *Helichrysum rosmarinifolium*; New South Wales, Victoria, Tasmania; shrub to 3m, branches erect, woolly at first; leaves crowded, to 4cm, linear, mucronate, woolly beneath, revolute; 'Silver Jubilee': leaves very silvery); *O.scutellifolius* (Tasmania; shrub to 2m, tomentose; leaves minute, scale-like, ovate, reflexed, revolute, glabrous to tomentose above, tomentose beneath); *O.secundiflorus* (CASCADE EVERLASTING; New South Wales, Victoria; shrub to 2m, branches white-woolly; leaves to 12mm, oblong to linear, sparsely downy above, densely woolly beneath); *O.selago* (syn. *Helichrysum selago*; New Zealand; shrub to 40cm with woolly, arching to pendulous branches; leaves minute, closely overlapping, leathery, woolly above, dark green beneath); *O.thyrsoideus* (SNOW IN SUMMER; New South Wales, Victoria, Tasmania; shrub to 3m, young shoots glutinous; leaves to 5cm, narrowly linear, adpressed, dark green, resinous above, paler beneath with fine down).

P

Pachycereus (from Greek *pachys*, thick, and *Cereus*). Cactaceae. Mexico. 9 species, tree-like or shrubby cacti, often massive with stout, erect stems and shortly tubular, funnelform or campanulate flowers. Grow in a neutral, gritty medium. Provide a minimum temperature of 10–15°C/50–60°F and shade in hot weather. Maintain low humidity. Keep dry from mid-autumn until early spring, except for light misting on warm days in late winter. Increase by stem cuttings.

P.pecten-aboriginum (W Mexico; resembles *P.pringlei*, but not glaucous when young and with fewer ribs (10–11) and spines (8–12 per areole), and less woolly flowers); *P.pringlei* (NW Mexico; massive tree to 15m, trunk short, branches erect, 25–50cm in diameter, blue-green; ribs 11–17; spines 20 or more per areole, 1–3cm, stout; flowers 6–8cm, densely woolly, white).

pachycaul thick-stemmed; sometimes applied to tree-like herbs or shrubs with thickly swollen stems or trunks stems (e.g. *Adenium*), and to simple-stemmed trees (e.g. *Carica* species).

Pachyphytum (from Greek, *pachys*, thick, and *phyton*, plant). Crassulaceae. Mexico. 12 species, succulent perennial herbs or subshrubs. The fleshy leaves are alternate or crowded in rosettes and may be flattened to almost circular in cross-section; they are usually glaucous and tinged purple or pink-red. Campanulate to cup-shaped, 5-petalled flowers are carried in erect to nodding racemes in spring and summer. Cultivate as for *Echeveria*.

P.compactum (Mexico; leaves 1.8–3.0 cm, fusiform, acute, glaucous, sometimes tinged red-purple; flowers orange-red with darker tips); *P.oviferum* (Mexico; leaves 3–5cm, thickly obovoid, purple-glaucous, with dense white bloom; flowers pale green-white).

Pachypodium (from Greek *pachys*, thick, and *pous*, foot, referring to the swollen stem bases). Apocynaceae. South Africa, Madagascar. 17 species, deciduous or semi-evergreen shrubs or small trees with greatly swollen stems, (usually) spiny branches, tough leaves and funnel-shaped flowers similar to those of *Adenium*. Cultivate as for *Adenium*.

P.lamerei (S Madagascar; tree to 6m with a tapered or cigar-shaped trunk covered in low spiralled tubercles, grossly swollen below and forking at its summit; flowers 3–6 cm, white); *P.succulentum* (C and S Cape to Orange Free State; low, spiny and twiggy shrublet arising from a turnip-like caudex; flowers1–4cm, red, pink or particoloured to white).

Pachysandra (from Greek *pachys*, thick, and *aner*, man, stamen). Buxaceae. E Asia, US. 4 species, evergreen or semi-evergreen, procumbent subshrubs or perennial herbs, with fleshy, rhizome-like stems and erect branches with leathery leaves clustered at the branch tips. Small, white flowers are carried in short spikes in spring and summer. *P.terminalis* is widely used as an evergreen groundcover, thriving beneath trees and shrubs. It responds well to a close shearing over after flowering. *P.procumbens* is less rampant. Both are fully hardy. They do not thrive in dry soils or if exposed to strong sun or winds. They are ideally suited to dappled shade. Increase by division in spring or by leafy shoot cuttings taken in late summer.

P.procumbens (SE US; semi-evergreen perennial herb, carpet-forming and with erect stems 5–30cm; leaves 5–10cm, grey-green with brown-green, mottling, ovate to round, coarsely toothed above); *P.terminalis* (Japan, north central China; evergreen subshrub, carpet-forming and with erect stems 5–20cm; leaves 5–10cm, oblong to rhombic or obovate, apex toothed, dark green, glossy; includes 'Green Carpet', erect, low and compact, with small, finely toothed, deep green leaves, 'Silver Edge', with leaves light green, narrowly edged silver-white, and 'Variegata', slow-growing, with leaves variegated white).

Pachystachys (from Greek *pachys*, thick, and *stachys*, spike). Acanthaceae. Tropical America. 12 species, evergreen, perennial herbs and shrubs with showy, tubular and two-lipped flowers in terminal spikes clothed with colourful bracts. Cultivate as for *Eranthemum*. *P.coccinea* (CARDINAL'S GUARD; West Indies, northern S America; shrub to 2m; inflorescence to 15cm, 4-sided, flowers scarlet); *P.lutea* (Peru; shrub to 1m; inflorescence to 10cm, 4-sided, bracts golden to amber, flowers white).

Paeonia (from Greek *paionia*, possibly derived from Paion, the physician of the gods, who used the plant medicinally.) Paeoniaceae. Europe, temperate Asia, NW America, China. PEONY. Some 33 species, perennial herbs or shrubs ('tree peonies') with ternately compound leaves and large cup- to bowl-shaped flowers in spring and summer. These long-lived perennials are divided into two distinct groups: herbs, which include the many garden forms and hybrids derived from *P. lactiflora*, the Chinese peony; and the shrubby tree peonies.

Herbaceous peonies are hardy to below –10°C/14°F to –20°C/–4°F, especially if mulched with ever-green branches or bracken. Established plants are best left undisturbed. Fork in generous amounts of well-rotted manure or garden compost, with bonemeal or super-phosphate, when planting. They are lime-tolerant and may be grown in any moist, fertile soil, with good drainage and a soil depth of not less than 30cm. Plant in sun or part shade, avoiding early morning sun on flowers in areas with late frosts. Mulch annually in spring with well-rotted manure or garden compost.

The tree peonies enjoy similar soil conditions with greater emphasis on good drainage, and a slight preference for lime. Because they start into growth earlier in spring than herbaceous peonies they are slightly more tender, needing a sheltered position. At winter temperatures below –10°C/14°F, protect when young with a covering of bracken; if cut back by frost, plants may still shoot from the base. No pruning is necessary, except to remove dead wood in spring. Rejuvenation is possible by cutting back to ground level in autumn.

Divide herbaceous peonies in early autumn, each division consisting of three stout shoots 10–15cm long, with 3–5 eyes. All peonies can be raised from seed but named cultivars and hybrids do not come true. Sow firm ripe seeds in autumn in a loam-based propagating medium in pots in a shaded cold frame, transplanting the seedlings to nursery bed 45cm apart when they are large enough to handle in the following spring. Germination is slow if dormancy is allowed to develop; with fresh seed, the root appears about six weeks after sowing, and the shoot the following spring. They will reach flowering size in 4–5 years, or slightly longer for tree peonies.

Propagation of tree peonies is usually by grafting in summer using *P. lactiflora* or *P. officinalis* as root-stock. A wedge-shaped scion, 4cm long, bearing a single leaf with a vegetative bud, is pushed into a slit in the rootstock, which should be 10cm long and 1.2cm in diameter. Plant in equal parts peat/sand, with the union covered, in a closed shaded frame, and leave to callus over. Admit air gradually to harden off. They should be ready to pot on by autumn; plunge the pots in a cold frame over winter, and spray against botrytis. As the pots fill with roots the following spring, plant out deeply in a frame to encourage scion rooting. Layering and 15–20cm hardwood cuttings in autumn are possible but success rates are low.

The major disease of peonies is peony grey mould blight (sometimes called peony wilt). Soft brown areas develop at leaf bases causing them to wilt. This is followed by a dark brown rot in both stem and leaf bases, and grey mould can subsequently be seen

P. officinalis   *P. mlokosewitschii*

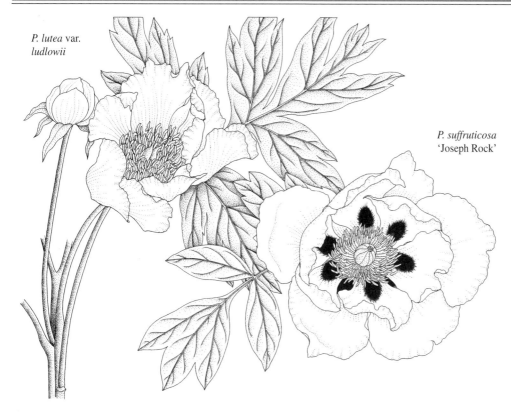

P. lutea var. *ludlowii*

P. suffruticosa 'Joseph Rock'

on stems at ground level. Young buds may blacken and wither, and grey mould develops on all damaged tissue. *B.cineria* causes similar symptoms later in the season. Both are best prevented by good hygiene, good drainage, and good air circulation.

P.cambessedesii (Balearic Islands; herb to 45cm, tinged red; leaves grey-green, tinted purple; flowers to 10cm diameter rose to pale magenta); *P.delavayi.* (China; tree peony to 1.6m leaves deeply and gracefully dissected; flowers to 9cm diameter, with very dark red to maroon golden anthers; 'Anne Rosse': *P.delavayi* × *P.lutea* variety *ludlowii*, flowers to 10cm diameter, lemon, streaked red); *P.emodi* (India; herb to 75cm flowers to 12cm diameter, white); *P.lactiflora* (Tibet, China, Siberia; herb to 60cm; flowers to 10cm diameter, fragrant, white; hybrids of *P.lactiflora* (Chinese Peonies) vary in flower colour from white to pink, deep red, crimson or maroon, and are single, double or semi-double in form); *P.* × *lemoinei* (*P.lutea* × *P.suffruticosa*; tree peony to 2m; flowers yellow or combinations of yellow and pink; includes 'Alice Harding': flowers double, lemon yellow; 'Argosy': flowers single, to 18cm diameter, petals primrose yellow, basal blotch carmine; 'Chromatella': flowers double, sulphur yellow; 'La Lorraine': flowers double, cream-yellow; 'L'Espérance': flowers 15–20cm diameter, petals pale yellow, basal blotch crimson; 'Souvenir de Maxime Cornu': flowers double, yellow to peach, tinged brown, orange and red); *P.lutea* (Tibet, China; as for *P.delavayi* except flowers to 7.5cm diameter, yellow; 'Superba': new growth bronze, flowers large, tinged pink at base; var. *ludlowii*: TIBETAN PEONY, to 2.5m, flowers to 12.5cm diameter produced

earlier); *P.mascula* (S Europe, Asia Minor; herb, 25–60cm flowers to 13cm diameter, red, pink or white; subspecies *mascula*: leaves smooth beneath flowers to 12cm diameter, purple-red; subspecies *arietina*: leaves grey-green, hairy beneath, flowers pink to red; 'Mother of Pearl': flowers pale pink; 'Northern Glory': leaves grey-green, flowers single, deep magenta-carmine; 'Purple Emperor': flowers single, rose-purple; 'Rose Gem': flowers single, bright blood-red); *P.mlokosewitschii* (Caucasus; herb to 1m; leaves broad, dark to silvery or blue-green above, glaucous beneath; flowers to 12cm diameter, cupped, yellow); *P.officinalis* (Europe; herb to 60cm flowers to 13cm diameter, red; 'Albaplena': flowers double, white; 'China Rose': flowers single, salmon pink, anthers orange-yellow; 'Crimson Globe': flowers single, garnet-red, petaloid stamens crimson and gold; 'Mutabilis Plena': flowers deep pink fading to blush pink; 'Roseaplena': flowers double, darker pink; 'Rosea Superba': flowers large double, pink; 'Rubra Plena': flowes double, crimson); *P.peregrina* (S Europe; herb to 50cm; flowers to 12cm diameter, cup-shaped, deep red; 'Otto Froebel' flowers vermilion, produced early); *P.potaninii* (China, Tibet; TREE PEONY; as for *P.delavayi* except stoloniferous, leaf segments narrower, flowers to 6.5m diameter, nodding, maroon to dark velvety red; 'Alba': flowers and stigmas white; variety *trollioides*: flowers pale yellow); *P.suffruticosa* (China, Tibet, Bhutan; MOUTAN PEONY; tree peony to 2m; leaves tinted purple-bronze at first; flowers to 15cm diameter, pink to white, each petal with a deep purple, red-bordered basal patch and finely scalloped margins; 'Banksii': flowers double, carmine; 'Godaishu': flowers semi-double, large, clear white; 'Koka-mon': flowers large, red-brown striped white; 'Reine Elizabeth': flowers fully double, large, salmon-pink tinged red, margins ruffled; 'Renkaku': flowers double, dense, white, ruffled, anthers long, deep yellow; 'Superba': flowers cherry red; subsp. *rockii*: flowers semi-double, spreading, petals very pale flesh pink to silvery-white blotched maroon at base, the best-known clone is 'Joseph Rock'); *P.tenuifolia* (S E Europe to Caucasus; herb to 60cm; leaves finely divided; flowers cup-shaped, to 8cm diameter, deep red; 'Early Bird': flowers single, deep red; 'Plena': flowers double, longer-lasting; 'Rosea': flowers pale pink); *P.veitchii* China; herb to 50cm; flowers nodding to 9cm diameter, pale to deep magenta; 'Alba': flowers cream-white); *P.wittmanniana* (Caucasus; herb to '120cm; flowers to 12.5cm diameter, pale yellow). *Cultivars*: SAUNDERS HYBRIDS: a very wide range of hybrid herbaceous peonies; they include 'Archangel': flowers single, white; 'Daystar': leaves slender, flowers yellow; 'Defender': flowers red, early; 'Early Windflower': leaves narrow, fern-like, flowers white, hanging; 'Golden Hind': flowers double, large, deep yellow; 'Renown': flowers single, strawberry red suffused yellow; 'Savage Splendour': flowers large, ivory splashed and edged purple, petals twisted; 'Thunderbolt': flowers single, black-crimson. 'Vesuvian': compact, leaves deep green, finely cut, flowers fully double, black-red; 'White Innocence': to 1.5m; flowers white. DAPHNIS HYBRIDS (tree peonies). 'Artemis': tall, tinged red, flowers single, large, silky yellow; 'Gauguin': flowers large, yellow veined red from bold red centre; 'Marie Laurencin': flowers semi-double, lavender, centre of darker shade).

paired usually applied to flowers or leaflets in opposite pairs; bi-, tri-, and multijugate describe compound leaves with two, three or many such pairs or leaflets.

palate the lower lip of a corolla, apparently closing its throat.

palea specifically the upper and generally slimmer of the two glumes enclosing the floret in most Gramineae; more generally, a small, dry bract or scale.

paleaceous bearing small, chaffy bracts or scales (paleae); more generally, chaffy in texture.

palisade a fence made up of spaced upright planks or railings fixed to horizontal members. It is sometimes used to describe a closely clipped hedge, grown as edging on narrow pathways or to conceal walls.

Paliurus (name for the genus used by Dioscorides.) Rhamnaceae. S Europe to E Asia. 8 species, spiny, deciduous or evergreen trees or shrubs with small yellow flowers in axillary clusters in summer, followed by dry, woody and winged fruits. Hardy to −15°C/5°F and suited to planting at the base of a warm, south- or southwest-facing wall or for dry

palmette

palms

Pests
aphids
mealy bug
scale insects

Disease
virus

borders, especially on limy soils. If the top growth is cut down by frost, it will resprout from the base so long as the roots are protected from damage. Grow in full sun in any well-drained fertile soil. Prune in winter to remove old and overcrowded growth. Cut overgrown specimens hard back to the base. Propagate by softwood cuttings in summer or by seed sown in a frame or under glass after a winter scarification. Alternatively, increase by root cuttings or simple layering. *P.spina-christi* (CHRIST'S THORN; Spain to C Asia and N China; deciduous, 3–7m, twigs flexuous, hairy, thorns paired; leaves 2–4cm, entire to crenate-serrate, hairy on veins beneath; fruit 1.8–3cm in diameter, woody, 3-lobed, subglobose, wings undulate).

palmate of leaves, with three or more free lobes or similar segments originating in a palm-like manner from the same basal point; of venation, with three or more veins arising from the point of attachment of the petiole.

palmatifid palmately cleft rather than lobed.

palmatisect palmately cut into segments almost to the base of the leaf.

palmette a trained form of fruit tree; similar to an espalier, with the tiers staggered and inclined upwards.

palms A group of monocotyledonous plants, mostly of tree and shrubby form, belonging to the family Palmae. They have special potential as garden and landscape features, and where dwarf species are used in their distinctive juvenile state, can be equally valuable as indoor container-grown specimens.

For outdoor cultivation, most palms need tropical or subtropical temperatures, besides which local climatic conditions and soils may affect their success rate. Palms suitable for outdoor planting in temperate areas are limited; those that thrive in favoured areas include the CHUSAN PALMS, *Trachycarpus fortunei*, and *T. wagnerianus*, *Chamaerops humilis*, *Butia capitata*, and *Jubaea chilensis*. Where varying degrees of protection can be provided during winter, species of borderline hardiness can be grown, including the COTTON PALM, *Washingtonia filifera*, the New Zealand FEATHER DUSTER PALM, *Rhopalostylis sapida*, and the CANARY DATE PALM, *Phoenix canariensis*. Several species are extremely cold-resistant but require summer heat to grow well. They include the DWARF PALMETTO, *Sabal minor*, and the BLUE HESPER PALM, *Brahea armata*.

Care should be exercised in the choice of species. A large number are extremely fast-growing and can rapidly dominate the landscape. Many are best grown as individual specimens, whereas others look good in groups of a single species. Palms require virtually no maintenance and the predictability of their mature size and shape is an advantage in garden design. Most palms succeed on a range of soil types.

The majority of palms grown in pots under cover benefit from high air humidity. In the greenhouse, this can be provided by damping down; in conservatories and the home, it can be achieved by means of an electric humidifier or by frequent misting. Grouping plants together and standing pots on a tray of pebbles, kept permanently moist, is helpful. A kitchen or bathroom is often a good place for palms because of the usually high humidity level.

Bright indirect light benefits most species. Many palms, including *Caryota*, *Phoenix roebelinii*, *Chamaedorea*, *Howea* and *Rhapis*, grow as seedlings on the jungle floor and thus tolerate low light levels. Sunlight filtered through net curtains and a north-facing window position are ideal, but direct sunlight through glass must be avoided. Dry air and insufficient levels of light in the greenhouse, conservatory or home, cause most palms to grow slowly; this can be an advantage, as long as the plant remains healthy, for a palm can then remain in the same container for a number of years. Conservatories are ideal locations.

Palms are most easily propagated from seed, although some such as *Chamaerops*, *Rhapis*, *Caryota*, *Ptychosperma*, and *Chamaedorea* species, can be increased by division or from suckers. The more tropical the plant's origin, the more essential it is that seed is fresh. Palms from temperate or arid zones have a considerably longer seed-viability period, for example, *Washingtonia*, *Trachycarpus*, *Chamaerops* and *Phoenix*.

Seed should be planted to its own depth in any moist proprietary potting medium that both retains moisture and is free-draining. Plant in pots or trays in a seed propagator or sealed plastic bag to retain humidity. Large quantitites of seed can be mixed with moistened medium and placed directly in a clear plastic bag and removed individually on germination. No light is required. Large seeds such as the COCONUT, *Cocos*, are half-buried in a seed-raising medium in individual pots.

Temperate-area palms will germinate well without additional heat but tropical species require 30–35°C/85–95°F. Fresh seed may sprout within a week or two, but most species take two or three months to germinate, and some take up to two years. During this time, the planting medium should not be allowed to dry out, and the temperature must be maintained. Germination-hastening practices include filing or cracking the hard seed coat, soaking in warm or hot water for long periods and soaking in acid.

'Sprouts' should be potted up individually as soon as the first leaf appears and placed in a warm and bright location out of direct sunlight. The seed should not be removed from the seedling at any stage and pots must be kept watered. Pot on as soon as roots fill the old pot, and feed regularly with a pot plant fertilizer.

In general, palms transplant successfully if moved when in active growth, generally in spring and early summer. Ideally, the tree should be prepared some months before being moved by trenching two or three spade-widths from its base to a depth of four or five spits, cutting through any roots encountered.

The trench is then filled in for at least three months, during which time new roots will grow from the base of the trunk.

Pamianthe (for Major Albert Pam (d. 1955), English horticulturist, and Greek *anthe*, flower). Amaryllidaceae. S America. 2 or 3 species, bulbous herbs with linear leaves and scapose umbels of large, fragrant flowers in spring. These consist of a long cylindrical tube, six lobes and six stamens, their filaments short, incurved and fused at the base into a campanulate corona with six short lobes. Maintain a winter minimum temperature of 10°C/50°F. Plant in a fibrous, loam-based medium in late summer/early autumn with the neck of the bulb at soil level, and water sparingly until growth commences. Grow in bright indirect light and water moderately when in growth. Propagate by offsets. Seeds take 12–15 months to mature on the plant, but germinate rapidly if sown when ripe in a warm and humid closed case. *P.peruviana* (Peru; scape to 60cm; flowers fragrant, perianth tube to 13cm, green, lobes to 13cm, white or flushed cream, the inner ones to 3cm wide, oblanceolate with a central green stripe; corona to 8cm, lobes bifid or mucronate).

pan (1) a compacted layer within a soil profile that is impermeable to water and air and thus impedes drainage and also root growth. It may arise naturally in soils of high iron content where the element accummulates at depth, or it may result from mechanical cultivation or treading of medium or heavy textured soil, especially when wet. The presence of a pan can be determined by digging a narrow pit into the profile, and the layer can be broken up by deep cultivation; (2) a plant container the diameter of which measures much more than its depth; it is useful for seed sowing, for growing plants with spreading or surface roots or for small bulbs. See *pots and potting*.

Pancratium (name in Dioscorides for a bulbous plant). Amaryllidaceae. Canary Islands, Mediterranean to tropical Asia, W Africa to Namibia. About 16 species, bulbous perennial herbs with 2-ranked, linear to lorate leaves. The large, white flowers are borne in summer in scapose umbels. The perianth is more or less funnelform with six spreading, linear-lanceolate lobes; the bases of the anther filaments are united in a conspicuous corona. They are suitable for the cool greenhouse or conservatory, or for outdoor cultivation in full sun at the base of a south-facing wall. *P.illyricum* is hardy to –5°C/23°F. *P.maritimum* is almost as hardy although the foliage is susceptible to frost, and in cool temperate gardens, the hot dry conditions necessary for bulb ripening are seldom achieved. Propagate by offsets or ripe seed.

P.illyricum (Corsica, Sardinia; scape to 40cm; flowers fragrant; perianth white, sometimes ivory or cream at base of segments and on corona, tube 8cm, lobes 5cm, linear-oblong to narrowly elliptic; corona much longer than perianth, teeth paired, long and narrow, alternating with stamens); *P.maritimum* (SEA

DAFFODIL, SEA LILY; Mediterranean, SW Europe; flowers highly fragrant; perianth tube to 7.5cm, very slender; corona two-thirds length of perianth, with short triangular teeth alternating with stamens).

Pandanus (from *pandang*, the Malayan name for these plants). Pandanaceae. SE Asia, especially Malaysia, Australia. SCREW PINE. 600 species, evergreen trees and shrubs with sparsely branching, stout stems bearing numerous stilt roots. Sword-shaped with saw-toothed margins, the leaves are inserted in three, twisted ranks, forming spirally arranged, terminal rosettes. The fruit is an oblong to globose syncarp of woody drupes, often pendulous and sometimes edible. Juveniles of *P.veitchii* and *P.sanderi*, are handsome foliage plants for the home or conservatory. Grow in sun or light shade in a sandy loam-based medium with additional leafmould and some charcoal; maintain a minimum temperature of 13°C/55°F and a humid, buoyant atmosphere. Water plentifully and feed fortnightly with dilute liquid feed when in growth, allowing the medium to dry slightly between waterings; reduce water during winter to keep almost dry. Avoid wetting the foliage in cool weather – an accumulation of water in the leaf axils will cause rot. As pot-grown specimens mature, they tend to heave themselves out of the pots. Propagate by offsets, suckers, by cuttings of lateral shoots in summer or by seed, pre-soaked for 24 hours before sowing.

P.sanderi (origin uncertain, possibly from Timor; leaves 75 × 5cm, striped yellow or golden and green, margins minutely spiny; Roehrsianus': leaves deep golden-yellow, later striped pale yellow); *P.veitchii* (origin uncertain, possibly from Polynesia; leaves 50–100 × 75cm, slightly drooping, margins spiny, dark green bordered with pure or silvery-white; 'Compactus': dwarf form).

Pandorea (for the Greek goddess *Pandora*). Bignoniaceae. Australia, Papuasia, E Malaysia, New Caledonia. 6 species, lianes with pinnate leaves. Produced in clusters between winter and summer, the flowers are cylindrical or funnelform with five, rounded lobes. Both *P.pandorana* and *P.jasminoides* withstand light frost and are drought-tolerant. Cultivate as for *Bignonia*.

P.jasminoides (BOWER OF BEAUTY, BOWER PLANT; NE Australia; to 5m; flowers 4–5cm, white, streaked or stained deep, rich pink within; 'Alba': flowers pure white; 'Lady Di': flowers white, with a cream throat; 'Rosea': flowers pink, with a darker throat; 'Rosea Superba': large, pink, with a darker, spotted purple throat); *P.pandorana* (WONGA-WONGA VINE; Australia, New Guinea, Pacific Islands; evergreen to 6m; flowers fragrant, to 4cm, creamy yellow, streaked and splashed red or purple; 'Rosea': pale pink).

pandurate fiddle-shaped; rounded at both ends.

panicle an indeterminate branched inflorescence; the branches generally racemose or corymbose.

paniculate resembling, or in the form of a panicle.

*Pandorea
jasminoides*

Panicum (from a Latin name for millet). Gramineae. Pantropical to temperate N America. PANIC GRASS. Some 470 species, annual or perennial grasses with thread-like to lanceolate leaves and flowers in panicles and racemes in spring and summer. The annual *P.capillare* is grown for its large, intricately branched flowerheads which are used for drying. *P.virgatum*, a hardy perennial, is noted for its generosity of bloom and the summer blue foliage colours of some cultivars, and the russet autumn tones. Grow in any moderately fertile well-drained soil in sun. Propagate *P.capillare* by seed sown in early spring under glass and set out after danger of frost is passed; increase *P.virgatum* by division.

P.capillare (OLD WITCH GRASS, WITCH GRASS; S Canada, US; annual, to 90cm; culms clumped, upright to spreading, slender to robust; leaves undulate, linear to narrow-lanceolate, to 30 × 1.4cm hairy; panicles open, to 45cm+, green to purple; branches very fine, spreading, resembling a cloud); *P.virgatum* (SWITCH GRASS; C America to S Canada; perennial, narrowly upright, to 180cm; culms clumped, flimsy to robust, purple to glaucous green; leaves linear, flat, erect, to 60cm × 1.4cm, usually glabrous, green, sometimes glaucous blue, becoming vivid yellow, bronze or burgundy in autumn; panicle open, to 50 × 25cm, with many fine branches and spikelets resembling a cloud or swarm lasting well into winter; branches spreading, feathery, stiff to drooping; 'Cloud Nine': to 2m, with blue-green leaves; 'Hänse Herms': RED SWITCH GRASS, to 1.25m, weeping, with plum to rich burgundy leaves in autumn and flowers suffused red; 'Heavy Metal': stiffly erect with pale metallic blue leaves, turning yellow in autumn; 'Rotbraun': to 80cm, with light brown leaves, flushed red at tips, with rich autumn colour; 'Rubrum': to 1m, with leaves flushed red, turning bright red in autumn, and rich brown seedheads in clouds; 'Squaw': leaves tinted red in autumn; 'Strictum': TALL SWITCH GRASS, narrowly upright, to 1.2m; 'Warrior': tall, strong-growing, with leaves tinted red-brown in autumn).

pannose felt-like in texture; being densely covered in woolly hairs.

Papaver (from Latin *pappa*, food or milk, an allusion to the milky latex). Papaveraceae. Europe, Asia, South Africa, W Australia, western N America. POPPY. 50 species, annual or perennial herbs with white or yellow latex. The leaves are generally oblong to lanceolate, pinnately lobed, incised and toothed. Produced in spring and summer on slender stalks, the short-lived flowers consist of four silky, obovate to rounded petals around numerous stamens. These in turn surround a top-shaped ovary capped with a sculpted, downy stigmatic disc. Poppies are easily grown in the mixed border, herbaceous border and in rock gardens.The annuals will usually self-seed freely unless deadheaded. Sow annuals *in situ*, in spring or late summer on any well-drained soil in a sunny position. Thin to 25–30cm apart and keep weed-free. Give late-sown plants cloche protection in winter. The biennial *P.triniifolium* is a superb foliage plant for silver and dry gardens. *P.nudicaule*, a short-lived perennial, can also be treated as an annual, but best results are obtained by growing it as a biennial. The blooms, 10–12cm across on wiry stems, make good cut flowers, lasting for up to a week if cut in bud and the stems tips scalded in boiling water before arranging. Sow thinly in late spring/early summer in a cold frame or outdoors. Transplant when large enough to handle, 15cm apart in rows 30cm apart, in a sunny well-drained bed. Transfer to the final site in full sun and fertile soil, in autumn or the following spring. *P.alpinum* and its allies are short-lived perennials for the rock garden, stone walls and raised beds. They need a gritty soil and full sun.

The oriental poppies, *P.orientale*, are among the loveliest hardy herbaceous perennials, although their flowering period is brief, and the leaves die back and leave gaps in the border. The seed pods are attractive when dried. Plant on any deep well-drained soil, in full sun. Propagate by division in early autumn or spring, or by root cuttings in late summer/early autumn. Place pieces of thick roots, crown end up, in boxes of sandy medium in a cold frame. Keep moist, not wet, and protect from frost. Transfer to a nursery bed in spring for a season before planting out. Propagation from seed is possible, but cultivars do not come true.

P.alpinum (Alps, Carpathians, Pyrenees; perennial to 25cm; leaves 5–20cm, 2–3-pinnate, lobes 6–8, linear to ovate-lanceolate, acute, occasionally pinnatisect, glaucous grey-green to green, glabrous to sparsely bristly, toothed; flowers solitary on scapes to 25cm; petals white, yellow or orange, to 2.5cm; a variable species – *P.burseri*, *P.kerneri*, *P.pyrenaicum* and *P.rhaeticum* are perhaps best treated as members of this complex); *P.atlanticum* (Morocco; perennial to 45cm; leaves to 15cm, oblong to lanceolate, jagged-toothed or pinnatisect, pilose; scapes to 45cm, simple or forked; petals to 2.5cm, buff to orange to red); *P.burseri* (C Europe; perennial to 25cm; leaves to 20cm, 2–3-pinnate, glaucous, segments 6–8, linear to lanceolate, acute; scapes to 25cm, setose; flowers solitary, petals to 2cm, usually white); *P.commutatum* (Caucasus, Asia Minor; annual to 40cm; stem erect, branching, sparsely hairy; leaves to 15cm, pinnatifid, downy, segments to 3cm, oblong to ovate, dentate to entire; scapes long, hairy, petals to 3cm, red with black basal spot; includes 'Lady Bird', with scarlet flowers splashed with black); *P.nudicaule* (ICELANDIC POPPY, ARCTIC POPPY; subarctic regions; perennial; leaves 3–30 cm, pinnatifid to pinnatisect, somewhat glaucous, hairy, segments 3–4, oblong, incised; flowers to 7.5cm in diameter, solitary on scapes to 60cm, white with yellow basal patch, yellow, orange, peach or pale red, ruffled; many cultivars and seed races are available, with large, long-stalked flowers in a wide range of pastel shades, bright yellows and warm oranges); *P.orientale* (ORIENTAL POPPY; SW Asia; robust perennial to 90cm; leaves to 25cm, bristly, pinnate to pinnatisect at apex, segments lanceolate or oblong; flowers solitary, to 10cm in diameter, red, orange or pale pink

usually with a purple to black basal blotch; includes many cultivars, ranging from dwarf and compact to tall, the flowers sometimes double, in a range of colours from white, pink and orange to deep red and purple, sometimes bicoloured, some with petals fringed or ruffled; among the finest are 'Cedric Morris' with grey hairy leaves and shell-pink petals with a large violet-black basal spot, 'Glowing Embers', to 110cm, robust, erect, with bright orange-red flowers, 10cm in diameter, with ruffled petals, 'Harvest Moon', to 1m, with semi-double, deep orange flowers, and 'Perry's White', to 80cm, with grey-white flowers with a purple-black centre); *P.rhoeas* (CORN POPPY, FIELD POPPY, FLANDERS POPPY; temperate Old World; bristly annual to 90cm; stem erect, branching; leaves pinnate or pinnatisect, segments lanceolate, to 15cm; flowers solitary, to 7.5cm across, brilliant red sometimes with a black basal spot; includes the Shirley Poppies, with medium-sized flowers, single or double, in a wide range of colours except yellow, and 'Valerie Finnis', with grey mauve, pink and white flowers); *P.somniferum* (OPIUM POPPY; SE Europe, W Asia; annual to 120cm; glaucous, grey-green; leaves to 12.5cm, obovate-oblong, deeply and jaggedly incised and toothed; flowers to 10cm across, pale white, pink, light mauve, purple or marked, frequently double, with erose petals, occasionally with a dark basal blotch; includes the Peony-flowered Hybrids, with double flowers); *P.triniifolium* (Asia Minor; biennial to 40cm; leaves silver-grey, bristly, finely divided and cut; flowers to 4cm across, pale orange-red).

papery bark a disorder occurring most frequently on trees of apple and ornamental *Malus,* where the bark peels in loose, papery and translucent sheets. In severe attacks, the affected area may be girdled, causing branch or trunk dieback. Papery bark commonly results from waterlogging, but is also a sign of infection with silver leaf (*Chondrostereum purpureum*) or apple canker (*Nectria galligena*). Where poor growing conditions persist, papery bark may recur, ultimately severely reducing fruit yield or even proving fatal. Badly damaged or dead shoots should be cut back to healthy wood. Papery bark also occurrs as a non-pathogenic ornamental feature on many trees, for example *Acer griseum, Betula utilis, Prunus serrula.*

Paphiopedilum (from Greek Paphos, an Aegean island with a temple to Aphrodite, and *pedilon*, slipper, describing the saccate lip formed by the third petal of each flower). Orchidaceae. SE Asia, India, Indonesia, SW China, New Guinea, Philippines, Solomon Islands. VENUS' SLIPPER. About 60 species, largely terrestrial, evergreen orchids. They lack stems and pseudobulbs. Borne in loose basal fans, the leaves are leathery, more or less strap-shaped and plain green or mottled with light or dark markings. The flowers are terminal, waxy, and carried one to several on a slender stalk; the dorsal sepal is large and erect, the lateral sepals fused to form a shield-like synsepalum behind the lip and the petals are

Papaver 'Cedric Morris'

Paphiopedilum niveum

horizontal or pendent. The lip is strongly saccate, forming a slipper-like pouch.

Compact orchids, with handsome, evergreen foliage and intriguing, long-lived flowers, usually from winter to spring. They are easy to grow in in the greenhouse and conservatory or in the home, in a growing case or stood on a tray of moist gravel. All need shade and a buoyant, humid atmosphere. Most need a night-minimum temperature of 15–18°C/ 60–65°F with a daytime increase of at least 5°C/9°F. Some of the most popular species, however, hail from cooler locations and will tolerate a night-minimum of 7°C/45°F. These include *P.hirsutissimum, P.insigne, P.fairrieanum, P.spicerianum,* and *P.venustum.* With the exception of *P.venustum,* it can generally be assumed that mottled-leaved species require warmer conditions, as do species and grexes with plain leaves and multiple-flowered inflorescences. Water through-out the year, aiming to drench the medium once it begins to dry out. Plants should not be allowed to dry out completely, but waterlogged conditions are disastrous. Although overhead misting is beneficial in optimum conditions, never allow water to stand on foliage or to settle for long periods in the centre of growths. Compost may be composed of a variety of mixtures using conifer-bark, perlag, perlite and charcoal with possible additions of sphagnum moss and/or coir. *P.bellatulum* benefits from the addition of broken chalk to the crocks. Feed fortnightly when the plants are growing. Propagate by division.

In addition to the species described below, there are many grexes and cultivars of *Paphiopedilum.* These vary widely in size, shape and colour. Some bear single flowers of great size and substance and almost perfectly circular in outline. Others are smaller and finer, closer to the species, with slender, some-times twisted petals, carried one to many per spike. In colour, they range from white to cream, yellow, ochre, green, rose, ruby red, garnet, plum purple, violet, maroon and deep, glossy chestnut. These colours may be combined—for example, a white dorsal sepal with a maroon lip and ochre petals. All parts may be spotted, mottled, striped or veined, some-times with hairy margins or hairy warts. One of the most popular grexes is *P.*Maudiae with attractively mottled leaves and lime green flowers with a white, green-striped dorsal sepal. In the clone *P.*Maudiae 'Coloratum', the flowers are large and wine red with a white-striped dorsal sepal and green-brown lip.

P.armeniacum (China; leaves marbled dark and light blue-green above, densely spotted purple beneath; flowers, bright golden yellow; dorsal sepal ovate, to 5cm, petals ovate, to 5cm; lip inflated, thin-textured, to 5 × 4cm, margins incurved, white-hairy and dotted purple inside at base); *P.barbatum* (Peninsular Malaysia and Penang Islands; leaves mottled pale and dark green above, pale green beneath; flowers 1-2; dorsal sepal ovate, to 5cm, white, green at base, veined purple; petals pale green beneath, purple above, veined darker, upper margin spotted dark maroon, deflexed, to 6 × 1.5cm, ciliate; lip to 4.5 × 2.5cm); *P.bellatulum* (W Burma, Thailand;

leaves dark green mottled pale green above, spotted purple beneath fls 1, white or cream, heavily spotted maroon; dorsal sepal to 3.5 × 4cm; petals to 6 × 4.5cm; lip narrowly ovoid, to 4 × 2cm, margins strongly incurved); *P.callosum* (Thailand, Cambodia, Laos; leaves tessellated pale and dark green above; flowers 1; sepals white flushed purple in lower half, veined purple and green, dorsal sepal broadly ovate, to 5.5 × 6cm, ciliate, recurved; petals curved downwards, to 6.5 × 2cm, white to yellow-green, apical third purple, maroon-spotted on upper margin and sometimes basal half; lip to 4.5 × 2.5cm, incurved lateral lobes warty, green, flushed maroon); *P.concolor* (SE Burma, SW China, Thailand, Indochina; leaves tessellated dark and pale green above, finely spotted purple beneath; flowers 1 (–3), yellow, rarely ivory or white, finely spotted purple; dorsal sepal broadly ovate, to 3.5 × 3.3cm; petals elliptic, rounded, to 4.5 × 2.5cm; lip ellipsoid, fleshy, margins incurved, to 3.8 × 1.5cm); *P.delenatii* (Vietnam; leaves mottled dark and pale green above, spotted purple beneath; flowers 1 (–2), pale pink with red and yellow markings on staminode, hairy; dorsal sepal ovate, to 3.5 × 2.5cm; petals broadly elliptic, rounded at apex, 4.3 × 5cm; lip ellip-soid to subglobose, 4 × 3cm, margins incurved, minutely hairy); *P.fairrieanum* (Sikkim, Bhutan, NE India; leaves very faintly mottled above; flowers 1, sepals white, veined green and purple, with a purple suffusion on dorsal sepal, petals similar; dorsal sepal elliptic, to 8 × 7cm, ciliate, apical margins recurved, lateral margins undulate; petals curved downwards with rising tips, 5 × 1.5cm, margins ciliate, undulate; lip deep, outcurved at rim, 4 × 2.5cm, olive to yellow-green, veined darker); *P.glaucophyllum* (leaves glau-cous; flowers to 20, produced in succession; dorsal sepal ovate to broad, to 3.3 × 3.2cm, shortly ciliate, outer surface pubescent, white or cream, centre yellow-green, veins flushed maroon; petals deflexed, linear, to 5 × 1 cm, apical half twisted, hairy on margins and at base, white, spotted purple; lip to 4 × 2cm, pink-purple, finely spotted darker, margins pale yellow); *P.godefroyae* (Peninsular Thailand, adjacent islands; leaves tessellated dark and pale green above, usually spotted purple beneath; flowers 1–2, white or ivory, all segments usually spotted purple; dorsal sepal concave, broadly ovate, to 3.3 × 3.7cm; petals oblong to elliptic, rounded, to 5.5 × 3cm, margins undulate; lip ellipsoid, to 3.5 × 1.5cm, margins strongly incurved); *P.haynaldianum* (Philippines; flowers 3–4; dorsal sepal obovate to elliptic, apex hooded, to 6 × 2.5cm, basal margin recurved, creamy white, sides flushed purple, centre green or yellow with basal half spotted maroon; petals arching, spathulate, half-twisted, to 8 × 1.4cm, ciliate, green or yellow, basal half spotted maroon, purple above; lip to 4.5 × 2.3cm, ochre-green, veined darker, purple-hairy within); *P.hirsutissimum* (NE India; leaves spotted purple beneath; flowers 1, sepals pale yellow to pale green with glossy dark brown suffusion almost to margins, petals pale yellow, lower half spotted purple-brown, apical half flushed rose-purple; dorsal sepal ovate to elliptic, 4.5 × 4cm, undulate, ciliate; petals horizontal to deflexed, spathulate, 7 × 2.2cm, half-twisted toward apex, strongly undulate, hairy, ciliate; lip 4.5 × 2cm); *P.insigne* (NE India, E Nepal; flowers 1; dorsal sepal obovate to elliptic, 6.4 × 4cm, apical margins incurved, pale green, inner surface with raised maroon spots, margin white; petals slightly oblong to spathulate, 6.3 × 1.8 cm, yellow-brown veined red-brown; lip 5 × 3cm, yellow, marked purple-brown); *P.lawrenceanum* (Borneo; leaves dark green mottled yellow-green above, pale green beneath; flowers 1; dorsal sepal broadly ovate to circular, to 6.2 × 6.2cm, lateral margins slightly reflexed, white, veined maroon above, green below; petals at right angles to dorsal sepal, ligulate, about 6 × 1 cm, green with purple apex, margins purple-ciliate and maroon-warted; lip 6.5 × 3.2cm, lateral lobes incurved and maroon-warted, green overlaid dull maroon, spotted maroon within); *P.lowii* (Peninsular Malaysia, Sumatra, Java, Borneo, Sulawesi; flowers 3–7; dorsal sepal elliptic to ovate, to 5.5 × 3.2cm, margins undulate, ciliate, pale green, basal half mottled purple, petals spathulate, often once-twisted in middle, to 9 × 2cm, ciliate, pale yellow, apical third purple, basal two-thirds spotted maroon; lip to 4 × 2.7cm, dull ochre brown); *P.malipoense* (China–SE Yunnan; flowers 1, sepals and petals green with purple stripes and spots; dorsal sepal elliptic to lanceolate, to 4.5 × 2.2cm, hairy; petals obovate, about 4 × 3.5cm; lip pale grey spotted purple within, deeply saccate, 4.5cm, subglobose, margins inrolled, hairy); *P.niveum* (N Malaysia and S Thailand; leaves mottled very dark and pale green above, dotted purple beneath; flowers 1–2, white, often dotted purple near base of segments and front of lip, hairy; dorsal sepal very broadly ovate, to 3 × 5cm; petals elliptic, rounded, to 4 × 3cm, ciliate; lip ovoid to ellipsoid, to 3 × 1.7cm, margins incurved); *P.parishii* (E Burma, Thailand, SW China; flowers to 9; sepals cream to green, veined darker, dorsal sepal elliptic, to 4.5 × 3cm, incurved on basal margins, recurved on apical margins; petals decurved to pendent, linear and finely tapering, to 10 × 1cm, tip spirally twisted, ciliate, basal margins undulate, green, spotted dark maroon below, margins and apical half dark maroon, maroon spots on lower basal margin; lip tapering to narrow apex, to 4.5 × 2cm, green, yellow-green or flushed purple); *P.primulinum* (Sumatra; flowers many, opening in succession, pale yellow with yellow-green sepals; dorsal sepal ovate, to 2.6 × 2.6cm, ciliate, hairy beneath; petals linear, finely tapering, spreading tip; twisted, to 3.2 × 0.8cm; lip to 3.5 × 2cm, bulbous toward apex); *P.rothschildianum* (Borneo; flowers 2–4; dorsal sepal ovate, to 6.6 × 4.1cm, ivory-white or yellow veined maroon; petals to 12 × 1.5cm, narrowly tapering to a rounded apex, yellow or ivory-white marked maroon; lip to 6 × 2cm, golden, heavily suffused purple); *P.spicerianum* (NE India, NW Burma; flowers 1; dorsal sepal curving forward, obovate to transversely elliptic, to 4.2 × 5cm, sides recurved, hairy, white with central maroon stripe and green-tinged base; petals falcate, linear-oblong, to 3.9 × 1.3cm, yellow-green with central brown-purple vein, flecking on other veins; lip to 5 × 3cm, glossy, pale

green, flushed brown with darker veins); *P.sukhakulii* (NE Thailand; leaves tessellated dark and yellow-green above; flowers 1; sepals white veined green, spotted purple at base, outer surface downy, dorsal sepal concave, to 4 × 3cm, ciliate; petals to 6.2 × 2cm, green heavily spotted maroon, margins ciliate; lip saccate, 5 × 2.3cm, green, veined and flushed maroon, lateral lobes warty); *P.venustum* (NE India, E Nepal, Sikkim, Bhutan; leaves tessellated dark green and grey green above, densely spotted purple beneath; flowers 1, sepals white veined green, petals white veined with green, warted maroon-black, flushed purple in apical half; dorsal sepal ovate, to 4 × 3cm, outer surface hirsute; petals oblanceolate, recurved, ciliate, to 5.5 × 1.5cm; lip to 4 × 3cm, yellow tinged purple and veined green, lateral lobes verrucose).

papilionaceous describes the pea-type corolla characteristic of papilionoid legumes.

Papilionanthe (from Latin *papilio*, butterfly and Greek *anthos*, flower). Orchidaceae. Himalaya to Malaysia. Some 11 species, evergreen epiphytic orchids. The stems are slender and erect to scrambling with many aerial roots. The leaves are narrow, terete, fleshy and tough. The flowers are borne at various times of year in short, axillary racemes or panicles. The sepals are obovate and wavy, the lateral sepals usually clawed. The petals are rounded, spreading and wavy. The lip is trilobed with a spurred, saccate base, a wedge-shaped, cleft midlobe and large, erect lateral lobes. Grow *P.teres* in full sun with high humidity and a minimum temperature of 15°C/60 °F. Plant in baskets or beds containing coarse bark mix. Syringe, feed and water freely throughout the year. *P.vandarum* requires lower temperatures (minimum 7°C/45°F) and drier conditions in winter. Propagate by stem cuttings or division. *P. teres* is an important garden plant and cut flower in the tropics and subtropics, notably in Singapore and Hawaii. Its clones and hybrid offspring, with butterfly-like flowers in brilliant pastel shades, are usually listed under *Vanda*.

P.teres (syn. *Vanda teres*; Thailand, Burma, Himalaya; stems to 1.75m; flowers 5–10cm across, tepals white or ivory deepening to rose or magenta, lip buff to golden, banded or dotted rose, blood-red or mauve); *P.vandarum* (syn. *Aerides vandarum*; India, Burma; stems to 2m; flowers nocturnally fragrant, lacy, to 5cm across, crystalline white, sometimes faintly tinted opal or lilac to pink).

papilla a nipple-like projection from an epidermal cell, often swollen and covered with wax.

papillate, papillose covered with papillae.

pappus a whorl or tuft of delicate bristles or scales in the place of the calyx, found in some flowers of Compositae.

papyraceous papery.

paradise garden a description of ancient gardens of the Near and Middle East. *Paradaisos* in Greek meant a park for a king, and was derived from the old Persian *pairidaeza*, an enclosure, and the Hebrew *pardes*, an enclosed garden or park. Its influence can be traced to India, China, Japan and into Italy and other parts of Europe.

Paradisea (for Count Giovanni Paradisi of Modena (1760–1826)). Liliaceae (Asphodelaceae). S Europe. 2 species, fleshy-rooted, perennial herbs with grassy leaves and, in spring and summer, 6-parted, funnelform to campanulate flowers in slender, lax racemes. Cultivate as for *Anthericum*. *P.liliastrum* (ST BRUNO'S LILY, PARADISE LILY; S Europe; to 60cm; leaves 3–5cm; flowers white tipped green, in a secund raceme; 'Major': robust, with larger flowers).

Parahebe (from Greek *para*, beside, i.e., close to, and *Hebe*). Scrophulariaceae. New Zealand, New Guinea, Australia. Some 30 species, prostrate or decumbent subshrubs, rarely herbs, similar to *Hebe* but with a less shrubby habit and with flowers in short axillary racemes. Hardy in zone 7. Grow on a gritty, free-draining soil in a sunny, sheltered position. Increase by semi-ripe cuttings, layering, and (*P.perfoliata*) by division. Cut away flowered stems of *P.perfoliata* when the new growth emerges in spring.
  P.catarractae (New Zealand; subshrub to 30cm, decumbent to ascending; leaves to 4cm, ovate to lanceolate, dark above, paler beneath, glabrous, acute, serrate; flowers white, veined pink or purple; includes 'Alba', with white flowers, 'Delight', to 15cm, bushy, with blue flowers, 'Diffusa', to 20cm, with small, toothed leaves and tiny, pink-flushed flowers, 'Miss Willmott', with rose-lilac flowers veined mauve, 'Porlock', to 25cm, with blue and white flowers, and 'Rosea', to 20cm, with pink flowers); *P.lyallii* (New Zealand; subshrub to 20cm, stems slender, prostrate to decumbent; leaves to 1cm, suborbicular to narrowly obovate, coriaceous, obtuse to rounded, crenate to serrate; flowers white to pink; 'Rosea': flowers pink); *P.perfoliata* (syn. *Veronica perfoliata*; DIGGER'S SPEEDWELL; Australia; woody-based herb, stems 20–100cm, clumped, arching; leaves to 5cm, ovate, amplexicaul to connate-perfoliate, blue-green, glaucous and thinly succulent; flowers blue to violet).

Paraquilegia (from Greek *para*, near, and *Aquilegia*, to which it is closely related). Ranunculaceae. C Asia to Himalaya. 4 species, low, tufted perennial herbs with ferny foliage and, in spring, solitary, cup-shaped flowers on delicate stalks. Hardy and with a preference for dry winters and cool climates. Grow in full sun on a well-drained alkaline soil. It is perhaps best cultivated in the alpine house, or in troughs in a gritty alpine mix. Sow fresh seed in autumn. *P.anemonoides* (syn. *P.grandiflora*; Himalaya; glabrous, to 18cm; leaves finely divided, glaucous; flowers 2–4cm across, white to lilac).

parasite an organism dependent for the whole or part of its nutrition on another living organism. An obligate parasite is wholly dependent on its host.

paring box a device for ensuring that grass turves (sods) are of the same thickness, consisting of a shallow wooden tray of the same dimensions and depth as that required of the turves. It has three sides, and the turves are laid grass side down into the box and pared to even depth with a sharp-bladed tool.

paripinnate where a pinnately compound leaf is not terminated by a single leaflet pinna or tendril.

Paris (from Latin *par*, *paris*, equal, on account of the numerical regularity of its leaves and flowers and their overall similarity). Liliaceae (Trilliaceae). Europe to E Asia. 4 species, deciduous herbaceous perennials with swollen creeping rhizomes. The stem is erect with ovate to lanceolate leaves in a whorl at its summit. Produced in spring and summer, the flowers are solitary and terminal, consisting of four to six green sepals somewhat like the leaves, and four to six linear, yellow-green sepals, often marked with purple-red. The fruit is a fleshy capsule. Suited to the woodland or wild garden, or for shaded parts of the rock garden; hardy to at least –15°C/5°F. Plant in a moisture-retentive, humus-soil enriched with leafmould. Propagate by division or by seed sown ripe in autumn. *P. polyphylla* (syn. *Daiswa polyphylla*; Himalaya, Burma, Thailand, China, Tibet, Taiwan; to 1m; leaves six to twelve, oblong to oblanceolate, tapering finely; sepals to 10cm, narrowly lanceolate to linear, green, spreading, petals to 10cm, thread-like, semi-erect, yellow).

parted, partite divided almost to the base into a determinate number of segments; a more general term for pinnate and palmate.

Parkinsonia (for John Parkinson (1567–1650), apothecary of London, court botanist and author of the great horticultural work *Paradisi in Sole Paradisus Terrestris* (1629)) Leguminosae. America, S and NE Africa. 12 or more species, shrubs or trees, usually armed, with green bark, finely pinnate or bipinnate leaves and fragrant, 5-petalled flowers in short racemes. Grow in perfectly drained soils in full sun. Provide a minimum temperature of 13°C/55°F. Water moderately when in full growth, very sparingly at other times. Under glass, maintain a dry, buoyant environment and avoid damp winter conditions. Propagate from seed in spring. *P.aculeata* (JERUSALEM THORN; tropical America; shrub or tree, deciduous or evergreen, to 10m, armed; branchlets weeping, yellow-green to dark green, smooth; leaflets to 0.5cm, ovate to oblong; flowers fragrant, to 2cm in diameter, dotted orange).

Parnassia (a shortened form of the ancient name *Gramen Parnassi*, grass of Mount Parnassus). Saxifragaceae. Northern temperate regions. GRASS OF PARNASSUS, BOG STAR. 15 species, small perennial, glabrous, usually evergreen herbs with slender-stalked leaves in basal rosettes and solitary, scapose, 5-petalled flowers in spring and early summer. Fully hardy. Plant in the rock or bog garden in damp but porous, alkaline soil and leave undisturbed. In the alpine house, plant in pans of high-fertility loam-based medium with added grit and leafmould and ground limestone. Water plentifully. Propagate from seed in late autumn or by division in spring. *P.palustris* (northern temperate regions; to 15cm; leaves ovate to 3cm, slender-stalked; flowers white, netted green to buff, 2.5cm indiameter).

Parochetus (from Greek *para*, near, and *ochetos*, brook, referring to the moist habitat). Leguminosae. Mountains of tropical Africa, Asia to Java. 1 species, *P. communis*, SHAMROCK PEA or BLUE OXALIS, a prostrate clover-like herb with slender, rooting stems, trifoliolate leaves and deep blue, pea-like flowers in summer. *Parochetus* is tolerant of temperates to –5°C/23°F and well-suited to the rock garden. In cool temperature zones, however, it is more commonly and safely grown in the cold greenhouse or alpine house, in pots and hanging baskets. Grow in a moist but gritty, well-drained medium, and give a sheltered position in full sun outdoors; protect with a cloche in winter. Propagate by seed or rooted runners.

Parodia (for Dr Domingo Parodi (1823–90), pharmacist and student of the flora of Paraguay). Cactaceae. Southern S America. 35–50 species, low-growing cacti, mostly with small, globose to shortly cylindric, ribbed or tuberculate stems and brightly coloured, funnel-shaped flowers. Grow in a cool frost-free greenhouse (minimum temperature 7°C/45°F) in a neutral to acid gritty medium. Provide full sun and low air-humidity; keep dry from mid-autumn until early spring, except for light misting on warm days in late winter. Increase by seed or offsets.

P.chrysacanthion (N Argentina; stem globose, to 12 × 10cm, apex woolly and tufted with erect spines, tubercles spiralled, spines 30–40, to 3cm, golden-yellow or paler; flowers yellow); *P.microsperma* (N Argentina; stem globose, sometimes elongate, 5–20 × 5–10cm, tubercles spiralled, spines 1–5cm, red, brown and white; flowers yellow or red); *P.nivosa* (N Argentina; stem globose to elongate, to 15 × 8cm, tubercles spiralled, spines to 2cm, white or dark; flowers fiery red).

Paronychia (from Greek *para*, beside, and *onyx*, nail, from the ancient use of the plant as a remedy for whitlows). Caryophyllaceae. Tropical and warmer temperate regions, common in the Mediterranean area. WHITLOW-WORT About 40 species, low-growing annuals or perennials, usually with very small leaves with silvery stipules and small flowers in dense heads surrounded by conspicuous silvery bracts. Hardy to –10°C/14°F. Grow in a warm open position in full sun, in sharply drained, sandy or gritty soils. Propagate by division in spring or from cuttings rooted in a closed case with bottom heat.

P.argentea (S Europe, N Africa, SW Asia; much-branched mat-forming perennial; leaves ovate to lanceolate; inflorescence dense, partially covered by silvery, bracts); *P.capitata* (Mediterranean; like *P.argentea* but with leaves linear to lanceolate and flowers in distinct, very silvery heads); *P.kapela* (Mediterranean; like *P.argentea* but with flowers in distinct, very silvery heads).

parrot a class of tulip with frilled and twisted petals, usually accompanied by splashes or streaks of contrasting colour.

Parrotia (for F.W. Parrot (1792–1841), German naturalist, traveller and professor of medicine, who climbed Mount Ararat in 1829). Hamamelidaceae. N Iran. IRONWOOD, IRONTREE. 1 species, *P.persica*, a deciduous, slow-growing, shrub or tree, to 10m. The bark is smooth and grey, ultimately flaking to leave paler patches. In wild colonies, the habit is erect with ascending branches. Clones in cultivation tend to be lower-growing with a broad, loose crown of spreading to arching branches. To 10cm long, the leaves are ovate to obovate, deep green above, paler beneath. In autumn, they colour richly, turning from wine red to flame and amber. Clusters of small, red flowers are produced before the leaves. This tree is fully hardy and should be grown in sun or part shade in deep, fertile, well-drained loamy soils. Acid soils are preferable, although it shows good tolerance of alkaline and chalky soils. Propagate by seed in autumn (germination may take up to 18 months), by softwood cuttings under mist in summer or by layering.

Parrotiopsis (from *Parrotia* and Greek *opsis*, appearance). Hamamelidaceae. Himalaya. 1 species, *P.jacqemontiana*, a deciduous, erect tree, to 6m. The leaves are ovate to orbicular, 5–9cm long, short-toothed, and usually turn yellow in autumn. Produced in spring and summer, the inflorescences are made up of densely packed heads of yellow stamens, subtended by white, petal-like bracts. Cultivate as for *Parrotia*.

parsley (*Petroselinum crispum*) a biennial herb, native of southern Europe and naturalized in many temperate countries; it is grown as an annual for garnishing and seasoning. Parsley is generally hardy but quality deteriorates without winter protection. There are two main forms, curled-leaved with thin roots, and French or flat-leaved with thicker roots; the latter is usually better flavoured and often easier to grow. Parsley struggles on soils that are light, acid or poorly drained.

For a summer crop sow *in situ* in spring, 15mm deep, in rows 20cm apart, thinning plants to 15cm apart; alternatively, raise in modules under protection. For an autumn and over-wintered crop, sow in mid-summer. Germination is slow and may take 5–6 weeks, with up to four months for plants to reach reasonable size. Seedlings may be potted up and grown on in large containers for convenient location

parsley

Pest
carrot fly

*Parthenocissus
quinquefolia*

indoors or outside. The crop needs to be kept well supplied with water. Plants can be cut to ground level for harvest once or twice during the year and protected with cloches to prolong quality.

Hamburg or turnip-rooted parsley (*P.crispum* var. *tuberosum*) is a dual-purpose form, producing smooth parsnip-like roots and usable leaves that are hardier than other types. Cultivate as for parsnip.

parsnip (*Pastinaca sativa*) a biennial grown as an annual winter root vegetable, valued for its hardiness and distinctive flavour. It is a native of Siberia and Europe, including Britain, and has been cultivated at least since Roman times. Root length varies from 13–20cm and cultivars are available of squat, rounded form with tapering roots, others with long narrow roots, and some with broad-shouldered wedged-shaped roots.

Parsnips grow best on a deep, well-cultivated soil free of stones. Seed rapidly loses viability so fresh seed must be used; germination is slow and erratic under cold conditions. Although the crop benefits from a long growing season, most success is achieved by April-May sowing as growing conditions improve. Sow *in situ*, 15–20mm deep in rows 30cm apart for large roots. Place 3–4 seeds at 15cm stations, later thinning to one plant: closer spacings will produce smaller roots. Canker-resistant cultivars are available. On poor soils, good-quality roots may be grown from seed sown directly into planting holes, about 90cm deep and 15cm across, made with a crowbar; the planting hole is filled with a fertile loam-based mix. This technique is also used for exhibition specimens.

Parsnips are relatively slow-growing and suitable for intercropping with faster-growing species such as radish or small lettuce. Roots can be lifted from late summer onwards but exposure to frost enhances flavour. They may be stored under cool conditions in boxes of sand.

parterre a level garden area of any size or shape containing ornamental flowerbeds. Parterres were highly developed in 16th-century France and Italy as separate garden areas set apart from the main garden by stone balustrades or hedges. The name superseded 'knot' in the 17th century, being of similar meaning. In the *parterre de broderie*, the main pattern is laid out in masses of low-growing flowers edged with sand or soil, in the style of carpet bedding. A water parterre is a shallow formal pool in which a knot-like pattern is laid out in stonework.

parthenocarpy the development of fruit without fertilization of its ovules, as occurs in greenhouse cucumbers and the cultivated banana.

Parthenocissus (from Greek *parthenos*, virgin, and *kissos*, ivy, a play on 'virginia creeper'). Vitaceae. N America, E Asia, Himalaya. VIRGINIA CREEPER. 10 species, generally deciduous woody vines attaching themselves by branched, twining tendrils usually with adhesive disks. The leaves are palmately lobed or parted. Small, yellow-green flowers in cymose panicles give rise to dark blue or black berries. Attractive and vigorous, self-clinging climbers invaluable for growing through sturdy trees, over pergolas, walls and fences and for covering buildings. Cold-tolerance varies: –25°C/–13°F for *P.quinquefolia* and at least –15°C/5°F for *P.tricuspidata* and its cultivars; *P.henryana* withstands cold to between –5 and –10°C/23–14°F. Plant in a moisture-retentive but well-drained fertile soil in shade or sun. Give initial support until plants produce adhesive pads. Prune annually, in autumn or early winter where it is necessary to limit growth – for example, to keep it away from roof eaves. For pergola-trained specimens, cut the current season's growth back to one or two viable buds from the main framework, thus creating a spur system, from which the following season's growth can cascade.

Propagate by seed. Remove pulp and sow fresh outdoors or in the cold frame, or stratify for six weeks, at or just below 5°C/40°F, and sow under glass in spring. Alternatively, increase by 10–12cm basal hardwood cuttings of the current season's growth, taken immediately after leaf fall; treat with rooting hormone and root in individual pots with a bottom heat of 18–20°C/65–70°F. Increase also by simple layering.

P.henryana (China; to 5m; leaves palmately 5-parted; leaflets 4–12 cm, obovate to oblanceolate or narrowly ovate, acute at apex, tapered to base, coarsely dentate except near base, dark velvety red- or brown-green with pink and silvery variegations along main veins, especially at first, becoming dark green then red in autumn); *P.quinquefolia* (VIRGINIA CREEPER, WOODBINE; US to Mexico; to 30m; leaves usually 5-parted; leaflets 2.5–10 cm, elliptic to obovate, acutely serrate particularly toward acuminate apex, cuneate at base, bright to dull green, rich crimson in autumn); *P.thomsonii* (syn. *Cayratia thomsonii*; W and C China; to 8m; leaves 5-parted; leaflets 2.5–10cm, rather thick-textured, glossy, ovate to obovate, acute at apex, tapered to base, shallowly serrate above; bright purple-red when young, dark shining green tinged purple when mature, ageing deep maroon); *P.tricuspidata* (syn. *Ampelopsis tricuspidata*; JAPANESE CREEPER, BOSTON IVY, VIRGINIA CREEPER; China, Japan; to 20m; leaves very variable, often so on the same plant, 6–20 cm, broadly ovate and variably 3-lobed with acuminate, serrate lobes or trifoliolate with stalked obovate leaflets, dull to glossy mid-green, turning crimson, scarlet or purple in fall; 'Atropurpurea': leaves large, green, tinged blue, purple-red in spring and autumn; 'Lowii': MINIATURE JAPANESE IVY, smaller and more slender, with small leaves, 2–3cm, 3–7-lobed, bright green tinged purple when young, colouring brilliant red in autumn; 'Veitchii': JAPANESE IVY, leaves small, simple or 3-foliate, and leaflets with 1–3 large teeth on each side, purple when young).

parthenogenesis the development of an embryo from an egg cell without fertilization, as occurs with dandelion (*Taraxacum officinale*) and certain aphids.

parsnip

Pests
carrot fly
celery fly

Disease
canker

Passiflora (from Latin *passio*, passion, and *flos*, flower, hence 'passion flower'. The name was given by the early Spanish missionaries in South America, in reference to a fancied representation in the flowers of the story of the Crucifixion; the corona represented the crown of thorns, the five anthers the five wounds, the three styles the three nails, the five sepals and five petals the apostles (less Peter and Judas), and the hand-like leaves and flagellate tendrils the hands and scourges of Christ's persecutors). Passifloraceae. Tropical Americas, Asia, Australia, Polynesia. PASSION FLOWER. Some 500 species, those here evergreen, woody climbers with tightly spiralling tendrils. The flowers are usually solitary, with five fleshy or membranous sepals and five petals (these sometimes absent). The filaments are arranged in several series, forming a showy corona between petals and stamens. Five stamens are inserted on a gynophore and the stigma is capitate with three styles. Sometimes edible, the fruit is usually a juicy, indehiscent, many-seeded berry.

P.caerulea is hardy in zone 6 and will thrive in any moderately fertile, well-drained and adequately moisture-retentive soil, in sun or part shade. Provide the support of wire or trellis. In less favoured areas, site with shelter from cold, drying winds and mulch the base, also protecting top growth with burlap or fleece if necessary. The remaining species need protected cultivation in cool temperate zones. *P.edulis, P.antioquiensis, P. × exoniensis, P.manicata* and *P.mollissima* need minimum temperatures of 10°C/50°F. Most of the remaining species require a minimum temperature of 15°C/60°F.

Grow all tender species in a freely draining, fibrous, loam-based mix with additional leafmould and sharp sand. Protect from fierce sun in summer. Water plentifully when in growth, sparingly at other times. Prune to prevent overcrowding by removing the weakest shoots in spring, and again, if necessary, after flowering; pinch out unwanted growth regularly during the growing season; spur back to an established framework in spring. Propagate by heel or nodal cuttings rooted in a sandy propagating medium in a closed case, or by seed. Some frost-tender species show greater tolerance of low soil temperatures when grafted on to *P.caerulea*.

PASSIFLORA

Name	Distribution	Leaves	Flowers
P.alata	NE Peru, E Brazil	to 15 cm, ovate to oblong, membranous, glabrous	to 12cm diam., fragrant; sepals pale crimson; petals brilliant carmine; filaments banded purple red and white
Comments: Spring–summer. Fruit edible. Z10			
P.× alato-caerulea (*P.alata × P.caerulea*)	garden origin	trilobed	to 10cm diam., pink to purple, white outside; sepals white inside; filaments blue-violet tipped white
Comments: 'Imperatrice Eugénie': flowers to 16cm diam., sepals white, petals lilac-pink, filaments white and mauve			
P.antioquiensis *P.van-volxemii* BANANA PASSION FRUIT	Colombia	to 15cm, ovate to lanceolate, or trilobed, serrate, hairy beneath	to 12.5cm diam.; sepals rose-red or magenta; petals often of a deeper shade than sepals; filaments violet
Comments: Summer. Z9.			
P.caerulea PASSION FLOWER; BLUE PASSION FLOWER	Brazil, Argentina	to 10cm, palmately 5-9-lobed, glabrous	to 10cm diam., faintly fragrant; sepals white or pink-white inside; petals of a slightly clearer colour than sepals; filaments banded blue at apex, white at centre, purple at base
Comments: Summer–autumn. 'Constance Eliott': flowers ivory white. 'Grandiflora': flowers to 15cm diam. Z7.			
P.capsularis	Nicaragua to C Brazil to Paraguay, Greater Antilles	to 7cm, bilobed, lobes downward-pointing, lanceolate, 3-veined	to 6cm diam.; sepals green-white; petals ivory; filaments yellow-white
Comments: Summer. Z10.			
P.coccinea *P.fulgens* RED PASSION FLOWER; RED GRANADILLA	Guianas, S Venezuela, Peru, Bolivia, Brazil.	to 14cm, oblong, serrate or crenate	to 12.5cm diam.; sepals scarlet; petals vivid scarlet, filaments pale pink to white at base, deep purple toward apex
Comments: Z10.			
P.coriacea BAT-LEAF PASSION FLOWER	Mexico to N Peru and N Bolivia, Guyana.	to 25cm across, usually lobed, broadly divergent, the whole leaf resembling a bat with outspread wings, coriaceous	to 3.5cm diam.; sepals yellow-green; petals absent; filaments ivory
Comments: Z10.			
P.edulis GRANADILLA; PURPLE GRANADILLA; PASSION FRUIT	Brazil	to 10cm, trilobed, dentate, shiny	to 7.5cm diam.; sepals white; petals paler; filaments white banded purple or indigo, strongly wavy towards tips
Comments: Summer. Z10.			

PASSIFLORA

Name	Distribution	Leaves	Flowers
P.× exoniensis (*P.antioquiensis* × *P.mollissima*) Comments: Z10.	garden origin	to 10cm, downy, trilobed	to 12.5cm diam., pendulous, cylindric, exterior brick red, interior rosy pink; petals bright pink with a violet tint in the throat; filaments white
P.foetida RUNNING POP; LOVE-IN-A-MIST; WILD WATER LEMON Comments: Summer–autumn. Z10.	South America, Puerto Rico, Jamaica, Lesser Antilles	hastate, 3–5-lobed, thin-textured, hairy	to 5cm diam., surrounded by showy, deeply and finely fringed bracts; sepals white to ivory streaked green; petals, ivory-lilac; filaments white banded violet
P.insignis *Tacsonia insignis* Comments: Summer–autumn. Z10.	Bolivia, Peru.	to 25cm, ovate to lanceolate, finely-toothed, coriaceous, shiny above, woolly beneath	pendulous to 9 cm, tubular, sepals violet-crimson; petals rose-purple; filaments white, mottled blue
P.kermesina Comments: Z10.	E Brazil	to 8cm, trilobed, lobes oblong to ovate, deep green above, somewhat glaucous and purple-tinted beneath	to 8cm diam.; sepals scarlet; petals similar to sepals, filaments violet-purple
P.× kewensis (*P.caerulea* × *P.kermesina*) Comments: Z8	garden origin		9cm diam., carmine, suffused blue
P.manicata *Tacsonia manicata* RED PASSION-FLOWER Comments: Z9.	Colombia to Peru.	to 10cm, trilobed, lobes ovate, serrate, downy beneath	vivid scarlet, tubular to bell-shaped, to 5 cm long, green-white inside; filaments blue and white
P.mixta *Tacsonia mixta* Comments: Z10.	C Venezuela, Colombia, Ecuador, Peru, Bolivia	to 10 cm, trilobed, lobes ovate-oblong	pendulous, to 11cm, tubular, sepals pink to orange-red; petals same colour on sepals; filaments deep lavender or purple
P.mollissima *Tacsonia mollissima* CURUBA; BANANA PASSION FRUIT Comments: Z9.	W Venezuela, Colombia, SE Peru, W Bolivia	to 10cm, trilobed, lobes ovate, downy	pendulous, to 8cm, tubular, sepals olive-green, soft pink; petals obtuse, same colour as sepals
P.morifolia Comments: Z10.	Guatemala, Peru, Paraguay, Argentina	to 11cm, trilobed, lobes ovate, deep green, minutely hairy	to 3cm diam.; sepals white and purple-mottled or uniformly green
P.organensis *P.maculifolia* Comments: Z10	E Brazil	to 10 cm, bilobed, shaped like a fishtail, dark green tinted purple to chocolate brown with regions of silver, cream, pink or lime-green variegation	to 3cm diam., cream to dull purple
P.quadrangularis *P.macrocarpa* GRANADILLA; GIANT GRANADILLA Comments: Fruit edible. 'Variegata': leaves blotched yellow. Z10.	Tropical America	to 20cm, ovate-lanceolate	to 12cm diam.; sepals pearly grey-green tinted flesh pink; petals fleshy, pale mauve-pink; filaments white, banded blue and red-purple, twisted
P.racemosa *P.princeps* RED PASSION FLOWER Comments: Z10	Brazil	to 10cm, ovate and simple, or trilobed, coriaceous	to 12cm diam., scarlet-red or white, in pendulous racemes; filaments white banded dark purple
P.rubra Comments: Autumn. Z10.	Colombia, Venezuela, Peru, Bolivia, E Brazil, W Indies	to 8cm, bilobed, membranous, downy	to 5cm diam., ivory; filaments red-purple or lavender
P.suberosa Comments: Summer. Stems becoming winged and corky. Z10	Tropical America	entire to deeply trilobed, lobes narrow	to 3cm diam., green-yellow to ivory
P.vitifolia Comments: Spring–summer. Z10.	C and S America	to 15cm, trilobed, toothed, shiny above, hairy beneath	to 9cm diam., scarlet, bright red or vermilion; filaments red to bright yellow

Passiflora caerulea

Passiflora quadrangularis

Passiflora coccinea

patching out the use of a small clump of seedlings for planting unthinned direct into a garden border. In the case of some plants which produce very delicate seedlings, such as bedding lobelia, the technique may be used at pricking-out stage instead of handling individual seedlings.

patellate, patelliform orbicular and thick, having a convex lower surface and a concave upper surface.

patent spreading.

Patersonia (for William Paterson (1755–1810), lieutenant-governor of New South Wales 1800–1810, an early botanical collector in Australia). Iridaceae. Australia, Borneo, New Guinea. About 20 species, perennial, rhizomatous herbs with a fan of linear leaves and short-lived flowers composed of three, broad and spreading outer tepals and three small, erect inner tepals. Grow in frost-free conditions in a light, fertile and well-drained medium in full sun, watering plentifully when in growth. Propagate by division or by seed in autumn. *P.umbrosa* (W Australia; to 50cm; flowers blue; 'Xanthina': flowers yellow).

pathogen a microorganism that is harmful to its host and causes disease. Most are parasitic but some are primarily saprophytic and only become pathogenic if the host is injured or stressed.

pathotype a subdivision of a species of pathogen based on common pathogenicity characteristics, especially host range.

pathovar (abbrev. pv.) a strain or pathotype of a bacterial species that infects only plants in a certain genus or species.

patio originally, the inner courtyard of a house which was built in a square to form an enclosed open space. Derived from Arab architecture, patios are characteristic of Spanish gardens and are usually associated with pools, fountains, and plants in beds and containers. The term is now generally used to describe a paved area close to a house, especially a sitting area, which may be better termed a terrace.

patio squares squares of wood used for paving.

Patrinia (from E.L.M. Patrin (1742–1814), French traveller in Siberia). Valerianaceae. Temperate Asia. Some 15 species, small, clump-forming herbaceous perennials with pinnately cut or lobed leaves. Produced in spring or summer in corymbose panicles, the small flowers consist of five spreading lobes. Hardy and suited to the herbaceous border, rock garden and woodland garden. Cultivate as for *Valeriana*. *P.triloba* (Japan; to 60cm; leaves to 5cm, 3–5-pinnately lobed, toothed; flowers golden-yellow, fragrant).

patte d'oie used to describe a set of paths or vistas radiating from a central point.

Paulownia (for Anna Paulowna, daughter of Paul I, Tsar of Russia (1795–1865)). Scrophulariaceae. E Asia. About 17 species, deciduous trees with stout shoots, large leaves and, in spring and summer, terminal panicles of showy, foxglove-like flowers. The pollution-tolerant *P.tomentosa* is grown as a street and park tree in cities of continental Europe and has become widely naturalized in the eastern states of the US, often as a roadside weed. *P.tomentosa* and its cultivar 'Lilacina' are also grown in the mixed border, for their juvenile foliage – when young plants are grown as stooled specimens, their downy heart-shaped leaves assume enormous proportions.

In temperate maritime climates, like Britain's, flowering may be poor and sporadic – the result of inadequate wood ripening and late frost. Once established, these trees are hardy to between –15 and –20°C/5 to –4°F. Although frost-tender when young, they will usually resprout from the base if given a protective mulch at the roots in winter.

Grow in any deep, moderately fertile, moisture-retentive but well-drained soil, in full sun with shelter from strong winds. For foliage effects, stool young plants back to the woody base in spring and pinch out any sideshoots. Propagate from seed sown at 15–20°C/60–70°F in spring, or when ripe; seed requires light for germination. Increase also from root cuttings in winter, or by semi-ripe, heeled cuttings in summer, in the cold frame. Overwinter under glass in the first year.

P.fortunei (China, Japan; to 20m; leaves 14–21 × 7–12cm, ovate to cordate, glabrous and shining above, densely hairy beneath; flowers 8–10cm, cream flushed with lilac); *P.tomentosa* (syn. *P.imperialis*; C and W China; to 20m; leaves 20–100 × 12–60cm, broadly ovate, hairy above, densely so beneath; flowers 5–6cm, violet with yellow stripes inside; 'Coreana': leaves tinted yellow, woolly beneath, flowers violet, with throats speckled yellow inside; 'Lilacina': flowers pale lilac).

paver a clay or concrete paving block, used especially for patios and terraces.

pavilion a substantial garden building, more ostentatious than a garden house, usually architect-designed and often highly decorated inside and out.

paving various solid materials used for making paths, terraces, and patios, mainly of real or composition stone, but including concrete slabs, bricks, tiles and sets.

Pavonia (for Jose Antonio Pavon (d. 1840), Spanish botanist and traveller). Malvaceae. Tropics and subtropics. About 150 species, herbs, subshrubs and shrubs with *Abutilon*-like flowers. Cultivate as for tender, shrubby *Hibiscus*. *P. × gledhillii* (*P.makoyana* × *P.multiflora*; shrub to 2m; epicalyx bright red, corolla to 3cm, tubular, dark purple, anthers exserted, filaments bright red, pollen chalky lilac-blue; 'Kermesina': dwarf, with carmine flowers; 'Rosea': flowers rose).

Paxistima (from Greek *pachys*, thick, and *stigma*, stigma). Celastraceae. N America. 2 species, dwarf, evergreen, glabrous shrubs with 4-angled, corky and warted branches and leathery leaves and, in summer, very small, 4-petalled flowers. Fully hardy, evergreen groundcover for exposed situations in the rock garden. Propagate by semi-ripe cuttings, taken in late summer and given a little bottom heat. *P.canbyi* (CLIFF GREEN, MOUNTAIN LOVER; N America; stem to 40cm, decumbent, rooting; leaves to 2cm, linear to oblong, obtuse, revolute, entire or finely serrate; flowers green, to 2.5mm; 'Compacta': hardy, dwarf, with dark green leaves later turning bronze).

peach

Pests
aphids
red spider mite
scale insects

Diseases
bacterial canker
  (*see* bacterial
  diseases)
brown rot
peach leaf curl
silver leaf

Disorder
gumming

peaches and nectarines the peach (*Prunus persica*) originated in China and is known to have been cultivated by at least 2000BC; the nectarine is a smooth-skinned form of the peach, common in parts of Turkestan.

With the demise of large estate gardens and consequent loss of their beneficial wall sites, peaches and nectarines are now little grown in British gardens. A limiting factor is the widespread incidence of peach leaf curl disease which severely debilitates trees.

By contrast, they are widely grown both domestically and commercially in warmer climates, notably southern Europe, parts of the US, Australia, New Zealand, South Africa and China. The peach flowers early and requires warm spring weather, but any tendency towards a tropical climate will not provide adequate winter chilling.

Successful seedlings are often raised by amateurs, but when propagated for growing elsewhere, they are seldom successful. Established named cultivars are propagated by budding on to either seedling peach or selected clonal plum rootstocks known to be compatible with the peach and nectarine; the former produce quite large trees. Of the plum rootstocks 'Brompton' (vigorous) and 'St. Julien A' (semi-vigorous) are widely used.

A wide range of cultivars is available and since they have different chilling requirements, local advice should be sought on which are best-suited for individual localities. Peaches may be classified as: (a) white-fleshed and predominantly free-stoned; (b) yellow-fleshed, of firm texture and free-stoned; (c) yellow-fleshed and cling-stoned. In Britain, the peaches 'Peregrine' and 'Rochester' and the nectarine 'Lord Napier' generally perform well.

Most peaches and nectarines are self-fertile, but satisfactory pollination is encouraged by providing sheltered conditions, and since flowering is very early, hand-pollination is advisable to supplement insect activity. Using a soft camel-hair brush or cotton wool, the centres of open flowers are gently brushed, in the middle of each day, to transfer pollen from anthers to stigma.

Plant on to well-drained sites only, by mid-winter to ensure good establishment before growth begins. Trees planted in the open ground should be supported with a stake for the first two seasons, and mulched. In Britain, two tree forms predominate, the bush

for open-ground cultivation and the fan for wall-training. Plant bushes 5–7m apart, and fans 3.5–6m apart, allowing greater distances for trees on peach seedlings and Brompton rootstocks. Shaping to achieve the desired form should be undertaken in the spring after planting.

With each form, it is important to establish a distinct branch system in order to produce annually a succession of young shoots for fruiting the following year; unlike apples and pears, peach and nectarine shoots carry fruit only once, after which they remain bare and unproductive.

Pruning is done while the tree is in active growth, not during winter when the risk of disease infection is greater. With bushes, prune a proportion of older shoots to young shoots, keeping the centre of the tree open and removing branches that are too low, crowded or crossing.

With fans, select two new shoots arising from existing growths, one from the base and one half-way up; remove any others near the base, together with any growing awkwardly, and shorten the remainder to one leaf. Tie the new shoots in where possible. After the fruit has been gathered, either cut the fruit-bearing lateral back to the replacement shoot(s) or, in younger trees, tie in all new shoots to fill existing space.

Regular applications of a compound fertilizer, higher in nitrogen and potash than phosphate, should be given at about $75g/m^2$ in late winter, followed by a good mulch of well-rotted manure or compost when the soil is moist. A liquid feed high in potash can also be given after flowering, at three week intervals, to induce quality and disease resistance.

Irrigation in summer is likely to be essential for wall-trained trees: apply approximately 5 litres around each tree every 10–14 days. Following successful pollination and fertilization, the number of fruitlets usually needs to be limited by thinning when approximately hazelnut size, reducing them to one per cluster and 8–12cm apart. Once at walnut size, thin again if necessary to about 22cm apart.

Peaches and nectarines in gardens should be picked when the flesh gives under gentle thumb pressure near the stalk end.

The use of fruit nets as bird protection is advisable against bullfinches damaging overwintering buds, and fruit damage by other birds. Frost protection of blossom on wall-trained trees can be provided by fixing hessian netting as a temporary cover.

Where climatic conditions are not conducive to regular cropping in the open, cultivation under glass is a practical although laborious alternative. A border of good deep soil is essential. A wall-trained fan tree is the best form to develop, hence a lean-to greenhouse is ideal. Heating is an advantage, and horizontal support wires fixed 7–8cm away from the wall will be needed. Pruning and training are as for fan-trained trees.

The greenhouse must be continuously ventilated once the crop has been cleared to allow adequate chilling. From late winter maintain a temperature of

8–10°C/46–50°F, raising this to approximately 20°C/68°F after ten days, with a night minimum of 4–8°C/39–46°F. Careful attention to watering is essential, and on bright days the trees must be syringed with tepid water mid-morning and early afternoon. Light top dressings of well-rotted manure or a compound fertilizer should be given in late winter, followed by liquid feeds high in potash every 10–14 days during summer. Peach leaf curl will not be a problem under glass, but mildew, aphids and red spider mite will require preventive measures.

Where there is no suitable greenhouse or garden space, cultivation in large containers is a possibility and trees can be moved to frost-free positions during the vulnerable periods of flowering, early fruitlet production and ripening. A miniature bush-type tree should be developed, firmly planted in a soil-based potting mix, watered regularly and fed every 10–14 days. Repotting in autumn is necessary, replacing some of the compost and removing any coarse roots in favour of fibrous ones. The trees should be kept in the open through autumn and winter to allow adequate chilling, with the rootball protected against freezing.

peach leaf curl (*Taphrina deformans*) a fungus that affects almonds, nectarines, peaches and, rarely, apricots, becoming apparent as young leaves emerge in spring. The leaves become puckered, curled, twisted and sometimes chlorotic, and as they mature, the thickened parts change from yellow to red. A white bloom of fungus spores can sometimes be seen on the surface of swollen areas, and severely attacked leaves eventually fall off. The fungus survives on shoots or in buds, and spores infect young leaves as the buds open. Cold and wet spring weather favours the disease. Affected leaves should be removed and burnt as soon as seen, and a copper-based fungicide spray can be applied during the winter, at the latest just before the buds begin to swell. Infection can be reduced on wall-trained fruit by protecting the plants from rain in spring with temporary plastic shelters. Leaf curl due to severe aphid infection can be confused with this fungal disease and close inspection is necessary. See *leaf blister.*

pea gravel small regular-shaped gravel, used especially by alpine gardeners for surface-mulching pots and sinks.

pea guard a narrow length of small-mesh wire netting formed into an inverted U-shape over wire supports to protect rows of seeds or seedlings from attack by birds.

peanut shells the fruit husks of the peanut, (*Arachis hypogaea*), used as a soil- and potting-medium conditioner. Physical properties vary with particle size, the coarser grade improving drainage and aeration, the finer ones improving water-holding. Structural stability is reasonable, although decomposition continues during use. Nitrogen depletion, salinity, and pathogens are potential problems.

pears (*Pyrus* species) hardy deciduous trees grown for their distinctive and variably shaped fruits. Freedom from frost and high summer temperature is essential for consistent quality cropping. The European pear has developed mainly from *Pyrus communis*, which is indigenous to Europe and much of northern Asia, particularly the Caucasus and Turkestan. It is among the most long-lived of fruit trees.

The shape of pears varies greatly with cultivar, and the classification usually adopted (from Bunyard's *Handbook of Fruits – Apples and Pears*, 1920) includes six groups: (1) Flat and round (e.g. 'Passe Crassanne'); (2) Bergamotte (flat, rounded/conical, e.g. 'Winter Nelis'; (3) Conical (e.g. 'Beurré Hardy'); (4) Pyriform, or waisted (e.g. 'Doyenné du Comice', which is widely grown and accepted as the finest flavoured pear); (5) Oval (e.g. 'Emile d'Heyst'); (6) Calabash, longer than pyriform (e.g. 'Conference', the most successfully grown pear in Britain).

Pears flower comparatively early and freedom from frost and summer warmth are essential for consistent cropping. Adequate shelter is necessary for cross-pollination, as is protection from cold winds, which can blacken leaves and fruitlets.

The pear usually succeeds on most soil types but calcareous sites may induce deficiencies of iron and manganese; it is intolerant of drought, so mulching and irrigation must be considered.

Most cultivars of European pear are not self-fertile and provision of suitable pollinator cultivars is imperative. 'Conference' is not self-fertile, as often erroneously stated, although it sometimes does produce parthenocarpic fruits under adverse conditions. Most cultivars are diploid, a few triploid or tetraploid, and flowering times vary so that it is essential to obtain detailed advice on pollination on purchase.

Propagation is by budding or grafting on to rootstocks of quince. 'Quince A' ('Angers Quince') is moderately vigorous and the most widely used; 'Quince C' is more dwarfing and having virus-free strains is useful for inducing early fruitfulness in vigorous cultivars such as 'Doyenné du Comice'. A major disadvantage of quince rootstock is that it is not compatible with some cultivars of pear, such as 'Williams Bon Chrétien', 'Jargonelle', 'Marie Louise' and 'Packham's Triumph', and double working is necessary.

Quince rootstock is insufficiently vigorous for the hot dry conditions often encountered in the fruit-growing areas of Australia, South Africa and the US, although it is used to produce small trees for gardens.

Plant in late autumn or early winter while the soil is still comparatively warm. Staking, mulching and irrigation are important. Tree forms for garden trees are similar to those used for apple.

Established cropping pear trees develop more intricate spur systems than do apples. Pruning is done between leaf-fall and late winter, the degree depending on the fruit bud to growth ratio, with less pruning being required for vigorous trees than for those of more moderate or weak habit. Some laterals can be shortened to three or four buds to encourage spur formation, and a few of the most vigorous

pear

Pests
birds
aphids
capsid bug
gall midge
winter moth

Diseases
brown rot
fireblight
canker
powdery mildew
scab

removed completely if overcrowded. Spur systems on older trees require shortening or selective removal. Branch shortening (dehorning) may be necessary when individual branches become too long, but in general tip-pruning should not be necessary. Where heavy crops develop, some fruit thinning is essential to enhance fruit size.

The harvesting of pears requires careful monitoring, particularly for earlier cultivars, which may be picked while still green yet parting readily from the spur; later cultivars must be allowed to mature fully.

Domestically, pears should be stored in cool, dark, slightly moist conditions, such as cellars or outhouses; the fruits are best not wrapped but laid out in single layers in trays. Alternatively they may be kept in a refrigerator, maintained above freezing level They should be conditioned in normal room temperature (at 18°C/65°F) for 2–3 days before consumption, to ensure full flavour and good eating texture. In Britain, the possibility of bullfinch damage to overwintering buds should be noted. Netting the trees is the only certain protection: without it, severe damage can render a tree worthless.

ASIAN PEARS (ORIENTAL, CHINESE, JAPANESE, NASHI) are derived from *P.pyrifolia* and *P.ussuriensis*. Until recently they were generally restricted to their regions of origin in China and Japan, where they have been known and used since ancient times, but cultivation has expanded worldwide. The chil-ling requirement is less than that for European pears, and so the potential for extended cultivation is considerable.

In Britain the climate is not usually conducive to success, the fruit in poor seasons remaining small, and even in more favoured regions the flavour is less rich and the texture crisper than that of European cultivars.

Cultivation requirements are similar to those of European pears. Rootstocks include Asian pear seedlings, *P.betulifolia* and *P.calleryana*; in Japan, *P.pyrifolia* seedlings are favoured. Growth can be vigorous but trees tend to commence fruiting quite early. Fruit is borne on two-year and older wood, with some on one-year laterals. Asian pears are partially self-fertile and flowering occurs at the same time as with European pears, which are suitable as pollinators. Heavy fruit set requires judicious thinning. The trees can be decorative, with attractive white blossom, glossy green foliage and striking red autumn colouring.

peas (*Pisum sativum*) an annual, mostly grown for its edible seeds, but including some cultivars in which the immature pod may be eaten. It is believed to have originated in central or southeast Asia, but is no longer found in the wild. Mature pea seeds were known as a food crop in Europe during the 15th century, and the cultivation of green peas is now widespread in temperate regions and highland areas of the tropics.

Garden peas are either round or wrinkle-seeded, the former being hardiest and the latter sweeter. Most cultivars have green pods but there are some with purple pods. Petit Pois peas produce a very small well-flavoured seed; Mangetout or Sugar type produce edible pods, some cultivars being dual purpose. Plant height varies from 45cm to 1.5m and supports are essential or desirable. Cultivars are classified as earlies, second earlies and main crop and the earlier-maturing type are the most suitable for small gardens.

Most pea cultivars succeed in a cool, humid climate with a temperature range of 13–18°C/55–65°F. A minimum soil temperature of 10°C/50°F is required for germination.

In mild areas, first sowings can be made in October or November, with protection from cold and mice. Otherwise, sow earlies from late February onwards, preferably protecting with cloches, and follow with successional sowings at four week intervals until midsummer. Sow 2.5cm deep, 5cm apart, either in single rows or 23cm-wide flat-bottomed drills. Leave a distance between lines of approximately the ultimate height of the cultivar.

Emerging plants need to be protected from birds and weed competition, and peas should be rotated around the vegetable garden to deter a build up of soil-borne disease. The roots produce nitrogen-fixing *Rhizobium* nodules, so that high nitrogen feeding should be avoided. Harvest regularly for prime quality and high yield.

peat incompletely decomposed plant remains, predominantly sedges, reed and moss arising in certain wetland areas, often in vast drifts. In its formation, natural decay processes are arrested by acid waterlogging and the exclusion of oxygen; the remains of succeeding bog plants accumulate and are compacted to form peat. It is a slow process and the average growth rate of peat layers in British bogs is estimated at no more than 1mm per annum. Wetland peat habitat, especially raised bogs, is rich in bird, insect and plant life, including butterflies, moths, dragonflies and damselflies, the bog asphodel and insectivorous plants. Occasionally, peatland holds important information of archaeological interest.

The first extensive use of peat in horticulture stemmed from its disposal as a waste product from cattle bedding, where it formed part of a useful bulky organic soil additive. The significant development of peat as an ingredient of standardized seed and potting composts dates back to development work at the John Innes Institute in the 1930s. The introduction of soilless compost in the 1960s led to the evolution of peat-based products, backed by intensive research and marketing to make multi-purpose peat compost widely available.

Peat offers the advantages of good air- and water-holding properties, general freedom from plant pathogens and low nutrient content. It is slow to decompose, odourless, light in weight and valued for its light brown colour. Its properties are however variable, depending on origin, age and processing; Sphagnum peat has excellent water-holding properties; sedge peat is more fully decomposed with less air-filled porosity and higher nitrogen content. Acidity varies with peat types and sources and the pH range is usually between 3.5 and 5.

peas

Pests
aphids
birds
weevils
pea moth
   (*see* moths)
thrips

The harvesting of peat initially despoils the landscape, and also destroys or threatens wildlife habitat. Its use for the improvement of soil structure and as a surface mulch is unnecessary in view of the wide range of alternative, mainly bulky organic materials, many of which contribute more benefits. Incorporating peat for the purpose of acidifying soil is not advisable in view of the large quantities actually required to effect a change. For gardeners who wish to avoid the use of peat in growing media, alternatives are available, with coconut fibre, leafmould and reduced wood products the most promising.

pectinate pinnately divided; the segments being many, slender and long, and close together, like the teeth of a comb.

ped an aggregate of soil particles; the basic unit of soil structure. A more general term than crumb.

pedate palmate except for the basal/lowest lobes on each side, which are themselves lobed.

pedicel the stalk of an individual flower or fruit.

pedicel necrosis a condition, particularly of roses or peonies, in which the pedicel just behind the flower bud shrivels and collapses. It is also known as bud disease or topple, and is usually the result of poor growth or sunscorch.

Pedilanthus (from Greek *pedilon*, a sandal or shoe, and *anthos*, flower, referring to the appearance of the flowers). Euphorbiaceae. Southern N America to tropical America. SLIPPER SPURGE, RED BIRD CACTUS, JEW BUSH. 14 species, succulent shrubs close to *Euphorbia* but with zygomorphic involucres composed of bracts resembling bird's heads. Provide a minimum temperature of 10°C/50°F. Grow in dappled sun or bright, indirect light in a dry, airy environment. Plant in a gritty, fast-draining medium. Water regularly in spring and summer, very sparingly in winter. Increase by stem cuttings in spring. *P.tithymaloides* (West Indies; evergreen or deciduous shrub, to 3m; leaves to 6cm, ovate to elliptic, keeled beneath; cyathia red above, yellow-green at base; subsp. *smallii*: JACOB'S LADDER, DEVIL'S BACKBONE, RIBBON CACTUS stems flexuous; 'Variegatus': leaves variegated white and red).

peg plant a young plant pulled from a seedbed to transplant into its growing position. It is especially used of brassica plants, the 'peg' alluding to the planting dibber.

Pelargonium (from Greek *pelargos*, stork: the fruit resembles the head and beak of a stork). Geraniaceae. Mostly from S Africa with a few from tropical Africa, Australia and Middle East; those here are S African. About 250 species, subshrubs, herbaceous perennials and annuals, sometimes with tubeous roots. Most are aromatic, glandular and hairy, many have somewhat succulent or swollen stems and leaves. The flowers are borne in stalked umbels. They consist of 5 petals, the 2 upper petals usually larger than the lower 3.

Pelargonium includes many of the garden plants commonly known as 'Geraniums'. Those described here are perennials, although some of the hybrids and cultivars are treated as half-hardy 'annuals'. They usually succumb to wet cold and prolonged frost and are therefore tender in most cool temperate zones. Many have a long flowering season and are ideal for greenhouse or house decoration, for summer bedding schemes, tubs, window boxes and hanging baskets. The garden hybrids may be roughly divided into the following groups – Angel, Zonal and Regal Pelargoniums, valued for their flowers and foliage, Ivy-Leaved Pelargoniums, derived from *P.peltatum* and valued for their trailing habit, foliage and flowers, and the diverse Scented-Leaved Pelargoniums, with a range of foliage colour and form, variously scented of apple, balsam, lemon, mint and spice. Under glass or in the home, they should be grown in a free-draining, gritty medium, in a brightly lit, dry and well-ventilated position. Water carefully at all times to avoid waterlogging and keep nearly dry during periods of low temperatures and low light intensity.

Tuberous and succulent species like *P.triste* and *P.carnosum* lose their foliage and should be kept dry while dormant. These species should be grown permanently under glass and associate well with cacti and other succulents.

Feed container-grown plants regularly when in growth. Pinch out the tips of young plants to encourage branching and prune shrubby, vigorous types like the scented-leaved pelargoniums to avoid legginess. Many of the more robust hybrids benefit from severe pruning in autumn; it also encourages growth suitable for use as cuttings. Overwinter species and hybrids in frost-free conditions. Plants used outdoors in summer need a sunny position and well-drained soil. Because their siting is not permanent, they may thrive for a season in conditions that would not suit them in the long term – for example, ivy-leaved pelargoniums in hanging baskets filled with a dense, soilless medium. Zonal pelargoniums may be overwintered as cuttings taken in mid to late summer. Alternatively, old plants may be dug up and stored dry in a cool, frost-free place. Cut back, removing any dead growth, and repot in late winter when the plants show signs of regrowth. Regal pelargoniums need a temperature slightly above 5°C/40°F in winter to continue growing and produce flowers for spring. Cuttings intended for bedding out should be hardened off before planting out after the last frosts. Deadheading will encourage a succession of blooms.

Increase by cuttings or (species) by seed. Small tubers and offsets of tuberous-rooted species may be detached and potted on. Whitefly is a serious pest under glass. The leaves of some species, especially the succulent types, may be scorched by insecticides. Grey mould will attack both cuttings and mature foliage. Cuttings may be killed by blackleg, a fungal disease that attacks the base of soft fleshy stems, blackening and rotting the tissue. It is encouraged by waterlogging and excess organic matter in the soil. *(see table overleaf)*

PELARGONIUM

Name	Habit	Leaves	Flowers
P.abrotanifolium SOUTHERNWOOD GERANIUM	erect and bushy, to 50cm, becoming woody	to 1.5cm, aromatic, grey-green, finely divided	to 1.5cm diam., white or pink, upper 2 petals narrow, marked maroon

Comments: Scented-leaved and species category

ANGEL PELARGONIUMS

Similar to Regal pelargoniums, but seldom exceeding 30cm and always with single flowers. Fine examples of this group include 'Catford Belle' with rose-red flowers with darker markings, and 'Mrs G.H. Smith', with off-white flowers marked pink.

P.capitatum ROSE-SCENTED GERANIUM	to 1m, woody-based, decumbent or weakly erect, aromatic, softly hairy	2-8cm, 3-5-lobed, rose-scented, velvety with crinkled margins	to 2cm diam., mauve-pink, upper 2 petals marked with darker veins

Comments: Scented-leaved and species category

P.carnosum	erect, to 30cm with a stout, succulent stem, gouty below, tapering above and covered in persistent leaf bases	to 15cm, succulent, grey-green, pinnately lobed	to 1cm diam., white to pale yellow-green, upper 2 petals marked red

P.crispum LEMON GERANIUM	erect, freely branched, aromatic subshrub to 70cm	to 1.5cm, lemon-scented, rounded to reniform with crisped to obscurely lobed margins, coarsely hairy	to 2cm diam., pink, upper petals marked deep pink

Comments: Scented-leaved and species category. 'Variegatum' has leaves edged with cream

P. Fragrans Group	erect, much-branched subshrubs to 40cm	usually small, cordate-ovate with three lobes and blunt teeth, grey-green, velvety, with a spicy scent	to 1.5cm diam., white, upper petals with red lines

Comments: Scented-leaved and species category. Cultivars include 'Old Spice', compact, with grey-green, crinkled leaves, and 'Variegatum' with leaves edged creamy yellow.

P.graveolens ROSE GERANIUM, SWEET-SCENTED GERANIUM	erect, downy subshrub to 150cm	to 4cm, triangular, finely pinnately lobed, grey-green, rose-scented	to 1.5cm diam., palest pink, upper petals veined purple

Comments: Scented-leaved and species category

P.peltatum IVY GERANIUM, HANGING GERANIUM	trailing or climbing with slender stems	3-7cm diam., thinly fleshy, rounded in outline with 5 triangular lobes	to 4cm diam., pale purple to pink, 2 upper petals with darker veins

Comments: This species is the parent of many excellent plants for hanging baskets, suspended pots, etc. Cultivars within the Ivy-leaved Group range from small- to large-leaved, with single to double flowers in shades of white, peach, salmon, orange-red, red, magenta and mauve. Some, such as 'Crocodile', have attractively variegated leaves

P.quercifolium OAK-LEAVED GERANIUM, ALMOND GERANIUM, VILLAGE OAK GERANIUM	erect, balsam-scented, clammy subshrub to 1.5m	triangular, deeply pinnately or palmately divided, the lobes themselves divided and toothed, coarsely hairy	to 1.5cm diam., purple-pink, the upper petals marked darker pink

Comments: Scented-leaved and species category. 'Royal Oak': shrubby, leaves with rounded lobes, with purple-brown blotch

REGAL PELARGONIUMS

Often grouped as *P.* × *domesticum*. A group of complex ancestry. To 50cm tall, the stems are usually thick and branching, giving a bushy habit well-furnished with foliage. The leaves are to 10cm diam., rounded to reniform, wavy and sharply toothed to obscurely 3-lobed and hairy. The flowers are large and showy (to 10cm diam.), carried in dense, straight-stalked umbels. Single or double, they consist of rounded and ruffled to lacy petals in a wide range of colours from white to rose, red, peach, salmon, orange, magenta, mauve and maroon-black and combinations thereof.

SCENTED-LEAVED PELARGONIUMS

A varied group of hybrids and selections drawn largely from the scented-leaved species described here and sharing qualities of one or the other. The foliage may be scented of roses, peppermint, lemon, balsam or apple. The leaves may be small or large, rounded and ruffled to narrowly palmately lobed, mid to grey-green and variegated white, yellow or marked chocolate brown. The flowers are generally small and white to pale pink, salmon or magenta.

P.tomentosum	low-growing and spreading to 50cm	to 6cm, base cordate, with 3-5, rounded lobes, velvety, peppermint-scented	large, paired, cream to pale pink, upper petals marked purple

Comments: Scented-leaved and species category

P.triste	low-growing deciduous herb to 20cm with a swollen, woody caudex and rootstock	to 15cm, ferny and very finely divided, softly downy	scented, petals narrow, cream to brown-purple overlaid with violet to maroon markings

UNIQUE PELARGONIUMS

Garden hybrids, these are sprawling to erect, woody-based plants with aromatic leaves, usually grey-green and palmately to pinnately lobed and toothed. The flowers are 2-5cm diam. and range in colour from white to pale pink, magenta, bright red and combinations thereof.

ZONAL PELARGONIUMS

Garden hybrids often grouped as *P.* × *hortorum*. These are the most popular members of the genus, widely grown for perennial display under glass or outdoors in favoured regions, and otherwise used for summer bedding. The plants are generally tall with succulent, jointed stems, these eventually become leggy and woody. The leaves are to 10cm diam., rounded to reniform and wavy to shallowly toothed; they may be plain green, variegated with white, cream, pink, purple-red and brown, or with a dark chocolate, horse-shoe-like marking (zoned). The flowers are freely produced in dense, long-stalked umbels. They are 2-4cm across, single or double, and composed, usually, of clawed obovate petals in shades of white, rose, pink, crimson, coral, peach, salmon, orange, flame and scarlet. Many cultivars have been named and further subcategories have emerged, namely *Cactus-flowered* (petals furled or quilled);*Irenes* (fast-growing and vigorous with large flowerheads used for cutting); *Stellar* (small plants with star-shaped, zoned leaves and flowers with cut and narrow petals); *Miniature* (no taller than 13cm, very free-flowering); *Rosebud* (flowers double with many small petals resembling a half-opened rosebud).

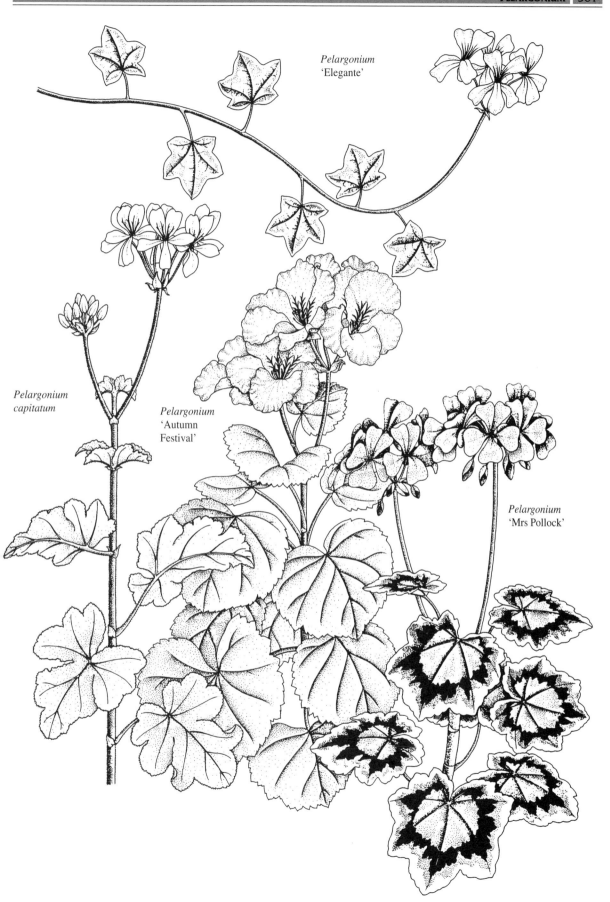

Pelargonium
'Elegante'

Pelargonium
capitatum

Pelargonium
'Autumn
Festival'

Pelargonium
'Mrs Pollock'

Pellaea (from Greek *pellos*, dusky, referring to the dark stipes). Pteridaceae (Adiantaceae). Tropical and warm temperate regions. 80 species, evergreen ferns with dark-stalked, pinnate to variously compound fronds. Cultivate as for tender *Adiantum* species.

P.atropurpurea (PURPLE ROCK BRAKE; North America; fronds 10–25 × 4–10cm, tufted, narrowly ovate-triangular to lanceolate, bipinnate, pinnules to 1.5cm, ovate to elliptic-lanceolate, auriculate at base, attenuate at tip, often glaucous-rusty-red when young); *P.rotundifolia* (BUTTON FERN; New Zealand, Australia; fronds 15–30cm, narrowly oblong pinnate, pinnae 1–2cm, dark glossy green, narrowly oblong to suborbicular, apiculate, minutely crenate).

pelleted (pelletized) seed seed coated with an inert material, often incorporating a pesticide, to make it spherical and facilitate accurate and more economical mechanical or hand sowing.

Pellionia (for Alphonse Pellion (1796–1868), a French naval officer who took part in Freycinet's second voyage around the world 1817–1820). Urticaceae. Tropical and subtropical Asia. Some 50 species, perennial herbs or subshrubs, sometimes with succulent stems. The leaves are alternate, 2-ranked, and entire or dentate. Inconspicuous flowers are borne in dense cymes. Cultivation as for *Pilea*.

P.pulchra (RAINBOW VINE; Vietnam; low spreading to creeping herb, stems fleshy, tinged purple; leaves 2–5cm, obliquely oblong to elliptic, marked dull black-green along midrib and veins above, purple beneath); *P.repens* (TRAILING WATERMELON BEGONIA; SE Asia; creeping herb to 60cm, stems succulent, tinged pink; leaves 1–6cm, oblong to orbicular, scalloped, bronze-green above, tinged violet with a broad central band of pale grey-green, tinged pink beneath, purple-edged; 'Argentea': leaves silver-white marked pale green).

peloria, peloric see *malformations*.

peltate a leaf whose stalk is attached inside its margin rather than at the edge, usually at the centre beneath.

pendent hanging downwards, more markedly than arching or nodding.

pendulous dependent.

penicillate brush-shaped, or like a tuft of hairs.

penjing the Chinese art of container cultivation, analogous to Japanese bonsai, and often a means of creating mixed miniature landscapes. In penjing, form is usually achieved by stricter pruning-based training than in bonsai.

Pennisetum (from Latin *penna*, feather, and *seta*, bristle, referring to the plume-like bristles of some species). Gramineae. Tropics, subtropics and warm temperate regions. Some 80 species, rhizomatous or stoloniferous, annual or perennial grasses with clumped, slender culms, narrow leaves and, in summer and autumn, showy, foxtail- or bottle-brush-like flowerheads with feathery to bristly spikelets. The following are valued for their foliage and graceful flowerheads produced from midsummer to autumn. Some of the cultivars are among the most colourful and elegant of all ornamental grasses. They are not reliably hardy, but may be treated as halfhardy annuals or overwintered in a cool greenhouse. They are excellent candidates for mixed container plantings for patios. Grow in full sun in a fertile, gritty medium. Keep moist when in growth, rather drier in winter. Increase by division in spring, or by stem cuttings in a closed case or under mist in summer

P.alopecuroides (CHINESE PENNISETUM, SWAMP FOX-TAIL GRASS, FOUNTAIN GRASS; E Asia to W Australia; perennial to 1.5m; leaves to 60 × 1.2cm, glabrous, scabrous; inflorescence cylindric to narrow-oblong, to 20 × 5cm, yellow-green to dark purple; 'Burgundy Giant': to 1.2m, with broad leaves, bronze tinted claret; 'Hameln': dwarf, to 50cm, with golden leaves in autumn and white flowers tinted green; 'Moudry': low, with wide, dark green, shiny leaves and dark purple to black flowerheads; 'Woodside': robust, with a dark purple inflorescence, early-flowering; var. *purpurascens*: dark purple spikelet bristles; var. *viridescens*: pale green spikelet bristles); *P.setaceum* (FOUNTAIN GRASS; tropical Africa, SW Asia, Arabia; perennial to 90cm; leaves to 30 × 0.3cm, rigid, very scabrous; inflorescence erect to inclined, plumed, to 30 × 30cm, tinged pink to purple; 'Purpureum': purple leaves and a deep crimson inflorescence; 'Rubrum': tall, with maroon leaves tinted bronze and deep burgundy plumes); *P.villosum* (FEATHERTOP; NE tropical Africa; perennial, to 60cm; leaves to 15cm × 0.6cm; inflorescence cylindric to subglobose, compact, plumed, to 11 × 5cm, tinged tawny brown to purple).

penninerved, penniveined, penniribbed with closely arranged veins extending from the midrib in a feather-like manner.

Penstemon (from Greek *pente*, five, and *stemon*, stamen, referring to the prominent, sterile fifth stamen). Scrophulariaceae. Western N America, 1 species in Kamchatka and N Japan. About 250 species, subshrubs or perennial herbs, with terminal racemes or panicles of tubular, two-lipped flowers. Although hardy in climate zone 7, the taller species and cultivars may be cut to the ground or killed by hard frosts. For this reason, they should be thickly mulched in autumn and a few rooted cuttings overwintered in frost-free conditions as an insurance policy. They are otherwise colourful and undemanding perennials for the sunny herbaceous and mixed border, thriving in all well-drained, fertile soils. Low-growing small shrubs or mat-formers, such as *P.davidsonii*, *P.menziesii*, *P.newberryi*, *P.pini-*

folius and *P.rupicola*, may be grown in full sun in rock gardens, alpine sinks, among paving stones or as front of border plants. They soon deteriorate in wet, cold conditions. Shear low, shrubby species back after flowering to maintain vigour, and cut back taller plants almost to ground level in spring after the last severe frosts. Propagate by semi-ripe cuttings of non-flowering side shoots in late summer, rooted in a sandy, free-draining medium in a closed frame and overwintered in their first year in frost-free conditions. Increase also from softwood cuttings in early summer. Herbaceous species forming basal tufts of foliage may be divided in spring.

P.barbatus (New Mexico to Utah, Arizona and N Mexico; to 1m; leaves lanceolate to ovate, to linear; flowers 3–4cm, red tinged pink to carmine, lower lobes yellow-hirsute at base; 'Albus': white; 'Carneus': pale pink; 'Coccineus': bright scarlet; 'Roseus': pink); *P.campanulatus* (Mexico and Guatemala; 30–60cm; leaves linear-lanceolate, serrate, to 7cm; flowers 2.5cm+, rosy purple or violet, glandular-pubescent; 'Pulchellus': violet or lilac; 'Purpureus': purple); *P.confertus* (YELLOW PENSTEMON; British Colombia to Alberta, Montana and Oregon; stem slender, 2–5cm; leaves 3–7cm, lanceolate to oblanceolate, thin; flowers 8–12mm, pale sulphur-yellow, palate brown-hirsute; 'Violaceus': violet); *P.davidsonii* (Washington to California; differs from *P.menziesii* in having entire and 18–35mm flowers; 'Albus': white; 'Broken Top': low, spreading, rich purple); *P.fructicosus* (SHRUBBY PENSTEMON; Washington to Oregon, east to Montana and Wyoming; shrubby, to 40cm; leaves 1–5cm, lanceolate to elliptic, entire or thinly coriaceous; flowers lavender-blue to pale purple; 'Albus': white; 'Major': flowers large); *P.hartwegii* (Mexico; stem 90–120cm; leaves to 10cm, lanceolate to ovate-lanceolate; flowers 5cm, deep scarlet, minutely viscid-pubescent); *P.heterophyllus* (FOOTHILL PENSTEMON; California; subshrub or shrub, 30–50cm; leaves 2–8cm, linear to lanceolate green or leaden-grey; flowers 2.5–3.5cm, rosy violet below, lobes blue or opalescent lilac; 'Blue Gem': dwarf; 'Heavenly Blue': blue to mauve); *P.hirsutus* (eastern N America; stem erect, 40–80cm, minutely glandular-pubescent; leaves 5–12cm, lanceolate to oblong, subentire to dentate; flowers 2.5cm, dull purple with white lobes; 'Caeruleus': blue-tinted; 'Purpureus': clear purple; 'Pygmaeus': to 15cm high, violet; 'Roseus': pink); *P.isophyllus* (Mexico; stem 70cm, decumbent then erect, simple, stout, purple; leaves 3–4cm, lanceolate, revolute, rather thick; flowers 4cm, red with a yellow throat); *P.menziesii* (British Colombia and Vancouver Island to Washington; forming creeping mats; flowering stems to 10cm; leaves 0.5–1.5cm, elliptic to orbicular, minutely serrate; flowers 2.5–3.5 cm, violet-purple; 'Microphyllus': very compact, with small leaves and lavender flowers); *P.newberryi* (MOUNTAIN PRIDE; California and adjacent Nevada; forming mats 15–30cm tall, woody below; leaves 1.5–4cm, elliptic to ovate, minutely serrate, coriaceous; flowers 2–3cm, rosy red); *P.pinifolius* (New

Mexico to Arizona and Mexico; stems to 40cm, numerous, woody below; leaves filiform, to 1cm; flowers 2.5–3cm, scarlet; 'Mersea Yellow': bright yellow); *P.procerus* (SMALL-FLOWERED PENSTEMON; northwestern N America; stem 10–40cm, slender; leaves 2–6cm lanceolate to oblanceolate or oblong, deep green, thin; flowers blue-purple); *P.rupicola* (ROCK PENSTEMON; Washington to California; mat-forming, woody below, to 10cm; leaves 0.8–2cm, elliptic to orbicular, minutely serrate-dentate, very glaucous, thick, glabrous or canescent; flowers 3cm, deep rose; 'Albus': white; 'Diamond Lake': large, rich pink; 'Pink Dragon': light salmon-pink; 'Roseus': pink); *P.serrulatus* (CASCADE PENSTEMON; S Alaska to Oregon; subshrub, 30–70cm; leaves 2–9cm, broadly lanceolate to spathulate, subentire to serrate or shallowly laciniate; flowers 1.5–2.5cm, deep blue to dark purple; 'Albus': white). *Cultivars*: 'Alice Hindley': mauve and white; 'Amethyst': amethyst-blue; 'Andenken an Friedrich Hahn': deep wine red; 'Apple Blossom': small white, tipped pink; 'Burgundy': deep wine-purple; 'Edithae': shrubby, prostrate, deep lilac; 'Evelyn': neat, with narrow leaves and slim rose-pink flowers with pale striped throats; 'Firebird': clear red; 'Garnet': garnet-red; 'Hidcote Pink': rose pink, throat white with dark pink streaks; 'Hidcote White': white; 'Hopley's Variegated': variegated form of 'Alice Hindley'; 'Maurice Gibb': rich cerise with a white throat; 'Mother of Pearl': white flushed purple; 'Old Candy Pink': vibrant pink; 'Osprey': pink, throat cream; 'Pink Endurance': candy-pink, with honey lines at throat; 'Prairie Fire': orange-red; 'Shoenholzeri': deep red; 'Six Hills': prostrate with fleshy grey-green leaves, lilac; 'Snow Storm': white, flushing pink; 'Sour Grapes': tinted purple; 'Weald Beacon': tinted blue.

Penstemon heterophyllus

Penstemon pinifolius 'Mersea Yellow'

pentamerous with parts in groups of five, or multiples of five; often written 5-merous.

pentaploid a polyploid with five sets of chromosomes.

Pentas (from Greek *pente*, five, referring to the characteristic number of floral parts). Rubiaceae. Tropical Arabia and Africa, Madagascar. Some 30 or 40 species, perennial or, rarely, biennial herbs or shrubs with ovate to lanceolate leaves. Produced at various times of year in crowded, much-branched cymes or flat-topped corymbs, the flowers are small, tubular to cylindric, with a spreading limb of five, ovate to oblong lobes. Grow in sun or light shade. Minimum temperature 10°C/45°F. Plant in a sandy, fertile mix. Water and feed freely in spring and summer. Keep moist in winter. Increase by soft stem-tip cuttings taken in late spring and inserted in a heated case. *P.lanceolata* (syn. *P.carnea*; STAR-CLUSTER, EGYPTIAN STAR-CLUSTER; Yemen to E Africa; herb or subshrub, to 2m, erect or prostrate; flowers pink or magenta to lilac or, occasionally, white to 4cm in dense, domed clusters).

Peperomia (from Greek *peperi*, pepper, and *homoios*, resembling, referring to its resemblance to the closely allied genus *Piper*). Piperaceae. Pantropical. RADIATOR PLANT. About 1000 species, small, generally evergreen, succulent herbs (those listed here are perennial). They are grown for their handsome, fleshy leaves, and, less commonly, for their narrow spikes of very small flowers. Suitable for the home, warm greenhouse or conservatory (minimum temperature 15°C/60°F). Grow in pots or hanging baskets in a porous but moisture-retentive medium rich in fibrous organic matter. Provide a moderately humid but buoyant atmosphere, with shade or bright indirect light. When in growth liquid-feed fortnightly and water by drenching thoroughly and allowing the medium to dry partially between waterings; in winter, keep just moist. Propagate by division, leaf or stem cuttings, or by seed.

P.argyreia (syn. *P.sandersii*; WATERMELON BEGONIA, WATERMELON PEPPER; northern S America to Brazil; stem to 20cm, erect, dark red; leaves to 7cm, broadly ovate, concave, acute at apex, rounded at base, peltate, silver-grey above with dark green stripes along main veins); *P.caperata* (EMERALD-RIPPLE PEPPER, GREEN-RIPPLE PEPPER, LITTLE-FANTASY PEPPER; Brazil; stem erect to 20cm; leaves 3.2cm, cordate, broadly acute to rounded at apex, rounded or auriculate at base, often peltate, dark glossy green, veins impressed, in rippling folds; 'Tricolor': leaves with a wide cream edge, marked pink at base; 'Variegata': edged cream); *P.clusiifolia* (West Indies, perhaps Venezuela; erect to 25cm; leaves 7.5cm, obovate to elliptic, broadly acute to rounded at apex, tapered to often clasping base, green or tinged purple, margin flushed maroon; 'Variegata': leaves light green, variegated cream towards edge, with a red margin); *P.fraseri* (FLOWERING PEPPER; Colombia and Ecuador; to 40cm, erect; leaves 3.5cm, broadly ovate to suborbicular, acute at apex, cordate at base, tinged purple on veins, pale green beneath with bright red to pink veins and pink spots; flowers bright white, fragrant, mignonette-like, in spikes to 4cm); *P.glabella* (C and S America, West Indies; to 20cm, erect or sprawling; leaves 3.8cm, broadly elliptic to obovate, rounded to obtusely acute at apex, broadly tapered to rounded at base, green, often black-spotted, fleshy; 'Variegata': leaves edged or variegated off-white); *P.griseo-argentea* (IVY-LEAF PEPPER, SILVER-LEAF PEPPER, PLATINUM PEPPER; Brazil; to 20cm, erect; leaves 3.8cm, cordate, broadly acute to rounded at apex, auriculate or rounded at base, often peltate, grey-green above, paler beneath, coriaceous, veins deeply impressed above; 'Nigra': leaves corrugated, metallic black); *P.magnoliifolia* (Panama, northern S America and West Indies; differs from *P.obtusifolia* in its more upright habit and leaves more contracted to rather acute apex; 'Greengold': leaves spattered green and yellow; 'Variegata': leaves variegated lime green); *P.marmorata* (S Brazil; to 30cm, stem erect; leaves 10cm, ovate, obtuse to acute at apex, deeply cordate to auriculate at base with overlapping lobes, green above with silver-grey or white patterns between veins, coriaceous; 'Silver Heart': leaves dappled silver between veins); *P.metallica* (Peru; to 15cm, erect; leaves 2.5cm, elliptic, acute at apex, broadly tapered to rounded at base, green tinged brown with silvery-green band above, tinged red beneath, succulent); *P.obtusifolia* (BABY RUBBER PLANT, AMERICAN RUBBER PLANT, PEPPER-FACE; Mexico to northern S America and West Indies; stoloniferous, stem rooting, ascending, to 15cm; leaves 10cm, elliptic to obovate, rounded to emarginate at apex, cuneate at base; 'Alba': cream new growth and stem and petioles dotted red; 'Minima': dwarf, dense, with shiny leaves; 'Variegata': leaves more pointed, variegated pale green and marked cream towards margin); *P.rotundifolia* (C and S America, West Indies; to 25cm, creeping; leaves orbicular to broadly elliptic, rounded or cordate at base, 1.5cm, green, paler beneath, sparsely hairy, fleshy; var. *pilosior*: more densely hairy, leaves with pale green reticulate pattern above); *P.rubella* (West Indies; erect to 15cm; leaves in whorls to 1cm, elliptic, broadly acute to rounded at apex, broadly tapered to rounded at base, light to dark green above, flushed pink, convex and sparsely hairy beneath); *P.scandens* (origin unknown; stem 60cm+, scandent, stout; leaves to 7.5cm, ovate to suborbicular, long-acuminate at apex, truncate to subcordate at base).

pepo a multicarpellate fruit with fleshy pulp and a hard exocarp derived from the receptacle.

pepper dust a repellent to cats and dogs, sprinkled on and around plants and beds.

peppers annual or short-lived perennials, natives of tropical America, grown for their edible, often highly decorative fruits; they include the SWEET PEPPER (*Capsicum annuum* Grossum Group), the CHILLI PEPPER (*Capsicum annuum* Longum Group) and the HOT PEPPER (*Capsicum frutescens*). Cultivars of sweet pepper may be red, yellow, orange or purple, ripening from green in warm conditions; they come in a range of shapes, round, tapering or irregularly oblong or bell-shaped and measuring 3–15cm long and 3–7cm in diameter.

Chilli peppers have narrow, pointed fruits, ripening to bright red, with a hot flavour which increases in intensity with maturity. Sweet and chilli peppers plants grow up to 75cm tall. Hot peppers are branching perennials, grown as annuals, which may reach 1.5m in height and bear small red, orange or yellow fruits that are highly pungent; they need a long growing season.

For best results, peppers require a minimum temperature of 20°C/70°F with 70–75% humidity, so greenhouse protection is desirable. Fruit set is adversely affected at 30°C/86°F and above.

Sow under protection at 20°C/70°F in mid-March for indoor cropping, one month later for outdoor cropping, either in trays for pricking out or direct into small pots or cell modules. Plant when 10cm

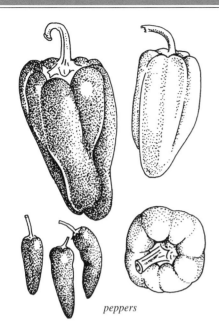

peppers

high, 38–45cm apart, for sweet and chilli peppers, 60cm apart for hot peppers. Plants normally branch naturally, but the terminal shoot may be pinched out to ensure a bushy habit. Most cultivars do not require staking, except those that are over 75cm tall. Mulch the plants, water sparingly but regularly, and apply liquid feed every 14 days; keep damped down where under protection.

Peppers can be picked when still green, and are ready when glossy and smooth. They may be left on the plant to turn colour, and generally are sweeter once ripened. Sweet peppers last well on the plant, and also freeze well. Chilli peppers may be left to shrivel on the plant and can be kept for long periods. See *Capsicum*.

perch see *rod*

perennial a plant that persists for more than two seasons. The term is most usually applied to herbaceous ornamentals, but, strictly, is applicable also to trees and woody shrubs.

Pereskia (for Nicholas Claude Fabry de Peiresc (1580–1637), councillor at Aix and patron of science). Cactaceae. S Mexico, C America, north and west S America, Brazil. 16 species, spiny trees, shrubs and woody climbers. They are true cacti, but lack obviously succulent stems; moreover, they bear 'leaf-like' leaves that are scarcely succulent. The flowers are rotate, showy and borne in panicles and corymbs. They are followed by berry- to pear-like fruit. Provide a minimum temperature of 10°C/50°F Plant in a mildly acidic to neutral, gritty medium; shade in hot weather and maintain low humidity. Water freely in summer; keep dry from mid-autumn until early spring, except for light misting on warm days in late winter. *P.aculeata* (tropical America; woody climber to 10m; leaves lanceolate to elliptic

or ovate, to 11cm, spines stipular or at areoles on older growth only; flowers 2.5–5cm in diameter, scented, white or nearly so; 'Godseffiana': leaves variegated yellow to peach-coloured and purple-tinted beneath).

perfect a bisexual flower or an organ with all its constituent members.

perfoliate of a sessile leaf with basal lobes united and surrounding the stem, which then passes through the blade.

pergola a tall structure upon which climbing and rambling plants are grown. It normally consists of vertical brick or timber pillars supporting horizontal beams. Double pergolas are often constructed one to each side of a pathway and linked by cross pieces to form a covered walk.

perianth the collective term for the floral envelopes, the corolla and calyx; especially when the two are not clearly differentiated.

Pericallis (from Greek *peri*, around, and *kallos*, beauty). Compositae. Canary Islands. About 14 species, perennial herbs, subshrubs or shrubs with daisy-like flowerheads. The florists' cineraria, *P. × hybrida*, is the most frequently seen in cultivation, flowering within about six months from seed, and valued for the exceptional profusion of bloom which sometimes covers the plant so densely as to obscure the foliage. In warmer climates with frost-free winters and warm dry springs, *P. × hybrida* is sometimes used as bedding but is more usually grown as a winter-flowering house or cool conservatory plant and discarded after blooming. Available in a wide range of cultivars and seed races, from compact Grandiflora and Multiflora types to the more graceful, branching 'Stellata' forms, which carry the blooms clear of the mound of foliage.

Surface sow seed in early to late summer in a sandy, soilless propagating medium and germinate in light shade at 13°C/55°F. Prick out individually into small pots containing a low-fertility loam-based mix, and pot on successively into a medium-fertility potting mix. Ensure excellent drainage and water freely but carefully; plants collapse rapidly if overwatered. Liquid feed fortnightly when roots have filled the pots. Shade in summer, maintaining cool, humid and well-ventilated conditions at temperatures of no more than 15°C/60°F. Overwinter at 7–10°C/45–50°F, increasing temperatures to about 16°C/60°F if necessary, to hasten flowering once buds have formed. The two species listed here need gritty soils in full sun with low humidity and a minimum temperature of 5°C/40°F. Increase by seed or stem cuttings in summer or autumn.

P.cruenta (Tenerife; white-woolly perennial herb to 1m; leaves ovate to triangular-cordate, 5–15cm wide, toothed, pink to purple-tomentose beneath; flowerheads 2.4–4cm in diameter, ray florets pink to maroon

peppers

Pests
aphids
red spider mite

Disease
grey mould

Disorder
blossom end rot

*Perovskia
atriplicifolia*

or purple, disc florets dark purple); *P.* × *hybrida* (FLORIST'S CINERARIA; a variable complex of hybrids between *P.lanata* and *P.cruenta* and possibly other species – perennial herbs, often grown as annuals, forming compact cushions or openly branched plants to 1m; leaves ovate to suborbicular, often light green; flowerheads to 5cm in diameter, florets white to pink, red, maroon, deep purple, violet, lilac and blue, ray florets sometimes bicoloured, with white base and coloured apex; numerous cultivars and seed races are available with flowers in a wide range of bright colours); *P.lanata* (S Tenerife; white-hairy subshrub to 1m, ascending to procumbent; leaves to 15cm, broadly ovate-cordate to suborbicular, toothed or pinnatifid; flowerheads 3–5cm in diameter, violet-scented, ray florets mauve, disc florets purple).

pericarp the wall of a ripened ovary or fruit, sometimes differentiated into exocarp, mesocarp and endocarp.

periderm an outer layer of protective tissue on stems and roots which may consist of cork, cork cambium and phelloderm.

perigone the perianth, especially when undifferentiated; or anything surrounding a reproductive structure.

perigynous of flowers in which the perianth and stamens are apparently basally united and borne on the margins of a cup-shaped rim, itself borne on the receptacle of a superior ovary.

Perilla (derivation of name obscure). Labiatae. India to Japan. Some 6 species, fast-growing, half hardy annuals, grown for their ornamental and often aromatic foliage. Grow in rich, well-drained but moisture-retentive soils in full sun. Propagate from seed sown under glass in early spring, to plant out in early summer. Pinch out young plants to encourage bushiness. *P.frutescens* (Himalaya to E Asia, naturalized Ukraine; to 1m, erect, pubescent; leaves 4.5–12cm, broadly ovate, acuminate, corrugated, deeply serrate, green sometimes speckled purple; 'Atropurpurea': leaves deep red-purple; var. *nankinensis*: leaves laciniate-dentate, dark bronze or purple with crisped, fringed margins; f. *rosea*: leaves crenate, variegated red pink).

perlite, perlag heat-expanded granules of volcanic alluminosilicate mineral, used as a potting-compost ingredient or separately as a medium for rooting cuttings or for hydroculture. It has a stable structure and high porosity, with negligible nutrient content and cation-exchange capacity, and is a useful additive for improving drainage and aeration. The coarser formulation is known as perlag.

permanent wilting point the level of soil moisture content at which transpiration ceases and plants remain wilted even in a humid environment.

Perovskia (for the l9th-century Turkestani statesman, B.A. Perovskii). Labiatae. Asia Minor, Iran, C Asia, Himalaya. 7 species, deciduous, perennial, aromatic shrubs or subshrubs with deeply toothed or laciniate, grey-green leaves and, in summer and autumn, panicles of many small, tubular flowers. All species are tolerant of dry chalk soils, suited to maritime situations and hardy to –15°C/5°F, although stems above ground may be killed by frost. The best foliage effects are obtained on new growth and plants should be cut back almost to the base in spring. Grow in perfectly drained soils, in full sun. Propagate by softwood cuttings in late spring, or from heeled cuttings of lateral shoots in summer.

P.abrotanoides (Afghanistan, W Himalaya; to 1m; leaves to 7cm, oval to oblong, bipinnatisect, grey-green; panicle to 40cm, flowers 4–6 per whorl, calyx violet, corolla pink); *P.atriplicifolia* (Afghanistan, Pakistan; to 150cm; leaves to 6cm, lanceolate to cuneate, grey-green crenate or coarsely toothed; panicle narrow, flowers soft blue; 'Blue Mist', earlier flowering with light blue flowers, and 'Blue Spire', deeply cut leaves, lavender blue flowers in a larger panicle); *P.* 'Hybrida' (*P.abrotanoides* × *P.atriplicifolia*; leaves to 5cm, ovate, pinnatisect to bipinnatisect, grey-green; inflorescence a long panicle of lavender-blue flowers).

perpetual used of plants that flower more or less continually over a long period, especially greenhouse carnations.

Persea (name used by Theophrastus for an Egyptian tree). Tropical and subtropical America, Macaronesia, SE Asia. Lauraceae. 150 species, evergreen trees or shrubs with large, entire leaves and small, yellow-green flowers massed in panicles. The fruit is a berry or drupe, ellipsoid to pyriform, with a leathery skin, richly oily flesh and a single, large nut. In cool temperate zones, *P.americana* is occasionally grown for interest, although plants are prone to leaf spotting, take up to seven years to flower, and, if grown from seed, are likely to produce fruit of inferior quality. Grow in a well-drained, medium-fertility, loam-based mix. Syringe daily in summer and water moderately when in growth, less in winter. Maintain a minimum temperature of 10°C/50°F. To germinate, plant the seed with the pointed end about lcm above the surface of the potting mix. Keep moist and shaded at 15–20°C/60–70°F. Move into full light following germination, which may take up to six months. Propagate also by greenwood cuttings in a closed case with bottom heat. *P.americana* (AVOCADO PEAR, AGUACATE, ALLIGATOR PEAR, PALTA; C America; to 20m; leaves 10–25cm, ovate to elliptic, subcoriaceous to papery, dull dark-green above, paler beneath; fruit 2–12cm, oblong-ovoid to pyriform, skin leathery, glossy dark green, pale-punctate, to dark purple-green and tuberculate, flesh lime-green to yellow, firm, smooth and oily; seed ovoid, 4cm; numerous cultivars are grown commercially).

persistent neither falling off nor withering.

perule a bud scale.

pest in its widest sense, any organism that causes harm or damage to humans, their animals, plants or possessions, even if only by annoyance. Invariably, the term is used more narrowly for animal organisms, as opposed to pathogens that cause disease, or weeds. Generally, a pest is considered to be of economic importance when it causes a reduction in yield of 5–10%; for gardeners, the relative importance of a particular pest is determined by circumstances. Methods of pest control can be broadly divided into those that are non-chemical and essentially preventive, and those that are chemical and mostly curative. Non-chemical methods include cultural, mechanical, physical, legislative and biological control.

pesticides substances that kill animal organisms such as insects, eelworms, mites, slugs and snails, and rodents; but also used more generally to include chemicals for the control of plant diseases and weeds. They can be naturally occurring materials, a purified concentrate of such materials, or manufactured substances not found in nature. The term covers acaricides, algicides, bactericides, fungicides, herbicides, insecticides, molluscicides and nematicides. Pesticides are prepared in different ways to ensure efficiency, such as sprays, aerosols, dusts, smokes, seed dressing, fogging, injection, and weed wiping. cf. *biological controls, cultural controls*

petal one of the modified leaves of the corolla, generally brightly coloured and often providing a place on which pollinators may alight.

petal fall also known as blossom fall. See *bud stages.*

petaloid, petalody used of flower parts that take the form of petals in colour, shape or texture, such as the calyces or stamens of double flowers.

Petasites (from Greek *petasos*, a broad-brimmed hat, referring to the shape of the leaves). Compositae. Northern temperate regions. BUTTERBUR, SWEET COLTSFOOT. Some 15 species, hardy, rhizomatous, perennial herbs with broad, stalked, basal leaves and button-like flowerheads in a dense panicle, produced usually before the leaves. Although invasive, some *Petasites* are not without value in the garden. *P.fragrans* (not reliably hardy where temperatures fall below –15°C/5°F) has strongly vanilla-scented flowers carried in mid-winter, with or just before the leaves. Good groundcover for wilder areas of the garden, it tolerates drier conditions than other species. In moist and shady areas of the wild or woodland garden the hardy *P.japonicus* makes stately and impressive colonies, especially at the waterside. Both species may be confined by planting in sunken tubs or barrels, although best effects are obtained *en masse*. Propagate by division or seed.

P.fragrans (WINTER HELIOTROPE; C Mediterranean; to 30cm; leaves reniform to cordate, lobed, glabrous above, pubescent beneath, toothed; flowerheads white-pink, vanilla-scented); *P.japonicus* (Korea, China, Japan, naturalized Europe; to 1m; leaves to 80cm in diameter, reniform to cordate, more or less lobed, glabrous above, pubescent beneath, toothed; flowerheads pale mauve to almost white; 'Variegatus': leaves spotted milky yellow to cream; var. *giganteus*: syn. *P.giganteus*, leaves 0.9–1.5m across on petioles to 2m).

petiolate, petioled furnished with a petiole.

petiole the leaf stalk.

petiolule the stalk of a leaflet.

Petrea (for Lord Robert James Petrie (1714–1743) of Thorndon, Essex, owner of one of the best private collections of exotics in Europe, supervised by Phillip Miller). Verbenaceae. Tropical America and Mexico; *P.volubilis* naturalized in India. 30 species, evergreen lianes, shrubs, or small trees with elongate racemes in summer. The flowers consist of a showy campanulate calyx with five linear-oblong lobes and a hypocrateriform corolla, darker than the calyx, with five rounded lobes. Under glass, grow in direct sun or bright filtered light with medium to high humidity and a winter minimum temperature of 10°C/50°F. Mist frequently in warm sunny weather. Plant in large tubs, open beds or under and through staging in a freely draining, lime-free medium enriched with leafmould. Propagate by layering (both simple and air) or semi-hardened nodal cuttings inserted in sand with bottom heat and misted regularly. *P.volubilis* (BLUEBIRD VINE, PURPLE WREATH, SAND PAPER VINE; C America and Lesser Antilles, introduced elsewhere; woody vine to undershrub, to 12m; leaves to 21cm, oblong to elliptic, subcoriaceous, scurfy to bumpy and bristly, deep green; inflorescence to 36cm, calyx lilac, lobes to 1.8cm, corolla to 0.8cm, indigo to amethyst).

Petrea volubilis

Petrocosmea (from Greek *petros*, rock, and *kosmos*, decoration, referring to the natural habitat). Gesneriaceae. Asia. Some 29 species, perennial, rhizomatous herbs with hairy leaves in basal rosettes. Produced in summer, the flowers are solitary or clustered on scapose inflorescences, and resemble those of *Saintpaulia*. Near-hardy in climate zone 8 and elsewhere suitable for the alpine house. Grow in light shade in a fibrous, gritty, acid mix. Keep moist and cool in summer; protect from damp and draughts from autumn till spring. Increase by seed or by leaf cuttings in late spring. *P.kerrii* (Thailand; leaves to 10cm, ovate to oblong; flowers white, lobes blotched yellow at base).

Petrorhagia (from Greek *petros*, rock, and *rhagas*, fissure referring to the habitat). Caryophyllaceae. Eurasia, especially in the E Mediterranean. 25–30 species, summer-flowering annual or perennial herbs, resembling *Gypsophila*. Suitable for rock gardens, dry walls, raised beds and for border edging. Hardy to −15°C/5°F. Grow in sun in a well-drained, sandy soil. Propagate by seed sown in late winter, or by soft stem cuttings of non-flowering shoots in early summer. *P.saxifraga* (S and C Europe; mat-forming perennial to 40cm; leaves linear; inflorescence loose delicate, petals 5–10mm, pale pink with deeper veins, limb indented; includes cultivars with double and single, pure white, pale rose and deep pink flowers).

Petroselinum (from Greek *petros*, a rock, and *selinon*, parsley or celery). Umbelliferae. Europe. 3 species, biennial herbs with aromatic, ternately compound leaves and small, white or green-yellow flowers in compound umbels. *P.crispum* is PARSLEY (q.v.).

Petunia (from the Brazilian *Petum*, a name applied to the closely related tobacco plant, *Nicotiana*). Solanaceae. Tropical S America. Some 35 species, downy annual or perennial herbs, subshrubs or shrubs with salverform to funnelform flowers. Colourful, half-hardy annuals used for bedding in borders, tubs and window boxes or cultivated in pots under glass for winter and early spring flowering. The dwarf types make particularly good edging plants, while the trailing cultivars (e.g. 'Cloud Mixed' and 'Balcony Blended Mixed') are suitable for hanging baskets and as summer groundcover in borders; a single plant may cover up to a metre of soil. Grandiflora Hybrids generally have fewer and larger flowers than the Multiflora types, but are less weather-resistant, spotting badly after rain: F_1 hybrids of both types show increased vigour and uniformity.

All species tolerate poor soils and maritime exposure. They require plenty of sunshine and settled, dry summer weather to make a good display, although recent developments such as the 'Resisto' and 'Plum Crazy' mixtures show greatly improved performance in windy, rainy conditions. Propagate from seed pressed lightly on to the surface of a finely sieved seed mix: sow from late summer to autumn for winter- to early spring-flowering pot plants, and from late winter to early spring for summer bedding. Germinate at 15–20°C/60–68°F and keep evenly moist, watering from below when necessary. Prick out both large and small seedlings in seed mixtures to retain the balance of colour. Harden off in the cold frame for planting out after the last frosts. Increase also from softwood cuttings in late summer or early autumn, or in spring from stock plants, cut back and top-dressed in late winter and then forced under gentle heat. Cuttings were the most common method of increasing cultivars until reliable seed races became available, and are still useful for the double-flowered and ruffle-edge sorts which do not always come true from seed. Affected by a range of viruses that cause leaf mosaic, mottling and failure to flower, including those viruses common to many of the Solanaceae, such as cucumber mosaic, tobacco mosaic and potato viruses X and Y.

P. × *hybrida* (PETUNIA; a complex group of hybrids, thought to be *P.axillaris* × *P.integrifolia* –

they are basically bushy plants, grown as annuals and covered with short, clammy hairs; the leaves are ovate to elliptic and the showy flowers funnel-shaped with a broad, spreading limb; among the most popular half-hardy bedding plants, petunias range in habit from the erect and robust to the small, low-growing or cascading; in flower size, from dwarf (to 5cm in diameter) to giant (to 15cm in diameter); in form, from single to semi-double, double and heavily crisped and ruffled; in flower colour, from white to cream, blush, pale pink to deep rose, magenta, deep crimson, dark red, plum, deep purple, mauve, lilac and blue, and many cultivars are bicoloured, with a contrasting throat or dark veins; two of the most successful new races are the Million Bells Series, miniatures with a low, bushy habit, neat foliage and flowers to 3cm across in shades of purple, magenta, and the Surfinia Series, with a cascading habit and masses of flowers to 6cm diam., in various colours).

pF a logarithmic scale used as a common basis of expressing water content in soil. The range varies from about pF 2 for a wet, well-drained soil to pF 7 for an over-dry soil.

pH a scale describing the degree of acidity of a substance; the negative index of the logarithm of hydrogen ion concentration. For horticultural purposes, of special relevance to soils. pH 7.0 represents the neutral condition; pH values below this indicate acidity, and those above it, alkalinity. Soil pH may be measured by simple colorimetric methods or with a pH meter. See *soil*

Phacelia (from Greek *phakelos*, a cluster, referring to the arrangement of the flowers). Hydrophyllaceae. Western N America, E US, S America. SCORPION WEED. 150 species, annual, biennial and perennial, downy to glandular-pubescent herbs, with pinnatifid, pinnatisect or entire leaves and 5-lobed, shortly tubular flowers in terminal cymes or racemes in summer. The following thrive in poor sandy soils, and, if overwintered in pots in the cool greenhouse at 5–7°C/40–45°F, are also useful for early spring flowers. Sow seed under glass in early spring or *in situ* in late spring on any moderately fertile well-drained soil in sun.

P.campanularia (CALIFORNIA BLUEBELL; S California; glandular-hispid annual, 15–40cm; flowers to 2.5cm, broadly campanulate-funnelform, dark blue spotted white at the base of each lobe, sometimes wholly white); *P.tanacetifolia* (FIDDLE-NECK; California to Mexico; hispidulous annual, 15–120cm; flowers to 1.5cm, deeply campanulate, blue to lilac or mauve).

Phaedranassa (from Greek *phaidros*, bright, and *anassa*, lady, referring to the beauty of the flowers). Amaryllidaceae. S America. QUEEN LILY. 6 species, bulbous, herbaceous perennials with stalked, narrow to broadly oblong leaves. Produced in spring and summer in scapose umbels, the flowers are drooping, narrow and funnel-shaped or nearly cylindric, with five lobes spreading from the apex. Plant in autumn, with the bulb neck at soil level, in a mix of equal parts loam, leafmould, and sharp sand; keep cool and just moist until growth begins, then water moderately and grow in full light at a minimum temperature of 7°C/45°F. Dry off as foliage dies down in late summer or autumn, and keep cool and dry over winter until growth resumes in early spring. Propagate by offsets and seed. *P.carnioli* (S America; to 60cm; flowers 6–10, to 5cm, glaucous crimson, tipped green with yellow fringe).

Phaius (from Greek *phaios*, grey, referring to the flowers which darken with age or damage). Orchidaceae. Indomalaya, S China, Tropical Australia. Some 50 species, large evergreen perennial herbs, with or without pseudobulbs. The leaves are large, elliptic to lanceolate and plicately ribbed. Carried in spring and summer on an axillary, erect raceme, the flowers consist of spreading, oblong to oblanceolate tepals and a funnelform lip. Grow in shaded, humid conditions with a minimum temperature of 15°C/60°F. Plant in a fast-draining mix of bark, leafmould, charcoal, garden compost and a little dried FYM. Water throughout the year, allowing a slight drying between waterings and avoiding wetting of foliage. Repot every third year and increase by division.

P.flavus (syn. *P.maculatus*; India, Malaysia, Java; leaves spotted pale yellow, 40–48cm; flowers 6–8cm in diameter, yellow, rarely white, lip streaked red-brown); *P.tankervilleae* (syn. *P.grandifolius*, *P.wallichii*; Himalaya to Australia; leaves 30–100cm, unspotted; flowers 10–12.5cm in diameter, tepals white, green or rose beneath, rusty red, yellow-brown, or white above, lip interior pink to burgundy, base yellow, exterior white, midlobe red-orange or white and pink).

Phalaenopsis (from Greek *phalaina*, moth, and *opsis*, appearance). Orchidaceae. Asia, Australasia. MOTH ORCHID. About 50 species, evergreen, perennial herbs. They are more or less stemless with a few, large and leathery, obovate to elliptic leaves in two loose ranks with, at their bases, numerous aerial roots. The flowers are borne in erect to arching racemes; the sepals are spreading, elliptic to broadly spathulate and usually smaller than petals; the lip is 3-lobed, the lateral lobes erect, the midlobe fleshy with an anchor- or arrow-shaped tip ending in horn-like projections or filaments.

Epiphytes for the intermediate or warm greenhouse, for growing cases (especially *P.cornu-cervi* and *P.violacea*) and for warm, shaded and humid positions in the home. They are valued for their beautiful moth-like flowers matched, in some species, by superbly marked foliage (*P.schilleriana* and *P.stuartiana*). Many grexes are now offered. Grow in pots, baskets or on rafts. The medium should consist of coarse bark, perlag and charcoal. Aerial roots are common and should be encouraged by frequent mistings (periodically, with a dilute feed); never mist at temperatures below 18°C/65°F, since rots will be encouraged by an accumulation of moisture in the growth axis. A humid, buoyant atmosphere is essential, as is shade. Drench pots throughout the year whenever the medium begins to dry out. Propagate by meristem culture or by plantlets sometimes produced on old inflorescences. The genus is also easily raised from seed.

P.amabilis (syn. *P.grandiflora*; East Indies, Australia; leaves glossy green; inflorescence to 1m; flowers to 10cm in diameter, fragrant, tepals white, often pink below, lip white, base red, margins yellow, midlobe cruciform, side projections triangular, with 2 yellow-tipped appendages, callus yellow dotted red); *P.cornu-cervi* (SE Asia; leaves olive green; inflorescence to 40cm, stalk compressed; flowers waxy yellow to yellow-green, sepals marked red-brown, blotches stripes and spots cinnamon, lip fleshy, white, midlobe anchor-shaped, lateral lobes red-brown or striped cinnamon); *P.equestris* (Philippines, Taiwan; leaves dark green above, often flushed rose-purple beneath; inflorescence arched, simple or branched, to 30cm; flowers to 4cm in diameter, tepals rose or white suffused rose, lip deep pink to purple, midlobe ovate, concave, apex fleshy, callus yellow, spotted red, lateral lobes marked yellow); *P.gigantea* (Borneo, Sabah; leaves glossy mid green; inflorescence to 40cm, flowers scented, cream to dull yellow, tepals blotched and lined maroon to dark purple, lip fleshy, white, striped, or lined magenta, midlobe ovate); *P.hieroglyphica* (Philippines; leaves dark green; inflorescence pendent or arched; tepals white lined red-purple, apex tinted green, lip to 2cm, midlobe truncate, apex jagged); *P.lueddemanniana* (Philippines; leaves dull olive green; inflorescence erect or pendent, to 30cm, branched or simple, flexuous, becoming horizontal; flowers to 6cm in diameter, tepals white, laterally or horizontally striped brown-purple, lip to 2.5cm, carmine, base yellow, midlobe oblong to ovate, apical callus white, downy); *P.parishii* (Himalaya, Vietnam; leaves dark green; inflorescence erect or arching, to 15cm; tepals white, lip with midlobe purple, triangular, lateral lobes white or yellow, spotted brown or purple); *P.sanderiana* (Philippines; leaves dark green above, marked silver-grey beneath; inflorescence to 80cm, axis purple, branched or simple, flowers to 7.5cm in diameter, colour and marking variable, tepals pink, dappled white, or wholly white, lip with midlobe triangular, white or yellow, striped purple or brown, apex with 2 filiform projections, callus horseshoe-shaped, yellow or white, spotted red, brown, or purple, lateral lobes white spotted pink); *P.schilleriana* (Philippines; leaves dark green, mottled silver-grey above, purple beneath; inflorescence branching, pendent; flowers to 8cm in diameter, fragrant, white to pink, mauve and rose-purple, tepals edged white, lateral sepals spotted carmine-purple at base, lip midlobe white to magenta, anchor-shaped, lateral lobes yellow at base, dotted red-brown); *P.stuartiana* (Philippines; leaves green blotched grey above, purple beneath; inflorescence branched, pendent, to 60cm; flowers to 6cm in diameter, fragrant, petals white, sepals white to yellow-

Phalaenopsis stuartiana

green, dotted red-brown at base, lip anchor-shaped, white stained green to yellow spotted brown to orange); *P.violacea* (Sumatra, Borneo, Malaysia; leaves glossy light green; inflorescence ascending or arched, thick, jointed, to 12.5cm, usually shorter; flowers few, to 4.5cm in diameter, often fragrant, tepals amethyst fading to white or lime green at apex, lip midlobe oblong, apiculate, violet tipped with white down, central crest yellow, lateral lobes yellow; column white or amethyst).

Phalaris (Classical Greek name for a kind of grass). Gramineae. Temperate regions. Some 15 species, annual or perennial grasses with slender culms, narrow leaves and silky, spike-like to ovoid panicles in summer. Hardy to between –15 and –20°C/5 to –4°F. Grow in sun or light shade. Tolerant of most soils and sites from pond margins to dry borders. Sometimes invasive; they can be confined where necessary by planting in a sunken, bottomless half barrel. Propagate by division. *P.arundinacea* (REED CANARY GRASS, RIBBON GRASS; Eurasia, N America, S Africa; reed-like, fast-spreading rhizomatous perennial, to 1.5m; leaves to 35 × 1.8cm, glabrous; inflorescence narrow, to 17cm; 'Dwarf's Garters': as 'Picta' but dwarf, to 30cm; 'Luteopicta': small, with leaves striped golden-yellow; 'Feesey': small, less invasive, with light green leaves, boldly striped white, and a pink-tinted inflorescence; 'Picta': GARDENER'S GARTERS, invasive, leaves striped white, usually predominantly on one side of leaf; 'Streamlined': leaves mainly green, edged white; 'Tricolor': leaves white-striped with a strong pink flush).

phanerogam a spermatophyte; a plant reproducing by means of seeds not spores; cf. *cryptogam*.

phanerophyte a plant whose latent buds are located more than 50cm above the soil surface.

Phaseolus (Latin diminutive of Greek *phaselos*, name used by Dioscorides for a kind of bean). Leguminosae. New World. BEAN. Over 20 species, annual or perennial, usually climbing herbs with trifoliolate leaves, pea-like flowers and – usually – edible pods and seeds; see *beans*. *P.coccineus* is the SCARLET RUNNER BEAN; *P.lunatus* the LIMA BEAN; *P.vulgaris* the KIDNEY BEAN, HARICOT, COMMON BEAN, FRENCH BEAN or DWARF STRING, FLAGEOLET, SNAP BEAN.

phellem see *cork*.

Phellodendron (from Greek *phellos*, cork, and *dendron*, tree, referring to the corky bark of some species). Rutaceae. Temperate and subtropical E Asia. 10 species, aromatic, deciduous trees, their bark often thick and corky. They are grown for their handsome, pinnate foliage and autumn colour. Small, yellow-green flowers are carried in terminal panicles in early summer, and give rise to pea-sized, black fruits. They perform best in zones where summers are long and hot. The pungently aromatic foliage gives fine, clear yellow autumn tints, and, on female trees, the black fruits often persist well into winter. Grow in deep, fertile, moisture-retentive but well-drained soils in an open position in full sun. Although hardy to at least –20°C/–4°F, the young growth is sometimes damaged by late spring frosts. Propagate by seed in autumn, or by softwood or heeled semi-ripe cuttings in summer rooted in a closed case with bottom heat. Alternatively, take root cuttings in late winter.

P.amurense (N China, Manchuria; to 15m; bark pale grey, corky; leaflets 5–10cm, broadly ovate to lanceolate, dark glossy green above, glaucous beneath, turning yellow in autumn); *P.chinense* (C China; to 10m, bark thin, dark grey-brown; leaflets 7–14cm, oblong to lanceolate, yellow-green above, light green and tomentose beneath).

phelloderm a layer of parenchymatous tissue found on the inside of the cork cambium.

phellogen see *cork cambium*.

Phenakospermum (Greek *phenax*, a cheat or impostor, and *sperma*, seed). Strelitziaceae. Brazil, Guyana. 1 species, *P.guyannense*, a giant herb to 10m, closely resembling *Ravenala* and *Strelitzia nicolai* except stems less prominent, leaf blades to 2m, dark glossy green, midrib often tinted coral-red, more elliptic to ovate than oblong to lanceolate. The inflorescence is terminal with white flowers amid carinate, deflexed bracts. Cultivate as for *Heliconia*.

phenomenal berry see *hybrid berries*.

phenotype the external appearance of an organism, which may vary according to its environment; for example, the foliage type of water crowsfoot (*Ranunculus aquatilis*) where growing in different depths and speeds of water; cf. *genotype*.

pheromone a chemical produced by an insect to induce a specific behavioural response, such as mating, in other insects of the same species. Pheromones are often useful in biological control or integrated pest management systems.

The principal types of pheromones are aggregation, sex, dispersal or spacing, alarm and trail pheromones. Aggregation pheromones stimulate individuals to aggregate for shelter, oviposition, colonization, feeding, or as a prelude to swarming or dispersal; they are widespread among insects such as cockroaches, bees, wasps and many beetles. Sex pheromones can be produced by either sex but they are most commonly produced by 'calling' females to attract males; these attractants or lures are typically found in butterflies and moths. Dispersal or spacing pheromones are used by insects to prevent over exploitation of a food source; for example, the apple fruit fly applies the pheromone to the surface of apples and other host fruits after depositing eggs in order to deter other females from ovipositing. Alarm pheromones are characteristic of social insects such as wasps, bees and ants and are used to alert the colony to the presence of intruders; aphids secrete alarm pheromones when they are being attacked by predators. Trail phero-mones occur commonly in social insects, particularly ants and termites; on locating a source of food, a worker ant lays down a pheromone trail back to the nest.

Once identified chemically, most pheromones can be prepared synthetically and the resulting compounds have a potential use in insect control for monitoring populations, mass trapping or mating disruption. Simple cardboard traps are available for pests including codling moth and other tortricids on apple, pea moth on peas, gypsy moth, fruit flies in citrus and peach orchards and Japanese beetle. By gathering information on populations, spraying can be more effectively timed and the total amount of insecticide required reduced.

Philadelphus (probably for Ptolemy Philadelphus (ruled 285–246 BC), patron of the arts and sciences). Hydrangeaceae (Philadelphaceae). C and N America, Caucasus, E Asia. MOCK ORANGE. 60 species, deciduous shrubs with peeling bark, ovate to lanceolate leaves and, in spring and summer, rounded, 4-petalled flowers with many stamens, usually fragrant and pale, in racemes, panicles, cymes, or solitary. They are easily grown in any moderately fertile soil, including thin, chalky soils. Most will tolerate winter temperatures of at least –15°C/5°F. They flower best in full sun. Cut back old shoots after flowering. Increase by softwood cuttings.

P.coronarius (S Europe, Caucasus; shrub to 3m, bark dark brown, slowly peeling; current growth downy becoming glabrous; leaves 4.5–9cm, ovate, mostly glabrous, irregularly and shallowly toothed, apex acuminate; flowers 5–9 in short terminal racemes, creamy white, strongly fragrant, 2.5–3cm across; cultivars include 'Aureus', compact, with golden yellow to lime green leaves; 'Bowles's variety', with leaves edged white; 'Dianthiflorus', dwarf, with double flowers and narrow petals; 'Maculiformis', with 3cm flowers, with petals red at base; 'Salicifolius', with narrower leaves and tips of petals bearing a few hairs – may be of hybrid origin); *P. × cymosus* (erect shrub to 2.5m; bark brown, not peeling; leaves ovate, sparsely toothed, hairy beneath; flowers white to cream, sometimes tinted rose, 4–10cm in diameter, cupped to spreading, fragrant in cymes of 1–5; some stamens petal-like; cultivars include 'Nuée Blanche', with single to semi-double, creamy, fragrant, profuse flowers; 'Perle Blanche', of compact habit, with single to semi-double, fragrant flowers; 'Voie Lactée', vigorous, with large leaves and flowers, to 5cm, freely produced and with reflexed petals); *P.delavayi* (SW China; shrub to 4m; bark grey-brown, grey or chestnut-brown, not peeling; current growth glaucous; leaves ovate-lanceolate or ovate-oblong, acuminate, 2–8cm, serrate or entire, sparsely bristly above, shaggy-hairy beneath; flowers in raceme of 5–9, 2.5–3.5cm across, disc-shaped, pure white, fragrant; calyx purple-tinted;

includes 'Nymans Variety', with plum-coloured calyx); *P. × lemoinei*. (*P.coronarius × P.microphyllus*; low compact shrub, bark peeling; leaves 1.5–2.5cm, ovate, glabrous above, sparsely bristly below, apex acuminate, toothed; flowers usually in threes, fragrant, cross-shaped, about 3cm across, petals notched; cultivars range from arching to upright in habit, with 2.5–1.5cm leaves, mottled and variegated to smooth dark green, and 2.5–6cm flowers, white to creamy, fragrant to very fragrant, and with petals sometimes waved, dentate or cut); *P.microphyllus* (SW US; erect shrub to 1m; new growth adpressed-pubescent, bark chestnut-brown, shiny, soon flaking; leaves 1–1.5cm, oval-elliptic or lanceolate, entire and ciliate, glabrate above, softly shaggy-hairy beneath; flowering shoots 1.5–4cm long, with 1 or rarely 2 pure white, very fragrant, cross-shaped, flowers about 3cm wide); *P. × polyanthus* (*P.insignis × P.lemoinei?*; erect shrub, bark dark brown, peeling; new growth sparsely shaggy; leaves 3.5 × 5cm, ovate, apex acuminate, entire or with a few sharp teeth, sparsely bristly beneath; flowers 3–5 in cymes or corymbs cross-shaped, about 3cm across; cultivars include 'Boule d'Argent', of compact habit, with 4cm, semi-double flowers in cymes of 5–7, scarcely fragrant; 'Favourite', to 2m, with large, cross-shaped, cupped flowers, serrate petals and yellow stamens; 'Mont Blanc', to 1m, with cross-shaped, pure white flowers of a strong fragrance; and 'Pavillon Blanc', a low bush, with arching branches and creamy-white, fragrant flowers). *P. × purpureo-maculatus* (*P. × lemoinei × P.coulteri*; shrub to 1.5m, bark black-brown, peeling in second year; new growth hairy; leaves 1–3.5cm, broadly ovate, tip acute, more or less entire, with a few scattered hairs beneath; flowers solitary or in threes or fives, fragrant, disc-shaped, 2.5–3cm across, white, purple at centre; cultivars include 'Beauclerk' to 2m, with pineapple-scented flowers to 8cm, with a purple centre and broad petals; 'Bicolore', dwarf, with 3cm leaves and creamy flowers with a purple centre; 'Etoile Rose', with flowers carmine at centre fading at apex, scarcely fragrant; 'Galathée', with arching branches and solitary flowers, with a pink centre; 'Nuage Rose', with wide-opening flowers, with a pink centre; 'Purpureo-Maculatus', with solitary flowers, with a pink-purple centre); *P. × virginalis* (stiffly upright to 2.5m; bark grey, peeling when old; new growth hairy; leaves 4–7cm, ovate, apex shortly acuminate, becoming glabrous above, shaggy beneath; flowers in raceme, usually double, pure white, very fragrant, 4–5cm wide; cultivars include 'Burfordiensis', to 3m, with 7cm flowers in raceme of 5–9, and conspicuous, yellow stamens, 'Fleur de Neige', to 1m, with 4cm, semi-double flowers and yellow stamens; 'Fraicheur', with large, very double, creamy, profuse flowers; 'Glacier', with creamy, very double, late flowers; 'Schneestrum', fast-growing, with pure white double flowers; and 'Virginal', with double flowers in loose heads). *Cultivars and hybrids*: 'Buckley's Quill' (Frosty Morn' × 'Bouquet Blanc'), also known as 'Faxonii', of arching habit, and with cross-shaped

Philadelphus delavayi

flowers; 'Frosty Morn', frost-tolerant, and with 3cm, double flowers; 'Mrs E. L. Robinson', with large, fragrant flowers; 'Patricia' (resembles *P.lewisii*), with leathery, dark green leaves and very fragrant flowers in raceme of 3–7; 'Slavinii', rounded bush to 3m, with 6.5–9.5cm leaves and 6cm, cross-shaped, abundant flowers; 'Splendor' (possibly *P.inodorus* var. *grandiflorus* × *P.lewisii* var. *gordonianus*), of full, rounded habit, with single, wide-opening, slightly fragrant flowers with rounded petals; 'Stenopetalus' (possibly a form of *P.pubescens*), with campanulate flowers; and 'Thelma' (origin unknown, possibly *P.purpurascens*), of low, graceful habit, with 2.5cm leaves and 1.5cm, campanulate flowers.

× Philageria *Philesia* × *Lapageria*) Liliaceae (Philesiaceae). One garden hybrid, × *P.veitchii* (*P.magellanica* × *L.rosea*), a scrambling shrub, similar in habit to *Lapageria* but more shrubby and less vigorously climbing. The flowers are drooping and consist of three outer tepals to 4cm long, fleshy, dull red deep or magenta with a glaucous bloom, and three inner tepals to 6cm long, bright rose and more faintly bloomed. Cultivate as for *Lapageria*. Propagate by simple layering.

Philesia (from Greek *phileo*, to love, from the beauty of the flower). Liliaceae (Philesiaceae). Chile. 1 species, *P.magellanica*, a branching, erect shrubby evergreen, 30–100cm, with a dense, rather box-like habit. The leaves are 1.5–3.5cm long, oblong to lanceolate, leathery and deep glossy green. Produced in summer and early autumn, the pendent flowers are narrowly campanulate with six tepals – these are 6cm long and pink to purple-red, sometimes faintly flecked orange-pink. Cultivate as for *Lapageria*. Propagate by removal of suckers or by greenwood or semi-ripe cuttings in summer in a closed shaded case.

Phillyrea (Greek name for the genus). Oleaceae. Mediterranean to Asia Minor. MOCK PRIVET. 4 species, evergreen shrubs, with glossy green, glabrous leaves and small, fragrant, green-white flowers, resembling those of *Osmanthus*. These are followed by fleshy, ovoid, blue-black fruits. Cultivate as for *Osmanthus*.

P.angustifolia (Mediterranean; shrub to 3m; leaves linear to lanceolate, usually entire; 'Rosmarinifolia': leaves narrower, smaller, grey-green, somewhat glaucous); *P.latifolia* (Mediterranean; shrub or small tree to 9m; leaves ovate to lanceolate, dentate or entire; 'Buxifolia': leaves small, obovate; 'Spinosa': leaves spiny-dentate, ovate).

Philodendron (from Greek *phileo*, to love, and *dendron*, tree, alluding to the climbing or epiphytic habit). Araceae. Tropical America. 350+ species, epiphytic or terrestrial evergreen perennials, some climbing with long, slender, rooting stems, some with leaves in a dense basal rosette, or along or crowning a stout, unbranched stem. The leaves are often large, entire or lobed to pinnatifid to pedate, oblong or ovate, cordate to sagittate, coriaceous and usually dark,

glossy green. They differ in size and shape in the juvenile and adult states. The inflorescence is *Monstera*-like and gives rise to white, orange or red berries.

Grow in an open, fertile, moisture-retentive mix of loam, coir or bark, leafmould and coarse sand. Maintain a day temperature of 20–27°C/70–80°F, falling to 15–20°C/60–70°F at night when in growth, with a winter minimum of 15°C/60°F. Site in part-shade or in filtered light, ensuring protection from bright summer sun; provide good air circulation, but avoid draughts. Mist frequently when in growth and water moderately, reducing as temperatures fall in winter. When plants are in full growth and roots have filled the pots, feed regularly with a 10:10:5 NPK liquid fertilizer. Train climbers on moss poles, wire cylinders filled with the potting medium, or canes. Smaller species will also trail, making good ground-cover and cascade plants for baskets and planters. The larger species may be pruned, or their stem tips repropagated, to create bushier, more or less free-standing specimens. Propagate by air layering, or by stem tip or leaf bud cuttings rooted in individual small pots in a closed case with a bottom heat of 20–25°C/70–77°F.

P.bipennifolium (syn. *P.panduraeforme* of gardens; HORSEHEAD PHILODENDRON, FIDDLE-LEAVED PHILODENDRON; SE Brazil; stems climbing, tall, internodes long; leaves to 45 × 15cm, reflexed, 5-lobed, glossy dark green, terminal lobe long, to 25 × 10cm, obovate, lateral lobes angular, obtuse, basal lobes broadly oblong-triangular); *P.cordatum* (HEART-LEAF PHILODENDRON; SE Brazil; stems tall, climbing, internodes short; leaves 45 × 25cm, reflexed, ovate-triangular, base sagittate, basal lobes round to angular, margins undulate); *P.domesticum* (syn. *P.hastatum* of gardens; SPADE-LEAF PHILODENDRON; origin unknown; stems climbing; leaves to 60 × 30cm, reflexed, elongate-triangular, sagittate, undulate, glossy bright green, basal lobes rounded-oblong to triangular; 'Variegatum': leaves splashed yellow, cream and green); *P.erubescens* (RED-LEAF PHILODENDRON, BLUSHING PHILODENDRON; Colombia; stems tall, climbing, purple or red-purple when young, with deep pink to pink-brown bracts; leaves to 40cm, reflexed, ovate-triangular, base shortly sagittate to cordate, somewhat coriaceous, glossy dark green above, coppery-purple beneath; 'Burgundy': leaves to 30cm, flushed red with burgundy veins, with a cordate to hastate base and claret stems; 'Golden Erubescens': climber, with thin, round stems and gold leaves, tinted pink beneath and when young; 'Imperial Red': leaves dark purple to red; and 'Red Emerald', vigorous, with claret stems and long-cordate leaves to 40cm, dark green with red ribs beneath, shiny, and with long, rich red petioles); *P.imbe* (syn. *P.sellowianum*; SE Brazil; stem climbing, red-purple; leaves 33 × 18cm, reflexed, ovate-oblong, cordate to sagittate, glossy above, basal lobes rounded; 'Goldiana': leaves long-ovate, rich green with gold speckles, red beneath, yellow when young, and with short petioles; 'Variegatum': leaves irregularly blotched green, dark green and cream; 'Weber's Selfheading': climber, with thick, very shiny, oblanceolate leaves, with a light green midrib, ribs red beneath, and petioles red spotted yellow-green); *P.melanochrysum* (syn. *P.andreanum*; BLACK-GOLD PHILODENDRON; Colombia; climbing; leaves 100 × 30cm, reflexed to pendent, oblong-lanceolate or narrow-ovate, sagittate, acuminate, velvety black-green above, veins pale green, copper-coloured when young); *P.pedatum* (syn. *P.laciniatum, P.laciniosum*; S Venezuela, Surinam to SE Brazil; stems tall, climbing; leaves reflexed, ovate in outline, irregularly pinnatifid, terminal lobe 45 × 30cm, median segments elliptic, obovate or rhombic, lateral segments to 5 per side, oblong, obtuse, basal lobes 27 × 17cm, widely spreading; 'Purple Green': leaves flushed red-purple); *P.pinnatilobum* (syn. *P.* 'Fernleaf'; Brazil; climbing; leaves to 50cm across, erect, ovate-orbicular, pinnatifid, lateral segments 1.5cm across, to 13 per side, narrow, acuminate, basal lobes bifid); *P.radiatum* (syn. *P.dubium* of gardens; DUBIA PHILODENDRON; C America; climbing; leaves to 90 × 70cm, reflexed, ovate in outline, cordate-sagittate, deeply pinnatifid, terminal lobe with 8 pairs oblong or further 3-parted segments, basal lobes 5-parted); *P.sagittifolium* (syn. *P.sagittatum* of gardens; SE Mexico; stems climbing; leaves to 60 × 30cm, long triangular-oblong, sagittate, subcoriaceous, glossy bright green, basal lobes to 12.5cm, triangular, obtuse); *P.scandens* (syn. *P.cordatum* of gardens; HEART-LEAF PHILODENDRON; Mexico and West Indies to SE Brazil; stems slender, climbing, becoming pendent; leaves 8(–30) × 6–23cm, reflexed, ovate-cordate, acuminate, glossy green above, green or red-purple beneath; f. *micans*: leaves bronze above, red to red brown beneath, and basal lobes larger and slightly overlapping; subsp. *oxycardium*: juvenile leaves glossy brown when immature, green when mature; 'Variegatum': leaves dark green marbled off-white and green-grey).

Phlebodium

Phlebodium (diminutive of Greek *phlebion*, vein). Polypodiaceae. Tropical America. Some 10 species, epiphytic ferns with thick and fleshy, creeping rhizomes and pinnately divided fronds bearing circular sori. Provide a minimum temperature of 7°C/45°F and bright, indirect light. Plant in an open, soilless medium with added coarse bark. Keep moist and humid in warm weather, rather dry in cooler conditions. Propagate by division of rhizomes or by spores – these are freely produced and often sow themselves. *P.aureum* (GOLDEN POLYPODY, RABBIT'S FOOT FERN, HARE'S FOOT FERN; tropical America; rhizome thick, hard and creeping; fronds 30–100cm, deeply pinnatifid, segments linear to oblong or strap-shaped, sea green to glaucous blue-green, sori golden yellow).

Phlomis

Phlomis (from Greek *phlomos*, mullein, alluding to the woolly stems and leaves of some species). Labiatae. Mediterranean to C Asia and China. About 100 species, downy, felted or woolly herbs or evergreen shrubs with tubular, 2-lipped and hooded flowers in summer; these arise from conspicuous and persistent calyces bunched in axillary whorls. The herbaceous species are hardy to –15°C/5°F. With a sheltered position and good drainage, The shrubby species will tolerate temperatures almost as low. Grow in full sun in well-drained soils, with shelter from cold winds. The large-leaved *P.russeliana* tolerates light shade. Cut back shrubby species after flowering. Herbaceous perennials may be cut down in autumn, although the dead flowering stems are sometimes left for winter interest. Propagate shrubby species by softwood cuttings in summer, or by semi-ripe cuttings in a sandy propagating mix in the cold frame in autumn; increase also by simple layering. Divide herbaceous species in autumn or spring. Sow seed in spring at 15–18°C/60–65°F.

P.cashmeriana (Kashmir, West Himalaya; to 90cm; stem densely woolly; leaves 13–23cm, ovate-lanceolate, obtuse, broadly rounded at base, downy, white beneath; flowers pale lilac); *P.chrysophylla* (Lebanon; low evergreen subshrub; leaves to 6cm, broadly elliptic to oblong-ovate, truncate to cordate at base, yellow-pubescent, golden and downy when young; flowers golden yellow); *P.fruticosa* (JERUSALEM SAGE; Mediterranean west to Sardinia; spreading subshrub or shrub to 130cm, tawny-hairy to felty; leaves 3–9cm, ovate to lanceolate, truncate or cuneate at base, dull sage green, rugose above, woolly beneath, entire or crenulate; flowers golden yellow); *P.italica* (Balearic Islands, *not* native to Italy; subshrub to 50cm; leaves to 10cm, oblong to bluntly scutate or lanceolate, obtuse, white- to golden-felted, shallow-crenate; flowers pink or pale lilac); *P.longifolia* (Anatolia, Syria, Lebanon, Cyprus; hairy shrub to 130cm; leaves 3–7cm, lanceolate to oblong or ovate, cordate or subcordate at base, crenulate or crenate-serrate, green tomentose above, grey- to yellow-tomentose beneath; flowers yellow); *P.russeliana* (W Syria; to 1m; leaves 6–20cm, broadly ovate, obtuse, crenate, cordate at base, grey-green, thinly stellate-tomentose above, hoary beneath, long-stalked; flowers golden yellow).

Phlox

Phlox (from Greek *phlox*, the name used for *Lychnis* by Theophrastus). Polemoniaceae. N America. PHLOX. 67 species, herbaceous to shrubby, evergreen or deciduous annuals or perennials. They vary widely in habit, from tufted and cushion-forming to tall and clumped. The leaves may be needle-like or ovate to lanceolate. The flowers are produced in late spring and summer in terminal corymbs or panicles (sometimes solitary in smaller species); they consist of a slender, short tube and a spreading limb of five, obovate lobes.

Grown for their tall, stately habit and generous masses of flowers, the taller herbaceous types, such as *P.paniculata* and *P.maculata,* have long been mainstays of the herbaceous and mixed border in summer, valued for their often intensely fragrant blooms in a wide range of colours from pure white through shades of pink to crimson and purple. They are also suitable for less formal situations, in the light dappled shade of the woodland edge. Grow in full sun or part shade in deep, moisture-retentive fertile soils that are rich in organic matter. Top-dress

Phlox douglasii

with a balanced fertilizer and mulch with well-rotted organic matter in spring. Staking is usually necessary only in exposed situations; deadhead to prevent self-sown seedlings which are usually of inferior quality. Divide in early spring. Alternatively, take stem cuttings in spring from forced rootstocks; increase also by root cuttings in late winter.

Most of the lower-growing types are suited to the rock garden and peat terrace, forming mats of foliage which become dense carpets of often brilliant colour when in flower; the more vigorous species such as *P.subulata* and the slightly less rampant *P. douglasii* make good groundcover in the rock garden or at path edges, and are especially useful for softening hard lines at the border front. The smaller cushion-forming species such as *P.caespitosa* are suitable in scale for sink and trough plantings. Grow in fertile, well-drained soils enriched with leafmould; *P.stolonifera*, *P.adsurgens* and *P. × procumbens* are calcifuge. Most species grow in full sun or in light shade in drier soils; *P.divaricata* prefers some shade. Clip over after flowering to remove dead flowerheads. In the alpine house, grow in pans in a mix of equal parts loam, leafmould and sharp sand. Propagate trailing species by removal of rooted sections in spring or early autumn, cushion types by cuttings of non-flowering shoots in spring. Increase also by seed. The annual species such as *P. drummondii* and its cultivars are useful as seasonal fillers in the border or for bedding. Sow under glass in spring and harden off before planting out after the last frost.

P.adsurgens (US (Oregon, N California); slender perennial; stem prostrate to ascending, to 30cm; leaves 1–2cm, rounded to narrowly ovate, glabrous; inflorescence lax, few-flowered, hairy; flowers to 2.5cm in diameter, pink or purple; includes cultivars with lavender to deep pink flowers some with grey-green leaves and a dense, creeping habit); *P.amoena* (SE US; perennial, decumbent to erect, to 60cm; leaves to 5cm, oblong to lanceolate, pilose; flowers magenta, red-purple, pink or white, crowded in a terminal cluster subtended by leafy bracts); *P.bifida* (SAND PHLOX, tufted perennial to 20cm; leaves to 3–6cm, distant, linear to narrowly elliptic, ciliate, pilose; flowers sweet-scented, 0.9–1.4cm, lavender to white, lobes cleft; includes cultivars with fragrant white to pale lavender flowers and a low, compact, mat-forming habit); *P.divaricata* (WILD SWEET WILLIAM, BLUE PHLOX; SE US; perennial to 45cm, spreading; leaves to 5cm, oblong to ovate or narrowly lanceolate; inflorescence a compound cyme, flowers 1.2–1.8cm, to 4cm in diameter, lavender to pale violet or white, tube sometimes darker inside, lobes cleft to erose; includes cultivars with white and clear blue flowers); *P.douglasii* (NW US; perennial to 20cm tall, usually less, loosely tufted, glandular-hairy; leaves 1–1.2cm, stiff, subulate to linear, pointed; inflorescence 2.5–7.5cm, 1–3-flowered; flowers strongly fragrant, white, lavender or pink, 1–1.3cm, lobes about 7.5mm, obovate, includes numerous cultivars with flowers in shades of white, pink, crimson, magenta, mauve and lavender, some with a darker 'eye' or tint);

P.drummondii (ANNUAL PHLOX, DRUMMOND PHLOX; US (E Texas); annual, 10–50cm tall, hairy, sometimes glandular; leaves narrowly oblanceolate to ovate; flowers grouped in clusters 1–2.2cm, purple, violet, pink, lavender, red or white, rarely pale yellow, often paler inside tube with markings around throat; cultivars and seed races include tall to dwarf plants with flowers in shades of white, pink, red and purple and combinations thereof, some with cut and fringed margins); *P.hoodii* (northwest N America; dwarf, tufted, mat-forming perennial to 6cm high; leaves to 1cm, subulate, tomentose; flowers solitary, white to pink, magenta or palest violet); *P.maculata* (WILD SWEET WILLIAM, MEADOW PHLOX; US; rhizomatous perennial, 35–70cm tall, glabrous to minutely hairy; stem often red-spotted; leaves 6.5–13cm, linear to ovate; inflorescence a many-flowered panicle, flowers often sweetly scented, 1.8–2.5cm, pink, purple or white, sometimes with a dark purple ring in throat; includes cultivars with white to pink or lilac flowers, often fragrant and with a dark ring or blotch in the throat); *P.paniculata* (PERENNIAL PHLOX; SUMMER PHLOX, AUTUMN PHLOX, FALL PHLOX; US; erect perennial 60–100cm tall, subglabrous to puberulent; leaves 1.2–12cm ovate or lanceolate to elliptic, toothed, ciliate; inflorescence a terminal, compound cyme, many-flowered, flowers 2–2.8cm, blue, lavender, pink or white; many cultivars ranging in habit from medium to tall with flowers in large, cylindrical to pyramidal heads and ranging in colour from white to pink, purple or red, often with darker throats; some, for example 'Harlequin', have variegated leaves); *P. × procumbens* (*P.stolonifera × P.subulata*; decumbent clump-forming perennial, 15–25cm tall; leaves to 2.5cm, oblanceolate to elliptic; inflorescence a lax, flat panicle, flowers bright purple; includes selections with a dense habit, dark green to variegated leaves and pale rose to lilac-pink flowers); *P.stolonifera* (CREEPING PHLOX; E US; creeping perennial, forming 15–25cm tall mats; leaves to 4.5cm, obovate and long-stalked to oblong, sessile; inflorescence lax, about 6-flowered, glandular-hairy, flowers to 2.5cm, violet to lavender or purple to lilac; includes cultivars with flowers in shades of white, rose, mauve, violet and blue, some with yellow 'eyes'); *P.subulata* (MOSS PHLOX, MOUNTAIN PHLOX, MOSS PINK; E US; downy perennial, forming dense mats or cushions to 50cm downy; leaves 0.5–2cm, rigid, needle-like to subulate; inflorescence a few-flowered, terminal, bracteate cyme; flowers to 1.5cm, white, pink or lavender; includes cultivars with white, rose, carmine and violet flowers, many with a darker 'eye', some fragrant).

Phoenix (Greek word for the date palm). Palmae. Africa, Asia. DATE PALM. Some 17 species, large, evergreen palms with solitary or clustered trunks clothed with persistent leaf bases or scars. The leaves are pinnate with tough-textured, narrow leaflets, the lowermost reduced to spines. Erect to arching, the inflorescences bear cream-yellow or pale orange flowers. These are followed by oblong-ellipsoid to ovoid, 1-seeded fruits, their skin waxy and shiny, the mesocarp fleshy or pasty and often sweetly edible.

P.dactylifera and *P.canariensis* withstand full sun and dry atmospheres, and are tolerant of a wide range of soil types. In frost-free zones, they are excellent trees for specimen and avenue plantings. In colder areas, younger examples can be used for dot planting in bedding schemes and for containers on terraces; overwinter in cool greenhouse conditions and move outdoors after the last frosts. In mild locations in climate zones 7 and 8, young plants may survive outside all year, especially if the base is lagged with thick sacking. *P.roebelinii* requires a minimum temperature of 13°C/55°F, bright light and a well-drained, fibrous, loam-based medium. Water and feed plentifully in warm weather; keep rather dry at other times. This is a graceful palm ideally suited to cultivation in the home when young. Older plants are a popular choice for large containers in public buildings, and for interior landcsaping. Propagate from seed, after soaking for 24 hours, with a bottom heat of 20–27°C/70–80°.

P.canariensis (CANARY ISLAND DATE, CANARY DATE PALM; Canary Islands; trunk to 15×0.9m, solitary with oblong scars, these wider than long; leaves to 6m, rachis sometimes twisted, pinnae crowded, regularly spaced; fruit 2cm, oblong-ellipsoid, yellow tinged red); *P.dactylifera* (DATE, DATE PALM; cultigen, probably originating in W Asia and N Africa; trunk to 30m, slender, suckering, scars as long as wide, or longer; leaves to 3m, rachis rigid, pinnae 30×2cm, to 80 on each side, regularly spaced or clustered, rigid; fruit 4–7cm, oblong-ellipsoid, yellow to brown, edible); *P.roebelinii* (MINIATURE DATE PALM, PYGMY DATE PALM, ROEBELIN PALM; Laos; stem to 2m, slender at base, expanding toward crown, often leaning, roots forming a basal mass; leaves to 1.2m, pinnae to 25×1cm, grey-green, regularly spaced, drooping, silver-scurfy; fruit to 1cm, ellipsoid, black).

Pholistoma (from Greek *pholis*, scale, and *stoma*, mouth). Hydrophyllaceae. Southwest N America. 3 species, prostrate or weakly climbing annual herbs closely related to *Nemophila*, from which the genus can be distinguished by the retrorse bristly spikes on the upper stems. A useful climber for temporary cover on fences and tree stumps. Cultivate as for *Nemophila*. *P.auritum* (FIESTA FLOWER; California; to 1.2m; leaves oblong, hirsute, lobes 7–11; flowers to 3cm in diameter, blue, lilac or violet marked with deeper streaks).

Phormium (from Greek *phormion* , a woven mat – referring to the use made of the fibre). Agavaceae. New Zealand. FLAX LILY. 2 species, large or giant evergreen, perennial herbs with basal clumps of tough-textured, keeled, sword-shaped leaves. Carried in summer, in tall, dark-stalked and tiered panicles, the flowers are tubular and consist of six tepals, their tips curved upwards and the stamens protruding. Hardy in climate zones 7 and over, especially if given a deep winter mulch of bracken litter or leafmould.

The typical plants and larger cultivars are unsurpassed in their dramatic evergreen foliage and the scale of their inflorescences. They do particularly well by the sea. Smaller cultivars like 'Bronze Baby' contrast strikingly with silver-grey foliage (*Artemisia* and *Stachys*, for example), provide a rich and exotic foil for the glowing, late summer tones of bronze-leaved *Ligularia* and the smaller *Kniphofia* cultivars, and sit well with such copper and bronze sedges as *Uncinia* and the New Zealand *Carex*. Grow in full sun in any moderately fertile, moisture-retentive but well drained soil. Propagate by division or by seed.

P.colensoi (MOUNTAIN FLAX; leaves to 150×6cm, flexible, dark olive green, base red-tinted; inflorescence to 2m, often inclined, flowers 2.5–4cm, green tinged orange or yellow; cultivars range in height from dwarf (30cm) to large (1.5m) with narrow to broad leaves, some solid plum-purple or bronze, others striped or flushed white, cream, yellow, lime, orange, copper, rose-pink and maroon). *P.tenax* (NEW ZEALAND FLAX, NEW ZEALAND HEMP; leaves to $3m \times 5$–12cm, stiff, olive green, margin red or orange, base pale; inflorescence to 5m, usually erect, flowers to 6cm, dull red; many cultivars including dwarf to tall plants (35–120cm) with narrow to broad leaves tinted, striped or solidly stained purple-bronze, wine red, red, pink, copper, orange, gold, yellow, cream, white and lime).

phosphate a salt of phosphoric acid; a form in which phosphorus is applied as a nutrient. See *phosphorus.*

phosphorus (abbreviation P) a major (macro-) nutrient, important in plant enzyme reactions, and a constitutent of cell nuclei; it is essential for cell division and the development of meristem tissue, and is commonly associated with healthy root growth. Plant requirements vary, legumes, for example, having a high demand.

Phosphorus naturally originates from the weathering of rocks, and some is held in organic matter from which it is slowly mineralised. It is stored in most soils and forms insoluble compounds under extreme acidity or alkalinity; availability is related to rooting capacity, young seedlings being more prone to shortage than mature plants. Symbiotic mycorrhiza often aid phosphorus uptake on deficient soils.

Due to its immobility, rarely more than 20–30% of added phosphorus is taken up from fertilizer application so that acute soil deficiency is rare on cultivated ground; long-established grazing land receiving minimal fertilizer dressing is often low in phosphorus.

Symptoms of deficiency are reduced growth rate, especially shortly after crop emergence during a cold spell; in some instances, there is a tendency to dull, blue/green leaves, often purple and with marginal scorch in older ones.

Phosphorus fertilizers include rock phosphate (26% P_2O_5), triple superphosphate (44% P_2O_5), bone meal (20% P_2O_5), fish, blood and bone (8% P_2O_5) and compound Growmore (7% P_2O_5). Forms of ammonium phosphate are very high in phosphorus and are used in concentrated fertilizer and liquid feed.

Photinia (from Greek *photeinos*, shining, referring to the leaves). Rosaceae. E and SE Asia north to Himalaya, W US. CHRISTMAS BERRY. To 60 species, deciduous and evergreen trees and shrubs with small, white to cream, 5-petalled flowers massed in terminal corymbs or short panicles in late spring and summer. These are followed by small, orange-red pomes. Hardy to −15°C/5°F. Grow in well-drained, fertile soils in sun or light shade. Protect evergreens from cold winds that may scorch foliage. Deciduous species tend to be calcifuge; evergreens are tolerant of calcareous soils. *P.davidiana* will grow in a range of soils including heavy clay, and will tolerate a degree of drought. Propagate deciduous species from seed sown in autumn, cultivars and evergreen species by semi-ripe basal cuttings rooted with bottom heat in a closed case.

P.davidiana (W China; evergreen shrub to 5m; leaves to 10cm, lanceolate to oblanceolate, entire, tough, smooth dark green; 'Palette': leaves blotched cream, dark green and lime, flushed or streaked pink at first; 'Fructuluteo': fruit bright yellow). *P.* × *fraseri* (*P.glauca* × *P.serratifolia*; vigorous evergreen shrub; leaves 7–9cm, elliptic to ovate, small-toothed, coppery red at first, becoming shiny dark green; 'Red Robin': leaves deep glowing red at first, becoming dark shiny green); *P.glabra* (JAPANESE PHOTINIA; Japan; evergreen shrub, 3–6m; leaves 5–8cm, elliptic to narrowly obovate, red at first, becoming dark shining green; 'Roseamarginata': leaves variegated green, white, grey and pink; 'Variegata': leaves pink at first, becoming green edged white); *P.* 'Redstart' (*P.davidiana* 'Fructuluteo' × *P.* × *fraseri* 'Robusta'; large shrub or small tree; leaves to 11cm, bright red at first, becoming dark green, finely toothed above); *P.serratifolia* (China, Taiwan; evergreen shrub to tree, 5–12m; leaves to 10cm, ovate to obovate, saw-toothed, becoming leathery, copper-red becoming glossy dark green); *P.villosa* (Japan, Korea, China; deciduous shrub or small tree to 5m; leaves 3–8cm, obovate to lanceolate dark ovate, apex long-tapered, leathery, teeth glandular-tipped).

photoperiodic lighting see *supplementary light*.

photoperiodism the influence of day length on plant growth, frequently concerning aspects of reproduction, especially flowering. By reason of this phenomenon, life cycles may be manipulated, and beneficial effects are extended to certain insect pollinators. Photoperiodic responses in plants are often related to the avoidance of seasonal stress factors, such as drought and cold, and are associated with leaf growth, dormancy, storage-organ formation and leaf fall.

Scientific knowledge of the detailed mechanism of photperiodism is incomplete, most of what is known concerning the effects of day length on flowering. In general, plants growing far from the equator respond to longer day lengths than those growing close to it. Response relative to latitude can occur within a species, sometimes quite precisely, as with soybean varieties grown in the United States which only achieve optimum performance within an approximately 50-mile range of latitutde.

True day-neutral plants where flowering is unaffected by day length are probably rare; cucumber (*Cucumis sativa*) and haricot (*Phaseolus vulgaris*) are two examples. Short-day plants (SDPs) flower in response to days shorter than a certain critical length. Cocklebur (*Xanthium strumarium*) is an example of an SDP, which will flower after just one short day. Spinach (*Spinacia oleracea*) is a long day-plant (LDP), responding to a single inductive long day. More common are species requring several cycles of the appropriate photoperiod for floral induction.

Among SDPs is *Kalanchoe blossfeldiana*, which has a precise requirement, and the sunflower (*Helianthus annuus*), which will eventually flower in days longer than the critical length. An example of a precise LDP is hibiscus (e.g. *Hibiscus syriacus*), while spring wheat (*Triticum aestivum*) is an LDP not totally dependent on a critical day length. There are sub-variations, for example, several species respond to various combinations of day lengths, for instance, long days followed by short days promote flowering in *Kalanchoe laxiflora*, while short days followed by long days are required by white clover (*Trifolium repens*).

The great majority of plants must reach a certain stage of development to respond to day length; for example, henbane (*Hyoscyamus niger*) must be 10–30 days old before it can respond to long days, and many trees will not flower until they are tens of years old.

For all the response types, it is the length of night or darkness, rather than day length, which is the critical factor and it appears that the plant pigment phytochrome inhibits flowering in SDPs, while promoting it in LDPs. This phenomenon makes possible the artifical manipulation of flowering time in practical horticulture, the best-known example of which in Britain is the supply of all-year-round chrysanthemums. Chrysanthemums are SDPs, which are naturally induced to flower by the shortening days of late summer. However, when grown in controlled conditions under glass, they can be prevented from flowering until required by using artificial lights to extend day lengths or, more efficiently, to provide one or several brief night breaks. When flowering is required, the night breaks are suspended and a few long nights allow flowering to be induced. In the long days (short nights) of summer, flowering can be induced by artificially lengthening the nights with blackouts in the greenhouse.

photosynthesis a complex chemical process, promoting the production in green plants of organic substances, such as sugars and starches, from water and atmospheric carbon dioxide. It is the source of energy for plant activity. The process involves the activation by sunlight of the green pigment chlorophyll. Photosynthesis, (carbon assimilation) takes place principally in the leaves, into which carbon dioxide diffuses through stomata, and free oxygen is discharged as a product of the reaction.

phototropism directional growth of a plant in response to light. See *tropism*.

Phragmipedium (from Greek *phragma*, partition, and *pedilon*, slipper, referring to the trilocular ovary and the slipper-shaped lip). Orchidaceae. C and S America. LADY-SLIPPER. Some 20 species, evergreen perennial herbs with narrowly strap-shaped leaves in a basal fan. The flowers are borne several to a stem in spring and summer. They resemble those of *Paphiopedilum* but have a more inflated lip, with the petals often narrower and drooping – in some species, greatly attenuated and spiralling. The genus was formerly included variously in *Cypripedium*, *Paphiopedilum*, and *Selenipedium*. Cultivate as for warm-growing *Paphiopedilum*.

P.besseae (Colombia, NE Peru; flowers to 6cm in diameter; dorsal sepal ovate-lanceolate, petals oblong-elliptic, larger, all bright scarlet; lip strongly pouched, fiery red to golden yellow or peach); *P.caudatum* (Mexico to Peru, Venezuela, Colombia, Ecuador; flowers to 125cm across; dorsal sepal white to yellow-green with dark green veins, to 15 × 3cm, lanceolate, acute, arched over lip, strongly undulate to spiralled, petals purple-brown to green-brown, to 60 × 1cm, linear-lanceolate, pendulous, spiralling, lip yellow near base, apex tinted pink or maroon, veined green, to 7 × 2cm, lateral lobes strongly incurved, spotted green); *P.grande* (*P.longifolium* × *P.caudatum*; flowers green-brown with very long petals); *P.lindenii* (Colombia, Peru, Ecuador; resembles *P.caudatum* except lip not pouched, similar to petals, wider at base); *P.lindleyanum* (Venezuela, Guyana; sepals pale green or yellow-green, veined red-brown, downy below, dorsal sepal to 3.5 × 2cm, elliptic, obtuse, concave, petals yellow-green at base, white-green toward apex, margins and veins flushed purple toward apex, to 5.5 × 1cm, linear-oblong, apex rounded, undulate; lip pale yellow-green with yellow-brown venation, to 3 × 1.5cm, lateral lobes spotted light purple); *P.longifolium* (Costa Rica, Panama, Colombia, Ecuador; dorsal sepal pale yellow-green, veined dark green or rose, edged white, to 6 × 2cm, lanceolate, acute, erect or curved forward, sometimes undulate, petals pale yellow-green, margins rose-purple, to 12 × 1cm, spreading, linear to lanceolate, twisted, lip yellow-green, to 6 × 1.5cm, margins spotted pale rose to magenta); *P.schlimii* (Colombia; tepals white flushed rose-pink, dorsal sepal to 2 × 1cm, ovate-oblong, obtuse, concave, petals slightly longer, spreading, spotted pink at base, elliptic, lip rose-pink, lateral lobes incurved, streaked white and rose-carmine); *P. × sedenii* (*P.longifolium* × *P.schlimii*; sepals ivory white, flushed pale rose, exterior rose-pink, petals white, margins tinged rose-pink, twisted, lip rose-pink, lobes white spotted rose).

Phragmites (from Greek *phragma*, fence or screen, referring to the screening effect of many plants growing together along streams). Gramineae. Cosmopolitan. REED. 4 species, rhizomatous perennial grasses; culms clumped, reedy; leaves linear, flat; inflorescence a large, terminal, plumed panicle produced in summer and early autumn. Occurring in marsh, fen and riverside habitats in temperate and tropical zones, *P.australis* is an elegant perennial, valued for the soft and showy flowering panicles – these retain their beautiful metallic sheen even on drying – and for the golden russet autumn colours. The species is invasive and hardy to –20°/–4°F and below. It thrives in deep, moisture-retentive soils, but is generally suitable only for the larger landscape, unless confined in containers, or used as a marginal or submerged aquatic. Propagate by division. *P.australis* (syn. *P.communis*; COMMON REED, CARRIZO; cosmopolitan; culms robust, to 3.5m; leaves to 60 × 5cm+, arching, mid to grey-green; inflorescence erect to nodding, silky grey tinged pearly pink-brown to purple, then brown, to 45cm; 'Rubra': inflorescence crimson-tinted; 'Striatopictus': leaves bright green striped pale yellow; 'Variegatus': leaves striped bright yellow, fading to white).

Phuopsis (from its resemblance to *Valeriana phu*). Rubiaceae. Caucasus. 1 species, *P.stylosa*, a slender mat-forming perennial herb to 30cm, with whorls of narrow-lanceolate, slender-pointed and spiny-ciliate leaves. Produced in summer, the clustered flowers are scented, small and pink, with a tubular corolla and five oblong to ovate, obtuse lobes. Suitable for groundcover in the rock garden, and on banks. Hardy to –20°C/–4°F. Grow in moist, gritty but well-drained soils in full sun, or light shade. Propagate by removal of rooted stems, by semi-ripe cuttings or from seed sown in autumn.

Phygelius (from Greek, phyge, flight – so-named because it long evaded the researches of botanists). Scrophulariaceae. South Africa. 2 species, evergreen or semi-evergreen shrubs and subshrubs with ovate to lanceolate, serrate leaves and, in summer, narrowly tubular flowers hung in terminal panicles. Hardy to –5°C/23°F, they are usually treated as herbaceous perennials in areas where temperatures fall much below freezing in winter. In warm-areas or if given the protection of a south-facing wall, *P.capensis* will thrive in its naturally shrubby form and may be trained as a wall shrub and tied into trellis supports. Plant in a sunny site with a moderately fertile, light soil. Where the ground freezes, provide a heavy winter mulch. Cut out dead wood in spring. Sow seed in spring at 15–18°C/60–65°F and overwinter young stock in frost-free conditions. Alternatively, take semi-ripe cuttings in late summer.

P.aequalis (to 1m; flowers dusky pink, throat orange, lobes crimson; 'Yellow Trumpet': flowers pale creamy yellow); *P.capensis* (CAPE FUCHSIA; to 3m, somewhat sprawling; flowers pale orange to deep red; 'Coccineus': flowers rich orange with orange-red lobes); *P. × rectus* (*P.aequalis* × *P.capensis*; varied group of hybrids from 50–150cm tall, erect to loosely scrambling with flowers ranging from pale cream, to lime green, pale pink, coral red, dusky pink, salmon, scarlet and orange).

Phygelius capensis

phyllary one of the bracts that subtends a flower or inflorescence, composing an involucre.

phylloclade a stem or branch, flattened and functioning as a leaf.

Phyllocladus (from Greek *phyllon*, leaf, and *klados*, young branch or shoot, referring to the leaf-like shoots). Phyllocladaceae. Malaysia, Indonesia, New Guinea, Tasmania, New Zealand. CELERY PINE. 5 species, evergreen, coniferous trees or shrubs with modified leaf-like shoots (phylloclades), which are leathery, incised and often deeply or shallowly toothed. Plant in a free-draining, acid to neutral medium with additional sand and grit. Grow in full sun in a cool, moist environment. This tree withstands mild frosts. It needs a cool, humid environment in full sun. Propagate by seed or semi-ripe cuttings. *P.trichomanoides* (TANEKAHA; New Zealand; tree to 20m; phylloclades, ovate to 30cm, dark green above, glaucous beneath, tough pinnately lobed with segments roughly rhombic and shallowly crenate).

phyllode an expanded petiole, taking on the function of a leaf blade, as in *Acacia*.

Phyllodoce (for a sea nymph of Greek mythology, one of Cyrene's attendants, mentioned by Virgil). Ericaceae. Arctic and Alpine regions of northern Hemisphere. 8 species, dwarf, evergreen shrubs with small, linear, revolute and leathery leaves. Produced in late spring and early summer, the flowers are pitcher- or bell-shaped and nod in racemes or clusters. Suitable for the peat terrace and for cool, damp, peaty niches in the rock garden. Hardy to −15°C/5°F. Grow in moist but well-drained, peaty, acid soils, in partial shade or in bright indirect light. Protect in areas that experience prolonged frosts. Trim over after blooming. Propagate by semi-ripe heel cuttings or softwood tip cuttings in late summer. Increase also by seed in spring, or by layering.
P.aleutica (Japan; dwarf, decumbent or scandent; flowers to 8mm, pale yellow-green, pitcher-shaped, downy, with pink anthers; subsp. *glanduliflora*: Alaska; YELLOW MOUNTAIN HEATHER; flowers fragrant, yellow to olive with purple anthers); *P.caerulea* (syn. *P.taxifolia*; Asia, Europe and US; 10–35cm, erect or scandent; flowers 7–12mm, lilac to purple-pink, pitcher-shaped); *P.empetriformis* (PINK MOUNTAIN HEATHER; British Colombia to California; low, diffuse, mat-forming shrub, 10–38cm; flowers 5–9mm, bell-shaped, rose-pink); *P. × intermedia.* (*P.empetriformis* × *P.aleutica* subsp. *glanduliflora*; western N America; dense, bushy subshrub, 15–23cm; flowers mauve to purple-red or yellow-pink, ovate-campanulate to pitcher-shaped; 'Drummondii': flat habit, dark purple flowers; 'Fred Stoker': flowers light purple); *P.nipponica* (North Japan; small, branched, suberect shrub, 7–23cm; flowers rose to white, bell-shaped; var. *amabilis*: calyx lobes red, corolla red- or pink-tipped).

Phyllostachys (from Greek *phyllon*, leaf, and *stachys*, spike, for the 'leafy' inflorescence). Gramineae. China, India, Burma. Some 80 species of medium and large bamboos; see *bamboos*.

phyllotaxy the arrangement of plant organs on an axis. Commonly used of leaves on a stem.

phylloxerids (Hemiptera: Phylloxeridae) a small family of sap-sucking insects, related to aphids, which are important pests of grapes, oak and pecans.

The VINE LOUSE or GRAPE PHYLLOXERID (*Daktulosphaira vitifolii*) is a North American species, introduced into Europe about 1863. Infested vines develop galls on the roots and leaves followed by rapid collapse and death. American vines have both foliar and root-feeding forms but in Europe the life cycle is restricted to subterranean forms. In the 19th-century, serious epidemics were eventually overcome by grafting European grape vines on to resistant rootstocks from North America.

The OAK PHYLLOXERIDS (*Phylloxera glabra* and other species.) infest the underside of oak leaves in summer, causing yellow and brown spots to develop; severe infestations cause extensive browning of foliage followed by premature leaf-fall. Older trees are seldom harmed but young trees may become seriously weakened and can be protected by contact or systemic insecticides recommended for aphid control.

Pecans are cultivated in the southern states of the US for their nuts. They are subject to infestation by the PECAN PHYLLOXERA (*P.devastatrix*), which causes extensive galling of leaves, shoots and nuts; the PECAN LEAF PHYLLOXERA (*P.notabilis*) is less serious on mature trees but can be damaging to young nursery trees.

Physalis (from Greek *physa*, bladder or bellows, referring to the inflated calyx). Solanaceae. Cosmopolitan, especially Americas. GROUND CHERRY, HUSK TOMATO. Some 80 species, annual or perennial herbs. Produced in summer, the flowers are solitary, white tinged violet or yellow, campanulate to rotate and 5-lobed. After pollination, the calyx develops into a colourful, inflated and ribbed bladder or lantern, enclosing the berry-like fruit. The hardy *P.alkekengi* is used for decoration and provides a vivid orange display in the autumn garden; it is also used for drying, with the stems cut when the calyces begin to colour and air-dried. *P.peruviana* yields the dessert fruit, Cape gooseberry; it is not reliably hardy beyond areas that are frost-free or almost so, but in temperate zones can be grown for fruit in a cool greenhouse. Both species are easily cultivated on any well-drained soil, in sun or light shade. Propagate by division or seed.
P.alkekengi (CHINESE LANTERN, WINTER CHERRY, BLADDER CHERRY; C and S Europe, W Asia to Japan; perennial to 60cm, thinly glandular-pubescent; calyx to 2cm, becoming inflated, lantern-like to 5cm, orange; fruit to 17mm, red to scarlet, inedible; 'Variegata': leaves deeply bordered cream and yellow-green); *P.peruviana* (CAPE GOOSEBERRY,

PURPLE GROUND CHERRY; tropical S America; perennial to 1m; calyx becoming ovoid, pubescent to 4cm, ultimately withered and thinly papery; fruit edible, globose, yellow to purple, to 2cm in diameter).

physic garden usually, an alternative name for a botanic garden; strictly, a garden primarily concerned with the collection, cultivation and study of medicinal plants, such as were established in Europe as adjuncts to faculties of medicine in the 16th and 17th centuries.

physiological disorder a condition harmful to growth, flowering or fruiting, not caused by pest or disease attack. It is often associated with defects in the growing conditions.

physiology the study of the functions of living organisms.

Physocarpus (from Greek *physa*, bladder, and *karpos*, fruit, referring to the inflated follicles). N America, NE Asia. NINEBARK. 10 species, deciduous shrubs with exfoliating bark and palmately lobed leaves. Produced in early summer in terminal corymbs, the flowers are small, white to light pink with five, rounded petals. Hardy to –25°C/–13°F. Grow in moist, moderately fertile, acid soils in full sun. Prune after flowering, thinning out old and overcrowded shoots by cutting back to ground level. Propagate by removal of suckers, by seed, or from softwood cuttings in a closed case with bottom heat. *P.opulifolius* (NINEBARK; central and east N America; to 3m; bark brown, shredded; leaves to 7.5cm, oval to rounded, cordate at base glabrous, double-toothed, usually 3-lobed; flowers often pale pink, or white tinged with rose, in many-flowered corymbs; 'Dart's Gold': bright gold leaves, flowers white washed pink; 'Luteus': leaves golden at first, later olive green or tinted bronze).

Physostegia (from Greek *physa*, a bladder, and *stege*, a covering). Labiatae. N America. OBEDIENT PLANT, FALSE DRAGON HEAD. 12 species, erect perennial herbs. They flower in summer and early autumn, producing spikes of tubular and hooded, 2-lipped flowers. If pushed to one side on the spike, a flower will remain there, hence the popular name 'obedient plant'. Fully hardy. Plant on rich fertile loamy soil with abundant water. Propagate by division when dormant. Sow seed in autumn in a cold frame. *P.virginiana* (central and south US and NE Mexico; erect to 180cm, often shorter; flowers red-violet, lavender or white, usually spotted and streaked purple; includes short and tall cultivars, some with variegated leaves and flowers ranging in colour from white to pale rose, deep pink, lilac, lavender or claret).

Phyteuma (name used by Dioscorides for an aphrodisiac plant, meaning simply 'the plant'). Campanulaceae. Europe, Asia. HORNED RAMPION. Some 40 species, summer-flowering perennial herbs. The flowers are produced in compact terminal spikes or heads, usually subtended by bracts. They consist of five, linear lobes, the whole head appearing spiky. Fully hardy. Plant in sun on fast-draining, gritty neutral or slightly alkaline soils in the rock garden. Propagate by division in spring or after flowering, or by seed in autumn. *P.scheuchzeri* (Alps; to 40cm; inflorescence spherical, bracts linear, leafy, flowers dark blue).

phytochrome a plant pigment involved in the developmental responses of plants to light. It exists in two interchangeable forms depending upon the ratio of red to far-red light. See *photoperiodism*.

Phytolacca (from Greek *phyton*, plant, and French *lac*, lake, a red pigment, from the crimson juice of the berries.) Phytolaccaceae. Temperate and warm regions. POKEWEED. 35 species, perennial herbs, shrubs and trees. In summer, they bear small, apetalous flowers in erect or drooping spike-like racemes. These are followed by wine-red to purple-black, berries. Fully hardy and suitable for the open woodland garden, for massed plantings and as a specimen in the shrub and mixed border. Its tinted, forked stems and garnet spikes of toxic fruits are especially striking in late summer. Grow in moisture-retentive, fertile soils in full sun or light shade. Propagate by seed sown in autumn or spring or by division. *P.americana* (POKE, POKEWEED, SCOKE, GARGET, PIGEON BERRY; N and C America; herbaceous perennial, to 4m; stem purple-red, dichotomously branching; leaves to 30cm, becoming purple-pink to yellow in autumn; fruit glossy, green becoming red then purple-black in erect persistent spikes).

phytophthora a genus of parasitic fungi allied to the downy mildews. Of wide importance is POTATO BLIGHT (*Phytophthora infestans*) which spreads rapidly in warm wet seasons. First signs are brown dead patches on the leaf tips or edges which enlarge to cause withering; haulm becomes affected with brownish black patches and all growth may collapse and rot. Tubers become infected by washed-down spores and develop rot before lifting and in store. Weather conditions conducive to infection are two consecutive 24-hour periods during which the minimum temperature is 10°C/50°F and for no less than 11 hours the relative humidity is at least 89%. Control is with protective sprays of Bordeaux mixture; and by the choice of resistant cultivars including 'Cara', 'Estima', 'Maris Peer' and 'Wilja'. Other *Phytophthora* diseases include RED CORE DISEASE of strawberry (*P.fragariae*), which is soil inhabiting and persistent, causing loss of root function; CROWN ROT of apple and strawberry (*P.cactorum*); STEM AND FRUIT ROT of red pepper (*P.capsici*); ROOT ROT of many woody plants (*P.cinnamomi*); DOWNY MILDEW of Lima bean (*P.phaseoli*) and LILAC BLIGHT and APPLE FRUIT ROT (*P.syringae*). The majority of phytophthoras are dispersed by water and are therefore important as soil-borne diseases, *P.infestans* and *P.phaseoli* are dispersed by air-borne spores, thereby causing rapidly spreading epidemics. See *root rot*.

Picea pungens

phytoplasmas (formerly known as MYCOPLASMAS, MYCOPLASMA-LIKE ORGANISMS, MLOs) are micro organisms which lack a true cell wall. Allied to bacteria, they attack plants to cause YELLOWS and other diseases transmitted by leafhoppers. They are morphologically similar to the Mycoplasmataceae (animal pathogens and saprophytes) which are related to the Spiroplasmataceae (which cause CITRUS STUBBORN DISEASE, etc), but phytoplasmas are not now thought to be closely related to these families. The true taxonomic position of phytoplasmas is not yet settled.

ASTER YELLOWS is a well-known disease in North America, Europe and Japan, formerly thought to be caused by a virus, but in 1967 shown to be an example of a group of similar plant diseases caused by phytoplasmas. These cannot be grown in artificial culture and their identification depends on symptoms, microscopic examination and transmission and inoculation studies. Infection is transmitted by the feeding of leafhoppers (e.g. *Macrosteles*) migrating from herbaceous perennials; such ornamentals particularly affected include the composites *Calendula*, *Callistephus*, *Chrysanthemum*, *Gaillardia*, *Gerbera* and *Tagetes*. Others are *Anemone*, *Delphinium*, *Gypsophila*, *Phlox* and *Primula*. ASTER YELLOWS is more common in warmer climates such as southern and eastern Europe, where insect vectors are numerous; and it causes yellowing, dwarfing, proliferation of shoots and malformation and greening of flowers, which can be lethal on some hosts whilst others recover. In North America the disease is considered to be caused by a different strain of phytoplasma to those occurring in Europe. Vegetable diseases caused by ASTER YELLOWS phytoplasma are LETTUCE PHYLLODY, ONION PROLIFERATION YELLOWS of beet, carrot, celery, courgette and spinach, and VIRESCENCE (GREENING) of brassicas. MAIZE BUSHY STUNT and STRAWBERRY WITCHES' BROOM occur in north and north west America but are not known in Europe.

HYDRANGEA VIRESCENCE (GREEN FLOWER) is an important disease in Europe, and *Lonicera*, *Jasminum* and *Vaccinium* species are affected by WITCHES' BROOM. 'GREEN PETAL' of strawberries and trifolium clover is a linked condition caused by a phytoplasma.

Other diseases thought to be caused by phytoplasma include APPLE PROLIFERATION, APPLE RUBBERY WOOD, APRICOT CHLOROTIC LEAF ROLL, BLACKCURRANT REVERSION, LITTLE CHERRY, CHERRY MOLIÉRE DISEASE, ELM PHLOEM NECROSIS, GRAPEVINE FLAVESCENCE DORÉE (transmitted by the N American leafhopper, *Scaphoideus litoralis*), PALM LETHAL YELLOWING, PEACH YELLOWS, PEACH ROSETTE, PEAR DECLINE, and RUBUS STUNT.

phytotoxicity the poisonous effects of a substance on plants, as may occur from adverse reaction to a pesticide. Stage and state of growth and environmental conditions, especially bright sunlight, may predispose plants to phytotoxicity.

Picea (from Latin *pix*, pitch, obtained from the resin of *P.abies*). Pinaceae. Northern Hemisphere except Africa. SPRUCE. Some 35 species, resinous evergreen, coniferous trees. The crown is conical in outline, often domed or columnar in older specimens. The branches are borne in more or less horizontal whorls. Needle-like leaves are carried in spirals, usually pressed forward on the branchlets. The cones are ovoid to cylindric with large, usually thin scales obscuring the very short bracts, green to purple, ripening pale to dark brown, becoming pendulous after pollination. Most species are hardy to –30°C/–22°F or lower. Cultivate as for *Abies*, except that spruces are more tolerant of exposure and less tolerant of shade; they should therefore be planted in the open. They are also more tolerant of poor, peaty soils. As with other conifers, propagation is better from seed than grafts; in *Picea breweriana* this is especially important, grafted plants never making good trees. Seed should be wild-collected or from extensive single-species plantings. Propagate dwarf cultivars by cuttings; if grafted on normal rootstocks, they grow too vigorously.

Several rust fungi, mostly species of *Chrysomyxa*, affect spruce, especially *P.abies*, but none is a serious problem. Those occurring in Britain are *Chrysomyxa abietis* and *C.ledi* var. *rhododendri*, both of which cause yellow bands on the needles. *C.abietis* is only known to infect spruce, on which it produces the teliospore stage, whereas *C.ledi* var. *rhododendri* produces uredio- and teliospores on rhododendron and aeciospores on spruce. The most important pests of spruce are common to Europe and North America. These include the green spruce aphid, which can cause complete defoliation of several species. This is the most serious pest of spruce in gardens, particularly in warmer areas and following mild winters. Spruce gall adelgids induce 1–3cm pineapple-shaped galls to develop. The European spruce sawfly and the spruce budworm are destructive pests of opening buds and young shoots in North America.

P.abies (NORWAY SPRUCE; N and C America; to 55m; bark red-brown to grey or purple, flaking in thin plates; crown conical, columnar with age, branches level or drooping and upcurved, branchlets pendulous; leaves 1–2.5cm, dark green, obtuse; cones cylindric, green or purplish, ripening brown, 8–18 × 3–4cm, to x5–6cm open; scales rhombic. Upright, conical selections include the slender, to 20m-tall 'Cupressina' and 'Viminalis', and the broadly conical 'Pyramidata'; sparse forms include the irregularly branched 'Virgata' (SNAKE-BRANCH SPRUCE) and 'Cranstonii'; weeping forms include the very pendulous 'Frohburg' and 'Inversa' and the semi-prostrate 'Lorely'; notable low-growers (to 5m) include the narrow-conic 'Concinna', the broad-conic 'Conica' and the profusely coning 'Acrocona'; dwarf selections include the very slow-growing 'Clanbrassiliana' (3m in 180 years), the irregular 'Pachyphylla' and 'Pygmaea', and the more regular 'Elegans' and 'Humilis'; globose dwarfs include the stoutly branched 'Nana Compacta' and 'Pyramidalis

Gracilis' and the finer 'Gregoryana' and 'Hystrix'; flat-topped dwarfs include the slow-growing 'Little Gem' (2cm per annum) and the mat-forming 'Repens'; cultivars noted for needle colour include the yellow to gold 'Aurea' and 'Aurescens', the steel-blue 'Coerulea' and the dwarf, gold-variegated 'Callensis' and 'Helen Cordes'); *P.breweriana* (BREWER'S SPRUCE; N California and S Oregon; to 35m, slow-growing; bark grey or purple-grey, becoming scaly; crown ovoid-conic; branches horizontal, upward-pointing, branchlets pendulous; leaves flattened, obtuse, 2.5–3.5cm, glossy deep green above, 2 white stomatal bands beneath, somewhat curved; cones 8–14 × 2–2.5cm, opening to 3–4cm across, cylindric, purple then red-brown, resinous; scales large, obovate, apex rounded); *P. × engelmannii* (ENGLEMANN SPRUCE; western N America; to 45m; bark thin, buff, splitting into small plates, resinous; crown conic to narrowly acute; branches in dense whorls, upward-pointing, branchlets pendulous; leaves straight or curved, blue-green or glaucous, soft, flexible 1.5–3cm, forward-pointing; cones 3–5 × 1cm; ovoid to cylindric, tapering, green to light brown; scales very thin, obtuse; 'Argentea', with grey leaves, tinged silver; 'Fendleri', with pendulous branches and leaves to 3cm; 'Glauca', upright, with vivid blue leaves; 'Microphylla', dwarf, bushy, compact; 'Snake', of highly irregular upright habit, tinged blue; f. *glauca*, with intense glaucous blue leaves); *P.glauca* (WHITE SPRUCE; Canada, far NE US; to 25m; bark light grey becoming darker with white cracks; crown conic; branches level or descending, with upward-growing tips; branchlets dense; leaves stiff, dull blue-green, 10–18mm, imbricate to spreading; cones 3–5 × 1cm, ovoid to cylindric, tapering, green to light brown; scales very thin obtuse; 'Aureospicata', with yellow young shoots, later green; 'Coerulea', to 2m, densely pyramidal, leaves tinged silver-blue; 'Hendersonii', similar to 'Coerulea', with young shoots lateral, later pendulous; 'Pendula', weeping, upright leader, with very pendulous red branches and leaves tinged blue; var. *albertiana*, from Canadian Rocky Mountains, to 45m, with pubescent shoots, leaves to 2.5cm, and ovoid cones to 4cm; 'Conica', to 4m, dense, conical, with permanently juvenile, very slender, curved leaves, 11–15 × 0.5mm; 'Gnom', slow-growing mutation of 'Conica' (to 5cm per annum); 'Laurin', small, dense, mutation of 'Conica'); *P.likiangensis* (LIJIANG SPRUCE; SW China, SE Tibet, E Assam; to 50m; bark grey, smooth to scaly or shallow-fissured; crown open, broadly conic; branches level to ascending; leaves 8–15mm, sharp, dark green to blue-green, flattened, keeled, 2 blue-white stomatal lines beneath, 2 faint bands above, forward-growing, loosely imbricate; cones oval-cylindric, 7–13 × 3cm, red to purple; scales thin, obovate, finely toothed); *P.mariana* (MOUNTAIN SPRUCE; northern N America; to 20m+; bark red-brown to grey-brown, exfoliating in thin flakes; crown conic; branches downswept; leaves densely arranged, 0.5–1.5cm, blue-green above, off-white stomatal stripes beneath, stiff, obtuse; cones 2–4 ×

1.5cm, fusiform, mauve, becoming grey-brown, persisting; scales woody, rounded, margins finely toothed; 'Argenteovariegata', with some leaves almost completely white; 'Aurea', with leaves tinged gold; 'Beissneri', dwarf, slow-growing to 5m, similar to 'Doumettii' but with broader leaves tinged blue; 'Doumettii', dwarf to 6m, globose-conic when young, becoming irregular, with a dense crown and bright blue-green leaves; 'Fastigiata', conical dwarf, with slender, ascending branches; 'Nana', very dwarf and globose to 50cm, slow-growing to 3cm per annum, more than one clone in cultivation; 'Pendula', upright leader with weeping branches); *P.morrisonicola* (TAIWAN SPRUCE; Taiwan; to 40m; shoots sparse, pendulous; leaves 10–20mm, slender, sharply acuminate, glossy dark green; cones 5–8 × 1.5–2cm, oblong-cylindric; scales obovate, obtuse); *P.omorika* (SERBIAN SPRUCE; Balkans; to 35m; bark orange-brown to purple-brown, becoming cracked and flaking in fine plates; crown conic to narrowly spire-like; branches downswept to upcurved; branchlets dense, more or less pendulous; leaves 10–20mm, flattened, keeled, dark to blue-green, with 2 silver-grey bands beneath; cones 3–7 × 2cm, fusiform, purple, ripening red-brown, scales closely adpressed, broadly obtuse, finely toothed; 'Expansa', vigorous procumbent dwarf, with branch ends slightly ascending; 'Gnom', broadly conical dwarf, slow-growing to 3cm per annum; 'Nana', globose to broadly conic-pyramidal dwarf; 'Pendula', slender and upright, with pendulous branches); *P.orientalis* (ORIENTAL SPRUCE, CAUCASIAN SPRUCE, Caucasus, NE Turkey; to 50m; bark smooth, pink-grey, becoming cracked; forming small raised plates; crown conic, dense, to ovoid-columnar; branches slightly ascending; leaves 0.5–0.8cm, slightly flattened, blunt, bevelled, dark green, adpressed to stamen above; cones 5 × 9–1.5cm, purple, ripening, brown, cylindric-conic to fusiform; scales obovate, obtuse, entire, to 15mm wide; 'Atrovirens', with very dark green leaves; 'Aurea', with gold young leaves, later dark green, with gold tint; 'Aurea Compacta', broadly pyramidal and compact semi-dwarf, with gold upper leaves and green lower leaves; 'Compacta', broadly conical dwarf; 'Gracilis', very dense dwarf, oval in growth, to 6m, slow-growing (to 7cm per annum); 'Nana', globose dwarf to 1m, very slow-growing (to 2.5cm per annum); 'Nutans', spreading, weeping and irregular in growth, with very dark green leaves; 'Pendula', compact and slow-growing, with nodding twigs); *P.pungens* (BLUE SPRUCE, COLORADO SPRUCE; US (S Rocky Mountains); to 40m or more; bark purple-grey, deeply grooved, forming thick scales; crown dense, conic to columnar-conic; lower branches downswept; leaves 2–2.5cm, radial, assurgent, grey-green to bright pale blue, thickly glaucous, stiff, pungent; cones 6–12 × 2.5–3cm, oblong-cylindric, bloomed violet, ripening light brown; scales thick-based with thin tips, flexible, emarginate, finely toothed and undulate; blue-leaved upright cultivars include the short-needled 'Microphylla', 'Endtz', 'Oldenburg', the very blue 'Koster' and 'Moerheim' and the compact 'Fat Albert'; notable blue-tinted dwarfs include the globose 'Glauca Globosa' and 'Pumila' and the slow-growing 'Montgomery'. 'Glauca Procumbens' is almost recumbent and 'Glauca Pendula' an irregular, graceful weeping form; yellow and green selections include the sulphur 'Aurea' and the dark green 'Viridis'; forma *glauca* is the group name for all plants with glaucous, blue-grey leaves); *P.sitchensis* (SITKA SPRUCE; N America (Pacific coast); to 90m; bark red-grey, exfoliating in coarse scales; crown broad, conic, branches straight, branchlets moderately pendulous; leaves 15–25mm, dark green, with 2 blue-white stomatal bands beneath, sharply pointed, flattened, slightly keeled; cones 3–10 × 1.5cm when closed, to 3.5cm broad, cylindric-oblong, pale green, ripening beige; scales very thin oblong to rhombic, undulate, serrate; 'Compacta', broadly conic, dense dwarf to 2m, with spreading branches and young shoots tinted yellow; 'Microphylla', dwarf, very slow-growing, to 25cm after 10 years, upright and narrowly conic; 'Nana', slow-growing dwarf, of open habit, with leaves tinted blue); *P.smithiana* (MORINDA SPRUCE, HIMALAYAN SPRUCE; W Himalaya; to 55m; bark dull grey-purple, grooved, slitting into round scales; crown conic, columnar with age; branches horizontal, shoots pendulous; leaves loosely radial, 3–5cm, dark green, apex acuminate; cones 10–16cm, cylindric, curved, tapered, shiny pale green, ripening brown, resinous; scales semi-circular, woody, entire, to 2.5cm wide).

pick, pickaxe a short-handled tool with a thick, centrally-fixed head fashioned as a narrow, curving blade with two working points. It is used from a stooping position to break up soil, rock or concrete. A pick has two pointed heads, a pickaxe one pointed and one chisel-shaped head.

pickfork see *fork.*

picotee describing flowers where the petals have a thin band of contrasting colour at their edges. It is used, for example, of carnations and sweetpeas.

picturesque a late 18th-century style of British Landscape gardening, fostered by the Reverend William Gilpin and the landscapers Richard Payne Knight and Uvedale Price. It sought to bring subtleties of texture, colour, and light and shade into designs, such as might be achieved by a painter.

Pieris (named for the *Pierides*, (Muses)). Ericaceae. E Asia, Himalaya, E US, W Indies. Some 7 species, evergreen trees or shrubs. The new growth is often tinged red, coral or bronze-pink. The leaves are ovate-elliptic to lanceolate and leathery. Produced in spring in erect or drooping, terminal or axillary racemes or panicles, the flowers are small, white, pitcher-shaped, 5-lobed and often scented. Hardy to −15°C/5°F. Grow in moisture-retentive, well-drained soils that are lime-free and humus-rich, in dappled shade or in sun. Give a position with protection to

the north and east, sheltered from strong winds and frost; the young foliage is particularly vulnerable to frost, although damaged specimens will usually regenerate giving a second, less spectacular display of coloured foliage. Flower buds formed in autumn may drop prematurely due to frost or insufficient soil moisture. Propagate by semi-ripe or softwood basal cuttings in summer; treat with 0.8% IBA and root in a closed case with bottom heat.

P.floribunda (FETTER BUSH; SE US; shrub to about 2m; bark grey to grey-brown; leaves 3–8cm, elliptic to ovate, serrate and ciliate, sharp-tipped, dull green above, sparsely glandular-hirsute; panicles 5–10cm; includes the later-flowering 'Elongata', with flowers in a long terminal panicle, and the early-flowering, compact 'Spring Snow'); *P.formosa* (SW China, Vietnam, Himalaya, Nepal; shrub or small tree, 2.5–5m; bark grey to grey-brown; leaves 2.5–10cm, elliptic or obovate, finely serrate, sharp-tipped or blunt, coriaceous, lustrous, sparsely glandular-hirsute; panicles or racemes erect to drooping to 15cm, flowers sometimes pink-tinged, urceolate to cylindric, hairy; var. *forrestii*: syn. *P.forrestii*, to 3m, with scarlet new growth, leaves elliptic-lanceolate, acuminate, finely serrate, 6–10cm-long, flowers fragrant, in a drooping terminal panicle; 'Wakehurst': to 5.4m, hardy, vigorous, with oblong-elliptic to oblanceolate, bright red leaves, fading to pink before turning deep green, with serrate margins and flowers in large clusters); *P.japonica* (LILY OF THE VALLEY BUSH; Japan, Taiwan, E China; shrub or small tree, 2.7–4m, bark grey to brown; leaves 2.5–10cm, obovate to lanceolate, serrate, coriaceous, sharp-tipped to blunt, emerging pink to red or bronze, hardening dark and lustrous green; racemes or panicles to 12cm, erect to drooping, axillary, flowers sometimes tinged pink; plants formerly known as *P.taiwanensis* usually have bronze-red young growth, tougher, matt green leaves with sparser teeth and a spreading to erect panicle – they include 'Crispa', a small, slow-growing shrub with curled or wavy leaves; 'Bert Chandler', with pale pink young leaves, turning glossy yellow-white then dark green; 'Compacta', compact, dense, to 1.8m, with small leaves; 'Christmas Cheer', early-flowering, with white and deep rose flowers; 'Firecrest', with bright red young growth; 'Flamingo', with bronze-red young growth and deep pink flowers; 'Forest Flame', compact, with oblanceolate finely serrate leaves, and red young growth changing to pink, ivory, then pale green; 'Snow Drift', with bronze-red young growth and abundant flowers on a long, upright panicle; and 'Variegata', with small leaves variegated cream to silver, flushed pink at first, and white flowers).

Pilea (from Greek *pilos*, cap, referring to the shape of the larger sepal, which in some species covers the achene). Urticaceae. Tropics (except Australia). Some 600 species, annual or perennial herbs grown for their ornamental foliage. The flowers are very small and inconspicuous. Attractive pot plants for the home or warm conservatory (minimum temperature 10°C/50°F). The trailing species are suitable for small hanging baskets and as groundcover in the greenhouse. Some of the smaller species are especially suited to the close conditions of the terrarium and bottle garden. Grow in bright indirect light or part shade in an open, porous and freely draining soilless medium, rich in organic matter, or in a medium-fertility loam-based mix with additional sharp sand. Keep evenly moist (not wet), and humid when in full growth, and liquid feed every two or three weeks. Reduce water in winter. Do not allow plants to dry out. Site them away from draughts. Repot annually in spring; plants often degenerate in their third or fourth year and should then be replaced. Propagate by stem cuttings in spring.

P.cadierei (Vietnam; spreading to erect perennial herb or subshrub to 50cm; branchlets green tinted pink; leaves to 8.5cm, obovate to oblong-oblanceolate, dentate, silver on a dark green ground or wholly metallic; 'Minima': dwarf, with small, elliptic, deep olive green leaves with raised patches of silver); *P.involucrata* (FRIENDSHIP PLANT; C and S South America; trailing to erect herb; leaves to 6cm, ovate to obovate, marked bronze, silver or red; 'Coral': stem tinged red; leaves long, ovate, glossy, crenate, copper above, tinged purple beneath; 'Liebmannii': leaves tinged silver; 'Moon Valley': leaves ovate, quilted, crenate, tinged bronze, with a broad silver band along middle and edges dotted silver; 'Norfolk': small, dense; leaves broad, oval, bronze to black-green, with raised silver bands; 'Silver Tree': with white, hairy stalks, ovate, quilted, crenate leaves, tinged bronze, with a broad silver band along middle and sides dotted silver); *P.microphylla* (ARTILLERY PLANT, GUNPOWDER PLANT, PISTOL PLANT; Mexico to Brazil; annual or short-lived perennial to 30cm; stems slender, translucent, densely branched; leaves very small, crowded, mossy, obovate to orbicular, pale green; 'Variegata': with leaves blotched white and pink); *P.repens* (BLACK-LEAF PANAMICA; creeping herb to 30cm; leaves to 3.5cm, obovate to suborbicular, obtuse or rounded, glabrous or pilose, crenate, dark; 'Black Magic: mat-forming, with small, round, wavy, green leaves shaded dark bronze).

Pileostegia (from Greek *pilos*, felt, and *stege*, roof, referring to the form of the corolla). Hydrangeaceae. E Asia. 4 species, climbing or prostrate evergreen shrubs, related to *Hydrangea* and *Schizophragma*, but distinguished in having flowers all alike in terminal, corymbose panicles. These are produced in summer and autumn and are small and pale with numerous showy stamens. A useful, hardy climber for a north- or east-facing wall, although it flowers more profusely in sun. Propagate by nodal stem and tip cuttings taken from young growth as the foliage expands; treat with rooting hormone and root in a closed case with bottom heat. Increase also by seed and layering. Otherwise, cultivate as for *Decumaria*. *P.viburnoides* (India; 6–10m; stem self-clinging; leaves 5–18cm, narrowly oblong to ovate-lanceolate, coriaceous, glossy dark green; panicle to 15cm; crowded, flowers, white, stamens conspicuous).

pillar an upright form of trained fruit tree.

pilose covered with diffuse, soft, slender hairs.

Pilosocereus (from Latin *pilosus*, hairy, and *Cereus*). Cactaceae. US (Florida), Mexico, Caribbean region, and tropical S America (especially E Brazil). About 45 species, shrubs and trees with succulent, ribbed stems often covered in long, woolly hairs. Opening at dusk, the flowers are tubular to campanulate with numerous stamens. Provide a minimum temperature of 10°C/50°F. Plant in an acid compost containing more than 50% grit. Grow in full sun and maintain low humidity. In winter, water only to avoid shrivelling. Increase by stem cuttings. *P.leucocephalus* (syn. *P.palmeri*; E Mexico, C America; eventually a tree to 6m; stem 5–10cm in diameter, ribbed, clothed with white hairs, spines pale brown or yellow at first; flowers 6–8cm, outer tepals purple-brown, inner tepals pale pink).

pilosulous slightly pilose.

Pimelea (from Greek *pimele*, fat, referring to the oily seeds). Thymelaeaceae. Australasia (New Zealand, Australia, Timor, Lord Howe Island). RICE FLOWER. 80 species, compact evergreen shrubs or herbs, often with decussate leaves. The flowers are tubular with four spreading lobes and silky-hairy; they are borne in terminal heads, to 5cm across and surrounded by coloured bracts. Hardy to 5°C/40°F. Grow in a neutral to slightly acid soil with perfect drainage. Water moderately when in growth, and allow to become rather dry when temperatures drop. Good direct light and ventilation are essential in winter. Flowers are borne on the tips of shoots made in the previous season; deadhead and prune after flowering. Propagate by heeled, semi-ripe cuttings in summer, or by softwood cuttings in spring, in a closed case. *P.ferruginea* (W Australia; erect shrub to 2m; leaves to 12mm, ovate or oblong, shiny green above, often hairy beneath, revolute; flowers rose-pink, in heads to 4cm across).

pin-eyed see *heterostyly*.

pinching the removal of the growing tip of a plant with the finger and thumb, to prevent growth extension from that point and to encourage the formation of side shoots. See *break, stopping*.

Pinellia (for Giovanni Vincenzo Pinelli (1535–1601), of the Botanic Garden, Naples). Araceae. China, Japan. 6 species of low, cormous, perennial herbs. The leaves are simple to compound. Produced in summer, the flowers are minute and packed on a slender spadix enclosed by a narrow, hooded spathe. Cultivate as for *Arisaema*. *P.ternata* (Japan, Korea, China; leaflets 3–12cm, ovate-elliptic to oblong; spathe to 7cm, green, tube 1.5–2cm, limb curved at apex, spadix appendix to 10cm, erect, purple below).

pine needles leaves of pine trees. These decay slowly and are suitable as a ground mulch where acid soil conditions are desirable, or for making a leafmould for use in acidic potting compost.

pinery a low greenhouse, heated by a hot bed and often built as a pit. Pineries were used throughout the 18th and 19th centuries for the growing of pine-apples, and sometimes strawberries.

pinetum an arboretum devoted to conifers.

Pinguicula (from Latin *pinguis*, fat, from the greasy appearance of leaves). Lentibulariaceae. BUTTER-WORT. Europe, Circumboreal, Americas to Antarctic. 46 species, evergreen or deciduous, carnivorous perennial herbs. Borne in a low, flat rosette, the leaves act as living flypaper, secreting sticky mucilage and digestive enzymes. Solitary, 2-lipped and spurred, the slender-stalked flowers are produced in spring and summer. All species require an acid mix of peat, sand and chopped sphagnum moss, a humid atmosphere and full sun. When in growth, keep them permanently wet by standing the containers in rainwater; when dormant, keep damp and cool (frost-tender species may grow continu-ously). Apart from the European species, all listed here require a minimum temperature of 10°C/50°F. The European species are essentially hardy in climate zone 7 and prefer cold winter conditions which they survive in the form of tight resting buds. With care, they may be established outdoors in wet, acid pockets of the rock garden or beside bog-garden pools. Increase by division of resting buds in early spring, divsion of leaf rosettes in late spring, or by surface-sown seed or leaf cuttings.

P.alpina (Arctic, mountains of Europe; leaves 2.5–4cm, elliptic to lanceolate-oblong, yellow-green; flowers white with yellow spots on palate); *P.fili-folia* (Cuba; leaves 8–15cm, erect, linear to filiform; flowers white, pink, blue, purple or pale lilac); *P.grandiflora* (W Europe; leaves 3–4cm, oblong to ovate, yellow-green; flowers purple to pink or white, white at throat); *P.longifolia* (mountains of S Europe; leaves 6–13cm, lowermost elliptic, others linear-lanceolate, somewhat undulate; flowers lilac to pale blue, spotted white at base of lower lip); *P.mora-nensis* (syn. *P.caudata*; Mexico; leaves 6–11cm, round to broadly and bluntly ovate; flowers crimson to magenta or pink, throat white with darker markings at base of lobes); *P.vulgaris* (NW and C Europe; leaves 2–4.5cm, oblong to ovate, yellow-green; flowers violet, throat white).

pink bud see *bud stages*.

pinks
Pest
thrips
Disease
rust

pinks the term garden pink describes a group of *Dianthus* descended from the COTTAGE PINK (*Dianthus plumarius*), which has been cultivated in Europe for more than 500 years. Garden pinks are represented by the cultivar 'Mrs Simkins', which blooms for about two weeks in early summer and must be allowed to grow naturally without stopping. Modern pinks are mainly of the *D.* × *allwoodii* type, derived from crossing garden pinks with perpetual flowering carnations. These should be pinched to induce a bushy habit and flushes of blooms. They are classified as SELFS, with one colour; BICOLOURS, with a central eye of contrasting colour; LACED or SCOTTISH PINKS, with a ring of contrasting colour besides a central eye; and FANCIES, with streaks or stripes of contrasting colour.

Plant in early spring spaced 30cm apart; they should survive for at least three years. Propagate by cuttings, taken at flowering time and comprising four pairs of leaves trimmed just below a joint; insert into a peat/sand or peat/perlite mix and maintain at 15°C/59°F. Annual and biennial pinks (*D.chinensis* and *D.* × *heddeweggii*) are seed-raised for bedding.

pinna (plural pinnae) the primary division of a pinnately compound leaf.

pinnate a feather-like arrangement of leaflets in two rows along the axis. Bipinnate refers to a leaf in which the leaflets are themselves pinnately divided.

pinnatifid pinnately cleft nearly to the midrib in broad divisions, but without separating into distinct leaflets or pinnae.

pinnatisect deeply and pinnately cut to, or near to, the midrib; the divisions, narrower than when pinnat-ifid, and not truly distinct segments.

pinnule the ultimate division of a (generally pinnate) compound leaf.

Pinus (Classical Latin name for this tree). Pinaceae. PINE. Cosmopolitan. About 110 species, evergreen coniferous trees and shrubs. The bark is furrowed or scaly, usually thick and resinous. Needle-like leaves are borne in bundles. The male cones are cylindric, catkin-like and produced in spirally arranged clus-ters. The female cones are solitary or whorled, ovoid or cylindric to subglobose and composed of spirally arranged, woody scales.

Pines are excellent trees for parks and gardens. The diversity of species and cultivars means that at least one pine can be found for most situations, from extreme cold and drought to coastal sites, temperate woodlands, stony banks and rock gardens. The needs and suitability of each species are described in the table below. Most listed here are fully hardy and show a preference for full sun and well-drained soils, including poor sandy and gravelly substrates. Increase by seed sown in autumn or spring, by grafting in late summer, winter or early spring, and by layering.

Many different rust fungi can affect pines and several are widespread. The white pine blister rust, *Cronartium ribicola*, causes a devastating disease of the American white pines, especially *Pinus strobus*. Other less serious rusts include the pine twisting rust, *Melampsora populnea*, which mainly affects *Pinus sylvestris*; the pine resin-top rust, *Endocronartium pini* also on *P.sylvestris*, may girdle and kill the branches. A needle rust is caused by *Coleosporium tussilaginis*, and needle cast by the fungus *Lophodermium* spp., the latter is worst in cool, damp areas on species adapted to drier conditions. Pines can also be affected by armillaria root rot and phytophthora root rot. In Britain the most serious disease of plantation conifers, including pines, is fomes root and butt rot *Heterobasidion annosum*, although this and many of the other diseases to which the trees are susceptible are unlikely to be a problem on isolated garden trees. Important pests include the pine shoot moth, pine adelgids, several species of sawflies including the fox-coloured or European pine sawfly and the conifer spinning mite.

P. aristata (ROCKY MOUNTAINS BRISTLECONE PINE; W US; to 18m; crown with ascending, whorled, dense branches; bark dark grey, becoming fissured, rust-brown; needles 5, 2.5–4cm x1mm, grooved, with flecks of white resin, bright green, blue-white on inner surface in first year of growth, darker later, densely arranged; cones dark purple, maturing cylindric-ovoid, brown, 5–10cm); *P. armandii* (CHINESE WHITE PINE; C & W China; to 40m; branches horizontal, spreading widely; bark thin, grey to green-grey, smooth, becoming cracked; needles 5, 10–18cm, bright glossy green, inner face white-green to glaucous blue, thin, flexible, spreading or pendulous, often kinked near the base, dentate; cones cylindric to oblong-conic, in groups of 1–3, 8–20 x 4–11cm, erect, pendulous in the second year, scales yellow-brown); *P. banksiana* (JACK PINE; N America; to 23m; crown regular, ovoid-conic; bark scaly, fissured, orange-grey to red-brown; shoots flexible; needles 2, twisted, spreading, 2–5cm x1.5mm, light green to yellow green, obscurely serrate; cones 3–6.5 x 2cm, ovoid-conic, often in pairs, yellow-buff fading to grey; 'Annae': needles tinted yellow; 'Compacta': dense and fast-growing dwarf; 'Tucker's Dwarf': denser than 'Compacta'; 'Uncle Fogy': prostrate, fast-growing); *P. bungeana* (LACEBARK PINE; C & N China; to 25m, often multi-stemmed, slow growing; bark white to grey-green, smooth, exfoliating to reveal cream or pale yellow; needles 3, 5–9cm x 2mm, hard, shiny, sharply acuminate, dark yellow-green on outer face, pale grey-green on inner face, smelling of turpentine when crushed; cones bluntly ovoid, 4–6.5 x 3–5cm); *P. cembra* (SWISS PINE, AROLLA PINE; C Europe; to 25m; crown dense, narrowly columnar to ovoid, often branched from ground; bark smooth, grey-green, becoming brown suffused grey, furrowed; branches densely twiggy; needles stiff, 5, 6–11cm, dark green, inner surface blue-white, serrulate; cones resinous, obtuse-ovate, 4–7.5 x 4cm, purple when young; seeds edible; 'Aurea': needles yellow; 'Compacta Glauca': compact; needles tinted blue; 'Jermyns': compact, conic, very slow-growing; 'Monophylla': slow-growing, irregular dwarf; 'Nana': tightly pyramidal dwarf; needles tinted blue; 'Pendula': pendulous; 'Pygmaea': dwarf; 'Variegata': needles stippled yellow, sometimes entirely yellow); *P. cembroides* (PINYON PINE, MEXICAN NUT PINE;

Mexico, US; to 15m; crown domed; bark thick, deeply rectangularly fissured, black-brown, branches outspread; needles 2–3, clustered at shoot tips, 3–6cm, entire, olive green; cones globose, 3–4cm, bright green when young, turning orange- or buff-brown; seeds edible; 'Blandsfortiana': very slow-growing dwarf); *P. contorta* (SHORE PINE; Coastal NW US; to 25m, shrubby on poor sites; crown domed, columnar or ovoid; young trees conical, bushy at base; bark red to yellow-brown, fissured into small squares; needles 2, densely arranged, twisted, yellow to dark green, 4–5cm x1.5mm; cones ovoid, 3–7cm, yellow-brown; 'Compacta': upright, dense; needles dark green; 'Frisian Gold': needles tinted gold; 'Pendula':weeping; subspecies *latifolia*: LODGEPOLE PINE; to 30m; bark thick, ridged, needles 8cm, more spreading, brighter green); *P. coulteri* (BIG-CONE PINE; Mexico, California; to 30m; crown broad conic, becoming ovoid; branches erect to spreading; bark brown to black with scaly ridges; needles 3, 20–32cm x 2mm, grey or blue-green, stiff, finely serrate; cones yellow-brown, ovoid-conic, 20–40 x 15cm, woody); *P. densiflora* (JAPANESE RED PINE; Japan, Korea; to 35m; crown conic at first, later wide-spreading, irregular, rounded when mature; bark rust-brown, scaly, fissured and grey at base; needles 2, 7–10cm, slender, bright green; cones buff to pale brown, ovoid, 4–6cm; 'Alice Verkade': habit dwarf, tightly globose; 'Aurea': needles with gold tint or spots; 'Globosa': slow-growing, hardy and hemispherical to 1m; 'Umbraculifera': very slow-growing umbrella form, needles bright green); *P. halepensis* (ALEPPO PINE; Mediterranean; to 20m; crown conic to umbrella-shaped, becoming globose, stem often bowed or twisted; bark silver-grey, becoming red-brown, fissured; needles 2, spreading, 6–10cm, stiff, very slender); *P. heldreichii* (syn. *P. leucodermis*; BOSNIAN PINE; Balkans, SE Italy, Greece; tree to 25m; crown ovoid conic; bark ash grey, splitting into furrows, exposing yellow-grey patches; needles 2, 6–9cm x 2mm, spiny-tipped, stiff, curved forward, glossy dark green, dentate; cones ovoid, yellow-brown; 'Aureopicta': needles tipped yellow; 'Compact Gem': compact and slow-growing dwarf); *P. x holfordiana* (*P. veitchii* x *P. wallichiana*; to 30m; differs from *P. wallichiana* in its downy young shoots and wider cones with scales not reflexed); *P. jeffreyi* (JEFFREY'S PINE; W US; to 55m; crown dome-shaped, conic; bark black-brown, splitting in large plates; branches stout, spreading; needles 3, 14–27cm x 2mm, matt grey-green, finely serrate; cones ovoid-conical, 10–24 x 6–8cm; buff to brown); *P. longaeva* (ANCIENT PINE; NW US; to 20m; crown conic, becoming rough, twisted; bark chocolate brown, in plates, branches erect to pendulous; needles 5, 2–4cm x1mm, green with white lines on inner faces; cones ovoid to cylindric, red brown, 5–11cm; differs from *P. aristata* in lacking resin drops on the foliage); *P. montezumae* (MONTEZUMA PINE, ROUGH-BARKED MEXICAN PINE; Mexico, Guatemala; to 35m; crown columnar to conic, becoming rounded; bark rust-brown, rough, fissured; needles 5, rarely 4–6, erect or pendent, 15–30cm x 2mm, green; cones 12–25cm, broadly cylindric-conical to ovoid-conical, yellow to rust-brown); *P. mugo* (MOUNTAIN PINE, DWARF MOUNTAIN PINE; C Europe, Balkans; shrub to 6m; crown conic; branches erect or decumbent; bark grey-brown, scaly; needles 2, often bowed, twisted, 3–7cm x 2–3mm, dark green; cones 2–6 x 3cm, ovoid to conic, grey-brown; 'Aurea': semi-dwarf to 1m; needles gold in winter; 'Compacta': dense and globose with ascending shoots; 'Hesse': cushion-forming dwarf, needles slightly twisted; 'Humpy': very compact, rounded, slow-growing dwarf; 'Mops': broadly upright dwarf to 1.5m, slow-growing, dense; 'Ophir': flattened-globose to 60cm, yellow in winter; 'Prostrata': prostrate; 'Slavinii': mat-forming with ascending branch tips, needles dense, tinted blue; 'Variegata': needles stippled yellow; 'Winter Gold': wide and low, needles twisted, tinted gold in winter); *P. muricata* (BISHOP PINE; California; to 40m; crown broadly conic, rounded when mature; bark red-brown, deeply splitting; branches spreading; needles 2 -3, 12-15cm x 2mm, twisted, green; cones glossy, 4-9 x 3-7cm, ovoid, scales with a thick, recurved thorn); *P. nigra* (BLACK PINE; AUSTRIAN PINE; SE Europe; to 40m+; crown ovoid-conic, becoming flat-topped or rounded, very dense; bark grey to dark grey-brown, deeply fissured; needles 2, dark green, stiff, 8-14cm x 2mm, straight or bowed, finely toothed; cones 5-8 x 2 - 5cm, yellow-grey to buff, glossy; 'Aurea': needles gold at first; 'Geant de Suisse': columnar-fastigiate; needles very long, to 18cm; 'Globosa': semi-dwarf, slow-growing, rounded; 'Hornibrookiana': compact, globose, mound-forming dwarf; 'Variegata': needles stippled gold; 'Wurstle': compactly globose; needles long, vivid green; witches' broom; var. *maritima*: CORSICAN PINE; to 55m in wild, 45m in cultivation, needles12-18cm, slender, grey-green); *P. parviflora* (JAPANESE WHITE PINE; Japan; to 20m; crown ovoid-conic, compact, usually spreading; bark grey-black, smooth, exfoliating in small plates; needles 5, 3-6cm, curved, twisted, stiff, finely toothed, glossy dark green with blue-white lines on inner surfaces; cones ovoid to cylindric, to 10 x 3cm, leathery-woody, wrinkled, rust-brown; 'Adcock's Dwarf': diminutive and slow-growing, to 75cm, forms a congested, grey-green bun; 'Blue Giant': vigorous, to 15m, ascending, needles tinted blue; 'Glauca Nana': semi-dwarf, to 1m, needles short, strongly tinted blue); *P. peuce* (MACEDONIAN PINE; BALKAN PINE; South Balkans, Greece, Albania; to 35m; crown ovoid-conic; bole often branched from base; bark smooth, grey, thick, deeply ridged; needles 5, 7–10cm x1mm, pendulous, blue-green, finely serrate; cones resinous, pendulous, 7–15 x 3–7cm, cylindric, pale brown); *P. pinaster* (MARITIME PINE; Atlantic Europe to Greece, Mediterranean; to 35m; crown ovoid-conic, becoming broad rounded or irregular, bark thick, rust-brown, deeply fissured; branches level to pendulous; needles 2, to 2.5mm wide, yellow- or grey-green, stiff, shiny, finely dentate; cones 8–18 x 4–11cm, ovoid-conic, red-brown and shiny); *P. pinea* (STONE PINE; Mediterranean; to 25m; crown domed, umbrella-shaped; bark scaly, orange, red to yellow-brown, with deep furrows; branches horizontal, upswept; needles 2, 12–18cm x 2mm, twisted, blue-grey to glossy green; cones ovoid to subglobose, 8–15 x 6–10cm; . 'Fragilis': widely cultivated for edible seed in South Europe); *P. ponderosa* (PONDEROSA PINE, WESTERN YELLOW PINE; W US, SW Canada; to 50m; bark thick, yellow-brown, deeply fissured; branches stout, whorled, spreading; needles 3, spreading straight or gently curved, 11–22cm x1.5–2mm, dull green; cones green or purple at first, later brown, 6–10 x 4–8cm); *P. pumila* (DWARF SIBERIAN PINE; NE Asia; prostrate shrub, to 3m, or small tree to 6m; needles 5, 4–8cm x1mm, twisted, densely arranged, glossy, green on outer face with blue-white lines on interior face, finely toothed or entire; cones clustered, ovoid, 3–6 x 3cm, violet-black when young, red to yellow-brown when mature; 'Blue Dwarf': dwarf, irregular, ascending, needles twisted, tinted blue; 'Globe': very dense globose form, needles tinted silver; 'Nana': dense and globose semi-dwarf); *P. radiata* (syn. *P. insignis*; MONTEREY PINE; US (California); to 45m; crown ovoid-conic, becoming rounded, dense; young bark purple-grey, becoming grey to dark brown, deeply fissured; needles 3, 10–16cm x1mm, bright green; cones 6-16 x 4-11cm, glossy yellow-brown, becoming grey); *P. rigida* (PITCH PINE, NORTHERN PITCH PINE; NE US, SE Canada tree to 25m; crown irregular, rounded, open, bark dark grey or red-brown, deeply furrowed, branches level; needles 3, yellow-green to pale green, stout, twisted, becoming dark grey-green, 7–14 cm x 2–2.5mm; cones ovoid-conic, 4–7 x 3–6cm, pale brown); *P. strobus* (WEYMOUTH PINE, EASTERN WHITE PINE; SE Canada, E US; to 50m; crown broadly conic; bark grey-green to brown, fissured; branches horizontal; needles 5, blue-green, 7–13cm, obtuse, serrate; cones pink-brown, pendulous, narrowly cylindric, 8–16 x 2cm; numerous cultivars - dwarfs include the blue-tinted 'Billaw', 'Blue Shag' and 'Pumila' and the Nana Group ('Nana', 'Pygmaea', 'Radiata', 'Umbraculifera') with light green needles, some dwarfs (e.g. 'Sea Urchin') developed from witches' brooms; upright pyramidal cultivars include the tortuous 'Fastigiata' and the gold-needled 'Hillside Winter Gold'; 'Pendula': weeping; 'Alba': new growth pure white); *P. sylvestris* (SCOTS PINE; Siberia to East Asia, Europe; to 30m, rarely 40m; crown ovoid to conic, becoming rounded and often limited to upper trunk as a loosely tiered, flat canopy; bark thin, red-brown, exfoliating to show rust-brown beneath, thick and fissured at base; needles 2, 4–10cm x 2mm, twisted, blue to pale grey-green, finely serrate; cones ovoid to conic, 3–7 x 2–4cm, buff to grey-brown; many cultivars - dwarfs include the blue-tinted conic 'Glauca Nana', and 'Compressa', and the vivid blue 'Doone Valley' ; 'Globosa Viridis' and the yellow-needled 'Moseri' are pyramidal dwarf forms; prostrate, creeping dwarfs include the blue-tinted 'Hillside Creeper' and the very dark green 'Repens' and 'Saxatilis'; 'Lodge Hill' and 'Oppdal' are irregular dwarfs derived from witches' brooms, 'Gold Coin' and 'Gold Medal' are richly gold in winter; notable cultivars of normal growth include the pyramidal 'Fastigiata' and the vivid blue of 'Mt. Vernon Blue'; weeping forms include 'Mitsch Weeping' and 'Pendula'. Tinted cultivars include the silver 'Argentea Compacta' and 'Alba', the gold-tinted 'Aurea', 'Beissneriana' and 'Nisbet's Gem', the soft white 'Nivea' and the stippled white 'Variegata'); *P. thunbergii* (JAPANESE BLACK PINE; Japan, South Korea to 30m; crown open, with few, long, level or sinuous branches; bark black-grey, furrowed; needles 2, 7–14cm x 2mm, densely arranged, spreading, twisted, dark green; cones 3–7 x 3cm, ovoid-conic, pink to buff; 'Majestic Beauty': dense and compact, resistant to salt and smog damage; 'Pygmaea': compact dwarf to 1.5m, needles long, rich green); *P. virginiana* (SCRUB PINE, VIRGINIA PINE; E US; tree to 15m or shrub, trunk to 50cm diameter; bark thin, furrowed; branches irregular, spreading, often twisted; needles 2, 4–7cm, yellow-green to dark green, twisted, stiff, sharp, finely serrated; cones oblong to conic, 3–7cm, yellow-buff to rust-brown); *P. wallichiana* (syn. *P. excelsa*, *P. griffithii*; HIMALAYAN PINE, BLUE PINE, BHUTAN PINE; Himalaya; to 50m; crown conic, or irregular; bark grey-brown, becoming fissured; upper branches whorled, ascending; needles 5, hanging downwards, flexible, grey-green to waxy blue, to 20cm x1mm; cones becoming pendulous, cylindric, 14–28 x 4-10cm, green, ripening buff or yellow-brown; 'Umbraculifera': low-growing, mushroom-headed dwarf).

pip (1) a single growth or stem of lily-of-the-valley (*Convallaria majalis*); (2) a florist's term for a single flower in a cluster or truss, especially used of auriculas and other primulas; (3) a popular name for the seeds of apples, pears, citrus fruits.

Piper (classical Latin name.) Piperaceae. Pantropical. PEPPER. More than 1000 species, erect, evergreen shrubs, climbers and small trees, rarely more herb-like, often with a pungent odour. They produce very small flowers in cylindric spikes, followed by brightly coloured berries harvested in some species for pepper. Those here are grown for their foliage and general habit. Suitable for the warm greenhouse or conservatory. Grow in humid, shady conditions in a fertile, moist medium. Propagate by seed, or by semi-ripe cuttings in sand in a closed case.

P.magnificum (LACQUERED PEPPER; Peru; shrub, erect, to 1m; leaves 15cm, ovate to broadly elliptic or suborbicular, rounded or broadly acute at apex, cordate to auriculate at base, glossy deep green, bright maroon beneath with white margin and veins, quilted); *P.ornatum* (CELEBES PEPPER; Sulawesi; shrub, spreading, creeping or weakly climbing, to 5m; leaves 9.5cm, broadly cordate to suborbicular, peltate, rounded or acute and attenuate at apex, rounded to cordate at base, finely mottled dark green, pink and silver above, flushed maroon beneath).

piping a method of taking cuttings of pinks, done in early summer by pulling a young vegetative shoot out at a node.

Piptanthus (from Greek *piptein*, to fall, and *anthos*, flower; the calyx, corolla and stamens fall off together, leaving the young pod without a calyx at base). Leguminosae. Himalaya. 2 species, evergreen or semi-deciduous shrubs or small trees, usually with green stems. The leaves are trifoliolate, with narrow leaflets, silky at first, later dark, shining green. Pealike, bright yellow flowers are carried in loose or congested, racemes in spring and summer. Hardy to –15°C/5°F. Where frosts are prolonged, these shrubs become deciduous and may suffer some damage to their rather tender hollow branches – given adequate protection at the roots, they will re-sprout from the base. Grow in full sun in any moderately fertile, well-drained soil. Prune after flowering to remove old and overcrowded growth at ground level. Sow seed under glass when ripe or in early spring.

P.nepalensis (Himalaya; leaflets subglabrous to puberulent beneath; flowers bright yellow, standard occasionally with purple-brown or grey markings); *P.tomentosus* (SW China; leaflets thickly silky-tomentose especially beneath; flowers lemon-yellow).

Piptanthus nepalensis

Pistacia (from the Greek name for the nut, *pistake*). Anacardiaceae. Mediterranean, C Asia to Japan, Malaysia, Mexico, S US. PISTACHIO. About 9 species, deciduous or evergreen, dioecious trees and shrubs, usually with pinnate leaves. Produced in spring and summer in axillary clusters, the flowers are inconspicuous but give rise to small, single-seeded drupes. These contain an edible nut in some species. *P.terebinthus* is hardy in climate zone 8. *P.lentiscus* needs wall protection in cool regions. Plant in a sunny position, in well-drained deep but light soil. Both are drought-tolerant. Sow seed in late winter early spring in a loam-based seed propagating medium at 25°C/77°F.

P.lentiscus (MASTIC, LENTISCO, CHIOS MASTIC; Mediterranean except NE Africa; tree or shrub to 4m, evergreen; leaves coriaceous, leaflets 4–6, rarely more, 1.5–5cm, ovate, oblong-lanceolate or elliptic; fruit red becoming black); *P.terebinthus* (CYPRUS TURPENTINE, TEREBINTE, TEREBINTHO; Iberia to Turkey, Morocco to Egypt; tree or shrub, 2–6m, deciduous; leaves glossy, aromatic, 10–20cm, leaflets 6–12, 3–5cm, ovate-lanceolate, to oblong; fruit purple, wrinkled).

Pistia (from Greek *pistos*, water, referring to the habitat). Araceae. Pantropical, now a widespread weed of rivers and lakes, first recorded on the Nile, possibly having originated from Lake Victoria. WATER LETTUCE; SHELL FLOWER. 1 species, *P.stratiotes*, an evergreen, floating aquatic, forming dense, freely offsetting rosettes. The roots are fine, feathery and hang in spreading bunches from their undersides. To 20 × 7cm, the leaves are broadly wedge-shaped with a rounded, truncate or retuse apex. They are blue-green and fluted above, pearly and prominently ribbed beneath, filled with spongy tissue toward the base and covered with fine, water-repellent hairs. The inflorescence is typical of the Araceae but very small, nestling

Pistia stratiotes

in the leaf axils. Grow as a free-floating aquatic in heated aquaria or greenhouse pools (minimum temperature 10°C/50°F) in full sun or bright, indirect light in soft water. These plants may be grown outdoors in summer on garden ponds. In autumn, bring a few clean and healthy rosettes under cover and overwinter them in bright, frost-free conditions on a pan of wet coir and sand. Propagate by removal of plantlets.

pistil one of the female reproductive organs of a flower, together making up the gynostemium, and usually composed of ovary, style and stigma. A simple pistil consists of one carpel, a compound pistil of two or more.

pistillate a unisexual, female flower bearing a pistil or pistils but no functional stamens.

pistillode a sterile, vestigial pistil remaining in a staminate flower.

pitch resinous exudate.

pitcher a tubular or cup-like vessel, usually a modified leaf. In several carnivorous genera, these are trapping adaptations.

Pisum (the classical Latin name). Leguminosae. Mediterranean, W Asia. PEA. 5 species, annual herbs, often climbing by means of tendrils. The flowers are typically papilionaceous – 'pea-like'; the fruit a legume containing subglobose seeds. See *peas. P.sativum* is the GARDEN PEA; its variety *macrocarpon*, is the EDIBLE-PODDED PEA, MANGE TOUT, SUGAR PEA, or SNOW PEA.

pit a type of frame or greenhouse constructed with the floor sunk below ground level so that the frame light or roof is at or just above soil level, thus providing wind protection and conserving heat. Pits are a feature of walled gardens in particular, and traditionally were maintained on large estates.

Pitcairnia (for Dr William Pitcairn (1711–1791), London doctor). Bromeliaceae. C and S America. 260 species, evergreen, terrestrial perennial herbs, occasionally lithophytic or epiphytic. Borne usually in loose basal clumps, the leaves are tough-textured but pliable, oblong-lanceolate to linear and slender-stalked. Showy, tubular flowers are carried in racemes or panicles, often amid colourful bracts. Cultivate as for *Billbergia. P.andreana* (Colombia; leaves to 35cm, scaly beneath; flowers orange, yellow near apex); *P.heterophylla* (Mexico to Peru; leaves to 70cm, flowers pink-red).

pitchfork see *fork.*

Pittosporum (from Greek *pitta*, pitch, and *sporos*, seed, referring to the sticky, resinous coating found on the seeds). Pittosporaceae. Australasia and South Africa to S and E Asia and Hawaii. About 200 species,

evergreen trees and shrubs, with thinly leathery to fleshy leaves and small, often sweetly scented flowers produced singly or in corymbs, umbels or clusters from late spring to early autumn. *P.tenuifolium* and *P.tobira* are among the hardiest species, tolerating temperatures as low as –10°C/14°F, and lower with wall shelter: *P.crassifolium* and *P.undulatum* are hardy to about –5°C/23°F. Grow in well-drained soils in full sun. All species need shelter from cold, drying winds and will benefit from the protection of a deep dry mulch of bracken litter at the root. Under glass, water moderately, maintain good ventilation and provide a winter minimum temperature at about 5°C/40°F. They regenerate freely from old wood and, if necessary, can be cut quite hard back in spring to remove frost-damaged growth. Propagate by seed sown in autumn or spring. Increase also by semi-ripe basal cuttings, rooted in a closed case with bottom heat at 15°-18°C/60–65°F in summer, or by basal ripewood cuttings in late autumn in the cold frame.

P.crassifolium (CARO, KARO, EVERGREEN PITTOSPORUM; New Zealand; shrub or small tree to 5(–10)m, crown columnar, young twigs tomentose; leaves 5–7cm, oblong to elliptic or obovate, leathery, dark green above, white- or buff-tomentose beneath when young, margins thickened; flowers dark crimson to purple, in terminal clusters; 'Variegatum': leaves variegated white); *P.tenuifolium* (TAWHIWHI, KOHUHU; New Zealand; tree to 10m, trunk slender; leaves 2.5–10cm, elliptic, obovate or oblong, glabrous, flat or undulate, pale green; inflorescence axillary, 1- to few-flowered, flowers small, dark maroon to purple-black, fragrant; 'Garnetii', with 4–6cm, ovate-elliptic, grey-green leaves with a white wavy margin sometimes with pink spots; 'Golden King', with 3–5cm, ovate to broadly ovate, pale golden-green leaves; 'Purpureum', with leaves green at first, dark purple-bronze when mature, very wavy; 'Tom Thumb', small, compact, with 3–6cm, oblong-elliptic, dark purple-bronze leaves with very wavy margins; 'Variegatum', leaves green with a cream margin). *P.tobira* (TOBIRA, MOCK ORANGE, JAPANESE PITTOSPORUM; China, Japan; shrub or small tree to 5m; leaves 3–10cm, obovate, apex rounded, base cuneate, leathery, dark green and glossy, revolute; inflorescence terminal, umbellate, 5–7.5cm in diameter, flowers orange-blossom-scented, 2.5cm in diameter, cream-white to lemon-yellow); *P.undulatum* (VICTORIAN BOX, ORANGE BERRY PITTOSPORUM, CHEESEWOOD; E Australia; tree, 9–14m; leaves 8–15cm, acuminate, laurel-like, shiny dark green above, pale beneath, margins wavy; flowers 1.2–1.8cm in diameter, fragrant, cream-white in clustered umbels; 'Variegatum': leaves with white margins).

Pityrogramma (from Greek *pityron*, chaff or bran, and *gramma*, line, referring to the scaly linear sori). Pteridaceae (Adiantaceae). Americas and Africa. About 40 species, evergreen or deciduous ferns. The rhizomes are short-creeping and scaly, giving rise to loose clumps or tufts of fronds. These have wiry, dark and glossy stipes and 2–3 times pinnately compound blades whose undersides are often covered in metallic

Pittosporum tenuifolium 'Variegatum'

farina. Grow in light shade in a humid, buoyant atmosphere with a minimum temperature of 10°C/50°F. Pot tightly in a soilless mix rich in coarse bark and grit. Water and syringe freely in summer; keep just moist in winter. Increase by division or spores.

P.calomelanos (SILVER FERN; tropical America, now pantropical; stipes 20–55cm, dark purple, fairnaceous when young; frond blade to 60 × 30cm, ovate, bipinnate or tripinnatifid at base, undersurface covered in silver-white farina); *P.chrysophylla* (GOLDEN FERN; West Indies and S America, widely naturalized as a subtropical weed; stipes to twice length of blade, dark brown, tinged black; frond blade 20–60cm, ovate to ovate-triangular, bipinnate or more compound, undersurface covered in golden farina).

Plantago (the Latin name for these plants). Plantaginaceae. Cosmopolitan. PLANTAIN. 200+ species, annual, biennial or perennial herbs and shrublets with basal rosettes of broadly spathulate leaves, or, in caulescent species, stems covered in finer, narrower leaves. Inconspicuous, grain-like flowers are carried in taper-like, stalked spikes. Although better known to most gardeners for the common and persistent lawn weeds it numbers *Plantago* includes some attractive alpine species whose slender leaves are covered in fine silvery down. These are hardy in climate zone 7 and are suitable for the rock garden, crevices, scree or warm, gritty soils in full sun. The more robust and familiar *P.major* is fully hardy and its cultivars are sometimes grown in the herbaceous border, in any moderately fertile soil in sun. Propagate by seed or division.

P.arborescens (Canary Islands; dwarf shrub to 60cm; stem woody, much-branched; leaves 2–4cm, crowded, linear to filiform, dark green, hairy); *P.major* (COMMON PLANTAIN, WHITE-MAN'S FOOT, CART-TRACK-PLANT; Eurasia, naturalized worldwide; perennial herb, sparingly pubescent; leaves 5–20cm, rosulate, spreading, broadly obovate, stalked, with prominent veins, tough, puckered; 'Atropurpurea': leaves purple tinged bronze to purple; 'Rosularis': ROSE PLANTAIN, monstrous in form, with a stout scape and apical coma of green leaves to 9cm in diameter, replacing flowers; 'Rubrifolia': large leaves, flushed wine red; 'Variegata': leaves variegated cream and green); *P.nivalis* (S Spain; perennial herb; leaves rosulate, linear to narrowly lanceolate, covered in silky white hairs).

plant breeder's rights (abbrev. PBR) the protection afforded plant breeders by the formal official registration and cataloguing of new cultigens. It prevents unauthorized propagation of the new plant and enables the breeder to collect royalties. The requirement is covered by plant variety protection laws.

planting and transplanting both terms describe the moving of plants from one growing position to another. Planting usually means the setting out of newly propagated trees, shrubs, herbaceous plants, bulbs and corms, and certain vegetables into a place where they will establish to maturity. Transplanting more often refers to the moving of established plants from one place to another, or the moving of nursery specimens, either in open ground or in containers, during the course of production for sale.

Most nursery stock is established in containers for sale, thus extending the planting season over that of more expensive bare root or root-balled plants. The system is particularly useful for genera that resent root disturbance, for example *Magnolia*, *Cytisus*, *Elaeagnus* and *Cistus*. Plants purchased in containers should be carefully inspected to ensure that their root systems have not become woody and constricted; it is also important to ascertain that they have not been too recently potted and may therefore suffer severe check on transplanting. Rapid establishment after planting will be best achieved with plants that are robust and evidently well-fed and whose roots fully explore the container compost without being pot-bound.

The year-round availability of containerized plants invites planting during unfavourable conditions, such as summer heat and drought, and thought should be given as to the wisdom of attempting to establish purchases during such periods unless the utmost aftercare can be provided. Field-grown plants are available for planting during periods of lowest growth activity but, as with those available in containers, this must not be attempted when the ground is frozen or waterlogged. Some plants, such as hollies and daphnes, are difficult to handle successfully as bare-root specimens and these, together with species that produce fibrous root systems, such as rhododendrons, are suitably prepared by careful lifting and ball-wrapping as an alternative to containerizing. Where a large number of plants is required, for example, for hedging, field-grown bare-rootstock is usually cheapest.

Mid- and late autumn is generally the best time to plant or transplant since the soil is relatively warm and usually sufficiently moist to encourage early rooting. Winter or spring planting may subject young plants to strong winds, frost damage and waterlogging, and allows for only minimal growth before a drought period in early summer. In very exposed locations, spring planting may nevertheless be the best course to avoid winter damage, with special precautions taken to aid establishment. Most evergreens are best spring-planted. The majority of herbaceous species can be planted in the autumn, but those susceptible to cold or wet should be held over until spring. Bearded irises should be planted mid- to late summer and many woodland plants, such as primulas, set out immediately after flowering.

Where bare root plants are received during periods unsuitable for planting, they should be heeled-in, if possible, or stored in a cool shed with care taken to ensure that the roots are not allowed to dry out. Leaving any packing intact is helpful.

Where moderately large trees or shrubs require transplanting in a garden, preparations should be made six to twelve months in advance by taking out a trench around half the root system and refilling it to encourage the production of additional fibrous roots. When the tree or shrub is finally undercut, the ground must be moist and a thick polythene sheet used to cover the root ball during transportation.

Plants being planted out from under protection must be thoroughly hardened off, especially in the case of frost-tender bedding and vegetable subjects.

Planting sites should be well-prepared to ensure good texture and fertility; this can be done by incorporating well-rotted organic matter throughout the cultivation depth well in advance of planting. A light dressing of general compound fertilizer applied just before planting is worthwhile, and removal of perennial weeds at planting stage is most important.

Ensure that stock is well-watered before planting, and that the planting hole is larger than the specimen it is to receive. Always remove plants from containers, gently teasing out compacted roots. Bare root plants should have soil settled around the roots; this can be achieved by continually shaking the specimen up and down as planting proceeds. Small plants are firmed by hand, trees and shrubs with the foot, and in all cases the plant should be planted to the same depth as at its previous site. A thorough watering-in should be provided.

For most garden situations, firm staking of trees is desirable, although there is evidence that unstaked trees develop stronger root systems on exposed sites. For all plants, it is worth considering the erection of a temporary shelter screen to aid establishment. Where rabbits are a threat, tree guards should be fitted. All planting and transplanting operations should include provision of a ground mulch to eliminate weed competition and conserve soil moisture. This can either be a layer of organic material, kept clear of the immediate base of the plant, or a thick black polythene sheet.

planting board usually, a device for accurately positioning a tree, consisting of a narrow piece of wood, up to 1m long, with one V-notch at the centre point and one at each end. The central notch is placed against a cane marking the required plant position, and canes then inserted into the ground within each of the two end notches. The board and central cane are then removed while the planting hole is prepared. The planting board is later repositioned within the two end canes in order to locate the original planting point, and to assist in restoring soil level.

The term is also sometimes used to refer to a wooden plank put down to reduce soil compaction during planting operations.

planting house, plant house a greenhouse fitted with benches for the cultivation of pot plants.

plantlet a small plant with leaves, and often roots, produced upon certain plants as a means of increase. Some plantlets detach naturally (e.g. bryophyllums), others need to be brought into contact with the soil (e.g. *Tolmiea)*.

plant lice a term often applied to aphids and related sap-sucking insects, such as adelgids and phylloxerids.

plant physiology the study of the life processes or functions of plants. It is one of the most important divisions of botanical science.

plantsman formerly, a nurseryman or florist; the term is now used to describe an expert gardener who is primarily a plant connoisseur, often one with a preference for growing individual plants or specific groups of plants over general gardening.

plant window a glazed structure built within a room, usually in front of a window or else with a source of artificial light, as a virtual indoor greenhouse. It is most often of narrow construction and must have facilities for ventilation and access, as provided by sliding panels. Akin to a Wardian case.

plash see *pleach.*

Platanus (classical name from the Greek *platys*, broad, in allusion to the broad, flat leaves). Platanaceae. Mostly N America; one in each of SE Europe to SW Asia, and Indochina. SYCAMORE, BUTTONWOOD, PLANE, PLANE TREE. 7 species, deciduous trees with thin bark, flaking in plates to reveal patches of paler new bark, large, palmately lobed leaves and bristly, brown-green flowers in stalked and pendulous, spherical heads. Planes are large, fast-growing and long-lived trees, with decorative bark and a remarkable tolerance of air pollution and compacted soil. *P. × acerifolia* is a much-planted street tree. One parent, *P.orientalis*, is a finer tree, with a more rugged crown and deeply cut leaves, but is slower-growing, while the other parent, *P.occidentalis*, is large and fast-growing but susceptible to disease. Hardy to zone 6. The natural habitat of all planes is on river gravels and silt with ready access to water, but, given access to ground water, they tolerate drought well, making large trees in dry southern Europe and California. The short, fine, stiff hairs, shed from the leaves in spring, and the fruit in autumn have been implicated in bronchial problems; effects are likely to be worse in dry climates. Propagate by seed, cuttings or layers; pre-treat moistened seed of the species at 2°C/36°F for two months and sow in early spring in a cold frame for frost protection. Seed of *P. × acerifolia* is often infertile. Heeled hardwood cuttings taken in autumn at leaf fall are easily rooted in a cold frame. Plane anthracnose causes sporadic dieback of shoots, occasionally spreading to make large cankers on branches. It is aggravated in cool, damp summers and on those species less adapted to such conditions–*P.occidentalis* is the most likely to be attacked by it in Britain, while *P.orientalis* is virtually immune. *P. × acerifolia* varies in susceptibility between clones.

P. × acerifolia (LONDON PLANE; *P.occidentalis × P.orientalis*; to 50m; bark peeling off in flakes, cream weathering to grey; leaves usually truncate to shallowly cordate at the base, 3- or 5-lobed, 12–25cm wide, lobes triangular, sinuately toothed to entire; fruit usually in groups of 2; 'Augustine Henry', with pendulous lower branches and 25–35cm wide, 5-

lobed, slightly blue-green leaves, tomentose beneath at first; 'Cantabrigiensis', with smaller, more deely lobed leaves, more delicate in all respects; 'Hispanica', with leaves to 30cm wide, normally 5–lobed, with lobes toothed, veins tomentose beneath, and fruit grouped 1–2; 'Kelseyana', with yellow-variegated leaves; 'Mirkovec', with leaves becoming red-tinged in summer, purple-red in autumn; 'Pyramidalis', of upright, conical habit, with rougher bark and leaves mostly 3-lobed, with lobes slightly toothed – close to *P.occidentalis*; 'Sutternii', with leaves white-blotched and speckled over the entire surface; and 'Tremonia', of quite narrow, conical habit, very fast-growing); *P.occidentalis* (AMERICAN SYCAMORE, BUTTONWOOD, AMERICAN PLANE; SE US; to 50m; bark exfoliating in small plates; leaves generally 3-lobed, 10–18cm wide, coarsely sinuate, occasionally entire, base obtuse to cuneate, tomentose beneath at first; fruit usually solitary, occasionally in pairs; *P. × acerifolia* was often grown and sold under this name; includes var. *glabrata*, with smaller, tougher, more deeply lobed leaves, with lobes long acuminate and often entire); *P.orientalis* (ORIENTAL PLANE; SE Europe to Asia Minor; to 30m; bark exfoliating in large plates; leaves deeply 5–7-lobed, occasionally 3-lobed on younger shoots, 10–20cm wide, lobes longer than wide, base cuneate or truncated; fruit in groups of 3–4 or more, occasionally in pairs; includes var. *insularis,* CYPRIAN PLANE, from Crete, with 5, deep leaf lobes, further lobed with ascending teeth, and long acuminate, and 'Cuneata', with leaves deeply incised and lobes entire or themselves lobed).

Platycarya (from Greek *platys*, broad, and *karya*, nut, alluding to the winged fruit). Juglandaceae. C and S China. 1 species, *P.strobilacea*, a deciduous tree or shrub to 12m. The leaves are pinnate and 15–30cm long; the leaflets are ovate to lanceolate, finely tapered, biserrate and 4-l0cm long. The male inflorescence is a slender, erect catkin, 5–8cm long; the female inflorescence is a cone-like catkin, ovoid to oblong, and about 3cm long. A small to medium-sized tree, which prefers warm dry continental climates where temperatures do not drop below about −12°C/10°F in winter. Propagate from ripe or stratified seed, by layering or by splice/veneer grafting on to *Carya* stock.

Platycerium (from Greek *platys*, broad, and *keras*, a horn, referring to the fronds). Polypodiaceae. Tropical Asia and Australasia. STAGHORN FERN, ELKHORN FERN, ANTELOPE EARS. Perhaps 18 species, epiphytic or lithophytic, evergreen ferns. The fronds are of two types: (1) sterile – mantle-like, stalkless, conspicuously veined and variously lobed on the upper margin, these fronds are produced in succession, overlapping each other, clasping the substrate and (the older ones) decaying to create a pocket of moist, fertile humus; (2) fertile – pendulous, stalked, forked and lobed like a elk's or stag's horn, with the spores carried in large, brown velvety patches on the undersides.

These are large, epiphytic ferns suitable for the home, greenhouse or conservatory (minimum temperature 10°C/50°F). All will grow in pots in a soilless medium rich in coarse bark. They will also thrive mounted on wood or bark slabs and stout branches, and tied in place until their spreading sterile fronds have clasped the surface. Encase the roots in an envelope of compost and sphagnum moss prior to mounting. *P.bifurcatum* and other clump-forming species can be grown in baskets and will colonize their outer surfaces. Grow in dappled sun or light shade. High humidity is ideal, but these ferns will tolerate the dry atmosphere of the home if they are frequently syringed, plunged and watered. Keep moist in warm weather, rather dry in winter. Propagate clump-formers by division, larger, non-clumping species (e.g. *P.grande*) by spores. Scraped from the undersides of the drooping fertile fronds and sown thinly on a sterile substrate, spores will soon develop prothalli; these may enter a rest period of up to one year before moving to the next generation.

P.bifurcatum (COMMON STAGHORN FERN, ELKHORN FERN; SE Asia, Polynesia, subtropical Australia; clump-forming; sterile fronds to 60 × 45cm, rounded, entire, wavy, or shallowly lobed, papery; fertile fronds to 90cm, pendent, base cuneate, 2 or 3 (occasionally to 5) × forked, leathery, segments strap-shaped, obtuse, stellate-pubescent); *P.grande* (STAGHORN FERN; Malaysia, Australia, Philippines; solitary; sterile fronds to 110 × 180cm, suborbicular to reniform, upper margin to 5 times deeply and irregularly dichotomously lobed, lower margin entire, papery, bronze to green, strongly veined; fertile fronds to 180cm, pendent, cuneate below, 2 × forked, leathery, dividing into strap-shaped segments each to 30cm); *P.hillii* (NORTHERN ELKHORN FERN; Australia (Queensland), New Guinea; clump-forming sterile fronds to 40 × 24cm, adpressed, rounded and shallowly lobed at upper margin; fertile fronds to 70cm+, erect or suberect, broadly cuneate to spathulate in lower part, irregularly forked or palmately lobed above, dark green, segments narrowly elliptic to obovate); *P.superbum* (STAGHORN FERN; Australia; solitary; sterile fronds to 160 × 150cm, spreading, cuneate, apex truncate, to 4 times forked or lobed, grey

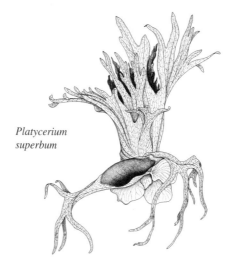

Platycerium superbum

to green; fertile fronds to 2m, spreading to pendent, cuneate below, to 5 times forked, segments to 30cm ribbon-like, often twisted).

Platycodon (from Greek *platys*, broad, and *kodon*, bell, alluding to the form of the flower). Campanulaceae. China, Manchuria, Japan. BALLOON FLOWER, CHINESE BELLFLOWER. 1 species, *P.grandiflorus*, a perennial herb with clumped, branching stems erect to 70cm. The leaves are elliptic-lanceolate, dentate and blue-green. Produced in summer singly or in few-flowered corymbs, the terminal flowers are to 4cm across and sky blue, azure, white or opal-pink. In bud, the flower is inflated like a grey to slate blue balloon; on opening, it is broadly campanulate, with five spreading, ovate-acuminate lobes. Suitable for the herbaceous border and rock garden and hardy to between –15 and –20°C/5 to –4°F; mulch with bracken or leafmould where low temperatures are prolonged. It tolerates a range of soils but fares best in deep, well-drained, loamy soil in sun or light, dappled shade. Propagate by seed sown in spring by basal cuttings with a piece of root attached; also by careful division.

Platystemon (from Greek *platys*, broad, and *stemon*, stamen, referring to the expanded filaments). Papaveraceae. N America (California). CREAMCUPS, CALIFORNIAN POPPY. 1 species, *P.californicus*, an annual herb to 30cm tall with erect or decumbent, loose, spreading branches. To 7.5cm long, the leaves are linear to oblong-lanceolate, entire, stem-clasping, grey-green and densely pubescent. Produced in summer, the flowers are terminal, solitary and to 2.5cm in diameter, with six, ovate to rounded, cream to yellow petals and numerous yellow stamens. Suitable for edging borders or for sowing in drifts in the rock garden. Cultivate as for annual *Papaver*.

pleach a term derived from plash or plait to describe hedge-making by laying, or, sometimes, the training of woody plants against a wall. The term is now used more specifically to describe the formation of a narrow screen or hedge by training new flexible branches along wires and pruning out surplus growth annually. In many cases, the pleaching is restricted to the upper parts of trees, especially lime, hornbeam, or yew, so that a dense green hedge is produced some distance from ground level supported on pillar-like trunks.

pleasance an enclosed park of the Middle Ages, intended for sport, recreation, and aesthetic pleasure, often including water and architectural features as well as trees.

pleasure garden in 18th- and 19th-century England, a public garden used, chiefly at night, for commercial concerts, firework displays, banquets, and other entertainments.

Plectranthus (from Greek *plektron*, spur, and *anthos*, flower: the corolla often has a basal spur). Labiatae. Africa, Asia, Australia. 350 species, annual or perennial herbs, subshrubs or shrubs. They are usually aromatic and minutely clammy-hairy. Small, tubular and 2-lipped, the flowers are whorled or clustered in spikes. Provide a minimum temperature of 10°C/50°F and bright light. The trailing species and cultivars are easily grown in large hanging baskets, or planted as groundcover under glass. Large plants can be pruned hard after flowering. Propagate by cuttings of new growth taken at any time of the year, by removal of rooted branches of the trailing species, or from seed sown under glass in spring.

P.amboinicus (SOUP MINT, MEXICAN MINT, INDIAN MINT, COUNTRY BORAGE, FRENCH THYME, SPANISH THYME; tropics to South Africa; decumbent, many-stemmed, aromatic, perennial herb to 1.5m; leaves 4.5cm, ovate, pubescent glandular-toothed, rounded, crenulate; flowers lilac, mauve or white; a form with white-edged leaves is in cultivation); *P.argentatus* (Australia; spreading, more or less hairy shrub to 1m; leaves 5–11cm, canescent, ovate, crenate, glandular; flowers pale blue-white; 'Green Silver' is a widely cultivated clone); *P.forsteri* (New Caledonia, Fiji, East Australia Island; decumbent, aromatic, perennial herb, stems straggling, to 1m; leaves 1.5–3.5cm, ovate to broad-ovate, pubescent, glandular beneath, crenate or serrate; flowers pale to mid-blue or mauve; includes 'Marginatus', syn. 'Variegatus', with leaves variegated cream and small, white flowers; a gold-variegated form is widely cultivated under the name *P.coleoides*); *P.madagascariensis* (South Africa, Mozambique and Madagascar; procumbent or decumbent semi-succulent herb, stem to 1m, rooting; leaves 1.5–4.5cm, slightly succulent, ovate to subrotund, strigose above, tomentose beneath, glandular-dotted, obtuse to rounded, crenate to dentate; flowers white or mauve, to purple, often dotted with red glands; a fragrant-leaved variegated form is grown as a trailing plant under the name 'Variegated Mintleaf' or, incorrectly, as *P.coleoides* 'Variegata'); *P.oertendahlii* (South Africa; freely branching, semi-succulent, perennial herb; stem decumbent, to 1m, rooting, glandular-tomentose; leaves 3–4cm, semi-succulent, ovate to suborbicular, sparingly villous, purple beneath, crenate, ciliate; flowers white or suffused with mauve; 'Variegatus': leaves to 4cm, green-bronze, variegated silver and off-white, particularly at edges).

Pleioblastus (from Greek *pleios*, many, and *blastos*, buds, alluding to the branches which are borne several per node). Gramineae. China, Japan. Some 20 species of dwarf to medium-sized bamboos with short or far-running rhizomes. See *bamboos*.

Pleione (for Pleione, mother of the Pleiades). Orchidaceae. India to Taiwan, Thailand. INDIAN CROCUS. Some 16 species, deciduous, epiphytic or terrestrial perennial herbs. They consist of one or two pseudobulbs only, these near-spherical to box- or flask-shaped and topped with one or two plicately veined and oblong-lanceolate leaves. Solitary and disproportionately large, the flowers appear in spring. The tepals are usually oblong to oblanceolate. The lip

*Pleione
formosana*

is broadly funnel-shaped and often fringed and crested.

These small and beautiful orchids will thrive in the alpine house and cold greenhouse. *P.formosana* has been grown successfully outdoors in zone 7 in pockets of perfectly drained peaty soil in the rockery, with minimal winter protection. Repot annually at the end of a cool dry rest in a fairly dense bark-based medium rich in leafmould. Commence watering to promote root and bud development. After flowering, keep moist and well-fed in bright, airy conditions to ensure full leaf and pseudobulb formation. As the leaves wither, reduce water and temperatures. Increase by division and by plantlets. (the latter may develop at the apex of the pseudobulb, especially where plants have beeen grown in overwarm, moist conditions).

P.bulbocodioides (syn. *P.pogonioides*; China; flowers pink to rose-purple or magenta, lip marked dark purple); *P.formosana* (syn. *P.pricei*; E China, Taiwan; flowers white, lilac or magenta, lip white, typically stained or marked yellow); *P.forrestii* (China, N Burma; flowers pale golden yellow or white, lip spotted brown or crimson); *P.hookeriana* (C Nepal to S China; sepals and petals lilac-pink to rose, sometimes dotted pale violet, rarely white, lip white, dotted yellow-brown or purple, lamellae and disc yellow); *P.humilis* (Burma, NE India; flowers white, lip spotted and streaked bronze or blood red, central zone pale yellow); *P. × lagenaria* (*P.maculata* × *P.praecox*; SW China; petals and sepals pink to rose-purple, lip white, central patch yellow, margins blotched purple). *P.limprichtii* (SW China, N Burma; flowers rose-pink to magenta, lip paler, spotted ochre or crimson, lamellae white); *P.praecox* (Indochina, Burma; flowers white to rose-purple, lamellae yellow); *P.yunnanensis* (N Burma, China; flowers pale lavender to rose pink, rarely white; lip flecked red or purple). In addition to the species and natural hybrid described above, numerous grexes and clones of *Pleione* are grown. Among the finest are: *P.*Shantung (flowers deep yellow or white flushed with pink; the best-known and most vigorous clone is the large-flowered, apricot 'Muriel Harberd'; 'Desert Sands' has creamy-yellow tepals blushing pink and a red-spotted, golden lip; 'Fieldfare' has large pale yellow flowers); *P.*Stromboli (flowers mostly dark red-pink; 'Fireball' is one of the most intensely coloured of all *Pleione* clones); *P.*Versailles (flowers very pale mauve-pink to deep rose pink; the clone 'Bucklebury' is particularly fine and very floriferous).

Pleiospilos (from Greek *pleios*, full, and *spilos*, dot, referring to the leaf surface, which is often covered in dots). Aizoaceae. South Africa. LIVING GRANITE, LIVING ROCK, STONE MIMICRY PLANT. 4 species, highly succulent, stemless, clump-forming perennials. Solid and similar in texture to lumps of granite, the leaves are usually carried in an opposite or decussate arrangement, one to two pairs per shoot. They are very thick, flat above, the apex obtuse to acute, the bases united and grey-green or dark green with translucent dots. Yellow to orange and often scented of coconut, the daisy-like flowers open in bright sun. Cultivate as for *Conophytum*.

P.bolusii (MIMICRY PLANT, LIVING ROCK; leaves, 4–7cm long, 3–3.5cm thick, upper surface broader than long, lower surfaces rounded, apex thickened and drawn over upper surface, red or brown-green with numerous dots; flowers golden yellow); *P.nelii* (SPLITROCK, CLEFTSTONE, MIMICRY PLANT; close to *P.bolusii* in size; leaves 2–4, rounded below, drawn over upper surface so that leaf is hemispherical, dark grey-green, densely dotted; flowers salmon pink to yellow to orange); *P.simulans* (leaves 6–8cm, usually paired, ovate-triangular, tip thickened beneath, not drawn over upper surface, red-, yellow-, or brown-green, spotted, undulating and tuberculate; flowers yellow, sometimes orange, fragrant).

plicate folded lengthwise; pleated, as a closed fan, a term usually applied to leaves for example those of *Veratrum*.

ploidy the number of chromosome sets of a nucleus, cell, or organism.

plug (1) a seedling raised in a single unit of a cellular seed tray. Also often called modules, plugs are a means of planting or potting up with minimal root disturbance; (2) a small piece of turf used for vegetative propagation of lawns; (3) of fruits, such as the raspberry, where the flesh can be detached, leaving the plug around which it is formed.

Plumbago (from Latin *plumbum*, lead; the plant was thought to be a cure for lead poisoning). Plumbaginaceae. Warm and tropical regions. LEADWORT. 15 species, shrubs and perennial or annual herbs. The flowers are produced at various times of year in branching, spicate racemes; the corolla tube is long and slender with a limb of five, spreading lobes. Grow *P.auriculata* in a well-drained, high-fertility, loam-based mix in direct sunlight, with a winter minimum of 7°C/45°F; ventilate freely when conditions allow at temperatures above 10°C/50°F. Water plentifully and liquid-feed fortnightly when in full growth. Keep just moist in winter. Tie to supports as growth proceeds. Prune hard in late winter, either to the base or by cutting lateral growth hard back to a more permanent framework of branches. Grow *P.indica* in bright filtered light or in sun, with some shade in summer, maintaining moderate to high humidity, with a minimum temperature of 15°C/60°F For flowers in summer, cut back to within 10–15cm of the base in spring; given some heat, unpruned specimens will bloom in late winter. Propagate *P.auriculata* by semi-ripe cuttings of non-flowering lateral growth with a heel. Propagate *P.indica* by basal cuttings and root cuttings, and also by seed in spring.

P.auriculata (syn. *P.capensis*; CAPE LEADWORT; South Africa; shrub, stems long-arching, somewhat scandent; spikes short, flowers to 4cm, pale blue); *P.indica* (syn. *P.rosea*; SE Asia; herb or subshrub, semi-scandent or erect; spikes 10–30cm, flowers to 2.5cm, deep rosy pink to pale red or purple).

Plumbago auriculata

Plumeria (for Charles Plumier (1646–1706), French traveller and writer on the flora of tropical America). Apocynaceae. Tropical Americas. FRANGIPANGI, TEMPLE TREE, NOSEGAY, WEST INDIAN JASMINE, PAGODA TREE. 8 species, deciduous shrubs and small trees with a sparse, candelabra-like crown of cylindrical, swollen, grey-green branches and plentiful milky sap. The leaves are oblong or lanceolate-elliptic. Borne usually on bare branches in terminal thyrses, the flowers are fragrant, salverform or funnelform, with five, spreading, ovate-elliptic lobes. Grow in full sun in a dry atmosphere with a minimum temperature of 10°C/50°F. Plant in a fertile, loam-based medium with additional grit and sand. Water and feed generously in spring and summer. Keep dry in winter when blooms may be produced on bare branches. The soft, thick stems will bleed profusely if broken. Propagate by stem tip cuttings before leaves emerge in spring. Allow the cut surface to dry for several days before inserting in a moist but well-drained medium.

P.alba (WEST INDIAN JASMINE; Puerto Rico, Lesser Antilles; to 6m tall; leaves to 30cm, lanceolate; flowers to 6cm in diameter, yellow with white centre); *P.rubra* (FRANGIPANGI, TEMPLE TREE, PAGODA TREE; Mexico to Panama; to 7m; leaves to 40cm, obovate; flowers typically rose-pink with a yellow throat, but highly variable in colour, to 10cm in diameter; includes forms and cultivars with flowers in shades and mixtures of white, yellow, gold, copper, peach, salmon, rose, cerise and deep red, often with a yellow or white throat).

plumose feather-like; with long, fine hairs, which themselves have fine secondary hairs.

plumule the axis (stem) of an embryo or seedling.

plum

Pests
aphids
birds
plum moth
  (*see* moth)
red spider mite
winter moth
bacterial canker
  (*see* bacterial
  diseases)
brown rot
silver leaf

Disorder
gumming

plums the fruits of selected *Prunus* species. The plum grown in Britain and Europe is classified as *P.* × *domestica*, and is considered to have arisen from hybridization between *P.cerasifera* and *P.spinosa* in southwest Asia/southeast Europe. The Romans introduced different types of plum into Britain, from which seedlings arose in the wild incorporating the indigenous *P.institia* and *P.spinosa*. Until the 16th century, the fruit was eaten dried and preserved.

The main groups of plums belong to *P.domestica*, with wide variations in shape, size and colour, within which several types are distinguishable, including greengage types, with green, round fruits of sweet flavour; transparent gage types, where the stone is visible; and prune types, which are purple-skinned, usually have a high sugar content and are suitable for drying; DAMSONS, which are selections from *P.institia*; see *damson*; CHERRY PLUMS, which are very early flowering, red- and yellow-fruited selections from *P.cerasifera*; of more indifferent cooking quality, some are grown in Britain for foliage effect (e.g. *P.cerasifera* 'Pissardii') and as rootstocks (e.g. 'Myrobalan B'); SALICINE PLUMS, which produce round, large fruits of red, yellow and purple; these are selections from *P.salicina*, the Japanese plum, and are widely grown in Japan, southern Europe, South Africa and California.

Other minor groups are identifiable, including the Mirabelle plums of France, which are selections of *P.institia* (see *bullace*) and gage-type plums such as 'Imperial Gage' (syn. 'Denniston's Superb').

In North America, the native species *P.americana*, *P.angustifolia*, *P.munsoniana*, *P.nigra* and *P.subcordata* have given rise to a number of disease-resistant cultivars suited to both cold and hot conditions.

Plums are widely grown in gardens and the cultivar 'Victoria' is generally successful; a sunny but sheltered site and reasonable freedom from spring frosts will encourage good pollination and improve fruiting potential. In Britain, the southern and particularly the drier southeast counties offer suitable climates for the widest range of cultivars, and wall culture is advisable for consistently good results with the best gages and gage types. For all plums, high lime soils should be avoided, and for longevity of trees, continuously good husbandry is necessary on sandy soils.

Propagation is by budding or grafting predominantly on to moderately vigorous 'St Julien A' and moderately dwarfing 'Pixy' rootstocks; the older vigorous rootstocks 'Myrobalan B' and 'Brompton' are occasionally used. In the US, seedling 'Myrobalan' is the normal rootstock, or 'Brompton' or 'St Julien' where incompatibility occurs; the European plums are suited especially to New York State and east coast regions, while the Japanese types are preferred on the west coast.

Fertility in European plums falls into three distinct categories. Class A: cultivars entirely failing to set with their own pollen; class B: cultivars setting a poor crop (2–5%) with their own pollen; class C: cultivars setting a full crop with their own pollen. Cultivars in classes A and B require at least one other suitable cultivar close by as a pollen source for cross-pollination; these requirements should be considered at the time of purchasing planting stock. The presence of bees and other pollinating insects is paramount and sheltered growing conditions are therefore desirable.

Fruit-thinning to 10–15cm apart is essential when a good set has occurred, but should be delayed until stoning has taken place.

The introduction of less vigorous rootstocks, producing smaller trees, has greatly increased the possibility of successful cultivation in gardens, and the best-suited tree forms are the bush, pyramid, or spindle, and, for walls, the fan. In the US, large vase-shaped bushes are favoured, and sometimes also centre-leader trees.

Since growth commences early, planting is best done in late autumn or no later than late winter; trees should be staked and tied. Planting distances in gardens should be approximately 3.5m for bushes, 3m for pyramids and 2.75m for spindles. Fan trees require a wall or fence height of at least 1.75m and a span of 3m minimum, with at least 3m between trees.

Pruning of established trees should always be done during the summer when there is less risk of infection from silver leaf; early summer is the ideal time, while fruitlets are still small. Wounds over 2.5cm in diameter should be sealed immediately with a suitable paint, the potential benefit outweighing the observation that painted wounds heal more slowly. Steady production of young growth is essential for fruit quality, and on bush trees and spindles, pruning consists of removing any low, crowded, crossing or damaged growths or branches, keeping the number of cuts to a minimum. On pyramids, summer pruning consists of cutting all laterals and leaders, other than the central one, to 20cm, and all sublaterals to 15cm, once the shoots are well-ripened; for fans, shoots are shortened to three leaves from their base, with any excessively vigorous or surplus growths pinched out as soon as noticed.

Watering and mulching are advantageous. Gather fruit when firmly ripe preferably picking a tree over two or three times for dessert samples rather than clearing a whole crop. Broken branches should be pruned off as as soon as noticed, and wounds painted.

plunge to sink potted plants, usually up to their rims, into a bed of sand, gravel or soil, known as a plunge bed, in order to protect them from extremes of temperature, to reduce water loss and to maintain stability in wind.

plur-, pluri- a prefix denoting 'many', as plurilocular, meaning many-celled.

pneumatophore a porous apogeotropic root, usually taking the function of a respiratory organ. It is found in plants of wet places, e.g. mangroves.

pod a general term for any dry, dehiscent fruit.

Podalyria (named for Podalyrius, son of Aesculapius, celebrated in Greek mythology as a skilful physician). Leguminosae. South Africa. 25 species, downy shrubs with simple leaves and fragrant, pea-like flowers. Grow in a medium-fertility loam-based mix with additional leafmould. Water moderately, giving a dilute liquid feed fortnightly when in growth. Maintain a winter minimum temperature of 7°C/45°F, with good ventilation. Propagate from seed, or by cuttings of short lateral shoots with a heel, in a sandy propagating mix in a closed case with bottom heat. *P.calyptrata* (to 3m; leaves to 5cm, elliptic or obovate, grey-green, thinly pubescent; flowers to 3cm, petals light pink to lavender-purple, purple at base, keel white).

Podocarpus (from Greek *pous*, foot, and *karpos*, fruit, referring to the fleshy receptacle at the base of the seed). Podocarpaceae. Mexico, C and S America, central and South Africa, Asia (Himalaya to Japan), Australasia. PODOCARP, YELLOW-WOOD. About 100 species, evergreen, coniferous shrubs and trees. They vary widely in habit from dwarf and densely bushy to tall and erect to weeping. The foliage ranges in size and shape from small to large, from needle-like to broadly rhombic or lanceolate-falcate, and in

colour from dull yellow-olive to deep, glaucous sea green. In the female, cones develop fleshy, bead-like arils. The following species are hardy in climate zone 7, but suffer in exposed sites and hard winters. For this reason, they tend to be grown as young, shrubby plants in sheltered rock gardens. *P.alpinus* and *P.nivalis* are especially suited to such positions. However, in mild, humid, maritime climates, some, like *P.salignus*, will attain considerable stature. Plant on moist, acid to neutral, gritty soils in full sun. Propagation is generally similar to that for *Taxus*, by seed or cuttings; seed may need to be stratified for up to a year. Cuttings should include an erect lead shoot if good shape is wanted; side shoots produce prostrate plants.

P.alpinus (TASMANIAN PODOCARP; SE Australia (northern southwest Victoria, Tasmania; shrub to 3m, branches level or upswept; leaves linear-oblong, 6–12mm, dark buff green above, tinged blue beneath, obtuse to apiculate, midrib keeled; receptacle vivid red); *P.latifolius* (YELLOWWOOD; Africa (S Sudan to S Natal); tree to 30m, or shrubby, bark dark grey, smooth, exfoliating in long strips; leaves linear-elliptic, 4–10cm on mature trees, longer on young trees, straight or falcate, rigid, dark glossy green to green tinged blue above, midrib distinct beneath; receptacle red tinged purple); *P.macrophyllus* (BIG-LEAF PODOCARP, KUSAMAKI; S China, Japan; tree to 15m, or shrub to 2m; leaves broadly linear-lanceolate, 8–10cm on mature trees, longer on young trees, green tinged yellow beneath, midrib, distinct, bluntly acute; receptacle red; 'Angustifolius': leaves narrower; 'Argenteus': leaves bordered white); *P.nivalis* (ALPINE TOTARA; S New Zealand; resembles *P.alpinus*, but leaves elliptic, wider; shrub, erect to prostrate or procumbent, to 2 × 3m wide; leaves 6–18mm, obtuse, rigid, margins thickened; receptacle red; 'Bronze': leaves tinged yellow-bronze, especially when young; 'County Park Fire': foliage creamy yellow to gold flushed bronze-red to pink); *P.salignus* (WILLOWLEAF PODOCARP, WILLOW PODOCARP; S Chile; tree, to 20m, or shrub in cold areas, bark fibrous; branches arching, foliage somewhat weeping; adult leaves narrow-lanceolate, more or less falcate, 8–12cm, shiny above, paler beneath, juvenile leaves straighter, 5–10cm; receptacle dark red to violet); *P.totara* (TOTARA; New Zealand; tree, to 30m; bark dark brown, to silver-grey, thick, exfoliating in strips; leaves 15–25cm, linear-lanceolate, straight to falcate, sharply spined, coriaceous, stiff, green tinged yellow-grey and ridged above, with raised midrib beneath; receptacle fleshy, orange-red to bright red; 'Aureus': leaves yellow-green).

Podophyllum (from Greek *pous*, foot and *phyllon*, leaf – the leaves were thought to resemble a duck's foot in outline). Berberidaceae. Eastern N America to E Asia and Himalaya. Around 5 species, perennial herbs with tuber-like rhizomes. The leaves are large, peltate and palmately lobed, standing over or beneath the terminal flowers like a parasol or ruff. Produced singly or a few together in spring and early summer, the flowers are cup- or bud-like with six to nine petals. They give rise to large, fleshy, ovoid berries. These are hardy plants for the woodland or wild garden, or for moist and shaded borders. Grow in deep, humus-rich, damp soil, in filtered light or shade. Propagate by division, from cuttings of the rhizome, or from fresh, ripe seed.

P.hexandrum (syn. *P.emodi*; W China, Himalaya; leaves to 25cm across, 3–5-lobed, toothed, often flushed red-bronze at first – marbled red-bronze or brown in cultivar Majus; flowers solitary, erect, appearing before leaves mature, white to rose pink; fruit red); *P.peltatum* (MAY APPLE, WILD MANDRAKE; eastern N America south to Texas; leaves 30cm across, lobes 5–9, cleft, finely hairy beneath; flowers nodding, fragrant, white to rose; fruit green-yellow, rarely red); *P.pleianthum* (syn. *Dysosma pleiantha*; C and SE China; differs from *P.versipelle* in having glossy mid-green leaves to 60cm in diameter, shallowly and bluntly lobed, and appallingly malodorous, dark amber flowers); *P.versipelle* (China, Tibet; leaves 40cm across, irregularly deeply divided, 5–8 lobed, finely toothed; flowers nodding, dull orange-bronze to deep crimson, malodorous).

Podranea (an anagram of *Pandorea*, a genus in which these vines were once included). Bignoniaceae. South Africa. 2 species, evergreen climbing shrubs with pinnate leaves and clusters of funnel- to bell-shaped flowers in summer. Cultivate as for *Bignonia*. *P.ricasoliana* (PINK TRUMPET VINE, PORT ST JOHN CLIMBER, RICASOL PODRANEA; flowers to 6cm long, pale pink striped red, fragrant).

pod-spot any of several fungal diseases that cause spots on the pods of peas and beans, and may also affect leaves, stems and flowers. *Aschochyta* and *Mycosphaerella* are commonly responsible, and damage may be severe in wet seasons. Infection of pods invariably leads to infection of seed, which may be disfigured, and this gives rise to diseased seedlings. Badly infected haulm should be burnt and the growing site rested.

Pogonatherum (Greek, meaning *bearded with bristles*). Gramineae. Tropical Asia. 3 species, low-growing, clump-forming to spreading, bamboo-like grasses with many very slender culms. A delightful miniature 'bamboo', *Pogonatherum* forms a mop of culms, dense with narrow, bright green leaves. It favours moist soils rich in garden compost, leafmould or composted bark, shade, a minimum temperature of 10°C/50°F and frequent misting. Plants will regenerate readily if cut hard back. Propagate by division. *P.saccharoideum* (BAMBOO GRASS; India, China, Malaysia; culms 15–40cm, simple or branched, tufted, slender, leafy; leaves to 2cm, linear-lanceolate, bright fresh green).

poisonous plants many plants contain chemicals that are poisonous to humans and animals. In some plants, every part is poisonous; in others, poison may be concentrated primarily in one type of organ – for example, rhizome or rootstocks (as in *Convallaria* and *Iris*), fruit (as in *Daphne*), or foliage (as in *Ruta*). In some cases, the poisonous parts of a plant can be rendered innocuous by cooking; with cassava or manioc (*Manihot esculenta*), this is achieved by combining cooking with crushing of the starchy roots to expel the prussic acid within them.

The lethal uses of some plant poisons are well-known, for example, the curare and wourate arrow poisons from *Strychnos toxicaria* in South America, and *S.nux-vomica* as the source of strychnine. Hemlock *Conium maculatum* and angel's trumpet *Brugmansia arborea* are recorded as intentionally administered poisons. However, some poisons, like the morphine of *Papaver somniferum*, and equally curare, have become useful drugs and anaesthetics.

A great many garden plants are poisonous, but incidents of serious effects are rare. It is important that possible risks are understood, but the subject should be seen in perspective to avoid unnecessary alarm and the spoiling of enjoyment of gardens and gardening.

Young children are most at risk from plant poisoning because they frequently put things into their mouths and colourful flowers or berries may be especially attractive to them. A sensible rule is that if a plant is not a recognized food it must not be eaten, and children must be taught this. Mushrooms and 'toadstools' should never be eaten unless definitely identified as harmless.

Adults are most at risk of poisoning from physical contact with plants and sap during weeding and pruning. This type of poisoning usually produces a skin reaction known as contact dermatitis, ranging from mild irritation to chronic blistering. The responsible mechanisms can be divided into five groups: (a) mechanical irritants, such as cactus spines; (b) stings, such as those produced by stinging nettles (*Urtica dioica*), where a toxin is mechanically injected; (c) phototoxins, which are chemicals activated by sunlight, are produced, for example, by rue (*Ruta graveolens*), and giant hogweed (*Heracleum mantegazzianum*), the sap of which can give rise to skin blisters and cause permanent scarring; (d) allergens inducing immediate or delayed sensitization, for example, chrysanthemum (*Dendranthema × grandiflorum*) *Primula obconica*, ivy (*Hedera helix*) and poison ivy (*Toxicodendron radicans*). Hayfever falls into the allergen-produced group of symptoms and is the result of allergy to wind-borne pollen; (e) direct irritants, as a result of which any person is likely to have a reaction if the concentration of toxin is sufficient, for example, irritation from the sap of *Daphne*, *Euphorbia* and *Ranunculus* species.

Contact with potentially dermatitic plants can be avoided by wearing protective clothing. Where poisoning by swallowing is suspected or evident, medical advice must be obtained, taking samples of the plants where known.

Young pets are notoriously adventurous and should be kept away from potentially hazardous plants. Poisoning of livestock is often more serious due to the large quantities ingested and the longer time spent on digestion. Thoughtless disposal of prunings into fields accessible to livestock is an important cause of sporadic poisoning. Particularly hazardous are hedge trimmings from plants such as box (*Buxus sempervirens*), laburnum (*Laburnum anagyroides*), cherry laurel (*Prunus laurocerasus*), privet (*Ligustrum* species), *Rhododendron* species and yew (*Taxus baccata*). Dried remains of many poisonous plants can have the same or a greater level of toxicity.

In the UK, there exists a voluntary Code of Recommended Retail Practice relating to the labelling of potentially harmful plants, which is intended to help gardeners purchase plants wisely. Individuals must always exercise proper responsibility for their own safety and that of children and animals within the garden environment. At the same time, gardeners have a duty to take reasonable steps to avoid exposure to risk.

The following is a list of the better-known poisonous species often found in gardens. The letters in parentheses indicate the types of potential risk. For external effects: (S) skin irritant; (E) eye irritant. For internal effects: (P) poisonous when ingested i.e. known to present *any* level of hazard from mild illness to, occasionally, severe poisoning. *Aconitum* (P/S); *Aesculus* (P); *Agrostemma githago* (P); *Alstroemeria* (S); *Aquilegia* (P); *Arum* (P/S/E); *Altropa* (P/S); *Brugmansia* (P); *Caladium* (P); *Caltha* (P/S/E); *Catharanthus roseus* (P); *Colchicum* (P); *Convallaria majalis* (P); × *Cupressocyparis leylandii* (S); *Daphne* (P/S); *Datura* (P/S); *Delphinium* (inc. *Consolida*) (P); *Dendranthema* (S); *Dictamnus* (S); *Dieffenbachia* (P/S/E); *Digitalis* (P); *Echium* (S); *Euonymus* (P); *Euphorbia* (P/S/E) –poinsettia does not present a serious hazard; *Ficus carica* (S/E); *Fremontodendron* (S/E); *Gaultheria* (P); *Gloriosa superba* (P); *Hedera* (P/S); *Helleborus* (P/S); *Heracleum mantegazzianum* (severe skin irritant in bright sunlight); *Hyacinthus* (S); *Hyoscyamus* (P); *Ipomoea* (P); *Iris* (P/S); *Juniperus sabina* (P); *Kalmia* (P); *Laburnum* (P); *Lantana* (P/S); *Ligustrum* (P); *Lobelia* (P/S/E); *Lupinus* (P); *Narcissus* (P/S); *Nerium oleander* (P); *Ornithogalum* (P/S); *Phytolacca* (P/S); *Polygonatum* (P); *Primula obconica* (S); *Prunus laurocerasus* (P); *Rhamnus* (inc. *Frangula*) (P/S); *Rhus verniciflua*, *R.radicans*, *R.succedanea* (P/S, severe); *Ricinus communis* (P); *Ruta* (severe skin irritant in bright sunlight); *Schefflera* (S); *Solandra* (P); *Solanum* (most species) (P); *Taxus* (P); *Thuja* (P/S); *Tulipa* (S); *Veratrum* (P); *Wisteria* (P). All of these plants are safe to grow provided they are treated with respect. They are ornamental plants not food plants.

pole see *rod*

Polemonium (name used by Dioscorides, possibly for the early Athenian philosopher Polemon, or perhaps from *polemos* war). Polemoniaceae. Temperate to Arctic regions, northern Hemisphere, southern S America. JACOB'S LADDER, SKY PILOT. 25 species, annuals or perennial herbs, often muskily scented. The leaves are usually pinnate. Borne in spring and summer in axillary or terminal cymes, the flowers are narrowly funnelform to rotate-campanulate and 5-petalled. Hardy in climate zone 6, the following species are suitable for the herbaceous and mixed border and the wild garden. Plant in sun or part shade in moist, well-drained and fertile soil. Deadhead to prevent self-seeding. Propagate species by seed in autumn, cultivars by division in spring.

P.brandegei (W US; erect perennial 10–30cm, densely glandular-pubescent and muskily aromatic; flowers bright yellow to pale gold or white, funnelform with a very narrow tube); *P.caeruleum* (JACOB'S LADDER, GREEK VALERIAN, CHARITY; N America, N Europe, N Asia; perennial 30–90cm tall, glandular-pubescent above; flowers rotate to campanulate, blue, rarely white); *P.carneum* (W US; erect perennial, 10–40cm; flowers pink or yellow, sometimes dark purple to lavender, rarely pink, white or blue); *P.foliosissimum* (LEAFY JACOB'S LADDER; erect perennial, 40–120cm, sparsely villous, glandular above; flowers campanulate, blue-violet, cream, white, or – var. *flavum* – yellow and tawny-red outside); *P.pulcherrimum* (NW America; perennial, 0.5–3m, erect; flowers campanulate, blue to violet or white, tube yellow inside). *Cultivars.* Hybrids and cultivars range in height from 20–100cm, in flower colour from white to pale rose, rich pink, lilac, mauve and pale blue to azure; there are also some striking white and yellow leaf variegations.

Polemonium caeruleum

polesaw a saw blade attached to a long pole, used to cut branches of tall trees.

Polianthes (from Greek *polios*, bright, and *anthos*, flower). Agavaceae. Mexico. 13 species, evergreen perennial herbs with thick, bulb-like bases, short rhizomes and thickened roots. Tough and thinly fleshy, the leaves are lanceolate or linear. Borne in terminal, bracteate, spike-like racemes, the flowers are cylindrical to narrowly funnel-shaped with six segments. Frost-tender, tuberous perennials, *Polianthes* usually flowers in midsummer, but, with successional plantings and given protection and a minimum temperature of 15°C/60°F, may be forced to bloom throughout the year; *P.tuberosa* has long been valued for its spikes of very strongly scented, waxen-white flowers. In cool temperate climates, *Polianthes* is grown in the warm, sheltered flower border, to be lifted and dried off in autumn and overwintered in sand in frost free-conditions. Otherwise, grow in bright direct sunlight in the greenhouse or conservatory. Plant in a mix of fibrous loam with sand, well-rotted manure and leafmould or equivalent; give bottom heat of about 15–18°C/60–65°F, keeping the potting mix just moist until the leaves appear. Water plentifully when in full growth, and feed fortnightly with liquid fertilizer. Dry off after the leaves fade in winter. Propagate by seed or offsets in spring. *P.geminiflora* (to 70cm; flowers to 2.5cm, bright orange red, nodding in a lax raceme); *P.tuberosa* (TUBEROSE; only known as a cultigen, having been cultivated in pre-Columbian Mexico; to 1.2m; flowers very fragrant, in a lax spike to 1m, pure waxy white, 3–6cm; 'Excelsior Double Pearl', an improved form of 'The Pearl'; 'Single Mexican', with single flowers with up to 5 spikes, long-lasting; 'The Pearl', with double, highly fragrant flowers; var. *gracilis*, of more slender habit, with narrower leaves, a perianth with a long slender tube and linear segments).

pollarding

pollard to cut back annually all branches of a tree to the main trunk, usually up to a height of about 2m. It is practised to contain tree spread, as with planes (*Platanus*), or to produce a decorative mophead effect, as with willows (*Salix*).

pollen the microspores, spores or grains producing male gametes and borne by an anther in a flowering plant.

pollen beetles (*Meligethes* species) (Coleoptera: Nitidulidae) small, shiny, black or green-bronze beetles, up to 3mm long, with clubbed antennae; both adults and larvae feed on pollen. The larvae of common species, such as *Meligethes aeneus* and *M.viridescens,* feed in the flower buds of brassicas and related plants, such as mustard and oil seed rape, causing a reduction in the seed crop. From spring onwards, adults invade the flowers of a wide variety of garden plants, including roses, sweet peas, marrows, runner beans and daffodils. Problems arise when infested cut flowers are brought indoors, and infested blooms are downmarked on the show bench. Cut flowers for the house should be tapped gently to dislodge as many beetles as possible, and then placed for a few hours during the day in a shed or outhouse where most of the remaining beetles will fly to windows or open doors. Blooms intended for exhibition can be enclosed in fine netting before opening to exclude beetles.

pollenizer, pollonizer a plant that provides pollen.

pollinating insects the most important insect pollinators are found in the order Hymenoptera, which includes sawflies, wasps, bees, ants and, particularly, the cosmopolitan honey bee (*Apis mellifera*) and bumble bees (*Bombus* species); some solitary bees, such as leaf-cutting bees (*Megachile* species) and mason bees (*Osmia* species), are also useful. Among the Diptera, blowflies and hoverflies play a significant role. Blowflies are easily reared in captivity and have been used in the breeding of crops, such as brassicas, onions and carrots. Some beetles (Coleoptera), for example, pollen beetles, and various species of thrips (Thysanoptera) pollinate flowers but any benefit can be offset by injury to the blooms. Butterflies and moths (Lepidoptera) are important pollinators of flowers with long tubular corollas.

pollination the transfer of pollen grains from an anther to a stigma, a process that precedes fertilization and the production of seed. It may involve either self-pollination, if pollen is transferred within the same flower, or cross-pollination where the transfer is between different flowers on the same plant or to flowers on another plant of the same species. Many flowers have processes and structures developed to ensure cross-pollination, such as nectar production to attract insects, mechanisms to ensure pollen distribution (as in *Viola* and *Salvia),* and profuse pollen production for wind distribution (as in many grasses). There are adaptations to ensure self-pollination, such as the maturing of anthers and stigmas at the same time. Intervention by gardeners can effect improved pollination, for example, dusting between open flowers of peach with a brush or rabbit's tail. Such action can also be a means of making controlled crosses in plant breeding.

pollinator (1) an agent or means of pollen transfer, for example insects, birds, bats, or wind; (2) used in fruit growing to describe a cultivar planted to ensure fruit set on another cultivar that is self-sterile or partially so.

pollinium a regular mass of more or less coherent pollen grains.

polycarbonate see *greenhouse.*

Polygala (name used by Dioscorides, from Greek *polys*, much, and *gala*, milk; these plants were believed to promote the secretion of milk). Polygalaceae. Widespread. MILKWORT, SENECA, SNAKEROOT. 500 species, annual or perennial herbs and shrubs or, rarely, trees. The pea-like flowers consist of two inner, petal-like sepals (wings) and three to five petals, the lowermost petal (keel) often crested. The hardy *P.calcarea* and *P.chamaebuxus* are suitable for the rock garden. Some alpine species, such as *P.vayredae*, do not thrive in mild damp winters and may be more successfully grown in the alpine house. Tender species, such as *P.myrtifolia,* need the protection of the cool greenhouse in cool temperate zones. Grow in part shade or in sun in moisture-retentive but well-drained soil. Under glass, grow in bright filtered light, providing shade from the hottest summer sun and good ventilation. Water plentifully and feed fortnightly with dilute liquid feed when in full growth. Keep just moist in winter with a minimum temperature of 5°C/40°F. Cut leggy specimens hard back in late winter. Propagate by seed in spring or by softwood or semi-ripe cuttings in a sandy propagating mix in a closed case with gentle bottom heat.

P.calcarea (W Europe; perennial herb, erect to 20cm, rosette leaves spathulate to obovate, stem leaves smaller, linear-lanceolate; flowers usually blue or white); *P.chamaebuxus* (C Europe to Italy; evergreen, low-growing shrub, 5–15cm; leaves 1.5–3cm, elliptic to obovate, leathery, somewhat glossy; flowers with wings cream-white to pink, rosy purple or yellow, keel yellow); *P.myrtifolia* (South Africa; erect, much-branched shrub, 1–2.5m; leaves 2.5–5cm, elliptic-oblong or obovate; flowers green-white veined purple, or wholly rich purple); *P.vayredae* (E Pyrenees; perennial to 5cm; leaves 2–2.5cm, linear-lanceolate; flowers pink-purple with a yellow-tinged keel).

polygamodioecious (1) a plant that is functionally dioecious but contains some perfect flowers in its inflorescence; (2) a species that has perfect and imperfect flowers on separate individuals.

polygamous describing a plant that has male (staminate), female (pistillate) and hermaphrodite flowers, on the same or separate plants of the same species, for example ash (*Fraxinus excelsior*).

Polygonatum (from Greek *polys*, many, and *gony* joint, referring to the many-jointed rhizome; name used by Dioscorides). Liliaceae (Convallariaceae). N US, Europe, Asia. Some 30 species, rhizomatous, perennial herbs with erect to arching stems clothed with alternate, opposite, or whorled, ovate to linear leaves. The flowers are tubular with six, erect to spreading segments, and usually nod or hang from the leaf axils. The berries are blue-black or red. The following species are hardy and tolerant of a range of conditions, excepting heat and drought. However, they are best grown in fertile, humus-rich, moist but well-drained soil in cool, semi-shade or shade. Leave undisturbed once planted and allow large clumps to establish for best effects. *P.hookeri* is suited to damp pockets in the rock garden and to alpine sinks and troughs. Propagate by division or by seed in autumn. Larvae of the saw fly will strip foliage in summer.

P.biflorum (syn. *P.canaliculatum*, *P.commutatum*; E US, south central Canada; stem 40cm–2m, erect or arched; leaves 4–18cm, alternate, narrow-lanceolate to broadly elliptic, glaucous beneath; flowers drooping, 1–2.3cm, green-white); *P.hookeri* (E Himalaya, China; stem 5–10cm, more or less erect; leaves crowded, alternate, 1.5–2cm, linear to narrow-elliptic; flowers solitary, erect, to 2cm, purple or lilac-pink); *P. × hybridum* (*P.multiflorum × P.odoratum*; intermediate between parents; common in cultivation, includes 'Florepleno', with double flowers, and 'Striatum' with creamy white striped leaves); *P.multiflorum* (Europe, Asia; stem to 90cm, arched; leaves 5–15cm, alternate, elliptic-oblong to ovate; flowers 1–2cm, drooping, white tipped green); *P.odoratum* (syn. *P.officinale*; Europe, Asia; stem to 85cm, arched; leaves 10–12cm lanceolate to ovate, alternate, ascending; flowers 2–4 per axil, drooping, fragrant 1–2cm, white tipped green; includes 'Gilt Edge', with leaves edged yellow, 'Grace Barker', with leaves striped creamy white, and 'Variegatum', with red stems when young and leaves narrowly edged creamy white); *P.verticillatum* (Europe, temperate Asia, Afghanistan; stem 20–100cm, erect; leaves 6.5–15cm, opposite or whorled, linear-lanceolate to narrowly ovate; flowers to 1.5cm, drooping, 1–3 per axil, green-white).

Polygonum (from Greek *polys* many, and *gony*, joint; the stems have conspicuously swollen nodes). Polygonaceae. Northern temperate regions. KNOTWEED, SMARTWEED, FLEECE VINE, SILVER LACE VINE. Some 150 species, annual or perennial herbs or shrubs, some aquatic or climbing. The stems appear jointed. Produced in spring and summer in axillary clusters or in terminal panicles or spikes, the flowers are small, funnel- or bell-shaped and consist usually of five segments. Small and angled, the fruits are enclosed by persistent perianth.

Polygonum is a diverse genus with species suited to a number of situations in the garden. Although most have invasive potential, this can often be used to advantage in larger landscape plantings. *P.japonicum*, introduced to Britain as an ornamental, has become widely naturalized as a pernicious and serious weed pest. Only the less invasive, variegated forms should be accommodated in (larger) gardens. The climbing species, such as *P.baldschuanicum* and *P.aubertii*, provide vigorous and rapid cover in even the most hostile situations. They may be confined by severe pruning in late winter, sprouting freely from older wood.

In cool, moist semi shade, *P.campanulatum* is a beautiful and robust perennial, bearing small elegant bell-shaped blooms over long periods in summer. In the herbaceous border, the vigorous clump-forming *P.bistorta* is valued for its large, weed-smothering foliage and dense spikes of soft pink bloom; *P.milletii* is similar, but more compact. These border species flower for longer periods on moist soils. *P.amplexicaule* is a stout, clumping perennial extending its period of interest from the first flush of bloom in midsummer often through until first frosts. *P.virginianum* 'Painter's Palette' is a remarkable foliage plant for cool, moist and semi-shaded placs. It associates especially well with hardy ferns, hostas, the smaller bamboos and grasses like *Hakonechloa macra* 'Albo-aurea'. Lower mat-forming species, such as *P.affine* and *P.vacciniifolium,* make good groundcover for the front of herbaceous borders and areas of the rock garden where their spread will not threaten less robust alpines. These last two are valuable for their late flowers, which dry to rich chestnut shades and persist into winter, creating a fine contrast with the rusty tints of the fading foliage. The attractively marked foliage of the creeping *P.capitatum* also provides good low cover. It may need protection in zones experiencing winter temperatures below about –5°C/23°F, or can be treated as a half-hardy annual. Most species are hardy to between –15 and –20°C/5 to –4°F. Propagate perennials by division; climbers in late winter by hardwood nodal cuttings of stem pieces with two nodes. Increase also by seed.

P.affine (syn. *Bistorta affinis*; Himalaya; low, mat-forming perennial to 25cm; leaves 3–10cm, mostly basal, elliptic to oblanceolate, dark green, red-bronze to brown in autumn; flowers in dense, erect spikes, 5–7.5cm, pale pink to crimson or deep salmon red, often becoming darker with age); *P.amplexicaule* (syn. *Bistorta amplexicaulis;* Himalaya; perennial to 1m, rootstock woody; leaves 8–25cm, ovate to lanceolate, acuminate, cordate at base, downy beneath, long-stalked or stem-clasping; flowers in loose spikes to 8cm, rose-red to purple or white); *P.aubertii* (syn. *Bilderdykia aubertii, Fallopia aubertii;* RUSSIAN VINE, CHINA FLEECE VINE, SILVER LACE

Polygonum japonicum 'Spectabile'

VINE; W China, Tibet, Tajikistan; rampant, twining, woody vine to 15m; leaves 3–10cm, ovate to ovate-oblong, apex acute, slightly cordate at base; flowers in lacy panicles, white or green-white to pink); *P.baldschuanicum* (syn. *Bilderdykia baldschuanica, Fallopia baldschuanica;* MILE-A-MINUTE VINE, RUSSIAN VINE; E. Europe, Iran; differs from *P.aubertii* in more woody habit; flowers larger in broader, drooping panicles, white tinged pink in time); *P.bistorta* (BISTORT, SNAKEWEED, EASTER LEDGES; Europe, N and W Asia; perennial to 60cm, rootstock stout; leaves 10–20cm, ovate to oblong, obtuse, truncate at base, wavy, long-stalked or sessile; flowers in dense, cylindrical spikes, rose, pale pink or white); *P.campanulatum* (LESSER KNOT-WEED; Himalaya; creeping perennial to 1m; leaves 3.5–12cm, lanceolate to elliptic, white- to pink-brown, hairy beneath; flowers in loosely branched, nodding panicles, pink to dark rosy red or white, fragrant); *P.capitatum* (Himalaya; perennial to 7.5cm with creeping stems to 30cm, rooting, glandular-hairy; leaves to 2.5cm, ovate to elliptic, green with a purple-bronze V-shaped marking; flowers pink in small, dense, rounded heads); *P.japonicum* (JAPAN-ESE KNOTWEED, MEXICAN BAMBOO; Japan; massive, freely suckering rhizomatous perennial to 2m; stems stout, cane-like, branched above; leaves broadly ovate, acuminate, base truncate, 6–12cm; panicles axillary; flowers cream-white; 'Spectabile': leaves red later marbled with yellow); *P.milletii* (Himalayas to SW China; stem to 50cm; leaves to 30cm, linear-lanceolate to oblong, petiole winged, upper leaves clasping; flowers crimson in broad cylindrical to rounded heads); *P.orientale* (PRINCE'S FEATHER, PRINCESS FEATHER, KISS-ME-OVER-THE-GARDEN-GATE; E and SE Asia, Australia, naturalized N America; pilose annual 1–1.5m, stem stout, branching; leaves 10–20cm, broadly ovate, base slightly cordate; flowers in dense, branched, nodding spikes, pink to rose-purple or white); *P.sachalinense* (GIANT KNOTWEED, SACALINE; Sakhalin Island; perennial resembling *P.japonicum,* but 2–4m, stems more robust, red-brown, forming a coarse thicket; leaves 15–30cm, ovate-oblong, base cordate; flowers in shorter, denser panicles, white-green); *P.vaccini-ifolium* (Himalaya; trailing perennial; stem woody, much-branched, to 30cm; leaves 1–2.5cm, ovate or elliptic, glaucous beneath, petiole short; flowers in loose, erect spikes, pink); *P.virginianum* (syn. *Tovara virginiana;* Japan, Himalaya, NE US; perennial to 120cm, glabrous to roughly hairy; leaves 8–15cm, ovate to elliptic, acuminate to rounded, glabrous to roughly pubescent; flowers in slender spikes, green-white or tinged pink and dark red bracts; 'Painter's Palette': leaves splashed and mottled with dark green, lime, grey, cream, pale yellow, salmon pink and maroon-black; 'Variegatum': leaves variegated ivory and primrose yellow).

polymorphic occurring in more than two distinct forms; possessing variety in some morphological feature.

polyploid a plant with more than the normal two sets of chromosomes as the genetic components of its cells. It arises from a multiplying of the chromosomes complement at the time of fertilization. See *triploid, tetraploid.*

Polypodium (from Greek *polys,* much, and *pous,* foot). Polypodiaceae. Mainly temperate northern Hemisphere. POLYPODY. Some 75 species, epiphytic, lithophytic or terrestrial ferns, evergreen or semi-deciduous with creeping, scaly rhizomes. The fronds are entire to pinnate or pinnatifid. Hardy in climate zone 7 and easily established in the fern border, mixed border, or (*P.vulgare*) on dry stone walls. Although they favour moist, lightly shaded cond-tions, they will tolerate drought and direct sunlight. Propagate by division or spores.

P.cambricum (WELSH POLYPODY; Europe; differs from *P.vulgare* in its broader, softer fronds; 13–50 × 7–10cm, pinnatifid, deltoid or oblong, base truncate, segments to 9mm wide; includes numerous cultivars with finely dissected, plumose, forked, lacerate and crested fronds); *P.glycyrrhiza* (LICORICE FERN; Alaska to California; fronds to 35 × 1.5cm, pinnate to pinnat-ifid, lanceolate to elliptic or oblong, caudate or atten-uate, thin-textured, segments to 6 × 1cm, alternate, falcate, linear, base dilated, notched; includes crested and lacerated and narrow, elongated cultivars); *P.poly-podioides* (RESURRECTION FERN; Americas, South Africa; fronds to 15 × 6cm, deltoid or oblong, apex attenuate to acute, margins rolling inwards during dry weather, somewhat leathery, segments to 25cm × 5mm, distant, spreading, linear or oblong, apex obtuse, base dilated, entire or notched, scaly); *P.scouleri* (COAST POLYPODY, LEATHERY POLYPODY; N America; fronds to 40 × 15cm, ovate to deltoid, thick-textured and rigid, segments to 14 pairs, spreading, linear to oblong, margin entire to notched or undulate and cartilaginous); *P.virginianum* (ROCK POLYPODY, AMERICAN WALL FERN; N America, E Asia; fronds to 25 × 7cm, arching or pendent, lance-olate or deltoid to oblong, leathery to thin-textured, segments to 4 × 1cm, to 25 pairs, alternate to subop-posite, lanceolate to linear or oblong, entire to notched and undulate; includes 'Bipinnatifidum', with regularly and deeply lacerated pinnae, plumose); *P.vulgare* (COMMON POLYPODY, ADDER'S FERN, WALL FERN, GOLDEN MAIDENHAIR; N America, Europe, Africa, E Asia; fronds to 30 × 15cm, ascending to erect, lanceolate to ovate or oblong or linear, glabrous, thin-textured to subcoriaceous, segments to 6cm × 7mm, close, spreading, horizontal to ascending, oblong to linear, obtuse, entire to dentate; includes cultivars with plumose, crested, forked, lacerated and pinnately cut fronds).

Polyscias (from Greek *polys,* many, and *skias,* shadow, i.e. sunshade or canopy, referring to the some-times large and many-branched, spreading inflores-cences). Araliaceae. Old World Tropics. About 100 species, evergreen shrubs or trees. Clustered toward the branch ends, the leaves are simple to trifoliolate or pinnate. Small, pale flowers are borne in umbels or heads in compound inflorescences. Cultivation as for *Schefflera*. *P.filicifolia* (FERN-LEAF ARALIA; probable cultigen originating in E Malaysia or the West Pacific, now widespread in warmer regions; large erect shrub; leaves 9–17-foliolate, leaflets narrowly elliptic to linear-lanceolate, stalked, deeply cut or pinnatifid, to 10cm, bright green, midribs tinted purple, shallowly toothed; white-variegated cultivars are grown); *P.guilfoylei* (GERANIUM ARALIA; probable cultigen originating in E Malaysia or the West Pacific, now widespread in warmer regions; erect shrub or treelet to 6m, little-branched; leaves to 60cm, leaflets 5–9, stalked, rotund to broadly ovate or oblong-elliptic, to 15cm, irregularly spiny-toothed often obscurely lobed, margins white or cream; includes cultivars with crisped, finely cut and variegated leaves); *P.scutellaria* (possible cultigen originating in E Malaysia or the West Pacific, now widespread in warmer regions; shrub or small tree to 6m; leaves to 30cm, leaflets 1–3(–5), broadly elliptic or orbicular, often shield-like, to 28cm wide, apex rounded, the base obtuse to cordate, entire or minutely spiny-toothed, sometimes lobed; includes cultivars with finely toothed or cut and variegated leaves).

Polystichum (from Greek *polys*, many, and *stix*, a row, referring to the regular rows of sori seen on many species). Dryopteridaceae. Cosmopolitan. HOLLY FERN. More than 175 species, small to medium-sized terrestrial ferns, evergreen to semi-deciduous, with shaggy-scaly rhizomes and 1–3-pinnate fronds. Most species are hardy to –30°C/ –22°F or lower, but where harsh winters are common, it may be wise to mulch their bases thickly and to cover the crown with dried leaves or bracken. *Polystichum* will tolerate part sun for up to six hours a day in moist soil. In full shade, it will tolerate quite dry conditions once the fronds are developed. Preferred pH is 6.5 to 7.5; in very acid soils, some species, such as *P.setiferum*, gradually lose vigour. Propagate by spores or division in spring. Increase also by bulbil-like plantlets, where produced.

P.acrostichoides (CHRISTMAS FERN; N America; fronds 20–50 × 5–12cm, linear-lanceolate, pinnae alternate, 20–35 each side, linear-oblong, acutely auriculate at base, minutely spinose-dentate, dark green, with hair-like scales beneath); *P.aculeatum* (Europe; fronds 30–90 × 5–22cm, lanceolate, rigid, pinnae to 50 per side, pinnate or pinnatifid, pinnules small, serrate, decurrent); *P.lonchitis* (NORTHERN HOLLY FERN; Europe; fronds 20–60 × 3–7cm, linear-lanceolate, rigid, coriaceous, pinnae 25–40 each side, lanceolate, slightly curved, auriculate on upper side at base, serrate, dark green above, scaly beneath, very shortly stalked); *P.proliferum* (MOTHER SHIELD FERN; Australia; rhizome trunk-like with age; fronds to 100 × 30cm, bipinnate to tripinnate, dark green, with proliferous buds near apex, pinnules spiny-serrate); *P.setiferum* (SOFT-SHIELD FERN; S, W and C Europe; fronds 30–120 × 10–25cm, lanceolate, soft, pinnae to 11–2.5cm, to 40 per side, pinnules serrate, stipitate; rachis bearing bulbils; includes

many cultivars with crested, finely dissected, narrowed and pointed, proliferous and densely congested feathery fronds); *P.tsussimense* (NE Asia; fronds 25–40 × 10–20cm, broadly lanceolate to oblong-ovate, acuminate, pinnules 7–15mm, ovate to oblong-ovate, mucronate, spinose-dentate, glabrous and glaucous above).

polythene a thermoplastic material of versatile use in gardening. It is available in various sheet thicknesses, opaque or coloured, and also in moulded form. Sheet polythene is used for cladding greenhouses and frames; types incorporating an ultraviolet light inhibitor will last longer. Low continuous polythene tunnels are suitable for advancing crops such as strawberries. Black sheet may be used as a weed-inhibiting soil mulch, and clear sheet as a ground mulch for raising soil temperature or for advancing maturity of crops, such as carrots and potatoes – in such instances, the sheet may be perforated. Polythene sheet is also an excellent means of protecting rooting cuttings and of insulating greenhouses in winter, especially in bubble form. Thick gauge sheet is used as a pool liner. It is also fabricated into flimsy or rigid plant containers. Polythene waste should be disposed of through approved collection agencies and never burnt because the process emits toxic gases.

polytunnel a greenhouse clad with polythene sheet.

pome a multiloculate fruit, formed by the fusion of an inferior ovary and the hypanthium, from which the flesh, which is tough but not woody or bony, is derived.

pome fruit the fruit of apple, pear, quince, medlar, and some other members of the family Rosaceae.

pomology the science and practice of fruit-growing.

pomona a treatise on fruits.

pompon applied especially to types of dahlia and chrysanthemum with small globular flowers composed of tight-packed florets. See *chrysanthemum, dahlia.*

Poncirus (Latinized form of French *poncire*, a type of citrus). Rutaceae. C and N China. 1 species, *P.trifoliata*, TRIFOLIATE ORANGE, BITTER ORANGE, a fast-growing, small, deciduous tree with many, interlacing, green twiggy branches armed with formidable thorns. The leaves are trifoliolate, the terminal leaflet 3–6 × 1.5–2.5cm and obovate-cuneate. Borne in early spring, the flowers are solitary and scented with waxy, white, oblong petals to 3cm. Roughly spherical and 3–5cm in diameter, the fruit is dull lemon-yellow and fragrant when ripe, with a thick, rough peel covered in copious oil glands; the pulp is very acidic. Hardy to –15°C/5°F, this shrub is valued for its interlacing branch pattern, orange blossom flowers and curious, inedible fruit. It

Polystichum tsussimense

is also used for hedging and as a frost-resistant root-stock for *Citrus*. Grow in any well-drained fertile soil in sun. Prune if necessary to remove any dead or damaged wood in spring; clip hedges in early summer. Propagate by seed, removed from flesh when ripe and sown in the cold frame in autumn. Increase also by semi-ripe cuttings in summer.

pond see *water gardening*.

Pontederia (for Giulio Pontedera (1688–1757), professor of botany at Padua). Pontederiaceae. N and S America, mostly eastern. PICKEREL WEED, WAMPEE. 5 species, perennial, aquatic or semi-aquatic, marginal herbs with emergent, lanceolate to sagittate or hastate leaves. Small and zygomorphic, with a 2-lipped tubular perianth, the flowers are borne in dense, scapose spikes in spring and summer. A hardy, marginal aquatic. Grow in full sun. Plant in the pond or lake bed, at its margins or in large, submerged baskets containing a fertile, loam-based medium. Remove dead foliage after the last frosts. Propagate by division in spring. *P.cordata* (PICKEREL WEED, WAMPEE; eastern N America to Caribbean; aquatic perennial, 40–100cm; leaves to 18cm, emergent, erect, ovate to lanceolate with a cordate to sagittate base and long stalks; flowers to 2cm across, blue, lilac or white, crowded in cylindric spikes to 18cm).

Populus (classical Latin vernacular name for the genus). Salicaceae. Europe, Asia, N Africa, N America. POPLAR, COTTONWOOD, ASPEN. Some 35 species, deciduous, dioecious trees with ovate to lanceolate or triangular leaves. The flowers are borne in pendulous catkins before leaves, the male catkins denser than the females. Tufted with white hairs, the minute seeds are released in billowing, cotton-down-like masses in late spring or early summer.

A large genus of northern Hemisphere natives, from a wide variety of habitats, *Populus* shares many characteristics with the closely related *Salix*, and requires careful selection and siting. Most species produce debris in the form of seed 'cotton', twigs and leaves, and most have greedy and extensive root systems. They are best suited to forestry and to the broader landscape, where they are undeniably useful as windbreaks and visual screens. One commonly used fastigiate poplar for this purpose, *P.nigra* 'Italica', is probably the least suitable, having fragile branches, and being particularly prone to basal rots that may cause sudden collapse; *P. nigra* 'Plantierensis' has stronger and lower branches, and a denser head. While selected cultivars may be suitable for large gardens, they should be sited at least 40 m from buildings, drains, walls and roads, especially on clay soils, since they are known to cause extensive damage to foundations and drainage systems.

Many *Populus* species are useful on difficult sites. *P.alba* and its cultivars and *P. × canescens* are tolerant of salt-laden winds (but not of saline conditions at the roots). *P.tremula* tolerates extremely alkaline soils. *P. × canadensis* 'Regenerata' is tolerant of urban pollution and may be pollarded, where it has outgrown its allotted space. *P. tremula* is sometimes used to improve heavy soils in neglected woodland. The young foliage of some, notably *P.balsamifera*, *P. × jackii* 'Gileadensis' (balm of Gilead), and *P.trichocarpa*, emits a pleasant, light balsam scent, especially after rain. Others colour well in autumn, among them *P. × canadensis*, *P.maximowiczii* and *P.tremuloides*. Grow on deep, fertile and well-drained, near neutral soil. Mound planting is advised on wet sites. Propagate by hardwood cuttings, suckers, layers or root cuttings. Sow newly ripened seed on moist silt. All species hybridize readily. The most serious disease is bacterial canker.

P.alba (WHITE POPLAR, SILVER-LEAVED POPLAR, ABELE; S, C and E Europe, N Africa to C Asia; to 30m, crown broad, bark smooth, grey suckering; leaves 6–12cm, ovate, base subcordate, those on long shoots lobed, coarsely toothed, on short shoots serrate ovate to elliptic-oblong, dark green and glabrous above, white-woolly beneath; 'Globosa', a tall shrub with a broadly rounded habit and pink young leaves; 'Intertexta', with dull white young leaves becoming yellow-speckled; 'Nivea', a juvenile form, with chalk white young twigs, leaf undersides and petioles, and deeply lobed leaves; 'Paletzkyana', with leaves deeply lobed; 'Pendula', with pendent branches; 'Pyramidalis', a tall, narrowly conical tree, with large, lobed leaves, often glabrous beneath; 'Raket', slender, columnar, with a grey stem and leaves glossy green above, silvery-grey beneath; 'Richardii', with leaves golden-yellow above; var. *subintegerrima*, with leaves almost entire); *P.balsamifera* (BALSAM POPLAR; N US, Canada, Russia; to 30m; suckering from base; young twigs glabrous; buds and new growth covered in a fragrant resin; leaves 7–12cm, oval to ovate-lanceolate, base rounded, cordate or broadly cuneate, apex acute, glossy mid-green, white or pale green beneath, subcoriaceous, more or less crenate, glabrous; 'Aurora', with yellow white leaves, spotted dark green); *P. × berolinensis* (*P.laurifolia × P.nigra* 'Italica'; BERLIN POPLAR; to 25m, slenderly columnar; leaves 7–12cm, broadly ovate or rhombic-ovate, long-acuminate, base rounded or cuneate, crenate to serrate, bright glossy green above, paler beneath, more or less glabrous); *P. × canadensis* (*P.deltoides × P.nigra*; CANADIAN POPLAR; tall, fast-growing tree to 30m, crown broad; leaves 7–10cm, triangular to ovate, long-acuminate, base truncate, sparsely toothed, and sometimes with 1 or 2 glands, margins crenate, initially ciliate, petioles tinged red; 'Aurea', with golden yellow leaves, becoming yellow-green, and red petioles; 'Eugenei', to 50m, columnar, with pale bark, 5–8cm, triangular to rhombic, short-acuminate leaves, coppery-brown when young; 'Marilandica', MAY POPLAR, with a short trunk, broad, rounded crown, spreading, often crooked branches, growing downwards, and leaves brown when young, soon pale green, to 10cm, rhombic to triangular-ovate, apex slenderly acute, entire; 'Robusta', columnar, with ascending, almost whorled branches, young twigs green, becoming red, finely hairy, and leaves red-brown when young, 10–12cm, glossy, tough, triangular, with to 2 basal glands, rounded teeth, and petioles becoming red; 'Serotina', LATE POPLAR, BLACK ITALIAN POPLAR, to 40m, with a broadly conical, open crown, brown, flexible, glabrous young twigs, with 7–10cm, ovate to triangular leaves, not borne until spring, with a truncate base, red-brown when young, matt dark green when mature, with petioles tinged red); *P.* × *canescens* (*P.alba* × *P.tremula*; GREY POPLAR; natural hybrid; Russia (Georgia), Iran to C Europe; to 45m, crown rounded, bark yellow-grey, scarred, young twigs grey-woolly, soon glabrous; leaves 6–12cm, triangular-ovate, base cordate, dark green above, grey-woolly beneath, ciliate, teeth glandular, rounded, petioles 1–7.5cm, woolly to glabrous; 'Aureovariegata', slow-growing, with leaves marbled yellow, and 'Pyramidalis', with a broadly conical crown, grey-green bark, upper branches steeply ascending, and lower branches spreading); *P.deltoides* (EASTERN COTTONWOOD, NECKLACE POPLAR; E and C US; to 30m, crown broad, bark pale green-yellow, young twigs ribbed, glabrous, buds sticky, balsam-scented; leaves 7–18cm, deltoid-ovate to rhombic, base cordate to truncate, often with 2–5 glands, apex acute, densely ciliate, teeth coarse, glandular, petioles tinged red; subsp. *wislizenii*, with 5–10cm, triangular to broadly ovate, short-acuminate leaves, with base slightly cuneate or rounded, lacking glands, coarsely crenate, and more or less glabrous); *P.lasiocarpa* (CHINESE NECKLACE POPLAR; SW China; to 25m, crown rounded; young twigs angular, densely woolly at first; leaves 15–35cm, ovate, acute, base cordate, glandular-crenate, glossy grey-green above, paler and downy beneath, veins red, petioles 5–10cm, red); *P.maximowiczii* (JAPANESE POPLAR; north central China, Japan, Korea; to 40m, crown broad, bark grey, deeply fissured; young twigs red, hairy; leaves 6–12cm, elliptic to oval-elliptic, apex abruptly acuminate to a twisted cusp, base slightly cordate to rounded, matt dark green and wrinkled above, pale green beneath, glandular-serrate and ciliate, veins hairy, somewhat leathery); *P.nigra* (BLACK POPLAR; W Europe, N Africa, Siberia; to 30m; crown broad, rounded, trunk often thickly knotted, bark deeply fissured; leaves 5–10cm, rhombic, triangular or ovate, sometimes wider than long, apex slenderly acuminate, base cuneate or truncate, green, paler beneath, finely crenate; 'Afghanica', columnar, with bark almost white with age, grey young twigs, initially hairy, and triangular-ovate leaves, with a broadly cuneate base; var. *betulifolia*, DOWNY BLACK POPLAR, MANCHESTER POPLAR, with brown-orange young twigs, initially hairy, smaller leaves, tapering gently to acute apex, hairy when young, and yellow-green petioles; 'Charkowensis', with oblong, broad, pyramidal crown, ascending branches, and sparsely pubescent young branches; 'Italica', LOMBARDY POPLAR, ITALIAN POPLAR, PYRAMIDAL POPLAR, columnar, with branches steeply ascending, young twigs brown at apex, later grey, small, rounded-rhombic leaves and petioles tinged red; 'Gigantea', as 'Italica' but broader, with orange winter twigs; 'Lombardy Gold', with golden leaves; 'Plantierensis', of columnar habit, but less so than 'Italica', with hairy young twigs, small leaves, and hairy petioles tinged red); *P.szechuanica* (W China; to 40m in wild; young twigs red-brown, angular; leaves 7–20cm, tinged red at first, oblong to broadly ovate, crenate, teeth with apical glands, apex acuminate, base rounded, or slightly cordate, bright green above, white-silver beneath, veins red); *P.tremula* (ASPEN, QUAKING ASPEN; NW Europe to N Africa, Siberia; to 20m, crown broad, much-branched, bark fissured and dark grey with age, suckering; leaves 3–12cm, oval to suborbicular, acute, base with 2 glands, truncate or cordate, undulate, crenate, grey-green above, very pale beneath, petioles slender; includes the narrow, upright 'Erecta', 'Pendula', a small tree, with pendulous twigs, and 'Purpurea', with leaves tinged red); *P.tremuloides* (AMERICAN ASPEN; Canada to Mexico; to 20m, trunk slender, bark very pale yellow-grey, suckering, young twigs red-brown; leaves 3–7cm, broadly oval to suborbicular, short-acuminate, base truncate to broadly cuneate, finely serrate, ciliate, dark glossy above, blue-green beneath); *P.trichocarpa* (BLACK COTTONWOOD, WESTERN BALSAM POPLAR; western N America; to 35m, slenderly pyramidal, crown open, bark smooth, yellow-grey, young twigs olive-brown, glabrous to hairy, slightly angular; leaves 8–25cm, ovate to rhombic-oblong, slenderly acuminate, base rounded to truncate, leathery, shallowly toothed, dark green and glabrous above, white or pale brown beneath; 'Fritz Pauley', a female clone, 'Pendula', with weeping branches, and 'Scott Pauley', a male clone).

Porana (native name in E Indies). Convolvulaceae. Tropical Asia, Australia. Some 20 species, slender, twining herbs or shrubs with cordate-ovate leaves and small, campanulate or funnelform flowers in terminal panicles or cymes in summer. Cultivate as for tender, perennial *Ipomoea* species. *P.paniculata* (BRIDAL-BOUQUET, CHRIST VINE, SNOW CREEPER, SNOW-IN-THE-JUNGLE, WHITE CORALLITA; N India, Upper Burma; liane to 9m; leaves cordate, to 15cm, white-pubescent beneath; panicles large, pendulous, flowers to 8mm, white).

poricidal a form of dehiscence in which the pollen is released through pores at the apex of the anther.

porous tube rubber or polythene tubing with micropores, buried in soil to provide subirrigation; see *irrigation*.

porrect extending or stretching outward and forward.

Portulaca (from *portilaca*, Latin name used by Pliny). Portulacaceae. Widely distributed in warm and tropical regions. PURSLANE, MOSS ROSE. 40

species, fleshy or trailing, mostly annual herbs. Opening in bright sunlight in spring and summer, the flowers consist of four to six, spreading, brightly coloured petals and many stamens. *P.grandiflora* is grown for its fleshy, moss-like foliage and for the profusion of individually short-lived, brightly coloured flowers carried over long periods in summer. Ideal as a low-growing seasonal filler in flowerbeds and borders, as edging and in window boxes and other containers, it is easily grown from seed sown *in situ* in spring, in any low-nutrient, freely draining, sandy soil in full sun. *P.grandiflora* (ROSE MOSS, SUN PLANT, ELEVEN-O'CLOCK; Brazil, Argentina, Uruguay; stem prostrate or ascending, to 30cm; leaves cylindrical, to 2.5cm, thick, fleshy, tinted red when grown in full sun; flowers single to double, rose, red, crimson, scarlet, orange, apricot, gold, yellow, white, sometimes striped or bicoloured, 2.5cm in diameter, opening in sunlight; numerous seed races and cultivars are available).

post-emergence used to describe herbicides that are applied after the emergence of a crop from the soil.

posterior at or towards the back or adaxial surface.

post-harvest, post-storage diseases various parasitic diseases and non-parasitic disorders that are liable to affect fresh fruit, vegetables and flower bulbs from the time of harvesting until they are used by the consumer. Diseases such as FRUIT BROWN ROT and POTATO BLIGHT are caused by pathogens which may affect plants during the growing season and also attack their fruit or tubers at maturity. Others, such as BLUE MOULD OF CITRUS and RHIZOPUS ROT of tomato, are diseases specific to harvested produce and their incidence is strongly influenced by damage during growth or handling and the conditions under which produce is stored.

Post-harvest disorders are caused by unfavourable environmental factors before and after harvest, of which extremes of temperature are probably the most important, resulting in chilling or overheating.

post-hole borer a large diameter auger-like tool, usually mechanically operated to facilitate the boring of holes for fence posts.

potager a kitchen garden of vegetables and fruits laid out for ornamental effect, often with the beds edged with low hedges, as in a parterre.

Potamogeton (from Greek *potamos*, river, *and geiton*, neighbour, referring to the natural habitat). Potamogetonaceae. Cosmopolitan. About 90 species, aquatic perennial herbs. The shoots arise from creeping rhizomes rooted in the bottom of the pond, stream or river. The leaves are submerged and/or floating, and variable in shape and size. Small, apetalous, green flowers are produced either above the water or submerged in fleshy spikes. Hardy aquatics suitable for large, cold water aquaria and outdoor pools. Plant the stems in sandy loam and plunge in neutral to alkaline water. Give full sun or strong light. Remove fading foliage and thin colonies when necessary. Propagate by stem cuttings in spring/summer or by scaly resting buds in spring where produced.

P.crispus (CURLED PONDWEED; Europe, naturalized E US and California; leaves submerged, to 10cm, narrow-oblong, crisped or undulate); *P.lucens* (SHINING PONDWEED; Europe, W Africa; leaves submerged, to 20cm, narrow-elliptic to obovate-elliptic, tips acuminate or cuspidate, translucent, undulate); *P.natans* (BROAD-LEAVED PONDWEED; Europe, N US; floating leaves to 12.5cm, broadly oval, with flexible joint at base, petioles often longer than blade; submerged leaves reduced, linear).

potassium (abbreviation K) a major (macro-) nutrient, important in plant metabolism. It is highly mobile in soils, and although effectively held by the cationic exchange capacity of most soils, it is leached on free-draining, acidic and sandy soils. On cropped sites, potassium is commonly in short supply and replenishment through fertilizer application is regularly required; as a base dressing, it should always be applied to balance applications of nitrogen.

Apples, gooseberries and redcurrants, potatoes, beans, spinach and tomatoes are notably affected by potassium deficiency, which is mostly seen well into the growing season. Symptoms of deficiency are restricted shoot growth, leading to die-back in fruit trees. In most crops, there is marginal scorch of older leaves, often with curling, and this may be preceded by spotting and chlorosis. Flowers and fruits drop excessively, and fruits are invariably small and of poor quality; in tomatoes, the fruit quality defects known as 'greenback' and 'blotchy ripening' are symptoms of potassium deficiency. Cure is by prevention only.

Potassium fertilizers include: potassium chloride (60% K_2O) which is the least expensive but potentially toxic where excessively applied, and some plants, especially tomatoes, gooseberries and redcurrants, are susceptible to chloride damage; potassium nitrate (44% K_2O, 13% N), which is expensive and mostly used in liquid feeds; and potassium sulphate (50% K_2O), which is more expensive than potassium chloride but widely used and available as sulphate of potash. Potassium is not stored in high quantity in organic matter.

potato (*Solanum tuberosum*) a short-lived herbaceous perennial cultivated as an annual for its swollen, underground stem tubers. The potato originated in the Andean highlands, being cultivated by the Incas for over 2000 years before the arrival of the Spanish in the 16th century. It was first recorded in Europe in 1587 and rapidly became a staple crop in Ireland and much of northern Europe. It was taken to North America by immigrants from Scotland and Ireland in the early 18th century.

Potatoes are widely cultivated in temperate and tropical regions throughout the world for human consumption and stock feed, and there is large

potato
Pests
cutworms
eelworm
slugs
wireworms

Diseases
blight
(see
  phytophthora)
scab
virus
wart disease

Disorders
hollow heart
internal rust spot

genetic variability within the species. It is a versatile food crop, amenable to processing. However, if developing tubers are exposed to light, they become green and poisonous, as are the green berries produced after flowering.

Tuber formation proceeds better under cooler conditions and potatoes are a diffciult crop to grow in hot humid lowlands. Temperatures around 22°C/72°F favour early growth; during later development, temperatures around 18°C/64°F are optimal. High soil temperature can retard stolon initiation and tuberization, which is generally best in the range 16–20°C/61–68°F. Potato foliage is very susceptible to damage by frost.

Propagation is through healthy 'seed' tubers, true seed generally producing unacceptable levels of variability in resulting tubers. The use of microtubers produced *in vitro* by tissue culture is a possibility.

In temperate regions, potatoes can be harvested throughout the summer months and tubers stored to extend the season; it is also possible to obtain earlier harvests of immature or 'new' potatoes, which have a superior flavour to the later crop.

Potatoes require a frost-free site, preferably with a slightly acid soil, and they should be grown on at least a three-year rotation to avoid the build-up of eelworm. Good quality 'seed' tubers, certified as virus free, should be chosen and growth can be started by the process of 'sprouting' or 'chitting', which is particularly useful for advancing development of early cultivars. During the latter part of winter, tubers are stood in shallow trays in a well-lit, cool but frost-free environment. The dormant buds or 'eyes' are concentrated at what is often referred to as the 'rose' end and these should be placed uppermost in the trays to encourage strong shoots about 2.5cm long prior to planting. Chitting will take about six weeks. For early potatoes of good size, the number of sprouts per tuber should be restricted to two or three by rubbing off the excess at planting; for later maincrops, all shoots should be retained for higher yield. Seed-tuber size can significantly affect final yield and tuber size, and ideally seed tubers should be approximately the size of a hen's egg.

Planting can commence in early spring, starting with early cultivars, the chitted tubers planted rose end uppermost either in drills 10–12cm deep or in individual planting holes, before covering with 2–3cm of soil. As a general guide, earlies should be planted closer together than later maincrops; typically, early potatoes are planted approximately 35cm apart in rows 35–45cm apart, whereas row spacing for later cultivars should be increased to 75cm.

Planting can be advanced by planting under cloches or by covering with perforated clear plastic film or non-woven polypropylene 'fleece'. An uncovered early crop can be protected from the occasional late frost by covering exposed shoots with straw or by drawing a thin layer of soil over them.

Potatoes should be earthed-up during growth to prevent greening of tubers; the process also serves to kill off weed seedlings. A single operation when plants are about 20cm high is usually sufficient, drawing soil to approximately 10cm depth around the stems.

Early cultivars in particular can be grown in large containers, of 30cm width and depth minimum. These can be placed on a patio or held in a cool greenhouse for early cropping. Chitted tubers are placed in the containers on 10–13cm depth of fertile garden soil or potting compost and covered to a similar depth, with further soil or compost added when the plants are about 15cm high. Crops can be raised in carefully managed, large, black polythene sacks with drainage holes.

Developing tubers should be kept well-supplied with water and, depending on growing conditions, will be ready to harvest 90–120 days from planting. Potatoes may be lifted as required; the entire crop is removed, including the smallest tubers, which may otherwise contribute to a weed and disease problem in following seasons. Potatoes for storage should be left in the ground for a couple of weeks after the haulm has been removed to allow the skins to harden; tubers should be allowed to dry thoroughly before storing in a cool dark environment between 4–10°C/39–50°F with high humidity.

Cultivars vary in time taken to reach maturity, disease resistance, flavour and skin colour. In the UK, they are commonly categorized as earlies (e.g. 'Epicure', 'Charlotte', 'Maris Bard'); second earlies (e.g. 'Estima', 'Nadine', 'Wilja'); and maincrop (e.g. 'King Edward', 'Maris Piper', 'Romana'). Some are often referred to as salad potatoes (e.g. 'Pink Fir Apple', 'Ratte'). Cultivars in the US are categorized as early (e.g. 'Red Norland', 'Irish Cobbler', 'Yukon Gold'); maincrop (e.g. 'All Blue', 'Chieftain', 'Russet Burbank'); and late season (e.g. 'Bintje', 'McNeilly Everbearing', 'Yellow Finn').

potato planter a tool for planting potatoes and other tubers or bulbs, which has two triangular scoop-shaped blades that hold the tuber or bulb in a V-shaped cradle whilst the tool is inserted into the soil. The tuber or bulb is released when the handles are pulled together and the blades open.

pot-bound used of a pot-grown plant whose roots have completely explored the growing medium and are densely packed into the container, often becoming woody and constricted. The condition should be avoided by timely potting-on or planting out, because affected plants are likely to be starved and their potential for producing a vigorous, strong root system after planting out is much reduced.

Potentilla (from Latin *potens*, powerful, referring to the plant's alleged medicinal properties). Rosaceae. Northern Hemisphere. CINQUEFOIL, FIVE-FINGER. About 500 species, mostly perennial but some annual and biennial herbs and shrubs, with palmate, pinnate or trifoliate leaves. Produced in spring and summer, the flowers are usually saucer-shaped, solitary or in terminal or axillary clusters, and 4-, 5-, or 6-parted.

Those listed here are hardy to –25°C/–13°F. The shrubby *P.fruticosa* and its numerous cultivars form loose and floriferous mounds, suitable for the front of the shrub border, for mass planting as groundcover, and also as low informal hedging. The summer and autumn blooms of the border perennials, largely derived from *P.nepalensis*, *P.atrosanguinea*, and *P.recta*, are held on slender stems well above the foliage; these are selected for their warm and brilliant shades of yellow, terracotta, vermilion, mahogany and deep crimson, and are well-suited to strong colour schemes and to cottage gardens. The alpine species are suitable for crevices and scree in the rock garden, or for growing on walls; some species, such as *P.alba*, form low mats of grey or silver foliage, carrying their flowers on short stems. *P.nitida* is small and neat and well-suited to sink or trough plantings.

Grow in well-drained soils in full sun to part shade; too rich a soil will produce soft foliar growth at the expense of flowers. Grow small alpines in poor gritty soil or on scree in full sun, or in the alpine house, using a low fertility loam-based mix, with added grit. Prune shrubby species in early spring, removing weak growth at ground level and cutting back strong growth by half or one third. Old neglected plants may be rejuvenated by cutting hard back, although this may be more safely done in two or more stages. Trim hedges in spring. Propagate herbaceous species by seed sown in autumn or early spring, or by division in spring or autumn. Remove rooted runners to increase alpine species. Propagate shrubby species by softwood cuttings in summer and, except for cultivars, by seed in autumn.

P.alba (C, S and E Europe; low-growing, spreading perennial herb to 10cm; leaves palmate, leaflets 5, 2–6cm, oblong to obovate-lanceolate, apex toothed, glabrous above, silvery silky beneath at first; flowers to 2.5cm in diameter, white); *P.atrosanguinea* (Himalaya; hairy perennial to 90cm; leaflets 3, to 7.5cm, elliptic-ovate to obovate, toothed, silky above, white-hairy beneath; flowers to 3cm in diameter, deep purple-red; includes var. *argyrophylla*, with yellow or yellow-orange flowers; hybrids derived from these and (in some cases) *P.nepalensis* include 'Etna', to 45cm, with leaves tinted silver and deep crimson flowers; 'Fire-dance', compact, with deep coral flowers; 'Gibson's Scarlet', with bright scarlet flowers; and 'Mons. Rouillard', with double, dark copper flowers); *P.aurea* (Alps, Pyrenees; mat-forming perennial to 30cm, silvery hairy; leaves digitate, leaflets 5, silver-hairy on margins and veins beneath, oblong, toothed at tip; flowers to 2cm in diameter, golden yellow, base deeper; 'Aurantiaca', with sunset yellow flowers, 'Florepleno', with double, light gold flowers, and 'Goldklumpen', with bright gold flowers with an orange ring); *P.crantzii* (ALPINE CINQUEFOIL; N US, N, C and S Europe; perennial herb to 20cm, rootstock woody; leaves digitate, leaflets 3 or 5, 2cm, obovate to cuneate, toothed at apex, green, hairless above, hairy beneath; flowers to 2.5cm in diameter, yellow, often orange spotted at base); *P.fruticosa* (SHRUBBY CINQUEFOIL, GOLDEN HARDHACK, WIDDY; northern Hemisphere; deciduous, much-branched shrub, to 150cm; leaves pinnate or trifoliate, leaflets 3 or 5, to 2.5cm, ovate to lanceolate, light to mid-green, silky; flowers to 4cm in diameter, bright yellow or white; var. *mandschurica*, low-growing, to 45cm, with grey silky hairy leaves and white flowers to 2.5cm in diameter; many cultivars and hybrids of *P.fruticosa* are available, ranging in habit from the dwarf and compact to the large and robust, from low and spreading to upright, in foliage from small- to large-leaved, bright green to lime, variegated or silver and silkily hairy, and in flower colour from white to cream, palest primrose, buttercup yellow, rich gold, tangerine, vermilion, fiery red, apricot, pink and bicolours); *P.megalantha* (Japan; softly hairy tufted perennial to 30cm; leaves to 8cm wide, leaflets 3, broad obovate, coarsely crenate, hairy beneath; flowers to 4cm in diameter, bright yellow); *P.nepalensis* (W Himalaya; perennial herb, stem leafy, erect, to 60cm; leaflets 5, 3–8cm, obovate or elliptic-obovate, coarsely toothed, hairy; flowers 2.5cm in diameter, purple-red or crimson, base darker; 'Flammenspiel', with red flowers narrowly edged yellow; 'Miss Willmott', with cherry red flowers; 'Roxana', with red buds and salmon pink flowers); *P.neumanniana* (N, W and C Europe; mat-forming, perennial to 10cm, stem woody, procumbent; leaves digitate, leaflets 5–7, to 4cm, obovate, dentate; flowers to 2.5cm in diameter, yellow; 'Goldrausch', to 10cm, with bright gold flowers; 'Nana', to 7.5cm, with vivid green leaves and gold flowers); *P.nitida* (SW and SE Alps; tufted, silver-grey, downy perennial to 10cm; leaflets 3, to 1cm, oblanceolate to obovate, apex usually 3-toothed, silvery-silky; flowers 2.5cm+ in diameter, white or pink, base deeper; 'Alannah', with pale pink flowers, 'Alba', with white flowers, 'Compacta', very dwarf, to 5cm, with large, gold flowers, 'Lissadel', with vivid pink flowers, and 'Rubra', with deep rose flowers); *P.recta* (Europe, Caucasus, Siberia, naturalized eastern N America; perennial; stem to 45cm, velvety hairy, with glandular hairs; leaves digitate, leaflets 5–7, to 3.5cm, oblong-lanceolate, green, serrate to pinnatisect; flowers yellow, to 2.5cm in diameter, many in flat corymbs; 'Macrantha', with large, bright yellow flowers in loose clusters); *P.* × *tonguei* (*P.anglica* × *P.nepalensis*; stem procumbent, not rooting; leaflets 3–5, obovate, dark green; flowers apricot with carmine-red eye).

pot herb used of herb-like plants that are edible rather than medicinal. See *herbs*.

× **Potinara** (*Brassavola* × *Cattleya* × *Laelia* × *Sophronitis*). Orchidaceae. Quadrigeneric hybrids made from species and hybrids of these genera and named after a French orchid grower, Monsieur Potin. The plants are small and compact with flowers in shades of yellow, orange and red.

pot-in-pot a method of displaying creeping or trailing plants, in which two or more pots of different sizes are filled with growing compost and partially embedded inside one another, with plants or cuttings grown in the smallest pot and around the edges of the larger one(s) to produce a cascade effect.

pot-pourri dried aromatic petals and leaves, often mixed with spices used to scent rooms or clothes.

pots and potting plants were grown in pots by the early Egyptians, Greeks and Chinese and traditionally made from baked clay. Old hand-made pots can be identified by the uneven finish and rather ridged surface inside the pots. Terracotta pots were produced from the late 1940s by mechanized manufacturing methods. Various plastics have replaced clay or terracotta, except in the case of decorative tubs and containers, and most plants grow equally well in either. *Pot types and uses* Conventionally shaped pots are usually classified by internal diameter at the rim, although commercial growers often describe containers by capacity (e.g. 1 or 2 litre), which relates to the volume of growing compost needed to fill them.

References are sometimes made to the practice of grading pots according to the number that could be made from a cast of clay. The largest size, made from one cast, was of 45cm in diameter and known as a 'number one'; 32 pots measuring 16cm in diameter could be produced from the same amount of clay and these were known as '32s', Very small sizes were often referred to as 'thumbs' (6cm) and 'thimbles' (5cm).

Special sizes and shapes are produced for particular uses. The pan (seed pan or mini-pot) is less than half the depth of a normal pot and used for seed-sowing or for small bulbs and shallow-rooting plants such as alpines. Half-pots are half the depth of a normal pot and often used for plants such as azaleas. Long toms are much deeper in proportion to width, and are used for plants with early developing tap roots, those that produce deep roots and resent restriction or disturbance, and those that demand perfect drainage, as can be provided with a deep layer of crocks. All pots must have adequate drainage holes. With terracotta sorts it is usual to place a layer of broken pots, known as crocks, over the drainage holes, to form a shallow layer at the bottom of the pot. Plastic pots are not normally crocked and are generally more suited to capillary watering.

Plastic pots are light, clean, and easy to handle but although the potting medium dries out less quickly than in clay pots, accurate assessment of water requirements is more difficult. Polypropylene is the plastic widely used, but for large container pots, a proportion of polyethylene (polythene) is used for added strength in cold weather. Polythene is used for inexpensive, collapsible, fold-flat pots.

Clay pots are attractive, and useful where their weight provides stability, for example for tall plants. Plants that need good drainage, such as alpines and many orchids, benefit from the porosity of clay pots.

Peat pots are available for plant raising, and are generally made of a mixture of sphagnum moss peat and wood fibre. They are useful for plants, such as *Salvia*, sweetpeas and runner beans. Roots eventually grow through the walls and the pots disintegrate when planted out. Expandable pellets of fertilized compressed peat held in a net are available for rooting cuttings and the direct sowing of seeds, which can be later planted into conventional pots intact. Care must be taken to keep these peat containers moist at all times. Durable containers made solely of compressed paper or wood fibre are available as alternatives to peat-based ones.

Bitumenized cardboard pots, known as 'whale-hides', are used for ring culture; these consist of bottomless tubes, approximately 23cm in diameter at the top and tapered towards the bottom to hold the potting medium. They are stood on an aggregate, especially for the ring culture of tomatoes, cucumbers, and peppers. Versions with bottoms are available for plants requiring a large pot, such as chrysanthemums. Both types are discarded after one season.

Potting The initial placing of seedling plants in pots is known as potting-off or potting-up, while moving potted plants to larger containers is known as potting-on or repotting. Young and rapidly growing plants need potting-on fairly frequently to avoid growth being checked, slow-growing plants less often. Plants that cannot be potted-on due to limits on the possible size and weight of container can often be maintained by replacing the top and bottom layers of the potting medium each year, and by regular feeding. Established plants should be potted-up or repotted when they are about to start growth or are growing actively; with young plants, the best time to move them is when the roots have fully explored the growing medium but have not yet started to spiral round the pot. Overpotting (using a pot that is too large) must be avoided because the unexplored potting medium may remain too wet. Normally, choose a pot one or two sizes larger than the existing pot.

A few plants succeed best if the roots are restricted, for example, sansevierias are best repotted only when the roots fill the pot and start to emerge from the compost; aspidistras are not repotted frequently, and the variegated form is usually better if slightly starved and pot-bound. Cacti can be repotted less frequently than most plants, and these, together with bromeliads, orchids and other epiphytic or partly epiphytic groups, require relatively small pots for their size. In general, large woody plants should be potted more firmly than those with soft growth, while soil-based potting media can be firmed much more than peat-based types. Newly potted plants will appreciate moister and shadier conditions than usual for a few days after moving.

pot-thick the arrangement of container-grown plants on a bed or greenhouse bench so that their pots are touching.

potting machine a device used in commercial nurseries to fill pots mechanically with potting medium and thus increase the speed of the potting operation.

potting media (singular: potting medium) used to describe mineral and organic substances suitable for growing containerised plants. In some European countries the term is synonymous with substrate. The word 'compost' is widely used in the UK as a general term for potting, seed and cutting media but in some other countries it refers only to the product of a compost heap. The term 'compost' traditionally describes all soil-based potting mixes, especially the John Innes composts. Lightweight media primarily based on peat are often referred to as 'mixes', particularly in the US.

Potting media are classified as either soil- or loam-based if they contain mineral soil, and soilless or loamless if they are based on lightweight ingredients, such as peat or composted bark. In the past, gardeners used their own special recipes, in which a wide range of materials were added to soil, including leafmould, animal manures, mortar, wood ashes, sand and grit. To overcome variable results, research in the 1930s by W.J.C. Lawrence and J. Newell at the John Innes Horticultural Research Institute, Hertford, England led to development of the standard John Innes composts; these are based on loam, sterilized to kill weed seeds and reduce the incidence of soil-borne pests and diseases, and generally produce reliable results.

Advances in producing containerized nursery stock in the late 1950s led to the development of soilless mixes, based on peat. These overcame the difficulties of obtaining bulk supplies of good-quality loam and reduced the labour costs of sterilizing and handling.

Soil-based composts, such as the John Innes range, confer the possible advantage of container wind-stability; they are also often considered more tolerant of over- and under-watering, and have better nutrient-storage capacity. However, batches are likely to be more variable than those of soilless mixes due to the difficulty of standardizing loam texture and organic content.

There are three basic formulations of John Innes composts (*see table*): seed compost for germinating seeds; cutting compost for rooting cuttings; and potting composts for growing on plants, designated JIP-1, JIP-2 and JIP-3. The potting composts contain increasing amounts of base fertilizers to accommodate varying degrees of plant vigour and seasonal changes in growth rate.

For the higher fertility JIP-2 and JIP-3, the quantities of fertilizer in JIP-1 are doubled and tripled, respectively. Hoof and horn, superphosphate and potassium sulphate can be obtained ready-mixed as 'John Innes Base' fertilizers.

Loams suitable for John Innes composts should be of clay texture, contain fibrous organic matter and have a pH value of between 5.5 and 6.5. It is essential that the loam is pasteurized at around 82°C /180°F, but this heat treatment, which is aimed at killing most harmful insects and diseases and weed seeds, must be carefully controlled to avoid risk of producing substances toxic to plants.

Sphagnum peats are the most suitable, since they retain water and provide aeration. Coconut fibre or coir appears to be a generally satisfactory substitute for peat as an ingredient of John Innes compost, where used at the same volume. The sand will improve drainage and should be a coarse grade and lime-free.

Ideally, John Innes composts should be used within four weeks of making since changes in the forms and amounts of available nitrogen and the pH can occur; successful storage will depend on moisture content, temperature and the amounts of organic nitrogen added. Mixing the fertilizers into the compost shortly before use will minimize storage problems.

The low-fertility seed compost is used for seed-sowing and pricking out and JIP-1, a medium fertility compost, for potting into 9cm pots. JIP-2 and JIP-3 are used for growing more vigorous plants and for potting on into larger pot sizes, and for large tubs and window boxes.

Lime-hating plants can be successfully grown in a compost termed JIS-A, which is the John Innes seed compost with ground limestone replaced by 0.6 kg/m3 of sulphur.

Soilless mixes are usually based on peat blended with materials such as sand, perlite and vermiculite. Some soilless media are mixes of different grades of peat. Peat is deficient in the principal plant nutrients and these are added as fertilizers; lime is also included to neutralize natural acidity. The use of peat is contentious due to actual and potential damage caused to the natural environment during harvest; of special concern is sphagnum peat taken from raised bogs.

The search for alternatives to peat has included developments with waste materials, such as straw, sewage sludge, municipal refuse, and animal wastes. However, the quality of different batches of media derived from these types of organic materials can be variable and they can deteriorate during storage. In addition, there are potential health risks if contaminated residues persist in the medium and it is advisable to handle any organic material with care; always wear gloves or wash hands when handling such mixes, and avoid dusty conditions by working in a well-ventilated area.

Composted bark and coir are well-tried peat alternatives, as are pine needles and leafmould. When using growing media based upon these materials, or ones in which they are used as additives, special attention needs to be given to avoid overwatering and early starvation of nutrients.

Coarse sands or grits can be added to peat to change the physical properties of the mix and to increase its weight in order to improve the stability of tall plants growing in pots; very fine sands can reduce aeration and drainage, and suitable types should also be lime-free. Perlite and vermiculite can be used as a lightweight substitute for sand in soilless mixes. Mixes containing 50% or more perlite are very useful for rooting cuttings, especially under mist. Unlike perlite, vermiculite contains some potassium and magnesium and can absorb mineral elements such as calcium, magnesium, potassium and ammonium nitrogen from added fertilizers, which will reduce the loss of these nutrients through leaching. Particles of vermiculite are less stable than perlite, but normally last long enough for plants that are regularly repotted. Also used in peat mixes are polystyrene granules, expanded aggregates made from clays, shales and pulverized fuel ash, and flocks of rockwool.

Lime is added variously to peat mixes as a mixture of ground chalk or limestone and ground magnesium limestone, both to reduce acidity and as a source of calcium and magnesium. Nitrogen, phosphorus and potassium are applied either as soluble inorganic fertilizers, such as potassium nitrate and superphosphate, which usually require frequent replenishment, or in a slow-release form. Trace elements are usually added in slow-release form as powdered glass frits or as soluble chemicals.

Composition of the John Innes range:

	parts/ volume	fertilizer	kg/m^3
SEED			
loam	2		
peat	1	grnd. limestone	0.6
sand	1	superphosphate	1.2
CUTTING			
loam	1		
peat	2	nil	
sand	1		
POTTING (JIP-1)			
loam	7	grnd. limestone	0.6
peat	3	hoof and horn	1.2
sand	2	superphosphate	1.2
		potass. sulphate	0.6

potting off, potting on, potting up see *pots and potting*.

potting stick a short, thick rounded stick used to firm potting media down the sides of large pots when potting-on or repotting specimens with a large root-ball.

pot worms see *enchytraeid worms*.

powdery mildew a type of pathogenic fungus characteristically showing as white mycelium on the surface of green tissues of a host plant, with powdery spores. The fungus is mostly external on its host and feeds by penetrating the epidermal cells. The mildews are obligate parasites, and may overwinter in the buds of deciduous and perennial plants or in resting structures on fallen leaves. Symptoms of powdery mildew infection may include chlorosis, stunting and distortion, all of which appear before obvious signs of the pathogen. As infection progresses, there can be loss of vigour and reduction in quality of foliage, flowers and fruit.

Unlike most other fungal pathogens of plants, the powdery mildews are capable of developing on their hosts in the absence of free water; their geographical distribution reflects their adaptation to drier conditions.

Species of powdery mildews are often restricted to certain host families, and within species there may be pathotypes affecting different cultivars.

Plants commonly affected by powdery mildews include apple (parasitized by *Podosphaera leucotricha*), crucifers (*Erysiphe cruciferarum*), cucurbits (*E.cichoracearum* and *Sphaerotheca fuliginea*), *Euonymus japonicus* (*Oidium euonymi-japonici*), gooseberry (*Microsphaera grossulariae* and *Sphaerotheca mors-uvae*), grapevine (*Uncinula necator*), hazelnut (*Phyllactinea guttata*), *Lagerstroemia* (*Erysiphe lagerstroemiae*), oak (*Microsphaera alphitoides*), pea (*Erysiphe pisi*), peach and rose (*Sphaerotheca pannosa*) and strawberry (*S.macularis*), as well as very many ornamental garden plants.

Despite their importance as pathogens of flowering plants, only one species of powdery mildew (*Erysiphe graminis*) is known to occur on monocotyledons, but this has many specialized pathotypes.

Control measures involve the removal of infected plant debris, pruning affected shoots and growing resistant varieties. Dusting or spraying with sulphur is a traditional remedy but can cause injury to some ornamental plants and certain cultivars of apples, gooseberries and vines. Some systemic fungicides are available to gardeners for use as preventative sprays on ornamentals.

praemorse raggedly or irregularly truncated, as if gnawed or bitten.

Pratia (for M.C. Prat-Bernon of the French navy who accompanied Freycinet on his second round-the-world voyage, but died in 1817, a few days after the expedition set sail). Campanulaceae. Australia, New Zealand, tropical Africa, S America. About 20 species, dwarf, perennial herbs with slender, creeping stems, very small leaves and, in summer, delicate, 2-lipped flowers followed by berries. Grow in a sheltered position (minimum temperature –13°C/9°F), in sun or light, part-day shade, in gritty, moist soil. In the alpine house, grow in pans in a mix of equal parts loam, leafmould and sharp sand. Water plentifully when in growth and keep just moist in winter. Propagate by seed in autumn in a sandy propagating mix, or by division.

P.angulata (New Zealand; leaves suborbicular, succulent, serrate; flowers to 1cm, subsessile, white with purple veins; 'Ohau': large, white flowers and bright red fruit); *P.pedunculata* (Australia; leaves ovate or orbicular, with few teeth; flowers to 7mm, blue, on pedicels longer than leaves; 'County Park': bright blue flowers); *P.perpusilla* (New Zealand; leaves minute, oblong or ovate-oblong, acute or obtuse, fleshy, dentate, glabrous or short-pubescent; flowers to 6mm, white, more or less sessile; 'Fragrant Carpet': white, fragrant flowers; 'Summer Meadows': bronze-tinted leaves, fragrant, white flowers).

precocious appearing or developing early, usually applied to flowers that open before the leaves emerge.

pre-emergence used to describe herbicides that are applied after sowing but before the emergence of a crop from the soil or, in the case of perennial crops, before shoot emergence. The herbicides may be of contact or selective residual action.

prepared used of bulbs that have been treated, usually with heat, whilst dormant, so as to advance growth and/or flowering, for example hyacinths, shallots.

pre-sowing a method of preparing seeds for sowing, whereby they are fixed to a sheet fabric, such as paper, which rots after being covered with soil, leaving the germinating seeds to develop; it is designed to ensure proper spacing and is used for grass, vegetables and annual flowers.

pricking lightly forking the soil among growing plants to stir and break up the surface for aeration and improved water penetration. It is also done to aerate lawns.

pricking out the operation of transferring very young seedlings from the site where they have germinated to other containers or beds, in order to provide more space for growth. Great care should be taken to handle only the seed leaves and to ensure that the roots are mostly intact and gently settled in with a dibber and by subsequent watering.

prilled granules small solid globules of fertilizer or herbicide. These are designed to aid even distribution, and may allow for slow release of the effective ingredient as the outer casing decomposes.

primary infection the first infection by a pathogen after a period of dormancy.

priming the preparation of seeds prior to sowing by promoting the partial imbibition of water to aid germination and establishment. Priming has been successful with crops such as carrot, parsnip and celery. Polyethylene glycol 600 is an example of a substance used to activate processes *leading* to germination, and after which seeds can be stored dry. Primed seeds germinate more quickly and evenly after sowing.

primocane the stem of a biennial cane fruit, such as raspberry, in its first, non-fruiting year. cf. *floricane*.

Primula (the diminutive form of the Latin *primus*, first; from the mediaeval Latin name for the cowslip, *Primula veris*, meaning 'first little thing of spring'). Primulaceae. PRIMROSE. Northern Hemisphere, Ethiopia, Tropical mts to Java and New Guinea, southern S America. 400 species, small or dwarf perennial herbs. The leaves are usually borne in basal clumps or rosettes. Produced at various times of year, the flowers may be solitary or clustered and short-stalked, or carried in scapose heads, spikes and umbels. The corolla is cup-like to tubular, with a limb of 5, spreading or incurved lobes.

Almost without exception, primulas like relatively cool, humid summers with a ready supply of soil moisture when in growth. Most need a well-aerated soil that allows excess moisture to drain freely away during the winter months to avoid risk of root rots. A fertile loam with additional organic matter and coarse sand is a good starting point for most primulas and can be adapted to suit individual requirements. The majority of European and American species enjoy or at least tolerate sun, whilst the Asiatics generally need light shade unless the soil remains constantly cool and moist. Propagation is by seed, division, cuttings, root cuttings, leafbud cuttings and, increasingly, by micropropagation. With the possible exceptions of *P.vialii* and *P.denticulata*, seed germination in primulas is inhibited at temperatures above 20°C/70°F, and seed of many species needs stratification. Primulas are relatively free from pests and diseases. The most damaging pests are vine weevil, red spider mite and aphid, particularly root aphid, and slugs. The main diseases, generally avoided by good hygiene and cultural conditions, are botrytis, root rots and viruses.

In terms of individual cultural requirements, primulas can be divided into nine groups -

1. full sun; permanently moist soil
2. full sun or part-shade; moist but fast-draining soil
3. part-shade; moist but well-drained soil
4. part-shade; well-drained, gritty loam
5. full sun or part-shade; gritty, alkaline soil
6. part-shade; gritty, alkaline soil
7. full or part shade; moist and humus-rich, neutral to acid soil
8. sun or part-shade; humus-rich, neutral to acid soil with grit
9. part-shade; humus-rich neutral to acid soil with grit

In addition to these groups, there are the Auricula Primulas (see *auriculas*), the Polyanthus Primulas or Pruhonicensis Hybrids derived from *P.vulgaris*, *P.veris*, *P.juliae* and related species, and those species such as *P.malacoides*, *P.obconica* and *P.sinensis*, (which, for all that they conform to cultivation group 4, are commonly grown under glass and in the home). Polyanthus Primulas are available in a number of named series with an extensive colour range. They are raised from seed and treated as biennials, used for bedding. They include the exceptionally large flowered Regal series; the Pacific series, both dwarf and giant, with fragrant, brilliant, velvety flowers, and the low growing Posy series, predominantly pastel with contrasting centres. Surface sow seed in late spring, keep constantly moist and germinate at temperatures no higher than 20°C/68°F; prick out and grow on to plant out in late autumn. *P.malacoides*, *P.obconica* and *P.sinensis* are decorative pot plants for winter and spring flowering. Sow seed between early summer and early autumn for a succession of bloom; use a soilless propagating mix and germinate at 15°C/60°F. Grow on in the shaded frame in a medium-fertility loam-based mix with additional leafmould and grit. Bring under glass in late autumn, or before first frost and maintain good light and ventilation with a minimum temperature range of 5–7°C/40–45°F. *P.* × *kewensis* is less commonly grown but requires similar treatment.

The group numbers given in each description refer to the cultivation requirements outlined above.

P. allionii (France and Italian Maritime Alps; leaves cushion-forming, oblanceolate to rounded, entire or finely toothed, fleshy, viscid, hairy; scapes very short ; flowers 1-5, tubular, 1.5–3cm diameter, pale pink to red-purple with white eye, or wholly white; group 5); *P. alpicola* (Tibet; leaves elliptic to spathulate, finely toothed; scape15–50cm, flowers to 2.5cm diameter, bell-shaped, pendent, white, yellow, purple or violet, eye farinose, fragrant; var. *luna*: flowers soft yellow; group 7); *P. aurantiaca* (SW China; leaves oblanceolate to obovate, finely toothed; scape to 30cm, tinged red; flowers whorled, 1.5cm diameter, tubular, red-orange; group 1 or 7); *P. aureata* (Nepal; leaves broadly spathulate to oblong, toothed, farinose; scapes very short; flowers to 10, to 4cm diameter, flat, cream to yellow, eye dark yellow; subspecies *fimbriata*: flowers pale cream-yellow, with orange eye, lobes deeply toothed; group 8); *P. auricula* (Alps, Carpathians, Apennines; leaves rounded to obovate, sometimes farinose, entire to dentate, fleshy; scapes 1–16cm; flowers 2–30, 1.5–2.5cm diameter, flat deep yellow, fragrant; group 5 or 6, see also *Auriculas*); *P. bulleyana* (SW China; leaves ovate to lanceolate, toothed; scapes to 60cm; flowers many per whorl, to 2cm diameter, tubular, deep orange; group 1 or 7); *P. chungensis* (China, Bhutan, Assam; leaves elliptic to oblong-obovate, toothed and shallowly lobed; scapes to 60cm; flowers whorled, to 1.8cm diameter, tubular, pale orange, tube red, fragrant; group 1 or 7); *P. clarkei* (Kashmir; leaves rounded to broadly ovate, toothed; scape usually absent; flowers to 2cm diameter, flat, rose-pink, eye yellow; group 8); *P. clusiana* (W Austrian Alps; leaves oblong to obovate, leathery, entire, ciliate, shiny above, grey-green beneath; scapes1–5cm, glandular-hairy; flowers1–4, 1.5–4cm diameter, tubular, bright rose, fading to lilac, throat glandular, eye white; group 8); *P. denticulata* (DRUMSTICK PRIMULA; Afghanistan to Tibet and Burma; leaves obovate to spathulate, finely toothed; scapes 10–30cm; flowers in a tight, rounded head, each to 1.7cm diameter, flat, pale purple to red-purple, sometimes white, eye yellow; group 1 or 3); *P. edgeworthii* (Himalaya; leaves spathulate to ovate, irregularly toothed or lobed, wavy; scapes absent; flowers solitary, to 3cm diameter, blue, lilac, pink or white, eye orange-yellow edged white; group 8); *P. elatior* (OXLIP; Europe to Near East; leaves ovate to oblong or elliptic, crenate to jaggedly toothed, downy beneath; scapes10–30cm; flowers many, 2–2.5cm, yellow, throat green-yellow to orange, tubular, fragrant; group 1,7 or 9); *P. farinosa* (Scotland and Central

Sweden to Central Spain to Bulgaria, North Asia and North Pacific; leaves oblanceolate to elliptic, entire or finely toothed, farinose; scape 3–20cm; flowers 2 to many, 0.8–1.6cm diameter, lilac-pink, rarely purple or white, throat yellow, tubular; group 8); *P. flaccida* (China; leaves elliptic or obovate, finely toothed, downy; scapes 20–40cm; flowers 5-15, to 2.5cm diameter, lavender to violet, tubular, farinose, pendent; group 7 or 8); *P. florindae* (Tibet; leaves broadly ovate, toothed; scapes to 90cm; flowers many, to 2cm diameter, yellow, creamy-farinose within and on calyx, fragrant, funnel-shaped; group 1 or 7); *P. forrestii* (China; leaves ovate to elliptic, toothed, rugose above, farinose beneath; scapes15–22cm; flowers10–25, to 2cm diameter, flat, yellow, eye orange; group 8); *P. frondosa* (Balkans; leaves spathulate or obovate, finely toothed, farinose; scapes 5–12cm; flowers 1–30, flat, rose-lilac to red-purple, to 1.5cm diameter, eye and tube yellow; group 8); *P. gracilipes* (Nepal, Tibet; leaves oblong-spathulate to elliptic, irregularly toothed; scapes absent; flowers solitary, 1-4cm diameter, bright pink-purple, eye small, green-to orange-yellow with narrow white border, tubular, lobes toothed; group 8); *P. hirsuta* (syn. *P. rubra*; Central Alps, Pyrenees; leaves obovate to rounded, toothed, fleshy, glandular-hairy, sticky; scape1–7cm; flowers1–15, to 2.5cm diameter, flat, pale lilac to deep purple-red, usually with white centre; group 8); *P. ioessa* (Tibet; leaves narrowly oblong to spathulate, toothed; scapes10–30cm; flowers 2–8, to 2.5cm diameter, funnel-shaped, pink-mauve to violet or white, fragrant, calyx almost black, farinose; group 7); *P. japonica* (CANDELABRA PRIMULA; Japan; leaves obovate to broadly spathulate, finely toothed; scapes to 45cm; flowers in tiered whorls, to 2cm diameter, tubular, purple-red, crimson, pink to white; Miller's Crimson': crimson; 'Postford White': white with a yellow eye; 'Valley Red': bright red; other candelabra-type primulas include *P. aurantiaca*, *P. bulleyana*, *P. chungensis*, *P. prolifera*, *P. pulverulenta*; group 1 or 7); *P.x kewensis.(P. floribunda* × *P. verticillata*; garden origin; leaves obovate to spathulate, finely toothed, thinly farinose; scapes to 30cm; flowers whorled, to 2cm diameter, tubular, yellow, fragrant; group 4); *P. malacoides* (FAIRY PRIMROSE, BABY PRIMROSE; China; leaves broadly oblong to ovate, shallowly lobed and toothed; scapes 20–30cm; flowers whorled, to 1.5cm diameter, flat, mauve to pink, red, or white; group 4); *P. marginata* (Alps; leaves obovate to oblong, toothed, leathery, fleshy, farinose on margins; scape 2–12cm; flowers 2–20, 1.5–3cm diameter, lilac to lavender or blue, sometimes violet or pink, shallow funnel-shaped, faintly fragrant, eye white-farinose; 'Prichard's Variety': lilac to purple, eye white-farinose; group 5); *P. melanops* (China; leaves lanceolate, toothed, farinose beneath; scape 20–35cm; flowers 6–12, to 2cm diameter, purple, eye black, to 2cm diameter, fragrant, narrowly funnel-shaped, pendent; group 1 or 7); *P. modesta* (Japan; leaves elliptic to spathulate, wavy-toothed, yellow-farinose; scapes 2.5–12.5cm; flowers 2–15, to 1.5cm diameter, pink-purple, tubu-

Primula Auricula

Primula bhutanica

Primula vialii

lar; var. *fauriae*: smaller, white-farinose flowers pink-purple with a yellow eye; group 8); *P. nutans* (Mountains of North Asia, North Russia, North Scandinavia, Alaska, South West Yukon; leaves oblong to ovate to orbicular, entire or finely toothed, fleshy; scapes 2–30cm; flowers 1–10,1–2cm diameter, lilac to pink-purple, eye yellow, pendent; group 7 or 8); *P. obconica* (GERMAN PRIMROSE, POISON PRIMROSE; China; leaves ovate to elliptic to oblong, finely toothed or lobed, hairy; scapes 15–17cm; flowers10–15, to 2.5cm diameter, flat, pale-lilac or purple, eye yellow; group 4); *P. palinuri* (S Italy; leaves broadly spathulate to oblong-ovate toothed, fleshy, viscid, aromatic, glandular-hairy; scapes 8–20cm; flowers 5–25, to 3cm diameter, narrowly funnel-shaped, deep yellow, throat with white, farinose ring; group 5); *P. petiolaris* (Himalaya; leaves spathulate, finely toothed; scapes 3–5cm, flowers many, 1.5cm diameter, tubular, pink, eye yellow with thin white border, lobes 3-toothed; group 8); *P. polyneura* (China; leaves ovate to orbicular, more or less downy; scapes to 23cm; flowers10–50 per umbel, each to 2.5cm diameter, tubular pale rose to rich rose-red to crimson or purple, eye yellow; group 7);*P.* Pruhonicensis Hybrids (POLYANTHUS; the name here is taken to cover several groups of hybrids between *P. juliae*, *P. veris*, *P. elatior* and P. vulgaris, all of which have interbred producing a vast range of plants in many colours and conforming to either the primrose-type habit or the cow-/oxlip with stalked umbels. *P.* x *pruhoniciana* (*P.juliae* × *P. vulgaris*) (*P.* × *helenae*) covers the early spring primulas with flowers in red, purple, pink and white, often with white or yellow eyes. *P.* x *margotae* (*P. juliae* × *P.* Elatior Hybrids) have leaves often flushed purple-bronze and pink, red or purple flowers in short-stalked umbels. For example 'Garryarde Guinevere' (leaves bronze; flowers lilac-pink). Elatior Hybrids (*P.* × *polyantha*) are strictly hybrids between members of *Primula* section Vernales. These are the polyanthuses with stalked umbels of flowers in shades of gold, pale yellow, bronze, brown and red. Included here are the gold and silver-laced forms with small, virtually black flowers edged in golden yellow and silver-white;). *P.* × *pubescens* (*P. auricula* × *P. hirsuta*; a group including many popular cultivars in a wide range of colours, among them shades of white, yellow, pink, red, purple and brown, sometimes with a differently coloured eye; they differ basically from *P. hirsuta* in having farinose, sometimes entire leaves and a slightly farinose calyx; group 2 or 8); *P. reidii* (Himalaya; leaves oblong or lanceolate, crenate or lobed, hairy; scapes 6–15cm; flowers 3–10, 2–2.5cm diameter, white, bell- to urn shaped, fragrant, pendent; var. williamsii: more robust, flowers pale blue to white; group8); *P. rosea* (Himalaya; leaves tinted red-bronze at first, emerging with or after flowers, , obovate to oblance-olate; scapes 3–10cm; flowers 4–12, 1–2cm diameter, flat, rose-pink to red; group1 or 7); *P. secundiflora* (W China; leaves oblong to obovate or oblanceolate, toothed, farinose beneath; scapes 30–60cm; flowers 10-20, nodding to one side, 1.5–2.5cm diameter, fun-nel-shaped, red-purple or deep rose-red; group 1 or 7); *P. sieboldii* (N E Asia; leaves ovate to oblong, with toothed lobes; scapes to 30cm; flowers 6–10, 2.5–3cm diameter, flat, white, pink or purple, eye white; includes numerous cultivars with flowers ranging in colour from white to bright cerise and wisteria blue; some of the best have cut or toothed corolla lobes, in the pure white 'Kuisakigarri', for example, they are so delicately fimbriate as to resemble snowflakes; group 7); *P. sikkimensis* (Nepal to SW China; leaves elliptic or oblong to oblanceolate, toothed; scapes 15–90cm; flowers many, to 2.5cm diameter, yellow or cream-white, pendent, funnel-shaped, variably fragrant, calyx yellow-farinose; group 1 or 7); *P. sinensis* (China; leaves broadly ovate to rounded, hairy, lobed and toothed, often red beneath; scapes10–15cm; flowers whorled, 2–5cm diameter, flat, purple, pink or white with a yellow eye and entire or incised lobes; group 4); *P. sonchifolia* (W China; leaves oblong to obovate, toothed and lobed; scapes to 30cm; flowers 3–20, to 2.5cm diameter, tubular, blue to purple with a yellow eye and white-edged, sometimes toothed lobes; group 8); *P. veris* (COWSLIP; Europe, W Asia; leaves ovate to oblong-ovate, crenate or entire; scapes 6–30cm; flowers 2–16, 1–1.5cm diameter, tubular, yellow, fragrant with an orange mark at base of each lobe; group 1 or 3); *P. verticillata* (SW Arabian Peninsula, NE Africa; leaves lanceolate to ovate-lanceolate, sharply and finely toothed, white-farinose beneath; scapes 10–60cm; flowers many, whorled, each to 2cm diameter, bell-shaped, yellow, fragrant; group 4); *P. vialii* (China; leaves broadly lanceolate to oblong, hairy, toothed; scape 30–40cm; flowers pointing downwards in a dense, narrowly conical spike, each to 1cm diameter, tubular, calyx, red in bud, later pink, corolla blue to violet; group 1 or 3); *P. vulgaris* (PRIMROSE; W & S Europe; leaves oblanceolate to obovate, toothed, downy; scapes absent; flowers 2.5–4cm diameter, solitary, short-stalked, flat, sometimes fragrant, pale yellow, throat more or less orange; cultivars include forms with single, double or hose-in-hose flowers, in a wide range of colours. 'Jack in the Green': abnormality with a leafy calyx; group 1 or 3); *P. warshenewskiana* (C Asia to Himalaya; leaves oblong to oblanceolate, finely toothed, dark green; scapes short; flowers 1-8, to 1.2cm diameter, flat, bright rose or pink with yellow eye surrounded by narrow white zone; group 8); *P. whitei* (syn. *P. bhutanica*; E Himalaya; leaves oblong to spathulate, finely toothed, slightly farinose; scapes short; flowers 5–10, to 2.5cm diameter, tubular to bell-shaped, blue to violet, eye white or yellow-green, lobes toothed ; *P. bhutanica*, syn. *P. whitei* 'Sheriff's Variety', differs from this species in its more conspicuously toothed flowers; group 8).

Proboscidea (from Greek *proboskis*, snout, referring to the beaked fruit). Pedaliaceae. S America north to central US. UNICORN PLANT, DEVIL'S-CLAW, ELEPHANT-TUSK, PROBOSCIS FLOWER. About 9 species, clammy-hairy, muskily scented, annual or perennial herbs. The flowers are funnelform to campanulate, with a spreading limb of five lobes. The fruit is a woody capsule with large, curving horns at its apex. Cultivate as for *Martynia*. *P.fragrans* (SWEET UNICORN PLANT; Mexico; flowers fragrant violet- to red-purple, upper lobes often blotched with darker purple, lower lobe with bright yellow band).

proliferation (1) the presence of an abnormal number of organs in a flower or of an abnormal number of flowers or flowerheads on a stem. See *hen and chickens, hose-in-hose;* (2) a stage in micropropagation, where new cells develop from tissue, such as an excised meristem; (3) a form of natural vegetative reproduction whereby plantlets or offsets (foliar embryos) are produced on leaves or stems of the mother plant, for example *Tolmiea menziesii, Asplenium bulbiferum.* Also known as vivipary; (4) a disease of apples and onions caused by *phytoplasma* (q.v.).

propagating frame a garden frame where used for a similar purpose to a propagating case but constructed at ground level outdoors. It is used for sowing and rooting hardy specimens. A propagating frame allows for progressive hardening off as the covering lights are variably raised or removed. It should be sited in a sheltered position out of full sun, and provided with shading and insulation against cold as necessary.

propagating house a greenhouse devoted to propagation, commonly including mist-propagation equipment and propagating cases.

propagation the increase of plants by seed or vegetative means. See *cuttings, division, grafting, layering, micropropagation, seed sowing.*

propagation case an enclosed box-like structure used for seed raising and the rooting of cuttings within a greenhouse, usually with wooden walls about 30–40cm high and a glass or sometimes plastic-clad detachable cover, fitted tightly to retain a humid environment. It is usually mounted on the greenhouse benching, and may feature an electrically heated warming cable; this is fitted beneath an overall layer of rooting medium or a covering of sand, upon which containers of seed or cuttings are stood. The term close case (sometimes closed case) is an alternative description.

propagator a purpose-built portable propagating unit, consisting of a stout plastic seed tray to which is fitted a moulded translucent plastic dome. There are small ventilators in the top of the dome, and some propagators have warming cables incorporated in their base.

propagule any part of a plant used for propagation.

propping a method of support for leaning trees or long low branches, in which the ground is used as a firm base for a timber or metal crutch.

Proserpinaca (possibly derived from the Latin *proserpo*, to creep, referring to the creeping stems of most species). Haloragidaceae. N and C America, West Indies. MERMAID WEED. Some 5 species, aquatic herbs, with long stems submerged or creeping in mud or shallow water; these are densely clothed in finely pinnately cut leaves, thus resembling feathery cylinders. The two best-known species are *P.palustris* and *P.pectinata*. Grow in a mix of coarse sand and sticky loam, in water of pH 7.0–8.0, with a minimum temperature of 14°C/57°F. Propagate by stem cuttings.

prickle a small, weak thorn-like outgrowth from the bark or outer layers, generally irregularly arranged.

procumbent trailing loosely or lying flat along the surface of the ground, without rooting.

proliferous producing buds or off-shoots; especially from unusual organs,for example plants or plantlets produced from stolons or runners.

prop root a root that arises adventitiously from the stem of a plant, descends to and roots in the ground and supports the stem.

Prostanthera (from Greek *prostheke*, appendage, and Latin *anthera*, anther, from the spur-like projections on the anthers). Labiatae. SE Australia, Tasmania. AUSTRALIAN MINT BUSH. About 50 species, viscid, usually aromatic evergreen shrubs or small trees. Produced in summer in leafy racemes or terminal panicles, the flowers are shortly tubular and 2-lipped, the upper lip 2-lobed and hooded, the lower lip 3-lobed. Free-flowering, dependable shrubs for the cool greenhouse and conservatory. Grow in a medium-fertility loam-based mix. Give direct light, maintain low humidity and water plentifully when in full growth. Keep just frost-free and almost dry in winter. Container-grown plants can be moved outdoors in late spring, making fine patio specimens. Prune out exhausted growth after flowering. Propagate from seed in spring or by semi-ripe cuttings in late summer.

P.ovalifolia (E Australia; erect shrub, to 4m; leaves to 1.5cm, oval to ovate, glaucous; flowers numerous in short terminal racemes, purple, mauve or white tinted lilac); *P.rotundifolia* (MINT BUSH; SE and S Australia; erect shrub to 3m; leaves orbicular to ovate, obscurely dentate or crenate, dark green above; flowers in short, loose, racemes, violet or lilac; 'Chelsea Pink': aromatic, grey-green leaves and pale rose flowers with mauve anthers).

prostrate lying flat on the ground; the Latin equivalent *prostratus, -a,-um* is frequently encountered as a botanical and cultivar name.

protandry the shedding of pollen before the stigma of the same flower is receptive, which discourages self-pollination; cf. *protogyny.*

Protea (for Proteus, the ancient Greek sea-god, famed for his ability to change shape, a reference to their diversity of form). Proteaceae. South Africa. 115 species, evergreen shrubs or small trees with leathery leaves and slender flowers packed in terminal, cone-like heads surrounded by tough, brightly coloured bracts. Grow in full light with low humidity and good air circulation. Use a perfectly drained, low-nutrient potting mix, comprising equal parts peat and grit, with additional charcoal; ensure pH is 6.5 or lower. Proteas are extremely sensitive to nitrates and phosphates, which can be toxic even at moderate levels. Conversely, they may suffer from magnesium deficiencies: apply dilute liquid feed of magnesium sulphate and urea in spring and autumn and water plentifully but judiciously when in flower and full growth. Maintain a winter minimum temperature of 5–7°C/40–45°F with a summer maximum at about 27°C/80°F. Propagate by seed sown fresh in autumn, in a 2:3 mix of peat and grit in individual small pots in the frost-free cold frame or cool greenhouse. Keep seedlings moist but not waterlogged and ensure full ventilation. Avoid root disturbance. Root semi-ripe cuttings in summer.

P.cynaroides (KING PROTEA; shrub to 2m; inflorescence 12–30cm, goblet-shaped resembling a globe artichoke, bracts to 12cm, ovate-lanceolate, deep crimson to pink or cream, silky); *P.magnifica* (small shrub; inflorescence turbinate-obovoid, to 15cm, bracts creamy pubescent to tomentose, ovate-lanceolate to linear-lanceolate, white-ciliate); *P.neriifolia* (shrub to 3m; inflorescence goblet-shaped, to 13cm, bracts 14cm, oblong to spathulate, incurved, pink to dark rose, tipped with long black hairs); *P.repens* (low shrub or small tree to 4m; inflorescence goblet-shaped, to 9cm in diameter, bracts creamy white, sometimes tinged dark red or pink, glabrous, resinous).

protection (1) any measure taken to make conditions tolerable for tender plants grown in a climate to which they are not adapted, or to guard hardier plants from vagaries of weather. Protected cultivation involves growing plants under some form of cover, such as a cloche, frame, or greenhouse; (2) used of the protection of plants against pests and diseases, as in 'crop protection'.

prothallus the gametophyte generation of ferns and other cryptogams, a delicate liverwort-like structure.

protogyny the development of a receptive stigma before pollen shedding by the same flower, which discourages self-pollination; cf. *protandry*.

protoplasm the contents of living cells. Protoplasm is composed of water, proteins, lipids, sugars, a variety of other organic compounds, and mineral salts.

proximal used of the part nearest to the axis, thus the base of a leaf is proximal; cf. *distal*.

proximate close together.

Prunella (from *Prunella*, pre-Linnaean name for the genus; the origin is obscure but may stem from *Die Breaune*, German for quinsy, a throat infection which these plants allegedly cured). Labiatae. Eurasia, N Africa, N America. SELF HEAL, HEAL ALL. 7 species, decumbent perennial herbs. Produced in spring and summer, whorled in dense spikelets or heads, the flowers are tubular and 2-lipped, the upper lip erect and somewhat hooded, the lower lip shorter, deflexed and 3-lobed. Fully hardy, grow in fully exposed areas or partial shade. It is also useful for groundcover and for meadow and other naturalistic plantings. Propagate from seed or lengths of rooted stem. *P.grandiflora* (Europe; sparsely pubescent, to 60cm; leaves to 10cm, ovate to lanceolate, entire or crenulate; flowers 18–30mm, lips deep mauve to violet, tube paler to off-white; 'Alba': white; 'Rosea': pink; 'Rotkäppchen': carmine; 'Loveliness': pale lilac; 'Loveliness White': white; 'Loveliness Pink': pink).

pruner usually, a pruning tool used for branch lopping, where a movable blade cuts against a fixed one. See *long arm pruner.*

pruning the practice of cutting out parts of plants for specific purposes, carried out almost entirely on woody plants. The main purposes of pruning are to remove dead, diseased or unsightly growth, to maintain a manageable and attractive shape, to prevent overcrowding of other plants, to improve the size or quality of flowers and fruits, and to enhance vigour.

Pruning requirements differ according to species, and special consideration should be given to timing.

Evergreens Cutting evergreens may be necessary or beneficial for rejuvenation, shaping and the removal of dead or diseased parts. Young growth is often more readily damaged by extremes of temperature and strong sunlight, and pruning of examples such as *Laurus*, *Rosmarinus* and evergreen *Cotoneaster* should therefore be undertaken in late spring.

Rhododendrons and camellias can be suitably pruned by removing flowering branches; special care should be taken to cut back to buds or shoots because there may be no regrowth from old bare wood.

Conifers have a natural tendency for lower branches to die as the trees age and pruning can improve the overall appearance. Occasionally, pruning to improve shape may be desirable, but otherwise this group requires little attention, except when grown as hedging; vigorous young specimens can have crowns removed. Conifers are best pruned when they are in a slow-growing phase.

Deciduous plants Deciduous woody ornamentals such as *Amelanchier*, *Hamamelis*, *Potentilla*, *Viburnum*, *Hibiscus* and *Chaenomeles*, which bear flowers on short shoots and spurs of an established framework, require no more than occasional cutting for shaping, tidying and renovation. This is best carried out in late winter or spring, outside the flowering period. The majority of deciduous shrubs which benefit from regular attention, especially as a means of encouraging a good supply of flowering shoots, may be considered in two main groups. The first group comprises those blooming from midsummer onwards on wood produced in the current season, for example, *Buddleja davidii*, *Caryopteris* × *clandonensis*, *Eccremocarpus scaber*, *Hydrangea paniculata* and *Spiraea japonica*; these should have their flowered shoots pruned in early spring, cutting back to within one or two buds of the previous season's growth. Within this group are those which may or may not form a woody framework, depending on locality, such as *Ceratostigma willmottianum*, *Perovskia atriplici-folia*, *Leycesteria formosa*, *Rubus cockburnianus*, *Cotinus coggygria* and hardy fuchsias; these should have all stems cut back to a few inches from ground level, a procedure known as stooling. Bright-stemmed *Salix* and *Cornus* may be stooled, or headed back on an upright stem or 'leg', using the procedure known as pollarding. *Paulownia tomentosa* produces leaves many times normal size if stooled or pollarded. Pruning should be delayed until the danger of frost is past.

The second group comprises the large number of summer-flowering woody plants, such as *Phila-delphus*, *Deutzia*, *Buddleja*, *Kerria*, *Hydrangea* × *macrophylla* and *Weigela*, which bear blooms on shoots produced in the previous season. These respond to the thinning out of old branches and crowded shoots, and the removal of faded flower trusses, immediately after blooming. Earlier-flow-ering subjects in this group, such as *Jasminum nudi-florum*, *Forsythia* and *Prunus triloba*, should have flowered shoots cut out soon after blooming to encourage vigorous replacement growth over the longest possible growing season. Deadheading is a suitable method of pruning heathers and lavender.

Roses Roses are pruned according to type, for example, shrub roses should have old, diseased or dead wood removed, with some trimming to main-tain shape, while modern bush roses, like hybrid teas and floribundas, need cutting back moderately hard, or sometimes severely, to produce good growth and flowers the next season. Prune when dormant, between leaf fall and budbreak; never in frosty condi-tions. See *roses*.

Climbers Pruning of woody climbers can be made difficult by tangled growth. With wisterias and orna-mental vines, it may be possible to spread out a permanent stout framework of growth and shorten all side growths to it. Climbers with weaker stems, such as *Clematis*, *Jasminum* and *Lonicera*, may require more drastic pruning of the previous season's flowered growth, which is borne on tangles of old bare stems, and severe thinning or pruning of a few selected leads to near ground level may be neces-sary. Timing is relative to flowering habit.

Fruit For fruit trees and bushes, the objective is to build a strong framework in the formative years, and to ensure a regular supply of replacement shoots or spurs capable of producing sufficient blossom for a full crop of good-quality fruit. The removal of diseased parts and the maintenance of controlled growth with branches open to air and sunlight, are also important reasons for pruning.

Pome and stone fruits flower on wood that is usually one or more years old, and pruning of branches, shoots and spurs of established trees should aim at producing a balance of old and new wood. Currants and gooseberries bear flowers mainly on shoots of the previous year, but also on older wood and spurs. In pruning established blackcurrants, about one third of the total number of branches is removed annually to encourage the production of young wood, which fruits best. Other currants and gooseberries should have leaders reduced to about half their length, and laterals cut back short. Summer-fruiting raspberries bear fruit on laterals arising on canes produced in the previous year, and these are cut to ground level after cropping. Autumn-fruiting raspberries bear fruit on canes of the current year, and are pruned to ground level in late winter. Blackberries and hybrid berries, although their canes are perennial, are pruned on the replacement system in the same way as summer-fruiting raspberries.

The pruning of apples, pears, gooseberries and currants is carried out during dormancy. Stone fruits are pruned in spring and summmer in order to avoid exposure to the active stages of silver leaf disease and bacterial canker. Additional summer pruning is an essential means of restricting vigorous growth with intensive tree forms, and of developing and maintain-ing good structure in gooseberries and red and white currants. See *renewal pruning*, *summer pruning*.

Pruning techniques All pruning cuts should be made cleanly, with sharp tools, directly above a growth bud or close to where another branch arises. It is good practice to make pruning cuts to shoots and small branches slope down from a bud at an angle of 45°. The wounds left by the removal of large branches are likely to heal more rapidly if the cut is made a little distance from the trunk, beyond the slight swelling that encircles its base. Care must be taken to ensure that branches which are being removed with a saw do not break and tear bark from the main limb. A branch should first be cut off a short distance from the trunk and two further cuts then made; with each cut, a shallow undercut from the lower surface of the branch should be followed by a second complete cut from above, just a little nearer the trunk or limb. The branch will snap off cleanly and the short remaining stub can be removed without risk of damage.

Wound dressing with proprietary paints may protect from infection, and this is an important requirement for stone fruits, but untreated wounds heal faster, and large wounds are only briefly protected since paint cover soon cracks.

Root pruning The roots of established fruit trees may be cut to reduce growth and encourage fruiting. This practice was more common before the advent of dwarfing rootstocks, but it is still an effective treatment for older trees, including shy flowering ornamentals, albeit less convenient than bark ringing. A trench 45–60cm deep should be dug encircling the tree beneath the branch extremities, and exposed thicker roots severed. The provision of subsequent support may be advisable as anchorage is weakened.

pruning knife

pruning knife a pocket knife used for light pruning, more usually by experts, with one folding blade that is slightly curved.

pruning saw a narrow bladed saw, used for pruning small branches. The blade and handle may be straight or angled, or more usually curved. See *Grecian saw*.

pruning shears see *secateurs*.

pruinose thickly frosted with white or blue-grey bloom, usually on a dark ground colour.

Prunus (the Classical Latin name for these plants). Rosaceae. PLUM, CHERRY, PEACH, ALMOND, APRICOT. N Temperate regions, S America. 430 species, deciduous and evergreen trees and shrubs. Produced between autumn and early summer (predominantly spring), the flowers consist of five rounded and spreading petals and numerous stamens; they are usually white, pink or red. They give rise to fleshy drupes with hard stones. In some cases these fruits are edible and important crops – see *almond, apricot, cherry, peaches and nectarines, plums*.

Grown for their attractive habit and form, their decorative bark, their fine autumn colours and for their unrivalled display of blossom which, with judicious selection of species and cultivars, can be extended almost throughout the year, from the late autumn and winter blooms of *P.subhirtella* 'Autumnalis', through late spring and early summer with the late-flowering Japanese cherries such as 'Fugenzo' and 'Shirofugen', and into summer with the intermittent flushes of blossom on *P.cerasus* 'Semperflorens'. Most flower in spring. *P.cerasifera* and its cultivars, and *P.davidiana* are among the earliest to flower: the former with spectacular combinations of bridal white blossom and wine red new growth, the latter blooming soon after New Year in a mild season. The shrubby species, *P.incisa*, *P.* × *cistena* and *P.spinosa*, are used for hedging, the last especially useful in cold, exposed and maritime areas. When grown as shrubs, *P.cerasifera* and its dark-leaved cultivars also make excellent hedging plants. The evergreen *P.laurocerasus* and *P.lusitanica* are often used as informal screens and in more manicured hedging. Both make good game cover and are often planted as understorey in woodland. *P.laurocerasus* is extremely tolerant of shade and tree drip. Species grown primarily for their beautiful peeling bark include *P.serrula*, a glossy deep mahogany-red, and *P.maackii*, a shining, translucent amber to gold. The Japanese Flowering Cherries (Sato-zakura Group), contribute a range of form and habit, from the upright narrow columns of 'Amanogawa' and stiffly ascending branches of 'Sekiyama', to the small weeping domes of 'Kiku-shidare'. The flowers may be large or small, single, semi-double or double, regular or ragged, sweetly fragrant or almond-scented, snow white, cream, pale rose, pink, pale red, pale lilac or purple-pink. This group includes many small trees for the confined spaces of urban gardens, where their tolerance of pollution is especially useful. Most colour well in autumn, in shades of yellow and tawny orange, and many also have bronzed new growth, in beautiful contrast to the emergent blossom.

The following are fully hardy. Grow them in well-drained moisture-retentive soils, siting deciduous species in full sun to ensure good flowering and autumn colour. *P.glandulosa* and *P.triloba* thrive when trained against a sheltered south-facing wall. Plant evergreens in sun to semi-shade. Stake standard specimen trees until established, especially in areas exposed to wind, and protect young stems from rabbits and hares. With the exception of *P.laurocerasus*, which becomes chlorotic on shallow chalk, most grow particularly well on calcareous soils. Prune if necessary in late summer, ensuring a long enough period for wounds to heal before the onset of winter. Cut back the flowered shoots of *P.glandulosa* and *P.triloba* to 2–3 buds from the base, especially if grown as wall or forcing shrubs. If grown as a flowering hedge, trim after flowering. Cut back *P.laurocerasus* and *P.lusitanica* with secateurs in spring or late summer; neglected specimens can be rejuvenated by cutting hard back into old wood in spring.

Propagate species by seed; they may hybridize and not come true, although seed-raised specimens are generally longer-lived. Collect seed when ripe, remove flesh and stratify in damp coir or sand for 10–14 weeks (3–4 weeks for *P.dulcis* and *P.armeniaca*) prior to sowing. Alternatively sow outdoors in the cold frame in autumn and protect from rodents. Propagate shrubby species such as *P.glandulosa*, *P.laurocerasus*, *P.pumila*, *P.tenella* and *P.triloba* by simple layering in early spring. Cultivars are propagated commercially by budding or grafting – the Japanese and large-flowered cherries on to *P.avium*; bird cherry types, such as *P.cornuta*, *P.serotina* and *P.virginiana*, on to *P.padus*. Species such as *P.cerasifera*, *P.incisa*, *P.spinosa* and *P.subhirtella*, can be propagated by semi-ripe heel cuttings in summer, in a closed case with bottom heat Most species can be propagated by softwood cuttings taken when plants are growing strongly, in spring and early summer, treated with hormone-rooting powder, and rooted with mist and bottom heat.

Ornamental Prunus species are affected to various degrees by the same pests and diseases that attack the edible species (q.v.).

P. x *amygdalo-persica* (*P. dulcis* x *P. persica*; tree or shrub; leaves similar to *P. dulcis*, , coarsely toothed; flowers to 5cm diameter, light pink; fruit peach-like, dry; 'Pollardii': flowers large, rich pink); *P. armeniaca* (APRICOT; North China; tree to 10m; leaves to 12cm, orbicular or ovate, glabrous, toothed; flowers to 4cm diameter, white or pink; fruit to 5cm, white to orange-red or yellow, downy); *P. avium* (BIRD CHERRY, SWEET CHERRY, WILD CHERRY; Europe, Asia; tree to 20m; leaves to 10cm, oblong-ovate, acuminate, toothed; flowers 2.5cm diameter, white; fruit 2cm, dark maroon; 'Plena': flowers double, pure white); *P.* x *blireana* (*P. cerasifera* 'Atropurpurea' x *P. mume*; garden origin; shrub or small tree to 4.5m; young leaves bronze-red at first ; flowers 3cm diameter, bright rose, double; 'Moseri': leaves light red flushed brown; flowers small, pale pink); *P. campanulata* (TAIWAN CHERRY, FORMOSAN, CHERRY, BELL-FLOWERED CHERRY; South Japan, Taiwan; small tree; bark purple-brown; leaves to 11cm, elliptic acuminate, toothed; flowers to 3cm diameter, dark rose-red, bell-shaped, nodding; fruit 1cm diameter, black-purple; 'Okame' (*P. campanulata* x *P. incisa*): leaves small, dark green, flaming orange in autumn; flowers shocking pink); *P. cerasifera* (CHERRY PLUM, MYROBALAN; Asia Minor, Caucasus; large shrub or tree to 9m, often spiny; leaves to 6.5cm, ovateto obovate, toothed; flowers to 2.5cm, pure white; fruit to 3cm diameter, round red to yellow, somewhat pruinose; 'Pissardii': leaves emerging dark red, later deep purple to red-bronze; flowers white, buds pink, profuse, appearing before foliage; fruit maroon); *P. cerasus* (SOUR CHERRY; SE Europe to W Asia; shrub or tree to 6m; leaves 6cm, narrowly ovate to elliptic, finely toothed, dark green, glossy; flowers to 2.5cm diameter, white; fruit 1.8cm diameter, dark red; var. *austera*: MORELLO CHERRY; tree to 9m, somewhat pendulous; flowers to 6cm diameter; fruit black-red; var. *caproniana* : AMARELLE CHERRY, KENTISH RED CHERRY, tree to 9m; fruit globose, light red; var. *marasca*: MARASCHINO CHERRY; arching tree; fruit black-red, small); *P.* x *cistena* (*P. cerasifera* 'Atropurpurea' x *P. pumila*; garden origin; shrub to 2.5m; leaves to 6cm, obovate to lanceolate, red-brown, toothed; flowers white, calyx red-brown; fruit black-purple; 'Crimson Dwarf': dense, upright, to 1.2m high; leaves red tinted bronze, crimson when young; flowers shell pink, calyx dark red); *P. davidiana* (DAVID'S PEACH; China; tree to 9m; leaves to 1cm, lanceolate, tapering finely, dark green and shiny, sharp-toothed; flowers 2.5cm diameter, white or pale pink; fruit 3cm diameter, yellow to red, downy); *P.* x *domestica* (*P. spinosa* x *P. cerasifera* subspecies *divaricata*; PLUM, COMMON PLUM; S Europe, Asia; tree to 12m, sometimes spiny; leaves to 10cm, elliptic or oblong, downy at first; flowers to 2.5cm diameter, white; fruit to 8cm, yellow or red to blue-black or violet-purple, pruinose, flesh green or yellow; subspecies *insititia*: BULLACE; branches often spiny; ruit smaller, usually dark purple); *P. dulcis* (ALMOND; Syria to N Africa; tree to 9m; leaves to 13cm, lanceolate, tapering narrowly, finely toothed; flowers palest pink to deep rose, to 5cm; fruit to 6.5cm, green, with thin, tough flesh and velvety skin; numerous cultivars with single or double flowers in tones of white, pale pink, rose and magenta); *P. glandulosa* (DWARF FLOWERING ALMOND; China, Japan; shrub to 2m; leaves to 7cm, oblong to lanceolate, finely toothed with glandular stipules; flowers to 2cm diameter, red becoming pink or white; fruit 1cm, dark red; 'Alba Plena': flowers double, large, white; 'Sinensis': flowers large, double, bright pink); *P.* x *hillieri* (*P. incisa* x *P. sargentii*; garden origin; densely branched tree to 9m; leaves bronze at first, colouring brilliantly in autumn, toothed; flowers 3cm diameter, pink, calyx bronze-red; 'Spire': habit conical, erect; leaves richly tinted in autumn; flowers single, pink, produced early); *P. incisa* (FUJI CHERRY; Japan; tree to 5m; leaves to 5cm, obovate or ovate, downy on veins, sharply toothed; flowers produced in late winter and

early spring, to 2cm diameter, white to rose, petals notched; fruit to 0.8cm, purple-black; 'February Pink': young growth red-tinted, sutumn colour good; flowers pale pink, produced early in the year); *P. laurocerasus* (CHERRY LAUREL, LAUREL; S E Europe, Asia Minor; evergreen shrub or tree to 6m; leaves to 25cm, oblong to elliptic, glossy and dark green, thick and rather rigid; flowers white, small, in dense racemes to 10cm; fruit 0.8cm, plum red to black; 'Aureovariegata': leaves tipped yellow; 'Castlewellan': leaves slightly contorted, speckled white; 'Marbled White': leaves marbled grey or white; 'Otto Luyken': compact and low-growing, spreading shrub to 1m; leaves narrow glossy dark green; free-flowering; 'Schipkaensis': broad, goblet shaped, to 2m; leaves large; very hardy); *P. lusitanica* (PORTUGUESE LAUREL CHERRY, PORTUGAL LAUREL; Iberian Peninsula; evergreen shrub of wide, bushy habit, or tree to 20m; branchlets red; leaves to 12cm, oblong to ovate, somewhat toothed, thinly coriaceous dark green, shiny above; flowers to 1cm diameter, white, erect, in racemes to 15cm; fruit 8mm, red to dark purple or black; subsp. *azorica*:AZORES LAUREL CHERRY; Azores; shrub or small tree to 4m; leaves to 10cm, ovate to elliptic; racemes shorter, fewer-flowered;'Variegata': leaves edged white, sometimes flushed pink in winter); *P. maackii* (MANCHURIAN CHERRY, AMUR CHERRY; Korea, Manchuria; tree to 16m; bark golden to russet, pellucid, papery, peeling; leaves 10cm, elliptic or oblong; flowers small, white, in dense racemes; fruit 4mm, dry, black;'Amber Beauty': bark golden); *P. mahaleb* (ST. LUCIE CHERRY; Europe, Asia Minor; tree to 9m; leaves 4.5cm, broadly ovate to rounded, toothed; flowers 1.4cm diameter, white, in racemes; fruit 0.6cm, black; 'Albomarginata': leaves edged white; 'Aurea': leaves splashed yellow); *P. mume* (JAPANESE APRICOT, MEI; Japan; tree to 9m; branchlets glossy; leaves to 10cm, broad, toothed; flowers to 3cm, pale rose; fruit to 3cm diameter, somewhat downy, yellow; 'Benishidori': flowers double, intense pink, fading later, fragrant; 'Omoi-no-wae': flowers semi-double, cup-shaped, white, occasional petal or even whole flower pink); *P. padus* (BIRD CHERRY, COMMON BIRD CHERRY; Europe, Asia; tree to 1.5m; leaves to 9cm, obovate to elliptic, toothed; flowers to 1.5cm diameter, white with toothed petals, in racemes to 12cm; fruit pea-sized, black; 'Plena': flowers semi-double, large, long-lasting; 'Watereri': flowers white, large, almond-scented in pendulous racemes; var. *commutata*: medium-sized tree; leaves coarser, crenate; racemes to 15cm); *P. pensylvanica* (BIRD CHERRY, PIN CHERRY, RED CHERRY; N America; tree to 9m; leaves 9cm, narrowly ovate, acuminate, with glandular teeth; flowers 1.5cm, white; fruit 0.6cm diameter, red); *P. persica* (PEACH; China; small tree to 8m; leaves to 15cm, elliptic to lanceolate, tapering narrowly; flowers pink to white, to 3.5cm diameter; fruit to 7cm diameter, downy, yellow blushed red, fragrant, very juicy; numerous cultivars grown for their fruit; the following are ornamental - 'Klara Mayer' : shrub-like; flowers double, to 4cm, strong red-pink; fruit pale green, tinted red; 'Russel's Red': habit dense; flowers double, striking red;

'Palace Peach' : flowers double, small, deepest red; 'Peppermint Stick': flowers double, white, striped red; 'Dianthifolia' : flowers very large, semi-double, petal narrow, striped deep red); *P. sargentii* (Japan; spreading tree to 18m; leaves to 10cm, broadly oblong-elliptic to obovate, tapering, sharply toothed, somewhat, glaucous beneath; flowers to 4cm diameter, blush-pink with cut petals; fruit to 1cm, glossy crimson; 'Columnaris': syn. 'Rancho'; columnar, to 10m; bark mahogany; leaves flaming red and gold in autumn; flowers single, pink; 'Accolade': *P. sargentii* x *P. sub-hirtella*; habit flat-topped, older plants somewhat pendulous; flowers semi-double, to 4cm diameter, shocking pink in pendulous clusters); *P.* Sato-zakura Group (JAPANESE FLOWERING CHERRIES; most probably derived from *P. serrulata*; 'Amanogawa': narrowly erect, to 6m; flowers single or semi-double, pale pink, fragrant ; 'Kanzan': syn. 'Sekiyama'; ascending to spreading, to 12m; leaves lightly tinted red-bronze at first, yellow and copper in autumn; flowers large, double, pink in hanging clusters; 'Kikushidare': syn. 'Cheal's Weeping'; small weeping tree; flowers double, clear deep pink ; 'Okumiyaku': syn. 'Shimidsu', 'Shogetsu'; small with a wide-spreading, flattened crown; flowers large, double, fringed, blush pink opening white; 'Shirofugen': wide and spreading, flat-crowned; flowers large, double, pink in bud, opening white; 'Shirotae': small, spreading to pendulous; flowers large, single to semi-double, white with a hawthorn-like fragrance; 'Ukon': habit erect to spreading; leaves colouring well in autumn; flowers large, semi-double, creamy white tinted primrose yellow; 'Uzuzakura': syn. 'Hokusai'; spreading, vigorous; autumn colour good; flowers large, semi-double, shell pink developing a dark stain in centre with age; 'Yaemarasakizakura': syn. 'Yae-Murasaki'; small, slow-growing; leaves brilliant orange in autumn; flowers semi-double, purple-red in bud, opening purple-pink, very freely produced); *P. serrula* (BIRCH-BARK CHERRY; SW China; tree to 15m; bark shiny mahogany-brown, peeling and birch-like; leaves to 7.5cm, lanceolate, finely toothed; flowers 2cm diameter, white; fruit 1cm, bright red); *P. serrulata* (ORIENTAL CHERRY; China; tree to 3m; leaves to 13cm, ovate, tapering narrowly, finely toothed; flowers 3.5cm diameter, pure white, double; fruit small, black; var. *spontanea*: HILL CHERRY, JAPANESE MOUNTAIN CHERRY; Japan; to 18m; bark brown or grey with conspicuous lenticels; leaves bronze when young; flowers to 2.5cm, diameter, white or pink, petals incised); *P. spinosa* (SLOE, BLACKTHORN; Europe, N Africa, Asia Minor; very spiny shrub or small tree to 8m; leaves to 5cm, elliptic to oblong or obovate, finely toothed; flowers to 2cm diameter, white; fruit to 1.5cm, black, pruinose); *P.* x *subhirtella* (WINTER FLOWERING CHERRY, SPRING CHERRY, HIGAN CHERRY, ROSEBUD CHERRY; Japan; tree to 18m with a flat to somewhat weeping crown; branchlets slender, cascading; leaves to 8cm, ovate to lanceolate, tapering, toothed, colouring richly in autumn; flowers to 1.8cm diameter, pale pink to white with cut petals, appearing at leaf fall and and often again before new leaves; fruit to 9mm, purple-black; 'Autumnalis':

habit spreading, to 5m; leaves deeply serrate, red and orange in autumn; flowers produced in autumn, late winter and early spring, semi-double, petals somewhat frilled, opening white, pink in bud, almond-scented); *P. tenella* (DWARF RUSSIAN ALMOND; C Europe to E Siberia; low, bushy shrub to 1.5m; leaves to 9cm, obovate or oblong, somewhat thick and glossy, coarsely toothed; flowers to 1.5cm across, rose-red; fruit 2.5cm, grey-yellow, velvety; 'Fire Hill': dwarf, to 75cm high, forms thickets of thin erect stem; flowers intense red-pink, profuse); *P. tomentosa* (DOWNY CHERRY; Kasmir, China, Tibet; shrub to 2.5m; young shoots downy; leaves to 5cm, obovate to oblong, slightly rugose, downy above, densely so beneath, toothed; flowers to 2.3cm diameter, white or pink; fruit 1.2cm, subglobose, hairy, red); *P. triloba* (FLOWERING ALMOND; China; shrub or small tree to 4.5m; branches brown, often rod-like and erect; leaves to 6cm, ovate or obovate, often trilobed, coarsely toothed; fowers to 2.5cm diameter, pink to white, often double; fruit 13mm diameter, red, downy; 'Multiplex': flowers double, pink); *P. virginiana* (VIRGINIAN BIRD CHERRY; N America; shrub to 3.5m, or rarely small tree; leaves to 8cm, broadly obovate or elliptic, very finely toothed; flowers 1cm diameter, white, in dense racemes; fruit dark red to black; 'Schubert': habit dense, spreading to 6m high; leaves becoming purple-brown in summer); *P.* x *yedoensis* (*P.* x *subhirtella* x *P. speciosa*; YOSHINO CHERRY, TOKYO CHERRY; small tree to 15m, broadly upright; bark smooth; leaves to 12cm, elliptic, tapering, toothed, stalk yellow with red hairs; flowers to 3.5cm diameter, pure white in racemes; fruit pea-sized, black; 'Moerheimii': shrub-like, to 3m, weeping; twigs pendulous, grey; flowers pink fading to white; to 2cm diameter; late-flowering; 'Shidare Yoshino': syn. 'Pendula'; habit weeping; flowers snow white, profuse).

Pseuderanthemum (from Greek *pseudo*, false, and *Eranthemum*, a closely related genus with which this one is much confused). Acanthaceae. Tropics. Some 60 species, evergreen herbs, subshrubs and shrubs, differing from *Eranthemum* in their inconspicuous floral bracts and overlapping, not twisted corolla lobes. Cultivate as for *Eranthemum*.

P.atropurpureum (Polynesia, naturalized tropical America; slender shrub to 1.5m; leaves 10–15cm, linear to ovate-elliptic, dark purple to deep metallic green; flowers to 2.5cm, white, spotted rose or purple at base; 'Variegatum': leaves bronze-purple, marked pink and variegated cream-yellow, flowers magenta marked red); *P.reticulatum* (Polynesia; subshrub to 1m; leaves to 27cm, ovate-lanceolate, undulate, dark green net-veined cream-yellow; flowers to 3cm, flushed and spotted damson).

pseudobulb the water-storing thickened bulb-like stems found in many sympodial orchids. Pseudobulbs vary in shape and size and usually grow actively for only one season, persisting thereafter as backbulbs.

pseudocarp see *accessory fruit*.

pseudogamy development of an embryo following pollination but not fertilization.

Pseudogynoxys (literally a 'false *Gynoxys*' – the name of another genus of South American Compositae). Compositae. Tropical S America. About 13 species, evergreen perennial shrubs or climbers with daisy-like flowerheads. A vigorous evergreen climber suited to the cool to the intermediate greenhouse or conservatory, where it will rapidly cover walls and pillars with deep green foliage and cheerful orange-red flowerheads. Plant in a moist, but not overwet, loam-based fertile mix in semi-shade or full sun. Feed fortnightly in spring and summer. Trim back shoots after flowering. Increase by seed or semi-ripe nodal cuttings. *P.chenopodioides* (syn. *Senecio chenopodioides*, *S. confusus*; MEXICAN FLAMEVINE, ORANGEGLOW VINE, TRAILING GROUNDSEL, MEXICAN FIREVINE, MEXICAN DAISY; Colombia; climbing shrub to 6m; leaves narrowly ovate, dentate, light green, glabrous; flowerheads to 5cm in diameter, bright orange to flame, usually fragrant).

Pseudolarix (from Greek *pseudo*, false, and *Larix*, which it resembles). Pinaceae. E China. GOLDEN LARCH. 1 species, *P.amabilis*, a deciduous, coniferous tree, to 40m. The crown is broadly conic with horizontal branches, the bole covered with rust-brown, narrow-ridged bark, becoming grey and fissured in old trees. The leaves are whorled or spiralled, needle-like, 3–7cm long and pale green becoming rich gold in autumn. A difficult tree to grow well in cool-temperate regions: if planted out when small, it is easily scorched and may be killed by –5°C/23°F. Good trees are found only in southern England yet the 30m trees in New Jersey and New York show clearly that very cold winters are no problem where summers are long and hot. The golden larch fares best in warm, sheltered sites on fertile, deep soils, around pH 5–6, with about 1000mm rainfall. Avoid exposure to dry winds, and fast-drying soils. Propagation as for *Larix*.

Pseudopanax (from Greek *pseudo*, false, and *Panax*.) Araliaceae. New Zealand and associated islands, Tasmania, Chile. LANCEWOOD. 12–20 species, evergreen trees or shrubs, some with distinct kinds of foliage in different life-stages. The leaves are simple or palmately compound, the blades entire or variously toothed. Small, pale flowers are borne in umbels on large, compound inflorescences. The following species tolerate temperatures down to at least –5°C/23°F, provided that they are given a warm, sheltered location. Cultivate in any fertile, well-drained soil, giving full sun or part shade. Large plants respond well to heavy pruning and to careful transplanting. In cold areas, grow in large containers in the conservatory or cool to intermediate greenhouse (winter minimum 7–10°C/45–50°F), giving strong, filtered light in summer and low humidity; ventilate whenever possible. Water moderately in summer, sparingly in winter. Propagate from seed in autumn,

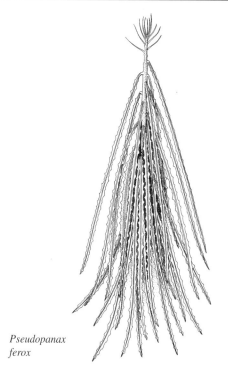

Pseudopanax ferox

by semi-ripe cuttings in summer or by air-layering.

P.arboreus (FIVE-FINGER, PUAHOU, WHAU-WHAUPAKU; New Zealand; round-headed tree to 8m, trunk slender, branches pointed upwards; leaflets 3–7, thick, narrowly oblong to oblong-obovate, to 20cm, coarsely toothed, dark glossy green above); *P.crassifolius* (HOROEKA, LANCEWOOD; New Zealand; tree to 15 m tall, usually erect, slender and unbranched when young; juvenile leaves to 100cm, linear, tough to rigid, sinuately toothed, downward pointing, olive to flesh pink mottled black to maroon or silver-grey, sometimes solid black-brown with an orange-pink midrib; adult-leaves simple or 3–5-foliate, upright, linear to linear-obovate, leathery, to 20cm, sinuate or coarsely toothed; *P.ferox* (TOOTHED LANCEWOOD; New Zealand; resembling *P.crassifolius* but to 7m; leaves always simple, in adults to 15cm or so; juvenile leaves to 45cm, jaggedly toothed); *P.laetus* (New Zealand; shrub or small tree, to 6m or so; leaves 5–7-foliolate, petioles purple-red, to 25cm, leaflets obovate, thick, to 30cm, stalked, toothed above).

Pseudosasa (from Greek *pseudos*, false, and *Sasa*). Gramineae. China, Japan, Korea. 6 species of tall or short bamboos; the rhizomes are usually spreading and running but in colder climates may become shorter and congested, giving the culms a clumped appearance; see *bamboos*.

pseudostem an erect aerial 'stem' apparently furnished with leaves but in fact composed of the packed or overlapping sheaths and stalks of essentially basal leaves. It is a common feature of the families Musaceae, Zingiberaceae, Heliconiaceae, Strelitziaceae and Cannaceae.

Pseudotsuga (from Greek *pseudo-*, false, and *Tsuga*). Pinaceae. Western N America south to Mexico; E Asia from S Japan to Taiwan and SW China. DOUGLAS FIR. 8 species, evergreen coniferous trees. The crown is conic, becoming irregular with age in exposure, the branches long, spreading or sweeping downwards, and the branchlets and shoots somewhat drooping. The bark is smooth, dark purple-grey, with resin blisters when young, becoming very thick, deeply fissured and corky red-brown with paler fissures on mature trees. The needle-like leaves are radially arranged with their short stalks twisted to position the leaves more or less pectinately. Ovoid to cylindric, the cones are pendulous, green, pink or purple, with a pale or violetbloom, ripening brown, often resinous. The following species is an important timber tree. Fully hardy, it should be planted on moist but well-drained, acid to neutral soils. Increase by seed, or (cultivars) by grafting on stocks of typical *P.menziesii*. Leaf-cast fungus can partly defoliate trees of *P.menziesii* subsp. *glauca*, and more particularly its var. *caesia*, but severe attacks are rare in gardens. Adelgids attack several species.

P.menziesii (syn. *P.douglasii*, *P.taxifolia*; DOUGLAS FIR, GREEN DOUGLAS FIR; SW British Colombia, to west central California; to 100m, bark green-brown with resin blisters, becoming red-brown, corky, fissured, crown dense, conic, more slender and open in vigorous trees, spire-like, branches level or gently downswept, branchlets drooping, shoots aromatic; leaves 1.5–4cm, bright green above, two dull white bands beneath, blunt, rounded; cone 6–12cm, green; subsp. *glauca*: BLUE DOUGLAS FIR, COLORADO DOUGLAS FIR, C and S US, to 40m, leaves strongly upswept, 1.5–2.5cm, grey-green to blue-green, with stomatal bands above as well as beneath, cultivars including dwarf and tall clones with blue to silver-white foliage; var. *caesia*: GREY-DOUGLAS FIR, FRASER RIVER DOUGLAS FIR, leaves dull grey-green with some stomata above, grey-white beneath).

Psychopsis (from Greek *psyche*, butterfly, and *opsis*, appearance referring to the flower shape). Orchidaceae. C and S America. BUTTERFLY ORCHID. 5 species, evergreen epiphytic perennial herbs, closely allied to and formerly included in *Oncidium*. The pseudobulbs are circular to ovoid and laterally compressed. Each bears a leathery, oblong to lanceolate leaf, often dotted or mottled red-brown. The flowers are produced at various times of year, singly in long succession atop a wiry raceme. They consist of a large, frilled and near-circular lip, broadly scimitar-shaped, with decurved lateral sepals and very slender petals and dorsal sepal, arising like antennae. Provide full sunlight and high summer temperatures (30–40°C/85–105°F). Pot in an open mixture of equal parts coarse bark and charcoal, surrounding the lead base with fresh sphagnum at first to promote rooting. Alternatively, mount on rafts or cork slabs. Suspend plants in a dry, airy position, water freely during the growing period and give a weak foliar feed every third week. Once pseudobulbs are completed, withhhold water, providing only a heavy misting every few days. Provide a minimum winter temperature of 18°C–65°F. Do not remove flower spikes, which may continue to produce flowers even though they appear spent. Propagation is by division, although plants seldom become large enough to warrant this.

P.krameriana (syn. *Oncidium kramerianum*, *O.nodosum*, *O.papilio* var. *kramerianum*; Costa Rica, Panama, Ecuador, Peru, Colombia; flowers to 12cm in diameter, showy; sepals deep red-brown with golden yellow margins or bands; lip deep red-brown); *P.limminghei* (syn. *Oncidium limminghei*; Brazil, Venezuela; flowers to 5cm in diameter; sepals dull red-brown; petals bright red-brown barred with pale yellow-brown; lip cream-yellow spotted orange-brown); *P.papilio* (syn. *Oncidium papilio* ; Trinidad, Venezuela, Colombia, Ecuador, Peru; flowers to 15cm in diameter; dorsal sepal and petals purple-brown mottled yellow-green; lateral sepals bright chestnut red, barred or blotched deep yellow; lip yellow spotted orange-brown, apex golden yellow bordered or heavily mottled red-brown at margins).

psyllids (Hemiptera: Psyllidae) small sap-feeding insects up to 3mm long, the adults of which have two pairs of wings and are usually very active. The flat-bodied nymphs are known as suckers, and produce drops of honeydew and conspicuous white waxy threads; some have toxic saliva which induces various forms of galling. Most species are host-specific. Psyllids common to Europe and North America include the APPLE SUCKER (*Psylla mali*), PEAR SUCKER (*P.pyricola*), BOX SUCKER (*P.buxi*) and BAY SUCKER (*Trioza alacris*).

Apple suckers overwinter as straw-coloured, elongate-oval eggs on the bark of apple. Hatching occurs in spring and nymphs crawl into the opening buds where they feed between developing leaves and blossoms. There is only one generation each year. In severe attacks, the blossoms turn brown; this can be mistaken for frost damage. Pear suckers differs from apple suckers in that they overwinter as adults and pass through three generations each year. Spring infestations are associated with death of blossoms and black sooty moulds growing on the excreted honeydew; later generations may cause premature defoliation in late summer and death of developing fruit buds in autumn.

Box suckers cause the leaves of box shoot tips to turn inwards forming a tight cluster, like a small cabbage, in which the developing nymphs may be found in the spring. The nymphs of the bay sucker live in curled over leaf edges of bay laurel. In severe attacks, leaves shrivel prematurely and shoots are killed; this type of damage is most serious on young trees under five years old.

Other psyllids include the ALDER SUCKER (*P.alni*); the BLUE-GREEN GUM SUCKER (*Ctenarytaina eucalypti*), introduced into Europe, South Africa and New Zealand from Australia, and sometimes a serious pest of nursery trees; and the CITRUS SUCKER (*Trioza erytreae*), prevalent on citrus in eastern and southern parts of Africa. In North America, the POTATO PSYLLID (*Paratnoza cockerelli*) is an important pest of potato and tomato, causing leaves to become curled and yellow, a condition known as 'psyllid yellows'. Severe attacks cause loss of yield.

Psyllids are best controlled by using systemic insecticides which are effective against aphids. Apple sucker eggs may also be controlled with winter washes applied to apple trees in the dormant season.

Ptelea (Greek name for the elm tree). Rutaceae. N America. HOP TREE, SHRUBBY TREFOIL. About 11 species, shrubs or small trees, with trifoliolate, muskily aromatic leaves, small, starry green flowers in summer and winged fruit. Hardy to at least −15°C/5°F. Grow in fertile, well-drained but moisture-retentive soil in full sun or light part-day shade. Sow seed outdoors in autumn, or in spring following three months of stratification at 5°C/40°F. Root softwood cuttings in early summer. *P.trifoliata* (HOP TREE, STINKING ASH, WATER ASH; E and C US; shrub or tree to 8m; leaflets 4–12cm, ovate to oblanceolate; fruit 2–2.5cm, suborbicular, rounded or notched at base; 'Aurea': leaves soft yellow, later lime; 'Glauca': leaves glaucous).

pteridophytes a division of non-flowering vascular plants. They reproduce by spores in the process of alternation of generations. Among these are ferns, club mosses, and horsetails. There are 365 genera organized in four classes comprising 2000 species.

Pteris (from the Greek word for fern, *pteron*, wing, referring to the feathery fronds). Pteridaceae. Subtropics and tropics. BRAKE, DISH FERN, TABLE FERN. About 280 species, terrestrial, rhizomatous ferns. The fronds are slender-stalked and loosely clumped, the blades pinnately divided with narrow segments, the lowermost segments often forked or themselves divided. The following species are widely grown as house plants and will tolerate neglect and low temperatures. For best results, grow in dappled light to deep shade, with a minimum temperature of 10°C/50°F and a humid but buoyant atmosphere. Plant in a porous, soilless, acid to neutral medium. Keep just moist in cool winters. Water, feed and syringe generously in spring and summer. Propagate by division or spores.

P.argyraea (tropics; fronds pinnate to pinnatisect, dark green with a broad silver-white central stripes, terminal pinna 15–30 x 3.5cm, segments linear-oblong); *P.cretica* (Old World tropics and subtropics; fronds with 1–5 pairs of simple or forked pinnae, pinnae 10–20cm, linear to lanceolate; 'Albolineata': fronds striped and edged white; 'Childsii': fronds bright green, lobed, waved or frilled; 'Wilsonii': fronds compact, bright green, lobed and heavily crested, giving a fan-like appearance); *P.ensiformis* (SWORD BRAKE; Himalaya to Japan, Philippines, Polynesia and tropical Australia; fronds bipinnate, 15–30 × 8–15cm, terminal pinna 5–10cm, slightly compound with central portion entire, lateral pinnae to 4–5 pairs, cut into 2–6 obovate-oblong pinnules to 12mm wide, dentate; 'Arguta': fronds dark green with central, silver-white markings; 'Victoriae': fronds with a white zone flanking the midrib on either side); *P.multifida* (SPIDER BRAKE, SPIDER FERN; E Asia, introduced America; fronds pedate, 2-pinnate at base, pinnatisect above, 20–50cm long, 10–25cm broad, pinnae elongate-pinnatisect, decurrent, to 5mm broad, long-attenuate; 'Cristata': compact, fronds crested); *P.tremula* (TENDER BRAKE, SHAKING BRAKE, TURAWERA, AUSTRALIAN BRACKEN, POORMAN'S CIBOTIUM; New Zealand, Australia, Fiji; fronds 3–4 pinnate at base, 30–90 × 20–70cm; pinnules to 2cm, narrowly oblong to linear, dentate, ultimate segments linear to 3.5 × 0.5cm, toothed).

Pterocarya (from Greek *pteron*, wing, and *karyon*, nut, referring to the winged fruit). Juglandaceae. Caucasus to E and SE Asia. WINGNUT. Some 10 species, deciduous trees with handsome pinnate leaves and long, green catkins in summer. Hardy to about −12°C/10°F and tolerant of soil compaction. The root system may be too invasive for smaller gardens. *P.fraxinifolia* grows well in damp soils. Most frost-resistant and vigorous is probably the hybrid wingnut, *P. × rehderiana*, which has a tendency to spread by suckering. Propagate from seed, by suckers or by layering; root cuttings are sometimes successful.

P.fraxinifolia (CAUCASIAN WALNUT; Caucasus to N Iraq; to 25m, often multi-stemmed, bark grey-black, deeply furrowed; young twigs olive-brown, slightly pubescent; leaves 20–40cm, leaflets 11–21, oval-oblong to oblong-lanceolate, 8–12cm; 'Albomaculata': young leaves speckled white); *P. × rehderiana* (*P.fraxinifolia* × *P.stenoptera*; to 30m, suckering; young twigs red-brown; leaves about 20cm, rachis partially winged, leaflets 11–21, narrow-oblong, 6–12cm); *P.stenoptera* (China; to 25m; young twigs tawny pubescent; leaves 20–40cm, rachis winged, sometimes serrate, leaflets 11–23, oval-oblong to narrow-oblong, serrulate).

Pteroceltis (from Greek *pteron*, wing, and *Celtis*). Ulmaceae. N and C China. 1 species, *P.tartarinowii*, a deciduous tree to 10m, with a broad crown and pale grey bark, peeling in flakes. To 10cm long, the leaves are ovate-oblong to ovate-lanceolate and serrate. The flowers are inconspicuous, the fruits circular, broad-winged, and to 2cm wide. Cultivate as for *Celtis*.

Pterocephalus (from Greek *pteron*, wing, and *kephale*, head). Dipsacaceae. Mediterranean to C and E Asia. Some 25 species, annual or perennial herbs, subshrubs and shrubs with entire or pinnately lobed leaves. Produced in summer, the flowerheads are rounded, flattened, long-stalked and subtended by narrow bracts. The flowers are pink to purple, the outermost conspicuously 2-lipped and

Pteris cretica
'Albolineata'

larger than those at the centre. The following is an undemanding, hardy mat-forming perennial with attractive foliage, flowers and seedheads. It is suitable for sunny well-drained situations in the rock garden, where it may self-seed when conditions suit. Propagate by seed, or by softwood or semi-ripe cuttings. *P.perennis* (Greece; cushion-forming perennial, to 10cm, usually shorter, tufted; leaves to 4cm, ovate to oblong, crenate to lyrate; flowers purple-pink in heads to 4cm in diameter, on 5–7cm stalks).

Pterostyrax (from Greek *pteron*, wing, and *Styrax*). Styracaceae. Japan, China, Burma. EPAULETTE TREE. 3 species, deciduous shrubs or trees with drooping racemes or panicles of bell-like flowers in summer. Hardy to –20°C/–4°F and fast-growing. The fragrant-flowered *P.hispida* is a small tree with shedding grey bark, which gives off a foetid odour when bruised. Grow in deep acid soil, in sun or semi-shade, allowing ample room for the branches to spread; if necessary, prune to shape after flowering. Sow ripe seed, after stratification at 5°C/40°F for three months. Increase also by semi-ripe cuttings in late summer. *P.hispida* (FRAGRANT EPAULETTE TREE; Japan, China; tree to 15m or shrub to 4.5–6m, bark grey, peeling, aromatic; leaves 7–17cm, oval to obovate; flowers white, in downy, pendulous panicles 12–25cm long).

puberulent, puberulous minutely pubescent; covered with minute, soft hairs.

pubescent generally hairy; more specifically, covered with short, fine, soft hairs.

puddling (1) the breakdown of soil surface structure by the impact of raindrops; (2) the practice of dipping the roots of transplants into a slurry of soil and water, traditionally held to aid establishment and discourage root pests; (3) sometimes used to describe the application of a heavy flow of water at the time of inserting a transplant; (4) the compression of a clay layer by treading or pounding, to form a natural waterproof lining to a pool.

Pueraria (for M.N. Puerari (1766–1845), professor at Copenhagen). Leguminosae. SE Asia, Japan. Some 20 species, herbaceous or woody twiners with trifoliolate leaves and pea-like flowers. Particularly useful as a rapidly growing screen, or, if unsupported, as groundcover. In good soils growth may exceed 15m during a season, as evinced by its performance in the SE US, where it has now become a notorious weed. In areas where winter temperatures fall much below –15°C/5°F, *P.montana* var. *lobata* may be grown as an annual; given adequate protection at the roots, however, it will re-sprout from the base in spring if cut down by frost. Grow in full sun on most well-drained soils, training the young stems to cover any support. Prune in spring, if necessary, to control spread. Propagate from seed

Pulmonaria
saccharata

in spring, sown singly in pots in the warm greenhouse; plant out when the danger of frost has passed. *P.montana* var. *lobata* (syn. *P.lobata*, *P.thunbergiana*; JAPANESE ARROWROOT, KUDZU VINE; China, Japan; deciduous, woody, hairy-stemmed vine to 20m; roots tuberous; leaflets to 18cm bright green turning yellow and copper in autumn; raceme erect, to 25cm; flowers to 1.5cm, purple fragrant, in summer and autumn).

pulhamite a type of artificial rock, noted for its adherence to geological detail, manufactured by James Pulham (1820–98) in the second half of the 19th century. It was made of clinkers and cement and moulded into strata and boulders for use in rock garden construction.

Pulmonaria (from Latin *pulmonarius*, relating to the lung, referring to the spotted leaves of *P.officinalis*, which resemble diseased lungs and were thought to be a cure for lung disease). Boraginaceae. Europe, Asia. LUNGWORT. Some 14 species, bristly-hairy, rhizomatous perennial herbs. The leaves are ovate to elliptic or lanceolate and often spotted or blotched. Produced from late winter to early spring in forked cymes, the flowers are bell-shaped to shortly tubular with five lobes. *Pulmonaria* provides hardy groundcover for open woodland and border edging, and is also suited to plantings in the sunless shade of buildings; in the wild garden, it is a valuable early nectar source for honey bees, especially in massed plantings. Grow in part to full shade in moist, humus-rich soils. In zones with cool summers, *P.saccharata* may be grown in full sun, provided that the soil remains adequately moist throughout summer. Increase by division in autumn or after flowering.

P.angustifolia (Europe; leaves to 40 × 5cm, typically unspotted, setose; flowers bright blue; includes cultivars with spotted or unspotted leaves, low-growing or tall, with flowers in shades of deep azure, clear blue, coral red and pink); *P.longifolia* (Europe; leaves to 50 × 6cm, usually spotted silver-white, setose and sparingly glandular above, setose and glandular beneath; flowers violet to blue-violet; includes cultivars with strongly spotted to solid silver leaves and flowers in tones of pale blue and deep azure, some tinted pink, also dusky mauve); *P.montana* (syn. *P.rubra*; Europe; leaves to 50 × 12.5cm, usually unspotted, setose and sparingly glandular-pubescent above; flowers violet to blue; includes cultivars with white, coral red and pink flowers); *P.officinalis* (JERUSALEM SAGE; Europe; leaves to 16 × 10cm, spotted white, setose; flowers red to rose-violet or blue; includes cultivars with white, pale blue, pink and opalescent flowers); *P.saccharata* (JERUSALEM SAGE; Europe; leaves to 27 × 10cm, green, usually spotted silver-white, long-setose, glandular-pubescent; flowers white or red-violet to dark violet; includes cultivars with white, pink- or blue-tinted, pale to deep blue, pink-tinted blue, mauve, rich indigo, pink and red flowers).

Pulsatilla pratensis 'Nigricans'

Pulsatilla (from the Latin *pulso*, strike – meaning obscure). Ranunculaceae. Eurasia, N America. 30 species, clump-forming, perennial herbs with finely dissected, downy leaves. Borne in spring and summer, the flowers are solitary and bowl- to bell-shaped. They consist (usually) of six oblong to elliptic segments with silky-hairy exteriors, and numerous golden stamens in a central boss surrounded by a ring of nectar-secreting staminodes. The globe-like fruiting heads are composed of feathery styles. Fine, spring-flowering perennials for the rock garden, alpine sinks and well-drained borders. The seed heads may be cut and used in dried-flower arrangements. All species here are fully hardy and tolerant of alkaline soils. Large plants have a deep woody rootstock and transplant poorly. Plant out young pot-grown specimens in full sun, in humus-rich, gritty, well-drained soils, and leave undisturbed. Increase by fresh seed or root cuttings.

P.alpina (ALPINE PASQUE FLOWER; C Europe; 20–45cm; flowers more or less erect, 4–6cm in diameter, white flushed blue-purple or palest creamy yellow, sericeous outside); *P.halleri* (C and SE Europe, Crimea; to 15cm; flowers erect or nearly so, 4–9cm in diameter, violet-purple to lavender blue; forms with lacinate petals are cultivated as well as others with semi-double flowers); *P.occidentalis* (N America; 10–60cm; flowers 4–6cm in diameter,

erect, white or cream-white, sometimes flushed purple or blue, exterior villous); *P.pratensis* (Central and Eastern Europe; flower stalks to 10cm, slender, hairy, flowers 2–4cm, nodding, campanulate to urceolate with recurved sepal tips, very hairy, pale to deep purple – black-violet in subsp. *nigricans*); *P.vernalis* (Europe to Siberia; to 15cm; flowers pendent, becoming erect, 4–6cm in diameter, with brown silky hairs outside and a white outer segment, usually flushed pink or violet-blue outside); *P.vulgaris* (PASQUE FLOWER; Great Britain and W France to Sweden, eastwards to Ukraine; 3–12cm; flowers 4–9cm in diameter, pale or dark violet, rarely white; many cultivars exist from dwarf to tall, woolly to almost smooth with coarsely to finely dissected leaves, and with flowers large or small, open or closed, erect or nodding, and in tones of white, lilac, mauve, violet, cobalt, plum, blood red and pink).

pulverulent powdery; covered in a fine bloom, dusty, or minutely freckled.

pulvinus a cushion of enlarged tissue, such as found on a stem at the insertion of a petiole.

pumice a light, porous aluminosilicate volcanic mineral, used in potting mixtures to improve drainage and aeration. It has low cation exchange capacity and contains some potassium and sodium.

pumpkins and squashes (*Cucurbitaceae*) loose collective terms for four species of *Cucurbita*: *C. maxima*, *C.argyrosperma* (*C.mixta*), *C.moschata* and *C.pepo*. They are annual herbs, which can be of bushy or more often trailing habit, and are usually separated into summer and autumn/winter harvesting groups. Summer squashes are usually boiled or fried when immature, before the rind hardens. Autumn/winter squashes and pumpkins are usually baked when mature, or stored for later use. Winter squashes are usually white-fleshed, whereas pumpkins are usually orange-fleshed, with types derived from each of the four species. Pumpkins embrace cultivars with coarse, strong-flavoured flesh, used for pies in the US, for Hallowe'en 'Jack o'lanterns' and for stock feed, and some grow very large. The acorns, buttercup and butternut groups are the most edible.

All pumpkins and squashes are frost-tender. They may be sown directly into the ground in appropriate conditions, otherwise in small pots or cell modules for planting out usually within two weeks of sowing. In frost-susceptible climates, seed must be sown in warm protected conditions for transplanting after the danger of frost is past. Spacing should suit the plant type and cultivar. Bush forms are typically sown on ridges or mounds, 60–90cm each way, with 90–120cm between rows, leaving one seedling per station when established; for trailers, leave 120–150cm between plants and 2–4m between rows. Planting into a black plastic ground mulch is a useful technique to suppress weed growth and reduce surface evaporation of soil moisture.

pumpkins and squashes

Pests
aphids
red spider mite
whiteflies

Diseases
virus
downy mildew
grey mould

Soil should be prepared with well-rotted manure or garden compost. Supplementary feeding with a balanced fertilizer at fortnightly intervals is valuable; foliar feed plants grown under plastic mulch. Irrigation is essential in dry periods.

Suitable cultivars are best stored over winter at temperatures of 10–16°C/50–61°F; curing fruits at 21–27°C/70–81°F for 2–3 weeks enhances sweetness in cultivars not subject to deterioration.

punctate dotted with minute, translucent impressions, pits or dark spots.

pungent ending in a rigid and sharp long point.

Punica (from the classical Latin name, *Malum punicum*). Punicaceae. E Mediterranean to Himalaya. 2 species, deciduous or semi-evergreen, densely branched shrubs or small trees. The flowers consist of a fleshy, tubular or more or less campanulate calyx with five to eight sepals, five to seven, silky petals, and numerous stamens. The fruit is a globose, pulpy berry packed with flesh-coated seeds inside a thick, leathery shell. Generally requiring long, hot continental summers for its fruit to ripen, *P.granatum* is nevertheless a handsome ornamental for warm south- or southwest-facing walls in regions where frosts are light and short-lived. In Mediterranean climates, it is sometimes used as hedging. Where size and fruit are not priorities, the dwarf cultivar 'Nana' is an altogether better garden plant, ideal in scale for patio tubs overwintered under cool glass. It is also hardier than the typical plant and may survive outdoors in sheltered rock gardens and against sun-drenched walls in climate zone 7. Grow in well-drained, fertile, moderately-retentive, loamy soils in full sun. Although it may be cut to the ground by severe frost, given wall protection and a deep mulch at the roots, *P.*'Nana' will re-sprout vigorously from the base. Prune in late spring or summer to remove old or weak wood. Shorten outward-growing shoots from trained specimens in spring at bud break – flowers are carried on the tips of the current year's growth. Propagate by suckers, by semi-ripe cuttings, by layering, or by grafting on to seedling understock. Increase also by seed in spring. *P.granatum* (POMEGRANATE; E Mediterranean to Himalaya; deciduous or semi-evergreen shrub to 2m, or small tree to 6m; leaves 2–8cm, obovate to oblong, pale green, glossy, subcoriaceous, tinted red in autumn; flowers to 3cm in diameter, silky, crumpled, scarlet; fruit to 12cm in diameter, globose, brown-yellow to purple-red, with a leathery rind and a chimney-like, persistent calyx; seeds with an edible fleshy crimson coat; includes 'Nana', a dwarf, intricately branched tree, with small leaves, flowers and fruit).

pupa see *chrysalis*.

Puschkinia (for Count Apollo Apollosovich Mussin-Pushkin (d. 1805), Russian chemist who collected plants in the Caucasus and Ararat). Liliaceae (Hyacinthaceae). Caucasus, Turkey, N Iran, N Iraq, Lebanon. 1 species, *P.scilloides*, a small, spring-flowering, bulbous perennial closely related to *Chionodoxa* and *Scilla*. Hardy to about –20°C/–4°F. Grow in sun or partial shade on gritty, well-drained, humus-rich soil. Flowering is usually better where bulbs can dry out in summer. Propagate by offsets when dormant in late summer, or by seed in autumn.

pustule a pimple-like or blister-like eruption.

Puya (from the Chilean vernacular name for the genus). Bromeliaceae. Andean S America, N Brazil, Guyana, Costa Rica. 168 species, perennial, terrestrial herbs, stemless or with long, stout, simple or branched stems. Packed in dense terminal rosettes, the leaves are tough to rigid, linear to narrowly triangular, the margins normally toothed and spinose. The tubular flowers are produced in great abundance in massive, spire-like pancles. Plant in a soil, rich in grit, sand and leafmould. Water liberally in hot weather; keep almost dry in winter. Under glass, grow in full sun with a buoyant, dry atmosphere and a minimum temperature of 5°C/40°F. Outdoors, light frosts may be tolerated where plants are kept dry and sheltered (see *Fascicularia*). Increase by seed or by offsets detached after flowering. *P.alpestris* (south central Chile; 1.2–1.5m in flower; leaves to 60cm, arching, covered in white scales beneath, hooked-spinose; inflorescence sparsely branched on a stout scape, loosely bipinnate, pyramidal; flowers petrol blue with orange-red anthers).

Pycnostachys (from Greek *pyknos*, dense, *stachys*, ear of corn, referring to the density of the flower spikes). Labiatae. Tropical and South Africa, Madagascar. About 40 species, erect perennial herbs or subshrubs. Borne in dense terminal spikes, the flowers consist of a spiny-toothed calyx and a 2-lipped, tubular corolla. Grow from seed in spring, at 13–15°C/55–60°F. Pot into a high-fertility loam-based potting mix. Give bright indirect light, a minimum temperature of 15°C/60°F and water plentifully. Propagate also by stem cuttings in early summer, or by softwood cuttings in autumn, after flowering.

P.dawei (tropical Central Africa; to 180cm; leaves to 30cm, linear to lanceolate, serrate, red-glandular beneath; flowers to 25mm, cobalt blue); *P.urticifolia* (tropical and South Africa; to 2.5m; leaves 4.5–12cm, ovate, crenate to deeply incised-dentate, smooth to hairy; flowers 12–20mm, gentian blue, rarely white tinted blue with a purple-red calyx).

Pyracantha (from Greek *pyr*, fire, referring to the colour of the berries, and *akanthos*, thorn). Rosaceae. SE Europe to China. FIRETHORN. 7 species, evergreen, viciously thorny shrubs, with obovate to oblong-lanceolate leaves and small, white to cream flowers in dense corymbs in late spring and summer. These give rise to small and spherical, orange to red fruits. Pollution- and exposure-tolerant evergreens for the border, where they form mounds of glossy green foliage, for training against walls, and for thorny and impenetrable hedging. Some, such as *P.atalantoides*, can be trained with a single stem to form small specimen trees. They are grown for the masses of white flowers in spring and early summer and for the profusion of long-lasting, brightly coloured berries. Given protection from cold winds, all species are tolerant of temperatures down to at least −15°C/5°F.

Grow in well-drained, moisture-retentive soils, in sun or part shade. Avoid very hot, sunny south-facing walls for wall-grown plants. Prune free-standing specimens after flowering to restrict size or to remove damaged wood. Clip back formal, wall-trained plants after flowering to give dense cover; otherwise, remove only outward-facing shoots, tying-in new growth in late summer. Cuttings may be taken from late spring onwards, but better growth is obtained from autumn cuttings: large, branched semi-hard cuttings (to 30cm), wounded and treated with hormone rooting compound and placed under mist, will root in 3–4 weeks, and will flower and berry the following season. Raise larger numbers of plants for hedging from fresh ripe seed; remove the flesh and sow in the cold frame in autumn. Susceptible to pyracantha scab which causes small scabby lesions on twigs, olive-brown spotting on leaves and flowers, and serious disfigurement of the fruit; it is often accompanied by premature leaf drop. Treat with systemic fungicide in spring and early summer. Leaves may also blister and discolour as a result of the larvae of the firethorn leaf miner moth, *Phyllonorycter leucographella*. As yet, there is no specific treatment for this pest, although control measures for other types of leaf miner may succeed.

P.angustifolia (SW China; shoots stiff, densely downy in the first year; leaves 1.5–5cm, linear-oblong to obovate, apex rounded and sometimes toothed; fruit to 1cm, felted at first, later yellow-orange; 'Gnome': very hardy, of erect, dense habit, with profuse flowers and orange fruit orange; 'Yukon Belle': of dense habit, with persistent orange fruit); *P.atalantoides* (SE to W China; branches olive-brown, downy only at first; leaves 3–7cm, lanceolate to elliptic or obovate, mostly entire, sometimes small-toothed; fruit 7–8mm, scarlet-crimson; 'Aurea': yellow fruit; 'Bakeri': red fruit; 'Nana': dwarf, to 1m, with red fruit); *P.coccinea* (PYRACANTH, FIRETHORN, BUISSON ARDENT; Italy to Asia Minor; shoots downy at first; leaves 2–4cm, elliptic to lanceolate; fruit 5–6mm, bright scarlet; 'Baker's Red': hardy, with red fruit; 'Kasan': compact, upright, fast-growing to 5m, with orange fruit; 'Lalandei': upright to 5m, hardy, with profuse, bright orange fruit; 'Lalandei Monrovia': upright, dense, with orange fruit; 'Sparkler': slow-growing, with leaves variegated white and red fruit); *P.rogersiana* (SW China; shoots pale downy at first becoming glabrous, red-brown; leaves 2–3.5cm, oblanceolate to narrow obovate, apex rounded and shallow toothed; fruit 8–9mm, yellow to orange-red; 'Flava': bright yellow fruit). *Cultivars*:

'Alexander Pendula', to 1.8m, with weeping branches and coral red fruit; 'Apache', forming a compact spreading mound 1.5 × 2m, with bright red fruit, resistant to scab and fireblight; 'Brilliant', with bright red fruit; 'Buttercup', spreading, with small, yellow fruit; 'Fiery Cascade', upright, 2.5 × 3m, with small, shining leaves and small, red, disease-resistant fruit; 'Golden Charmer', to 2.4m, with large, golden yellow fruit, resistant to scab; 'Harlequin', with variegated leaves and red fruit; 'Mohave', upright, to 4 × 5m, with bright orange fruit, unaffected by bird feeding and resistant to scab and fireblight; 'Navajo', low, dense, with orange-red fruit, resistant to scab; 'Orange Giant', with bright orange fruit; 'Pueblo', spreading, compact, with orange red fruit, resistant to scab and fireblight; 'Red Column', upright, with small, red fruit; 'Red Cushion', dense, to 1 × 2m, with mid-red fruit; 'Red Elf', dwarf, compact, mounding, with bright red fruit; 'Red Pillar', vigorous, upright, with orange-red fruit; 'Renault d'Or', with yellow fruit; 'Ruby Mound', low, dense, with orange-red fruit, resistant to scab; 'Shawnee', dense, spreading to 3.5m, with abundant, yellow-orange fruit, colouring early, resistant to scab and fireblight; 'Soleil d'Or', spreading, with pale green leaves and yellow fruit; 'Sunshine', open, orange fruit; 'Tiny Tim', with dwarf habit, to 1m, densely leafy, with profuse fruit; 'Watereri', compact, vigorous to 2.5m, with profuse, orange fruit.

pyramid see *dwarf pyramid*.

pyrene the nutlet of a drupe or drupelet; a seed and the bony endocarp that surrounds it.

pyriform pear-shaped.

Pyrola (diminutive of the Latin *Pyrus*, pear, alluding to the supposed similarity of the leaves). Pyrolaceae. Northern temperate regions. WINTERGREEN, SHIN-LEAF. 15 species, perennial, dwarf herbs with creeping rhizomes and slender-stalked leaves in basal clusters. Borne in summer in scapose, loose racemes, the flowers consist of five, broad, cupped petals. Fully hardy and generally grown in wild gardens; it is also suitable for cool positions in the rock garden, peat terrace and at the woodland edge. Plant in shade on damp, sandy acid soils containing leafmould and pine needles. Avoid disturbance – any damage to the wide-spreading fibrous-feeding roots may cause either death or prevent any further increase in size. Increase by division, or by seed, which germinates infrequently and should be sown on moist sphagnum moss.

P.asarifolia (BOG WINTERGREEN; Canada, E US, Asia; to 65cm; leaves broadly elliptic to reniform, to 10cm, entire or crenate, often purple beneath; flowers pink to crimson; includes var. *purpurea*, from N America, with leaves sometimes red beneath and large, fragrant, bright violet or rose flowers); *P.rotundifolia* (Europe, N America; to 25cm; leaves round to oval; flowers white).

Pyrostegia (from Greek *pyr*, fire, and *stege*, covering, referring to the fiery corolla). Bignoniaceae. Americas. 3–4 species, lianes climbing by tendrils. The leaves consist of paired leaflets with a tendril between them. Borne in terminal sprays usually in summer, the flowers are showy, tubular to club-shaped, with the curved corolla tube swollen and expanded towards its apex. Grow in sun in a fertile, well-drained medium. Maintain a humid atmosphere and a minimum temperature of about 10°C/50°F. Water plentifully in summer; sparingly in winter. Prune in late winter early spring, cutting back flowered stems to within 30cm of the base. *P.venusta* works well if trained to a single overhead stem, with flowered shoots cut back to 2–3 buds each spring. Propagate by semi-ripe cuttings of 2–3 nodes in summer. *P.venusta* (syn. *Bignonia ignea, B. venusta, P.ignea*; GOLDEN SHOWER, FLAME VINE, FLAMING TRUMPET, ORANGE-FLOWERED STEPHANOTIS, TANGO; Brazil, Paraguay, Bolivia, NE Argentina; flowers 3.5–6 × 0.3cm, fiery orange).

Pyrus (classical Latin name for the pear). Rosaceae. Europe to E Asia and N Africa. PEAR. About 30 species, deciduous trees and shrubs, occasionally thorny. They are grown for their habit and foliage, their white to cream, 5-petalled flowers borne in spring, and for their fruit, in some cases ornamental, in others, edible (see *pears*).

Most species are hardy to at least –15°C/5°F and are tolerant of atmospheric pollution and a range of soil types. The genus includes a number of attractive foliage trees, notably *P.salicifolia* 'Pendula'. *P.communis*, is especially beautiful in flower, and is worth growing for this alone, even in regions where frosts make fruit set unreliable. *P.calleryana* 'Bradford' flowers profusely in spring and colours well in autumn, in shades of russet and red. It is commonly grown as a street tree in the US. The narrowly conical *P.calleryana* 'Chanticleer' is more suitable for the smaller garden. Grow in full sun. Propagate cultivars by budding or grafting on to seedling rootstocks of *P.communis*. Propagate species by seed, although progeny may be variable; stratify for 8–10 weeks, at 1°C/34°F, and sow in spring. Temperatures above 15–20°C/60–70°F in the sowing medium may induce a secondary dormancy. Most of the pests and diseases are as for *pears* (q.v.).

P.amygdaliformis (S Europe, Asia Minor; shrub or small tree to 6m; shoots slender, thorny; leaves 2.5–7cm, ovate-obovate, pointed or blunt, coriaceous, entire or finely scalloped, grey-felted at first becoming shiny green, glabrous; fruit 2–3cm, rounded, yellow-brown; includes var. *oblongifolia*, with elliptic to oblong leaves, with a blunt apex and rounded base, and larger fruit, yellow flushed red); *P.calleryana* (CALLERY PEAR; China; tree to 10m; shoots thorny; leaves 4–8cm, ovate, short tapered, undulate-scalloped, shiny, glabrous; fruit 1cm, rounded, brown, pitted; 'Aristocrat'. fast-growing to 13m, broadly pyramidal, with thornless branches, wavy-edged, glossy leaves, brilliant red in autumn, plum when young, and red to yellow fruit; 'Autumn Blaze', fast-growing, to 12m, of asymmetrical and open habit, with glossy leaves and crimson autumn colour; 'Bradford', to 13m, with a broadly ovate, dense crown and red to maroon leaves in autumn; 'Capital', to 12m, loosely pyramidal, dense, with rich glossy green leaves, purple in autumn; 'Chanticleer', to 13m, with a narrow-conical crown, evenly branched, with glossy green leaves, becoming carmine scarlet in autumn; 'Redspire', to 12m, with a moderately pyramidal crown, upright, well-branched, with long, thick, glossy leaves, coloured crimson to purple in autumn, and profuse flowers; 'Stone Hill', to 12m, upright, with an oval crown, deep green, shiny leaves, brilliant orange in autumn; 'Trinity', to 10m, with a rounded crown, light green, somewhat serrate leaves, orange to red in autumn; and 'White House', to 10m, upright, compact, with crown ovate in outline and red to plum leaves in early autumn); *P.communis* (COMMON PEAR; Europe, Asia Minor; tree to 15m, crown conical, broad, bark dark, fissured to flaking; branches often thorny; branchlets, crabbed; leaves 2–8cm, oval-elliptic, pointed, scalloped, soon glabrous, autumn colour red-yellow; fruit 2.5–5cm, pyriform or sub-globose, yellow green, flesh bitter; var. *sativa* is the collective name for all edible pear cultivars); *P.salicifolia* (WEEPING PEAR, SILVER PEAR; SE Europe, Asia Minor, Caucasus; tree 5–8m, shoots slender, arching to pendulous; leaves 3–9cm, narrowly oblong-elliptic, apex acuminate, entire or sparsely toothed, silvery grey-downy, becoming more or less glabrous above; flowers cream; fruit 2–3cm, pyriform, green; 'Pendula', with weeping branches and narrow, silvery leaves; 'Silfrozam', with a broadly weeping crown and willow-like, silvery grey leaves).

pyxis a capsule exhibiting circumscissile dehiscence.

Q

quadrate more or less square.

quadripinnate pinnately divided into four pinnae, or groups of four pinnae.

quassia a bitter extract produced by boiling wood fragments of the tree *Picrasma excelsa.* It is used as a non-toxic deterrent to dogs, cats, rabbits and birds, applied repeatedly in the form of sprays. When used as an insect repellent, it is sometimes combined with derris.

quenouille a form of ornamental wall-trained fruit tree, with a central stem and tiers of successively shorter horizontal branches forming a triangle in outline.

quercetum a plantation of oak trees.

Quercus (Latin name for this tree). Fagaceae. N America to western tropical S America, N Africa, Europe and Asia, in temperate and subtropical zones, and tropics at high altitudes. OAK. 600 species, evergreen or deciduous trees or shrubs. The male flowers are inconspicuous and crowded in slender pendulous catkins. The female flowers are solitary or two to many in a spike. They give rise to a single-seeded nut (acorn). This is ovoid to globose and partly or almost wholly enclosed by a cup-shaped, scale-covered involucre (cupule), composed of adpressed scales.

Oaks are excellent trees for parks and large gardens, having interesting foliage and often good crown shape; most are long-lived and many are large trees. All oaks like warm summers. The constraints of low summer heat affect the range and health of species which can be grown in Britain; oaks from continental E US, China and Japan are generally of poor vigour and many suffer autumn frost damage to the unripened shoots. Of these, the most tolerant is *Q.rubra,* almost as vigorous as in its native E US. *Q.alba* has very rarely been successful in the UK except in the warmest parts of SE England. Conversely, the Mediterranean oaks, adapted to rela-

tively cooler summers during the ice ages, thrive in Britain's climate, responding to the higher rainfall by growing faster in S England than in their native areas.

In the eastern US, all oaks can be grown in either the northeast, or (the less winter-hardy species) in the southeast. None reacts adversely to the hot, humid summer. The European oaks retain considerable winter hardiness from ice age adaptations: *Q.petraea* from N England is hardy in Wisconsin and also enjoys the warmer summers there. *Q.cerris* from S Europe is hardy to zone 6. No oak shows extreme cold tolerance, and the genus is absent from most of Canada and Siberia.

Growth is best on fertile, deep soils with medium drainage and 400–1500mm rainfall, but in such a large genus, different oaks can be found to tolerate almost all conditions, from poor dry acid sands, freely drained mountain screes, through heavy damp clays to ill-drained marshy areas and dry chalk. Only acid deep peats are consistently avoided. Most are sensitive to salt, but *Q.ilex* is often planted in maritime gardens. Most tolerate exposure only moderately, surviving best in a woodland site, among or not far from other similarly sizes trees or shrubs.

Propagation is by seed, or by grafting where seed is unavailable and for desirable cultivars. For grafting, side graft sucker shoots (preferably erect) on seedlings of a closely related species and bind with tape. Normal branch shoot scions should be avoided if possible, as they have poor apical dominance and may not develop into good trees.

As many oaks hybridize freely in cultivation, wild-collected seed should be used whenever possible. Even in the wild hybrids are frequent. Acorn production is usually cyclical: there is low production in most years to defeat parasites and predators, with, more sporadically, a massive crop in the 'mast' years, where a high proportion of seed escapes damage. If possible, seed collection should be made in such years; crop assessment can be made from midsummer. Collect the acorns in mid-autumn after falling.

quince

quince

Diseases
brown rot
fireblight

Acorns must be stored moist and cold; they will be killed if they drop below 60% fresh weight. They cannot normally be stored for more than one winter, except in the case of very hardy species (zone 6 origin or colder); these can be deep-frozen. For freezing, the acorns, which must not be showing any sign of germination, should be *slightly* dried to about 80% fresh weight (to prevent ice damage to the cells) and acclimatized first for a few days at 0°C/32°F, then –5°C/41°F and finally to –20°C/–4°F. The seed should be stratified for 3–5 months in damp sand refrigerated at 0–2°C/32–36°F before sowing, stored in plastic bags.

Acorns commonly start to germinate in early winter (even those stored at 0°C/32°F), establishing a root system (though not a shoot). These sprouting acorns must be planted immediately, the others in early spring. Lie the acorn horizontal at its own thickness under soil surface. Place in a cold frame in deep narrow pots to accommodate the long tap-root. If the seed supply is plentiful and high losses to rodents are acceptable, it is always preferable to avoid checks to growth by planting the acorns directly where the tree is to grow.

Oaks are best planted as small trees, one to two years old and 10–40cm tall. Larger, older plants require special nursery treatment; for these, sever the tap-root early and regularly (every few weeks) either by cutting under the seedbed with a spade, or pruning roots that appear at the bottom of the pot; this will encourage the seedling to develop a more fibrous root system. Plant when 1–2m tall; lengthy check can occur. Water well, protect from rabbits, and keep weeds clear until established.

Oaks are affected by numerous pests and diseases. though only a few are serious. In North America, oak wilt has caused heavy losses. The red oaks are most affected; white oaks are more resistant and suffer only scattered shoot death and only rarely whole tree death. *Q.robur* is commonly defoliated by tortrix moths, but the tree merely grows a second flush of leaves; it is also attacked by numerous gall-wasps, but these also do no real harm. *Q.coccifera* is host to the Kermes insect. Acorns are commonly infested by weevils and gall wasps, including the conspicuous 'knopper gall' on *Q.robur* and *Q.petraea;* they may account for most of the crop in years of low seed production, but in mast years most of the very large seed crops will escape their attention.

quicklime see *lime.*

quiescence the inability of a seed to germinate, particularly where due to insufficient hydration. It is also used of other plant parts that fail to grow due to suboptimal environmental conditions or natural dormancy.

quilled applied to petals that are inrolled into a narrow tube for much of their length, as in certain dahlias and chrysanthemums.

quinate possessing five leaflets emanating from the same point of attachment.

quince *(Cydonia oblonga)* A small deciduous tree belonging to the family Rosaceae. Grown for its golden-yellow fruits, which are downy, strongly aromatic, and of gritty texture, it is used for preserves, in cooking and as a herbal medicine. The seeds are poisonous. Quince is of uncertain origin but grows wild in Iran, Anatolia, Turkestan and Transcaucasia; it was highly regarded by the ancient Greeks.

Quince makes an attractive small specimen tree, with white or pale pink flowers in spring and yellow foliage in autumn. It may be grown as a standard, bush, fan or espalier. Selections have been grown as rootstocks for pears for many centuries and quince is propagated by chip budding on to quince rootstock or by hardwood cuttings or suckers. Trees should be planted about 4m apart. The flowers are self-fertile and fruits are best picked when turned golden yellow. Store in a dark well-ventilated place, unwrapped and separated from other fruits to avoid aroma taint; they will keep for 2–3 months under suitable conditions. Quince bears its fruits on spurs and on shoot tips of the previous year's growth. Apart from the need to remove congested growth, pruning required for established trees is minimal.

quincunx an arrangement of five trees with one at each corner of a square and one in the centre, favoured from ancient times to the 17th century. The term is also often loosely applied to a square planting of trees.

Quisqualis (from Latin *quis*, who, and *qualis*, what kind – there was at first some uncertainty as to the family to which this plant belonged.) Combretaceae. Tropical and South Africa, tropical Indomalaysia. Some 17 species of scandent shrubs often climbing by woody hooks (persistent petioles). Borne in spring and summer in racemes or panicles, the flowers are narrowly tubular, with five spreading, oblong lobes. Grow in sun or part-shade in well-drained and not too fertile soil. Provide a minimum temperature of 10°C/50°F and bright filtered light or full sun with shade during the hottest summer months. Water plentifully when in growth. Reduce water in winter to allow a period of almost dry rest. In late winter, thin congested growth and shorten flowered stems back to the main framework of growth. Propagate by heeled softwood cuttings or by seed. *Q.indica* (DRUNKEN SAILOR, RANGOON CREEPER); Old World Tropics; rampant climber; flowers fragrant to 6cm, white becoming pink to red on upper surfaces and within throat over a period of days).

Quisqualis indica

QUERCUS

Name	Distribution	Habit	Bark	Leaves	Fruit
Q.acutissima SAWTHORN OAK	China, Korea, Japan	deciduous tree, seldom exceeding 25m	ashen to black corky, fissured	6–20cm, lanceolate to obovate, apex long acuminate, base cuneate to rounded, teeth broadly triangular, bristle-tipped, lustrous green above, paler and glabrous beneath	acorns to 2.5cm, solitary, ovoid; cupule to 3.5cm diameter, concealing two-thirds of acorn, covered with long, hairy scales
Q.agrifolia CALIFORNIA LIVE OAK	California	evergreen tree to 25m, often shrubby in cultivation; crown broad	thick, smooth to grooved, grey, red or black	2.5–7cm, ovate-elliptic or broadly elliptic, apex acute, base rounded, spiny-toothed, deep glossy green and rather concave above, paler beneath, glabrous except for tufted hairs in axils of veins	acorns 2–3.5cm, ovoid, acute, solitary; cupule enclosing one-quarter to one-third of acorn, sericeous
Q.alba WHITE OAK	E US	deciduous tree in 45m; crown rounded	pale grey to brown, exfoliating in plates	10–20cm, obovate, oblong or elliptic, apex abruptly acute, base cuneate, lobes 3–4 per side, entire or sparsely toothed, initially pubescent, dull mid-green above, orange to burgundy in fall, glaucous beneath	acorn 1–3cm, ovoid-oblong; cupule enclosing quarter to one-third of acorn, grey-white, composed of adpressed, compacted, hairy scales

Comments: f. *elongata*: leaves slender, to 25cm, orange to purple red in fall. f. *pinnatifida*: leaves deeply pinnatisect, lobes slender, dentate. f. *repanda*: leaves very shallowly lobed. Hybrids between *Q.alba* and *Q.bicolor*. are referred to as *Q. ×jackiana.*

Name	Distribution	Habit	Bark	Leaves	Fruit
Q.aliena	Japan, Korea	deciduous tree to 25m		10–20cm, obovate to oblong, apex acute to rounded, base cuneate, sinuately dentate, teeth coarse, blunt, 10–15 per side, yellow-green glabrous above, grey-white to blue-green and lightly hairy beneath	acorns 2–2.5cm, ovoid, solitary or grouped 2–3; cupule enclosing one third of acorn, grey-tomentose with scales adpressed

Comments: var. *acuteserrata*: leaves smaller and narrower, teeth acute, rather incurved, gland-tipped.

Name	Distribution	Habit	Bark	Leaves	Fruit
Q.alnifolia GOLDEN OAK	Cyprus	evergreen shrub to 2m or tree to 8m	coarsely textured, grey	2.5–6cm, broadly obovate to suborbicular, apex and base rounded to subtruncate, denticulate above, glossy dark or livid green above, with ochre or ash-coloured felty hairs beneath	acorns 2.5–3.5cm, obovoid, mucronate, solitary or paired; cupule small, enclosing to half acorn, scales spreading, long-pubescent, arranged in a broad band
Q.canariensis ALGERIAN OAK, MIRBECK'S OAK	N Africa, Iberian peninsula (not Canaries)	deciduous tree, to 40m tall in habitat	thick, rugged, black	5–18cm, oval or obovate, apex rounded, base tapering, rounded or subcordate, margins obtusely lobed or coarsely toothed, coriaceous, rusty-floccose at first, becoming dark green and glabrous above, paler beneath, glabrous except for rusty residual hairs on veins	acorns to 2.5cm, solitary, paired or in small clusters; cupule with adpressed downy scales, enclosing one-third of acorn
Q.castaneifolia	Caucasus, Iran, N Africa	deciduous tree to 35m; crown large, rounded	brown, rough-textured and corky	leaves 6–16cm, narrowly elliptic to oblong-lanceolate, apex acute, base cuneate or rounded, with coarse triangular aristate saw-teeth, deep glossy green above, dull grey-green beneath, glabrous or minutely downy	acorns 2–3cm, ovoid, often dorsally compressed, solitary or to 5 on a stout stalk; cupule clothed with slender scales (sometimes reflexed), enclosing one-third to half of acorn
Q.cerris TURKEY OAK	C and S Europe, Asia Minor	deciduous tree to 43m; crown conical	grey-white, splitting into thick plates	5–12cm, oblong-lanceolate, tapered to an acute apex and a rounded or truncate base, lobes or teeth triangular to lobulate, mucronulate, dark green and stellate-pubescent initially above, lighter and pubescent to tomentose beneath, yellow-brown in fall	acorns in groups of 1–4, to 3.5cm, ellisoid, apex subtruncate and mucronate; cupule of subulate often spreading or reflexed scales, enclosing one-third of acorn

Comments: 'Argenteovariegata': leaf margins variegated cream-white. 'Aureovariegata': leaves yellow-variegated. 'Pendula': branchlets pendulous.

Name	Distribution	Habit	Bark	Leaves	Fruit
Q.coccifera KERMES OAK, GRAIN OAK	W Mediterranean,S Europe, NW Africa	evergreen bushy shrub, usually 0.25–1.5m, rarely a tree 4.5m	smooth and grey at first, cracked with age	1.2–3.5cm, oval, elliptic or oblong, apex acute or rounded, mucronate, base rounded or cordate, becoming leathery, dark and glabrous above, paler and glabrous or with a few hairs in vein axils beneath, margin undulate, 4–6 pairs of spreading, spine-like teeth	acorns to 1.5×3cm, ovoid or oblong-ovoid, apex attenuate with an apical spine light brown striped darker; cupule enclosing one-half to two-thirds of acorn, scales spiny, often reflexed, lightly puberulent.

Comments: ssp. *calliprinos*: PALESTINE OAK, SIND OAK; tree to 12m; leaves larger and more oblong, to 5cm; acorn to three-quarters enclosed by cup. East Mediterranean, South West Asia.

Name	Distribution	Habit	Bark	Leaves	Fruit
Q.coccinea SCARLET OAK	E US, S Canada	deciduous tree to 30m	pale grey-brown, cracked into irregular, scaly plates	8–12cm, oval or obovate, rarely oblong, base truncate, rarely cuneate, silky at first, becoming glabrous and glossy, paler and with tufts of hair in vein axils beneath, scarlet in fall, margin deeply pinnately sinuate-dentate or serate, lobes usually 7	acorns 1.5–2.5cm, ovoid to subspherical, apex and base rounded, with thin red-brown tomentum; cupule turbinate, scales yellow-brown, oval, subglabrous, enclosing one-third to one-half of acorn

Comments: 'Splendens': leaves brilliant scarlet in fall.

QUERCUS Name	Distribution	Habit	Bark	Leaves	Fruit
Q.dentata DAIMIO OAK, JAPANESE EMPEROR OAK	NE Asia,Taiwan	deciduous tree, fast-growing to 20–25m; crown large, rounded	brown, fissured and split into grey-scaly plates	15–50cm, orbicular-obovate to oblong-obovate, base attenuate to cordate, apex attenuate or rounded, sinuately lobed, lobes large, rounded, often mucronulate, pale yellow tomentose at first, becoming dark green and glabrous except on nerves above, paler and densely tomentose beneath	acorns 1.2–2.4cm, ovoid to subglobose, glabrous, apex rounded, mucronate; cupule scales adpressed, upper scales erect to recurved, enclosing more than half of acorn

Comments: 'Pinnatifida': leaves crispate, deeply and narrowly lobed

Q.ellipsoidalis NORTHERN PIN OAK	NEast US	deciduous tree to 20m	grey-brown, fissured into narrow plates	6–15cm, obovate or elliptic, base truncate or cuneate, margin deeply lobed, lobes 5–7, oblong, with 1–3 teeth, silky-tomentose at first, becoming glabrous except in vein axils beneath, dark green and shining above, paler beneath, yellow to light brown with red markings in fall	acorns 1.2–2cm, solitary or paired, ellipsoid, mucronate, puberulent, chestnut-brown with darker lines; cupule light red-brown, scales adpressed, finely pubescent, enclosing one-third to one-half of acorn
Q.frainetto HUNGARIAN OAK	S Italy, Turkey, Balkans	deciduous tree to 30m	smooth with small scale-like plates, dark grey; crown large; branches sometimes pendulous	8–25cm, obovate or oblong-obovate, apex rounded, base subcordate or truncate, margins pinnatifid, lobes 6–10(–12) pairs, sometimes lobulate or sinuate, initially white-yellow-tomentose, becoming glabrous or stellate-pubescent above, paler glaucescent beneath	acorns 1.25–3.5cm, oblong-ellipsoid or ovoid-oblong, apex rounded, mucronate; cupule densely tomentose outside, scales adpressed, enclosing up to one half of acorn
Q.garryana OREGON WHITE OAK	WN America	deciduous tree 10–30m; crown rounded, branches crooked, ascending or pendulous	pale grey, shallowly cracked	leaves 10–15cm, obovate or oblong-obovate, apex rounded, base cuneate, rounded, or subcordate, margins slightly revolute, lobes 3–5 per side, entire or dentate, dark green and glabrous above, paler and soft pubescent beneath	acorns 2–2.5cm, ovoid, base truncate, apex rounded, mucronate, glabrous; cupule shallow, puberulent inside, scales adpressed, pubescent
Q. × heterophylla (*Q.phellos* × *Q.rubra*)	US	deciduous tree 20–25m	smooth, pale-grey or brown.	10–18cm, oblong-lanceolate to obovate, apex mucronate, base cuneate, entire or with few shallow teeth, or with 4–6 shallow or deep, often mucronate lobes each side, ultimately glabrous or slightly hairy on veins beneath	acorns closely resembling those of *Q.rubra*
Q. × hispanica (*Q.cerris* × *Q.suber*)	S France to Portugal and Italy, Balkans	semi-evergreen tree or shrub to 30m	less corky than in *Q.suber*	4–10cm, oblong-elliptic, acute, margin sinuate-dentate, teeth small, mucronulate, dark green and sparsely hairy above, densely yellow-white tomentose beneath	acorns 3–4cm, oblong-ovoid, mucronate, mostly glabrous; cupule hemispherical to ovoid, tomentose scales erect to reflexed, enclosing half or more of acorn

Comments: 'Ambrozyana': leaves 6–10cm, usually rather small, oblong-obovate, lobes subulate-mucronate, dark green and shining above, grey-tomentose beneath. 'Crispa': compact shrub or small tree; bark more corky; leaves 5–8cm, crispate, densely white tomentose beneath. 'Dentata': bark corky; leaves more closely resembling *Q.cerris*, coarsely dentate, dark green above, grey-tomentose beneath; sometimes confused with 'Fulhamensis'. 'Diversifolia': small tree; bark very corky; leaves 5×2cm, ovate, each side with conspicuous sinus, thus resembling *Q.cerris* leaves, margin lobed in lower half, entire to denticulate above; acorn cup hemispherical, scales more adpressed than in hybrid type. 'Fulhamensis': graceful tree; branch more slender and bark less corky; leaves 7.5–9×3.5cm, ovate, apex acute, base rounded to subcordate, margin dentate, teeth 5–8 each side, apex acute, tomentum white; often confused with and grown as 'Dentata'. 'Fulhamensis Latifolia': leaves to 8×6cm, larger than 'Fulhamensis', apex obtuse, margin teeth wider and shallower than 'Fulhamensis', tomentum grey. 'Heterophylla': closely resembles *Q.cerris*. var. **haliphloeos**: leaves oblong, irregularly lobed, sometimes with conspicuous broad sinus each side of mid-rib. 'Lucombeana': crown conical; bark only slightly corky; leavs resemble those of *Q.cerris*, 6–12×2.5–4cm, oblong, coarsely dentate, teeth 6–7(9) each side, green, glabrous, shining above, paler and tomentose beneath; acorn to 2.5cm, cupule with lower scales reflexed, upper scales erect, enclosing more than half of acorn.

Q.ilex HOLM OAK, EVERGREEN OAK, HOLLY-LEAVED OAK	Mediterranean	evergreen tree to 20m; crown broad, rounded	smooth, grey, becoming shallowly split	2–9cm, sometimes wider, narrowly elliptic, ovate-lanceolate or suborbicular, slightly concave beneath, apex acute or obtuse, base attenuate or slightly cordate, entire or with mucronate, acuminate teeth, leathery, subglabrous, above, densely pubescent beneath	acorns in groups of 1–3, 1.5–3cm, oblong-ovoid to subglobose, mucronate, grey-brown with darker lines, lightly lanate becoming pulverulent; cupule scales adpressed, triangular, obtuse, tomentose, enclosing about half of acorn

Comments: 'Rotundifolia': leaves suborbicular. 'Crispa': leaves 1.5cm, suborbicular, margins slightly revolute. 'Fordii': crown narrow. Leaves 2.5–3.75cm, oblong, apex and base attenuate, margin undulate, entire or dentate. forma **microphylla**: leaves 2–2.7cm, elliptic, margins spiny-dentate.

Q.imbricaria SHINGLE OAK	E and C US	deciduous tree 15–20cm; crown conical to rounded	light brown, smooth, split by narrow cracks into plates	10–17cm, oblong-lanceolate to ovate, apex acuminate, apiculate, base attenuate or cuneate, entire, often crispate undulate, revolute, occasionally lobed, initially red-puberulent above and thick white-tomentose beneath, becoming dark green, glabrous above, soft-tomentose beneath	acorns 1–1.5cm, subglobose, mucronate, subglabrous to silky-pubescent; cupule scales adpressed, hairy, enclosing up to half of acorn

QUERCUS

Name	Distribution	Habit	Bark	Leaves	Fruit
Q.ithaburensis VALLONEA OAK	Syria, Palestine	semi-evergreen to deciduous tree to 25m	bark furrowed, dark brown	4–9cm, ovate to lanceolate-elliptic, apex acuminate, base subcordate, teeth 5–9 each side, triangular, aristate, grey-tomentose becoming glabrous above, persistent grey-tomentose beneath	acorns clustered 1–3, 2.5–4.5cm, ovoid, apex obtuse or depressed; cupule to 5×5cm, with broad, thick, flattened, woody scales enclosing two thirds to nearly all of the acorn

Comments: Subspecies *macrolepis*: leaves more deeply incised; acorn cupule scales more flexible and less woody

Q.laurifolia LAUREL OAK	SE US	deciduous to semi-evergreen tree to 30m	smooth, dark brown to black, fissured	6–13cm, oblong to obovate, apex acute to rounded, base tapered, acute or rounded, margin entire or with to 3 shallow teeth at apex, glossy above, paler and initially pubescent beneath, midvein and petiole yellow	acorns solitary or paired, 1–1.5cm, ovoid-globose, black-brown; cupule scales adpressed, obtuse, enclosing only the base of acorn, subsessile
Q.macranthera CAUCASIAN OAK, PERSIAN OAK	Caucasus, N Iran	deciduous tree 25–30m		8–20cm, obovate, apex rounded or blunt, base rounded, subcordate or subcuneate, serrate, teeth 7–11 each side, rounded or obtuse, rarely mucronate, dark green, glabrous above, densely yellow or red-brown tomentose beneath	acorns 1.6–2.5cm, ovoid-ellipsoid, almost glabrous; cupule scales erect to spreading, lanceolate, pubescent, enclosing half of acorn
Q.macrocarpa BURR OAK, MOSSY CUP OAK	C and NE US, SE Canada	deciduous tree to 50m; crown rounded	rugose, split into irregular plates, light grey-brown	10–45cm, obovate to oblong, base attenuate, cuneate or cordate, lyrate, lobes 5–7 per side, dark green and glabrous above, paler, glaucous or white-tomentose beneath, pale-yellow to brown in fall	acorns 2.5–4cm, ovoid to hemispherical, mucronate, softly tomentose; cupule hairy, lower scales adpressed, upper scales with free, spreading apices, enclosing half or more of acorn
Q.marilandica BLACKJACK OAK	C and SE US	small deciduous, slow-growing tree, 6–15m	rough, black-brown, split into square plates	10–17cm, broadly obovate, apex obtuse, often divided into 3 mucronate lobes, base rounded to cordate, dark green, shining above, paler and red-brown tomentose beneath, yellow to brown in fall	acorns solitary or paired, 1–2cm, ovoid; cupule hairy, made up of broad, adpressed scales, enclosing one- to two-thirds of acorn
Q.mongolica	NE Asia	deciduous tree to 30m		10–20cm, obovate, apex obtuse, base attenuate, cordate, teeth rounded, glabrous above, glabrous or pubescent on veins beneath	acorns 2cm, ovoid; cupule made up of tuberculate scales, uppermost scales acuminate, ciliate, enclosing one-third of acorn

Comments: Largely represented in cultivation by var. *grosseserrata*: leaves 10–20cm, teeth acute, sometimes themselves dentate; acorn cupule made of closely adpressed scales, these are not ciliate.; native to Japan, Sakhalin and the Kuriles

Q.muehlenbergii YELLOW CHESTNUT OAK, CHINKAPIN OAK	E US	deciduous tree to 20m, rarely 45m	grey, split into thin, flaky scales	10–18cm, oblong, obovate or lanceolate, apex acute or acuminate, base cuneate, serrate, teeth acute, pale green flushed bronze at first, subglabrous above, tomentose beneath, becoming glabrous above, white silky-pubescent beneath, orange or scarlet in fall	acorns solitary, 1.3–3cm, ovoid, mucronate, silky-pubescent; cupule made up of tomentose scales, upper most scales acute, forming fring, enclosing half of acorn
Q.myrsinifolia	Japan, China, Laos	evergreen tree 12–25m		5–15cm, lanceolate or lanceolate-elliptic, apex long-acuminate, base rounded or shortly acuminate, leathery, dark green above, glaucescent, slightly papillose beneath, margin with fine teeth in upper half	acorns 17–25×7–10mm, narrowly ovoid, puberulent at first, becoming glabrous, crowned with persistent styles; cupule slightly glaucous, ashy-puberulent outside, silky inside, with scales forming 7–9 rings
Q.nigra WATER OAK	SE US	deciduous tree 20–35m	brown, smooth initially, becoming black-brown deeply channelled.	3–15cm, obovate with shallowly lobed apex, lobe tips bristle-like, or oblong and entire, leathery, matt blue-green and glabrous above, glabrous but for tufts of hair in axils of veins beneath	acorns solitary, 1–1.5cm, globose to ovoid, black; cupule made up of short, adpressed scales, enclosing up to half of acorn
Q.palustris PIN OAK	NE US, SE Canada	deciduous tree, 20–25(–35)m; crown dense, ovoid-conical	smooth, grey-brown becoming fissured and ridged	8–15cm, obovate, apex acute, base attenuate or truncate, pinnately lobed, lobes oblong to triangular, tips slightly lobed and bristle-like, glabrous and glossy above, paler and glabrous except for tufts of green-white hair in vein axils beneath	acorns 1.2–1.7cm, mucronate; cupule saucer-shaped, puberulent, red-brown, enclosing only base of acorn

Comments: 'Crownright': crown narrowly conical. 'Pendula': branches pendent. 'Reichenbachii': young leaves and shoots flushed red. 'Umbraculifera': crown rounded, symmetrical; leaves shining green, colouring red in autumn.

QUERCUS

Name	Distribution	Habit	Bark	Leaves	Fruit
Q.petraea	Europe to W Russia	deciduous tree to 45m; crown regular, trunk reaching far into crown	bark grey to black-brown, furrowed	6–17cm, broadly or narrowly obovate, base attenuate or truncate to subcordate but not auriculate is as *Q.robur*, apex rounded, margin with 4–6 pairs of even, rounded lobes, glossy, glabrous above, paler, glaucous, glabrous to pubescent beneath	acorns clustered, 2–3cm, ovoid to oblong-ovoid; cupule made up of closely adpressed, pubescent scales

Comments: 'Albovariegata': leaves variegated with white. 'Aurea': young branchlets yellow; leaves initially yellow, becoming green but petiole and nerves persistently yellow. 'Aureovariegata': leaves yellow-green variegated. 'Cochleata': leaves more leathery, slightly convex. 'Columna': crown columnar; leaves narrower, more oblong, margins irregularly lobed, grey-green. 'Laciniata': leaves laciniate. 'Pendula': branches pendent. 'Pinnata': leaves pinnatisect. ' Purpurea': leaves crimson-purple becoming grey-green flushed red, veins red.

Name	Distribution	Habit	Bark	Leaves	Fruit
Q.phellos WILLOW OAK	SE US	deciduous tree to 40m; crown rounded to columnar	bark grey tinted red-brown, glabrous	7–12cm, oblong-lanceolate, apex acute, base attenuate, entire, often undulate, dark green above, paler beneath, almost glabrous	acorns to 1cm, subglobose, pale yellow-brown; cupule shallow, scales adpressed, grey-tomentose, enclosing only base of acorn.
Q.pontica ARMENIAN OAK	NE Turkey, Caucasus	deciduous shrub or small tree, to 6m, rarely 10m		15–25cm, obovate to elliptic, apex acute, base attenuate, margin shallowly and irregularly dentate, teeth mucronate, dark green and shining above, paler with sparse hair on veins beneath, midrib yellow	acorns 2cm, ovoid; cupule of triangular, pubescent, adpressed scales, enclosing to half of acorn
Q.robur ENGLISH OAK, COMMON OAK, PEDUNCULATE OAK	Europe to W Russia	deciduous tree 20–30m; crown broad, spreading, irregular	grey-brown, deeply fissured	5–14cm, oblong to obovate, apex rounded, base attenuate, auriculate, margin with 3–6 pairs of deep, rounded lobes, glabrous dark green above, paler blue-green beneath	acorns solitary or clustered, 1.5–2.5cm, ovoid to oblong-ovoid, apex mucronate; cupule subhemispherical, made up of tightly adpressed, velvety scales, enclosing one-quarter to one-third of fruit

Comments: Cultivars vary in habit, from dwarf to columnar, in colour from variegated to yellow or purple throughout, and in leaf from flat to convex, entire to deeply incised and crispate. *Habit* - 'Contorta': dwarf; leaf lobes twisted. 'Cupressoides': columnar tree; leaves smaller than in 'Fastigiata'. 'Fastigiata': columnar, branchlets erect. 'Pendula': vigorous; branches pendent. 'Tortuosa': dwarf, with twisted branches. 'Umbraculifera': crown globose. *Leaf colouring* - 'Argenteomarginata': leaf margins white. 'Argenteovariegata': leaves white-variegated, flushed red. 'Atropurpurea': to 10m; leaves plum purple. 'Aureobicolor': first leaves sparsely yellow-dotted, late leaves yellow-variegated, flushed red. 'Concordia': leaves golden. 'Nigra': leaves deep purple. *Leaf form* - 'Cucullata': leaves elongate, convex, margin shallowly divided. 'Cristata': leaves small, curled and twisted. 'Fennessii': leaves flat or convex, margins deeply incised, lobes irregular, narrow. 'Filicifolia': slow-growing tree; leaves often incised to midrib, lobes regular, acute sometimes crispate. 'Strypemonde': slow-growing; leaves irregularly incised, lobes sharp or 0, mottled yellow. 'Salicifolia': slow-growing; leaves elliptic, entire.

Name	Distribution	Habit	Bark	Leaves	Fruit
Q.rubra RED OAK, NORTHERN RED OAK	EN America	deciduous tree to 45m; crown rounded		10–22cm, oblong to obovate, base attenuate or rounded, margin with 3–5 pairs of lobes, lobes triangular or ovate, acute, sinuses run to midrib, margins irregularly toothed, dark green, glabrous except for red-brown hairs in vein axils, colouring dull red or yellow-brown in fall	acorns 2–3cm, ovoid, cupule made up of tightly adpressed, short scales, enclosing one-third of acorn

Comments: 'Aurea': leaves golden yellow becoming slightly green towards autumn. 'Heterophylla': leaves oblong-ovate to linear-lanceolate, often falcate, margin with a few shallow teeth. 'Schrefeldii': leaf margins deeply divided, lobes often overlapping.

Name	Distribution	Habit	Bark	Leaves	Fruit
Q.suber CORK OAK	NAfrica, S Europe	evergreen tree to 26m	to 15cm thick, very corky, deeply fissured	to 3.5cm, ovate to lanceolate, apex acute, base sometimes slightly cordate, margin dentate, teeth 5–7 each side, mucronate, dark green, shining, glabrous above, grey-tomentose beneath	acorns solitary or paired, 2–4.5cm, ovoid or ellipsoid, apex slightly mucronate; cupule subhemispherical, lower scales closely adpressed, upper scales spreading enclosing half or more of acorn
Q. × turneri (*Q.ilex × Q.robur*)		semi-evergreen tree to 25m		obovate or elliptic, apex acute or obtuse, base rounded to subcordate, margin sinuately-lobed, lobes 5–6, obtuse, leathery, dark green and glabrous above, paler and subglabrous but pubescent on veins beneath	acorns, where produced, in clusters of 3–7, to 2cm, ovoid; cupule subhemispherical, pubescent, enclosing half of acorn

Comments: 'Pseudoturneri': leaves narrowly oblong-ovate, lobes narrower, more glabrous beneath, remaining green for longer

Name	Distribution	Habit	Bark	Leaves	Fruit
Q.velutina BLACK OAK	E US, SE Canada	deciduous tree to 30m	black-brown, deeply fissured, fissures orange at centre	6–25cm, narrowly ovate to obovate, often misshapen, apex acute, base truncate, deeply 5- or 7-lobed, lobes ovate to triangular, 1–3 bristle-tipped teeth, dark green, glabrous and glossy above, paler, densely tomentose at first beneath	acorns solitary or paired, 1.5–2.5cm, ovoid to subglobose, pale brown; cupule made up of loosely overlapping, hairy scales, enclosing half of acorn

Comments: 'Macrophylla': leaves with 4 pairs of wide lobes, densely pubescent becoming glabrous above, veins red. 'Magnifica': leaves to 25cm, apex rounded, mucronate, sinuses rounded. 'Nobilis': leaves with 2–4 pairs of rounded lobes, grey-pubescent at first, becoming glabrous, dark green above, paler and usually persistently tomentose beneath, veins red-brown.

R

rabbits (*Oryctolagus cuniculus*) small burrowing mammals of the family Leporidae that feed on a wide range of wild and cultivated plants, and may graze them to ground level; they also cause damage in gardens by gnawing the bark of young trees, frequently girdling them, and by digging holes in lawns and borders. Rabbits generally live in colonies underground with interconnecting tunnels, but some spend all of their adult lives above ground. They usually breed all the year round, but especially from January to July. There may be 2–5 litters per year, consisting of 3–6 young, which are at first blind and naked of fur but able to leave the nest after 3–4 weeks.

Foxes, stoats, cats, badgers and some large birds of prey are natural enemies. Deliberate introduction of the the virus disease myxomatosis, spread by fleas and mosquitoes, initially reduced numbers but rabbits have become partially immune to it. Control by shooting or gassing is not practicable for most gardeners, and humane trapping requires skill.

Fencing is an alternative, effective means of control. In order to be rabbit-proof, it needs to be constructed out of 2.5cm wire mesh, 1.2–1.4m high, with the bottom 30cm sunk below ground level; of the underground section, the lower 15cm should be bent outwards to discourage burrowing. Trees and shrubs may be protected from bark stripping by netting or spiral tree protectors. Proprietary chemical repellents are variably effective.

Hares cause similar damage to rabbits but they are much more erratic in their attacks since they usually exist only in small numbers. Fences against hares need to be 1.5–1.75m high.

raccoon an omnivorous, largely nocturnal climbing animal which in the US can cause severe damage to fruit and vegetables, and also to water lilies. Shooting and trapping are the only effective controls and local advice on procedures should be sought.

raceme an indeterminate, unbranched and, usually, elongate inflorescence composed of pedicelled flowers.

rachilla a secondary axis of a compound leaf or inflorescence.

rachis (plural rachises, rachides) the axis of a compound leaf or a compound inflorescence, as an extension of the petiole or peduncle, respectively.

raddichio see *chicory*.

Radermachera (for J.C.M. Radermacher (1757–83), Dutch resident in Java who published a list of Javanese plants). Bignoniaceae. Indomalaysia, S China, Hainan, Taiwan, Ryukyu Islands. About 15 species, trees or shrubs, mostly evergreen with 2–3-pinnate leaves. Tubular, 5-lobed flowers are carried in terminal thyrses in summer. Young examples of the following species are often treated as pot plants, valued for their evergreen, lustrous foliage. Eventually, they will make handsome flowering trees. Grow in sun or part shade. Water moderately when in growth; maintain a winter minimum temperature of 15°C/60°F. Propagate by seed, cuttings or air layering. *R.sinica* (S and E Asia; shrub or tree; leaves to 75cm, triangular to deltoid in outline, 2-pinnate; segments ovate to rhomboid, somewhat lobed or toothed, glossy, deep green; flowers scented, to 7.5cm, deep yellow).

radiate (1) spreading outward from a common centre; (2) possessing ray florets, as in the typically daisy-like capitula of many Compositae.

radical arising directly from the root, rootstock or root-like stems; usually applied to leaves that are basal rather than cauline.

radicle the rudimentary root of the embryo.

radish

Pests
cabbage root fly
flea beetle

radish (*Raphanus sativus*) an annual or biennial species, variable in size, colour and pungency, grown primarily as an annual for its enlarged taproot, which is usually eaten raw; the immature leaves and seed pods are also edible.

The radish probably originated in the eastern Mediterranean region and was important as a food crop in Egypt by 2000 BC. It is widely distributed and adapted to various climatic conditions.

Roots can be available year round from two main types. Small-rooted cultivars are suitable for summer harvest and are globose or long in shape, with red, pink, bicoloured or white skins; selections are available for autumn/winter sowing with cold or warm protection, and these usually produce less leaf and are slower to mature. Large-rooted types include the white mooli radishes, which are about 20cm long and suitable for summer use, and the hardy Chinese and Spanish types, which may be white-, red-, black- or violet-skinned, round or long, in some cases up to 60cm long, and capable of producing roots up to 500g in weight. All of the large-rooted types may be eaten raw or cooked.

Small radishes, such as 'French Breakfast' and 'Crystal Ball', can be sown at 10 day intervals from early spring until autumn, and the season extended with cloches or other forms of protection. Sow in rows 15cm apart with seeds 2.5cm apart, or sow broadcast. Too dense a stand will result in lanky unproductive plants. Mooli types should not be sown before June to avoid bolting, and these take 10–12 weeks to mature. Large radishes for winter use should be sown in July or August in rows 25cm apart, aiming at a final spacing between plants of 15cm. Sow all types 13–20mm deep.

An open site is preferable, although summer crops tolerate light shade; crop rotation is necessary because, like other members of the family Cruciferae, radishes are liable to club root infection.

Small-rooted types can be sown thinly along the prepared drills of slower germinating vegetables, such as parsnips and parsley, to mark the rows, and these may be progressively pulled as they mature; alternatively, they may be grown as an intercrop in rows between other vegetables. Rapid germination and establishment is essential and summer radishes must be pulled early since they soon become woody; small-rooted types may be ready within a month from sowing. Winter types can be left in the ground until required, or lifted for indoor storage.

raffia a flat, fibrous material used for tying plants, including shoot and bud grafts. It is derived from the leaves of certain palms, especially the RAFFIA PALM (*Raphia farinifera*). The term is often used interchangeably with *bass* or *bast* q.v.

raft (1) an piece of bark, osmunda fibre or tree fern fibre on to which epiphytic plants are fixed and hung – this method of growing is particularly suitable for orchids with abundant clinging aerial roots and rambling stems and rhizomes e.g. *Bulbophyllum*; (2) a form of training used in bonsai where the principal trunk of the tree is laid on its side and the lateral branches along the exposed side are trained up as if they were themselves individual trees.

rain shadow the area immediately to the leeward of a wall, fence, hedge, screen or shelter belt, which receives much less rainfall than surrounding areas because of its protection from prevailing wind. It is an important consideration in planting and in the provision of irrigation.

raised bed any growing bed formed above ground level, which may range in height from a very shallow construction up to almost 1 metre. All raised beds offer the advantages of good drainage, a growing medium of relatively high temperature, and free root run. Beds up to 30cm high are used in the vegetable garden and for growing strawberries, in a system referred to as the deep bed method. Taller raised beds are appropriate for sites where there is no soil or where its composition or profile is unsuitable for satisfactory plant growth; they are often used for growing alpines, either direct planted or as plunged container-grown specimens.

Raised beds provide opportunities for gardeners who use a wheelchair or otherwise have restricted mobility, and it is possible to grow a wide range of plants, including ornamentals, vegetables and some forms of fruit in this way.

Beds have walls made of brick or cement blocks, or of timber such as old railway sleepers; these may have planting holes left during construction in which to establish ornamental trailing plants and to provide side drainage. For most plant use, a foundation layer of rubble is desirable as an aid to drainage. The growing medium can be mixed according to intended use, a soil-based mix being the best choice.

rake a hand implement consisting of a crossbar with short prongs, or 'teeth', fixed to a long handle. Most commonly used, for levelling and preparing seed-beds, is the steel-headed rake, which consists of a head 30–38cm wide, with prongs 6mm long, mounted on a wooden handle 150cm long. It is sometimes used for forming seed drills, where with the head turned so that the prongs are facing upwards a continuous depression can be made in a fine tilth; this is a suitable method where less precise drilling is required, as may be acceptable, for example, in the sowing of annual flowerbeds. Held at the same angle, the rake may then be used for pulling soil back over the sown drills.

The wooden rake consists of a head about 76cm wide, bearing prongs 8cm long, fixed to a 177cm-long handle, and is particularly useful for levelling ground and pulling off stones and litter. The spring-tined rake has a special purpose in lawn maintenance; it comprises a fan-shaped head about 30cm long, with a number of thin-gauge tensile wire prongs, spring mounted over a central wire bearer, with the end of each prong bent to a common shallow angle. Plastic and bamboo versions are available. It is used for raking out thatch or gathering mowings and, like every other type of rake is useful for general tidying, such as leaf gathering; for some purposes, a rubber-headed rake is suitable. A very small, short-handled rake is sometimes used as a cultivator in much the same manner as a hand fork.

rambler a climbing plant that trails or scrambles in an informal habit; used especially of a type of rose.

ramentaceous stems or leaves possessing small, loose, brownish scales.

ramet (1) an individual member of a clone; (2) an underground tree system giving rise to large suckering colonies.

ramiflorous bearing flowers directly on large branches and leafless twigs, but not the trunk.

ramiform branched, branch-like.

rammer see *potting stick*.

Ramonda (for L.F. Ramond (d. 1827), French botanist and traveller). Gesneriaceae. S Europe. 3 species, stemless, more or less evergreen perennial herbs. Borne in a low, basal rosette, the leaves are rugose and white-hairy. Produced in spring and summer, the flowers are solitary or few together in scapose umbels; they are shallowly tubular with a broad, spreading limb of four to six lobes. Plant on the rock garden and in wall crevices, especially those with a northerly aspect. Kept free of damp in winter, these plants may be hardy to about –15°C/5°F. They may also be grown in the alpine house or cold frame in shallow pans, with a well-drained neutral or slightly acidic, loam-based potting mix. Outdoors, establish in pockets of gritty and freely draining soil, and position so that water is unable to settle in the leaf rosette. Plants may shrivel under dry conditions but will readily resurrect given water. In winter, keep barely moist. Propagate by seed or by leaf cuttings as for *Saintpaulia. R.myconi* (syn. *R.pyrenaica*; Pyrenees; flowers to 4cm in diameter, 1–6 on glandular-hairy scapes to 12cm, violet to lilac, rosy pink or white with a yellow centre).

rank (1) a vertical row or column, for example of leaves; (2) in taxonomy, the position of a taxon in the hierarchy.

Ranunculus (name used by Pliny, derived from Latin *rana*, frog: several species are aquatic or grow in marshes). Ranunculaceae. BUTTERCUP, CROWFOOT. Cosmopolitan. Some 400 species, annual, biennial and perennial herbs (those here perennial).

They vary greatly in size and habit, but most produce simple to palmately lobed leaves in basal clumps. Solitary or carried in cymose panicles, the flowers are bowl-shaped to starry with five or more petals and numerous stamens.

Some of those commonly grown are selections of garden weeds (for example, cultivars of *R.acris*), fully hardy and suited to damp places in sun or part shade in informal borders, woodland or water gardens. Where they prove invasive, cut back the running rhizomes with a spade. Cultivars of the lesser celandine, *R.ficaria*, are charming additions to the woodland garden and equally suited to underplanting in shrubberies and mixed borders. They prefer some shade and a moist, humus-rich soil, at least when in growth. They flower in early to mid spring and their foliage (often their finest feature) will disappear by summer, when their pea-sized tubers will withstand drought. This species may also become invasive and is easily propagated by tubers taken in summer and autumn. Taller and sadly non-invasive, *R.gouanii* has similar requirements. *R.lingua* is an excellent plant for water margins, pools and bog gardens. *R.asiaticus* should be grown as for the tuberous-rooted De Caen and St Brigid anemones. Its cultivars make beautiful pot plants and cut flowers. Alpine species such as *R.alpestris* and *R.montanus* are hardy and will thrive in sun on the rock garden given a moist, gritty soil. *R.crenatus* and *R.glacialis* are harder to please, needing an acid medium and careful watering. They are perhaps more easily cultivated in the alpine house. *R.lyallii. R.amplexicaulis, R.bullatus, R.calandrinioides* and *R.gramineus* need a fast-draining, sandy soil in full sun. They grow well in the more sheltered reaches of the rock garden, gravel garden and silver border. Some of them may become dormant in summer and dislike wet winter conditions. Increase by seed or division.

Ranunculus ficaria 'Salmon's White'

RANUNCULUS

Name	Distribution	Habit	Leaves	Flowers	
R.aconitifolius	W to C Europe	perennial to 60cm	dark green, palmately, 3–5-lobed	few to many, white, to 2cm, across; sepals red or purple beneath	spring

Comments: 'Florepleno': WHITE BACHELOR'S BUTTONS, FAIR MAIDS OF FRANCE, FAIR MAIDS OF KENT; flowers fully double, long-lasting. 'Luteus Plenus': flowers double, yellow.

Name	Distribution	Habit	Leaves	Flowers	
R.acris MEADOW BUTTERCUP	Europe, Asia, naturalized elsewhere	perennial to 1m	with 3–7 deep lobes simple or further divided, dentate, hairy, sometimes marked black	several, glossy golden yellow, to 2.5cm across	spring

Comments: 'Florepleno': YELLOW BACHELOR'S BUTTONS; to 90cm; flowers double. 'Hedgehog': leaves marked dark brown; flowers pale.

Name	Distribution	Habit	Leaves	Flowers	
R.alpestris	Europe	perennial, 3–12cm	glossy dark green, orbicular, lobes 3–5, deeply crenate	1–3, white to 2cm across; sepals tinged red-brown 3–5	spring-summer
R.amplexicaulis	Pyrenees, N Spain	perennial 8–30cm	ovate-lanceolate, glaucous, blue-grey; stem leaves amplexicaul	several, white, occasionally pink, 2–2.5cm diameter; petals sometimes more than 5	spring
R.asiaticus	SE Europe, SW Asia	pubescent perennial to 45cm	3-lobed, lobes stalked, divided and toothed	few to several, red, pink, purple, yellow to white, 3–5cm across; anthers purple-black	spring-summer

Comments: Bloomingdale Hybrids: dwarf; flowers double in shades of pink, yellow, rose, red and white. 'Color Carnival': tall, robust; flowers large, double, many colours. 'Picotee': compact; flowers double, very large, ruffled, in pink, red, orange, white and yellow, picotee edge. 'Pot Dwarf': flowers semi-double and double, in scarlet, pink, salmon, orange and white. 'Superbissimus': tall; flowers large. Tecolote Hybrids: robust; flowers exceptionally large. Victoria Hybrids: flowers large, fully double in wide range of colours.

Name	Distribution	Habit	Leaves	Flowers	
R.bullatus	Mediterranean to NW Spain and Portugal	perennial; roots tuberous; stem to 30cm	ovate, crenate, bullate	1–2, yellow, 2.5cm across, violet-scented; petals 5–12	autumn-spring
R.calandrinioides	Morocco	perennial to 15cm	to 6.5cm, long- petiolate, lanceolate to ovate, entire, undulate, grey-green, glaucous	1–3, white, sometimes flushed pink, to 5cm across	winter-spring
R.crenatus	E Alps, Apennines to Balkans and Carpathians	perennial to 15cm	orbicular, weakly cordate, crenate or 3-lobed at apex	1–2, white, to 2.5cm across	spring
R.ficaria LESSER CELANDINE, PILEWORT	Europe, NW Africa, W Asia, naturalized N America	perennial 5–30cm with small, clustered tubers	cordate, angled or crenate, 1–4cm, dark green, long-petiolate	solitary to few, brilliant golden-yellow, fading to white, 2–3cm across; petals 8–12	spring

Comments: 'Albus': leaves with a dark marks; flowers very pale yellow, blue-green below. 'Bowles Double': flowers double, green centre on opening, later paler yellow. 'Brazen Hussy': leaves chocolate brown; flowers golden. 'Collarette': anthers petaloid, forming a tight anemone centre. 'Cupreus': leaves heavily marked coppery-bronze; flowers coppery-orange. 'Double Bronze': flowers double, yellow, with bronze backs to petals. 'Double Cream': flowers double, upper surface of petals creamy, lower surface tinged grey. 'E.A. Bowles': flowers golden yellow, anemone-centred. 'Florepleno': flowers double, yellow, petals tinged green beneath. 'Green Petal': flowers with many narrow, wavy-edged green and yellow staminodes; sepals 0. 'Hoskin's Miniature': small; leaves slightly marked silver. 'Lemon Queen': flowers simple, pale yellow, petals tinged bronze beneath. 'Major': large in all parts. 'Randall's White': upper surface of petals pale cream, lower surface slate-blue. 'Salmon's White': leaves with dark markings; upper surface of petals very pale cream, lower surface slate-blue. 'Whiskey Double Yellow': flowers double, yellow, petals tinged yellow and bronze beneath.

Name	Distribution	Habit	Leaves	Flowers	
R.glacialis	Alps, Pyrenees, Sierra Nevada, Arctic, Iceland, Greenland	perennial, 4–25cm	dark-green, somewhat fleshy, lobes 3, stalked, deeply segmented	1–3, white or pale pink, becoming red after pollination, 2–3cm across	spring-summer
R.gouanii	Pyrenees	pubescent perennial to 30cm	leaves 3–5-lobed, lobes obovate, dentate	1–5 per stem, rich yellow, to 4cm across	spring

Comments: close to *R.montanus* . *R.gouanii* 'Plenus', also known as *R.speciosus* 'Plenus' is, properly, *R.bulbosus* 'Pleniflorus', with double flowers to 4cm across in chrome yellow with a green centre.

Name	Distribution	Habit	Leaves	Flowers	
R.gramineus	S Europe, N Africa	perennial, 20–50cm	linear to lanceolate, flat, glaucous	1–3 per stem, citron yellow, to 2cm across	spring-summer

Comments: 'Florepleno': flowers double

Name	Distribution	Habit	Leaves	Flowers	
R.lingua GREATER SPEARWORT	Europe to Siberia	semi-aquatic stoloniferous perennial; stem stout, 50–200cm	to 20cm, ovate to lanceolate, usually toothed	in lax cyme, bright yellow, to 5cm across	summer

Comments: 'Grandiflorus': flowers large, fleshy

Name	Distribution	Habit	Leaves	Flowers	
R.lyallii MOUNT COOK LILY, MOUNTAIN LILY	New Zealand	perennial; rootstock thick, roots fleshy; stem stout, to 1.5m	peltate, 12–40cm broad, crenate, leathery, dark green, long-petiolate	5–15 in a panicle, white, 5–8cm across	summer
R.montanus	Europe (mountains)	glabrous to pubescent rhizomatous perennial, 16–30cm	3–5-lobed, lobes obovate, dentate.	1–3, glossy golden-yellow, 2–3cm across	spring-summer

Comments: 'Molten Gold': dwarf, robust; leaves dark; flowers large, rounded, gold

Ranzania (for Ono Ranzan, (1729–1810) Japanese naturalist). Berberidaceae. Japan (Honshu). 1 species, *R.japonica*, a perennial, rhizomatous herb, 15–40cm tall. The compound leaves consist of three leaflets, each ovate to round, 3-lobed and 8–12cm across. To 2.5cm across, the nodding flowers appear in early summer. They are pale lavender-violet and composed of six petaloid sepals and six, small petals each with two nectaries at its base. For cool, humus-rich, moist soils in light woodland, for the shaded rock garden. Hardy to –10°C/14°F. Propagate from ripe seed in a cold frame – this can take up to 18 months to germinate and a further 3–5 years to flower.

Raoulia (for Edouard F.L. Raoul (1825–52), French botanist). Compositae. New Zealand. About 20 species, dwarf, evergreen perennial herbs or subshrubs, often forming dense cushions or mats. The leaves are very small and usually densely packed or overlapping. Button-like, stalkless flowerheads with yellow florets appear in spring and summer. Silvery cushion-formers for the rock garden, fine scree and raised beds. Grow in sun on moist, gritty, humus-rich soils. Protect from excessive wet in winter. Hardy to –15°C/5°F. In the alpine house, grow in pans with a mix of equal parts loam, leafmould and sharp sand with a top dressing of grit. Water freely when in growth; keep almost dry in winter. Propagate by division of rooted tufts.

R.australis (leaves to 2mm, spathulate, closely overlapping, silver-pubescent; flowerheads to 5mm in diameter, phyllaries tipped bright yellow); *R.haastii* (leaves to 5mm, ovate to linear-oblong, loosely overlapping, pale sea green; flowerheads to 9mm in diameter, phyllaries tipped white or yellow); *R.hookeri* (leaves to 2mm, narrowly obovate to spathulate, closely overlapping, silvery white-tomentose toward apex; phyllaries green or straw-coloured).

rape see *green manuring*.

Raphanus (name used by Theophrastus for this plant). Cruciferae. Europe, temperate Asia. 8 species, annual, biennial or perennial herbs with pungently flavoured and crisply succulent taproots and lyrate to pinnatifid leaves. *R. sativus* is the RADISH (q.v.)

raphide an elongated, needle-like crystal of calcium oxalate, sometimes in a bundle and found in the vegetative parts of many plants.

raspberry (*Rubus* species) a deciduous woody perennial, with fruits generally red, but also yellow, black or purple. The temperate climate species is the red-fruited *R.idaeus*, named after Mount Ida in Asia Minor, the possible origin of the raspberry. European and North American raspberries are classified respectively into subspecies *vulgatus* (native to much of Europe and northern Asia to Japan), and *strigosus* (which grows wild in North America, in the mountains of Georgia to Pennsylvania, through Maine and Dakota to Canada). The North American subspecies is hardier and has brighter coloured fruits.

The AMERICAN BLACK RASPBERRY, (*R.occidentalis,*) has a similar natural range to the AMERICAN RED RASPBERRY (*R.idaeus* subspecies *strigosus.*), but is not found so far north and also further south; and it is less hardy. A selection of the HILL RASPBERRY (*R.niveus* 'Mysore') is grown commercially in Florida and harvested from March to May.

Domestication of the raspberry in Europe was mentioned by Roman writers in the 4th century, the fruits being eaten or used for flavouring drinks and the leaves used for making tea. By the 16th century, raspberries were cultivated in gardens, and during the next century, different forms were described, including red, yellow and white fruited and thornless cultivars. Selection for larger fruits had made considerable progress by the end of the 18th century, and in 1823 the Horticultural Society of London reported that it had 23 cultivars in its collection. Over the next hundred years, nurserymen introduced more selections and chance seedlings.

Some European cultivars were introduced to North America from the end of the 18th century onwards, but due to temperature extremes they did not flourish. They were combined with American forms to produce improved cultivars like 'Cuthbert' and 'Willamette'.

Virus disease developed as a serious problem of raspberries in the first half of the 20th century, but the work of breeders in both North America and Europe has greatly reduced its incidence with the introduction of the virus-tolerant cultivar 'Malling Jewel' and a series of others that show resistance to aphid vectors. 'Glen Clova' and 'Glen Prosen' are widely grown as early and late red-fruited cultivars.

The season is lengthened with autumn-fruiting or primocane cultivars, such as 'Autumn Bliss', 'Zeva' and 'Heritage'. These fruit on the current season's canes, from late summer until the first autumn frosts, and can be grown in areas where severe winter temperature would normally damage growth.

Yellow-fruited raspberries have been known for centuries – 'Yellow Antwerp' was one of the first to be selected; more recent cultivars include 'Amber' and 'Golden Queen'. Black raspberries, such as 'Ohio Everbearer' and 'Black Hawk', are grown in North America, but in the UK are susceptible to *Phytophthora* root rots. Purple raspberries, hybrids of red and black types, grow more vigorously than black raspberries and give higher yields, and cultivars such as 'Brandywine' are grown commercially in a limited area in New York State.

Raspberries are self-fertile, so that single cultivars may be planted.

Raspberries are best grown in cool temperate conditions, sheltered from spring frosts and wind; soil moisture content is crucial, and should be adequate to supply water in summer for growing canes and swelling fruit, but not so high in winter as to asphyxiate roots. Raspberries show iron deficiency on calcareous soils.

Plant canes of certified stock, with a good fibrous root system, during the dormant period, with plants

raspberry

Pests
aphids
birds
raspberry cane borer
  (see longhorn beetles)
raspberry beetle

Diseases
cane spot
virus

spaced 40cm apart in rows 2m apart. In temperate climates, planting should preferably be undertaken in late autumn to encourage early root growth; in countries with very cold winters, spring planting is usual to avoid winter losses. Plant the canes with the roots 5–7cm deep, and cut back each cane to 20cm; autumn-fruiting canes are cut back to ground level.

In gardens, raspberries are usually grown in upright rows, running north-south for uniform light distribution, with post and wire support. For red raspberries, three wires are fixed, along single rows of 2.3m long posts driven 45cm into the ground, at 0.7m, 1.0m and 1.6m above ground level and the canes are tied to the wires. An alternative system is to erect two parallel rows of support posts, with 60cm between each row, and the line of raspberry canes is planted down the centre. Wires are attached along each row of posts at 0.75cm and 1.5m above ground level, and twine or wire ties fixed across each tier of parallel wires. The raspberry canes grow up through the structure with no tying. Canes of autumn-fruiting raspberries are shorter and do not need permanent support. Black and purple raspberries require more space than red ones, and are planted 75–120cm apart in rows about 2.5m apart; they are usually grown without support or attached to a single wire running 45–60cm above soil level along a row of posts.

In cold areas such as the central US and east Canada, canes are not tied in until the spring since the winter snowfall protects them from very low winter temperatures.

Summer-fruiting red raspberries are pruned by cutting out the fruited canes immediately after harvesting, together with any weak ones, to leave about 8–10 canes per metre of row; winter-damaged tips are cut out in late winter. Autumn-fruiting raspberries are cut to the ground in winter. New canes of black and purple raspberries are tipped in summer to encourage the growth of laterals, which are then pruned back in spring, according to climate and varietal vigour, to increase fruit size. The old fruited canes are usually cut out each winter.

The main nutrients required by raspberries are potassium for fruiting and nitrogen for cane production; $15–30g/m^2$ of sulphate of ammonia applied in spring is sufficient to meet requirements, together with a mid-winter dressing of sulphate of potash at $30g/m^2$. Phosphates are needed every third year, applied as superphosphate at $60g/m^2$. An annual mulch of organic matter will help conserve soil moisture and control weeds.

Watering in dry spells as the fruits start to ripen is beneficial. It is also recommended for autumn-fruiting raspberries to maintain fruit size and number.

Fruits should be harvested when dry, pulled from the stalk, and kept as cool as possible.

raspberry beetle (*Byturus tomentosus*) a small brown beetle, about 4mm long, which overwinters in the soil as an adult to emerge in late April to early June. The beetles feed upon the petals and stamens of cane fruit flowers, and also apple and hawthorn. Eggs are laid on the fertilised flowers of blackberry, loganberry, raspberry and related fruits. The brownish-white maggots, up to 8mm long, feed on the outside of developing fruits; as the berries ripen they tunnel into the plug and feed inside the fruit. Fully fed larvae move to pupate for one month in the soil. There is one generation. Control by spraying a recommended insecticide when the larvae are feeding on the outside of the fruits.

Ratibida (derivation of name obscure). Compositae. N America, Mexico. PRAIRIE CONEFLOWER, MEXICAN HAT. 6 species, erect, roughly hairy biennials or perennial herbs. The leaves are pinnate to pinnatifid. Large and long-stalked, the daisy-like flowerheads are produced in summer and early autumn. The ray florets are long, colourful and somewhat shaggy, the disc florets are small, dark and packed into a prominent rounded or cone-like receptacle. Tolerant of drought, heat and high humidity, the following species are hardy and suitable for use in the herbaceous border, as cut flowers, in native plant collections and in other naturalistic plantings. Propagate by seed. The deep taproots make division difficult. *R.columnifera* is often treated as an annual.

R.columnifera (LONG-HEAD CONEFLOWER, PRAIRIE CONEFLOWER; British Colombia to New Mexico; receptacle cone-shaped, ray florets to 2.5cm, yellow to red or partly to entirely brown-purple); *R.pinnata* (GREY-HEAD CONEFLOWER, DROOPING CONEFLOWER; central North America; receptacle globose, ray florets to 5cm, yellow).

Ravenala (the native name in Madagascar). Strelitziaceae. Madagascar. TRAVELLER'S TREE. 1 species, *R.madagascariensis*, a giant, evergreen tree-like herb, to 16m. The columnar stems are unbranched, ring-scarred and olive green or grey. The paddle-shaped leaves are to 4m long, 2 ranked, dull green and become torn along the veins. The petiole bases overlap to form a distinctive fan at the summit of the stem. The flowers resemble those of the larger *Strelitzia* species, but are duller. Cultivate as for the larger *Strelitzia* species.

ray (1) the primary division of an umbel or umbelliform inflorescence; (2) a ray flower, or a circle of ray flowers, or the corolla of a ray flower.

ray flower, ray floret a small flower with a tubular corolla and the limb expanded and flattened in a strap-like blade (ligule); usually occupying the peripheral rings of a radiate Compositae capitulum; cf. *disc floret*.

Rebutia (for P. Rebut, wine-grower and cactus dealer of Chazay d'Azergues). Cactaceae. Bolivia, NW Argentina. To 40 species, low-growing cacti with small, globose to shortly cylindric stems – these are tuberculate or weakly ribbed and covered in rather weak spines. Small, funnelform flowers usually appear near the stem-base. Grow in a cool greenhouse (minimum temperature 7°C/45°F) in a gritty, mildly acid to neutral medium in full sun; ensure a dry atmosphere. Keep dry from mid-autumn until early spring, except for light misting on warm days in later winter.

R.aureiflora (NW Argentina; stems depressed-globose to globose, to 6cm in diameter, often red-tinged; spines to 1cm, bristly; flowers about 4cm in diameter, yellow with white throat, sometimes orange, red or purple); *R.fiebrigii* (syn. *R.muscula*; Bolivia, NW Argentina; stem globose, depressed at apex, 5cm tall; spines to 2cm, brown-tipped; flower 2–3cm, bright orange to red); *R.krainziana* (probably garden origin; stem depressed-globose, 4cm in diameter; spines about 1–2mm, bristly, white; flowers about 3 × 4cm, bright red with violet sheen and violet throat); *R.minuscula* (syn. *R.violaciflora*; N Argentina; stem globose, to about 5cm in diameter; spines 2–3mm, bristly; flowers to 4cm, variable in colour scarlet to purple-red); *R.pygmaea* (NW Argentina; stem ovoid to short-cylindric, 1–3 × 1.2–2cm; spines 2–3mm; flowers 18–25mm, rose-purple); *R.senilis* (N Argentina; stem to 7cm in diameter, deep green; spines to 3cm, chalky white or tinged yellow; flowers 3.5cm, carmine with white throat or citron yellow).

receptacle the enlarged or elongated, and either flat, concave or convex, end of the stem from which the floral parts derive.

recessive used of a particular hereditable characteristic which is masked by the dominance of another characteristic-transmitting component of a gene. For example, dwarfness being recessive may be carried as a latent characteristic, whilst tallness being dominant is expressed in offspring following fertilization; cf. *dominant.*

recurved curved backwards and downwards.

red spider mites (Acarina: Tetranychidae) small, spider-like mites, up to 1mm long, red, green or yellow and with four pairs of legs; they are divided into two subfamilies: red spider mites and bryobia mites. Colonies develop on plant foliage and throughout the summer all stages are usually present; overwintering is as eggs or adults depending upon species. Young and adult mites suck sap from plant tissues, producing a characteristic pale mottling of the upper surface of leaves; in severe attacks, leaves become chlorotic, desiccated, bronzed or silvered and usually covered with spun silk webbing.

The FRUIT TREE RED SPIDER MITE *(Panonychus ulmi),* known as the EUROPEAN RED MITE in America, is an important pest, especially of apples and plums, and it also attacks other deciduous trees and shrubs. Severe infestations cause leaves to become bronzed; premature defoliation follows and on fruit trees loss of yield. Eggs overwinter in bark crevices of woody plants. The CITRUS RED MITE *(P.citri),* is a worldwide pest of citrus. The GREENHOUSE RED SPIDER MITE *(Tetranychus urticae),* known as the TWO-SPOTTED

MITE in North America, is common and very troublesome, attacking a wide range of greenhouse and outdoor plants, including strawberries, peaches and nectarines, cucumbers, beans, carnations, chrysanthemums, fuchsias, orchids and roses. The closely-related CARMINE MITE *(T.cinnabarinus)* is found in similar situations, but is mostly restricted to greenhouses in temperate countries. Both species overwinter as adults. Other injurious species in North America include the PACIFIC SPIDER MITE *(T.pacificus),* SCHOENE SPIDER MITE *(T.schoenei),* the FOUR-SPOTTED MITE *(T.canadensis),* the STRAWBERRY SPIDER MITE *(T.turkestani)* and the MCDANIEL SPIDER MITE *(T.mcdanieli).* All the mites attack a wide range of edible and ornamental plants. The LIME MITE *(Eotetranychus tiliarius)* in Europe attracts attention because, in severe infestations, mites vacate the leaves in late summer and autumn and swarm over the trunk, coating it with glistening silk webbing. Other species occur in North America, including the YELLOW SPIDER MITE *(E.carpani borealis)* and the YUMA SPIDER MITE *(E.yumensis).* The CONIFER SPINNING MITE *(Oligonychus ununguis),* known as the SPRUCE SPIDER MITE in North America, is an important pest of conifers; affected foliage turns yellow during the summer and in severe attacks needles are shed. Other North American species include the AVOCADO BROWN MITE *(O.punicae),* the SOUTHERN RED MITE *(O.ilicis),* the BANKS GRASS MITE *(O.pratensis)* and the AVOCADO RED MITE *(O.yothersi).*

Natural enemies play an important part in the control of red spider mites and the red spider predator *(Phytoseiulus persimilis)* is widely used in Europe and North America to control *Tetranychus* species in greenhouses. Spraying greenhouse plants with a fine mist of water at frequent intervals helps check populations. The choice of acaricides is restricted for amateur gardeners and red spider mites readily become resistant to the main groups of chemicals. See *mites.*

Rebutia minuscula

redcurrant *(Ribes rubrum* Red Currant Group) a deciduous shrubby plant grown for its translucent berries borne in pendant trusses. The main difference between the red- and whitecurrant is the absence of cyanidin glycosides in the whitecurrant. Each is cultivated in the same manner. Cultivars of both fruits are developed mainly from *Ribes rubrum* and *R.sativum,* and also *R.petraeum, R.vulgare* and *R.multiflorum,* all of which are native to Europe. They are first recorded in cultivation in the late 15th century, and were introduced into North America in the early 17th century.

The red- and whitecurrant grows easily in cool moist climates, requiring a sheltered site that is not readily prone to frost at flowering time. In gardens, it is commonly grown as a bush on a short leg planted 1.5m², although commercially it is often trained as a higher yielding stooled bush. Redcurrants and whitecurrants may also be grown as standards, or as cordons, espaliers or fans which may be trained against a fence or on a wired structure as both space-saving and potentially attractive garden features.

redcurrant

Pests
birds
capsid bug
sawflies

Diseases
grey mould
leaf spot

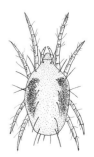
red spider mite

Single cordons are conveniently spaced 45cm apart. Two year old bushes from a certified source are a good choice for planting in autumn.

Redcurrants depend on potash for healthy growth; this should be applied as $25g/m^2$ of sulphate of potash in late winter. Nitrogen dressing is advisable only on poor soils. Organic mulching and early summer irrigation are highly beneficial. Fruit is produced from buds borne at the base of one-year-old shoots, and the objective is to form a goblet-shaped bush of 8–10 main branches on which fruiting spurs are carried. In the formative years, prune leading shoots to about half their length. On established bushes, laterals should be cut back to two or three buds each winter, with old less fruitful branches being selectively removed. Trained forms should have spur growth encouraged by shortening laterals to one or two buds each winter; all current season growth should be shortened to five leaves during early summer. New plants can be established from hardwood cuttings, 30–38cm long, taken in autumn or early winter, and from which all but the top 3 or 4 buds are removed. 'Jonkheer van Tets', 'Redlake' and 'Rondom' are amongst well-liked cultivars of redcurrant which will provide some harvest succession; 'White Versailles' is a notable cultivar of whitecurrant.

red thread *(Laetisaria fuciformis)* a common disease of turf, especially where conditions of low fertility are combined with hot weather; it is also known as PINK PATCH. RED FESCUE *(Festuca rubra)* and PERENNIAL RYE-GRASS *(Lolium perenne)* are most often affected. Patches of diseased grass, up to 35cm in diameter, appear pink or red due to needle-like threads of fungus sticking out from the leaves. Fungicide and nitrogen fertilizer applications can control the disease. See *lawns.*

reel mower a cylinder mower.

reflexed abruptly deflexed at more than a 90° angle.

regular see *actinomorphic.*

regulated pruning a system of pruning involving thinning out and spacing branches and removing weak growth, mostly with a saw, without attention being given to leaders, laterals or spurs. It is applied primarily to large vigorous apple trees, and sometimes referred to as saw pruning.

Rehderodendron (for Alfred Rehder (1863–1949), German botanist, dendrologist and gardener, and Greek *dendron,* tree). Styracaceae. S and W China. 9 species, deciduous shrubs or trees with serrulate leaves and white, 5-petalled flowers in axillary leafless panicles or racemes in spring. The fruit is a woody capsule. *R.macrocarpum* is a beautiful small tree, valued for its lemon-scented flowers in mid- to late spring; these are followed by bright red fruits, which resemble miniature coconuts. The glossy leaves often have attractive autumn tints. Hardy to –10°C/14°F. Plant in a sheltered

position. Sow seed when ripe, in midsummer; take semi-ripe cuttings, and root in a closed case with bottom heat. Otherwise, cultivate as for *Styrax. R.macrocarpum* (W China; tree to 9m flowers to 5cm in diameter, finely downy, fragrant, appearing before leaves).

Rehmannia (for Joseph Rehmann (1799–1831), German physician active in St. Petersburg). Gesneriaceae. China. 9 species, hairy perennial herbs with 2-lipped, tubular flowers in terminal racemes in spring and summer. Suitable for permanent outdoor cultivation, in the lightly shaded herbaceous border only where frosts are short-lived. In cooler zones they are best treated as half-hardy, overwintered in frost-free conditions. Alternatively, grow all year in the alpine house in an open, humus-rich loam. Propagate from seed, root cuttings or cuttings of basal shoots.

R.angulata (as for *R.glutinosa,* except leaves deeply lobed and toothed); *R.elata* (to 1.5m, usually smaller; leaves lobed, sometimes toothed; flowers 7–10cm, bright rose-purple, throat yellow, spotted red); *R.glutinosa* (to 30cm, with sticky purple hairs; leaves obovate, coarsely toothed; flowers to 7.5cm, red-brown to yellow, striped purple, lobes pale yellow).

Reinwardtia (for K.G.K. Reinwardt (1773–1822), director of the Botanic Garden, Leiden). Linaceae. N India, China. 1 species, *R.indica,* YELLOW FLAX, a glabrous, evergreen shrub or subshrub, to 90cm. Produced in late spring and summer, the flowers are bright yellow, 2.5–5cm in diameter, broadly funnel-shaped to bowl-shaped and consist of five, rounded petals. Grow in full sun or bright filtered light; water plentifully and feed fortnightly when in full growth. Provide a minimum winter temperature of 7–10°C/45–50°F. Cut back after flowering. Propagate by seed or by softwood cuttings.

rejuvenation pruning the heading back of non-productive branches on old fruit trees, which results in a mass of new shoots from which a new fruiting framework can be developed.

remontant used of plants that flower and/or fruit more than once in a growing season.

rendzina a very shallow, humus-rich soil overlying chalk or limestone.

renewal pruning an important system of pruning apples and pears, carried out in order to ensure a successional supply of one-year-old (growth), two-year-old (bud) and three-year-old (fruiting) shoots, thereby restraining the tendency of most cultivars to produce too much fruit bud. On established trees, about one third of the number of one-year-old laterals is left uncut, and the remainder cut back to stubs or completely removed. Budding and fruiting laterals are either left uncut, cut back to stubs or removed completely, according to the balance of growth desired. Leaders are pruned lightly or heavily, depending on general vigour, and always with a view to providing major replacement shoots. See *pruning*.

reniform kidney-shaped in outline.

repand sinuate, with less pronounced undulations.

repellents materials intended to deter animals from attacking plants, while not killing or injuring them. Mechanical methods, especially those intended to repel birds, include devices such as scarecrows, distress-call recordings, mimic birds of prey, and explosives. Chemical repellents have limited use in the garden, except against birds and mammal pests. Lawn seed may be dressed to deter seed-eating birds and several proprietary substances are available to repel deer, rabbits, dogs and cats, although most remain effective only for relatively short periods.

replant disease a reduction in growth or yield, often with associated root lesions, seen when plants of the same species are replanted on the same site. In many cases, replant disease is associated with pathogenic soil fungi, for example *Pythium sylvaticum* in specific apple replant disease (SARD). It is also known as soil sickness.

Reseda (from Latin *resedo,* to heal or calm: the Romans applied the plant to external bruises). Resedaceae. Mediterranean, E Africa, SW Asia to NW India. MIGNONETTE. 55 species, herbs, producing erect racemes crowded with small, often fragrant flowers in summer; the flowers are composed of four to eight, entire to cut petals and many stamens. *R.odorata,* is valued for its intense, spicy fragrance. It is grown in the annual and cut-flower border (the flowers are remarkably long lived, persisting for many weeks in water in cool conditions) and as a pot plant for the cool greenhouse or conservatory in winter. Sow seed *in situ* in spring, or in autumn where temperatures do not fall much below –10°C/14°F; alternatively, sow in cellular trays and introduce young plants as plugs. For winter/early spring blooms under glass, make successional sowings at monthly intervals from late summer; sow several seeds in small pots, germinate at about 13°C/55°F and thin to the strongest, potting on into a medium fertility, loam-based medium. Pinch to encourage bushy plants. Grow in full light and maintain good ventilation with a temperature range between 7–13°C/45–55°F; high temperatures will prevent flowering. *R.odorata* (MIGNONETTE, SWEET RESEDA, BASTARD ROCKET; Mediterranean; annual, erect to ascending, to 80cm; flowers white, sometimes tinged green or yellow or buff, intensely fragrant).

reserve border see *nursery bed*.

residual used of herbicides that persist in soil for some time after application and continue to kill germinating weeds.

resistance the ability of one organism to ward off the harmful effects of another; for example, some plant species or cultivars may be resistant to disease infection by reason of having thicker surface leaf cells, which prevent ready penetration of a fungus, or as a result of gum barriers or antibodies formed after infection. Certain tomato cultivars may be resistant to attack by root eelworm. These are important factors in reducing reliance on pesticides in the garden. Another context is where pests and diseases may become resistant to a particular chemical control treatment; this points to the importance of using alternative substances from time to time. See *immunity, insecticides.*

respiration the breakdown of organic molecules in plant cells, during which process energy is released and carbon dioxide produced. It takes place following the absorption of oxygen by a plant cell. Respiration can proceed independent of light intensity. Diffusion of the gases involved occurs through stomata and lenticels.

resting period a period when a plant makes little or no growth, usually in response to environmental conditions of temperature, light or moisture. See *dormancy, quiescence.*

resupinate, resupine a leaf or flower inverted by a twisting of 180° of the petiole or stalk.

retaining wall a supporting wall between different levels of a terraced slope.

retarding the process of artificially delaying growth and/or flowering in plants. The retarding method most commonly used is that of storing at low temperature, as may be practised with lifted stock of lily-of-the-valley, daffodils and flowering shrubs. It can also be achieved with chemicals (e.g. maleic hydrazide, which is used to suppress the growth of hedges) and by mechanical means (e.g. deblossoming strawberries in order to encourage a later flush).

reticulate netted, i.e. with a close or open network of anastomosing veins, ribs or colouring.

retrorse turned, curved or bent downwards or backwards, away from the apex; cf. *introrse*.

reversion used of (1) alteration in a plant's growth back to a less desirable form, as indicated by changes in colour or other characteristics. It is commonly encountered in variegated clones where leaves or part of the plant are found to be wholly green or yellow, representing the nature of the plant from which the variegation is a sport or mutant. Such reverted shoots should be cut out. Suckering, seeding and virus infection sometimes produce forms that are mistaken for reverted forms, while circumstances of breeding occasionally result in the expression of normally recessive characteristics, which may be misinterpreted as true genetic reversion to an ancestral species; (2) a disease of blackcurrants, affecting the size and shape of the leaves and causing a reduction in the number of flowers and therefore crop yield. It is caused by an unknown, graft-transmissible agent, which may be a mycoplasma-like organism, probably transmitted by the BLACKCURRANT GALL MITE *(Cecidophyopsis ribis).*

revolute with margins rolled under, i.e. toward the dorsal surface.

reworking changing the scion cultivar of a fruit tree by grafting onto existing branches. See *frameworking, grafting, top working.*

Rhamnus (from the Greek name for the genus, *Rhamnos).* Rhamnaceae. Widely distributed in northern temperate areas, with a few species in E and South Africa and Brazil. BUCKTHORN. 150 species, deciduous or evergreen trees or shrubs. Produced in summer, the small, yellow-green flowers are followed by small berries, red at first and ripening black. Fully hardy, the following will thrive on most fertile soils in full sun, but benefit from some protection from harsh winds. Prune in late winter/early spring only to remove dead and overcrowded wood. Propagate by seed, stratified for 60–90 days at 5°C/40°F and sown in spring, by semi-ripe cuttings or by simple layering.

R.alaternus (S Portugal to N Africa and SE Russia; evergreen shrub, 1–5m; leaves 3–7cm, ovate to elliptic, coriaceous, more or less serrate, glabrous; 'Argenteovariegata': leaves elliptic to oblong-lanceolate with broad creamy-white margins on leaden green); *R.imeretina* (E Turkey, Georgia, Armenia; deciduous shrub, 2–4m; leaves 8–25cm, elliptic to broadly lanceolate, finely toothed, glabrous and dull green above, pale, velvety beneath, bronze-purple in autumn).

Rhaphidophora (from Greek *raphis,* needle, and *phora,* to bear; the cells contain raphides.) (Sometimes spelt *Raphidophora).* Araceae. SE Asia. 70 species, evergreen climbers or creepers intermediate in appearance between the smaller *Monstera* species and *Philodendron.* Cultivate as for the climbing *Philodendron* species and cultivars. *R.celatocaulis* (syn. *Pothos celatocaulis;* SHINGLE PLANT; Borneo; tall climber; leaves of juvenile plants 8.5–10cm, in 2 ranks, closely adpressed to substrate, overlapping and obscuring stem, entire, elliptic to ovate, coriaceous, blue-green; adult leaves to 40cm, entire, pinnatifid or pinnatisect, occasionally perforate).

Rhaphiolepis (from Greek *rhaphis,* needle and *lepis,* scale, referring to the extremely narrow persistent bracteoles on the inflorescence). Rosaceae. Subtropical E Asia. Some 15 species, evergreen shrubs or small trees with tough, glossy dark green leaves. Produced mostly in spring and summer in terminal racemes or panicles, the flowers resemble apple blossom and consist of five obovate petals and numerous stamens. They are followed by small, subglobose pomes, often black or purple-black. Hardy to −15°C/5°F. Grow in a warm sheltered position in full sun. Prune if necessary, after flowering. Propagate by semi-ripe cuttings, layering, or seed.

R. × delacourii (R.indica × R.umbellata; garden origin; to 2m; flowers to 1.8cm in diameter, rose-pink; includes 'Coates' Crimson', with crimson flowers, 'Majestic Beauty', with large, dark green leaves and milky pink, scented flowers, in clusters to 25cm across, and 'Springtime', compact, vigorous, with new growth flushed pink-brown and broad, candy pink flowers); *R.indica* (S China; to 1m; flowers to 1.6cm in diameter, white tinted pale pink toward centre, stamens bright red-pink; cultivars include 'Charisma', small tightly branched, with double, soft pink, abundant flowers in late winter-spring, 'Enchantress', dwarf, with small, rose flowers in late winter-early summer, 'Indian Princess', broadly mounding, compact, with bright pink flowers fading to white, in broad clusters, in spring, 'Rosea', with deep pink flowers, 'Snow White', with white flowers in early spring, and 'White Enchantress', dwarf, compact, with white flowers, in spring); *R.umbellata* (Japan, Korea; flowers fragrant to 2cm in diameter, white, stamens bright pink-red; 'Minor': dwarf, small, white, scented flowers, in early spring).

Rhapis (from Greek *rhaphis,* a needle, probably referring to the narrow leaf segments). Palmae. China, Japan. LADY PALM. 12 species, small to medium-sized palms. Ultimately, they make a clump of slender, cane-like stems, clothed in their upper portions with the remains of fibrous leaf sheaths. The leaves are palmate, dark green and slender-stalked, the leaflets, finger-like, ribbed and folded. Suitable for pot and tub cultivation in the home, greenhouse or conservatory (minimum temperature 10°C/50°F). Plant in a fast-draining, loamy medium, rich in leafmould. Water, syringe and feed generously in warm conditions. Shade from direct sunlight. Propagate by seed, suckers or division.

R.excelsa (syn. *R.flabelliformis;* GROUND RATTAN, BAMBOO PALM, LADY PALM, MINIATURE FAN PALM, FERN RHAPIS; probably originally from China but introduced from Japan; stem to 5m; leaf segments 3–10, 20–30 × 1.5cm, tips truncate; includes 'Variegata', with leaves striped ivory to white; many other variegated forms are cultivated and highly prized, particularly in Japan); *R.humilis* (REED RHAPIS, SLENDER LADY PALM; S China; differs from *R.excelsa* in leaf segments numbering at least 9, and 15–18 × to 2cm, with more acuminate tips).

rheophyte a plant that grows in running water.

Rheum (from Greek *rheon,* rhubarb). Polygonaceae. Asia. RHUBARB. 50 species, deciduous perennial herbs with tough or woody rhizomes. Produced in basal clumps, the leaves are large and sinuately toothed to palmately lobed, their stalks often thick and differently coloured to the leaf blades. Small flowers are massed amid large bracts on tall, panicles in late spring and early summer. Impressive specimens for focal plantings in moist borders, at the woodland edge, bog gardens and the margins of pools and streams. Hardy to about −15°C/5°F. Grow in sun to part shade in humus-rich, moisture-retentive soils. Propagate by division in spring, cutting through the tough fleshy rootstock with a sharp knife or spade. Alternatively, sow seed in autumn in the shaded cold frame. *R.* × *hybridum*, RHUBARB is cultivated for its edible leafstalks.

R.alexandrae (W China, Tibet; to 1.5m; leaves to 22 × 13cm, ovate to oblong, base cordate, entire, dark green in neat rosettes, glabrous, petiole slender; inflorescence bracts, yellow-green, subcordate, flowers green-yellow); *R.nobile* (SIKKIM RHUBARB; Nepal to SE Tibet; to 2m; leaves to 30cm in diameter, rounded, base cuneate to rounded, glossy, leathery, dark green veined red, entire, edged red, petiole stout, red; inflorescence bracts cream; flowers green); *R.palmatum* (NW China; to 1.5m; leaves 50–90 × 50–70cm, more or less rounded, palmately lobed, toothed; inflorescence bracts tinted pink to bronze-red, flowers brick red to deep red; includes various forms with leaves deeply and sharply to broadly cut, many flushed bronze-red at first, later coppery olive; two popular cultivars are 'Atrosanguineum', the whole of which is suffused blood-red, later bronze-red, and 'Rubrum', with leaves flushed dark maroon-red when young).

Rhipsalis (from Greek *rhips*, wickerwork, and *-alis*, alluding to the very slender stems). Cactaceae. Tropical America, with one species extending to tropical Africa, Madagascar and Sri Lanka. Up to 50 species, epiphytic or lithophytic cacti. The stems are cylindric, ribbed, angled, winged or flat. They are usually pendulous and segmented, lacking spines and with the younger segments arising singly or in whorls. The flowers are small, rotate and usually white; they are followed by berry-like fruit. Grow in an intermediate greenhouse (minimum temperature 10°C/50°F). Plant in an acid medium rich in coarse bark and perlite. Shade in summer. Maintain high humidity. Reduce watering in winter. Increase by stem cuttings.

R.baccifera (tropical America, Africa, Madagascar, Ceylon; stems cylindric, pendent, 1–4m, 4–6mm in diameter; fruit white or pink); *R.gracilis* (E Brazil; stems 10–30cm long, 1–3mm in diameter; fruit white to deep red;) For *R.gaertneri*, see *Hatiora gaertneri*.

rhizoid a structure resembling a root in appearance and function, as found in *Selaginella*.

rhizome a specialized stem, slender or swollen, branching or simple, subterranean or lying close to the soil surface (except in epiphytes), and producing roots, stems, leaves and inflorescences along its length and at its apex. Rhizomes are common to many perennial herbs.

rhizomorph a root-like aggregation of fungal threads, or hyphae, often referred to as a 'bootlace' or 'shoestring', of *Armillaria* species, by means of which the fungus survives and spreads. See *armillaria root rot*

rhizosphere the zone of soil immediately surrounding plant roots.

Rhodanthemum (from Greek, *rhodo-*, red and *anthemon*, flower) Compositae, N Africa. 5 species. Semi-hardy perennial, evergreen subshrubs or tufted, woody-based herbs, usually with silver, finely divided leaves and long-stalked daisy-like flowerheads in summer. Hardy in climate zone 7, if planted in a sheltered, warm site and protected from winter wet; grow in full sun on a free-draining, gritty soil. Propagate by semi-ripe cuttings in summer.

R.hosmariense (syn. *Chrysanthemopsis hosmariensis*, *Chrysanthemumhosmariense*,*Leucanthemopsis hosmariensis*; Morocco; low subshrub to 30 × 50cm; leaves finely-divided, silver-grey, flowerheads to 4cm, ray florets white and disc florets yellow); *R.maresii* (syn. *Chrysanthemopsis maresii*; Algeria; woody-based, bushy herb or subshrub to 30cm tall, with silver-grey, finely dissected or simple leaves and yellow flowerheads ultimately tinged purple to purple-black).

Rhodiola (from Greek *rhodon*, rose, referring to the rose-scented roots of the type species). Crassulaceae. Himalaya, NW China, C Asia, N America, Europe. Around 50 species, perennial herbs with thick, fleshy and scaly rhizomes. The flowering stems arise annually bearing fleshy leaves. Produced in late spring and summer in terminal heads, the small flowers consist of four or five petals and twice as many rather longer stamens. The following species is very hardy and favours a fast-draining position in full sun. It grows well on the rock garden and in wall crevices. Propagation as for *Sedum*. *R.rosea* (syn. *Sedum rosea*; ROSE-ROOT; northern Hemisphere; 5–30cm; leaves 1–4cm, glaucous, grey-green oblong, apex obtuse to acuminate, entire or toothed, often tinged red, base round or slightly cordate; flowers green-yellow, anthers deep yellow to copper).

Rhodochiton (from Greek *rhodon*, rose, and *chiton*, cloak, referring to the large, rosy calyx). Scrophulariaceae. Mexico. Some 3 species, perennial, climbing herbs with pendulous solitary flowers in summer; these have a decorative parasol-like calyx and a tubular corolla with five, blunt lobes. In zones 7 and under, it may be grown outdoors as an annual (sometimes surviving as a herbaceous perennial in favoured situations) or treated as a perennial in cool greenhouses or conservatories. Grow in a sunny, sheltered spot, on rich, well-drained soil. Provide support while plants are still very young. Under glass, water plentifully during the growing season, sparingly in winter, and give a minimum temperature of 5°C/40°F; ventilate freely and shade from direct summer sun. Cut back and re-pot in spring. Sow seed under glass in early spring or as soon as ripe. *R.atrosanguineum* (to 6m, slender-stemmed; calyx to 3cm in diameter, 5-lobed, shallowly-campanulate, rose-pink or mauve, corolla dark blood-red to maroon-black, to 6.5cm, minutely hairy, anthers ash-white).

Rhododendron (the Ancient Greek name for the unrelated rose-flowered oleander, *Nerium oleander*, from *rhodon*, rose and *dendron*, tree). Ericaceae. Cosmopolitan. Some 800 species, trees and shrubs, evergreen or deciduous with simple leaves. Produced in racemes in spring and summer, the flowers are tubular, funnel-shaped or campanulate with 5, more or less spreading lobes.

The following are hardy in climate zones 6 or 7, but prefer a mild and sheltered site. Plant in dappled sun or light shade on a fertile, humus-rich and acid to neutral soil. Ensure that the root-run is cool and moist, especially in summer. Prune only to keep within bounds (after flowering), or to remove dead or exhausted growth (in early spring). Increase by layer-ing, grafting, or semi-ripe cuttings in a case in late summer. Affected by powdery mildew, honey fungus, phytophthora root rot, rusts and chlorosis (caused by poor drainage and unsuitable soil pH). Attacked by rodents, weevils and leatherjackets.

Several of the small, evergreen hybrids and cultivars popularly known as azaleas grow well in containers in the cool greenhouse. They can be brought indoors at flowering time. In addition to the species listed below, hundreds of hybrids and cultivars have been named. Varying greatly in size and shape, the flowers range in colour from pure white to cream, yellow, gold, apricot, salmon, rose, deep pink, cerise, crimson, magenta, mauve, lilac, scarlet, orange and coral red, often spotted, stained or suffused a different colour.

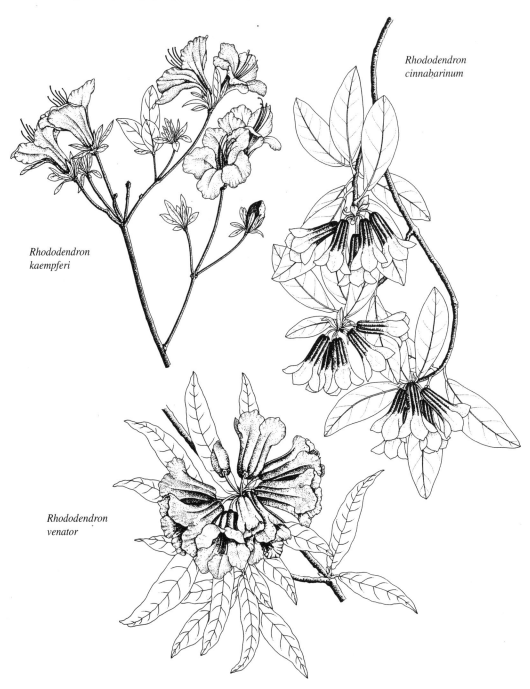

Rhododendron cinnabarinum

Rhododendron kaempferi

Rhododendron venator

RHODODENDRON

Name	Distribution	Habit	Leaves	Flowers
R.aberconwayi	SW China	evergreen, to 2m	3–6cm, thick, elliptic.	2.8–3.5cm, open-campanulate, white to pale rose with purple spots

Comments: 'His Lordship': flowers white with crimson streaks

R.albrechtii	Japan	deciduous, to 2m	4–12cm, obovate, ciliate, finely toothed, densely grey-tomentose beneath	to 5cm, open-campanulate, red-purple

Comments: 'Michael McLaren': flowers purple, spotted with yellow-green

R.arboreum	Himalaya, China to Thailand, S India and Sri Lanka	evergreen tree to 50m	6.5–19cm, broadly oblong-lanceolate, glabrous and wrinkled above, densely hairy beneath	3–5cm, tubular-campanulate, fleshy, pink to deep crimson or rarely white, often with dark spots

Comments: ssp. *arboreum*: leaves 8–19cm, with silvery or white hairs beneath. 'Goat Fell': flowers cherry red with some spots in throat. 'Rubaiyat': flowers red with darker spots. ssp. *cinnamomeum* var. *cinnamoneum*: leaves with off-white or fawn hairs beneath; flowers pink to red, rarely white. 'Sir Charles Lemon': flowers creamy white; leaves dark green above with bright orange-brown hairs beneath. var. *roseum*: similar to var. *cinnamoneum*, but leaves with a 1-layered, fawn to off-white compacted indumentum, flowers tinged red. 'Tony Schilling': flowers deep pink with darker spots. ssp. *delavayi*; leaves 7–15.5cm, off-white to fawn – hairy beneath; flowers red.

R.argyrophyllum	SW China	evergreen, spreading, to 12m	1.8–6cm, elliptic to oblanceolate, glabrous above, with thin, silvery to fawn hairs beneath.	3–5.5cm, funnelform-campanulate or open-campanulate, white to pale pink with purple spots

Comments: 'Henry Wood': flowers white, flushed and spotted with purple, opening from pink buds

R.augustinii	China	evergreen, bushy shrub or small tree to 10m	narrowly elliptic to elliptic, usually evergreen, 4–11cm, glabrous above, golden scaly beneath	2.8–4cm, openly funnel-shaped, purple, lavender or almost blue, rarely white, with greenish or brownish spots inside

Comments: 'Electra': flowers large, deep blue. 'Tower Court Form': flowers large, lavender-blue, veined red . ssp. *augustinii*: leaves mostly evergreen; flowers lavender to blue. ssp. *chasmanthum*: leaves mostly evergreen; flowers lavender to blue. Exbury form: flowers blue-purple with some yellow spots.

R.auriculatum	W China	evergreen shrub or tree to 6m	15–30cm, oblong-oblanceolate, base auriculate, margin fringed with small glands, villous and glandular beneath.	8–11cm, fragrant, funnel-shaped, white or cream to pink, tinged green at the base

R.barbatum	Himalaya	evergreen shrub or tree to 6m	9–19cm, elliptic to obovate, glabrous above, with scattered hairs and glands beneath	3–3.5cm, fleshy, tubular-campanulate, crimson to blood-red or, rarely, pure white

R.calendulaceum FLAME AZALEA	SE US	much-branched, deciduous shrub to 3m	3.5–9cm, elliptic to broadly elliptic or ovate-oblong, finely pubescent above, more densely so beneath	to 5cm, funnel-shaped, exterior pubescent, orange, red or yellow, more rarely pink

Comments: Involved in the parentage of the deciduous azalea hybrids. 'Alhambra': flowers bright orange-red. 'Burning Light': flowers coral red, orange in throat. 'Camp's Red': flowers rich red. 'Croceum': flowers rich yellow. 'Luteum': flowers yellow. 'Roseum': flowers pink. 'Smoky Mountaineer': flowers almost scarlet

R.calophytum	SW China	evergreen tree to 12m	14–30cm, oblong-oblanceolate, glabrous when mature	4–6cm, open-campanulate, glabrous, pink-white with purple spots and a basal blotch

R.calostrotum	NE India, N Burma, W China	evergreen, low to prostrate, matted or erect shrub, 5–150cm	1–3.3cm, oblong-ovate to almost circular, upper surface usually matt with persistent, dried-out scales, margins bristly, lower surface with dense overlapping scales	1.8–2.8cm, magenta, rarely pink or purple, often with darker spots on the upper lobes, exterior hairy and occasionally scaly

Comments: 'Gigha': compact, very free-flowering flowers large, rosy-crimson. 'Rock's form': flowers large, blue-purple; foliage purple-tinged. 'Calciphilum': leaves smaller, flowers pink. 'Nitens': pink-purple flowers in June-July, later than other cultivars.

R.catawbiense	E US	evergreen shrub to 3m	6.5–11.5cm, broadly elliptic to obovate, glabrous when mature	3–4.5cm, funnelform to campanulate, usually lilac-purple with faint spots

Comments: 'Album': flowers large, white, opening from lilac buds. 'English Roseum': flowers bright pink, faintly tinged with lilac. 'Lavender Queen': flowers pale lavender. 'Lee's Dark Purple': flowers deep purple. 'Powell Glass': flowers pure white. 'Roseum Elegans': flowers purple-pink. Some of the above forms, heat-resistant and widely grown in the United States, may be hybrids.

R.cinnabarinum	Himalaya to N Burma	evergreen straggling shrub to 7m.	3–9cm, broadly to narrowly elliptic, scaly or scaleless above, with fleshy scales beneath often tinged red or purple	2.5–3.5cm, tubula to campanulate, yellow, orange, red, red and yellow or purple, usually with a waxy bloom

Comments: 'Aestivale': a late-flowering form (July in Great Britain) with narrow leaves. 'Mount Everest': flowers apricot, more open than usual. 'Nepal': flowers red at base, yellow towards margin. Blandfordiiflorum Group: flowers narrowly tubular, red on outside, yellow, apricot or green inside; late-flowering, hardy. 'Bodnant': flowers vermilion at base on outside, paler towards rim. Roylei Group: flowers rose-red to purple-red, relatively open; leaves glaucous blue-green. 'Crarae Crimson': flowers plum-crimson. 'Magnificum': flowers large, red-purple tinged with orange. 'Vin Rosé': flowers red with a waxy bloom. ssp. *xanthocodon*: flowers campanulate, yellow, orange, orange with a purple flush or purple. Concatenans Group: flowers campanulate, apricot to coral; leaves glaucous-white beneath, blue-green above. 'Cooper': flowers coral-red flushed with orange and red. Pallidum Group: flowers pale pink-purple, broadly funnel-shaped. Purpurellum Group: flowers pale pink-purple, broadly funnel-shaped. Purpureum Group: flowers bright mauve-pink or plum-purple, campanulate. *R.cinnabarinum* has been widely used in hybridization; some of the hybrids, previously thought to be interspecific, are now within the species, e.g. 'Conroy' (*R.cinnabarinum* ssp. *cinnabarinum* 'Roylei' × ssp. *xanthocodon*); others involve other species, e.g. 'Cinnkeys' (× *R.keysii*); 'Royal Flush' and 'Rose Mangles' (both × *R.maddenii* ssp. *maddenii*) and 'Yunncinn' (× *R.yunnanense*).

RHODODENDRON

Name	Distribution	Habit	Leaves	Flowers
R.dauricum	E Siberia, Mongolia, N China, Japan	straggling, semi-evergreen shrub to 1.5m	leathery, densely scaly beneath, 1–3.5cm, elliptic.	1.4–2cm, pink or violet-pink

Comments: 'Arctic Pearl': flowers white. 'Hokkaido': flowers large, white, later-flowering. 'Midwinter': flowers bright rose-purple. 'Nanum': dwarf form. Sempervirens Group: plants evergreen. Sometimes split into 3 species: *R.dauricum* – leaves dark green above, rusty brown beneath, most overwintering. *R.ledebourii* – leaves pale above and beneath, mostly falling in autumn, with a few scales on the upper surface; *R.sichotense* – leaves without scales on the upper leaf surface.

R.davidsonianum	W China	evergreen shrub to 5m	3–6.5cm, often V-shaped in section, lower surface covered in small brown scales	2.3–2.7cm, pink, pink-lavender or lavender

Comments: 'Caerhays Blotched': flowers pale pink with darker spots. 'Exbury': flowers pink. 'Ruth Lyons': flowers rich pink with few spots, becoming deeper coloured with age. 'Serenade': leaves small, rounded; flowers clear pink

R.decorum	W China, NE Burma, Laos	evergreen shrub or small tree to 6m.	6–20cm, oblanceolate to elliptic, glabrous when mature with punctate hair bases beneath	4.5–11cm, funnelform-campanulate, sparsely glandular outside, densely so within, white to pale pink, with or without green or crimson spots

Comments: ssp. *decorum*: flowers 4–6.5cm, and leaves to 12cm. 'Cox's Uranium Green': flowers chartreuse green. ssp. *diaprepes*: leaves and flowers larger. 'Gargantua': flowers very large, white, flushed green below.

R.degronianum	Japan	evergreen shrub to 2.5m	8–14cm, elliptic to oblanceolate, with dense grey to fawn hairs beneath	3–4.5cm, widely funnelform-campanulate, pink with conspicuous darker spots

Comments: 'Gerald Loder': flowers white, spotted and flushed with red-purple. ssp. *yakushimanum*: syn. *R.yakushimanum*; leaves silvery then dark green, covered in thick, dense, brown-felty hairs beneath; flowers broad, pink fading to white, speckled green within- a ssp. much used in recent hybridization as a parent of a range of striking very hardy hybrids. 'Ken Janeck': plants and leaves larger than usual; flowers pink. 'Koichiro Wada': flowers white opening from pink buds, in dense trusses, freely produced. 'Overstreet': flowers white opening from pink buds; leaves very narrow. 'Tremeer': tall.

R.fortunei	China	evergreen shrub or tree to 10m	8–18cm, broadly oblanceolate to obovate, glabrous when mature except for persistent hair-bases beneath	5.5–7cm, open-campanulate or funnelform-campanulate, pale pink to almost pure white, exterior glandular or glabrous

Comments: 'Mrs Butler': flowers pink. ssp. *discolor*: syn. *R.discolor*; leaves oblanceolate; flowers fragrant, pink, funnelform. 'John R. Elcock': flowers purple-pink, yellow in throat.

R.fulvum	W China, NE Burma	evergreen shrub or small tree to 8m	8–22cm, oblanceolate to elliptic, glabrous above when mature, velvety brown-hairy beneath, appearing granular	2.5–4.5cm, campanulate, white to pink usually with a dark basal blotch, with or without crimson spots

R.glaucophyllum	Himalaya, W China	shrub to 1.5m	3.5–6cm, elliptic, upper surface dark brown-green, lower surface white, with pale and dark scales	1.8–3.2cm, tubular to campanulate, pink or white flushed with pink, scaly outside

Comments: Plants grown under the name *R.shweliense* are usually this species.'Branklyn': flowers pale pink, bell-shaped. 'Prostratum': low-growing.

R.glischrum	W China, NE Burma	shrub or small tree to 8m	11–30cm, obovate to elliptic, ciliate, upper surface smooth to wrinkled, with some glandular bristles, lower surface with a dense covering of bristles sometimes also with a thin brown covering of hairs	3–5cm, campanulate, pink to scarlet or occasionally white flushed pink, with purple spots and usually also a basal blotch

Comments: 'Frank Kingdon Ward': flowers purple-pink, spotted purple

R.hippophaeoides	W China	erect, evergreen openly branched shrub to 1.2m	5–12mm, narrowly elliptic, oblong or narrowly obovate, aromatic, grey-green above, yellow-green or grey-green with numerous scales beneath	broadly funnel-shaped, pale lavender to deep purple, rarely white

Comments: 'Bei-ma-shan': tall, with lavender-blue flowers . 'Haba-shan': flowers lavender in large trusses.

R.impeditum	W China	compact, much-branched, evergreen shrub to 1m	4–14mm, aromatic, lower surface pale grey-green speckled with brown or more uniformly brown.	8–15mm, broadly funnel-shaped, violet to purple, rarely lavender

Comments: Parent of a number of low, blue-flowered hybrids, e.g. 'Blue Tit' (× *R.augustinii*).

R.kaempferi	Japan	semi-evergreen shrub to 3m	1–5cm, entire, strigose	2–3cm, funnel-shaped, pink or red

Comments: Very similar to *R.indicum*, but with broader leaves. 'Mikado': flowers red with darker spotting in throat.

R.kiusianum	Japan	semi-evergreen to deciduous, dwarf shrub to 1m	0.5–2cm, obovate, red-brown strigose	1.5–2cm, funnel-shaped, usually rose-pink, occasionally purple

Comments: The Obtusum Group of cultivars and the Kurume Azaleas had their origin in this species and its hybrids with *R.kaempferi*. 'Album': flowers white. 'Benichidori': flowers salmon orange. 'Betty Muir': flowers bright pink. 'Chidori': flowers white. 'Hanejiro': flowers white. 'Harunokikari': flowers lavender-blue. 'Harusame': flowers bright pink. 'Hillier's Pink': flowers pink. "Komokul-shan": flowers pale pink, tipped rose pink. 'Muraski Shikibu': flowers hose-in-hose, red-purple.

R.lacteum	W China	shrub or small tree to 7.5m	8–17cm, elliptic to obovate, with grey-brown hairs beneath	4–5cm, widely campanulate, pure yellow, without spots through sometimes with a purple basal blotch

Comments: 'Blackhills': flowers clear yellow, unmarked.

RHODODENDRON

Name	Distribution	Habit	Leaves	Flowers
R.lutescens	W China	semi-evergreen, straggling shrub to 6m	5–9cm, lanceolate to oblong, bronze-red at first, variably scaly above, with golden scales beneath	1.8–2.5cm, pale yellow with green spots in the inside of the upper lobes, funnel-shaped
Comments: Flowers in late winter and early spring. 'Bagshot Sands': hardy and free-flowering; flowers primrose yellow with darker spots. 'Exbury': flowers clear lemon yellow; new growth red, rather tender.				
R.luteum	E Europe to Caucasus	deciduous shrub to 4m	5–10cm, oblong to lanceolate, sticky with adpressed bristles	to 3.5cm, funnel-shaped, yellow, fragrant, exterior glandular-sticky
R.macabeanum	NE India	evergreen tree to 15m	14–25cm, broadly ovate to broadly elliptic, glabrous above when mature, with dense woolly hairs beneath	to 5cm, tubular-campanulate, lemon yellow with a purple blotch at the base
R.mallotum	W China, NE Burma	evergreen shrub or small tree to 6.5m	10–13cm, broadly oblanceolate to obovate, glabrous above, with dense, red, woolly hairs beneath	4–4.5cm, fleshy, tubular-campanulate, crimson
R.maximum	EN America	shrub or small tree to 3.5m	10–16cm, oblanceolate to elliptic, glabrous above, thinly hairy beneath	2.5–3cm, campanulate, white to pink-purple, with yellow-green spots
Comments: 'Summertime': flowers white, flushed red-purple at tips; yellow-green spots in throat				
R.moupinense	W China	evergreen, compact shrub to 1m	3–4cm, narrowly ovate, elliptic or obovate, green or tinged brown beneath with rather dense scales, margins bristly	3–4cm, white, often flushed pink and with dark red spots on the upper part of the tube inside
Comments: *R.× cilpinense*, a hybrid between *R.moupinense* and *R.ciliatum*, is a semi-evergreen, compact shrub to 1.5m tall with pink flowers produced in early spring.				
R.occidentale	W US	deciduous shrub to 3m	3–9cm, elliptic to oblong-lanceolate, ciliate, thinly pubescent	3.5–5cm, funnel-shaped, white or sometimes pink, with a yellow blotch, glandular-hairy outside
Comments: 'Crescent City Double': flowers semi-double, cream flushed with apricot. 'Humboldt Picotee': flowers white, edged with rose carmine. 'Leonard Frisbee': flowers white with broad, frilly petals with pink mid-line, upper petal with yellow flare. 'Miniskirt': dwarf; flowers white, very small, with long red style. 'Stagecoach Frills': flowers frilly, white flushed with pink, with orange-yellow flare.				
R.orbiculare	W China	evergreen shrub or tree to 15m	7–12.5cm, orbicular to ovate, glabrous	3.5–4cm, campanulate, rose-pink
R.oreotrephes	W China	evergreen shrub to 8m	2–8cm, mostly evergreen, circular to elliptic or oblong, blunt, lower surface with dense purple, red-brown or grey scales	2.5–3.4cm, rose to rose-lavender
Comments: 'Exquisetum': flowers pale mauve-pink spotted red. 'James's White Form': flowers white.				
R.ponticum	W and E Mediterranean, widely naturalized in Britain	evergreen shrub to 8m, young shoots glabrous	6–18cm, oblanceolate to broadly elliptic, glabrous when mature	3.5–5cm, campanulate, lilac-pink to purple, usually with green-yellow spots
Comments: 'Aucubifolium': leaves spotted yellow. 'Cheiranthifolium': leaves very narrow, undulate; flowers small. 'Foliis-purpureis': leaves coppery-purple in winter. 'Variegatum': leaves bordered with cream-white.				
R.racemosum	W China	evergreen shrub to 3m	1.5–5cm, broadly obovate to oblong-elliptic, glabrous above, shining white-papillose beneath, with dense scales	1–2cm, openly funnel-shaped, white to pale or deep pink
Comments: 'Forrest': dwarf; flowers pink. 'Glendoick': taller, with deep pink flowers. 'Rock Rose': flowers bright purple-pink. 'White Lace': flowers white.				
R.rex	W China, NE Burma	evergreen, large shrub or small tree to 12m	2–37cm, obovate to oblanceolate, glabrous above, smooth or finely wrinkled, lower surface with a dense fawn to red hairs	campanulate, white, pale yellow or pink, with a crimson basal blotch and spots.
Comments: 'Quartz': flowers pink with crimson blotch and spots. 'Roseum': flowers pale rose pink with small dark blotch, opening from deeper pink buds. ssp. *fictolacteum*: syn. *R.fictolacteum*; leaves brown, flowers white or pink. 'Cherry Tip': flowers white, edged with pink, and with crimson blotch spots.				
R.rubiginosum	W China, NE Burma	evergreen shrubs or small trees to 10m or more	5–11.5cm, narrowly elliptic to elliptic or almost lanceolate, glabrous and scaleless above, scaly beneath	1.7–3.8cm, openly funnel-shaped, pink, mauve-pink or rarely white, flushed with pink-purple, scaly outside
Comments: 'Wakehurst': flowers mallow purple with crimson spots. Desquamatum Group: flowers usually larger and flatter; plants often more compact with slightly larger leaves.				
R.schlippenbachii	Korea, E Russia	deciduous shrub to 5m	2.5–11cm, obovate or broadly ovate, truncate, rounded or notched at apex, sparsely pubescent at first	3–6cm, very openly funnel-shaped, pale pink, upper lobes spotted with red-brown
Comments: 'Prince Charming': flowers rose, flushed with deeper pink and spotted with crimson.				
R.simsii INDIAN AZALEA	NE Burma, China, Taiwan	evergreen to semi-evergreen shrub to 3m	1–7cm, elliptic, ovate to obovate or oblanceolate, sparsely strigose above, more densely so beneath	2.5–5cm, broadly funnel-shaped, red, spotted
R.sinogrande	W China, NE Burma	evergreen tree to 10m.	20–90cm, oblong-oblanceolate to oblong-elliptic, glabrous and wrinkled above when mature, with silvery or fawn hairs beneath	4–6cm, ventricose-campanulate, pale cream to ivory, blotched crimson at base

RHODODENDRON

Name	Distribution	Habit	Leaves	Flowers
R.souliei	W China	evergreen shrub to 5m	5.5–8cm, broadly ovate, glabrous	2.5–4cm, openly saucer-shaped, pale purple-pink

Comments: 'Exbury Pink': flowers deep rose pink. 'Windsor Park': flowers white, edged with pink, the 3 upper petal with small red blotch at base.

R.sutchuenense	W China	evergreen shrub to 5m	11–25cm, oblong-lanceolate, glabrous above, floccose along midrib beneath	5–7.5cm, widely campanulate, pink with darker spots

Comments: 'Seventh Heaven': flowers red-purple, white in centre

R.thomsonii	Himalaya, W China	evergreen shrub or small tree to 3.5m	3–11cm, orbicular to obovate or elliptic, glabrous, glaucous beneath, sometimes with some red glands	3.5–5cm, campanulate, fleshy, deep crimson, usually without darker spots
R.trichostomum	W China	small, intricately branched shrub, 30–150cm, often forming rounded bushes	1.2–3cm, linear, oblong or oblanceolate, upper surface green; scaly or not, margins usually revolute, lower surface pale brown with golden scales	0.6–2cm, white or pink, tube glabrous outside, with a few scales on exterior

Comments: Ledoides Group: corolla not scaly. 'Collingwood Ingram': flowers red-purple, paler in throat. 'Lakeside': flowers white, flushed with red-purple. 'Quarry Wood': flowers white, flushed with red-purple. 'Rae Berry': flowers clear, pale pink. 'Sweet Bay': flowers white flushed with rose.

R.venator	W China	shrub to 3m	8-14cm, elliptic to lanceolate, glabrous except for hairs on midrib beneath	to 3.5cm, fleshy, tubular to campanulate, crimson with darker nectar pouches
R.wardii	W China	evergreen shrub or small tree to 8m	6–11cm, narrowly obovate to broadly ovate, glabrous, somewhat glaucous beneath	2.5–4cm, saucer-shaped, white to pale yellow, sometimes with a purple basal blotch

Comments: 'Ellestee': flowers lemon yellow with crimson blotch. Litiense Group: leaves oblong, wavy, glaucous on underside. 'Meadow Pond': flowers primrose yellow with crimson blotch. var. *puralbum*: flowers white.

R.williamsianum	W China	dwarf, evergreen shrub to 1.5m	2–4.5cm, ovate-orbicular, more or less glabrous	3–4cm, campanulate, pale pink with darker spots

Comments: 'Exbury White': flowers white.

R.yunnanense	Burma, W China	shrub to 6m	3–7cm, evergreen or deciduous, narrowly elliptic or elliptic, scaly beneath, margins and upper surface bristly	2–3cm, white, pink or lavender, usually densely spotted above with red or yellow

Rhodohypoxis (from Greek *rhodon,* rose, and *Hypoxis*). Hypoxidaceae. SE Africa. 6 species, small perennial herbs with fibrous and fleshy roots and hairy, grass-like leaves in a basal clump. Produced in spring and summer on slender stalks, the solitary flowers consist of a short tube and six, spreading, oblong to obovate lobes. Plant in peaty pockets in the rock garden, or in sinks and troughs. Although hardy to at least –5°C/23°F, *Rhodohypoxis* is intolerant of winter wet. When grown outside, protect in winter with a propped pane of glass. Otherwise, keep in the alpine house or bulb frame. Grow in sun in a free-draining, acid to neutral medium, with additional sharp sand and leafmould; water plentifully when in growth, avoiding the foliage; keep almost completely dry in winter. Propagate by offsets, by division of established clumps, or by seed. *R.baurii* (RED STAR; leaves to 15cm, usually shorter, flowers 1–2, white, pink or red, to 4cm diam., lobes hairy).

Rhodoleia (from Greek *rhodon,* rose, and *leios,* smooth, an allusion to the rose-like flowers and spineless stems). Hamamelidaceae. S China to Sumatra. About 7 species, evergreen small trees with oblong leaves; these are thickly leathery and glaucous beneath. Produced in spring, the small, flowers are grouped in stalked, nodding heads, surrounded by coloured bracts. Plant in a slightly acidic, moisture-retentive but well-drained medium. Provide a minimum temperature of 7°C/45°F. Water plentifully when in growth, moderately at other times. Propagate from seed in autumn/spring or from semi-ripe cuttings rooted with bottom heat in summer. *R.championii* (Tibet, China, Hong Kong; leaves to 9cm; outer bracts brown, silky, inner bracts tomentose, petals dark pink).

Rhodothamnus (from Greek *rhodon,* rose, and *thamnos,* a shrub). Ericaceae. E Alps. 1 species, *R. chamaecistus,* an evergreen, dwarf shrub to 40cm. The leaves are 0.5–1.5 cm long, elliptic to oblanceolate, leathery and dark green. Produced in late spring and summer, the flowers are 2–3cm in diameter, rose pink, with five spreading lobes and slender stamens tipped with purple-red anthers. For the rock or peat garden; hardy to –15°C/5°F. Plant in sun or part shade in moist, humus-rich, acid soils with a cool root run. Propagate from seed, by careful layering, or by semi-ripe cuttings in late summer.

Rhodotypos (from Greek *rhodon,* rose, and *typos,* type, referring to its resemblance to a rose). Rosaceae. China, Japan. 1 species, *R.scandens,* a spreading, deciduous shrub to 5m. To 6cm long, the leaves are ovate, acuminate, sharply serrate with corrugated veins. Produced in spring and summer, the flowers are solitary, white, to 4cm in diameter and composed of four, spreading, rounded petals and numerous stamens. They are followed by shiny black berries. Suitable for the border and for woodland garden or other informal plantings. It will tolerate a wide range of soil conditions, shade, and temperatures as low as –20°C/–4°F. For best results, grow in dappled shade on a well-drained, moisture-retentive soil. Thin out old flowered shoots to ground level. Propagate by seed sown in autumn, or by softwood or semi-ripe cuttings in a closed case with bottom heat

Rhoicissus (from Greek *rhoia*, pomegranate, and *Cissus*). Vitaceae. Tropical and southern Africa. 12 species, more or less woody vines climbing by tendrils. The leaves are simple or 3-foliolate, rarely digitately 5-foliolate. Inconspicuous flowers are produced in axillary cymes and followed by small, grape-like fruit. The following species is a useful house or conservatory plant, grown for its attractive, leathery foliage. Cultivate as for the more resilient *Cissus* species, with a minimum temperature of 7°C/45°F. *R.capensis* (syn. *Cissus capensis*, *Vitis capensis*; South Africa; tuberous-rooted evergreen vine; leaves simple, orbicular to reniform, broadly cordate at base, obtuse at apex, bluntly 5-angled, toothed, 8–20cm across, leathery, rusty-tomentose when young; fruit very dark red tinged black, glossy).

rhomboid of leaves and tepals, diamond-shaped, angularly oval, the base and apex forming acute angles, and both sides forming obtuse angles.

Rhombophyllum (from Greek *rhombos,* lozenge, *and phyllon,* leaf, referring to the leaf shape). Aizoaceae. South Africa (Cape Province). 3 species, compact succulent perennials, sometimes small shrubs. The leaves are more or less united at the base and semicylindric, keeled below, with the underside pulled forward and chin-like. Their margins are entire or with 1-2 short teeth. The surface is smooth, more or less glossy and deep green with pale translucent dots. Golden-yellow, daisy-like flowers are produced a few per stalk from summer to autumn. Cultivate as for *Pleiospilos*. *R.rhomboideum* (leaves 4–5 pairs resting on soil, 2.5–5 × 1–2cm, rhombic above, dark grey-green with white dots and pale margins).

rhubarb *(Rheum × hybridum)* a leafy perennial grown for its thick succulent petioles. It is probably a hybrid of *R.rhaponticum* and *R.palmatum* but its origins are unclear. The two likely parents and other *Rheum* species are common in China and elsewhere in the east; *R.officinale* and *R.palmatum* were described in a Chinese herbal of 2700 BC for their strong laxative properties. The culinary use of rhubarb in Europe dates from the 18th century.

The crop is more frequently cultivated in temperate regions, but cultivars have been selected for high elevations in tropical areas. With shortening day length in autumn, rhubarb becomes dormant and leaf production is only resumed following exposure to low temperature; the period of exposure required varies between cultivars. Those requiring the shortest cold treatment to break dormancy produce the earliest crops.

Harvesting takes place from the spring, but plants can be forced to produce an early crop of blanched stalks during late winter. On no account should leaves be eaten since they contain a poisonous glycoside. Blanched and natural stalks may be cooked by boiling or baked in pies and tarts, with sugar usually being necessary to reduce acidity.

Where grown on a well-prepared site, rhubarb plants may extend to 2m. Soil must be made moisture-retentive through the incoporation of rotted organic matter; good drainage is essential and plants do not thrive in shade.

Rhubarb plants should be re-propagated every five years or so by dividing crowns, with a spade, into wedge-shaped pieces, each bearing at least one strong bud; vigorous healthy mother plants of named cultivars should be chosen because long-standing stocks are prone to degeneration from virus infection. Plant during late autumn, 75–90cm apart, with the buds just showing at the soil surface. Propagation from seed is possible from spring-sowing but the resulting plants are likely to be variable. Plants should be kept well-watered and free from weeds, and provided with a thick layer of organic mulch each autumn or spring.

To enable crowns to build up reserves, harvesting should not be carried out during the first growing season. Always retain three or four leaves per plant to sustain the crown. Flowering stems should be removed since they weaken the plant. Early harvests of outdoor crops can be obtained by forcing plants, either by covering with a thick layer of straw during mid-winter or by covering individual crowns with a rhubarb pot or similar to provide protection and to exclude light, thereby blanching the developing leaves. Crowns can also be lifted during autumn and early winter and forced under protection in a greenhouse or shed; they must first be exposed to frost in order to break dormancy. The lifted crowns should be replanted in soil in large pots or boxes and brought under protection, where growth will resume at a temperature between 7–16°C/45–61°F. The containers should be positioned where light can be excluded, for example, under a greenhouse bench with an improvized curtain attached. Plants forced *in situ* will take time to recover and are best left uncropped the following season; those forced indoors are best discarded.

Recommended cultivars include: (early) 'Hawkes Champagne', 'Timperley Early'; (mid-season) 'Prince Albert', 'Canada Red'; (late) 'The Sutton', 'Victoria'. Notable in the US are 'Linneus' and 'Martha Washington'.

rhubarb pot a tall, upright earthenware pot with a removable lid, specially made for forcing rhubarb.

Rhus (from Greek name for this plant, *rhous*). Anacardiaceae. Temperate and subtropical N America, southern Africa, subtropical E Asia, NE Australia. SUMAC, SUMACH, TAAIBOS. About 200 species, trees and shrubs, deciduous or evergreen, with simple to pinnate or palmatifid leaves and very small flowers massed in panicles or spikes. The species listed here are deciduous with pinnately compound leaves. The autumn foliage colour is

Rhoicissus capensis

rhubarb

Diseases
crown rot
virus

Ribes speciosum

superb, and the ornamental flower and fruiting spikes persist until winter. Hardy to at least –25°C/–13°F. Plant on a well-drained, fertile soil in full sun. Propagate by removal of suckers from autumn onwards, also by 10cm cuttings of semi-ripe shoots with a heel, taken in late summer and inserted in a sandy medium in a propagating frame with bottom heat. Root cuttings will succeed potted vertically in a sandy medium in winter, under glass. Prune only to keep in bounds or to remove damaged or ungainly wood; the brittle wood of the taller shrubs and tree types is prone to storm damage.

R.aromatica (FRAGRANT SUMAC, LEMON SUMAC, POLECAT BUSH; SE Canada to S and E US; shrub to 1m, prostrate; leaflets 3, to 6.5cm, ovate, serrate; flower spikes to 2cm, yellow; 'Gro-low': very low and spreading, with golden yellow, fragrant flowers); *R.glabra* (SMOOTH SUMAC, SCARLET SUMAC, VINEGAR TREE; NE US to S Canada; deciduous shrub or tree, to 4m; leaflets 11–31, to 11cm, oblong to lanceolate, serrate, glabrous, dark green above, blue-green beneath, fiery red in autumn; flower spikes dense, to 25cm, green turning deep rusty red; 'Flavescens': leaves yellow in autumn; 'Laciniata': leaflets deeply and jaggedly cut); *R.typhina* (STAGHORN SUMAC, VELVET SUMAC, VIRGINIAN SUMAC; eastern N America; spreading deciduous tree or shrub to 10m; twigs bristly-velvety; leaflets 11–31, to 11cm, oblong to lanceolate, serrate, coarsely hairy, fiery red to orange and gold in autumn; flower spikes to 30cm, dense rusty-red; 'Dissecta': leaflets pinnately dissected; 'Laciniata': leaflets deeply and jaggedly cut).

Rhyncholaelia (Greek *rhynchos*, snout, and *Laelia*, the name of a closely allied genus). Orchidaceae. Mexico, C America. 2 species, robust epiphytes or lithophytes, formerly included in *Brassavola*, from which they differ in their habit, (closer to the squatter, tougher-leaved *Cattleya* species), and in their large flowers with very showy lips. Cultivate as for the larger *Cattleya* species but with maximum sunlight at all times.

R.digbyana (syn. *Brassavola digbyana*; Mexico, Belize; flowers to 17cm in diameter, strongly scented, long-lived, sepals and petals pale yellow-green, oblong to lanceolate, lip white to cream, tinged green, margin finely cut into many beard-like filaments; *R.glauca* (syn. *Brassavola glauca*; Mexico, Guatemala, Honduras; flowers to 12cm in diameter, fragrant, long-lived, sepals and petals white, olive-green or very pale lavender, oblong to lanceolate, lip white to yellow-cream, with a rose-pink or purple mark at base, margin slightly wavy).

rhytidome the periderm and tissues external to it; bark.

ribbed possessing one or more prominent veins or nerves.

Ribes (from the Arabic *ribas* 'sharp-tasting'). Grossulariaceae. Northern temperate regions. CURRANT, GOOSEBERRY. 150 species of small to medium-sized shrubs, mainly deciduous, often thorny. They bear small, 4–5-petalled flowers singly or in racemes in spring. These are followed by small, juicy berries, which, in some cases, are edible. The ornamental *Ribes* species are grown primarily for their flowers, although some also have attractive foliage as in the golden-leaved *R.sanguineum* 'Brocklebankii'. The low-growing evergreen *R.laurifolium* is particularly valuable in the winter garden, bearing dense panicles of green-yellow flowers in late winter/early spring. The most commonly grown species, the spice-scented *R.odoratum* and the rather feline *R.sanguineum,* are reliably hardy where temperatures fall to –25°C/-13°F. *R.speciosum,* with fuchsia-like flowers and fine glossy foliage, will tolerate temperatures to –15°C/5°F, and is suited to wall or pillar training, especially in zones at the limits of hardiness. *R.laurifolium* shows similar cold tolerance, given shelter from cold drying winds. *R.viburnifolium* does not tolerate temperatures much below –5°C/23°F.

Grow in any moderately fertile, moisture-retentive but well-drained soil in sun. *R.sanguineum* 'Brocklebankii' performs better in partial shade. *R.laurifolium* is also shade-tolerant. Trim over immediately after flowering to maintain a compact habit and remove a proportion (a quarter to one third) of flowered wood at the base to maintain a supply of younger wood: flowers are carried on the previous season's growth. Older wood in wall-trained specimens of *R.speciosum* should be removed in late summer and new basal growth tied in. Propagate deciduous species by nodal hardwood cuttings in late autumn/early winter. Increase *R.laurifolium* by semi-ripe cuttings in summer.

R.laurifolium (W China; evergreen shrub to 1.5m; leaves 5–10cm, oval or ovate, coarsely toothed, leathery, smooth; flowers green-yellow; fruit red to black, downy); *R.odoratum* (CLOVE CURRANT, BUFFALO CURRANT, MISSOURI CURRANT; C US; shrub to 2m; deciduous or semi-evergreen; leaves 3–8cm, oval to rounded, 3–5-lobed, toothed; flowers, sweetly scented, yellow; fruit globose, black); *R.sanguineum* (WINTER CURRANT; W US to California; to 4m, deciduous shrub, glandular, mustily aromatic; leaves 2.5–7cm, rounded, base cordate, 3–5-lobed, dark green, downy; flowers red or deep rosy pink; fruit glandular, blue-black with white bloom; 'Albescens': flowers white flushed pink, appearing earlier than other cultivars; 'Brocklebankii': leaves yellow; 'Plenum': double, red flowers; 'Pulborough Scarlet': bright mid-pink flowers; 'Pulborough Scarlet Variegated': leaves variegated yellow; 'Tydemans White': bright white flowers); *R.speciosum* (FUCHSIA-FLOWERED GOOSEBERRY; California; evergreen shrub to 4m, bristly–thorny; leaves 1–4cm, rounded, 3–5-lobed, bluntly toothed, smooth, glossy; flowers pendulous, red, glandular-bristly, with long, red, downward-pointing stamens; fruit glandular-bristly, red); *R.viburnifolium* (California; evergreen shrub to 1.5m, stems low often rooting; leaves 2–4cm, broadly ovate to oval, entire or slightly dentate, glandular, turpentine-scented; flowers pink; fruit red).

rice husks a by-product of the processing of rice, useful for ground mulching and as an ingredient of potting mixtures. They provide good aeration and drainage and are especially beneficial where incorporated into mixes for epiphytes.

Richea (for the French naturalist Colonel A.G. Riche, who died in 1791 during the Australian voyage of Admiral d'Entrecasteaux in search of La Perouse). Epacridaceae. Australia, Tasmania. 11 species, evergreen trees or shrubs. Crowded along the stems or tufted at their tips, the leaves are narrow and usually sword-shaped with sheathing bases. The flowers do not open fully, but are waxy and colourful and crowded in simple terminal clusters, spikes or panicles. Hardy to –15°C/5°F. Grow in moist, humus-rich, acid soils in partial or dappled shade, or with part-day sun where soils remain moist throughout the summer. Propagate from semi-ripe cuttings or by seed in autumn. *R.scoparia* (Tasmania; bushy, erect shrub to 1.5m; flowers white, pink or orange in dense cylindric racemes to 30cm long).

Ricinus (from Latin *ricinus*, a tick, which the seeds resemble). Euphorbiaceae. NE Africa to Middle East, naturalized throughout tropics. 1 species, *R. communis*, CASTOR OIL PLANT, a glabrous shrub to 12m, seldom exceeding 2m in cultivation. The leaves are up to 60cm across, rounded in outline and palmately lobed with toothed margins. In summer, red-green flowers are borne in terminal inflorescences to 15cm. These are followed by prickly capsules which split to reveal highly toxic marbled, bead-like seeds. Several cultivars are grown with leaves variously veined, flushed and marbled with yellow, lime green, bronze, purple-red, orange and fiery red. One of the finest is 'Impala' with metallic purple-bronze leaves veined with scarlet, and scarlet flowers. The castor oil plant is often treated as a half-hardy annual, grown each year from seed and planted out after the last frosts. It is valued for its bold foliage, and traditionally used as a dot plant in subtropical bedding. It is also suitable for the cool greenhouse or conservatory. Grow in sun with shelter from wind, in a fertile, moisture-retentive but well-drained soil enriched with well-rotted organic matter. Sow seed in spring in individual pots at 20°C/70°F.

riddle a wide-meshed sieve used for sifting soil or coarse materials.

ridger a hand tool shaped like a miniature plough. It is used to produce a furrow into which seeds can be sown.

ridging see *digging*.

rigid paving paving units bedded on mortar, and with the joints sealed with mortar.

ring barking see *bark-ringing*.

ring culture a method of cultivating tomatoes, also used satisfactorily for chrysanthemums and carnations, in which young plants are set into a growing medium contained in bottomless whalehide (q.v.) rings; these measure 22cm in depth and diameter and are stood on a 15cm deep layer of gravel, weathered clinker or similar aggregate. The aggregate is sealed at the base with a polythene sheet in order to prevent root penetration into disease- or pest-infested soil. All water is subsequently applied to the aggregate, and liquid feed to the medium. In this way, nutrition is concentrated within the ring and mainly water-extracting roots develop throughout the aggregate. The method is especially useful on unsuitable soils, and also on solid garden surfaces.

ringspot an area of visually normal leaf tissue surrounded by one or more rings, which may be darker green, yellow, chlorotic or necrotic. It is usually a virus-disease symptom.

ripeness, ripening used to describe (1) the stage at which fruits reach physiological maturity and are in a condition favourable for eating. In apple fruits, it coincides more or less with maximum respiration rate, at a phase known as the climacteric. Carbon dioxide, water and volatile products, notably ethylene, are evolved at this time; (2) the maturation of young woody shoots on trees and shrubs, or of bulbs, especially onions, which occurs during high summer temperature.

Robinia (for Jean Robin (d. 1629), herbalist to Henry IV of France). Leguminosae. US. Some 20 species, deciduous trees and shrubs with pinnate leaves and pea-like, fragrant flowers in pendulous racemes in late spring and early summer. The following will tolerate temperatures to –15°C/5°F and atmospheric pollution. Good drainage and shelter from strong winds are necessary. *Robinia* regenerates strongly after pruning, which should be done in late summer to reduce risk of bleeding. Most sucker freely and felled trees will produce coppices. Propagate by seed in spring, scarified or soaked in hot water for 12 hours before sowing, or by removal of rooted suckers. Cultivars can be side-grafted on to seedling rootstock of the species in winter, although suckering may be a problem. Take root cuttings of plants grown on their own roots. Locust borer and locust leaf miner are serious pests in North America.

R. × *ambigua* (*R.pseudoacacia* × *R.viscosa*; differs from *R.pseudoacacia*, in the somewhat glutinous branches with small thorns; flowers pale pink to lavender); *R.hispida* (ROSE ACACIA, MOSS LOCUST, BRISTLY LOCUST; SE US; shrub to 2.25m, bristly, suckering; flowers 3–5 in short raceme, pale purple or rose; includes 'Monument', to 4m, compact, conical, sparsely bristled, with racemes to 10cm and large, delicate lilac-pink flowers); *R.kelseyi* (ALLEGHANY MOSS; SE US; shrub to 3m, branches bristly at first; racemes to 6cm, crowded, glandular-hairy, flowers lilac to rose); *R.pseudoacacia* (BLACK LOCUST, YELLOW LOCUST; E and C US, widely naturalized in temperate regions; tree, to 27m, nearly glabrous; young branches thorny; racemes dense, nodding, to 20cm, flowers white to ivory, fragrant, blotched with yellow at base; 'Frisia': a fine small tree with bright golden new foliage, turning a cool lime by midsummer).

robin's pincushion see *bedeguar.*

rocambole (*Allium sativum* var. *ophioscorodon*) a perennial grown for its bulbs, which are used in flavouring in a similar way to garlic. It is propagated vegetatively from both offsets and bulblets produced at the top of its stem. These should be planted 5cm deep and about 20cm apart during early spring, and lifted during the summer months after the foliage dies back. They may be stored in a cool dry place in a similar way to onions.

rock phosphate an inorganic fertilizer made of finely ground phosphate-containing rock (25–40% phosphoric acid). Its usefulness varies according to the source of the rock and processing treatment, and it produces best results on acid soils.

rock plant a herbaceous, shrubby, or bulbous plant that is suitable for growing in a rock garden. See *alpine*

rock plant gardening the cultivation of alpine and other rock plants in a garden area specially constructed to simulate natural rock formation. The construction of rock gardens dates from 1772, when the director of what is now the Chelsea Physic Garden in London had placed there about forty tons of stone from the Tower of London, together with lava, brought from Iceland by Sir Joseph Banks, and a topping of flints and chalk. This was the precursor of many such collections of stones, known as 'stoneries' or 'lapideums', built largely for their own sake, although an attempt was sometimes made to grow plants on them.

One of the earliest natural-looking rock gardens in England was built in the early 1800s by William Beckford in a quarry at Fonthill, Wiltshire. In the UK and North America interest and achievements in rock gardening evolved throughout the 19th century. Today, rock gardeners in the UK, US, and Canada cultivate rock plants with great seriousness under the aegis of The Alpine Garden Society and the American Rock Garden Society. There are various ways in which rock plants can be grown in the garden, accomodating interest on a large or small scale.

Rock gardens This is the commonest form of rock plant growing. The cheapest and most natural-looking features are those made by exposing existing rock, but the incorporation of any existing mound, slope or change of level is an opportunity to be looked out for. Artificially created mounds in a flat scene look contrived, especially in a small garden, but this effect can be minimized by making gentle slopes or by creating a surrounding pavement of flat rocks. A south or west outlook is best, but a northwest-facing rock garden still provides good west-facing planting pockets; alternatively, a mound of smaller rocks can be gently stepped on both sides, offering a variety of growing conditions, perhaps with one short, sharp slope and a second long, gentle one. In all cases, the site must be open to good light.

Arrange the stones as they would be found in a natural outcrop, but with some positioned as if they have rolled to the bottom of the slope. Incorporate some flat stones for standing upon while tending plants, and, in the larger rock garden, plan discreet gravel or stone-based paths. Ensure good drainage, and at each level set the largest stones into the bank first, deeply enough to be stable to stand upon and tilted slightly backwards so that rain will run off the top surface to the plants' roots. Smaller stones are placed on either side, their tops more or less level, but always with the strata lying in straight lines. The second layer of rocks uses some of the first layer as a base, together with an excavated section of bank. As each stone is positioned, soil must be packed firmly at its back and in any vertical and horizontal crevices. Adjacent rocks should be butted up close but the distance between each cluster of rocks can be varied. It is easiest to plant alpines in the crevices as building progresses.

All true alpines need a well-drained growing medium, which can be packed into the rock crevices; a generally suitable mixture is three parts loam to two parts grit and one part coir or leafmould. Plants need very little feeding and some thrive in very dry and poor soil.

MATERIALS Wherever possible, local stone should be chosen since it will harmonize with the landscape, as well as being relatively cheap to transport. Where this is not possible, matching the colours of existing stone or brick used for house and garden may be a consideration. A mixture of sizes is best, with several bold blocks and some large, flat pieces included.

Limestone is an interesting rock, which varies in consistency and colour, producing flowing lines where it has been worn by running water; it is also quite easy to cut. A disadvantage of its weathering is that the surrounding soil becomes alkaline, thus precluding the cultivation of some desirable plants. In Britain, conservation measures justifiably restrict the supply of the choicest rock, which was formerly obtained at the cost of severe damage to surface limestone pavements, but quarried limestone is available. Tufa is a water-retentive form of limestone, deposited by dripping water or mineral springs; it has no strata, is soft to cut and although it hardens on exposure to the air can be easily penetrated by the roots of plants. It is generally used for special situations, including sinks or troughs, where holes 2.5–5cm wide and 5cm deep are drilled at a 45° angle and filled with equal parts tufa debris, soil and leafmould.

Sandstone is a more angular rock, available in a wide range of colours, including soft pinks and yellows; it has obvious strata and can be split easily. Its main advantage is that it absorbs moisture, which it will release into surrounding soil in dry spells and which makes it relatively cool. As sandstone weathers, it produces acidic residues, and mosses and lichens will establish on its surface.

Slate can be used to dramatic effect, echoing natural outcrops, where strata point almost vertically upwards; it consists of relatively thin layers of easily split, grey rock.

Granite is least satisfactory because it is heavy, hard to cut or split, and does not weather. It has few strata lines, and if crevices develop, they tend to be vertical and unsuitable for planting.

Scree and moraine beds These are made to simulate particular natural rocky habitats. Scree is a sloping area of stone fragments at the base of a rock face, formed by deposits of weathered rock and notable for its perfect drainage and lack of moisture. Moraine is an accumulation of rock fragments, with sand or silt particles, which arises from the grinding action of a glacier on bedrock; it is well-drained but retains moisture, becoming a damp habitat.

Scree beds may be sited either running between large rocks or as a separate feature. They can be made with a deep layer of stone fragments and chippings laid over 20cm of a very gritty soil, which in turn tops a well-drained foundation. Moraine can be constructed by digging out a site to about 60cm deep and placing coarse drainage material to a depth of 30cm. Inverted turves are placed on top of this layer as a moisture reservoir, and a very gritty soil mix is added to a depth of 20cm. The site is finished with 5cm of coarse gravel into which is laid a 3cm-deep subsurface watering system comprising plastic pipes with small widely spaced perforations.

Scree depends on rainfall or overhead watering; moraine relies mainly on its subterranean watering system which should be used in hot dry weather. Both scree and moraine benefit from occasional overhead fine spray in very hot weather.

Raised beds These provide a more formal way of growing scree plants, allowing a large number to be grown in a small space. They can be any size or shape, built against a wall or in the form of an island bed, and may be partitioned to provide both acid and alkaline growing conditions. Dry-stone retaining walls, built to a height up to 1m, are best since plants can be grown in narrow cracks between the stones. A gritty soil mix is used to fill the bed, topped with a 5cm layer of coarse gravel. A few pieces of rock protruding from the finished level provide an authentic-looking miniature landscape.

Walls A variety of trailing rock plants can be grown in the gaps between the stones of a dry wall. The top of a brick wall can be used if it has a coping at least 30cm wide and a narrow trough specially built on top of it. The trough is filled with a mixture of two parts fibrous loam to one part leafmould and one part sharp sand; to this is added slow-release fertilizer.

Paving Alpine plants can be grown where soil rather than cement is used between paving slabs in courtyards, terraces, patios, and in paths laid on sand. Planting should be done during construction, or if the paving is already in place, holes can be made in the cement and gritty soil trickled into the opening.

Sink or trough gardens Old stone sinks and animal drinking troughs provide excellent foundations for growing alpine plants, but original ones are scarce and new ones very expensive. Artificial troughs can be made using redundant glazed white sinks, simulating natural stone with a coating of hypertufa; the mix is plastered over a layer of bonding agent, such as tile cement. The same material can be used to construct a trough using cardboard boxes as moulds, a smaller sized box placed inside another. See *hypertufa*

Alpine houses Alpine houses are a means of growing mountain plants with protection from winter rain, severe frost and icy winds. They permit the cultivation of slightly tender species and the propagation, growing on and establishment of young plants. They can also be used for the display of spring-flowering alpines and dwarf bulbs, which often flower two or three weeks earlier in containers than in the garden outside. Another use in summer is for drying and ripening many dwarf bulbs, which need summer heat when dormant to ensure regular flowering.

The alpine house should be situated in an open site and, where possible, orientated north/south. Typically, it is unheated, very well-ventilated but draught-free; wooden structures are preferable to aluminium, which gives rise to condensation and drips. Thermostatically controlled heating will provide plants with the protection against very low temperatures afforded by snow cover in their natural habitat, and prevent freezing of pots. Protection in summer with external blinds or a layer of green plastic netting is desirable, any covering being easily removable on dull days. Plants are best kept on sturdy benches covered with shingle.

Frames Frames allow perfect water control and protection from winter damage for plants starting growth in autumn, and can be used for the propagation, growing on and storage of all types of rock plants. They can also be used to hold plants not currently on display; these are trimmed over after flowering, repotted if necessary and stood out until flower buds reappear.

Containers in frames are best plunged in a well-drained bed of ashes, sand or bark fibre, to reduce water loss and protect roots and pots from frost. By watering the bed in which pots are planted, rather than the pots themselves, even plants drying off over the summer can be given the minute amount of moisture they require for optimum growth in the next season. Plants within frames need plenty of air movement, so lights should only be closed in periods of frost or fog; they will also need shading in high summer.

rockwool a lightweight material manufactured from spun mineral fibres, with high air-filled porosity and low nutrient content or nutrient-holding capacity. There are water-absorbent and water-repellent forms, both available in granules and suitable for use as growing-media additives; the water-absorbent form is widely used for soilless cultivation of greenhouse crops and for propagation. Rockwool is also available as slabs or blocks, and as small modules for rooting cuttings.

rod an outmoded linear measure of land, referred to in old gardening literature, and equivalent to about 5m. It is often loosely used to mean an area about 5m^2. It is also known as a pole or perch.

Rodgersia (for US admiral John Rodgers (1812–1882), commander of the expedition on which *R. podophylla* was discovered). Saxifragaceae. Nepal, China, Japan. 6 species, rhizomatous perennial herbs. The leaves are large and basal, with stout stalks, palmate or pinnate blades and leathery, toothed leaflets. Numerous, small, white, red or purple flowers are carried in dense panicles in summer. Hardy foliage plants for damp borders, bog gardens and pond margins, their leaves often emerging bronze or becoming so later in the season. The plume-like flowers are fine and the seedheads become red-tinted in autumn. Plant in moist humus-rich soils, in sun or part shade, with shelter from strong winds. Mulch annually to conserve moisture and give a balanced slow-release fertilizer. Water plentifully in drought. Propagate by division in early spring.

R.aesculifolia (China; to 2m; leaves palmate, leaflets acute and coarsely toothed at apex, veins rusty-tomentose; flowers white); *R.pinnata* (SW China; to 120cm; leaves pinnate, leaflets obovate-lanceolate, flowers red; 'Superba': tall, with leaves tinted bronze to purple and rose flowers); *R.podophylla* (Japan, Korea; to 120cm; leaves palmate, turning bronze in autumn, leaflets cleft and toothed at apex; flowers cream; 'Rotlaub': leaves tinted red); *R.sambucifolia* (China; to 90cm; leaves pinnate, leaflets acute; flowers white or pink).

rogue any plant with characteristics which deviate from those desired or expected. Rogueing is the culling of such plants, used especially for the removal of atypical or diseased bulbs.

Rohdea (for Michael Rohde (1782–1812), physician and botanist of Bremen). Liliaceae. SW China, Japan. 1 species, *R.japonica*, a rhizomatous perennial herb. The leaves are 30–40cm long, dark green, thick and strap-shaped and loosely 2-ranked in a basal rosette. Fleshy, white or yellow-green flowers emerge in a short dense spike in early spring. They are followed by red or yellow berries. Numerous variegated cultivars of this plant are available with leaves variously striped or edged gold, cream and white. They are highly prized in Japan. Grow in cool but frost-free, airy conditions in bright, indirect light. Plant in a loamy mix rich in leafmould and garden compost. Keep evenly moist. Propagate by seed or division.

roller a large garden hand-tool, comprising a wide, smooth-surfaced wheel mounted on an axle and fixed to a handle; used as an aid in the preparation of level sites, especially prior to laying paving. Modern garden rollers come in various weights, are made of metal or of plastic (when they may be filled with water) and are usually a little longer in width than diameter. Rollers are constructed so as to be pushed or pulled.

Rolling lawns, traditionally done to disperse worm casts and level minor surface undulations, is not recommended because of the deleterious effects of consolidation, particularly on heavy soil; the small roller built into a cylinder mower is adequate for levelling, and certain types of roller fitted with spikes are effective for aerating the surface layers. Light rollers have a place in aiding seedbed preparation, after initial cultivation, but treading is invariably a more effective technique; such rollers may prove more effective for lawn seedbed preparation.

Romneya (for Rev. T. Romney Robinson (1792–1882), astronomer in Armagh, Ireland). Papaveraceae. California, Mexico. MATILIJA POPPY; CALIFORNIAN TREE POPPY. 1 species, *R.coulteri*, a woody-based, suckering perennial herb 20–200cm tall. The leaves are to 15cm long, pinnately lobed and toothed, grey-blue and glaucous. Produced in summer, the flowers are to 15cm in diameter, white and fragrant, with six obovate, silky petals and many golden stamens. The flowers unfurl from tough, pod-like calyces. In plants sometimes named var. *trichocalyx*, these are covered in sharp bristles. Such plants also tend to have more finely divided, steely blue foliage. Hardy to −10°C/14°F if given sharp drainage and a warm, sunny, sheltered site. Cut back in spring before the appearance of the new growth. This plant may become invasive. Propagate by root cuttings in late winter or by rooted suckers in spring.

Romulea (for Romulus, founder and first King of Rome). Iridaceae. Mediterranean north to SW England; South Africa, and at high altitudes in C and E Africa north to the Mediterranean. Some 80 species, spring-flowering, crocus-like, perennial cormous herbs. Where they can be kept dry in summer (when dormant), the European and Asiatic species are hardy to −10°C/14°F; they are best planted at the base of a south-facing wall or niche in the rock garden. *R.bulbocodium* is slightly more cold-tolerant. The South African species, with the possible exception of the alpine *R.macowanii*, are probably more safely grown under glass or in the bulb frame in cool temperate zones. Grow in full sun, in a sandy soil. Under glass, use a loam-based mix with additional sharp sand and leafmould; grow in direct sunlight, with good ventilation, and water moderately when in growth. Dry off gradually as the foliage dies down. Propagate by seed in autumn, or in spring for *R.macowanii*. Increase also by offsets.

R.bulbocodium (syn. *R.grandiflora*, *R.clusiana*; Mediterranean, Portugal, NW Spain, Bulgaria; flowers 1–6 per scape; perianth 2.25–5cm, bright violet, rarely white, tube 0.3–0.8cm, yellow, orange or white within or at throat; var. *crocea*: flowers yellow; var. *leichtliniana*: flowers white with a yellow throat; var. *subpalustris*: flowers tinted purple-blue, with a white throat); *R.flava* (syn. *R.bulbocodioides*; South Africa; flowers 1–4 per scape ; perianth 2.25–4cm, yellow-green, occasionally white or tinted blue or pink, tube yellow within and at throat); *R.macowanii* (syn. *R.longituba*; South Africa; flowers 1–3 per scape; perianth 3–10cm, golden yellow, tube deep orange-yellow within and at throat); *R.nivalis* (Lebanon; flowers 1–3 per scape; perianth to 2.5cm, lilac to mauve, tube to 0.3cm, yellow within and at throat); *R.sabulosa* (South Africa; flowers 1–4 per scape; perianth to 3–5.5cm, bright scarlet, tube with a dark white- or pink-edge blotch within and at throat).

Rondeletia (for W.G. Rondelet (1507–1566), French scholar, physician, and specialist in fish and algae). Rubiaceae. C to S America, Polynesia. Some 150 species, shrubs or trees with axillary or terminal panicles, cymes or corymbs of funnel- to salverform, 4–6-lobed flowers in summer. Grow in the intermediate greenhouse (minimum temperature 13°C/55°F), in a loam-based medium, in bright filtered light. Water plentifully when in growth, sparingly in winter. Propagate by semi-ripe cuttings in a closed case with bottom heat in autumn. *R.amoena* (Mexico, Panama, Guatemala; shrub to 1.5m, or tree to 14m; flowers white to pink, to 1.5cm in many-flowered cymes or panicles).

roof gardens the practice of constructing gardens on the rooftops of buildings dates back to ancient times, the Hanging Gardens of Babylon, built about 600 BC, being terraced gardens arranged in a stepped pyramid over elaborate state rooms. There were rooftop gardens in Roman cities such as Pompeii, balcony gardens in Constantinople, and roof gardens were made by the Aztecs in Mexico. On a smaller scale, roof terraces with potted plants have probably always been a common feature of private houses in the Mediterranean and other dry mild regions. In the 20th century, Le Corbusier and Frank Lloyd Wright are among architects to have promoted rooftop gardens.

Ecologically, roof gardens contribute to an increase in air humidity, and help to purify the air and filter dust in towns. However, proper consideration must be given during the planning stage to factors such as access, weight of soil and containers, roof waterproofing and durability, drainage and also local by-laws.

Plants can be grown on roofs in large tubs, boxes and other containers, or, in some instances, in beds. Growing media should be fertile, soil-based and with good moisture-retention properties since rapid drying out is inevitable; an adequate water supply is necessary. Screening from wind is essential for the protection of plants and users – panels of fine wire mesh or similar material will reduce wind force and sheets of strengthened glass are often used.

A very wide range of shrubs and small trees can be grown, provided large containers are available. Species with strong branch structure and natural adaptation to hot dry conditions are a good choice; rock-garden plants, heathers and low-growing shrubs can be effective. Even vegetables and soft fruit can be grown with adequate protection and watering facilities, and growing bags are suitable in some instances. Maximum use should be made of climbing plants against walls, chimney stacks, and, if possible, on pergola-like structures.

Similar considerations apply to gardening on balconies, where long narrow window boxes, growing bags and smaller containers are a suitable choice.

rond-point a circular open space in a large formal garden, from which alleys or avenues radiate and several vistas are presented.

root the axis of a plant that serves to anchor it and absorb nutrients from the soil; usually geotropic, subterranean and derived from the radicle. Roots may also be adventitious or aerial.

root aphids the roots of various plants are subject to attack by aphids, invariably attended by ants. In severe infestations, plant growth is checked and wilting occurs. Most root aphids are light-coloured and covered with white powdery wax. Some, such as the ARTICHOKE TUBER APHID *(Trama troglodytes)* live on roots continuously, but for most species root-feeding represents one phase in their annual migratory life cycle; for example, the LETTUCE ROOT APHID *(Pemphigus bursarius)* migrates from leaf petiole galls on poplars to the roots of lettuce. Other species injurious to garden plants include the AURICULA ROOT APHID *(P.auriculae)*, which infests the roots of auriculas and primulas, especially in pots; the ELDER APHID *(Aphis sambuci)*, which migrates from elder to *Dianthus* species, *Lychnis* species, London pride *(Saxifraga umbrosa)* and many weeds; and the HAWTHORN-CARROT APHID *(Dysaphis crataegi)*, which attacks the roots of carrots and parsnips.

Romulea bulbocodium

In North America, the WOOLLY APHID (*Eriosoma lanigerum*) has a bark-infesting form as well as one that attacks the roots of hosts, such as apple, *Cotoneaster* and *Pyracantha*; in Europe only the bark feeding form occurs. Another injurious pest in North America is the STRAWBERRY ROOT APHID *(Aphis forbesi)*; several other species of root aphids attack ornamentals, such as asters, calendulas, cosmos, dahlias, primulas and sweeet peas.

Root aphids may be controlled by applying soil drenches of a recommended aphicide. Pot plants may be treated by washing the roots and dipping them in similar insecticides. Some lettuce cultivars are resistant to the lettuce root aphid, for example the butterheads 'Avondefiance' and 'Sabine'. See *woolly aphid*.

rootball the mass of roots and growing medium formed by a plant, especially a container-grown one, but also applicable to plants lifted from the open ground.

root-bound see *pot-bound*.

root crop a vegetable crop grown for its edible roots or tubers, such as carrot, parsnip and potato.

root cuttings see *cuttings*.

root dipping the practice of immersing the roots of transplants into an approved pesticide solution. It is also applied to the dipping of bare-root trees or shrubs into a water-retentive polymer or alginate in order to prevent their desiccation. See *puddling*.

root flies, maggots Common to Europe and North America, root flies are some of the most serious vegetable pests and include CABBAGE ROOT FLY (*Delia radicum*), ONION FLY (*D. antiqua*), BEAN SEED FLY (*D. platura*) and CARROT FLY (*Psila rosae*). All the *Delia* species have adult forms about 6mm long, which are of similar appearance to the HOUSEFLY (*Musca domestica*).

Root fly attacks are best prevented by treating the soil around the plants with granules of recommended insecticides; correct timing is important and varies with different species. On a small scale, physical barriers are effective against cabbage root fly and carrot fly, and peas and beans may be protected from bean seed fly by seed dressing. See *bean seed fly, cabbage root fly, carrot fly, chrysanthemum stool miner, onion fly*.

root grafting see *grafting*.

root hairs specialized epidermal cells situated a short distance behind the root tip, which increase the root's capacity to absorb water and mineral salts. Root hairs are short-lived and continually being produced as root growth proceeds.

root house an ornamental garden feature of Victorian gardens, formed from the bases of tree trunks, and often thatched.

rooting hormone see *cuttings* (synthetic hormones).

rootlet a small or secondary root.

root nodules small galls of different shapes that occur on the roots of certain plants, in particular members of the Leguminosae, and are produced by the reaction of the host to infection by various nitrogen-fixing microorganisms; they are beneficial. See *nitrogen*.

root pruning see *pruning*.

root restriction reducing the root run of a plant by planting it in a confined space in order to contain vegetative growth and induce flowering. Root restriction is practised, for example, in fig cultivation.

root rot decay of plant roots encouraged primarily by disease organisms, often in unfavourable soil conditions such as waterlogging. Fleshy roots, such as beet, carrot and turnip, may develop soft rots, caused by the bacterium *Erwinia carotovora*, or hard dry rots, such as those caused by the fungi *Leptosphaeria maculans* on swede and *Alternaria radicina* on carrot (BLACK ROT). Many different fungi, including *Colletotrichum, Fusarium, Phoma, Phytophthora, Pythium, Rhizoctonia* and *Verticillium,* cause root rots of herbaceous plants and some are also responsible for damping-off in seedlings. Root rots caused by *Phytophthora* species commonly affect fruit trees and ornamental trees and shrubs. HONEY FUNGUS (*Armillaria mellea*), WHITE ROOT ROT FUNGUS (*Rosellinia necatrix*) and VIOLET ROOT ROT (*Helicobasidium brebissonii*) are other root rots with a wide host range. See *armillaria root rot, black rot, violet root rot*.

root run the area of soil explored by a plant's roots.

root spiralization the root configuration developed by plants left too long in a pot or container. See *pot-bound*.

rootstock (1) a woody or fleshy plant upon which is grafted a shoot or bud from a chosen specimen. The rootstock thereby provides the root system of the new plant. Besides facilitating propagation, rootstocks may influence growth and other characteristics in the new individual, such as ultimate size and early bearing in fruit trees and resistance to soilborne pest infestations in tomatoes. Widely used examples of clonal rootstocks are the EMLA selections for apple, Quince for pear and Myrobolan for plum. Many ornamental plants are grafted on to seedlings of the same or closely related species, see *EMLA*; (2) botanically, a rhizome, as in some ferns; (3) loose descriptive term for the crown and root system of a compact plant.

root vegetable see *root crop*.

roridulous covered with small, translucent prominences, giving the appearance of dewdrops.

Rosa (the Classical Latin name for the rose). Rosaceae. ROSE. N Hemisphere. Some 150 species, mostly deciduous shrubs, erect, arching or scrambling, with prickly and/or bristly stems. The leaves are pinnate, trifoliate or simple. Solitary or corymbose, the flowers consist of 4 or 5 (-7), usually obovate and spreading petals (8–14 in semi-double, and 15–30+ in double flowers), and numerous stamens. In some species and many cultivars, flowering is remontant (occurs at least twice per season). Known as hips or heps, the fruits are usually waxy, flesh, brightly coloured and crowned with remnants of the sepals.

The following are fully hardy. Plant in sun or light shade on a well-drained, fertile soil. Avoid sites where other roses have been grown and have deteriorated. Feed generously in late winter and mulch with garden compost. Continue to feed at regular intervals between mid-spring and late summer. Remove spent flower-heads unless the hips are decorative. Prune in late winter or early spring. Pruning requirements vary, but can be broadly described as follows – remove any dead, damaged, weak or spent growth for all species and hybrids. Old Garden and Groundcover Roses usually need only a light trim. Modern Bush and Miniature Roses can be divested of as much as two thirds of the previous season's growth. Modern Shrub and Climbing Roses, ramblers and species usually need only a light prune. Increase by budding (bud grafting) in summer or by hardwood cuttings in autumn. Affected by powdery mildew, black spot and rose rust; also canker and crown gall, rose anthracnose and die-back. Pests include aphids, Japanses beetles and chafers, rose curculio, fuller beetles, froghoppers and leafhoppers, leaf miners, leaf-rolling sawflies, rose scale, rose slugworms and shoot or pith borers.

Roses are divided into three principal groups, the first includes the species described in detail in the table below; the other two groups comprise hybrids and cultivars so numerous that only brief generalisations can be made here. Each group is further divided.

SPECIES ROSES

Species – all rose species and their varieties and culti-vars, including shrub roses and climbers

Species Hybrids – hybrids between species and clearly showing their influence. Flowers usually single, borne in one phase. Hips often handsome.

OLD GARDEN ROSES

Alba – large shrubs. Foliage grey-green. Flowers semi-double or double, 5–7 per cluster in midsummer. Hardy. Good for border and specimen planting.

Bourbon – shrubs, loosely branched and open. Remontant – flowers usually double, produced in threes in summer and autumn. Can be trained to climb: good for borders and for growing on fences, walls and stakes and pillars.

China – sparsely branched shrubs with pointed, glossy leaflets. Remontant – flowers single or double, soli-tary or in clusters of 2–13 in summer and autumn. Good for sheltered borders and walls.

Damask – lax, open shrubs. Flowers very fragrant, semi-double to double, solitary or in 5–7 in loose clus-ters in summer. Good for borders.

Gallica – dense, freely branching shrubs. Leaves dull green. Flowers single to double, richly coloured, usually in threes, in summer. Good for borders and hedging.

Hybrid Perpetual – vigorous, freely branching shrubs. Leaves usually olive green. Remontant – flowers double, solitary or in threes in summer and autumn. Good for beds and borders.

Moss – loosely branched shrubs with moss-like, fuzzy growth on stems and calyces. Flowers double in summer.

Noisette – climbing shrubs with smooth stems and glossy leaves. Remontant – flowers usually double, somewhat spicily scented, to 9 per cluster in summer and autumn. Good for sheltered, warm walls.

Portland – erect shrubs. Remontant – flowers semi-double to double, solitary or in threes in summer and autumn. Good for borders.

Provence (Centifolia) – loose, thorny shrubs. Leaves dark green. Flowers scented, usually double, solitary or in threes in summer. Good for borders.

Sempervirens – semi-evergreen climbers. Leaves glossy, bright pale to mid-green. Flowers semi-double to double, many, in late summer. Good for pergolas and fences, also for naturalizing.

Tea – shrubs and climbers. Leaves glossy, pale green. Remontant – flowers semi-double to double, pointed, spicily scented, slender-stemmed, solitary or in threes in summer and autumn. Good for sheltered borders.

MODERN ROSES

Shrub – varied category of bush roses, mostly remon-tant and attaining 1–2m in height. Flowers single or double, solitary or in sprays, in summer and autumn. Good for beds and specimen planting.

Large-flowered Bush (*Hybrid Tea*) – vigorous shrubs. Remontant – flowers usually double, large, and pointed, solitary or in threes, in summer and autumn. Good for borders and cutting.

Cluster-flowered Bush (*Floribunda*) – vigorous, free-flowering shrub. Remontant – flowers single or double, 3–25 per spray in summer and autumn. Good for borders and hedges.

Dwarf Cluster-flowered Bush (*Patio*) – small, compact shrubs (to 60 × 60cm). Remontant – flowers single to double, 3–11 per spray in summer and autumn. Good for beds, hedging and containers.

Miniature Bush – dwarf bush (to 45 × 40cm). Leaves very small. Remontant – flowers small, single to double, 3–11 per spray. Good for containers and small gardens.

Polyantha – compact, resilient shrubs. Remontant – flowers small, single or double, 7–15 per spray in summer and autumn. Good for borders.

Groundcover – stems long, trailing or spreading, more or less prostrate. Often remontant – flowers single to double, in clusters of 3–11 in summer and autumn. Good for low to medium groundcover, threading through other low shrubs and for beds, banks and walls.

Climbing – vigorous climbers with rather rigid stems. Sometimes remontant – flowers single to double, solitary or clustered, from late spring to autumn. Good for walls, fences and pergolas.

Rambler – climbers with robust, lax stems. Flowers single to double, in clusters of 3–21 in summer. Good for walls, pergolas and trees. (*see table overleaf*)

Rosa 'Canary Bird'

Rosa glauca

Rosa moyesii

Rosa 'Verschuren'

Rosa sericea subsp. omeiensis f. pteracantha

ROSA Name	Distribution	Habit	Leaves	Flowers	Fruit
R. × alba (*R.gallica × R.arvensis* or *R.corymbifera* or *R.canina × R.damascena*) WHITE ROSE, WHITE ROSE OF YORK	garden origin	stout, arching, 1.8–2.5m, prickles scattered, hooked, often mixed with bristles	leaflets 5(-7), dull green, ovate to round, shortly acuminate to obtuse, glabrous above, downy beneath, toothed	1–3 per cluster, semi-double or double, 6–8cm across, fragrant; sepals with lobes and leafy tips, glandular-bristly, reflexed; petals white to pale pink	2–2.5cm, more or less spherical, red

Comments: 'Incarnata': MAIDEN'S BLUSH; stems with few prickles but densely bristly below inflorescence; flowers double, pale pink. 'Maxima': GREAT DOUBLE WHITE ROSE, JACOBITE ROSE, CHESHIRE ROSE; flowers double, pink fading to creamy white. 'Semiplena': flowers semi-double, white.

R.banksiae BANKSIAN ROSE, LADY BANKS'S ROSE	W and C China	strong, climbing to 12m, more or less unarmed	leaflets 2–6.5cm, 3–7, oblong-lanceolate to elliptic-ovate, acute or obtuse, glossy above, sometimes downy beneath, margin wavy, toothed	many per umbel-like cyme, single or double, fragrant, 2.5–3cm across; sepals entire, reflexed; petals creamy white to lemon or pale primrose yellow	0.7cm, spherical, dull red.

Comments: var. *banksiae*: flowers double, white, violet-scented. var. *normalis*: flowers single, white, very fragrant, and stem usually prickly. 'Lutea': stem generally unarmed; leaflets usually 5; flowers double, yellow, slightly fragrant. 'Lutescens': flowers single, yellow, highly scented.

R.californica	W US to Mexico (Baja California)	erect, 1.5–3m tall, red-brown, prickles stout, broad-based, recurved in pairs, with bristles on young shoots.	leaflets 1–3.5cm, 5–7, ovate to broadly elliptic, obtuse, hairless or downy above, downy and often glandular beneath, toothed	in clusters, single, fragrant, 3.7–4cm across; sepals entire, hairy on the back, erect; petals deep pink to bright crimson	to 1.5cm, spherical with a neck, smooth

Comments: 'Plena': flowers semi-double. The rose usually cultivated under this name is *R.nutkana* (NOOTKA ROSE), native to W N America, with single, bright red to purple-pink or white flowers to 6cm across.

R. × centifolia PROVENCE ROSE, CABBAGE ROSE, HOLLAND ROSE	garden origin	to 2m, prickles many small, almost straight and larger, hooked	leaflets 5–7, dull green, broadly ovate to almost round, glabrous above, downy beneath, glandular-toothed	1 to few, double, globose, very fragrant, 6–8cm across; sepals lobed, glandular; spreading; petals usually pink, more rarely white or dark red	ellipsoidal or spherical, red

Comments: A complex hybrid involving *R.gallica*, *R.moschata*, *R.canina* and *R.damascena*. 'Bullata': leaflets crinkled, tinged brown above when young; flowers pink. 'Cristata': sepals divided, providing each pink flower with a green fringe. 'Muscosa': COMMON MOSS ROSE; calyx and flower-stalks bearing a mass of many-branched, green, scented glands forming the so-called 'moss'. 'Parvifolia': BURGUNDY ROSE; to 1m; leaflets small, dark green; flowers fragrant, deep pink suffused with purple and with a paler centre – possibly not related to *R. × centifolia*.

R.chinensis CHINA ROSE, BENGAL ROSE	garden origin	dwarf to more or less climbing to 6m, prickles absent or scattered, hooked, flattened	leaflets 2.5–6cm, 3–5, lanceolate to broadly ovate, acuminate, glossy and hairless above, hairless beneath except for downy midrib, margins with simple teeth	solitary or in clusters, single or semi-double, 5cm across, often fragrant; sepals entire or lobed, smooth or glandular, reflexed; petals pale pink to scarlet or crimson	1.5–2cm, ovoid to pear-shaped, green-brown to scarlet

Comments: Repeat-flowering, contributing this feature to European roses when it was introduced around 1800. 'Minima': FAIRY ROSE, PYGMY ROSE; 20–50cm; flowers *circa* 3cm across, rose red, single or double, petals pointed. 'Mutabilis': 1–1.7m; leaflets tinged purple or coppery when young; flowers single, fragrant, 4.5–6cm across, petals yellow with an orange back, turning coppery salmon-pink and eventually deep pink. 'Pallida': to 1m or more; flowers in clusters, semi-double, fragrant, blush-pink. 'Semperflorens': 1–1.5m; flowers semi-double, deep pink or crimson-scarlet, delicately scented. 'Viridiflora': GREEN ROSE; petals green streaked with purple-red, stamens and pistils mutated to leafy, narrow, toothed segments. var. *spontanea*: climber or bush usually 1–2.5m tall; flowers single, pink, turning red, 5–6cm across; fruit orange.

R. × damascena (*R.gallica × R.moschata*) summer damask rose	Asia Minor	to 2.2m, prickles dense, stout, curved, bristles stiff	leaflets 5(-7), grey-green, ovate to elliptic, acute to obtuse, downy beneath, toothed	in clusters of up to 12, semi-double, fragrant; sepals lobed, with slender tips, glandular and hairy, reflexed; petals pink	2.5cm, turbinate, red, bristly

Comments: var. *semperflorens*: AUTUMN DAMASK ROSE, FOUR SEASONS ROSE, QUATRE SAISONS ROSE, MONTHLY ROSE; flowers produced in autumn. 'Versicolor': YORK AND LANCASTER ROSE; flowers loosely double, petals deep pink or very pale pink or partly deep pink and partly pale pink. 'Trigintipetala': KAZANLIK ROSE; flowers semi-double, *circa* 8cm across, red. 'Portlandica': PORTLAND ROSE; flowers semi-double, bright red, faintly scented, produced from midsummer to autumn.

R. 'Dupontii' DUPONT ROSE; SNOW-BUSH ROSE	garden origin				

Comments: Differs from *R.moschata* in its smaller size (2–3m tall), leaflets with compound teeth, flowers single, 6–7.5cm across, creamy pink with sepals glandular beneath

R.ecae	NE Afghanistan, NW Pakistan and adjacent Russia, N China	much-branched, erect, suckering, to 1.5m, prickles dense, straight, flattened, red-tinged	aromatic, leaflets 0.4–0.8cm, 5–9, elliptic to obovate or more or less orbicular, obtuse, glandular beneath, teeth often glandular	solitary, single, 2–3m across; sepals entire, hairless, spreading or deflexed; petals deep yellow	0.5–1cm, spherical, shiny red-brown, smooth and hairless

Comments: 'Golden Chersonese': *R.ecae* × *R.* 'Canary Bird'; to 2m, with the aromatic leaves and deep yellow flowers of *R.ecae*. 'Helen Knight': *R.ecae* × *R.pimpinellifolia* 'Grandiflora'; stems red-brown to 3m with support; flowers large, yellow.

ROSA

Name	Distribution	Habit	Leaves	Flowers	Fruit
R.filipes	W China	arching and climbing to 9m, purple when young, prickles few, small, hooked	coppery when young; leaflets 3.5–8cm, 5–7, narrowly ovate to narrowly elliptic, acuminate, hairless or downy on veins beneath, toothed	to 100 or more in large clusters, single, fragrant, 2–2.5cm across; sepals lobed, glandular and slightly downy or hairless, reflexed; petals creamy or white	0.8–1.5cm, spherical to ellipsoid, orange to crimson-scarlet

Comments: 'Kiftsgate': fast-growing climber; very free-flowering, inflorescence to 45cm across, flowers white to cream.

Name	Distribution	Habit	Leaves	Flowers	Fruit
R.foetida AUSTRIAN BRIAR, AUSTRIAN YELLOW ROSE	SW and WC Asia	erect or arching, 1–3m, then grey-brown, prickles few, straight or curved, mixed with bristles	leaflets 1.5–4cm, 5–9, elliptic to obovate, obtuse to subacute, bright green and subglabrous above, dull green, downy and somewhat glandular beneath, teeth few, compound, glandular	solitary or 2–4, single, 5–7.5cm across, malodorous; sepals entire or lobed, apex expanded, hairless or glandular-bristly, erect; petals deep yellow	to 1cm, spherical, red, smooth or bristly

Comments: 'Bicolor': AUSTRIAN COPPER BRIAR; petals coppery red inside and yellow-buff on the back; branches occasionally revert to the yellow-flowered form. 'Persiana': PERSIAN DOUBLE YELLOW; flowers very double, yellow, freely produced, smaller than those of *R.foetida*

Name	Distribution	Habit	Leaves	Flowers	Fruit
R.gallica FRENCH ROSE, RED ROSE	S and C Europe to Caucasus	erect, shrubby, suckering, 50–200cm, green to dull red, prickles slender, curved to hooked, bristles glandular	leaflets 3–7, leathery, broadly elliptic to almost orbicular, dark green above, hairy and glandular beneath, teeth compound, usually glandular	solitary, or 2–4, single or semi-double, fragrant, 4–8cm across; sepals lobed, glandular, reflexed; petals rose-pink or crimson	to 1cm, spherical to ellipsoid, brick red, glandular-bristly

Comments: 'Officinalis': syn. *R.officinalis*; APOTHECARIES' ROSE, OFFICINAL ROSE, RED ROSE OF LANCASTER; flowers semi-double, crimson, very fragrant. 'Versicolor' flowers semi-double 7–9cm across, striped white, pink and red; often reverts; often confused with *R.damascena* 'Versicolor'.

Name	Distribution	Habit	Leaves	Flowers	Fruit
R.glauca *R.rubrifolia*	C and S Europe	erect to arching 1.5–3m tall, dark red and bloomy when young, prickles sparse, straight or decurved, broad-based, bristles sometimes present	leaflets 2–4.5cm, 5–9, leaden grey-green to blue-green, often tinged pink to purple-red, glaucous, ovate to narrowly elliptic, acute, hairless, toothed	solitary or 2–12, single, 2.5–4cm across; sepals entire or lobed, smooth or glandular-bristly, spreading; petals pale to deep pink, white at base	1.3–1.5cm, ovoid to subglobose, brown-red, smooth or with sparse glandular bristles

Name	Distribution	Habit	Leaves	Flowers	Fruit
R.laevigata CHEROKEE ROSE	S China, Taiwan, Vietnam, Laos, Cambodia, Burma,, naturalized S US	green, climbing to 10m or more, prickles scattered, stout, red-brown, hooked	leaflets 3–9cm, 3–5, lanceolate to elliptic or ovate, leathery, acute or acuminate, glossy above, midrib sometimes prickly beneath, toothed	solitary, single, fragrant, 5–10cm across; sepals entire, bristly, erect; petals white or creamy white	3.5–4cm, pear-shaped, orange-red to red, bristly

Name	Distribution	Habit	Leaves	Flowers	Fruit
R.longicuspis	NE India, W China, Burma	scrambling and climbing to 6m, tinged red when young, prickles short, curved or hooked or absent	tinged red when young; leaflets 5–10cm 3–7, narrowly ovate to elliptic, leathery, acuminate, occasionally downy on midrib beneath, toothed	to 15 in a loose cluster, single, smelling of bananas, to 5cm across; sepals lobed, hairy and glandular; petals white, silky beneath.	1.5–2cm, broadly ellipsoid to spherical, red to orange, often hairy and glandular

Name	Distribution	Habit	Leaves	Flowers	Fruit
R.moschata MUSK ROSE	unknown in the wild; widely cultivated	arching or semi-climbing, purple or red-tinged 3–10m, prickles few, scattered, straight or slighly curved	leaflets 3–7cm, 5–7, broadly ovate to broadly elliptic, acute or acuminate, shiny and hairless above, downy or not on veins beneath, toothed	in few-flowered loose clusters, single, with a musky scent, 3–5.5cm across; sepals entire lobed, hairy, reflexed; petals white or cream, reflexing with age	to 1.5cm, subglobose to ovoid, orange-red, usually downy and sometimes glandular

Comments: var. *nastarana*: leaflets smaller, always hairless beneath; flowers larger, more numerous, tinged pink

Name	Distribution	Habit	Leaves	Flowers	Fruit
R.moyesii	W China	erect, bristly below, stout, red-brown 2–3.5m, prickles pale, scattered, stout, broad-based	leaflets 1–4cm, 7–13, broadly elliptic to ovate, acute, dark green above, somewhat glaucous beneath, hairless except for midrib beneath, toothed	1–2(–4), single, 4–6.5cm across; sepals entire, expanded at tip, hairy and sparsely glandular; petals pink to blood-red	3.8–6cm, bottle-shaped, with a neck, orange-red, glandular-bristly at least towards the base

Comments: 'Fargesii': flowers pink to rose red; leaflets shorter and wider, obtuse. 'Geranium': compact; leaflets paler green; flowers scarlet, crimson with a shorter neck. 'Sealing Wax': flowers pink; fruit large, bright red.

Name	Distribution	Habit	Leaves	Flowers	Fruit
R.multiflora JAPANESE ROSE, BABY ROSE	Japan, Korea,	arching, trailing or sometimes climbing 3–5m, prickles many small, stout, decurved	leaflets 1.5–5cm, 5–11, obovate or elliptic, acute, acuminate or obtuse, sometimes downy beneath, toothed	few to many in branched clusters, single, with a fruity scent, 1.5–3cm across; sepals lobed, glandular-bristly; petals cream, fading to white, occasionally pink	0.6–0.7cm, ovoid to spherical, red

Comments: 'Carnea': flowers double, flesh-pink. 'Grevillei': SEVEN SISTERS ROSE; flowers in clusters of 25 to 50, usually double, deep pink-purple fading to white. 'Wilsonii': flowers single, white. var. *cathayensis*: flowers single, rosy pink, to 4cm across, in rather flat clusters. 'Nana': dwarf; flowers single, fragrant, pale pink or cream.

ROSA

Name	Distribution	Habit	Leaves	Flowers	Fruit
R.pimpinellifolia BURNET ROSE, SCOTCH ROSE	Europe, Asia	suckering, 90–200cm, purple-brown, prickles dense straight, slender, bristles stiff	leaflets 0.6–2cm, 5–11, broadly elliptic to broadly obovate to orbicular, obtuse, midvein sometimes downy beneath, teeth glandular	solitary, single or double, 3.8–6cm across; sepals entire, narrow, margin woolly, erect; petals creamy white	0.7–1.5cm, spherical or almost so, dark brown becoming black, smooth

Comments: 'Andrewsii': flowers double, rose-pink. 'Grandiflora': flowers 7–7.5cm across. 'Hispida': flowers 5–7cm across, opening pale yellow, fading to

Name	Distribution	Habit	Leaves	Flowers	Fruit
R.primula	C Asia to N China	erect, slender, to 3m, red-brown when young, prickles stout, straight, somewhat compressed, broad-based	very aromatic; leaflets 0.6–2cm, 7–13, elliptic to obovate or oblanceolate, acute or obtuse, with large glands beneath, teeth compound, glandular	solitary, single, 2.5–4.5cm across; sepals entire, hairless, erect; petals primrose-yellow	1–1.5cm, spherical to turbinate, brown-red to maroon, smooth
R.roxburghii CHESTNUT ROSE, CHINQUAPIN ROSE	China			double, pink, darker in the centre	

Comments: The typical plant was originally introduced from Chinese gardens and is rarely grown. More commonly seen is f. *normalis*, with spreading, rather stiff stems to 5m+ with sparse, straight or hooked prickles. The leaflets are 1–2.5cm long, narrowly ovate to obovate, hairless and toothed. Solitary, single and fragrant, the flowers are 5–7.5cm across with lobed, downy and prickly sepals and mid- to deep pink petals. The fruit is 3–4cm long, flattened spherical, yellow-green and prickly.

Name	Distribution	Habit	Leaves	Flowers	Fruit
R.rubiginosa **R.eglanteria** SWEET BRIER, EGLANTINE	Europe, N Africa, W Asia	branches arching, 2–3m, dense, bearing many, stout, hooked prickles and stiff bristles	aromatic; leaflets 1–3cm, 5–9, dark green, ovate to rounded, glandular and downy beneath, teeth compound	1–7, single, fragrant, 2.5–5cm across; sepals lobed, glandular-bristly, erect; petals pale to deep pink	1–2.5cm, ovoid to subglobose, red or orange, smooth or glandular-bristly
R.rugosa JAPANESE ROSE, TURKESTAN ROSE	E Russia, Korea, Japan, N China	stems erect, stout, 1.2–2.5m, downy when young, densely prickly and bristly	leaflets 2.5–5cm, 5–9, dark green, oblong to elliptic, acute, wrinkled above, downy beneath with conspicuous veins, teeth shallow	1 to few, single, fragrant, 6–9cm across; sepals entire, expanded at tip, downy, erect; petals purple-pink	2–2.5cm, subglobose, with a neck at the top, red to orange-red, smooth

Comments: 'Alba': flowers single, white, opening from pale pink buds; very free-fruiting. var. *alboplena*: flowers double, white. var. *rosea*: flowers single, rose-pink

Name	Distribution	Habit	Leaves	Flowers	Fruit
R.sericea	Himalaya, NEast India, N Bhutan, SW and C China	upright or somewhat spreading 2–4m, grey or brown, prickles straight or curved, often upward-pointing, red-tinged with broad bases, mixed with slender bristles	leaflets 0.6–3cm, 7–11, elliptic, oblong or obovate, obtuse or acute, hairy or not beneath, toothed above	solitary, on short lateral shoots, single, 2.5–6cm across; sepals entire, hairless or silky; petals usually 5, white or cream, notched	0.8–1.5cm, spherical to pear-shaped, dark crimson, scarlet, orange or yellow, smooth

Comments: ssp. *omeiensis*: f. *pteracantha*: stem with very broad-based prickles forming interrupted wings along stems, bright red at first, hardening grey-brown.

Name	Distribution	Habit	Leaves	Flowers	Fruit
R.wichuraiana MEMORIAL ROSE	Japan, Korea, E China, Taiwan	procumbent, trailing or climbing 3–6m, prickles strong, curved	leaflets 5–9, dark green, elliptic to broadly ovate or rounded, obtuse, hairless except for midrib beneath.	6–10 in loose clusters, toothed single, fragrant, 2.5–5cm across; sepals entire or lobed, often downy or slightly glandular; petals white.	1–1.5cm, ovoid or spherical, orange-red to dark red

Comments: 'Variegata': erect to clambering, leaflets cream with pink tips when young, turning green, marked with cream and tinted red

Name	Distribution	Habit	Leaves	Flowers	Fruit
R.willmottiae	W and NW China	erect or arching to 3m, glaucous-pruinose, becoming red-brown, prickles slender, straight, mostly in pairs	leaflets 0.6–1.5cm, 7–9, obovate to oblong or almost round, obtuse, hairless, toothed	usually solitary, single, slightly fragrant, 2.5–3.8cm across; sepals entire, hairless; petals purple-pink	1–1.8cm, ovoid, pear-shaped or more or less spherical, orange-red

Comments: 'Wisley': flowers deeper pink with narrower petals and the sepals persistent in fruit

Name	Distribution	Habit	Leaves	Flowers	Fruit
R. × wintoniensis (*R.moyesii* × *R.setipoda*)	garden origin				

Comments: Differs from *R.moyesii* in clusters of 7–10 flowers with petal crimson, white at base, and aromatic leaves.

Name	Distribution	Habit	Leaves	Flowers	Fruit
R.xanthina	N China, Korea	1.5–3.5m, erect, brown to grey-brown, prickles straight or slightly curved, broad-based, sometimes much flattened on non-flowering shoots	leaflets 0.8–2cm, 7–13, broadly elliptic to obovate to orbicular, obtuse, hairy beneath, toothed	solitary, rarely 2, semi-double, 3.8–5cm across; sepals entire, acuminate, leafy and toothed, more or less hairless; petals bright yellow	1.2–1.5cm, spherical or broadly ellipsoid, brown red or maroon, smooth and hairless.

Comments: f. *hugonis*: leaflets elliptic to obovate, flower-stalks hairless bearing single flowers 4–6cm across. f. *spontanea*: flowers single, 5–6cm across – one of the parents of 'Canary Bird', a cultivar with neat foliage and large, freely produced bright yellow flowers, the other parent being f. *hugonis*.

Roscoea (for William Roscoe (1753–1831), founder of the Liverpool Botanic Garden). Zingiberaceae. China, Himalaya. Some 17 species, perennial, rhizomatous herbs with lanceolate to strap-shaped leaves, their bases enveloping each other to form a pseudostem. Short terminal spikes of showy flowers are produced in summer and early autumn. The corolla is tubular with three petals, the uppermost erect and hooded, the lateral petals hanging either side of a showy, 2-lobed lip. Hardy to –20°C/-4°F if mulched in winter. Grow in part shade and in rich, moist but well-drained acid, peaty loam. Propagate by division in spring.

R.alpina (Kashmir, Nepal; to 30cm; flowers pink or mauve, to 3cm in diameter); *R.cautleoides* (W China; to 55cm; flowers pale yellow. to 5cm diam.); *R.humeana* (SW China; to 34cm; flowers purple or lilac, to 5cm diam.); *R.purpurea* (Himalaya, Sikkim; to 30cm; flowers pale purple or white with dark purple markings, to 6cm diam.).

rose a perforated cap attached to the spout of a watering can in order to produce a spray of water. Roses may be round or oval in surface outline; and are fitted with the distribution holes uppermost for a fine spray when watering newly sown seed or seedlings, and pointed downwards for a more drenching application when watering newly potted or established open-ground plants.

rosemary (*Rosmarinus officinalis*) an evergreen perennial, up to 2m in height, with procumbent forms. A native of the Mediterranean region, it has been cultivated for centuries for the aromatic oils distilled from the flowers and leaves; the fresh or dried leaves are used for flavouring meat. Grow in a sheltered sunny site, in pots in areas subject to severe cold. Prune after flowering to contain habit. Propagate from softwood or semi-ripe cuttings in summer.

rosette leaves radiating from a common crown or centre.

Rosmarinus (from Latin *ros,* dew, and *marinus,* of the sea, from the maritime habitat). Labiatae. S Europe, N Africa. 2 species, very aromatic evergreen shrubs with crowded, linear to oblong, leathery leaves with revolute margins. The flowers are 2-lipped, small and borne in short axillary racemes at various times of year, but usually in early spring and late summer. *R.officinalis*, ROSEMARY (q.v.) has been an important constituent of herb gardens for centuries, and is an attractive evergreen shrub for hot, dry, sheltered positions, particularly on stony, calcareous soils. It is hardy in climate zone 6. *R.officinalis* (Mediterranean; to 2m, erect or procumbent; leaves 1.5–4cm, dark green, leathery, white-tomentose beneath, margins strongly revolute; flowers pale blue to lilac; includes cultivars with white, pink and deep blue flowers, prostrate to tall and narrowly erect, some with white – or yellow – variegated leaves).

Rossioglossum (for J. Ross, who collected orchids in Mexico in the 1830s). Orchidaceae. Mexico to Panama. 6 species, epiphytic orchids with clumped, ovoid pseudobulbs and broadly elliptic to ovate, leathery leaves. Carried on basal racemes, the flowers are large and waxy, with spreading tepals, usually wavy and with tiger-like colouring. The lip is smaller with a bumpy callus. Provide a minimum temperature of 10°C/50°F. Pot every two years in a dense, bark-based medium containing some leafmould, coir and sphagnum. Place in a cool, well-ventilated and brightly lit situation, avoiding sunscorch and wetting of the foliage. Water and feed copiously when in growth to encourage the plumpest possible pseudobulbs. On their completion, withhold water, except where it is necessary to prevent excessive shrivelling. Flower spikes will develop in late summer and autumn, blooming in winter. Repot and recommence watering on first appearance of new growth. Propagate by division when repotting. *R.grande* (syn. *Odontoglossum grande*; CLOWN ORCHID; Guatemala, Mexico; raceme 14–30cm, 2–8-flowered; flowers to 15cm in diameter, long-lived, waxy; sepals linear-lanceolate, yellow, barred and flecked rust to chestnut brown, narrow, wavy; petals oblanceolate, yellow, basal half solidly deep red-brown, broader than sepals; lip cream-white or cream-yellow, banded red-brown, fan-shaped callus yellow, protuberances eye-like, ringed with brick red).

rostellum (1) any small beak-like structure; (2) the tissue separating anther from stigma in Orchidaceae.

rosulate (of leaves) arranged in a rosette.

rotate wheel-shaped.

rotation the practice of changing the position of annual crops in a garden each season for a period of several years. Rotation reduces the likelihood of a build-up of soil-borne pests and diseases, and helps make the most efficient use of nutrients some different crops absorbing surpluses left by others. It also aids weed control where some dense canopy crops smother seedlings and through the continual cultivation required for crops such as potatoes. It is especially advocated for the vegetable garden, where crops from the same family are likely to suffer from the same pests and diseases; legumes, onions and miscellaneous crops are grouped to follow potatoes which in turn follow brassicas over three years.

Strict adherence to such rotation is often impractical in a small garden where there may be limited areas of particular crops and where extended cropping may be practised using covering methods and a range of cultivars. Moreover, clubroot disease and potato cyst eelworm, for example, can persist in the soil for much longer than three years. The general principle of rotation is advisable, although also worthy of consideration is the view that crops may be grown continually on the same site until pest or disease problems arise and at that point moved elsewhere; this, however, obviates other sound advantages of rotation.

rosemary

rotavator a powered cultivator with rotating blades, originally a trade name but now used generally to describe this type of implement. Rotavators are suitable for cultivating large areas but are capable of producing a semi-permeable pan at operating depth when used continually on heavy soils, especially under wet conditions.

Rothmannia (for Dr Georgius Rothmann (1739–1778), explorer, botanist and pupil of Linnaeus). Rubiaceae. Tropical Africa, Madagascar, Asia. 30 species, shrubs or small trees grown for their *Gardenia*-like flowers produced in summer. Cultivate as for *Gardenia*. *R.capensis* (CANDLEWOOD; South Africa; evergreen tree to 14m; leaves to 10cm, lustrous dark green; flowers white or cream to yellow, fragrant, to 8cm, funnel-shaped).

rotund rounded, curved like the arc of a circle.

rough leaves the first true leaves produced by a seedling, used particularly when describing a suitable stage of growth at which to prick out or apply pesticide treatment.

rove beetles (Coleoptera: Staphylinidae) a worldwide family of about 20,000 species. Most are black or brown, but some are brightly coloured; slender and elongate beetles varying in length from 1–30mm; their most characteristic feature is their very short wing cases, which give a flexibility unusual in beetles. Many fly readily and they are found in a wide variety of habitats, including moss, dung, carrion, decaying vegetable matter, flowers, under bark and in birds' nests.

Many species are predacious as both adults and larvae, exerting natural control of various pests, and several species have been recorded feeding on the eggs and larvae of the cabbage root fly; especially active in this way are *Aleochara* species *Oligoa flavicornis,* with small, shiny black adults up to 1mm in length, is a predator of the fruit tree red spider mite. Familiar in Europe and North America is the DEVIL'S COACH HORSE *(Staphylinus olens)*, which is up to 30mm long.

Rudbeckia fulgida

Roystonea (for General Roy Stone (1836–1905), US military engineer in Puerto Rico). Palmae. Caribbean Islands and continental coasts. ROYAL PALM. 12 species, palms with columnar trunks and pinnate leaves, their bases forming a distinct crownshaft. Used in landscaping in the lowland tropics, and for avenue and specimen plantings. Although they are sometimes grown in sheltered subtropical gardens, they seldom achieve their full stature. *R.regia* is a slender and elegant palm when young, suitable for pot and tub cultivation in the warm greenhouse (minimum temperature 15°C/60°F). Propagate by seed. *R.regia* (CUBAN ROYAL; Cuba; trunk to 25m, thickened in middle, often swollen at base, sometimes leaning or crooked; lower leaves curved downwards, covering the crownshaft, pinnae to 1m × 2–4cm in 2 rows and differing planes).

Rubus (the Classical Latin name). Rosaceae. BLACKBERRY, BRAMBLE, RASPBERRY. Cosmopolitan. Some 250 species, evergreen or deciduous, erect, scrambling, trailing or prostrate shrubs and shrublets, usually bristly and prickly. The leaves are simple, lobed or pinnately to palmately divided. Produced in spring and summer in clusters, racemes and panicles, the flowers consist of 4 or 5 broad and spreading petals and numerous stamens. They are followed by aggregate fruits composed of many, fleshy drupelets. The following are hardy to –25°C/13°F, with the exception of *R.squarrosus*, which needs a warm situation in zones 7 and over. All prefer a fertile, well-drained soil in sun or light shade. Train climbers on walls, fences and trellis or among larger shrubs and over tree stumps. *R.tricolor* is an excellent groundcover. *R.cockburnianus* and *R.thibetanus* are striking silver shrubs for the winter garden. Cut out old and exhausted stems of all species in early spring. Increase by seed. Propagate evergreens by semi-ripe cuttings in a heated case; deciduous species by softwood or hardwood cuttings; also by layering. See *blackberry, hybrid berries* and *raspberry.*

Rudbeckia (for Olof Rudbeck the elder (1630–1702), and Olof Rudbeck the younger (1660–1740), the former a Swedish physician/botanist and founder of the Uppsala botanic garden, the latter also a professor at Uppsala – both were friends, patrons and colleagues of Linnaeus). Compositae. N America, now widespread as garden escape. CONEFLOWER. About 15 species, usually perennial, rarely annual or biennial herbs with simple or branched stems and large, daisy-like flowerheads in summer and autumn. Fully hardy. Grow in any moderately fertile and moisture-retentive garden soil in sun or part-day shade. Propagate by division or removal of rooted basal portions of stem; also (cultivars) by seed sown under glass in early spring. Plant out seedlings in late spring. The seed races are often treated as annuals, flowering in their first year.

R.fulgida (SE US; perennial herb to 1m; flowerheads to 7cm in diameter, receptacle hemispherical to conic, ray florets yellow to orange, disc florets purple-brown); *R.hirta* (BLACK-EYED SUSAN, MARGUERITE JAUNE; central US; biennial or short-lived perennial herb to 2m; flowerheads to 9cm in diameter, receptacle conic, ray florets to 4cm, pale yellow, darker at base, disc florets purple-brown; includes cultivars and seed races with single and double flowerheads with large ray florets in shades of gold, some stained or solidly overlaid with brown, bronze or coppery red; the disc florets range in colour from bright green to mahogany and black-brown); *R.laciniata* (N America; perennial herb to 3m, mostly glabrous and glaucous; flowerheads solitary, receptacle hemispheric to conic, ray florets to 6cm, yellow, disc florets yellow-green or grey-green; includes cultivars with short (60cm) to tall (2m) habits and large flowerheads in shades of yellow and gold).

ruderal a plant that grows in waste materials or on waste land.

rudimentary fragmentary, imperfectly developed.

RUBUS

Name	Distribution	Habit	Stems	Leaves	Flowers	Fruit
R.'Benenden' R. 'Tridel'; (R.deliciosus × R.trilobus)	garden origin	erect, vigorous deciduous shrub to 3m	branches arching; bark peeling	3–5-lobed, broad, bright green, downy	to 5cm diameter, solitary, axillary; petals broad, bright white	
R.biflorus	Himalaya	deciduous shrub to 3m	stems erect, white-glaucous, with straight prickles	to 25cm; petiole and rachis bristly; leaflets 3–5, ovate to elliptic, to 10cm, acute or acuminate, sharply toothed, dark green above, white-tomentose beneath	white, 1–1.7cm diameter, terminal and axillary	yellow
R.cockburnianus	N and C China	deciduous shrub to 3m.	stems erect, much-branched, strongly white-glaucous, sparsely prickly; branches arching	pinnate, to 20cm; petioles white-woolly, with hooked prickles; leaflets 9, to 6cm, ovate, or rhomboid or oval-lanceolate, acuminate, serrate, white-tomentose beneath	purple, small, in terminal panicles to 12cm	black
R.henryi	C and W China	scrambling evergreen shrub to 6m	stems slender, with a few prickles, tomentose-floccose when young	trilobed, to 15cm, glabrous above, white-tomentose beneath, lobes to 2.5cm, narrow, long-acuminate, serrate	pink, to 2cm diameter, in 6–10-flowered, glandular raceme to 7.5cm	shiny black

Comments: var. *bambusarum*: leaflets 3, narrowly lanceolate, to 12.5cm

Name	Distribution	Habit	Stems	Leaves	Flowers	Fruit
R.idaeus RASPBERRY, WILD RASPBERRY, RED RASPBERRY, EUROPEAN RASPBERRY, FRAMBOISE	Europe, N Asia, Japan	to 1.5m, erect	stems erect, bloomed, bristly and softly hairy, usually with weak prickles	usually pinnate; leaflets 3–7 ovate or oblong, occasionally somewhat lobed, all shortly acuminate, cordate at the base, glabrescent above, white-tomentose beneath	white, to 1cm diameter, in few-flowered, leafy, terminal and axillary racemes	red or orange

Comments: 'Aureus': low-growing; leaves clear bright yellow. 'Phyllanthus': flowers replaced by tasselled green shoots, inflorescence divided. var. *strigosus*: AMERICAN RASPBERRY; densely glandular-bristly. 'Albus': fruit orange-yellow.

Name	Distribution	Habit	Stems	Leaves	Flowers	Fruit
R.laciniatus CUT-LEAVED BRAMBLE, CUT-LEAF BLACKBERRY, PARSLEY-LEAVED BRAMBLE	origin unknown	scrambling, deciduous shrub	stems robust, angled, pubescent with stout, recurved prickles	leaflets 3–5, divided into pairs of laciniate segments, glabrous or pubescent beneath	pink-white, in large terminal panicles, petals toothed	black

Comments: 'Elegans': leaves small, upper leaf surface hairy; flowers and fruit rare.

Name	Distribution	Habit	Stems	Leaves	Flowers	Fruit
R.odoratus FLOWERING RASPBERRY, PURPLE-FLOWERING RASPBERRY, THIMBLEBERRY	E N America	vigorous, deciduous shrub to 3m	stems woody, erect, with pale brown, peeling bark, glandular, pubescent, without prickles	to 25cm, 5-lobed, cordate at base, serrate, pubescent beneath	fragrant, pink-purple, to 5cm diameter, in many-flowered panicles	red or orange

Comments: 'Albus': bark light; leaves pale; flowers white

Name	Distribution	Habit	Stems	Leaves	Flowers	Fruit
R.pentalobus	Taiwan	spreading or prostrate shrub	stems brown-pubescent, sparsely prickly	wrinkled, ovate to orbicular, to 20cm, cordate, lobes 3–5, sharply serrate, pilose to glabrous above, tomentose or subglabrous beneath	white, solitary	red

Comments: 'Emerald Carpet': carpet-forming

Name	Distribution	Habit	Stems	Leaves	Flowers	Fruit
R.spectabilis SALMONBERRY	W N America	deciduous shrub to 2m	stems erect, glabrous, with fine prickles	to 15cm; leaflets 3, ovate, to 10cm, long-acuminate, serrate, glabrous	pink to maroon, 2–3.5cm diameter, usually solitary	orange-yellow

Comments: 'Florepleno': flowers large, double

Name	Distribution	Habit	Stems	Leaves	Flowers	Fruit
R.squarrosus	New Zealand	sprawling evergreen shrub	stems slender, glossy, red-brown, fiercely prickly	to 15cm, slender-stalked; leaflets 3–5, linear-lanceolate, toothed, often reduced to midrib only or a vestigial blade terminating midrib, tinted red-brown	yellow-white, small in panicles	orange-red
R.thibetanus GHOST BRAMBLE	W China	deciduous shrub to 2m, resembling a shorter, finer R.cockburnianus.	stems slender, silver to purple-pruinose, bristly	to 23cm; leaflets 7–13, oval or ovate, to 5cm, coarsely serrate, mid- to grey-green to silvery above, bright, white-tomentose beneath; terminal leaflet pinnately lobed	purple, to 1.2cm diameter, solitary or in few-flowered terminal racemes	black, pruinose

Comments: 'Silver Fern': twigs and leaves silver-grey, elegant

Name	Distribution	Habit	Stems	Leaves	Flowers	Fruit
R.tricolor	W China	prostrate, semi-evergreen shrub	stems with dense, buff bristles	to 10cm, acute, cordate at base, serrate, dark green above, tomentose and bristly beneath	white, to 2.5cm diameter, solitary or in small, few-flowered terminal panicles	bright red

Comments: 'Betty Ashburner' : *R.tricolor × R.fockeanus*; leaves large, rounded, with crinkled edges, shiny. 'Kenneth Ashburner': stems long; leaves pointed, shiny; vigorous

Name	Distribution	Habit	Stems	Leaves	Flowers	Fruit
R.ulmifolius BRAMBLE	W & C Europe	vigorously scrambling shrub	stems robust, arching, bloomed, with straight or hooked prickles	leaflets 3–5, white-hairy beneath	inflorescence raceme-like; petals pink or white, crumpled	red to purple-black

Comments: 'Bellidiflorus': flowers double. 'Variegatus': leaves variegated.

Ruellia (for Jean de la Ruelle of Soissons (1474–1537), botanist and physician to Francois I, author of *De Natura Plantarum* (1536)). Acanthaceae. Tropical America, Africa, Asia and temperate N America. Some 150 species, perennial herbs, subshrubs or shrubs with 5-lobed, tubular to funnelform flowers in spring and summer. Cultivate as for *Crossandra*. *R.devosiana* (Brazil; pubescent subshrub to 45cm, leaves with silvery white veins above purple-tinted beneath; flowers white flushed lavender, tube slender, lobes veined purple); *R.graecizans* (S America; bushy subshrub to 60cm or more, spreading; flowers to 2.5cm, scarlet); *R.macrantha* (CHRISTMAS PRIDE; Brazil; bushy subshrub, erect to 2m; flowers to 7.5cm, crimson to lavender-rose, throat paler with red veins); *R.makoyana* (MONKEY PLANT, TRAILING VELVET PLANT; Brazil; spreading herb to 60cm; leaves tinted violet above and veined silvery grey, purple beneath; flowers to 5cm, rosy carmine).

rufescent, rufous reddish-brown.

rugose wrinkled by irregular lines and veins.

rugulose finely rugose.

Rumex (name used by Pliny). Polygonaceae. Northern temperate regions. DOCK, SORREL. 200 species, annual, biennial or perennial herbs, rarely shrubs, with variously shaped leaves and panicles or racemes of very small, rusty-red-green flowers. COMMON SORREL, *R.acetosa*, and the more drought-tolerant FRENCH SORREL, *R. scutatus*, are valued for their sharply flavoured leaves, used raw in salads and cooked as for spinach. Treated as annuals or short-term perennials (replace every third or fourth year), they are easily grown and have a long harvesting season. Grow in sun on any moderately fertile and well-drained soil. Sow seed in spring and thin to 10cm, where the whole plant is to be harvested, or to 25cm for perennial crops. Leaves can be harvested about eight weeks from sowing. *R.sanguineus* is the ornamental BLOODY DOCK, a native of Europe, Asia and N Africa. It grows to 1m tall, with bold, oblong and puckered, glossy olive leaves, the veins stained with dark blood red. Fully hardy, it prefers a damp or wet situation and full sun. Increase by seed or division. Allowed to set seed, it may become invasive.

ruminate appearing as if chewed.

runcinate a leaf, petal or petal-like structure, usually oblanceolate in outline and with sharp, prominent teeth or broad, incised lobes pointing backwards towards the base, away from a generally acute apex.

runner a common term for a stolon; especially used of strawberries.

run off (1) rain or irrigation water that is neither absorbed by soil nor evaporated, and therefore a potential cause of erosion; (2) used of the stage of surface saturation of a plant that results from spraying.

Russelia equisetiformis

Ruschia (for Ernest Rusch of Lichtenstein Farm, Namaqualand). Aizoaceae. South Africa (Cape Province, Namibia. 360 species, evergreen shrubs, shrublets and perennial herbs with succulent leaves and daisy-like flowers in summer. Cultivate as for *Lampranthus*.

R.acuminata (shrublet to 20cm; leaves 1.5–2.5cm, tinged blue, spotted, roughly papillose; flowers 3cm in diameter, white to pale pink); *R.crassa* (shrub ascending or prostrate; leaves blue-green, covered in short hairs, 1–2cm long, apiculate; flowers 2.2cm in diameter, white); *R.macowanii* (shrublet, 15–20cm, prostrate; leaves 2–3.5cm, swollen; flowers 2.2cm in diameter, pink with darker stripes).

Ruscus (name used by Virgil). Liliaceae (Asparagaceae). Azores, N Africa, Madeira, W Europe to Caspian Sea. BUTCHER'S BROOM. 6 species, evergreen perennial herbs with clumped shrub-like stems bearing cladophylls. Fleshy at first, these later become hard and rigid. The true leaves are minute and scale-like. The cladophylls are rhombic to ovate-lanceolate, usually acute to sharply pointed and crowded. Very small, pale flowers are borne on the midrib of the cladophylls. Where plants of both sexes are grown together, bright red berries may be produced. *Ruscus* tolerates a range of soil conditions, including chalk and heavy clay, and particularly dry shade. Dried stems are used in floral arrangements, taking on red tints when dried in glycerine. They are often dyed dark green. Hardy to –20°C/–4°F. Propagate by division or by seed, which may take up to 18 months to germinate. *R.aculeatus* (BUTCHER'S BROOM, BOX HOLLY, JEW'S MYRTLE; Mediterranean and Black Sea north to Great Britain, west to Azores; 30–125cm, erect, densely clumped and branched, branches green, cladophylls to 2.5cm, ovate, sharply pointed, crowded, dark glossy green at first, becoming dry and matt, rigid).

Russelia (for Dr. Alexander Russell, physician, traveller and author of *Natural History of Aleppo* (1756)). Scrophulariaceae. Mexico, Cuba to Colombia. 52 species, evergreen shrubs and subshrubs with many erect to pendent, slender green stems, the leaves sometimes reduced to scales. In summer, they bear masses of tubular, 2-lipped flowers. A superb basket plant for the cool to intermediate greenhouse (minimum temperature 10°C/50°F). Grow in full sun in a fast-draining loam-based mix with additional sand. Water very sparingly in cold weather. Propagate from seed and by division or layering (each drooping tip of *R.equisetiformis* that touches damp ground is likely to root); alternatively, increase by greenwood cuttings. *R.equisetiformis* (syn. *R.juncea*; CORAL PLANT, FOUNTAIN PLANT, FIRECRACKER PLANT; Mexico; shrub to 1.5m with a weeping habit; branches many, green, slender; leaves very small, scale-like, short-lived; flowers bright coral red to scarlet, to 3cm).

russeting the formation of rough brown patches of cork on the surface of fruits, mainly apples and pears, resulting from genetic characteristic, environmental conditions or disease. Apple varieties which naturally have a rough brown skin are called russets.

rust a fungal plant disease organism typified by orange-red, yellow or white powdery spore masses on affected plants. The term is often confusingly used to describe any red/brown discolouration of foliage.

Morphologically identical rusts are sometimes restricted to one host genus when each is referred to as *formae speciales* (f.sp.). Within a special form of a rust, there may be races which only attack certain cultivars of the host and these are known as pathogenic races or pathotypes.

Rusts are parasites on seed plants and ferns and include many of the most important diseases of cereals, vegetables and trees. Well-known garden rusts include those of *Antirrhinum,* apple (cedar-apple rust), leek, bean, chrysanthemum, carnation, hollyhock, mint, *Pelargonium,* and rose.

Rust fungi are most commonly found on the green, aerial parts of plants, the foliage and stems; infection is often localized to where pustules containing spore masses of the fungus have burst through the epidermis of the plant. Spores may be orange, whitish, yellow, brown or black, and occasionally gelatinous, depending on species. Affected plants are always weakened and heavy infection can cause stunted growth. Some rusts may spread internally in their hosts causing systemic infections that persist from one season to another, for example, mint rust.

A feature of many rusts is the dependence of their reproductive phase on infection of two different types of plants, which may be of a different botanical family; these are called alternate hosts. Such heteroecious rusts include WHEAT/BARBERRY RUST *(Puccinia graminis)* and JUNIPER/PEAR RUST *(Gymnosporangium asiaticum).* It is sometimes possible to prevent infection of a cultivated host plant by eradicating non-crop alternate hosts in the vicinity; for example, the control of PINE BLISTER RUST can be achieved by the eradication of *Ribes.* Where the whole life cycle occurs on one kind of host, the rust is termed autoecious, for example, ROSE RUSTS *(Phragmidium* species) and MINT RUST *(Puccinia menthae).*

During the development of their host plants, rusts can produce up to five different types of spore, and life cycles vary according to which ones are present. CHRYSANTHEMUM WHITE RUST *(Puccinia horiana),* which originated in Japan, has caused serious damage in Europe to its confined host range of Asiatic *Dendranthema,* including cultivars of florist's chrysanthemums. It was introduced into England in 1963 on infected cuttings and from around that time also spread to Argentina, Australia, New Zealand, South Africa and the US. The name is descriptive of the rust pustules formed mainly on the undersides of leaves, but any green tissues and flowers are liable to infection. It is spread by air-borne spores and is highly contagious in cool, moist conditions. Quarantine measures in the UK were successful until the late 1980s, by which time the disease had become common on the continent of Europe. Strict control of propagation premises prevents outbreaks in new areas of cultivation, but once established on outdoor chrysanthemums, there is little chance of eradicating the disease and nearby greenhouse specimens will be at risk. The hot water treatment of lifted stools will kill infected tissue. There are several different strains of chrysanthemum white rust fungus.

CHRYSANTHEMUM BROWN RUST *(Puccinia chrysanthemi),* sometimes called BLACK RUST because of the very dark-coloured spore pustules, is also of Japanese origin and was first found in the UK nearly a century ago. It has no alternate host and occurs worldwide, although it is not usually common or serious enough to warrant control measures.

The fact that some *Berberis* species are alternate hosts for WHEAT RUST *(Puccinia graminis)* has led to barberry being the subject of quarantine measures in several countries, legislation requiring the destruction of certain species of *Berberis* and restrictions on their importation and interstate movement. The rust forms cup-shaped orange bodies on the underside of the leaves and sometimes on the fruits. Some species, such as *B.thunbergii,* are immune from infection by wheat rust and have been widely planted in N America. *Mahonia* species can be infected by wheat rust, but *M.aquifolium* and *M.bealei* are highly resistant and the western North American *M.repens* is immune from infection.

SEMPERVIVUM RUST *(Endophyllum sempervivi)* is systemic and infects the leaves and grows down into the roots, the first symptoms being a stunting or abnormal elongation of the leaves. It is not possible to cure infected plants and they should be destroyed as soon as possible. Neighbouring healthy plants of *Sempervivum* and *Echeveria* should be given a protective spray of a rust-specific fungicide.

On junipers, the *Gymnosporangium* rusts cause a range of symptoms, from relatively minor needle infections to witches' brooms and large galls, but their importance as garden diseases mainly depends on their effect on their rosaceous hosts, which is most damaging when fruits are infected. The most economically important is AMERICAN CEDAR-APPLE RUST *(G. juniperae-virginianae),* for the damage it causes to apples rather than any harm to its natural hosts, the red cedar *(J. virginiana),* the southern red cedar *(J. silicicola),* and the Rocky Mountain juniper *(J. scopulorum).*

Another important NORTH AMERICAN JUNIPER RUST is *G.clavipes,* quince rust, which attacks junipers of the *J. oxycedrus* and *J. sabina* alliances, in particular *J. virginiana* and *J. horizontalis,* and has a wide range of alternate host genera including *Amelanchier, Chaenomeles, Cotoneaster, Crataegus, Cydonia, Malus, Photinia* and *Pyrus.* A similar host range of junipers and Rosaceae is infected in North

America by *G. globosum,* known mainly as a serious disease of *Crataegus,* but also found on *Malus, Pyrus* and *Sorbus.*

PEAR RUST (*G. fuscum*) and MEDLAR RUST (*G. confusum*) are Eurasian species which have now spread, probably in the form of latent infection of their host, *J. sabina,* to North America. The case of JAPANESE PEAR RUST (*G. asiaticum*) in recent years illustrates the potential for intercontinental movement of plant diseases in commerce. It was found on *Juniperus chinensis* bonsai imported into England from Japan and investigations showed that the fungus was capable of infecting the EUROPEAN PEAR (*Pyrus communis*).

HOLLYHOCK RUST (*Puccinia malvacearum*) is found worldwide, and although HOLLYHOCK (*Alcea rosea*) is most commonly infected, other members of Malvacae are susceptible, including species growing wild. All green parts are liable to infection, which is characterized by bright yellow or orange spots with reddish centres. On the underside of leaves, infected areas bear circular pin-head-sized pustules; these bear spores in the presence of moisture, which are distributed by air current. Where affected areas coalesce, portions of the leaf may fall out. The disease overwinters in crowns and on plant debris. There are no resistant cultivars, but vigorously growing seedlings are less susceptible than old plants, and regular re-sowing is advisable, as is the removal of the first leaves. Recommended protectant fungicide spraying from 12.5–15.0 cm high is worthwhile, and severely diseased plants should be removed and burnt.

PELARGONIUM RUST (*Puccinia pelargonii-zonalis*) was first observed in Britain in 1965 and is now established nationally. In the UK, it only affects zonal pelargoniums, first showing as white spots on the underside of leaves and yellow spots on the uppersides. Brown masses of spores later develop on the undersurfaces in the presence of moisture, and are spread by air current, especially in a humid environment. Control measures include purchasing healthy stock, good greenhouse ventilation and careful watering, and spraying or dusting with an approved fungicide.

ROSE RUST (*Phragmidium* species) infection is most commonly by *P.tuberculatum,* which attacks all sorts of roses. Although distributed throughout Britain, it mostly occurs in southern and eastern England, were it weakens the plant especially in wet seasons. On rose species, the first sign of infection may be bright orange rust patches up to 2.5cm-long on stems and leaf stalks in early spring; on cultivars, leaf infection in May is more usual as a first sign. Rose rust overwinters as special spores on fallen leaves, plant debris and mulching material, and on soil and other surfaces. Regular pruning, with removal of infected shoots, and the gathering of and burning of fallen leaves are useful means of reducing infection level. Spraying with a recommended fungicide in mid-April before overwintered spores germinate is advisable, and where the disease is prevalent, subsequent summer treatment should be applied.

LEEK RUST (*Puccinia porrii*) is common on leeks and occasionally severe. Diseased plants should be destroyed, and raising and growing on carried out on a different site each year. Avoidance of high-nitrogen feeding and very close spacing will reduce the incidence of infection.

MINT RUST (*Puccinia menthae*) occurs widely on cultivated and wild mint, the rhizomes and stems becoming infected in late autumn and winter; affected shoots are stunted and bear yellow leaves. Healthy mother plants should be identified for propagation; where suspect plants must be used, they should be washed and then immersed in water at 44°C/112°F for 10 minutes, before plunging in cold water.

Apart from the destruction of alternate hosts, the enforcement of quarantine and hygiene measures, hot water treatment, and cultivar resistance, the control of rusts is possible with systemic chemicals, and those recommended for garden use should be sought.

rustication a style of stonework in which surfaces are left rough and the joints are deeply recessed.

rustic work used of ornamental garden features made of timber in the form of branches, with bark intact or removed. The term may be used of garden furniture, fencing, pergolas and archways.

Ruta (the Classical Latin name). Rutaceae. Macaronesia, E Europe, SW Asia. RUE. 8 species, pungently aromatic, woody-based perennial herbs and small shrubs with finely 2–3× pinnately cut leaves. Produced in spring and summer, the small flowers consist of four, outspread, yellow petals with incurved tips. Rue has long been cultivated as a bitter medicinal herb, and is often held to have magical properties, particularly if obtained by theft. It is also highly decorative, with glaucous, deeply divided leaves and curious yellow flowers. When handled in sunlight, all parts may cause blistering or dermatitis. Hardy in climate zone 6. Grow in full sun or part-shade on a well-drained soil. Cut back damaged or exhausted growth in spring. Propagate from semi-ripe cuttings taken in late summer, or from seed sown as soon as ripe. *R.graveolens* (RUE, HERB OF GRACE; Balkan Peninsula, SE Europe; intensely aromatic, glaucous sea green to blue subshrub or shrub 20–50cm; leaves finely divided, segments usually obovate, to 1cm long; includes 'Blue Curl', dwarf, bushy, with lacy and curling, ice-blue leaves, 'Blue Mound', mound-forming, with tinted blue leaves, 'Jackman's Blue', with leaves distinctly glaucous blue-green, and 'Variegata', with leaves edged with creamy white).

S

Sabal (from a native American name). Palmae. PALMETTO. N Carolina through Florida and West Indies to E Texas and Colombia. Some 14 species, small to medium-sized palms with palmate leaves. They are usually stemless and clump-forming, but may produce short trunks. Grow in full sun or light shade in a free-draining, sandy medium. In winter keep frost-free and rather dry, and in summer warm and moist. Propagate by seed or by rooted suckers. *S.minor* (DWARF PALMETTO, SCRUB PALMETTO, BUSH PALMETTO; SE US; leaves to 1.5m, blue-green, segments to 150 × 5cm, filaments few or none); *S.palmetto* (BLUE PALMETTO, CABBAGE PALMETTO, CABBAGE TREE, COMMON PALMETTO; SE US; to 30m; leaves to 2m, green, segments to 80 × 4cm, with many filaments).

sac see *locule*.

sage (*Salvia officinalis*) a perennial subshrub, native of the Mediterranean region; it is widely cultivated for its aromatic leaves, which are used in cooking and in medicines. The plant grows to about 60cm high, and bears grey-green foliage with predominantly purple-blue flowers in summer. Plant 45–60cm apart. To avoid legginess, propagate every four or five years by softwood cuttings taken in early summer, or by semi-ripe cuttings with a heel taken in late summer–early autumn; earthing up of established plants produces rooted shoots. Leaves are used fresh but they also dry well. Trim shoots back after flowering to retain a compact bush. There are variegated and coloured foliage selections, many of which are less hardy, but all sages can be grown as ornamental plants.

Sagina (Latin meaning 'fodder', referring originally to *Spergula arvensis*, a plant cultivated for fodder). Caryophyllaceae. Widespread in northern temperate regions, also occurring on tropical mountains. PEARL-WORT. About 20 species, small, low-growing, annual or perennial herbs, carpeting or hummock-forming, with narrow leaves and small, green to white flowers. Suitable for the interstices of paving, pans in the alpine house, open sites in the rock garden, scree and trough plantings. Hardy to 15°C/5°F. Grow in sun on a gritty, moist soil. Propagate by seed in autumn or by division in spring. *S.boydii* (NW Europe; densely cushion-forming, glabrous perennial; leaves crowded, rigid, recurved, glossy dark green).

Sagittaria (from Latin *sagittarius*, armed with arrows, referring to the leaf shape). Alismataceae. Temperate and tropical regions. 20 species, aquatic, perennial herbs with a swollen rootstock, sagittate leaves and white, 3-petalled flowers in racemes or panicles in spring and summer. Hardy aquatics with cultural requirements as for *Alisma*. Propagate by division in spring. *S.latifolia* (WAPATO, DUCK POTATO; US; leaves to 30cm; scape to 1.2m; flowers 3–4cm in diameter; 'Florepleno': double-flowered); *S.sagittifolia* (OLD WORLD ARROWHEAD; Eurasia; leaves to 25cm; scape to 1m; flowers to 2.5cm in diameter, spotted purple; 'Florepleno': double-flowered).

sagittate arrow- or spear-shaped; where the equal and approximately triangular basal lobes point downward or toward the stalk; cf. *hastate*.

Saintpaulia (for Baron Walter von Saint Paul-Illaire (1860–1910), who discovered *S.ionantha*). Gesneriaceae. Tropical E Africa. AFRICAN VIOLET. 20 species, small perennial herbs with thinly fleshy, hairy leaves in low rosettes, their blades suborbicular to elliptic, the petioles slender and succulent. Produced in cymes at various times of year, the flowers are 2-lipped and consist of five, spreading, oblong to rounded lobes and bright yellow anthers.

Saintpaulias require a temperature range of 17–25°C/63–77°F all year round, with an overnight drop of no more than 5°C/9°F for active growth. Prolonged temperatures below 10°C/50°F or above 30°C/85°F may prove fatal. Grow in shade, indirect light or bright filtered light.

When in active growth, water plentifully. If room temperatures fall below 13°C/55°F during winter, water only moderately. Saintpaulias require medium to high humidity. Stand plants over wet pebbles in trays or

sage

surround pots with continuously damp leafmould or coir. Give soluble or liquid balanced fertilizers while the plant is in active growth. Plant in an open, well-drained, soil-less potting medium, with a pH range of 6.5–7.0.

Saintpaulias prefer to be slightly pot-bound. As a plant matures, a 'neck' develops when spent leaves are removed from the main stem. Repot the plant into a clean pot of the same size by cutting away the bottom third or half of the rootball, thus lowering the plant into the fresh potting medium for new roots to grow from the main stem. When the 'neck' is too long for this treatment, cut through the main stem 2cm below the leaves, trim the stem below the leaves into a cone shape and pot into a small pot with the leaf-stalks just clear of the potting medium. Place the pot in a covered, heated propagator.

To propagate from leaves, break a mature leaf away cleanly from the main stem using a sharp side-ways tug. Cut the leafstalk to 5cm and insert at a slight angle into potting medium in a 5cm pot; place in bright indirect light in a covered, heated propa-gator. The leaf will root in 3–4 weeks and plantlets will appear 8–10 weeks after potting. Separate the plantlets from the mother leaf when they are 5cm tall and have at least four leaves, and pot into indi-vidual 5cm pots. The trimmed leaf will also root in water, but must be potted into potting medium when the roots are no longer than 1.5cm.

Sow fresh seed on the surface of moist potting medium, cover with clear plastic film, and place in bright filtered or bright indirect light at a tempera-ture of 25°C/77°F. On germination, reduce the temperature to 20°C/70°F. When seedlings have four leaves, prick out into a small tray for growing on. Cultivars do not come true from seed.

S.ionantha (AFRICAN VIOLET, USAMBARA VIOLET; Tanzania; leaves to 5cm, ovate or oblong-ovate, crenate, dark green often tinted red-purple above, pale or purple-tinted beneath, hairy; flowers to 2.5cm in diameter, light blue to blue-violet, or white with a violet throat). *S.* cultivars (*Standard*) 'Dupont Blue': violet-blue; 'Glacier': single to semi-double, star-shaped, white with pale cream stamens; 'Hawaii': single, star-shaped, deep royal-blue, fringed silver-white; 'Marguerite': large, pale pink, wavy; 'Porcelain': single, white with blue blotches. (*Dwarf and minia-ture*) "Love Bug": semi-double, wine-red; "Pip Squeak': pale pink; 'Wee Hope': semi-double, white with a blue centre. (*Trailing*) 'Ding Dong Trail': campanulate, pale pink with deeper shading in throat; 'Magic Trail': mid-green, quilted leaves; 'Snowy Trail': semi-dwarf, semi-double, white flowers; 'Trail's Delight': leaves, edged and blotched white; 'Winding Trail': dwarf, flowers double, blue and white.

salad plant any vegetable or fruit eaten raw in salads; e.g. lettuce, tomato, cucumber, radish, celery and salad onion, but also soft green vegetables, especially where grown as 'cut and come again' crops.

salicetum a collection of willows (*Salix* species); known popularly as a sally garden in the 19th century.

Salix (Classical Latin name for Willow; also Celtic *sal*, near, and *lis*, water). Salicaceae. WILLOW, OSIER, SALLOW. Cosmopolitan except Australia, chiefly temperate regions. Around 300 species, deciduous trees and shrubs. They vary greatly in habit, from dwarf, creeping shrubs to medium-sized, weeping trees. Minute flowers are carried in dense, hairy catkins usually appearing before or with the foliage in spring. There are willows suited to a variety of soils and situations, and to a number of uses in the garden. Although generally undemanding in cultiva-tion, they should be chosen with care – the larger-growing trees may cause problems in drainage systems, and the rapid growth rate of many makes them weak-branched and prone to wind and storm damage. Debris is shed throughout the year, and the genus is predisposed to a range of pests and diseases, many of which reduce the natural life span. That said, willows have strong ornamental and functional values in the garden, which may be categorized as follows -

(a) Weeping willows. Especially beautiful at the waterside; they include *S.babylonica*, *S.* × *sepulcralis* and *S.* × *pendulina*. For smaller spaces, *S.caprea* 'Pendula', *S.purpurea* 'Pendula' are excellent.

(b) Ornamental catkins. Usually medium to large shrubs, flowering in spring. This group includes *S.aegyptiaca*, *S.caprea*, *S.daphnoides* 'Aglaia', *S.far-gesii*, *S.gracilistyla*, *S.hastata*, *S.lanata*, *S.magnifica*, *S.pentandra* and *S.purpurea*.

(c) Coloured stems. The young growth has distinc-tive colouring, and is particularly useful in the winter garden or beside lakes. Members of this group should be grown as stools or low pollards (stubs), cut back hard in late winter/early spring every one to three years. They include cultivars of *S.alba*, *S.daphnoides* and S.*irrorata*.

(d) Ornamental foliage. This group includes grey-leaved plants such as *S.alba* var. *sericea*, *S.elae-agnos*, *S.helvetica*, *S.lanata*, and those with scented leaves such as *S.pentandra*.

(e) Dwarf and alpine species. Suitable for the rock garden, including low-growing or creeping plants that make good groundcover; among them, *S.apoda*, *S.arctica*, *S.* × *boydii* (a suitable trough plant), *S.glau-cosericea*, *S.* × *grahamii* (vigorous), *S.helvetica*, *S.herbacea* (one of smallest), *S.hylematica*, *S.lanata*, *S.reticulata* (will grow in full shade, tolerant of dry conditions).

Willows tolerate a wide range of climates, and many will thrive where temperatures regularly fall to -25°C/-13°F. *S.magnifica* may suffer below -10°C/14°F. Most require ample moisture and rarely thrive on shallow chalk. Grow in full sun. Propagate from cuttings: with a few exceptions (e.g. *S.caprea*) 30cm/12in. hardwood cuttings of one-year-old wood root easily at any time (usually late autumn/early spring) and in any soil. Take semi-ripe cuttings of dwarf creepers during summer; these species can also be divided or increased by spontaneous layers. Pendulous cultivars are usually grafted. Seed is frequently sterile; fertile seed rapidly loses viability and there is a strong risk that the progeny will in any case be hybrid.

The large willow aphid infests willows in Europe; the giant bark aphid infests trees in North America. Willows across the northern hemisphere can become seriously infested by the willow scale and those in North America may also be invaded by several other species of scale insects. The willow bean sawfly and hazel sawfly commonly infest willows in Europe. Various leaf beetles including flea beetles also attack willows in both continents.

Anthracnose is particularly serious on weeping willows especially in wet seasons. Small brown spots occur on the leaves which fall prematurely, and black cankers on the shoots result in dieback. The disease is spread by spores produced on the leaves and shoots, and although it may be useful to collect and burn fallen leaves the cankers on the shoots are often too numerous for pruning to be an effective control.

Regular sprays of Bordeaux mixture or other copper fungicide commencing when the leaves unfold should have some effect. Watermark disease, a name referring to the characteristic discoloration of the wood, is caused by the bacterium *Erwinia salicis* and is especially important on cultivars of *Salix alba*. The first symptom is the sudden appearance of bright red leaves during hot summer weather. The affected branches die back to produce a 'stag's head' and a sticky liquid containing bacteria oozes from the bark. The only control method is to fell severely affected trees to prevent such spread occurring. The rust fungus *Melampsora* also affects willows, causing stem cankers: control by spraying with Bordeaux mixture. Willow heart rot is caused by a bracket fungus which gains entry through wounded branches, mostly on old trees. Willows can also be affected by armillaria, root rot, crown gall, silver leaf and tar spot.

SALIX

Name	Distribution	Habit	Shoots	Leaves	Catkins
S.aegyptiaca	Turkey, Armenia, Iran	tree or shrub to 4m	red, grey-hairy at first	to 15cm, oblong, toothed, glaucous beneath, hairy at first	to 8cm
S.alba WHITE WILLOW	Europe, N Africa, C Asia	tree to 25m	light grey-pink to olive, grey-hairy at first	to 10cm, lanceolate, serrulate, blue-green beneath, silky white-hairy at first	to 6cm

Comments: 'Britzensis': stems bright orange-red in winter; var. *caerulea*: CRICKET BAT WILLOW; leaves to 10cm, hairless when mature, blue green above, glaucous beneath; var. *sericea*: SILVER WILLOW: leaves to 10cm, silver-white; var. *vitellina*: GOLDEN WILLOW; twigs bright orange-yellow in winter

Name	Distribution	Habit	Shoots	Leaves	Catkins
S.arbuscula	Scotland, Scandinavia, Russia	twiggy shrub to 50 cm		0.5-4cm, elliptic to lanceolate, toothed, glossy above, glaucous beneath	1-2cm, with red anthers
S.babylonica BABYLON WEEPING WILLOW	Asia	tree to 12m with long, weeping branches	tinted green	to 16cm, narrowly lanceolate, serrulate, deep green above, grey-green beneath	to 2cm

Comments: 'Crispa': leaves twisted into rings, slow-growing; 'Pendula': gracefully weeping, scab-resistant. The PEKING WILLOW, *S.matsudana* (syn. *S.babylonica* var. *pekinensis*) differs from this species in its erect to arching branches and leaves with blue-green undersides. This in turn contains the cultivar 'Tortuosa' (DRAGON'S CLAW WILLOW), an erect shrub or tree to 4m with corkscrew-like stems and branches and much-twisted leaves – a popular florist's plant.

Name	Distribution	Habit	Shoots	Leaves	Catkins
S.× boydii (*S.lapponum* × *S.reticulata*)	Great Britain	dwarf shrub	hairy	more or less round, grey-hairy with prominent veins	
S.caprea GOAT WILLOW, PUSSY WILLOW, FLORIST'S WILLOW, SALLOW	Europe to NE Asia	tree or shrub to 10m	stout smooth, yellow-brown	to 12cm, broadly oval, toothed, deep green above, grey-green and downy beneath	to 2.5cm, ovoid, with soft, silver hairs and golden anthers

Comments: Includes several weeping cultivars; the best-known, 'Pendula', is grown as a small, grafted standard with a dome of weeping branches. Produced before the leaves, the catkins are a beautiful sight in late winter and often used for floral decoration.

Name	Distribution	Habit	Shoots	Leaves	Catkins
S.daphnoides VIOLET WILLOW	Europe to Asia	tree or tall shrub to 10m	red- or purple-brown to violet and thickly glaucous	to 10cm, oblong to lanceolate, serrate, dark green above, glaucous beneath	to 3cm, silver and silky, produced before leaves

Comments: Cut hard back each spring to encourage heads of frosted, violet wands for the following winter

Name	Distribution	Habit	Shoots	Leaves	Catkins
S.elaeagnos HOARY WILLOW, ROSEMARY WILLOW	C Europe to Asia Minor	tall shrub or small tree to 6m	twigs wand-like, grey-velvety then smooth and red-brown	to 15cm, narrowly lanceolate, silver at first becoming dark, glossy green above, white-felted beneath	to 6cm, narrow, appearing before or with leaves
S.fargesii	China	shrub to 3m	glossy purple to red-brown with large, red winter buds	to 16cm, elliptic, glossy and dark with sunken veins above, silky beneath	to 12cm

SALIX

Name	Distribution	Habit	Shoots	Leaves	Catkins
S.fragilis CRACK WILLOW, BRITTLE WILLOW	Europe, N Asia, naturalized N America	tree to 25m	smooth, olive-brown, easily broken at their joints	to 15cm, lanceolate, dark green above, more or less glaucous beneath	to 7cm, drooping, appearing with leaves
S.gracilistyla ROSEGOLD PUSSY WILLOW	Japan, Korea, Manchuria	erect shrub to 3m	white-hairy at first	to 12cm, narrowly oblong, covered in grey silky hairs at first, becoming smooth above, persisting well into autumn	to 5cm, appearing before leaves, grey- silky with red then bright yellow anthers

Comments: var. *melanostachys* is highly distinctive, with smooth twigs and leaves and black-brown catkins with rust then golden anthers.

Name	Distribution	Habit	Shoots	Leaves	Catkins
S.hastata HALBERD WILLOW	C Europe to NE Asia	erect, dense shrub to 1.5m	smooth, sometimes red- or purple-tinted	to 8cm, oval to broadly oblong, base cordate, dull green above with a yellow- tinted midrib, glaucous beneath	to 6cm

Comments: 'Wehrhahnii': slow-growing, spreading, small to medium shrub with many silver to golden yellow catkins in early spring. *S.apoda* differs from this species in its bright green, obovate leaves.

Name	Distribution	Habit	Shoots	Leaves	Catkins
S.helvetica SWISS WILLOW	Alps	low, and densely branched shrub	grey-downy becoming more or less smooth, red- brown	3-6cm, obovate to oblanceolate, thickly grey- downy at first becoming smooth and glossy above	to 4cm, before or with leaves

Comments: A fine silver-leaved shrub for the large rock garden

Name	Distribution	Habit	Shoots	Leaves	Catkins
S.herbacea DWARF WILLOW	Arctic Circle, Europe, N Asia	dwarf shrub with very slender, prostrate or buried stems		to 2cm, broadly ovate, serrate and emarginate, smooth, bright green	to 1.5cm, erect

Comments: One of the smallest true shrubs, it will form dense mats in the rock or peat garden.

Name	Distribution	Habit	Shoots	Leaves	Catkins
S.hylematica	W Himalaya	dwarf shrub	short and stout, shaggy at first, later red-brown	to 1cm, lanceolate to oblanceolate	to 1.5cm

Comments: Suitable for the rock garden

Name	Distribution	Habit	Shoots	Leaves	Catkins
S.integra	Korea, Japan	very similar to *S.purpurea*, with slender, slightly weeping branches		oblong, bright pale to mid green thus differing from *S.purpurea*	

Comments: 'Hakuro Nikishi' to 2m, somewhat weeping especially if grafted as a standard; leaves pink and white at first becoming bright green boldly splashed with pure white

Name	Distribution	Habit	Shoots	Leaves	Catkins
S.irrorata	SW US	vigorous shrub to 3m	long, wand-like, yellow- green at first becoming plum purple in winter and covered in a thick silver-white bloom	to 10cm, oblong to lanceolate, glossy above, glaucous beneath	to 3cm, anthers brick red becoming golden

Comments: A striking, leaden-stemmed shrub for the winter garden

Name	Distribution	Habit	Shoots	Leaves	Catkins
S.lanata WOOLLY WILLOW	Arctic and Subarctic Europe and Asia	slow-growing shrub to 1m	ascending, woolly at first becoming gnarled	2-7cm, broadly oval to obovate, densely grey- to white-hairy, becoming smoother above	to 5cm, with golden anthers

Comments: 'Stuartii': dwarf , gnarled shrub with yellow twigs and orange-tinted buds, its leaves smaller but catkins larger than in typical *S.lanata*.

Name	Distribution	Habit	Shoots	Leaves	Catkins
S.lindleyana	Nepal, Yunnan	dwarf, procumbent shrub		0.5-1.2cm, ovate to spathulate, leathery, glossy above, paler beneath, with a glossy, yellow-brown petiole	to 1 cm

Comments: Differs from *S.hylematica* in being glabrous throughout

Name	Distribution	Habit	Shoots	Leaves	Catkins
S.magnifica	China	large shrub or tree to 6m	purple to red	to 25cm, oval to obovate with a cordate base, sage green above, glaucous beneath with red veins	to 20cm, narrow and ascending, ivory to yellow deepening to brick red and maroon at the tip

Comments: A remarkable species with *Magnolia*-like foliage and catkins that resemble glowing tapers

SALIX

Name	Distribution	Habit	Shoots	Leaves	Catkins
S.pentandra BAY WILLOW, LAUREL WILLOW	Europe, naturalized E US	tree or shrub to 10m	glossy brown-green, buds yellow	5-12cm, elliptic, base cordate or rounded, finely toothed, leathery, glossy dark green above, pale beneath	to 5cm
Comments: The foliage is aromatic when young or crushed					
S.purpurea PURPLE WILLOW, PURPLE OSIER, BASKET WILLOW	Europe, N Africa to E Asia	shrub or tree to 5m with a somewhat arching habit	usually purple when young, downy at first then glossy	to 10cm, lanceolate to oblong, serrulate, dull blue-green above, paler and glaucous beneath	to 3cm, narrow, curving, anthers red becoming purple-black
Comments: The inner wood of young shoots is usually yellow. Includes 'Pendula', with long, weeping branches; often trained as a standard.					
S.repens CREEPING WILLOW	Europe, Asia Minor, Siberia	shrub to 1.5m with creeping stems and ascending branches		1-3.5cm, narrowly elliptic to ovate or oblong, covered in white silky hairs at first, often becoming glabrous and dull grey-green above	to 2.5cm
Comments: var. *argentea*: syn. *S.arenaria*, *S.repens* var. *nitida*; dwarf shrub with leaves densely coated in silver hairs; native to coastal regions, it grows well in maritime gardens and on sandy soils.					
S.reticulata	N Europe, N America, N Asia	dwarf, creeping and mat-forming shrublet		1–4cm, rounded, wrinkled to reticulate and dark green above, white-felted beneath	1–3cm
Comments: Suitable for damper places in the rock garden.					
S.× rubens (*S.alba × S.fragilis*)	C Europe	fast-growing shrub or tree to 10m	olive tinted yellow or red	to 16cm, lanceolate, toothed, more or less glaucous beneath	
Comments: Includes 'Basfordiana': branches or stems rod-like, orange-red in winter, especially if stooled in spring; 'Sanguinea': branches or stems sealing wax red in winter, especially if stooled in spring.					
S. × sepulcralis WEEPING WILLOW (*S.alba* 'Tristis' × *S.babylonica*)	garden origin	tree to 10m, slightly less weeping than *S.babylonica*	bark more fissured than in *S.babylonica*	with a glandular apex to the petiole, thus differing from *S.babylonica*	
Comments: var. *chrysocoma*: strongly weeping, twigs yellow, leaves bright green – the most commonly cultivated weeping willow.					
S.udensis *S. sachalinensis*	Japan, E Russia	tree or shrub to 5m	hoary at first then smooth shiny and chestnut brown	to 10cm, narrowly lanceolate, dark green above, blue-green and slightly hairy beneath	to 3cm
Comments: 'Sekka': a fasciated form with sporadic flattened and curling stems, used in flower arranging					

salinity the soluble salt concentration of a soil or compost, usually measured by electrical conductivity. See *conductivity*.

Salpiglossis (from Greek *salpinx*, tube, and *glossa*, tongue, from the resemblance of the style to a tongue in the corolla throat). Solanaceae. S Andes. 2 species, erect, clammy-pubescent annual, biennial or perennial herbs with 5-lobed, funnel-shaped flowers in summer. Used for bedding schemes, as late winter- and spring-flowering pot plants and for cut flowers. If sown *in situ*, thin to 22–30cm apart. Pot plants should be allowed to dry slightly between waterings. *S.sinuata* (PAINTED TONGUE; to 60cm; flowers to 6 × 5cm, yellow to ochre, mauve-scarlet or violet-blue with darker purple veins and marking). *Cultivars* Bolero Hybrids, dwarf, vigorous, with flowers in a range of rich colours; Casino Hybrids, dwarf, compact, with flowers veined in shades of red, yellow, orange, rose and purple; 'Kew Blue', dwarf, with abundant, rich blue flowers veined golden-yellow; and Splash Hybrids, compact, tolerant, with abundant flowers in a range of colours.

salsify (*Tragopogon porrifolius*) VEGETABLE OYSTER, OYSTER PLANT a biennial grown primarily for its creamy-white taproots, which are used as a winter vegetable and eaten hot or cold after boiling. Sow fresh seed March-May, 10mm deep, in rows 15cm apart; thin to 10cm apart. Lift for use from October onwards, or store in boxes of sand kept in a cool building.

Also edible are the young shoots, or chards, of overwintered plants and the mauve flower buds and flowering shoots. Chards can be produced by cutting the plants down to ground level in autumn, and earthing up to 13cm deep or covering with a similar depth of straw; use in salads or lightly cooked. Pick the flower buds just before opening, with about 10cm length of stem, and cook as for asparagus.

salt 1) the chemical compound formed as the reaction product of an acid and a base. Most inorganic fertilizers are salts. The name of the salt of a particular acid is derived from the name of the acid: nitric acid forms nitrates; sulphuric acid, sulphates; phosphoric acid, phosphates; carbonic acid, carbonates;

salsify

Disease
white blister

Salvia fulgens

Salvia leucantha

silicic acid, silicates. The first part of the the full name of a salt indicates the base in the reaction, for example, *potassium* sulphate. 2) Salt commonly refers to sodium chloride, which has a number of horticultural associations: in deflocculating clay soils; as a nutrient supplement (at very low dosage) for crops such as beet, asparagus and seakale; as a partial weedkiller; and as a phytotoxic compound from sea spray or from winter road dressings.

Salvia (Classical Latin name, from *salvare*, to save or heal, alluding to the supposed medicinal properties of certain species). Labiatae. SAGE. Cosmopolitan. Some 900 or more species, perennial, annual or biennial herbs, shrubs or subshrubs. The flowers are tubular and two-lipped, the upper lip arching and hood-like, the lower lip spreading and 3-lobed. They emerge from bell-shaped calyces, disposed in whorls arranged in spikes, racemes and panicles.

A number of species are annuals or treated as annuals in cool temperate gardens, including *S.coccinea* and *S.splendens*, used in summer bedding. *S.viridis* provides blooms for cutting to use fresh or air dried, as does *S.sclarea*. Several species usually grown as annuals, (notably *S.farinacea* and *S.patens*), are almost hardy and will behave as perennials given a warm, sheltered and well-drained position. Other Mexican and Central American species (e.g.*S.blepharophylla*, *S.involucrata* and *S.leucantha*) might be tried outdoors in similar conditions. In cooler regions they are amenable to large pot cultivation in the cool glasshouse or conservatory. Some

such as the 'black'-flowered *S.discolor* may be cut to the base in winter but will re-sprout in milder climates if given a deep protective mulch. Shrubby species that will survive temperatures down to –10°C/14°F unharmed include *S.blancoana* and *S.officinalis*. Several other shrubby species will survive to –5°C/23°F, e.g. *S.greggii*, *S.microphylla* and *S.guaranitica*. All shrubby salvias prefer a well-drained, loamy soil in a sunny, sheltered position. The more tender species may be kept alive outdoors in climate zone 7 through careful siting against south-facing walls, mulching, wrapping, etc.

Cuttings root readily and soon make flowering-size plants – where winters are harsh, it is an easy matter to keep a few rooted cuttings alive under glass. In winter, the herbaceous species are more likely to suffer from excess moisture than from low temperatures. They prefer well-drained situations on lighter, sandier soils which benefit from a mulch of organic matter each spring. Several species are nonetheless reliable and easy perennials for any soil; these include *S.nemorosa* and *S.pratensis*. Biennials can be sown *in situ* with a view to their eventual size at flowering-some, like *S.sclarea* and *S.argentea* form rosettes of very large leaves. The latter needs a very fast-draining, gritty soil. Annuals can also be sown outside *in situ* in late spring after the last frosts, although it is more usual to sow *S.splendens* under glass in early spring. All shrubby and most herbaceous salvias are propagated easily from softwood cuttings taken at any stage of the growing season. All can be grown from seed sown under glass in spring.

SALVIA

Name	Distribution	Habit	Leaves	Flowers
S.argentea	Mediterranean	short-lived perennial or biennial herb, more or less stemless	to 20cm, low-lying, ovate to oblong, tip rounded, base tapering, margins scalloped to lobed, thick-textured, rugose and thickly covered in shaggy, silver-white hairs	white to yellow flecked pink to violet in a spreading panicle to 1m tall
Comments: Zone 7, but very intolerant of winter wet				
S.blancoana	Morocco, Spain	evergreen shrub to 1m	to 9cm, oblong-elliptic, silver-hairy	pale blue to violet, sometimes streaked white, on leafless stalks
Comments: Zone 7				
S.blepharophylla	Mexico	perennial herb or subshrub to 40cm	to 5cm, ovate to deltoid, glossy and more or less hairless	red, protruding from purple to maroon calyces in loose racemes to 12cm
Comments: Zone 9				
S.bulleyana	China	more or less stemless perennial herb	to 12cm, ovate, base cordate to hastate, margins crenate, dark green covered in sparse hairs	yellow with purple to maroon lips in racemes to 60cm tall
Comments: Zone 6				
S.discolor	Peru	strongly aromatic and clammy perennial herb or subshrub to 1.5m	to 6cm, oblong-ovate, leathery, shiny dark green above, silver-white beneath	indigo to black, protruding from silver calyces in narrow terminal racemes
Comments: Hardy in zone 7 if planted in a sheltered, sunny site, thickly mulched in winter and treated as an herbaceous perennial				
S.farinacea MEALY SAGE	Texas, Mexico	erect, mealy-downy, perennial herb to 60cm tall	to 8cm, ovate to oblong or lanceolate, margins wavy to notched, minutely hairy	blue to lavender or deep purple, emerging from similarly coloured, hairy calyces in an interrupted, terminal raceme
Comments: Hardy in climate zone 9, but usually treated as a half-hardy annual in cultivation. Includes cultivars with white, opal, pale blue, dark blue and purple flowers				
S.fulgens	Mexico	freely branching perennial herb or evergreen subshrub to 1m tall	6-12cm, ovate to deltoid, toothed or notched, white-hairy beneath	bright red to deep scarlet, in terminal racemes in late summer
Comments: Zone 8 with protection from frosts and harsh winds				

SALVIA

Name	Distribution	Habit	Leaves	Flowers
S.greggii AUTUMN SAGE	Texas, Mexico	evergreen, glandular-hairy subshrub or perennial herb erect to 1m	to 3cm, ovate to elliptic or oblong, entire, tough	red to purple or magenta pink (more rarely, yellow or white) in racemes to 20cm
Comments: Zone 7 with protection. Flowers in late summer and autumn; has hybridized with *S.microphylla*				
S.involucrata	Mexico	glandular-hairy, woody-based perennial herb to 1m	to 14cm, ovate, base usually cordate, glabrous to hairy	pink to red or purple, in racemes to 30cm long in late summer and autumn
Comments: Frost-tender and treated as half-hardy. Includes 'Bethellii': flowers deep cerise with paler lips, amid pink to red bracts. 'Deschampsiana': flowers rose pink, from red calyces and bracts				
S.jurisicii	Macedonia	white-hairy, perennial herb to 60cm	ovate to oblong, pinnately cut into narrow lobes	white to pink or deep violet in loosely branched racemes to 20cm
Comments: Zone 6				
S.leucantha MEXICAN BUSH SAGE	Mexico to Tropical America	evergreen, erect perennial subshrub or herb, to 1m	to 10cm, linear-lanceolate to oblong and finely tapering, rugose above, white-downy beneath	white, woolly, from downy mauve to violet calyces in racemes to 24cm
Comments: Frost-tender				
S.microphylla *S.grahamii*	Mexico	evergreen subshrub or shrub to 120cm	1.5-4cm, ovate to elliptic, fresh green, glabrous to thinly hairy	deep crimson, pink, magenta or purple, emerging from purple-tinted calyces in racemes to 15cm in late summer
Comments: Half hardy. Includes var. *neurepia* with cerise flowers, and 'James Compton' (a cross with *S.greggii*) with large, dark crimson flowers.				
S.nemorosa	Europe to C Asia	clump-forming, perennial herb, erect to 1m	to 10cm, ovate to lanceolate or oblong, rugose, glandular-hairy	violet to purple or pink to white amid purple bracts in dense, branched spikes to 40cm
S.officinalis SAGE, COMMON SAGE	Mediterranean, N Africa	bushy, intensely aromatic subshrub or shrub to 1m	to 6cm, oblong to elliptic, finely rugose and minutely hairy, typically grey-green	violet-blue to mauve or, rarely, pink to white, amid grey-green to purple-tinted bracts in terminal racemes to 10cm
Comments: Zone 6. The herb sage, widespread in cultivation and with numerous cultivars, varying mostly in leaf colouring - from deep purple-tinted to silver-white, to pure gold, or moss green and gold-variegated, to cream-variegated or grey-green zoned yellow to cream and pink to purple-red				
S.patens	Mexico	glandular-hairy, perennial herb, erect to 1m	to 20cm, ovate to deltoid, base hastate to cordate, downy, mid-green	large, deep to pale blue, lilac or white in loose, erect and branching racemes to 40cm
Comments: Zone 8				
S.pratensis MEADOW CLARY *S.haematodes*	Europe	glandular-hairy perennial herb, often short-lived, to 1m	to 15cm, ovate to oblong, apex blunt, base cordate, margins toothed to scalloped, rugose, hairy	violet, lavender, white or pink in panicles to 45cm
Comments: Fully hardy				
S.rutilans PINEAPPLE-SCENTED SAGE	known only in cultivation	strongly pineapple-scented, bushy perennial herb or subshrub to 80cm	to 10cm, ovate to oblong-lanceolate, more or less toothed, fresh green	pink to scarlet, usually produced in autumn or winter in erect racemes
Comments: Frost-tender. Probably a form of the Central American *S.elegans*, and often listed as a cultivar of this species under the name 'Scarlet Pineapple'.				
S.sclarea CLARY	Europe to C Asia	erect, aromatic glandular-hairy, biennial to perennial herb to 1m tall	to 23cm, ovate to oblong, base cordate, toothed to erose, rugose	cream to pink, lilac, blue or mauve, amid large, white, mauve or pink bracts in tall branching spikes in summer
Comments: Usually grown as an annual; var. *turkestanica*: stems and bracts purple-pink, flowers white flecked with pink				
S.splendens SCARLET SAGE	Brazil	erect, sparsely branched annual to perennial herb, 20-100cm tall	to 15cm, ovate-lanceolate, mid-green, more or less glabrous	large, scarlet, from scarlet calyces amid red to maroon bracts in terminal, unbranched or branched spikes to 18cm
Comments: Frost-tender and widely cultivated as a half-hardy annual. Available in tall, slender or short, bushy forms with flowers in shades of white, pink, crimson, scarlet, blood red and dark violet-purple				
S.uliginosa BOG SAGE	Brazil, Uruguay, Argentina	branched, perennial herb, erect to 2m	to 9cm, lanceolate-oblong, entire to serrate, mid-green	bright blue in long racemes in autumn
Comments: Zone 9. Needs a moist soil				
S.viridis *S.horminum*	Mediterranean	erect, bushy annual herb, 20-100cm, more or less glandular-hairy	to 5cm, ovate to oblong	small, white to lilac or purple, from green to purple calyces amid showy, papery bracts in shades of white, pink, lilac or purple, sometimes with contrasting veins, in erect spikes in summer
Comments: A popular annual				

salverform of a corolla with a long, slim tube and an abruptly expanded, flattened limb.

Salvinia (for Antonio Mario Salvini (1633–1729), professor of Greek at Florence). Salviniaceae. Widespread in tropical and warm-temperate zones. 10 species, free-floating, aquatic ferns with oblong, hairy or papillose leaves arranged along thread-like stems. The individual plants are small, but soon proliferate to form large colonies. The best-known species is *Salvinia natans*. Cultivate as for *Azolla*, but in frost-free conditions.

samara a one-seeded, indehiscent, winged fruit, as found in maple or ash.

samara

Sambucus (the Latin name, perhaps from *sambuca*, a kind of harp, made from elder wood). Caprifoliaceae. Temperate and subtropical northern Hemisphere, E and NW Africa, S America, E Australia, Tasmania. ELDER, ELDERBERRY. Some 25 species, deciduous shrubs or small trees with pinnately compound leaves and flat-topped cymose corymbs or panicles of small cream-white flowers in spring and summer. These are followed by glossy, dark berries.

Sambucus species tolerate a wide range of soils (including chalk), atmospheric pollution, coastal situations and shade, the temperate species withstanding temperatures to at least –25°C/–13°F; *S.canadensis* grows where winter temperatures drop to –34°C/–29°F. Plant in sun or part shade. Golden-leaved cultivars colour earlier in sun, but retain their colour better in cool moist shade. Propagate from hardwood cuttings taken in winter, with a heel so as not to expose the pith. For best foliage effects on plants grown as medium-sized shrubs, prune to the ground in late winter.

S.canadensis (AMERICAN ELDER, SWEET ELDER; eastern N America; shrub to 3.5m, often suckering; leaflets (5–)7(–11), serrate, lowermost often lobed, glabrous or lanuginose beneath; inflorescence to 20cm in diameter, flowers ivory; fruit purple-black; cultivars include 'Acutiloba', with finely dissected, deep green leaves; 'Argenteomarginata', with leaves edged silver; 'Aurea', with pale green shoots, golden yellow leaves, and red, sweet fruit; 'Chlorocarpa', with pale yellow-green leaves and light green fruit; 'Maxima', robust, tall-growing, with an inflorescence to 40cm in diameter and tinted red; and 'Rubra', with pale red fruit); *S.nigra* (BLACK ELDER, BOURTREE, COMMON ELDER, ELDERBERRY, EUROPEAN ELDER; Europe, N Africa, SW Asia; shrub or small tree to 10m; leaflets 3–9, to 12 × 6cm, ovate to elliptic, serrate, sparsely pubescent on veins beneath, dark green above; flowers cream, muskily scented, in flat-topped inflorescences to 20cm in diameter; fruit shiny purple-black; cultivars include 'Alba', with pellucid chalky white fruit; 'Albopunctata', with leaves splashed white, especially at margins; 'Albovariegata', with leaves coarsely speckled white; 'Argentea', with white leaves, reverse-variegated green; 'Asplenifolia', with fern-like leaves; 'Aurea', with golden to lime green leaves; 'Aureomarginata', with leaves edged gold to yellow-

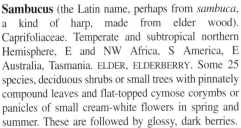

green; 'Guincho Purple': leaves leaden purple-black to wine-red; flowers bright white with purple stamens; f. *laciniata*: leaflets finely laciniate; 'Linearis': leaflets irregularly cut to midrib and blade often reduced to midrib; 'Marginata': leaflets broadly edged gold at first, later margined pale yellow to cream; 'Pulverentula': leaves striped, spotted and splashed pure white, appearing wholly white at first; 'Purpurea': leaves metallic maroon-bronze); *S.racemosa* (EUROPEAN RED ELDER, RED-BERRIED ELDER; Europe, Asia Minor, Siberia, W Asia; shrub to 3.5m; leaves to 23cm, leaflets 5, to 10 × 4.5cm, oval or ovate, acuminate, scabrous, serrate; flowers pale green to yellow-white in bluntly pyramidal panicles to 7.5cm; fruit scarlet; cultivars include 'Aurea', with bright yellow leaves; 'Goldenlocks', with golden-yellow, deeply cut leaves, lasting well in bright sun; 'Laciniata', with green leaves and leaflets deeply slashed, in narrow segments; 'Plumosa', with purple new growth and deeply incised leaflets with fine, slender teeth; 'Plumosa Aurea', resembling 'Plumosa', but with foliage striking gold throughout and new growth often tinted purple-red; 'Sutherland Gold': leaves gold, finely cut; and 'Tenuifolia', dwarf suckering shrub, with finely incised leaves and filiform leaflets, curling in frothy masses).

Sanchezia (for Josef Sanchez, professor of botany at Cadiz (18th century)). Acanthaceae. C and S America. Some 20 species, perennial herbs and soft-stemmed shrubs with handsome foliage and 5-lobed, tubular flowers amid colourful bracts in summer. Provide a minimum temperature of 10°C/50°F, bright, indirect light and medium to high humidity. Water and feed freely in spring and summer, sparingly in winter. Prune after flowering. Increase by semi-ripe cuttings. *S.speciosa* (syn. *S.nobilis*; Ecuador, Peru; evergreen, glabrous shrub to 2.25m; leaves to 30cm, glossy dark green, veins sunken and outlined white or yellow; panicles tinted purple, bracts bright red, flowers yellow).

sand soil particles ranging in diameter between 2mm (very coarse) and 0.05mm (very fine), mainly derived from solid rocks, such as quartz and sandstone; see *soil*. In gardening parlance, sharp sand is the coarse, angular grade around 3mm diameter, used in potting media and for cutting mixes; fine sand may be used on capillary benches and as a top dressing for lawns.

sand-bed see *capillary bed*.

Sandersonia (for John Sanderson (1820–1891), honorary secretary of the Horticultural Society of Natal). Liliaceae (Colchicaceae). South Africa. 1 species, *S.aurantiaca* (CHINESE LANTERN LILY, CHRISTMAS BELLS), a summer-flowering perennial herb with tuberous roots. To 75cm tall, the stems are erect and clothed with tendril-tipped leaves. Solitary and axillary, the 2cm-long flowers are broadly campanulate to urceolate and warm orange. Cultivate as for *Gloriosa*.

sandstone a type of sedimentary rock, of which the harder kinds are valuable for rock garden construction. It contains no lime, and so is particularly useful for calcifuges. See *rock plant gardening*.

Sanguinaria (from Latin *sanguis*, blood, referring to the orange-red latex). Papaveraceae. Eastern N America. BLOODROOT, RED PUCCOON. 1 species, *S. canadensis*, a perennial herb 10–60cm tall. The leaves are long-stalked and grey-green, rounded to reniform with much-scalloped margins. Produced in spring, the flowers are solitary, to 7.5cm in diameter and cup- to bowl-shaped with 8–12, oblong to obovate, white petals (more in the double 'Florepleno'). Suitable for the rock garden, woodland garden, peaty borders and for pot culture in the alpine house. It will tolerate at least –20°C/14°F and a soil pH range from 5–7. Plant in sun or part shade, on a well-drained soil, containing leafmould, composted bark and peat. Increase by division of well-established clumps in early spring, or by seed.

Sanguisorba (from Latin *sanguis*, blood, and *sorbeo*, to soak up, referring to the plant's alleged styptic qualities and its former use as an infusion to prevent bleeding). Rosaceae. Temperate northern Hemisphere. Some 18 species, rhizomatous perennial herbs or small shrubs with pinnate, saw-toothed leaves and small, white, green or pink flowers in dense, bottle-brush-like spikes in summer. Fully hardy. Grow in damp, fertile soils in sun to part shade. Propagate by division in spring or by seed in spring or autumn.

S.canadensis (CANADIAN BURNET; N America; to 2m; flowers white, in cylindric spikes to 20cm); *S.obtusa* (Japan; to 50cm; flowers pale rose, in nodding spikes to 7cm; 'Alba': flowers white); *S.officinalis* (GREAT BURNET, BURNET BLOODWORT; W Europe, Mongolia, Japan, China, N America; to 1m; flowers dark red to black-purple, in oval or cylindric heads to 3cm).

Sansevieria (after the Prince of Sanseviero (1710–1771)). Agavaceae. Tropical and subtropical Africa, Madagascar, India, East Indies. BOWSTRING HEMP, SNAKE PLANT, MOTHER-IN-LAW'S TONGUE. Around 70 species, evergreen, perennial herbs with tough and succulent, linear-cylindrical to ovate or lanceolate leaves in loose tufts or low rosettes. Carried in narrow racemes, the flowers are pale, with narrow perianth segments, and are infrequently produced in cultivation. They may be followed by colourful berries. Maintained at a minimum temperature of 10°C/50°F, with plenty of light, *Sansevieria* is extremely tolerant of dry atmospheres and neglect. Pot in a porous, well-drained medium of pH 6.0–7.0. Water sparingly in winter. Propagate by division or by leaf cuttings in spring, placing 5cm deep, transverse sections of leaf in 2:1 sand/coir in a propagating case with bottom heat of 18°C/65°F. Variegated plants usually revert to green when produced from leaf cuttings.

S.cylindrica (Angola; leaves 3–4 to a shoot, in 2 ranks, 75cm+, 25mm thick, semi-terete, straight or slightly arching, banded when young); *S.trifasciata* (MOTHER-IN-LAW'S TONGUE, SNAKE PLANT; Nigeria; leaves to 75+ × 7cm, erect, stiff, linear to lanceolate, in clusters of 1–2 or more per shoot, sometimes forming rosettes, flat, with a yellow-green sharp tip and margins, banded pale and dark green or yellow; 'Hahnii': dwarf with shorter, broader leaves in open rosettes; 'Golden Hahnii': leaves banded grey and margins banded cream; 'Silver Hahnii': leaves pale green tinted silver; 'Laurentii': leaves as for type but with bright yellow margins).

Santolina (from Latin *sanctum linum*, holy flax, old name for *S.virens*). Compositae. Mediterranean. 18 species, small, aromatic evergreen shrubs with silky to felty, entire to finely dissected leaves and long-stalked button-like flowerheads in late spring and summer. Hardy in climate zone 6. Plant in full sun on a fast-draining, medium to low fertility soil, with shelter from harsh winds. Trim over after flowering to give shape and rejuvenate. Prune out exhausted and frost-damaged wood in spring. These shrubs tolerate shearing and may be used for edging the herb garden and in parterres and knot gardens. Propagate by semi-ripe cuttings in a sandy propagating mix in the cold frame.

S.chamaecyparissus (LAVENDER COTTON; 10–50cm; leaves 2–4cm, narrow, much divided and feathery with short, tight segments, grey- to white-felted; flowerheads bright yellow; includes very dwarf, compact, silver and pale-flowered cultivars); *S.pinnata* (40–80cm; leaves 2–4cm, narrow, pinnately divided, green; flowerheads white to cream or pale butter yellow; includes cultivars with silver leaves).

sap the fluid contained in the cells and vascular system of plants. It is a variable solution of water, salts, and complex organic molecules.

sap beetles (Coleoptera: Nitidulidae) small beetles that feed on the sap of tree wounds, damaged fruits and vegetables, and fermenting plant fluids; they are closely related to pollen beetles. In North America, the DUSTY SAP BEETLE (*Carpophilus lugubris*), the CORN SAP BEETLE (*C. dimidiatus*) and certain other species feed on sweetcorn, while *Glischrochilus quadrisignatus* feeds in large numbers on vegetable and fruit debris. Some related species occur in the UK but are of no great significance.

sap-drawer (1) a shoot left unpruned on the side or underside of a branch during formative pruning, in order to provide nourishment for the branch without competing with it; (2) a branch retained during top-working to maintain a flow of sap.

saprophyte a plant deriving nutrition directly from dead or decayed organic matter and usually lacking chlorophyll.

Sansevieria trifasciata 'Laurentii'

Sarcococca hookeriana var. *humilis*

sapling a young tree at any stage of growth between a strongly developed seedling and the stage at which the heartwood becomes hard.

Saponaria (from medieval Latin *sapo*, soap – the roots of *S.officinalis* produce lather if rubbed under water and were used as a soap substitute). Caryophyllaceae. Widespread in Europe and SW Asia, with most species in the mountains of S Europe. SOAPWORT. About 20 species, perennial herbs, rarely annuals. Those listed here are tall and erect or low and cushion-forming with flat, 5-parted flowers in summer. *Saponaria* includes robust perennials, such as *S.officinalis* and *S.ocymoides*, together with dwarf, tufted species like *S.caespitosa*. *S. officinalis*, especially in its double forms, is grown in cottage gardens and other informal and naturalistic plantings, in sun or light shade, in any moderately fertile well-drained soil. *S.ocymoides* is well-suited to the border front, path edge and dry walls. Other compact mat- and cushion-forming species are grown in the rock garden, in troughs and in the alpine house. Grow alpine species in sun in a gritty, well-drained but moisture-retentive soil. Trim over after flowering. Propagate by seed in spring or autumn or by softwood cuttings taken after flowering and rooted in a closed case with gentle bottom heat. Increase also by division in spring or autumn.

S.caespitosa (C Pyrenees; densely tufted, slightly hairy perennial to 15cm, with woody stock; flowers pink-purple); *S.ocymoides* (S Europe, Spain to Balkans; mat-forming, hairy perennial to 15cm; flowers red, pink or white); *S.officinalis* (SOAPWORT; widespread in Europe; robust, glabrous, perennial to 60cm; flowers pink, red or white; some cultivars with double flowers and variegated leaves); *S.pumilio* (E Alps, SE Carpathians; densely tufted, glabrous perennial with woody stock; flowers pinkpurple or white; the hybrid *S.* × *olivana*, with *S.pumilio*, *S.caespitosa* and *S.ocymoides* in its parentage, is easier to grow – it has the attractive cushion habit of *S.pumilio*, and large, pale pink flowers).

sap-rot see *wood rots*.

sapwood the outer, functional xylem layers of a woody stem; cf. *heartwood*. The term is also used, more loosely, to describe young, non-woody shoots.

Sarcococca (from Greek *sarx*, flesh, and *kokkos*, berry, an allusion to the fleshy fruits). Buxaceae. SE Asia, Himalaya, W China. SWEET BOX. 11 species, small, evergreen shrubs with suckering stems, tough and shiny, neat foliage, small, pale and fragrant winter flowers and glossy berries. They grow best in partial shade, but will tolerate full sun given sufficient moisture. All are hardy in climate zone 6, but prefer a sheltered site and some winter protection where frosts are hard and prolonged. Propagate by division of rooted suckers or by hardwood cuttings taken in autumn and inserted into a cold frame.

S.confusa (origin unknown; 0.25–2m; leaves 3–6.5cm, elliptic to lanceolate or ovate, thinly leathery, dark green above, pale green beneath; male flowers with cream-coloured anthers; fruit ripening from red to black); *S.hookeriana* (Himalaya to W China; 0.25–1.8m; leaves to 5.5cm, coriaceous, rugose, dull green, ovate to lanceolate; male flowers with deep pink anthers; fruit black; var. *digyna*: spreading and suckering, 0.3–1m, with 4–8cm, narrowly elliptic to oblong leaves and purple-black fruit; var. *humilis*: 0.2–0.5m tall, with 3.5–7cm, elliptic leaves); *S.ruscifolia* (W China, Tibet; to 1m; leaves 3–5.5cm, dark green, broadly ovate and tapering; male flowers with cream-coloured anthers; fruit red).

sarmentose producing long, slender stolons.

Sarracenia (for Dr. M. Sarrazin de l'Etang (1659–1734), physician at the court of Quebec, who sent plants of *S. purpurea* to Tournefort). Sarraceniaceae. Eastern N America. PITCHER PLANT. 8 species, carnivorous perennial herbs. Prey is trapped in much the same way as it is for *Cephalotus*, *Darlingtonia* and *Nepenthes*. The leaves are developed as pitcher-like traps and are erect to decumbent, elongate ('trumpet-shaped') or squat, with a broad lateral wing and a terminal lid or hood. The pitcher mouth has a distinct thickened margin. The whole may be green or yellow-green, often variously marked red or brown, especially on veins, sometimes with many translucent white spots toward the apex. The colour tends to intensify when plants are grown in full sun. Lantern-like flowers nod on slender scapes, usually in spring.

Grow in full sun in buoyant, humid conditions with a minimum temperature of 7°C/45°F. Plant in an acid, soiless medium high in sphagnum, leafmould and sand. Keep moist in winter, when the leaves are often lost. From spring to autumn, keep constantly wet, standing the containers in water and misting or watering overhead frequently. Always use rainwater. *S.purpurea* is certainly hardy outdoors in climate zone 6, as may be other species and hybrids. It succeeds in a sunny, sheltered postion in wet areas of the bog garden, or stood, containerized, on the marginal shelves of pools (again, it is essential that the water should be soft – plants in small ponds filled by tap water will fail). Increase by division in late winter.

S.flava (YELLOW TRUMPET, YELLOW PITCHER PLANT, TRUMPETS; Virginia to Florida, Louisiana; pitchers 30–120cm, erect, trumpet-shaped, wide at mouth, lid erect, rounded, yellow green, often veined red on lid, especially at base, sometimes wholly red or maroon; flowers yellow); *S.leucophylla* (WHITE TRUMPET; SE and C US; pitchers to 120cm, erect, trumpet-shaped, slender, green below, white netted with green, maroon or red above, lid erect, undulate, hairy within; flowers red-purple); *S.minor* (HOODED PITCHER PLANT; S Carolina to Florida; pitchers 25–45(–100)cm, erect, trumpet-shaped, green with purple veins and translucent white patches above, wings broad, lid concave, hooded, bending over mouth; flowers yellow); *S.psittacina* (LOBSTER-POT PITCHER PLANT; SE US; pitchers to 15cm, prostrate, apex curved inwards, green, veined and marked red-purple above with distinct translucent patches, wings broad, lid forming an inflated, round hood; flowers red-purple); *S.purpurea* (COMMON PITCHER PLANT, HUNTSMAN'S CUP; New Jersey north to Canadian Arctic, naturalized Europe, especially Eire; pitchers 10–30cm, decumbent, squat, curved and erect above, inflated above, narrowed below, green with red, or purple veins and markings, or entirely green, wing broad, lid entire, erect; flowers dark red or yellow; subsp. *venosa*: pitchers sqatter, heavily veined and stained red); *S.rubra* (SWEET TRUMPET PLANT, SWEET PITCHER PLANT; SE US; pitchers 10–45cm, erect to sprawling, trumpet-shaped, narrow, green, becoming bronze- to copper-coloured with maroon veins above, wing broad, lid bending over pitcher mouth, elliptic, rim and throat glistening with nectar; flowers red to maroon).

Sasa (from *zasa*, Japanese name for the smaller bamboos). Gramineae. Japan, Korea, China. About 40 small or medium-sized bamboos with running rhizomes; see *bamboos*.

Sasaella (diminutive of *Sasa*). Gramineae. Japan. Some 12 species of small, running bamboos; see *bamboos*.

Sasamorpha (from its resemblance to the related *Sasa*). Gramineae. Far East. About 6 species of medium-sized, usually glabrous bamboos; see *bamboos*.

Satureja (from *Satureja*, name used by Pliny). Labiatae. Temperate and warm temperate northern Hemisphere. SAVORY. 30 species, annual herbs or perennial subshrubs with tubular, 2-lipped flowers. Among the most fragrant of herbs, savory is usually grown in the herb garden, but is also suitable for edging in the herbaceous border or for hot, dry and sunny situations on the rock garden or terrace. It is hardy to –10°C/14°F and needs full sun and a well-drained soil. Propagate by division in spring, or by softwood cuttings in summer. *S.montana* (WINTER SAVORY; S Europe; aromatic shrublet to 50cm; leaves small, tough, grey-green; flowers white to pale violet; 'Nana': dwarf; 'Procumbens': creeping).

Sauromatum (from the Greek *sauros*, lizard, a reference to the spathe and the petioles, which resemble a mottled lizard skin). Araceae. E and W Africa, E Asia. 2 species, perennial herbs with circular tubers and long-stalked, pedately compound leaves. The malodorous inflorescence usually appears before the foliage – numerous minute cream flowers cover the base of a long, tapering spadix, the whole arrangement more or less enveloped by a narrow spathe, bulbous at the base, reflexed and ribbon-like at its tip. Tubers are sometimes grown as curios in the home, with neither soil nor water,

their spathes developing with astonishing speed. They should be planted and fed immediately after flowering. Grow in a fertile and well-drained medium, rich in organic matter. Water as the new leaf appears and withhold water as it dies down. Keep dry but not arid when dormant in winter, maintaining a minimum temperature of 5–7°C/40–45°F. They might equally be attempted outdoors in sheltered sites in zones 7 and over. Propagate by removal of offsets or by seed. *S.venosum* (syn. *S.guttatum*; MONARCH OF THE EAST, VOODOO LILY, RED CALLA; Himalaya, S India; leafstalk to 70cm, mottled maroon; spathe tube 5–10cm, blade 30–70cm, soon becoming reflexed, twisted and undulate, yellow-green to ochre spotted and mottled blood-red to dark maroon, spadix to 35cm, narrowly cylindric, maroon or bronze).

sawdust wood particles resulting from sawing, used as a soil improver or surface mulch. It is a variable product and should be applied only after composting or weathering to prevent nitrogen deficiency in soils into which it is incorporated, and in order to neutralise possible toxins. The carbon:nitrogen ratio is higher than that of bark. The physical properties of sawdust depend on grade, and very fine samples can adversely affect drainage and aeration. Its cation exchange capacity is moderately high, but it is low in nutrients.

sawflies (Hymenoptera: Tenthredinoidea) flies up to 20mm long, mainly black or yellow-brown, sometimes brightly coloured with yellow and black predominating; females have a saw-edged ovipositor for inserting eggs into plant tissues. All sawfly larvae feed on plants, many being host specific. Most feed openly on the foliage, others are leaf miners, leaf rollers, stem borers, fruit feeders, or live inside galls. Unlike most other hymenopterous insects, there is no constriction between the thorax and abdomen. The larvae resemble caterpillars of moths and butterflies, but they have more abdominal legs, from six to eight pairs compared to five pairs or fewer.

Roses are host to several open-feeding sawflies. The BANDED SAWFLY (*Allantus cinctus*), known as the CURLED ROSE SAWFLY in the US, has yellow-brown larvae, up to 12mm long, which feed with the hind part of their body raised in the air and, when at rest, remain curled up on the underside of leaves. The EUROPEAN LARGE ROSE SAWFLY (*Arge ochropus*) has blue-green larvae with black spots. The ANTLER SAWFLY (*Cladius difformis*), known as the BRISTLY ROSE SLUGWORM in the US, has yellow-green larvae, up to 12mm long, with bodies covered in stiff hairs; the adult males have horn-like projections on the antennae.

The COMMON GOOSEBERRY SAWFLY (*Nematus ribesii*), known as the IMPORTED CURRANT WORM in the US, has green larvae, with black heads and black spots, and can cause complete defoliation of the host plants. A related European species, the SPIRAEA SAWFLY (*N. spiraea*), has pale green larvae up to

20mm long, which can completely skeletonize the leaves of its only host *Aruncus dioicus*.

Pines may be attacked by many species of sawfly, including the LARGE PINE SAWFLY (*Diprion pini*) in Europe, with yellow-green larvae, up to 25mm long, which feed in groups on the needles; the PINE SAWFLY (*D. simile*), imported into North America from Europe; and the FOX-COLOURED SAWFLY (*Neodiprion sertifer*), known as the EUROPEAN PINE SAWFLY in the US, which has grey-green larvae with a pale line along the back and a black line along each side.

Raspberry foliage is attacked by the RASPBERRY SAWFLY (*Monophadnoides geniculatus*), which has light green larvae, up to 12mm long, with prominent bristles. Other defoliating European species include the HAZEL SAWFLY (*Croesus septentrionalis*), with black-spotted green-blue larvae, up to 25mm long, which attack birch, hazel, alders, poplars, and willows; the IRIS SAWFLY (*Rhadinoceraea micans*), with blue-grey larvae, up to 20mm long, which attack wild and cultivated water-side irises, especially *Iris pseudocorus* and *I. laevigata*; and the SOLOMON'S SEAL SAWFLY (*Phymatocera aterrima*), with blue-grey larvae, up to 20mm long, which feed on polygonatums during the summer.

Examples of sawflies with leaf-mining larvae include the BIRCH LEAFMINER (*Fenusa pusilla*) in Europe and North America, causing blister-like mines, and the raspberry leaf-mining sawflies (*Metallus* species), in Europe, which live in irregularly-shaped mines on raspberry and blackberry. The ROSE LEAF-ROLLING SAWFLY (*Blennocampa pusilla*) causes tight rolling of rose leaves, but a leaf rolling habit is not common among sawflies. Some European sawflies cause galling, for example, the WILLOW BEAN-GALL SAWFLY (*Pontania proxima*), known as the WILLOW RED GALL SAWFLY in North America, which causes galls up to 6mm long – these resemble bean seeds and protrude from both sides of the leaves of willows; a related species, the WILLOW PEA-GALL SAWFLY (*P.viminalis*), is responsible for pea-shaped galls on the same host.

Several sawflies are of considerable economic importance as pests of fruit. These include the APPLE SAWFLY (*Hoplocampa testudinea*) in Europe and North America, with cream-white larvae, up to 15mm long, which tunnel into apple fruitlets; in Europe, the PLUM SAWFLY (*H. flava*) has similar larvae which bore into plums and damsons in spring and early summer. In North America, the CHERRY FRUIT SAWFLY (*H. cookei*) has yellow-white larvae, up to 6mm long, which enter the fruits of cherries and plums and occasionally those of peaches and apricots.

Small infestations of sawflies can be removed by hand-picking. Otherwise, caterpillars of open-feeding species should be controlled by applying recommended insecticides. Species with larvae that become hidden must be killed at the adult stage before egg-laying; for example, preventative sprays should be applied against the leaf-rolling sawfly in late spring. See *slugworms*.

apple sawfly
and larva

Saxegothaea (after Prince Albert of Saxe-Coburg-Gotha (1819–61), Prince Consort). Podocarpaceae. Chilean Andes, Patagonia. 1 species, *S.conspicua*, PRINCE ALBERT'S YEW, a coniferous evergreen tree or shrub to 20m or more. The bark is brown, suffused grey and flaking; the branches are whorled with pendulous tips. The leaves are linear to lanceolate, 1.5–3cm long, coriaceous and dark green above with blue-white bands beneath, and a spiny tip. To 1cm long, the mature female cones are blue-grey to glaucous brown. Hardy to –15°C/5°F. Plant on a moist, acid soil in a sheltered, humid environment.

saxicolous growing on or among rocks.

Saxifraga (from Latin *saxum*, rock, and *fragere*, to break: under the doctrine of the signatures, saxifrages with kidney-shaped leaves were used as a cure for kidney or bladder stones). Saxifragaceae. SAXIFRAGE. Europe, Asia, N America, mainly in arctic and mountainous; extending into S America via the Andes. Some 370 species, dwarf or small annual to perennial herbs (those here perennial or monocarpic). Many are alpines. They range widely in habit from miniature cushion plants to carpeters, rosette-formers and loose clumps. The leaves may be needle- or scale-like, or toothed and lobed, tongue- or strap-shaped, or larger and broader. The flowers are solitary or carried in cymes, racemes or panicles. They consist usually of 5 spreading, rounded petals.

All of the following are hardy in climate zone 6. For many, the greatest threat is not cold, but overwet conditions. One exception to this is *S.fortunei*, a handsome foliage plant that favours mild climates and damp, acid soils. It associates well with sedges, hostas and ferns. Plant form and habit reflect natural habitat: the cushion habit generally relates to species occurring in the arctic and high mountain ranges, whilst those with softer, wide-spreading foliage mats are typical of the more sheltered montane conditions that prevail in areas of late snow melt. Taller, leafy-stemmed species are generally found in meadow, in open woodland at woodland margin. The presence of lime-encrustation, secreted by ducts or hydathodes on the leaf surface, usually indicates a plant with a preference for very rocky or gritty, rather dry and alkaline conditions. In terms of cultural requirements saxifrages be can be broken down into four groups -1. plants needing protection from fierce sun and a moist soil; 2. plants suitable for rock gardens and scree, needing some shade and a well-drained soil; 3. plants suitable for well-drained pockets in the rock garden, sink gardens, alpine troughs and pans in the alpine house, needing shade from fierce sun and moist but never wet soil; 4. plants suitable for the rock garden, sink gardens, alpine troughs and pans in the alpine house, needing full sun and impeccably drained, alkaline soil.

Increase by division, by cuttings of non-flowering rosettes, by soft tip cuttings of *S.oppositifolia* and by seed (the only practicable means for monocarpic species such as *S.longifolia*). Seed is surface sown when ripe, or in spring after stratification.

SAXIFRAGA

Name	Distribution	Habit	Leaves	Flowers
S.aizoides YELLOW MOUNTAIN SAXIFRAGE, YELLOW SAXIFRAGE	Iceland, Scandinavia, N Britain, Alps, Apennines	7–20cm, forming loose cushions	4–22mm, linear to oblong, fleshy, ciliate, covered in a thin, chalky crust	yellow, often spotted orange-red, sometimes orange or brick red, in hairy, stalked racemes

Comments: Z5. Group 1. Grow in moist soil; protect from harsh sunlight. var. *atrorubens*: flowers dark red.

S. × apiculata (*S.marginata × S.sancta*)	garden origin	densely cushion-forming, to 10cm.	to 12mm, linear to lanceolate, bright green	yellow in small clusters on short, erect stems

Comments: Z6. Group 2. Grow in a moist but well-drained soil on the rock garden, in pans or troughs and sinks. Protect from burning sun. 'Alba': flowers white; 'Gregor Mendel': leaves glossy pale green; flowers pale yellow. 'Primrose Bee': flowers large, pale yellow.

S. × arco-valleyi (*S.lilacina × S.marginata*)	garden origin	forms small, tight and domed cushions	to 10mm, oblong to linear, silver-grey	cup-like, pale pink to bright red, solitary, short-stalked

Comments: Z6. Group 3. Grow on the rock garden and in alpine sinks and pans, on a moist, gritty soil; protect from scorching sun. 'Arco': pale pink. 'Dainty Dame': bright rose lilac fading to white flushed pink. 'Ophelia': white. 'Silver Edge': leaves grey-green edged white; flowers pale lilac.

S. × borisii (*S.ferdinandi-coburgi ×S.marginata*)	garden origin	cushion-forming, to 8cm high	to 10mm, linear, grey-green or dark green	usually pale yellow, sometimes white or rich yellow, large, in red-tinted cymes

Comments: Z6. Group 3. Grow in a gritty soil on the rock garden or in alpine sinks and pans; keep damp and protect from burning sun

S. × boydii (*S.aretioides × S.burseriana*)	garden origin	low-growing, forming tight cushions to 8cm	linear to lanceolate, silver-green	vivid yellow, solitary or a few on short, red-tinted stems

Comments: Z6. Group 3. For the rock garden or alpine sink. 'Valerie Finnis' syn. 'Aretiastrum'; cushions small, dense; flowers deep sulphur-yellow.

S.brunonis *S.brunoniana*	Himalaya	deciduous or semi-evergreen, mat-forming, with thread-like, red runners	lanceolate, rigid, soft green, in rosettes.	yellow, in panicles to 15cm

Comments: Z6. Group 1. Grow in a moist soil; protect from harsh sun.

S.bryoides MOSSY SAXIFRAGE	Europe (mountains)	forms dense mats of spherical leaf rosettes	3.5–5mm, oblong to lanceolate, pointed, hairy	white spotted orange, with orange stain at centre, solitary on short stems

Comments: Z6. Group 4. Rock gardens and alpine sinks; gritty soil, full sun

S.burseriana BURSER'S SAXIFRAGE	E Alps	stems numerous, woody-based, forming dense cushions	6–12mm, packed in low cones, narrowly lanceolate, silver-grey, with a thin, chalky crust	solitary, large, white on short, erect, hairy red stems

Comments: Z6. Group 3. Rock gardens, alpine sinks and pans on a moist, gritty soil with full sun. 'Crenata': petals crenate. 'Gloria': flowers large, white with yellow centres on bright red stalks.'His Majesty': rosettes silver-grey; flowers large, rounded, pink, flushed white.

S.callosa *S.lingulata* LIMESTONE SAXIFRAGE	NE Spain, SW Alps, Apennines	forms tight leaf rosettes to 16cm diameter, each dying after flowering but replaced by new growths on short stolons	4–7cm, linear, covered in a chalky grey-white crust	white, sometimes spotted red, many borne together in erect to arching panicles to 40cm long

Comments: Z6. Group 4. Grow in crevices and ledges on the rock garden and in alpine sinks; needs full sun and a fast-draining, alkaline soil. 'Alberti': rosettes silver, inflorescence large, white. 'Leichtlinii': flowers rose-red. 'Rosea': flowers rose-red. 'Superba': flowers large, in arching, plume-like inflorescence; cream-white.

S.cochlearis	Maritime Alps	leaf rosettes numerous, 1.5–8cm diameter, forming a dense cushion	to 4.5cm, spathulate to oblanceolate, fleshy, tough, grey-green, chalk-encrusted	white, sometimes spotted red, many together in loose panicles to 30cm

Comments: Z6. Group 4. For the rock garden and alpine sinks and troughs; full sun on an alkaline, gritty soil. 'Major': flowers larger. 'Minor': tight silvery hummocks

S.cotyledon GREAT ALPINE ROCKFOIL, GREATER EVERGREEN SAXIFRAGE	Scandinavia and Iceland, Pyrenees, Alps	leaf rosettes 7–12cm diameter, with daughter-rosettes on short runners	2–8cm, oblong to oblanceolate, margins toothed, covered in a chalky crust	white, sometimes spotted or veined red, many in broadly pyramidal panicles

Comments: Z6. Group 2. For the rock, scree or gravel garden

S.cuneifolia SHIELD-LEAVED SAXIFRAGE	Mountains of S and SC Europe	stems prostrate, loosely carpeting with leaves in rosettes	1–3cm, broadly obovate to cuneate, fleshy, glossy bright green above, red-purple beneath, toothed	white, often with yellow basal patch and sometimes red spots, in loose panicles to 25cm

Comments: Z5. Group 1. Will thrive in most situations but prefers a moist, fertile soil and protection from harsh sun. 'Variegata': leaves variegated creamy white to yellow

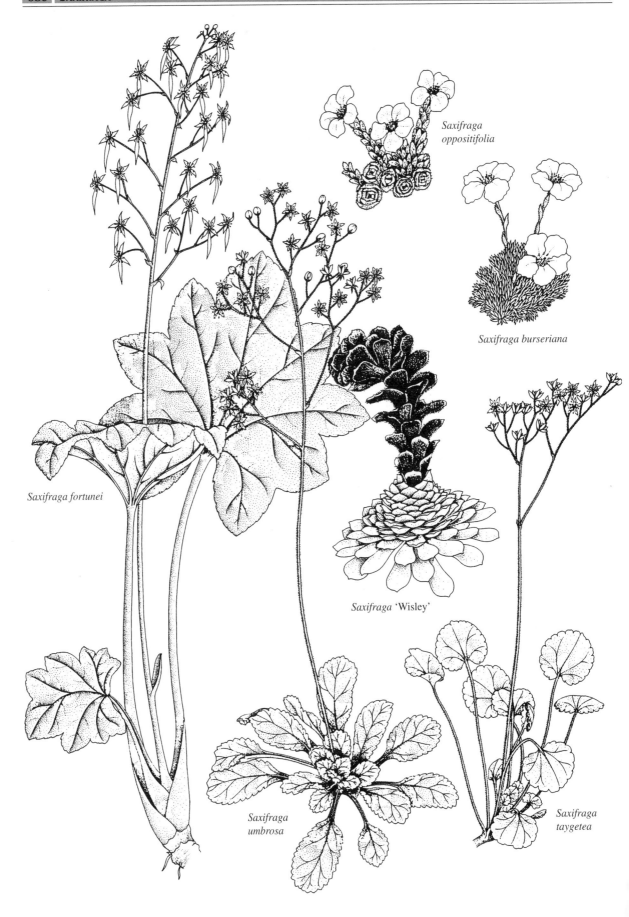

Saxifraga oppositifolia

Saxifraga burseriana

Saxifraga fortunei

Saxifraga 'Wisley'

Saxifraga umbrosa

Saxifraga taygetea

SAXIFRAGA

Name	Distribution	Habit	Leaves	Flowers
S.exarata FURROWED SAXIFRAGE	C & S Europe, Caucasus	shoots leafy, to 4cm, forming dense, spreading cushions	to 10mm, reniform, deeply divided into 3–5, pointed and furrowed lobes, bright green	white to dull cream, sometimes tinted red, 2–5 together on short, erect stalks

Comments: Z6. Group 1. Grow on a moist but well-drained soil with protection from harsh sun. In ssp. ***moschata*** (syn. ***S.moschata***), the leaf segments are not furrowed and the flowers are cream, sometimes tinted red. 'Cloth of Gold': leaves solid, bright gold – best in light shade

S.ferdinandi-coburgi	E Macedonia	leaves crowded, forming dense cushions	4–8mm, linear, pointed, covered with thick chalk crust	bright yellow, 3–5 together on a hairy, red-tinted stem 3–12cm tall

Comments: Z6. Group 3. For the rock garden and alpine sinks and pans in moist, gritty soil with protection from scorching sun

S.fortunei	Japan	loosely clump-forming	15cm, rounded, toothed, glossy and thinly fleshy, apple green tinted bronze or copper and slender stalks to 20cm long	white with one or two lower petals far longer than upper petals, in a loose, airy, red-tinted panicle to 50cm tall

Comments: Z7. Group 1. Grow in dappled sun to shade on a moist, neutral to acid soil rich in leafmould – a fine companion for sedges, ferns, small bamboos and hostas in the damp or woodland garden. 'Rubrifolia': leaves tinted coppery red; 'Wada's Variety': leaves beetroot red.

S.frederici-augusti	Balkan Peninsula	forms dense rosettes of spreading leaves	12–35mm, oblanceolate, grey-green covered in chalky crust	bright red to deep maroon, more or less concealed by pink to wine red sepals and leafy bracts in erect then nodding, hairy raceme to 10cm tall

Comments: Z7. Group 4. For the rock garden or alpine trough in gritty, alkaline soil. ssp. ***grisebachii***. (syn. ***S.grisebachii***) inflorescence cherry red to maroon. 'Wisley': more vigorous and brightly coloured with the inflorescence bracts deep purple red and tipped bright green.

S. × geum (***S.hirsuta*** × ***S.umbrosa***)	garden origin	forms loose mats of spreading leaf rosettes	2–6cm, broadly oblong to spathulate, crenulate, hairy	small, pink in bud, opening white spotted pink, in loose, frothy, red-tinted panicles to 18cm tall

Comments: Z6. Group 1. Grow in dappled sun or part-shade in a moist, fertile soil

S.granulata MEADOW SAXIFRAGE, BULBOUS SAXIFRAGE, FAIR MAIDS OF FRANCE	Europe	rosette-forming, bulbiferous	0.6–2cm, reniform to cordate, toothed, glandular-hairy, slightly fleshy, short-stalked	white, sometimes red-veined in few-branched panicles 10–30cm tall

Comments: Z5. Group 1. Grow in a moist soil; protect from harsh sun. 'Florepleno': flowers sterile with many petals.

S.hirsuta KIDNEY SAXIFRAGE	SW Europe and Iceland	loosely mat-forming with sprawling and rooting stems	1.5–4cm, reniform to rounded, crenate to bluntly toothed, soft pale green, stained purple-red beneath, slender-stalked	small, white, faintly spotted pink in loose, narrow, red-tinted panicles 12–40cm

Comments: Z5. Group 1. Grow in sun or shade on a moist, fertile soil.

S.juniperifolia	Bulgaria	stems short, columnar with densely packed leaves, forming dense cushions	9–13mm, linear, pointed, persisting long after death	bright yellow, 3–8 on hairy stems 3–6cm long

Comments: Z6. Group 3/4. For the rock garden and alpine sink, on a gritty soil in sun

S. × kelleri	garden origin	leaf rosettes cushion-forming	17mm, cuneate, grey-green, tightly packed	pink, 1–3 together on short stalks

Comments: Z6. Group 3. For the rock garden or alpine sink in sun on a gritty soil

S.longifolia PYRENEAN SAXIFRAGE	Pyrenees	leaves form a single, broad rosette, dying after flowering and producing no daughter rosettes	6–11cm, linear, fleshy, pale grey-green, with a thick chalk crust	small, white, sometimes spotted red, packed in a cylindrical to conical panicle to 60cm

Comments: Z6. Group 4. For the rock garden or alpine sink in full sun on a gritty, alkaline soil. Increase by seed. 'Magnifica': rosettes to 15cm diameter; panicle very large and showy. 'Walpole's Variety': smaller, blue-grey rosettes; flowers in short white spikes.

S.oppositifolia PURPLE SAXIFRAGE	Arctic, high mountains of Europe, Asia and N America	stems short, woody, freely branched, forming dense mats and cushions	0.5–0.8cm, oblong, blunt, bristly, rigid, dark dull green, chalky	solitary, short-stalked, pink to deep purple fading to white

Comments: Z5. Group 1. For the rock garden and alpine sinks, in an open, sunny position on a gritty, acid soil. Includes cultivars with white, pale rose, pink, crimson and deep magenta flowers, some with silvery foliage

S.paniculata ***S.aizoon*** LIFELONG SAXIFRAGE	Pyrenees, SW Alps north to Lapland and Iceland	mat-forming	to 4cm, oblong to lanceolate, in hemispherical rosettes, margins inrolled, chalky	creamy-white, sometimes red-spotted in panicles to 30cm

Comments: Z5. Group 4. For the rock garden and alpine troughs; grow in full sun on a gritty, alkaline soil. Includes cultivars with pure white, pink-spotted, rose and cream to lemon yellow flowers, some with fine, silvery foliage

SAXIFRAGA

Name	Distribution	Habit	Leaves	Flowers
S.rosacea	NW and C Europe	variable, low-growing, stems short and forming dense, compact cushions or longer and trailing in mats with dead leaves persistent	1–2cm, obovate to cuneate, bright green to red-tinted, usually cut into 5 or more pointed segments	white to pink or dark red, 2–5 on erect, slender scapes to 6cm.
Comments: Z5. Group 1. A very popular garden plant, tolerant of most conditions; it includes many forms and hybrids. Similar to *S.exarata*.				
S.rotundifolia ROUND-LEAVED SAXIFRAGE	Europe (SW France to Black Sea and Aegean), Caucasus	rhizomatous, forming loose, spreading rosettes	1.5–4.5cm, rounded, base deeply cordate, smooth to hairy, crenate to sharp-toothed or palmately lobed, slender-stalked	white, usually spotted crimson-purple, shading to orange-yellow at base in loose, spreading panicles 15–70cm
Comments: Z6. Group 1. Grow in light shade or dappled sun on a moist, fertile soil				
S.sancta	NE Greece	differs from *S.juniperifolia* in its shorter leaves (to 10mm), and glabrous inflorescence		
Comments: Z6. Group 3/4.				
S.stolonifera MOTHER OF THOUSANDS, STRAWBERRY GERANIUM	China, Japan	rosette-forming, spreading by long, thin, red stolons	to 6cm, round, mid-green with silvery veins above, flushed red beneath, bristly, long-stalked	white, lower petals longer than upper, in loose panicles to 40cm
Comments: Z6. Group 1. Grow in shade or dappled sun in a moist, fertile soil. A good groundcover plant more commonly used as a houseplant and for hanging baskets. *S.*'Tricolor' ('Magic Carpet'): leaves variegated green, pink, red and silver				
S.taygetea	SW Greece, Balkans	forms loose, spreading rosettes, similar to *S.rotundifolia*	5-13mm, rounded, crenate, smooth to downy, mid-green, slender-stalked	white spotted pink or purple-red, in loose panicles to 20cm
Comments: Z6. Group 1.				
S. 'Tumbling Waters'	garden origin	mat-forming with tight low rosettes	to 8cm, narrow, with a chalky crust	white, crowded in an arching, conical panicle to 50cmlong
Comments: Z6. Group 4. For ledges and crevices in the rock garden or alpine sink in sun and on gritty alkaline soil. A spectacular cultivar, each rosette dies after flowering but is replaced by offsets				
S.umbrosa	Pyrenees	stems short, bearing leaf rosettes forming loose cushions	to 3cm, oblong to obovate, crenate, tough to thinly fleshy, smooth, ciliate, short-stalked	white spotted crimson in loose, red-tinted panicles to 35cm
Comments: Z5. Group 1. Will grow in sun or shade on most soil. *S.*'Clarence Elliott': dwarf compact; flowers very small, rose-pink. 'Primuloides': to 15cm; leaves primrose-like; flowers rose-pink				
S. × *urbium* (*S.umbrosa* × *S.hirsuta*) LONDON PRIDE	garden origin	like *S.umbrosa*, but petioles longer and leaves more strongly toothed	white spotted red-pink or flushed pink, in panicles to 25cm	
Comments: Z5. Group 1. Good low groundcover, tolerant of most conditions.				

scab used of various fungal diseases, the most damaging of which occur on apple, pear and potato. Scab characteristically shows as a discrete superficial lesion, involving localized severe roughening, pitting or cracking, and abnormal thickening of the surface layers, with or without the development of cork.

APPLE SCAB (*Venturia inaequalis*) and PEAR SCAB (*Venturia pirina*) are restricted to their individual fruit hosts but are very similar in life cycle and the damage caused. The skin of infected young fruits develops blackish brown blotches, and in severe cases fruits may be completely covered, with cracks developing and resultant misshapen young apples; the disease is superficial. Infected leaves show brown or olive green spots and may fall prematurely, reducing growth and also fruiting in the following season; ornamental *Malus* leaves are particularly affected. The fungi overwinter in scabby shoot pustules and in special fruiting bodies on fallen leaves. Disposing of fallen leaves contributes to control, as does regular pruning for improved air circulation and the removal of infected shoots. Regular sprays of recommended fungicide provide effective control. Amongst resistant apple cultivars are 'Discovery', 'Lane's Prince Albert' and 'Sunset'; resistant pear cultivars include 'Jargonelle' and 'Catillac'.

COMMON POTATO SCAB (*Streptomyces scabies*) causes unsightly, generally raised, scabby patches on the surface of tubers; different strains probably cause different forms of scabbing. The corky scabs largely prevent infection spreading into the tuber flesh, and because the skin can be peeled for cooking, the disease is tolerable. Common scab occurs predominantly on light soils, especially when these remain dry during tuber initiation, and is encouraged by high lime content. Fungicide treatment is not appropriate, and attention should be paid to maintaining soil organic-matter content and to irrigation. Amongst resistant cultivars are 'Arran Pilot' 'King Edward' and 'Maris Peer', while those susceptible include 'Desiree', 'Majestic' and 'Maris Piper'.

scaberulous finely or minutely scabrous.

Scabiosa (from Latin *scabies*, itch, a disease these plants were thought to cure). Dipsacaceae. Europe, Asia, Africa, mainly Mediterranean. PINCUSHION FLOWER, SCABIOUS. 80 species, annual or perennial herbs with flowers in slender-stalked hemispherical, heads subtended by involucral bracts. The following species are hardy in climate zone 6. Grow in sun in any moderately fertile, well-drained soil, preferably circumneutral or alkaline. Deadhead to prolong flowering and lift and divide perennials every third or fourth year to maintain vigour. *S.graminifolia* should be given a warm, sunny position in the rock garden with excellent drainage. Propagate by seed. Perennials may also be increased by division or by basal cuttings in spring.

S.atropurpurea (MOURNFUL WIDOW, SWEET SCABIOUS, PINCUSHION FLOWER, EGYPTIAN ROSE; South Africa; annual 60–90cm; flowers dark purple to deep crimson, fragrant, in heads to 5cm in diameter; includes 'Blue Cockade', to 90cm, with double, globe-shaped, rich blue, fragrant flowers, 'Chile Black', with darkest purple-black flowers, 'Grandiflora', with large flowerheads, and 'Nana', slightly dwarf and much-branched); *S.caucasica* (Caucasus; perennial to 60cm; flowers pale blue, in heads to 7.5cm in diameter; includes wide range of cultivars and seed races – tall or short, with flowers in shades of white, cream, lavender and blue, good for cutting); *S.columbaria* var. *ochroleuca* (syn. *S.ochroleuca;* Europe, W Asia; biennial or perennial to 75cm; flowers primrose yellow to creamy white in heads to 6cm in diameter); *S.graminifolia* (S Europe; perennial to 40cm forming loose mats; leaves grass-like; flowers blue-violet to rose in heads to 4cm in diameter); *S.lucida* (C Europe; perennial to 30cm; flowers rose-lilac to pale pink in dense heads to 4cm in diameter); For *S.rumelica*, see *Knautia macedonica*.

scabrid a little scabrous, or rough.

scabridulous minutely scabrid.

scabrous rough, harsh to the touch because of minute projections, scales, tiny teeth or bristles.

Scadoxus (name derived from Greek *skiadion*, parasol (i.e. umbel), and *doxa*, glory). Amaryllidaceae. Tropical Arabia and Africa to S Namibia and E Cape. 9 species bulbous or rhizomatous herbs differing from

Haemanthus in their leaves, which are ovate to lanceolate and not 2-ranked. Cultivate as for *Haemanthus*, maintaining slightly higher temperatures and humidity. *S.multiflorus* (syn. *Haemanthus coccineus*, *H. multiflorus*; tropical and South Africa, Yemen; scape to 75cm, spathe valves colourless or tinged red, soon withering, umbel more less spherical, flowers scarlet, fading pink; subsp. *katherinae*: syn. *Haemanthus katherinae*, flowers larger); *S.puniceus* (syn. *Haemanthus magnificus*, *H. natalensis*, *H.puniceus*; ROYAL PAINT BRUSH; E and South Africa; scape to 40cm; spathe valves red, umbel conical, flowers scarlet to green-yellow).

scaffold branch a main structural branch of a tree used to support renewable fruiting laterals or spurs.

scald damage caused to plant tissues by unsatisfactory atmospheric temperature or humidity, or toxicity from pollutants produced by paraffin or gas burners. Sun scald is a variation where leaves, stems and fruits may be damaged as a result of sudden exposure to hot sunshine, especially in cases where the plant surface is wet. Damage takes the form of chlorotic or brown blotches or spots on leaves and shoots, discolouration of bark, shrivelling of fruits, as occurs with grapes (cf. *shanking*), and pale areas on fruits of tomato and sometimes apple and plum. The condition can be particularly troublesome in plants grown in greenhouses, and attention should be paid to shading and watering.

Sun scorch is separately identified, referring more to the withering of shoots as a result of exposure to hot sunshine.

scale see *trichome*.

scaling see *cuttings*.

scale leaves specialized, scale-like leaves, including those covering buds or forming bulbs.

scale insects (Hemiptera: Coccoidea) sap-sucking insects, some of which excrete copious amounts of honeydew, which encourages the growth of sooty moulds. Adult males are winged, with well-developed antennae and legs; the females are wingless, having scale-like bodies without well-defined divisions, and legs and antennae absent or reduced. The female lays eggs once in her lifetime; these are protected by a white, cottony, waxy egg sac (as in mealybugs), by the dead body of the female, or, more typically, by a waxy scale. The first-stage nymph, known as a crawler, has functional legs and is often the only means of natural dispersal of the species; once its stylets are inserted, the subsequent nymphal stages and the adult female move infrequently (mealybugs, however, retain their mobility).

The GREENHOUSE ORTHEZIA (*Orthezia insignis*) attacks a wide range of greenhouse plants in Europe, North America and Australia, and the mobile nymphs and adults resemble mealybugs; the females carry conspicuous white egg sacs attached to their abdomens. The COTTONY CUSHION or FLUTED SCALE (*Icerya purchasi*), so-called because of its conspicuous longitudinally grooved white egg sacs, which can be up to 10mm long, is a serious pest of citrus and also attacks many other plants, including acacia, cytisus, mimosas and roses. The BEECH SCALE (*Cryptococcus fagisuga*) in Europe and North America is a widespread pest of beech, where it can be identified by the conspicuous, white, waxy powder on the trunks and branches; it can predispose the trees to infection by fungi (*Nectria* species).

Other scale insects of importance have either soft or armoured scales. The soft scales have a waxy covering, which remains part of the body wall and cannot be separated from it; some species are large and rounded, others are broad and flattened. The soft scales common to Europe and North America include BROWN SCALE (*Parthenolecanium corni*), known as the EUROPEAN FRUIT SCALE in the US, which is up to 5mm long, with brown oval convex scales. It infests peach, nectarine, currants, grape vines and many ornamental plants. The SOFT SCALE (*Coccus hesperidum*) is up to 4mm long, with flat, elongate-oval scales. It usually lies alongside the main vein on the underside of the leaves of abutilons, bay laurel, camellias, *Citrus*, *Clematis*, escallonias, ferns, *Ficus*, *Hibiscus*, hippeastrums, hollies, ivies, oleanders, poinsettias, *Stephanotis* and many other ornamental plants. The HEMISPHERICAL SCALE (*Saissetia coffeae*), up to 4mm long, with a hemispherical shell, infests asparagus ferns, begonias, carnations, clerodendrons, codiaeums, cycads, ferns, *Ficus*, oleanders, orchids, *Stephanotis* and other ornamental plants. Other unarmoured scale insects in North America include the TERRAPIN SCALE (*Mesolecanium nigrofasciatum*), a dark red-brown, hemispherical scale, which attacks fruits and shade trees; the BLACK SCALE (*Saissetia oleae*), up to 6mm, with large black or dark brown rounded scales, which is a serious pest of citrus, olive, grapefruit and greenhouse plants; and the CITRICOLA SCALE (*Coccus pseudomagnoliarum*), with flattish scales, at first transparent but later darkening, which also attacks *Citrus*.

Some species of unarmoured scales are conspicuous because of their 'woolly' egg sacs; these scale insects are known as WOOLLY SCALES (*Pulvinaria* species). The most widespread in Europe are the HORSE CHESTNUT SCALE (*P. regalis*), on horse chestnuts, limes, elms, sycamores, magnolias, maples, *Cornus* and other ornamental plants; the WOOLLY VINE SCALE (*P. vitis*) on currants and gooseberries; and the CUSHION SCALE (*Chloropulvinaria floccifera*) on camellia and holly. The COTTONY PEACH SCALE (*P. amygdali*), on orchard trees, and the COTTONY MAPLE SCALE (*P. innumerabilis*), on maples and other trees and shrubs, are widespread in North America.

The armoured scales have a hard, waxy scale-like covering, which can be separated from the body, and several species are common to Europe and North America. The MUSSEL SCALE (*Lepidosaphes ulmi*),

known as the OYSTERSHELL SCALE in the US, is up to 3mm long, with mussel-shaped blackish-brown scales. It infests the woody stems and branches of *Ceanothus*, cotoneasters, ornamental and fruiting apples and pears, roses, and various other trees and shrubs. The OYSTERSHELL SCALE (*Quadraspidiotus ostreaeformis*), known as the EUROPEAN FRUIT SCALE in the US, is found on the bark of a wide range of trees and shrubs, especially apple, pear, peach, plum, birch, poplar and horse chestnut. The CAMELLIA SCALE (*Hemiberlesia rapax*), known as the GREEDY SCALE in the US, has circular scales, up to 2mm across. It infests a wide range of plants, including begonias, camellias, citrus, deciduous fruits, cacti and orchids. The OLEANDER SCALE (*Aspidiotus nerii*) has circular scales, up to 2mm across with a yellow centre and pale edges. It infests the underside of leaves of many plants, including acacias, ornamental asparagus, aucubas, azaleas, pot *Cyclamen*, dracaenas, ericas, fatsias, *Ficus*, laurels, oleanders and palms. The EUONYMUS SCALE (*Unaspis euonymi*), with female scales similar to those of mussel scales, infests the woody stems of *Euonymus japonica*; the SCURFY SCALE (*Aulacaspis rosae*), known as the ROSE SCALE in the US, has round, flat, opaque white scales, up to 3mm across, and forms dense colonies on the woody stems of certain species of rose. In North America, there are several other species of armoured scales, particularly on citrus and deciduous fruits.

Scale insects are relatively immobile and most garden infestations are introduced on plant material; all newly acquired plants should therefore be inspected carefully. Recommended systemic insecticides control most infestations, while scale insects on deciduous trees and shrubs can also be controlled by applying appropriate winter washes in the dormant season.

scandent climbing.

scape an erect, leafless stalk arising at or near the base of the plant and bearing a terminal inflorescence or solitary flower; such inflorescences are described as *scapose*.

scarification (1) the removal of thatch and moss from a lawn using a spring-tined rake or a special tool, known as a scarifier, which has short metal, crescent-shaped teeth or sturdy brushes; see *lawns*; (2) mechanical or chemical treatment of hard-coated seeds, such as those of the family Leguminosae, before sowing to permit imbibition and improve germination. See *seeds and seed sowing*.

scarious thin, dry and shrivelled; consisting of more or less translucent tissue.

Schefflera (for J.G. Scheffler (1722–1811), physician in Danzig). Araliaceae. Warmer regions of the world, with the greatest numbers of species in the Americas and S and E Asia to the Pacific. IVY TREE, SCHEFFLERA, UMBRELLA TREE. Over 700 species, evergreen trees and shrubs, some climbers, some epiphytic. The leaves are usually palmately compound; the flowers are small, green to creamy white or red and borne in umbels on massive panicles. Popular foliage plants cultivated in the intermediate to hot greenhouse or as house plants. Large specimens of *S.actinophylla* are common in offices and public buildings, where they adapt well to hydroponic cultivation. *S.elliptica* and *S.pueckleri* are sold as young rooted leads trained on moss poles. *S.elegantissima* (usually sold in its juvenile form) is the well-known houseplant *Dizygotheca*.

Maintain a minimum winter temperature of 13°C/55°F. Give strong, filtered light all year round. Under glass, maintain medium to high humidity, syringing plants 2–3 times daily in hot weather. When grown as house plants, sponge the foliage occasionally to remove dust and stand in trays of damp pebbles or in groups to create a humid microclimate. Water moderately, allowing the medium to become slightly dry between waterings. Feed monthly during the growing season with a dilute liquid fertilizer. Plants tolerate hard cutting back. Sow ripe seed in a humid atmosphere at 20–24°C/ 70–75°F in autumn. Alternatively, propagate by air-layering or by sectional stem cuttings and softwood cuttings rooted with bottom heat in spring and summer.

S.actinophylla (AUSTRALIAN IVY-PALM, OCTOPUS-TREE, UMBRELLA-TREE; Australia, New Guinea; erect, glabrous shrub or small tree to 12m; leaflets to 7–16, to 30 × 10cm, oblong, entire, stalked, bright green and glossy above, obtuse; juvenile leaflets fewer, smaller, toothed, acute; includes variegated cultivars); *S.arboricola* (Taiwan; glabrous epiphytic shrub or liane; leaflets 7–11, obovate, entire, to 11 × 4.5cm, stalked, bright green and semi-glossy above, obtuse; juveniles differ in leaflets with short, widely spaced teeth; includes variegated cultivars); *S.digitata* (New Zealand; spreading shrub or tree to 8m, glabrescent; young branches red-purple; leaflets 7–10, thin, to 20 × 6cm, narrowly obovate, shortly stalked, sharply toothed, leaflets in juvenile plants 5, more or less pinnatifid, lobes toothed, pink beneath); *S.elegantissima* (syn. *Dizygotheca elegantissima*; FALSE ARALIA, PETIT BOUX CALEDONIEN; New Caledonia; glabrous, sparsely branching trees to 15m with dark bark; juvenile stage to 2m, unbranched; stem and petioles dark green spotted white; leaflets 7–11, linear, to 23 × 3cm, coarsely serrate, dark green to chocolate brown above, dark brown-green beneath, midrib white; adults with leaflets oblong, to 25 × 8cm, stalked, obtuse, broadly toothed, sinuate or entire); *S.pueckleri* (S and SE Asia; erect, glabrescent, spreading tree to 18 × 0.6m or scandent; leaflets 5–10, narrowly oblong to 25 × 8cm, resembling *S.actinophylla* but upper surface darker and glossier, yellow-green beneath, always entire; variegated cultivars are grown).

Schima (name said to be derived from the Arabic). Theaceae. S Asia. 1 species, *S.wallichii* (syn. *S.argentea*, *S.superba*), a frost-tender, summer-flow-ering evergreen tree to 40m tall. The leaves are 7–24 cm long, elliptic to oblong, entire or wavy. Red-purple at first, they become dark green and thinly coriaceous. Scarlet in bud, the fragrant flowers are to 6.5cm across and open purple-cream to white, with small, circular petals and many yellow stamens. Grow in a freely draining, humus-rich, lime-free soil, in full sun or light dappled shade, in a position sheltered from cold drying winds. Under glass, water plentifully when in full growth, at other times moderately; maintain a winter minimum temperature of 3–5°C/37–40°F. Propagate by seed sown ripe, or by semi-ripe cuttings in summer, rooted in a closed case with bottom heat.

Schinus (from Greek *schinos*, name used for *Pistacia lentiscus*). Anacardiaceae. S America, Uruguay to Mexico, naturalized in S US, Canary Islands, China. Around 30 species, evergreen, resinous shrubs and trees with pinnate or simple leaves and small flowers massed in panicles in summer. Where plants of both sexes are grown together, the females sometimes bear colourful, pea-sized berries. The graceful, weeping *S.molle* is widely planted in more or less frost-free climates for shade and ornament. *S.terebinthifolius*, used as a street tree and lawn specimen in warm climates, is of more rigid aspect, and makes a roundheaded tree, its fruit clusters showy in winter. Plant both in well-drained soil and full sun with wall protection. Increase by seed in spring or by semi-ripe cuttings.

S.molle (PEPPER TREE, PERUVIAN MASTIC TREE, MOLLE; Brazil, Uruguay, Paraguay, N Argentina; tree to 15m; branches slender, pendulous; leaves 10–30cm, leaflets lanceolate, serrate; flowers pale yellow; fruit rose-pink); *S.terebinthifolius* (BRAZIL-IAN PEPPER TREE, CHRISTMAS BERRY TREE, AROEIRA; Venezuela to Argentina; shrub or small tree to 7m; leaves 10–17cm, leaflets 3–6cm, oval-lanceolate to obovate; flowers white; fruit red).

Schisandra (from Greek *schizo*, to split, and *aner*, man, referring to the distinctly separated anther cells). Schisandraceae. E Asia, Eastern N America. 25 species, deciduous or evergreen climbing shrubs with twining stems and oblong to lanceolate leaves. Produced in spring, the flowers are small, cup-shaped and composed of numerous tepals. Where plants of both sexes are cultivated together, the female flowers develop into slender, pendulous spikes bearing bright red, berry-like fruit. The following are hardy in climate zone 7, but may need the protection of a south- or west-facing wall, or of larger trees and shrubs in areas that experience prolonged frosts and harsh winds. Grow in any fertile, well-drained but moisture-retentive soil with some protection from intense sunlight; slightly acid soils are preferred. Topdress annually with organic material, and water plentifully during dry periods, particularly wall-grown specimens. Walls and trellis apart, these shrubs will also scramble through trees and shrubs and over large rocks. Prune in late winter or spring to remove dead wood and to train or limit to the allocated space. Sow seed in cold frames

Schisandra grandiflora

in the autumn or root semi-ripe cuttings in summer. Alternatively, layer long shoots during the autumn, severing plants from parent material the following year.

S.chinensis (China; flowers to 1.25cm in diameter, pale rose to bright pink, fragrant); *S.coccinea* (BAY STAR VINE, WILD SARSPARILLA; SE US; flowers to 1cm in diameter, crimson); *S.grandiflora* (N India, Bhutan, Nepal; flowers 2.5cm in diameter, fragrant, white, cream or pale rose; var. *cathayensis*: flowers rose pink; var. *rubriflora*: syn. *S.rubriflora*, flowers deep scarlet); *S.henryi* (C and S China; flowers white); *S.propinqua* (C and W China, Himalaya; flowers to 1.5cm in diameter, orange-red; var. *sinensis*: flowers smaller, somewhat yellow, cold-resistant).

schist rock composed of layers of different minerals, which splits into thin plates.

Schizanthus (from Greek *schizo*, to cut, and *anthos*, flower, referring to the deeply cut corolla). Solanaceae. Chile. POOR MAN'S ORCHID, BUTTERFLY FLOWER. 12 species, glandular-hairy annual herbs with pinnately cut leaves and cymes of showy, 5-lobed and 2-lipped flowers. For outdoor cultivation, sow *in situ* in spring in a sunny, sheltered spot on light, well-drained, fertile soils. Alternatively, sow under glass 7–8 weeks before planting out; harden off before planting out. For pot culture, sow at 13–15°C/55–60°F from autumn to spring. Grow on in a medium-fertility, loam-based mix, potting on regularly. Give cool, airy conditions in full sun and ventilate when temperatures rise above 10°C/50°F. Water sparingly in winter, plentifully in spring and summer. *Schizanthus* is prone to a number of virus diseases, in particular tomato spotted wilt, which causes severe bronzing and wilting of the upper leaves; control aphid infestations and destroy affected plants. *S.pinnatus* (annual 20–50cm; flowers violet to pink, middle upper lip with yellow throat with violet markings; there are many cultivars, hybrids and seed races tall or dwarf plants with large flowers in a range of colours: white, flesh-pink, golden-yellow centred, shades of crimson, amber, pink and mauve to mauve-blue).

schizocarp a dry, usually dehiscent syncarpous ovary which splits into separate, one-seeded halves, at maturity.

Schizopetalon (from Greek *schizo*, to cut, and *petalon*, petal, referring to the lobed petals). Cruciferae. Chile. 5 species, hairy herbs with pinnately lobed leaves and, in summer, racemes of fragrant flowers each composed of four clawed and deeply cut petals. Grow in a well-drained fertile soil in full sun. Sow under glass at 13–15°C/55–60°F; prick out into individual pots in order to minimize root disturbance on setting out. Plant out after the last frost. For pot plants, sow in early autumn in a well-drained, medium-fertility loam-based mix. Maintain a minimum temperature of 10°C/50°F. Water sparingly, increasing quantities when temperatures rise in spring; apply dilute liquid feed when pots are well-

filled with roots. *S.walkeri* (perennial treated as an annual to 45cm; leaves to 14cm, deeply pinnatifid; flowers almond-scented, petals white, fimbriate).

Schizophragma (from Greek *schizo*, to split, and *phragma*, fence, referring to the fruits, which split and fragment between the ribs). Hydrangeaceae. Himalaya to Japan, Taiwan. 4 species, deciduous climbing and creeping shrubs, attaching themselves by means of short, adhesive, aerial roots. The leaves are basically ovate with a tapering tip and a rounded to heart-shaped base. The flowers are carried in flat-topped, branching terminal cymes in summer. Each inflorescence branch bears a long-stalked peripheral enlarged sepal. Ovate to rhombic and white to cream, these sepals appear petal-like and form a showy outer ring encompassing the smaller white fertile flowers. These magnificent climbing shrubs resemble a larger, showier *Hydrangea petiolaris*, whose cultural requirements they share.

S.hydrangeoides (Japan, Korea; to 10m ; leaves 8–12.5cm, coarsely toothed; inflorescence to 20cm in diameter, sterile outer sepals 2.5–3.5cm, white to ivory; 'Roseum': outer sepals tinted rose); *S.integrifolium* (C and W China; 10–15m; leaves 10–18cm, entire or sparsely toothed; inflorescence to 30cm in diameter, sterile outer sepals 6.5–7.25cm, white). For *S.viburnoides*, see *Pileostegia viburnoides*.

Schizostylis (from Greek *schizo*, to split or cut and *stylis*, style). Iridaceae. South Africa. KAFFIR LILY. 1 species, *S.coccinea*, a perennial, cormous herb, 20–60cm tall, with narrowly sword-shaped leaves and 2–ranked spikes of flowers in late summer and autumn. These are scarlet, pink, salmon or white, the perianth tube to 3cm, slender at the base and abruptly widened at throat, the lobes ovate to lanceolate and about 3cm long. Plant in a sunny position on any moist soil, including sodden, heavy clay. Although hardy to between –5 and –10°C/14°–23°F, kaffir lilies are also grown in the cool greenhouse, for long-lasting cut flowers. Numerous cultivars are available. Increase by division in spring.

Schlumbergera (for Frédéric Schlumberger, a well-known collector of cacti at Chateau des Anthieux, near Rouen, and a patron of horticultural botany). Cactaceae. SE Brazil. 6 species, epiphytic or lithophytic cacti. The stems are branched and arching, composed of numerous flattened segments, sometimes with small, bristly spines. Borne usually in winter and spring at the stem tips, the flowers are tubular with several ranks of overlapping and reflexed tepals, an expanded limb and prominent stamens. These cacti, and especially the hybrid *S.* × *buckleyi*, are popular as flowering house plants, with more than 200 cultivars named. Maintain a minimum temperature of 10°C/50°F. Grow in an open acid to neutral, soilless medium. In spring and summer, shade from direct sun, provide high humidity and water frequently. From late summer until spring, admit full sun, keep cool and draught-free and water sparingly. Propagate by stem cuttings.

Schizophragma hydrangeoides (left) and *S. integrifolium (right)*

S. × buckleyi (*S.truncata* × *S.russelliana*; syn. *S.bridgesii*; CHRISTMAS CACTUS; garden origin; stem segments to 5cm, crenate; flowers purple, magenta, pink, white, red, orange or yellow, to 6.5cm); *S.truncata* (SE Brazil; stem segments to 8cm, serrate; flowers to 7cm, magenta to rose or red).

Schoenoplectus (from Greek *schoinos*, rush, and *pleko*, plait, alluding to the mat-forming rhizomes of some species). Cyperaceae. Cosmopolitan. BULRUSH. 80 species, annual or perennial, grass-like, rhizomatous herbs with reduced leaves, rush-like stems and small, brown flowers in dense heads. *S.lacustris* (syn. *Scirpus lacustris*) is the widespread and well-known bulrush, a perennial with clumps of terete stems 1–3m tall. Fully hardy, it soons establishes itself in damp or wet places in sun or part-shade. The white-banded cultivar 'Zebrinus' is an especially attractive addition to the pond or streamside. Increase by division in spring.

Schwantesia (for Dr M.H.G. Schwantes (1881–1960), professor at the University of Kiel). Aizoaceae. South Africa. 10 species, many-headed, cushion-forming perennial succulents with fleshy, toothed leaves and daisy-like flowers opening on bright days in summer and autumn. Cultivate as for *Lithops*. *S.ruedebuschii* (leaves erect, to 5cm, navicular, margins more or less rounded, suffused blue-green with white mottling, tip expanded and blunt, with broad teeth; flowers light yellow).

Sciadopitys (from Greek *skiadeion*, parasol, and *pitys*, a pine tree: the leaves resemble the spokes of an umbrella). Sciadopityaceae. Japan. UMBRELLA PINE, JAPANESE UMBRELLA PINE. 1 species, *S.verticillata*, an evergreen coniferous tree to 40m tall. The crown is conical with tiered branches. Distinctly whorled along the branchlets and at their tips, the glossy midgreen needles are 7–12cm long and linear. Hardy to –30°C/–22°F but best in mild, damp areas with warm, humid summers. Grow on acid to neutral soils. Increase by seed or cuttings.

sciarid flies see *fungus gnats*

Scilla (classical name for *Urginea maritima*). Liliaceae (Hyacinthaceae). South Africa, Asia, Europe. SQUILL. Some 90 species, bulbous herbs – those listed below are perennial and small to medium-sized, with scapose spikes of flowers, generally blue to mauve, star- or bell-like and produced in spring, summer and autumn.

Most are hardy to at least –10°C/14°F. *S.peruviana* needs protection in regions experiencing temperatures below about –5°C/23°F. The South African species need cool greenhouse protection in cool temperate zones. Grow in freely draining but humus-rich soils, with adequate moisture when in growth, in sun or dappled shade. Propagate by offsets, by seed when ripe or by division of established clumps as the foliage dies back.

Scilla bifolia

scorzonera

Disease
white blister

S.autumnalis (S, W and C Europe, NW Africa to C Asia, Iran, Iraq; autumn-flowering; scapes 5–30cm; flowers with tepals 3–5mm, spreading, lilac to pink; 'Alba': white; 'Praecox': large, purple-blue; and 'Rosea': pink); *S.bifolia* (C and S Europe, Turkey; early spring-flowering; scape 7.5–15cm; flowers with tepals 5–10mm, ovate to elliptic, tips hooded, blue to purple-blue); *S.litardieri* (syn. *S.pratensis*, *S.amethystina*; Balkans; early summer-flowering; scape 5–15cm; flowers with tepals 4–6mm, ovate, blue-violet or pink); *S.mischtschenkoana* (syn. *S.tubergeniana*; Iran, Russia; early spring-flowering; scapes 1–3, 5–10cm; flowers with tepals 1–1.5cm, oblong to elliptic, obtuse, white-blue with a darker stripe); *S.natalensis* (South Africa; summer-flowering; scape to 40cm; flowers with tepals 6–10mm, elliptic to oblong, obtuse, spreading, light violet-blue or occasionally pink or white); *S.peruviana* (SW Europe, W Africa; early summer-flowering; scape 15–25cm; flowers with tepals 8–15mm, deep violet-blue to dull purple-brown, or white); *S.scilloides* (syn. *S.chinensis*; China, Korea, Taiwan, Japan, Ryukyu Islands; late summer- or autumn-flowering; scape 20–40cm; flowers with tepals 3–4mm, narrowly oblong, acute, spreading, pale pink or mauve-pink); *S.siberica* (SIBERIAN SQUILL; S Russia, naturalized C Europe; early spring-flowering; scapes 10–20cm; flowers rotate or broadly bowl-shaped, pendent, with tepals 1–2cm, elliptic to oblong, obtuse, bright-blue with a darker median band; 'Alba': white flowers; 'Spring Beauty': deep blue, long-lasting, scented). See also *Ledebouria* and *Hyacinthoides*.

Scindapsus (name used in ancient Greece for ivy-like plant) Araceae. SE Asia, Pacific, Brazil. About 40 species, evergreen trailers and climbers with adhesive adventitious roots. They resemble *Epipremnum* in most respects. Cultivate as for the smaller *Philodendron* species. *S.pictus* (Java to Borneo; stems to 12m+, slender; leaves 6–18cm, broadly ovate to lanceolate, coriaceous, dull green; 'Argyraeus': SATIN POTHOS, with 7–10 cm, ovate to cordate, silky deep green leaves with silver spots). For *S.aureus*, see *Epipremnum aureum*.

scion a shoot or bud cut from a chosen specimen for the purpose of grafting or budding on to a rootstock, in order to produce a new individual.

scion-rooting the production of roots by a scion, as may occur when a grafted plant is established with the graft union below soil level. Scion-rooting is undesirable where the rootstock is required to transmit a characteristic such as dwarfness, so depth of planting can be critical.

sclerotium (plural: sclerotia) a resistant, usually rounded, fungal resting body, composed of a compact mass of mycelia. Most sclerotium are dark-coloured. Diameter varies with type from less than 0.1mm (e.g. *Verticillium*) to several centimetres in diameter (e.g. *Sclerotinia*).

scooping a method of bulb propagation, especially suitable for hyacinths, involving scooping out the inner portion of the basal plate of the parent bulb with a knife or sharpened spoon to leave a sound outer rim, dusting the cut surface with fungicide and then placing the bulb base upwards in a container of moist sand for storage in a dark, warm place. Bulblets arise around the edge of the scooped area and are removed when of handleable size to grow on. See *scoring*.

scorch, scorching the discolouration and death of plant tissue, especially leaf margins, resulting from disease or any of a wide range of environmental factors, including strong sunlight, contact with chemical substances, salt spray and strong wind.

scoria porous volcanic mineral with properties similar to perlite, used in crushed form in potting media or as a substrate in hydroculture.

scoring a method of propagating bulbs, similar in principle to scooping, but with the base plate scored with two shallow cuts, which are made at right angles to each other across the diameter of the base of the bulb. See *scooping*.

scorzonera (*Scorzonera hispanica*) COMMON VIPER'S GRASS, a hardy perennial grown for its black skinned roots. It has broader leaves and yellow flowers, and may be raised and grown in a similar manner to *salsify* (q.v.), to harvest selectively over two seasons. The flower buds are plump and succulent.

scrambler a long stemmed plant, often equipped with prickles, which climbs through shrubs and trees.

scree see *rock garden plants*.

screen a barrier designed to provide privacy, as an ornamental feature, to mask unwanted views, or to protect from sound, or wind. Visual screens may consist of tree belts, hedges, tall individual plant specimens, fences, or walls. Aural screens are effective only where large areas are available to accommodate several rows of tall, dense evergreen trees. Screens used for ornament can be developed from plantings, or made of materials such as wire or wooden lattice work, trellis or pierced wall blocks, any of which may be covered with plants. Screen walls composed of relatively thick blocks of concrete, earthenware, or other materials, incorporating pierced patterns, allow a partial view while reducing wind flow. See *windbreaks*.

scrim a fine cotton mesh used as a shading fabric for greenhouses. It has now largely been superseded by more durable materials.

Scrophularia (from medieval Latin *scrofula*, referring to its supposed ability to cure scrofula – the roots of some species resemble scrofulous tumours). Scrophulariaceae. Northern temperate regions. FIG-

WORT. Some 200 species, foetid herbs and subshrubs with 4-angled, erect stems and small, squatly tubular and 2-lipped flowers in terminal, paniculate cymes in summer. *S.auriculata* 'Variegata' is a bold, perennial foliage plant for waterside plantings or damp borders in sun or part-shade. It is fully hardy. Propagate by division in spring. Attacked by figwort weevil, which may skeletonize foliage. *S.auriculata* (syn. *S.aquatica*; WATER BETONY, WATER FIGWORT; W Europe; perennial to 1m; leaves 5–25cm, ovate to oblong, sometimes lobed at base; flowers red-brown; 'Variegata': leaves dark green marked cream).

scutate shaped like a shield.

Scutellaria (from Latin *scutella*, a small shield or dish, descriptive of the crest or pouch on the upper calyx lip). Labiatae. Cosmopolitan, through temperate regions and tropical montane, with the exception of South Africa. SKULLCAP, HELMET FLOWER. Some 300 species, largely summer-flowering annual or perennial herbs, erect or sprawling with terminal spikes or racemes of tubular and hooded, 2–lipped flowers. The following are hardy in climate zone 7. They need a sunny position. The shorter species are ideal for the front of a border or the rock garden. Propagate by division or seed sown outside in late spring.

S.indica (Japan, Korea, China; slender, very white-hairy, procumbent perennial, 20–40cm; flowers 1.8–2cm, pale purple-blue; 'Japonica': larger flowers, lilac to blue); *S.orientalis* (SE Europe, mountains of SE Spain; sprawling, subshrubby perennial to 45cm, grey-woolly; flowers 1.5–3cm, yellow, sometimes spotted red or with lip red, occasionally pink or bright red to purple); *S.scordiifolia* (E Europe, Siberia; erect, basally branched perennial, 10–30cm, hairy; flowers 1.8–2cm, violet-blue).

scythe a large hand tool for cutting rough grass and newly sown lawns, consisting of a curved, sharpened blade attached to a handle, with two handgrips. The tool is held with both hands and cutting is achieved with a swinging movement, keeping the blade parallel to the ground. The grass hook is in effect a scythe with a short handle, the blade being slightly curved and the tool used with one hand.

seakale (*Crambe maritima*) a deep-rooting perennial herbaceous plant, native to northern Europe, especially seaboard areas. It makes a handsome ornamental plant in gardens, reaching 75cm in height, and is grown especially as a vegetable crop for its edible stems, which have a characteristic flavour when eaten raw or cooked (after blanching). The young flowerheads and young leaves may also be eaten raw.

Grow on an organically rich, open site, raised either from seed or from root cuttings or thongs. Plants from spring-sown seed may be variable and germination can be very slow. Thongs, selected from plants at least three-years-old, are lifted in November and pencil-thick side roots cut into portions 7.5–15cm long with a sloping cut at the lower end.

These should be stored in moist sand in a cool building until March, when buds will appear at the upper cut end. The buds are reduced to the strongest one and the thongs planted with a dibber 2.5cm below soil level, with 38cm between plants.

Sea kale can be forced after its third season. This can be done *in situ* between November and January. Cover the crowns first with dry litter and then with a special earthenware forcing pot (which should be about 38cm high and fitted with a lid) or any upturned container that is large enough and effectively excludes light. Stems are ready after about 12 weeks and should be cut with a knife. Alternatively, roots can be lifted after the first frost, lightly trimmed and packed into boxes or large pots and moved to a dark place, within a building or greenhouse, at a temperature of 16–20°C/61–70°F. The planted containers must be kept moist and stems are useable within a few weeks.

seaweed a useful bulky manure, which can be incorporated fresh, or after composting, or dried. There are liquid preparations for use as supplementary feeds. An average analysis of the nutrient content of fresh seaweed is around 0.5% nitrogen, 0.1% phosphate and 1.0% potash, together with a content of trace elements and alginates, which have a beneficial effect on soil structure. The strap-shaped kelps (*Laminaria* species) are richer in plant nutrients than bladderwrack (*Fucus vesiculosus*). Calcified seaweeds concentrate calcium carbonate in their cell walls, and are dried and crushed to produce an organic liming agent and soil conditioner with nutrient content.

secateurs (pruning shears in the US) small shears used for pruning. They are available in three designs: anvil, bypass and parrot-beak. Anvil secateurs have a straight-edged blade that cuts against a thickened bar or anvil; these must be kept sharp to avoid crush damage to stems. Bypass and parrot-beak types have a scissor-like action which makes a clean cut. All are available in light or heavy duty grade, the latter being suitable for woody pruning.

secondary growth (1) new or extension shoots produced by a tree or shrub after summer pruning. The term is applied especially to fruit trees; (2) new growth from the tip of a shoot that had apparently stopped developing earlier in the same season; (3) used botanically to describe the extension of vascular systems, through the formation of new tissue, that arises from activity of the cambium. It results in stem thickening, also known as secondary thickening.

secondary organism an organism that invades diseased tissues after they have been killed by a primary pathogen. Secondary organisms include bacteria, fungi, nematodes and mites.

sectio, section in plant classification, the rank of taxa between subgenus and series, for example *Rhododendron* subgenus *Rhododendron* section *Pogonanthum*.

seakale

Pest
flea beetle

Disease
clubroot

secund with all the parts along one side of the axis.

sedge peat peat formed from the remains of sedges, reeds, mosses and heathers. It is usually coarser and darker than sphagnum moss peat and less tractable for use in potting media; it is also generally less acidic and has a lower water-holding capacity. cf. *sphagnum peat.*

Sedum (from Latin *sedere*, to sit, a reference to the way some species attach themselves to rocks and walls). Crassulaceae. Northern temperate and tropical mountain regions. Over 300 species, annual to perennial herbs and subshrubs. They vary widely in habit and size, but usually have clumped stems with succulent leaves and terminal, branching cymes of small, starry flowers in spring and summer.

The hardy perennials *S.spectabile* and *S.telephium*, together with their hybrids and cultivars, are among the finest plants for late summer and autumn borders. Most species withstand frost to at least –15°C/5°F; *S.spurium* tolerates temperatures to –5°C/23°F. A number of species, for example those native to the hot dry Americas, need protection from frost and are best-grown as house or greenhouse plants with a winter minimum of 5°C/40°F. Among the frost-tender species are several fine foliage plants; for example, *S.morganianum*, which is suitable for hanging baskets. Most species are drought-tolerant. Plant open-ground species in fertile, well-drained soils, in full sun. Grow tender species in a gritty mix in direct light and with good ventilation. Propagate by cuttings, by division in spring and early summer or by seed.

S.acre (STONE CROP, WALL PEPPER; widespread Europe, Turkey and N Africa, naturalized E US; loosely tufted or mat-forming perennial, 5–12cm; stem slender, erect or trailing; leaves 3–6mm, overlapping, triangular, blunt, thick; flowers yellow; 'Aureum': leaves variegated pale yellow; 'Elegans': shoots striped silver at tips); *S.aizoon* (N Asia, naturalized N and C Europe; herbaceous perennial, 30–40cm; stems few, erect, mostly unbranched; leaves 50–80mm, oblong to lanceolate, flowers golden yellow; 'Euphorbioides': more compact and robust than type, with flowers of a stronger yellow); *S.caeruleum* (W Mediterranean Islands; bushy annual, 5–20cm; stems erect, hairy above; leaves about 10mm, oblong to linear, subterete, suffused red; flowers white at base, blue above); *S.kamtschaticum* (Japan; similar to *S.aizoon* except 5–30cm; rhizome stout, stem branched below; leaves 20–40mm, obovate to lanceolate, apex coarsely toothed; 'Variegatum': leaves edged

Sedum 'Cape Blanco'

white and tinged pink, flowers yellow, turning crimson); *S.lydium* (W and C Turkey; differs from *S.tenellum* in having rooting stems; leaves linear, to 6mm, tipped red; flowers many); *S.morganianum* (BURRO'S TAIL, DONKEY'S TAIL; origin obscure, probably Mexico; evergreen perennial to 6cm; stems numerous, prostrate or pendulous, sparsely branched; leaves to 18mm, very succulent, glaucous, overlapping, oblong to lanceolate, subcylindric, tips pointed incurved; flowers deep pink); *S.obtusatum* (California; loose hummock-forming perennial, 3–30cm; leaves 5–25mm, in loose rosettes at stem tips, spathulate, glaucous, blunt to rounded, tinged red; flowers pale yellow-orange); *S.palmeri* (Mexico; sprawling, evergreen perennial, 15–22cm; stems arching, branching, naked below, scarred; leaves 2.5–15mm, spathulate, glaucous, in loose rosettes; flowers deep orange); *S.reflexum* (C and W Europe; mat-forming, evergreen perennial, 15–35cm; stems horizontal, woody; leaves 12mm, linear, terete; flowers bright yellow; 'Cristatum': leaves crested; 'Minus': dwarf; 'Monstrosum Cristatum': a fasciated form, with stem thin at base, widening towards the top edge, along which grey-green leaves are borne; 'Viride': green leaves and yellow flowers); *S.sempervivoides* (Armenia, Georgia, Caucasus; erect, hairy biennial, to 20cm; leaves 10–30mm, ovate, pointed, tinged purple, in a basal rosette; flowers bright red); *S.spathulifolium* (western N America; clump-forming perennial, 9–15cm, stems short, often forming long runners; leaves about 18mm, scattered or in compact rosettes at stem tips, spathulate, blunt, grey-glaucous; flowers yellow; 'Aureum': leaves grey, tinged pink and flecked yellow; 'Cape Blanco': compact, with pale silver-grey leaves; 'Carneum': leaves tinged crimson; 'Purpureum': glaucous grey leaves flushed purple; 'Roseum': flowers rose); *S.spectabile* (syn. *Hylotelephium spectabile*; ICE PLANT; China, Korea; glaucous, pale sea green perennial to 50cm; stems clumped, erect; leaves to 10cm, ovate-elliptic, crenate; flowers pale pink to deep rusty red; includes cultivars with white, deep pink and pale rose flowers, some with leaves very pale and glaucous or variegated); *S.spurium* (Caucasus, N Iran, naturalized Europe; evergreen mat-forming, perennial to 15cm; stem much-branched, finely hairy, creeping; leaves 25mm, obovate, toothed, fringed; flowers white, pink or red purple; 'Atropurpureum': leaves burgundy; flowers rose-red; 'Bronze Carpet': leaves bronze; 'Coccineum': scarlet-red flowers; 'Green Mantle': leaves forming fresh green groundcover, flowerless; 'Purpurteppich': compact, leaves plum purple; 'Roseum': flowers rose-pink; 'Schorbuser Blut': leaves bronze with age, flowers large, very dark red; 'Variegatum': leaves edged pink and cream; var. *album*: flowers white); *S.telephium* (syn. *Hylotelephium telephium*; ORPINE, LIVE FOREVER; to 60cm, erect; leaves to 7cm, oblong or ovate, slightly toothed; flowers purple-red; includes cultivars with burgundy-red leaves and deep red flowers); *S.tenellum* (N Iran, Caucasus; ascending, tufted perennial, 5–8cm; stem woody, bare below, creeping and ascending; leaves 3mm, overlapping, oblong, terete, smooth; flowers white suffused red). *Cultivars* 'Autumn Joy': flowers bright salmon pink becoming bronze; 'Ruby Glow':

leaves purple-grey, flowers rich ruby red; 'Vera Jameson': leaves deep purple; flowers pale pink. See also *Rhodiola*.

seed a ripened, fertilized ovule; an embryonic plant.

seedbed an area of soil prepared for seed sowing. Firming and raking provide a level, fine tilth, ensuring good aeration and water penetration and facilitating the sowing of seeds at an even depth. A stale seedbed is a prepared area left unsown for a period to allow weed seeds to germinate. The seedling weeds are destroyed with a minimum of disturbance to the bed, possibly using a contact herbicide. The crop is then sown with the benefit of a reduced level of competition from weed growth, and less need for early cultivations,

seed box, seed tray a container used for sowing seeds, pricking out seedlings and for the insertion of cuttings; of standard dimension $35 \times 17 \times 5$–7.5cm deep depending on purpose; half trays are available. Plastic has largely replaced wood as the preferred manufacturing material, being easier to clean and stack and more durable; available thicknesses of plastic vary, depending on price and intended use. Some trays are made as cellular units, which may or may not need supporting in sturdier trays; the cells may be of a size suitable for direct sowing or cutting insertion, or large enough for growing on individual plants to planting size. All types must have drainage holes in the base.

seed dressing a pesticide applied as a coating to seeds prior to sowing, to control seed-borne diseases and to protect against certain pests and diseases that attack germinating seeds or young seedlings. Formerly, home treatments by the gardener were advocated, but, in the interests of safety, it is now recommended that seed is purchased ready-dressed, where appropriate, and instructions on safe handling followed.

seed leaf see *cotyledon*.

seedling a young plant that develops from a seed.

seed-raised used of plants grown from seed, as distinct from those grown by vegetative methods.

seeds and seed sowing seeds are the means of sexual reproduction in plants, as opposed to vegetative or asexual means. By allowing for recombination of genes following fertilization, they facilitate variation and selection of characteristics. A seed contains an embryonic plant, with root and shoot, and food reserves to support it during the periods of germination and initial growth. Viable seeds are living organisms that maintain a species through a period of dormancy, and offer a means of dispersal over time and distance and protection during unfavourable conditions.

The term seed is used loosely to describe all forms of dispersal unit. Strictly speaking, the only true seeds are those in which the ovary wall is dehiscent, for example *Delphinium*, *Lupinus*, *Aquilegia*, *Allium*. In many cases, the dispersal unit is a dry indehiscent fruit, for example *Hibiscus, Lavatera, Malva*, or a succulent indehiscent fruit, for example *Prunus, Malus, Ribes*. Sometimes the dispersal unit is a false fruit, where a part other than the ovary wall develops as a result of fertilization, for example *Fragaria*. To germinate naturally, seeds must be mature and provided with suitable environmental conditions.

Environmental control of germination The four main environmental factors that influence germination are water, light, temperature and gases.

WATER For the majority of seeds, the first step in germination is rehydration of the seed tissues. Since water availability can be unpredictable, its shortage is the most common cause of seed or seedling death. Seeds are naturally hygroscopic and the rate of water uptake or imbibition is mainly determined by properties of the seed coat. If the growing medium becomes waterlogged, germination may be suppressed, due to reduced oxygen availability, a risk increased if the germination period is prolonged, for example, by sub-optimal temperatures.

LIGHT Most seeds germinate equally well in the light or dark. However, some will only germinate if given a light stimulus and a very few will germinate only in the dark. The photo reversible pigment phytochrome allows plants to detect information about their light environment. While the seeds of certain species respond in a particular way to light – for example, small-seeded annual weeds are usually promoted by light – the response is often dependent on other factors, especially temperature. Thus, an individual seed may be insensitive to light at one temperature, but will only germinate at at a different temperature if first given a light stimulus. In the garden situation, sown seed is usually covered, with the primary objectives of moisture conservation and protection from predators, and the exclusion of light rarely presents difficulties.

TEMPERATURE For all seeds, at any given time, there is a maximum and minimum temperature above and below which germination cannot occur. At a point between these temperatures, there is a specific temperature or narrow band of temperatures which allows for maximum germination in the shortest time, and this is the so-called optimum temperature. Temperate species generally have lower optimum germination temperatures than tropical species, but the range of temperature over which germination may occur can vary considerably, even within a single batch of stored seeds. Seeds of wild plants often fail to germinate under constant temperature conditions. GASES As a rule, seeds germinate best in air, and germination is suppressed if the carbon dioxide level is raised or the oxygen level lowered; the gaseous atmosphere that surrounds seeds probably plays a significant role in controlling the germination of buried seeds in the natural environment. The seeds of certain aquatic species, such as North American wild rice (*Zizania* species), germinate best if completely submerged in water under more or less anaerobic conditions.

Dormancy Dormancy is a highly variable character controlled by the interaction of genetic and environmental factors:

(a) innate or primary dormancy is expressed when seeds are freshly harvested, and develops during seed development and maturation. It may be enforced, where germination fails because of unsuitable environmental conditions, or physiological where seeds fail to germinate without a prolonged treatment, such as growth-regulator application or chilling. Seed coat impermeability is a true form of innate dormancy;

(b) induced or secondary dormancy describes the condition where seeds able to germinate are subsequently switched into a dormant state; for example, a combination of high temperature and darkness can often induce a state of dormancy.

Overcoming innate seed dormancy

SCARIFICATION Various methods exist for penetrating hard seed coats:

(1) chipping – using a scalpel and forceps, a small shallow cut is made in the seed coat to allow the entry of water when the seed is sown. To avoid risk of damage, the cut should be located at a site on the seed surface well away from the axis;

(2) filing – used as an alternative to chipping for large-seeded species with particularly thick seed coats;

(3) acid treatment – concentrated sulphuric acid is an effective agent for degrading hard impermeable testas but it is only a suitable treatment for commercial operations, where suitable safety precautions can be taken;

(4) abrasives – rubbing seeds between sheets of sandpaper is effective in some cases, as is tumbling seeds in rotating drums lined with abrasive;

(5) hot water – some species respond positively to simple hot water treatment, in which a volume of seed is plunged into an equal volume of recently boiled water, and left to stand until cool. The proportion of seeds that have swollen are assessed.

STRATIFICATION Very many species of trees and shrubs, as well as many herbaceous species, have seeds with coats that are mechanically resistant to germination. The seed coats will allow water to penetrate but the embryo will not germinate until the hard seed coat splits and the embryo has undergone a period of cold stratification. In the case of a few species, splitting can be achieved by exposure for two weeks at temperatures of between 20–25°C/68–77°F. Many other species require exposure for up to twelve weeks at 15–20°C/59–68°F, followed by a period of 12–18 weeks' exposure at between 0–10°C/32–50°F.

The treatment is known as stratification because of the traditional practice of carrying it out in strata of sand and seed in deep pits in the garden, and this remains a suitable method. Alternatively, very small quantitites of seed can be mixed with a well-aerated, free-draining medium and placed in a nylon net bag; this is tied loosely but securely and buried just below the soil surface outdoors, carefully marked. Another method is to place the seeds in large pots, in two levels separated by 2–3cm layers of sand or a sand/peat mixture, and plunge well below the soil surface. Similar benefit can be achieved by storing containerised seeds within a refridgerator.

Viability and storage Seeds of the majority of species used in horticulture are tolerant of desiccation, and it is important to reduce moisture content to a low level as soon as possible after harvest. The optimum moisture content (expressed on a fresh weight basis) for survival of most desiccation-tolerant seeds is in the range of 5–8%.

Desiccation-intolerant seeds will die if dried, for example, SYCAMORE (*Acer pseudoplatanus*), oak (*Quercus* species) and SWEET CHESTNUT (*Castanea sativa*). Such seeds must be maintained at a high-moisture content, equivalent to a relative humidity of not less than 98%. Oxygen must be available when seeds are stored at a high-moisture content.

Lowering the temperature tends to improve storage life. As a rule of thumb, a 5° reduction in the ambient range (say, 10–29°C/50–85°F) results in an approximate doubling of lifespan; lower limits apply to tropical and moist seeds.

The realization that desiccation-tolerant seeds may be stored for anything between several decades to several thousand years, providing that moisture content and temperature are properly controlled, has led to the widespread use of seed banks.

Seed quality Many types of seeds are now covered by statutory regulations detailing minimum germination standards. Storage conditions in warehouses or on display racks in shops or warm houses may significantly reduce viability. For seeds stored in such condition, especially where self-gathered or held over from previous years, a quick germination test can often be justified. A weighed or counted sample from the measured bulk is set to germinate in containers covered with glass, plastic or cling film; these are placed in a propagator maintained at 20–24°C/70–75°F. For seeds the size of cereal grains or smaller, a bedding of moist absorbent paper should be laid flat in the container; for larger seeds, sand should be used.

After ten days, the number of seeds that have germinated or begun to germinate is counted and the likely total viable number in the batch calculated.

Outdoor bed sowing Sowing direct in the open ground requires preparation of a level seedbed, which is achieved by raking and cross raking ground previously cultivated by digging or forking, and by weathering. Except on wet soils, treading is a helpful means of breaking down clods, but great care must be made not to compact the soil, which impedes drainage and aeration. A wide wooden rake is a valuable aid to levelling and removing large debris. The finished seedbed should be moist and friable to a depth of at least 5cm.

There are two methods of sowing seed outdoors, in drills or broadcast. Drills can best be marked out with the aid of a strong garden line pulled taught between two canes. Suitable depth is achieved by pulling the edge of a swan-necked or similar hoe along the line; the same effect can be obtained with a thick cane. For short rows in, for example, narrow beds or frames, a board can be used instead of a garden line. For informal areas, such as an annual flowerbed, suitable drills can be made by pushing the reverse side of a metal rake head into the bed and spacing the row intervals by eye. In the garden, seed may be sown from the hand, packet or from a specially designed dispenser; a wheeled drilling device is suitable for larger undertakings. Seed should be sown thinly under most circumstances, and in some cases it can be sown at spaced stations for economy.

Container sowing Sowing in containers requires the tray, pot, pan or cellular-tray modules to be over-filled with compost, which is then struck off level and settled by gently tapping it upon the bench. The medium is then gently pressed down with a shaped firmer so that the surface is level, and the seed sown broadcast by the methods described. Cell- and thimble-sized plugs may be sown direct to overcome the inevitable damage and check caused by pricking out. It is usual to cover sown seeds with sieved compost to the depth of the seed, but this is rarely essential, and some subjects, such as begonias, coleus, *Impatiens* and primulas, should not be covered. The containers should be lightly watered, either with a fine overhead spray, or, where appropriate, by steeping the container in a larger vessel of water so that moisture is absorbed by capillary action. In all cases, very fine seed, such as that of *Begonia*, may be mixed with sand to aid even distribution. A sheet of glass or clear plastic placed over the container will enhance humidity, and covering with a sheet of newspaper reduces water loss by evaporation. Small numbers of seed containers can be accommodated in a proprietary seed-tray-sized propagator, which should have a built-in heating element; optimum germination temperature varies according to species.

seep hose, soaker hose small-bore, flexible irrigation tubing from which water leaks or oozes to produce a continuous wet band of soil. It may consist of plastic tubing with durable stitching along one side, or compressed rubber tubing where water oozes from pores in the tube wall. See *irrigation*.

segment a discrete portion of an organ or appendage which is deeply divided, or compound.

Selaginella (diminutive of *Selago*, the old name for another club moss). Selaginellaceae. Tropics and subtropics worldwide, with a few species also in temperate zones, extending rarely to Arctic. LITTLE CLUB MOSS, SPIKE MOSS. Over 700 species, evergreen moss-like plants with bushy and creeping to slender and scrambling stems, clothed with overlapping, scale-like leaves. *Selaginella* species occur in a range of habitats but most – with the exception of specialized desert natives, such as *S.lepidophylla* – are found in moist soils in humid shaded forest, both tropical and temperate. They are grown for their beautiful foliage, usually in a range of rich vibrant greens; several have golden or variegated cultivars and some, such as *S.uncinata* and the climbing *S.willdenovii*, have foliage of luminescent electric blue. Compact cultivars, for example, *S.kraussiana* 'Brownii', may be grown in bottle gardens and terraria. Plant in fast-draining, soilless potting mixes that are open, porous and rich in organic matter. Keep the medium evenly moist but not constantly wet; maintain high humidity. Shade from direct sun. Provide a minimum temperature of 10°C/50°F. *S.lepidophylla* will grow in similar conditions, although it is usually sold as a dry, dead or half-dead curio, unfurling quickly if made wet. It is difficult to maintain in cultivation. Propagate by division or layers.

S.kraussiana (SPREADING CLUBMOSS, TRAILING SPIKE MOSS, MAT SPIKE MOSS; South Africa; stem 15–30cm, trailing, jointed, copiously pinnate, branches erect to outspread, compound; leaves bright green; includes 'Aurea', with bright golden green foliage, and 'Variegata', with foliage splashed ivory to pale yellow); *S.lepidophylla* (RESURRECTION PLANT, ROSE-OF-JERICHO; Texas and Arizona south to Peru; stems 5–10cm, frond-like, in a rosette; leaves dark green, drying or ageing rusty brown; dried, the fronds will roll into a tight, brown ball, unfurling if wet); *S.martensii* (Mexico; stem 15–30cm, trailing, branches decompound, fan-shaped to pinnate; leaves bright green; includes 'Albolineata', with some leaves wholly or partially white, and 'Albovariegata', with white-tipped leaves); *S.uncinata* (PEACOCK MOSS, RAINBOW FERN, BLUE SPIKE MOSS; China; stem 30–60cm, trailing, rooting, slender, branches alternate, pinnate, short, leaves iridescent blue); *S.willdenovii* (PEACOCK FERN; Old World tropics; stems to 4m, scandent, branched from base; pinnae 30–60cm, deltoid, and decompound; leaves bronze to kingfisher blue with an iridescent sheen).

selection see *cultivar*.

selective used of herbicides and other pesticides that suppress or kill some weeds or pests without seriously affecting others.

Selenicereus (from Greek *selene*, the moon, and *Cereus*, referring to the nocturnal flowers). Cactaceae. Tropical America and the Caribbean region. Some 20 species, lithophytic or epiphytic cacti, with slender and ribbed, spiny and clambering stems. The flowers are very large and funnelform, opening at night. In spring and summer, shade from direct sun and maintain high humidity and good water supplies. In autumn and winter, admit full light, reduce water and provide a minimum temperature range of 10–15°C/50–60°F. Pot in a fast-draining, acid mix rich in coarse bark and sand. Provide support. Increase by stem cuttings. *S.grandiflorus* (QUEEN OF THE NIGHT; Jamaica, Cuba; stem to 4m × 3cm; flowers 17–30 × 12–17cm, outer tepals pale yellow or brown-tinged, inner white).

Sempervivum
'Kramers Spinrad'

self a loose term used by some specialist growers to describe flowers of one colour.

self-compatible, self-fertile a flower or plant that is capable of self-fertilization.

self-fertilization fertilization resulting from self-pollination; cf. *cross-fertilization*.

self-incompatible, self-sterile a flower or plant that is incapable of self-fertilization.

selfing the act of artificially pollinating a plant with its own pollen.

self-pollination the transfer of pollen from the anthers to the stigma of the same flower or a different flower of the same plant; a process often used to imply self-fertilization; cf. *cross-pollination*.

self-seeding the establishment of a succeeding generation from the germination of naturally sown seed shed by the spent flowers of the mother plant.

self-sterile incapable of producing viable seed after self-fertilization.

Semele (named for the daughter of Kadmos and mother of Dionysos). Liliaceae (Asparagaceae). Canary Islands. CLIMBING BUTCHER'S BROOM. 1 species, *S.androgyna*, a robust, climbing, evergreen, perennial herb with clumped stems to 7m tall. The true leaves are scale-like and soon wither, becoming dry and prickly. The stem branches are clothed with leaf-like cladophylls; these are dark green, rigid to semi-pliable, 2.5–7cm long and broadly ovate to lanceolate. Small cream to buff flowers are carried along the margins of the cladophylls and give rise to orange-red berries Grow in a gritty but fertile medium in bright, filtered light in frost-free conditions; provide strong support and water moderately. Flowers are carried on year-old growth and fruit will not ripen for a further twelve months, therefore pruning should be restricted, with shoots removed only where necessary to confine to allotted space. Propagate by division.

Semiaquilegia (from Latin *semi-*, half, and *Aquilegia*, to which it is closely related). Ranunculaceae. E Asia. 7 species, low perennial herbs closely related to *Aquilegia*, from which they differ in their petals – these are gibbous or slightly saccate at the base, not spurred. Hardy and suitable for the rock garden. Cultivate as for the smaller *Aquilegia* species. *S.ecalcarata* (W China; to 45cm; flowers wine-red to purple).

Semiarundinaria (from its similarity to *Arundinaria*.). Gramineae. Far East. 10 or more species of tall bamboos; see *bamboos*.

semiterete a semi-cylindrical form; terete on one side, but flattened on the other, as some petioles.

semi-double see *double*.

semi-ripe see *cuttings*.

Sempervivum (from Latin *semper*, forever, and *vivere*, to live). Crassulaceae. Europe, N Africa, W Asia. HOUSE LEEK. 42 species, small, stoloniferous perennial herbs with thickly fleshy, short leaves in low and dense basal rosettes. The flowers are bell- to star-shaped and carried in terminal, long-stemmed cymes. *Sempervivum* species are suitable for the rock garden, scree and alpine house. Drought-tolerant, they thrive in rock crevices, walls, rooftops and paths. All species are hardy to -15°C/5°F and will tolerate snow cover. Most will not tolerate warm, high-humidity climates like the North American Gulf Coast. Small and hairy species, such as *S.arachnoideum* may need protection from winter wet. Grow in well-drained gritty soils in sun. Propagate by division, stem cuttings, offsets or seed. Affected by rust.

S.arachnoideum (Pyrenees to Carpathians; rosettes around 5cm in diameter, tight, with dense white arachnoid hairs; leaves 7–12mm, obovate-oblong or oblanceolate; flowering stem to 12cm, flowers bright rose-red); *S.ciliosum* (Bulgaria; rosettes 2–5cm in diameter, somewhat closed; leaves to 2.4cm, grey-green, hairy, incurved, convex on both sur-faces; flowering stem to 10cm, flowers yellow); *S.erythraeum* (SW Bulgaria; rosettes flat, 3–6cm in diameter; leaves about 2cm, spathulate-obovate, with a short point, glandular-hairy, grey-purple; flowers red-purple dashed white above, red on back; 'Red Velvet': velvety, pale purple-pink leaves; 'Glasnevin': blue-purple leaves, tinged green, pink, and grey). *S. × giuseppii* (Spain; rosettes 2cm in diameter, compact; leaves 1.6cm, obovate, hairy, green, apex brown; flowers red, edged white); *S.grandifolium* (W and C Alps; rosettes 5–20cm in diameter, lax, flat; leaves oblong-triangular, abruptly pointed, hairy, dark green, tinged red at tip, smelling of resin; flowering stem 15–30cm, flowers yellow, stained purple at base); *S.montanum* (C Europe; rosettes dense, open, clustered, 1.5–3.5cm in diameter; leaves about 1cm, dark green, dull, very fleshy, finely hairy, oblanceolate, pointed; flowers red-purple, white, hairy below); *S.tectorum* (C Europe; rosettes large, 5–10cm in diameter, open, rather flat, offsets numerous; leaves 2–4cm, apex with distinct dark markings, glabrous, oblanceolate, thick, bristle-tipped; flowering stem to 30cm, hairy, inflorescence large, many-flowered, flowers dull rose; 'Atropurpureum': dark violet leaves; 'Nigrum': with leaves with conspicuous red-purple tips; 'Red Flush': leaves flushed red; 'Sunset': leaves orange-red; 'Triste': leaves red-brown). See also *Jovibarba*.

Senecio (from Latin *senex*, old man, referring to the usually white or grey, hairy pappus). Compositae. Cosmopolitan. About 1000 species, trees, shrubs, lianes and herbs, varying greatly in size and habit, but generally hairy and with daisy-like, usually yellow, flowerheads. Some of the more commonly seen species are grown for their white or silver

heathers and conifers

Is it any wonder that dwarf conifers and heathers (*right*) should be planted together so often when they have so much in common? Evergreen, dense and compact in habit, presenting a wide range of foliage colour, from dark green to lime, gold, bronze, cream and silver-grey. Suitable for smaller gardens and needing little maintenance, these beds come into their own in winter: gently modulated miniature landscapes, intriguing in form, subtle in colouring. Not that heathers need always be subtle – for example, this brilliant carpet of lime-tolerant *Erica* x *darleyensis* cultivars (*left*).

foliage. *S.cineraria*, for example, is frequently used as an annual in summer bedding; although, given a warm sunny situation in well-drained soils, many of its cultivars will tolerate several degrees of frost, becoming perennial subshrubs. Grow them in sunny positions in moderately fertile and well-drained soil. Propagate by seed or by semi-ripe cuttings in coarse sharp sand in the greenhouse or cold frame. True annuals, such as *S.elegans*, are valued for their magenta or purple daisy flowerheads, an uncommon colour in the genus, but shared with the slightly frost-tender perennial *S.pulcher*. Their requirements are similar to those of *S.cineraria*, but they do not flower well where summers are very hot and/or humid.

Species that require greenhouse protection in cool temperate zones can be divided into the succulents and non-succulents. Non-succulents include the evergreen vining species, *S.macroglossus*, which thrives outdoors in warm, dry, essentially frost-free climates. Under glass, grow in a freely draining medium-fertility potting mix in full sun, with good ventilation and a minimum temperature of 7°C/45°F. Water moderately and feed fortnightly with dilute liquid feed when in full growth. Allow a dry rest, cut back as necessary and re-pot after flowering.

The succulent species show great diversity of form and habit, and are grown primarily for their range of beautiful pastel-coloured and often chalky or waxy bloomed foliage. They include trailing species with long strands of curious, globose leaves, such as *S.rowleyanus*, well-suited to hanging baskets in the home or under glass, and taller species with sausage- or candelabra-like stems, such as *S.articulatus*. Grow them in a medium rich in sand and grit. Provide full sun and good ventilation; water moderately when in full growth, allowing the medium to become almost dry between waterings. When dormant, they need water only to prevent shrivelling. Maintain a minimum temperature of 7°C/45°F. Propagate by seed, division and cuttings.

S.articulatus (CANDLE PLANT, HOT-DOG CACTUS, SAUSAGE CRASSULA; South Africa; perennial succulent to 60cm; stem with cylindric, swollen joints to 2 × 1.2cm, glaucous blue with dark lines; leaves to 5cm, deeply 3–5–lobed, fleshy, glaucous; flowerheads white); *S.cineraria* (DUSTY MILLER; W and C Mediterranean; perennial shrub to 50cm+, much-branched, densely white-tomentose; leaves to 15cm, ovate or lanceolate, dentate to pinnate; flowerheads typically yellow; includes cultivars selected for their habit and intensely silver and finely cut foliage); *S.elegans* (South Africa; annual to 60cm, clammy-hairy, erect or diffuse, branched; leaves to 8cm, oblong, subentire or toothed, lyrate, pinnatifid; flowerheads purple, rarely white). *S.grandiflorus* (South Africa; perennial herb to 1.5m; leaves to 15cm, pinnately divided, lobes linear, entire or dentate; flowerheads purple); *S.macroglossus* (NATAL IVY, WAX VINE; eastern South Africa to Mozambique and Zimbabwe; slender, twining perennial herb to 2m; stem somewhat succulent; leaves to 8cm, deltoid, hastate, 3-lobed, somewhat succulent; flowerheads

cream or pale yellow; includes 'Variegatus', with leaves variegated cream); *S.pulcher* (South Brazil, Uruguay, Argentina; perennial herb to 60cm, lanate at first; leaves to 20cm, in a rosette, elliptic to lanceolate, lobed to toothed; flowerheads large, violet or purple); *S.rowleyanus* (STRING-OF-BEADS; Namibia; succulent perennial to 20cm, forming creeping mats; stems prostrate, slender, rooting; leaves 8mm in diameter, succulent, globose, with a narrow translucent line; flowerheads white); *S.scandens* (E Asia; climbing perennial to 5m, woody at base; leaves to 10cm, deltoid to hastate, incised to toothed, pubescent; flowerheads yellow); *S.serpens* (BLUE-CHALK-STICKS; South Africa; perennial succulent shrub to 30cm, stems glaucous blue, branched below; leaves linear to lanceolate, more or less cylindric, obtuse, grooved above; flowerheads white); *S.smithii* (Chile, Falkland Islands; perennial to 1.2m; stem erect, stout, usually branched above, woolly; basal leaves to 35cm, oblong to cordate, dentate, woolly; flowerheads white). See also *Brachyglottis*, *Delairea*, *Pericallis* and *Pseudogynoxis*.

senescence the physiological changes that occur following maturity of a plant or plant organ, ultimately resulting in tissue death.

Senna (from Arabic *sanā*). Leguminosae. Tropics and subtropics, few temperate. Some 260 species, shrubs, trees and herbs. The leaves are pinnate and the flowers racemose, with five, broad, clawed and, usually, yellow petals. Treatment as for *Cassia*. *S.artemisioides* (syn. *Cassia artemisioides*; WORMWOOD SENNA, SHOWER CASSIA, FEATHERY CASSIA; Australia; shrub 1–3m; leaves 3–6cm, grey-green, leaflets narrow, downy; flowers fragrant, rich yellow); *S.corymbosa* (syn. *Cassia corymbosa*; S US, Uruguay, Argentina; shrub or small tree to 4m, erect or producing procumbent runners; leaves 40–95cm, dull yellow-green, glabrous, leaflets oblong to obovate; flowers golden-yellow); *S.didymobotrya* (syn. *Cassia didymobotrya*; S India, Sri Lanka, Malaysia, tropical Africa, naturalized neotropics; diffuse shrub 60cm-5m; leaves 10–50cm, leaflets elliptic to oblong, downy; flowers yellow); *S.siamea* (syn. *Cassia siamea*; KASSOD TREE; C and S America; diffusely branched, swift-growing tree; leaves 10–35cm, leaflets ovate, oblong to elliptic, dark yellow-green and glabrous above, paler and thinly hairy beneath; flowers yellow with dark-hued sepal).

sensitive (1) used of plants that respond to the stimulus of touch, such as certain carnivorous plants, and *Mimosa pudica*, in which all the leaflets fold together following contact with an object. This phenomenon is also seen in some flowers where stamens or stigmas may move on contact with visiting insects to the benefit of pollination, for example in the families Berberidaceae and Scrophulariaceae; see *haptonasty*, *haptotropism*. (2) used of a plant that develops severe disease when exposed to a pathogen.

sensu in the sense of; often used before the first author cited in a homonym where two or more authors are contrasted by 'non' and 'nec'; for example *Mimulus luteus* sensu Greene non L., where Linnaeus's taxon is the one to which the name should correctly be applied.

sensu lato in the broad sense; used of a taxon when it is cited in such a way as to signify its broadest interpretation, i.e. embracing variants.

sensu stricto in the narrow sense; used of a taxon when it is cited in such a way as to signify only its type.

sepal one of the members of the outer floral envelope, enclosing and protecting the inner floral parts; the segments composing the calyx, sometimes leafy, sometimes bract- or scale-like, sometimes petaloid.

septate see *locular*.

sequestrol, sequestrene a complex organic compound used to bind an element tightly, altering its typical properties so that it is not subject to reactions in the soil which normally make it unavailable for plant uptake. Sequestrol treatment is an effective means of making trace elements available, especially manganese and iron, which on high-lime soils are insoluble and cause so-called lime-induced chlorosis of leaves. The compounds are also known as chelates, and manufacturers' instructions usually recommend several applications to overcome a nutrient deficiency.

Sequoia (after Sequoyah, son of an English trader and a Cherokee mother, inventor of a syllabary for transcribing the Cherokee language). Cupressaceae. W US (Oregon to California, within 30km of the Pacific). CALIFORNIA REDWOOD, COAST REDWOOD. 1 species, *S.sempervirens*, an evergreen coniferous tree to 110m tall in habitat. The crown is conic becoming columnar; the bole is to 8.5m in diameter and the bark red-brown, thick, fibrous and spongy. The leaves are scale-like or 1–2cm long, 2-ranked, linear to oblong and deep green to blue-grey. Cultivars include the weeping 'Pendula' and the dwarf, sprawling to weeping 'Nana Pendula' (syn. 'Prostrata'). Treat as for *Abies*, although this tree is more sensitive to exposure to dry winds. It does best in cool humid areas; it may scorch in cold, dry weather but without any lasting harm. Most soils are suitable. It is, intolerant of pollution.

Sequoiadendron (from *Sequoia* and Greek *dendron*, tree). Cupressaceae. W US (western slopes of Sierra Nevada, California). GIANT SEQUOIA, BIG TREE, SIERRA REDWOOD. 1 species, *S.giganteum*, an evergreen, coniferous tree to 95m tall; its bole is to 10m in diameter and covered in thick, spongy, rust brown bark. The crown is conic and irregularly spreading with age; the branches are curved downward with upturned tips. The leaves are to 0.6cm

long, scale-like to subulate, acute and green suffused grey. It includes pendulous, glaucous, variegated and dwarf cultivars. Hardy to about –30°C/–22°F; soil type, altitude and exposure seem to have little effect on growth, although it does not flourish on shallow chalky soils in exposed areas, or thrive in polluted cities. Light shade is acceptable in early years only. Increase by seed or cuttings.

seriate in a row or whorl. The term is generally prefixed by a number, for example monoseriate, in one row.

sericeous silky; covered with fine, soft, adpressed hairs.

series in plant classification, the category of taxa intermediate in rank between section and species, for example *Saxifraga* section *Porphyrion*, series *Juniperifoliae*.

Seriphidium (diminutive of Greek *seriphon*, a type of wormwood). Compositae. North temperate regions. Some 60 species of aromatic annual to perennial herbs and shrubs resembling *Artemisia*. Cultivate as for *Artemisia*. *S.canum* (syn. *Artemisia cana*; western N America; shrub to 1m; leaves silvery grey-pubescent, to 6cm, linear to oblanceolate, acute or acuminate, entire or occasionally irregularly 3-dentate or pinnatifid, sometimes viscid); *S.maritimum* (syn. *Artemisia maritima*; W and N European coasts; perennial herb to 1.5m, strongly aromatic, pale grey- to white-tomentose, occasionally glabrescent; leaves to 4.5 × 3cm, 2–3-pinnatisect, segments to 15mm, spathulate to linear); *S.palmeri* (California, Baja California; subshrub to 3m; leaves to 12cm, pinnatifid, lobes 3–9, linear to lanceolate, revolute, densely tomentose beneath); *S.vallesiacum* (syn. *Artemisia vallesiaca*; S Switzerland and S France to N Italy; strongly aromatic perennial to 2m, densely grey- to white-tomentose; leaves 3–4 pinnatisect, lobes to 5mm, linear, often pinnatisect).

Serissa (from the old Indian name). Rubiaceae. SE Asia. 1 species, *S.foetida* (syn. *S.japonica*), an evergreen, densely branched shrub to 60cm tall. The branches are initially pubescent and the leaves to 0.5–2cm long, ovate, tough, dark glossy green and foetid. Produced in spring and summer, the flowers are to 1cm long, white and funnel-shaped. It includes cultivars with dwarf habit and minute leaves, double flowers, pink flowers and variegated leaves. Grow in the cool greenhouse (minimum temperature 7°C /45°F), in direct sunlight or with bright filtered light. Use a pH-neutral mix. Water moderately when in growth, less at other times. *Serissa* responds to clipping into formal shapes, after flowering or in winter. Propagate from softwood cuttings in spring or from semi-ripe cuttings in autumn.

serpentine wall a wall built in a regularly curving line, creating sheltered positions for fruit trees and tender shrubs. It usually consists of only one row of bricks with no buttresses. It is sometimes called crinkle-crankle.

serrate essentially dentate, but with forward-pointing teeth resembling those of a saw. Biserrate is when the teeth are themselves serrate.

Serratula (from Latin *serratula*, a little saw, referring to the leaf margins). Compositae. Europe to N Africa and Japan. About 70 species, erect, perennial herbs with pinnately divided leaves and small thistle-like flowerheads in late summer and autumn. Fully hardy. Grow in any well-drained soil in sun. Propagate by seed or division. *S.seoanei* (syn. *S.shawii*; SW Europe; stem to 1m; leaves dark green and scurfy with narrow, serrate segments; flowerheads 1.5–2cm in diameter, purple-pink).

serrulate minutely or finely serrate.

Sesleria (for Leonardo Sesler (d. 1785), Venetian physician and owner of a botanic garden). Gramineae. Europe. MOOR GRASS. 23 species, tufted perennial grasses with narrow, flat leaves and spike-like, cylindrical to globose inflorescences, often tinted blue, grey or white. These small, hardy grasses are grown for their attractive foliage and short flower spikes. They tolerate a wide range of soils and exposures, but prefer in full sunlight and work especially well planted in rockeries, among paving or at the edges of borders. Propagate by division.

　　S.albicans (BLUE MOOR GRASS; W and C Europe; leaves (1.5–)2.5–3(–5)mm wide, grey-green or green, flat or plicate); *S.autumnalis* (syn. *S.elongata*; AUTUMN MOOR GRASS; N and E Italy to C Albania; leaves up to 4mm wide, usually flat, glabrous, glaucescent); *S.caerulea* (syn. *S.uliginosa*, BLUE MOOR GRASS; C Sweden and NW Russia southwards to Bulgaria; leaves lying rather flat in a loose, spreading tussock, bright blue and glaucous above, usually green beneath, 2–8 × 0.2–0.5cm, flat, tip blunt); *S.heufleriana* (BLUE-GREEN MOOR GRASS; eastern C Europe; leaves 2–3(–7)mm wide, strongly glaucous above when young, the uppermost 1.5–3(–5)cm); *S.nitida* (GREY MOOR GRASS; C and S Italy and Sicily; leaves 3.5–6mm wide, flat, rarely convolute, glabrous, pale blue-green, glaucescent, with rough margins, the uppermost 3.5–7.5cm).

sessile stalkless, usually of a leaf lacking a petiole.

set (1) formerly, any cutting, sucker, or graft material, or a young plant intended for bedding out; nowadays, reserved for potato and other tubers, and small onion or shallot bulbs, grown for planting; (2) used of flowers that have been fertilized and are starting to develop fruit or seed. See *dry set*.

seta a bristle, or bristle-shaped organ.

setaceous either bearing bristles, or bristle-like.

Setaria (from Latin *seta*, a bristle, referring to the bristles of the inflorescence). Gramineae. Tropics, subtropics, warm temperate zones. Some 100 species, annual or perennial grasses. In summer and autumn they produce densely cylindrical and bristly, nodding panicles. Among the most ornamental of grasses, grown for their large arching flower spikes. For dried-flower arrangements, these should be cut when green. Sow seed *in situ* in spring on any well-drained soil in full sun. *S.italica* (syn. *Panicum italicum*; FOXTAIL MILLET, ITALIAN MILLET, JAPANESE MILLET; warm temperate Asia; annual, to 1.5m; inflorescence erect to pendent, cylindric, to 30 × 3cm, spikelets bristled, white, cream, yellow, purple-red, brown to black).

setiform bristle-shaped.

setose covered with sharply pointed bristles.

set(t)s small rectangles of shaped natural or artificial stone, used for making paths.

setulose finely or minutely setose.

sewage sludge processed human excrement and industrial effluent from sewage works; a potential source of organic matter for the improvement of soil structure and the provision of nutrients. Raw sludge is rich in nitrogen and phosphate, but should not be considered for garden use. Treated sludge is sometimes offered as an odourless powder, which contains up to 7.0% nitrogen and 4.0% phosphate. It is variable according to source and processing method and is difficult to re-wet; there is some risk of heavy metal contamination, which can cause long-term soil pollution, and also, where treatment is deficient, risk of transmission of human pathogens.

sexual propagation propagation by means of seed (except, in the case of apomixis), or spores.

shade house any structure used for the cultivation of plants requiring partial shade for acclimatization. Shade houses may be clad with narrow wooden laths, plastic strapping or netting made of extruded plastic or woven nylon. See *lathhouse*.

shading the provision of shade in greenhouses and frames; an important factor in preventing excessive temperature rise and water loss, and in protecting against the effects of sunscorch. Movable blinds are the best choice, fixed so that shading can be selectively applied. Roller systems are ideal, made of wooden lath or nylon, but cheaper methods can be devised using translucent fabric, especially for small greenhouses. Wattle hurdles are suitable for covering frames. A useful, alternative method of shading is to apply a proprietary paint to glass for the duration of the brightest summer months. Shading the south side of a greenhouse is most important. See *lathhouse, shade house*.

shallot see *onion*.

shanking a disorder of grapes where the short stalk of each fruit withers and the fruits fail to develop. It is caused by water imbalance in the soil or inadequate nutrition.

shards see *crocks*.

sharp (1) of soils, sometimes used to describe one that is free-draining, with a high sand or stone content; (2) of drainage, denoting fast and free drainage achieved either by crocking or the incorporation of sand, chippings or rubble; (3) of sand, particles with angular edges, as opposed to those with rounded edges.

shears a two-bladed tool, of scissor action, with short or long handles in varying planes to the blades, used for trimming grass and hedges. Pruning shears in the US are secateurs in Europe.

sheath a tubular structure surrounding a plant part; most often, the basal part of a leaf surrounding the stem in Gramineae and other monocot families, either as a tube, or with overlapping edges and therefore tube-like and distinct from the leaf blade.

sheet composting see *green manuring*.

shell house a garden house with walls lined with shells, occasionally created during the 18th and 19th centuries.

shells crushed sea shells are sometimes used to raise the pH of growing media and to improve aeration. They can also be used mixed with grit to cover greenhouse staging. See *oyster shells*.

shelter belt a row of trees or shrubs, over 4.5m high, grown principally for wind protection, but also suitable for screening, noise reduction and, incidentally, as a wildlife habitat. Plantings less than 4.5m high are better described as hedges, which are more likely to be subject to regular pruning or trimming. Due to shadow and competition effects, shelter belts are only suitable around the perimeter of very large gardens, although small gardens can gain much benefit from shelter belts established on adjacent land. For maximum effect, shelter belts should consist of two or three stagger-planted rows. Suitable evergreen species include the Monterey pine (*Pinus radiata*), especially for milder coastal areas, and the Austrian pine (*P. nigra*); suitable deciduous species include the Italian alder (*Alnus cordata*) and the Swedish whitebeam (*Sorbus intermedia*). Mixed plantings of suitable species are visually the most acceptable and extend opportunity for wildlife.

Shepherdia. (for John Shepherd (1764–1836), British botanist). Elaeagnaceae. N America. BUFFALO BERRY. 3 species, deciduous or evergreen, dioecious shrubs with scaly twigs and foliage and inconspicuous flowers followed by brightly coloured berries. Fully hardy, tolerant of poor dry soils and saline conditions, the following species is grown for its silver foliage and attractive fruits. Cultivate as for *Hippophaë*. *S.argentea* (SILVERBERRY, BUFFALO BERRY; N America; deciduous shrub to 4m; leaves 3–5cm, oblong, silvery; flowers small, white-yellow; fruit pea-sized, bright red, glossy with silvery scales, edible; 'Xanthocarpa': fruit yellow).

Shibataea (for Keita Shibata (1897–1949), Japanese botanist and biochemist). Gramineae. Far East. 8 species, squat, shrubby bamboos; see *bamboos*.

shingle (1) small sea-worn pebbles, suitable for covering greenhouse staging; (2) a small rectangular piece of thin wood, used for roofing and fencing.

shoddy waste from wool mills, used as an organic manure for its humus content and slowly released nitrogen. It is not widely available to amateur gardeners. Pure wool shoddy contains 12–15% nitrogen, but the inclusion of cotton and dirt in poor grades can lower the nitrogen content to 2.0%.

shoot (1) the first vertical growth of a seedling or, more loosely, a young stem or lateral growth; (2) applied equally to side growths, twigs, or branches.

short-day plant see *photoperiodism*.

Shortia (for Dr Charles W. Short (1794–1863), botanist of Kentucky). Diapensiaceae. N America, E Asia. 6 species, evergreen perennial herbs with long-stalked, rounded and toothed leaves and solitary, bell-shaped and fringed flowers in spring. Hardy to at least −15°C/5°F, suitable for the rock garden, for an acid border or for a moist, open woodland garden. Each spring provide an acidic mulch such as pine needles. The glossy evergreen foliage assumes bronze tints in winter. Propagate by division in spring.

S.galacifolia (OCONEE BELLS; N America; to 12cm, flowers solitary, to 2.5cm in diameter, white and red ultimately flushed rose, or pink or blue, funnelform, lobes crenate); *S.soldanelloides* (FRINGED GALAX, FRINGE-BELL; Japan; 6.5–10cm, flowers to 2.5cm in diameter, broadly campanulate, deep rose at centre fading to white on lobes, lobes lacerate).

shot-hole a symptom of several bacterial or fungal diseases, in which affected fragments of leaves fall out, leaving small holes.

shot-hole borer see *bark beetles*.

shredder a powered machine with cutting blades that reduces woody and fibrous prunings and similar garden waste to shreds or chips. These may be added to a compost heap or used as a mulch.

shrub a woody stemmed plant with a number of branches arising at or a little above ground level, but lacking a conspicuous trunk.

shrubbery a garden area devoted to shrubs, traditionally evergreens. Today, shrubberies are less fashionable than shrub borders – mixed plantings of shrubs, bulbs and perennials.

shrublet a small shrub or a dwarf, woody-based and closely branched plant.

Sibiraea (for the Siberian native distribution of *S.altaiensis*, the type species). Rosaceae. SE Europe, Siberia, W China. 2 species, deciduous shrubs with small, white to cream flowers in terminal, long-branched racemes in spring and summer. Hardy to at least –20°C/–4°F. Grow in sun on well-drained soil. Cut old or weak shoots back to the base after flowering. Propagate by seed, softwood cuttings or by simple layering. *S.altaiensis* (syn. *S.laevigata*; Siberia, W China, Balkans; erect to 1.5m; leaves to 10cm, oblong to lanceolate, soft blue-green; flowers white in slender panicle to 12cm).

sickle a hand tool consisting of a virtually semi-circular blade fixed to a short handle, used for cutting rough grass and weeds, especially in awkward places.

Sidalcea (from *Sida*, a genus of the Malvaceae, and *Alcea*). Malvaceae. Western N America. PRAIRIE MALLOW, CHECKER MALLOW. 20 species, annual and perennial herbs with rounded, palmately lobed or dissected leaves (both types often on the same plant) and mallow-like flowers in terminal spikes or racemes in summer. Hardy to at least –15°C/5°F. It will grow in full sun on most soils, heavy clay included. Cut back immediately after flowering unless seed is needed. Propagate by division or seed in spring. *S.malviflora* (CHECKERBLOOM; Oregon, California, Baja California; perennial erect to 80cm; petals 1–2.5cm, pink or lilac, usually white-veined; includes many cultivars, low-growing or to 1m or more, with spikes of flowers in shades of white, pale pink, rose, salmon, red and purple-red).

side dressing see *top dressing*.

Sideritis (name used by Discorides, derived from Greek *sideros*, iron, referring to its supposed medicinal properties in the treatment of iron-inflicted injuries). Labiatae. Mediterranean, Atlantic Islands. Some 100 species, hairy annual or perennial herbs, subshrubs, or shrubs bearing terminal spikes of whorled, tubular and 2-lipped flowers. The following frost-tender species is very drought-tolerant and can be used in hot, dry and sunny situations and for landscape plantings in warm, semi-arid climates. Elsewhere, grow in the cool, well-ventilated greenhouse with full sun and in a perfectly drained, gritty and humus-rich medium. Root softwood cuttings of new growth in early summer in a sandy propagating mix. *S.candicans* (Canary Islands, Maderia; shrub, to 90cm, much-branched, white-tomentose; flowers yellow, tipped red to brown).

side shooting cutting or rubbing out side shoots, especially those arising between the leaf stalks and stem of tomatoes.

sieve a usually circular device, consisting of a raised rim across which is fixed a wire-mesh covering with 3–6mm perforations (much smaller than in a riddle). It is used to separate coarse particles and stones from growing-media ingredients, and for covering newly sown seed.

sigmoid, sigmoidal S-shaped, curving in one direction and then in the other.

Silene (a classical plant name of obscure origin, used by medieval botanists more or less in our modern sense). Caryophyllaceae. Widespread in northern Hemisphere, also in South Africa and S America. CAMPION, CATCHFLY. About 500 species, annual, biennial or perennial herbs, often woody at base. Produced in spring and summer the flowers have five, spreading petals each with an entire, notched or cleft limb and a narrow claw. The low-growing species are suited to rock garden, scree and trough; most are hardy to at least –15°C/5°F, although the tightly tufted species are more safely grown in the alpine house. *S.uniflora* is an excellent edging plant for paths, terraces and dry gardens. It thrives in maritime conditions. Shear over after flowering. Others, such as *S.armeria*, *S.coeli-rosa* and the long-blooming *S.pendula*, are hardy annuals. Grow in moisture-retentive but well-drained fertile soils in sun. Propagate annuals by seed sown *in situ*; perennials by division or basal cuttings.

S.acaulis (MOSS CAMPION; circumpolar arctic regions, mountains of W and C Europe and N America; dwarf perennial forming moss-like tufts or mats; flowers solitary, small deep pink or white); *S.alpestris* (E Alps, Balkans; loosely tufted, hairy perennial to 20cm; flowers small, fringed, white flushed pale pink); *S.armeria* (C and S Europe; annual or biennial to 40cm, glaucous; flowers to 1.5cm across, bright pink); *S.coeli-rosa* (Mediterranean, naturalized elsewhere in Europe; glabrous annual to 50cm; flowers to 3cm diameter, pink-purple with white 'eye' and deeply notched petals; includes 'Blue Angel', with clear blue flowers with dark eye, 'Candida', with white flowers, 'Kermesina', with red flowers, 'Nana', dwarf, 'Oculata', with light pink flowers with purple eye, and 'Rose Angel', with bright rose flowers with dark eye). *S.dioica* (RED CAMPION; Europe; perennial with spreading leafy shoots and erect flowering stem to 80cm; flowers to 2cm across, red with deeply cleft petals; hybridizes freely with the white-flowered *S.latifolia*; includes 'Compacta', neat, low and dense, 'Florepleno', with double flowers, 'Graham's Delight', tall, to 90cm, with leaves striped cream,

sickle

'Minikin', compact, with bright dusky pink flowers, 'Richmond', with deep rose pink flowers, 'Roseaplena', tall, with double, large, dusky pink flowers, 'Tresevern Gold', with leaves variegated gold, and 'Variegata', with leaves striped white and pale pink flowers); *S.elisabethae* (Italian Alps; tufted, hairy woody-based perennial to 25cm; flowers white with deeply cut petals); *S.hookeri* (Oregon, California; tufted, grey-hairy prostrate perennial; flowers pink or purple with white rays and deeply cut petals); *S.pendula* (Mediterranean, often naturalized elsewhere; glandular-hairy annual with weak, branching, ascending stem to 20cm; flowers to 1,8cm across, petals cut, usually pink; includes 'Compacta', dense, to 10cm high, 'Ruberrima', compact, with deep pink flowers, 'Ruberrima Bonnettii', stem glabrous, flushed purple, with bright pink flowers, and 'Triumph', to 20cm, with double, red flowers); *S.schafta* (Caucasus; mat-forming, hairy perennial with ascending flowering stem to 25cm; flowers to 2cm diameter, red-purple with cleft petals; includes 'Deep Rose', with rosy magenta flowers, 'Shelly Pink', with palest pink flowers, and 'Splendens', tall, with rose pink flowers); *S.uniflora* (syn. *S.maritima*; coasts of Atlantic and arctic Europe; mat-forming perennial; leaves to 2cm, lanceolate-elliptic, grey-green, glaucous; flowers to 2cm across with a large, inflated calyx, petals deeply cleft, usually white, anthers dark; includes 'Robin Whitebreast', with large white double flowers, resembling a small carnation, 'Rosea', with flowers palest pink, and 'Silver Lining', prostrate, to 20cm, with leaves finely edged silver).

silicle, silicula, silicule a short siliqua, no more than twice as long as broad.

siliqua, silique a thin, 2-carpelled fruit, three or more times as long as wide, dehiscing longitudinally and leaving a central frame of tissue (septum). Siliquas are peculiar to the Cruciferae.

Silphium (from Greek *silphion*, name used by Hippocrates for a plant which produced gum-resin, perhaps asafoetida). Compositae. Eastern N America. ROSINWEED, PRAIRIE DOCK. About 23 species, coarse, hairy perennial herbs with pinnately lobed leaves and daisy-like flowerheads in late summer and autumn. Fully hardy. Grow in any deep, moisture-retentive, moderately fertile soil, which is not too nitrogen-rich, in sun or dappled shade. Propagate by seed or by division in early spring. *S.laciniatum* (COMPASS-PLANT; C South US; 1–3m, glandular, hairy; leaves broadly and deeply pinnately cut, fern-like, standing erect and orienting themselves with the margins pointing north/south and the flat surfaces east/west, thus avoiding full exposure to midday sun; flowerheads to 12cm across, ray florets pale gold, disc florets yellow-orange.

silt fine soil particles 0.05 to 0.002mm in diameter. See *soils*.

Sinningia speciosa

silvanberry see *hybrid berries*.

silver leaf (*Chondrostereum purpureum*) a fungal disease of rosaceous trees, especially plum and its cultivar 'Victoria', characterized by a silvery sheen on leaves and die-back of branches with discolouration of the wood. It is a wound parasite, the spores entering only through cuts and breakages. Silvering is due to air accumulation, with the toxins causing a separation of tissue layers. Flat, overlapping bracket-shaped fruiting bodies develop on dead wood, on which the fungus can live as a saprophyte. The spores are wind-borne and new infections occur most readily between September and May; hence, it is recommended that necessary pruning on susceptible trees is carried out during high summer when natural gums resist infection. Dying branches should be cut back into sound wood with no sign of discolouration. However, if all branches show silvering, the tree should be dug out and burned.

silviculture the practice and science of woodland management for commercial purposes and for wildlife conservation.

Silybum (name used by Dioscorides). Compositae. Mediterranean, Europe and mountains of E Africa. 2 species, annual or biennial herbs with pinnately lobed, white-veined or variegated and spiny leaves and thistle-like flowerheads in summer. Sow in summer or early autumn. It is hardy to –15°C/5°F. Grow in any well-drained fertile soil in an open site in full sun. *S.marianum* (BLESSED THISTLE, HOLY THISTLE, OUR LADY'S MILK THISTLE; Mediterranean, SW Europe to Afghanistan; biennial to 1.5m; leaves dark green, glossy, marbled white, 25–50cm, deeply lobed and spiny; flowerheads purple).

simple not divded or branched.

Sinarundinaria (from *sino*, Chinese, and *Arundinaria*). Gramineae. China, Himalaya. 12 or more species of clumping bamboos; see *bamboos*.

singling the removal of seedlings from a sown row to leave plants at their final spacing. Particularly used of vegetable crops, where it is also referred to as thinning.

single applied to flowers with the normal number of petals for the species; not double or semi-double.

sink garden see *rock plant gardening*.

Sinningia (for Wilhelm Sinning (1792–1874), Prussian horticulturist and botanist). Gesneriaceae. Mexico to Argentina and Brazil. Some 40 species, tuberous or rhizomatous perennial herbs and shrubs (those listed here are squatly tuberous, downy, deciduous herbs). Typically large, thinly fleshy and densely hairy, the leaves are carried in a basal rosette or on annual stems. Produced in summer, the flowers are narrowly tubular to broadly bell-shaped with a spreading, 2-lipped limb, the upper lip 2-lobed, the lower lip 3-lobed.

Pot individual tubers into shallow containers with a fibrous potting medium, covering them to a depth of 2–3cm. Apply water sparingly at first, increasing the amount as growth progresses and watering copiously once plants are in full growth. Maintain a temperature of 20–23°C/70–75°F until flowering, then move to cooler temperatures for display. Apply a balanced liquid feed monthly throughout the growing season. Frequent damping-down during hot weather is beneficial, although the foliage is easily marked by water and syringing must be carried out cautiously; full sunlight will also damage the leaves. Gradually withhold water after flowering until foliage dies back and the root ball is quite dry; store tubers in a cool, dry place (at 7°C/45°F) until the following season. *S.speciosa* and its many hybrids and cultivars are the florist's gloxinias, dependable and free-flowering houseplants. *S.canescens* is grown in collections of caudiciform succulents (i.e. alongside such plants as *Adenium*), and is altogether more tolerant of bright sunlight and arid conditions than the other species. Propagate by seed sown on a finely graded propagating mix, maintaining high humidity and a temperature of 18–20°C/65–70°F; germination will occur within 2–3 weeks. Increase temperatures to 20–23°C/70–75°F after germination and prick out into seed trays or individual pots. Maintain a constant supply of water to avoid initiation of dormancy – for example, by using capillary matting. Seed and seedlings are very small. Tubers may also be cut up, each segment with a bud. Cuttings of emerging shoots may be rooted in a closed case with bottom heat. Leaf cuttings may be made; tubers are produced at the cut edge, the old leaf decaying in the process.

S.aggregata (syn. *Rechsteineria aggregata*; Brazil; to 60cm; flowers to 3cm, cylindric, red-orange, throat yellow-orange, spotted with red); *S.barbata* (syn. *S.carolinae*; Brazil; to 60cm; flowers to 3cm, white, cylindric, inflated at base); *S.canescens* (syn. *S.leucotricha, Rechsteineria canescens*; Brazil; tuber large, exposed, gouty; stems to 25cm; leaves to 15cm, whorled, oblong-obovate, grey-green, densely white-pubescent; flowers to 3cm, cylindric, pink to orange to red, hairy, lobes maroon-violet, marked brown-black); *S.cardinalis* (syn. *Rechsteineria cardinalis*; Brazil; to 30cm; flowers to 5cm, bright red, pubescent; 'Feuerschein': pale tangerine-red to scarlet flowers; 'Innocence': compact, with bright green leaves and snow white flowers; 'Splendens': large, bright scarlet flowers); *S.concinna* (Brazil; flowers trumpet-shaped, lilac above, pale mauve to yellow-brown below, purple-spotted inside, upper lobes deep purple, lower lobes paler). *S.speciosa* (syn. *Gloxinia speciosa, S.regina*; FLORISTS' GLOXINIA, GLOXINIA; Brazil; to 30cm; flowers 4–8cm, stoutly bell-shaped with a broad, spreading limb, purple, lavender, red or white in wild, variously coloured in cultivation; hybrids and cultivars include Fyfiana Group, with large, campanulate, erect flowers, white, through pink, yellow and orange, to red and violet, variously blotched and striped; 'Blanche de Meru', white, fringed pink; 'Boonwood Yellow Bird', yellow; 'Chic', flame-red;

'Mont Blanc', pure white; 'Violacea', deep violet-blue; Maxima Group, with large, nodding flowers; 'Buell's Blue Slipper', trumpet-shaped, curved, mauve-blue flowers; 'Buell's Queen Bee', white with a bright pink blotch on either side of throat; 'Pink Slipper', rose-pink, with a dark centre and spotted throat).

Sinofranchetia (for Adrien René Franchet (1834–1900), French botanist). C and W China. Lardizabalaceae. 1 species, *S.chinensis*, a fast-growing, deciduous, dioecious twining shrub to 10m. The leaves consist of three leaflets, each to 14cm long, ovate, entire, dark green above, blue-green beneath. Small, dull white flowers are carried in pendulous racemes, 10–30cm long, in late spring. They are followed by globose, lavender-purple berries. Cultivate as for *Holboellia*.

sinuate of the outline of a margin, wavy, alternatively concave and convex. cf. *undulate*.

sinus the indentation or space between two divisions, for example between lobes.

siphunculi a pair of horn-like structures on the rear end of the abdomen of aphids. They differ in shape and size and so are useful characteristics in determining species. Siphunculi function as anti-predatory devices, secreting a fluid that physically interferes with predators and also produces alarm pheromones.

Sisyrinchium (name used by Theophrastus for a plant related to *Iris*). Iridaceae. N, C and S America, with 1 species in Ireland; also Australia, New Zealand and Hawaiian Islands, probably naturalized in these countries. About 90 species, perennial herbs with clumped fans of 2-ranked, linear to sword-shaped leaves. Short-lived, starry to cup-shaped flowers are carried in clusters on slender stalks from spring to summer. Those described here are hardy to between –10 and –15°C/14–5°F. Grow in freely draining, sandy loam soils with additional organic matter. *S.californicum, S.bellum* and *S.angustifolium* prefer moist soils, but still need good drainage. Position in full sun or part-day shade. Most species will self-seed if conditions suit. Propagate by seed in spring or autumn or by division in early spring.

S.angustifolium (syn. *S.anceps, S.graminoides, S.iridoides*; BLUE-EYED GRASS; SE US; 15–45cm; flowers about 1.5cm in diameter, blue, yellow in centre); *S.bermudianum* (BLUE-EYED GRASS; West Indies; 15–45cm; flowers 1.5cm in diameter, blue, yellow in centre); *S.californicum* (syn. *S.boreale, S.brachypus*; W US; to 40cm; leaves grey-green; flowers to 3.5cm in diameter, bright yellow, sometimes turning orange on drying); *S. idahoense* var. *bellum* (syn. *S.bellum, S.eastwoodiae*; CALIFORNIAN BLUE-EYED GRASS; California; 10–50cm; flowers to 2cm in diameter, violet-blue with purple veins); *S.striatum* (syn. *S.lutescens*; Argentina, Chile; 40–80cm; flowers to 3cm in diameter, cream or pale yellow, veined with brown; 'Aunt May': leaves green tinged grey, striped cream). For *S.douglasii*, see *Olsynium douglasii*.

Skimmia (from *shikimi*, the Japanese name). Rutaceae. Himalaya through E and SE Asia. 4 species, evergreen dioecious shrubs or small trees with leathery, aromatic leaves, dense panicles of small, white flowers from winter to summer and, where plants of both sexes are present, brightly coloured berries. Fully hardy and tolerant of a wide range of conditions, including heavy atmospheric pollution. They thrive in partial shade on free-draining rich soils. Most will grow fairly well in fully exposed sunny places, but the leaves may yellow, fading and scorching at their margins. Propagate from cuttings of new growth taken in early summer and given bottom heat, or from semi-ripe wood placed in a cold frame in late summer. *Skimmia* will also grow from seed sown outside in spring.

S.anquetilia (Afghanistan, Pakistan, India, Nepal; creeping or erect, shrub, 30cm–2m; leaves to 18cm, oblanceolate to oblong or elliptic, pale green or yellow-green; inflorescence globose, 5cm in diameter, flowers green-yellow to yellow, unpleasantly scented; fruit orange or red); *S. × confusa.* (*S.anquetilia × S.japonica*; garden origin; spreading shrub, 0.5–3m; leaves 7–15cm, oblanceolate to narrowly elliptic; inflorescence to 15cm, pyramidal; flowers sweetly scented, creamy white to white; fruit red; includes 'Kew Green', with all-male, creamy-yellow flowers, and 'Isabella', with female, white flowers and red berries); *S.japonica* (Japan, China; erect or low shrub, 0.5–7m; leaves oblanceolate to elliptic, entire to obscurely crenate, dark to yellow-green; male inflorescence well-developed, female and hermaphrodite inflorescences 1–5-flowered, flowers very sweetly scented white, sometimes flushed pink or red; fruit red; 'Bowles' Dwarf Male': to 30cm, with wide leaves, tinted red in winter, and abundant flowers; 'Bronze Knight': male, of open habit, tall, with long leaves, tinted red in winter, and flowers red in bud; 'Fortunei': angular habit, slow-growing, with abundant fruit; 'Nana Femina': female, of stiff, dense, mound-forming habit, with large, broad, thick, twisted, deep green leaves, flowers in large, round heads and abundant fruit; 'Nana Mascula': as 'Nana Femina', but male and more erect, smaller, slow-growing, with small, less twisted leaves and white flowers in dense heads; 'Rubella': male, with bright red leaves in winter, especially upper leaves, and flowers red in bud in a loose panicle); *S.laureola* (Nepal, Burma, China; low, creeping shrub or erect tree, 0.5–13m; leaves oblanceolate to narrowly elliptic, very dark green; inflorescence globose to pyramidal, flowers sweetly scented, creamy or green-white; fruit black; 'Fragrant Cloud': dwarf, wide-spreading, to 45cm, with more pointed, bright green leaves and abundant flowers).

skirret (*Sium sisarum*) a moisture-loving perennial of the Umbelliferae family, grown for its edible tuberous roots, which are slender, grey-skinned and with a flavour characteristic of carrot. Sow in spring for autumn/winter use or in late summer for the following spring; alternatively, prepare root cuttings in early spring. Plant on a moisture-retentive site, with plants spaced 20cm apart in rows 45cm apart; irrigate if necessary. Skirret roots should be eaten young before they become woody. They may be lifted and stored in boxes of moist sand, kept within a building.

slasher a long handled billhook.

sleepy disease used of some kinds of wilt disease, especially that affecting tomatoes.

slime moulds lower fungi whose vegetative bodies consist of a plasmodium or a multinucleate mass of protoplasm. They are non-parasitic, but may be unsightly on low-growing plants, such as turf grasses, strawberries, and small ornamentals.

slip a colloquial term for a shoot, especially a cutting. The term is sometimes used more specifically to mean a cutting with a heel, or a shoot separated with roots attached, otherwise known as an Irishman's cutting.

slit drains rectangular plastic pipes capable of absorbing water along their length, layed to a relatively shallow depth in drainage systems.

slitting (1) making short horizontal cuts in a lawn with a sharp-bladed slitting machine, in order to improve aeration; (2) making vertical incisions in the bark of a fruit tree where the bark has become so hardened as to prevent an increase in girth, thereby restricting overall growth; (3) a form of wounding treatment applied to semi-ripe or hardwood cuttings to encourage growth. See *cuttings.*

slow-release used of fertilizer formulations that break down slowly in the soil, releasing nutrients over a long period. This may occur naturally as with a number of organic fertilizers, such as hoof and horn, that depend upon bacterial action for their activity, or it may be manufactured by coating individual granules with resin or sulphur. Slow-release substances are particularly useful for plants growing over a long period in containers. Also known as controlled release.

slugs and snails (Mollusca: Gastropoda) soft-bodied, unsegmented animals that secrete a slimy mucus from epidermal glands and a large slime gland opening below the mouth. The body is divided into a head, bearing tentacles and eyes, a muscular foot and a visceral hump. In snails, the hump is spiralled and protected by a shell; most slugs have either lost the shell completely or have a small internal shell, a notable exception being *Testacella* species, which have a small cap-like external shell at the rear end of the body (they are unusual in being carnivorous, preying mostly on earthworms). Slugs feed mostly at night and, providing the temperature is above freezing, will continue to do so throughout the winter. The main period of activity is in spring, summer and autumn, although they remain inactive during conditions of drought, drying winds, very low temperatures and heavy rain, sheltering under stones, logs, leaf litter or in the soil.

Breeding can occur throughout the year, but most eggs are laid during late summer and autumn. The eggs are opaque white or transparent, and are deposited in clusters of 10–50 in small cavities in the soil. Young slugs resemble the adults in shape; at first, the young feed mainly on rotting vegetable material but as they mature they eat an increasing amount of living plant material. Plants particularly susceptible to attack include tulips, *Dianthus*, *Gladiolus*, *Iris*, *Hosta*, hyacinth, lily, *Delphinium*, sweetpea seedlings, ripening strawberry fruits and many annuals and low-growing plants. Among vegetables especially prone to attack are peas, beans, lettuce, celery and potato tubers.

The GARDEN SLUG (*Arion hortensis*), which is dark grey to black, up to 40mm long, and with a yellow-orange foot, is an important European pest, attacking plants both above and below ground; it is typical of types that have no keel.

Some slugs have a keel only on the posterior half of the dorsal surface. Included in this group are the GREAT SLUG (*Limax maximus*), known as the SPOTTED GARDEN SLUG in the US, a black-grey slug up to 200mm long; the CELLAR SLUG (*L. flavus*), known as the TAWNY GARDEN SLUG in the US, which is yellow-grey and up to 100mm long, producing yellow slime; and the FIELD SLUG (*Agriolimax reticulatus*), known as the GREY GARDEN SLUG in the US, which is mottled light grey-brown in colour and up to 50mm long, with a chalky white slime.

Slugs with a well-developed keel down the back are known as KEELED SLUGS (*Milax* species); many of these are subterranean and attack the underground parts of plants, such as the tubers of potatoes and the bulbs of tulips. *M. budapestensis*, a dark brown or black slug up to 100mm long with a tripartite coloured foot, is a serious pest in northern Europe, whereas *M. gagates*, although present in Europe, appears to be more important in North America, where it is known as the GREENHOUSE SLUG. Specimens of *M. gagates* have a pale foot and can reach a length of 60mm.

Slugs and snails are abundant in most gardens, particularly if the soil is moist with a high organic content. Numbers may be reduced by good garden hygiene, including the removal of weeds and dead vegetation. Measures may be taken to encourage useful predators, such as thrushes, frogs, shrews, hedgehogs and ground and rove beetles. Cultivation of the soil and disturbance of places where slugs are sheltering will help to expose them to these predators. Where slugs are persistent pests, the application of mulches should be reviewed and dressings of manure or organic composts might usefully be replaced by inorganic fertilizers. Susceptible plants or crops should be protected from slugs and snails by using liquid or pelleted forms of recommended molluscicides. Pellets are poisonous to cats, dogs and birds and should be covered by netting or hidden under raised covers. As an alternative to poisonous baits, slugs may be trapped by placing cabbage leaves, grapefruit skins or hollowed-out potato tubers on the ground. Slugs shelter in such places during the day, and can be collected and killed by dropping them into a strong salt solution. A non-chemical treatment, which employs a microscopic eelworm *Phas-marhabditis hermaphrodita*, is available; it infects slugs to release a destructive bacterium. Treatment should be applied in the evening during March–April or September–October by watering. Earthworms, birds, insects or mammals are not affected.

Small black keeled slugs can damage maincrop potatoes, and may be treated with molluscicide when they surface during warm moist weather. Less susceptible cultivars include 'Wilja', 'Charlotte' and 'Estima'.

slugworms (Hymenoptera: Tenthredinidae) used of certain sawfly larvae bearing a superficial resemblance to slugs. The PEAR AND CHERRY SLUGWORM (*Caliroa cerasi*) is a cosmopolitan species, occurring in Australasia, North and South America and Europe. The larvae, which taper from front to rear end, are shiny black, up to 15mm long and covered with slime. They feed on the upper surface of leaves, removing all the tissue except the main veins and the lower epidermis. Plants attacked include ornamental and fruiting varieties of cherry, pear, plum and, occasionally, apple, almond, hawthorn, quince and rowan. The ROSE SLUG SAWFLY (*Endelomyia aethiops*), a widespread pest in Europe and North America, has soft, pale green larvae, up to 15mm long. They infest mostly the underside of rose leaves, leaving the upper epidermis and main veins intact; when plants are grown in deep shade, the larvae may feed on both leaf surfaces. Also damaged by species of slugworm are oak, lime, willow, poplar, birch and beech. Slugworms may be controlled with recommended contact insecticides when the larvae are first noticed. See *sawflies*.

Smilacina (diminutive of *Smilax*). Liliaceae (Convallariaceae). N America, C America, Asia. FALSE SOLOMON'S SEAL, SOLOMON'S FEATHERS, SOLOMON'S PLUME. 25 species, perennial herbs, distinguished from the related *Polygonatum* by the racemose or paniculate terminal inflorescence and free, spreading tepals, and from *Maianthemum* by its floral parts in threes, not twos. Suited to the woodland garden and other shaded situations, and hardy to at least −20°C/−4°F. Grow in a deep, fertile, humus-rich and moisture-retentive soil that is neutral to slightly acid and does not dry out when plants are in leaf. Propagate by division. *S.racemosa* (FALSE SPIKENARD, SOLOMON'S ZIGZAG, TREACLEBERRY; N America, Mexico; stems to 90cm, arching; leaves to 15cm, lanceolate to elliptic or ovate; panicles to 15cm, many-flowered, flowers small, white, sometimes tinged green in frothy terminal panicles; fruit green to red, sometimes mottled red or purple).

Smilax (Greek name for this plant). Liliaceae (Smilacaceae). Tropics, temperate Asia, US. GREENBRIER, CATBRIER. Some 200 species, dioecious, perennial, evergreen or deciduous, climbing or scrambling vines. The stems are often prickly; the leaves are leathery, smooth and conspicuously veined. The flowers are green to white, small and carried in clusters and racemes in spring and summer. They give rise to spherical berries. Florists' smilax is usually *Asparagus asparagoides*. Grow in semi-wooded or woodland locations on damp, fertile soils rich in leafmould. All are hardy to –20°C/–4°F and should be allowed to scramble through trees and shrubs and over tree stumps, or be trained on walls, in sun or part to deep shade. Propagate by division or by seed.

S.aspera (Canary Islands, S Europe to Ethiopia and India; evergreen to 15m; stems zigzag, angled, spiny; leaves 4–11 cm, narrowly to broadly lanceolate, triangular, ovate, oblong or kidney-shaped, abruptly narrowed above the usually cordate base, 5–9-veined, shiny on both surfaces with margins and veins prickly; fruit black or red; 'Maculata': with leaves blotched white); *S.china* (Korea, Japan; deciduous climber to 5m, stems sparsely prickly or unarmed; leaves to 8cm, broadly ovate to orbicular, sometimes wider than long, apex abruptly acuminate, base somewhat cordate, leathery or papery, 5–7-veined; fruit red); *S.glauca* (SAWBRIER, WILD SARSPARILLA; SE US; climbing deciduous or semi-evergreen shrub, stems spiny, often glaucous; leaves 4–13cm, ovate to lanceolate, broadly tapered or cordate at base, glaucous beneath, 7-veined, often mottled or streaked in a paler or darker silvery tone; fruit black, glaucous); *S.laurifolia* (LAUREL-LEAVED GREENBRIER, BLASPHEME VINE, BAMBOO VINE; SE US; evergreen climber, stems prickly below; leaves 5–20cm, narrowly ovate to oblong-lanceolate, thick-textured, 3-veined; fruit black); *S.rotundifolia* (COMMON GREENBRIER, COMMON CATBRIER, BULLBRIER, HORSE BRIER; E US; high-climbing, thinly woody, deciduous or partly evergreen vine to 10m, stems 4–angled, prickly; leaves 5–15cm, broadly ovate to circular, base round to cordate, thick-textured, 5-veined, dark green; fruit blue to black).

smoke a pesticide applied as fine dry particles which are produced and activated by igniting a mixture of the pesticide, an oxidant and combustible material. Smokes are most commonly used in greenhouses, which must be kept shut during treatment.

smother crop a crop that makes rapid dense growth and suppresses germinating weed seeds beneath its canopy.

smuts a group of diseases characterized by the formation of conspicuous masses of very dark soot-like spores. Smut fungi are particularly common on grasses, sedges and cereal crops, but only of relatively minor importance in horticulture, occurring on annuals, herbaceous plants, bulbs and vegetables; they do not attack shrubs and trees, apart from unusual forms that occur on palms. The fungi usually spread within the tissues of the infected plant before spore production occurs. They rarely kill their hosts but infected plants can be severely stunted. The most serious smut diseases of garden plants are those affecting bulbs and corms; for example, ONION SMUT, which leads to loss in yield and the formation of poor quality bulbs, and flower smuts, such as CARNATION ANTHER SMUT and SCILLA SMUT, in which the flowers are disfigured by masses of dark spores. The best control is to destroy smutted plants and dispose of any debris, thereby reducing infection in the following season. Also, avoid growing related plants in the same place continually. Spores may be very long-lived in the soil.

Smyrnium (name used by Dioscorides, from Greek *smyrna*, myrrh, referring to the smell of the plants). Umbelliferae. W Europe, Mediterranean. 7 species, glabrous biennial or monocarpic herbs with ternately compound leaves and small, yellow flowers in compound umbels on branching stems, sometimes subtended by colourful leaves and bracts. The following species is ideal for naturalizing in light, open woodland, or among shrubs. A hardy biennial, it prefers a moist, fertile site with filtered sun. Sow seed *in situ* in autumn or spring. *S.perfoliatum* (S Europe to Czech Republic; 40–150cm; leaves bright green; upper leaves lime green to golden, stem clasping; umbels golden yellow).

snag (1) a short stump of a shoot or branch left after accidental breakage or faulty pruning; (2) the short piece of rootstock stem retained after beheading of the stock (carried out once a grafted scion bud has established), and to which the developing scion shoot is tied for early support.

snails see *slugs*.

snow mould (*Monographella nivalis*) a common disease of lawns, especially where a high proportion of ANNUAL MEADOW GRASS (*Poa annua*) exists. It can occur at any time of year but is most severe in October and during mild spells in winter, and may appear following a thaw in places where snow has been trodden upon. Small patches of yellowing grass turn brown on death and gradually enlarge and coalesce; in moist weather, the patches become covered with white or pink cotton-like fungal growth. The disease is encouraged by lack of aeration and excessive nitrogen feeding after August. Scarifying, spiking and regular mowing mitigate against the disease, and it is worthwhile dispersing heavy dew on susceptible sites. Treatment with a recommended fungicide, applied as a preventative each month from August, offers some protection.

soakaway a pit filled with rubble or stones and, ideally, lined with brick, placed at the lowest point of a drainage system to collect excess water.

soap sometimes used as an insecticide against aphids and red spider mite, especially in the form of soft soap made from vegetable oil and potash. It is also used as an adjuvant or spreader. For both purposes more effective substances are available.

soboliferous clump-forming; producing suckering lateral shoots from below ground, usually applied to shrubs and small trees, for example *Aesculus parviflora*.

Sobralia (for the Spanish botanist Dr Francisco Sobral, fl. 1790). Orchidaceae. C and S America. Some 100 species, evergreen perennial herbs. The stems are tall, clumped and reed-like, and clothed with plicate, lanceolate leaves. Produced in spring and summer in terminal clusters, the flowers are large and showy, basically resembling those of *Cattleya* in shape. Pot in a coarse bark mix, with leafmould, garden compost and dried FYM. Maintain a minimum temperature of 15°C/60°F. Provide full sunlight or, when in bloom, light shade. Water and feed freely when in growth (which is often continuous); at other times, keep just moist. Propagate by division after flowering. *S.macrantha* (Mexico to Costa Rica; stems 1–2m; flowers to 25cm in diameter, fleshy, short-lived, fragrant, tepals rose-purple, lip rose-purple, white at base, centre tinged yellow, apex rounded, broad and frilly; many clones have been recorded, among them white, deep magenta and crimson forms).

sodding US term for turfing lawns.

sod-lifter see *turfing iron*.

soft fruit a group name for bush and cane fruits, such as blackcurrants and raspberries, and also strawberries; cf. *top fruit*.

soft rot a general term for those bacterial and fungal diseases that cause a soft and slimy decay of plant tissues. Storage rots of root crops and other vegetables, fruits, bulbs and corms can be caused by bacteria, mostly *Erwinia* species, or fungi such as *Rhizopus*. Soft rots can occur on growing plants, especially in very humid conditions under glass or on the sappy growth of garden plants caused by excess nitrogen. They often follow infection by the grey mould fungus (*Botrytis cinerea*). See *grey mould*.

soft water water that is relatively free of dissolved mineral salts, especially calcium. It is of relevance to gardening as collected rainwater.

softwood (1) a common name for coniferous trees that produce relatively softwood when compared to broad-leaved deciduous trees; also descriptive of their timber; cf. *hardwood*; (2) used of soft, green, young plant shoots that contain no woody tissue and which may be induced to root as cuttings.

soil the outermost part of the earth's crust, formed from weathering of rock, the growth of vegetation and the decomposition of plant and animal remains. It comprises a complex mixture of inert minerals, chemicals and organic materials intimately colonized by living organisms.

The deeper a soil is excavated, the lower is the percentage of organic matter present, and the more the material resembles its mineral origins; the horticultural terms 'topsoil' and 'subsoil' relate to this distribution.

Soil classification Soils can vary enormously in make-up and behaviour, depending on their location, history and husbandry. Extensive survey work in many countries has led to the recognition and naming of numerous natural soil types based on consistent characteristics, such as the presence and depth of different layers and soil geology. Maps showing the distribution of soil types or 'soil series' can be used to assess the nature of a soil in a given area, and to make inferences about qualities, such as the likely ease of working, fertility and drainage.

An alternative approach to soil classification is the 'land use capability index' based on direct assessment of general physical qualities only, such as aspect, stoniness, drainage and topography. This system classifies land in seven appraisal grades, covering a range of potential productivity under good management.

Of more relevance to gardening is the classification of soil texture.

Soil texture Soil texture describes the make-up of soil, based on the relative proportions of variously-sized mineral particles in the root zone. Particles larger than 2.0mm in size are classified as stones; those between 0.05mm and 2.0mm as sand; those between 0.002mm and 0.05mm as silt; and those less than 0.002mm as clay.

Soils composed of a very high proportion of sand particles are referred to as light soils. They do not aggregate well and are therefore susceptible to erosion; they are also drought-prone and poor in nutrient retention. They have the advantages of being relatively early-warming, and are usually freely draining and easily cultivated.

Soils in which clay particles predominate are referred to as heavy soils. They are potentially more fertile than sandy soils, showing greater facility in holding certain nutrients in chemical combination, in binding together particles, and in moisture retention. However, they are relatively slow to warm, susceptible to compaction and poor drainage, and are liable to bake hard in summer. Special care is needed in cultivation.

Silt soils are made up of intermediate-sized particles and are of smooth and silky texture. Although potentially fertile and more moisture-retentive than sand, they are unstable and liable to compaction.

Peat soils are formed from incompletely decomposed sedges or mosses. They are naturally wet and usually acidic, but where drained and dressed with fertilizer, can become highly fertile. Limestone or chalk soils are typically shallow and free-draining, alkaline and of moderate fertility. See *loam*.

Soil structure The structure of a soil relates to how well the soil particles are aggregated into crumbs, with a network of pores of different size connecting them, through which water, nutrients and air circulate. Soil structure is influenced by a great variety of factors – the presence of lime, nutrient fertilizers, and organic matter derived from animal and plant residues; the physical effects of frost, wetting and drying; the activity of plant roots, earthworms and microorganisms; and cultivation practices. Structure is at its most vulnerable when a soil is wet and liable to compaction; timely cultivation is important in producing a good tilth.

Soil nutrients Mineral nutrients required for plant growth are normally absorbed in solution from the soil. Those needed in relatively large quantities and known as macronutrients are nitrogen, potassium, phosphorus, magnesium, calcium and sulphur. Micronutrients or trace elements, which are also essential but only required in small amounts, are iron, manganese, copper, zinc, boron, molybdenum and chlorine. Nitrogen and potassium are the most readily depleted by intensive cropping, phosphorus rather less so. Soil nutrients may be supplemented by the application of organic or inorganic fertilizers. Dressings of bulky organic manure, and the practice of green manuring, are valuable means of maintaining nutrient fertility in well-managed gardens.

Soil pH A term commonly used in horticulture to describe levels of acidity, especially in soil and other growing media. pH values refer to the negative logarithm of the hydrogen ion concentration of an aqueous suspension. Soil acidity is recorded on a scale between pH 0 and pH 14, where the pH value increases as acidity decreases. pH 7 represents a neutral condition and scale points above it a less acidic, or more alkaline, condition. Acidity has several direct and indirect effects on the relationship between soil and plant growth. It affects the solubility and availability of soil minerals; the existence of soil microorganisms, which may be involved in the breakdown of organic matter or be present as plant pathogens; the state of soil structure; and the well-being of plant roots. A pH value of 7 indicates a neutral condition. For many horticultural purposes, a pH of 6.5 is recommended as optimum. Whereas most plants are tolerant of a wide pH range, some genera, such as *Erica* and *Rhododendron*, consist of species that thrive on acid soil, and these are known as calcifuges. Plants that do well on soil of high lime content, for example, *Betula* and *Delphinium*, are known as calcicoles.

Soil pH can be raised by the application of lime. The lowering of soil pH is more challenging and best attempted by applying flowers of sulphur, or iron or aluminium sulphate, in line with recommendations resulting from soil testing.

Soil water Water is held in soils as a film around individual soil particles and in the spaces between them. In a well-balanced soil, water is held in fine pores and air in large ones. A soil that is holding the maximum amount of water possible after drainage is said to be at field capacity. Winter rain replenishes the water lost in the growing season and brings soil up to field capacity. Waterlogging occurs when the amount of rain or irrigation exceeds drainage rate. It commonly occurs where a compacted layer exists beneath the topsoil, either due to a naturally-occurring impervious seam or due to compression resulting from surface treading or machinery traffic; a naturally high ground water table can also give rise to waterlogging. Under such conditions, oxygen in the soil is displaced, causing damage to and eventual death of roots where the situation is prolonged. Extreme fluctuation in water level can be detrimental to root growth. To maintain satisfactory plant growth, the gardener must take care to produce a balance between good water-holding capacity and drainage in the soil, as well as taking measures to replenish water that is lost during the growing season through uptake by plants and evaporation.

Salt levels Soil water contains many dissolved mineral salts, some of which are plant nutrients. Where these are present in too high concentrations, they can be damaging to plants and such a soil is termed 'saline'. Soil salinity can be determined by measuring the electrical conductivity of the soil; the less resistance there is to a current, the higher the salt levels. The term conductivity is therefore often used instead of salinity.

High salt levels make it harder for plant roots to extract water, causing drought stress; symptoms are a blueing of foliage, small stunted leaves and poor growth. Some salts are more harmful than others and sodium is regarded as the most detrimental.

Heavy dressing of soluble fertilizers can increase soil salinity to damaging levels, particularly in the case of plants with shallow, poorly developed roots and germinating seedlings. Soil salt levels quickly fall as a result of leaching, achieved by flooding treatment in greenhouses where crops are continually grown and fed in the border soil, but resulting naturally from rainfall outdoors.

Soil temperature The temperature of the soil is an important consideration for both the hardiness and growth rate of plants. In summer, soils are generally cooler than the air, and in winter, warmer. Soil temperature is affected by depth, texture and the amount of water held; wet soils tend to be comparatively cold and slow to warm up in spring. Thick ground mulches influence soil temperature, making the soil warmer in winter and colder in summer and slowing the rate of temperature change; their use favours the growing of warm-climate plants in areas with cold winters, and cold-climate plants in areas with hot summers. Thin mulches do not behave in the same way, although black polythene, which absorbs heat from the sun, can raise soil temperatures in summer and is helpful in speeding the growth of crops that need warmth, such as sweetcorn.

Soil life Soil is intimately colonized by many forms of life feeding off organic remains, chemical reactions, and each other. These organisms are involved in many essential functions, such as release of plant nutrients and improvement of soil structure. Soil organisms may be categorized as soil microflora (bacteria and actinomycetes), soil flora (fungi and algae), soil microfauna (protozoa) and soil fauna (worms and arthropods).

Bacteria and actinomycetes are present in vast numbers in most soils. Some of these form symbiotic relationships with plants, for example, with nitrogen-fixers such as alder (*Alnus*) and members of the family Leguminosae. Many types of bacteria play essential roles in both the breakdown and build-up of organic compounds within the soil that are required for plant growth.

Algae need minerals and light, and therefore occur near the soil surface. They play an important role in colonizing bare ground, in initiating the build-up of organic matter and in binding soil particles together.

Fungi colonize the soil with their growth strands and are important for the breakdown of complex organic matter. They prefer well-aerated soil and usually grow better in acidic conditions.

Mycorrhizae are a special group of symbiotic fungi that live in and extend a plant's root system. They extract nutrients that roots cannot extract and, in return, receive carbohydrates. Mycorrhizae are particularly useful in improving phosphorus uptake on deficient soils.

Most of the soil fauna prefer open, aerated soil and many live in litter above soil. Their food supply consists largely of dead and living plants and other soil organisms, and they play a vital part in the breakdown of organic matter, which eventually becomes humus.

Small soil animals indirectly influence the production of organic glues that bind soil particles together; large soil animals are important in improving aeration and helping to distribute organic matter and other organisms, such as fungi.

Worms are important soil organisms. Nematodes, which are usually too small to see with the eye, live in soil water films and, in dry conditions, form cysts in which they can survive for many years. Some feed on bacteria, fungi and insects, while others attack plants. Earthworms have a profound effect on soil formation, fertility, structure and aeration; they are most common in loam soils, and in undisturbed pasture rather than cultivated land. They are intolerant of drought, frost, and dry or sandy soils, and need reasonable aeration. Most require calcium and are unable to survive on very acid soils; they feed upon dead and decaying plant remains and are encouraged by the addition of organic waste. Earthworm action aids drainage and water infiltration, helps bind soil together and decomposes organic matter into mineral nutrients. Through their casts, they help transfer nutrients and fine soil particles from depth to the topsoil.

soil block a small block of compressed soil-based growing medium, into which seeds may be directly sown or seedlings pricked out. Soil blocks are formed into modules in a mould. They have the disadvantage of drying out rapidly and, for best results, a proprietary block-maker should be used and care taken in judging the correct moisture content of the medium used.

soil conditioner, soil ameliorant any material that improves soil structure, including animal manure, garden compost and leafmould, and inert materials, such as lime, gypsum, coarse sand or grit. Synthetic soil stabilizers such as polyvinyl alcohol, cellulose acetate, and polyacrylamide may be used to prevent soil capping.

soilless cultivation methods of growing plants without soil, particularly using peat/sand, coir, or bark mixes, but also including hydroponics, rock-wool and straw-bale culture.

soil mark the usually noticeable mark on the stem of a plant that indicates the soil level before it was lifted from the open ground or removed from a container. It is an essential guide to correct replanting depth.

soil moisture deficit the amount of water required to return a soil to field capacity.

soil profile a vertical section of soil at a particular site, running from the surface to the underlying parent rock.

soil sickness see *replant disease*.

soil sterilization the destruction of plant-damaging or competitive soil-borne organisms, such as insects, pathogenic fungi, some viruses and weeds, through the application of heat, usually in the form of steam, or chemicals. The process is, strictly, partial sterilization.

Heat treatment is generally more effective than chemicals for sterilizing loam for seed and potting composts and proprietary electrical soil sterilizers or steaming systems improvised by the gardener may be used. The effectiveness of soil sterilization by these methods depends on the level and duration of temperature; a useful treatment is to heat moist soil to between 71°C/160°F and 79°C/175°F for 10 minutes, thereby achieving pasteurization.

Soils with extreme pH levels or high organic-matter content should not be treated. Chemical sterilants vary in their range of effectiveness, and the most useful are not suitable for garden use. To overcome pest or disease problems in greenhouse border soil, the best course for the amateur is to resort to container growing or re-soiling.

soil structure see *soil*.

soil warming see *bottom heat*.

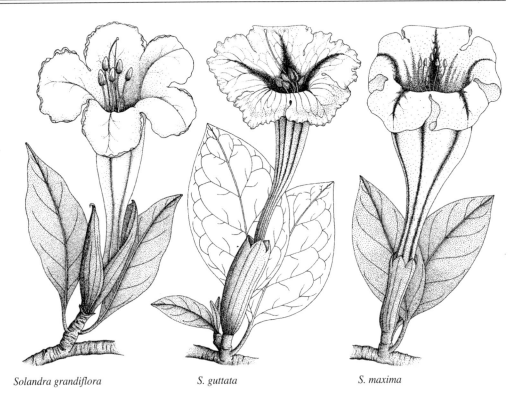

Solandra grandiflora   *S. guttata*   *S. maxima*

Solandra (for Daniel Carlsson Solander (1733–1782), botanist). Solanaceae. Tropical America. CHALICE VINE. 8 species, shrubby climbers with leathery leaves and very large and waxy, funnel-shaped and 5–lobed flowers. These are produced usually in spring and summer and are headily scented, especially at night. Provide a minimum temperature of 10°C/50°F. Plant in a well-drained, fertile soil, in full sun or part shade. To induce flowering, stimulate an initial period of shoot growth, then reduce watering until plants almost reach wilting point. Shorten long, vigorous shoots annually in late winter/early spring, removing weak growth altogether (flowering occurs on the current year's wood). Water plentifully during the summer months, sparingly in winter. Propagate by seed in spring and from greenwood cuttings.

S.grandiflora (Jamaica, Puerto Rico, Lesser Antilles; evergreen shrub, to 5m+; flowers white, becoming yellow to pink tinged yellow, to 23cm, lobes undulate to crenate); *S.guttata* (Mexico; evergreen climbing shrub, to 2m+; flowers to 26cm, throat pale yellow, spotted or striped purple, lobes crispate; many plants labelled as *S.guttata* are *S.maxima*); *S.maxima* (syn. *S.hartwegii*; Mexico, C America, Colombia, Venezuela; evergreen to deciduous shrub, to 4m; flowers yellow, to 20cm, tube with five purple veins, lobes reflexed).

Solanecio (Latin *sol*, sun, and *Senecio*). Compositae. Tropical Africa, Madagascar, Yemen. 15 species, evergreen, tuberous-rooted perennial herbs and shrubs, some climbing, most with thinly succulent foliage. Daisy-like flowerheads are produced from spring to autumn. Grow in full sun in a rather dry, well-ventilated atmosphere with a minimum temperature of 10°C/50°F. Plant in a fast-draining, loam-based medium. Provide support with canes or trellis. Water freely when in full growth, sparingly at other times. Increase by cuttings. *S.angulatus* (tropical Africa; perennial herb climbing to 3m; leaves to 12cm, deeply dissected; flowerheads yellow).

Solanum (Latin name for this plant). Solanaceae. Cosmopolitan, particularly tropical America. NIGHT-SHADE. About 1400 species, herbs, shrubs, trees or vines. They vary greatly in habit, but are for the most part toxic, rank and clammy to hairy. The flowers are star- to bell-shaped with five lobes. The fruit is a berry.

The hardiest climbers include the semi-evergreen, fast growing *S.crispum* (to at least –5°C/23°F) and its slightly tougher cultivar 'Glasnevin' (–10°C/14°F), along with the more tender *S.jasminoides* and its handsome cultivar 'Album' (although this last is known to have survived to –10°C/14°F). Grow in full sun against walls or fences on any fertile soil. Tie in to trellis or wires. *S.wendlandii* and *S.seaforthianum* are handsome climbers, requiring greenhouse conditions in temperate climates. Protect from the fiercest summer sun. Grow in the greenhouse border or in large pots in a medium-fertility, loam-based mix, providing a minimum temperature of 10°C/50°F. Ventilate freely as temperatures rise above 15°C/60°F; water plentifully and feed established plants when in full growth, reducing water in winter. Tie in periodically as growth progresses. Prune all climbers immediately after flowering, or in spring

Clockwise from top left: *Solanum crispum, S. jasminoides, S. wendlandii, S. seaforthianum*

before growth begins, shortening strong stems and removing weak and crowded shoots. Propagate by seed in spring or by softwood or greenwood cuttings in a closed case with bottom heat or under mist.

The poisonous fruits of the frost-tender shrublets *S.capsicastrum* and *S.pseudocapsicum* (WINTER CHERRIES) give a colourful display over the winter period in cool rooms or in window boxes where temperatures do not fall far below freezing. Sow seed in early spring at 18°C/65°F, pot on into a medium-fertility loam-based mix, pinching out for bushy plants. Grow on in the cool greenhouse or stand out for the summer, feeding fortnightly during the summer months. Bring indoors before the first frost into a minimum temperature of 10°C/50°F. Good light and humidity (a daily syringe) will prevent premature fruit drop. After berry drop, water sparingly and cut back in spring; these plants are, however, generally produced from seed each year.

S.quitoense, a large shrub with magnificent mauve foliage and yellow-orange fruit, is cultivated in the high altitiude tropics for fruit and juice production. *S.muricatum*, has larger fruit and can grow at lower altitudes. Grow under glass in cool-temperate regions and give minimum night temperatures of 7–10°C/45–50°F: ventilate freely and water carefully as both overwatering and underwatering will soon cause foliage wilt. Propagate from seed, by cuttings and by grafting. The hardy *S.melanocerasum* and *S. × burbankii* are grown for their edible berries, which resemble blackcurrants. Grow in sun in a fertile and well-drained soil. Sow seed *in situ* in spring, thinning to a final spacing of 60cm, in rows 75–90cm apart. Irrigate in dry periods. See also *aubergines* (*S.melongena), potatoes (S.tuberosum).* For *S.rantonnetii,* see *Lycianthes rantonnetii.*

S.capsicastrum (FALSE JERUSALEM CHERRY; Brazil; shrublet to 60cm, downy; leaves 4–7cm, oblong to lanceolate, entire to undulate; flowers small, white; fruit 1–2.5cm, ovoid, pointed, orange to scarlet or creamy yellow); *S.crispum* (Chile; coarse semi-evergreen vigorous shrub, climbing to 4m, branches stout and glandular; flowers fragrant, to 3cm in diameter, lilac-blue, anthers yellow; includes 'Glasnevin', of vigorous habit, with deep blue flowers with golden-yellow stamens in large clusters); *S.jasminoides* (POTATO VINE; Brazil; slender, deciduous vine, flowers star-shaped, 2–2.5cm in diameter, white tinted lilac-blue, stamens lemon-yellow; includes 'Album', with dark green, purple-tinted leaves and pure white, long-lasting flowers); *S.melanocerasum* (GARDEN HUCKLE-BERRY; origin unknown; widely cultivated in west tropical Africa; annual herb, to 75cm; flowers to 1cm in diameter, thick, white, anthers brown; fruit 1–1.5cm in diameter, black); *S.muricatum* (PEPINO, MELON PEAR, MELON SHRUB; Andes; perennial herb or shrub, erect or ascending, to 1m+; flowers to 4cm in diameter, violet-purple or white with purple markings; fruit pendent, to 10cm, ovoid to ellipsoid, white or pale green marked purple, flesh yellow, aromatic, edible); *S.pseudocapsicum* (WINTER CHERRY, JERSULAM CHERRY, MADEIRA WINTER CHERRY; Madeira, widely

naturalized; small, bushy shrub, 1–2m, thinly tomentose when young; leaves neat, dark green to mid-green; flowers star-shaped, 1cm in diameter, white; fruit 1–1.5cm in diameter, globose, bright orange, when ripe, succulent, toxic; includes 'Cherry Jubilee', with white, yellow and orange fruit, 'Giant Red Cherry', with orange-scarlet fruit, to 2.5cm, and 'Patersonii', of dwarf, spreading habit, with abundant fruit); *S.quitoense* (NARANJILLA, LULO; S America; shrubby scrambler or shrub, to 2m, densely stellate-pubescent, sometimes armed, all parts flushed purple and purple-hairy; flowers 6mm in diameter, white; fruit tomato-shaped, 4+cm in diameter, orange, fleshy juicy, green); *S.seaforthianum* (ST VINCENT LILAC, GLYCINE, ITALIAN JASMINE; S America, widely cultivated and naturalized in subtropics and tropics; vine to 6m, branches slender, glabrous or sparsely pubescent; flowers campanulate to rotate, to 2cm in diameter, blue, purple, pink or white); *S.wendlandii* (POTATO VINE, GIANT POTATO CREEPER, PARADISE FLOWER; Costa Rica; vine to 5m+, branches thick, sparsely armed with short hook-like spines; flowers rotate, 5cm in diameter, lilac-blue with paler interstices).

Soldanella (from Italian *soldo*, a coin, referring to the shape of the leaf). Primulaceae. Alps, Carpathians, Balkans. 10 species, dwarf, alpine perennial herbs with very small slender-stalked, round and smooth leaves and solitary flowers. These are nodding and bell-shaped with toothed to fringed margins. They appear in spring, often penetrating the snow blanket in their native areas. Grow on a sharply drained humus-rich soil with a neutral to slightly acid pH. Give an open position with a cool aspect, avoiding strong midday sun. Drier winter conditions may be provided by a propped pane of glass. Otherwise, grow in the alpine house or cold frame, in a gritty medium with added leafmould, giving bright, filtered light, cool temperatures and good ventilation; water plentifully in the summer months. Propagate from fresh seed or by division in spring.

S.alpina (Pyrenees, Alps to the Tirol; flowers violet or blue-violet, with crimson marks inside, to 15mm, cut to middle or beyond, lobes each cut into 4–5 lobules); *S.minima* (S Alps; flowers to 15mm, campanulate, divided to quarter of its length, pale violet or lilac to almost white, tinted with lilac streaks); *S.montana* (N Alps, Carpathians, Balkans peninsula; flowers to 15mm, divided to three-quarters of its length, blue to lilac, lobes deeply, divided into 3–4 narrow lobules); *S.villosa* (Pyrenees; flowers to 15mm, violet, divided to three-quarters length, lobes deeply cut into 3–5 lobules).

Soleirolia (for M. Soleirol, a military man, who collected extensively in Corsica (early 19th century)). Urticaceae. W Mediterranean Islands, naturalized W Europe. 1 species, *S.soleirolii* (MIND YOUR OWN BUSINESS, BABY'S TEARS, IRISH MOSS, CORSICAN CURSE), a dwarf creeping, mat-forming, evergreen, perennial herb. The stems are delicate,

Soldanella alpina

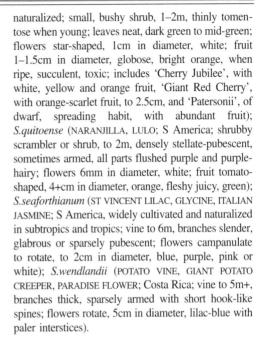

intricately branching, translucent and rooting. The leaves are 2–4mm long and rounded; the flowers minute. It includes cultivars with white-variegated, silver-grey, lime and golden leaves. Low, compact groundcover for under the greenhouse bench, terraria, and pots. It can become invasive. It is generally considered frost-tender, but may survive temperatures of –13°C/9°F, especially in damp, equable climates. Grow in a porous, humus-rich medium, with additional grit if necessary; keep evenly moist when in full growth, but reduce water as temperatures and light levels drop in winter. Grow in bright, filtered light or part shade. Feed container-grown plants occasionally with dilute liquid fertilizer. Propagate by division.

Solenostemon (from Greek *solen*, tube and *stemon*, stamen – the stamens are united into the corolla tube). Labiatae. Tropical Africa and Asia. 60 species, shrubby perennial herbs with rather succulent stems and softly hairy, ovate and toothed leaves, often beautifully marked and coloured. Small, 2-lipped flowers are borne in slender, branched spikes. *S.scutellarioides* and its many colourful cultivars and hybrids have long been familiar to gardeners as 'coleus' – striking foliage plants for year-round indoor display, for permanent outdoor bedding in frost-free climates and for half-hardy bedding use in zones 8 and under. Although easily raised from seed, to perpetuate true leaf colour, vegetative propagation is necessary. Root softwood cuttings of young side shoots in early summer under mist or in a closed frame. Pinch back young plants two or three times throughout the season to enhance their shape and foliage. They prefer a winter minimum temperature of 10°C/50°F, a rich, loam-based potting medium, plentiful water and bright, indirect light. They thrive in the home.

S.scutellarioides (syn. *Coleus blumei, C.hybridus, C.pumilus, C.scutellarioides, C.verschaffeltii*; COLEUS, PAINTED NETTLE; Malaysia, SE Asia; perennial herb or subshrub, procumbent or erect to 1m; stems semi-succulent, 4-angled; leaves ovate to deltoid or oblong, crenate, dentate or incised, minutely downy; flowers small, blue-violet, in spikes; the 'coleus' of florists, with many cultivars and seed races ranging in habit from dwarf or trailing to tall; height from 20–60cm; leaves 3–15cm long, oak-leafed to heart-shaped or saber-like, edges sometimes deeply serrated or frilled, in a wide range of brilliant colours including bright pink, rose, scarlet, ivory, dark green, purple and black with edges and markings golden, white, lilac, salmon, lemon, cream, burgundy, copper, lime and mixtures thereof).

Solidago (from Latin *solidus*, whole, from its traditional healing properties). Compositae. Northern Hemisphere, particularly N America; a few in S America, Eurasia and Macaronesia. GOLDENROD. About 100 species, hardy perennial herbs with clumped, erect stems, clothed with narrow, toothed leaves and terminating in spreading panicles of numerous, small, button-like and golden flowerheads in late summer. Suitable for the larger herbaceous border; its hybrids and cultivars are greatly valued for their tolerance of a range of conditions in cultivation and for their late summer colour. Grow in any moderately fertile and retentive soil in sun or part shade. Propagate by division. *S.virgaurea* (Europe; to 1m; flowerheads to 1cm in diameter, golden, many in a thyrse or panicle with ascending racemose branches; subsp. *alpestris*: to 20cm, with few, large flowerheads in compact, spike-like racemes). Cultivars smaller cultivars include the 30cm yellow 'Golden Dwarf', the 30cm 'Queenie' ('Golden Thumb') with variegated gold and green leaves, the 40cm light gold 'Laurin', the deep gold 50cm semi-dwarf 'Golden Gate', and the early-flowering 50cm 'Praecox'; taller cultivars include the 70cm buttercup-yellow 'Leraft', the 80cm deep yellow 'Ledsham' and the paler gold 80cm 'Goldschleier'.

× **Solidaster** (*Aster* × *Solidago*). Compositae. Garden origin. An intergeneric hybrid more or less intermediate between its parents, × *S.luteus* is a perennial to 1m tall, with daisy-like flowerheads to 1cm in diameter in corymbose panicles, the ray florets canary yellow, becoming creamy yellow with age, and the disc florets golden yellow. Lacking the vigour and coarseness of its *Solidago* parent, and with flowers that mature to a softer, more creamy yellow, × *Solidaster* is suitable for the sunny herbaceous or cut-flower border. Propagate by division or basal cuttings.

solitary used either of a single (i.e. lone) flower that constitutes a whole inflorescence, or a single flower in an axil.

solitary bees (Hymenoptera: Apoidea) a group of mainly beneficial garden insects differing from social bees in that each female constructs and provisions her own nest without the assistance of her offspring. Nest sites vary according to the type of bee, but include tunnels which they excavate in soil or rotten wood, hollow plant stems, and empty snail shells. In most species, the female dies once the nest is complete and the next generation of bees may not emerge until the following year. Solitary bees are useful pollinating insects of fruit trees and bushes; those most likely to be seen in gardens are the BURROWING BEES (*Andrena* species), which sometimes nest in lawns and paths. Solitary bees do little damage to lawns and soon disappear once the breeding season is over, so they should be left unmolested. See *leaf-cutter bees*.

Sollya (for R.H. Solly (1778–1858), botanical anatomist and physiologist). Pittosporaceae. Australia. 3 species, evergreen, slender-stemmed climbing shrubs with narrow, dark green leaves and cymes of nodding, bell-shaped flowers with five obovate petals, produced from late spring to autumn. *Sollya* will tolerate light frosts, provided the wood

Sollya heterophylla

is well-ripened. In cold regions, it may be grown in the cool greenhouse or conservatory. Plant in a humus-rich, moisture-retentive but well-drained medium in full sun or part shade. Provide support. Under glass, shade from the hottest sun; water moderately when in full growth, less in winter; maintain a minimum temperature of 5°C/40°F. Propagate by seed sown at 20°C/70°F, by greenwood cuttings in early summer. *S.heterophylla* (BLUEBELL CREEPER, AUSTRALIAN BLUEBELL; W Australia; 0.5–1.5m; flowers to 2cm across, bright blue).

soluble fertilizer a type of fertilizer that readily dissolves in water; used in solution for liquid feeding, including foliar application, and also in solid form as a base or top dressing.

Sonerila (from the native Malabar name for one species). Melastomataceae. SE Asia, S China. Some 100 species, tender herbs and subshrubs, often fleshy-stemmed with deeply veined and marked leaves and spikes of rounded, rose-pink flowers. Grow in a coarse, open, soilless medium with leaf-mould or other organic matter, sharp sand and charcoal. Admit bright filtered light and maintain high humidity. Keep evenly moist, but never saturated. Provide a minimum temperature of 15°C/60°F. Propagate by cuttings in late winter/early spring. *S.margaritacea* (Java to Burma; procumbent, red-stemmed herb to 30cm; leaves to 12cm, oblong to lanceolate, claret to dark green above with pearly-white spots or zones in lines between veins, pale with red-purple veins beneath; flowers to 1.5cm in diameter, rose).

soot residue collected during the sweeping of domestic chimneys; useful as a minor nitrogenous fertilizer (containing 2–4% nitrogen) and for improving soil structure provided it has been weathered for at least three months to remove toxic ingredients, such as sulphur, and then stored dry. It can be prepared as a liquid feed by filling a porous bag, which is then suspended in a tank of water. Traditionally, soot is suggested as a slug and snail deterrent for sprinkling around susceptible plants.

sooty bark (*Cryptostroma corticale*) a fungal disease of SYCAMORE (*Acer pseudoplatanus*) and other *Acer* species, the leaves of affected trees wilting and the bark lifting to reveal masses of black pores. The fate of a tree depends on the extent of bark infection. Although widespread in Europe and N America, the disease is erratic in appearance, occurring in hot summers when trees are under water stress. Infection probably takes place through wounds, such as those inflicted by squirrels, and these animals may be vectors.

sooty blotch a fungal disease (*Gloeodes pomigena*) of top fruits, characterized by black spots on affected fruits.

sooty mould a superficial black fungal growth that sometimes develops on plant surfaces, particularly those of evergreen shrubs in warm climates. It is formed by various fungi, especially species of *Alternaria*, *Capnodium* and *Cladosporium*, that grow on the honeydew excreted by aphids, mealybugs, whiteflies and other sap-sucking insects. The sooty moulds are not parasitic and prevention is by control of the honeydew-producing insects. On house plants and greenhouse plants, the mould can be cleaned off.

Sophora (from Arabic *sophera*, a tree with pea flowers). Leguminosae. Cosmopolitan. Some 52 species, deciduous or evergreen trees, shrubs or robust woody-based herbs with pinnate leaves and clusters of pea-like flowers in spring and summer. The late-flowering, fragrant *S.japonica* tolerates heat and drought, poor soils and city conditions; when mature, it will withstand temperatures as low as –25°C/–13°C. *S.microphylla* and *S.tetraptera* need the shelter of a south-facing wall and excellent drainage. They are hardy to about –10°C/14°F. Grow all in full sun. Propagate from seed sown fresh or in early spring under glass; pre-soaking in hot water may improve germination. Cultivars of *S.japonica* are grafted on seedling understock of the type. Take softwood or semi-ripe cuttings of semi-evergreens. Increase also by air-layering.

S.japonica (JAPANESE PAGODA TREE, CHINESE SCHOLAR TREE; China, Korea; deciduous tree to 25m; leaflets 2.5–5cm, in 3–8 pairs, ovate to lanceolate, dark green; panicle terminal, 15–25cm, flowers to 1.3cm, creamy-white fragrant; 'Columnaris'; columnar; 'Pendula': weeping; 'Regent': heat-tolerant, vigorous, oval-crowned, to 18m tall, with large, glossy dark green leaves and white flowers, produced from an early age; 'Variegata': leaves edged pale cream; 'Violacea': flowers flushed violet); *S.microphylla* (New Zealand; evergreen tree to 9m, branches flexuous, tangled; leaflets to 0.8cm, to 30 pairs, closely opposite, ovate to oblong, dark green; flowers to 4.5cm, golden-yellow, rather tubular, clustered); *S.tetraptera* (KOWHAI; New Zealand; evergreen or semi-deciduous shrub or tree, 4.5–12m; branchlets yellow-hairy, slender; leaflets 1.5–3.5cm, 10–20 pairs, ovate to oblong, sericeous; flowers 2.5–5cm, somewhat tubular, golden-yellow, pendulous).

× **Sophrolaeliocattleya** (*Sophronitis* × *Laelia* × *Cattleya*). Orchidaceae. Trigeneric hybrids. The plants are usually smaller than those of other *Cattleya* hybrids and tolerate slightly cooler conditions. The flowers are mostly intense, glowing colours of various shades of red.

Sophronitis (from Greek *sophron*, modest, referring to the small flowers and habit). Orchidaceae. E Brazil, Paraguay. 9 species, dwarf, evergreen, perennial epiphytic herbs. The pseudobulbs are small and clumped, each bearing a leathery, oblong to elliptic leaf. Produced from autumn to spring, either singly or in terminal clusters, the flowers are disproportionately

Sophronitis coccinea

large, brightly coloured and basically *Cattleya*-like in form. An exquisite dwarf orchid for well-ventilated situations in light shade in the cool house or growing cases (minimum temperature 10°C/50°F). Pot in small clay pans in a fine-bark mix with added sphagnum moss, charcoal and sterilized leafmould. Alternatively, attach to rafts or cork slabs. Water throughout the year, allowing a slight drying between each application; mist growing plants in warm weather. Propagate by division. *S.coccinea* (syn. *S.grandiflora*, *S.militaris*; E Brazil; pseudobulbs to 4cm; leaves to 6cm; flowers to 7.5cm in diameter, usually vivid scarlet, the lip marked yellow at the base).

Sorbaria

Sorbaria (from Latin, supposedly resembling *Sorbus*). Rosaceae. E Asia. FALSE SPIRAEA. 4 species, deciduous shrubs with freely suckering, slender stems, fine, pinnate leaves and foamy panicles of small, creamy white flowers in summer. Hardy to at least −15°C/5°F. Grow in sun or dappled shade, in fertile, moisture-retentive soils, with protection from cold winds. Prune in early spring, cutting back the previous year's growth to 2–3 buds, and removing old and weak branches at ground level. This species suckers very freely and may prove invasive. Propagate by removal of suckers in autumn or winter. *S.sorbifolia* (E Siberia, Manchuria, N China, Korea, Japan; to 3m; leaves to 25cm, leaflets to 9cm, lanceolate or oblong, tapering finely).

Sorbus

Sorbus (from Latin *sorbum*, the fruit of *Sorbus domestica*, a name used by Pliny). Rosaceae. MOUNTAIN ASH. N Hemisphere. Some 100 species, deciduous trees or shrubs with simple to pinnately lobed or pinnate leaves and small, white to cream or, rarely, pink flowers borne in spring on terminal corymbs. These are followed by small, usually rounded pomes, often brightly coloured and persisting well into autumn and winter. The following are fully hardy and valued for foliage, flowers and fruit. They will thrive on most fertile, well-drained soils – smaller species and those with pinnate leaves tend to suffer in dry positions. All prefer full sun. Propagate cultivars by T-budding or chip-budding in summer, or by whip and tongue grafting in winter. Occasionally species are grafted on to *Crataegus* rootstock – a practice to be discouraged since such plants almost invariably lean. Increase also by softwood cuttings in spring and early summer, treated with rooting hormone and in a closed case with bottom heat or under mist. Propagation from seed is complicated by dormancy. Collect berries as they begin to colour, remove flesh and sow seed immediately, in boxes or pots in a north-facing cold frame. Bring under glass as germination commences, and prick out on development of the second leaf. Return pots containing ungerminated seeds to the frame for a second winter: there may be a second crop of seedlings the following spring. Affected by fireblight.

SORBUS				
Name	**Distribution**	**Habit**	**Leaves**	**Fruit**
S.alnifolia KOREAN MOUNTAIN ASH	Japan, Korea	tree, to 20m; crown erect at first, becoming conical to rounded with age, young stems red-brown, shiny, sometimes hairy at first	5–10cm, ovate-elliptic, acuminate, saw-toothed, bright green, glabrous beneath, scarlet to orange in autumn	pea-sized, red and yellow
S.americana AMERICAN MOUNTAIN ASH	C & E US	tree to 10m or shrubby, branches red-brown, buds large, glutinous	to 25cm, pinnate with 11-17 oblong-lanceolate leaflets, serrate, bright green, slightly downy when young, grey-green beneath, turning golden-yellow in autumn	globose, to 0.5cm, scarlet-red
S.aria WHITEBEAM	Europe	tree 6-12m, often with multiple stems; crown broadly conical, young shoots and buds grey-felted	simple, elliptic-ovate, 8–12cm, coriaceous, silver-white at first, then dull sea green above and white-felted beneath	to 1.2cm, oval to rounded, orange-red, with powdery coating
Comments: 'Aurea': leaves white at first becoming golden yellow. 'Lutescens': young leaves silvery, slightly yellow above becoming silver-green, lower leaf surfaces remaining white. 'Magnifica': leaves large, dark green waxy above, white-felted beneath. 'Majestica': crown spreading, leaves large, dull green above, white beneath becoming green , fruit dark orange-red.				
S.aucuparia COMMON MOUNTAIN ASH, ROWAN	Europe to Asia Minor, Siberia	tree, 5–15m, occasionally with multiple stems; young shoots soon glabrous, grey-brown	to 20cm, leaflets 9-15 to 6cm, oblong-lanceolate, pointed, roughly senate, dull green, glabrous, grey-green, slightly hairy beneath	0.75cm, bright red
Comments: 'Rossica Major': vigorous; branches tightly ascending; leaflets 8 × 2.5cm, petiole red, woolly. 'Xanthocarpa': fruit orange-yellow.				
S.cashmeriana	Himalaya, Kashmir	tree to 10m; shoots glabrous, red.	leaflets 15–19, 2–3cm, oval-elliptic to oblong, long-tapered, rough, margins serrate, dark green glabrous above, pale green beneath, midrib and rachis brown-hairy at first becoming red	to 1.8cm, white
S.commixta	Korea, Japan	tree to 7m+, upright at first; buds glabrous, gummy; shoots red brown	leaflets 11-15, 4–8cm, elliptic-lanceolate, long-tapered, coarsely serrate, early foliage brown, becoming pale green, blue green beneath becoming yellow-red with age	0.75cm, globose, scarlet
Comments: 'Ethel's Gold': fruit golden-amber, long-lasting				
S.cuspidata	Himalaya	broadly conical tree to 12m; branches glabrous purple-brown	to 22 × 16cm or larger, elliptic-oblong, tapered, serrate, coriaceous, grey-green above, white-felted beneath	to 2cm, globose, orange-red

SORBUS Name	Distribution	Habit	Leaves	Fruit
S.decora	NE US	shrub or tree to 10m	leaflets usually 15, to 7cm, elliptic to oval-lanceolate, blunt or slightly tapered, dark blue-green, glabrous above, paler beneath, soon glabrous, petiole and midrib red	to 1cm, subglobose, red
S.esserteauana	W China	tree, 8-15m; buds red, white-hairy	18-29cm; leaflets 11–13, 6–10cm, oblonglanceolate, coriaceous, bright green above, densely downy beneath, turning red in autumn	0.8cm, globose, bright red
Comments: 'Flava': fruit orange-yellow. 'Winter-Cheer': fruit yellow later orange-red, long-lasting				
S.hupehensis	C & W China	tree to 10m; buds glabrous, shoots soon becoming so	leaflets 13-17, 2–5cm, elliptic-oblong, tips finely serrate,blue-green above, becoming pink- to red-tinted in autumn, grey-green beneath	0.8cm, globose, white becoming pink
Comments: 'Coral Fire': leaves red in autumn, fruit coral red; hardy. 'November Pink': vigorous small tree, fruit white-pink becoming purple-red, persistent. 'Rosea': leaves pale blue-green tinted pink then flame in autumn, fruit pink, persistent.				
S.insignis	Himalaya, Sikkim	small tree	to 30cm, leaflets 9-13, 5–10cm, elliptic-oblong, the central leaflets larger, lightly toothed; buds covered in rust-coloured silky hairs	pink
Comments:'Bellona': shrubby, conical; leaflets to 5 pairs, oblong-obovate, to 9cm, glabrous, grey beneath; fruit coral red. 'Ghose': small, erect; leaflets flushed blue beneath, oblong-lanceolate; fruit red to pink, abundant; hardy.				
S.intermedia	Scandinavia	tree to 12m, occasionally shrubby; young shoots densely tomentose, becoming glabrous	to 10m broadly ovate, blunt, thin, base rounded to wedge-shaped, lobed or occasionally pinnatisect below, serrate, waxy above, grey-woolly beneath	1.3cm, ellipsoid, orange-red
S. 'Joseph Rock'		erect, to 9m	pinnate, to 6 × 16cm, vibrant orange to plum in autumn, leaflets 15-19, narrowly oblong to serrate	cream to gold
S.latifolia SERVICE TREE OF FONTAINEBLEAU	Europe	vigorous tree to 15m+; crown broadly conical, branches shiny olive-brown	7–10cm, round-ovate, pinnately lobed, serrate, dull dark green above, yellow-grey-tomentose beneath	1.5cm, ellipsoid, brown-speckled
Comments: 'Henk Vink': crown narrowly pyramidal, fast-growing; leaves not lobed, wide, light green, grey tomentose beneath; fruit rusty red. 'Red Tip': upright, crown pyramidal; leaves small; fruit white spotted red.				
S.pohuashanensis	N China	small tree, twigs and buds hairy	leaflets 11-15, 3–6cm, elliptic to oblong-lanceolate, acute, tips sharply toothed, grey-green downy beneath; stipules very large, ovate	6-8mm, rounded, orange-red
Comments:'Chinese Lace': small tree; leaflets deeply toothed giving lace-like effect, rich autumn colour; fruit red, glossy, in wide clusters. 'Kewensis': (*S.pohuashanensis × S.esserteauana*) ; low, leaflets long, deeply cut, rich green, purple in autumn, grey beneath; fruit oval, bright red				
S.prattii	W China	tree to 8m; twigs glabrous	leaflets 21-27, 2–3cm, blue-green and downy beneath, tips toothed	to 0.8cm, white
Comments: var. *subarachnoidea* : leaves covered in rusty web-like down beneath.				
S.reducta	W China, Burma	dwarf, suckering shrub 15-40cm; shoots slender, thinly bristled	7–10cm, leaflets 9-15, 1.5–2.5cm, ovate to elliptic serrate, slightly downy above, glabrous beneath turning bronze-red in autumn	0.6cm, globose, carmine
Comments: Suitable for the large rock garden or wild garden				
S.sargentiana	W China	tree 7–10m, buds gummy, slightly downy; shoots stout	large, 20-25cm, leaflets 7-11, 8–13cm, oblong-lanceolate, serrate, downy beneath, turning brilliant orange to red in autumn	0.6cm, red
Comments: 'Warleyensis': leaflets smaller, petiole red; fruit smaller.				
S.scalaris	China	large shrub or small tree to 6m, shoots downy	10–20cm, leaflets 21-37, 2–3cm, narrowly-oblong, apex slightly toothed, deep green above, grey-felted beneath, turning red-purple in autumn	to 0.6cm, globose, bright red
S.scopulina	W US	shrub to 4m, erect, rigid; buds black, downy	leaflets 11-15, 3–6cm, oblong-lanceolate, serrate, glabrous, green above, blue-green beneath	0.5cm, globose, glossy red
S.thibetica	China	conical tree to 20m	large, oval, grey-white at first, later deep green above and silvery-white beneath	rounded, red-brown
Comments: 'John Mitchell': leaves to 15 × 15cm, rounded, silvery-white beneath; fruit pear-shaped, orange tinted red.				
S. × *thuringiaca* (*S.aria* × *S.aucuparia*)	Germany	broadly conical tree to 12m	dark green, oval, serrate, with 1-4 lobes or leaflets at base	small, red
Comments: 'Fastigiata': crown narrow, upright at first, later spreading.				
S.vilmorinü	China	fine shrub or small tree to 6m, branches spreading, buds and shoots downy, red-brown	8–12cm, leaflets 11-31, 1.5–2.5cm, elliptic, upper margins sharply serrate, glabrous above, grey-green beneath	0.8cm, globose, red becoming pink
Comments: 'Pearly King': small; leaves elegant, feathery, glossy green, yellow to red in autumn; fruit large, white flushed pink				

sorus in ferns, a cluster of sporangia, generally found on the underside of the leaf blade.

sour a colloquial expression for soil in poor condition because of strong acidity but more usually waterlogging, giving rise to poor structure.

southern stem rot (*Corticium rolfsii*), CROWN ROT, SOUTHERN WILT, SOUTHERN BLIGHT, is a fungal disease which in warm climate gardens may attack vegetables, herbaceous and bulbous plants, and, occasionally, greenhouse crops in cooler climates. The characteristic symptom of the disease is rotting of the stem at soil level, leading to yellowing and wilting of shoots and eventual collapse of the entire plant; it can also cause post-harvest rotting. Infection is encouraged by wet soil that is low in nitrogen and organic matter, with undecomposed plant remains are present in the top layers. Infected plants should be removed as soon as they are seen, together with the surrounding soil over a 15cm stretch beyond the diseased area. Long-term control involves rotation, deep cultivation and maintenance of high fertility.

spade the principal hand tool for digging, trenching and planting large specimens, consisting of a rectangular metal blade fixed to a long handle. The standard spade has a slightly dished blade of dimensions 32×20cm; medium versions have blades 25×16.5cm, while the border or ladies' spade has a blade 23×14cm. There may be a thickened tread moulded along the blade's top edges. In the UK, the blade is usually fashioned at a slight angle to the handle, which is no more than 1m long with a T-, Y- or D- shaped handgrip; elsewhere in Europe and in the US, the blade is fashioned to lie in line with a longer handle. The best quality blades are made of stainless steel.

There are specialized spades for lifting nursery stock, drainage and post-hole digging, and in some localities, spade work is undertaken with a long-handled pointed shovel. See *graft*.

spadix the fleshy axis of a spike; often surrounded or subtended by a spathe, most familiar as the club- or tail-like inflorescence of Araceae.

Spanish garden a style of design derived from the Islamic garden. The major features are courtyards or patios, often with arcaded sides, narrow canals, rectangular pools, fountains, and formal plantings.

Sparaxis (from Greek *sparasso*, to tear, referring to the bracts). Iridaceae. South Africa. 6 species, cormous perennial herbs; those listed here are 10–30cm tall with narrow, ribbed leaves. In spring and summer, they produce spikes of funnel-shaped flowers, each 4–6cm across with six, spreading segments. Cultivate as for *Ixia*.

S.elegans (syn. *Streptanthera elegans*; flowers with a yellow throat, and vermilion segments fading to pink, each with a violet band, sometimes marked with yellow at edge inside at base; 'Coccinea': flowers orange-red with near-black centre); *S.grandiflora* (flowers usually red-purple but sometimes white with purple marks, the tube yellow inside, purple outside); *S.tricolor* (flowers with a yellow throat and vermilion or salmon segments; 'Alba': flowers white, yellow segments at throat with dark central stripe; 'Honneur de Haarlem': flowers large, deep crimson, petals blotched black in middle with yellow markings; Magic Border Hybrids: hardy, with large flowers in a rich selection of colours, often four colours in one bloom).

Sparganium (name used by Dioscorides for a waterside plant probably *Butomus*). Sparganiaceae. Northern and southern temperate regions. 21 species, aquatic or marginal perennial herbs with reed-like stems, ribbon-like leaves and small, green flowers arranged tightly in globose heads on simple or branched racemes. Several species of BURREEDS are grown, notably *S.natans*. They are hardy to at least –15°C/5°F. Plant in full sun or part shade at the water's edge or in submerged containers. Remove dead foliage and cut back regularly to confine to bounds. Propagate by seed or division.

Sparmannia (for Dr Anders Sparmann (1748–1820), Swedish traveller, member of Captain Cook's second expedition). Tiliaceae. Tropics, South Africa. Some 4 species, hairy, evergreen shrubs or trees with large leaves. Borne in umbels on long stalks, the flowers consist of four obovate petals with a boss of long and colourful stamens. If touched, these move rapidly outwards. Provide a minimum temperature of 10°C/50°F, bright, filtered light and airy conditions. Pot in a fertile, loam-based mix in large containers or borders. Water and feed liberally in spring and summer; in winter, keep rather dry unless in warm conditions. Prune to restrict size in early spring. Increase by soft cuttings in spring, or by seed. *S.africana* (AFRICAN HEMP; South Africa; to 6m; leaves 15–30cm, cordate to ovate, irregularly toothed, bright green, hairy; flowers to 5cm in diameter, white, stamens yellow with red to purple tips).

Spartina (from Greek *spartion*, a grass used for cordage). Gramineae. W and S Europe, NW and South Africa, America, S Atlantic Islands. MARSH GRASS, CORD GRASS. Some 15 species, rhizomatous, perennial grasses. An adaptable, hardy grass found in freshwater swamps, salt marshes and wet prairies. It prefers damp or wet soils, thrives in full sun and will quickly colonize the banks of ponds and streams, flowering and fruiting over the summer and autumn. Propagate by division. *S.pectinata* (PRAIRIE CORD GRASS, FRESHWATER CORD GRASS, SLOUGH GRASS; N America; to 2m, stems tufted, robust; leaves to 120×1.5cm, narrow and arching, scabrous, turning yellow in autumn; inflorescence narrow, compact, erect, to 10cm; 'Aureomarginata': leaves edged golden yellow).

Spartium (from Greek *sparton*, the rush, which this plant resembles in its wand-like branches and its use as a fibre source). Leguminosae. Mediterranean regions and SW Europe; naturalized elsewhere. 1 species, *S.junceum* (SPANISH BROOM, WEAVER'S BROOM), a deciduous shrub to 3m tall with erect, rush-like, green branches and small, slender and short-lived leaves. Fragrant, golden yellow, pea-like flowers are borne in terminal racemes from spring to autumn. Hardy to –10°C/14°F. Grow in full sun in any well-drained but not too fertile soil. Propagate by seed.

Spathiphyllum (from Greek *spatha*, spathe, and *phyllon*, leaf, alluding to the leaf-like spathe). Araceae. Tropical America, Philippines, Indonesia. PEACE LILY. 36 species, evergreen, perennial herbs with glossy green, elliptic to lanceolate leaves and showy, long-lasting inflorescences – these consist of a club-like spadix backed by a sail-like spathe. These are ideal plants for the home and interior landscapes. They need growing conditions similar to those of *Dieffenbachia*, but are stemless and should be propagated by division.

S.floribundum (SNOWFLOWER; Colombia; leaves to 20cm, velvety dark green; spathe recurved, spreading, lanceolate to oblong-elliptic, to 8cm, white or green; includes 'Mauna Loa', compact, with deep green leaves and a white, fragrant inflorescence, and 'Mauna Loa Supreme', upright, tall, with rich green leaves); *S.wallisii* (Panama, Costa Rica; leaves 24–36cm, bright green; spathe ovate to oblong or elliptic, concave, to 17cm, white, becoming green, scented; the plant usually grown is the cultivar 'Clevelandii' (syn. *S.clevelandii*) with glossy green leaves to 40cm long and erect to slightly spreading, ovate, acuminate spathes, to 18cm and white).

Spathodea (from Greek *spathe*, a sheath, and *-odes*, resembling, referring to calyx form). Bignoniaceae. Tropical Africa. 1 species, *S.campanulata*, (AFRICAN TULIP), a tree to 20m with large pinnate leaves and terminal racemes of flowers; these are each to 12cm deep, scarlet to blood-red and campanulate with crisped lobes. Widely planted in tropical and sub-tropical regions as a park specimen and street tree. In cool temperate zones, it can be grown in the warm greenhouse or conservatory in a well-drained but moisture-retentive, high-fertility medium with full sun and a buoyant atmosphere. Water plentifully when in full growth and maintain a minimum temperature of about 15°C/60°F. Propagate by seed sown in spring or by semi-ripe cuttings rooted in a closed case with bottom heat.

spathulate, spatulate spatula-shaped, essentially oblong, but attenuated at the base and rounded at the apex.

spawn (1) the mycelium of cultivated mushrooms; (2) offsets of bulbs and corms, particularly of gladiolus; (3) used of raspberry suckers or primocane.

species the category of taxa of the lowest principal taxonomic rank, occurring below genus level and usually containing closely related, morphologically similar individuals often found within a distinct geographical range. The species may further be divided into subspecies, variety and forma. It is the basic unit of naming, forming a binomial in combination with the generic name. The species itself is signified by a specific epithet, or trivial name.

specimen a sample of plant material, living, dried and pressed or pickled, taken for purposes of identification, record-making and demonstration.

specimen plant (1) used of any plant, but usually a tree or shrub, which is typical of its species or cultivar and growing in prime condition; (2) used of a plant placed singly as a conspicuous feature.

spermatophyte a seed-bearing plant.

Sphaeralcea (from Greek *sphaira*, a globe, and *akaia*, mallow, referring to the globose fruits). Malvaceae. N and S America. GLOBE MALLOW, FALSE MALLOW. 60 species, annual or perennial, hairy herbs and subshrubs with racemes and panicles of mallow-like flowers in summer. Grow in a gritty medium in a frost-free greenhouse or overwinter in same and plant outdoors in late spring in a warm dry position in full sun. Propagate by seed, by division, or by softwood cuttings.

S.ambigua (DESERT MALLOW; SW US; white-hairy, somewhat shrubby perennial to 1m; flowers 3–6cm diameter, orange); *S.munroana* (western US; grey-hairy, erect perennial to 1m; flowers 3–5cm diameter, apricot-pink to pale red or orange).

sphagnum peat peat formed from the remains of sphagnum moss. It has excellent physical properties and is slow to decompose, with a pH level of 3.5–4.0, a high cation exchange capacity and a low nutrient content. Also known as moss peat; cf. *sedge peat*.

spicate spike-like, or borne in a spike-like inflorescence.

spiciform spike-shaped.

spiculate furnished with fine, fleshy points.

spike an indeterminate inflorescence bearing sessile flowers on an unbranched axis.

spikelet a small spike.

spiking a method of aerating lawns, carried out either with an ordinary digging fork, a hollow-tined fork, or a wheeled spiking machine. See *lawns, slitting*.

spiling making conical holes in soil by inserting a crowbar and rotating it to make a narrow shaft

45–60cm deep, which is filled with specially prepared compost. Taprooted vegetables, such as carrots and parsnips, are sown in each hole with the objective of producing excellent specimens for the show bench. The technique is also suitable for raising such vegetables on very stony sites.

spinach (*Spinacia oleracea*) an annual grown for its edible leaves. Believed to have originated in Central Asia, spinach was cultivated in China as early as the 7th century and in Ethiopia by the 14th century. Young leaves can be eaten raw in salads but older leaves are usually served cooked.

It grows best where average temperatures are around 15–18°C/60–65°F, but will withstand much lower temperatures, with young plants tolerating frost down to –9°C/16°F. It is liable to run to seed in dry weather, especially on soils low in organic matter. There are two forms of spinach, one with round seeds, used for a summer crop, the other a hardier type with prickly seeds, more suited for an overwintered crop; improved modern cultivars are often dual-purpose. With appropriate cultivars and sowing dates, it is possible in most temperate areas to maintain supply throughout the year.

Sow summer spinach at three-weekly intervals from February to May, the overwintered crop in August or September. Sow 2cm deep, in rows about 30cm apart, thinning to 15cm apart in the case of summer spinach and to 22cm apart in the case of the winter crop. Harvesting can commence 10–12 weeks from spring-sowing through to late autumn; the overwintered crop can be harvested throughout winter until spring, and cloches will improve its quality. Cut-and-come-again is possible under favourable conditions.

NEW ZEALAND SPINACH (*Tetragonia tetragonioides*) is spikier-leaved and half-hardy. A sprawling perennial, cultivated for its edible young shoots and succulent leaves, it was discovered growing wild in New Zealand by James Cook in 1770. It is indigenous to the coastal region of south and Western Australia and Tasmania. It spreads over the ground, producing smaller leaves and being less dependent on moist conditions than true spinach, or leaf beets. The 'seed' consists of a hard, dry, angular fruit, 8–19mm long, containing several true seeds; these are best soaked in water for 24 hours before sowing. For early summer use, raise singly in pots under protection for planting out, at 60–90cm² spacings, after any risk of frost has passed.

For SPINACH BEET (PERPETUAL SPINACH), see *swiss chard*.

spindlebush a training system for fruit trees, consisting of a cone-shaped framework around a single vertical stem, grafted on to a dwarfing rootstock, and supported by a permanent stake. The side shoots are tied down to induce fruit-bud formation and early cropping. The system is used especially for apples.

spine a modified stem or reduced branch, stiff and sharp pointed.

spinescent (1) bearing or capable of developing spines; (2) terminating in, or modified to, a spine-like tip.

spin trimmer a brush cutter, commonly known as a strimmer; a powered cutter with a spinning nylon cord or metal blades, used for cutting down weeds and grass, especially lawn edges or around the base of trees.

spinule a small spine.

spinulose bearing small or sparsely distributed spines.

Spiraea (from Greek *speiraia*, a plant used for garlands). Rosaceae. Asia, Europe, N America, Mexico. SPIREA, SPIRAEA, BRIDAL-WREATH. Some 80 species, deciduous shrubs with small, 4–5-petalled, white, pink or carmine flowers in racemes, panicles or corymbs in spring and summer. The following are fully hardy. Grow in full sun on a well-drained, moisture-retentive and moderately fertile soil. With those that flower in spring and early summer on the previous season's wood, prune immediately after flowering, thinning out old and weak growth by cutting back to within a few centimetres of old wood. Those that flower on the current season's wood can be pruned in early spring, and cut back to ground level. Propagate from softwood or semi-ripe nodal cuttings, or take cuttings of ripe wood in autumn and root in the cold frame. Thicket-forming species can be increased by rooted suckers in early spring.

S. × *arguta*. (*S.* × *multiflora* × *S. thunbergii*; GARLAND SPIRAEA; erect, to 2m, rounded, bushy; leaves to 4cm, oblanceolate, light green, serrate, initially downy beneath; flowers snow-white, in crowded corymbs); *S.* × *billardii* (*S.douglasii* × *S.salicifolia*; closer to *S.douglasii* but with leaves acute at apex and base, downy grey beneath; to 2m; leaves 5–8cm, oblong to lanceolate; flowers bright pink in narrow, dense panicle to 20cm; 'Alba': flowers white); *S.canescens* (Himalaya; to 4.5m, habit domed; branches thin, arching or pendulous; leaves 1–2.5cm, oval or obovate, apex usually blunt and dentate, dull green and sparsely lanuginose above, grey and densely lanuginose beneath; flowers white or dull cream-white, in hemispherical corymbs to 5cm in diameter). *S.douglasii* (western N America; erect, to 2.5m, compact, freely suckering; leaves 4–10cm, oblong, apex usually obtuse and serrate, dark green above, felty white beneath; flowers deep pink, crowded in erect, tomentose narrowly conical panicles to 20cm); *S.japonica* (JAPANESE SPIRAEA; Japan, China; to 1.5m, erect; leaves 2–8cm, lanceolate to ovate, acute to acuminate, biserrate, glaucescent with downy veins beneath; flowers rose-pink to carmine, rarely white, crowded in flat corymbs; 'Albiflora': flowers white; 'Allgold': leaves clear gold; 'Alpina': mound-forming, with rose flowers, in tiny heads; 'Anthony Waterer': upright, dwarf, with deep green leaves, occasionally cream and pink

spinach

Diseases
downy mildew
leaf spot
spinach blight
   (*see* blight)

spindle bush

variegated; 'Bullata': dwarf, with dark olive-green, coarsely serrate, crinkled leaves and crimson flowers; 'Bumalda': dwarf, with toothed leaves, often cream and pink variegated, and dark pink flowers; 'Goldflame': leaves bronze to copper turning pale gold to lime; 'Golden Princess': leaves neat, emerging copper turning gold to lime); *S.prunifolia* (BRIDALWREATH SPIRAEA; China, naturalized Japan; upright to 2m, dense, bushy; branches slender, arching; leaves 2.5–4.5cm, ovate, obtuse to acute, denticulate, bright green above, downy grey beneath; flowers pure white, double, in clusters along branches); *S.thunbergii* (China, Japan; densely branching shrub to 1m; branchlets wiry, angled; leaves 2.5–4cm, linear to lanceolate, acuminate to acute, sparsely and sharply serrulate, pale green, glabrous; flowers pure white, in clusters closely set on slender, nodding branches); *S.trichocarpa* (Korea; 1–2m; branches rigid, spreading, young branchlets angled, glabrous; leaves 3–6.5cm, oblong to lance-olate, few-toothed or entire, vivid green above, some-what glaucous beneath; flowers white, in rounded corymbs at ends of short, leafy twigs, forming a graceful spray; 'Snow White': bushy, to 1.5m, twigs drooping, leaves large, light green, flowers large, creamy white, in umbrella-shaped heads); *S.trilobata* (N Siberia, Turkestan to N China; intricately branching to 1.2m, broad, compact; stems often flex-uous, young branchlets slender; leaves 1.5–2.75cm, more or less circular, sometimes cordate, serrate, occasionally 3–5-lobed, blue-green; flowers white, crowded in umbels to 4cm in diameter at the ends of short, leafy twigs); *S. × vanhouttei* (*S.trilobata* × *S.cantoniensis*; to 2m; stems slender, rod-like, arching; leaves 3–4cm, rhomboid or obovate, occa-sionally 3–5-lobed, apex coarsely serrate, dark green above, somewhat glaucous beneath, glabrous; flowers white, in umbellate clusters to 5cm in diam-eter, borne profusely along branches); *S.veitchii* (W and C China; vigorous, to 3.5m; shoots arching, young branches tinted red, lanuginose, striate; leaves to 5cm, oblong or obovate, entire, sometimes slightly lanuginose beneath; flowers white, in crowded, dense corymbs to 6.5cm in diameter).

spit a spade's depth of soil, usually 25–30cm. The terms top spit, second spit, and third spit refer to the first three 25–30cm layers of soil found when working from the surface downwards. See *digging*.

spittlebug see *froghoppers*.

split stone a physiological condition of stone fruits, especially peaches, in which calcium deficiency and irregular water supply are implicated.

spoon-type applied to chrysanthemums, especially those in which the quilled florets expand at their ends into a spoon shape.

sporangium a sac or body that contains or produces spores; used especially of cryptogams.

spore the propagule of fungi, ferns and mosses, often specialized for a particular method of dispersal, and, in effect, similar to seed in higher plants.

sporophyll a leaf, or modified leaf, or leaf-like organ, which bears spores.

sporophyte in cryptogams, the vegetative, spore-bearing generation, as opposed to the gametophyte generation.

sport see *mutation*.

spraing a disordered condition of the flesh of potatoes characterized, in cut tubers, by rust-brown arc-shaped streaks arising in concentric or parallel formation. It is most likely to be caused by virus infection.

spray (1) a group of flowers on a single stem, used especially of chrysanthemums and carnations; (2) a pesticide applied mixed with water in the form of an emulsion, suspension, or solution using various types of spraying equipment. See *spraying and spraying equipment*.

spraying and spraying equipment insecticides, fungicides, and herbicides should be sprayed only where really necessary: when there is a rapid increase in numbers of a particular pest, danger of spread of disease, or competition from weeds that cannot be dealt with by cultivation methods. Cultural controls, such as pruning, removal of plant residue, rotations and manual weeding are alternative measures, as is encouraging natural enemies to provide biological control. Where spraying can be justified, care must be taken to avoid excessive use since this will reduce the survival rate of beneficial insects and also because the target organism may become resistant to a chemical that is regularly used; treatments should also be localized as spot applications in order to avoid contamination of a large area with pesticide. Generally, spraying upwards through a plant canopy achieves better distribution of a pesticide than drenching from overhead. Pests hidden deep in plant foliage may be controlled by systemic insecticides, while translocated herbicides, such as glyphosate, will move to underground parts of a target weed. It is very important to identify the cause of a problem correctly and to select the most appropriate chem-ical substance in each instance.

A pesticide is usually formulated with other chem-icals, such as surfactants, emulsifiers and 'stickers', so that it mixes with water easily, will spread over the plant surface and will resist removal by rain. Some pesticide products are sold as liquids to be poured and measured out by the user, whilst others are only available as wettable powders or dispersible grains that break up and form a suspension in water; with the user's safety in view, the latter are prefer-able since there is less risk of small particles being inhaled during preparation.

With most sprayers, the diluted spray liquid is applied to surfaces by forcing it, under pressure, through a small hole or orifice in a nozzle so that it breaks up into droplets. The droplets can vary in size, the largest tending to fall vertically downwards due to gravity while the smaller ones may be wind blown some distance as spray drift. Particular care is needed when applying herbicides to avoid drift which may accidentally damage susceptible valued plants.

Many small garden sprayers are fitted with a variable cone nozzle, which can be screwed clockwise to provide a wide-angled fine spray with mostly small droplets, or anti-clockwise to produce a coarser spray or to form a narrow jet of liquid. Inside the cap with the orifice is the swirl plate, which has one or more tangential slots directing the liquid flow so that it rotates through the orifice; the distance between the swirl plate and the orifice defines the angle of spray. Ideally, the position of the cap should be fixed whilst spraying with plain water, so that no later adjustments are necessary which could cause the operator to come into contact with the pesticide.

Other sprayers have a nozzle body and cap into which a variety of fittings can be inserted, enabling a particular volume of liquid to be applied or different spray patterns to be obtained. Deflector and fan nozzles are used to apply herbicides; cone nozzles are more suitable for pesticides and fungicides. The nozzles should always be held to the side and directed down wind to avoid the operator walking into the spray.

All types of sprayer consist of a container, one or more nozzles and a pump or other alternative method of liquid transfer to the nozzle, e.g. gravity. Filters are necessary on all but the most basic equipment, since spray liquid is readily contaminated with dust or other foreign matter.

Hand-operated trigger sprayer Here, a hand-operated trigger is used to draw liquid via a dip tube from a container, of usually less than one-litre capacity, and force it through a cone nozzle to produce an intermittent spray. Some pesticides can be purchased pre-packed in such sprayers.

Pressure pack This type of applicator, often referred to as an 'aerosol', consists of a lightweight metal can, containing a pesticide dissolved in a solvent, together with a gas propellent, such as compressed air or butane. When a plastic valve is opened, the pressure forces the liquid to the orifice, from where it issues as a fine spray, the process of atomization being assisted by volatilization of the liquid.

Syringe-type sprayer This is a cylindrical hand-operated pump, with a hose at one end connected to a container, and a nozzle at the other end. By holding the pump firmly in one hand and sliding the handle in and out, liquid is drawn from the container and forced through the nozzle on the compression stroke. This equipment is mainly suitable for spot treatment in gardens.

Compression sprayer This is the most widely used type of garden sprayer, requiring no pumping whilst spraying. Models have a robust plastic container of 1–10 litres capacity, with a removable hand-operated pump, which also provides a lid for the container. Air above the liquid is pressurized using the pump; this forces the liquid through a flexible delivery hose to a trigger valve on a hand-held lance, at the end of which is usually a variable cone nozzle. Pressure-indicating devices and safety valves avoid over-pressurization. Large sprayers of this type can be carried as a knapsack. Sprayers where compression is achieved by the need for continuous pumping during spraying are less convenient.

Calibration For correct dosage, it is desirable to know the output from a sprayer over a given time and area. This can be obtained by measuring the quantity of water sprayed into a receptacle for one minute, the distance of travel per minute and the width of the nozzle's spray pattern. The output in litres per square metre will equal the output per minute (litres per minute) divided by the speed (metres per minute) times the swath (metres).

Maintenance Spraying equipment should always be washed after use and kept dry while in store. Nozzles, washers and other parts are subject to wear and must be replaced periodically.

It is essential that appropriate protective clothing is worn during spraying operations. Always read the label to check what is advised for the product being used.

spreader see *adjuvant*.

Sprekelia (for J.H. von Sprekelsen (d. 1764), German botanist). Amaryllidaceae. Mexico. 1 species, *S.formosissima*, (JACOBEAN LILY, ST JAMES LILY, AZTEC LILY, ORCHID LILY), a bulbous, perennial herb with narrow leaves and solitary, scapose flowers produced in spring and summer. These are typically scarlet (sometimes deep crimson or white) with six perianth segments, the lower three overlapping and partly inrolled at their bases into a loose, waisted cylinder, their free parts narrowly lanceolate and drooping, the upper three more or less erect, outspread or reflexed, the central one to 2.5cm across and broader than the lateral segments. Within the cylinder, attached to the bases of the segments, are long stamens – these project down, outwards and upwards and end in large versatile anthers. Provide a minimum temperature of 7°C/45°F. Cultivation otherwise as for *Hippeastrum*.

Sprekelia formosissima

springtails (Collembola) small wingless insects, 1–5mm long, with biting mouthparts; typically, they have a forked tail, which when at rest is held under the body, but when released propels the insect into the air. Springtails abound in leaf litter and commonly occur in soil, especially soils that are moist and rich in the organic matter upon which they feed; infrequently, plant leaves may be damaged, especially at the seedling stage, by the biting of pin-sized holes. Soft stems may also be attacked, and damage to roots and root hairs of seedlings can cause plants to become stunted. Damage has been recorded on plants such as beans, carrots, and other root crops, cucumber, lettuce, tomatoes, cyclamen, cinerarias, chrysanthemums, conifers, orchids, violets and zantedeschias. Springtails are also often found in the compost of house plants. Some springtails attack mushrooms, making pits and tunnels in the stalk and caps. In most situations, limiting measures are unnecessary, predatory mites and insects providing adequate natural control. Where damage is persistent, springtails can be destroyed by soil application of recommended contact insecticides.

sprinkler see *irrigation*.

sprinkle bar see *dribble bar*.

sprouts (1) the small leafy axillary buds of Brussels sprouts; (2) any new shoot, such as the first growth of a potato tuber or the first aerial growth of a seedling, especially the first shoots of vegetables eaten in the seedling stage, of which there is a considerable variety. The Chinese traditionally used MUNG and SOYA BEANS (*Vigna radiata* and *Glycine max*); other suitable legumes include the ADZUKI BEAN (*Vigna angularis*), a small-seeded variety of BROAD BEAN (*Vicia faba*), known as SILK BEAN, ALFALFA (*Medicago sativa*), LENTILS (*Lens culinaris*), many kinds of PEA (*Pisum sativum*), ranging from normal GARDEN PEAS to RED-COATED PEAS and YELLOW FODDER PEAS, and the herb FENUGREEK (*Trigonella foenum-graecum*). Several cruciferous plants can be used, including most brassicas, especially CHINESE CABBAGE, SPINACH MUSTARDS, mild-flavoured TURNIP, ORIENTAL MUSTARDS, RADISH (*Raphanus sativus*), and the familiar MUSTARD AND CRESS. Other candidates include *Perilla frutescens*, WATER PEPPER (*Persicaria hydropiper*) and its relative BUCKWHEAT (*Polygonum fagopyrum*). ONIONS and LEEKS (*Allium* species) and grains of the grass family, including BARLEY (*Hordeum vulgare*) and RYE (*Secale cereale*) may also be used.

There are two kinds of sprouted seeds: seed sprouts which are grown in the dark and used between germination and the first signs of seed leaves, such as those of most beans; seedling sprouts which are allowed to grow into seedlings in the light and are green, for example, mustard and cress. The easiest domestic method of growing seed sprouts is in a jam jar, with the lid replaced by a fine mesh cover held in place by an elastic band. A dessert-spoonful of seed – which will expand to 7 or 10

times its bulk when sprouted – is placed at the bottom; it should be rinsed thoroughly by shaking in water and straining well. The lid is secured, and each morning and evening the jar is filled with tepid water, then shaken and drained. A temperature of 13–24°C/55–75°F must be maintained; sprouting is slower at the lower temperatures but the quality of sprouts is better.

Seed sprouts are usually boiled, steamed or stir-fried, or added to soup, but they can be eaten raw in salads, as are most seedling sprouts when green. Some leguminous seed sprouts, such as those of broad beans, contain toxins that may be harmful if consumed raw in quantity; cooking greatly lessens the toxicity. Ensure that seeds purchased have not been treated with pesticide for sowing outdoors. See *cress, mustard*.

spud a weeding tool with a narrow chisel-shaped blade, which is either straight, as in a thistle spud, or made with a hook-like projection on one side for extracting severed weeds. It may be long- or short handled.

spur (1) a tubular or sac-like basal extension of the perianth, generally projecting backwards and nectariferous; (2) a short branch or branchlet, bearing whorls of leaves and fascicled flowers and fruits from closely spaced nodes; often a short shoot terminating in a fruit bud, which may branch to form a multiple spur or spur system. Some cultivars of apple produce fruit spurs freely at the expense of lateral growths, and pears similarly.

spurred calcarate, furnished with a spur or spurs.

squamate, squamose, squamous covered with scales.

squamulose covered or furnished with small scales.

squarrose rough or hostile as a result of the outward projection of scales, prickles or bracts with reflexed tips perpendicular to the axis; also used of parts spreading sharply at right angles from a common axis.

squash see *pumpkin*.

squirrels (*Sciurus* species) arboreal rodents, of which only the grey squirrel (*S.carolinensis*) of North America, subsequently introduced into Europe, is likely to be a pest in gardens, especially those adjoining woodland. They may make holes in lawns, attack hanging baskets, dig and eat various bulbs, such as those of crocuses, lilies and tulips, attack nuts and fruits, eat the young shoots of various trees and gnaw bark, especially on beech and sycamore. Commercial control measures are not appropriate in gardens. Fruit cages help to deter attacks on fruits. Dormant bulbs can be protected by wire netting, laid over the soil surface and then removed once the bulb shoots break through. See *gophers*.

Stachys (Greek word for spike, used by Dioscorides in compound words descriptive of flowering spikes). Labiatae. Widespread, especially in temperate zones and eastern Hemisphere; also Australasia and tropical/subtropical mountains. BETONY, HEDGE NETTLE, WOUNDWORT. Some 300 species herbs, subshrubs or shrubs, usually hairy and with 2–lipped flowers whorled in erect spikes in summer. Fully hardy and among the most tolerant of silver-leaved plants, especially suitable for low edging. It is easily grown on any well-drained soil in full sun. Propagate by division. *S.byzantina* (syn. *S.lanata*, *S.olympica*; LAMB'S TONGUE, LAMB'S TAILS, LAMB'S EARS, WOOLLY BETONY; Caucasus to Iran; low-growing, perennial, densely white-woolly; leaves thick, 3–10cm, narrowed at base, oblong to elliptic; flowering stem erect, 20–90cm, flowers 15–25mm, purple or pink; includes 'Big Ears', with large, grey leaves, 'Cotton Ball', with fine foliage and no flowers, 'Primrose Heron', with bright gold leaves in spring, 'Silver Carpet', low, with grey, woolly leaves and flowers absent, and 'Striped Phantom', with leaves broadly striped cream).

Stachyurus (from Greek *stachys*, a spike, and *oura*, tail, referring to the shape of the racemes). Stachyuraceae. E Asia, Himalaya. 10 species, deciduous shrubs or small trees with small, campanulate to rounded flowers in narrow, drooping racemes appearing before the leaves in early spring. Both are hardy to about –15°C/5°F (lower with shelter). Grow in deep, fertile, well-drained leafy soils, in sun or light, dappled shade, with shelter from cold drying winds. Humus-rich acid soils are preferred, although *Stachyurus* will tolerate some lime. Propagate by fresh seed, by layering or by heeled semi-ripe cuttings in a closed case or under mist with gentle bottom heat.

S.chinensis (China; to 2.5m; leaves 3–15cm, ovate to oblong, crenate; racemes 5–10cm, flowers pale yellow, more spreading than in *S.praecox* and opening later; 'Magpie': leaves grey-green above, spotted pale green or pink, and with an irregular ivory edge); *S.praecox* (Japan; to 2m; leaves to 14cm, ovate to broadly lanceolate, serrate; racemes 5–8cm, flowers to 8mm, yellow).

stage, staging (1) see *bench, benching*. (2) a table or trestle on which exhibits are arranged at a flower show.

stag-headed describing trees suffering from a condition in which the ends of branches die, leaving leafless, woody projections beyond the leafy crown. The condition may be a symptom of disease, poor root condition or old age.

staking the provision of vertical stakes or poles to which the single stem of a plant is secured for support; used more generally for the provision of diverse plant supports. Except in situations of extreme exposure to wind, or on sites prone to vandalism, most trees require staking, at most, only for the period after planting when a specimen is establishing. Where stakes are left for more than about 18 months or, in the case of large specimens or mature transplants, 30 months, trees may become weaker and less supple in roots and stem than where not staked. Single stakes should preferably be inserted before planting, on the windward side of the stem, and with the tree secured with ties that do not chafe and which can be adjusted as girth expansion occurs to avoid constriction. See *tree ties*. For most trees, short single stakes are advisable; in some circumstances, however, support may be desirable up to the head of the tree and this can be achieved with a single tall stake, two parallel stakes with a cross member or by guying branches to three short stakes.

Stakes should be driven into the ground to no less than one quarter of their length. Trained fruit trees, such as cordons, fans and espaliers, require stake provision for longer periods of their life to aid development of tree structure.

Herbaceous plants, such as delphiniums, may require individual spikes to be staked, while other plants can be contained by inserting three stakes at regular intervals around the plant's circumference and holding growths in with a soft string tie (proprietary metal hoop supports serve the same purpose). Twiggy sticks are suitable for many clump-forming herbaceous plants. Whatever type of staking is used, it should be installed early and with the aim of it becoming unobtrusive. Careful choice of herbaceous plants will greatly reduce the need for staking.

Ornamental pot plants, such as *Stephanotis*, *Jasminum*, and *Hedera*, can be attractively staked with a tripod of canes or on hoops. For small pot plants, split canes and raffia ties are best; for large flowering specimens of chrysanthemums, bamboo canes are necessary.

In the vegetable garden, maximum yield from peas and beans is achieved with rows that are supported in some way. Although netting is effective, twiggy pea sticks and coppiced bean rods are visually pleasing. Climbing beans can be trained up sticks or bamboo canes, arranged in parallel rows, with opposing sticks or canes tied together near their tops and a line of horizontal canes placed along the points of contact, linking all pairs. An alternative means of staking involves drawing together the tops of four stakes set out in a square, to form a wigwam. Vegetable crops such as tomatoes and Brussels sprouts benefit from staking.

stale seedbed see *seedbed*.

stamen the male floral organ, bearing an anther, generally on a filament, and producing pollen.

staminate of the male; a unisexual, male flower, bearing stamens and no functional pistils.

staminode a sterile stamen or stamen-like structure, either rudimentary or modified and petal-like.

standard tree

standard (1) a plant grown with a bare stem or leg surmounted by a head of branches. The length of stem varies according to species grown, from 2m for a standard fruit or ornamental tree to 1–1.3m for plants such as a standard rose or fuchsia. A half standard tree has a leg about 1m high. (2) the large, uppermost petal in flowers of plants belonging to the Leguminosae sub-family Papilionoideae. (3) an erect or ascending unit of the inner whorl of an *Iris* flower.

standing ground a site on which plants in containers are kept to harden off or grow on before planting out or moving into a display greenhouse. Standing grounds usually take the form of beds covered with gravel, sand, weathered ashes or special types of irrigation matting.

stand out, stand down to place newly containerized plants on to a standing ground, bed or greenhouse bench on which they are to grow on.

Stanhopea (for the Earl of Stanhope, president of the London Medico-Botanical Society 1829–1837). Orchidaceae. C and S America. 55 species, epiphytic evergreen perennial herbs. The pseubobulbs are clumped and flask-shaped, each bearing one or two, lanceolate to elliptic leaves. Produced in spring and summer in strongly pendulous racemes, the flowers are large, thickly waxy, richly scented and short-lived. They consist of reflexed, wavy and curling, ovate to lanceolate tepals and a downward-pointing, fleshy lip, complex in structure and heavily marked. The large column arches forward over the lip. Suitable for the cool or intermediate house (minimum temperature 10°C/50°F). Plant in wooden baskets, lined with sphagnum and filled with a coarse bark mix with added leafmould and dried manure. Inflorescences tend to bury themselves and wither where pots are used; baskets will permit the racemes to escape. Water and syringe freely when in growth and suspend in a humid, airy place in bright, filtered light. Apply a weak foliar feed every second week when in growth. Once the new pseudobulbs are fully developed, reduce water and temperature and remove all shading. Until the emergence of new growths and roots, water only to prevent excessive shrivelling. Propagate by division.

S.eburnea (syn. *S.grandiflora*; Brazil, Guyana, Venezuela, Trinidad, Colombia; flowers to 15 × 12cm, waxy, ivory-white variously dotted purple); *S.tigrina* (Mexico to Brazil; flowers large, to 20cm in diameter, deep yellow, heavily barred or blotched purple-brown, strongly fragrant); *S.wardii* (Venezuela to Peru, Mexico to Panama; flowers overpoweringly scented of spice or chocolate, to 14cm in diameter, green-white to pale yellow or peach, variously dotted red-purple, orange-yellow at base).

Stapelia (named for J.B. von Stapel (d. 1636), Amsterdam doctor and author on the botanical works of Theophrastus). Asclepiadaceae. South, C and E Africa to E India. CARRION FLOWER, STARFISH FLOWER. 99 species, perennial herbs with clump-forming and sprawling, succulent stems. These are leafless and angled, the ridges flattened and coarsely toothed. Produced at various times of year, the starfish-like flowers are deeply 5-lobed. They are malodorous, fleshy and rugose or swollen with a lobed corona. Grow in a cool greenhouse (minimum temperature 10°C/50°F), with low humidity, direct sunlight from autumn to spring and bright, filtered light in the summer. Plant in a loam-based medium high in sand and grit. Keep more or less dry from mid-autumn to late winter; water sparingly at other times. Liquid feed with a low-nitrogen fertilizer from late spring to late summer. Increase by stem cuttings, layers and division.

S.flavirostris (Cape Province, Lesotho; stem to 17 × 2–3cm; flowers 13–16cm in diameter, deeply cleft, tube short, flat, lobes lanceolate, exterior pubescent, interior dull purple-red, lower part hairy with pale yellow or dull purple lines, tips dull purple, margins with red or white hairs); *S.gigantea* (syn. *S.nobilis*; GIANT STAPELIA; South Africa to Tanzania; stem 15–20 × 3cm; flowers very malodorous, 25–35cm in diameter, flat to campanulate, tube short, dark red, rugose, lobes deep, ovate-acuminate, attenuate, pale ochre-yellow with tiny crimson wrinkles, sparsely red-pubescent, margins white-ciliate, corona dark purple-brown). For *S.variegata*, see *Orbea variegata*.

Staphylea (from Greek *staphyle*, cluster, alluding to the arrangement of the flowers). Staphyleaceae. North temperate regions. BLADDERNUT. 11 species, hardy deciduous shrubs or small trees with pinnate leaves and 5-petalled, bell-shaped flowers in terminal panicles in late spring. These give rise to thin-walled, lobed and inflated capsules. Hardy to at least −15°C/5°F and tolerant of a range of soil types in sun or part shade. Prune to restrict size and shape, immediately after flowering; overgrown specimens can be rejuvenated by cutting hard back in winter.

Sow ripe seed in autumn in the cold frame. Root softwood or greenwood cuttings in summer.

S.colchica (Caucasus; shrub to 4m; flowers fragrant, to 1.5cm, in panicle 5–10cm long and wide, white); *S.holocarpa* (China; shrub to 5m or tree to 10m; flowers 0.8cm, rose in bud becoming pure white, in drooping axillary panicles to 10cm; includes var. *rosea*, with rose flowers); *S.pinnata* (BLADDER-NUT; Europe, Asia Minor; shrub to 5m; about 1cm, in panicles to 12cm, white, sepals tipped red).

Stauntonia (for Sir George Staunton (1737–1801), Irish physician, diplomat and naturalist; he accompanied Lord Macartney on his Embassy to China in 1793). Lardizabalaceae. E Asia (Burma to Taiwan and Japan). 16 species, dioecious, evergreen, twining shrubs with palmately compound leaves and, in spring and summer, racemes of small, cup-shaped flowers. These give rise to fleshy berries. Hardy to −5°C/23°F; plant in a sheltered site and cultivate as for *Akebia*. *S.hexaphylla* (S Korea, Japan, Ryukyu Islands; to 10m; leaflets 3–7, to 14cm, ovate to elliptic, dark green, glossyflowers to 2cm, white to ivory, with a violet tint, fragrant; fruit 2.5–5cm, globose to ovoid, purple).

steckling a young plant of a root vegetable, for example beetroot or carrot; referring particularly to a transplant, such as those specially raised by late sowing for overwintering and seed production.

steeping (1) the practice of soaking seeds or corms with very hard coats before sowing or planting, to ensure rapid germination or sprouting; (2) the preparation of liquid manure by suspending a porous bag of animal manure in a large container of water.

stellate star-like; commonly used of hairs with branches that radiate in a star-like manner from a common point.

stem and bulb eelworm (*Ditylenchus dipsaci*) a species of eelworm recorded on more than 450 hosts, worldwide. Garden plants most frequently attacked are narcissus, hyacinths, tulips, scillas and snowdrops, and also strawberries, potatoes, onions and phlox. This eelworm exists in various biolgical races, each of which shows a preference for particular groups of plants, although some hosts are common to more than one race.

The adults are microscopic, slender and just over 1mm long. Damage generally involves tissues becoming blistered or bloated, with a growth check leading to stunting, twisting of stems and leaves and distortion of flowers. On narcissus, for example, small pale yellow swellings known as 'spickels' appear on the leaves, and if the bulb is cut transversely, concentric brown rings can be seen. On tulips, flowers typically bend over, part of the bloom remaining green, and the foliage becomes ragged and tears easily. Infested onions become soft, flabby and swollen in the neck region, a condition known as 'onion bloat'. On annual and perennial phlox, the upper leaves become narrowed and crinkled and the stems brittle and stunted. See *eelworms*, *hot water treatment*.

stem rot a general term for disease affecting plant stems. There may not always be a clear distinction between a stem rot and other types of rot, such as collar rot, foot rot, leaf rot, and root rot, since the same disease organism may attack several different parts of the plant. Basal stem rot is sometimes used to describe a disease that originates from infection at soil level and spreads upwards.

stem-rooting applied to plants with stems that produce aerial roots, such as *Hydrangea petiolaris*, and to those with roots on subterranean stems, for example, some lilies.

Stenotaphrum (from Greek *stenos*, narrow, and *taphros*, trench, referring to the depressions in the raceme axis). Gramineae. New and Old World tropics. Some 7 species, annual or perennial grasses with creeping or ascending stems, rooting at nodes. The leaves are linear to oblong, with the blades held perpendicular to the stem and the broad sheaths compressed and overlapping. Used for lawns in the tropics and subtropics. In cooler regions, it is used as a cascading basket plant or as fast-spreading ground-cover in the warm greenhouse or in subtropical bedding. Grow in rich, moisture-retentive soil in sun. Propagate by plugs rooted in moist sand with bottom heat. *S.secundatum* (ST AUGUSTINE GRASS, BUFFALO GRASS; tropical America, W Africa, Pacific Islands; stems creeping and rooting; leaves to 15 × 1.5cm; 'Variegatum': leaves striped pale green and ivory).

Stephanandra (from Greek *stephanos*, crown, and *aner*, male, referring to the disposition of the persistent, numerous, short stamens in the shape of a crown). Rosaceae. E Asia. 4 species, deciduous shrubs, multi-stemmed and graceful with incised to serrate to shallow-lobed leaves, the veins conspicuous and impressed above. Small, star-shaped flowers are produced in corymbose panicles in late spring and early summer. Hardy to −20°C/-4°F. Grow in full sun or semi-shade, in well-drained moisture-retentive soil; mulch with well-rotted organic matter annually. Prune after flowering to remove old and overcrowded branches; overgrown plants can be cut hard back almost to ground level in winter. Apply a general fertilizer in early spring. Propagate by division when dormant, or by softwood or semi-ripe cuttings in summer.

S.incisa (Japan, Korea, Taiwan; much-branched shrub to 2.5m; branchlets slender, flexuous, terete; leaves to 5cm, ovate or triangular, apex caudate to acute, base cordate, lobes incised or serrate, pilose to glabrescent, pale green beneath; flowers cream-white or yellow, in short, terminal racemes; 'Crispa': dwarf, mound-forming, with arching branches and small, crinkled, ferny-like leaves, tinted maroon in autumn);

S.tanakae (Japan; shrub to 2m; shoots slender, arching, smooth, branchlets terete or angled; leaves to 9cm, ovate, shallowly 3– to several-lobed, apex caudate to acuminate, base subcordate, orange and scarlet or bright yellow in autumn, scabrous, biserrate, glabrous above, somewhat pubescent on veins beneath; flowers off-white in terminal panicles to 10cm).

Stephanotis (from Greek *stephanotis*, a name given to the myrtle used in Ancient Greece for making crowns or garlands (a use to which these flowers are put); equally, the name here derives from *stephanos*, crown, and *otos*, ear, referring to the auricled staminal corona). Asclepiadaceae. Old World tropics. 5 species, twining, glabrous, evergreen shrubs with leathery leaves and large, white, waxy flowers in umbel-like axillary cymes. They are salverform to funnelform, the tube cylindrical, often slightly inflated at base, the limb with 5 spreading lobes. Train on trellis or wire hoops. Grow in a freely draining medium, fertile, moisture-retentive, well-aerated and rich in organic matter. Maintain a minimum temperature of 15°C/60°F and high humidity. Admit bright filtered light. Water plenti-fully and feed fortnightly with dilute liquid feed when in full growth, misting frequently in warm weather. In winter keep just moist. Blooms are carried on young shoots, so prune in winter before growth commences, thinning out congested shoots. Propagate by seed, layering or semi-ripe cuttings. *S.floribunda* (MADAGASCAR JASMINE, BRIDAL WREATH, CHAPLET FLOWER, WAX FLOWER, FLORADORA; Madagascar; stems to 4m; leaves to 15cm, oval to oblong, dark green; flowers highly fragrant, to 6cm, pure white to ivory).

*Stephanotis
floribunda*

sterile of sex organs, not producing viable seed and non-functional, barren; of shoots and branches, not bearing flowers; of plants, without functional sex organs, or not producing fruit.

Sternbergia (for Count Kaspar Moritz von Sternberg of Prague (1761–1838), botanist, author of *Revisio Saxifragum*, 1819). Amaryllidaceae. Turkey, west to Spain and east to Kashmir. AUTUMN DAFFODIL. 8 species of perennial, bulbous herbs with narrow leaves borne with or after flowers. The flowers are funnel- to goblet-shaped with six segments; they are solitary and scapose, the scapes sometimes below ground at flowering time. With the exception of the winter and spring-flowering *S.fischeriana* and *S.candida*, these are autumn-flowering bulbs. *S.lutea* and *S.sicula* are the most reliable species for cultivation in the open, well-suited to the rock garden or the border front, in sunny positions and on well-drained soils. All are hardy but dislike overwet winter conditions. Plant deeply in full sun on sharply draining soils. Allow to dry out in summer. Propagate by division of offsets when dormant, or from ripe seed.

S.candida (SW Turkey; like *S.fischeriana*, but scape 12–20cm; perianth segments 4.3–5 × 0.9–

Sternbergia lutea

1.8cm, obovate, apex rounded, white; winter–spring-flowering); *S.clusiana* (syn. *S.macrantha*; Turkey, Jordan, Israel, Iran; scape very short, below ground at flowering; perianth segments 3.7–7.5 × 1.1–3.3cm, obovate to oblanceolate, bright yellow to yellow-green; autumn-flowering); *S.colchiciflora* (SE Spain, Italy, west to Iran; scape below ground at flowering; perianth segments 2–3.4 × 0.4–1.2cm, linear, pale yellow; autumn-flowering); *S.fischeriana* (Caucasus to Kashmir; scape 3–15cm; perianth segments 2–3.5 × 0.5–0.8cm, oblanceolate, bright yellow; spring-flowering); *S.lutea* (GOLDEN CROCUS; Spain to Iran and C Asia; scape 2.5–20cm; perianth segments 3–5.5 × 1–2cm, oblanceolate to obovate, deep yel-low; autumn-flowering); *S.sicula* (syn. *S.lutea* subsp. *sicula*, *S.lutea* var. *gracea*, *S.lutea* var. *angustifolia*; Italy, Greece, Aegean Islands, W Turkey; similar to *S.lutea*; scape to 7cm; perianth segments 2–3.4 × 0.4–1.2cm, yellow; autumn-flowering).

Stewartia (for John Stuart, Earl of Bute (1713–1792)). Theaceae. Eastern N America, E Asia. 9 species, deciduous trees and shrubs with colourful, exfoliating bark, fine autumn colour and large flowers in summer. These consist of five to eight, spreading, silken petals with a boss of numerous stamens. The genus is also spelt *Stuartia*. Handsome specimens for the woodland garden, or for lawns and shrubberies in cool shady sites. They offer an excep-tionally long season of interest, displaying their beau-tiful branchwork and peeling bark in winter, creamy flowers over long periods in mid- to late summer, and attractive foliage from spring through to the wine reds and fiery orange before leaf fall in autumn. Most species tolerate temperatures to –15°C/5°F. Grow in deep, humus-rich soils that are lime-free and moist but not boggy, ensuring a cool root run and light shade at least from the hottest summer sun. Propagate by seed sown fresh in autumn in an outdoor seedbed; seed may not germinate until the second spring, since breaking of dormancy requires some months of warmth followed by three months at temperatures of about 5°C/40°F. Take softwood cuttings in summer or heeled semi-ripe cuttings in late summer; root these in a sandy propagating mix in the cold frame. Alternatively, increase by simple layering.

S.malacodendron (SILKY CAMELLIA; SE US; shrub or small tree to 5m; flowers to 10cm in diameter, petals white, fringed, filaments purple, anthers blue-grey); *S.ovata* (MOUNTAIN CAMELLIA; SE US; shrub to 5m; flowers 6–8cm across, cup-shaped, petals creamy white, wavy-edged, one often deformed, fila-ments yellow-white, anthers white, orange or tinted purple; var. *grandiflora*: flowers to 12cm in diam-eter with purple filaments and orange anthers); *S.pseudocamellia* (JAPANESE STEWARTIA; Japan; tree to 20m, usually a shrub to 4m in cultivation; flowers 5–6cm in diameter, broadly cup-shaped, petals white, suborbicular, concave with jagged edges, filaments white, anthers orange-yellow); *S.sinensis* (China; tree to 10m; flowers 4–5cm in diameter, cup-shaped, fragrant, petals broadly ovate, white, anthers yellow).

stigma the apical unit of a pistil that receives the pollen and normally differs in texture from the rest of the style.

stigmatose having especially well-developed or conspicuous stigmas.

stilt roots the oblique adventitious support roots of the mangrove, screw pine and other woody plants of warm coastal and flooded places.

stimulus movements see *geotropism, haptonasty, phototropism, sensitive.*

Stipa (from Latin *stipa*, tow or oakum, from the feathery inflorescences). Gramineae. Temperate regions. NEEDLE GRASS, SPEAR GRASS, FEATHER GRASS. Some 300 species, perennial grasses with clumps of short or tall culms, slender leaves and, in summer, large, finely branched panicles. These are large grasses, making specimen features in herbaceous borders and grass gardens. Four are particularly noteworthy: *S.arundinacea*, an Australasian native, acquires rusty orange and yellow-brown leaf tints at the end of the season, lasting well into winter; *S.calamagrostis* carries windswept plumes high above its foliage, late into the year; *S.gigantea* produces spectacular open heads of long-awned, shimmering spikelets; the awns of *S.pennata* are longer still and feathery. They are hardy to at least –15°C/5°F. The European species sometimes suffer in intensely hot, humid summers. They prefer full sun, a rather dry situation and a gritty soil. *S.arundinacea* needs a moist, slightly acid and porous soil and will grow in sun or partial shade. Propagate by seed or division.

S.arundinacea (NEW ZEALAND WIND GRASS, PHEASANT'S TAIL GRASS; E Australia, New Zealand; to 1.5m, culms slender, clumped, upright or arching; leaves to 30 × 0.6cm, dark bronze-green, acquiring orange-red to rusty spots and streaks often with black eye-like dots from late summer through winter; panicle pendent; spikelets sparse, tinged purple, awns to 8mm); *S.calamagrostis* (syn. *Achnatherum calamagrostis*; SILVER SPIKE GRASS; C and S Europe; to 120cm, culms clumped, robust; leaves to 30 × 0.5cm, attenuate; panicle to 80cm, loose, appearing one-sided and windswept, feathery; spikelets tinged purple, awns to 1cm, straight or curved); *S.gigantea* (GIANT FEATHER GRASS; C and S Spain, Portugal, NW Africa; to 250cm, culms short or absent; leaves to 70 × 0.3cm, involute when dry, smooth to slightly rough; panicle to 50cm, very long-stalked, loose, open and arching, held high above foliage, very spectacular; spikelets shimmering straw-yellow, awns to 12cm, rough); *S.pennata* (EUROPEAN FEATHER GRASS; S and C Europe to Himalaya; to 60cm, culms short or absent; leaves erect to weeping in a tight bunch, to 60 × 0.6cm, smooth, glabrous above; panicle very loose, rather sparse, spikelets yellow, awn to 28cm, plumose).

stipe the stalk of a pistil or a similar small organ; also, the petiole of a fern or palm frond.

stipel, stipella, stipellae the secondary stipule of a compound leaf, i.e. at the base of a leaflet or petiolule.

stipitate provided with, or borne on, a stipe or small stalk or stalk-like base.

stipulate possessing stipules.

stipule a leafy or bractlike appendage at the base of a petiole, usually occurring in pairs and soon shed.

stock (1) used of a group of mother plants, usually selected and maintained for particular qualities, such as virus freedom (hence *certified stock*), ease of propagation, or flower production; (2) see *rootstock*.

stock plant a plant maintained as a source of propagating material.

Stokesia (for the English botanist Jonathon Stokes, MD (1755–1831)). Compositae. SE US. STOKES' ASTER. 1 species, *S.laevis* (syn. *S.cyanea*), an erect, evergreen perennial herb to 1m. To 20cm long and borne in basal rosettes, the leaves are elliptic to lanceolate, dark green and entire or spiny toward the base. Daisy-like flowerheads to 10cm across appear in summer, consisting of white, yellow to pale lavender to deep indigo florets. Cultivars include forms with flowerheads of varying size, with colours ranging from white ('Alba', 'Silver Moon') and yellow to blue (the deep blue of 'Blue Danube' and the deep dark blue of 'Wyoming'), lilac ('Blue Star') and purple; some are early-flowering. Where winter temperatures fall as low as –20 to –25°C/–4 to –13°F, mulch with straw or dry bracken litter. Grow in a warm, sunny position in light, fertile and freely draining soils. Propagate by seed in autumn, division in spring or from root cuttings in late winter.

stolon an above-ground stem, either aerial or procumbent, that gives rise to new plants at its tip or at intermediate nodes.

stoloniferous producing stolons.

stoloniform resembling a stolon.

stoma (plural stomata) an aperture or pore in the epidermis of a leaf or stem, allowing gaseous exchange.

stomach poison an insecticide which, in order to be effective, must be ingested by an insect feeding upon treated plant parts.

stone the seed(s) with surrounding hardened endocarp in a drupe.

stone fruits a group name for fruits such as apricot, cherry, peach, and plum that contain a hard stone, or endocarp, enclosing and protecting the seed.

Stipa gigantea

stoning the period of grape development when the seeds start to harden.

stony pit a disease of pears, probably caused by a virus, in which leaves become discoloured and fruits are deformed with deep irregular pits and internal stony pockets. Inedibility and premature drop result. The disease is also known as dimpling.

stool a group of shoots emerging from the base of a single plant. The term is applied to plants used for propagation, such as chrysanthemums, and to those cut back to produce attractive young woody shoots, for example, *Cornus alba* cultivars.

stooling see *coppicing, layering.*

stopping see *break, pinching.*

storage organ a swollen stem or root, usually subterranean, in which food and moisture are stored, for example, a parsnip root, crocus corm, or dahlia tuber.

storage rot decay in stored plant produce caused by microorganisms, especially bacteria and fungi. Starchy root crops, fruits, vegetables, bulbs and corms have a high moisture content and are therefore particularly susceptible. SOFT ROT BACTERIA (*Erwinia* species) are common on vegetables, while fungi are the most frequent cause of damage to fruits and root crops. Initial infection may occur while the plant is still growing, through mechanical damage during harvest and transit, or through wounds at the point of attachment to the parent plant. Exposure to temperature extremes can predispose plants to rotting. Often initial infection by a specific pathogen allows invasion by non-specific organisms, such as *Botrytis, Fusarium* and *Rhizopus*, all of which develop rapidly in storage.

Prevention involves controlling pests and diseases on growing crops, choosing correct harvest times, careful handling at all stages and the provision of correct storage conditions.

stove house, stove a greenhouse where the temperature minimum is maintained at 18°C/65°, used for the cultivation of tropical plants. It is also known as a hot house or warm house.

straight fertilizer a fertilizer containing one major plant nutrient, such as sulphate of potash; cf. *compound fertilizer.*

strawberry

Pests
aphids
eelworm
red spider mite
slugs

Diseases
grey mould
powdery mildew
virus

strain in traditional usage, a maintained selection of a plant, especially used of selections raised from seed. The term is invalid under the International Code of Nomenclature for Cultivated Plants, being considered too imprecise.

straining bolt a device used to maintain tension in wires on fences, and on post and wire structures used for supporting plants, especially fruits. A simple form is the threaded bolt which has a looped head to which the wire is fixed by twisting; the bolt is progressively eased through a narrow hole made in the end post by the tightening of a nut, which tensions the wire against its other, fixed, end.

stramineous straw-like, in colour, texture or shape.

strangles a form of heat injury of vegetables caused by contact with hot soil.

stratification see *seeds and seed sowing.*

straw the dry stalks of barley, wheat, and oats, which can prove beneficial to the garden in several ways. As a soil improver, straw is best chopped, wetted, and added in small amounts to the compost heap, or composted separately by wetting and adding an activator to loose layers. Provided an ample dressing of nitrogen fertilizer is applied, it may be dug into the soil directly. In either of these forms of application, straw is a useful conditioner improving soil structure and water- and nutrient-holding. Used in a dry, untreated form it is also valuable for protecting plants from frost damage, for vegetable clamping, and as a ground mulch. Where specially treated, it is a useful ingredient of growing media; and it has useful effect in controlling algal growth in large ponds. See *blanketweed, straw bale culture.*

straw bale culture a specialized system for growing tomatoes, cucumbers and peppers in a greenhouse. It is a suitable means of overcoming soil-borne disease infection, but is more demanding of initial effort than ring, pot, or growing-bag culture. Straw bales are placed end to end on a plastic sheet laid over the greenhouse border soil and then drenched with water for a period of three weeks at a minimum temperature of 13°C/55°F. High nitrogen fertilizer, such as ammonium nitrate with lime is applied at a rate of 680gm per bale and watered in. A further nitrogenous dressing is applied four days later at a rate of 453gm per bale, and again, after another four days, at a rate of 340gm, together with 680gm of a general fertilizer. The fermenting bales are allowed to cool before a topping of a general compost is added into which the crop is planted.

strawberry (*Fragaria* × *ananassa*) a herbaceous and comparatively short-term crop, which is the quickest fruit to come into bearing after planting. The large-fruited hybrids grown today are of relatively recent origin. Previously, over many centuries, fruits of a few native European species were collected from the wild for eating, including the WILD STRAWBERRY (*F. vesca*), the ALPINE STRAWBERRY (*F. vesca* var. *semperflorens*), *F. moschata* and *F. viridis*.

Modern cultivars are the progeny of very variable species, so that hybrids can be found to suit a wide range of conditions, from the equator to the Arctic Circle, and from sea level to about 3000m. Selection of cultivars with the right response to local daylengths and temperatures is the key to successful cultivation, whether on a large or small scale.

Plants are susceptible to very low temperatures in winters of cool-temperate climates and may need protection. Flowers are susceptible to injury by spring frosts, and windy conditions are known to reduce growth, so a sunny sheltered site is best for strawberries. Infection by virus diseases is a serious problem and it is therefore essential to select and plant healthy stock, which has been certified following expert inspection.

Strawberries may be planted in either autumn or spring, but spring is preferable in areas with very cold winters. If runners can be obtained for late summer planting, a crop can be taken in the first year after planting; delays reduce the eventual yield, and with later planting flowers are removed in the first season to allow the plant to build up for the following year. Autumn-fruiting cultivars can be picked in their first season after planting, although they are generally deblossomed until just after midsummer. Large-fruited cultivars start to initiate flowers in daylengths of about ten hours, continuing until they become dormant, after which a chilling period is needed before flowering when the temperature rises; an ideal time for planting these cultivars in the UK is the first week in August. With everbearers, flowers are initiated at days longer than 12 hours, but once in a flowering phase, flower development will continue even in short days of less than 8 hours. Remontants initiate flowers in short days and crop in summer after autumn or spring planting; they then have the ability to produce a second crop in autumn. Almost all strawberry cultivars are self-fertile.

Plant in rows 75–90cm apart, with individual plants spaced 30–45cm apart. Crowns must be positioned at soil level: if placed above it, the roots will dry out, and if placed below, the plants may rot. Runners on large-fruited cultivars are usually removed as they appear, but if left and laid into the row, then a 'matted row' develops, producing a greater number of fruits per unit area, but ones that are smaller and more susceptible to rotting.

Planting on ridges is a satisfactory system, which encourages good establishment by reason of early-warming and good drainage. Planting through slits in a thin-gauge black-polythene ground mulch is of further temperature benefit and serves to conserve moisture and suppress weeds; it is advisable to fit low-level irrigation lines beneath the polythene, along each plant row. A polythene mulch also serves to keep fruits clean of soil; in the UK, a straw mulch is traditionally used for this purpose, being placed round the plants as the fruits start to swell. Both types of mulch may encourage slugs and this should be borne in mind.

Strawberries need adequate water at three particular stages: after planting, at the end of flowering, and during fruit swelling. In dry conditions, irrigate every seven days to make up for any shortfall in rain. Low level application is advisable since overhead watering encourages fruit rots.

Fruit is picked with the calyx attached, using the thumbnail to sever the stalk, without touching the fruit, which is easily marked by any pressure. In the garden, keep picked fruits in the shade and use as soon as possible. Strawberries are less suitable for freezing than other soft fruits since they collapse on thawing.

Unless a matted bed is required, runners should be cut off after cropping: removal of runners promotes a multi-crowned plant. Immediately harvesting is complete, the old leaves are cut off at about 10cm above ground level. They are then removed and burnt to destroy pathogens and allow new growth to develop in healthy conditions. This treatment should not be applied to remontant cultivars, from which only the oldest leaves should be removed by hand.

Strawberries propagate naturally by runners. These are produced in quantity in late summer, and, if healthy, may be used for new plantations. However, because of the susceptibility to virus infection, it is generally preferable to buy in plants certified as virus-tested.

Fruiting can be advanced by covering plants with cloches or low plastic tunnels or by growing them in greenhouses. All of these methods give sufficient protection from low temperature in spring encouraging early ripening. One-year-old plants are most suitable since they tend to crop earlier anyway and are less leafy for the limited space under covers. July planting from cold-stored runners guarantees an even earlier crop.

Strawberry plants should be replaced more frequently than those of other soft fruits, in pursuit of fruit quality and the prevention of pest and disease build up. Cultivars with very large fruits and red flesh include 'Gorella' (early) and 'Hapil' (mid-season). Older English cultivars include the Cambridge series of which 'Cambridge Favourite' is a reliable cropper. 'Honeoye' is a US-raised cultivar of good performance. Among everbearers 'Ostara' and 'Aromel' are notably good in Europe.

strawberry barrel, strawberry pot a wooden or plastic barrel or earthenware pot, with many planting holes in its sides, into which strawberry plants are set for cropping in restricted areas or on patios.

streak a condition most usually caused by virus infection, in which plant stems, leaves and occasionally fruits show discoloured streaks and sometimes sunken spots. In tomatoes, the condition is often due to a combination of virus disease and growing conditions; in healthy plants, it may be caused by magnesium or potash deficiency.

Strelitzia (for Charlotte of Mecklenburg Strelitz (1744–1818), Queen to George III). Strelitziaceae. South Africa. BIRD OF PARADISE. 4 species, large or giant perennial, evergreen herbs, more or less stemless and clump-forming or bearing multiple, stout, unbranched stems. The leaves are large, paddle-

strawberry pot

shaped and 2-ranked. The inflorescence is axillary and long- or short-stalked, the flowers emerging in succession along the upper edge of waxy, rigid beak or boat-like spathes. Each flower consists of three lanceolate, colourful calyx segments and two corolla lobes; the lobes are arrow-shaped and paired to form a projecting 'tongue' enclosing the stamens and style.

In tropical and subtropical areas, plant out into a rich deep, loamy soil in full sun, incorporating organic matter at planting to ensure against drying of the root zone. Under glass, pot into a high-fertility loam-based mix, moisture-retentive but well-drained. If planted directly into greenhouse borders, *Strelitzia* will thrive and flower more freely. Water plentifully in summer, shading against the fiercest sun and ventilating whenever temperatures rise above 20°C/70°F. Feed pot-grown plants with a dilute liquid fertilizer fortnightly when in growth. In winter, water sparingly and maintain a minimum temperature of 10°C/50°F. Plants may also be grown in the home, given large containers, good light and moderate humidity. Container-grown plants benefit from being placed outdoors on sheltered terraces in summer. Propagate by seed sown into a moist, sandy medium at 20°C/70°F in spring. Increase also by removal of rooted offsets.

S. × *kewensis* (*S.alba* × *S.reginae*; garden origin; trunk to 1.5m or absent; spathe long-stalked, calyx pale yellow, corolla blue, lilac-pink at base); *S.nicolai* (Natal, NE Cape; stems to 10m, stout, clumped; leaf blades to 2 × 0.8m, oblong or ovate-oblong, petiole to 2m; inflorescence short-stalked, spathes chestnut-red, pruinose, 40–45cm, calyx 20cm, white; corolla purple-blue, very gummy); *S.reginae* (BIRD OF PARADISE, CRANE FLOWER; South Africa; more or less stemless; leaf blades 25–100 × 10–30cm, glaucescent, oblong-lanceolate, petiole, about 1m; inflorescence long-stalked, equalling leaves, spathe about 12cm, green tinted purple and orange, glaucous, calyx to 10cm, orange, corolla dark; cultivars include 'Humilis', dwarf (to 80cm), closely clump-forming, with ovate-oblong, short-petioled leaves and disproportionately large, short-stalked flowers; 'Kirstenbosch Gold', with pale orange-gold flowers; 'Rutilans', with leaves with red or purple midrib, and brilliantly coloured flowers; var. *juncea*, with leaf blades absent or much reduced – the plant appearing spiky and rush-like).

Streptocarpus (from Greek *streptos*, twisted, and *karpos*, fruit). Gesneriaceae. Tropical, C and E Africa, Madagascar, Thailand, SW China to East Indies. CAPE PRIMROSE. 130 species, annual, perennial or monocarpic herbs, rarely subshrubs, often hairy. They range widely in habit – some are shrubby with small, rather fleshy leaves; others may produce a clump of basal, oblong to oblanceolate leaves, or one massive and wrinkled leaf only. Tubular and 2–lipped, the flowers are produced throughout the year in branched cymes. Provide a minimum temperature of 15°C/60°F, moderate to high humidity, shade from direct sunlight and protection from draughts. Pot in a porous, soilless medium. Water and feed

generously during the growing period, taking care not to wet the foliage. During cooler, resting periods, keep barely moist. Increase by seed, leaf cuttings and stem cuttings.

S.caulescens (Tanzania, Kenya; perennial; stem upright, branched, fleshy, to 60cm; leaves elliptic or ovate, softly hairy, to 6.5cm; flowers to 2cm in diameter, violet or pale lilac; includes var. *pallescens*, with flowers violet, or striped violet on white); *S.* × *hybridus* (a group of popular garden hybrids, in habit generally like *S.rexii*, with puckered, oblanceolate to oblong leaves and large, mauve-purple to rosy crimson flowers; 'Constant Nymph', with pale blue-mauve flowers with darker lines in throat; 'Merton Blue', with blue-mauve flowers with a white throat white with yellow patch; 'Eira', with small, semi-double to double, white flowers, tinged mauve on lower lip; 'Lisa', with shell-pink flowers with a white centre; 'Nicola', with small, semi-double, deep pink flowers; 'Falling Stars', with small, numerous, sky-blue flowers; 'Wiesmoor', with trumpet-shaped, bright ruby flowers); *S.rexii* (South Africa; rosette- or clump-forming perennial; leaves obovate to strap-shaped, obtuse, to 30cm, short-pubescent, bullate, wavy-crenate; inflorescence 1–2–flowered, rarely more, to 20cm, flowers to 7.5cm, white or pale mauve, palate and lower lip striped purple); *S.saxorum* (Kenya, Tanzania; perennial; stem low to prostrate, hairy, few-branched, fleshy; leaves opposite, elliptic to ovate, 1.5–3cm, fleshy, hairy; flowers to 3.5cm, tube white, lobes pale lilac, carried in sparse axillary cymes); *S.wendlandii* (South Africa; monocarpic and stemless; leaf solitary, to 75 × 60cm, massive and somewhat pendulous, dark purple-green above, red-purple beneath, hairy, crenate, deeply puckered; inflorescence to 30cm, arising from base of leaf, flowers many, 1.8 × 3.5cm, flushed blue-purple, interior white, limb blue-purple with 2 deep purple patches on palate).

Streptosolen (from Greek *streptos* twisted, and *solon*, tube, referring to the twisted corolla tube). Solanaceae. Colombia, Peru. 1 species, *S.jamesonii* (ORANGE BROWALLIA, MARMALADE BUSH, FIREBUSH), an erect to weakly climbing or sprawling evergreen, hairy shrub, to 2.5m. The leaves are to 5cm long and ovate to elliptic. Borne in spring and summer in terminal corymbs, the flowers are funnel-shaped with a long, yellow to orange and spirally twisted tube, to 2cm long, and a 5-lobed, orange-red limb. Grow under intermediate greenhouse conditions (minimum temperature 7°C/45°F); conservatory plants may be trained to form standards or attractive specimen shrubs for patios during the summer months. Plant in a well-drained, high-fertility loam-based medium. Provide bright filtered light and medium humidity. Water moderately in summer, sparingly in winter. Feed established plants monthly with a dilute liquid fertilizer and ventilate when temperatures rise above 20°C/70°F. Propagate by heeled greenwood cuttings in a closed case with bottom heat in spring/summer.

stress any environmental factor impairing the normal functioning of plants. Drought, waterlogging, extreme cold or heat, unsuitable light conditions, salinity, and mechanical factors are counted among potential major stresses. Any of these can cause loss of plant performance or even death, and may also make plants more susceptible to pest and disease attack.

striate striped with fine longitudinal lines, grooves or ridges.

strig a single cluster of currants.

strigose covered with sharp, stiff, adpressed hairs.

strigulose minutely or finely strigose.

strike to cause a cutting to root, or to increase a plant by means of cuttings. A struck cutting is one that has formed roots.

strimmer see *spin trimmer*.

stringfellowing severe root-pruning of a bare-root fruit tree prior to planting. A system developed in commercial practice, having been found to speed up planting, stimulate stronger roots and obviate the need for staking. Roots are pruned to 2–5cm from the rootstock.

strip cropping a system of protected growing in which crops are grown in narrow beds or strips with pathways in between. It enables maximum use to be made of cloches and low polythene tunnels, since these can be moved from strip to strip during the year according to the development stage of the crops grown.

strobile, strobilus a cone, or dense cone-like cluster of sporophylls.

strobiloid, strobiliform shaped like a cone, as in the closely bracteate inflorescence of some Acanthaceae.

Stromanthe (from Greek *stroma*, bed, and *anthos*, flower, describing the inflorescence). Marantaceae. C America to S Brazil. Some 15 species, evergreen perennial herbs. They resemble *Calathea* but bear their small flowers among brightly coloured, waxy bracts. Cultivate as for *Calathea*. *S.sanguinea* (Brazil; leaves to 50cm, elliptic-oblong, dark green above, purple-red beneath; inflorescence bracts red, calyx orange-red, corolla white).

Strombocactus (from Greek *strombos*, a spinning top, and *cactus*). Cactaceae. Mexico. 1 species, *S.disciformis*, a low-growing cactus with a disc- to turnip-shaped stem to 8 × 17cm. The tubercles are spiralled, rhomboid and pale grey-green, the spines weak, to 15mm and off-white with darker tips. The flowers are funnelform, 3cm long and pale yellow to almost white with a red throat. Provide a minimum temperature of 7°C/45°F and plant in a gritty, neutral medium. Admit full sun and maintain low humidity. Keep dry in winter except for a light misting on warm days in late winter.

Strongylodon (from Greek *strongylos*, round, and *odous*, tooth, an allusion to the rounded teeth of the calyx). Leguminosae (Papilionoideae). SE Asia to Polynesia. Some 20 species, evergreen, twining shrubs with pendulous racemes of pea-like flowers with long, sharp beak-like keels. Grow in light shade. Plant in tubs or greenhouse borders in well-drained, neutral to acid soil, into which leafmould and well-rotted farmyard manure have been incorporated. Water and syringe plentifully during the growing season (avoid wetting flowers); at other times, allow a slight drying between waterings to discourage continuous weak growth. It is essential to maintain a winter minimum of 15°C/60°F, and to train carefully (along wires, but not too near the glass) and prune lightly, as for young plants of *Wisteria*. Sow in high temperatures (27–30°C/80–85°F) in a closed case; seed is quick to germinate but slow to reach maturity. Air-layering and stem cuttings taken in early summer will usually produce better results. *S.macrobotrys* (JADE VINE, EMERALD CREEPER; Philippines; vigorous evergreen twining liane to 13m; leaves trifoliolate; raceme cylindrical, to 90cm, flowers to 7.5cm, waxy aquamarine to luminous jade green).

Strongylodon macrobotrys

stub (1) the remnant of wood left after a tree shoot is cut back hard, or any short piece of cut or broken branch projecting from a main stem; (2) sometimes used of the stump of a tree left in the ground after felling. Stubbing out refers to the complete removal of a plant.

stump the base of a felled or fallen tree, complete with roots, left in the ground; also, sometimes, used of the base of a tree still in growth.

stumpery a collection of tree stumps, often interplanted with ferns or woodland plants. Stumperies were an occasional feature of Victorian gardens.

stunt used of disordered growth in plants, such as dwarfing, contorted shoots and leaves and abnormal flowers, all of which prevent a plant from developing and functioning normally. Stunted growth may be due to environmental conditions, pest infestation or disease, especially virus infection.

style the elongated and narrow part of the pistil between the ovary and stigma.

Stylophorum (from Greek *stylos*, style, and *phoros*, bearing, referring to the columnar style). Papaveraceae. E Asia, eastern N America. 3 species, perennial herbs with yellow or red sap, pinnatifid, incised leaves and poppy-like flowers in spring and summer. Suitable for shady places, open woodland or a peat terrace, requiring some shade, since it may suffer sun scorch in direct sunlight; otherwise, cultivate and propagate as for *Chelidonium*. *S.diphyllum* (CELANDINE POPPY, WOOD POPPY; E US; to 50cm, downy; leaves 20–55cm, deeply incised, scalloped and toothed; flowers ochre in terminal clusters).

Styrax (Greek name for this plant). Styracaceae. Americas, Asia, Europe. SNOWBELL, SILVERBELLS, STORAX. Some 100 species, deciduous or evergreen shrubs and small trees with pendulous, white, bell-shaped flowers. Hardy to –15°C/5°F, especially with the shelter of a south-facing wall, where the wood will ripen during summer. Grow in moist but well-drained, lime-free, loamy soils, in sun or part shade. Propagate from softwood cuttings in summer; treat with hormone rooting powder and root under mist or in a closed case. Sow ripe seed and keep at 15–24°C/60–75°F for 3–5 months, then cold-stratify at 0.5–5°C/33–40°F for three months, and bring into germination.
S.japonicum (China, Japan; shrub or small tree to 10m; flowers 3–6 on short lateral shoots, pendulous on glabrous stalks to 3.5cm, corolla lobes elliptic-oblong, 1.5cm, pubescent; 'Benibana': a name applied in Japan to all pink-flowered, seed-raised forms); *S.obassia* (Japan; to 10m; flowers in terminal racemes, 10–20cm, stalks 8–10mm, pubescent, corolla lobes 2cm); *S.wilsonii* (W China; shrub or small tree to 3m; flowers 3–5, 1.5–1.8cm across, short-stalked; corolla lobes ovate-oblong, 0.8cm).

subalpine describing plants that inhabit the zone between foothills and alpine slopes; usually referring to relatively tall herbaceous subjects that are not clearly adapted as a result of their high altitude habitat. cf. *alpine*.

sub- a prefix meaning (1) (in conjunction with a descriptive term) slightly, somewhat, nearly, as in subacute, meaning somewhat or slightly acute; (2) (in conjunction with an anatomical term) below or under, as in subapical, meaning below the apex.

sub-base the basic foundation on which a path or paved area is laid. It may be composed of hard-core, stones or firm soil, topped with sand or gravel, where appropriate.

suberous, suberose corky.

subgenus the category of supplementary taxa intermediate between genus and order.

sub-irrigation methods of irrigating, in which water is delivered below the soil or substrate surface or on to a standing area; as with buried seephose, and systems used in moraine beds and on capillary and flood benches. See *irrigation*, *rock plant gardening*.

submerged, submersed beneath the water.

subsessile with a partial or minute stalk.

subshrub (1) a very small but truly woody shrub; (2) a perennial herb with a woody base and partially soft stems.

subsoil the layer of the soil profile directly beneath the topsoil; generally of lower organic matter content, inferior fertility and structure. It should not be mixed with the topsoil during cultivation.

subsoiling trenching or digging deeply in order to loosen the subsoil, especially to break up a pan or generally improve drainage.

substrate a material in which plants can be grown or propagated.

subulate awl-shaped, tapering from a narrow base to a fine, sharp point.

successional applied to sowings made at regular intervals or to the planting of cultivars that fruit at different times, to ensure continuity of supply of a crop.

succulent plant a plant with thick, fleshy leaves or stems developed for water storage, usually as an adaptation to arid conditions, although some succulents grow as epiphytes in humid climates and others by the sea. They are mostly tender plants, such as cacti, but include hardy genera, for example, sedums and sempervivums.

sucker (1) a shoot that arises from plant roots or underground stems and, in the case of grafted plants, may be untypical of the scion. Suckers often occur on plums and roses. (2) see *pysllid*.

sucker cutter, suckering iron a tool used to remove unwanted suckers from the roots of grafted trees and shrubs, consisting of a V-shaped blade fixed to a handle. Older versions of the tool take the form a very narrow spade with a concave sharp blade.

suffrutex see *subshrub*.

suffruticose, suffrutescent of a subshrub; perennial, with a shrubby, woody, persistent base, and soft stem summits.

sulcate lined with deep longitudinal grooves or channels.

sulphate of ammonia see *ammonium sulphate*.

sulphate of magnesium see *magnesium sulphate*.

sulphate of potash see *potassium sulphate*.

sulphur a major (macro-)nutrient; an essential constituent of the protein contained in living cells, and found in the seed oils of Brassicae. Deficiency occurs only very rarely on soils low in organic matter, in areas distant from industrial sulphur dioxide; it is not known in the UK. Refined sulphur dust, in the form of flowers of sulphur, can be used to acidify non-calcareous soils. Sulphur is also used as a broad spectrum inorganic fungicide, especially against powdery mildews; it has some beneficial effect as a foliar feed.

summer pruning pruning carried out mostly between July and August, particularly on trained apple and pear trees, as a means of restricting growth and encouraging fruit bud. Lateral growth of the current season arising from a main stem is cut back to 3 leaves, while that from spurs or existing side shoots is cut back to 1 leaf. Shoots are suitable for pruning when woody and dark brown at their base. Summer pruning is appropriate for trees that fruit on one-year-old wood, for example, fan-trained peaches or plums.

summerhouse a garden house, usually a simple, utilitarian timber building, used as a retreat from the dwelling house, particularly in summer.

sunberry see *hybrid berries*.

sunflower husks milled sunflower husks can be used in potting mixtures to improve drainage and aeration. Nitrogen depletion and toxins are potential problems.

sunken garden a part of a larger, usually formal, garden sunk below ground level. About 60cm deep, with edges and walls of stone, it is accessed by steps, and often occupied by a pool, ferns or a knot garden.

sun scald see *scald*.

superior above, uppermost. A superior ovary is one borne above the insertion of the floral envelope and stamens, or of a hypanthium from which it is distinct.

superphosphate, superphosphate of lime an inorganic phosphate fertilizer containing 18–21% P_2O_5, made by treating insoluble rock phosphate with sulphuric acid; the suffix lime refers to gypsum, which comprises about half its bulk, but which will not raise soil pH. Triple superphosphate is a more concentrated form containing 47% P_2O_5. Both types are water soluble and may be used in compound liquid feed. See *phosphate*.

supplementary light artificial light used, in addition to natural light, in greenhouses. Mercury vapour, sodium vapour, or mercury halide lamps are used to accelerate growth (photosynthetic lighting), while tungsten filament lamps are employed for controlling flowering times (photoperiodic lighting). See *photoperiodism*.

Surrey School a term sometimes applied to English gardeners of the latter part of the 19th century, such as William Robinson and Gertrude Jekyll, who dis-

approved of elaborate Victorian bedding schemes and advocated more natural-looking plantings, including the herbaceous border and woodland gardens.

suspension finely divided particles of a solid or liquid dispersed in a liquid.

Sutera (for Johann Rudolf Suter (1766–1827), Swiss botanist, professor at Berne, author of *Flora Helvetica*, 1802). Scrophulariaceae. Macaronesia, South Africa. 130 species, annual or perennial herbs, subshrubs and small shrubs with tubular, 2–lipped flowers in summer. A charming, compact perennial, suitable for bedding schemes in cool regions or for permanent edging where temperatures do not fall below 5°C/40°F. Plant in warm light soils in full sun. Sow seed or divide in spring; strike cuttings of shoot tips with 3–4 nodes in summer. Plants used for bedding will require yearly repropagation. *S.grandiflora* (PURPLE GLORY PLANT; South Africa; woody-based glandular-pubescent perennial herb to 1m; flowers lavender to deep purple, tube to 2.5cm, slender, widening and curved upwards toward white throat, limb to 2.5cm in diameter).

Sutherlandia (for James Sutherland (d. 1719), superintendent of the Edinburgh Botanic Garden and author of *Hortus Medicus Edinburgensis* (1683)). Leguminosae. South Africa. Some 5 species, evergreen or deciduous shrubs with pinnate leaves, racemes of showy, pea-like flowers in summer and bladder-like seed pods. Given optimum conditions, *S.frutescens* may tolerate temperatures to –5°C/23°F. Grow in a sheltered position on any perfectly drained soil in full sun. Propagate by seed in spring or by semi-ripe cuttings in a closed case with bottom heat. *S.frutescens* (BALLOON PEA, DUCK PLANT; to 1.25m; branches wand-like; leaves thinly downy; flowers to 3.2cm, scarlet to deep purple; fruit inflated, to 5 × 3.2cm).

swap hook see *sickle*.

sward stretch of turf or lawn.

swath, swathe the width of the area covered by a sprayer or duster, or the width of an area of grass cut with a scythe.

swede (*Brassica napus*, Napobrassica Group) RUTABAGA, SWEDISH TURNIP a biennial grown for its large edible root; it is similar to the turnip but with a milder, sweeter flavour and better storage properties. Swede is one of the hardiest of root crops and can be harvested from autumn through the winter months, or lifted and stored either outdoors in clamps or in a cool shed. The flesh is usually yellow, and purple-topped roots are best for the garden. They are peeled and sliced for boiling.

Sow in late spring 2cm deep, in rows 40–45cm apart; thin individual plants to 25cm apart. Swedes are often challenging to grow successfully: the season from sowing to maturity is up to 26 weeks

swede

Pests
flea beetle
cabbage root fly

Diseases
club root
turnip gall weevil
  (*see* weevils)

and the crop must be kept growing steadily, which demands an open fertile site, ideally of light texture and with access to irrigation.

sweetcorn

Pest
frit fly

sweet corn (*Zea mays*) an important cereal crop worldwide, of which primitive forms were cultivated over 5000 years ago. It is grown for its seed-bearing cobs; the seeds are eaten after boiling, either direct from the cob or separated. It is suitable for freezing.

Sweetcorn is a warm season crop that will not withstand frost, usually requiring 70–110 frost-free days from planting to harvest. Optimum temperature requirements are 20–27°C/70–80°F for germination, and 20–29°C/70–85°F air temperature for growth. Flowering is promoted by short days, although adaptable cultivars are available. Sweetcorn needs a long, warm growing season to succeed, and later maturing cultivars are generally of better quality; the sugar content decreases with maturity and the supersweet cultivars are notable.

Choose a moisture-retentive soil in a sheltered site, on which *in situ* sowings can be made in late May or June. Protect with cloches or upturned jam jars for advantage, or with clear polythene or fine net film, which can be either slit or removed on crop establishment – net must be removed by the five-leaf stage. Sow seeds 2.5–4cm deep, with two or three seeds per station, either in block formation with approximately 35cm between stations, or in rows 60cm apart with 30cm between stations; thin germinated groups to one plant. For superior results, multisow in 7.5cm pots or similar-sized modules in April, in gentle heat, planting out after the risk of frost has passed. Sweetcorn is suitable for intercropping with faster-maturing vegetables or low-growing ones, such as marrows.

The plants are wind-pollinated by transference of pollen from the male flowers or 'tassels' to the female flowers or 'silks'. Earthing up for stability is advisable, and ground mulching helps to suppress weeds and conserve moisture; irrigation during seed swelling is beneficial. Each plant yields one or two cobs, which are ready from July-October when the tassels and silks die away and the cob assumes a 45° angle to the stem. Mature seeds yield a milky juice when squeezed.

Mini cobs can be obtained by planting early maturing cultivars 15cm apart and harvesting the cobs when about 7.5cm long.

sweetpea

Pests
aphids
slugs and snails

Diseases
grey mould
foot and root rots

sweetpea (*Lathyrus odoratus*) a tender, climbing annual with winged stems, originating in Crete, Italy and Sicily. Grown for its highly ornamental and fragrant flowers, it has been subjected to much horticultural breeding and selection on both sides of the Atlantic. There is a range of types: (1) tall forms, which reach 2–3m and depend on support, producing long stemmed blooms primarily for cutting or exhibition. These include the Grandiflora type or Old-fashioned sweetpeas, mostly raised before 1900, with dainty well-scented flowers, and now largely superceded by the Spencer types, originating from the early 1900s, with larger frilled or waved stan-

dards in great variety. Other tall forms are the Early Multiflora Giganteas, Cuthbertson Floribundas and Galaxy and Mammoth series, which are variously earlier, more tolerant of hot weather and productive of more blooms per stem; (2) dwarf forms, which grow to a maximum of about 15–30cm, are of bushy habit and need little or no support, producing short-stemmed flowers ideal for a border display. These include the Little Sweetheart series and the prostrate Cupid and Patio series. The Snoopea series is prostrate and without tendrils, reaching a height of about 30cm; similar is the Bijou series, suitable for baskets and tubs; (3) semi-dwarf or hedge forms, such as the Jet Set, Knee-hi and Continental series, can reach 1m in height, with flowers larger than the dwarf forms; in most situations they benefit from support.

Sow seed under protection in late September in mild districts, overwintering in a cold, airy frame, or in late winter or spring in cold districts. Nick darker-coloured seeds opposite the original point of attachment to aid germination. Sow in trays, or one to three seeds per small pot or cellular-tray module, or in special 5cm-diameter sweetpea tubes, which are 15cm deep; maintain at 15°C/60°F, before transferring to a frame. Prick out tray-sown seedlings when 3.5cm high. Pinch out the tips of all autumn-sown plants when 10cm high; stop spring-sown plants at the first pair of leaves. Plant out after progressive hardening off.

Set out autumn-sown plants in mid-spring and those from spring-sowings in late spring, 23cm apart. Bush-grown plants are allowed to develop naturally, supported where appropriate with pea-sticks or nets, or with canes arranged as four-legged tents or in parallel rows. These plants are initially tied loosely with wire rings and the side shoots left to grow. Cordon-grown plants produce the best quality blooms and this system is mainly suited to growing for exhibition. Plants are supported on a structure consisting of stout 2.5m long posts driven in at each end of the row rising 2m out of the ground. To the top of each post is fixed a 45cm long cross piece. Two parallel wires are run between the ends of the cross-pieces and to these are secured 2.2m canes, spaced 23cm apart at a slight angle. After two weeks establishment, the plants have the weakest shoots pinched out, leaving one strong shoot to be tied to each cane with raffia, a sweetpea ring or tape. Side shoots and tendrils are removed as the plants grow, as are flower stems with less than four buds, and the plants are progressively tied in. When plants reach the top wire, they can be untied, laid down and trained 30–40cm up a cane further along the row, thus facilitating new stem growth to prolong flower production. A supplementary feed of a high potash fertilizer is recommended, and an organic ground mulch keeps the rows weed-free and retentive of moisture.

Early blooms can be obtained by growing autumn-sown plants in a cold greenhouse, in pots or border soil, maintaining a minimum temperature of 5–10°C/40–50°F. Early flowering peach or pink cultivars are particularly suitable for this method. Special watch must be kept for attack by aphids, thrips, slugs and snails.

sweet potato (*Ipomoea batatas*) a vigorous, climbing herbaceous perennial, grown as an annual for its edible fleshy, tuberous rootstock. It is of tropical origin, producing an optimum yield in areas with 750–1250mm rainfall, a temperature range of 22–25°C/72–77°F, and good light levels. Although naturally a short-day plant for flower production, tuber initiation and development are not inhibited by variation in daylength.

In temperate climates, sweet potatoes should be grown under cover or in a sheltered place outdoors in very favourable locations. Plant tubers in spring 5–7cm deep in ridges 75cm apart. Alternatively, raise from stem cuttings, inserted to about half their length below a ridge top, or from seed sown at 24°C/75°F in trays; potted-up seedlings are planted out when 10–15cm high. Under cover, the crop can be grown in pots, growing bags, or border soil, provided it is watered regularly. Harvest sweet potatoes grown from tubers 12–16 weeks after planting, when the foliage turns yellow.

swift moths (Lepidoptera: Hepialidae) In Europe, two species are pests in gardens: the GHOST SWIFT MOTH (*Hepialus humuli*), which has forewings up to 50mm, silvery-white in males and yellow-brown in females; and the GARDEN SWIFT MOTH (*H.lupulinus*), which is smaller, with a wingspan of up to 40mm and yellow-brown forewings with white markings (these are less conspicuous in the females).

The moths fly in early summer and fertilized females drop their eggs during flight over suitable vegetation; resulting larvae enter the soil and feed on the roots and underground organs of a wide range of plants. The larvae of the garden swift moth reach maturity and pupate the following year, those of the ghost swift moth a year later. The root-feeding larvae are white, with shiny brown heads, and up to 65mm long. They are particularly prevalent in herbaceous borders, soft fruit plantations and vegetable gardens. Corms, bulbs, tubers, rhizomes and the roots of many plants are frequently attacked, including anemones, asters, carrots, chrysanthemums, dahlias, delphiniums, gladioli, grasses, lettuces, *Narcissus* and other bulbs, onions, peonies, parsnips, *Phlox*, potatoes, pyrethrums, scabious, solidagos, strawberries, raspberries and rudbeckias.

The caterpillars are often spotted when old herbaceous borders, strawberry beds, weedy vegetable plots or grassland are dug during the winter. Such areas should be well-cultivated and if necessary treated with an appropriate soil insecticide before being resown or replanted. Rigorous weed elimination and routine cultivation can help prevent damage by these pests, since eggs are rarely dropped on to bare or sparsely planted soil.

swiss chard (*Beta vulgaris* subsp. *cicla*) SILVER CHARD, SEAKALE BEET, SILVER BEET, SPINACH BEET, PERPETUAL SPINACH; a biennial, related to the beetroot, occurring in a range of forms collectively known as leaf beets. These are cultivated for their edible leaves, which are larger than those of spinach, with swollen mid-ribs and stems; both leaf blades and stems may be used as separate vegetables. Leaf stalks are white, pink, red or silver-coloured, according to form; they are sometimes grown as garden ornamentals, notably 'Ruby Chard'. Swiss chard is hardy and less susceptible to drought than spinach and year-round harvesting is possible; crop quality is improved by covering with cloches during winter months.

The main sowings are made *in situ*, 2cm deep, in March-April in rows 30–38cm apart, thinning to 30cm apart; these will supply a crop from summer until flowering the following spring. A July-August sowing will produce a lower-yielding crop, lasting until well into the following summer, and quality is improved if this is module-raised and grown with unheated protection.

switch a long pliable rod used to disperse wormcasts and dew on turf before mowing.

switch plant a plant with long green photosynthetic shoots and greatly reduced or absent leaves; adapted for arid conditions. For example some *Cytisus* species.

swoe a trade name, now in common use, for a hoe, with an angled blade sharpened on three edges, which can be used with both forward and backward motion.

syconium a fruit-like inflorescence characteristic of *Ficus* where the interior of the globose, fleshy, hollow receptacle is lined with flowers.

Sycopsis (from Greek *syke*, fig-tree and *opsis*, appearance). Hamamelidaceae. Himalaya, China, Malaysia. Some 7 species, evergreen shrubs or small trees. They bear heads of flowers in late winter and early spring – these lack petals but have clusters of showy stamens. *Sycopsis* is tolerant of temperatures down to about –15°C/5°F. Cultivate as for *Hamamelis* and propagate from semi-ripe cuttings in summer. *S.sinensis* (C China; to 6m; leaves to 12cm, leathery, ovate to elliptic-lanceolate, tips occasionally serrulate, dark green above, pale beneath; flowers in thick clusters to 2.5cm, bracts red-brown, woolly, filaments yellow, anthers red).

symbiosis a mutually beneficial relationship between two or more organisms, for example, mycorrhizal fungi and heaths, mycorrhiza, fig and the fig wasp.

sympetalous see *gamopetalous*.

Symphoricarpos (from Greek *symphorein*, bear together, and, *karpos*, fruit, referring to the arrangement of the berries, sometimes so tightly clustered as to resemble a compound fruit). Caprifoliaceae. N and C America, China. SNOWBERRY, CORALBERRY. Some 17 species, deciduous, suckering shrubs with

swiss chard

Diseases
downy mildew
leaf spot

small, pink or white, bell-shaped flowers in summer. Persisiting through winter, the fruits are globose, white, pink, red or purple-black, and juicy with mealy flesh. *Symphoricarpos* withstands a wide range of soil types, urban pollution and maritime conditions. The following species are hardy to at least –25°C/–13°F. They make useful screens and informal hedges, in cold climates, especially on poor soils, even thriving among the roots and under the drip of trees. The suckering habit of species such as *S.albus* renders them suitable for soil stabilization on banks and slopes; the procumbent rooting branches of *S.* × *chenaultii* make dense groundcover. Propagate by division in autumn, or by softwood cuttings in summer. Alternatively, place hardwood cuttings in open ground in winter. Prune out weak, dead or crowded shoots each or every other year, in late winter or early spring.

S.albus (SNOWBERRY, COMMON SNOWBERRY, WAXBERRY; N America; to 120cm; shoots slender, erect, slightly lanuginose; leaves to 5cm, oval to oval-oblong, blunt, rounded at base, sometimes lacerate, dark green above, lighter and finely downy beneath; flowers pink, in spikes or clusters; fruit 12.5mm in diameter, globose or ovoid, snow-white; includes variegated cultivars, and var. *laevigatus*, to 180cm, forming dense, erect thickets, with glabrous shoots, leaves to 7.5cm and glabrous, and larger fruit); *S.* × *chenaultii* (*S.microphyllus* × *S.orbiculatus*; erect to spreading and low, to 2m+; shoots densely downy; leaves to 2cm, ovate, dark green above, blue-green beneath, densely soft-pubescent; flowers pink, in short, terminal spikes; fruit red and white-spotted, or white stippled red on exposed side, or wholly pink); *S.* × *doorenbosii* (*S.albus* var. *laevigatus* × *S.* × *chenaultii*; also known as the Doorenbos Hybrids; very vigorous to 2m; shoots downy; leaves to 4cm, elliptic to broadly ovate, obtuse, dark green and glabrous above, paler and pubescent beneath; flowers white flushing pink, in short racemes; fruit to 1.3cm in diameter, globose, white flushed pink on exposed side; includes numerous cultivars suitable for hedging); *S.orbiculatus* (CORALBERRY, INDIAN CURRANT; E US, Mexico; to 2m, erect, branches thin, pubescent at first; leaves to 3cm, oval to ovate, rounded at base, full dark green above, pubescent, grey-green beneath, tinted red in autumn; flowers ivory flushing dull pink, in short clusters; fruit to 0.6cm in diameter, ovoid-globose, grey-white becoming opaque wine-red; includes variegated cultivars).

Symphyandra (from Greek *symphyo*, to grow together and *aner*, anther, the anthers being connate). Campanulaceae. E Mediterranean. RING BELL-FLOWER. Some 12 species, perennial herbs, sometimes treated as biennials, grown for their bell-shaped flowers produced in summer. Attractive specimens for the rock garden and herbaceous border, valued for their large, narrowly campanulate flowers, carried over long periods in summer. They are generally short-lived as perennials – *S.armena* is best treated as a biennial. Most species are, however, hardy to about -15°C/5°F and, in favourable conditions, will self-seed freely. Grow in sun or light shade in fertile, well-drained, sandy loam soils. Propagate by careful division of the fleshy roots or by seed in autumn.

S.armena (Caucasus; erect or sprawling herb to 60cm, tomentose; leaves to 25cm, cordate-acuminate, incised; flowers erect or pendulous, solitary or terminal, corymbose, white sometimes flushed blue); *S.pendula* (Caucasus; stems to 60cm, pendulous, pubescent; leaves cordate-lanceolate to linear, crenate, velutinous; inflorescence racemose, flowers cream); *S.wanneri* (Alps; erect to 35cm, pubescent; leaves narrowly elliptic to linear-oblong, irregularly dentate, with a winged petiole or sessile; inflorescence paniculate, flowers violet).

symphylids (Symphyla) small, white, delicate-looking arthropods, up to 10mm in length, with long antennae and 12 pairs of legs; they resemble miniature centipedes in their general appearance and high activity. They are found in the soil at various depths down to 2m. Although unable to make burrows themselves, they are adept at moving up and down cracks and fissures, through spaces between soil particles and along earthworm tunnels. They may be abundant in greenhouses, and although outdoor populations are generally lower, numbers may build up to injurious levels on moist soils with a high organic content and open texture.

The most important species is the greenhouse symphylid (*Scutigerella immaculata*), known as the garden symphylan in North America. It feeds on roots, devouring root hairs and smaller roots, its feeding causing rust-coloured lesions. Affected plants make little new growth and readily wilt in dry weather. Damaging attacks in amateur greenhouses are infrequent, but susceptible plants include tomatoes, chrysanthemums and lettuce; outdoor susceptible crops include anemones, asparagus, beans, brassicas, celery, marrows, parsley, garden peas, potatoes, strawberries and sweetpeas. To confirm the presence of symphylids, plants should be lifted and their roots immersed in a bucket of water – most of the symphylids will float to the top within a few minutes. Symphylids may be controlled by drenching the soil with recommended contact insecticides.

Symphytum (from Greek *symphyo*, to grow together, and *phyton*, plant, from the apparent healing powers of the plants). Boraginaceae. Europe to Caucasus and Iran. COMFREY. 35 species, bristly-hairy perennial herbs, with 5–lobed, tubular flowers in arching cymes in spring. Fully hardy, *Symphytum* species make fast-growing groundcover for shaded borders and woodland. *S.officinalis* is also grown as green manure. The leaves also make a valuable high-bulk addition to the compost heap. *S.officinalis* is rich in potassium, phosphorus, calcium and nitrogen, and is used in organic gardens as a high-potash tomato feed or for pot plants. Comfrey water is prepared by packing a water butt or other large container with leaves and steeping in water until decayed. Propagate by seed in autumn, by division, or by root cuttings.

S.caucasicum (Caucasus, Iran; stem to 60cm, branched, pilose; leaves to 20cm, ovate-lanceolate to oblong-lanceolate; cymes paired, scorpioid, flowers to 14mm, red-purple becoming blue); *S.grandiflorum* (Europe, Caucasus; stem to 40cm, ascending-decumbent, shortly pubescent; leaves elliptic or ovate to ovate-lanceolate, base rounded or subcordate; inflorescence many-flowered, flowers to 16mm, pale yellow or cream; includes 'Lilacinum', with flowers white, tinted pink and blue); *S.* × *uplandicum* (*S.asperum* × *S.officinale*; Caucasus; stem to 2m; leaves oblong to elliptic-lanceolate, pubescent; flowers to 2cm, rose becoming blue or purple). *Cultivars* 'Goldsmith', to 30cm, with leaves edged and splashed gold and cream, flowers blue, white and pink; 'Hidcote Blue', spreading, vigorous, to 45cm high, with rough leaves and soft blue and white flowers, red in bud, later fading; 'Hidcote Pink', to 45cm, with pink and white flowers; 'Langthorn's Pink' ('Roseum'), to 1.2m, vigorous, with pink flowers, in clusters on high stem (possibly a hybrid of *S.* 'Rubrum' and *S.* × *uplandicum*); 'Pink Robins', to 45cm, with dark green leaves and narrowly tubular, strong pink flowers, in clusters; and 'Rubrum', to 30cm, with deep red flowers in early summer.

sympodial a form of growth in which the terminal bud dies or terminates in an inflorescence, and growth is continued by successive secondary axes growing from lateral buds; cf. *monopodial*.

symptom a visible or measurable expression of a disease or disorder, which may be characteristic and an aid to diagnosis.

Synadenium (from Greek syn, with, and *aden*, gland). Euphorbiaceae. Tropical America to Mascarenes. 19 species, shrubs or small trees, with fleshy branches containing milky latex and handsome, thinly succulent leaves. The inflorescences are inconspicuous. Cultivate as for the tender, succulent species of *Euphorbia*. *S.grantii* (Uganda to Zimbabwe, Mozambique; succulent shrub, to 3m; leaves to 18cm, oblanceolate to obovate, blunt to apiculate; 'Rubrum': leaves red above, purple-red beneath, and finely toothed, and red flowers).

synanthous with leaves appearing alongside the flowers; cf. *hysteranthous*.

syncarp an aggregate or multiple fruit produced from the connate or coherent pistils of one or more flowers, and composed of massed and more or less coalescent fruits.

syndrome the totality of symptoms characteristic of a disease or pest infestation.

synergism (1) the combined effect of two or more organisms, which is greater than their separate effects; (2) the increased activity that may result from mixing pesticides, and which is more than the sum of their individual effects.

synflorescence a compound inflorescence, possessing a terminal inflorescence and lateral inflorescences.

syngamy see *fertilization*.

Syngonium (from Greek *syn*, together or joined, and *gone*, womb, alluding to united ovaries). Araceae. Tropical America. 33 species, evergreen epiphytic or terrestrial perennials with climbing or trailing, rooting stems and sagittate to ovate or pedately lobed leaves. These are attractive foliage plants for pots and hanging baskets, suited to the heated conservatory and intermediate to warm greenhouse, but also tolerant of drier atmospheres in the home and other interiors. Grow in an open and well-drained medium, rich in organic matter, and shade from bright direct sunlight; provide a moist moss pole for support and a minimum temperature of 16–18°C/61–65°F. Water moderately, allowing the medium to dry a little between waterings. Pinch out stem tips to encourage branching. Repot in spring, if necessary. Propagate by leaf bud or stem tip cuttings, taken in summer and rooted in a closed case, or by division.

S.auritum (Jamaica, Cuba, Hispaniola; juvenile leaves ovate or sagittate to hastate, acute; adult leaves 3–5–pedatisect, lobes 10–30cm, broadly elliptic, acuminate, somewhat glossy above; includes 'Fantasy', with thick, deep green, shiny, compound leaves, mottled white, with a midlobe with smaller lobes each side, and petiole streaked cream); *S.erythrophyllum* (Panama; juvenile leaves 3–9cm, ovate, cordate, lobes rounded, black-green above, pale green, soon becoming deep violet-purple beneath; adult leaves trisect, segments 10–22cm, lanceolate-elliptic to ovate, lateral segments to 10cm, glossy dark green above, pale green or violet-purple beneath); *S.hoffmannii* (Costa Rica to Panama; juvenile leaves sagittate, 4–18cm, green with grey veins; adult leaves trisect, segments 9–28cm, oblong-elliptic to ovate-oblong or lanceolate, acuminate, somewhat glossy above); *S.podophyllum* (syn. *S.vellozianum*, *S.gracile*, *S.riedelianum*, *S.xanthophilum*; Mexico to Brazil and Bolivia; juvenile leaves 7–14cm, simple, cordate, acuminate at apex, becoming sagittate or hastate as plant ages, adult leaves pedatisect, segments 3–11, obovate to broad-elliptic, acuminate, dark green above, pale or sometimes glaucescent beneath). *Cultivars*: 'Albolineatum', with leaves heart-shaped with white centre and veins when young, later palmate and green; 'Emerald Gem', with shiny dark green leaves, fleshy and sagittate when young; 'Emerald Gem Variegated', with white to pale grey leaves with irregular variegation; 'Imperial White', with leaves tinted blue and white veins; 'Roxanne', of rosette form, later creeping, with pinky brown petioles and glossy dull green leaves with muddy green centre and shading along the white ribs, hastate when young; 'Silver Knight', with silver-green leaves; 'Tricolor', with dark green leaves with light green and white variegation, hastate and trilobed when young; and 'Trileaf Wonder', with leaves marked ash green along lateral veins and midrib).

synonym (abbrev. syn.) one of two or more names used for the same taxon. Synonyms fall into two types: *nomenclatural* synonyms are different names based on the same type; *taxonomic* synonyms are different names based on different types that are later judged to belong to the same taxon.

synsepalum a calyx cup or tube; a discrete structure formed by the fusion of two or more sepals.

synthecology the study of plant communities.

Synthyris (from Greek *syn*, with, and *thyris*, a window, referring to the valves of the capsule). Scrophulariaceae. US. Some 14 species, low-growing, spring-flowering perennial herbs with 4–parted, campanulate flowers in narrow erect racemes. The following species are hardy woodland denizens, tolerating very poor soils under deciduous trees. The leathery clumps of evergreen foliage will, however, be larger and freer-flowering where the growing medium is a gritty, richer mix. Plant in light shade. Propagate from seed in spring and by division after flowering.

S.reniformis (SNOW QUEEN, ROUND-LEAVED SYNTHYRIS; to 10cm; leaves with shallow, rounded lobes; flowers campanulate); *S.stellata* (COLUMBIA SYNTHYRIS; to 20cm; leaves deeply toothed or cut; flowers starry).

Syringa (from Greek *syrinx*, the reed, via *Syrinx*, a nymph transformed into a reed, thus an Arcadian reed pipe: a reference to the hollow stems). Oleaceae. E Asia, SE Europe. LILAC. 20 or more species, deciduous trees and shrubs with simple to pinnately compound leaves and, in spring and summer, terminal panicles of small, scented, tubular and 4–lobed flowers.

Hardy, sweetly scented spring-flowering shrubs and small trees, lilacs make good informal hedging or screens and may be used as a spring backdrop to a summer-flowering border. *S.josikaea* and its hybrids stand shearing well and make good hedging plants. Smaller types, such as *S.meyeri* 'Palibin', are much neater in habit, slowly forming rounded, small-leafed shrubs that look well as single lawn or patio specimens, or in the rock garden. Plant on a sunny, open site, in fertile, neutral to alkaline loams. Prune initially to one leader. In later years, prune after flowering to thin out old and dead wood. Late-flowering tree lilacs, such as *S.reticulata*, may grow as single stems into small trees. Lax, vigorous plants, like *S.* × *hyacinthiflora* and its cultivars, should have stems

shortened by one third, since they do not form a terminal bud and naturally fork each year. Old plants of shrubby lilacs respond well to hard cutting back. Mulch and feed well, particularly on poorer, dry soils. Remove flowerheads on small plants directly after flowering to increase vigour. On grafted plants, suckers that overtake the scion material may be a problem, especially where *S.vulgaris* is used as stock. Plants grafted on *Ligustrum* tend to be short-lived unless planted with the graft union 7–10cm below soil level, encouraging the lilac to root. Layers are a better method, producing own root plants. Propagate also from softwood cuttings in mist in summer, by semi-ripe cuttings later in a cold frame, or (for species) from seed or suckers in spring.

Lilacs are susceptible to leaf spotting and blotching (*Ascochyta* and *Phyllosticta* species, *Heterosporium syringae*); and blight (*Pseudomonas syringae*), which produces blackened young shoots, often confused with frost damage and chlorosis. Insect pests include the lilac leafminer, whose caterpillars live in blister-shaped mines. Other pests prevalent in Europe include the privet thrip which causes silvering of the foliage but is only serious in young plants. In North America, the lilac borer (*Podosesia syringae*), a clearwing moth with white caterpillars up to 35mm long, bores into the branches, causing them to wilt.

S. × *chinensis* (*S.* × *persica* × *S.vulgaris*; CHINESE LILAC; garden origin; to 3m; branches dense, slender, arching; leaves to 8cm, ovate, apex acuminate, base rounded or broad-cuneate; flowers in long, nodding axillary panicles, fragrant, lilac, corolla tube to 8mm, lobes obtuse or acute; includes 'Alba', with white flowers white, 'Bicolor', with slate-grey flowers with a violet-tinted throat, 'Duplex', with double, lilac-purple flowers, 'Metensis', with opalescent flowers, 'Nana', dwarf, and 'Saugeana', with lilac flowers flushed wine-red – paler reversions may appear); *S.emodi* (HIMALAYAN LILAC; Afghanistan, Himalaya; robust shrub, upright to 5m; branches very erect, thick, warty; leaves to 15cm, oblong-elliptic, dark above, pale grey beneath, glabrous; flowers in columnar panicles to 15cm, terminal, pale lilac, malodorous, corolla 5mm; includes 'Aurea', with soft yellow leaves, and 'Variegata', with large leaves, bordered gold-green); *S.* × *henryi* (*S.josikaea* × *S.villosa*; garden origin; shrub or tree to 4m; leaves hairy beneath; flowers pale mauve to red in large, downy, pyramidal panicles; includes 'Alba', with white flowers, 'Lutèce', with mauve to white flowers, and 'Prairial', with pale lavender flowers in large panicle); *S.* × *hyacinthiflora* (*S.oblata* × *S.vulgaris*; garden origin; leaves broad-ovate, emerging bronze as in *S.oblata*, turning purple in autumn; flowers scented as in *S.vulgaris*, single or double with lobes somewhat incurved; *S.* × *hyacinthiflora* embraces the Early-Flowering Hybrids or Praecox Hybrids, where one parent is *S.oblata* var. *giraldii*; these produce looser panicles of bloom a fortnight before *S.vulgaris*; there are single and double flowers with colours ranging

Syringa × *chinensis*

cylindric, terminal, paired panicles to 15cm, corolla to 10mm, slender, lilac, white within, lobes narrow, reflexed); *S.* × *persica* (PERSIAN LILAC; W Asia, naturalized Asia Minor; thought to be an ancient backcross between *S.* × *laciniata* and *S.vulgaris*; to 2m; branches rigid, glabrous; leaves to 6cm, lanceolate, rarely pinnate or 3–lobed; flowers in panicles to 5 × 5cm, lilac, fragrant; includes 'Alba', with white to rose pink flowers, 'Laciniata' (see *S.* × *laciniata*), and 'Rosea', with light to dark pink flowers); *S.* × *prestoniae* (*S.reflexa* × *S.villosa*; garden origin; flowers rose-white to deep lilac in dense, nodding panicles; cultivars include 'Audrey', with pink, single flowers; 'Desdemona', with magenta to blue, single flowers; 'Donald Wyman', sturdy, dense and upright, with flowers in large pyramidial panicles, lavender tinted purple; 'Elinor', with pale lavender flowers in erect panicle, tinted purple in bud; 'James MacFarlane', very hardy and adaptable, free-flowering, with clear pink flowers; 'Hecla', magenta, single; 'Hiawatha', mauve-red flowers, opening rose pink; 'Isabella', purple-pink flowers in erect panicle; 'Regan', rose-mauve, single flowers; and 'Redwine', tinted claret); *S.reticulata* (JAPANESE TREE LILAC; N Japan; small tree with low, domed crown, to 10m; branches shaggy with exfoliating bark; leaves to 15cm, ovate, narrow-acuminate, shiny light green above, initially downy beneath; flowers cream, strongly scented of musk in panicle to 30cm; cultivars and varieties include 'Ivory Silk', sturdy and compact, with bark with cherry-pink tint and cream flowers; 'Miss Kim', hardy and compact, dwarf, with glossy dark green leaves, burgundy in autumn, and late, pale lilac flowers, and var. *mandschurica*, AMUR LILAC, a large shrub to 4m, with ovate-acuminate, glabrous leaves and white, unscented flowers, in shorter panicles); *S.vulgaris* (COMMON LILAC; SE Europe; tree or shrub to 7m; leaves ovate, acuminate, to 12cm, glabrous; flowers in pyramidal, terminal panicles to 12cm, lilac, highly fragrant, corolla tube to 10mm, lobes concave; includes var. *alba*, with white flowers, and 'Aurea', with leaves opening yellow-green; other cultivars include early- and late-flowering, single- and double-flowered whites, pinks, magentas, lilac-mauves and blues, some with white throats or a different colour tone in bud, and flowers and panicle of varying sizes and shapes); *S.yunnanensis* (SW China; deciduous shrub to 3m, erect, narrow; branches slender, new growth hairy; leaves to 8cm, oval or narrowly obovate, acuminate, tapering to base, olive green above, glaucous beneath, ciliate; flowers in terminal pubescent panicles, to 15cm, fragrant, shell-pink, fading; includes 'Alba', with white flowers, and 'Rosea', with pink flowers in long, slender panicle). For *S.palibiniana* of gardens, see *S.meyeri* 'Palibin'; for *S.palibiniana*, see *S.patula*.

through white, lilac, and powder blue to garnet, magenta and pink, including bicolors, with forms changing colour as the buds open); *S.josikaea* (HUNGARIAN LILAC; Hungary, Galicia; erect shrub to 4m; branches rigid, warty; leaves to 12cm, elliptic, ciliate, glossy above, glaucous beneath; flowers in slender panicles to 15cm, pubescent, carried erect, deep violet, corolla to 15mm, lobes held forward; includes 'Eximia', with large, light red flowers and panicle, later tinted pink. 'H. Zabel', compact, with red flowers fading to white, 'Pallida', with pale violet flowers, 'Rosea', with pink flowers, and 'Rubra', with violet flowers, tinted red); *S.* × *laciniata* (*S.protolaciniata* × *S.vulgaris*; CUT LEAF LILAC; SW Asia; shrub to 2m; first leaves pinnately cleft or lobed ×3–9, later leaves entire; flowers deep lilac in bud, opening faded mauve with a violet centre, in loose axillary panicle to 7cm, fragrant); *S.meyeri* (N China; compact shrub to 1.5m; shoots 4–sided, downy when young; leaves to 4cm, elliptic, veins pubescent beneath; flowers in broad panicles to 8cm, corolla tube to 15mm, very narrow, purple, faintly scented, sometimes flowering twice in a season; includes 'Palibin', compact, with flowers fading from violet to rose pink on profuse short panicles); *S.microphylla* (China; bushy, upright shrub, 1–1.5m; branches thinly pubescent; leaves 1–4cm, rounded to elliptic-ovate, slightly hairy and ciliate; panicles 4–7cm, finely pubescent, flowers very fragrant, corolla to 1cm, lilac-pink, exterior darker, tube narrow; flowers in early summer and early autumn; includes 'Superba', fine, free-flowering form, with rose-pink flowers produced from spring to autumn); *S.patula* (Korea, N China; deciduous shrub to 3m; young branches flushed purple, sparsely pubescent or glabrous; leaves to 8cm, ovate to rhombic or lanceolate, acuminate, glabrous or minutely downy above, paler, pubescent beneath; flowers fragrant in

systemic (1) used of a pathogen not restricted to the surface of a plant or a point close to the place of infection; (2) used of a pesticide that is absorbed by a plant and translocated through its vascular system.

T

Tabebuia (from the vernacular Brazilian name for *Tabebuia uliginosa*). Bignoniaceae. C and S America, West Indies. Some 100 species, trees or shrubs with simple or digitately compound leaves and panicles of 2-lipped, tubular to campanulate flowers. Beautiful lawn and park specimens in subtropical and tropical climates, *Tabebuia* species are valued for their showy and often fragrant flowers (on deciduous species, these are carried when the leaves have fallen). Many flower sporadically throughout the year and some, such as *T.rosea*, require a dry resting period to induce flowering. In cool temperate areas they are grown under glass as for *Tecoma*. Propagate by fresh seed, by semi-ripe cuttings or by air-layering.

T.chrysotricha (Colombia, Brazil; small tree; flowers to 6cm, yellow, with red down); *T.rosea* (Mexico to Colombia and N Venezuela; tree to 27m; flowers to 10cm, white, pink or lilac with a yellow to white eye).

Tacca (from the Indonesian name for these plants). Taccaceae. SE Asia, W Africa. 10 species, evergreen perennial herbs with starchy rhizomes or tubers and clumped, lanceolate-ovate to decompound leaves. Fleshy, bell-shaped flowers hang in scapose umbels subtended by heavily veined bracts and long, whisker-like filaments. Grow in filtered light in a humid greenhouse or conservatory, with a minimum temperature of 13–15°C/55–60°F. Pot in a leafmould and coarse bark mix with a slow-release fertilizer. Water and syringe continuously, giving a monthly dilute foliar feed. Propagate by division, or sections of old rhizomes, each with an eye. Alternatively, surface-sow seed at 22–27°C/72–80°F.

T.chantrieri (syn. *T.macrantha*, *T.lancifolia*, *T.roxburghii*; DEVIL FLOWER, BAT FLOWER, CAT'S WHISKERS; Thailand; like *T.integrifolia*, but to 63cm, outer bracts green to black, filaments to 25cm, olive green to maroon or violet-black); *T.integrifolia* (*T.aspera*; BAT PLANT, BAT FLOWER; E India to S China, south to Sumatra, Borneo, W Java; to 100cm, dark purple or red, sometimes brown, outer bracts to 14cm, green to dark purple, stained and veined black, filaments to 20cm, pale green shaded violet, darkening as flowering continues to dark brown or purple-black, flowers nodding, purple-black).

Tacca integrifolia

tachinid flies (Diptera: Tachinidae) a diverse family of stout, bristled flies, distributed worldwide, ranging in size from about half as big as a housefly to a wing span of 35mm. The adults are free-living, often frequenting flowers, but the larvae are internal parasites of various other arthropods, especially insects, including larvae and nymphs of beetles, grasshoppers, plant bugs, sawflies, occasionally other flies and most commonly moths and butterflies. Tachinids are important as natural regulators of insect populations, but have been used only to a limited extent in biological control.

Tagetes (from Tages, an Etruscan deity, grandson of Jupiter, said to have taught the Etruscans soothsaying). Compositae. Tropical and warm America, one species from Africa. MARIGOLD. About 50 species, annual or perennial herbs with glandular, pinnatifid or pinnate leaves and showy, daisy-like flowerheads in summer. Grow in sun in any moderately fertile and well-drained soil. Deadhead to prolong blooming. Sow seed *in situ* in spring for informal effects or earlier under glass for bedding out and containers. Plants begin to bloom about 12–14 weeks from sowing.

T.erecta (AFRICAN MARIGOLD, AZTEC MARIGOLD, BIG MARIGOLD; C America; annual to 1m; flowerheads 5–12cm in diameter, ray florets orange to lemon yellow; includes many seed races, tall to dwarf plants with single to double flowerheads in shades of orange and yellow); *T.patula* (FRENCH MARIGOLD; Mexico, Guatemala; annual to 50cm; flowerheads to 5cm in diameter, ray florets red-brown, orange, yellow or particoloured; includes numerous seed races, among them dwarf and bushy to tall plants with single or double flowerheads in shades and combinations of gold, bronze and rusty red).

Talinum (from the native African name, meaning unknown). Portulacaceae. Tropics and subtropics, western N America. FAMEFLOWER. About 50 species, woody-based, succulent annual or perennial herbs with showy, short-lived flowers consisting of five or more silky petals. Hardy to –15°C/5°F, given perfect drainage. It is well-suited to sunny, perfectly drained

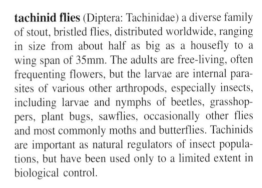

niches in the rock garden and in troughs. Cultivate as for *Lewisia*. *T.okanoganense* (western N America; cushion-forming perennial; leaves cylindrical, fleshy, grey-green; flowers to 2cm in diameter, white).

Tamarix (Classical Latin name for tamarisk). Tamaricaceae. W Europe and the Mediterranean to E Asia and India. TAMARISK, SALT CEDAR. 54 species, deciduous shrubs or small trees with many slender branches, their branchlets feathery and clothed with very small, green, needle- or scale-like leaves. In summer, they are clothed in powdery masses of tiny pink or white flowers. Most *Tamarix* species occur in coastal habitats, often on saline soils; they are particularly useful in exposed maritime gardens, as hedging and windbreaks or as border specimens. Grow in any well-drained soil in full sun. Cut back after flowering. Propagate by hardwood cuttings of the current year's growth, taken in winter and rooted in a sandy propagating mix in the open frame, or by semi-ripe cuttings in summer. Increase also by seed.

T.gallica (Mediterranean; shrub or small tree, 2–10m tall with spreading to erect, slender branches; flowers pink in dense, cylindric racemes 3–5cm long); *T.pentandra* (SE Europe to C Asia; shrub or small tree 3–5m tall; flowers pink to red in small spikes grouped in large panicles); *T.ramosissima* (E Europe to Asia; shrub or small tree to 6m tall; flowers pink in slender racemes to 3cm long; includes cultivars with pale pink to deep rose flowers, some such as 'Summer Glow' with very feathery, grey-tinted foliage and bright pink flowers).

Tanacetum (from the Medieval Latin name, *tanazita*, itself derived from the Greek *athanasia*, immortality – the herb tansy was scattered in shrouds to discourage worms and maggots). Compositae. N temperate regions. Some 70 species, mostly scented, annual or perennial herbs, occasionally slightly woody, with entire, toothed or dissected leaves often covered in hairs and sometimes silvery. The flowerheads are daisy- or button-like and white, yellow or both.

A number of low-growing, mound-forming species have extremely attractive white or silver-downy, finely divided foliage, including *T.densum*, *T.haradjanii* and *T.argenteum*. These are suited to the rock garden, raised bed and warm, south-facing, dry walls and banks, or other situations characterized by their open, sunny nature and perfect drainage. *T.parthenium*, feverfew, is more commonly grown in the herb garden; it has been of recent interest for its role in the treatment of migraine. *T.coccineum* is grown in the herbaceous and cut-flower border. It tolerates a range of soils, excepting those that are very heavy or wet, and will thrive in sun (for better flowering) or light dappled shade; it requires the support of pea sticks in the border and, if cut back after flowering, may give a second flush of bloom later in the season. Propagate by seed, and in the case of perennials, by division or from basal cuttings in spring. *T.parthenium* will self-seed, often to the point of nuisance, although self-sown seedlings will transplant readily if watered in well. As with the smaller alpine species, it can also be increased by soft cuttings with a heel in summer.

T.argenteum (Mediterranean; perennial to 30cm, white-tomentose; leaves 2–7cm, ovate to rounded, 1–2-pinnate to entire; flowerheads yellow); *T.coccineum* (PYRETHRUM, PAINTED DAISY; W Asia, Caucasus; glabrous perennial to 60cm; leaves elliptic to oblong, pinnatisect, dark green; flowerheads to 7cm in diameter, white to pink or purple-red; cultivars include large to dwarf plants with single to double flowerheads in tones of pink, rose, scarlet and dark red); *T.densum* (Turkey; perennial subshrub to 30cm; leaves 2–5cm, ovate, 2-pinnatisect, white-tomentose; flowerheads small, yellow); *T.haradjanii* (Syria, Turkey; silvery white-tomentose perennial subshrub to 30cm; leaves to 5cm, oblong to elliptic to ovate, finely 2–3-pinnatisect; flowerheads yellow); *T.parthenium* (FEVERFEW; SE Europe, Caucasus; aromatic, bushy perennial to 40cm; leaves 3–8cm, ovate, pinnatisect, cut and toothed, glandular, mid-green; flowerheads to 2cm in diameter, ray florets white, disc florets yellow; includes cultivars with gold to lime green leaves, dwarf and compact to tall, and with single to pompon flowerheads).

Tapeinochilus (from Greek *tapeinos*, *low*, and *cheilos*, lip, referring to the small labellum). Zingiberaceae. Malaysia, Indonesia, New Guinea, NE Australia. Some 15 species, perennial herbs resembling *Costus*. The flowers are yellow to brown with small lips. They are carried in showy, cone-like spikes composed of waxy, overlapping and colourful bracts. Cultivate as for *Costus*. *T.ananassae* (PINE-APPLE GINGER; Moluccas; 1–2m tall; inflorescence to 20cm, bracts scarlet to crimson, closely overlapping with their tips recurved, the whole resembling a pineapple, flowers yellow).

tapis vert an expanse of turf; but also applied to knots or parterres created from dwarf box (*Buxus* species).

taproot the primary root, serving to convey nutrition and to anchor the plant.

tar oils tar distillates containing hydrocarbon and phenolic oils; used mainly as insecticidal sprays during winter, especially on dormant fruit trees, against hibernating insects or eggs, including those of aphids, scale insects and winter moths. They are also effective as eradicants of moss and lichens on tree trunks.

tar spot (*Rhytisma acerinum*) a fungal disease of *Acer* species, causing pale yellow leaf spots on leaves in spring, which develop into conspicuous, large, shiny black blotches with yellow halos by mid-summer. It is not usually the cause of serious damage but premature leaf fall can result from heavy infection. The fungus overwinters on fallen leaves and young foliage is infected in spring.

The fungus is highly susceptible to sulphur dioxide in the air, and clean air legislation has resulted in an increased incidence in urban areas.

For small ornamental trees that are regularly disfigured, it is worth collecting and burning infected leaves and spraying expanding leaves with a copper fungicide in the spring. A similar disease, *Rhytisma salicinum*, affects willows.

tarsonemid mites (Acari: Tarsonemidae) minute mites, up to 0.25mm long, with four pairs of legs and glistening white or caramel-coloured bodies. Males are smaller than females and the last pair of legs is held erect to enable them to carry pre-adult female nymphs prior to mating. Females lay relatively large eggs, from which emerge active larvae with only three pairs of legs; the larval stage is followed by non-feeding pre-adult nymphs enclosed in their inflated larval skins. The life cycle may be completed in 10–14 days under warm conditions and breeding may continue all year round.

The BROAD MITE (*Polyphagotarsonemus latus*) is of worldwide distribution and in warmer regions is an important pest of castor oil, citrus, cotton, pepper, potato, rubber and tea; in temperate countries, it infests greenhouse plants, such as aubergine, begonias, browallias, capsicums, chrysanthemums, *Cissus, Cyclamen*, dahlias, fatshederas, fuchsias, gerberas, *Impatiens*, pelargoniums and tomatoes. The foliage of infested plants becomes stunted, brittle and puckered, often accompanied by a downward roll of the leaf margin. The blooms of some plants, including gerberas and chrysanthemums, become malformed and lopsided. Begonias are particularly susceptible to infestation by these mites and typically develop a rust-coloured scarring on the underside of leaves.

The STRAWBERY MITE (*Tarsonemus pallidus*), also known as the CYCLAMEN MITE, occurs in Europe and North America. Outdoors, it attacks strawberries and some perennial asters; suscepible greenhouse plants include aphelandras, azaleas, chrysanthemums, *Cissus*, fatshederas, *Ficus*, fuchsias, gloxinias, heliotropes, ivies, pelargoniums, saintpaulias and verbenas, and *Cyclamen*. Symptoms include leaf puckering, distorted flowers, severe stunting of growth and extensive bronzing of young leaves; the leaf margins of *Cyclamen* and strawberry are often turned inwards.

The BULB SCALE MITE (*Steneotarsonemus laticeps*) attacks *Narcissus*, especially those forced for early flowering, and hippeastrums. The mites congregate in the neck of the bulb, feeding in the spaces between the bulb scales. Emerging leaves first appear a brighter green than those of healthy plants, with yellow or brown streaks at the leaf margins and edges of the flower stem; these later develop into longitudinal scars, which are brown in *Narcissus* and red in hippeastrums. Leaves become curved and both leaves and flower stems are stunted, with buds either failing to open or producing distorted blooms. Heavily-infested plants have the appearance of being dusted with a fine white powder.

The FERN MITE (*Hemitarsonemus tepidariorum*), which infests greenhouse ferns causing stunting and distortion, and the MUSHROOM MITE (*Tarsonemus myceliophagus*), are other species of tarsonemid mite.

Infested plants should be destroyed, but move carefully since mites readily fall off, spreading infestation. Only healthy plants should be introduced; mites on strawberry runners, *Narcissus* bulbs and some other plants can be controlled by *hot water treatment* (q.v.).

Taxodium (from *Taxus*, the yew, and Greek *eidos*, resemblance: the genus superficially resembles *Taxus*). Cupressaceae. US, Mexico. SWAMP CYPRESS, BALD CYPRESS. 3 species, deciduous coniferous trees to 45m, commonly found in sodden or submerged habitats. The trunk is often buttressed and, when growing in wet conditions, it may be surrounded by exposed, conical roots or *knees*. The bark is stringy, fissured, brown-grey to rusty red and peels in strips. Broadly conic at first, the crown becomes irregular with age. The leaves are scale-like (on the persistent new growths, or throughout in *T.ascendens*) or oblong to lanceolate to linear and 2-ranked along feathery, deciduous branchlets. *T.distichum* and *T.ascendens* are both trees of flooding river bottoms and coastal swamps. In cultivation, neither needs waterlogged soils. They can, nonetheless, be kept in up to 60cm of permanent water, planted in the pond or lake bed, at pond or lake margins, or on islands. Both respond to warm summers, and frost damage can occur below about –10°C/14°F. *T.distichum* is very tolerant of city air and the high pH of chalk streams; its autumn colour is superb. The more southerly *T.ascendens* is more tender. Propagate by seed or by cuttings in late summer.

T.ascendens (POND CYPRESS; SE US; to 25m, 'knees' rarely produced, crown columnar; leaves to 1cm, subulate, bright green turning fox red in autumn, closely adpressed to slender, erect branchlets; 'Nutans': branchlets pendulous); *T.distichum* (SWAMP CYPRESS, BALD CYPRESS; SE US; to 40m, 'knees' often produced in wet sites, crown conic; leaves to 2cm, linear to lanceolate-oblong, pale green turning rust red in autumn, 2-ranked and spreading on branchlets).

taxon (plural taxa) a general term for a taxonomic unit or group of any category or rank. Thus, *Buddleja* is a taxon at genus level, *Buddleja davidii* is a taxon at species level and *Buddleja* 'Black Knight' is a taxon at cultivar level.

taxonomy identification, nomenclature and classification of living things.

Taxus (Latin name for Yew.) Taxaceae. Northern temperate zones to Mexico, C Malesia. YEW. 10 species, normally dioecious, slow-growing, evergreen coniferous trees or shrubs. The bark is thin, red-brown, scaly and exfoliates in plates on older stems. The leaves are narrow, linear to oblong or lanceolate to falcate and glossy dark green with a distinct midrib. The leaves and seeds are highly toxic, the latter enclosed by a colourful, fleshy aril open at the apex.

T.baccata is very tolerant of cold and heat, shade and sun, wet and dry exposure, and any pH; it is, however, sometimes damaged, or even killed when temperatures fall to –25°C/–13°F or lower. It withstands most types of urban pollution, but is sensitive to soil compaction by roads; damage has also been associated with acid rain. It is ideal for shelter, screens and as a dark background for pale flowers, and is the traditional choice for backing the herbaceous border, garden seats and marble statuary, as well as an ideal material for topiary. Old and overgrown specimens can be pruned back savagely to the pole of the central stem. Where *T.baccata* is not hardy, for example, in the NE US and E Canada, its place in gardens is taken by *T.cuspidata* and *T.* × *media*; these are hardy to at least –35°C/–30°F, but also more demanding of summer heat and humidity. *T.mairei* needs very hot humid summers to grow well.

Seed may require lengthy stratification. However, good seedlings spring up wherever birds void the seeds; when they are first found at about 10cm tall (they form very little fibrous root and will not move easily when larger), put them in pots or line them out. The cultivars are propagated by cuttings, taken as for *Chamaecyparis*, or by grafting on to seedlings; scions must include an erect leading shoot if the upright growth pattern is to be maintained. Side shoots will not develop a leader and remain prostrate.

T.baccata (YEW; Europe to Asia Minor; tree to 15m, crown ovoid-conic, becoming domed, much-branched; leaves 1.5–3 × 0.2–0.25cm, linear-oblong, glossy dark green; aril red; includes many cultivars, among them dwarf to tall clones with spreading and low to tall and fastigiate habit, small to long leaves, sometimes marked cream or wholly yellow to lime, with red to orange or glowing yellow arils); *T.cuspidata* (JAPANESE YEW; Japan; to 20m, often a shrub in cultivation; leaves 1.5–2.5 × 0.2–0.3cm, linear, with a pointed tip, banded yellow-green beneath; aril scarlet; range of cultivars as for *T.baccata*); *T.mairei* (S China to SE Asia; to 15m, or bushy in cultivation; leaves to 1.5–4 × 0.2–0.3cm, linear to lanceolate, falcate, yellow-green above; aril red); *T.* × *media* (*T.baccata* × *T.cuspidata*; garden origin; pyramidal, spreading bush, leaves as for *T.cuspidata*, aril scarlet; range of cultivars as for *T.baccata*).

tayberry see *hybrid berries*.

Tecoma (from the Mexican name for these plants, *tecomaxochitl*). Bignoniaceae. Arizona and Florida to Argentina. TRUMPET BUSH, YELLOW BELLS. Some 12 species, evergreen or semi-deciduous trees or shrubs with pinnate leaves and tubular flowers. Grow in full light but with protection from direct sunlight in summer. Maintain a a minimum temperature of 10–13°C/50–55°F. Water moderately in summer, hardly at all in winter. Prune after flowering. Propagate by seed sown in spring, by semi-ripe cuttings, or by simple layering. *T.stans* (US, C and S America; to 3m; flowers 3–6cm long, yellow lined or stained red).

Tecomaria (from the genus *Tecoma*, to which these plants are closely related). Bignoniaceae. South Africa. 1 species, *T.capensis* (CAPE HONEYSUCKLE), an erect or scrambling shrub or tree, usually evergreen, to 7m tall with pinnate leaves and, in spring and summer, clusters or racemes of yellow, orange or scarlet, tubular flowers. Grow in full sun on walls and pergolas in more or less frost-free conditions. Water and feed generously when in growth, sparingly in winter. Prune after flowering. Propagate by air-layering or by semi-ripe cuttings.

Tecophilaea (for Tecofila Billiotti, daughter of Professor Colla of Turin, *c.*1830). Liliaceae (Tecophilaeaceae). Andes of Chile. 2 species, small cormous perennial herbs with narrow leaves and solitary, salverform to broadly funnel-shaped flowers in spring. Grow in the alpine or cold greenhouse. Plant in a fertile, sandy potting mix, and position in sun; water only when in growth, reducing water gradually after flowering to allow a warm, dry dormancy in summer. Propagate by offsets or by seed, which will develop more reliably following hand-pollination. *T.cyanocrocus* (Chile; flowers 3–3.5cm, royal blue, veined and often tinted white in neck, sometimes with a white margin; 'Leichtlinii': flowers paler blue with broad white central zone; 'Violacea': flowers deep purple-blue).

*Tecophilaea
cyanocrocus
'Leichtlinii'*

Telekia (for Teleki de Szek (fl. 1816). who founded a great library in Hungary). Compositae. C Europe to Caucasia. 2 species, tall, coarse perennial herbs with heart-shaped, stalked leaves and loose racemes of large, daisy-like flowerheads in late summer. Fully hardy and suited to waterside or woodland plantings in large gardens, where it will make huge mounds of foliage, and bear long, branched stems of fine-rayed daisies. Grow on any moist soil in sun or part shade, and with some shelter from cold winds. Propagate by seed or division. *T.speciosa* (C Europe, Caucasus, Balkans, C and S Russia; to 2m; leaves to 30cm, aromatic, coarsely toothed, hairy beneath; flowerheads yellow).

Tellima (an anagram of *Mitella*, a closely related genus). Saxifragaceae. Western N America. FRINGE CUPS. 1 species, *T.grandiflora*, a semi-evergreen, glandular-hairy, perennial herb. The leaves are low-lying, cordate to reniform, 3–10cm broad and nearly as long, the margins 5–7 lobed and toothed. Produced in late spring, the flowering stem is 20–80cm tall and carries a raceme of small, campanulate flowers, their petals fringed and white tinged green to deep red. Fully hardy and a useful groundcover for moist borders, woodland and rock gardens, on humus-rich soils in shady spots. The cultivar 'Rubra' (syn. 'Purpurea') is strongly tinted purple-red throughout. Cultivate as for *Heuchera*.

Telopea (from Greek *telopos*, seen at a distance, referring to the conspicuous crimson flowers). Proteaceae. Australia. 4 species, evergreen shrubs with leathery leaves and slender flowers packed in terminal heads and surrounded by showy bracts. Grow in a well-drained, slightly acid sandy loam in full sun or light shade, with protection from strong winds and with ample summer moisture. Once established, *Telopea* benefits from regular light pruning and will stand being cut hard back to rejuvenate. It tolerates light frosts and may be grown in sheltered temperate gardens where winter temperatures seldom fall much below –5°C/23°F. Increase by seed.

T.speciosissima (WARATAH; E Australia; straggling shrub to 3m; flowers red amid red bracts in a dense, rounded head to 15cm across); *T.truncata* (TASMANIAN WARATAH; low spreading shrub or a tree to 8m; flowers rich crimson or yellow in heads to 8cm across).

temperate house a greenhouse where the temperature is maintained at about 13°C/55°F minimum.

tender used of plants liable to injury or death when subjected to low temperature; cf. *hardy*.

tendril a slender modified branch, leaf or axis, capable of attaching itself to a support either by twining or adhesion.

tent caterpillars (Lepidoptera) the caterpillars of some moths and butterflies that feed gregariously, for all or part of their lives, in extensive silk webs or tents; they are also known as WEBBERS or WEBWORMS. They include the larvae of the LACKEY MOTH (*Malacosoma neustria*), a common European species, which are blue-grey and hairy, up to 50mm long, with white and orange red longitudinal stripes; they feed on the foliage of many trees and shrubs, including apple, cherry, pear, plum, alder, blackthorn, elm, hawthorn, hazel, oak, rose and willow. Several closely related North American species such as the EASTERN TENT CATERPILLAR (*M. americanum*), the WESTERN TENT CATERPILLAR (*M. californicum*), the PACIFIC TENT CATERPILLAR (*M. constrictum*), and the SOUTH WESTERN TENT CATERPILLAR (*M. incurvum*), attack a wide range of fruiting and ornamental trees and shrubs.

SMALL ERMINE MOTHS (*Yponomeuta* species) occur in Europe and North America and all have dusky grey caterpillars spotted with black, up to 12mm long, that feed on plants such as blackthorn, hawthorn, apple, *Euonymus*, bird cherry and willow.

Other common tent caterpillars include the larvae of the HAWTHORN WEBBER (*Scythropia crataegella*) in Europe, which are brown, up to 15mm long, and also attack *Cotoneaster*; the larvae of the JUNIPER WEBBER (*Dichomeris marginella*), which are red-brown with two white stripes, up to 12mm long, and feed on junipers in Europe and North America; and the larvae of the BROWN-TAIL MOTH (*Euproctis chrysorrhoea*), found in both continents, which are black and hairy, up to 37mm long, with brown warts bearing tufts of hair and two conspicuous red spots at the posterior end. Brown-tail moth larvae feed on fruit trees, roses, hawthorn, willow and other trees and

lackey moth

terrarium

shrubs, and bear hairs which may cause a painful skin rash. Two other pests of garden crops in North America are the GARDEN WEBWORM (*Achyra rantalis*), which has green-yellow caterpillars, up to 18mm long, with black dots on each segment, and the BEET WEBWORM (*Loxostege sticticalis*), which has green to yellow larvae, up to 25mm long, with a black stripe.

Tent caterpillars can be controlled by removing and burning infested branches or spraying the young caterpillars with a recommended contact insecticide, applying sufficient force to penetrate the webbing.

tepal a unit of a perianth, which cannot be clearly distinguished as a sepal or petal; for example, the flower segments of *Tulipa*. In orchids, the term is applied to the sepals and petals minus the labellum and was originally coined for this purpose.

teratology the study of plant abnormalities or monstrosities. See *malformations*, *proliferation*.

terete cylindrical, and smoothly circular in cross-section.

ternate leaves, sepals or petals in threes. A ternate leaf is a compound leaf divided into three leaflets; in a biternate leaf, these divisions are themselves divided into three parts; in a triternate leaf, these parts are further divided into three.

Ternstroemia (for Christopher Ternstroem (d. 1748), Swedish naturalist and traveller in China). Theaceae. Asia, Americas, Africa. 85 species, evergreen trees or shrubs with leathery leaves and 5-petalled flowers with numerous stamens in summer. These are followed by small, red fruit. *T.japonica* may tolerate temperatures to about –10°C/14°F. Grow in fertile, humus-rich, lime-free soils in part shade. Pinch shoot tips to encourage branching and prune to shape in spring before bud break. Propagate by fresh seed or by semi-ripe cuttings in late summer. *T.japonica* (syn. *T.gymnanthera*; (Japan; shrub or small tree to 3.5m; leaves 4–7cm, cuneate to obovate, dark green; flowers white with obovate to cuneate petals; fruit to 1cm in diameter, red; includes 'Variegata', with leaves grey-green edged cream becoming red-tinted).

terrace a flat area raised above main ground level; more generally, any level area treated formally, for example, with paving, and often separated from the rest of the garden by a balustrade or low wall; especially such an area adjacent to a dwelling, acting as a transition between house and garden and which is frequently referred to as a patio. A terrace is also a level area created artificially on a slope, with retaining walls.

terrarium a small enclosed glass container in which ornamental plants are grown, *see* also Wardian case.

terrestrial growing in the soil; a land plant.

Tetranema (from Greek *tetra*, four, and *nema*, thread, referring to the stamens). Scrophulariaceae. Mexico, Guatemala. 2 species, low perennial herbs with long-stalked clusters of small, tubular and 2-lipped flowers in spring and summer. Provide a minimum temperature of 13°C/55°F. Water moderately (avoiding waterlogging) and feed mature specimens regularly with a dilute liquid fertilizer. Provide strong filtered light. Grown as house plants, *Tetranema* species will flower sporadically throughout the year on a well-lit windowsill. Propagate from fresh seed (which may take 1–2 months to germinate), or by division in spring. *T.roseum* (syn. *T.mexicanum*; MEXICAN FOXGLOVE, MEXICAN VIOLET; Mexico; leaves to 12cm, obovate, dark green, somewhat leathery; flowers to 1.5cm, lilac to mauve with darker markings, on stalks to 20cm tall).

Tetrapanax (from Greek *tetra*, four, and *Panax*, a related genus, in reference to the presence of four petals in the plant originally described). Araliaceae. Asia (Taiwan). RICE-PAPER PLANT, CHINESE RICE-PAPER PLANT. 1 species, *T.papyrifer*, a suckering shrub or small tree to 7m tall with slender branches. To 50cm or more across, the leaves are palmately compound, toothed and covered in scurfy pale felt at first. In the cultivar 'Variegata', the leaves are patterned with cream, white and various shades of green. Produced at various times of year in massive, umbellate panicles, the flowers are small, cream and invested with large, woolly bracts. Hardy to –5°C/23°F. In colder areas, the top growth may be cut back to ground level during the winter but the base will sprout in spring. Plant in most soils and situations, but allow plenty of room since *Tetrapanax* often creates running and suckering thickets. Where space is restricted, it may be confined to containers on patios or plunged in the border. Plants grown in the open ground may be regularly chopped back. Increase by suckers.

tetraploid a plant that has four sets of chromosomes, instead of two sets, as in the majority of plant species; it is annotated 4×. Examples of natural tetraploids are *Cyclamen persicum* and the potato.

Tetrastigma (from Greek *tetra*, four, and stigma, referring to the 4-lobed stigma). Vitaceae. Indomalaysia to tropical Australia. 90 species, deciduous or evergreen shrubs climbing vigorously by tendrils. The leaves are palmately or pedately compound; the flowers are inconspicuous and followed by inedible grape-like fruits. Suitable for the intermediate greenhouse or conservatory, or for large spaces in the home or in public buildings (minimum temperature 10°C/50°F). Train on strong wires, posts and rafters. Grow in pots, tubs or beds in a high-fertility loam-based mix with added garden compost and leafmould. Give bright, filtered light or partial shade. In summer, maintain high humidity by syringing or hosing once the foliage is no longer in bright sunshine. Water plentifully when in growth; reduce

watering in winter. Cut out overcrowded growth in spring. Increase by stem cuttings in summer, by simple layers, or by detached stem tips. *T.voinierianum* (syn. *Cissus voinieriana*, *Vitis voinieriana*; CHESTNUT VINE, LIZARD PLANT; Laos; very vigorous, covered in dense, tawny down, especially at first; stems stout, becoming woody, tendrils simple; leaflets 3–5, tough, tawny-hairy beneath, 10–25 cm, obovate to broadly rhombic, toothed).

tessellated chequered; marked with a grid of small squares.

testa the outer coating of a seed.

tetradynamous having four long stamens and two short as in Cruciferae.

tetramerous with parts in fours, or groups of four; often written as 4-merous.

Teucrium (from Greek *teukrion*, used in Dioscorides and Theophrastus, perhaps derived from the Trojan king Teucer, supposed to have used the plant medicinally). Labiatae. Cosmopolitan, especially Mediterranean to W Asia. GERMANDER, WOOD SAGE. Some 300 species, evergreen, usually aromatic perennial herbs, shrubs and subshrubs with tough leaves and spikes or heads of tubular, 2-lipped flowers in summer. *T.aroanium* and *T.polium* are suited to the rock garden, raised bed and trough, *T.chamaedrys* to border edging, where it will stand light clipping. All three are hardy in climate zone 7. *T.fruticans* is more sensitive to cold and needs the shelter of a warm, sunny wall on a sharply draining soil. This species should be mulched in hard winters and the frost-damaged top growth pruned back hard in mid-spring. It is one of the finest silver-leaved shrubs for dry, sheltered gardens. Grow all in full sun in well-drained soils. Protect from excessive wet in winter. Propagate by softwood or semi-ripe cuttings in a sandy propagating mix.

T.aroanium (Greece; suckering subshrub to 10cm; leaves to 2cm, tomentose; flowers mauve); *T.chamaedrys* (WALL GERMANDER; Europe to Caucasus; perennial herb or subshrub, 10–50cm tall; leaves to 4cm, dark green, toothed; flowers pink, red, purple or white; includes cultivars with cream-variegated leaves and compact habit); *T.fruticans* (TREE GERMANDER; Iberia; evergreen shrub to 2m; leaves to 2.5cm, silver-hairy at least beneath; flowers large, blue to lilac or white); *T.polium* (Mediterranean to W Asia; perennial procumbent subshrub to 45cm; leaves 1–3cm, rugose white-woolly; flowers white, yellow, purple or pink).

Thalia (for Johann Thal (1542–1583), German botanist). Marantaceae. Tropical and sub-tropical Americas, tropical Africa. About 12 species, very large, evergreen perennial herbs of wet places. The leaves arise in two ranks from a basal clump, their petioles long, erect and terminating in a knee-like joint, the blades ovate to lanceolate and held more or less horizontally. Bell-like, silken flowers hang from the flexuous branches of tall, slender panicles in summer. Suitable for subtropical gardens, where the bold, grey-green foliage and graceful inflorescences make a dramatic impact in bog gardens or at pond margins. If its roots do not freeze, *T.dealbata* will withstand winter minimums of –7°C/20°F. Plant in sun or partial shade on moist, fertile soils with plenty of organic matter. *Thalia* grows well with the soil surface either at water level or covered by up to 45cm of water. In cool-temperate areas, grow in the greenhouse or conservatory (minimum temperature 10°C/45°F); stand in shallow water or submerge in indoor pools. In summer, these plants can be moved outdoors to small ponds or to water-filled half barrels. Propagate from seed in spring or by division in spring or summer.

T.dealbata (S US, Mexico; 1.5–3m; leaves to 50cm, covered with mealy lead-white bloom; flowers to 8cm, violet); *T.geniculata* (tropical America, tropical Africa; differs from last in its height, 3–4m, its longer, narrower and less mealy leaves and larger flowers in a more strongly pendulous panicle).

Thalictrum aquilegiifolium

Thalictrum (name used by Dioscorides for a plant with coriander-like leaves, possibly a member of this genus). Ranunculaceae. Northern temperate zones, tropical S America, tropical and South Africa, Indonesia. MEADOW RUE. 130 species, perennial herbs with ferny, ternately compound foliage and delicate panicles of small, brush-like flowers in spring and summer. These lack petals, consisting of petal-like sepals and numerous showy stamens. Fully hardy. Plant in dappled shade or, where a cool, damp rootrun can be guaranteed, in full sun. Increase by division in spring or from fresh seed sown in summer and autumn.

T.aquilegiifolium (Europe to Asia; 30–150cm; sepals green-white, filaments mauve to white, pink or deep purple); *T.chelidonii* (Himalaya; 30–250cm; sepals lilac to white, filaments mauve to white); *T.delavayi* (W China to Tibet; to 120cm; sepals lilac to white); *T.diffusiflorum* (Tibet; to 100cm; sepals mauve); *T.flavum* (YELLOW MEADOW RUE; S W Europe to N Africa; to 100cm; sepals pale yellow); *T.kiusianum* (Japan; to 15cm; sepals white to purple, stamens lilac); *T.lucidum* (Europe to temperate Asia; to 100cm; sepals yellow-green); *T.orientale* (Greece to Asia Minor; to 30cm; sepals white to lilac).

thallophyte a cryptogam consisting principally of a thallus.

thallus an undifferentiated vegetative growth.

Thamnocalamus (from Greek *thamnos*, bush, and *kalamos*, reed). Gramineae. Himalaya, South Africa. Some 6 species of clumping bamboos. See *bamboos*.

thatch organic debris that accumulates at the soil surface in lawns, as a result of mowing or natural senescence or disease of grass plants. It is more prevalent on compacted or poorly drained lawns. See *lawns*.

theca a case; specifically, the pollen sac in flowering plants or the capsule in bryophytes.

Thelocactus (from Greek *thele*, nipple, and *cactus*). Cactaceae. C and N Mexico, SW US. 11 species, low-growing cacti with globose to short-cylindric, ribbed or tuberculate, spiny stems and funnelform flowers. Provide a minimum temperature of 7°C/45°F. Grow in a neutral, gritty compost, in full sun, with low humidity. Keep dry from mid-autumn until early spring, except for light misting on warm days in late winter. Increase by seed or offsets.

T.bicolor (SW US; stem depressed-globose, ovoid to cylindric, 5–38 × 5–14cm, ribbed, spines tinged red, yellow or white; flowers 3–7cm, pink to magenta, throat red); *T.leucacanthus* (Mexico; stem globose to cylindric, 4.5–15 × 2.5–5cm, ribbed and tuberculate, spines tinged yellow; flowers to 5cm in diameter, magenta); *T.setispinus* (syn. *Ferocactus setispinus*; SW US, N E Mexico; stems globose to cylindric, 7–12 × 6–9cm, ribbed but not tuberculate, spines pale yellow to red-tinged, hooked; flowers to 5cm, yellow with a red throat).

Thelypteris (Theophrastean name, from *thelys*, female, and *pteris*, fern). Thelypteridaceae. Temperate regions. 2 species, rhizomatous ferns, with softly scaly rhizomes and lanceolate to deltoid, bipinnatifid fronds, often scaly and grey-hairy above and glandular beneath. Fully hardy. Plant in a moist to wet soil with a pH of 5.0 to 7.0, in sun or shade. Increase by division or spores. *T.palustris* (MARSH FERN; cosmopolitan; rhizomes long, creeping; sterile fronds 8–38cm, thin-textured, lanceolate with to 25 pairs of linear to lanceolate, pinnately divided pinnae, fertile fronds taller, stouter and stiffer with narrower pinnae and stipe usually longer than blade).

Thermopsis (from Greek *thermos*, lupin, and *opsis*, appearance). Leguminosae. S and W US, Siberia, N India, E Asia. 23 species, perennial herbs with trifoliolate leaves and erect racemes of pea-like flowers in summer. Suited to the large herbaceous and mixed border, they are hardy to at least –15°C/5°F. Grow in any moderately fertile well-drained soil in sun. Propagate by seed or division.

T.rhombifolia (syn. *T.montana*; Rocky Mountains to New Mexico; 20–100cm; leaflets hairy beneath; flowers yellow in downy or glabrous racemes); *T.villosa* (syn. *T.caroliniana*; CAROLINA LUPIN; SE US; 50–100cm; leaflets somewhat glaucous and downy beneath; flowers yellow in downy racemes).

thermostat a device for regulating temperature in greenhouses and propagators. It incorporates a temperature sensor, by which means heating and ventilating equipment can be operated at chosen settings.

therophyte a plant that completes its growth cycle in one season and survives adverse conditions by means of protective seed.

Thespesia (from Greek *thespesios*, divine). Malvaceae. Tropics. PORTIA TREE. 17 species, evergreen trees and shrubs, usually evergreen with large leaves and showy, 5-petalled and cup-shaped flowers. Cultivate as for *Pavonia*. The following species is well-suited to seaside locations in tropical and subtropical regions. *T.populnea* (PORTIA TREE; tropics; tree to 12m; leaves cordate; flowers 6–10cm across, yellow with a maroon centre).

Thevetia (for Andre Thevet (1502–1592), French monk and traveller in Brazil and Guyana). Apocynaceae. Tropical America, West Indies. 8 species, evergreen shrubs and trees. The leavesresemble those of *Nerium*, but are darker and more glossy. The flowers are funnel-shaped with a cylindric tube dilating sharply to a campanulate throat, and broad, spirally overlapping lobes. All parts are highly poisonous. Cultivate as for *Nerium*, but with a minimum temperature of 15°C/60°F. *T.peruviana* (syn. *T.neriifolia*; YELLOW OLEANDER; shrub or small tree to 8m; flowers fragrant, to 7cm, yellow to orange or peach, rarely white).

thigmonasty, thigmotropism see *haptonasty, haptotropism*.

thin used of soil that is poorly structured and prone to drying, erosion or capping, or otherwise impoverished. Such soils are sometimes alternatively referred to as lean.

thinning (1) the selective removal of plants from a bed, sown row, container, orchard or other plantation, in order to reduce competition for nutrients, moisture and light. In the case of seedling plants, thinning is best carried out when they are very young to allow for typical and steady development of the remaining plants. Thinning outdoor rows of vegetable plants is often referred to as singling; (2) the removal of overcrowded shoots, especially during the pruning of trees and shrubs; (3) the reduction of flower buds, and more usually young fruits, on top fruit plants, in order to improve fruit quality, to prevent branch breakage, or to influence cropping pattern.

thistles a group of handsome plants, including the genera *Carduus*, *Onopordum*, and *Silybum*, suitable for carefully chosen places in the garden. The CREEPING THISTLE (*Circium arvensis*) is, however, unwelcome because of its invasive habit; when young, it produces a taproot from which laterals develop sending up shoots from underground buds. The roots are brittle and the plant is readily propagated during cultivating operations; it also perpetuates itself through seed.

A population of creeping thistle can be reduced by repeated cutting of shoots, over a number of years, just before flowering. Forking out requires concentrated effort so that as many root fragments as possible are removed. Where space can be afforded, a long term mulch with an opaque, durable fabric will reduce a stand of the weed. Most effective as a means of control is herbicide treatment with glyphosate or glyphosate trimesium, applied according to product recommendations; the weedkillers are non-selective and therefore potentially damaging to valued plants, and more than one application is likely to be necessary.

Thlaspi (from Greek *thlao*, to flatten, and *aspis*, shield, alluding to the fruits). Cruciferae. Northern temperate Regions. 60 species, annual to perennial, low-growing herbs, those here cushion- or mat-forming with small leaves and racemes of 4-petalled flowers in spring and summer. Grow on scree and moraine in the rock garden, or in sinks and troughs. They will tolerate winter temperatures to –15°C/5°F, given a gritty and well-drained growing medium. Small plants may exhaust themselves in flowering. Frequent propagation is recommended. Increase by seed or by division in autumn.

T.alpinum (SW and C Europe; biennial or perennial; flowers white to purple with violet anthers); *T.bulbosum* (Greece, Aegean Islands; biennial or perennial with a swollen rootstock; flowers violet); *T.montanum* (MOUNTAIN PENNYCRESS; C Europe; perennial; flowers white).

thong a root cutting taken from fleshy rooted plants, such as horseradish and seakale.

thrips (Thysanoptera) small brown, yellow or black insects, up to 2–3mm long, with slender bodies. Most species have wings when adult, which are very narrow and fringed with long hairs; when the insects are in repose, the wings lie more or less parallel along the back. Wingless thrips live in the surface of soils, while winged types are abundant in flowers, many species causing considerable damage. Thrips have rasping and sucking mouthparts, which are used to penetrate plant tissue and imbibe the sap; emptied plant cells become filled with air, resulting in characteristic silvering of foliage. In severe infestations, whole leaves may wither and fall prematurely and the seedlings of many plants are particularly vulnerable.

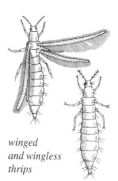

winged and wingless thrips

Some thrips attack fruits, scarring the surface; others invade flowers, discolouring them with blotches or streaks, and causing blooms to become malformed or withered. The number of generations varies with species and locations, for example, the onion thrips (*Thrips tabaci*) passes through two generations in cold northern climates, but under glass and in warmer regions the number of generations increases to twelve. Thrips often swarm in vast numbers, encouraged by dry, settled weather and a temperature of at least 20°C/68°F.

Of the 5000 or more species distributed worldwide, most are plant feeders but some feed on mites, aphids and other small, soft-bodied insects, or on fungal spores.

The ONION THRIPS (*Thrips tabaci*), of which the adults are yellow to dark brown, attacks plants outdoors and under glass, especially onions, brassicas, begonias, cucumber, cinerarias, chrysanthemums, *Cyclamen*, dahlias, gerberas, gloxinias, tomatoes and sweetpeas. The GREENHOUSE THRIPS (*Heliothrips haemorrhoidalis*), infests greenhouse plants in temperate countries, including arum lily, azaleas, chrysanthemums, crotons, ferns, fuchsias, orchids, nicotianas, and tomatoes; the adults are dark brown with an orange-coloured apex to the abdomen. The WESTERN FLOWER THRIPS (*Frankliniella occidentalis*) infests a similar range of plants, and also attacks seed heads and scars the surface of fruits, especially nectarine. The GLADIOLUS THRIPS (*Thrips simplex*) has dark brown adults, and also infests freesias. The PEAR THRIPS (*Taeniothrips inconsequens*), with dark brown adults, attacks the leaf and blossom buds of pears, plums, cherries, apricots, apples and almond, and causes russetting on the fruit surfaces. The PRIVET THRIPS (*Dendrothrips ornatus*) infests privet and lilac, causing severe silvering of the foliage; the adults are dark brown with alternate light and dark bands on the forewings, giving the insect's back a banded appearance. The LILY THRIPS (*Liothrips vaneeckei*) congregates between the scales of lily bulbs, which develop brown, sunken areas; the shiny black adults are relatively large at up to 3mm long.

Other important species in Europe include the PEA THRIPS (*Kakothrips robustus*) and the ROSE THRIPS (*Thrips fuscipennis*). The dark brown pea thrips infests the young leaves, flowers and pods of peas and, to a lesser extent, broad beans, causing loss of yield and undersized, distorted pods, scarred with silvery brown areas. The rose thrips injures the apex of rose blooms in the bud stage, causing the tips of petals to turn brown and wither, with flecking of blooms and silvering of foliage; the adults are yellow to dark brown.

Other North American species that attack garden flowers and vegetable crops include the FLOWER THRIPS (*Frankliniella tritici*); the TOBACCO THRIPS (*F. fusci*); the BEAN THRIPS (*Caliothrips fasciatus*); and the IRIS THRIPS (*F. iridis*). Another species, the CITRUS THRIPS (*Scirtothrips citri*), is a serious pest of citrus, avocado, nectarines and grapes, causing leaf distortion, blindness of buds and scabby fruits.

Some species of thrips, notably onion thrips (*T. tabaci*), tobacco thrips (*F. fusa*) and the western flower thrips (*F. occidentalis*), are vectors of tomato spotted wilt virus.

Thrips populations build up quickly in hot dry weather, so under-glass infestations can be discouraged by increasing humidity and growing under cooler conditions. Most species can be controlled by recommended contact insecticides. Infested gladioli corms or lily bulbs should be dusted with recommended contact insecticides before storing.

thrum-eyed see *heterostyly*.

Thuja (from Greek *thyia*, referring to a scented gum). Cupressaceae. N America, E Asia. THUJA, RED CEDAR. 5 species of evergreen coniferous trees, ultimately with open, conical crowns, bark exfoliating in long strips and flattened, fan-like sprays of scale-like, aromatic leaves. All species are from cool woodlands with high rainfall or cold, wet coastal or low plains. *T.plicata* grows to very large sizes in areas with cool wet summers; it is smaller and sparser in drier and warmer regions, and does not survive in the hot summers of the eastern US. *T.koraiensis* and *T.standishii* need more sheltered, humid sites. *T.occidentalis* is considerably more cold-tolerant, and can survive in boggy sites too wet for other species. Although completely hardy, *T.koraiensis* is not seen in the far north, and frost hollows are to be avoided for this species and *T.standishii*. All species tolerate clipping and make good hedges. Propagate as for *Chamaecyparis*. The hybrid *T.plicata* × *T.standishii* has been bred in Denmark to introduce resistance to the fungus *Didymascella thujina* to *T.plicata*. Young seedlings of *T.plicata* can be killed by this fungus, but it can easily be controlled by the use of fungicides on the seedbed. Damage to older plants, and the other species, is negligible.

T.koraiensis (KOREAN THUJA; Korea; tree to 15m, crown conic with trailing then ascending branchlets; leaves green and glandular above with silvery bands beneath); *T.occidentalis* (eastern N America; tree to 20m, crown narrowly conic with horizontal branches and spreading, flattened branchlets; leaves yellow-green above, grey-green beneath, becoming yellow-brown; many cultivars available, including dwarf to tall, domed, compact and weeping habits); *T.plicata* (WESTERN RED CEDAR; western N America; to 70 m, crown conic to columnar, branches horizontal to drooping with erect tips; leaves in flattened sprays, dark green above, grey-white beneath; includes dwarf to tall cultivars, erect and narrowly conic to low and spreading with fan- to thread-like foliage in shades of olive to bronze or with white to yellow variegation); *T.standishii* (JAPANESE THUJA; Japan; to 30m, crown broadly conic and open, branchlets spreading to ascending; leaves glandular, dark green to yellow-green).

Thujopsis (from *Thuja* and Greek *opsis*, appearance, referring to the similarity between these two genera). Cupressaceae. C and S Japan. 1 species, *T.dolobrata*, HIBA, an evergreen conifer to 30m with red-brown bark exfoliating in long stringy strips, a conic crown with numerous, layering stems and horizontal to upswept branches. The leaves are scale-like, thick, waxy and glossy green above and white beneath. Raise from cuttings or by seed and plant in a sheltered site where the soil remains damp, preferably in or beside woodland. Although hardly affected by cold or pH, this tree does demand high humidity.

Thunbergia (for Carl Peter Thunberg (1743–1828), Swedish botanist. A student of Linnaeus, he travelled for the Dutch East India Company as a doctor and sent back plants from Japan, Java and South Africa. He became professor of botany at Uppsala). Acanthaceae. South and tropical Africa, Madagascar, warm Asia. 100 species, perennial shrubs and herbs, most climbing or scrambling; some are treated as annuals in cultivation. Produced at various times of

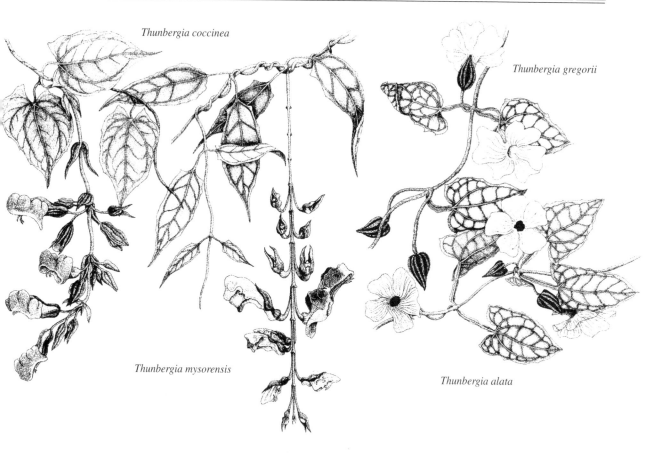

Thunbergia coccinea

Thunbergia gregorii

Thunbergia mysorensis

Thunbergia alata

year, the flowers are large and funnel-shaped with a broad, 2-lipped rim of five lobes. *Thunbergia* includes some of the most beautiful climbers for the arches and pergolas of tropical and subtropical gardens; *T.erecta* may be used as a flowering hedge in such climates. For cooler zones, the genus contains many elegant and vigorous specimens for borders and large containers in the warm greenhouse or conservatory. Some species, such *T.alata* and *T.gregorii*, flower in their first year from seed and are commonly grown as frost-tender annuals. The latter may overwinter at the base of a warm south-facing wall in mild winter areas, especially if cut back and mulched at the end of the season. Grow in a fertile loam-based mix with additional sharp sand and screened manure, in direct sun but with shade from the strongest summer sun. Train on wires or trellis. Maintain a minimum temperature of 10°C/50°F and moderate humidity with good air movement. Water plentifully when in growth. Reduce water as light and temperature levels fall, keeping the plants almost dry in winter. Prune lightly in late winter or early spring before growth resumes; hard pruning will severely reduce the production of flowers. Easily propagated by most types of cuttings and by seed.

T.alata (BLACK-EYED SUSAN; tropical Africa; twining herb, often grown as an annual; flowers solitary, to 3.75 × 4cm, creamy-white or yellow-orange usually with a dark purple-brown throat; includes 'Alba', with white flowers with a dark centre,

'Aurantiaca', with orange flowers with a dark centre, 'Bakeri', with snow white flowers, and Suzie Hybrids, with orange, yellow or white flowers with a dark eye); *T.capensis* (South Africa; sprawling herb or subshrub, sometimes grown as an annual; flowers 4–10 together, to 3cm, creamy yellow); *T.coccinea* (India, Nepal, Burma; woody-stemmed climber, 3–8m; flowers in pendent racemes 15–45cm long, each to 2.5cm, orange-red); *T.erecta* (KING'S MANTLE, BUSH CLOCK VINE; tropical W to South Africa; erect or twining shrub to 2m; flowers solitary, 7cm, tube yellow-cream, lobes dark blue-violet; includes 'Alba', with white flowers); *T.fragrans* (India, Sri Lanka, Australasia; woody-stemmed, aromatic climber; leaves 5–7.5cm, triangular-ovate to oblong, base cordate to hastate, entire to toothed; flowers solitary, to 3 × 5cm, white, lobes spreading; includes 'Angel Wings', with snow white, lightly scented flowers); *T.grandiflora* (BLUE TRUMPET VINE, CLOCK VINE, BENGAL CLOCK VINE, SKY VINE, SKYFLOWER, BLUE SKYFLOWER; N India; large twining shrub; flowers usually solitary, to 7.5 × 7.5cm, blue-violet, sometimes with a paler, or white to golden, throat; includes 'Alba', with white flowers); *T.gregorii* (tropical Africa; twining herb, often cultivated as an annual; flowers solitary, 4.5 × 4.5cm, orange); *T.mysorensis* (India; twining shrub to 6m; flowers in pendent racemes to 40cm long, each to 3.75 × 5cm, rich golden yellow with red-brown lobes).

Thymus serpyllum

thunderflies see *thrips*.

thunderworms (*Mermis* species) types of eelworm with thin bodies up to 160mm long. They are most frequently seen in early summer after heavy rain, when the thin, creamy white or blackish worms can be found writhing on the foliage of various low-growing plants. Thunderworms are harmless to plants and animals other than certain plant-feeding insects, for example, caterpillars, grasshoppers or earwigs, in the bodies of which eggs hatch to develop as parasites.

Thunia (for Franz, Graf von Thun (1786–1873)). Orchidaceae. Himalaya to Burma. Some 6 species, terrestrial orchids with clumped, cane-like pseudobulbs and terminal clusters of showy flowers. These have oblong to lanceolate tepals and a funnel-shaped lip with a frilly to lacy margin. Pot canes annually in a mix of turfy loam, leafmould and bark. Position in a brightly lit, well-ventilated location in the cool or intermediate house (minimum temperature 10°C/50°F), away from direct sunlight, draughts and dripping water. Water and feed copiously when in growth. After flowering, dry canes entirely (foliage is usually lost) and remove to a cool dry place in full sunlight. Increase by division, stem cuttings placed on warm sphagnum or perlite, or by plantlets developing on badly grown plants (i.e. where they have been given too much heat and water and too little rest). *T.alba* (syn. *T.marshalliana*; India, Burma, China; stems to 1m, erect, tufted; flowers to 14cm in diameter, often not opening fully, crystalline white, the lip orange or yellow, striped purple-red).

thyme (*Thymus* species) a perennial, aromatic, low-growing herb or subshrub, grown for food flavouring and as a source of thymol for antiseptics and deodorants. Thyme is also useful for ornamental planting and as a bee plant. There are many different types. *T.vulgaris* retains its flavour well in slow cooking, and is a useful ingredient of bouquet garni; *T × citriodorus*, the LEMON-SCENTED THYME, and *T.pulegioides* are commonly grown, and *T.herba-barona* has a distinctive caraway flavour. Thyme becomes straggly after a few years and should be re-propagated either from spring-sown seed or from cuttings or division in April or May which will produce established plants sooner. Plant 15–30cm apart on a sunny, dry, well-drained site. Leaves can be picked year-round to use fresh, or just before flowering for drying.

Thymus vulgaris

Thymus (from Greek *thyo*, to perfume: *thymon* or *thymos* used by Dioscorides, probably for this plant). Labiatae. Eurasia. THYME. 350 species, dwarf or low-growing perennial, aromatic herbs and subshrubs with small leaves and tubular, 2-lipped flowers at various times of year.

Thymes are culinary herbs and important bee plants. All are strongly aromatic in favourable locations. They retain their fragrance well on drying, and are used in pot-pourri, sweet sachets and herb pillows. Plant in dry sunny borders, open scree, troughs, rock gardens, herb gardens and between paving stones, where low-growing species will tolerate treading. Most species are hardy to at least −15°C/5°F, but, in very cold climates, they should be protected with a dry mulch of bracken litter or evergreen prunings. Plant on well-drained, preferably calcareous soils, in full sun. Clip over after flowering to maintain vigour and compactness. Propagate by division, by simple layering, and from semi-ripe heel cuttings in spring or summer.

T.caespititius (SW Europe; dwarf, mat-forming shrublet; to 5cm tall; leaves to 0.6cm, narrowly spathulate, ciliate; flowers rose, lilac or white; includes 'Aureus', with pale gold leaves); *T.carnosus* (Portugal; woody, erect, to 40cm; leaves ovate, light green to grey-green, fleshy, hairy beneath; flowers white, lilac or pink); *T. × citriodorus* (small shrublet, seldom exceeding 20cm with very small, smooth, dark green leaves strongly scented of lemon; includes cultivars with white to cream or silver-marked leaves and solid lime to gold foliage); *T.herba-barona* (CARAWAY THYME; Corsica, Sardinia; dwarf, wiry subshrub to 20cm; leaves very small, hairy, caraway-scented; flowers pink to mauve; includes strongly caraway- and lemon-scented forms); *T.leucotrichus* (Greece, Turkey; creeping, dwarf shrub to 10cm with very small, pale-hairy, narrow leaves; flowers pink to purple); *T.praecox* (S, W and C Europe; mat-forming subshrub to 5cm; leaves lanceolate to obovate, ciliate; flowers white, mauve or purple); *T.pseudolanuginosus* (origin unknown; stems prostrate; leaves small, grey-green, hairy; flowers pale pink); *T.pulegioides* (Europe; highly aromatic dwarf shrub, ascending to 20cm; leaves oval to elliptic, hairy beneath; flowers pink to purple); *T.serpyllum* (WILD THYME, N Europe; very aromatic dwarf shrublet with small, ovate, ciliate leaves and heads of pink to purple flowers; includes many cultivars, creeping to erect with white to silver- or gold-variegated leaves); *T.vulgaris* (Mediterranean; very aromatic, woody, erect to 30cm; leaves small, narrow, revolute; flowers white to purple; includes variegated and camphor-scented cultivars).

thyrse, thyrsus an indeterminate, paniculate or racemose primary axis, with determinate cymose or dichasially compound lateral axes.

Tiarella (from Latin *tiara*, little diadem, referring to the shape of the fruit). Saxifragaceae. FALSE MITREWORT, SUGAR SCOOP. Eastern and northwestern N America, 1 species Asia. 5 species, rhizomatous, evergreen perennial herbs with low clumps of long-stalked, rounded to palmately lobed leaves and, in late spring and early summer, erect racemes of small, star-like flowers. Hardy ground-cover with excellent autumn colour. Its cultivation requirements are as for *Heuchera*, except that it requires more shade and a moist, humus-rich soil. Remove decayed leaves in winter and protect the rhizomes with a mulch of garden compost. *T.cordifolia* (FOAM FLOWER; N America; stoloniferous; leaves to 10 cm, cordate, 3–5-lobed, toothed, hairy bright green, usually with some maroon staining around veins; racemes to 30cm tall; includes cultivars with pink to wine red flowers and bronze to purple-marbled or maroon leaves; *T.wherryi* differs in lacking stolons and its bronze-pink cast).

Tibouchina (native name in Guyana). Melastomataceae. GLORY BUSH. Tropical S America, mostly Brazil. Some 350 species, evergreen shrubs or subshrubs, usually hairy with rounded, 5-petalled flowers in spring and summer. Plant in freely draining, neutral or (preferably) acid, moisture-retentive, fertile soils. Admit bright light with some shade from the hottest midsummer sun; water plentifully when in full growth; Keep almost dry in winter, with a minimum temperature of 7–10°C/45–50°F. Propagate by greenwood or semi-ripe cuttings. *T.urvilleana* (syn. *T.semidecandra*; GLORY BUSH, PRINCESS FLOWER; Brazil; shrub 2–5m; leaves 5–10cm, ovate, deep green and finely bristly above, paler and downy beneath; flowers 7–10cm across, silky, mauve, indigo or violet, stamens pink to purple).

Tigridia (from Latin *tigris*, tiger, in this case referring to the jaguar, whose spotted pelt these flowers resemble). Iridaceae. Mexico, Guatemala. PEACOCK FLOWER, TIGER FLOWER, FLOWER OF TIGRIS. 23 species, perennial bulbous herbs with sword-shaped, plicate leaves. Borne in summer, the flowers are large, showy and shallowly cupped, the three outer segments broadly obovate and spreading, the inner segments smaller and ovate to pandurate. In frost-free regions, these bulbs will grow outdoors all year round in a sunny position on free-draining soil. Elsewhere, plant them outdoors in spring, lifting in autumn to be dried off and stored at 8–12°C/46–54°F in sand. Otherwise, plant in pots or in the border of the cool greenhouse or conservatory. Grow in a warm, sheltered site in full sun, in a fertile, well-drained, sandy or gritty soil. Plant in late spring, 10–12.5cm deep in light soils, more shallowly in heavier soils or if bulbs are small. Propagate from seed, or by offsets. *T.pavonia* (Mexico; 60–125cm; outer perianth segments 6–10cm, orange, bright pink, red, yellow or white, variously spotted red, brown or maroon at base, inner segments of similar ground colour but more distinctly marked).

Tilia (from Latin *tilia*, lime, derived from Greek *ptilon*, feather, referring to the flower bract). Tiliaceae. Eastern and central N America, Europe, W, C and E Asia. LIME, LINDEN, BASSWOOD. About 45 species, deciduous trees, usually with cordate, toothed leaves. Produced in summer, the flowers are small, 5-petalled and fragrant, carried in short, pendulous cymes fused for half their length to an oblong to lanceolate, yellow-green bract.

The following are fully hardy and grow best on fertile, alkaline to neutral soils with medium to good drainage. They suffer when exposed to harsh winds and are best among or not far from other trees. Propagate by root suckers, layering, grafting or from wild-collected seed. Dig up suckers with as good a root as possible, and replant. Layers respond well to rooting hormone, dusted into the wound; cut from the parent tree when well rooted. Grafts should only be used where suckers are too high up for easy layering; side-graft erect suckers on to seedlings of *T.tomentosa* or *T.platyphyllos*. Stocks of *T. × vulgaris* should not be used, since they produce dense basal suckers below the graft, and normal branch-shoot scions must not be used, since they lack apical dominance.

Root and stem suckers are not attractive features on specimen trees, and the best limes for gardens are those species, like *T.tomentosa* and *T.platyphyllos*, that rarely produce them. These should be grown from seed. Collect seed in mid-autumn from wild trees or extensive single-species plantings; stratify for 3–5(–16) months in damp sand at 0–3°C/32–37°F before sowing in early spring in a cold frame. Germination takes up to 18 months (particularly if the stratification period is short). Garden-collected seed will probably produce hybrids of dubious parentage and should not be used, except in the search for new cultivars.

T.americana (AMERICAN BASSWOOD, AMERICAN LIME; N America; to 40m, crown broad, ovoid to rounded; leaves 6–16cm, broadly ovate to orbicular, base truncate, apex acuminate, serrate, dark green and glabrous above, paler beneath with prominent veins and tufts of hairs in their axils); *T.cordata* (SMALL-LEAVED LIME, LITTLELEAF LINDEN; Europe; to 40m, usually smaller, crown outspread, suckering; leaves 3–7cm, rounded, apex abruptly acuminate, base cordate, serrate, glossy dark green above, somewhat glaucous beneath with tufts of hairs in the vein axils beneath); *T. × euchlora* (CAUCASIAN LIME, CRIMEAN LINDEN; to 20m, crown rounded, branches arching to pendulous; leaves 5–10cm, rounded to ovate, apex abruptly acuminate, base cordate, margins finely serrate, glossy dark green above, paler beneath with tawny tufts in the vein axils); *T.henryana* (China; to 25m; leaves emerging bronze-red, 5–12cm, broadly ovate, apex acuminate, base cordate to truncate with bristly teeth, veins hairy above, with tufts of hairs in veins beneath); *T.mongolica* (MONGOLIAN LIME; Russia, Mongolia, China; to 20m, usually shorter, branches arching; leaves 4–7cm, broadly ovate, apex abruptly acuminate, base truncate to cordate, often 3–5-lobed, finely toothed, flushed red at first, hardening glossy dark green above, somewhat glaucous beneath); *T.oliveri* (China; to 25m; leaves 7–11cm, broadly ovate, apex abruptly acuminate, base cordate or truncate with sparse, glandular teeth, dark green and glabrous above, white-tomentose beneath); *T.* 'Petiolaris' (PENDENT WHITE LIME, WEEPING LIME; Europe to Asia Minor; to 30m, crown domed, branches pendulous; leaves 5–10cm, oval to rounded, apex acute, base cordate, serrate, dark green and slightly downy above, white-tomentose beneath); *T.platyphyllos* (BROAD-LEAVED LIME, LARGE-LEAVED LIME; Europe to W Asia; to 40m, crown conical to broadly columnar; leaves 4.5–11cm, rounded to ovate, apex abruptly acuminate, base cordate, serrate, dull dark green above, light green and somewhat downy beneath; includes 'Rubra', with branches coral to sealing wax red in winter); *T.tomentosa* (SILVER LIME; S Europe to Asia Minor; to 35m, crown dense, ovoid-conical becoming domed with age, branches erect; leaves 5–11cm, rounded, apex abruptly and finely acuminate, base cordate, serrate and sometimes lobed, dark green above, grey-tomentose beneath); *T. × vulgaris* (syn. *T. × europaea*; COMMON LIME, EUROPEAN LIME; to 40m, crown broadly conical; leaves 5–9cm, broadly ovate, apex abruptly acuminate, base cordate, dark green, glabrous above, paler with tufts of hairs in vein axils beneath).

Tillandsia (for Dr Elias Tillands (1640–1693), Swedish botanist and physician). Bromeliaceae. S US, C and S America, W Indies. AIR PLANT. Some 400 species, mostly epiphytic, evergreen perennial herbs. They vary greatly in habit and size, from the miniature and mossy to large rosette-formers. Many species are very sparsely rooting, taking much of their moisture and nourishment from the air through grey, absorbent scales that cover the foliage. The leaves may be thread-like or sword-shaped. Each growth produces a terminal spike of showy, short-lived flowers, often surrounded by vividly coloured bracts.

Tillandsia argentea

T.cyanea and *T.lindenii* need a minimum temperature of 10°C/50°F, with high humidity and protection from full sun. They grow best planted in small pots containing an open, bark-based medium. Water throughout the year, allowing the medium to dry between waterings. Given dry, airy conditions, the remaining species listed here will survive barely frost-free conditions. They prefer, however, a humid and buoyant atmosphere in full sun and temperatures of 7–23°C/45–75°F. They are grown exclusively as epiphytes – wired, pinned or even glued to branches, cork slabs, bromeliad 'trees' and other objects. Given a daily misting with soft water in warm weather, a weekly misting at cold times, an occasional very dilute foliar feed and adequate light, they may thrive for many years as house plants. Under glass, their culture is easier, but avoid keeping these species too wet during cool, dull spells. *T.usneoides*, a signature plant of the Deep South of the US, is useful for providing light shade in sections of the greenhouse – merely hang it in swags from the roof. It is also a good, live lining material for orchid baskets. Increase all species by division; *T.usneoides* may also be increased by lengths of stem.

T.aeranthos (S Brazil, Paraguay, Uruguay, Argentina; to 30cm , with a distinct stem; leaves 6–14cm, narrowly triangular, densely grey-scaly; flowers dark blue in an ovoid to cylindric spike with pink-purple bracts); *T.argentea* (C America, West Indies; to 25cm in flower, stem more or less absent; leaves 3–9cm, filiform, silvery-scaly in a dense rosette; flower spike narrow, flowers bright cerise amid red-green bracts); *T.bulbosa* (Mexico to Brazil, Ecuador, W Indies; to 22cm in flower, stem absent but swollen leaf bases form a 'bulb'; leaves terete, narrow, snaking, dark green to olive marked maroon; flowers violet amid red bracts in compound spikes); *T.caput-medusae* (C America; to 25cm in flower, stem absent but base bulbous; leaves linear to triangular with inrolled margins, snaking, silver-scaly; inflorescence branched with red to pink bracts and violet flowers); *T.cyanea* (Ecuador, Peru; to 30cm in flower, stem absent; leaves linear to triangular, dark green in a low, spreading rostte; flowers large, dark violet, emerging from closely overlapping, 2-ranked, waxy pink bracts in a flattened, elliptic spike); *T.ionantha* (Mexico to Nicaragua; to 10cm, stems short, clumped; leaves to 6cm, narrowly triangular, mid-green, grey-scaly, the innermost flushed red at flowering; flowers violet-blue in a short, nestling spike amid rose to scarlet bracts); *T.lindenii* (Peru; differs from *T.cyanea* in its pink- to purple-tinted, green bracts and royal blue flowers with a white base); *T.recurvata* (BALL MOSS; SE US to S America; dwarf, clump-forming with linear, terete, curving, olive and scaly leaves to 2cm long and white to violet flowers in a very short inflorescence); *T.streptophylla* (Mexico to Honduras; to 25cm in flower; leaves strap-shaped to narrowly triangular, silvery-scaly, twisted and curling in a low bulbous-based rosette; flowers purple amid red bracts in branching spikes); *T.usneoides* (SPANISH MOSS, OLD MAN'S BEARD; SE US to S America; individual growths composed of three or four, thread-like, silvery-scaly leaves to 4cm long and connected by wire-like rhizomes in dense, mossy festoons to 3m long; flowers very small, stalk-less, pale blue and yellow-green).

tiller a shoot issuing from the base or roots of a plant; used particularly of basal side shoots of grasses.

tilth a fine crumbly structure to the surface layer of a soil with good moisture status, produced by careful preparation, and especially suitable for seed sowing.

tine an individual prong, spike, or tooth on a cultivation tool, such as a fork, rake, or harrow.

tip bearer used of apple and pear cultivars that produce fruit buds at the tips of one-year-old shoots. During pruning, only shoots longer than 22cm are cut back. The apple cultivars 'Worcester Pearmain' and 'Bramley's Seedling' are partial tip bearers, and the pear 'Jargonelle' is a tip bearer.

tipburn a disorder of lettuces in which the leaf margins become brown and the plant disfigured, with arrested growth and associated vulnerability to

Tillandsia lindenii

disease infection. The condition arises particularly on light soils outdoors, where there is shortage of water, and under greenhouse conditions of high humidity during bright sunshine.

tipping removing the tip of a shoot during pruning, often used specifically to describe the light pruning of leaders.

Tipuana (referring to the Tipuani Valley in Bolivia, where there are large populations of this tree). Leguminosae. S America. 1 species, *T.tipu* (syn. *T.speciosa*), the TIPU TREE, YELLOW JACARANDA, ROSE WOOD, PRIDE OF BOLIVIA, an evergreen tree to 35m tall with pinnate leaves and yellow to apricot or orange, pea-like flowers in pendulous panicles. A fast-growing, spring-flowering tree, used in the tropics and subtropics for park and avenue plantings. Cultivate as for tender *Bauhinia*.

tissue culture see *micropropagation*.

Titanopsis (from Greek *titanos*, chalk, and *opsis*, resemblance, referring to the chalk-like surface of the leaves of some species). Aizoaceae. South Africa, Namibia. 6 species, highly succulent, rosette- and clump-forming perennial herbs, with thickly fleshy leaves, their tips covered with chalky tubercles resembling stone. Yellow or orange, daisy-like flowers open in bright weather. Cultivate as for *Pleiospilos*.

T.calcarea (leaves to 2.5cm, spathulate, tinged white or green-blue, tip truncate with green-red to grey-white tubercles; flowers golden to orange-yellow); *T.schwantesii* (leaves to 3cm, spathulate, grey to blue-green sometimes tinged red, tip rounded to triangular with yellow-brown tubercles; flowers pale yellow).

Tithonia (for Tithonus, companion of Aurora, goddess of the dawn). Compositae. C America. MEXICAN SUNFLOWER. 10 species, erect, robust annual or perennial herbs and shrubs. Produced in summer and autumn, the daisy-like flowerheads are solitary and long-stalked with golden-yellow to orange-scarlet ray florets and yellow disc florets. Grow in any well-drained, moderately fertile soil in full sun. Provide the support of stakes, irrigate in dry periods and deadhead regularly. Sow seed *in situ* in spring or earlier under glass. *T.rotundifolia* (Mexico to Panama; annual, 1–4m tall; flowerheads to 7cm in diameter, ray florets orange to scarlet; seed races are available with single or double flowerheads, in shades of pale yellow, gold, orange, burnt orange and flame, the plants seldom exceeding 1m, and sometimes with finely cut leaves).

Todea (for Henry Julius Tode, of Mecklenburg (1733–1797), mycologist.) Osmundaceae. Australasia, South Africa. 2 species, large ferns with stout, erect rhizomes crowned with bipinnate fronds. Provide a minimum temperature of 10°C/50°F, a humid atmosphere and filtered light or shade. Plant in an acid to neutral mix rich in bark, leafmould and garden compost. Keep moist at all times. Syringe in summer. *T.barbara* (syn. *T.africana*; CREPE FERN, KING FERN; Australasia, South Africa; rhizome to 1m tall, stout, black and fibrous; fronds to 1m, pale to deep, shining green, pinnules to 4cm, lanceolate, serrate).

toggle lopper a lopper, or long-arm pruner, with a double pivot for extra leverage, used for cutting thick branches.

tolerance (1) used of a plant that shows few or no signs of disease when infected by a pathogen; (2) used of a plant capable of adapting to an environmental stress by altered metabolism or growth; (3) the amount of pesticide permitted by legal regulations to remain on a crop at or after harvesting, expressed in parts per million.

Tolmiea (for Dr. William Fraser Tolmie (1812–1886), surgeon for the Hudson's Bay Company at Fort Vancouver). Saxifragaceae. Western N America. PICKABACK PLANT. 1 species, *T.menziesii*, a viviparous, hairy perennial herb. The leaves are long-stalked, 5–10cm long, reniform to cordate, shallowly palmately lobed and toothed. Plantlets develop at the junction of the petiole and leaf blade. Small, green-brown flowers are carried in erect racemes in spring and summer. The popular cultivar 'Taff's Gold' (syn. 'Maculata', 'Variegata') has fresh lime green leaves finely spotted and mottled cream to gold. Hardy to –17°C/1°F, although often grown as a house plant. Cultivate as for *Heuchera*, but provide shade. Propagate by potting plantlets in a sandy, low-fertility, loam-based medium. Feed and grow on in intermediate greenhouse conditions for planting out the following spring, or peg leaves to surrounding soil and detach young plantlets when rooted.

tomato (*Lycopersicon esculentum*) a short-lived perennial cultivated as an annual for its usually bright red, fleshy fruit. It is a variable species with a growth habit ranging from an indeterminate vine, attaining a potential length of around 10m in one season, to compact, determinate bush types. Fruits are borne in clusters and may be round, lobed, or pear-shaped, varying in size from 2–12cm in diameter; cultivated forms are usually red but also orange, yellow and striped. There is a wide choice of cultivars available suited to local conditions. In section, the fruits contain cavities or locules filled with seed-containing gel. Most of the smaller, round-fruited cultivars are bilocular in contrast to the larger-fruited beefsteak types which are multi-locular, producing a more lobed fruit. The cherry tomato, *L.esculentum* var. *cerasiforme*, has smaller and more sweetly flavoured fruit.

The tomato originated mainly in Ecuador and Peru and, following domestication in Mexico and Central America, was introduced into Europe during the early 16th century. It is adapted to a wide range of conditions but is frost-tender; optimum temperature range for growth is 20–24°C/70–75°F, temperatures below 13°C/55°F and greater than 27°C/80°F

tomato

Pests
aphids
eelworm
red spider mite
whiteflies

Diseases
grey mould
leaf mould
tomato blight
(see phytophthora)

Disorder
blossom end rot

adversely affects growth. High light levels are required for maximum fruit set and high yields.

Optimum soil temperature for seed germination is 29°C/85°F. Air temperature greater than 38°C/100°F can destroy both pollen and developing embryos, and setting is poor at high night temperatures of 25–27°C/77–80°F, except with specially selected cultivars, whilst temperatures below 10°C/50°F increases the rate of flower abortion. Fruit ripening is optimum at 18–24°C/65–75°F and poor between 13°C/55°F and 10°C/50°F. Fresh fruit, picked at an early stage of ripening, can be stored for several days, optimally at a temperature of 13°C/55°F in conditions of high humidity.

In temperate regions, fresh outdoor tomatoes are only harvestable during the summer months. Commercially, heated greenhouse production serves most of the year, although poor winter light in northern latitudes makes all-year-round production difficult.

Plants are seed-raised in early spring, under protection at 15–18°C/60–65°F, either sown thinly in seed trays and pricked out into 7.5cm pots, or direct into pots, two or three seeds per pot, and thinned to one. Plants are ready to plant 6–8 weeks after sowing, when 15–20cm high with the first flowering truss is visible.

For growing outdoors, plants must be hardened off, and bush (determinate) types are the best choice. These have no prominent leading shoot but a number of side branches, giving rise to a bushy plant; they require a minimum of support. Compact forms are available, suitable for growing in pots, hanging baskets or window boxes. For open-ground sites, a sheltered, sunny position is essential, the soil made fertile and moisture-retentive by the incorporation of organic matter, and ground mulch. Plants are set out 45–60cm apart.

Better quality and earlier fruit is obtainable under greenhouse protection, where tall (indeterminate) types are usually grown, with cultivars selected for growing either with or without supplementary heating. Plants may be planted direct into the border soil or grown in pots, by ring culture, or in growing bags, set 40–45cm apart. They may be trained up tall canes but, more usually, the plants are progressively twisted clockwise around strong soft twine, which is tied to horizontal wires fixed along the roof of the structure and secured to the base of each plant or to a wire peg.

Indeterminate types require the constant removal of side shoots that arise in axils between leaf stalks and the main stem. Thorough and regular watering and supplementary feeding are necessary from about the time of first fruit set, with a proprietary feed high in potash. Pollination is encouraged by shaking the trusses, and plants are stopped at one leaf above the last required truss. In the greenhouse, wide fluctuation of temperature should be avoided; summer ventilation and shading are essential to keep the temperature as near as possible to a maximum of 20°C/70°F.

tomentose with densely woolly, short, rigid hairs, perceptible to the touch.

tomentum hair or, more specifically, a tomentose pubescence.

Toona Meliaceae. E Asia to Australasia. 6 species, evergreen or deciduous trees formerly included in *Cedrela* (a genus now confined to the neotropics). Hardy to –25°C/–13°F. Plant on fertile, well-drained soils in full sun. Sow seed in autumn or strike root cuttings in late winter in a cold frame. *T.sinensis* (syn. *Cedrela sinensis*; China; deciduous tree to 20m; leaves 3–60cm long, pinnate, leaflets 6–14cm, ovate to lanceolate, emerging pink to bronze, hardening deep green, turning gold in autumn; flowers small, white to cream, fragrant in pendulous panicles; 'Flamingo': leaves emerging vivid pink, turning creamy yellow then bright green).

top draining the use of shallow channels or ditches to drain off surface water.

top dressing the application of quick-acting fertilizers to the soil surface around growing plants to stimulate growth; usually carried out in spring; cf. *base dressing*. Care must be taken to avoid leaf contact, which can cause scorching, and to protect against root damage and pollution of ground water. Top dressing of a lawn involves applying sharp sand and organic matter, such as sieved compost, peat, or spent hops, to level the surface and encourage root growth. Sometimes known as side dressing.

top fruit a descriptive term for tree fruits, such as apples, cherries, peaches, pears, and plums; cf. *soft fruit*.

topiary the art of cutting or training trees and shrubs into specified shapes, probably dating from the end of the 1st century BC. Topiary was widely used in Italian gardens of the 16th and 17th centuries and spread throughout Europe. A wide range of plants was used, including rosemary, juniper, phillyreas, myrtle and privet.

To create either an individual specimen or a structured topiary garden can take from between six to ten years. Good soil preparation and a well-drained site in full sunlight are required for vigorous and even growth. Wind shelter is essential, and plenty of space should be left around specimen topiary to allow for maintenance. Topiary looks best on level ground and, if possible, should be viewed from above. Shaping should begin when plants are very young. It is done by temporarily tying shoots down and clipping with secateurs or shears against a simple cane template, or against a shaped wire framework, which may be incorporated into the plant as it grows. Shoots must be tied in whilst young and supple, pinching them back to encourage branching and training new shoots to fill gaps. Yew, holly, bay, privet and box are some of the best candidates for topiary; when established all need clipping once a year, holly and box around midsummer and yew in early autumn. Precise clipping is required, geometric and bird or animal designs always with the aid of guiding templates, and working from the top down. Intricate designs require frequent clipping during the formative years. A dressing of

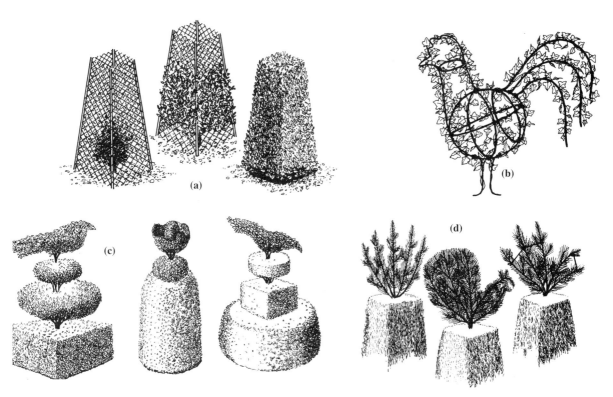

Topiary (a) *Forming an obelisk with wire formers* (b) *Forming a rooster on a wire frame* (c) *Four-tier sculptural forms* (d) *Stages in forming a peacock*

slow-release fertilizer applied in spring is beneficial.

Cones These follow the natural shape of many trees and have the advantage that they can later be developed into a more complex shape, such as a spiral. Plants are first cut to shape by eye before setting up a structure of canes as a template.

Square-based styles Geometric shapes, such as pyramids, obelisks and cubes, are easy to commence, but attention must be paid to maintaining strict verticals and horizontals. They can be used for the formal garden, as part of a structural and proportioned plan. Rounded shapes are easier to maintain.

Crowns For these, it is necessary to select plants with four equally vigorous shoots growing out from the central stem at or near a coincident point. The shoots should be bent downwards so as to use the tension of the tree itself and give a more resilient curve. Yew or box are most suitable.

Birds The traditional topiary shapes of birds are most satisfactorily produced from box or yew.

topping the removal of the growing point of a plant, other than where done purposely to induce the growth of side shoots (which is referred to as stopping). Topping may, for example, be carried out to restrict the height of a hedge, or to remove aphid-infested shoots of broad beans.

topsoil soil that, in gardens, is usually fertile and suitable for the cultivation of plants, in contrast to subsoil, which is less weathered and of lower organic-matter content. Topsoil varies in depth, and its natural position in the profile should be preserved during digging and levelling operations.

top working see *grafting*.

Torenia (for Olaf Toren, 18th-century Swedish clergyman). Scrophulariaceae. Africa, Asia; some species naturalized in US. Some 40 species, perennial or annual herbs with short racemes of broadly tubular, 2-lipped flowers in summer. Provide a position in moist soil in part shade. Cultivate as for *Alonsoa*. *T.fournieri* (WISHBONE FLOWER, BLUE-WINGS; Asia; to 30cm, bushy; flowers 2–5cm across, tube violet, upper lip pale blue, throat lilac to purple marked yellow, lower lobes velvety, dark violet to indigo).

Torreya (for Dr John Torrey (1796–1873), American botanist). Cephalotaxaceae. N America (California, Florida) and Asia (Japan to SW China). NUTMEG-YEW. 7 species, evergreen coniferous shrubs or trees with a broadly oval to conic crown and red-brown, finely shredding bark. Lying flat in two ranks, the leaves are narrowly lanceolate and sharp-tipped. Pollinated cones develop as fleshy drupe-like fruits. Hardy in zone 7. Provide shelter and high humidity or a moist soil – they tolerate woodland shade very well but not exposure to wind; undemanding as to soil pH. Propagate as for *Taxus*. *T.californica* (CALIFORNIA NUTMEG-YEW; California; tree to 20m, crown broadly conic, branches more or less horizontal with pendulous branchlets; leaves to 5cm, dark green to olive, aromatic if crushed; fruit 3–4cm, purple-green; 'Spreadeagle': low-growing, branching horizontally; 'Variegata': shoots blotched yellow-green).

tortrix moths (Lepidoptera: Tortricidae) small moths with broad, leaf-like wings; their larvae, which are usually green with brown or black heads, wriggle actively backwards if removed from their concealed sites. Most tortrix moth caterpillars feed within the turned-over edge of a single leaf, or between adjacent leaves held together by silken threads; others bore into fruits or buds, or tunnel into stems.

Several species are common to Europe and North America. The larvae of the CODLING MOTH (*Cydia pomonella*) burrow into apples and pears; the PEA MOTH (*C. nigricana*) causes similar damage to edible peas. The PINE SHOOT MOTH (*Rhyacionia buoliana*) has red-brown larvae, up to 20mm long, that tunnel into the buds and young shoots of pines, causing them to turn brown and die. The CARNATION TORTRIX (*Cacoecimorpha pronubana*) has yellow to olive green larvae, up to 20mm long; they feed between leaves fastened together with webbing, on the buds and blooms of carnations, and many other green-house plants, including acacias, *Asparagus*, crassulas, *Cyclamen*, euphorbias, grevilleas, as well as outdoors on *Cytisus*, daphnes, *Euonymus*, ivies, laurel and privet. The BUD MOTH (*Spilonota ocellana*), known in the US as the eye spotted bud moth, has red-brown caterpillars that attack buds in spring and later feed on the foliage, tying together leaves of various woody plants, including cotoneasters and fruit trees and bushes. The STRAWBERRY TORTRIX MOTH (*Acleris comariana*) has pale green larvae with brown heads. They are up to 12mm long, and feed in folded or rolled leaves of strawberries, causing them to turn brown and wither. Common to both continents is the FRUIT TREE TORTRIX MOTH (*Archips podana*), which is one of a complex of species that attack the fruits of apple and to a lesser extent those of pears, plums and cherries. The caterpillars are green with a darker line along the back; typically, they feed between two leaves fastened together with webbing, but more serious damage is caused when they browse the surface of fruits by feeding beneath a leaf attached to the fruit by webbing. They may also attack blackberries, hawthorns, ivies, oaks, raspberries, rhododendrons and roses.

Other important species in Europe are the FLAX TORTRIX (*Cnephasia interjectana*) and the GREEN OAK TORTRIX (*Tortrix viridana*). The flax tortrix has also been recorded in the Canary Islands and Newfoundland; the green-grey to black caterpillars are commonly found in gardens, attacking in particular heleniums, lettuces, phlox, rudbeckias and solidagos. The green oak tortrix larvae sometimes completely defoliates oaks and later pupates in rolled leaves of underlying plants. The adults are green and the larvae grey-green and up to 20mm long.

Other important North American species include the ORIENTAL FRUIT MOTH (*Grapholitha molesta*), the GRAPE BERRY MOTH (*Endopiza viteana*) and the omnivorous LEAF ROLLER (*Platynota stultana*). Larvae of the oriental fruit moth attacks shoots, causing blindness, before tunnelling into fruits of peach, quince and apple, plum and pear; the larvae are white, up to 12mm long. The grape berry moth has two generations of grey-green caterpillars, up to 12mm long; the first generation attacks blossoms and small berries, and the second bores into maturing grapes.

Pheromone traps are available for some species, including codling moth, plum moth, carnation tortrix, and pea moth, allowing contact insecticides to be applied at the correct time. On a small garden scale the trapping may reduce numbers of the pests to an acceptable level without the need to resort to spraying. Recommended contact insecticides may be used to control other species. The bacterial spray containing *Bacillus thuringiensis* has been used successfully against many of these pests. In the US, the oriental fruit moth has been adequately controlled by other biological control methods employing hymenopterous parasites. See *codling moth, leaf rollers*.

tortuous irregularly bent and twisted.

torulose of a cylindrical, ellipsoid or terete body, swollen and constricted at intervals but less markedly and regularly so than when moniliform.

torus a receptacle, the region bearing the floral parts.

total weedkiller a herbicide that destroys all vegetation with which it comes into contact, for example, sodium chlorate.

town refuse municipal waste, originating from household garbage and street cleanings, is sorted by some local authorities, and segregated organic matter may be composted. It has the disadvantages of variability, low nutrient content, some possibility of heavy metal contamination, and usually glass shard content. It is however in principle an environmentally friendly option, and the processed material is sometimes offered mixed with other organic materials. Some local authorities provide collection points for plant waste, woody forms of which may be shredded before composting; this is referred to as GREENWASTE.

trace elements see *nutrients and plant nutrition*.

Trachelium (from Greek *trachelos*, neck; plants reputedly effective against tracheal diseases). Campanulaceae. Mediterranean. 7 species, perennial herbs with tubular, 5-parted flowers in spring and summer. *T.caeruleum*, suitable for the annual and cut-flower border, is doubtfully hardy, but will bloom in its first year from seed if sown under glass in early spring. Grow in sun on a moderately fertile, well-drained soil. The tiny, cushion-forming *T.asperuloides* is suitable for crevice plantings in the rock garden and scree, for trough gardens and for small

*Trachelospermum
jasminoides*

*Trachelospermum
asiaticum*

pans in the alpine house. Although hardy to
−15°C/5°F, it needs some protection from winter wet.
Grow in perfectly drained, alkaline soils in sun or
with light shade at the hottest part of the day.
Propagate by seed or by softwood cuttings in sand
in a closed case with gentle bottom heat.

T.asperuloides (syn. *Diosphaera asperuloides*;
Greece; tufted, mat-forming, to 3cm; leaves very
small, glossy; flowers disproportionately large, solitary or clustered, white suffused mauve to pink); *T.
caeruleum* (W and C Mediterranean; erect to 1m;
flowers small, blue, lilac, mauve or white, in dense,
terminal umbels).

Trachelospermum (from Greek *trachelos*, neck, and
sperma, seed). Apocynaceae. India to Japan. 20 species,
evergreen climbing shrubs with milky sap. Produced
in summer in terminal and axillary cymes, the flowers
are tubular with five spreading lobes twisting and overlapping to the right. Hardy to –15°C/5°F with good
drainage and wall shelter. Grow in any well-drained,
moderately retentive soil in sun or with light part-day
shade. Prune to remove deadwood and weak growths,
and cut back excessively long stems almost to flowering wood. Propagate by seed, by simple layering or
by semi-ripe cuttings in summer.

T.asiaticum (Japan, Korea; 5–7m; leaves 2–5cm,
dark glossy green; flowers fragrant, yellow-white,
tube 0.7–0.8cm, 2cm across limb, lobes obovate);
T.jasminoides (China; to 7m; leaves 4–7.5cm, dark
shining green; flowers intensely fragrant, white, tube
0.75–1cm, 2.5cm across limb, lobes narrow, wavy;
'Japonicum': leaves veined white, turning bronze in
autumn; 'Minimum': dwarf, with mottled leaves;
'Variegatum': leaves striped white, often tinged pink
to red-bronze; 'Wilsonii': leaves flushed red-bronze
to maroon in winter).

Trachycarpus (from Greek *trachys*, rough, and
karpos, fruit, referring to the fruits of some species).
Palmae. Subtropical Asia. 6 species, medium-sized
to large palms developing columnar trunks clothed
with shaggy, brown fibres and petiole bases. The
leaves are large, glossy and palmate with many
narrow segments. Golden to cream inflorescences
are produced by mature plants. One of the few palms
suitable for outdoor cultivation in cool temperate
zones, *T.fortunei* will withstand frosts to about
–10°C/14°F, with protection from wind. It is an
excellent plant, at least when young, for city gardens
and container planting. Plant on any well-drained
but moist, fertile soil in sun or shade. Propagate by
seed. *T.fortunei* (FAN PALM, CHINESE WINDMILL
PALM, CHUSAN PALM; N Burma, China, naturalized
Japan; trunk ultimately to 20m; petioles to 1m,
sharply toothed, leaf blades to 1 × 1.25m, more or
less circular, dark green with many, linear-lanceolate segments, their tips drooping; also hardy,
T.wagnerianus differs in its smaller, more rigid
leaves with narrower, finger-like segments).

Trachymene (from Greek *trachys*, rough, and
mene, moon, referring to the appearance of the fruit).
Umbelliferae. Australia, W Pacific. 12 or more
species, annual, biennial or perennial herbs with
ternately divided leaves and small flowers in long-
stalked umbels in summer. Grow on fertile, well-
drained soils, in a sheltered position in sun. Sow
seed under glass in early spring or *in situ* in spring.
Alternatively, grow in pots of medium-fertility loam-
based mix in the cool greenhouse, and provide light
shade from strong sun to preserve flower colour.
T.coerulea (BLUE LACE FLOWER; W Australia; annual
or biennial to 70cm; flowers sky blue to pale
lavender, slightly fragrant).

*Tradescantia
virginiana*

Tradescantia (for John Tradescant (d. 1638), and his son, also John Tradescant (d. 1662), British naturalists, travellers, collectors and gardeners). Commelinaceae. Americas. SPIDER-LILY, SPIDERWORT. 70 species of perennial or very rarely annual or short-lived herbs. They vary greatly in habit, from erect clump-formers to slender creepers and trailers. Most have succulent stems, thinly fleshy leaves and terminal clusters of 3-petalled flowers in spring and summer.

T.virginiana is hardy and suitable for the herbaceous border. Plant in any fertile, well-drained and moisture-retentive soil in sun or light dappled shade. Propagate by division. The remaining, tender species make good dense groundcover for greenhouse borders and subtropical bedding. They are also commonly grown as house plants. The trailing species (all except *T.spathacea*) are especially useful for hanging baskets. Grow in a freely draining and moderately fertile soil, kept moist when the plant is in active growth, in sun or bright filtered light with a minimum temperature of about 15°C/60°F. Propagate by division or cuttings.

T.cerinthoides (syn. *T.blossfeldiana*; FLOWERING INCH PLANT; SE Brazil; stems decumbent, rooting; leaves to 15cm, elliptic to oblong, more or less fleshy, glossy dark green and hairy to smooth above, green to purple and hairy beneath; flowers purple-pink to white; 'Variegata': leaves variegated; *T.fluminensis* (WANDERING JEW; SE Brazil; stems ascending and decumbent or sprawling to hanging and rooting; leaves 1.5–12cm, broadly ovate to lanceolate, smooth, green, sometimes flushed purple beneath; flowers white; includes several cultivars very popular as house plants, with leaves striped or zoned white, silver, yellow-cream and pink, sometimes tinted purple); *T.pallida* (syn. *Setcreasea purpurea*; Mexico; stem to 40cm, ascending to decumbent; leaves 7–15cm, oblong-lanceolate to elliptic, trough-shaped, thinly fleshy, glaucous, tinted purple-red, smooth to weakly hairy; flowers pink; 'Purple Heart': syn. 'Purpurea', a very popular cultivar with leaves solidly coloured deep purple); *T.sillamontana* (syn. *T.pexata*, *T.velutina*; Mexico; stems ascending or decumbent, to 30cm; leaves 3–7cm, 2-ranked with clasping bases, elliptic to broadly ovate, somewhat succulent, green flushed purple-red, densely woolly; flowers purple-pink); *T.spathacea* (syn. *T.discolor*, *Rhoeo discolor*; OYSTER PLANT, BOAT LILY, MOSES IN HIS CRADLE, MOSES IN THE BULRUSHES, MEN-IN-A-BOAT; Mexico, Belize, Guatemala; stem short, stout, erect; leaves 20–35cm, lanceolate, semi-succulent, glossy, dark green above, purple-tinted beneath, in a loose rosette; inflorescences axillary, each with a pair of purple-green, fleshy bracts arranged like a boat and containing small white flowers; 'Vittata': leaves striped cream to pale yellow or lime); *T.virginiana* (SPIDERWORT; E US; stems slender, clumped, purple-tinted, erect to 40cm; leaves to 30cm, linear to lanceolate, dark green, arching, thinly hairy; flowers in terminal clusters, dark violet, mauve, magenta, pink, indigo, royal blue, sky blue or white; popular hardy perennials – many cultivars are available, some of them hybrids involving several other US species, these are sometimes included in *T.virginiana* or grouped under the name *T.* × *andersoniana*); *T.zebrina* (syn. *Zebrina pendula*; Mexico; stems decumbent to creeping, rooting; leaves 2.5–10cm, ovate to oblong, slightly succulent, dark green usually striped silver above, purple-tinted beneath; flowers pink, magenta or violet; includes cultivars with leaves stained purple or striped and splashed green, pink, red and white).

trailers plants with stems that grow over the soil surface, often rooting as they do so. They include plants such as trailing *Lobelia*, used at the edges of hanging baskets and other containers. Trailing climbers are scrambling plants that grow upwards and horizontally, loosely attached to other plants or to structures; they do not have tendrils or aerial roots, neither do they twine.

translocation (1) the movement of water and solutes within the vascular system of a plant; (2) used of a pesticide which is absorbed into a plant's translocation system and distributed to points distant from where it was applied; thus having widespread effect throughout the plant; (3) the movement of a chromosome segment to another location on the same or another chromosome.

transpiration loss of water by evaporation from plant surfaces, occurring chiefly from open stomata. It is a process accentuated by low atmospheric humidity and temperature rise, and can lead to wilting in dry conditions. Anti-transpirant sprays are available to reduce growth check on woody transplants and are especially relevant to evergreens.

transplant (1) (verb) to move a seedling or well-developed plant from one position to another; (2) (noun) a young plant grown from seed or cuttings, intended for direct planting outdoors or for growing on in a container or a nursery bed.

transplanter see *bulb planter*.

Trapa (a compressed form of the term *calcitrapa*, the caltrop, a 4-spiked iron ball used to maim horses in battle and which the fruit of this genus is said to resemble). Trapaceae. C Europe to E Asia, Africa. WATER CHESTNUT. 15 species, aquatic, perennial herbs with rosettes of leaves floating by means of inflated petioles, and inconspicuous flowers giving rise to nuts. The raw nuts, rich in carbohydrate and fat, contain toxins which must be destroyed by boiling. They are eaten whole or ground as flour. The following species is sometimes cultivated in aquaria and ponds. They are hardy in zone 6. Give full sun, slightly acidic water rich in nutrients. Seeds lose viability if allowed to dry and should be stored in water or wet moss. Propagate also by offsets. *T.natans* (WATER CHESTNUT, WATER CALTROP, JESUIT'S NUT; widespread; leaf blades to 3cm, broadly ovate to rhombic, toothed, glossy dark green sometimes turning bronze, petioles to 5cm, spindle-shaped, pale, radiating in floating rosettes; fruit a 2–4-horned nut).

trapeziform asymmetrical and four-sided, as a trapezium.

treading the compression of soil in a growing area by using one's feet. Treading is an effective aid for improving soil structure in the preparation of a seedbed, where walking or shuffling over the surface breaks down clods (especially on heavy soils) and serves to firm a soft surface (especially on light soils). Treading is also done to firm the soil around trees and shrubs immediately after planting, and where newly planted specimens are subject to 'lifting' out of the ground due to frost action. Treading; should never be carried out on wet soils because compaction can cause impeded drainage and the exclusion of air.

tree a woody, perennial plant with a single main stem, generally branching at some distance from the ground and possessing a more or less distinct, elevated crown.

tree banding the use of bands of grease, hay, sacking or corrugated paper around tree trunks to trap migrating insects, such as winter moths and certain weevils; in the US, it is known as barrier banding. See *grease banding*.

tree guards devices and constructions used to protect trees from bark-gnawing animals, especially rabbits, hares, cattle and deer. They may take the form of metal or wooden enclosures, to protect specimen trees from larger animals. Wire netting or perforated spiral plastic collars are effective around small trees against rabbit and hare damage. Some tall tree guards, in the form of tubes of translucent plastic material, have the added beneficial effect of providing an ideal growing environment for establishing tree seedlings.

tree house platforms, balconies, seats, and small houses constructed in trees were once popular in both Eastern and Western gardens, where they were sometimes described as roosts, and used for viewing from or as a place for relaxation. A modern US equivalent is the platform or deck that may surround the base of a tree or lead from house level into the crown of a nearby tree. Tree houses provide garden interest for children.

tree lopper, tree pruner see *long-arm pruner*.

tree surgery the removal of large tree branches that are damaged, diseased or overcrowded and the treatment of wounds resulting from branch breakage; especially used of professional operations.

tree ties any means of securing trees to their supports. Tree ties must allow for increasing trunk girth and also aim to prevent chafing on the stake. They can be made of soft rope, or improvized with strong wire threaded through lengths of old water hose. Proprietary ties, consisting of plastic straps, are a good choice; the strap passes through a thick buffering band between the tree and stake and is secured with a buckle.

treillage used of trellis and also structures made of trellis, such as arbours, columns, domes and even garden houses and temples.

trellis a construction of diagonally intersecting wooden laths, metal latticework or plastic bars, usually forming a 10cm mesh and mounted as variously-sized panels. Trellis has been a feature of garden decoration at least since Roman times, and is useful for creating screens or visually attractive supports for climbing plants, placed either free-standing or against a wall.

trench a vertical-sided furrow or trough made in soil with, depending on the depth required, a spade or draw hoe. Planting in trenches is recommended for certain vegetable crops, such as celery and leeks, and trenches filled with organic matter are also prepared for growing runner beans. Trenches also provide a means of increasing headroom for plants grown under cloches.

trenching see *digging*.

triangulation a method for determining a position, whereby measurements are made from two selected fixed points on a plan to a third point that is being established.

tribe a category of taxa intermediate between subfamily and genus.

Trichoderma a genus of soil-inhabiting fungi containing several species, including *T. viride* and *T. harzianum*, that have antagonistic effects on certain pathogenic fungi.

Trichodiadema (from Greek *thrix*, hair, and *diadema*, crown, referring to the spiny hair-clusters at the tips of the leaves). Aizoaceae. South Africa, Namibia. Some 36 species, succulent shrubs with woody or tuberous roots and terete leaves terminating in tufts of bristle-like spines. Daisy-like flowers open on bright days from spring to late autumn. Cultivate as for *Pleiospilos*.
    T.densum (caudex fleshy, thickened, stems short, mat-forming; leaves to 2cm, papillose, with long, white bristles; flowers crimson); *T.mirabile* (shrublet to 8cm, stems white-bristly; leaves to 2.5cm, papillose with stiff, dark brown bristles; flowers white).

trichome an epidermal hair or scale, formed of one or more specialized epidermal cells, often glandular-tipped.

Trichopilia (from Greek *thrix*, hair, and *pilion*, a cap, the like of which conceals the anther). Orchidaceae. Tropical C and S America. Some 30 species, epiphytic orchids with laterally compressed, cylindrical to ovoid pseudobulbs and oblong to lanceolate, coriaceous leaves. The flowers are held low or hang down, and consist of linear to lanceolate tepals – often strongly twisted – and a funnel-shaped lip. Grow in the intermediate greenhouse (minimum temperature 12°C/54°F). Pot in pans or baskets in a medium-grade bark mix. Position in good light and high humidity. Impose a short, dry winter rest. Propagate by division.

T.suavis (Costa Rica, Panama, Colombia; flowers to 10cm in diameter, delicately fragrant, tepals to 5.5 × 1cm, white or cream-white, sometimes spotted pale violet-rose or red, usually undulate but *not* twisted, lip to 6.5 × 5cm, white or cream-white, often densely spotted pale violet-rose or spotted and lined yellow, throat yellow or yellow-orange, undulate-crisped); *T.tortilis* (Mexico, Guatemala, Honduras, El Salvador; flowers to 15cm in diameter, fragrant, waxy, tepals to 8 × 1cm, off-white tinted pale lavender to fleshy purple-brown or yellow-grey with livid blotches, margins much paler, strongly and spirally twisted, lip to 6.5 × 4.5cm, white to pale yellow, throat yellow spotted brown or crimson, margin crisped, undulate).

trichotomous branching regularly by dividing regularly into three.

trickle irrigation, trickle tape see *irrigation*.

Tricyrtis (from Greek *treis*, three, and *kyrtos*, convex; the three outer petals are bag-like at the base). Liliaceae (Tricyrtidaceae). E Himalaya to Japan and Taiwan. TOAD LILY. 16 species, deciduous perennial herbs with creeping and tangled rhizomes and clumped, erect to arching stems. The leaves are lanceolate to elliptic. Borne in late summer and autumn in axillary clusters or terminal branching cymes, the flowers are cup- to star-shaped, and composed of six, oblong to lanceolate tepals with swollen bases. Hardy in climate zone 6, especially if thickly mulched in winter. Grow in dappled sun or shade in a sheltered position. Plant in a cool, porous soil, acid to neutral and rich in leafmould and garden compost. Keep permanently moist, never wet. Increase by division in spring, or by stem cuttings in late summer (plantlets will develop in the leaf axils).

T.affinis (differs from *T.macropoda* in its flowers borne on separate stalks); *T.formosana* (syn. *T.stolonifera*; stems erect; leaves sometimes spotted; inflorescence branched, terminal, flowers to 3cm across, widely funnelform to starry, held erect, tepals white to lilac-pink with crimson to mauve spots inside; cultivars include 'Amethystina', with white flowers densely spotted darkest violet, and 'Tojen', a cross with *T.hirta*, with spreading to erect, thinly hairy stems and white to palest rose flowers, yellow and finely purple-spotted at the centre and edged with a purple-pink flush); *T.hirta* (syn. *T.japonica*; Japan; stems thickly hairy, erect to 80cm; leaves

Tricyrtis 'Tojen'

Tricyrtis formosana (dark form)

Tricyrtis formosana

minutely hairy; flowers to 4cm across, erect, funnelform, solitary or clustered in leaf axils, white with dark purple spots or blotches; includes 'Alba', with white flowers flushed green, 'Lilac Towers', with lilac flowers, 'Miyazaki', with white flowers spotted lilac with sporadic blotches of dark wine red; 'Miyazaki Gold', as 'Miyazaki', but with leaves edged gold, 'Variegata', with leaves edged gold and lavender flowers, and 'White Towers', to 60cm, with arching stems and white flowers in most leaf-axils); *T.macrantha* (Japan; stems to 90cm, arched, drooping, with coarse brown hairs; flowers 3–4cm, axillary, pendulous, campanulate, yellow with chocolate spots inside); *T.macranthopsis* (Japan; stems 40–80cm, arching; flowers pendulous, terminal or axillary, to 5cm, tubular to campanulate, clear yellow with fine brown spots within); *T.macropoda* (China; stem erect, to 70cm, occasionally minutely hairy above; flowers to 3cm, erect, borne in branching cymes, mostly terminal, occasionally axillary, white-purple with small purple spots); *T.ohsumiensis* (Japan; stems 20–50cm, erect, subglabrous; flowers erect, axillary and terminal, 2–5cm, broadly funnel-shaped, primrose-yellow, faintly spotted red-brown).

tridentate when the teeth of a margin have teeth which are themselves toothed.

tridynamous when three stamens out of six are longer than the rest.

trifoliate three-leaved, or loosely, with three leaflets.

trifoliolate having three leaflets.

Trifolium (from Latin *tri*, three, and *folium*, leaf). Leguminosae. Temperate and subtropical regions excluding Australasia. CLOVER. Some 230 species, annual, biennial or perennial herbs with creeping to ascending stems, palmately compound leaves and stalked heads of small, pea-like flowers in spring and summer. Fully hardy. Grow in moist but well-drained, more or less neutral soils, in full sun. The variegated cultivars listed here are ideal for planting among small to medium-sized perennials, contrasting well with grey, lime green and bronze foliage. They tend, however, to become bare and exhausted and should be repropagated frequently. They are also suitable for hanging baskets. Propagate by division or rooted stem sections. *T. pratense* (RED CLOVER, PURPLE CLOVER; Europe, naturalized US; erect to decumbent short-lived perennial, 5–20cm tall; leaflets 1–3cm, obovate to oblong or rounded, thinly hairy beneath; flowers to 1.5cm, red-purple to pink or cream; 'Susan Smith': syn. 'Gold Net', pale green leaves, laced and veined with gold); *T.repens* (WHITE CLOVER, DUTCH CLOVER, SHAMROCK; Europe, naturalized US; creeping to erect perennial, 5–30cm tall; leaflets 0.5–2cm, broadly obovate to rounded, tips notched or truncate, finely toothed; flowers 0.5–1cm, white or pink; 'Atropurpureum': dwarf, with dark purple-red leaflets edged green; 'Aureum': leaflets veined gold; 'Purpurascens': large leaflets, almost solid maroon; forms with a high proportion of leaves with 4 or 5 leaflets are available – 'Quinquefolium' has 5 leaflets overlaid with bronze and yellow-veined).

trifurcate with three branches or forked in three.

trigonous a solid body which is triangular but obtusely angled in cross-section.

Trillium (from Greek *tris*, thrice: the leaves and floral parts occur in threes). Liliaceae (Trilliaceae). N America, W Himalaya, NE Asia. WOOD LILY, BIRTHROOT, WAKE ROBIN, STINKING BENJAMIN. 30 species, deciduous perennial herbs with short, stout rhizomes. The stems arise in spring, unbranched and bearing an apical whorl of three ovate to elliptic leaves. The flowers are solitary, stalked or unstalked and carried above the leaves at the stem's tip. Bell-shaped to narrowly cupped, they consist of three leafy sepals and three broader, colourful petals. Hardy in climate zone 5. They need a sheltered position in dappled sun or shade, and are ideal for the woodland and peat gardens. Plant in an open, acid to neutral soil enriched with leafmould and garden compost. Keep the roots cool and moist at all times. Mulch in winter. Increase by division or seed.

T.cernuum (eastern N America; flowers on pendent pedicels to 2cm, held among or below leaves; petals to 2cm, undulate, white sometimes flushed rose); *T.chloropetalum* (California; leaves mottled maroon; flowers sessile, fragrant, held erect, green to yellow to purple); *T.erectum* (eastern N America; flowers borne upright or obliquely on pedicel to 10cm, malodorous; sepals light green suffused red-purple, petals to 8cm, spreading or incurved, dark garnet to white; includes 'Ochroleucum', with flowers tinged green. f. *albiflorum*, with white flowers, occasionally flushed rose, and f. *luteum*, with green-yellow flowers stained blood red); *T.grandiflorum* (eastern N America; flowers erect on pedicels to 5cm, petals to 8cm, undulate, opening pure white, usually flushing pale pink with age; includes double-flowered forms and 'Roseum', with flowers flushing pink almost immediately and ageing damask); *T.luteum* (SE US; leaves mottled; flowers erect, sessile, sweetly scented, petals to 9cm, erect, cupped, golden or bronzy green); *T.nivale* (SE US; ovate; flowers stalked, erect at first, nodding with age, petals to 4cm, brilliant white; calcicole, flourishing in exposed situations); *T.ovatum* (western N America; flowers stalked, borne erect, muskily fragrant; petals spreading rather than erect to spreading as in *T.grandiflorum*, white becoming pink or red); *T.rivale* (W US; flowers stalked, erect; petals to 2.5cm, white flushed pink, spotted purple toward base); *T.sessile* (NE US; leaves with dark mottling; flowers sessile with pungent odour, petals to 4.5cm, erect, maroon; includes 'Rubrum', with crimson-purple flowers, 'Snow Queen', with white flowers, and f. *viridiflorum*, with green-bronze to yellow flowers); *T.undulatum* (eastern N America; flowers erect, stalked; sepals dark green bordered maroon, petals to 3cm, undulate, white or pale rose pink, with a dark maroon basal blotch suffusing the veins).

Trillium grandiflorum (left) and *T. sessile* (right)

trilocular having three locules.

trimerous with parts in threes; often written as 3-merous.

trimming knife see *pruning knife*.

tripinnate when the ultimate divisions of a bipinnate leaf are themselves pinnate.

triploid a plant that has three sets of chromosomes, instead of two sets as in the majority of plant species; it is annotated 3×. An example of a triploid is the apple cultivar 'Bramley's Seedling'.

Tripterygium (from Greek *tri-*, three, and *pteryx*, wing, in reference to the 3-winged fruit). Celastraceae. E Asia. 2 species, deciduous climbing shrubs with ovate to elliptic leaves to 15cm long and small flowers in terminal panicles; these are followed by 3-winged fruit in autumn. Both species are well-adapted to climbing through sturdy trees or over unsightly walls. They flower best in full sun; thrive in a moist loamy soil and will tolerate chalk. *T.regelii* will tolerate cold –15°C/5°F, but *T.wilfordii* needs the protection of a south-facing wall where temperatures fall below –5°C/23°F. Propagate from seed in autumn.
T.regelii (Japan, Korea, Manchuria; to 10m; leaves sometimes papillose-pilose on nerves beneath; fruit to 1.8cm in diameter, pale green); *T.wilfordii* (Taiwan, E China; to 13m; leaves glabrous, usually glaucous beneath; fruit to 1.2cm in diameter, purple-red).

triquetrous triangular in cross-section.

Triteleia (from Greek *tri-*, three and *telos*, end; the parts of the flower are in threes). Liliaceae (Alliaceae). W US. 15 species, perennial, cormous herbs with grass-like leaves and funnel-shaped, 6-lobed flowers borne on slender stalks in scapose umbels. Cultivate as for *Brodiaea*.
T.hyacintha (syn. *Brodiaea hyacintha*, *Brodiaea lactea*; to 70cm, flowers usually white, occasionally blue or lilac); *T.ixioides* (syn. *Brodiaea ixioides*, *Brodiaea lutea*; to 60cm, flowers golden-yellow); *T.laxa* (syn. *Brodiaea laxa*, *Brodiaea candida*, *T.candida*; GRÁSSNUT, TRIPLET LILY; to 75cm; flowers deep violet-blue, rarely white; 'Queen Fabiola': flowers pale violet blue); *T.peduncularis* (syn. *Brodiaea peduncularis*; stems to 40cm, flexuous, flowers white, blue, or tinged lavender).

Tritonia (from Greek *triton*, referring to the variable direction of the stamens of different species). Iridaceae. Tropical and South Africa. 28 species, cormous perennial herbs with fans of sword-shaped, plicate leaves and, in late summer, slender sprays of tubular to cup-shaped flowers. Cultivate as for *Ixia*. *T.crocata* (South Africa; 20–50cm tall, flowers bright orange, orange-red or pink-orange, the 3 lower lobes with a yellow mid-line).

Trochodendron (from Greek *trochos*, wheel, and *dendron*, tree, alluding to the stamens which spread like spokes of a wheel). Trochodendraceae. Japan, Korea, Taiwan. 1 species, *T.aralioides*, an evergreen tree or shrub to 20m tall. To 14cm long, the leaves are ovate to oblong, crenate, with a tapering tip, and dark green to olive and glossy. Circular, petal-less, green flowers appear in spring and early summer. Hardy to about –10°C/14°F, given shelter from cold drying winds. Grow in a moisture-retentive, moderately fertile, circumneutral soil in sun or light, part-day shade. Propagate from seed or semi-ripe cuttings in a closed case with gentle bottom heat or under mist.

Trollius (from German *Trollblume*, globe flower). Ranunculaceae. Northern temperate regions. GLOBE FLOWER. 31 species, perennial herbs with palmately lobed or divided, toothed leaves. Produced in spring and summer, the buttercup-like flowers are large and rounded with many staminodes and stamens. Fully hardy and suitable for the damp border, bog garden and water margin. Plant in sun on a fertile soil that never dries out. Increase by division.
T. × *cultorum* (garden hybrids between *T.europaeus*, *T.asiaticus* and *T.chinensis*, 30–90cm tall with large, globe-shaped flowers in shades of yellow, gold, orange and flame); *T.europaeus* (GLOBE-FLOWER; Europe, Russia, N America; to 80cm tall; flowers 1.5–5cm in diameter, globose, lemon-yellow to deep gold); *T.pumilus* (Himalaya, Tibet; 6–30cm; flowers 2–3.5cm in diameter, globose to cup-like, interior yellow-orange, exterior dark wine red or crimson); *T.yunnanensis* (W China; to 70cm; flowers 4–8cm in diameter, globose to bowl-shaped, golden yellow to pale orange).

trompe-l'oeil a garden feature intended to deceive the eye, usually by wall paintings, mirrors, ha-has, sunken pools and special plantings.

Tropaeolum (from Greek *tropaion*, Latin *tropaeum*, trophy, i.e. a sign of victory consisting originally of a wigwam of spears hung with the helmets and shields of the vanquished; gardeners used to grow *T.majus* up pyramids of poles and Linnaeus compared its rounded leaves to shields and the flowers to blood-stained helmets). Tropaeolaceae. S Mexico to Brazil and Patagonia. NASTURTIUM, INDIAN CRESS, CANARY BIRD VINE, CANARY BIRD FLOWER, FLAME FLOWER. Some 86 species, perennial or annual herbs, some tuberous-rooted, most climbing or scrambling. The leaves are rounded and angled to palmately lobed. Produced in summer, the flowers are spurred, with five, clawed petals, the upper two usually smaller than the lower three.
The nasturtium is derived mainly from *T.majus*, although its many cultivars reflect hybridity between this and other species. They range from low-growing dwarf cultivars with a wide colour range, for edging and for the flower border, through semi-trailing types good for hanging baskets, to taller-climbing types, suited to trellis, fence and other supports. These

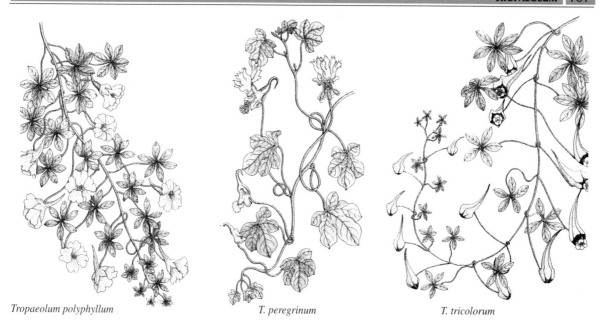

Tropaeolum polyphyllum　　　　*T. peregrinum*　　　　*T. tricolorum*

annuals grow well in sun on any well-drained but moisture-retentive soil. Among the perennial species, *T.polyphyllum* is perhaps the hardiest: given full sun and the perfect drainage of a raised bed, it is reliable to about −15°C/5°F. *T.speciosum* shows similar cold-tolerance. *T.peregrinum* is usually treated as an annual, as for *T.majus*. *T.tricolorum* and *T.tuberosum* will survive short-lived, light frost, given a warm sheltered situation with good drainage. *T.tuberosum* is a short day plant, and in cool temperate gardens, the late flowers are likely to be spoiled by frost; its large-tubered cultivar 'Ken Aslet' begins to bloom much earlier, in midsummer, and is more desirable for cooler climates. The tubers of these and other slightly tender species, such as *T.pentaphyllum*, may be lifted in autumn and stored in cool, dry conditions for replanting in spring. *T.azureum* and *T.tricolorum* work very well in the cool greenhouse or conservatory, trained on pea sticks. Grow in good light, in a neutral or slightly acidic mix of loam, leafmould and sharp sand. Water plentifully when in growth; withdraw moisture gradually as foliage fades; store the plants more or less dry in their pots over winter. *T.speciosum* differs from other species in that it performs best in areas with cool, moist summers. It favours relatively high rainfall and an acid soil. It is hardy in zone 6, but can be difficult to establish: one of the fundamental requirements for success is that the roots and lower stems be cool and shaded. Sow seed of annual species *in situ* or for earlier blooms under glass in late winter/early spring at 13–15°C/55–60°F. Propagate perennials with fleshy rhizomes, such as *T.polyphyllum* and *T.speciosum*, by careful division; for tuberous species, separate small tubers when repotting or take basal stem cuttings in spring.

T.azureum (Chile; climber to 1.2m, with small tubers; leaves 5-lobed; flowers purple-blue, 1–2cm in diameter, petals emarginate); *T.majus* (NASTURTIUM,

INDIAN CRESS; Colombia to Bolivia; vigorous annual climber to 2m; leaves orbicular, rarely lobed; flowers large, to 6cm, very variable in colour, in all shades of red, orange, or yellow (orange in wild plant), petals rounded, claws fimbriate; cultivars include plants ranging in habit from dwarf and bushy to tall and scrambling, some with variegated or deep blue-grey foliage; flowers single to double in shades of cream, yellow, peach, apricot, salmon, pink, orange, scarlet and dark velvety red); *T.pentaphyllum* (S America; perennial climber to 6m with long, beaded tubers, stems and leaf-stalks purple; leaves 5-lobed; flowers 2–3cm, in pendent masses, upper sepals spotted, red, petals scarlet); *T.peregrinum* (syn. *T.canariense*; CANARY CREEPER; Peru, Ecuador; perennial climber to 2.5m, treated as a half-hardy annual; leaves usually 5-lobed; flowers to 2.5cm in diameter, petals sulphur or lemon-yellow, upper petals fimbriate-lacerate); *T.polyphyllum* (Chile, Argentina; grey-green herbaceous annual or perennial with long rhizome and prostrate or climbing leafy stems to 3m; leaves 5- to 7- or more lobed, glaucous; flowers freely produced, in long masses, petals yellow, orange or ochre); *T.speciosum* (FLAME NASTURTIUM, SCOTTISH FLAME FLOWER; Chile; perennial climber to 3m; leaves 5–7-lobed; flowers to 3cm across, scarlet, upper petals narrowly wedge-shaped, lower rounded to square, notched; fruit blue); *T.tricolorum* (Bolivia, Chile; perennial climber to 2m, with small tubers; leaves 5–7-lobed; flowers of various colours, calyx urn-shaped, often orange-scarlet, rimmed black, but may be blue, violet, red or yellow with green margin, and yellow within, petals held mostly within calyx, orange, spur red to yellow with a blue or green tip); *T.tuberosum* (Peru, Bolivia, Colombia and Ecuador; perennial climber, 2–3m, with large yellow tubers marbled with purple ; leaves 3–5-lobed, grey-green; flowers cup-shaped, calyx red, upper petals rounded, all orange or scarlet; 'Ken Aslet': orange flowers).

tropism growth movement of a plant or organ in the direction of an environmental stimulus, for example, light (phototropism); cf. *nastic movements*.

trough garden see *rock plant gardening*.

trowel a small hand tool with a concave blade about 15cm long with a pointed or rounded end, mounted on a handle of similar length or of 30cm. Trowels are used for planting or lifting small plants, and are available in various forms, including some with very narrow blades.

true-breeding used of plants which when self-pollinated give rise to progeny similar to their parent; also known as homozygotic.

trug a shallow oblong basket, usually made with overlapping strips of wood, and having a curved, centrally mounted handle. Also made in plastic, trugs are useful for carrying flowers and produce, plants and small tools. They are sometimes referred to as Sussex trugs.

trullate, trulliform trowel-shaped.

trumpet (1) a description of flowers of tubular shape with a flared opening; (2) a name for the corona of *Narcissus* flowers, particularly used to define one of its classifying divisions. See *daffodil*.

truncate where the organ is abruptly terminated as if cut cleanly straight across, perpendicular to the midrib.

truncheon an archaic term describing a thick piece of stem used as a cutting or for grafting.

trunk the main stem of a tree; also, the hard, thickened stem of certain primulas and cyclamens, and of tree ferns.

trunk slitting see *bark-bound, slitting*.

truss a compact cluster of flowers or fruits, arising from one position on a stem. The term is applied, for example, to the flowerheads of tomatoes, rhododendrons and auriculas.

tryma drupaceous nut with dehiscent exocarp, for example, the walnut.

Tsuga (the Japanese name). Pinaceae. Himalaya, China, Japan and the west and east coasts of N America. HEMLOCK, HEMLOCK SPRUCE. 10 species, evergreen, coniferous trees with horizontal branches, slender branchlets arranged in fine sprays and 2-ranked, linear to needle-like leaves. Cultivation largely as for *Abies*. They will tolerate prolonged dense shade for long periods. Most species require hot, humid summers for full ripening of the shoots (where wood is not fully hardened, trees become shrubby and multi-stemmed following snow damage to weak stems). All are thin and poor in dry and exposed conditions. Most make excellent informal or formal hedges and screens, especially in semi-shade.

T.canadensis (EASTERN HEMLOCK; N America; to 40m, crown conic; leaves 0.6–2cm, dull green and grooved above, with two grey-white bands beneath; includes dwarf, prostrate, globose, conical and fastigiate cultivars, some with weeping foliage, some variegated); *T.caroliniana* (CAROLINA HEMLOCK; SE US; to 40m, crown densely conic; leaves 0.8–1.5cm, dark green, grooved above, entire or sometimes notched; includes pyramidal, dwarf, rounded and weeping cultivars); *T.diversifolia* (NORTH JAPANESE HEMLOCK; Japan; to 25m tall, but usually a shrub in cultivation; crown ovoid to conical; leaves 0.5–1.5cm, linear to oblong, tip often hooded, dark green, shiny and grooved above, with two white bands beneath); *T.heterophylla* (WESTERN HEMLOCK; Alaska to California; to 70m, crown conic; leaves 0.5–2cm, grooved, matt bright green above with two white bands beneath, tip blunt, toothed; includes weeping, narrowly columnar, dwarf, conical and grey-leaved cultivars); *T.mertensiana* (MOUNTAIN HEMLOCK; western N America; to 45m, crown narrowly conic; leaves 1.5–2.5cm, thick, keeled or slightly grooved above, often terete, grey-green to blue-green with pale bands on both surfaces; includes dwarf, blue- and silver-leaved cultivars); *T.sieboldii* (SOUTHERN JAPANESE HEMLOCK; S Japan; to 30m or a shrub in cultivation, crown conical to ovoid; leaves 0.7–2cm, yellow-green above, apple green or dull white beneath, notched).

tuber a swollen, generally subterranean stem, branch or root used for storage.

tubercle a small, warty, conical to spherical excrescence.

tuberous bearing tubers, or resembling a tuber.

tufa see *rock plant gardening, hypertufa*.

Tulbaghia (for Ryk Tulbagh (d. 1771), governor of the Cape of Good Hope). Liliaceae (Alliaceae). South Africa. WILD GARLIC, SOCIETY GARLIC. 26 species, perennial herbs, bulbous or fleshy-rooted, with clumps of onion-scented, narrowly strap-shaped leaves. Produced in spring and summer in scapose umbels, the flowers are star-like with an urn-shaped perianth tube, six spreading lobes and a fleshy corona. The species below will tolerate temperatures of –5°C/23°F to –10°C/14°F; they are sometimes given front positions in the herbaceous border, or grown at the base of a warm wall, on well-drained soils in sun, with a dry mulch of bracken litter in winter. In containers, plant in a gritty, loam-based mix in direct sunlight; water plentifully when in active growth, less as plants enter dormancy. Propagate by division or by seed sown when ripe or in spring. *T.capensis* (leaves mid-green; scapes to 60cm, flowers olive green with a maroon corona); *T.natalensis* (leaves light green; scape 30cm; flowers fragrant, white sometimes tinged lilac with a yellow-orange or green-white corona); *T.violacea* (leaves grey-green; scape to 60cm, flowers sweetly scented, bright lilac, corona tinged purple, red or white; includes 'Variegata', with grey-green leaves, striped cream to white).

Tulipa (From Turkish *tulbend*, turban, which this flower resembles). Liliaceae (Liliaceae). TULIP. N temperate Old World, especially C Asia. Some 100 species, perennial bulbous herbs with linear to broadly ovate leaves. Produced in spring and early summer, the flowers are solitary and scapose or carried in loose racemes. Bowl-shaped to narrowly bell-like, they are usually held erect and consist of six tepals in two whorls, of which the inner three at least are often blotched or stained in a different colour. The introduction of garden tulips into Europe from Turkey dates from the mid sixteenth century. The plant became the subject of extravagant enthusiasm, and great financial speculation took place in the trade for bulbs throughout the following century, especially in Holland. Commercial production remains important there, and similarly in Lincolnshire in England, in the Western United States and in British Columbia.

The derivation of today's garden cultivars cannot be traced directly to a particular species, although it is probable that many arose from *Tulipa gesneriana* and *T.suaveolens*. There are many early selections of *T.fosteriana*, *T.kauffmanniana* and *T.greigii*.

Horticulturally, tulips may be grouped into 15 divisions:

(1) *Single early*, cup-shaped, single flowers, often opening wider in sunshine, from early to mid spring. (2) *Double early*, long lasting double flowers, opening wide in early to mid spring. (3) *Triumph*, conical, single flowers borne on strong stems, and opening to a rounded shape in mid to late spring. (4) *Darwin hybrids*, large single flowers of variable shape, borne on strong stems in mid to late spring. (5) *Single late*, single flowers of variable shape, often ovoid or squarish, usually with pointed petals, flowering in late spring and very early summer. (6) *Lily-flowered*, narrow-waisted, single flowers with long pointed petals often reflexed at the tips, borne on strong stems in late spring. (7) *Fringed*, similar to lily-flowered but with fringed petals. (8) *Viridiflora*, single, variably shaped flowers with partly greenish petals, borne in late spring. (9) *Rembrandt*, mostly old cultivars, similar to Div. 6 with the colours in striped or feathered patterns due to virus infection; flowering in late spring. (10) *Parrot*, large single flowers of variable shape, the petals frilled or fringed and usually twisted; blooming in late spring. (11) *Double late* (Peony flowered), double bowl-shaped flowers in late spring. (12) *Kauffmanniana hybrids*, large single flowers which open wide in sunshine from early to mid spring. The leaves are often mottled or striped. (13) *Fosteriana hybrids*, large single flowers which open wide in sunshine from early to mid spring. The leaves are often mottled or striped. (14) *Greigii hybrids*, large single flowers borne in mid to late spring. The leaves are generally wavy-edged, and always mottled or striped. (15) *Miscellaneous*, other species, their varieties and hybrids, flowering in spring and early summer.

Tulips grow well on sheltered but sunny, well-drained sites, rich in humus. Some types including the species and the Kauffmanniana, Fosteriana and Greigii groups may be left in the ground to flower from year to year, but most hybrid garden tulips are best grown for annual bedding out; after which they may be lifted to discard or replant in the autumn for growing on to flowering size in about two years. Tulips may be planted much later in the autumn than other spring-flowering bulbs, up to mid November in suitable conditions, setting them out 10–15cm apart and at least 15cm deep. Tulips should not be continually planted on the same ground.

Tulips can be flowered early in 20cm wide pots or pans, using top quality bulbs or suitable cultivars, planted early. The containers should be subjected to 10–12 weeks in a cool, dark and damp environment, such as can be provided by burying them outdoors under 6–8cm of soil or sand in a shady site. They should subsequently be removed to a situation out of direct sunlight, and can be introduced gradually to a moderate heating regime under protection for flower development.

tulip

Pests
aphids
eelworms

Diseases
tulip fire
(see botrytis)

TULIPA

Name	Distribution	Stem	Leaves	Flowers
T.acuminata T.cornuta	known only in cultivation (Turkey, 18th century)	glabrous, 30–45cm	to 20×5cm, 3, lanceolate, slightly undulate, glaucous	yellow, sometimes streaked red; tepals 7.5–13cm, narrowly tapering, apex often convoluted; filaments yellow or white, anthers red-brown
Comments: Closely related to *T.gesneriana* and possibly a variety of that species. Z5.				
T.armena T.schrenkii; T.suaveolens	Turkey, NW Iran, Iraq, Transcaucasia	to 25cm, upper part coarsely hairy	to 16×2.5cm, 3–6, lanceolate, glaucous, sometimes downy above, recurved, undulate, margins ciliate	solitary, variable, cup-shaped to bowl-shaped, crimson, vermilion, yellow or multi-coloured; tepals to 6×3cm, rhombic to long-obovate, basal blotch black, navy or yellow-green, sometimes fan-shaped or bordered yellow; stamens yellow or black, filaments narrow-triangular
Comments: Z7				
T.aucheriana	Iran			
Comments: Close to *T.humilis* but flowers opening later, star-shaped, pink, tepals 3×1cm, basal blotch yellow. Z5.				
T.bakeri	Crete			
Comments: Differs from *T.saxatilis* in the deeper lilac to purple flowers. Z6.				
T.batalinii	C Asia			
Comments: Tepals pale yellow, golden to bronze at centre, stamens yellow, pollen yellow, otherwise very close to *T.linifolia*. 'Apricot Jewel': apricot-orange, interior yellow. 'Bright Gem': sulphur yellow, flushed orange. 'Bronze Charm': sulphur, feathered apricot-bronze. 'Red Gem': bright red, long-lasting. 'Yellow Jewel': pale lemon tinged pink. Z5.				

TULIPA

Name	Distribution	Stem	Leaves	Flowers
T.biflora *T.polychroma*	Balkans, SE Russia	to 10cm, glabrous, brown-green, glaucous	to 15×1cm, 2, widely spaced, linear-lanceolate, glaucous, margins and tips tinted claret, sometimes ciliate	fragrant, 1–2, broad-campanulate; tepals to 3×1cm, narrowly rhombic, cream to ivory, outside grey-green or green-violet, inner tepals with green midrib and yellow basal blotch fringed with yellow cilia; filaments yellow, anthers yellow, tip purple-black
Comments: Late winter–spring. Z5.				
T.clusiana *T.aitchisonii* LADY TULIP	Iran to Afghanistan, naturalized S Europe	to 30cm, glabrous	to 30×1cm, 2–5, linear, glaucous	1, rarely 2, opening to form a star; tepals tapering, acute, white to cream, outer tepals to 6×1.5cm, exterior carmine, edged white, elliptic, acuminate, basal blotch purple or red; filaments smooth, anthers purple
Comments: 'Cynthia': cream, exterior red with green edge, base purple. 'Tubergen's Gem': yellow, exterior red. var. ***chrysantha*** : golden-yellow, exterior of tepals stained red or purple-brown, basal blotch absent; stamens yellow. var. ***stellata*** : starry, white with a yellow centre.				
T.dasystemon	C Asia	to 5cm, glabrous	to 10×1cm, 2, semi-subterranean at flowering, narrowly lanceolate, blue-green, glabrous	solitary; tepals to 2×0.7cm, narrowly lanceolate, bright yellow, outer tepals with a broad, brown-claret or green band along midrib on the outside, inside pure yellow, inner tepals streaked brown-green along midrib; anthers yellow
Comments: In gardens, the name *T.dasystemon* has been misapplied to *T.tarda*. Z5.				
T.edulis *Amana edulis*	S Japan, NE China, Korea	to 15cm, glabrous, with 2–3 narrow bracts beneath flowers	to 25×1cm, to 6	1–2; tepals to 3×0.6cm, ivory, exterior veined claret or mauve, basal blotch purple-black edged yellow; filaments glabrous
Comments: Late winter–spring.				
T.fosteriana	C Asia	15–50cm, tinted pink, sometimes hairy	to 30×16cm, 3–5, widely spaced, oblong to broadly ovate, glossy green, downy above	solitary, faintly scented; tepals to 18×8.5cm, lustrous vivid red, long-ovate to narrowly rhombic to broadly lanceolate, tip downy, basal blotch black, fan-shaped or 2–3-pointed, edged yellow; filaments black, triangular, anthers black-violet, pollen purple-brown or yellow
Comments: Fosteriana Hybrids (*T.fosteriana* × *T.greigii* or *T.kaufmanniana*): flowers large, white or yellow. 'Princeps': resembles 'Red Emperor' but stem shorter, flowers later. 'Red Emperor' ('Mme Lefeber'): flowers large, tepals to 15cm, brilliant red, glossy, basal blotch black with irregular yellow border. Crossed with Darwin tulips to produce Darwin Hybrids, which include Mendel tulips (flowers large, stem stout). Z5.				
T.gesneriana	E Europe, Asia Minor	glabrous or finely hairy, to 60cm	to 15cm, 2–7, lanceolate to ovate-lanceolate, glaucous, margin ciliate toward apex	solitary, cup-shaped, opening to form a star; tepals 4–8cm (inner tepals wider than outer tepals), purple to dull crimson to yellow, sometimes variegated ('broken'), sometimes with yellow or dark olive basal blotch which may be edged lemon-yellow; filaments glabrous, yellow or purple, anthers yellow or purple
Comments: A complex and variable species, from which most late-flowering cultivars are derived. Z5.				
T.greigii	C Asia	to 45cm, often tinged pink or brown, hairy	to 32×16cm, 3–5, usually closely set and reflexed, lanceolate, glaucous, stained, veined and streaked maroon above	solitary; tepals to 16×10cm, usually vermilion but sometimes claret, orange, yellow, cream or multi-coloured, rhombic to oblong-obovate, angles rounded, acute, tip downy, basal blotch rhombic, black on red forms, red on yellow forms; filaments black or yellow, anthers black
Comments: Z5				
T.hageri				
Comments: Close to *T.orphanidea*. in range and appearance, except flowers dull red, exterior tinted or marked green. 'Splendens': bronze tinted red, exterior dark red. Z5.				
T.humilis	SE Turkey, N & W Iran, N Iraq, Azerbaidjan	to 20cm	1–3 - 10–15×1cm, 2–5, channelled, somewhat glaucous	solitary, or to 3, cup-shaped opening to a star, pale pink, yellow at centre; tepals 2–5×1–2cm, acute inner tepals longer than outer; filaments yellow or purple, anthers yellow
Comments: 'Eastern Star': rose, flamed bronze green on outer tepals, base yellow. 'Magenta Queen': lilac with yellow centre, exterior green and flame. 'Odalisque': light purple, base yellow. 'Persian Pearl': cyclamen-purple, base yellow, exterior light magenta. 'Violacea': deep violet, centre yellow. Z7.				
T.kaufmanniana	C Asia	to 50cm, upper part hairy, sometimes glabrous, often tinged red	2–20cm wide, 2–5, closely set, often rosulate, lanceolate to oblanceolate, slightly undulate, pale grey-green, veins darker, glabrous except minutely ciliate margins	1–5, star-shaped to campanulate to cup-shaped, often fragrant; tepals to 11×5cm, white or cream, sometimes shades from yellow to brick-red, outer tepals lanceolate, often recurved, basal blotch yellow, outside red or pink along midrib, inner tepals often erect, long-elliptic or long-rhombic, obtuse, sometimes multi-coloured, midrib often green or red in lighter-coloured forms, basal blotch fan-shaped, bright yellow; filaments yellow, anthers yellow
Comments: Often crossed with *T.fosteriana* and *T.greigii*. 'Lady Killer': white, exterior with a crimson flame, centre purple.				
T.linifolia	C Asia, N Iran, Afghanistan	to 30cm, glabrous	to 8×1cm, 3–8, closely set, linear, falcate, margin often wavy, usually ciliate and pink	solitary, star-shaped; tepals to 6×3.5cm, rhombic to subovate, scarlet, basal blotch blue-black, truncate, often edged cream to lemon-yellow; filaments yellow or black, anthers grey-yellow
Comments: Z5.				

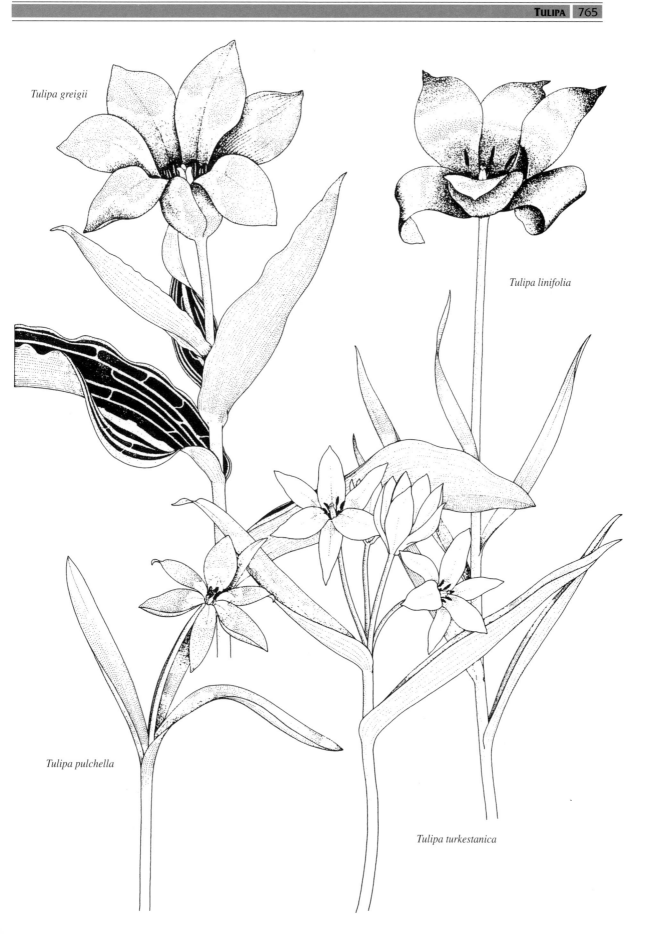

Tulipa greigii

Tulipa linifolia

Tulipa pulchella

Tulipa turkestanica

TULIPA

Name	Distribution	Stem	Leaves	Flowers
T.marjolettii	SE France	to 50cm	close to *T.gesneriana*.	tepals pale yellow to cream edged pink, exterior flushed rose-mauve, especially along midrib; filaments black, anthers pale yellow

Comments: Z6.

| *T.maximowiczii* | | | | |

Comments: Differs from *T. linifolia* in its tepals with a black blotch edged white, and black filaments fading to white at apex, anthers mauve or yellow. Z5.

| *T.orphanidea* | E Mediterranean | 10–35cm, glabrous or hairy | to 30×1.5cm, 2–7, lanceolate, glabrous, margin often claret | 1–4, globose; tepals 3–5×1–2cm, elliptic, vermilion to brick-red, outside of outer tepals buff stained green or purple, basal blotch olive to black, sometimes edged yellow; stamens brown or olive |

Comments: Closely related to *T.sylvestris*, producing many natural hybrids. *T.orphanidea* is a variable species and may encompass such colourful and valued plants as *T.hageri* and *T.whittallii* . The plant usually offered as *T.orphanidea* has dull orange-brown flowers, the exterior suffused green and mauve. 'Flava': yellow, flushed red; almost indistinguishable from *T.sylvestris*. Z5.

| *T.praestans* | C Asia | 10–60cm, finely downy above | 3–6, widely spaced, oblong or lanceolate, pale grey-green, midrib thick, distinctly keeled, margins sometimes undulate, ciliate | 1–5, cupped; tepals orange-red, outer tepals 7×2.5cm, broadly elliptic or subovate, tinged yellow toward base, inner tepals shorter, lanceolate, elliptic or obovate; filaments red, anthers yellow, violet or claret |

Comments: 'Fusilier': flowers glowing orange-red, to 4 per stem. 'Unicum': leaves broadly edged white; flowers to 5 per stem, red, base small, yellow, anthers black. 'Van Tubergen's Variety': flowers red, several per stem. 'Zwanenburg': flowers large, striking red. Z5.

| *T.pulchella* | Turkey | | | |

Comments: As for *T.humilis* but tepals to 3×1.5cm, strongly cupped, mauve,basal blotch navy edged white. var. **albocaerulea**: flowers white tinted grey-mauve with a dark indigo centre, or , in the finest selections, pure white with a steel blue centre. Z5.

| *T.saxatilis* CANDIA TULIP | Crete, W Turkey | 15–45cm, glabrous | 10–30×2–5cm, 2–4, glabrous, lustrous or, rarely, glaucous | 1–4, fragrant; tepals 4–5×1.5–3cm, elliptic, acute, pink to mauve, basal blotch yellow edged white; stamens yellow |

Comments: See also *T.bakeri* – plants offered as *T.saxatilis* will usually have lilac flowers, those named *T.bakeri* have deeper mauve tones. 'Lilac Wonder': rosy lilac, base lemon-yellow. Z6.

| *T.sprengeri* | N Turkey | to 30cm | to 25×3cm, 3–6, linear, glabrous | solitary; tepals brilliant red, acute, tapering toward base, outer tepals 6×1.5cm, buff-coloured beneath, inner tepals to 2.5cm wide, basal blotch absent; filaments bright red, anthers yellow |

Comments: Z5. Late-flowering (i.e. late spring-early summer), easily raised from seed and excellent for naturalizing.

| *T.sylvestris* *T.florentina*; *T.grisebachiana* | origin unknown; naturalized from Europe and N Africa to C Asia and Siberia | to 45cm, glabrous, sometimes tinted red at base | to 24×2.5cm, 2–4, widely spaced, declinate, linear, acuminate, channelled, dark green, glabrous, glaucous | 1–2, starry; tepals to 7×2.5cm, lanceolate to subrhombic, acuminate, golden, midrib bordered green, inner tepals convex, sometimes tinged pink; filaments yellow, anthers orange |

Comments: 'Major': flowers to 3 per stem, gold, abundant, larger, with 8 petals. 'Tabriz': tall; flowers large, lemon-yellow, sweetly scented. Z5.

| *T.tarda* | C Asia | to 5–11cm | to 12×1.5cm, 3–7, closely set, lanceolate, recurved, bright green, glabrous, margins claret, often ciliate | 4–15, broadly star-shaped, fragrant; outer tepals to 3.5×1cm, broadly lanceolate, acute, white, the exterior midrib edged with a broad green stripe shading to purple and yellow, inner tepals 3.5×2cm, spoon-shaped, acute, interior often yellow toward base; filaments yellow, anthers yellow |

Comments: Z5.

| *T.turkestanica* | C Asia | to 30cm, very finely white-hairy | to 20cm, 2-4, ovate-lanceolate, smooth, grey-green, glaucous | to 12 per stem, to 2cm, sometimes malodorous, ivory to white, base yellow or orange within, exterior of outer tepals stained olive green to grey or slate-mauve; anthers purple, brown or yellow tipped purple |

Comments: Very similar to *T. biflora*. Z5.

| *T.undulatifolia* *T.baeotica*; *T.eichleri* | Balkans, Greece, Turkey, Iran, C Asia | to 50cm, hairy | to 19×5.5cm, 3–4, widely spaced, linear to lanceolate, reflexed, glaucous, downy, margins crispate or undulate, ciliate | solitary, cup-shaped to broadly campanulate; tepals to 7×3.5cm, crimson to dark red, usually paler below, downy, broadly lanceolate to obovate, tapering to downy tip, basal blotch elliptic to rhombic, black to purple, often edged yellow; filaments black, anthers black to yellow |

Comments: 'Clare Benedict': bright red, base black, edged yellow, early-flowering. 'Excelsa': large, scarlet, base black, edged yellow

| *T.urumiensis* | NW Iran | 1–2, to 20cm, mostly subterranean, glabrous | 10–12×1cm, 2–4 in a flat rosette, plicate, glabrous, glaucous | 1–2, cup-shaped, opening to a star in sunlight; tepals to 4×1cm, yellow, outside streaked green or red; stamens and pollen yellow |

Comments: Z5.

| *T.violacea* | N Iran, SE Turkey | | | |

Comments: As for *T.humilis* but flowers more rounded, tepals 3–5×2–3cm, violet-pink, basal blotch black, edged yellow. var. **pallida** : paler mauve with a dark navy central blotch.

| *T.whittallii* | | | | |

Comments: Vivid orange-bronze, otherwise very similar to *T.orphanidea*. Z5.

tummelberry see *hybrid berries*.

tunic a bulb or corm covering, usually papery, membranous or fibrous.

tunicate (1) enclosed in a tunic; (2) having concentrically layered coats, the outermost being a tunic, like an onion.

tunnel cloche, tunnel house see *cloche, greenhouse*.

turbinate top-shaped; inversely conical, but contracted near the apex.

turf, sod (1) an area of grass; (2) a piece of lifted grass, primarily used for lawn making (turfing).

turfing iron, sod lifter (US) a spade with a heart-shaped blade, about 25cm long and 20cm wide, set at an angle and fixed to a long handle. Turfing irons are used for cutting and lifting turves.

turf plugger a bulb planter.

turf seat a raised seat covered with turf, camomile or thyme, usually built against a wall, or sometimes a grass mound. Turf seats were common in medieval gardens.

turgid filled out or inflated, usually by fluid.

turgor the distension of plant cells by water taken in by osmosis, which results in outward pressure exerted by the protoplasts against the cell walls.

turion a detached winter bud, often thick and fleshy, by which aquatic plants perennate; also, an adventitious shoot or sucker.

turnip (*Brassica rapa* Rapifera Group) a biennial grown as an annual primarily for the swollen edible root, which is flat, round or long in shape, with white or yellow flesh. Native to central and southern Europe, it is widely cultivated in temperate and highland tropical regions; it grows best under cool moist conditions. The root freezes well, either raw or cooked, and turnip tops are a useful substitute for spring greens.

The crop should be grown in rotation, like other brassicas, on a moisture-retentive soil of high organic-matter content, in a lightly shaded area. Turnip is a fast-maturing crop, harvestable in 6–10 weeks from sowing (although some selections may be ready as small roots even sooner), and it is therefore suitable for catch cropping or inter-cropping.

Sow at 3-week intervals from early spring onwards. Very early crops may be obtained from February sowing under cloches. Sow thinly, 2.5cm deep: earlies in rows 20cm apart, thinning individual seedlings to 10cm apart, and later cultivars in rows 30cm apart, thinning to 15cm apart. Sow turnips grown for their tops in autumn and early spring. Cut when 15cm high as required.

The crop requires adequate water throughout growth, and summer sowings will bolt if allowed to dry out. Early cultivars are ready when about 5cm in diameter, and quickly deteriorate in quality if left; later ones, sown July-August, are more hardy and can be left in the soil in winter, or lifted and stored in boxes in a cool shed.

Tweedia (for J. Tweedie (1775–1862), of Glasgow and Argentina). Asclepiadaceae. S Brazil, Uruguay. 1 species, *T.caerulea* (syn. *Oxypetalum caeruleum*), a twining subshrub to 1m tall. Produced in summer, the flowers are to 2.5cm in diameter, powder-blue becoming lilac, with five, outspread and narrow petals. Grow under glass with a minimum temperature of 5°C/40°F in any well-drained, moderately fertile soil in sun. Alternatively, treat as an annual, planted out after the last frosts. Propagate from seed sown under glass in spring.

Tweedia caerulea

twiner a climbing plant that twines around a support.

twin-scaling see *cuttings*.

Tylecodon (an anagram of *Cotyledon*). Crassulaceae. South Africa, Namibia. 28 species, succulent perennial subshrubs, usually with swollen stems covered in papery bark and crowned with spreading branches bearing fleshy leaves and panicles of 5-petalled, tubular flowers. These gouty-stemmed plants grow throughout the winter in the northern hemisphere, resting in a deciduous state in summer. Cultivate as for *Cotyledon*. *T.paniculata* (syn. *Cotyledon paniculata*; to 1.5m; flowers orange to red marked yellow and spotted red); *T.reticulata* (syn. *Cotyledon reticulata*; to 1m; flowers yellow-green tipped yellow); *T.wallichii* (syn. *Cotyledon wallichii*; to 1.4m; flowers yellow-green).

type one of several categories of specimen or, less commonly, image or description, used by a botanist as the basis of a taxon; the proof of its existence.

Typha (name used by Theophrastus for these or possibly other aquatic plants). Typhaceae. Cosmopolitan. BULRUSH, REEDMACE, CAT TAIL. 10 species, aquatic and marginal perennial herbs, with two-ranked, narrowly strap-shaped leaves and scapose inflorescences. Produced in summer and autumn, these are composed of many tightly packed and minute flowers, the whole forming a brown, velvety club with a spike at its summit. Suitable for larger water features and lakesides, except for *T.minima*, a charming dwarf ideal for small ponds. They are useful bank stabilizers. For floral decorations, pick the spikes early in the season, while male flowers are still in bloom, and air-dry. Spray with hair lacquer or aerosol fixative.

T.angustifolia (LESSER BULRUSH, NARROW-LEAVED REEDMACE; N Hemisphere; to 2m; leaves 3–6mm wide); *T.latifolia* (CAT'S TAIL, BULRUSH, GREAT REEDMACE; N Hemisphere; to 2m, robust; leaves 8–20mm wide); *T.minima* (DWARF CAT TAIL, DWARF REED-MACE; Eurasia; 40–80cm, slender; leaves to 2mm wide).

turnip

Pests
aphids
flea beetle
cabbage root fly
turnip gall weevil
(see weevils)

Diseases
canker
clubroot
virus

U

U-cordon a cordon formed of two parallel upright or oblique growths branching from near the base of a plant, often used as a training method for redcurrants and gooseberries.

Ulex (name used by Pliny for a type of heather). Leguminosae. W Europe, N Africa. GORSE. 20 species, shrubs, ultimately leafless but with dense branching spines that are green and perform photosynthesis. Fragrant, pea-like flowers are produced throughout the year. Fully hardy, *U.europaeus* is useful for plantings on land too dry and barren to accommodate other shrubs; more fertile soils will result in lank growth and poor flowering. It is also a valuable soil stabilizer on sandy substrates, an impenetrable barrier and a useful windbreak in coastal areas and on heathlands. Propagate from seed; also by semi-ripe cuttings in a sandy propagating mix in the cold frame. *U.europaeus* (IRISH GORSE, COMMON GORSE, FURZE, WHIN; W Europe; 0.6–2.5m, erect or ascending, densely branched, spines to 1.5cm, dark olive; flowers to 2cm, golden yellow marked red at base of keel, coconut-scented; 'Aureus': foliage golden yellow to lime; 'Florepleno': flowers double).

Ulmus (Classical Latin name). Ulmaceae. Nothern temperate regions to Mexico and C Asia. ELM. 45 species, deciduous or semi-evergreen trees and shrubs (those listed here are hardy deciduous trees), grown for their habit and foliage. Dutch Elm disease (q.v.) has destroyed almost all native elms in both Europe and North America since about 1965. Specimens remain only on the northern fringes of these areas. None of the European or North American indigenous elms shows the slightest degree of tolerance to the disease. The disease originated in E Asia and the elms there are resistant; there is less evidence of resistance for Himalayan species, some of which have been diseased in Britain. Asian species grow well in the eastern US, but they do not thrive in the cool summers of northwest Europe or the Pacific northwest. Elms can be hybridized easily in cultivation and some hybrids will resist the disease to a useful extent.

Propagation is by seed for some species, by layers, grafting or root suckers for others. Seed can be collected fresh on ripening in early summer and sown immediately without drying, germination then occurs within a few days; alternatively, the seed can be dried gently and refrigerated for planting the next spring, but germination rates are then lower. The autumn-flowering species are treated by the latter method. Layers are pegged from suckers or coppiced plants and, like root suckers, are cut from the parent when a good root system is formed. For grafting, use *U.glabra* or related Asian species for the rootstock; these will not produce suckers from the base.

In addition to Dutch Elm disease, this genus is attacked by elm phloem necrosis, a phytoplasma disease with symptoms similar to Dutch elm disease, but in which the inner bark is stained yellow, not brown, and smells of oil of wintergreen. It is spread by a leafhopper. Serious epidemics occurred in North America, but it has now declined quite simply through lack of elms. It has not spread to Europe. The disease is not easily controlled. The use of insecticidal sprays to kill the vectors has been tried, but the volume required to treat large trees is impractical and also causes severe environmental disturbance.

U.americana (AMERICAN ELM, WHITE ELM, WATER ELM; eastern N America; to 35m; branches pendulous toward tips, bark ash grey, furrowed, twigs downy at first; leaves 7–15cm, oblong to ovate to elliptic, apex acuminate, base unequal, glabrous or scabrous above, downy beneath; includes cultivars with columnar, narrowly upright, vase-shaped and pyramidal crowns, and dark, golden or incised leaves); *U.angustifolia* (GOODYER'S ELM; England, France; to 30m, crown rounded with age; leaves 4–7cm, broadly ovate, glossy above with tufted hairs in vein axils beneath); *U.carpinifolia* (EUROPEAN FIELD ELM, SMOOTH-LEAVED ELM; S and C Europe; to 30m, pyramidal or narrowly upright; leaves

4–10cm, obliquely oval to ovate, biserrate, glabrous above, downy on veins beneath; includes numerous cultivars, tall to dwarf, erect to weeping, narrow to rounded, corky and densely twiggy, with purple, yellow, and white-marked leaves; also includes var. *cornubiensis*, the CORNISH ELM, a narrowly conical tree to 30m, with ovate leaves, glossy dark green and smooth above with conspicuous tufts beneath); *U.glabra* (WYCH ELM, SCOTCH ELM; N and C Europe to Asia Minor; to 40m, crown open and spreading, bark fissured with age, young shoots hairy; leaves 5–16cm, oval to obovate, apex acuminate, occasionally 3-lobed, margins toothed, rough above, downy beneath; includes numerous cultivars, among the most popular are 'Camperdownii', compact and dome-shaped with drooping branches, and 'Exoniensis', erect and conical, to 7m tall, with irregularly serrate leaves); *U. × hollandica* (*U.glabra × U.carpinifolia*; DUTCH ELM; spreading, upright tree to 30m; leaves ovate, serrate, glossy dark green; includes 'Jacqeline Hillier', a densely branched, slow-growing shrub to 2m, with elliptic to lanceolate, 2-ranked leaves to 3cm); *U.parvifolia* (CHINESE ELM, LACEBARK; China, Korea, Japan; semi-evergreen, to 25m, crown rounded or globose, branchlets slender, bark smooth, flaking in rounded pieces; leaves 2–5cm, elliptic to obovate, acute, serrate, coriaceous; includes dwarf, shrubby, slow-growing, pendulous and variegated cultivars); *U.procera* (ENGLISH ELM; England; tree to 35m, trunk long and straight, crown tall with rounded heads of foliage on a few large branches, bark with rectangular fissures; leaves 3–7cm, rounded to ovate, toothed, scabrous above, bristly beneath, dark green, yellow in autumn; includes variegated cultivars); *U.pumila* (SIBERIAN ELM; E Siberia, N China, Turkestan; tree or shrub to 10m; leaves 2–7cm, ovate, serrate, acute to acuminate, deep green and glabrous above, becoming glabrous beneath; includes cultivars with small, rounded and bushy crowns, some with resistance to Dutch Elm disease); *U.* 'Sarniensis' (*U.carpinifolia × U. × hollandica*; JERSEY ELM, WHEATELY ELM; differs from *U.carpinifolia* var. *cornubiensis* in its tapering crown with erect branches and broader, biserrate leaves); *U.* 'Vegeta' (HUNTINGDON ELM, CHICHESTER ELM; to 40m, trunk thick, short, branching low down, crown open, bark rough; leaves 8–12cm, elliptic, acuminate, biserrate, glabrous above, downy in vein axils beneath).

umbel a flat-topped inflorescence like a corymb, but with all the flowered pedicels (rays) arising from the same point at the apex of the main axis.

umbellate, umbelliform borne in or furnished with an umbel; resembling an umbel.

Umbellularia (from Latin *umbella*, navel, referring to the inflorescence form). Lauraceae. California, Oregon. 1 species, *U.californica*, CALIFORNIA LAUREL, PEPPERWOOD, HEADACHE TREE, an ever-green tree to 25m. The leaves are 5–12cm long, ovate to oblong or lanceolate, glossy dark green. Produced in stalked umbels in summer, the small flowers give rise to purple berries. The leaves are intensely aromatic when bruised, and on warm days the whole plant exudes a fragrance which may cause dizziness, headaches and nausea. Given protection from frost when young and some shelter from cold drying winds, it will tolerate temperatures to –15°C/5°F. Grow in moisture-retentive but well-drained lime-free soils with full sun. Propagate from seed sown when ripe in autumn, by simple layering or by semi-ripe cuttings in a shaded closed case with bottom heat.

umbellule the umbel that terminates a compound umbel.

umbilicate with a more or less central depression, like a navel.

umbo a rough, conical projection, arising from the centre of a surface, as in the raised part of the scale in a pine cone.

umbonate bossed; orbicular, with a point projecting from the centre.

umbraculiferous shaped like a parasol.

uncinate hooked.

Uncinia (from Latin *uncinatus*, hooked, referring to the seedheads). Cyperaceae. Southern Hemisphere. 35 species, evergreen, tufted grass-like herbs with finely hooked seed-heads. The following species resembles the New Zealand *Carex*, whose cultivation requirements it shares. *U.rubra* (New Zealand; leaves linear, finely tapered, arching in dense tussocks to 30cm tall, glossy dark maroon-bronze to blood red).

Uncinia rubra

undercutting severing the roots of lined out trees and shrubs in order to encourage the development of a fibrous root system, carried out especially in advance of lifting for transplanting.

underplanting the establishment of low-growing plants among distinctly taller ones.

undershrub a low shrub.

understorey a population of lower-growing trees and shrubs established under a canopy of tall trees in deciduous woodland or tropical rainforest.

undulate wavy, but of a margin's surface rather than outline. cf. *sinuate*.

unguiculate petals and sepals bearing a basal claw.

unijugate see *paired*.

union the point where a scion and rootstock are joined by grafting. It may become swollen, where the rate of growth of one component is faster than the other. It is necessary to keep the union above soil level in situations where scion rooting could be a disadvantage.

unisexual either staminate or pistillate; with flowers of one sex only.

united parts usually free or separate held closely or clinging together so as to be scarcely distinguishable.

urceolate an urn-shaped corolla.

urea an organic chemical found in animal urine; it is synthesized as a fertilizer, containing about 45% nitrogen, and used as a liquid feed diluted at 25g per 27 litres of water. Urea formaldehyde, the product of a chemical reaction between urea and formaldehyde, is an insoluble, slow-release nitrogenous fertilizer.

Urginea (for the Afro-Arab tribe, the Urgin). Liliaceae (Hyacinthaceae). Africa, Mediterranean. 100 species, bulbous perennial herbs with 6-tepalled flowers in scapose racemes in summer and autumn. Occasionally successful out of doors in northern Europe, although it is rarely free-flowering in cooler climates. Plant with the bulb neck exposed in very free-draining gritty or sandy soils, in full sunlight; protect from winter wet with a loose mulch of dry bracken or straw. Increase by offsets or seed. *U.maritima* (SEA ONION, SQUILL; Mediterranean, Portugal; bulb 5–15cm in diameter; leaves to 100cm, narrow, glaucous; scape to 100cm, tinted red, flowers white lined with green or purple-red).

urine liquid waste excreted by animals, which is a quick-acting fertilizer, containing nitrogen and some potassium. It can be diluted with water for use as a liquid plant food or used as an accelerator in compost heaps.

Ursinia (for Johann Ursinus of Regensburg (1608–1666), author of *Arboretum Biblicum*). Compositae. Ethiopia, Namibia, South Africa. 40 species, annual and perennial herbs, subshrubs and shrubs with pinnately cut leaves and dasiy-like flowerheads in summer. Cultivate as for *Gazania*. *U.anthemoides* (South Africa; thinly hairy annual to 40cm; flowerheads 1.5–6cm in diameter, ray florets yellow or dark purple towards base or coppery yellow beneath, disc florets yellow or purple at

apex); *U.chrysanthemoides* (South Africa; hairy annual to shrubby perennial to 1m; flowerheads 2.5–6cm in diameter, ray florets yellow or red throughout or yellow or white above and coppery beneath, disc florets yellow or dark purple at the apex); *U.sericea* (S Africa; silvery-hairy procumbent perennial shrub to 70cm; flowerheads 1–3cm in diameter, yellow).

Utricularia (from Latin *utriculus*, little bottle or bladder, alluding to the bladder-like traps). Lentibulariaceae. Cosmopolitan. BLADDERWORT. 214 species, perennial carnivorous herbs. They may be aquatic, terrestrial or epiphytic, with a range of habit and foliage types and 2-lipped, spurred flowers. Prey is caught in very small, bladder-like traps carried on the main growth axis or on root-like, specialized stems. *U.longifolia* and *U.renifomis* require high humidity, light shade and a minimum temperature of 10°C/50°F. The compost should be very open – two parts coarse bark, two parts moss peat, one part perlite and a little sand. Keep permanently moist, with soft water only. Propagate by division in early summer. They grow well in baskets and shallow pans. *U.vulgaris* is hardy and may be grown outside in still, acidic pools. In aquaria and other containers, it needs to float in soft water out of direct sunlight. It will feed on water fleas.

U.longifolia (Brazil; leaves dark green, strap-shaped; flowers to 3cm, violet blotched orange-yellow); *U.renifomis* (Brazil; leaves kidney-shaped; flowers to 4cm, violet with yellow markings); *U.vulgaris* (temperate Old World; aquatic perennial with floating stems bearing many small, dark green leaves divided into fine segments; flowers to 2cm, yellow marked red-brown).

utricle a round, inflated sheath or appendage.

Uvularia (from Latin *uvula*, the soft palate, referring to the drooping flowers). Liliaceae (Uvulariaceae). US. BELLWORT, MERRY-BELLS. 5 species, rhizomatous, hardy herbaceous perennials with erect to arching stems clothed with oblong-lanceolate to elliptic leaves and hung with narrowly bell-shaped flowers in spring and early summer. These are pendulous and composed of 6 lanceolate and twisted tepals. Suitable for the woodland or rock garden, or at the waterside above water level. Propagate by division of established clumps, or from ripe seed sown in the cold frame.

U.grandiflora (to 75cm; flowers to 2cm, yellow sometimes tinted green); *U.perfoliata* (to 60cm; flowers to 3.5cm, yellow, tepals papillose).

Uvularia grandiflora

V

Vaccinium (ancient Latin name of disputed origin, used by Virgil and Pliny, perhaps a corruption of *Hyacinthus* or from *bacca*, berry). Ericaceae. Northern Hemisphere and some tropical highlands. 450 species, shrubs and small trees, evergreen or deciduous, with small, urn- to bell-shaped flowers and spherical berries. Cultivate as for *Arctostaphylos*. See also *blueberry* and *cranberry*.

V.angustifolium (LOW-BUSH BLUEBERRY, LATE SWEET BLUEBERRY; northeast N America; deciduous shrublet 8–20cm tall; leaves 1–3cm, lanceolate; flowers white tinged green, often streaked red; fruit to 1.2cm, blue-black, pruinose, edible); *V.arctostaphylos* (CAUCASIAN WHORTLEBERRY; N Asia Minor, W Caucasus; deciduous shrub 1–3m tall; leaves 3–10cm, elliptic to oblong, carmine in autumn; flowers white tinged green, faintly flushed red; fruit to 1cm, purple); *V.corymbosum* (BLUEBERRY, AMERICAN BLUEBERRY, HIGH-BUSH BLUEBERRY; E US; deciduous shrub to 2m tall; leaves 3–8cm, ovate to lanceolate, scarlet in autumn; flowers white, sometimes slightly red-tinted; fruit 0.8–1.5cm, blue-black, pruinose, edible); *V.glauco-album* (Sikkim; evergreen shrub to 1m tall; leaves 3–8cm, ovate to oblong; flowers white tinged pink; fruit to 0.6cm, black, pruinose); *V.myrtillus* (BILBERRY, WHORTLEBERRY, BLAEBERRY; semi-deciduous, erect shrub to 50cm; leaves 1–3cm, oval to elliptic, finely toothed; flowers tinged green, later red; fruit to 1cm, blue-black, pruinose, edible; 'Leucocarpum': fruit off-white); *V.nummularium* (Sikkim, Bhutan; evergreen shrub to 40cm; leaves 1–2.5cm, rounded to elliptic; flowers pink; fruit to 0.6cm, black, edible); *V.parvifolium* (RED BILBERRY, RED HUCKLEBERRY; western N America; deciduous shrub, erect to 2m or more; leaves 1–2cm, ovate to elliptic, red in autumn; flowers tinted red-green; fruit to 1.2cm, coral red, slightly translucent, edible); *V.vitis-idaea* (COWBERRY, CRANBERRY; northern temperate regions; evergreen, creeping shrub 10–30cm tall; leaves 0.6–1.5cm, obovate to ovate; flowers white to pink; fruit to 0.6cm, bright red, edible).

vaginate possessing or enclosed by a sheath.

Valeriana (Medieval name, perhaps derived from *valere*, to be healthy, referring to its powerful medicinal qualities). Valerianaceae. Widespread. VALERIAN. Some 200 species, perennial herbs and subshrubs with simple to pinnately lobed leaves and small, rotate to campanulate flowers in cymose panicles from spring to autumn. The following are fully hardy and will grow on most soils in sun or part shade. Increase by division or seed.

V.officinalis (CAT VALERIAN, COMMON VALERIAN, GARDEN HELIOTROPE; Europe, W Asia; herb to 1.5m, strongly attractive to cats; leaves pinnate or pinnatisect; flowers pink or white); *V.phu* (Caucasus; differs from *V.officinalis* in its undivided or irregularly and sparsely divided lower leaves, the leaf divisions tend to be broader, the flowers are white; includes 'Aurea', with brilliant golden yellow leaves, turning lime green by midsummer). Red valerian is *Centranthus ruber*, (syn. *Valeriana rubra*).

Vallea (for Felice Valle (d. 1747), author of *Florula Corsicae* (1761)). Elaeocarpaceae. Andes. 1 species, *V.stipularis*, an evergreen shrub to 5m tall. The heart-shaped leaves are 3–12cm long. Borne in short cymes in late summer, the flowers are cup-shaped, to 1.5cm long and pale rose with darker veins. Hardy to about –5°C/23°F. Cultivate as for *Crinodendron*.

Vallisneria (for Antonio Vallinieri de Vallisnera (1661–1730), professor at the University of Padua). Hydrocharitaceae. Widespread. EEL GRASS, VALLIS. 2 aquatic perennial herbs, evergreen and bottom-rooting, with a clump of translucent green, ribbon-like leaves and inconspicuous flowers. Suitable for warm or cold aquaria. Plant the base in the substrate; grow in neutral to acid water; give good light, especially from above. Propagate by division of the runners from parent plants.

V.americana (syn. *V.gigantea*; E US, E Asia to Australia; leaves to 2m × 2cm; male flowers with stamens partially or wholly united); *V.spiralis* (EEL GRASS, TAPE GRASS; cosmopolitan; leaves to 2m × 1cm; male flowers with free stamens; 'Torta': leaves spirally twisted).

Valeriana phu 'Aurea'

valvate (1) describing parts with touching but not overlapping margins; (2) opening by or pertaining to valves.

valve one of the parts into which a dehiscent fruit or capsule splits at maturity.

Vancouveria (for Captain George Vancouver, Royal Navy (1758–1798), British navigator and explorer). Berberidaceae. W US. 3 species, perennial herbs resembling a smaller and more delicate *Epimedium*. The foliage is finely divided and often assumes attractive purple-red tints. Cultivate as for *Epimedium*.

V.chrysantha (SW Oregon; differs from *V.hexandra* in its yellow flowers); *V.hexandra* (Washington to California; deciduous, 10–40cm tall; flowers white with red-spotted sepals).

Vanda (Sanskrit name for *Vanda tessellata*.) Orchidaceae. Asia. 35 species, evergreen, epiphytic perennials with long stems clothed by 2-ranked, strap-shaped leaves and bearing abundant aerial roots. Produced at various times of year in axillary racemes, the flowers consist of five obovate tepals and a smaller, tongue-like lip. *V.coerulea* is suitable for the cool house or conservatory (minimum temperature 10°C/50 °F). The remainder need a minimum temperature of 15°C/60°F. Pot in baskets or orchid pots, containing the coarsest bark-based mix and with perfect drainage. Rooting into the substrate serves largely to anchor the plant: a healthy growth of aerial roots is essential. Older specimens may have bare stem bases, in which case cut away the lower portion of stem altogether, replanting the healthy, leafy, rooting summit. The old base may then produce new shoots. In summer, provide light shade in warm, humid conditions; water and mist frequently, applying a dilute foliar feed when in full growth. In winter, reduce temperatures and humidity, admitting maximum light. Increase by stem cuttings and offsets.

V.coerulea (India, Burma, Thailand; flowers 7–10cm, pale to deep lilac-blue, obscurely chequered, lip darker); *V.denisoniana* (Burma, Thailand; flowers fragrant, to 5cm in diameter, lime green to ivory, often mottled or spotted ginger, lip marked cinnamon to orange at base); *V. Rothschildiana* (a garden hybrid between *V.coerulea* and *Euanthe sanderiana* with large, broad-tepalled flowers in deep lavender blue with darker, chequered markings); *V.tessellata* (syn. *V.roxburghii*; India, Burma to Malaysia; flowers 4–5cm in diameter, cream, yellow-green or very pale lilac-blue, chequered brown, lip violet to blue edged white); *V.tricolor* (syn. *V.suavis*; Java, Laos; flowers fragrant, 5–7.5cm in diameter, usually pale yellow or cream dotted red-brown or purple-red, lip magenta to mauve). For *V.sanderiana*, see *Euanthe sanderiana*; for *V.teres*, see *Papilionanthe teres*.

variability, variation the deviation of plant characteristics from a type as a result of alteration to hereditary material which may arise as adaptations to, for example, environmental change or pest or disease influence. It is the process by which evolution occurs.

variegation the occurrence of patterns in plant leaves and, in some cases, stems and flowers, due especially to differences in distribution of the amount or compostion of chlorophyll; in the case of multi-coloured leaf patterns, other pigments, such as anthocyanin, carotenoids and xanthophylls, are involved. Variegation is usually an abnormal condition, due to mutation or, less frequently, infection by viruses. It can, however, be a normal character of a species, breeding true from one generation to the next, as in the case of the white-edged leaves and bracts of the summer annual *Euphorbia marginata*. Species of *Maranta* and *Calathea*, *Pilea cadierei* and *Silybum marianum* are examples of natural variegation.

Mutations that inhibit chlorophyll production are the most likely cause where areas of white tissue occur sporadically during development of otherwise normal green foliage. Such mutations may be under genetic control and repeated as new leaves develop, or they may arise as a single event directly affecting one chlorophyll-producing cell component. Genetically-controlled repeat mutations are evident in *Acer pseudoplatanus* 'Leopoldii', a form of common sycamore which has leaves striped and flecked pink-white, and also in the white and green marbled leaves of the annual *Tropaeolum majus* 'Alaska'. This kind of variegation may be transmitted by seed, providing the parent flowers are self-pollinated.

Chlorophyll mutations not genetically controlled may dominate, through cell divisions, individual seams of a multi-layered shoot apex, resulting in layers of white and normal green tissues; these are known as chimaeras. Shoots with white layers overlying green will produce leaves with white 'skins' and green 'cores', and because the skin tissues develop a broad band of cells around the leaf margins, the margins will be white whilst the central areas of the leaves remain green; the reverse arrangement of green-over-white shoots produces leaves with green margins and white centres. Variegations of this kind are usually stable, although, occasionally, individual layers will separate and hence entirely green or white shoots and leaves will result. Shoots with entirely green leaves are said to have 'reverted' and, since they are more vigorous than the variegated kind, they should be pruned out at the earliest opportunity. All-white shoots will not survive on their own if used as cuttings. Variegated plants of the layered or chimaera kind include the Norway maple, *Acer platanoides* 'Drummondii', *Philadelphus coronarius* 'Variegatus', *Cornus mas* 'Variegata' and *Aralia elata* 'Variegata'. Perennial herbaceous plants with layered variegation include *Symphytum × uplandicum* 'Variegatum', and many cultivars of *Hosta*. Greenhouse plants, such as *Pelargonium*, *Ficus* and *Codiaeum*, have equally attractive leaf variegations.

Sometimes, apples and chrysanthemums have sectorial chimaeras of a different colour, in which one kind of tissue lies alongside another and not over or under it. Such chimaeras, with more regular patterning, flecking or colour alteration, may be perpetuated as ornamentals, as in *Camellia* 'Donckelaeri' and certain rhododendrons. Chimaeras with layered kinds of variegation cannot be propagated by seed. Vegetative propagation sometimes gives rise to curious results as in the sword plant, *Sansevieria trifasciata*, where the popular yellow-edged form *S.trifasciata laurentii* is a chimaera and will not reproduce from cuttings, new plants reverting to the plain form.

Some chimaeras are produced by grafting as in the case of + *Laburnocytisus adamii*, created by grafting laburnum *(Laburnum anagyroides)* on to the broom *(Chamaecytisus purpureus)*. The stock tissue becomes mingled with that of the scion, so that the tree produces, at random, laburnum growths and flowers, broom growth and flowers, and intermediate growth with laburnum-like flowers tinged purple like those of the broom.

Mutations that affect the chemical composition of chlorophyll are generally responsible for plants that produce golden-coloured foliage. The golden tissues may occur alone and breed true from seed as in *Tanacetum parthenium* 'Aureum' and Bowles' golden grass, *Milium effusum* 'Aureum', but in many kinds of conifers, for instance *Thuja occidentalis* 'Rheingold' and *Cupressus macrocarpa* 'Goldcrest', it must be passed from one generation to the next by vegetative propagation. Golden tissues also occur in combination with layers of green or even white to produce chimaeras. *Elaeagnus pungens* 'Maculata' is an example of an evergreen shrub whose leaves have golden centres and green margins. The reverse arrangement, green centres and golden margins, is present in *Elaeagnus pungens* 'Dicksonii'. Similar reciprocal golden variegations are found in cultivars of *Euonymus* and *Hedera*.

Viruses are responsible for relatively few plant variegations; for example, abutilon mosaic virus causes the striking golden-mosaic leaves in *Abutilon striatum* 'Thomsonii' and other abutilons, and the greater periwinkle, *Vinca major*, and common bugle, *Ajuga reptans*, may be infected with vein-clearing virus to produce variegated forms.

Viruses are also responsible for certain kinds of flower variegation where the normal base colour of the petals is broken or flecked with different colour pigments, as in the Bizarre Group of tulips and forms of the common wallflower and Algerian iris, *Iris unguicularis*. Some flower variegations, however, such as the Damask rose 'York and Lancaster', *Antirrhinum* 'Bizarre Hybrids' and *Zinnia* 'Peppermint Stick', are known to be the visible expression of mutant genes. Propagation of virus-transmitted variegation needs to be by vegetative means.

Abnormal leaf variegations may be induced by deficiency of minerals, such as magnesium or iron, or by herbicide action, but such variegations are transient with no ornamental value.

variety, varietas (abbrev. var.) the category of taxa intermediate between subspecies and forma.

vascular plant a plant with strands of cells that form a system for conducting fluids throughout its roots and aerial parts, for example ferns, horsetails and seed-producing plants, as distinct from algae, fungi, liverworts and mosses.

vascular wilt a disease in which the pathogen is almost entirely confined to the vascular system. Wilting is a characteristic symptom, but may not always occur. *Verticillium* species often cause typical wilt diseases.

vase see *goblet*.

vector an organism that transmits a pathogen to a plant or animal.

vegetable (1) a general description of plant forms, distinguishing them from animal forms; (2) generally, a plant grown for its edible leaves, stems, roots, bulbs or tubers. However, tomatoes, cucumbers, marrows, peppers, aubergines and sweet corn cobs are strictly fruits; rhubarb is also usually regarded as a fruit, because of the manner of its culinary use, although it is commonly grown in the vegetable garden.

vegetative bud a bud that normally produces only a shoot, as opposed to a fruit bud.

vegetative propagation, vegetative reproduction the increase of a plant by asexual means, from vegetative parts of the plant; a process which normally results in a population of genetically identical individuals. It occurs by natural means in the case of apomicts, bulbils, cormlets, offsets, plantlets and runners, or may be brought about by artificial means through cuttings, division, budding, grafting and layering.

vein an externally visible strand of vascular tissues.

veitchberry see *hybrid berries*.

velamen the corky epidermis of aerial roots in some epiphytes, through which atmospheric moisture is absorbed.

Veltheimia (for August Ferdinand, Graf von Veltheim (1741–1801), German patron of botany). Liliaceae (Hyacinthaceae). South Africa. 2 species, bulbous perennial herbs. The bulbs are large, bearing a rosette of thinly fleshy, broadly lanceolate to strap-shaped, undulate leaves. Tubular flowers hang in a dense, torch-like raceme at the top of a stout scape from autumn to spring. Plant the bulbs with their necks just above soil level. Grow in sun or bright indirect light, in a freely draining medium, with additional coarse sharp sand, and in a minimum temperature of 5°C/40°F. Water moderately and regularly when in growth, allowing the surface of the medium to dry between waterings. Give an occasional dilute low-nitrogen liquid feed; reduce water gradually as foliage fades to allow near-dry dormancy in summer. Propagate by offsets when dormant, by seed or by leaf cuttings; mature leaves will form bulbs at their bases if inserted in damp sharp sand.

V.bracteata (syn. *V.viridifolia*; leaves glossy green; scape to 45cm, purple spotted yellow-green; flowers to 4cm, pink-purple sometimes tinged yellow); *V.capensis* (syn. *V.glauca*, *V.roodeae*; leaves glaucous; scape to 30cm or more, flecked purple, glaucous, flowers to 3cm, white spotted red to dull pink, tipped cream or pink with a green or claret apex, to deep and solid coral red).

velutinous coated with fine soft hairs giving the appearance and texture of velvet.

venation the arrangement or disposition of veins on the surface of an organ.

ventilator a hinged window or louvre unit used to admit air to a greenhouse in order to control temperature and humidity. See *greenhouse*.

ventral attached or relating to the inner or adaxial surface or part of an organ; sometimes wrongly applied to the undersurface.

ventricose swollen or inflated on one side.

Veratrum (from Latin *vere*, truly, and *ater*, black, from the colour of the root). Liliaceae (Melanthiaceae). Northern temperate regions. 20 species, perennial herbs with erect stems clothed with broad, plicate leaves and terminating in dense panicles of small, 6-parted flowers in summer. Hardy to about –20°C/–4°F, especially if given a protective mulch when dormant in winter. Grow in deep, fertile, moisture-retentive soils, rich in well-rotted organic matter. In moist soils, *Veratrum* will tolerate full sun. However, best foliage effects are obtained in partial shade with shelter from drying winds. Propagate by division. Alternatively, take 6mm root sections, each with a bud, and root in a sandy propagating mix in the cold frame. Seed germination may be erratic and seedlings plants may take up to ten years to reach maturity.

V.album (Europe, N Africa, N Asia; 1–2m; flowers green-white); *V.nigrum* (S Europe, Asia, Siberia; 60–120cm; flowers purple-black).

Verbascum (Latin name used by Pliny). Scrophulariaceae. Europe, N Africa, W and C Asia. MULLEIN. 300 species, annual, biennial or perennial herbs and subshrubs, usually hairy, often glandular. The flowers are produced from spring to autumn in erect spikes or panicles; they consist of five, rounded and rather silky petals and hairy anthers.

They are grown for their erect spikes of flower in summer; a number also have white- or golden-felted foliage, and although these are more commonly used to give height at the back of the herbaceous border, their beautiful rosettes are worthy of positions to the fore of borders where they will not be overgrown or overlooked. One of the most attractive of the rosette formers, *V.olympicum*, may take two years to achieve its full stature, forming a silver-white rosette up to a metre across, before producing majestic white-stemmed candelabra. Most *Verbascum* species, particularly the alpine *V.dumulosum*, are sun-loving and favour dry soils; they will not tolerate wet soils in winter. Given good drainage, most are hardy to –15°C/5°F. Many are not reliably perennial, but those that can usually be counted on include *V.nigrum*, *V.chaixii*, and subshrubby species, such as *V.dumulosum*. Grow alpine species in a sharply drained medium and protect from winter wet with a propped pane or open cloche; alternatively, grow in the alpine house in a gritty low-fertility alpine mix. Remove spent flowering spikes to encourage a second flush of flowers. Sow seed in late spring/early summer in the cold frame; if seed is sown earlier, biennial species may flower and die in the same season. Increase cultivars by root cuttings in late winter. The subshrubby species may be propagated by semi-ripe cuttings of lateral shoots in late summer in the cold frame.

V.bombyciferum (Asia Minor; white-felty biennial to 180cm; leaves to 50cm, forming large rosettes; flowers sulphur yellow with white-hairy stamens, in much-branched, woolly spikes); *V.chaixii* (NETTLE-LEAVED MULLEIN; S Europe, E Europe, Russia; grey-tomentose perennial to 100cm; leaves to 30cm, sometimes lobed; inflorescence usually branched, flowers yellow with violet-hairy stamens); *V.densiflorum* (syn. *V.thapsiforme*; Europe, Siberia; biennial to 2m, grey- or yellow-tomentose; leaves to 45cm; inflorescence dense, usually unbranched, flowers yellow with white-yellow-hairy stamens); *V.dumulosum* (SW Turkey; perennial, low-growing, much-branched subshrub to 30cm, grey-green and covered with white, yellow or creamy-brown felt; leaves 1–5cm; inflorescence 10–30-flowered, flowers lemon yellow with a central ring of coral or brick red). *V.lychnitis* (WHITE MULLEIN; Europe, W Asia; sparsely grey-tomentose biennial to 150cm; leaves to 30cm, green above, white beneath; flowers in a narrow panicle, bright yellow or white, stamens with white or yellow hairs); *V.nigrum* (DARK MULLEIN; Europe, Asia; sparingly hairy perennial to 100cm tall; leaves to 40cm, hairy beneath; flowers yellow with violet-hairy stamens, in slender, branched or unbranched spikes); *V.olympicum* (Turkey; densely white-woolly perennial or biennial to 180cm tall; leaves to 70cm; flowers golden yellow with white-hairy stamens in candelabra-like branching spikes). Numerous hybrids and cultivars are available. Among the finest are: 'Cotswold Beauty', to 1.2m, with pale buff to amber flowers with purple-hairy anthers; 'Gainsborough', to 1.4m, with silver-grey leaves and sulphur to primrose yellow flowers; 'Helen Johnson', to 1m, with grey leaves and peach to pale rust flowers; 'Mont Blanc', to 1m, with silky white leaves and pure white flowers; and 'Pink Domino', to 1m, with rosy pink flowers with a darker eye.

Verbena (Latin *verbena*, classical name for certain sacred herbs, probably of *V.officinalis* of Europe; common name vervain said to be derived from the Celtic words *fer*, to remove, and *faen*, stone, referring to its use in treating bladder stones). Verbenaceae. Tropical and subtropical America. VERVAIN. 250 species, annual or perennial herbs and subshrubs. The leaves are usually toothed to pinnately lobed. Produced in summer in corymbs or branching spikes, the flowers are small and tubular with a limb of five lobes. *V. × hybrida*, *V.rigida* and their cultivars are valued for their tight heads of often brilliantly coloured flowers, used in bedding, in the annual border and in hanging baskets or other containers. *V.bonariensis* will bloom in its first year from seed and is often treated as an annual, although it is well-suited to the herbaceous border where, given good drainage, it will tolerate temperatures to between –5 and –10°C/23–14°F. Grow in full sun in any moderately fertile and moisture-retentive soil. Propagate by seed in autumn or spring, and grow on with a minimum night temperature of 7–10°C/45–50°F; increase also by stem cuttings in late summer, rooted in a closed case with gentle bottom heat. Increase perennials also by division.

V.bonariensis (syn. *V.patagonica*; PURPLE TOP, SOUTH AMERICAN VERVAIN, TALL VERBENA; S America; slender, erect annual or perennial herb to 2m tall; leaves to 13cm, oblong to lanceolate, toothed, rugose above, white-hairy beneath; flowers blue, violet, lavender or mauve in short spikes terminating the branches of a long-stalked, forking panicle); *V. × hybrida* (COMMON GARDEN VERBENA, FLORIST'S VERBENA; garden origin; perennial herb usually grown as an annual, 20–100cm, erect or low to trailing; leaves to 8cm, ovate to lanceolate, toothed to incised, dark green, rugose; flowers fragrant, white to purple, blue, vermilion, scarlet crimson, red, pink or yellow, sometimes with a white or yellow mouth, in elongated spikes or large, flat corymbs; numerous cultivars and seed races are available, popular for half-hardy bedding and covering the wide range of habits and colours described above); *V.peruviana* (syn. *V.chamaedrifolia*, *V.chamaedrioides*, *V.coccinea;*; Argentina to S Brazil; creeping, small perennial; leaves to 5cm, ovate to oblong, toothed, rough; flowers bright scarlet or crimson, sometimes pink or striped white); *V.rigida* (syn. *V.venosa*; VEINED VERBENA; S Brazil, Argentina; erect or spreading, rigid perennial herb to 60cm; leaves to 7.5cm, oblong, toothed, rugose, dark green; flowers purple, magenta or lavender-blue in short spikes, usually terminating the three, candelabra-like branches of a long-stalked panicle); *V.tenera* (Brazil to Argentina; low, shrubby perennial with tufted, decumbent stems; leaves to 2.5cm, oval to ovate, hairy, finely pinnately divided; flowers rose-violet edged white in elongated spikes).

verge the edge of a lawn; sometimes, also used of a narrow strip of turf alongside a bed, border or path.

vermicide a pesticide active against earthworms.

vermicompost casts from worms (especially those fed on organic waste material), which can be used as a potting medium ingredient. Vermicompost has good structural stability and useful nutrient content.

vermiculate worm-shaped, with a pattern of impressed, close, wavy lines.

vermiculite an aluminosilicate mineral, heat-treated to form expanded or 'exfoliated' granules with very low bulk density and a laminated structure that allows for good aeration and water retention; it is used in potting media or for rooting cuttings. The cation exchange capacity is high and there are useful levels of magnesium and potassium. Industrial grades should be avoided for horticultural use due to possible high alkalinity.

vernal appearing in spring.

vernalization the treatment of seed, bulbs, or plants in order to break dormancy and induce or hasten growth or flowering, most commonly involving low-temperature storage or the application of the growth-influencing plant hormones known as gibberellins.

vernation the arrangement of leaves in the bud.

Veronica (perhaps named for St Veronica). Scrophulariaceae. Northern temperate regions. SPEEDWELL, BIRD'S-EYE. 250 species, annual or perennial herbs and subshrubs. Small, bell-shaped to rotate and 4-parted the flowers are produced singly in the leaf axils or in slender racemes from spring to summer. Hardy in climate zone 6 and suited to a range of situations, from the rock garden to the herbaceous border. Increase by division or seed.

V.austriaca (Europe; perennial, stems 25–50cm, erect or decumbent, more or less hairy; leaves linear to rounded, pinnately cut; inflorescence axillary, flowers bright blue; includes subsp. *teucrium*, syn. *V.teucrium*, with royal blue and sky blue flowers); *V.cinerea* (E Mediterranean, Asia Minor; tufted to spreading, white-tomentose perennial subshrub; leaves to 1.5cm, linear, silvery-tomentose; raceme to 3cm, flowers dark blue sometimes tinged purple, white in centre); *V.fruticans* (ROCK SPEEDWELL; Europe; woody-based perennial, erect to 15cm; leaves 1–2cm, obovate to oblong, entire to serrate; flowers deep blue stained red in centre in short, capitate racemes); *V.gentianoides* (Caucasus, SW Asia; erect perennial to 80cm; leaves to 10cm, linear to broadly ovate; flowers white or pale blue to lilac with dark blue veins, or wholly dark blue, or white faintly tinted blue, in terminal racemes to 30cm; a variegated cultivar is also grown); *V.longifolia* (syn. *V.exaltata*; Europe, naturalized eastern N America;

Veronica gentianoides

erect perennial to 100cm; leaves 3–12cm, linear to lanceolate, saw-toothed; flowers pale blue to lilac, bright blue to rich mauve in some cultivars, in long, slender spikes); *V.pectinata* (E Balkans, Asia Minor; evergreen, perennial, mat-forming subshrub to 15cm tall; leaves to 2cm, elliptic to oblong, saw-toothed to incised; flowers deep blue with a white eye, or pink, in axillary racemes to 15cm long); *V.prostrata* (Europe; mat-forming perennial; leaves 1–2.5cm, linear to ovate, entire to toothed; flowers pale to deep blue in dense erect spikes; includes cultivars with flowers in blue, lavender and pink, some with yellow leaves); *V.spicata* (Europe; erect or ascending perennial to 60cm tall; leaves 2–8cm, linear to ovate, crenate to entire, mid-green, hairy; flowers clear blue with purple stamens in dense, terminal racemes to 30cm; includes white, dark blue, pink, dark rose and dwarf cultivars; closely related is *V.incana*, SILVER SPEEDWELL, which differs in being covered in fine, silvery white hairs). For *V.perfoliata*, see *Parahebe perfoliata*; for *V.virginica*, see *Veronicastrum virginicum*.

Veronicastrum (somewhat similar to *Veronica*). Scrophulariaceae. NE Asia, eastern N America. 2 species, erect perennial herbs with whorled leaves and *Veronica*-like flowers produced in summer. Cultivate as for *Veronica*. *V.virginicum* (syn. *Veronica virginica*; eastern N America; CULVER'S ROOT, BLACKROOT, BOWMAN'S ROOT; 60–180cm tall; leaves lanceolate to oblanceolate, serrate, glabrous; flowers blue, white or pink with protruding stamens, crowded in slender spikes to 30cm long).

verrucose warty.

verticil a ring or whorl of three or more parts at one node.

verticillate forming or appearing to form a whorl.

verticillium wilt see *wilt*.

Vestia (for L.C. de Vest (1776–1840), professor at Graz, Austria). Solanaceae. Chile. 1 species, *V.foetida* (syn. *V.lycioides*), a musky scented evergreen, glabrous shrub 1–3m tall with pendent, tubular to funnel-shaped and yellow-green flowers in summer). In climate zones 8 and under, plant it outdoors against a warm, sheltered wall or grow in the cool greenhouse. Site in sun or part-day shade on any well-drained, fertile soil. Propagate from seed in spring and by greenwood cuttings rooted with bottom heat in summer.

vestigial of a part or organ that was functional and fully developed in ancestral forms, but is now reduced and obsolete.

viability the ability to live and grow, used especially of the capacity of seeds to germinate. See *seeds and seed sowing*.

Viburnum (Classical Latin name of *V.lantana*). Caprifoliaceae. ARROW-WOOD, WAYFARING TREE. Temperate northern hemisphere, extending into Malaysia and S America. At least 150 species, deciduous or evergreen shrubs or small trees. The flowers are small, white or cream, and sometimes flushed pink or wholly pink. They are borne at various times of year in corymbs, sometimes with sterile and showy ray flowers, or in umbels or axillary and terminal panicles. The corolla is rotate, campanulate, or cylindric and 5-lobed. The fruit is a drupe, usually small and glossy, and coloured red, blue or black.

Viburnum includes some of the most choice and valuable shrubs in cultivation. They are grown for their beautiful flowers, usually white, sometimes in showy 'lacecap' corymbs and often very fragrant especially in the winter-flowering species and hybrids. Some bear highly ornamental fruits. Almost all species grown for their fruits are self-incompatible and require pollination by seedlings of a different clone, or by related species flowering at the same time (*V.opulus* is a notable exception). Many of the deciduous species have exceptional autumn tints, among them *V.acerifolium*, *V.lantana*, and *V.opulus*. The evergreens, with the possible exception of *V.rhytidophyllum*, generally require a milder climate than the deciduous species (zones 6 and over), but include a number of species with handsome foliage, such as *V.davidii* and *V.cinnamomifolium*.

They tend to be used in shrubberies and mixed borders, but many are worthy of specimen plantings, especially those *V.plicatum* cultivars with tiered habit and 'lace-cap' flowers, for example, *V.*'Mariesii'. The evergreen *V.tinus* flowers in winter, but it is the deciduous or semi-deciduous winter-flowerers that are outstanding – beautiful *and* fragrant. Among these are *V.farreri*, *V.grandiflorum* and their hybrid offspring, *V. × bodnantense*. In some years, they will bloom from early autumn till spring, the flowers withstanding hard frosts without damage. They are followed chronologically by the semi-evergreen *V. × burkwoodii*, with slightly more frost-tender flowers but equally sweetly scented, flowering from early winter, and *V. × juddii*, *V.carlesii* and *V. × carlcephalum* flowering in spring, all hardy to –29°C/–20°F.

Plant in any deep, rich, moisture-retentive soil, in sun or semi-shade. Shelter winter-flowering species from north and east winds, and position them so that they avoid early morning sun on flowers and young growth after frost. Propagate deciduous species by softwood cuttings in early summer, also by hardwood cuttings in early autumn, in the cold frame, or by french layering. Propagate evergreens from semi-ripe cuttings in summer. Prune evergreens in late spring, deciduous species after flowering, to maintain shape, prevent overcrowding and remove dead or damaged wood. Viburnum aphids cause tight curling of the leaves. The eggs overwinter on the plant: treat with a tar-oil winter wash before buds break. Viburnum whitefly infests the underside of the leaves of *V.tinus* and may cause sooty moulds.

VIBURNUM

Name	Distribution	Habit	Leaves	Flowers	Fruit
V.acerifolium DOCKMACKIE	Eastern N America	erect, deciduous shrub to 1.8m	to 10cm, maple-like, rounded or cordate at base, downy beneath, carmine in autumn, 3-lobed, lateral lobes narrowly tapering, toothed	white, in long-stalked terminal cymes to 7.5cm diam. in summer	ovoid, 8.5mm, red, later purple-black
V betulifolium	W & C China	deciduous shrub to 3.5m	to 10cm, ovate to rhomboid or elliptic-oblong, finely tapering, coarsely serrate, glabrous and rich green above, paler beneath	white, in cymes to 10cm diam. in summer	to 7mm, rounded, bright red
V.bitchiuense	Japan, Korea	deciduous shrub to 3m, habit open upright	to 7cm, ovate or oblong, usually obtuse, subcordate at base, slightly dentate, hairy, densely so beneath	pink, later white, in lax infl. to 7cm diam. in spring	black
V. × bodnantense (*V.farreri* × *V.grandiflorum*)	garden origin	deciduous shrub to 3m, of upright habit	to 10cm, lanceolate to ovate or obovate, acute, serrate, soon glabrous	sweetly fragrant, rich rose red, later fading to white with a strong flush of pink, in dense 7cm-diam. clusters from autumn to spring	
Comments: 'Dawn': the type of the cross. 'Pink Dawn': flowers pink, rose in bud, intensely fragrant					
V. × burkwoodii (*V.carlesii* × *V.utile*)	garden origin	semi-evergreen shrub to 2.5m, of open and divaricate habit	to 10cm, ovate to elliptic, acute, rounded or somewhat cordate at base, obscurely toothed, glossy dark green and scabrous above, pale brown-hairy beneath	fragrant, pink-white, later pure white, in rounded, terminal, many-flowered clusters to 9cm diam. in spring	
Comments: 'Anne Russell': *V. × burkwoodii* × *V.carlesii* ; flowers in rounded corymbs, pink in bud, opening white, fragrant. 'Chenault': more compact; leaves more persistent, often bronze-brown in autumn; flowers pale pink, later white. 'Fulbrook': as 'Anne Russell' but lvs pale green, infl. larger, conical, later-flowering. 'Park Farm': habit broader ; leaves more ovate to lanceolate, to 10cm, entire to slightly serrate, hairy beneath; flowers pink at first later pure white, in cymes to 12cm diam.					
V. × carlcephalum (*V.carlesii* × *V.macrocephalum*)	garden origin	vigorous, deciduous shrub to 2.5m	close to *V.carlesii* but larger, to 12cm, somewhat glossy, paler beneath, tinged red in autumn	pink, later white and pink-flushed, in dense trusses to 15cm diam. in spring	
V.carlesii	Korea, Japan	rounded deciduous shrub to 2.5m	to 9cm, broad-ovate, acute, somewhat cordate at base, irregular-serrate, dull green above, grey-green, downy beneath	fragrant, pink, later white, in terminal, rounded clusters to 7.5cm diam. in spring	to 7mm, ovoid, compressed, black
Comments: 'Diana': vigorous, compact; leaves tinged chocolate when young; flowers red, then pink, somewhat purple-flushed, later white					
V.cinnamomifolium	China	evergreen shrub, sometimes a tree, to 6m	to 15cm, elliptic-oblong, tapering, more or less entire, leathery and dark green with 3 sunken veins above	cream-white, in lax cymes to 15cm diam. in summer	to 4mm, ovoid, blue-black, shiny
Comments: Differs from the next species in its thinner-textured, entire leaves and the fruit colour. It needs more shelter than *V.davidii* and thrives in moist shade.					
V.davidii	W China	densely branched evergreen shrub to 1.5m	to 15cm, leathery, narrowly oval to obovate, obscurely toothed toward apex, dark green above, paler beneath, more or less glabrous with 3 distinct veins	dull white, in dense, stiff cymes to 7.5cm diam.in summer	to 6.5mm, narrow-ovoid, metallic blue
Comments: As with many *Viburnum* species, plants of both sexes are needed if the beautiful fruits are to form					
V.dilatatum	Japan, China	erect deciduous shrub to 3m+	to 12cm, broadly ovate, rounded or obovate, apex abruptly short-acuminate, base rounded to somewhat cordate, toothed, hairy	pure white, very abundant, in downy cymes to 12cm diam. in summer	8.5mm, rounded-ovoid, bright red
Comments: 'Catskill': to 1.8m, broad, compact ; leaves to 14cm, orange-yellow to red in autumn; fruit dark red					
V.farreri **V.fragrans**	N China	upright deciduous shrub to 3m+	to 10cm, obovate or ovate, acute, coarsely toothed, tough, glabrous except for downy vein axils beneath	fragrant, white or pink-tinged, in dense terminal and axillary panicles to 5cm diam. in winter	brilliant red, later black
Comments: 'Candidissimum': flowers pure white. 'Nanum': dwarf; leaves small; flowers pink, scented, sparse					
V.foetens	Himalaya, Kashmir, Korea	closely related to *V.grandiflorum*, but to 1.5m, of wider and more open habit	large, glabrous or subglabrous beneath	white, in lax, 5cm panicles in winter	red, later black
V.grandiflorum	Himalaya, W China	deciduous shrub or small tree, of broad and somewhat stocky habit	to 10cm, firm, narrowly oval, acute, tapered at base, finely toothed, woolly beneath	fragrant, bright rose-pink at first, later nearly pure white, in dense corymbs to 7.5cm diam. in winter and spring	to 2cm, ovoid, black-purple
Comments: 'Snow White': flowers white, sometimes becoming pink					

VIBURNUM

Name	Distribution	Habit	Leaves	Flowers	Fruit
V. × juddii (*V.bitchiuense × V.carlesii*)	garden origin	intermediate between parents – broad, deciduous shrub to 1.5m	more oblong than in *V.carlesii*	more numerous than in *V.carlesii*, in somewhat laxer and broader corymbs in spring	
V.lantana WAYFARING TREE, TWISTWOOD	Europe, N Africa, Asia Minor, Caucasus, NW Iran	upright, deciduous shrub to 4.5m, occasionally tree-like	to 12cm, broadly ovate to oblong, apex acute or somewhat blunt, cordate at base, finely toothed, velvety-hairy above, at least at first, densely hairy beneath	white, in cymes to 10cm diam. in spring and summer	oblong, 8.5mm, red later black, glossy
V.lentago SHEEPBERRY	Eastern N America	robust deciduous shrub or small tree to 9m	to 10cm, ovate to obovate, apex tapering, base cuneate or rounded, sharply toothed, glossy dark green, lighter beneath, glabrous except for scurfy veins, bright red in autumn; petiole winged	creamy white, in terminal cymes to 11cm diam. in spring and summer	ovoid, to 16mm, blue-black, pruinose

Comments: 'Pink Beauty': fruit pink, later almost violet

Name	Distribution	Habit	Leaves	Flowers	Fruit
V.odoratissimum	India, Bunma, China, Japan, Philippines	shrub to 10m, evergreen	to 14cm, coriaceous, glabrous, ovate to obovate, glossy green	sweetly fragrant, white, in terminal cymes to 10cm in spring and summer	to 10mm, oblong-ovoid, slightly compressed, red ripening black

Comments: 'Nanum': habit dwarf, somewhat tender

Name	Distribution	Habit	Leaves	Flowers	Fruit
V.opulus GUELDER ROSE, EUROPEAN CRANBERRY BUSH, CRAMPBARK	Europe, NW Africa, Asia Minor, Caucasus, C Asia	deciduous shrub to 4.5m, thicket-forming; stems erect, grey; bark thin	to 10cm, maple-like, lobes 3-5, acute, coarsely and irregularly toothed, dark green and glabrous above, lanuginose beneath, turning wine-red to carmine in autumn	creamy white, in two forms – small and fertile, and sterile and showy, to 2cm diam., in flat cymes to 7.5cm diam. in summer	to 8.5mm diam., globose, bright red

Comments: 'Aureum': leaves emerging bronze, becoming deep yellow yellow then lime green. 'Compactum': dense and compact, to 1.5m high; leaves small, glossy, maroon in autumn; flowers and fruit abundant. 'Roseum' : syn. 'Sterile', 'Snowball'; habit rounded, to 4m; leaves tinted purple in autumn; flowers large, double, sterile, in 'snowball' heads, white or green tinted, sometimes turning pink. 'Xanthocarpum': leaves pale; fruit bright green at first, later pale ochre.

Name	Distribution	Habit	Leaves	Flowers	Fruit
V.plicatum	Japan, China	deciduous bushy shrub to 3m; branches mostly horizontal	to 10cm, broadly ovate to elliptic-obovate, base round to broadly cuneate, crenate, plicately veined, deep green above, downy beneath, dark red to violet-brown in autumn	white, small and fertile or sterile and showy (to 4cm diam.), in flat umbels to 10cm diam. in summer	rounded-ovoid, coral red, later blue-black. Summer.

Comments: f. *tomentosum* : branches horizontally spreading and overlapping; leaves wine red in autumn; fruit blue-black. 'Mariesii': low, spreading; branches tiered; inflorescences large, resembling a lacecap *Hydrangea* with outer flowers larger and sterile. 'Nanum Semperflorens' : syn. 'Watanabe'; dwarf, to 0.5m; leaves richly coloured in autumn; flowers white, a fine 'lacecap' form

Name	Distribution	Habit	Leaves	Flowers	Fruit
V. × pragense (*V.rhytidophyllum × V.utile*)	garden origin	evergreen shrub to 2.5m; shoots somewhat thin, spreading to slightly nodding	to 10cm, elliptic to lanceolate, shiny green and somewhat rugose above, densely tomentose beneath		
V.rhytidophyllum	C &W China	evergreen shrub to 6m	to 20cm, ovate to oblong, tough, shiny and finely rugose above, grey- to tawny-felted beneath	dull yellow-white in tawny-felted, umbel-like trusses to 20cm diam., in spring and summer	ovoid, red ripening black
V.sargentii V.opulus var. sargentii	NE Asia	deciduous shrub to 3m	to 12cm, 3-lobed, maple-like, emerging bronze becoming rich yellow-green, turning yellow to red in autumn	white to cream, sterile and fertile in lacecap cymes to 10cm diam. in spring and summer	subglobose, light red

Comments: Differs from *V.opulus* in flowers with purple, not yellow, anthers. Includes 'Onondaga', with leaves emerging bronze and turning purple-red in autumn, blush pink flowerbuds and pure white, larger, outer sterile flowers

Name	Distribution	Habit	Leaves	Flowers	Fruit
V.sieboldii	Japan	deciduous shrub or small tree to 3m	to 12cm, obovate to oblong, serrate, glossy and smooth above	creamy white in long-stalked cymes to 10cm diam. in spring and summer	ovoid, pink later blue-black
V.tinus LAURUSTINUS	S Europe, N Africa	evergreen shrub to 3.5m	to 10cm, ovate to ovate-oblong, entire, shiny dark green above, paler beneath	pink-brown in bud, opening white to palest pink in domed, terminal cymes to 10cm diam., in winter and spring	ovoid, indigo to black

Comments: Includes numerous cultivars, some compact and bushy, some with white- to cream- or yellow-variegated leaves, others with pink-to red-tinted flowers. One of the most popular is 'Eve Price', dense and compact with red-stained flowerbuds, pink-tinted flowers and blue fruits

Vigna (for Dominico Vigna (d. 1647), professor of botany at Pisa). Leguminosae. Tropics, subtropics. 150 species, erect, twining or creeping herbs with tri-foliolate leaves and pea-like flowers in summer. *V.caracalla* is suitable for outdoor cultivation in frost-free zones on any fertile, well-drained soil in sun; otherwise, grow in the warm greenhouse. Propagate by seed. *V.caracalla* (SNAIL FLOWER, CORKSCREW FLOWER; S America; twining perennial to 6m; flowers to 5cm, fragrant, white or yellow with purple-pink wings, the keel coiled like a snail's shell).

villous with shaggy pubescence.

Vinca (contraction of ancient Latin name *vinca pervinca*, from *vincio*, bind, wind about, in reference to making wreaths, whence late Middle English *per wynke* and Modern English *periwinkle*). Apocynaceae. Europe, W Asia. PERIWINKLE. 6 species, low, evergreen perennial herbs and subshrubs with long and slender, arching to creeping stems and smooth, dark green leaves. Solitary flowers are produced in the leaf axils from winter to summer; they are funnel-shaped with 5 spreading lobes. Fully hardy, trailing groundcover. They will grow in almost any soil in sun or shade and, once established, spread rapidly. They may be propagated by division of rooted shoots, or by cuttings. The over-wintering shoots of *V.major* can be cut away in late winter before new growth begins. These plants are poisonous.

V.major (GREATER PERIWINKLE; fertile shoots arching, to 30cm; flowers 3–5cm across, blue-purple or violet); *V.minor* (LESSER PERIWINKLE; fertile shoots more or less erect, to 20cm; flowers to 3cm across, blue, violet or rich mauve). Both the above species include cultivars with variegated leaves and single and double flowers in shades of pale blue, dark blue, indigo and violet; those derived from *V.minor* may also show strong mauves and purple-reds. For *Vinca rosea*, see *Catharanthus roseus*.

vine a climbing plant.

vine eye (1) a broad, flat, wedge-shaped nail with a hole through which wire can be passed; used to support climbing plants on brick walls. It is driven into the mortar at suitable spacing; (2) a single-node cutting of a grape vine, traditionally taken with the bud in the middle of a short length of internode and inserted horizontally into the rooting medium.

vine eye cutting

vine weevil *(Otiorhynchus sulcatus)* a widespread pest of many garden plants, especially those grown in pots, or other containers. Under greenhouse conditions, corms of *Cyclamen, Begonia* and *Sinningia* are particularly vulnerable to the feeding grubs, with coleus, crassulas, ferns, fuchsias, *Impatiens*, pelargoniums, primulas and saintpaulias also attacked. In gardens, primroses, polyanthus, strawberries, saxifrages and sedums may be severely damaged. Fully grown grubs are about 10mm long but curled; they are plump, legless and creamy-white with a brown head. Vine weevil larvae are active mainly from late summer to spring, when they feed on roots and burrow into corms; on woody plants, they remove outer tissue from the stem base and large roots. The first indication of infestation is usually a collapse of the plant.

The adult beetles are flightless, about 9mm long and dull black with speckles. They move slowly and eat notches in the leaf margins of a wide range of plants; especially rhododendrons and camellias, where they feed mostly amongst the lower branches at night. Outdoors they are active from spring until autumn, with egg-laying occurring mainly in the period June-August. They gain access to greenhouses through doors or ventilators or on pot plants moved from outside standing areas; they are proficient climbers.

Vine weevil grubs are tolerant of many insecticide treatments, particularly once they are large enough to be most damaging. Good results can, however, be achieved by drenching the growing medium with a recommended product in early July and early August. On a small scale, control vine weevil by picking out and destroying grubs found in the rootballs of container-grown plants (which will then need repotting).

Biological control with the nematodes *Heterorhabditis megidis* and *Steinernema carpocapsae* is most worthwhile in gardens. These microscopic eelworms are applied in a water suspension, as a drench in late summer when the growing medium temperature is in the range 14–20°C/57–70°F; moist, open compost affords best results. The nematodes enter the bodies of larvae and transmit a fatal bacteria. Proprietary packs are available.

Adult weevils can be controlled by trapping in corrugated cardboard rolls, placed amongst greenhouse plants, or by shaking them from the plants on to newspaper at night. Recommended insecticide sprays can be applied to the foliage and growing medium surface, at night, as soon as damage is first noticed.

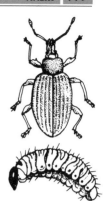

vine weevil and grub

vinery a traditional greenhouse designed for grape vines, with a wide, airy span and low side-walls of often less than 1m in height. Often constructed as a lean-to, a vinery allows relatively easy access to all parts of the vine, with minimal need for ladders. Vinery-type greenhouses have been used for commercial tomato growing, in spans 8.5–9.2m wide, with heights of 1.4–1.7m high to gutters and about 3.9m to the ridge.

Viola (Classical Latin name, variant of the Greek *ion*, a name applied to a number of plants, including violets). Violaceae. VIOLET. Temperate regions, especially N Temperate & Andes, Sandwich Is. Some 500 species, annual or perennial herbs and small shrubs. Axillary and solitary, the flowers nod on erect stalks. They consist of five petals, the two uppermost often larger and outspread, the two outer, lower petals smaller, the central, lower petal often differently coloured and shaped. The flowers are usually spurred.

With the exception of the Parma violets, violets are quite frost-hardy (to at least –10°C/14°F). Grow them in cool, moist but well-drained soil, in partial or dappled shade, with protection from scorching winds. An acid to neutral pH is ideal. Propagate from rooted runners, by division immediately after flowering or in autumn, or by cuttings. For winter flowers, protect open-ground plants with cloches, plant in the cold frame in early autumn, or in pots in the cold greenhouse, ensuring maximum light and ventilation but with protection from frost and cold winds. Violets will not thrive in high temperatures, and the range should be 5–10°C/40–50°F, never above 15°C/60°F. Parma violets should be given such winter treatment as a matter of course in all but the mildest districts.

"Pansy", from the French *pensées*, referred originally to *V.tricolor*, which has a long history in cultivation. Several modern violettas have been derived from crosses between named violas and *V.cornuta* and *V.c.* 'Alba', by using colchicine. All have the characteristic compact, tufted foliage and delicate fragrant blooms of the true violetta. All pansies have a single-stemmed root system, whilst violas originating from crosses between *V.lutea*, *V.altaica* and *V.cornuta* have either a multi-stemmed root system, or an underground stolon system. Elongated leaves with a sprawling habit, flowers with a short spur or 'horn' and little or no scent, are signs of a pansy; rounded leaves, short-jointed stems, long flower spurs, fragrance, and a neat compact habit are sure signs of a true bedding viola.

The garden pansies, *V. × wittrockiana*, usually treated as annuals or biennials, are commonly used for bedding and edging. Some selections, such as the Ice Queen, Floral Dance, and Universal series, are useful for colour throughout the winter in mild weather; they also make colourful and attractive plants for window boxes, pots and tubs. The garden violas, of compact form and perennial habit, make neat edging plants, or can be given foreground positions in herbaceous beds and borders. They can be used to good effect in the underplanting of the shrub border. The species violas, and violettas are well-suited in scale to the rock garden, or in window boxes and pots, violettas also being useful for bedding in the smaller garden. Those blooming early in the season make a good carpet under taller spring-flowering bulbs. With hybrids, the strongest and most floriferous plants are obtained if treated as biennials, sown in autumn for spring planting after overwintering in a well-ventilated cold frame or in summer for autumn plantings. For one-year plants, sow at any time from midwinter onwards, thinly in pans of sandy propagating medium, covering lightly. Germinate at 20–25°C/68–77°F – higher temperatures will cause failure; grow on at 10°C/50°F, keeping the potting medium moist at all times. Show and exhibition cultivars, or other particularly attractive plants, may be propagated from cuttings as for named violas, or by simple layering, new roots appearing at nodes. Where Violas and violettas are treated as perennials, they require a well-drained but moisture-retentive soil, with a pH between 5.5 and 7, and a cool root run. Position them in sun or, in warmer areas, in part shade. Full sun can be tolerated with sufficient moisture at the root; too-deep shade will result in straggling, etiolated plants. When necessary, cut back; flowering will resume in 2–3 weeks. For container culture, grow in a medium fertility loam-based mix. Cultivars can be propagated from cuttings, preferably taken in late summer or early autumn. Mature growth is hollow-stemmed and difficult to root: cut back hard to generate good cutting material about one month before propagating. Take nodal cuttings of young growth and insert in a soilless propagating medium in a shaded and well-ventilated cold frame. Violettas and those species of compact habit and a fibrous root system can be divided in late spring or in early autumn after cutting back to ensure short, strong shoots.

The rust fungus *Puccinia violae* is more common on species than on hybrids. The fungus mycelium is perennial in the rootstocks of pansies, so infected plants are best destroyed. The smut *Urocystis violae* produces black spore masses in large swellings on the leaves and leaf stalks; affected plants should be destroyed and the site kept free of violas for several years. The violet leaf midge, *Dasineura affinis* causes abnormal thickening and rolling of the leaves and can be severely debilitating. Pick off and destroy affected leaves and treat with insecticide.

Viola
'Bowles' Black'

VIOLA

Name	Distribution	Habit	Leaves	Flowers
V.aetolica	Balkans	perennial herb 5–40cm	to 2cm, ovate to lanceolate, crenate	to 2cm, yellow, the upper petals sometimes tipped violet
V.biflora TWIN-FLOWERED VIOLET	Europe to N Asia, N America	creeping to weakly erect perennial herb to 20cm	reniform to cordate, toothed	to 1.5cm, yellow, lower petals veined or streaked brown to purple-black
V.calcarata	C Europe	perennial herb to 6cm	to 4cm, elliptic to lanceolate, base cuneate, crenate	large and slender-spurred, blue to mauve, lilac or yellow, with a yellow centre
V.canadensis CANADA VIOLET, TALL WHITE VIOLET	N America	erect perennial herb, 5–30cm	ovate, sparsely hairy, toothed	white with a yellow centre and purple back, the lower petals sometimes flushed purple throughout
Comments: Pure white and dwarf forms are also grown				
V.cornuta VIOLA, HORNED VIOLET, BEDDING PANSY		perennial, 10–30cm, prostrate to ascending	ovate with a truncate to cordate base and crenate margins	2–4cm diam., the petals broadly wedge-shaped and violet to pale lilac-blue
Comments: Many cultivars are available including dwarf and compact to larger and rather sprawling plants, some with variegated leaves, and flowers ranging from white to opal, silvery lilac, pale blue, lilac, mauve and dark violet				
V.elatior	C Europe, W Asia	erect perennial herb, 20–50cm	to 9cm, broadly lanceolate, toothed	to 2.5cm, pale blue with a white centre
V.glabella STREAM VIOLET	NE Asia, W N America	perennial to 30cm, forming a clump from a horizontal, scaly rootstock	reniform to ovate-cordate, toothed, bright green	to 3cm across, yellow veined purple
V.gracilis *V.olympica*	Balkans, Greece to Asia Minor	low-perennial to 10cm	oblong to broadly ovate, more or less toothed, with large, pinnately divided stipules	to 4cm across, deep violet with a yellow centre
Comments: Cultivars with larger and white flowers are available				
V.hederacea AUSTRALIAN VIOLET, TRAILING VIOLET *Erpetion reniforme*; *V.reniformis*	Australia	stemless or creeping and mat-forming perennial herb seldom more than 7cm high	to 3cm across, ovate-cordate to reniform, entire to toothed	to 2cm, blue-violet or white
Comments: Needs protection in winter and is well-suited to a shady, cool and moist situation in the cold or alpine house				
V.labradorica LABRADOR VIOLET	N N America, Greenland	clump-forming, spreading perennial to 5cm	ovate-cordate to reniform, dark green	small, typically mauve with a paler throat overlaid with darker veins, rarely blue, lilac or white
Comments: 'Purpurea' is a popular cultivar with dark purple-tinted leaves and pale to mid-mauve flowers				
V.lutea MOUNTAIN PANSY	W & C Europe	creeping to mat-forming perennial herb to 10cm tall	to 3cm, ovate-cordate to lanceolate	to 3cm, similar to *V.tricolor* but wholly yellow, or with upper petals sometimes purple, or the whole flower violet
V.obliqua MARSH BLUE VIOLET *V.cucullata*		spreading perennial herb to 15cm	to 9cm wide, ovate-cordate to reniform, crenate	to 4cm, blue-violet with a white eye, or wholly white, or pale blue
Comments: Includes cultivars with flowers red-eyed and white, white with violet or sky blue eyes, and rose to red				
V.odorata SWEET VIOLET, GARDEN VIOLET, ENGLISH VIOLET	Europe	spreading perennial herb to 15cm	broadly ovate-cordate to reniform, crenate	fragrant, to 2cm, dark violet, lilac, white or yellow
Comments: Includes numerous cultivars ranging in flower size, form and colour				
V.palmata EARLY BLUE VIOLET, WILD OKRA	US	spreading perennial herb to 10cm	palmately lobed with 5–11, toothed segments	pale violet with a white eye overlaid with darker veins
Comments: Plant in a dry, fast-draining situation				
V.pedata BIRD'S FOOT VIOLET, PANSY VIOLET, CROWFOOT VIOLET	E N America	thick-rooted, clump-forming perennial to 15cm	palmately divided into 3-5, cleft, narrow segments	deep, velvety purple, the lower petals paler than the upper with dark veins
Comments: White-flowered forms are grown, likewise 'Bicolor', with flowers half dark violet, half pale lilac. Although hardy, this species fares better in the alpine house grown in an acid, sandy medium				
V.sororia WOOLLY BLUE VIOLET	E N America	perennial herb, erect or spreading, to 8cm	5-10cm across, ovate to rounded, crenate, hairy beneath, pale to mid-green	to 2.5cm, deep violet-blue with a white eye overlaid with darker veins
Comments: 'Albiflora': flowers pure white. 'Freckles': flowers white flecked and spotted pale blue				

VIOLA

Name	Distribution	Habit	Leaves	Flowers
V.tricolor WILD PANSY, HEART'S EASE, LOVE-IN-IDLENESS, JOHNNY JUMP UP	Europe, Asia	annual, biennial or perennial herb, erect and bushy or sprawling to spreading, 5–30cm	ovate-cordate to lanceolate, crenate	2–4cm, upper 2 petals typically violet to mauve, lateral petals white striped violet-black, lowest petals white with yellow bases overlaid with violet-black; forms with violet, pale mauve, yellow and bicoloured flowers are common

Comments: Many cultivars are known, one of the finest is 'Bowles' Black', with velvety black flowers

Name	Distribution	Habit	Leaves	Flowers
V. × wittrockiana PANSY, LADIES'-DELIGHT, HEART'S-EASE, STEPMOTHER'S FLOWER	garden origin	erect and bushy to sprawling annual, biennial or short-lived perennial herb 6–20cm	5–12cm, ovate-cordate to spathulate-lanceolate, crenate	to 8cm across, the petals rounded and overlapping, in a wide range of colours

Comments: One of the most popular hardy bedding plants with many cultivars and seed races, ranging in flowering time (winter to summer), longevity, size and colour, including pure tones and mixtures of white, pale blue, azure, indigo, violet, mauve, lilac, opal, pale rose, red, orange and yellow

violet root rot (*Helicobasidium brebissonii*, syn. *H. purpureum*) a fungal disease, in which strands of violet- or purple-coloured mycelium cover the roots, crown and stem bases of plants, in which small dark resting bodies occur as well as in the surrounding soil. Plants become yellow, stunted and may wilt and die. A wide range of plants is affected, including asparagus, beetroot, carrot, celery, clover, parsnip, potato, strawberry, sugarbeet, swede and turnip, as well as ornamentals, weeds and young trees. The disease may perpetuate in the soil for many years, especially in acid waterlogged conditions. Control by rotation, removing crop debris and burning affected roots.

virescence greening of tissue that is normally devoid of chlorophyll; often a symptom of *phytoplasma* (q.v.) infection.

virgate long, slim and straight, like a rod or wand.

viridarium a mediaeval garden or park planted with trees.

viroid one of the smallest pathogens known, akin to viruses and transmitted mainly by grafting and other vegetative propagation methods. Viroids are the cause of a number of diseases with symptoms similar to those of viruses, including CITRUS EXOCORTIS, CHRYSANTHEMUM STUNT, COCONUT CADANG-CADANG, CUCUMBER PALE FRUIT, HOP STUNT, and POTATO SPINDLE TUBER.

viruses non-cellular obligately parasitic pathogens, visible only under the electron microscope, responsible for a wide range of plant diseases. Viruses multiply only in living cells, being dependent on the host's processes; they disrupt cells, metabolic activity, and usually cause a variety of symptoms. Almost any plant may be attacked. Some viruses can infect a wide host range, causing very different symptoms, of varying severity, in different plants.

Reduction in growth is the most common symptom, and may be associated with stunting, loss of vigour, decreased yield, loss of flower potential, or crop failure. Where fruit trees and ornamentals such as pelargoniums

are hosts, infection may not be obvious, and it is then termed 'latent'. Most conspicuous is colour change; this is particularly noticeable in leaves, which develop chlorotic and yellow patterns, such as mosaics, mottles, blotches, spots, ringspots, vein-banding and vein-yellowing. Flower-breaking, for example, the light streaks on dark petals apparent in 'broken' tulips, may also occur. Leaf variegation of some garden cultivars is viral in origin, for example *Lonicera japonica* 'Aureoreticulata' and *Jasminum officinale* 'Aureum'. Flower malformations, leaf-rolling, leaf-curling, leaf-narrowing, abortion of reproductive organs, abnormal branching, rosetting, and enations are malformations caused by viruses. Tissue death may occur as local lesions, or it may affect phloem, stem tips or buds, or the internal tissue of organs such as tubers. Wilting or withering may result. Many of these symptoms are similar to those caused by fungal or bacterial diseases; environmental effects and mineral deficiencies can also produce comparable symptoms.

The three major plant viruses are TOBACCO MOSAIC VIRUS, CUCUMBER MOSAIC VIRUS and CYMBIDIUM MOSAIC VIRUS.

TOBACCO MOSAIC VIRUS (TMV) and the closely related tomato mosaic virus occur worldwide and are capable of infecting species from more than 150 genera in 30 families, being economically important chiefly in Solanaceae. Leaf mottling and malformation, stunting, and reduced yield are the usual symptoms of infection.

TMV is a highly infectious virus, which can survive in plant debris in soil for many years and infect roots directly; it is spread by contact between plants, particularly during cultural operations, and in seed. Use of fresh or steam-sterilized soil for susceptible crops is advisable, as is thorough hand-washing and sterilization of tools before and after working on plants. Resistant tomato cultivars are available.

CUCUMBER MOSAIC VIRUS (CMV) also occurs worldwide and can infect more than 775 species in 86 families, especially cucurbits, tomato, pepper, spinach, many other vegetables, and ornamentals, including gladioli, lilies, delphiniums and petunias. Leaf mosaic, leaf malformation, stunting, colour-breaking and sterility of flowers and severe reduction in yield are the major symptoms, and are most severe on outdoor plants in summer. Aphid vectors are the most important means of transmission, but spread by seed and mechanical means also occurs during cultural operations, and the virus can overwinter in many weeds and crop plants. Isolation of susceptible crops, aphid control, and weed elimination are standard control measures, together with the use of resistant or tolerant cultivars.

CYMBIDIUM MOSAIC VIRUS, which has spread worldwide as a result of interest in orchid growing, has no known natural vector and is typically transmitted on cutting implements, and probably also on hands during cultural operations. Typical symptoms are long, well-defined yellow and black streaks on leaves and a marked reduction in size and number of flowers. It affects many other orchids and is sometimes associated with other orchid viruses.

Aphids, leafhoppers, mites, nematodes, fungi, and parasitic plants can all serve as virus vectors. Because of their intimate relationship with plant cells, viruses are virtually impossible to eliminate by chemical means; they are not affected by antibiotics and no viricides are commercially available. Insecticides, nematicides, and fungicides can reduce spread by vectors; disinfection of tools, hands and greenhouse wires are additional useful cultural measures. Avoidance of sources of infection, by the use of certified seed or planting material, is the most important measure of control. Also vital is the prompt removal and destruction of infected plants, and the elimination of weeds that may act as a reservoir of infection.

Virus-free plants can be produced by heat treatment or meristem culture.

viscid covered in a sticky or gelatinous exudation.

Vitaliana (for Donati Vitaliano (1717–1762), professor of botany at Turin). Primulaceae. Mountains of S and WC Europe. 1 species, *V.primuliflora* (syn. *Douglasia vitaliana*), a mat-forming, tufted, evergreen perennial herb. The leaves are small, narrow and silvery. Produced in spring, the flowers are solitary, stalkless and bright yellow to rich gold; they consist of a short tube and a limb to 2.5cm across, of 5 segments. Cultivate as for *Androsace*.

Vitex (the Latin name used by Pliny for *V.agnus-castus*). Verbenaceae. Mostly tropics, some temperate. 250 species, trees and shrubs (those here deciduous small aromatic trees and shrubs) with digitate leaves and small, tubular and 2-lipped flowers in racemes and panicles in late summer and autumn. Given the shelter of a warm wall with good drainage and protection from cold drying winds, they will tolerate temperatures to between –5 and –10°C/23–14°F. Increase by semi-ripe cuttings.

V.agnus-castus (CHASTE TREE; S Europe; to 5m; leaflets 5–9, with grey-white hairs beneath; flowers fragrant, lavender, lilac or white); *V.negundo* (E Africa, E Asia; to 8.5m; leaflets 3–5, white-hairy beneath; flowers lilac to lavender).

viticulture the practice and science of cultivating grape vines.

Vitis (name used by the Romans for the grapevine). Vitaceae. Northern Hemisphere, especially N America. VINE, GRAPE. 65 species, deciduous shrubs mostly climbing by tendrils. The leaves are simple to palmately lobed and toothed. Small, green-white flowers in spring and summer give rise to succulent berries, usually pruinose and sometimes edible. The ornamental species are usually found in rich, damp woodland soils. They are hardy vigorous vines, ideally suited to clothing pergolas, trellis and fences and for creating covered walkways. They will also climb and cascade through sturdy trees and, with support, make good wall plants. They are grown primarily for their luxuriant and often very attractive foliage, although some, such as *V.davidii*, with distinctive prickly stems, also bear edible fruits. The edible grape, *V.vinifera*, has given rise to a number of clones with beautiful foliage. Many other species are valued for their brilliance in autumn, including the spectacular, large-leaved *V.coignetiae*, and *V.davidii*, *V.amurensis* and *V.thunbergii*.

Grow in deep, moist, well-drained, moderately fertile soils in full sun or partial shade (sun and warmth are needed for fruit ripening and good autumn colours). Prune in mid-winter, before the sap begins to rise. In a restricted space, most species can be trained to a permanent framework and cut back to within one or two buds to form a spur system that becomes increasingly picturesque with age. Excessive new growth is removed (i.e. new shoots stopped at 5–6 leaves in midsummer as the basal growth begins to ripen), frequently giving rise to attractive contrasts between new and mature foliage. Propagate by hardwood cuttings in late winter, rooted in a closed case with bottom heat at 18–21°C/60–65°F; also by simple layers, or by seed, following a 6 week stratification period at or below –5°C/23°F.

V.amurensis (AMUR GRAPE; NE Asia, vigorous vine to 15m; leaves 10–25cm, 5-lobed, midlobe broadly ovate, acuminate, slightly hairy beneath, turning rich crimson and purple in autumn; fruit black, glaucous, bitter); *V.coignetiae* (CRIMSON GLORY VINE; Japan, Korea; vigorous vine to 25m; leaves to 30cm, rounded, obscurely 3–5-lobed, toothed, thick-textured, rather puckered with impressed veins but more or less smooth above, brown-scurfy beneath, turning metallic bronze then fiery red to glowing scarlet in autumn; fruit purple-black, glaucous, scarcely edible); *V.davidii* (China; vigorous vine to 15m, young shoots with spine-like bristles; leaves to 25cm, ovate to cordate, toothed, tip tapered, glossy dark green above, glaucous beneath, brilliant red in autumn; fruit black, edible; includes 'Veitchii', with leaves tinged bronze in summer, and var. *cyanocarpa*, with fruit thickly blue-pruinose); *V.thunbergii* (temperate E Asia; moderately vigorous vine; leaves 7–15cm, cordate at base with 3–5, ovate, toothed lobes, dull dark green turning rich crimson in autumn, woolly beneath; fruit purple-black, glaucous); *V.vinifera* (COMMON GRAPE VINE; S and C Europe, W Asia; vigorous vine to 20m; leaves 5–15cm, rounded

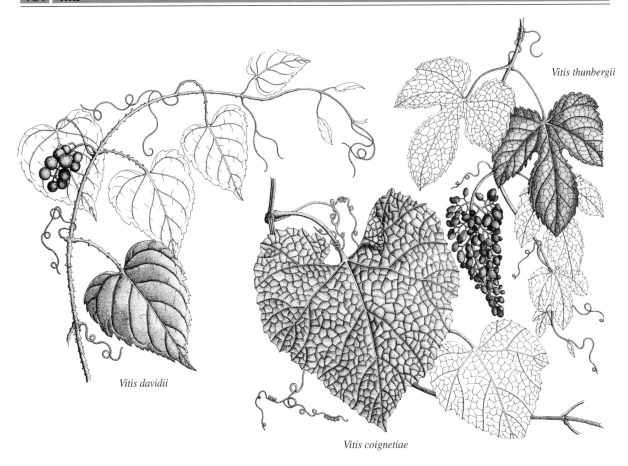

Vitis thunbergii

Vitis davidii

Vitis coignetiae

in outline, palmately 3–7-lobed, toothed, base cordate, becoming glabrous above, hairy beneath; fruit large, sweet and edible green, yellow, purple-red or purple-black; a widespread and much-developed fruit crop (see *grapes*), with many hybrids and cultivars; ornamental cultivars include 'Apiifolia', with finely cut and deeply lobed leaves, 'Brant', a fruiting hybrid with large, deeply lobed leaves turning deep purple-red in autumn except for yellow-green veins and margins, and 'Purpurea', with plum leaves turning a rich dark purple to garnet in autumn).

viviparous developing plantlets.

vivipary (1) the development of plantlets on non-floral organs, for example, plantlets on the leaves of *Kalanchoe daigremontiana* see *foliar embryo, proliferation*; (2) the germination of seeds within a fruit still attached to the parent plant, as occurs in some *Poa* species; (3) the production of living offspring, as occurs in *aphids* (q.v.).

Vriesea (for W. de Vriese (1807–1862), Dutch botanist). Bromeliaceae. C and S America. 249 species, evergreen perennial herbs with strap-shaped, glossy leaves, their bases overlapping and forming a funnel or vase. From its centre, a raceme or spike arises clothed with brightly coloured bracts and bearing small flowers. Cultivate as for *Aechmea*.

V.fenestralis (SE Brazil; leaves to 50cm, pale green netted with dark green lines, sometimes spotted red-brown and scaly beneath; floral bracts green, spotted, flowers white-green); *V.fosteriana* (Brazil; leaves to 70cm, green, banded maroon beneath; floral bracts yellow, flowers pale green tipped purple); *V.hieroglyphica* (Brazil; leaves to 80cm, broadly banded dark green or purple; floral bracts yellow-green, flowers yellow); *V.psittacina* (leaves to 50cm, green covered in pale scales; floral bracts yellow tipped green or red tipped yellow to wholly red or yellow-green, flowers yellow sometimes tipped green); *V.splendens* (E Venezuela to Surinam; FLAMING SWORD; leaves to 80cm, dark green with broad, dark purple-brown bands; floral bracts flaming scarlet to orange, flowers yellow).

× **Vuylstekeara** Orchidaceae. Garden hybrids between the genera *Cochlioda*, *Miltonia* and *Odontoglossum*. They have laterally flattened pseudobulbs, strap-shaped leaves and arching racemes of large, frilly flowers in shades of white, yellow, rose and red, usually spotted or mottled darker red. The most typical coloration is white heavily overlaid with deep velvety red, as in the very popular × *V.* Cambria 'Plush'. Cultivate as for *Odontoglossum*, although some, especially the clone mentioned here, will take higher temperatures and will thrive in the home if kept moderately moist.

W

Wahlenbergia (for Georg Wahlenberg (1780–1851) of Uppsala, author of *Flora Lapponica*). Campanulaceae. ROCK BELL. Cosmopolitan. 150 species, annual or perennial herbs with bell-shaped, 5-lobed flowers in summer. Hardy in climate zone 7. Plant in full sun on a moist, sandy and acid soil. *W.albomarginata* is more safely grown in the alpine house in damp, cool temperate gardens. Propagate by seed or division.

W.albomarginata (New Zealand; tufted perennial to 30cm; leaves to 2cm, hairy; flowers white to pale blue or pale slate blue); *W.congesta* (New Zealand; mat-forming perennial; leaves to 2.5cm, glabrous; flowers pale blue to white).

Waldsteinia (for Count Franz Adam Waldstein von Wartenburg (1759–1823), Austrian botanist, who published in cooperation with the Hungarian botanist Kitaibel an outstanding work on Hungarian flora in 1799–1812). Rosaceae. Northern temperate regions. 6 species resembling yellow-flowered *Fragaria*. Suitable for the rock garden, the front of the flower border and the woodland edge; *W.ternata* will cover banks. Hardy to at least –15°C/5°F. Grow in part shade, or sun where soils remain moist in summer. Propagate by division of rooted runners in early spring, or by seed.

W.fragarioides (BARREN STRAWBERRY; E US; leaves bronze; flowers to 2cm in diameter); *W.ternata* (leaves hairy; flowers to 1.5cm in diameter).

walk a path created primarily as a feature, possibly between trees, under a pergola or arbour, or linking dominant features in a garden. In Renaissance gardens, walks were aggrandized into alleys and avenues, and usually terminated in some focal point.

wall case a very narrow lean-to greenhouse used to protect wall-trained fruit trees, such as peaches and apricots.

wall nail a nail with a strip of soft lead attached to its head. The lead strip can be bent around wires to support climbing plants on brick walls, with the nail being driven into the mortar. It is also known as a leadheaded wall nail.

walnut (*Juglans regia*) ENGLISH WALNUT, PERSIAN WALNUT) a large tree reaching 18m or more, grown as a broad-crowned, attractive ornamental tree and for its nuts; its native area stretches from the Balkan peninsula to Kashmir. It is self-fertile and wind-pollinated, but male and female flowers may mature at different times and pollination is therefore more likely where two cultivars are established; flowers and shoots are susceptible to damage at temperatures between –2 and –3°C/28–26°F. Fruit ripening is variable throughout Britain, but immature nuts are suitable for pickling.

Plant young trees 12–18m apart, taking care to keep the taproot intact and to reject pot-bound specimens. Establishment is slow, and trees may be trained as centre leader standards for best ornamental effect, or contained, for example, as a goblet form, in smaller gardens. Pruning must only be carried out in the completely dormant season, and minimal attention is required for established trees. Walnuts for pickling are picked in summer before the shells harden; mature nuts are harvested in early autumn on release from their casings and should be stored dry in a cool, airy, slightly humid environment.

Propagation is possible from seed but is best done by chip budding or grafting cultivar scions on to BLACK WALNUT (*J. nigra*) seedlings.

Wardian case originally a wooden-sided, glazed container with a pitched roof, developed by Nathaniel Bagshaw Ward in the early 19th century for transporting actively growing plants over long distances. Widely used by Victorian plant hunters for transporting collections from overseas, it was subsequently adopted in various forms for the indoor cultivation of plants, especially those requiring high humidity; modern versions are bottle gardens and terrariums. It is also known as a fern case.

warm house see *stove house*.

walnut

Pests
aphids
caterpillars

wart disease *(Synchytrium endobioticum)* a destructive fungal disease of potato, highly localized worldwide. It was first formally recorded in 1869 in Czechoslovakia and England, and has since been recognised in most of north and central Europe, the Indian subcontinent, New Zealand and South Africa. Wart disease probably originated in South America, where it occurs. It has been successfully eradicated in the eastern US, following outbreaks from 1918 onwards, and in North America is only established in Newfoundland.

The fungus is soil-borne and can survive there for up to 30 years; resting bodies germinate in wet conditions giving rise to infection of the 'eyes' and young sprouts of potato tubers early in the growing season. Cells near to the site of infection enlarge and irregular cauliflower-like galls develop on the seed tuber, stems, young tubers and sometimes aerial stems and foliage. Underground, the galls are whitish and turn green if exposed to light. The disease flourishes in wet seasons, when it can result in total loss of crop; it is also spread in soil on machinery, tools, and footwear, and the manure of animals fed with diseased tubers can also be infective.

In Britain, the control of wart disease has been effected by the growing of resistant cultivars, available through the development of government-certified seed programmes; the majority of cultivars now grown are of the immune type, selected during breeding by official laboratory testing. The disease is rare in commercial crops and only of limited occurrence in private gardens and allotments. Of non-immune cultivars, only 'King Edward' and 'Red King Edward' are grown to any extent in Britain, but these are of limited susceptibility. Although permitted, it is not advisable to grow non-immune cultivars. Extreme vigilance is necessary, especially as regards the appearance of new races of the pathogen; these exist on the continent of Europe, in the ex-Soviet republics and in Newfoundland, and can infect immune cultivars. The Department of Agriculture should be notified of all outbreaks.

Most countries impose strict import regulations on soil and, especially, on seed potatoes.

Washingtonia (for George Washington (1732–1799), first president of the US). Palmae. SW US, N Mexico. 2 species, tall palms with columnar stems, scarred below and clothed above with coarse, rusty fibres. The leaves are large and palmate, composed of numerous, dark green, linear segments. They are frequently seen as avenue trees in the dry tropics and subtropics. They show great tolerance of transplanting, despite their deep root system, and grow rapidly even when young. Grow in frost-free conditions with full sun and a free-draining soil. Water container-grown plants frequently in spring and summer. Propagate by fresh seed.

W.filifera (COTTON PALM, DESERT FAN PALM, CALIFORNIAN WASHINGTONIA; S California, Arizona, N Mexico; to 15m; petioles to 2m, toothed at base, leaf blades to 2m, segments tipped with white filaments); *W.robusta* (SOUTHERN WASHINGTONIA, MEXICAN WASHINGTONIA; NW Mexico; to 25m; petioles toothed throughout, leaf blades to 1m, segments drooping, with white filaments when young).

wasps winged insects of the order Hymenoptera, frequently troublesome in gardens. The common wasp *(Vespula vulgaris)* is the best known of seven species of social wasp, which include the hornet *(V. crabro)*; the adult is approximately 15mm long and brightly banded in yellow and black. Wasps perform a beneficial function in spring and early summer, feeding their grubs mainly on insects, many of which are pests of garden plants. However, they cause serious damage to top fruits and grapes in late summer, usually following primary damage by birds; dahlias are sometimes attacked by gnawing the stems. In some seasons, wasps are a serious pest of honey bee hives.

The overwintered queen leaves hibernation in spring and builds a nest of a paper-like material, made by chewing small pieces of wood; the nest is constructed as an outer covering protecting combs of downward-facing cells, into each of which an egg is laid. During mid- to late summer, males and young queens are produced and the fertilized queens fly off to hibernate. A reduced number of female workers and males are active until killed by frosts in late autumn. Nests are not recolonized the following year.

Control of wasps in and around gardens is by destroying nests with recommended chemicals, applied strictly according to instructions; where possible, trained operators should be engaged.

watercress *(Nasturtium officinale)* a hardy aquatic and wet-soil perennial, native throughout Europe, especially in limestone areas that give rise to pure springs running at a constant temperature of around 10°C/50°F. It is grown as a salad vegetable for its pungent tasting leaves, which are rich in vitamins and minerals, particularly iron.

The availability of suitable free-flowing uncontaminated water in gardens is rare. If special care is taken to keep the crop always moist, watercress can be successfully grown in a trench, 30cm deep and 60cm wide; prepare with a 10cm layer of rotted organic matter topped with a similar depth of fertile soil. Cuttings from purchased bunches root easily in a container of water, and these are set out 15cm apart; pinching out the growing tips encourages bushy growth.

Watercress can also be grown in well-crocked 20cm pots, filled with loam-based growing compost; plant a number of well-rooted cuttings per pot. Keep thoroughly watered by standing container in a shallow tray of water and replenishing daily. Protect all crops from wind and frost to maintain good quality.

water gardening there are records of decorative pools and water features dating from 3000 BC in ancient Egypt. The Chinese grew water lilies in pools, lakes, urns and vases by the 10th century AD, and water was a vital element in the design of Persian and Indian gardens. In succeeding centuries, water was

watercress

Pests
aphids
flea beetle

Diseases
virus

Washingtonia filifera

used primarily for the effect of stillness and its mirroring qualities; fountains and waterfalls were used to add movement to gardens. Plants seldom featured until the early 18th century, and the cultivation of tropical and temperate water lilies followed. The type of water gardening in which cultivation of plants and care of fish predominate did not develop in Europe and North America until the end of the 19th century.

Siting A pool should be sited in a sheltered position, where cold air does not accumulate, and away from trees so that aquatic plants and fish can benefit from full sunshine. Even a shallow pool can present danger to small children, and should be sited within visible range to enable supervision. For good effect, construct the pool at a naturally low point, carefully distributing excavated soil to emphasise a low spot.

Types of pool An informal pool may take any irregular non-angular shape; a few sweeping curves are more attractive than a number of smaller ones. It should be edged with natural materials, such as stones or turf, with marginal moisture-loving plants established to soften the outline. Blending with a rock garden creates an impressive ornamental feature.

A formal pool may be raised or sunken, incorporating coping or paving stones and made to a geometric shape. Such features usually look best with fewer plants, in particular species of more rigid form; fountains are valuable complements. Formal pools make good focal points, and can be attractively situated near to a wall where water spouts and sculptures can be added.

In small areas, such as a patio, it is possible to construct a water feature by utilizing a large ornamental container and planting it with a miniature water lily and other dwarf species, such as *Typha minima*, the DWARF REED MACE, and *Mimulus ringens*, together with floating plants that will need reducing from time to time.

On spacious sites, a bog garden can be created adjacent to a pool by laying impermeable sheeting shallowly beneath an absorbent, highly organic growing medium. This should be kept constantly moist, either using the pool or by means of buried irrigation lines.

Choice of plants Plants for water gardens may be considered under group headings. **Oxygenators** are submerged fast-growing species, usually established from a bunch of cuttings planted in an individual pot container. They help keep a pool clear of algae and are essential if fish are stocked, for which purpose *Myriophyllum* and *Lagarosiphon* species are particularly effective. **Deep water plants** succeed in water depths from 30–90cm, and are most commonly represented by the water lilies (*Nymphaea* species). *Aponogeton*, *Nymphoides* and *Orontium* species are also suitable. **Surface floaters**, such as *Lemna trisulca* and *Trapa natans*, add character and balance but must be prevented from completely covering the water surface. Marginals include the aquatic irises and dwarf rushes. Bog plants, such as *Caltha* and *Myosotis* species, thrive in waterlogged soil.

Construction Pools can be made with flexible liners of synthetic rubber or plastic; butyl rubber is more durable than polythene or PVC and its long life justifies its expense. To obtain the correct liner dimensions, take the maximum width of the intended pool plus twice the maximum depth, by the maximum length plus twice the maximum depth. To allow for a turned-over flap at the finished edges, an extra 15cm should be left round the edge of the liner sheet.

The pool site may be marked out with a rope or hosepipe. The hole should then be excavated with sloping sides and with shelves constructed at various depths to accommodate a range of plants; ensure that the finished edges are perfectly level all round to prevent seepage. The completed hole must be lined with fibreglass wadding or plastered with fine sand to protect the sheet from puncturing. The liner is then draped over the hole, anchored all around with large stones and slowly filled with water. Surplus sheet should be trimmed and the pool edged with stones protruding above the water by 5cm or so. Pre-formed rigid fibreglass or plastic liners are available, and care must be taken to ensure that these are bedded level all round.

By incorporating moving water and fish and carefully maintaining a balanced population of plants of various, contrasting forms, a special garden feature can be created that is both ornamental and restful to observe, as well as being attractive to wildlife.

water-holding polymers polyacrylamides, polyvinyl alcohols and starches, specifically manufactured to increase water retention in soil or substrates. They are of particular value in hanging baskets and patio containers.

watering can a spouted can for watering plants, made either of galvanized iron or plastic. See *irrigation*, *rose*.

waterlogging the state of growing media when saturated with water. It may result from compaction, natural defects in the soil profile, the lie of the land or other factors which permit no drainage to occur. It is a common cause of root death. See *drainage*.

water shoot a vigorous upright shoot growing from an adventitious bud on the upper surface of a tree or shrub branch, especially near a large pruning wound. In fruit trees they are unproductive and often congesting, but are sometimes used for bending to encourage fruit bud formation, or as replacement shoots in some pruning systems.

water soaked describing plant lesions that are soft and watery, such as those caused by *Sclerotinia* or bacterial soft rots.

waterworks various systems for displaying water motion in lakes, pools, canals and formal basins. Most commonly employed are fountains and cascades, but there is also a wide range of other contrivances, including the modern bubbler.

pea and bean weevil

Watsonia (for Sir William Watson (1715–1758), British scientist). Iridaceae. South Africa. 52 species, cormous perennial herbs with sword-shaped leaves. Produced in summer in 2-ranked spikes, the flowers are tubular and curving with six tepals. Slightly frost-tender and more safely grown in the cool greenhouse in cool temperate zones; however, in areas where temperatures only rarely fall to –5°C/23°F, most species can be grown outside with a protective mulch of bracken litter. They are generally best left undisturbed, but spring- and early summer-flowering species that rest during late summer are sometimes lifted and stored in a cool, frost-free place in winter. The late summer and autumn-flowering species are more or less evergreen and should not be dried out. Grow in a sunny position on a light, well-drained, sandy loam. Provide plentiful water when in growth. Apply a slow-release fertilizer in summer. Propagate by seed or offsets.

W.borbonica (syn. *W.pyramidata*; 1–2m; flowers to 6cm across, slightly scented, pale to deep pink or purple with white lines at the base of each tepal, rarely wholly white); *W.fourcadei* (1–2m; flowers to 6cm across, pink, orange or vermilion, rarely purple, sometimes with a paler tube); *W.meriana* (0.5–2m; flowers to 4cm across, orange, red, pink, purple, rarely yellow); *W.pillansii* (syn. *W.beatricis*; 0.5–1.2m; flowers to 6cm across, orange or orange-red).

wattle a construction of pliable, woody plant stems, for example, young willow or hazel branches, mostly used in the form of hurdles for fencing.

weaning the gradual acclimatization of vegetatively propagated plants from a rooting environment, especially used of cuttings rooted in a mist unit and of micropropagated plants raised by tissue culture. It is sometimes referred to as acclimatization.

webbers, webworms see *tent caterpillars*.

weed a plant growing where it is not wanted. Weeds are found in habitats disturbed by human activity and are usually native species, although some garden plants and lawn grasses can become weeds because of their free-seeding habit or invasive growth. Annual weeds complete their life cycle in one year, and many produce several generations during that time; perennial weeds perpetuate themselves from year to year by means of rhizomes, bulbs, tubers and taproots.

weed wiper a device for smearing individual weeds with herbicide, often in the form of a purpose-made glove. It is particularly suitable for applying translocated herbicides on a small scale.

weep holes small round drainage holes set into the base of a brick or block retaining wall to provide drainage.

weeping (1) describing the growth habit of trees and shrubs with pendulous branches or twigs; (2) used of the phenomena of bleeding and gumming.

weevils (Coleoptera: mostly Curculionidae) the largest family in the animal kingdom, with over 40,000 described species. The head of a weevil is elongated at the front to form a snout, with the mouthparts at the tip and with the antennae, which are often elbowed and with a terminal club, in varying positions along its length. Some weevils have long snouts that facilitate feeding; females also use their snouts to make deep incisions in plant tissue in which they deposit eggs. The larvae are all plant feeders; they are remarkably uniform, the fleshy grubs being off-white in colour, with a well-developed yellow-brown head capsule, and devoid of legs. Some weevil larvae feed openly on foliage, or in rolled leaf portions or as leaf miners; many are root feeders; still others are wood or shoot borers, seed or fruit eaters, or live in galls on roots or stems. When fully fed the larvae pupate, usually in the soil, but sometimes within the tissues of the host plant.

The BOLL WEEVIL (*Anthonomus grandis grandis*) is the most serious pest of cotton in the US, while other *Anthonomus* species attack garden plants. The larvae of the STRAWBERRY BLOSSOM WEEVIL (*A. rubi*), a European species, feed in wilted buds; the adults are black and up to 3mm long. The STRAWBERRY BUD WEEVIL (*A. signatus*) of North America attacks strawberries in a similar way. The larvae of another European species, the APPLE BLOSSOM WEEVIL (*A. pomorum*), feed in apple blossom buds; the adults are up to 4mm long, with a white V-shaped mark on their backs. In the US, the larvae of the PEPPER WEEVIL (*A. eugenii*) destroy pepper buds; the adults are black and up to 3mm long. Also important in North America is the PLUM CURCULIO (*Conotrachelus nenuphar*), whose brown weevils, up to 6mm long, feed on foliage and puncture the fruit of stone fruits, apples and pears, while the larvae bore into fruit. Larvae of the BLACK WALNUT CURCULIO (*C. retentus*) tunnel into walnuts; the BUTTERNUT CURCULIO (*C. juglandis*) causes similar damage to walnuts and butternuts; and the larvae of the QUINCE CURCULIO (*C. crataegi*) invade the fruits of quince. Other North American weevil pests include the CARROT WEEVIL (*Listornotus oregonensis*), whose larvae bore into the roots and stems of carrots, celery, parsley and parsnips, and the VEGETABLE WEEVIL (*Listorderes costirostris obliquus*), whose adults and larvae feed on the foliage and roots of various vegetables including carrots, turnips, potatoes and tomatoes.

Species prevalent in Europe include the APPLE TWIG CUTTER (*Rhynchites caeruleus*) whose bright blue weevils, up to 4mm long, attack apple and other fruit trees, depositing eggs in shoots that are then partially or completely severed. The APPLE FRUIT RHYNCHITES (*R.aequatus*) are small, red-brown weevils that deposit eggs in apple fruitlets into which the larvae then tunnel. The STRAWBERRY RHYNCHITES (*R.germanicus*) are metallic blue-green weevils, up to 3mm long, which sever the leaves and blossoms of strawberries and other plants, such as raspberry, blackberry, geums, potentillas and helianthemums. The larvae of the NUT WEEVIL (*Curculio nucum*) feed in the kernels of hazel nuts. The TURNIP GALL WEEVIL (*Ceuthorrhynchus pleurostigma*), a dull black weevil up to 3mm long, has larvae that live in marble-shaped galls just below soil level on the taproots of brassicas, especially cabbages, savoys, turnips and swede.

PEA AND BEAN WEEVILS (*Sitona species*) are common in both Europe and North America. The adults are black with yellow-brown scales that give a striped appearance. They feed on the leaf margins, removing symmetrical U-shaped notches from broad beans and, to a lesser extent, from garden peas and sweetpeas; the larvae feed in the root nodules. Other species of weevil are *leaf miners* (q.v.), *leaf weevils* (q.v.), leaf rovers and wingless weevils.

Wingless weevils Several medium to large weevils are flightless, with absent or vestigial wings, and with a comparatively short snout. The most important are *Otiorhynchus* species, whose largely nocturnal adults feed on leaves, buds, and shoots, while their larvae attack underground parts of plants. The larvae have slightly curved, creamy white, legless bodies, with red-brown heads, and may be up to 12mm long, according to species; when fully fed, they pupate in the soil. The principal pest species common to Europe and North America are the VINE, or YEW, WEEVIL (*O. sulcatus*), with black adults up to 9mm long; the CLAY-COLOURED WEEVIL (*O. singularis*), with brown adults up to 7mm long; and the STRAWBERRY ROOT WEEVILS (*O. nugostriatus* and *O. ovatus*), with adults similar to the vine weevil but smaller. Other major European species include another STRAWBERRY ROOT WEEVIL (*O. rugifrons*), and the RED-LEGGED WEEVIL (*O. clavipes*), which has black adults up to 13mm long, with red legs.

Clay-coloured weevil adults attack a wide range of plants, including apples, currants, raspberries, rhododendrons, roses, spruce and other conifers, causing ring barking and damage to buds and new growths. The larvae are less injurious than those of the vine weevil, but they are difficult to distinguish from one another. The strawberry root weevils are common in woodlands and hedgerows but seldom damage garden plants except strawberries, whose roots the larvae may sever. Red-legged weevil adults attack soft and top fruits and roses, and their larvae damage the roots of strawberry and pot plants.

The STRAWBERRY CROWN BORER (*Tyloderma fragariae*), with dark brown adults up to 4mm long, is an important pest of strawberries in the eastern US; its larvae bore into and may destroy the crowns of the plants.

Methods of weevil control include clearing plant debris, in which adults shelter, trapping with corrugated paper or sacking, grease banding the trunks of young tees, and, for strawberries, long rotation and burning off foliage after cropping. Recommended persistent contact insecticides can be used in potting composts, or as a spray when leaf damage is first noticed. The parasitic eelworms *Heterorhabditis heliothis* and *Steinernema bibionis* are useful biological control agents, and the fungus *Metarhizium anisopliae*, applied as a spore drench, may prove to be effective. See *vine weevil*.

Weigela (for Christian Ehrenfried van Weigel (1748–1831), German botanist, professor at Greifswald). Caprifoliaceae. E Asia. 10 species, deciduous shrubs with corymbs of bell- to funnel-shaped, 5-lobed flowers in spring and summer. Hardy to –23°C/–10°F. Plant in fertile, moist but well-drained soil, preferably in sun, or in partial shade. Prune after flowering to maintain vigour, thinning out exhausted shoots to ground level, and shorten the strongest shoots to within a few centimetres of old wood. Propagate in early summer from softwood basal cuttings rooted in a sandy propagating mix in the cold frame, or in summer by semi-ripe cuttings with a heel, at 15°C/60°F. Alternatively, take 20cm hardwood cuttings in autumn after leaf fall, and place in a sheltered nursery bed.

W.florida (Korea, N China, Japan; to 3m; flowers to 4cm, dark rose red, paler within; includes numerous hybrids and cultivars, small or large with variegated, dark green or purple-red-tinted leaves and white, pink, or dark carmine flowers); *W.middendorfiana* (Japan, N China; to 1.5m; flowers to 4cm, yellow); *W.praecox* (Japan, Korea; to 2m; flowers to 4cm, purple-pink to carmine; includes tall and short cultivars, some with variegated leaves, early-flowering with white, pink, carmine and red flowers, the exteriors often darker).

Weigela florida

Weldenia (for L. von Welden (1780–1853), Austrian soldier and student of the Alpine flora). Commelinaceae. Mexico, Guatemala. 1 species, *W.candida*, a tuberous-rooted, perennial herb, stemless with a basal rosette of linear to lanceolate leaves 5–20cm long. Produced in clusters in spring and summer, the flowers are 5–6cm long and white or blue with ovate segments. A choice perennial for the alpine house, plant in a deep pot of freely draining loam-based medium. Keep dry but not arid during winter rest. Resume watering in spring and water freely in summer, reducing supplies as the leaves wither after flowering. Propagate by seed or division.

well-head the structure built at the top of a water well, often finely carved and used as a garden ornament.

wettable powder a pesticide formulation in which the active chemical constituent is adsorbed on to an inert carrier, ground to a fine powder and a surfactant added; the resultant product is insoluble and forms a suspension when vigorously stirred in water to form a substance for spraying.

wetter, wetting agent see *adjuvant*.

whalehide a trade name, now in common use for the bitumenized cardboard used to make plant containers, especially bottomless pots for the ring culture of tomatoes. Some forms of whalehide pots degrade more quickly than others and are suitable for leaving intact at planting time.

wheelbarrow a large open-topped container, with sloping sides, mounted on a frame, at one end of which is an axle usually bearing one wheel. At the opposite end the frame is fashioned to form two handles, below which are two legs for supporting the container when stationery. It is used for transporting soil, bulky manure or rubbish. See *ball-barrow*.

whetstone a shaped stone for sharpening cutting tools, especially scythes and sickles.

whip a young seedling or grafted tree bearing no lateral growths.

white blister (*Albugo* species) a fungal disease confusingly sometimes called white rust, which is the accepted common name of the true rust, *Puccinia horiana*, on chrysanthemum. *Albugo* is closely related to the downy mildews, and similarly encouraged by damp weather. It can overwinter in the remains of infected plants by means of resting bodies.

Albugo candida infects members of the Cruciferae family and has strains restricted to certain hosts. Incidence of the disease varies from year to year on cultivated plants, but the following vegetables are prone to infection: brassicas in general, horseradish, radish and watercress. Ornamental hosts include alyssums, aubretias, honesty and wallflower, and the disease is also commonly seen on the weed shepherd's purse *(Capsella bursa-pastoris)*. The first signs of infection are small raised white pustules on the foliage, which spread in concentric rings and become powdery; the edge of the infected area may turn blackish as the infection progresses and, in severe infections, distortion of the host occurs. It may be necessary to destroy affected plants to check spread, and frequent protective treatment with a recommended fungicide may be warranted.

A. tragopogonis infects members of the Compositae, on which it causes yellow spotting followed by white pustules and stunting of plants. It is the most important disease of salsify and scorzonera and there are no effective control measures. Gerberas, feverfew, ragwort and sunflower are also hosts.

white bud see *bud stages*.

whitecurrant see *redcurrant*.

whiteflies (Hemiptera: Aleyrodidae) small, sap-sucking, moth-like insects, up to 2mm long, with two pairs of rounded wings covered with a white, powdery wax. All stages are found on the underside of leaves, the elongated, slightly curved eggs suspended by a short pedicel and, in several species, deposited in complete or partial circles. In all

whitefly

species, both adults and larvae excrete large quantities of honeydew that fall on underlying leaves; a heavy infestation weakens affected plants, which become coated with unsightly black sooty moulds. Some species of whitefly cause leaf spotting.

The GREENHOUSE WHITEFLY (*Trialeurodes vaporariorum*) is a cosmopolitan species living under glass and, in warmer climates, also outdoors. It attacks a very wide range of plants. The adults live for about a month, and the length of life cycle varies according to temperature, being about three weeks at 20°C/70°F. Other species common to Europe and North America include the RHODODENDRON WHITEFLY (*Dialeurodes chittendeni*), which attacks smooth-leaved rhododendrons, and the AZALEA WHITEFLY (*Pealius azaleae*), which infests hairy-leaved evergreen azaleas. In Europe, brassicas may become infested with the CABBAGE WHITEFLY (*Aleyrodes proletella*), and *Viburnum tinus* with the VIBURNUM WHITEFLY (*Aleurotuba jelinekii*), which overwinters as conspicuous black pupae fringed with a frill of white wax. Several other species are present in North America, including the TOBACCO WHITEFLY (*Bemisia tabaci*); the CITRUS WHITEFLY (*Dialeurodes citri*), which is one of several species that attack citrus.

Infestations should be controlled with recommended contact or systemic insecticides, applied as soon as symptoms are first noticed. To guard against resistance, from time to time use sprays from different insecticide groups. The GREENHOUSE WHITEFLY (*T. vaporariorum*) may be controlled biologically by introducing the parasite *Encarsia formosa*, a minute, winged, chalcid wasp now widely available to gardeners. The parasite is self-perpetuating at temperatures above 24°C/75°F, and it is likely to die out over the winter period.

whorl three or more organs arranged in a circle at one node or around the same axis.

wide-span house a greenhouse with a span of 18m or more.

white rot (*Sclerotium cepivorum*) affects onions, garlic and to a lesser extent shallots and leeks. First symptoms are yellowing of the foliage and wilting, due to root destruction. White, fluffy mycelium develops around the neck of the bulb, and the bulb eventually shrivels. Resting bodies of the fungus form to perpetuate the disease in the soil. Contaminated land must be avoided, and care taken to destroy infected tissue and to prevent movement of the soil from an infected site.

widger a small proprietary tool shaped like a tapering spatula, for use in pricking out seedlings.

wild and woodland gardening a form of gardening based on the concept of using indigenous or appropriate exotic plants in an informal way, the aim frequently being to create, under a controlled regime of good husbandry, an image of the surrounding natural environment. It was popularized from the late 19th century by William Robinson, Gertrude Jekyll and others, but its progenitors can be traced back to medieval times, when informally managed areas were allowed to flourish within the ramparts of castles and fortified dwellings. In the US, many woodland gardens were started in the late 20th century.

Site selection is of fundamental importance in the creation of the successful woodland garden, and consideration must be given to the composition and maturity of the existing area. In Britain and mainland Europe, oaks *(Quercus robur* and *Q. petraea)* provide a good environment due to their longevity, deep-rootedness and relatively open canopy, which creates an ideal balance of shelter and shade. Occasional trees of beech or ash will not be detrimental but large numbers are likely to result in too dense a canopy, while the shallow root systems of beech and the spreading roots of ash can suppress understorey plantings. Excellent woodland gardens can be created in pine woods, but these may prove too dark in late winter and early spring when many plants are growing vigorously; judicious thinning is advisable.

The native woodlands of eastern North America offer a suitable selection of canopy species, with a wide range of oaks as well as black walnut, butternut and forest species of hickory. In the west, the available native woodland is largely dominated by coniferous species, which can provide excellent growing conditions.

Mature woodland is always to be preferred for the creation of a woodland garden, and in its design, informality is paramount. The aim should be to develop three tiers of vegetation: a canopy of large trees; a woody understorey of shade-demanding shrubs and small trees, such as rhododendrons; and hardy herbaceous and bulbous plants. Existing trees must be healthy and structurally sound, and any necessary removal or thinning undertaken as a first step. Crown-thinning, where appropriate, will increase the levels of light without substantially altering the appearance of the woodland. There must be a balance between wooded and open areas, since not all plants require the same degree of shade; glades provide a vital change of environment, which adds to the woodland garden experience. Bold groups of herbaceous plants, added in drifts amongst the woody plantings, extend interest, as well as providing weed-suppressing groundcover. Suitable species include primroses *(Primula vulgaris),* snowdrops *(Galanthus nivalis),* wood anemones *(Anemone nemorosa),* solomon's seal *(Polygonatum),* lungwort *(Pulmonaria),* comfrey *(Symphytum)* and hellebores *(Helleborus).*

Before undertaking bed preparation, efforts must be made to eradicate perennial weeds. If the soil is poorly structured and impoverished, as is frequently the case where coniferous species predominate, ground preparation must be thorough. On a small scale, single digging is adequate, but where larger beds are required, it is advisable to use machinery, such as a pedestrian-operated rotary cultivator. In all cases, generous quantities of organic matter must be incorporated into the soil. After planting, the beds should be mulched to a minimum depth of 50mm. Leafmould is particularly suitable as a mulch if available in sufficient quantity; alternatively, very well-rotted manure, garden compost, bark or woodchips may be used. During establishment, new plantings will need regular watering, with repeated applications to beds of herbaceous plants and bulbs during spring and summer.

wilderness an area of a garden allowed to develop with minimum control, often contrived by dense planting of shrubs and trees crossed by a maze of intersecting paths. Wildernesses were popular from the mid 17th to the 18th century.

wilding a self-sown plant.

× **Wilsonara** Orchidaceae. Garden hybrids between *Cochlioda, Odontoglossum* and *Oncidium.* They have laterally compressed pseudobulbs, more or less strap-shaped leaves and tall, branching sprays of small flowers with crisped tepals in shades of yellow, cream and white marked brown to deep mauve. Cultivate as for *Odontoglossum.*

wilt used to describe a collapsed condition in a plant evidently due to disease. Various species of plant pathogens are described as wilt, and the following are commonly of concern to gardeners:

ASTER WILT *(Fusarium oxysporum* f.sp. *callistephi)* China asters are susceptible to this rapid wilting disease, where the stems are found to be blackened, frequently covered with white or pinkish fungal growth. The disease is most prevalent on poorly-drained soils and is persistent. Control is by prevention through the avoidance of planting infected stock or planting on contaminated sites, rotation, and destruction of infected plants. Resistant cultivars are available.

BLOSSOM WILT a symptom of fungal infection on top fruits, notably by brown rot fungus *(Monilinia* species).

CLEMATIS WILT *(Aschochyta clematidina)* Clematis, especially Jackmanii types, some other large-flowered cultivars and *C.montana,* may be affected by wilting or die-back of shoots. Although cultural effects are possible fungal infection is the major cause. The fungus enters lower parts of the plant and symptoms develop from a flagging of upper shoots, blackening of leaf stalks to general collapse, although the plant is rarely killed outright in one season. Wilted shoots should be cut back to remove apparently infected tissue, and new shoots should be sprayed with a recommended fungicide.

FUSARIUM WILT *(Fusarium bulbigenum* var *lycopersicum)* used to describe a troublesome disease of tomatoes, the fungus entering roots from the soil and giving rise to brown discolouration of the root vascular system. Yellowing and wilting of the aerial parts follow. Soil sterilization or container growing are control methods.

PAEONY WILT (*Botrytis paeoniae*) a common cause of die-back and often death if the plants are sufficiently weakened: Shoots turn brown from spring or early summer, in wet seasons especially and under some conditions fluffy fungal growth grows profusely at the stem bases. Resting fungal bodies may be formed to perpetuate infection in the soil. Infected tissue should be cut out as soon as noticed, and destroyed. In herbaceous paeonies this requires cutting below soil level. Sprays of a recommended fungicide may be worthwhile.

VERTICILLIUM WILT a soil-borne fungal disease most frequently caused by *Verticillium albo-atrum* and *V.dahliae*, which has a very wide host range and is most troublesome on young plants. *V.dahliae*, usually included in *V.albo-atrum* in the US, is mostly responsible for damage to woody plants and resting bodies may remain viable in the soil for several years even in the absence of host plants. Symptoms appear first on older leaves as yellowed or dull foliage, later becoming necrotic and desiccated and the leaves may curl and bend downwards. The wilting symptom is initially reversible, but as the disease takes hold permanent wilting occurs and the vascular strands of plants become stained, and some plants, such as tomato, produce adventitious roots on the stem. Damage to trees and shrubs can be divided into two: acute wilt where there is a sudden wilting and die-back of the foliage, with leaves remaining on the plant, and chronic wilt which is slower or progressive, with often one-sided, wilting of the leaves, crown die-back over several years and early leaf fall. *Acer* is particularly susceptible and other commonly affected woody plants include ash, beech, *Cercis*, *Cotinus*, *Cornus*, elm, laburnum, *Rhus*, *Robinia*, rose and *Rubus*; also herbaceous plants such as *Antirrhinum*, aubergine, *Aster*, chrysanthemums, *Dahlia*, *Fuchsia*, mint and tomato.

Complete removal of all affected debris and root material is vital, together with good weed control as several broad-leafed weed hosts allow it to persist. There is no reliable chemical control and affected plants are best removed and burned. The soil should be changed or sterilized and susceptible plants avoided.

wilting the loss of turgor in plant cells that results in drooping and possibly eventual death of plant parts. It occurs when transpiration exceeds water uptake, and can result from lack of water, or from the withdrawal of water by parasites. Wilting may be a symptom of disease or disorder caused by damage to roots and/or blocking of the vascular tissue.

wind-rocking root disturbance caused by the movement of plants in wind; a potential cause of long term instability and root damage.

wind-throw the uprooting of plants, particularly trees, by wind.

windbreak, windscreen a barrier established to reduce the force of strong wind, thereby preventing physical damage to plants and structures; it also reduces water loss through transpiration and evaporation. A windbreak may consist of living plants or take the form of a fabricated screen. The most effective barriers are those that are 60% permeable, since damaging eddies occur to leeward and windward of a solid barrier. A windbreak is most beneficial for a distance to leeward of up to 10 times its height, and it affects wind speeds up to 1½ times its height.

windfall fruit that drops to the ground because of wind or ripeness; occasionally, also used of trees or branches brought down by wind.

window box a plant display container, primarily designed for use on external window ledges or narrow balconies and therefore normally narrowly oblong. Window boxes are also suitable for the sides of paved terraces, or for placing within or on top of porches. For an attractive display, they can be planted in a similar way to hanging baskets. Regular attention should be paid to deadheading, watering and supplementary feeding.

wing (1) a thin, flat, often membranous extension or appendage or an organ; (2) a lateral petal of a papilionaceous flower.

winged pyramid a decorative form of fruit tree, trained on a metal framework to form a conical shape.

winter annual an annual plant that germinates in the autumn and overwinters and dies during the following summer.

winter garden (1) an outdoor garden planned for attractiveness in winter, utilizing flowering plants, including many fragrant sorts, as well as plants with coloured stems or conspicuous bark or fruit and evergreen foliage; (2) an entire garden under shelter, taking the form of a large-scale conservatory, with paths between extensive beds and trees sited among other plants. Winter gardens were popular in the 19th century.

winter moths (Lepidoptera: Geometridae) a group of moths, active between September and March, in which females are wingless and males winged. Female moths emerge from pupal cases in the soil, crawl up the trunks of host plants and deposit eggs either singly or in groups, depending upon species. The caterpillars are of the looper type, known in North America as cankerworms or inchworms. They feed openly within loose silk webbing, on the buds, flower trusses and foliage of various trees and shrubs. As well as causing extensive defoliation and damage to flowers buds, they also bite holes in the fruitlets of fruit trees; these callous over and, on mature fruits, appear as corky areas, often developing into scarred clefts. Apples, pears, plums, cherries and their ornamental cultivars are frequently attacked, as well as many ornamental trees and shrubs, especially beech, *Cotoneaster*, dogwood, elm, hawthorn, hazel, hornbeam, lime, rhododendrons, rose, sycamore and willow. Fully-fed caterpillars drop to the ground in summer on silken threads and pupate in the soil to begin the cycle again.

One of the most prevalent species in Europe and North America is the WINTER MOTH, *Operophtera brumata*, whose green looper caterpillars are up to 25mm long, with a light green or yellow stripe along each side and one along the back. The MARCH MOTH, *Alsophila aescularia*, is a European species with more slender caterpillars; the closely-related FALL CANKERWORM (*A. pometaria*) occurs in North America. Both species lay eggs as a bracelet of about 100 eggs encircling a twig. March moths appear in late winter to early spring and the fall cankerworm in winter. The MOTTLED UMBER MOTH (*Erannis defoliaria*) is a common European species with conspicuous red-brown and yellow caterpillars, up to 30mm long, which feed more openly than the other species. A close relative, the LINDEN LOOPER (*E. tiliaria*), occurs in North America. Also prevalent there is the SPRING CANKERWORM (*Paleacrita vernala*), the moths of which appear in spring followed by slender larvae coloured light green, brown or black with white lines down the back.

Winter moths can be controlled, where feasible, by grease banding tree trunks in autumn to prevent wingless females ascending to deposit eggs. Alternatively, apply recommended contact insecticides just as the buds open.

winter wash a tar distillate spray, manufactured from hydrocarbon and phenolic oils, applied to dormant deciduous plants in winter to kill hibernating insects and overwintering eggs, and to clean off lichen and other superficial growths.

wirestem used to describe the damping-off of brassicas caused by the fungus *Rhizoctonia solani*, which shows as a characteristic constricted red-brown stem lesion at soil level. See *damping-off*.

wireworms (Coleoptera: Elateridae) the larvae of click beetles, or skipjacks, which, when turned on their backs, right themselves using a mechanism under the thorax that throws them into the air with an audible click. The beetles are dark brown or black, elongate and oval in shape, tapering towards the hind end, and seldom exceed 13mm in length. *Agriotes* species are the most important pests in Europe. The COMMON CLICK BEETLE (*A. lineatus*) and the GARDEN CLICK BEETLE (*Athous haemorrhoidalis*) are often the most prevalent species in gardens. In North America, the WHEAT WIREWORM (*Agriotes mancus*) and *Limonius, Ctenicera,* and *Conoderus* species are the principal pests.

Their natural habitat is permanent grassland, but females also lay eggs in cereal crops, and weedy soil. The larvae are shiny golden-brown, cylindrical,

hard-bodied grubs, up to 25mm long, with strong mandibles, three small pairs of thoracic legs, and a characteristic projection on the underside of the terminal segment. Larvae of *Agriotes* species are distinguished by a dark spot on each side of the terminal segment. The larvae feed in the soil for up to five years, depending on species, and attack a wide range of plants, including anemones, carnations, chrysanthemums, dahlias, gladioli, primulas, brassicas, beans, beet, carrots, lettuce, onions, potatoes, strawberries, sweet corn, tomatoes, and turnips. Feeding occurs in bouts, peaking in spring and autumn. Seedlings and young plants are particularly vulnerable, the wireworms attacking roots and stems at or just below soil level, causing wilting and death. Potato tubers are tunnelled deeply and the base of tomato stems bored into and the pith invaded; the fibrous roots of fruit trees and nursery stock can also be damaged.

Infestation is most troublesome in the first few years after the cultivation of grassland, and also in weedy arable land. Such areas should be ploughed or dug early in the season and cultivated several times. Repeated, clean cultivation of vegetable and fruit areas and rigorous weed control can deter egg-laying. Drenches, sprays, or granules of recommended insecticide applied to the soil before replanting or sowing can be effective, and suitable seed dressing will protect some vegetable crops during early growth. Potatoes and carrots should be lifted early from affected sites.

Wisteria (for Caspar Wistar (1761–1818), Professor of Anatomy at the University of Pennsylvania). Leguminosae. E Asia, E US. 10 species, deciduous, twining shrubs with pinnate leaves composed of ovate to oblong leaflets, and pendulous racemes of pea-like flowers in spring and summer. Hardy climbers suited to walls, arches, arbours and pergolas. With careful pruning and a little support, *W.floribunda* and *W.sinensis* can also be grown as 'free-standing' lawn specimens. Equally, they may be left unhindered to scramble through tall trees. Plant in a sunny position, but with protection from early morning sun in frosty periods and with shelter from cold winds. All species are hardy to at least –15°C/5°F, but are more reliably so given the additional warmth of a wall. Grow in a deep, fertile, well-drained but moisture-retentive soil.

Prune to encourage the formation of flowering spurs close to the main framework of the plant. Once the desired framework is achieved cut back all shoots to 5–6 buds in midsummer, about two months after flowering; any subsequent twining growth and late-flowering shoots are again cut back to 2–3 buds in late autumn or winter to form flowering spurs for the following season. The flower buds are quite apparent in winter, being rounder and plumper than the vegetative buds. Any basal shoots not required for propagation by layering should be removed.

Propagate by simple layering, or by basal cuttings of side shoots in early to midsummer, in a closed case with bottom heat. Increase also by bench

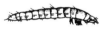

wireworm

Wisteria floribunda

Wisteria sinensis

grafting, making an apical wedge graft on to seedling stock of *W.sinensis* in late winter. Plunge the pot in a closed case with bottom heat at 18–20°C/65–70°F. When established, pot on, ensuring that the graft union is below the soil surface to encourage the scion to root. Seed-raised plants are variable and may, in some cases, take up to twenty years to bear flowers of poor quality that are then obscured by foliage. If this method is attempted, sow seed in spring at 10–13°C/50–55°F, after pre-soaking in hot water for about 18 hours. Seeds are poisonous. The aphid-borne wisteria vein mosaic virus causes yellow mottling or spotting of the leaves. Late frosts will kill the developing racemes, but these will often be replaced by a second flush.

W. floribunda (JAPANESE WISTERIA; Japan; stem twining clockwise, to 8m; racemes 40–120cm, flowering gradually from base, flowers to 2cm across, fragrant, typically lilac-blue, also darkest violet, blue, lilac, pink, red or white; includes numerous cultivars with large or small flowers, some double, in a wide range of colours, some with very long racemes); *W. × formosa* (*W. floribunda × W.sinensis*; stem twining clockwise; raceme to 25cm, flowers strongly fragrant, blooming all at once, to 2cm across, standard pale violet, wings and keel darker); *W.japonica* (Japan, Korea; racemes 15–30cm, often branching, flowers to 1.3cm across, white or cream to rose); *W.sinensis* (CHINESE WISTERIA; China; stems to 10m, twining anti-clockwise; racemes 15–30cm, flowers to 2.5cm across, faintly scented, blue-violet, includes white, pale blue, dark blue, lilac, mauve and purple-black cultivars with single and double flowers); *W.venusta* (SILKY WISTERIA; Japan; stems twining anti-clockwise, to 9m; racemes 10–15cm, flowers to 2.5cm across, white or purple, highly fragrant, on silky hairy stalks).

witches' broom the development of an abnormally large number of shoots, mainly on woody plants, often growing in the form of a broom or, on trees in particular, as a tangled mass of branching stems, usually with reduced or abnormally shaped leaves. They are mainly produced following infection by fungi, including rusts, but mycoplasma-like organisms may also be responsible. Mite infestation of the growing points of plants can cause witches' brooms, as can damage by other factors, such as low temperature. Dwarf mistletoes (*Arceuthobium* species.) attack many conifers in the west of North America, causing damaging witches' brooms.

wood, woody lignified plant tissue that forms rigid stems and branches, giving the plant a robust structure.

wood ash the product of burning wood or other vegetable material which, although variable, is useful in containing 4–15% potash, as well as phosphate, iron, magnesium, and manganese in small amounts. Ash from herbaceous material or young woody growths is richest in potassium. Wood ash should be applied immediately or stored dry. It slightly increases the alkalinity of soil and should be well-incorporated, especially on heavy soils, to avoid damage to the surface soil structure.

wood bud see *vegetative bud*.

woodchips, wood shavings waste wood fragments produced during milling or from the shredding of prunings. It may be used satisfactorily as a degradable path covering or a mulch around plants. If incorporated into the topsoil, it can cause nitrogen depletion as it decomposes. It is unlikely to transmit, or encourage the development of, honey fungus.

woodchuck (*Marmota monax*) a North American mammal of the marmot tribe that devours salads, other vegetables and flowers in the garden; it is also known as a ground hog. It is usually deterred by construction of low wire fencing, buried 15cm below ground level and extending 1m above it. Shooting, trapping or gassing can be carried out in badly infested areas, but advice on appropriate local procedures should be taken.

woodland garden see *wild and woodland gardening*.

woodlice (Crustacea: Isopoda) crustaceans, oval in shape, with a convex body, one pair of antennae and seven pairs of walking legs. Also known as sowbugs, pillbugs or slaters, they are restricted to damp places and mainly nocturnal activity. Immature woodlice resemble adults, except in size; during the breeding season, females have four pairs of plates that project inwards between the rear legs to form a brood chamber in which up to 50 eggs, and then newly hatched young, are carried.

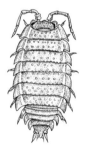

woodlouse

Woodlice feed mostly on decaying vegetable matter and are generally beneficial in accelerating the breakdown of plant remains. They sometimes attack living plants, especially seedlings under glass, which may be severed at soil level, and they may bite irregular-shaped holes in the leaves, stems, flowers and fruit of older plants. Susceptible plants include cacti, chrysanthemums, cucumbers, *Cyclamen*, delphiniums, lupins, pansies, petunias and the aerial roots of orchids; mushrooms grown on floor beds in greenhouses are also attacked.

Some species roll into a ball for protection, including the COMMON PILLBUG *(Armadillidium vulgare)* in Europe and North America, which has light grey to black adults up to 12mm long.

Woodlice are numerous where daytime shelter is available, and where plant damage occurs, these sites should be reduced. On a small scale, woodlice may be successfully trapped in inverted flower pots or scooped-out potato tubers, providing the traps are examined daily and all the woodlice destroyed. Recommended contact insecticides applied to the soil surface are effective, as are certain slug baits, but both methods of control should only be employed where essential.

X

Xanthoceras (from Greek *xanthos*, yellow, and *keras*, horn, referring to the horn-like growths between the petals). Sapindaceae. N China. 2 species, deciduous shrubs and small trees with fissured bark, yellow wood and pinnate leaves. Borne in spring and summer in erect panicles, the flowers consist of five broad and spreading petals. Hardy in zone 6. Plant in any fertile, well-drained soil in sun and with shelter from harsh winds. To flower well, this tree needs a long, hot summer; it is often grown against south-facing walls in cooler, duller areas. Increase by seed in autumn, or by rooted suckers in early spring. *X.sorbifolium* (erect shrub to 7m; flowers to 2cm wide, white with the centre stained green to yellow then red).

Xanthorhiza (from Greek *xanthos*, yellow, and *rhiza*, root). Ranunculaceae. E N America. 1 species, *X.simplicissima*, a deciduous, running and suckering woody-based perennial herb or shrub to 80cm. The stems are thin and sparsely branched with yellow wood. The leaves are to 18cm long, 1–2-pinnate with narrowly elliptic to lanceolate, toothed leaflets. They emerge purple, become mid-green and turn bronze in autumn. Small, star-shaped and purple-brown, the flowers are produced in slender, drooping panicles in spring. Excellent, medium-height groundcover hardy in climate zone 5. Plant in sun or shade on a moist, fertile soil. Older plants can be rejuvenated by shearing to ground level in spring. Increase by division.

Xanthorrhoea (from Greek *xanthos*, yellow, and *rheo*, to flow, referring to the resin produced by the plants). Liliaceae (Xanthorrhoeaceae). Australia. BLACK BOY, GRASS TREE. 15 species, large evergreen perennial herbs and shrubs with thick, dark and fibrous cylindrical trunks composed of countless, hard, compacted leaf bases. Carried in a dense, terminal crown the leaves are long and needle-like. Small, white flowers are carried in tapering, club-like spikes. Several species are grown, all superficially very similar, including *X. australis, X.arborea* and *X.preisii*. They may reach a height of 4 metres with leaves to 1.5m long, but are painfully slow-growing. Grow these rare and remarkable plants in a fast-draining, acid to neutral medium. Provide a minimum temperature of 10°C/50°F, full sun and a buoyant, dry atmosphere. Water freely in hot weather, sparingly at other times. Increase by offsets or seed.

Xanthosoma (Greek *xanthos*, yellow, and *soma*, body, referring to yellow tissues of some species). Araceae. Tropical America. TANNIA. 50 species, large, tuberous perennial herbs with stout-stalked, cordate to sagittate leaves dull, *Arum*-like inflorescences. Cultivate as for *Alocasia*.

X.sagittifolium (TANNIA; native range unknown, widespread through cultivation in tropics; leaf blades to 90cm, broadly arrow-shaped, glaucous above, petiole to 1m; includes cultivars with leaves veined, edged and splashed with white and gold); *X.violaceum* (BLUE TANNIA, BLUE TARO; native range unknown, widespread through cultivation in tropics; leaf blades to 70cm, broadly arrow-shaped, glaucous, margins and veins flushed dark purple, petiole to 2m).

Xeranthemum (from Greek *xeros*, dry, and *anthos*, flower). Compositae. Mediterranean to W Asia. 6 species, erect, annual summer-flowering herbs with daisy-like flowerheads, papery and pearly in texture and lasting well when dried. Sow seed *in situ* in spring on a fertile, well-drained soil in sun. *X.annuum* (IMMORTELLE; SE Europe to W Asia, naturalized elsewhere; stem 25–75cm; leaves to 6cm, linear to oblong, white-tomentose; flowerheads 3–5cm in diameter, bright pink or white; seed races include plants with white, pink, rose, red and purple flowerheads).

Xerophyllum (from Greek *xeros*, dry, and *phyllon*, leaf; the leaves are dry and grass-like). Liliaceae (Melanthiaceae). N America. 3 species, rhizomatous perennial herbs with tufts of linear, rough-edged leaves and pyramidal racemes of long-stalked, star-shaped flowers in summer. Fully hardy. Grow in an acid soil, permanently moist and enriched with leafmould, in full sun. Increase by seed or division. *X.tenax* (ELK GRASS, BEAR GRASS, INDIAN BASKET GRASS, SQUAW GRASS; western N America; to 180cm; flowers to 1.5cm across, white to cream with violet stamens).

xerophyte a plant adapted to survival in an arid habitat.

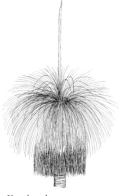

*Xanthorrhoea
preisii*

Y

yard, yard gardening in the US, 'yard' is the usual name for what in the UK is known as a back garden; 'yard gardening' is occasionally used to denote cultivation in courtyard situations.

yellows disease characterized by a general yellowing of foliage or other yellow-coloured symptoms, caused by various fungi, bacteria and phytoplasma-like organisms. *Fusarium oxysporum* f.sp. *conglutinans* causes CABBAGE YELLOWS, a form of the disease that is particularly destructive to cabbage and other brassicas in the US. Plant roots are infected from the soil, and at high temperatures the fungus progresses rapidly through the vascular system causing the leaves to turn yellow-green with the lower ones curling and dying. The pathogen can survive many years in the soil and spreads in crop debris, soil and drainage water or on implements. The best control is by the choice of resistant cultivars.

F. oxysporum f.sp. *gladioli* causes a VASCULAR WILT DISEASE, gladiolus yellow, in which the plants become yellow and die during the growing season, and there is rotting of corms. Freesias are also susceptible, and the symptoms on corms of these and gladiolus are seen when the tunic is removed to reveal circular, black sunken areas around the base, with irregularly concentric ridges. The disease is encouraged by heavy manuring, and insufficient ripening before storage. Yellowed plants should be disposed of, together with a spadeful of the surrounding soil, as soon as noticed.

Xanthomonas campestris pv. *hyacinthi* is a bacterium causing yellows of hyacinths. It is an occasional problem where affected plants fail to sprout, the growth is stunted or the bulbs rotten; further signs are brown striping and rotting of the leaves and flower stalks. The disease spreads from plant to plant and diseased ones should be carefully removed.

Aster yellows, palm lethal yellowing and peach yellows are phytoplasma diseases. See *phytoplasma*.

yostaberry see *jostaberry*.

youngberry see *hybrid berries*.

Yucca (from *yucca*, the native name for *Manihot esculenta*, named by Gerard, who mistook the genus for *Manihot*). Agavaceae. N and C America, West Indies. 40 species, large evergreen perennial herbs, shrubs or trees. Where present, the stems are stout, cylindrical, sparsely branched and scarred by the traces of dead leaves (these may also persist in a dense collar). In age, they may become massive and freely branched above. The leaves are sword-shaped to needle-like, thick, and tough but pliable, or slender and rigid; they are commonly glaucous and grey-green, clothing the stems or arranged in dense, terminal rosettes. The waxy, bell-shaped flowers are usually fragrant at night and hang in stately spire-like panicles arising from the crown in summer.

All described below prefer full sun, a sheltered position and a fast-draining, gritty soil with minimal winter damp. *Y.filamentosa*, *Y.filifera*, *Y. glauca*, *Y.gloriosa* and *Y.recurvifolia* are fully hardy and superb plants for border and outdoor container plantings. *Y.gloriosa* soon forms a large and free-flowering stand and is perhaps better used for specimen plantings in lawns. In cool temperate cultivation, *Y. rostrata* and *Y.whipplei* are smaller, finer plants, ideal for sheltered, dry rockeries, for the base of a south-facing wall or for the silver border. Although hardy in climate zone 7, they will both suffer unless planted on a very lean, sandy soil and protected from prolonged cold and wet. Of the remaining species, most will only thrive in climate zone 9, or in the cool greenhouse or conservatory (minimum temperature 7°C/45°F) where they associate well with cacti and succulents. *Y.elephantipes* is a popular and durable house plant, sold in the form of rooted and sprouting stem sections. It tolerates more shade and moisture than the other tender species. Increase by seed sown in heat, by division, by removal of rooted suckers and, where possible, by stem cuttings.

Y.aloifolia (SPANISH BAYONET, DAGGER PLANT; SE US, West Indies; stems to 8m, slender, sometimes branching; leaves to 40 × 6cm, rigid, flat, sharply pointed and minutely toothed; flowers to 5 × 10cm, cream tinged purple or lined green at base; includes 'Marginata', with leaves edged yellow, 'Quadricolor', with leaves striped yellow and white, and 'Tricolor', with leaves striped yellow, sometimes with white central stripe); *Y.brevifolia* (JOSHUA TREE; California to SW Utah; stem to 9m or more, often branching, bark deeply fissured; leaves to 40 × 4cm, stiletto-like, rigid, minutely toothed; flowers to 7cm, green tinged yellow to cream, malodorous); *Y.elephantipes* (Mexico, Guatemala; differs from *Y.aloifolia* in its taller, freely branching and rather stouter stems; leaves to 100 × 7cm, semi-rigid, base swollen, minutely toothed; flowers to 4cm, white or cream); *Y.filamentosa* (SPOONLEAF YUCCA, ADAM'S NEEDLE, NEEDLE PALM; US; stems short or absent, clump-forming and suckering; leaves to 75 × 10cm, sword-shaped, stiff, margins bearing wavy to curling threads and fibres; inflorescence 1–4m, flowers to 5cm, white tinged yellow; includes 'Bright Edge', to 60cm, with leaves broadly edged yellow,

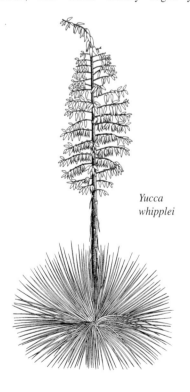

Yucca whipplei

'Elegantissima', erect to 1.2m, with white flowers, 'Rosenglocken', with flowers tinted rose, 'Starburst', forming low rosettes with cream-edged leaves, and 'Variegata', with white leaf margins becoming pink-tinged, and centre of leaves glaucous green); *Y.filifera* (syn. *Y.flaccida*; much-branched shrub to 10m; leaves to 55 × 4cm, linear-lanceolate, glaucous, somewhat flexible and reflexed above middle, margins with straight, not curling, threads; flowers to 5cm, creamy white; includes 'Golden Sword', with leaves edged yellow); *Y.glauca* (western C US; stems absent or very short; leaves to 60 x 5cm, linear, rigid, smooth and glaucous, margins white to grey, bearing sparse straight threads; flowers to 6cm, creamy white tinged green or brown; includes 'Rosea', with flowers tinged rose); *Y.gloriosa* (SPANISH DAGGER, ROMAN CANDLE, PALM LILY; SE US; stem to 2.5m, thick, rarely branching but suckering freely; leaves to 60 × 7cm, narrowly lanceolate, smooth, stiff, glaucous, entire; flowers to 7cm, creamy white sometimes tinged red or green; cultivars include: ' Garland's Gold', with leaves with a central gold stripe, similar to 'Mediostriata'; 'Nobilis', with large and very glaucous leaves, a tall and massive inflorescence, and flowers tinted red; ' Variegata', with leaves striped creamy yellow; and ' Vittorio Emmanuel II', low and compact, with flowers strongly tinted red in bud); *Y.recurvifolia* (SE US; similar to *Y. gloriosa* but with stems often branching, leaves to 90cm long and recurved, and a more open inflorescence; includes 'Marginata', with leaves edged yellow, and 'Variegata', with leaves with a central yellow stripe); *Y.rostrata* (N Mexico, S US; stem absent or to 3m, solitary or several in a clump, leaf rosettes dense; leaves 20–60 × 0.5–1.3cm, linear, sharply pointed, rigid, somewhat glaucous with a thin pale yellow margin, entire or minutely toothed; flowers to 6cm, white); *Y.smalliana* (ADAM'S NEEDLE, BEAR GRASS; S California to Florida; resembles *Y.filamentosa* but with narrower and flatter leaves, the marginal threads distinctly curled, a finely hairy inflorescence axis and white flowers; includes 'Maxima', tall, with an inflorescence with large, leafy bracts, and 'Variegata', with leaves striped yellow); *Y.whipplei* (OUR LORD'S CANDLE; W California, Baja California, Mexico, Arizona; stem usually absent, leaf rosettes dense; leaves 30–90 × 1–7cm, linear, glaucous, rigid, pungent, margins minutely toothed; flowers to 6.5cm, creamy white often tipped green; young plants resemble *Y.rostrata*).

Yushania (for the Yushan mountains in Taiwan). Gramineae. E Asia. 2 species, clump-forming bamboos. See *bamboos*.

Z

Zaluzianskya (Scrophulariaceae). S Africa. 35 species, sticky, hairy annual or perennial herbs or subshrubs with short-stalked flowers in late spring and summer; these are tubular with a spreading limb of 5 broad lobes. Hardy in zone 7 in a sheltered, sunny position. Grow on a gritty, fast-draining soil. Avoid wet conditions. Where winters are harsh, overwinter young plants in an unheated greenhouse. Increase by stem tip cuttings in spring. *Z.ovata* (bushy, low-growing evergreen perennial to 10 × 30cm; leaves to 3cm, ovate to oblong, toothed, midgreen with short, grey hairs; flowers to 2cm diam., pure white with deep wine red reverses to lobes).

Zamia (from Greek *zamia*, loss, echoing Pliny's description of the withered appearance of fir cones). Zamiaceae. Americas. Some 30 species, evergreen cycads, with swollen subterranean stems or small, cylindrical trunks. The leaves are pinnately compound with tough leaflets. Male cones tend to be narrow and scurfy, the females squatter, resembling a hand grenade. Where plants of both sexes cone together, the females may produce small, hard nuts with a brightly coloured, fleshy coating. Grow in frost-free conditions with bright indirect light to full sun. Plant in a free-draining, loam-based mix with additional sand, leafmould and composted bark;

water, feed and syringe freely in warm weather, at other times keep rather dry. Both are excellent foliage plants for the home and cool greenhouse. In climate zones 9 and 10, they are resilient candidates for landscaping. Propagate by seed sown in warmth with the fleshy coat removed, or by rooted offsets.

Z.furfuracea (CARDBOARD PALM; Mexico; stem subterranean or stout, short; leaves to 1m, leaflets to 12cm, obovate to oblong, tip broad, bluntly toothed, rigid, olive green, scurfy); *Z.pumila* (syn. *Z.angustifolia*, *Z.integrifolia*; COONTIE, FLORIDA ARROWROOT, SEMINOLE BREAD; stem subterranean or short, swollen and sometimes multi-headed; leaves to 1m, leaflets 5–12cm, linear to oblanceolate or oblong, tip broad and blunt to slender and pointed, minutely toothed, somewhat pliable, dark green, glossy).

Zamioculcas (a combination of *Zamia*, which this aroid very broadly resembles, and *culcas*, the Arabic name for *Colocasia*). Araceae. Tropical Africa. 1 species, *Z.zamiifolia*, an evergreen or partially deciduous perennial herb. The pinnate leaves are 30–100cm long and stand erect in tight clumps. The petioles are succulent and swollen at their bases. The leaflets are 5–15cm long, broadly ovate to elliptic or lanceolate, closely set, glossy dark green and somewhat fleshy. A handsome foliage plant, tolerant of temperatures to 10°C/50°F and dry conditions. Where cool drought continues, however, the leaves may drop, leaving only their swollen bases as storage vessels. For luxuriant, year-round growth, plant in a fertile soilless medium; position in bright, indirect light, keep evenly moist and maintain a minimum temperature of 15°C/60°F. Increase by inserting mature leaflets in a moist, soilless mix with bottom heat, or by division.

Zantedeschia (for Giovanni Zantedeschi (1773–1846), Italian botanist and physician). Araceae. South Africa, some widely naturalized elsewhere. ARUM LILY, CALLA LILY. 6 species, perennial herbs usually evergreen and with short, thick rhizomes. The leaves arise in a basal clump and are usually heart- or arrow-shaped or, less commonly,

Zamia furfuracea

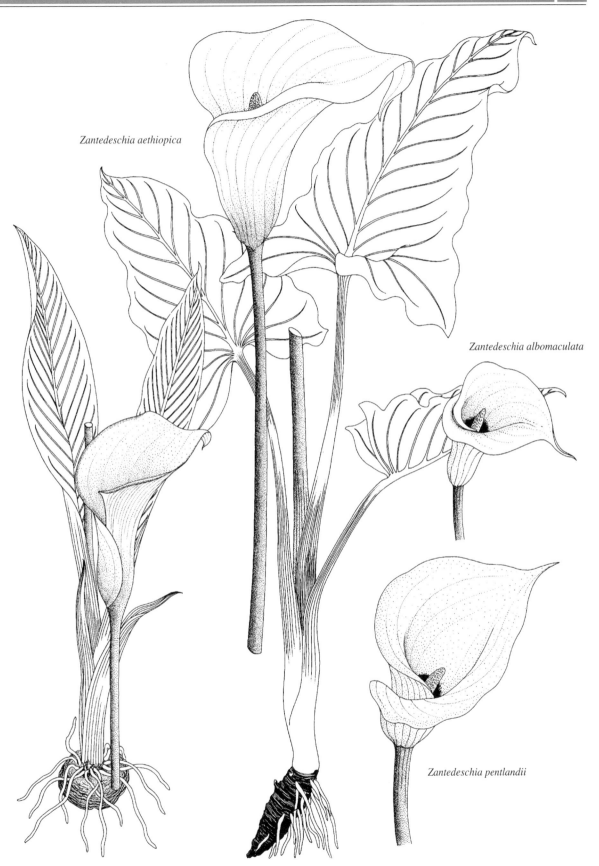

Zantedeschia aethiopica

Zantedeschia albomaculata

Zantedeschia pentlandii

Zantedeschia rehmannii

lanceolate. Produced in summer, the inflorescence consists of numerous minute flowers packed at the base of a club-shaped spadix, the whole surrounded by a showy, broadly funnel-shaped spathe. *Z.aethiopica* is hardy in climate zone 7, especially the cultivar 'Crowborough'. It will grow in sun or part shade with year-round moisture on most fertile soils, and will also thrive with its roots fully or partially submerged in water in ponds, lakes and half barrels. The remaining species and cultivars require bright, frost-free conditions. Plant them in a porous, fertile and preferably soilless medium. As new growth emerges, increase water; feed and water liberally when in full leaf. After flowering, decrease water supplies. Dry off as the foliage withers and keep virtually dry over winter. Increase by offsets.

Z.*aethiopica* (ARUM LILY; leaf blades to 40cm; spathe to 25cm, white, spadix yellow; includes the large and hardy, pure white 'Crowborough', and 'Green Goddess' with large spathes stained and streaked with green); Z.*albomaculata* (leaf blades to 45cm, with white spots or blotches or plain; spathe to 12cm, white to ivory, pale yellow or pink); Z.*elliotiana* (leaves densely spotted; spathe golden yellow); Z.*rehmannii* (leaves to 40cm, lanceolate; spathe to 12cm, white, pink or plum, sometimes with a darker throat). Numerous hybrids and cultivars have been developed within this genus, including plants tall and dwarf with leaves spotted and unspotted and spathes in tones of white, cream, yellow, peach, pink, red and plum. In some, the throat is far darker (e.g. purple-black) or paler.

Zanthoxylum (from Greek *xanthos*, yellow, and *xylon*, wood). Rutaceae. Widespread in warmer areas. PRICKLY ASH. Some 250 species, deciduous or evergreen prickly trees and shrubs with aromatic, usually pinnate foliage, small, cymose flowers and fleshy or leathery fruit. Hardy to at least −15°C/5°F. Plant in full sun or semi-shade on a deep, moist and fertile soil. Increase by seed in autumn, by semi-ripe cuttings in autumn, or by root cuttings in late winter. Z.*piperitum* (JAPAN PEPPER; N China, Korea, Japan; shrub or tree to 6m; leaflets 11–23, to 3.5cm, ovate, tip toothed); Z.*simulans* (China, Taiwan; spreading shrub or small tree to 3m; leaflets 7–11, to 5cm, ovate to oblong, serrate).

Zauschneria (Onagraceae). SW US, Mexico. 4 species, low-growing, evergreen perennial subshrubs with small leaves covered in grey down. Produced in summer, the flowers are tubular to trumpet-shaped with a limb of 4, flared lobes. Hardy in zone 7. Plant on a gritty, free-draining soil in full sun. Protect from hard frosts and wet weather with fleece or a dry mulch. Increase by stem cuttings in spring. Z.*californica* (syn. Z.*cana*, *Epilobium canum*; HUMMINGBIRD'S TRUMPET, CALIFORNIA FUCHSIA; SW US to Mexico; 20–60cm, covered in grey to silver down; flowers vermilion to scarlet; 'Album': flowers white; 'Dublin', syn. 'Glasnevin': flowers long, flame-coloured).

Zea mays 'Quadricolor'

Zea (from Greek *zea*, a type of cereal). Gramineae. C America, native origin unclear, now widespread through cultivation. 4 species, large annual or, rarely, perennial grasses. The stem is usually tall, thick and unbranched, sheathed by the overlapping bases of arching strap-shaped leaves. The male inflorescence consists of several, tassel-like racemes in a terminal panicle. The female inflorescence is axillary and enveloped by tough spathes (husks). It develops as a stoutly cylindrical 'cone' of tightly packed, juicy grains, tipped with a plume of long, silky styles. The ornamental cultivars are treated as half-hardy annuals. Sow seed under glass in late winter or early spring and plant out after the last frost. In regions with long, hot summers, seed may be sown *in situ* in late spring. Grow in sunny, warm and sheltered conditions on a moist, fertile soil. As the cobs near maturity, plants prefer hotter, drier weather. See *sweet corn*. Z.*mays* (MAIZE, CORN, MEALIE, SWEET CORN; 1–4m; leaves to 90cm; female inflorescence to 20cm, grains to 1cm, yellow; ornamental cultivars include plants with pink and white-striped leaves, others with small cobs, some with blue-black or wine red grains or grains in a mixture of colours (the highly decorative indian corn) and is used in autumnal dried-flower arrangements).

Zelkova (from the local name for *Z.carpinifolia*). Ulmaceae. Crete, Caucasus, E Asia. 5 species, deciduous trees with smooth to scaly, grey bark and many branches usually drawn gently upwards in a rounded or goblet-like crown atop a short, clean trunk. Ovate to elliptic, saw-toothed and distinctly veined, the leaves turn gold, amber or red in autumn. Inconspicuous, green flowers give rise to dry, nut-like drupes. The following are fully hardy but prefer a sheltered situation. Plant in sun on a deep, fertile soil with plentiful moisture. Z.*serrata* makes a beautiful bonsai. Increase by seed sown in autumn.

Z. abelicea (Crete; to 15m, branchlets white-hairy; leaves 1.5–5cm, ovate to oblong, obtuse, cordate to rounded at base, triangular-toothed, somewhat hairy at first); *Z.carpinifolia* (CAUCASIAN ELM; Caucasus; to 35m, branchlets densely downy; leaves 2–7cm, elliptic to oblong, acute, rounded or subcordate at base, teeth obtuse, ciliate, with scattered hairs above, paler and downy beneath); *Z.serrata* (JAPANESE ZELKOVA, SAW LEAF ZELKOVA; Japan, Taiwan, E China; to 35m, branchlets tinged purple-grey, slightly hairy; leaves 3–7cm, oblong-elliptic, acuminate, rounded or somewhat cordate at base, teeth acute, ciliate, sometimes thinly hairy above, pale green and glossy beneath with hairy veins; includes dwarf, compact, shapely and variegated cultivars).

Zenobia (for Zenobia, Queen of Palmyra, *c* AD 266). Ericaceae. SE US. 1 species, *Z.pulverulenta*, a deciduous or semi-evergreen shrub, 50–180cm tall. The leaves are 2–7cm long, ovate to broadly elliptic, entire or toothed, blue-green and somewhat glaucous. Bell-shaped, sparkling white and sweetly fragrant, the flowers hang on slender stalks in summer. Cultivate as for *Leucothoë*.

Zephyranthes (from Greek *zephyros*, the west wind, and *anthos*, flower – the genus originates in the western Hemisphere). Amaryllidaceae. Americas. ZEPHYR FLOWER, FAIRY LILY, RAIN LILY. 71 species, bulbous perennial herbs with narrowly strap-shaped to grassy leaves and solitary, funnel-shaped flowers usually appearing before or with the leaves in late spring or autumn. In favoured locations in climate zones 7 and over, the following species may grow outdoors on fast-draining soils at the base of a south-facing wall. Otherwise, grow under glass with a minimum temperature of 7°C/45°F. Pot in a sandy, loam-based mix and position in full sun. Keep dry and warm once the leaves have faded. After a period of dry rest, resume watering and lower temperatures to induce flowering and new growth. Increase by seed or division.

Z.atamasca (ATAMASCO LILY; SE US; to 30cm, flowers to 8cm, pure white or tinged purple); *Z.candida* (Argentina, Uruguay; to 25cm, flowers to 5cm, pure white or slightly rose-tinged on the exterior); *Z.citrina* (tropical America; to 13cm, flowers to 5cm, bright yellow); *Z.grandiflora* (Mexico; to 30cm, flowers 8cm, bright pink with a white throat); *Z.rosea* (Cuba; to 15cm, flowers to 6cm, bright rose).

Zigadenus (from Greek *zygon*, yoke, and *aden*, gland – a pair of glands is typically present at the base of the tepals). Liliaceae (Melanthiaceae). N America, N Asia. DEATH CAMAS, ZYGADENE. 18 species, bulbous or thickly rhizomatous perennial herbs with narrowly strap-shaped leaves and erect spikes of star-shaped flowers in summer. Cultivate as for *Camassia.*

Z.elegans (WHITE CAMAS, ALAKALI GRASS; N America; to 90cm, flowers 1.6–2cm diam. green-white); *Z.fremontii* (STAR ZYGADENE, STAR LILY; SW US; to 90cm, flowers 2–3cm diam., off-white to ivory).

zinc an essential plant micronutrient, rarely deficient on garden soils of the UK, although its lack can cause shoot dwarfing and die-back in fruit with small, narrow leaves. It is most common in high sunlight climates, and on alkaline soils or those rich in phosphorus; citrus fruits are susceptible.

Zinc toxicity is a possibility where sewage-based organic manure is contaminated with so-called 'heavy metals'; severe chlorosis and interveinal necrosis are likely symptoms, possibly through induced iron deficiency.

Zinnia (for Johann Gottfried Zinn (1727–1759), professor of botany at Gottingen). Compositae. S US to Argentina, particularly Mexico. Some 20 species, annual or perennial herbs or low shrubs with brightly coloured daisy-like flowerheads in summer. Half-hardy annuals, they thrive on most soils in full sun. Avoid damp conditions. Deadhead frequently. Sow seed under glass in early spring, and plant out after the last frosts. *Z.elegans* (Mexico; erect annual to 1m; flowerheads 4–8cm across, ray florets red, white, yellow, orange, pink, scarlet, lilac or purple, disc florets yellow and black; a highly developed and popular garden annual with many seed races ranging from tall (80cm) to dwarf (20cm) with large or small flowerheads, single or double, ruffled or dahlia-like, in a wide spectrum of colours).

zones see *hardiness*.

zygomorphic bilaterally symmetrical, having only one plane of symmetry by which it can be divided into two equal halves.

Zygopetalum (from Greek, *zygon*, yoke, and *petalon*, petal, referring to the fusion of the tepals at the base of the column). Orchidaceae. C and S America. 20 species, evergreen epiphytic or terrestrial orchids. Those listed here have clumped, ovoid to conical pseudobulbs and lanceolate to strap-shaped leaves. Produced in spring and summer on tall, erect racemes, the flowers are waxy, often fragrant with spreading oblong to lanceolate tepals and an obovate to fan-shaped lip with a fleshy, ridged callus. Cultivate as for *Odontoglossum* but at slightly higher temperatures.

Z.crinitum (E Brazil; inflorescence to 50cm, flowers strongly fragrant, tepals to 2.7cm, grey-green or yellow-green streaked and spotted chestnut brown, lip to 5cm, white veined red or purple, veins branching, hairy, callus white); *Z.intermedium* (Peru, Bolivia, Brazil; inflorescence to 50cm, flowers hyacinth-scented, tepals to 3.7cm, equal, green or yellow-green blotched red-brown or purple to maroon, lip to 3.5cm, white with radiating, broken lilac to deep purple-violet lines, callus mauve); *Z.mackayi* (Brazil; differs from *Z.intermedium* in its taller inflorescence and unequal petals and sepals).

zygote the cell resulting from the union of gametes at fertilization, from which an embryo develops.

SOME COMMON PESTS AND DISEASES OF GARDEN PLANTS
Diagnostic Chart

Plants growing strongly are less likely to succumb to pests, diseases and disorders than those which struggle. Good performance can be encouraged by careful choice of plants for the location and climate, by the establishment of a well planned and prepared site, and by continuous maintenance with watchfulness and good technique. Pests and diseases thrive on neglected plants. See *cultural controls*.

Problems are bound to arise even under good management, but in the amateur garden some loss of quantity and quality in edible crops can be tolerated. Where pest and disease control is essential because of competition, threat or unsightliness, the first consideration should be the use of so-called mechanical methods like protection with physical barriers, trapping and hand-picking - all of which are perfectly feasible and effective on a small scale. The next most suitable means of control is with the use of predators and parasites, a good selection of which are now available to the gardener for reducing the numbers of some specific pests. See *biological controls*.

The last resort should be to chemical substances, most usually applied as sprays, dusts or smokes. The range is wide but the authorised use of specific substances is under constant review in the interests of safety and protection of the natural environment. For these and other reasons the retail availability of generic substances, and of commercial products and formulations, is continually changing; and the sensible and practical course for the gardener is to refer to what is currently offered and recommended by retailers. See *pesticides*.

In all of these considerations the correct identification of pests and diseases is vital. The following selective groups of descriptions provide a means of deciding the most likely cause of some more commonly observed symptoms.

FRUITS (TREE)

Symptoms	possible causes
Apple	
Ripening fruit with maggot holes and tunnelling.	codling moth larvae
Leaves and shoots distorted.	aphids
Swellings on woody stems with white woolly masses.	woolly aphids
Leaves eaten; blossom and young fruits damaged.	winter moth larvae
Leaves with white powdery fungal growth, mostly on the upper sides and especially at shoot tips; leaves may curl inwards.	powdery mildew
Leaves, stems and fruits with brownish felty patches; early defoliation. Black scabby patches on fruits, which may be cracked and distorted.	scab
Bark areas sunken and discoloured, often near to a bud, joint or wound. Rings of loose flaky tissue bearing white pustules in summer, red pustules in winter. Infected shoots die.	canker
Fruitlets tunnelled and fall early in midsummer. Some damaged fruits remain to maturity, but are misshapen, with a long ribbon scar.	sawfly larvae
Fruits developing soft brown areas penetrating the flesh. Concentric rings of cream pustules. Fruit fall or shrivel on the tree.	brown rot
Cherries	
Leaves and shoots severely distorted.	aphids
Leaves eaten; blossom and young fruits damaged.	winter moth larvae
Leaves with a silver sheen. Tree parts die progressively.	silver leaf
Bark oozes resin-coloured gum. Small brown spots on leaves with holes developing.	bacterial canker
Peaches	
Leaves thicken and blister and turn red. White fungal bloom on the leaves. Early defoliation.	peach leaf curl
Leaves and shoots severely distorted.	aphids
Leaf surfaces show progressive pale mottling. Premature leaf fall. Sometimes a fine silk webbing apparent.	red spider mite

FRUIT (TREE)

Symptoms	possible causes
Pear	
Leaves and shoots distorted.	aphids
Fruitlets turn black and fall.	gall midge larvae
Leaves, stems and fruits with brownish felty patches; early defoliation. Black scabby patches on fruits which may be cracked and distorted.	scab
Fruits developing soft brown areas penetrating the flesh. Concentric rings of cream pustules. Fruits fall or shrivel on the tree.	brown rot
Leaves turn brown to black, wither and die, but remain attached. Individual limbs affected; oozing cankers appear. Progressive decline of the tree.	fireblight
Plum	
Leaves with a silver sheen. Tree parts die progressively.	silver leaf
Bark oozes resin-coloured gum. Small brown spots on leaves with holes developing.	bacterial canker
Leaves and shoots severely distorted; sticky foliage. Black sooty mould.	aphids
Fruits ripen prematurely and fall. Pink maggot within the fruit.	plum moth larvae
Buds destroyed between autumn and spring.	bullfinches

FRUIT (SOFT)

Symptoms	possible causes
Currants	
Leaves, especially at shoot tips, coated with white powdery fungus. Shoots and leaves wither and die.	American gooseberry mildew (see powdery mildew)
Leaf upper surfaces with dark brown spots; premature defoliation. Loss of vigour.	leaf spot
Bark lifted. Small fuzzy grey fungal patches. Stems die back.	*Botrytis* grey mould
Leaves distorted, with reddening in some cases.	aphids
Buds much enlarged, and fail to open.	big bud mite
Gooseberries	
Leaves and shoots distorted.	aphids
Leaves eaten; bushes defoliated, especially from their centres.	sawfly larvae
Buds destroyed between autumn and spring.	bullfinches
Leaves, especially at shoot tips, and developing fruits, coated with white powdery fungus. Shoots and leaves wither and die.	American gooseberry mildew (see powdery mildew)
Bark lifted. Small fuzzy grey fungal patches. Stems die back.	*Botrytis* grey mould
Raspberries	
Leaves distorted.	aphids
Ripened fruits, dried at the stalk end. Small maggot within the flesh.	raspberry beetle larvae
Foliage yellowed, mottled and streaked. Stunted/dwarfed growth. Poor cropping.	virus
Purple spots with silvery central areas on stems, leaves and flower stalks. Die-back of canes may occur.	cane spot

FRUIT (SOFT)

Symptoms	possible causes
Strawberries	
Leaves distorted; growth checked.	aphids
Leaf surfaces show progressive pale mottling. Foliage withered and fine silk webbing may be apparent.	red spider mite
Foliage yellowed and puckered. Stunted growth, and poor cropping.	virus
Fruits rotted, with fuzzy grey fungal growth.	*Botrytis* grey mould

VEGETABLES

Symptoms	possible causes
Brassicas	
Foliage eaten; hearts bored into.	cabbage caterpillars
Foliage yellow and puckered, with dense colonies of grey/white insects.	mealy cabbage aphids, treat as aphids
Leaves with yellow spots becoming brown or grey; concentric rings from which holes develop.	leaf spot
Plants discoloured, wilted and stunted; grossly swollen roots.	clubroot
Plants wilted and collapsed; taproots tunnelled, fine roots destroyed.	cabbage rootfly
Seedling leaves with neat round holes.	flea beetle
Peas and beans	
Leaf margins conspicuously scalloped with U-shaped notches.	pea and been weevil adults
Leaves and pods of peas silver-flecked and finely mottled.	thrips
Leaf undersides and shoot tips with dense colonies of black insects.	aphids
Leaves of broad beans with brown spots and streaks.	*Botrytis* chocolate spot
Leaves and stems of runner and kidney beans with water-soaked patches and angular spots each with a pale halo	bacterial halo blight
Pea pods with dark sunken blotches.	pod spot
Pea seeds damaged as they develop in the pod.	pea moth larvae
Roots of runner and kidney beans infested with colonies of white-wax-covered insects.	root aphids
Seeds of runner and kidney beans fail to germinate.	bean seed fly

VEGETABLES (BULBS, ROOTS AND TUBERS)

Symptoms	possible causes
Carrots	
Leaves distorted; growth checked.	aphids
Seedlings wilt and die; plants stunted, foliage reddened. Tap roots tunnelled.	carrot fly
Leeks	
Young plants abnormally swollen, stunted and thick-necked; rot in later stages.	stem and bulb eelworm
Leaves yellow in places with bright orange spots and blotches.	rust

VEGETABLES (BULBS, ROOTS AND TUBERS)

Symptoms	possible causes
Onions	
Bulbs developing grey rot on the necks in store.	neck rot
Plants abnormally swollen, stunted and thick-necked.	stem and bulb eelworm
Leaves with pale, rounded areas; tips discoloured and withered.	downy mildew
Leaves yellow; white fluffy fungus at stem base and roots.	white rot
Young plants collapse, usually in groups, with yellow drooping leaves.	onion fly larvae
Parsnips	
Roots cracked and blackened especially at the shoulders; rot developing.	canker
Potatoes	
Lower leaves discolour and die; growth stunted and weak; plants die down prematurely; tubers small.	potato cyst eelworm
Leaves with yellow, brown patches; haulm dies progressively; tubers rot.	potato blight

ORNAMENTALS

Symptoms	possible causes
Seeds	
developing seeds damaged	moth larvae
tunnelled	beetle larvae
germinating seed damaged	millipedes; wireworms
Bulbs corms and tubers	
eaten/tunnelled	bulb fly larvae; cutworms; millipedes; slugs; vine weevil larvae; wireworms
internal discolouration	bulb scale mite; stem and bulb eelworm
rotted	daffodil basal rot; *Phytophthora* root rots; violet root rot
Roots	
constricted/rotted	*Pythium*/*Phytophthora* damping off; foot and root rots; *Fusarium* wilt; *Verticillium* wilt; waterlogging
eaten/tunnelled	cockchafers (*see* chafers); cutworms; leatherjackets; millipedes; slugs; symphylids; vine weevil larvae; wireworms; woodlice
fungal fruiting bodies/ compacted black strands	honey fungus
galled	clubroot; crown gall; *Phylloxerid* vine louse; rootknot eelworm; woolly aphids
woolly masses	root aphids
Stems and shoots	
distorted	aphids; capsids; froghoppers; hormone weedkiller; mechanical wounding; rust; stem and bulb eelworm
galled	clubroot; crown gall; gall mites; leafy gall
gnawed	beetle adults; cockchafers (*see* chafers); cutworms; deer; millipedes; rabbits/hares; slugs/snails; springtails; voles; weevil adults and larvae; woodlice
sticky/black mould	aphids; whitefly; scale insects
wilted/tunnelled/withered/dead	bark beetle; canker; drought; Dutch elm disease; fireblight; fly larvae; *Botrytis* grey mould; moth larvae; waterlogging

ORNAMENTALS
Symptoms	possible causes
Foliage	
eaten- (clean holes, perforations, skeltonising, notching)	birds; beetle adults; butterfly/moth larvae; deer; earwig; leaf-cutter bees; millipedes; rabbits/hares; sawfly larvae; slugs/snails; slugworms; weevil adults; woodlice
finely mottled surface	thrips
finely mottled surface with defoliation and fine webbing	red spider mite
tunnelled	leaf miner
fuzzy grey fungal growth	*Botrytis* grey mould
marginal yellowing/scorch	drought; high temperature; nutrient deficiency; pesticide damage; waterlogging
overall or interveinal yellowing	drought; nutrient deficiency; waterlogging
puckered/blistered/distorted	aphids; hormone weedkiller; peach leaf curl; sucker; stem and bulb eelworm
rolled	aphids; sawfly larvae; tortrix larvae
silvered	silver leaf
spots/discoloured patches	leaf spot; leaf and bud eelworm; rust; scab
sticky/black mould	aphids; scale insect; whitefly
white fungal growth	powdery mildew; downy mildew
woolly masses	adelgids; mealy bugs; woolly aphids
yellow variegation/mottle/mosaic/ streaking/ring-spotting	virus
Flowers	
failure to open	drought; gall midge larvae; high temperature
buds blackened	frost
petals and flower buds lacerated/eaten	birds; beetle adults; earwigs; weevil adults
small/distorted/streaked	virus
spotted/withered	*Botrytis* grey mould;
uneven opening/distortion	moth larvae; capsid bug; leaf and bud eelworm; gall midge larvae
white flecks on petals	*Botrytis* grey mould; red spider mite; thrips

Leaf Arrangements 1

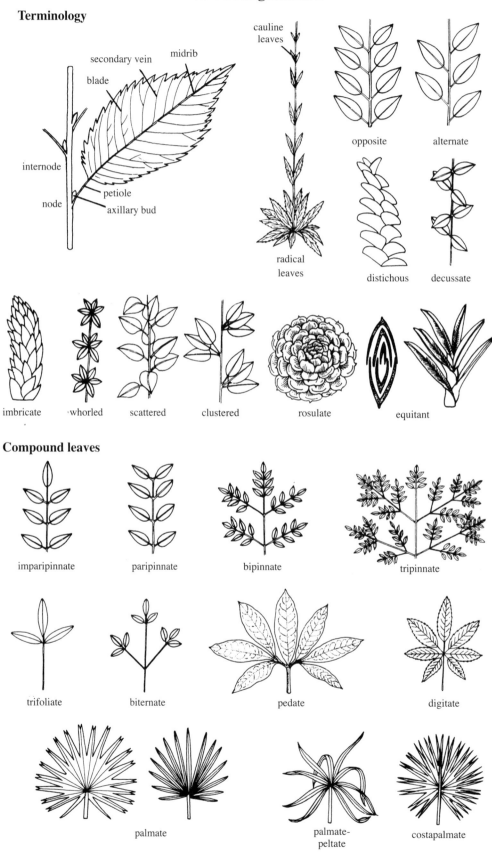

Terminology

cauline leaves

secondary vein

midrib

blade

internode

node

petiole

axillary bud

opposite

alternate

radical leaves

distichous

decussate

imbricate · whorled scattered clustered rosulate equitant

Compound leaves

imparipinnate paripinnate bipinnate tripinnate

trifoliate biternate pedate digitate

palmate palmate-peltate costapalmate

Leaf Arrangements 2

Insertion

petiolate

sessile

amplexicaul

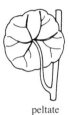

peltate

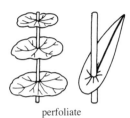

perfoliate

Ligules

connate-
perfoliate

sheathing

decurrent

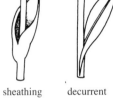

auricled

hairy

Stipules

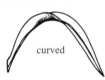

leafy   scale-like   glandular   spinose
or filiform   sheathing   protective

Leaf attitudes

recurved

curved

revolute

carinate

involute

plicate

reflexed

deflexed

conduplicate

Leaf-like organs

phyllode

leafy petiole

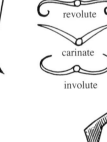

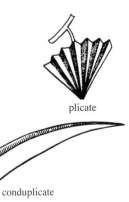

phylloclades

Leaf Shapes 1

Whole leaves

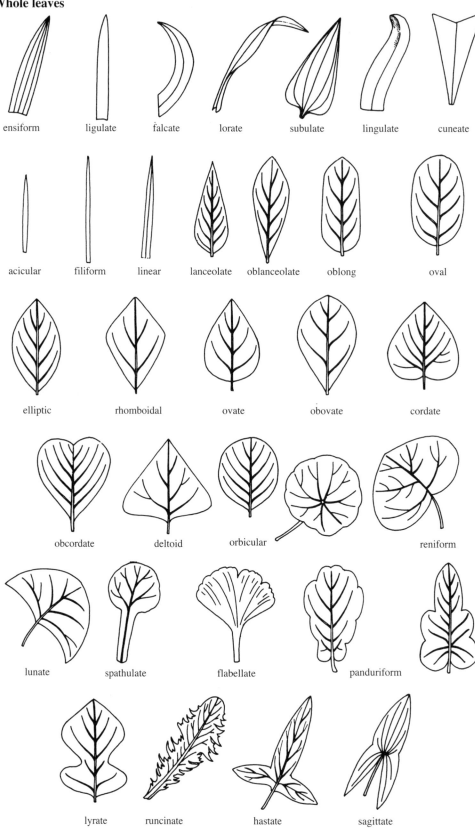

ensiform

ligulate

falcate

lorate

subulate

lingulate

cuneate

acicular

filiform

linear

lanceolate

oblanceolate

oblong

oval

elliptic

rhomboidal

ovate

obovate

cordate

obcordate

deltoid

orbicular

reniform

lunate

spathulate

flabellate

panduriform

lyrate

runcinate

hastate

sagittate

Leaf Shapes 2

Leaf tips

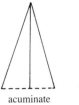
acuminate

acute

abruptly acute

apiculate

aristate

caudate

cirrhose

cupsidate

pungent

mucronate

mucronulate

obtuse

rounded

truncate

praemorse

retuse

emarginate

cleft

Leaf bases

acute

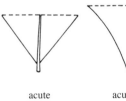
acuminate

attenuate

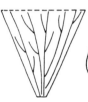
cuneate

cordate

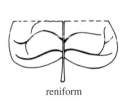
reniform

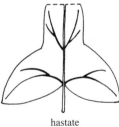
hastate

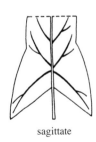
sagittate

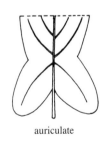
auriculate

rounded

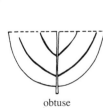
obtuse

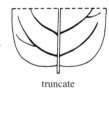
truncate

unequal

oblique

Leaf Margins

Margins

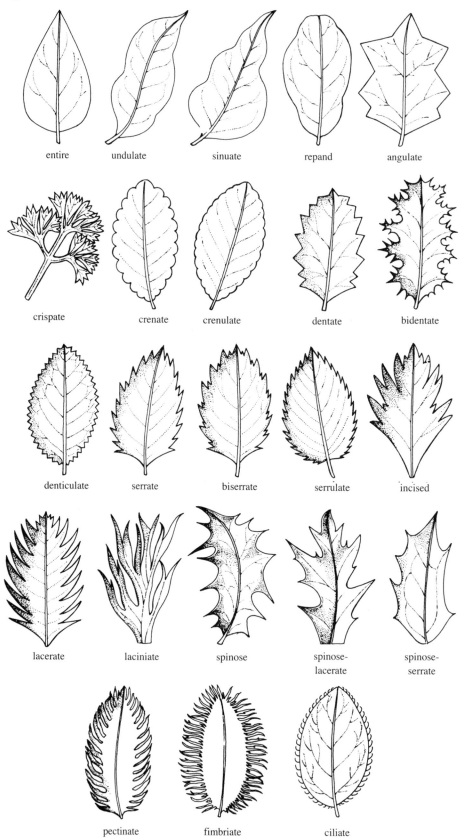

entire undulate sinuate repand angulate

crispate crenate crenulate dentate bidentate

denticulate serrate biserrate serrulate incised

lacerate laciniate spinose spinose-lacerate spinose-serrate

pectinate fimbriate ciliate

Leaf Lobing and Veining

Lobing

lobed

cleft

parted

pinnatifid

pinnatisect

pedately
lobed

palmately
lobed

Venation

secondary
vein

tertiary
veins

midrib

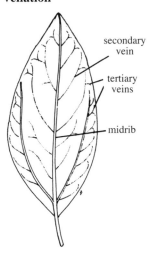

parallel

laciniate

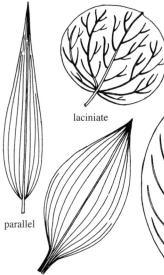

camptodrome

tesselate

reticulate

pinnate

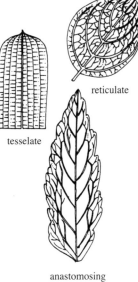

anastomosing

ribbed

conspicuous

prominent

furrowed

Flower Forms and Arrangements

Parts of the flower

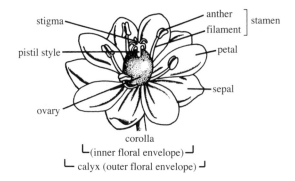

stigma

anther
filament } stamen

pistil style

petal

ovary

sepal

corolla
⌞(inner floral envelope)⌟
⌞ calyx (outer floral envelope) ⌟

polypetalous flower

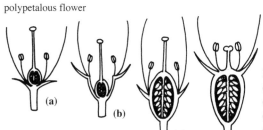

(a)

(b)

(c)

(d)

Position of ovary
(a) ovary superior, perianth and stamens hypogynous
(b) ovary superior, petals and stamens perigynous
(c) ovary partly inferior, petals and stamens epigynous
(d) ovary inferior, petals and stamens epigynous

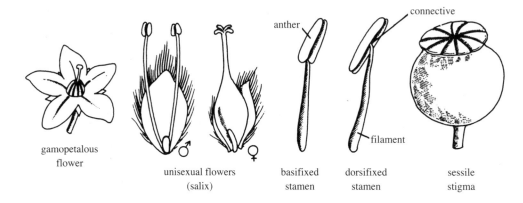

connective

anther

filament

gamopetalous
flower

unisexual flowers
(salix)

basifixed
stamen

dorsifixed
stamen

sessile
stigma

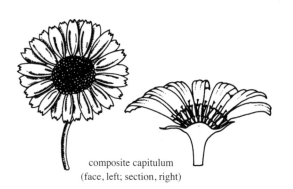

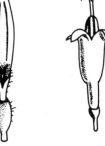

composite capitulum
(face, left; section, right)

ray floret

disc floret

Flower Forms and Arrangements 2

Perianth forms

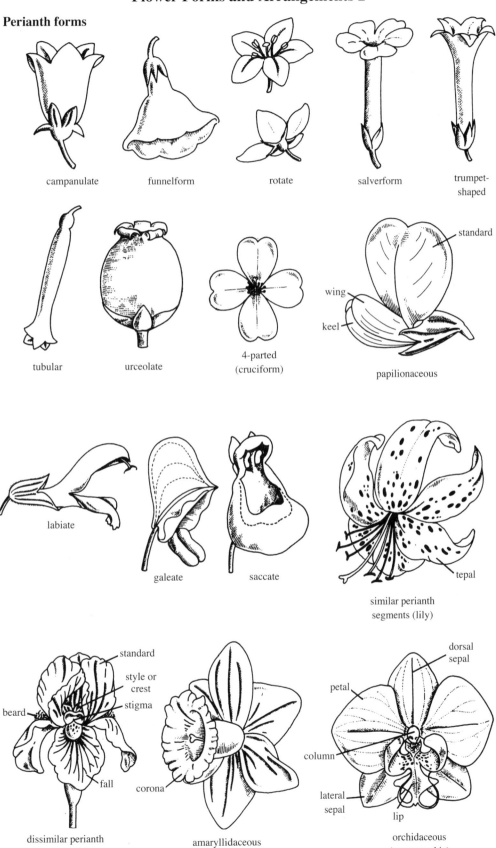

campanulate

funnelform

rotate

salverform

trumpet-shaped

tubular

urceolate

4-parted
(cruciform)

papilionaceous

standard

wing

keel

labiate

galeate

saccate

similar perianth
segments (lily)

tepal

dissimilar perianth
segments (iris)

standard

style or
crest

stigma

beard

fall

amaryllidaceous

corona

orchidaceous
(zygomorphic)

dorsal
sepal

petal

column

lateral
sepal

lip

Fruits 1

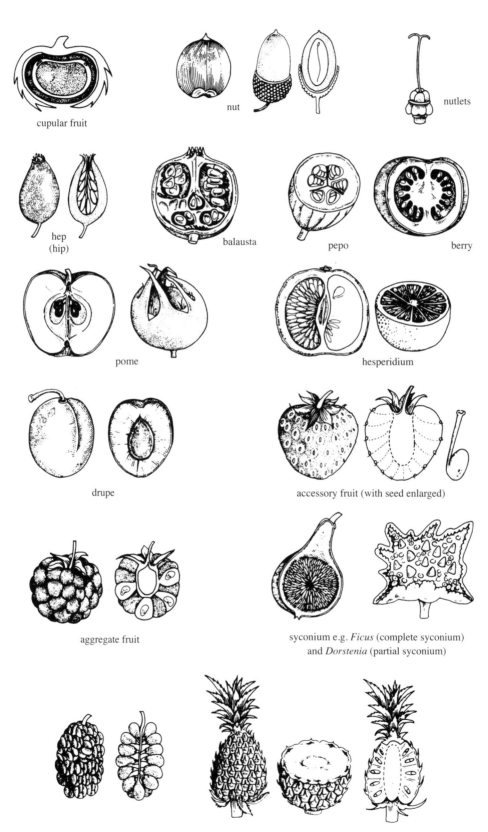

cupular fruit

nut

nutlets

hep
(hip)

balausta

pepo

berry

pome

hesperidium

drupe

accessory fruit (with seed enlarged)

aggregate fruit

syconium e.g. *Ficus* (complete syconium)
and *Dorstenia* (partial syconium)

sororsis (multiple fruit e.g. *Morus*, *Ananas*)

Fruits 2

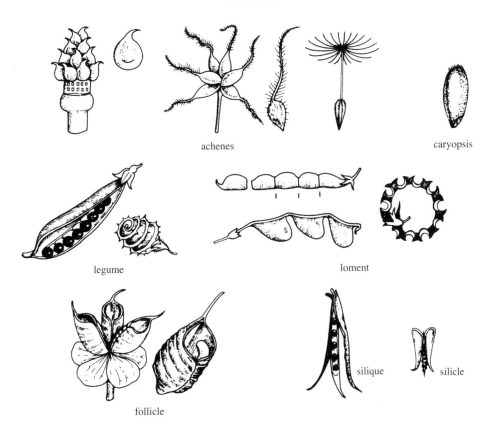

achenes

caryopsis

legume

loment

follicle

silique   silicle

Capsules (showing different modes of dehiscence)

loculicidal   septicidal   septifragal   poricidal   valvate

circumscissile
(pyxis)

utricle

schizocarp

samara

strobilus (conifer)

strobilus (angiosperm
e.g. *Humulus*)

Inflorescences

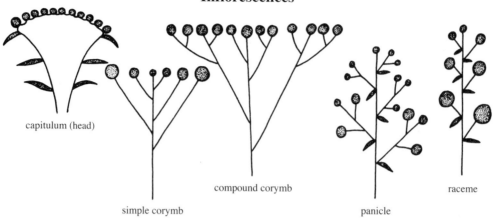

capitulum (head)

simple corymb

compound corymb

panicle

raceme

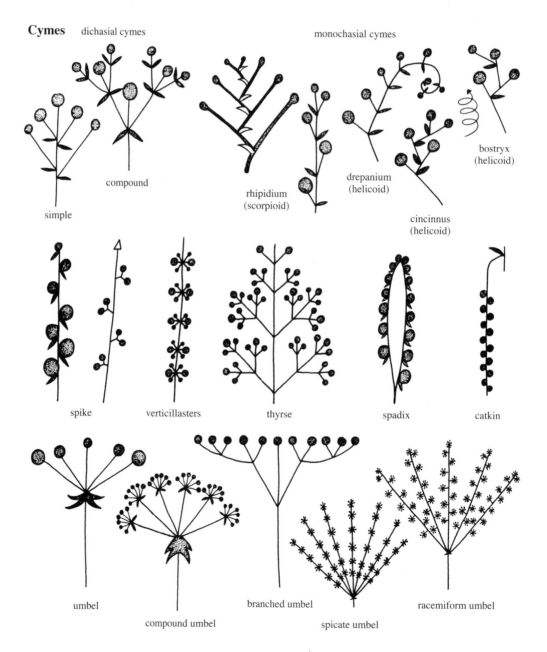

Cymes   dichasial cymes

monochasial cymes

simple

compound

rhipidium
(scorpioid)

drepanium
(helicoid)

cincinnus
(helicoid)

bostryx
(helicoid)

spike

verticillasters

thyrse

spadix

catkin

umbel

compound umbel

branched umbel

spicate umbel

racemiform umbel

INDEX OF COMMON NAMES

This index excludes common names of edible plants that appear as main text entries (e.g. apple) unless that name also happens to cover some ornamentals (e.g. onion *Allium*). Where a common name applies to an entire genus, its specific variants are not usually listed. Thus 'pine' is given below as *Pinus*. Readers seeking the Bhutan pine (*Pinus wallichiana*) should look under *Pinus* in the main text, where the name stands after its Latin equivalent. Exceptions to this rule do appear below (e.g. screw pine *Pandanus*). Where a name applies to more than one plant, the more common application of the name is given first (e.g. mother-in-law's tongue *Sansevieria, Dieffenbachia*).

Aaron's beard *Hypericum calycinum*
Abele *Populus alba*
Aconite *Aconitum*
Adam's needle *Yucca filamentosa*
Adder's tongue *Erythronium*
Aeroplane propeller *Crassula perfoliata* var. *falcata*
African daisy *Arctotis*
African harebell *Dierama*
African hemp *Sparmannia africana*
African lily *Agapanthus*
African milkbush *Synadenium grantii*
African tulip tree *Spathodea*
African violet *Saintpaulia*
Air plant *Tillandsia*
Alder *Alnus*
Alexandrian laurel *Danae racemosa*
Alkanet *Anchusa*
Allegheny vine *Adlumia fungosa*
Allspice *Calycanthus*
Almond *Prunus*
Alpine anemone *Pulsatilla alpina*
Alpine avens *Geum montanum*
Alpine catchfly *Lychnis alpina*
Alpine forget-me-not *Eritrichium*
Alpine strawberry *Fragaria vesca* var. *semperflorens*
Alpine thistle *Carlina acaulis*
Aluminium plant *Pilea cadierei*
Alum root *Heuchera*
Amaryllis *Hippeastrum*
American bittersweet *Celastrus scandens*
American black raspberry *Rubus occidentalis*
American cowslip *Dodecatheon*
American lime *Tilia americana*
American white oak *Quercus alba*
Amethyst violet *Browallia*
Amur cherry *Prunus maackii*
Amur cork tree *Phellodendron amurense*
Angelica *Angelica archangelica*
Angel's fishing rod *Dierama*
Angel's tears *Narcissus triandrus*; *Soleirolia soleirolii*;
    Billbergia x *windii*; *nutans*
Angel's trumpet *Brugmansia, Datura*
Angel's wings *Caladium*
Anise tree *Illicium*
Annual phlox *Phlox drummondii*
Antarctic beech *Nothofagus antarctica*
Antelope ears *Platycerium*
Apache plume *Fallugia*
Apple of Peru *Nicandra*
Arabian primrose *Arnebia*
Arrowhead *Sagittaria*
Arum lily *Zantedeschia*

Ash *Fraxinus*
Asarabacca *Asarum europaeum*
Asiatic poppy *Meconopsis*
Aspen *Populus*
Asphodel *Asphodelus*
Atlas cedar *Cedrus libani* subsp. *libani*
Aubrietia *Aubrieta*
August lily *Hosta plantaginea*
Australian fuchsia *Correa*
Australian heath *Epacris*
Australian honeysuckle *Banksia*
Australian mint bush *Prostanthera*
Australian pine *Casuarina*
Australian pitcher plant *Cephalotus*
Australian tree fern *Cyathea australis*; *Dicksonia antarctica*
Autumn crocus *Colchicum autumnale*; *Crocus nudiflorus*
Autumn daffodil *Sternbergia*
Autumn snowdrop *Galanthus reginae-olgae*
Autumn snowflake *Leucojum autumnale*
Avalanche lily *Erythronium grandiflorum*; *E. montanum*
Avens *Geum*
Aztec lily *Sprekelia formosissima*
Baboon flower *Babiana*
Baby blue eyes *Nemophila menziesii*
Baby's breath *Gypsophila*
Baby rubber plant *Peperomia obtusifolia*
Baby's tears *Soleirolia*
Baby's toes *Fenestraria rhopalophylla*
Bald cypress *Taxodium*
Balloon flower *Platycodon*
Balloon vine *Cardiospermum halicacabum*
Balm of Gilead *Populus* x *jackii*
Balsam *Impatiens*
Banana *Musa*
Banana passion fruit *Passiflora antioquiensis*; *P. mollissima*
Baneberry *Actaea*
Banyan *Ficus benghalensis*
Barbados gooseberry *Pereskia aculeata*
Barberry *Berberis*
Barberton daisy *Gerbera*
Barley *Hordeum*
Barrel cactus *Ferocactus*
Barrenwort *Epimedium*
Bastard balm *Melittis*
Bastard jasmine *Cestrum*
Bat flower *Tacca*
Bay laurel *Laurus nobilis*
Bay tree *Laurus*
Bayonet plant *Aciphylla squarrosa*
Bead fern *Onoclea*
Bead plant *Nertera granadensis*
Bead tree *Melia azedarach*

Elder *Sambucus*
Elephant's ear *Caladium*
Elephant's ear plant *Alocasia*
Elephant's foot *Dioscorea elephantipes*; *Nolina recurvata*
Elephant's tusk *Proboscidea*
Elkhorn fern *Platycerium*
Elm *Ulmus*
Epaulette tree *Pterostyrax*
Eryngoe *Eryngium*
European fan palm *Chamaerops humilis*
Evening primrose *Oenothera*
Everlasting *Antennaria*
Everlasting flower *Helichrysum*
Everlasting pea *Lathyrus grandiflorus*
Everlastings *Helipterum*
Fair maids of France *Ranunculus aconitifolius*; *Saxifraga granulata*
Fair maids of Kent *Ranunculus aconitifolius*
Fairy duster *Calliandra eriophylla*
Fairy foxglove *Erinus*
Fairy lily *Zephyranthes*
Fairy moss *Azolla*
False acacia *Robinia pseudoacacia*
False anemone *Anemonopsis*
False bird of paradise *Heliconia*
False dragon head *Physostegia*
False indigo *Baptisia australis*
False mallow *Sphaeralcea*
False mitrewort *Tiarella*
False rue anemone *Isopyrum*
False sago *Cycas*
False Solomon's seal *Smilacina*
False spiraea *Sorbaria*
Fameflower *Talinum*
Farewell to spring *Clarkia*
Fawn lily *Erythronium*
Feather grass *Stipa*
Fennel *Foeniculum vulgare*
Fern palm *Cycas*
Fescue *Festuca*
Feverfew *Tanacetum parthenium*
Fig *Ficus*
Figwort *Scrophularia*
Filbert *Corylus maxima*
Fir *Abies*
Firecracker vine *Manettia luteorubra*; *M. cordifolia*
Fire lily *Cyrtanthus*
Firethorn *Pyracantha*
Fishbone cactus *Epiphyllum anguliger*
Fishpole bamboo *Phyllostachys aurea*
Fishtail palm *Caryota*
Five-finger *Potentilla*
Five-spot baby *Nemophila maculata*
Flag *Iris*
Flame coral tree *Erythrina*
Flame creeper *Tropaeolum speciosum*
Flame flower *Pyrostegia venusta*
Flame pea *Chorizema*
Flame vine *Pyrostegia venusta*
Flame violet *Episcia cupreata*
Flaming Katy *Kalanchoe blossfeldiana*
Flamingo flower *Anthurium*
Flannel bush; flannel flower *Fremontodendron*

Flax *Linum*
Flax lily *Dianella*; *Phormium*
Fleabane *Erigeron*
Fleur-de-lis *Iris*
Floating fern *Ceratopteris*
Floating heart *Nymphoides*
Floating stag's horn fern *Ceratopteris*
Flower of Tigris *Tigridia*
Flowering maple *Abutilon*
Flowering quince *Chaenomeles*
Flowering rush *Butomus umbellatus*
Flowering stones *Lithops*
Foamflower *Tiarella*
Forget-me-not *Myosotis*
Four o'clock flower *Mirabilis*
Foxglove *Digitalis*
Foxglove tree *Paulownia tomentosa*
Foxtail barley *Hordeum jubatum*
Foxtail grass *Alopecurus*
Foxtail lily *Eremurus*
Frangipani *Plumeria*
Freckle face *Hypoestes phyllostachya*
French honeysuckle *Hedysarum coronarium*
Friendship tree *Crassula ovata*
Fringe cups *Tellima*
Fringe tree *Chionanthus*
Fritillary *Fritillaria*
Frogbit *Hydrocharis*
Galaxy *Galax*
Gardener's garters *Phalaris arundinacea* 'Picta'
Garland flower *Hedychium*
Gay feather *Liatris*
Gentian *Gentiana*
Geraldton waxflower *Chamelaucium uncinatum*
Geranium *Pelargonium*
Germander *Teucrium*
Ghost tree *Davidia*
Giant bellflower *Ostrowskia*
Giant fennel *Ferula*
Giant hyssop *Agastache*
Giant lily *Cardiocrinum*
Giant mallow *Hibiscus*
Gillyflower *Matthiola*
Ginger lily *Hedychium*
Gingham golfball *Euphorbia obesa*
Gladwyn *Iris foetidissima*
Gland bellflower *Adenophora*
Glastonbury thorn *Crataegus monogyna* 'Biflora'
Globe amaranth *Gomphrena globosa*
Globe daisy *Globularia*
Globe flower *Trollius*
Globe lily *Calochortus*
Globe mallow *Sphaeralcea*
Globe thistle *Echinops*
Glory bush *Tibouchina*
Glory flower; glory vine *Eccremocarpus*
Glory lily *Gloriosa*
Glory-of-the-snow *Chionodoxa*
Glory-of-the-sun *Leucocoryne ixioides*
Goat's beard *Aruncus*
Goat's rue *Galega*
Godetia *Clarkia*
Gold guinea plant *Hibbertia*

Red orach *Atriplex hortensis* var. *rubra*
Red passion flower *Passiflora coccinea*
Red puccoon *Sanguinaria*
Red spider lily *Lycoris radiata*
Redwood *Sequoia sempervirens*
Reed *Phragmites*
Reed grass *Calamagrostis*
Reedmace *Typha*
Regal lily *Lilium regale*
Resurrection plant *Selaginella lepidopylla*
Rest-harrow *Ononis*
Rhubarb *Rheum*
Ribbon wood *Hoheria sexstylosa*
Rice flower *Pimelea*
Rice-paper plant *Tetrapanax*
Ring bellflower *Symphyandra*
Roast beef plant *Iris foetidissima*
Rock brake *Cryptogramma*
Rock cress *Arabis*
Rocket *Eruca sativa*
Rock jasmine *Androsace*
Rock lily *Arthropodium cirrhatum*
Rock rose *Helianthemum*; *Cistus*
Rock spiraea *Petrophytum*
Rose bay *Nerium*
Rose mallow *Hibiscus*
Rose of Sharon *Hypericum calycinum*
Rose periwinkle *Catharanthus roseus*
Rose-root *Rhodiola*
Rosemary *Rosmarinus*
Rosin-weed *Grindelia*
Rosinweed *Silphium*
Rowan *Sorbus aucuparia*
Royal fern *Osmunda regalis*
Royal jasmine *Jasminum officinale* f. *grandiflorum*
Royal palm *Roystonea*
Royal red bugler *Aeschynanthus pulcher*
Rubber plant *Ficus elastica*
Rue *Ruta*
Russian comfrey *Symphytum* x *uplandicum*
Russian vine *Polygonum baldshuanicum*; *P. aubertii*
Sacred bamboo *Nandina domestica*
Sacred fig tree *Ficus religiosa*
Sacred lotus *Nelumbo nucifera*
Saffron *Crocus sativus*
Sage *Salvia*
Sagebrush *Artemisia*
Sago *Cycas*
Sago fern *Cyathea*
Saguaro *Carnegiea gigantea*
St Bruno's lily *Paradisea*
St Catherine's lace *Eriogonum*
St Dabeoc's heath *Daboecia*
Salsify *Tragopogon*
Saltbush *Atriplex*
Sand verbena *Abronia*
Sandpaper vine *Petrea*
Sandwort *Arenaria*
Sarviceberry *Amelanchier*
Satin poppy *Meconopsis napaulensis*
Savin *Juniperus sabina*
Savory *Micromeria*; *Satureja*
Sawara cypress *Chamaecyparis pisifera*

Saxifrage *Saxifraga*
Scabious *Scabiosa*; *Knautia*
Scorpion grass *Myosotis*
Scorpion orchid *Arachnis*
Scorpion weed *Phacelia*
Scotch heather *Calluna vulgaris*
Scotch laburnum *Laburnum alpinum*
Scotch thistle *Onopordum*
Scots pine *Pinus sylvestris*
Screw pine *Pandanus*
Sea buckthorn *Hippophaë*
Sea daffodil *Pancratium maritimum*
Sea holly *Eryngium maritimum*
Sea kale *Crambe maritima*
Sea lavender *Limonium*
Sea lily *Pancratium maritimum*
Sea onion *Urginea maritima*
Sea pink *Armeria*
Sea squill *Urginea maritima*
Sedge *Carex*
Self heal *Prunella*
Semi bean *Lablab*
Seneca *Polygala*
Sensitive fern *Onoclea sensibilis*
Sensitive plant *Mimosa pudica*
Sentry palm *Howea*
Serviceberry *Amelanchier*
Service tree of Fontainebleau *Sorbus latifolia*
Shadbush *Amelanchier*
Shagbark hickory *Carya ovata*
Shallon *Gaultheria shallon*
Shamrock *Oxalis*
Shasta daisy *Leucanthemum* x *superbum*
Sheep's burrs *Acaena*
Shell flower *Chelone*
Shell ginger *Alpinia zerumbet*
Shield fern *Polystichum*; *Dryopteris*
Shinleaf *Pyrola*
Shoo-fly *Nicandra*
Shooting stars *Dodecatheon*
Shower tree *Cassia*
Shrimp plant *Justicia brandegeana*
Shrimp tree *Koelreuteria*
Shrubby germander *Teucrium fruticans*
Shrubby trefoil *Ptelea*
Shrub verbena *Lantana*
Siberian bugloss *Brunnera macrophylla*
Siberian crab *Malus baccata*
Siberian dogwood *Cornus alba*
Siberian elm *Ulmus pumila*
Siberian iris *Iris sibirica*
Siberian squill *Scilla siberica*
Silk tassel bush *Garrya elliptica*
Silk tree *Albizia julibrissin*
Silkweed *Asclepias*
Silky oak *Grevillea robusta*
Silky wisteria *Wisteria venusta*
Silverballs *Styrax*
Silver beech *Nothofagus menziesii*
Silverbell tree *Halesia*
Silver birch *Betula pendula*
Silver dollar cactus *Astrophytum asterias*
Silver fir *Abies*

INDEX OF SYNONYMS

Picture Acknowledgements

The publishers, editors and the Royal Horticultural Society wish to thank the following persons and institutions, whose gardens illustrate this work. All photographs were taken by Andrew Lawson.

Spring

Abbotswood, Stow-on-the-Wold, Gloucestershire (Robin Scully) **viii** (bottom); **ix** (top); **x** (bottom)
Barnsley House, Gloucestershire (Rosemary Verey) **viii** (top)
Exbury Gardens, Hampshire (Edmund de Rothschild) **iv**; **v**; **xii** (bottom)
The Garden House, Buckland Monachorum, Devon (Fortescue Garden Trust) **v** (bottom)
Marwood Hill, Devon (Dr J A Smart) **iii** (left and bottom); **xi**
Regent's Park, London (Royal Parks) **ix** (bottom)
Royal Botanic Gardens, Kew (Trustees of the Royal Botanic Gardens) **iii** (top)
Rupert Golby, Oxfordshire **i**
Steeple Aston, Oxfordshire (Connie Franks) **viii** (right)
Waterperry Gardens, Oxfordshire (School of Economic Science) **vii** (top)

Summer

Ashtree Cottage, Wiltshire (Wendy Lauderdale) **vi** (top)
Bosvigo House, Truro, Cornwall (Wendy and Michael Perry) **iv** (top); **v** (bottom)
Bourton House, Gloucestershire (Mr & Mrs Richard Paice) **iii** (top)
Brook Cottage, Alkerton, Oxfordshire (Mr & Mrs D M Hodges) **x** (top)
Cottage Garden Roses, Staffordshire **viii** (top)
Giverny, France **xi** (top)
Gothic House, Oxfordshire (Andrew Lawson) **iv** (bottom left and right)
Hadspen Gardens, Somerset (Nori and Sandra Pope) **vii** (top)
Hermannshof, Weinheim, Germany **vi** (bottom)
Lower Hopton Farm, Hereford (Giles and Veronica Cross) **vii** (bottom right)
Powis Castle, Powys (The National Trust) **ii** (top left)
The Priory, Kemerton, Hereford & Worcester (The Hon Mrs Peter Kealing) **v** (top right)
Rupert Golby, Oxfordshire **ii** (top right)

Autumn

Barnsley House, Gloucestershire (Rosemary Verey) **vii** (top right)
Castle Ashby, Northamptonshire (Marqess of Northampton) **iv** (top)
The Old Rectory, Burghfield, Berkshire (R Merton) **iv** (bottom left)
Powis Castle, Powys (The National Trust) **i**
Sticky Wicket, Buckland Newton, Dorset (Peter and Pam Lewis) **xii** (bottom left)
Westonbirt Arboretum, Gloucestershire (Forestry Commission) **v** (bottom); **vi** (bottom right)

Winter

Barnsley House, Gloucestershire (Rosemary Verey) **iii** (top left and bottom)
St John's College, Oxford (President and Fellows of St John's College) **vii** (top)
University Botanic Garden, Cambridge (University of Cambridge) **ix** (left)
Waterperry Gardens, Oxfordshire (School of Economic Science) **i** (right); **ii**